Solar System Data

Body	Mass (kg)	Mean Radius (m)	Period (s)	Distance from Sun (m)
Mercury	3.18×10^{23}	2.43×10^6	7.60×10^6	5.79×10^{10}
Venus	4.88×10^{24}	6.06×10^6	1.94×10^7	1.08×10^{11}
Earth	5.98×10^{24}	6.37×10^6	3.156×10^7	1.496×10^{11}
Mars	6.42×10^{23}	3.37×10^6	5.94×10^7	2.28×10^{11}
Jupiter	1.90×10^{27}	6.99×10^7	3.74×10^8	7.78×10^{11}
Saturn	5.68×10^{26}	5.85×10^7	9.35×10^8	1.43×10^{12}
Uranus	8.68×10^{25}	2.33×10^7	2.64×10^9	2.87×10^{12}
Neptune	1.03×10^{26}	2.21×10^7	5.22×10^9	4.50×10^{12}
Pluto	$\approx 1.4 \times 10^{22}$	$\approx 1.5 \times 10^6$	7.82×10^9	5.91×10^{12}
Moon	7.36×10^{22}	1.74×10^6	—	—
Sun	1.991×10^{30}	6.96×10^8	—	—

Physical Data Often Used[a]

Average Earth-Moon distance	3.84×10^8 m
Average Earth-Sun distance	1.496×10^{11} m
Average radius of the Earth	6.37×10^6 m
Density of air (20°C and 1 atm)	1.20 kg/m^3
Density of water (20°C and 1 atm)	1.00×10^3 kg/m^3
Free-fall acceleration	9.80 m/s^2
Mass of the Earth	5.98×10^{24} kg
Mass of the Moon	7.36×10^{22} kg
Mass of the Sun	1.99×10^{30} kg
Standard atmospheric pressure	1.013×10^5 Pa

[a] These are the values of the constants as used in the text.

Some Prefixes for Powers of Ten

Power	Prefix	Abbreviation	Power	Prefix	Abbreviation
10^{-18}	atto	a	10^1	deka	da
10^{-15}	femto	f	10^2	hecto	h
10^{-12}	pico	p	10^3	kilo	k
10^{-9}	nano	n	10^6	mega	M
10^{-6}	micro	μ	10^9	giga	G
10^{-3}	milli	m	10^{12}	tera	T
10^{-2}	centi	c	10^{15}	peta	P
10^{-1}	deci	d	10^{18}	exa	E

PHYSICS

For Scientists & Engineers
with Modern Physics

Fourth Edition

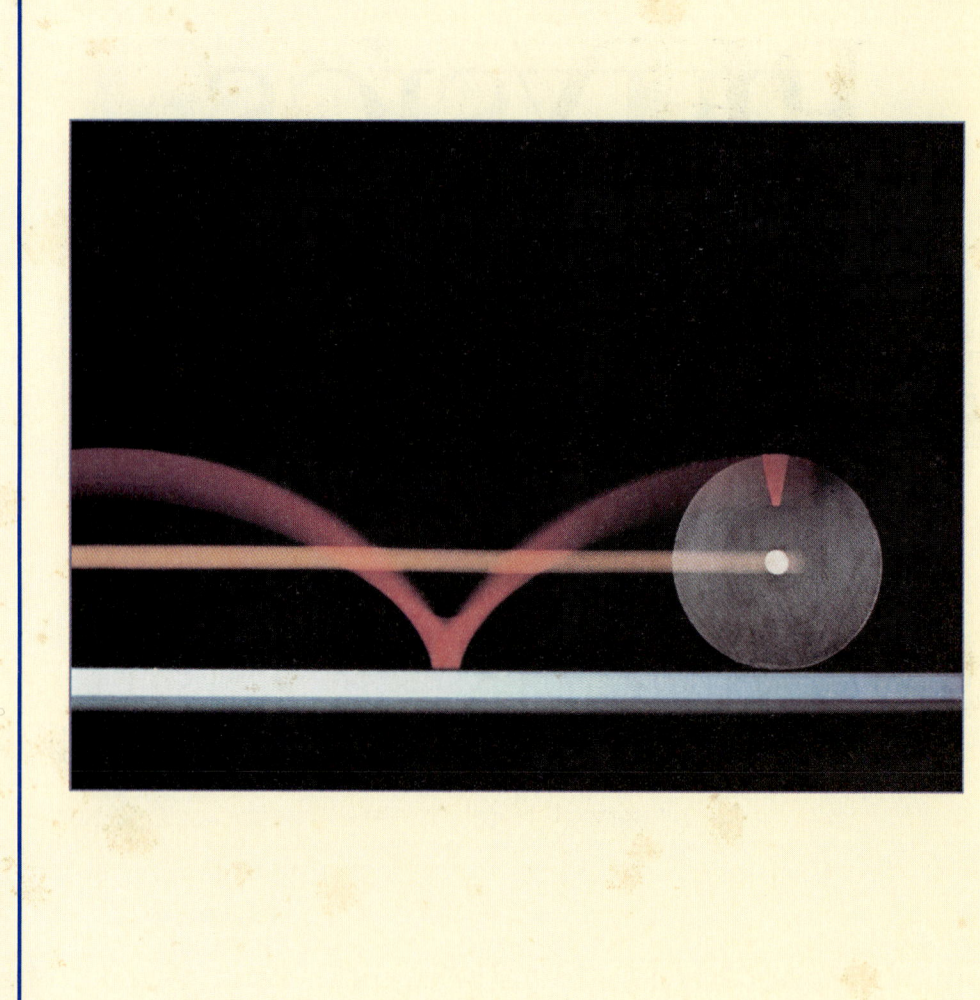

PHYSICS

For Scientists & Engineers
with Modern Physics

Fourth Edition

Raymond A. Serway

James Madison University

SAUNDERS GOLDEN SUNBURST SERIES

SAUNDERS COLLEGE PUBLISHING

Philadelphia   Fort Worth   Chicago   San Francisco
Montreal   Toronto   London   Sydney   Tokyo

Requests for permission to make copies of any part of the work should be mailed to: Permissions Department, Harcourt Brace & Company, 6277 Sea Harbor Drive, Orlando, Florida 32887-6777.

Text Typeface: New Baskerville
Composition and Layout: Progressive Information Technologies
Publisher: John Vondeling
Developmental Editor: Laura Maier
Senior Project Editor: Sally Kusch
Copy Editor: Charlotte Nelson
Managing Editor: Carol Field
Manager of Art and Design: Carol Bleistine
Associate Art Director: Sue Kinney
Art and Design Coordinator: Kathleen Flanagan
Text Designer: Rebecca Lemna
Cover Designer: Lawrence R. Didona
Text Artwork: Rolin Graphics
Photo Researcher: Sue Howard
Director of EDP: Tim Frelick
Production Manager: Charlene Squibb
Director of Marketing: Marjorie Waldron

Cover and title page: Rolling motion of a rigid body: disk by Richard Megna, © Fundamental Photographs, NYC

Printed in the United States of America

Physics for Scientists and Engineers with Modern Physics, Fourth edition

ISBN 0-03-015654-8

Library of Congress Catalog Card Number: 95-069620

89012345 032 10 9876

The publisher and author have gone to extreme measures in attempting to ensure the publication of an error-free text. The manuscript, galleys, and page proofs have been carefully checked by the author, the editors, and a battery of reviewers. While we realize that a 100% error-free text may not be humanly possible, Serway's *Physics for Scientists and Engineers* is very close. Confirmed in this belief, we are offering $5.00 for any first-time error you may find. (Note that we will only pay for each error the first time it is brought to our attention.) Please write to John Vondeling, Publisher.

Preface

Physics for Scientists and Engineers has been used successfully at over 700 colleges and universities over the course of three editions. This fourth edition has many new pedagogical features, and a major effort was made to improve clarity of presentation, precision of language, and accuracy throughout. Based on comments from users of the third edition and reviewers' suggestions, refinements have been added such as an increased emphasis on teaching concepts. The fourth edition has also integrated several new interactive software products that will be useful in courses using computer-assisted instruction.

This textbook is intended for a course in introductory physics for students majoring in science or engineering. The book is an extended version of *Physics for Scientists and Engineers* in that it includes eight additional chapters covering selected topics in modern physics. This material on modern physics has been added to meet the needs of those universities that choose to cover the basic concepts of quantum physics and its application to atomic, molecular, solid state, and nuclear physics as part of their curriculum.

The entire contents of the text could be covered in a three-semester course, but it is possible to use the material in shorter sequences with the omission of selected chapters and sections. The mathematical background of the student taking this course should ideally include one semester of calculus. If that is not possible, the student should be enrolled in a concurrent course in introduction to calculus.

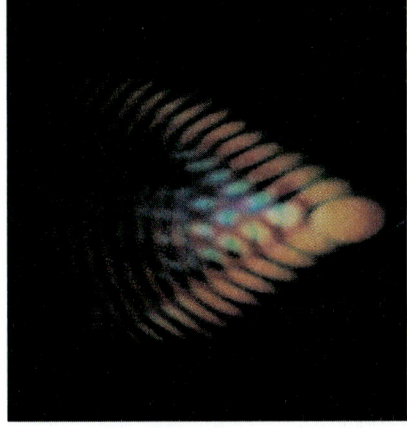

(Norman Goldberg)

OBJECTIVES

The main objectives of this introductory physics textbook are twofold: to provide the student with a clear and logical presentation of the basic concepts and principles of physics, and to strengthen an understanding of the concepts and principles through a broad range of interesting applications to the real world. In order to meet these objectives, emphasis is placed on sound physical arguments. At the same time, I have attempted to motivate the student through practical examples that demonstrate the role of physics in other disciplines including engineering, chemistry, and medicine.

CHANGES TO THE FOURTH EDITION

A number of changes and improvements have been made in preparing the fourth edition of this text. Many changes are in response to comments and suggestions offered by instructors and students using the third edition and by reviewers of the manuscript. The following represent the major changes in the fourth edition:

• **Line-by-Line Revision** The entire text has been carefully edited to improve clarity of presentation and precision of language. We hope that the result is a book that is both accurate and enjoyable to read.

• **Organization** The organization of the textbook is essentially the same as that of the third edition with one exception. Chapters 2 and 3 have been interchanged, so that the treatment of vectors precedes the discussion of motion in two dimensions, where vectors and their components are first used. Many sections have been streamlined or combined with other sections to allow for a more balanced presentation.

• **Problems**   A substantial revision of the end-of-chapter problems and questions was made in an effort to provide a greater variety and to reduce repetition. Approximately 25 percent of the problems (approximately 800), most of which are at the intermediate level, are new. The remaining problems have been carefully edited and reworded where necessary. All new problems are marked with an asterisk in the Instructors Manual. Solutions to approximately 25 percent of the problems are included in the Student Solutions Manual and Study Guide. These problems are identified by boxes around their numbers.

• **Significant Figures**   Significant figures in both worked examples and end-of-chapter problems have been handled with care. Most numerical examples and problems are worked out to either two or three significant figures, depending on the accuracy of the data provided.

• **Visual Presentation**   Most of the line art and many of the color photographs have been replaced or modified to improve the clarity of presentation, pedagogy, and visual appeal of the text. As in the third edition, color is used primarily for pedagogical purposes. A chart explaining the pedagogical use of color is included after the To the Student section following the preface.

NEW FEATURES IN THE FOURTH EDITION

• **Integrated Software**   The textbook is accompanied by two interactive software packages. *SD2000* is a self-contained software package of physics simulations and demonstrations that have been developed exclusively to accompany this textbook. Concepts and examples are presented and explained in an interactive format. Simulations developed for the Interactive Physics II™ program are keyed to appropriate worked-example problems and to selected end-of-chapter problems. Both packages are provided on disks and are described in more detail in the section dealing with ancillaries.

• **Conceptual Examples**   Approximately 150 conceptual examples have been included in this edition. These examples, which include reasoning statements, provide students with a means of reviewing the concepts presented in that section. The examples could also serve as models when students are asked to respond to end-of-chapter questions, which are largely conceptual in nature.

• **Review Problems**   Many chapters now include a multi-part review problem located prior to the list of end-of-chapter problems. The review problem requires the student to draw on numerous concepts covered in the chapter as well as those discussed in previous chapters. These problems can be used by students in preparing for tests, and by instructors in classroom discussions and review.

• **Paired Problems**   Several end-of-chapter problems have been paired with the same problem in symbolic form. For example, numerical Problem 9 may be followed by symbolic Problem 9A. If Problem 9 is assigned, Problem 9A can be used to test the student's understanding of the concepts used in solving the problem.

• **Spreadsheet Problems**   Most chapters will include several spreadsheet problems following the end-of-chapter problem sets. Spreadsheet modeling of physical phenomena enables the student to obtain graphical representations of physical quantities and perform numeri-

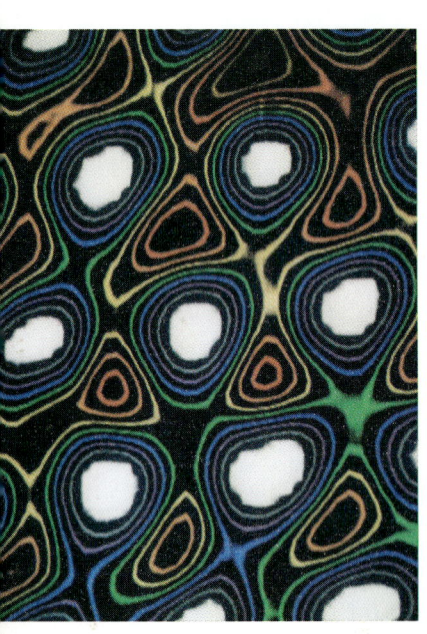

(Prof. R. V. Coleman, University of Virginia)

cal analyses of problems without the burden of having to learn a high-level computer language. Spreadsheets are particularly valuable in exploratory investigations; "what if" questions can be addressed easily and depicted graphically.

Level of difficulty in the spreadsheet problems, as with all end-of-chapter problems, is indicated by the color of the problem number. For the most straightforward problems (black) a disk with spreadsheet templates is provided. The student must enter the pertinent data, vary the parameters, and interpret the results. Intermediate level problems (blue) usually require students to modify an existing template to perform the required analysis. The more challenging problems (magenta) require students to develop their own spreadsheet templates. Brief instuctions on using the templates are provided in Appendix F.

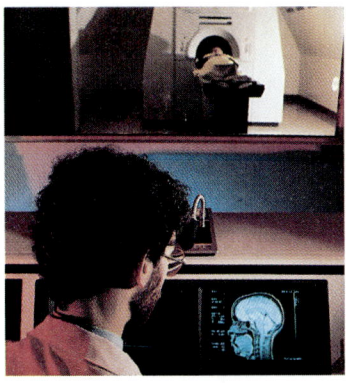

(Hank Morgan/Science Source)

COVERAGE

The material covered in this book is concerned with fundamental topics in classical physics and an introduction to modern physics. The book is divided into six parts. Part I (Chapters 1–15) deals with the fundamentals of Newtonian mechanics and the physics of fluids; Part II (Chapters 16–18) covers wave motion and sound; Part III (Chapters 19–22) is concerned with heat and thermodynamics; Part IV (Chapters 23–34) treats electricity and magnetism; Part V (Chapters 35–38) covers light and optics; and Part VI (Chapters 39–47) deals with relativity, quantum physics, and selected topics in modern physics. Each part opener includes an overview of the subject matter to be covered in that part and some historical perspectives.

TEXT FEATURES

Most instructors would agree that the textbook selected for a course should be the student's primary "guide" for understanding and learning the subject matter. Furthermore, a textbook should be easily accessible and should be styled and written for ease in instruction. With these points in mind, I have included many pedagogic features in the textbook which are intended to enhance its usefulness to both the student and instructor. These are as follows:

Style   As an aid for rapid comprehension, I have attempted to write the book in a style that is clear, logical, and engaging. The writing style is somewhat informal and relaxed, which I hope students will find appealing and enjoyable to read. New terms are carefully defined, and I have tried to avoid jargon.

Previews   Most chapters begin with a chapter preview, which includes a brief discussion of chapter objectives and content.

Important Statements and Equations   Most important statements and definitions are set in bold print for added emphasis and ease of review. Important equations are highlighted with a tan screen for review or reference.

Problem-Solving Strategies and Hints   I have included general strategies for solving the types of problems featured in both the examples and in the end-of-chapter problems. This feature will help students identify necessary steps in solving problems and eliminate any uncertainty they might have. Problem-solving strategies are highlighted by a light color screen for emphasis and ease of location.

Marginal Notes   Comments and marginal notes are used to locate important statements, equations, and concepts in the text.

Illustrations   The readability and effectiveness of the text material and worked examples are enhanced by the large number of figures, diagrams, photographs, and tables. Full color is used to add clarity to the artwork and to make it as realistic as possible. For example, vectors are color-coded, and curves in *xy*-plots are drawn in color. Three-dimensional effects are produced with the use of color airbrushed areas, where appropriate. The color photographs have been carefully selected, and their accompanying captions have been written to serve as an added instructional tool. Several chapter-opening photographs, particularly in the chapters on mechanics, include color-coded vector overlays that illustrate and present physical principles more clearly and apply them to real-world situations.

Mathematical Level   Calculus is introduced gradually, keeping in mind that a course in calculus is often taken concurrently. Most steps are shown when basic equations are developed, and reference is often made to mathematical appendices at the end of the text. Vector products are introduced later in the text where they are needed in physical applications. The dot product is introduced in Chapter 7, "Work and Energy." The cross product is introduced in Chapter 11, which deals with rotational dynamics.

Worked Examples   A large number of worked examples of varying difficulty are presented as an aid in understanding concepts. In many cases, these examples serve as models for solving the end-of-chapter problems. The examples are set off in a box, and the solution answers are highlighted with a light blue screen. Most examples are given titles to describe their content.

Worked-Example Exercises   Many of the worked examples are followed immediately by exercises with answers. These exercises are intended to make the textbook more interactive with the student and to immediately reinforce the student's understanding of concepts and problem-solving techniques. The exercises represent extensions of the worked examples.

Units   The international system of units (SI) is used throughout the text. The British engineering system of units (conventional system) is used only to a limited extent in the chapters on mechanics, heat, and thermodynamics.

Biographical Sketches   Throughout the text I have included short biographies of important scientists to add more historical emphasis and show the human side of the lives of scientists.

Optional Topics   Many chapters include special topic sections which are intended to expose the student to various practical and interesting applications of physical principles. These optional sections are labeled with an asterisk (*).

(*Joseph Brignolo/The Image Bank*)

Summaries   Each chapter contains a summary which reviews the important concepts and equations discussed in that chapter.

Thought Questions   Questions requiring verbal answers are provided at the end of each chapter. Some questions provide the student with a means of self-testing the concepts presented in the chapter. Others could serve as a basis for initiating classroom discussions.

Problems   An extensive set of problems is included at the end of each chapter. Answers to odd-numbered problems are given at the end of the book; these pages have colored edges for ease of location. For the convenience of both the student and the instructor, about two thirds of the problems are keyed to specific sections of the chapter. The remaining problems, labeled "Additional Problems," are not keyed to specific sections. In my opinion, assignments should consist mainly of the keyed problems to help build self-confidence in students.

   Usually, the problems within a given section are presented so that the straightforward problems (numbered in black print) are first, followed by problems of increasing difficulty.

For ease in identifying the intermediate-level problems, the problem number is printed in blue. I have also included a small number of challenging problems, which are indicated by a problem number printed in magenta.

Appendices and Endpapers   Several appendices are provided at the end of the text, including the new appendix with instructions for problem-solving with spreadsheets. Most of the appendix material represents a review of mathematical techniques used in the text, including scientific notation, algebra, geometry, trigonometry, differential calculus, and integral calculus. Reference to these appendices is made throughout the text. Most mathematical review sections include worked examples and exercises with answers. In addition to the mathematical reviews, the appendices contain tables of physical data, conversion factors, atomic masses, and the SI units of physical quantities, as well as a periodic chart. Other useful information, including fundamental constants and physical data, planetary data, a list of standard prefixes, mathematical symbols, the Greek alphabet, and a table of standard abbreviations and symbols of units appears on the endpapers.

ANCILLARIES

The ancillary package has been updated and expanded in response to suggestions from users of the third edition. The most essential changes are an expansive set of interactive software, an updated test bank with greater emphasis on conceptual questions and open-ended problems, a new Student Solutions Manual and Study Guide with complete solutions to 25 percent of the text problems, a student's Pocket Guide, and a new spreadsheet supplement.

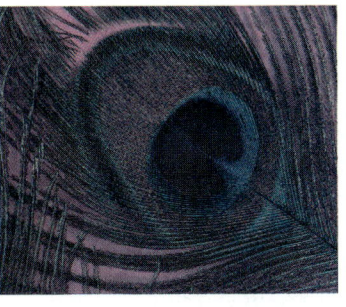

(Werner H. Muller/Peter Arnold, Inc.)

Interactive Software

Interactive Homework System

The World Wide Web (WWW) is the platform for an interactive homework system developed out of the University of Texas at Austin. This system, developed to coordinate with *Physics for Scientists and Engineers*, uses WWW, telnet, telephone, and Scantron submission of student work. The system has been class-tested at the University of Texas with over 1800 students participating each semester. Over 100,000 questions are answered electronically per month. Instructors at any university using Serway's *Physics for Scientists and Engineers* may establish access to this system by providing a class roster and making problem selections. Over 2000 algorithm-based problems are available; problem parameters vary from student to student, so that each student must do original work. All grading is done by computer, with results automatically posted on WWW. Students receive immediate right/wrong feedback, with multiple tries allowed for incorrect answers. When students answer incorrectly, they are automatically linked into text from the appropriate section of the fourth edition of Serway's *Physics for Scientists and Engineers with Modern Physics*.

A demo using the WWW interface is available at the URL **http://hw.ph.utexas.edu:80** by clicking on the demo link. Further information for instructors interested in importing the system to their institutions is available from **see@physics.utexas.edu**. The fourth edition of *Physics for Scientists and Engineers* will be linked into the system by January 1996.

SD2000 Interactive Software

This learning environment of physics simulations and demonstrations has been developed by Future Graph, Inc., exclusively to accompany this textbook. Its applications span all of the basic topics treated in the textbook. SD2000 is available on computer disk or CD-ROM in Macintosh and IBM Windows formats. The icon identifies examples and sections for which a simulation or demonstration exists.

Simulations   A collection of 11 powerful simulators allows students to model and bring to life an infinite number of physics problems. Students can model systems that include kinetic motion, collisions, geometric optics, and electric and magnetic fields, as well as laboratory tools such as Fourier synthesizers, wave form generators, and oscilloscopes. SD2000 boxes throughout the text identify how these simulators can be used to reinforce the concepts presented in the text. In modeling individualized simulations, students may investigate how varying the components of a situation will affect the outcome.

- Chapter 4: Motion, Section 4.4
- Chapter 9: Collisions, Section 9.5
- Chapter 16: Wave Motion, Section 16.4
- Chapter 18: Complex Waves—The Fourier Synthesizer, Section 18.8
- Chapter 21: Systems of Particles, Section 21.1
- Chapter 23: Motion in an Electric Field, Section 23.7
- Chapter 25: Mapping the Electric Field, Section 25.5
- Chapter 29: Motion of Charged Particles in Electric and Magnetic Fields, Section 29.5
- Chapter 33: The Oscilloscope, Section 33.5
- Chapter 36: Optical Instruments, Section 36.10
- Chapter 41: Experiments in Modern Physics

Demonstrations   Lessons derived from worked examples in the text of *Physics for Scientists and Engineers* allow students to investigate the results of changing parameters within the context of the example. Students can interactively explore physics through equations, calculations, graphs, tables, animations, and simulations. A complete list of demonstrations follows:

Chapter 2
Section 2.4
Example 2.3
Example 2.15

Chapter 3
Example 3.8

Chapter 4
Section 4.2
Section 4.3
Example 4.7

Chapter 5
Example 5.6
Example 5.9
Example 5.15

Chapter 6
Example 6.4
Example 6.5
Example 6.6
Example 6.10

Chapter 7
Example 7.3
Example 7.5
Example 7.18

Chapter 8
Example 8.10

Chapter 9
Example 9.7
Example 9.10
Example 9.11
Example 9.12
Example 9.15
Section 9.6
Example 9.16
Example 9.23

Chapter 10
Example 10.4

Chapter 11
Example 11.8
Example 11.13

Chapter 13
Example 13.2
Section 13.4
Section 13.6
Section 13.7

Chapter 14
Example 14.2
Example 14.4
Example 14.7
Example 14.11

Chapter 16
Example 16.1
Example 16.3

Chapter 18
Example 18.2
Section 18.7

Chapter 20
Example 20.4
Example 20.6

Chapter 21
Section 21.6

Chapter 22
Example 22.3

Chapter 23
Example 23.2
Example 23.3
Example 23.14
Section 23.7

Chapter 24
Example 24.5
Example 24.6

Chapter 25
Example 25.5
Example 25.7
Example 25.11

Chapter 26
Example 26.1
Example 26.4

Chapter 27
Example 27.1
Example 27.3
Example 27.4

Chapter 28
Example 28.3
Example 28.8
Example 28.11

Chapter 30
Example 30.1
Example 30.3
Example 30.4

Chapter 31
Example 31.1
Example 31.6
Example 31.11

Chapter 32
Example 32.1
Example 32.3
Section 32.6
Example 32.7

Chapter 33

Example 33.1
Example 33.2
Example 33.4
Example 33.5
Section 33.8

Chapter 34

Example 34.1

Chapter 35

Example 35.2
Example 35.3
Example 35.6
Example 35.7

Chapter 36

Example 36.10

Chapter 37

Example 37.1
Example 37.4

Chapter 38

Example 38.1
Example 38.4

Chapter 39

Example 39.3
Example 39.8

Chapter 40

Example 40.1
Example 40.2
Example 40.3
Example 40.4

(Ken Sakomoto, Black Star)

Interactive-Physics Simulations

Approximately 100 simulations developed by Ray Serway and Knowledge Revolution are available on computer disk in either Macintosh or IBM format to be used in conjunction with the highly acclaimed program *Interactive Physics II* from Knowledge Revolution. Most of these simulations are keyed to appropriate worked-example problems and to selected end-of-chapter problems. The remainder are demonstrations that complement concepts or applications discussed in the text. Simulations can be used in the classroom or laboratory to help students understand physics concepts by developing better visualization skills. The simulation is started by simply clicking the RUN button. The simulation engine calculates the motion of the defined system and displays it in smooth animation. The results can be displayed in graphical, digital, tabular, and bar-graph formats. The acquired data can also be exported to a spreadsheet of your choice for other types of analyses. The Interactive Physics Icon   identifies the examples, problems, and figures for which a simulation exists. A complete list of physics simulations follows.

List of Interactive Physics Simulations

Chapter 2

Example 2.10
Example 2.12
Example 2.14
Example 2.15
Problem 2.46
Problem 2.49
Problem 2.72
Problem 2.76
Problem 2.80
Problem 2.81

Chapter 3

Example 3.8
Problem 3.50

Chapter 4

Example 4.2
Example 4.5
Example 4.6
Example 4.7
Example 4.11
Figure 4.5
Problem 4.17
Problem 4.55
Problem 4.58
Problem 4.66
Problem 4.82
Problem 4.84

Chapter 5

Example 5.8
Example 5.9

Example 5.12
Example 5.13
Example 5.14
Problem 5.18
Problem 5.37
Problem 5.38
Problem 5.42
Problem 5.47
Problem 5.55
Problem 5.70
Problem 5.73
Problem 5.74
Problem 5.76
Problem 5.83
Problem 5.84
Problem 5.87
Problem 5.88

Chapter 6

Example 6.1
Example 6.3
Problem 6.5
Problem 6.21
Problem 6.30

Chapter 7

Example 7.7
Example 7.8
Example 7.12
Figure 7.8
Problem 7.37
Problem 7.43

Problem 7.44
Problem 7.82
Problem 7.89

Chapter 8

Example 8.1
Example 8.3
Example 8.6
Example 8.8
Example 8.9
Problem 8.10
Problem 8.11
Problem 8.17
Problem 8.19
Problem 8.33
Problem 8.35
Problem 8.59
Problem 8.64
Problem 8.67

Chapter 9

Example 9.7
Example 9.11
Example 9.13
Example 9.14
Problem 9.66
Problem 9.72
Problem 9.83
Problem 9.87

Chapter 10

Example 10.11
Example 10.12

Example 10.15
Problem 10.51

Chapter 11

Problem 11.51
Problem 11.65

Chapter 12

Example 12.1
Example 12.3
Example 12.4
Problem 12.36
Problem 12.40
Problem 12.51

Chapter 13

Example 13.4
Example 13.5
Example 13.8
Figure 13.9
Problem 13.18
Problem 13.57
Problem 13.63

Chapter 23

Example 23.3
Example 23.8
Example 23.14
Problem 23.54

Chapter 29

Problem 29.43
Problem 29.71

f(g) Scholar — Spreadsheet/Graphing, Calculator/Graphing Software

f(g) Scholar is a powerful, scientific/engineering spreadsheet software program with over 300 built-in math functions, developed by Future Graph, Inc. It uniquely integrates graphing calculator, spreadsheet, and graphing applications into one, and allows for quick and easy movement between the applications. Students will find many uses for *f(g) Scholar* across their science, math and engineering courses, including working through their laboratories from start to finished reports. Other features include a programming language for defining math functions, curve fitting, three-dimensional graphing and equation displaying. When bookstores order *f(g) Scholar* through Saunders College Publishing they can pass on our exclusive low price to the student.

Student Ancillaries

Student Solutions Manual and Study Guide by John R. Gordon, Ralph McGrew, Steve Van Wyk, and Ray Serway   The manual features detailed solutions to 25 percent of the end-of-chapter problems from the text. These are indicated in the text with boxed problem numbers. The manual also features a skills section that reviews mathematical concepts and important notes from key sections of the text and provides a list of important equations and concepts.

Pocket Guide by V. Gordon Lind   This 5″ × 7″ notebook is a section-by-section capsule of the textbook that provides a handy guide for looking up important concepts, formulas, and problem-solving hints.

Discovery Exercises for Interactive Physics by Jon Staib   This workbook is designed to be used in conjunction with the Interactive Physics simulations previously described. The workbook consists of a set of exercises in which the student is required to fill in blanks, answer questions, construct graphs, predict results, and perform simple calculations. Each exercise is designed to teach at least one physical principle and/or to develop student's physical intuition. The workbook and templates can be used either as stand-alone tutorials, or in a laboratory setting.

Spreadsheet Templates   The Spreadsheet Template Disk contains spreadsheet files designed to be used with the end-of-chapter problems entitled Spreadsheet Problems. The files have been developed in Lotus 1-2-3 using the WK1 format. These can be used with most spreadsheet programs including all the recent versions of Lotus 1-2-3, Excel for Windows and Macintosh, Quattro Pro, and f(g) scholar. Over 30 templates are provided for the student.

Spreadsheet Investigations in Physics by Lawrence B. Golden and James R. Klein   This workbook with the accompanying disk illustrates how spreadsheets can be used for solving many physics problems and when spreadsheet analysis is useful. The workbook is divided into two parts. The first part consists of spreadsheet tutorials, while the second part is a short introduction to numerical methods. The tutorials include basic spreadsheet techniques emphasizing navigating the spreadsheet, entering data, constructing formulas, and graphing. The numerical methods include differentiation, integration, interpolation, and the solution of differential equations. Many examples and exercises are provided. Step-by-step instructions are given for constructing numerical models of selected physics problems. The exercises and examples used to illustrate the numerical methods are chosen from introductory physics and mathematics. The spreadsheet material is presented using Lotus 1-2-3 Release 2.x features, with specific sections devoted to features of other spreadsheet programs, including recent versions of Lotus 1-2-3 for Windows, Excel for Windows and the Macintosh, Quattro Pro, and f(g) Scholar.

Mathematical Methods for Introductory Physics with Calculus by Ronald C. Davidson, Princeton University   This brief book is designed for students who find themselves unable to keep pace in their physics class because of a lack of familiarity with the necessary mathe-

(D.O.E./Science Source/Photo Researchers)

matical tools. *Mathematical Methods* provides a brief overview of all the various types of mathematical topics that may be needed in an introductory-level physics course through the use of many worked examples and exercises.

So You Want to Take Physics: A Preparation for Scientists and Engineers by Rodney Cole   This text is useful to those students who need additional preparation before or during a course in physics. The book includes typical problems with worked-out solutions, and a review of techniques in mathematics and physics. The friendly, straightforward style makes it easier to understand how mathematics is used in the context of physics.

Practice Problems with Solutions   This collection of more than 500 level-1 problems taken from the third edition of *Physics for Scientists and Engineers* is available with full solutions. These problems can be used for homework assignments or student practice and drill exercises.

Challenging Problems in Physics by Boris Korsunsky   This set of 600 thought-provoking problems is meant to test the student's understanding of basic concepts and help them develop general approaches to solving physics problems.

Life Science Applications for Physics   This supplement, compiled by Jerry Faughn, provides examples, readings, and problems from the biological sciences as they relate to physics. Topics include "Friction in Human Joints," "Physics of the Human Circulatory System," "Physics of the Nervous System," and "Ultrasound and Its Applications." This supplement is useful in those courses having a significant number of pre-med students.

Physics Laboratory Manual by David Loyd   To supplement the learning of basic physical principles while introducing laboratory procedures and equipment, each chapter of the laboratory manual includes a pre-laboratory assignment, objectives, an equipment list, the theory behind the experiment, experimental procedure, calculations, graphs, and questions. In addition, a laboratory report is provided for each experiment so the student can record data, calculations, and experimental results.

Instructor's Ancillaries

Instructor's Manual with Solutions by Steve Van Wyk, Ralph McGrew, Ray Serway, and Louis Cadwell   This manual consists of complete, worked-out solutions to all the problems in the text and answers to even-numbered problems. The solutions to the new problems in the fourth edition are marked so the instructor can identify them. All solutions have been carefully reviewed for accuracy.

Computerized Test Bank by Louis H. Cadwell and Michael Carchidi   Available for the IBM PC and Macintosh computers, this test bank contains over 2300 multiple-choice and open-ended problems and questions, representing every chapter of the text. The test bank enables the instructor to customize tests by rearranging, editing, and adding new questions. The software program prints each answer on a separate grading key.

Printed Test Bank   This test bank is the printed version of the computerized test bank; it contains all of the multiple-choice questions and open-ended problems and questions from the software disk. Answers are also provided.

Interactive Physics Demonstrations by Ray Serway   A set of physics computer simulations that use the Interactive Physics II program is available for use in classroom presentations. These simulations are very useful to show animations of motion, and most are keyed to specific sections or examples in the textbook.

Saunders Physics Videodisc   Contains animations derived from SD2000 software and Interactive Physics II software, video clips demonstrating real-world applications of physics, and still images from the text of *Physics for Scientists and Engineer with Modern Physics*, fourth edition. The still images include most of the line art from the text with enlarged labels for better classroom viewing.

Physics Demonstration Videotape by J. C. Sprott of the University of Wisconsin, Madison   A unique two-hour video-cassette divided into 12 primary topics. Each topic contains between four and nine demonstrations for a total of 70 physics demonstrations.

Selected Solutions Transparency Masters   Selected worked-out solutions are identical to those included in the Student Solutions Manual and Study Guide. These can be used in the classroom when transferred to acetates.

Overhead Transparency Acetates   This collection of transparencies consists of more than 200 full-color figures from the text and features large print for easy viewing in the classroom.

Instructor's Manual to Accompany Challenging Problems for Physics by Boris Korsunsky   This book contains the answers and solutions to all 600 problems that appear in *Challenging Problems for Physics*. All problems are restated for convenience, along with the necessary diagrams.

Instructor's Manual for Physics Laboratory Manual by David Loyd   Each chapter contains a discussion of the experiment, teaching hints, answers to selected questions, and a post-laboratory-quiz with short answer and essay questions. A list of the suppliers of scientific equipment and a summary of the equipment needed for all the laboratory experiments in the manual are also included.

TEACHING OPTIONS

This book is structured in the following sequence of topics: classical mechanics, matter waves, heat and thermodynamics, electricity and magnetism, light waves, optics, relativity, and modern physics. This presentation is a more traditional sequence, with the subject of matter waves presented before electricity and magnetism. Some instructors may prefer to cover this material after completing electricity and magnetism (after Chapter 34). The chapter on relativity was placed at the end of the text because this topic is often treated as an introduction to the era of "modern physics." If time permits, instructors may choose to cover Chapter 39 in Volume II after completing Chapter 14, which concludes the material on Newtonian mechanics.

For those instructors teaching a two-semester sequence, some sections and chapters could be deleted without any loss in continuity. I have labeled these with asterisks (*) in the Table of Contents and in the appropriate sections of the text. For student enrichment, some of these sections or chapters could be given as extra reading assignments.

ACKNOWLEDGMENTS

In preparing the fourth edition of this textbook, I have been guided by the expertise of the many people who reviewed part or all of the manuscript. Robert Bauman was instrumental in the reviewing process, checking the entire text for accuracy and offering numerous suggestions to improve clarity. I appreciate the assistance of Irene Nunes for skillfully editing and refining the language in the text. In addition, I would like to acknowledge the following scholars and express my sincere appreciation for their helpful suggestions, criticisms, and encouragement:

Edward Adelson, Ohio State University
Joel Berlinghieri, The Citadel
Ronald E. Brown, California Polytechnic State University–San Luis Obispo
Lt. Commander Charles Edmonson, U.S. Naval Academy
Phil Fraundorf, University of Missouri–St. Louis
Ken Ganeger, California State University–Dominiquez Hills
Alfonso Diaz-Jimínez, ADJOIN Research Center, Colombia
George Kottowar, Texas A & M
Raymond L. Kozub, Tennessee Technological University
Tom Moon, Montana Tech

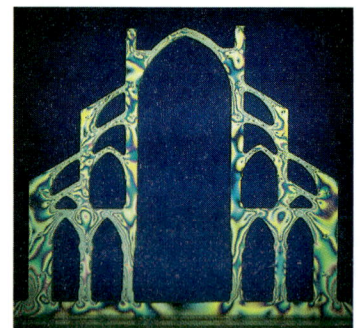

(Peter Aprahamian/Science Photo Library)

Edward Mooney, Youngstown State University
Marvin Payne, Georgia Southern University
Sama'an Salem, California State University–Long Beach
John Sheldon, Florida International University
J. C. Sprott, University of Wisconsin–Madison
Larry Sudduth, Georgia Institute of Technology
David Taylor, Northwestern University
George Williams, University of Utah

I would also like to thank the following professors for their suggestions during the development of the prior editions of this textbook:

Reviewers

George Alexandrakis, University of Miami
Elmer E. Anderson, University of Alabama
Wallace Arthur, Fairleigh Dickinson
Duane Aston, California State University at Sacramento
Stephen Baker, Rice University
Richard Barnes, Iowa State University
Albert A. Bartlett
Stanley Bashkin, University of Arizona
Marvin Blecher, Virginia Polytechnic Institute and State University
Jeffrey J. Braun, University of Evansville
Kenneth Brownstein, University of Maine
William A. Butler, Eastern Illinois University
Louis H. Cadwell, Providence College
Bo Casserberg, University of Minnesota
Ron Canterna, University of Wyoming
Soumya Chakravarti, California State Polytechnic University
C. H. Chan, The University of Alabama in Huntsville
Edward Chang, University of Massachusetts, Amherst
Don Chodrow, James Madison University
Clifton Bob Clark, University of North Carolina at Greensboro
Walter C. Connolly, Appalachian State University
Hans Courant, University of Minnesota
David R. Currot
Lance E. De Long, University of Kentucky
James L. DuBard, Birmingham-Southern College
F. Paul Esposito, University of Cincinnati
Jerry S. Faughn, Eastern Kentucky University
Paul Feldker, Florissant Valley Community College
Joe L. Ferguson, Mississippi State University
R. H. Garstang, University of Colorado at Boulder
James B. Gerhart, University of Washington
John R. Gordon, James Madison University
Clark D. Hamilton, National Bureau of Standards
Mark Heald, Swarthmore College
Herb Helbig, Clarkson University

Howard Herzog, Broome Community College
Larry Hmurcik, University of Bridgeport
Paul Holoday, Henry Ford Community College
Jerome W. Hosken, City College of San Francisco
William Ingham, James Madison University
Mario Iona, University of Denver
Karen L. Johnston, North Carolina State University
Brij M. Khorana, Rose-Hulman Institute of Technology
Larry Kirkpatrick, Montana State University
Carl Kocher, Oregon State University
Robert E. Kribel, Jacksonville State University
Barry Kunz, Michigan Technological University
Douglas A. Kurtze, Clarkson University
Fred Lipschultz, University of Connecticut
Chelcie Liu
Francis A. Liuima, Boston College
Robert Long, Worcester Polytechnic Institute
Roger Ludin, California Polytechnic State University
Nolen G. Massey, University of Texas at Arlington
Howard McAllister
Charles E. McFarland, University of Missouri at Rolla
Ralph V. McGrew, Broome Community College
James Monroe, The Pennsylvania State University, Beaver Campus
Bruce Morgan, U.S. Naval Academy
Clem Moses, Utica College
Curt Moyer, Clarkson University
David Murdock, Tennessee Technological College
A. Wilson Nolle, The University of Texas at Austin
Thomas L. O'Kuma, San Jacinto College North
Fred A. Otter, University of Connecticut
George Parker, North Carolina State University
William F. Parks, University of Missouri, Rolla
Philip B. Peters, Virginia Military Institute
Eric Peterson, Highland Community College
Richard Reimann, Boise State University
Joseph W. Rudmin, James Madison University

(Courtesy Jeanne Maier)

Jill Rugare, DeVry Institute of Technology
C. W. Scherr, University of Texas at Austin
Eric Sheldon, University of Massachusetts–Lowell
John Shelton, College of Lake County
Stan Shepard, The Pennsylvania State University
A. J. Slavin
James H. Smith, University of Illinois at Urbana-Champaign
Richard R. Sommerfield, Foothill College
Kervork Spartalian, University of Vermont
J. C. Sprott
Robert W. Stewart, University of Victoria
James Stith, United States Military Academy
Charles D. Teague, Eastern Kentucky University
Edward W. Thomas, Georgia Institute of Technology

Carl T. Tomizuka, University of Arizona
Herman Trivilino, San Jacinto College North
Som Tyagi, Drexel University
Steve Van Wyk, Chapman College
Joseph Veit, Western Washington University
T. S. Venkataraman, Drexel University
Noboru Wada, Colorado School of Mines
James Walker, Washington State University
Gary Williams, University of California, Los Angeles
George Williams, University of Utah
William W. Wood
Edward Zimmerman, University of Nebraska, Lincoln
Earl Zwicker, Illinois Institute of Technology

I would like to thank Michael Carchidi for coordinating and contributing to the end-of-chapter problems in the fourth edition. I am very grateful to the following individuals for contributing many creative and interesting new problems to the text: Barry Gilbert, Rhode Island College; Boris Korsunsky, Northfield Mound Hermon School; Bo Lou, Ferris State University; and Roger Ludin, California Polytechnic State University–San Luis Obispo.

I appreciate the assistance of Steve Van Wyk and Ralph McGrew for their careful review of all new end-of-chapter problems and for the preparation of the answer section in the text and the Instructor's Manual. I am indebted to my colleague and friend John R. Gordon for his many contributions during the development of this text, for his continued encouragement and support, and for his expertise in preparing the Student Solutions Manual and Study Guide with the assistance of Ralph McGrew and Steve Van Wyk. Linda Miller is to be thanked for assisting in the preparation of the manuscript and for typesetting and proofreading the Student Solutions Manual and Study Guide. My thanks to Michael Rudmin for the contribution of illustrations to the Student Solutions Manual and Study Guide. Thanks to Sue Howard for locating many excellent photographs and to Jim Lehman and the late Henry Leap for providing numerous photographs of physics demonstrations.

My thanks to Larry Golden and James Klein for their development of end-of-chapter spreadsheet problems and the accompanying templates, as well as the supplement, Numerical Analysis: Spreadsheet Investigations in Physics.

The support package is becoming an ever more essential component of a textbook. I would like to thank the following individuals for authoring the ancillaries that accompany this text: Jorge Cossio for thoroughly reviewing and updating the test bank; Jon Staib for developing the Discovery Exercises workbook that accompanies the Interactive Physics simulations and for reviewing and fine tuning many of the simulations; Evelyn Patterson of the U.S. Air Force Academy for her insightful review and improvements to the Interactive Physics simulations; John Minnerly for converting the Interactive Physics files into an IBM Windows format; Bob Blitshtein and the staff at Future Graph, Inc., for their creation of the SD2000 software package to accompany this text; V. Gordon Lind for concepting and authoring the Pocket Guide; David Loyd for preparing the lab manual and accompanying instructor's manual; Boris Korsunsky for preparing a supplement of Challenging Problems for Physics; Jerry Faughn for compiling *Life Science Applications for Physics;* Ron Davidson for authoring *Mathematical Methods for Introductory Physics with Calculus;* and Rodney Cole for preparing the preparatory manual for physics, *So You Want to Take Physics.*

During the development of this textbook, I benefited from valuable discussions and communications with many people including Subash Antani, Gabe Anton, Randall Caton, Don Chodrow, Jerry Faughn, John R. Gordon, Herb Helbig, Lawrence Hmurcik, William Ingham, David Kaup, Len Ketelsen, Alfonso Diaz-Jiménez, Henry Leap, H. Kent Moore, Charles McFarland, Frank Moore, Clem Moses, Curt Moyer, William Parks, Dorn Peterson, Joe Rudmin, Joe Scaturro, Alex Serway, John Serway, Georgio Vianson, and Harold Zimmerman. Special recognition is due to my mentor and friend, Sam Marshall, a gifted teacher and scientist who helped me sharpen my writing skills while I was a graduate student.

Special thanks and recognition go to the professional staff at Saunders College Publishing for their fine work during the development and production of this text, especially Laura Maier, Developmental Editor; Sally Kusch, Senior Project Editor; Charlene Squibb, Production Manager; and Carol Bleistine, Manager of Art and Design. Thanks also to Tim Frelick, VP/Director of Editorial, Design, and Production, and to Margie Waldron, VP/Marketing, for their continued support of this project. I thank John Vondeling, Vice President/Publisher, for his great enthusiasm for the project, his friendship, and his confidence in me as an author. I am most appreciative of the intelligent copyediting by Charlotte Nelson, the excellent artwork by Rolin Graphics, Inc., and the attractive design by Rebecca Lemna.

A special note of appreciation goes to the hundreds of students at Clarkson University who used the first edition of this text in manuscript form during its development. I also wish to thank the many users of the second and third editions who submitted suggestions and pointed out errors. With the help of such cooperative efforts, I hope to have achieved my main objective; that is, to provide an effective textbook for the student.

And last, I thank my wonderful family for continuing to support and understand my commitment to physics education.

Raymond A. Serway

James Madison University

August 1995

To The Student

I feel it is appropriate to offer some words of advice which should be of benefit to you, the student. Before doing so, I will assume that you have read the preface, which describes the various features of the text that will help you through the course.

HOW TO STUDY

Very often instructors are asked "How should I study physics and prepare for examinations?" There is no simple answer to this question, but I would like to offer some suggestions based on my own experiences in learning and teaching over the years.

First and foremost, maintain a positive attitude towards the subject matter, keeping in mind that physics is the most fundamental of all natural sciences. Other science courses that follow will use the same physical principles, so it is important that you understand and be able to apply the various concepts and theories discussed in the text.

CONCEPTS AND PRINCIPLES

It is essential that you understand the basic concepts and principles before attempting to solve assigned problems. This is best accomplished through a careful reading of the textbook before attending your lecture on that material. In the process, it is useful to jot down certain points which are not clear to you. Take careful notes in class, and then ask questions pertaining to those ideas that require clarification. Keep in mind that few people are able to absorb the full meaning of scientific material after one reading. Several readings of the text and notes may be necessary. Your lectures and laboratory work should supplement the text and clarify some of the more difficult material. You should reduce memorization of material to a minimum. Memorizing passages from a text, equations, and derivations does not necessarily mean you understand the material. Your understanding of the material will be enhanced through a combination of efficient study habits, discussions with other students and instructors, and your ability to solve the problems in the text. Ask questions whenever you feel it is necessary.

STUDY SCHEDULE

It is important to set up a regular study schedule, preferably on a daily basis. Make sure to read the syllabus for the course and adhere to the schedule set by your instructor. The lectures will be much more meaningful if you read the corresponding textual material before attending the lecture. As a general rule, you should devote about two hours of study time for every hour in class. If you are having trouble with the course, seek the advice of the instructor or students who have taken the course. You may find it necessary to seek further instruction from experienced students. Very often, instructors will offer review sessions in addition to regular class periods. It is important that you avoid the practice of delaying study until a day or two before an exam. More often than not, this will lead to disastrous results. Rather than staying up for an all-night session, it is better to review the basic concepts and equations briefly, followed by a good night's rest. If you feel in need of additional help in understanding the concepts, in preparing for exams, or in problem-solving, we suggest that you acquire a copy of the Student Solutions Manual and Study Guide which accompanies the text and should be available at your college bookstore.

USE THE FEATURES

You should make full use of the various features of the text discussed in the preface. For example, marginal notes are useful for locating and describing important equations and concepts, while important statements and definitions are highlighted in color. Many useful tables are contained in appendices, but most are incorporated in the text where they are used most often. Appendix B is a convenient review of mathematical techniques.

Answers to odd-numbered problems are given at the end of the text, and answers to end-of-chapter questions are provided in the study guide. Exercises (with answers), which follow some worked examples, represent extensions of those examples, and in most cases you are expected to perform a simple calculation. Their purpose is to test your problem-solving skills as you read through the text.

Problem-Solving Strategies and Hints are included in selected chapters throughout the text to give you additional information to help you solve problems. An overview of the entire text is given in the table of contents, while the index will enable you to locate specific material quickly. Footnotes are sometimes used to supplement the discussion or to cite other references on the subject. Most chapters include several problems that make use of spreadsheets. These are intended for those courses that place some emphasis on numerical methods. In some cases, spreadsheet templates are provided, while others require the modification of these templates or the creation of new templates.

After reading a chapter, you should be able to define any new quantities introduced in that chapter and to discuss the principles and assumptions that were used to arrive at certain key relations. The chapter summaries and the review sections of the study guide should help you in this regard. In some cases, it will be necessary to refer to the index of the text to locate certain topics. You should be able to correctly associate with each physical quantity a symbol used to represent that quantity and the unit in which the quantity is specified. Furthermore, you should be able to express each important relation in a concise and accurate prose statement.

(Ken Kay/Fundamental Photographs)

PROBLEM SOLVING

R. P. Feynman, Nobel laureate in physics, once said, "You do not know anything until you have practiced." In keeping with this statement, I strongly advise that you develop the skills necessary to solve a wide range of problems. Your ability to solve problems will be one of the main tests of your knowledge of physics, and therefore you should try to solve as many problems as possible. It is essential that you understand basic concepts and principles before attempting to solve problems. It is good practice to try to find alternate solutions to the same problem. For example, problems in mechanics can be solved using Newton's laws, but very often an alternative method using energy considerations is more direct. You should not deceive yourself into thinking you understand the problem after seeing its solution in class. You must be able to solve the problem and similar problems on your own.

The method of solving problems should be carefully planned. A systematic plan is especially important when a problem involves several concepts. First, read the problem several times until you are confident you understand what is being asked. Look for any key words that will help you interpret the problem, and perhaps allow you to make certain assumptions. Your ability to interpret the question properly is an integral part of problem solving. You should acquire the habit of writing down the information given in a problem and deciding what quantities need to be found. You might want to construct a table listing quantities given and quantities to be found. This procedure is sometimes used in the worked examples of the text. After you have decided on the method you feel is appropriate for the situation, proceed with your solution. General problem-solving strategies of this type are included in the text and are highlighted by a light color screen.

I often find that students fail to recognize the limitations of certain formulas or physical laws in a particular situation. It is very important that you understand and remember the assumptions which underlie a particular theory or formalism. For example, certain equations in kinematics apply only to a particle moving with constant acceleration. These equations are not valid for situations in which the acceleration is not constant, such as the motion of an object connected to a spring or the motion of an object through a fluid.

General Problem-Solving Strategy

Most courses in general physics require the student to learn the skills of problem solving, and examinations are largely composed of problems that test such skills. This brief section describes some useful ideas which will enable you to increase your accuracy in solving problems, enhance your understanding of physical concepts, eliminate initial panic or lack of direction in approaching a problem, and organize your work. One way to help accomplish these goals is to adopt a problem-solving strategy. Many chapters in this text will include a section labeled "Problem-Solving Strategies and Hints" which should help you through the "rough spots."

In developing problem-solving strategies, five basic steps are commonly used.

- Draw a suitable diagram with appropriate labels and coordinate axes if needed.
- As you examine what is being asked in the problem, identify the basic physical principle (or principles) that are involved, listing the knowns and unknowns.
- Select a basic relationship or derive an equation that can be used to find the unknown, and solve the equation for the unknown symbolically.
- Substitute the given values along with the appropriate units into the equation.
- Obtain a numerical value for the unknown. The problem is verified and receives a check mark if the following questions can be answered properly: Do the units match? Is the answer reasonable? Is the plus or minus sign proper or meaningful?

One of the purposes of this strategy is to promote accuracy. Properly drawn diagrams can eliminate many sign errors. Diagrams also help to isolate the physical principles of the problem. Symbolic solutions and carefully labeled knowns and unknowns will help eliminate other careless errors. The use of symbolic solutions should help you think in terms of the physics of the problem. A check of units at the end of the problem can indicate a possible algebraic error. The physical layout and organization of your problem will make

---

EXAMPLE

A person driving in a car at a speed of 20 m/s applies the brakes and stops in a distance of 100 m. What was the acceleration of the car?

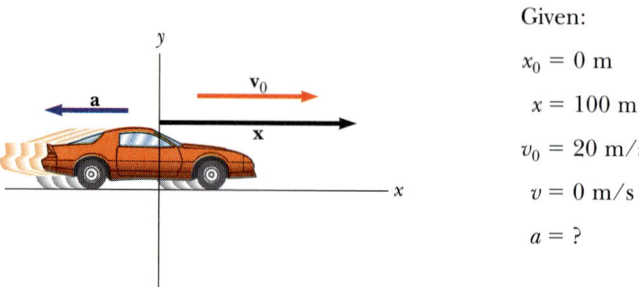

Given:

$x_0 = 0$ m

$x = 100$ m

$v_0 = 20$ m/s

$v = 0$ m/s

$a = ?$

$v^2 = v_0{}^2 + 2a(x - x_0)$

$v^2 = v_0{}^2 + 2a(x - x_0)$

$a = \dfrac{v^2 - v_0{}^2}{2(x - x_0)}$

$a = \dfrac{(0 \text{ m/s})^2 - (20 \text{ m/s})^2}{2(100 \text{ m})} = -2 \text{ m/s}^2$

$\dfrac{\text{m}^2/\text{s}^2}{\text{m}} = \dfrac{\text{m}}{\text{s}^2}$

(Guy Sauvage/Photo Researchers)

the final product more understandable and easier to follow. Once you have developed an organized system for examining problems and extracting relevant information, you will become a more confident problem solver.

EXPERIMENTS

Physics is a science based upon experimental observations. In view of this fact, I recommend that you try to supplement the text through various types of "hands-on" experiments, either at home or in the laboratory. These can be used to test ideas and models discussed in class or in the text. For example, the common "Slinky" toy is excellent for studying traveling waves; a ball swinging on the end of a long string can be used to investigate pendulum motion; various masses attached to the end of a vertical spring or rubber band can be used to determine their elastic nature; an old pair of Polaroid sunglasses and some discarded lenses and a magnifying glass are the components of various experiments in optics; you can get an approximate measure of the free-fall acceleration by dropping a ball from a known height and measuring the time of its fall with a stopwatch. The list is endless. When physical models are not available, be imaginative and try to develop models of your own.

PEDAGOGICAL USE OF COLOR

The various colors that you will see in the illustrations of this text are used to improve clarity and understanding. Many figures with three-dimensional representations are air-brushed in various colors to make them as realistic as possible.

Most graphs are presented with curves plotted in either rust or blue, and coordinate axes are in black. Several colors are used in those graphs where many physical quantities are plotted simultaneously, or in those cases where different processes may be occurring and need to be distinguished.

Motional shading effects have been incorporated in many figures to remind the readers that they are dealing with a dynamic system rather than a static system. These figures will appear to be somewhat like a "multiflash" photograph of a moving system, with faint images of the "past history" of the system's path. In some figures, a broad, colored arrow is used to indicate the direction of motion of the system.

Color coding has been used in various parts of the book to identify specific physical quantities. The chart on p. xxiii should be a good reference when examining illustrations.

AN INVITATION TO PHYSICS

It is my sincere hope that you will find physics an exciting and enjoyable experience, and that you will profit from this experience, regardless of your chosen profession. Welcome to the exciting world of physics.

> The scientist does not study nature because it is useful; he studies it because he delights in it, and he delights in it because it is beautiful. If nature were not beautiful, it would not be worth knowing, and if nature were not worth knowing, life would not be worth living.
>
> *Henri Poincaré*

PEDAGOGICAL COLOR CHART

Part I (Chapters 1–15) : Mechanics

Displacement and
position vectors

Velocity vectors (**v**)
Velocity component vectors

Force vectors (**F**)
Force component vectors

Acceleration vectors (**a**)
Acceleration component vectors

Torque (τ) and
angular momentum
(**L**) vectors

Linear or rotational
motion directions

Springs

Pulleys

Part IV (Chapters 23–34) : Electricity and Magnetism

Electric fields

Magnetic fields

Positive charges

Negative charges

Resistors

Batteries and other
dc power supplies

Switches

Capacitors

Inductors (coils)

Voltmeters

Ammeters

Galvanometers

ac generators

Ground symbol

Part V (Chapters 35–38) : Light and Optics

Light rays

Lenses and prisms

Mirrors

Objects

Images

Contents Overview

Part I   Mechanics   1

1. Physics and Measurement   3
2. Motion in One Dimension   23
3. Vectors   53
4. Motion in Two Dimensions   71
5. The Laws of Motion   104
6. Circular Motion and Other Applications of Newton's Laws   144
7. Work and Energy   171
8. Potential Energy and Conservation of Energy   202
9. Linear Momentum and Collisions   235
10. Rotation of a Rigid Object About a Fixed Axis   276
11. Rolling Motion, Angular Momentum, and Torque   306
12. Static Equilibrium and Elasticity   337
13. Oscillatory Motion   360
14. The Law of Gravity   391
15. Fluid Mechanics   421

Part II   Mechanical Waves   453

16. Wave Motion   454
17. Sound Waves   479
18. Superposition and Standing Waves   499

Part III   Thermodynamics   529

19. Temperature   531
20. Heat and the First Law of Thermodynamics   551
21. The Kinetic Theory of Gases   586
22. Heat Engines, Entropy, and the Second Law of Thermodynamics   615

Part IV   Electricity and Magnetism   647

23. Electric Fields   649
24. Gauss's Law   684
25. Electric Potential   707
26. Capacitance and Dielectrics   741
27. Current and Resistance   772
28. Direct Current Circuits   797
29. Magnetic Fields   832
30. Sources of the Magnetic Field   864
31. Faraday's Law   905
32. Inductance   939
33. Alternating Current Circuits   964
34. Electromagnetic Waves   994

(Courtesy of Central Scientific Co.)

Part V   Light and Optics   1020

35. The Nature of Light and the Laws of Geometric Optics   1023
36. Geometric Optics   1052
37. Interference of Light Waves   1091
38. Diffraction and Polarization   1117

Part VI   Modern Physics   1146

39. Relativity   1147
40. Introduction to Quantum Physics   1189
41. Quantum Mechanics   1216
42. Atomic Physics   1252
43. Molecules and Solids   1286
44. Superconductivity   1317
45. Nuclear Structure   1346
46. Fission and Fusion   1382
47. Particle Physics and Cosmology   1413

Appendices   A.1
Answers to Odd-Numbered Problems   A.43
Index   I.1

Contents

Part I Mechanics 1

1 Physics and Measurement 3

  1.1 Standards of Length, Mass, and Time 4
  1.2 The Building Blocks of Matter 8
  1.3 Density and Atomic Mass 9
  1.4 Dimensional Analysis 10
  1.5 Conversion of Units 13
  1.6 Order-of-Magnitude Calculations 13
  1.7 Significant Figures 15
  1.8 Mathematical Notation 16
  Summary 17

2 Motion in One Dimension 23

  2.1 Displacement, Velocity, and Speed 24
  2.2 Instantaneous Velocity and Speed 26
  2.3 Acceleration 28
  2.4 One-Dimensional Motion with Constant Acceleration 32
  2.5 Freely Falling Objects 36
  * 2.6 Kinematic Equations Derived from Calculus 41
  Summary 43

3 Vectors 53

  3.1 Coordinate Systems and Frames of Reference 54
  3.2 Vector and Scalar Quantities 55
  3.3 Some Properties of Vectors 56
  3.4 Components of a Vector and Unit Vectors 59
  Summary 64

4 Motion in Two Dimensions 71

  4.1 The Displacement, Velocity, and Acceleration Vectors 71
  4.2 Two-Dimensional Motion with Constant Acceleration 74
  4.3 Projectile Motion 77
  4.4 Uniform Circular Motion 85
  4.5 Tangential and Radial Acceleration 87
  4.6 Relative Velocity and Relative Acceleration 89
  * 4.7 Relative Motion at High Speeds 92
  Summary 93

5 The Laws of Motion 104

  5.1 The Concept of Force 105
  5.2 Newton's First Law and Inertial Frames 107
  5.3 Inertial Mass 110
  5.4 Newton's Second Law 111
  5.5 Weight 113
  5.6 Newton's Third Law 114
  5.7 Some Applications of Newton's Law 116
  5.8 Forces of Friction 124
  Summary 129

6 Circular Motion and Other Applications of Newton's Law 144

  6.1 Newton's Second Law Applied to Uniform Circular Motion 144
  6.2 Nonuniform Circular Motion 150
  * 6.3 Motion in Accelerated Frames 151
  * 6.4 Motion in the Presence of Resistive Forces 154
  * 6.5 Numerical Modeling in Particle Dynamics 158
  6.6 The Fundamental Forces of Nature 160
  Summary 162

7 Work and Energy 171

  7.1 Work Done by a Constant Force 172
  7.2 The Scalar Product of Two Vectors 174
  7.3 Work Done by a Varying Force 176
  7.4 Kinetic Energy and the Work-Energy Theorem 180
  7.5 Power 186
  * 7.6 Energy and the Automobile 188
  * 7.7 Kinetic Energy at High Speeds 191
  Summary 192

8 Potential Energy and Conservation of Energy 202

  8.1 Potential Energy 202
  8.2 Conservative and Nonconservative Forces 204
  8.3 Conservative Forces and Potential Energy 206
  8.4 Conservation of Energy 207
  8.5 Changes in Mechanical Energy When Nonconservative Forces Are Present 210
  8.6 Relationship Between Conservative Forces and Potential Energy 217

(David Madison, Tony Stone Images)

* 8.7   Energy Diagrams and the Equilibrium
        of a System   218
  8.8   Conservation of Energy in General   219
* 8.9   Mass-Energy Equivalence   220
* 8.10 Quantization of Energy   220
  Summary   222

9   **Linear Momentum and Collisions   235**
   9.1   Linear Momentum and its Conservation   236
   9.2   Impulse and Momentum   239
   9.3   Collisions   242
   9.4   Elastic and Inelastic Collisions in
         One Dimension   243
   9.5   Two-Dimensional Collisions   248
   9.6   The Center of Mass   251
   9.7   Motion of a System of Particles   255
* 9.8   Rocket Propulsion   259
   Summary   261

10   **Rotation of a Rigid Object About a
     Fixed Axis   276**
   10.1 Angular Velocity and Angular Acceleration   277
   10.2 Rotational Kinematics: Rotational Motion with
         Constant Angular Acceleration   278
   10.3 Relationships Between Angular and
         Linear Quantities   279
   10.4 Rotational Energy   281

   10.5 Calculation of Moments of Inertia   283
   10.6 Torque   287
   10.7 Relationship Between Torque and
         Angular Acceleration   289
   10.8 Work, Power, and Energy in
         Rotational Motion   293
   Summary   296

11   **Rolling Motion, Angular Momentum,
     and Torque   306**
   11.1 Rolling Motion of a Rigid Body   307
   11.2 The Vector Product and Torque   309
   11.3 Angular Momentum of a Particle   312
   11.4 Rotation of a Rigid Body About a Fixed
         Axis   315
   11.5 Conservation of Angular Momentum   317
 *11.6 The Motion of Gyroscopes and Tops   321
 *11.7 Angular Momentum as a Fundamental
         Quantity   324
   Summary   325

12   **Static Equilibrium and Elasticity   337**
   12.1 The Conditions of Equilibrium of a
         Rigid Object   338
   12.2 More on the Center of Gravity   340
   12.3 Examples of Rigid Objects in
         Static Equilibrium   341
   12.4 Elastic Properties of Solids   345
   Summary   349

13   **Oscillatory Motion   360**
   13.1 Simple Harmonic Motion   361
   13.2 Mass Attached to a Spring   364
   13.3 Energy of the Simple Harmonic Oscillator   368
   13.4 The Pendulum   371
 *13.5 Comparing Simple Harmonic Motion with
         Uniform Circular Motion   375
 *13.6 Damped Oscillations   377
 *13.7 Forced Oscillations   379
   Summary   380

14   **The Law of Gravity   391**
   14.1 Newton's Law of Gravity   392
   14.2 Measurement of the Gravitational Constant   393
   14.3 Weight and Gravitational Force   394
   14.4 Kepler's Laws   395
   14.5 The Law of Gravity and the
         Motion of Planets   396
   14.6 The Gravitational Field   400
   14.7 Gravitational Potential Energy   401

14.8 Energy Considerations in Planetary and Satellite Motion 404
*14.9 The Gravitational Force Between an Extended Object and a Particle 407
*14.10 Gravitational Force Between a Particle and a Spherical Mass 408
  Summary 410

15 Fluid Mechanics 421

15.1 Pressure 422
15.2 Variation of Pressure with Depth 424
15.3 Pressure Measurements 427
15.4 Buoyant Forces and Archimedes' Principle 427
15.5 Fluid Dynamics 431
15.6 Streamlines and the Equation of Continuity 432
15.7 Bernoulli's Equation 433
*15.8 Other Applications of Bernoulli's Equation 436
*15.9 Energy from the Wind 437
*15.10 Viscosity 439
  Summary 441

Part II Mechanical Waves 453

16 Wave Motion 454

16.1 Introduction 455
16.2 Types of Waves 456
16.3 One-Dimensional Traveling Waves 457
16.4 Superposition and Interference of Waves 459
16.5 The Speed of Waves on Strings 462
16.6 Reflection and Transmission of Waves 464
16.7 Sinusoidal Waves 465
16.8 Energy Transmitted by Sinusoidal Waves on Strings 469
*16.9 The Linear Wave Equation 470
  Summary 472

17 Sound Waves 479

17.1 Speed of Sound Waves 480
17.2 Periodic Sound Waves 481
17.3 Intensity of Periodic Sound Waves 482
17.4 Spherical and Plane Waves 484
*17.5 The Doppler Effect 487
  Summary 491

18 Superposition and Standing Waves 499

18.1 Superposition and Interference of Sinusoidal Waves 500
18.2 Standing Waves 502
18.3 Standing Waves in a String Fixed at Both Ends 505
18.4 Resonance 508
18.5 Standing Waves in Air Columns 510
*18.6 Standing Waves in Rods and Plates 512

*18.7 Beats: Interference in Time 513
*18.8 Complex Waves 516
  Summary 518

Part III Thermodynamics 529

19 Temperature 531

19.1 Temperature and the Zeroth Law of Thermodynamics 532
19.2 Thermometers and Temperature Scales 533
19.3 The Constant-Volume Gas Thermometer and the Kelvin Scale 534
19.4 Thermal Expansion of Solids and Liquids 536
19.5 Macroscopic Description of an Ideal Gas 541
  Summary 543

20 Heat and the First Law of Thermodynamics 551

20.1 Heat and Thermal Energy 552
20.2 Heat Capacity and Specific Heat 554
20.3 Latent Heat 557
20.4 Work and Heat in Thermodynamic Processes 561
20.5 The First Law of Thermodynamics 563
20.6 Some Applications of the First Law of Thermodynamics 565
20.7 Heat Transfer 568
  Summary 575

21 The Kinetic Theory of Gases 586

21.1 Molecular Model of an Ideal Gas 587
21.2 Specific Heat of an Ideal Gas 591
21.3 Adiabatic Processes for an Ideal Gas 594
21.4 The Equipartition of Energy 596
*21.5 The Boltzmann Distribution Law 599
*21.6 Distribution of Molecular Speeds 602
*21.7 Mean Free Path 604
*21.8 Van der Waals' Equation of State 606
  Summary 607

22 Heat Engines, Entropy, and the Second Law of Thermodynamics 615

22.1 Heat Engines and the Second Law of Thermodynamics 616
22.2 Reversible and Irreversible Processes 619
22.3 The Carnot Engine 620
22.4 The Absolute Temperature Scale 624
22.5 The Gasoline Engine 625
22.6 Heat Pumps and Refrigerators 627
22.7 Entropy 628
22.8 Entropy Changes in Irreversible Processes 631
*22.9 Entropy on a Microscopic Scale 635
  Summary 637

(E. R. Degginger/H. Armstrong Roberts)

Part IV   Electricity and Magnetism   647

23   Electric Fields   649

23.1   Properties of Electric Charges   649
23.2   Insulators and Conductors   651
23.3   Coulomb's Law   653
23.4   The Electric Field   658
23.5   Electric Field of a Continuous Charge Distribution   661
23.6   Electric Field Lines   665
23.7   Motion of Charged Particles in a Uniform Electric Field   667
*23.8   The Oscilloscope   669
Summary   671

24   Gauss's Law   684

24.1   Electric Flux   684
24.2   Gauss's Law   687
24.3   Application of Gauss's Law to Charged Insulators   690
24.4   Conductors in Electrostatic Equilibrium   693
*24.5   Experimental Proof of Gauss's Law and Coulomb's Law   696
*24.6   Derivation of Gauss's Law   696
Summary   697

25   Electric Potential   707

25.1   Potential Difference and Electric Potential   708
25.2   Potential Differences in a Uniform Electric Field   709
25.3   Electric Potential and Potential Energy Due to Point Charges   712
25.4   Obtaining $\mathbf{E}$ from the Electric Potential   715
25.5   Electric Potential Due to Continuous Charge Distributions   717
25.6   Potential of a Charged Conductor   720
*25.7   The Millikan Oil Drop Experiment   724
*25.8   Applications of Electrostatics   725
Summary   728

26   Capacitance and Dielectrics   741

26.1   Definition of Capacitance   741
26.2   Calculation of Capacitance   742
26.3   Combinations of Capacitors   745
26.4   Energy Stored in a Charged Capacitor   749
26.5   Capacitors with Dielectrics   751
*26.6   Electric Dipole in an External Electric Field   756
*26.7   An Atomic Description of Dielectrics   757
Summary   760

27   Current and Resistance   772

27.1   Electric Current   773
27.2   Resistance and Ohm's Law   775
27.3   Resistance and Temperature   780
27.4   Superconductors   782
27.5   A Model for Electrical Conduction   783
27.6   Electrical Energy and Power   786
Summary   789

28   Direct Current Circuits   797

28.1   Electromotive Force   797
28.2   Resistors in Series and in Parallel   799
28.3   Kirchhoff's Rules   804
28.4   *RC* Circuits   808
*28.5   Electrical Instruments   813
*28.6   Household Wiring and Electrical Safety   817
Summary   818

29   Magnetic Fields   832

29.1   The Magnetic Field   833
29.2   Magnetic Force on a Current-Carrying Conductor   837
29.3   Torque on a Current Loop in a Uniform Magnetic Field   841
29.4   Motion of a Charged Particle in a Magnetic Field   843
*29.5   Applications of the Motion of Charged Particles in a Magnetic Field   847

*29.6   The Hall Effect   851
*29.7   The Quantum Hall Effect   853
Summary   853

30   Sources of the Magnetic Field   864

30.1   The Biot-Savart Law   865
30.2   The Magnetic Force Between Two Parallel Conductors   869
30.3   Ampère's Law   870
30.4   The Magnetic Field of a Solenoid   875
*30.5   The Magnetic Field Along the Axis of a Solenoid   876
30.6   Magnetic Flux   877
30.7   Gauss's Law in Magnetism   878
30.8   Displacement Current and the Generalized Ampère's Law   879
*30.9   Magnetism in Matter   881
*30.10   Magnetic Field of the Earth   888
Summary   890

31   Faraday's Law   905

31.1   Faraday's Law of Induction   906
31.2   Motional emf   911
31.3   Lenz's Law   914
31.4   Induced emfs and Electric Fields   917
*31.5   Generators and Motors   919
*31.6   Eddy Currents   922
31.7   Maxwell's Wonderful Equations   923
Summary   925

32   Inductance   939

32.1   Self-Inductance   940
32.2   *RL* Circuits   941
32.3   Energy in a Magnetic Field   944
*32.4   Mutual Inductance   947
32.5   Oscillations in an *LC* Circuit   948
*32.6   The *RLC* Circuit   952
Summary   954

33   Alternating Current Circuits   964

33.1   ac Sources and Phasors   965
33.2   Resistors in an ac Circuit   965
33.3   Inductors in an ac Circuit   967
33.4   Capacitors in an ac Circuit   969
33.5   The *RLC* Series Circuit   971
33.6   Power in an ac Circuit   975
33.7   Resonance in a Series *RLC* Circuit   977
*33.8   Filter Circuits   979
*33.9   The Transformer and Power Transmission   981
Summary   983

34   Electromagnetic Waves   994

34.1   Maxwell's Equations and Hertz's Discoveries   995
34.2   Plane Electromagnetic Waves   997
34.3   Energy Carried by Electromagnetic Waves   1002
34.4   Momentum and Radiation Pressure   1004
*34.5   Radiation from an Infinite Current Sheet   1006
*34.6   The Production of Electromagnetic Waves by an Antenna   1008
34.7   The Spectrum of Electromagnetic Waves   1011
Summary   1013

Part V   Light and Optics   1020

35   The Nature of Light and the Laws of Geometric Optics   1023

35.1   The Nature of Light   1023
35.2   Measurements of the Speed of Light   1025
35.3   The Ray Approximation in Geometric Optics   1026
35.4   Reflection and Refraction   1027
*35.5   Dispersion and Prisms   1033
35.6   Huygens' Principle   1036
35.7   Total Internal Reflection   1039
*35.8   Fermat's Principle   1042
Summary   1042

36   Geometric Optics   1052

36.1   Images Formed by Flat Mirrors   1052
36.2   Images Formed by Spherical Mirrors   1055
36.3   Images Formed by Refraction   1061
36.4   Thin Lenses   1064
*36.5   Lens Aberrations   1070
*36.6   The Camera   1072
*36.7   The Eye   1073
*36.8   The Simple Magnifier   1077
*36.9   The Compound Microscope   1078
*36.10   The Telescope   1080
Summary   1082

37   Interference of Light Waves   1091

37.1   Conditions for Interference   1091
37.2   Young's Double-Slit Experiment   1092
37.3   Intensity Distribution of the Double-Slit Interference Pattern   1096
37.4   Phasor Addition of Waves   1098
37.5   Change of Phase Due to Reflection   1101
37.6   Interference in Thin Films   1103
*37.7   The Michelson Interferometer   1107
Summary   1108

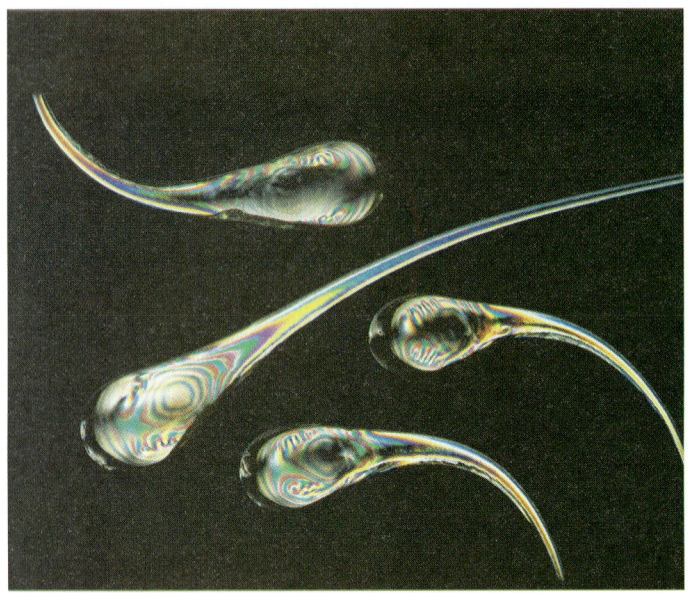

(*James L. Amos/Peter Arnold, Inc.*)

38 Diffraction and Polarization 1117

38.1   Introduction to Diffraction 1117
38.2   Single-Slit Diffraction 1119
38.3   Resolution of Single-Slit and Circular Apertures 1124
38.4   The Diffraction Grating 1127
*38.5 Diffraction of X-Rays by Crystals 1130
38.6   Polarization of Light Waves 1132
Summary 1138

Part VI Modern Physics 1146

39 Relativity 1148

39.1   The Principle of Newtonian Relativity 1149
39.2   The Michelson-Morley Experiment 1154
39.3   Einstein's Principle of Relativity 1156
39.4   Consequences of Special Relativity 1157
39.5   The Lorentz Transformation Equations 1167
39.6   Relativistic Momentum and the Relativistic Form of Newton's Laws 1171
39.7   Relativistic Energy 1173
39.8   Equivalence of Mass and Energy 1176
39.9   Relativity and Electromagnetism 1178
*39.10 General Relativity 1179
Summary 1182

40 Introduction to Quantum Physics 1189

40.1   Blackbody Radiation and Planck's Hypothesis 1190
40.2   The Photoelectric Effect 1194
40.3   Applications of the Photoelectric Effect 1196
40.4   The Compton Effect 1197
40.5   Atomic Spectra 1200
40.6   Bohr's Quantum Model of the Atom 1202
Summary 1208

41 Quantum Mechanics 1216

41.1   Photons and Electromagnetic Waves 1216
41.2   The Wave Properties of Particles 1218
41.3   The Double-Slit Experiment Revisited 1221
41.4   The Uncertainty Principle 1224
41.5   Introduction to Quantum Mechanics 1228
41.6   A Particle in a Box 1230
41.7   The Schrödinger Equation 1234
*41.8 A Particle in a Well of Finite Height 1236
*41.9 Tunneling Through a Barrier 1238
*41.10 The Scanning Tunneling Microscope 1240
*41.11 The Simple Harmonic Oscillator 1241
Summary 1243

42 Atomic Physics 1252

42.1   Early Models of the Atom 1253
42.2   The Hydrogen Atom Revisited 1255
42.3   The Spin Magnetic Quantum Number 1256
42.4   The Wave Functions for Hydrogen 1257
42.5   The "Other" Quantum Numbers 1261
42.6   The Exclusion Principle and the Periodic Table 1266

(*Philippe Plailly/SPL/Photo Researchers*)

42.7 Atomic Spectra: Visible and X-Ray 1272

42.8 Atomic Transitions 1275

*42.9 Lasers and Holography 1276

Summary 1280

43 Molecules and Solids 1286

43.1 Molecular Bonds 1287

43.2 The Energy and Spectra of Molecules 1290

43.3 Bonding in Solids 1296

43.4 Band Theory of Solids 1300

43.5 Free-Electron Theory of Metals 1301

43.6 Electrical Conduction in Metals, Insulators, and Semiconductors 1304

*43.7 Semiconductor Devices 1307

Summary 1311

44 Superconductivity 1317

44.1 Brief Historical Review 1318

44.2 Some Properties of Type I Superconductors 1319

44.3 Type II Superconductors 1324

44.4 Other Properties of Superconductors 1326

44.5 Electronic Specific Heat 1327

44.6 The BCS Theory 1328

44.7 Energy Gap Measurements 1332

44.8 Josephson Tunneling 1333

44.9 High-Temperature Superconductivity 1336

*44.10 Applications 1339

Summary 1340

45 Nuclear Structure 1346

45.1 Some Properties of Nuclei 1347

45.2 Nuclear Magnetic Resonance and MRI 1352

45.3 Binding Energy and Nuclear Forces 1354

45.4 Nuclear Models 1357

45.5 Radioactivity 1359

45.6 The Decay Processes 1363

45.7 Natural Radioactivity 1370

45.8 Nuclear Reactions 1370

Summary 1372

46 Fission and Fusion 1382

46.1 Interactions Involving Neutrons 1382

46.2 Nuclear Fission 1383

46.3 Nuclear Reactors 1386

46.4 Nuclear Fusion 1389

*46.5 Recent Fusion Energy Developments 1397

*46.6 Radiation Damage in Matter 1400

*46.7 Radiation Detectors 1402

*46.8 Uses of Radiation 1406

Summary 1408

47 Particle Physics and Cosmology 1413

47.1 The Fundamental Forces in Nature 1414

47.2 Positrons and Other Antiparticles 1415

47.3 Mesons and the Beginning of Particle Physics 1416

47.4 Classification of Particles 1420

47.5 Conservation Laws 1420

47.6 Strange Particles and Strangeness 1423

47.7 The Eightfold Way 1424

47.8 Quarks—Finally 1425

47.9 The Standard Model 1429

47.10 The Cosmic Connection 1432

47.11 Problems and Perspectives 1437

Summary 1438

Appendix A Tables

Table A.1 Conversion Factors A.1

Table A.2 Symbols, Dimensions, and Units of Physical Quantities A.3

Table A.3 Table of Selected Atomic Masses A.4

Appendix B Mathematics Review A.14

B.1 Scientific Notation A.14

B.2 Algebra A.15

B.3 Geometry A.20

B.4 Trigonometry A.22

B.5 Series Expansions A.24

B.6 Differential Calculus A.24

B.7 Integral Calculus A.27

Appendix C The Periodic Table A.32

Appendix D SI Units A.34

Appendix E Nobel Prize Winners A.35

Appendix F Spreadsheet Problems A.40

Answers to Odd-Numbered Problems A.43

Index I.1

(Paul Hanny/Gamma Liaison)

Stonehenge is a circle of stone built by the ancient Britons on Salisbury Plain, England. Its orientation marks the seasonal rising and setting points of the Sun. For example, on the summer solstice, an observer at the center of the stone circle sees the Sun rising directly over a marker stone. (©John Serafin, Peter Arnold, Inc.)

Mechanics

Philosophy is written in that great book which ever lies before our gaze — I mean the Universe — but we cannot understand if we do not first learn the language and grasp the symbols in which it is written. The book is written in the mathematical language, and the symbols are triangles, circles, and other geometrical figures, without the help of which it is impossible to conceive a single word of it, and without which one wanders in vain through a dark labyrinth.

GALILEO GALILEI

Physics, the most fundamental physical science, is concerned with the basic principles of the Universe. It is the foundation upon which the other physical sciences—astronomy, chemistry, and geology—are based. The beauty of physics lies in the simplicity of the fundamental physical theories and in the manner in which just a small number of fundamental concepts, equations, and assumptions can alter and expand our view of the world around us.

The myriad physical phenomena in our world are a part of one or more of the following five areas of physics:

1. Classical mechanics, which is concerned with the motion of objects moving at speeds that are low compared to the speed of light
2. Relativity, which is a theory describing objects moving at any speed, even those whose speeds approach the speed of light
3. Thermodynamics, which deals with heat, work, temperature, and the statistical behavior of a large number of particles
4. Electromagnetism, which involves the theory of electricity, magnetism, and electromagnetic fields
5. Quantum mechanics, a theory dealing with the behavior of particles at the submicroscopic level as well as the macroscopic world

The first part of this textbook deals with mechanics, sometimes referred to as either classical mechanics or Newtonian mechanics. This is an appropriate place to begin an introductory text since many of the basic principles used to understand mechanical systems can later be used to describe such natural phenomena as waves and heat transfer. Furthermore, the laws of conservation of energy and momentum introduced in mechanics retain their importance in the fundamental theories that follow, including the theories of modern physics.

Today, mechanics is of vital importance to students from all disciplines. It is highly successful in describing the motions of material bodies, such as planets, rockets, and baseballs. In the first part of the text, we shall describe the laws of mechanics and examine a wide range of phenomena that can be understood with these fundamental ideas.

CHAPTER 1

Physics and Measurement

A modern pocket calculator together with a Chinese abacus. The abacus represents one of the first technological approaches to counting and calculating. There are evidences of its use as early as 3000 BC. *(Sheila Terry/Photo Researchers)*

Physics is a fundamental science concerned with understanding the natural phenomena that occur in our Universe. Like all sciences, physics is based on experimental observations and quantitative measurements. The main objective of physics is to use the limited number of fundamental laws that govern natural phenomena to develop theories that can predict the results of future experiments. The fundamental laws used in developing theories are expressed in the language of mathematics, the tool that provides a bridge between theory and experiment.

When a discrepancy between theory and experiment arises, new theories and experiments must be formulated to remove the discrepancy. Many times a theory is satisfactory only under limited conditions; a more general theory might be satisfactory without such limitations. A classic example is Newton's laws of motion, which accurately describe the motion of bodies at normal speeds but do not apply to objects moving at speeds comparable to the speed of light. The special theory of relativity developed by Einstein successfully predicts the motion of objects at low speed and at speeds approaching the speed of light and hence is a more general theory of motion.

3

Classical physics, which means all of the physics developed prior to 1900, includes the theories, concepts, laws, and experiments in classical mechanics, thermodynamics, and electromagnetism. Galileo Galilei (1564–1642) made significant contributions to classical mechanics through his work on the motion of objects having constant acceleration. In the same era, Johannes Kepler (1571–1630) analyzed astronomical data to develop empirical laws for the motion of planetary bodies.

The most important contributions to classical mechanics were provided by Isaac Newton (1642–1727), who developed classical mechanics as a systematic theory and was one of the originators of the calculus as a mathematical tool. Although major developments in classical physics continued in the 18th century, thermodynamics and electricity and magnetism were not developed until the latter part of the 19th century, principally because the apparatus for controlled experiments was either too crude or unavailable until then. In this text we shall treat the various disciplines of classical physics in separate sections; however, we will see that the disciplines of mechanics and electromagnetism are basic to all the branches of classical and modern physics.

A new era in physics, usually referred to as *modern physics,* began near the end of the 19th century. Modern physics developed mainly because of the discovery that many physical phenomena could not be explained by classical physics. The two most important developments in this modern era were the theories of relativity and quantum mechanics. Einstein's theory of relativity completely revolutionized the traditional concepts of space, time, and energy. Among other things, Einstein's theory corrected Newton's laws of motion for describing the motion of objects moving at speeds comparable to the speed of light. The theory of relativity also assumes that the speed of light is the upper limit of the speed of an object or signal and shows the relationship between mass and energy. Quantum mechanics was formulated by a number of distinguished scientists to provide descriptions of physical phenomena at the atomic level.

Scientists are constantly working at improving our understanding of fundamental laws, and new discoveries are being made every day. In many research areas, there is a great deal of overlap between physics, chemistry, and biology, as well as engineering. Some of the most notable developments are (1) numerous space missions and the landing of astronauts on the Moon, (2) microcircuitry and high-speed computers, and (3) sophisticated imaging techniques used in scientific research and medicine. The impacts of such developments and discoveries on our society have indeed been great, and it is very likely that future discoveries and developments will be just as exciting and challenging and of great benefit to humanity.

1.1   STANDARDS OF LENGTH, MASS, AND TIME

The laws of physics are expressed in terms of basic quantities that require a clear definition. For example, such physical quantities as force, velocity, volume, and acceleration can be described in terms of more basic quantities that in themselves are defined in terms of measurements or comparison with established standards. In mechanics, the three basic quantities are length (L), mass (M), and time (T). All other physical quantities in mechanics can be expressed in terms of these three quantities.

Obviously, if we are to report the results of a measurement to someone who wishes to reproduce this measurement, a standard must be defined. It would be meaningless if a visitor from another planet were to talk to us about a length of 8 ''glitches'' if we do not know the meaning of the unit glitch. On the other hand, if someone familiar with our system of measurement reports that a wall is 2 meters high and our unit of length is defined as 1 meter, we then know that the height of the wall is twice our fundamental unit of length. Likewise, if we are told that a person has a mass of 75 kilograms and our unit of mass is defined as 1.0 kilogram, then that person is 75 times as massive as our basic unit of mass.[1]

In 1960, an international committee established a set of standards for these fundamental quantities. The system that was established is an adaptation of the metric system, and it is called the **International System (SI)** of units. The abbreviation SI comes from its French name ''Système International.'' In this system, the units of length, mass, and time are the meter, kilogram, and second, respectively. (The SI system is closely related to the *mks* system which preceded it.) Other standard SI units established by the committee are those for temperature (the *kelvin*), electric current (the *ampere*), luminous intensity (the *candela*) and for the amount of substance (the *mole*, see Sec. 1.3). These seven units are the basic SI units. In the study of mechanics, however, we will be concerned only with the units of length, mass, and time. Definitions of units are under constant review and are changed from time to time.

Length

In A.D. 1120 the king of England decreed that the standard of length in his country would be the yard and that the yard would be precisely equal to the distance from the tip of his nose to the end of his outstretched arm. Similarly, the original standard for the foot adopted by the French was the length of the royal foot of King Louis XIV. This standard prevailed until 1799, when the legal standard of length in France became the meter, defined as one ten-millionth the distance from the equator to the North Pole along a longitudinal line that passes through Paris.

Many other systems have been developed in addition to those just discussed, but the advantages of the French system have caused it to prevail in most countries and in scientific circles everywhere. As recently as 1960, the length of the meter was defined as the distance between two lines on a specific platinum-iridium bar stored under controlled conditions. This standard was abandoned for several reasons, a principal one being that the limited accuracy with which the separation between the lines on the bar can be determined does not meet the current requirements of science and technology. Until recently, the meter was defined as 1 650 763.73 wavelengths of orange-red light emitted from a krypton-86 lamp. However, in October 1983, the **meter** was redefined as the distance traveled by light in vacuum during a time of 1/299 792 458 second. In effect, this latest definition establishes that the speed of light in vacuum is 299 792 458 meters per second.

Mass

The basic SI unit of mass, the **kilogram,** is defined as the mass of a specific platinum-iridium alloy cylinder kept at the International Bureau of Weights and Measures at Sèvres, France. This mass standard was established in 1887, and there has

---

[1] The need for assigning numerical values to various physical quantities through experimentation was expressed by Lord Kelvin (William Thomson) as follows: ''I often say that when you can measure what you are speaking about, and express it in numbers, you should know something about it, but when you cannot express it in numbers, your knowledge is of a meagre and unsatisfactory kind. It may be the beginning of knowledge but you have scarcely in your thoughts advanced to the state of science.''

(Left) The National Standard Kilogram No. 20, an accurate copy of the International Standard Kilogram kept at Sèvres, France, is housed under a double bell jar in a vault at the National Institute of Standards and Technology (NIST). *(Right)* The primary frequency standard (an atomic clock) at the NIST. This device keeps time with an accuracy of about 3 millionths of a second per year.   *(Photos courtesy of National Institute of Standards and Technology (NIST), U.S. Department of Commerce)*

been no change since that time because platinum-iridium is an unusually stable alloy. A duplicate is kept at the National Institute of Standards and Technology (NIST) in Gaithersburg, Md.

Time

Before 1960, the standard of time was defined in terms of the *mean solar day* for the year 1900.[2] The *mean solar second,* representing the basic unit of time, was originally defined as $(\frac{1}{60})(\frac{1}{60})(\frac{1}{24})$ of a mean solar day. The rotation of the Earth is now known to vary substantially with time, however, and therefore this motion is not a good one to use for defining a standard.

In 1967, consequently, the second was redefined to take advantage of the high precision obtainable in a device known as an *atomic clock.* In this device, the frequencies associated with certain atomic transitions (which are extremely stable and insensitive to the clock's environment) can be measured to a precision of one part in 10^{12}. This is equivalent to an uncertainty of less than one second every 30 000 years. Thus, in 1967 the SI unit of time, the *second,* was redefined using the characteristic frequency of a particular kind of cesium atom as the "reference clock": The basic SI unit of time, the **second,** is defined as 9 192 631 770 periods of the radiation from cesium-133 atoms.

---

[2] A solar day is the time interval between successive appearances of the Sun at the highest point it reaches in the sky each day.

TABLE 1.1 **Approximate Values of Some Measured Lengths**	
	Length (m)
Distance from Earth to most remote known quasar	1.4×10^{26}
Distance from Earth to most remote known normal galaxies	4×10^{25}
Distance from Earth to nearest large galaxy (M 31 in Andromeda)	2×10^{22}
Distance from Earth to nearest star (Proxima Centauri)	4×10^{16}
One lightyear	9.46×10^{15}
Mean orbit radius of the Earth about the Sun	1.5×10^{11}
Mean distance from Earth to Moon	3.8×10^{8}
Distance from the equator to the North Pole	1×10^{7}
Mean radius of the Earth	6.4×10^{6}
Typical altitude of satellite orbiting Earth	2×10^{5}
Length of a football field	9.1×10^{1}
Length of a housefly	5×10^{-3}
Size of smallest dust particles	1×10^{-4}
Size of cells of most living organisms	1×10^{-5}
Diameter of a hydrogen atom	1×10^{-10}
Diameter of an atomic nucleus	1×10^{-14}
Diameter of a proton	1×10^{-15}

Approximate Values for Length, Mass, and Time

Approximate values of various lengths, masses, and time intervals are presented in Tables 1.1 to 1.3. Note the wide range of values.[3] You should study these tables and get a feel for what is meant by a kilogram of mass (for example, the mass of an adult human is about 70 kilograms) or by a time interval of 10^8 seconds (one year is about 3×10^7 seconds). In addition to SI, there are two other systems of units you might run across in the literature. The *cgs system* was used in Europe before SI, and the *British engineering system* (sometimes called the conventional system) is still used in the United States despite acceptance of SI by the rest of the world. In the cgs system, the units of length, mass, and time are the centimeter (cm), gram (g), and second (s), respectively; in the British engineering system, the units of length, mass, and time are the foot (ft), slug, and second, respectively. Throughout most of this text we shall use SI units since they are almost universally accepted in science and industry. We will make some limited use of British engineering units in the study of classical mechanics.

TABLE 1.2 **Masses of Various Bodies (Approximate Values)**	
	Mass (kg)
Universe	1×10^{52}
Milky Way Galaxy	7×10^{41}
Sun	2×10^{30}
Earth	6×10^{24}
Moon	7×10^{22}
Horse	1×10^{3}
Human	7×10^{1}
Frog	1×10^{-1}
Mosquito	1×10^{-5}
Bacterium	1×10^{-15}
Hydrogen atom	1.67×10^{-27}
Electron	9.11×10^{-31}

CONCEPTUAL EXAMPLE 1.1

What types of natural phenomena could serve as alternative time standards?

Reasoning We could imagine any process with a beginning and an end to define the unit of time. For example, the duration of any biological process or the time it takes a fluid to empty from a container depends on many factors. To make the definition reproducible, each factor would have to be specified (such as the composition of the fluid, its temperature, and the dimensions and tilt of the container). Alternatively, a naturally isolated object like a flashing pulsar repeats the same flashing process in very precise time intervals. A "simpler" system is preferred as a time standard, such as a particular species of isolated atom or its nucleus.

---

[3] If you are unfamiliar with the use of powers scientific notation, you should review Section B.1 of the mathematical appendix at the back of this book.

TABLE 1.3   Approximate Values of Some Time Intervals

	Interval (s)
Age of the Universe	5×10^{17}
Age of the Earth	1.3×10^{17}
Average age of a college student	6.3×10^{8}
One year	3.2×10^{7}
One day (time for one revolution of Earth about its axis)	8.6×10^{4}
Time between normal heartbeats	8×10^{-1}
Period[a] of audible sound waves	1×10^{-3}
Period of typical radio waves	1×10^{-6}
Period of vibration of an atom in a solid	1×10^{-13}
Period of visible light waves	2×10^{-15}
Duration of a nuclear collision	1×10^{-22}
Time for light to cross a proton	3.3×10^{-24}

[a] Period is defined as the time interval of one complete vibration.

TABLE 1.4   Some Prefixes for SI Units

Power	Prefix	Abbreviation
10^{-24}	yocto	y
10^{-21}	zepto	z
10^{-18}	atto	a
10^{-15}	femto	f
10^{-12}	pico	p
10^{-9}	nano	n
10^{-6}	micro	μ
10^{-3}	milli	m
10^{-2}	centi	c
10^{-1}	deci	d
10^{1}	deka	da
10^{3}	kilo	k
10^{6}	mega	M
10^{9}	giga	G
10^{12}	tera	T
10^{15}	peta	P
10^{18}	exa	E
10^{21}	Zetta	Z
10^{24}	Yotta	Y

Some of the most frequently used prefixes for the various powers of ten and their abbreviations are listed in Table 1.4. For example, 10^{-3} m is equivalent to 1 millimeter (mm), and 10^{3} m is 1 kilometer (km). Likewise, 1 kg is 10^{3} g, and 1 megavolt (MV) is 10^{6} volts.

1.2   THE BUILDING BLOCKS OF MATTER

A 1-kg cube of solid gold has a length of approximately 3.73 cm on a side. Is this cube nothing but wall-to-wall gold, with no empty space? If the cube is cut in half, the two resulting pieces still retain their chemical identity as solid gold. But what if the pieces are cut again and again, indefinitely? Will the smaller and smaller pieces always be the same substance, gold? Questions such as these can be traced back to early Greek philosophers. Two of them—Leucippus and Democritus—could not accept the idea that such cuttings could go on forever. They speculated that the process ultimately must end when it produces a particle that can no longer be cut. In Greek, *atomos* means "not sliceable." From this comes our English word *atom* for the smallest, ultimate particle of matter. Elementary-particle physicists still engage in speculation and experimentation concerning the ultimate building blocks of matter.

Let us review briefly what is known about the structure of the world around us. It is useful to view the atom as a miniature Solar System with a dense positively charged nucleus occupying the position of the Sun and negatively charged electrons orbiting like the planets. This model of the atom enables us to understand some properties of the simpler atoms, such as hydrogen, but fails to explain many fine details of atomic structure.

Following the discovery of the nucleus in the early 1900s, the question arose: Does it have structure? That is, is the nucleus a single particle or a collection of particles? The exact composition of the nucleus is not known completely even today, but by the early 1930s a model evolved that helped us understand how the nucleus behaves. Specifically, scientists determined that occupying the nucleus are two basic entities, protons and neutrons. The *proton* carries a positive charge, and a specific element is identified by the number of protons in its nucleus. For instance, the nucleus of a hydrogen atom contains one proton, the nucleus of a helium atom contains two protons, and the nucleus of a uranium atom contains ninety-two protons.

The existence of *neutrons* was verified conclusively in 1932. A neutron has no charge and a mass about equal to that of a proton. One of its primary purposes is to act as a "glue" to hold the nucleus together. If neutrons were not present in the nucleus, the repulsive force between the positively charged particles would cause the nucleus to break apart.

But is this where the breaking down stops? Protons, neutrons, and a host of other exotic particles are now known to be composed of six different varieties of particles called **quarks,** which have been given the names of *up, down, strange, charmed, bottom,* and *top* (Fig. 1.1). The up, charmed, and top quarks have charges of + 2/3 that of the proton, whereas the down, strange, and bottom quarks have charges of − 1/3 that of the proton. The proton consists of two up quarks and one down quark, which you can easily show leads to the correct charge for the proton. Likewise, the neutron consists of two down quarks and one up quark, giving a net charge of zero.

1.3   DENSITY AND ATOMIC MASS

A property of any substance is its **density** ρ (Greek letter rho), defined as *mass per unit volume* (a table of the letters in the Greek alphabet is provided on the back endsheet of the textbook):

$$\rho \equiv \frac{m}{V} \tag{1.1}$$

For example, aluminum has a density of 2.70 g/cm³, and lead has a density of 11.3 g/cm³. Therefore, a piece of aluminum of volume 10.0 cm³ has a mass of 27.0 g, while an equivalent volume of lead would have a mass of 113 g. A list of densities for various substances is given in Table 1.5.

The difference in density between aluminum and lead is due, in part, to their different *atomic masses;* the atomic mass of lead is 207 atomic mass units and that of aluminum is 27.0 atomic mass units. However, the ratio of atomic masses, 207/27.0 = 7.67, does not correspond to the ratio of densities, 11.3/2.70 = 4.19. The discrepancy is due to the difference in atomic spacings and atomic arrangements in the crystal structure of these two substances.

TABLE 1.5 **Densities of Various Substances**	
Substance	**Density ρ (kg/m³)**
Gold	19.3×10^3
Uranium	18.7×10^3
Lead	11.3×10^3
Copper	8.93×10^3
Iron	7.86×10^3
Aluminum	2.70×10^3
Magnesium	1.75×10^3
Water	1.00×10^3
Air	0.0012×10^3

FIGURE 1.1   Distances at the frontier of nuclear physics are astonishingly short. An atom is so small that a single-file line of 250 000 of them would fit within the thickness of aluminum foil. The nucleus at the atom's center is a cluster of nucleons, each 100 000 times smaller than the atom itself. The three quarks inside each nucleon are even smaller.   *(Courtesy of SURA, Inc.)*

All ordinary matter consists of atoms, and each atom is made up of electrons and a nucleus. Practically all of the mass of an atom is contained in the nucleus, which consists of protons and neutrons. Because different elements contain different numbers of protons and neutrons, the atomic masses of the various elements differ. The mass of a nucleus is measured relative to the mass of an atom of the carbon-12 isotope (this isotope of carbon has six protons and six neutrons).

The mass of ^{12}C is defined to be exactly 12 atomic mass units (u), where 1 u = 1.660 540 2 × 10^{-27} kg. In these units, the proton and neutron have masses of about 1 u. More precisely,

$$\text{mass of proton} = 1.0073 \text{ u}$$

$$\text{mass of neutron} = 1.0087 \text{ u}$$

One **mole** (mol) of a substance is that amount of it that consists of Avogadro's number, N_A, of molecules. Avogadro's number is defined so that one mole of carbon-12 atoms has a mass of exactly 12 g. Its value has been found to be $N_A = 6.02 \times 10^{23}$ molecules/mol. For example, one mole of aluminum has a mass of 27 g, and one mole of lead has a mass of 207 g. But one mole of aluminum contains the same number of atoms as one mole of lead, since there are 6.02×10^{23} atoms in one mole of *any* element. The mass per atom for a given element is then given by

The mass of an atom

$$m_{\text{atom}} = \frac{\text{atomic mass of the element}}{N_A} \tag{1.2}$$

For example, the mass of an aluminum atom is

$$m_{\text{Al}} = \frac{27 \text{ g/mol}}{6.02 \times 10^{23} \text{ atoms/mol}} = 4.5 \times 10^{-23} \text{ g/atom}$$

Note that 1 u is equal to N_A^{-1} g.

EXAMPLE 1.2   How Many Atoms in the Cube?

A solid cube of aluminum (density 2.7 g/cm^3) has a volume of 0.20 cm^3. How many aluminum atoms are contained in the cube?

Solution   Since density equals mass per unit volume, the mass of the cube is

$$m = \rho V = (2.7 \text{ g/cm}^3)(0.20 \text{ cm}^3) = 0.54 \text{ g}$$

To find the number of atoms, N, we can set up a proportion using the fact that one mole of aluminum (27 g) contains 6.02×10^{23} atoms:

$$\frac{6.02 \times 10^{23} \text{ atoms}}{27 \text{ g}} = \frac{N}{0.54 \text{ g}}$$

$$N = \frac{(0.54 \text{ g})(6.02 \times 10^{23} \text{ atoms})}{27 \text{ g}} = 1.2 \times 10^{22} \text{ atoms}$$

1.4   DIMENSIONAL ANALYSIS

The word *dimension* has a special meaning in physics. It usually denotes the physical nature of a quantity. Whether a distance is measured in units of feet or meters or furlongs, it is a distance. We say its dimension is *length*.

The symbols we use in this book to specify length, mass, and time are L, M, and T, respectively. We shall often use brackets [ ] to denote the dimensions of a physical quantity. For example, the symbol we use for speed in this book is v, and in our notation the dimensions of speed are written $[v] = L/T$. As another exam-

TABLE 1.6   Dimensions of Area, Volume, Speed, and Acceleration

System	Area (L^2)	Volume (L^3)	Speed (L/T)	Acceleration (L/T^2)
SI	m^2	m^3	m/s	m/s^2
cgs	cm^2	cm^3	cm/s	cm/s^2
British engineering	ft^2	ft^3	ft/s	ft/s^2

ple, the dimensions of area, A, are $[A] = L^2$. The dimensions of area, volume, speed, and acceleration are listed in Table 1.6, along with their units in the three common systems. The dimensions of other quantities, such as force and energy, will be described as they are introduced in the text.

In many situations, you may have to derive or check a specific formula. Although you may have forgotten the details of the derivation, there is a useful and powerful procedure called *dimensional analysis* that can be used to assist in the derivation or to check your final expression. This procedure should always be used and should help minimize the rote memorization of equations. Dimensional analysis makes use of the fact that *dimensions can be treated as algebraic quantities*. That is, quantities can be added or subtracted only if they have the same dimensions. Furthermore, the terms on both sides of an equation must have the same dimensions. By following these simple rules, you can use dimensional analysis to help determine whether or not an expression has the correct form because the relationship can be correct only if the dimensions on each side of the equation are the same.

To illustrate this procedure, suppose you wish to derive a formula for the distance x traveled by a car in a time t if the car starts from rest and moves with constant acceleration a. In Chapter 2, we shall find that the correct expression is $x = \frac{1}{2}at^2$. Let us use dimensional analysis to check the validity of this expression.

The quantity x on the left side has the dimension of length. In order for the equation to be dimensionally correct, the quantity on the right side must also have the dimension of length. We can perform a dimensional check by substituting the dimensions for acceleration, L/T^2, and time, T, into the equation. That is, the dimensional form of the equation $x = \frac{1}{2}at^2$ is

$$L = \frac{L}{T^2} \cdot T^2 = L$$

The units of time cancel as shown, leaving the unit of length.

A more general procedure using dimensional analysis is to set up an expression of the form

$$x \propto a^n t^m$$

when n and m are exponents that must be determined and the symbol $\propto$ indicates a proportionality. This relationship is correct only if the dimensions of both sides are the same. Since the dimension of the left side is length, the dimension of the right side must also be length. That is,

$$[a^n t^m] = L = LT^0$$

Since the dimensions of acceleration are L/T^2 and the dimension of time is T, we have

$$(L/T^2)^n T^m = L$$

or

$$L^n T^{m-2n} = L$$

Since the exponents of L and T must be the same on both sides, we see that $n = 1$ and $m = 2$. Therefore, we conclude that

$$x \propto at^2$$

This result differs by a factor of 2 from the correct expression, which is $x = \frac{1}{2}at^2$. Because the factor $1/2$ is dimensionless, there is no way of determining this via dimensional analysis.

CONCEPTUAL EXAMPLE 1.3

Does dimensional analysis give any information on constants of proportionality that may appear in an algebraic expression? Explain.

Reasoning Dimensional analysis gives the units of the proportionality constant but gives no information about its numerical value. For example, experiments show that doubling or tripling the radius of a spherical water balloon makes its mass get eight or twenty-seven times larger. Its mass is proportional to the cube of its radius. Because $m \propto r^3$, we can write $m = kr^3$. Dimensional analysis shows that the proportionality constant k must have units kg/m^3, but to determine its value requires experimental data or geometrical reasoning.

EXAMPLE 1.4 Analysis of an Equation

Show that the expression $v = v_0 + at$ is dimensionally correct, where v and v_0 represent speeds, a is acceleration, and t is a time interval.

Solution For the speed terms, we have from Table 1.6

$$[v] = [v_0] = L/T$$

The same table gives us L/T^2 for the dimensions of acceleration, and so the dimensions of at are

$$[at] = (L/T^2)(T) = L/T$$

Therefore the expression is dimensionally correct. (If the expression were given as $v = v_0 + at^2$, it would be dimensionally *incorrect*. Try it and see!)

EXAMPLE 1.5 Analysis of a Power Law

Suppose we are told that the acceleration of a particle moving with uniform speed v in a circle of radius r is proportional to some power of r, say r^n, and some power of v, say v^m. How can we determine the powers of r and v?

Solution Let us take a to be

$$a = kr^n v^m$$

where k is a dimensionless constant. Knowing the dimensions of a, r, and v, we see that the dimensional equation must be

$$L/T^2 = L^n (L/T)^m = L^{n+m}/T^m$$

This dimensional equation is balanced under the conditions

$$n + m = 1 \quad \text{and} \quad m = 2$$

Therefore, $n = -1$, and we can write the acceleration

$$a = kr^{-1}v^2 = k\frac{v^2}{r}$$

When we discuss uniform circular motion later, we shall see that $k = 1$ if a consistent set of units is used. The constant k would not equal 1 if, for example, v were in km/h and you wanted a in m/s^2.

(Left) Conversion of miles to kilometers. *(Right)* This vehicle's speedometer gives speed readings in miles per hour and in kilometers per hour. Try confirming the conversion between the two sets of units for a few readings of the dial.   *(Paul Silverman, Fundamental Photographs)*

1.5   CONVERSION OF UNITS

Sometimes it is necessary to convert units from one system to another. Conversion factors between the SI and conventional units of length are as follows:

1 mile = 1609 m = 1.609 km	1 ft = 0.3048 m = 30.48 cm
1 m = 39.37 in. = 3.281 ft	1 in. = 0.0254 m = 2.54 cm

A more complete list of conversion factors can be found in Appendix A.

Units can be treated as algebraic quantities that can cancel each other. For example, suppose we wish to convert 15.0 in. to centimeters. Since 1 in. = 2.54 cm (exactly), we find that

$$15.0 \text{ in.} = (15.0 \text{ in.})\left(2.54 \, \frac{\text{cm}}{\text{in.}}\right) = 38.1 \text{ cm}$$

EXAMPLE 1.6   The Density of a Cube

The mass of a solid cube is 856 g, and each edge has a length of 5.35 cm. Determine the density ρ of the cube in basic SI units.

Solution   Since $1 \text{ g} = 10^{-3}$ kg and $1 \text{ cm} = 10^{-2}$ m, the mass, m, and volume, V, in basic SI units are given by

$$m = 856 \text{ g} \times 10^{-3} \text{ kg/g} = 0.856 \text{ kg}$$

$$V = L^3 = (5.35 \text{ cm} \times 10^{-2} \text{ m/cm})^3$$
$$= (5.35)^3 \times 10^{-6} \text{ m}^3 = 1.53 \times 10^{-4} \text{ m}^3$$

Therefore

$$\rho = \frac{m}{V} = \frac{0.856 \text{ kg}}{1.53 \times 10^{-4} \text{ m}^3} = \boxed{5.59 \times 10^3 \text{ kg/m}^3}$$

1.6   ORDER-OF-MAGNITUDE CALCULATIONS

It is often useful to compute an approximate answer to a physical problem even where little information is available. Such results can then be used to determine whether or not a more precise calculation is necessary. These approximations are usually based on certain assumptions, which must be modified if more precision is needed. Thus, we shall sometimes refer to the order of magnitude of a certain quantity as the power of ten of the number that describes that quantity. If, for example, we say that a quantity increases in value by three orders of magnitude, this means that its value is increased by a factor of $10^3 = 1000$.

The spirit of order-of-magnitude calculations, sometimes referred to as "guesstimates" or "ball-park figures," is given in the following quotation: "Make an estimate before every calculation, try a simple physical argument . . . before every derivation, guess the answer to every puzzle. Courage: no one else needs to know what the guess is."[4]

EXAMPLE 1.7 Breaths in a Lifetime

Estimate the number of breaths taken during an average life span of 70 years.

Solution The only estimate we must make in this example is the average number of breaths that a person takes in 1 min. This number varies, depending on whether the person is exercising, sleeping, angry, serene, and so forth. We shall choose 8 breaths per minute as our estimate of the average.

The number of minutes in a year is

$$1 \text{ year} \times 365 \frac{\text{days}}{\text{year}} \times 24 \frac{\text{h}}{\text{day}} \times 60 \frac{\text{min}}{\text{h}} = 5.26 \times 10^5 \text{ min}$$

Thus, in 70 years there will be $(70)(5.26 \times 10^5 \text{ min}) = 3.68 \times 10^7$ min. At a rate of 8 breaths/min the individual would take about 3×10^8 breaths. You may want to check your own rate and repeat the above calculation.

EXAMPLE 1.8 How Many Atoms?

Estimate the number of atoms in 1 cm³ of a solid.

Solution From Table 1.2 we note that the diameter of an atom is about 10^{-10} m. Thus, if in our model we assume that the atoms in the solid are solid spheres of this diameter, then the volume of each sphere is about 10^{-30} m³ (more precisely, volume $= 4\pi r^3/3 = \pi d^3/6$, where $r = d/2$). Therefore,

since 1 cm³ $= 10^{-6}$ m³, the number of atoms in the solid is of the order of $10^{-6}/10^{-30} = 10^{24}$ atoms.

A more precise calculation would require knowledge of the density of the solid and the mass of each atom. However, our estimate agrees with the more precise calculation to within a factor of 10.

EXAMPLE 1.9 How Much Gas Do We Use?

Estimate the number of gallons of gasoline used by all U.S. cars each year.

Solution Since there are about 240 million people in the United States, an estimate of the number of cars in the country is 120 million (assuming two cars and four people per family). We also estimate that the average distance traveled per year is 10 000 miles. If we assume a gasoline consumption of 0.05 gal/mi, each car uses about 500 gal/year. Multiplying this by the total number of cars in the United States gives an estimated total consumption of 6×10^{10} gal. This number of gallons corresponds to a yearly consumer expenditure of over 60 billion dollars and is probably a low estimate because we haven't accounted for commercial consumption.

---

[4] E. Taylor and J. A. Wheeler, *Spacetime Physics*, San Francisco, W. H. Freeman, 1966, p. 60.

1.7 SIGNIFICANT FIGURES

When certain quantities are measured, the measured values are known only to within the limits of the experimental uncertainty. The value of the uncertainty can depend on various factors, such as the quality of the apparatus, the skill of the experimenter, and the number of measurements performed.

Suppose that in a laboratory experiment we are asked to measure the area of a rectangular plate using a meter stick as a measuring instrument. Let us assume that the accuracy to which we can measure a particular dimension of the plate is ± 0.1 cm. If the length of the plate is measured to be 16.3 cm, we can claim only that its length lies somewhere between 16.2 cm and 16.4 cm. In this case, we say that the measured value has three significant figures. Likewise, if its width is measured to be 4.5 cm, the actual value lies between 4.4 cm and 4.6 cm. This measured value has only two significant figures. Note that the significant figures include the first estimated digit. Thus, we could write the measured values as 16.3 ± 0.1 cm and 4.5 ± 0.1 cm.

Suppose now that we would like to find the area of the plate by multiplying the two measured values. If we were to claim that the area is $(16.3 \text{ cm})(4.5 \text{ cm}) = 73.35 \text{ cm}^2$, our answer would be unjustifiable because it contains four significant figures, which is greater than the number of significant figures in either of the measured lengths. A good rule of thumb to use as a guide in determining the number of significant figures that can be claimed is as follows:

> When **multiplying** several quantities, the number of significant figures in the final answer is the same as the number of significant figures in the *least* accurate of the quantities being multiplied, where "least accurate" means "having the lowest number of significant figures." The same rule applies to division.

Applying this rule to the multiplication example above, we see that the answer for the area can have only two significant figures because the length of 4.5 cm has only two significant figures. Thus, all we can claim is that the area is 73 cm^2, realizing that the value can range between $(16.2 \text{ cm})(4.4 \text{ cm}) = 71 \text{ cm}^2$ and $(16.4 \text{ cm})(4.6 \text{ cm}) = 75 \text{ cm}^2$.

Zeros may or may not be significant figures. Those used to position the decimal point in such numbers as 0.03 and 0.0075 are not significant. Thus there are one and two significant figures, respectively, in these two values. When the positioning of zeros comes after other digits, however, there is the possibility of misinterpretation. For example, suppose the mass of an object is given as 1500 g. This value is ambiguous because we do not know whether the last two zeros are being used to locate the decimal point or whether they represent significant figures in the measurement. In order to remove this ambiguity, it is common to use scientific notation to indicate the number of significant figures. In this case, we would express the mass as 1.5×10^3 g if there are two significant figures in the measured value, 1.50×10^3 g if there are three significant figures, and 1.500×10^3 g if there are four. Likewise, 0.000 15 should be expressed in scientific notation as 1.5×10^{-4} if it has two significant figures or as 1.50×10^{-4} if it has three significant figures. The three zeros between the decimal point and the digit 1 in the number 0.000 15 are not counted as significant figures because they are present only to locate the decimal point. In general, a **significant figure** is a reliably known digit (other than a zero used to locate the decimal point).

For addition and subtraction, the number of decimal places must be considered when you are determining how many significant figures to report.

> When numbers are **added** or **subtracted,** the number of decimal places in the result should equal the smallest number of decimal places of any term in the sum.

For example, if we wish to compute $123 + 5.35$, the answer would be 128 and not 128.35. If we compute the sum $1.0001 + 0.0003 = 1.0004$, the result has five significant figures, even though one of the terms in the sum, 0.0003, has only one significant figure. Likewise, if we perform the subtraction $1.002 - 0.998 = 0.004$, the result has only one significant figure even though one term has four significant figures and the other has three. In this book, *most of the numerical examples and end-of-chapter problems will yield answers having either two or three significant figures.*

EXAMPLE 1.10   The Area of a Rectangle

A rectangular plate has a length of (21.3 ± 0.2) cm and a width of (9.80 ± 0.10) cm. Find the area of the plate and the uncertainty in the calculated area.

Solution

$$\text{Area} = \ell w = (21.3 \pm 0.2) \text{ cm} \times (9.80 \pm 0.10) \text{ cm}$$
$$\approx (21.3 \times 9.80 \pm 21.3 \times 0.10 \pm 0.2 \times 9.80) \text{ cm}^2$$
$$\approx (209 \pm 4) \text{ cm}^2$$

Because the input data were given only to three significant figures, we cannot claim any more in our result. Furthermore, you should realize that the uncertainty in the product (2%) is *approximately* equal to the sum of the uncertainties in the length and width (each uncertainty is about 1%). Note that this approach only gives *bounds* on the true error. The relative error in the area is actually the square root of the sum of the squares of the individual relative errors in ℓ and w.

EXAMPLE 1.11   Installing a Carpet

A carpet is to be installed in a room whose length is measured to be 12.71 m (four significant figures) and whose width is measured to be 3.46 m (three significant figures). Find the area of the room.

Solution   If you multiply 12.71 m by 3.46 m on your calculator, you will get an answer of 43.9766 m². How many of these numbers should you claim? Our rule of thumb for multiplication tells us that you can claim only the number of significant figures in the least accurate of the quantities being measured. In this example, we have only three significant figures in our least accurate measurement, so we should express our final answer as 44.0 m². Note that in reducing 43.9766 to three significant figures for our answer, we used a general rule for rounding off numbers that states that the last digit retained is increased by 1 if the first digit dropped equals 5 or greater. Furthermore, if the last digit is 5, the result should be rounded to the nearest even number. (This helps avoid accumulation of errors.)

1.8   MATHEMATICAL NOTATION

Many mathematical symbols are used throughout this book, some of which you are surely aware of, such as the symbol $=$ to denote the equality of two quantities.

The symbol $\propto$ is used to denote a proportionality. For example, $y \propto x^2$ means that y is proportional to the square of x.

The symbol $<$ means *less than,* and $>$ means *greater than.* For example, $x > y$ means x is greater than y.

The symbol $\ll$ means *much less than,* and $\gg$ means *much greater than.*

The symbol $\approx$ is used to indicate that two quantities are *approximately equal* to each other.

The symbol $\equiv$ means *is defined as.* This is a stronger statement than a simple $=$.

It is convenient to use a symbol to indicate the change in a quantity. For example, Δx (read delta x) means the *change in the quantity x.* (It does not mean the product of Δ and x.) For example, if x_i is the initial position of a particle and x_f is its final position, then the *change in position* is written

$$\Delta x = x_f - x_i$$

We shall often have occasion to sum several quantities. A useful abbreviation for representing such a sum is the Greek letter Σ (capital sigma). Suppose we wish to sum a set of five numbers represented by x_1, x_2, x_3, x_4, and x_5. In the abbreviated notation, we would write the sum

$$x_1 + x_2 + x_3 + x_4 + x_5 \equiv \sum_{i=1}^{5} x_i$$

where the subscript i on a particular x represents any one of the numbers in the set. For example, if there are five masses in a system, m_1, m_2, m_3, m_4, and m_5, the total mass of the system $M = m_1 + m_2 + m_3 + m_4 + m_5$ could be expressed

$$M = \sum_{i=1}^{5} m_i$$

Finally, the *magnitude* of a quantity x, written $|x|$, is simply the absolute value of that quantity. The magnitude of x is *always positive,* regardless of the sign of x. For example, if $x = -5$, $|x| = 5$; if $x = 8$, $|x| = 8$.

A list of these symbols and their meanings is given on the back endsheet.

SUMMARY

The physical quantities we shall encounter in our study of mechanics can be expressed in terms of three fundamental quantities, length, mass, and time, which in the SI system have the units meters (m), kilograms (kg), and seconds (s), respectively.

The **density** of a substance is defined as its *mass per unit volume.* Different substances have different densities mainly because of differences in their atomic masses and atomic arrangements.

The number of atoms in one mole of any element or compound, called **Avogadro's number,** N_A, is 6.02×10^{23}.

It is often useful to use the method of *dimensional analysis* to check equations and to assist in deriving expressions.

QUESTIONS

1. In this chapter we described how the Earth's daily rotation on its axis was once used to define the standard unit of time. What other types of natural phenomena could serve as alternative time standards?

2. A hand is defined as 4 inches; a foot is defined as 12 inches. Why should the former be any less acceptable than the latter, which we use all the time?

3. Express the following quantities using the prefixes given in Table 1.4: (a) 3×10^{-4} m, (b) 5×10^{-5} s, (c) 72×10^2 g.

4. Suppose that two quantities A and B have different dimensions. Determine which of the following arithmetic operations *could* be physically meaningful: (a) $A + B$, (b) A/B, (c) $B - A$, (d) AB.

5. What level of accuracy is implied in an order-of-magnitude calculation?

6. Do an order-of-magnitude calculation for an everyday situation you might encounter. For example, how far do you walk or drive each day?

7. Estimate your age in seconds.

8. Estimate the masses of various objects around you in grams or in kilograms. If a scale is available, check your estimates.

9. Is it possible to use length, density, and time as three fundamental units rather than length, mass, and time? If so, what could be used as a standard of density?

10. An automobile tire is rated to last for 50 000 miles. Estimate the number of revolutions the tire will make in its lifetime.

11. Estimate the total length of all McDonald's french fries (laid end to end) sold in the United States in one year. How many round trips to the Moon would this equal?

PROBLEMS

Section 1.3   Density and Atomic Mass

1. Calculate the density of a solid cube that measures 5.00 cm on each side and has a mass of 350 g.

2. The mass of the planet Saturn is 5.64×10^{26} kg and its radius is 6.00×10^7 m. (a) Calculate its density. (b) If this planet were placed in a large enough ocean of water, would it float? Explain.

3. How many grams of copper are required to make a hollow spherical shell with an inner radius of 5.70 cm and an outer radius of 5.75 cm? The density of copper is 8.93 g/cm³.

3A. How many grams of copper are required to make a hollow spherical shell having an inner radius r_1 (in cm) and an outer radius r_2 (in cm)? The density of copper is ρ (in g/cm³).

4. The planet Jupiter has an average radius 10.95 times that of the average radius of the Earth and a mass 317.4 times that of the Earth. Calculate the ratio of Jupiter's mass density to the mass density of the Earth.

5. Calculate the mass of an atom of (a) helium, (b) iron, and (c) lead. Give your answers in atomic mass units and in grams. The atomic masses are 4, 56, and 207, respectively, for the atoms given.

6. A small cube of iron is observed under a microscope. The edge of the cube is 5.00×10^{-6} cm. Find (a) the mass of the cube and (b) the number of iron atoms in the cube. The atomic mass of iron is 56 u, and its density is 7.86 g/cm³.

7. A structural I beam is made of steel. A view of its cross-section and its dimensions is shown in Figure P1.7. (a) What is the mass of a section 1.5 m long? (b) How many atoms are there in this section? The density of steel is 7.56×10^3 kg/m³.

8. A flat circular plate of copper has a radius of 0.243 m and a mass of 62.0 kg. What is the thickness of the plate?

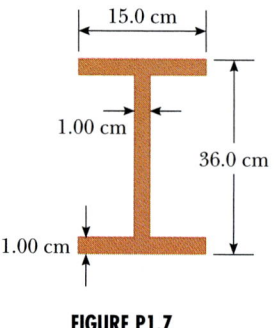

FIGURE P1.7

Section 1.4   Dimensional Analysis

9. Show that the expression $x = vt + \frac{1}{2}at^2$ is dimensionally correct, where x is a coordinate and has units of length, v is speed, a is acceleration, and t is time.

10. The displacement of a particle when moving under uniform acceleration is some function of the elapsed time and the acceleration. Suppose we write this displacement $s = ka^m t^n$, where k is a dimensionless constant. Show by dimensional analysis that this expression is satisfied if $m = 1$ and $n = 2$. Can this analysis give the value of k?

11. The square of the speed of an object undergoing a uniform acceleration a is some function of a and the displacement s, according to the expression $v^2 = ka^m s^n$, where k is a dimensionless constant. Show by dimensional analysis that this expression is satisfied only if $m = n = 1$.

12. The radius r of a circle inscribed in any triangle whose sides are a, b, and c is given by $r = [(s - a)(s - b)(s - c)/s]^{1/2}$, where s is an abbreviation for $(a + b + c)/2$. Check this formula for dimensional consistency.

13. Which of the equations below is dimensionally correct?

☐ indicates problems that have full solutions available in the Student Solutions Manual and Study Guide.

(a) $v = v_0 + ax$

(b) $y = (2 \text{ m})\cos(kx)$, where $k = 2 \text{ m}^{-1}$

14. The period T of a simple pendulum is measured in time units and is

$$T = 2\pi \sqrt{\frac{\ell}{g}}$$

where ℓ is the length of the pendulum and g is the free-fall acceleration in units of length divided by the square of time. Show that this equation is dimensionally correct.

15. The volume of an object as a function of time is calculated by $V = At^3 + B/t$, where t is time measured in seconds and V is in cubic meters. Determine the dimension of the constants A and B.

16. The consumption of natural gas by a company satisfies the empirical equation $V = 1.5t + 0.0080t^2$, where V is the volume in millions of cubic feet and t the time in months. Express this equation in units of cubic feet and seconds. Put the proper units on the coefficients. Assume a month is 30 days.

17. Newton's law of universal gravitation is

$$F = G\frac{Mm}{r^2}$$

Here F is the force of gravity, M and m are masses, and r is a length. Force has the SI units $\text{kg}\cdot\text{m}/\text{s}^2$. What are the SI units of the constant G?

Section 1.5 Conversion of Units

18. Convert the volume 8.50 in.^3 to m^3, recalling that $1 \text{ in.} = 2.54 \text{ cm}$ and $1 \text{ cm} = 10^{-2} \text{ m}$.

19. A rectangular building lot is 100.0 ft by 150.0 ft. Determine the area of this lot in m^2.

20. A classroom measures $40.0 \text{ m} \times 20.0 \text{ m} \times 12.0 \text{ m}$. The density of air is 1.29 kg/m^3. What are (a) the volume of the room in cubic feet, and (b) the weight of air in the room in pounds?

21. A creature moves at a speed of 5.0 furlongs per fortnight (not a very common unit of speed). Given that 1.0 furlong = 220 yards and 1 fortnight = 14 days, determine the speed of the creature in m/s. (The creature is probably a snail.)

22. A section of land has an area of 1 square mile and contains 640 acres. Determine the number of square meters there are in 1 acre.

23. A solid piece of lead ·has a mass of 23.94 g and a volume of 2.10 cm^3. From these data, calculate the density of lead in SI units (kg/m^3).

24. A quart container of ice cream is to be made in the form of a cube. What should be the length of a side in cm? (Use the conversion 1 gallon = 3.786 liters.)

25. An astronomical unit (AU) is defined as the average distance between the Earth and Sun. (a) How many astronomical units are there in one lightyear?

(b) Determine the distance from Earth to the Andromeda galaxy in astronomical units.

26. The mass of the Sun is about 1.99×10^{30} kg, and the mass of a hydrogen atom, of which the Sun is mostly composed, is 1.67×10^{-27} kg. How many atoms are there in the Sun?

27. At the time of this book's printing, the U.S. national debt is about \$4 trillion. (a) If payments were made at the rate of \$1000/sec, how many years would it take to pay off the debt, assuming no interest were charged? (b) A dollar bill is about 15.5 cm long. If 4 trillion dollar bills were laid end to end around the Earth's equator, how many times would they encircle the Earth? Take the radius of the Earth at the equator to be 6378 km. (*Note:* Before doing any of these calculations, try to guess at the answers. You may be very surprised.)

28. A room measures $4.0 \text{ m} \times 4.0 \text{ m}$, and its ceiling is 2.5 m high. Is it possible to completely wallpaper the walls of this room with the pages of this book? Explain.

29. (a) Find a conversion factor to convert from mi/h to km/h. (b) Until recently, federal law mandated that highway speeds would be 55 mi/h. Use the conversion factor of part (a) to find this speed in km/h. (c) The maximum highway speed has been raised to 65 mi/h in some places. In km/h, how much increase is this over the 55 mi/h limit?

30. (a) How many seconds are there in a year? (b) If one micrometeorite (a sphere with a diameter of 1.00×10^{-6} m) strikes each square meter of the Moon each second, how many years would it take to cover the Moon to a depth of 1.00 m? (*Hint:* Consider a cubic box on the Moon 1.00 m on a side, and find how long it will take to fill the box.)

31. One gallon of paint (volume = $3.78 \times 10^{-3} \text{ m}^3$) covers an area of 25.0 m^2. What is the thickness of the paint on the wall?

32. A pyramid has a height of 481 ft and its base covers an area of 13.0 acres (1 acre = $43\,560 \text{ ft}^2$). If the volume of a pyramid is given by the expression $V = (1/3)Bh$, where B is the area of the base and h is the height, find the volume of this pyramid in cubic meters.

33. The pyramid described in Problem 32 contains approximately two million stone blocks that average 2.50 tons each. Find the weight of this pyramid in pounds.

34. Assuming that 70 percent of the Earth's surface is covered with water at an average depth of 1 mile, estimate the mass of the water on Earth in kilograms.

35. The diameter of our disk-shaped galaxy, the Milky Way, is about 1.0×10^5 lightyears. The distance to Andromeda, the galaxy nearest our own Milky Way, is about 2.0 million lightyears. If we represent the Milky Way by a dinner plate 25 cm in diameter, determine the distance to the next dinner plate.

36. The mean radius of the Earth is 6.37×10^6 m, and that of the Moon is 1.74×10^8 cm. From these data calculate (a) the ratio of the Earth's surface area to that of the Moon and (b) the ratio of the Earth's volume to that of the Moon. Recall that the surface area of a sphere is $4\pi r^2$ and the volume of a sphere is $\frac{4}{3}\pi r^3$.

37. From the fact that the average density of the Earth is 5.5 g/cm^3 and its mean radius is 6.37×10^6 m, compute the mass of the Earth.

38. Assume that an oil slick consists of a single layer of molecules and that each molecule occupies a cube 1.0 nm on a side. Determine the area of an oil slick formed by 1.0 m^3 of oil.

39. One cubic meter (1.00 m^3) of aluminum has a mass of 2.70×10^3 kg, and 1.00 m^3 of iron has a mass of 7.86×10^3 kg. Find the radius of a solid aluminum sphere that will balance a solid iron sphere of radius 2.00 cm on an equal-arm balance.

39A. Let ρ_{Al} represent the density of aluminum and ρ_{Fe} that of iron. Find the radius of a solid aluminum sphere that balances a solid iron sphere of radius r_{Fe} on an equal-arm balance.

Section 1.6   Order-of-Magnitude Calculations

40. Assuming 60 heartbeats a minute, estimate the total number of times the heart of a human beats in an average lifetime of 70 years.

41. Estimate the number of Ping-Pong balls that would fit into an average-size room (without being crushed).

42. Estimate the amount of motor oil used by all cars in the United States each year and its cost to the consumers.

43. When a droplet of oil spreads out on a smooth water surface, the resulting "oil slick" is approximately one molecule thick. An oil droplet of mass 9.00×10^{-7} kg and density 918 kg/m^3 spreads out into a circle of radius 41.8 cm on the water surface. What is the diameter of an oil molecule?

44. Approximately how many raindrops fall on a 1.0-acre lot during a 1.0-in. rainfall?

45. Army engineers in 1946 determined the distance from the Earth to the Moon by using radar. If the round trip from Earth to Moon and back again took the radar beam 2.56 s, what is the distance from the Earth to the Moon? (The speed of radar waves is 3.00×10^8 m/s.)

46. A billionaire offers to give you $1 billion if you can count it out using only one-dollar bills. Should you accept her offer? Assume you can count one bill every second, and be sure to allow for the fact that you need about 8 hours a day for sleeping and eating, and that right now you are probably at least 18 years old.

47. A high fountain of water is located at the center of a circular pool as in Figure P1.47. A student walks around the pool and estimates its circumference to be 150 m. Next, the student stands at the edge of the pool and uses a protractor to gauge the angle of elevation of the top of the fountain to be 55°. How high is the fountain?

FIGURE P1.47

48. Assume that you watch every pitch of every game of a 162-game major league baseball season. Approximately how many pitches would you see thrown?

49. Estimate the number of piano tuners living in New York City. This problem was posed by the physicist Enrico Fermi, who was well known for making order-of-magnitude calculations.

Section 1.7   Significant Figures

50. Determine the number of significant figures in the following numbers: (a) 23 cm, (b) 3.589 s, (c) 4.67×10^3 m/s, (d) 0.0032 m.

51. The radius of a circle is measured to be 10.5 ± 0.2 m. Calculate the (a) area and (b) circumference of the circle, and give the uncertainty in each value.

52. Carry out the following arithmetic operations: (a) the sum of the numbers 756, 37.2, 0.83, and 2.5; (b) the product 3.2×3.563; (c) the product $5.6 \times \pi$.

53. If the length and width of a rectangular plate are measured to be (15.30 ± 0.05) cm and (12.80 ± 0.05) cm, respectively, find the area of the plate and the approximate uncertainty in the calculated area.

54. The radius of a solid sphere is measured to be (6.50 ± 0.20) cm, and its mass is measured to be (1.85 ± 0.02) kg. Determine the density of the sphere in kg/m^3 and the uncertainty in the density.

55. How many significant figures are there in (a) 78.9 ± 0.2, (b) 3.788×10^9, (c) 2.46×10^{-6}, (d) 0.0053?

56. A farmer measures the distance around a rectangular field. The length of the long sides of the rectangle is found to be 38.44 m, and the length of the short sides is found to be 19.5 m. What is the total distance around the field?

57. A sidewalk is to be constructed around a swimming pool that measures (10.0 ± 0.1) m wide by (17.0 ± 0.1) m long. If the sidewalk is to measure (1.00 ± 0.01) m wide by (9.0 ± 0.1) cm thick, what volume of concrete is needed, and what is the approximate uncertainty of this volume?

Section 1.8 Mathematical Notation

58. Compute the value of $\Sigma_{i=1}^{4}\, x_i$ if $x_i = (2i + 1)$.

59. Determine whether the expression $\Sigma_{i=1}^{4}\, i^2$ is equal to $(\Sigma_{i=1}^{4}\, i)^2$.

ADDITIONAL PROBLEMS

60. In physics it is important to use mathematical approximations. Demonstrate for yourself that for small angles (< 20°)

$$\tan \alpha \approx \sin \alpha \approx \alpha = \frac{\pi \alpha'}{180°}$$

where α is in radians and α' is in degrees. Use a calculator to find the largest angle for which $\tan \alpha$ may be approximated by $\sin \alpha$ if the error is to be less than 10%.

61. A useful fact is that there are about $\pi \times 10^7$ s in one year. Use a calculator to find the percentage error in this approximation. *Note:* "percentage error" is defined as (difference/true value) × 100%.

62. Assume that there are 50 million passenger cars in the United States and that the average fuel consumption is 20 mi/gal of gasoline. If the average distance traveled by each car is 10 000 mi/yr, how much gasoline would be saved per year if average fuel consumption could be increased to 25 mi/gal?

63. One cubic centimeter (1.0 cm³) of water has a mass of 1.0×10^{-3} kg. (a) Determine the mass of 1.0 m³ of water. (b) Assuming biological substances are 98% water, estimate the mass of a cell that has a diameter of 1.0 μm, a human kidney, and a fly. Assume a kidney is roughly a sphere with a radius of 4.0 cm and a fly is roughly a cylinder 4.0 mm long and 2.0 mm in diameter.

64. The data in the following table represent measurements of the masses and dimensions of solid cylinders of aluminum, copper, brass, tin, and iron. Use these data to calculate the densities of these substances. Compare your results for aluminum, copper, and iron with those given in Table 1.5.

Substance	Mass (g)	Diameter (cm)	Length (cm)
Aluminum	51.5	2.52	3.75
Copper	56.3	1.23	5.06
Brass	94.4	1.54	5.69
Tin	69.1	1.75	3.74
Iron	216.1	1.89	9.77

65. Determine an order-of-magnitude estimation of (a) the number of words in this textbook, (b) the number of symbols (letters, numbers, Greek letters, mathematical operators, and so on) in this textbook, (c) the number of problems in this textbook.

SPREADSHEET PROBLEMS

S1. "On a Clear Day You Can See Forever" is the title song from a Broadway musical. How far can you really see? Assuming that light travels in straight lines, you could see the whole Earth from any point above the Earth's surface if the Earth were perfectly

THE WIZARD OF ID   By Parker and Hart

flat. Because the Earth is round, however, there is a maximum distance you can see from any given height. The location of the most distant visible point is called the *horizon*. Suppose your eyes are located at point P in Figure S1.1, where P is a height h above the Earth's surface. Assuming a perfectly smooth Earth of radius R, the horizon is located at point H. Point C is the center of the Earth, d is the straight-line distance from P to H, and s is the distance measured along the Earth's surface from a point directly beneath P to H. The Pythagorean theorem tells us that

$$(R + h)^2 = R^2 + d^2$$

from which we find

$$d = \sqrt{h^2 + 2hR}$$

From Figure S1.1, we see that

$$\cos \phi = \frac{R}{R + h}$$

$$s = R\phi$$

In the limit $h \ll R$, both d and s approach $\sqrt{2hR}$, which we can call L.

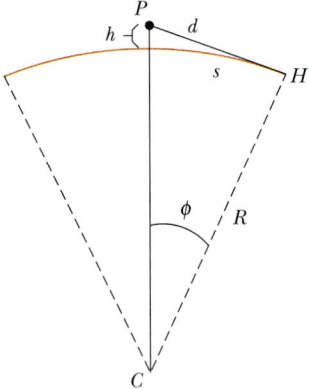

FIGURE S1.1

   Spreadsheet 1.1 uses these equations to calculate h/R, d, ϕ, s, and L, for any given h and R. The angle ϕ is given in both radians and degrees. Use the spreadsheet to answer the following questions: (a) Suppose Jud and Spud are standing on the Earth and are gazing off into the distance. Jud's eyes are 2 m above the ground, and Spud's are 1.5 m above the ground. How far away are Jud's and Spud's horizons? Does it make any difference whether you use d, s, or L to answer this question? (b) If Jud and Spud were on the Moon ($R = 1740$ km), where would their horizons be?

S2. Suppose you are standing on top of New York's World Trade Center, 417 m high, looking out toward the Atlantic Ocean. (a) How far is your horizon? Use Spreadsheet 1.1. How high would you have to be above the Earth's surface to see from New York City to Paris? (Estimate the New York–Paris distance using a world map.) (b) How high would you have to be for d and s to differ by 1 percent? 10 percent? (c) Choose h such that $h = NR$. Set N successively equal to 10, 100, 1000, . . . , and observe what happens to s and ϕ. Explain your observations.

S3. Spreadsheet programs are useful in examining data graphically. Most spreadsheet programs can fit the best straight line (a regression line) to a set of data. (See Example 4 in Chapters 1 or 2 of *Computer Investigations* for instructions on applying the least-squares method to a set of data.) The following table gives experimental results for the measurement of the period T of a pendulum of length L. These data are consistent with an equation of the form $T = CL^n$, where C and n are constants and n is not necessarily an integer.

Length L(m)	Period T(s)
0.25	1.00
0.50	1.40
0.75	1.75
1.00	2.00
1.50	2.50
2.00	2.80

(a) Use the least-squares or the regression-line procedure of your spreadsheet program to find the best fit to the data. Since $T = CL^n$, we see that

$$\log T = \log C + n \log L$$

First calculate a column of log T values and another of log L values. Use log L as the independent variable and log T as the dependent variable. The least-squares fit finds the slope n and the intercept log C. (b) Which data points deviate most from a straight-line plot of T versus L^n? (c) Is the experimental value of n found in part (a) consistent with a dimensional analysis of $T = CL^n$.

S4. The period T and orbit radius R for the motions of four moons of Jupiter are:

	Io	Europa	Ganymede	Callisto
Period, T(days)	1.77	3.55	7.16	16.69
R(km)	422 000	671 000	1 070 000	1 880 000

(a) These data can be fitted by the formula $T = CR^n$. Follow the procedures in Problem S3a to find C and n. (b) A fifth satellite, Amalthea, has a period of 0.50 days. Use $I = CR^n$ to find the radius for its orbit.

Motion in One Dimension

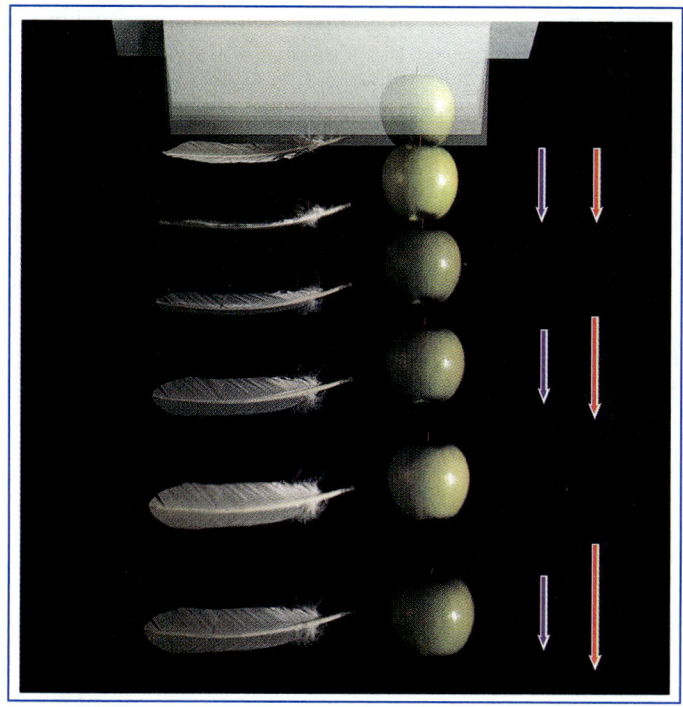

An apple and a feather released from rest in a vacuum chamber fall at the same rate regardless of their mass difference. If we neglect air resistance, we can say that all objects fall to the Earth with the same constant acceleration of magnitude 9.80 m/s^2, as indicated by the violet arrows in this multiflash photograph. The velocity of the two objects increases linearly with time, as indicated by the red arrows. (© 1993 James Sugar/Black Star)

A s a first step in studying mechanics, it is convenient to describe motion in terms of space and time, ignoring for the present the agents that caused that motion. This portion of mechanics is called *kinematics*. In this chapter we consider motion along a straight line, that is, one-dimensional motion. Starting with the concept of displacement discussed in the previous chapter, we first define velocity and acceleration. Then, using these concepts, we study the motion of objects traveling in one dimension under a constant acceleration.

From everyday experience we recognize that motion represents the continuous change in the position of an object. In physics we are concerned with three types of motion: translational, rotational, and vibrational. A car moving down a highway is undergoing translational motion, the Earth's daily spin on its axis is an example of rotational motion, and the back-and-forth motion of a pendulum is an example of vibrational motion.

In this and the next few chapters, we are concerned only with translational motion. In many situations, we can treat the moving object as a particle, which in mathematics is defined as a point having no size. For example, if we wish to describe the motion of the Earth around the Sun, we can treat the Earth as a particle

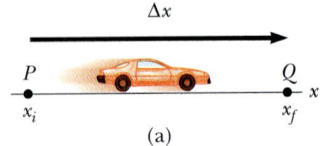

(a)

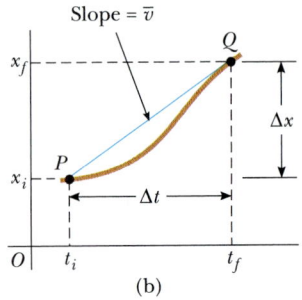

(b)

FIGURE 2.1 (a) A car moving to the right along a straight line taken to be the x axis. Because we are interested only in the car's translational motion, we can treat it as a particle. (b) Position-time graph for the motion of the "particle."

Displacement

Average velocity

---

and obtain reasonable accuracy in a prediction of the Earth's orbit. This approximation is justified because the radius of the Earth's orbit is large compared with the dimensions of the Earth and Sun. As an example, on a much smaller scale, it is possible to explain the pressure exerted by a gas on the walls of a container by treating the gas molecules as particles. Of course, the particle approach doesn't work in *all* situations. For instance, we cannot treat the Earth as a particle when we are examining its internal structure or when we are studying such phenomena as tides, earthquakes, and volcanic activity. On the other hand, we cannot consider gas molecules to be particles when we are studying properties that depend on molecular rotation and vibration. In general, we shall often find it both valid and convenient to treat a moving object as a particle.

2.1   DISPLACEMENT, VELOCITY, AND SPEED

The motion of a particle is completely known if the particle's position in space is known at all times. Consider a car (which we treat as a particle) moving along the x axis from a point P to a point Q as in Figure 2.1a. Let its position at point P be x_i at some time t_i, and let its position at point Q be x_f at time t_f. (The indices i and f refer to the initial and final values.) At times other than t_i and t_f, the position of the particle between these two points may vary as in Figure 2.1b. Such a plot is called a *position-time graph*. As the particle moves from position x_i to position x_f, its displacement is given by $x_f - x_i$. As mentioned in Chapter 1, we use the Greek letter delta (Δ) to denote the *change* in a quantity. Therefore, we write the change in the position of the particle (the displacement)

$$\Delta x \equiv x_f - x_i \tag{2.1}$$

From this definition, we see that Δx is positive if x_f is greater than x_i and negative if x_f is less than x_i.

> The **average velocity** $\bar{v}$ of the particle is defined as the ratio of its displacement Δx and the time interval Δt:
>
> $$\bar{v} \equiv \frac{\Delta x}{\Delta t} = \frac{x_f - x_i}{t_f - t_i} \tag{2.2}$$

From this definition,[1] we see that average velocity has dimensions of length divided by time (L/T)—m/s in SI units and ft/s in British engineering units. The average velocity is *independent* of the path taken between the points P and Q. This is true because the average velocity is proportional to the displacement, Δx, which depends only on the initial and final coordinates of the particle. It therefore follows that if a particle starts at some point and returns to the same point via any path, its average velocity for this trip is zero, because its displacement is zero.

Displacement should not be confused with the distance traveled, since the distance traveled for any motion is clearly nonzero. For instance, when a baseball player hits a home run as in Figure 2.2, he travels a distance of 360 ft in his trip

---

[1] In one-dimensional motion, all quantities describing the motion are scalars (having size only; see Chapter 3). Nevertheless, we introduce the term *velocity* here, rather than only *speed*, where the sign of the rate of motion is important.

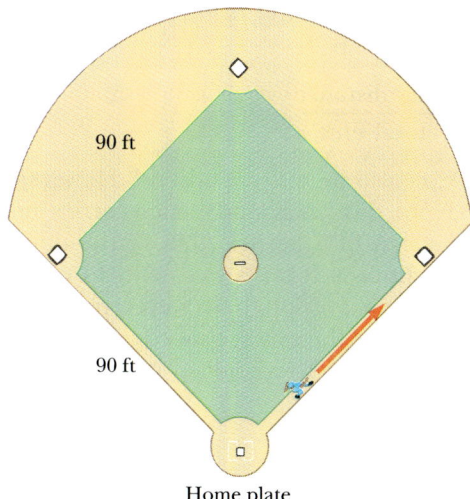

90 ft

90 ft

Home plate

FIGURE 2.2 Bird's-eye view of a baseball diamond. A batter who hits a home run travels 360 ft as he rounds the bases, but his displacement for the round trip is zero.

around the bases; however, his displacement is 0 ft because the player's final and initial positions are identical.

Average velocity gives us no details of the motion between points P and Q in Figure 2.1b. (How we evaluate the velocity at some instant in time is discussed in the next section.) The average velocity of a particle in one dimension can be positive or negative, depending on the sign of the displacement. (The time interval, Δt, is always positive.) If the coordinate of the particle increases in time (that is, if $x_f > x_i$), then Δx is positive and $\bar{v}$ is positive. This case corresponds to motion in the positive x direction. If the coordinate decreases in time ($x_f < x_i$), Δx is negative and hence $\bar{v}$ is negative. This case corresponds to motion in the negative x direction.

The average velocity can also be interpreted geometrically by drawing a straight line between the points P and Q in Figure 2.1b. This line forms the hypotenuse of a triangle of height Δx and base Δt. The slope of this line is the ratio $\Delta x / \Delta t$. Therefore, we see that the *average* velocity of the particle during the time interval t_i to t_f is equal to the slope of the straight line joining the initial and final points on the space-time graph.[2]

EXAMPLE 2.1 Calculate the Average Velocity

A particle moving along the x axis is located at $x_i = 12$ m at $t_i = 1$ s and at $x_f = 4$ m at $t_f = 3$ s. Find its displacement, average velocity, and average speed during this time interval.

Solution The displacement is given by Equation 2.1:

$$\Delta x = x_f - x_i = 4 \text{ m} - 12 \text{ m} = \boxed{-8 \text{ m}}$$

The average velocity is, from Equation 2.2,

$$\bar{v} = \frac{\Delta x}{\Delta t} = \frac{x_f - x_i}{t_f - t_i} = \frac{4 \text{ m} - 12 \text{ m}}{3 \text{ s} - 1 \text{ s}} = -\frac{8 \text{ m}}{2 \text{ s}} = \boxed{-4 \text{ m/s}}$$

The displacement and average velocity are negative for this time interval because the particle has moved to the left, toward decreasing values of x. Its average speed for this trip is 4 m/s.

---

[2] Slope represents the ratio of the change in the quantity represented on the vertical axis to the change in the quantity represented on the horizontal axis.

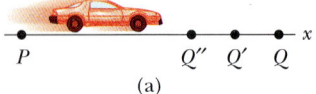

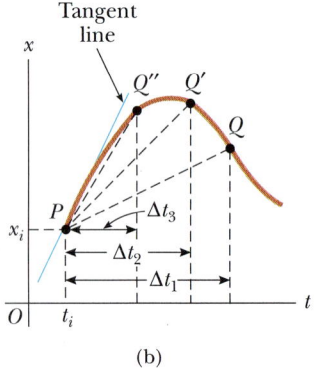

FIGURE 2.3 (a) As the car moves along the *x* axis, and *Q* is brought closer to *P*, the time it takes to travel the distance decreases. (b) Position-time graph for the "particle." As the time intervals get smaller and smaller, the average velocity for that interval, equal to the slope of the dashed line connecting *P* and the appropriate *Q*, approaches the slope of the line tangent at *P*. The instantaneous velocity at *P* is the slope of the blue tangent line at the time t_1.

Definition of instantaneous velocity

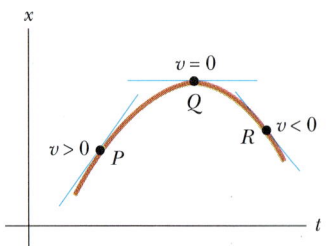

FIGURE 2.4 In this position-time graph, the velocity is positive at *P*, where the slope of the tangent line is positive, zero at *Q*, where the slope of the tangent line is zero, and negative at *R*, where the slope of the tangent line is negative.

The **average speed** of a particle is defined as the ratio of the total distance traveled to the total time it takes to travel that distance:

$$\text{Average speed} = \frac{\text{total distance}}{\text{total time}}$$

The SI unit of average speed, like velocity, is also meters per second. However, unlike average velocity, average speed has no direction, and hence carries no algebraic sign. Knowledge of the average speed of a particle tells you nothing about the details of the trip. For example, suppose it takes you 8.0 h to travel 280 km in your car. The average speed for your trip is 35 km/h. However, you most likely traveled at various speeds during the trip, and the average speed of 35 km/h could result from an infinite number of possible speed variations.

2.2 INSTANTANEOUS VELOCITY AND SPEED

We would like to be able to define the velocity of a particle at a particular instant of time, rather than just over a finite interval of time. The velocity of a particle at any instant of time—in other words, at some point on a space-time graph—is called the **instantaneous velocity**. This concept is especially important when the average velocity in different time intervals is *not* constant.

Consider the straight-line motion of a particle between two points *P* and *Q* on the *x* axis as in Figure 2.3a. As *Q* is brought closer and closer to *P*, the time needed to travel the distance gets progressively smaller. The average velocity for each time interval is the slope of the appropriate dotted line in the space-time graph shown in Figure 2.3b. As *Q* approaches *P*, the time interval approaches zero and the slope of the dotted line approaches that of the blue line tangent to the curve at *P*. The slope of this line is defined to be the *instantaneous velocity* at the time t_i. In other words,

> the instantaneous velocity, *v*, equals the limiting value of the ratio $\Delta x/\Delta t$ as Δt approaches zero[3]:
>
> $$v \equiv \lim_{\Delta t \to 0} \frac{\Delta x}{\Delta t} \tag{2.3}$$

In the calculus notation, this limit is called the *derivative* of *x* with respect to *t*, written *dx/dt*:

$$v \equiv \lim_{\Delta t \to 0} \frac{\Delta x}{\Delta t} = \frac{dx}{dt} \tag{2.4}$$

The instantaneous velocity can be positive, negative, or zero. When the slope of the position-time graph is positive, such as at *P* in Figure 2.4, *v* is positive. At point *R*, *v* is negative since the slope is negative. Finally, the instantaneous velocity is zero at the peak *Q* (the turning point), where the slope is zero.

---

[3] Note that the displacement, Δx, also approaches zero as Δt approaches zero. As Δx and Δt become smaller and smaller, the ratio $\Delta x/\Delta t$ approaches a value equal to the slope of the line tangent to the *x* versus *t* curve.

From here on, we use the word *velocity* to designate instantaneous velocity. When it is *average velocity* we are interested in, we always use the adjective *average*.

The **speed** of a particle is defined to be equal to the magnitude of its velocity. Speed has no direction associated with it and, hence, carries no algebraic sign. For example, if one particle has a velocity of $+25$ m/s, and another particle has a velocity of -25 m/s, both have a speed of 25 m/s. A car's speedometer indicates the speed of the car, not its velocity.

It is also possible to use a mathematical technique called integration to find the displacement of a particle if its velocity is known as a function of time. Because integration procedures may not be familiar to many students, the topic is treated in (optional) Section 2.6 for general interest and for those courses that cover this material.

Speed

EXAMPLE 2.2   Average and Instantaneous Velocity

A particle moves along the x axis. Its x coordinate varies with time according to the expression $x = -4t + 2t^2$, where x is in meters and t is in seconds. The position-time graph for this motion is shown in Figure 2.5. Note that the particle moves in the negative x direction for the first second of motion, is at rest at the moment $t = 1$ s, and then heads back in the positive x direction for $t > 1$ s. (a) Determine the displacement of the particle in the time intervals $t = 0$ to $t = 1$ s and $t = 1$ s to $t = 3$ s.

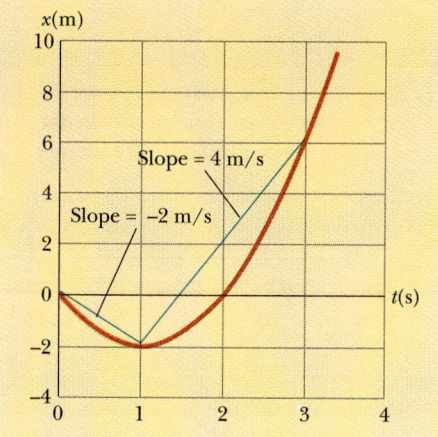

FIGURE 2.5   (Example 2.2) Position-time graph for a particle having an x coordinate that varies in time according to the expression $x = -4t + 2t^2$.

Solution   In the first time interval, we set $t_i = 0$ and $t_f = 1$ s. Since $x = -4t + 2t^2$, and using Equation 2.1, we get for the first displacement

$$\Delta x_{01} = x_f - x_i$$
$$= [-4(1) + 2(1)^2] - [-4(0) + 2(0)^2]$$

$$= -2 \text{ m}$$

In the second time interval we can set $t_i = 1$ s and $t_f = 3$ s. Therefore, the displacement in this interval is

$$\Delta x_{13} = x_f - x_i$$
$$= [-4(3) + 2(3)^2] - [-4(1) + 2(1)^2]$$

$$= +8 \text{ m}$$

These displacements can also be read directly from the position-time graph (Fig. 2.5).

(b) Calculate the average velocity in the time intervals $t = 0$ to $t = 1$ s and $t = 1$ s to $t = 3$ s.

Solution   In the first time interval, $\Delta t = t_f - t_i = 1$ s. Therefore, using Equation 2.2 and the results from (a) gives

$$\bar{v}_{01} = \frac{\Delta x_{01}}{\Delta t} = \frac{-2 \text{ m}}{1 \text{ s}} = -2 \text{ m/s}$$

In the second time interval, $\Delta t = 2$ s; therefore,

$$\bar{v}_{13} = \frac{\Delta x_{13}}{\Delta t} = \frac{8 \text{ m}}{2 \text{ s}} = +4 \text{ m/s}$$

These values agree with the slopes of the lines joining these points in Figure 2.5.

(c) Find the instantaneous velocity of the particle at $t = 2.5$ s.

Solution   By measuring the slope of the position-time graph at $t = 2.5$ s, we find that $v = +6$ m/s. We could also use Equation 2.4 and the rules of differential calculus to find the velocity from the displacement:

$$v = \frac{dx}{dt} = \frac{d}{dt}(-4t + 2t^2) = 4(-1 + t) \text{ m/s}$$

Therefore, at $t = 2.5$ s,

$$v = 4(-1 + 2.5) = +6 \text{ m/s}$$

A review of basic operations in the calculus is provided in Appendix B.6.

Exercise   Do you see any symmetry in the motion? For example, does the speed ever repeat itself?

EXAMPLE 2.3   The Limiting Process

The position of a particle moving along the x axis varies in time according to the expression $x = (3 \text{ m/s}^2)t^2$, where x is in meters, and t is in seconds. Find the velocity at any time.

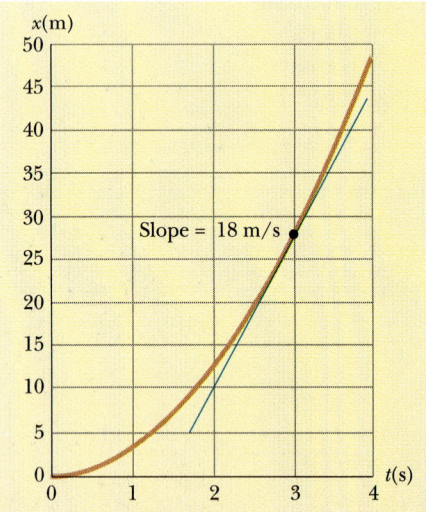

FIGURE 2.6   (Example 2.3) Position-time graph for a particle having an x coordinate that varies in time according to $x = 3t^2$. Note that the instantaneous velocity at $t = 3$ s equals the slope of the blue line tangent to the curve at this instant.

Reasoning and Solution   The position-time graph for this motion is shown in Figure 2.6. We can compute the velocity at any time t by using the definition of the instantaneous velocity (Eq. 2.3). If the initial coordinate of the particle at time t is $x_i = 3t^2$, then the coordinate at a later time $t + \Delta t$ is

$$x_f = 3(t + \Delta t)^2 = 3[t^2 + 2t\,\Delta t + (\Delta t)^2]$$
$$= 3t^2 + 6t\,\Delta t + 3(\Delta t)^2$$

Therefore, the displacement in the time interval Δt is

$$\Delta x = x_f - x_i = 3t^2 + 6t\,\Delta t + 3(\Delta t)^2 - 3t^2$$
$$= 6t\,\Delta t + 3(\Delta t)^2$$

The average velocity in this time interval is

$$\bar{v} = \frac{\Delta x}{\Delta t} = 6t + 3\,\Delta t$$

To find the instantaneous velocity, we take the limit of this expression as Δt approaches zero, as shown by Equation 2.3. In doing so, we see that the term $3\,\Delta t$ goes to zero, therefore

$$v = \lim_{\Delta t \to 0} \frac{\Delta x}{\Delta t} = \boxed{(6 \text{ m/s}^2)t}$$

Notice that this expression gives us the velocity at *any* general time t. It tells us that v is increasing linearly in time. It is then a straightforward matter to find the velocity at some specific time from the expression $v = (6 \text{ m/s}^2)t$. For example, at $t = 3.0$ s, the velocity is $v = (6 \text{ m/s}^2)(3.0 \text{ s}) = +18$ m/s. Again, this can be checked from the slope of the graph (the blue line) at $t = 3.0$ s.

The limiting process can also be examined numerically. For example, we can compute the displacement and average velocity for various time intervals beginning at $t = 3.0$ s, using the expressions for Δx and $\bar{v}$. The results of such calculations are given in Table 2.1. As the time intervals get smaller and smaller, the average velocity approaches the value of the instantaneous velocity at $t = 3.0$ s, namely, $+18$ m/s.

TABLE 2.1	Displacement and Average Velocity for Various Time Intervals for the Function $x = (3 \text{ m/s}^2)t^2$ (the intervals begin at $t = 3.00$ s)	
$\Delta t(s)$	$\Delta x(m)$	$\bar{v} = \Delta x/\Delta t(\text{m/s})$
1.00	21	21
0.50	9.75	19.5
0.25	4.69	18.8
0.10	1.83	18.3
0.05	0.9075	18.15
0.01	0.1803	18.03
0.001	0.018003	18.003

2.3   ACCELERATION

When the velocity of a particle changes with time, the particle is said to be *accelerating*. For example, the speed of a car will increase when you "step on the gas" and decreases when you apply the brakes. However, we need a more precise definition of acceleration than this.

Suppose a particle moving along the x axis has a velocity v_i at time t_i and a velocity v_f at time t_f, as in Figure 2.7a.

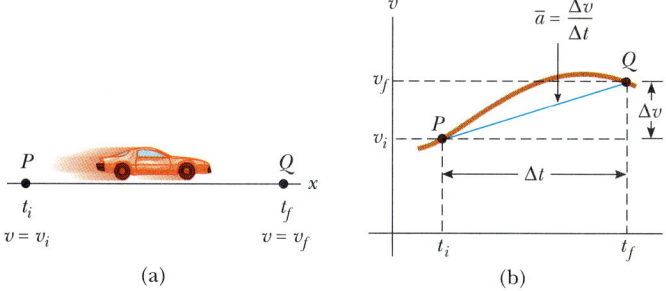

FIGURE 2.7   (a) A ''particle'' moving from P to Q has velocity v_i at $t = t_i$ and velocity v_f at $t = t_f$. (b) Velocity-time graph for the particle moving in a straight line. The slope of the blue straight line connecting P and Q is the average acceleration in the time interval $\Delta t = t_f - t_i$.

The **average acceleration** of the particle in the time interval $\Delta t = t_f - t_i$ is defined as the ratio $\Delta v / \Delta t$, where $\Delta v = v_f - v_i$ is the *change* in velocity in this time interval:

$$\bar{a} \equiv \frac{\Delta v}{\Delta t} = \frac{v_f - v_i}{t_f - t_i} \tag{2.5}$$

Average acceleration

Acceleration has dimensions of length divided by (time)^2, or L/T^2. Some of the common units of acceleration are meters per second per second (m/s^2) and feet per second per second (ft/s^2). As with velocity, we can use positive and negative signs to indicate the direction of the acceleration when the motion being analyzed is one dimensional.

In some situations, the value of the average acceleration may be different over different time intervals. It is therefore useful to define the **instantaneous acceleration** as the limit of the average acceleration as Δt approaches zero. This concept is analogous to the definition of instantaneous velocity discussed in the previous section. If we imagine that the point Q is brought closer and closer to the point P in Figure 2.7a, and take the limit of $\Delta v / \Delta t$ as Δt approaches zero, we get the *instantaneous acceleration:*

$$a \equiv \lim_{\Delta t \to 0} \frac{\Delta v}{\Delta t} = \frac{dv}{dt} \tag{2.6}$$

Instantaneous acceleration

That is, the instantaneous acceleration equals the derivative of the velocity with respect to time, which by definition is the slope of the velocity-time graph (Fig. 2.7b). One can interpret the derivative of the velocity with respect to time as the *time rate of change of velocity.* If a is positive, the acceleration is in the positive x direction, whereas negative a indicates acceleration in the negative x direction.

From now on we shall use the term *acceleration* to mean instantaneous acceleration.

Since $v = dx/dt$, the acceleration can also be written

$$a = \frac{dv}{dt} = \frac{d}{dt}\left(\frac{dx}{dt}\right) = \frac{d^2x}{dt^2} \tag{2.7}$$

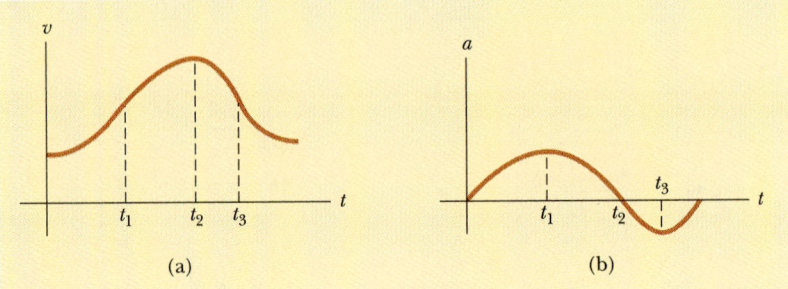

FIGURE 2.8 The instantaneous acceleration can be obtained from the velocity-time graph. (a) At each instant, the acceleration in the *a* versus *t* graph (b) equals the slope of the line tangent to the *v* versus *t* curve.

That is, in one-dimensional motion, the acceleration equals the *second derivative* of the *x* coordinate with respect to time.

Figure 2.8 shows how the acceleration-time curve can be derived from the velocity-time curve. The acceleration at any time is the slope of the velocity-time graph at that time. Positive values of the acceleration correspond to those points in Figure 2.8a where the velocity is increasing in the positive *x* direction. The acceler-

---

EXAMPLE 2.4   Graphical Relations Between *x*, *v*, and *a*

The position of an object moving along the *x* axis varies with time as in Figure 2.9a. Let us use graphical procedures to obtain graphs of the velocity versus time and acceleration versus time for the object.

Reasoning and Solution   The velocity at any instant is the slope of the tangent to the *x-t* graph at that instant. Between $t = 0$ and $t = t_1$, the slope of the *x-t* graph increases uniformly, so the velocity increases linearly as in Figure 2.9b. Between t_1 and t_2, the slope of the *x-t* graph is constant, so the velocity remains constant. At t_4, the slope of the *x-t* graph is zero, so the velocity is zero at that instant. Between t_4 and t_5, the slope of the *x-t* graph is negative and decreases uniformly; therefore the velocity is negative and constant in this interval. In the interval t_5 to t_6, the slope of the *x-t* graph is still negative, and goes to zero at t_6. Finally, after t_6, the slope of the *x-t* graph is zero, so the object is at rest.

Similarly, the acceleration at any instant is the slope of the tangent to the *v-t* graph at that instant. The graph of acceleration versus time for this object is shown in Figure 2.9c. Note that the acceleration is constant and positive between 0 and t_1, where the slope of the *v-t* graph is positive; the acceleration is zero between t_1 and t_2 and for $t \geq t_6$, because the slope of the *v-t* graph is zero.

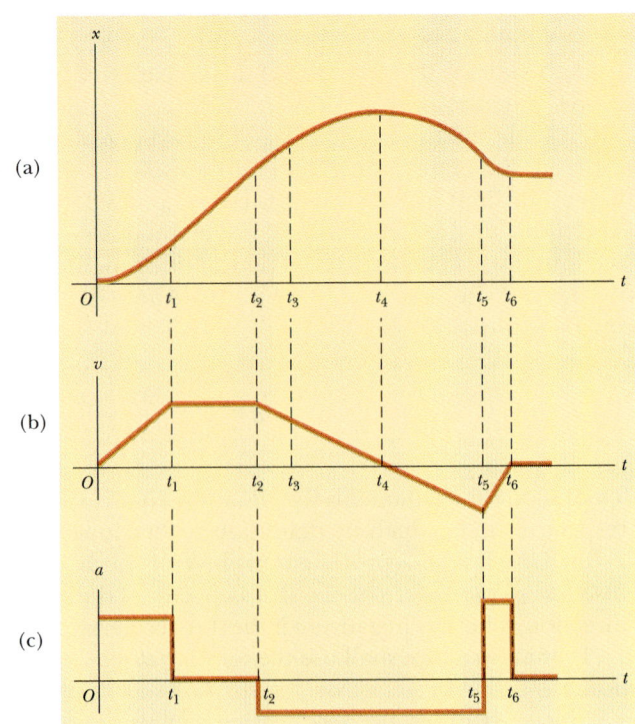

FIGURE 2.9 (Example 2.4) (a) Position-time graph for an object moving along the *x* axis. (b) The velocity versus time for the object is obtained by measuring the slope of the position-time graph at each instant. (c) The acceleration versus time for the object is obtained by measuring the slope of the velocity-time graph at each instant.

ation reaches a maximum at time t_1, when the slope of the velocity-time graph is a maximum. The acceleration then goes to zero at time t_2, when the velocity is a maximum (that is, when the velocity is momentarily not changing and the slope of the v versus t graph is zero). The acceleration is negative when the velocity in the positive x direction is decreasing in time, and it reaches its minimum value at time t_3.

CONCEPTUAL EXAMPLE 2.5

(a) If a car is traveling eastward, can its acceleration be westward? Explain. If a car is slowing down, can its acceleration be positive?

Reasoning   (a) Yes. This occurs when a car is slowing down, so that the direction of its acceleration is opposite to its direction of motion. (b) Yes. If the motion is in the direction chosen as negative, a positive acceleration causes a decrease in speed, often called a deceleration.

EXAMPLE 2.6   Average and Instantaneous Acceleration

The velocity of a particle moving along the x axis varies in time according to the expression $v = (40 - 5t^2)$ m/s, where t is in seconds. (a) Find the average acceleration in the time interval $t = 0$ to $t = 2.0$ s.

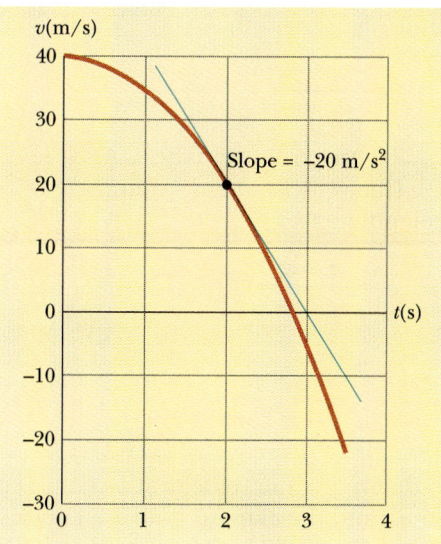

FIGURE 2.10   (Example 2.6) The velocity-time graph for a particle moving along the x axis according to the relation $v = (40 - 5t^2)$ m/s. Note that the acceleration at $t = 2$ s is equal to the slope of the blue tangent line at that time.

Solution   The velocity-time graph for this function is given in Figure 2.10. The velocities at $t_i = 0$ and $t_f = 2.0$ s are found by substituting these values of t into the expression given for the velocity:

$$v_i = (40 - 5t_i^2) \text{ m/s} = [40 - 5(0)^2] \text{ m/s} = +40 \text{ m/s}$$

$$v_f = (40 - 5t_f^2) \text{ m/s} = [40 - 5(2.0)^2] \text{ m/s} = +20 \text{ m/s}$$

Therefore, the average acceleration in the specified time interval $\Delta t = t_f - t_i = 2.0$ s is

$$\bar{a} = \frac{v_f - v_i}{t_f - t_i} = \frac{(20 - 40) \text{ m/s}}{(2.0 - 0) \text{ s}} = \boxed{-10 \text{ m/s}^2}$$

The negative sign is consistent with the fact that the slope of the line joining the initial and final points on the velocity-time graph is negative.

(b) Determine the acceleration at $t = 2.0$ s.

Solution   The velocity at time t is $v_i = (40 - 5t^2)$ m/s, and the velocity at time $t + \Delta t$ is

$$v_f = 40 - 5(t + \Delta t)^2 = 40 - 5t^2 - 10t \, \Delta t - 5(\Delta t)^2$$

Therefore, the change in velocity over the time interval Δt is

$$\Delta v = v_f - v_i = [-10t \, \Delta t - 5(\Delta t)^2] \text{ m/s}$$

Dividing this expression by Δt and taking the limit of the result as Δt approaches zero gives the acceleration at *any* time t:

$$a = \lim_{\Delta t \to 0} \frac{\Delta v}{\Delta t} = \lim_{\Delta t \to 0} (-10t - 5 \, \Delta t) = -10t \text{ m/s}^2$$

Therefore, at $t = 2.0$ s, we find that

$$a = (-10)(2.0) \text{ m/s}^2 = \boxed{-20 \text{ m/s}^2}$$

This result can also be obtained by measuring the slope of the velocity-time graph at $t = 2.0$ s. Note that the acceleration is not constant in this example. Situations involving constant acceleration are treated in the next section.

So far we have evaluated the derivatives of a function by starting with the definition of the function and then taking the limit of a specific ratio. Those of you familiar with the calculus should recognize that there are specific rules for taking the derivatives of various functions. These rules, which are listed in Appendix B.6, enable us to evaluate derivatives quickly.

Suppose x is proportional to some power of t, such as

$$x = At^n$$

where A and n are constants. (This is a very common functional form.) The derivative of x with respect to t is given by

$$\frac{dx}{dt} = nAt^{n-1}$$

Applying this rule to Example 2.3, where $x = 3t^2$, we see that $v = dx/dt = 6t$, in agreement with our result when we took the limit explicitly. Likewise, in Example 2.6, where $v = 40 - 5t^2$, we find that $a = dv/dt = -10t$. (Note that the rate of change of any constant quantity is zero.)

2.4   ONE-DIMENSIONAL MOTION WITH CONSTANT ACCELERATION

If the acceleration of a particle varies in time, the motion can be complex and difficult to analyze. However, a very common and simple type of one-dimensional motion occurs when the acceleration is constant, or uniform. When the acceleration is constant, the average acceleration equals the instantaneous acceleration. Consequently, the velocity increases or decreases at the same rate throughout the motion.

If we replace $\bar{a}$ by a in Equation 2.5, we find that

$$a = \frac{v_f - v_i}{t_f - t_i}$$

For convenience, let $t_i = 0$ and t_f be any arbitrary time t. Also, let $v_i = v_0$ (the initial velocity at $t = 0$) and $v_f = v$ (the velocity at any arbitrary time t). With this notation, we can express the acceleration as

$$a = \frac{v - v_0}{t}$$

or

Velocity as a function of time

$$v = v_0 + at \qquad \text{(for constant } a) \qquad (2.8)$$

This expression enables us to determine the velocity at *any* time t if the initial velocity, the (constant) acceleration, and the elapsed time are known. A velocity-time graph for this motion is shown in Figure 2.11a. The graph is a straight line the slope of which is the acceleration, a, consistent with the fact that $a = dv/dt$ is a constant. Note that if the acceleration were negative, the slope of Figure 2.11a would be negative. If the acceleration is in the direction opposite to the velocity, then the particle is slowing down. From this graph and from Equation 2.8, we see that the velocity at any time t is the sum of the initial velocity, v_0, and the change in velocity, at.

The graph of acceleration versus time (Fig. 2.11b) is a straight line with a slope of zero, since the acceleration is constant.

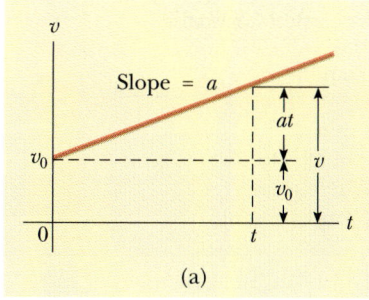

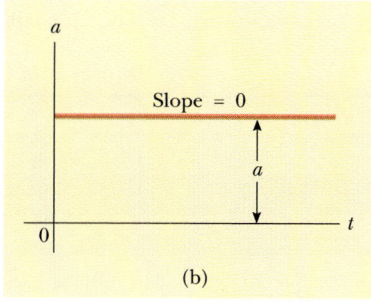

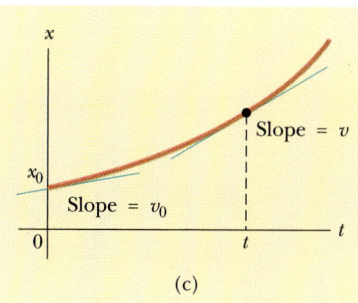

FIGURE 2.11 A particle moving along the *x* axis with constant acceleration *a*; (a) the velocity-time graph, (b) the acceleration-time graph, and (c) the position-time graph.

Because the velocity varies linearly in time according to Equation 2.8, we can express the average velocity in any time interval as the arithmetic mean of the initial velocity, v_0, and the final velocity, v:

$$\bar{v} = \frac{v_0 + v}{2} \quad \text{(for constant } a\text{)} \tag{2.9}$$

Note that this expression is useful only when the acceleration is constant, that is, when the velocity varies linearly with time.

We can now use Equations 2.2 and 2.9 to obtain the displacement as a function of time. Again, we choose $t_i = 0$, at which time the initial position is $x_i = x_0$. This gives

$$\Delta x = \bar{v}\,\Delta t = \left(\frac{v_0 + v}{2}\right) t$$

$$x - x_0 = \tfrac{1}{2}(v + v_0)t \quad \text{(for constant } a\text{)} \tag{2.10}$$

Displacement as a function of velocity and time

We can obtain another useful expression for the displacement by substituting Equation 2.8 into Equation 2.10:

$$x - x_0 = \tfrac{1}{2}(v_0 + v_0 + at)t$$

$$x - x_0 = v_0 t + \tfrac{1}{2}at^2 \quad \text{(for constant } a\text{)} \tag{2.11}$$

Displacement as a function of time

The validity of this expression can be checked by differentiating it with respect to time:

$$v = \frac{dx}{dt} = \frac{d}{dt}(x_0 + v_0 t + \tfrac{1}{2}at^2) = v_0 + at$$

Finally, we can obtain an expression that does not contain the time by substituting the value of *t* from Equation 2.8 into Equation 2.10:

$$x - x_0 = \tfrac{1}{2}(v_0 + v)\left(\frac{v - v_0}{a}\right) = \frac{v^2 - v_0^2}{2a}$$

$$v^2 = v_0^2 + 2a(x - x_0) \quad \text{(for constant } a\text{)} \tag{2.12}$$

Velocity as a function of displacement

TABLE 2.2 Kinematic Equations for Motion in a Straight Line Under Constant Acceleration	
Equation	**Information Given by Equation**
$v = v_0 + at$	Velocity as a function of time
$x - x_0 = \frac{1}{2}(v + v_0)t$	Displacement as a function of velocity and time
$x - x_0 = v_0 t + \frac{1}{2}at^2$	Displacement as a function of time
$v^2 = v_0^2 + 2a(x - x_0)$	Velocity as a function of displacement

Note: Motion is along the x axis. At $t = 0$, the position of the particle is x_0 and its velocity is v_0.

A position-time graph for motion under constant acceleration assuming positive a is shown in Figure 2.11c. Note that the curve representing Equation 2.11 is a parabola. The slope of the tangent to this curve at $t = 0$ equals the initial velocity, v_0, and the slope of the tangent line at any time t equals the velocity at that time.

If motion occurs in which the acceleration is *zero,* then we see that

$$\left. \begin{array}{l} v = v_0 \\ x - x_0 = vt \end{array} \right\} \qquad \text{when } a = 0$$

That is, when the acceleration is zero, the velocity is a constant and the displacement changes linearly with time.

Equations 2.8 through 2.12 are *kinematic expressions that may be used to solve any problem in one-dimensional motion with constant acceleration.* Keep in mind that these relationships were derived from the definition of velocity and acceleration, together with some simple algebraic manipulations and the requirement that the acceleration be constant. It is often convenient to choose the initial position of the particle as the origin of the motion, so that $x_0 = 0$ at $t = 0$. In such a case, the displacement is simply x.

The four kinematic equations that are used most often are listed in Table 2.2 for convenience.

The choice of which kinematic equation or equations you should use in a given situation depends on what is known beforehand. Sometimes it is necessary to use two of these equations to solve for two unknowns, such as the displacement and velocity at some instant. For example, suppose the initial velocity, v_0, and acceleration, a, are given. You can then find (1) the velocity after a time t has elapsed, using $v = v_0 + at$, and (2) the displacement after a time t has elapsed, using $x - x_0 = v_0 t + \frac{1}{2}at^2$. You should recognize that the quantities that vary during the motion are velocity, displacement, and time.

You will get a great deal of practice in the use of these equations by solving a number of exercises and problems. Many times you will discover that there is more than one method for obtaining a solution.

CONCEPTUAL EXAMPLE 2.7

Can these equations of kinematics be used in a situation where the acceleration varies with time? Can they be used when the acceleration is zero?

Reasoning   The equations of kinematics cannot be used if the acceleration varies continuously. However, if the acceleration changes in steps, having one constant value for a while and then another constant value for some later time interval, the equations for constant-acceleration motion can be used to follow each section of the motion separately. If the acceleration is zero during some time interval, the velocity is constant, and the kinematic equations can be used; in this case, because $a = 0$, the equations become $v = v_0$, and $x - x_0 = vt$.

EXAMPLE 2.8   The Supercharged Sports Car

A certain automobile manufacturer claims that its super-deluxe sports car will accelerate from rest to a speed of 42.0 m/s in 8.00 s. Under the improbable assumption that the acceleration is constant: (a) determine the acceleration of the car in m/s².

Solution   First note that $v_0 = 0$ and the velocity after 8.00 s is $v = 42.0$ m/s. Because we are given v_0, v, and t, we can use $v = v_0 + at$ to find the acceleration:

$$a = \frac{v - v_0}{t} = \frac{42.0 \text{ m/s}}{8.00 \text{ s}} = \boxed{+5.25 \text{ m/s}^2}$$

In reality, this is an average acceleration, since it is unlikely that a car accelerates uniformly.

(b) Find the distance the car travels in the first 8.00 s.

Solution   Let the origin be at the original position of the car, so that $x_0 = 0$. Using Equation 2.10 we find that

$$x = \tfrac{1}{2}(v_0 + v)t = \tfrac{1}{2}(42.0 \text{ m/s})(8.00 \text{ s}) = \boxed{168 \text{ m}}$$

(c) What is the speed of the car 10.0 s after it begins its motion, assuming it continues to accelerate at the average rate of 5.25 m/s²?

Solution   Again we can use $v = v_0 + at$, this time with $v_0 = 0$, $t = 10.0$ s, and $a = 5.25$ m/s²:

$$v = v_0 + at = 0 + (5.25 \text{ m/s}^2)(10.0 \text{ s}) = \boxed{52.5 \text{ m/s}}$$

EXAMPLE 2.9   Accelerating an Electron

An electron in the cathode ray tube of a television set enters a region where it accelerates uniformly from a speed of 3.00×10^4 m/s to a speed of 5.00×10^6 m/s in a distance of 2.00 cm. (a) For what length of time is the electron in this region where it accelerates?

Solution   Taking the direction of motion to be along the x axis, we can use Equation 2.10 to find t, since the displacement and velocities are known:

$$x - x_0 = \tfrac{1}{2}(v_0 + v)t$$

$$t = \frac{2(x - x_0)}{v_0 + v} = \frac{2(2.00 \times 10^{-2} \text{ m})}{(3.00 \times 10^4 + 5.00 \times 10^6) \text{ m/s}}$$

$$= \boxed{7.95 \times 10^{-9} \text{ s}}$$

(b) What is the acceleration of the electron in this region?

Solution   To find the acceleration, we can use $v = v_0 + at$ and the results from (a):

$$a = \frac{v - v_0}{t} = \frac{(5.00 \times 10^6 - 3.00 \times 10^4) \text{ m/s}}{7.95 \times 10^{-9} \text{ s}}$$

$$= \boxed{6.25 \times 10^{14} \text{ m/s}^2}$$

We also could have used Equation 2.12 to obtain the acceleration, since the velocities and displacement are known. Try it! Although the acceleration is very large in this example, it occurs over a very short time interval and is a typical value for charged particles in accelerators.

EXAMPLE 2.10   Watch Out for the Speed Limit

A car traveling at a constant speed of 30.0 m/s (≈ 67 mi/h) passes a trooper hidden behind a billboard. One second after the speeding car passes the billboard, the trooper sets in chase after the car with a constant acceleration of 3.00 m/s². How long does it take the trooper to overtake the speeding car?

Reasoning   To solve this problem algebraically, we write expressions for the position of each vehicle as a function of time. It is convenient to choose the origin at the position of the billboard and take $t = 0$ as the time the trooper begins moving. At that instant, the speeding car has already traveled a distance of 30.0 m because it travels at a constant speed of

30.0 m/s. Thus, the initial position of the speeding car is $x_0 = 30.0$ m.

Solution   Because the car moves with constant speed, its acceleration is zero, and applying Equation 2.11 gives

$$x_C = 30.0 \text{ m} + (30.0 \text{ m/s})t$$

Note that at $t = 0$, this expression does give the car's correct initial position $x_C = x_0 = 30.0$ m.

For the trooper, who starts from the origin at $t = 0$, we have $x_0 = 0$, $v_0 = 0$, and $a = 3.00$ m/s^2. Hence, the position of the trooper as a function of time is

$$x_T = \tfrac{1}{2}at^2 = \tfrac{1}{2}(3.00 \text{ m/s}^2)t^2$$

The trooper overtakes the car at the instant that $x_T = x_C$, or

$$\tfrac{1}{2}(3.00 \text{ m/s}^2)t^2 = 30.0 \text{ m} + (30.0 \text{ m/s})t$$

This gives the quadratic equation

$$1.50t^2 - 30.0t - 30.0 = 0$$

whose positive solution is $t = 21.0$ s. (For help in solving quadratic equations, see Appendix B.2.) Note that in this time interval, the trooper travels a distance of about 660 m.

Exercise   This problem can also be solved graphically. On the same graph, plot the position versus time for each vehicle, and from the intersection of the two curves determine the time at which the trooper overtakes the speeding car.

2.5   FREELY FALLING OBJECTS

It is well known that all objects, when dropped, fall toward the Earth with nearly constant acceleration. There is a legend that Galileo first discovered this fact by observing that two different weights dropped simultaneously from the Leaning Tower of Pisa hit the ground at approximately the same time. Although there is some doubt that this particular experiment was carried out, it is well established that Galileo did perform many systematic experiments on objects moving on inclined planes. Through careful measurements of distances and time intervals, he was able to show that the displacement of an object starting from rest is proportional to the square of the time the object is in motion. This observation is consistent with one of the kinematic equations we derived for motion under constant acceleration (Eq. 2.11). Galileo's achievements in the science of mechanics paved the way for Newton in his development of the laws of motion.

You might want to try the following experiment. Drop a coin and a crumpled-up piece of paper simultaneously from the same height. In the absence of air resistance, both will experience the same motion and hit the floor at the same

Galileo performing demonstrations of balls rolling down a grooved inclined plane. As a ball rolled down the incline, Galileo carefully measured its position at the end of equal time intervals and showed that the displacement was proportional to the square of the elapsed time. This painting by Giuseppe Bezzouli is located in the Zoological Museum in Florence, Italy.   *(Art Resource)*

time. In a real (non-ideal) experiment, air resistance cannot be neglected. In the idealized case, where air resistance *is* neglected, such motion is referred to as *free-fall*. If this same experiment could be conducted in a good vacuum, where air friction is truly negligible, the paper and coin would fall with the same acceleration, regardless of the shape of the paper. This point is illustrated very convincingly on page 23, in the photograph of the apple and feather falling in a vacuum. On August 2, 1971, such an experiment was conducted by astronaut David Scott on the Moon (where air resistance is negligible). He simultaneously released a geologist's hammer and a falcon's feather, and in unison they fell to the lunar surface. This demonstration would have surely pleased Galileo!

We shall denote the *free-fall acceleration* by the symbol g. The value of g on Earth decreases with increasing altitude. Furthermore, there are slight variations in g with latitude. The free-fall acceleration is directed downward toward the center of the Earth. At the Earth's surface, the value of g is approximately 9.80 m/s², or 980 cm/s², or 32 ft/s². Unless stated otherwise, we shall use this value for g when doing calculations.

When we use the expression *freely falling object,* we do not necessarily refer to an object dropped from rest. A freely falling object is any object moving freely under the influence of gravity, *regardless* of its initial motion. Objects thrown upward or downward and those released from rest are all falling freely once they are released. Furthermore, it is important to recognize that any freely falling object experiences an acceleration directed *downward*. This is true regardless of the initial motion of the object.

> An object thrown upward and one thrown downward will both experience the same acceleration as an object released from rest. Once they are in free-fall, all objects have an acceleration downward, equal to the free-fall acceleration.

If we neglect air resistance and assume that the free-fall acceleration does not vary with altitude, then the vertical motion of a freely falling object is equivalent to motion in one dimension under constant acceleration. Therefore, our kinematic equations for constant acceleration can be applied. We shall take the vertical direction to be the y axis and call y positive upward. With this choice of coordinates, we can replace x by y in Equations 2.10, 2.11, and 2.12. Furthermore, since positive y is upward, the acceleration is negative (downward) and given by $a = -g$. The negative sign simply indicates that the acceleration is downward. With these substitutions, we get the following expressions[4]:

$$v = v_0 - gt \tag{2.13}$$

$$y - y_0 = \tfrac{1}{2}(v + v_0)t \qquad \text{(for constant } a = -g) \tag{2.14}$$

$$y - y_0 = v_0 t - \tfrac{1}{2} gt^2 \tag{2.15}$$

$$v^2 = v_0{}^2 - 2g(y - y_0) \tag{2.16}$$

You should note that the *negative sign for the acceleration is already included in these expressions.* Therefore, when using these equations in any free-fall problem, you should simply substitute $g = 9.80$ m/s².

[4] One can also take y positive downward, in which case $a = +g$. The results will be the same, regardless of the convention chosen.

Free-fall acceleration $g = 9.80$ m/s²

Definition of free-fall

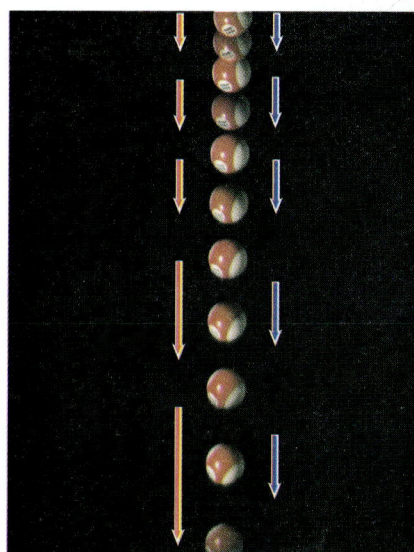

A multiflash photograph of a falling billiard ball. As the ball falls, the spacing between successive images increases, indicating that the ball accelerates downward. The motion diagram shows that the ball's velocity (red arrows) increases with time while its acceleration (violet arrows) remains constant. *(Richard Megna/Fundamental Photographs)*

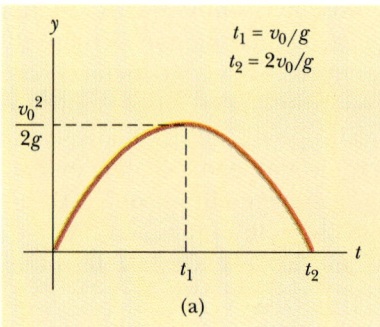
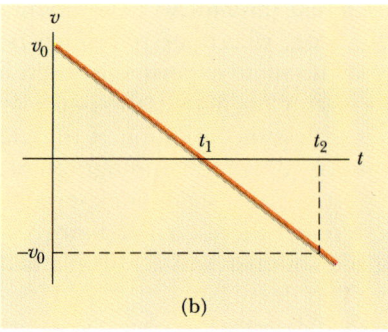

FIGURE 2.12   Graphs of (a) the displacement versus time and (b) the velocity versus time for a freely falling particle, where y and v are taken to be positive upward. Note the symmetry in (a) about $t = t_1$.

Consider the case of a particle thrown vertically upward from the origin with a velocity v_0. In this case, v_0 is positive and $y_0 = 0$. Graphs of the displacement and velocity as functions of time are shown in Figure 2.12. Note that the velocity is initially positive but decreases in time and goes to zero at the peak of the path. From Equation 2.13, we see that this peak occurs at the time $t_1 = v_0/g$. At this time, the displacement has its largest positive value, which can be calculated from Equation 2.15 with $t = t_1 = v_0/g$. This gives $y_{max} = v_0^2/2g$.

At the time $t_2 = 2t_1 = 2v_0/g$, we see from Equation 2.15 that the displacement is again zero, that is, the particle has returned to its starting point. Furthermore, at time t_2 the velocity is $v = -v_0$. (This follows directly from Eq. 2.13.) Hence, there is symmetry in the motion. In other words, both the displacement and the magnitude of the velocity repeat themselves in the time interval $t = 0$ to $t = 2v_0/g$.

In the examples that follow, we shall, for convenience, assume that $y_0 = 0$ at $t = 0$. This choice does not affect the solution to the problem. If y_0 is nonzero, then the graph of y versus t (Fig. 2.12a) is simply shifted upward or downward by an amount y_0, while the graph of v versus t (Fig. 2.12b) remains unchanged.

CONCEPTUAL EXAMPLE 2.11

A child throws a marble into the air with some initial velocity. Another child drops a ball at the same instant. Compare the accelerations of the two objects while they are in the air.

Reasoning   Once the objects leave the hand, both are freely falling objects, and hence both experience the same downward acceleration equal to the free-fall acceleration, $g = 9.80$ m/s^2.

CONCEPTUAL EXAMPLE 2.12

A ball is thrown upward. While the ball is in the air, (a) what happens to its velocity? (b) does its acceleration increase, decrease, or remain constant?

Reasoning   (a) The velocity of the ball changes continuously. As it travels upward, its speed decreases by 9.80 m/s during each second of its motion. When it reaches the peak of its motion, its speed becomes zero. As the ball moves downward, its speed increases by 9.80 m/s each second. (b) The acceleration of the ball remains constant while it is in the air, from the instant it leaves the hand until the instant before it strikes the ground. Its magnitude is the free-fall acceleration, $g = 9.80$ m/s^2. (If the acceleration were zero at the peak when the velocity is zero, this would say that there would be no change in velocity thereafter, so the ball would stop at the peak, and remain there, which is not the case.)

CONCEPTUAL EXAMPLE 2.13   Try to Catch the Dollar Bill!

Emily challenges her best friend David to catch a dollar bill as follows. She holds the bill vertically, as in Figure 2.13, with the center of the bill between David's index finger and thumb. David must catch the bill after Emily releases it without moving his hand downward. Who would you bet on?

Reasoning   Place your bets on Emily. There is a time delay between the instant Emily releases the bill and the time David reacts and closes his fingers. The reaction time of most people is at best about 0.2 s. Since the bill is in free-fall, and undergoes a downward acceleration of 9.80 m/s^2, in 0.2 s it falls a distance of $\frac{1}{2}gt^2 \cong 0.2$ m = 20 cm. This distance is about twice the distance between the center of the bill and its top edge ($\cong 8$ cm). Thus, David will be unsuccessful. You might want to try this on one of your friends.

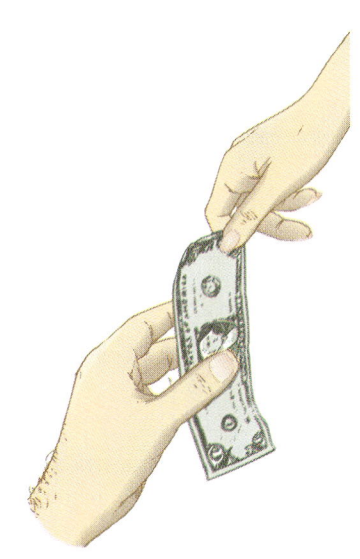

FIGURE 2.13   (Example 2.13) David is challenged to catch the dollar bill between his index finger and thumb after the bill is released by Emily. Is he able to meet the challenge?

EXAMPLE 2.14   Look Out Below!

A golf ball is released from rest from the top of a very tall building. Neglecting air resistance, calculate the position and velocity of the ball after 1.00, 2.00, and 3.00 s.

Solution   We choose our coordinates such that the starting point of the ball is at the origin ($y_0 = 0$ at $t = 0$) and remember that we have defined y to be positive upward. Since $v_0 = 0$, Equations 2.13 and 2.15 become

$$v = -gt = -(9.80 \text{ m/s}^2)t$$

$$y = -\tfrac{1}{2}gt^2 = -\tfrac{1}{2}(9.80 \text{ m/s}^2)t^2$$

where t is in seconds, v is in meters per second, and y is in meters. These expressions give the velocity and displacement at any time t after the ball is released. Therefore, at $t = 1.00$ s,

$$v = -(9.80 \text{ m/s}^2)(1.00 \text{ s}) = \boxed{-9.80 \text{ m/s}}$$

$$y = -\tfrac{1}{2}(9.80 \text{ m/s}^2)(1.00 \text{ s})^2 = \boxed{-4.90 \text{ m}}$$

At $t = 2.00$ s, we find that $v = -19.6$ m/s and $y = -19.6$ m. At $t = 3.00$ s, $v = -29.4$ m/s and $y = -44.1$ m. The minus signs for v indicate that the velocity is directed downward, and the minus signs for y indicate displacement in the negative y direction. (Because of air resistance, the actual speed of the ball is limited to approximately 30 m/s.)

Exercise   Calculate the position and velocity of the ball after 4.00 s.

Answer -78.4 m, -39.2 m/s.

EXAMPLE 2.15   Not a Bad Throw for a Rookie

A stone thrown from the top of a building is given an initial velocity of 20.0 m/s straight upward. The building is 50.0 m high, and the stone just misses the edge of the roof on its way down, as in Figure 2.14. Determine (a) the time needed for the stone to reach its maximum height, (b) the maximum height, (c) the time needed for the stone to return to the top of the building, (d) the velocity of the stone at this instant, and (e) the velocity and position of the stone at $t = 5.00$ s.

Solution   (a) To find the time necessary to reach the maximum height, use Equation 2.13, $v = v_0 - gt$, noting that $v = 0$ at maximum height:

$$20.0 \text{ m/s} - (9.80 \text{ m/s}^2)t_1 = 0$$

$$t_1 = \frac{20.0 \text{ m/s}}{9.80 \text{ m/s}^2} = \boxed{2.04 \text{ s}}$$

(b) This value of time can be substituted into Equation 2.15, to give the maximum height as measured from the position of the thrower:

$$y = v_0 t - \tfrac{1}{2}gt^2$$

$$y_{max} = (20.0 \text{ m/s})(2.04 \text{ s}) - \tfrac{1}{2}(9.80 \text{ m/s}^2)(2.04 \text{ s})^2$$

$$= \boxed{20.4 \text{ m}}$$

(c) When the stone is back at the height of the building top, the y coordinate is zero. Using Equation 2.15, with $y = 0$, we obtain

$$y = v_0 t - \tfrac{1}{2}gt^2$$

$$0 = (20.0 \text{ m/s})t - (4.90 \text{ m/s}^2)t^2$$

This is a quadratic equation and has two solutions for t. The equation can be factored to give

$$t(20.0 - 4.90t) = 0$$

One solution is $t = 0$, corresponding to the time the stone starts its motion. The other solution is $t = 4.08$ s, which is the solution we are after.

(d) The value for t found in (c) can be inserted into Equation 2.13 to give

$$v = v_0 - gt = 20.0 \text{ m/s} - (9.80 \text{ m/s}^2)(4.08 \text{ s})$$

$$= \boxed{-20.0 \text{ m/s}}$$

Note that the velocity of the stone when it arrives back at its original height is equal in magnitude to its initial velocity but opposite in direction. This indicates that the motion is symmetric.

(e) From Equation 2.13, the velocity after 5.00 s is

$$v = v_0 - gt = 20.0 \text{ m/s} - (9.80 \text{ m/s}^2)(5.00 \text{ s})$$

$$= \boxed{-29.0 \text{ m/s}}$$

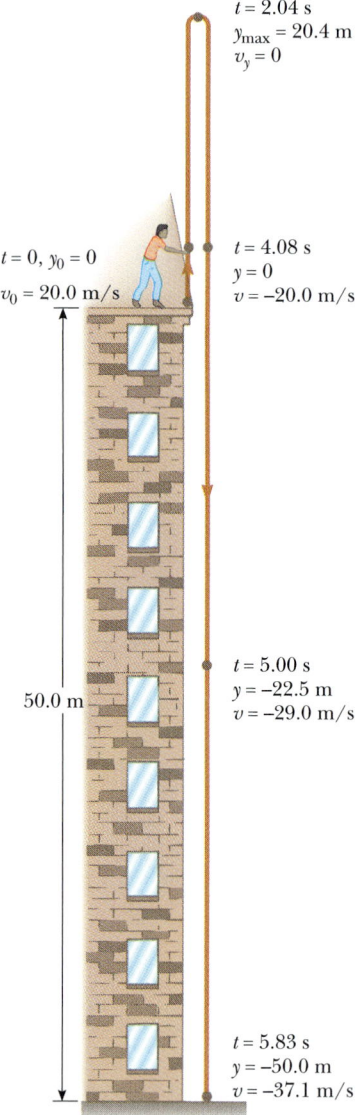

FIGURE 2.14   (Example 2.15) Position and velocity versus time for a freely falling particle thrown initially upward with a velocity of $v_0 = 20$ m/s.

We can use Equation 2.15 to find the position of the particle at $t = 5.00$ s:

$$y = v_0 t - \tfrac{1}{2}gt^2$$

$$= (20.0 \text{ m/s})(5.00 \text{ s}) - \tfrac{1}{2}(9.80 \text{ m/s}^2)(5.00 \text{ s})^2 = \boxed{-22.5 \text{ m}}$$

Exercise   Find (a) the velocity of the stone just before it hits the ground and (b) the total time the stone is in the air.

Answer   (a) -37.1 m/s (b) 5.83 s.

*2.6 KINEMATIC EQUATIONS DERIVED FROM CALCULUS

This is an optional section that assumes the reader is familiar with the techniques of integral calculus. If you have not studied integration in your calculus course as yet, this section should be skipped or covered at some later time after you become familiar with integration.

The velocity of a particle moving in a straight line can be obtained from a knowledge of its position as a function of time. Mathematically, the velocity equals the derivative of the coordinate with respect to time. It is also possible to find the displacement of a particle if its velocity is known as a function of time. In the calculus, this procedure is referred to as integration, or finding the antiderivative. Graphically, it is equivalent to finding the area under a curve.

Suppose the velocity versus time plot for a particle moving along the x axis is as shown in Figure 2.15. Let us divide the time interval $t_f - t_i$ into many small intervals of duration Δt_n. From the definition of average velocity, we see that the displacement during any small interval, such as the shaded one in Figure 2.15, is given by $\Delta x_n = \bar{v}_n \Delta t_n$, where $\bar{v}_n$ is the average velocity in that interval. Therefore, the displacement during this small interval is simply the area of the shaded rectangle. The total displacement for the interval $t_f - t_i$ is the sum of the areas of all the rectangles:

$$\Delta x = \sum_n \bar{v}_n \Delta t_n$$

where the sum is taken over all the rectangles from t_i to t_f. Now, as each interval is made smaller and smaller, the number of terms in the sum increases and the sum approaches a value equal to the area under the velocity-time graph. Therefore, in the limit $n \rightarrow \infty$, or $\Delta t_n \rightarrow 0$, we see that the displacement is given by

$$\Delta x = \lim_{\Delta t_n \rightarrow 0} \sum_n v_n \Delta t_n \tag{2.17}$$

or

Displacement = area under the velocity-time graph

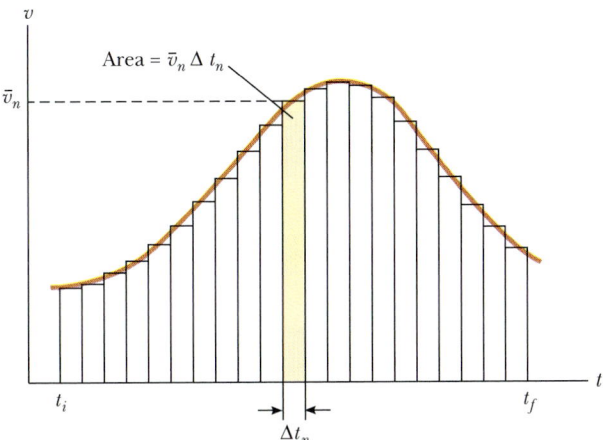

FIGURE 2.15 Velocity versus time for a particle moving along the x axis. The area of the shaded rectangle is equal to the displacement Δx in the time interval Δt_n, while the total area under the curve is the total displacement of the particle.

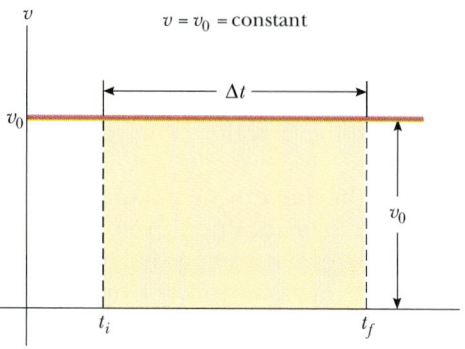

FIGURE 2.16 The velocity versus time curve for a particle moving with constant velocity v_0. The displacement of the particle during the time interval $t_f - t_i$ is equal to the area of the shaded rectangle.

Note that we have replaced the average velocity $\bar{v}_n$ by the instantaneous velocity v_n in the sum. As you can see from Figure 2.15, this approximation is clearly valid in the limit of very small intervals.

We conclude that if the velocity-time graph for motion along a straight line is known, the displacement during any time interval can be obtained by measuring the area under the curve corresponding to that time interval.

The limit of the sum in Equation 2.17 is called a **definite integral** and is written

Definite integral

$$\lim_{\Delta t_n \to 0} \sum_n v_n \, \Delta t_n = \int_{t_i}^{t_f} v(t) \, dt \qquad (2.18)$$

where $v(t)$ denotes the velocity at any time t. If the explicit functional form of $v(t)$ is known, and the limits are given, the integral can be evaluated.

If a particle moves with a constant velocity v_0 as in Figure 2.16, its displacement during the time interval Δt is simply the area of the shaded rectangle, that is,

$$\Delta x = v_0 \, \Delta t \qquad \text{(when } v = v_0 = \text{constant)}$$

As another example, consider a particle moving with a velocity that is proportional to t, as in Figure 2.17. Taking $v = at$, where a is the constant of proportionality (the acceleration), we find that the displacement of the particle during the time interval $t = 0$ to $t = t_1$ is the area of the shaded triangle in Figure 2.17:

$$\Delta x = \tfrac{1}{2}(t_1)(at_1) = \tfrac{1}{2} a t_1^2$$

Kinematic Equations

We now make use of the defining equations for acceleration and velocity to derive two of our kinematic equations.

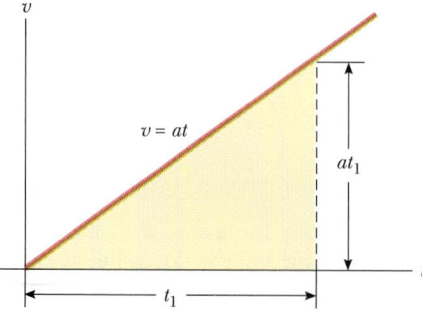

FIGURE 2.17 The velocity versus time curve for a particle moving with a velocity that is proportional to the time.

The defining equation for acceleration (Eq. 2.6),

$$a = \frac{dv}{dt}$$

may also be written in terms of an integral (or antiderivative) as

$$v = \int a \, dt + C_1$$

where C_1 is a constant of integration. For the special case where the acceleration is constant, this reduces to

$$v = at + C_1$$

The value of C_1 depends on the initial conditions of the motion. If we take $v = v_0$ when $t = 0$ and substitute these values into the last equation, we have

$$v_0 = a(0) + C_1$$

$$C_1 = v_0$$

Hence, we obtain the first kinematic equation (Eq. 2.8):

$$v = v_0 + at \qquad \text{(for constant } a\text{)}$$

Now let us consider the defining equation for velocity (Eq. 2.3):

$$v = \frac{dx}{dt}$$

We can write this in integral form as

$$x = \int v \, dt + C_2$$

where C_2 is another constant of integration. Since $v = v_0 + at$, this expression becomes

$$x = \int (v_0 + at) \, dt + C_2$$

$$x = \int v_0 \, dt + \int at \, dt + C_2$$

$$x = v_0 t + \tfrac{1}{2}at^2 + C_2$$

To find C_2, we make use of the initial condition that $x = x_0$ when $t = 0$. This gives $C_2 = x_0$. Therefore, we have

$$x - x_0 = v_0 t + \tfrac{1}{2} at^2 \qquad \text{(for constant } a\text{)}$$

This is a second equation of kinematics (Eq. 2.11). Recall that $x - x_0$ is equal to the displacement of the object, where x_0 is its initial position.

SUMMARY

As a particle moves along the x axis from some initial position x_i to some final position x_f, its **displacement** is

$$\Delta x \equiv x_f - x_i \tag{2.1}$$

The **average velocity** of a particle during some time interval is equal to the ratio of the displacement, Δx, and the time interval, Δt:

$$\bar{v} \equiv \frac{\Delta x}{\Delta t} \tag{2.2}$$

The **average speed** of a particle is equal to the ratio of total distance it travels to the total time it takes to travel that distance.

The **instantaneous velocity** of a particle is defined as the limit of the ratio $\Delta x / \Delta t$ as Δt approaches zero. By definition, this limit equals the derivative of x with respect to t, or the time rate of change of the position:

$$v \equiv \lim_{\Delta t \to 0} \frac{\Delta x}{\Delta t} = \frac{dx}{dt} \tag{2.3, 2.4}$$

The **speed** of a particle is defined to be equal to the magnitude of its velocity.

The **average acceleration** of a particle during some time interval is defined as the ratio of the change in its velocity, Δv, and the time interval, Δt:

$$\bar{a} \equiv \frac{\Delta v}{\Delta t} \tag{2.5}$$

The **instantaneous acceleration** is equal to the limit of the ratio $\Delta v / \Delta t$ as Δt approaches 0. By definition, this limit equals the derivative of v with respect to t, or the time rate of change of the velocity:

$$a \equiv \lim_{\Delta t \to 0} \frac{\Delta v}{\Delta t} = \frac{dv}{dt} \tag{2.6}$$

The slope of the tangent to the v versus t curve equals the instantaneous acceleration of the particle.

The **equations of kinematics** for a particle moving along the x axis with uniform acceleration a (constant in magnitude and direction) are

$$v = v_0 + at \tag{2.8}$$

$$x - x_0 = \tfrac{1}{2}(v_0 + v)t \tag{2.10}$$

$$x - x_0 = v_0 t + \tfrac{1}{2}at^2 \tag{2.11}$$

$$v^2 = v_0{}^2 + 2a(x - x_0) \tag{2.12}$$

An object falling freely in the presence of the Earth's gravity experiences a free-fall acceleration directed toward the center of the Earth. If air friction is neglected, and if the altitude of the motion is small compared with the Earth's radius, then one can assume that the free-fall acceleration, g, is constant over the range of motion, where g is equal to 9.80 m/s², or 32 ft/s². Assuming y is positive upward, the acceleration is given by $-g$ and the equations of kinematics for an object in free fall are the same as those given above, with the substitutions $x \to y$ and $a \to -g$.

QUESTIONS

1. Average velocity and instantaneous velocity are generally different quantities. Can they ever be equal for a specific type of motion? Explain.

2. If the average velocity is nonzero for some time interval, does this mean that the instantaneous velocity is never zero during this interval? Explain.

3. If the average velocity equals zero for some time interval Δt and if $v(t)$ is a continuous function, show that the instantaneous velocity must go to zero some time in this interval. (A sketch of x versus t might be useful in your proof.)

4. Is it possible to have a situation in which the velocity and

acceleration have opposite signs? If so, sketch a velocity-time graph to prove your point.

5. If the velocity of a particle is nonzero, can its acceleration ever be zero? Explain.

6. If the velocity of a particle is zero, can its acceleration ever be nonzero? Explain.

7. A stone is thrown vertically upward from the top of a building. Does the stone's displacement depend on the location of the origin of the coordinate system? Does the stone's velocity depend on the origin? (Assume that the coordinate system is stationary with respect to the building.) Explain.

8. A student at the top of a building of height h throws one ball upward with an initial speed v_0 and then throws a second ball downward with the same initial speed. How do the final velocities of the balls compare when they reach the ground?

9. Can the magnitude of the instantaneous velocity of an object ever be greater than the magnitude of its average velocity? Can it ever be less?

10. If the average velocity of an object is zero in some time interval, what can you say about the displacement of the object for that interval?

11. A rapidly growing plant doubles in height each week. At the end of the 25th day, the plant reaches the height of a building. At what time was the plant one-fourth the height of the building?

12. Two cars are moving in the same direction in parallel lanes along a highway. At some instant, the velocity of car A exceeds the velocity of car B. Does this mean that the acceleration of A is greater than that of B? Explain.

13. An apple is dropped from some height above the Earth's surface. Neglecting air resistance, how much does its speed increase each second during its fall?

PROBLEMS

Section 2.1   Displacement, Velocity, and Speed

1. The position of a car coasting down a hill was observed at various times and the results are summarized in the table below. Find the average velocity of the car during (a) the first second, (b) the last three seconds, and (c) the entire period of observation.

x(m)	0	2.3	9.2	20.7	36.8	57.5
t(s)	0	1.0	2.0	3.0	4.0	5.0

2. A motorist drives north for 35 min at 85 km/h and then stops for 15 min. He then continues north, traveling 130 km in 2.0 h. (a) What is his total displacement? (b) What is his average velocity?

3. The displacement versus time graph for a certain particle moving along the x axis is shown in Figure P2.3. Find the average velocity in the time intervals (a) 0 to 2 s, (b) 0 to 4 s, (c) 2 s to 4 s, (d) 4 s to 7 s, (e) 0 to 8 s.

4. A jogger runs in a straight line with an average velocity of $+5.00$ m/s for 4.00 min, and then with an average velocity of $+4.00$ m/s for 3.00 min. (a) What is her total displacement? (b) What is her average velocity during this time?

5. A person walks first at a constant speed of 5.0 m/s along a straight line from point A to point B and then back along the line from B to A at a constant speed of 3.0 m/s. (a) What is her average speed over the entire trip? (b) Her average velocity over the entire trip?

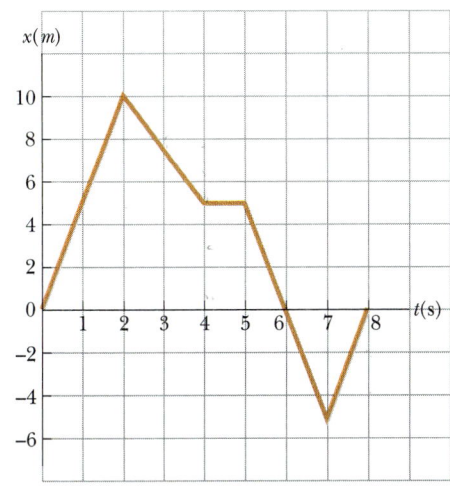

FIGURE P2.3

5A. A person walks first at a constant speed v_1 along a straight line from point A to point B, and then back along the line from B to A at a constant speed v_2. (a) What is her average speed over the entire trip? (b) Her average velocity over the entire trip?

6. A particle moves according to the equation $x = 10t^2$ where x is in meters and t is in seconds. (a) Find the average velocity for the time interval from 2.0 s to 3.0 s. (b) Find the average velocity for the time interval from 2.0 s to 2.1 s.

7. A car makes a 200-km trip at an average speed of 40 km/h. A second car starting 1.0 h later arrives at

their mutual destination at the same time. What was the average speed of the second car for the period that it was in motion?

Section 2.2   Instantaneous Velocity and Speed

8. A speedy tortoise can run at 10.0 cm/s, and a hare can run 20 times as fast. In a race, they start at the same time, but the hare stops to rest for 2.0 min, and so the tortoise wins by a shell (20 cm). (a) How long does the race take? (b) What is the length of the race?

9. The position-time graph for a particle moving along the *x* axis is as shown in Figure P2.9. (a) Find the average velocity in the time interval $t = 1.5$ s to $t = 4.0$ s. (b) Determine the instantaneous velocity at $t = 2.0$ s by measuring the slope of the tangent line shown in the graph. (c) At what value of *t* is the velocity zero?

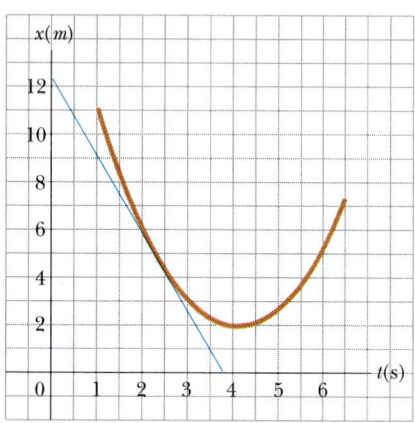

FIGURE P2.9

10. Two cars travel in the same direction along a straight highway, one at 55 mi/h and the other at 70 mi/h. (a) Assuming they start at the same point, how much sooner does the faster car arrive at a destination 10 miles away? (b) How far must the faster car travel before it has a 15-min lead on the slower car?

11. At $t = 1.0$ s, a particle moving with constant velocity is located at $x = -3.0$ m, and at $t = 6.0$ s, the particle is located at $x = 5.0$ m. (a) From this information, plot the position as a function of time. (b) Determine the velocity of the particle from the slope of this graph.

12. (a) Use the data in Problem 1 to construct a graph of position versus time. (b) By constructing tangents to the $x(t)$ curve, find the instantaneous velocity of the car at several instants. (c) Plot the instantaneous velocity versus time and, from this, determine the average acceleration of the car. (d) What is the initial velocity of the car?

13. Find the instantaneous velocity of the particle described in Figure P2.3 at the following times: (a) $t = 1.0$ s, (b) $t = 3.0$ s, (c) $t = 4.5$ s, and (d) $t = 7.5$ s.

14. The position-time graph for a particle moving along the *z* axis is as shown in Figure P2.14. Determine whether the velocity is positive, negative, or zero at times (a) t_1, (b) t_2, (c) t_3, (d) t_4.

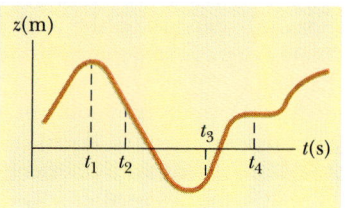

FIGURE P2.14

Section 2.3   Acceleration

15. A particle is moving with a velocity $v_0 = 60$ m/s at $t = 0$. Between $t = 0$ and $t = 15$ s, the velocity decreases uniformly to zero. What is the average acceleration during this 15-s interval? What is the significance of the sign of your answer?

16. An object moves along the *x* axis according to the equation $x(t) = (3.0t^2 - 2.0t + 3.0)$ m. Determine (a) the average speed between $t = 2.0$ s and $t = 3.0$ s, (b) the instantaneous speed at $t = 2.0$ s and at $t = 3.0$ s, (c) the average acceleration between $t = 2.0$ s and $t = 3.0$ s, and (d) the instantaneous acceleration at $t = 2.0$ s and $t = 3.0$ s.

17. A particle moves along the *x* axis according to the equation $x = 2.0t + 3.0t^2$, where *x* is in meters and *t* is in seconds. Calculate the instantaneous velocity and instantaneous acceleration at $t = 3.0$ s.

18. A particle moving in a straight line has a velocity of 8.0 m/s at $t = 0$. Its velocity at $t = 20$ s is 20.0 m/s. (a) What is its average acceleration in this time interval? (b) Can the average velocity be obtained from the information presented? Explain.

19. A particle starts from rest and accelerates as shown in Figure P2.19. Determine (a) the particle's speed at $t = 10$ s and at $t = 20$ s and (b) the distance traveled in the first 20 s.

20. The velocity of a particle as a function of time is shown in Figure P2.20. At $t = 0$, the particle is at $x = 0$. (a) Sketch the acceleration as a function of time. (b) Determine the average acceleration of the particle in the time interval $t = 2.0$ s to $t = 8.0$ s. (c) Determine the instantaneous acceleration of the particle at $t = 4.0$ s.

21. A particle moves along the *x* axis according to the equation $x = 2.0 + 3.0t - 1.0t^2$, where *x* is in meters and *t* is in seconds. At $t = 3.00$ s, find (a) the position of the particle, (b) its velocity, and (c) its acceleration.

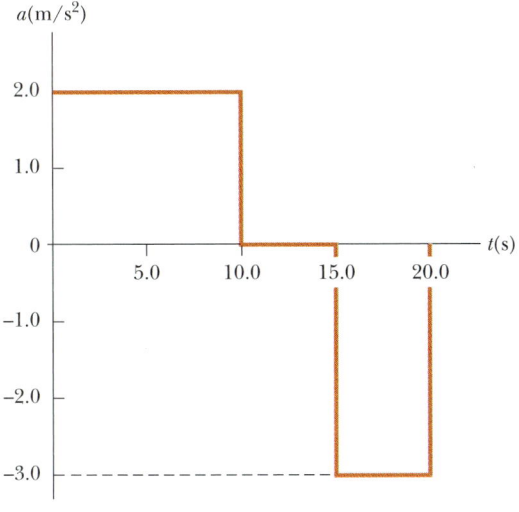

FIGURE P2.19

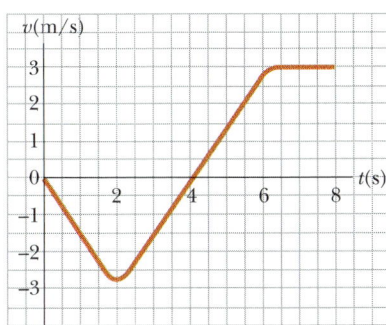

FIGURE P2.20

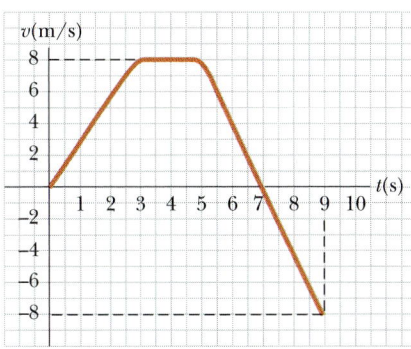

FIGURE P2.22

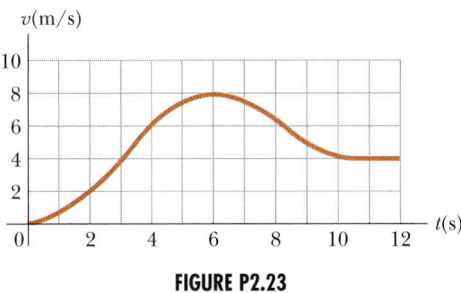

FIGURE P2.23

22. A student drives a moped along a straight road as described by the speed-versus-time graph in Figure P2.22. Sketch this graph in the middle of a sheet of graph paper. (a) Directly above this graph, sketch a graph of the position versus time, aligning the time coordinates of the two graphs. (b) Sketch a graph of the acceleration versus time directly below the v-t graph, again aligning the time coordinates. On each graph, show the numerical values of x and a for all points of inflection. (c) What is the acceleration at $t = 6$ s? (d) Find the position (relative to the starting point) at $t = 6$ s. (e) What is the moped's final position at $t = 9$ s?

23. Figure P2.23 shows a graph of v versus t for the motion of a motorcyclist as she starts from rest and moves along the road in a straight line. (a) Find the average acceleration for the time interval $t_0 = 0$ to $t_1 = 6.0$ s. (b) Estimate the time at which the acceleration has its greatest positive value and the value of the acceleration at this instant. (c) When is the acceleration zero? (d) Estimate the maximum negative value of the acceleration and the time at which it occurs.

Section 2.4 One-Dimensional Motion with Constant Acceleration

24. A particle travels in the positive x direction for 10 s at a constant speed of 50 m/s. It then accelerates uniformly to a speed of 80 m/s in the next 5 s. Find (a) the average acceleration of the particle in the first 10 s, (b) its average acceleration in the interval $t = 10$ s to $t = 15$ s, (c) the total displacement of the particle between $t = 0$ and $t = 15$ s, and (d) its average speed in the interval $t = 10$ s to $t = 15$ s.

25. A body moving with uniform acceleration has a velocity of 12.0 cm/s when its x coordinate is 3.00 cm. If its x coordinate 2.00 s later is -5.00 cm, what is the magnitude of its acceleration?

26. The new BMW M3 can accelerate from zero to 60 mi/h in 5.6 s. (a) What is the resulting acceleration in m/s²? (b) How long would it take for the BMW to go from 60 mi/h to 130 mi/h at this rate?

27. The minimum distance required to stop a car moving at 35 mi/h is 40 ft. What is the minimum stopping distance for the same car moving at 70 mi/h, assuming the same rate of acceleration?

28. Figure P2.28 represents part of the performance data of a car owned by a proud physics student. (a) Calculate from the graph the total distance traveled. (b) What distance does the car travel between the

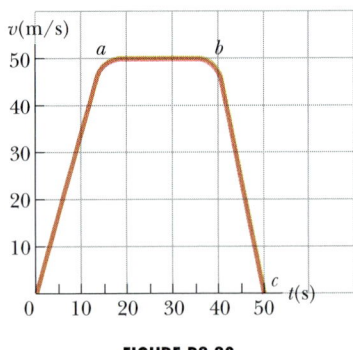

FIGURE P2.28

times $t = 10$ s and $t = 40$ s? (c) Draw a graph of its acceleration versus time between $t = 0$ and $t = 50$ s. (d) Write an equation for x as a function of time for each phase of the motion, represented by (i) Oa, (ii) ab, (iii) bc. (e) What is the average velocity of the car between $t = 0$ and $t = 50$ s?

29. The initial speed of a body is 5.20 m/s. What is its speed after 2.50 s if it accelerates uniformly at (a) 3.00 m/s² and (b) at -3.00 m/s²?

30. A hockey puck sliding on a frozen lake comes to rest after traveling 200 m. If its initial velocity is 3.00 m/s, (a) what is its acceleration if that acceleration is assumed constant, (b) how long is it in motion, and (c) what is its speed after traveling 150 m?

31. A jet plane lands with a velocity of 100 m/s and can accelerate at a maximum rate of -5.0 m/s² as it comes to rest. (a) From the instant it touches the runway, what is the minimum time needed before it stops? (b) Can this plane land at a small airport where the runway is 0.80 km long?

32. A car and train move together along parallel paths at 25.0 m/s. The car then undergoes a uniform acceleration of -2.50 m/s² because of a red light and comes to rest. It remains at rest for 45.0 s, then accelerates back to a speed of 25.0 m/s at a rate of 2.50 m/s². How far behind the train is the car when it reaches the speed of 25.0 m/s, assuming that the train speed has remained at 25.0 m/s?

33. A drag racer starts her car from rest and accelerates at 10.0 m/s² for the entire distance of 400 m ($\frac{1}{4}$ mile). (a) How long did it take the car to travel this distance? (b) What is its speed at the end of the run?

34. An electron in a cathode ray tube (CRT) accelerates from 2.0×10^4 m/s to 6.0×10^6 m/s over 1.5 cm. (a) How long does the electron take to travel this distance? (b) What is its acceleration?

35. A particle starts from rest from the top of an inclined plane and slides down with constant acceleration. The inclined plane is 2.00 m long, and it takes 3.00 s for the particle to reach the bottom. Find (a) the acceleration of the particle, (b) its speed at the bottom of the incline, (c) the time it takes the particle to reach the middle of the incline, and (d) its speed at the midpoint.

36. Two express trains started 5 min apart. Starting from rest, each is capable of a maximum speed of 160 km/h after uniformly accelerating over a distance of 2.0 km. (a) What is the acceleration of each train? (b) How far ahead is the first train when the second one starts? (c) How far apart are they when they are both traveling at maximum speed?

37. A teenager has a car that accelerates at 3.0 m/s² and decelerates at -4.5 m/s². On a trip to the store, he accelerates from rest to 12 m/s, drives at a constant speed for 5.0 s, and then comes to a momentary stop at the corner. He then accelerates to 18 m/s, drives at a constant speed for 20 s, decelerates for 8/3 s, continues for 4.0 s at this speed, and then comes to a stop. (a) How long does the trip take? (b) How far has he traveled? (c) What is his average speed for the trip? (d) How long would it take to walk to the store and back if he walked at 1.5 m/s?

38. A ball accelerates at 0.5 m/s² while moving down an inclined plane 9.0 m long. When it reaches the bottom, the ball rolls up another plane, where, after moving 15 m, it comes to rest. (a) What is the speed of the ball at the bottom of the first plane? (b) How long does it take to roll down the first plane? (c) What is the acceleration along the second plane? (d) What is the ball's speed 8.0 m along the second plane?

39. A car moving at a constant speed of 30.0 m/s suddenly stalls at the bottom of a hill. The car undergoes a constant acceleration of -2.00 m/s² (opposite its motion) while ascending the hill. (a) Write equations for the position and the velocity as functions of time, taking $x = 0$ at the bottom of the hill, where $v_0 = 30.0$ m/s. (b) Determine the maximum distance traveled by the car after it stalls.

40. An electron has an initial speed of 3.0×10^5 m/s. If it undergoes an acceleration of 8.0×10^{14} m/s², (a) how long will it take to reach a speed of 5.4×10^5 m/s and (b) how far has it traveled in this time?

41. Speedy Sue driving at 30 m/s enters a one-lane tunnel. She then observes a slow-moving van 155 m ahead traveling at 5.0 m/s. Sue applies her brakes but can decelerate only at 2.0 m/s² because the road is wet. Will there be a collision? If yes, determine how far into the tunnel and at what time the collision occurs. If no, determine the distance of closest approach between Sue's car and the van.

42. An indestructible bullet 2.00 cm long is fired straight through a board that is 10.0 cm thick. The bullet strikes the board with a speed of 420 m/s and emerges with a speed of 280 m/s. (a) What is the average acceleration of the bullet through the board? (b) What is the total time that the bullet is in contact with the board? (c) What thickness of boards

(calculated to 0.1 cm) would it take to stop the bullet?

43. Until recently, the world's land speed record was held by Colonel John P. Stapp, USAF. On March 19, 1954, he rode a rocket-propelled sled that moved down the track at 632 mi/h. He and the sled were safely brought to rest in 1.4 s. Determine (a) the negative acceleration he experienced and (b) the distance he traveled during this negative acceleration.

44. A hockey player is standing on his skates on a frozen pond when an opposing player skates by with the puck, moving with a uniform speed of 12.0 m/s. After 3.00 s, the first player makes up his mind to chase his opponent. If the first player accelerates uniformly at 4.00 m/s², (a) how long does it take him to catch the opponent? (b) How far has the first player traveled in this time? (Assume the opponent moves at constant speed.)

Section 2.5 Freely Falling Bodies

45. A woman is reported to have fallen 144 ft from the 17th floor of a building, landing on a metal ventilator box, which she crushed to a depth of 18.0 in. She suffered only minor injuries. Neglecting air resistance, calculate (a) the speed of the woman just before she collided with the ventilator, (b) her average acceleration while in contact with the box, and (c) the time it took to crush the box.

46. A ball is thrown directly downward with an initial speed of 8.00 m/s from a height of 30.0 m. When does the ball strike the ground?

47. A student throws a set of keys vertically upward to her sorority sister in a window 4.00 m above. The keys are caught 1.50 s later by the sister's outstretched hand. (a) With what initial velocity were the keys thrown? (b) What was the velocity of the keys just before they were caught?

48. A hot air balloon is traveling vertically upward at a constant speed of 5.00 m/s. When it is 21.0 m above the ground, a package is released from the balloon. (a) After it is released, for how long is the package in the air? (b) What is its velocity just before impact with the ground? (c) Repeat (a) and (b) for the case of the balloon descending at 5.00 m/s.

49. A ball is thrown vertically upward from the ground with an initial speed of 15.0 m/s. (a) How long does it take the ball to reach its maximum altitude? (b) What is its maximum altitude? (c) Determine the velocity and acceleration of the ball at $t = 2.00$ s.

50. A ball thrown vertically upward is caught by the thrower after 20.0 s. Find (a) the initial velocity of the ball and (b) the maximum height it reaches.

51. A baseball is hit such that it travels straight upward after being struck by the bat. A fan observes that it requires 3.00 s for the ball to reach its maximum height. Find (a) its initial velocity and (b) its maximum height. Ignore the effects of air resistance.

52. An astronaut standing on the Moon drops a hammer, letting it fall 1.00 m to the surface. The lunar gravity produces a constant acceleration of magnitude 1.62 m/s². Upon returning to Earth, the astronaut again drops the hammer, letting it fall to the ground from a height of 1.00 m with an acceleration of 9.80 m/s². Compare the times of fall in the two situations.

53. The height of a helicopter above the ground is given by $h = 3.00t^3$, where h is in meters and t is in seconds. After 2.00 s, the helicopter releases a small mailbag. How long after its release does the mailbag reach the ground?

54. A stone falls from rest from the top of a high cliff. A second stone is thrown downward from the same height 2.00 s later with an initial speed of 30.0 m/s. If both stones hit the ground simultaneously, how high is the cliff?

55. A daring stunt woman sitting on a tree limb wishes to drop vertically onto a horse galloping under the tree. The speed of the horse is 10.0 m/s, and the distance from the limb to the saddle is 3.00 m. (a) What must be the horizontal distance between the saddle and limb when the woman makes her move? (b) How long is she in the air?

*Section 2.6 Kinematic Equations Derived from Calculus

56. The speed of a bullet shot from a gun is given by $v = (-5.0 \times 10^7)t^2 + (3.0 \times 10^5)t$, where v is in meters/second and t is in seconds. The acceleration of the bullet just as it leaves the barrel is zero. (a) Determine the acceleration and position of the bullet as a function of time when the bullet is in the barrel. (b) Determine the length of time the bullet is accelerated while in the barrel. (c) Find the speed at which the bullet leaves the barrel. (d) What is the length of the barrel?

57. The position of a softball tossed vertically upward is described by the equation $y = 7.00t - 4.90t^2$, where y is in meters and t in seconds. Find (a) the ball's initial speed v_0 at $t_0 = 0$, (b) its velocity at $t = 1.26$ s, and (c) its acceleration.

58. A rocket sled for testing equipment under large accelerations starts at rest and accelerates according to the expression $a = (3 \text{ m/s}^3)t + 5.00 \text{ m/s}^2$. How far does the sled move in the time interval $t = 0$ to $t = 2.00$ s?

59. Automotive engineers refer to the time rate of change of acceleration as the "jerk." If an object moves in one dimension such that its jerk J is constant, (a) determine expressions for its acceleration $a(t)$, speed $v(t)$, and position $x(t)$, given that its initial acceleration, speed, and position are a_0, v_0,

and x_0, respectively. (b) Show that $a^2 = a_0{}^2 + 2J(v - v_0)$.

60. The acceleration of a marble in a certain fluid is proportional to the speed of the marble squared, and is given (in SI units) by $a = -3.00v^2$ for $v > 0$. If the marble enters this fluid with a speed of 1.50 m/s, how long will it take before the marble's speed is reduced to half of its initial value?

ADDITIONAL PROBLEMS

61. Another scheme to catch the roadrunner has failed, and a safe falls from rest from the top of a 25-m-high cliff toward Wiley Coyote, who is standing at the base. Wiley first notices the safe after it has fallen 15 m. How long does he have to get out of the way?

62. A motorist is traveling at 18.0 m/s when he sees a deer in the road 38.0 m ahead. (a) If the maximum negative acceleration of the vehicle is -4.50 m/s², what is the maximum reaction time Δt of the motorist that will allow him to avoid hitting the deer? (b) If his reaction time is 0.300 s, how fast will he be traveling when he reaches the deer?

63. An inquisitive physics student climbs a 50.0-m cliff that overhangs a calm pool of water. She throws two stones vertically downward 1.00 s apart and observes that they cause a single splash. The first stone has an initial velocity of 2.00 m/s. (a) At what time after release of the first stone do the two stones hit the water? (b) What initial velocity must the second stone have if they are to hit simultaneously? (c) What is the velocity of each stone at the instant they hit the water?

64. In a 100-m linear accelerator, an electron is accelerated to 1.0 percent of the speed of light in 40 m before it coasts 60 m to a target. (a) What is the electron's acceleration during the first 40 m? (b) How long does the total flight take?

65. A "superball" is dropped from a height of 2.00 m above the ground. On the first bounce, the ball reaches a height of 1.85 m, where it is caught. Find the velocity of the ball (a) just as it makes contact with the ground and (b) just as it leaves the ground on the bounce. (c) Neglecting the time the ball spends in contact with the ground, find the total time required for the ball to go from the dropping point to the point where it is caught.

66. A Cessna 150 aircraft has a lift-off speed of approximately 125 km/h. (a) What minimum constant acceleration does this require if the aircraft is to be airborne after a take-off run of 250 m? (b) What is the corresponding take-off time? (c) If the aircraft continues to accelerate at this rate, what speed will it reach 25.0 s after it begins to roll?

67. One runner covered the 100-m dash in 10.3 s. Another runner came in second at a time of 10.8 s. Assuming that the runners traveled at their average speeds for the entire distance, determine the separation between them when the winner crossed the finish line.

68. A falling object requires 1.50 s to travel the last 30.0 m before hitting the ground. From what height above the ground did it fall?

69. A young woman named Kathy Kool buys a super-deluxe sports car that can accelerate at the rate of 4.90 m/s². She decides to test the car by dragging with another speedster, Stan Speedy. Both start from rest, but experienced Stan leaves 1.00 s before Kathy. If Stan moves with a constant acceleration of 3.50 m/s² and Kathy maintains an acceleration of 4.90 m/s², find (a) the time it takes Kathy to overtake Stan, (b) the distance she travels before she catches him, and (c) the velocities of both cars at the instant she overtakes him.

70. A hockey player takes a slap shot at a puck at rest on the ice. The puck glides over the ice for 10.0 ft without friction, at which point it runs over a concrete surface. The puck then accelerates opposite its motion at a uniform rate of -20.0 ft/s². If the velocity of the puck is 40.0 ft/s after traveling 100 ft from the point of impact, (a) what is the average acceleration imparted to the puck as it is struck by the hockey stick? (Assume that the time of contact is 0.0100 s.) (b) How far does the puck travel before stopping? (c) What is the total time the puck is in motion, neglecting contact time?

71. Two cars are traveling along a straight line in the same direction, the lead car at 25 m/s and the other at 30 m/s. At the moment the cars are 40 m apart, the lead driver applies the brakes so that her car accelerates at -2.0 m/s². (a) How long does it take for the lead car to stop? (b) Assuming that the chasing car brakes at the same time as the lead car, what must be the chasing car's minimum negative acceleration so as not to hit the lead car? (c) How long does it take for the chasing car to stop?

72. A motorist drives along a straight road at a constant speed of 15.0 m/s. Just as she passes a parked motorcycle police officer, the officer starts to accelerate at 2.00 m/s² to overtake her. Assuming the officer maintains this acceleration, (a) determine the time it takes the police officer to reach the motorist. Find (b) the speed and (c) the total displacement of the officer as he overtakes the motorist.

73. In 1987, Art Boileau won the Los Angeles Marathon, 26 mi and 385 yd, in 2 h, 13 min, and 9 s. (a) Find his average speed in meters per second and in miles per hour. (b) At the 21-mi marker, Boileau had a 2.50-min lead on the second-place winner, who later crossed the finish line 30.0 s after Boileau. Assume that Boileau maintained his constant average speed and that both runners were running at the same speed when Boileau passed the 21-mi marker. Find

the average acceleration (in meters per second squared) that the second-place runner had during the remainder of the race after Boileau passed the 21-mi marker.

74. A rock is dropped from rest into a well. (a) If the sound of the splash is heard 2.40 s later, how far below the top of the well is the water surface? The speed of sound in air (for the air temperature that day) was 336 m/s. (b) If the travel time for the sound is neglected, what percentage error is introduced when the depth of the well is calculated?

75. A train travels along a straight track between Stations 1 and 2 as shown in Figure P2.75. The engineer is instructed to start from rest at Station 1, accelerate uniformly between *A* and *B*, coast with a uniform speed between *B* and *C*, and then decelerate uniformly between *C* and *D* (at the same rate as between *A* and *B*) until the train stops at Station 2. If the distances *AB*, *BC*, and *CD* are all equal, and if it takes 5.00 min to travel between the two stations, determine how much of this 5.00-min period the train spends between points (i) *A* and *B*, (ii) *B* and *C*, and (iii) *C* and *D*.

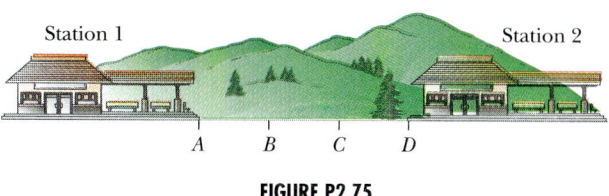

FIGURE P2.75

76. A rocket is fired vertically upward with an initial velocity of 80.0 m/s. It accelerates upward at 4.00 m/s² until it reaches an altitude of 1000 m. At that point, its engines fail and the rocket goes into free-fall with acceleration -9.80 m/s². (a) How long is the rocket in motion? (b) What is its maximum altitude? (c) What is its velocity just before it collides with the Earth? (*Hint:* Consider the motion while the engine is operating separate from the free-fall motion.)

77. In a 100-m race, Maggie and Judy cross the finish line in a dead heat, both taking 10.2 s. Accelerating uniformly, Maggie takes 2.00 s and Judy 3.00 s to attain maximum speed, which they maintain for the rest of the race. (a) What is the acceleration of each sprinter? (b) What are their respective maximum speeds? (c) Which sprinter is ahead at the 6.00-s mark, and by how much?

78. A train travels in time in the following manner. In the first 60 min, it travels with a speed v; in the next 30 min it has a speed $3v$; in the next 90 min it travels with a speed $v/2$; in the final 120 min, it travels with a speed $v/3$. (a) Plot the speed-time graph for this trip.

(b) How far does the train travel? (c) What is the average speed of the train over the entire trip?

79. A commuter train can minimize the time t between two stations by accelerating at a rate $a_1 = 0.100$ m/s² for a time t_1 and then undergoing a negative acceleration $a_2 = -0.500$ m/s² as the engineer uses the brakes for a time t_2. Since the stations are only 1.00 km apart, the train never reaches its maximum speed. Find the minimum time of travel t and the time t_1.

80. In order to protect his food from hungry bears, a boy scout raises his food pack, mass m, with a rope that is thrown over a tree limb of height h above his hands. He walks away from the vertical rope with constant speed v_s while holding the free end of the rope in his hands (Fig. P2.80). (a) Show that the speed v_p of the food pack is $x(x^2 + h^2)^{-1/2}v_s$, where x is the distance he has walked away from the vertical rope. (b) Show that the acceleration a_p of the food pack is $h^2(x^2 + h^2)^{-3/2}v_s^2$. (c) What values do the acceleration and speed have shortly after he leaves the vertical rope? (d) What values do the speed and acceleration approach as the distance x continues to increase?

80A. In Problem 80, let the height h equal 6.00 m and the speed v_s equal 2.00 m/s. Assume that the food pack starts from rest. (a) Tabulate and graph the speed-time graph. (b) Tabulate and graph the acceleration-time graph. (Let the range of time be from 0 s to 6.00 s and the time intervals be 0.50 s.)

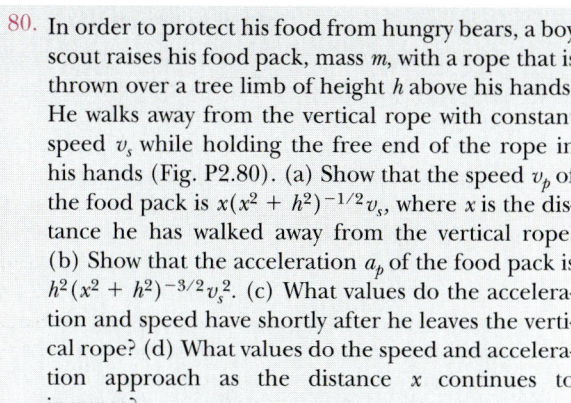

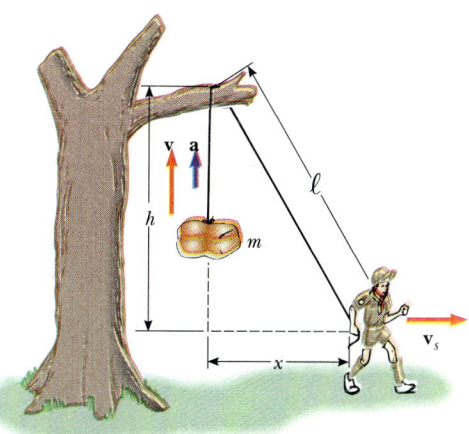

FIGURE P2.80

81. Two objects *A* and *B* are connected by a rigid rod that has a length *L*. The objects slide along perpendicular guide rails, as shown in Figure P2.81. If *A* slides to the left with a constant speed v, find the velocity of *B* when $\alpha = 60°$.

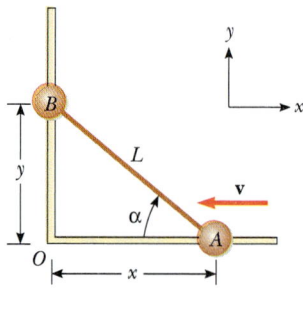

FIGURE P2.81

SPREADSHEET PROBLEMS

S1. Use Spreadsheet 2.1 to plot position and velocity as functions of time for one object traveling at constant velocity and another traveling with constant acceleration. Choose a variety of velocity and acceleration values, and view the graphs. Be sure to investigate zero and negative values. *Note:* There are two graphs associated with Spreadsheet 2.1, one for velocity versus time and one for position versus time.

S2. Spreadsheet 2.2 models the motion of a two-person 1.5-km race. For simplicity, we will assume that the runners can change speed instantaneously. Each runner can set his or her strategy. For instance, Racer 1 decides to lead initially, so his original speed, $V11 = 7$ m/s, is a little greater than Racer 2's, $V21 = 6.5$ m/s. Racer 2 realizes that she has fallen behind more than she wants, so she starts her "kick," at $X2 = 250$ m before the finish line. She is able to increase her speed, $V22$, to 10 m/s for 14 s. She then slows down to $V23 = 8$ m/s. During this time Racer 1 sees that his opponent is gaining, so at $X1 = 150$ m from the finish, he starts to "kick" at $V12 = 11$ m/s for 10 s. He then slows down to $V13 = 6$ m/s. (a) Who wins the race? (b) Devise a strategy so that Racer 1 beats the world's record of 3 min, 15 s, while Racer 2 finishes 0.2 s behind Racer 1. Your input data must be reasonable; neither runner is Superman.

S3. Spreadsheet 2.3 models the sport of drag racing. Two dragsters can have different accelerations A1 and A2, as well as different maximum speeds V1 and V2. Both cars start from rest at the same starting position. However, a time delay t' in starting times may be introduced if the cars are quite different. Enter the following data:

	Acceleration (ft/s^2)	Maximum Speed (ft/s)	Delay Time t' (s)
Car 1	38.6	175	
Car 2	43.5	170	0.1

(a) Which car wins a 1/4-mile race? (b) If the acceleration of Car 1 is increased to 44.0 ft/s^2, which car now wins?

S4. Modify Spreadsheet 2.1 to solve this problem: The State Police have set up a "speed trap" on the interstate highway. From a police car hidden behind a billboard, an officer with a radar gun measures a motorist's speed to be 35.0 m/s. Three seconds later, she alerts her partner, who is in another police car 100 m down the road. The second police car starts from rest accelerating at 2.00 m/s^2 in pursuit of the speeder 2.00 s after receiving the alert. (a) How much time elapses before the speeder is overtaken? (b) What is the police car's speed when it overtakes the speeder? (c) What is the distance from where the second police car was sitting to the point where the speeder is overtaken?

S5. As a demonstration, astronauts on a distant planet toss a rock into the air. With the aid of a high-speed camera, they record the height of the rock as a function of time as given in the table below: (a) Find the speed over each time interval by using a difference equation to approximate the differential. (b) Find the acceleration during each time interval by the same type of approximation.

Height of a Rock Versus Time for Problem S5

Time (s)	0.00	0.25	0.50	0.75	1.00	1.25	1.50	1.75	2.00	2.25	2.50
Height (m)	5.00	5.75	6.40	6.94	7.38	7.72	7.96	8.10	8.13	8.07	7.90
Time (s)	2.75	3.00	3.25	3.50	3.75	4.00	4.25	4.50	4.75	5.00	
Height (m)	7.62	7.25	6.77	6.20	5.52	4.73	3.85	2.86	1.77	0.58	

CHAPTER 3

Vectors

This interesting set of directional arrows was photographed in Estes Park, Colorado. A particular location relative to the position of the park is specified by the direction of the arrow and the distance to that location. This is one type of vector quantity. We examine it and other vector quantities in this chapter. *(Ray Serway)*

Physical quantities that have both numerical and directional properties are represented by vectors. Some examples of vector quantities are force, displacement, velocity, and acceleration. This chapter is primarily concerned with vector algebra and with some general properties of vector quantities. The addition and subtraction of vector quantities are discussed, together with some common applications to physical situations. Discussion of the products of vector quantities shall be delayed until these operations are needed.[1]

Vector quantities are used throughout this text, and it is therefore imperative that you master both their graphical and their algebraic properties.

---

[1] The dot, or scalar, product is discussed in Section 7.2, and the cross, or vector, product is introduced in Section 11.2.

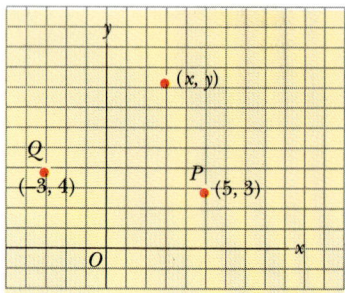

FIGURE 3.1   Designation of points in a cartesian coordinate system. Every point is labeled with coordinates (x, y).

3.1   COORDINATE SYSTEMS AND FRAMES OF REFERENCE

Many aspects of physics deal in some form or other with locations in space. For example, the mathematical description of the motion of an object requires a method for describing the position of the object at various times. Thus, it is fitting that we first discuss how to describe the position of a point in space. This description is accomplished by means of coordinates.

A point on a line can be described with one coordinate. A point in a plane is located with two coordinates, and three coordinates are required to locate a point in space. In general, a coordinate system used to specify locations in space consists of

• A fixed reference point O, called the origin
• A set of specified axes with appropriate scales and labels on the axes
• Instructions that tell us how to label a point in space relative to the origin and axes.

One convenient coordinate system that we shall use frequently is the *cartesian coordinate system*, sometimes called the *rectangular coordinate system*. Such a system in two dimensions is illustrated in Figure 3.1. An arbitrary point in this system is labeled with the coordinates (x, y). The positive x direction is arbitrarily defined to be to the right of the origin, and the positive y direction is arbitrarily defined to be upward from the origin. The negative x direction is to the left of the origin, and the negative y direction is downward from the origin. For example, the point P, which has coordinates $(5, 3)$, may be reached by first going 5 units to the right of the origin and then going 3 units above the origin. Similarly, the point Q has coordinates $(-3, 4)$, corresponding to going 3 units to the left of the origin and 4 units above the origin.

Sometimes it is more convenient to represent a point in a plane by its *plane polar coordinates*, (r, θ), as in Figure 3.2a. In this coordinate system, r is the distance from the origin to the point having cartesian coordinates (x, y) and θ is the angle between r and a fixed axis. This fixed axis is usually the positive x axis and θ is usually measured counterclockwise from it. From the right triangle in Figure 3.2b, we find $\sin \theta = y/r$ and $\cos \theta = x/r$. (A review of trigonometric functions is given in Appendix B.4.) Therefore, starting with plane polar coordinates, the cartesian coordinates can be obtained through the equations

$$x = r \cos \theta \tag{3.1}$$

$$y = r \sin \theta \tag{3.2}$$

Furthermore, the definitions of trigonometry tell us that

$$\tan \theta = \frac{y}{x} \tag{3.3}$$

and

$$r = \sqrt{x^2 + y^2} \tag{3.4}$$

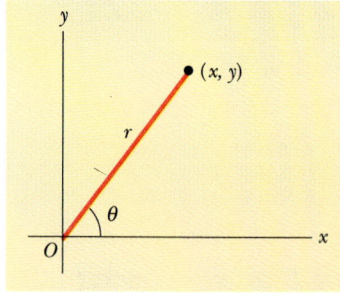

(a)

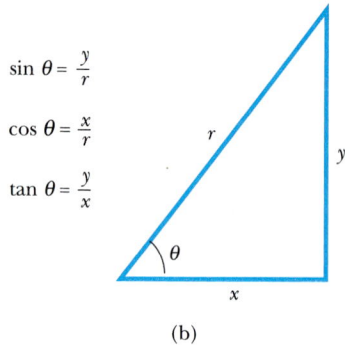

$\sin \theta = \dfrac{y}{r}$

$\cos \theta = \dfrac{x}{r}$

$\tan \theta = \dfrac{y}{x}$

(b)

FIGURE 3.2   (a) The plane polar coordinates of a point are represented by the distance r and the angle θ, where θ is measured counterclockwise from the positive x axis. (b) The right triangle used to relate (x, y) to (r, θ).

These four expressions relating the coordinates (x, y) to the coordinates (r, θ) apply only when θ is defined as in Figure 3.2a — in other words, where positive θ is an angle measured *counterclockwise* from the positive x axis. If the reference axis for the polar angle θ is chosen to be other than the positive x axis or if the sense of increasing θ is chosen differently, then the corresponding expressions relating the two sets of coordinates will change. Scientific calculators provide conversions between cartesian and polar coordinates based on these standard conventions.

EXAMPLE 3.1 **Polar Coordinates**

The cartesian coordinates of a point in the xy plane are $(x, y) = (-3.50, -2.50)$ m, as in Figure 3.3. Find the polar coordinates of this point.

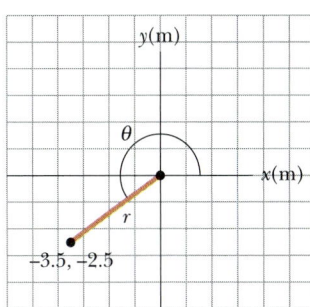

FIGURE 3.3 (Example 3.1).

Solution

$$r = \sqrt{x^2 + y^2} = \sqrt{(-3.50 \text{ m})^2 + (-2.50 \text{ m})^2} = \boxed{4.30 \text{ m}}$$

$$\tan \theta = \frac{y}{x} = \frac{-2.50 \text{ m}}{-3.50 \text{ m}} = 0.714$$

$$\theta = \boxed{216°}$$

Note that you must use the signs of x and y to find that the point lies in the third quadrant of the coordinate system. That is, $\theta = 216°$ and not $36°$.

3.2 VECTOR AND SCALAR QUANTITIES

The physical quantities we shall encounter in this text may be treated as either scalar quantities or vector quantities. A scalar quantity is one that is completely specified by a number with appropriate units. That is,

> A **scalar quantity** has only magnitude and no direction.

On the other hand, a vector quantity is a physical quantity that is completely specified by a number with appropriate units plus a direction. That is,

> A **vector quantity** has both magnitude and direction.

(a)

(b)

(a) The number of apples in the basket is one example of a scalar quantity. Can you think of other examples? *(Superstock)* (b) Jennifer pointing in the right direction. A vector is a physical quantity that must be specified by both magnitude and direction. *(Photo by Ray Serway)*

The number of apples in a basket is an example of a scalar quantity. If you are told there are 38 apples in the basket, this completes the required information; no specification of direction is required. Other examples of scalar quantities are temperature, volume, mass, and time intervals. The rules of ordinary arithmetic are used to manipulate scalar quantities.

Force is one example of a vector quantity. To describe completely the force on an object, we must specify both the direction of the applied force, a number to indicate the magnitude of the force, and often the line or point of application of the force. Velocity is another example of a vector quantity. If we wish to describe the velocity of a moving car, we must specify both its speed (say, 25 m/s) and the direction in which the car is moving (say, southwest). The rules of ordinary arithmetic cannot be used to manipulate vector quantities. Instead, we combine vectors according to special rules that are discussed in Sections 3.3 and 3.4.

A third example of a vector quantity is the displacement of a particle. Suppose the particle moves from some point O to the point P along a straight path, as in Figure 3.4. We represent this displacement by drawing an arrow from O to P, where the tip of the arrow represents the direction of the displacement and the length of the arrow represents the magnitude of the displacement. If the particle travels along some other path from O to P, such as the broken line in Figure 3.4, its displacement is still the arrow drawn from O to P. If the particle travels along any indirect path from O and P, its displacement is defined as being equivalent to the displacement for the direct path from O to P.

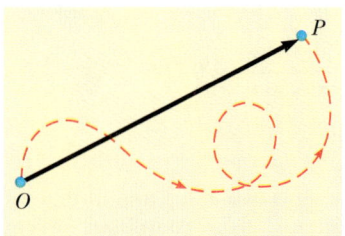

FIGURE 3.4 As a particle moves from O to P along an arbitrary path represented by the broken line, its displacement is a vector quantity shown by the arrow drawn from O to P.

If a particle moves along the x axis from position x_i to position x_f, as in Figure 3.5, its displacement is given by $\Delta x = x_f - x_i$. (The indices i and f refer to the initial and final values.) We use the Greek letter delta (Δ) to denote the *change* in a quantity.

It is important to remember that the distance traveled by a particle is distinctly different from its displacement. The distance traveled (a scalar quantity) is the length of the path, which can be much greater than the magnitude of the displacement (see Fig. 3.4). The magnitude of any displacement is the shortest distance between the end points of the displacement vector.

In this text, we use boldface letters, such as **A**, to represent a vector quantity. Another common method for vector notation that you should be aware of is to use an arrow over the letter $\vec{A}$. The magnitude of the vector **A** is written A or, alternatively, $|\mathbf{A}|$. The magnitude of a vector has physical units, such as meters for displacement or meters per second for velocity.

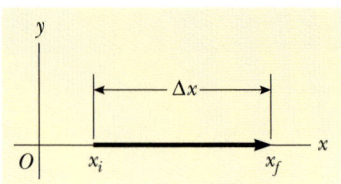

FIGURE 3.5 A particle moving along the x axis from x_i to x_f undergoes a displacement $\Delta x = x_f - x_i$.

3.3 SOME PROPERTIES OF VECTORS

Equality of Two Vectors

For many purposes, two vectors **A** and **B** may be defined to be equal if they have the same magnitude and point in the same direction. That is $\mathbf{A} = \mathbf{B}$, only if $A = B$ *and* they act along parallel directions. For example, all the vectors in Figure 3.6 are equal even though they have different starting points. This property allows us to translate a vector parallel to itself in a diagram without affecting the vector.

Addition

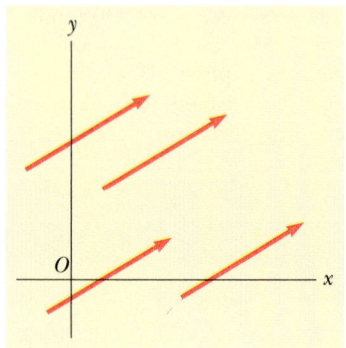

FIGURE 3.6 These four vectors are all equal since they have equal lengths and point in the same direction.

When two or more vectors are added together, *all* of them must have the same units. For example, it would be meaningless to add a velocity vector to a displacement vector since they are different physical quantities. Scalars also obey the same

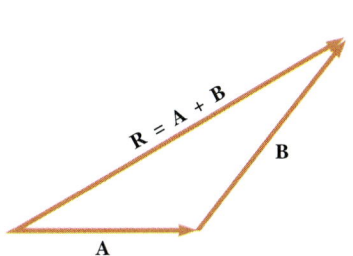

FIGURE 3.7 When vector **A** is added to vector **B**, the resultant **R** is the vector that runs from the tail of **A** to the tip of **B**.

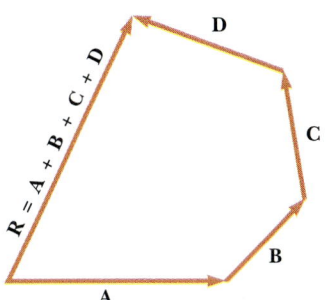

FIGURE 3.8 Geometric construction for summing four vectors. The resultant vector **R** is by definition the one that completes the polygon.

rule. For example, it would be meaningless to add time intervals and temperatures.

The rules for vector sums are conveniently described by geometric methods. To add vector **B** to vector **A**, first draw vector **A**, with its magnitude represented by a convenient scale, on graph paper and then draw vector **B** to the same scale with its tail starting from the tip of **A**, as in Figure 3.7. The *resultant* vector $R = A + B$ is the vector drawn from the tail of **A** to the tip of **B**. This is known as the *triangle method of addition.*

Geometric constructions can also be used to add more than two vectors. This is shown in Figure 3.8 for the case of four vectors. The resultant vector sum $R = A + B + C + D$ is *the vector that completes the polygon.* In other words, **R** is *the vector drawn from the tail of the first vector to the tip of the last vector.* The order of the summation is unimportant.

An alternative graphical procedure for adding two vectors, known as the *parallelogram rule of addition,* is shown in Figure 3.9a. In this construction, the tails of the two vectors **A** and **B** are together and the resultant vector **R** is the diagonal of a parallelogram formed with **A** and **B** as its sides.

When two vectors are added, the sum is independent of the order of the addition. This can be seen from the geometric construction in Figure 3.9b and is known as the **commutative law of addition:**

$$A + B = B + A \qquad (3.5)$$

Commutative law

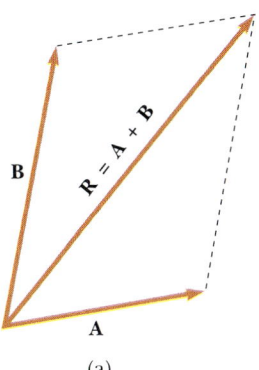

(a)

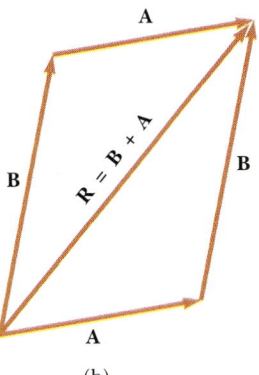

(b)

FIGURE 3.9 (a) In this construction, the resultant **R** is the diagonal of a parallelogram with sides **A** and **B**. (b) This construction shows that $A + B = B + A$. These geometric constructions verify the commutative law of addition for vectors.

If three or more vectors are added, their sum is independent of the way in which the individual vectors are grouped together. A geometric proof of this rule for three vectors is given in Figure 3.10. This is called the **associative law of addition:**

Associative law

$$A + (B + C) = (A + B) + C \tag{3.6}$$

Thus we conclude that *a vector quantity has both magnitude and direction and also obeys the laws of vector addition* as described in Figures 3.7 to 3.10.

Negative of a Vector

The negative of the vector **A** is defined as the vector that when added to **A** gives zero for the vector sum. That is, $A + (-A) = 0$. The vectors **A** and $-A$ have the same magnitude but point in opposite directions.

Subtraction of Vectors

The operation of vector subtraction makes use of the definition of the negative of a vector. We define the operation $A - B$ as vector $-B$ added to vector **A**:

$$A - B = A + (-B) \tag{3.7}$$

The geometric construction for subtracting two vectors is shown in Figure 3.11.

Multiplication of a Vector by a Scalar

If vector **A** is multiplied by a positive scalar quantity m, the product mA is a vector that has the same direction as **A** and magnitude mA. If m is a negative scalar quantity, the vector mA is directed opposite **A**. For example, the vector 5A is five times as long as **A** and points in the same direction as **A**; the vector $-\frac{1}{3}$A is one third the length of **A** and points in the direction opposite **A**.

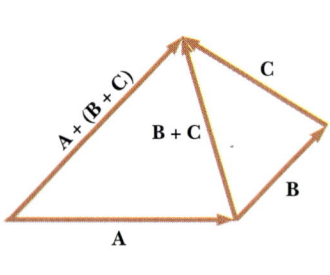

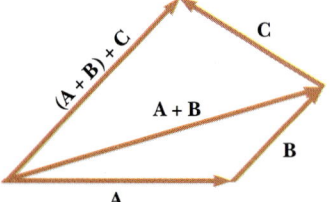

FIGURE 3.10 Geometric constructions for verifying the associative law of addition.

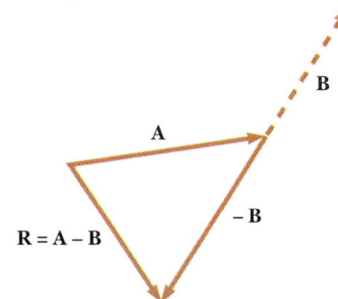

FIGURE 3.11 This construction shows how to subtract vector **B** from vector **A**. The vector $-B$ is equal in magnitude to vector **B** and points in the opposite direction. To subtract **B** and **A**, apply the rule of vector addition to the combination of **A** and $-B$: draw **A** along some convenient axis, place the tail of $-B$ at the tip of **A**, and the resultant is the difference $A - B$.

CONCEPTUAL EXAMPLE 3.2

If **B** is added to **A**, under what condition does the resultant vector **A** + **B** have a magnitude equal to $A + B$? Under what conditions is the resultant vector equal to zero?

Reasoning The resultant has a magnitude $A + B$ when **A** is oriented in the same direction as **B**. The resultant vector **A** + **B** = 0 when **A** is oriented in the direction opposite to **B**, and when $A = B$.

EXAMPLE 3.3 A Vacation Trip

A car travels 20.0 km due north and then 35.0 km in a direction 60.0° west of north, as in Figure 3.12. Find the magnitude and direction of the car's resultant displacement.

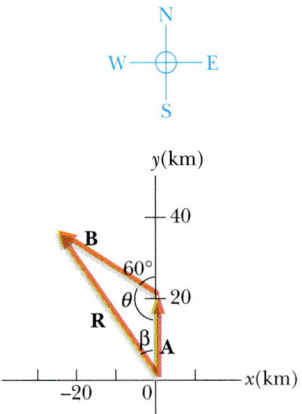

FIGURE 3.12 (Example 3.3) Graphical method for finding the resultant displacement vector **R** = **A** + **B**.

Reasoning The problem can be solved geometrically using graph paper and a protractor, as shown in Figure 3.12.

The resultant displacement **R** is the sum of the two individual displacements **A** and **B**. The following is an algebraic solution.

Solution The magnitude of **R** can be obtained using the law of cosines from trigonometry as applied to the obtuse triangle (Appendix B.4). Since $\theta = 180° - 60° = 120°$ and $R^2 = A^2 + B^2 - 2AB \cos \theta$, we find that

$$R = \sqrt{A^2 + B^2 - 2AB \cos \theta}$$
$$= \sqrt{(20.0)^2 + (35.0)^2 - 2(20.0)(35.0) \cos 120°} \text{ km}$$

$$= \boxed{48.2 \text{ km}}$$

The direction of **R** measured from the northerly direction can be obtained from the law of sines from trigonometry:

$$\frac{\sin \beta}{B} = \frac{\sin \theta}{R}$$

$$\sin \beta = \frac{B}{R} \sin \theta = \frac{35.0 \text{ km}}{48.2 \text{ km}} \sin 120° = 0.629$$

$$\beta = \boxed{38.9°}$$

Therefore, the resultant displacement of the car is 48.2 km in a direction 38.9° west of north.

3.4 COMPONENTS OF A VECTOR AND UNIT VECTORS

The geometric method of adding vectors is not the recommended procedure in situations where high precision is required or in three-dimensional problems. In this section, we describe a method of adding vectors that makes use of the *projections* of a vector along the axes of a rectangular coordinate system. These projections are called the **components** of the vector. Any vector can be completely described by its components.

Consider a vector **A** lying in the *xy* plane and making an arbitrary angle θ with the positive *x* axis, as in Figure 3.13. This vector can be expressed as the sum of two other vectors A_x and A_y. From Figure 3.13, we see that the three vectors form a right triangle and $A = A_x + A_y$. We shall often refer to the components of a vector **A**, written as A_x and A_y (without the boldface notation). The component A_x represents the projection of **A** along the *x* axis, and A_y represents the projection of **A**

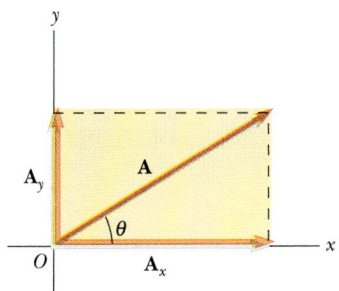

FIGURE 3.13 Any vector **A** lying in the *xy* plane can be represented by a vector lying along the *x* axis, A_x, and a vector lying along the *y* axis, A_y, where $A = A_x + A_y$.

along the y axis. These components can be positive or negative. The component A_x is positive if $\mathbf{A}_x$ points along the positive x axis and is negative if $\mathbf{A}_x$ points along the negative x axis. The same is true for the component A_y.

From Figure 3.13 and the definition of sine and cosine, we see that $\cos \theta = A_x/A$ and $\sin \theta = A_y/A$. Hence, the components of $\mathbf{A}$ are

Components of the vector A

$$A_x = A \cos \theta \tag{3.8}$$

$$A_y = A \sin \theta \tag{3.9}$$

These components form two sides of a right triangle the hypotenuse of which has a magnitude A. Thus, it follows that the magnitude of $\mathbf{A}$ and its direction are related to its components through the expressions

Magnitude of A

$$A = \sqrt{A_x^{\,2} + A_y^{\,2}} \tag{3.10}$$

and

Direction of A

$$\tan \theta = \frac{A_y}{A_x} \tag{3.11}$$

To solve for θ, we can write $\theta = \tan^{-1}(A_y/A_x)$, which is read "$\theta$ equals the angle whose tangent is the ratio A_y/A_x." *Note that the signs of the components A_x and A_y depend on the angle θ.* For example, if $\theta = 120°$, A_x is negative and A_y is positive. If $\theta = 225°$, both A_x and A_y are negative. Figure 3.14 summarizes the signs of the components when $\mathbf{A}$ lies in the various quadrants. When solving problems, you can specify a vector $\mathbf{A}$ with *either* the notation A_x, A_y or its magnitude and direction, A, θ.

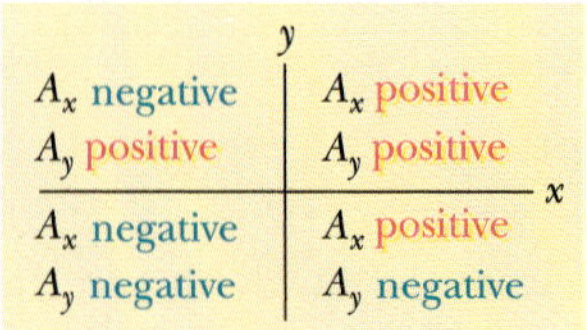

Suppose you are working a physics problem that requires resolving vectors into its components. If you choose reference axes or an angle other than the axes and angle shown in Figure 3.13, the components must be modified accordingly. In many applications, for example, it is convenient to express the components in a coordinate system having axes that are not horizontal and vertical but still perpendicular to each other. Suppose a vector $\mathbf{B}$ makes an angle θ' with the x' axis defined in Figure 3.15. The components of $\mathbf{B}$ along these axes are $B_{x}' = B \cos \theta'$ and $B_{y}' = B \sin \theta'$, as in Equations 3.8 and 3.9. The magnitude and direction of $\mathbf{B}$ are obtained from expressions equivalent to Equations 3.10 and 3.11. Thus, we can express the components of a vector in *any* coordinate system that is convenient for a particular situation.

The components of a vector are different when viewed from different coordinate systems. Furthermore, components can change with respect to a fixed coordinate system if the vector changes in magnitude, orientation, or both.

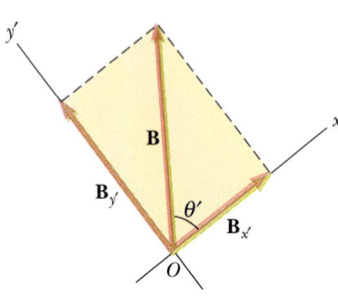

A_x negative A_y positive	A_x positive A_y positive
A_x negative A_y negative	A_x positive A_y negative

FIGURE 3.14 The signs of the components of a vector $\mathbf{A}$ depend on the quadrant in which the vector is located.

FIGURE 3.15 The component vectors of $\mathbf{B}$ in a coordinate system that is tilted.

Unit Vectors

Vector quantities are often expressed in terms of unit vectors. A **unit vector** is a dimensionless vector having a magnitude of exactly one. Unit vectors are used to specify a given direction and have no other physical significance. They are used solely as a convenience in describing a direction in space. We shall use the symbols $\mathbf{i}$, $\mathbf{j}$, and $\mathbf{k}$ to represent unit vectors pointing in the positive x, y, and z directions, respectively. The unit vectors $\mathbf{i}$, $\mathbf{j}$, and $\mathbf{k}$ form a set of mutually perpendicular vectors in a right-handed coordinate system as shown in Figure 3.16a. The magnitude of each unit vector equals unity; that is, $|\mathbf{i}| = |\mathbf{j}| = |\mathbf{k}| = 1$.

Consider a vector $\mathbf{A}$ lying in the xy plane, as in Figure 3.16b. The product of the component A_x and the unit vector $\mathbf{i}$ is the vector $A_x\mathbf{i}$, which is parallel to the x axis and has magnitude $|A_x|$. (The vector $A_x\mathbf{i}$ is an alternative and more common way of representing $\mathbf{A}_x$.) Likewise, $A_y\mathbf{j}$ is a vector of magnitude $|A_y|$ parallel to the y axis. (Again, $A_y\mathbf{j}$ is an alternative way of representing $\mathbf{A}_y$.) Thus, the unit-vector notation for the vector $\mathbf{A}$ is written

$$\mathbf{A} = A_x\mathbf{i} + A_y\mathbf{j} \tag{3.12}$$

For example, consider a point lying in the xy plane having cartesian coordinates (x, y) as in Figure 3.17. The point can be specified by the **position vector r**, which in unit-vector form is given by

$$\mathbf{r} = x\mathbf{i} + y\mathbf{j} \tag{3.13}$$

That is, the components of $\mathbf{r}$ are the coordinates x and y.

Now let us see how components are used to add vectors when the geometric method described in the preceding section is not appropriate. Suppose we wish to add vector $\mathbf{B}$ to vector $\mathbf{A}$, where $\mathbf{B}$ has components B_x and B_y. The procedure for performing this sum via the component method is to simply add the x and y components separately. The resultant vector $\mathbf{R} = \mathbf{A} + \mathbf{B}$ is therefore

$$\mathbf{R} = (A_x + B_x)\mathbf{i} + (A_y + B_y)\mathbf{j} \tag{3.14}$$

Since $\mathbf{R} = R_x\mathbf{i} + R_y\mathbf{j}$, we see that the components of the resultant vector are

$$R_x = A_x + B_x$$
$$R_y = A_y + B_y \tag{3.15}$$

The magnitude of $\mathbf{R}$ and the angle it makes with the x axis can then be obtained from its components using the relationships

$$R = \sqrt{R_x{}^2 + R_y{}^2} = \sqrt{(A_x + B_x)^2 + (A_y + B_y)^2} \tag{3.16}$$

and

$$\tan\theta = \frac{R_y}{R_x} = \frac{A_y + B_y}{A_x + B_x} \tag{3.17}$$

The procedure just described for adding two vectors $\mathbf{A}$ and $\mathbf{B}$ using the component method can be checked using a geometric construction, as in Figure 3.18.

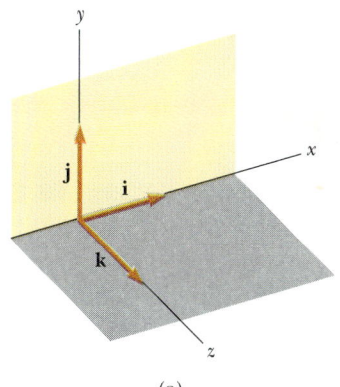

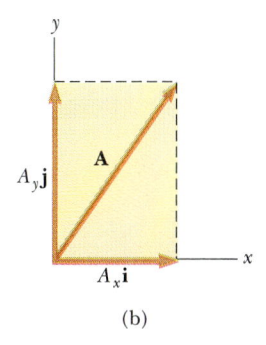

FIGURE 3.16   (a) The unit vectors i, j, and k are directed along the x, y, and z axes, respectively. (b) Vector $\mathbf{A} = A_x\mathbf{i} + A_y\mathbf{j}$ lying in the xy plane has components A_x and A_y.

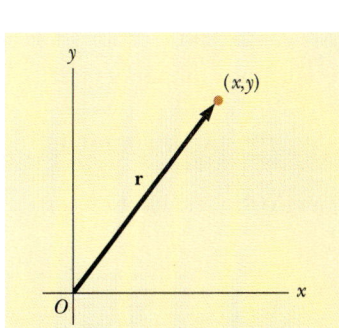

FIGURE 3.17   The point whose cartesian coordinates are (x, y) can be represented by the position vector $\mathbf{r} = x\mathbf{i} + y\mathbf{j}$.

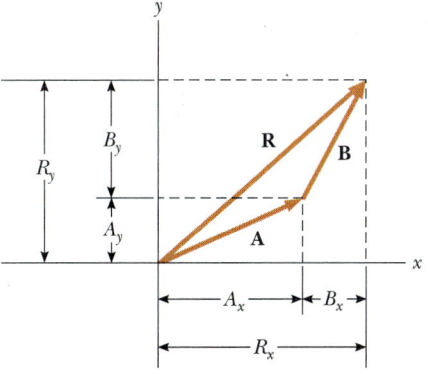

FIGURE 3.18   Geometric construction for the sum of two vectors showing the relationship between the components of the resultant $\mathbf{R}$ and the components of the individual vectors.

Again, you must take note of the *signs* of the components when using either the algebraic or the geometric method.

The extension of these methods to three-dimensional vectors is straight forward. If **A** and **B** both have x, y, and z components, we express them in the form

$$\mathbf{A} = A_x\mathbf{i} + A_y\mathbf{j} + A_z\mathbf{k} \tag{3.18}$$

$$\mathbf{B} = B_x\mathbf{i} + B_y\mathbf{j} + B_z\mathbf{k} \tag{3.19}$$

The sum of **A** and **B** is

$$\mathbf{R} = \mathbf{A} + \mathbf{B} = (A_x + B_x)\mathbf{i} + (A_y + B_y)\mathbf{j} + (A_z + B_z)\mathbf{k} \tag{3.20}$$

Thus, the resultant vector also has a z component $R_z = A_z + B_z$.

The component procedure just described can be used to sum up three or more vectors.

CONCEPTUAL EXAMPLE 3.4

If one component of a vector is not zero, can its magnitude be zero? Explain.

Reasoning   No. The magnitude of a vector **A** is equal to $\sqrt{A_x^2 + A_y^2 + A_z^2}$. Therefore, if any component is nonzero, **A** cannot be zero. Proof of this generalization of the Pythagorean theorem is left to Problem 52.

CONCEPTUAL EXAMPLE 3.5

If $\mathbf{A} + \mathbf{B} = 0$, what can you say about the components of the two vectors?

Reasoning $\mathbf{A} = -\mathbf{B}$, therefore the components of the two vectors must have opposite signs and equal magnitudes.

Problem-Solving Strategy
Adding Vectors

When two or more vectors are to be added, the following step-by-step procedure is recommended:

- Select a coordinate system.
- Draw a sketch of the vectors and label each one.
- Find the x and y components of all vectors.
- Find the resultant components (the algebraic sum of the components) in the x and y directions.
- Use the Pythagorean theorem to find the magnitude of the resultant vector.
- Use a suitable trigonometric function to find the angle the resultant vector makes with the x axis.

EXAMPLE 3.6   The Sum of Two Vectors

Find the sum of two vectors **A** and **B** lying in the xy plane and given by

$$\mathbf{A} = 2.0\mathbf{i} + 2.0\mathbf{j} \quad \text{and} \quad \mathbf{B} = 2.0\mathbf{i} - 4.0\mathbf{j}$$

Solution   Comparing the above expression for **A** with the general relation $\mathbf{A} = A_x\mathbf{i} + A_y\mathbf{j}$, we see that $A_x = 2.0$ and $A_y = 2.0$. Likewise, $B_x = 2.0$, and $B_y = -4.0$. Therefore, the resultant vector **R** is obtained by using Equation 3.20:

$\mathbf{R} = \mathbf{A} + \mathbf{B} = (2.0 + 2.0)\mathbf{i} + (2.0 - 4.0)\mathbf{j} = 4.0\mathbf{i} - 2.0\mathbf{j}$

or

$$R_x = 4.0 \qquad R_y = -2.0$$

The magnitude of $\mathbf{R}$ is given by Equation 3.16:

$R = \sqrt{R_x^2 + R_y^2} = \sqrt{(4.0)^2 + (-2.0)^2} = \sqrt{20} = \boxed{4.5}$

Exercise Find the angle θ that $\mathbf{R}$ makes with the positive x axis.

Answer $330°$.

EXAMPLE 3.7 The Resultant Displacement

A particle undergoes three consecutive displacements: $\mathbf{d}_1 = (1.5\mathbf{i} + 3.0\mathbf{j} - 1.2\mathbf{k})$ cm, $\mathbf{d}_2 = (2.3\mathbf{i} - 1.4\mathbf{j} - 3.6\mathbf{k})$ cm, and $\mathbf{d}_3 = (-1.3\mathbf{i} + 1.5\mathbf{j})$ cm. Find the components of the resultant displacement and its magnitude.

Solution

$\mathbf{R} = \mathbf{d}_1 + \mathbf{d}_2 + \mathbf{d}_3$

$\quad = (1.5 + 2.3 - 1.3)\mathbf{i} + (3.0 - 1.4 + 1.5)\mathbf{j}$

$\quad\quad + (-1.2 - 3.6 + 0)\mathbf{k}$

$\quad = (2.5\mathbf{i} + 3.1\mathbf{j} - 4.8\mathbf{k})$ cm

That is, the resultant displacement has components

$R_x = \boxed{2.5 \text{ cm}} \quad R_y = \boxed{3.1 \text{ cm}} \quad \text{and} \quad R_z = \boxed{-4.8 \text{ cm}}$

Its magnitude is

$$R = \sqrt{R_x^2 + R_y^2 + R_z^2}$$
$$= \sqrt{(2.5 \text{ cm})^2 + (3.1 \text{ cm})^2 + (-4.8 \text{ cm})^2}$$
$$= \boxed{6.2 \text{ cm}}$$

EXAMPLE 3.8 Taking a Hike

A hiker begins a trip by first walking 25.0 km southeast from her base camp. On the second day, she walks 40.0 km in a direction 60.0° north of east, at which point she discovers a forest ranger's tower. (a) Determine the components of the hiker's displacement for each day.

Solution If we denote the displacement vectors on the first and second days by $\mathbf{A}$ and $\mathbf{B}$, respectively, and use the camp as the origin of coordinates, we get the vectors shown in Figure 3.19. Displacement $\mathbf{A}$ has a magnitude of 25.0 km and is 45.0° southeast. From Equations 3.8 and 3.9, its components are

$A_x = A \cos(-45.0°) = (25.0 \text{ km})(0.707) = \boxed{17.7 \text{ km}}$

$A_y = A \sin(-45.0°) = -(25.0 \text{ km})(0.707) = \boxed{-17.7 \text{ km}}$

The negative value of A_y indicates that the y coordinate had decreased for this displacement—in other words, the hiker has walked in the negative y direction. The signs of A_x and A_y are also evident from Figure 3.19. The second displacement, $\mathbf{B}$, has a magnitude of 40.0 km and is 60.0° north of east. Its rectangular components are

$B_x = B \cos 60.0° = (40.0 \text{ km})(0.500) = \boxed{20.0 \text{ km}}$

$B_y = B \sin 60.0° = (40.0 \text{ km})(0.866) = \boxed{34.6 \text{ km}}$

(b) Determine the components of the hiker's resultant displacement $\mathbf{R}$ for the trip. Find an expression for $\mathbf{R}$ in terms of unit vectors.

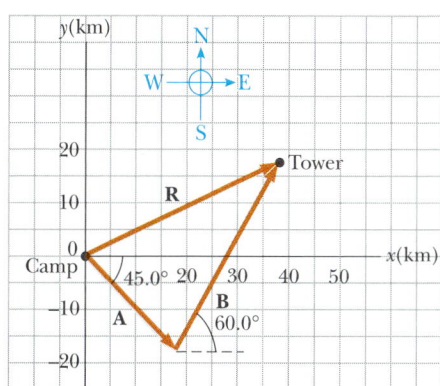

FIGURE 3.19 (Example 3.8) The total displacement of the hiker is the vector $\mathbf{R} = \mathbf{A} + \mathbf{B}$.

Solution The resultant displacement for the trip, $\mathbf{R} = \mathbf{A} + \mathbf{B}$, has components given by Equation 3.15:

$R_x = A_x + B_x = 17.7 \text{ km} + 20.0 \text{ km} = \boxed{37.7 \text{ km}}$

$R_y = A_y + B_y = -17.7 \text{ km} + 34.6 \text{ km} = \boxed{16.9 \text{ km}}$

In unit-vector form, we can write the total displacement as $\mathbf{R} = (37.7\mathbf{i} + 16.9\mathbf{j})$ km.

Exercise Determine the magnitude and direction of the total displacement.

Answer 41.3 km, 24.1° north of east from the base camp.

EXAMPLE 3.9   Let's Fly Away

A commuter airplane starts from an airport and takes the route shown in Figure 3.20. First, it flies to city A located 175 km in a direction 30.0° north of east. Next, it flies 150 km 20.0° west of north to city B. Finally, it flies 190 km west to city C. Find the location of city C relative to the location of the starting point.

Solution   As in the previous example, it is convenient to choose the coordinate system shown in Figure 3.20, where the *x* axis points to the east and the *y* axis points to the north. Let us denote the three consecutive displacements by the vectors **a**, **b**, and **c**. The first displacement **a** has a magnitude of 175 km and has components

$$a_x = a\cos(30.0°) = (175\text{ km})(0.866) = 152\text{ km}$$

$$a_y = a\sin(30.0°) = (175\text{ km})(0.500) = 87.5\text{ km}$$

Displacement **b**, whose magnitude is 150 km, has components

$$b_x = b\cos(110°) = (150\text{ km})(-0.342) = -51.3\text{ km}$$

$$b_y = b\sin(110°) = (150\text{ km})(0.940) = 141\text{ km}$$

Finally, displacement **c**, whose magnitude is 190 km, has components

$$c_x = c\cos(180°) = (190\text{ km})(-1) = -190\text{ km}$$

$$c_y = c\sin(180°) = 0$$

Therefore, the components of the position vector **R** from the starting point to city C are

$$R_x = a_x + b_x + c_x = 152\text{ km} - 51.3\text{ km} - 190\text{ km}$$

$$= -89.7\text{ km}$$

$$R_y = a_y + b_y + c_y = 87.5\text{ km} + 141\text{ km} + 0$$

$$= 228\text{ km}$$

In unit-vector notation, **R** = (−89.7**i** + 228**j**) km. That is, city C can be reached from the starting point by first traveling 89.7 km due west and then traveling 228 km due north.

Exercise   Find the magnitude and direction of **R**.

Answer   245 km, 21.4° west of north.

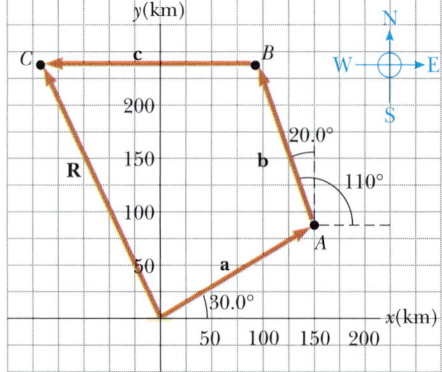

FIGURE 3.20   (Example 3.9) The airplane starts at the origin, flies first to *A*, then to *B*, and finally to *C*.

SUMMARY

Vector quantities have both magnitude and direction and obey the vector law of addition. **Scalar quantities** have only magnitude and obey the laws of ordinary arithmetic.

Two vectors **A** and **B** can be added graphically using either the triangle method or the parallelogram rule. In the triangle method (Fig. 3.21a), the resultant vector

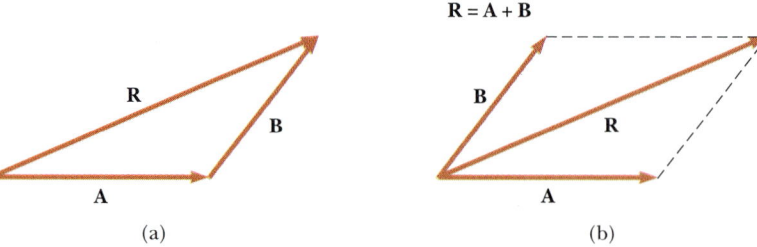

FIGURE 3.21   (a) Vector addition using the triangle method. (b) Vector addition using the parallelogram rule.

$\mathbf{R} = \mathbf{A} + \mathbf{B}$ runs from the tail of $\mathbf{A}$ to the tip of $\mathbf{B}$. In the parallelogram method (Fig. 3.21b), $\mathbf{R}$ is the diagonal of a parallelogram having $\mathbf{A}$ and $\mathbf{B}$ as its sides.

The x component, A_x, of the vector $\mathbf{A}$ is equal to its projection along the x axis of a coordinate system as in Figure 3.22, where $A_x = A \cos \theta$. Likewise, the y component, A_y, of $\mathbf{A}$ is its projection along the y axis, where $A_y = A \sin \theta$.

If a vector $\mathbf{A}$ has an x component equal to A_x and a y component equal to A_y, the vector can be expressed in unit-vector form as $\mathbf{A} = A_x\mathbf{i} + A_y\mathbf{j}$. In this notation, $\mathbf{i}$ is a unit vector pointing in the positive x direction and $\mathbf{j}$ is a unit vector in the positive y direction. Since $\mathbf{i}$ and $\mathbf{j}$ are unit vectors, $|\mathbf{i}| = |\mathbf{j}| = 1$.

The resultant of two or more vectors can be found by resolving all vectors into their x and y components, adding their resultant x and y components, and then using the Pythagorean theorem to find the magnitude of the resultant vector. The angle that the resultant vector makes with respect to the x axis can be found by use of a suitable trigonometric function.

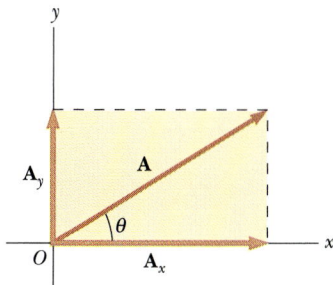

FIGURE 3.22 The x and y vector components of $\mathbf{A}$ are $\mathbf{A}_x$ and $\mathbf{A}_y$. The x and y rectangular components of $\mathbf{A}$ are $A_x = A \cos \theta$ and $A_y = A \sin \theta$.

QUESTIONS

1. A book is moved once around the perimeter of a rectangular tabletop of dimensions 1.0 m × 2.0 m. If the book ends up at its initial position, what is its displacement? What is the distance traveled?
2. Can the magnitude of a particle's displacement be greater than the distance traveled? Explain.
3. The magnitudes of two vectors $\mathbf{A}$ and $\mathbf{B}$ are $A = 5$ units and $B = 2$ units. Find the largest and smallest values possible for the resultant vector $\mathbf{R} = \mathbf{A} + \mathbf{B}$.
4. Vector $\mathbf{A}$ lies in the xy plane. For what orientations of $\mathbf{A}$ will both of its components be negative? For what orientations will its components have opposite signs?
5. If the component of vector $\mathbf{A}$ along the direction of vector $\mathbf{B}$ is zero, what can you conclude about the two vectors?
6. Can the magnitude of a vector have a negative value? Explain.
7. Which of the following are vectors and which are not: force, temperature, volume, ratings of a television show, height, velocity, age?
8. Under what circumstances would a nonzero vector lying in the xy plane have components that are equal in magnitude?
9. Is it possible to add a vector quantity to a scalar quantity? Explain.
10. Two vectors have unequal magnitudes. Can their sum be zero? Explain.

PROBLEMS

Section 3.1   Coordinate Systems and Frames of Reference

1. Two points in the xy plane have cartesian coordinates (2.00, −4.00) m and (−3.00, 3.00) m. Determine (a) the distance between these points and (b) their polar coordinates.
2. If the rectangular and polar coordinates of a point are (2, y) and (r, 30°), respectively, determine y and r.
3. The polar coordinates of a point are $r = 5.50$ m and $\theta = 240.0°$. What are the cartesian coordinates of this point?
4. Two points in a plane have polar coordinates (2.50 m, 30.0°) and (3.80 m, 120.0°). Determine (a) the cartesian coordinates of these points and (b) the distance between them.
5. A certain corner of a room is selected as the origin of a rectangular coordinate system. A fly is crawling on a wall adjacent to one of the axes. If the fly is located at a point having coordinates (2.00, 1.00) m, (a) how far is it from the corner of the room? (b) what is its location in polar coordinates?
6. If the polar coordinates of the point (x, y) are (r, θ), determine the polar coordinates for the points: (a) (−x, y), (b) (−2x, −2y), and (c) (3x, −3y).
7. A point is located in a polar coordinate system by the coordinates $r = 2.50$ m and $\theta = 35.0°$. Find the cartesian coordinates of this point, assuming the two coordinate systems have the same origin.

☐ indicates problems that have full solutions available in the Student Solutions Manual and Study Guide.

Section 3.2   Vector and Scalar Quantities
Section 3.3   Some Properties of Vectors

8. An airplane flies 200 km due west from city A to city B and then 300 km in the direction of 30° north of west from city B to city C. (a) In straight-line distance, how far is city C from city A? (b) Relative to city A, in what direction is city C?

9. A surveyor estimates the distance across a river by the following method: standing directly across from a tree on the opposite bank, she walks 100 m along the riverbank, then sights across to the tree. The angle from her baseline to the tree is 35.0°. How wide is the river?

10. A pedestrian moves 6.00 km east and then 13.0 km north. Find the magnitude and direction of the resultant displacement vector using the graphical method.

11. A plane flies from base camp to lake A, a distance of 280 km at a direction of 20.0° north of east. After dropping off supplies, it flies to lake B, which is 190 km and 30.0° west of north from lake A. Graphically determine the distance and direction from lake B to the base camp.

12. Vector **A** has a magnitude of 8.00 units and makes an angle of 45.0° with the positive *x* axis. Vector **B** also has a magnitude of 8.00 units and is directed along the negative *x* axis. Using graphical methods, find (a) the vector sum **A** + **B** and (b) the vector difference **A** − **B**.

13. A person walks along a circular path of radius 5.00 m, around one half of the circle. (a) Find the magnitude of the displacement vector. (b) How far did the person walk? (c) What is the magnitude of the displacement if the person walks all the way around the circle?

14. A force **F**$_1$ of magnitude 6.00 units acts at the origin in a direction 30.0° above the positive *x* axis. A second force **F**$_2$ of magnitude 5.00 units acts at the origin in the direction of the positive *y* axis. Find graphically the magnitude and direction of the resultant force **F**$_1$ + **F**$_2$.

15. Each of the displacement vectors **A** and **B** shown in Figure P3.15 has a magnitude of 3.00 m. Find graphically (a) **A** + **B**, (b) **A** − **B**, (c) **B** − **A**, (d) **A** − 2**B**.

16. A dog searching for a bone walks 3.5 m south, then 8.2 m at an angle 30° north of east, and finally 15 m west. Find the dog's resultant displacement vector using graphical techniques.

17. A roller coaster moves 200 ft horizontally, then travels 135 ft at an angle of 30.0° above the horizontal. It then travels 135 ft at an angle of 40.0° below the horizontal. What is its displacement from its starting point? Use graphical techniques.

18. The driver of a car drives 3.00 km north, 2.00 km northeast (45.0° east of north), 4.00 km west, and then 3.00 km southeast (45.0° east of south). Where

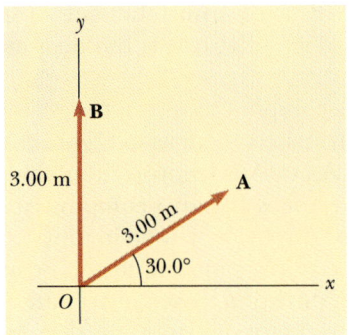

FIGURE P3.15

does he end up relative to his starting point? Work out your answer graphically. Check by using components. (The car is not near the North Pole or the South Pole.)

19. Find the horizontal and vertical components of the 100-m displacement of a superhero who flies from the top of a tall building following the path shown in Figure P3.19.

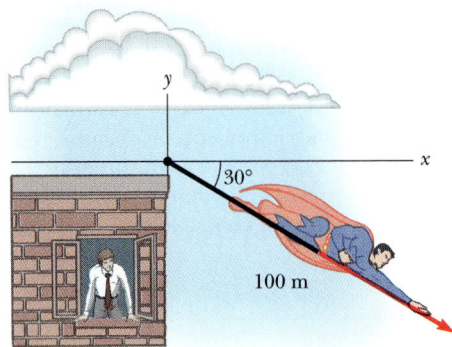

FIGURE P3.19

20. A person walks 25.0° north of east for 3.10 km. How far would she have to walk due north and due east to arrive at the same location?

21. Indiana Jones is trapped in a maze. To find his way out, he walks 10 m, makes a 90° right turn, walks 5.0 m, makes another 90° right turn, and walks 7.0 m. What is his displacement from his initial position?

22. While exploring a cave, a spelunker starts at the entrance and moves the following distances. She goes 75.0 m north, 250 m east, 125 m at an angle 30.0° north of east, and 150 m south. Find the resultant displacement from the cave entrance.

Section 3.4   Components of a Vector and Unit Vectors

23. A vector has an x component of -25.0 units and a y component of 40.0 units. Find the magnitude and direction of this vector.

24. Vector **B** has x, y, and z components of 4.00, 6.00, and 3.00 units, respectively. Calculate the magnitude of **B** and the angles that **B** makes with the coordinate axes.

25. Given the vectors $\mathbf{A} = 2.0\mathbf{i} + 6.0\mathbf{j}$ and $\mathbf{B} = 3.0\mathbf{i} - 2.0\mathbf{j}$, (a) sketch the vector sum $\mathbf{C} = \mathbf{A} + \mathbf{B}$ and the vector subtraction $\mathbf{D} = \mathbf{A} - \mathbf{B}$. (b) Find analytical solutions for **C** and **D** first in terms of unit vectors and then in terms of polar coordinates, with angles measured with respect to the $+x$ axis.

26. A displacement vector lying in the xy plane has a magnitude of 50.0 m and is directed at an angle of 120.0° to the positive x axis. What are the rectangular components of this vector?

27. Find the magnitude and direction of the resultant of three displacements having rectangular components (3.00, 2.00) m, $(-5.00, 3.00)$ m, and (6.00, 1.00) m.

28. Vector **A** has x and y components of -8.7 cm and 15 cm, respectively; vector **B** has x and y components of 13.2 cm and -6.6 cm, respectively. If $\mathbf{A} - \mathbf{B} + 3\mathbf{C} = 0$, what are the components of **C**?

29. Consider two vectors $\mathbf{A} = 3\mathbf{i} - 2\mathbf{j}$ and $\mathbf{B} = -\mathbf{i} - 4\mathbf{j}$. Calculate (a) $\mathbf{A} + \mathbf{B}$, (b) $\mathbf{A} - \mathbf{B}$, (c) $|\mathbf{A} + \mathbf{B}|$, (d) $|\mathbf{A} - \mathbf{B}|$, (e) the direction of $\mathbf{A} + \mathbf{B}$ and $\mathbf{A} - \mathbf{B}$.

30. A boy runs 3.0 blocks north, 4.0 blocks northeast, and 5.0 blocks west. Determine the length and direction of the displacement vector that goes from the starting point to his final position.

31. Obtain expressions for the position vectors having polar coordinates (a) 12.8 m, 150°; (b) 3.30 cm, 60.0°; (c) 22.0 in., 215°.

32. Consider the displacement vectors $\mathbf{A} = (3\mathbf{i} + 3\mathbf{j})$ m, $\mathbf{B} = (\mathbf{i} - 4\mathbf{j})$ m, and $\mathbf{C} = (-2\mathbf{i} + 5\mathbf{j})$ m. Use the component method to determine (a) the magnitude and direction of the vector $\mathbf{D} = \mathbf{A} + \mathbf{B} + \mathbf{C}$, (b) the magnitude and direction of $\mathbf{E} = -\mathbf{A} - \mathbf{B} + \mathbf{C}$.

33. A particle undergoes the following consecutive displacements: 3.50 m south, 8.20 m northeast, and 15.0 m west. What is the resultant displacement?

34. A quarterback takes the ball from the line of scrimmage, runs backward for 10 yards, then sideways parallel to the line of scrimmage for 15 yards. At this point, he throws a forward pass 50 yards straight downfield perpendicular to the line of scrimmage. What is the magnitude of the football's resultant displacement?

35. A jet airliner moving initially at 300 mph to the east moves into a region where the wind is blowing at 100 mph in a direction 30.0° north of east. What are the new speed and direction of the aircraft?

36. A novice golfer on the green takes three strokes to sink the ball. The successive displacements are 4.00 m to the north, 2.00 m northeast, and 1.00 m 30.0° west of south. Starting at the same initial point, an expert golfer could make the hole in what single displacement?

37. Find the x and y components of the vectors **A** and **B** shown in Figure P3.15. Derive an expression for the resultant vector $\mathbf{A} + \mathbf{B}$ in unit-vector notation.

38. A particle undergoes two displacements. The first has a magnitude of 150 cm and makes an angle of 120.0° with the positive x axis. The *resultant* displacement has a magnitude of 140 cm and is directed at an angle of 35.0° to the positive x axis. Find the magnitude and direction of the second displacement.

39. The vector **A** has x, y, and z components of 8, 12, and -4 units, respectively. (a) Write a vector expression for **A** in unit-vector notation. (b) Obtain a unit-vector expression for a vector **B** one fourth the length of **A** pointing in the same direction as **A**. (c) Obtain a unit-vector expression for a vector **C** three times the length of **A** pointing in the direction opposite the direction of **A**.

40. Instructions for finding a buried treasure include the following: Go 75 paces at 240°, turn to 135° and walk 125 paces, then travel 100 paces at 160°. Determine the resultant displacement from the starting point.

41. Given the displacement vectors $\mathbf{A} = (3.0\mathbf{i} - 4.0\mathbf{j} + 4.0\mathbf{k})$ m and $\mathbf{B} = (2.0\mathbf{i} + 3.0\mathbf{j} - 7.0\mathbf{k})$ m, find the magnitudes of the vectors (a) $\mathbf{C} = \mathbf{A} + \mathbf{B}$ and (b) $\mathbf{D} = 2\mathbf{A} - \mathbf{B}$, also expressing each in terms of its rectangular components.

42. As it passes over Grand Bahama Island, the eye of a hurricane is moving in a direction 60.0° north of west with a speed of 41.0 km/h. Three hours later, it shifts due north, and its speed slows to 25.0 km/h. How far from Grand Bahama is the eye 4.50 h after it passes over the island?

43. Vector **A** has a negative x component 3.00 units in length and a positive y component 2.00 units in length. (a) Determine an expression for **A** in unit-vector notation. (b) Determine the magnitude and direction of **A**. (c) What vector **B** when added to **A** gives a resultant vector with no x component and a negative y component 4.00 units in length?

44. An airplane starting from airport A flies 300 km east, then 350 km 30.0° west of north, and then 150 km north to arrive finally at airport B. There is no wind on this day. (a) The next day, another plane flies directly from A to B in a straight line. In what direction should the pilot travel in this direct flight? (b) How far will the pilot travel in this direct flight?

45. Point A in Figure P3.45 is an arbitrary point along the line connecting the two points (x_2, y_2). Show that the coordinates of A are $(1 - f)x_1 + fx_2$, $(1 - f)y_1 + fy_2$.

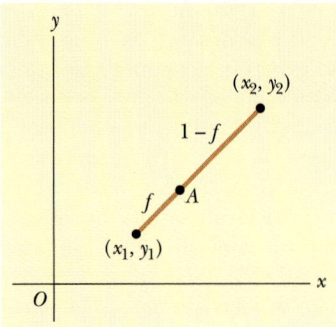

FIGURE P3.45

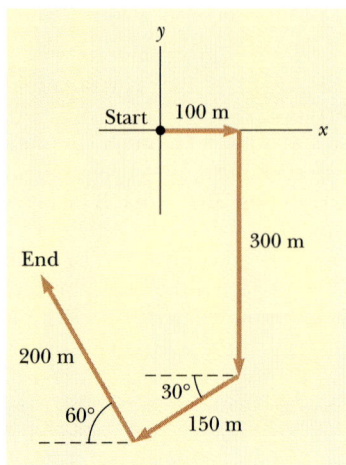

FIGURE P3.49

46. If $\mathbf{A} = (6.0\mathbf{i} - 8.0\mathbf{j})$ units, $\mathbf{B} = (-8.0\mathbf{i} + 3.0\mathbf{j})$ units, and $\mathbf{C} = (26.0\mathbf{i} + 19.0\mathbf{j})$ units, determine a and b so that $a\mathbf{A} + b\mathbf{B} + \mathbf{C} = 0$.

47. Three vectors are oriented as shown in Figure P3.47, where $|\mathbf{A}| = 20.0$ units, $|\mathbf{B}| = 40.0$ units, and $|\mathbf{C}| = 30.0$ units. Find (a) the x and y components of the resultant vector and (b) the magnitude and direction of the resultant vector.

 50. The helicopter view in Figure P3.50 shows two people pulling on a stubborn mule. Find (a) the single force that is equivalent to the two forces shown, and (b) the force that a third person would have to exert on the mule to make the resultant force equal to zero.

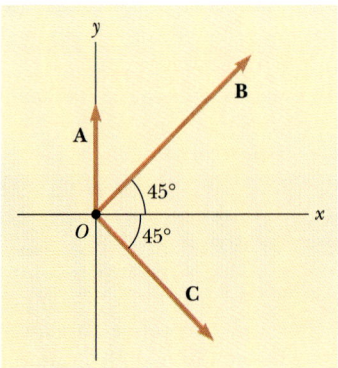

FIGURE P3.47

ADDITIONAL PROBLEMS

48. A vector is given by $\mathbf{R} = 2\mathbf{i} + \mathbf{j} + 3\mathbf{k}$. Find (a) the magnitudes of the x, y, and z components, (b) the magnitude of $\mathbf{R}$, and (c) the angles between $\mathbf{R}$ and the x, y, and z axes.

49. A person going for a walk follows the path shown in Figure P3.49. The total trip consists of four straight-line paths. At the end of the walk, what is the person's resultant displacement measured from the starting point?

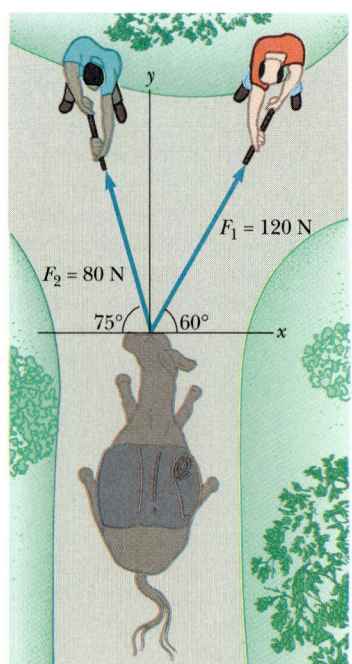

FIGURE P3.50

51. A pirate has buried his treasure on an island on which grow five trees located at the following points: $A(30\,\text{m}, -20\,\text{m})$, $B(60\,\text{m}, 80\,\text{m})$, $C(-10\,\text{m}, -10\,\text{m})$, $D(40\,\text{m}, -30\,\text{m})$, and $E(-70\,\text{m}, 60\,\text{m})$, all mea-

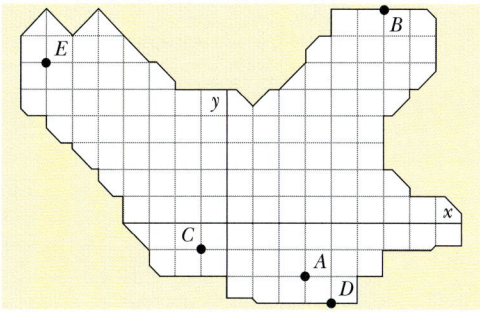

FIGURE P3.51

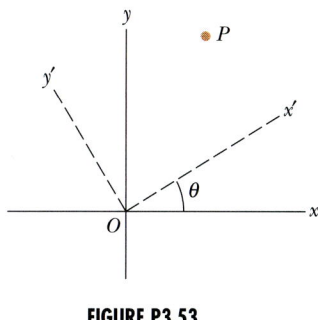

FIGURE P3.53

sured relative to some origin, as in Figure P3.51. His map instructs him to start at A and move toward B, but cover only one-half the distance between A and B. Then move toward C, covering one-third the distance between your current location and C. Then move toward D, covering one-fourth the distance between where you are and D. Finally move toward E, covering one-fifth the distance between you and E, stop and dig. (a) What are the coordinates of the point where his treasure is buried? (b) Rearrange the order of the trees (for instance, $B(30$ m, -20 m), $A(60$ m, 80 m), $E(-10$ m, -10 m), $C(40$ m, -30 m), and $D(-70$ m, 60 m)), and repeat the calculation to show that the answer does not depend on the order of the trees. (*Hint:* See Problem 45.)

52. A rectangular parallelepiped has dimensions a, b, and c, as in Figure P3.52. (a) Obtain a vector expression for the face diagonal vector $\mathbf{R}_1$. What is the magnitude of this vector? (b) Obtain a vector expression for the body diagonal vector $\mathbf{R}_2$. Note that $\mathbf{R}_1$, $c\mathbf{k}$, and $\mathbf{R}_2$ make a right triangle and prove that the magnitude of $\mathbf{R}_2$ is $\sqrt{a^2 + b^2 + c^2}$.

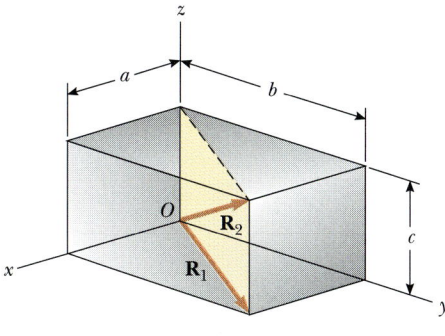

FIGURE P3.52

53. A point P is described by the coordinates (x, y) with respect to the normal cartesian coordinate system shown in Figure P3.53. Show that (x', y'), the coordinates of this point in the rotated coordinate system,

are related to (x, y) and the rotation angle θ by the expressions

$$x' = x \cos \theta + y \sin \theta$$

$$y' = -x \sin \theta + y \cos \theta$$

54. A point lying in the xy plane and having coordinates (x, y) can be described by the position vector $\mathbf{r} = x\mathbf{i} + y\mathbf{j}$. (a) Show that the displacement vector for a particle moving from (x_1, y_1) to (x_2, y_2) is given by $\mathbf{d} = (x_2 - x_1)\mathbf{i} + (y_2 - y_1)\mathbf{j}$. (b) Plot the position vectors $\mathbf{r}_1$ and $\mathbf{r}_2$ and the displacement vector $\mathbf{d}$, and verify by the graphical method that $\mathbf{d} = \mathbf{r}_2 - \mathbf{r}_1$.

SPREADSHEET PROBLEMS

S1. Use Spreadsheet 3.1 to find the total displacement corresponding to the sum of the three displacement vectors: $\mathbf{A} = 3.0\mathbf{i} + 4.0\mathbf{j}$, $\mathbf{B} = -2.3\mathbf{i} - 7.8\mathbf{j}$, and $\mathbf{C} = 5.0\mathbf{i} - 2.0\mathbf{j}$, where the units are in meters. (a) Give your answer in component form. (b) Give your answer in polar form. (c) Find a fourth displacement that returns you to the origin.

S2. A vendor must call on four customers once in every sales period. The four customers are in different cities, and the salesman wants to visit all customers in the least time. What path should he take?

To solve this problem, suppose you start at the origin and visit four different points (A, B, C, D) once before returning to the origin, always traveling in a straight line between points. The distance traveled depends upon the route taken. Spreadsheet 3.2 enables you to pick any four points A, B, C, D lying in a plane; the spreadsheet calculates the magnitude of the vector displacement for each straight-line portion of the trip and the total distance traveled. Find the path that gives the shortest total distance traveled.

Brute force can be used to find this path: try every path possible. To use Spreadsheet 3.2, enter the X and Y coordinates in the columns corresponding to the points A, B, C, D. The first site, the starting point, is labeled O, and the final site, which is also the starting position, is also labeled O. Note the total distance traveled and examine the graph that displays the path. To change the order in which the sites are visited, change the index order of the four non-origin sites and sort the block. For example, to visit *B* first, enter a "1" in the order column to the left of *B*; to visit *D* next, enter a "2" in the order column next to *D*, and so on; then sort the block containing the rows in which the points *A*, *B*, *C*, *D* appear in ascending order according to the order column. Note the total distance traveled and again examine the graph to see your new path. Repeat this process until you are convinced you have found the shortest path. (a) Find the shortest possible distance for visiting four points, when the locations are $(-10, 5)$, $(-8,$ $-7)$, $(1, 11)$, and $(12, 9)$; find the minimum distance necessary to visit these points. How many trials did you use to find this shortest distance? In what order should the vendor visit these four customers? (b) Choose any other four points and repeat part (a).

S3. Modify Spreadsheet 3.2 to include two more points. Choose any six points and use your spreadsheet to find the shortest path for visiting all six sites. How many trials did you need this time? This brute-force method is not practical when there are more than a handful of customers. For *N* sites, the method requires $(N-1)!/2$ trials. For example, if $N = 10$, you would need over 180 000 trials.

CHAPTER 4

Motion in Two Dimensions

The circus stuntmen being shot out of cannons are human projectiles. Neglecting air resistance, they move in parabolic paths until they land in a strategically placed net. What initial condition(s) will determine where the catching net should be placed? *(Ringling Brothers Circus)*

I n this chapter we deal with the kinematics of a particle moving in a plane, which is two-dimensional motion. Some common examples of motion in a plane are the motion of projectiles and satellites and the motion of charged particles in uniform electric fields. We begin by showing that displacement, velocity, and acceleration are vector quantities. As in the case of one-dimensional motion, we derive the kinematic equations for two-dimensional motion from the fundamental definitions of displacement, velocity, and acceleration. As special cases of motion in two dimensions, we then treat constant-acceleration motion in a plane as well as uniform circular motion.

4.1 THE DISPLACEMENT, VELOCITY, AND ACCELERATION VECTORS

In Chapter 2 we found that the motion of a particle moving along a straight line is completely known if its coordinate is known as a function of time. Now let us extend this idea to motion in the *xy* plane. We begin by describing the position of a

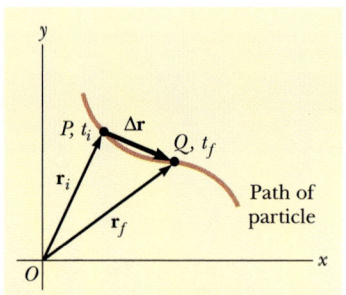

FIGURE 4.1 A particle moving in the *xy* plane is located with the position vector **r** drawn from the origin to the particle. The displacement of the particle as it moves from *P* to *Q* in the time interval $\Delta t = t_f - t_i$ is equal to the vector $\Delta \mathbf{r} = \mathbf{r}_f - \mathbf{r}_i$.

Average velocity

particle with a *position vector* **r**, drawn from the origin of some reference frame to the particle located in the *xy* plane, as in Figure 4.1. At time t_i, the particle is at point *P*, and at some later time t_f, the particle is at *Q*, where the indices *i* and *f* refer to initial and final values. As the particle moves from *P* to *Q* in the time interval $\Delta t = t_f - t_i$, the position vector changes from $\mathbf{r}_i$ to $\mathbf{r}_f$. As we learned in Chapter 2, the displacement of a particle is the difference between its final position and initial position. Therefore, the **displacement vector** for the particle of Figure 4.1 equals the difference between its final position vector and its initial position vector:

$$\Delta \mathbf{r} \equiv \mathbf{r}_f - \mathbf{r}_i \qquad (4.1)$$

The direction of $\Delta \mathbf{r}$ is indicated in Figure 4.1. As we see from Figure 4.1, the magnitude of the displacement vector is *less* than the distance traveled along the curved path.

We define the **average velocity** of the particle during the time interval Δt as the ratio of the displacement to that time interval:

$$\bar{\mathbf{v}} \equiv \frac{\Delta \mathbf{r}}{\Delta t} \qquad (4.2)$$

Since displacement is a vector quantity and the time interval is a scalar quantity, we conclude that the average velocity is a vector quantity directed along $\Delta \mathbf{r}$. Note that the average velocity between points *P* and *Q* is *independent of the path* between the two points. This is because the average velocity is proportional to the displacement, which in turn depends only on the initial and final position vectors and not on the path taken between those two points. As we did with one-dimensional motion, we conclude that if a particle starts its motion at some point and returns to this point via any path, its average velocity is zero for this trip since its displacement is zero.

Consider again the motion of a particle between two points in the *xy* plane, as in Figure 4.2. As the time intervals over which we observe the motion become smaller and smaller, the direction of the displacement approaches that of the line tangent to the path at the point *P*.

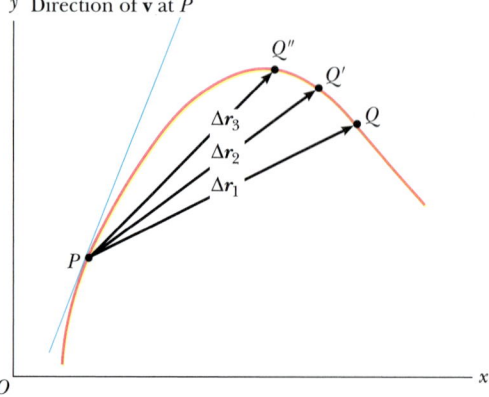

FIGURE 4.2 As a particle moves between two points, its average velocity is in the direction of the displacement vector $\Delta \mathbf{r}$. As the end point of the path is moved from *Q* to *Q'* to *Q''*, the respective displacements and corresponding time intervals become smaller and smaller. In the limit that the end point approaches *P*, Δt approaches zero and the direction of $\Delta \mathbf{r}$ approaches that of the line tangent to the curve at *P*. By definition, the instantaneous velocity at *P* is in the direction of this tangent line.

The **instantaneous velocity, v,** is defined as the limit of the average velocity, $\Delta \mathbf{r}/\Delta t$, as Δt approaches zero:

$$\mathbf{v} \equiv \lim_{\Delta t \to 0} \frac{\Delta \mathbf{r}}{\Delta t} = \frac{d\mathbf{r}}{dt} \tag{4.3}$$

Instantaneous velocity

That is, the instantaneous velocity equals the derivative of the position vector with respect to time. The direction of the instantaneous velocity vector at any point in a particle's path is along a line that is tangent to the path at that point and in the direction of motion; this is illustrated in Figure 4.3. The magnitude of the instantaneous velocity vector is called the *speed.*

As a particle moves from one point to another along some path as in Figure 4.3, its instantaneous velocity vector changes from $\mathbf{v}_i$ at time t_i to $\mathbf{v}_f$ at time t_f.

The **average acceleration** of a particle as it moves from P to Q is defined as the ratio of the change in the instantaneous velocity vector, $\Delta \mathbf{v}$, to the elapsed time, Δt:

$$\bar{\mathbf{a}} \equiv \frac{\mathbf{v}_f - \mathbf{v}_i}{t_f - t_i} = \frac{\Delta \mathbf{v}}{\Delta t} \tag{4.4}$$

Average acceleration

Since the average acceleration is the ratio of a vector quantity, $\Delta \mathbf{v}$, and a scalar quantity, Δt, we conclude that $\bar{\mathbf{a}}$ is a vector quantity directed along $\Delta \mathbf{v}$. As is indicated in Figure 4.3, the direction of $\Delta \mathbf{v}$ is found by adding the vector $-\mathbf{v}_i$ (the negative of $\mathbf{v}_i$) to the vector $\mathbf{v}_f$, since by definition $\Delta \mathbf{v} = \mathbf{v}_f - \mathbf{v}_i$.

The **instantaneous acceleration, a,** is defined as the limiting value of the ratio $\Delta \mathbf{v}/\Delta t$ as Δt approaches zero:

$$\mathbf{a} \equiv \lim_{\Delta t \to 0} \frac{\Delta \mathbf{v}}{\Delta t} = \frac{d\mathbf{v}}{dt} \tag{4.5}$$

Instantaneous acceleration

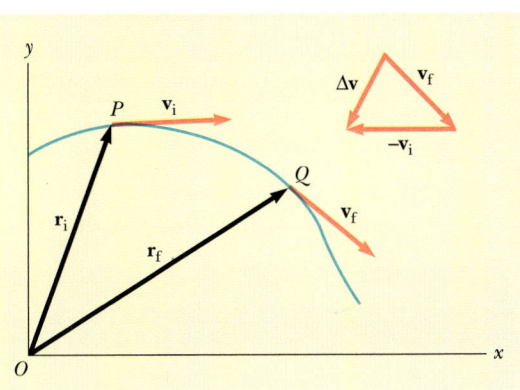

FIGURE 4.3   The average acceleration vector $\bar{\mathbf{a}}$ for a particle moving from P to Q is in the direction of the change in velocity, $\Delta \mathbf{v} = \mathbf{v}_f - \mathbf{v}_i$.

In other words, the instantaneous acceleration equals the derivative of the velocity vector with respect to time.

It is important to recognize that various changes may constitute acceleration of a particle. First, the magnitude of the velocity vector (the speed) may change with time as in straight-line (one-dimensional) motion. Second, only the direction of the velocity vector may change with time as its magnitude (speed) remains constant, as in curved-path (two-dimensional) motion. Finally, both the magnitude and the direction of the velocity vector may change.

CONCEPTUAL EXAMPLE 4.1

(a) Can an object accelerate if its speed is constant? (b) Can an object accelerate if its velocity is constant?

Reasoning   (a) Yes. Although its speed may be constant, the direction of its motion (that is, the direction of **v**) may change, causing an acceleration. For example, an object moving in a circle with constant speed has an acceleration directed toward the center of the circle. (b) No. An object that moves with constant velocity has zero acceleration. Note that constant velocity means that both the direction and magnitude of **v** remain constant.

4.2   TWO-DIMENSIONAL MOTION WITH CONSTANT ACCELERATION

Let us consider two-dimensional motion during which the acceleration remains constant. That is, we assume that the magnitude and direction of the acceleration remain unchanged during the motion.

As we learned in Chapter 3, the motion of a particle can be determined by its position vector **r**. The position vector for a particle moving in the *xy* plane can be written

> Position vector

$$\mathbf{r} = x\mathbf{i} + y\mathbf{j} \tag{4.6}$$

where *x*, *y*, and **r** change with time as the particle moves. If the position vector is known, the velocity of the particle can be obtained from Equations 4.3 and 4.6, which give

$$\mathbf{v} = \frac{d\mathbf{r}}{dt} = \frac{dx}{dt}\mathbf{i} + \frac{dy}{dt}\mathbf{j}$$

$$\mathbf{v} = v_x\mathbf{i} + v_y\mathbf{j} \tag{4.7}$$

Because **a** is assumed constant, its components a_x and a_y are also constants. Therefore, we can apply the equations of kinematics to the *x* and *y* components of the velocity vector. Substituting $v_x = v_{x0} + a_x t$ and $v_y = v_{y0} + a_y t$ into Equation 4.7 gives

$$\mathbf{v} = (v_{x0} + a_x t)\mathbf{i} + (v_{y0} + a_y t)\mathbf{j}$$
$$= (v_{x0}\mathbf{i} + v_{y0}\mathbf{j}) + (a_x\mathbf{i} + a_y\mathbf{j})t$$

> Velocity vector as a function of time

$$\mathbf{v} = \mathbf{v}_0 + \mathbf{a}t \tag{4.8}$$

This result states that the velocity of a particle at some time *t* equals the vector sum of its initial velocity, $\mathbf{v}_0$, and the additional velocity **a**t acquired in the time *t* as a result of its constant acceleration.

Similarly, from Equation 2.9 we know that the x and y coordinates of a particle moving with constant acceleration are

$$x = x_0 + v_{x0}t + \tfrac{1}{2}a_x t^2 \qquad \text{and} \qquad y = y_0 + v_{y0}t + \tfrac{1}{2}a_y t^2$$

Substituting these expressions into Equation 4.6 gives

$$\mathbf{r} = (x_0 + v_{x0}t + \tfrac{1}{2}a_x t^2)\mathbf{i} + (y_0 + v_{y0}t + \tfrac{1}{2}a_y t^2)\mathbf{j}$$
$$= (x_0\mathbf{i} + y_0\mathbf{j}) + (v_{x0}\mathbf{i} + v_{y0}\mathbf{j})t + \tfrac{1}{2}(a_x\mathbf{i} + a_y\mathbf{j})t^2$$

$$\boxed{\mathbf{r} = \mathbf{r}_0 + \mathbf{v}_0 t + \tfrac{1}{2}\mathbf{a}t^2} \tag{4.9}$$

Position vector as a function of time

This equation implies that the displacement vector $\mathbf{r} - \mathbf{r}_0$ is the vector sum of a displacement $\mathbf{v}_0 t$, arising from the initial velocity of the particle, and a displacement $\tfrac{1}{2}\mathbf{a}t^2$, resulting from the uniform acceleration of the particle. Graphical representations of Equations 4.8 and 4.9 are shown in Figures 4.4a and 4.4b. For simplicity in drawing the figure, we have taken $\mathbf{r}_0 = 0$ in Figure 4.4b. That is, we assume that the particle is at the origin at $t = 0$. Note from Figure 4.4b that $\mathbf{r}$ is generally not along the direction of $\mathbf{v}_0$ or $\mathbf{a}$, because the relationship between these quantities is a vector expression. For the same reason, from Figure 4.4a we see that $\mathbf{v}$ is generally not along the direction of $\mathbf{v}_0$ or $\mathbf{a}$. Finally, if we compare the two figures, we see that $\mathbf{v}$ and $\mathbf{r}$ are not in the same direction.

Because Equations 4.8 and 4.9 are *vector* expressions, we may write their x and y component forms with $\mathbf{r}_0 = 0$:

$$\mathbf{v} = \mathbf{v}_0 + \mathbf{a}t \qquad \begin{cases} v_x = v_{x0} + a_x t \\ v_y = v_{y0} + a_y t \end{cases}$$

$$\mathbf{r} = \mathbf{v}_0 t + \tfrac{1}{2}\mathbf{a}t^2 \qquad \begin{cases} x = v_{x0}t + \tfrac{1}{2}a_x t^2 \\ y = v_{y0}t + \tfrac{1}{2}a_y t^2 \end{cases}$$

These components are illustrated in Figure 4.4. In other words, two-dimensional motion having constant acceleration is equivalent to two independent motions in the x and y directions having constant accelerations a_x and a_y.

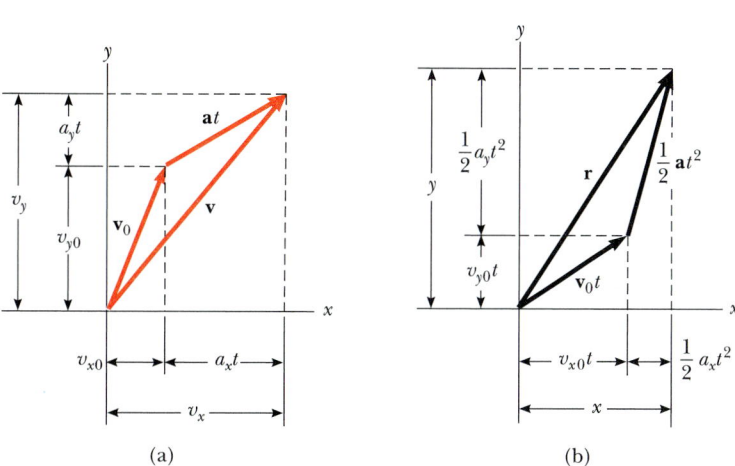

(a)   (b)

FIGURE 4.4   Vector representations and rectangular components of (a) the velocity and (b) the displacement of a particle moving with a uniform acceleration **a**.

Multiflash exposure of a tennis player executing a backhand swing. Note that the ball follows a parabolic path characteristic of a projectile. Such photographs can be used to study the quality of sports equipment and the performance of an athlete. (© *Zimmerman, FPG International*)

EXAMPLE 4.2   Motion in a Plane

A particle starts from the origin at $t = 0$ with an initial velocity having an x component of 20 m/s and a y component of -15 m/s. The particle moves in the xy plane with an x component of acceleration only, given by $a_x = 4.0$ m/s². (a) Determine the components of velocity as a function of time and the total velocity vector at any time.

Solution   With $v_{x0} = 20$ m/s and $a_x = 4.0$ m/s², the equations of kinematics give

$$v_x = v_{x0} + a_x t = (20 + 4t) \text{ m/s}$$

Also, with $v_{y0} = -15$ m/s and $a_y = 0$,

$$v_y = v_{y0} = -15 \text{ m/s}$$

Therefore, using these results and noting that the velocity vector $\mathbf{v}$ has two components, we get

$$\mathbf{v} = v_x \mathbf{i} + v_y \mathbf{j} = \boxed{[(20 + 4.0t)\mathbf{i} - 15\mathbf{j}] \text{ m/s}}$$

We could also obtain this result using Equation 4.8 directly, noting that $\mathbf{a} = 4.0\mathbf{i}$ m/s² and $\mathbf{v_0} = (20\mathbf{i} - 15\mathbf{j})$ m/s. Try it!

   (b) Calculate the velocity and speed of the particle at $t = 5.0$ s.

Solution   With $t = 5.0$ s, the result from (a) gives

$$\mathbf{v} = \{[20 + 4(5.0)]\mathbf{i} - 15\mathbf{j}\} \text{ m/s} = \boxed{(40\mathbf{i} - 15\mathbf{j}) \text{ m/s}}$$

That is, at $t = 5.0$ s, $v_x = 40$ m/s and $v_y = -15$ m/s. Knowing these two components for this two-dimensional motion, we know the numerical value of the velocity vector. To determine the angle θ that $\mathbf{v}$ makes with the x axis, we use the fact that $\tan \theta = v_y/v_x$, or

$$\theta = \tan^{-1}\left(\frac{v_y}{v_x}\right) = \tan^{-1}\left(\frac{-15 \text{ m/s}}{40 \text{ m/s}}\right) = \boxed{-21°}$$

The speed is the magnitude of $\mathbf{v}$:

$$v = |\mathbf{v}| = \sqrt{v_x^2 + v_y^2} = \sqrt{(40)^2 + (-15)^2} \text{ m/s} = \boxed{43 \text{ m/s}}$$

(*Note:* If you calculate v_0 from the x and y components of $\mathbf{v_0}$ you will find that $v > v_0$. Why?)

   (c) Determine the x and y coordinates of the particle at any time t and the displacement vector at this time.

Solution   Since at $t = 0$, $x_0 = y_0 = 0$, Equation 2.11 gives

$$x = v_{x0}t + \tfrac{1}{2}a_x t^2 = \boxed{(20t + 2.0t^2) \text{ m}}$$

$$y = v_{y0}t = \boxed{(-15t) \text{ m}}$$

Therefore, the displacement vector at any time t is

$$\mathbf{r} = x\mathbf{i} + y\mathbf{j} = [(20t + 2.0t^2)\mathbf{i} - 15t\mathbf{j}] \text{ m}$$

Alternatively, we could obtain $\mathbf{r}$ by applying Equation 4.9 directly, with $\mathbf{v}_0 = (20\mathbf{i} - 15\mathbf{j})$ m/s and $\mathbf{a} = 4\mathbf{i}$ m/s^2. Try it!

Thus, for example, at $t = 5.0$ s, $x = 150$ m and $y = -75$ m, or $\mathbf{r} = (150\mathbf{i} - 75\mathbf{j})$ m. It follows that the distance of

the particle from the origin to this point is the magnitude of the displacement:

$$|\mathbf{r}| = r = \sqrt{(150)^2 + (-75)^2} \text{ m} = 170 \text{ m}$$

Note that this is *not* the distance that the particle travels in this time! Can you determine this distance from the available data?

4.3   PROJECTILE MOTION

Anyone who has observed a baseball in motion (or, for that matter, any object thrown into the air) has observed projectile motion. The ball moves in a curved path when thrown at some angle with respect to the Earth's surface. This very common form of motion is surprisingly simple to analyze if the following two assumptions are made: (1) the free-fall acceleration, g, is constant over the range of motion and is directed downward,[1] and (2) the effect of air resistance is negligible.[2] With these assumptions, we find that the path of a projectile, which we call its *trajectory*, is *always* a parabola. *We use these assumptions throughout this chapter.*

If we choose our reference frame such that the y direction is vertical and positive upward, then $a_y = -g$ (as in one-dimensional free-fall) and $a_x = 0$ (because air friction is neglected). Furthermore, let us assume that at $t = 0$, the projectile leaves the origin ($x_0 = y_0 = 0$) with speed v_0, as in Figure 4.5. If the vector $\mathbf{v}_0$ makes

Assumptions of projectile motion

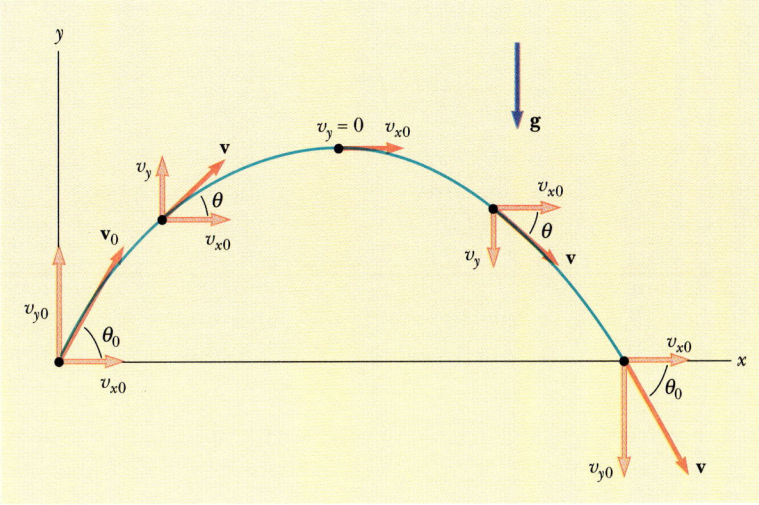

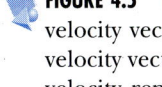
FIGURE 4.5   The parabolic path of a projectile that leaves the origin with a velocity $\mathbf{v}_0$. The velocity vector $\mathbf{v}$ changes with time in both magnitude and direction. The change in the velocity vector is the result of acceleration in the negative y direction. The x component of velocity remains constant in time because there is no acceleration along the horizontal direction. Also, the y component of velocity is zero at the peak of the path.

---

[1] This approximation is reasonable as long as the range of motion is small compared with the radius of the Earth (6.4×10^6 m). In effect, this approximation is equivalent to assuming that the Earth is flat over the range of motion considered.

[2] This approximation is generally *not* justified, especially at high speeds. In addition, any spin imparted to a projectile, such as happens when a pitcher throws a curve ball, can give rise to some very interesting effects associated with aerodynamic forces.

an angle θ_0 with the horizontal, where θ_0 is the angle at which the projectile leaves the origin as in Figure 4.5, then from the definitions of the cosine and sine functions we have

$$\cos \theta_0 = v_{x0}/v_0 \quad \text{and} \quad \sin \theta_0 = v_{y0}/v_0$$

Therefore, the initial x and y components of velocity are

$$v_{x0} = v_0 \cos \theta_0 \quad \text{and} \quad v_{y0} = v_0 \sin \theta_0$$

Substituting these expressions into Equations 4.8 and 4.9 with $a_x = 0$ and $a_y = -g$ gives the velocity components and coordinates for the projectile at any time t:

Horizontal velocity component

$$v_x = v_{x0} = v_0 \cos \theta_0 = \text{constant} \tag{4.10}$$

Vertical velocity component

$$v_y = v_{y0} - gt = v_0 \sin \theta_0 - gt \tag{4.11}$$

Horizontal position component

$$x = v_{x0}t = (v_0 \cos \theta_0)t \tag{4.12}$$

Vertical position component

$$y = v_{y0}t - \tfrac{1}{2}gt^2 = (v_0 \sin \theta_0)t - \tfrac{1}{2}gt^2 \tag{4.13}$$

From Equation 4.10 we see that v_x remains constant in time and is equal to v_{x0}; there is no horizontal component of acceleration. For the y motion note that v_y and y are similar to Equations 2.12 and 2.14 for freely falling bodies. In fact, *all* of the equations of kinematics developed in Chapter 2 are applicable to projectile motion.

If we solve for t in Equation 4.12 and substitute this expression for t into Equation 4.13, we find that

$$y = (\tan \theta_0)x - \left(\frac{g}{2v_0^2 \cos^2 \theta_0} \right)x^2 \tag{4.14}$$

which is valid for the angles in the range $0 < \theta_0 < \pi/2$. This expression is of the form $y = ax - bx^2$, which is the equation of a parabola that passes through the origin. Thus, we have proved that the trajectory of a projectile is a parabola. Note that the trajectory is *completely* specified if v_0 and θ_0 are known.

One can obtain the speed, v, of the projectile as a function of time by noting that Equations 4.10 and 4.11 give the x and y components of velocity at any instant. Therefore, by definition, since v is equal to the magnitude of **v**,

$$v = \sqrt{v_x^2 + v_y^2} \tag{4.15}$$

Also, since the velocity vector is tangent to the path at any instant, as shown in Figure 4.5, the angle θ that v makes with the horizontal can be obtained from v_x and v_y through the expression

$$\tan \theta = \frac{v_y}{v_x} \tag{4.16}$$

Multiflash photograph of a tennis ball undergoing several bounces off a hard surface is an example of projectile motion. Note the parabolic path of the ball following each bounce. *(Richard Megna, Fundamental Photographs)*

The vector expression for the position vector of the projectile as a function of time follows directly from Equation 4.9, with **a** = **g**:

$$\mathbf{r} = \mathbf{v}_0 t + \tfrac{1}{2}\mathbf{g}t^2$$

This expression gives the same information as the combination of Equations 4.12 and 4.13 and is plotted in Figure 4.6. Note that this expression for **r** is consistent with Equation 4.13, since the expression for **r** is a vector equation and **a** = **g** = $-g\mathbf{j}$ when the upward direction is taken to be positive.

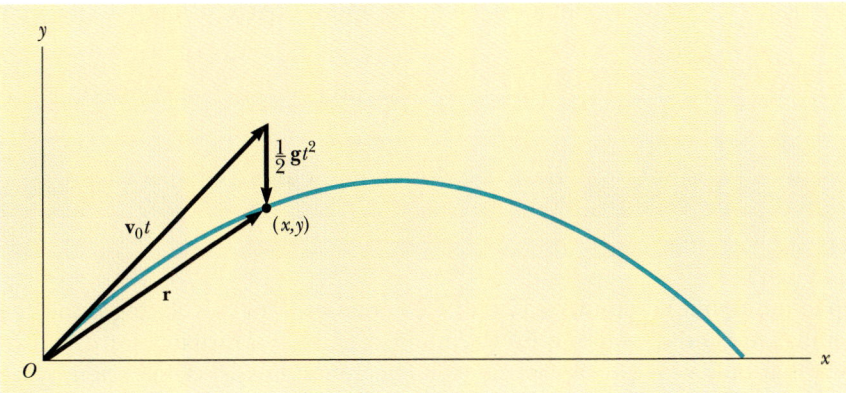

FIGURE 4.6 The displacement vector **r** of a projectile whose initial velocity at the origin is $\mathbf{v}_0$. The vector $\mathbf{v}_0 t$ would be the displacement of the projectile if gravity were absent, and the vector $\frac{1}{2}\mathbf{g}t^2$ is its vertical displacement due to its downward gravitational acceleration.

It is interesting to note that the motion of a particle can be considered the superposition of the term $\mathbf{v}_0 t$, which would be the displacement if no acceleration were present, and the term $\frac{1}{2}\mathbf{g}t^2$, which arises from the acceleration due to gravity. In other words, if there were no gravitational acceleration, the particle would continue to move along a straight path in the direction of $\mathbf{v}_0$. Therefore, the vertical distance $\frac{1}{2}\mathbf{g}t^2$ through which the particle "falls" off the straight path line is the same distance a freely falling body would fall during the same time interval. *We conclude that projectile motion is the superposition of two motions: (1) constant velocity motion in the initial direction and (2) the motion of a particle freely falling in the vertical direction under constant acceleration.*

Horizontal Range and Maximum Height of a Projectile

Let us assume that a projectile is fired from the origin at $t = 0$ with a positive v_y component, as in Figure 4.7. There are two special points that are interesting to analyze: the peak that has cartesian coordinates $(R/2, h)$, and the point having coordinates $(R, 0)$. The distance R is called the *horizontal range* of the projectile, and h is its *maximum height*. Let us find h and R in terms of v_0, θ_0, and g.

We can determine h by noting that at the peak, $v_y = 0$. Therefore, Equation 4.11 can be used to determine the time t_1 it takes to reach the peak:

$$t_1 = \frac{v_0 \sin \theta_0}{g}$$

Substituting this expression for t_1 into Equation 4.13 and replacing y with h gives h in terms of v_0 and θ_0:

$$h = (v_0 \sin \theta_0)\frac{v_0 \sin \theta_0}{g} - \frac{1}{2}g\left(\frac{v_0 \sin \theta_0}{g}\right)^2$$

$$h = \frac{v_0^2 \sin^2 \theta_0}{2g} \tag{4.17}$$

The range, R, is the horizontal distance traveled in twice the time it takes to reach the peak, that is, in a time $2t_1$. (This can be seen by setting $y = 0$ in Equation 4.13 and solving the quadratic for t. One solution of this quadratic is $t = 0$, and the

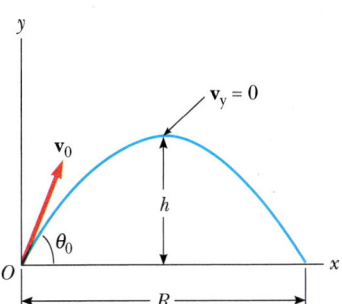

FIGURE 4.7 A projectile fired from the origin at $t = 0$ with an initial velocity $\mathbf{v}_0$. The maximum height of the projectile is h, and the horizontal range is R. At the peak of the trajectory, the particle has coordinates $(R/2, h)$.

second is $t = 2t_1$.) Using Equation 4.12 and noting that $x = R$ at $t = 2t_1$, we find that

$$R = (v_0 \cos \theta_0)2t_1 = (v_0 \cos \theta_0)\frac{2v_0 \sin \theta_0}{g} = \frac{2v_0^2 \sin \theta_0 \cos \theta_0}{g}$$

Since $\sin 2\theta = 2 \sin \theta \cos \theta$, R can be written in the more compact form

Range of projectile

$$R = \frac{v_0^2 \sin 2\theta_0}{g} \tag{4.18}$$

Keep in mind that Equations 4.17 and 4.18 are useful for calculating h and R only if v_0 and θ_0 are known and only for a symmetric path, as shown in Figure 4.7 (which means that only v_0 has to be specified). The general expressions given by Equations 4.10 through 4.13 are the *more important* results, because they give the coordinates and velocity components of the projectile at *any* time t.

You should note that the maximum value of R from Equation 4.18 is given by $R_{max} = v_0^2/g$. This result follows from the fact that the maximum value of $\sin 2\theta_0$ is unity, which occurs when $2\theta_0 = 90°$. Therefore, we see that R is a maximum when $\theta_0 = 45°$, as you would expect if air friction is neglected.

Figure 4.8 illustrates various trajectories for a projectile of a given initial speed. As you can see, the range is a maximum for $\theta_0 = 45°$. In addition, for any θ_0 other than $45°$, a point with coordinates $(R, 0)$ can be reached by using either one of two complementary values of θ_0 such as $75°$ and $15°$. Of course, the maximum height and the time of flight will be different for these two values of θ_0.

CONCEPTUAL EXAMPLE 4.3

As a projectile moves in its parabolic path, is there any point along its path where the velocity and acceleration vectors are (a) perpendicular to each other? (b) parallel to each other?

Reasoning (a) At the top of its flight, **v** is horizontal and **a** is vertical. This is the only point where the velocity and acceleration vectors are perpendicular. (b) If the object is thrown straight up or down, then **v** and **a** will be parallel throughout the downward motion. Otherwise, the velocity and acceleration vectors are never parallel.

FIGURE 4.8 A projectile fired from the origin with an initial speed of 50 m/s at various angles of projection. Note that complementary values of θ_0 will result in the same value of x (range of the projectile).

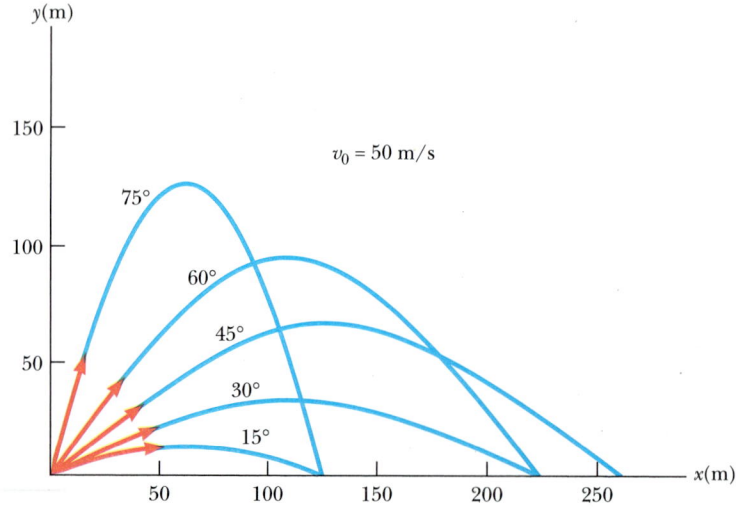

EXAMPLE 4.4   The Long-Jump

A long-jumper leaves the ground at an angle of 20.0° to the horizontal and at a speed of 11.0 m/s. (a) How far does he jump? (Assume that his motion is equivalent to that of a particle.)

Solution   The horizontal motion is described by Equation 4.12:

$$x = (v_0 \cos \theta_0)t = (11.0 \text{ m/s})(\cos 20.0°)t$$

The value of x can be found if t, the total time of the jump, is known. We are able to find t using Equation 4.11, $v_y = v_0 \sin \theta_0 - gt$, and by noting that at the top of the jump the vertical component of velocity goes to zero:

$$v_y = v_0 \sin \theta_0 - gt$$

$$0 = (11.0 \text{ m/s}) \sin 20.0° - (9.80 \text{ m/s}^2)t_1$$

$$t_1 = 0.384 \text{ s}$$

Note that t_1 is the time interval to reach the *top* of the jump. Because of the symmetry of the vertical motion, an identical time interval passes before the jumper returns to the ground. Therefore, the *total* time in the air is $t = 2t_1 = 0.768$ s. Substituting this value into the above expression for x gives

$$x = (11.0 \text{ m/s})(\cos 20.0°)(0.768 \text{ s}) = \boxed{7.94 \text{ m}}$$

(b) What is the maximum height reached?

Solution   The maximum height reached is found using Equation 4.13 with $t = t_1 = 0.384$ s:

$$y_{max} = (v_0 \sin \theta_0)t_1 - \tfrac{1}{2}gt_1^2$$

(Example 4.4) In a long-jump event, Willie Banks can leap horizontal distances of at least 8 meters.   *(© R. Mackson/FPG)*

$$= (11.0 \text{ m/s})(\sin 20.0°)(0.384 \text{ s})$$
$$- \tfrac{1}{2}(9.80 \text{ m/s}^2)(0.384 \text{ s})^2$$

$$= \boxed{0.722 \text{ m}}$$

The assumption that the motion of the long-jumper is that of a particle is an oversimplification. Nevertheless, the values obtained are reasonable. Note that we also could have used Equations 4.17 and 4.18 to find the maximum height and horizontal range. However, the method used in our solution is more instructive.

EXAMPLE 4.5   It's a Bull's-Eye Every Time

In a very popular lecture demonstration, a projectile is fired at a target in such a way that the projectile leaves the gun at the same time the target is dropped from rest, as in Figure 4.9. Let us show that if the gun is initially aimed at the stationary target, the projectile hits the target.

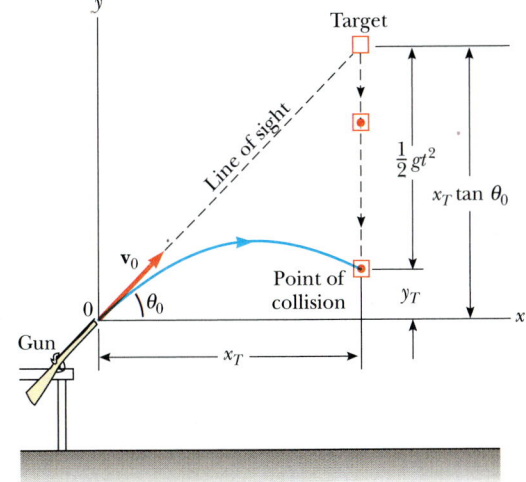

FIGURE 4.9   (Example 4.5) Schematic diagram of the projectile-and-target demonstration. If the gun is aimed directly at the stationary target and is fired at the same instant the target begins to fall, the projectile will hit the target. Both fall through the same vertical distance in a time t, because both experience the same acceleration, $a_y = -g$.

Reasoning and Solution   We can argue that a collision will result under the conditions stated by noting that both the projectile and the target experience the same acceleration, $a_y = -g$, as soon as they are released. First, note from Figure 4.9 that the initial y coordinate of the target is $x_T \tan \theta_0$ and that it falls through a distance $\frac{1}{2}gt^2$ in a time t. Therefore, the y coordinate of the target as a function of time is, from Equation 4.14,

$$y_T = x_T \tan \theta_0 - \tfrac{1}{2}gt^2$$

Now if we write equations for x and y for the projectile path over time, using Equations 4.12 and 4.13 simultaneously, we get

$$y_p = x_p \tan \theta_0 - \tfrac{1}{2}gt^2$$

Thus, we see by comparing the two equations above that when $x_p = x_T$, $y_p = y_T$ and a collision results.

The result could also be arrived at with vector methods, using expressions for the position vectors for the projectile and target.

You should also note that a collision will *not* always take place. There is the further restriction that a collision results only when $v_0 \sin \theta_0 \geqslant \sqrt{gd/2}$, where d is the initial elevation of the target above the *floor*. If $v_0 \sin \theta_0$ is less than this value, the projectile will strike the floor before reaching the target.

EXAMPLE 4.6   That's Quite an Arm

A stone is thrown from the top of a building upward at an angle of 30.0° to the horizontal and with an initial speed of 20.0 m/s, as in Figure 4.10. If the height of the building is 45.0 m, (a) how long is the stone "in flight"?

Solution   The initial x and y components of the velocity are

$$v_{x0} = v_0 \cos \theta_0 = (20.0 \text{ m/s})(\cos 30.0°) = 17.3 \text{ m/s}$$

$$v_{y0} = v_0 \sin \theta_0 = (20.0 \text{ m/s})(\sin 30.0°) = 10.0 \text{ m/s}$$

To find t, we can use $y = v_{y0}t - \frac{1}{2}gt^2$ (Eq. 4.13) with $y = -45.0$ m and $v_{y0} = 10.0$ m/s (we have chosen the top of the building as the origin, as in Figure 4.10):

$$-45.0 \text{ m} = (10.0 \text{ m/s})t - \tfrac{1}{2}(9.80 \text{ m/s}^2)t^2$$

Solving the quadratic equation for t gives, for the positive root, $t = 4.22$ s. Does the negative root have any physical meaning? (Can you think of another way of finding t from the information given?)

(b) What is the speed of the stone just before it strikes the ground?

Solution   The y component of the velocity just before the stone strikes the ground can be obtained using the equation $v_y = v_{y0} - gt$ (Eq. 4.11) with $t = 4.22$ s:

$$v_y = 10.0 \text{ m/s} - (9.80 \text{ m/s}^2)(4.22 \text{ s}) = -31.4 \text{ m/s}$$

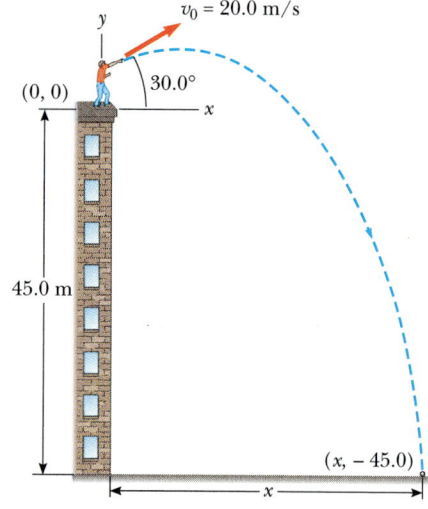

FIGURE 4.10   (Example 4.6)

Since $v_x = v_{x0} = 17.3$ m/s, the required speed is

$$v = \sqrt{v_x^2 + v_y^2} = \sqrt{(17.3)^2 + (-31.4)^2} \text{ m/s} = \boxed{35.9 \text{ m/s}}$$

Exercise   Where does the stone strike the ground?

Answer   73.0 m from the base of the building.

EXAMPLE 4.7   The Stranded Explorers

An Alaskan rescue plane drops a package of emergency rations to a stranded party of explorers, as shown in Figure 4.11. If the plane is traveling horizontally at 40.0 m/s at a height of 100 m above the ground, where does the package strike the ground relative to the point at which it is released?

Reasoning   The coordinate system for this problem is selected as shown in Figure 4.11, with the positive x direction to the right and the positive y direction upward.

Consider first the horizontal motion of the package. The only equation available to us is $x = v_{x0}t$ (Eq. 4.12).

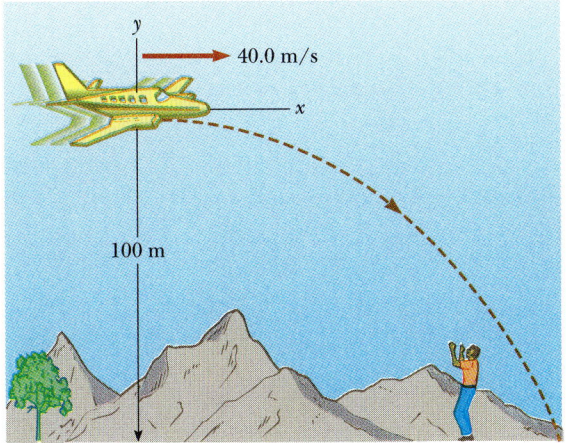

FIGURE 4.11 (Example 4.7) According to a ground observer, a package released from the rescue plane travels along the path shown. How does the path followed by the package appear to an observer on the plane (assumed to be moving at constant speed)?

The initial x component of the package velocity is the same as that of the plane when the package is released, 40.0 m/s. Thus, we have

$$x = (40.0 \text{ m/s})\, t$$

If we know t, the length of time the package is in the air, we can determine x, the distance traveled by the package in the horizontal direction. To find t, we move to the equations for the vertical motion of the package. We know that at the instant the package hits the ground, its y coordinate is -100 m. We also know that the initial component of velocity of the package in the vertical direction, v_{y0}, is zero because the package was released with only a horizontal component of velocity.

Solution From Equation 4.13, we have

$$y = -\tfrac{1}{2}gt^2$$
$$-100 \text{ m} = -\tfrac{1}{2}(9.80 \text{ m/s}^2)\, t^2$$
$$t^2 = 20.4 \text{ s}^2$$
$$t = 4.51 \text{ s}$$

The value for the time of flight substituted into the equation for the x coordinate gives

$$x = (40.0 \text{ m/s})(4.51 \text{ s}) = \boxed{180 \text{ m}}$$

The package hits the ground not directly under the drop point but 180 m to the right of that point.

Exercise What are the horizontal and vertical components of the velocity of the package just before it hits the ground?

Answer $v_x = 40.0$ m/s; $v_y = -44.1$ m/s.

EXAMPLE 4.8 The End of the Ski Jump

A ski jumper travels down a slope and leaves the ski track moving in the horizontal direction with a speed of 25.0 m/s as in Figure 4.12. The landing incline below her falls off with a slope of 35.0°. (a) Where does she land on the incline?

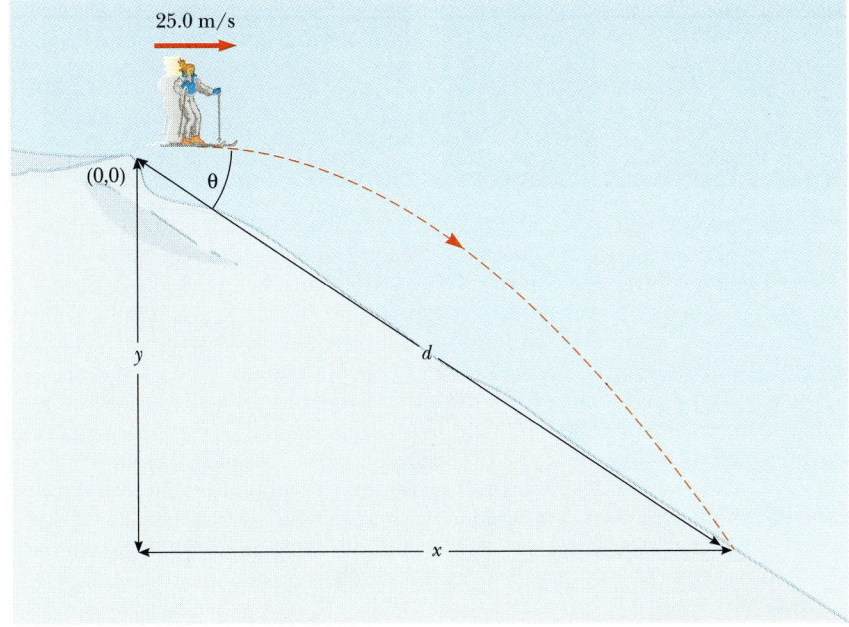

FIGURE 4.12 (Example 4.8)

Solution   It is convenient to select the origin ($x = y = 0$) at the beginning of the jump. Since $v_{x0} = 25.0$ m/s, and $v_{y0} = 0$ in this case, Equations 4.12 and 4.13 give

$$(1) \qquad x = v_{x0}t = (25.0 \text{ m/s})t$$

$$(2) \qquad y = v_{y0}t - \tfrac{1}{2}gt^2 = -\tfrac{1}{2}(9.80 \text{ m/s}^2)t^2$$

Taking d to be the distance she travels along the incline before landing, then from the right triangle in Figure 4.12, we see that her x and y coordinates at the point of landing are $x = d \cos 35.0°$ and $y = -d \sin 35.0°$. Substituting these relationships into (1) and (2) gives

$$(3) \qquad d \cos 35.0° = (25.0 \text{ m/s})t$$

$$(4) \qquad -d \sin 35.0° = -\tfrac{1}{2}(9.80 \text{ m/s}^2)t^2$$

Eliminating t from these equations gives $d = 109$ m. Hence, the x and y coordinates of the point at which she lands are

$$x = d \cos 35.0° = (109 \text{ m}) \cos 35.0° = \boxed{89.3 \text{ m}}$$

$$y = -d \sin 35.0° = -(109 \text{ m}) \sin 35° = \boxed{-62.5 \text{ m}}$$

Exercise   Determine how long the ski jumper is airborne and her vertical component of velocity just before she lands.

Answer   3.57 s; $v_y = -35.0$ m/s.

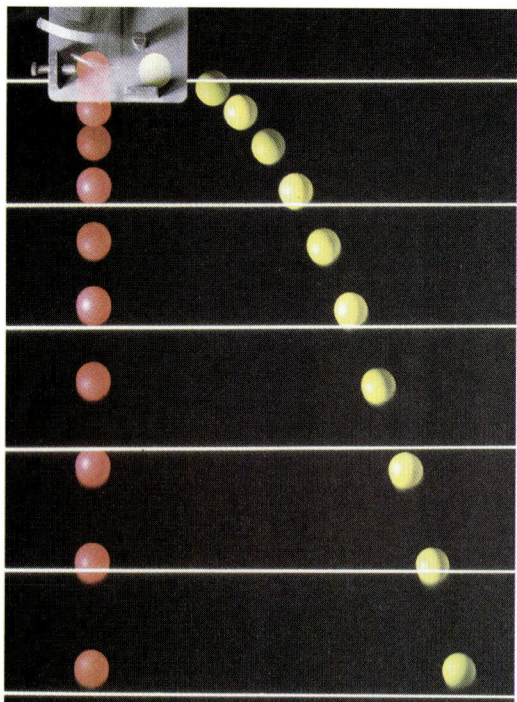
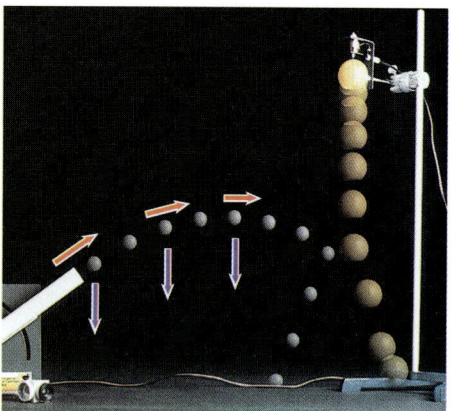

(Left) This multiflash photograph of two balls released simultaneously illustrates both free fall (red ball) and projectile motion (yellow ball). The yellow ball was projected horizontally, while the red ball was released from rest. Can you explain why both balls reach the floor simultaneously?   *(Richard Megna, Fundamental Photographs) (Right)* A multiflash photograph of a popular lecture demonstration in which a projectile is fired at a target that is being held by a magnet in the device at the top right of the photograph. The conditions of the experiment are that the gun is aimed at the target and the projectile leaves the gun at the instant the target is released from rest. Under these conditions the projectile will hit the target, independent of the initial speed of the projectile. The reason is that they both experience the same downward acceleration and hence the velocities of the projectile and target change by the same amount in the same time interval. Note that the velocity of the projectile (red arrows) changes in direction and magnitude, while its downward acceleration (violet arrows) remains constant.   *(Central Scientific Co.)*

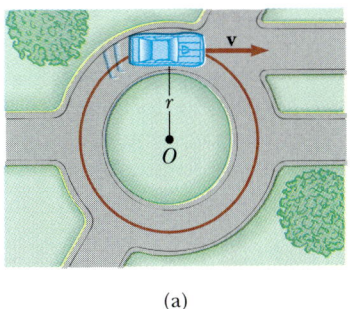
(a)

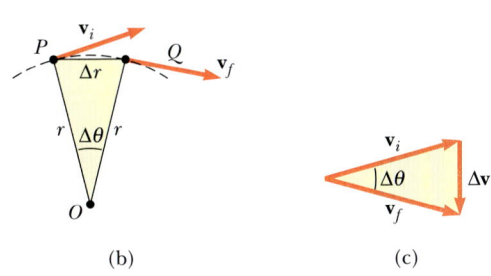
(b)

(c)

FIGURE 4.13   (a) An object moving along a circular path at constant speed experiences uniform circular motion. (b) As the particle moves from P to Q, the direction of its velocity vector changes from $\mathbf{v}_i$ to $\mathbf{v}_f$. (c) The construction for determining the direction of the change in velocity $\Delta\mathbf{v}$ that is toward the center of the circle.

4.4   UNIFORM CIRCULAR MOTION

Figure 4.13a shows an object moving in a circular path with constant linear speed v. Such motion is called **uniform circular motion**. It is often surprising to students to find that *even though an object moves at a constant speed, it still has an acceleration.* To see why, consider the defining equation for average acceleration, $\overline{\mathbf{a}} = \Delta\mathbf{v}/\Delta t$ (Eq. 4.4).

Note that the acceleration depends on the change in the velocity vector. Because velocity is a vector quantity, there are two ways in which an acceleration can be produced, as mentioned in Section 4.1: by a change in the *magnitude* of the velocity and by a change in the *direction* of the velocity. It is the latter situation that is occurring for an object moving with constant speed in a circular path. The velocity vector is always tangent to the path of the object and perpendicular to the radius r of the circular path. We now show that the acceleration vector in uniform circular motion is always perpendicular to the path and always points toward the center of the circle. An acceleration of this nature is called a *centripetal acceleration* (center seeking), and its magnitude is

$$a_r = \frac{v^2}{r} \tag{4.19}$$

| Magnitude of centripetal acceleration

where r is the radius of the circle.

To derive Equation 4.19, consider Figure 4.13b. Here an object is seen first at point P and then at point Q. The particle is at P at time t_i, and its velocity at that time is $\mathbf{v}_i$; it is at Q at some later time t_f, and its velocity at that time is $\mathbf{v}_f$. Let us also assume here that $\mathbf{v}_i$ and $\mathbf{v}_f$ differ only in direction; their magnitudes are the same (that is, $v_i = v_f = v$). In order to calculate the acceleration of the particle, let us begin with the defining equation for average acceleration (Eq. 4.4):

$$\overline{\mathbf{a}} = \frac{\mathbf{v}_f - \mathbf{v}_i}{t_f - t_i} = \frac{\Delta\mathbf{v}}{\Delta t}$$

This equation indicates that we must vectorially subtract $\mathbf{v}_i$ from $\mathbf{v}_f$, where $\Delta\mathbf{v} = \mathbf{v}_f - \mathbf{v}_i$ is the change in the velocity. Since $\mathbf{v}_i + \Delta\mathbf{v} = \mathbf{v}_f$, the vector $\Delta\mathbf{v}$ can be found using the vector triangle in Figure 4.13c. When Δt is very small, Δr and $\Delta\theta$ are also

Motion Simulator

This simulator allows you to model a wide range of problems dealing with motion, including linear motion under constant acceleration and projectile motion. You will have the ability to vary an object's mass, as well as its initial position and initial velocity. In addition, you will be able to observe changes in the motion of an object in the presence or absence of a gravitational field and/or friction.

very small. In this case, $\mathbf{v}_f$ is almost parallel to $\mathbf{v}_i$ and the vector $\Delta\mathbf{v}$ is approximately perpendicular to them, pointing toward the center of the circle.

Now consider the triangle in Figure 4.13b, which has sides Δr and r. This triangle and the one in Figure 4.13c, which has sides Δv and v, are similar. (Two triangles are similar if the angle between any two sides is the same for both triangles and if the ratio of lengths of these sides is the same.) This enables us to write a relationship between the lengths of the sides:

$$\frac{\Delta v}{v} = \frac{\Delta r}{r}$$

This equation can be solved for Δv and the expression so obtained can be substituted into $\bar{a} = \Delta v/\Delta t$ (Eq. 4.4) to give

$$\bar{a} = \frac{v\,\Delta r}{r\,\Delta t}$$

Now imagine that points P and Q in Figure 4.13b become extremely close together. In this case $\Delta\mathbf{v}$ would point toward the center of the circular path, and because the acceleration is in the direction of $\Delta\mathbf{v}$, it too is toward the center. Furthermore, as P and Q approach each other, Δt approaches zero, and the ratio $\Delta r/\Delta t$ approaches the speed v. Hence, in the limit $\Delta t \rightarrow 0$, the magnitude of the acceleration is

$$a_r = \frac{v^2}{r}$$

Thus we conclude that in uniform circular motion, the acceleration is directed inward toward the center of the circle and has a magnitude given by v^2/r, where v is the speed of the particle and r is the radius of the circle. You should show that the dimensions of a_r are L/T^2. We shall return to the discussion of circular motion in Section 6.1.

4.5 TANGENTIAL AND RADIAL ACCELERATION

Let us consider the motion of a particle along a curved path where the velocity changes both in direction and in magnitude, as described in Figure 4.14. The velocity vector is always tangent to the path; however, in this situation, the acceleration vector **a** is at some angle to the path.

As the particle moves along the curved path in Figure 4.14, we see that the direction of the total acceleration vector, **a**, changes from point to point. This vector can be resolved into two component vectors: a radial component vector, $\mathbf{a}_r$, and a tangential component vector, $\mathbf{a}_t$. That is, the total acceleration vector can be written as the vector sum of these component vectors:

$$\mathbf{a} = \mathbf{a}_r + \mathbf{a}_t \qquad (4.20)$$

Total acceleration

The tangential acceleration arises from the change in the speed of the particle, and the projection of the acceleration along the direction of the velocity is

$$a_t = \frac{d|\mathbf{v}|}{dt} \qquad (4.21)$$

Tangential acceleration

The radial acceleration is due to the change in direction of the velocity vector and has an absolute magnitude given by

$$a_r = \frac{v^2}{r} \qquad (4.22)$$

Centripetal acceleration

where r is the radius of curvature of the path at the point in question. Since $\mathbf{a}_r$ and $\mathbf{a}_t$ are perpendicular component vectors of **a**, it follows that $a = \sqrt{a_r^2 + a_t^2}$. As in the case of uniform circular motion, $\mathbf{a}_r$ always points toward the center of curvature, as shown in Figure 4.14. Also, at a given speed, a_r is large when the radius of curvature is small (as at points P and Q in Fig. 4.14) and small when r is large (such

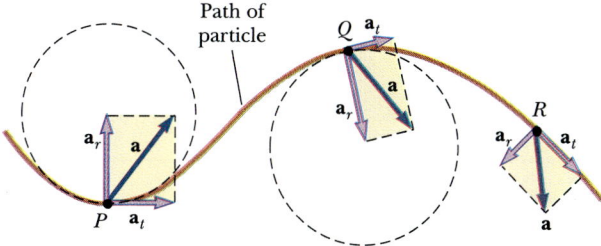

FIGURE 4.14   The motion of a particle along an arbitrary curved path lying in the *xy* plane. If the velocity vector **v** (always tangent to the path) changes in direction and magnitude, the component vectors of the acceleration **a** are a tangential vector $\mathbf{a}_t$ and a radial vector $\mathbf{a}_r$. The acceleration vectors are not necessarily to scale.

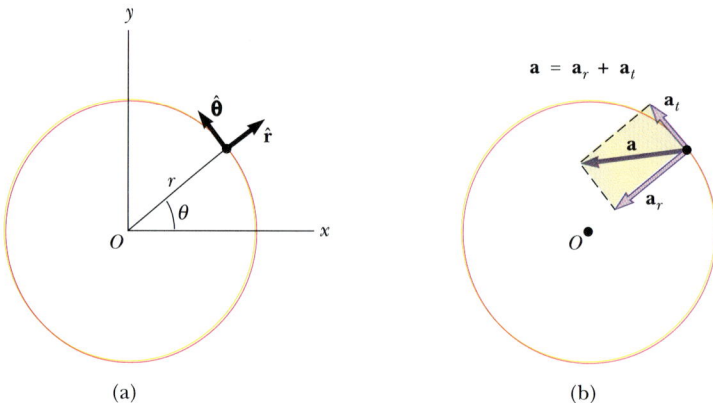

FIGURE 4.15   (a) Descriptions of the unit vectors $\hat{\mathbf{r}}$ and $\hat{\boldsymbol{\theta}}$. (b) The total acceleration **a** of a particle moving along a curved path (which at any instant is part of a circle of radius r) consists of a radial component $\mathbf{a}_r$ directed toward the center of rotation, and a tangential component $\mathbf{a}_t$. The tangential component vector $\mathbf{a}_t$ is zero if the speed is constant.

as at point R). The direction of $\mathbf{a}_t$ is either in the same direction as **v** (if v is increasing) or opposite **v** (if v is decreasing).

Note that in the case of uniform circular motion, where v is constant, $a_t = 0$ and the acceleration is always radial, as we described in Section 4.4. (Note that Eq. 4.22 is identical to Eq. 4.19.) In other words, uniform circular motion is a special case of motion along a curved path. Furthermore, if the direction of **v** doesn't change, then there is no radial acceleration and the motion is one dimensional ($a_r = 0$, $a_t \neq 0$).

It is convenient to write the acceleration of a particle moving in a circular path in terms of unit vectors. We do this by defining the unit vectors $\hat{\mathbf{r}}$ and $\hat{\boldsymbol{\theta}}$, where $\hat{\mathbf{r}}$ is a *unit vector along the radius vector directed radially outward* from the center of the circle, and $\hat{\boldsymbol{\theta}}$ is a *unit vector tangent to the circular path*, as in Figure 4.15a. The direction of $\hat{\boldsymbol{\theta}}$ is in the direction of increasing θ where θ is measured counterclockwise from the positive x axis. Note that both $\hat{\mathbf{r}}$ and $\hat{\boldsymbol{\theta}}$ "move along with the particle" and so vary in time relative to a stationary observer. Using this notation, we can express the total acceleration as

$$\mathbf{a} = \mathbf{a}_t + \mathbf{a}_r = \frac{d|\mathbf{v}|}{dt}\hat{\boldsymbol{\theta}} - \frac{v^2}{r}\hat{\mathbf{r}} \tag{4.23}$$

These vectors are described in Figure 4.15b. The negative sign on the v^2/r term for $\mathbf{a}_r$ indicates that the radial acceleration is always directed radially inward, *opposite* the unit vector $\hat{\mathbf{r}}$.

EXAMPLE 4.9   The Swinging Ball

A ball tied to the end of a string 0.50 m in length swings in a vertical circle under the influence of gravity, as in Figure 4.16. When the string makes an angle of $\theta = 20°$ with the vertical, the ball has a speed of 1.5 m/s. (a) Find the magnitude of the radial component of acceleration at this instant.

Solution   Since $v = 1.5$ m/s and $r = 0.50$ m, we find that

$$a_r = \frac{v^2}{r} = \frac{(1.5 \text{ m/s})^2}{0.50 \text{ m}} = \boxed{4.5 \text{ m/s}^2}$$

(b) When the ball is at an angle θ to the vertical, it has a tangential acceleration of magnitude $g \sin \theta$ (produced by the tangential component of the force $m\textbf{g}$.). Therefore, at $\theta = 20°$, $a_t = g \sin 20° = 3.4$ m/s². Find the magnitude and direction of the *total* acceleration at $\theta = 20°$.

Solution   Since $\textbf{a} = \textbf{a}_r + \textbf{a}_t$, the magnitude of $\textbf{a}$ at $\theta = 20°$ is

$$a = \sqrt{a_r^2 + a_t^2} = \sqrt{(4.5)^2 + (3.4)^2} \text{ m/s}^2 = \boxed{5.6 \text{ m/s}^2}$$

If ϕ is the angle between $\textbf{a}$ and the string, then

$$\phi = \tan^{-1}\frac{a_t}{a_r} = \tan^{-1}\left(\frac{3.4 \text{ m/s}^2}{4.5 \text{ m/s}^2}\right) = \boxed{36°}$$

Note that all of the vectors—$\textbf{a}$, $\textbf{a}_t$, and $\textbf{a}_r$—change in direction *and* magnitude as the ball swings through the circle. When the ball is at its lowest elevation ($\theta = 0$), $a_t = 0$, because there is no tangential component of $\textbf{g}$ at this angle and a_r is a *maximum* because v is a maximum. If the ball has enough speed to reach its highest position ($\theta = 180°$), a_t is again zero but a_r is a minimum because v is now a minimum. Finally, in the two horizontal positions ($\theta = 90°$ and $270°$), $|a_t| = g$ and a_r has a value between its minimum and maximum values.

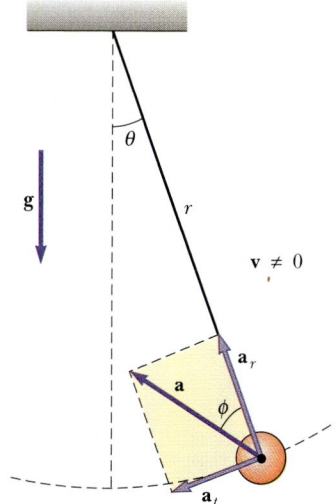

FIGURE 4.16   (Example 4.9) Motion of a ball suspended by a string of length r. The ball swings with nonuniform circular motion in a vertical plane, and its acceleration $\textbf{a}$ has a radial component $\textbf{a}_r$ and a tangential component $\textbf{a}_t$.

4.6   RELATIVE VELOCITY AND RELATIVE ACCELERATION

In this section, we describe how observations made by different observers in different frames of reference are related to each other. We find that observers in different frames of reference may measure different displacements, velocities, and accelerations for a given particle. That is, two observers moving relative to each other generally do not agree on the outcome of a measurement.

For example, if two cars are moving in the same direction with speeds of 50 mi/h and 60 mi/h, a passenger in the slower car will measure the speed of the faster car relative to that of the slower car to be 10 mi/h. Of course, a stationary observer will measure the speed of the faster car to be 60 mi/h. This simple example demonstrates that velocity measurements differ in different frames of reference.

Next, suppose a person riding on a moving vehicle (observer A) throws a ball in such a way that it appears, in his frame of reference, to move first straight upward and then straight downward along the same vertical line, as in Figure 4.17a. However, a stationary observer (B) sees the path of the ball as a parabola, as illustrated in Figure 4.17b. Relative to the observer B, the ball has a vertical component of velocity (resulting from the initial upward velocity and the downward acceleration of gravity) *and* a horizontal component of velocity.

Another simple example is to imagine a package being dropped from an airplane flying parallel to the Earth with a constant velocity, which is the situation we studied in Example 4.7. An observer on the airplane would describe the motion of the package as a straight line toward the Earth. The stranded explorer on the ground, however, would view the trajectory of the package as a parabola. If the airplane continues to move horizontally with the same velocity, the package will hit the ground directly beneath the airplane (assuming that friction is neglected)!

In a more general situation, consider a particle located at the point P in Figure 4.18. Imagine that the motion of this particle is being described by two observers,

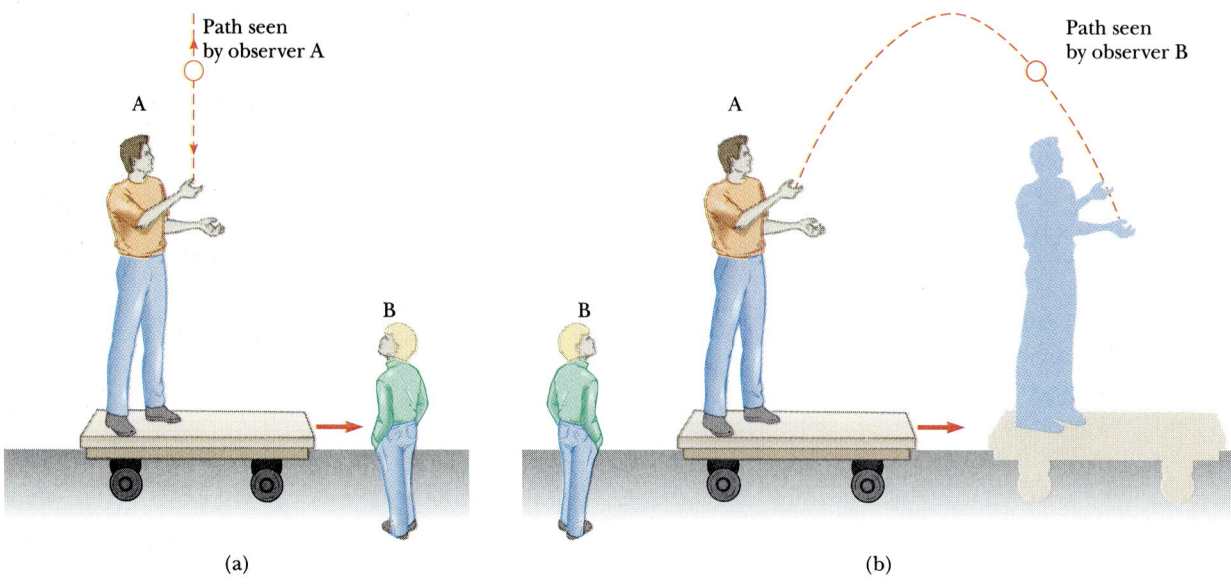

Path seen
by observer A

A

B

(a)

Path seen
by observer B

A

B

(b)

FIGURE 4.17 (a) Observer A in a moving vehicle throws a ball upward and sees it rise and fall in a straight-line path. (b) Stationary observer B sees a parabolic path for the same ball.

one in reference frame S, fixed relative to the Earth, and another in reference frame S', moving to the right relative to S (and therefore relative to Earth) with a constant velocity $\mathbf{u}$. (Relative to an observer in S', S moves to the left with a velocity $-\mathbf{u}$.) Where an observer stands in a reference frame is irrelevant in this discussion, but to be definite let us place both observers at the origin.

We label the position of the particle relative to the S frame with the position vector $\mathbf{r}$ and label its position relative to the S' frame with the vector $\mathbf{r}'$ both at some time t. If the origins of the two reference frames coincide at $t = 0$, then the vectors $\mathbf{r}$ and $\mathbf{r}'$ are related to each other through the expression $\mathbf{r} = \mathbf{r}' + \mathbf{u}t$, or

Galilean coordinate transformation

$$\mathbf{r}' = \mathbf{r} - \mathbf{u}t \tag{4.24}$$

That is, after a time t, the S' frame is displaced to the right by an amount $\mathbf{u}t$.

If we differentiate Equation 4.24 with respect to time and note that $\mathbf{u}$ is constant, we get

$$\frac{d\mathbf{r}'}{dt} = \frac{d\mathbf{r}}{dt} - \mathbf{u}$$

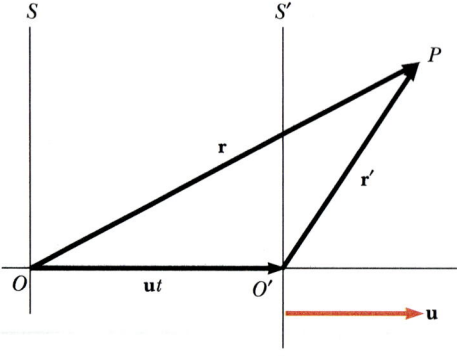

FIGURE 4.18 A particle located at P is described by two observers, one in the fixed frame of reference, S, the other in the frame S', which moves with a constant velocity $\mathbf{u}$ to the right. The vector $\mathbf{r}$ is the particle's position vector relative to S, and $\mathbf{r}'$ is its position vector relative to S'.

$$\mathbf{v}' = \mathbf{v} - \mathbf{u} \qquad\qquad (4.25)$$

Galilean velocity transformation

where $\mathbf{v}'$ is the velocity of the particle observed in the S' frame and $\mathbf{v}$ is its velocity observed in the S frame. Equations 4.24 and 4.25 are known as *Galilean transformation equations*. The equations relate the coordinates and velocity of a particle as measured, for example, in a frame fixed relative to the Earth to those measured in a frame moving with uniform motion relative to the Earth.

Although observers in the different reference frames measure different velocities for the particles, they measure the *same acceleration* when $\mathbf{u}$ is constant. This can be seen by taking the time derivative of Equation 4.25:

$$\frac{d\mathbf{v}'}{dt} = \frac{d\mathbf{v}}{dt} - \frac{d\mathbf{u}}{dt}$$

But $d\mathbf{u}/dt = 0$ because $\mathbf{u}$ is constant. Therefore, we conclude that $\mathbf{a}' = \mathbf{a}$ because $\mathbf{a}' = d\mathbf{v}'/dt$ and $\mathbf{a} = d\mathbf{v}/dt$. That is, *the acceleration of the particle measured by an observer in the Earth's frame of reference is the same as that measured by any other observer moving with constant velocity relative to the Earth frame.*

CONCEPTUAL EXAMPLE 4.10

A ball is thrown upward in the air by a passenger on a train that is moving with constant velocity. (a) Describe the path of the ball as seen by the passenger. Describe the path as seen by a stationary observer outside the train. (b) How would these observations change if the train were accelerating along the track?

Reasoning   (a) The passenger sees the ball moving vertically up and then down with a velocity that changes with time. The outside observer sees the ball moving in a parabolic path, with constant horizontal velocity equal to that of the train, as well as a continuously changing vertical velocity. (b) If the train were accelerating horizontally forward, the passenger would see the ball accelerating horizontally backward as well as vertically downward. Relative to the passenger, the ball would move in a parabola with its axis inclined to the vertical. The outside observer would see pure parabolic motion for the ball, with a horizontal acceleration of zero, a vertical acceleration of $-g$, and a horizontal velocity equal to that of the train at the moment the ball was released.

EXAMPLE 4.11   A Boat Crossing a River

A boat heading due north crosses a wide river with a speed of 10.0 km/h relative to the water. The river has a uniform speed of 5.00 km/h due east relative to Earth. Determine the velocity of the boat relative to a stationary ground observer.

Solution   We know

$\mathbf{v}_{br}$ = the velocity of the boat, *b*, relative to the river, *r*

$\mathbf{v}_{re}$ = the velocity of the river, *r*, relative to Earth, *e*

and we want $\mathbf{v}_{be}$, the velocity of the *boat* relative to *Earth*. The relationship between these three quantities is

$$\mathbf{v}_{be} = \mathbf{v}_{br} + \mathbf{v}_{re}$$

The terms in the equation must be manipulated as vector quantities; the vectors are shown in Figure 4.19. The quantity $\mathbf{v}_{br}$ is due north, $\mathbf{v}_{re}$ is due east, and the vector sum of the two, $\mathbf{v}_{be}$, is at an angle θ, as defined in Figure 4.19. Thus, the speed

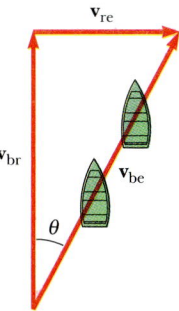

FIGURE 4.19   (Example 4.11)

of the boat relative to Earth can be found from the Pythagorean theorem:

$$v_{be} = \sqrt{v_{br}^2 + v_{re}^2} = \sqrt{(10.0)^2 + (5.00)^2} \text{ km/h} = \boxed{11.2 \text{ km/h}}$$

The direction of $\mathbf{v}_{be}$ is

$$\theta = \tan^{-1}\left(\frac{v_{re}}{v_{br}}\right) = \tan^{-1}\left(\frac{5.00}{10.00}\right) = 26.6°$$

Therefore, the boat will be traveling at a speed of 11.2 km/h in the direction 63.4° north of east relative to Earth.

Exercise   If the width of the river is 3.0 km, find the time it takes the boat to cross the river.

Answer   18 min.

EXAMPLE 4.12   Which Way Should We Head?

If the boat of the preceding example travels with the same speed of 10.0 km/h relative to the water and is to travel due north, as in Figure 4.20, what should be its heading?

Solution   As in the previous example, we know $\mathbf{v}_{br}$ and $\mathbf{v}_{re}$, and we want to find $\mathbf{v}_{be}$. The relationship between these three quantities, $\mathbf{v}_{be} = \mathbf{v}_{br} + \mathbf{v}_{re}$, is shown in Figure 4.20. That is, the boat must head upstream in order to be pushed directly northward across the river. The speed v_{be} can be found from the Pythagorean theorem:

$$v_{be} = \sqrt{v_{br}^2 - v_{re}^2} = \sqrt{(10.0)^2 - (5.00)^2} \text{ km/h} = 8.66 \text{ km/h}$$

The direction of $\mathbf{v}_{be}$ is

$$\theta = \tan^{-1}\left(\frac{v_{re}}{v_{be}}\right) = \tan^{-1}\left(\frac{5.00}{8.66}\right) = \boxed{30.0°}$$

The boat must steer a course 30.0° west of north.

Exercise   If the width of the river is 3.0 km, how long does it take the boat to cross the river?

Answer   21 min.

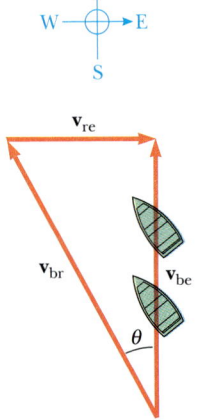

FIGURE 4.20   (Example 4.12)

---

*4.7   RELATIVE MOTION AT HIGH SPEEDS

The Galilean transformation equations, Equations 4.24 and 4.25, are valid *only* at particle speeds (relative to both observers) that are small compared with the speed of light. When the particle speed relative to either observer approaches the speed of light, these transformation equations must be replaced by the equations used by Einstein in his special theory of relativity.

Although we shall discuss the theory of relativity in Chapter 39, a forecast of some of its predictions is in order. As it turns out, the relativistic transformation equations reduce to the Galilean transformations when the particle speed is small compared with the speed of light. This is in keeping with the **correspondence principle** first proposed by Niels Bohr, which in effect states that, if an old theory accurately describes a number of physical phenomena, then any new theory *must* account for the same phenomena over the range of validity of the old theory.

You might wonder how one can test the validity of the transformation equations. The Galilean transformations are used in Newtonian mechanics, while Einstein's theory uses relativistic transformations. From experiments on high-speed particles, one finds that Newtonian mechanics *fails* at particle speeds approaching the speed of light. On the other hand, Einstein's theory of special relativity is in agreement with experiment at *all* speeds. Finally, Newtonian mechanics places no upper limit on speed of a particle. In contrast, the relativistic velocity transformation equation predicts that *no particle speed can exceed the speed of light*. Electrons and protons accelerated through very high voltages can acquire speeds close to the speed of light, but their speeds never reach this limiting value. Hence, experimental results are in complete agreement with the theory of relativity.

SUMMARY

If a particle moves with *constant* acceleration a and has velocity $\mathbf{v}_0$ and position $\mathbf{r}_0$ at $t = 0$, its velocity and position vectors at some later time t are

$$\mathbf{v} = \mathbf{v}_0 + \mathbf{a}t \tag{4.8}$$

$$\mathbf{r} = \mathbf{r}_0 + \mathbf{v}_0 t + \tfrac{1}{2}\mathbf{a}t^2 \tag{4.9}$$

For two-dimensional motion in the xy plane under constant acceleration, these vector expressions are equivalent to two-component expressions, one for the motion along x and one for the motion along y.

 Projectile motion is two-dimensional motion under constant acceleration, where $a_x = 0$ and $a_y = -g$. In this case, if $x_0 = y_0 = 0$, the components of Equations 4.8 and 4.9 reduce to

$$v_x = v_{x0} = \text{constant} \tag{4.10}$$

$$v_y = v_{y0} - gt \tag{4.11}$$

$$x = v_{x0}t \tag{4.12}$$

$$y = v_{y0}t - \tfrac{1}{2}gt^2 \tag{4.13}$$

where $v_{x0} = v_0 \cos\theta_0$, $v_{y0} = v_0 \sin\theta_0$, v_0 is the initial speed of the projectile, and θ_0 is the angle $\mathbf{v}_0$ makes with the positive x axis. Note that these expressions give the velocity components (and hence the velocity vector) and the coordinates (and hence the position vector) at *any* time t that the projectile is in motion.

 It is useful to think of projectile motion as the superposition of two motions: (1) uniform motion in the x direction, and (2) motion in the vertical direction subject to a constant downward acceleration of magnitude $g = 9.80 \text{ m/s}^2$. Hence, one can analyze the motion in terms of separate horizontal and vertical components of velocity, as in Figure 4.21.

 A particle moving in a circle of radius r with constant speed v undergoes a

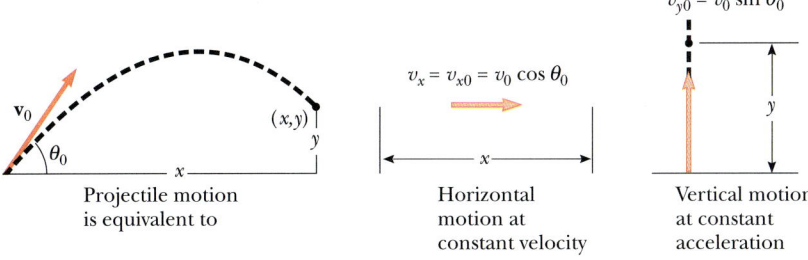

FIGURE 4.21 Analyzing projectile motion in terms of the horizontal and vertical components of velocity.

centripetal (or radial) acceleration, $\mathbf{a}_r$, because the direction of $\mathbf{v}$ changes in time. The magnitude of $\mathbf{a}_r$ is

$$a_r = \frac{v^2}{r} \qquad (4.19)$$

and its direction is always toward the center of the circle.

If a particle moves along a curved path in such a way that the magnitude and direction of $\mathbf{v}$ change in time, the particle has an acceleration vector that can be described by two component vectors: (1) a radial component vector, $\mathbf{a}_r$, arising from the change in direction of $\mathbf{v}$, and (2) a tangential component vector, $\mathbf{a}_t$, arising from the change in magnitude of $\mathbf{v}$. The magnitude of $\mathbf{a}_r$ is v^2/r, and the magnitude of $\mathbf{a}_t$ is $d|\mathbf{v}|/dt$.

The velocity of a particle, $\mathbf{v}$, measured in a fixed frame of reference, S, is related to the velocity of the same particle, $\mathbf{v}'$, measured in a moving frame of reference, S', by

$$\mathbf{v}' = \mathbf{v} - \mathbf{u} \qquad (4.25)$$

where $\mathbf{u}$ is the velocity of S' relative to S.

QUESTIONS

1. If the average velocity of a particle is zero in some time interval, what can you say about the displacement of the particle for that interval?

2. If you know the position vectors of a particle at two points along its path and also know the time it took to get from one point to the other, can you determine the particle's instantaneous velocity? its average velocity? Explain.

3. Describe a situation in which the velocity of a particle is perpendicular to the position vector.

4. Explain whether or not the following particles have an acceleration: (a) a particle moving in a straight line with constant speed and (b) a particle moving around a curve with constant speed.

5. Correct the following statement: "The racing car rounds the turn at a constant velocity of 90 miles per hour."

6. Determine which of the following moving objects have an approximate parabolic trajectory: (a) a ball thrown in an arbitrary direction, (b) a jet airplane, (c) a rocket leaving the launching pad, (d) a rocket a few minutes after launch with failed engines, (e) a tossed stone moving to the bottom of a pond.

7. A student argues that as a satellite orbits the Earth in a circular path, the satellite moves with a constant velocity and therefore has no acceleration. The professor claims that the student is wrong because the satellite must have a centripetal acceleration as it moves in its circular orbit. What is wrong with the student's argument?

8. What is the fundamental difference between the unit vectors $\hat{\mathbf{r}}$ and $\hat{\boldsymbol{\theta}}$ defined in Figure 4.15 and the unit vectors $\mathbf{i}$ and $\mathbf{j}$?

9. At the end of its arc, the velocity of a pendulum is zero. Is its acceleration also zero at this point?

10. If a rock is dropped from the top of a sailboat's mast, will it hit the deck at the same point whether the boat is at rest or in motion at constant velocity?

11. A stone is thrown upward from the top of a building. Does the stone's displacement depend on the location of the origin of the coordinate system? Does the stone's velocity depend on the location of the origin?

12. Is it possible for a vehicle to travel around a curve without accelerating? Explain.

13. A baseball is thrown with an initial velocity of $(10\mathbf{i} + 15\mathbf{j})$ m/s. When it reaches the top of its trajectory, what are (a) its velocity and (b) its acceleration? Neglect air resistance.

14. An object moves in a circular path with constant speed v. (a) Is the velocity of the object constant? (b) Is its acceleration constant? Explain.

15. A projectile is fired at some angle to the horizontal with some initial speed v_0, and air resistance is neglected. Is the projectile a freely falling body? What is its acceleration in the vertical direction? What is its acceleration in the horizontal direction?

16. A projectile is fired at an angle of 30° to the horizontal with some initial speed. What other projectile angle gives the same range if the initial speed is the same in both cases? Neglect air resistance.

17. A projectile is fired on the Earth with some initial velocity. Another projectile is fired on the Moon with the same initial velocity. Neglecting air resistance, which projectile has the greater range? Which reaches the greater altitude? (Note that the free-fall acceleration on the Moon is about 1.6 m/s².)

18. As a projectile moves through its parabolic trajectory,

which of these quantities, if any, remain constant: (a) speed, (b) acceleration, (c) horizontal component of velocity, (d) vertical component of velocity?

19. The maximum range of a projectile occurs when it is launched at an angle of 45° with the horizontal if air re-sistance is neglected. If air resistance is not neglected, will this optimum angle be greater or less than 45°? Explain.

20. A passenger on a train that is moving with constant veloc-ity drops a spoon. What is the acceleration of the spoon relative to (a) the train and (b) the Earth?

PROBLEMS

Review Problem

A ball is thrown with an initial speed v_0 at an angle θ_0 with the horizontal. The range of the ball is R, and the ball reaches a maximum height $R/6$. In terms of R and g, find (a) the time the ball is in motion, (b) the ball's speed at the peak of its path, (c) the initial vertical component of its velocity, (d) its initial speed, (e) the angle θ_0, (f) the maximum height that can be reached if the ball is thrown at the appropriate angle and at the speed found in (d), (g) the maximum range that can be reached if the ball is thrown at the appropriate angle and at the speed found in (d). (h) Suppose that two rocks are thrown from the same point at the same moment as in the figure below. Find the distance between them as a function of time. Assume that v_0 and θ_0 are given.

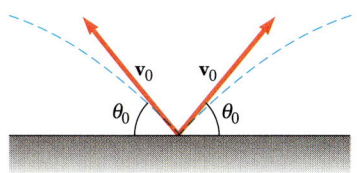

Section 4.1 The Displacement, Velocity, and Acceleration Vectors

1. Suppose that the trajectory of a particle is given by $r(t) = x(t)\mathbf{i} + y(t)\mathbf{j}$ with $x(t) = at^2 + bt$ and $y(t) = ct + d$, where a, b, c, and d are constants that have appropriate dimensions. What displacement does the particle undergo between $t = 1$ s and $t = 3$ s?

2. Suppose that the position vector for a particle is given as $r(t) = x(t)\mathbf{i} + y(t)\mathbf{j}$, with $x(t) = at + b$ and $y(t) = ct^2 + d$, where $a = 1.00$ m/s, $b = 1.00$ m, $c = 0.125$ m/s^2, and $d = 1.00$ m. (a) Calculate the aver-age velocity during the time interval from $t = 2.00$ s to $t = 4.00$ s. (b) Determine the velocity and the speed at $t = 2.00$ s.

3. A motorist drives south at 20.0 m/s for 3.00 min, then turns west and travels at 25.0 m/s for 2.00 min, and finally travels northwest at 30.0 m/s for 1.00 min. For this 6.00-min trip, find (a) the result-ant vector displacement, (b) the average speed, and (c) the average velocity.

4. A golf ball is hit off a tee at the edge of a cliff. The x and y coordinates of the golf ball versus time are given by the expressions $x = (18.0 \text{ m/s})t$ and $y = (4.00 \text{ m/s})t - (4.90 \text{ m/s}^2)t^2$. (a) Write a vector ex-pression for the position $\mathbf{r}$ as a function of time t using the unit vectors $\mathbf{i}$ and $\mathbf{j}$. By taking derivatives, repeat for (b) the velocity vector $\mathbf{v}(t)$ and (c) the ac-celeration vector $\mathbf{a}(t)$. (d) Find the x and y coordi-nates of the ball at $t = 3.00$ s. Using the unit vectors $\mathbf{i}$ and $\mathbf{j}$, write expressions for (e) the velocity $\mathbf{v}$ and (f) the acceleration $\mathbf{a}$ at the instant $t = 3.00$ s.

Section 4.2 Two-Dimensional Motion with Constant Acceleration

5. At $t = 0$, a particle moving in the xy plane with con-stant acceleration has a velocity of $\mathbf{v}_0 = (3\mathbf{i} - 2\mathbf{j})$ m/s at the origin. At $t = 3$ s, its velocity is given by $\mathbf{v} = (9\mathbf{i} + 7\mathbf{j})$ m/s. Find (a) the acceleration of the particle and (b) its coordinates at any time t.

6. A particle starts from rest at $t = 0$ at the origin and moves in the xy plane with a constant acceleration of $\mathbf{a} = (2\mathbf{i} + 4\mathbf{j})$ m/s^2. After a time t has elapsed, deter-mine (a) the x and y components of velocity, (b) the coordinates of the particle, and (c) the speed of the particle.

7. A fish swimming in a horizontal plane has velocity $\mathbf{v}_0 = (4.0\mathbf{i} + 1.0\mathbf{j})$ m/s at a point in the ocean whose position vector is $\mathbf{r}_0 = (10.0\mathbf{i} - 4.0\mathbf{j})$ m relative to a stationary rock at the shore. After the fish swims with constant acceleration for 20.0 s, its velocity is $\mathbf{v} = (20.0\mathbf{i} - 5.0\mathbf{j})$ m/s. (a) What are the components of the acceleration? (b) What is the direction of the ac-celeration with respect to the fixed x axis? (c) Where is the fish at $t = 25$ s and in what direction is it mov-ing?

8. The position of a particle varies in time according to the expression $\mathbf{r} = (3.00\mathbf{i} - 6.00t^2\mathbf{j})$ m. (a) Find ex-pressions for the velocity and acceleration as func-tions of time. (b) Determine the particle's position and velocity at $t = 1.00$ s.

9. The coordinates of an object moving in the xy plane vary with time according to the equations $x =$

$-(5.0 \text{ m})\sin(t)$ and $y = (4.0 \text{ m}) - (5.0 \text{ m})\cos(t)$, where t is in seconds. (a) Determine the components of the velocity and those of the acceleration at $t = 0$ s. (b) Write expressions for the position vector, the velocity vector, and the acceleration vector at any time $t > 0$. (c) Describe the path of the object in an xy plot.

Section 4.3   Projectile Motion

(Neglect air resistance in all problems.)

10. Jimmy is at the base of a hill, while Billy is 30 m up the hill. Jimmy is at the origin of an xy coordinate system, and the line that follows the slope of the hill is given by the equation, $y = 0.4x$, as shown in Figure P4.10. If Jimmy throws an apple to Billy at an angle of $50°$ with respect to the horizontal, with what speed must he throw the apple if it is to reach Billy?

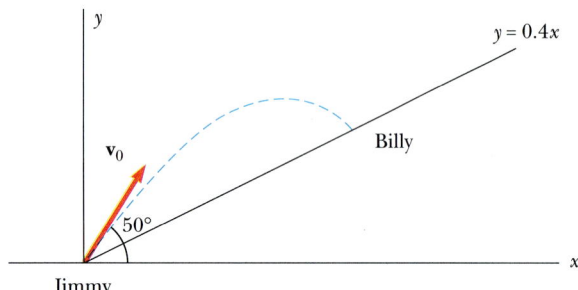

FIGURE P4.10

11. In a local bar, a customer slides an empty beer mug on the counter for a refill. The bartender is momentarily distracted and does not see the mug, which slides off the counter and strikes the floor 1.40 m from the base of the counter. If the height of the counter is 0.860 m, (a) with what speed did the mug leave the counter and (b) what was the direction of the mug's velocity just before it hit the floor?

12. A student decides to measure the muzzle speed of the pellets from her BB gun. She points the gun horizontally at a target placed on a vertical wall a distance x away from the gun. The shots hit the target a vertical distance y below the gun. (a) Show that the position of the pellet when traveling through the air is $y = Ax^2$, where A is a constant. (b) Express the constant A in terms of the initial speed and the free-fall acceleration. (c) If $x = 3.0$ m and $y = 0.21$ m, what is the speed of the BB?

13. A ball is thrown horizontally from the top of a building 35 m high. The ball strikes the ground at a point 80 m from the base of the building. Find (a) the time the ball is in flight, (b) its initial velocity, and (c) the x and y components of velocity just before the ball strikes the ground.

14. Superman is flying at treetop level near Paris when he sees the Eiffel Tower elevator start to fall (the cable snapped). His x-ray vision tells him Lois Lane is inside. If Superman is 1.00 km away from the tower, and the elevator falls from a height of 240 m, how long does he have to save Lois, and what must his average speed be?

14A. Superman is flying at treetop level near Paris when he sees the Eiffel Tower elevator start to fall (the cable snapped). His x-ray vision tells him Lois Lane is inside. If Superman is a distance d away from the tower and the elevator falls from a height h, how long does he have to save Lois, and what must his average speed be?

15. A soccer player kicks a rock horizontally off the edge of a 40.0-m-high cliff into a pool of water. If the player hears the sound of the splash 3.0 s after the kick, what was the initial speed? Assume the speed of sound in air is 343 m/s.

16. A baseball player throwing the ball in from the outfield usually allows it to take one bounce on the theory that the ball arrives sooner this way. Suppose the ball hits the ground at an angle θ and then rebounds at the same angle after the bounce but loses half its speed. (a) Assuming the ball is always thrown with the same initial speed, at what angle θ should it be thrown in order to go the same distance D with one bounce (solid path in Fig. P4.16) as one thrown upward at $45°$ and reaches its target without bouncing (dashed path in Fig. P4.16)? (b) Determine the ratio of the times for the one-bounce and the no-bounce throws.

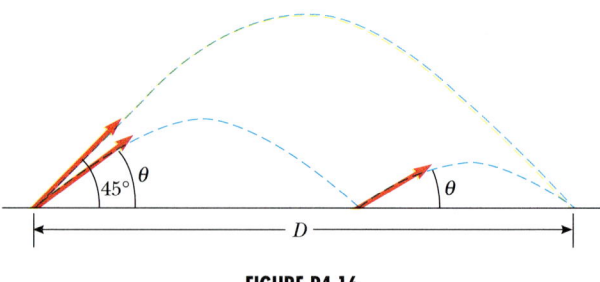

FIGURE P4.16

17. A place kicker must kick a football from a point 36.0 m (about 40 yards) from the goal and the ball must clear the crossbar, which is 3.05 m high. When kicked, the ball leaves the ground with a speed of 20.0 m/s at an angle of $53.0°$ to the horizontal. (a) By how much does the ball clear or fall short of clearing the crossbar? (b) Does the ball approach the crossbar while still rising or while falling?

18. A firefighter, 50.0 m away from a burning building, directs a stream of water from a firehose at an angle

of 30.0° above the horizontal as in Figure P4.18. If the initial speed of the stream is 40.0 m/s, at what height does the water strike the building?

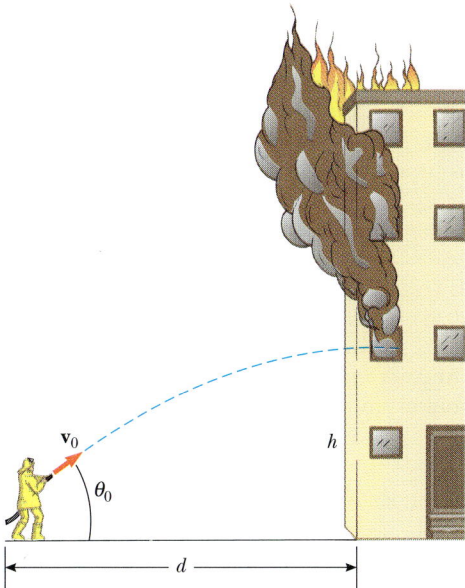

FIGURE P4.18

18A. A firefighter, a distance d from a burning building, directs a stream of water from a firehose at an angle of θ_0 above the horizontal as in Figure P4.18. If the initial speed of the stream is v_0, at what height h does the water strike the building?

19. An astronaut standing on the Moon fires a gun so that the bullet leaves the barrel initially moving in a horizontal direction. (a) What must the muzzle speed be if the bullet is to travel completely around the Moon and return to its original location? (b) How long is the bullet in flight? Assume that the free-fall acceleration on the Moon is one-sixth that on the Earth.

20. A rifle is aimed horizontally at the center of a large target 200 m away. The initial speed of the bullet is 500 m/s. (a) Where does the bullet strike the target? (b) To hit the center of the target, the barrel must be at an angle above the line of sight. Find the angle of elevation of the barrel.

21. During World War I, the Germans had a gun called Big Bertha that was used to shell Paris. The shell had an initial speed of 1.70 km/s at an initial inclination of 55.0° to the horizontal. In order to hit the target, adjustments were made for air resistance and other effects. If we ignore those effects, (a) how far away did the shell hit? (b) How long was it in the air?

22. One strategy in a snowball fight is to throw a snowball at a high angle over level ground. While your oppo-

nent is watching this first snowball, you throw a second one at a low angle timed to arrive at your opponent either before or at the same time as the first one. Assume both snowballs are thrown with a speed of 25 m/s. The first one is thrown at an angle of 70° with respect to the horizontal. (a) At what angle should the second snowball be thrown in order to arrive at the same point as the first? (b) How many seconds later should the second snowball be thrown after the first to arrive at your target at the same time?

23. A projectile is fired in such a way that its horizontal range is equal to three times its maximum height. What is the angle of projection?

24. A flea can jump a vertical height h. (a) What is the maximum horizontal distance it can jump? (b) What is the time in the air in both cases?

25. A cannon having a muzzle speed of 1000 m/s is used to destroy a target on a mountaintop. The target is 2000 m from the cannon horizontally and 800 m above the ground. At what angle, relative to the ground, should the cannon be fired? Ignore air friction.

26. A ball is tossed from an upper-story window of a building. The ball is given an initial velocity of 8.00 m/s at an angle of 20.0° below the horizontal. It strikes the ground 3.00 s later. (a) How far horizontally from the base of the building does the ball strike the ground? (b) Find the height from which the ball was thrown. (c) How long does it take the ball to reach a point 10.0 m below the level of launching?

Section 4.4 Uniform Circular Motion

27. If the rotation of the Earth increased to the point that the centripetal acceleration was equal to the gravitational acceleration at the equator, (a) what would be the tangential speed of a person standing at the equator, and (b) how long would a day be?

28. Young David, who slew Goliath, experimented with slings before tackling the giant. He found that with a sling of length 0.60 m, he could revolve the sling at the rate of 8.0 rev/s. If he increased the length to 0.90 m, he could revolve the sling only 6.0 times per second. (a) Which rate of rotation gives the larger linear speed? (b) What is the centripetal acceleration at 8.0 rev/s? (c) What is the centripetal acceleration at 6.0 rev/s?

29. An athlete rotates a 1.00-kg discus along a circular path of radius 1.06 m. The maximum speed of the discus is 20.0 m/s. Determine the magnitude of its maximum radial acceleration.

30. From information in the endsheets of this book, compute the radial acceleration of a point on the surface of the Earth at the equator.

31. The orbit of the Moon about the Earth is approximately circular, with a mean radius of 3.84×10^8 m.

It takes 27.3 days for the Moon to complete one revolution about the Earth. Find (a) the mean orbital speed of the Moon and (b) its centripetal acceleration.

32. In the spin cycle of a washing machine, the tub of radius 0.300 m rotates at a constant rate of 630 rev/min. What is the maximum linear speed with which water leaves the machine?

33. A ball on the end of a string is whirled around in a horizontal circle of radius 0.30 m. The plane of the circle is 1.2 m above the ground. The string breaks and the ball lands 2.0 m away from the point on the ground directly beneath the ball's location when the string breaks. Find the centripetal acceleration of the ball during its circular motion.

34. A tire 0.500 m in radius rotates at a constant rate of 200 rev/min. Find the speed and acceleration of a small stone lodged in the tread on the outer edge of the tire.

Section 4.5   Tangential and Radial Acceleration

35. Figure P4.35 represents, at a given instant, the total acceleration of a particle moving clockwise in a circle of radius 2.50 m. At this instant of time, find (a) the centripetal acceleration, (b) the speed of the particle, and (c) its tangential acceleration.

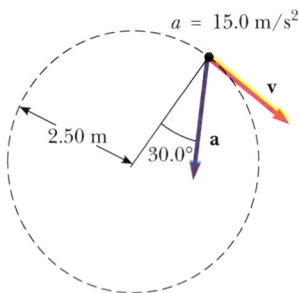

$a = 15.0 \text{ m/s}^2$

2.50 m

30.0°

FIGURE P4.35

36. A point on a rotating turntable 20.0 cm from the center accelerates from rest to 0.700 m/s in 1.75 s. At $t = 1.25$ s, find the magnitude and direction of (a) the centripetal acceleration, (b) the tangential acceleration, and (c) the total acceleration of the point.

37. A train slows down as it rounds a sharp, level turn, slowing from 90.0 km/h to 50.0 km/h in the 15.0 s that it takes to round the bend. The radius of the curve is 150 m. Compute the acceleration at the moment the train speed reaches 50.0 km/h.

38. A pendulum of length 1.00 m swings in a vertical plane (Fig. 4.16). When the pendulum is in the two horizontal positions $\theta = 90°$ and $\theta = 270°$, its speed

is 5.00 m/s. (a) Find the magnitude of the centripetal acceleration and tangential acceleration for these positions. (b) Draw vector diagrams to determine the direction of the total acceleration for these two positions. (c) Calculate the magnitude and direction of the total acceleration.

39. A student attaches a ball to the end of a string 0.600 m in length and then swings the ball in a vertical circle. The speed of the ball is 4.30 m/s at its highest point and 6.50 m/s at its lowest point. Find its acceleration at (a) its highest point and (b) its lowest point.

40. A ball swings in a vertical circle at the end of a rope 1.50 m long. When it is 36.9° past the lowest point on its way up, the ball's total acceleration is $(-22.5\mathbf{i} + 20.2\mathbf{j})$ m/s². For that instant, (a) sketch a vector diagram showing the components of its acceleration, (b) determine the magnitude of its centripetal acceleration, and (c) determine the magnitude and direction of its velocity.

Section 4.6   Relative Velocity and Relative Acceleration

41. Heather in her Corvette accelerates at the rate of $(3.0\mathbf{i} - 2.0\mathbf{j})$ m/s², while Jill in her Jaguar accelerates at $(1.0\mathbf{i} + 3.0\mathbf{j})$ m/s². They both start from rest at the origin of an xy coordinate system. After 5.0 s, (a) what is Heather's speed with respect to Jill, (b) how far apart are they, and (c) what is Heather's acceleration relative to Jill?

42. A motorist traveling west at 80.0 km/h is being chased by a police car traveling at 95.0 km/h. What is the velocity of (a) the motorist relative to the police car and (b) the police car relative to the motorist?

43. A river has a steady speed of 0.500 m/s. A student swims upstream a distance of 1.00 km and returns to the starting point. If the student can swim at a speed of 1.20 m/s in still water, how long does the trip take? Compare this with the time the trip would take if the water were still.

44. How long does it take an automobile traveling in the left lane at 60.0 km/h to overtake (pull alongside) a car traveling in the right lane at 40.0 km/h, if the cars' front bumpers are initially 100 m apart?

45. When the Sun is directly overhead, a hawk dives toward the ground at a speed of 5.00 m/s. If the direction of his motion is at an angle of 60.0° below the horizontal, calculate the speed of his shadow moving along the ground.

46. A boat crosses a river of width $w = 160$ m in which the current has a uniform speed of 1.50 m/s. The pilot maintains a bearing (i.e., the direction in which the boat points) perpendicular to the river and a throttle setting to give a constant speed of 2.00 m/s relative to the water. (a) What is the speed of the boat relative to a stationary shore observer? (b) How far

downstream from the initial position is the boat when it reaches the opposite shore?

47. The pilot of an airplane notes that the compass indicates a heading due west. The airplane's speed relative to the air is 150 km/h. If there is a wind of 30.0 km/h toward the north, find the velocity of the airplane relative to the ground.

48. Two swimmers, A and B, start at the same point in a stream that flows with a speed v. Both move at the same speed c relative to the stream, where $c > v$. A swims downstream a distance L and then upstream the same distance, while B swims directly perpendicular to the stream's flow a distance L and then back the same distance, so that both swimmers return back to the starting point. Which swimmer returns first? (*Note:* First guess at the answer.)

49. A car travels due east with a speed of 50.0 km/h. Rain is falling vertically relative to the Earth. The traces of the rain on the side windows of the car make an angle of 60.0° with the vertical. Find the velocity of the rain relative to (a) the car and (b) the Earth.

50. A child in danger of drowning in a river is being carried downstream by a current that has a speed of 2.50 km/h. The child is 0.600 km from shore and 0.800 km upstream of a boat landing when a rescue boat sets out. (a) If the boat proceeds at its maximum speed of 20.0 km/h relative to the water, what heading relative to the shore should the pilot take? (b) What angle does the boat velocity make with the shore? (c) How long does it take the boat to reach the child?

51. A bolt drops from the ceiling of a train car that is accelerating northward at a rate of 2.50 m/s². What is the acceleration of the bolt relative to (a) the train car? (b) the Earth?

52. A science student is riding on a flatcar of a train traveling along a straight horizontal track at a constant speed of 10.0 m/s. The student throws a ball into the air along a path that he judges to make an initial angle of 60.0° with the horizontal and to be in line with the track. The student's professor, who is standing on the ground nearby, observes the ball to rise vertically. How high does she see the ball rise?

ADDITIONAL PROBLEMS

53. At $t = 0$ a particle leaves the origin with a velocity of 6.00 m/s in the positive y direction. Its acceleration is given by $\mathbf{a} = (2.00\mathbf{i} - 3.00\mathbf{j})$ m/s². When the particle reaches its maximum y coordinate, its y component of velocity is zero. At this instant, find (a) the velocity of the particle and (b) its x and y coordinates.

54. The speed of a projectile when it reaches its maximum height is one half the speed when the projectile is at half its maximum height. What is the initial projection angle?

55. A car is parked overlooking the ocean on an incline that makes an angle of 37.0° with the horizontal. The distance from where the car is parked to the bottom of the incline is 50.0 m, and the incline terminates at a cliff that is 30.0 m above the ocean surface. The negligent driver leaves the car in neutral, and the parking brakes are defective. If the car rolls from rest down the incline with a constant acceleration of 4.00 m/s², find (a) the speed of the car just as it reaches the cliff and the time it takes to get there, (b) the velocity of the car just as it lands in the ocean, (c) the total time the car is in motion, and (d) the position of the car relative to the base of the cliff just as it lands in the ocean.

56. A projectile is fired up an incline (incline angle ϕ) with an initial speed v_0 at an angle θ_0 with respect to the horizontal ($\theta_0 > \phi$), as shown in Figure P4.56. (a) Show that the projectile travels a distance d up the incline, where

$$d = \frac{2v_0^2 \cos \theta_0 \sin(\theta_0 - \phi)}{g \cos^2 \phi}$$

(b) For what value of θ_0 is d a maximum, and what is the maximum value?

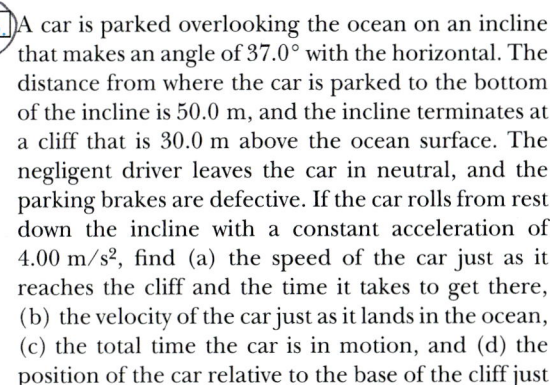

Path of the projectile

FIGURE P4.56

57. A batter hits a pitched baseball 1.00 m above the ground, imparting to the ball a speed of 40.0 m/s. The resulting line drive is caught on the fly by the left fielder 60.0 m from home plate with his glove 1.00 m above the ground. If the shortstop, 45.0 m from home plate and in line with the drive, were to jump straight up to make the catch instead of allowing the left fielder to make the play, how high above the ground would his glove have to be?

58. A basketball player 2.00 m tall throws to the basket from a horizontal distance of 10.0 m, as in Figure P4.58. If he shoots at a 40° angle with the horizontal, at what initial speed must he throw so that the ball goes through the hoop without striking the backboard?

59. A boy can throw a ball a maximum horizontal distance of 40.0 m on a level field. How far can he throw the same ball vertically upward? Assume that his muscles give the ball the same speed in each case.

59A. A boy can throw a ball a maximum horizontal dis-

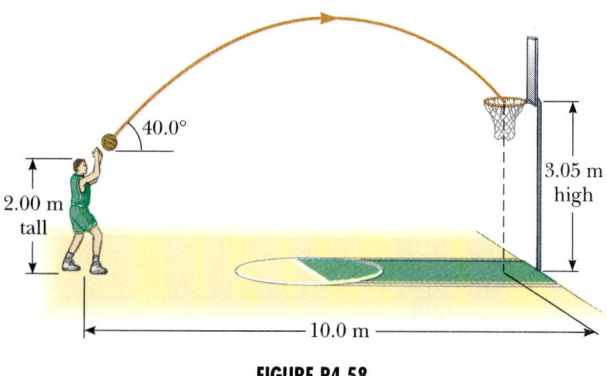

FIGURE P4.58

tance R on a level field. How far can he throw the same ball vertically upward? Assume that his muscles give the ball the same speed in each case.

60. The x and y coordinates of a particle are given by

$$x = 2.00 \text{ m} + (3.00 \text{ m/s})t \qquad y = x - (5.00 \text{ m/s}^2)t^2$$

How far from the origin is the particle at (a) $t = 0$; (b) $t = 2.00$ s?

61. A stone at the end of a sling is whirled in a vertical circle of radius 1.20 m at a constant speed $v_0 =$ 1.50 m/s as in Figure P4.61. The center of the string is 1.50 m above the ground. What is the range of the stone if it is released when the sling is inclined at 30.0° with the horizontal (a) at A? (b) at B? What is the acceleration of the stone (c) just before it is released at A? (d) just after it is released at A?

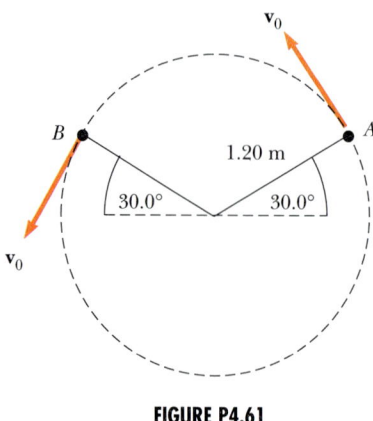

FIGURE P4.61

62. A truck is moving north with a constant speed of 10.0 m/s on a horizontal stretch of road. A boy riding on the back of the truck wishes to throw a ball while the truck is moving and to catch the ball after the truck has gone 20.0 m. (a) Neglecting air resistance, at what angle to the vertical should the ball be thrown? (b) What should be the initial speed of the

ball? (c) What is the shape of the path of the ball as seen by the boy? (d) An observer on the ground watches the boy throw the ball up and catch it. In this observer's fixed frame of reference, determine the general shape of the ball's path and the initial velocity of the ball.

63. A dart gun is fired while being held horizontally at a height of 1.00 m above ground level. With the gun at rest relative to the ground, the dart from the gun travels a horizontal distance of 5.00 m. A child holds the same gun in a horizontal position while sliding down a 45.0° incline at a constant speed of 2.00 m/s. How far will the dart travel if the gun is fired when it is 1.00 m above the ground?

64. A rocket is launched at an angle of 53.0° to the horizontal with an initial speed of 100 m/s. It moves along its initial line of motion with an acceleration of 30.0 m/s² for 3.00 s. At this time its engines fail and the rocket proceeds to move as a free body. Find (a) the maximum altitude reached by the rocket, (b) its total time of flight, and (c) its horizontal range.

65. A person standing at the top of a hemispherical rock of radius R kicks a ball (initially at rest on the top of the rock) so that its initial velocity is horizontal as in Figure P4.65. (a) What must its minimum initial speed be if the ball is never to hit the rock after it is kicked? (b) With this initial speed, how far from the base of the rock does the ball hit the ground?

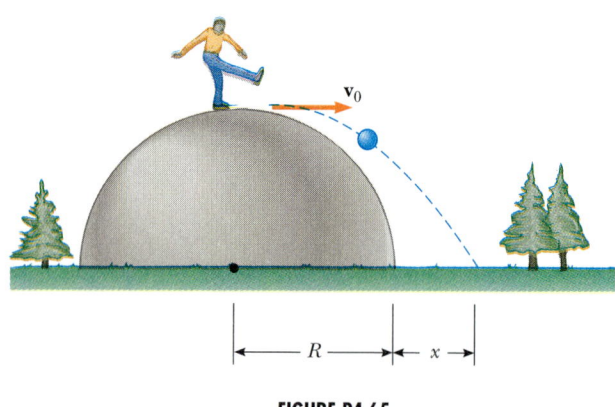

FIGURE P4.65

66. A home run in a baseball game is hit in such a way that the ball just clears a wall 21.0 m high, located 130 m from home plate. The ball is hit at an angle of 35.0° to the horizontal, and air resistance is negligible. Find (a) the initial speed of the ball, (b) the time it takes the ball to reach the wall, and (c) the velocity components and the speed of the ball when it reaches the wall. (Assume the ball is hit at a height of 1.00 m above the ground.)

67. A daredevil is shot out of a cannon at 45.0° to the horizontal with an initial speed of 25.0 m/s. A net is

located at a horizontal distance of 50.0 m from the cannon. At what height above the cannon should the net be placed in order to catch the daredevil?

68. The position of a particle as a function of time is described by

$$\mathbf{r} = (bt)\mathbf{i} + (c - dt^2)\mathbf{j} \qquad b = 2.00 \text{ m/s}$$

$$c = 5.00 \text{ m} \qquad d = 1.00 \text{ m/s}^2$$

(a) Express y in terms of x and sketch the trajectory of the particle. What is the shape of the trajectory? (b) Derive a vector relationship for the velocity. (c) At what time ($t > 0$) is the velocity vector perpendicular to the position vector?

69. A bomber flies horizontally with a speed of 275 m/s relative to the ground. The altitude of the bomber is 3000 m and the terrain is level. Neglect the effects of air resistance. (a) How far from the point vertically under the point of release does a bomb hit the ground? (b) If the plane maintains its original course and speed, where is it when the bomb hits the ground? (c) At what angle from the vertical at the point of release must the telescopic bomb sight be set so that the bomb hits the target seen in the sight at the time of release?

70. A football is thrown toward a receiver with an initial speed of 20.0 m/s at an angle of 30.0° above the horizontal. At that instant, the receiver is 20.0 m from the quarterback. In what direction and with what constant speed should the receiver run in order to catch the football at the level at which it was thrown?

71. A flea is at point A on a horizontal turntable 10.0 cm from the center. The turntable is rotating at $33\frac{1}{3}$ rev/min in the clockwise direction. The flea jumps vertically upward to a height of 5.00 cm and lands on the turntable at point B. Place the coordinate origin at the center of the turntable with the positive x axis fixed in space and passing through A. (a) Find the linear displacement of the flea. (b) Find the position of point A when the flea lands. (c) Find the position of point B when the flea lands.

72. A student who is able to swim at a speed of 1.50 m/s in still water wishes to cross a river that has a current of velocity 1.20 m/s toward the south. The width of the river is 50.0 m. (a) If the student starts from the west bank, in what direction should she head in order to swim directly across the river? How long will this trip take? (b) If she heads due east, how long will it take to cross the river? (*Note:* The student travels farther than 50.0 m in this case.)

73. A rifle has a maximum range of 500 m. (a) For what angles of elevation would the range be 350 m? What is the range when the bullet leaves the rifle (b) at 14.0°? (c) at 76.0°?

74. A river flows with a uniform velocity **v**. A person in a motorboat travels 1.00 km upstream, at which time a log is seen floating by. The person continues to travel upstream for 60.0 min at the same speed and then returns downstream to the starting point, where the same log is seen again. Find the velocity of the river. (*Hint:* The time of travel of the boat after it meets the log equals the time of travel of the log.)

75. An airplane has a velocity of 400 km/h due east relative to the moving air. At the same time, a wind blows northward with a speed of 75.0 km/h relative to Earth. (a) Find the airplane's velocity relative to Earth. (b) In what direction must the airplane head in order to move east relative to Earth?

76. A sailor aims a rowboat toward an island located 2.00 km east and 3.00 km north of her starting position. After an hour she sees the island due west. She then aims the boat in the opposite direction from which she was rowing, rows for another hour, and ends up 4.00 km east of her starting position. She correctly deduces that the current is from west to east. (a) What is the speed of the current? (b) Show that the boat's velocity relative to the shore for the first hour can be expressed as $\mathbf{u} = (4.00 \text{ km/h})\mathbf{i} + (3.00 \text{ km/h})\mathbf{j}$, where $\mathbf{i}$ is directed east and $\mathbf{j}$ is directed north.

77. Two soccer players, Mary and Jane, begin running from approximately the same point at the same time. Mary runs in an easterly direction at 4.0 m/s, while Jane takes off in a direction 60° north of east at 5.4 m/s. (a) How long is it before they are 25 m apart? (b) What is the velocity of Jane relative to Mary? (c) How far apart are they after 4.0 s?

78. After delivering his toys in the usual manner, Santa decides to have some fun and slide down an icy roof, as in Figure P4.78. He starts from rest at the top of the roof, which is 8.00 m in length, and accelerates at the rate of 5.00 m/s². The edge of the roof is 6.00 m above a soft snow bank, which Santa lands on. Find (a) Santa's velocity components when he reaches the snow bank, (b) the total time he is in motion, and (c) the distance d between the house and the point where he lands in the snow.

FIGURE P4.78

79. A skier leaves the ramp of a ski jump with a velocity of 10 m/s, 15° above the horizontal, as in Figure P4.79.

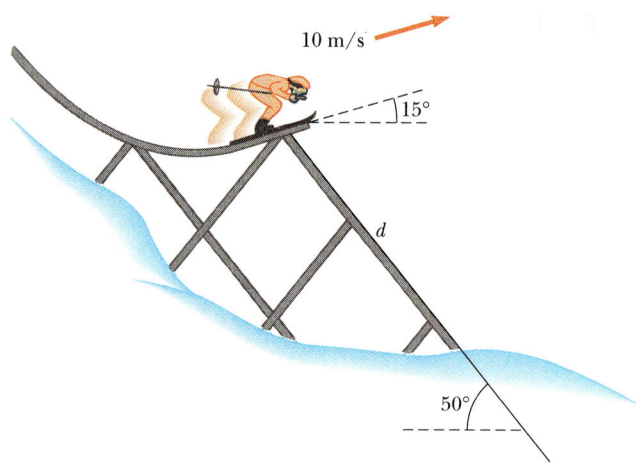

FIGURE P4.79

FIGURE P4.81

The slope is inclined at 50°, and air resistance is negligible. Find (a) the distance that the jumper lands down the slope and (b) the velocity components just before landing. (How do you think the results might be affected if air resistance were included? Note that jumpers lean forward in the shape of an airfoil with their hands at their sides to increase their distance. Why does this work?)

80. A golf ball leaves the ground at an angle θ and hits a tree while moving horizontally at height h above the ground. If the tree is a horizontal distance of b from the point of projection, show that (a) $\tan \theta = 2\, h/b$. (b) What is the initial velocity of the ball in terms of b and h?

81. A truck loaded with cannonball watermelons stops suddenly to avoid running over the edge of a washed-out bridge (Fig. P4.81). The quick stop causes a number of melons to fly off the truck. One melon rolls over the edge with an initial speed $v_0 = 10$ m/s in the horizontal direction. What are the x and y coordinates of the melon when it splatters on the bank, if a cross-section of the bank has the shape of a parabola ($y^2 = 16x$ where x and y are measured in meters) with its vertex at the edge of the road?

82. An enemy ship is on the east side of a mountain island as shown in Figure P4.82. The enemy ship can maneuver to within 2500 m of the 1800-m-high mountain peak and can shoot projectiles with an initial speed of 250 m/s. If the western shoreline is horizontally 300 m from the peak, what are the distances from the western shore at which a ship can be safe from the bombardment of the enemy ship?

83. A hawk is flying horizontally at 10.0 m/s in a straight line 200 m above the ground. A mouse it was carrying is released from its grasp. The hawk continues on its path at the same speed for two seconds before attempting to retrieve its prey. To accomplish the retrieval, it dives in a straight line at constant speed and recaptures the mouse 3.0 m above the ground. Assuming no air resistance (a) find the diving speed of the hawk. (b) What angle did the hawk make with the horizontal during its descent? (c) For how long did the mouse "enjoy" free flight?

84. The determined coyote is out once more to try to capture the elusive road runner. The coyote wears a pair of Acme jet-powered roller skates, which provide a constant horizontal acceleration of 15 m/s² (Fig. P4.84). The coyote starts off at rest 70 m from the edge of a cliff at the instant the road runner zips by in the direction of the cliff. (a) If the road runner moves with constant speed, determine the minimum speed he must have in order to reach the cliff before the coyote. (b) If the cliff is 100 m above the base of a canyon, determine where the coyote lands in the

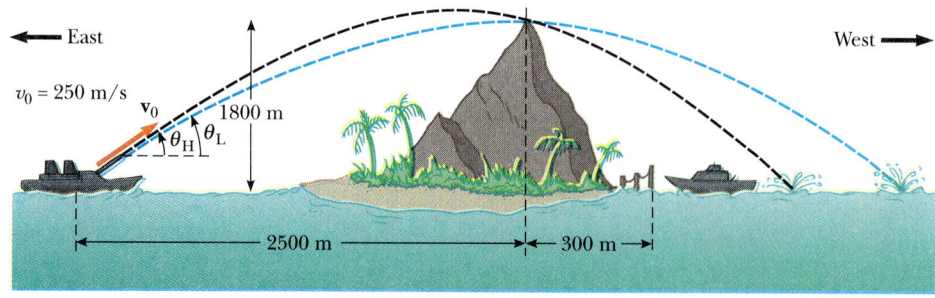

FIGURE P4.82

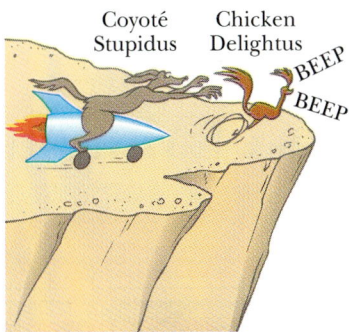

Coyoté Stupidus   Chicken Delightus

BEEP BEEP

FIGURE P4.84

canyon (assume his skates are still in operation when he is in "flight"). (c) Determine the coyote's velocity components just before he lands in the canyon. (As usual, the road runner is saved by making a sudden turn at the cliff.)

85. The Earth is 1.50×10^{11} m from the Sun and makes one revolution around the Sun in 3.16×10^7 s. The Moon is 3.84×10^8 m from the Earth and makes one revolution around the Earth in 2.36×10^6 s. Determine the velocity of the Moon relative to the Sun at the instant the Moon is heading directly toward the Sun, as in Figure P4.85.

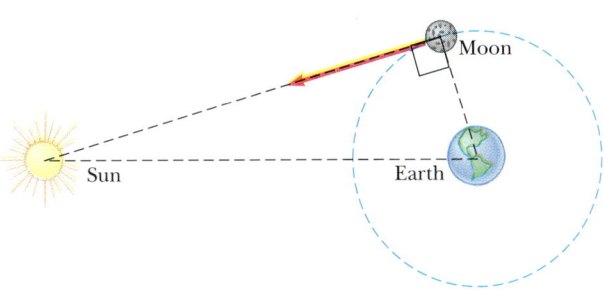

Moon

Sun   Earth

FIGURE P4.85

SPREADSHEET PROBLEMS

S1. The punter for the St. Louis Rams needs to kick the football 40 to 45 yd (36.6 to 41.1 m) to keep the opposing team inside the 5-yd line. Use Spreadsheet 4.1 to estimate the angle at which he should kick the ball. *Note:* He must keep the ball in the air as long as possible to help prevent a runback (in other words, he wants a long "hang time"). The initial speed of the football should be in the range of 20 to 30 m/s.

S2. Modify Spreadsheet 4.1 by adding two columns to calculate the maximum height reached by the football in Problem S4.1 and the total time it is in the air.

S3. The x and y coordinates of a projectile are

$$x = x_0 + v_{x0}t$$

$$y = y_0 + v_{y0}t - \tfrac{1}{2}gt^2$$

where $v_{x0} = v_0 \cos \theta_0$ and $v_{y0} = v_0 \sin \theta_0$. Using these equations, set up a spreadsheet to calculate the x and y coordinates of a projectile and the components of its velocity v_x and v_y as functions of time. The initial speed and initial angle of the projectile should be input parameters. Use the graph function of the spreadsheet to plot the x and y coordinates as functions of time. Also graph v_x and v_y as functions of time. Use this spreadsheet to solve the following problem:

   The New York Giants are tied with the Chicago Bears with only a few seconds left in the game. The Giants have the football, and call the place kicker into the game. He must kick the ball 52 yd (47.5 m) for a field goal. If the crossbar of the goal is 10 ft (3.05 m) high, at what angle and speed should he kick the ball to score the winning points? Is there only one solution to this problem? Solutions can be found by varying the parameters and studying the graph of y versus x.

The Wizard of Id   by Parker and Hart

The Laws of Motion

This calf-roping scene, taken in Steamboat, Colorado, is a standard rodeo event. The external forces acting on the horse are the force of friction between the horse and ground, the force of gravity, the tension force of the rope attached to the calf, the force of the cowboy on the horse, and the upward force of the ground. Can you identify the forces acting on the calf? *(FourbyFive, Inc.)*

I n Chapters 2 and 4, we described the motion of particles based on the definitions of displacement, velocity, and acceleration. However, we would like to be able to answer specific questions related to the causes of motion, such as "What mechanism causes motion?" and "Why do some objects accelerate at a higher rate than others?" In this chapter, we use the concepts of force and mass to describe the change in motion of particles. We then discuss the three basic laws of motion, which are based on experimental observations and were formulated more than three centuries ago by Newton.

Classical mechanics describes the relationship between the motion of a body and the forces acting on that body. Classical mechanics deals only with objects that (a) are large compared with the dimensions of atoms ($\approx 10^{-10}$ m) and (b) move at speeds that are much less than the speed of light (3.00×10^8 m/s).

We learn in this chapter how it is possible to describe an object's acceleration in terms of the resultant force acting on the object and its mass. This force represents the interaction of the object with its environment. Mass is a measure of the object's inertia, that is, its tendency to resist an acceleration when a force acts on it.

We also discuss force laws, which describe the quantitative method of calculating the force on an object if its environment is known. We see that, although the force laws are simple in form, they successfully explain a wide variety of phenomena and experimental observations. These force laws, together with the laws of motion, are the foundations of classical mechanics.

5.1 THE CONCEPT OF FORCE

Everyone has a basic understanding of the concept of force from everyday experiences. When you push or pull an object, you exert a force on it. You exert a force when you throw or kick a ball. In these examples, the word *force* is associated with the result of muscular activity and some change in the state of motion of an object. Forces do not always cause an object to move, however. For example, as you sit reading this book, the force of gravity acts on your body, and yet you remain stationary. As a second example, you can push on a large block of stone and not be able to move it.

What force (if any) causes a distant star to drift freely through space? Newton answered such questions by stating that the change in velocity of an object is caused by forces. Therefore, if an object moves with uniform motion (constant velocity), no force is required to maintain the motion. Since only a force can cause a change in velocity, we can think of force as that which causes a body to accelerate.

Now consider a situation in which several forces act simultaneously on an object. In this case, the object accelerates only if the *net force* acting on it is not equal to zero. (We sometimes refer to the net force as either the *resultant force* or the *unbalanced force*.) *If the net force is zero, the acceleration is zero and the velocity of the object remains constant.* That is, if the net force acting on the object is zero, the object either remains at rest or continues to move with constant velocity. *When the velocity of a body is constant or when the body is at rest, it is said to be in equilibrium.*

Whenever a force is exerted on an object, the shape of the object can change. For example, when you squeeze a rubber ball or strike a punching bag with your fist, the objects are deformed to some extent. Even more rigid objects, such as an automobile, are deformed under the action of external forces. The deformations can be permanent if the forces are large enough, as in the case of a collision between vehicles.

In this chapter, we are concerned with the relationship between the force exerted on an object and the acceleration of that object. If you pull on a coiled spring, as in Figure 5.1a, the spring stretches. If the spring is calibrated, the distance it stretches can be used to measure the strength of the force. If you pull hard enough on a cart to overcome friction, as in Figure 5.1b, the cart moves. When a football is kicked, as in Figure 5.1c, the football is both deformed and set in motion. These are all examples of a class of forces called *contact forces.* That is, they represent the result of physical contact between two objects. Other examples of contact forces include the force exerted by gas molecules on the walls of a container and the force exerted by our feet on the floor.

Another class of forces, which do not involve physical contact between two objects but act through empty space, are known as *field forces.* The force of gravitational attraction between two objects is an example of this class of force, illustrated in Figure 5.1d. This gravitational force keeps objects bound to the Earth and gives rise to what we commonly call the *weight* of an object. The planets of our Solar System are bound under the action of gravitational forces. Another common example of a field force is the electric force that one electric charge exerts on another electric charge, as in Figure 5.1e. These charges might be an electron and proton forming a hydrogen atom. A third example of a field force is the force that a bar magnet exerts on a piece of iron, as shown in Figure 5.1f. The forces holding

A body accelerates due to an external force

Definition of equilibrium

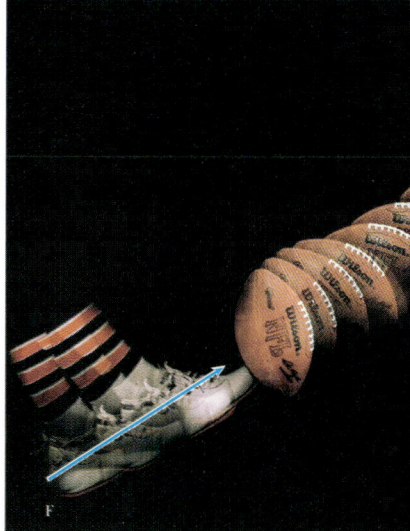

A football is set in motion by the contact force, **F**, on it due to the kicker's foot. The ball is deformed during the short time in contact with the foot. *(Ralph Cowan, Tony Stone Worldwide)*

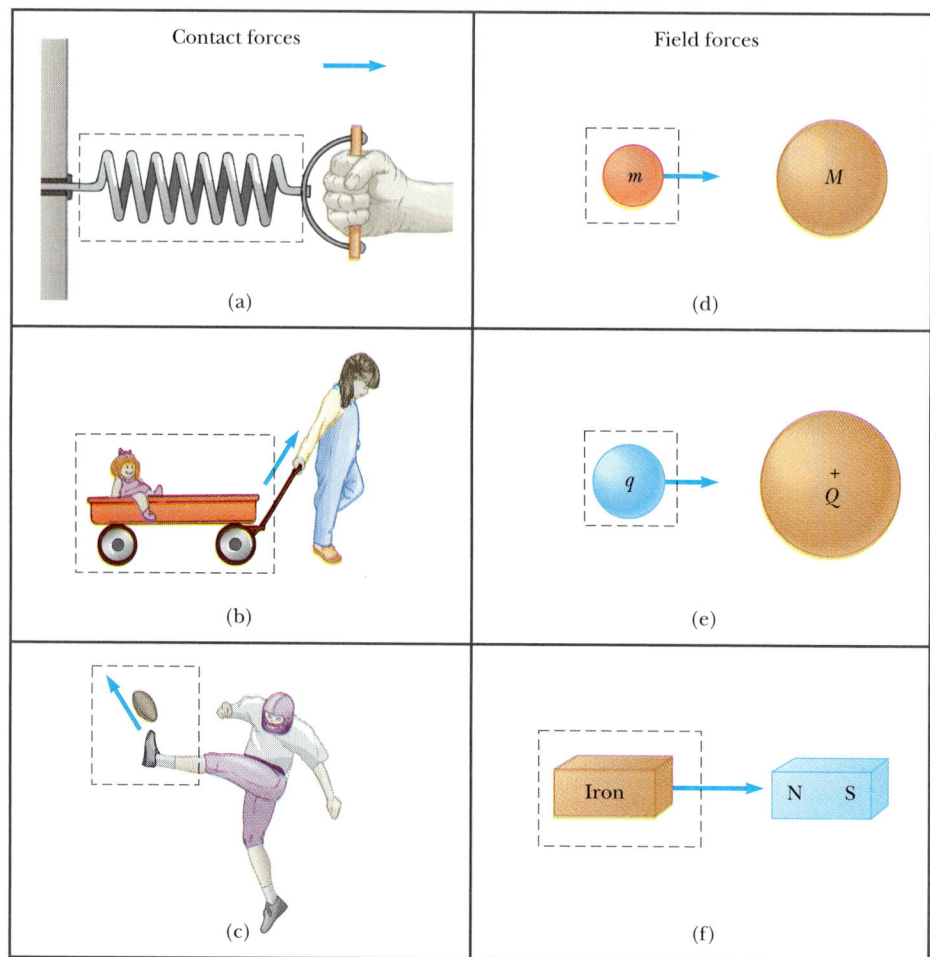

FIGURE 5.1 Some examples of forces applied to various objects. In each case a force is exerted on the object within the boxed area. Some agent in the environment external to the boxed area exerts the force on the object.

an atomic nucleus together are also field forces but are usually very short-range. They are the dominating interaction for particle separations of the order of 10^{-15} m.

Early scientists, including Newton, were uneasy with the concept of a force acting between two disconnected objects. To overcome this conceptual problem, Michael Faraday (1791–1867) introduced the concept of a *field*. According to this approach, when a mass m_1 is placed at some point P near a mass m_2, one can say that m_1 interacts with m_2 by virtue of the gravitational field that exists at P. The field at P is created by mass m_2. Likewise, a field exists at the position of m_2 created by m_1. In fact, all objects create a gravitational field in the space around them.

The distinction between contact forces and field forces is not as sharp as you may have been led to believe by the above discussion. When examined at the atomic level, all the forces we classify as contact forces turn out to be due to repulsive electrical (field) forces of the type illustrated in Figure 5.1e. Nevertheless, in developing models for macroscopic phenomena, it is convenient to use both classifications of forces. However, the only known *fundamental* forces in nature are all field forces: (1) gravitational attraction between objects, (2) electro-

Fundamental forces in nature

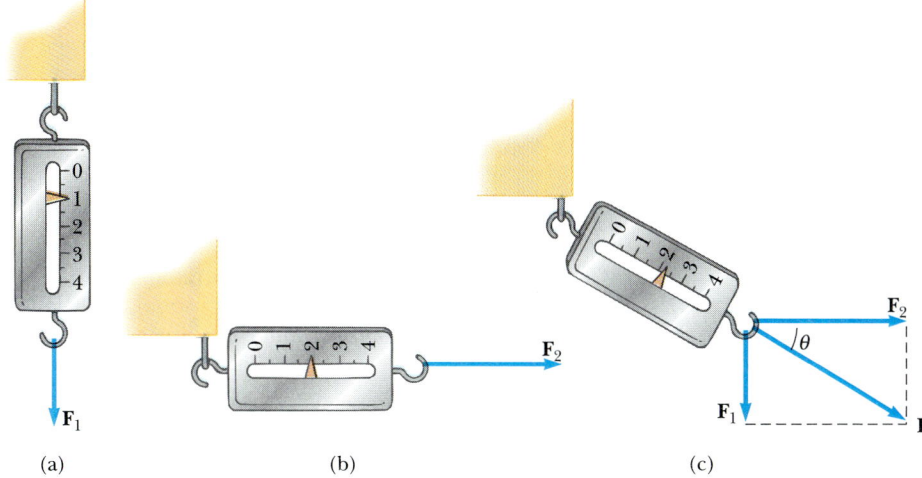

FIGURE 5.2   The vector nature of a force is tested with a spring scale. (a) The downward force F_1 elongates the spring 1 unit. (b) A horizontal force F_2 elongates the spring 2 units. (c) When F_2 is applied horizontally, and F_1 is downward, the combination of the two forces elongates the spring $\sqrt{1^2 + 2^2} = \sqrt{5}$ units.

magnetic forces between electric charges, (3) strong nuclear forces between sub-atomic particles, and (4) weak nuclear forces that arise in certain radioactive decay processes. In classical physics, we are concerned only with gravitational and elec-tromagnetic forces.

It is convenient to use the deformation of a spring to measure force. Suppose a force is applied vertically to a spring that has a fixed upper end, as in Figure 5.2a. We can calibrate the spring by defining the unit force, F_1, as the force that pro-duces an elongation of 1.00 cm. (Because force is a vector quantity, we use the bold-faced symbol F.) If a second downward force F_2 is applied whose magnitude is 2 units, as in Figure 5.2b, the scale elongation is 2.00 cm. This shows that the combined effect of the two forces that are collinear is the sum of the effect of the individual forces. Now suppose that the two forces are applied simultaneously with F_1 downward and F_2 horizontal, as in Figure 5.2c. In this case, the elongation of the spring is found to be $\sqrt{5} = 2.24$ cm. The single force, F, that would produce this same elongation is the vector sum of F_1 and F_2, as described in Figure 5.2c. That is, $|F| = \sqrt{F_1^2 + F_2^2} = \sqrt{5}$ units, and its direction is $\theta = \arctan(-0.500) = -26.6°$. *Because forces are vectors, you must use the rules of vector addition to get the resultant force on a body.*

Springs that elongate in proportion to an applied force are said to obey *Hooke's law.* Such springs can be constructed and calibrated to measure the magnitude of unknown forces.

5.2   NEWTON'S FIRST LAW AND INERTIAL FRAMES

Before we state Newton's first law, consider the following simple experiment. Sup-pose a book is lying on a table. Obviously, the book remains at rest in the absence of any influences. Now imagine that you push the book with a horizontal force large enough to overcome the force of friction between book and table. The book can then be kept in motion with constant velocity if the force you apply is equal in magnitude to the force of friction and in the direction opposite the friction force.

Isaac Newton, a British physicist and mathematician, is regarded as one of the greatest scientists in history. Before the age of 30 he formulated the basic concepts and laws of motion, discovered the universal law of gravitation, and invented the calculus. Newton was able to explain the motions of the planets, the ebb and flow of the tides, and many special features of the motion of the Moon and the Earth. He also made many important discoveries in optics, showing, for example, that white light is composed of a spectrum of colors. His contributions to physical theories dominated scientific thought for two centuries and remain important today.

Newton was born prematurely on Christmas Day in 1642, shortly after his father's death. When he was three, his mother remarried and he was left in his grandmother's care. Because he was small in stature as a child, he was bullied by other children and took refuge in such solitary activities as the building of water clocks, kites carrying fiery lanterns, sundials, and model windmills powered by mice. His mother withdrew him from school at the age of 12 with the intention of turning him into a farmer. Fortu-

Isaac Newton

| 1 6 4 2 – 1 7 2 7 |

nately for later generations, his uncle recognized his scientific and mathematical abilities and helped send him to Trinity College in Cambridge.

In 1665, the year Newton completed his Bachelor of Arts degree, the university was closed because of the bubonic plague that was raging

through England. Newton returned to the family farm at Woolsthorpe to study. During this especially creative period, he laid the foundations of his work in mathematics, optics, motion, celestial mechanics, and gravity.

Newton was a very private person who studied alone and labored day and night in his laboratory, conducting experiments, performing calculations, and immersing himself in theological studies. His greatest single work, *Mathematical Principles of Natural Philosophy,* was published in 1687. In his later years he spent much of his time quarreling with other eminent minds, including the mathematician Gottfried Leibnitz, who worked independently on the development of calculus; Christian Huygens, who developed the wave theory of light; and Robert Hooke, who supported Huygens' theory. These disputes, the strain of his studies, and his work in alchemy, which involved mercury (a poison), caused him in 1692 to suffer severe depression. He was elected president of the Royal Society in 1703, and he retained that office until his death in 1727.

(Giraudon/Art Resource)

If the applied force exceeds the force of friction, the book accelerates. If you stop pushing, the book stops sliding after moving a short distance because the force of friction retards its motion. Now imagine pushing the book across a smooth, highly waxed floor. The book again comes to rest after you stop pushing, but not as quickly as before. Now imagine a floor so highly polished that friction is absent; in this case, the book, once set in motion, slides until it hits the wall. The forces that we have described are *external* forces, that is, forces exerted on the object by other objects.

Before about 1600, scientists felt that the natural state of matter was the state of rest. Galileo was the first to take a different approach to motion and the natural state of matter. He devised thought experiments, such as the one we just discussed for a book on a frictionless surface, and concluded that it is not the nature of an object to stop once set in motion: rather, it is its nature to resist changes in its motion. In his words, "Any velocity once imparted to a moving body will be rigidly maintained as long as the external causes of retardation are removed."

This new approach to motion was later formalized by Newton in a form that has come to be known as **Newton's first law of motion:**

An object at rest will remain at rest and an object in motion will continue in motion with constant velocity unless a net external force acts on the object. In this case, the wall of the building did not exert a large enough force on the moving train to stop it. *(Roger Viollet, Mill Valley, CA, University Science Books, 1982)*

An object at rest remains at rest and an object in motion will continue in motion with a constant velocity (that is, constant speed in a straight line) unless it experiences a net external force.

Newton's first law

In simpler terms, we can say that *when the net force on a body is zero, its acceleration is zero.* That is, when $\Sigma\mathbf{F} = 0$, then $\mathbf{a} = 0$. From the first law, we conclude that an isolated body (a body that does not interact with its environment) is either at rest or moving with constant velocity.

Another example of uniform (constant velocity) motion on a nearly frictionless plane is the motion of a light disk on a column of air (the lubricant), as in Figure 5.3. If the disk is given an initial velocity, it coasts a great distance before stopping. This idea is used in the game of air hockey, where the disk makes many collisions with the walls before coming to rest.

Finally, consider a spaceship traveling in space and far removed from any planets or other matter. The spaceship requires some propulsion system to *change* its velocity. However, if the propulsion system is turned off when the spaceship reaches a velocity v, the spaceship coasts in space at a constant velocity and the astronauts get a free ride (that is, no propulsion system is required to keep them moving at the velocity v).

Inertial Frames

Newton's first law, sometimes called the *law of inertia,* defines a special set of reference frames called inertial frames. An **inertial frame** of reference is one in which Newton's first law is valid. Thus, an inertial frame of reference is an unaccel-

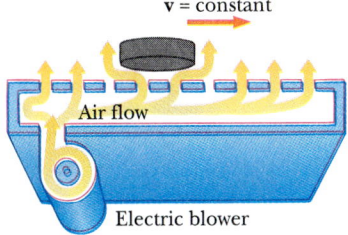

FIGURE 5.3   A disk moving on a column of air is an example of uniform motion, that is, motion in which the acceleration is zero and the velocity remains constant.

Inertial frame

erated frame. Any reference frame that moves with constant velocity relative to an inertial frame is itself an inertial frame. A reference frame that moves with constant velocity relative to the distant stars is the best approximation of an inertial frame. The Earth is not an inertial frame because of its orbital motion about the Sun and rotational motion about its own axis. As the Earth travels in its nearly circular orbit around the Sun, it experiences a centripetal acceleration of about 4.4×10^{-3} m/s^2 toward the Sun. In addition, since the Earth rotates about its own axis once every 24 hours, a point on the equator experiences an additional centripetal acceleration of 3.37×10^{-2} m/s^2 toward the center of the Earth. However, these accelerations are small compared with g and can often be neglected. In most situations *we shall assume that a set of nearby points on the Earth's surface constitutes an inertial frame.*

Thus, if an object is moving with constant velocity, an observer in one inertial frame (say, one at rest relative to the object) will claim that the acceleration and the resultant force on the object are zero. An observer in *any other* inertial frame will also find that $\mathbf{a} = 0$ and $\mathbf{F} = 0$ for the object. According to the first law, a body at rest and one moving with constant velocity are equivalent. In other words, the absolute motion of an object has no effect on its behavior. Unless stated otherwise, we shall write the laws of motion relative to an observer "at rest" in an inertial frame.

CONCEPTUAL EXAMPLE 5.1

Is it possible to have motion in the absence of a force?

Reasoning   Motion requires no force. Newton's first law says that motion needs no cause but continues by itself. That is, an object in motion continues to move by itself in the absence of external forces. As an example, witness the motion of a meteoroid in outer space (imitated by a glider on an air track).

5.3   INERTIAL MASS

Inertia

If you attempt to change the velocity of an object, the object resists this change. **Inertia** is solely a property of an individual object; it is a measure of the response of an object to an external force. For instance, consider two large, solid cylinders of equal size, one balsa wood and the other steel. If you were to push the cylinders along a horizontal surface, the force required to give the steel cylinder some acceleration would be larger than the force needed to give the balsa wood cylinder the same acceleration. Therefore, we say that the steel cylinder has more inertia than the balsa wood cylinder.

Mass is used to measure inertia, and the SI unit of mass is the kilogram. The greater the mass of a body, the less that body accelerates (changes its state of motion) under the action of an applied force. For example, if a given force acting on a 3-kg mass produces an acceleration of 4 m/s^2, the same force when applied to a 6-kg mass will produce an acceleration of 2 m/s^2.

Let us now use this idea to obtain a quantitative description of the concept of mass. We begin by comparing the accelerations that a given force produces on different bodies. Suppose a force acting on a body of mass m_1 produces an acceleration $\mathbf{a}_1$, and the *same force* acting on a body of mass m_2 produces an acceleration

a_2. The ratio of the two masses is defined as the *inverse* ratio of the magnitudes of the accelerations produced by the same force:

$$\frac{m_1}{m_2} \equiv \frac{a_2}{a_1} \qquad (5.1)$$

If one of these is a standard known mass—of say, 1 kg—the mass of an unknown can be obtained from acceleration measurements. For example, if the standard 1-kg mass undergoes an acceleration of 6 m/s² under the influence of some force, a 2-kg mass will undergo an acceleration of 3 m/s² under the action of the same force.

   Mass is an inherent property of a body and is independent of the body's surroundings and of the method used to measure it. It is an experimental fact that *mass is a scalar quantity.* Finally, *mass is a quantity that obeys the rules of ordinary arithmetic.* That is, several masses can be combined in a simple numerical fashion. For example, if you combine a 3-kg mass with a 5-kg mass, their total mass is 8 kg. This result can be verified experimentally by comparing the acceleration of each object produced by a known force with the acceleration of the combined system using the same force.

   Mass should not be confused with weight. *Mass and weight are two different quantities.* As we shall see below, the weight of a body is equal to the magnitude of the force exerted by the Earth on the body and varies with location. For example, a person who weighs 180 lb on Earth weighs only about 30 lb on the Moon. On the other hand, the mass of a body is the same everywhere, regardless of location. An object having a mass of 2 kg on Earth also has a mass of 2 kg on the Moon.

Mass and weight are different quantities

5.4   NEWTON'S SECOND LAW

Newton's first law explains what happens to an object when the resultant of all external forces on it is zero: it either remains at rest or moves in a straight line with constant speed. Newton's second law answers the question of what happens to an object that has a nonzero resultant force acting on it.

   Imagine you are pushing a block of ice across a frictionless horizontal surface. When you exert some horizontal force **F**, the block moves with some acceleration **a**. If you apply a force twice as large, the acceleration doubles. Likewise, if the applied force is increased to 3**F**, the acceleration is tripled, and so on. From such observations, we conclude that *the acceleration of an object is directly proportional to the resultant force acting on it.*

   As stated in the preceding section, the acceleration of an object also depends on its mass. This can be understood by considering the following set of experiments. If you apply a force **F** to a block of ice on a frictionless surface, the block undergoes some acceleration **a**. If the mass of the block is doubled, the same applied force produces an acceleration a/2. If the mass is tripled, the same applied force produces an acceleration a/3, and so on. According to this observation, we conclude that *the acceleration of an object is inversely proportional to its mass.*

   These observations are summarized in **Newton's second law:**

> The acceleration of an object is directly proportional to the net force acting on it and inversely proportional to its mass.

Newton's second law

Thus we can relate mass and force through the following mathematical statement of Newton's second law[1]:

$$\sum \mathbf{F} = m\mathbf{a} \tag{5.2}$$

You should note that Equation 5.2 is a *vector* expression and hence is equivalent to the following three component equations:

$$\sum F_x = ma_x \qquad \sum F_y = ma_y \qquad \sum F_z = ma_z \tag{5.3}$$

Units of Force and Mass

The SI unit of force is the **newton,** which is defined as the force that, when acting on a 1-kg mass, produces an acceleration of 1 m/s^2. From this definition and Newton's second law, we see that the newton can be expressed in terms of the following fundamental units of mass, length, and time:

Definition of newton

$$1 \text{ N} \equiv 1 \text{ kg} \cdot \text{m/s}^2 \tag{5.4}$$

The unit of force in the cgs system is called the **dyne** and is defined as that force that, when acting on a 1-g mass, produces an acceleration of 1 cm/s^2:

Definition of dyne

$$1 \text{ dyne} \equiv 1 \text{ g} \cdot \text{cm/s}^2 \tag{5.5}$$

In the British engineering system, the unit of force is the **pound,** defined as the force that, when acting on a 1-slug mass,[2] produces an acceleration of 1 ft/s^2:

Definition of pound

$$1 \text{ lb} \equiv 1 \text{ slug} \cdot \text{ft/s}^2 \tag{5.6}$$

Since $1 \text{ kg} = 10^3 \text{ g}$ and $1 \text{ m} = 10^2 \text{ cm}$, it follows that $1 \text{ N} = 10^5$ dynes. It is left as a problem to show that $1 \text{ N} = 0.225$ lb. The units of force, mass, and acceleration are summarized in Table 5.1.

TABLE 5.1   Units of Force, Mass, and Acceleration[a]

System of Units	Mass	Acceleration	Force
SI	kg	m/s^2	$\text{N} = \text{kg} \cdot \text{m/s}^2$
cgs	g	cm/s^2	$\text{dyne} = \text{g} \cdot \text{cm/s}^2$
British engineering	slug	ft/s^2	$\text{lb} = \text{slug} \cdot \text{ft/s}^2$

[a] $1 \text{ N} = 10^5 \text{ dyne} = 0.225 \text{ lb.}$

---

[1] Equation 5.2 is valid only when the speed of the particle is much less than the speed of light. We shall treat the relativistic situation in Chapter 39.

[2] The *slug* is the *unit of mass* in the British engineering system and is that system's counterpart of the SI *kilogram.* When we speak of going on a diet to lose a few pounds, we really mean that we want to lose a few slugs, that is, we want to reduce our mass. When we lose those few slugs, the force of gravity (pounds) on our reduced mass decreases and that is how we "lose a few pounds." Since most of the calculations in our study of classical mechanics are in SI units, the slug is seldom used in this text.

EXAMPLE 5.2   **An Accelerating Hockey Puck**

A hockey puck with a mass of 0.30 kg slides on the horizontal frictionless surface of an ice rink. Two forces act on the puck as shown in Figure 5.4. The force F_1 has a magnitude of 5.0 N, and F_2 has a magnitude of 8.0 N. Determine the magnitude and direction of the puck's acceleration.

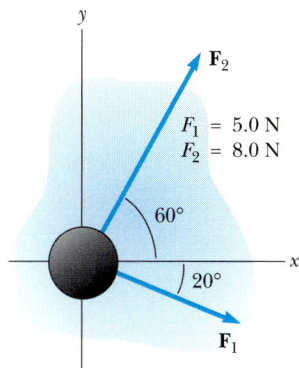

FIGURE 5.4   (Example 5.2) A hockey puck moving on a frictionless surface accelerates in the direction of the *resultant* force, $F_1 + F_2$.

Solution   The resultant force in the *x* direction is

$$\sum F_x = F_{1x} + F_{2x} = F_1 \cos 20° + F_2 \cos 60°$$
$$= (5.0 \text{ N})(0.940) + (8.0 \text{ N})(0.500) = 8.7 \text{ N}$$

The resultant force in the *y* direction is

$$\sum F_y = F_{1y} + F_{2y} = -F_1 \sin 20° + F_2 \sin 60°$$
$$= -(5.0 \text{ N})(0.342) + (8.0 \text{ N})(0.866) = 5.2 \text{ N}$$

Now we use Newton's second law in component form to find the *x* and *y* components of acceleration:

$$a_x = \frac{\sum F_x}{m} = \frac{8.7 \text{ N}}{0.30 \text{ kg}} = 29 \text{ m/s}^2$$

$$a_y = \frac{\sum F_y}{m} = \frac{5.2 \text{ N}}{0.30 \text{ kg}} = 17 \text{ m/s}^2$$

The acceleration has a magnitude of

$$a = \sqrt{(29)^2 + (17)^2} \text{ m/s}^2 = \boxed{34 \text{ m/s}^2}$$

and its direction relative to the positive *x* axis is

$$\theta = \tan^{-1}(a_y/a_x) = \tan^{-1}(17/29) = \boxed{31°}$$

Exercise   Determine the components of a third force that, when applied to the puck, causes it to be in equilibrium.

Answer $F_x = -8.7 \text{ N}, F_y = -5.2 \text{ N}$.

5.5   WEIGHT

We are well aware of the fact that all objects are attracted to the Earth. The force exerted by the Earth on an object is called the **weight** of the object w. This force is directed toward the center of the Earth.[3]

We have seen that a freely falling body experiences an acceleration g acting toward the center of the Earth. Applying Newton's second law to the freely falling body of mass *m*, we have $F = ma$. Because $F = mg$ and also $F = ma$, it follows that $a = g$ and $F = w$, or

$$\mathbf{w} = m\mathbf{g} \qquad (5.7)$$

Since it depends on g, weight varies with geographic location. Bodies weigh less at higher altitudes than at sea level. (You should not confuse the italicized symbol *g* that we use for gravitational acceleration with the symbol g that is used as a symbol for grams.) This is because *g* decreases with increasing distance from the center of the Earth. Hence, weight, unlike mass, is not an inherent property of a body. For example, if a body has a mass of 70 kg, then its weight in a location where $g = 9.80 \text{ m/s}^2$ is $mg = 686 \text{ N}$ (about 154 lb). At the top of a mountain where

---

[3] This statement ignores the fact that the mass distribution of the Earth is not perfectly spherical.

Astronaut Edwin E. Aldrin, Jr., walking on the Moon after the Apollo 11 lunar landing. The weight of the astronaut on the Moon is less than it is on Earth, but his mass remains the same. *(Courtesy of NASA)*

$g = 9.76$ m/s², this weight is 683 N. Therefore, if you want to lose weight without going on a diet, climb a mountain or weigh yourself at 30 000 ft during an airplane flight.

Because $\mathbf{w} = m\mathbf{g}$, we can compare the masses of two bodies by measuring their weights on a spring scale or balance. At a given location, the ratio of the weights of two bodies equals the ratio of their masses.

CONCEPTUAL EXAMPLE 5.3

A baseball of mass m is thrown upward with some initial speed. If air resistance is neglected, what is the force on the ball (a) when it reaches half its maximum height and (b) when it reaches its peak?

Reasoning   The only external force on the ball at *all* points in its trajectory is the downward force of gravity. The magnitude of this force is $w = mg$. For example, if the mass of the baseball is 0.15 kg, $w = 1.47$ N.

5.6   NEWTON'S THIRD LAW

Newton's third law states that if two bodies interact, the force exerted on body 1 by body 2 is equal to and opposite the force exerted on body 2 by body 1:

$$\mathbf{F}_{12} = -\mathbf{F}_{21} \tag{5.8}$$

This law, which is illustrated in Figure 5.5a, is equivalent to stating that *forces always occur in pairs* or that *a single isolated force cannot exist*. The force that body 1 exerts on body 2 is sometimes called the *action force*, while the force body 2 exerts on body 1 is called the *reaction force*. In reality, either force can be labeled the action or reaction force. *The action force is equal in magnitude to the reaction force and opposite in direction. In all cases, the action and reaction forces act on different objects.* For example, the force acting on a freely falling projectile is its weight, $\mathbf{w} = m\mathbf{g}$, which is the force exerted by the Earth on the projectile. The reaction to this force is the force

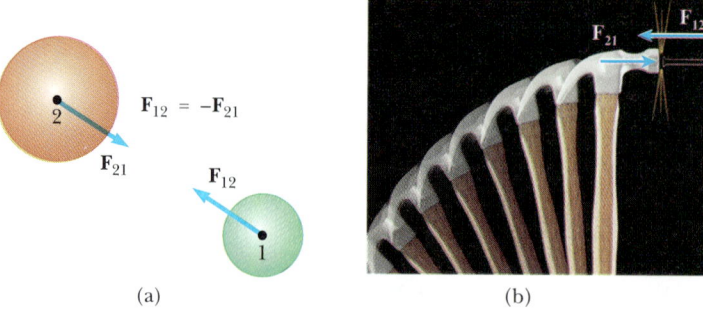

FIGURE 5.5   Newton's third law. (a) The force exerted by object 1 on object 2 is equal to and opposite the force exerted by object 2 on object 1. (b) The force exerted by the hammer on the nail is equal to and opposite the force exerted by the nail on the hammer. *(John Gillmoure, The Stock Market)*

of the projectile on the Earth, $\mathbf{w'} = -\mathbf{w}$. The reaction force, $\mathbf{w'}$, must accelerate the Earth toward the projectile just as the action force, $\mathbf{w}$, accelerates the projectile toward the Earth. However, since the Earth has such a large mass, its acceleration due to this reaction force is negligibly small.

Another example of Newton's third law is shown in Figure 5.5b. The force exerted by the hammer on the nail (the action) is equal to and opposite the force exerted by the nail on the hammer (the reaction). You experience Newton's third law directly whenever you slam your fist against a wall or kick a football. You should be able to identify the action and reaction forces in these cases.

The weight of an object, $\mathbf{w}$, is defined as the force the Earth exerts on the object. If the object is a TV at rest on a table, as in Figure 5.6a, the reaction force to $\mathbf{w}$ is the force the TV exerts on the Earth, $\mathbf{w'}$. The TV does not accelerate because it is held up by the table. The table, therefore, exerts an upward action force, $\mathbf{n}$, on the TV, called the **normal force**.[4] The normal force is the force that prevents the TV from falling through the table and can have any value needed, up to the point of breaking the table. The normal force balances the force of gravity acting on the TV and provides equilibrium. The reaction to $\mathbf{n}$ is the force exerted by the TV on the table, $\mathbf{n'}$. Therefore, we conclude that

$$\mathbf{w} = -\mathbf{w'} \qquad \text{and} \qquad \mathbf{n} = -\mathbf{n'}$$

The forces $\mathbf{n}$ and $\mathbf{n'}$ have the same magnitude, which is the same as $\mathbf{w}$ unless the table has broken. Note that the forces acting on the TV are $\mathbf{w}$ and $\mathbf{n}$, as shown in Figure 5.6b. The two reaction forces, $\mathbf{w'}$ and $\mathbf{n'}$, are exerted on objects other than

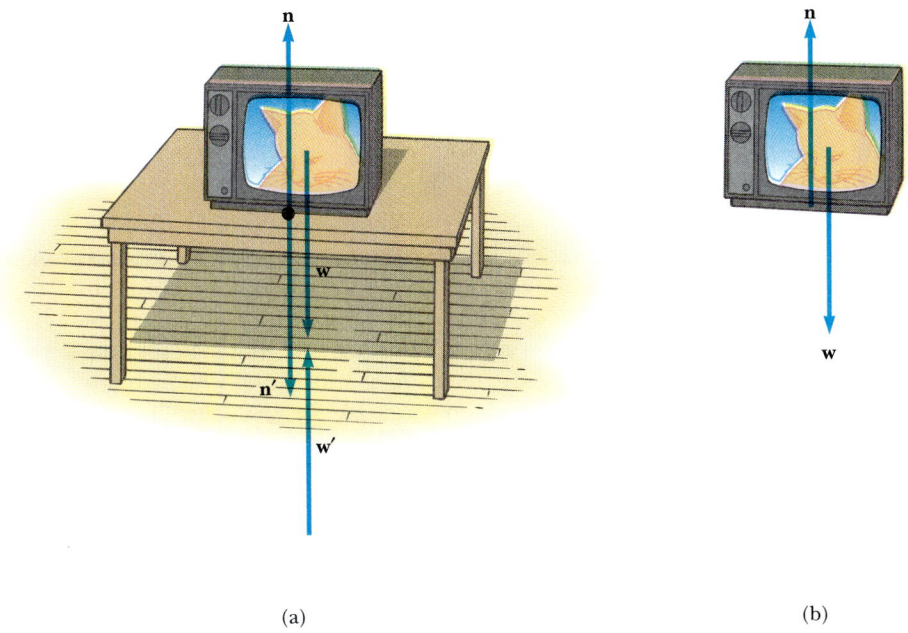

(a)   (b)

FIGURE 5.6   When a TV is at rest on a table, the forces acting on the TV are the normal force, $\mathbf{n}$, and the force of gravity, $\mathbf{w}$, as illustrated in (b). The reaction to $\mathbf{n}$ is the force exerted by the TV on the table, $\mathbf{n'}$. The reaction to $\mathbf{w}$ is the force exerted by the TV on the Earth, $\mathbf{w'}$.

---

[4] *Normal* in this context means *perpendicular*.

the TV. Remember, the two forces in an action-reaction pair always act on two different objects.

From the second law, we see that, because the TV is in equilibrium ($a = 0$), it follows that $w = n = mg$.

CONCEPTUAL EXAMPLE 5.4

If a small sports car collides head-on with a massive truck, which vehicle experiences the greater impact force? Which vehicle experiences the greater acceleration?

Reasoning   The car and truck experience forces that are equal in magnitude but in opposite directions. A calibrated spring scale placed between the colliding vehicles reads the same whichever way it faces. Because the car has the smaller mass, it stops with much greater acceleration.

CONCEPTUAL EXAMPLE 5.5

Is there any relation between the total force acting on an object and the direction in which it moves?

Reasoning   There is no relation between the total force on an object and the direction of its motion before or after the force acts. The force does determine the direction of the *change in motion*. Force describes what the rest of the Universe does to the object. The environment can push the object forward, backward, sideways, or not at all.

5.7   SOME APPLICATIONS OF NEWTON'S LAWS

In this section we apply Newton's laws to objects that are either in equilibrium ($a = 0$) or moving linearly under the action of constant external forces. We assume that the objects behave as particles so that we need not worry about rotational motions. We also neglect the effects of friction for those problems involving motion; this is equivalent to stating that the surfaces are *frictionless*. Finally, we usually neglect the mass of any ropes involved. In this approximation, the magnitude of the force exerted at any point along a rope is the same at all points along the rope. In problem statements, the terms *light* and *of negligible mass* are used to indicate that a mass is to be ignored when you work the problem. These two terms are synonymous in this context.

When we apply Newton's laws to a body, we are interested only in those external forces that *act on the body*. For example, in Figure 5.6 the only external forces acting on the TV are **n** and **w**. The reactions to these forces, **n**′ and **w**′, act on the table and on the Earth, respectively, and therefore do not appear in Newton's second law as applied to the TV.

When an object is being pulled by a rope attached to it, the rope exerts a force on the object. In general, **tension** is a scalar and is defined as the magnitude of the force that the rope exerts on whatever is attached to it.

Consider a crate being pulled to the right on the frictionless, horizontal surface, as in Figure 5.7a. Suppose you are asked to find the acceleration of the crate and the force the floor exerts on it. First, note that the horizontal force being applied to the crate acts through the rope. The force that the rope exerts on the crate is denoted **T**. The magnitude of **T** is equal to the tension in the rope.

Tension

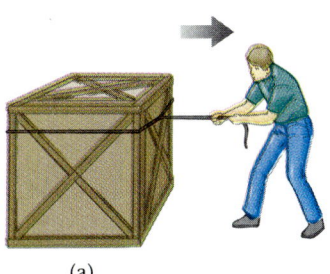

(a)

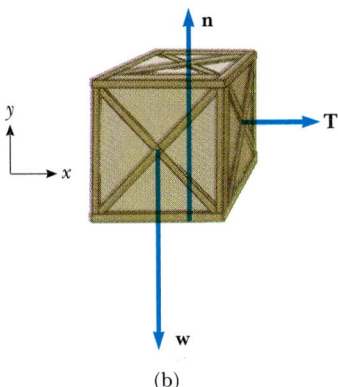

(b)

FIGURE 5.7   (a) A crate being pulled to the right on a frictionless surface. (b) The free-body diagram that represents the external forces acting on the crate.

The forces acting on the crate are illustrated in Figure 5.7b. In addition to the force T, this force diagram for the crate includes the force of gravity, **w**, and the normal force, **n**, exerted by the table on the crate. Such a force diagram is referred to as a **free-body diagram.** The construction of a correct free-body diagram is an important step in applying Newton's laws. The reactions to the forces we have listed, namely, the force exerted by the rope on the hand, the force exerted by the crate on the Earth, and the force exerted by the crate on the floor, are not included in the free-body diagram because they act on *other* bodies and not on the crate.

Free-body diagrams are important when applying Newton's laws

We can now apply Newton's second law in component form to the system. The only force acting in the x direction is T. Applying $\sum F_x = ma_x$ to the horizontal motion gives

$$\sum F_x = T = ma_x \qquad \text{or} \qquad a_x = \frac{T}{m}$$

There is no acceleration in the y direction. Applying $\sum F_y = ma_y$ with $a_y = 0$ gives

$$n - w = 0 \qquad \text{or} \qquad n = w$$

That is, the normal force is equal to and opposite the force of gravity.

If T is a constant force, then the acceleration, $a_x = T/m$, is also a constant. Hence, the equations of kinematics from Chapter 2 can be used to obtain the displacement, Δx, and velocity, v, as functions of time. Since $a_x = T/m = \text{constant}$, these expressions can be written

$$\Delta x = v_0 t + \tfrac{1}{2}\left(\frac{T}{m}\right) t^2$$

$$v = v_0 + \left(\frac{T}{m}\right) t$$

where v_0 is the speed of the crate at $t = 0$.

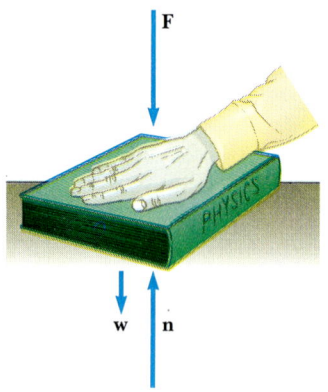

FIGURE 5.8 When one object pushes downward on another object with a force **F**, the normal force **n** is greater than the force of gravity. That is, $n = w + F$.

In the example just presented, the normal force **n** is equal in magnitude and opposite the force of gravity **w**. *This is not always the case.* For example, suppose you were to push down on a book with a force **F** as in Figure 5.8. In this case, the book has no motion in the y direction. Therefore $\Sigma F_y = 0$, which gives $n - w - F = 0$, or $n = w + F$. Other examples in which $n \neq w$ are presented later.

Consider a lamp of weight w suspended from a chain of negligible weight fastened to the ceiling, as in Figure 5.9a. The free-body diagram for the lamp (Fig. 5.9b) shows that the forces on it are the force of gravity **w** acting downward, and the force exerted by the chain on the lamp **T** acting upward.

If we apply the second law to the lamp, noting that $\mathbf{a} = 0$, we see that because there are no forces in the x direction, $\Sigma F_x = 0$ provides no helpful information. The condition $\Sigma F_y = 0$ gives

$$\sum F_y = T - w = 0 \qquad \text{or} \qquad T = w$$

Note that **T** and **w** are *not* an action-reaction pair. The reaction force to **T** is **T′**, the downward force exerted by the lamp on the chain, as in Figure 5.9c. The ceiling exerts an equal and opposite force, **T″**, on the chain.

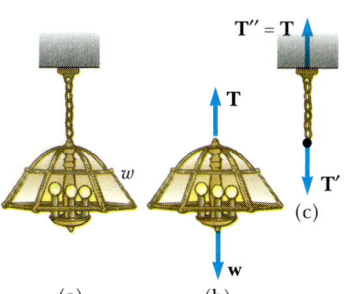

FIGURE 5.9 (a) A lamp of weight w suspended from a ceiling by a chain of negligible mass. (b) The forces acting on the lamp are the force of gravity, **w**, and the force exerted by the chain, **T**. (c) The forces acting on the chain are the force exerted by the lamp, **T′**, and the force exerted by the ceiling, **T″**.

Problem-Solving Strategy
Applying Newton's Laws

The following procedure is recommended when dealing with problems involving the application of Newton's laws:

- Draw a simple, neat diagram of the system.
- Isolate the object whose motion is being analyzed. Draw a free-body diagram for this object, that is, a diagram showing *all external forces acting on the object*. For systems containing more than one object, draw separate free-body diagrams for each object. Do *not* include in the free-body diagram forces that the object exerts on its surroundings.
- Establish convenient coordinate axes for each object and find the components of the forces along these axes. Apply Newton's second law, $\Sigma \mathbf{F} = m\mathbf{a}$, in *component* form. Check your dimensions to make sure that all terms have units of force.
- Solve the component equations for the unknowns. Remember that you must have as many independent equations as you have unknowns in order to obtain a complete solution.
- Check the predictions of your solutions for extreme values of the variables. You can often detect errors in your results by doing so.

EXAMPLE 5.6   A Traffic Light at Rest

A traffic light weighing 125 N hangs from a cable tied to two other cables fastened to a support, as in Figure 5.10a. The upper cables make angles of 37.0° and 53.0° with the horizontal. Find the tension in the three cables.

Reasoning We must construct two free-body diagrams in order to work this problem. The first of these is for the traffic light, shown in Figure 5.10b; the second is for the knot that holds the three cables together, as in Figure 5.10c. This knot is a convenient point to choose because all the forces we are interested in act through this point. Because the acceleration of the system is zero, we can use the condition that the net force on the light is zero, and the net force on the knot is zero.

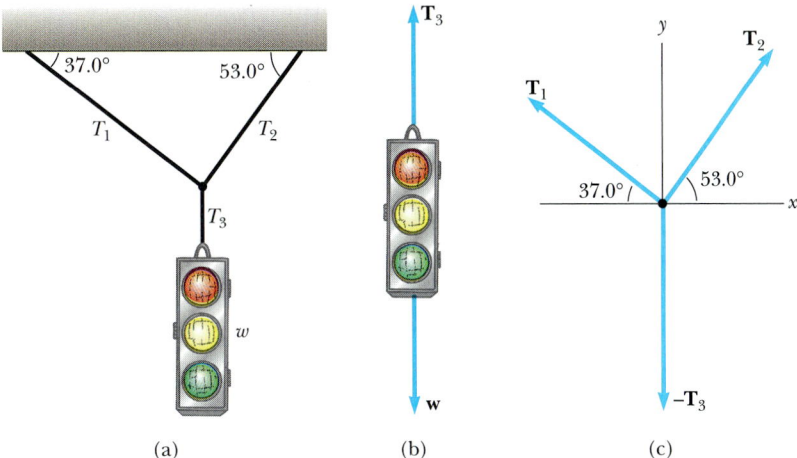

FIGURE 5.10 (Example 5.6) (a) A traffic light suspended by cables. (b) Free-body diagram for the traffic light. (c) Free-body diagram for the knot where the three cables are joined.

Solution First we construct a free-body diagram for the traffic light as in Figure 5.10b. The force exerted by the vertical cable, T_3, supports the light, and so $T_3 = w = 125$ N. Next, we choose the coordinate axes as shown in Figure 5.10c and resolve the forces into their x and y components:

Force	x Component	y Component
T_1	$-T_1 \cos 37.0°$	$T_1 \sin 37.0°$
T_2	$T_2 \cos 53.0°$	$T_2 \sin 53.0°$
T_3	0	-125 N

The first condition for equilibrium gives us the equations

(1) $\quad \sum F_x = T_2 \cos 53.0° - T_1 \cos 37.0° = 0$

(2) $\quad \sum F_y = T_1 \sin 37.0° + T_2 \sin 53.0° - 125 \text{ N} = 0$

From (1) we see that the horizontal components of $\mathbf{T}_1$ and $\mathbf{T}_2$ must be equal in magnitude, and from (2) we see that the sum of the vertical components of $\mathbf{T}_1$ and $\mathbf{T}_2$ must balance the weight of the light. We can solve (1) for T_2 in terms of T_1 to give

$$T_2 = T_1 \left(\frac{\cos 37.0°}{\cos 53.0°} \right) = 1.33 T_1$$

This value for T_2 can be substituted into (2) to give

$$T_1 \sin 37.0° + (1.33 T_1)(\sin 53.0°) - 125 \text{ N} = 0$$

$$T_1 = \boxed{75.1 \text{ N}}$$

$$T_2 = 1.33 T_1 = \boxed{99.9 \text{ N}}$$

Exercise In what situation will $T_1 = T_2$?

Answer When the supporting cables make equal angles with the horizontal support.

EXAMPLE 5.7 Crate on a Frictionless Incline

A crate of mass m is placed on a frictionless, inclined plane of angle θ, as in Figure 5.11a. (a) Determine the acceleration of the crate after it is released.

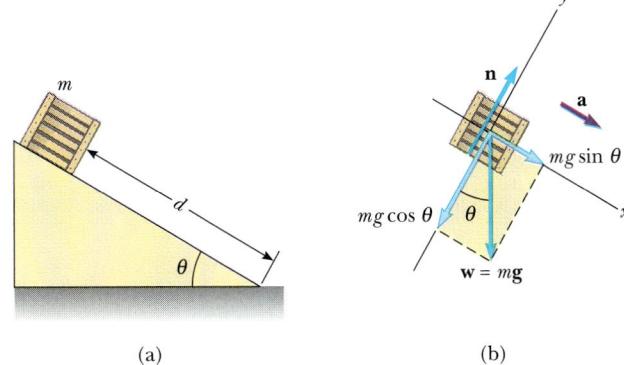

FIGURE 5.11 (Example 5.7) (a) A crate of mass m sliding down a frictionless incline. (b) The free-body diagram for the crate. Note that its acceleration along the incline is $g \sin \theta$.

Reasoning Because the forces acting on the crate are known, Newton's second law can be used to determine its acceleration. First, we construct the free-body diagram for the crate as in Figure 5.11b. The only forces on the crate are the normal force, **n**, acting perpendicular to the plane and the weight, **w**, acting vertically downward. For problems of this type involving inclined planes, *it is convenient to choose the coordinate axes with x along the incline and y perpendicular to it.* Then, we replace the weight vector by a component of magnitude $mg \sin \theta$ along the positive x axis and one of the magnitude $mg \cos \theta$ in the negative y direction.

Solution Applying Newton's second law to the crate in component form while noting that $a_y = 0$ gives

$$(1) \qquad \sum F_x = mg \sin \theta = ma_x$$

$$(2) \qquad \sum F_y = n - mg \cos \theta = 0$$

From (1) we see that the acceleration along the incline is provided by the component of weight directed down the incline:

$$(3) \qquad a_x = g \sin \theta$$

From (2) we conclude that the component of weight perpendicular to the incline is balanced by the normal force; that is, $n = mg \cos \theta$.

Note that the acceleration given by (3) is independent of the mass of the crate! It depends only on the angle of inclination and on g!

Special Cases When $\theta = 90°$, $a = g$ and $n = 0$. This condition corresponds to the crate in free-fall. When $\theta = 0$, $a_x = 0$ and $n = mg$ (its maximum value).

(b) Suppose the crate is released from rest at the top, and the distance from the crate to the bottom is d. How long does it take the crate to reach the bottom, and what is its speed just as it gets there?

Solution Since $a_x = $ constant, we can apply Equation 2.10, $x - x_0 = v_{x0}t + \frac{1}{2}a_xt^2$, to the crate. Since the displacement $x - x_0 = d$ and $v_{x0} = 0$, we get

$$d = \tfrac{1}{2}a_xt^2$$

or

$$(4) \qquad t = \sqrt{\frac{2d}{a_x}} = \sqrt{\frac{2d}{g \sin \theta}}$$

Using Equation 2.12, $v_x^2 = v_{x0}^2 + 2a_x(x - x_0)$, with $v_{x0} = 0$, we find that

$$v_x^2 = 2a_xd$$

or

$$(5) \qquad v_x = \sqrt{2a_xd} = \sqrt{2gd \sin \theta}$$

Again, t and v_x are independent of the mass of the crate. This suggests a simple method of measuring g using an inclined air track or some other frictionless incline. Simply measure the angle of inclination, the distance traveled by the crate, and the time it takes to reach the bottom. The value of g can then be calculated from (4) and (5).

EXAMPLE 5.8 Atwood's Machine

When two unequal masses are hung vertically over a frictionless pulley of negligible mass as in Figure 5.12a, the arrangement is called *Atwood's machine*. The device is sometimes used in the laboratory to measure the gravitational field strength. Determine the magnitude of the acceleration of the two masses and the tension in the string.

Reasoning The free-body diagrams for the two masses are shown in Figure 5.12b. Two forces act on each block: the upward force exerted by the string, **T**, and the downward force of gravity. Thus, the magnitude of the net force exerted on m_1 is $T - m_1g$, while the magnitude of the net force exerted on m_2 is $T - m_2g$. Because the blocks are connected by a string, their accelerations must be equal in magnitude. If we assume that $m_2 > m_1$, then m_1 must accelerate upward, while m_2 must accelerate downward.

Solution When Newton's second law is applied to m_1, with a upward for this mass (because $m_2 > m_1$), we find (taking upward to be the positive y direction)

$$(1) \qquad \sum F_y = T - m_1g = m_1a$$

Similarly, for m_2 we find

$$(2) \qquad \sum F_y = T - m_2g = -m_2a$$

The negative sign on the right-hand side of (2) indicates that m_2 accelerates downward, in the negative y direction.

When (2) is subtracted from (1), T drops out and we get

$$-m_1g + m_2g = m_1a + m_2a$$

or

$$(3) \qquad a = \left(\frac{m_2 - m_1}{m_1 + m_2}\right)g$$

When (3) is substituted into (1), we get

$$(4) \qquad T = \left(\frac{2m_1m_2}{m_1 + m_2}\right)g$$

The result for the acceleration, (3), can be interpreted as the ratio of the unbalanced force on the system to the total mass of the system.

Special Cases When $m_1 = m_2$, $a = 0$ and $T = m_1 g = m_2 g$, as we would expect for this balanced case. If $m_2 \gg m_1$, $a \approx g$ (a freely falling body) and $T \approx 2 m_1 g$.

Exercise Find the magnitude of the acceleration and tension of an Atwood's machine in which $m_1 = 2.00$ kg and $m_2 = 4.00$ kg.

Answer $a = 3.27$ m/s², $T = 26.1$ N.

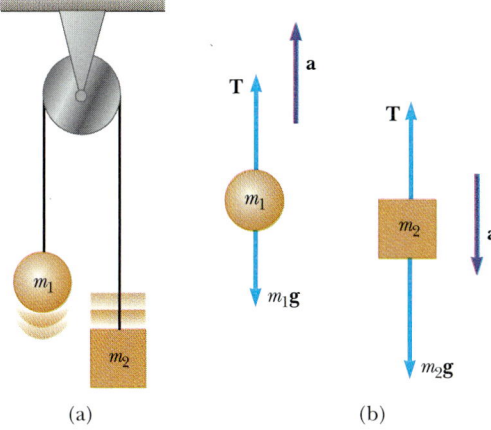

(a)   (b)

FIGURE 5.12 (Example 5.8) Atwood's machine. (a) Two masses ($m_2 > m_1$) connected by a massless string over a frictionless pulley. (b) Free-body diagrams for m_1 and m_2.

EXAMPLE 5.9 Two Connected Objects

Two unequal masses are attached by a lightweight string that passes over a frictionless pulley of negligible mass as in Figure 5.13a. The block of mass m_2 lies on a frictionless incline of angle θ. Find the magnitude of the acceleration of the two masses and the tension in the string.

Reasoning and Solution Since the two masses are connected by a string (which we assume doesn't stretch), they both have accelerations of the same magnitude. The free-body diagrams for the two masses are shown in Figures 5.13b and 5.13c. Applying Newton's second law in component

form to m_1 while assuming that a is upward for this mass (and taking upward to be the positive y direction) gives

$$(1) \qquad \sum F_x = 0$$

$$(2) \qquad \sum F_y = T - m_1 g = m_1 a$$

Note that in order for this mass to accelerate upward, it is necessary that $T > m_1 g$.

For m_2 it is convenient to choose the positive x' axis along the incline as in Figure 5.13c. Here we choose positive acceleration to be down the incline, in the $+x'$ direction. Applying Newton's second law in component form to m_2 gives

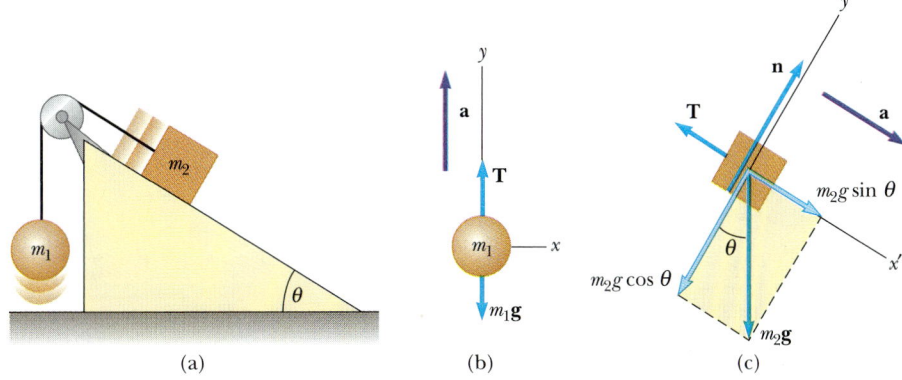

(a)   (b)   (c)

FIGURE 5.13 (Example 5.9) (a) Two masses connected by a massless string over a frictionless pulley. (b) Free-body diagram for m_1. (c) Free-body diagram for m_2 (the incline is frictionless).

(3) $\quad \sum F_{x'} = m_2 g \sin \theta - T = m_2 a$

(4) $\quad \sum F_{y'} = n - m_2 g \cos \theta = 0$

Expressions (1) and (4) provide no information regarding the acceleration. However, if we solve (2) and (3) simultaneously for a and T, we get

(5) $\qquad a = \dfrac{m_2 g \sin \theta - m_1 g}{m_1 + m_2}$

When this is substituted into (2), we find

(6) $\qquad T = \dfrac{m_1 m_2 g (1 + \sin \theta)}{m_1 + m_2}$

Note that m_2 accelerates down the incline only if $m_2 \sin \theta > m_1$ (that is, if a is in the direction we assumed). If $m_1 > m_2 \sin \theta$, the acceleration of m_2 is up the incline and downward for m_1. You should also note that the result for the acceleration, (5), can be interpreted as the resultant unbalanced force on the system divided by the total mass of the system.

Exercise   If $m_1 = 10.0$ kg, $m_2 = 5.00$ kg, and $\theta = 45.0°$, find the acceleration.

Answer $a = -4.22$ m/s², where the negative sign indicates that m_2 accelerates up the incline and m_1 accelerates downward.

EXAMPLE 5.10   One Block Pushes Another

Two blocks of masses m_1 and m_2 are placed in contact with each other on a frictionless, horizontal surface as in Figure 5.14a. A constant horizontal force **F** is applied to m_1 as shown. (a) Find the magnitude of the acceleration of the system.

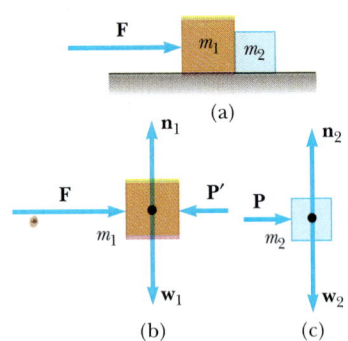

(a)

(b)   (c)

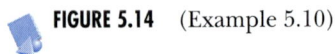

FIGURE 5.14   (Example 5.10).

Reasoning and Solution   Both blocks must experience the same acceleration because they are in contact with each other. Because **F** is the only horizontal force on the system (the two blocks), we have

$$\sum F_x \text{ (system)} = F = (m_1 + m_2) a$$

(1) $\qquad a = \dfrac{F}{m_1 + m_2}$

(b) Determine the magnitude of the contact force between the two blocks.

Reasoning and Solution   To solve this part of the problem, it is necessary to first construct a free-body diagram for

each block, as shown in Figures 5.14b and 5.14c, where the contact force is denoted by **P**. From Figure 5.14c, we see that the only horizontal force acting on m_2 is the contact force **P** (the force exerted by m_1 on m_2), which is directed to the right. Applying Newton's second law to m_2 gives

(2) $\qquad \sum F_x = P = m_2 a$

Substituting the value of a given by (1) into (2) gives

(3) $\qquad P = m_2 a = \left(\dfrac{m_2}{m_1 + m_2} \right) F$

From this result, we see that the contact force **P** is less than the applied force **F**. This is consistent with the fact that the force required to accelerate m_2 alone must be less than the force required to produce the same acceleration for the system of two blocks.

It is instructive to check this expression for P by considering the forces acting on m_1, as shown in Figure 15.14b. In this case, the horizontal forces acting on m_1 are the applied force **F** to the right, and the contact force **P'** to the left (the force exerted by m_2 on m_1). From Newton's third law, **P'** is the reaction to **P**, so that $|\mathbf{P'}| = |\mathbf{P}|$. Applying Newton's second law to m_1 gives

(4) $\qquad \sum F_x = F - P' = F - P = m_1 a$

Substituting the value of a from (1) into (4) gives

$$P = F - m_1 a = F - \dfrac{m_1 F}{m_1 + m_2} = \left(\dfrac{m_2}{m_1 + m_2} \right) F$$

This agrees with (3), as it must.

Exercise   If $m_1 = 4.00$ kg, $m_2 = 3.00$ kg, and $F = 9.00$ N, find the magnitude of the acceleration of the system and that of the contact force.

Answer $a = 1.29$ m/s², $P = 3.86$ N.

EXAMPLE 5.11   Weighing a Fish in an Elevator

A person weighs a fish of mass m on a spring scale attached to the ceiling of an elevator, as shown in Figure 5.15. Show that if the elevator accelerates in either direction, the spring scale gives a reading different from the weight of the fish.

Reasoning   The external forces acting on the fish are the downward force of gravity, $\mathbf{w}$, and the upward force, $\mathbf{T}$, exerted on it by the scale. By Newton's third law, the tension T is also the reading of the spring scale. If the elevator is either at rest or moving at constant velocity, then the fish is not accelerating and $T = w = mg$. However, if the elevator accelerates in either direction, the tension is no longer equal to the weight of the fish.

Solution   If the elevator accelerates upward with an acceleration $\mathbf{a}$ relative to an observer outside the elevator in an inertial frame (Fig. 5.15a), then the second law applied to the fish gives the total force on the fish:

$$(1) \qquad \sum F = T - w = ma \qquad \text{(if } \mathbf{a} \text{ is upward)}$$

If the elevator accelerates downward as in Figure 5.15b, Newton's second law applied to the fish becomes

$$(2) \qquad \sum F = T - w = -ma \qquad \text{(if } \mathbf{a} \text{ is downward)}$$

Thus, we conclude from (1) that the scale reading T is greater than the weight, w, if $\mathbf{a}$ is upward. From (2) we see that T is less than w if $\mathbf{a}$ is downward.

For example, if the weight of the fish is 40.0 N, and a is 2.00 m/s^2 *upward*, then the scale reading is

$$T = ma + mg = mg\left(\frac{a}{g} + 1\right)$$

$$= W\left(\frac{a}{g} + 1\right) = (40.0 \text{ N})\left(\frac{2.00 \text{ m/s}^2}{9.80 \text{ m/s}^2} + 1\right)$$

$$= \boxed{48.2 \text{ N}}$$

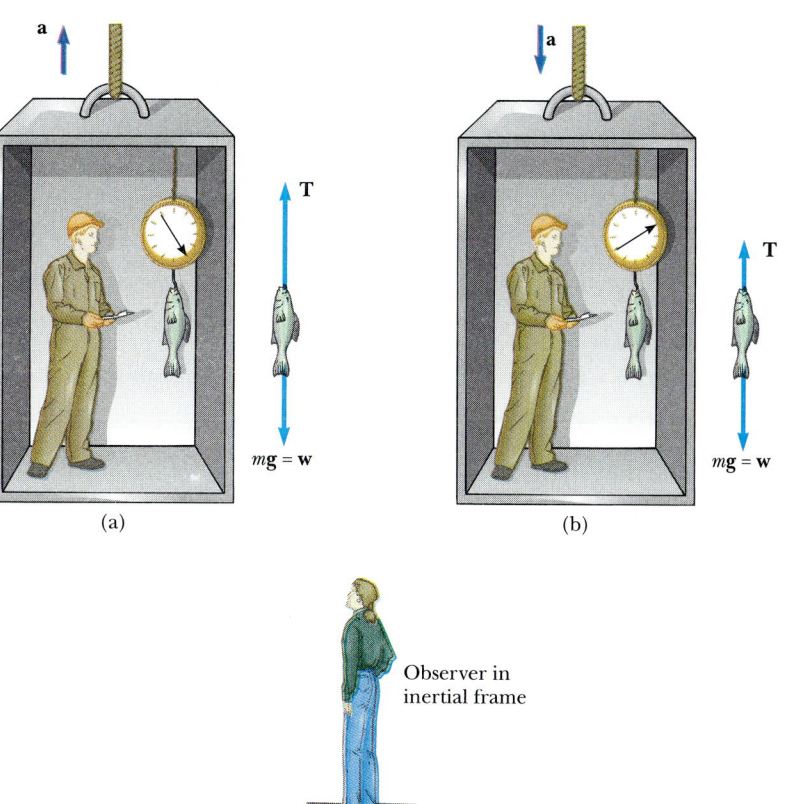

(a)   (b)

Observer in
inertial frame

FIGURE 5.15   (Example 5.11) Apparent weight versus true weight. (a) When the elevator accelerates *upward* the spring scale reads a value *greater* than the weight of the fish. (b) When the elevator accelerates *downward* the spring scale reads a value *less* than the weight of the fish. When the elevator is accelerating, the spring scale reads the *apparent weight*.

If a is 2.00 m/s² *downward,* then

$$T = -ma + mg = mg\left(1 - \frac{a}{g}\right)$$

$$= W\left(1 - \frac{a}{g}\right) = (40.0 \text{ N})\left(1 - \frac{2.00 \text{ m/s}^2}{9.80 \text{ m/s}^2}\right)$$

$$= \boxed{31.8 \text{ N}}$$

Hence, if you buy a fish by weight in an elevator, make sure the fish is weighed while the elevator is at rest or acceler-

ating downward! Furthermore, note that from the information given here, one cannot determine the direction of motion of the elevator.

Special Cases   If the elevator cable breaks, then the elevator falls freely and $a = g$. Since $w = mg$, we see from (1) that the scale reading, T, is zero in this case; that is, the fish appears to be weightless. If the elevator accelerates downward with an acceleration greater than g, the fish (along with the person in the elevator) eventually hits the ceiling since the acceleration of the fish and person is still that of a freely falling body relative to an outside observer.

5.8   FORCES OF FRICTION

When a body is in motion either on a surface or through a viscous medium such as air or water, there is resistance to the motion because the body interacts with its surroundings. We call such resistance a **force of friction.** Forces of friction are very important in our everyday lives. They allow us to walk or run and are necessary for the motion of wheeled vehicles.

Consider a block on a horizontal table, as in Figure 5.16a. If we apply an external horizontal force **F** to the block, acting to the right, the block remains stationary if **F** is not too large. The force that counteracts **F** and keeps the block from moving acts to the left and is called the *frictional force,* **f**. As long as the block is not moving, $f = F$. Since the block is stationary, we call this frictional force the *force of static friction,* $\mathbf{f}_s$. Experiments show that this force arises from contacting points that protrude beyond the general level of the surfaces, even for surfaces that are apparently very smooth, as in Figure 5.16a. (If the surfaces are clean and smooth at the atomic level, they are likely to weld together when contact is made.) The frictional force arises in part from one peak physically blocking the motion of a peak from the opposing surface and in part from chemical bonding of opposing points as they come into contact. If the surfaces are rough, bouncing is likely to occur, further complicating the analysis. Although the details of friction are quite complex at the atomic level, it ultimately involves the electrostatic force between atoms or molecules.

If we increase the magnitude of **F**, as in Figure 5.16b, the block eventually slips. When the block is on the verge of slipping, f_s is a maximum as shown by the graph in Figure 5.16c. When F exceeds $f_{s,\text{max}}$, the block moves and accelerates to the right. When the block is in motion, the retarding frictional force becomes less than $f_{s,\text{max}}$ (Fig. 5.16c). When the block is in motion, we call the retarding force the *force of kinetic friction,* $\mathbf{f}_k$. The unbalanced force in the $+x$ direction, $F - f_k$, accelerates the block to the right. If $F = f_k$, the block moves to the right with constant speed. If the applied force **F** is removed, then the frictional force **f** acting to the left accelerates the block in the $-x$ direction and eventually brings it to rest.

Experimentally, one finds that, to a good approximation, both $f_{s,\text{max}}$ and f_k are *proportional to the normal force acting on the block.* The experimental observations can be summarized by the following empirical laws of friction:

- The direction of the force of static friction between any two surfaces in contact is opposite the direction of any applied force and can have values

Force of static friction

Force of kinetic friction

$$f_s \leq \mu_s n \tag{5.9}$$

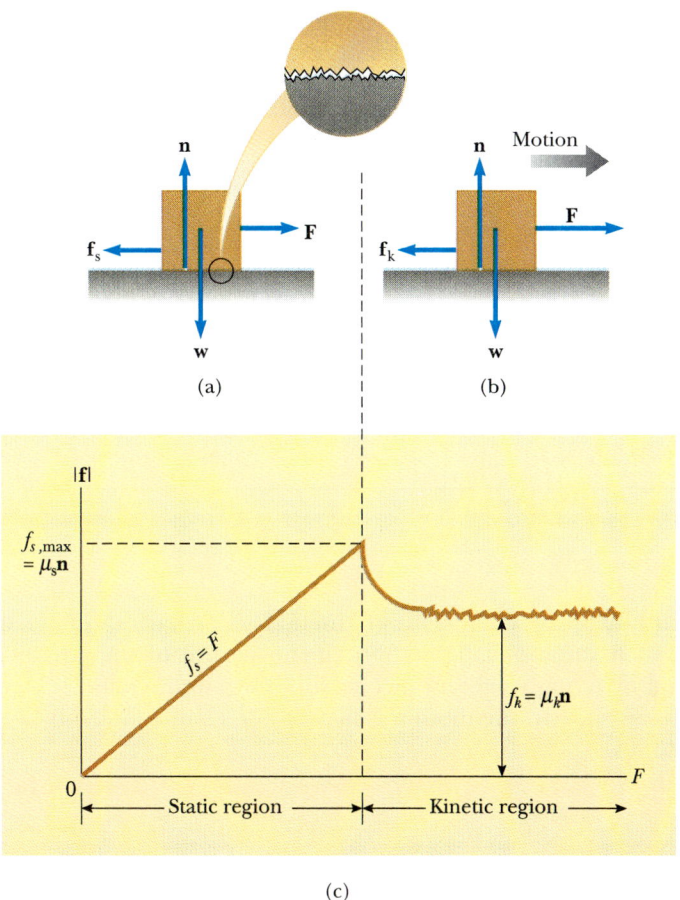

FIGURE 5.16   The direction of the force of friction, **f**, between a block and a rough surface is opposite the direction of the applied force, **F**. Due to the roughness of the two surfaces, contact is made only at a few points as illustrated in the "magnified" view. (a) The magnitude of the force of static friction equals the applied force. (b) When the magnitude of the applied force exceeds the force of kinetic friction, the block accelerates to the right. (c) A graph of the frictional force versus the applied force. Note that $f_{s,\text{max}} > f_k$.

where the dimensionless constant μ_s is called the **coefficient of static friction** and n is the magnitude of the normal force. The equality in Equation 5.9 holds when the block is on the verge of slipping, that is, when $f_s = f_{s,\text{max}} = \mu_s n$. The inequality holds when the applied force is less than this value.

- The direction of the force of kinetic friction acting on an object is opposite the direction of its motion and is given by

$$f_k = \mu_k n \tag{5.10}$$

where μ_k is the **coefficient of kinetic friction**.

- The values of μ_k and μ_s depend on the nature of the surfaces, but μ_k is generally less than μ_s. Typical values of μ range from around 0.05 to 1.5. Table 5.2 lists some reported values.
- The coefficients of friction are nearly independent of the area of contact between the surfaces.

TABLE 5.2   Coefficients of Friction[a]

	μ_s	μ_k
Steel on steel	0.74	0.57
Aluminum on steel	0.61	0.47
Copper on steel	0.53	0.36
Rubber on concrete	1.0	0.8
Wood on wood	0.25–0.5	0.2
Glass on glass	0.94	0.4
Waxed wood on wet snow	0.14	0.1
Waxed wood on dry snow	—	0.04
Metal on metal (lubricated)	0.15	0.06
Ice on ice	0.1	0.03
Teflon on Teflon	0.04	0.04
Synovial joints in humans	0.01	0.003

[a] All values are approximate.

Finally, although the coefficient of kinetic friction varies with speed, we shall neglect any such variations. The approximate nature of the equations is easily demonstrated by trying to get a block to slip down an incline at constant speed. Especially at low speeds, the motion is likely to be characterized by alternate *stick* and *slip* episodes.

CONCEPTUAL EXAMPLE 5.12

A horse pulls a sled with a horizontal force, causing it to accelerate as in Figure 5.17a. Newton's third law says that the sled exerts an equal and opposite force on the horse. In view of this, how can the sled accelerate? Under what condition does the system (horse plus sled) move with constant velocity?

Reasoning The motion of any object is determined by the external forces that act on it. In this situation, the horizontal forces exerted on the sled are the forward force exerted by the horse and the backward force of friction between sled and surface (Fig. 5.17b). When the forward force exerted on the sled exceeds the backward force, the resultant force on it is in the forward direction. This resultant force causes the sled to accelerate to the right. The horizontal forces that act on the horse are the forward force of friction between horse and surface and the backward force of the sled (Fig. 5.17c). The resultant of these two forces causes the horse to accelerate. When the forward force of friction acting on the horse balances the backward force of the sled, the system moves with constant velocity.

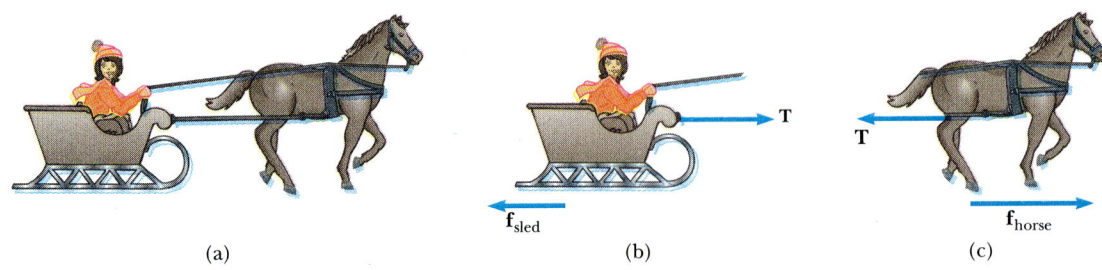

(a)   (b)   (c)

FIGURE 5.17 (Conceptual Example 5.12).

EXAMPLE 5.13   Experimental Determination of μ_s and μ_k

In this example we describe a simple method of measuring coefficients of friction. Suppose a block is placed on a rough surface inclined relative to the horizontal, as in Figure 5.18. The angle of the inclined plane is increased until the block slips. Let us show that by measuring the critical angle θ_c at which this slipping just occurs, we can obtain μ_s directly.

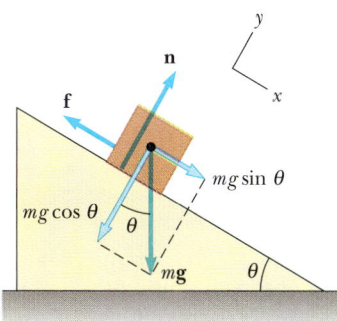

FIGURE 5.18   (Example 5.13) The external forces exerted on a block lying on a rough incline are the force of gravity, mg, the normal force, n, and the force of friction, f. For convenience, the force of gravity is resolved into a component along the incline, $mg \sin \theta$, and a component perpendicular to the incline, $mg \cos \theta$.

Solution   The only forces acting on the block are the force of gravity, mg, the normal force, n, and the force of static friction, f_s. When we take x parallel to the plane and y per-

pendicular to it, Newton's second law applied to the block gives

Static case: (1) $\sum F_x = mg \sin \theta - f_s = 0$

                (2) $\sum F_y = n - mg \cos \theta = 0$

We can eliminate mg by substituting $mg = n/\cos \theta$ from (2) into (1) to get

(3) $f_s = mg \sin \theta$

$$= \left(\frac{n}{\cos \theta} \right) \sin \theta = n \tan \theta$$

When the inclined plane is at the critical angle, θ_c, $f_s = f_{s,\,max} = \mu_s n$, and so at this angle, (3) becomes

$$\mu_s n = n \tan \theta_c$$

Static case: $\mu_s = \tan \theta_c$

For example, if the block just slips at $\theta_c = 20°$, then $\mu_s = \tan 20° = 0.364$. Once the crate starts to move at $\theta \geq \theta_c$, it accelerates down the incline and the force of friction is $f_k = \mu_k n$. However, if θ is reduced below θ_c, it may be possible to find an angle θ_c' such that the block moves down the incline with constant speed ($a_x = 0$). In this case, using (1) and (2) with f_s replaced by f_k gives

Kinetic case: $\mu_k = \tan \theta_c'$

where $\theta_c' < \theta_c$.

You should try this simple experiment using a coin as the block and a notebook as the inclined plane. Also, you can try taping two coins together to prove that you still get the same critical angles as with one coin.

EXAMPLE 5.14   The Sliding Hockey Puck

A hockey puck on a frozen pond is hit and given an initial speed of 20.0 m/s. If the puck always remains on the ice and slides 115 m before coming to rest, determine the coefficient of kinetic friction between the puck and the ice.

Reasoning   The forces acting on the puck after it is in motion are shown in Figure 5.19. If we assume that the force of friction, f_k, remains constant, then this force produces a uniform negative acceleration of the puck. First, we find the acceleration using Newton's second law. Knowing the acceleration of the puck and the distance it travels, we can then use kinematics to find the coefficient of kinetic friction.

Solution   Applying Newton's second law in component form to the puck gives

(1) $\sum F_x = -f_k = ma$

(2) $\sum F_y = n - mg = 0$ ($a_y = 0$)

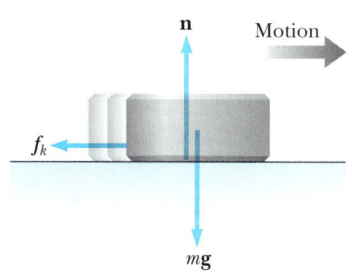

FIGURE 5.19   (Example 5.14) *After* the puck is given an initial velocity to the right, the only external forces acting on it are the force of gravity, mg, the normal force, n, and the force of kinetic friction, f_k.

But $f_k = \mu_k n$, and from (2) we see that $n = mg$. Therefore, (1) becomes

$$-\mu_k n = -\mu_k mg = ma$$

$$a = -\mu_k g$$

The negative sign means that the acceleration is to the left, corresponding to a negative acceleration of the puck. The acceleration is independent of the mass of the puck and is constant because we assume that μ_k remains constant.

Since the acceleration is constant, we can use Equation 2.12, $v^2 = v_0^2 + 2ax$, with the final speed $v = 0$.

$$v_0^2 + 2ax = v_0^2 - 2\mu_k gx = 0$$

$$\mu_k = \frac{v_0^2}{2gx}$$

$$\mu_k = \frac{(20.0 \text{ m/s})^2}{2(9.80 \text{ m/s}^2)(115 \text{ m})} = 0.177$$

Note that μ_k has no dimensions.

EXAMPLE 5.15   Connected Objects with Friction

A mass m_1 on a rough, horizontal surface is connected to a second mass m_2 by a lightweight cord over a lightweight, frictionless pulley as in Figure 5.20a. A force of magnitude F at an angle θ with the horizontal is applied to m_1 as shown. The coefficient of kinetic friction between m_1 and the surface is μ. Determine the magnitude of the acceleration of the masses and the tension in the cord.

Reasoning   First we draw the free-body diagrams of m_1 and m_2 as in Figures 5.20b and 5.20c. Next, we apply Newton's second law in component form to each block, and make use of the fact that the magnitude of the force of kinetic friction is proportional to the normal force according to $f_k = \mu n$. Finally we solve for the acceleration in terms of the parameters given.

Solution   The applied force F has components $F_x = F \cos \theta$ and $F_y = F \sin \theta$. Applying Newton's second law to both masses and assuming the motion of m_1 is to the right, we get

Motion of m_1: $\sum F_x = F \cos \theta - f_k - T = m_1 a$

(1) $\sum F_y = n + F \sin \theta - m_1 g = 0$

Motion of m_2: $\sum F_x = 0$

(2) $\sum F_y = T - m_2 g = m_2 a$

But $f_k = \mu n$, and from (1), $n = m_1 g - F \sin \theta$ (note that in this case n is *not* equal to $m_1 g$); therefore

(3) $f_k = \mu(m_1 g - F \sin \theta)$

That is, the frictional force is reduced because of the positive y component of $\mathbf{F}$. Substituting (3) and the value of T from (2) into (1) gives

$$F \cos \theta - \mu(m_1 g - F \sin \theta) - m_2(a + g) = m_1 a$$

Solving for a, we get

(4) $$a = \frac{F(\cos \theta + \mu \sin \theta) - g(m_2 + \mu m_1)}{m_1 + m_2}$$

We can find T by substituting this value of a into (2).

Note that the acceleration for m_1 can be either to the right or to the left,[5] depending on the sign of the numerator in (4). If the motion of m_1 is to the left, we must reverse the sign of f_k because the frictional force must oppose the motion. In this case, the value of a is the same as in (4) with μ replaced by $-\mu$.

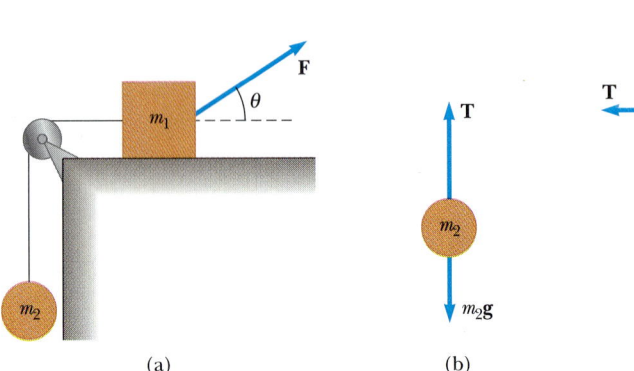

(a)   (b)   (c)

FIGURE 5.20   (Example 5.15) (a) The external force, $\mathbf{F}$, applied as shown can cause m_1 to accelerate to the right. (b) and (c) The free-body diagrams assuming that m_1 accelerates to the right while m_2 accelerates upward. The magnitude of the force of kinetic friction in this case is given by $f_k = \mu_k n = \mu_k(m_1 g - F \sin \theta)$.

[5] Equation (4) shows that, when $\mu m_1 > m_2$, there is a range of values of F for which no motion occurs at a given angle θ.

SUMMARY

Newton's first law states that a body at rest remains at rest and a body in uniform motion in a straight line maintains that motion unless an external resultant force acts on it. An **inertial frame** is one that is not accelerated.

Newton's second law states that the acceleration of an object is directly proportional to the resultant force acting on it and inversely proportional to its mass. If the mass of the body is constant, the net force equals the product of the mass and its acceleration, or $\Sigma \mathbf{F} = m\mathbf{a}$.

The **weight** of a body is equal to the product of its mass (a scalar quantity) and the free-fall acceleration, or $\mathbf{w} = m\mathbf{g}$.

Newton's third law states that if two bodies interact, the force exerted on body 1 by body 2 is equal to and opposite the force exerted on body 2 by body 1. Thus, an isolated force cannot exist in nature.

The **maximum force of static friction**, $f_{s,\max}$ between an object and a surface is proportional to the normal force acting on the object. In general, $f_s \leq \mu_s n$, where μ_s is the coefficient of static friction and $\mathbf{n}$ is the normal force. When an object slides over a surface, the force of kinetic friction, $\mathbf{f}_k$, is opposite the motion and is also proportional to the normal force. The magnitude of this force is given by $f_k = \mu_k n$, where μ_k is the coefficient of kinetic friction. Usually, $\mu_k < \mu_s$.

More on Free-Body Diagrams

In order to be successful in applying Newton's second law to a mechanical system you must first be able to recognize all the forces acting on the system. That is, you must be able to construct the correct free-body diagram. The importance of constructing the free-body diagram cannot be overemphasized. In Figure 5.21 a number of mechanical systems are presented together with their corresponding free-body diagrams. You should examine these carefully and then construct free-body diagrams for other systems described in the problems. When a system contains more than one element, it is important that you construct a free-body diagram for *each* element.

As usual, F denotes some applied force, $\mathbf{w} = m\mathbf{g}$ is the force of gravity, $\mathbf{n}$ denotes a normal force, $\mathbf{f}$ is the force of friction, and $\mathbf{T}$ is the force of the string on the object.

QUESTIONS

1. If gold were sold by weight, would you rather buy it in Denver or in Death Valley? If sold by mass, at which of the two locations would you prefer to buy it? Why?

2. A passenger sitting in the rear of a bus claims that he was injured when the driver slammed on the brakes, causing a suitcase to come flying toward the passenger from the front of the bus. If you were the judge in this case, what disposition would you make? Why?

3. A space explorer is in a spaceship moving through space far from any planet or star. She notices a large rock, taken as a specimen from an alien planet, floating around the cabin of the spaceship. Should she push it gently toward a storage compartment or kick it toward the compartment? Why?

4. How much does an astronaut weigh out in space, far from any planet?

5. A massive metal object on a rough metal surface may undergo contact welding to that surface. Discuss how this affects the frictional force between the object and the surface.

6. The observer in the elevator of Example 5.11 would claim that the "weight" of the fish is T, the scale reading. This

(continued on page 131)

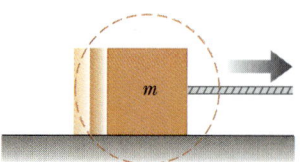

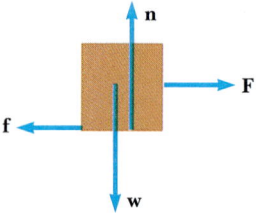

A block pulled to the right on a
rough, horizontal surface

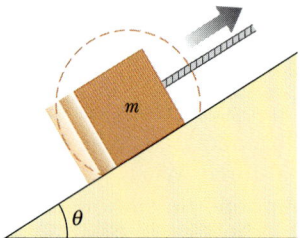

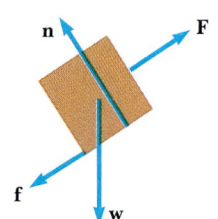

A block pulled up a rough incline

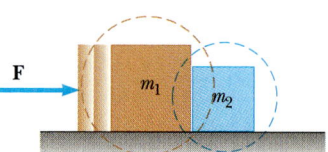

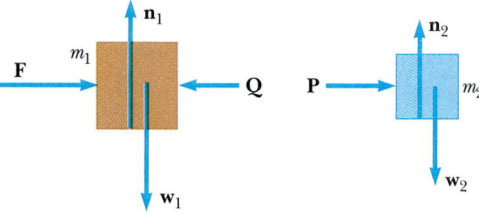

Two blocks in contact, pushed to the
right on a frictionless surface

Note: **P** = −**Q** because they are an action-reaction pair

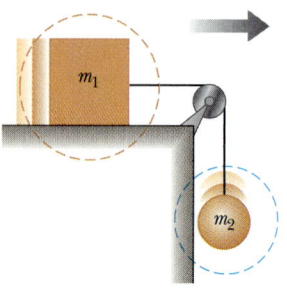

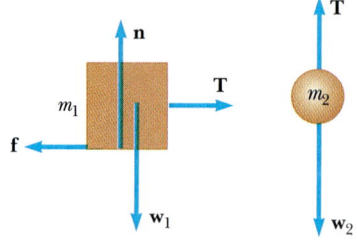

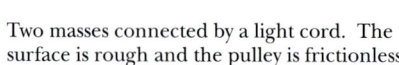

Two masses connected by a light cord. The
surface is rough and the pulley is frictionless

FIGURE 5.21   Various mechanical configurations *(left)* and the corresponding free-body diagrams *(right)*. The term *rough* here means only that the surface is not frictionless.

claim is obviously wrong. Why does this observation differ from that of a person outside the elevator at rest relative to the elevator?

7. Identify the action-reaction pairs in the following situations: a man takes a step; a snowball hits a woman in the back; a baseball player catches a ball; a gust of wind strikes a window.

8. A ball is held in a person's hand. (a) Identify all the external forces acting on the ball and the reaction to each. (b) If the ball is dropped, what force is exerted on it while it is falling? Identify the reaction force in this case. (Neglect air resistance.)

9. If a car is traveling westward with a constant speed of 20 m/s, what is the resultant force acting on it?

10. A large crate is placed on the bed of a truck without being tied to the truck. (a) As the truck accelerates forward, the crate remains at rest relative to the truck. What force causes the crate to accelerate? (b) If the truck driver slams on the brakes, what could happen to the crate?

11. A rubber ball is dropped onto the floor. What force causes the ball to bounce?

12. What is wrong with the statement, "Since the car is at rest, there are no forces acting on it"? How would you correct this sentence?

13. Suppose you are driving a car along a highway at a high speed. Why should you avoid slamming on your brakes if you want to stop in the shortest distance?

14. If you have ever taken a ride in an elevator of a high-rise building, you may have experienced the nauseating sensation of "heaviness" and "lightness" depending on the direction of acceleration. Explain these sensations. Are we truly weightless in free-fall?

15. The driver of a speeding empty truck slams on the brakes and skids to a stop through a distance *d*. (a) If the truck carried a heavy load such that its mass were doubled, what would be its skidding distance? (b) If the initial speed of the truck is halved, what would be its skidding distance?

16. Does it make sense to say that an object possesses force? Explain.

17. In an attempt to define Newton's third law, a student states that the action and reaction forces are equal to and opposite each other. If this is the case, how can there ever be a net force on an object?

18. In a tug-of-war between two athletes, each pulls on the rope with a force of 200 N. What is the tension in the rope?

19. If you push on a heavy box that is at rest, it requires some force F to start its motion. However, once it is sliding, it requires a smaller force to maintain that motion. Why?

20. What causes a rotary lawn sprinkler to turn?

21. The force of gravity is twice as great on a 20-N rock as on a 10-N rock. Why doesn't the 20-N rock have a greater free-fall acceleration?

PROBLEMS

Review Problem

Consider the three connected blocks shown in the diagram. If the inclined plane is frictionless, and the system is in equilibrium, find (in terms of m, g and θ) (a) the mass M and (b) the tensions T_1 and T_2. If the suspended mass is double that value found in part (a), find (c) the acceleration of each block, and (d) the tensions T_1 and T_2. If the coefficient of static friction between m and $2m$ and the inclined plane is μ_s, and the sys-

tem is in equilibrium, find (e) the minimum value of M and (f) the maximum value of M. (g) Compare the values of T_2 when M has its minimum and maximum values.

Section 5.2 through Section 5.6

1. A force, **F**, applied to an object of mass m_1 produces an acceleration of 3.00 m/s². The same force applied to an object of mass m_2 produces an acceleration of 1.00 m/s². (a) What is the value of the ratio m_1/m_2? (b) If m_1 and m_2 are combined, find their acceleration under the action of **F**.

2. Three forces, given by $F_1 = (-2.00i + 2.00j)$ N, $F_2 = (5.00i - 3.00j)$ N, and $F_3 = (-45.0i)$ N act on an object to give it an acceleration of magnitude 3.75 m/s². (a) What is the direction of the acceleration? (b) What is the mass of the object? (c) If the object is initially at rest, what is its speed after 10.0 s? (d) What are the velocity components of the object after 10.0 s?

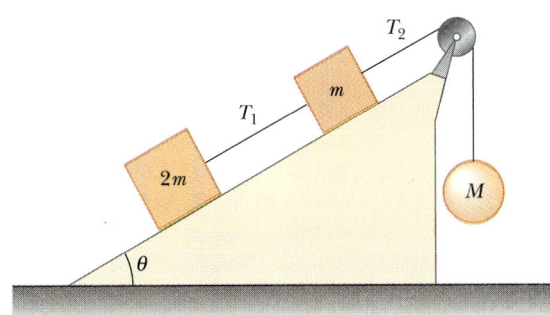

☐ indicates problems that have full solutions available in the Student Solution Manual and Study Guide.

3. A time-dependent force, $F = (8.00i - 4.00tj)$ N (where t is in seconds), is applied to a 2.00-kg object initially at rest. (a) At what time will the object be moving with a speed of 15.0 m/s? (b) How far is the object from its initial position when its speed is 15.0 m/s? (c) Through what total displacement has the object traveled at this time?

4. A 3.00-kg particle starts from rest and moves a distance of 4.00 m in 2.00 s under the action of a single, constant force. Find the magnitude of the force.

5. A 5.0-g bullet leaves the muzzle of a rifle with a speed of 320 m/s. What average force is exerted on the bullet while it is traveling down the 0.82-m-long barrel of the rifle?

6. A pitcher releases a baseball of weight 1.4 N with a speed of 32 m/s by uniformly accelerating his arm for 0.090 s. If the ball starts from rest, (a) through what distance does it accelerate before release? (b) What average force is exerted on it to produce this acceleration?

6A. A pitcher releases a baseball of weight w with a speed of v by uniformly accelerating his arm for a time t. If the ball starts from rest, (a) through what distance does it accelerate before release? (b) What average force is exerted on it to produce this acceleration?

7. A 3.0-kg mass undergoes an acceleration given by $a = (2.0i + 5.0j)$ m/s². Find the resultant force, F, and its magnitude.

8. A freight train has mass of 1.5×10^7 kg. If the locomotive can exert a constant pull of 7.5×10^5 N, how long does it take to increase the speed of the train from rest to 80 km/h?

9. A person weighs 125 lb. Determine (a) her weight in newtons and (b) her mass in kilograms.

10. If the Earth's gravitational force causes a falling 60-kg student to accelerate downward at 9.8 m/s², determine the upward acceleration of the Earth during the student's fall. Take the mass of the Earth to be 5.98×10^{24} kg.

11. The average speed of a nitrogen molecule in air is about 6.7×10^2 m/s, and its mass is about 4.68×10^{-26} kg. (a) If it takes 3.0×10^{-13} s for a nitrogen molecule to hit a wall and rebound with the same speed but in an opposite direction, what is the average acceleration of the molecule during this time interval? (b) What average force does the molecule exert on the wall?

12. If a man weighs 875 N on Earth, what would he weigh on Jupiter, where the free-fall acceleration is 25.9 m/s²?

13. On planet X, an object weighs 12 N. On planet B where the magnitude of the free-fall acceleration is $1.6g$, the object weighs 27 N. What is the mass of the object and what is the free-fall acceleration (in m/s²) on planet X?

14. One or more external forces are exerted on each object enclosed in a dashed box shown in Figure 5.1. Identify the reaction to each of these forces.

15. A brick of weight w rests on top of a vertical spring of weight w_s. The spring rests on a table. (a) Draw a free-body diagram of the brick and label all forces acting on it. (b) Repeat (a) for the spring. (c) Identify all action-reaction pairs in the brick-spring-table-Earth system.

16. Forces of 10.0 N north, 20.0 N east, and 15.0 N south are simultaneously applied to a 4.00-kg mass. Obtain its acceleration.

17. A fire helicopter carries a 620-kg bucket of water at the end of a 20-m-long cable. Flying back from a fire at a constant speed of 40 m/s, the cable makes an angle of 40.0° with respect to the vertical. (a) Determine the force of air resistance on the bucket. (b) After filling the bucket with sea water, the helicopter returns to the fire at the same speed with the bucket now making an angle of 7.0° with the vertical. What is the mass of the water in the bucket?

18. Two forces F_1 and F_2 act on a 5.00-kg mass. If $F_1 = 20.0$ N and $F_2 = 15.0$ N, find the acceleration in (a) and (b) of Figure P5.18.

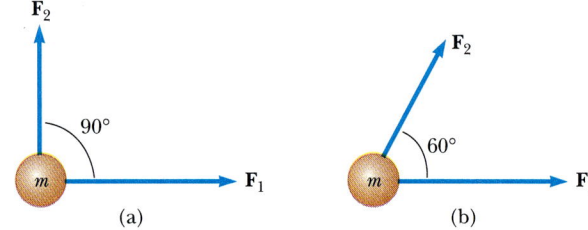

FIGURE P5.18

19. A constant force changes the speed of an 85-kg sprinter from 3.0 m/s to 4.0 m/s in 0.50 s. Calculate (a) the magnitude of the acceleration of the sprinter, (b) the magnitude of the force, and (c) the magnitude of the acceleration of a 58-kg sprinter experiencing the same force. (Assume linear motion.)

20. Besides its weight, a 2.80-kg object is subjected to one other constant force. The object starts from rest and in 1.20 s experiences a displacement of $(4.20 \text{ m})i - (3.30 \text{ m})j$, where the direction of j is the upward vertical direction. Determine the other force.

21. A 4.0-kg object has a velocity of $3.0i$ m/s at one instant. Eight seconds later, its velocity is $(8.0i + 10.0j)$ m/s. Assuming the object was subject to a constant net force, find (a) the components of the force and (b) its magnitude.

22. A barefoot field-goal kicker imparts a speed of 35 m/s to a football initially at rest. If the football has a mass of 0.50 kg and the time of contact with the ball

is 0.025 s, what is the force exerted by the ball on the foot?

23. A 2.0-ton truck provides an acceleration of 3.0 ft/s² to a 5.0-ton trailer. If the truck exerts the same force on the road while pulling a 15.0-ton trailer, what acceleration results?

24. An electron of mass 9.1×10^{-31} kg has an initial speed of 3.0×10^5 m/s. It travels in a straight line, and its speed increases to 7.0×10^5 m/s in a distance of 5.0 cm. Assuming its acceleration is constant, (a) determine the force on the electron and (b) compare this force with the weight of the electron.

25. Figure P5.25 shows the speed of a person's body, during a chin-up. Assuming the motion is vertical and the mass of the person (excluding the arms) is 64.0 kg, determine the magnitude of the force exerted on the body by the arms at various stages of the motion.

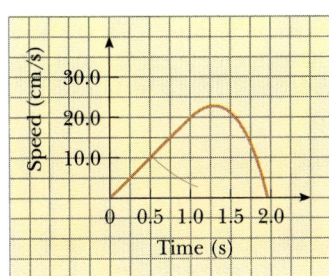

FIGURE P5.25

Section 5.7   Some Applications of Newton's Laws

26. Find the tension in each cord for the systems shown in Figure P5.26. (Neglect the mass of the cords.)

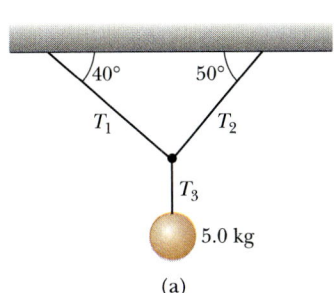

(a)

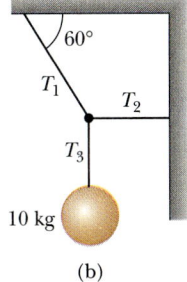

(b)

FIGURE P5.26

27. A 2.0-kg mass accelerates at 11 m/s² in a direction 30.0° north of east (Fig. P5.27). One of the two forces acting on the mass has a magnitude of 11 N

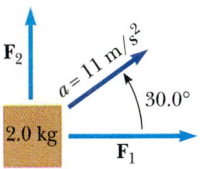

FIGURE P5.27

and is directed north. Determine the magnitude of the second force.

28. A 225-N weight is tied to the middle of a strong rope, and two people pull at opposite ends of the rope in an attempt to lift the weight. (a) What is the magnitude F of the force that each person must apply in order to suspend the weight as shown in Figure P5.28? (b) Can they pull in such a way as to make the rope horizontal? Explain.

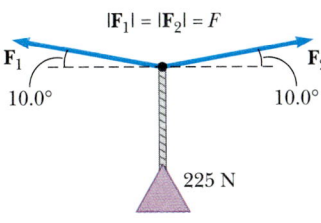

FIGURE P5.28

29. The distance between two telephone poles is 45 m. After a 1.0-kg bird lands on the telephone wire midway between the poles, the wire sags 0.18 m. What is the tension in the wire? Ignore the weight of the wire.

30. The systems shown in Figure P5.30 are in equilibrium. If the spring scales are calibrated in newtons, what do they read in each case? (Neglect the mass of the pulleys and strings, and assume the incline is frictionless.)

31. A bag of cement hangs from three wires as shown in Figure P5.31. Two of the wires make angles θ_1 and θ_2 with the horizontal. If the system is in equilibrium, (a) show that

$$T_1 = \frac{w \cos \theta_2}{\sin(\theta_1 + \theta_2)}$$

(b) Given that $w = 325$ N, $\theta_1 = 10°$, and $\theta_2 = 25°$, find the tensions T_1, T_2, and T_3 in the wires.

32. A woman is pulling her 25-kg suitcase at constant speed by pulling on a strap at an angle θ above the horizontal (Fig. P5.32). She pulls on the strap with a force of magnitude 35 N. A horizontal retarding force of 22 N also acts on the suitcase. (a) What is the

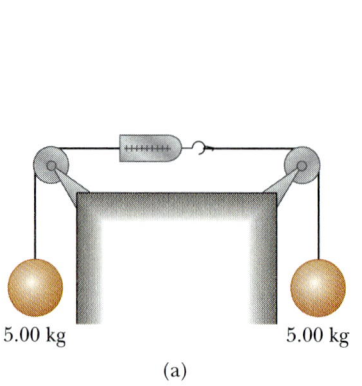

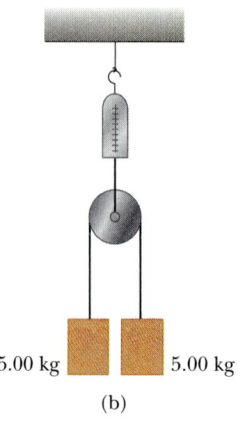

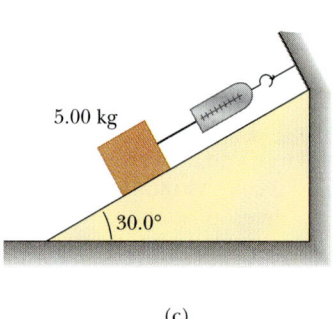

5.00 kg

5.00 kg   5.00 kg

5.00 kg   5.00 kg

30.0°

(a)   (b)   (c)

FIGURE P5.30

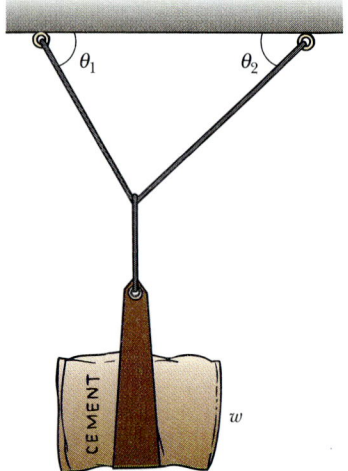

θ_1 θ_2

CEMENT

w

FIGURE P5.31

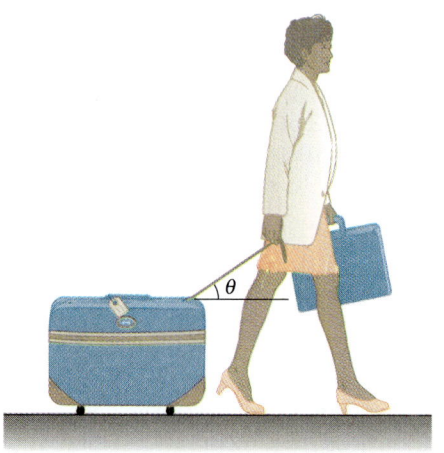

θ

FIGURE P5.32

value of θ? (b) What normal force does the ground exert on the suitcase?

33. A block of mass $m = 2.0$ kg is held in equilibrium on an incline of angle $\theta = 60°$ by the horizontal force F, as shown in Figure P5.33. (a) Determine the value of F, the magnitude of F. (b) Determine the normal force exerted by the incline on the block (ignore friction).

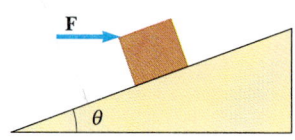

F

θ

FIGURE P5.33

34. A rifle bullet with a mass of 12 g, traveling with a speed of 400 m/s, strikes a large wooden block, which it penetrates to a depth of 15 cm. Determine the magnitude of the retarding force (assumed constant) that acts on the bullet.

35. A simple accelerometer is constructed by suspending a mass m from a string of length L that is tied to the top of a cart. As the cart is accelerated the string system makes an angle of θ with the vertical. (a) Assuming that the mass of the string is negligible compared to m, derive an expression for the cart's acceleration in terms of θ and show that it is independent of the mass m and the length L. (b) Determine the acceleration of the cart when $\theta = 23°$.

36. The force of the wind on the sails of a sailboat is 390 N north. The water exerts a force of 180 N east. If the boat including crew has a mass of 270 kg, what are the magnitude and direction of its acceleration?

37. In the system shown in Figure P5.37, a horizontal force F_x acts on the 8.00-kg mass. (a) For what values of F_x does the 2.00-kg mass accelerate upward? (b) For what values of F_x is the tension in the cord

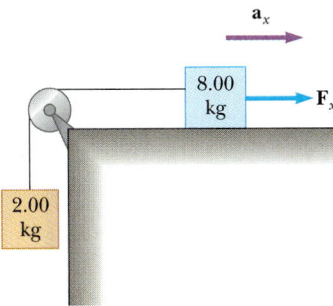

FIGURE P5.37

zero? (c) Plot the acceleration of the 8.00-kg mass versus F_x. Include values of F_x from -100 N to $+100$ N.

 38. Two masses, m_1 and m_2, situated on a frictionless, horizontal surface are connected by a massless string. A force, **F**, is exerted on one of the masses to the right (Fig. P5.38). Determine the acceleration of the system and the tension, T, in the string.

FIGURE P5.38

39. A small bug is placed between two blocks of masses m_1 and m_2 ($m_1 > m_2$) on a frictionless table. A horizontal force, **F**, can be applied to either m_1, as in Figure P5.39a, or m_2, as in Figure P5.39b. For which of these two cases does the bug have a greater chance of surviving? Explain. (*Hint:* Determine the contact force between the blocks in each case.)

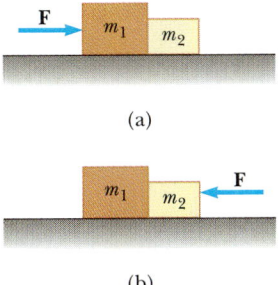

(a)

(b)

FIGURE P5.39

40. A block slides down a frictionless plane having an inclination of $\theta = 15°$. If the block starts from rest at the top and the length of the incline is 2.0 m, find

(a) the magnitude of the acceleration of the block and (b) its speed when it reaches the bottom of the incline.

41. A block of mass $m = 2.0$ kg is released from rest $h = 0.5$ m from the surface of a table, at the top of a $\theta = 30°$ incline as shown in Figure P5.41. The incline is fixed on a table of height $H = 2.0$ m, and the incline is frictionless. (a) Determine the acceleration of the block as it slides down the incline. (b) What is the speed of the block as it leaves the incline? (c) How far from the table will the block hit the floor? (d) How much time has elapsed between when the block is released and when it hits the floor? (e) Does the mass of the block affect any of the above calculations?

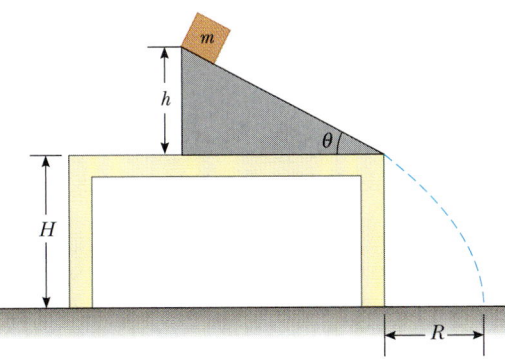

FIGURE P5.41

 42. Two masses are connected by a massless string that passes over a massless pulley as in Figure 5.13. If the incline is also frictionless and if $m_1 = 2.00$ kg, $m_2 = 6.00$ kg, and $\theta = 55.0°$, find (a) the magnitude of the acceleration of the masses, (b) the tension in the string, and (c) the speed of each mass 2.00 s after they are released from rest.

43. A 72-kg man stands on a spring scale in an elevator. Starting from rest, the elevator ascends, attaining its maximum speed of 1.2 m/s in 0.80 s. It travels with this constant speed for the next 5.0 s. The elevator then undergoes a uniform acceleration in the negative y direction for 1.5 s and comes to rest. What does the spring scale register (a) before the elevator starts to move? (b) during the first 0.80 s? (c) while the elevator is traveling at constant speed? (d) during the time it is slowing down?

44. A ball of mass m is dropped (from rest) at the top of a building having height h. If a wind blowing along the side of the building exerts a constant horizontal force of magnitude F on the ball as it drops (Fig. P5.44), (a) show that the ball follows a straight-line path. (b) Does this mean that the ball falls with constant velocity? Explain. (c) If the ball is dropped with

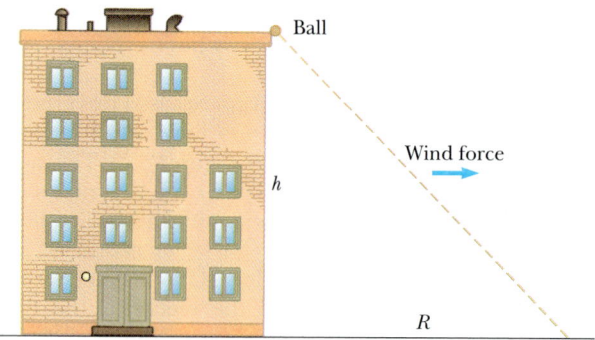

FIGURE P5.44

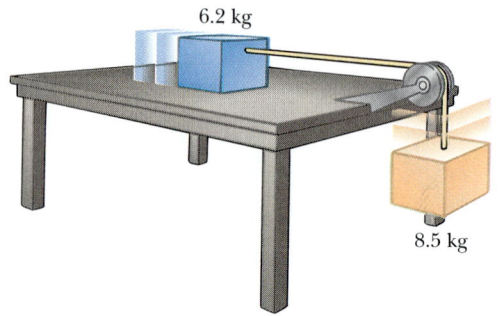

FIGURE P5.47

an initial nonzero vertical velocity of v_0, will it still follow the path of a straight line? Explain. (d) Using $m = 10.0$ kg, $h = 10.0$ m, $F = 20.0$ N, and $v_0 = 4.00$ m/s downward, how far from the building will the ball hit the ground?

45. A net horizontal force $F = A + Bt^3$ acts on a 3.5-kg object, where $A = 8.6$ N and $B = 2.5$ N/s^3. What is the horizontal speed of this object 3.0 s after it starts from rest?

46. Mass m_1 on a frictionless horizontal table is connected to mass m_2 through a massless pulley P_1 and a massless fixed pulley P_2 as shown in Figure P5.46. (a) If a_1 and a_2 are the magnitudes of the accelerations of m_1 and m_2, respectively, what is the relationship between these accelerations? Find expressions for (b) the tensions in the strings and (c) the accelerations a_1 and a_2 in terms of m_1, m_2, and g.

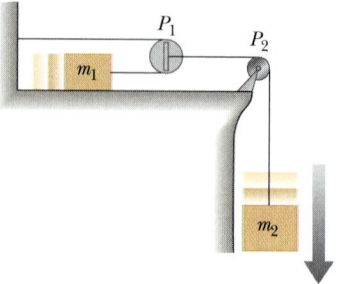

FIGURE P5.46

Section 5.8   Forces of Friction

47. An 8.5-kg hanging block is connected by a string over a pulley to a 6.2-kg block sliding on a flat table (Fig. P5.47). If the coefficient of sliding friction is 0.20, find the tension in the string.

48. A 25-kg block is initially at rest on a horizontal surface. A horizontal force of 75 N is required to set the

block in motion. After it is in motion, a horizontal force of 60 N is required to keep the block moving with constant speed. Find the coefficients of static and kinetic friction from this information.

49. Assume the coefficient of friction between the wheels of a race car and the track is 1.00. If the car starts from rest and accelerates at a constant rate for 335 m, what is its speed at the end of the race?

50. What force must be exerted on block A in order for block B not to fall (Fig. P5.50). The coefficient of static friction between blocks A and B is 0.55, and the horizontal surface is frictionless.

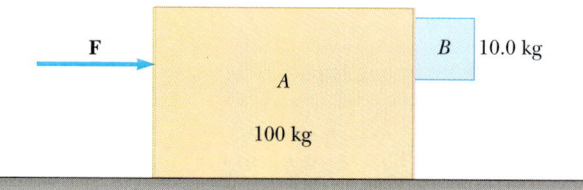

FIGURE P5.50

51. An ice skater moving at 12 m/s coasts to a halt in 95 m on an ice surface. What is the coefficient of friction between ice and skates?

52. A car is traveling at 50.0 mi/h on a horizontal highway. (a) If the coefficient of friction between road and tires on a rainy day is 0.10, what is the minimum distance in which the car will stop? (b) What is the stopping distance when the surface is dry and $\mu = 0.60$?

53. A boy drags his 60.0-N sled at constant speed up a 15° hill. He does so by pulling with a 25-N force on a rope attached to the sled. If the rope is inclined at 35° to the horizontal, (a) what is the coefficient of kinetic friction between sled and snow? (b) At the top of the hill, he jumps on the sled and slides down the hill. What is the magnitude of his acceleration down the slope?

54. A block moves up a 45° incline with constant speed under the action of a force of 15 N applied *parallel* to the incline. If the coefficient of kinetic friction is 0.30, determine (a) the weight of the block and (b) the minimum force required to allow it to move *down* the incline at constant speed.

55. Two blocks connected by a massless rope are being dragged by a horizontal force **F** (Fig. P5.38). Suppose that $F = 68$ N, $m_1 = 12$ kg, $m_2 = 18$ kg, and the coefficient of kinetic friction between each block and the surface is 0.10. (a) Draw a free-body diagram for each block. (b) Determine the tension, T, and the magnitude of the acceleration of the system.

56. A mass $M = 2.2$ kg is accelerated across a horizontal surface by a rope passing over a pulley, as shown in Figure P5.56. The tension in the rope is 10.0 N and the pulley is 10.0 cm above the top of the block. The coefficient of sliding friction is 0.40. (a) Determine the acceleration of the block when $x = 0.40$ m. (b) Find the value of x at which the acceleration becomes zero.

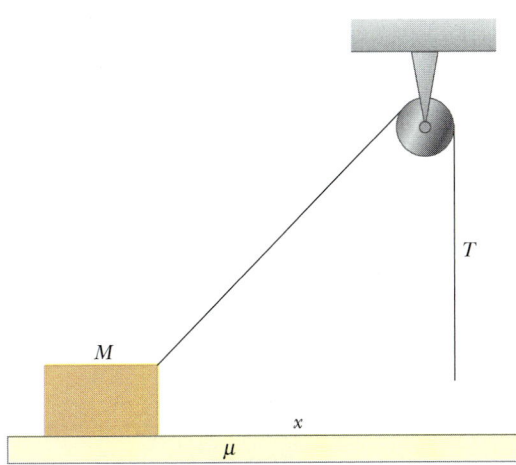

FIGURE P5.56

57. A 3.0-kg block starts from rest at the top of a 30.0° incline and slides 2.0 m down the incline in 1.5 s. Find (a) the magnitude of the acceleration of the block, (b) the coefficient of kinetic friction between block and plane, (c) the frictional force acting on the block, and (d) the speed of the block after it has slid 2.0 m.

58. A block slides on an incline having an inclination of θ with the horizontal. The coefficient of kinetic friction between block and plane is μ_k. (a) If the block accelerates down the incline, show that the magnitude of its acceleration is given by $a = g(\sin \theta - \mu_k \cos \theta)$. (b) If the block is projected up the incline, show that the magnitude of its acceleration is $a = -g(\sin \theta + \mu_k \cos \theta)$.

59. Three masses are connected on the table as shown in Figure P5.59. The table has a coefficient of sliding friction of 0.35. The three masses are 4.0 kg, 1.0 kg, and 2.0 kg, respectively, and the pulleys are frictionless. (a) Determine the acceleration of each block and their directions. (b) Determine the tensions in the two cords.

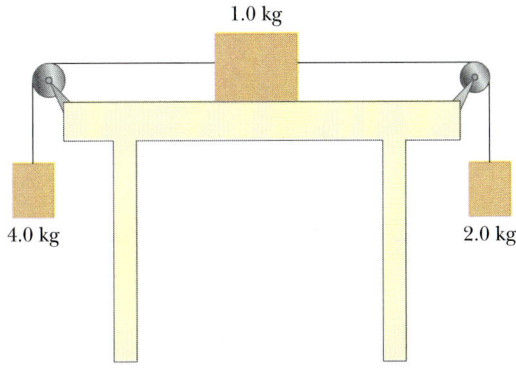

FIGURE P5.59

60. A crate of weight w is pushed by a force **F** on a horizontal floor. If the coefficient of static friction is μ_s and **F** is directed at angle ϕ below the horizontal, (a) show that the minimum value of F that will move the crate is

$$F = \frac{\mu_s w \sec \phi}{1 - \mu_s \tan \phi}$$

(b) Find the minimum value of F that can produce motion when $\mu_s = 0.40$, $w = 100$ N, and $\phi = 0°$, 15°, 30°, 45°, and 60°.

61. A block is placed on a plane inclined at 35° relative to the horizontal. If the block slides down the plane with an acceleration of magnitude $g/3$, determine the coefficient of kinetic friction between block and plane.

62. A crate is carried in a truck traveling horizontally at 15 m/s. If the coefficient of static friction between crate and truck is 0.40, determine the minimum stopping distance for the truck so that the crate will not slide. 28.7 m

63. An Olympic skier moving at 25 m/s down a 20° slope encounters a region of wet snow of coefficient of friction $\mu_k = 0.55$. How far down the slope does she travel before coming to a halt?

ADDITIONAL PROBLEMS

64. A car moving at 20.0 m/s brakes to a stop without skidding. The driver in the car behind the first, moving at 30.0 m/s, sees the brake lights, applies his

brakes after a 0.10-s delay, and brakes to a stop without skidding. Assume that $\mu_s = 0.75$ for both cars. Calculate the minimum distance between the cars at the instant the driver of the lead car applies the brakes if a rear-end collision is to be avoided. (Assume constant accelerations.)

65. A mass M is held in place by an applied force **F** and a pulley system as shown in Figure P5.65. The pulleys are massless and frictionless. Find (a) the tension in each section of rope, T_1, T_2, T_3, T_4, and T_5, and (b) the magnitude of **F**.

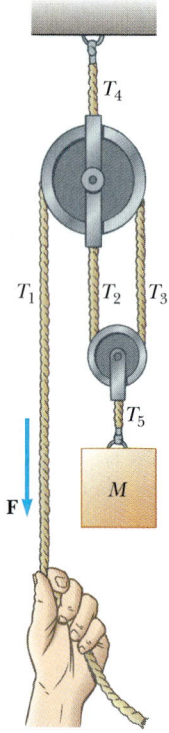

FIGURE P5.65

66. Alex remembered from high-school physics that pulleys can be used to aid in lifting heavy objects. Alex designed the frictionless pulley system shown in Figure P5.66 to lift a safe to a second-floor office. The safe weighs 400 lb, and Alex can pull with a force of 240 lb. (a) Will he be able to raise the safe? (b) What is the maximum weight he can lift using his pulley system? (*Note:* The large pulley is fastened by a yoke to the rope that Alex is pulling.)

67. As part of a laboratory investigation, a student wishes to measure the coefficients of friction between a block of metal and a wooden board. The board has length L, and the block is placed at one end of it. This end of the board is raised, and the block begins to slide when it is a distance of h above the lower end of the board, as in Figure P5.67. At this angle, the

FIGURE P5.66

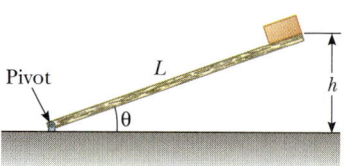

FIGURE P5.67

block slides down the length of the board in the time t. Determine (a) the coefficient of static friction between block and board, (b) the acceleration of the block, (c) the smallest angle that causes the block to move, and (d) the coefficient of kinetic friction between block and board.

68. About 200 years ago, Charles Coulomb invented the tribometer, a device used to investigate static friction. The instrument is represented schematically in Figure P5.68. To determine the coefficient of static friction, the hanging mass M is increased or decreased as necessary until m is on the verge of sliding. Prove that $\mu_s = M/m$.

69. A 2.00-kg aluminum block and a 6.00-kg copper block are connected by a light string over a frictionless pulley. They are allowed to move on a fixed steel block-wedge (of angle $\theta = 30.0°$) as shown in Figure P5.69. Considering there is friction between the blocks and wedge, determine (a) the acceleration of the two blocks and (b) the tension in the string.

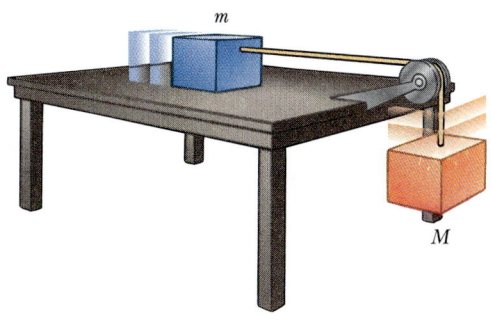

FIGURE P5.68

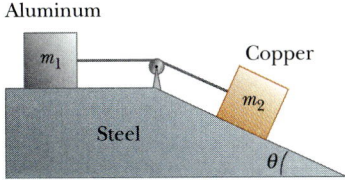

FIGURE P5.69

70. A block of mass $m = 2.00$ kg rests on the left edge of a block of length $L = 3.00$ m and mass $M = 8.00$ kg. The coefficient of kinetic friction between the two blocks is 0.300, and the surface on which the 8.00-kg block rests is frictionless. A constant horizontal force of magnitude $F = 10.0$ N is applied to the 2.00-kg block, setting it in motion as shown in Figure P5.70a. (a) How long will it take before this block makes it to the right side of the 8.00-kg block, as shown in Figure P5.70b? (*Note:* Both blocks are set in motion when **F** is applied.) (b) How far does the 8.00-kg block move in the process?

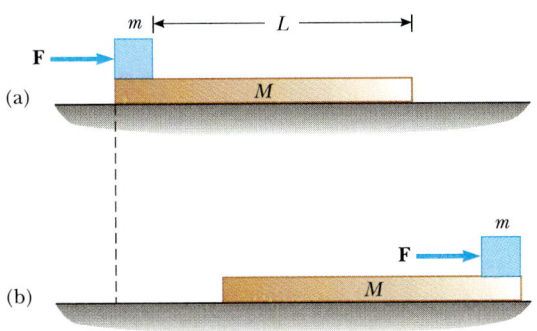

FIGURE P5.70

71. Three baggage carts of masses m_1, m_2, and m_3 are towed by a tractor of mass M along an airport apron. The wheels of the tractor exert a total frictional force **F** on the ground as shown (Fig. P5.71). In the following, express your answers in terms of F, M, m_1, m_2, m_3, and g. (a) What are the magnitude and direction of the horizontal force exerted on the tractor by the ground? (b) What is the smallest value of the coefficient of static friction that will prevent the wheels from slipping? Assume that each of the two drive wheels on the tractor bears $1/3$ of the tractor's weight. (c) What is the acceleration, a, of the system (tractor plus baggage carts)? (d) What are the tensions T_1, T_2, and T_3 in the connecting cables? (e) What is the net force on the cart of mass m_2?

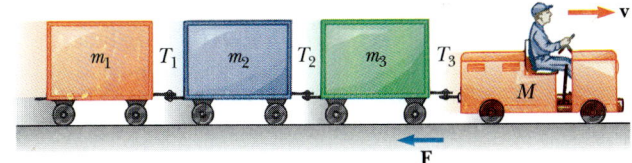

FIGURE P5.71

72. In Figure P5.72, the man and the platform together weigh 750 N. Determine how hard the man would have to pull to hold himself off the ground. (Or is it impossible? If so, explain why.)

FIGURE P5.72

73. A 2.0-kg block is placed on top of a 5.0-kg block as in Figure P5.73. The coefficient of kinetic friction between the 5.0-kg block and the surface is 0.20. A horizontal force **F** is applied to the 5.0-kg block. (a) Draw a free-body diagram for each block. What force accel-

erates the 2.0-kg block? (b) Calculate the magnitude of the force necessary to pull both blocks to the right with an acceleration of 3.0 m/s². (c) Find the minimum coefficient of static friction between the blocks such that the 2.0-kg block does not slip under an acceleration of 3.0 m/s².

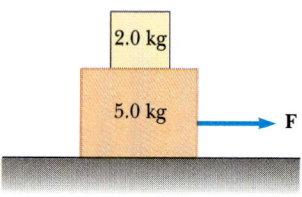

FIGURE P5.73

74. A 5.0-kg block is placed on top of a 10-kg block (Fig. P5.74). A horizontal force of 45 N is applied to the 10-kg block, and the 5.0-kg block is tied to the wall. The coefficient of kinetic friction between the moving surfaces is 0.20. (a) Draw a free-body diagram for each block and identify the action-reaction forces between the blocks. (b) Determine the tension in the string and the magnitude of the acceleration of the 10-kg block.

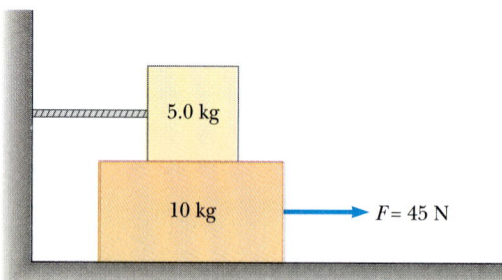

FIGURE P5.74

75. An inventive child named Brian wants to reach an apple in a tree without climbing the tree. Sitting in a chair connected to a rope that passes over a frictionless pulley (Fig. P5.75), he pulls on the loose end of the rope with such a force that the spring scale reads 250 N. His true weight is 320 N, and the chair weighs 160 N. (a) Draw free-body diagrams for Brian and the chair considered as separate systems, and another diagram for Brian and the chair considered as one system. (b) Show that the acceleration of the system is upward and find its magnitude. (c) Find the force that Brian exerts on the chair.

FIGURE P5.75

76. In Figure P5.76, a 500-kg horse pulls a sledge of mass 100 kg. The system (horse plus sledge) has a forward acceleration of 1.00 m/s² when the frictional force on the sledge is 500 N. Find (a) the tension in the connecting rope and (b) the magnitude and direction of the force of friction exerted on the horse. (c) Verify that the total forces of friction the Earth exerts on the system will give to the total system an acceleration of 1.00 m/s².

FIGURE P5.76

77. A block is released from rest at the top of a plane inclined at an angle of 45°. The coefficient of kinetic friction varies along the plane according to the relation $\mu_k = \sigma x$, where x is the distance along the plane measured in meters from the top and where $\sigma = 0.50$ m⁻¹. Determine (a) how far the block slides before coming to rest and (b) the maximum speed it attains.

78. A small block of mass m is initially at the bottom of an incline of mass M, angle θ, and length L, as shown in Figure P5.78a. Assume that all surfaces are friction-

less, and that a constant horizontal force of magnitude F is applied to the block so as to set it, *and the incline,* in motion. (a) Show that the mass m will reach the top of the incline (Fig. P5.78b) in time

$$t = \sqrt{\frac{2L[1 + (m/M)\sin^2\theta]}{(F/m)\cos\theta - g(1 + m/M)\sin\theta}}$$

(*Hint:* The block must always lie on the incline.) (b) How far does the incline travel in the process? (c) Does the expression in part (a) reduce to the expected result when $M \gg m$? Explain.

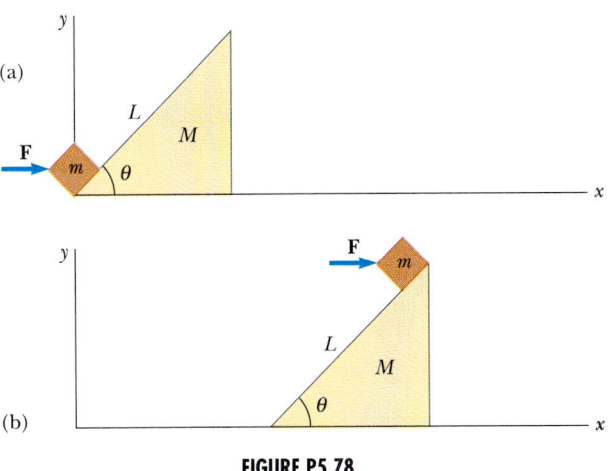

(a)

(b)

FIGURE P5.78

79. A wire ABC supports a body of weight w as shown in Figure P5.79. The wire passes over a fixed pulley at B and is firmly attached to a vertical wall at A. The line AB makes an angle ϕ with the vertical, and the pulley at B exerts on the wire a force of magnitude F inclined at angle θ with the horizontal. (a) Show that if the system is in equilibrium, $\theta = \phi/2$. (b) Show that $F = 2w\sin(\phi/2)$. (c) Sketch a graph of F as ϕ increases from $0°$ to $180°$.

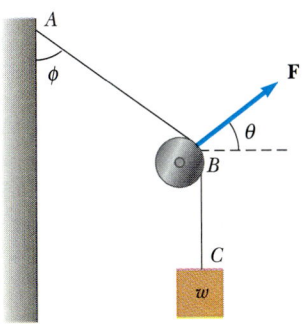

FIGURE P5.79

80. A mobile is formed by supporting four metal butterflies of equal mass m from a string of length L. The points of support are evenly spaced a distance ℓ apart as shown in Figure P5.80. The string forms an angle θ_1 with the ceiling at each end point. The center section of string is horizontal. (a) Find the tension in each section of string in terms of θ_1, m, and g. (b) Find the angle θ_2, in terms of θ_1, that the sections of string between the outside butterflies and the inside butterflies form with the horizontal. (c) Show that the distance D between the end points of the string is

$$D = \frac{L}{5}\left(2\cos\theta_1 + 2\cos\left[\tan^{-1}\left(\frac{1}{2}\tan\theta_1\right)\right] + 1\right)$$

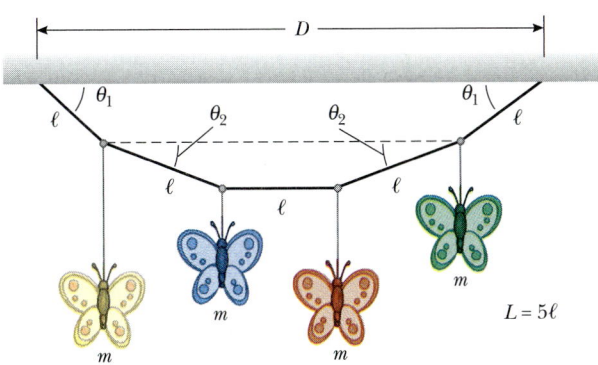

FIGURE P5.80

81. Forces $\mathbf{F}_1 = (-6\mathbf{i} - 4\mathbf{j})$ N and $\mathbf{F}_2 = (-3\mathbf{i} + 7\mathbf{j})$ N act on a 2-kg particle initially at rest at coordinates $(-2\text{ m}, +4\text{ m})$. (a) What are the components of the particle's velocity at $t = 10$ s? (b) In what direction is the particle moving at $t = 10$ s? (c) What displacement does the particle undergo during the first 10 s? (d) What are the coordinates of the particle at $t = 10$ s?

82. A student is asked to measure the acceleration of a crate on a frictionless inclined plane as in Figure 5.11 using an air track, a stopwatch, and a meter stick. The vertical height of the incline is 1.774 cm, and the total length of the incline is $d = 127.1$ cm. Hence, the angle of inclination θ is determined from the relationship $\sin\theta = 1.774/127.1$. The crate is released from rest at the top of the incline, and its displacement along the incline, x, is measured versus time, where $x = 0$ refers to the initial position of the crate. For x values of 10.0, 20.0, 35.0, 50.0, 75.0, and 100 cm, the measured times to undergo these displacements (averaged over six runs) are 1.02, 1.53, 2.01, 2.64, 3.30, and 3.75 s, respectively. Construct a graph of x versus t^2, and perform a linear least-

squares fit to the data. Determine the magnitude of the acceleration of the crate from the slope of this graph, and compare it with the value you would get using $a' = g \sin \theta$.

 83. What horizontal force must be applied to the cart shown in Figure P5.83 in order that the blocks remain stationary relative to the cart? Assume all surfaces, wheels, and pulley are frictionless. (*Hint:* Note that the force exerted by the string accelerates m_1.)

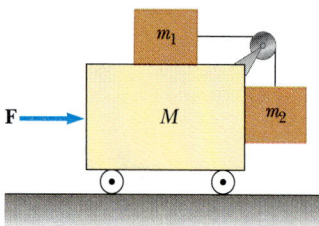

FIGURE P5.83

84. Initially the system of masses shown in Figure P5.83 is held motionless. All surfaces, pulley, and wheels are frictionless. Let the force **F** be zero and assume that m_2 can move only vertically. At the instant after the system of masses is released, find (a) the tension T in the string, (b) the acceleration of m_2, (c) the acceleration of M, and (d) the acceleration of m_1. (*Note:* The pulley accelerates along with the cart.)

85. The three blocks in Figure P5.85 are connected by massless strings that pass over frictionless pulleys. The acceleration of the system is 2.35 m/s² to the left and the surfaces are rough. Find (a) the tensions in the strings and (b) the coefficient of kinetic friction between blocks and surfaces. (Assume the same μ for both blocks.)

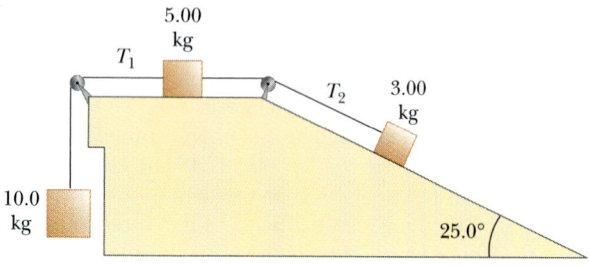

FIGURE P5.85

86. In Figure P5.86, the coefficient of kinetic friction between the 2.00-kg and 3.00-kg blocks is 0.300. The

horizontal surface and the pulleys are frictionless, and the masses are released from rest. (a) Draw a free-body diagram for each block. (b) Determine the acceleration of each block. (c) Find the tension in the strings.

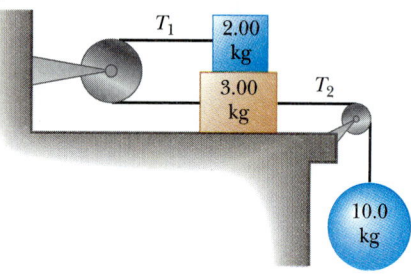

FIGURE P5.86

 87. Two blocks of mass 3.50 kg and 8.00 kg are connected by a massless string that passes over a frictionless pulley (Fig. P5.87). The inclines are frictionless. Find (a) the magnitude of the acceleration of each block and (b) the tension in the string.

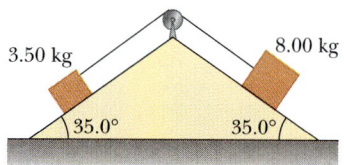

FIGURE P5.87

 88. The system shown in Figure P5.87 has an acceleration of magnitude 1.5 m/s². Assume the coefficient of kinetic friction between block and incline is the same for both inclines. Find (a) the coefficient of kinetic friction and (b) the tension in the string.

89. A van accelerates down a hill (Fig. P5.89), going from rest to 30.0 m/s in 6.00 s. During the acceleration, a toy ($m = 100$ g) hangs by a string from the ceiling. The acceleration is such that the string remains perpendicular to the ceiling. Determine (a) the angle θ and (b) the tension in the string.

90. Before 1960 it was believed that the maximum attainable coefficient of static friction of an automobile tire was less than 1. Then about 1962, three companies independently developed racing tires with coefficients of 1.6. Since then, tires have improved, as illustrated in the following problem. According to the 1990 Guinness Book of Records, the fastest $\frac{1}{4}$ mile covered by a piston-engine car from a standing start is 4.96 s. This record elapsed time was set by

FIGURE P5.89

Shirley Muldowney in September 1989. (a) Assuming that the rear wheels nearly lifted the front wheels off the pavement, what minimum value of μ is necessary to achieve the record time? (b) Suppose Muldowney were able to double her engine power, keeping other things equal. How would this change affect the elapsed time?

91. A magician attempts to pull a tablecloth from under a 200-g mug located 30 cm from the edge of the cloth. If there is a frictional force of 0.10 N exerted on the mug by the cloth, and the cloth is pulled with a constant acceleration of magnitude 3.0 m/s², how far does the mug move on the tabletop before the cloth is completely out from under it? (*Hint:* The cloth moves more than 30 cm before it is out from under the mug!)

SPREADSHEET PROBLEMS

S1. An 8.4-kg mass slides down a fixed, frictionless inclined plane. Design and write a spreadsheet to determine the normal force exerted on the mass and its acceleration for a series of incline angles (measured from the horizontal) ranging from 0° to 90° in 5° increments. Use the graphing capability of your spreadsheet data to plot the normal force and the acceleration as functions of the incline angle. In the limiting cases of 0° and 90°, are your results consistent with the known behavior?

S2. A person must move a 65-kg crate resting on a level floor. The coefficient of static friction between crate and floor is 0.48. A force of magnitude F is applied at an angle θ with the horizontal. (a) Construct a spreadsheet that will enable you to calculate the force necessary to move the box for a sequence of angles. Take the angle θ as positive if the force has an upward component, and negative if the force has a downward component. Plot F versus θ, and from this graph determine the minimum force required to move the box. At what angle should the force be applied? (b) Investigate what happens when you change the coefficient of friction.

B.C.   **By John Hart**

CHAPTER 6

Circular Motion and Other Applications of Newton's Laws

The passengers on this corkscrew rollercoaster experience the thrill of various forces as they travel along the curved track. The forces on one of the passenger cars include the force exerted by the track, the force of gravity, and the force of air resistance. *(Robin Smith/Tony Stone Images)*

I n the previous chapter we introduced Newton's laws of motion and applied them to situations involving linear motion. In this chapter we apply Newton's laws of motion to circular motion. We also discuss the motion of an object when observed in an accelerated, or noninertial, frame of reference and the motion of an object through a viscous medium. Finally, we conclude this chapter with a brief discussion of the fundamental forces in nature.

6.1 NEWTON'S SECOND LAW APPLIED TO UNIFORM CIRCULAR MOTION

In Section 4.4 we found that a particle moving in a circular path of radius r with uniform speed v experiences an acceleration that has a magnitude

$$a_r = \frac{v^2}{r}$$

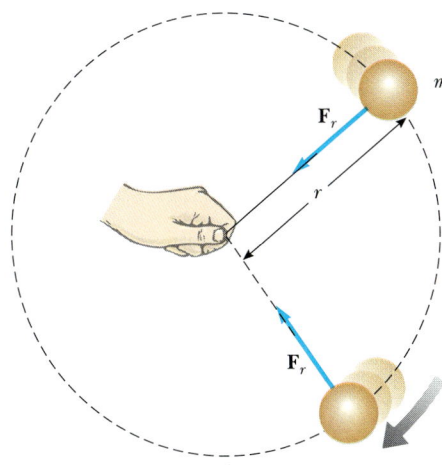

FIGURE 6.1 An overhead view of a ball moving in a circular path in a horizontal plane. A force $\mathbf{F}_r$ directed toward the center of the circle keeps the ball moving in the circle with constant speed.

Because the velocity vector $\mathbf{v}$ changes its direction continuously during the motion, the acceleration vector $\mathbf{a}_r$ is directed toward the center of the circle and hence is called **centripetal acceleration.** Furthermore, $\mathbf{a}_r$ is *always* perpendicular to $\mathbf{v}$.

Consider a ball of mass m tied to a string of length r and being whirled in a horizontal circular path on a table top as in Figure 6.1. Let us assume that the ball moves with constant speed. The ball tends to maintain the motion in a straight-line path; however, the string prevents this motion along a straight line by exerting a force on the ball to make it follow its circular path. This force is directed along the length of the string toward the center of the circle, as shown in Figure 6.1, and is an example of a class of forces called **central forces.** If we apply Newton's second law along the radial direction, we find that the required central force is

$$F_r = ma_r = m\frac{v^2}{r} \qquad (6.1)$$

Central force

Like the centripetal acceleration, the central force acts toward the center of the circular path followed by the particle. Because they act toward the center of rotation, central forces cause a change in the direction of the velocity. Central forces are no different from any other forces we have encountered. The term *central* is used simply to indicate that *the force is directed toward the center of a circle.* In the case of a ball rotating at the end of a string, the force exerted by the string on the ball is the central force. For a satellite in a circular orbit around the Earth, the force of gravity is the central force. The central force acting on a car rounding a curve on a flat road is the force of friction between the tires and the pavement, and so forth. In general, a body can move in a circular path under the influence of such forces as friction, the gravitational force, or a combination of forces.

Regardless of the example used, if the central force acting on an object should vanish, the object would no longer move in its circular path; instead, it would move along a straight-line path tangent to the circle. This idea is illustrated in Figure 6.2 for the case of the ball whirling in a circle at the end of a string. If the string breaks at some instant, the ball will move along the straight-line path tangent to the circle at the point where the string broke.

Let us consider some examples of uniform circular motion. In each case, be sure to recognize the external force (or forces) that causes the body to move in its circular path.

An athlete in the process of throwing the hammer. The force exerted by the chain is the central force. Only when the athlete releases the hammer will it move along a straight-line path tangent to the circle. *(Focus on Sports)*

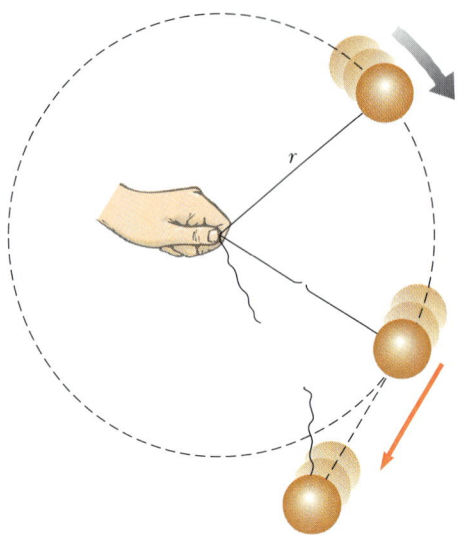

FIGURE 6.2 When the string breaks, the ball moves in the direction tangent to the circular path.

EXAMPLE 6.1 **How Fast Can It Spin?**

A ball of mass 0.500 kg is attached to the end of a cord whose length is 1.50 m. The ball is whirled in a horizontal circle as in Figure 6.2. If the cord can withstand a maximum tension of 50.0 N, what is the maximum speed the ball can have before the cord breaks?

Solution Because the central force in this case is the force T exerted by the cord on the ball, Equation 6.1 gives

$$T = m\,\frac{v^2}{r}$$

Solving for v, we have

$$v = \sqrt{\frac{Tr}{m}}$$

The maximum speed that the ball can have corresponds to the maximum tension. Hence, we find

$$v_{max} = \sqrt{\frac{T_{max}r}{m}} = \sqrt{\frac{(50.0\ \text{N})(1.50\ \text{m})}{0.500\ \text{kg}}} = \boxed{12.2\ \text{m/s}}$$

Exercise Calculate the tension in the cord if the speed of the ball is 5.00 m/s.

Answer 8.33 N.

EXAMPLE 6.2 **The Conical Pendulum**

A small body of mass m is suspended from a string of length L. The body revolves in a horizontal circle of radius r with constant speed v, as in Figure 6.3. (Since the string sweeps out the surface of a cone, the system is known as a *conical pendulum*.) Find the speed of the body and the period of revolution, T_p, defined as the time needed to complete one revolution.

Solution The free-body diagram for the mass m is shown in Figure 6.3, where the force exerted by the string, T, has been resolved into a vertical component, $T\cos\theta$, and a component $T\sin\theta$ acting toward the center of rotation. Since the body does not accelerate in the vertical direction, the vertical component of T must balance the weight. Therefore,

(1) $T\cos\theta = mg$

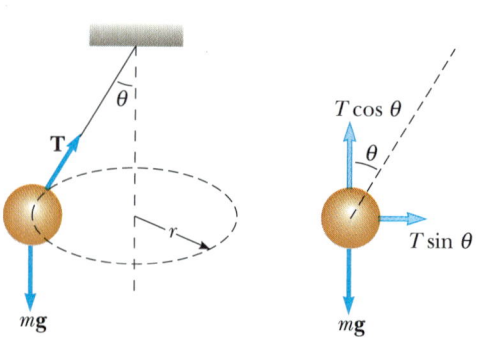

FIGURE 6.3 (Example 6.2) The conical pendulum and its free-body diagram.

Since the central force in this example is provided by the component $T \sin \theta$, from Newton's second law we get

$$(2) \qquad T \sin \theta = ma_r = \frac{mv^2}{r}$$

By dividing (2) by (1), we eliminate T and find that

$$\tan \theta = \frac{v^2}{rg}$$

But from the geometry, we note that $r = L \sin \theta$; therefore

$$v = \sqrt{rg \tan \theta} = \boxed{\sqrt{Lg \sin \theta \tan \theta}}$$

Since the ball travels a distance of $2\pi r$ (the circumference

of the circular path) in a time equal to the period of revolution, T_p (not to be confused with the force T), we find

$$(3) \qquad T_p = \frac{2\pi r}{v} = \frac{2\pi r}{\sqrt{rg \tan \theta}} = \boxed{2\pi \sqrt{\frac{L \cos \theta}{g}}}$$

The intermediate algebraic steps used in obtaining (3) are left to the reader. Note that T_p is independent of m! If we take $L = 1.00$ m and $\theta = 20.0°$, we find using (3) that

$$T_p = 2\pi \sqrt{\frac{(1.00 \text{ m})(\cos 20.0°)}{9.80 \text{ m/s}^2}} = 1.95 \text{ s}$$

Is it physically possible to have a conical pendulum with $\theta = 90°$?

EXAMPLE 6.3   What Is the Maximum Speed of the Car?

A 1500-kg car moving on a flat, horizontal road negotiates a curve whose radius is 35.0 m as in Figure 6.4. If the coefficient of static friction between the tires and the dry pavement is 0.500, find the maximum speed the car can have in order to make the turn successfully.

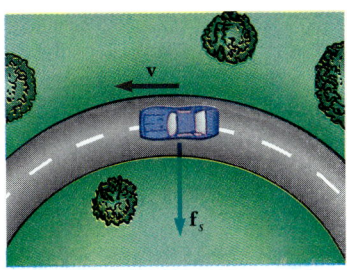

FIGURE 6.4   (Example 6.3) The force of static friction directed toward the center of the arc keeps the car moving in a circle.

Solution   In this case, the central force that enables the car to remain in its circular path is the force of static friction. Hence, from Equation 6.1 we have

$$(1) \qquad f_s = m\frac{v^2}{r}$$

The maximum speed that the car can have around the curve corresponds to the speed at which it is on the verge of skidding outwards. At this point, the friction force has its maximum value

$$f_{s\,\text{max}} = \mu n$$

Because the normal force equals the weight in this case, we find

$$f_{s\,\text{max}} = \mu mg = (0.500)(1500 \text{ kg})(9.80 \text{ m/s}^2) = 7350 \text{ N}$$

Substituting this value into (1), we find that the maximum speed is

$$v_{\text{max}} = \sqrt{\frac{f_{s\,\text{max}}r}{m}} = \sqrt{\frac{(7350 \text{ N})(35.0 \text{ m})}{1500 \text{ kg}}} = \boxed{13.1 \text{ m/s}}$$

Exercise   On a wet day, the car described in this example begins to skid on the curve when its speed reaches 8.00 m/s. What is the coefficient of static friction in this case?

Answer   0.187.

EXAMPLE 6.4   The Banked Exit Ramp

An engineer wishes to design a curved exit ramp for a toll road in such a way that a car will not have to rely on friction to round the curve without skidding. Suppose that a typical car rounds the curve with a speed of 30.0 mi/h (13.4 m/s) and that the radius of the curve is 50.0 m. At what angle should the curve be banked?

Reasoning   On a level road, the central force must be provided by a force of friction between car and road. However, if

the road is banked at an angle θ, as in Figure 6.5, the normal force, n, has a horizontal component $n \sin \theta$ pointing toward the center of the circular path followed by the car. We assume that only the component $n \sin \theta$ furnishes the central force. Therefore, the banking angle we calculate will be one for which *no frictional force is required.* In other words, a car moving at the correct speed (13.4 m/s) can negotiate the curve even on an icy surface.

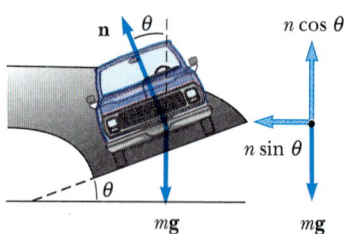

FIGURE 6.5 (Example 6.4) End view of a car rounding a curve on a road banked at an angle θ to the horizontal. The central force is provided by the horizontal component of the normal force when friction is neglected. Note that **n** is the *sum* of the forces that the road exerts on the wheels of the car.

Solution Newton's second law written for the radial direction gives

$$(1) \qquad n \sin \theta = \frac{mv^2}{r}$$

The car is in equilibrium in the vertical direction. Thus, from $\Sigma F_y = 0$, we have

$$(2) \qquad n \cos \theta = mg$$

Dividing (1) by (2) gives

$$\tan \theta = \frac{v^2}{rg}$$

$$\theta = \tan^{-1}\left[\frac{(13.4 \text{ m/s})^2}{(50.0 \text{ m})(9.80 \text{ m/s}^2)}\right] = \boxed{20.1°}$$

If a car rounds the curve at a speed lower than 13.4 m/s, the driver will have to rely on friction to keep from sliding down the incline. A driver who attempts to negotiate the curve at a speed higher than 13.4 m/s will have to depend on friction to keep from sliding up the ramp.

Exercise Write Newton's second law applied to the radial direction for the car in a situation in which a frictional force **f** is directed *down* the slope of the banked road.

Answer $n \sin \theta + f \cos \theta = \dfrac{mv^2}{r}$.

EXAMPLE 6.5 Satellite Motion

This example treats the problem of a satellite moving in a circular orbit around the Earth. In order to understand this problem, we must first note that the gravitational force between two particles having masses m_1 and m_2, separated by a distance r, is attractive and has a magnitude

$$F = G\frac{m_1 m_2}{r^2}$$

where $G = 6.672 \times 10^{-11}$ N·m²/kg². This is Newton's law of gravity, which we shall discuss in detail in Chapter 14.

Now consider a satellite of mass m moving in a circular orbit around the Earth at a constant speed v and at an altitude h above the Earth's surface as in Figure 6.6. (a) Determine the speed of the satellite in terms of G, h, R_E (the radius of the Earth), and M_E (the mass of the Earth).

Solution Because the only external force on the satellite is the force of gravity, which acts toward the center of the Earth, we have

$$F_r = G\frac{M_E m}{r^2}$$

From Newton's second law we get

$$G\frac{M_E m}{r^2} = m\frac{v^2}{r}$$

Solving for v and remembering that $r = R_E + h$, we get

$$(1) \qquad v = \sqrt{\frac{GM_E}{r}} = \boxed{\sqrt{\frac{GM_E}{R_e + h}}}$$

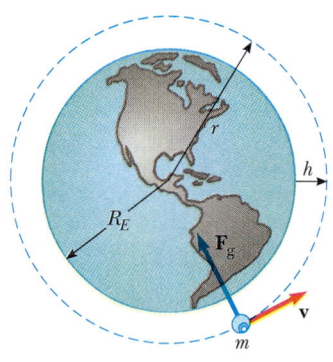

FIGURE 6.6 (Example 6.5) A satellite of mass m moving in a circular orbit of radius r and with constant speed v around the Earth. The central force **F** is provided by the gravitational force exerted by the Earth on the satellite.

(b) Determine the satellite's period of revolution, T_p (the time for one revolution about the Earth).

Solution Since the satellite travels a distance of $2\pi r$ (the circumference of the circle) in a time T_p, we find using (1) that

$$(2) \qquad T_p = \frac{2\pi r}{v} = \frac{2\pi r}{\sqrt{GM_E/r}} = \boxed{\left(\frac{2\pi}{\sqrt{GM_E}}\right) r^{3/2}}$$

The planets move around the Sun in approximately circular orbits. The radii of these orbits can be calculated from (2) with M_E replaced by the mass of the Sun. If (2) is squared, we see that $T_p^2 \propto r^3$. The fact that the square of the period is proportional to the cube of the radius of the orbit was first recognized as an empirical relation based on planetary data. We shall return to this topic in Chapter 14.

Exercise   A satellite is in a circular orbit at an altitude of 1000 km. The radius of the Earth is 6.37×10^6 m. Find the speed of the satellite and the period of its orbit.

Answer 7.35×10^3 m/s; 6.31×10^3 s = 105 min.

EXAMPLE 6.6   Let's Go Loop-the-Loop

A pilot of mass m in a jet aircraft executes a "loop-the-loop" maneuver as illustrated in Figure 6.7a. In this flying pattern, the aircraft moves in a vertical circle of radius 2.70 km at a *constant speed* of 225 m/s. Determine the force exerted by the seat on the pilot at (a) the bottom of the loop and (b) the top of the loop. Express the answers in terms of the weight of the pilot, mg.

Solution   (a) The free-body diagram for the pilot at the bottom of the loop is shown in Figure 6.7b. The only forces acting on the pilot are the downward force of gravity, mg, and the upward force n_{bot} exerted by the seat. Since the net upward force that provides the centripetal acceleration has a magnitude $n_{bot} - mg$, Newton's second law for the radial direction gives

$$n_{bot} - mg = m\frac{v^2}{r}$$

$$n_{bot} = mg + m\frac{v^2}{r} = mg\left[1 + \frac{v^2}{rg}\right]$$

Substituting the values given for the speed and radius gives

$$n_{bot} = mg\left[1 + \frac{(225 \text{ m/s})^2}{(2.70 \times 10^3 \text{ m})(9.80 \text{ m/s}^2)}\right] = \boxed{2.91\,mg}$$

Hence, the force exerted by the seat on the pilot is *greater* than his weight by a factor of 2.91. The pilot experiences an

apparent weight that is greater than his true weight by the factor 2.91. This is discussed further in Section 6.4.

(b) The free-body diagram for the pilot at the top of the loop is shown in Figure 6.7c. At this point, both the weight and the force exerted by the seat on the pilot, n_{top}, act *downward*, so the net force downwards which provides the centripetal acceleration has a magnitude $n_{top} + mg$. Applying Newton's second law gives

$$n_{top} + mg = m\frac{v^2}{r}$$

$$n_{top} = m\frac{v^2}{r} - mg = mg\left[\frac{v^2}{rg} - 1\right]$$

$$n_{top} = mg\left[\frac{(225 \text{ m/s})^2}{(2.70 \times 10^3 \text{ m})(9.80 \text{ m/s}^2)} - 1\right]$$

$$= \boxed{0.911\,mg}$$

In this case, the force exerted by the seat on the pilot is *less* than the true weight by a factor of 0.911. Hence, the pilot will feel lighter at the top of the loop.

Exercise   Calculate the central force on the pilot if the aircraft is at point A in Figure 6.7a, midway up the loop.

Answer $n_A = 1.911\,mg$ directed to the right.

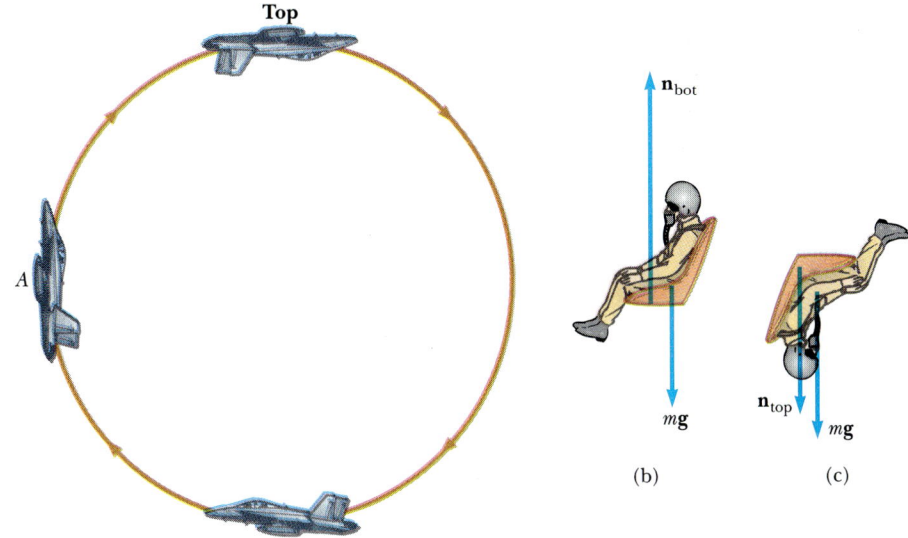

FIGURE 6.7   (a) An aircraft executes a "loop-the-loop" maneuver as it moves in a vertical circle at constant speed. (b) Free-body diagram for the pilot at the bottom of the loop. In this position the pilot experiences an apparent weight that is greater than his true weight. (c) Free-body diagram for the pilot at the top of the loop.

Some examples of central forces acting during circular motion. (*Left*) Cyclists in the Tour de France negotiate a curve on a flat racing track. (*Right*) Passengers in a roller coaster at Knott's Berry Farm. What are the origins of the central forces in these two examples?   *(Left: Michel Gouverneur, Photo News, Gamma Sport; Right: Superstock)*

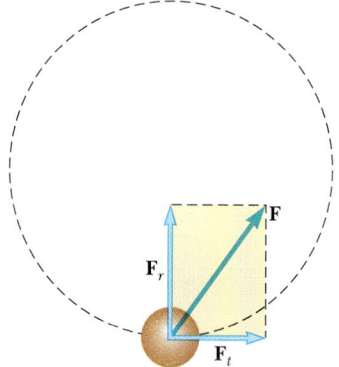

FIGURE 6.8   When the force acting on a particle moving in a circular path has a tangential component F_t, its speed changes. The total force on the particle in this case is the vector sum of the tangential force and the central force. That is, $\mathbf{F} = \mathbf{F}_t + \mathbf{F}_r$.

6.2   NONUNIFORM CIRCULAR MOTION

In Chapter 4 we found that if a particle moves with varying speed in a circular path, there is, in addition to the centripetal component of acceleration, a tangential component of magnitude dv/dt. Therefore, the force acting on the particle must also have a tangential and a radial component. That is, since the total acceleration is $\mathbf{a} = \mathbf{a}_r + \mathbf{a}_t$, the total force exerted on the particle is $\mathbf{F} = \mathbf{F}_r + \mathbf{F}_t$, as shown in Figure 6.8. The vector $\mathbf{F}_r$ is directed toward the center of the circle and is responsible for the centripetal acceleration. The vector $\mathbf{F}_t$ tangent to the circle is responsible for the tangential acceleration, which causes the speed of the particle to change with time. The following example demonstrates this type of motion.

EXAMPLE 6.7   Follow the Rotating Ball

A small sphere of mass m is attached to the end of a cord of length R, which rotates in a *vertical* circle about a fixed point O, as in Figure 6.9a. Let us determine the tension in the cord at any instant when the speed of the sphere is v and the cord makes an angle θ with the vertical.

Solution   First we note that the speed is *not* uniform because there is a tangential component of acceleration arising from the weight of the sphere. From the free-body diagram in Figure 6.9a, we see that the only forces acting on the sphere are the weight, $m\mathbf{g}$, and the force exerted by the cord, $\mathbf{T}$. Now we resolve $m\mathbf{g}$ into a tangential component, $mg \sin \theta$, and a radial component, $mg \cos \theta$. Applying Newton's second law to the forces in the tangential direction gives

$$\sum F_t = mg \sin \theta = ma_t$$

$$a_t = g \sin \theta$$

This component causes v to change in time, since $a_t = dv/dt$.

Applying Newton's second law to the forces in the radial direction and noting that both $\mathbf{T}$ and $\mathbf{a}_r$ are directed toward O, we get

$$\sum F_{r_i} = T - mg \cos \theta = \frac{mv^2}{R}$$

$$T = m\left(\frac{v^2}{R} + g \cos \theta\right)$$

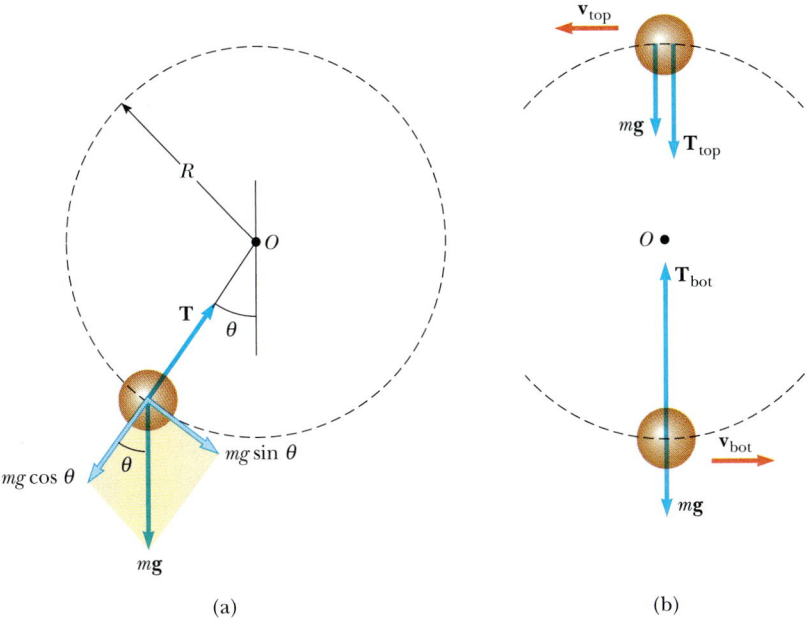

FIGURE 6.9   (Example 6.7) (a) Forces acting on a mass m connected to a string of length R and rotating in a vertical circle centered at O. (b) Forces acting on m when it is at the top and bottom of the circle. Note that the tension at the bottom is a maximum and the tension at the top is a minimum.

Limiting Cases   At the top of the path, where $\theta = 180°$, we have $\cos 180° = -1$, and the tension equation becomes

$$T_{\text{top}} = m \left(\frac{v_{\text{top}}^2}{R} - g \right)$$

This is the *minimum* value of T. Note that at this point $a_t = 0$ and therefore the acceleration is radial and directed downward, as in Figure 6.9b.

At the bottom of the path, where $\theta = 0$, we see that since $\cos 0 = 1$,

$$T_{\text{bot}} = m \left(\frac{v_{\text{bot}}^2}{R} + g \right)$$

This is the *maximum* value of T. Again, at this point $a_t = 0$ and the acceleration is radial and directed upward.

Exercise   At what orientation of the system would the cord most likely break if the average speed increased?

Answer   At the bottom of the path, where T has its maximum value.

*6.3   MOTION IN ACCELERATED FRAMES

When Newton's laws of motion were introduced in Chapter 5, we emphasized that the laws are valid only when observations are made in an inertial frame of reference. In this section, we analyze how an observer in a noninertial frame of reference (one that is accelerating) would attempt to apply Newton's second law.

If a particle moves with an acceleration **a** relative to an observer in an inertial frame, then the inertial observer may use Newton's second law and correctly claim that $\Sigma \mathbf{F}_i = m\mathbf{a}$. If an observer in an accelerated frame (the noninertial observer) tries to apply Newton's second law to the motion of the particle, she or he must introduce **fictitious forces** to make Newton's second law work in that frame. These forces "invented" by the noninertial observer appear to be real forces in the accelerating frame. However, we emphasize that these fictitious forces do not exist when the motion is observed in an inertial frame. The fictitious forces are used only in an accelerating frame but do not represent "real" forces on the body. (By

Fictitious forces

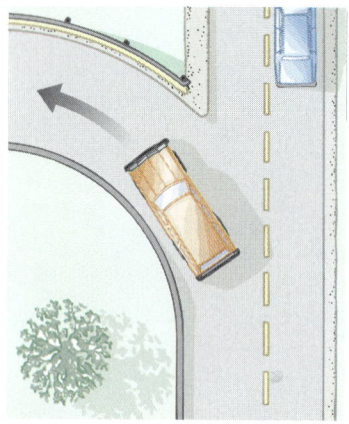

FIGURE 6.10   A car approaching a curved exit ramp. What causes the passenger on the right side to move toward the door?

real forces, we mean the interaction of the body with its environment.) If the fictitious forces are properly defined in the accelerating frame, then the description of motion in this frame will be equivalent to the description by an inertial observer who considers only real forces. Usually, motions are analyzed using inertial reference frames, but there are cases in which an accelerating frame is more convenient.

Suppose you are a passenger in a car accelerating along a straight line down a highway. What causes a building to appear to accelerate in the opposite direction? From this observation, you would have to imagine a force acting on the (stationary) building to cause its apparent acceleration.

In order to understand the motion of a system that is noninertial due to rotation, consider a car traveling along a highway at a high speed and approaching a curved exit ramp, as in Figure 6.10. As the car takes the sharp left turn onto the ramp, a person sitting in the passenger seat slides to the right across the seat and hits the door. At that point, the force exerted by the door keeps her from being ejected from the car. What causes the passenger to move toward the door? A popular, but improper, explanation is that some mysterious force pushes her outward. (This is often called the "centrifugal" force, but we shall not use this term since it often creates confusion.) The passenger invents this fictitious force in order to explain what is going on in her accelerated frame of reference. The driver of the car also experiences this effect, but holds on to the steering wheel to keep from sliding across the seat.

The phenomenon is correctly explained as follows. Before the car enters the ramp, the passenger is moving in a straight-line path. As the car enters the ramp and travels a curved path, the passenger tends to move along the original straight-line path. This is in accordance with Newton's first law: The natural tendency of a body is to continue moving in a straight line. However, if a sufficiently large central force (toward the center of curvature) acts on the passenger, she will move in a curved path along with the car. The origin of this central force is the force of friction between the passenger and the car seat. If this frictional force is not large enough, the passenger will slide across the seat as the car turns under her. Eventually, the passenger encounters the door, which provides a large enough central force to enable the passenger to follow the same curved path as the car. The passenger slides toward the door not because of some mysterious outward force but because *there is no central force large enough to allow her to travel along the circular path followed by the car.*

In summary, one must be very careful to distinguish real forces from fictitious ones in describing motion in an accelerating frame. An observer in a car rounding a curve is in an accelerating frame and invents a fictitious outward force to explain why he or she is thrown outward. A stationary observer outside the car, however, considers only real forces on the passenger. To this observer, the mysterious outward force *does not exist!* The only real external force on the passenger is the central (inward) force due either to friction or to the normal force exerted by the door.

EXAMPLE 6.8   Fictitious Forces in Linear Motion

A small sphere of mass m is hung by a cord from the ceiling of an accelerating boxcar, as in Figure 6.11. (When the boxcar is not accelerating, the cord is vertical, or $\theta = 0$.) According to the inertial observer at rest (Figure 6.11a), the forces on the sphere are force exerted by the cord T and the force of

gravity mg. The inertial observer concludes that the acceleration of the sphere is the same as that of the boxcar and that this acceleration is provided by the horizontal component of T. Also, the vertical component of T balances the weight. Therefore, the inertial observer writes Newton's second law

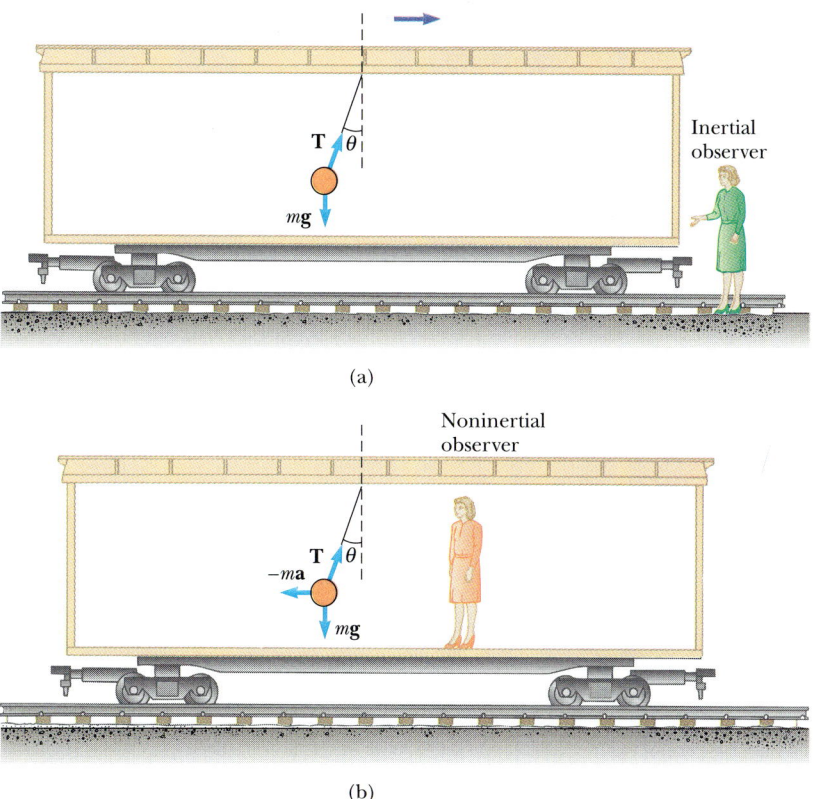

(a)

(b)

FIGURE 6.11 (Example 6.8) (a) A small sphere suspended from the ceiling of a boxcar accelerating to the right is deflected as shown. The inertial observer at rest outside the car claims that the acceleration of the sphere is provided by the horizontal component of **T**. (b) A noninertial observer riding in the car says that the net force on the sphere is zero and that the deflection of the string from the vertical is due to a fictitious force, $-m\mathbf{a}$, which balances the horizontal component of **T**.

as $\mathbf{T} + m\mathbf{g} = m\mathbf{a}$, which in component form becomes

$$\text{Inertial observer} \begin{cases} (1) & \sum F_x = T \sin\theta = ma \\ (2) & \sum F_y = T \cos\theta - mg = 0 \end{cases}$$

Thus, by solving (1) and (2) simultaneously, the inertial observer can determine the acceleration of the car through the relationship

$$a = g \tan\theta$$

Therefore, since the deflection of the string from the vertical serves as a measure of the acceleration of the car, *a simple pendulum can be used as an accelerometer.*

According to the noninertial observer riding in the car, described in Figure 6.11b, the cord still makes an angle θ with the vertical, the sphere is at rest, and its acceleration is zero. Therefore, the noninertial observer introduces a fictitious force, $-m\mathbf{a}$, to balance the horizontal component of **T** and claims that the net force on the sphere is *zero!* In this noninertial frame of reference, Newton's second law in component form gives

$$\text{Noninertial observer} \begin{cases} \sum F'_x = T \sin\theta - ma = 0 \\ \sum F'_y = T \cos\theta - mg = 0 \end{cases}$$

These expressions are equivalent to (1) and (2); therefore, the noninertial observer gets the same mathematical results as the inertial observer. However, the physical interpretation of the deflection of the string differs in the two frames of reference. Note that even though a pendulum is used, it does not oscillate in this application.

EXAMPLE 6.9   Fictitious Force in a Rotating System

An observer in a rotating system is another example of a noninertial observer. Suppose a block of mass m lying on a horizontal, frictionless turntable is connected to a string as in Figure 6.12. According to an inertial observer, if the block rotates uniformly, it undergoes an acceleration of magnitude v^2/r, where v is its tangential speed. The inertial observer concludes that this centripetal acceleration is provided by the force exerted by the string, T, and writes Newton's second law $T = mv^2/r$.

According to a noninertial observer attached to the turn-table, the block is at rest. Therefore, in applying Newton's second law, this observer introduces a fictitious outward force of magnitude mv^2/r. According to the noninertial observer, this outward force balances the force exerted by the string and therefore $T - mv^2/r = 0$.

You should be careful when using fictitious forces to describe physical phenomena. Remember that fictitious forces are used *only* in noninertial frames of reference. When solving problems, it is often best to use an inertial frame.

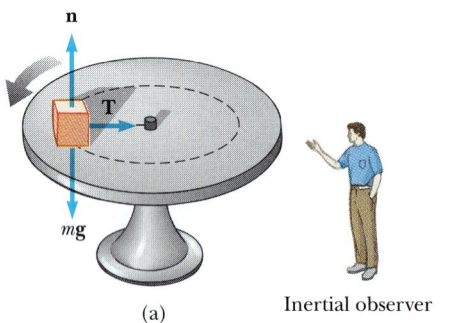
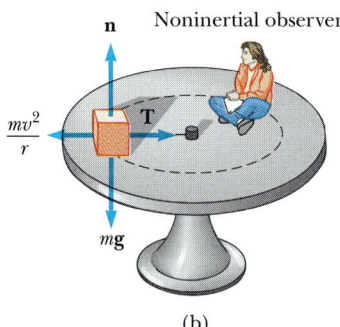

(a)   Inertial observer   (b)

FIGURE 6.12   (Example 6.9) A block of mass m connected to a string tied to the center of a rotating turntable. (a) The inertial observer claims that the central force is provided by the force T exerted by the string on the block. (b) The noninertial observer claims that the block is not accelerating and therefore he introduces a fictitious force of magnitude mv^2/r, which acts outward and balances the force T.

*6.4   MOTION IN THE PRESENCE OF RESISTIVE FORCES

In the previous chapter we described the interaction between a moving object and the surface along which it moves. We completely ignored any interaction between the object and the medium through which it moves. Now let us consider the effect of a medium such as a liquid or gas. The medium exerts a **resistive force R** on the object moving through it. The magnitude of this force depends on such factors as the speed of the object, and the direction of **R** is always opposite the direction of motion of the object relative to the medium. Generally, the magnitude of the resistive force increases with increasing speed. Some examples are the air resistance associated with moving vehicles (sometimes called air drag) and the viscous forces that act on objects moving through a liquid.

The resistive force can have a complicated speed dependence. In the following discussions, we consider two situations. First, we assume that the resistive force is proportional to the speed; this is the case for objects that fall through a liquid with low speed and for very small objects, such as dust particles, that move through air. Second, we treat situations for which the resistive force is proportional to the square of the speed of the object; large objects, such as a skydiver moving through air in free-fall, experience such a force.

Resistive Force Proportional to Object Speed

If we assume that the resistive force acting on an object that is moving through a viscous medium is proportional to the object's velocity, then the resistive force can be expressed as

$$\mathbf{R} = -b\mathbf{v} \tag{6.2}$$

where $\mathbf{v}$ is the velocity of the object and b is a constant that depends on the properties of the medium and on the shape and dimensions of the object. If the object is a sphere of radius r, then b is proportional to r.

Consider a sphere of mass m released from rest in a liquid, as in Figure 6.13a. Assuming the only forces acting on the sphere are the resistive force, $-b\mathbf{v}$, and the weight, $m\mathbf{g}$, let us describe its motion.[1]

Applying Newton's second law to the vertical motion, choosing the downward direction to be positive, and noting that $\Sigma F_y = mg - bv$, we get

$$mg - bv = m\frac{dv}{dt}$$

where the acceleration is downward. Simplifying this expression gives

$$\frac{dv}{dt} = g - \frac{b}{m}v \tag{6.3}$$

Equation 6.3 is called a *differential equation,* and the methods of solving such an equation may not be familiar to you as yet. However, note that initially, when $v = 0$, the resistive force is zero and the acceleration, dv/dt, is simply g. As t increases, the resistive force increases and the acceleration decreases. Eventually, the acceleration becomes zero when the resistive force equals the weight. At this point, the object reaches its **terminal speed,** v_t, and from then on it continues to move with zero acceleration. The terminal speed can be obtained from Equation 6.3 by setting $a = dv/dt = 0$. This gives

$$mg - bv_t = 0 \qquad \text{or} \qquad v_t = mg/b$$

The expression for v that satisfies Equation 6.3 with $v = 0$ at $t = 0$ is

$$v = \frac{mg}{b}(1 - e^{-bt/m}) = v_t(1 - e^{-t/\tau}) \tag{6.4}$$

This function is plotted in Figure 6.13b. The time constant $\tau = m/b$ is the time it takes the object to reach 63.2% of its terminal speed. This can be seen by noting that when $t = \tau$, Equation 6.4 gives $v = 0.632v_t$. We can check that Equation 6.4 is a solution to Equation 6.3 by direct differentiation:

$$\frac{dv}{dt} = \frac{d}{dt}\left(\frac{mg}{b} - \frac{mg}{b}e^{-bt/m}\right) = -\frac{mg}{b}\frac{d}{dt}e^{-bt/m} = ge^{-bt/m}$$

Substituting this expression and Equation 6.4 into Equation 6.3 shows that our solution satisfies the differential equation.

---

[1] There is also a *buoyant* force acting on the submerged object, and this force is constant and equal to the weight of the displaced fluid. This force will only change the apparent weight of the sphere by a constant factor. We shall discuss such buoyant forces in Chapter 15.

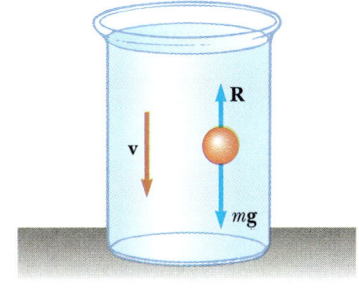

(a)

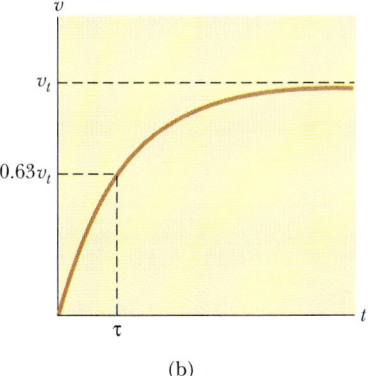

(b)

FIGURE 6.13   (a) A small sphere falling through a viscous fluid. (b) The speed-time graph for an object falling through a viscous medium. The object reaches a maximum, or terminal speed v_t, and τ is the time it takes to reach 0.63 v_t.

Aerodynamic car. Streamlined bodies are used for sports cars and other vehicles to reduce air drag and increase fuel efficiency.   *(© 1992 Dick Kelley)*

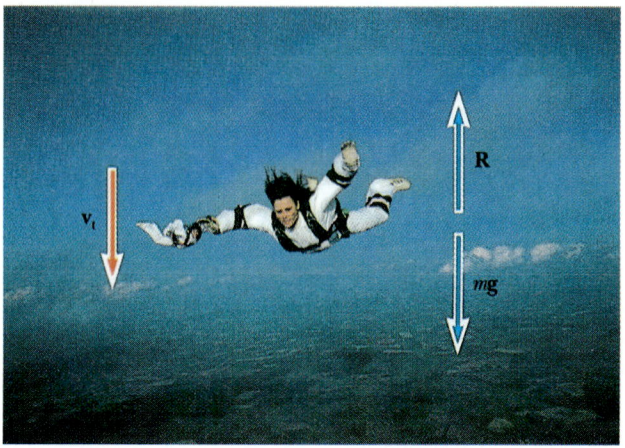

By spreading their arms and legs and by keeping their bodies parallel to the ground, sky divers experience maximum air resistance resulting in a minimum terminal speed.   *(Guy Sauvage, Photo Researchers, Inc.)*

EXAMPLE 6.10   Sphere Falling in Oil

A small sphere of mass 2.00 g is released from rest in a large vessel filled with oil. The sphere reaches a terminal speed of 5.00 cm/s. Determine the time constant τ and the time it takes the sphere to reach 90% of its terminal velocity.

Solution   Since the terminal speed is given by $v_t = mg/b$, the coefficient b is

$$b = \frac{mg}{v_t} = \frac{(2.00 \text{ g})(980 \text{ cm/s}^2)}{5.00 \text{ cm/s}} = 392 \text{ g/s}$$

Therefore, the time τ is given by

$$\tau = \frac{m}{b} = \frac{2.00 \text{ g}}{392 \text{ g/s}} = \boxed{5.10 \times 10^{-3} \text{ s}}$$

The speed of the sphere as a function of time is given by Equation 6.4. To find the time t it takes the sphere to reach a speed of $0.900 v_t$, we set $v = 0.900 v_t$ into this expression and solve for t:

$$0.900 v_t = v_t(1 - e^{-t/\tau})$$

$$1 - e^{-t/\tau} = 0.900$$

$$e^{-t/\tau} = 0.100$$

$$-\frac{t}{\tau} = -2.30$$

$$t = 2.30\tau = 2.30(5.10 \times 10^{-3} \text{ s})$$

$$= 11.7 \times 10^{-3} \text{ s} = \boxed{11.7 \text{ ms}}$$

Exercise   What is the sphere's speed through the oil at $t = 11.7$ ms? Compare this value with the speed the sphere would have if it were falling in a vacuum and so influenced only by gravity.

Answer   4.50 cm/s in oil versus 11.5 cm/s in free-fall.

Air Drag at High Speeds

For large objects moving at high speeds through air, such as airplanes, sky divers, and baseballs, the resistive force is approximately proportional to the square of the speed. In these situations, the magnitude of the resistive force can be expressed as

$$R = \tfrac{1}{2}D\rho A v^2 \tag{6.5}$$

where ρ is the density of air, A is the cross-sectional area of the falling object measured in a plane perpendicular to its motion, and D is a dimensionless empirical quantity called the *drag coefficient*. The drag coefficient has a value of about 0.5 for spherical objects but can be as high as 2 for irregularly shaped objects.

Consider an airplane in flight that experiences such a resistive force. Equation 6.5 shows that the force is proportional to the density of air and hence decreases with decreasing air density. Since air density decreases with increasing altitude, the resistive force on a jet airplane flying at a given speed must also decrease with increasing altitude. However, if at a given altitude the plane's speed is doubled, the resistive force increases by a factor of 4. In order to maintain this increased speed, the propulsive force must also increase by a factor of 4 and the power required (force times speed) must increase by a factor of 8.

Now let us analyze the motion of a mass in free-fall subject to an upward air resistive force whose magnitude is $R = \frac{1}{2}D\rho Av^2$. Suppose a mass m is released from rest from the position $y = 0$ as in Figure 6.14. The mass experiences two external forces: the downward force of gravity, mg, and the resistive force, $\mathbf{R}$, upward. (There is also an upward buoyant force that we neglect.) Hence, the magnitude of the net force is

$$F_{net} = mg - \tfrac{1}{2}D\rho Av^2 \tag{6.6}$$

Substituting $F_{net} = ma$ into Equation 6.6, we find that the mass has a downward acceleration of magnitude

$$a = g - \left(\frac{D\rho A}{2m}\right)v^2 \tag{6.7}$$

Again, we can calculate the terminal speed, v_t, using the fact that when the weight is balanced by the resistive force, the net force is zero and therefore the acceleration is zero. Setting $a = 0$ in Equation 6.7 gives

$$g - \left(\frac{D\rho A}{2m}\right)v_t^2 = 0$$

$$v_t = \sqrt{\frac{2mg}{D\rho A}} \tag{6.8}$$

Using this expression, we can determine how the terminal speed depends on the dimensions of the object. Suppose the object is a sphere of radius r. In this case, $A \propto r^2$ and $m \propto r^3$ (since the mass is proportional to the volume). Therefore, $v_t \propto \sqrt{r}$.

Table 6.1 lists the terminal speeds for several objects falling through air.

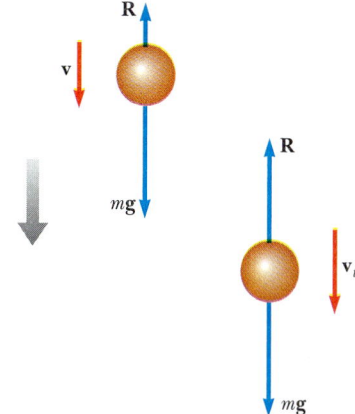

FIGURE 6.14   An object falling through air experiences a resistive drag force $\mathbf{R}$ and a gravitational force $m\mathbf{g}$. The object reaches terminal speed (on the right) when the net force is zero, that is, when $\mathbf{R} = -m\mathbf{g}$, or $R = mg$. Before this occurs, the acceleration varies with speed according to Equation 6.7.

TABLE 6.1   Terminal Speed for Various Objects Falling Through Air

Object	Mass (kg)	Surface Area (m²)	v_t(m/s)
Sky diver	75	0.70	60
Baseball (radius 3.7 cm)	0.145	4.2×10^{-3}	43
Golf ball (radius 2.1 cm)	0.046	1.4×10^{-3}	44
Hailstone (radius 0.50 cm)	4.8×10^{-4}	7.9×10^{-5}	14
Raindrop (radius 0.20 cm)	3.4×10^{-5}	1.3×10^{-5}	9.0

CONCEPTUAL EXAMPLE 6.11

Consider a sky diver falling through air before reaching her terminal speed. As the speed of the sky diver increases, what happens to her acceleration? What is her acceleration once she reaches terminal speed?

Reasoning   The forces exerted on the sky diver are the downward force of gravity (her weight) and an upward force of air resistance, which is less than her weight before she reaches terminal speed. As her downward speed increases, the force of air resistance increases. The vector sum of the force of gravity and the force of air resistance gives a total force which decreases with time, so her acceleration decreases. Once she reaches terminal speed, the two forces balance each other, the total force is zero, and her acceleration is zero.

*6.5 NUMERICAL MODELING IN PARTICLE DYNAMICS[2]

As we have seen in this and the preceding chapter, the study of dynamics of a particle focuses on describing the position, velocity, and acceleration as functions of time. Cause-and-effect relationships exist between these quantities: A velocity causes position to change, an acceleration causes velocity to change, and an acceleration is the direct result of applied forces. Therefore, the study of motion usually begins with an evaluation of the net force on the particle.

In this section, we confine our discussion to one-dimensional motion, so boldface notation will not be used for vector quantities. If a particle of mass m moves under the influence of a net force F, Newton's second law tells us that the acceleration of the particle is given by $a = F/m$. In general, we can then obtain the solution to a dynamics problem by using the following procedure:

1. Sum all the forces on the particle to get the net force F.
2. Use this force to determine the acceleration, using $a = F/m$.
3. Use this acceleration to determine the velocity from $dv/dt = a$.
4. Use this velocity to determine the position from $dx/dt = v$.

EXAMPLE 6.12   An Object Falling in a Vacuum

We can illustrate the procedure just described by considering a particle falling in a vacuum under the influence of the force of gravity as in Figure 6.15.

$$mg$$

FIGURE 6.15   An object falling in vacuum under the influence of gravity.

Reasoning and Solution   Applying Newton's second law, we set the sum of the external forces equal to the mass of the particle times its acceleration (taking upward to be the positive direction):

$$F = ma = -mg$$

Thus, $a = -g$, a constant. Because $dv/dt = a$, the resulting differential equation for velocity is $dv/dt = -g$, which may be integrated to give

$$v(t) = v_0 - gt$$

Then, since $dx/dt = v$, the position of the particle is obtained from another integration, which yields the well-known result

$$x(t) = x_0 + v_0 t - \tfrac{1}{2}at^2$$

In this expression, x_0 and v_0 represent the position and speed of the particle at $t = 0$.

The procedure just described is straightforward for some physical situations, such as the one described in the previous example. In the "real world," however, complications often arise that make analytical solutions for many practical situations difficult and perhaps beyond the mathematical abilities of most students

---

[2] The author is most grateful to Colonel James Head of the U.S. Air Force Academy for preparing this section. A more extensive treatment of this topic by Colonel Head is included in a supplement entitled *Numerical Methods in Physics*.

taking introductory physics. For example, the force may depend on the position of the particle, as in cases where variation in gravitational acceleration with height must be taken into account. The force may also vary with velocity, as we have seen in cases of resistive forces caused by motion through a liquid or gas. The force may depend on both position and velocity, as in the case of an object falling through air where the resistive force depends on velocity and on height (air density). In rocket motion, the mass changes with time, so even if the force is constant, the acceleration is not.

Another complication arises because the equations relating acceleration, velocity, position, and time are differential, not algebraic, equations. Differential equations are usually solved using integral calculus and other special techniques that introductory students may not have mastered.

So how does one proceed to solve real-world problems without advanced mathematics? One answer is to solve such problems on personal computers, using elementary numerical methods. The simplest of these is the Euler method, named after the Swiss mathematician Leonhard Euler (1707–1783).

The Euler Method

In the **Euler method** of solving differential equations, derivatives are approximated as finite differences. Considering a small increment of time, Δt, the relationship between speed and acceleration may be approximated as

$$a(t) = \frac{\Delta v}{\Delta t} = \frac{v(t + \Delta t) - v(t)}{\Delta t}$$

Then the speed of the particle at the end of the period Δt is approximately equal to the speed at the beginning of the period, plus the acceleration during the interval multiplied by Δt:

$$v(t + \Delta t) = v(t) + a(t) \, \Delta t \tag{6.9}$$

Since the acceleration is a function of time, this estimate of $v(t + \Delta t)$ will be accurate only if the time interval Δt is short enough that the change in acceleration during it is very small (as will be discussed later).

The position can be found in the same manner:

$$v(t) = \frac{x(t + \Delta t) - x(t)}{\Delta t}$$

so

$$x(t + \Delta t) = x(t) + v(t) \, \Delta t \tag{6.10}$$

It may be tempting to add the term $\frac{1}{2} a (\Delta t)^2$ to this result to make it look like the familiar kinematics equation, but this term is not included in the Euler method of integration because Δt is assumed to be so small that $(\Delta t)^2$ is nearly zero.

If the acceleration at any instant t is known, the particle's velocity and position at $(t + \Delta t)$ can be calculated from Equations 6.9 and 6.10. The calculation can then proceed in a series of finite steps to determine the velocity and position at any later time. The acceleration is determined by the net force acting on the object,

$$a(x, v, t) = \frac{F(x, v, t)}{m} \tag{6.11}$$

which may depend explicitly on the position, velocity, or time.

TABLE 6.2 **The Euler Method for Solving Dynamics Problems**

Step	Time	Position	Velocity	Acceleration
0	t_0	x_0	v_0	$a_0 = F(x_0, v_0, t_0)/m$
1	$t_1 = t_0 + \Delta t$	$x_1 = x_0 + v_0\,\Delta t$	$v_1 = v_0 + a_0\,\Delta t$	$a_1 = F(x_1, v_1, t_1)/m$
2	$t_2 = t_1 + \Delta t$	$x_2 = x_1 + v_1\,\Delta t$	$v_2 = v_1 + a_1\,\Delta t$	$a_2 = F(x_2, v_2, t_2)/m$
3	$t_3 = t_2 + \Delta t$	$x_3 = x_2 + v_2\,\Delta t$	$v_3 = v_2 + a_2\,\Delta t$	$a_3 = F(x_3, v_3, t_3)/m$
.	.	.	.	.
.	.	.	.	.
.	.	.	.	.
n	t_n	x_n	v_n	a_n

It is convenient to set up the numerical solution to this kind of problem by numbering the steps and entering the calculations in a table. Table 6.2 illustrates how to do this in an orderly way.

The equations provided in the table can be entered into a spreadsheet and the calculations performed row by row to determine the velocity, position, and acceleration as functions of time. The calculations can also be carried out using a program written in BASIC, PASCAL, or FORTRAN or with commercially available mathematics packages for personal computers. Many small increments can be taken, and accurate results can usually be obtained with the help of a computer. Graphs of velocity versus time or position versus time can be displayed to help you visualize the motion.

The Euler method has the advantage that the dynamics is not obscured — the fundamental relationships of acceleration to force, velocity to acceleration, and position to velocity are clearly evident. Indeed, these relationships form the heart of the calculations. There is no need to use advanced mathematics, and the basic physics governs the dynamics.

The Euler method is completely reliable for infinitesimally small time increments, but for practical reasons a finite increment size must be chosen. In order for the finite difference approximation of Equation 6.9 to be valid, the time increment must be small enough that the velocity can be approximated as being constant during the increment. We can determine an appropriate size for the time increment by examining the particular problem that is being investigated. The criterion for the size of the time increment may need to be changed during the course of the motion. In practice, however, we usually choose a time increment appropriate to the initial conditions and use the same value throughout the calculations.

The size of the time increment influences the accuracy of the result, but unfortunately it is not easy to determine the accuracy of a solution by the Euler method without a knowledge of the correct analytical solution. One method of determining the accuracy of the numerical solution is to repeat the calculations with a smaller time increment and compare results. If the two calculations agree to a certain number of significant figures, you can assume that the results are correct to that precision.

6.6 THE FUNDAMENTAL FORCES OF NATURE

In the previous chapter, we described a variety of forces that are experienced in our everyday activities, such as the force of gravity that acts on all bodies at or near the Earth's surface and the force of friction as one surface slides over another.

Other forces we have encountered are the force exerted by a rope on an object, the normal force acting on an object in contact with some other object, and the resistive force as an object moves through a liquid or gas. Other forces we shall encounter include the restoring force in a deformed spring, the electrostatic force between two charged objects, and the magnetic force between a magnet and a piece of iron.

Forces also act in the atomic and subatomic world. For example, atomic forces within the atom are responsible for holding its constituents together, and nuclear forces act on different parts of the nucleus to keep its parts from separating.

Until recently, physicists believed that there were four fundamental forces in nature: the gravitational force, the electromagnetic force, the strong nuclear force, and the weak nuclear force.

The **gravitational force** mentioned in Example 6.5 is the mutual force of attraction between all masses. We have already encountered the gravitational force when describing the weight of an object. Although the magnitude of gravitational forces can be significant between macroscopic objects, they are the weakest of the four fundamental forces. For example, the gravitational force between the electron and proton in the hydrogen atom is only about 10^{-47} N, whereas the electrostatic force between the two particles is about 10^{-7} N. We shall learn more about the nature of the gravitational force in Chapter 14.

The **electromagnetic force** is an attraction or repulsion between two charged particles that may be in relative motion. Although much stronger than the gravitational force, the electromagnetic force between two charged elementary particles is of medium strength. The force that causes a rubbed comb to attract bits of paper and the force that a magnet exerts on an iron nail are examples of electromagnetic forces. It is interesting to note that essentially all forces in our macroscopic world (apart from the gravitational force) are manifestations of the electromagnetic force when examined closely. For example, friction forces, contact forces, tension forces, and forces in stretched springs or other deformed bodies are essentially the consequence of electromagnetic forces between charged particles in close proximity.

The **strong nuclear force** is responsible for the stability of nuclei. This force represents the ''glue'' that holds the nuclear constituents (called nucleons) together. It is the strongest of all the fundamental forces. For separations of about 10^{-15} m (a typical nuclear dimension), the strong nuclear force is one to two orders of magnitude stronger than the electromagnetic force. However, the strong nuclear force decreases rapidly with increasing separation and is negligible for separations greater than about 10^{-14} m.

Finally, the **weak nuclear force** is a short-range nuclear force that tends to produce instability in certain nuclei. Most radioactive decay reactions are caused by the weak nuclear force, which is about 12 orders of magnitude weaker than the electromagnetic force.

In 1979, physicists predicted that the electromagnetic force and the weak force are manifestations of one and the same force called the **electroweak** force. This prediction was confirmed experimentally in 1984.

Physicists and cosmologists believe that the fundamental forces of nature are closely related to the origin of the Universe. The Big Bang theory of the creation of the Universe states that the Universe erupted from a point-like singularity 15 to 20 billion years ago. According to this theory, the first moments ($\approx 10^{-10}$ s) after the Big Bang saw such extremes of energy that all four fundamental forces were unified. Scientists continue in their search for other possible connections among the four fundamental forces.

SUMMARY

Newton's second law applied to a particle moving in uniform circular motion states that the net force, which is a **central force**, is

$$F_r = ma_r = \frac{mv^2}{r} \tag{6.1}$$

The central force acting on an object that provides the centripetal acceleration could be the force of gravity (as in satellite motion), a force of friction, or a force exerted by a string.

A particle moving in nonuniform circular motion has both a centripetal acceleration and a nonzero tangential component of acceleration. In the case of a particle rotating in a vertical circle, the force of gravity provides the tangential acceleration and part or all of the centripetal acceleration.

An observer in a noninertial frame of reference must introduce **fictitious forces** when applying Newton's second law in that frame. If these fictitious forces are properly defined, the description of motion in the noninertial frame is equivalent to that made by an observer in an inertial frame. However, the observers in the two different frames will not agree on the causes of the motion.

A body moving through a liquid or gas experiences a **resistive force** that is speed dependent. This resistive force, which opposes the motion, generally increases with speed. The force depends on the shape of the body and the properties of the medium through which the body is moving. In the limiting case for a falling body, when the magnitude of the resistive force equals the weight ($a = 0$), the body reaches its **terminal speed**.

There are four fundamental forces in nature: the strong nuclear force, the electromagnetic force, the gravitational force, and the weak force.

QUESTIONS

1. Because the Earth rotates about its axis and revolves around the Sun, it is a noninertial frame of reference. Assuming the Earth is a uniform sphere, why would the apparent weight of an object be greater at the poles than at the equator?

2. Explain why the Earth bulges at the equator.

3. How would you explain the force that pushes a rider toward the side of a car as the car rounds a corner?

4. When an airplane does an inside "loop-the-loop" in a vertical plane, at what point does the pilot appear to be heaviest? What is the constraint force acting on the pilot?

5. A sky diver in free-fall reaches terminal speed. After the parachute is opened, what parameters change to decrease this terminal speed?

6. Why is it that an astronaut in a space capsule orbiting the Earth experiences a feeling of weightlessness?

7. Why does mud fly off a rapidly turning wheel?

8. A pail of water can be whirled in a vertical path such that none is spilled. Why does the water stay in, even when the pail is above your head?

9. Imagine that you attach a heavy object to one end of a spring and then whirl the spring and object in a horizon-

tal circle (by holding the free end of the spring). Does the spring stretch? If so, why? Discuss this in terms of central force.

10. It has been suggested that rotating cylinders about 10 mi in length and 5 mi in diameter be placed in space and used as colonies. The purpose of the rotation is to simulate gravity for the inhabitants. Explain this concept for producing an effective gravity.

11. Why does a pilot tend to black out when pulling out of a steep dive?

12. Describe a situation in which an automobile driver can have a centripetal acceleration but no tangential acceleration.

13. Is it possible for a car to move in a circular path in such a way that it has a tangential acceleration but no centripetal acceleration?

14. Analyze the motion of a rock dropped into water in terms of its speed and acceleration as it falls. Assume that there is a resistive force acting on the rock that increases as the speed increases.

15. Centrifuges are often used in dairies to separate the cream from the milk. Which remains on the inside?

16. We often think of the brakes and the gas pedal on a car as

the devices that accelerate the car. Could a steering wheel also fall into this category? Explain.

17. Consider a small raindrop and a large raindrop falling through the atmosphere. Compare their terminal speeds. What are their accelerations when they reach terminal speed?

PROBLEMS

Section 6.1   Newton's Second Law Applied to Uniform Circular Motion

1. A toy car moving at constant speed completes one lap around a circular track (a distance of 200 m) in 25.0 s. (a) What is the average speed? (b) If the mass of the car is 1.50 kg, what is the magnitude of the central force that keeps it in a circle?

2. In a cyclotron (one type of particle accelerator), a deuteron (of atomic mass 2 u) reaches a final speed of 10% of the speed of light while moving in a circular path of radius 0.48 m. The deuteron is maintained in the circular path by a magnetic force. What magnitude of force is required?

3. The wheels of a certain roller coaster are both above and below the rails as shown in Figure P6.3 so that the car will not leave the rails. If the mass supported by this particular wheel is 320 kg and the radius of this section of track is 15 m, (a) what is the magnitude and direction of the force that the track exerts on the wheel when the speed of the car is 20 m/s? (b) What would be the net force exerted on a 60-kg person riding in the car? (c) What supplies this force?

FIGURE P6.3

4. In a hydrogen atom, the electron in orbit around the proton feels an attractive force of about 8.20×10^{-8} N. If the radius of the orbit is 5.30×10^{-11} m, what is the frequency in revolutions per second? See the front cover for additional data.

5. A 3.00-kg mass attached to a light string rotates on a horizontal, frictionless table. The radius of the circle is 0.800 m, and the string can support a mass of 25.0 kg before breaking. What range of speeds can the mass have before the string breaks?

6. A satellite of mass 300 kg is in a circular orbit around the Earth at an altitude equal to the Earth's mean radius (see Example 6.5). Find (a) the satellite's orbital speed, (b) the period of its revolution, and (c) the gravitational force acting on it.

7. While two astronauts were on the surface of the Moon, a third astronaut orbited the Moon. Assume

(Problem 6) *(NASA/Peter Arnold, Inc.)*

the orbit to be circular and 100 km above the surface of the Moon. If the mass and radius of the Moon are 7.40×10^{22} kg and 1.70×10^{6} m, respectively, determine (a) the orbiting astronaut's acceleration, (b) his orbital speed, and (c) the period of the orbit.

8. An automobile moves at constant speed over the crest of a hill. The driver moves in a vertical circle of radius 18.0 m. At the top of the hill, she notices that she barely remains in contact with the seat. Find the speed of the vehicle.

8A. An automobile moves at constant speed over the crest of a hill. The driver moves in a vertical circle of radius R. At the top of the hill, she notices that she barely remains in contact with the seat. Find the speed of the vehicle.

9. A crate of eggs is located in the middle of the flat bed of a pickup truck as the truck negotiates an unbanked curve in the road. The curve may be regarded as an arc of a circle of radius 35 m. If the coefficient of static friction between crate and truck is 0.60, what must be the maximum speed of the truck if the crate is not to slide?

10. A 55-kg ice skater is moving at 4.0 m/s when she grabs the loose end of a rope the opposite end of which is tied to a pole. She then moves in a circle of radius 0.80 m around the pole. (a) Determine the

☐ indicates problems that have full solutions available in the Student Solutions Manual and Study Guide.

force exerted by the rope on her arms. (b) Compare this force with her weight.

11. An air puck of mass 0.250 kg is tied to a string and allowed to revolve in a circle of radius 1.00 m on a horizontal, frictionless table. The other end of the string passes through a hole in the center of the table and a mass of 1.00 kg is tied to it. The suspended mass remains in equilibrium while the puck revolves. (a) What is the tension in the string? (b) What is the central force acting on the puck? (c) What is the speed of the puck?

12. The speed of the tip of the minute hand on a town clock is 1.75×10^{-3} m/s. (a) What is the speed of the tip of the second hand of the same length? (b) What is the centripetal acceleration of the tip of the second hand?

13. A coin placed 30.0 cm from the center of a rotating, horizontal turntable slips when its speed is 50.0 cm/s. (a) What provides the central force when the coin is stationary relative to the turntable? (b) What is the coefficient of static friction between coin and turntable?

Section 6.2 Nonuniform Circular Motion

14. A car traveling on a straight road at 9.00 m/s goes over a hump in the road. The hump may be regarded as an arc of a circle of radius 11.0 m. (a) What is the apparent weight of a 600-N woman in the car as she rides over the hump? (b) What must be the speed of the car over the hump if she is to experience weightlessness? (That is, her apparent weight is zero.)

15. A pail of water is rotated in a vertical circle of radius 1.00 m. What is the minimum speed of the pail at the top of the circle if no water is to spill out?

16. A hawk flies in a horizontal arc of radius 12.0 m at a constant speed of 4.00 m/s. (a) Find its centripetal acceleration. (b) It continues to fly along the same horizontal arc but increases its speed at the rate of 1.20 m/s². Find the acceleration (magnitude and direction) under these conditions.

17. A 40.0-kg child sits in a swing supported by two chains, each 3.00 m long. If the tension in each chain at the lowest point is 350 N, find (a) the child's speed at the lowest point and (b) the force of the seat on the child at the lowest point. (Neglect the mass of the seat.)

17A. A child of mass m sits in a swing supported by two chains, each of length R. If the tension in each chain at the lowest point is T, find (a) the child's speed at the lowest point and (b) the force of the seat on the child at the lowest point. (Neglect the mass of the seat.)

18. A 0.40-kg object is swung in a vertical circular path on a string 0.50 m long. If a constant speed of 4.0 m/s is maintained, what is the tension in the string when the object is at the top of the circle?

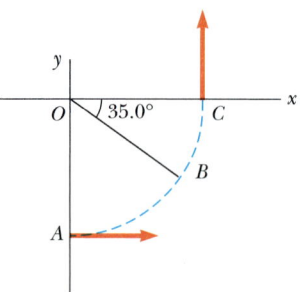

FIGURE P6.19

19. A car initially traveling eastward turns north by traveling in a circular path at uniform speed as in Figure P6.19. The length of the arc *ABC* is 235 m, and the car completes the turn in 36.0 s. (a) What is the acceleration when the car is at *B* located at an angle of 35.0°? Express your answer in terms of the unit vectors **i** and **j**. Determine (b) the car's average speed and (c) its average acceleration during the 36.0-s interval.

20. A roller-coaster vehicle has a mass of 500 kg when fully loaded with passengers (Fig. P6.20). (a) If the vehicle has a speed of 20.0 m/s at point *A*, what is the force exerted by the track on the vehicle at this point? (b) What is the maximum speed the vehicle can have at *B* and still remain on the track?

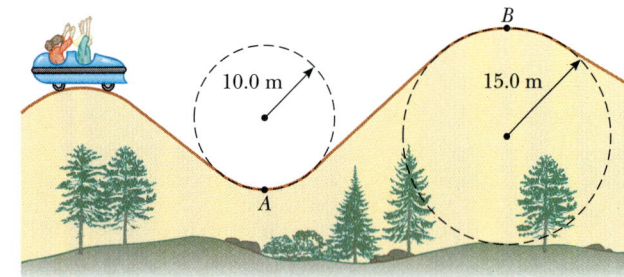

FIGURE P6.20

21. Tarzan ($m = 85.0$ kg) tries to cross a river by swinging from a vine. The vine is 10.0 m long, and his speed at the bottom of the swing (as he just clears the water) is 8.00 m/s. Tarzan doesn't know that the vine has a breaking strength of 1000 N. Does he make it safely across the river?

22. At the Six Flags Great America amusement park in Gurnee, Illinois, there is a roller coaster that incorporates some of the latest design technology and some basic physics. Each vertical loop, instead of being circular, is shaped like a teardrop (Fig. P6.22). The cars ride on the inside of the loop at the top, and the speeds are high enough to ensure that the cars remain on the track. The biggest loop is 40.0 m high,

FIGURE P6.22   *(Frank Cezus/FPG International)*

with a maximum speed of 31.0 m/s (nearly 70 mph) at the bottom. Suppose the speed at the top is 13.0 m/s and the corresponding centripetal acceleration is $2g$. (a) What is the radius of the arc of the teardrop at the top? (b) If the total mass of the cars plus people is M, what force does the rail exert on it at the top? (c) Suppose the roller coaster had a loop of radius 20.0 m. If the cars have the same speed, 13.0 m/s at the top, what is the centripetal acceleration at the top? Comment on the normal force at the top in this situation.

*Section 6.3   Motion in Accelerated Frames

23. A merry-go-round makes one complete revolution in 12.0 s. If a 45.0-kg child sits on the horizontal floor of the merry-go-round 3.00 m from the center, find (a) the child's acceleration and (b) the horizontal force of friction that acts on the child. (c) What minimum coefficient of static friction is necessary to keep the child from slipping?

24. The Earth rotates about its axis with a period of 24 h. Imagine that the rotational speed can be increased. If an object at the equator is to have zero apparent weight, (a) what must the new period be? (b) By what factor would the speed of the object be increased when the planet is rotating at the higher speed? (*Hint:* See Problem 39 and note that the apparent weight of the object becomes zero when the normal force exerted on it is zero. Also, the distance traveled during one period of rotation is $2\pi R$, where R is the Earth's radius.)

25. A 0.500-kg object is suspended from the ceiling of

an accelerating boxcar as in Figure 6.11. If $a = 3.00$ m/s², find (a) the angle that the string makes with the vertical and (b) the tension in the string.

26. A 5.00-kg mass attached to a spring scale rests on a frictionless, horizontal surface as in Figure P6.26. The spring scale, attached to the front end of a boxcar, reads 18.0 N when the car is in motion. (a) If the spring scale reads zero when the car is at rest, determine the acceleration of the car while it is in motion. (b) What will the spring scale read if the car moves with constant velocity? (c) Describe the forces on the mass as observed by someone in the car and by someone at rest outside the car.

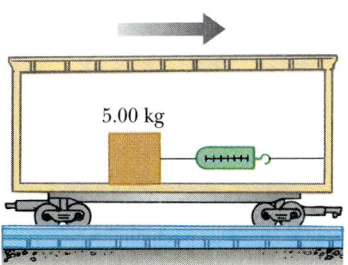

5.00 kg

FIGURE P6.26

27. A person stands on a scale in an elevator. The maximum and minimum scale readings are 591 N and 391 N, respectively. Assume the magnitude of the acceleration is the same during starting and stopping, and determine (a) the weight of the person, (b) the person's mass, and (c) the acceleration of the elevator.

28. A plumb bob does not hang exactly along a line directed to the center of the Earth's rotation. How much does the plumb bob deviate from a radial line at 35° north latitude? Assume that the Earth is spherical.

*Section 6.4   Motion in the Presence of Resistive Forces

29. Assume that the resistive force exerted on a speed skater is $f = -kmv^2$, where k is a constant and m is the skater's mass. The skater crosses the finish line of a straight-line race with a speed v_f and then slows down by coasting without skating. Show that the skater's speed at any time t after crossing the finish line is $v(t) = v_f/(1 + ktv_f)$.

30. A small piece of Styrofoam packing material is dropped from a height of 2.00 m above the ground. Until the terminal speed is reached, the acceleration is given by $a = g - cv$. After falling 0.500 m, it reaches terminal speed, and the Styrofoam takes an extra 5.00 s to reach the ground. (a) What is the value of the constant c? (b) What is the acceleration at $t = 0$? (c) What is the acceleration when the speed is 0.150 m/s?

31. A motor boat cuts its engine when its speed is 10.0 m/s and coasts to rest. The equation governing the motion of the motorboat during this period is

$v = v_0 e^{-ct}$, where v is the speed at time t, v_0 is the initial speed, and c is a constant. At $t = 20.0$ s, the speed is 5.00 m/s. (a) Find the constant c. (b) What is the speed at $t = 40.0$ s? (c) Differentiate the expression for $v(t)$ and thus show that the acceleration of the boat is proportional to the speed at any time.

32. A fire helicopter carries a 620-kg bucket at the end of a cable 20.0 m long as in Figure P6.32. As the helicopter flies to a fire at a constant speed of 40.0 m/s, the cable makes an angle of 40.0° with respect to the vertical. The bucket presents a cross-sectional area of 3.80 m² in a plane perpendicular to the air moving past it. Determine the drag coefficient assuming that the resistive force is proportional to the square of the bucket's speed.

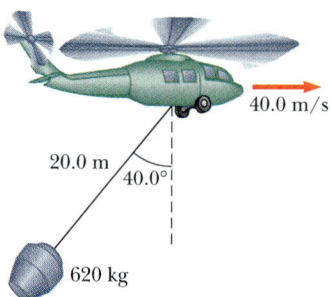

20.0 m
40.0°
40.0 m/s
620 kg

FIGURE P6.32

33. A small, spherical bead of mass 3.00 g is released from rest at $t = 0$ in a bottle of liquid shampoo. The terminal speed is observed to be 2.00 cm/s. Find (a) the value of the constant b in Equation 6.4, (b) the time, τ, it takes to reach $0.630v_t$, and (c) the value of the resistive force when the bead reaches terminal speed.

ADDITIONAL PROBLEMS

34. Suppose the boxcar of Figure 6.11 is moving with constant acceleration a up a hill that makes an angle ϕ with the horizontal. If the plumb bob makes an angle of θ with the perpendicular to the ceiling, what is a?

35. In the Bohr model of the hydrogen atom, the speed of the electron is approximately 2.2×10^6 m/s. Find (a) the central force acting on the electron as it revolves in a circular orbit of radius 0.53×10^{-10} m and (b) the centripetal acceleration of the electron.

36. A 9.00-kg object starting from rest moves through a viscous medium and experiences a resistive force $\mathbf{R} = -b\mathbf{v}$, where $\mathbf{v}$ is the velocity of the object. If the object's speed reaches one-half its terminal speed in 5.54 s, (a) determine the terminal speed. (b) At what time is the speed of the object three-fourths the ter-

minal speed? (c) How far has the object traveled in the first 5.54 s of motion?

37. Consider a conical pendulum with an 80.0-kg bob on a 10.0-m wire making an angle of 5.00° with the vertical (Fig. 6.3). Determine (a) the horizontal and vertical components of the force exerted by the wire on the pendulum and (b) the radial acceleration of the bob.

38. An air puck of mass 0.25 kg is tied to a string and allowed to revolve in a circle of radius 1.0 m on a frictionless horizontal table. The other end of the string passes through a hole in the center of the table, and a mass of 1.0 kg is tied to it (Fig. P6.38). The suspended mass remains in equilibrium while the puck on the tabletop revolves. What are (a) the tension in the string, (b) the central force exerted on the puck, and (c) the speed of the puck?

38A. An air puck of mass m_1 is tied to a string and allowed to revolve in a circle of radius R on a frictionless horizontal table. The other end of the string passes through a hole in the center of the table, and a mass m_2 is tied to it (Fig. P6.38). The suspended mass remains in equilibrium while the puck on the tabletop revolves. (a) What is the tension in the string? (b) What is the central force acting on the puck? (c) What is the speed of the puck?

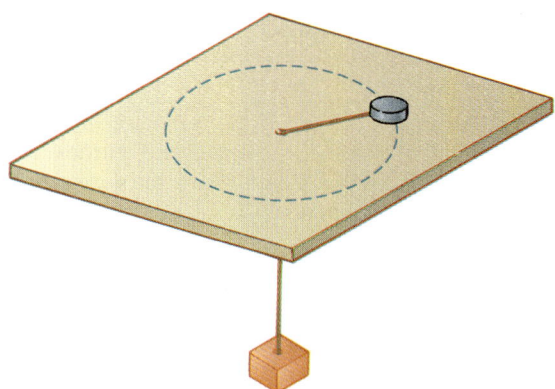

FIGURE P6.38

39. Because the Earth rotates about its axis, a point on the equator experiences a centripetal acceleration of 0.0340 m/s², while a point at the poles experiences no centripetal acceleration. (a) Show that at the equator the gravitational force on an object (the true weight) must exceed the object's apparent weight. (b) What is the apparent weight at the equator and at the poles of a person having a mass of 75.0 kg? (Assume the Earth is a uniform sphere and take $g = 9.800$ N/kg.)

40. A piece of putty is initially located at point A on the rim of a grinding wheel rotating about a horizontal

axis. The putty is dislodged from point A when the diameter through A is horizontal. It then rises vertically and returns to A the instant the wheel completes one revolution. (a) Find the speed of a point on the rim of the wheel in terms of the free-fall acceleration and the radius R of the wheel. (b) If the mass of the putty is m, what is the magnitude of the force that held it to the wheel?

41. A string under a tension of 50.0 N is used to whirl a rock in a horizontal circle of radius 2.50 m at a speed of 20.4 m/s. The string is pulled in and the speed of the rock increases. When the string is 1.00 m long and the speed of the rock is 51.0 m/s, the string breaks. What is the breaking strength (in newtons) of the string?

42. A car rounds a banked curve as in Figure 6.5. The radius of curvature of the road is R, the banking angle is θ, and the coefficient of static friction is μ. (a) Determine the range of speeds the car can have without slipping up or down the road. (b) Find the minimum value for μ such that the minimum speed is zero. (c) What is the range of speeds possible if $R = 100$ m, $\theta = 10°$, and $\mu = 0.10$ (slippery conditions)?

43. A model airplane of mass 0.75 kg flies in a horizontal circle at the end of a 60-m control wire, with a speed of 35 m/s. Compute the tension in the wire if it makes a constant angle of $20°$ with the horizontal. The forces exerted on the airplane are the pull of the control wire, its own weight, and aerodynamic lift, which acts at $20°$ inward from the vertical as shown in Figure P6.43.

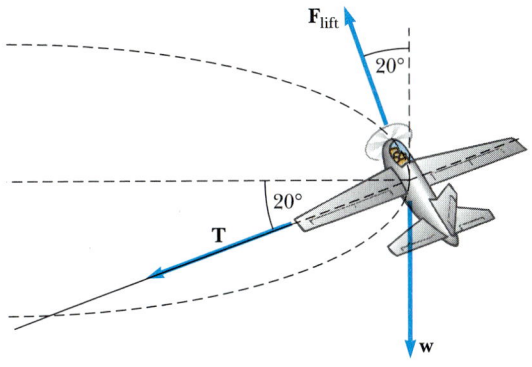

FIGURE P6.43

44. A child's toy consists of a small wedge that has an acute angle θ (Fig. P6.44). The sloping side of the wedge is frictionless, and the wedge is spun at a constant speed by rotating a rod that is firmly attached to it at one end. Show that, when the mass m rises up the wedge a distance L, the speed of the mass is

$$v = \sqrt{gL \sin \theta}$$

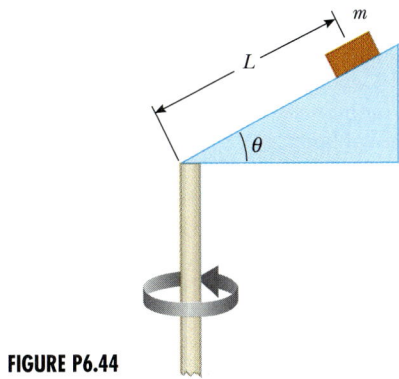

FIGURE P6.44

45. The pilot of an airplane executes a constant-speed loop-the-loop maneuver in a vertical plane. The speed of the airplane is 300 mi/h, and the radius of the circle is 1200 ft. (a) What is the pilot's apparent weight at the lowest point if his true weight is 160 lb? (b) What is his apparent weight at the highest point? (c) Describe how the pilot could experience weightlessness if both the radius and the speed can be varied. (*Note:* His apparent weight is equal to the force that the seat exerts on his body.)

46. In order for a satellite to move in a stable circular orbit at a constant speed, its centripetal acceleration must be inversely proportional to the square of the radius r of the orbit. (a) Show that the tangential speed of a satellite is proportional to $r^{-1/2}$. (b) Show that the time required to complete one orbit is proportional to $r^{3/2}$.

47. An 1800-kg car passes over a bump in a road that follows the arc of a circle of radius 42 m as in Figure P6.47. (a) What force does the road exert on the car as the car passes the highest point of the bump if the car travels at 16 m/s? (b) What is the maximum speed the car can have as it passes the highest point before losing contact with the road?

47A. A car of mass m passes over a bump in a road that follows the arc of a circle of radius R as in Figure P6.47. (a) What force does the road exert on the car as the car passes the highest point of the bump if the car travels at a speed v? (b) What is the maximum speed the car can have as it passes this point before losing contact with the road?

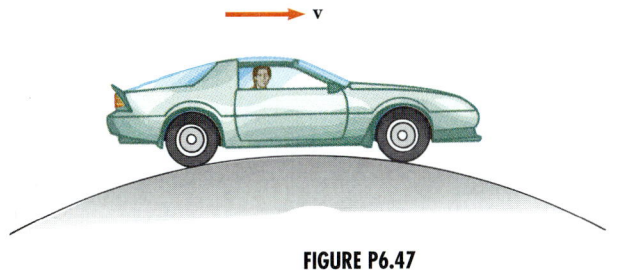

FIGURE P6.47

48. A student builds and calibrates an accelerometer, which she uses to determine the speed of her car around a certain highway curve. The accelerometer is a plumb bob with a protractor that she attaches to the roof of her car. A friend riding in the car with her observes that the plumb bob hangs at an angle of 15.0° from the vertical when the car has a speed of 23.0 m/s. (a) What is the centripetal acceleration of the car rounding the curve? (b) What is the radius of the curve? (c) What is the speed of the car if the plumb bob deflection is 9.0° while rounding the same curve?

49. An amusement park ride consists of a large vertical cylinder that spins about its axis fast enough that any person inside is held up against the wall when the floor drops away (Fig. P6.49). The coefficient of static friction between person and wall is μ_s, and the radius of the cylinder is R. (a) Show that the maximum period of revolution necessary to keep the person from falling is $T = (4\pi^2 R \mu_s / g)^{1/2}$. (b) Obtain a numerical value for T if $R = 4.00$ m and $\mu_s = 0.400$. How many revolutions per minute does the cylinder make?

FIGURE P6.49

50. A penny of mass 3.1 g rests on a small 20.0-g block supported by a spinning disk (Fig. P6.50). If the coefficients of friction between block and disk are 0.75 (static) and 0.64 (kinetic) while those for the penny

and block are 0.45 (kinetic) and 0.52 (static), what is the maximum speed of the disk in revolutions per minute without the block or penny sliding on the disk?

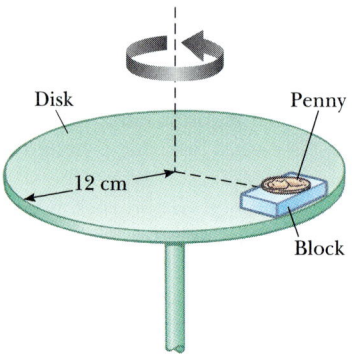

FIGURE P6.50

51. Figure P6.51 shows a Ferris wheel that rotates four times each minute and has a diameter of 18.0 m. (a) What is the centripetal acceleration of a rider? What force does the seat exert on a 40.0-kg rider (b) at the lowest point of the ride and (c) at the highest point of the ride? (d) What force (magnitude and direction) does the seat exert on a rider when the rider is halfway between top and bottom?

FIGURE P6.51   *(Color Box/FPG)*

52. An amusement park ride consists of a rotating circular platform 8.00 m in diameter from which 10.0-kg seats are suspended at the end of 2.50-m massless chains (Fig. P6.52). When the system rotates, the chains make an angle $\theta = 28.0°$ with the vertical. (a) What is the speed of each seat? (b) If a child of mass 40.0 kg sits in a seat, what is the tension in the chain?

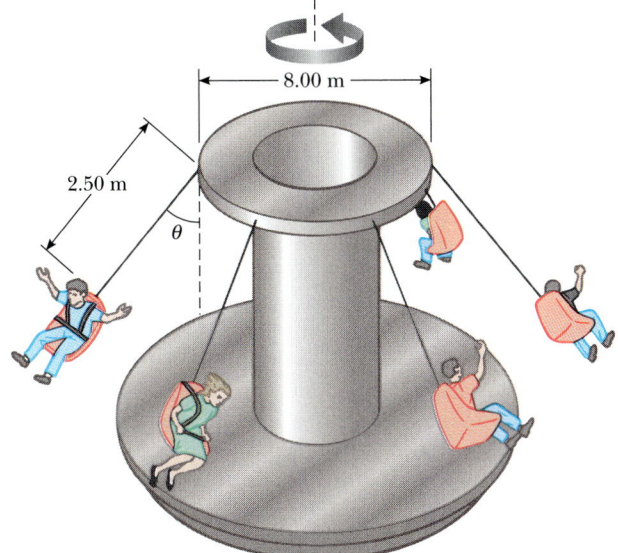

FIGURE P6.52

53. A stream of air moving at speed v exerts a resistive force on a sphere of radius r. The magnitude of the resistive force (in newtons) is $F = arv + br^2v^2$, where v is in meters per second, r is in meters, and a and b are constants with appropriate SI units. Their numerical values are $a = 3.10 \times 10^{-4}$ and $b = 0.870$. Using this formula, find the terminal speed for water droplets falling under their own weight in air, taking these values for the drop radii: (a) 10.0 μm, (b) 100 μm, and (c) 1.00 mm. Note that for (a) and (c) you can obtain accurate answers without solving a quadratic equation by considering which of the two contributions to the air resistance is dominant and ignoring the lesser contribution.

SPREADSHEET PROBLEMS

S1. In studying the drag forces that act on a falling object, it is often assumed that the magnitude of the drag force is $R = bv^n$. The net force acting on the object in this case is

$$F_{net} = mg - bv^n$$

where m is the mass of the object, and b and n are constants. When the net force goes to zero, the object reaches its terminal speed v_t:

$$v_t = \left(\frac{mg}{b}\right)^{\frac{1}{n}}$$

Spreadsheet 6.1 numerically integrates the force equation for $n = 1$ and calculates the position and speed of the falling object. In addition, it integrates the force equation without air resistance. Choose the

initial position and initial speed to be zero, and enter these values in the spreadsheet. (a) Set $b = 0.200$ and use $m = 0.100$, 1.00, and 10.0 kg to find v_T numerically in each case as follows. Plot v as a function of time, t. For t large enough, v approaches v_t. Compare your results for v_t obtained this way with the corresponding results calculated from the formula for v_t. *Note:* You may need to change the time increment to reach the terminal speed. (b) Change b to 0.500 and repeat part (a).

S2. Use Spreadsheet 6.1 with $m = 1$ kg, $v_0 = 0$, $x_0 = 100$ m, and $b = 0.2$. (a) Compare the position of the object with air resistance and without air resistance at times $t = 0$ s, 0.5 s, 1.0 s, 5.0 s, and 10 s. (b) Do the same for $m = 2$ kg and 10 kg. What is the effect of changing the mass?

S3. Members of a skydiving club were given the following data to use in planning their jumps. In the table, d is the distance fallen from rest by a sky diver in a "free-fall stable spread position" versus the time of fall t. (a) Enter the data into a spreadsheet and convert the distances in feet to distances in meters. (b) Graph d (in meters) versus t. (c) Determine the value of the terminal speed v_t by finding the slope of the linear portion of the curve. Use a least-squares fit to determine the slope.

t (s)	d (ft)
1	16
2	62
3	138
4	242
5	366
6	504
7	652
8	808
9	971
10	1138
11	1309
12	1483
13	1657
14	1831
15	2005
16	2179
17	2353
18	2527
19	2701
20	2875

S4. A 50.0-kg box at rest on a level floor must be moved. The coefficient of static friction between box and floor is 0.600. A force of magnitude F is applied at an angle θ with the horizontal. (a) Using a spreadsheet, calculate the force F necessary to move the box for a sequence of angles θ. Take the angle θ as positive if the force has an upward component, negative if it has a downward component. Plot F versus θ. From the graph, find the minimum force necessary to move the box. At what angle should it be applied? (b) Investigate what happens when you change the coefficient of friction.

S5. It is "common knowledge" that when air resistance is *not* negligible, a heavier object always hits the ground before a lighter object when both are dropped simultaneously. (a) Use Spreadsheet 6.1 to test this prediction, with $m = 0.100$ kg, 1.00 kg, and 5.00 kg and $b = 0.200$. This corresponds to dropping objects of the same size and shape but with each having a different mass, such as a cork ball, a wooden ball, and a lead ball. (b) Investigate what happens if the objects have an initial velocity. Use different values (both positive and negative) for the initial velocity. Make a table of the times it takes the objects to hit the ground when they are thrown from a height of 200 m. In the spreadsheet, $x = 0$ is the ground. Does the heavier object always hit first?

S6. Modify Spreadsheet 6.1 so that $n = 2$ (see Problem S1). Then the net force in Newton's second law has the form

$$F_{net} = mg - b\mathbf{v} \cdot |\mathbf{v}|$$

since the drag force term must always be in the opposite direction from the velocity. Compare your results for both position and speed versus time for $n = 1$ and $n = 2$. Note that the value of b should also change as n is changed. Pick values for b such that the terminal speeds for $n = 1$ and $n = 2$ are the same for each mass. Plot the $n = 1$ and $n = 2$ solutions on the same graphs and compare them to the case where air resistance can be neglected. Vary x_0, v_0, and m, and recompare.

S7. If the drag force due to air resistance for the sky diver in Problem S3 can be represented as $R = bv^n$, the acceleration of the falling sky diver is

$$a = g \left[1 - \left(\frac{v}{v_t} \right)^n \right]$$

The most commonly used empirical values of n are $n = 1$ and $n = 2$. Which choice best fits the sky diver data in Problem S3? To answer this question, use your value of the terminal speed from Problem S3 and solve numerically for the speed and position of the sky diver as functions of time for both cases, $n = 1$ and $n = 2$. Plot the position-versus-time curves for both cases along with the sky diver data on the same graph.

S8. A large artillery shell with a mass of 44.0 kg is fired with a muzzle speed $v_0 = 570$ m/s. It is reported that the maximum horizontal range is 15.0 km. However, the maximum range if air resistance were negligible would be 33.0 km according to

$$R_{max} = \frac{v_0^2}{g}$$

Clearly, air resistance is not negligible even for such a heavy projectile. Assuming the force of air resistance on the shell is $\mathbf{f} = -b\mathbf{v}|\mathbf{v}|$, do a numerical study to find an appropriate value for b such that $R_{max} = 15.0$ km. (You could modify Spreadsheet 6.1 to include two-dimensional motion.) *Note:* The maximum range when air resistance is included does not necessarily occur at $\theta_0 = 45°$. You will have to vary θ_0 for each value of b to find R_{max} as a function of b.

S9. A 0.142-kg baseball has a terminal speed of 42.5 m/s (95 mph). (a) If a baseball experiences a drag force of magnitude $R = Kv^2$, what is the value of K? (b) What is the magnitude of the drag force when the speed of the baseball is 36.0 m/s? (c) Set up a spreadsheet (or modify Spreadsheet 6.1) to determine the motion of a baseball thrown vertically upward at an initial speed of 36.0 m/s. What maximum height does the ball reach? How long is it in the air? What is its speed just before it hits the ground?

S10. A 50.0-kg parachutist jumps from an airplane and falls to Earth with a drag force of magnitude $R = Kv^2$. Take $K = 0.200$ kg/m with the parachute closed and $K = 20.0$ kg/m with the parachute open. (a) Determine the terminal speed of the parachutist before and after the parachute is opened. (b) Set up a spreadsheet to determine the position and speed of the parachutist as functions of time. Assume the jumper began the descent at an altitude of 1000 m, and fell in free fall for 10 s before opening the parachute. (*Hint:* When the parachute is opened there is a sudden large acceleration; a smaller time step may be necessary in this region.)

S11. A 10.0-kg projectile is launched with an initial speed of 150 m/s at an elevation angle of 35.0° with the horizontal. The resistive force acting on the projectile is $\mathbf{R} = b\mathbf{v}$, where $b = 15.0$ kg/s. (a) Set up a spreadsheet to determine the horizontal and vertical positions of the projectile as functions of time. (b) Find the range of this projectile. (c) Determine the elevation angle that gives the maximum range for this projectile. (*Hint:* Adjust the elevation angle by trial and error to find the greatest range.)

Work and Energy

These cyclists are working hard and expending energy as they pedal uphill in Marin County, California.   *(David Madison/Tony Stone Images)*

The concept of energy is one of the most important concepts in both contemporary science and engineering practice. In everyday usage, we think of energy in terms of fuel for transportation and heating, electricity for lights and appliances, and the foods that we consume. However, these ideas do not really define energy. They tell us only that fuels are needed to do a job and that those fuels provide us with something we call energy.

Energy is present in the Universe in various forms, including mechanical energy, electromagnetic energy, chemical energy, thermal energy, and nuclear energy. Furthermore, one form of energy can be converted to another. For example, when an electric motor is connected to a battery, chemical energy is converted to electrical energy, which in turn is converted to mechanical energy. The transformation of energy from one form to another is an essential part of the study of physics, engineering, chemistry, biology, geology, and astronomy. When energy is changed from one form to another, its total amount remains the same. Conservation of energy says that although the form of energy may be changed, if an object (or system) loses energy, that same amount of energy appears in another object (or in the surroundings).

In this chapter, we shall first introduce the concept of work. Work is done by a force acting on an object when the point of application of that force moves through some distance and the force has a component along the line of motion. We then define kinetic energy, which is energy associated with the motion of an object. We shall see that the concepts of work and energy can be applied to the dynamics of a mechanical system without resorting to Newton's laws. However, it is important to note that the work-energy concepts are based upon Newton's laws and therefore do not involve any new physical principles.

Although the approach we shall use provides the same results as Newton's laws in describing the motion of a mechanical system, the general ideas of the work-energy concept can be applied to a wide range of phenomena in the fields of electromagnetism and atomic and nuclear physics. In addition, in a complex situation the "energy approach" can often provide a much simpler analysis than the direct application of Newton's second law.

This alternative method of describing motion is especially useful when the force acting on a particle is not constant. In this case, the acceleration is not constant, and we cannot apply the simple kinematic equations we developed in Chapter 2. Often, a particle in nature is subject to a force that varies with the position of the particle. Such forces include gravitational forces and the force exerted on an object attached to a spring. We shall describe techniques for treating such systems with the help of an extremely important development called the *work-energy theorem*, which is the central topic of this chapter.

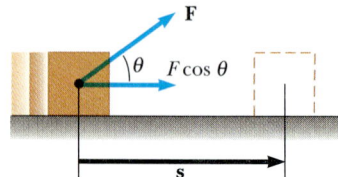

FIGURE 7.1   If a force acting on an object undergoes a displacement **s**, the work done by the force **F** is $(F \cos \theta)s$.

Work done by a constant force

7.1   WORK DONE BY A CONSTANT FORCE

Almost all of the terms we have used thus far—velocity, acceleration, force, and so on—have conveyed the same meaning in physics as they do in everyday life. Now, however, we encounter a term whose meaning in physics is distinctly different from its everyday meaning. That new term is **work**. Consider a particle that undergoes a displacement **s** along a straight line while acted on by a constant force **F**, which makes an angle θ with **s**, as in Figure 7.1.

> The work W done by an agent exerting a constant force is the product of the component of the force in the direction of the displacement and the magnitude of the displacement of the force.
>
> $$W = Fs \cos \theta \qquad (7.1)$$

From this definition, we see that a force does no work on a particle if the particle does not move. That is, if $s = 0$, Equation 7.1 gives $W = 0$. For example, if a person pushes against a brick wall, a force is exerted on the wall but the person does no work on the wall provided the point of application of the force does not move. Also note from Equation 7.1 that the work done by a force is zero when the force is perpendicular to the displacement. That is, if $\theta = 90°$, then $W = 0$ and $\cos 90° = 0$. For example, in Figure 7.2, the work done by the normal force and the work done by the force of gravity are zero because both forces are perpendicular to the displacement and have zero components in the direction of **s**.

The sign of the work also depends on the direction of **F** relative to **s**. The work done by the applied force is positive when the vector associated with the component $F \cos \theta$ is in the same direction as the displacement. For example, when an object is lifted, the work done by the applied force is positive because the lifting force is upward, that is, in the same direction as the displacement. In this situation, the work done by the gravitational force is negative.

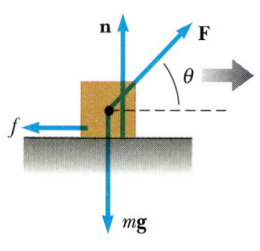

FIGURE 7.2   When an object is displaced horizontally, the normal force **n** and the force of gravity, *m***g**, do no work. The force of friction (box on floor) does some work on the floor.

Does the weight lifter do any work as he holds the weight on his shoulders? Does he do any work as he raises the weight?   *(Gerard Vandystadt/Photo Researchers)*

When the vector associated with the component $F \cos \theta$ is in the direction opposite the displacement, W is negative. The factor $\cos \theta$ that appears in the definition of W (Eq. 7.1) automatically takes care of the sign. In the case of the object being lifted, for instance, the work done by the gravitational force is negative. It is important to note that work is an energy transfer; if energy is transferred *to* the system (object), W is positive; if energy is transferred *from* the system, W is negative.

If an applied force **F** acts along the direction of the displacement, then $\theta = 0$, and $\cos 0 = 1$. In this case, Equation 7.1 gives

$$W = Fs$$

Work is a scalar quantity, and its units are force multiplied by length. Therefore, the SI unit of work is the **newton · meter** (N · m). Another name for the newton · meter is the **joule** (J). The unit of work in the cgs system is the **dyne · cm,** also called the **erg**; the unit in the British engineering system is the **ft · lb.** These are summarized in Table 7.1. Note that $1 \text{ J} = 10^7$ ergs.

In general, a particle may be moving with a constant or varying velocity under the influence of several forces. In that case, since work is a scalar quantity, the total work done as the particle undergoes some displacement is the algebraic sum of the amounts of work done by each of the forces.

TABLE 7.1   Units of Work in the Three Common Systems of Measurement

System	Unit	Alternate Name
SI	newton · meter (N · m)	joule (J)
cgs	dyne · centimeter (dyne · cm)	erg
British engineering	foot · pound (ft · lb)	foot · pound (ft · lb)

EXAMPLE 7.1   Mr. Clean

A man cleaning his apartment pulls a vacuum cleaner with a force of magnitude $F = 50$ N. The force makes an angle of $30°$ with the horizontal as shown in Figure 7.3. The vacuum cleaner is displaced 3.0 m to the right. Calculate the work done by the 50-N force.

Solution   Using the definition of work (Equation 7.1), we have

$$W_F = (F \cos \theta) s = (50 \text{ N})(\cos 30°)(3.0 \text{ m})$$

$$= 130 \text{ N} \cdot \text{m} = \boxed{130 \text{ J}}$$

Note that the normal force **n**, the weight $m\mathbf{g}$, and the upward component of the applied force, $(50 \text{ N})\sin 30°$, do no work because they are perpendicular to the displacement.

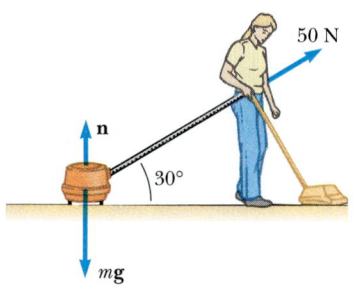

FIGURE 7.3   (Example 7.1) A vacuum cleaner being pulled at an angle of $30°$ with the horizontal.

Exercise   Find the work done by the man on the vacuum cleaner if he pulls it 3.0 m with a horizontal force of 32 N.

Answer   96 J.

CONCEPTUAL EXAMPLE 7.2

A person lifts a cement block of mass m a vertical height h, and then walks horizontally a distance d while holding the block, as in Figure 7.4. Determine the work done by the person and by the force of gravity in this process.

Reasoning   Assuming that the person lifts the block with a force of magnitude equal to the weight of the block, mg, the work done by the person during the vertical displacement is mgh, since the force in this case is in the direction of the displacement. The work done by the person during the horizontal displacement of the block is zero since the applied force in this process is perpendicular to the displacement. Thus, the net work done by the person is mgh. The work done by the force of gravity during the vertical displacement of the block is $- mgh$, because this force is opposite the displacement. The work done by the force of gravity is zero during the horizontal displacement because this force is also perpendicular to the displacement. Hence, the net work done by the force of gravity is $- mgh$. The net work done on the block is zero $(+ mgh - mgh = 0)$. The kinetic energy of the block does not change.

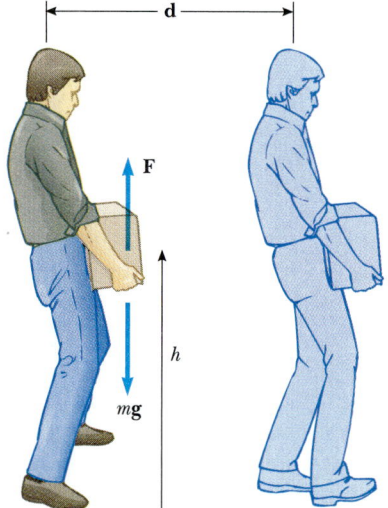

FIGURE 7.4   (Conceptual Example 7.2) A person lifts a cement block of mass m a vertical height h and then walks horizontally a distance d.

7.2   THE SCALAR PRODUCT OF TWO VECTORS

It is convenient to express Equation 7.1 in terms of a **scalar product** of the two vectors **F** and **s**. We write this scalar product $\mathbf{F} \cdot \mathbf{s}$. Because of the dot symbol, the scalar product is often called the *dot product*. Thus, we can express Equation 7.1 as a scalar product:

$$W = \mathbf{F} \cdot \mathbf{s} = Fs \cos \theta \qquad (7.2)$$

In other words, $\mathbf{F} \cdot \mathbf{s}$ (read "F dot s") is a shorthand notation for $Fs \cos \theta$.

Work expressed as a dot product

> In general, the scalar product of any two vectors **A** and **B** is a scalar quantity equal to the product of the magnitudes of the two vectors and the cosine of the angle θ between them:
>
> $$\mathbf{A} \cdot \mathbf{B} \equiv AB \cos \theta \qquad (7.3)$$

Scalar product of any two vectors **A** and **B**

where A is the magnitude of **A**, B is the magnitude of **B**, and θ is the smaller angle between **A** and **B**, as in Figure 7.5. Note that **A** and **B** need not have the same units.

In Figure 7.5, $B \cos \theta$ is the projection of **B** onto **A**. Therefore, Equation 7.3 says that $\mathbf{A} \cdot \mathbf{B}$ is the product of the magnitude of **A** and the projection of **B** onto **A**.[1]

From Equation 7.3 we also see that the scalar product is *commutative*. That is,

$$\mathbf{A} \cdot \mathbf{B} = \mathbf{B} \cdot \mathbf{A}$$

The order of the dot product can be reversed

Finally, the scalar product obeys the *distributive law of multiplication,* so that

$$\mathbf{A} \cdot (\mathbf{B} + \mathbf{C}) = \mathbf{A} \cdot \mathbf{B} + \mathbf{A} \cdot \mathbf{C}$$

The dot product is simple to evaluate from Equation 7.3 when **A** is either perpendicular or parallel to **B**. If **A** is perpendicular to **B** ($\theta = 90°$), then $\mathbf{A} \cdot \mathbf{B} = 0$. (The equality $\mathbf{A} \cdot \mathbf{B} = 0$ also holds in the more trivial case when either **A** or **B** is zero.) If **A** and **B** point in the same direction ($\theta = 0°$), then $\mathbf{A} \cdot \mathbf{B} = AB$. If **A** and **B** point in opposite directions ($\theta = 180°$), then $\mathbf{A} \cdot \mathbf{B} = -AB$. The scalar product is negative when $90° < \theta < 180°$.

The unit vectors **i, j,** and **k,** which were defined in Chapter 3, lie in the positive $x, y,$ and z directions, respectively, of a right-handed coordinate system. Therefore, it follows from the definition of $\mathbf{A} \cdot \mathbf{B}$ that the scalar products of these unit vectors are

$$\mathbf{i} \cdot \mathbf{i} = \mathbf{j} \cdot \mathbf{j} = \mathbf{k} \cdot \mathbf{k} = 1 \qquad (7.4)$$

$$\mathbf{i} \cdot \mathbf{j} = \mathbf{i} \cdot \mathbf{k} = \mathbf{j} \cdot \mathbf{k} = 0 \qquad (7.5)$$

Dot products of unit vectors

Two vectors **A** and **B** can be expressed in component vector form as

$$\mathbf{A} = A_x\mathbf{i} + A_y\mathbf{j} + A_z\mathbf{k}$$

$$\mathbf{B} = B_x\mathbf{i} + B_y\mathbf{j} + B_z\mathbf{k}$$

Therefore Equations 7.4 and 7.5 reduce the scalar product of **A** and **B** to

$$\mathbf{A} \cdot \mathbf{B} = A_xB_x + A_yB_y + A_zB_z \qquad (7.6)$$

In the special case where **A** = **B**, we see that

$$\mathbf{A} \cdot \mathbf{A} = A_x^2 + A_y^2 + A_z^2 = A^2$$

[1] This is equivalent to stating that $\mathbf{A} \cdot \mathbf{B}$ equals the product of the magnitude of **B** and the projection of **A** onto **B**.

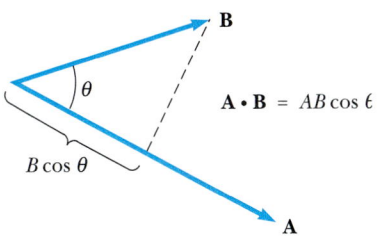

FIGURE 7.5 The scalar product $\mathbf{A} \cdot \mathbf{B}$ equals the magnitude of **A** multiplied by the projection of **B** onto **A**.

EXAMPLE 7.3   The Scalar Product

The vectors **A** and **B** are given by $\mathbf{A} = 2\mathbf{i} + 3\mathbf{j}$ and $\mathbf{B} = -\mathbf{i} + 2\mathbf{j}$. (a) Determine the scalar product $\mathbf{A} \cdot \mathbf{B}$.

Solution

$$\mathbf{A} \cdot \mathbf{B} = (2\mathbf{i} + 3\mathbf{j}) \cdot (-\mathbf{i} + 2\mathbf{j})$$

$$= -2\mathbf{i} \cdot \mathbf{i} + 2\mathbf{i} \cdot 2\mathbf{j} - 3\mathbf{j} \cdot \mathbf{i} + 3\mathbf{j} \cdot 2\mathbf{j}$$

$$= -2 + 6 = \boxed{4}$$

where we have used the facts that $\mathbf{i} \cdot \mathbf{i} = \mathbf{j} \cdot \mathbf{j} = 1$ and $\mathbf{i} \cdot \mathbf{j} = \mathbf{j} \cdot \mathbf{i} = 0$. The same result is obtained using Equation 7.6 directly, where $A_x = 2$, $A_y = 3$, $B_x = -1$, and $B_y = 2$.

(b) Find the angle θ between **A** and **B**.

Solution   The magnitudes of **A** and **B** are given by

$$A = \sqrt{A_x^2 + A_y^2} = \sqrt{(2)^2 + (3)^2} = \sqrt{13}$$

$$B = \sqrt{B_x^2 + B_y^2} = \sqrt{(-1)^2 + (2)^2} = \sqrt{5}$$

Using Equation 7.3 and the result from (a) gives

$$\cos \theta = \frac{\mathbf{A} \cdot \mathbf{B}}{AB} = \frac{4}{\sqrt{13}\sqrt{5}} = \frac{4}{\sqrt{65}}$$

$$\theta = \cos^{-1} \frac{4}{8.06} = \boxed{60°}$$

EXAMPLE 7.4   Work Done by a Constant Force

A particle moving in the xy plane undergoes a displacement $\mathbf{s} = (2.0\mathbf{i} + 3.0\mathbf{j})$ m while a constant force $\mathbf{F} = (5.0\mathbf{i} + 2.0\mathbf{j})$ N acts on the particle. (a) Calculate the magnitude of the displacement and that of the force.

Solution

$$s = \sqrt{x^2 + y^2} = \sqrt{(2.0)^2 + (3.0)^2} = \boxed{3.6 \text{ m}}$$

$$F = \sqrt{F_x^2 + F_y^2} = \sqrt{(5.0)^2 + (2.0)^2} = \boxed{5.4 \text{ N}}$$

(b) Calculate the work done by **F**.

Solution   Substituting the expressions for **F** and **s** into Equation 7.2 and using Equations 7.4 and 7.5, we get

$$W = \mathbf{F} \cdot \mathbf{s} = (5.0\mathbf{i} + 2.0\mathbf{j}) \cdot (2.0\mathbf{i} + 3.0\mathbf{j}) \text{ N} \cdot \text{m}$$

$$= 5.0\mathbf{i} \cdot 2.0\mathbf{i} + 5.0\mathbf{i} \cdot 3.0\mathbf{j} + 2.0\mathbf{j} \cdot 2.0\mathbf{i} + 2.0\mathbf{j} \cdot 3.0\mathbf{j}$$

$$= 10 + 0 + 0 + 6 = 16 \text{ N} \cdot \text{m} = \boxed{16 \text{ J}}$$

Exercise   Calculate the angle between **F** and **s**.

Answer   35°.

7.3   WORK DONE BY A VARYING FORCE

Consider a particle being displaced along the x axis under the action of a varying force, as in Figure 7.6. The particle is displaced in the direction of increasing x from $x = x_i$ to $x = x_f$. In such a situation, we cannot use $W = (F \cos \theta)s$ to calculate

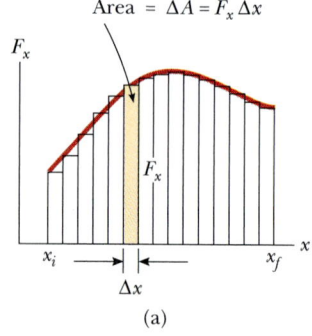
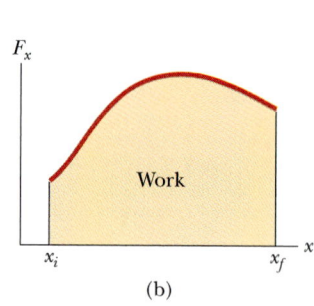

FIGURE 7.6   (a) The work done by the force F_x for the small displacement Δx is $F_x \Delta x$, which equals the area of the shaded rectangle. The total work done for the displacement from x_i to x_f is approximately equal to the sum of the areas of all the rectangles. (b) The work done by the variable force F_x as the particle moves from x_i to x_f is *exactly* equal to the area under this curve.

the work done by the force, because this relationship applies only when **F** is constant in magnitude and direction. However, if we imagine that the particle undergoes a very small displacement Δx, shown in Figure 7.6a, then the x component of the force, F_x, is approximately constant over this interval, and we can express the work done by the force for this small displacement as

$$W_1 = F_x \, \Delta x$$

This is just the area of the shaded rectangle in Figure 7.6a. If we imagine that the F_x versus x curve is divided into a large number of such intervals, then the total work done for the displacement from x_i to x_f is approximately equal to the sum of a large number of such terms:

$$W \cong \sum_{x_i}^{x_f} F_x \, \Delta x$$

If the displacements are allowed to approach zero, then the number of terms in the sum increases without limit but the value of the sum approaches a definite value equal to the area under the curve bounded by F_x and the x axis:

$$\lim_{\Delta x \to 0} \sum_{x_i}^{x_f} F_x \, \Delta x = \int_{x_i}^{x_f} F_x \, dx$$

This definite integral is numerically equal to the area under the F_x versus x curve between x_i and x_f. Therefore, we can express the work done by F_x for the displacement of the object from x_i to x_f as

$$W = \int_{x_i}^{x_f} F_x \, dx \qquad (7.7)$$

Work done by a varying force

This equation reduces to Equation 7.1 when $F_x = F \cos \theta$ is constant.

If more than one force acts on a particle, the total work done is just the work done by the resultant force. For systems that do not act as particles, work must be found for each force separately. If we express the resultant force in the x direction as ΣF_x, then the *net work* done as the particle moves from x_i to x_f is

$$W_{\text{net}} = \int_{x_i}^{x_f} \left(\sum F_x \right) dx \qquad (7.8)$$

EXAMPLE 7.5   Calculating Total Work Done from a Graph

A force acting on a particle varies with x as shown in Figure 7.7. Calculate the work done by the force as the particle moves from $x = 0$ to $x = 6.0$ m.

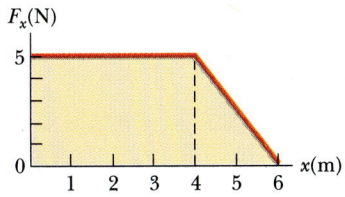

FIGURE 7.7 (Example 7.5) The force acting on a particle is constant for the first 4.0 m of motion and then decreases linearly with x from $x = 4.0$ m to $x = 6.0$ m. The net work done by this force is the area under this curve.

Solution   The work done by the force is equal to the area under the curve from $x = 0$ to $x = 6.0$ m. This area is equal to the area of the rectangular section from $x = 0$ to $x = 4.0$ m plus the area of the triangular section from $x = 4.0$ m to $x = 6.0$ m. The area of the rectangle is $(4.0)(5.0)$ N·m $= 20$ J, and the area of the triangle is $\frac{1}{2}(2.0)(5.0)$ N·m $= 5.0$ J. Therefore, the total work done is 25 J.

Work Done by a Spring

A common physical system for which the force varies with position is shown in Figure 7.8. A block on a horizontal, frictionless surface is connected to a spring. If the spring is stretched or compressed a small distance from its unstretched,

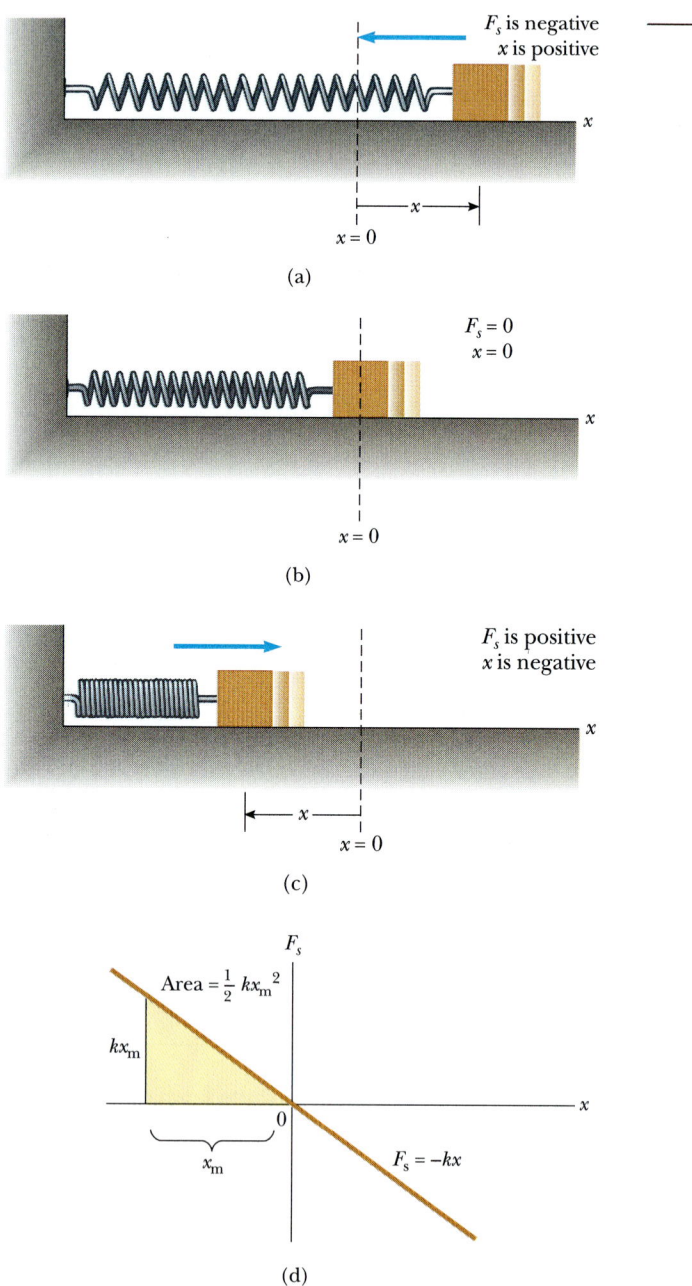

FIGURE 7.8 The force exerted by a spring on a block varies with the block's displacement from the equilibrium position $x = 0$. (a) When x is positive (stretched spring), the spring force is to the left. (b) When x is zero, the spring force is zero (natural length of the spring). (c) When x is negative (compressed spring), the spring force is to the right. (d) Graph of F_s versus x for the mass-spring system. The work done by the spring force as the block moves from $-x_m$ to 0 is the area of the shaded triangle, $\frac{1}{2}kx_m^2$.

or equilibrium, configuration, the spring will exert a force on the block given by

$$F_s = -kx \qquad (7.9)$$

where x is the displacement of the block from its unstretched ($x = 0$) position and k is a positive constant called the *force constant* of the spring. As we learned in Chapter 5, this force law for springs is known as **Hooke's law.** Note that Hooke's law is valid only in the limiting case of small displacements. The value of k is a measure of the stiffness of the spring. Stiff springs have large k values, and soft springs have small k values.

The negative sign in Equation 7.9 signifies that the force exerted by the spring is always directed *opposite* the displacement. For example, when $x > 0$ as in Figure 7.8a, the spring force is to the left, or negative. When $x < 0$ as in Figure 7.8c, the spring force is to the right, or positive. Of course, when $x = 0$ as in Figure 7.8b, the spring is unstretched and $F_s = 0$. Since the spring force always acts toward the equilibrium position, it is sometimes called a *restoring force.* Suppose the spring is compressed so that the block is displaced a distance $-x_m$ from equilibrium. After it is released, the block moves from $-x_m$ through zero to $+x_m$. If the spring is stretched until the block is at x_m and then released, the block moves from $+x_m$ through zero to $-x_m$. The details of the ensuing oscillating motion will be given in Chapter 13.

Suppose that the block is pushed to the left a distance x_m from equilibrium and then released. Let us calculate the work done by the spring force as the block moves from $x_i = -x_m$ to $x_f = 0$. Applying Equation 7.7 assuming the block may be treated as a particle, we get

$$W_s = \int_{x_i}^{x_f} F_s \, dx = \int_{-x_m}^{0} (-kx) \, dx = \tfrac{1}{2} kx_m^2 \qquad (7.10)$$

where we have used the indefinite integral $\int x \, dx = x^2/2$. That is, the work done by the spring force is positive because the spring force is in the same direction as the displacement (both are to the right). The positive value of W_s confirms that energy is transferred from the spring to the block. However, if we consider the work done by the spring force as the block moves from $x_i = 0$ to $x_f = x_m$, we find that $W_s = -\tfrac{1}{2} kx_m^2$, since for this part of the motion the displacement is to the right and the spring force is to the left. Therefore, the *net* work done by the spring force as the block moves from $x_i = -x_m$ to $x_f = x_m$ is *zero.*

Figure 7.8d is a plot of F_s versus x. The work calculated in Equation 7.10 is the area of the shaded triangle in Figure 7.8d, corresponding to the displacement from $-x_m$ to 0. Because the triangle has base x_m and height kx_m, its area is $\tfrac{1}{2} kx_m^2$, the work done by the spring as given by Equation 7.10.

If the block undergoes an arbitrary displacement from $x = x_i$ to $x = x_f$, the work done by the spring force is

$$W_s = \int_{x_i}^{x_f} (-kx) \, dx = \frac{1}{2} kx_i^2 - \frac{1}{2} kx_f^2 \qquad (7.11)$$

From this equation we see that the work done by the spring force is zero for any motion that ends where it began ($x_i = x_f$). We shall make use of this important result in Chapter 8, where we describe the motion of this system in more detail.

Equations 7.10 and 7.11 describe the work done by the spring force exerted on the block. Now let us consider the work done on the spring by an *external agent* as the spring is stretched very slowly from $x_i = 0$ to $x_f = x_m$, as in Figure 7.9. This

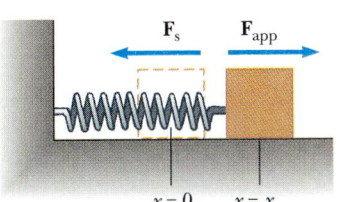

FIGURE 7.9   A block being pulled from $x = 0$ to $x = x_m$ on a frictionless surface by a force F_{app}. If the process is carried out very slowly, the applied force is equal to and opposite the spring force at all times.

work can be easily calculated by noting that the *applied force*, F_{app}, is equal to and opposite the spring force, F_s, at any value of the displacement, so that $F_{app} = -(-kx) = kx$. Therefore, the work done by this applied force (the external agent) is

$$W_{F_{app}} = \int_0^x F_{app} \, dx = \int_0^x kx \, dx = \frac{1}{2}kx^2$$

You should note that this work is equal to the negative of the work done by the spring force for this displacement. For example, if a spring of force constant 80 N/m is compressed 3.0 cm from equilibrium, the work done by the spring force as the block moves from $x_i = -3.0$ cm to its unstretched position, $x_f = 0$, is 3.6×10^{-2} J.

EXAMPLE 7.6   Measuring *k* for a Spring

A common technique used to measure the force constant of a spring is described in Figure 7.10. The spring is hung vertically and then a mass *m* is attached to the lower end of the spring. The spring stretches a distance *d* from its equilibrium position under the action of the "load" *mg*. Since the spring force is upward, it must balance the weight *mg* downward when the system is at rest. In this case, we can apply Hooke's law to give $|F_s| = kd = mg$, or

$$k = \frac{mg}{d}$$

For example, if a spring is stretched 2.0 cm by a suspended mass of 0.55 kg, the force constant of the spring is

$$k = \frac{mg}{d} = \frac{(0.55 \text{ kg})(9.80 \text{ m/s}^2)}{2.0 \times 10^{-2} \text{ m}} = 2.7 \times 10^2 \text{ N/m}$$

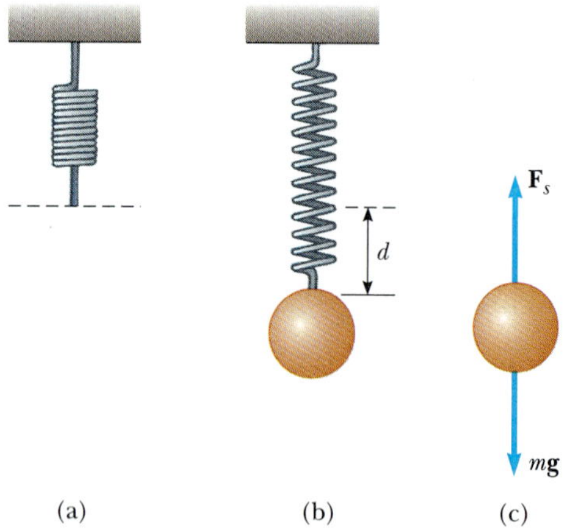

FIGURE 7.10   (Example 7.6) Determining the force constant *k* of a helical spring. The elongation *d* of the spring is due to the attached weight *mg*. Because the spring force balances the weight, it follows that $k = mg/d$.

7.4   KINETIC ENERGY AND THE WORK-ENERGY THEOREM

Solutions using Newton's second law can be difficult if the forces in the problem are complex. An alternative approach that enables us to understand and solve such motion problems is to relate the speed of the particle to its displacement under the influence of some net force. As we shall see in this section, if the work done by the net force on a particle can be calculated for a given displacement, the change in the particle's speed will be easy to evaluate.

Figure 7.11 shows a particle of mass *m* moving to the right under the action of a constant net force **F**. Because the force is constant, we know from Newton's second law that the particle will move with a constant acceleration **a**. If the particle is displaced a distance *s*, the net work done by the force **F** is

$$W_{net} = Fs = (ma)s \tag{7.12}$$

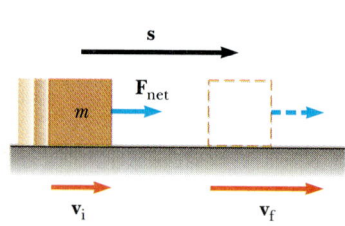

FIGURE 7.11   A particle undergoing a displacement and change in velocity under the action of a constant net force **F**.

In Chapter 2 we found that the following relationships are valid when a particle undergoes constant acceleration:

$$s = \tfrac{1}{2}(v_i + v_f)t \qquad a = \frac{v_f - v_i}{t}$$

where v_i is the speed at $t = 0$ and v_f is the speed at time t. Substituting these expressions into Equation 7.12 gives

$$W_{\text{net}} = m\left(\frac{v_f - v_i}{t}\right)\tfrac{1}{2}(v_i + v_f)t$$

$$W_{\text{net}} = \tfrac{1}{2}\,mv_f^2 - \tfrac{1}{2}\,mv_i^2 \tag{7.13}$$

The quantity $\tfrac{1}{2}mv^2$ represents the energy associated with the motion of a particle. It is so important that it has been given a special name — **kinetic energy**. The kinetic energy, K, of a particle of mass m moving with a speed v is defined as

$$K \equiv \tfrac{1}{2}mv^2 \tag{7.14}$$

Kinetic energy is a scalar quantity and has the same units as work. For example, a 2.0-kg mass moving with a speed of 4.0 m/s has a kinetic energy of 16 J. Table 7.2 lists the kinetic energies for various objects. We can think of kinetic energy as energy associated with the motion of a body. It is often convenient to write Equation 7.14 in the form

$$W_{\text{net}} = K_f - K_i = \Delta K \tag{7.15}$$

That is,

> the work done by the constant net force F_{net} in displacing a particle equals the change in kinetic energy of the particle.

Equation 7.15 is an important result known as the **work-energy theorem**. For convenience, it was derived under the assumption that the net force acting on the particle was constant. You should note that when energy is transferred to a particle

Kinetic energy is energy associated with the motion of a body

Work-energy theorem

TABLE 7.2   Kinetic Energies for Various Objects

Object	Mass (kg)	Speed (m/s)	Kinetic Energy (J)
Earth orbiting the Sun	5.98×10^{24}	2.98×10^4	2.65×10^{33}
Moon orbiting the Earth	7.35×10^{22}	1.02×10^3	3.82×10^{28}
Rocket moving at escape speed[a]	500	1.12×10^4	3.14×10^{10}
Automobile at 55 mi/h	2000	25	6.3×10^5
Running athlete	70	10	3.5×10^3
Stone dropped from 10 m	1.0	14	9.8×10^1
Golf ball at terminal speed	0.046	44	4.5×10^1
Raindrop at terminal speed	3.5×10^{-5}	9.0	1.4×10^{-3}
Oxygen molecule in air	5.3×10^{-26}	500	6.6×10^{-21}

[a] Escape speed is the minimum speed an object must reach near the Earth's surface in order to escape the Earth's gravitational force.

(as work), it must appear as kinetic energy of the particle because that is the *only* form of energy that can change for a particle.

Now we shall show that the work-energy theorem is valid even when the force is varying. If the resultant force acting on a body in the *x* direction is ΣF_x, then Newton's second law states that $\Sigma F_x = ma$. Thus, we can use Equation 7.8 and express the net work done as

$$W_{net} = \int_{x_i}^{x_f} \left(\sum F_x \right) dx = \int_{x_i}^{x_f} ma \, dx$$

Because the resultant force varies with *x*, the acceleration and speed also depend on *x*. We can now use the following chain rule to evaluate W_{net}:

$$a = \frac{dv}{dt} = \frac{dv}{dx}\frac{dx}{dt} = v\frac{dv}{dx}$$

Substituting this into the expression for W_{net} gives

$$W_{net} = \int_{x_i}^{x_f} mv\frac{dv}{dx}dx = \int_{v_i}^{v_f} mv \, dv$$

Work done on a particle equals the change in its kinetic energy

$$W_{net} = \tfrac{1}{2}mv_f^2 - \tfrac{1}{2}mv_i^2 \qquad (7.16)$$

where the limits of the integration were changed because the variable was changed from *x* to *v*. Thus, we conclude that the work done on a particle by the net force acting on it is equal to the change in the kinetic energy of the particle.

The work-energy theorem also says that the speed of the particle will increase if the net work done on it is positive, because the final kinetic energy will be greater than the initial kinetic energy. The speed will decrease if the net work is negative because the final kinetic energy will be less than the initial kinetic energy. The speed and kinetic energy of a particle change only if work is done on the particle by some external force.

Consider the relationship between the work done on a particle and the change in its kinetic energy as expressed by Equation 7.15. Because of this connection, we can also think of kinetic energy as the work the particle can do in coming to rest. For example, suppose a hammer is on the verge of striking a nail, as in Figure 7.12. The moving hammer has kinetic energy and can do work on the nail. The work done on the nail appears as the product *Fs*, where *F* is the average force exerted on the nail by the hammer and *s* is the distance the nail is driven into the wall.

Situations Involving Kinetic Friction

Suppose that an object of mass *m* sliding on a horizontal surface is pulled with a constant horizontal external force **F** to the right and a kinetic frictional force **f** acts to the left, where **F** > **f**. In this case, the net force is to the right as in Figure 7.13,

FIGURE 7.12   The hammer has kinetic energy associated with its motion and can do work on the nail, driving it into the wall.

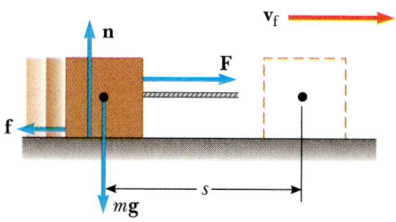

FIGURE 7.13   A mass *m* on a horizontal surface undergoes a displacement *s*. **F** is a constant horizontal external force and **f** is the force of kinetic friction directed toward the left.

and we might believe that we could find the net work done on the object as it undergoes a displacement **s** to the right by evaluating

$$W_{net} = (\mathbf{F} - \mathbf{f}) \cdot \mathbf{s} = Fs - fs$$

However, the object is not a particle, and it is incorrect to say that $-fs$ is the work done by the frictional force on the object. The frictional force acts on the object at the microscopic level through a series of microscopic displacements. The work done by kinetic friction depends on both the displacement of the object and on the details of the motion between the initial and final positions. In fact, the work done by kinetic friction on an extended object cannot be explicitly evaluated because friction forces and their individual displacements are very complex.

Now suppose that a block moving on a horizontal surface and given an initial horizontal velocity $\mathbf{v}_i$ slides a distance s before reaching a final velocity $\mathbf{v}_f$ as in Figure 7.14. The external force that causes the block to undergo an acceleration in the negative x direction is the force of kinetic friction $\mathbf{f}$ acting to the left, opposite the motion. The initial kinetic energy of the block is $\frac{1}{2}mv_i^2$ and its final kinetic energy is $\frac{1}{2}mv_f^2$. The change in kinetic energy of the block is equal to $-fs$. This can be shown by applying Newton's second law to the block. (Newton's second law gives the acceleration of the center of mass of any object regardless of how or where the forces act.) Since the net force on the block in the x direction is the friction force, Newton's second law gives $-f = ma$. Multiplying both sides of this expression by s, and using the equation $v_f^2 - v_i^2 = 2as$ for motion under constant acceleration gives $-fs = (ma)s = \frac{1}{2}mv_f^2 - \frac{1}{2}mv_i^2$ or

$$\Delta K = -fs \tag{7.17}$$

Loss in kinetic energy due to friction

This result says that the loss in kinetic energy of the block is equal to $-fs$, which corresponds to the energy dissipated by the force of kinetic friction. Part of this energy is transferred to internal energy of the block, and part is transferred from the block to the surface.[2] In effect, the loss in kinetic energy of the block results in an increase in internal energy of both the block and surface in the form of thermal (heat) energy. For example, if the loss in kinetic energy of the block is 300 J, and 100 J appears as an increase in internal energy of the block, then the remaining 200 J must have been transferred from the block to the surface.

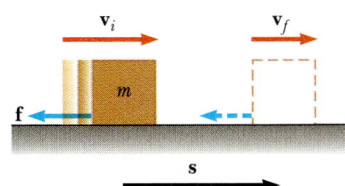

FIGURE 7.14   A block sliding to the right on a horizontal surface slows down in the presence of a force of kinetic friction acting to the left. The initial velocity of the block is $\mathbf{v}_i$, and its final velocity is $\mathbf{v}_f$. The normal force and force of gravity are not included in the diagram because they are perpendicular to the direction of motion and therefore do not influence the change in velocity of the block.

EXAMPLE 7.7   A Block Pulled on a Frictionless Surface

A 6.0-kg block initially at rest is pulled to the right along a horizontal, frictionless surface by a constant, horizontal force of 12 N, as in Figure 7.15a. Find the speed of the block after it has moved 3.0 m.

Solution   The weight of the block is balanced by the normal force, and neither of these forces does work since the displacement is horizontal. Since there is no friction, the resultant external force is the 12-N force. The work done by this force is

---

[2] For more details on energy transfer situations involving forces of kinetic friction, see B. A. Sherwood and W. H. Bernard, *American Journal of Physics*, 52:1001, 1984, and R. P. Bauman, *The Physics Teacher,* 30:264, 1992.

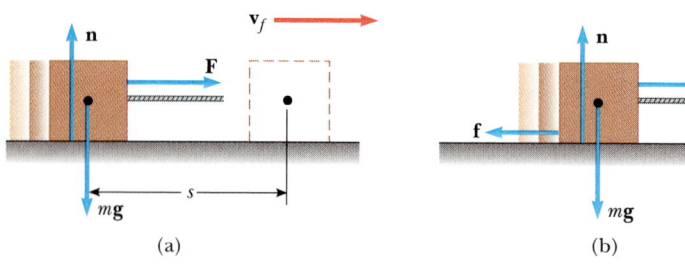

FIGURE 7.15   (a) Example 7.7. (b) Example 7.8.

$$W = Fs = (12 \text{ N})(3.0 \text{ m}) = 36 \text{ N} \cdot \text{m} = 36 \text{ J}$$

Using the work-energy theorem and noting that the initial kinetic energy is zero, we get

$$W = K_f - K_i = \tfrac{1}{2} mv_f^2 - 0$$

$$v_f^2 = \frac{2W}{m} = \frac{2(36 \text{ J})}{6.0 \text{ kg}} = 12 \text{ m}^2/\text{s}^2$$

$$v_f = \boxed{3.5 \text{ m/s}}$$

Exercise   Find the acceleration of the block, and determine its final speed using the kinematic equation $v_f^2 = v_i^2 + 2as$.

Answer $a = 2.0 \text{ m/s}^2$; $v_f = 3.5 \text{ m/s}$.

EXAMPLE 7.8   A Block Pulled on a Rough Surface

Find the final speed of the block described in Example 7.7 if the surface is rough and the coefficient of kinetic friction is 0.15.

Reasoning   In this case, we must use Equation 7.17 to calculate the change in kinetic energy, ΔK. The net force exerted on the block is the sum of the applied 12-N force and the frictional force, as in Figure 7.15b. Since the frictional force is in the direction opposite the displacement, it must be subtracted.

Solution   The magnitude of the frictional force is $f = \mu n = \mu mg$. Therefore the net force acting on the block is

$$F_{net} = F - \mu mg = 12 \text{ N} - (0.15)(6.0 \text{ kg})(9.80 \text{ m/s}^2)$$
$$= 12 \text{ N} - 8.82 \text{ N} = 3.18 \text{ N}$$

Multiplying this constant force by the displacement gives

$$\Delta K = F_{net} s = (3.18 \text{ N})(3.0 \text{ m}) = 9.54 \text{ J} = \tfrac{1}{2} mv_f^2$$

using the information that $v_i = 0$. Therefore,

$$v_f^2 = \frac{2(9.54 \text{ J})}{6.0 \text{ kg}} = 3.18 \text{ m}^2/\text{s}^2$$

$$v_f = \boxed{1.8 \text{ m/s}}$$

Exercise   Find the acceleration of the block from Newton's second law, and determine its final speed using kinematics.

Answer $a = 0.53 \text{ m/s}^2$; $v_f = 1.8 \text{ m/s}$.

CONCEPTUAL EXAMPLE 7.9

A team of furniture movers wishes to load a truck using a ramp from the ground to the rear of the truck. One of the movers claims that less work would be required to load the truck if the length of the ramp were increased, reducing the angle of the ramp with respect to the horizontal. Is his claim valid? Explain.

Reasoning   His claim is not valid. Although less force is required with a longer ramp, the force must act over a larger distance and do the same amount of work. Suppose a refrigerator attached to a frictionless wheeled dolly is rolled up a ramp at constant speed as in Figure 7.16. The normal force acting 90° to the motion does no work. Because $\Delta K = 0$ in

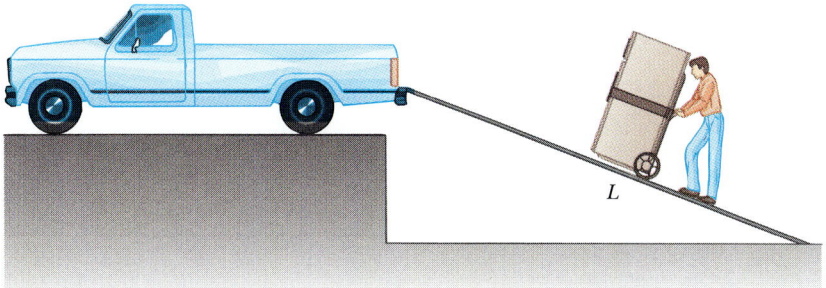

FIGURE 7.16   (Conceptual Example 7.9) A refrigerator attached to a frictionless wheeled dolly is moved up a ramp at constant speed. Does the amount of work required depend on the length of the ramp?

this case, the work-energy theorem gives

$$W_{\text{net}} = W_{\text{by movers}} + W_{\text{by gravity}} = 0$$

The work done by gravity equals the weight of the refrigera-tor times the vertical height through which it is displaced times cos 180°, or $W_{\text{by gravity}} = -mgh$. Thus, the movers, how-ever long the ramp, must do work mgh on the refrigerator.

CONCEPTUAL EXAMPLE 7.10

A car traveling at a speed v skids a distance d after its brakes lock. Estimate how far it will skid if its brakes lock when its initial speed is $2v$. What happens to the car's kinetic energy as it stops?

Reasoning   Let us assume that the force of kinetic friction between car and road surface is constant and the same in both cases. The net force times the displacement of the center of mass is equal to the initial kinetic energy of the car. If the speed is doubled as in this example, the kinetic energy of the car is quadrupled. For a given applied force (in this case, the frictional force), the distance traveled is four times as great when the initial speed is doubled, so the estimated distance it skids is $4d$. The kinetic energy of the car is changed into internal energy associated with the tires, brake pads, and road as they heat up.

CONCEPTUAL EXAMPLE 7.11

In most situations we have encountered in this chapter, fric-tional forces tend to reduce the kinetic energy of an object. However, frictional forces can sometimes increase an object's kinetic energy. Describe a few situations in which friction causes an increase in kinetic energy.

Reasoning   If a crate is located on the bed of a truck, and the truck accelerates to the east, the static friction force ex-erted on the crate by the truck acts to the east to give the crate the same acceleration as the truck (assuming the crate doesn't slip). Another example is a car that accelerates be-cause of the frictional forces exerted on the car's tires by the road. These forces act in the direction of the car's motion, and the sum of these forces causes an increase in the car's kinetic energy.

EXAMPLE 7.12   **A Mass-Spring System**

A block of mass 1.6 kg is attached to a spring that has a force constant of 1.0×10^3 N/m as in Figure 7.8. The spring is compressed a distance of 2.0 cm, and the block is released from rest. (a) Calculate the speed of the block as it passes through the equilibrium position $x = 0$ if the surface is fric-tionless.

Solution   We use Equation 7.10 to find the work done by the spring with $x_m = -2.0$ cm $= -2.0 \times 10^{-2}$ m:

$$W_s = \tfrac{1}{2} kx_m{}^2 = \tfrac{1}{2}(1.0 \times 10^3 \text{ N/m})(-2.0 \times 10^{-2}\text{ m})^2 = 0.20 \text{ J}$$

Using the work-energy theorem with $v_i = 0$ gives

$$W_s = \tfrac{1}{2} mv_f^2 - \tfrac{1}{2} mv_i^2$$

$$0.20 \text{ J} = \tfrac{1}{2}(1.6 \text{ kg}) v_f^2 - 0$$

$$v_f^2 = \frac{0.40 \text{ J}}{1.6 \text{ kg}} = 0.25 \text{ m}^2/\text{s}^2$$

$$v_f = \boxed{0.50 \text{ m/s}}$$

(b) Calculate the speed of the block as it passes through the equilibrium position if a constant frictional force of 4.0 N retards its motion.

Solution   We use Equation 7.17 to calculate the kinetic energy lost due to friction and add this to the kinetic energy found in the absence of friction. Considering only the frictional force, the kinetic energy lost due to friction is

$$-fs = -(4.0 \text{ N})(2.0 \times 10^{-2}\text{ m}) = -0.08 \text{ J}$$

The final kinetic energy, without this loss, was found in part (a) to be 0.20 J. Therefore, the final kinetic energy in the presence of friction is

$$K_f = 0.20 \text{ J} - 0.08 \text{ J} = 0.12 \text{ J} = \tfrac{1}{2} mv_f^2$$

$$\tfrac{1}{2}(1.6 \text{ kg}) v_f^2 = 0.12 \text{ J}$$

$$v_f^2 = \frac{0.24 \text{ J}}{1.6 \text{ kg}} = 0.15 \text{ m}^2/\text{s}^2$$

$$v_f = \boxed{0.39 \text{ m/s}}$$

CONCEPTUAL EXAMPLE 7.13

An Earth satellite is in a circular orbit at an altitude of 500 km. Explain why the work done by the gravitational force acting on the satellite is zero. Using the work-energy theorem, what can you conclude about the speed of the satellite?

Reasoning   As the satellite moves in a circular orbit about the Earth as in Figure 7.17, its velocity is tangent to the circular path. Its incremental displacement ds in any small time interval is always at 90° to the inward gravitational force, which always acts toward the center of the Earth. Now $\cos 90° = 0$, therefore $\mathbf{F} \cdot d\mathbf{s} = 0$. So as the satellite turns through any angle or through many revolutions, the total work done on it by the gravitational force is always zero. The work-energy theorem says that the net work done on a particle during any displacement is equal to the change in its kinetic energy. Because the net work done on the satellite is zero, the change in its kinetic energy is zero, and its speed remains constant.

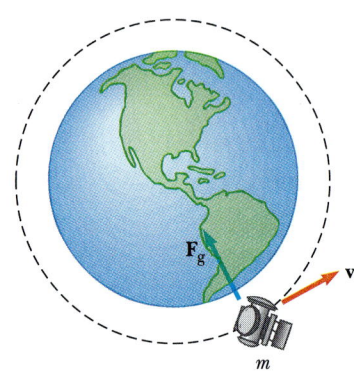

FIGURE 7.17   (Conceptual Example 7.13) An Earth satellite in a circular orbit. What is the work done by the gravitational force acting on the satellite?

7.5   POWER

From a practical viewpoint, it is interesting to know not only the work done on an object but also the rate at which the work is being done. The time rate of doing work is called **power.**

If an external force is applied to an object (which we assume acts as a particle), and if the work done by this force is W in the time interval Δt, then the **average power** during this interval is defined as

Average power

$$\bar{P} \equiv \frac{W}{\Delta t}$$

The work done on the object contributes to increasing the energy of the object. A more general definition of power is the *time rate of energy transfer.* The **instantaneous** power is the limiting value of the average power as Δt approaches zero:

$$P \equiv \lim_{\Delta t \to 0} \frac{W}{\Delta t} = \frac{dW}{dt}$$

where we have represented the infinitesimal value of the work done by dW (even though it is not a change and therefore not a differential). We find from Equation 7.2 that $dW = \mathbf{F} \cdot d\mathbf{s}$. Therefore, the instantaneous power can be written

$$P = \frac{dW}{dt} = \mathbf{F} \cdot \frac{d\mathbf{s}}{dt} = \mathbf{F} \cdot \mathbf{v} \qquad (7.18)$$

Instantaneous power

where we have used the fact that $\mathbf{v} = d\mathbf{s}/dt$.

The SI unit of power is joules per second (J/s), also called a *watt* (W) (after James Watt):

$$1 \text{ W} = 1 \text{ J/s} = 1 \text{ kg} \cdot \text{m}^2/\text{s}^3$$

The watt

The symbol W for watt should not be confused with the symbol W for work.

A unit of power in the British engineering system is the horsepower (hp):

$$1 \text{ hp} = 550 \text{ ft} \cdot \text{lb/s} = 746 \text{ W}$$

A unit of energy (or work) can now be defined in terms of the unit of power. One kilowatt hour (kWh) is the energy converted or consumed in 1 h at the constant rate of 1 kW. The numerical value of 1 kWh is

$$1 \text{ kWh} = (10^3 \text{ W})(3600 \text{ s}) = 3.60 \times 10^6 \text{ J}$$

The kilowatt hour is a unit of energy

It is important to realize that a kilowatt hour is a unit of energy, not power. When you pay your electric bill, you are buying energy, and the amount of electricity used by an appliance is usually expressed in kilowatt hours. For example, an electric bulb rated at 100 W would "consume" 3.60×10^5 J of energy in 1 h.

EXAMPLE 7.14 Power Delivered by an Elevator Motor

An elevator has a mass of 1000 kg and carries a maximum load of 800 kg. A constant frictional force of 4000 N retards its motion upward, as in Figure 7.18. (a) What must be the minimum power delivered by the motor to lift the elevator at a constant speed of 3.00 m/s?

Solution The motor must supply the force $\mathbf{T}$ that pulls the elevator upward. From Newton's second law and from the fact that $a = 0$ since v is constant, we get

$$T - f - Mg = 0$$

where M is the *total* mass (elevator plus load), equal to 1800 kg. Therefore,

$$\begin{aligned} T &= f + Mg \\ &= 4.00 \times 10^3 \text{ N} + (1.80 \times 10^3 \text{ kg})(9.80 \text{ m/s}^2) \\ &= 2.16 \times 10^4 \text{ N} \end{aligned}$$

Using Equation 7.18 and the fact that $\mathbf{T}$ is in the same direction as $\mathbf{v}$ gives

$$\begin{aligned} P &= \mathbf{T} \cdot \mathbf{v} = Tv \\ &= (2.16 \times 10^4 \text{ N})(3.00 \text{ m/s}) = 6.49 \times 10^4 \text{ W} \\ &= 64.9 \text{ kW} = 87.0 \text{ hp} \end{aligned}$$

(b) What power must the motor deliver at any instant if it is designed to provide an upward acceleration of 1.00 m/s²?

Solution Applying Newton's second law to the elevator gives

$$\begin{aligned} T - f - Mg &= Ma \\ T &= M(a + g) + f \\ &= (1.80 \times 10^3 \text{ kg})(1.00 + 9.80)\text{m/s}^2 + 4.00 \times 10^3 \text{ N} \\ &= 2.34 \times 10^4 \text{ N} \end{aligned}$$

Therefore, using Equation 7.18 we get for the required power

$$P = Tv = \boxed{(2.34 \times 10^4 v)}\ \text{W}$$

where v is the instantaneous speed of the elevator in meters per second. Hence, the power required increases with increasing speed.

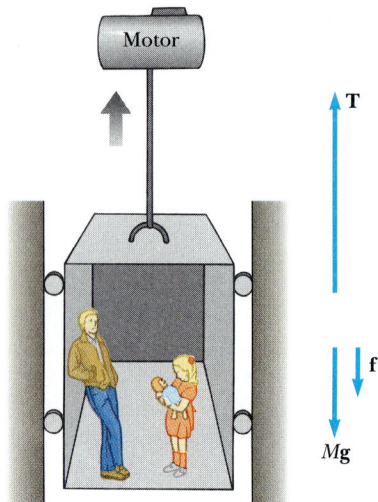

FIGURE 7.18   (Example 7.14) A motor exerts an upward force, equal in magnitude to the tension T, on the elevator. A frictional force **f** and the force of gravity $M\mathbf{g}$ act downward.

CONCEPTUAL EXAMPLE 7.15

In the first part of the previous example, the motor delivers power to lift the elevator, yet the elevator moves at constant speed. A student analyzing this situation claims that according to the work-energy theorem, if the speed of the elevator remains constant, the work done on it is zero. The student concludes that the power that must be delivered by the motor must also be zero. How would you explain this apparent paradox?

Reasoning   The student has attempted to apply the work-energy theorem to a system that is not acting as a particle (there is significant friction). Applying Newton's law, it is the *net* force on the system, times the displacement (of the center of mass), that is equal to the change in the kinetic energy of the system. In this case, there are three forces acting on the elevator: the upward force T exerted by the cable, the downward force of gravity, and the downward frictional force (see Fig. 7.18). The elevator moves at constant speed (zero acceleration) when the upward force is balanced by the sum of the two downward forces ($T = Mg + f$). The power that is supplied by the motor is equal to Tv, which is *not* zero. Part of the energy supplied by the motor in some time interval is used to increase its potential energy and part is lost due to the frictional force. So there is no paradox.

*7.6   ENERGY AND THE AUTOMOBILE

Automobiles powered by gasoline engines are very inefficient machines. Even under ideal conditions, less than 15% of the available energy in the fuel is used to power the vehicle. The situation is much worse under stop-and-go driving in the city. This section uses the concepts of energy, power, and friction to analyze automobile fuel consumption.

Many mechanisms contribute to energy loss in a typical automobile. About two thirds of the energy available from the fuel is lost in the engine. This energy ends up in the atmosphere, partly via the exhaust system and partly via the cooling system. As we shall see in Chapter 22, the large power loss in the exhaust and cooling system is not easy to overcome because of some fundamental laws of thermodynamics. Approximately 10% of the available energy is lost to friction in the transmission, drive shaft, wheel and axle bearings, and differential. Friction in other moving parts dissipates approximately 6% of the energy, and 4% is used to operate fuel and oil pumps and such accessories as power steering and air conditioning. Finally, approximately 13% of the available energy is used to propel the automobile. This energy is used mainly to overcome road friction and air resistance.

TABLE 7.3   Frictional Forces and Power Requirements for a Typical Car

$v\,(\text{m/s})$	$n\,(\text{N})$	$f_r\,(\text{N})$	$f_a\,(\text{N})$	$f_t\,(\text{N})$	$P = f_t v\,(\text{kW})$
0	14 200	227	0	227	0
8.9	14 100	226	51	277	2.5
17.8	13 900	222	204	426	7.6
26.8	13 600	218	465	683	18.3
35.9	13 200	211	830	1041	37.3
44.8	12 600	202	1293	1495	66.8

In this table, n is the normal force, f_r is road friction, f_a is air friction, f_t is total friction, and P is the power delivered to the wheels.

Let us examine the power required to overcome road friction and air drag. The coefficient of rolling friction μ between tires and road is about 0.016. For a 1450-kg car, the weight is 14 200 N and the force of rolling friction is $\mu n = \mu w = 227$ N. As the speed of the car increases, there is a small reduction in the normal force as a result of a reduction in air pressure as air flows over the top of the car. This causes a slight reduction in the force of rolling friction f_r with increasing speed.

Now let us consider the effect of the resistive force that results from air moving past the various surfaces of the car. For large objects, the resistive force associated with air friction is proportional to the square of the speed (in meters per second) (Section 6.4) and is given by Equation 6.5:

$$f_a = \tfrac{1}{2}DA\rho v^2$$

where D is the drag coefficient, A is the cross-sectional area of the moving object, and ρ is the density of air. This expression can be used to calculate the f_a values in Table 7.3 using $D = 0.50$, $\rho = 1.293\ \text{kg/m}^3$, and $A \approx 2\ \text{m}^2$.

The magnitude of the total frictional force, f_t, is the sum of the rolling friction force and the air resistive force:

$$f_t = f_r + f_a \approx \text{constant} + \tfrac{1}{2}DA\rho v^2 \tag{7.19}$$

At low speeds, road resistance is the predominant resistive force, but at high speeds air drag predominates, as shown in Table 7.3. Road friction can be reduced by reducing tire flexing (increase the air pressure slightly above recommended values) and using radial tires. Air drag can be reduced by using a smaller cross-sectional area and streamlining the car. Although driving a car with the windows open does create more air drag, resulting in a 3% decrease in mileage, driving with the windows closed and the air conditioner running results in a 12% decrease in mileage.

The total power needed to maintain a constant speed v is $f_t v$, and this is the power that must be delivered to the wheels. For example, from Table 7.3 we see that at $v = 26.8$ m/s, the required power is

$$P = f_t v = (683\ \text{N})\left(26.8\ \frac{\text{m}}{\text{s}}\right) = 18.3\ \text{kW}$$

This can be broken into two parts: (1) the power needed to overcome road friction, $f_r v$, and (2) the power needed to overcome air drag, $f_a v$. At $v = 26.8$ m/s,

these have the values

$$P_r = f_r v = (218 \text{ N}) \left(26.8 \, \frac{\text{m}}{\text{s}} \right) = 5.84 \text{ kW}$$

$$P_a = f_a v = (465 \text{ N}) \left(26.8 \, \frac{\text{m}}{\text{s}} \right) = 12.5 \text{ kW}$$

Note that $P = P_r + P_a$.

On the other hand, at $v = 44.8$ m/s (100 mi/h), we find that $P_r = 9.05$ kW, $P_a = 57.9$ kW, and $P = 67.0$ kW. This shows the importance of air drag at high speeds.

EXAMPLE 7.16   Gas Consumed by a Compact Car

A compact car has a mass of 800 kg, and its efficiency is rated at 18%. (That is, 18% of the available fuel energy is delivered to the wheels.) Find the amount of gasoline used to accelerate the car from rest to 60 mi/h (27 m/s). Use the fact that the energy equivalent of one gallon of gasoline is 1.3×10^8 J.

Solution   The energy required to accelerate the car from rest to a speed v is its kinetic energy, $\frac{1}{2}mv^2$. For this case,

$$E = \tfrac{1}{2}mv^2 = \tfrac{1}{2}(800 \text{ kg}) \left(27 \, \frac{\text{m}}{\text{s}} \right)^2 = 2.9 \times 10^5 \text{ J}$$

If the engine were 100% efficient, each gallon of gasoline would supply 1.3×10^8 J of energy. Since the engine is only 18% efficient, each gallon delivers only $(0.18)(1.3 \times 10^8 \text{ J}) = 2.3 \times 10^7$ J. Hence, the number of gallons used to accelerate the car is

$$\text{Number of gal} = \frac{2.9 \times 10^5 \ \text{J}}{2.3 \times 10^7 \ \text{J/gal}} = \boxed{0.013 \text{ gal}}$$

At this rate, a gallon of gas would be used after 77 such accelerations. This demonstrates the severe energy requirements for extreme stop-and-start driving.

EXAMPLE 7.17   Power Delivered to Wheels

Suppose the car described in Example 7.16 has a mileage rating of 35 mi/gal when traveling at 60 mi/h. How much power is delivered to the wheels?

Solution   The car consumes 60/35 = 1.7 gal/h. Using the fact that each gallon is equivalent to 1.3×10^8 J, we find that the total power used is

$$P = \frac{(1.7 \text{ gal/h})(1.3 \times 10^8 \text{ J/gal})}{3.6 \times 10^3 \text{ s/h}}$$

$$= \frac{2.2 \times 10^8 \text{ J}}{3.6 \times 10^3 \text{ s}} = 62 \text{ kW}$$

Since 18% of the available power is used to propel the car, the power delivered to the wheels is $(0.18)(62 \text{ kW}) = 11$ kW. This is about one-half the value obtained for the large 1450-kg car discussed in the text. Size is clearly an important factor in power-loss mechanisms.

EXAMPLE 7.18   Car Accelerating Up a Hill

Consider a car of mass m accelerating up a hill, as in Figure 7.19. Assume that the magnitude of the resistive force is

$$|\mathbf{f}| = (218 + 0.70v^2) \text{ N}$$

where v is the speed in meters per second. Calculate the power the engine must deliver to the wheels.

Solution   The forces on the car are shown in Figure 7.19, where **F** is the force of static friction that propels the car and the remaining forces have their usual meaning. Newton's

second law applied to the motion along the road surface gives

$$\sum F_x = F - |\mathbf{f}| - mg \sin \theta = ma$$

$$F = ma + mg \sin \theta + |\mathbf{f}|$$

$$= ma + mg \sin \theta + (218 + 0.70v^2)$$

Therefore, the power required for propulsion is

$$P = Fv = mva + mvg \sin \theta + 218v + 0.70v^3$$

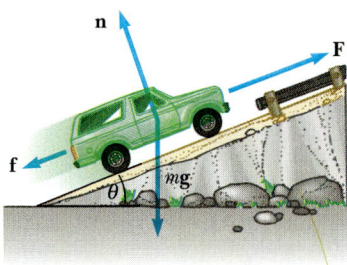

FIGURE 7.19   (Example 7.18)

where *mva* represents the power the engine must deliver to accelerate the car. If the car moves at constant speed, this term is zero and the power requirement is reduced. The term $mvg \sin \theta$ is the power required to overcome the force of gravity as the car moves up the incline. This term would be zero for motion on a horizontal surface. The term $218v$ is the power required to counterbalance rolling friction. Fi-

nally, the term $0.70v^3$ is the power needed to overcome air drag.

If we take $m = 1450$ kg, $v = 27$ m/s $(= 60$ mi/h$)$, $a = 1.0$ m/s^2, and $\theta = 10°$, the various terms in P are calculated to be

$$mva = (1450 \text{ kg})(27 \text{ m/s})(1.0 \text{ m/s}^2)$$
$$= 39 \text{ kW} = 52 \text{ hp}$$

$$mvg \sin \theta = (1450 \text{ kg})(27 \text{ m/s})(9.80 \text{ m/s}^2)(\sin 10°)$$
$$= 67 \text{ kW} = 89 \text{ hp}$$

$$218v = 218(27) = 5.9 \text{ kW} = 7.9 \text{ hp}$$

$$0.70v^3 = 0.70(27)^3 = 14 \text{ kW} = 18 \text{ hp}$$

Hence, the total power required is 126 kW, or 167 hp. Note that the power requirements for traveling at constant speed on a horizontal surface are only 20 kW, or 26 hp (the sum of the last two terms). Furthermore, if the mass is halved (as in compact cars), the power required is also reduced by almost the same factor.

*7.7   KINETIC ENERGY AT HIGH SPEEDS

The laws of Newtonian mechanics are valid only for describing the motion of particles moving at speeds that are small compared with the speed of light in a vacuum, c ($\approx 3 \times 10^8$ m/s). When the particle speeds are comparable to c, the equations of Newtonian mechanics must be replaced by the more general equations predicted by the theory of relativity. One consequence of the theory of relativity is that the kinetic energy of a particle of mass m moving with a speed v is no longer given by $K = mv^2/2$. Instead, one must use the relativistic form of the kinetic energy:

$$K = mc^2 \left(\frac{1}{\sqrt{1 - (v/c)^2}} - 1 \right) \qquad (7.20)$$

Relativistic kinetic energy

According to this expression, speeds greater than c are not allowed because K would be imaginary for $v > c$. Furthermore, as v approaches c, K approaches ∞. This is consistent with experimental observations on subatomic particles, which have shown that no particles travel at speeds greater than c. (That is, c is the ultimate speed.) From the point of view of the work-energy theorem, v can only approach c, since it would take an infinite amount of work to attain the speed $v = c$.

All formulas in the theory of relativity must reduce to those in Newtonian mechanics at low particle speeds. It is instructive to show that this is the case for the kinetic energy relationship by analyzing Equation 7.20 when v is small compared to c. In this case, we expect that K should reduce to the Newtonian expression. We can check this by using the binomial expansion applied to the quantity $[1 - (v/c)^2]^{-1/2}$, with $v/c \ll 1$. If we let $x = (v/c)^2$, the expansion gives

$$\frac{1}{(1 - x)^{1/2}} = 1 + \frac{x}{2} + \frac{3}{8}x^2 + \cdots$$

Making use of this expansion in Equation 7.20 gives

$$K = mc^2 \left(1 + \frac{v^2}{2c^2} + \frac{3}{8}\frac{v^4}{c^4} + \cdots - 1 \right)$$

$$= \frac{1}{2}mv^2 + \frac{3}{8}m\frac{v^4}{c^2} + \cdots$$

$$\approx \frac{1}{2}mv^2 \quad \text{for} \quad \frac{v}{c} \ll 1$$

Thus, we see that the relativistic kinetic energy expression does indeed reduce to the Newtonian expression for speeds that are small compared with c. We shall return to the subject of relativity in more depth in Chapter 39.

SUMMARY

The **work** done by a *constant* force $\mathbf{F}$ acting on a particle is defined as the product of the component of the force in the direction of the particle's displacement and the magnitude of the displacement. If $\mathbf{F}$ makes an angle θ with the displacement $\mathbf{s}$, the work done by $\mathbf{F}$ is

$$W \equiv Fs \cos \theta \tag{7.1}$$

The **scalar**, or **dot**, **product** of two vectors $\mathbf{A}$ and $\mathbf{B}$ is defined by the relationship

$$\mathbf{A} \cdot \mathbf{B} \equiv AB \cos \theta \tag{7.3}$$

where the result is a scalar quantity and θ is the angle between the directions of the two vectors. The scalar product obeys the commutative and distributive laws.

The **work** done by a *varying* force acting on a particle moving along the x axis from x_i to x_f is

$$W \equiv \int_{x_i}^{x_f} F_x \, dx \tag{7.7}$$

where F_x is the component of force in the x direction. If there are several forces acting on the particle, the net work done by all forces is the sum of the individual amounts of work done by each force.

The **kinetic energy** of a particle of mass m moving with a speed v (where v is small compared with the speed of light) is

$$K \equiv \tfrac{1}{2}mv^2 \tag{7.14}$$

The **work–energy theorem** states that the net work done on a particle by external forces equals the change in kinetic energy of the particle:

$$W_{\text{net}} = K_f - K_i = \tfrac{1}{2}mv_f^2 - \tfrac{1}{2}mv_i^2 \tag{7.16}$$

The **instantaneous power** is defined as the time rate of energy transfer. If an agent applies a force $\mathbf{F}$ to an object moving with a velocity $\mathbf{v}$, the power delivered by that agent is

$$P \equiv \frac{dW}{dt} = \mathbf{F} \cdot \mathbf{v} \tag{7.18}$$

QUESTIONS

1. When a particle rotates in a circle, a central force acts on it directed toward the center of rotation. Why is it that this force does no work on the particle?
2. Is there any direction associated with the dot product of two vectors?
3. If the dot product of two vectors is positive, does this imply that the vectors must have positive rectangular components?
4. As the load on a spring hung vertically is increased, one would not expect the F_s versus x curve to always remain linear as in Figure 7.8d. Explain qualitatively what you would expect for this curve as m is increased.
5. Can kinetic energy be negative? Explain.
6. If the speed of a particle is doubled, what happens to its kinetic energy?
7. What can be said about the speed of a particle if the net work done on it is zero?
8. Can the average power ever equal the instantaneous power? Explain.
9. In Example 7.18, does the required power increase or decrease as the force of friction is reduced?
10. An automobile sales representative claims that a "souped-up" 300-hp engine is a necessary option in a compact car (instead of a conventional 130-hp engine). Suppose you intend to drive the car within speed limits ($\leqslant 55$ mi/h) and on flat terrain. How would you counter this sales pitch?
11. One bullet has twice the mass of a second bullet. If both are fired so that they have the same speed, which has more kinetic energy? What is the ratio of the kinetic energies of the two bullets?

12. When a punter kicks a football, is he doing any work on the ball while his toe is in contact with it? Is he doing any work on the ball after it loses contact with his toe? Are there any forces doing work on the ball while it is in flight?
13. Discuss the work done by a pitcher throwing a baseball. What is the approximate distance through which the force acts as the ball is thrown?
14. Estimate the time it takes you to climb a flight of stairs. Then approximate the power required to perform this task. Express your answer in horsepower.
15. Cite two examples in which a force is exerted on an object without doing any work on the object.
16. Two sharpshooters fire 0.30-caliber rifles using identical shells. The barrel of rifle A is 2.00 cm longer than that of rifle B. Which rifle will have the higher muzzle speed? (*Hint:* The force of the expanding gases in the barrel accelerates the bullets.)
17. As a simple pendulum swings back and forth, the forces acting on the suspended mass are the force of gravity, the tension in the supporting cord, and air resistance. (a) Which of these forces, if any, does no work on the pendulum? (b) Which of these forces does negative work at all times during its motion? (c) Describe the work done by the force of gravity while the pendulum is swinging.
18. The kinetic energy of an object depends on the frame of reference in which its motion is measured. Give an example to illustrate this point.

PROBLEMS

Review Problem

Two constant forces act on a 5.0-kg object moving in the xy plane as shown in the figure. The force $\mathbf{F}_1$ has a magnitude of 25 N and makes an angle of $35°$ with the x axis, while the force $\mathbf{F}_2$ has a magnitude of 42 N and makes an angle of $150°$ with the x axis. At $t = 0$, the object is at the origin and has a velocity of $(4.0\mathbf{i} + 2.5\mathbf{j})$ m/s. Find (a) the components of the applied forces and expressions for the applied forces in unit vector notation; (b) a graphical solution for the resultant force on the object; (c) the components of the resultant force and an expression for the resultant force in unit vector notation; (d) the magnitude and direction of the object's acceleration, and an expression for the acceleration in unit vector notation; (e) the components of the velocity and an expression for the velocity in unit vector notation at $t = 3.0$ s; (f) the coordi-

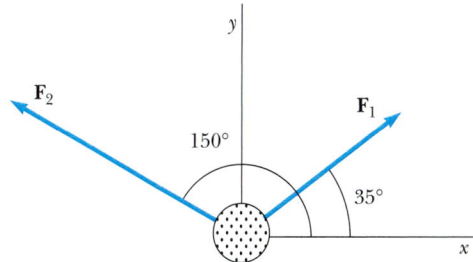

nates and position vector of the object at $t = 3.0$ s; (g) the kinetic energy of the object at $t = 3.0$ s; and (h) the net work done on the object treating it as a particle as it moves from the origin to its location at $t = 5.0$ s.

Section 7.1   Work Done by a Constant Force

1. If a person lifts a 20.0-kg bucket from a well and does 6.00 kJ of work, how deep is the well? Assume the speed of the bucket remains constant as it is lifted.

2. A raindrop ($m = 3.35 \times 10^{-5}$ kg) falls vertically at constant speed under the influence of gravity and air resistance. After the drop has fallen 100 m, what is (a) the work done by gravity and (b) the energy dissipated by air resistance?

3. A block of mass 2.5 kg is pushed 2.2 m along a frictionless horizontal table by a constant 16.0-N force directed 25° below the horizontal. Determine the work done by (a) the applied force, (b) the normal force exerted by the table, (c) the force of gravity, and (d) the net force on the block.

4. Two objects having masses $m_1 = 10.0$ kg and $m_2 = 8.0$ kg hang from a frictionless pulley, as in Figure P7.4. (a) Determine the work done by the force of gravity on each object separately as the 10.0-kg mass is displaced downward by 0.50 m. (b) What is the total work done on each object, including the work done by the force of the string? (c) Comment on any relationship you have discovered connecting these quantities.

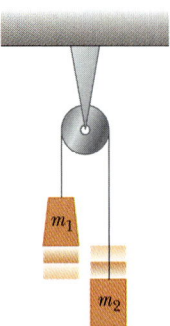

FIGURE P7.4

5. A cheerleader lifts his 50.0-kg partner straight up off the ground a distance of 0.60 m before releasing her. If he does this 20 times, how much work has he done?

6. A team of dogs drags a 100-kg sled 2.0 km over a horizontal surface at a constant speed. If the coefficient of friction between sled and snow is 0.15, find (a) the work done by the dogs and (b) the energy lost due to friction.

7. A horizontal force of 150 N is used to push a 40.0-kg box 6.00 m on a rough, horizontal surface. If the box moves at constant speed, find (a) the work done by the 150-N force, (b) the energy lost due to friction, and (c) the coefficient of kinetic friction.

8. A 15-kg block is dragged over a rough, horizontal surface by a 70-N force acting 20° above the horizontal. The block is displaced 5.0 m, and the coefficient of kinetic friction is 0.30. Find the work done by (a) the 70-N force, (b) the normal force, and (c) the force of gravity. (d) What is the energy lost due to friction?

8A. A block of mass m is dragged over a rough, horizontal surface by a force F acting at an angle θ above the horizontal. The block is displaced a distance d, and the coefficient of kinetic friction is μ_k. Find the work done by (a) the force F, (b) the normal force, and (c) the force of gravity. (d) What is the energy lost due to friction?

9. If you push a 40.0-kg crate at a constant speed of 1.40 m/s across a horizontal floor ($\mu_k = 0.25$), at what rate (a) is work being done on the crate by you and (b) is energy dissipated by the frictional force?

10. Batman, whose mass is 80.0 kg, is holding on to the free end of a 12.0-m rope, the other end of which is fixed to a tree limb above. He is able to get the rope in motion as only Batman knows how, eventually getting it to swing enough that he can reach a ledge when the rope makes a 60° angle with the vertical. How much work was done against the force of gravity in this maneuver?

11. A cart loaded with bricks has a total mass of 18.0 kg and is pulled at constant speed by a rope. The rope is inclined at 20.0° above the horizontal, and the cart moves 20.0 m on a horizontal surface. The coefficient of kinetic friction between ground and cart is 0.500. (a) What is the tension in the rope? (b) How much work is done on the cart by the rope? (c) What is the energy lost due to friction?

11A. A cart loaded with bricks has a total mass m and is pulled at constant speed by a rope. The rope is inclined at an angle θ above the horizontal, and the cart moves a distance d on a horizontal surface. The coefficient of kinetic friction between ground and cart is μ_k. (a) What is the tension in the rope? (b) How much work is done on the cart by the rope? (c) What is the energy lost due to friction?

Section 7.2   The Scalar Product of Two Vectors

12. For $\mathbf{A} = 4\mathbf{i} + 3\mathbf{j}$ and $\mathbf{B} = -\mathbf{i} + 3\mathbf{j}$, find (a) $\mathbf{A} \cdot \mathbf{B}$ and (b) the angle between $\mathbf{A}$ and $\mathbf{B}$.

13. Vector $\mathbf{A}$ extends from the origin to a point having polar coordinates (7, 70°) and vector $\mathbf{B}$ extends from the origin to a point having polar coordinates (4, 130°). Find $\mathbf{A} \cdot \mathbf{B}$.

□ indicates problems that have full solutions available in the Student Solutions Manual and Study Guide.

13A. Vector **A** extends from the origin to a point having polar coordinates (r_1, θ_1), and vector **B** extends from the origin to a point having polar coordinates (r_2, θ_2). Find **A** · **B**.

14. Vector **A** has a magnitude of 5.00 units, and **B** has a magnitude of 9.00 units. The two vectors make an angle of 50.0° with each other. Find **A** · **B**.

15. Show that **A** · **B** $= A_x B_x + A_y B_y + A_z B_z$. (*Hint:* Write **A** and **B** in unit vector form and use Eqs. 7.4 and 7.5.)

16. For **A** $= 3\mathbf{i} + \mathbf{j} - \mathbf{k}$, **B** $= -\mathbf{i} + 2\mathbf{j} + 5\mathbf{k}$, and **C** $= 2\mathbf{j} - 3\mathbf{k}$, find **C** · (**A** − **B**).

17. A force **F** $= (6\mathbf{i} - 2\mathbf{j})$ N acts on a particle that undergoes a displacement **s** $= (3\mathbf{i} + \mathbf{j})$ m. Find (a) the work done by the force on the particle and (b) the angle between **F** and **s**.

18. Vector **A** is 2.0 units long and points in the positive *y* direction. Vector **B** has a negative *x* component 5.0 units long, a positive *y* component 3 units long, and no *z* component. Find **A** · **B** and the angle between the vectors.

19. A force **F** $= (3.00\mathbf{i} + 4.00\mathbf{j})$ N acts on a particle. The angle between **F** and the displacement vector **s** is 32.0°, and 100.0 J of work is done by **F**. Find **s**.

20. Find the angle between **A** $= -5\mathbf{i} - 3\mathbf{j} + 2\mathbf{k}$ and **B** $= -2\mathbf{j} - 2\mathbf{k}$.

21. Using the definition of the scalar product, find the angles between (a) **A** $= 3\mathbf{i} - 2\mathbf{j}$ and **B** $= 4\mathbf{i} - 4\mathbf{j}$; (b) **A** $= -2\mathbf{i} + 4\mathbf{j}$ and **B** $= 3\mathbf{i} - 4\mathbf{j} + 2\mathbf{k}$; (c) **A** $= \mathbf{i} - 2\mathbf{j} + 2\mathbf{k}$ and **B** $= 3\mathbf{j} + 4\mathbf{k}$.

Section 7.3   Work Done by a Varying Force

22. A force **F** $= (4.0x\mathbf{i} + 3.0y\mathbf{j})$ N acts on a particle as the object moves in the *x* direction from the origin to $x = 5.0$ m. Find the work done on the object by the force.

23. A particle is subject to a force F_x that varies with position as in Figure P7.23. Find the work done by the force on the body as it moves (a) from $x = 0$ to $x = 5.0$ m, (b) from $x = 5.0$ m to $x = 10$ m, and (c) from $x = 10$ m to $x = 15$ m. (d) What is the total work done by the force over the distance $x = 0$ to $x = 15$ m?

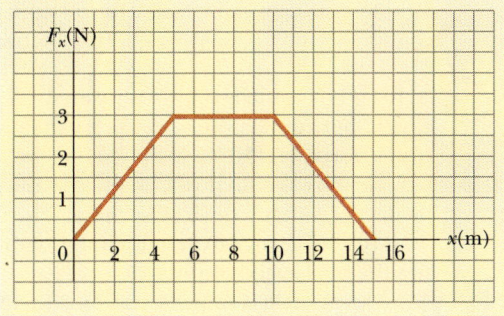

FIGURE P7.23

24. The force acting on a particle varies as in Figure P7.24. Find the work done by the force as the particle moves (a) from $x = 0$ to $x = 8.0$ m, (b) from $x = 8.0$ m to $x = 10$ m, and (c) from $x = 0$ to $x = 10$ m.

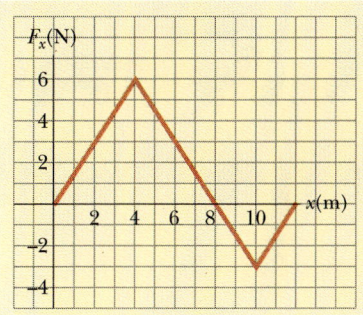

FIGURE P7.24

25. An archer pulls her bow string back 0.400 m by exerting a force that increases uniformly from zero to 230 N. (a) What is the equivalent spring constant of the bow? (b) How much work is done in pulling the bow?

25A. An archer pulls her bow string back a distance *d* by exerting a force that increases uniformly from zero to *F*. (a) What is the equivalent spring constant of the bow? (b) How much work is done in pulling the bow?

26. A 100-g bullet is fired from a rifle having a barrel 0.60 m long. Assuming the origin is placed where the bullet begins to move, the force (in newtons) exerted on the bullet by the expanding gas is $15\,000 + 10\,000x - 25\,000x^2$ where *x* is in meters. (a) Determine the work done by the gas on the bullet as the bullet travels the length of the barrel. (b) If the barrel is 1.00 m long, how much work is done, and how does this value compare to the work calculated in (a)?

27. A 6000-kg freight car rolls along rails with negligible friction. The car is brought to rest by a combination of two coiled springs, as illustrated in Figure P7.27. Both springs obey Hooke's law, with $k_1 = 1600$ N/m and $k_2 = 3400$ N/m. After the first spring compresses a distance of 30.0 cm, the second spring (acting with the first) increases the force so that there is additional compression, as shown in the graph. If the car is brought to rest 50.0 cm after first contacting the two-spring system, find the car's initial speed.

28. A marine in the jungle finds himself in the middle of a swamp. The force F_x he must exert in the *x* direction as he struggles to get out is $F_x = (1000 - 50.0x)$ N, where *x* is in meters. (a) Sketch the graph of F_x versus *x*. (b) What is the average force he exerts in traveling from 0 to *x*? (c) If he travels $x = 20.0$ m to

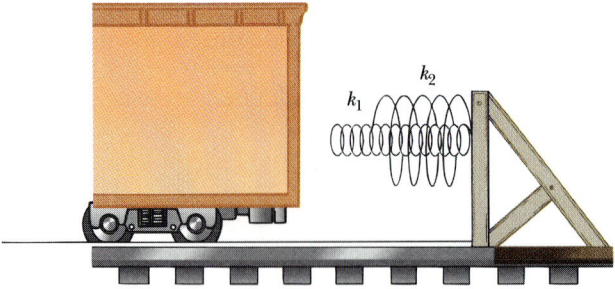

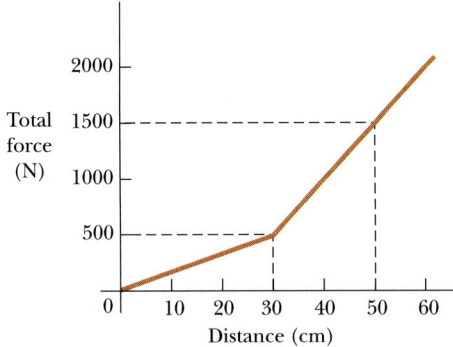

FIGURE P7.27

get completely free of the swamp, how much energy does he expend against the swamp?

29. A small mass m is pulled to the top of a frictionless half-cylinder (of radius R) by a cord passing over the top of the cylinder, as illustrated in Figure P7.29. (a) If the mass moves at a constant speed, show that $F = mg \cos \theta$. (*Hint*: If the mass moves at constant speed, the component of its acceleration tangent to the cylinder must be zero at all times.) (b) By directly integrating $W = \int \mathbf{F} \cdot d\mathbf{s}$, find the work done in moving the mass at constant speed from the bottom to the top of the half-cylinder.

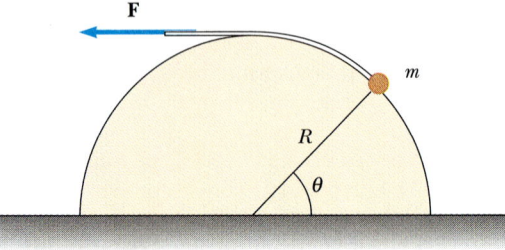

FIGURE P7.29

30. The force required to stretch a Hooke's-law spring varies from zero to 50.0 N as we stretch the spring by

moving one end 12.0 cm from its unstressed position. (a) Find the force constant of the spring. (b) Find the work done in stretching the spring.

31. If it takes 4.00 J of work to stretch a Hooke's-law spring 10.0 cm from its unstressed length, determine the extra work required to stretch it an additional 10.0 cm.

31A. If it takes work W to stretch a Hooke's-law spring a distance d from its unstressed length, determine the extra work required to stretch it an additional distance d.

Section 7.4   Kinetic Energy and the Work-Energy Theorem

32. A 0.600-kg particle has a speed of 2.00 m/s at point A and kinetic energy of 7.50 J at B. What is (a) its kinetic energy at A? (b) its speed at B? (c) the total work done on the particle as it moves from A to B?

33. A picture tube in a certain television set is 36 cm long. The electrical force accelerates an electron in the tube from rest to 1% the speed of light over this distance. Determine (a) the kinetic energy of the electron as it strikes the screen at the end of the tube, (b) the magnitude of the average electrical force acting on the electron over this distance, (c) the magnitude of the average acceleration of the electron over this distance, and (d) the time of flight.

34. A 7.00-kg bowling ball moves at 3.00 m/s. How fast must a 46-g golf ball move so that the two balls have the same kinetic energy?

35. A mechanic pushes a 2500-kg car from rest to a speed v, doing 5000 J of work in the process. During this time, the car moves 25.0 m. Neglecting friction between car and road, (a) what is the final speed, v, of the car? (b) What is the horizontal force exerted on the car?

35A. A mechanic pushes a car of mass m from rest to a speed v, doing work W in the process. During this time, the car moves a distance d. Neglecting friction between car and road, (a) what is the final speed, v, of the car? (b) What is the horizontal force exerted on the car?

36. A 3.0-kg mass has an initial velocity $\mathbf{v}_0 = (6.0\mathbf{i} - 2.0\mathbf{j})$ m/s. (a) What is its kinetic energy at this time? (b) Find the change in its kinetic energy if its velocity changes to $(8.0\mathbf{i} + 4.0\mathbf{j})$ m/s. (*Hint*: Remember that $v^2 = \mathbf{v} \cdot \mathbf{v}$.)

37. A 40-kg box initially at rest is pushed 5.0 m along a rough, horizontal floor with a constant applied horizontal force of 130 N. If the coefficient of friction between box and floor is 0.30, find (a) the work done by the applied force, (b) the energy lost due to friction, (c) the change in kinetic energy of the box, and (d) the final speed of the box.

38. A 4.0-kg particle is subject to a force that varies with

position as shown in Figure P7.23. The particle starts from rest at $x = 0$. What is its speed at (a) $x = 5.0$ m, (b) $x = 10$ m, (c) $x = 15$ m?

39. A 15.0-g bullet is accelerated in a rifle barrel 72.0 cm long to a speed of 780 m/s. Use the work-energy theorem to find the average force exerted on the bullet while it is being accelerated.

40. A bullet with a mass of 5.00 g and a speed of 600 m/s penetrates a tree to a depth of 4.00 cm. (a) Use energy considerations to find the average frictional force that stops the bullet. (b) Assuming that the frictional force is constant, determine how much time elapsed between the moment the bullet entered the tree and the moment it stopped.

40A. A bullet with a mass m and speed v penetrates a tree to a depth d. (a) Use energy considerations to find the average frictional force that stops the bullet. (b) Assuming that the frictional force is constant, determine how much time elapsed between the moment the bullet entered the tree and the moment it stopped.

41. A block of mass m hangs on the end of a cord and is connected to a block of mass M by the pulley arrangement shown in Figure P7.41. Using energy considerations, (a) find an expression for the speed of m as a function of the distance it has fallen. Assume that the blocks are initially at rest and there is no friction. (b) Repeat (a) assuming sliding friction (coefficient μ_k) between M and the table. (c) Show that the result obtained in (b) reduces to that obtained in (a) in the limit as μ_k goes to zero.

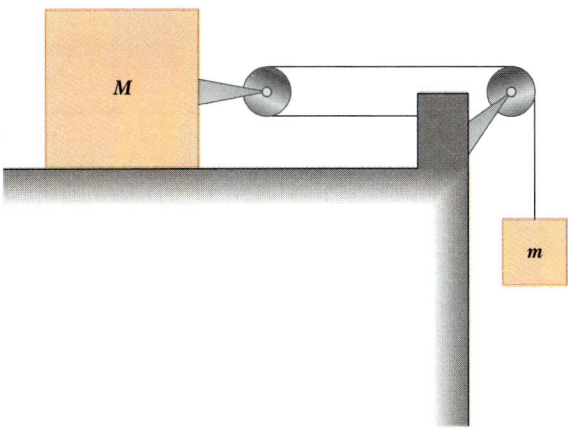

FIGURE P7.41

42. A 5.0-kg steel ball is dropped onto a copper plate from a height of 10.0 m. If the ball leaves a dent 0.32 cm deep, what is the average force exerted on the ball by the plate during the impact?

43. An Atwood's machine (Fig. 5.12) supports masses of 0.20 kg and 0.30 kg. The masses are held at rest beside each other and then released. Neglecting friction, what is the speed of each mass the instant they have both moved 0.40 m?

44. A 2.0-kg block is attached to a spring of force constant 500 N/m as in Figure 7.8. The block is pulled 5.0 cm to the right of equilibrium and released from rest. Find the speed of the block as it passes through equilibrium if (a) the horizontal surface is frictionless and (b) the coefficient of friction between block and surface is 0.35.

45. A sled of mass m is given a kick on a frozen pond, imparting to it an initial speed $v_i = 2$ m/s. The coefficient of kinetic friction between sled and ice is $\mu_k = 0.10$. Use energy considerations to find the distance the sled moves before stopping.

46. A block of mass 12.0 kg slides from rest down a frictionless 35.0° incline and is stopped by a strong spring with $k = 3.00 \times 10^4$ N/m. The block slides 3.00 m from the point of release to the point where it comes to rest against the spring. When the block comes to rest, how far has the spring been compressed?

47. A crate of mass 10.0 kg is pulled up a rough incline with an initial speed of 1.50 m/s. The pulling force is 100 N parallel to the incline, which makes an angle of 20.0° with the horizontal. The coefficient of kinetic friction is 0.400, and the crate is pulled 5.00 m. (a) How much work is done by gravity? (b) How much energy is lost due to friction? (c) How much work is done by the 100-N force? (d) What is the change in kinetic energy of the crate? (e) What is the speed of the crate after being pulled 5.00 m?

48. A block of mass 0.60 kg slides 6.0 m down a frictionless ramp inclined at 20° to the horizontal. It then travels on a rough horizontal surface where $\mu_k = 0.50$. (a) What is the speed of the block at the end of the incline? (b) What is its speed after traveling 1.00 m on the rough surface? (c) What distance does it travel on this horizontal surface before stopping?

49. A 4.00-kg block is given an initial speed of 8.00 m/s at the bottom of a 20.0° incline. The frictional force that retards its motion is 15.0 N. (a) If the block is directed up the incline, how far does it move before stopping? (b) Will it slide back down the incline?

50. A time-varying net force acting on a 4.0-kg particle causes the particle to have a displacement given by $x = 2.0t - 3.0t^2 + 1.0t^3$, where x is in meters and t is in seconds. Find the work done on the particle in the first 3.0 s of motion.

51. A 3.00-kg block is moved up a 37.0° incline under the action of a constant horizontal force of 40.0 N. The coefficient of kinetic friction is 0.100, and the block is displaced 2.00 m up the incline. Calculate (a) the work done by the 40.0-N force, (b) the work done by gravity, (c) the energy lost due to friction,

and (d) the change in kinetic energy of the block. (*Note:* The applied force is *not* parallel to the incline.)

52. A 4.0-kg block attached to a string 2.0 m in length rotates in a circle on a horizontal surface. (a) If the surface is frictionless, identify all the forces on the block and show that the work done by each force is zero for any displacement of the block. (b) If the coefficient of friction between block and surface is 0.25, find the energy lost due to friction in each revolution.

Section 7.5   Power

53. A 700-N marine in basic training climbs a 10.0-m vertical rope at a constant speed in 8.00 s. What is his power output?

54. Water flows over a section of Niagara Falls at a rate of 1.2×10^6 kg/s and falls 50 m. How many 60-W bulbs can be lit with this power?

55. A 650-kg elevator starts from rest. It moves upward for 3.00 s with constant acceleration until it reaches its cruising speed, 1.75 m/s. (a) What is the average power of the elevator motor during this period? (b) How does this power compare with its power while it moves at its cruising speed?

56. A 200-kg crate is pulled along a level surface by an engine. The coefficient of friction between crate and surface is 0.40. (a) How much power must the engine deliver to move the crate at 5.0 m/s? (b) How much work is done by the engine in 3.0 min?

57. A 1500-kg car accelerates uniformly from rest to 10 m/s in 3.0 s. Find (a) the work done on the car in this time, (b) the average power delivered by the engine in the first 3.0 s, and (c) the instantaneous power delivered by the engine at $t = 2.0$ s.

57A. A car of mass m accelerates uniformly from rest to a speed v in a time t. Find (a) the work done on the car in this time, (b) the average power delivered by the engine during this time, and (c) the instantaneous power delivered by the engine at some time less than t, neglecting drag.

58. A certain automobile engine delivers 30.0 hp $(2.24 \times 10^4$ W) to its wheels when moving at 27.0 m/s $(\approx 60$ mi/h). What is the resistive force acting on the automobile at that speed?

59. An outboard motor propels a boat through the water at 10.0 mi/h. The water resists the forward motion of the boat with a force of 15.0 lb. How much power is delivered through the propeller?

60. A car of weight 2500 N operating at a rate of 130 kW develops a maximum speed of 31 m/s on a level, horizontal road. Assuming that the resistive force (due to friction and air resistance) remains constant, (a) what is the car's maximum speed on an incline of 1 in 20 (i.e., if θ is the angle of the incline with the horizontal, $\sin \theta = 1/20$)? (b) What is its power output on a 1-in-10 incline if the car is traveling at 10 m/s?

61. If a certain horse can maintain 1.0 hp of output for 2.0 h, how many 70.0-kg bundles of shingles can the horse hoist (via some pulley arrangement) to the roof of a house 8.0 m tall, assuming 70% efficiency?

62. A force F acts on a particle of mass m. The particle starts at rest at $t = 0$. (a) Show that the instantaneous power delivered by the force at any time t is $(F^2/m)t$. (b) If $F = 20.0$ N and $m = 5.00$ kg, what is the power delivered at $t = 3.00$ s?

*Section 7.6   Energy and the Automobile

63. The horsepower required to keep an aerodynamic car of mass 1500 kg moving at 30.0 mi/h is about 10.0 hp. Assuming that the total retarding force due to friction, air resistance, and so forth is proportional to the square of the car's speed, what is the horsepower required to keep the car moving at 60.0 mi/h?

64. A passenger car carrying two people has a fuel economy of 25 mi/gal. It travels 3000 miles. A jet airplane making the same trip with 150 passengers has a fuel economy of 1.0 mi/gal. Compare the fuel consumed per passenger for the two modes of transportation.

65. A compact car of mass 900 kg has an overall motor efficiency of 15%. (That is, 15% of the energy supplied by the fuel is transformed into kinetic energy of the car.) (a) If burning one gallon of gasoline supplies 1.34×10^8 J of energy, find the amount of gasoline used in accelerating the car from rest to 55 mph. (b) How many such accelerations will one gallon provide? (c) The mileage claimed for the car is 38 mi/gal at 55 mph. What power is delivered to the wheels (to overcome frictional effects) when the car is driven at this speed?

66. Suppose the empty car described in Table 7.3 has a fuel economy of 6.40 km/liter (15 mi/gal) when traveling at 26.8 m/s (60 mi/h). Assuming constant efficiency, determine the fuel economy of the car if the total mass of passengers plus driver is 350 kg.

67. When an air conditioner is added to the car described in Problem 66, the additional output power required to operate the air conditioner is 1.54 kW. If the fuel economy is 6.40 km/liter without the air conditioner, what is it when the air conditioner is operating?

*Section 7.7   Kinetic Energy at High Speeds

68. An electron moves with a speed of $0.995c$. (a) What is its kinetic energy? (b) If you use the classical expression to calculate its kinetic energy, what percentage error would result?

69. A proton in a high-energy accelerator moves with a

speed of $c/2$. Use the work-energy theorem to find the work required to increase its speed to (a) $0.75c$, (b) $0.995c$.

ADDITIONAL PROBLEMS

70. Diatomic molecules exert attractive forces on each other at large distances and replusive forces at short distances. For many molecules the Lennard-Jones law is a good approximation to the magnitude of these intermolecular forces:

$$ F = F_0 \left[2 \left(\frac{\sigma}{r} \right)^{13} - \left(\frac{\sigma}{r} \right)^{7} \right] $$

where r is the center-to-center distance between the atoms in the molecule, σ is a length parameter, and F_0 is the force when $r = \sigma$. For an oxygen molecule, $F_0 = 9.6 \times 10^{-11}$ N and $\sigma = 3.5 \times 10^{-10}$ m. Determine the work done by this force from $r = 4.0 \times 10^{-10}$ m to $r = 9.0 \times 10^{-10}$ m.

71. A bartender slides a bottle of rye on the horizontal counter to a customer 7.0 m away. With what speed did she release the bottle if the coefficient of sliding friction is 0.10 and the bottle comes to rest in front of the customer?

72. When a spring is stretched near its elastic limit, the spring force satisfies the equation $F = -kx + \beta x^3$. If $k = 10$ N/m and $\beta = 100$ N/m³, calculate the work done by this force when the spring is stretched 0.10 m.

73. A particle of mass m moves with a constant acceleration $\mathbf{a}$. If the initial position vector and velocity of the particle are $\mathbf{r}_0$ and $\mathbf{v}_0$, respectively, show that its speed v at any time satisfies the equation

$$ v^2 = v_0{}^2 + 2\mathbf{a} \cdot (\mathbf{r} - \mathbf{r}_0) $$

where $\mathbf{r}$ is the position vector of the particle at that same time.

74. The direction of an arbitrary vector $\mathbf{A}$ can be completely specified with the angles α, β, γ that the vector makes with the x, y, and z axes, respectively. If $\mathbf{A} = A_x\mathbf{i} + A_y\mathbf{j} + A_z\mathbf{k}$, (a) find expressions for $\cos \alpha$, $\cos \beta$, and $\cos \gamma$ (these are known as *direction cosines*) and (b) show that these angles satisfy the relationship $\cos^2 \alpha + \cos^2 \beta + \cos^2 \gamma = 1$. (*Hint:* Take the scalar product of $\mathbf{A}$ with $\mathbf{i}, \mathbf{j},$ and $\mathbf{k}$ separately.)

75. A 4.0-kg particle moves along the x axis. Its position varies with time according to $x = t + 2.0t^3$, where x is in meters and t is in seconds. Find (a) the kinetic energy at any time t, (b) the acceleration of the particle and the force acting on it at time t, (c) the power being delivered to the particle at time t, and (d) the work done on the particle in the interval $t = 0$ to $t = 2.0$ s.

76. An Atwood's machine has a 3.00-kg mass and a 2.00-kg mass at the ends of the string (Fig. 5.12). The

2.00-kg mass is released from rest on the floor, 4.00 m below the 3.00-kg mass. (a) If the pulley is frictionless, what will be the speed of the masses when they pass each other? (b) Suppose that the pulley does not rotate and the string must slide over it. If the total frictional force between the pulley and string is 5.00 N, what are their speeds when the masses pass each other?

77. A 2100-kg pile driver is used to drive a steel I beam into the ground. The pile driver falls 5.0 m before contacting the beam, and it drives the beam 12 cm into the ground before coming to rest. Using energy considerations, calculate the average force the beam exerts on the pile driver while the pile driver is brought to rest.

78. A particle of mass m is attached to two identical springs on a horizontal frictionless table as in Figure P7.78. Both springs have spring constant k and an unstretched length L. (a) If the particle is pulled a distance x along a direction perpendicular to the initial configuration of the springs, show that the force exerted on the particle due to the springs is

$$ \mathbf{F} = -2kx \left(1 - \frac{L}{\sqrt{x^2 + L^2}} \right) \mathbf{i} $$

(b) Determine the amount of work done by this force in moving the particle from $x = A$ to $x = 0$.

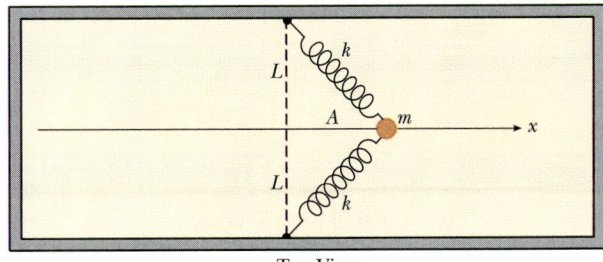

Top View

FIGURE P7.78

79. A 200-g block is pressed against a spring of force constant 1.40 kN/m until the block compresses the spring 10.0 cm. The spring rests at the bottom of a ramp inclined at 60.0° to the horizontal. Use energy considerations to determine how far up the incline the block moves before it stops (a) if there is no friction between block and ramp and (b) if the coefficient of kinetic friction is 0.400.

80. A block of mass m is attached to a massless spring of force constant k as in Figure 7.8. The spring is compressed a distance d from its position and released from rest. (a) If the block comes to rest when it first reaches the equilibrium position, what is the coeffi-

cient of friction between block and surface? (b) If the block first comes to rest when the spring is stretched a distance $d/2$ from equilibrium, what is μ?

81. A 0.400-kg particle slides on a horizontal, circular track 1.50 m in radius. It is given an initial speed of 8.00 m/s. After one revolution, its speed drops to 6.00 m/s because of friction. (a) Find the energy lost due to friction in one revolution. (b) Calculate the coefficient of kinetic friction. (c) What is the total number of revolutions the particle makes before stopping?

82. A horizontal string is attached to a 0.25-kg mass lying on a rough, horizontal table. The string passes over a light frictionless pulley and a 0.40-kg mass is then attached to its free end. The coefficient of sliding friction between the 0.25-kg mass and table is 0.20. Use energy considerations to determine (a) the speed of the masses after each has moved 20 m from rest and (b) the mass that must be added to the 0.25-kg mass so that, given an initial velocity, the masses continue to move at a constant speed. (c) What mass must be removed from the 0.40-kg mass to accomplish the same thing as in (b)?

82A. A string is attached to a mass m_1 lying on a rough, horizontal table. The string passes over a light frictionless pulley and a mass m_2 is then attached to its free end. The coefficient of sliding friction between m_1 and table is μ_k. Use energy considerations to determine (a) the speed of the masses after each has moved a distance d from rest and (b) the mass that must be added to m_1 so that, given an initial velocity, the masses continue to move at a constant speed. (c) What mass must be removed from m_2 to accomplish the same thing as in (b)?

83. A projectile of mass m is shot horizontally with initial velocity v_i from a height h above a flat desert floor. The instant before the projectile hits the desert floor find (a) the work done on the projectile by gravity, (b) the change in kinetic energy of the projectile since it was fired, and (c) the kinetic energy of the projectile.

84. A cyclist and her bicycle have a combined mass of 75 kg. She coasts down a road inclined at 2.0° with the horizontal at 4.0 m/s and coasts down another road inclined at 4.0° at 8.0 m/s. She then holds on to a moving vehicle and coasts on a level road. What power must the vehicle expend to maintain her speed at 3.0 m/s? Assume that the force of air resistance is proportional to her speed, and assume that other frictional forces remain constant.

85. A 60.0-kg load is raised by a two-pulley arrangement as shown in Figure P7.85. How much work is done by the force F to raise the load 3.00 m if there is a frictional force of 20.0 N in each pulley? (The pulleys do not rotate, but the rope slides across each surface.)

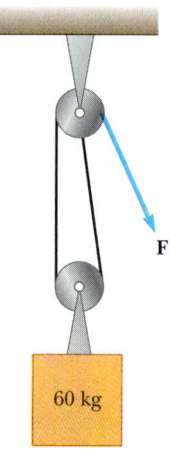

FIGURE P7.85

86. A small sphere of mass m hangs from a string of length L as in Figure P7.86. A variable horizontal force F is applied to the sphere in such a way that it moves slowly from the vertical position until the string makes an angle θ with the vertical. Assuming the sphere is always in equilibrium, (a) show that $F = mg \tan \theta$. (b) Show that the work done by F is mgL $(1 - \cos \theta)$. (*Hint:* Note that $s = L\theta$, and so $ds = L \, d\theta$.)

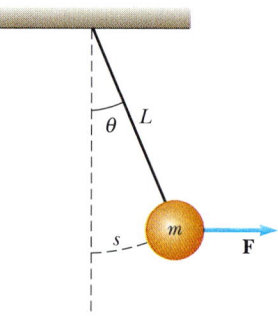

FIGURE P7.86

87. To operate one of the generators in the Grand Coulee Dam in Washington, 7.24×10^8 W of mechanical power is required. This power is provided by gravity as it performs work on the water while it falls 87 m to the generator. The kinetic energy acquired during the fall is given up turning the generator. (a) Prove that the available water power is mgh/t, where m is the mass of water falling through the height h during the time t. (b) A maximum of about 60% can be extracted and keep the water flowing. Calculate the flow rate in kilograms per second. (c) Determine the volume of water needed to power this generator for one day. (d) If the water for one day were stored in a circular lake 10 m deep, what would the radius of the lake be?

88. Power windmills turn in response to the force of high-speed drag. For a sphere moving through a fluid, the resistive force F_R is proportional to r^2v^2 where r is the radius of the sphere and v is the fluid speed. The power developed, $P = F_R v$, is proportional to r^2v^3. The power developed by a windmill can be expressed as $P = ar^2v^3$ where r is the windmill radius, v wind speed, and $a = 2.00$ W·s³/m⁵. For a home windmill with $r = 1.50$ m, calculate the power delivered to the generator if (a) $v = 8.00$ m/s and (b) $v = 24.0$ m/s. For comparison, a typical home needs about 3.0 kW of electric power. (*Note:* This representation ignores system efficiency—which is about 25%.)

89. The ball launcher in a pinball machine has a spring that has a force constant of 1.20 N/cm (Fig. P7.89). The surface on which the ball moves is inclined 10.0° with respect to the horizontal. If the spring is initially compressed 5.00 cm, find the launching speed of a 100-g ball when the plunger is released. Friction and the mass of the plunger are negligible.

FIGURE P7.89

90. Suppose a car is modeled as a cylinder moving with a speed v, as in Figure P7.90. In a time Δt, a column of air of mass Δm must be moved a distance $v\Delta t$ and hence must be given a kinetic energy $\frac{1}{2}(\Delta m)v^2$. Using this model, show that the power loss due to air resistance is $\frac{1}{2}\rho Av^3$ and the resistive force is $\frac{1}{2}\rho Av^2$, where ρ is the density of air.

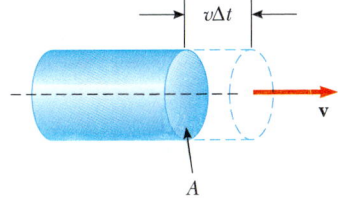

FIGURE P7.90

SPREADSHEET PROBLEMS

S1. When different weights are hung on a spring, the spring stretches to different lengths, as shown in the table below. (a) Use a spreadsheet to plot the length of the spring versus applied force. Use the spreadsheet's least-squares fitting features to determine the straight line that best fits the data. (You may not want to use all the data points.) (b) From the slope of the

least-squares fit line, find the spring constant k. (c) If the spring is extended to 105 mm, what force does it exert on the suspended weight?

F(N)	L(m)
2.0	15
4.0	32
6.0	49
8.0	64
10	79
12	98
14	112
16	126
18	149
20	175
22	190

S2. A 0.178-kg particle moves along the x axis from $x = 12.8$ m to $x = 23.7$ m under the influence of a force

$$F = \frac{375}{x^3 + 3.75x}$$

where F is in newtons and x is in meters. Use numerical integration and set up a spreadsheet to determine the total work done by this force during this displacement. Your calculations should have an accuracy of at least 2%.

S3. A box of mass m is pushed up a curved inclined plane as in Figure PS7.3. The equation of the curve is $y = 5 - x^2$. The applied force $\mathbf{F}$ remains constant in magnitude but changes direction so that it is always parallel to the curve. (a) Show that the work done by the applied force is

$$W = \int_a^b \mathbf{F} \cdot d\mathbf{s} = \int_a^b \frac{Fdx}{\cos\theta} = \int_a^b \frac{Fdx}{\cos[\tan^{-1}(-2x)]}$$

(b) Use a rectangular trapezoidal numerical integration scheme to calculate the work done by the applied force as the box is moved from $x = -2$ m to $x = 0$ m if $F = 100$ N.

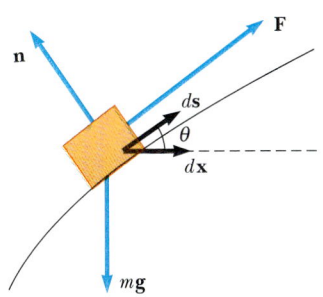

FIGURE PS7.3

CHAPTER 8

Potential Energy and Conservation of Energy

Twin Falls on the Island of Kauai, Hawaii. The gravitional potential energy of the water at the top of the falls is converted to kinetic energy at the bottom. In many locations, this mechanical energy is used to produce electrical energy.   *(Bruce Byers, FPG)*

In Chapter 7 we introduced the concept of kinetic energy, which is the energy associated with the motion of an object. In this chapter we introduce another form of mechanical energy, *potential energy*, which is the energy associated with the position or configuration of an object. Potential energy can be thought of as stored energy that can be converted to kinetic energy or other forms of energy.

The potential energy concept can be used only when dealing with a special class of forces called *conservative forces*. When only internal conservative forces, such as gravitational or spring forces, act within a system, the kinetic energy gained (or lost) by the system as its members change their relative positions is compensated by an equal energy loss (or gain) in potential energy.

8.1 POTENTIAL ENERGY

An object with kinetic energy can do work on another object, as illustrated by a moving hammer driving a nail into a wall. Now we shall see that an object can also do work because of the energy resulting from its *position* in space.

As an object falls in a gravitational field, the field exerts a force on it in the direction of its motion, doing work on it, and thereby increasing its kinetic energy. Consider a brick dropped from rest directly above a nail in a board that is lying horizontally on the ground. When the brick is released, it falls toward the ground gaining speed and therefore gaining kinetic energy. Due to its position in space, the brick has potential energy (it has the *potential* to do work), which is converted into kinetic energy as it falls. When the brick reaches the ground, it does work on the nail, driving it into the board. The energy that an object has due to its position in space is called **gravitational potential energy.** It is energy held by the gravitational field and transferred to the object as it falls.

Let us now derive an expression for the gravitational potential energy of an object at a given location. To do this, consider a block of mass m at an initial height y_i above the ground, as in Figure 8.1. Neglecting air resistance, as the block falls, the only force that does work on it is the gravitational force, $m\mathbf{g}$. The work done by the gravitational force as the block undergoes a downward displacement $\mathbf{s}$ is the product of the downward force times the displacement, or

$$W_g = (m\mathbf{g}) \cdot \mathbf{s} = (-mg\mathbf{j}) \cdot (y_f - y_i)\mathbf{j} = mgy_i - mgy_f$$

We now define the quantity mgy to be the gravitational potential energy, U_g:

$$U_g \equiv mgy \qquad (8.1)$$

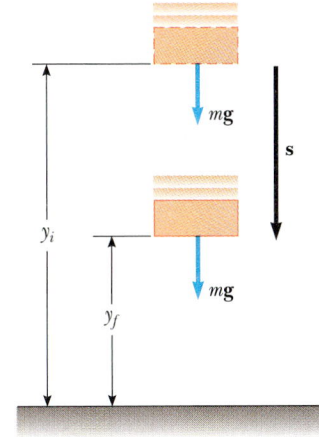

FIGURE 8.1 The work done by the gravitational force as the block falls from y_i to y_f is equal to $mgy_i - mgy_f$.

Gravitational potential energy

Thus, the gravitational potential energy associated with an object at any point in space is the product of the object's weight and its vertical coordinate. The origin of the coordinate system could be located at the surface of the Earth or at any other convenient point.

If we substitute U for the mgy terms in the expression for W_g, we have

$$W_g = U_i - U_f \qquad (8.2)$$

From this result, we see that the work done on any object by the gravitational force — that is, the energy transferred to the object from the gravitational field — is equal to the initial value of the potential energy minus the final value of potential energy. For convenience, the potential energy is often "assigned to" the object, which includes the gravitational field as part of the system. However, when this choice is made, one must be careful *not* to include the work done by the gravitational field on the object; because it is part of the system, the field no longer exerts an external force.

The units of gravitational potential energy are the same as those of work. That is, potential energy may be expressed in joules, ergs, or foot-pounds. Potential energy, like work and kinetic energy, is a scalar quantity.

Note that the gravitational potential energy associated with an object depends only on the vertical height of the object above the surface of the Earth. From this result, we see that the work done by the force of gravity on an object as it falls vertically to the Earth is the same as if it starts at the same point and slides down a frictionless incline to the Earth. Also note that Equation 8.1 is valid only for objects near the surface of the Earth, where $\mathbf{g}$ is approximately constant.[1]

In working problems involving gravitational potential energy, it is always necessary to set the gravitational potential energy equal to zero at some location. The choice of zero level is completely arbitrary, because the important quantity is the

---

[1] The assumption that the force of gravity is constant is a good one as long as the vertical displacement is small compared with the Earth's radius.

difference in potential energy, and this difference is independent of the choice of zero level.

It is often convenient to choose the surface of the Earth as the reference position for zero potential energy, but again this is not essential. Often, the statement of the problem suggests a convenient level to choose.

8.2   CONSERVATIVE AND NONCONSERVATIVE FORCES

Forces found in nature can be divided into two categories: conservative and non-conservative. We shall describe the properties of conservative and nonconservative forces separately in this section.

Conservative Forces

Properties of a conservative force

A force is conservative if the work it does on a particle moving between any two points is independent of the path taken by the particle. Furthermore, the work done by a conservative force exerted on a particle moving through any closed path is zero.

The force of gravity is conservative. As we learned in the previous section, the work done by the gravitational force on an object moving between any two points near the Earth's surface is

$$W_g = mgy_i - mgy_f$$

From this we see that W_g depends only on the initial and final coordinates of the object and, hence, is independent of path. Furthermore, W_g is zero when the object moves over any closed path (where $y_i = y_f$).

We can associate a potential energy function with any conservative force. In the previous section, the potential energy function associated with the gravitational force was found to be

$$U_g \equiv mgy$$

Gravitational potential energy is the energy stored in the gravitational field when the object is lifted against the field.

Potential energy functions can be defined only for conservative forces. In general, the work, W_c, done on an object by a conservative force is equal to the initial value of the potential energy associated with the object minus the final value:

Work done by a conservative force

$$W_c = U_i - U_f \tag{8.3}$$

Another example of a conservative force is the force of a spring on an object attached to the spring, where the spring force is given by $F_s = -kx$. In the previous chapter, we learned that the work done by the spring is

$$W_s = \tfrac{1}{2}kx_i^2 - \tfrac{1}{2}kx_f^2$$

where the initial and final coordinates of the block are measured from its equilibrium position, $x = 0$. Again we see that W_s depends only on the initial and final x coordinates of the object and is zero for any closed path. The **elastic potential** energy function associated with the spring force is defined by

$$U_s \equiv \tfrac{1}{2}kx^2 \qquad\qquad (8.4)$$

Elastic potential energy

The elastic potential energy can be thought of as the energy stored in the deformed spring (one that is either compressed or stretched from its equilibrium position). To visualize this, consider Figure 8.2, which shows an undeformed spring on a frictionless, horizontal surface. When the block is pushed against the spring (Fig. 8.2b), compressing the spring a distance x, the elastic potential energy stored in the spring is $kx^2/2$. When the block is released from rest, the spring snaps back to its original length and the stored elastic potential energy is transformed into kinetic energy of the block (Fig. 8.2c). The elastic potential energy stored in the spring is zero whenever the spring is undeformed ($x = 0$). Energy is stored in the spring only when the spring is either stretched or compressed. Furthermore, the elastic potential energy is a maximum when the spring has reached its maximum compression or extension (that is, when $|x|$ is a maximum). Finally, since the elastic potential energy is proportional to x^2, we see that U_s is always positive in a deformed spring. If the spring and object connected to it are taken together as the system, then no work is done as the spring changes length because the forces are internal.

Nonconservative Forces

A force is **nonconservative** if it causes a change in mechanical energy. For example, if you move an object on a horizontal surface, returning to the same location and same state of motion, but you found that it was necessary to do a net

Properties of a nonconservative force

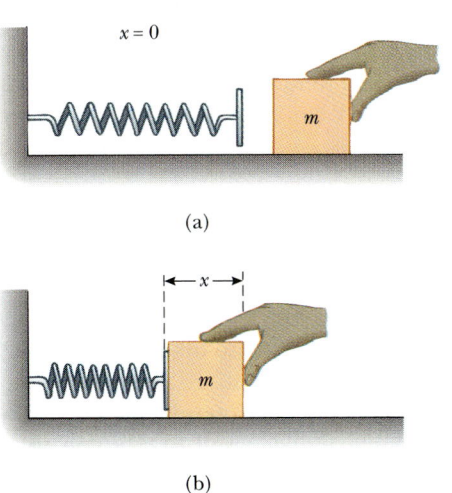

(a)

(b)

(c)

FIGURE 8.2   (a) An undeformed spring on a frictionless horizontal surface. (b) A block of mass m is pushed against the spring, compressing the spring a distance x. (c) The block is released from rest and the elastic potential energy stored in the spring is transferred to the block in the form of kinetic energy.

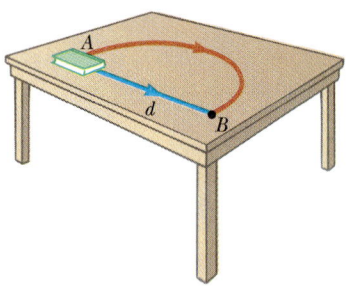

FIGURE 8.3   The loss in mechanical energy due to the force of friction depends on the path taken as the book is moved from *A* to *B*. The loss in mechanical energy is greater along the red path compared with the blue path.

amount of work on the object, then something must have dissipated that energy transferred to the object. That dissipative force is recognized as friction between the surface and object. Friction is a dissipative or "nonconservative" force. By contrast, if the object is lifted, work is required, but that energy is recovered when the object is lowered. The gravitational force is a nondissipative or "conservative" force.

Suppose you displace a book between two points on a table. If the book is displaced in a straight line between two points *A* and *B* in Figure 8.3, the loss in mechanical energy due to friction is simply $-fd$, where *d* is the distance between the points. However, if you move the book along *any other* path on the table between the two points, the loss in mechanical energy due to friction is *greater* (in absolute value) than $-fd$. For example, the loss in mechanical energy due to friction along the semicircular path in Figure 8.3 is equal to $-f(\pi d/2)$, where *d* is the diameter of the circle.

8.3   CONSERVATIVE FORCES AND POTENTIAL ENERGY

Because work done is a function only of a particle's initial and final coordinates, we can define a **potential energy function** *U* such that the work done by a conservative force equals the decrease in the potential energy of the particle. The work done by a conservative force **F** as a particle moves along the *x* axis is[2]

$$W_c = \int_{x_i}^{x_f} F_x \, dx = -\Delta U = U_i - U_f \tag{8.5}$$

where F_x is the component of **F** in the direction of the displacement. That is, *the work done by a conservative force equals the negative of the change in the potential energy associated with that force*, where the change in the potential energy is defined as $\Delta U = U_f - U_i$. For example, the work done by the gravitational field on an object, as the object is lowered in the field, is $U_i - U_f$, which is positive, showing that energy has been transferred from the gravitational field to the object. This energy may appear as kinetic energy of the falling object or may be transferred to something else. We can also express Equation 8.5 as

$$\Delta U = U_f - U_i = -\int_{x_i}^{x_f} F_x \, dx \tag{8.6}$$

Therefore ΔU is negative when F_x and dx are in the same direction, as when an object is lowered in a gravitational field or a spring pushes an object toward equilibrium.

The term *potential energy* implies that the object has the potential, or capability, of either gaining kinetic energy or doing work when released from some point under the influence of gravity. It is often convenient to establish some particular location, x_i, as a reference point and measure all potential energy differences with respect to it. We can then define the potential energy function as

$$U_f(x) = -\int_{x_i}^{x_f} F_x \, dx + U_i \tag{8.7}$$

[2] For a general displacement, the work done in two or three dimensions also equals $U_i - U_f$, where $U = U(x, y, z)$. We write this formally as $W = \int_i^f \mathbf{F} \cdot d\mathbf{s} = U_i - U_f$.

Furthermore, as we discussed earlier, the value of U_i is often taken to be zero at some arbitrary reference point. It really doesn't matter what value we assign to U_i, since any nonzero value only shifts $U_f(x)$ by a constant, and only the *change* in potential energy is physically meaningful. If the conservative force is known as a function of position, we can use Equation 8.7 to calculate the change in potential energy of a body as it moves from x_i to x_f. It is interesting to note that in the one-dimensional case, a force is always conservative if it is a function of position only. This is generally not the case for motion involving two- or three-dimensional displacements.

The amount of mechanical energy dissipated by a nonconservative force depends on the path as an object moves from one position to another and can also depend on the object's speed or on other quantities. Because the work done by a nonconservative force is not simply a function of the initial and final coordinates, there is no potential energy function associated with a nonconservative force.

8.4   CONSERVATION OF ENERGY

An object held at some height h above the floor has no kinetic energy, but as we learned earlier, there is an associated gravitational potential energy equal to *mgh* relative to the floor if the gravitational field is included as part of the system. If the object is dropped, it falls to the floor; as it falls, its speed and thus its kinetic energy increase, while the potential energy decreases. If factors such as air resistance are ignored, whatever potential energy the object loses as it moves downward appears as kinetic energy. In other words, the sum of the kinetic and potential energies, called the *mechanical energy E*, remains constant in time. This is an example of the principle of **conservation of energy.** For the case of an object in free-fall, this principle tells us that any increase (or decrease) in potential energy is accompanied by an equal decrease (or increase) in kinetic energy.

Since the total mechanical energy E is defined as the sum of the kinetic and potential energies, we can write

$$E = K + U \qquad (8.8)$$

Total mechanical energy

Therefore, we can apply conservation of energy in the form $E_i = E_f$, or

$$K_i + U_i = K_f + U_f \qquad (8.9)$$

The mechanical energy of an isolated system remains constant

Conservation of energy requires that the total mechanical energy of a system remains constant in any isolated system of objects that interact only through conservative forces.

It is important to note that Equation 8.9 is valid provided no energy is added to or removed from the system. Furthermore, there must be no nonconservative forces within the system.

If more than one conservative force acts on the object, then a potential energy function is associated with each force. In such a case, we can apply the principle of conservation of mechanical energy for the system as

$$K_i + \sum U_i = K_f + \sum U_f \qquad (8.10)$$

where the number of terms in the sums equals the number of conservative forces acting on the system. For example, if a mass connected to a spring oscillates vertically, two conservative forces act on it: the spring force and the force of gravity. (We discuss this situation later, in a worked example.)

If the force of gravity is the *only* force acting on an object, then the total mechanical energy of the object is constant. Therefore, the principle of conservation of energy for an object in free-fall can be written

$$\tfrac{1}{2}mv_i^2 + mgy_i = \tfrac{1}{2}mv_f^2 + mgy_f \qquad (8.11)$$

Likewise, if the only force acting on an object is the conservative spring force for which the potential energy is $U_s = \tfrac{1}{2}kx^2$, the principle of conservation of energy gives

$$\tfrac{1}{2}mv_i^2 + \tfrac{1}{2}kx_i^2 = \tfrac{1}{2}mv_f^2 + \tfrac{1}{2}kx_f^2 \qquad (8.12)$$

The mechanical energy for a freely falling body remains constant

The mechanical energy for a mass-spring system remains constant

---

CONCEPTUAL EXAMPLE 8.1

A mass is connected to a massless spring that is suspended vertically from the ceiling as in Figure 8.4. If the mass is displaced downward from its equilibrium position and released, it will oscillate up and down. If air resistance is neglected, will the total mechanical energy of the system (mass plus spring) be conserved? How many forms of potential energy are there for this situation?

Reasoning   Yes, the total mechanical energy of the system is conserved because the only forces acting are conservative: the force of gravity and the spring force. There are two forms of potential energy in this case, gravitational potential energy and elastic potential energy stored in the spring. The sum of the kinetic energy, the gravitational potential energy, and the elastic potential energy remains constant for the system.

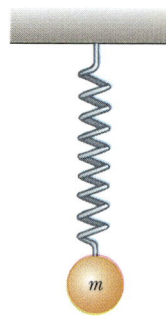

FIGURE 8.4   (Conceptual Example 8.1) A mass m connected to a massless spring suspended vertically from a support. What forms of potential energy are associated with the mass-spring system when the mass is displaced downward?

CONCEPTUAL EXAMPLE 8.2

Discuss the energy transformations that occur during the pole vault event pictured in the multiflash photograph. Ignore rotational motion.

Reasoning   As an athlete runs, chemical energy in the body is converted mostly into thermal energy but also into kinetic energy. The kinetic energy of the athlete just before ascending becomes elastic potential energy stored in the bent pole at the bottom, then gravitational potential energy in the elevated body of the vaulter, and then kinetic energy as he falls. When the vaulter lands on the pad, this energy turns into additional thermal energy, mostly in the pad.

(Conceptual Example 8.2) Multiflash photograph of a pole vault event. How many forms of energy can you identify in this picture?   (© *Harold E. Edgerton, Courtesy of Palm Press Inc.*)

CONCEPTUAL EXAMPLE 8.3

Three identical balls are thrown from the top of a building, all with the same initial speed. The first ball is thrown horizontally, the second at some angle above the horizontal, and the third at some angle below the horizontal as in Figure 8.5. Neglecting air resistance, describe their motions and compare the speeds of the balls as they reach the ground.

Reasoning   The first and third balls speed up after they are thrown, while the second ball first slows down and then speeds up after reaching its peak. The paths of all three are parabolas. The three take different times to reach the ground. However, all have the same impact speed because all start with the same kinetic energy and undergo the same change in gravitational potential energy. In other words, $E_{\text{total}} = \frac{1}{2}mv^2 + mgh$ is the same for all three balls.

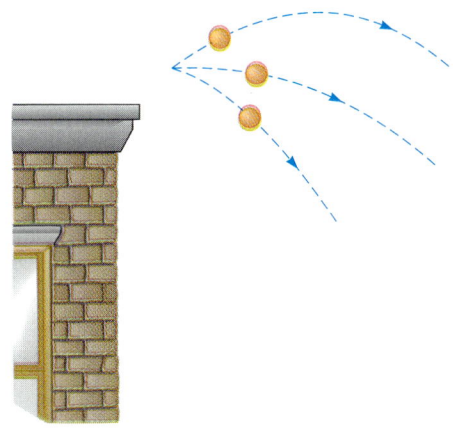

FIGURE 8.5   (Conceptual Example 8.3) Three identical balls are thrown with the same initial speed from the top of a building.

EXAMPLE 8.4   Ball in Free-Fall

A ball of mass m is dropped from a height h above the ground as in Figure 8.6. (a) Neglecting air resistance, determine the speed of the ball when it is at a height y above the ground.

Reasoning   Since the ball is in free-fall, the only force acting on it is the gravitational force. Therefore, we can use the principle of constancy of mechanical energy. Initially, the ball has potential energy and no kinetic energy. As it falls, its total energy (the sum of kinetic and potential energies) remains constant and equal to its initial potential energy.

Solution   When the ball is released from rest at a height h above the ground, its kinetic energy is $K_i = 0$ and its potential energy is $U_i = mgh$, where the y coordinate is measured from ground level. When the ball is at a distance y above the ground, its kinetic energy is $K_f = \frac{1}{2}mv_f^2$ and its potential energy relative to the ground is $U_f = mgy$. Applying Equation 8.9, we get

$$K_i + U_i = K_f + U_f$$
$$0 + mgh = \tfrac{1}{2}mv_f^2 + mgy$$
$$v_f^2 = 2g(h - y)$$
$$v_f = \boxed{\sqrt{2g(h - y)}}$$

(b) Determine the speed of the ball at y if it is given an initial speed v_i at the initial altitude h.

Solution   In this case, the initial energy includes kinetic energy equal to $\frac{1}{2}mv_i^2$ and Equation 8.11 gives

$$\tfrac{1}{2}mv_i^2 + mgh = \tfrac{1}{2}mv_f^2 + mgy$$

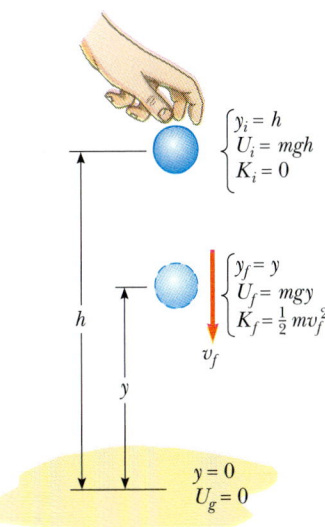

FIGURE 8.6   (Example 8.4) A ball is dropped from a height h above the ground. Initially, its total energy is potential energy, equal to mgh relative to the ground. At the elevation y, its total energy is the sum of kinetic and potential energies.

$$v_f^2 = v_i^2 + 2g(h - y)$$
$$v_f = \boxed{\sqrt{v_i^2 + 2g(h - y)}}$$

This result is consistent with the expression $v_y^2 = v_{y0}^2 - 2g(y - y_0)$, from kinematics, where $y_0 = h$. Furthermore, this result is valid even if the initial velocity is at an angle to the horizontal (the projectile situation).

EXAMPLE 8.5   The Pendulum

A pendulum consists of a sphere of mass m attached to a light cord of length L as in Figure 8.7. The sphere is released from rest when the cord makes an angle θ_0 with the vertical, and the pivot at P is frictionless. (a) Find the speed of the sphere when it is at the lowest point, b.

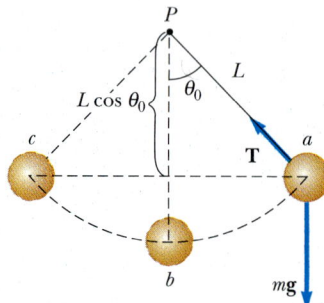

FIGURE 8.7   (Example 8.5) If the sphere is released from rest at the angle θ_0, it will never swing above this position during its motion. At the start of the motion, position a, its energy is entirely potential. This initial potential energy is all transformed into kinetic energy at the lowest elevation, position b. As the sphere continues to move along the arc, the energy again becomes entirely potential energy at position c.

Reasoning   The only force that does work on m is the force of gravity, since the force of tension is always perpendicular to each element of the displacement and hence does no work. Since the force of gravity is a conservative force, the total mechanical energy is constant. Therefore, as the pendulum swings, there is a continuous transfer between potential and kinetic energy. At the instant the pendulum is re-

leased, the energy is entirely potential energy. At point b, the pendulum has kinetic energy but has lost some potential energy. At point c, the pendulum has regained its initial potential energy and its kinetic energy is again zero.

Solution   If we measure the y coordinates from the center of rotation, then $y_a = -L \cos \theta_0$ and $y_b = -L$. Therefore, $U_a = -mgL \cos \theta_0$ and $U_b = -mgL$. Applying the principle of constancy of mechanical energy gives

$$K_a + U_a = K_b + U_b$$

$$0 - mgL \cos \theta_0 = \tfrac{1}{2}mv_b^2 - mgL$$

$$(1) \qquad v_b = \boxed{\sqrt{2gL(1 - \cos \theta_0)}}$$

(b) What is the tension T in the cord at b?

Solution   Since the force of tension does no work, it cannot be determined using the energy method. To find T_b, we can apply Newton's second law to the radial direction. First, recall that the centripetal acceleration of a particle moving in a circle is equal to v^2/r directed toward the center of rotation. Since $r = L$ in this example, we get

$$(2) \qquad \sum F_r = T_b - mg = ma_r = mv_b^2/L$$

Substituting (1) into (2) gives for the tension at point b

$$(3) \qquad T_b = mg + 2mg(1 - \cos \theta_0) = \boxed{mg(3 - 2\cos \theta_0)}$$

Exercise   A pendulum of length 2.00 m and mass 0.500 kg is released from rest when the cord makes an angle of 30.0° with the vertical. Find the speed of the sphere and the tension in the cord when the sphere is at its lowest point.

Answer   2.29 m/s; 6.21 N.

8.5   CHANGES IN MECHANICAL ENERGY WHEN NONCONSERVATIVE FORCES ARE PRESENT

In real physical systems, nonconservative forces, such as friction, are usually present. Such forces remove mechanical energy from the system. Therefore, the total mechanical energy is not constant. In general, we cannot calculate the work done by nonconservative forces, but we can find the change in kinetic energy of the system, acted upon by a net force, using the net-force equation:

$$\int \mathbf{F}_{net} \cdot d\mathbf{x} = \Delta K \tag{8.13}$$

Since the change in kinetic energy can be the result of many types of force, it is convenient to separate ΔK into three parts:

1. The change in kinetic energy due to internal conservative forces, $\Delta K_{int\text{-}c}$
2. The change in kinetic energy due to internal nonconservative forces, $\Delta K_{int\text{-}nc}$

3. The change in kinetic energy due to external forces (conservative or nonconservative), ΔK_{ext}.

The first of these is $\Delta K_{int\text{-}c} = -\Delta U$. This is simply an exchange of potential energy into kinetic energy within the system. The second term, $\Delta K_{int\text{-}nc}$, may be positive or negative: if it represents internal friction, it will be negative; however, it can be positive (as in the case of a muscle operating on biochemical energy). The last term, ΔK_{ext}, is the change in kinetic energy of the system due to external forces. Thus, we have

$$\Delta K = \Delta K_{int\text{-}c} + \Delta K_{int\text{-}nc} + \Delta K_{ext}$$

or

$$(K + U)_i + \Delta K_{int\text{-}nc} + \Delta K_{ext} = (K + U)_f \qquad (8.14)$$

From this we see that if there are no internal nonconservative forces and no external forces acting on the system, then the right side of Equation 8.14 is zero, and the sum $K + U$ is constant, which is consistent with Equation 8.9. If internal nonconservative forces are present, they may increase or decrease the kinetic energy. Similarly, external forces may either increase or decrease the kinetic energy of the system.

Problem-Solving Strategies
Conservation of Energy

Many problems in physics can be solved using the principle of conservation of energy. In this chapter we applied the principle to special cases in which the total energy of the system is constant and any loss of mechanical energy is determined as loss of kinetic energy from the net-force equation. The following procedure should be used when you apply this principle:

- Define your system, which may consist of more than one object and may or may not include fields, springs, or other sources of potential energy. Choose initial and final points.
- Select a reference position for the zero point of potential energy (both gravitational and elastic), and use this position throughout your analysis. If there is more than one conservative force, write an expression for the potential energy associated with each force.
- Remember that if friction or air resistance is present, mechanical energy *is not constant*.
- If mechanical energy is constant, write the total initial energy, E_i, at some point as the sum of the kinetic and potential energies at that point. Then write an expression for the total final energy, $E_f = K_f + U_f$, at the final point. Since mechanical energy is constant, equate the two total energies and solve for the unknown.
- If frictional forces are present (and thus mechanical energy is not constant), first write expressions for the total initial and total final energies. In this case, however, the total final energy differs from the total initial energy, the difference being the amount of energy dissipated by nonconservative forces. That is, apply Equation 8.14.

EXAMPLE 8.6   Crate Sliding Down a Ramp

A 3.00-kg crate slides down a ramp at a loading dock. The ramp is 1.00 m in length and inclined at an angle of 30.0°, as shown in Figure 8.8. The crate starts from rest at the top, experiences a constant frictional force of magnitude 5.00 N, and continues to move a short distance on the flat floor. Use energy methods to determine the speed of the crate when it reaches the bottom of the ramp.

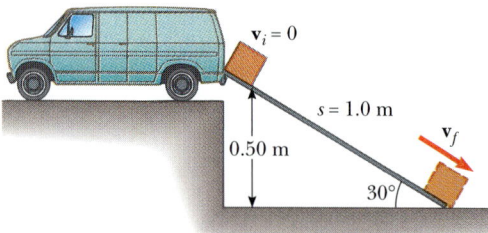

FIGURE 8.8   (Example 8.6) A crate slides down a ramp under the influence of gravity. Its potential energy decreases while its kinetic energy increases.

Solution   Since $v_i = 0$, the initial kinetic energy is zero. If the y coordinate is measured from the bottom of the ramp, then $y_i = 0.500$ m. Therefore, the total mechanical energy of the system at the top is all potential energy:

$$U_i = mgy_i = (3.00 \text{ kg}) \left(9.80 \, \frac{\text{m}}{\text{s}^2} \right) (0.500 \text{ m}) = 14.7 \text{ J}$$

When the crate reaches the bottom, the potential energy is *zero* because the elevation of the crate is $y_f = 0$. Therefore,

the total mechanical energy at the bottom is all kinetic energy,

$$K_f = \tfrac{1}{2} m v_f^2$$

However, we cannot say that $U_i = K_f$ in this case, because there is an external nonconservative force that removes mechanical energy from the system: the force of friction. In this case, $\Delta K_{\text{ext}} = -fs$, where s is the displacement along the ramp. (Remember that the forces normal to the ramp do no work on the crate because they are perpendicular to the displacement.) With $f = 5.00$ N and $s = 1.00$ m, we have

$$\Delta K_{\text{ext}} = -fs = (-5.00 \text{ N})(1.00 \text{ m}) = -5.00 \text{ J}$$

This says that some mechanical energy is lost because of the presence of the retarding frictional force. Applying Equation 8.13 gives

$$-fs = \tfrac{1}{2} m v_f^2 - mgy_i$$

$$\tfrac{1}{2} m v_f^2 = 14.7 \text{ J} - 5.00 \text{ J} = 9.70 \text{ J}$$

$$v_f^2 = \frac{19.4 \text{ J}}{3.00 \text{ kg}} = 6.47 \text{ m}^2/\text{s}^2$$

$$v_f = \boxed{2.54 \text{ m/s}}$$

Exercise   Use Newton's second law to find the acceleration of the crate along the ramp and the equations of kinematics to determine the final speed of the crate.

Answer   3.23 m/s²; 2.54 m/s.

Exercise   If the ramp is assumed to be frictionless, find the final speed of the crate and its acceleration along the ramp.

Answer   3.13 m/s; 4.90 m/s².

EXAMPLE 8.7   Motion on a Curved Track

A child of mass m takes a ride on an irregularly curved slide of height $h = 6.00$ m, as in Figure 8.9. The child starts from rest at the top. (a) Determine the speed of the child at the bottom, assuming no friction is present.

Reasoning   The normal force, **n**, does no work on the child since this force is always perpendicular to each element of the displacement. Furthermore, since there is no friction, mechanical energy is constant; that is, $K + U = \text{constant}$.

Solution   If we measure the y coordinate from the bottom of the slide, then $y_i = h$, $y_f = 0$, and we get

$$K_i + U_i = K_f + U_f$$

$$0 + mgh = \tfrac{1}{2} m v_f^2 + 0$$

$$v_f = \sqrt{2gh}$$

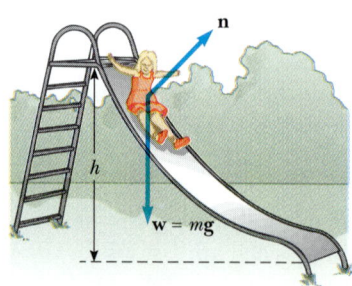

FIGURE 8.9   (Example 8.7) If the slide is frictionless, the speed of the child at the bottom depends only on the height of the slide.

Note that the result is the same as it would be if the child fell vertically through a distance h! In this example, $h = 6.00$ m, giving

$$v_f = \sqrt{2gh} = \sqrt{2\left(9.80\,\frac{m}{s^2}\right)(6.00\text{ m})} = \boxed{10.8\text{ m/s}}$$

(b) If a frictional force acts on the child, how much mechanical energy is dissipated by this force? Assume that $v_f = 8.00$ m/s and $m = 20.0$ kg.

Solution   In this case $\Delta K_{ext} \neq 0$ and mechanical energy is *not* constant. We can use Equation 8.13 to find the loss of kinetic energy due to friction, assuming the final speed at the bottom is known:

$$\Delta K_{ext} = E_f - E_i = \tfrac{1}{2}mv_f^2 - mgh$$
$$\Delta K_{ext} = \tfrac{1}{2}(20.0\text{ kg})(8.00\text{ m/s})^2$$
$$- (20.0\text{ kg})\left(9.80\,\frac{m}{s^2}\right)(6.00\text{ m})$$
$$= \boxed{-536\text{ J}}$$

Again, ΔK_{ext} is negative since friction is removing kinetic energy from the system. Note, however, that because the slide is curved, the normal force changes in magnitude and direction during the motion. Therefore, the frictional force, which is proportional to n, also changes during the motion. Do you think it would be possible to determine μ from these data?

EXAMPLE 8.8   Let's Go Skiing

A skier starts from rest at the top of a frictionless incline of height 20.0 m as in Figure 8.10. At the bottom of the incline, the skier encounters a horizontal surface where the coefficient of kinetic friction between the skis and snow is 0.210. How far does the skier travel on the horizontal surface before coming to rest?

Solution   First, let us calculate the speed of the skier at the bottom of the incline. Since the incline is frictionless, the mechanical energy remains constant and we find

$$v = \sqrt{2gh} = \sqrt{2(9.80\text{ m/s}^2)(20.0\text{ m})} = 19.8\text{ m/s}$$

Now we apply Equation 8.13 as the skier moves along the rough horizontal surface. The change in kinetic energy along the horizontal is $\Delta K_{ext} = -fs$, where s is the horizontal displacement. Therefore,

$$\Delta K_{ext} = -fs = K_f - K_i$$

To find the distance the skier travels before coming to rest, we take $K_f = 0$. Since $v_i = 19.8$ m/s, and the friction force is $f = \mu n = \mu mg$, we get

$$-\mu mgs = -\tfrac{1}{2}mv_i^2$$

or

$$s = \frac{v_i^2}{2\mu g} = \frac{(19.8\text{ m/s})^2}{2(0.210)(9.80\text{ m/s}^2)} = \boxed{95.2\text{ m}}$$

Exercise   Find the horizontal distance the skier travels before coming to rest if the incline also has a coefficient of kinetic friction equal to 0.210.

Answer   40.3 m.

20.0 m

20.0°

FIGURE 8.10   (Example 8.8).

EXAMPLE 8.9   The Spring-Loaded Popgun

The launching mechanism of a toy gun consists of a spring of unknown spring constant (Fig. 8.11a). When the spring is compressed 0.120 m, the gun is able to launch a 35.0-g pro-

jectile to a maximum height of 20.0 m when fired vertically from rest. (a) Neglecting all resistive forces, determine the spring constant.

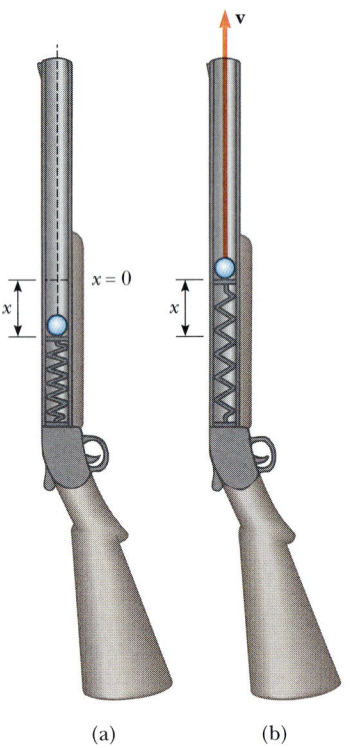

FIGURE 8.11 (Example 8.9).

Reasoning   Since the projectile starts from rest, the initial kinetic energy in the system is zero. If the zero point for the gravitational potential energy is taken at the lowest position of the projectile, then the initial gravitational potential energy is also zero. The mechanical energy of this system remains constant because there are no nonconservative forces present.

Solution   The total initial energy of the system is the elastic potential energy stored in the spring, which is $kx^2/2$. Since

the projectile rises to a maximum height $h = 20.0$ m, the final gravitational potential energy is mgh, its final kinetic energy is zero, and the final elastic potential energy is zero. Because the mechanical energy of the system is constant, we find

$$\tfrac{1}{2}kx^2 = mgh$$

$$\tfrac{1}{2}k(0.120 \text{ m})^2 = (0.0350 \text{ kg})(9.80 \text{ m/s}^2)(20.0 \text{ m})$$

$$k = \boxed{953 \text{ N/m}}$$

(b) Find the speed of the projectile as it moves through the equilibrium position of the spring (where $x = 0$) as shown in Figure 8.11b.

Solution   Using the same reference level for the gravitational potential energy as in part (a), we see that the initial energy of the system is still the elastic potential energy $kx^2/2$. The energy of the system when the projectile moves through the unstretched position of the spring consists of the kinetic energy of the projectile, $mv^2/2$, and the gravitational potential energy of the projectile, mgx. Hence, conservation of energy in this case gives

$$\tfrac{1}{2}kx^2 = \tfrac{1}{2}mv^2 + mgx$$

Solving for v gives

$$v = \sqrt{\frac{kx^2}{m} - 2gx}$$

$$v = \sqrt{\frac{(953 \text{ N/m})(0.120 \text{ m})^2}{(0.0350 \text{ kg})} - 2(9.80 \text{ m/s}^2)(0.120 \text{ m})}$$

$$= \boxed{19.7 \text{ m/s}}$$

Exercise   What is the speed of the projectile when it is at a height of 10.0 m?

Answer   14.0 m/s.

EXAMPLE 8.10   Mass-Spring Collision

A mass of 0.80 kg is given an initial velocity $v_i = 1.2$ m/s to the right and collides with a light spring of force constant $k = 50$ N/m, as in Figure 8.12. (a) If the surface is frictionless, calculate the initial maximum compression of the spring after the collision.

Reasoning   Before the collision, the mass has kinetic energy and the spring is uncompressed, so the energy stored in the spring is zero. Thus, the total energy of the system (mass plus spring) before the collision is $\tfrac{1}{2}mv_i^2$. After the collision, and when the spring is fully compressed, the mass is at rest and has zero kinetic energy, while the energy stored in the spring has its maximum value, $\tfrac{1}{2}kx_f^2$. The total mechanical

energy of the system is constant since no nonconservative forces act on the system.

Solution   Because mechanical energy is constant, the kinetic energy of the mass before the collision must equal the maximum energy stored in the spring when it is fully compressed, or

$$\tfrac{1}{2}mv_i^2 = \tfrac{1}{2}kx_f^2$$

$$x_f = \sqrt{\frac{m}{k}}\, v_i = \sqrt{\frac{0.80 \text{ kg}}{50 \text{ N/m}}}\,(1.2 \text{ m/s}) = \boxed{0.15 \text{ m}}$$

(b) If a constant force of friction acts between block and surface with $\mu = 0.50$ and if the speed of the block just as it

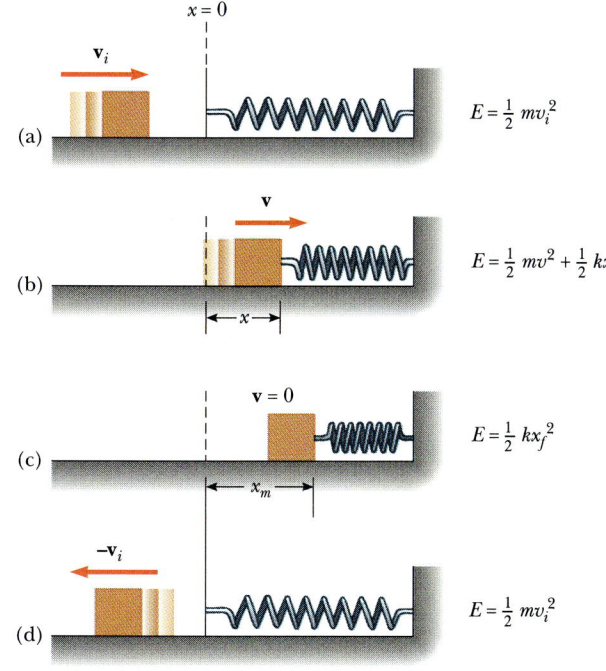

FIGURE 8.12 (Example 8.10) A block sliding on a smooth, horizontal surface collides with a light spring. (a) Initially the mechanical energy is all kinetic energy. (b) The mechanical energy is the sum of the kinetic energy of the block and the elastic potential energy in the spring. (c) The energy is entirely potential energy. (d) The energy is transformed back to the kinetic energy of the block. The total energy remains constant throughout the motion.

collides with the spring is $v_i = 1.2$ m/s, what is the maximum compression in the spring?

Solution In this case, mechanical energy is *not* conserved because of friction. The magnitude of the frictional force is

$$f = \mu n = \mu m g = 0.50(0.80 \text{ kg})\left(9.80 \, \frac{\text{m}}{\text{s}^2}\right) = 3.9 \text{ N}$$

Therefore, the loss in kinetic energy due to friction as the block is displaced from $x_i = 0$ to $x_f = x$ is

$$\Delta K_{\text{ext}} = -fx = (-3.92x) \text{ J}$$

Substituting this into Equation 8.14 gives

$$\Delta K_{\text{ext}} = (0 + \tfrac{1}{2}kx^2) - (\tfrac{1}{2}mv_i^2 + 0)$$

$$-3.92x = \tfrac{1}{2}(50)x^2 - \tfrac{1}{2}(0.80)(1.2)^2$$

$$25x^2 + 3.92x - 0.576 = 0$$

Solving the quadratic equation for x gives $x = 0.092$ m and $x = -0.25$ m. The physically meaningful root is $x = 0.092$ m $= 9.2$ cm. The negative root is meaningless because the block must be to the right of the origin when it comes to rest. Note that 9.2 cm is less than the distance obtained in the frictionless case (a). This result is what we expect because friction retards the motion of the system.

EXAMPLE 8.11 **Connected Blocks in Motion**

Two blocks are connected by a light string that passes over a frictionless pulley as in Figure 8.13. The block of mass m_1 lies on a horizontal surface and is connected to a spring of force constant k. The system is released from rest when the spring is unstretched. If m_2 falls a distance h before coming to rest, calculate the coefficient of kinetic friction between m_1 and the surface.

Reasoning and Solution In this situation there are two forms of potential energy to consider: gravitational and elastic. We can write Equation 8.14 as

$$(1) \qquad \Delta K_{\text{ext}} = \Delta K + \Delta U_g + \Delta U_s$$

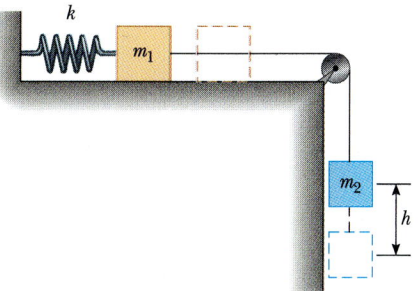

FIGURE 8.13 (Example 8.11) As m_2 moves from its highest elevation to its lowest, the system loses gravitational potential energy but gains elastic potential energy in the spring. Some mechanical energy is lost because of friction between m_1 and the surface.

where ΔU_g is the change in the gravitational potential energy and ΔU_s is the change in the elastic potential energy of the system. In this situation, $\Delta K = 0$ because the initial and final speeds of the system are zero. Also, the loss in kinetic energy due to friction is

$$(2) \qquad \Delta K_{ext} = -fh = -\mu m_1 gh$$

The change in the gravitational potential energy is associated only with m_2 since the vertical coordinate of m_1 does not change. Therefore, we get

$$(3) \qquad \Delta U_g = U_f - U_i = -m_2 gh$$

where the coordinates have been measured from the lowest position of m_2.

The change in the elastic potential energy in the spring is

$$(4) \qquad \Delta U_s = U_f - U_i = \tfrac{1}{2}kh^2 - 0$$

Substituting (2), (3), and (4) into (1) gives

$$-\mu m_1 gh = -m_2 gh + \tfrac{1}{2}kh^2$$

$$\mu = \frac{m_2 g - \tfrac{1}{2}kh}{m_1 g}$$

This set-up represents a way of measuring the coefficient of kinetic friction between an object and some surface.

EXAMPLE 8.12   One Way to Lift an Object

Two blocks are connected by a massless cord that passes over a frictionless pulley and a frictionless peg as in Figure 8.14. One end of the cord is attached to a mass $m_1 = 3.00$ kg that is a distance $R = 1.20$ m from the peg. The other end of the cord is connected to a block of mass $m_2 = 6.00$ kg resting on a table. From what angle θ (measured from the vertical) must the 3.00-kg mass be released in order to just begin to lift the 6.00-kg block off the table?

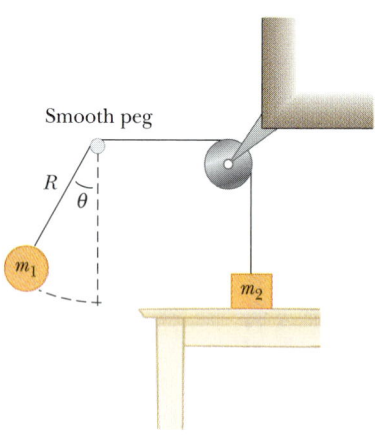

FIGURE 8.14   Example 8.12.

Reasoning   It is necessary to use several concepts to solve this problem. First, we use conservation of energy to find the speed of the 3.00-kg mass at the bottom of the circular path as a function of θ and the radius of the path. Next, we apply Newton's second law to the 3.00-kg mass at the bottom of its path to find the tension as a function of the given parameters. Finally, we note that the 6.00-kg block lifts off the table when the upward force exerted on it by the cord exceeds the force of gravity acting on the block. This procedure enables us to find the required angle.

Solution   Applying conservation of energy to the 3.00-kg mass gives

$$K_i + U_i = K_f + U_f$$

$$(1) \qquad 0 + m_1 gy_i = \tfrac{1}{2}m_1 v^2 + 0$$

where v is the speed of the 3.00-kg mass at the bottom of its path. (Note that $K_i = 0$ since the 3.00-kg mass starts from rest and $U_f = 0$ because the bottom of the circle is the zero level of potential energy.) From the geometry in Figure 8.14, we see that $y_i = R - R\cos\theta = R(1 - \cos\theta)$. Using this relation in (1) gives

$$(2) \qquad v^2 = 2gR(1 - \cos\theta)$$

Now we apply Newton's second law to the 3.00-kg mass when it is at the bottom of the circular path:

$$T - m_1 g = m_1 \frac{v^2}{R}$$

$$(3) \qquad T = m_1 g + m_1 \frac{v^2}{R}$$

This same force is transmitted to the 6.00-kg block, and if it is to be just lifted off the table, the normal force on it becomes zero, and we require that $T = m_2 g$. Using this condition, together with (2) and (3), gives

$$m_2 g = m_1 g + m_1 \frac{2gR(1 - \cos\theta)}{R}$$

Solving for θ, and substituting in the given parameters, we get

$$\cos\theta = \frac{3m_1 - m_2}{2m_1} = \frac{3(3.00 \text{ kg}) - 6.00 \text{ kg}}{2(3.00 \text{ kg})} = \frac{1}{2}$$

$$\theta = 60.0°$$

Exercise   If the initial angle is $\theta = 40.0°$, find the speed of the 3.00-kg mass and the tension in the cord when the 3.00-kg mass is at the bottom of its circular path.

Answer   2.35 m/s; 43.2 N.

8.6   RELATIONSHIP BETWEEN CONSERVATIVE FORCES AND POTENTIAL ENERGY

In the previous sections we saw that one mode of energy storage is as potential energy, which is related to the configuration, or coordinates, of a system. Potential energy functions are associated only with conservative forces. If an object or field does work on some external object, energy is transferred from the object or field to the external object. The work done is

$$\int \mathbf{F} \cdot d\mathbf{x} = -\Delta U$$

That is, the energy transferred as work decreases the potential energy of the system from which the energy came. (In this expression, $\mathbf{F}$ is the force exerted by the object or field on the external object, and $d\mathbf{x}$ is the displacement of the point of application of the force.)

If there is an infinitesimal displacement, dx, we can express the infinitesimal change in potential energy of the system, dU, as

$$dU = -F_x \, dx$$

Therefore, the conservative force is related to the potential energy function through the relationship[3]

$$F_x = -\frac{dU}{dx} \qquad (8.15)$$

Relationship between force and potential energy

That is, *the conservative force equals the negative derivative of the potential energy with respect to x.*

We can easily check this relationship for the two examples already discussed. In the case of the deformed spring, $U_s = \frac{1}{2}kx^2$, and therefore

$$F_s = -\frac{dU_s}{dx} = -\frac{d}{dx}\left(\tfrac{1}{2}kx^2\right) = -kx$$

which corresponds to the restoring force in the spring. Since the gravitational potential energy function is $U_g = mgy$, it follows from Equation 8.15 that $F_g = -mg$.

We now see that U is an important function because the conservative force can be derived from it. Furthermore, Equation 8.15 should clarify the fact that adding a constant to the potential energy is unimportant.

Often we define our system to include all the interacting parts (for example, a spring and the mass on which it acts, or a gravitational field and a mass). Then no energy is transferred to or from the system, so

$$\Delta E = \Delta K + \Delta U = 0$$

$$\Delta K = -\Delta U$$

---

[3] In three dimensions, the expression is $\mathbf{F} = -\mathbf{i}\dfrac{\partial U}{\partial x} - \mathbf{j}\dfrac{\partial U}{\partial y} - \mathbf{k}\dfrac{\partial U}{\partial z}$, where $\dfrac{\partial U}{\partial x}$, etc., are partial derivatives. In the language of vector calculus, $\mathbf{F}$ equals the negative of the gradient of the scalar quantity $U(x, y, z)$.

*8.7   ENERGY DIAGRAMS AND THE EQUILIBRIUM OF A SYSTEM

The motion of a system can often be understood qualitatively through its potential energy curve. Consider the potential energy function for the mass-spring system, given by $U_s = \frac{1}{2}kx^2$. This function is plotted versus x in Figure 8.15a. The force is related to U through Equation 8.15:

$$F_s = -\frac{dU_s}{dx} = -kx$$

That is, the force is equal to the negative of the slope of the U versus x curve. When the mass is placed at rest at the equilibrium position ($x = 0$), where $F = 0$, it will remain there unless some external force acts on it. If the spring is stretched from equilibrium, x is positive and the slope dU/dx is positive; therefore F_s is negative and the mass accelerates back toward $x = 0$. If the spring is compressed, x is negative and the slope is negative; therefore F_s is positive and again the mass accelerates toward $x = 0$.

From this analysis, we conclude that the $x = 0$ position is one of **stable equilibrium**. That is, any movement away from this position results in a force directed back toward $x = 0$. In general, *positions of stable equilibrium correspond to those points for which U(x) has a minimum value.*

From Figure 8.15 we see that if the mass is given an initial displacement x_m and released from rest, its total energy initially is the potential energy stored in the spring, $\frac{1}{2}kx_m^2$. As motion commences, the system acquires kinetic energy and loses an equal amount of potential energy. Because the total energy must remain constant, the mass oscillates between the two points $x = \pm x_m$, called the *turning points*. In fact, because there is no energy loss (no friction), the mass will oscillate between $-x_m$ and $+x_m$ forever. (We discuss these oscillations further in Chapter 13.) From an energy viewpoint, the energy of the system cannot exceed $\frac{1}{2}kx_m^2$; therefore the

Stable equilibrium	

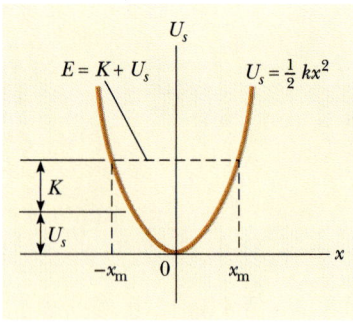

(a)

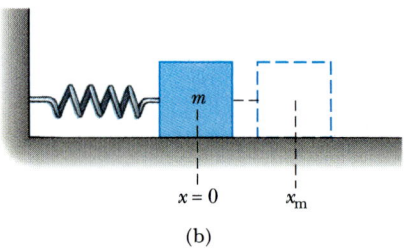

(b)

FIGURE 8.15   (a) Potential energy as a function of x for the mass-spring system shown in (b). The mass oscillates between the turning points, which have the coordinates $x = \pm x_m$. Note that the restoring force of the spring always acts toward $x = 0$, the position of stable equilibrium.

mass must stop at these points and, because of the spring force, accelerate toward $x = 0$.

Another simple mechanical system that has a position of stable equilibrium is a ball rolling about in the bottom of a bowl. If the ball is displaced from its lowest position, it will always tend to return to that position when it is released.

Now consider an example where the U versus x curve is as shown in Figure 8.16. Once again, $F_x = 0$ at $x = 0$, and so the particle is in equilibrium at this point. However, this is a position of **unstable equilibrium** for the following reason. Suppose the particle is displaced to the right ($x > 0$). Because the slope is negative for $x > 0$, $F_x = -dU/dx$ is positive and the particle accelerates away from $x = 0$. Now suppose that the particle is displaced to the left ($x < 0$). In this case, the force is negative because the slope is positive for $x < 0$, and the particle again accelerates away from the equilibrium position. The $x = 0$ position in this situation is one of unstable equilibrium because for any displacement from this point, the force pushes the particle farther away from equilibrium. In fact, the force pushes the particle toward a position of lower potential energy. A ball placed on the top of an inverted bowl in the shape of a hemisphere is in a position of unstable equilibrium. If the ball is displaced slightly from the top and released, it will surely roll off the bowl. In general, *positions of unstable equilibrium correspond to those points for which U(x) has a maximum value.*

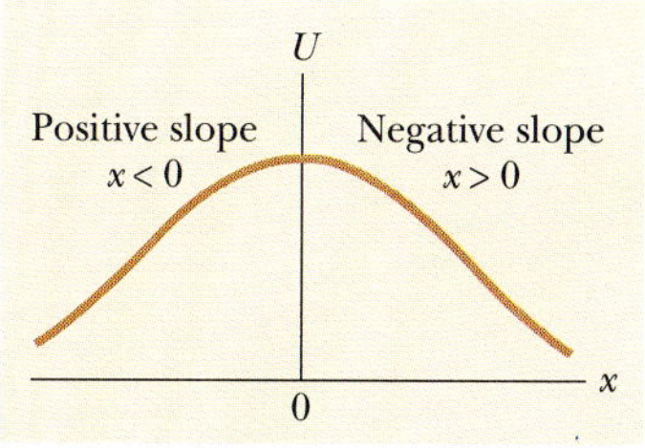

Finally, a situation may arise where U is constant over some region and hence $F = 0$. This is called a position of **neutral equilibrium.** Small displacements from this position produce neither restoring nor disrupting forces. A ball lying on a flat horizontal surface is an example of an object in neutral equilibrium.

FIGURE 8.16   A plot of U versus x for a particle that has a position of unstable equilibrium, located at $x = 0$. For any finite displacement of the particle, the force on the particle is directed away from $x = 0$.

Neutral equilibrium

8.8   CONSERVATION OF ENERGY IN GENERAL

We have seen that the total mechanical energy of a system is constant when only conservative forces act within the system. Furthermore, we can associate a potential energy function with each conservative force. On the other hand, as we saw in Section 8.5, mechanical energy is lost when nonconservative forces such as friction are present.

In the study of thermodynamics we shall find that mechanical energy can be transformed to internal energy of the system. For example, when a block slides over a rough surface, the mechanical energy lost is transformed into internal energy temporarily stored in the block and the surface, as evidenced by a measurable increase in the block's temperature. We shall see that on a submicroscopic scale, this internal energy is associated with the vibration of atoms about their equilibrium positions. Such internal atomic motion has kinetic and potential energy. Therefore, if we include this increase in the internal energy of the system in our energy expression, the total energy is conserved.

This is just one example of how you can analyze an isolated system and always find that its total energy does not change, as long as you account for all forms of energy. That is, *energy can never be created or destroyed. Energy may be transformed from one form to another, but the total energy of an isolated system is always constant.* From a universal point of view, we can say that the *total energy of the Universe is constant:* If one part of the Universe gains energy in some form, another part must lose an equal amount of energy. No violation of this principle has been found.

Total energy is always conserved

*8.9 MASS-ENERGY EQUIVALENCE

This chapter has been concerned with the important principle of energy conservation and its application to various physical phenomena. Another important principle, **conservation of mass,** says that, in any kind of physical or chemical process, *mass is neither created nor destroyed.* That is, the mass before the process equals the mass after the process.

For centuries, it appeared to scientists that energy and mass were two quantities that were separately conserved. However, in 1905 Einstein made the incredible discovery that mass, or inertia, of any system is a measure of the total energy of the system. Hence, *energy and mass are related concepts.* The relationship between the two is given by Einstein's most famous formula

$$E = mc^2 \tag{8.16}$$

where c is the speed of light and E is the energy equivalent of a mass m. As we shall see in Chapter 39, mass increases with speed; however, this dependence is insignificant for $v \ll c$. Hence, the masses we use for describing situations in our everyday experiences are always taken to be rest masses.

The energy associated with even a small amount of matter is enormous. For example, the energy of 1 kg of any substance is

$$E = mc^2 = (1 \text{ kg})(3 \times 10^8 \text{ m/s})^2 = 9 \times 10^{16} \text{ J}$$

This is equivalent to the energy content of about 15 million barrels of crude oil (about one day's consumption in the United States)! If this energy could easily be released, as useful work, our energy resources would be unlimited.

In reality, only a small fraction of the energy contained in a material sample can be released through chemical or nuclear processes. The effects are largest in nuclear reactions, where fractional changes in energy, and hence mass, of approximately 10^{-3} are routinely observed. A good example is the enormous energy released when the uranium-235 nucleus splits into two smaller nuclei. This happens because the ^{235}U nucleus is more massive than the sum of the masses of the product nuclei. The awesome nature of the energy released in such reactions is vividly demonstrated in the explosion of a nuclear weapon.

Expressed briefly, Equation 8.16 tells us that *energy has mass.* Whenever the energy of an object changes in any way, its mass changes. If ΔE is the change in energy of an object, its change in mass is

$$\Delta m = \frac{\Delta E}{c^2} \tag{8.17}$$

Any time energy ΔE in any form is supplied to an object, the change in the mass of the object is $\Delta m = \Delta E/c^2$. Because c^2 is so large, however, the changes in mass in any ordinary mechanical experiment or chemical reaction are too small to be detected.

*8.10 QUANTIZATION OF ENERGY

As you may have learned in your chemistry course, all ordinary matter consists of atoms, and each atom consists of a collection of electrons and a nucleus, which in turn contains smaller particles, each with a distinctive electric charge. Thus, on the fine scale of the atomic world, mass comes in discrete quantities corresponding to the atomic masses. In the language of modern physics, we say that certain quantities such as electric charge are *quantized*.

As we shall learn in the extended version of this text, many more physical quantities are quantized. The quantized nature of energy in bound systems is especially important in the atomic and subatomic world. For example, let us consider the energy levels of the hydrogen atom (an electron orbiting around a proton). The atom can occupy only certain energy levels, called *quantum states,* as shown in Figure 8.17a. The atom cannot have any energy values lying between these quantum states. The lowest energy level, E_0, is called the *ground state* of the atom. The ground state corresponds to the state an isolated atom usually occupies. The atom can move to higher energy states by absorbing energy from some external source or by colliding with other atoms. The highest energy on the scale shown in Figure 8.17a, E_∞, corresponds to the energy of the atom when the electron is completely removed from the proton and is called the *ionization energy*. Note that the energy levels get closer together at the high end of the scale. Above this energy level, the energies are continuous.

Next, consider a satellite in orbit about the Earth. If you were asked to describe the possible energies the satellite could have, it would be reasonable (but incorrect) to say it could have any arbitrary energy value. Just like the hydrogen atom, however, *the energy of the satellite is also quantized*. If you were to construct an energy level diagram for the satellite showing its allowed energies, the levels would be so close to one another, as in Figure 8.17b, that it would be impossible to tell that they were not continuous. In other words, we have no way of experiencing quantization of energy in the macroscopic world; hence we can ignore it in describing everyday experiences.

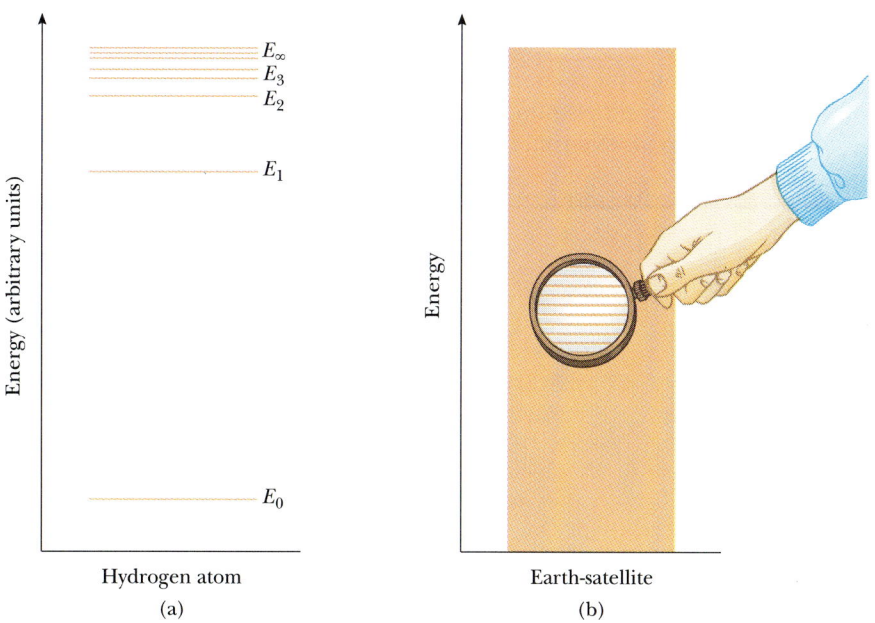

FIGURE 8.17   Energy-level diagrams: (a) Quantum states of the hydrogen atom. The lowest state, E_0, is the ground state. (b) The energy levels of an Earth satellite are also quantized but are so close together that they cannot be distinguished from each other.

SUMMARY

The **gravitational potential energy** of a particle of mass m that is elevated a distance y near the Earth's surface is

$$U_g \equiv mgy \qquad (8.1)$$

The **elastic potential energy** stored in a spring of force constant k is

$$U_s \equiv \tfrac{1}{2} kx^2 \qquad (8.4)$$

A force is **conservative** if the work it does on a particle is independent of the path the particle takes between two points. Alternatively, a force is conservative if the work it does is zero when the particle moves through an arbitrary closed path and returns to its initial position. A force that does not meet these criteria is said to be **nonconservative.**

A **potential energy** function U can be associated only with a conservative force. If a conservative force $\mathbf{F}$ acts on a particle that moves along the x axis from x_i to x_f, the change in the potential energy of the particle equals the negative of the work done by that force:

$$U_f - U_i = -\int_{x_i}^{x_f} F_x \, dx \qquad (8.6)$$

The **total mechanical energy of a system** is defined as the sum of the kinetic energy and potential energy:

$$E \equiv K + U \qquad (8.8)$$

If no external forces do work on a system, and there are no nonconservative forces, the total mechanical energy of the system is constant:

$$K_i + U_i = K_f + U_f \qquad (8.9)$$

The change in total mechanical energy of a system equals the change in the kinetic energy due to internal nonconservative forces, $\Delta K_{\text{int-nc}}$, plus the change in kinetic energy due to all external forces, ΔK_{ext}:

$$K_i + U_i + \Delta K_{\text{int-nc}} + \Delta K_{\text{ext}} = K_f + U_f \qquad (8.14)$$

QUESTIONS

1. A bowling ball is suspended from the ceiling of a lecture hall by a strong cord. The bowling ball is drawn away from its equilibrium position and released from rest at the tip of the demonstrator's nose. If the demonstrator remains stationary, explain why she will not be struck by the ball on its return swing. Would the demonstrator be safe if she pushed the ball as she released it?

2. Can gravitational potential energy ever be negative? Explain.

3. One person drops a ball from the top of a building, while another person at the bottom observes its motion. Will these two people agree on the value of the ball's potential energy? on its change in potential energy? on its kinetic energy?

4. When a person runs in a track event at constant velocity, is any work done? (*Note:* Although the runner moves with constant velocity, the legs and arms accelerate.) How does air resistance enter into the picture? Does the center of mass of the runner move horizontally?

5. Our body muscles exert forces when we lift, push, run, jump, and so forth. Are these forces conservative?

6. If three conservative forces and one nonconservative force act on a system, how many potential energy terms appear in the problem?

7. Consider a ball fixed to one end of a rigid rod with the other end pivoted on a horizontal axis so that the rod can rotate in a vertical plane. What are the positions of stable and unstable equilibrium?

8. Is it physically possible to have a situation where $E - U < 0$?

9. What would the curve of U versus x look like if a particle were in a region of neutral equilibrium?

10. Explain the energy transformations that occur during (a) the pole vault, (b) the shot put, (c) the high jump. What is the source of energy in each case?

11. Discuss all the energy transformations that occur during the operation of an automobile.

12. A ball is thrown straight up into the air. At what position is its kinetic energy a maximum? At what position is its gravitational potential energy a maximum?

13. Consider the Earth to be a perfect sphere. By how much does your potential energy change when you (a) walk from the North Pole to the equator? (b) drop through a

"tunnel" from the North Pole to the South Pole passing through the center of the Earth?

14. Does a single external force acting on a particle necessarily change (a) its kinetic energy? (b) its velocity?

15. In the high jump, is any portion of the athlete's kinetic energy converted to potential energy during the jump?

16. In the pole vault or high jump, why does the athlete attempt to keep his or her center of gravity as low as possible (consistent with passing over the bar) near the top of the jump?

17. A right circular cone can be balanced on a horizontal surface in three ways. Sketch these three equilibrium configurations and identify them as being stable, unstable, or neutral.

PROBLEMS

Review Problem

A block of mass m rests on top of a frictionless inclined plane of mass $3m$ and height h as shown below. The block slides down the plane and moves along a horizontal surface until it stops. The coefficient of kinetic friction between the block and horizontal surface is μ_k. At the same time, the plane slides to the left without friction and collides with a spring of force constant k. Find (a) the initial potential energy of the block, (b) the kinetic energy of the system when the block leaves the plane, (c) the speed of the block when it leaves the plane, (d) the speed of the plane when the block leaves the plane, (e) the distance the block travels before it stops, (f) the time it takes the block to stop, (g) the maximum compression of the spring caused by the inclined plane, and (h) the maximum value of the force exerted by the spring on the inclined plane. (i) Find the distance the block travels before it stops if the inclined plane is attached to the table.

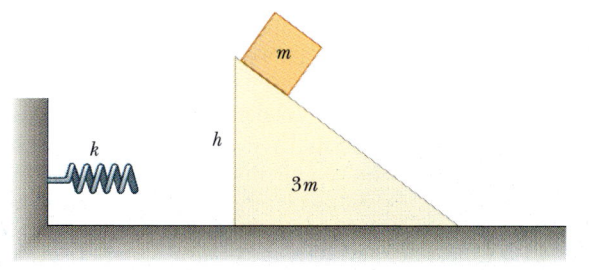

Section 8.2 Conservative and Nonconservative Forces

1. A 4.00-kg particle moves from the origin to the position having coordinates $x = 5.00$ m and $y = 5.00$ m under the influence of gravity acting in the negative y direction (Fig. P8.1). Using Equation 7.2, calculate the work done by gravity in going from O to C along (a) OAC, (b) OBC, (c) OC. Your results should all be identical. Why?

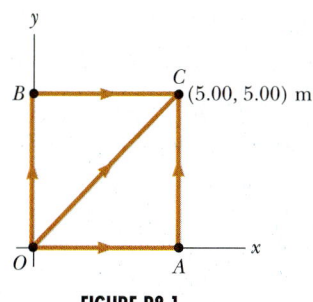

FIGURE P8.1

2. (a) Starting with Equation 7.2 for the definition of work done by a constant force, show that any constant force is conservative. (b) As a special case, suppose a particle of mass m is under the influence of force $\mathbf{F} = (3\mathbf{i} + 4\mathbf{j})$ N and moves from O to C in Figure P8.1. Calculate the work done by $\mathbf{F}$ along the three paths OAC, OBC, and OC. (Again, your three answers should be identical.)

3. Under the influence of gravity, a block of mass m

□ indicates problems that have full solutions available in the Student Solutions Manual and Study Guide.

slides down a quarter-circular track, concave upward, so slowly that the centripetal force is negligible. The coefficient of friction is μ_k and the track radius is R. (a) Show that the change in kinetic energy of the block is $mgR(1 - \mu_k)$. (b) If this change is 42.0 J when $m = 2.00$ kg and $R = 3.20$ m, determine the work done by the conservative forces and the energy dissipated by the nonconservative forces. (c) What is μ_k?

4. A single conservative force acting on a particle varies as $\mathbf{F} = (-Ax + Bx^2)\mathbf{i}$ N, where A and B are constants and x is in meters. (a) Calculate the potential energy associated with this force, taking $U = 0$ at $x = 0$. (b) Find the change in potential energy and change in kinetic energy as the particle moves from $x = 2.0$ m to $x = 3.0$ m.

5. A force acting on a particle moving in the xy plane is $\mathbf{F} = (2y\mathbf{i} + x^2\mathbf{j})$ N, where x and y are in meters. The particle moves from the origin to a final position having coordinates $x = 5.0$ m and $y = 5.0$ m, as in Figure P8.1. Calculate the work done by $\mathbf{F}$ along (a) OAC, (b) OBC, (c) OC. (d) Is $\mathbf{F}$ conservative or nonconservative? Explain.

Section 8.3   Conservative Forces and Potential Energy
Section 8.4   Conservation of Energy

6. A 4.0-kg particle moves along the x axis under the influence of a single conservative force. If the work done on the particle is 80.0 J as it moves from the point $x = 2.0$ m to $x = 5.0$ m, find (a) the change in its kinetic energy, (b) the change in its potential energy, and (c) its speed at $x = 5.0$ m if it starts at rest at $x = 2.0$ m.

7. A single conservative force $F_x = (2.0x + 4.0)$ N acts on a 5.0-kg particle, where x is in meters. As the particle moves along the x axis from $x = 1.0$ m to $x = 5.0$ m, calculate (a) the work done by this force, (b) the change in the potential energy of the particle, and (c) its kinetic energy at $x = 5.0$ m if its speed at $x = 1.0$ m is 3.0 m/s.

8. At time t_i, the kinetic energy of a particle is 30 J and its potential energy is 10 J. At some later time t_f, its kinetic energy is 18 J. (a) If only conservative forces act on the particle, what are its potential energy and its total energy at time t_f? (b) If the potential energy at time t_f is 5 J, are there any nonconservative forces acting on the particle? Explain.

9. A single constant force $\mathbf{F} = (3.0\mathbf{i} + 5.0\mathbf{j})$ N acts on a 4.0-kg particle. (a) Calculate the work done by this force if the particle moves from the origin to the point having the vector position $\mathbf{r} = (2.0\mathbf{i} - 3.0\mathbf{j})$ m. Does this result depend on the path? Explain. (b) What is the speed of the particle at $\mathbf{r}$ if its speed at the origin is 4.0 m/s? (c) What is the change in its potential energy?

10. A 5.0-kg mass is attached to a light cord that passes over a massless, frictionless pulley. The other end of the cord is attached to a 3.5-kg mass as in Figure

P8.10. Use conservation of energy to determine the final speed of the 5.0-kg mass after it has fallen (starting from rest) 2.5 m.

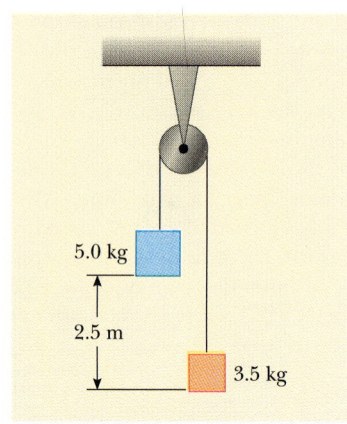

5.0 kg

2.5 m

3.5 kg

FIGURE P8.10

11. A bead slides without friction around a loop-the-loop (Fig. P8.11). If the bead is released from a height $h = 3.50R$, what is its speed at point A? How large is the normal force on it if its mass is 5.00 g?

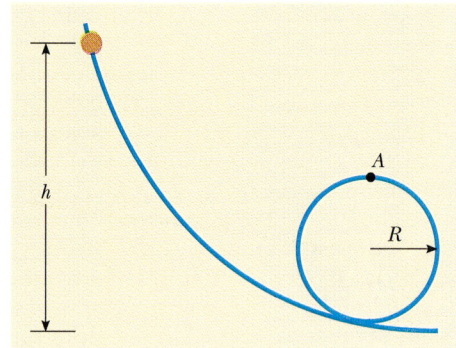

h

A

R

FIGURE P8.11

12. A particle of mass 0.500 kg is shot from P as shown in Figure P8.12 with an initial velocity $\mathbf{v}_0$ having a horizontal component of 30.0 m/s. The particle rises to a maximum height of 20.0 m above P. Using conservation of energy, determine (a) the vertical component of $\mathbf{v}_0$, (b) the work done by the gravitational force on the particle during its motion from P to B, and (c) the horizontal and the vertical components of the velocity vector when the particle reaches B.

13. A rocket is launched at an angle of 53° to the horizontal from an altitude h with a speed v_0. (a) Use energy methods to find its speed when its altitude is $h/2$. (b) Find the x and y components of velocity when the rocket's altitude is $h/2$, using the fact that $v_x = v_{x0} = $ constant (since $a_x = 0$) and the results to part (a).

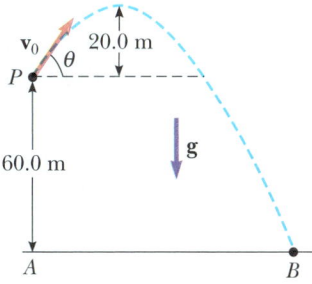

FIGURE P8.12

14. A 2.00-kg ball is attached to a 10-lb (44.5 N) fish line. The ball is released from rest at the horizontal position ($\theta = 90°$). At what angle θ (measured from the vertical), does the line break?

15. A 20.0-kg cannon ball is fired from a cannon at muzzle speed of 1000 m/s and at an angle of $37.0°$ with the horizontal. A second ball is fired at an angle of $90.0°$. Use the conservation of mechanical energy to find, for each ball, (a) the maximum height reached and (b) the total mechanical energy at the maximum height.

16. A ball of mass m is spun in a vertical circle having radius R. The ball has a speed v_0 at its highest point. Take zero potential energy at the lowest point and use the angle θ measured with respect to the vertical as shown in Figure P8.16. (a) Derive an expression for the speed v at any time as a function of R, θ, v_0, and g. (b) What minimum speed v_0 is required to keep the ball moving in a circle? (c) Does the equation you derived in part (a) account for the result found in part (b)? Explain.

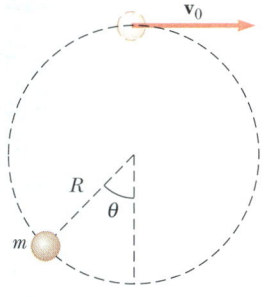

FIGURE P8.16

17. Two masses are connected by a light string passing over a light frictionless pulley as shown in Figure P8.17. The 5.0-kg mass is released from rest. Using the law of conservation of energy, (a) determine the speed of the 3.0-kg mass just as the 5.0-kg mass hits the ground. (b) Find the maximum height to which the 3.0-kg mass rises.

17A. Two masses are connected by a light string passing over a light frictionless pulley as in Figure P8.17. The mass m_1 is released from rest. Using the law of conservation of energy, (a) determine the speed of m_1 just as m_2 hits the ground. (b) Find the maximum height to which m_1 rises.

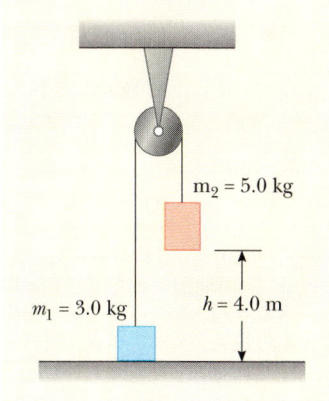

FIGURE P8.17

18. A child slides down the frictionless slide shown in Figure P8.18. In terms of R and H, at what height h will he lose contact with the section of radius R?

FIGURE P8.18

Section 8.5 Changes in Mechanical Energy When Nonconservative Forces Are Present

19. A 5.0-kg block is set into motion up an inclined plane with an initial speed of 8.0 m/s (Fig. P8.19). The block comes to rest after traveling 3.0 m along the plane, which is inclined at an angle of $30°$ to the horizontal. Determine (a) the change in the block's kinetic energy, (b) the change in its potential energy, (c) the frictional force exerted on it (assumed to be constant). (d) What is the coefficient of kinetic friction?

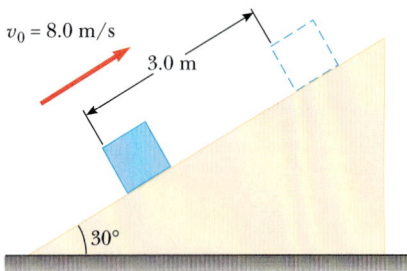

FIGURE P8.19

20. A 3.0-kg block starts at a height $h = 60$ cm on a plane that has an inclination angle of 30° as in Figure P8.20. Upon reaching the bottom, the block slides along a horizontal surface. If the coefficient of friction on both surfaces is $\mu_k = 0.20$, how far does the block slide on the horizontal surface before coming to rest? (*Hint:* Divide the path into two straight-line parts.)

20A. A block of mass m starts at a height h on a plane that has an inclination angle θ as in Figure P8.20. Upon reaching the bottom, the block slides along a horizontal surface. If the coefficient of friction on both surfaces is μ_k, how far does the block slide on the horizontal surface before coming to rest? (*Hint:* Divide the path into two straight-line parts.)

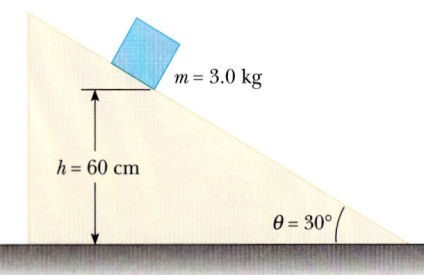

FIGURE P8.20

21. A parachutist of mass 50.0 kg jumps out of an airplane at a height of 1000 m and lands on the ground with a speed of 5.00 m/s. How much energy was lost to air friction during this jump?

22. A 0.500-kg bead slides on a curved wire, starting from rest at point A in Figure P8.22. The segment from A to B is frictionless, and the segment from B to C is rough. (a) Find the speed of the bead at B. (b) If the bead comes to rest at C, find the energy lost due to friction as it moves from B to C.

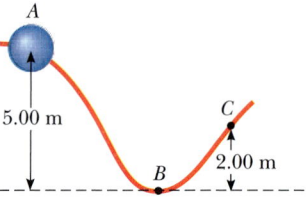

FIGURE P8.22

23. In the 1800's French engineer Hector Horeau proposed a design for a tunnel across the English Channel. Railroad cars would be able to roll freely through the tunnel until they ran out of kinetic energy, and then a steam engine would pull them the rest of the way (Fig. P8.23). Assume that the 32.0-km-long tunnel was designed with a 6.0° slope for the first 1.00 km at each end and ran level the rest of the way. Let the coefficient of rolling friction be 0.1 (a small value). (a) What is the change in potential energy of a 4000-kg railroad car as it rolls down the incline? (b) If the car starts from rest, what is its speed when it reaches the level track? (c) How far into the tunnel does the car get before it stops? (d) How feasible was this design?

24. An 80.0-kg skydiver jumps out of an airplane at an altitude of 1000 m and opens the parachute at an altitude of 200 m. (a) Assuming that the total retarding force on the diver is constant at 50.0 N with the parachute closed and constant at 3600 N with the parachute open, what is her speed when she lands? (b) Do you think the skydiver will get hurt? Explain. (c) At what height should the parachute be opened so that the speed of the diver just as she hits the ground is 5.00 m/s? (d) How realistic is the assumption that the total retarding force is constant? Explain.

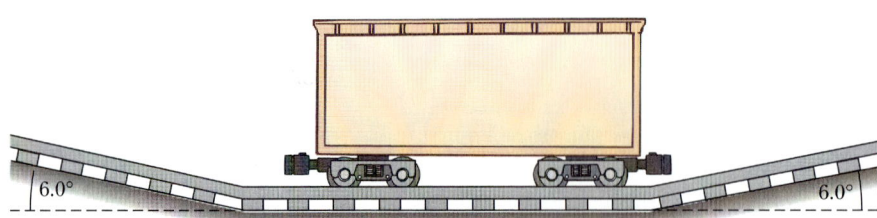

FIGURE P8.23

25. The coefficient of friction between the 3.0-kg mass and surface in Figure P8.25 is 0.40. The system starts from rest. What is the speed of the 5.0-kg mass when it has fallen 1.5 m?

25A. The coefficient of friction between m_1 and surface in Figure P8.25 is μ. The system starts from rest. What is the speed of m_2 when it has fallen a distance h?

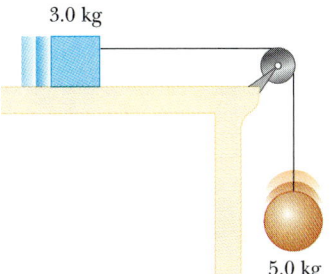

3.0 kg

5.0 kg

FIGURE P8.25

26. A toy gun uses a spring to project a 5.3-g soft rubber sphere. The spring constant is 8.0 N/m, the barrel of the gun is 15 cm long, and there is a constant frictional force of 0.032 N between barrel and projectile. With what speed is the projectile launched from the barrel if the spring is compressed 5.0 cm?

27. A block slides down a curved frictionless track and then up an inclined plane as in Figure P8.27. The coefficient of kinetic friction between block and incline is μ_k. Use energy methods to show that the maximum height reached by the block is

$$y_{max} = \frac{h}{1 + \mu_k \cot(\theta)}$$

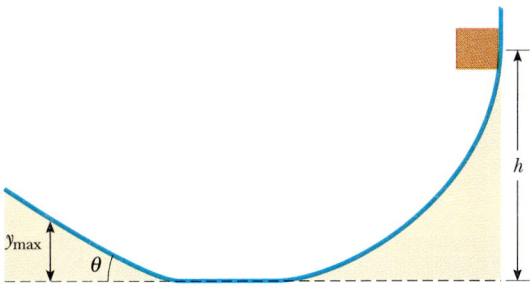

y_{max}

θ

h

FIGURE P8.27

28. A 1.5-kg mass is first held 1.2 m above a relaxed massless spring that has a spring constant of 320 N/m and then dropped onto the spring. (a) How far does the spring compress? (b) The same experiment is repeated on the Moon, where $g = 1.63$ m/s². (c) Repeat part (a), but this time assume that a constant

0.70-N air-resistance acts on the mass during the fall.

29. In the dangerous "sport" of bungee-jumping, a daring student jumps from a balloon with a specially designed elastic cord attached to his ankles. The unstretched length of the cord is 25.0 m, the student weighs 700 N, and the balloon is 36.0 m above the surface of a river. Calculate the required force constant of the cord if the student is to stop safely 4.00 m above the river.

30. A 3.0-kg mass starts from rest and slides a distance d down a frictionless 30° incline, where it contacts an unstressed spring of negligible mass as in Figure P8.30. The mass slides an additional 0.20 m as it is brought momentarily to rest by compressing the spring ($k = 400$ N/m). Find the initial separation d between mass and spring.

30A. A mass m starts from rest and slides a distance d down a frictionless incline of angle θ, where it contacts an unstressed spring of negligible mass as in Figure P8.30. The mass slides an additional distance x as it is brought momentarily to rest by compressing the spring (force constant k). Find the initial separation d between mass and spring.

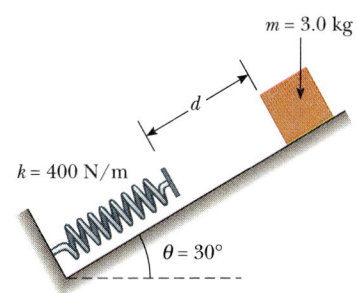

$m = 3.0$ kg

d

$k = 400$ N/m

$\theta = 30°$

FIGURE P8.30

31. An 8.00-kg block travels on a rough, horizontal surface and collides with a spring as in Figure 8.12. The speed of the block *just before* the collision is 4.00 m/s. As the block rebounds to the left with the spring uncompressed, its speed as it leaves the spring is 3.00 m/s. If the coefficient of kinetic friction between block and surface is 0.400, determine (a) the energy lost due to friction while the block is in contact with the spring and (b) the maximum distance the spring is compressed.

32. A child's pogo stick (Fig. P8.32) stores energy in a spring ($k = 2.5 \times 10^4$ N/m). At position A ($x_1 = -0.10$ m) the spring compression is a maximum and the child is momentarily at rest. At position B ($x = 0$) the spring is relaxed and the child is moving upward. At position C, the child is again momentarily at rest at the top of the jump. Assuming that the combined

mass of child and pogo stick is 25 kg, (a) calculate the total energy of the system if both potential energies are zero at $x = 0$, (b) determine x_2, (c) calculate the speed of the child at $x = 0$, (d) determine the value of x for which the kinetic energy of the system is a maximum, and (e) obtain the child's maximum upward speed.

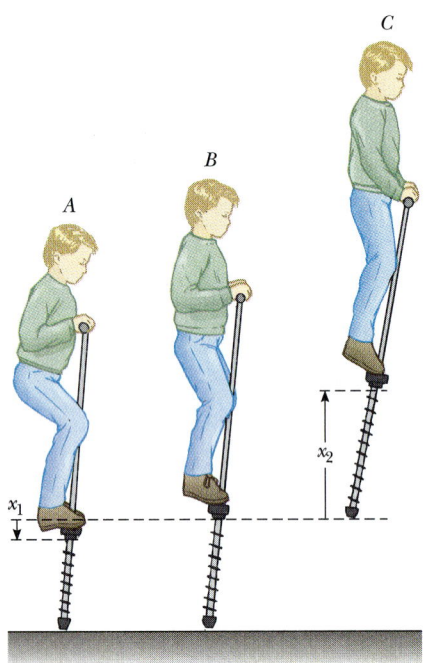

FIGURE P8.32

33. A block of mass 0.250 kg is placed on top of a vertical spring of constant $k = 5000$ N/m and pushed downward, compressing the spring 0.100 m. After the block is released it travels upward and then leaves the spring. To what maximum height above the point of release does it rise?

33A. A block of mass m is placed on top of a vertical spring of constant k and pushed downward, compressing the spring a distance d. After the block is released it travels upward and then leaves the spring. To what maximum height h above the point of release does it rise?

34. A block of mass 2.0 kg is kept at rest as it compresses a horizontal massless spring ($k = 100$ N/m) by 10 cm. As the block is released, it travels 0.25 m on a rough horizontal surface before stopping. Calculate the coefficient of kinetic friction between surface and block.

35. A 10.0-kg block is released from point A in Figure P8.35. The track is frictionless except for the portion BC, of length 6.00 m. The block travels down the track, hits a spring of force constant $k = 2250$ N/m, and compresses it 0.300 m from its equilibrium position before coming to rest momentarily. Determine the coefficient of kinetic friction between surface BC and block.

36. A 120-g mass is attached to the end of an unstressed vertical spring ($k = 40$ N/m) and then dropped. (a) What is its maximum speed? (b) How far does it drop before coming to rest momentarily?

36A. A mass m is attached to the end of an unstressed vertical spring with force constant k and then dropped. (a) What is its maximum speed? (b) How far does it drop before coming to rest momentarily?

Section 8.6   Relationship Between Conservative Forces and Potential Energy

37. The potential energy of a two-particle system separated by a distance r is $U(r) = A/r$, where A is a constant. Find the radial force F_r in terms of A and r.

38. A 3.00-kg block moving along the x axis is acted upon by a single force that varies with the block's position according to the equation $F_x = ax^2 + b$, where $a = 5.00$ N/m^2 and $b = -2.50$ N. At $x = 1.0$ m, the block is moving to the right at 4.0 m/s. Determine its speed at $x = 2.0$ m.

39. A potential energy function for a two-dimensional force is of the form $U = 3x^3y - 7x$. Find the force that acts at the point (x, y).

*Section 8.7   Energy Diagrams and the Equilibrium of a System

40. For the potential energy curve shown in Figure P8.40, (a) determine whether the force F_x is positive, negative, or zero at the five points indicated. (b) In-

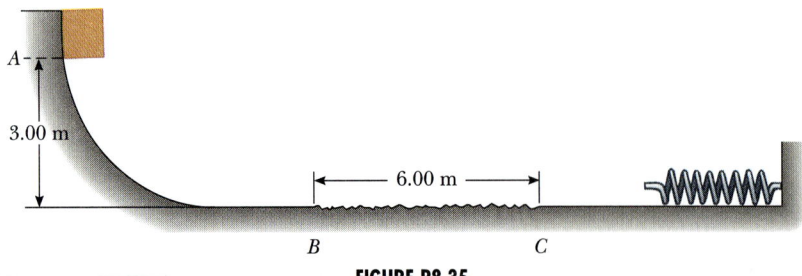

FIGURE P8.35

dicate points of stable, unstable, and neutral equilibrium. (c) Sketch the curve of F_x versus x from $x = 0$ to
$x = 8.0$ m.

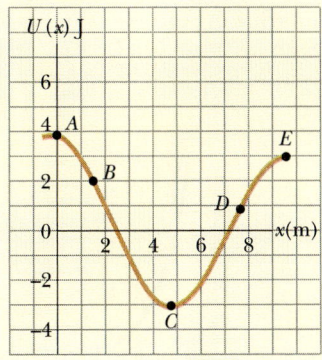

FIGURE P8.40

41. A particle of mass m is suspended between two identical springs on a horizontal frictionless tabletop as
shown in Figure P8.41. Both springs have spring constant k. (a) If the particle is pulled a distance x along
a direction perpendicular to the initial configuration
of the springs, show that its potential energy due to
the springs is

$$U(x) = kx^2 + 2kL(L - \sqrt{x^2 + L^2})$$

(*Hint:* See Problem 78 of Chapter 7.) (b) Plot $U(x)$
versus x and identify all equilibrium points. Assume
that $L = 1.20$ m and $k = 40.0$ N/m. (c) If the particle is pulled 0.500 m to the right and then released,
what is its speed when it reaches $x = 0$?

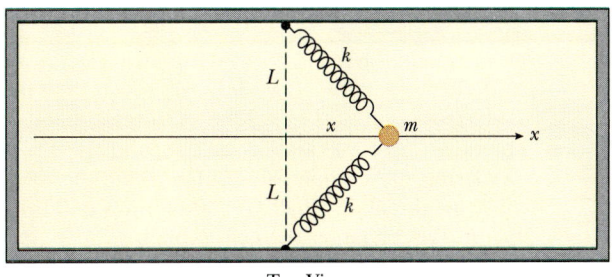

Top View

FIGURE P8.41

42. A hollow pipe has one or two weights attached to its
inner surface as shown in Figure P8.42. Characterize
each configuration as being stable, unstable, or neutral equilibrium and explain each of your choices.

43. A particle of mass $m = 5.00$ kg is released from point
A on the frictionless track shown in Figure P8.43. Determine (a) the particle's speed at points B and C and
(b) the net work done by the force of gravity in moving the particle from A to C.

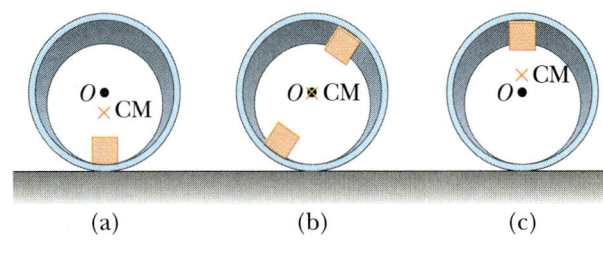

FIGURE P8.42

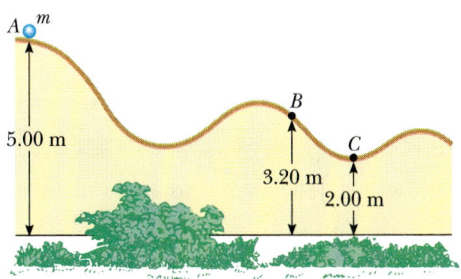

FIGURE P8.43

*Section 8.9 Mass-Energy Equivalence

44. Find the energy equivalence of (a) an electron of
mass 9.11×10^{-31} kg, (b) a uranium atom of mass
4.0×10^{-25} kg, (c) a paper clip of mass 2.0 g, and
(d) the Earth of mass 5.99×10^{24} kg.

45. The expression for the kinetic energy of a particle
moving with speed v is given by Equation 7.20, which
can be written as $K = \gamma mc^2 - mc^2$, where $\gamma = [1 - (v/c)^2]^{-1/2}$. The term γmc^2 is the total energy
of the particle, and the term mc^2 is its rest energy. A
proton moves with a speed of $0.990c$, where c is the
speed of light. Find (a) its rest energy, (b) its total
energy, and (c) its kinetic energy.

ADDITIONAL PROBLEMS

46. A 200-g particle is released from rest at point A along
the horizontal diameter on the inside of a frictionless, hemispherical bowl of radius $R = 30.0$ cm (Fig.
P8.46). Calculate (a) its gravitational potential energy at point A relative to point B, (b) its kinetic energy at point B, (c) its speed at point B, and (d) its
kinetic energy and potential energy at point C.

47. The particle described in Problem 46 (Fig. P8.46) is
released from rest at A, and the surface of the bowl is
rough. The speed of the particle at B is 1.50 m/s.
(a) What is its kinetic energy at B? (b) How much
energy is lost due to friction as the particle moves
from A to B? (c) Is it possible to determine μ from
these results in any simple manner? Explain.

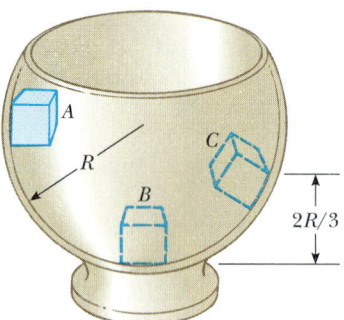

FIGURE P8.46

48. A child's toy consists of a piece of plastic attached to a spring (Fig. P8.48). The spring is compressed 2.0 cm, and the toy is released. If the mass of the toy is 100 g and it rises to a maximum height of 60 cm, estimate the force constant of the spring.

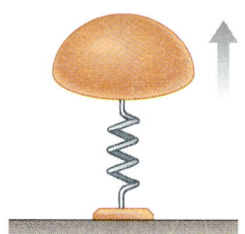

FIGURE P8.48

49. A child slides without friction from a height *h* along a curved water slide (Fig. P8.49). She is launched from a height *h*/5 into the pool. Determine her maximum airborne height *y* in terms of *h* and *θ*.

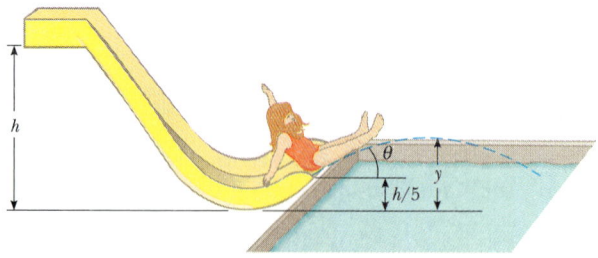

FIGURE P8.49

50. A particle of mass *m* starts from rest and slides down a frictionless track as in Figure P8.50. It leaves the track horizontally, striking the ground as indicated in the sketch. Determine *h*.

51. The masses of the javelin, the discus, and the shot are 0.80 kg, 2.0 kg, and 7.2 kg, respectively, and record throws in the track events using these objects are

about 89 m, 69 m, and 21 m, respectively. Neglecting air resistance, (a) calculate the minimum initial kinetic energies that produce these throws and (b) estimate the average force exerted on each object during the throw, assuming the force acts over a distance of 2.0 m. (c) Do your results suggest that air resistance is an important factor?

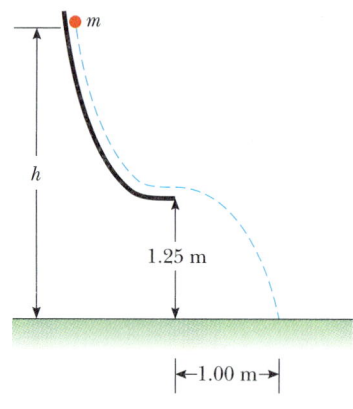

FIGURE P8.50

52. A ball having mass *m* is connected by a string of length *L* to a pivot point and held in place in a vertical position. A constant wind force of magnitude *F* is blowing from left to right as in Figure P8.52a. (a) If the ball is released from rest, show that the maximum height *H* it reaches, as measured from its initial height, is

$$H = \frac{2L}{1 + (mg/F)^2}$$

Assume that the string does not break in the process, and check that the above formula is valid for both $0 \leq H \leq L$ and $L \leq H < 2L$. (b) Compute *H* using the values *m* = 2.00 kg, *L* = 2.00 m, and *F* = 14.7 N. (c) Using these same values, determine the *equilibrium* height of the ball. (d) Can the equilibrium height ever be larger than *L*? Explain.

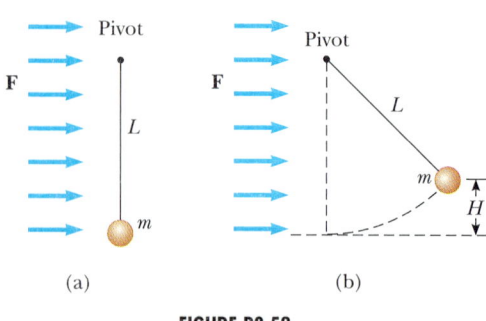

FIGURE P8.52

53. Prove that the following forces are conservative and find the change in potential energy corresponding to each, taking $x_i = 0$ and $x_f = x$: (a) $F_x = ax + bx^2$, (b) $F_x = Ae^{\alpha x}$. (a, b, A, and α are all constants.)

54. A bobsled makes a run down an ice track starting at a point on the track that is a vertical distance of 150 m above ground level. If friction is neglected, what is its speed at the bottom of the hill?

55. A 2.00-kg block situated on a rough incline is connected to a spring of negligible mass having a spring constant of 100 N/m (Fig. P8.55). The block is released from rest when the spring is unstretched, and the pulley is frictionless. The block moves 20.0 cm down the incline before coming to rest. Find the coefficient of kinetic friction between block and incline.

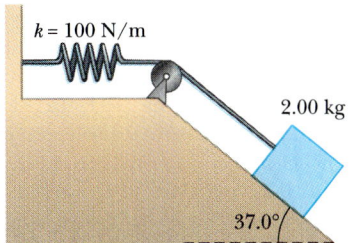

$k = 100$ N/m

2.00 kg

37.0°

FIGURE P8.55

56. Suppose the incline is frictionless for the system described in Problem 55 (Fig. P8.55). The block is released from rest with the spring initially unstretched. (a) How far does it move down the incline before coming to rest? (b) What is its acceleration at its lowest point? Is the acceleration constant? (c) Describe the energy transformations that occur during the descent.

57. A ball whirls around in a vertical circle at the end of a string. If the ball's total energy remains constant, show that the tension in the string at the bottom is greater than the tension at the top by six times the weight of the ball.

58. A pendulum made of a string of length L and a sphere swing in the vertical plane. The string hits a peg located a distance d below the point of suspension (Fig. P8.58). (a) Show that if the pendulum is

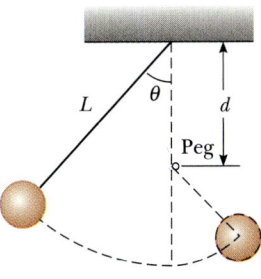

L

θ

d

Peg

FIGURE P8.58

released from a height below that of the peg, it will return to this height after striking the peg. (b) Show that if the pendulum is released from the horizontal position ($\theta = 90°$) and is to swing in a complete circle centered on the peg, then the minimum value of d must be $3L/5$.

59. A 20.0-kg block is connected to a 30.0-kg block by a string that passes over a frictionless pulley. The 30.0-kg block is connected to a spring that has negligible mass and a force constant of 250 N/m, as in Figure P8.59. The spring is unstretched when the system is as shown in the figure, and the incline is frictionless. The 20.0-kg block is pulled 20.0 cm down the incline (so that the 30.0-kg block is 40.0 cm above the floor) and released from rest. Find the speed of each block when the 30.0-kg block is 20.0 cm above the floor (that is, when the spring is unstretched).

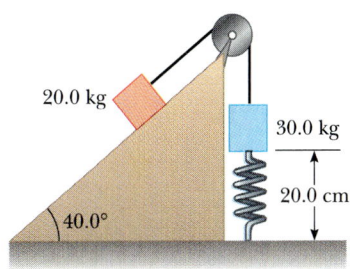

20.0 kg

30.0 kg

20.0 cm

40.0°

FIGURE P8.59

60. Consider a ball swinging in a vertical plane, with speed $v_0 = \sqrt{Rg}$ at the top of the circle, as in Figure P.860. At what angle θ should the string be cut so that the ball travels through the center of the circle?

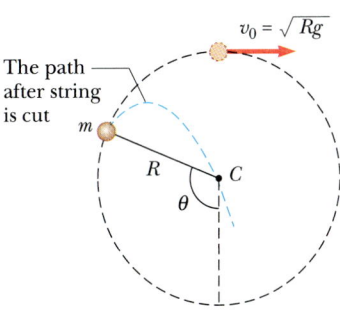

$v_0 = \sqrt{Rg}$

The path after string is cut

m

R

C

θ

FIGURE P8.60

61. A uniform chain of length 8.0 m initially lies stretched out on a horizontal table. (a) If the coefficient of static friction between chain and table is 0.60, show that the chain begins to slide off the table when 3.0 m of it hangs over the edge. (b) Determine the speed of the chain as all of it leaves the table,

given that the coefficient of kinetic friction between chain and table is 0.40.

62. Jane, whose mass is 50.0 kg, needs to swing across a river (of width D) filled with crocodiles in order to save Tarzan from danger. However, she must swing into a constant horizontal wind force **F** on a vine having length L and initially making an angle θ with the vertical (Fig. P8.62). Taking $D = 50.0$ m, $F = 110$ N, $L = 40.0$ m, and $\theta = 50.0°$, (a) with what minimum speed must Jane begin her swing in order to just make it to the other side? (b) Once the rescue is complete, Tarzan and Jane must swing back across the river. With what minimum speed must they begin their swing? Assume that Tarzan has a mass of 80.0 kg.

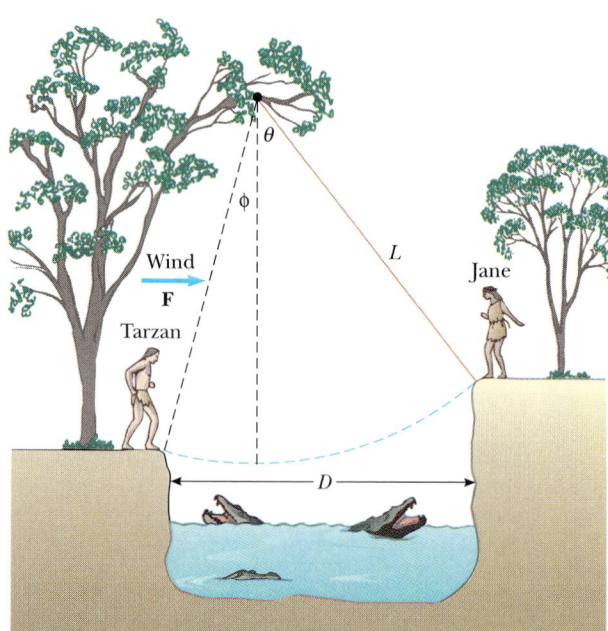

FIGURE P8.62

63. A 3.50-kg particle moves along the x direction under the influence of a force described by the potential energy function $U = (4.70 \text{ J/m}) |x|$, where x is the position of the particle in meters measured from the origin as in Figure P8.63. The total energy of the particle is 15.0 J. (a) Determine the distance it travels from the origin before reversing direction. (b) Find its maximum speed.

64. A 5.0-kg block free to move on a horizontal, frictionless surface is attached to a spring. The spring is compressed 0.10 m from equilibrium and released. The speed of the block is 1.2 m/s when it passes the equilibrium position of the spring. The same experiment is now repeated with the frictionless surface replaced

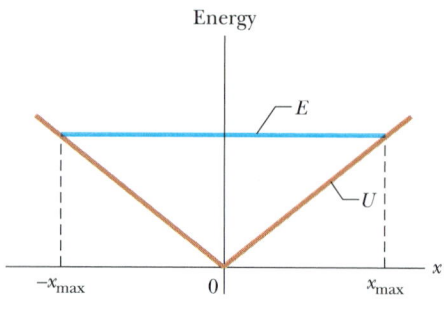

FIGURE P8.63

by a surface for which $\mu_k = 0.30$. Determine the speed of the block at the equilibrium position of the spring.

65. A block of mass 0.500 kg is pushed against a horizontal spring of negligible mass, compressing the spring a distance of Δx (Fig. P8.65). The spring constant is 450 N/m. When released, the block travels along a frictionless, horizontal surface to point B, the bottom of a vertical circular track of radius $R = 1.00$ m, and continues to move up the track. The speed of the block at the bottom of the track is $v_B = 12$ m/s, and the block experiences an average frictional force of 7.00 N while sliding up the track. (a) What is Δx? (b) What is the speed of the block at the top of the track? (c) Does the block reach the top of the track, or does it fall off before reaching the top?

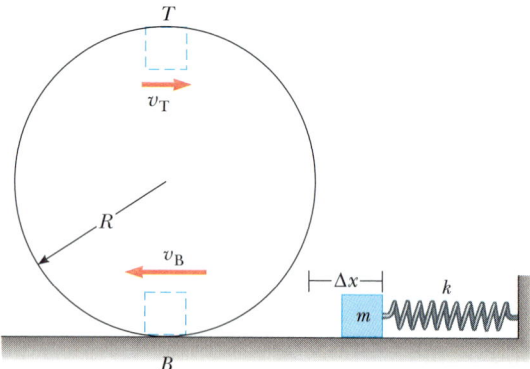

FIGURE P8.65

66. Two identical massless springs, both of constant $k = 200$ N/m, are fixed at opposite ends of a level track. A 5.00-kg block is pressed against the left spring, compressing it by 0.150 m. The block (initially at rest) is then released, as shown in Figure P8.66a. The entire track is frictionless *except* for the section between A and B. Given that the coefficient of kinetic friction between block and track along AB is $\mu_k = 0.080$, and given that the length of AB is 0.250 m,

(a) determine the maximum compression of the spring on the right (Fig. P8.66b). (b) Determine where the block eventually comes to rest, as measured from A (Fig. P8.66c).

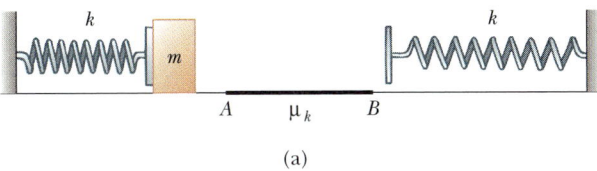

(a)

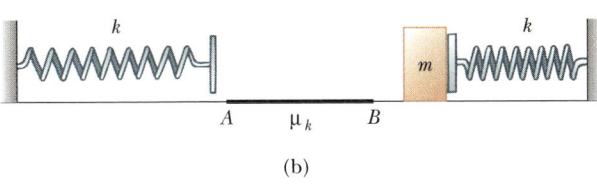

(b)

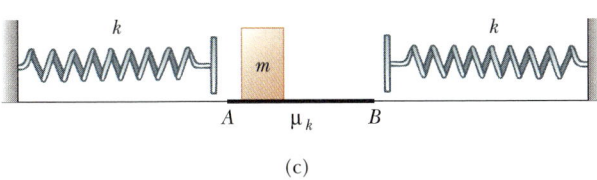

(c)

FIGURE P8.66

 67. A 50-kg block and 100-kg block are connected by a string as in Figure P8.67. The pulley is frictionless and of negligible mass. The coefficient of kinetic friction between the 50-kg block and incline is $\mu_k = 0.25$. Determine the change in the kinetic energy of the 50-kg block as it moves from C to D, a distance of 20 m.

67A. A block of mass m_1 and a block of mass m_2 are connected by a string as in Figure P8.67. The pulley is frictionless and of negligible mass and the incline angle is θ. The coefficient of kinetic friction between m_1 and incline is μ_k. Determine the change in the kinetic energy of m_1 as it moves from C to D, a distance d.

68. A pinball machine launches a 100-g ball with a spring-driven plunger (Fig. P8.68). The game board is inclined at $8°$ above the horizontal. Find the force constant k of the spring that will give the ball a speed of 80 cm/s when the plunger is released from rest with the spring compressed 5.0 cm from its relaxed position. Assume that the plunger's mass and frictional effects are negligible.

69. A 1.0-kg mass slides to the right on a surface having a coefficient of friction $\mu = 0.25$ (Fig. P8.69). The mass has a speed of $v_i = 3.0$ m/s when contact is made with a spring that has a spring constant $k = 50$ N/m. The mass comes to rest after the spring has

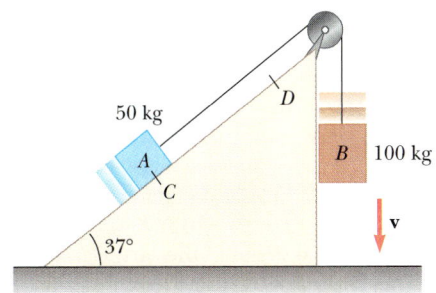

FIGURE P8.67

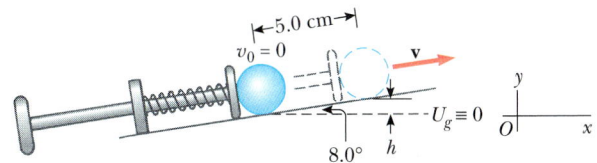

FIGURE P8.68

been compressed a distance d. The mass is then forced toward the left by the spring and continues to move in that direction beyond the unstretched position. Finally the mass comes to rest a distance D to the left of the unstretched spring. Find (a) the compressed distance d, (b) the speed v at the unstretched position when the system is moving to the left, and (c) the distance D where the mass comes to rest.

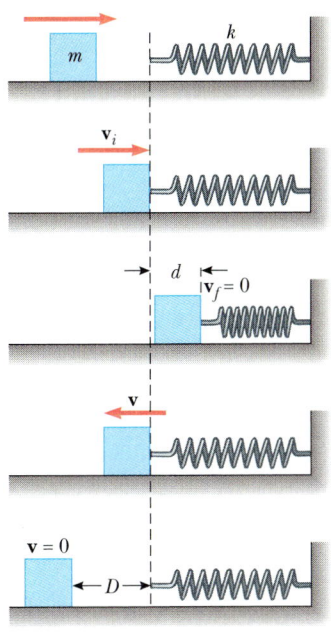

FIGURE P8.69

70. An object of mass m is suspended from the top of a cart by a string of length L, as in Figure P8.70a. The cart and object are initially moving to the right at constant speed v_0. The cart comes to rest after colliding with and sticking to a bumper, as in Figure P8.70b, and the suspended object swings through an angle θ. (a) Show that the cart speed is $v_0 = \sqrt{2gL(1 - \cos\theta)}$. (b) If $L = 1.2$ m and $\theta = 35°$, find the initial speed of the cart. (*Hint:* The force exerted by the string on the object does no work on it.)

from $x = -5.00$ m to $x = +15.0$ m. Describe the general motion of a particle under the influence of this potential energy function. How does the motion change when the initial energy of the object is increased?

S2. Using Spreadsheet 8.1, numerically differentiate the potential energy function given in Problem S8.1. Find the equilibrium points. Are they stable or unstable?

S3. The potential energy function associated with the force between two atoms is modeled by the Lennard-Jones potential

$$U(x) = 4\varepsilon\left[\left(\frac{\sigma}{x}\right)^{12} - \left(\frac{\sigma}{x}\right)^6\right]$$

In this model, there are two adjustable parameters, σ and ε, that are determined from experiments (σ is a range parameter and ε is the well depth). Modify Spreadsheet 8.1 to plot $U(x)$ versus x for $\sigma = 0.263$ nm and $\varepsilon = 1.51 \times 10^{-22}$ J. (These are typical values for helium-helium interactions.) Numerically differentiate $U(x)$ to find the force F_x.

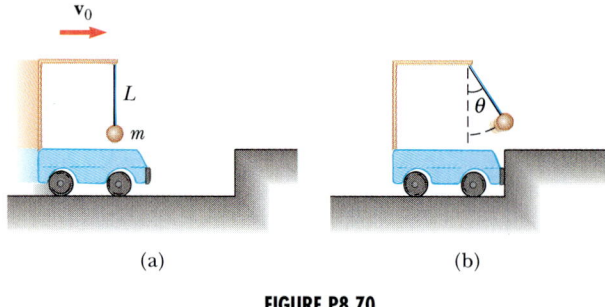

(a)   (b)

FIGURE P8.70

SPREADSHEET PROBLEMS

S1. The potential energy function of a particle is

$$U(x) = \tfrac{1}{2}kx^2 + bx^3 + c$$

where $k = 300$ N/m, $b = -12.0$ N/m², and $c = -1000$ J. Use Spreadsheet 8.1 to plot the function

B.C.   By John Hart

CHAPTER 9

Linear Momentum and Collisions

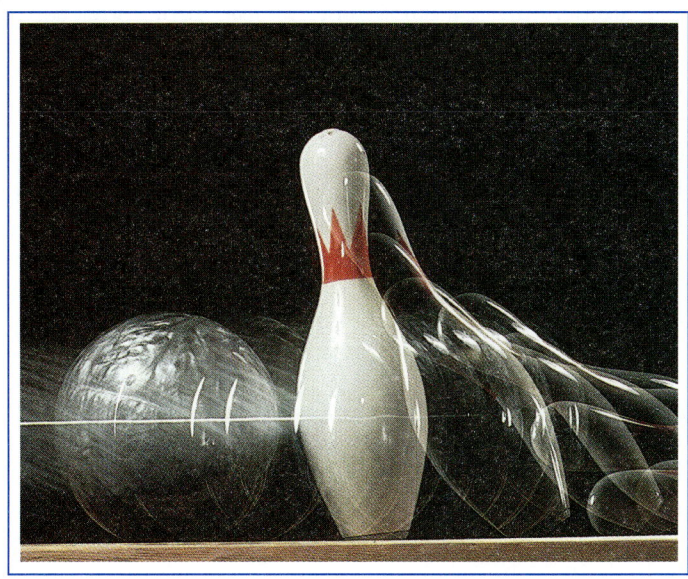

As a result of the collision between the bowling ball and pin, part of the ball's momentum is transferred to the pin. Consequently, the pin acquires momentum and kinetic energy, while the ball loses momentum and kinetic energy. However, the total momentum of the system (ball and pin) remains constant. *(Ben Rose / The Image Bank)*

Consider what happens when a golf ball is struck by a club. The ball is given a very large initial velocity as a result of the collision; consequently, it is able to travel more than a hundred meters through the air. The ball experiences a large change in velocity and a correspondingly large acceleration. Furthermore, because the ball experiences this acceleration over a very short time interval, the average force on it during the collision is very large. By Newton's third law, the club experiences a reaction force that is equal to and opposite the force on the ball. This reaction force produces a change in velocity of the club. Because the club is much more massive than the ball, however, the change in velocity of the club is much less than the change in velocity of the ball.

One of the main objectives of this chapter is to enable you to understand and analyze such events. As a first step, we shall introduce the concept of *momentum*, a term that is used in describing objects in motion. For example, a very massive football player is often said to have a great deal of momentum as he runs down the field. A much less massive player, such as a halfback, can have equal or greater momentum if his speed is greater than that of the more massive player. This follows from the fact that momentum is defined as the product of mass and velocity.

The concept of momentum leads us to a second conservation law, that of conservation of momentum. This law is especially useful for treating problems that involve collisions between objects and for analyzing rocket propulsion. The concept of the center of mass of a system of particles is also introduced, and we shall see that the motion of a system of particles can be represented by the motion of one representative particle located at the center of mass.

9.1   LINEAR MOMENTUM AND ITS CONSERVATION

> The linear momentum of a particle of mass m moving with a velocity $\mathbf{v}$ is defined to be the product of the mass and velocity:
>
> $$\mathbf{p} \equiv m\mathbf{v} \tag{9.1}$$

Linear momentum is a vector quantity since it equals the product of a scalar, m, and a vector, $\mathbf{v}$. Its direction is along $\mathbf{v}$, and it has dimensions of ML/T. The SI unit for linear momentum is the $kg \cdot m/s$.

If a particle is moving in an arbitrary direction, $\mathbf{p}$ will have three components and Equation 9.1 is equivalent to the component equations

$$p_x = mv_x \qquad p_y = mv_y \qquad p_z = mv_z \tag{9.2}$$

As you can see from its definition, the concept of momentum provides a quantitative distinction between heavy and light particles moving at the same velocity. For example, the momentum of a bowling ball moving at 10 m/s is much greater than that of a tennis ball moving at the same speed. Newton called the product $m\mathbf{v}$ *quantity of motion*, perhaps a more graphic description than *momentum*, which comes from the Latin word for movement.

Using Newton's second law of motion, we can relate the linear momentum of a particle to the resultant force acting on the particle: *The time rate of change of the linear momentum of a particle is equal to the resultant force acting on the particle.*[1] That is,

$$\mathbf{F} = \frac{d\mathbf{p}}{dt} \tag{9.3}$$

From Equation 9.3 we see that if the resultant force is zero, the time derivative of the momentum is zero, and therefore the linear momentum[2] of the particle must be constant. In other words, the linear momentum of a particle is *constant* when $\mathbf{F} = 0$. Of course, if the particle is *isolated*, then by necessity, $\mathbf{F} = 0$ and $\mathbf{p}$ remains unchanged.

Conservation of Momentum for a Two-Particle System

Consider two particles that can interact with each other but are isolated from their surroundings (Fig. 9.1). That is, the particles may exert a force on each other, but no external forces are present. It is important to note the impact of Newton's third

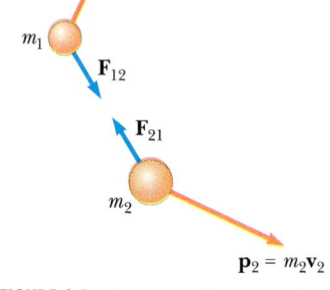

FIGURE 9.1   At some instant, the momentum of m_1 is $\mathbf{p}_1 = m_1\mathbf{v}_1$ and the momentum of m_2 is $\mathbf{p}_2 = m_2\mathbf{v}_2$. Note that $\mathbf{F}_{12} = -\mathbf{F}_{21}$.

[1] The formula $\mathbf{F} = d\mathbf{p}/dt$ is valid in relativity provided we use the relationship $\mathbf{p} = m\mathbf{v}/(1 - v^2/c^2)^{1/2}$ for the momentum. We shall return to the relativistic treatment of motion in Chapter 39.

[2] In this chapter, the terms "momentum" and "linear momentum" mean the same thing. Later, in Chapter 11, we will use the term "angular momentum" when dealing with rotational motion.

law on this analysis. Recall from Chapter 5 that Newton's third law states that the forces on these two particles are equal in magnitude and opposite in direction. Thus, if an internal force (say a gravitational force) acts on particle 1, then there must be a second internal force, equal in magnitude but opposite in direction, which acts on particle 2.

Suppose that at some instant, the momentum of particle 1 is $\mathbf{p}_1$ and that of particle 2 is $\mathbf{p}_2$. Applying Newton's second law to each particle, we can write

$$\mathbf{F}_{12} = \frac{d\mathbf{p}_1}{dt} \quad \text{and} \quad \mathbf{F}_{21} = \frac{d\mathbf{p}_2}{dt}$$

where $\mathbf{F}_{12}$ is the force exerted on particle 1 by particle 2 and $\mathbf{F}_{21}$ is the force exerted on particle 2 by particle 1. (These forces could be gravitational forces, or have some other origin. This really isn't important for the present discussion.) Newton's third law tells us that $\mathbf{F}_{12}$ and $\mathbf{F}_{21}$ are equal in magnitude and opposite in direction. That is, they form an action-reaction pair, $\mathbf{F}_{12} = -\mathbf{F}_{21}$. We can express this condition as

$$\mathbf{F}_{12} + \mathbf{F}_{21} = 0$$

or as

$$\frac{d\mathbf{p}_1}{dt} + \frac{d\mathbf{p}_2}{dt} = \frac{d}{dt}(\mathbf{p}_1 + \mathbf{p}_2) = 0$$

The force from a nitrogen-propelled, hand-controlled device allows an astronaut to move about freely in space without restrictive tethers. *(Courtesy of NASA)*

Since the time derivative of the total momentum ($\mathbf{p}_{\text{tot}} = \mathbf{p}_1 + \mathbf{p}_2$) is *zero*, we conclude that the *total* momentum, $\mathbf{p}_{\text{tot}}$, of the system must remain constant:

$$\mathbf{p}_{\text{tot}} = \mathbf{p}_1 + \mathbf{p}_2 = \text{constant} \tag{9.4}$$

or, equivalently,

$$\mathbf{p}_{1i} + \mathbf{p}_{2i} = \mathbf{p}_{1f} + \mathbf{p}_{2f} \tag{9.5}$$

where $\mathbf{p}_{1i}$ and $\mathbf{p}_{2i}$ are initial values and $\mathbf{p}_{1f}$ and $\mathbf{p}_{2f}$ are final values of the momentum during a time period, *dt*, over which the reaction pair interacts. Equation 9.5 in component form says that the total momenta in the *x, y,* and *z* directions are all independently conserved; that is,

$$\mathbf{p}_{ix} = \mathbf{p}_{fx} \quad \mathbf{p}_{iy} = \mathbf{p}_{fy} \quad \mathbf{p}_{iz} = \mathbf{p}_{fz} \tag{9.6}$$

This result is known as the **law of conservation of linear momentum.** It is considered as one of the most important laws of mechanics. We can state it as follows:

> Whenever two isolated, uncharged particles interact with each other, their total momentum remains constant.

Conservation of momentum

That is, *the total momentum of an isolated system at all times equals its initial momentum.*

We can also describe the law of conservation of momentum in another way. Since we require that the system be isolated, no external forces are present, and the total momentum of the system remains constant.

Notice that we have made no statement concerning the nature of the forces acting on the system. The only requirement was that the forces must be *internal* to the system. Thus, momentum is constant for an isolated two-particle system *regardless* of the nature of the internal forces. One can use a similar and equivalent argument to show that the law of conservation of linear momentum also applies to an isolated system of many particles.

EXAMPLE 9.1   The Recoiling Pitching Machine

A baseball player uses a pitching machine to help him improve his batting average. He places the 50-kg machine on a frozen pond as in Figure 9.2. The machine fires a 0.15-kg baseball horizontally with a velocity of 36i m/s. What is the recoil velocity of the machine?

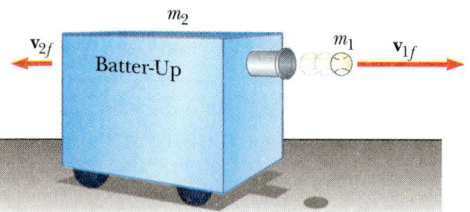

FIGURE 9.2   (Example 9.1) When the baseball is fired horizontally to the right, the pitching machine recoils to the left. The total momentum of the system before and after firing is zero.

Reasoning   We take the system to consist of the baseball and the pitching machine. Because of the force of gravity and the normal force, the system is not really isolated. However, both of these forces are directed perpendicularly to the mo-

tion of the system. Therefore, momentum is constant in the x direction because there are no external forces in this direction (assuming the surface is frictionless).

Solution   The total momentum of the system before firing is zero ($m_1 v_{1i} + m_2 v_{2i} = 0$). Therefore, the total momentum after firing must be zero; that is,

$$m_1 v_{1f} + m_2 v_{2f} = 0$$

With $m_1 = 0.15$ kg, $v_{1f} = 36i$ m/s, and $m_2 = 50$ kg, solving for v_{2f}, we find the recoil velocity of the pitching machine to be

$$v_{2f} = -\frac{m_1}{m_2} v_{1f} = -\left(\frac{0.15 \text{ kg}}{50 \text{ kg}}\right)(36i \text{ m/s}) = \boxed{-0.11i \text{ m/s}}$$

The negative sign for v_{2f} indicates that the pitching machine is moving to the left after firing, in the direction opposite the direction of motion of the cannon. In the words of Newton's third law, for every force (to the left) on the pitching machine, there is an equal but opposite force (to the right) on the ball. Because the pitching machine is much more massive than the ball, the acceleration and consequent speed of the pitching machine are much smaller than the acceleration and speed of the ball.

EXAMPLE 9.2   Decay of the Kaon at Rest

A meson is a nuclear particle that is more massive than an electron but less massive than a proton or neutron. One type of meson, called the neutral kaon (K^0), decays into a pair of charged pions (π^+ and π^-) that are oppositely charged but equal in mass, as in Figure 9.3. A pion is a particle associated with the strong nuclear force that binds the protons and neutrons together in the nucleus. Assuming the kaon is initially at rest, prove that after the decay, the two pions must have momenta that are equal in magnitude and opposite in direction.

Solution   The decay of the kaon, represented in Figure 9.3, can be written

$$K^0 \longrightarrow \pi^+ + \pi^-$$

If we let p^+ be the momentum of the positive pion and p^- be the momentum of the negative pion after the decay, then the final momentum of the system can be written

$$p_f = p^+ + p^-$$

Because the pion is at rest before the decay, we know that $p_i = 0$. Furthermore, because momentum is conserved, $p_i =$

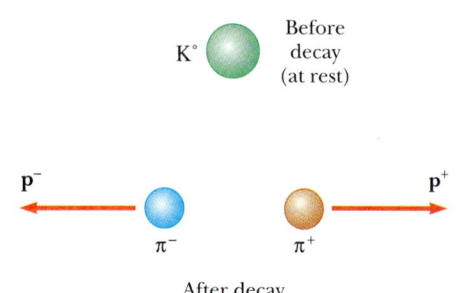

FIGURE 9.3   (Example 9.2) A kaon at rest decays spontaneously into a pair of oppositely charged pions. The pions move apart with momenta that are equal in magnitude but opposite in direction.

$p_f = 0$, so that $p^+ + p^- = 0$, or

$$p^+ = -p^-$$

Thus, we see that the two momentum vectors of the pions are equal in magnitude and opposite in direction.

9.2 IMPULSE AND MOMENTUM

As we have seen, the momentum of a particle changes if a net force acts on the particle. Let us assume that a single force $\mathbf{F}$ acts on a particle and that this force may vary with time. According to Newton's second law, $\mathbf{F} = d\mathbf{p}/dt$, or

$$d\mathbf{p} = \mathbf{F}\, dt \tag{9.7}$$

We can integrate this expression to find the change in the momentum of a particle. If the momentum of the particle changes from $\mathbf{p}_i$ at time t_i to $\mathbf{p}_f$ at time t_f, then integrating Equation 9.7 gives

$$\Delta\mathbf{p} = \mathbf{p}_f - \mathbf{p}_i = \int_{t_i}^{t_f} \mathbf{F}\, dt \tag{9.8}$$

The quantity on the right side of Equation 9.8 is called the *impulse* of the force $\mathbf{F}$ for the time interval $\Delta t = t_f - t_i$. Impulse is a vector defined by

$$\mathbf{I} \equiv \int_{t_i}^{t_f} \mathbf{F}\, dt = \Delta\mathbf{p} \tag{9.9}$$

Impulse of a force

That is, the impulse of the force $\mathbf{F}$ equals the change in the momentum of the particle.

Impulse-momentum theorem

This statement, known as the **impulse-momentum theorem**, is equivalent to Newton's second law. From this definition, we see that impulse is a vector quantity having a magnitude equal to the area under the force-time curve, as described in Figure 9.4a. In this figure, it is assumed that the force varies in time in the general manner shown and is nonzero in the time interval $\Delta t = t_f - t_i$. The direction of the impulse vector is the same as the direction of the change in momentum. Impulse has the dimensions of momentum, that is, M L /T. Note that impulse is *not* a property of the particle itself; rather, it is a measure of the degree to which an external force changes the momentum of the particle. Therefore, when we say that an impulse is given to a particle, it is implied that momentum is transferred from an external agent to that particle.

Since the force can generally vary in time as in Figure 9.4a, it is convenient to define a time-averaged force $\overline{\mathbf{F}}$ as

$$\overline{\mathbf{F}} \equiv \frac{1}{\Delta t} \int_{t_i}^{t_f} \mathbf{F}\, dt \tag{9.10}$$

where $\Delta t = t_f - t_i$. Therefore, we can express Equation 9.9 as

$$\mathbf{I} = \Delta\mathbf{p} = \overline{\mathbf{F}}\, \Delta t \tag{9.11}$$

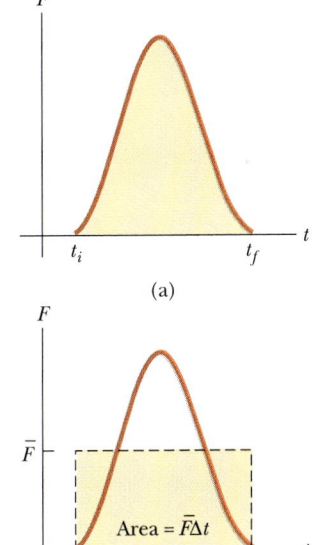

FIGURE 9.4 (a) A force acting on a particle may vary in time. The impulse is the area under the force versus time curve. (b) The average force (horizontal dashed line) gives the same impulse to the particle in the time Δt as the time-varying force described in (a).

This average force, described in Figure 9.4b, can be thought of as the constant force that would give the same impulse to the particle in the time interval Δt as the actual time-varying force gives over this same interval.

In principle, if $\mathbf{F}$ is known as a function of time, the impulse can be calculated from Equation 9.9. The calculation becomes especially simple if the force acting on the particle is constant. In this case, $\overline{\mathbf{F}} = \mathbf{F}$ and Equation 9.11 becomes

$$\mathbf{I} = \Delta\mathbf{p} = \mathbf{F}\, \Delta t \tag{9.12}$$

During the brief time the club is in contact with the ball, the ball gains momentum as a result of the collision, and the club loses the same amount of momentum. *(Courtesy Michel Hans/Photo Researchers)*

In many physical situations, we shall use the so-called **impulse approximation:** *Assume that one of the forces exerted on a particle acts for a short time but is much larger than any other force present.* This approximation is especially useful in treating collisions, where the duration of the collision is very short. When this approximation is made, we refer to the force as an *impulsive force.* For example, when a baseball is struck with a bat, the time of the collision is about 0.01 s, and the average force the bat exerts on the ball in this time is typically several thousand newtons. This is much greater than the force of gravity, so the impulse approximation is justified. When we use this approximation, it is important to remember that $\mathbf{p}_i$ and $\mathbf{p}_f$ represent the momenta *immediately* before and after the collision, respectively. Therefore, in the impulse approximation very little motion of the particle takes place during the collision.

EXAMPLE 9.3   Teeing Off

A golf ball of mass 50 g is struck with a club (Fig. 9.5). The force on the ball varies from zero when contact is made up to some maximum value (where the ball is deformed) back to zero when the ball leaves the club. Thus, the force-time curve is qualitatively described by Figure 9.4. Assuming that the ball travels 200 m, estimate the magnitude of the impulse due to the collision.

Solution   Neglecting air resistance, we can use Equation 4.18 for the range of a projectile:

$$R = \frac{v_0{}^2}{g} \sin 2\theta_0$$

Let us assume that the launch angle is 45°, the angle that provides the maximum range for any given launch speed. The initial speed of the ball is then estimated to be

$$v_0 = \sqrt{Rg} = \sqrt{(200 \text{ m})(9.80 \text{ m/s}^2)} = 44 \text{ m/s}$$

FIGURE 9.5   A golf ball being struck by a club.   *(© Harold E. Edgerton. Courtesy of Palm Press, Inc.)*

Since $v_i = 0$ and $v_f = v_0$ for the ball, the magnitude of the impulse imparted to the ball is

$$I = \Delta p = mv_0 = (50 \times 10^{-3} \text{ kg})\left(44 \frac{\text{m}}{\text{s}}\right) = \boxed{2.2 \text{ kg} \cdot \text{m/s}}$$

Exercise If the club is in contact with the ball 4.5×10^{-4} s, estimate the magnitude of the average force exerted by the club on the ball.

Answer 4.9×10^3 N. This force is extremely large compared with the weight (gravity force) of the ball, which is only 0.49 N.

EXAMPLE 9.4   How Good Are the Bumpers?

In a particular crash test, an automobile of mass 1500 kg collides with a wall as in Figure 9.6a. The initial and final velocities of the automobile are $v_i = -15.0i$ m/s and $v_f = 2.6i$ m/s. If the collision lasts for 0.150 s, find the impulse due to the collision and the average force exerted on the automobile.

Solution The initial and final momenta of the automobile are

$$\mathbf{p}_i = m\mathbf{v}_i = (1500 \text{ kg})(-15.0i \text{ m/s})$$
$$= -2.25 \times 10^4 i \text{ kg} \cdot \text{m/s}$$

$$\mathbf{p}_f = m\mathbf{v}_f = (1500 \text{ kg})(2.6i \text{ m/s})$$
$$= 0.39 \times 10^4 i \text{ kg} \cdot \text{m/s}$$

Hence, the impulse is

$$I = \Delta \mathbf{p} = \mathbf{p}_f - \mathbf{p}_i = 0.39 \times 10^4 i \text{ kg} \cdot \text{m/s}$$
$$- (-2.25 \times 10^4 i \text{ kg} \cdot \text{m/s})$$

$$I = \boxed{2.64 \times 10^4 i \text{ kg} \cdot \text{m/s}}$$

The average force exerted on the automobile is

$$\overline{\mathbf{F}} = \frac{\Delta \mathbf{p}}{\Delta t} = \frac{2.64 \times 10^4 i \text{ kg} \cdot \text{m/s}}{0.150 \text{ s}} = \boxed{1.76 \times 10^5 i \text{ N}}$$

Before

−15.0 m/s

After

2.6 m/s

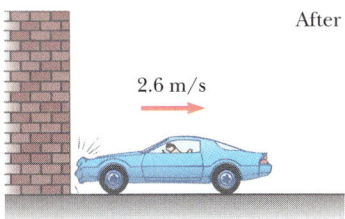

FIGURE 9.6 (Example 9.4).

CONCEPTUAL EXAMPLE 9.5

A boxer wisely moves his head backward just before receiving a punch. How does this maneuver help reduce the force of impact?

Reasoning As the boxer moves away from the moving fist, the time his head is in contact with the fist is increased. The impulse-momentum theorem tells us that the average force exerted on an object multiplied by the time during which the force acts equals the change in momentum of the object. The average force exerted on the boxer's head is reduced when he extends the time of contact by moving away from his opponent's fist. For the same reason, it is better to stop a truck out of control by hitting a haystack (long collision time and small force), rather than a brick wall (short collision time and large force).

(Conceptual Example 9.5) Boxer moving away from a punch. *(Focus on Sports)*

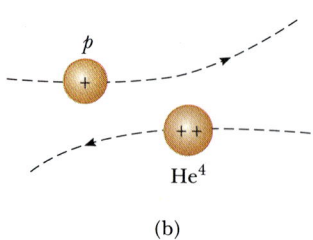

(a)

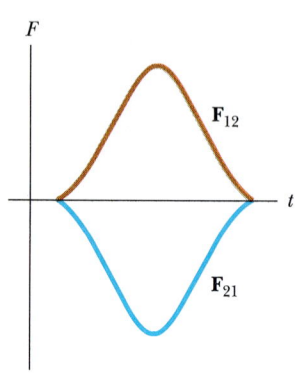

(b)

FIGURE 9.7 (a) The collision be-tween two objects as the result of direct physical contact. (b) The "collision" between two charged particles which results in a change in direction for each particle. In this case the action-reaction forces are elec-trostatic, and the two particles do not experience physical contact.

CONCEPTUAL EXAMPLE 9.6

A karate expert is able to break a stack of boards with a swift blow with the side of his bare hand. How is this possible?

Reasoning   The arm and hand have a large momentum before the collision with the boards. This momentum is reduced quickly as the hand is in contact with the boards for a short time. Since the contact time is very small, the impact force (the force of the hand on the boards) is very large. Thus, a karate expert obtains the best result by delivering the blow in a short time.

9.3   COLLISIONS

In this section we use the law of conservation of linear momentum to describe what happens when two particles collide. We use the term **collision** to represent the event of two particles coming together for a short time and thereby producing impulsive forces on each other. *The force due to the collision is assumed to be much larger than any external forces present.*

A collision may entail physical contact between two macroscopic objects, as described in Figure 9.7a, but the notion of what we mean by collision must be generalized because "physical contact" on a submicroscopic scale is ill-defined and hence meaningless. To understand this, consider a collision on an atomic scale (Fig. 9.7b), such as the collision of a proton with an alpha particle (the nucleus of the helium atom). Because the two particles are positively charged, they never come into physical contact with each other; instead, they repel each other because of the strong electrostatic force between them at close separations. When two particles of masses m_1 and m_2 collide as in Figure 9.7, the impulse forces may vary in time in a complicated way, one of which is described in Figure 9.8. If $\mathbf{F}_{12}$ is the force exerted on m_1 by m_2, and if we assume that no external forces act on the particles, then the change in momentum of m_1 due to the collision is given by Equation 9.8:

$$\Delta \mathbf{p}_1 = \int_{t_i}^{t_f} \mathbf{F}_{12}\, dt$$

Likewise, if $\mathbf{F}_{21}$ is the force exerted on m_2 by m_1, the change in momentum of m_2 is

$$\Delta \mathbf{p}_2 = \int_{t_i}^{t_f} \mathbf{F}_{21}\, dt$$

Newton's third law states that the force exerted on m_1 by m_2 is equal to and oppo-site the force exerted on m_2 by m_1. Hence, we conclude that

$$\Delta \mathbf{p}_1 = - \Delta \mathbf{p}_2$$

$$\Delta \mathbf{p}_1 + \Delta \mathbf{p}_2 = 0$$

Since the total momentum of the system is $\mathbf{p}_{tot} = \mathbf{p}_1 + \mathbf{p}_2$, we conclude that the *change* in the momentum of the system due to the collision is zero:

$$\mathbf{p}_{tot} = \mathbf{p}_1 + \mathbf{p}_2 = \text{constant}$$

This is precisely what we expect as an overall result because there are no external forces acting on the system (Section 9.2). Since the impulsive forces due to the

FIGURE 9.8 The impulse forces as a function of time for the two colliding particles described in Figure 9.7a. Note that $\mathbf{F}_{12} = -\mathbf{F}_{21}$.

collision are internal, they do not change the total momentum of the system (only external forces can do that). Therefore, we conclude that *the total momentum of a system just before the collision equals the total momentum of a system just after the collision.*

Momentum is conserved for any collision

EXAMPLE 9.7 Carry Collision Insurance

An 1800-kg car stopped at a traffic light is struck from the rear by a 900-kg car and the two become entangled. If the smaller car was moving at 20 m/s before the collision, what is the speed of the entangled mass after the collision?

Reasoning The total momentum of the system (the two cars) before the collision equals the total momentum of the system after the collision because momentum is conserved for any type of collision.

Solution The magnitude of the total momentum of the system before the collision is equal to that of the smaller car because the larger car is initially at rest:

$$p_i = m_i v_i = (900 \text{ kg})(20 \text{ m/s}) = 1.80 \times 10^4 \text{ kg} \cdot \text{m/s}$$

After the collision, the mass that moves is the sum of the masses of the cars. The magnitude of the momentum of the combination is

$$p_f = (m_1 + m_2)v_f = (2700 \text{ kg})v_f$$

Equating the momentum before to the momentum after and solving for v_f, the speed of the combined mass, we have

$$v_f = \frac{p_i}{m_1 + m_2} = \frac{1.80 \times 10^4 \text{ kg} \cdot \text{m/s}}{2700 \text{ kg}} = \boxed{6.67 \text{ m/s}}$$

CONCEPTUAL EXAMPLE 9.8

As a ball falls toward the Earth, its momentum increases. How would you reconcile this fact with the law of conservation of momentum?

Reasoning The momentum of the ball increases as it accelerates downward. A large change in its momentum occurs when it strikes the floor and reverses its direction of motion. The external forces exerted on the ball are the downward force of gravity and the upward normal force that acts while the ball is in contact with the floor to change its momentum. However, if we think of the ball and Earth together as our system, these forces are internal and do not change the total momentum of the system. That is, the total momentum of the system (ball and Earth) is constant. As the ball falls, the Earth moves up to meet it, on the order of 10^{25} times more slowly. Because the Earth's mass is so large, its upward motion is negligibly small.

CONCEPTUAL EXAMPLE 9.9

An open box slides across an icy (frictionless) surface of a frozen lake. What happens to the speed of the box as water from a rain shower collects in it, assuming that the rain falls vertically downward into the box? Explain.

Reasoning The rain initially has no horizontal component of momentum. The rain drops acquire a horizontal component of momentum as they collide with the box. Because the momentum of the system (box plus rain) must be conserved, the horizontal component of momentum of the box must decrease. Hence, the speed of the box decreases continuously as it collects water.

9.4 ELASTIC AND INELASTIC COLLISIONS IN ONE DIMENSION

As we have seen, momentum is conserved in any type of collision. Kinetic energy, however, is generally *not* constant in a collision because some of it is converted to thermal energy, to internal elastic potential energy when the objects are deformed, and to rotational energy.

Inelastic collision

We define various types of collisions on the basis of whether or not kinetic energy is constant. An **inelastic collision** is one in which *total kinetic energy is not constant (even though momentum is constant)*. The collision of a rubber ball with a hard surface is inelastic since some of the kinetic energy of the ball is lost when it is deformed while in contact with the surface. When two objects collide and stick together after the collision, some kinetic energy is lost, and the collision is called **perfectly inelastic.** For example, if two vehicles collide and become entangled, as in Example 9.7, they move with some common velocity after the perfectly inelastic collision. If a meteorite collides with the Earth, it becomes buried, and the collision is perfectly inelastic.

An **elastic collision** is one in which *total kinetic energy is constant (as well as momentum)*. Billiard-ball collisions and the collisions of air molecules with the walls of a container at ordinary temperatures are highly elastic. Real collisions in the macroscopic world are only approximately elastic because some deformation and loss of kinetic energy take place. Collisions between atomic and subatomic particles may also be inelastic, but are often elastic on the average. Elastic and perfectly inelastic collisions are limiting cases; most collisions fall in a category between them.

Elastic collision

Perfectly Inelastic Collisions

Consider two particles of masses m_1 and m_2 moving with initial velocities $\mathbf{v}_{1i}$ and $\mathbf{v}_{2i}$ along a straight line, as in Figure 9.9. If the two particles collide head-on, stick together, and move with some common velocity $\mathbf{v}_f$ after the collision, the collision is perfectly inelastic. Therefore, we can say that the total momentum before the collision equals the total momentum of the composite system after the collision:

$$m_1\mathbf{v}_{1i} + m_2\mathbf{v}_{2i} = (m_1 + m_2)\mathbf{v}_f \tag{9.13}$$

$$\mathbf{v}_f = \frac{m_1\mathbf{v}_{1i} + m_2\mathbf{v}_{2i}}{m_1 + m_2} \tag{9.14}$$

Elastic Collisions

Now consider two particles that undergo an elastic head-on collision (Fig. 9.10). In this case, both momentum and kinetic energy are constant; therefore, we can write

$$m_1\mathbf{v}_{1i} + m_2\mathbf{v}_{2i} = m_1\mathbf{v}_{1f} + m_2\mathbf{v}_{2f} \tag{9.15}$$

$$\tfrac{1}{2}m_1 v_{1i}^2 + \tfrac{1}{2}m_2 v_{2i}^2 = \tfrac{1}{2}m_1 v_{1f}^2 + \tfrac{1}{2}m_2 v_{2f}^2 \tag{9.16}$$

Because all velocities in Figure 9.10 are to the left or right, they can be represented by the corresponding speeds, where v is taken to be positive if a particle moves to the right and negative if it moves to the left.

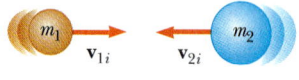

Before collision

(a)

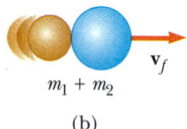

After collision

$m_1 + m_2$

(b)

FIGURE 9.9   Schematic representation of a perfectly inelastic head-on collision between two particles.

Before collision   After collision

m_1 (a) m_2 (b)

FIGURE 9.10   Schematic representation of an elastic head-on collision between two particles.

In a typical problem involving elastic collisions, there are two unknown quantities, and Equations 9.15 and 9.16 can be solved simultaneously to find these. However, an alternative approach, one that involves a little mathematical manipulation of Equation 9.16, often simplifies this process. To see this, let's cancel the factor of $\frac{1}{2}$ in Equation 9.16 and rewrite it as

$$m_1(v_{1i}^2 - v_{1f}^2) = m_2(v_{2f}^2 - v_{2i}^2)$$

Here we have moved the terms containing m_1 to one side of the equation and those containing m_2 to the other. Next, let us factor both sides:

$$m_1(v_{1i} - v_{1f})(v_{1i} + v_{1f}) = m_2(v_{2f} - v_{2i})(v_{2f} + v_{2i}) \tag{9.17}$$

We now separate the terms containing m_1 and m_2 in the equation for the conservation of momentum (Eq. 9.15) to get

$$m_1(v_{1i} - v_{1f}) = m_2(v_{2f} - v_{2i}) \tag{9.18}$$

To obtain our final result, we divide Equation 9.17 by Equation 9.18 and get

$$v_{1i} + v_{1f} = v_{2f} + v_{2i}$$

$$v_{1i} - v_{2i} = -(v_{1f} - v_{2f}) \tag{9.19}$$

This equation, in combination with Equation 9.15, can be used to solve problems dealing with perfectly elastic collisions. According to Equation 9.19, the relative speed of the two objects before the collision, $v_{1i} - v_{2i}$, equals the negative of their relative speed after the collision, $-(v_{1f} - v_{2f})$.

Suppose that the masses and the initial velocities of both particles are known. Equations 9.15 and 9.16 can be solved for the final speeds in terms of the initial speeds, since there are two equations and two unknowns:

$$v_{1f} = \left(\frac{m_1 - m_2}{m_1 + m_2}\right) v_{1i} + \left(\frac{2m_2}{m_1 + m_2}\right) v_{2i} \tag{9.20}$$

$$v_{2f} = \left(\frac{2m_1}{m_1 + m_2}\right) v_{1i} + \left(\frac{m_2 - m_1}{m_1 + m_2}\right) v_{2i} \tag{9.21}$$

Elastic collision: relations between final and initial speeds

It is important to remember that the appropriate signs for v_{1i} and v_{2i} must be included in Equations 9.20 and 9.21. For example, if m_2 is moving to the left initially, as we will see in Example 9.11, then v_{2i} is negative.

Let us consider some special cases: If $m_1 = m_2$, then $v_{1f} = v_{2i}$ and $v_{2f} = v_{1i}$. That is, the particles exchange speeds if they have equal masses. This is nearly what one observes in billiard ball collisions.

If m_2 is initially at rest, $v_{2i} = 0$, and Equations 9.20 and 9.21 become

$$v_{1f} = \left(\frac{m_1 - m_2}{m_1 + m_2}\right) v_{1i} \tag{9.22}$$

$$v_{2f} = \left(\frac{2m_1}{m_1 + m_2}\right) v_{1i} \tag{9.23}$$

If m_1 is very large compared with m_2, we see from Equations 9.22 and 9.23 that $v_{1f} \approx v_{1i}$ and $v_{2f} \approx 2v_{1i}$. That is, when a very heavy particle collides head-on with a very light one initially at rest, the heavy particle continues its motion unaltered after the collision, while the light particle rebounds with a speed equal to about twice the initial speed of the heavy particle. An example of such a collision would be that of a moving heavy atom, such as uranium, with a light atom, such as hydrogen.

If m_2 is much larger than m_1 and m_2 is initially at rest, then $v_{1f} \approx -v_{1i}$ and $v_{2f} \approx v_{2i} = 0$. That is, when a very light particle collides head-on with a very heavy particle initially at rest, the light particle has its velocity reversed, while the heavy particle remains approximately at rest. For example, imagine what happens when a marble hits a stationary bowling ball.

EXAMPLE 9.10   The Ballistic Pendulum

The ballistic pendulum (Fig. 9.11) is a system used to measure the speed of a fast-moving projectile, such as a bullet. The bullet is fired into a large block of wood suspended from some light wires. The bullet is stopped by the block, and the entire system swings through a height h. Because the collision is perfectly inelastic and momentum is conserved, Equation 9.14 gives the speed of the system right after the collision when we assume the impulse approximation. The kinetic energy right after the collision is

$$(1) \qquad K = \tfrac{1}{2}(m_1 + m_2)v_f^2$$

With $v_{2i} = 0$, Equation 9.14 becomes

$$(2) \qquad v_f = \frac{m_1 v_{1i}}{m_1 + m_2}$$

Substituting this value of v_f into (1) gives

$$K = \frac{m_1^2 v_{1i}^2}{2(m_1 + m_2)}$$

where v_{1i} is the initial speed of the bullet. Note that this kinetic energy is less than the initial kinetic energy of the bullet. In all the energy changes that take place after the collision, however, energy is constant; the kinetic energy at the bottom is transformed to potential energy at the height h:

$$\frac{m_1^2 v_{1i}^2}{2(m_1 + m_2)} = (m_1 + m_2)gh$$

$$v_{1i} = \left(\frac{m_1 + m_2}{m_1} \right) \sqrt{2gh}$$

Hence, it is possible to obtain the initial speed of the bullet by measuring h and the two masses. Why would it be incorrect to equate the initial kinetic energy of the incoming bullet to the final gravitational energy of the bullet-block combination?

Exercise   In a ballistic pendulum experiment, suppose that $h = 5.00$ cm, $m_1 = 5.00$ g, and $m_2 = 1.00$ kg. Find (a) the

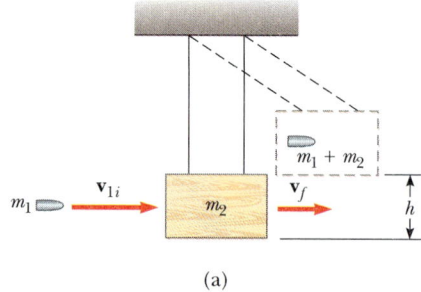

(a)

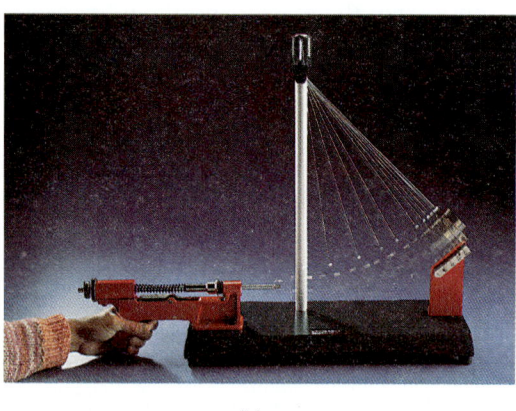

(b)

FIGURE 9.11   (Example 9.10) (a) Diagram of a ballistic pendulum. Note that $\mathbf{v}_f$ is the velocity of the system right after the perfectly inelastic collision. (b) Multiflash photo of a ballistic pendulum used in the laboratory.   *(Courtesy of Central Scientific Co.)*

initial speed of the projectile, and (b) the loss in energy due to the collision.

Answer   199 m/s; 98.5 J.

EXAMPLE 9.11 A Two-Body Collision with Spring

A block of mass $m_1 = 1.60$ kg initially moving to the right with a speed of 4.00 m/s on a frictionless horizontal track collides with a spring attached to a second block of mass $m_2 = 2.10$ kg moving to the left with a speed of 2.50 m/s, as in Figure 9.12a. The spring has a spring constant of 600 N/m. (a) At the instant when m_1 is moving to the right with a speed of 3.00 m/s, as in Figure 9.12b, determine the speed of m_2.

Solution   First, note that the initial velocity of m_2 is -2.50 m/s because its direction is to the left. Since the total momentum is conserved, we have

$$m_1 v_{1i} + m_2 v_{2i} = m_1 v_{1f} + m_2 v_{2f}$$

$$(1.60 \text{ kg})(4.00 \text{ m/s}) + (2.10 \text{ kg})(-2.50 \text{ m/s})$$
$$= (1.60 \text{ kg})(3.00 \text{ m/s}) + (2.10 \text{ kg}) v_{2f}$$

$$v_{2f} = \boxed{-1.74 \text{ m/s}}$$

The negative value for v_{2f} means that m_2 is still moving to the left at the instant we are considering.

(b) Determine the distance the spring is compressed at that instant.

Solution   To determine the compression in the spring, x, shown in Figure 9.12b, we use conservation of energy since there are no friction or other nonconservative forces acting on the system. Thus, we have

$$\tfrac{1}{2} m_1 v_{1i}^2 + \tfrac{1}{2} m_2 v_{2i}^2 = \tfrac{1}{2} m_1 v_{1f}^2 + \tfrac{1}{2} m_2 v_{2f}^2 + \tfrac{1}{2} kx^2$$

Substituting the given values and the result to part (a) into this expression gives

$$x = \boxed{0.173 \text{ m}}$$

Exercise   Find the velocity of m_1 and the compression in the spring at the instant that m_2 is at rest.

Answer   0.719 m/s to the right; 0.251 m.

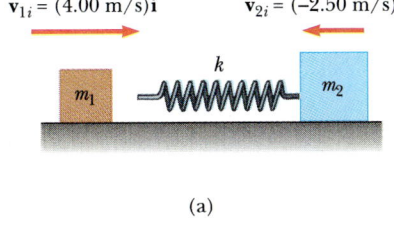
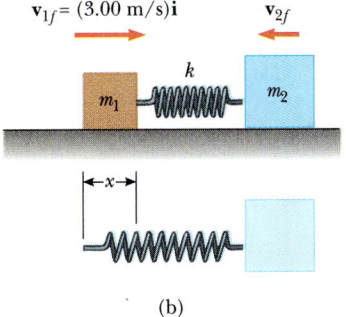

FIGURE 9.12   (Example 9.11).

EXAMPLE 9.12 Slowing Down Neutrons by Collisions

In a nuclear reactor, neutrons are produced when a $^{235}_{92}$U atom splits in a process called fission. These neutrons are moving at about 10^7 m/s and must be slowed down to about 10^3 m/s before they take part in another fission event. They are slowed down by being passed through a solid or liquid material called a *moderator*. The slowing-down process involves elastic collisions. Let us show that a neutron can lose most of its kinetic energy if it collides elastically with a moderator containing light nuclei, such as deuterium (in "heavy water," D_2O) or carbon (in graphite).

Reasoning   Because momentum and energy are constant, Equations 9.22 and 9.23 can be applied to the head-on collision of a neutron with the moderator nucleus.

Solution   Let us assume that the moderator nucleus of mass m_m is at rest initially and that the neutron of mass m_n and initial speed v_{ni} collides head-on with it. The initial kinetic energy of the neutron is

$$K_{ni} = \tfrac{1}{2} m_n v_{ni}^2$$

After the collision, the neutron has a kinetic energy $\tfrac{1}{2} m_n v_{nf}^2$, where v_{nf} is given by Equation 9.22:

$$K_{nf} = \tfrac{1}{2} m_n v_{nf}^2 = \frac{m_n}{2} \left(\frac{m_n - m_m}{m_n + m_m} \right)^2 v_{ni}^2$$

Therefore, the fraction of the total kinetic energy possessed by the neutron after the collision is

$$(1) \quad f_n = \frac{K_{nf}}{K_{ni}} = \left(\frac{m_n - m_m}{m_n + m_m}\right)^2$$

$$(2) \quad f_m = \frac{K_{mf}}{K_{ni}} = \frac{4 m_n m_m}{(m_n + m_m)^2}$$

From this result, we see that the final kinetic energy of the neutron is small when m_m is close to m_n and is zero when $m_n = m_m$.

We can calculate the kinetic energy of the moderator nucleus after the collision using Equation 9.23:

$$K_{mf} = \tfrac{1}{2} m_m v_{mf}^2 = \frac{2 m_n^2 m_m}{(m_n + m_m)^2} v_{ni}^2$$

Hence, the fraction of the total kinetic energy transferred to the moderator nucleus is

Since the total energy is constant, (2) can also be obtained from (1) with the condition that $f_n + f_m = 1$, so that $f_m = 1 - f_n$.

Suppose that heavy water is used for the moderator. For collisions of the neutrons with deuterium nuclei in D_2O ($m_m = 2 m_n$), $f_n = 1/9$ and $f_m = 8/9$. That is, 89% of the neutron's kinetic energy is transferred to the deuterium nucleus. In practice, the moderator efficiency is reduced because head-on collisions are very unlikely to occur.

How do the results differ with graphite as the moderator?

9.5   TWO-DIMENSIONAL COLLISIONS

In Sections 9.1 and 9.3, we showed that the total momentum of a system of two particles is constant when the system is isolated. For a general collision of two particles, this result implies that the total momentum in each of the directions x, y, and z is constant. However, an important subset of collisions takes place in a plane. The game of billiards is a familiar example involving multiple collisions of objects moving on a two-dimensional surface. For such two-dimensional collisions, we obtain two component equations for conservation of momentum:

$$m_1 v_{1ix} + m_2 v_{2ix} = m_1 v_{1fx} + m_2 v_{2fx}$$

$$m_1 v_{1iy} + m_2 v_{2iy} = m_1 v_{1fy} + m_2 v_{2fy}$$

Let us consider a two-dimensional problem in which a particle of mass m_1 collides with a particle of mass m_2, where m_2 is initially at rest, as in Figure 9.13. After the collision, m_1 moves at an angle θ with respect to the horizontal and m_2 moves at an angle ϕ with respect to the horizontal. This is called a *glancing* collision. Applying the law of conservation of momentum in component form, and noting that the total y component of momentum is zero, we get

$$m_1 v_{1i} = m_1 v_{1f} \cos \theta + m_2 v_{2f} \cos \phi \tag{9.24}$$

$$0 = m_1 v_{1f} \sin \theta - m_2 v_{2f} \sin \phi \tag{9.25}$$

We now have two independent equations. So long as only two of the preceding quantities are unknown, we can solve the problem completely.

Because the collision is elastic, we can use Equation 9.16 (conservation of

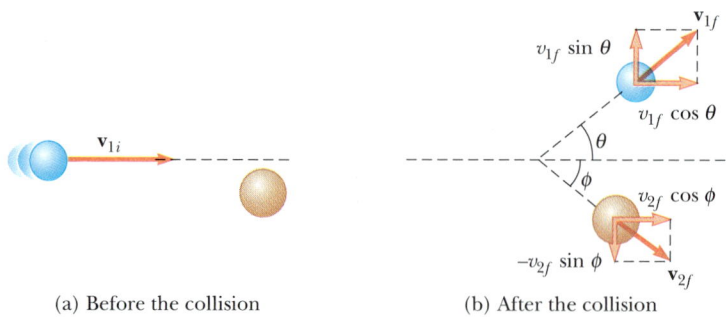

(a) Before the collision   (b) After the collision

FIGURE 9.13   An elastic glancing collision between two particles.

energy), with $v_{2i} = 0$, to give

$$\tfrac{1}{2}m_1 v_{1i}^2 = \tfrac{1}{2}m_1 v_{1f}^2 + \tfrac{1}{2}m_2 v_{2f}^2 \qquad (9.26)$$

Knowing the initial speed of the moving particle, and the masses, we are left with four unknowns. Since we only have three equations, one of the four remaining quantities (v_{1f}, v_{2f}, θ, or ϕ) must be given to determine the motion after the collision from conservation principles alone.

If the collision is inelastic, kinetic energy is *not* constant, and Equation 9.26 does *not* apply.

Problem-Solving Strategy
Collisions

The following procedure is recommended when dealing with problems involving collisions between two objects.

- Set up a coordinate system and define your velocities with respect to that system. It is convenient to have the *x* axis coincide with one of the initial velocities.
- In your sketch of the coordinate system, draw and label all velocity vectors and include all the given information.
- Write expressions for the *x* and *y* components of the momentum of each object before and after the collision. Remember to include the appropriate signs for the components of the velocity vectors.

This simulator enables you to model and investigate a wide variety of collisions such as collisions between moving objects and fixed boundaries as well as collisions between two moving objects in one and two dimensions (elastic and inelastic). For example, you will be able to make an animated simulation of a 1200-kg car traveling at 30 m/s colliding with a 1000-kg sports car traveling at 50 m/s in the opposite direction.

Collisions

- Write expressions for the total momentum in the *x* direction *before* and *after* the collision and equate the two. Repeat this procedure for the total momentum in the *y* direction. These steps follow from the fact that, because the momentum of the *system* is constant in any collision, the total momentum along any direction must be constant. Remember, it is the momentum of the *system* that is constant, not the momenta of the individual objects.
- If the collision is inelastic, kinetic energy is not constant, and additional information is probably required. If the collision is totally inelastic, the final velocities of the two objects are equal. Proceed to solve the momentum equations for the unknown quantities.
- If the collision is elastic, kinetic energy is constant, and you can equate the total kinetic energy before the collision to the total kinetic energy after the collision to get an additional relationship between the velocities.

EXAMPLE 9.13   Collision at an Intersection

A 1500-kg car traveling east with a speed of 25.0 m/s collides at an intersection with a 2500-kg van traveling north at a speed of 20.0 m/s, as shown in Figure 9.14. Find the direction and magnitude of the velocity of the wreckage after the collision, assuming that the vehicles undergo a perfectly inelastic collision (that is, they stick together).

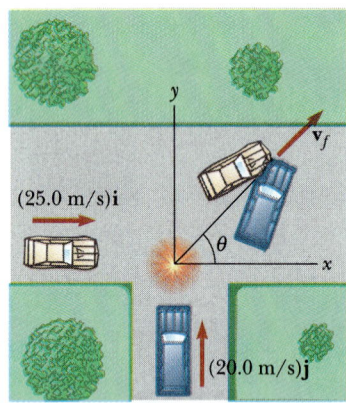

FIGURE 9.14   (Example 9.13) Top view of an eastbound car colliding with a northbound van.

Solution   Let us choose east to be along the positive *x* direction and north to be along the positive *y* direction, as in Figure 9.14. Before the collision, the only object having momentum in the *x* direction is the car. Thus, the magnitude of the total initial momentum of the system (car plus van) in the *x*

direction is

$$\sum p_{xi} = (1500 \text{ kg})(25.0 \text{ m/s}) = 3.75 \times 10^4 \text{ kg} \cdot \text{m/s}$$

Now let us assume that the wreckage moves at an angle θ and speed v after the collision. The magnitude of the total momentum in the *x* direction after the collision is

$$\sum p_{xf} = (4000 \text{ kg})\, v \cos \theta$$

Because the total momentum in the *x* direction is constant, we can equate these two equations to get

$$(1) \qquad (3.75 \times 10^4) \text{ kg} \cdot \text{m/s} = (4000 \text{ kg})\, v \cos \theta$$

Similarly, the total initial momentum of the system in the *y* direction is that of the van, whose magnitude is equal to (2500 kg)(20.0 m/s). Applying conservation of momentum to the *y* direction, we have

$$\sum p_{yi} = \sum p_{yf}$$

$$(2500 \text{ kg})(20.0 \text{ m/s}) = (4000 \text{ kg})\, v \sin \theta$$

$$(2) \qquad 5.00 \times 10^4 \text{ kg} \cdot \text{m/s} = (4000 \text{ kg})\, v \sin \theta$$

If we divide (2) by (1), we get

$$\tan \theta = \frac{5.00 \times 10^4}{3.75 \times 10^4} = 1.33$$

$$\theta = \boxed{53.1°}$$

When this angle is substituted into (2), the value of v is

$$v = \frac{5.00 \times 10^4 \text{ kg} \cdot \text{m/s}}{(4000 \text{ kg}) \sin 53°} = \boxed{15.6 \text{ m/s}}$$

EXAMPLE 9.14   Proton-Proton Collision

A proton collides in a perfectly elastic fashion with another proton initially at rest. The incoming proton has an initial speed of 3.50×10^5 m/s and makes a glancing "collision" with the second proton, as in Figure 9.13. After the collision, one proton moves at an angle of 37.0° to the original direction of motion, and the second deflects at an angle ϕ to the same axis. Find the final speeds of the two protons and the angle ϕ.

Solution   Since $m_1 = m_2$, $\theta = 37.0°$, and we are given $v_{1i} = 3.50 \times 10^5$ m/s, Equations 9.24, 9.25 and 9.26 become

$$v_{1f}\cos 37.0° + v_{2f}\cos \phi = 3.50 \times 10^5$$

$$v_{1f}\sin 37.0° - v_{2f}\sin \phi = 0$$

$$v_{1f}^2 + v_{2f}^2 = (3.50 \times 10^5)^2$$

Solving these three equations with three unknowns simulta-

neously gives

$$v_{1f} = \boxed{2.80 \times 10^5 \text{ m/s}} \qquad v_{2f} = \boxed{2.11 \times 10^5 \text{ m/s}}$$

$$\phi = \boxed{53.0°}$$

Note that $\theta + \phi = 90°$. This result is not accidental. *Whenever two equal masses collide elastically in a glancing collision and one of them is initially at rest, their final velocities are always at right angles to each other.* The next example illustrates this point in more detail.

EXAMPLE 9.15   Billiard Ball Collision

In a game of billiards, a player wishes to sink the target ball in the corner pocket, as shown in Figure 9.15. If the angle to the corner pocket is 35°, at what angle θ is the cue ball deflected? Assume that friction and rotational motion are unimportant, and assume the collision is elastic.

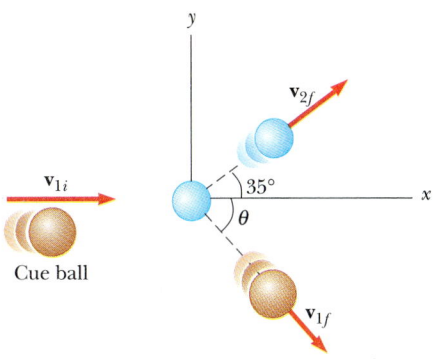

FIGURE 9.15   (Example 9.15).

Solution   Since the target is initially at rest, $v_{2i} = 0$, conservation of energy (Eq. 9.16) gives

$$\tfrac{1}{2}m_1 v_{1i}^2 = \tfrac{1}{2}m_1 v_{1f}^2 + \tfrac{1}{2}m_2 v_{2f}^2$$

But $m_1 = m_2$, so that

$$(1) \qquad v_{1i}^2 = v_{1f}^2 + v_{2f}^2$$

Applying conservation of momentum to the two-dimensional collision gives

$$(2) \qquad \mathbf{v}_{1i} = \mathbf{v}_{1f} + \mathbf{v}_{2f}$$

If we square both sides of (2), we get

$$v_{1i}^2 = (\mathbf{v}_{1f} + \mathbf{v}_{2f}) \cdot (\mathbf{v}_{1f} + \mathbf{v}_{2f}) = v_{1f}^2 + v_{2f}^2 + 2\mathbf{v}_{1f}\cdot\mathbf{v}_{2f}$$

But $\mathbf{v}_{1f}\cdot\mathbf{v}_{2f} = v_{1f}v_{2f}\cos(\theta + 35°)$, and so

$$(3) \qquad v_{1i}^2 = v_{1f}^2 + v_{2f}^2 + 2v_{1f}v_{2f}\cos(\theta + 35°)$$

Subtracting (1) from (3) gives

$$2v_{1f}v_{2f}\cos(\theta + 35°) = 0$$

$$\cos(\theta + 35°) = 0$$

$$\theta + 35° = 90° \qquad \text{or} \qquad \theta = \boxed{55°}$$

Again, this result shows that whenever two equal masses undergo a glancing elastic collision and one of them is initially at rest, they move at right angles to each other after the collision.

9.6   THE CENTER OF MASS

In this section we describe the overall motion of a mechanical system in terms of a very special point called the **center of mass** of the system. The mechanical system can be either a system of particles, such as a collection of atoms in a container, or an extended object, such as a gymnast leaping through the air. We shall see that the mechanical system moves as if all its mass were concentrated at the center of mass. Furthermore, if the resultant external force on the system is **F** and the total mass of the system is M, the center of mass moves with an acceleration given by $\mathbf{a} = \mathbf{F}/M$. That is, the system moves as if the resultant external force were applied to a single particle of mass M located at the center of mass. This behavior is independent of other motion, such as rotation or vibration of the system. This result was implicitly assumed in earlier chapters since nearly all examples referred to the motion of extended objects.

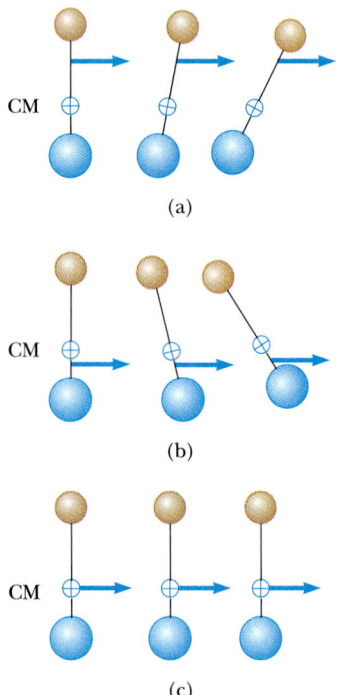

(a)

(b)

(c)

FIGURE 9.16 Two unequal masses are connected by a light, rigid rod. (a) The system rotates clockwise when a force is applied between the smaller mass and the center of mass. (b) The system rotates counterclockwise when the force is applied between the larger mass and the center of mass. (c) The system moves in the direction of the force without rotating when the force is applied at the center of mass.

Consider a mechanical system consisting of a pair of particles connected by a light, rigid rod (Fig. 9.16). The center of mass is located somewhere on the line joining the particles and is closer to the larger mass. If a single force is applied at some point on the rod somewhere between the center of mass and the smaller mass, the system rotates clockwise (Fig. 9.16a). If the force is applied at a point on the rod somewhere between the center of mass and the larger mass, the system rotates counterclockwise (Fig. 9.16b). If the force is applied at the center of mass, the system moves in the direction of **F** without rotating (Fig. 9.16c). Thus, the center of mass can be easily located.

One can describe the position of the center of mass of a system as being the weighted average position of the system's mass. For example, the center of mass of the pair of particles described in Figure 9.17 is located on the *x* axis and lies somewhere between the particles. Its *x* coordinate is

$$x_{CM} \equiv \frac{m_1 x_1 + m_2 x_2}{m_1 + m_2} \tag{9.27}$$

For example, if $x_1 = 0$, $x_2 = d$, and $m_2 = 2 m_1$, we find that $x_{CM} = \frac{2}{3}d$. That is, the center of mass lies closer to the more massive particle. If the two masses are equal, the center of mass lies midway between the particles.

We can extend the center of mass concept to a system of many particles in

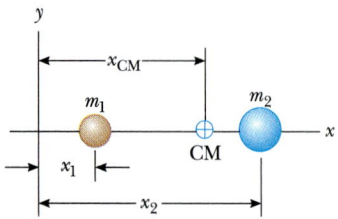

FIGURE 9.17 The center of mass of two particles on the *x* axis is located at x_{CM}, a point between the particles, closer to the larger mass.

This multiflash photograph shows that as the diver executes a back flip, his center of mass follows a parabolic path, the same path that a particle would follow. (*© The Harold E. Edgerton 1992 Trust. Courtesy of Palm Press, Inc.*)

three dimensions. The x coordinate of the center of mass of n particles is defined to be

$$x_{CM} \equiv \frac{m_1 x_1 + m_2 x_2 + m_3 x_3 + \cdots + m_n x_n}{m_1 + m_2 + m_3 + \cdots + m_n} = \frac{\Sigma m_i x_i}{\Sigma m_i} \qquad (9.28)$$

where x_i is the x coordinate of the ith particle and Σm_i is the *total mass* of the system. For convenience, we shall express the total mass as $M = \Sigma m_i$, where the sum runs over all n particles. The y and z coordinates of the center of mass are similarly defined by the equations

$$y_{CM} \equiv \frac{\Sigma m_i y_i}{M} \quad \text{and} \quad z_{CM} \equiv \frac{\Sigma m_i z_i}{M} \qquad (9.29)$$

The center of mass can also be located by its position vector, $\mathbf{r}_{CM}$. The rectangular coordinates of this vector are x_{CM}, y_{CM}, and z_{CM}, defined in Equations 9.28 and 9.29. Therefore,

$$\mathbf{r}_{CM} = x_{CM}\mathbf{i} + y_{CM}\mathbf{j} + z_{CM}\mathbf{k}$$
$$= \frac{\Sigma m_i x_i \mathbf{i} + \Sigma m_i y_i \mathbf{j} + \Sigma m_i z_i \mathbf{k}}{M}$$

$$\boxed{\mathbf{r}_{CM} \equiv \frac{\Sigma m_i \mathbf{r}_i}{M}} \qquad (9.30)$$

where $\mathbf{r}_i$ is the position vector of the ith particle, defined by

$$\mathbf{r}_i \equiv x_i \mathbf{i} + y_i \mathbf{j} + z_i \mathbf{k}$$

Although locating the center of mass for an extended object is somewhat more cumbersome, the basic ideas we have discussed still apply. We can think of an extended object as a system of a large number of particles (Fig. 9.18). The particle separation is very small, and so the object can be considered to have a continuous mass distribution. By dividing the object into elements of mass Δm_i, with coordinates x_i, y_i, z_i, we see that the x coordinate of the center of mass is approximately

$$x_{CM} \approx \frac{\Sigma x_i \Delta m_i}{M}$$

with similar expressions for y_{CM} and z_{CM}. If we let the number of elements, n, approach infinity, then x_{CM} is given precisely. In this limit, we replace the sum by an integral and replace Δm_i by the differential element dm, so that

$$x_{CM} = \lim_{\Delta m_i \to 0} \frac{\Sigma x_i \Delta m_i}{M} = \frac{1}{M} \int x \, dm \qquad (9.31)$$

Likewise, for y_{CM} and z_{CM} we get

$$y_{CM} = \frac{1}{M} \int y \, dm \quad \text{and} \quad z_{CM} = \frac{1}{M} \int z \, dm \qquad (9.32)$$

We can express the vector position of the center of mass of an extended object in the form

$$\boxed{\mathbf{r}_{CM} = \frac{1}{M} \int \mathbf{r} \, dm} \qquad (9.33)$$

which is equivalent to the three expressions given by Equations 9.31 and 9.32.

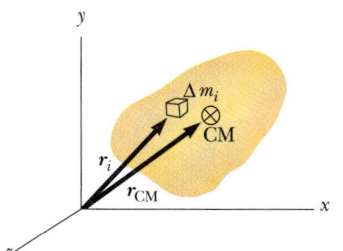

FIGURE 9.18 An extended object can be considered a distribution of many small elements of mass Δm_i. The center of mass is located at the vector position $\mathbf{r}_{CM}$, which has coordinates x_{CM}, y_{CM}, and z_{CM}.

Vector position of the center of mass for a system of particles

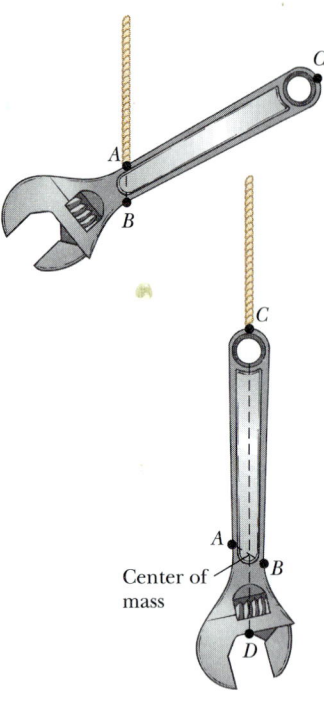

FIGURE 9.19   An experimental technique for determining the center of mass of a wrench. The wrench is hung freely from two different pivots, A and C. The intersection of the two vertical lines AB and CD locates the center of mass.

The center of mass of any symmetric object lies on an axis of symmetry and on any plane of symmetry.[3] For example, the center of mass of a rod lies in the rod, midway between its ends. The center of mass of a sphere or a cube lies at its geometric center.

One can determine the center of mass of an irregularly shaped object, such as a wrench, by suspending the wrench first from one point and then from another (Fig. 9.19). The wrench is first hung from point A, and a vertical line AB (which can be established with a plumb bob) is drawn when the wrench has stopped swinging. The wrench is then hung from point C, and a second vertical line, CD, is drawn. The center of mass coincides with the intersection of these two lines. In general, if the wrench is hung freely from any point, the vertical line through this point must pass through the center of mass.

Since an extended object is a continuous distribution of mass, each small mass element is acted upon by the force of gravity. The net effect of all of these forces is equivalent to the effect of a single force, $M\mathbf{g}$, acting through a special point, called the **center of gravity**. If $\mathbf{g}$ is constant over the mass distribution, then the center of gravity coincides with the center of mass. If an extended object is pivoted at its center of gravity, it balances in any orientation.

EXAMPLE 9.16   The Center of Mass of Three Particles

A system consists of three particles located at the corners of a right triangle as in Figure 9.20. Find the center of mass of the system.

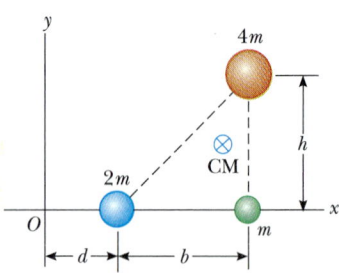

FIGURE 9.20   (Example 9.16) The center of mass of the three particles is located inside the triangle.

Solution   Using the basic defining equations for the coordinates of the center of mass, and noting that $z_{CM} = 0$, we get

$$x_{CM} = \frac{\Sigma m_i x_i}{M} = \frac{2md + m(d+b) + 4m(d+b)}{7m} = d + \frac{5}{7}b$$

$$y_{CM} = \frac{\Sigma m_i y_i}{M} = \frac{2m(0) + m(0) + 4mh}{7m} = \frac{4}{7}h$$

Therefore, we can express the position vector to the center of mass measured from the origin as

$$\mathbf{r}_{CM} = x_{CM}\mathbf{i} + y_{CM}\mathbf{j} = \boxed{(d + \tfrac{5}{7}b)\mathbf{i} + \tfrac{4}{7}h\mathbf{j}}$$

EXAMPLE 9.17   The Center of Mass of a Rod

(a) Show that the center of mass of a rod of mass M and length L lies midway between its ends, assuming the rod has a uniform mass per unit length (Fig. 9.21).

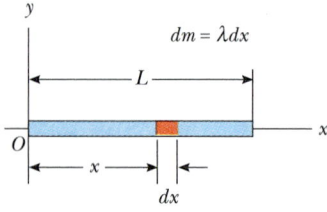

FIGURE 9.21   (Example 9.17) The center of mass of a uniform rod of length L is located at $x_{CM} = L/2$.

Solution   By symmetry, we see that $y_{CM} = z_{CM} = 0$ if the rod is placed along the x axis. Furthermore, if we call the mass per unit length λ (the linear mass density), then $\lambda = M/L$ for

---

[3] This statement is valid only for objects that have a uniform mass-per-unit volume.

a uniform rod. If we divide the rod into elements of length dx, then the mass of each element is $dm = \lambda\, dx$. Since an arbitrary element is at a distance x from the origin, Equation 9.31 gives

$$x_{CM} = \frac{1}{M}\int_0^L x\, dm = \frac{1}{M}\int_0^L x\lambda\, dx = \frac{\lambda}{M}\frac{x^2}{2}\Big]_0^L = \frac{\lambda L^2}{2M}$$

Because $\lambda = M/L$, this reduces to

$$x_{CM} = \frac{L^2}{2M}\left(\frac{M}{L}\right) = \boxed{\frac{L}{2}}$$

One can also argue that by symmetry, $x_{CM} = L/2$.

(b) Suppose a rod is *nonuniform* such that its mass per unit length varies linearly with x according to the expression $\lambda = \alpha x$, where α is a constant. Find the x coordinate of the center of mass as a fraction of L.

Solution   In this case, we replace dm by $\lambda\, dx$, where λ is not constant. Therefore, x_{CM} is

$$x_{CM} = \frac{1}{M}\int_0^L x\, dm = \frac{1}{M}\int_0^L x\lambda\, dx = \frac{\alpha}{M}\int_0^L x^2\, dx = \frac{\alpha L^3}{3M}$$

We can eliminate α by noting that the total mass of the rod is related to α through the relationship

$$M = \int dm = \int_0^L \lambda\, dx = \int_0^L \alpha x\, dx = \frac{\alpha L^2}{2}$$

Substituting this into the expression for x_{CM} gives

$$x_{CM} = \frac{\alpha L^3}{3\alpha L^2/2} = \boxed{\frac{2}{3}L}$$

EXAMPLE 9.18   The Center of Mass of a Right Triangle

An object of mass M is in the shape of a right triangle whose dimensions are shown in Figure 9.22. Locate the coordinates of the center of mass, assuming the object has a uniform mass per unit area.

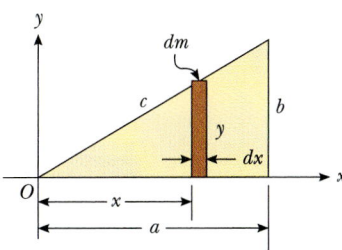

FIGURE 9.22   (Example 9.18).

Solution   To evaluate the x coordinate of the center of mass, we divide the triangle into narrow strips of width dx and height y as in Figure 9.22. The mass dm of each strip is

$$dm = \frac{\text{total mass}}{\text{total area}}\times \text{area of strip}$$

$$= \frac{M}{\frac{1}{2}ab}\,(y\, dx) = \left(\frac{2M}{ab}\right)y\, dx$$

Therefore, the x coordinate of the center of mass is

$$x_{CM} = \frac{1}{M}\int x\, dm = \frac{1}{M}\int_0^a x\left(\frac{2M}{ab}\right)y\, dx = \frac{2}{ab}\int_0^a xy\, dx$$

In order to evaluate this integral, we must express the variable y in terms of the variable x. From similar triangles in Figure 9.22, we see that

$$\frac{y}{x} = \frac{b}{a}\qquad\text{or}\qquad y = \frac{b}{a}x$$

With this substitution, x_{CM} becomes

$$x_{CM} = \frac{2}{ab}\int_0^a x\left(\frac{b}{a}x\right)dx = \frac{2}{a^2}\int_0^a x^2\, dx = \frac{2}{a^2}\left[\frac{x^3}{3}\right]_0^a = \boxed{\frac{2}{3}a}$$

By a similar calculation, we get for the y coordinate of the center of mass

$$y_{CM} = \boxed{\frac{1}{3}b}$$

9.7   MOTION OF A SYSTEM OF PARTICLES

We can begin to understand the physical significance and utility of the center of mass concept by taking the time derivative of the position vector given by Equation 9.30. Assuming that M remains constant for a system of particles, that is, no particles enter or leave the system, we get the following expression for the **velocity of the** center of mass of the system:

$$\mathbf{v}_{CM} = \frac{d\mathbf{r}_{CM}}{dt} = \frac{1}{M} \sum m_i \frac{d\mathbf{r}_i}{dt}$$

Velocity of the center of mass

$$\mathbf{v}_{CM} = \frac{\sum m_i \mathbf{v}_i}{M} \tag{9.34}$$

where $\mathbf{v}_i$ is the velocity of the ith particle. Rearranging Equation 9.34 gives

Total momentum of a system of particles

$$M\mathbf{v}_{CM} = \sum m_i \mathbf{v}_i = \sum \mathbf{p}_i = \mathbf{p}_{tot} \tag{9.35}$$

Therefore, we conclude that the **total linear momentum of the system** equals the total mass multiplied by the velocity of the center of mass. In other words, the total linear momentum of the system is equal to that of a single particle of mass M moving with a velocity $\mathbf{v}_{CM}$.

If we now differentiate Equation 9.35 with respect to time, we get the acceleration of the center of mass of the system:

Acceleration of the center of mass

$$\mathbf{a}_{CM} = \frac{d\mathbf{v}_{CM}}{dt} = \frac{1}{M} \sum m_i \frac{d\mathbf{v}_i}{dt} = \frac{1}{M} \sum m_i \mathbf{a}_i \tag{9.36}$$

Rearranging this expression and using Newton's second law, we get

$$M\mathbf{a}_{CM} = \sum m_i \mathbf{a}_i = \sum \mathbf{F}_i \tag{9.37}$$

where $\mathbf{F}_i$ is the force on particle i.

The forces on any particle in the system may include both external forces (from outside the system) and internal forces (from within the system). However, by Newton's third law, the force exerted by particle 1 on particle 2, for example, is equal to and opposite the force exerted by particle 2 on particle 1. Thus, when we sum over all internal forces in Equation 9.37, they cancel in pairs and the net force on the system is due *only* to external forces. Thus, we can write Equation 9.37 in the form

Newton's second law for a system of particles

$$\sum \mathbf{F}_{ext} = M\mathbf{a}_{CM} = \frac{d\mathbf{p}_{tot}}{dt} \tag{9.38}$$

That is, the resultant external force on a system of particles equals the total mass of the system multiplied by the acceleration of the center of mass. If we compare this to Newton's second law for a single particle, we see that

the center of mass moves like an imaginary particle of mass M under the influence of the resultant external force on the system.

In the absence of external forces, the center of mass moves with uniform velocity, as in the case of the rotating wrench shown in Figure 9.23.

Finally, we see that if the resultant external force is zero, then from Equation 9.38 it follows that

$$\frac{d\mathbf{p}_{tot}}{dt} = M\mathbf{a}_{CM} = 0$$

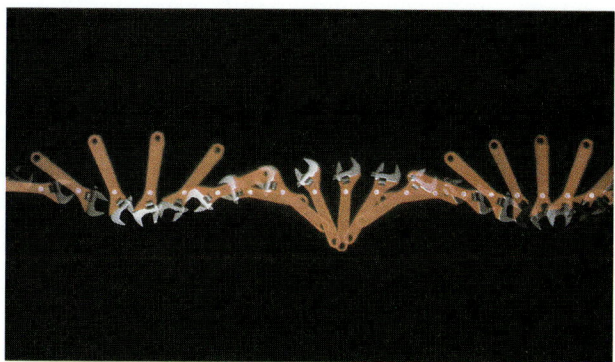

FIGURE 9.23   Multiflash photograph of a wrench moving on a horizontal surface. The center of mass of the wrench moves in a straight line as the wrench rotates about this point, shown by the white dot.   *(Education Development Center, Newton, Mass.)*

so that

$$\mathbf{p}_{tot} = M\mathbf{v}_{CM} = \text{constant} \qquad \left(\text{when } \sum \mathbf{F}_{ext} = 0 \right) \qquad (9.39)$$

That is, the total linear momentum of a system of particles is constant if there are no external forces acting on the system. Therefore, it follows that for an isolated system of particles, both the total momentum and the velocity of the center of mass are constant in time. This is a generalization to a many-particle system of the law of conservation of momentum discussed in Section 9.1 for a two-particle system.

Suppose an isolated system consisting of two or more members is at rest. The center of mass of such a system remains at rest unless acted upon by an external force. For example, consider a system made up of a swimmer and a raft, with the system initially at rest. When the swimmer dives off the raft, the center of mass of the system remains at rest (if we neglect friction between raft and water). Furthermore, the linear momentum of the diver is equal in magnitude to that of the raft but opposite in direction.

As another example, suppose an unstable atom initially at rest suddenly decays into two fragments of masses M_1 and M_2, with velocities $\mathbf{v}_1$ and $\mathbf{v}_2$, respectively. Since the total momentum of the system before the decay is zero, the total momentum of the system after the decay must also be zero. Therefore, we see that $M_1\mathbf{v}_1 + M_2\mathbf{v}_2 = 0$. If the velocity of one of the fragments after the decay is known, the recoil velocity of the other fragment can be calculated.

CONCEPTUAL EXAMPLE 9.19

A boy stands at one end of a floating raft that is stationary relative to the shore. He then walks to the opposite end of the raft, away from the shore. Does the raft move? Explain.

Reasoning   Yes, the raft moves toward the shore. Neglecting friction between the raft and water, there are no horizontal forces acting on the system consisting of the boy and raft. Therefore, the center of mass of the system remains fixed relative to the shore (or any stationary point). As the boy moves away from the shore, the boat must move toward the shore such that the center of mass of the system remains constant. An alternative explanation is that the momentum of the system remains constant if friction is neglected. As the boy acquires a momentum away from the shore, the boat must acquire an equal momentum toward the shore such that the total momentum of the system is always zero.

CONCEPTUAL EXAMPLE 9.20

Suppose you tranquilize a polar bear on a glacier as part of a research effort (Fig. 9.24). How might you be able to estimate the weight of the polar bear using a measuring tape, a rope, and a knowledge of your own weight?

Reasoning Tie one end of the rope around the bear. Lay out the tape measure on the ice between the bear's original position and yours, as you hold the opposite end of the rope.

Take off your spiked shoes and pull on the rope. Both you and the bear will slide over the ice until you meet. From the tape observe how far you have slid, x_y, and how far the bear has slid, x_b. The point where you meet the bear is the constant location of the center of mass of the system (bear plus you), so you can determine the mass of the bear from $m_b x_b = m_y x_y$. (Unfortunately, if the bear now wakes up you cannot get back to your spiked shoes.)

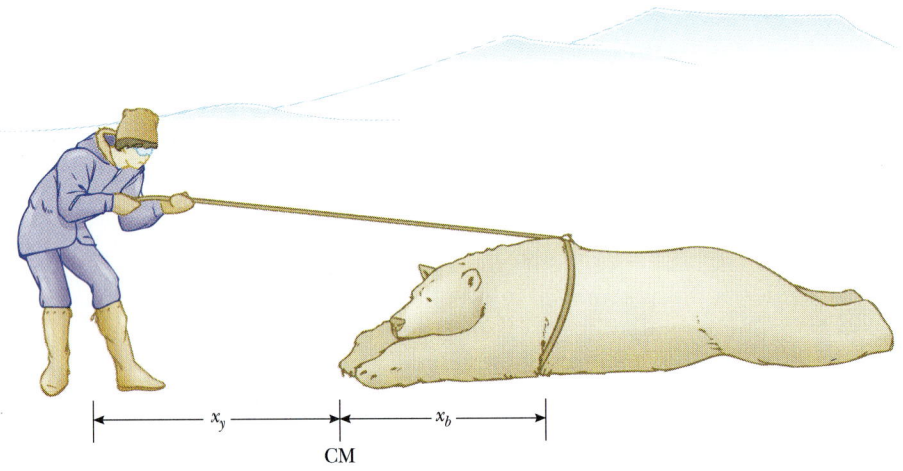

FIGURE 9.24 (Conceptual Example 9.20) The center of mass of an isolated system remains at rest unless acted on by an external force. How can you determine the weight of the polar bear?

CONCEPTUAL EXAMPLE 9.21 Exploding Projectile

A projectile is fired into the air and suddenly explodes into several fragments (Fig. 9.25). What can be said about the motion of the center of mass of the fragments after the explosion?

Reasoning Neglecting air resistance, the only external force on the projectile is the force of gravity. Thus, the projectile follows a parabolic path. If the projectile did not explode, it would continue to move along the parabolic path indicated by the broken line in Figure 9.25. Since the forces due to the explosion are internal, they do not affect the motion of the center of mass. Thus, after the explosion the center of mass of the fragments follows the same parabolic path the projectile would have followed if there had been no explosion.

FIGURE 9.25 (Conceptual Example 9.21) When a projectile explodes into several fragments, the center of mass of the fragments follows the same parabolic path the projectile would have taken had there been no explosion.

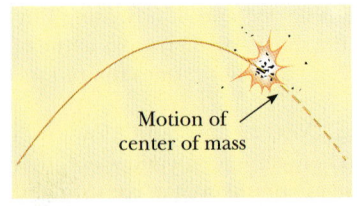

Motion of center of mass

EXAMPLE 9.22   The Exploding Rocket

A rocket is fired vertically upward. At the instant it reaches an altitude of 1000 m and a speed of 300 m/s, it explodes into three equal fragments. One fragment continues to move upward with a speed of 450 m/s following the explosion. The second fragment has a speed of 240 m/s and is moving east right after the explosion. What is the velocity of the third fragment right after the explosion?

Solution   Let us call the total mass of the rocket M; hence, the mass of each fragment is $M/3$. The total momentum just before the explosion must equal the total momentum of the fragments right after the explosion since the forces of the explosion are internal to the system and cannot affect its total momentum.

Before the explosion:

$$\mathbf{p}_{tot} = M\mathbf{v}_0 = 300M\mathbf{j}$$

After the explosion:

$$\mathbf{p}_{tot} = 240\left(\frac{M}{3}\right)\mathbf{i} + 450\left(\frac{M}{3}\right)\mathbf{j} + \frac{M}{3}\mathbf{v}$$

where $\mathbf{v}$ is the unknown velocity of the third fragment. Equating these two expressions gives

$$M\frac{\mathbf{v}}{3} + 80M\mathbf{i} + 150M\mathbf{j} = 300M\mathbf{j}$$

$$\mathbf{v} = \boxed{(-240\mathbf{i} + 450\mathbf{j})\,\text{m/s}}$$

Exercise   Find the position of the center of mass relative to the ground 3.00 s after the explosion. (Assume the rocket engine is nonoperative after the explosion.)

Answer   The x coordinate of the center of mass doesn't change. The y coordinate of the center of mass at this instant is 1.86 km.

*9.8   ROCKET PROPULSION

When ordinary vehicles, such as automobiles and locomotives, are propelled, the driving force for the motion is one of friction. In the case of the automobile, the driving force is the force exerted by the road on the car. A locomotive "pushes" against the tracks; hence, the driving force is the force exerted by the tracks on the locomotive. However, a rocket moving in space has no air or tracks to push against. Therefore, the source of the propulsion of a rocket must be different. Figure 9.26

FIGURE 9.26   Lift-off of the space shuttle Columbia. Enormous thrust is generated by the shuttle's liquid-fueled engines, aided by the two solid-fuel boosters. Many physical principles from mechanics, thermodynamics, and electricity and magnetism are involved in such a launch. *(Courtesy of NASA)*

is a dramatic photograph of a spacecraft at lift-off. *The operation of a rocket depends upon the law of conservation of linear momentum as applied to a system of particles, where the system is the rocket plus its ejected fuel.*

Rocket propulsion can be understood by first considering the mechanical system consisting of a machine gun mounted on a cart on wheels. As the gun is fired, each bullet receives a momentum $m\mathbf{v}$ in some direction, where $\mathbf{v}$ is measured with respect to a stationary Earth frame. For each bullet that is fired, the gun and cart must receive a compensating momentum in the opposite direction. That is, the reaction force exerted by the bullet on the gun accelerates the cart and gun. If there are n bullets fired each second, then the average force exerted on the gun is $\mathbf{F}_{av} = nm\mathbf{v}$.

In a similar manner, as a rocket moves in free space, *its linear momentum changes when some of its mass is released in the form of ejected gases. Since the ejected gases acquire some momentum, the rocket receives a compensating momentum in the opposite direction. Therefore, the rocket is accelerated as a result of the "push," or thrust, from the exhaust gases.* In free space, the center of mass of the system moves uniformly, independent of the propulsion process. It is interesting to note that the rocket and machine gun represent cases of the inverse of an inelastic collision; that is, momentum is conserved, but the kinetic energy of the system is increased (at the expense of internal energy).

Suppose that at some time t, the momentum of the rocket plus the fuel is $(M + \Delta m)v$ (Fig. 9.27a). At some short time later, Δt, the rocket ejects fuel of mass Δm and the rocket's speed therefore increases to $v + \Delta v$ (Fig. 9.27b). If the fuel is ejected with a speed v_e relative to the rocket, the speed of the fuel relative to a stationary frame of reference is $v - v_e$. Thus, if we equate the total initial momentum of the system to the total final momentum, we get

$$(M + \Delta m)v = M(v + \Delta v) + \Delta m(v - v_e)$$

Simplifying this expression gives

$$M\,\Delta v = \Delta m(v_e)$$

We also could have arrived at this result by considering the system in the center-of-mass frame of reference, that is, a frame whose velocity equals the center-of-mass velocity. In this frame, the total momentum is zero; therefore, if the rocket gains a momentum $M\,\Delta v$ by ejecting some fuel, the exhaust gases obtain a momentum $\Delta m(v_e)$ in the *opposite* direction, and so $M\,\Delta v - \Delta m(v_e) = 0$. If we now take the limit as Δt goes to zero, then $\Delta v \rightarrow dv$ and $\Delta m \rightarrow dm$. Furthermore, the increase in the exhaust mass, dm, corresponds to an equal decrease in the rocket mass, so that $dm = -dM$. Note that dM is given a negative sign because it represents a decrease in mass. Using this fact, we get

$$M\,dv = -v_e\,dM \tag{9.40}$$

Integrating this equation, and taking the initial mass of the rocket plus fuel to be M_i and the final mass of the rocket plus its remaining fuel to be M_f, we get

$$\int_{v_i}^{v_f} dv = -v_e \int_{M_i}^{M_f} \frac{dM}{M}$$

$$v_f - v_i = v_e \ln\left(\frac{M_i}{M_f}\right) \tag{9.41}$$

This is the basic expression of rocket propulsion. First, it tells us that the increase in speed is proportional to the exhaust speed, v_e. Therefore, the exhaust

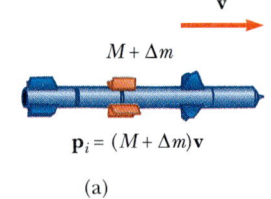

$$\mathbf{v}$$
$$M + \Delta m$$

$$\mathbf{p}_i = (M + \Delta m)\mathbf{v}$$

(a)

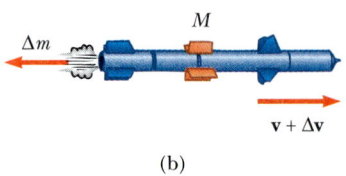

$$M$$
$$\Delta m$$
$$\mathbf{v} + \Delta\mathbf{v}$$

(b)

FIGURE 9.27 Rocket propulsion. (a) The initial mass of the rocket is $M + \Delta m$ at a time t, and its speed is v. (b) At a time $t + \Delta t$, the rocket's mass has reduced to M, and an amount of fuel Δm has been ejected. The rocket's speed increases by an amount Δv.

Expression for rocket propulsion

speed should be very high. Second, the increase in speed is proportional to the logarithm of the ratio M_i/M_f. Therefore, this ratio should be as large as possible, which means that the mass of the rocket without its fuel should be as small as possible and the rocket should carry as much fuel as possible.

The **thrust** on the rocket is the force exerted on the rocket by the ejected exhaust gases. We can obtain an expression for the thrust from Equation 9.40:

$$\text{Thrust} = M\frac{dv}{dt} = \left| v_e \frac{dM}{dt} \right| \qquad (9.42)$$

Here we see that the thrust increases as the exhaust speed increases and as the rate of change of mass (burn rate) increases.

EXAMPLE 9.23   A Rocket in Space

A rocket moving in free space has a speed of 3.0×10^3 m/s relative to Earth. Its engines are turned on, and fuel is ejected in a direction opposite the rocket's motion at a speed of 5.0×10^3 m/s relative to the rocket. (a) What is the speed of the rocket relative to Earth once its mass is reduced to one half its mass before ignition?

Solution   Applying Equation 9.41, we get

$$v_f = v_i + v_e \ln\left(\frac{M_i}{M_f}\right)$$

$$= 3.0 \times 10^3 + 5.0 \times 10^3 \ln\left(\frac{M_i}{0.5M_i}\right)$$

$$= 6.5 \times 10^3 \text{ m/s}$$

(b) What is the thrust on the rocket if it burns fuel at the rate of 50 kg/s?

Solution

$$\text{Thrust} = \left| v_e \frac{dM}{dt} \right| = \left(5.0 \times 10^3 \frac{\text{m}}{\text{s}} \right)\left(50 \frac{\text{kg}}{\text{s}} \right)$$

$$= 2.5 \times 10^5 \text{ N}$$

SUMMARY

The **linear momentum** of a particle of mass m moving with a velocity $\mathbf{v}$ is

$$\mathbf{p} \equiv m\mathbf{v} \qquad (9.1)$$

The law of **conservation of linear momentum** says that the total momentum of an isolated system is constant. If two particles form an isolated system, their total momentum is constant regardless of the nature of the force between them. Therefore, the total momentum of the system at all times equals its initial total momentum, or

$$\mathbf{p}_{1i} + \mathbf{p}_{2i} = \mathbf{p}_{1f} + \mathbf{p}_{2f} \qquad (9.5)$$

The **impulse** of a force $\mathbf{F}$ on a particle is equal to the change in the momentum of the particle:

$$\mathbf{I} \equiv \int_{t_i}^{t_f} \mathbf{F}\, dt = \Delta \mathbf{p} \qquad (9.9)$$

This is known as the **impulse–momentum theorem.**

Impulsive forces are often very strong compared with other forces on the system, and usually act for a very short time, as in the case of collisions.

When two particles collide, the total momentum of the system before the collision always equals the total momentum after the collision, regardless of the nature of the collision. An **inelastic collision** is one for which mechanical energy is not constant. A perfectly inelastic collision corresponds to the situation where the colliding bodies stick together after the collision. An **elastic collision** is one in which kinetic energy is constant.

In a two- or three-dimensional collision, the components of momentum in each of the three directions (*x*, *y*, and *z*) are conserved independently.

The **vector position of the center of mass of a system of particles** is defined as

$$\mathbf{r}_{CM} \equiv \frac{\sum m_i \mathbf{r}_i}{M} \tag{9.30}$$

where $M = \sum m_i$ is the total mass of the system and $\mathbf{r}_i$ is the vector position of the *i*th particle.

The **vector position of the center of mass of a rigid body** can be obtained from the integral expression

$$\mathbf{r}_{CM} = \frac{1}{M} \int \mathbf{r} \, dm \tag{9.33}$$

The **velocity of the center of mass for a system of particles** is

$$\mathbf{v}_{CM} = \frac{\sum m_i \mathbf{v}_i}{M} \tag{9.34}$$

The total momentum of a system of particles equals the total mass multiplied by the velocity of the center of mass, that is, $\mathbf{p}_{tot} = M\mathbf{v}_{CM}$.

Newton's second law applied to a system of particles is

$$\sum \mathbf{F}_{ext} = M\mathbf{a}_{CM} = \frac{d\mathbf{p}_{tot}}{dt} \tag{9.38}$$

where $\mathbf{a}_{CM}$ is the acceleration of the center of mass and the sum is over all external forces. Therefore, the center of mass moves like an imaginary particle of mass M under the influence of the resultant external force on the system. It follows from Equation 9.38 that the total momentum of the system is constant if there are no external forces acting on it.

QUESTIONS

1. If the kinetic energy of a particle is zero, what is its linear momentum? If the total energy of a particle is zero, is its linear momentum necessarily zero? Explain.

2. If the speed of a particle is doubled, by what factor is its momentum changed? By what factor is its kinetic energy changed?

3. If two particles have equal kinetic energies, are their momenta necessarily equal? Explain.

4. Does a large force always produce a larger impulse on a body than a small force? Explain.

5. An isolated system is initially at rest. Is it possible for parts of the system to be in motion at some later time? If so, explain how this might occur.

6. If two objects collide and one is initially at rest, is it possible for both to be at rest after the collision? Is it possible for one to be at rest after the collision? Explain.

7. Explain how linear momentum is conserved when a ball bounces from a floor.

8. Is it possible to have a collision in which all of the kinetic energy is lost? If so, cite an example.

9. In a perfectly elastic collision between two particles, does the kinetic energy of each particle change as a result of the collision?

10. When a ball rolls down an incline, its linear momentum increases. Does this imply that momentum is not conserved? Explain.

11. Consider a perfectly inelastic collision between a car and a large truck. Which vehicle loses more kinetic energy as a result of the collision?

12. Can the center of mass of a body lie outside the body? If so, give examples.

13. Three balls are thrown into the air simultaneously. What

is the acceleration of their center of mass while they are in motion?

14. A meter stick is balanced in a horizontal position with the index fingers of the right and left hands. If the two fingers are brought together, the stick remains balanced and the two fingers always meet at the 50-cm mark regardless of their original positions (try it!). Explain.

15. A sharpshooter fires a rifle while standing with the butt of the gun against his shoulder. If the forward momentum of a bullet is the same as the backward momentum of the gun, why isn't it as dangerous to be hit by the gun as by the bullet?

16. A piece of mud is thrown against a brick wall and sticks to the wall. What happens to the momentum of the mud? Is momentum conserved? Explain.

17. Early in this century, Robert Goddard proposed sending a rocket to the Moon. Critics took the position that in a vacuum, such as exists between Earth and Moon, the gases emitted by the rocket would have nothing to push against to propel the rocket. According to *Scientific American* (January 1975), Goddard placed a gun in a vacuum and fired a blank cartridge from it. (A blank cartridge fires only the wadding and hot gases of the burning gunpowder.) What happened when the gun was fired?

18. A pole-vaulter falls from a height of 6.0 m onto a foam rubber pad. Can you calculate his speed just before he reaches the pad? Can you calculate the force exerted on him due to the collision? Explain.

19. As a ball falls toward the Earth, the momentum of the ball increases. Reconcile this fact with the law of conservation of momentum.

20. Explain how you would use a balloon to demonstrate the mechanism responsible for rocket propulsion.

21. Explain the maneuver of decelerating a spacecraft. What other maneuvers are possible?

22. Does the center of mass of a rocket in free space accelerate? Explain. Can the speed of a rocket exceed the exhaust speed of the fuel? Explain.

23. A ball is dropped from a tall building. Identify the system for which linear momentum is constant.

24. A bomb, initially at rest, explodes into several pieces. Is linear momentum constant? (b) Is kinetic energy constant? Explain.

25. Is kinetic energy always lost in an inelastic collision? Explain.

26. A skater is standing still on a frictionless ice rink. Her friend throws a Frisbee straight at her. In which of the following cases is the largest momentum transferred to the skater? (a) She catches the Frisbee and holds it, (b) she catches it momentarily but drops it, (c) she catches it and then throws it back to her friend.

27. The Moon revolves around the Earth. Is the Moon's linear momentum constant? Is its kinetic energy constant? Assume the orbit is circular.

28. A large bedsheet is held vertically by two students. A third student, who happens to be the star pitcher on the baseball team, throws a raw egg at the sheet. Explain why the egg does not break when it hits the sheet, regardless of its initial speed. (If you try this one, make sure the pitcher hits the sheet near its center, and do not allow the egg to fall on the floor after being caught.)

29. A raw egg dropped to the floor breaks apart upon impact. However, a raw egg dropped onto a thick foam rubber cushion from a height of about 1 m rebounds without breaking. Why is this possible? (In this demonstration, be sure to catch the egg after the first bounce.)

PROBLEMS

Review Problem

A 60-kg person running at an initial speed of 4.0 m/s jumps onto a 120-kg cart initially at rest. The person slides on the cart's surface, and finally comes to rest relative to the cart. The coefficient of kinetic friction between the person and cart is 0.40 and the friction between the cart and ground can be neglected. (a) Find the final velocity of the person and cart relative to the Earth. (b) Find the frictional force exerted on the person while sliding on the surface of the cart. (c) How long does the frictional force act on the person? (d) Find the change in momentum of the person and the change in momentum of the cart. (e) Determine the displacement of the person relative to the Earth while sliding on the surface of the cart. (f) Determine the displacement of the cart relative to the Earth while the person slides on its surface. (g) Find the change in kinetic energy of the person. (i) Find the change in kinetic energy of the cart. (j) Explain why the answers to (g) and (i) differ. (What kind of collision is this, and what accounts for the loss in mechanical energy?)

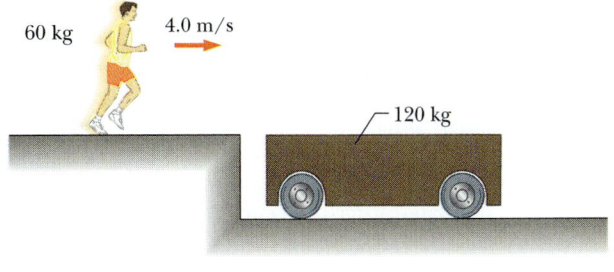

60 kg   4.0 m/s

120 kg

Section 9.1   Linear Momentum and Its Conservation
Section 9.2   Impulse and Momentum

1. A 3.0-kg particle has a velocity of $(3.0\mathbf{i} - 4.0\mathbf{j})$ m/s. Find its x and y components of momentum and the magnitude of its total momentum.

2. A 7.00-kg bowling ball moves in a straight line at 3.00 m/s. How fast must a 2.45-g Ping-Pong ball move in a straight line so that the two balls have the same momentum?

3. A child bounces a superball on the sidewalk. The linear impulse delivered by the sidewalk to the ball is 2.00 N·s during the 1/800 s of contact. What is the magnitude of the average force exerted on the ball by the sidewalk?

4. A superball with a mass of 60 g is dropped from a height of 2.0 m. It rebounds to a height of 1.8 m. What is the change in its linear momentum during the collision with the floor?

5. The force $\mathbf{F}_x$ acting on a 2.0-kg particle varies in time as shown in Figure P9.5. Find (a) the impulse of the force, (b) the final velocity of the particle if it is initially at rest, (c) its final velocity if it is initially moving along the x axis with a velocity of -2.0 m/s, and (d) the average force exerted on the particle for the time interval $t_i = 0$ to $t_f = 5.0$ s.

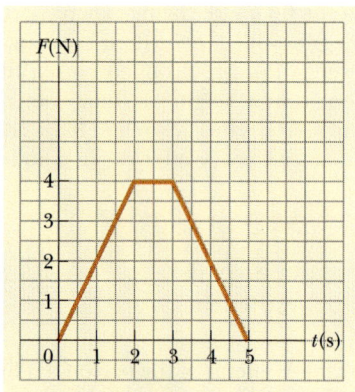

FIGURE P9.5

6. In a slow-pitch softball game, a 0.200-kg softball crossed the plate at 15.0 m/s at an angle of 45.0° below the horizontal. The ball was hit at 40.0 m/s, 30.0° above the horizontal. (a) Determine the impulse applied to the ball. (b) If the force on the ball increased linearly for 4.00 ms, held constant for 20.0 ms, then decreased to zero linearly in another 4.00 ms, find the maximum force on the ball.

7. An estimated force-time curve for a baseball struck by a bat is shown in Figure P9.7. From this curve,

determine (a) the impulse delivered to the ball, (b) the average force exerted on the ball, and (c) the peak force exerted on the ball.

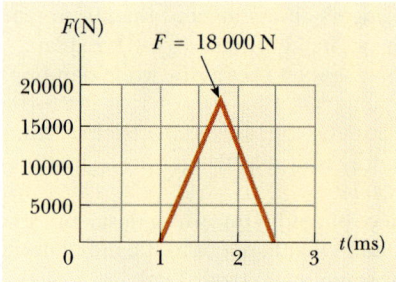

FIGURE P9.7

8. A garden hose is held in the manner shown in Figure P9.8. What force is necessary to hold the nozzle stationary if the discharge rate is 0.60 kg/s with a speed of 25 m/s?

FIGURE P9.8

9. A machine gun fires 35.0 g bullets at a speed of 750.0 m/s. If the gun can fire 200 bullets/min, what is the average force the shooter must exert to keep the gun from moving?

10. (a) If the momentum of an object is doubled in magnitude, what happens to its kinetic energy? (b) If the kinetic energy of an object is tripled, what happens to its momentum?

11. A 0.50-kg football is thrown with a speed of 15 m/s. A stationary receiver catches the ball and brings it to

□ indicates problems that have full solutions available in the Student Solutions Manual and Study Guide.

rest in 0.020 s. (a) What is the impulse delivered to the ball? (b) What is the average force exerted on the receiver?

12. A car is stopped for a traffic signal. When the light turns green, the car accelerates, increasing its speed from zero to 5.20 m/s in 0.832 s. What linear impulse and average force does a 70.0-kg passenger in the car experience?

13. A 0.15-kg baseball is thrown with a speed of 40 m/s. It is hit straight back at the pitcher with a speed of 50 m/s. (a) What is the impulse delivered to the baseball? (b) Find the average force exerted by the bat on the ball if the two are in contact for 2.0×10^{-3} s. Compare this with the weight of the ball and determine whether or not the impulse approximation is valid in the situation.

14. A tennis player receives a shot with the ball (0.060 kg) traveling horizontally at 50 m/s and returns the shot with the ball traveling horizontally at 40 m/s in the opposite direction. What is the impulse delivered to the ball by the racket?

15. A 3.0-kg steel ball strikes a wall with a speed of 10 m/s at an angle of 60° with the surface. It bounces off with the same speed and angle (Fig. P9.15). If the ball is in contact with the wall for 0.20 s, what is the average force exerted on the ball by the wall?

15A. A steel ball of mass m strikes a wall with a speed of v at an angle θ with the surface. It bounces off with the same speed and angle (Fig. P9.15). If the ball is in contact with the wall for a time t, what is the average force exerted on the ball by the wall?

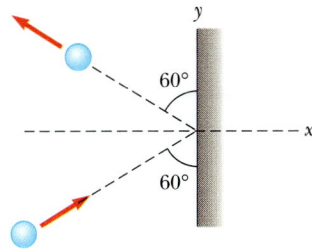

FIGURE P9.15

16. Water falls without splashing at a rate of 0.25 liter/s from a height of 60 m into a 0.75-kg bucket on a scale. If the bucket is originally empty, what does the scale read after 3.0 s?

Section 9.3 Collisions
Section 9.4 Elastic and Inelastic Collisions in One Dimension

17. A 79.5-kg man holding a 0.500-kg ball stands on a frozen pond next to a wall. He throws the ball at the wall with a speed of 10.0 m/s (relative to the ground) and then catches the ball after it rebounds from the

wall. (a) How fast is he moving after he catches the ball? (Ignore the projectile motion of the ball, and assume that it loses no energy in its collision with the wall.) (b) How many times does the man have to go through this process before his speed reaches 1.00 m/s relative to the ground?

18. Two blocks of masses M and $3M$ are placed on a horizontal, frictionless surface. A light spring is attached to one of them, and the blocks are pushed together with the spring between them (Fig. P9.18). A cord holding them together is burned, after which the block of mass $3M$ moves to the right with a speed of 2.00 m/s. What is the speed of the block of mass M?

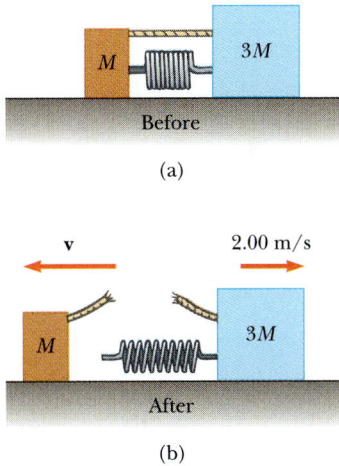

FIGURE P9.18

19. A 60.0-kg astronaut is on a space walk away from the shuttle when her tether line breaks. She is able to throw her 10.0-kg oxygen tank away from the shuttle with a speed of 12.0 m/s to propel herself back to the shuttle (Fig. P9.19). Assuming that she starts from rest (relative to the shuttle), determine the maximum distance she can be from the craft when the line breaks and still return within 60.0 s (the amount of time she can hold her breath).

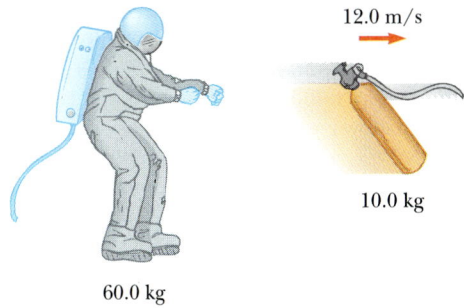

FIGURE P9.19

20. Identical air cars ($m = 200$ g) are equipped with identical springs ($k = 3000$ N/m). The cars move toward each other with speeds of 3.00 m/s on a horizontal air track and collide, compressing the springs (Fig. P9.20). Find the maximum compression of each spring.

20A. Identical air cars each of mass m are equipped with identical springs each having a force constant k. The cars move toward each other with speeds v on a horizontal air track and collide, compressing the springs (Fig. P9.20). Find the maximum compression of each spring.

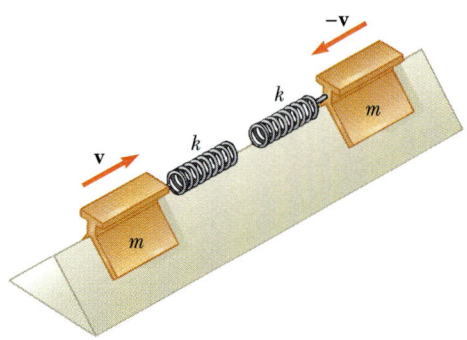

FIGURE P9.20

21. A 45.0-kg girl is standing on a plank that has a mass of 150 kg. The plank, originally at rest, is free to slide on a frozen lake, which is a flat, frictionless supporting surface. The girl begins to walk along the plank at a constant speed of 1.5 m/s relative to the plank. (a) What is her speed relative to the ice surface? (b) What is the speed of the plank relative to the ice surface?

22. A 7.00-kg bowling ball initially at rest is dropped from a height of 3.00 m. (a) What is the speed of the Earth coming up to meet the ball just before the ball hits the ground? Use 5.98×10^{24} kg as the mass of the Earth. (b) Use your answer to part (a) to justify ignoring the motion of the Earth when dealing with the motions of terrestrial objects.

23. A 2000-kg meteorite has a speed of 120 m/s just before colliding head-on with the Earth. Determine the recoil speed of the Earth (mass 5.98×10^{24} kg).

24. Gayle runs at a speed of 4.0 m/s and dives onto a sled that is initially at rest on the top of a frictionless snow-covered hill. After she and the sled have descended a vertical distance of 5.0 m, her brother, who is initially at rest, hops on her back and together they continue down the hill. What is their speed at the bottom if the total vertical drop is 15.0 m? Gayle's mass is 50.0 kg, the sled's is 5.0 kg, and her brother's is 30.0 kg.

25. A 10.0-g bullet is stopped in a block of wood ($m =$

5.00 kg). The speed of the bullet-plus-wood combination immediately after the collision is 0.600 m/s. What was the original speed of the bullet?

26. A 90-kg halfback running north with a speed of 10 m/s is tackled by a 120-kg opponent running south with a speed of 4.0 m/s. If the collision is perfectly inelastic and head-on, (a) calculate the speed and direction of the players just after the tackle and (b) determine the energy lost as a result of the collision. Account for the missing energy.

27. A 1200-kg car traveling initially with a speed of 25.0 m/s in an easterly direction crashes into the rear end of a 9000-kg truck moving in the same direction at 20.0 m/s (Fig. P9.27). The velocity of the car right after the collision is 18.0 m/s to the east. (a) What is the velocity of the truck right after the collision? (b) How much mechanical energy is lost in the collision? Account for this loss in energy.

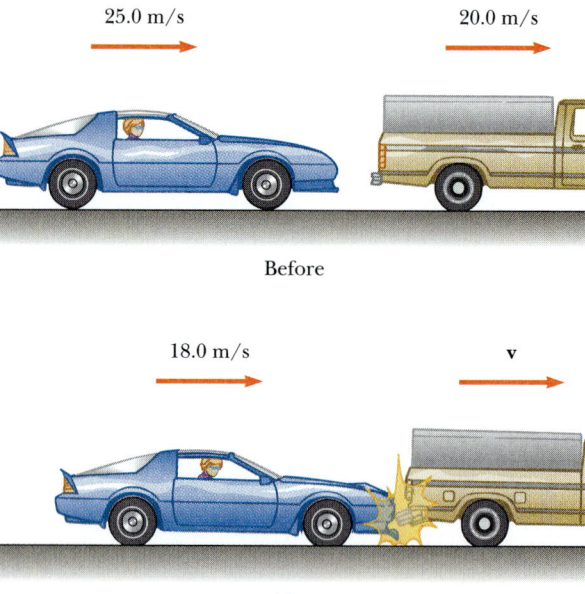

Before

After

FIGURE P9.27

28. A railroad car of mass 2.5×10^4 kg moving with a speed of 4.0 m/s collides and couples with three other coupled railroad cars, each of the same mass as the single car and moving in the same direction with an initial speed of 2.0 m/s. (a) What is the speed of the four cars after the collision? (b) How much energy is lost in the collision?

29. A neutron in a reactor makes an elastic head-on collision with the nucleus of a carbon atom initially at rest. (a) What fraction of the neutron's kinetic energy is transferred to the carbon nucleus? (b) If the initial kinetic energy of the neutron is 1.6×10^{-13} J,

find its final kinetic energy and the kinetic energy of the carbon nucleus after the collision. (The mass of the carbon nucleus is about 12 times the mass of the neutron.)

30. A ball of mass m is suspended by a string of length L above a block standing on end, as shown in Figure P9.30. The ball is pulled back an angle θ and released. In trial A, it rebounds elastically off the block. In trial B, two-sided tape causes the ball to stick to the block in a totally inelastic collision. In which case is the ball more likely to knock the block over? Explain.

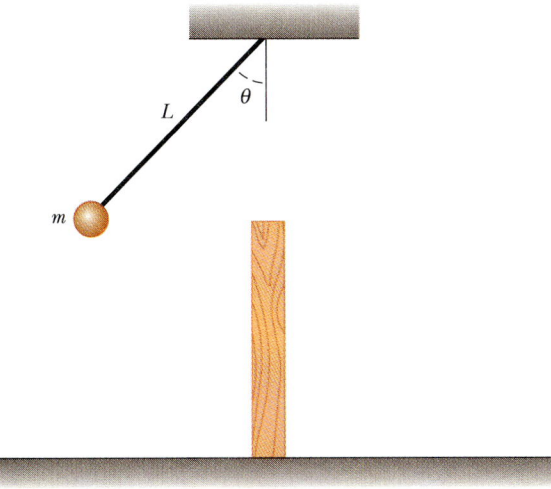

FIGURE P9.30

31. A 75-kg ice skater moving at 10 m/s crashes into a stationary skater of equal mass. After the collision, the two skaters move as a unit at 5.0 m/s. The average force a human skater can experience without breaking a bone is 4500 N. If the impact time is 0.10 s, does a bone break?

32. A 0.10-kg block is released from rest from the top of a 40.0° frictionless incline. When it has fallen a vertical distance of 1.5 m, a 0.015-kg bullet is fired into the block along a path parallel to the slope of the incline and momentarily brings the block to rest. (a) Find the speed of the bullet just before impact. (b) What bullet speed is needed to send the block up the incline to its initial position?

33. A 75.0-kg man stands in a 100.0-kg rowboat at rest in still water. He faces the back of the boat and throws a 5.00-kg rock out of the back at a speed of 20.0 m/s. The boat recoils forward and comes to rest 4.2 m from its original position. Calculate (a) the initial recoil speed of the boat, (b) the loss in mechanical energy due to the frictional force exerted by the water, and (c) the effective coefficient of friction between boat and water.

34. As shown in Figure P9.34, a bullet of mass m and speed v passes completely through a pendulum bob of mass M. The bullet emerges with a speed $v/2$. The pendulum bob is suspended by a stiff rod of length ℓ and negligible mass. What is the minimum value of v such that the pendulum bob will barely swing through a complete vertical circle?

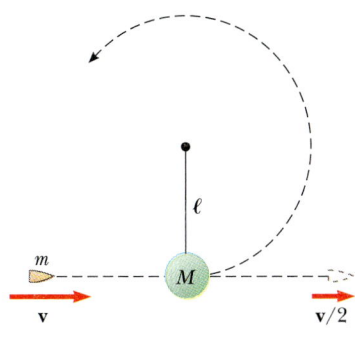

FIGURE P9.34

35. A 12-g bullet is fired into a 100-g wooden block initially at rest on a horizontal surface. After impact, the block slides 7.5 m before coming to rest. If the coefficient of friction between block and surface is 0.65, what was the speed of the bullet immediately before impact?

35A. A bullet of mass m_1 is fired into a wooden block of mass m_2 initially at rest on a horizontal surface. After impact, the block slides a distance d before coming to rest. If the coefficient of friction between block and surface is μ, what was the speed of the bullet immediately before impact?

36. A 7.00-g bullet fired into a 1.00-kg block of wood held in a vise penetrates to a depth of 8.00 cm. Then the vise is removed, the block of wood is placed on a frictionless horizontal surface, and another 7.00-g bullet is fired from the gun into it. To what depth does this second bullet penetrate?

37. Consider a frictionless track ABC as shown in Figure P9.37. A block of mass $m_1 = 5.001$ kg is released from A. It makes a head-on elastic collision with a block of mass $m_2 = 10.0$ kg at B, initially at rest. Cal-

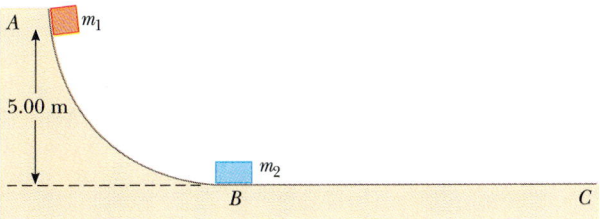

FIGURE P9.37

culate the maximum height to which m_1 rises after the collision.

38. A block of mass $m_1 = 2.0$ kg starts from rest on a surface inclined at $53°$ to the horizontal. The coefficient of kinetic friction between surface and block is $\mu_k = 0.25$. (a) If the speed of the block at the foot of the incline is 8.0 m/s to the right, determine the height from which the block was released. (b) Another block of mass $m_2 = 6.0$ kg is at rest on the smooth horizontal surface. Block m_1 collides with block m_2. After collision, the blocks stick together and move to the right. Determine the speed of the blocks after collision.

39. A 20.0-g bullet is fired horizontally into a 1.0-kg wooden block resting on a horizontal surface ($\mu_k = 0.25$). The bullet goes through the block and comes out with a speed of 250 m/s. If the block travels 5.0 m before coming to rest, what was the initial speed of the bullet?

40. Two blocks of mass $m_1 = 2.00$ kg and $m_2 = 4.00$ kg are released from a height of 5.00 m on a frictionless track as shown in Figure P9.40. The blocks undergo an elastic head-on collision. (a) Determine the two velocities just before collision. (b) Using Equations 9.20 and 9.21 determine the two velocities immediately after collision. (c) Determine the maximum height to which each block rises after collision.

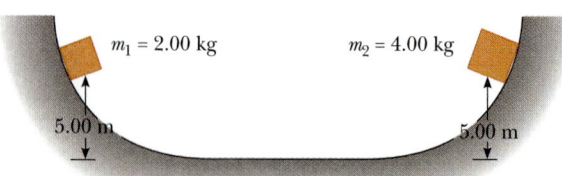

$m_1 = 2.00$ kg $m_2 = 4.00$ kg

5.00 m   5.00 m

FIGURE P9.40

Section 9.5   Two-Dimensional Collisions

41. A 3.00-kg mass with an initial velocity of $5.00\mathbf{i}$ m/s collides with and sticks to a 2.00-kg mass with an initial velocity of $-3.00\mathbf{j}$ m/s. Find the final velocity of the composite mass.

42. During the battle of Gettysburg, the gunfire was so intense that several bullets collided in midair and fused together. Assume a 5.0-g Union musket ball moving to the right at 250 m/s, and $20.0°$ above the horizontal, and a 3.0-g Confederate ball moving to the left at 280 m/s, and $15°$ above the horizontal. Immediately after they fuse together, what is their velocity?

43. An unstable nucleus of mass 17×10^{-27} kg initially at rest disintegrates into three particles. One of the particles, of mass 5.0×10^{-27} kg, moves along the y

axis with a speed of 6.0×10^6 m/s. Another particle, of mass 8.4×10^{-27} kg, moves along the x axis with a speed of 4.0×10^6 m/s. Find (a) the velocity of the third particle and (b) the total energy given off in the process.

44. A 0.30-kg puck, initially at rest on a horizontal, frictionless surface, is struck by a 0.20-kg puck moving initially along the x axis with a speed of 2.0 m/s. After the collision, the 0.20-kg puck has a speed of 1.0 m/s at an angle of $\theta = 53°$ to the positive x axis (Fig. 9.13). (a) Determine the velocity of the 0.30-kg puck after the collision. (b) Find the fraction of kinetic energy lost in the collision.

45. Two shuffleboard disks of equal mass, one orange and the other yellow, are involved in a perfectly elastic, glancing collision. The yellow disk is initially at rest and is struck by the orange disk moving with a speed of 5.00 m/s. After the collision the orange disk moves along a direction that makes an angle of $37.0°$ with its initial direction of motion, and the velocity of the yellow disk is perpendicular to that of the orange disk (after the collision). Determine the final speed of each disk.

45A. Two shuffleboard disks of equal mass, one orange and the other yellow, are involved in a perfectly elastic, glancing collision. The yellow disk is initially at rest and is struck by the orange disk moving with a speed v_0. After the collision the orange disk moves along a direction that makes an angle θ with its initial direction of motion, and the velocity of the yellow disk is perpendicular to that of the orange disk (after the collision). Determine the final speed of each disk.

46. A block of mass m_1 moves east on a tabletop, traveling at a speed v_0 toward a block of mass m_2, which is at rest. After the collision, the first block moves south with a speed v. (a) Show that

$$v \leq \sqrt{\frac{m_2 - m_1}{m_2 + m_1}}\, v_0$$

(*Hint:* Assume $K_{after} \leq K_{before}$.) (b) What does this expression for v tell you about m_1 and m_2?

47. A billiard ball moving at 5.00 m/s strikes a stationary ball of the same mass. After the collision, the first ball moves at 4.33 m/s at an angle of $30.0°$ with respect to the original line of motion. Assuming an elastic collision (and ignoring friction and rotational motion), find the struck ball's velocity.

48. A 200-g cart moves on a horizontal, frictionless surface with a constant speed of 25.0 cm/s. A 50.0-g piece of modeling clay is dropped vertically onto the cart. (a) If the clay sticks to the cart, find the final speed of the system. (b) After the collision, the clay has no momentum in the vertical direction. Does this mean that the law of conservation of momentum is violated?

49. A particle of mass m, moving with speed v, collides obliquely with an identical particle initially at rest. Show that if the collision is elastic, the two particles move at 90° from each other after collision. (*Hint:* $(\mathbf{A} + \mathbf{B})^2 = A^2 + B^2 + 2AB \cos \theta$.)

50. A mass m_1 having an initial velocity $\mathbf{v}_1$ collides with a stationary mass m_2. After the collision, m_1 and m_2 are deflected as shown in Figure P9.50. The velocity of m_1 after collision is $\mathbf{v}'_1$. Show that

$$\tan \theta_2 = \frac{v'_1 \sin \theta_1}{v_1 - v_1 \cos \theta_2}$$

From the information given and the result you derived, can you make the assumption that the collision is perfectly elastic?

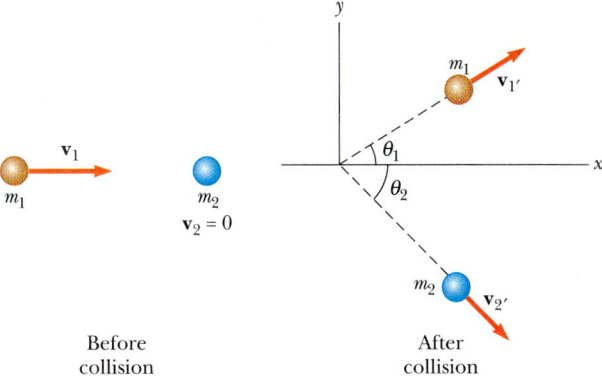

Before collision   After collision

FIGURE P9.50

Section 9.6 The Center of Mass

51. A 3.00-kg particle is located on the x axis at $x = -5.00$ m, and a 4.00-kg particle is on the x axis at $x = 3.00$ m. Find the center of mass of this two-particle system.

52. A water molecule consists of an oxygen atom with two hydrogen atoms bound to it (Fig. P9.52). The angle between the two bonds is 106°. If each bond is 0.100 nm long, where is the center of mass of the molecule?

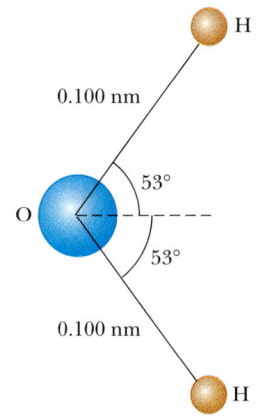

FIGURE P9.52

53. The mass of the Sun is 329 390 Earth masses, and the mean distance from the center of the Sun to the center of the Earth is 1.496×10^8 km. Treating the Earth and Sun as particles, with each mass concentrated at its respective geometric center, how far from the center of the Sun is the center of mass of the Earth-Sun system? Compare this distance with the mean radius of the Sun (6.960×10^5 km).

54. The separation between the hydrogen and chlorine atoms of the HCl molecule is about 1.30×10^{-10} m. Determine the location of the center of mass of the molecule as measured from the hydrogen atom. (Chlorine is 35 times more massive than hydrogen.)

55. Figure P9.55 shows three uniform objects—rod, right triangle, and square—with their masses given, along with their coordinates. Determine the center of mass for this three-object system.

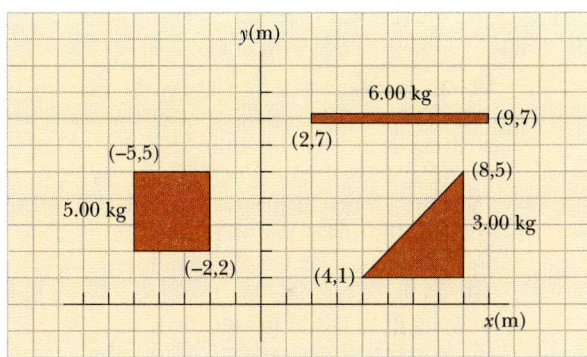

FIGURE P9.55

56. A circular hole of diameter a is cut out of a uniform square of sheet metal having sides $2a$, as in Figure P9.56. Where is the center of mass for the remaining portion?

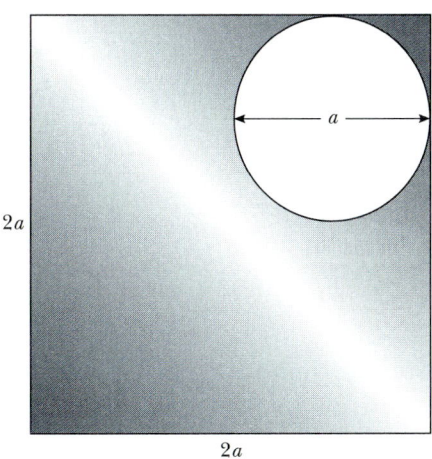

FIGURE P9.56

57. A uniform piece of sheet steel is shaped as in Figure P9.57. Compute the *x* and *y* coordinates of the center of mass of the piece.

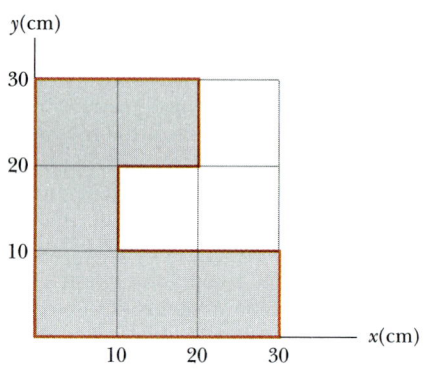

FIGURE P9.57

Section 9.7   Motion of a System of Particles

58. A block of mass *m* = 2.0 kg is placed on the top of an incline of mass *M* = 8.0 kg, height *h* = 2.0 m, and base length *L* = 6.0 m. If the block is released from rest (Fig. P9.58a), how far has the incline moved when the block reaches the bottom (Fig. P9.58b)? Assume all surfaces are frictionless. (*Hint:* The *x* coordinate for the center of mass of the block-incline system is fixed (why?) — see Example 9.18 for finding the center of mass of the incline.)

59. A 2.0-kg particle has a velocity $(2.0\mathbf{i} - 3.0\mathbf{j})$ m/s, and a 3.0-kg particle has a velocity $(1.0\mathbf{i} + 6.0\mathbf{j})$ m/s. Find (a) the velocity of the center of mass and (b) the total momentum of the system.

60. A 2.0-kg particle has a velocity of $\mathbf{v}_1 = (2.0\mathbf{i} - 10t\mathbf{j})$ m/s, where *t* is in seconds. A 3.0-kg particle moves with a constant velocity of $\mathbf{v}_2 = 4.0\mathbf{i}$ m/s. At *t* = 0.50 s, find (a) the velocity of the center of mass, (b) the acceleration of the center of mass, and (c) the total momentum of the system.

61. A 3.00-g particle is moving at 3.00 m/s toward a stationary 7.00-g particle. (a) With what speed does each approach the center of mass? (b) What is the momentum of each particle, relative to the center of mass?

61A. A particle of mass m_1 is moving at a speed v_1 toward a stationary particle of mass m_2. (a) With what speed does each approach the center of mass? (b) What is the momentum of each particle, relative to the center of mass?

62. Romeo entertains Juliet by playing his guitar from the rear of their boat in still water. After the serenade, Juliet carefully moves to the rear of the boat (away from shore) to plant a kiss on Romeo's cheek. If the 80-kg boat is facing shore and the 55-kg Juliet moves 2.7 m toward the 77-kg Romeo, how far does the boat move toward shore?

62A. Romeo entertains Juliet by playing his guitar from the rear of their boat in still water. After the serenade, Juliet carefully moves to the rear of the boat (away from shore) to plant a kiss on Romeo's cheek. If the boat (mass m_B) is facing shore and Juliet (mass m_J) moves a distance *d* toward Romeo (mass m_R), how far does the boat move toward shore?

*Section 9.8   Rocket Propulsion

63. A rocket engine consumes 80 kg of fuel per second. If the exhaust speed is 2.5×10^3 m/s, calculate the thrust on the rocket.

64. The first stage of a Saturn V space vehicle consumes fuel at the rate of 1.5×10^4 kg/s, with an exhaust speed of 2.6×10^3 m/s. (a) Calculate the thrust produced by these engines. (b) Find the initial acceleration of the vehicle on the launch pad if its initial mass is 3.0×10^6 kg. [*Hint:* You must include the force of gravity to solve part (b).]

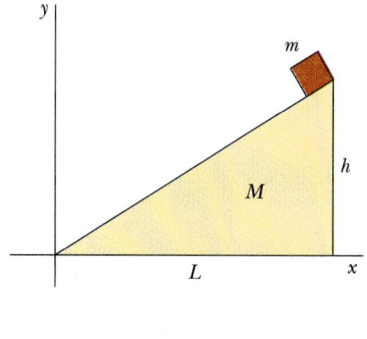

(a)

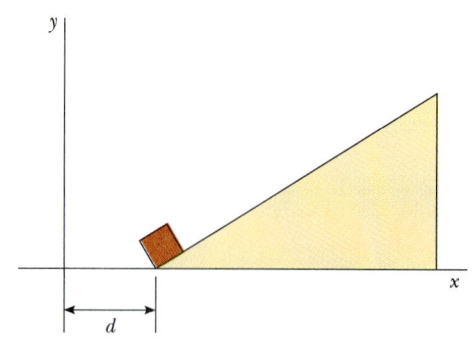

(b)

FIGURE P9.58

65. A 3000-kg rocket has 4000 kg of fuel on board. The rocket is coasting through space at 100.0 m/s and needs to boost its speed to 300.0 m/s. It does this by firing its engines and ejecting fuel at a relative speed of 650.0 m/s until the desired speed is reached. How much fuel is left on board after this maneuver?

 66. A rocket for use in deep space is to have the capability of boosting a payload (plus the rocket frame and engine) of 3.0 metric tons to a speed of 10 000 m/s with an engine and fuel designed to produce an exhaust speed of 2000 m/s. (a) How much fuel plus oxidizer is required? (b) If a different fuel and engine design could give an exhaust speed of 5000 m/s, what amount of fuel and oxidizer would be required for the same task?

67. Fuel aboard a rocket has a density of 1.4×10^3 kg/m^3 and is ejected with a speed of 3.0×10^3 m/s. If the engine is to provide a thrust of 2.5×10^6 N, what volume of fuel must be burned per second?

67A. Fuel aboard a rocket has a density ρ and is ejected with a speed v. If the engine is to provide a thrust F, what volume of fuel must be burned per second?

68. A rocket with an initial total mass M_i is launched vertically from the Earth's surface. When the launch fuel has been completely burned, the rocket has reached an altitude small compared to the Earth's radius (so that the gravitational field strength may be considered constant during the burn). Show that the final speed is $v = -v_e \ln(M_i/M_f) - gt_1$, where the time of burn t_1 is $t_1 = (M_i - M_f)(dm/dt)^{-1}$. ($M_f$ is the final total mass of the rocket, v_e is the exhaust gas speed, and dm/dt is the constant rate of fuel consumption.)

ADDITIONAL PROBLEMS

69. A golf ball ($m = 46$ g) is struck a blow that makes an angle of 45° with the horizontal. The drive lands 200 m away on a flat fairway. If the golf club and ball are in contact for 7.0 ms, what is the average force of impact? (Neglect air resistance.)

70. A 12.0-g bullet is fired horizontally into a 100-g wooden block that is initially at rest on a rough horizontal surface and connected to a massless spring of constant 150 N/m. If the bullet-block system compresses the spring by 0.800 m, what was the speed of the bullet just as it enters the block? Assume that the coefficient of kinetic friction between block and surface is 0.60.

71. A 30-06 caliber hunting rifle fires a bullet of mass 0.012 kg with a muzzle velocity of 600 m/s to the right. The rifle has a mass of 4.0 kg. (a) What is the recoil velocity of the rifle as the bullet leaves the rifle? (b) If the rifle is stopped by the hunter's shoulder in

a distance of 2.5 cm, what is the average force exerted on the shoulder by the rifle? (c) If the hunter's shoulder is partially restricted from recoiling, would the force exerted on the shoulder be the same as in part (b)? Explain.

72. An 8.00-g bullet is fired into a 2.50-kg block initially at rest at the edge of a frictionless table of height 1.00 m (Fig. P9.72). The bullet remains in the block, and after impact the block lands 2.00 m from the bottom of the table. Determine the initial speed of the bullet.

72A. A bullet of mass m is fired into a block of mass M initially at rest at the edge of a frictionless table of height h (Fig. P9.72). The bullet remains in the block, and after impact the block lands a distance d from the bottom of the table. Determine the initial speed of the bullet.

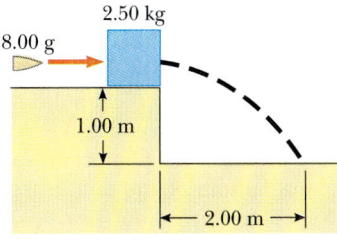

FIGURE P9.72

73. An attack helicopter is equipped with a 20-mm cannon that fires 130-g shells in the forward direction with a muzzle speed of 800 m/s. The fully loaded helicopter has a mass of 4000 kg. A burst of 160 shells is fired in a 4.0-s interval. What is the resulting average force on the helicopter and by what amount is its forward speed reduced?

74. A small block of mass $m_1 = 0.500$ kg is released from rest at the top of a curved frictionless wedge of mass $m_2 = 3.00$ kg, which sits on a frictionless horizontal surface as in Figure P9.74a. When it leaves the wedge, the block's velocity is 4.00 m/s to the right, as in Figure P9.74b. (a) What is the velocity of the wedge after the block reaches the horizontal surface? (b) What is the height h of the wedge?

74A. A small block of mass m_1 is released from rest at the top of a curved frictionless wedge of mass m_2, which sits on a frictionless horizontal surface as in Figure P9.74a. When it leaves the wedge, the block's velocity is v_1 to the right, as in Figure P9.74b. (a) What is the velocity of the wedge after the block reaches the horizontal surface? (b) What is the height h of the wedge?

75. A jet aircraft is traveling at 500 mi/h (223 m/s) in horizontal flight. The engine takes in air at a rate of

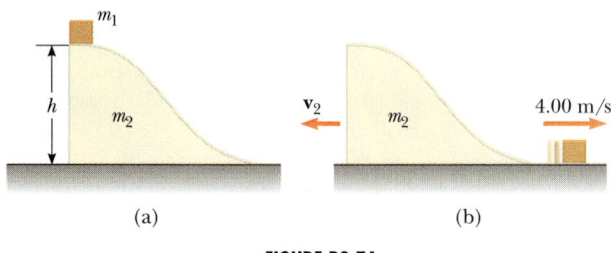

(a)   (b)

FIGURE P9.74

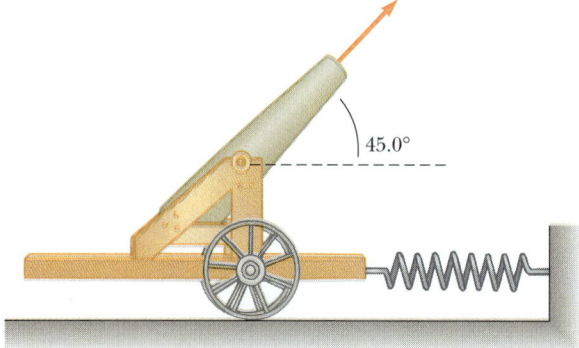

FIGURE P9.80

80.0 kg/s and burns fuel at a rate of 3.00 kg/s. If the exhaust gases are ejected at 600 m/s relative to the aircraft, find the thrust of the jet engine and the delivered horsepower.

76. Two ice skaters approach each other at right angles. Skater A has a mass of 50.0 kg and travels in the $+x$ direction at 2.00 m/s. Skater B has a mass of 70.0 kg and is moving in the $+y$ direction at 1.5 m/s. They collide and cling together. Find (a) the final velocity of the couple and (b) the loss in kinetic energy due to the collision.

77. A 75.0-kg firefighter slides down a pole while a constant frictional force of 300 N retards her motion. A horizontal 20.0-kg platform is supported by a spring at the bottom of the pole to cushion the fall. The firefighter starts from rest 4.00 m above the platform, and the spring constant is 4000 N/m. Find (a) the firefighter's speed just before she collides with the platform and (b) the maximum distance the spring is compressed. (Assume the frictional force acts during the entire motion.)

78. A 70.0-kg baseball player jumps vertically up to catch a baseball of mass 0.160 kg traveling horizontally with a speed of 35.0 m/s. If the vertical speed of the player at the instant of catching the ball is 0.200 m/s, determine the player's velocity just after the catch.

79. An 80.0-kg astronaut is working on the engines of his ship, which is drifting through space with a constant velocity. The astronaut, wishing to get a better view of the Universe, pushes against the ship and later finds himself 30.0 m behind the ship. Without a thruster, the only way to return to the ship is to throw his 0.500-kg wrench directly away from the ship. If he throws the wrench with a speed of 20.0 m/s relative to the ship, how long does it take the astronaut to reach the ship?

80. A cannon is rigidly attached to a carriage, which can move along horizontal rails but is connected to a post by a large spring of force constant $k = 2.00 \times 10^4$ N/m (Fig. P9.80). The cannon fires a 200-kg projectile at a velocity of 125 m/s directed 45.0° above the horizontal. (a) If the mass of the cannon and its carriage is 5000 kg, find the recoil speed of the cannon. (b) Determine the maximum extension of the spring. (c) Consider the system to consist of the cannon, carriage, and shell. Is the momentum of this system constant during the firing? Why or why not?

81. A chain of length L and total mass M is released from rest with its lower end just touching the top of a table, as in Figure P9.81a. Find the force exerted by the table on the chain after the chain has fallen through a distance x, as in Figure P9.81b. (Assume each link comes to rest the instant it reaches the table.)

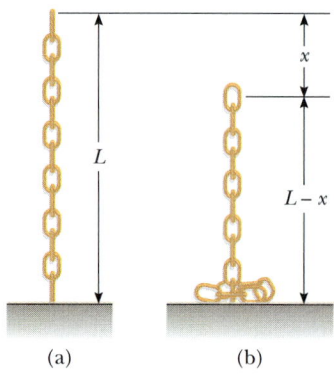

(a)   (b)

FIGURE P9.81

82. Two gliders are set in motion on an air track. A spring of force constant k is attached to the near side of one glider. The first glider of mass m_1 has velocity v_1 and the second glider of mass m_2 has velocity v_2, as in Figure P9.82 ($v_1 > v_2$). When m_1 collides with the spring attached to m_2 and compresses the spring to its maximum compression x_m, the velocity of the gliders is v. In terms of v_1, v_2, m_1, m_2, and k, find

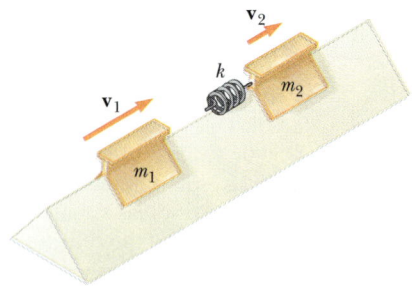

FIGURE P9.82

(a) the velocity v at maximum compression, (b) the maximum compression x_m, and (c) the velocities of each glider after m_1 has lost contact with the spring.

83. A 40.0-kg child stands at one end of a 70.0-kg boat that is 4.00 m in length (Fig. P9.83). The boat is initially 3.00 m from the pier. The child notices a turtle on a rock at the far end of the boat and proceeds to walk to that end to catch the turtle. Neglecting friction between boat and water, (a) describe the subsequent motion of the system (child plus boat). (b) Where is the child *relative to the pier* when he reaches the far end of the boat? (c) Will he catch the turtle? (Assume he can reach out 1.00 m from the end of the boat.)

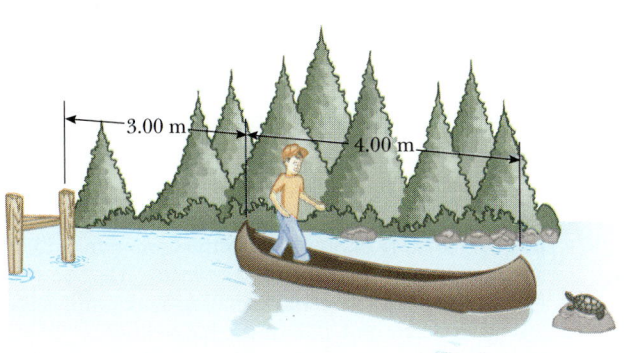

FIGURE P9.83

84. Two carts of equal mass, $m = 0.250$ kg, are placed on a frictionless track that has a light spring of force constant $k = 50.0$ N/m attached to one end of it, as in Figure P9.84. The red cart is given an initial velocity of $v_0 = 3.00$ m/s to the right, and the blue cart is initially at rest. If the carts collide elastically, find (a) the velocity of the carts just after the first collision and (b) the maximum compression in the spring.

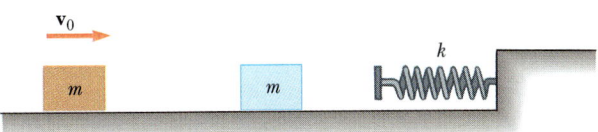

FIGURE P9.84

85. A 5.00-g bullet moving with an initial speed of 400 m/s is fired into and passes through a 1.00-kg block, as in Figure P9.85. The block, initially at rest on a

frictionless, horizontal surface, is connected to a spring of force constant 900 N/m. If the block moves 5.00 cm to the right after impact, find (a) the speed at which the bullet emerges from the block and (b) the energy lost in the collision.

85A. A bullet of mass m moving with an initial speed v_0 is fired into and passes through a block of mass M, as in Figure P9.85. The block, initially at rest on a frictionless, horizontal surface, is connected to a spring of force constant k. If the block moves a distance d to the right after impact, find (a) the speed at which the bullet emerges from the block and (b) the energy lost in the collision.

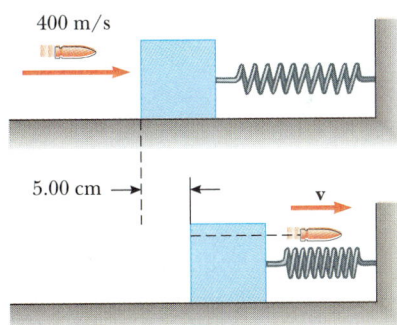

FIGURE P9.85

86. A student performs a ballistic pendulum experiment using an apparatus similar to that shown in Figure 9.11b. She obtains the following average data; $h = 8.68$ cm, $m_1 = 68.8$ g, and $m_2 = 263$ g. (a) Determine the initial speed v_{1i} of the projectile. (b) The second part of her experiment is to obtain v_{1i} by firing the same projectile horizontally (with the pendulum removed from the path), by measuring its horizontal displacement, x, and vertical displacement, y (Fig. P9.86). Show that the initial speed of the projectile is

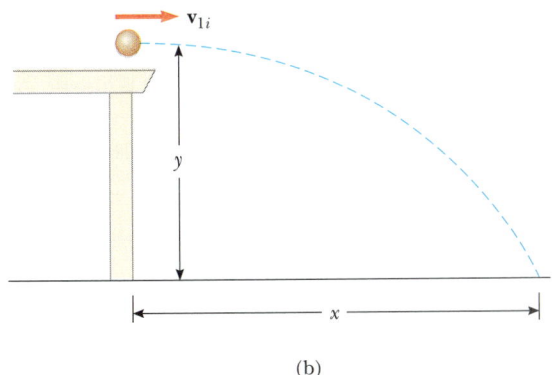

(b)

FIGURE P9.86

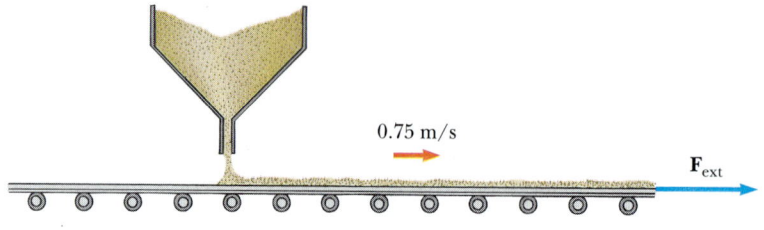

FIGURE P9.89

related to x and y through the relation

$$v_{1i} = \frac{x}{\sqrt{2y/g}}$$

What numerical value does she obtain for v_{1i} based on her measured values of $x = 257$ cm and $y = 85.3$ cm? What factors might account for the difference in this value compared to that obtained in part (a)?

87. Two particles, of masses m and $3m$, are moving toward each other along the x axis with the same initial speeds v_0. Mass m is traveling to the left, and mass $3m$ is traveling to the right. They undergo a head-on elastic collision, and each rebounds along the same line as it approached. Find the final speeds of the particles.

88. Two particles, of masses m and $3m$, are moving toward each other along the x axis with the same initial speeds v_0. Mass m is traveling to the left, and mass $3m$ is traveling to the right. They undergo an elastic glancing collision such that m is moving downward after the collision at a right angle to its initial direction. (a) Find the final speeds of the two masses. (b) What is the angle θ at which $3m$ is scattered?

89. Sand from a stationary hopper falls on a moving conveyor belt at the rate of 5.0 kg/s, as in Figure P9.89. The conveyor belt is supported by frictionless rollers and moves at 0.75 m/s under the action of a horizontal external force F_{ext} supplied by the motor that drives the belt. Find (a) the sand's rate of change of momentum in the horizontal direction, (b) the frictional force exerted by the belt on the sand, (c) the magnitude of F_{ext}, (d) the work done by F_{ext} in one second, and (e) the kinetic energy acquired by the falling sand each second due to the change in its horizontal motion. (f) Why is the answer to (d) different from the answer to (e)?

SPREADSHEET PROBLEMS

S1. A rocket has an initial mass of 20 000 kg, of which 20% is the payload. The rocket burns fuel at a rate of 200 kg/s and exhausts gas at a relative speed of 2.00 km/s. Its acceleration dv/dt is determined from the

equation of motion

$$M\frac{dv}{dt} = v_e \left|\frac{dM}{dt}\right| + F_{ext}$$

Assume that there are no external forces and that the rocket's initial velocity is zero. When $F_{ext} = 0$, the rocket's velocity is $v(t) = v_e \ln(M_i/M)$, where M is the mass at time t and M_i is the rocket's initial mass. Spreadsheet 9.1 calculates the acceleration and velocity of the rocket and plots the velocity as a function of time. (a) Using the given parameters, find the maximum acceleration and velocity. (b) At what time is the velocity equal to half its maximum value? Why isn't this time half of the burn time?

S2. Modify Spreadsheet 9.1 to calculate the distance traveled by the rocket in Problem S9.1. Add a new column to the spreadsheet to find the new position x_{i+1}. Estimate the new position of the rocket by

$$x_{i+1} = x_i + \tfrac{1}{2}(v_{i+1} + v_i)\Delta t,$$

where x_i is the old position, v_i is the old velocity, and v_{i+1} is the new velocity.

S3. If the rocket in the previous problem burns its fuel twice as fast (400 kg/s) while maintaining the same exhaust speed, how far does it travel before it runs out of fuel? The velocity reached by the rocket is the same in both cases; that is

$$v_f - v_i = v_e \ln\left(\frac{M_i}{M_f}\right)$$

Is there an advantage in burning the fuel faster? Is there a disadvantage? *Hint:* It may be useful to plot the accelerations, velocities, and positions for the same time intervals in both cases. In which case does the rocket travel farther? In which case does the rocket have to withstand the largest acceleration? By how much?

S4. Applying conservation of momentum to the motion of a rocket, we find:

$$M\frac{dv}{dt} = v_e \left|\frac{dM}{dt}\right| + F_{ext},$$

where $v_e |dM/dt|$ is the thrust of the rocket and M is the mass of the rocket. If the rocket is fired vertically upward from Earth, we have $F_{ext} = -mg$. Dividing the equation of motion by M gives

$$\frac{dv}{dt} = \frac{v_e}{M}\left|\frac{dM}{dt}\right| - g$$

If the rocket gets far enough away from Earth during its "burn" so that the variation in the Earth's gravitational field must be taken into account, then the constant g in the equation above must be replaced by $g[R_E/(R_E + x)]^2$, where R_E is the radius of the Earth and x is the distance above the Earth. Write your own program or modify Spreadsheet 6.1 to calculate and plot the velocity and acceleration of a rocket during its burn time, taking into account the variation of gravity with the distance x from the Earth's surface. Use the following data: $M_i = 2.50 \times 10^6$ kg; burn rate $|dM/dt| = 1.35 \times 10^4$ kg/s; and thrust = 25×10^6 N. The payload is 30% of the original mass. With these data, $v_e = 1.85$ km/s. How does the final speed of the rocket at burnout compare to the final speed if g were constant?

S5. To design a better golf club driver, a mechanical engineer equips a test club with a device that measures the impact forces generated when the driver strikes the ball. The force data as a function of time are recorded in the following table:

t (ms)	F (N)
0.45	0
0.50	20
0.55	210
0.60	1350
0.65	4350
0.70	12755
0.75	14430
0.80	11675
0.85	3005
0.90	1130
0.95	400
1.00	40
1.05	0
1.10	0

Using these data, calculate the impulse imparted to the ball by numerically integrating the force-versus-time data. Find the average force acting on the ball. (a) Use a simple rectangular integration approximation. Compare the impulse and average force you calculate with Example 9.3. (b) Repeat part (a) using a trapezoidal integration approximation.

GARFIELD

JIM DAVIS

Rotation of a Rigid Object About a Fixed Axis

The pair of skaters in this figure-skating routine is executing the beautiful "death spiral." In this circular motion, the female skater is spun in a circular arc by her partner with her body nearly parallel to the ice. *(© Duomo/Steven E. Sutton, 1994)*

W hen an extended object, such as a wheel, rotates about its axis, the motion cannot be analyzed by treating the object as a particle, since at any given time different parts of the object have different velocities and accelerations. For this reason, it is convenient to consider an extended object as a large number of particles, each with its own velocity and acceleration.

In dealing with the rotation of an object, analysis is greatly simplified by assuming the object to be rigid. A **rigid object** is defined as one that is nondeformable or, to say the same thing another way, one in which the separations between all pairs of particles remain constant. All real bodies are deformable to some extent; however, our rigid-object model is useful in many situations where deformation is negligible. In this chapter, we treat the rotation of a rigid object about a fixed axis, commonly referred to as *pure rotational motion.*

When the rigid-body approximation is not adequate, it often suffices as a first approximation. Thus, a rotating galaxy is far from a rigid body, but for many purposes, the internal motions may be neglected when treating the entire galaxy. A rotating molecule is likely to expand because of the rotation, but may be adequately approximated as a rigid body with the longer bond lengths.

Rigid object

10.1   ANGULAR VELOCITY AND ANGULAR ACCELERATION

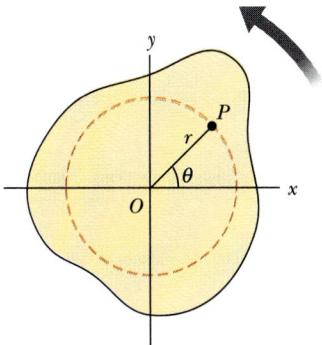

Figure 10.1 illustrates a planar rigid object of arbitrary shape confined to the xy plane and rotating about a fixed axis through O perpendicular to the plane of the figure. A particle on the object at P is at a fixed distance r from the origin and rotates in a circle of radius r about O. In fact, *every* particle on the object undergoes circular motion about O. It is convenient to represent the position of P with its polar coordinates (r, θ). In this representation, the only coordinate that changes in time is the angle θ; r remains constant. (In rectangular coordinates, both x and y vary in time.) As the particle moves along the circle from the positive x axis ($\theta = 0$) to P, it moves through an arc length s, which is related to the angular position θ through the relationship

$$s = r\theta \qquad (10.1a)$$

$$\theta = \frac{s}{r} \qquad (10.1b)$$

FIGURE 10.1   Rotation of a rigid object about a fixed axis through O perpendicular to the plane of the figure. (In other words, the axis of rotation is the z axis.) Note that a particle at P rotates in a circle of radius r centered at O.

It is important to note the units of θ in Equation 10.1b. The angle θ is the ratio of an arc length and the radius of the circle and, hence, is a pure number. However, we commonly give θ the artificial unit **radian** (rad), where

> one radian is the angle subtended by an arc length equal to the radius of the arc.

Radian

Since the circumference of a circle is $2\pi r$, it follows from Equation 10.1b that $360°$ corresponds to an angle of $2\pi r/r$ rad or 2π rad (one revolution). Hence, 1 rad $= 360°/2\pi \approx 57.3°$. To convert an angle in degrees to an angle in radians, we use the fact that 2π radians $= 360°$:

$$\theta \ (\text{rad}) = \frac{\pi}{180°} \theta \ (\text{deg})$$

For example, $60°$ equals $\pi/3$ rad, and $45°$ equals $\pi/4$ rad.

As the particle in question on our rigid object travels from P to Q in a time Δt, the radius vector sweeps out an angle $\Delta\theta = \theta_2 - \theta_1$ (Fig. 10.2), which equals the **angular displacement**. We define the **average angular speed** $\bar\omega$ (omega) as the ratio of this angular displacement to the time interval Δt:

$$\bar\omega \equiv \frac{\theta_2 - \theta_1}{t_2 - t_1} = \frac{\Delta\theta}{\Delta t} \qquad (10.2)$$

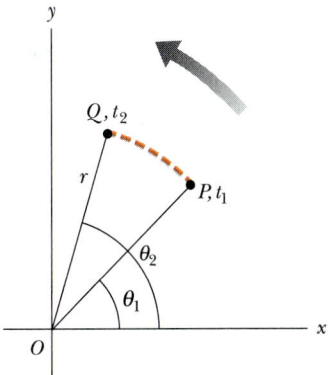

FIGURE 10.2   A particle on a rotating rigid object moves from P to Q along the arc of a circle. In the time interval $\Delta t = t_2 - t_1$, the radius vector sweeps out an angle $\Delta\theta = \theta_2 - \theta_1$.

In analogy to linear speed, the **instantaneous angular speed,** ω, is defined as the limit of the ratio in Equation 10.2 as Δt approaches zero:

$$\omega \equiv \lim_{\Delta t \to 0} \frac{\Delta\theta}{\Delta t} = \frac{d\theta}{dt} \qquad (10.3)$$

Instantaneous angular speed

Angular speed has units of radians per second, or s^{-1}, since radians are not dimensional. Let us adopt the convention that the fixed axis of rotation for the rigid object is the z axis, as in Figure 10.1. We take ω to be positive when θ is increasing (counterclockwise motion) and negative when θ is decreasing (clockwise motion).

If the instantaneous angular speed of an object changes from ω_1 to ω_2 in the time interval Δt, the object has an angular acceleration. The **average angular acceleration** $\bar{\alpha}$ (alpha) of a rotating object is defined as the ratio of the change in the angular speed to the time interval Δt:

Average angular acceleration

$$\bar{\alpha} \equiv \frac{\omega_2 - \omega_1}{t_2 - t_1} = \frac{\Delta\omega}{\Delta t} \tag{10.4}$$

In analogy to linear acceleration, the **instantaneous angular acceleration** is defined as the limit of the ratio $\Delta\omega/\Delta t$ as Δt approaches zero:

Instantaneous angular acceleration

$$\alpha \equiv \lim_{\Delta t \to 0} \frac{\Delta\omega}{\Delta t} = \frac{d\omega}{dt} \tag{10.5}$$

Angular acceleration has units of rad/s^2, or s^{-2}. Note that α is positive when ω is increasing in time and negative when ω is decreasing in time.

When rotating about a fixed axis, every particle on a rigid object has the same angular speed and the same angular acceleration. That is, the quantities ω and α characterize the rotational motion of the entire rigid object. Using these quantities, we can greatly simplify the analysis of rigid-body rotation. The angular displacement (θ), angular speed (ω), and angular acceleration (α) are analogous to linear displacement (x), linear speed (v), and linear acceleration (a), respectively, for the corresponding motion discussed in Chapter 2. The variables θ, ω, and α differ dimensionally from the variables x, v, and a, only by a length factor.

We have already indicated how the signs for ω and α are determined; however, we have not specified any direction in space associated with these vector quantities.[1] For rotation about a fixed axis, the only direction in space that uniquely specifies the rotational motion is the direction along the axis of rotation. Therefore, the directions $\boldsymbol{\omega}$ and $\boldsymbol{\alpha}$ are along this axis. If an object rotates in the xy plane as in Figure 10.1, the direction of $\boldsymbol{\omega}$ is out of the plane of the diagram when the rotation is counterclockwise and into the plane of the diagram when the rotation is clockwise. To illustrate this convention, it is convenient to use the *right-hand rule* shown in Figure 10.3a. The four fingers of the right hand are wrapped in the direction of the rotation. The extended right thumb points in the direction of $\boldsymbol{\omega}$. Figure 10.3b illustrates that $\boldsymbol{\omega}$ is also in the direction of advance of a similarly rotating right-handed screw. Finally, the sense of **a** follows from its definition as $d\boldsymbol{\omega}/dt$. It is the same as $\boldsymbol{\omega}$ if the angular speed (the magnitude of $\boldsymbol{\omega}$) is increasing in time and antiparallel to $\boldsymbol{\omega}$ if the angular speed is decreasing in time.

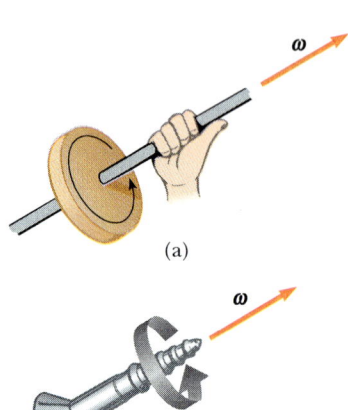

(a)

(b)

FIGURE 10.3 (a) The right-hand rule for determining the direction of the angular velocity vector. (b) The direction of $\boldsymbol{\omega}$ is in the direction of advance of a right-handed screw.

10.2 ROTATIONAL KINEMATICS: ROTATIONAL MOTION WITH CONSTANT ANGULAR ACCELERATION

In the study of linear motion, we found that the simplest form of accelerated motion to analyze is motion under constant linear acceleration (Chapter 2). Likewise, for rotational motion about a fixed axis, the simplest accelerated motion to analyze is motion under constant angular acceleration. Therefore, we next develop kinematic relationships for rotational motion under constant angular acceleration. If we write Equation 10.5 in the form $d\omega = \alpha\, dt$ and let $\omega = \omega_0$ at $t_0 = 0$,

---

[1] Although we do not verify it here, the instantaneous angular velocity and instantaneous angular acceleration are vector quantities, but the corresponding average values are not. This is because angular displacement is not a vector quantity for finite rotations.

TABLE 10.1	A Comparison of Kinematic Equations for Rotational and Linear Motion Under Constant Acceleration

Rotational Motion About a Fixed Axis with α = Constant Variables: θ and ω	Linear Motion with a = Constant Variables: x and v
$\omega = \omega_0 + \alpha t$	$v = v_0 + at$
$\theta = \theta_0 + \frac{1}{2}(\omega_0 + \omega)t$	$x = x_0 + \frac{1}{2}(v_0 + v)t$
$\theta = \theta_0 + \omega_0 t + \frac{1}{2}\alpha t^2$	$x = x_0 + v_0 t + \frac{1}{2}at^2$
$\omega^2 = \omega_0^2 + 2\alpha(\theta - \theta_0)$	$v^2 = v_0^2 + 2a(x - x_0)$

we can integrate this expression directly:

$$\omega = \omega_0 + \alpha t \qquad (\alpha = \text{constant}) \qquad (10.6)$$

Rotational kinematic equations

Likewise, substituting Equation 10.6 into Equation 10.3 and integrating once more (with $\theta = \theta_0$ at $t_0 = 0$), we get

$$\theta = \theta_0 + \omega_0 t + \tfrac{1}{2}\alpha t^2 \qquad (10.7)$$

If we eliminate t from Equations 10.6 and 10.7, we get

$$\omega^2 = \omega_0^2 + 2\alpha(\theta - \theta_0) \qquad (10.8)$$

Notice that these kinematic expressions for rotational motion under constant angular acceleration are of the same form as those for linear motion under constant linear acceleration with the substitutions $x \rightarrow \theta$, $v \rightarrow \omega$, and $a \rightarrow \alpha$. Table 10.1 compares the kinematic equations for rotational and linear motion. Furthermore, the expressions are valid for both rigid-body rotation and particle motion about a *fixed* axis.

EXAMPLE 10.1   Rotating Wheel

A wheel rotates with a constant angular acceleration of 3.50 rad/s². If the angular speed of the wheel is 2.00 rad/s at $t_0 = 0$, (a) what angle does the wheel rotate through in 2.00 s?

Solution

$$\theta - \theta_0 = \omega_0 t + \tfrac{1}{2}\alpha t^2 = \left(2.00\ \frac{\text{rad}}{\text{s}}\right)(2.00\ \text{s})$$

$$+ \tfrac{1}{2}\left(3.50\ \frac{\text{rad}}{\text{s}^2}\right)(2.00\ \text{s})^2$$

$$= 11.0\ \text{rad} = 630° = 1.75\ \text{rev}$$

(b) What is the angular speed at $t = 2.00$ s?

$$\omega = \omega_0 + \alpha t = 2.00\ \text{rad/s} + \left(3.50\ \frac{\text{rad}}{\text{s}^2}\right)(2.00\ \text{s})$$

$$= 9.00\ \text{rad/s}$$

We could also obtain this result using Equation 10.8 and the results of part (a). Try it!

Exercise   Find the angle that the wheel rotates through between $t = 2.00$ s and $t = 3.00$ s.

Answer   10.8 rad.

10.3   RELATIONSHIPS BETWEEN ANGULAR AND LINEAR QUANTITIES

In this section we derive some useful relationships between the angular speed and acceleration of a rotating rigid object and the linear speed and acceleration of an arbitrary point in the object. In order to do so, we must keep in mind that when a rigid object rotates about a fixed axis as in Figure 10.4, every particle of the object moves in a circle the center of which is the axis of rotation.

We can relate the angular speed of the rotating object to the tangential speed of a point *P* on the object. Since *P* moves in a circle, the linear velocity vector **v** is always tangent to the circular path, and hence the phrase *tangential velocity*. The magnitude of the tangential velocity of the point *P* is, by definition, ds/dt, where *s* is the distance traveled by this point measured along the circular path. Recalling that $s = r\theta$ (Eq. 10.1a) and noting that *r* is constant, we get

$$v = \frac{ds}{dt} = r\frac{d\theta}{dt}$$

Relationship between linear and angular speed

$$v = r\omega \qquad (10.9)$$

That is, the tangential speed of a point on a rotating rigid object equals the distance of that point from the axis of rotation multiplied by the angular speed. Therefore, although every point on the rigid object has the same *angular* speed, not every point has the same *linear* speed. In fact, Equation 10.9 shows that the linear speed of a point on the rotating object increases as one moves outward from the center of rotation, as you intuitively expect.

We can relate the angular acceleration of the rotating rigid object to the tangential acceleration of the point *P* by taking the time derivative of *v*:

$$a_t = \frac{dv}{dt} = r\frac{d\omega}{dt}$$

Relationship between linear and angular acceleration

$$a_t = r\alpha \qquad (10.10)$$

That is, the magnitude of the tangential component of the linear acceleration of a point on a rotating rigid object equals the distance of that point from the axis of rotation multiplied by the angular acceleration.

In Chapter 4 we found that a point rotating in a circular path undergoes a centripetal, or radial, acceleration of magnitude v^2/r directed toward the center of rotation (Fig. 10.5). Since $v = r\omega$ for the point *P* on the rotating object, we can

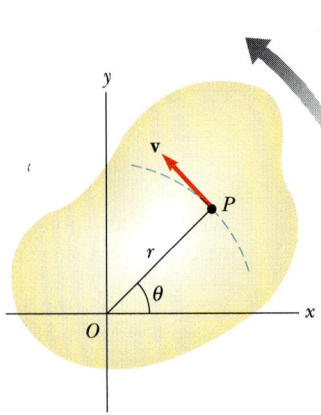

FIGURE 10.4   As a rigid object rotates about the fixed axis through *O*, the point *P* has a linear velocity **v** that is always tangent to the circular path of radius *r*.

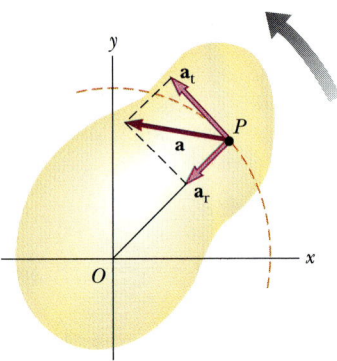

FIGURE 10.5   As a rigid object rotates about a fixed axis through *O*, the point *P* experiences a tangential component of linear acceleration, **a**$_t$, and a centripetal component of linear acceleration, **a**$_r$. The total linear acceleration of this point is **a** = **a**$_t$ + **a**$_r$.

express the magnitude of the centripetal acceleration as

$$a_r = \frac{v^2}{r} = r\omega^2 \qquad (10.11)$$

The total linear acceleration of the point is $\mathbf{a} = \mathbf{a}_t + \mathbf{a}_r$. Therefore, the magnitude of the total linear acceleration of the point P on the rotating rigid object is

$$a = \sqrt{a_t^2 + a_r^2} = \sqrt{r^2\alpha^2 + r^2\omega^4} = r\sqrt{\alpha^2 + \omega^4} \qquad (10.12)$$

CONCEPTUAL EXAMPLE 10.2

When a wheel of radius R rotates about a fixed axis as in Figure 10.3, do all points on the wheel have the same angular speed? Do they all have the same linear speed? If the angular speed is constant and equal to ω, describe the linear speeds and linear accelerations of the points located at $r = 0$, $r = R/2$, and $r = R$, where the points are measured from the center of the wheel.

Reasoning   Yes, all points on the wheel have the same angular speed. This is why we use angular quantities to describe

rotational motion. Not all points on the wheel have the same linear speed. The point at $r = 0$ has zero linear speed and zero linear acceleration; a point at $r = R/2$ has a linear speed $v = R\omega/2$ and a linear acceleration equal to the centripetal acceleration $v^2/(R/2) = R\omega^2/2$. (The tangential acceleration is zero at all points since ω is constant.) A point on the rim at $r = R$ has a linear speed $v = R\omega$ and a linear acceleration $R\omega^2$.

EXAMPLE 10.3   A Rotating Turntable

The turntable of a record player rotates initially at the rate 33 rev/min and takes 20.0 s to come to rest. (a) What is the angular acceleration of the turntable, assuming the acceleration is uniform?

Solution   Recalling that 1 rev $= 2\pi$ rad, we see that the initial angular speed is

$$\omega_0 = \left(33.0 \frac{\text{rev}}{\text{min}}\right)\left(2\pi \frac{\text{rad}}{\text{rev}}\right)\left(\frac{1}{60} \frac{\text{min}}{\text{s}}\right) = 3.46 \text{ rad/s}$$

Using $\omega = \omega_0 + \alpha t$ and the fact that $\omega = 0$ at $t = 20.0$ s, we get

$$\alpha = -\frac{\omega_0}{t} = -\frac{3.46 \text{ rad/s}}{20.0 \text{ s}} = \boxed{-0.173 \text{ rad/s}^2}$$

where the negative sign indicates that ω is decreasing.

   (b) How many rotations does the turntable make before coming to rest?

Solution   Using Equation 10.7, we find that the angular displacement in 20.0 s is

$$\Delta\theta = \theta - \theta_0 = \omega_0 t + \tfrac{1}{2}\alpha t^2$$

$$= [3.46(20.0) + \tfrac{1}{2}(-0.173)(20.0)^2]\text{rad} = \boxed{34.6 \text{ rad}}$$

This corresponds to $34.6/2\pi$ rev, or 5.50 rev.

   (c) If the radius of the turntable is 14.0 cm, what are the magnitudes of the radial and tangential components of the linear acceleration of a point on the rim at $t = 0$?

Solution   We can use $a_t = r\alpha$ and $a_r = r\omega^2$, which give

$$a_t = r\alpha = (14.0 \text{ cm})\left(-0.173 \frac{\text{rad}}{\text{s}^2}\right) = \boxed{-2.42 \text{ cm/s}^2}$$

$$a_r = r\omega_0^2 = (14.0 \text{ cm})\left(3.46 \frac{\text{rad}}{\text{s}}\right)^2 = \boxed{168 \text{ cm/s}^2}$$

Exercise   What is the initial linear speed of a point on the rim of the turntable?

Answer   48.4 cm/s.

10.4   ROTATIONAL ENERGY

Let us consider a rigid object as a collection of small particles and let us assume that the object rotates about the fixed z axis with an angular speed ω (Fig. 10.6). Each particle has kinetic energy determined by its mass and speed. If the mass of

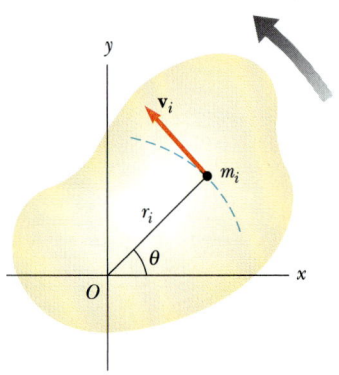

FIGURE 10.6 A rigid object rotating about the *z* axis with angular speed ω. The kinetic energy of the particle of mass m_i is $\frac{1}{2}m_i v_i^2$. The total energy of the object is $\frac{1}{2}I\omega^2$.

the *i*th particle is m_i and its speed is v_i, its kinetic energy is

$$K_i = \tfrac{1}{2}m_i v_i^2$$

To proceed further, we must recall that although every particle in the rigid object has the same angular speed, ω, the individual linear speeds depend on the perpendicular distance r_i from the axis of rotation according to the expression $v_i = r_i \omega$ (Eq. 10.9). The *total* energy of the rotating rigid object is the sum of the kinetic energies of the individual particles:

$$K_R = \sum K_i = \sum \tfrac{1}{2} m_i v_i^2 = \tfrac{1}{2}\sum m_i r_i^2 \omega^2$$

$$K_R = \tfrac{1}{2}\left(\sum m_i r_i^2\right)\omega^2 \tag{10.13}$$

where we have factored ω^2 from the sum since it is common to every particle. The quantity in parentheses is called the **moment of inertia, I:**

$$I = \sum m_i r_i^2 \tag{10.14}$$

Using this notation, we can express the rotational energy of the rotating rigid object (Eq. 10.13) as

$$K_R = \tfrac{1}{2} I\omega^2 \tag{10.15}$$

From the definition of moment of inertia, we see that it has dimensions of ML^2 ($kg \cdot m^2$ in SI units and $g \cdot cm^2$ in cgs units). Although we commonly refer to the quantity $\frac{1}{2}I\omega^2$ as the **rotational energy**, it is not a new form of energy. It is ordinary kinetic energy, because it was derived from a sum over individual kinetic energies of the particles contained in the rigid object. However, the form of the energy given by Equation 10.15 is a convenient one when dealing with rotational motion, providing we know how to calculate I. It is important that you recognize the analogy between kinetic energy associated with linear motion, $\frac{1}{2}mv^2$, and rotational energy, $\frac{1}{2}I\omega^2$. The quantities I and ω in rotational motion are analogous to m and v in linear motion, respectively. (In fact, I takes the place of m every time we compare a linear-motion equation to its rotational counterpart.)

EXAMPLE 10.4   The Oxygen Molecule

Consider the diatomic molecule oxygen, O_2, which is rotating in the *xy* plane about the *z* axis passing through its center, perpendicular to its length. The mass of each oxygen atom is 2.66×10^{-26} kg, and at room temperature, the average separation between the two oxygen atoms is $d = 1.21 \times 10^{-10}$ m (the atoms are treated as point masses). (a) Calculate the moment of inertia of the molecule about the *z* axis.

Solution   Since the distance of each atom from the *z* axis is $d/2$, the moment of inertia about the *z* axis is

$$I = \sum m_i r_i^2 = m\left(\frac{d}{2}\right)^2 + m\left(\frac{d}{2}\right)^2 = \frac{md^2}{2}$$

$$= \left(\frac{2.66 \times 10^{-26}}{2}\text{ kg}\right)(1.21 \times 10^{-10}\text{ m})^2$$

$$= 1.95 \times 10^{-46}\text{ kg} \cdot \text{m}^2$$

(b) If the angular speed of the molecule about the *z* axis is 4.60×10^{12} rad/s, what is its rotational kinetic energy?

Solution

$$K_R = \tfrac{1}{2}I\omega^2$$

$$= \tfrac{1}{2}(1.95 \times 10^{-46}\text{ kg} \cdot \text{m}^2)\left(4.60 \times 10^{12}\frac{\text{rad}}{\text{s}}\right)^2$$

$$= 2.06 \times 10^{-21}\text{ J}$$

This is approximately equal to the average kinetic energy associated with the linear motion of the molecule at room temperature, which is about 6.2×10^{-21} J.

EXAMPLE 10.5   Four Rotating Masses

Four point masses are fastened to the corners of a frame of negligible mass lying in the xy plane (Fig. 10.7). (a) If the rotation of the system occurs about the y axis with an angular speed ω, find the moment of inertia about the y axis and the rotational kinetic energy about this axis.

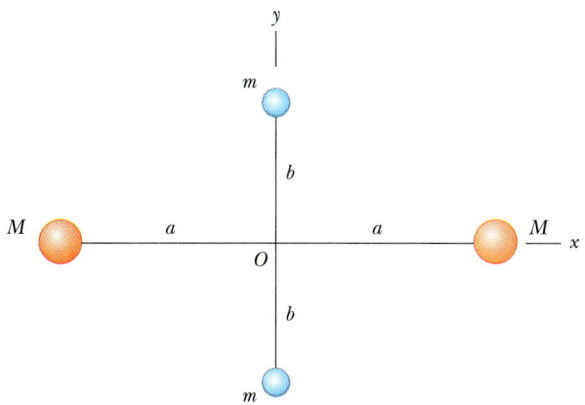

FIGURE 10.7   (Example 10.5) All four masses are at a fixed separation as shown. The moment of inertia depends on the axis about which it is evaluated.

Solution   First, note that the two masses m that lie on the y axis do not contribute to I_y (that is, $r_i = 0$ for these masses about this axis). Applying Equation 10.14, we get

$$I_y = \sum m_i r_i^2 = Ma^2 + Ma^2 = 2Ma^2$$

Therefore, the rotational energy about the y axis is

$$K_R = \tfrac{1}{2}I_y\omega^2 = \tfrac{1}{2}(2Ma^2)\omega^2 = \boxed{Ma^2\omega^2}$$

The fact that the masses m do not enter into this result makes sense, since they have no motion about the chosen axis of rotation; hence, they have no kinetic energy.

(b) Suppose the system rotates in the xy plane about an axis through O (the z axis). Calculate the moment of inertia about the z axis and the rotational energy about this axis.

Solution   Since r_i in Equation 10.14 is the *perpendicular* distance to the axis of rotation, we get

$$I_z = \sum m_i r_i^2 = Ma^2 + Ma^2 + mb^2 + mb^2 = \boxed{2Ma^2 + 2mb^2}$$

$$K_R = \tfrac{1}{2}I_z\omega^2 = \tfrac{1}{2}(2Ma^2 + 2mb^2)\omega^2 = \boxed{(Ma^2 + mb^2)\omega^2}$$

Comparing the results for (a) and (b), we conclude that the moment of inertia and, therefore, the rotational energy associated with a given angular speed depend on the axis of rotation. In (b), we expect the result to include all masses and distances because all four masses are in motion for rotation in the xy plane. Furthermore, the fact that the rotational energy in (a) is smaller than in (b) indicates that it would take less effort (work) to set the system into rotation about the y axis than about the z axis.

10.5   CALCULATION OF MOMENTS OF INERTIA

We can evaluate the moment of inertia of an extended object by imagining that the object is divided into many small volume elements, each of mass Δm. We use the definition $I = \sum r_i^2 \, \Delta m_i$ and take the limit of this sum as $\Delta m \rightarrow 0$. In this limit, the sum becomes an integral over the whole object:

$$I = \lim_{\Delta m_i \rightarrow 0} \sum r_i^2 \, \Delta m_i = \int r^2 \, dm \qquad (10.16)$$

To evaluate the moment of inertia using Equation 10.16, it is necessary to express each volume element (of mass dm) in terms of its coordinates. It is common to define a mass density in various forms. For a three-dimensional object, it is appropriate to use the *volume density*, that is, *mass per unit volume:*

$$\rho = \lim_{\Delta V \rightarrow 0} \frac{\Delta m}{\Delta V} = \frac{dm}{dV}$$

$$dm = \rho \, dV$$

Therefore, the moment of inertia for a three-dimensional object can be expressed in the form

$$I = \int \rho r^2 \, dV$$

If the object is homogeneous, then ρ is constant and the integral can be evaluated for a known geometry. If ρ is not constant, then its variation with position must be specified.

When dealing with a sheet of uniform thickness t, rather than with a three-dimensional object, it is convenient to define a *surface density* $\sigma = \rho t$, which signifies *mass per unit area*.

Finally, when mass is distributed along a uniform rod of cross-sectional area A, we sometimes use *linear density*, $\lambda = \rho A$, where λ is defined as *mass per unit length*.

EXAMPLE 10.6   Uniform Hoop

Find the moment of inertia of a uniform hoop of mass M and radius R about an axis perpendicular to the plane of the hoop, through its center (Fig. 10.8).

Solution   All mass elements are the same distance $r = R$ from the axis, and so, applying Equation 10.16, we get for the moment of inertia about the z axis through O:

$$I_z = \int r^2 \, dm = R^2 \int dm = \boxed{MR^2}$$

FIGURE 10.8   (Example 10.6) The mass elements of a uniform hoop are all the same distance from O.

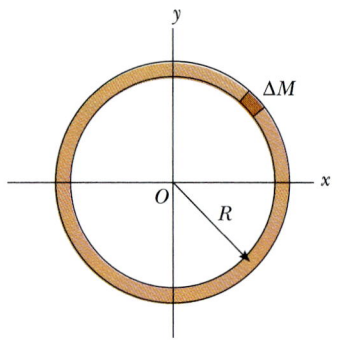

EXAMPLE 10.7   Uniform Rigid Rod

Calculate the moment of inertia of a uniform rigid rod of length L and mass M (Fig. 10.9) about an axis perpendicular to the rod (the y axis) and passing through its center of mass.

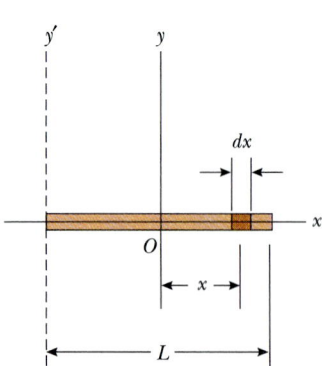

FIGURE 10.9   (Example 10.7) A uniform rigid rod of length L. The moment of inertia about the y axis is less than that about the y' axis.

Solution   The shaded length element dx has a mass dm equal to the mass per unit length multiplied by dx:

$$dm = \frac{M}{L} \, dx$$

Substituting this expression for dm into Equation 10.16, with $r = x$, we get

$$I_y = \int r^2 \, dm = \int_{-L/2}^{L/2} x^2 \, \frac{M}{L} \, dx = \frac{M}{L} \int_{-L/2}^{L/2} x^2 \, dx$$

$$= \frac{M}{L} \left[\frac{x^3}{3} \right]_{-L/2}^{L/2} = \boxed{\tfrac{1}{12} M L^2}$$

Exercise   Calculate the moment of inertia of a uniform rigid rod about an axis perpendicular to the rod and passing through one end (the y' axis). Note that the calculation requires that the limits of integration be from $x = 0$ to $x = L$.

Answer $\tfrac{1}{3} M L^2$.

EXAMPLE 10.8   Uniform Solid Cylinder

A uniform solid cylinder has a radius R, mass M, and length L. Calculate its moment of inertia about its central axis (the z axis in Fig. 10.10).

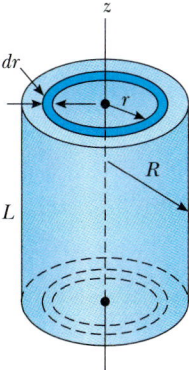

FIGURE 10.10   (Example 10.8) Calculating I about the z axis for a uniform solid cylinder.

Reasoning   It is convenient to divide the cylinder into many cylindrical shells each of radius r, thickness dr, and length L, as in Figure 10.10. (Cylindrical shells are chosen because we want all mass elements dm to have a single value for r, which makes the calculation of I more straightforward.) The volume of each shell is its cross-sectional area multiplied by the length, or $dV = dA \cdot L = (2\pi r\, dr)\, L$. If the mass per unit volume is ρ, then the mass of this differential volume element is $dm = \rho\, dV = \rho 2\pi rL\, dr$.

Solution   Substituting this expression for dm into Equation 10.16, we get

$$I_z = \int r^2\, dm = 2\pi\rho L \int_0^R r^3\, dr = \frac{\pi\rho LR^4}{2}$$

Because the total volume of the cylinder is $\pi R^2 L$, we see that $\rho = M/V = M/\pi R^2 L$. Substituting this value into the above result gives

$$I_z = \tfrac{1}{2}MR^2$$

As we saw in the previous examples, the moments of inertia of rigid bodies with simple geometry (high symmetry) are relatively easy to calculate provided the rotation axis coincides with an axis of symmetry. Table 10.2 gives the moments of inertia for a number of bodies about specific axes.[2]

The calculation of moments of inertia about an arbitrary axis can be somewhat cumbersome, even for a highly symmetric object. There is an important theorem, however, called the *parallel-axis theorem,* that often simplifies the calculation. Suppose the moment of inertia about any axis through the center of mass is I_{CM}. The parallel-axis theorem states that the moment of inertia about any axis that is parallel to and a distance D away from the axis that passes through the center of mass is

$$I = I_{CM} + MD^2 \qquad (10.17)$$

Parallel-axis theorem

Proof of the Parallel-Axis Theorem   Suppose an object rotates in the xy plane about the z axis as in Figure 10.11 and the coordinates of the center of mass are x_{CM}, y_{CM}. Let the mass element Δm have coordinates x, y. Since this element is at a distance $r = \sqrt{x^2 + y^2}$ from the z axis, the moment of inertia about the z axis is

$$I = \int r^2\, dm = \int (x^2 + y^2)\, dm$$

However, we can relate the coordinates x, y of the mass element Δm to the coordinates of the mass element relative to the center of mass, x', y'. If the coordinates of

[2] Civil engineers use the moment of inertia concept to characterize the elastic properties (rigidity) of such structures as loaded beams. Hence, this concept is often useful even in a nonrotational context.

TABLE 10.2   **Moments of Inertia of Some Rigid Objects**

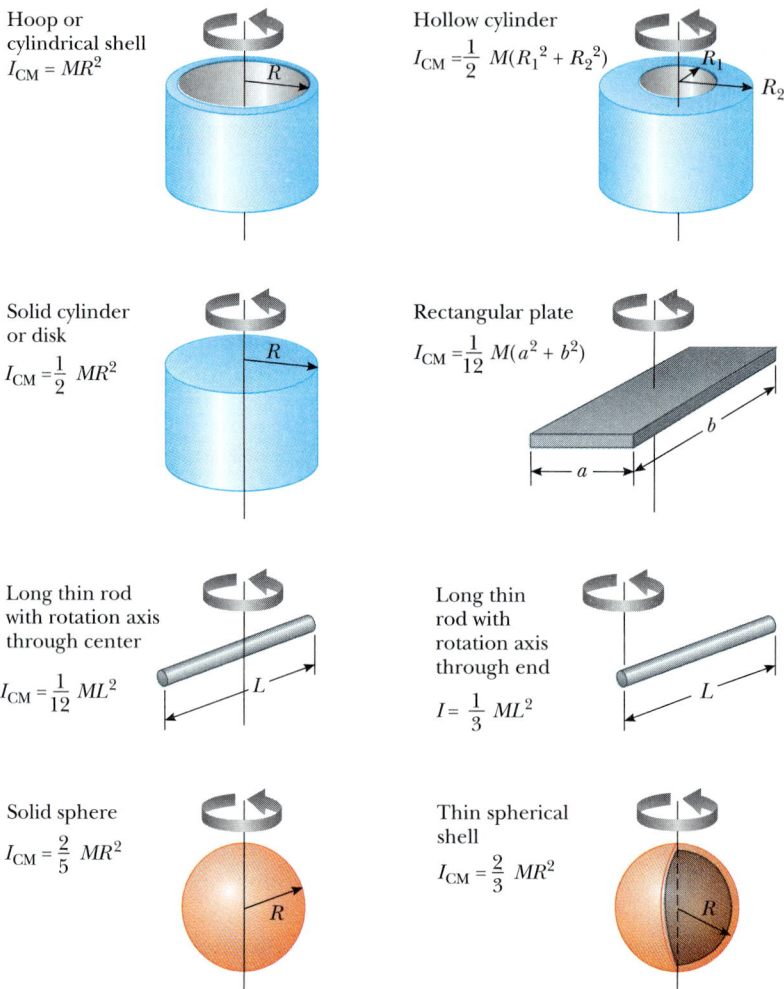

Hoop or cylindrical shell
$I_{CM} = MR^2$

Hollow cylinder
$I_{CM} = \frac{1}{2} M(R_1^2 + R_2^2)$

Solid cylinder or disk
$I_{CM} = \frac{1}{2} MR^2$

Rectangular plate
$I_{CM} = \frac{1}{12} M(a^2 + b^2)$

Long thin rod with rotation axis through center
$I_{CM} = \frac{1}{12} ML^2$

Long thin rod with rotation axis through end
$I = \frac{1}{3} ML^2$

Solid sphere
$I_{CM} = \frac{2}{5} MR^2$

Thin spherical shell
$I_{CM} = \frac{2}{3} MR^2$

the center of mass are x_{CM}, y_{CM}, then from Figure 10.11 we see that the relationships between the unprimed and primed coordinates are $x = x' + x_{CM}$ and $y = y' + y_{CM}$. Therefore,

$$I = \int [(x' + x_{CM})^2 + (y' + y_{CM})^2] \, dm$$

$$= \int [(x')^2 + (y')^2] \, dm + 2x_{CM} \int x' \, dm + 2y_{CM} \int y' \, dm + (x_{CM}^2 + y_{CM}^2) \int dm$$

The first term on the right is, by definition, the moment of inertia about an axis that is parallel to the z axis and passes through the center of mass. The second two terms on the right are zero because by definition of the center of mass, $\int x' \, dm = \int y' \, dm = 0$. Finally, the last term on the right is simply MD^2, since $\int dm = M$ and $D^2 = x_{CM}^2 + y_{CM}^2$. Therefore, we conclude that

$$I = I_{CM} + MD^2$$

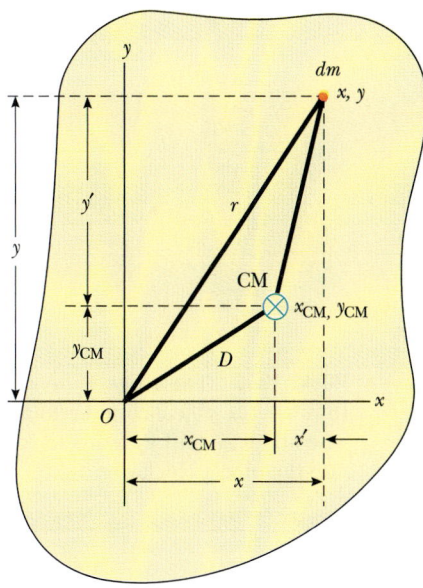

FIGURE 10.11 The parallel-axis theorem: If the moment of inertia about an axis perpendicular to the figure through the center of mass is I_{CM}, then the moment of inertia about the z axis is $I_z = I_{CM} + MD^2$.

EXAMPLE 10.9   Applying the Parallel-Axis Theorem

Consider once again a uniform rigid rod of mass M and length L as in Figure 10.9. Find the moment of inertia of the rod about an axis perpendicular to the rod through one end (the y' axis in Fig. 10.9).

Solution   Since the moment of inertia of the rod about an axis perpendicular to the rod passing through its center of mass is $ML^2/12$ and the distance between this axis and the parallel axis through its end is $D = L/2$, the parallel-axis

theorem gives

$$I = I_{CM} + MD^2 = \tfrac{1}{12}ML^2 + M\left(\frac{L}{2}\right)^2 = \tfrac{1}{3}ML^2$$

Exercise   Calculate the moment of inertia of the rod about a perpendicular axis through the point $x = L/4$.

Answer $I = \tfrac{7}{48}ML^2$.

10.6   TORQUE

When a force is exerted on a rigid object pivoted about an axis, the object tends to rotate about that axis. The tendency of a force to rotate an object about some axis is measured by a quantity called **torque** τ (Greek letter tau). Consider the wrench pivoted on the axis through O in Figure 10.12. The applied force **F** acts at a general angle ϕ (Greek letter phi) to the horizontal. We define the magnitude of the torque associated with the force **F** by the expression

$$\tau \equiv rF \sin \phi = Fd \tag{10.18}$$

Definition of torque

where r is the distance between the pivot point and the point of application of **F** and d is the perpendicular distance from the pivot point to the line of action of **F**. (The *line of action* of a force is an imaginary line extending out both ends of the vector representing the force. The dashed line extending from the tail of **F** in Figure 10.12 is part of the line of action of **F**.) From the right triangle in Figure 10.12, we see that $d = r \sin \phi$. This quantity d is called the **moment arm** (or *lever arm*) of **F**. The moment arm represents the perpendicular distance from the rotation axis to the line of action of **F**.

Moment arm

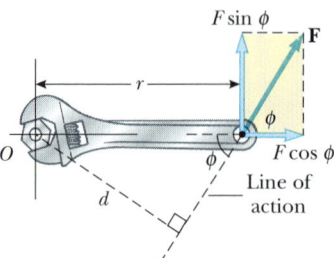

FIGURE 10.12 The force F has a greater rotating tendency about O as F increases and as the moment arm, d, increases. It is the component $F \sin \phi$ that tends to rotate the system about O.

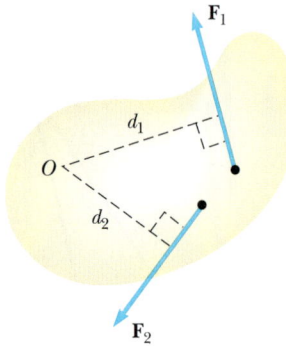

FIGURE 10.13 The force $\mathbf{F}_1$ tends to rotate the object counterclockwise about O, and $\mathbf{F}_2$ tends to rotate the object clockwise.

It is very important that you recognize that *torque is defined only when a reference axis is specified.* Torque is the product of a force and the moment arm of that force, and moment arm is defined only in terms of the axis of rotation.

The only component of **F** that tends to cause rotation is $F \sin \phi$, the component perpendicular to r. The horizontal component, $F \cos \phi$, because it passes through O, has no tendency to produce rotation. From the definition of torque, we see that the rotating tendency increases as **F** increases and as d increases. For example, it is easier to close a door if we push at the doorknob rather than at a point close to the hinge.

If there are two or more forces acting on a rigid object, as in Figure 10.13, each has a tendency to produce rotation about the pivot at O. In this example, $\mathbf{F}_2$ has a tendency to rotate the object clockwise, and $\mathbf{F}_1$ has a tendency to rotate the object counterclockwise. We use the convention that the sign of the torque resulting from a force is positive if its turning tendency of the force is counterclockwise and negative if the turning tendency is clockwise. For example, in Figure 10.13, the torque resulting from $\mathbf{F}_1$, which has a moment arm of d_1, is positive and equal to $+ F_1 d_1$; the torque from $\mathbf{F}_2$ is negative and equal to $- F_2 d_2$. Hence, the net torque about O is

$$\tau_{\text{net}} = \tau_1 + \tau_2 = F_1 d_1 - F_2 d_2$$

Torque should not be confused with force. Torque has units of force times length, newton-meters in SI units, and should be reported in these units.

The torque exerted on the nail by the hammer increases as the length of the handle (lever arm) increases. (© *Richard Megna, 1991, Fundamental Photographs*)

EXAMPLE 10.10   The Net Torque on a Cylinder

A one-piece cylinder is shaped as in Figure 10.14, with a core section protruding from the larger drum. The cylinder is free to rotate around the central axis shown in the drawing. A rope wrapped around the drum, of radius R_1, exerts a force F_1 to the right on the cylinder. A rope wrapped around the core, of radius R_2, exerts a force $\mathbf{F}_2$ downward on the cylinder. (a) What is the net torque acting on the cylinder about the rotation axis (which is the z axis in Fig. 10.14)?

Solution   The torque due to $\mathbf{F}_1$ is $-R_1F_1$ and is negative because it tends to produce a clockwise rotation. The torque due to $\mathbf{F}_2$ is $+R_2F_2$ and is positive because it tends to produce a counterclockwise rotation. Therefore, the net torque about the rotation axis is

$$\tau_{\text{net}} = \tau_1 + \tau_2 = \boxed{R_2F_2 - R_1F_1}$$

(b) Suppose $F_1 = 5.0$ N, $R_1 = 1.0$ m, $F_2 = 6.0$ N, and $R_2 = 0.50$ m. What is the net torque about the rotation axis and which way does the cylinder rotate?

$$\tau_{\text{net}} = (6.0\text{ N})(0.50\text{ m}) - (5.0\text{ N})(1.0\text{ m}) = \boxed{-2.0\text{ N}\cdot\text{m}}$$

Since the net torque is negative, the cylinder rotates clockwise.

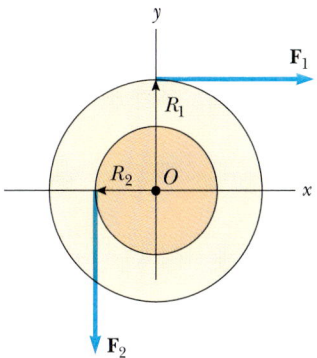

FIGURE 10.14   (Example 10.10) A solid cylinder pivoted about the z axis through O. The moment arm of $\mathbf{F}_1$ is R_1, and the moment arm of $\mathbf{F}_2$ is R_2.

10.7 RELATIONSHIP BETWEEN TORQUE AND ANGULAR ACCELERATION

In this section we show that the angular acceleration of a rigid object rotating about a fixed axis is proportional to the net torque acting about that axis. Before discussing the more complex case of rigid-body rotation, however, it is instructive first to discuss the case of a particle rotating about some fixed point under the influence of an external force.

Consider a particle of mass m rotating in a circle of radius r under the influence of a tangential force $\mathbf{F}_t$ as in Figure 10.15 and a central (radial) force $\mathbf{F}_c$ not shown in the figure. (The central force *must* be present to keep the particle moving in its circular path.) The tangential force provides a tangential acceleration $\mathbf{a}_t$, and

$$F_t = ma_t$$

The torque about the center of the circle due to $\mathbf{F}_t$ is

$$\tau = F_t r = (ma_t)r$$

Since the tangential acceleration is related to the angular acceleration through the relation $a_t = r\alpha$ (Eq. 10.10), the torque can be expressed as

$$\tau = (mr\alpha)r = (mr^2)\alpha$$

Recall from Equation 10.14 that mr^2 is the moment of inertia of the rotating mass about the z axis passing through the origin, so that

$$\tau = I\alpha \qquad (10.19)$$

That is, *the torque acting on the particle is proportional to its angular acceleration,* and the proportionality constant is the moment of inertia. It is important to note that $\tau = I\alpha$ is the rotational analogue of Newton's second law of motion, $F = ma$.

Now let us extend this discussion to a rigid object of arbitrary shape rotating about a fixed axis as in Figure 10.16. The object can be regarded as an infinite number of mass elements dm of infinitesimal size. If we impose a cartesian coordinate system on the object, then each mass element rotates in a circle about the origin, and each has a tangential acceleration $\mathbf{a}_t$ produced by an external tangential force $d\mathbf{F}_t$. For any given element, we know from Newton's second law that

$$dF_t = (dm)a_t$$

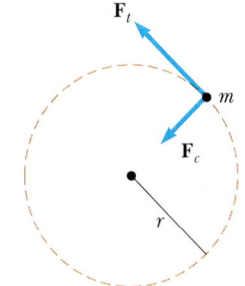

FIGURE 10.15   A particle rotating in a circle under the influence of a tangential force $\mathbf{F}_t$. A central force $\mathbf{F}_c$ must also be present to maintain the circular motion.

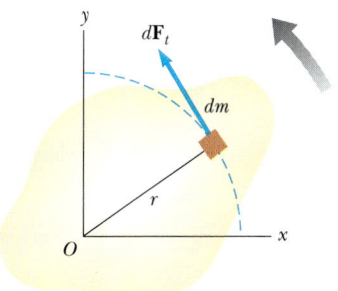

FIGURE 10.16   A rigid object rotating about an axis through O. Each mass element dm rotates about O with the same angular acceleration α, and the net torque on the object is proportional to α.

The torque $d\tau$ associated with the force $d\mathbf{F}_t$ acting about the origin is

$$d\tau = r\, dF_t = (r\, dm)\, a_t$$

Since $a_t = r\alpha$, the expression for $d\tau$ becomes

$$d\tau = (r\, dm)\, r\alpha = (r^2\, dm)\, \alpha$$

It is important to recognize that although each mass element of the rigid object may have a different linear acceleration $\mathbf{a}_t$, they all have the *same* angular acceleration, α. With this in mind, the above expression can be integrated to obtain the net torque of the external forces about O:

$$\tau_{\text{net}} = \int (r^2\, dm)\, \alpha = \alpha \int r^2\, dm$$

where α can be taken outside the integral because it is common to all mass elements. Since the moment of inertia of the object about the rotation axis through O is defined by $I = \int r^2\, dm$ (Eq. 10.16), the expression for τ_{net} becomes

Torque is proportional to angular acceleration

$$\tau_{\text{net}} = I\alpha \tag{10.20}$$

Again we see that the net torque about the rotation axis is proportional to the angular acceleration of the object with the proportionality factor being I, a quantity that depends upon the axis of rotation and upon the size and shape of the object.

In view of the complex nature of the system, it is interesting to note that $\tau_{\text{net}} = I\alpha$ is strikingly simple and in complete agreement with experimental observations. The simplicity of the result lies in the manner in which the motion is described.

Every point has the same ω and α

> Although each point on a rigid object rotating about a fixed axis may not experience the same force, linear acceleration, or linear velocity, each point experiences the same angular acceleration and angular velocity at any instant. Therefore, at any instant the rotating rigid object as a whole is characterized by specific values for angular acceleration, net torque, and angular velocity.

Finally, note that the result $\tau_{\text{net}} = I\alpha$ also applies when the forces acting on the mass elements have radial components as well as tangential components. This is because the line of action of all radial components must pass through the axis of rotation, and hence all radial components produce zero torque about that axis.

EXAMPLE 10.11   Rotating Rod

A uniform rod of length L and mass M is free to rotate about a frictionless pivot at one end in a vertical plane, as in Figure 10.17. The rod is released from rest in the horizontal position. What is the initial angular acceleration of the rod and the initial linear acceleration of the right end of the rod?

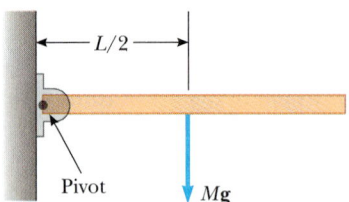

FIGURE 10.17   (Example 10.11) The uniform rod is pivoted at the left end.

Reasoning   We cannot use our kinematics equations to find α or a because they are not constant in this situation. We have enough information to find the torque, however, which we can then use in the torque-angular acceleration relationship (Eq. 10.19) to find α and then a.

$$\alpha = \frac{Mg(L/2)}{\frac{1}{3}ML^2} = \boxed{\frac{3g}{2L}}$$

Solution   The only force contributing to torque about an axis through the pivot is the force of gravity Mg exerted on the rod. (The force exerted by the pivot on the rod has zero torque about the pivot because its moment arm is zero.) To compute the torque on the rod, we can assume that the force of gravity is located at the geometric center and, thus, acts at the center of mass as shown in Figure 10.17. The magnitude of the torque due to this force about an axis through the pivot is

$$\tau = \frac{MgL}{2}$$

Since $\tau = I\alpha$, where $I = \frac{1}{3}ML^2$ for this axis of rotation (Table 10.2), we get

$$I\alpha = Mg\frac{L}{2}$$

This angular acceleration is common to all points on the rod.

To find the linear acceleration of the right end of the rod, we use the relationship $a_t = R\alpha$, with $R = L$. This gives

$$a_t = L\alpha = \boxed{\tfrac{3}{2}g}$$

This result, that $a_t > g$, is rather interesting. It means that if a coin were placed at the end of the rod, the end of the rod would fall faster than the coin when released.

Other points on the rod have a linear acceleration less than $\frac{3}{2}g$. For example, the middle of the rod has an acceleration $\frac{3}{4}g$.

Exercise   Show that in Figure 10.17, the force of gravity can be treated as a single force acting at the center of the rod.

EXAMPLE 10.12   Angular Acceleration of a Wheel

A wheel of radius R, mass M, and moment of inertia I is mounted on a frictionless, horizontal axle as in Figure 10.18. A light cord wrapped around the wheel supports an object of mass m. Calculate the linear acceleration of the object, the angular acceleration of the wheel, and the tension in the cord.

Reasoning   The torque acting on the wheel about its axis of rotation is $\tau = TR$, where T is the force exerted by the cord on the rim of the wheel. (The weight of the wheel and the normal force exerted by the axle on the wheel pass through the axis of rotation and produce no torque.) Since $\tau = I\alpha$, we get

$$\tau = I\alpha = TR$$

$$(1) \qquad \alpha = \frac{TR}{I}$$

Now let us apply Newton's second law to the motion of the suspended object, making use of the free-body diagram in Figure 10.18, taking the upward direction to be positive:

$$\sum F_y = T - mg = -ma$$

$$(2) \qquad a = \frac{mg - T}{m}$$

The linear acceleration of the suspended object is equal to the tangential acceleration of a point on the rim of the wheel. Therefore, the angular acceleration of the wheel and this linear acceleration are related by $a = R\alpha$. Using this fact to-

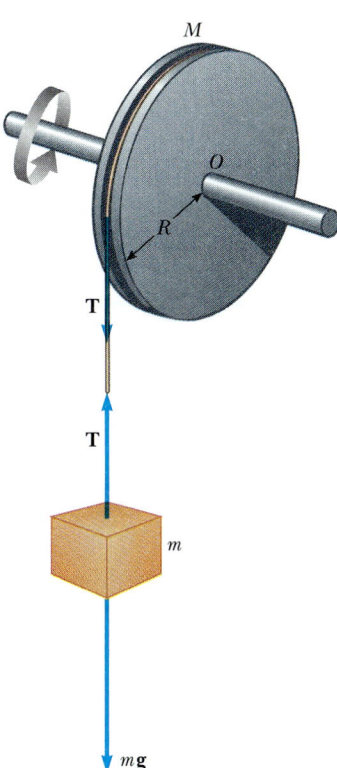

FIGURE 10.18   (Example 10.12) The cord attached to an object of mass m is wrapped around the wheel, and the tension in the cord produces a torque about the axle through O.

gether with (1) and (2) gives

(3) $a = R\alpha = \dfrac{TR^2}{I} = \dfrac{mg - T}{m}$

(4) $T = \dfrac{mg}{1 + mR^2/I}$

Substituting (4) into (3), and solving for a and α gives

$$a = \dfrac{g}{1 + I/mR^2}$$

$$\alpha = \dfrac{a}{R} = \dfrac{g}{R + I/mR}$$

Exercise   The wheel in Figure 10.18 is a solid disk of $M = 2.00$ kg, $R = 30.0$ cm, and $I = 0.0900$ kg·m². The suspended object has a mass of $m = 0.500$ kg. Find the tension in the cord and the angular acceleration of the wheel.

Answer   3.27 N; 10.9 rad/s².

EXAMPLE 10.13   The Atwood Machine Revisited

Two masses m_1 and m_2 are connected to each other by a light cord that passes over two identical pulleys, each having a moment of inertia I as in Figure 10.19a. Find the acceleration of each mass and the tensions T_1, T_2, and T_3 in the cord. (Assume no slipping between cord and pulleys.)

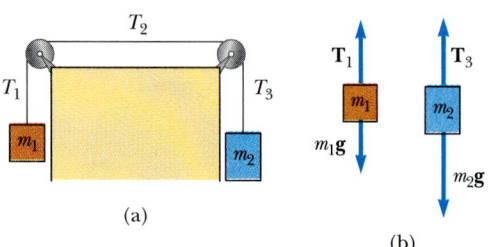

(a)

(b)

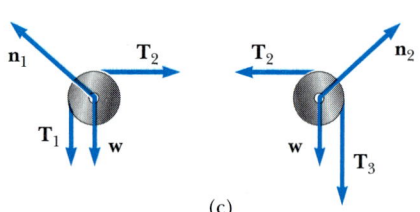

(c)

FIGURE 10.19   (Example 10.13).

Reasoning   First, let us write Newton's second law of motion as applied to each mass, taking $m_2 > m_1$. (The free-body diagrams for the two masses are shown in Figure 10.19b, and the upward direction is taken to be positive.)

(1) $T_1 - m_1 g = m_1 a$

(2) $T_3 - m_2 g = - m_2 a$

Next, we must include the effect of the pulleys on the motion. Free-body diagrams for the pulleys are shown in Fig-

ure 10.19c. The net torque about the axle for the pulley on the left is $(T_2 - T_1)R$, while the net torque for the pulley on the right is $(T_3 - T_2)R$. Using the relation $\tau_{\text{net}} = I\alpha$ for each pulley, and noting that each pulley has the same α, we get

(3) $(T_2 - T_1)R = I\alpha$

(4) $(T_3 - T_2)R = I\alpha$

We now have four equations with four unknowns: a, T_1, T_2, and T_3. These can be solved simultaneously. Adding Equations (3) and (4) gives

(5) $(T_3 - T_1)R = 2I\alpha$

Subtracting Equation (2) from Equation (1) gives

$T_1 - T_3 + m_2 g - m_1 g = (m_1 + m_2)a$

or

(6) $T_3 - T_1 = (m_2 - m_1)g - (m_1 + m_2)a$

Substituting Equation (6) into Equation (5), we have

$[(m_2 - m_1)g - (m_1 + m_2)a]R = 2I\alpha$

Since $\alpha = a/R$, this can be simplified as follows:

$(m_2 - m_1)g - (m_1 + m_2)a = 2I\dfrac{a}{R^2}$

or

(7) $a = \dfrac{(m_2 - m_1)g}{m_1 + m_2 + 2\dfrac{I}{R^2}}$

This value of a can then be substituted into Equations (1) and (2) to give T_1 and T_3. Finally, T_2 can be found from Equation (3) or Equation (4).

10.8   WORK, POWER, AND ENERGY IN ROTATIONAL MOTION

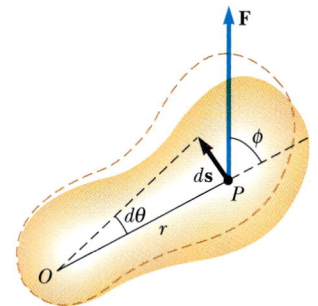

The description of a rotating rigid object would not be complete without a discussion of the rotational kinetic energy and how its change is related to the work done by external forces. In this section, we shall see that the important relationship $\tau_{\text{net}} = I\alpha$ can also be obtained by considering the rate at which energy is changing with time.

We again restrict our discussion to rotation about a fixed axis. Consider a rigid object pivoted at O in Figure 10.20. Suppose a single external force $\mathbf{F}$ is applied at P. The work done by $\mathbf{F}$ as the object rotates through an infinitesimal distance $ds = r\, d\theta$ in a time dt is

$$dW = \mathbf{F} \cdot d\mathbf{s} = (F \sin \phi) r\, d\theta$$

where $F \sin \phi$ is the tangential component of $\mathbf{F}$, or, in other words, the component of the force along the displacement. Note from Figure 10.20 that *the radial component of* $\mathbf{F}$ *does no work because it is perpendicular to the displacement.*

Since the magnitude of the torque due to $\mathbf{F}$ about O is defined to be $rF \sin \phi$ by Equation 10.18, we can write the work done for the infinitesimal rotation as

$$dW = \tau\, d\theta \qquad (10.21)$$

The rate at which work is being done by $\mathbf{F}$ as the object rotates about the fixed axis is

$$\frac{dW}{dt} = \tau \frac{d\theta}{dt}$$

Since dW/dt is the instantaneous power P (Section 7.5) delivered by the force, and since $d\theta/dt = \omega$, this expression reduces to

$$P = \frac{dW}{dt} = \tau\omega \qquad (10.22)$$

FIGURE 10.20   A rigid object rotates about an axis through O under the action of an external force F applied at P.

Power delivered to a rigid object rotating about a fixed axis

This expression is analogous to $P = Fv$ in the case of linear motion, and the expression $dW = \tau\, d\theta$ is analogous to $dW = F_x\, dx$.

The Work and Energy in Rotational Motion

In linear motion, we found the energy concept, and in particular the work-energy theorem, extremely useful in describing the motion of a system. The energy concept can be equally useful in describing rotational motion. From what we learned of linear motion, we expect that for rotation of a symmetric object about a fixed axis, the work done by external forces equals the change in the rotational energy.

To show that this is in fact the case, let us begin with $\tau = I\alpha$. Using the chain rule from the calculus, we can express the torque as

$$\tau = I\alpha = I\frac{d\omega}{dt} = I\frac{d\omega}{d\theta}\frac{d\theta}{dt} = I\frac{d\omega}{d\theta}\omega$$

Rearranging this expression and noting that $\tau\, d\theta = dW$, we get

$$\tau\, d\theta = dW = I\omega\, d\omega$$

Integrating this expression, we get for the total work done

$$W = \int_{\theta_0}^{\theta} \tau \, d\theta = \int_{\omega_0}^{\omega} I\omega \, d\omega = \tfrac{1}{2}I\omega^2 - \tfrac{1}{2}I\omega_0^2 \qquad (10.23)$$

where the angular speed changes from ω_0 to ω as the angular displacement changes from θ_0 to θ. That is,

> the net work done by external forces in rotating a symmetric rigid object about a fixed axis equals the change in the object's rotational energy.

Table 10.3 lists the various equations we have discussed pertaining to rotational motion, together with the analogous expressions for linear motion. The last two equations in Table 10.3, involving angular momentum L, are discussed in Chapter 11 and are included here only for the sake of completeness.

TABLE 10.3   A Comparison of Useful Equations in Rotational and Translational Motion

Rotational Motion About a Fixed Axis	Linear Motion
Angular speed $\omega = d\theta/dt$	Linear speed $v = dx/dt$
Angular acceleration $\alpha = d\omega/dt$	Linear acceleration $a = dv/dt$
Resultant torque $\Sigma\tau = I\alpha$	Resultant force $\Sigma F = Ma$
If $\alpha = $ constant $\begin{cases} \omega = \omega_0 + \alpha t \\ \theta - \theta_0 = \omega_0 t + \tfrac{1}{2}\alpha t^2 \\ \omega^2 = \omega_0^2 + 2\alpha(\theta - \theta_0) \end{cases}$	If $a = $ constant $\begin{cases} v = v_0 + at \\ x - x_0 = v_0 t + \tfrac{1}{2}at^2 \\ v^2 = v_0^2 + 2a(x - x_0) \end{cases}$
Work $W = \displaystyle\int_{\theta_0}^{\theta} \tau \, d\theta$	Work $W = \displaystyle\int_{x_0}^{x} F_x \, dx$
Rotational energy $K_R = \tfrac{1}{2}I\omega^2$	Translational energy $K = \tfrac{1}{2}mv^2$
Power $P = \tau\omega$	Power $P = Fv$
Angular momentum $L = I\omega$	Linear momentum $p = mv$
Resultant torque $\tau = dL/dt$	Resultant force $F = dp/dt$

CONCEPTUAL EXAMPLE 10.14

Consider an object in the shape of a hoop lying in the xy plane with all of its mass concentrated on its rim. In two separate experiments, the hoop is rotated by an external agent from rest to an angular speed ω. In one experiment, the rotation occurs about the z axis passing through the center of the hoop. In the other experiment, the rotation occurs about an axis parallel to z passing through a point P on the rim of the hoop. Which rotation requires more work?

Reasoning   Rotation about the axis through P requires more work. The moment of inertia of the hoop about the axis

through the center of the hoop is $I_{CM} = MR^2$, whereas by the parallel axis theorem, the moment of inertia about the axis through P is $I_P = I_{CM} + MR^2 = MR^2 + MR^2 = 2MR^2$. That is, $I_P = 2I_{CM}$. Using Equation 10.23, taking the initial angular speed to be zero, we see that the work required by the external agent for rotation about the axis through P is twice the work required for rotation about the axis through the center of the hoop.

EXAMPLE 10.15   Rotating Rod Revisited

A uniform rod of length L and mass M is free to rotate on a frictionless pin through one end (Fig. 10.21). The rod is released from rest in the horizontal position. (a) What is the angular speed of the rod at its lowest position?

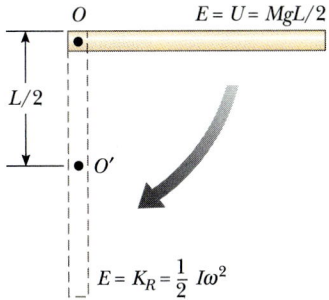

FIGURE 10.21   (Example 10.15) A uniform rigid rod pivoted at O rotates in a vertical plane under the action of gravity.

Reasoning and Solution   The question can be answered by considering the mechanical energy of the system. When the rod is horizontal, it has no rotational energy. The potential energy relative to the lowest position of its center of mass (O') is $MgL/2$. When it reaches its lowest position, the energy is entirely rotational energy, $\frac{1}{2}I\omega^2$, where I is the moment of inertia about the pivot. Since $I = \frac{1}{3}ML^2$ (Table 10.2) and since mechanical energy is constant, we have

$$\tfrac{1}{2}MgL = \tfrac{1}{2}I\omega^2 = \tfrac{1}{2}(\tfrac{1}{3}ML^2)\omega^2$$

$$\omega = \boxed{\sqrt{\dfrac{3g}{L}}}$$

(b) Determine the linear speed of the center of mass and the linear speed of the lowest point on the rod in the vertical position.

Solution

$$v_{\text{CM}} = r\omega = \frac{L}{2}\,\omega = \boxed{\tfrac{1}{2}\sqrt{3gL}}$$

The lowest point on the rod has a linear speed equal to $2v_{\text{CM}} = \sqrt{3gL}$.

EXAMPLE 10.16   Connected Masses

Consider two masses connected by a string passing over a pulley having a moment of inertia I about its axis of rotation, as in Figure 10.22. The string does not slip on the pulley, and the system is released from rest. Find the linear speeds of the masses after m_2 descends through a distance h, and the angular speed of the pulley at this time.

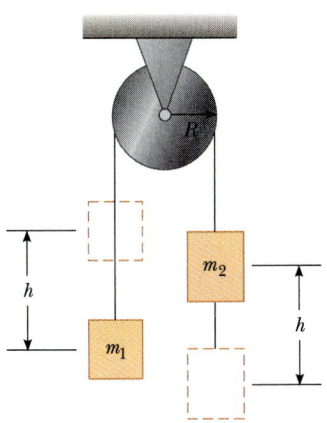

FIGURE 10.22   (Example 10.16).

Reasoning and Solution   If we neglect friction in the system, then mechanical energy is constant and we can state that the increase in kinetic energy equals the decrease in potential energy. Since $K_i = 0$ (the system is initially at rest), we have

$$\Delta K = K_f - K_i = \tfrac{1}{2}m_1 v^2 + \tfrac{1}{2}m_2 v^2 + \tfrac{1}{2}I\omega^2$$

where m_1 and m_2 have a common speed. Because $v = R\omega$, this expression becomes

$$\Delta K = \tfrac{1}{2}\left(m_1 + m_2 + \frac{I}{R^2} \right)v^2$$

From Figure 10.22, we see that m_2 loses potential energy while m_1 gains potential energy. That is, $\Delta U_2 = -m_2 gh$ and $\Delta U_1 = m_1 gh$. Applying the principle of conservation of energy in the form $\Delta K + \Delta U_1 + \Delta U_2 = 0$ gives

$$\tfrac{1}{2}\left(m_1 + m_2 + \frac{I}{R^2} \right)v^2 + m_1 gh - m_2 gh = 0$$

$$\boxed{v = \left[\dfrac{2(m_2 - m_1)\,gh}{\left(m_1 + m_2 + \dfrac{I}{R^2} \right)} \right]^{1/2}}$$

Since $v = R\omega$, the angular speed of the pulley at this instant is $\omega = v/R$.

Exercise   Repeat the calculation of v in Example 10.16 using $\tau_{\text{net}} = I\alpha$ applied to the pulley and Newton's second law applied to m_1 and m_2. Make use of the procedure presented in Examples 10.12 and 10.13.

SUMMARY

If a particle rotates in a circle of radius r through an angle θ (measured in radians), the arc length it moves through is $s = r\theta$.

The **instantaneous angular speed** of a particle rotating in a circle or of a rigid object rotating about a fixed axis is

$$\omega = \frac{d\theta}{dt} \tag{10.3}$$

where ω is in rad/s, or s^{-1}.

The **instantaneous angular acceleration** of a rotating object is

$$\alpha = \frac{d\omega}{dt} \tag{10.5}$$

and has units of rad/s², or s^{-2}. When a rigid object rotates about a fixed axis, every part of the object has the same angular speed and the same angular acceleration.

If a particle or object rotates about a fixed axis under constant angular acceleration α, one can apply equations of kinematics in analogy with kinematic equations for linear motion under constant linear acceleration:

$$\omega = \omega_0 + \alpha t \tag{10.6}$$

$$\theta = \theta_0 + \omega_0 t + \tfrac{1}{2}\alpha t^2 \tag{10.7}$$

$$\omega^2 = \omega_0{}^2 + 2\alpha(\theta - \theta_0) \tag{10.8}$$

$$\theta = \theta_0 + \tfrac{1}{2}(\omega_0 + \omega)t$$

When a rigid object rotates about a fixed axis, the angular speed and angular acceleration are related to the linear speed and tangential linear acceleration through the relationships

$$v = r\omega \tag{10.9}$$

$$a_t = r\alpha \tag{10.10}$$

The **moment of inertia of a system of particles** is

$$I = \sum m_i r_i^2 \tag{10.14}$$

If a rigid object rotates about a fixed axis with angular speed ω, its **rotational energy** can be written

$$K_R = \tfrac{1}{2}I\omega^2 \tag{10.15}$$

where I is the moment of inertia about the axis of rotation.

The **moment of inertia of a rigid object** is

$$I = \int r^2 \, dm \tag{10.16}$$

where r is the distance from the mass element dm to the axis of rotation.

The magnitude of the **torque** associated with a force **F** acting on an object is

$$\tau = Fd \tag{10.18}$$

where d is the moment arm of the force, which is the perpendicular distance from some origin to the line of action of the force. Torque is a measure of the tendency of the force to rotate the object about some axis.

If a rigid object free to rotate about a fixed axis has a **net external torque** acting on it, the object undergoes an angular acceleration α, where

$$\tau_{\text{net}} = I\alpha \tag{10.20}$$

The rate at which work is done by external forces in rotating a rigid object about a fixed axis, or the **power** delivered, is

$$P = \tau\omega \qquad\qquad (10.22)$$

The net work done by external forces in rotating a rigid object about a fixed axis equals the change in the rotational energy of the object:

$$W = \tfrac{1}{2}I\omega^2 - \tfrac{1}{2}I\omega_0{}^2 \qquad\qquad (10.23)$$

QUESTIONS

1. What is the magnitude of the angular velocity, ω, of the second hand of a clock? What is the direction of ω as you view a clock hanging vertically? What is the magnitude of the angular acceleration α of the second hand?
2. A wheel rotates counterclockwise in the xy plane. What is the direction of ω? What is the direction of α if the angular velocity is decreasing in time?
3. Are the kinematic expressions for θ, ω, and α valid when the angular displacement is measured in degrees instead of in radians?
4. A turntable rotates at a constant rate of 45 rev/min. What is its angular speed in radians per second? What is the magnitude of its angular acceleration?
5. Suppose $a = b$ and $M > m$ for the system of particles described in Figure 10.7. About what axis (x, y, or z) does the moment of inertia have the smallest value? the largest value?
6. Suppose the rod in Figure 10.9 has a nonuniform mass distribution. In general, would the moment of inertia about the y axis still equal $\frac{1}{12}ML^2$? If not, could the moment of inertia be calculated without knowledge of the manner in which the mass is distributed?
7. Suppose that only two external forces act on a rigid body, and the two forces are equal in magnitude but opposite in direction. Under what conditions does the body rotate?
8. Explain how you might use the apparatus described in Example 10.12 to determine the moment of inertia of the wheel. (If the wheel does not have a uniform mass density, the moment of inertia is not necessarily equal to $\frac{1}{2}MR^2$.)
9. Using the results from Example 10.12, how would you

calculate the angular speed of the wheel and the linear speed of the suspended mass at $t = 2$ s, if the system is released from rest at $t = 0$? Is the expression $v = R\omega$ valid in this situation?
10. If a small sphere of mass M were placed at the end of the rod in Figure 10.21, would the result for ω be greater than, less than, or equal to the value obtained in Example 10.15?
11. Explain why changing the axis of rotation of an object changes its moment of inertia.
12. Is it possible to change the translational kinetic energy of an object without changing its rotational energy?
13. Two cylinders having the same dimensions are set into rotation about their long axes with the same angular speed. One is hollow, and the other if filled with water. Which cylinder will be easier to stop rotating?
14. Must an object be rotating to have a nonzero moment of inertia?
15. If you see an object rotating, is there necessarily a net torque acting on it?
16. Can a (momentarily) stationary object have a nonzero angular acceleration?
17. The polar diameter of the Earth is slightly less than the equatorial diameter. How would the moment of inertia of the Earth change if some mass from near the equator were removed and transferred to the polar regions to make the Earth a perfect sphere?
18. During a wrecking operation, a tall chimney is toppled by an explosive charge at its base. The chimney ruptures on its lower half while toppling, so that the bottom part reaches the ground before the top part. Explain why this occurs.

PROBLEMS

Section 10.2   Rotational Kinematics: Rotational Motion with Constant Angular Acceleration

1. A wheel starts from rest and rotates with constant angular acceleration to an angular speed of 12.0

rad/s in 3.00 s. Find (a) the magnitude of the angular acceleration of the wheel and (b) the angle in radians through which it rotates in this time.
2. The turntable of a record player rotates at a rate of $33\frac{1}{3}$ rev/min and takes 60.0 s to come to rest when

□ indicates problems that have full solutions available in the Student Solutions Manual and Study Guide.

switched off. Calculate (a) the magnitude of its angular acceleration and (b) the number of revolutions it makes before coming to rest.

3. What is the angular speed in radians per second of (a) the Earth in its orbit about the Sun and (b) the Moon in its orbit about the Earth?

4. (a) The hour and minute hands on a clock coincide at 12 o'clock. Determine all the other times (up to the second) when these two hands coincide. (b) If the clock has a second hand, determine all the times when the three hands coincide, given that they all coincide at 12 o'clock.

5. An electric motor rotating a grinding wheel at 100 rev/min is switched off. Assuming constant negative angular acceleration of magnitude 2.00 rad/s^2, (a) how long does it take the wheel to stop? (b) Through how many radians does it turn during the time found in (a)?

6. The angular position of a point on a wheel is described by $\theta = 5.0 + 10t + 2.0t^2$ rad. Determine the angular position, speed, and acceleration of the point at $t = 0$ and $t = 3.0$ s.

7. A rotating wheel requires 3.0 s to rotate 37 rev. Its angular speed at the end of the 3.0 s interval is 98 rad/s. What is its constant angular acceleration?

8. A car accelerates uniformly from rest and reaches a speed of 22 m/s in 9 s. If the diameter of a tire is 58 cm, find (a) the number of revolutions the tire makes during this motion, assuming no slipping. (b) What is the final rotational speed of a tire in revolutions per second?

Section 10.3   Relationships Between Angular and Linear Quantities

9. A racing car travels on a circular track of radius 250 m. If the car moves with a constant linear speed of 45.0 m/s, find (a) its angular speed and (b) the magnitude and direction of its acceleration.

9A. A racing car travels on a circular track of radius R. If the car moves with a constant linear speed v, find (a) its angular speed and (b) the magnitude and direction of its acceleration.

10. A car traveling on a flat (unbanked) circular track accelerates uniformly from rest with a tangential acceleration of 1.70 m/s^2. The car makes it one quarter of the way around the circle before skidding off the track. Determine the coefficient of static friction between car and track.

11. A wheel 2.00 m in diameter rotates with a constant angular acceleration of 4.00 rad/s^2. The wheel starts at rest at $t = 0$, and the radius vector at point P on the rim makes an angle of 57.3° with the horizontal at this time. At $t = 2.00$ s, find (a) the angular speed of the wheel, (b) the linear speed and acceleration of the point P, and (c) the position of the point P.

12. A discus thrower accelerates a discus from rest to a speed of 25.0 m/s by whirling it through 1.25 rev. Assume the discus moves on the arc of a circle 1.00 m in radius. (a) Calculate the final angular speed of the discus. (b) Determine the magnitude of the angular acceleration of the discus, assuming it to be constant. (c) Calculate the acceleration time.

13. A disk 8.00 cm in radius rotates at a constant rate of 1200 rev/min about its central axis. Determine (a) its angular speed, (b) the linear speed at a point 3.00 cm from its center, (c) the radial acceleration of a point on the rim, and (d) the total distance a point on the rim moves in 2.00 s.

14. A car is traveling at 36 km/h on a straight road. The radius of the tires is 25 cm. Find the angular speed of one of the tires with its axle taken as the axis of rotation.

15. A 6.00-kg block is released from A on a frictionless track shown in Figure P10.15. Determine the radial and tangential components of acceleration for the block at P.

15A. A block of mass m is released from A on a frictionless track shown in Figure P10.15. Determine the radial and tangential components of acceleration for the block at P.

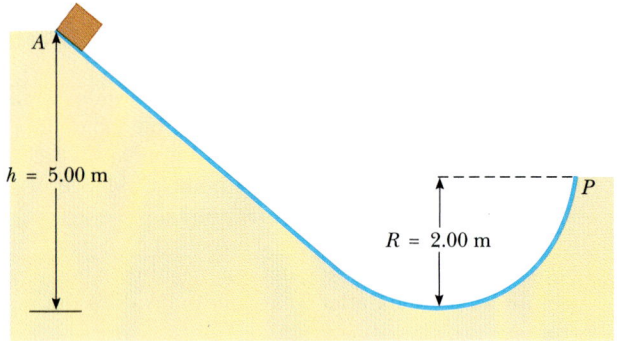

FIGURE P10.15

Section 10.4   Rotational Energy

16. A can of soup has a mass 215 g, height 10.8 cm, and diameter 6.38 cm. It is placed at rest on the top of an incline that is 3.00 m long and at 25.0° to the horizontal. Using energy methods, calculate the moment of inertia of the can if it takes 1.50 s to reach the bottom of the incline.

17. The four particles in Figure P10.17 are connected by rigid rods of negligible mass. The origin is at the center of the rectangle. If the system rotates in the xy plane about the z axis with an angular speed of 6.00 rad/s, calculate (a) the moment of inertia of the system about the z axis and (b) the rotational energy of the system.

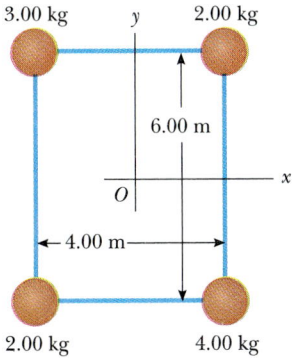

3.00 kg   *y*   2.00 kg

6.00 m

O — *x*

←4.00 m→

2.00 kg   4.00 kg

FIGURE P10.17

18. The center of mass of a pitched baseball (radius = 3.8 cm) moves at 38 m/s. The ball spins about an axis through its center of mass with an angular speed of 125 rad/s. Calculate the ratio of the rotational energy to the translational kinetic energy. Treat the ball as a uniform sphere.

18A. The center of mass of a pitched baseball of radius R moves at a speed v. The ball spins about an axis through its center of mass with an angular speed ω. Calculate the ratio of the rotational energy to the translational kinetic energy. Treat the ball as a uniform sphere.

19. Three particles are connected by rigid rods of negligible mass lying along the y axis (Fig. P10.19). If the system rotates about the x axis with an angular speed of 2.00 rad/s, find (a) the moment of inertia about the x axis and the total rotational energy evaluated from $\frac{1}{2}I\omega^2$ and (b) the linear speed of each particle and the total energy evaluated from $\Sigma\frac{1}{2}m_iv_i^2$.

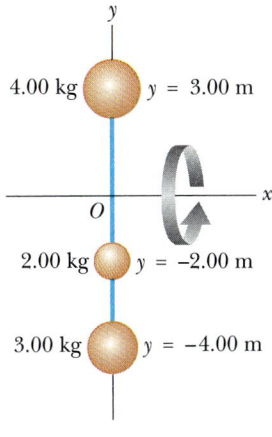

y

4.00 kg $y = 3.00$ m

O — *x*

2.00 kg $y = -2.00$ m

3.00 kg $y = -4.00$ m

FIGURE P10.19

20. The hour and minute hands of Big Ben in London are 2.7 m and 4.5 m long and have masses of 60 kg

and 100 kg, respectively. Calculate the total rotational kinetic energy of the two hands about the axis of rotation. (Model the hands as long, thin rods.)

21. Two masses M and m are connected by a rigid rod of length L and negligible mass as in Figure P10.21. For an axis perpendicular to the rod, show that the system has the minimum moment of inertia when the axis passes through the center of mass. Show that this moment of inertia is $I = \mu L^2$, where $\mu = mM/(m + M)$.

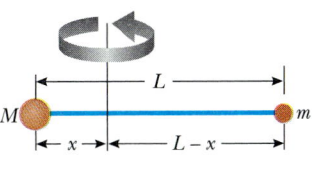

L

M *m*

←*x*→←*L − x*→

FIGURE P10.21

Section 10.5   Calculation of Moments of Inertia

22. Three identical thin rods of length L and mass m are placed perpendicular to each other as shown in Figure P10.22. The setup is rotated about an axis that passes through the end of one rod and is parallel to another. Determine the moment of inertia of this arrangement.

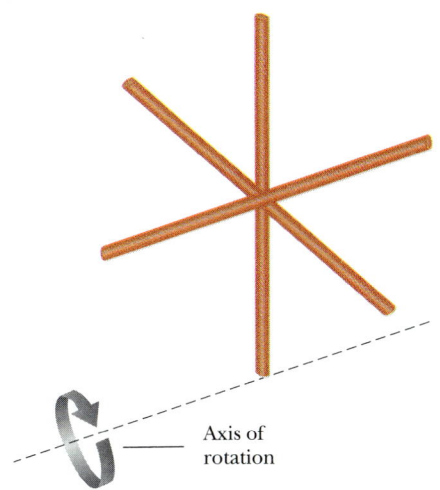

Axis of rotation

FIGURE P10.22

23. Use the parallel-axis theorem and Table 10.2 to find the moments of inertia of (a) a solid cylinder about an axis parallel to the center-of-mass axis and passing through the edge of the cylinder and (b) a solid sphere about an axis tangent to its surface.

Section 10.6   Torque

24. Using the lengths and masses given in Problem 20, (a) determine the total torque due to the weight of Big Ben's hands about the axis of rotation when the

time reads (i) 3:00, (ii) 5:15, (iii) 6:00, (iv) 8:20, (v) 9:45. (Model the hands as long, thin rods.) (b) Determine all times (up to the second) when the total torque about the axis of rotation is zero.

25. Find the net torque on the wheel in Figure P10.25 about the axle through O if $a = 10$ cm and $b = 25$ cm.

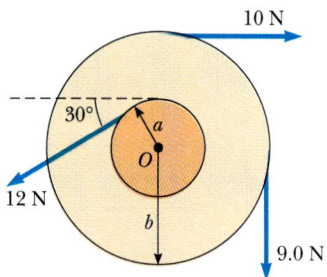

FIGURE P10.25

26. Find the mass m needed to balance the 1500-kg truck on the incline shown in Figure P10.26. Assume all pulleys are frictionless and massless.

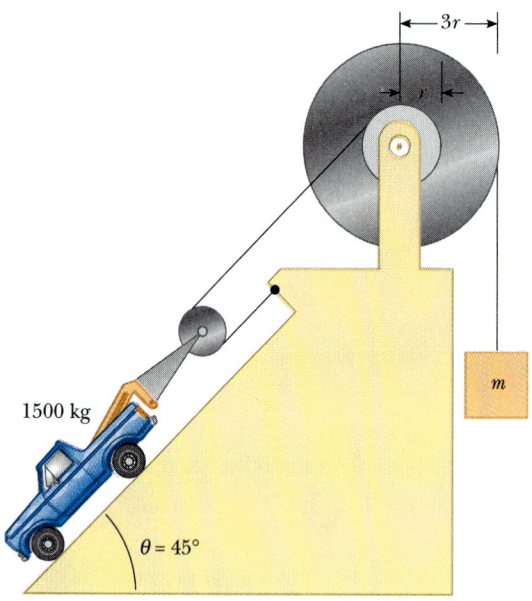

FIGURE P10.26

27. A flywheel in the shape of a solid cylinder of radius $R = 0.60$ m and mass $M = 15$ kg can be brought to an angular speed of 12 rad/s in 0.60 s by a motor exerting a constant torque. After the motor is turned off, the flywheel makes 20 rev before coming to rest because of friction (assumed constant during rotation). What percentage of the power generated by the motor is used to overcome friction?

Section 10.7 Relationship Between Torque and Angular Acceleration

28. A bicycle wheel has a diameter of 64.0 cm and a mass of 1.80 kg. The bicycle is placed on a stationary stand on rollers, and a resistive force of 120 N is applied to the rim of the tire. Assume all the mass of the wheel is concentrated on the outside radius. To give the wheel an acceleration of 4.5 rad/s², what force must be applied by a chain passing over (a) a 9.0-cm-diameter sprocket and (b) a 5.6-cm-diameter sprocket?

29. A block of mass $m_1 = 2.00$ kg and one of mass $m_2 = 6.00$ kg are connected by a massless string over a pulley that is in the shape of a disk having radius $R = 0.25$ m and mass $M = 10.0$ kg. In addition, the blocks are allowed to move on a fixed block-wedge of angle $\theta = 30.0°$ as in Figure P10.29. The coefficient of kinetic friction is 0.36 for both blocks. Determine (a) the acceleration of the two blocks and (b) the tensions in the string on both sides of the pulley.

29A. A block of mass m_1 and one of mass m_2 are connected by a massless string over a pulley that is in the shape of a disk having radius R and mass M. In addition, the blocks are allowed to move on a fixed block-wedge of angle θ as in Figure P10.29. Determine (a) the acceleration of the two blocks and (b) the tensions in the string on both sides of the pulley.

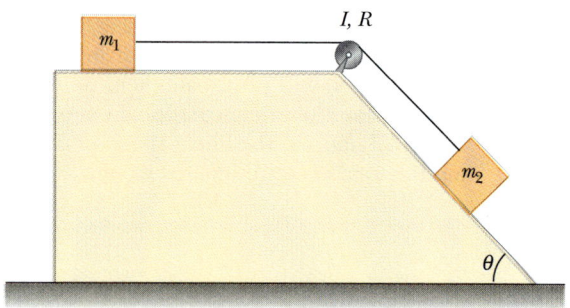

FIGURE P10.29

30. A model airplane whose mass is 0.75 kg is tethered by a wire so that it flies in a circle 30 m in radius. The airplane engine provides a net thrust of 0.80 N perpendicular to the tethering wire. (a) Find the torque the net thrust produces about the center of the circle. (b) Find the angular acceleration of the airplane when it is in level flight. (c) Find the linear acceleration of the airplane tangent to its flight path.

Section 10.8 Work, Power, and Energy in Rotational Motion

31. A cylindrical rod 24 cm long has mass 1.2 kg and radius 1.5 cm. A 20-kg ball of diameter 8.0 cm is attached to one end. The arrangement is originally ver-

tical with the ball at the top and is free to pivot about the other end. After the ball-rod system falls a quarter turn, what are (a) its rotational kinetic energy, (b) its angular speed, and (c) the linear speed of the ball? (d) How does this linear ball speed compare with the speed if the ball had fallen freely from a distance equal to the radius (28 cm)?

32. A 15-kg mass and a 10-kg mass are suspended by a pulley that has a radius of 10 cm and a mass of 3.0 kg (Fig. P10.32). The cord has a negligible mass and causes the pulley to rotate without slipping. The pulley rotates without friction. The masses start from rest 3.0 m apart. Treat the pulley as a uniform disk, and determine the speeds of the two masses as they pass each other.

32A. A mass m_1 and a mass m_2 are suspended by a pulley that has a radius R and a mass m_3 (Fig. P10.32). The cord has a negligible mass and causes the pulley to rotate without slipping. The pulley rotates without friction. The masses start from rest a distance d apart. Treat the pulley as a uniform disk, and determine the speeds of the two masses as they pass each other.

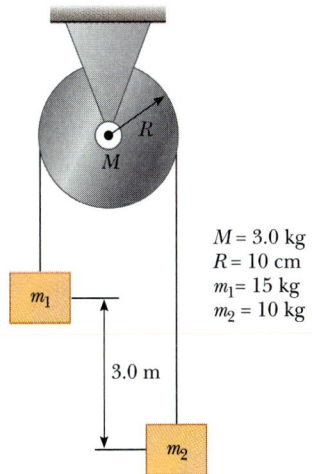

$M = 3.0$ kg
$R = 10$ cm
$m_1 = 15$ kg
$m_2 = 10$ kg

FIGURE P10.32

33. (a) A uniform solid disk of radius R and mass M is free to rotate on a frictionless pivot through a point on its rim (Fig. P10.33). If the disk is released from rest in the position shown by the green circle, what is the speed of its center of mass when the disk reaches the position indicated by the dashed circle? (b) What is the speed of the lowest point on the disk in the dashed position? (c) Repeat part (a) for a uniform hoop.

34. A potter's wheel, a thick stone disk of radius 0.50 m and mass 100 kg, is freely rotating at 50 rev/min. The potter can stop the wheel in 6.0 s by pressing a

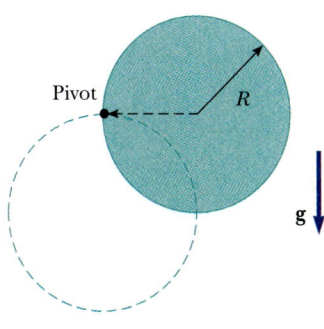

FIGURE P10.33

wet rag against the rim and exerting a radially inward force of 70 N. Find the effective coefficient of kinetic friction between wheel and wet rag.

35. A weight of 50.0 N is attached to the free end of a light string wrapped around a pulley of radius 0.250 m and mass 3.00 kg. The pulley is free to rotate in a vertical plane about the horizontal axis passing through its center. The weight is released 6.00 m above the floor. (a) Determine the tension in the string, the acceleration of the mass, and the speed with which the weight hits the floor. (b) Find the speed calculated in part (a) by using the principle of conservation of energy.

36. A bus is designed to draw its power from a rotating flywheel that is brought up to its maximum speed (3000 rpm) by an electric motor. The flywheel is a solid cylinder of mass 1000 kg and diameter 1.00 m. If the bus requires an average power of 10.0 kW, how long does the flywheel rotate?

ADDITIONAL PROBLEMS

37. A grinding wheel is in the form of a uniform solid disk of radius 7.00 cm and mass 2.00 kg. It starts from rest and accelerates uniformly under the action of the constant torque of 0.600 N·m that the motor exerts on the wheel. (a) How long does the wheel take to reach its final speed of 1200 rev/min? (b) Through how many revolutions does it turn while accelerating?

38. Two identical disks having moment of inertia 0.0080 kg·m² and radius 10 cm are free to pivot on frictionless axles. One disk has a 2.0-N weight attached to it by a cord wrapped around the circumference, while the other has a force of 2.0 N applied to its cord. Which disk is turning faster after 1.0 m of cord has been unwrapped? Explain.

39. The fishing pole in Figure P10.39 makes an angle of 20° with the horizontal. What is the torque exerted by the fish about an axis perpendicular to the page and passing through the fisher's hand.

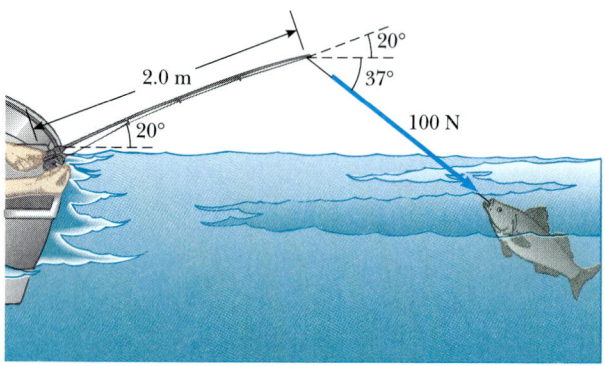

FIGURE P10.39

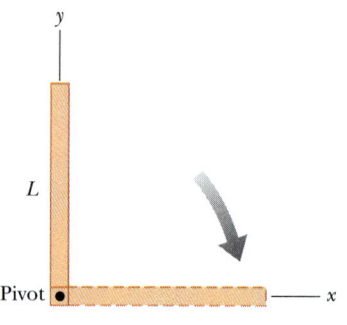

FIGURE P10.43

40. The density of the Earth, at any distance r from its center, is approximately

$$\rho = \left[14.2 - 11.6 \left(\frac{r}{R} \right) \right] \times 10^3 \ \text{kg/m}^3$$

where R is the radius of the Earth. Show that this density leads to a moment of inertia $I = 0.330MR^2$ about an axis through the center, where M is the mass of the Earth.

41. A 4.00-m length of light nylon cord is wound around a uniform cylindrical spool of radius 0.500 m and mass 1.00 kg. The spool is mounted on a frictionless axle and is initially at rest. The cord is pulled from the spool with a constant acceleration of magnitude 2.50 m/s². (a) How much work has been done on the spool when it reaches an angular speed of 8.00 rad/s? (b) Assuming there is enough cord on the spool, how long does it take the spool to reach this angular speed? (c) Is there enough cord on the spool?

42. A flywheel in the form of a heavy circular disk of diameter 0.600 m and mass 200 kg is mounted on a frictionless bearing. A motor connected to the flywheel accelerates it from rest to 1000 rev/min. (a) What is the moment of inertia of the flywheel? (b) How much work is done on it during this acceleration? (c) After 1000 rev/min is achieved, the motor is disengaged. A friction brake is used to slow the rotational rate to 500 rev/min. How much energy is dissipated as heat in the friction brake?

43. A long uniform rod of length L and mass M is pivoted about a horizontal, frictionless pin through one end. The rod is released from rest in a vertical position as in Figure P10.43. At the instant the rod is horizontal, find (a) its angular speed, (b) the magnitude of its angular acceleration, (c) the x and y components of the acceleration of its center of mass, and (d) the components of the reaction force at the pivot.

44. A bicycle is turned upside down while its owner repairs a flat tire. A friend spins the other wheel of radius 0.381 m and observes that drops of water fly off tangentially. She measures the height reached by drops moving vertically (Fig. P10.44). A drop that breaks loose from the tire on one turn rises $h = 54.0$ cm above the tangent point. A drop that breaks loose on the next turn rises 51.0 cm above the tangent point. The height to which the drops rise decreases because the angular speed of the wheel decreases. From this information, determine the magnitude of the average angular acceleration of the wheel.

44A. A bicycle is turned upside down while its owner repairs a flat tire. A friend spins the other wheel of radius R and observes that drops of water fly off tangentially. She measures the height reached by drops moving vertically (Fig. P10.44). A drop that breaks loose from the tire on one turn rises a distance h_1 above the tangent point. A drop that breaks loose on the next turn rises a distance $h_2 < h_1$ above the tangent point. The height to which the drops rise decreases because the angular speed of the wheel decreases. From this information, determine the magnitude of the average angular acceleration of the wheel.

45. *Radius of gyration:* For any given rotational axis, the radius of gyration, K, of a rigid body is defined by the expression $K^2 = I/M$, where M is the total mass of the body and I is the moment of inertia about the given axis. In other words, the radius of gyration is the distance between an imaginary point mass M, and the axis of rotation with I for the point mass about that axis is the same as for the rigid body. Find the radius of gyration of (a) a solid disk of radius R, (b) a uniform rod of length L, and (c) a solid sphere of radius R, all three rotating about a central axis.

46. A bright physics student purchases a wind vane for her father's garage. The vane consists of a rooster sitting on top of an arrow. The vane is fixed to a vertical shaft of radius r and mass m that is free to

FIGURE P10.44

turn in its roof mount as shown in Figure P10.46. The student sets up an experiment to measure the rotational inertia of the rooster and arrow. String wound about the shaft passes over a pulley and is connected to a mass M hanging over the edge of the roof. When the mass M is released, the student determines the time t that the mass takes to fall through a distance h. From these data, she is able to find the rotational inertia I of the rooster and arrow. Find the expression for I in terms of m, M, r, g, h, and t.

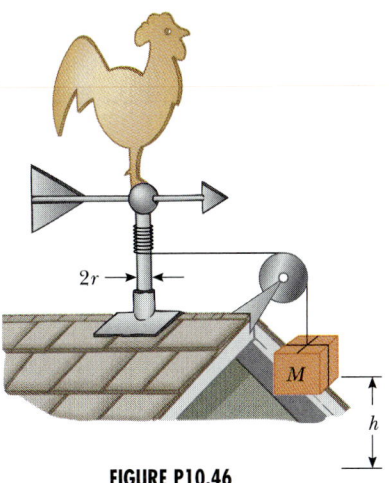

FIGURE P10.46

47. The top in Figure P10.47 has moment of inertia 4.00×10^{-4} kg·m² and is initially at rest. It is free to rotate about the stationary axis AA'. A string wrapped around a peg along the axis of the top is

pulled in such a manner as to maintain a constant tension of 5.57 N. If the string does not slip while being unwound, what is the angular speed of the top after 80.0 cm of string has been pulled off the peg?

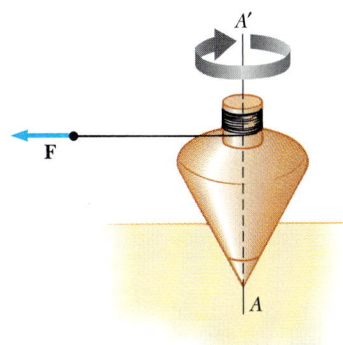

FIGURE P10.47

48. A cord is wrapped around a pulley of mass m and radius r. The free end of the cord is connected to a block of mass M. The block starts from rest and then slides down an incline that makes an angle θ with the horizontal. The coefficient of kinetic friction between block and incline is μ. (a) Use energy methods to show that the block's speed as a function of displacement d down the incline is

$$v = \left[4gd\left(\frac{M}{m + 2M}\right)(\sin\theta - \mu\cos\theta) \right]^{1/2}$$

(b) Find the magnitude of the acceleration of the block in terms of μ, m, M, g, and θ.

49. (a) What is the rotational energy of the Earth about its spin axis? The radius of the Earth is 6370 km and its mass is 5.98×10^{24} kg. Treat the Earth as a sphere of moment of inertia $\frac{2}{5}MR^2$. (b) The rotational energy of the Earth is decreasing steadily because of tidal friction. Estimate the change in rotational energy in one day, given that the rotational period increases by about 10 μs each year.

50. The speed of a moving bullet can be determined by allowing the bullet to pass through two rotating paper disks mounted a distance d apart on the same axle (Fig. P10.50). From the angular displacement $\Delta\theta$ of the two bullet holes in the disks and the rotational speed of the disks, we can determine the speed v of the bullet. Find the bullet speed for the following data: $d = 80$ cm, $\omega = 900$ rev/min, and $\Delta\theta = 31°$.

51. The blocks shown in Figure 10.51 are connected by a string of negligible mass passing over a pulley of radius $R = 0.250$ m and moment of inertia I. The

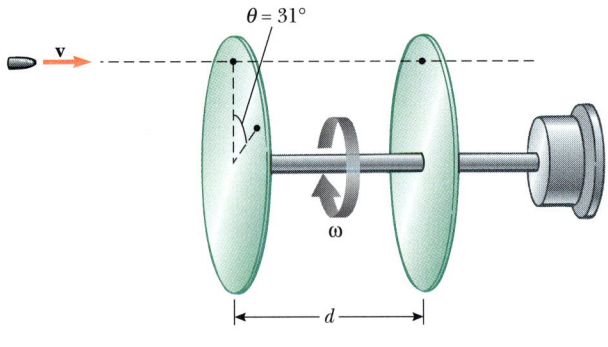

FIGURE P10.50

block on the frictionless incline is moving up with a constant acceleration of magnitude $a = 2.00 \text{ m/s}^2$. (a) Determine T_1 and T_2, the tensions in the two parts of the string, and (b) find the moment of inertia of the pulley.

51A. The blocks shown in Figure 10.51 are connected by a string of negligible mass passing over a pulley of radius R and moment of inertia I. The block on the incline is moving up with a constant acceleration of magnitude a. (a) Determine T_1 and T_2, the tensions in the two parts of the string, and (b) find the moment of inertia of the pulley.

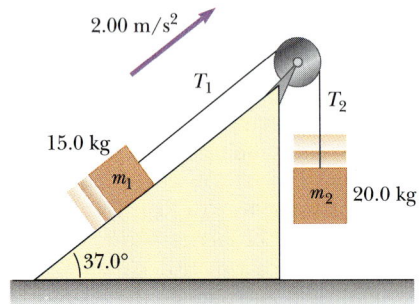

FIGURE P10.51

52. The pulley shown in Figure P10.52 has a radius R and moment of inertia I. One end of the mass m is connected to a spring of force constant k, and the other end is fastened to a cord wrapped around the pulley. The pulley axle and the incline are frictionless. If the pulley is wound counterclockwise so as to stretch the spring a distance d from its unstretched position and then released from rest, find (a) the angular speed of the pulley when the spring is again unstretched and (b) a numerical value for the angular speed at this point if $I = 1.0 \text{ kg} \cdot \text{m}^2$, $R = 0.30 \text{ m}$, $k = 50 \text{ N/m}$, $m = 0.50 \text{ kg}$, $d = 0.20 \text{ m}$, and $\theta = 37°$.

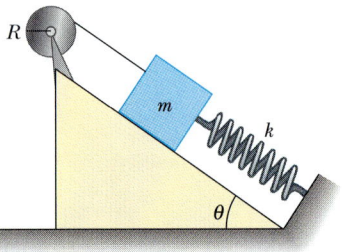

FIGURE P10.52

53. As a result of friction, the angular speed of a wheel changes with time according to

$$\frac{d\theta}{dt} = \omega_0 e^{-\sigma t}$$

where ω_0 and σ are constants. The angular speed changes from 3.50 rad/s at $t = 0$ to 2.00 rad/s at $t = 9.30$ s. Use this information to determine σ and ω_0. Then determine (a) the magnitude of the angular acceleration at $t = 3.00$ s, (b) the number of revolutions the wheel makes in the first 2.50 s, and (c) the number of revolutions it makes before coming to rest.

54. A wheel is formed from a hoop and n equally spaced spokes. The mass of the hoop is M and its radius (and hence length of each spoke) is R. If the mass of each spoke is m, determine the moment of inertia of the wheel (a) about an axis through its center and perpendicular to the plane of the wheel and (b) about an axis through the hoop and perpendicular to the plane of the wheel.

55. A uniform, hollow, cylindrical spool has inside radius $R/2$, outside radius R, and mass M (Fig. P10.55). It is mounted so as to rotate on a fixed horizontal axle. A mass m is connected to the end of a string wound around the spool. The mass m falls from rest through a distance y in time t. Show that the torque due to the frictional forces between spool and axle is

$$\tau_f = R\left[m\left(g - \frac{2y}{t^2} \right) - \frac{5}{4} M\left(\frac{y}{t^2} \right) \right]$$

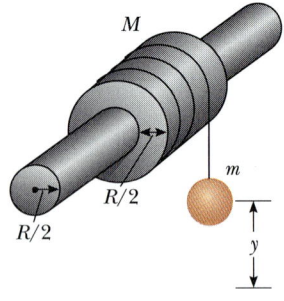

FIGURE P10.55

56. An electric motor can accelerate a Ferris wheel of moment of inertia $I = 20\,000$ kg·m² from rest to 10 rev/min in 12 s. When the motor is turned off, friction causes the wheel to slow down from 10 to 8.0 rev/min in 10 s. Determine (a) the torque generated by the motor to bring the wheel to 10 rev/min and (b) the power needed to maintain this rotational speed.

SPREADSHEET PROBLEM

S1. A disk having a moment of inertia of 100 kg·m² is free to rotate about a fixed axis through its center as in Figure 10.18. A tangential force whose magnitude ranges from $F = 0$ to $F = 50.0$ N can be applied at any distance ranging from $R = 0$ to $R = 3.00$ m measured from the axis of rotation. Using Spreadsheet 10.1, find values of F and R that causes the disk to complete 2 revolutions in 10.0 s. Are there unique values for F and R?

Rolling Motion, Angular Momentum, and Torque

Derek Swinson, professor of physics at the University of New Mexico, demonstrating the "gyro-ski" technique. The skier initiates a turn by lifting the axle of the rotating bicycle wheel. The direction of the turn depends on whether the left or right hand is used to lift the axle from the horizontal. Ignoring friction and gravity, the angular momentum of the system (the skier and the bicycle wheel) remains constant. *(Courtesy of Derek Swinson)*

I n the previous chapter we learned how to treat the rotation of a rigid body about a fixed axis. This chapter deals in part with the more general case, where the axis of rotation is not fixed in space. We begin by describing the rolling motion of an object. Next, we define a vector product, a convenient mathematical tool for expressing such quantities as torque and angular momentum. The central point of this chapter is to develop the concept of the angular momentum of a system of particles, a quantity that plays a key role in rotational dynamics. In analogy to the conservation of linear momentum, we find that angular momentum is always conserved. Like the law of conservation of linear momentum, the conservation of angular momentum is a fundamental law of physics, equally valid for relativistic and quantum systems.

11.1 ROLLING MOTION OF A RIGID BODY

In this section we treat the motion of a rigid body that is rotating about a moving axis. The general motion of a rigid body in space is very complex. However, we can simplify matters by restricting our discussion to a homogeneous rigid body having a high degree of symmetry, such as a cylinder, sphere, or hoop. Furthermore, we assume that the body undergoes rolling motion in a plane.

Suppose a cylinder is rolling on a straight path as in Figure 11.1. The center of mass moves in a straight line, while a point on the rim moves in a more complex path, which corresponds to the path of a cycloid. As we shall see later in this chapter, it is convenient to view the rolling motion as a combination of rotation about the center of mass and translation of the center of mass.

Now consider a uniform cylinder of radius R rolling without slipping on a horizontal surface (Fig. 11.2). As the cylinder rotates through an angle θ, its center of mass moves a distance $s = R\theta$. Therefore, the speed and acceleration magnitude of the center of mass for *pure rolling motion* are

$$v_{\text{CM}} = \frac{ds}{dt} = R\frac{d\theta}{dt} = R\omega \tag{11.1}$$

$$a_{\text{CM}} = \frac{dv_{\text{CM}}}{dt} = R\frac{d\omega}{dt} = R\alpha \tag{11.2}$$

The linear velocities of various points on the rolling cylinder are illustrated in Figure 11.3. Note that the linear velocity of any point is in a direction perpendicular to the line from that point to the contact point. At any instant, the point P is at rest relative to the surface since sliding does not occur.

A general point on the cylinder, such as Q, has both horizontal and vertical components of velocity. However, the points P and P' and the point at the center of mass are unique and of special interest. Relative to the surface on which the cylinder is moving, the center of mass moves with a speed $v_{\text{CM}} = R\omega$, whereas the contact point P has zero speed. The point P' has a speed $2v_{\text{CM}} = 2R\omega$, since all points on the cylinder have the same angular speed.

We can express the total energy of the rolling cylinder as

$$K = \tfrac{1}{2}I_P\omega^2 \tag{11.3}$$

where I_P is the moment of inertia about the axis through P. Applying the parallel-

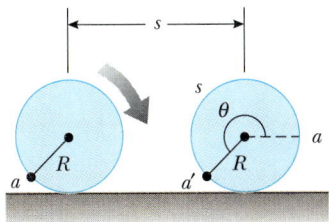

FIGURE 11.2 For pure rolling motion, as the cylinder rotates through an angle θ, its center moves a distance $s = R\theta$.

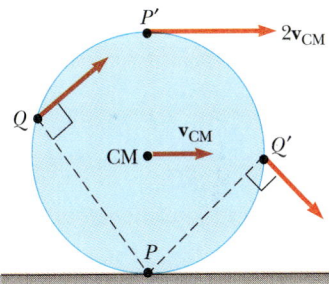

FIGURE 11.3 All points on a rolling body move in a direction perpendicular to an axis through the contact point P. The center of the body moves with a velocity $\mathbf{v}_{\text{CM}}$, while the point P' moves with a velocity $2\mathbf{v}_{\text{CM}}$.

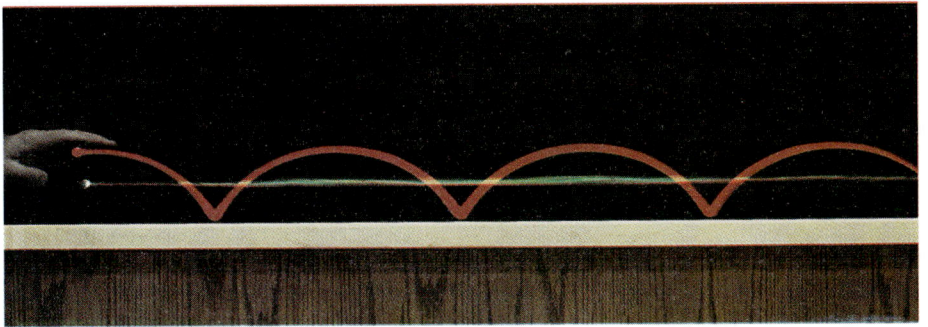

FIGURE 11.1 Light sources at the center and rim of a rolling cylinder illustrate the different paths these points take. The center moves in a straight line, as indicated by the green line, while a point on the rim moves in the path of a cycloid, as indicated by the red curve. *(Courtesy of Henry Leap and Jim Lehman)*

axis theorem, we can substitute $I_P = I_{CM} + MR^2$ into Equation 11.3 to get

$$K = \tfrac{1}{2}I_{CM}\omega^2 + \tfrac{1}{2}MR^2\omega^2$$

$$K = \tfrac{1}{2}I_{CM}\omega^2 + \tfrac{1}{2}Mv_{CM}^2 \qquad (11.4)$$

Total kinetic energy of a rolling body

where we have used the fact that $v_{CM} = R\omega$.

We can think of Equation 11.4 as follows: The term $\tfrac{1}{2}I_{CM}\omega^2$ represents the rotational kinetic energy about the center of mass, and the term $\tfrac{1}{2}Mv_{CM}^2$ represents the kinetic energy the cylinder would have if it were just translating through space without rotating. Thus, we can say that

> the total kinetic energy of an object undergoing rolling motion is the sum of the rotational kinetic energy about the center of mass and the translational kinetic energy of the center of mass.

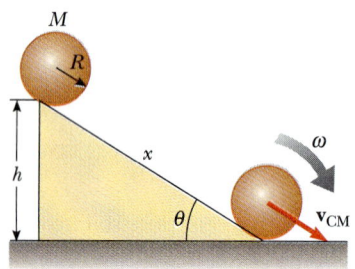

FIGURE 11.4   A round object rolling down an incline. Mechanical energy is conserved if no slipping occurs and there is no rolling friction.

We can use energy methods to treat a class of problems concerning the rolling motion of a rigid object down a rough incline. We assume that the object in Figure 11.4 does not slip and is released from rest at the top of the incline. Note that rolling motion is possible only if a frictional force is present between the object and the incline to produce a net torque about the center of mass. Despite the presence of friction, there is no loss of mechanical energy since the contact point is at rest relative to the surface at any instant. On the other hand, if the object were to slide, mechanical energy would be lost as motion progressed.

Using the fact that $v_{CM} = R\omega$ for pure rolling motion, we can express Equation 11.4 as

$$K = \tfrac{1}{2}I_{CM}\left(\frac{v_{CM}}{R}\right)^2 + \tfrac{1}{2}Mv_{CM}^2$$

$$K = \tfrac{1}{2}\left(\frac{I_{CM}}{R^2} + M\right)v_{CM}^2 \qquad (11.5)$$

When the rolling object reaches the bottom of the incline, work equal to Mgh has been done on it by the gravitational field, where h is the height of the incline. Because the body starts from rest at the top, its kinetic energy at the bottom, given by Equation 11.5, must equal this energy gain. Therefore, the speed of the center of mass at the bottom can be obtained by equating these two quantities:

$$\tfrac{1}{2}\left(\frac{I_{CM}}{R^2} + M\right)v_{CM}^2 = Mgh$$

$$v_{CM} = \left(\frac{2gh}{1 + I_{CM}/MR^2}\right)^{1/2} \qquad (11.6)$$

EXAMPLE 11.1   Sphere Rolling Down an Incline

If the object in Figure 11.4 is a solid sphere, calculate the speed of its center of mass at the bottom and determine the magnitude of the linear acceleration of the center of mass.

Solution   For a uniform solid sphere, $I_{CM} = \tfrac{2}{5}MR^2$, and therefore Equation 11.6 gives

$$v_{CM} = \left(\frac{2gh}{1 + \dfrac{\frac{2}{5}MR^2}{MR^2}}\right)^{1/2} = \left(\frac{10}{7}gh\right)^{1/2}$$

The vertical displacement is related to the displacement x along the incline through the relationship $h = x \sin\theta$.

Hence, after squaring both sides, we can express the equation above as

$$v_{CM}^2 = \frac{10}{7} gx \sin \theta$$

Comparing this with the familiar expression from kinematics, $v_{CM}^2 = 2a_{CM}x$, we see that the acceleration of the center of mass is

$$a_{CM} = \tfrac{5}{7}g \sin \theta$$

These results are quite interesting in that both the speed

and the acceleration of the center of mass are *independent* of the mass and radius of the sphere! That is, *all homogeneous solid spheres experience the same speed and acceleration on a given incline.*

If we repeated the calculations for a hollow sphere, a solid cylinder, or a hoop, we would obtain similar results. The constant factors that appear in the expressions for v_c and a_c depend only on the moment of inertia about the center of mass for the specific body. In all cases, the acceleration of the center of mass is *less* than $g \sin \theta$, the value it would have if the plane were frictionless and no rolling occurred.

EXAMPLE 11.2 Another Look at the Rolling Sphere

In this example, let us consider the solid sphere rolling down an incline and verify the results of Example 11.1 using dynamic methods. The free-body diagram for the sphere is illustrated in Figure 11.5.

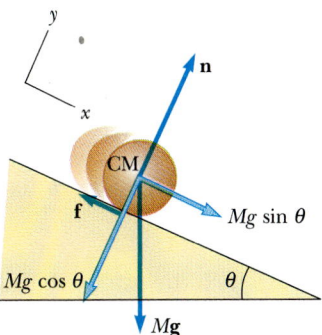

FIGURE 11.5 (Example 11.2) Free-body diagram for a solid sphere rolling down an incline.

Solution Newton's second law applied to the motion of the center of mass gives

$$(1) \qquad \sum F_x = Mg \sin \theta - f = Ma_{CM}$$
$$\sum F_y = n - Mg \cos \theta = 0$$

where x is measured downward along the inclined plane. Now let us write an expression for the torque acting on the sphere. A convenient axis to choose is an axis through the center of the sphere, perpendicular to the plane of the figure.[1] Since $\mathbf{n}$ and $M\mathbf{g}$ go through this origin, they have zero moment arms and do not contribute to the torque. However, the force of friction produces a torque about this axis equal to fR in the clockwise direction; therefore

$$\tau_{CM} = fR = I_{CM}\alpha$$

Since $I_{CM} = \tfrac{2}{5}MR^2$ and $\alpha = a_{CM}/R$, we get

$$(2) \qquad f = \frac{I_{CM}\alpha}{R} = \left(\frac{\tfrac{2}{5}MR^2}{R}\right)\frac{a_{CM}}{R} = \tfrac{2}{5}Ma_{CM}$$

Substituting (2) into (1) gives

$$a_{CM} = \tfrac{5}{7}g \sin \theta$$

which agrees with the result of Example 11.1. Note that $F_{net} = ma$ applies rigorously if F_{net} is the net force and a is the acceleration of the center of mass. Hence, in this case, although the frictional force does no work, it contributes to F_{net} and thus decreases the acceleration of the center of mass.

---

[1] You should note that although the point at the center of mass is not an inertial frame, the expression $\tau_{CM} = I\alpha$ still applies in the center-of-mass frame.

11.2 THE VECTOR PRODUCT AND TORQUE

Consider a force **F** acting on a rigid body at the vector position **r** (Fig. 11.6). *The origin O is assumed to be in an inertial frame, so that Newton's second law is valid.* The *magnitude* of the torque due to this force relative to the origin is, by definition, $rF \sin \phi$, where ϕ is the angle between **r** and **F**. The axis about which **F** tends to produce rotation is perpendicular to the plane formed by **r** and **F**. If the force lies in the *xy* plane as in Figure 11.6, then the torque, **τ**, is represented by a vector parallel to the *z* axis. The force in Figure 11.6 creates a torque that tends to rotate the body counterclockwise looking down the *z* axis, and so the sense of **τ** is toward

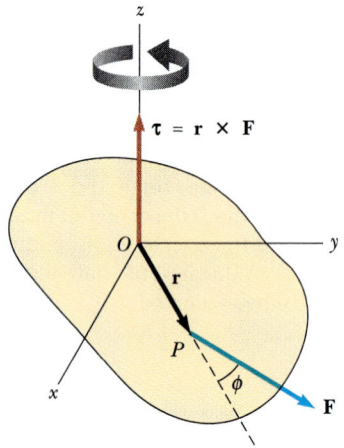

FIGURE 11.6   The torque vector τ lies in a direction perpendicular to the plane formed by the position vector **r** and the applied force **F**.

increasing z and τ is therefore in the positive z direction. If we reversed the direction of **F** in Figure 11.6, τ would then be in the negative z direction. The torque involves two vectors, **r** and **F**, and is in fact defined to be equal to the *vector product,* or *cross product,* of **r** and **F**:

$$\tau \equiv \mathbf{r} \times \mathbf{F} \tag{11.7}$$

We must now give a formal definition of the vector product. Given any two vectors **A** and **B**, the vector product $\mathbf{A} \times \mathbf{B}$ is defined as a third vector **C**, the *magnitude* of which is $AB \sin \theta$, where θ is the angle included between **A** and **B**. That is, if **C** is given by

$$\mathbf{C} = \mathbf{A} \times \mathbf{B} \tag{11.8}$$

then its magnitude is

$$C \equiv AB \sin \theta \tag{11.9}$$

Note that the quantity $AB \sin \theta$ is equal to the area of the parallelogram formed by **A** and **B**, as shown in Figure 11.7. The *direction* of $\mathbf{A} \times \mathbf{B}$ is perpendicular to the plane formed by **A** and **B**, as in Figure 11.7, and its sense is determined by the advance of a right-handed screw when turned from **A** to **B** through the angle θ. A more convenient rule to use for the direction of $\mathbf{A} \times \mathbf{B}$ is the right-hand rule illustrated in Figure 11.7. The four fingers of the right hand are pointed along **A** and then "wrapped" into **B** through the angle θ. The direction of the erect right thumb is the direction of $\mathbf{A} \times \mathbf{B}$. Because of the notation, $\mathbf{A} \times \mathbf{B}$ is often read "A cross B"; hence the term *cross product.*

Some properties of the vector product that follow from its definition are as follows:

Properties of the vector product

1. Unlike the scalar product, the order in which the two vectors are multiplied in a cross product is important, that is,

$$\mathbf{A} \times \mathbf{B} = -(\mathbf{B} \times \mathbf{A}) \tag{11.10}$$

Therefore, if you change the order of the cross product, you must change the sign. You could easily verify this relationship with the right-hand rule (Fig. 11.7).

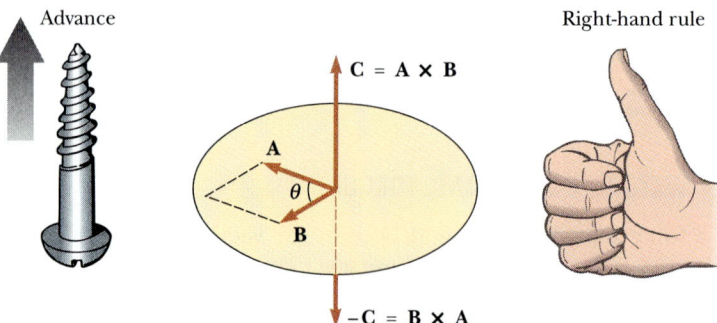

FIGURE 11.7   The vector product $\mathbf{A} \times \mathbf{B}$ is a third vector **C** having a magnitude $AB \sin \theta$ equal to the area of the parallelogram shown. The direction of **C** is perpendicular to the plane formed by **A** and **B**, and its sense is determined by the right-hand rule.

2. If **A** is parallel to **B** ($\theta = 0°$ or $180°$), then **A** × **B** = 0; therefore, it follows that **A** × **A** = 0.
3. If **A** is perpendicular to **B**, then $|\mathbf{A} \times \mathbf{B}| = AB$.
4. The vector product obeys the *distributive law*, that is,

$$\mathbf{A} \times (\mathbf{B} + \mathbf{C}) = \mathbf{A} \times \mathbf{B} + \mathbf{A} \times \mathbf{C} \qquad (11.11)$$

5. The derivative of the cross product with respect to some variable such as t is

$$\frac{d}{dt}(\mathbf{A} \times \mathbf{B}) = \mathbf{A} \times \frac{d\mathbf{B}}{dt} + \frac{d\mathbf{A}}{dt} \times \mathbf{B} \qquad (11.12)$$

where it is important to preserve the multiplicative order of **A** and **B**, in view of Equation 11.10.

It is left as an exercise to show from Equations 11.8 and 11.9 and the definition of unit vectors that the cross products of the rectangular unit vectors **i**, **j**, and **k** obey the following expressions:

$$\mathbf{i} \times \mathbf{i} = \mathbf{j} \times \mathbf{j} = \mathbf{k} \times \mathbf{k} = 0 \qquad (11.13a)$$

$$\mathbf{i} \times \mathbf{j} = -\mathbf{j} \times \mathbf{i} = \mathbf{k} \qquad (11.13b)$$

$$\mathbf{j} \times \mathbf{k} = -\mathbf{k} \times \mathbf{j} = \mathbf{i} \qquad (11.13c)$$

$$\mathbf{k} \times \mathbf{i} = -\mathbf{i} \times \mathbf{k} = \mathbf{j} \qquad (11.13d)$$

Cross products of unit vectors

Signs are interchangeable. For example, $\mathbf{i} \times (-\mathbf{j}) = -\mathbf{i} \times \mathbf{j} = -\mathbf{k}$.

The cross product of *any* two vectors **A** and **B** can be expressed in the following determinant form:

$$\mathbf{A} \times \mathbf{B} = \begin{vmatrix} \mathbf{i} & \mathbf{j} & \mathbf{k} \\ A_x & A_y & A_z \\ B_x & B_y & B_z \end{vmatrix}$$

Expanding this determinant gives the result

$$\mathbf{A} \times \mathbf{B} = (A_y B_z - A_z B_y)\mathbf{i} + (A_z B_x - A_x B_z)\mathbf{j} + (A_x B_y - A_y B_x)\mathbf{k} \qquad (11.14)$$

EXAMPLE 11.3 The Cross Product

Two vectors lying in the xy plane are given by the equations **A** = 2**i** + 3**j** and **B** = −**i** + 2**j**. Find **A** × **B**, and verify explicitly that **A** × **B** = −**B** × **A**.

Solution Using Equations 11.13a through 11.13d for the cross product of unit vectors gives

$$\mathbf{A} \times \mathbf{B} = (2\mathbf{i} + 3\mathbf{j}) \times (-\mathbf{i} + 2\mathbf{j})$$

$$= 2\mathbf{i} \times 2\mathbf{j} + 3\mathbf{j} \times (-\mathbf{i}) = 4\mathbf{k} + 3\mathbf{k} = \boxed{7\mathbf{k}}$$

(We have omitted the terms with $\mathbf{i} \times \mathbf{i}$ and $\mathbf{j} \times \mathbf{j}$, because they are zero.)

$$\mathbf{B} \times \mathbf{A} = (-\mathbf{i} + 2\mathbf{j}) \times (2\mathbf{i} + 3\mathbf{j})$$

$$= -\mathbf{i} \times 3\mathbf{j} + 2\mathbf{j} \times 2\mathbf{i} = -3\mathbf{k} - 4\mathbf{k} = \boxed{-7\mathbf{k}}$$

Therefore, **A** × **B** = −**B** × **A**.

As an alternative method for finding **A** × **B**, we could use Equation 11.8, with $A_x = 2$, $A_y = 3$, $A_z = 0$ and $B_x = -1$, $B_y = 2$, $B_z = 0$:

$$\mathbf{A} \times \mathbf{B} = (0)\mathbf{i} + (0)\mathbf{j} + [2 \times 2 - 3 \times (-1)]\mathbf{k} = 7\mathbf{k}$$

Exercise Use the results to this example and Equation 11.9 to find the angle between **A** and **B**.

Answer 60.3°.

11.3 ANGULAR MOMENTUM OF A PARTICLE

A particle of mass *m*, located at the vector position **r**, moves with a velocity **v** (Fig. 11.8).

> The instantaneous angular momentum **L** of the particle relative to the origin *O* is defined by the cross product of the instantaneous vector position of the particle and its instantaneous linear momentum **p**:

Angular momentum of a particle

$$\mathbf{L} \equiv \mathbf{r} \times \mathbf{p} \qquad (11.15)$$

The SI units of angular momentum are kg·m²/s. It is important to note that both the magnitude and direction of **L** depend on the choice of origin. The direction of **L** is perpendicular to the plane formed by **r** and **p**, and its sense is governed by the right-hand rule. For example, in Figure 11.8, **r** and **p** are in the *xy* plane, so that **L** points in the *z* direction. Since **p** = *m***v**, the magnitude of **L** is

$$L = mvr \sin \phi \qquad (11.16)$$

where ϕ is the angle between **r** and **p**. It follows that *L* is zero when **r** is parallel to **p** ($\phi = 0$ or $180°$). In other words, when the particle moves along a line that passes through the origin, it has zero angular momentum with respect to the origin. On the other hand, if **r** is perpendicular to **p** ($\phi = 90°$), then *L* = *mvr*. At that instant the particle moves exactly as though it were on the rim of a wheel rotating about the origin in a plane defined by **r** and **p**.

Alternatively, one should note that a particle has nonzero angular momentum about some point if its position measured from that point is nonzero and the velocity vector would not carry it through that point, even if the position vector does not rotate about the point. On the other hand, if the position vector simply increases or decreases in length, the particle moves along a line passing through the origin and, therefore, has zero angular momentum with respect to that origin.

In linear motion, we found that the resultant force on a particle equals the time rate of change of its linear momentum (Eq. 9.3). We now show that the resultant torque acting on a particle equals the time rate of change of its angular momentum. Let us start by writing the torque on the particle in the form

$$\boldsymbol{\tau} = \mathbf{r} \times \mathbf{F} = \mathbf{r} \times \frac{d\mathbf{p}}{dt} \qquad (11.17)$$

where we have used Newton's second law in the form **F** = *d***p**/*dt*. Now let us differentiate Equation 11.15 with respect to time using the rule given by Equation 11.12:

$$\frac{d\mathbf{L}}{dt} = \frac{d}{dt}(\mathbf{r} \times \mathbf{p}) = \mathbf{r} \times \frac{d\mathbf{p}}{dt} + \frac{d\mathbf{r}}{dt} \times \mathbf{p}$$

It is important to adhere to the order of terms since **A** × **B** = − **B** × **A**.

The last term on the right in the above equation is zero, because **v** = *d***r**/*dt* is parallel to **p** (property 2 of the vector product). Therefore,

$$\frac{d\mathbf{L}}{dt} = \mathbf{r} \times \frac{d\mathbf{p}}{dt} \qquad (11.18)$$

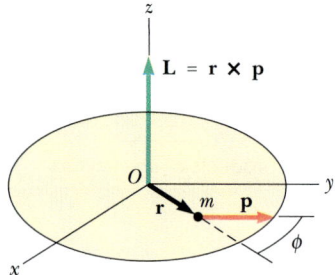

FIGURE 11.8 The angular momentum **L** of a particle of mass *m* and momentum **p** located at the position **r** is a vector given by **L** = **r** × **p**. The value of **L** depends on the origin and is a vector perpendicular to both **r** and **p**.

Comparing Equations 11.17 and 11.18, we see that

$$\boldsymbol{\tau} = \frac{d\mathbf{L}}{dt} \qquad (11.19)$$

which is the rotational analog of Newton's second law, $\mathbf{F} = d\mathbf{p}/dt$. This rotational result says that

the torque acting on a particle is equal to the time rate of change of the particle's angular momentum.

It is important to note that Equation 11.19 is valid only if the origins of $\boldsymbol{\tau}$ and L are the *same*. Equation 11.19 is also valid when there are several forces acting on the particle, in which case $\boldsymbol{\tau}$ is the net torque on the particle. *Furthermore, the expression is valid for any origin fixed in an inertial frame.* Of course, the same origin must be used in calculating all torques as well as the angular momentum.

A System of Particles

The total angular momentum, L, of a system of particles about some point is defined as the vector sum of the angular momenta of the individual particles:

$$\mathbf{L} = \mathbf{L}_1 + \mathbf{L}_2 + \cdots + \mathbf{L}_n = \sum \mathbf{L}_i$$

where the vector sum is over all n particles in the system.

Since the individual momenta of the particles may change in time, the total angular momentum may do so also. In fact, from Equations 11.17 and 11.18, we find that the time rate of change of the total angular momentum equals the vector sum of all torques, both those associated with internal forces between particles and those associated with external forces. However, the net torque associated with internal forces is zero. To understand this, recall that Newton's third law tells us that the internal forces occur in equal and opposite pairs. If we assume that these forces lie along the line of separation of each pair of particles, then the torque due to each action-reaction force pair is zero. By summation, we see that the net internal torque vanishes. Finally, we conclude that the total angular momentum can vary with time only if there is a net external torque on the system, so that we have

$$\sum \boldsymbol{\tau}_{\text{ext}} = \sum \frac{d\mathbf{L}_i}{dt} = \frac{d}{dt} \sum \mathbf{L}_i = \frac{d\mathbf{L}}{dt} \qquad (11.20)$$

That is,

the time rate of change of the total angular momentum of the system about some origin in an inertial frame equals the net external torque acting on the system about that origin.

Note that Equation 11.20 is the rotational analog of Equation 9.3, $\mathbf{F}_{\text{ext}} = d\mathbf{p}/dt$, for a system of particles.

CONCEPTUAL EXAMPLE 11.4

Can a particle moving in a straight line have nonzero angular momentum?

Reasoning   A particle has angular momentum about an origin when moving in a straight line as long as its line of motion does not pass through the origin. Its angular momentum is zero only if the line of motion passes through the origin, in which case **r** is parallel to **v**, and $L = r \times p = r \times mv = 0$.

EXAMPLE 11.5   Linear Motion

A particle of mass m moves in the xy plane with a velocity v along a straight line (Fig. 11.9). What are the magnitude and direction of its angular momentum (a) with respect to the origin O and (b) with respect to the origin O'?

Solution   (a) From the definition of angular momentum, $L = r \times p = rmv \sin \phi(-k)$. Therefore the magnitude of **L** is

$$L = mvr \sin \phi = \boxed{mvd}$$

where $d = r \sin \phi$ is the distance of closest approach of the particle from the origin. The direction of **L** from the right-hand rule is into the diagram, and we can write the vector expression $L = -(mvd)k$.

(b) Because the direction of **v** passes through O', the angular momentum relative to O' is zero.

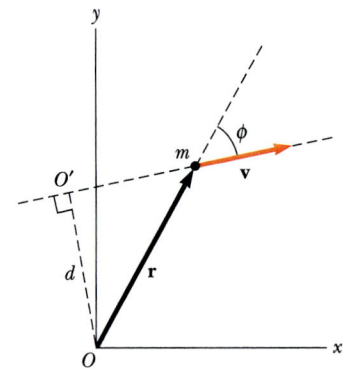

FIGURE 11.9   (Example 11.5) A particle moving in a straight line with a velocity **v** has an angular momentum equal in magnitude to mvd relative to O, where $d = r \sin \phi$ is the distance of closest approach to the origin. The vector $L = r \times p$ points *into* the diagram in this case.

EXAMPLE 11.6   Circular Motion

A particle moves in the xy plane in a circular path of radius r, as in Figure 11.10. (a) Find the magnitude and direction of its angular momentum relative to O when its velocity is **v**.

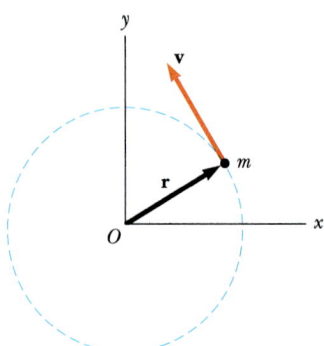

FIGURE 11.10   (Example 11.6) A particle moving in a circle of radius r has an angular momentum equal in magnitude to mvr relative to the center. The vector $L = r \times p$ points *out* of the diagram.

Solution   Since **r** is perpendicular to **v**, $\phi = 90°$ and the magnitude of **L** is simply

$$L = mvr \sin 90° = \boxed{mvr} \qquad \text{(for r perpendicular to v)}$$

The direction of **L** is perpendicular to the plane of the circle, and its sense depends on the direction of **v**. If the sense of the rotation is counterclockwise, as in Figure 11.10, then by the right-hand rule, the direction of $L = r \times p$ is *out* of the paper. Hence, we can write the vector expression $L = (mvr)k$. If the particle were to move clockwise, **L** would point into the paper.

(b) Find an alternative expression for L in terms of the angular speed, ω.

Solution   Since $v = r\omega$ for a particle rotating in a circle, we can express L as

$$L = mvr = mr^2\omega = \boxed{I\omega}$$

where I is the moment of inertia of the particle about the z axis through O. In this case the angular momentum is in the *same* direction as the angular velocity vector, ω (see Section 10.1), and so we can write $\mathbf{L} = I\boldsymbol{\omega} = I\omega\mathbf{k}$.

Exercise   A car of mass 1500 kg moves on a circular race track of radius 50 m with a speed of 40 m/s. What is the magnitude of its angular momentum relative to the center of the track?

Answer 3.0×10^6 kg·m²/s.

11.4   ROTATION OF A RIGID BODY ABOUT A FIXED AXIS

Consider a rigid body rotating about an axis that is fixed in direction. We assume that the z axis coincides with the axis of rotation, as in Figure 11.11. Each particle of the rigid body rotates in the xy plane about the z axis with an angular speed ω. The magnitude of the angular momentum of the particle of mass m_i is $m_i v_i r_i$ about the origin O. Because $v_i = r_i\omega$, we can express the magnitude of the angular momentum of the ith particle as

$$L_i = m_i r_i^2 \omega$$

The vector $\mathbf{L}_i$ is directed along the z axis, corresponding to the direction of ω.

We can now find the z component of the angular momentum of the rigid body by taking the sum of L_i over all particles of the body:

$$L_z = \sum m_i r_i^2 \omega = \left(\sum m_i r_i^2\right)\omega$$

or

$$L_z = I\omega \tag{11.21}$$

where I is the moment of inertia of the rigid body about the z axis.

Now let us differentiate Equation 11.21 with respect to time, noting that I is constant for a rigid body:

$$\frac{dL_z}{dt} = I\frac{d\omega}{dt} = I\alpha \tag{11.22}$$

where α is the angular acceleration relative to the axis of rotation. Because dL_z/dt is equal to the net torque (Eq. 11.20), we can express Equation 11.22 as

$$\sum \tau_{\text{ext}} = \frac{dL_z}{dt} = I\alpha \tag{11.23}$$

That is, the net external torque acting on an object rotating about a fixed axis equals the moment of inertia about the axis of rotation multiplied by the angular acceleration of the object relative to that axis.

You should note that if a symmetrical object rotates about a fixed axis passing through its center of mass, you can write Equation 11.21 in vector form, $\mathbf{L} = I\boldsymbol{\omega}$, where $\mathbf{L}$ is the total angular momentum of the object measured with respect to the axis of rotation. Furthermore, the expression is valid for any object, regardless of

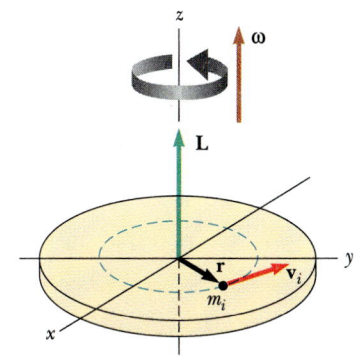

FIGURE 11.11   When a rigid body rotates about an axis, the angular momentum **L** is in the same direction as the angular velocity ω, according to the expression $\mathbf{L} = I\boldsymbol{\omega}$.

its symmetry, if **L** stands for the component of angular momentum along the axis of rotation.[2]

EXAMPLE 11.7   Rotating Sphere

A uniform solid sphere of radius $R = 0.50$ m and mass 15 kg rotates about the z axis through its center, as in Figure 11.12. Find the magnitude of its angular momentum when the angular speed is 3.0 rad/s.

Solution   The moment of inertia of the sphere about an axis through its center is, from Table 10.2,

$$I = \tfrac{2}{5}MR^2 = \tfrac{2}{5}(15 \text{ kg})(0.50 \text{ m})^2 = 1.5 \text{ kg} \cdot \text{m}^2$$

Therefore, the magnitude of the angular momentum is

$$L = I\omega = (1.5 \text{ kg} \cdot \text{m}^2)(3.0 \text{ rad/s}) = \boxed{4.5 \text{ kg} \cdot \text{m}^2/\text{s}}$$

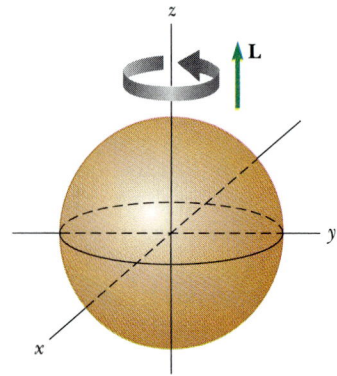

FIGURE 11.12   (Example 11.7) A sphere that rotates about the z axis in the direction shown has an angular momentum **L** in the positive z direction. If the direction of rotation is reversed, **L** will point in the negative z direction.

EXAMPLE 11.8   Rotating Rod

A rigid rod of mass M and length ℓ rotates in a vertical plane about a frictionless pivot through its center (Fig. 11.13). Particles of masses m_1 and m_2 are attached at the ends of the rod. (a) Determine the magnitude of the angular momentum of the system when the angular speed is ω.

Solution   The moment of inertia of the system equals the sum of the moments of inertia of the three components: the rod, m_1, and m_2. Using Table 10.2, we find that the total moment of inertia about the z axis through O is

$$I = \frac{1}{12}M\ell^2 + m_1\left(\frac{\ell}{2}\right)^2 + m_2\left(\frac{\ell}{2}\right)^2 = \frac{\ell^2}{4}\left(\frac{M}{3} + m_1 + m_2\right)$$

Therefore, when the angular speed is ω, the magnitude of the angular momentum is

$$L = I\omega = \frac{\ell^2}{4}\left(\frac{M}{3} + m_1 + m_2\right)\omega$$

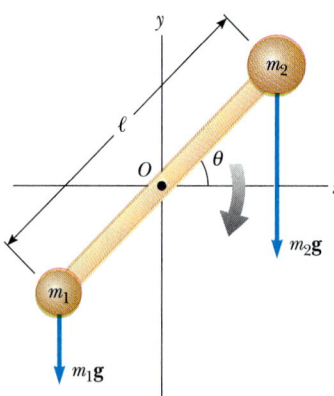

FIGURE 11.13   (Example 11.8) Since gravitational forces act on the system rotating in a vertical plane, there is in general a net nonzero torque about O when $m_1 \neq m_2$, which in turn produces an angular acceleration according to $\tau_{\text{net}} = I\alpha$.

---

[2] In general, the expression **L** = $I\boldsymbol{\omega}$ is not always valid. If a rigid body rotates about an arbitrary axis, **L** and $\boldsymbol{\omega}$ may point in different directions. In fact, in this case, the moment of inertia cannot be treated as a scalar. Strictly speaking, **L** = $I\boldsymbol{\omega}$ applies only to rigid bodies of any shape that rotate about one of three mutually perpendicular axes (called *principal axes*) through the center of mass. This is discussed in more advanced texts on mechanics.

(b) Determine the magnitude of the angular acceleration of the system when the rod makes an angle θ with the horizontal.

Solution   The torque due to the force $m_1 g$ about the pivot is

$$\tau_1 = m_1 g \frac{\ell}{2} \cos \theta \qquad (\tau_1 \text{ is out of the plane})$$

The torque due to the force $m_2 g$ about the pivot is

$$\tau_2 = - m_2 g \frac{\ell}{2} \cos \theta \qquad (\tau_2 \text{ is into the plane})$$

Hence, the net torque about O is

$$\tau_{\text{net}} = \tau_1 + \tau_2 = \tfrac{1}{2}(m_1 - m_2) g \ell \cos \theta$$

You should note that the direction of τ_{net} is out of the plane if $m_1 > m_2$ and is into the plane if $m_2 > m_1$.

To find α, we use $\tau_{\text{net}} = I\alpha$, where I was obtained in (a):

$$\alpha = \frac{\tau_{\text{net}}}{I} = \frac{2(m_1 - m_2) g \cos \theta}{\ell \left(\dfrac{M}{3} + m_1 + m_2 \right)}$$

Note that α is zero when θ is $\pi/2$ or $-\pi/2$ (vertical position) and a maximum when θ is 0 or π (horizontal position). Furthermore, the angular speed of the system changes because α is not zero.

Exercise   If $m_1 > m_2$, at what value of θ is ω a maximum? Knowing the angular speed at some instant, how would you calculate the linear speed of m_1 and m_2?

EXAMPLE 11.9   Two Connected Masses

Two masses, m_1 and m_2, are connected by a light cord that passes over a pulley of radius R and moment of inertia I about its axle, as in Figure 11.14. The mass m_2 slides on a frictionless, horizontal surface. Determine the acceleration of the two masses using the concepts of angular momentum and torque.

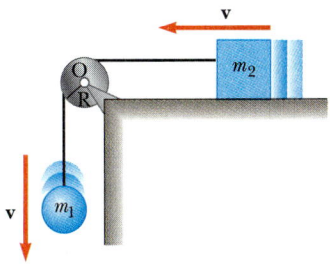

FIGURE 11.14   (Example 11.9).

Reasoning and Solution   First, calculate the angular momentum of the system, which consists of the two masses plus the pulley. Then calculate the torque about an axis along the axle of the pulley through O. At the instant m_1 and m_2 have a speed v, the angular momentum of m_1 is $m_1 vR$, and that of m_2 is $m_2 vR$. At the same instant, the angular momentum of the pulley is $I\omega = Iv/R$. Therefore, the total angular momentum of the system is

$$(1) \qquad L = m_1 vR + m_2 vR + I \frac{v}{R}$$

Now let us evaluate the total external torque on the system about the axle. Because it has zero moment arm, the force exerted by the axle on the pulley does not contribute to the torque. Furthermore, the normal force acting on m_2 is balanced by its weight $m_2 g$, and so these forces do not contribute to the torque. The external force $m_1 g$ produces a torque about the axle equal in magnitude to $m_1 gR$, where R is the moment arm of the force about the axle. This is the total external torque about O; that is, $\tau_{\text{ext}} = m_1 gR$. Using this result, together with (1) and Equation 11.23 gives

$$\tau_{\text{ext}} = \frac{dL}{dt}$$

$$m_1 gR = \frac{d}{dt} \left[(m_1 + m_2) Rv + I \frac{v}{R} \right]$$

$$(2) \qquad m_1 gR = (m_1 + m_2) R \frac{dv}{dt} + \frac{I}{R} \frac{dv}{dt}$$

Because $dv/dt = a$, we can solve this for a to get

$$a = \frac{m_1 g}{(m_1 + m_2) + I/R^2}$$

You may wonder why we did not include the forces that the cord exerts on the objects in evaluating the net torque about the axle. The reason is that these forces are internal to the system under consideration. Only the external torques contribute to the change in angular momentum.

11.5   CONSERVATION OF ANGULAR MOMENTUM

In Chapter 9 we found that the total linear momentum of a system of particles remains constant when the resultant external force acting on the system is zero. We have an analogous conservation law in rotational motion:

Conservation of angular
momentum .

> The total angular momentum of a system is constant if the resultant external torque acting on the system is zero.

This follows directly from Equation 11.20, where we see that if

$$\sum \tau_{\text{ext}} = \frac{d\mathbf{L}}{dt} = 0 \tag{11.24}$$

then

$$\mathbf{L} = \text{constant} \tag{11.25}$$

For a system of particles, we write this conservation law as $\Sigma \mathbf{L}_n$ = constant. If a body undergoes a redistribution of its mass, then its moment of inertia changes and we express this conservation of angular momentum in the form

$$\mathbf{L}_i = \mathbf{L}_f = \text{constant} \tag{11.26}$$

If the system is a body rotating about a *fixed* axis, such as the z axis, then we can write $L_z = I\omega$, where L_z is the component of $\mathbf{L}$ along the axis of rotation and I is the moment of inertia about this axis. In this case, we can express the conservation of angular momentum as

$$I_i\omega_i = I_f\omega_f = \text{constant} \tag{11.27}$$

This expression is valid for rotations either about a fixed axis or about an axis through the center of mass of the system as long as the axis remains parallel to itself. We require only that the net external torque be zero.

Although we do not prove it here, there is an important theorem concerning the angular momentum relative to the center of mass:

> The resultant torque acting on a body about an axis through the center of mass equals the time rate of change of angular momentum regardless of the motion of the center of mass.

This theorem applies even if the center of mass is accelerating, provided τ and $\mathbf{L}$ are evaluated relative to the center of mass.

In Equation 11.27 we have a third conservation law to add to our list. We can now state that the energy, linear momentum, and angular momentum of an isolated system all remain constant.

There are many examples that demonstrate conservation of angular momentum. You may have observed a figure skater undergoing a spin motion in the finale of an act. The angular speed of the skater increases upon pulling his or her hands and feet close to the body. Neglecting friction between skates and ice, we see that there are no external torques on the skater. The change in angular speed is due to the fact that, because angular momentum is conserved, the product $I\omega$ remains constant and a decrease in the moment of inertia of the skater causes an increase in the angular speed. Similarly, when divers (or acrobats) wish to make several somersaults, they pull their hands and feet close to their bodies in order to rotate at a higher rate. In these cases, the external force due to gravity acts through the center of mass and, hence, exerts no torque about this point. Therefore, the

Nancy Kerrigan, winner of a Silver Medal in the 1994 Winter Olympics. In this spin her angular speed increases when she pulls her arms in close to her body, demonstrating that angular momentum is conserved.
(© Duomo/Steven E. Sutton, 1994)

angular momentum about the center of mass must be constant, or $I_i\omega_i = I_f\omega_f$. For example, when divers wish to double their angular speed, they must reduce their moment of inertia to half its initial value.

CONCEPTUAL EXAMPLE 11.10

A particle moves in a straight line, and you are told that the net torque acting on it is zero about some unspecified origin. Does this necessarily imply that the net force on the particle is zero? Can you conclude that its velocity is constant?

Reasoning The net force is not necessarily zero. If the line of action of the net force passes through the origin, the net torque about an axis passing through that origin is zero, even though the net force is not zero. Because the net force is not necessarily zero, you cannot conclude that its velocity is constant.

EXAMPLE 11.11 Formation of a Neutron Star

A star of radius 1.0×10^4 km rotates about its axis with a period of 30 days. The star undergoes a supernova explosion, whereby its core collapses into a neutron star of radius 3.0 km. Estimate the period of the neutron star.

Solution Let us assume that during the collapse of the star, (1) no torque acts on it, (2) it remains spherical, and (3) its

mass remains constant. Since I is proportional to r^2, and $\omega = 2\pi/T$, conservation of angular momentum (Eq. 11.27) gives $T_f = T_i(r_f/r_i)^2 = (30 \text{ days})(3.0 \text{ km}/1.0 \times 10^4 \text{ km})^2 = 2.7 \times 10^{-6}$ days $= 0.23$ s. Thus the neutron star rotates about four times each second.

EXAMPLE 11.12 A Projectile–Cylinder Collision

A projectile of mass m and velocity $\mathbf{v}_0$ is fired at a solid cylinder of mass M and radius R (Fig. 11.15). The cylinder is initially at rest and is mounted on a fixed horizontal axle that runs through the center of mass. The line of motion of the projectile is perpendicular to the axle and at a distance $d < R$ from the center. Find the angular speed of the system after the projectile strikes and adheres to the surface of the cylinder.

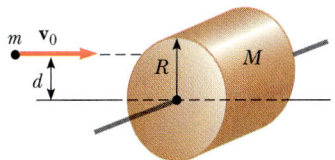

FIGURE 11.15 (Example 11.12) The angular momentum of the system before the collision equals the angular momentum right after the collision with respect to the center of mass if we take the center of mass to be along the cylinder axis (an approximation).

Reasoning Let us evaluate the angular momentum of the system (projectile + cylinder) about the axle of the cylinder.

The net external torque on the system is zero about this axle. Hence, the angular momentum of the system is the same before and after the collision.

Solution Before the collision, only the projectile has angular momentum with respect to a point on the axle. The magnitude of this angular momentum is $mv_0 d$, and it is directed along the axle into the paper. After the collision, the total angular momentum of the system is $I\omega$, where I is the total moment of inertia about the axle (projectile + cylinder). Since the total angular momentum is constant, we get

$$mv_0 d = I\omega = \left(\tfrac{1}{2}MR^2 + mR^2\right)\omega$$

$$\omega = \frac{mv_0 d}{\tfrac{1}{2}MR^2 + mR^2}$$

This suggests another technique for measuring the speed of a bullet.

Exercise In this example, mechanical energy is not conserved since the collision is inelastic. Show that $\tfrac{1}{2}I\omega^2 < \tfrac{1}{2}mv_0^2$. What do you suppose accounts for the energy loss?

EXAMPLE 11.13   The Merry-Go-Round

A horizontal platform in the shape of a circular disk rotates in a horizontal plane about a frictionless vertical axle (Fig. 11.16). The platform has a mass $M = 100$ kg and a radius $R = 2.0$ m. A student whose mass is $m = 60$ kg walks slowly from the rim of the platform toward the center. If the angular speed of the system is 2.0 rad/s when the student is at the rim, (a) calculate the angular speed when the student has reached a point 0.50 m from the center.

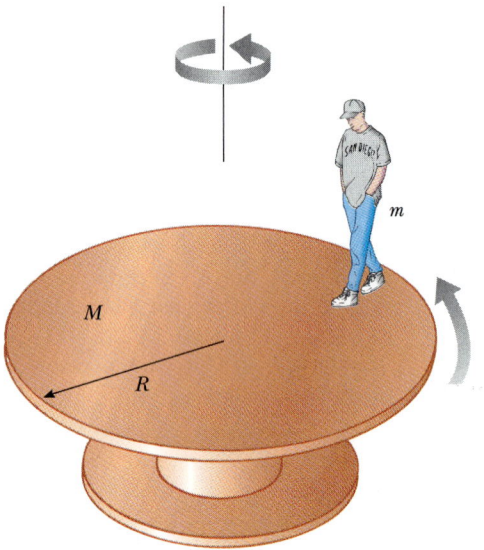

FIGURE 11.16   (Example 11.13) As the student walks toward the center of the rotating platform, the angular speed of the system increases since the angular momentum must remain constant.

Solution   Let us call the moment of inertia of the platform I_p and the moment of inertia of the student I_s. Treating the

student as a point mass m, we can write the initial moment of inertia of the system about the axis of rotation

$$I_i = I_p + I_s = \tfrac{1}{2}MR^2 + mR^2$$

When the student has walked to the position $r < R$, the moment of inertia of the system reduces to

$$I_f = \tfrac{1}{2}MR^2 + mr^2$$

Since there are no external torques on the system (student + platform) about the axis of rotation, we can apply the law of conservation of angular momentum:

$$I_i\omega_i = I_f\omega_f$$

$$(\tfrac{1}{2}MR^2 + mR^2)\,\omega_i = (\tfrac{1}{2}MR^2 + mr^2)\,\omega_f$$

$$\omega_f = \left(\frac{\tfrac{1}{2}MR^2 + mR^2}{\tfrac{1}{2}MR^2 + mr^2}\right)\omega_i$$

$$\omega_f = \left(\frac{200 + 240}{200 + 15}\right)(2.0 \text{ rad/s})$$

$$= \boxed{4.1 \text{ rad/s}}$$

(b) Calculate the initial and final rotational energies of the system.

Solution

$$K_i = \tfrac{1}{2}I_i\omega_i^2 = \tfrac{1}{2}(440 \text{ kg}\cdot\text{m}^2)\left(2.0\,\frac{\text{rad}}{\text{s}}\right)^2 = \boxed{880 \text{ J}}$$

$$K_f = \tfrac{1}{2}I_f\omega_f^2 = \tfrac{1}{2}(215 \text{ kg}\cdot\text{m}^2)\left(4.1\,\frac{\text{rad}}{\text{s}}\right)^2 = \boxed{1.8 \times 10^3 \text{ J}}$$

Note that the rotational energy of the system increases! What accounts for this increase in energy?

CONCEPTUAL EXAMPLE 11.14

A student sits on a pivoted stool while holding a pair of weights, as in Figure 11.17. The stool is free to rotate about a vertical axis with negligible friction. The student is set in rotating motion with the weights outstretched. Why does the angular speed of the system increase as the weights are pulled inward?

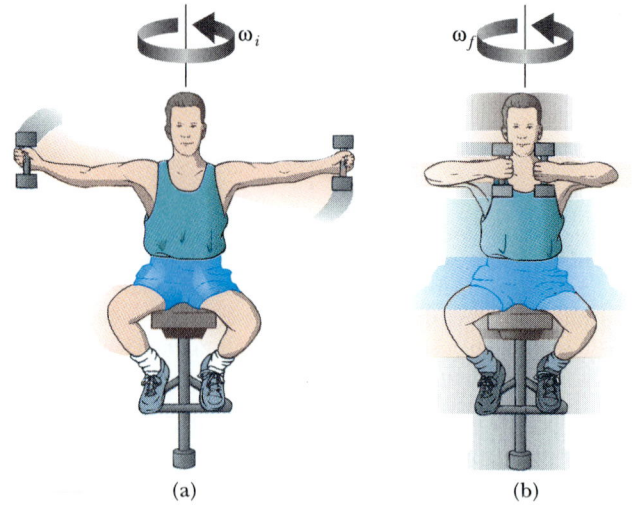

FIGURE 11.17   (Conceptual Example 11.14) (a) The student is given an initial angular speed while holding two masses as shown. (b) When the masses are pulled in close to the body, the angular speed of the system increases. Why?

(a)   (b)

Reasoning The initial angular momentum of the system (student + weights + stool) is $I_i\omega_i$. After the weights are pulled in, the angular momentum of the system is $I_f\omega_f$. Note that $I_f < I_i$ because the weights are now closer to the axis of rotation, reducing the moment of inertia. Since the net external torque on the system is zero, angular momentum is constant, and so $I_i\omega_i = I_f\omega_f$. Therefore, $\omega_f > \omega_i$.

As in the previous example, the kinetic energy of the system increases as the weights are pulled inward. The increase in rotational energy arises from the fact that the student does work in pulling the weights toward the axis of rotation.

EXAMPLE 11.15 The Spinning Bicycle Wheel

In another favorite classroom demonstration, a student holds the axle of a spinning bicycle wheel while seated on a pivoted stool (Fig. 11.18). The student and stool are initially at rest while the wheel is spinning in a horizontal plane with an initial angular momentum L_0 pointing upward. Explain what happens when the wheel is inverted about its center by 180°.

FIGURE 11.18 (Example 11.15) The wheel is initially spinning when the student is at rest. What happens when the wheel is inverted?

Solution In this situation, the system consists of the student, wheel, and stool. Initially, the total angular momentum of the system is L_0, corresponding to the contribution from the spinning wheel. As the wheel is inverted, a torque is supplied by the student, but this is internal to the system. There is no external torque acting on the system about the vertical axis. Therefore, *the angular momentum of the system about the vertical axis is constant.*

Initially, we have

$$L_{system} = L_0 \quad \text{(upward)}$$

After the wheel is inverted,

$$L_{system} = L_0 = L_{student+stool} + L_{wheel}$$

In this case, $L_{wheel} = -L_0$ because the wheel is now rotating in the opposite sense. Therefore

$$L_0 = L_{student+stool} - L_0$$

$$L_{student+stool} = 2L_0$$

This result tells us that, as the wheel is inverted, the student and stool will start to turn, acquiring an angular momentum having a magnitude twice that of the spinning wheel and directed upward.

This student is holding the axle of a spinning bicycle wheel while seated on a pivoted stool. The student and stool are initially at rest while the wheel is spinning in a horizontal plane. When the wheel is inverted about its center by 180°, the student and stool begin to rotate because angular momentum is conserved. *(Courtesy of Central Scientific Co.)*

---

*11.6 THE MOTION OF GYROSCOPES AND TOPS

A very unusual and fascinating type of motion that you probably have observed is that of a top spinning about its axis of symmetry as in Fig. 11.19a. If the top spins about its axis very rapidly, the axis will rotate about the vertical direction as indicated, thereby sweeping out a cone. The motion of the axis of the top about the

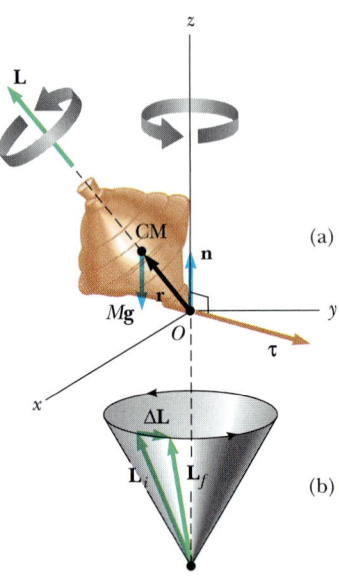

(a)

(b)

FIGURE 11.19 Precessional motion of a top spinning about its axis of symmetry. The only external forces acting on the top are the normal force **n**, and the force of gravity, *M***g**. The direction of the angular momentum, **L**, is along the axis of symmetry.

vertical, known as **precessional motion,** is usually slow compared with the spin motion of the top. It is quite natural to wonder why the top doesn't fall over. Since the center of mass is not directly above the pivot point *O*, there is clearly a net torque acting on the top about *O* due to the weight force *M***g**. From this description, it is easy to see that the top would certainly fall if it were not spinning. However, because the top is spinning, it has an angular momentum **L** directed along its axis of symmetry. As we shall show, the motion of the rotation axis about the *z* axis (the precessional motion) arises from the fact that the torque produces a change in the *direction* of the rotation axis. This is an excellent example of the importance of the directional nature of angular momentum.

The two forces acting on the top are the downward force of gravity, *M***g**, and the normal force, **n**, acting upward at the pivot point *O*. The normal force produces no torque about the pivot since its moment arm is zero. However, the force of gravity produces a torque $\boldsymbol{\tau} = \mathbf{r} \times M\mathbf{g}$ about *O*, where the direction of $\boldsymbol{\tau}$ is perpendicular to the plane formed by **r** and *M***g**. By necessity, the vector $\boldsymbol{\tau}$ lies in a horizontal plane perpendicular to the angular momentum vector. The net torque and angular momentum of the body are related through Equation 11.19:

$$\boldsymbol{\tau} = \frac{d\mathbf{L}}{dt}$$

From this expression, we see that the nonzero torque produces a change in angular momentum $d\mathbf{L}$, which is in the same direction as $\boldsymbol{\tau}$. Therefore, like the torque vector, $d\mathbf{L}$ must also be at right angles to **L**. Figure 11.19b illustrates the resulting precessional motion of the axis of the top. In a time Δt, the change in angular momentum $\Delta \mathbf{L} = \mathbf{L}_f - \mathbf{L}_i = \boldsymbol{\tau} \, \Delta t$. Because $\Delta \mathbf{L}$ is perpendicular to **L**, the magnitude of **L** doesn't change ($|\mathbf{L}_i| = |\mathbf{L}_f|$). Rather, what is changing is the *direction* of **L**. Since the change in angular momentum is in the direction of $\boldsymbol{\tau}$, which lies in the *xy* plane, the top undergoes precessional motion. Thus, the effect of the torque is to deflect the angular momentum of the top in a direction perpendicular to its spin axis.

The essential features of precessional motion can be illustrated by considering the simple gyroscope shown in Fig. 11.20a. This device consists of a wheel free to spin about an axle that is pivoted at a distance *h* from the center of mass of the wheel. When given an angular velocity $\boldsymbol{\omega}$ about its axis, the wheel will have a spin angular momentum $\mathbf{L} = I\boldsymbol{\omega}$ directed along the axle as shown. Let us consider the torque acting on the wheel about the pivot *O*. Again, the force **n** of the support on the axle produces no torque about *O*. On the other hand, the weight *M***g** produces a torque of magnitude *Mgh* about *O*. The direction of this torque is perpendicular to the axle (and perpendicular to **L**), as shown in Figure 11.20. This torque causes the angular momentum to change in the direction perpendicular to the axle. Hence, the axle moves in the direction of the torque, that is, in the horizontal plane.

There is an assumption we must make in order to simplify the description of the system. The total angular momentum of the precessing wheel is the sum of the spin angular momentum, $I\boldsymbol{\omega}$, and the angular momentum due to the motion of the center of mass about the pivot. In our treatment, we shall neglect the contribution from the center-of-mass motion and take the total angular momentum to be just $I\boldsymbol{\omega}$. In practice, this is a good approximation if $\boldsymbol{\omega}$ is made very large.

In a time dt the torque due to the weight force adds to the system angular momentum equal to $dL = \tau \, dt = (Mgh) \, dt$. When added vectorially to the original total angular momentum, $I\boldsymbol{\omega}$, this additional angular momentum causes a shift in the direction of the total angular momentum.

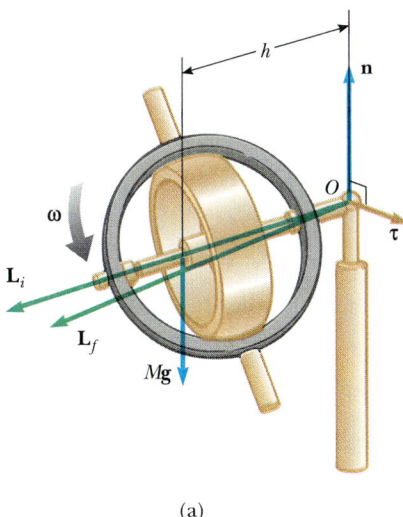

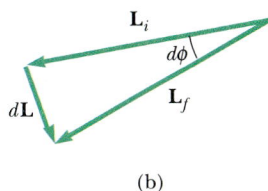

(a)   (b)

FIGURE 11.20   (a) The motion of a simple gyroscope pivoted a distance h from its center of mass. Note that the weight Mg produces about the pivot a torque that is perpendicular to the axle. (b) This torque results in a change in angular momentum $d\mathbf{L}$ in the direction perpendicular to the axle. The axle sweeps out an angle $d\phi$ in a time dt.

The vector diagram in Fig. 11.20b shows that in the time dt, the angular momentum vector rotates through an angle $d\phi$, which is also the angle through which the axle rotates. From the vector triangle formed by the vectors $\mathbf{L}_i$, $\mathbf{L}_f$, and $d\mathbf{L}$, we see that

$$d\phi = \frac{dL}{L} = \frac{(Mgh)\,dt}{L}$$

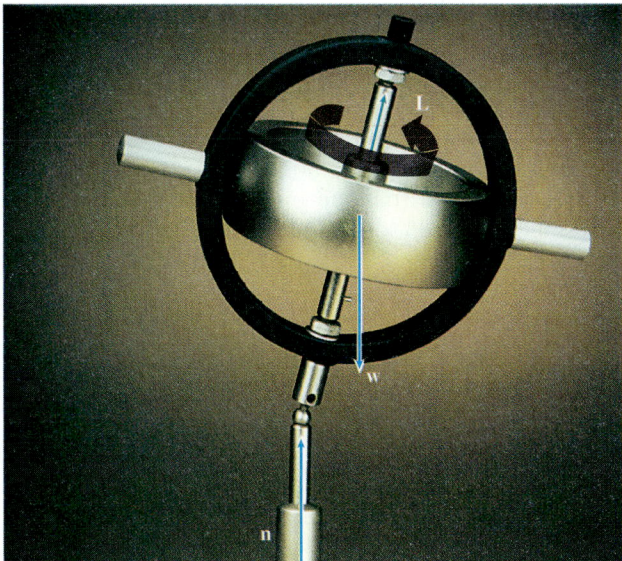

This toy gyroscope undergoes precessional motion about the vertical axis as it spins about its axis of symmetry. The only forces acting on it are the force of gravity, $\mathbf{w}$, and the upward force of the pivot, $\mathbf{n}$. The direction of its angular momentum, $\mathbf{L}$, is along the axis of symmetry. *(Courtesy of Central Scientific Company)*

Using the relationship $L = I\omega$, we find that the rate at which the axle rotates about the vertical axis is

$$\omega_p = \frac{d\phi}{dt} = \frac{Mgh}{I\omega} \qquad (11.28)$$

The angular frequency ω_p is called the **precessional frequency**. This result is valid only when $\omega_p \ll \omega$. Otherwise, a much more complicated motion is involved. As you can see from Equation 11.28, the condition that $\omega_p \ll \omega$ is met when $I\omega$ is large compared with Mgh. Furthermore, note that the precessional frequency decreases as ω increases, that is, as the wheel spins faster about its axis of symmetry.

*11.7   ANGULAR MOMENTUM AS A FUNDAMENTAL QUANTITY

We have seen that the concept of angular momentum is very useful for describing the motion of macroscopic systems. However, the concept is also valid on a submicroscopic scale and has been used extensively in the development of modern theories of atomic, molecular, and nuclear physics. In these developments, it was found that the angular momentum of a system is a fundamental quantity. The word *fundamental* in this context implies that angular momentum is an intrinsic property of atoms, molecules, and their constituents.

In order to explain the results of a variety of experiments on atomic and molecular systems, it is necessary to assign discrete values to the angular momentum. These discrete values are some multiple of a fundamental unit of angular momentum, which equals $\hbar = h/2\pi$, where h is called Planck's constant:

$$\text{Fundamental unit of angular momentum} = \hbar = 1.054 \times 10^{-34} \frac{\text{kg} \cdot \text{m}^2}{\text{s}^2}$$

Let us accept this postulate without proof for the time being and show how it can be used to estimate the frequency of a diatomic molecule. Consider the O_2 molecule as a rigid rotor, that is, two atoms separated by a fixed distance d and rotating about the center of mass (Fig. 11.21). Equating the rotational angular momentum to the fundamental unit $\hbar$, we can estimate the lowest frequency:

$$I_{CM}\omega \approx \hbar \qquad \text{or} \qquad \omega \approx \frac{\hbar}{I_{CM}}$$

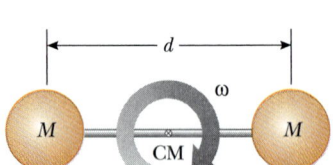

FIGURE 11.21   The rigid-rotor model of the diatomic molecule. The rotation occurs about the center of mass in the plane of the diagram.

In Example 10.4, we found that the moment of inertia of the O_2 molecule about this axis of rotation is 2.03×10^{-46} kg $\cdot$ m^2. Therefore,

$$\omega \approx \frac{\hbar}{I_{CM}} = \frac{1.054 \times 10^{-34} \text{ kg} \cdot \text{m}^2/\text{s}}{1.95 \times 10^{-46} \text{ kg} \cdot \text{m}^2} = 5.41 \times 10^{11} \text{ rad/s}$$

Actual frequencies are a multiple of this smallest possible value.

This simple example shows that certain classical concepts and models might be useful in describing some features of atomic and molecular systems. However, a wide variety of phenomena on the submicroscopic scale can be explained only if we assume discrete values of the angular momentum associated with a particular type of motion.

The Danish physicist Niels Bohr (1885–1962) accepted and adapted this radical idea of discrete momentum values in his theory of the hydrogen atom. Strictly classical models were unsuccessful in describing many properties of the hydrogen atom. Bohr postulated that the electron could occupy only those circular orbits about the proton for which the orbital angular momentum was equal to $n\hbar$, where

n is an integer. That is, he made the bold assumption that the orbital angular momentum is quantized. From this simple model, the rotational frequencies of the electron in the various orbits can be estimated (Problem 35).

SUMMARY

The **total kinetic energy** of a rigid body, such as a cylinder, that is rolling on a rough surface without slipping equals the rotational kinetic energy about its center of mass, $\frac{1}{2}I_{CM}\omega^2$, plus the translational kinetic energy of the center of mass, $\frac{1}{2}Mv_{CM}^2$:

$$K = \tfrac{1}{2}I_{CM}\omega^2 + \tfrac{1}{2}Mv_{CM}^2 \tag{11.4}$$

In this expression, v_{CM} is the velocity of the center of mass and $v_{CM} = R\omega$ for pure rolling motion.

The **torque** $\boldsymbol\tau$ due to a force **F** about an origin in an inertial frame is defined to be

$$\boldsymbol\tau \equiv \mathbf{r} \times \mathbf{F} \tag{11.7}$$

Given two vectors **A** and **B**, their **cross product** $\mathbf{A} \times \mathbf{B}$ is a vector **C** having a magnitude

$$C \equiv AB \sin\theta \tag{11.9}$$

where θ is the angle included between **A** and **B**. The direction of the vector $\mathbf{C} = \mathbf{A} \times \mathbf{B}$ is perpendicular to the plane formed by **A** and **B**, and its sense is determined by the right-hand rule. Some properties of the cross product include the facts that $\mathbf{A} \times \mathbf{B} = -\mathbf{B} \times \mathbf{A}$ and $\mathbf{A} \times \mathbf{A} = 0$.

The **angular momentum L** of a particle of linear momentum $\mathbf{p} = m\mathbf{v}$ is

$$\mathbf{L} \equiv \mathbf{r} \times \mathbf{p} = m\mathbf{r} \times \mathbf{v} \tag{11.15}$$

where **r** is the vector position of the particle relative to an origin in an inertial frame.

The **net external torque** acting on a particle or rigid body is equal to the time rate of change of its angular momentum:

$$\sum \boldsymbol\tau_{ext} = \frac{d\mathbf{L}}{dt} \tag{11.20}$$

The z *component* of **angular momentum of a rigid body rotating about a fixed axis** (the z axis) is

$$L_z = I\omega \tag{11.21}$$

where I is the moment of inertia about the axis of rotation, and ω is its angular velocity.

The **net external torque** acting on a rigid body equals the product of its moment of inertia about the axis of rotation and its angular acceleration:

$$\sum \tau_{ext} = I\alpha \tag{11.23}$$

If the net external torque acting on a system is zero, the total angular momentum of the system is constant. Applying this **conservation of angular momentum** law to a body whose moment of inertia changes gives

$$I_i\omega_i = I_f\omega_f = \text{constant} \tag{11.27}$$

QUESTIONS

1. Is it possible to calculate the torque acting on a rigid body without specifying a center of rotation? Is the torque independent of the location of the center of rotation?
2. Is the triple product defined by $\mathbf{A} \cdot (\mathbf{B} \times \mathbf{C})$ a scalar or vector quantity? Explain why the operation $(\mathbf{A} \cdot \mathbf{B}) \times \mathbf{C}$ has no meaning.
3. In the expression for torque, $\boldsymbol{\tau} = \mathbf{r} \times \mathbf{F}$, is r equal to the moment arm? Explain.
4. If the torque acting on a particle about an arbitrary origin is zero, what can you say about its angular momentum about that origin?
5. Suppose that the velocity vector of a particle is completely specified. What can you conclude about the direction of its angular momentum vector with respect to the direction of motion?
6. If a single force acts on an object, and the torque due to that force is nonzero about some point, is there any other point about which the torque is zero?
7. If a system of particles is in motion, is it possible for the total angular momentum to be zero about some origin? Explain.
8. A ball is thrown in such a way that it does not spin about its own axis. Does this mean that the angular momentum is zero about an arbitrary origin? Explain.
9. Why is it easier to keep your balance on a moving bicycle than on a bicycle at rest?
10. A scientist at a hotel sought assistance from a bellhop to carry a mysterious suitcase. When the unaware bellhop rounded a corner carrying the suitcase, it suddenly moved away from him for some unknown reason. At this point, the alarmed bellhop dropped the suitcase and ran off. What do you suppose might have been in the suitcase?
11. When a cylinder rolls on a horizontal surface as in Figure 11.3, are there any points on the cylinder that have only a vertical component of velocity at some instant? If so, where are they?
12. Three objects of uniform density—a solid sphere, a solid cylinder, and a hollow cylinder—are placed at the top of an incline (Fig. 11.22). If they all are released from rest at

the same elevation and roll without slipping, which reaches the bottom first? Which reaches last? You should try this at home and note that the result is independent of the masses and radii.

FIGURE 11.22

13. A mouse is initially at rest on a horizontal turntable mounted on a frictionless vertical axle. If the mouse begins to walk around the perimeter, what happens to the turntable? Explain.
14. Stars originate as large bodies of slowly rotating gas. Because of gravity, these regions of gas slowly decrease in size. What happens to the angular velocity of a star as it shrinks? Explain.
15. Often when a high diver wants to turn a flip in midair, she will draw her legs up against her chest. Why does this make her rotate faster? What should she do when she wants to come out of her flip?
16. As a tether ball winds around a pole, what happens to its angular velocity? Explain.
17. For a particle undergoing uniform circular motion, how are its linear momentum p and the angular momentum L oriented with respect to each other?
18. Why do tightrope walkers carry a long pole to help balance themselves?
19. Two balls have the same size and mass. One is hollow while the other is solid. How could you decide which is which without breaking them apart?
20. A particle is moving in a circle with constant speed. Locate one point about which the particle's angular momentum is constant and another about which it changes with time.

PROBLEMS

Review Problem

A rigid, massless rod has three equal masses attached to it as shown in the figure. The rod is free to rotate about a frictionless axle perpendicular to the rod through the point P and is released from rest in the horizontal position at $t = 0$. Assuming m and d are known, find (a) the moment of inertia of the system (rod plus masses) about the pivot, (b) the

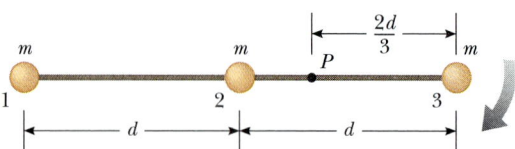

torque acting on the system at $t = 0$, (c) the angular acceleration of the system at $t = 0$, (d) the linear acceleration of the mass labeled 3 at $t = 0$, (e) the maximum kinetic energy of the system, (f) the maximum angular speed reached by the rod, (g) the maximum angular momentum of the system, and (h) the maximum speed reached by the mass labeled 2.

Section 11.1 Rolling Motion of a Rigid Body

1. A cylinder of mass 10.0 kg rolls without slipping on a horizontal surface. At the instant its center of mass has a speed of 10.0 m/s, determine (a) the translational kinetic energy of its center of mass, (b) the rotational energy about its center of mass, and (c) its total energy.

2. A solid sphere has a radius of 0.200 m and a mass of 150 kg. How much work is required to get the sphere rolling with an angular speed of 50.0 rad/s on a horizontal surface? (Assume the sphere starts from rest and rolls without slipping.)

3. (a) Determine the acceleration of the center of mass of a uniform solid disk rolling down an incline and compare this acceleration with that of a uniform hoop. (b) What is the minimum coefficient of friction required to maintain pure rolling motion for the disk?

4. A uniform solid disk and a uniform hoop are placed side by side at the top of an incline of height h. If they are released from rest and roll without slipping, determine their speeds when they reach the bottom. Which object reaches the bottom first?

5. A bowling ball has a mass of 4.0 kg, a moment of inertia of 1.6×10^{-2} kg·m², and a radius of 0.10 m. If it rolls down the lane without slipping at a linear speed of 4.0 m/s, what is its total energy?

5A. A bowling ball has a mass M, radius R, and a moment of inertia of $\frac{2}{5}MR^2$. If it rolls down the lane without slipping at a linear speed v, what is its total energy in terms of M and v?

6. A ring of mass 2.4 kg, inner radius 6.0 cm, and outer radius 8.0 cm is rolling (without slipping) up an inclined plane that makes an angle of $\theta = 36.9°$ with the horizontal (Fig. P11.6). At the moment the ring is $x = 2.0$ m up the plane its speed is 2.8 m/s. The ring continues up the plane for some additional distance and then rolls back down. Assuming that the plane is long enough so that the ring does not roll off the top end, how far up the plane does it go?

Section 11.2 The Vector Product and Torque

7. Two vectors are given by $\mathbf{A} = -3\mathbf{i} + 4\mathbf{j}$ and $\mathbf{B} = 2\mathbf{i} + 3\mathbf{j}$. Find (a) $\mathbf{A} \times \mathbf{B}$ and (b) the angle between $\mathbf{A}$ and $\mathbf{B}$.

8. A student claims to have found a vector $\mathbf{A}$ such that $(2\mathbf{i} - 3\mathbf{j} + 4\mathbf{k}) \times \mathbf{A} = (4\mathbf{i} + 3\mathbf{j} - \mathbf{k})$. Do you believe this claim? Explain.

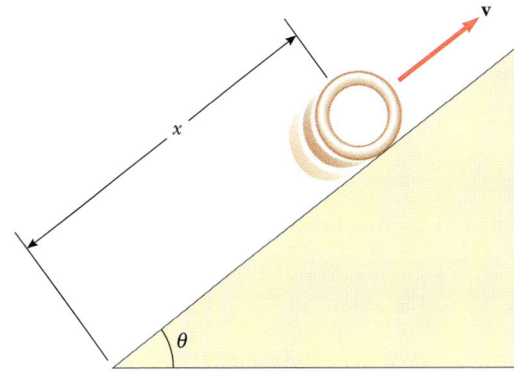

FIGURE P11.6

9. Vector $\mathbf{A}$ is in the negative y direction, and vector $\mathbf{B}$ is in the negative x direction. What are the directions of (a) $\mathbf{A} \times \mathbf{B}$ and (b) $\mathbf{B} \times \mathbf{A}$?

10. A particle is located at the vector position $\mathbf{r} = (\mathbf{i} + 3\mathbf{j})$ m, and the force acting on it equals $(3\mathbf{i} + 2\mathbf{j})$ N. What is the torque about (a) the origin and (b) the point having coordinates $(0, 6)$ m?

11. If $|\mathbf{A} \times \mathbf{B}| = \mathbf{A} \cdot \mathbf{B}$, what is the angle between $\mathbf{A}$ and $\mathbf{B}$?

12. Verify Equation 11.14 and show that the cross product may be written

$$\mathbf{A} \times \mathbf{B} = \begin{vmatrix} \mathbf{i} & \mathbf{j} & \mathbf{k} \\ A_x & A_y & A_z \\ B_x & B_y & B_z \end{vmatrix}$$

13. Two forces $\mathbf{F}_1$ and $\mathbf{F}_2$ act along the two sides of an equilateral triangle as shown in Figure P11.13. Find a third force $\mathbf{F}_3$ to be applied at B and along BC that will make the net torque about the point of intersection of the altitudes zero. Will the net torque change if $\mathbf{F}_3$ is applied not at B but at any other point along BC?

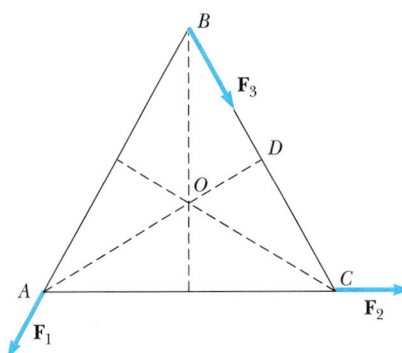

FIGURE P11.13

---

14. A force $\mathbf{F} = (2.0\mathbf{i} + 3.0\mathbf{j})$ N is applied to an object that is pivoted about a fixed axis aligned along the z coordinate axis. If the force is applied at the point $\mathbf{r} = (4.0\mathbf{i} + 5.0\mathbf{j} + 0\mathbf{k})$ m, find (a) the magnitude of the net torque about the z axis and (b) the direction of the torque vector $\boldsymbol{\tau}$.

Section 11.3   Angular Momentum of a Particle

15. A light rigid rod 1.00 m in length rotates in the xy plane about a pivot through the rod's center. Two particles of masses 4.00 kg and 3.00 kg are connected to its ends (Fig. P11.15). Determine the angular momentum of the system about the origin at the instant the speed of each particle is 5.00 m/s.

15A. A light rigid rod of length d rotates in the xy plane about a pivot through the rod's center. Two particles of masses m_1 and m_2 are connected to its ends (Fig. P11.15). Determine the angular momentum of the system about the origin at the instant the speed of each particle is v.

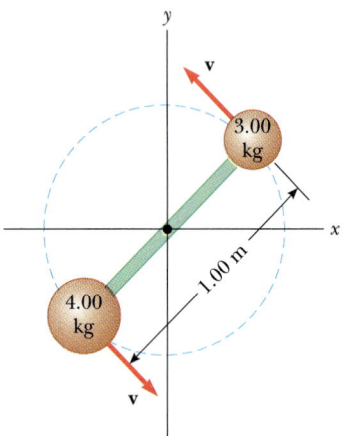

FIGURE P11.15

16. At a certain instant the position of a stone in a sling is given by $\mathbf{r} = (1.7\mathbf{i})$ m. The linear momentum $\mathbf{p}$ of the stone is $(12\mathbf{j})$ kg·m/s. Calculate its angular momentum $\mathbf{L} = \mathbf{r} \times \mathbf{p}$.

17. The position vector of a particle of mass 2.0 kg is given as a function of time by $\mathbf{r} = (6.0\mathbf{i} + 5.0t\mathbf{j})$ m. Determine the angular momentum of the particle as a function of time.

18. A conical pendulum consists of a bob of mass m moving in a circular path in a horizontal plane as shown in Figure P11.18. During the motion, the supporting wire of length ℓ maintains a constant angle θ with the vertical. Show that the magnitude of the angular momentum of the bob about the support point is

$$L = \sqrt{\frac{m^2 g\ell^3 \sin^4 \theta}{\cos \theta}}$$

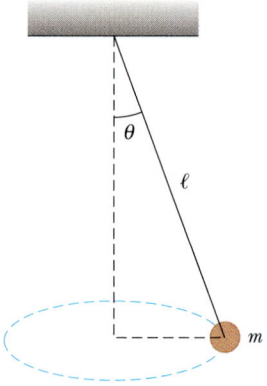

FIGURE P11.18

19. A particle of mass m moves in a circle of radius R at a constant speed v, as shown in Figure P11.19. If the motion begins at point Q, determine the angular momentum of the particle about point P as a function of time.

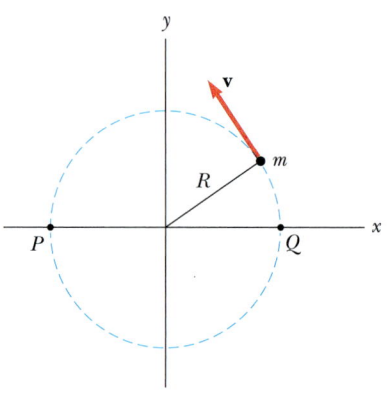

FIGURE P11.19

20. An airplane of mass 12 000 kg flies level to the ground at an altitude of 10.0 km with a constant speed of 175 m/s relative to the Earth. (a) What is the magnitude of the airplane's angular momentum relative to a ground observer directly below the airplane? (b) Does this value change as the airplane continues its motion along a straight line?

21. A ball having mass m is fastened at the end of a flagpole that is connected to the side of a tall building at point P shown in Figure P11.21. The length of the flagpole is ℓ, and it makes an angle θ with the horizontal. If the ball becomes loose and starts to fall, determine its angular momentum as a function of time about P. Neglect air resistance.

22. A 4.0-kg mass is attached to a light cord, which is wound around a pulley (Fig. 10.18). The pulley is a uniform solid cylinder of radius 8.0 cm and mass 2.0 kg. (a) What is the net torque on the system

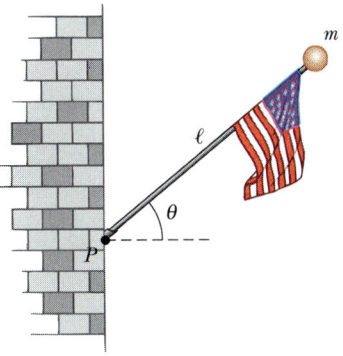

FIGURE P11.21

about the point O? (b) When the mass has a speed v, the pulley has an angular speed $\omega = v/R$. Determine the total angular momentum of the system about O. (c) Using the fact that $\tau = dL/dt$ and your result from (b), calculate the acceleration of the mass.

23. A particle of mass m is shot with an initial velocity $\mathbf{v}_0$ making an angle θ with the horizontal as shown in Figure P11.23. The particle moves in the gravitational field of the Earth. Find the angular momentum of the particle about the origin when the particle is (a) at the origin, (b) at the highest point of its trajectory, and (c) just before it hits the ground. (d) What torque causes its angular momentum to change?

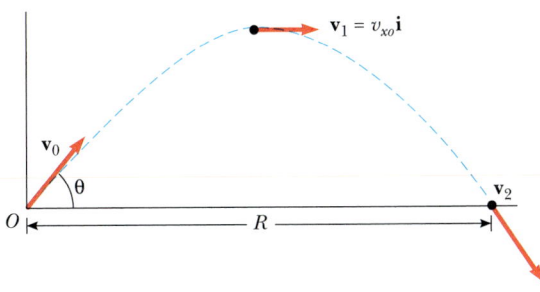

FIGURE P11.23

Section 11.4   Rotation of a Rigid Body About a Fixed Axis

24. A uniform solid disk of mass 3.00 kg and radius 0.200 m rotates about a fixed axis perpendicular to its face. If the angular frequency of rotation is 6.00 rad/s, calculate the angular momentum of the disk when the axis of rotation (a) passes through its center of mass and (b) passes through a point midway between the center and the rim.

24A. A uniform solid disk of mass M and radius R rotates about a fixed axis perpendicular to its face. If the angular frequency of rotation is ω, calculate the angular momentum of the disk when the axis of rotation (a) passes through its center of mass and (b) passes through a point midway between the center and the rim.

25. A particle of mass 0.400 kg is attached to the 100-cm mark of a meter stick of mass 0.100 kg. The meter stick rotates on a horizontal, frictionless table with an angular speed of 4.00 rad/s. Calculate the angular momentum of the system when the stick is pivoted about an axis (a) perpendicular to the table through the 50.0-cm mark and (b) perpendicular to the table through the 0-cm mark.

26. The hour and minute hands of Big Ben in London are 2.7 m and 4.5 m long and have masses of 60 kg and 100 kg, respectively. Calculate their total angular momentum about the center point. Treat the hands as long, thin rods.

Section 11.5   Conservation of Angular Momentum

27. A cylinder for which the moment of inertia is I_1 rotates about a vertical, frictionless axle with angular velocity ω_0. A second cylinder, this one having moment of inertia I_2 and initially not rotating, drops onto the first cylinder (Fig. P11.27). Since the surfaces are not frictionless, the two eventually reach the same angular velocity ω. (a) Calculate ω. (b) Show that energy is lost in this situation and calculate the ratio of the final to the initial rotational energy.

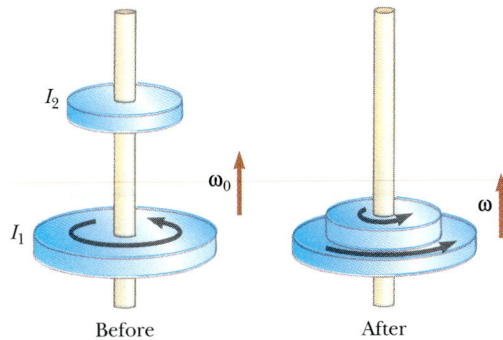

Before   After

FIGURE P11.27

28. A merry-go-round of radius $R = 2.0$ m has a moment of inertia $I = 250$ kg·m² and is rotating at 10 rev/min. A 25-kg child jumps onto the edge of the merry-go-round. What is the new angular speed of the merry-go-round?

29. A 60-kg woman stands at the rim of a horizontal turntable having a moment of inertia of 500 kg·m² and a radius of 2.00 m. The turntable is initially at rest and is free to rotate about a frictionless, vertical axle through its center. The woman then starts walking around the rim clockwise (as viewed from above the system) at a constant speed of 1.50 m/s relative to

the Earth. (a) In what direction and with what angular speed does the turntable rotate? (b) How much work does the woman do to set the turntable into motion?

30. A uniform rod of mass 100 g and length 50.0 cm rotates in a horizontal plane about a fixed, vertical, frictionless pin through its center. Two small beads, each of mass 30.0 g, are mounted on the rod such that they are able to slide without friction along its length. Initially the beads are held by catches at positions 10.0 cm on each side of center, at which time the system rotates at an angular speed of 20.0 rad/s. Suddenly, the catches are released and the small beads slide outward along the rod. Find (a) the angular speed of the system at the instant the beads reach the ends of the rod and (b) the angular speed of the rod after the beads fly off the ends.

30A. A uniform rod of mass M and length d rotates in a horizontal plane about a fixed, vertical, frictionless pin through its center. Two small beads, each of mass m, are mounted on the rod such that they are able to slide without friction along its length. Initially the beads are held by catches at positions x (where $x < d/2$) on each side of center, at which time the system rotates at an angular speed ω. Suddenly, the catches are released and the small beads slide outward along the rod. Find (a) the angular speed of the system at the instant the beads reach the ends of the rod and (b) the angular speed of the rod after the beads fly off the ends.

31. The student in Figure 11.17 holds two weights, each of mass 10.0 kg. When his arms are extended horizontally, the weights are 1.00 m from the axis of rotation and he rotates with an angular speed of 2.00 rad/s. The moment of inertia of student plus stool is 8.00 kg·m² and is assumed to be constant. If the student pulls the weights horizontally to 0.250 m from the rotation axis, calculate (a) the final angular speed of the system and (b) the change in the mechanical energy of the system.

32. A puck of mass 80.0 g and radius 4.00 cm slides along an air table at 1.5 m/s as shown in Figure P11.32a. It makes a glancing collision with a second puck of radius 6.00 cm and mass 120 g (initially at

rest) such that their rims just touch. The pucks stick together and spin after the collision (Fig. P11.32b). What are (a) the angular momentum of the system relative to the center of mass and (b) its angular speed about the center of mass?

33. A wooden block of mass M resting on a frictionless horizontal surface is attached to a rigid rod of length ℓ and of negligible mass (Fig. P11.33). The rod is pivoted at the other end. A bullet of mass m traveling parallel to the horizontal surface and normal to the rod with speed v hits the block and gets embedded in it. (a) What is the angular momentum of the bullet-block system? (b) What fraction of the original kinetic energy is lost in the collision?

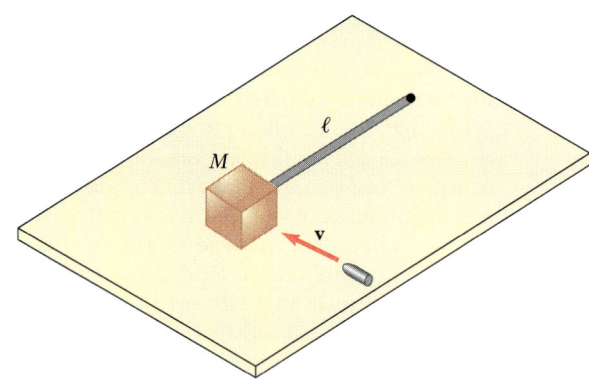

FIGURE P11.33

34. A wheel-shaped space station has a radius of 100 m and a moment of inertia of 5.00×10^8 kg·m². A crew of 150 live on the rim, and the station is rotated so that they experience an apparent gravity of g (Fig. P11.34). If 100 people move to the center, the angular rotation speed changes. What apparent gravity is experienced by those remaining at the rim? Assume an average mass of 65.0 kg per crew member.

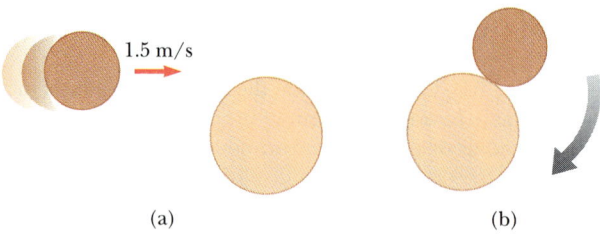

1.5 m/s

(a)   (b)

FIGURE P11.32

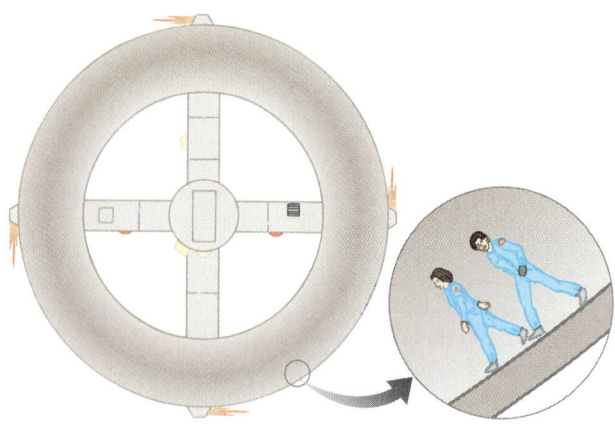

FIGURE P11.34

*Section 11.7 Angular Momentum as a Fundamental Quantity

35. In the Bohr model of the hydrogen atom, the electron moves in a circular orbit of radius 0.529×10^{-10} m around the proton. Assuming the orbital angular momentum of the electron is equal to h, calculate (a) the orbital speed of the electron, (b) the kinetic energy of the electron, and (c) the angular frequency of the electron's motion.

ADDITIONAL PROBLEMS

36. A uniform solid sphere of radius r is placed on the inside surface of a hemispherical bowl of radius R. The sphere is released from rest at an angle θ to the vertical and rolls without slipping (Fig. P11.36). Determine the angular speed of the sphere when it reaches the bottom of the bowl.

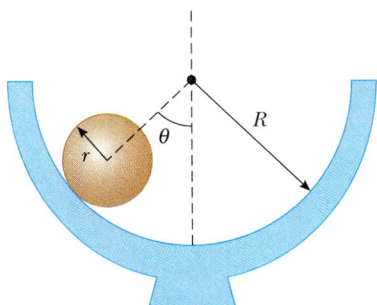

FIGURE P11.36

37. Halley's comet moves about the Sun in an elliptical orbit, with its closest approach to the Sun being 0.59 AU and its greatest distance from the Sun being 35 AU (1 AU = the Earth-Sun distance). If the comet's speed at closest approach is 54 km/s, what is its speed when it is farthest from the Sun, assuming that its angular momentum about the Sun is constant?

38. A thin uniform cylindrical turntable of radius 2.00 m and mass 30.0 kg rotates in a horizontal plane with an initial angular speed of 4π rad/s. The turntable bearing is frictionless. A small clump of clay of mass 0.250 kg is dropped onto the turntable and sticks at a point 1.80 m from the center of rotation. (a) Find the final angular speed of the clay and turntable. (Treat the clay as a point mass.) (b) Is mechanical energy constant in this collision? Explain and use numerical results to verify your answer.

39. A string is wound around a uniform disk of radius R and mass M. The disk is released from rest with the string vertical and its top end tied to a fixed support (Fig. P11.39). As the disk descends, show that (a) the tension in the string is one-third the weight of the disk, (b) the magnitude of the acceleration of the

center of mass is $2g/3$, and (c) the speed of the center of mass is $(4gh/3)^{1/2}$. Verify your answer to (c) using the energy approach.

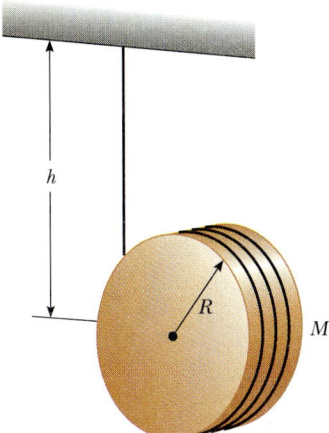

FIGURE P11.39

40. A constant horizontal force F is applied to a lawn roller in the form of a uniform solid cylinder of radius R and mass M (Fig. P11.40). If the roller rolls without slipping on the horizontal surface, show that (a) the acceleration of the center of mass is $2F/3M$ and (b) the minimum coefficient of friction necessary to prevent slipping is $F/3Mg$. (*Hint:* Take the torque with respect to the center of mass.)

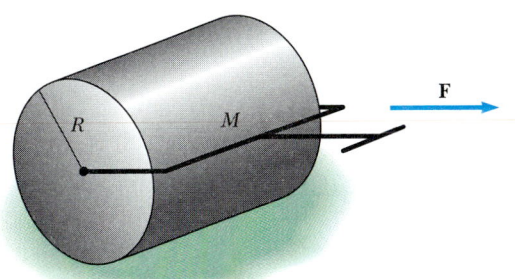

FIGURE P11.40

41. A light rope passes over a light, frictionless pulley. One end is fastened to a bunch of bananas of mass M, and a monkey of mass M clings to the other end (Fig. P11.41). The monkey climbs the rope in an attempt to reach the bananas. (a) Treating the system as consisting of the monkey, bananas, rope, and pulley, evaluate the net torque about the pulley axis. (b) Using the results to (a), determine the total angular momentum about the pulley axis and describe the motion of the system. Will the monkey reach the bananas?

FIGURE P11.41

42. A small, solid sphere of mass m and radius r rolls without slipping along the track shown in Figure P11.42. If it starts from rest at the top of the track at a height h, where h is large compared to r, (a) what is the minimum value of h (in terms of the radius of the loop R) such that the sphere completes the loop? (b) What are the force components on the sphere at the point P if $h = 3R$?

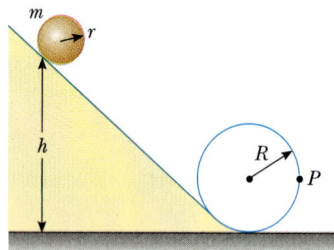

FIGURE P11.42

43. This problem describes a method of determining the moment of inertia of an irregularly shaped object such as the payload for a satellite. Figure P11.43 shows one method of determining I experimentally. A mass m is suspended by a cord wound around the inner shaft (radius r) of a turntable supporting the object. When the mass is released from rest, it descends with constant acceleration a distance h, acquiring a speed v. Show that the moment of inertia I of the equipment (including the turntable) is $mr^2(2gh/v^2 - 1)$.

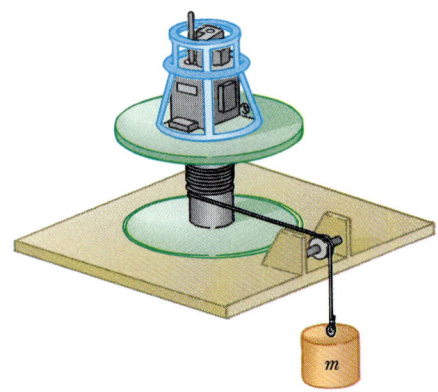

FIGURE P11.43

44. Consider the problem of the solid sphere rolling down an incline as described in Example 11.1. (a) Choose the instantaneous axis through the contact point P as the axis of the origin for the torque equation and show that the acceleration of the center of mass is $a_{CM} = \frac{5}{7}g\sin\theta$. (b) Show that the minimum coefficient of friction such that the sphere rolls without slipping is $\mu_{\min} = \frac{2}{7}\tan\theta$.

45. A common physics demonstration (Fig. P11.45) consists of a ball resting r meters away from the hinged end of a board of length ℓ that is elevated at an angle θ with the horizontal. A cup attached to the board at r_c is to catch the ball when the support stick is suddenly removed. Show that (a) the ball, placed at the very end, will lag behind the stick in falling when $\theta < 35.3°$ and (b) the cup should be placed at

$$r_c = \frac{2\ell}{3\cos\theta}$$

for this limiting angle. (c) If a ball is at the end of a 1.0-m stick at this critical angle, show that the cup must be 18.4 cm from the moving (free) end.

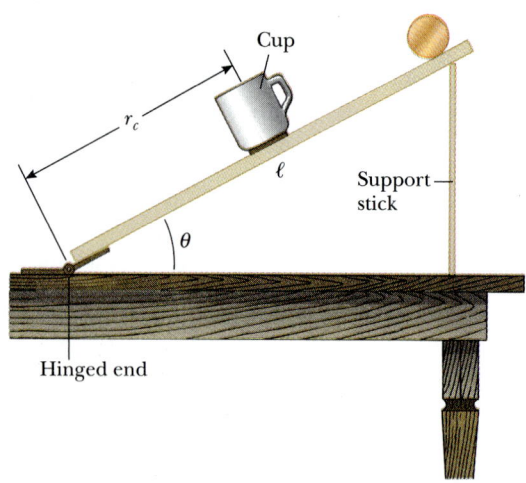

FIGURE P11.45

46. A bowling ball is both sliding and spinning on a horizontal surface such that its rotational kinetic energy equals its translational kinetic energy. What is the ratio of the ball's center-of-mass speed to the tangential speed of a point on its surface?

47. A constant torque of 25.0 N·m is applied to a grindstone for which the moment of inertia is 0.130 kg·m². Using energy principles, find the angular speed after the grindstone has made 15.0 rev. (Neglect friction.)

48. A projectile of mass m moves to the right with speed v_0 (Fig. P11.48a). The projectile strikes and sticks to the end of a stationary rod of mass M and length d that is pivoted about a frictionless axle through its center (Fig. P11.48b). (a) Find the angular speed of the system right after the collision. (b) Determine the fractional loss in mechanical energy due to the collision.

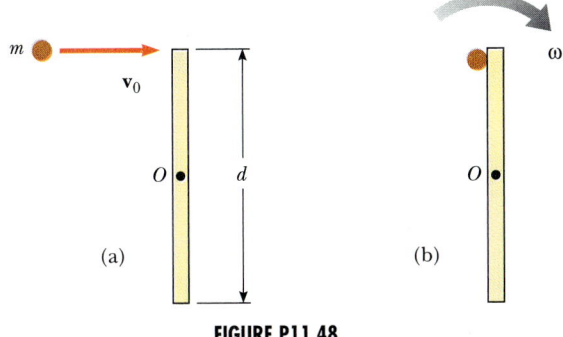

FIGURE P11.48

49. A mass m is attached to a cord passing through a small hole in a frictionless, horizontal surface (Fig. P11.49). The mass is initially orbiting with speed v_0 in a circle of radius r_0. The cord is then slowly pulled from below, decreasing the radius of the circle to r. (a) What is the speed of the mass when the radius is r? (b) Find the tension in the cord as a function of r. (c) How much work W is done in moving m from r_0 to r? (*Note:* The tension depends on r.) (d) Obtain numerical values for v, T, and W when $r = 0.100$ m, $m = 50.0$ g, $r_0 = 0.300$ m, and $v_0 = 1.50$ m/s.

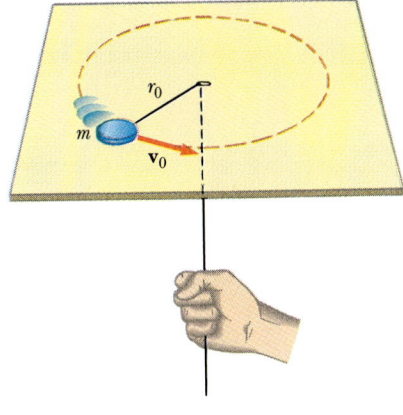

FIGURE P11.49

50. A bowling ball is given an initial speed v_0 on an alley such that it initially slides without rolling. The coefficient of friction between ball and alley is μ. Show that at the time pure rolling motion occurs, (a) the speed of the ball's center of mass is $5v_0/7$ and (b) the distance it has traveled is $12v_0^2/49\mu g$. (*Hint:* When pure rolling motion occurs, $v_{CM} = R\omega$. Since the frictional force provides the deceleration, from Newton's second law it follows that $a_{CM} = \mu g$.)

51. A trailer with loaded weight w is being pulled by a vehicle with a force F, as in Figure P11.51. The trailer is loaded such that its center of mass is located as shown. Neglect the force of rolling friction and assume the trailer has an acceleration of magnitude a. (a) Find the vertical component of F in terms of the given parameters. (b) If $a = 2.00$ m/s² and $h = 1.50$ m, what must be the value of d in order that $F_y = 0$ (no vertical load on the vehicle)? (c) Find F_x and F_y given that $w = 1500$ N, $d = 0.800$ m, $L = 3.00$ m, $h = 1.50$ m, and $a = -2.00$ m/s².

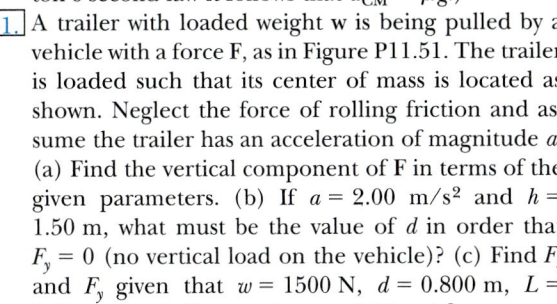

FIGURE P11.51

52. (a) A thin rod of length h and mass M is held vertically with its lower end resting on a frictionless horizontal surface. The rod is then let go to fall freely. Determine the speed of its center of mass just before it hits the horizontal surface. (b) Suppose the rod were pivoted at its lower end. Determine the speed of the rod's center of mass just before it hits the surface.

53. Two astronauts (Fig. P11.53), each having a mass of 75 kg, are connected by a 10-m rope of negligible mass. They are isolated in space, orbiting their center of mass at speeds of 5.0 m/s. Calculate (a) the magnitude of the angular momentum of the system by treating the astronauts as particles and (b) the rotational energy of the system. By pulling on the rope, the astronauts shorten the distance between them to 5.0 m. (c) What is the new angular momentum of the system? (d) What are their new speeds? (e) What is the new rotational energy of the system? (f) How much work is done by the astronauts in shortening the rope?

53A. Two astronauts (Fig. P11.53), each having a mass M, are connected by a rope of length d having negligible mass. They are isolated in space, orbiting their center of mass at speeds v. Calculate (a) the magnitude of the angular momentum of the system by treating the astronauts as particles and (b) the rotational energy of the system. By pulling on the rope, the astronauts shorten the distance between them to $d/2$. (c) What is the new angular momentum of the system? (d) What are their new speeds? (e) What is the new rotational energy of the system? (f) How much work is done by the astronauts in shortening the rope?

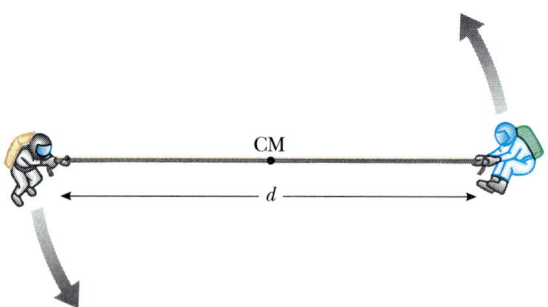

FIGURE P11.53

54. A solid cube of wood of side $2a$ and mass M is resting on a horizontal suface. The cube is constrained to rotate about an axis AB (Fig. P11.54). A bullet of mass m and speed v is shot at the face opposite $ABCD$ at a height of $4a/3$. The bullet gets embedded in the cube. Find the minimum value of v required to tip the cube so that it falls on face $ABCD$. Assume $m < M$.

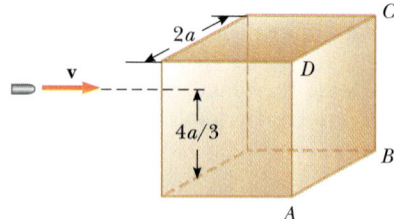

FIGURE P11.54

55. Toppling chimneys often break apart in mid-fall because the mortar between the bricks cannot withstand much tension force. As the chimney falls, this tension supplies the centripetal forces on the topmost segments that they need to keep them traveling in an arc. For simplicity, let us model the chimney as a uniform rod of length ℓ pivoted at the lower end. The rod starts at rest in a vertical position (with the pivot at the bottom) and falls over under the influ-

ence of gravity. What fraction of the length of the rod has a tangential acceleration greater than $g \sin \theta$, where θ is the angle the chimney makes with the vertical axis?

56. A solid sphere is positioned at the top of an incline that makes an angle θ with the horizontal. This initial position of the sphere is a vertical distance h above the ground. The sphere is released and moves down the incline. Calculate the speed of the sphere when it reaches the bottom of the incline in the case where (a) it rolls without slipping and (b) it slips frictionlessly without rolling. Compare the times required to reach the bottom in cases (a) and (b).

57. A spool of wire of mass M and radius R is unwound under a constant force **F** (Fig. P11.57). Assuming the spool is a uniform solid cylinder that doesn't slip, show that (a) the acceleration of the center of mass is $4F/3M$ and (b) the force of friction is to the *right* and equal in magnitude to $F/3$. (c) If the cylinder starts from rest and rolls without slipping, what is the speed of its center of mass after it has rolled through a distance d?

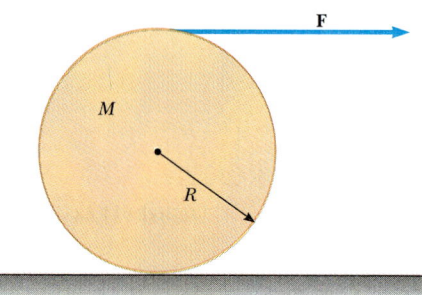

FIGURE P11.57

58. A uniform solid disk is set into rotation with an angular speed ω_0 about an axis through its center. While still rotating at this speed, the disk is placed into contact with a horizontal surface and released as in Figure P11.58. (a) What is the angular speed of the disk

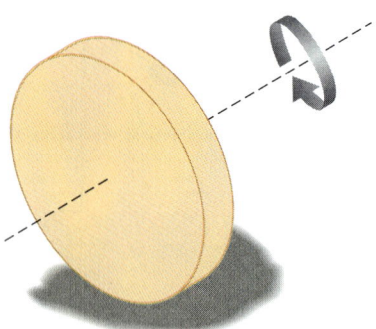

FIGURE P11.58

once pure rolling takes place? (b) Find the fractional loss in kinetic energy from the time the disk is released until pure rolling occurs. (*Hint:* Consider torques about the center of mass.)

59. Suppose a solid disk of radius R is given an angular speed ω_0 about an axis through its center and then lowered to a horizontal surface and released, as in Problem 58 (Fig. P11.58). Furthermore, assume that the coefficient of friction between disk and surface is μ. (a) Show that the time it takes pure rolling motion to occur is $R\omega_0/3\mu g$. (b) Show that the distance the disk travels before pure rolling occurs is $R^2\omega_0^2/18\mu g$.

60. A large, cylindrical roll of tissue paper of initial radius R lies on a long, horizontal surface with the open end of the paper nailed to the surface. The roll is given a slight shove ($v_0 \approx 0$) and commences to unroll. (a) Determine the speed of the center of mass of the roll when its radius has diminished to r. (b) Calculate a numerical value for this speed at $r = 1.0$ mm, assuming $R = 6.0$ m. (c) What happens to the energy of the system when the paper is completely unrolled? (*Hint:* Assume the roll has a uniform density and apply energy methods.)

61. A solid cube of side $2a$ and mass M is sliding on a frictionless surface with uniform velocity v_0 as in Figure P11.61a. It hits a small obstacle at the end of the table, which causes the cube to tilt as in Figure P11.61b. Find the minimum value of v_0 such that the cube falls off the table. Note that the moment of inertia of the cube about an axis along one of its edges is $8Ma^2/3$. (*Hint:* The cube undergoes an inelastic collision at the edge.)

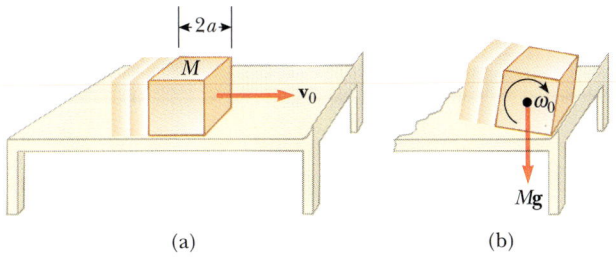

(a)   (b)

FIGURE P11.61

 62. In a demonstration known as the ballistics cart, a ball is projected vertically upward from a cart moving with constant velocity along the horizontal direction. The ball lands in the catching cup of the cart because both the cart and ball have the same horizontal component of velocity. Consider a ballistics cart moving on an incline making an angle θ with the horizontal as in Figure P11.62. The cart (including wheels) has a mass M and the moment of inertia of each of the two wheels is $mR^2/2$. (a) Using conservation of energy (assuming no friction between cart and axle),

and assuming pure rolling motion (no slipping), show that the acceleration of the cart along the incline is

$$a_x = \left(\frac{M}{M + 2m}\right) g \sin\theta$$

(b) Note that the x component of acceleration of the ball released by the cart is $g \sin\theta$. Thus, the x component of the cart's acceleration is *smaller* than that of the ball by the factor $M/(M + 2m)$. Use this fact and kinematic equations to show that the ball overshoots the cart by an amount Δx, where

$$\Delta x = \left(\frac{4m}{M + 2m}\right)\left(\frac{\sin\theta}{\cos^2\theta}\right)\frac{v_{y0}^2}{g}$$

and v_{y0} is the initial speed of the ball imparted to it by the spring in the cart. (c) Show that the distance d that the ball travels measured along the incline is

$$d = \frac{2v_{y0}^2}{g}\frac{\sin\theta}{\cos^2\theta}$$

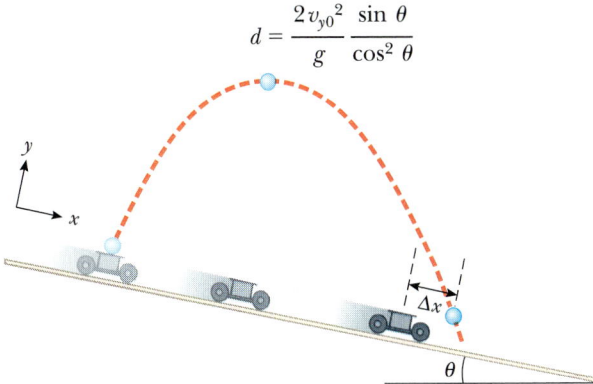

FIGURE P11.62

63. A spool of wire rests on a horizontal surface as in Figure P11.63, and, when pulled, does not slip at the contact point P. The spool is pulled in the directions indicated by the vectors $\mathbf{F}_1$, $\mathbf{F}_2$, $\mathbf{F}_3$, and $\mathbf{F}_4$. For each force, determine the direction the spool rolls. Note that the line of action of $\mathbf{F}_2$ passes through P.

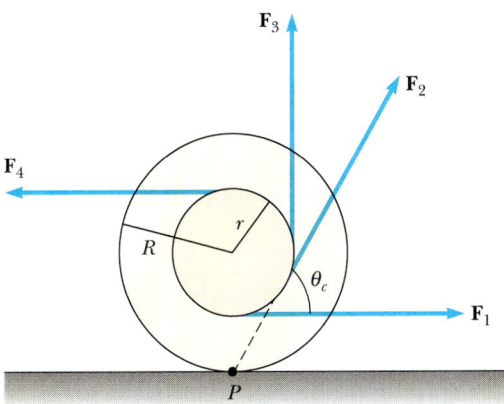

FIGURE P11.63

64. The spool shown in Figure P11.63 has inner radius r and outer radius R. The angle θ between the applied force and the horizontal can be varied. Show that the critical angle for which the spool does not roll and remains stationary is given by $\cos \theta_c = r/R$. (*Hint:* At the critical angle, the line of action of the applied force passes through the contact point.)

65. A plank having mass $M = 6.0$ kg rides on top of two identical solid cylindrical rollers each having radius $R = 5.0$ cm and mass $m = 2.0$ kg (Fig. P11.65). The plank is pulled by a constant horizontal force $F = 6.0$ N applied to its end and perpendicular to the axes of the cylinders (which are parallel). The cylin-

ders roll without slipping on a flat surface. There is also no slipping between cylinders and plank. (a) Find the acceleration of the plank and of the rollers. (b) What frictional forces are acting?

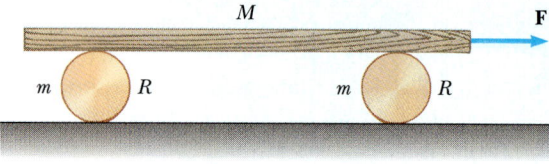

FIGURE P11.65

CHAPTER 12

Static Equilibrium and Elasticity

The Chinese acrobats in this difficult formation represent a balanced system. The external forces acting on the system, as shown by the blue vectors, are the weights of the acrobats, $\mathbf{w}_1$ and $\mathbf{w}_2$, and the upward force of the support, $\mathbf{n}$, on the lower acrobat. The vector sum of these external forces must be zero in such a balanced system. The net external torque acting on such a balanced system must also be zero. *(J.P. Lafont, Sygma)*

I n Chapters 10 and 11 we studied the dynamics of a rigid object, that is, one whose parts remain at a fixed separation with respect to each other when subjected to external forces. Part of this chapter is concerned with the conditions under which a rigid object is in equilibrium. The term *equilibrium* implies either that the object is at rest or that its center of mass moves with constant velocity. We deal here only with the former, which are referred to as objects in *static equilibrium*. Static equilibrium represents a common situation in engineering practice, and the principles involved are of special interest to civil engineers, architects, and mechanical engineers. Those of you who are engineering students will undoubtedly take an intensified course in statics in the future.

In Chapter 5 we stated that one necessary condition for equilibrium is that the net force on an object be zero. If the object is treated as a particle, this is the only condition that must be satisfied for equilibrium. That is, if the net force on the particle is zero, the particle remains at rest (if originally at rest) or moves with constant velocity (if originally in motion).

The situation with real (extended) objects is more complex when the objects cannot be treated as particles. In order for an extended object to be in static equilibrium, the net force on it must be zero *and* the net torque about any origin must be zero. In order to establish whether or not an object is in equilibrium, we must know its size and shape, the forces acting on different parts of it, and the points of application of the various forces.

The last section of this chapter deals with the realistic situation of objects that deform under load conditions. Such deformations are usually elastic in nature and do not affect the conditions of equilibrium. By *elastic* we mean that when the deforming forces are removed, the object returns to its original shape. Several elastic constants are defined, each corresponding to a different type of deformation.

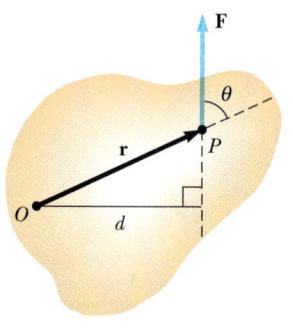

FIGURE 12.1   A single force **F** acts on a rigid object at the point *P*. The moment arm of **F** relative to *O* is the perpendicular distance *d* from **O** to the line of action of **F**.

12.1   THE CONDITIONS OF EQUILIBRIUM OF A RIGID OBJECT

Consider a single force **F** acting on a rigid object as in Figure 12.1. The effect of the force depends on its point of application, *P*. If **r** is the position vector of this point relative to *O*, the torque associated with the force **F** about *O* is given by Equation 11.7:

$$\boldsymbol{\tau} = \mathbf{r} \times \mathbf{F}$$

Recall from Section 11.2 that the vector $\boldsymbol{\tau}$ is perpendicular to the plane formed by **r** and **F**. Furthermore, the sense of $\boldsymbol{\tau}$ is determined by the sense of the rotation that **F** tends to give to the object. The right-hand rule can be used to determine the direction of $\boldsymbol{\tau}$: Close your right hand such that your four fingers wrap in the direction of rotation that **F** tends to give the object; your thumb then points in the direction of $\boldsymbol{\tau}$. Hence, in Figure 12.1, $\boldsymbol{\tau}$ is directed out of the paper.

As you can see from Figure 12.1, the tendency of **F** to make the object rotate about an axis through *O* depends on the moment arm *d* as well as on the magnitude of **F**. By definition, the magnitude of $\boldsymbol{\tau}$ is *Fd*.

Now suppose either of two forces, $\mathbf{F}_1$ and $\mathbf{F}_2$, act on a rigid object. The two forces will have the same effect on the object only if they have the same magnitude, the same direction, and the same line of action. In other words,

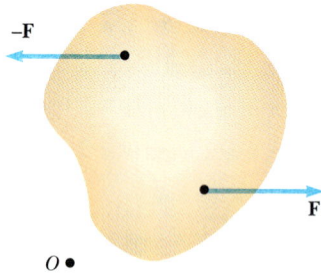

FIGURE 12.2   The two forces acting on the object are equal in magnitude and opposite in direction, yet the object is not in equilibrium.

Equivalent forces

> two forces $\mathbf{F}_1$ and $\mathbf{F}_2$ are equivalent if and only if $F_1 = F_2$ and if the two produce the same torque about any given point.

Two equal and opposite forces that are *not* equivalent are shown in Figure 12.2. The force directed to the right tends to rotate the object clockwise about an axis perpendicular to the diagram through *O*, whereas the force directed to the left tends to rotate it counterclockwise about that axis.

When pivoted about an axis through its center of mass, an object undergoes an angular acceleration about this axis if there is a nonzero torque acting. As an example, suppose an object is pivoted about an axis through its center of mass as in

Figure 12.3. Two equal and opposite forces act in the directions shown, such that their lines of action do not pass through the center of mass. A pair of forces acting in this manner form what is called a **couple.** (The two forces shown in Figure 12.2 also form a couple.) Since each force produces the same torque, *Fd*, the net torque has a magnitude 2*Fd*. Clearly, the object rotates clockwise and undergoes an angular acceleration about the axis. This is a nonequilibrium situation as far as the rotational motion is concerned. That is, the "unbalanced," or net, torque on the object gives rise to an angular acceleration α according to the relationship $\tau_{\text{net}} = 2Fd = I\alpha$ (Eq. 10.20).

In general, an object is in rotational equilibrium only if its angular acceleration $\alpha = 0$. Since $\tau_{\text{net}} = I\alpha$ for rotation about a fixed axis, a necessary condition of equilibrium is that *the net torque about any origin must be zero.* We now have *two necessary conditions for equilibrium of an object,* which can be stated as follows:

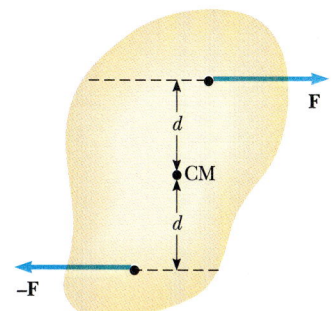

FIGURE 12.3   Two equal and opposite forces acting on the object form a couple. In this case, the object rotates clockwise. The net torque about the center of mass is 2*Fd*.

- The resultant external force must equal zero. $\quad \sum \mathbf{F} = 0 \qquad (12.1)$

- The resultant external torque must be zero about *any* origin. $\quad \sum \boldsymbol{\tau} = 0 \qquad (12.2)$

The first condition is a statement of translational equilibrium; it tells us that the linear acceleration of the center of mass of the object must be zero when viewed from an inertial reference frame. The second condition is a statement of rotational equilibrium and tells us that the angular acceleration about any axis must be zero. In the special case of **static equilibrium,** which is the main subject of this chapter, the object is at rest so that it has no linear or angular speed (that is, $v_{\text{CM}} = 0$ and $\omega = 0$).

The two vector expressions given by Equations 12.1 and 12.2 are equivalent, in general, to six scalar equations, three from the first condition of equilibrium, and three from the second (corresponding to *x*, *y*, and *z* components). Hence, in a complex system involving several forces acting in various directions, you would be faced with solving a set of equations with many unknowns. Here, we restrict our discussion to situations in which all the forces lie in the *xy* plane. (Forces whose vector representations are in the same plane are said to be *coplanar.*) With this restriction, we shall have to deal with only three scalar equations. Two of these come from balancing the forces in the *x* and *y* directions. The third comes from the torque equation, namely, that the net torque about *any* point in the *xy* plane must be zero. Hence, the two conditions of equilibrium provide the equations

$$\sum F_x = 0 \qquad \sum F_y = 0 \qquad \sum \tau_z = 0 \qquad (12.3)$$

where the axis of the torque equation is arbitrary, as we show later.

There are two cases of equilibrium that are often encountered. The first deals with an object subjected to only two forces, and the second is concerned with an object subjected to three forces.

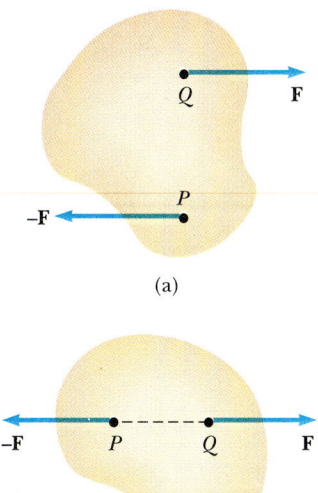

(a)

(b)

FIGURE 12.4   (a) The object is not in equilibrium because the two forces do not have the same line of action. (b) The object is in equilibrium because the two forces act along the same line.

Case I   *An object subjected to two forces is in equilibrium if and only if the two forces are equal in magnitude, opposite in direction, and have the same line of action.* Figure 12.4a shows a situation in which the object is not in equilibrium because the two forces are not along the same line. Note that the torque about any axis, such as one through *P*, is not zero, which violates the second condition of equilibrium. In Figure 12.4b, the object is in equilibrium because the forces have the same line of action. In this situation, it is easy to see that the net torque about any axis is zero.

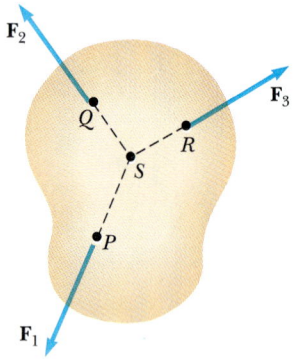

FIGURE 12.5 If three forces act on an object in equilibrium, their lines of action must intersect at a point S (or they must be parallel to each other).

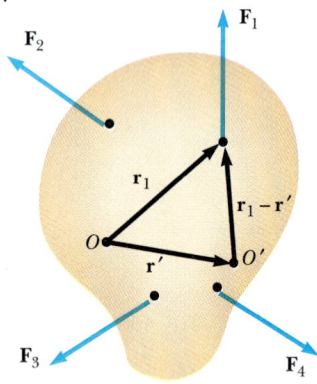

FIGURE 12.6 Construction for showing that if the net torque about origin O is zero, the net torque about any other origin, such as O', is also zero.

Case II *If an object subjected to three forces is in equilibrium, the lines of action of the three forces must intersect at a common point.* That is, the forces must be *concurrent*. (One exception to this rule is the situation in which none of the lines of action intersect. In this situation, the forces must all be parallel to each other and lie in the same plane.) Figure 12.5 illustrates the general rule. The lines of action of the three forces pass through the point S. The conditions of equilibrium require that $F_1 + F_2 + F_3 = 0$ and that the net torque about any axis be zero. Note that as long as the forces are concurrent, the net torque about an axis through S must be zero.

We can easily show that, regardless of the number of forces acting, if an object is in translational equilibrium and if the net torque is zero with respect to one point, it must also be zero about any other point. The point can be inside or outside the boundaries of the object. Consider an object under the action of several forces such that the resultant force $\Sigma \mathbf{F} = \mathbf{F}_1 + \mathbf{F}_2 + \mathbf{F}_3 + \cdots = 0$. Figure 12.6 describes this situation (for clarity, only four forces are shown). The point of application of $\mathbf{F}_1$ relative to O is specified by the position vector $\mathbf{r}_1$. Similarly, the points of application of $\mathbf{F}_2, \mathbf{F}_3, \ldots$ are specified by $\mathbf{r}_2, \mathbf{r}_3, \ldots$ (not shown). The net torque about O is

$$\sum \boldsymbol{\tau}_0 = \mathbf{r}_1 \times \mathbf{F}_1 + \mathbf{r}_2 \times \mathbf{F}_2 + \mathbf{r}_3 \times \mathbf{F}_3 + \cdots$$

Now consider another arbitrary point, O', having a position vector $\mathbf{r}'$ relative to O. The point of application of $\mathbf{F}_1$ relative to O' is identified by the vector $\mathbf{r}_1 - \mathbf{r}'$. Likewise, the point of application of $\mathbf{F}_2$ relative to O' is $\mathbf{r}_2 - \mathbf{r}'$, and so forth. Therefore, the torque about O' is

$$\begin{aligned} \sum \boldsymbol{\tau}_{O'} &= (\mathbf{r}_1 - \mathbf{r}') \times \mathbf{F}_1 + (\mathbf{r}_2 - \mathbf{r}') \times \mathbf{F}_2 + (\mathbf{r}_3 - \mathbf{r}') \times \mathbf{F}_3 + \cdots \\ &= \mathbf{r}_1 \times \mathbf{F}_1 + \mathbf{r}_2 \times \mathbf{F}_2 + \mathbf{r}_3 \times \mathbf{F}_3 + \cdots - \mathbf{r}' \times (\mathbf{F}_1 + \mathbf{F}_2 + \mathbf{F}_3 + \cdots) \end{aligned}$$

Since the net force is assumed to be zero, the last term in this last expression vanishes and we see that the torque about O' is equal to the torque about O. Hence,

> if an object is in translational equilibrium and the net torque is zero about one point, it is also zero about any other point.

12.2 MORE ON THE CENTER OF GRAVITY

Whenever we deal with rigid objects, one of the forces we must consider is the weight of the object, that is, the force of gravity acting on it. In order to compute the torque due to the weight force, all of the weight can be considered as being concentrated at a single point called the *center of gravity*. As we shall see, the center of gravity of an object coincides with its center of mass if the object is in a uniform gravitational field.

Consider an object of arbitrary shape lying in the xy plane, as in Figure 12.7. Suppose the object is divided into a large number of very small particles of masses $m_1, m_2, m_3, \ldots$ having coordinates $(x_1, y_1), (x_2, y_2), (x_3, y_3), \ldots$. In Chapter 9 we defined the x coordinate of the center of mass of such an object to be

$$x_{\mathrm{CM}} = \frac{m_1 x_1 + m_2 x_2 + m_3 x_3 + \cdots}{m_1 + m_2 + m_3 + \cdots} = \frac{\Sigma m_i x_i}{\Sigma m_i}$$

The y coordinate of the center of mass is similar to this, with x_{CM} replaced by y_{CM}.

Let us now examine the situation from another point of view by considering the weight of each part of the object, as in Figure 12.8. Each particle contributes a torque about the origin equal to the particle's weight multiplied by its moment arm. For example, the torque due to the weight $m_1\mathbf{g}_1$ is $m_1 g_1 x_1$, and so forth. We wish to locate the center of gravity, the one position of the single force **w** (the total weight of the object) whose effect on the rotation of the object is the same as that of the individual particles. Equating the torque exerted by **w** acting at the center of gravity to the sum of the torques acting on the individual particles gives

$$(m_1 g_1 + m_2 g_2 + m_3 g_3 + \cdots)x_{cg} = m_1 g_1 x_1 + m_2 g_2 x_2 + m_3 g_3 x_3 + \cdots$$

This expression accounts for the fact that the gravitational field strength, g, can in general vary over the object. If we assume that g is uniform over the object (as is usually the case), then the g terms cancel and we get

$$x_{cg} = \frac{m_1 x_1 + m_2 x_2 + m_3 x_3 + \cdots}{m_1 + m_2 + m_3 + \cdots} \qquad (12.4)$$

In other words, *the center of gravity is located at the center of mass as long as the object is in a uniform gravitational field.*

In several examples presented in the next section, we shall be concerned with homogeneous, symmetric objects for which the center of gravity coincides with the geometric center of the object. A rigid object in a uniform gravitational field can be balanced by a single force equal in magnitude to the weight of the object, as long as the force is directed upward through the center of gravity.

12.3 EXAMPLES OF RIGID OBJECTS IN STATIC EQUILIBRIUM

In working static equilibrium problems, it is important to recognize external forces acting on the object. Failure to do so will result in an incorrect analysis. The following procedure is recommended when analyzing an object in equilibrium under the action of several external forces:

Problem-Solving Strategy
Objects in Equilibrium

- Make a sketch of the object under consideration.
- Draw a free-body diagram and label all external forces acting on the object. Try to guess the correct direction for each force. If you select a direction that leads to a negative sign in your solution for a force, do not be alarmed; this merely means that the direction of the force is the opposite of what you guessed.
- Resolve all forces into rectangular components, choosing a convenient coordinate system. Then apply the first condition for equilibrium. Remember to keep track of the signs of the various force components.
- Choose a convenient axis for calculating the net torque on the object. Remember that the choice of the origin for the torque equation is arbitrary; therefore, choose an origin that will simplify your calculation as much as possible. Becoming adept at this is a matter of practice.
- The first and second conditions of equilibrium give a set of linear equations with several unknowns. All that is left is to solve the simultaneous equations for the unknowns in terms of the known quantities.

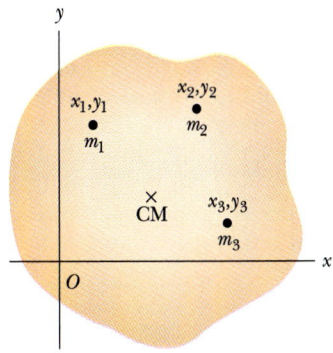

FIGURE 12.7 An object can be divided into many small particles each having a specific mass and coordinates. These particles can be used to locate the center of mass.

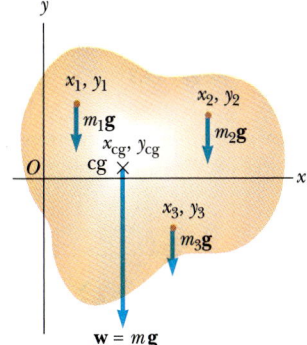

FIGURE 12.8 The center of gravity of the object is located at the center of mass if the value of **g** is constant over the object.

EXAMPLE 12.1   The Seesaw

A uniform board of weight 40.0 N supports two children weighing 500 N and 350 N, respectively, as shown in Figure 12.9. If the support (called the *fulcrum*) is under the center of gravity of the board and if the 500-N child is 1.50 m from the center, (a) determine the upward force n exerted on the board by the support.

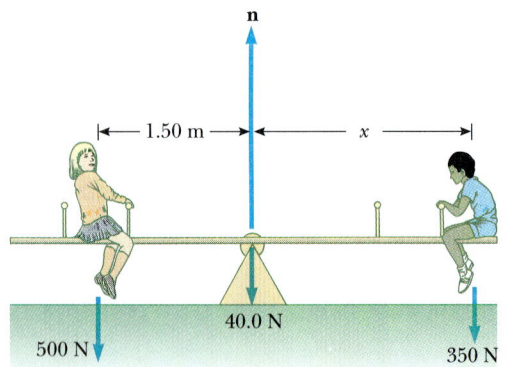

FIGURE 12.9   (Example 12.1) A balanced system.

Solution   First note that, in addition to n, the external forces acting on the board are the weights of the children and the weight of the board, all of which act downward. We can assume that the center of gravity of the board is at its geometric center because we were told that the board is uniform. Since the system is in equilibrium, the upward force n must balance all the downward forces. From $\Sigma F_y = 0$, we have

$$n - 500\ \text{N} - 350\ \text{N} - 40.0\ \text{N} = 0 \quad \text{or} \quad n = \boxed{890\ \text{N}}$$

(The equation $\Sigma F_x = 0$ also applies to this situation, but it is unnecessary to consider this equation because we have no forces acting horizontally on the board.)

(b) Determine where the 350-N child should sit to balance the system.

Solution   To find this position, we must invoke the second condition for equilibrium. Taking the center of gravity of the board as the axis for our torque equation, we see from $\Sigma \tau = 0$ that

$$(500\ \text{N})(1.50\ \text{m}) - (350\ \text{N})x = 0$$

$$x = \boxed{2.14\ \text{m}}$$

Exercise   If the fulcrum did not lie under the center of gravity of the board, what other information would you need to solve the problem?

EXAMPLE 12.2   A Weighted Hand

A 50.0-N weight is held in the hand with the forearm in the horizontal position, as in Figure 12.10a. The biceps muscle is attached 3.00 cm from the joint, and the weight is 35.0 cm from the joint. Find the upward force that the biceps exerts on the forearm and the downward force exerted by the upper arm on the forearm and acting at the joint. Neglect the weight of the forearm.

Solution   The forces acting on the forearm are equivalent to those acting on a bar as shown in Figure 12.10b, where F is the upward force exerted by the biceps and R is the downward force exerted by the upper arm at the joint. From the first condition for equilibrium, we have

$$(1) \quad \sum F_y = F - R - 50.0\ \text{N} = 0$$

From the second condition for equilibrium, we know that the sum of the torques about any point must be zero. With the joint O as the axis, we have

$$Fd - w\ell = 0$$

$$F(3.00\ \text{cm}) - (50.0\ \text{N})(35.0\ \text{cm}) = 0$$

$$F = \boxed{583\ \text{N}}$$

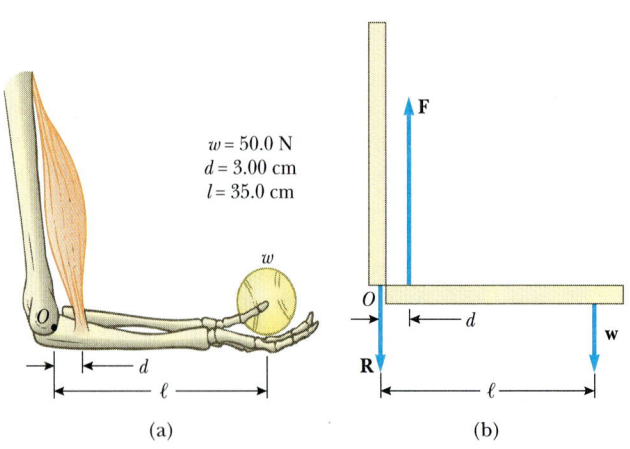

FIGURE 12.10   (Example 12.2) (a) In this skeletal view, the biceps muscle is pulling upward with a force F (essentially) at right angles to the forearm. (b) The mechanical model for the system described in (a).

This value for F can be substituted into (1) to give $R = 533$ N. Hence, the forces at joints and in muscles can be extremely large.

Exercise In reality, the biceps makes an angle of 15.0° with the vertical, so that **F** has both a vertical and a horizontal component. Find the value of **F** and the components of **R** including this fact in your analysis.

Answer $F = 604$ N, $R_x = 156$ N, $R_y = 533$ N.

EXAMPLE 12.3 Standing on a Horizontal Beam

A uniform horizontal beam of length 8.00 m and weight 200 N is attached to a wall by a pin connection. Its far end is supported by a cable that makes an angle of 53.0° with the horizontal (Fig. 12.11a). If a 600-N person stands 2.00 m from the wall, find the tension in the cable and the force exerted by the wall on the beam.

Solution First we must identify all the external forces acting on the beam. These are its weight, the force **T** exerted by the cable, the force **R** exerted by the wall at the pivot (the direction of this force is unknown), and the weight of the person on the beam. These are all indicated in the free-body diagram for the beam (Fig. 12.11b). If we resolve **T** and **R** into horizontal and vertical components and apply the first condition for equilibrium, we get

$$(1) \qquad \sum F_x = R \cos \theta - T \cos 53.0° = 0$$

$$(2) \qquad \sum F_y = R \sin \theta + T \sin 53.0° - 600 \text{ N} - 200 \text{ N} = 0$$

Because R, T, and θ are all unknown, we cannot obtain a solution from these expressions alone. (The number of simultaneous equations must equal the number of unknowns in order for us to be able to solve for the unknowns.)

Now let us invoke the condition for rotational equilibrium. A convenient axis to choose for our torque equation is the one that passes through the pivot at O. The feature that makes this point so convenient is that the force **R** and the horizontal component of **T** both have a lever arm of zero and, hence, zero torque about this pivot. Recalling our convention for the sign of the torque about an axis and noting that the lever arms of the 600-N, 200-N, and $T \sin 53°$ forces are 2.00 m, 4.00 m, and 8.00 m, respectively, we get

$$\sum \tau_O = (T \sin 53.0°)(8.00 \text{ m}) - (600 \text{ N})(2.00 \text{ m})$$
$$- (200 \text{ N})(4.00 \text{ m}) = 0$$

$$T = \boxed{313 \text{ N}}$$

Thus the torque equation with this axis gives us one of the unknowns directly! This value is substituted into (1) and (2) to give

$$R \cos \theta = 188 \text{ N}$$

$$R \sin \theta = 550 \text{ N}$$

We divide these two equations and recall the trigonometric

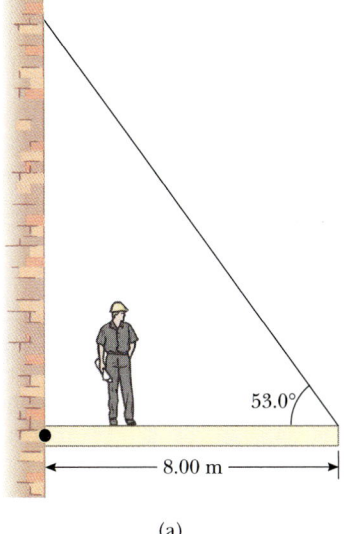

(a)

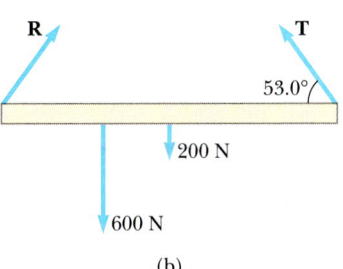

(b)

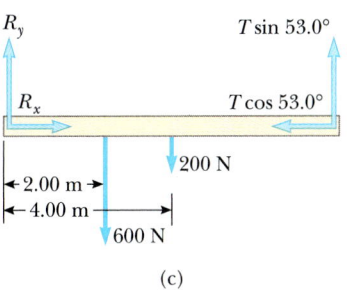

(c)

FIGURE 12.11 (Example 12.3) (a) A uniform beam supported by a cable. (b) The free-body diagram for the beam.

identity $\sin\theta/\cos\theta = \tan\theta$ to get

$$\tan\theta = \frac{550\ \text{N}}{188\ \text{N}} = 2.93$$

$$\theta = 71.1°$$

Finally,

$$R = \frac{188\ \text{N}}{\cos\theta} = \frac{188\ \text{N}}{\cos 71.1°} = \boxed{581\ \text{N}}$$

If we had selected some other axis for the torque equation, the solution would have been the same. For example, if we had chosen to have the axis pass through the center of gravity of the beam, the torque equation would involve both T and R. However, this equation, coupled with (1) and (2), could still be solved for the unknowns. Try it!

When many forces are involved in a problem of this na-

ture, it is convenient to set up a table of forces, lever arms, and torques. For instance, in the example just given, we would construct the following table. Setting the sum of the terms in the last column equal to zero represents the condition of rotational equilibrium.

Force Component	Lever Arm Relative to O (m)	Torque About O (N·m)
$T\sin 53.0°$	8.00	$(8.00)\,T\sin 53°$
$T\cos 53.0°$	0	0
200 N	4.00	$-(4.00)(200)$
600 N	2.00	$-(2.00)(600)$
$R\sin\theta$	0	0
$R\cos\theta$	0	0

EXAMPLE 12.4   The Leaning Ladder

A uniform ladder of length ℓ and weight $w = 50$ N rests against a smooth, vertical wall (Fig. 12.12a). If the coefficient of static friction between ladder and ground is $\mu_s = 0.40$, find the minimum angle $\theta_{\min}$ such that the ladder does not slip.

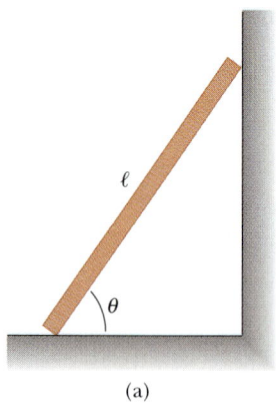

(a)

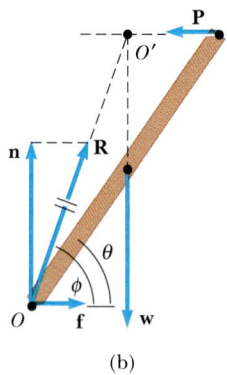

(b)

FIGURE 12.12   (Example 12.4) (a) A uniform ladder at rest, leaning against a smooth wall. The floor is rough. (b) The free-body diagram for the ladder. Note that the forces **R**, **w**, and **P** pass through a common point O'.

Solution   The free-body diagram showing all the external forces acting on the ladder is illustrated in Figure 12.12b. The reaction **R** exerted by the ground on the ladder is the vector sum of a normal force, **n**, and the force of friction, **f**. The reaction force **P** exerted by the wall on the ladder is horizontal, since the wall is frictionless. From the first condition of equilibrium applied to the ladder, we have

$$\sum F_x = f - P = 0$$

$$\sum F_y = n - w = 0$$

Since $w = 50$ N, we see from the second equation here that $n = w = 50$ N. Furthermore, *when the ladder is on the verge of slipping, the force of friction must be a maximum,* given by $f_{\max} = \mu_s n = 0.40(50\ \text{N}) = 20$ N. (Recall Eq. 5.9: $f_s \le \mu_s n$.) Thus, at this angle, $P = 20$ N.

To find θ, we must use the second condition of equilibrium. When the torques are taken about the origin O at the bottom of the ladder, we get

$$\sum \tau_O = P\ell \sin\theta - w\frac{\ell}{2}\cos\theta = 0$$

But $P = 20$ N when the ladder is about to slip and $w = 50$ N, so that this expression gives

$$\tan\theta_{\min} = \frac{w}{2P} = \frac{50\ \text{N}}{40\ \text{N}} = 1.25$$

$$\theta_{\min} = \boxed{51°}$$

It is interesting to note that the result does not depend on ℓ or w. The answer depends only on μ_s.

An alternative approach to analyzing this problem is to consider the intersection O' of the lines of action of forces **w** and **P**. Since the torque about any origin must be zero, the torque about O' must be zero. This requires that the line of action of **R** (the resultant of **n** and **f**) pass through O'. That is, since three forces act on this stationary object, the forces must be concurrent. With this condition, one could then obtain the angle ϕ that **R** makes with the horizontal (where ϕ is greater than θ), assuming the length of the ladder is known.

Exercise   For the angles labeled in Figure 12.12, show that $\tan\phi = 2\tan\theta$.

EXAMPLE 12.5 Raising a Cylinder

A cylinder of weight w and radius R is to be raised onto a step of height h as shown in Figure 12.13. A rope is wrapped around the cylinder and pulled horizontally. Assuming the cylinder doesn't slip on the step, find the minimum force **F** necessary to raise the cylinder and the reaction force at P exerted by the step on the cylinder.

Solution When the cylinder is just about to be raised, the reaction force at Q goes to zero. Hence, at this time there are only three forces on the cylinder, as shown in Figure 12.13b. From the dotted triangle in Figure 12.13a, we see that the moment arm d of the weight relative to the point P is

$$d = \sqrt{R^2 - (R - h)^2} = \sqrt{2Rh - h^2}$$

The moment arm of **F** relative to P is $2R - h$. Therefore, the net torque acting on the cylinder about P is

$$wd - F(2R - h) = 0$$

$$w\sqrt{2Rh - h^2} - F(2R - h) = 0$$

$$F = \frac{w\sqrt{2Rh - h^2}}{2R - h}$$

We can determine the components of **n** by using the first condition of equilibrium:

$$\sum F_x = F - n\cos\theta = 0$$

$$\sum F_y = n\sin\theta - w = 0$$

Dividing gives

$$(1) \qquad \tan\theta = \frac{w}{F}$$

and solving for n gives

$$(2) \qquad n = \sqrt{w^2 + F^2}$$

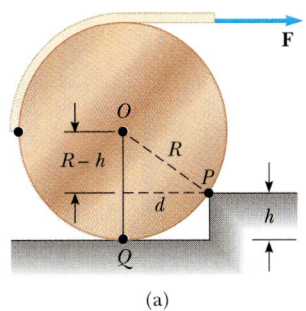

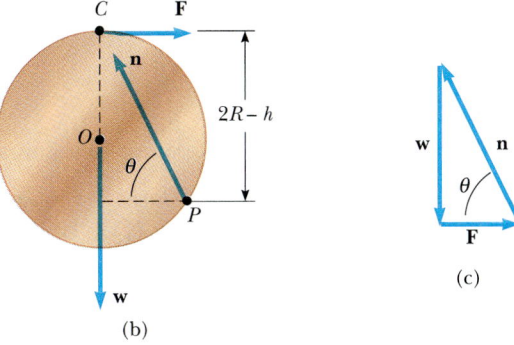

FIGURE 12.13 (Example 12.5) (a) A cylinder of weight w being pulled by a force **F** over a step. (b) The free-body diagram for the cylinder when it is just about to be raised. (c) The vector sum of the three external forces is zero.

Exercise Solve this problem by noting that the three forces acting on the cylinder are concurrent and therefore all pass through the point C. The three forces form the sides of the triangle shown in Figure 12.13c.

12.4 ELASTIC PROPERTIES OF SOLIDS

In our study of mechanics thus far, we have assumed that objects remain undeformed when external forces act on them. In reality, all objects are deformable. That is, it is possible to change the shape or size of an object (or both) by applying external forces. Although these changes are observed as large-scale deformations, the internal forces that resist the deformation are due to short-range forces between atoms.

We shall discuss the deformation of solids in terms of the concepts of stress and strain. **Stress** is a quantity that is proportional to the force causing a deformation; more specifically, stress is the external force per unit cross-sectional area acting on an object. **Strain** is a measure of the degree of deformation. It is found that, for sufficiently small stresses, the stress is proportional to the strain; the constant of proportionality depends on the material being deformed and on the nature of the deformation. We call this proportionality constant the **elastic modulus.** The elastic modulus is therefore the ratio of stress to strain:

Elastic modulus

$$\text{Elastic modulus} \equiv \frac{\text{stress}}{\text{strain}} \tag{12.5}$$

We consider three types of deformation and define an elastic modulus for each:

- **Young's modulus,** which measures the resistance of a solid to a change in its length.
- **Shear modulus,** which measures the resistance to motion of the planes of a solid sliding past each other
- **Bulk modulus,** which measures the resistance that solids or liquids offer to changes in their volume

A plastic model of an arch structure under load conditions observed between two crossed polarizers. The stress pattern is produced in those regions where the stresses are greatest. Such patterns and models are useful in finding the optimum design for architectural components. *(Peter Aprahamian/Science Photo Library)*

Young's Modulus: Elasticity in Length

Consider a long bar of cross-sectional area A and length L_0 that is clamped at one end (Fig. 12.14). When an external force $\mathbf{F}$ is applied along the bar and perpendicular to the cross-section, internal forces in the bar resist distortion ("stretching"), but the bar attains an equilibrium in which its length is greater than L_0 and in which the external force is exactly balanced by internal forces. In such a situation, the bar is said to be stressed. We define the **tensile stress** as the ratio of the magnitude of the external force F to the cross-sectional area A. The **tensile strain** in this case is defined as the ratio of the change in length, ΔL, to the original length, L_0, and is therefore a dimensionless quantity. Thus, we can use Equation 12.6 to define **Young's modulus:**

Young's modulus

$$Y \equiv \frac{\text{tensile stress}}{\text{tensile strain}} = \frac{F/A}{\Delta L / L_0} \tag{12.6}$$

This quantity is typically used to characterize a rod or wire stressed under either tension or compression. Note that because strain is a dimensionless quantity, Y has units of force per unit area. Typical values are given in Table 12.1. Experiments

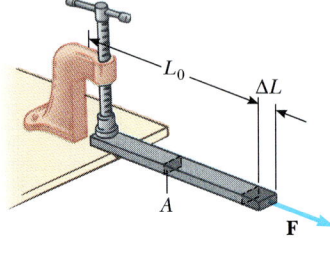

FIGURE 12.14 A long bar clamped at one end is stretched by an amount ΔL under the action of a force **F**.

TABLE 12.1   Typical Values for Elastic Modulus

Substance	Young's Modulus (N/m²)	Shear Modulus (N/m²)	Bulk Modulus (N/m²)
Aluminum	7.0×10^{10}	2.5×10^{10}	7.0×10^{10}
Brass	9.1×10^{10}	3.5×10^{10}	6.1×10^{10}
Copper	11×10^{10}	4.2×10^{10}	14×10^{10}
Steel	20×10^{10}	8.4×10^{10}	16×10^{10}
Tungsten	35×10^{10}	14×10^{10}	20×10^{10}
Glass	$6.5–7.8 \times 10^{10}$	$2.6–3.2 \times 10^{10}$	$5.0–5.5 \times 10^{10}$
Quartz	5.6×10^{10}	2.6×10^{10}	2.7×10^{10}
Water	—	—	0.21×10^{10}
Mercury	—	—	2.8×10^{10}

show that (a) for a fixed applied force, the change in length is proportional to the original length and (b) the force necessary to produce a given strain is proportional to the cross-sectional area. Both of these observations are in accord with Equation 12.6.

The **elastic limit** of a substance is defined as the maximum stress that can be applied to the substance before it becomes permanently deformed. It is possible to exceed the elastic limit of a substance by applying a sufficiently large stress (Fig. 12.15). When the stress exceeds the elastic limit, the object is permanently distorted and does not return to its original shape after the stress is removed. Hence, the shape of the object is permanently changed. As the stress is increased even further, the material will ultimately break.

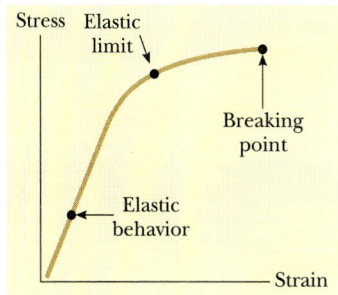

FIGURE 12.15 Stress-versus-strain curve for an elastic solid.

Shear Modulus: Elasticity of Shape

Another type of deformation occurs when an object is subjected to a force **F** tangential to one of its faces while the opposite face is held fixed by a force such as the force of friction, f_s (Fig. 12.16a). The stress in this case is called a shear stress. If the object is originally a rectangular block, a shear stress results in a shape whose cross-section is a parallelogram. As the stress increases, the stress-strain curve is no longer a straight line. A book pushed sideways as in Figure 12.16b is an example of an object under a shear stress. There is no change in volume under this deformation to a first approximation (for small distortions).

We define the **shear stress** as F/A, the ratio of the tangential force to the area, A, of the face being sheared. The **shear strain** is defined as the ratio $\Delta x/h$, where Δx is the horizontal distance the sheared face moves and h is the height of the object. In terms of these quantities, the **shear modulus** is

$$S \equiv \frac{\text{shear stress}}{\text{shear strain}} = \frac{F/A}{\Delta x/h} \qquad (12.7)$$

Values of the shear modulus for some representative materials are given in Table 12.1. The units of shear modulus are force per unit area.

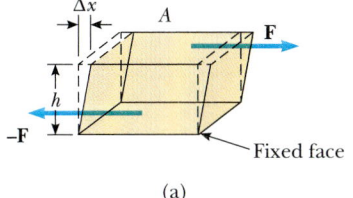

(a)

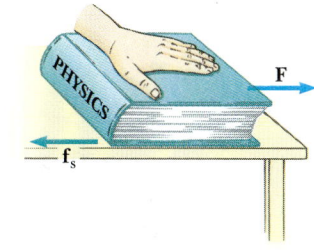

(b)

FIGURE 12.16 (a) A shear deformation in which a rectangular block is distorted by two forces of equal magnitude but opposite directions applied to two parallel faces. (b) A book under shear stress.

Bulk Modulus: Volume Elasticity

Bulk modulus characterizes the response of a substance to uniform squeezing. Suppose that the external forces acting on an object are at right angles to all its faces (Fig. 12.17) and distributed uniformly over all the faces. As we shall see in Chapter 15, such uniformly distributed forces occur when an object is immersed in a fluid. An object subject to this type of deformation undergoes a change in volume but no change in shape. The **volume stress**, ΔP, is defined as the ratio of the magnitude of the normal force, F, to the area, A. The quantity $\Delta P = F/A$ is called the **pressure**. The volume strain is equal to the change in volume, ΔV, divided by the original volume, V. Thus, from Equation 12.6 we can characterize a volume compression in terms of the **bulk modulus**, defined as

$$B \equiv \frac{\text{volume stress}}{\text{volume strain}} = -\frac{F/A}{\Delta V/V} = -\frac{\Delta P}{\Delta V/V} \qquad (12.8)$$

Bulk modulus

A negative sign is inserted in this defining equation so that B is a positive number. This maneuver is necessary because an increase in pressure (positive ΔP) causes a decrease in volume (negative ΔV) and vice versa.

Table 12.1 lists bulk moduli for some materials. If you look up such values in a different source, you will often find that the reciprocal of the bulk modulus is

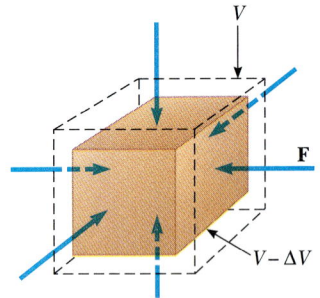

FIGURE 12.17 When a solid is under uniform pressure, it undergoes a change in volume but no change in shape. This cube is compressed on all sides by forces normal to its six faces.

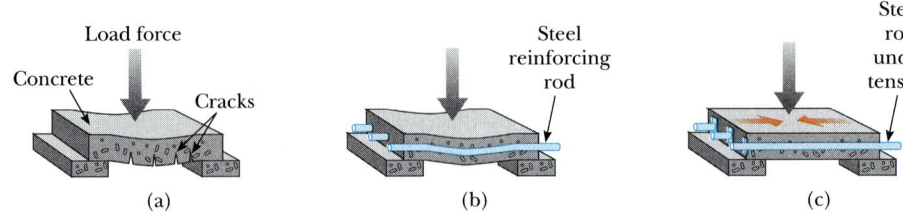

FIGURE 12.18   (a) A concrete slab with no reinforcement tends to crack under a heavy load. (b) The strength of the concrete is increased by using steel tensile reinforcement rods. (c) The concrete is further strengthened by prestressing it with steel rods under tension.

listed. The reciprocal of the bulk modulus is called the **compressibility** of the material.

Note from Table 12.1 that both solids and liquids have a bulk modulus. However, there is no shear modulus and no Young's modulus for liquids because a liquid does not sustain a shearing stress or a tensile stress (it flows instead).

Prestressed Concrete

If the stress on a solid object exceeds a certain value, the object will fracture. The maximum stress that can be applied before fracture occurs depends on the nature of the material and the type of applied stress. For example, concrete has a tensile strength of about $2 \times 10^6 \ N/m^2$, a compressive strength of $20 \times 10^6 \ N/m^2$, and a shear strength of $2 \times 10^6 \ N/m^2$. If the applied stress exceeds these values, the concrete fractures. It is common practice to use large safety factors to prevent failure in concrete structures.

Concrete is normally very brittle when cast in thin sections. Thus, concrete slabs tend to sag and crack at unsupported areas, as in Figure 12.18a. The slab can be strengthened by using steel rods to reinforce the concrete, as in Figure 12.18b. Because concrete is much stronger under compression than under tension, vertical columns of concrete that are under compression can support very heavy loads, whereas horizontal beams of concrete tend to sag and crack because of their smaller shear strength. A significant increase in shear strength is achieved, however, by prestressing the reinforced concrete, as in Figure 12.18c. As the concrete is being poured, the steel rods are held under tension by external forces. The external forces are released after the concrete cures, which results in a permanent tension in the steel and hence a compressive stress on the concrete. This enables the concrete slab to support a much heavier load.

EXAMPLE 12.6   Measuring Young's Modulus

A load of 102 kg is supported by a wire of length 2.0 m and cross-sectional area 0.10 cm². The wire is stretched by 0.22 cm. Find the tensile stress, tensile strain, and Young's modulus for the wire.

Solution

$$\text{Tensile stress} = \frac{F}{A} = \frac{Mg}{A} = \frac{(102 \ \text{kg})(9.80 \ \text{m/s}^2)}{0.10 \times 10^{-4} \ \text{m}^2}$$

$$= \boxed{1.0 \times 10^8 \ N/m^2}$$

$$\text{Tensile strain} = \frac{\Delta L}{L_0} = \frac{0.22 \times 10^{-2} \ \text{m}}{2.0 \ \text{m}}$$

$$= \boxed{0.11 \times 10^{-2}}$$

$$Y = \frac{\text{tensile stress}}{\text{tensile strain}} = \frac{1.0 \times 10^8 \ N/m^2}{0.11 \times 10^{-2}}$$

$$= \boxed{9.1 \times 10^{10} \ N/m^2}$$

Comparing this value for Y with the values in Table 12.1, we conclude that the wire is probably made of brass.

EXAMPLE 12.7 Squeezing a Lead Sphere

A solid lead sphere of volume 0.50 m³ is lowered to a depth in the ocean where the water pressure is equal to 2.0×10^7 N/m². The bulk modulus of lead is equal to 7.7×10^9 N/m². What is the change in volume of the sphere?

Solution From the definition of bulk modulus, we have

$$B = -\frac{\Delta P}{\Delta V / V}$$

$$\Delta V = -\frac{V \Delta P}{B}$$

In this case, the change in pressure, ΔP, is 2.0×10^7 N/m². (This is large relative to atmospheric pressure, 1.01×10^5 N/m².) Therefore,

$$\Delta V = -\frac{(0.50 \text{ m}^3)(2.0 \times 10^7 \text{ N/m}^2)}{7.7 \times 10^9 \text{ N/m}^2} = -1.3 \times 10^{-3} \text{ m}^3$$

The negative sign indicates a decrease in volume.

SUMMARY

A rigid object is in **equilibrium** if and only if *the resultant external force on it is zero* and *the resultant external torque on it is zero about any origin:*

$$\sum \mathbf{F} = 0 \tag{12.1}$$

$$\sum \boldsymbol{\tau} = 0 \tag{12.2}$$

The first condition is the *condition of translational equilibrium,* and the second is the *condition of rotational equilibrium.*

If two forces act on a rigid object, the object is in equilibrium if and only if the forces are equal in magnitude and opposite in direction and have the same line of action.

When three forces act on a rigid object that is in equilibrium, the three forces must be concurrent, that is, their lines of action must intersect at a common point.

The force of gravity exerted on an object can be considered to act at a single point called the **center of gravity.** The center of gravity of an object coincides with the center of mass if the object is in a uniform gravitational field.

The elastic properties of a substance can be described using the concepts of stress and strain. **Stress** is a quantity proportional to the force producing a deformation; **strain** is a measure of the degree of deformation. Stress is proportional to strain, and the constant of proportionality is the **elastic modulus:**

$$\text{Elastic modulus} \equiv \frac{\text{stress}}{\text{strain}} \tag{12.5}$$

Three common types of deformation are (1) the resistance of a solid to elongation under a load, characterized by **Young's modulus,** Y; (2) the resistance of a solid to the motion of internal planes sliding past each other, characterized by the **shear modulus,** S; (3) the resistance of a solid (or a liquid) to a volume change, characterized by the **bulk modulus,** B.

A large balanced rock at the Garden of the Gods in Colorado Springs, Colorado, an example of stable equilibrium. *(Photo by David Serway)*

QUESTIONS

1. Can a body be in equilibrium if only one external force acts on it? Explain.
2. Can a body be in equilibrium if it is in motion? Explain.
3. Locate the center of gravity for the following uniform objects: (a) sphere, (b) cube, (c) right circular cylinder.
4. The center of gravity of an object may be located outside the object. Give a few examples for which this is the case.
5. You are given an arbitrarily shaped piece of plywood, together with a hammer, nail, and plumb bob. How could

you use these items to locate the center of gravity of the plywood? (*Hint:* Use the nail to suspend the plywood.)

6. In order for a chair to be balanced on one leg, where must the center of gravity of the chair be located?

7. Give an example in which the net torque acting on an object is zero and yet the net force is nonzero.

8. Give an example in which the net force acting on an object is zero and yet the net torque is nonzero.

9. Can an object be in equilibrium if the only torques acting on it produce clockwise rotation?

10. A tall crate and a short crate of equal mass are placed side by side on an incline (without touching each other). As the incline angle is increased, which crate will topple first? Explain.

11. When lifting a heavy object, why is it recommended to keep the back as vertical as possible, lifting from the knees, rather than bending over and lifting from the waist?

12. Would you expect the center of gravity and the center of mass of the Empire State Building to coincide precisely? Explain.

13. Give several examples where several forces are acting on a system in such a way that their sum is zero but the system is not in equilibrium.

14. If you measure the net torque and the net force on a system to be zero, (a) could the system still be rotating with respect to you? (b) Could it be translating with respect to you?

15. A ladder is resting inclined against a wall. Would you feel safer climbing up the ladder if you were told that the ground is frictionless but the wall is rough or that the wall is frictionless but the ground is rough? Justify your answer.

16. What kind of deformation does a cube of Jello exhibit when it "jiggles"?

PROBLEMS

Section 12.1   The Conditions of Equilibrium of a Rigid Object

1. A baseball player holds a 36-oz bat (weight = 10.0 N) with one hand at the point O (Fig. P12.1). The bat is in equilibrium. The force of gravity acts along a line 60 cm to the right of O. Determine the force exerted on the bat by the player and the torque exerted on the bat by the force of gravity.

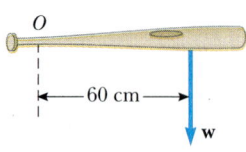

FIGURE P12.1

2. Write the necessary conditions of equilibrium for the body shown in Figure P12.2. Take the origin of the torque equation at the point O.

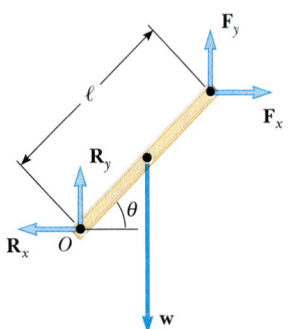

FIGURE P12.2

3. A uniform beam of weight w and length ℓ has weights w_1 and w_2 at two positions, as in Figure P12.3. The beam is resting at two points. For what value of x will the beam be balanced at P such that the normal force at O is zero?

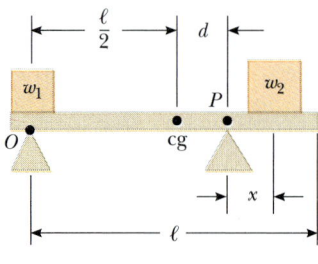

FIGURE P12.3

4. A letter "A" is formed from two uniform pieces of metal each of weight 26.0 N and length 1.00 m, hinged at the top and held together by a horizontal wire of length 1.20 m (Fig. P12.4). The structure rests on a frictionless surface. If the wire is connected at points a distance of 0.65 m from the top of the letter, determine the tension in the wire.

5. A ladder of weight 400 N and length 10.0 m is placed against a frictionless vertical wall. A person weighing 800 N stands on the ladder 2.00 m from the bottom as measured along the ladder. The foot of the ladder is 8.00 m from the bottom of the wall. Calculate the

□ indicates problems that have full solutions available in the Student Solutions Manual and Study Guide.

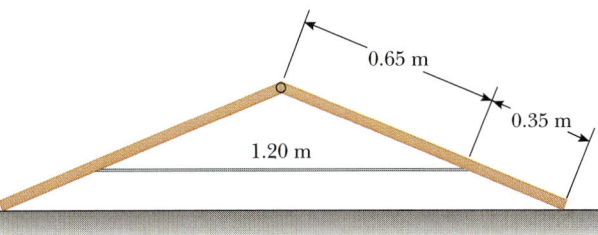

FIGURE P12.4

force exerted by the wall, and the normal force exerted by the floor on the ladder.

5A. A ladder of weight w_1 and length L is placed against a frictionless vertical wall. A person weighing w_2 stands on the ladder a distance x from the bottom as measured along the ladder. The foot of the ladder is a distance d from the bottom of the wall. Find expressions for the force exerted by the wall, and the normal force exerted by the floor on the ladder.

6. A student gets his car stuck in a snow drift. Not at a loss, having studied physics, he attaches one end of a stout rope to the vehicle and the other end to the trunk of a nearby tree, allowing for a small amount of slack. The student then exerts a force F on the center of the rope in the direction perpendicular to the car-tree line, as shown in Figure P12.6. If the rope is inextensible and if the magnitude of the applied force is 500 N, what is the force on the car? (Assume equilibrium conditions.)

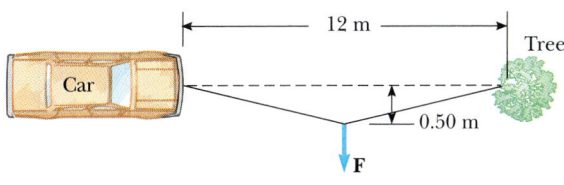

FIGURE P12.6

Section 12.2   More on the Center of Gravity

7. A circular pizza of radius R has a circular piece of radius $R/2$ removed from one side as shown in Figure P12.7. Clearly the center of gravity has moved from C to C' along the x axis. Show that the distance

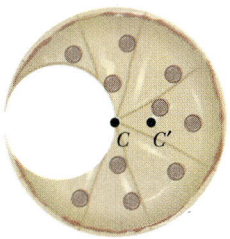

FIGURE P12.7

from C to C' is $R/6$. (Assume the thickness and density of the pizza are uniform throughout.)

8. A carpenter's square has the shape of an "L," as in Figure P12.8. Locate its center of gravity.

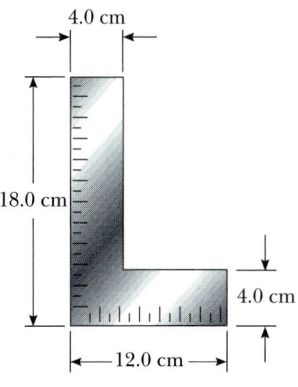

FIGURE P12.8

9. Consider the following mass distribution: 5.0 kg at $(0, 0)$ m, 3.0 kg at $(0, 4.0)$ m, and 4.0 kg at $(3.0, 0)$ m. Where should a fourth mass of 8.0 kg be placed so that the center of gravity of the four-mass arrangement will be at $(0, 0)$?

10. Pat builds a track for his model car out of wood (Fig. P12.10). The track is 5.0 cm wide, 1.0 m high, and 3.0 m long. The runway is cut such that it forms a parabola, $y = (x - 3)^2/9$. Locate the horizontal position of the center of gravity of this track.

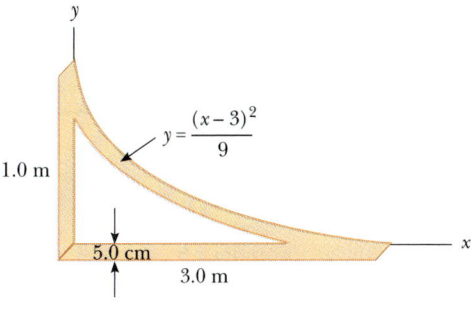

FIGURE P12.10

Section 12.3   Examples of Rigid Objects in Static Equilibrium

11. Chris is pushing his sister Nicole in a wheelbarrow when it is stopped by a brick 8.0 cm high (Fig. P12.11). The handles make an angle of 15.0° with the horizontal. A downward force of 400 N is exerted on the wheel, which has a radius of 20.0 cm. (a) What force must Chris apply along the handles in order to just start the wheel over the brick? (b) What is the force (magnitude and direction) that the brick exerts on the wheel just as the wheel begins to lift

FIGURE P12.11

over the brick? Assume in both parts that the brick remains fixed and does not slide along the ground.

12. Two pans of a balance are 50.0 cm apart. The fulcrum of the balance has been shifted 1.0 cm away from the center by a dishonest shopkeeper. By what percentage is the true weight of the goods being marked up by the shopkeeper? (Assume the balance has negligible mass.)

13. A ladder having a uniform density and a mass m rests against a frictionless vertical wall at an angle of 60°. The lower end rests on a flat surface where the coefficient of static friction is $\mu_s = 0.40$. A student with a mass $M = 2m$ attempts to climb the ladder. What fraction of the length L of the ladder will the student have reached when the ladder begins to slip?

14. A uniform plank of length 6.0 m and mass 30 kg rests horizontally on a scaffold, with 1.5 m of the plank hanging over one end of the scaffold. How far can a painter of mass 70 kg walk on the overhanging part of the plank before it tips?

14A. A uniform plank of length L and mass m_1 rests horizontally on a scaffold, with a length d of the plank hanging over one end of the scaffold. How far can a painter of mass m_2 walk on the overhanging part of the plank before it tips?

15. A 1500-kg automobile has a wheel base (the distance between the axles) of 3.0 m. The center of mass of the automobile is on the center line at a point 1.2 m behind the front axle. Find the force exerted by the ground on each wheel.

16. A uniform rod of weight w and length L is supported at its ends by a frictionless trough as shown in Figure P12.16. (a) Show that the center of gravity of the rod is directly over point O when the rod is in equilibrium. (b) Determine the equilibrium value of the angle θ.

17. A cue stick strikes a cue ball and delivers a horizontal impulse in such a way that the ball rolls without slipping as it starts to move. At what height above the ball's center (in terms of the radius of the ball) was the blow struck?

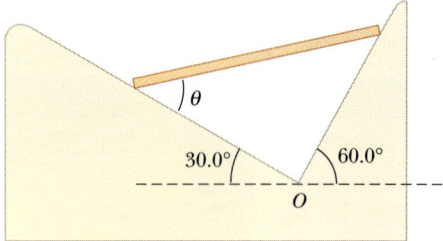

FIGURE P12.16

18. A flexible chain weighing 40 N hangs between two hooks located at the same height (Fig. P12.18). At each hook, the tangent to the chain makes an angle $\theta = 42°$ with the horizontal. Find (a) the magnitude of the force each hook exerts on the chain and (b) the tension in the chain at its midpoint. (*Hint:* For part (b), make a free-body diagram for half the chain.)

FIGURE P12.18

19. A hemispherical sign 1.0 m in diameter and of uniform mass density is supported by two strings as shown in Figure P12.19. What fraction of the sign's weight is supported by each string?

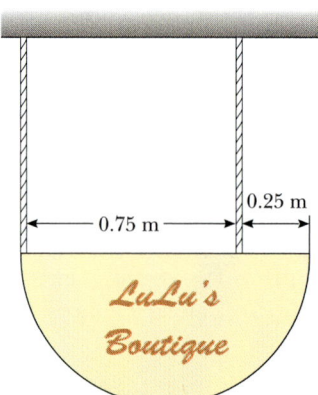

FIGURE P12.19

20. Sir Lost dons his armor and sets out from the castle on his trusty steed in his quest to rescue fair damsels from dragons (Fig. P12.20). Unfortunately his aide lowered the drawbridge too far and finally stopped it 20.0° below the horizontal. Sir Lost and his steed

FIGURE P12.20

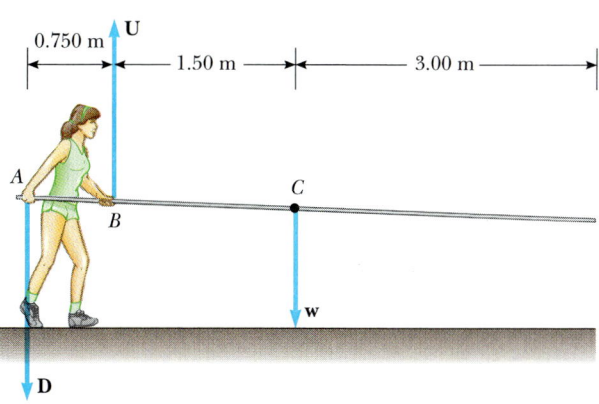

FIGURE P12.22

stop when their combined center of mass is 1.0 m from the end of the bridge. The bridge is 8.0 m long and has a mass of 2000 kg; the lift cable is attached to the bridge 5.0 m from the castle end and to a point 12.0 m above the bridge. Sir Lost's mass combined with his armor and steed is 1000 kg. (a) Determine the tension in the cable and (b) the horizontal and vertical force components acting on the bridge at the castle end.

21. Two identical uniform bricks of length L are placed in a stack over the edge of a horizontal surface with the maximum overhang possible without falling, as in Figure P12.21. Find the distance x.

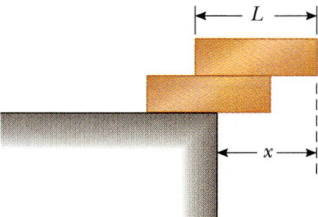

FIGURE P12.21

22. A vaulter holds a 29.4-N pole in equilibrium by exerting an upward force, **U**, with her leading hand, and a downward force, **D**, with her trailing hand, as shown in Figure P12.22. If we assume that the weight of the pole acts at its midpoint, what are the magnitudes of **U** and **D**?

Section 12.4   Elastic Properties of Solids

23. A 200-kg load is hung on a wire of length 4.0 m, cross-sectional area 0.20×10^{-4} m², and Young's modulus 8.0×10^{10} N/m². What is its increase in length?

24. A steel piano wire 1.12 m long has a cross-sectional area of 6.0×10^{-3} cm². When under a tension of 115 N, how much does it stretch?

25. Assume that Young's modulus for bone is 1.5×10^{10} N/m² and that the bone will fracture if a shear stress of more than 1.5×10^{8} N/m² is exerted. (a) What is the maximum force that can be exerted on the femur bone in the leg if it has a minimum effective diameter of 2.5 cm? (b) If this much force is applied compressively, by how much does the 25.0-cm-long bone shorten?

26. If the elastic limit of copper is 1.5×10^{8} N/m², determine the minimum diameter a copper wire can have under a load of 10 kg if its elastic limit is not to be exceeded.

27. A 2.0-m-long cylindrical steel wire with a cross-sectional diameter of 4.0 mm is placed over a frictionless pulley, with one end of the wire connected to a 5.00-kg mass and the other end connected to a 3.00-kg mass. By how much does the wire stretch while the masses are in motion?

27A. A cylindrical steel wire of length L with a cross-sectional diameter d is placed over a frictionless pulley, with one end of the wire connected to a mass m_1 and the other end connected to a mass m_2. By how much does the wire stretch while the masses are in motion?

28. Calculate the density of sea water at a depth of 1000 m where the hydraulic pressure is approximately 1.000×10^{7} N/m². (The density of sea water at the surface is 1.030×10^{3} kg/m³.)

29. If the shear stress in steel exceeds about 4.0×10^{8} N/m², the steel ruptures. Determine the shearing force necessary to (a) shear a steel bolt 1.0 cm in diameter and (b) punch a 1.0-cm-diameter hole in a 0.50-cm-thick steel plate.

30. (a) Find the minimum diameter of a steel wire 18 m long that will elongate no more than 9.0 mm when a load of 380 kg is hung on the lower end. (b) If the elastic limit for this steel is 3.0×10^{8} N/m², will permanent deformation occur with this load?

31. When water freezes, it expands about 9%. What

would be the pressure increase inside your automobile engine block if the water in it froze? (The bulk modulus of ice is 2.0×10^9 N/m².)

32. For safety in climbing, a mountaineer uses a 50-m nylon rope that is 10 mm in diameter. When supporting the 90-kg climber on one end, the rope elongates 1.6 m. Find Young's modulus for the rope material.

ADDITIONAL PROBLEMS

33. A bridge of length 50 m and mass 8.0×10^4 kg is supported at each end as in Figure P12.33. A truck of mass 3.0×10^4 kg is located 15 m from one end. What are the forces on the bridge at the points of support?

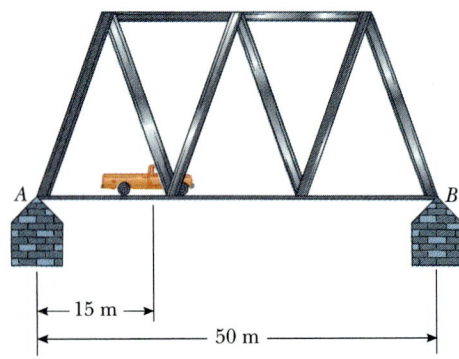

FIGURE P12.33

34. A solid sphere of radius R and mass M is placed in a wedge as shown in Figure P12.34. The inner surfaces of the wedge are frictionless. Determine the forces exerted by the wedge on the sphere at the two contact points.

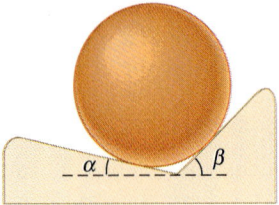

FIGURE P12.34

35. A 10-kg monkey climbs up a 120-N uniform ladder of length L, as in Figure P12.35. The upper and lower ends of the ladder rest on frictionless surfaces. The lower end is fastened to the wall by a horizontal rope that can support a maximum tension of 110 N. (a) Draw a free-body diagram for the ladder. (b) Find the tension in the rope when the monkey is one third the way up the ladder. (c) Find the maximum distance d the monkey can walk up the ladder

before the rope breaks, expressing your answer as a fraction of L.

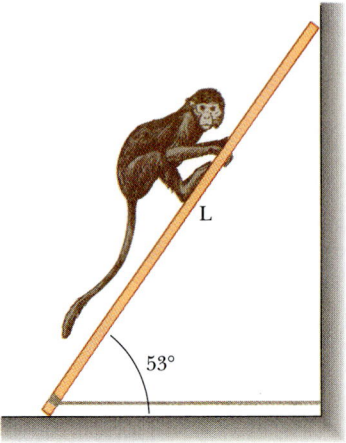

FIGURE P12.35

36. A hungry bear weighing 700 N walks out on a beam in an attempt to retrieve a basket of food hanging at the end of the beam (Fig. P12.36). The beam is uniform, weighs 200 N, and is 6.00 m long; the basket weighs 80.0 N. (a) Draw a free-body diagram for the beam. (b) When the bear is at $x = 1.00$ m, find the tension in the wire and the components of the force exerted by the wall on the left end of the beam. (c) If the wire can withstand a maximum tension of 900 N, what is the maximum distance the bear can walk before the wire breaks?

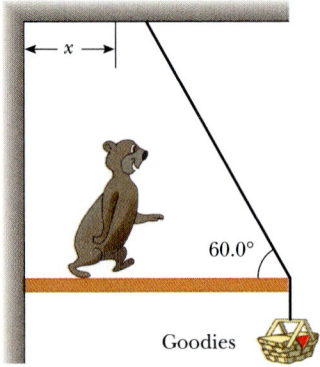

FIGURE P12.36

37. Old MacDonald had a farm, and on that farm he had a gate (Fig. P12.37). The gate is 3.0 m long and 1.8 m tall with hinges attached to the top and bottom. The guy wire makes an angle of 30.0° with the top of the gate and is tightened by turnbuckle to a tension of 200 N. The mass of the gate is 40.0 kg. (a) Determine the horizontal force exerted on the gate by the bot-

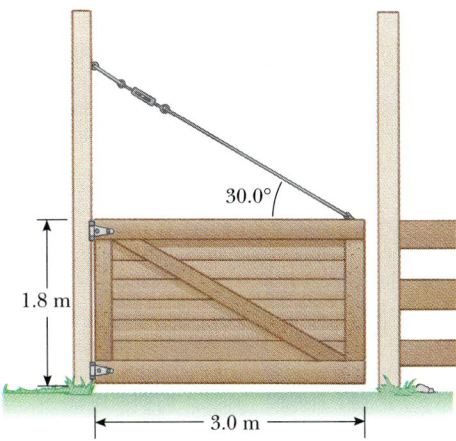

FIGURE P12.37

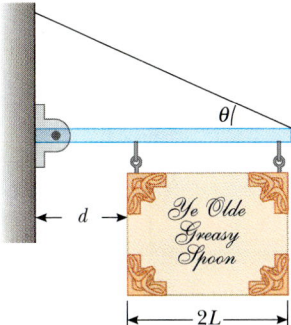

FIGURE P12.39

tom hinge. (b) Find the horizontal force exerted by the upper hinge. (c) Determine the combined vertical force exerted by both hinges. (d) What must be the tension in the guy wire so that the horizontal force exerted by the upper hinge is zero?

38. A 1200-N uniform boom is supported by a cable as in Figure P12.38. The boom is pivoted at the bottom, and a 2000-N object hangs from its top. Find the tension in the cable and the components of the reaction force on the boom by the floor.

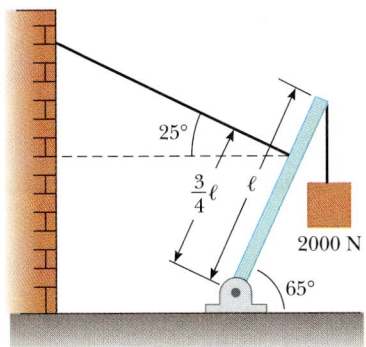

FIGURE P12.38

39. A uniform sign of weight w and width $2L$ hangs from a light, horizontal beam, hinged at the wall and supported by a cable (Fig. P12.39). Determine (a) the tension in the cable and (b) the components of the reaction force exerted by the wall on the beam in terms of w, d, L, and θ.

 40. A crane of mass 3000 kg supports a load of 10 000 kg as in Figure P12.40. The crane is pivoted with a smooth pin at A and rests against a smooth support at B. Find the reaction forces at A and B.

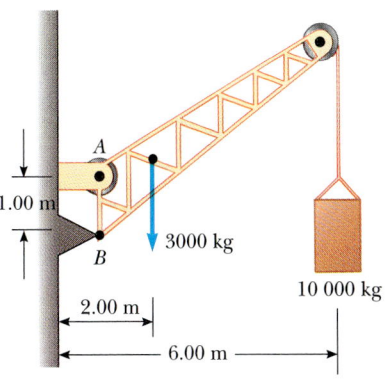

FIGURE P12.40

41. A 15-m uniform ladder weighing 500 N rests against a frictionless wall. The ladder makes a 60.0° angle with the horizontal. (a) Find the horizontal and vertical forces the ground exerts on the base of the ladder when an 800-N firefighter is 4.00 m from the bottom. (b) If the ladder is just on the verge of slipping when the firefighter is 9.00 m up, what is the coefficient of static friction between ladder and ground?

41A. A uniform ladder of length L and mass m_1 rests against a frictionless wall. The ladder makes an angle θ with the horizontal. (a) Find the horizontal and vertical forces the ground exerts on the base of the ladder when a firefighter of mass m_2 is a distance x from the bottom. (b) If the ladder is just on the verge of slipping when the firefighter is a distance d from the bottom, what is the coefficient of static friction between ladder and ground?

42. A uniform ladder weighing 200 N is leaning against a wall (Fig. 12.12). The ladder slips when θ is 60°. Assuming the coefficients of static friction at the wall and the ground are the same, obtain a value for μ_s.

43. A 10 000-N shark is supported by a cable attached to a 4.00-m rod that can pivot at the base. Calculate the cable tension needed to hold the system in the posi-

tion shown in Figure P12.43. Find the horizontal and vertical forces exerted on the base of the rod. (Neglect the weight of the rod.)

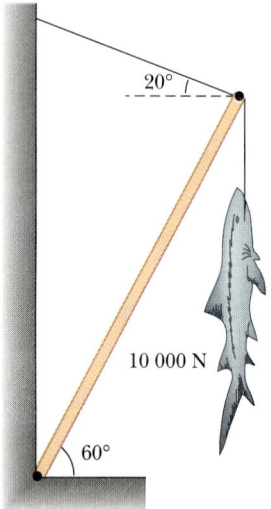

FIGURE P12.43

44. When a person stands on tiptoe (a strenuous position), the position of the foot is as shown in Figure P12.44a. The total weight of the body w is supported by the force **n** exerted by the floor on the toe. A mechanical model for the situation is shown in Figure P12.44b, where **T** is the force exerted by the Achilles tendon on the foot and **R** is the force exerted by the tibia on the foot. Find the values of T, R, and θ when $w = 700$ N.

45. A person bends over and lifts a 200-N object as in Figure P12.45a, with the back in the horizontal position (a terrible way to lift an object). The back muscle attached at a point two thirds up the spine maintains the position of the back, where the angle between the spine and this muscle is 12.0°. Using the mechanical model shown in Figure P12.45b and taking the weight of the upper body to be 350 N, find the tension in the back muscle and the compressional force in the spine.

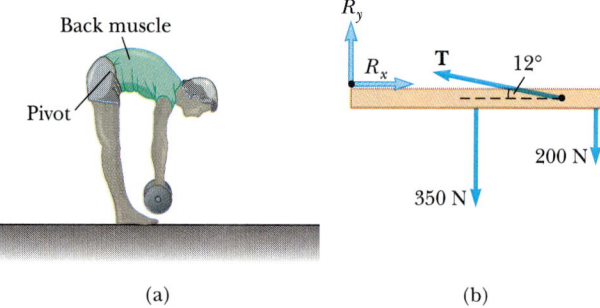

(a)   (b)

FIGURE P12.45

46. Two 200-N traffic lights are suspended from a single cable as shown in Figure P12.46. Neglect the cable weight and (a) prove that if $\theta_1 = \theta_2$, then $T_1 = T_2$. (b) Determine the three tensions if $\theta_1 = \theta_2 = 8.0°$.

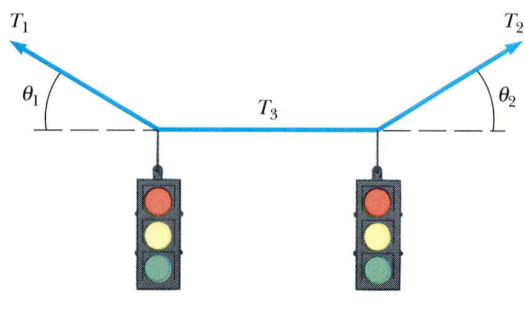

FIGURE P12.46

47. A force acts on a rectangular block weighing 400 N as in Figure P12.47. (a) If the block slides with constant speed when $F = 200$ N and $h = 0.400$ m, find the coefficient of sliding friction and the position of the resultant normal force. (b) If $F = 300$ N, find the value of h for which the block just begins to tip.

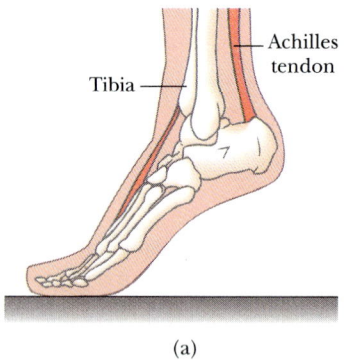

(a)

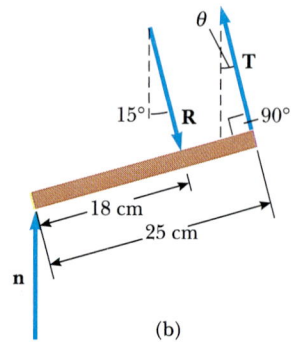

(b)

FIGURE P12.44

47A. A force *F* acts on a rectangular block of mass *m* as in Figure P12.47. (a) If the block slides with constant speed, find the coefficient of sliding friction and the position of the resultant normal force. (b) Find the value of *h* for which the block just begins to tip.

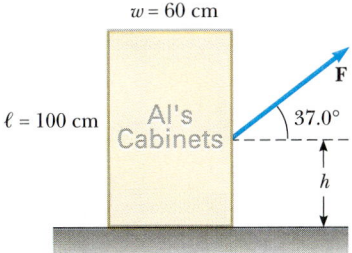

FIGURE P12.47

48. Consider the rectangular block of Problem 47. A force **F** is applied horizontally at the upper edge. (a) What is the minimum force required to start to tip the block? (b) What is the minimum coefficient of static friction required for the block to tip with the application of a force of this magnitude? (c) Find the magnitude and direction of the minimum force required to tip the block if the point of application can be chosen anywhere on the block.

49. A uniform beam of weight *w* is inclined at an angle θ to the horizontal with its upper end supported by a horizontal rope tied to a wall and its lower end resting on a rough floor (Fig. P12.49). (a) If the coefficient of static friction between beam and floor is μ_s, determine an expression for the maximum weight *W* that can be suspended from the top before the beam slips. (b) Determine the magnitude of the reaction force at the floor and the magnitude of the force exerted by the beam on the rope at *P* in terms of *w*, *W*, and μ_s.

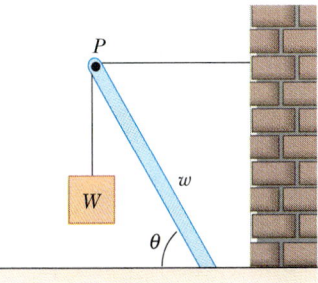

FIGURE P12.49

50. Figure P12.50 shows a truss that supports a downward force of 1000 N applied at the point *B*. Neglect-ing the weight of the truss, apply the conditions of equilibrium to prove that $n_A = 366$ N and $n_C = 634$ N.

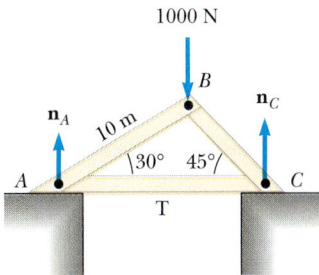

FIGURE P12.50

51. A stepladder of negligible weight is constructed as shown in Figure P12.51. A painter of mass 70.0 kg stands on the ladder 3.00 m from the bottom. Assuming the floor is frictionless, find (a) the tension in the horizontal bar connecting the two halves of the ladder, (b) the normal forces at *A* and *B*, and (c) the components of the reaction force at the hinge *C* that the left half of the ladder exerts on the right half. (*Hint:* Treat each half of the ladder separately.)

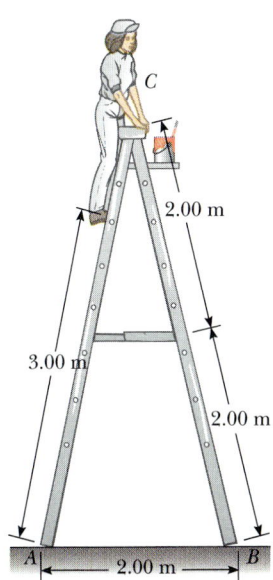

FIGURE P12.51

52. A flat dance floor of dimensions 20.0 m by 20.0 m has a mass of 1000 kg. Three dance couples, each of mass 125 kg, start in the top left, top right, and bottom left corners. (a) Where is the initial center of gravity? (b) The couple in the bottom left corner moves 10.0 m to the right. Where is the new center of gravity? (c) What was the speed of the center of gravity if it took that couple 8.00 s to change positions?

53. A shelf bracket is mounted on a vertical wall by a single screw, as shown in Figure P12.53. Neglecting the weight of the bracket, find the horizontal component of the force that the screw exerts on the bracket when an 80.0-N vertical force is applied as shown. (*Hint:* Imagine that the bracket is slightly loose.)

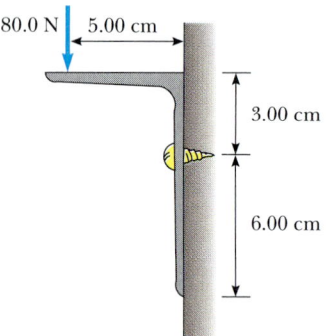

FIGURE P12.53

54. Figure P12.54 shows a claw hammer as it pulls a nail out of a horizontal surface. If a force of 150 N is exerted horizontally as shown, find (a) the force exerted by the hammer claws on the nail and (b) the force exerted by the surface on the point of contact with the hammer head. Assume that the force the hammer exerts on the nail is parallel to the nail.

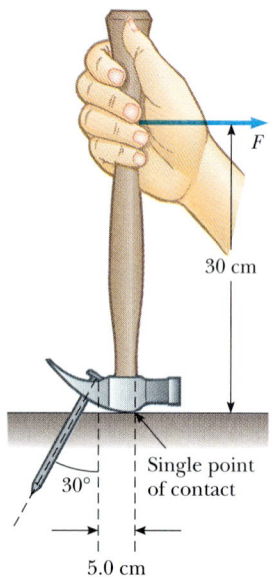

FIGURE P12.54

55. Figure P12.55 shows a vertical force applied tangentially to a uniform cylinder of weight *w*. The coeffi-

cient of static friction between the cylinder and all surfaces is 0.50. Find, in terms of *w*, the maximum force F that can be applied without causing the cylinder to rotate. (*Hint:* When the cylinder is on the verge of slipping, both friction forces are at their maximum values. Why?)

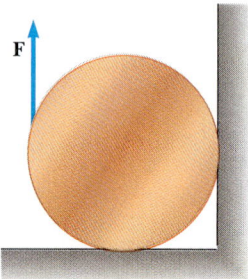

FIGURE P12.55

56. A wire of length *L*, Young's modulus *Y*, and cross-sectional area *A* is stretched elastically by an amount ΔL. By Hooke's law, the restoring force is $-k\,\Delta L$. (a) Show that $k = YA/L$. (b) Show that the work done in stretching the wire by an amount ΔL is

$$\text{Work} = \frac{1}{2}\frac{YA}{L}(\Delta L)^2$$

57. Two racquetballs are placed in a glass as shown in Figure P12.57. Their centers and the point *A* lie on a straight line. (a) Assume that the walls are frictionless, and determine P_1, P_2, and P_3. (b) Determine the magnitude of the force exerted on the right ball by the left ball. Assume each ball has a mass of 170 g.

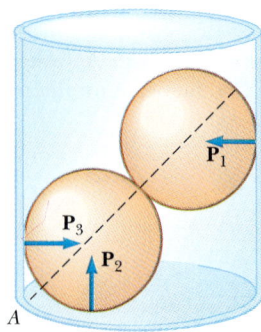

FIGURE P12.57

58. In Figure P12.58, the scales read $w_1 = 380$ N and $w_2 = 320$ N. Neglecting the weight of the supporting plank, how far from the woman's feet is her center of mass, given that her height is 2.0 m?

59. (a) Estimate the force with which a karate master strikes a board if the hand's speed at time of impact is 10.0 m/s, decreasing to 1.0 m/s during a 0.0020 s

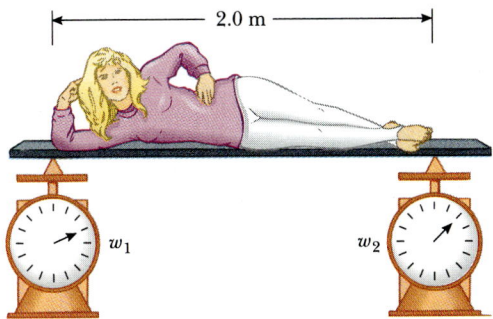

FIGURE P12.58

time-of-contact with the board. The mass of coordinated hand and arm is 1.0 kg. (b) Estimate the shear stress if this force is exerted on a 1.0-cm-thick pine board that is 10 cm wide. (c) If the maximum shear stress a pine board can receive before breaking is $3.6 \times 10^6 \ N/m^2$, will the board break?

60. A steel cable 3.0 cm² in cross-sectional area has a mass of 2.4 kg per meter of length. If 500 m of the cable is hung over a vertical cliff, how much does the cable stretch under its own weight? $Y_{steel} = 2.0 \times 10^{11} \ N/m^2$.

61. The bottom and top of a bucket have radii of 25.0 cm and 35.0 cm, respectively. The bucket is 30.0 cm high and filled with water. Where is the center of gravity? (Ignore the weight of the bucket itself.)

Oscillatory Motion

Time exposure of a simple pendulum, which consists of a small metal sphere suspended by a light string. The time it takes the pendulum to undergo one complete oscillation is the period of its motion, while the maximum displacement of the pendulum from the vertical position is called the amplitude.

As we shall see in this chapter, the period of its motion for small amplitudes depends only on the length of the pendulum and the value of the free-fall acceleration. *(James Stevenson/SPL/Photo Researchers)*

A very special kind of motion occurs when the force on a body is proportional to the displacement of the body from equilibrium. If this force always acts toward the equilibrium position of the body, there is a repetitive back-and-forth motion about this position. Such motion is an example of what is called *periodic* or *oscillatory* motion.

You are most likely familiar with several examples of periodic motion, such as the oscillations of a mass on a spring, the motion of a pendulum, and the vibrations of a stringed musical instrument. Numerous systems exhibit oscillatory motion. For example, the molecules in a solid oscillate about their equilibrium positions; electromagnetic waves, such as light waves, radar, and radio waves, are characterized by oscillating electric and magnetic field vectors; and in alternating-current circuits, voltage, current, and electrical charge vary periodically with time.

Most of the material in this chapter deals with *simple harmonic motion*. In this type of motion, an object oscillates between two spatial positions for an indefinite period of time with no loss in mechanical energy. In real mechanical systems, retarding (frictional) forces are always present and these forces are considered in an optional section at the end of the chapter.

13.1 SIMPLE HARMONIC MOTION

A particle moving along the x axis is said to exhibit **simple harmonic motion** when x, its displacement from equilibrium, varies in time according to the relationship

$$x = A \cos(\omega t + \phi) \tag{13.1}$$

Displacement versus time for simple harmonic motion

where A, ω, and ϕ are constants of the motion. In order to give physical significance to these constants, it is convenient to plot x as a function of t, as in Figure 13.1. First, note that A, called the **amplitude** of the motion, is the maximum displacement of the particle in either the positive or negative x direction. The constant angle ϕ is called the **phase constant** (or phase angle) and along with the amplitude A is determined uniquely by the initial displacement and initial velocity of the particle. The constants ϕ and A tell us what the displacement was at time $t = 0$. The quantity $(\omega t + \phi)$ is called the **phase** of the motion and is useful in comparing the motions of two systems of particles. Note that the function x is periodic and repeats itself when ωt increases by 2π rad.

The **period**, T, of the motion is the time it takes the particle to go through one full cycle. That is, the value of x at time t equals the value of x at time $t + T$. We can show that $T = 2\pi/\omega$ by using the fact that the phase increases by 2π rad in a time T:

$$\omega t + \phi + 2\pi = \omega(t + T) + \phi$$

Hence, $\omega T = 2\pi$, or

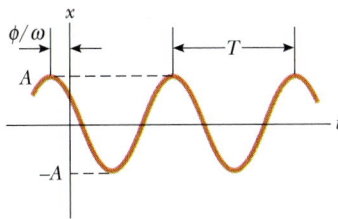

FIGURE 13.1 Displacement versus time for a particle undergoing simple harmonic motion. The amplitude of the motion is A, and the period is T.

$$T = \frac{2\pi}{\omega} \tag{13.2}$$

Period

The inverse of the period is called the **frequency** of the motion, f. The frequency represents the *number of oscillations the particle makes per unit time*:

$$f = \frac{1}{T} = \frac{\omega}{2\pi} \tag{13.3}$$

Frequency

The units of f are cycles/s, or hertz (Hz).

Rearranging Equation 13.3 gives

$$\omega = 2\pi f = \frac{2\pi}{T} \tag{13.4}$$

Angular frequency

The constant ω is called the **angular frequency** and has units of radians per second. We shall discuss the geometric significance of ω in Section 13.4.

We can obtain the speed of a particle undergoing simple harmonic motion by differentiating Equation 13.1 with respect to time:

$$v = \frac{dx}{dt} = -\omega A \sin(\omega t + \phi) \tag{13.5}$$

Speed in simple harmonic motion

The acceleration of the particle is

$$a = \frac{dv}{dt} = -\omega^2 A \cos(\omega t + \phi) \tag{13.6}$$

Since $x = A \cos(\omega t + \phi)$, we can express Equation 13.6 in the form

$$a = -\omega^2 x \tag{13.7}$$

From Equation 13.5 we see that since the sine and cosine functions oscillate between ± 1, the extreme values of v are $\pm \omega A$. Equation 13.6 tells us that the extreme values of the acceleration are $\pm \omega^2 A$. Therefore, the maximum values of the speed and acceleration are

$$v_{max} = \omega A \tag{13.8}$$

$$a_{max} = \omega^2 A \tag{13.9}$$

Figure 13.2a represents the displacement versus time for an arbitrary value of the phase constant. The velocity and acceleration versus time curves are illustrated in Figures 13.2b and 13.2c. These curves show that the phase of the velocity differs from the phase of the displacement by $\pi/2$ rad, or 90°. That is, when x is a maximum or a minimum, the velocity is zero. Likewise, when x is zero, the speed is a maximum. Furthermore, note that the phase of the acceleration differs from the phase of the displacement by π rad, or 180°. That is, when x is a maximum, a is a maximum in the opposite direction.

The phase constant ϕ is important when comparing the motion of two or more oscillating particles. Suppose that the initial position x_0 and initial speed v_0 of a single oscillator are given, that is, at $t = 0$, $x = x_0$ and $v = v_0$. Under these conditions, Equations 13.1 and 13.5 give

$$x_0 = A \cos \phi \quad \text{and} \quad v_0 = -\omega A \sin \phi \tag{13.10}$$

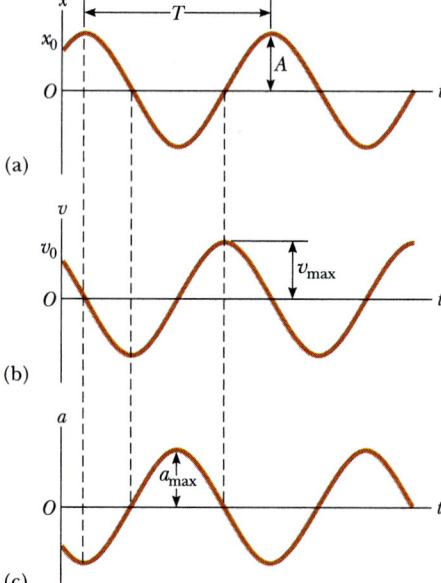

(a)

(b)

(c)

FIGURE 13.2 Graphical representation of simple harmonic motion: (a) displacement versus time, (b) velocity versus time, and (c) acceleration versus time. Note that at any specified time the velocity is 90° out of phase with the displacement and the acceleration is 180° out of phase with the displacement.

Dividing the second of these equations by the first eliminates A, giving $v_0/x_0 = -\omega \tan \phi$, or

$$\tan \phi = -\frac{v_0}{\omega x_0} \tag{13.11}$$

Furthermore, if we square Equations 13.10 and add terms, we get $x_0{}^2 + \left(\dfrac{v_0}{\omega}\right)^2 = A^2 \cos^2 \phi + A^2 \sin^2 \phi$. Solving for A, we find

$$A = \sqrt{x_0{}^2 + \left(\frac{v_0}{\omega}\right)^2} \tag{13.12}$$

The phase angle ϕ and amplitude A can be obtained from the initial conditions

Thus, we see that ϕ and A are known if x_0, ω, and v_0 are given.

The following are important properties of a particle moving in simple harmonic motion:

- The displacement, velocity, and acceleration all vary sinusoidally with time but are not in phase, as shown in Figure 13.2.
- The acceleration of the particle is proportional to the displacement but in the opposite direction.
- The frequency and the period of the motion are independent of the amplitude.

Properties of simple harmonic motion

CONCEPTUAL EXAMPLE 13.1

Does the acceleration of a simple harmonic oscillator remain constant during its motion? Is the acceleration ever zero? Explain.

Reasoning In simple harmonic motion, the acceleration is not constant. The acceleration is zero whenever the object passes through its equilibrium position, to the right whenever the object is to the left of its equilibrium position, and to the left whenever the object is to the right of its equilibrium position. In general, the acceleration is proportional to the displacement but oppositely directed.

EXAMPLE 13.2 An Oscillating Body

A body oscillates with simple harmonic motion along the x axis. Its displacement varies with time according to the equation

$$x = (4.00 \text{ m}) \cos\left(\pi t + \frac{\pi}{4}\right)$$

where t is in seconds and the angles in the second parentheses are in radians. (a) Determine the amplitude, frequency, and period of the motion.

Solution By comparing this equation with the general equation for simple harmonic motion, $x = A \cos(\omega t + \phi)$, we see that $A = 4.00$ m and $\omega = \pi$ rad/s; therefore we find $f = \omega/2\pi = \pi/2\pi = 0.500 \text{ s}^{-1}$ and $T = 1/f = 2.00$ s.

(b) Calculate the velocity and acceleration of the body at any time t.

Solution

$$v = \frac{dx}{dt} = -(4.00 \text{ m/s}) \sin\left(\pi t + \frac{\pi}{4}\right) \frac{d}{dt}(\pi t)$$

$$= -(4.00\pi \text{ m/s}) \sin\left(\pi t + \frac{\pi}{4}\right)$$

$$a = \frac{dv}{dt} = -(4.00\pi \text{ m/s}^2) \cos\left(\pi t + \frac{\pi}{4}\right) \frac{d}{dt}(\pi t)$$

$$= -(4.00\pi^2 \text{ m/s}^2) \cos\left(\pi t + \frac{\pi}{4}\right)$$

(c) Using the results to part (b), determine the position, velocity, and acceleration of the body at $t = 1.00$ s.

Solution   Noting that the angles in the trigonometric functions are in radians, we get at $t = 1.00$ s

$$x = (4.00 \text{ m}) \cos\left(\pi + \frac{\pi}{4}\right) = (4.00 \text{ m}) \cos\left(\frac{5\pi}{4}\right)$$

$$= (4.00 \text{ m})(-0.707) = -2.83 \text{ m}$$

$$v = -(4.00\pi \text{ m/s}) \sin\left(\frac{5\pi}{4}\right)$$

$$= -(4.00\pi \text{ m/s})(-0.707) = 8.89 \text{ m/s}$$

$$a = -(4.00\pi^2 \text{ m/s}^2) \cos\left(\frac{5\pi}{4}\right)$$

$$= -(4.00\pi^2 \text{ m/s}^2)(-0.707) = 27.9 \text{ m/s}^2$$

(d) Determine the maximum speed and maximum acceleration of the body.

Solution   From the general expressions for v and a found in part (b), we see that the maximum values of the sine and cosine functions are unity. Therefore, v varies between $\pm 4.00\pi$ m/s, and a varies between $\pm 4.00\pi^2$ m/s^2. Thus,

$v_{max} = 4.00\pi$ m/s and $a_{max} = 4.00\pi^2$ m/s^2. The same results are obtained using $v_{max} = \omega A$ and $a_{max} = \omega^2 A$, where $A = 4.00$ m and $\omega = \pi$ rad/s.

(e) Find the displacement of the body between $t = 0$ and $t = 1.00$ s.

Solution   The x coordinate at $t = 0$ is

$$x_0 = (4.00 \text{ m}) \cos\left(0 + \frac{\pi}{4}\right) = (4.00 \text{ m})(0.707) = 2.83 \text{ m}$$

In part (c), we found that the coordinate at $t = 1.00$ s is -2.83 m; therefore, the displacement between $t = 0$ and $t = 1.00$ s is

$$\Delta x = x - x_0 = -2.83 \text{ m} - 2.83 \text{ m} = -5.66 \text{ m}$$

Because the particle's velocity changes sign during the first second, the magnitude of Δx is not the same as the distance traveled in the first second.

Exercise   What is the phase of the motion at $t = 2.00$ s?

Answer $9\pi/4$ rad.

13.2   MASS ATTACHED TO A SPRING

Consider a physical system consisting of a mass attached to the end of a spring, where the mass is free to move on a horizontal, frictionless track (Fig. 13.3). When the spring is neither stretched nor compressed, the mass is at the position $x = 0$, called the *equilibrium position* of the system. We know from experience that such a system will oscillate back and forth if disturbed from the equilibrium position. Because the surface is frictionless, the mass moves in simple harmonic motion. An experimental arrangement that clearly demonstrates that such a system moves in simple harmonic motion is illustrated in Figure 13.4, in which a mass oscillating vertically on a spring has a pen attached to it. While the mass is in motion, a sheet of paper is moved horizontally, and the pen traces out a sinusoidal pattern.

We can understand this motion qualitatively by first recalling that, when the mass is displaced a small distance x from equilibrium, the spring exerts a force on m given by Hooke's law, as expressed by Equation 7.9:

$$F = -kx$$

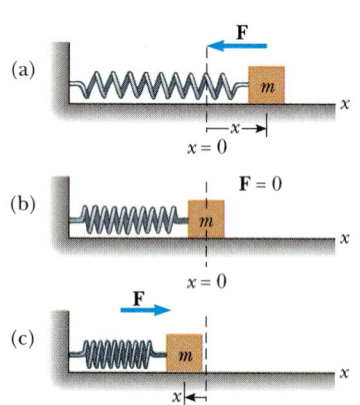

FIGURE 13.3   A mass attached to a spring on a frictionless track moves in simple harmonic motion. (a) When the mass is displaced to the right of equilibrium, the displacement is positive and the acceleration is negative. (b) At the equilibrium position, $x = 0$, the acceleration is zero but the speed is a maximum. (c) When the displacement is negative, the acceleration is positive.

As we learned in Section 7.3, we call this a linear restoring force because it is linearly proportional to the displacement and is always directed toward the equilibrium position and therefore *opposite* the displacement. That is, when the mass is displaced to the right in Figure 13.3, x is positive and the restoring force is to the left. When the mass is displaced to the left of $x = 0$, then x is negative and $\mathbf{F}$ is to the right.

If we apply Newton's second law to the motion of the mass in the x direction, we get

$$F = -kx = ma$$

$$a = -\frac{k}{m} x \tag{13.13}$$

That is, *the acceleration is proportional to the displacement of the mass from equilibrium and is in the opposite direction.* If the mass is displaced a maximum distance $x = A$ at some initial time and released from rest, its initial acceleration is $-kA/m$ (its extreme negative value). When the mass passes through the equilibrium position, $x = 0$ and its acceleration is zero. At this instant, its speed is a maximum. The mass then continues to travel to the left of equilibrium and finally reaches $x = -A$, at which time its acceleration is kA/m (maximum positive) and its speed is again zero. Thus, we see that the mass oscillates between the turning points $x = \pm A$. In one full cycle of its motion, the mass travels a distance $4A$.

We now describe the motion in a quantitative fashion. Recall that $a = dv/dt = d^2x/dt^2$, and so we can express Equation 13.13 as

$$\frac{d^2x}{dt^2} = -\frac{k}{m}x$$

If we denote the ratio k/m by the symbol ω^2, this equation becomes

$$\frac{d^2x}{dt^2} = -\omega^2 x \qquad (13.14)$$

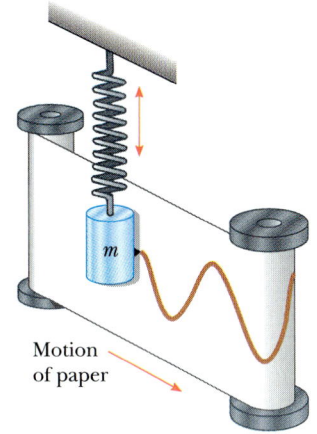

FIGURE 13.4 An experimental apparatus for demonstrating simple harmonic motion. A pen attached to the oscillating mass traces out a sine wave on the moving chart paper.

What we now require is a solution to Equation 13.14—that is, a function $x(t)$ that satisfies this second-order differential equation. Since Equations 13.14 and 13.7 are equivalent, we see that the solution must be that of simple harmonic motion:

$$x(t) = A\cos(\omega t + \phi)$$

To see this explicitly, note that if $x(t) = A\cos(\omega t + \phi)$ then

$$\frac{dx}{dt} = A\frac{d}{dt}\cos(\omega t + \phi) = -\omega A\sin(\omega t + \phi)$$

$$\frac{d^2x}{dt^2} = -\omega A\frac{d}{dt}\sin(\omega t + \phi) = -\omega^2 A\cos(\omega t + \phi)$$

Comparing the expressions for x and d^2x/dt^2, we see that $d^2x/dt^2 = -\omega^2 x$ and Equation 13.14 is satisfied.

The following general statement can be made based on the foregoing discussion:

> Whenever the force acting on a particle is linearly proportional to the displacement and in the opposite direction, the particle moves in simple harmonic motion.

Since the period is $T = 2\pi/\omega$ and the frequency is the inverse of the period, we can express the period and frequency of the motion for this system as

$$T = \frac{2\pi}{\omega} = 2\pi\sqrt{\frac{m}{k}} \qquad (13.15)$$

$$f = \frac{1}{T} = \frac{1}{2\pi}\sqrt{\frac{k}{m}} \qquad (13.16)$$

Period and frequency for mass-spring system

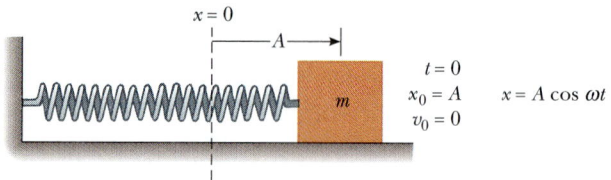

FIGURE 13.5   A mass-spring system that starts from rest at $x_0 = A$. In this case, $\phi = 0$, and so $x = A \cos \omega t$.

That is, the period and frequency depend only on the mass and on the force constant of the spring. As we might expect, the frequency is larger for a stiffer spring (the stiffer the spring, the higher the value of k) and decreases with increasing mass.

Special Case I   In order to better understand the physical significance of our solution of the equation of motion, let us consider the following special case. Suppose we pull the mass a distance A from equilibrium and release it from rest from this stretched position, as in Figure 13.5. We must then require that our solution for $x(t)$ obey the initial conditions that at $t = 0$, $x_0 = A$, and $v_0 = 0$. These conditions are met if we choose $\phi = 0$, giving $x = A \cos \omega t$ as our solution. To check this solution, we note that it satisfies the condition that $x_0 = A$ at $t = 0$, since $\cos 0 = 1$. Thus, we see that A and ϕ contain the information on initial conditions.

Now let us investigate the behavior of the velocity and acceleration for this special case. Since $x = A \cos \omega t$,

$$v = \frac{dx}{dt} = -\omega A \sin \omega t$$

and

$$a = \frac{dv}{dt} = -\omega^2 A \cos \omega t$$

From the preceding velocity expression, we see that at $t = 0$, $v_0 = 0$, as we require. The expression for the acceleration tells us that at $t = 0$, $a = -\omega^2 A$. Physically this negative acceleration makes sense, because the force on the mass is directed to the left when the displacement is positive. In fact, at the extreme position shown in Figure 13.5, $F = -kA$ (to the left), and the initial acceleration is $-kA/m$.

We could also use a more formal approach to show that $x = A \cos \omega t$ is the correct solution by using the relationship $\tan \phi = -v_0/\omega x_0$ (Eq. 13.11). Since $v_0 = 0$ at $t = 0$, $\tan \phi = 0$ and so $\phi = 0$ or π. Only $\phi = 0$ yields the correct sign for x_0.

The displacement, velocity, and acceleration versus time are plotted in Figure 13.6 for this special case. Note that the acceleration reaches extreme values of $\pm \omega^2 A$ when the displacement has extreme values of $\mp A$. Furthermore, the velocity has extreme values of $\pm \omega A$, which both occur at $x = 0$. Hence, the quantitative solution agrees with our qualitative description of this system.

Special Case II   Now suppose that the mass is given an initial velocity v_0 to the right while the system is in the equilibrium position, so that at $t = 0$, $x_0 = 0$, and $v = v_0$ (Fig. 13.7). Our solution must now satisfy these initial conditions. Since the mass is moving toward positive x values at $t = 0$ and since $x_0 = 0$ at $t = 0$, the solution has the form $x = A \sin \omega t$.

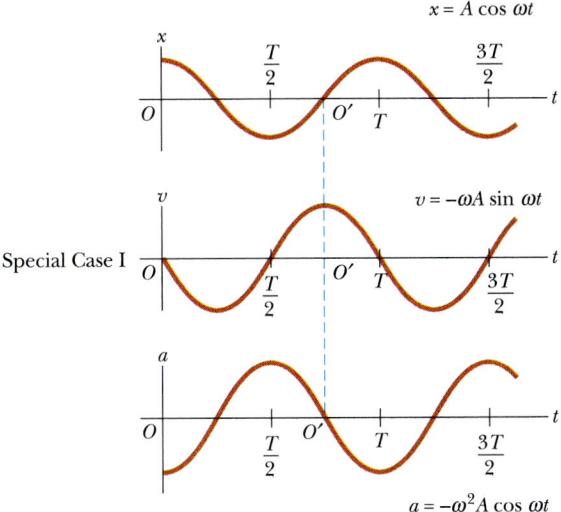

FIGURE 13.6   Displacement, velocity, and acceleration versus time for a particle undergoing simple harmonic motion under the initial conditions that at $t = 0$, $x = A$, and $v = 0$.

Applying Equation 13.11 and the initial condition that $x_0 = 0$ at $t = 0$ gives $\tan \phi = -\infty$ and $\phi = -\pi/2$. Hence, the solution is $x = A \cos(\omega t - \pi/2)$, which can be written $x = A \sin \omega t$. Furthermore, from Equation 13.11 we see that $A = v_0/\omega$; therefore, we can express our solution as

$$x = \frac{v_0}{\omega} \sin \omega t$$

The velocity and acceleration in this case are

$$v = \frac{dx}{dt} = v_0 \cos \omega t$$

$$a = \frac{dv}{dt} = -\omega v_0 \sin \omega t$$

These results are consistent with the fact that the mass always has a maximum speed at $x = 0$, and the force and acceleration are zero at this position. The graphs of these functions versus time in Figure 13.6 correspond to the origin at O'. What is the solution for x if the mass is initially moving to the left in Figure 13.7?

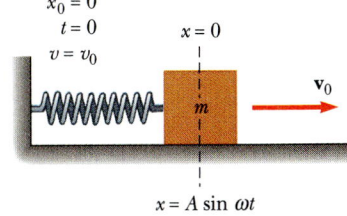

FIGURE 13.7   The mass-spring system starts its motion at the equilibrium position, $x = 0$ at $t = 0$. If its initial velocity is v_0 to the right, its x coordinate varies as

$$x = \frac{v_0}{\omega} \sin \omega t.$$

EXAMPLE 13.3   Watch Out for Potholes

A car of mass 1300 kg is constructed using a frame supported by four springs. Each spring has a force constant of 20 000 N/m. If two people riding in the car have a combined mass of 160 kg, find the frequency of vibration of the car when it is driven over a pothole in the road.

Solution   We assume the weight is evenly distributed. Thus, each spring supports one fourth of the load. The total mass supported by the springs is 1460 kg, and therefore each

spring supports 365 kg. Hence, the frequency of vibration is, from Equation 13.16,

$$f = \frac{1}{2\pi} \sqrt{\frac{k}{m}} = \frac{1}{2\pi} \sqrt{\frac{20\ 000\ \text{N/m}}{365\ \text{kg}}} = \boxed{1.18\ \text{Hz}}$$

Exercise   How long does it take the car to execute two complete vibrations?

Answer   1.70 s.

EXAMPLE 13.4   A Mass-Spring System

A mass of 200 g is connected to a light spring of force constant 5.00 N/m and is free to oscillate on a horizontal, frictionless track. If the mass is displaced 5.00 cm from equilibrium and released from rest, as in Figure 13.5, (a) find the period of its motion.

Solution   This situation corresponds to Special Case I, where $x = A \cos \omega t$ and $A = 5.00 \times 10^{-2}$ m. Therefore,

$$\omega = \sqrt{\frac{k}{m}} = \sqrt{\frac{5.00 \text{ N/m}}{200 \times 10^{-3} \text{ kg}}} = 5.00 \text{ rad/s}$$

and

$$T = \frac{2\pi}{\omega} = \frac{2\pi}{5} = \boxed{1.26 \text{ s}}$$

(b) Determine the maximum speed of the mass.

Solution

$$v_{max} = \omega A = (5.00 \text{ rad/s})(5.00 \times 10^{-2} \text{ m}) = \boxed{0.250 \text{ m/s}}$$

(c) What is the maximum acceleration of the mass?

Solution

$$a_{max} = \omega^2 A = (5.00 \text{ rad/s})^2 (5.00 \times 10^{-2} \text{ m}) = \boxed{1.25 \text{ m/s}^2}$$

(d) Express the displacement, speed, and acceleration as functions of time.

Solution   The expression $x = A \cos \omega t$ is our solution for Special Case I, and so we can use the results from (a), (b), and (c) to get

$$x = A \cos \omega t = \boxed{(0.0500 \text{ m}) \cos 5.00 t}$$

$$v = -\omega A \sin \omega t = \boxed{-(0.250 \text{ m/s}) \sin 5.00 t}$$

$$a = -\omega^2 A \cos \omega t = \boxed{-(1.25 \text{ m/s}^2) \cos 5.00 t}$$

13.3   ENERGY OF THE SIMPLE HARMONIC OSCILLATOR

Let us examine the mechanical energy of the mass-spring system described in Figure 13.5. Since the surface is frictionless, we expect that the total mechanical energy is constant, as was shown in Chapter 8. We can use Equation 13.5 to express the kinetic energy as

Kinetic energy of a simple harmonic oscillator

$$K = \tfrac{1}{2}mv^2 = \tfrac{1}{2}m\omega^2 A^2 \sin^2(\omega t + \phi) \tag{13.17}$$

The elastic potential energy stored in the spring for any elongation x is given by $\tfrac{1}{2}kx^2$. Using Equation 13.1, we get

Potential energy of a simple harmonic oscillator

$$U = \tfrac{1}{2}kx^2 = \tfrac{1}{2}kA^2 \cos^2(\omega t + \phi) \tag{13.18}$$

We see that K and U are *always* positive quantities. Since $\omega^2 = k/m$, we can express the total energy of the simple harmonic oscillator as

$$E = K + U = \tfrac{1}{2}kA^2 [\sin^2(\omega t + \phi) + \cos^2(\omega t + \phi)]$$

But $\sin^2\theta + \cos^2\theta = 1$, where $\theta = \omega t + \phi$; therefore, this equation reduces to

Total energy of a simple harmonic oscillator

$$E = \tfrac{1}{2}kA^2 \tag{13.19}$$

That is, the energy of a simple harmonic oscillator is a constant of the motion and proportional to the square of the amplitude. In fact, the total mechanical energy is equal to the maximum potential energy stored in the spring when $x = \pm A$. At these points, $v = 0$ and there is no kinetic energy. At the equilibrium position, $x = 0$ and $U = 0$, so that the total energy is all in the form of kinetic energy. That is, at $x = 0$, $E = \tfrac{1}{2}mv_{max}^2 = \tfrac{1}{2}m\omega^2 A^2$.

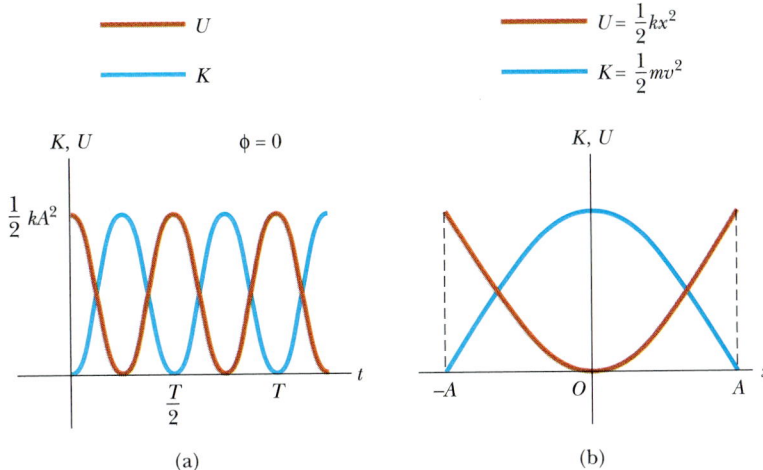

FIGURE 13.8 (a) Kinetic energy and potential energy versus time for a simple harmonic oscillator with $\phi = 0$. (b) Kinetic energy and potential energy versus displacement for a simple harmonic oscillator. In either plot, note that $K + U =$ constant.

Plots of the kinetic and potential energies versus time are shown in Figure 13.8a, where we have taken $\phi = 0$. As mentioned above, both K and U are always positive and their sum at all times is a constant equal to $\frac{1}{2}kA^2$, the total energy of the system. The variations of K and U with displacement are plotted in Figure 13.8b. Energy is continuously being transformed between potential energy stored in the spring and the kinetic energy of the mass.

EXAMPLE 13.5 **Oscillations on a Horizontal Surface**

A 0.500-kg mass connected to a light spring of force constant 20.0 N/m oscillates on a horizontal, frictionless track. (a) Calculate the total energy of the system and the maximum speed of the mass if the amplitude of the motion is 3.00 cm.

Solution Using Equation 13.19, we get

$$E = \tfrac{1}{2}kA^2 = \tfrac{1}{2}\left(20.0\ \frac{\text{N}}{\text{m}}\right)(3.00 \times 10^{-2}\ \text{m})^2 = 9.00 \times 10^{-3}\ \text{J}$$

When the mass is at $x = 0$, $U = 0$ and $E = \frac{1}{2}mv_{\text{max}}^2$; therefore,

$$\tfrac{1}{2}mv_{\text{max}}^2 = \boxed{9.00 \times 10^{-3}\ \text{J}}$$

$$v_{\text{max}} = \sqrt{\frac{18.0 \times 10^{-3}\ \text{J}}{0.500\ \text{kg}}} = \boxed{0.190\ \text{m/s}}$$

(b) What is the velocity of the mass when the displacement is equal to 2.00 cm?

Solution We can apply Equation 13.20 directly:

$$v = \pm\sqrt{\frac{k}{m}(A^2 - x^2)} = \pm\sqrt{\frac{20.0}{0.500}(3.00^2 - 2.00^2) \times 10^{-4}}$$

$$= \boxed{\pm 0.141\ \text{m/s}}$$

The positive and negative signs indicate that the mass could be moving to the right or left at this instant.

(c) Compute the kinetic and potential energies of the system when the displacement equals 2.00 cm.

Solution Using the result to part (b), we get

$$K = \tfrac{1}{2}mv^2 = \tfrac{1}{2}(0.500\ \text{kg})(0.141\ \text{m/s})^2$$

$$= \boxed{4.97 \times 10^{-3}\ \text{J}}$$

$$U = \tfrac{1}{2}kx^2 = \tfrac{1}{2}\left(20.0\ \frac{\text{N}}{\text{m}}\right)(2.00 \times 10^{-2}\ \text{m})^2$$

$$= \boxed{4.00 \times 10^{-3}\ \text{J}}$$

Note that the sum $K + U$ equals the total energy, E.

Exercise For what values of x does the speed of the mass equal 0.100 m/s?

Answer ± 2.55 cm.

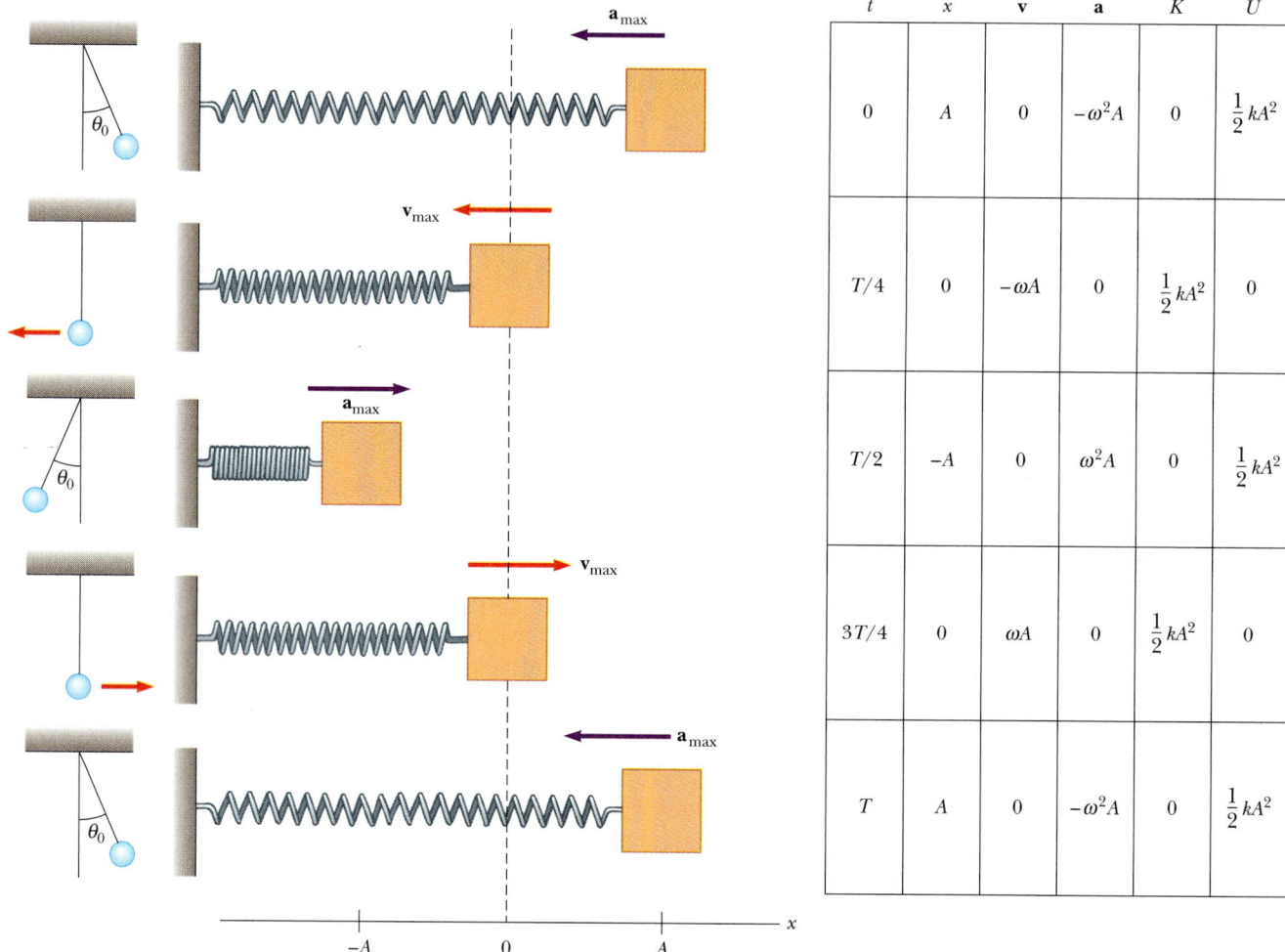

t	x	v	a	K	U
0	A	0	$-\omega^2 A$	0	$\frac{1}{2}kA^2$
$T/4$	0	$-\omega A$	0	$\frac{1}{2}kA^2$	0
$T/2$	$-A$	0	$\omega^2 A$	0	$\frac{1}{2}kA^2$
$3T/4$	0	ωA	0	$\frac{1}{2}kA^2$	0
T	A	0	$-\omega^2 A$	0	$\frac{1}{2}kA^2$

FIGURE 13.9   Simple harmonic motion for a mass-spring system and its analogy to the motion of a simple pendulum. The parameters in the table at the right refer to the mass-spring system, assuming that at $t = 0$, $x = A$ so that $x = A \cos \omega t$ (Special Case I).

Figure 13.9 illustrates the position, velocity, acceleration, kinetic energy, and potential energy of the mass-spring system for one full period of the motion. Most of the ideas discussed so far are incorporated in this important figure. Study it carefully.

Finally, we can use energy conservation to obtain the velocity for an arbitrary displacement x by expressing the total energy at some arbitrary position as

$$E = K + U = \tfrac{1}{2}mv^2 + \tfrac{1}{2}kx^2 = \tfrac{1}{2}kA^2$$

Velocity as a function of position for a simple harmonic oscillator

$$v = \pm \sqrt{\frac{k}{m}(A^2 - x^2)} = \pm \omega\sqrt{A^2 - x^2} \tag{13.20}$$

Again, this expression substantiates the fact that the speed is a maximum at $x = 0$ and is zero at the turning points, $x = \pm A$.

13.4   THE PENDULUM

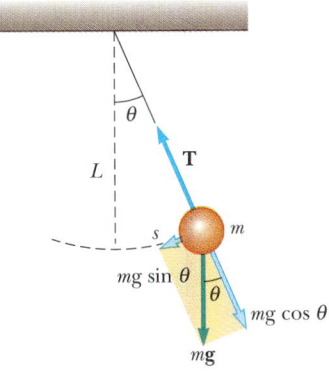

The **simple pendulum** is another mechanical system that moves in oscillatory motion. It consists of a point mass m suspended by a light string of length L, where the upper end of the string is fixed as in Figure 13.10. The motion occurs in a vertical plane and is driven by the gravitational force. We shall show that, provided the angle θ is small, the motion is that of a simple harmonic oscillator. The forces acting on the mass are the force exerted by the string T and the gravitational force mg. The tangential component of the gravitational force, $mg \sin \theta$, always acts toward $\theta = 0$, opposite the displacement. Therefore, the tangential force is a restoring force, and we can write the equation of motion in the tangential direction:

$$F_t = -mg \sin \theta = m \frac{d^2 s}{dt^2}$$

FIGURE 13.10   When θ is small, the simple pendulum oscillates in simple harmonic motion about the equilibrium position ($\theta = 0$). The restoring force is $mg \sin \theta$, the component of the weight tangent to the circle.

where s is the displacement measured along the arc and the minus sign indicates that F_t acts toward the equilibrium position. Since $s = L\theta$ and L is constant, this equation reduces to

$$\frac{d^2 \theta}{dt^2} = -\frac{g}{L} \sin \theta$$

The right side is proportional to $\sin \theta$ rather than to θ; hence, we conclude that the motion is not simple harmonic motion since it is not of the form of Equation 13.14. However, if we assume that θ is small, we can use the approximation

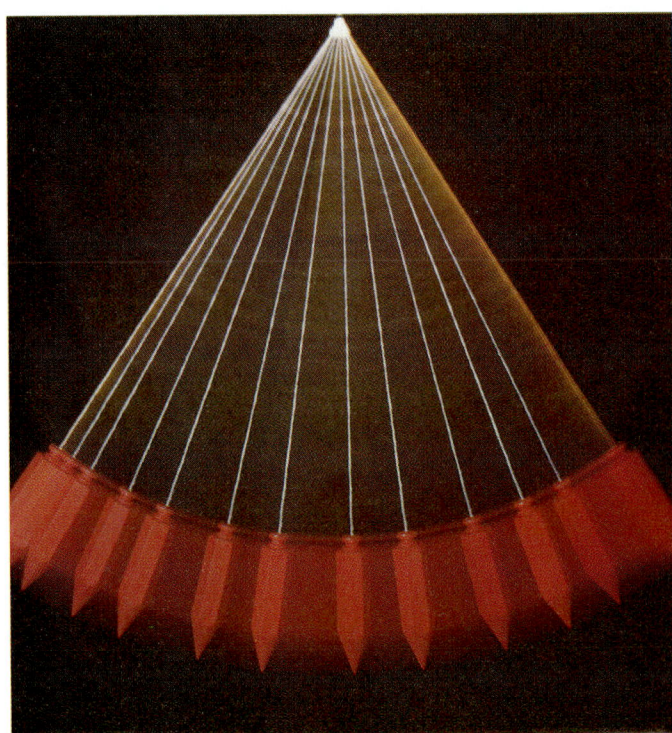

The motion of a simple pendulum captured with multiflash photography. Is the motion simple harmonic in this case?   *(Paul Silverman/Fundamental Photographs)*

The Foucault pendulum at the Smithsonian Institution in Washington, D.C. This type of pendulum was first used by the French physicist Jean Foucault to verify the Earth's rotation experimentally. During its swinging motion, the pendulum's plane of oscillation appears to rotate, as the bob successively knocks over the red indicators arranged in a horizontal circle.   *(Courtesy of the Smithsonian Institution)*

$\sin \theta \approx \theta$, where θ is measured in radians.[1] Therefore, the equation of motion becomes

Equation of motion for the simple pendulum (small θ)

$$\frac{d^2\theta}{dt^2} = -\frac{g}{L}\theta \qquad (13.21)$$

Now we have an expression that is of exactly the same form as Equation 13.14, and so we conclude that the motion is simple harmonic motion. Therefore, θ can be written as $\theta = \theta_0 \cos(\omega t + \phi)$, where θ_0 is the *maximum angular displacement* and the angular frequency ω is

Angular frequency of motion for the simple pendulum

$$\omega = \sqrt{\frac{g}{L}} \qquad (13.22)$$

The period of the motion is

Period of motion for the simple pendulum

$$T = \frac{2\pi}{\omega} = 2\pi\sqrt{\frac{L}{g}} \qquad (13.23)$$

In other words, *the period and frequency of a simple pendulum depend only on the length of the string and the value of g.* Since the period is independent of the mass, we conclude that all simple pendulums of equal length at the same location oscillate with

---

[1] This approximation can be understood by examining the series expansion for $\sin \theta$, which is $\sin \theta = \theta - \theta^3/3! + \cdots$. For small values of θ, we see that $\sin \theta \approx \theta$. The difference between θ (in radians) and $\sin \theta$ for $\theta = 15°$ is only about 1%.

equal periods.[2] The analogy between the motion of a simple pendulum and the mass-spring system is illustrated in Figure 13.9.

The simple pendulum can be used as a timekeeper. It is also a convenient device for making precise measurements of the free-fall acceleration. Such measurements are important since variations in local values of g can provide information on the location of oil and other valuable underground resources.

CONCEPTUAL EXAMPLE 13.6

A pendulum bob is made with a sphere filled with water. What would happen to the frequency of vibration of this pendulum if there were a hole in the sphere that allowed the water to leak out slowly? Neglect both the mass of the string and air resistance.

Reasoning   The frequency of a pendulum is equal to the inverse of its period. From Equation 13.23, we see that $f = (1/2\pi)\sqrt{g/L}$. Because the frequency depends only on the length of the pendulum and the free-fall acceleration, and

not on the mass, the frequency will not change appreciably if the sphere is small compared to the length of the suspension. However, as the water leaks out of the sphere, the frequency first decreases (as the distance from the pivot to the center of mass of the sphere increases). After the water level in the sphere reaches the half-way point, the frequency begins to increase again until the sphere is empty. At that point, the frequency is the same as it was when the sphere was completely filled with water.

EXAMPLE 13.7   A Measure of Height

A man enters a tall tower, needing to know its height. He notes that a long pendulum extends from the ceiling almost to the floor and that its period is 12.0 s. How tall is the tower?

Solution   If we use $T = 2\pi\sqrt{L/g}$ and solve for L, we get

$$L = \frac{gT^2}{4\pi^2} = \frac{(9.80 \text{ m/s}^2)(12.0 \text{ s})^2}{4\pi^2} = \boxed{35.7 \text{ m}}$$

Exercise   If the pendulum described in this example is taken to the Moon, where the free-fall acceleration is 1.67 m/s², what is the period there?

Answer   29.1 s.

The Physical Pendulum

If a hanging object oscillates about a fixed axis that does not pass through its center of mass, and the object cannot be accurately approximated as a point mass, then it must be treated as a physical, or compound, pendulum. Consider a rigid object pivoted at a point O that is a distance d from the center of mass (Fig. 13.11). The torque about O is provided by the weight of the object, and the magnitude of the torque is $mgd \sin \theta$. Using the fact that $\tau = I\alpha$, where I is the moment of inertia about the axis through O, we get

$$-mgd \sin \theta = I\frac{d^2\theta}{dt^2}$$

The minus sign indicates that the torque about O tends to decrease θ. That is, the weight of the object produces a restoring torque.

---

[2] The period of oscillation for the simple pendulum with arbitrary amplitude is

$$T = 2\pi\sqrt{\frac{L}{g}}\left(1 + \frac{1}{4}\sin^2\frac{\theta_0}{2} + \frac{9}{64}\sin^4\frac{\theta_0}{2} + \cdots\right)$$

where θ_0 is the maximum angular displacement in radians.

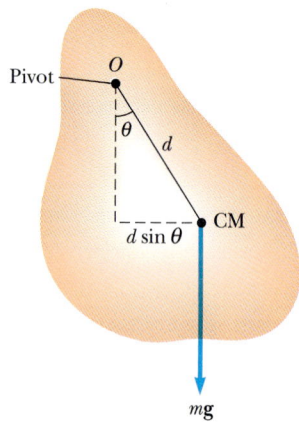

FIGURE 13.11   The physical pendulum consists of a rigid body pivoted at the point O, which is not at the center of mass. At equilibrium, the weight vector passes through O, corresponding to $\theta = 0$. The restoring torque about O when the system is displaced through an angle θ is $mgd \sin \theta$.

If we again assume that θ is small, then the approximation $\sin \theta \approx \theta$ is valid and the equation of motion reduces to

$$\frac{d^2\theta}{dt^2} = -\left(\frac{mgd}{I}\right)\theta = -\omega^2\theta \qquad (13.24)$$

This equation is of the same form as Equation 13.14, and so the motion is simple harmonic motion. That is, the solution of Equation 13.24 is $\theta = \theta_0 \cos(\omega t + \phi)$, where θ_0 is the maximum angular displacement and

$$\omega = \sqrt{\frac{mgd}{I}}$$

The period is

$$T = \frac{2\pi}{\omega} = 2\pi\sqrt{\frac{I}{mgd}} \qquad (13.25)$$

You can use this result to measure the moment of inertia of a planar rigid body. If the location of the center of mass and hence of d is known, the moment of inertia can be obtained by measuring the period. Finally, note that Equation 13.25 reduces to the period of a simple pendulum (Eq. 13.23) when $I = md^2$, that is, when all the mass is concentrated at the center of mass.

EXAMPLE 13.8   A Swinging Rod

A uniform rod of mass M and length L is pivoted about one end and oscillates in a vertical plane (Fig. 13.12). Find the period of oscillation if the amplitude of the motion is small.

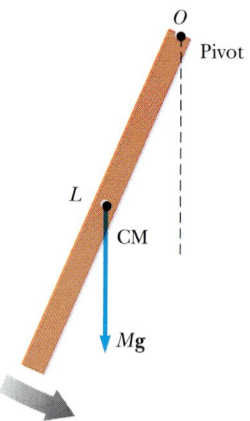

FIGURE 13.12   (Example 13.8) A rigid rod oscillating about a pivot through one end is a physical pendulum with $d = L/2$ and, from Table 10.2, $I_0 = \frac{1}{3}ML^2$.

Solution   In Chapter 10 we found that the moment of inertia of a uniform rod about an axis through one end is $\frac{1}{3}ML^2$. The distance d from the pivot to the center of mass is $L/2$. Substituting these quantities into Equation 13.25 gives

$$T = 2\pi\sqrt{\frac{\frac{1}{3}ML^2}{Mg\frac{L}{2}}} = 2\pi\sqrt{\frac{2L}{3g}}$$

Comment   In one of the Moon landings, an astronaut walking on the Moon's surface had a belt hanging from his space suit, and the belt oscillated as a compound pendulum. A scientist on Earth observed this motion on TV and from it was able to estimate the free-fall acceleration on the Moon. How do you suppose this calculation was done?

Exercise   Calculate the period of a meter stick pivoted about one end and oscillating in a vertical plane as in Figure 13.12.

Answer   1.64 s.

Torsional Pendulum

Figure 13.13 shows a rigid body suspended by a wire attached at the top of a fixed support. When the body is twisted through some small angle θ, the twisted wire exerts a restoring torque on the body proportional to the angular displacement.

That is,

$$\tau = -\kappa\theta$$

where κ (the Greek letter kappa) is called the *torsion constant* of the support wire. The value of κ can be obtained by applying a known torque to twist the wire through a measurable angle θ. Applying Newton's second law for rotational motion gives

$$\tau = -\kappa\theta = I\frac{d^2\theta}{dt^2}$$

$$\frac{d^2\theta}{dt^2} = -\frac{\kappa}{I}\theta \qquad (13.26)$$

Again, this is the equation of motion for a simple harmonic oscillator, with $\omega = \sqrt{\kappa/I}$ and a period

$$T = 2\pi\sqrt{\frac{I}{\kappa}} \qquad (13.27)$$

This system is called a *torsional pendulum*. There is no small-angle restriction in this situation, as long as the response of the wire remains linear. The balance wheel of a watch oscillates as a torsional pendulum, energized by the mainspring. Torsional pendulums are also used in laboratory galvanometers and the Cavendish torsional balance.

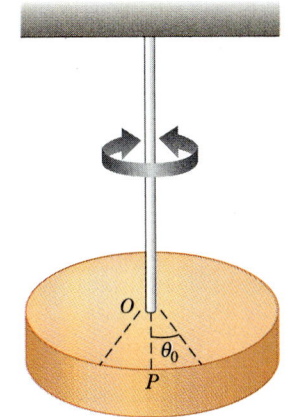

FIGURE 13.13 A torsional pendulum consists of a rigid body suspended by a wire attached to a rigid support. The body oscillates about the line *OP* with an amplitude θ_0.

*13.5 COMPARING SIMPLE HARMONIC MOTION WITH UNIFORM CIRCULAR MOTION

We can better understand and visualize many aspects of simple harmonic motion by studying its relationship to uniform circular motion. Figure 13.14 shows an experimental arrangement useful for developing this concept. Shadows are cast on a screen by a peg attached to the rim of a rotating disk and a mass attached to a spring. The two shadows move together when the period of the rotating disk equals the period of the oscillating mass-spring system.

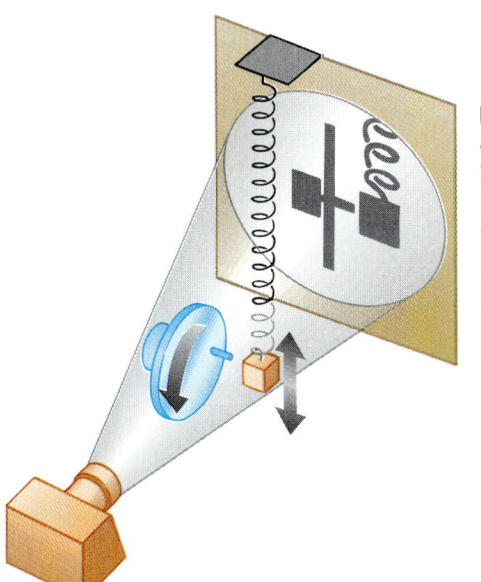

FIGURE 13.14 Experimental setup for demonstrating the connection between simple harmonic motion and uniform circular motion. The shadow of a peg on a turntable is projected on the screen along with the shadow of a mass on a spring. When the period of rotation of the turntable equals the period of oscillation of the mass, the shadows move together.

Consider a particle at point P moving in a circle of radius A with constant angular speed ω (Fig. 13.15a). We refer to this circle as the *reference circle* for the motion. As the particle rotates, its position vector rotates about the origin, O. Consider a particle located on the circumference of the circle of radius A, as in Figure 13.15a, with the line OP making an angle ϕ with the x axis at $t = 0$. We call this circle a reference circle for comparison of simple harmonic motion and uniform circular motion and take the position of P at $t = 0$ as our reference position. If the particle moves along the circle with constant angular speed ω until OP makes an angle θ with the x axis as in Figure 13.15b, then at some time $t > 0$, the angle between OP and the x axis is $\theta = \omega t + \phi$. As the particle rotates on the reference circle, the angle that OP makes with the x axis changes with time. Furthermore, the projection of P onto the x axis, labeled point Q, moves back and forth along a line superimposed on the horizontal diameter of the reference circle, between the limits $x = \pm A$.

Note that points P and Q always have the same x coordinate. From the right triangle OPQ, we see that this x coordinate is

$$x = A \cos(\omega t + \phi) \tag{13.28}$$

This expression shows that the point Q moves with simple harmonic motion along the x axis. Therefore, we conclude that

> simple harmonic motion along a straight line can be represented by the projection of uniform circular motion along a diameter of a reference circle.

By a similar argument, you can see from Figure 13.15b that the projection of P along the y axis also exhibits harmonic motion. Therefore, *uniform circular motion can be considered a combination of two simple harmonic motions,* one along x and one along y, where the two differ in phase by $90°$.

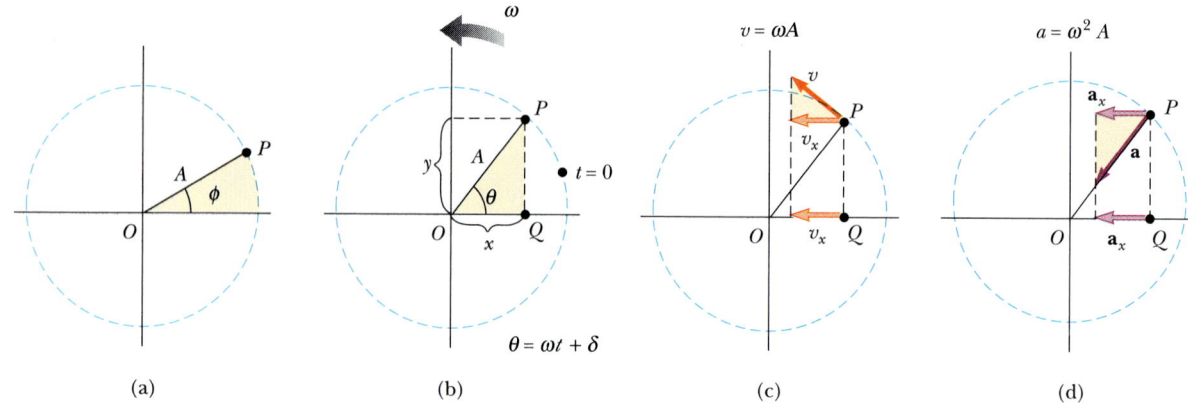

FIGURE 13.15   Relationship between the uniform circular motion of a point P and the simple harmonic motion of the point Q. A particle at P moves in a circle of radius A with constant angular frequency ω. (a) Reference circle showing position of P at $t = 0$. (b) The x components of the points P and Q are equal and vary in time as $x = A \cos(\omega t + \phi)$. (c) The x component of velocity of P equals the velocity of Q. (d) The x component of the acceleration of P equals the acceleration of Q.

This geometric interpretation shows that the time for one complete revolution of the point P on the reference circle is equal to the period of motion, T, for simple harmonic motion between $x = \pm A$. That is, the angular speed of P is the same as the angular frequency, ω, of simple harmonic motion along the x axis. The phase constant ϕ for simple harmonic motion corresponds to the initial angle that OP makes with the x axis. The radius of the reference circle, A, equals the amplitude of the simple harmonic motion.

Since the relationship between linear and angular speed for circular motion is $v = r\omega$ (Eq. 10.9), the particle moving on the reference circle of radius A has a velocity of magnitude ωA. From the geometry in Figure 13.15c, we see that the x component of this velocity is $-\omega A \sin(\omega t + \phi)$. By definition, the point Q has a velocity given by dx/dt. Differentiating Equation 13.28 with respect to time, we find that the velocity of Q is the same as the x component of the velocity of P.

The acceleration of P on the reference circle is directed radially inward toward O and has a magnitude $v^2/A = \omega^2 A$. From the geometry in Figure 13.15d, we see that the x component of this acceleration is $-\omega^2 A \cos(\omega t + \phi)$. This value is also the acceleration of the projected point Q along the x axis, as you can verify from Equation 13.28.

EXAMPLE 13.9 Circular Motion with Constant Speed

A particle rotates counterclockwise in a circle of radius 3.00 m with a constant angular speed of 8.00 rad/s. At $t = 0$, the particle has an x coordinate of 2.00 m. (a) Determine the x coordinate as a function of time.

Solution Since the amplitude of the particle's motion equals the radius of the circle and $\omega = 8.00$ rad/s, we have

$$x = A \cos(\omega t + \phi) = (3.00 \text{ m}) \cos(8.00t + \phi)$$

We can evaluate ϕ using the initial condition that $x = 2.00$ m at $t = 0$:

$$2.00 \text{ m} = (3.00 \text{ m}) \cos(0 + \phi)$$

$$\phi = \cos^{-1}\left(\tfrac{2.00}{3.00}\right) = 48.2° = 0.841 \text{ rad}$$

Therefore, the x coordinate versus time is of the form

$$x = (3.00 \text{ m}) \cos(8.00t + 0.841)$$

Note that the angles in the cosine function are in radians.

(b) Find the x components of the particle's velocity and acceleration at any time t.

Solution

$$v_x = \frac{dx}{dt} = (-3.00)(8.00) \sin(8.00t + 0.841)$$

$$= -(24.0 \text{ m/s}) \sin(8.00t + 0.841)$$

$$a_x = \frac{dv_x}{dt} = (-24.0)(8.00) \cos(8.00t + 0.841)$$

$$= -(192 \text{ m/s}^2) \cos(8.00t + 0.841)$$

From these results, we conclude that $v_{max} = 24$ m/s and $a_{max} = 192$ m/s^2. Note that these values also equal the tangential velocity, ωA, and centripetal acceleration, $\omega^2 A$.

---

*13.6 DAMPED OSCILLATIONS

The oscillatory motions we have considered so far have been for ideal systems, that is, systems that oscillate indefinitely under the action of a linear restoring force. In real systems, dissipative forces, such as friction, are present and retard the motion. Consequently, the mechanical energy of the system diminishes in time, and the motion is said to be *damped*.

One common type of retarding force, which we discussed in Chapter 6, is proportional to the speed and acts in the direction opposite the motion. This retarding force is often observed when an object moves through a gas. Because the

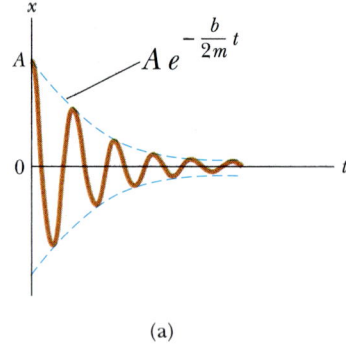

(a)

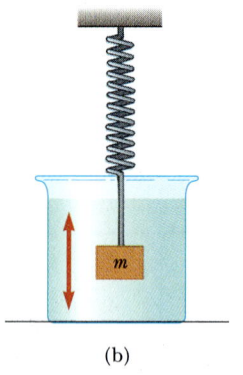

(b)

FIGURE 13.16 (a) Graph of the displacement versus time for a damped oscillator. Note the decrease in amplitude with time. (b) One example of a damped oscillator is a mass on a spring submersed in a liquid.

retarding force can be expressed as $\mathbf{R} = -b\mathbf{v}$, where b is a constant, and the restoring force of the system is $-kx$, we can write Newton's second law as

$$\sum F_x = -kx - bv = ma_x$$

$$-kx - b\frac{dx}{dt} = m\frac{d^2x}{dt^2} \tag{13.29}$$

The solution of this equation requires mathematics that may not be familiar to you as yet, and so we simply state it here without proof. When the retarding force is small compared with kx—that is, when b is small—the solution to Equation 13.29 is

$$x = Ae^{-\frac{b}{2m}t}\cos(\omega t + \phi) \tag{13.30}$$

where the angular frequency of motion is

$$\omega = \sqrt{\frac{k}{m} - \left(\frac{b}{2m}\right)^2} \tag{13.31}$$

This result can be verified by substituting Equation 13.30 into Equation 13.29. Figure 13.16a shows the displacement as a function of time in this case. We see that *when the retarding force is small compared with the restoring force, the oscillatory character of the motion is preserved but the amplitude decreases in time,* and the motion ultimately ceases. Any system that behaves in this way is known as a **damped oscillator.** The dashed blue lines in Figure 13.16a, which define what is called the *envelope* of the oscillatory curve, represent the exponential factor that appears in Equation 13.30. This envelope shows that *the amplitude decays exponentially with time.* For motion with a given spring constant and particle mass, the oscillations dampen more rapidly as the maximum value of the retarding force approaches the maximum value of the restoring force. One example of a damped harmonic oscillator is a mass immersed in a fluid as in Figure 13.16b.

It is convenient to express the angular frequency of a damped oscillator in the form

$$\omega = \sqrt{\omega_0{}^2 - \left(\frac{b}{2m}\right)^2}$$

where $\omega_0 = \sqrt{k/m}$ represents the angular frequency in the absence of a retarding force (the undamped oscillator). In other words, when $b = 0$, the retarding force is zero and the system oscillates with its natural frequency, ω_0. As the magnitude of the retarding force approaches the magnitude of the restoring force in the spring, the oscillations dampen more rapidly. When b reaches a critical value b_c such that $b_c/2m = \omega_0$, the system does not oscillate and is said to be **critically damped.** In this case, once released from rest at some nonequilibrium position, the system returns to equilibrium and then stays there. The displacement versus time graph for this case is the red curve in Figure 13.17.

If the medium is so viscous that the retarding force is greater than the restoring force—that is, if $b/2m > \omega_0$—the system is **overdamped.** Again, the displaced system does not oscillate but simply returns to its equilibrium position. As the damping increases, the time it takes the displacement to reach equilibrium also increases, as indicated in Figure 13.17. In any case, when friction is present, the energy of the oscillator eventually falls to zero. The lost mechanical energy dissipates into thermal energy in the retarding medium.

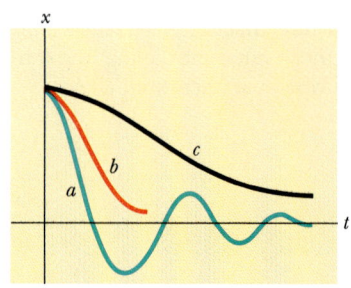

FIGURE 13.17 Plots of displacement versus time for an underdamped oscillator (a), a critically damped oscillator (b), and an overdamped oscillator (c).

*13.7   FORCED OSCILLATIONS

It is possible to compensate for the energy loss in a damped system by applying an external force that does positive work on the system. At any instant, energy can be put into the system by an applied force that acts in the direction of motion of the oscillator. For example, a child on a swing can be kept in motion by appropriately timed pushes. The amplitude of motion remains constant if the energy input per cycle of motion exactly equals the energy lost as a result of friction.

A common example of a forced oscillator is a damped oscillator driven by an external force that varies periodically, such as $F = F_0 \cos \omega t$, where ω is the angular frequency of the force and F_0 is a constant. Adding this driving force to the left side of Equation 13.29 gives

$$F_0 \cos \omega t - b\frac{dx}{dt} - kx = m\frac{d^2x}{dt^2} \qquad (13.32)$$

As earlier, the solution of this equation is not presented. After a sufficiently long period of time, when the energy input per cycle equals the energy lost per cycle, a steady-state condition is reached in which the oscillations proceed with constant amplitude. At this time, when the system is in a steady state, the solution of Equation 13.32 is

$$x = A \cos(\omega t + \phi) \qquad (13.33)$$

where

$$A = \frac{F_0/m}{\sqrt{(\omega^2 - \omega_0{}^2)^2 + \left(\dfrac{b\omega}{m}\right)^2}} \qquad (13.34)$$

and where $\omega_0 = \sqrt{k/m}$ is the angular frequency of the undamped oscillator $(b = 0)$. One can argue that, in steady state, the oscillator must physically have the same frequency as the driving force, and so the solution given by Equation 13.33 is expected. In fact, when this solution is substituted into Equation 13.32, one finds that it is indeed a solution, provided the amplitude is given by Equation 13.34.

Equation 13.34 shows that the motion of the forced oscillator is not damped since it is being driven by an external force. That is, the external agent provides the necessary energy to overcome the losses due to the retarding force. Note that the mass oscillates at the angular frequency of the driving force, ω. For small damping, the amplitude becomes large when the frequency of the driving force is near the natural frequency of oscillation. The dramatic increase in amplitude near the natural frequency ω_0 is called **resonance**, and ω_0 is called the **resonance frequency** of the system.

Physically, the reason for large-amplitude oscillations at the resonance frequency is that energy is being transferred to the system under the most favorable conditions. This can be better understood by taking the first time derivative of x, which gives an expression for the velocity of the oscillator. In doing so, one finds that v is proportional to $\sin(\omega t + \phi)$. When the applied force is in phase with the velocity, the rate at which work is done on the oscillator by the force $\mathbf{F}$ (or the power) equals Fv. Since the quantity Fv is always positive when $\mathbf{F}$ and $\mathbf{v}$ are in phase, we conclude that *at resonance the applied force is in phase with the velocity and the power transferred to the oscillator is a maximum.*

Figure 13.18 is a graph of the amplitude as a function of frequency for the forced oscillator, with and without a retarding force. Note that the amplitude

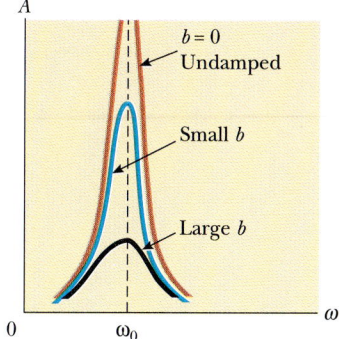

FIGURE 13.18   Graph of amplitude versus frequency for a damped oscillator when a periodic driving force is present. When the frequency of the driving force equals the natural frequency, ω_0, resonance occurs. Note that the shape of the resonance curve depends on the size of the damping coefficient, b.

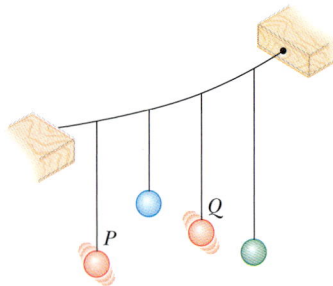

FIGURE 13.19 If pendulum P is set into oscillation, pendulum Q will eventually oscillate with the greatest amplitude because its length is equal to that of P and so they have the same natural frequency of vibration. The pendulums are oscillating in a direction perpendicular to the plane formed by the stationary strings.

increases with decreasing damping ($b \to 0$) and that the resonance curve broadens as the damping increases. Under steady-state conditions and at any driving frequency, the energy transferred into the system equals the energy lost because of the damping force; hence, the average total energy of the oscillator remains constant. In the absence of a damping force ($b = 0$), we see from Equation 13.35 that the steady-state amplitude approaches infinity as $\omega \to \omega_0$. In other words, if there are no losses in the system and if we continue to drive an initially motionless oscillator with a periodic force that is in phase with the velocity, the amplitude of motion builds without limit (Fig. 13.18). This limitless building does not occur in practice because some damping is always present. That is, at resonance the amplitude is large but finite for small damping.

One experiment that demonstrates a resonance phenomenon is illustrated in Figure 13.19. Several pendulums of different lengths are suspended from a stretched string. If one of them, such as P, is set in sideways motion, the others begin to oscillate because they are coupled by the stretched string. Pendulum Q, whose length is the same as that of P (and hence the two pendulums have the same natural frequency), oscillates with the greatest amplitude.

Later in the text we shall see that resonance appears in other areas of physics. For example, certain electrical circuits have natural frequencies. A bridge has natural frequencies that can be set into resonance by an appropriate driving force. A striking example of such resonance occurred in 1940, when the Tacoma Narrows bridge in Washington was destroyed by resonant vibrations. Although the winds were not particularly strong on that occasion, the bridge ultimately collapsed because turbulences generated by the wind blowing through the bridge structure occurred at a frequency that matched the natural frequency of the structure.

Many other examples of resonant vibrations can be cited. A resonant vibration you may have experienced is the "singing" of telephone wires in the wind. Machines often break if one vibrating part is at resonance with some other moving part. Finally, soldiers marching in cadence across bridges have been known to set up resonant vibrations in the structure, causing it to collapse. In one famous accident, which occurred in France in 1850, a collapsed suspension bridge resulted in the death of 226 soldiers.

SUMMARY

The position of a simple harmonic oscillator varies periodically in time according to the expression

$$x = A \cos(\omega t + \phi) \tag{13.1}$$

where A is the amplitude of the motion, ω is the angular frequency, and ϕ is the phase constant. The value of ϕ depends on the initial position and velocity of the oscillator.

The time T for one complete vibration is called the **period** of the motion:

$$T = \frac{2\pi}{\omega} \tag{13.2}$$

The inverse of the period is the **frequency** of the motion, which equals the number of oscillations per second.

The **velocity** and **acceleration** of a simple harmonic oscillator are

$$v = \frac{dx}{dt} = -\omega A \sin(\omega t + \phi) \tag{13.5}$$

$$a = \frac{dv}{dt} = -\omega^2 A \cos(\omega t + \phi) \tag{13.6}$$

Thus, the maximum speed is ωA, and the maximum acceleration is $\omega^2 A$. The speed is zero when the oscillator is at its turning points, $x = \pm A$, and a maximum at the equilibrium position, $x = 0$. The magnitude of the acceleration is a maximum at the turning points and zero at the equilibrium position.

A mass-spring system moves in simple harmonic motion on a frictionless surface, with a period

$$T = \frac{2\pi}{\omega} = 2\pi\sqrt{\frac{m}{k}} \tag{13.15}$$

The **kinetic energy** and **potential energy** for a simple harmonic oscillator vary with time and are given respectively by

$$K = \tfrac{1}{2}mv^2 = \tfrac{1}{2}m\omega^2 A^2 \sin^2(\omega t + \phi) \tag{13.17}$$

$$U = \tfrac{1}{2}kx^2 = \tfrac{1}{2}kA^2 \cos^2(\omega t + \phi) \tag{13.18}$$

The **total energy** of a simple harmonic oscillator is a constant of the motion and is

$$E = \tfrac{1}{2}kA^2 \tag{13.19}$$

The potential energy of a simple harmonic oscillator is a maximum when the particle is at its turning points (maximum displacement from equilibrium) and zero at the equilibrium position. The kinetic energy is zero at the turning points and a maximum at the equilibrium position.

A **simple pendulum** of length L moves in simple harmonic motion for small angular displacements from the vertical, with a period

$$T = 2\pi\sqrt{\frac{L}{g}} \tag{13.23}$$

A **physical pendulum** moves in simple harmonic motion about a pivot that does not go through the center of mass. The period of this motion is

$$T = 2\pi\sqrt{\frac{I}{mgd}} \tag{13.25}$$

where I is the moment of inertia about an axis through the pivot and d is the distance from the pivot to the center of mass.

QUESTIONS

1. What is the total distance traveled by a body executing simple harmonic motion in a time equal to its period if its amplitude is A?

2. If the coordinate of a particle varies as $x = -A \cos \omega t$, what is the phase constant in Equation 13.1? At what position does the particle begin its motion?

3. Does the displacement of an oscillating particle between $t = 0$ and a later time t necessarily equal the position of the particle at time t? Explain.

4. Determine whether or not the following quantities can be in the same direction for a simple harmonic oscillator: (a) displacement and velocity, (b) velocity and acceleration, (c) displacement and acceleration.

5. Can the amplitude A and phase constant ϕ be deter-

mined for an oscillator if only the position is specified at $t = 0$? Explain.

6. Describe qualitatively the motion of a mass-spring system when the mass of the spring is not neglected.

7. If a mass-spring system is hung vertically and set into oscillation, why does the motion eventually stop?

8. Explain why the kinetic and potential energies of a mass-spring system can never be negative.

9. A mass-spring system undergoes simple harmonic motion with an amplitude A. Does the total energy change if the mass is doubled but the amplitude is not changed? Do the kinetic and potential energies depend on the mass? Explain.

10. What happens to the period of a simple pendulum if the pendulum's length is doubled? What happens to the period if the mass that is suspended is doubled?

11. A simple pendulum is suspended from the ceiling of a stationary elevator, and the period is determined. Describe the changes, if any, in the period when the elevator (a) accelerates upward, (b) accelerates downward, and (c) moves with constant velocity.

12. A simple pendulum undergoes simple harmonic motion when θ is small. Is the motion periodic when θ is large? How does the period of motion change as θ increases?

13. Give a few examples of damped oscillations that are commonly observed.

14. Will damped oscillations occur for any values of b and k? Explain.

15. Is it possible to have damped oscillations when a system is at resonance? Explain.

16. At resonance, what does the phase constant ϕ equal in Equation 13.33? (*Hint:* Compare this equation with the expression for the driving force, which must be in phase with the velocity at resonance.)

17. A platoon of soldiers marches in step along a road. Why are they ordered to break step when crossing a bridge?

18. Give as many examples as you can in the workings of an automobile where the motion is simple harmonic or damped.

19. If a grandfather clock were running slow, how could we adjust the length of the pendulum to correct the time?

PROBLEMS

Section 13.1 Simple Harmonic Motion

1. The displacement of a particle at $t = 0.25$ s is given by the expression $x = (4.0 \text{ m}) \cos(3.0\pi t + \pi)$, where x is in meters and t is in seconds. Determine (a) the frequency and period of the motion, (b) the amplitude of the motion, (c) the phase constant, and (d) the displacement of the particle at $t = 0.25$ s.

2. A small ball is set in horizontal motion by rolling it with a speed of 3.00 m/s across a room 12.0 m long, between two walls. Assume that the collisions made with each wall are perfectly elastic and that the motion is perpendicular to the two walls. (a) Show that the motion is periodic, and (b) determine its period. Is this motion simple harmonic? Explain.

3. A ball dropped from a height of 4.00 m makes a perfectly elastic collision with the ground. Assuming no energy lost due to air resistance, (a) show that the motion is periodic and (b) determine the period of the motion. (c) Is the motion simple harmonic? Explain.

4. If the initial position and velocity of an object moving in simple harmonic motion are x_0, v_0, and a_0, and if the angular frequency of oscillation is ω, (a) show that the position and speed of the object for all time can be written as

$$x(t) = x_0 \cos \omega t + \left(\frac{v_0}{\omega}\right) \sin \omega t$$

$$v(t) = -x_0 \omega \sin \omega t + v_0 \cos \omega t$$

(b) If the amplitude of the motion is A, show that

$$v^2 - ax = v_0^2 - a_0 x_0 = A^2 \omega^2$$

5. The displacement of an object is $x = (8.0 \text{ cm}) \cos(2.0t + \pi/3)$, where x is in centimeters and t is in seconds. Calculate (a) the speed and acceleration at $t = \pi/2$ s, (b) the maximum speed and the earliest time ($t > 0$) at which the particle has this speed, and (c) the maximum acceleration and the earliest time ($t > 0$) at which the particle has this acceleration.

6. A 20-g particle moves in simple harmonic motion with a frequency of 3.0 oscillations/s and an amplitude of 5.0 cm. (a) Through what total distance does the particle move during one cycle of its motion? (b) What is its maximum speed? Where does this occur? (c) Find the maximum acceleration of the particle. Where in the motion does the maximum acceleration occur?

7. A particle moving along the x axis in simple harmonic motion starts from the origin at $t = 0$ and moves to the right. If the amplitude of its motion is 2.00 cm and the frequency is 1.50 Hz, (a) show that its displacement is given by $x = (2.00 \text{ cm}) \sin(3.00\pi t)$. Determine (b) the maximum speed and the earliest time ($t > 0$) at which the particle has this speed, (c) the maximum acceleration and the earliest time ($t > 0$) at which the particle has this acceleration, and (d) the total distance traveled between $t = 0$ and $t = 1.00$ s.

□ indicates problems that have full solutions available in the Student Solutions Manual and Study Guide.

8. A piston in an automobile engine is in simple harmonic motion. If its amplitude of oscillation from centerline is ±5.0 cm and its mass is 2.0 kg, find the maximum velocity and acceleration of the piston when the auto engine is running at the rate of 3600 rev/min.

Section 13.2 Mass Attached to a Spring

(*Note:* Neglect the mass of the spring in all of these problems.)

9. A block of unknown mass is attached to a spring of spring constant 6.50 N/m and undergoes simple harmonic motion with an amplitude of 10.0 cm. When the mass is halfway between its equilibrium position and the endpoint, its speed is measured to be + 30.0 cm/s. Calculate (a) the mass of the block, (b) the period of the motion, and (c) the maximum acceleration of the block.

10. A spring stretches by 3.9 cm when a 10-g mass is hung from it. If a 25-g mass attached to this spring oscillates in simple harmonic motion, calculate the period of motion.

11. A 7.00-kg mass is hung from the bottom end of a vertical spring fastened to an overhead beam. The mass is set into vertical oscillations having a period of 2.60 s. Find the force constant of the spring.

12. A 1.0-kg mass attached to a spring of force constant 25 N/m oscillates on a horizontal, frictionless track. At $t = 0$, the mass is released from rest at $x = -3.0$ cm. (That is, the spring is compressed by 3.0 cm.) Find (a) the period of its motion, (b) the maximum values of its speed and acceleration, and (c) the displacement, velocity, and acceleration as functions of time.

13. A simple harmonic oscillator takes 12.0 s to undergo five complete vibrations. Find (a) the period of its motion, (b) the frequency in Hz, and (c) the angular frequency in rad/s.

14. A 1.0-kg mass is attached to a horizontal spring. The spring is initially stretched by 0.10 m and the mass is released from rest there. After 0.50 s, the speed of the mass is zero for the first time. What is the maximum speed of the mass?

15. A 0.50-kg mass attached to a spring of force constant 8.0 N/m vibrates in simple harmonic motion with an amplitude of 10 cm. Calculate (a) the maximum value of its speed and acceleration, (b) the speed and acceleration when the mass is 6.0 cm from the equilibrium position, and (c) the time it takes the mass to move from $x = 0$ to $x = 8.0$ cm.

16. (a) A 100-g block is placed on top of a 200-g block as shown in Figure P13.16. The coefficient of static friction between the blocks is 0.20. The lower block is now moved back and forth horizontally in simple harmonic motion having an amplitude of 6.0 cm. Keeping the amplitude constant, what is the highest frequency for which the upper block will not slip relative to the lower block? (b) Suppose the lower block is moved vertically in simple harmonic motion rather than horizontally. The frequency is held constant at 2.0 oscillations/s while the amplitude is gradually increased. Determine the amplitude at which the upper block will no longer maintain contact with the lower block.

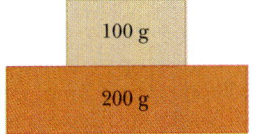

FIGURE P13.16

17. A particle that hangs from a spring oscillates with an angular frequency of 2.00 rad/s. The spring is suspended from the ceiling of an elevator car and hangs motionless (relative to the elevator car) as the car descends at a constant speed of 1.50 m/s. The car then stops suddenly. (a) With what amplitude does the particle oscillate? (b) What is the equation of motion for the particle? (Choose the upward direction to be positive.)

17A. A particle that hangs from a spring oscillates with an angular frequency ω. The spring is suspended from the ceiling of an elevator car and hangs motionless (relative to the elevator car) as the car descends at a constant speed v. The car then stops suddenly. (a) With what amplitude does the particle oscillate? (b) What is the equation of motion for the particle? (Choose the upward direction to be positive.)

Section 13.3 Energy of the Simple Harmonic Oscillator

(*Note:* Neglect the mass of the spring in all these problems.)

18. A 1.5-kg block at rest on a tabletop is attached to a horizontal spring having constant 19.6 N/m. The spring is initially unstretched. A constant 20.0-N horizontal force is applied to the object causing the spring to stretch. (a) Determine the speed of the block after it has moved 0.30 m from equilibrium if the surface between the block and tabletop is frictionless. (b) Answer part (a) if the coefficient of kinetic friction between block and tabletop is 0.20.

18A. A block of mass m, at rest on a tabletop is attached to a horizontal spring having constant k. The spring is initially unstretched. A constant horizontal force F is applied to the object causing the spring to stretch. (a) Determine the speed of the block after it has moved a distance d from equilibrium if the surface between the block and tabletop is frictionless. (b) Answer part (a) if the coefficient of kinetic friction between block and tabletop is μ.

19. An automobile having a mass of 1000 kg is driven into a brick wall in a safety test. The bumper behaves

like a spring of constant 5.0×10^6 N/m and compresses 3.16 cm as the car is brought to rest. What was the speed of the car before impact, assuming no energy is lost during impact with the wall?

20. A bullet of mass 10.0 g is fired into and embeds in a 2.00-kg block attached to a spring with constant 19.6 N/m. (a) How far will the spring be compressed if the speed of the bullet just before striking the block is 300 m/s and the block slides along a frictionless track? (b) Answer part (a) if the coefficient of kinetic friction between track and block is 0.200.

21. The amplitude of a system moving in simple harmonic motion is doubled. Determine the change in (a) the total energy, (b) the maximum speed, (c) the maximum acceleration, and (d) the period.

22. A 50-g mass connected to a spring of force constant 35 N/m oscillates on a horizontal, frictionless surface with an amplitude of 4.0 cm. Find (a) the total energy of the system and (b) the speed of the mass when the displacement is 1.0 cm. When the displacement is 3.0 cm, find (c) the kinetic energy and (d) the potential energy.

23. A particle executes simple harmonic motion with an amplitude of 3.00 cm. At what displacement from the midpoint of its motion does its speed equal one half of its maximum speed?

24. A 2.00-kg mass is attached to a spring and placed on a horizontal smooth surface. A horizontal force of 20.0 N is required to hold the mass at rest when it is pulled 0.200 m from its equilibrium position (the origin of the x axis). The mass is now released from rest with an initial displacement of $x_0 = 0.200$ m, and it subsequently undergoes simple harmonic oscillations. Find (a) the force constant of the spring, (b) the frequency of the oscillations, and (c) the maximum speed of the mass. Where does this maximum speed occur? (d) Find the maximum acceleration of the mass. Where does it occur? (e) Find the total energy of the oscillating system. When the displacement equals one-third the maximum value, find (f) the speed and (g) the acceleration.

Section 13.4   The Pendulum

25. A simple pendulum has a period of 2.50 s. (a) What is its length? (b) What would its period be on the Moon, where $g_{Moon} = 1.67$ m/s^2?

26. A light rod extends rigidly 0.50 m out from one end of a meterstick. The stick is suspended at the far end of the rod and set into oscillation. (a) Determine the period of small oscillations. (b) By what percentage does this differ from a 1.0-m-long simple pendulum?

27. A simple pendulum is 5.0 m long. (a) What is the period of simple harmonic motion for this pendulum if it is located in an elevator accelerating upward at 5.0 m/s^2? (b) What is the answer to part (a) if the elevator is accelerating downward at 5.0 m/s^2?

(c) What is the period of simple harmonic motion for this pendulum if it is placed in a truck that is accelerating horizontally at 5.0 m/s^2?

27A. A simple pendulum has a length L. (a) What is the period of simple harmonic motion for this pendulum if it is located in an elevator accelerating upward with an acceleration a? (b) What is the answer to part (a) if the elevator is accelerating downward with an acceleration a? (c) What is the period of simple harmonic motion for this pendulum if it is placed in a truck that is accelerating horizontally with an acceleration a?

28. A mass is attached to the end of a string to form a simple pendulum. The period of its harmonic motion is measured for small angular displacements and three lengths, each time clocking the motion with a stopwatch for 50 oscillations. For lengths of 1.00 m, 0.75 m, and 0.50 m, total times of 99.8 s, 86.6 s, and 71.1 s are measured for 50 oscillations. (a) Determine the period of motion for each length. (b) Determine the mean value of g obtained from these three independent measurements, and compare it with the accepted value. (c) Plot T^2 versus L, and obtain a value for g from the slope of your best-fit straight-line graph. Compare this value with that obtained in part (b).

29. A simple pendulum has a mass of 0.250 kg and a length of 1.00 m. It is displaced through an angle of 15.0° and then released. What are (a) the maximum speed? (b) the maximum angular acceleration? (c) the maximum restoring force?

30. Consider the physical pendulum of Figure 13.11. (a) If its moment of inertia about an axis passing through its center of mass and parallel to the axis passing through its pivot point is I_{CM}, show that its period is

$$T = 2\pi\sqrt{\frac{I_{CM} + md^2}{mgd}}$$

where d is the distance between the pivot point and center of mass. (b) Show that the period has a minimum value when d satisfies $md^2 = I_{CM}$.

31. A simple pendulum has a length of 3.00 m. Determine the change in its period if it is taken from a point where $g = 9.80$ m/s^2 to an elevation where the free-fall acceleration decreases to 9.79 m/s^2.

32. A horizontal rod 1.0 m long, with mass 2.0 kg, is suspended from a wire at its center, to form a torsional pendulum. If the resulting period is 3.0 minutes, what is the torsion constant for the wire?

33. A physical pendulum in the form of a planar body moves in simple harmonic motion with a frequency of 0.450 Hz. If the pendulum has a mass of 2.20 kg and the pivot is located 0.350 m from the center of mass, determine the moment of inertia of the pendulum.

34. The angular displacement of a pendulum is represented by the equation $\theta = 0.32 \cos \omega t$, where θ is in radians and $\omega = 4.43$ rad/s. Determine the period and length of the pendulum.

35. A clock balance wheel has a period of oscillation of 0.250 s. The wheel is constructed so that 20.0 g of mass is concentrated around a rim of radius 0.500 cm. What are (a) the wheel's moment of inertia and (b) the torsion constant of the attached spring?

*Section 13.5   Comparing Simple Harmonic Motion with Uniform Circular Motion

36. While riding behind a car traveling at 3.0 m/s, you notice that one of the car's tires has a small hemispherical boss on its rim, as in Figure P13.36. (a) Explain why the boss, from your viewpoint behind the car, executes simple harmonic motion. (b) If the radii of the car's tires are 0.30 m, what is the boss's period of oscillation?

Boss

FIGURE P13.36

37. Consider the simplified single-piston engine in Figure P13.37. If the wheel rotates at a constant angular speed ω, explain why the piston rod oscillates in simple harmonic motion.

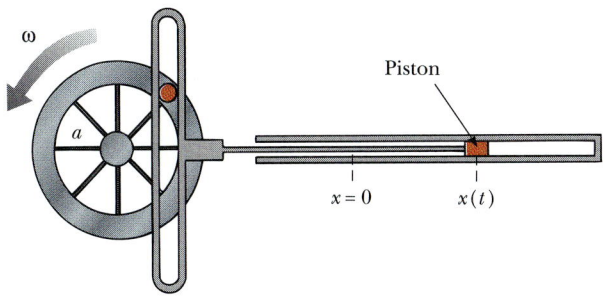

ω

Piston

a

$x = 0$ $x(t)$

FIGURE P13.37

*Section 13.6   Damped Oscillations

38. Show that the time rate of change of mechanical energy for a damped, undriven oscillator is given by $dE/dt = -bv^2$ and hence is always negative. (*Hint:* Differentiate the expression for the mechanical energy of an oscillator, $E = \frac{1}{2}mv^2 + \frac{1}{2}kx^2$, and use Eq. 13.29.)

39. A pendulum of length 1.00 m is released from an initial angle of 15.0°. After 1000 s, its amplitude is reduced by friction to 5.5°. What is the value of $b/2m$?

*Section 13.7   Forced Oscillations

40. A 2.00-kg mass attached to a spring is driven by an external force $F = (3.00 \text{ N}) \cos(2\pi t)$. If the force constant of the spring is 20.0 N/m, determine (a) the period and (b) the amplitude of the motion. (*Hint:* Assume there is no damping, that is, $b = 0$, and use Eq. 13.34.)

41. Calculate the resonant frequencies of (a) a 3.00-kg mass attached to a spring of force constant 240 N/m and (b) a simple pendulum 1.50 m in length.

42. Consider an undamped forced oscillator at resonance so that $\omega = \omega_0 = \sqrt{k/m}$. The equation of motion is

$$\frac{d^2x}{dt^2} + \omega^2 x = \left(\frac{F_0}{m}\right) \cos \omega t$$

Show by direct substitution that the solution of this equation is

$$x(t) = x_0 \cos \omega t + \left(\frac{v_0}{\omega}\right) \sin \omega t + \left(\frac{F_0}{2m\omega}\right) t \sin \omega t$$

where x_0 and v_0 are its initial position and velocity.

43. A weight of 40.0 N is suspended from a spring that has a force constant of 200 N/m. The system is undamped and is subjected to a harmonic force of frequency 10.0 Hz, resulting in a forced-motion amplitude of 2.00 cm. Determine the maximum value of the force.

ADDITIONAL PROBLEMS

44. A car with bad shock absorbers bounces up and down with a period of 1.5 s after hitting a bump. The car has a mass of 1500 kg and is supported by four springs of equal force constant k. Determine a value for k.

45. A large passenger of mass 150 kg sits in the car of Problem 44. What is the new period of oscillation?

46. A block rests on a flat plate that executes vertical simple harmonic motion with a period of 1.2 s. What is the maximum amplitude of the motion for which the block does not separate from the plate?

47. When the simple pendulum illustrated in Figure

P13.47 makes an angle with the vertical, its speed is v. (a) Calculate the total mechanical energy of the pendulum as a function of v and θ. (b) Show that when θ is small, the potential energy can be expressed as $\frac{1}{2}mgL\theta^2 = \frac{1}{2}m\omega^2 s^2$. (*Hint:* In part (b), approximate $\cos\theta$ by $\cos\theta \approx 1 - \theta^2/2$.)

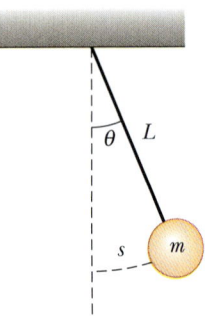

FIGURE P13.47

48. A horizontal platform vibrates with simple harmonic motion in the horizontal direction with a period of 2.0 s. A body on the platform starts to slide when the amplitude of vibration reaches 0.30 m. Find the coefficient of static friction between body and platform.

49. A particle of mass m slides inside a hemispherical bowl of radius R. Show that, for small displacements from equilibrium, the particle moves in simple harmonic motion with an angular frequency equal to that of a simple pendulum of length R. That is, $\omega = \sqrt{g/R}$.

50. A horizontal plank of mass m and length L is pivoted at one end, and the opposite end is attached to a spring of force constant k (Fig. P13.50). The moment of inertia of the plank about the pivot is $\frac{1}{3}mL^2$. When the plank is displaced a small angle θ from the horizontal and released, show that it moves with simple harmonic motion with an angular frequency $\omega = \sqrt{3k/m}$.

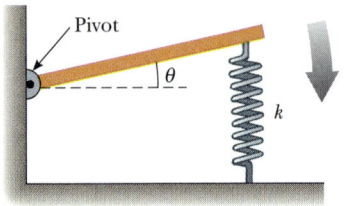

FIGURE P13.50

51. A mass M is attached to the end of a uniform rod of mass M and length L that is pivoted at the top (Fig. P13.51). (a) Determine the tensions in the rod at the pivot and at the point P when the system is stationary. (b) Calculate the period of oscillation for small dis-

placements from equilibrium, and determine this period for $L = 2.00$ m. (*Hint:* Assume the mass at the end of the rod is a point mass and use Eq. 13.25.)

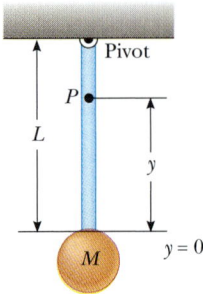

FIGURE P13.51

52. Consider a damped oscillator as illustrated in Figure 13.16. Assume the mass is 375 g, the spring constant is 100 N/m, and $b = 0.100$ kg/s. (a) How long does it take for the amplitude to drop to half its initial value? (b) How long does it take for the mechanical energy to drop to half its initial value? (c) Show that, in general, the rate at which the amplitude decreases in a damped harmonic oscillator is one-half the rate at which the mechanical energy decreases.

53. A small, thin disk of radius r and mass m is attached rigidly to the face of a second thin disk of radius R and mass M as shown in Figure P13.53. The center of the small disk is located at the edge of the large disk. The large disk is mounted at its center on a frictionless axle. The assembly is rotated through an angle θ from its equilibrium position and released. (a) Show that the speed of the center of the small disk as it passes through the equilibrium position is

$$v = 2\left[\frac{Rg(1-\cos\theta)}{(M/m) + (r/R)^2 + 2}\right]^{1/2}$$

(b) Show that the period of the motion is

$$T = 2\pi\left[\frac{(M + 2m)R^2 + mr^2}{2mgR}\right]^{1/2}$$

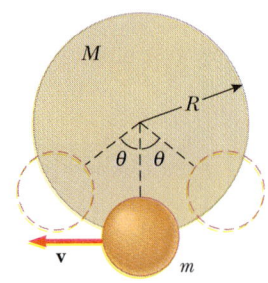

FIGURE P13.53

54. A mass m is connected to two springs of force constants k_1 and k_2 as in Figures P13.54a and P13.54b. In each case, the mass moves on a frictionless table and is displaced from equilibrium and released. Show that in each case the mass exhibits simple harmonic motion with periods

(a) $\quad T = 2\pi\sqrt{\dfrac{m(k_1 + k_2)}{k_1 k_2}}$

(b) $\quad T = 2\pi\sqrt{\dfrac{m}{k_1 + k_2}}$

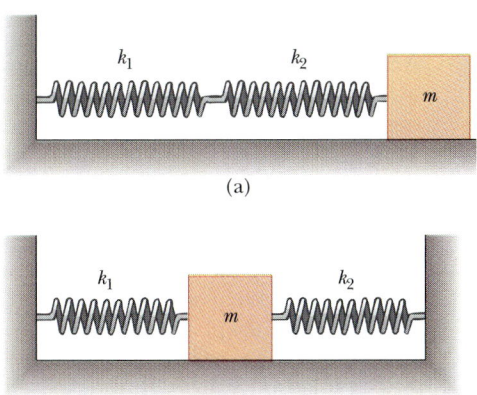

(a)

(b)

FIGURE P13.54

55. A pendulum of length L and mass M has a spring of force constant k connected to it at a distance h below its point of suspension (Fig. P13.55). Find the frequency of vibration of the system for small values of the amplitude (small θ). (Assume the vertical suspension of length L is rigid, but neglect its mass.)

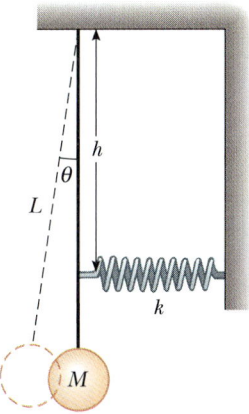

FIGURE P13.55

56. A mass m is oscillating freely on a vertical spring (Fig. P13.56). When $m = 0.810$ kg, the period is 0.910 s. An unknown mass on the same spring has a period of 1.16 s. Determine (a) the spring constant k and (b) the unknown mass.

56A. A mass m is oscillating freely on a vertical spring with a period T (Fig. P13.56). An unknown mass m' on the same spring oscillates with a period T'. Determine (a) the spring constant k and (b) the unknown mass m'.

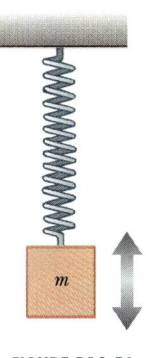

FIGURE P13.56

57. A large block P executes horizontal simple harmonic motion by sliding across a frictionless surface with a frequency $f = 1.5$ Hz. Block B rests on it, as shown in Figure P13.57, and the coefficient of static friction between the two is $\mu_s = 0.60$. What maximum amplitude of oscillation can the system have if the block is not to slip?

57A. A large block P executes horizontal simple harmonic motion by sliding across a frictionless surface with a frequency f. Block B rests on it, as shown in Figure P13.57, and the coefficient of static friction between the two is μ_s. What maximum amplitude of oscillation can the system have if the block is not to slip?

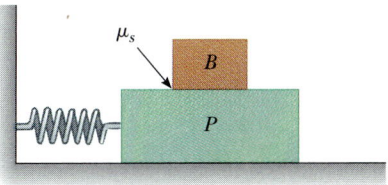

FIGURE P13.57

58. A long, thin rod of mass M and length L oscillates about its center on a cylinder of radius R (Fig. P13.58). Show that small displacements give rise to simple harmonic motion with a period $\pi L/\sqrt{3gR}$.

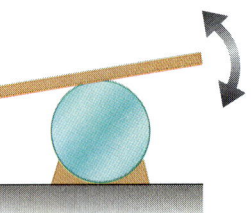

FIGURE P13.58

59. A simple pendulum having a length of 2.23 m and a mass of 6.74 kg is given an initial speed of 2.06 m/s at its equilibrium position. Assume it undergoes simple harmonic motion and determine its (a) period, (b) total energy, and (c) maximum angular displacement.

60. A mass, $m_1 = 9.0$ kg, is in equilibrium while connected to a light spring of constant $k = 100$ N/m that is fastened to a wall as in Figure P13.60a. A second mass, $m_2 = 7.0$ kg, is slowly pushed up against mass m_1, compressing the spring by the amount $A = 0.2$ m, as shown in Figure P13.60b. The system is then released, causing both masses to start moving to the right on the frictionless surface. (a) When m_1 reaches the equilibrium point, m_2 loses contact with m_1 (Fig. P13.60c) and moves to the right with speed v. Determine the value of v. (b) How far apart are the masses when the spring is fully stretched for the first time (D in Fig. P13.60d)? (*Hint:* First determine the period of oscillation and the amplitude of the m_1-spring system after m_2 loses contact with m_1.)

61. The mass of the deuterium molecule (D_2) is twice that of the hydrogen molecule (H_2). If the vibrational frequency of H_2 is 1.30×10^{14} Hz, what is the vibrational frequency of D_2, assuming that the "spring constant" of attracting forces is the same for the two molecules?

62. Show that if a torsional pendulum is twisted through an angle and then held the potential energy is $U = \frac{1}{2}\kappa\theta^2$.

63. A 2.00-kg block hangs without vibrating at the end of a spring ($k = 500$ N/m) that is attached to the ceiling of an elevator car. The car is rising with an upward acceleration of $g/3$ when the acceleration suddenly ceases (at $t = 0$). (a) What is the angular frequency of oscillation of the block after the acceleration ceases? (b) By what amount is the spring stretched during the time that the elevator car is accelerating? (c) What are the amplitude of the oscillation and the initial phase angle observed by a rider in the car? Take the upward direction to be positive.

64. A solid sphere (radius $= R$) rolls without slipping in a cylindrical trough (radius $= 5R$) as shown in Figure P13.64. Show that, for small displacements from equilibrium perpendicular to the length of the trough, the sphere executes simple harmonic motion with a period $T = 2\pi\sqrt{28R/5g}$.

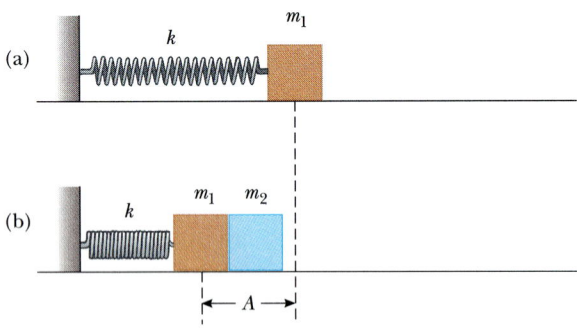

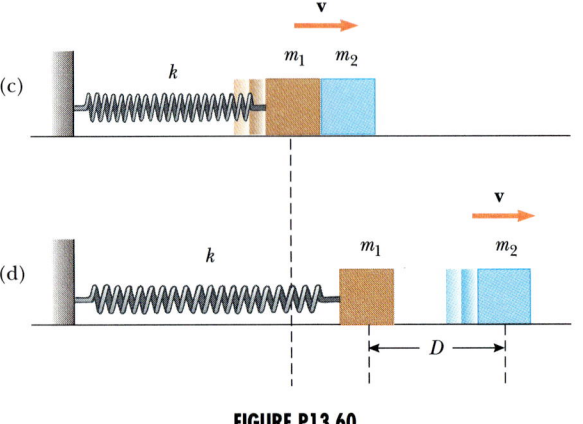

FIGURE P13.60

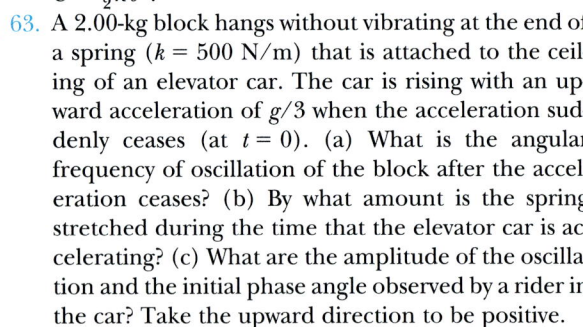

FIGURE P13.64

65. A mass m is connected to two rubber bands of length L, each under tension T, as in Figure P13.65. The mass is displaced vertically by a small distance y. Assuming the tension does not change, show that (a) the restoring force is $-(2T/L)y$ and (b) the sys-

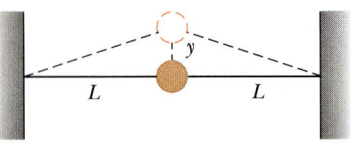

FIGURE P13.65

tem exhibits simple harmonic motion with an angular frequency $\omega = \sqrt{2T/mL}$.

66. A light, cubical container of volume a^3 is initially filled with a liquid of mass density ρ. The cube is initially supported by a light string to form a pendulum of length L_0 measured from the center of mass of the filled container. The liquid is allowed to flow from the bottom of the container at a constant rate (dM/dt). At any time t, the level of the fluid in the container is h and the length of the pendulum is L (measured relative to the instantaneous center of mass). (a) Sketch the apparatus and label the dimensions a, h, L_0, and L. (b) Find the time rate of change of the period as a function of time t. (c) Find the period as a function of time.

67. When a mass M connected to the end of a spring of mass $m_s = 7.40$ g and force constant k is set into simple harmonic motion, the period of its motion is

$$T = 2\pi\sqrt{\frac{M + (m_s/3)}{k}}$$

A two-part experiment is conducted using various masses suspended vertically from the spring. (a) Displacements of 17.0, 29.3, 35.3, 41.3, 47.1, and 49.3 cm are measured for M values of 20.0, 40.0, 50.0, 60.0, 70.0, and 80.0 g, respectively. Construct a graph of Mg versus x, and perform a linear least-squares fit to the data. From the slope of your graph, determine a value for k for this spring. (b) The system is now set into simple harmonic motion, and periods are measured with a stopwatch. With $M = 80.0$ g, the total time for 10 oscillations is measured to be 13.41 s. The experiment is repeated with M values of 70.0, 60.0, 50.0, 40.0, and 20.0 g, with corresponding times for 10 oscillations of 12.52, 11.67, 10.67, 9.62, and 7.03 s. Obtain experimental values for T for each of these M values. Plot a graph of T^2 versus M and determine a value for k from the slope of the linear least-squares fit through the data points. Compare this value of k with that obtained in part (a). (c) Obtain a value for m_s from your graph and compare it with the given value of 7.40 g.

SPREADSHEET PROBLEMS

S1. Use a spreadsheet to plot the position x of an object undergoing simple harmonic motion as a function of time. (a) Use $A = 2.00$ m, $\omega = 5.00$ rad/s, and $\phi = 0$. On the same graph, plot the function for $\phi = \pi/4$, $\pi/2$, and π. Explain how the different phase constants affect the position of the object. (b) Repeat part (a) with $\omega = 7.50$ rad/s. How does the change in ω affect the plots?

S2. The potential and kinetic energies of a particle of mass m oscillating on a spring are given by Equations 13.17 and 13.18. Use a spreadsheet to plot $U(t)$,

$K(t)$, and their sum as functions of time on the same graph. The input variables should be k, A, m, and ϕ. What happens to the sum of $U(t)$ and $K(t)$ as the input variables are changed?

S3. Consider a simple pendulum of length L that is displaced from the vertical by an angle θ_0 and released. In this problem, you are to determine the angular displacement for any value of the initial angular displacement. Modify Spreadsheet 6.1 or write your own program to integrate the equation of motion of a simple pendulum:

$$\frac{d^2\theta}{dt^2} = -\frac{g}{L}\sin\theta$$

Take the initial conditions to be $\theta = \theta_0$ and $d\theta/dt = 0$ at $t = 0$. Choose values for θ_0 and L and find the angular displacement θ as a function of time. Using the same values of θ_0, compare your results for θ with those obtained from $\theta(t) = \theta_0\cos\omega_0 t$ where $\omega_0 = \sqrt{g/L}$. Repeat for different values of θ_0. Be sure to include large initial angular displacements. Does the period change when the initial angular displacement is changed? How do the periods for large values of θ_0 compare to those for small values of θ_0? (*Note*: Using the modified Euler's method to solve this differential equation, you may find that the amplitude tends to increase with time. The fourth-order Runge-Kutta method would be a better choice to solve the differential equation. However, if you chose dt small enough, the solution using the modified Euler's method will still be good.)

S4. In the real world friction is always present. Modify your spreadsheet for Problem S3 to include the force of air resistance: $F_{air} = -bv$. In this situation, the equation of motion is

$$\frac{d^2\theta}{dt^2} = -\frac{g}{L}\sin\theta - \frac{b}{m}\frac{d\theta}{dt}$$

where m is the mass of the pendulum and b is the drag coefficient. Integrate this equation and calculate the sum of the kinetic and potential energies as a function of time using your calculated values of θ and $d\theta/dt$. Since energy is not constant, when in the pendulum's cycle is energy dissipated the fastest?

S5. The differential equation for the position x of a driven, damped harmonic oscillator of mass m is

$$m\frac{d^2x}{dt^2} + b\frac{dx}{dt} + m\omega_0^2 x = F_0\cos\omega t$$

where ω_0 and b are constants, F_0 is the amplitude of the driving force, and ω is the driving angular frequency. For the case of a mass on a spring, $\omega_0 = \sqrt{k/m}$, where k is the force constant. Write a computer program or spreadsheet to integrate this equation of motion. Your input variables are the damping coefficient b, the mass m, ω_0 (the natural frequency of the system), and F_0. Choose values for the initial velocity and position and calculate the position x as a function of time. Vary the driving angular frequency ω. Pick some values near ω_0. Does resonance occur? (*Note*: Your program may "crash" or blow up near resonance. The amplitude of the motion may have very large values when $\omega = \omega_0$. Also, the Euler method usually tends to overestimate the solution. Other methods such as the 4th-order Runge-Kutta method discussed in the Supplement give a more accurate solution.)

CHAPTER 14

The Law of Gravity

Astronauts F. Story Musgrave and Jeffrey A. Hoffman complete the final of five space walks needed to repair the Hubble Space Telescope in December 1993. *(Courtesy NASA)*

P rior to 1686, a large amount of data had been collected on the motions of the Moon and the planets but a clear understanding of the forces that caused these celestial bodies to move the way they did was not available. In that year, however, Newton provided the key that unlocked the secrets of the heavens. He knew, from his first law, that a net force had to be acting on the Moon because without such a force the Moon would move in a straight-line path rather than in its almost circular orbit. Newton reasoned that this force was the gravitational attraction that the Earth exerts on the Moon. He also concluded that there could be nothing special about the Earth-Moon system or about the Sun and its planets that would cause gravitational forces to act on them alone. In other words, he saw that the same force of attraction that causes the Moon to follow its path also causes an apple to fall to Earth from a tree. He wrote, "I deduced that the forces which keep the planets in their orbs must be reciprocally as the squares of their distances from the centers about which they revolve; and thereby compared the force requisite to keep the Moon in her orb with force of gravity at the surface of the Earth; and found them answer pretty nearly."

In this chapter we study the law of gravity. Emphasis is placed on describing the motion of the planets, because astronomical data provide an important test of the validity of the law of gravity. We show that the laws of planetary motion developed by Johannes Kepler follow from the law of gravity and the concept of the conservation of angular momentum. A general expression for the gravitational potential energy is derived, and the energetics of planetary and satellite motion are treated. The law of gravity is also used to determine the force between a particle and an extended body.

14.1   NEWTON'S LAW OF GRAVITY

It has been said that Newton was struck on the head by a falling apple while napping under a tree (or some variation of this legend). This accident supposedly prompted him to imagine that perhaps all bodies in the Universe are attracted to each other in the same way the apple was attracted to the Earth. Newton then analyzed astronomical data on the motion of the Moon around the Earth. From this analysis, he made the bold statement that the force law governing the motion of planets has the *same* mathematical form as the force law that attracts a falling apple to the Earth.

In 1686 Newton published his work on the law of gravity in his *Mathematical Principles of Natural Philosophy*. **Newton's law of gravity** states that

> every particle in the Universe attracts every other particle with a force that is directly proportional to the product of their masses and inversely proportional to the square of the distance between them.

If the particles have masses m_1 and m_2 and are separated by a distance r, the magnitude of this gravitational force is

The law of gravity

$$F_g = G \frac{m_1 m_2}{r^2} \tag{14.1}$$

where G is a universal constant called the *universal gravitational constant*, which has been measured experimentally. Its value in SI units is

$$G = 6.672 \times 10^{-11} \frac{\text{N} \cdot \text{m}^2}{\text{kg}^2} \tag{14.2}$$

The force law given by Equation 14.1 is often referred to as an **inverse-square law** because the magnitude of the force varies as the inverse square of the separation of the particles. We can express this force in vector form by defining a unit vector $\hat{\mathbf{r}}_{12}$ (Fig. 14.1). Because this unit vector is in the direction of the displacement vector $\mathbf{r}_{12}$ directed from m_1 to m_2, the force exerted on m_2 by m_1 is

$$\mathbf{F}_{21} = -G \frac{m_1 m_2}{r_{12}^2} \hat{\mathbf{r}}_{12} \tag{14.3}$$

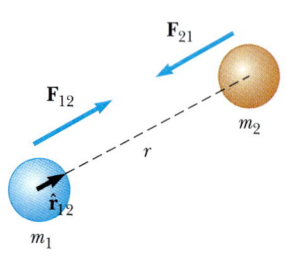

FIGURE 14.1   The gravitational force between two particles is attractive. The unit vector $\hat{\mathbf{r}}_{12}$ is directed from m_1 to m_2. Note that $\mathbf{F}_{12} = -\mathbf{F}_{21}$.

where the minus sign indicates that m_2 is attracted to m_1, and so the force must be directed toward m_1. Likewise, by Newton's third law the force exerted on m_1 by m_2, designated $\mathbf{F}_{12}$, is equal in magnitude to $\mathbf{F}_{21}$ and in the opposite direction. That is, these forces form an action-reaction pair, and $\mathbf{F}_{12} = -\mathbf{F}_{21}$.

There are several features of the inverse-square law that deserve some attention. The gravitational force is a field force that always exists between two particles, regardless of the medium that separates them. The force varies as the inverse square of the distance between the particles and therefore decreases rapidly with increasing separation. Finally, the gravitational force is proportional to the mass of each particle.

Properties of the gravitational force

Another important fact is that *the gravitational force exerted by a finite-size, spherically symmetric mass distribution on a particle outside the sphere is the same as if the entire mass of the sphere were concentrated at its center.* For example, the force exerted by the Earth on a particle of mass m at the Earth's surface has the magnitude

$$F_g = G\frac{M_E m}{R_E^{2}}$$

where M_E is the Earth's mass and R_E is the Earth's radius. This force is directed toward the center of the Earth.

14.2   MEASUREMENT OF THE GRAVITATIONAL CONSTANT

The universal gravitational constant, G, was measured in an important experiment by Henry Cavendish in 1798. The Cavendish apparatus consists of two small spheres each of mass m fixed to the ends of a light horizontal rod suspended by a fine fiber or thin metal wire, as in Figure 14.2. Two large spheres each of mass M are then placed near the smaller spheres. The attractive force between the smaller and larger spheres causes the rod to rotate and twist the wire suspension. If the system is oriented as shown in Figure 14.2, the rod rotates clockwise when viewed from above. The angle through which it rotates is measured by the deflection of a light beam reflected from a mirror attached to the vertical suspension. The deflected spot of light is an effective technique for amplifying the motion. The experiment is carefully repeated with different masses at various separations. In addition to providing a value for G, the results show that the force is attractive, proportional to the product mM, and inversely proportional to the square of the distance r.

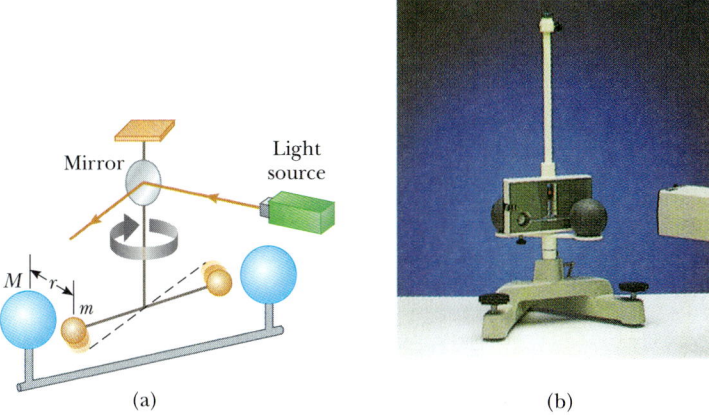

(a)   (b)

FIGURE 14.2   (a) Schematic diagram of the Cavendish apparatus for measuring G. The smaller spheres of mass m are attracted to the large spheres of mass M, and the rod rotates through a small angle. A light beam reflected from a mirror on the rotating apparatus measures the angle of rotation. The dashed line represents the original position of the rod. (b) Photograph of a student Cavendish apparatus.   *(Courtesy of PASCO Scientific)*

EXAMPLE 14.1   Three Interacting Masses

Three uniform spheres of mass 2.00 kg, 4.00 kg, and 6.00 kg are placed at the corners of a right triangle as in Figure 14.3. Calculate the resultant gravitational force on the 4.00-kg mass, assuming the spheres are isolated from the rest of the Universe.

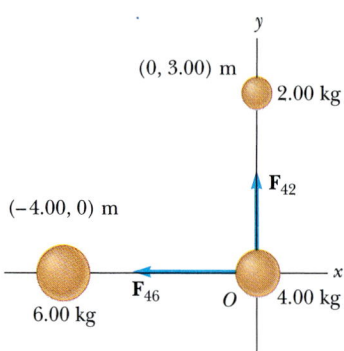

FIGURE 14.3   (Example 14.1) The resultant gravitational force on the 4.00-kg mass is the vector sum $\mathbf{F}_{46} + \mathbf{F}_{42}$.

Reasoning   First we calculate the individual forces on the 4.00-kg mass due to the 2.00-kg and 6.00-kg masses separately, and then we find the vector sum to get the resultant force.

Solution   The force exerted on the 4.00-kg mass by the 2.00-kg mass is directed upward and given by

$$\mathbf{F}_{42} = G\frac{m_4 m_2}{r_{42}^2}\mathbf{j}$$

$$= \left(6.67 \times 10^{-11}\ \frac{\mathrm{N\cdot m^2}}{\mathrm{kg^2}}\right)\frac{(4.00\ \mathrm{kg})(2.00\ \mathrm{kg})}{(3.00\ \mathrm{m})^2}\mathbf{j}$$

$$= 5.93 \times 10^{-11}\mathbf{j}\ \mathrm{N}$$

The force exerted on the 4.00-kg mass by the 6.00-kg mass is directed to the left:

$$\mathbf{F}_{46} = G\frac{m_4 m_6}{r_{46}^2}(-\mathbf{i})$$

$$= \left(-6.67 \times 10^{-11}\ \frac{\mathrm{N\cdot m^2}}{\mathrm{kg^2}}\right)\frac{(4.00\ \mathrm{kg})(6.00\ \mathrm{kg})}{(4.00\ \mathrm{m})^2}\mathbf{i}$$

$$= -10.0 \times 10^{-11}\mathbf{i}\ \mathrm{N}$$

Therefore, the resultant force on the 4.00-kg mass is

$$\mathbf{F}_4 = \mathbf{F}_{42} + \mathbf{F}_{46} = (-10.0\mathbf{i} + 5.93\mathbf{j}) \times 10^{-11}\ \mathrm{N}$$

Exercise   Find the magnitude and direction of the force $\mathbf{F}_4$.

Answer 11.6×10^{-11} N at an angle of 149° with the positive *x* axis.

14.3   WEIGHT AND GRAVITATIONAL FORCE

In Chapter 5 when defining the weight of a body of mass *m* as *mg*, we referred to *g* simply as the magnitude of the free-fall acceleration. Now we are in a position to obtain a more fundamental description of *g*. Since the force on a freely falling body of mass *m* near the surface of the Earth is given by Equation 14.1, we can equate *mg* to this expression to give

$$mg = G\frac{M_E m}{R_E^2}$$

Free-fall acceleration near the Earth's surface

$$g = G\frac{M_E}{R_E^2} \tag{14.4}$$

where M_E is the mass of the Earth and R_E is the Earth's radius. Using the facts that $g = 9.80$ m/s² at the Earth's surface and $R_E \approx 6.38 \times 10^6$ m, we find from Equation 14.4 that $M_E = 5.98 \times 10^{24}$ kg. From this result, the average density of the Earth is calculated to be

$$\rho_E = \frac{M_E}{V_E} = \frac{M_E}{\frac{4}{3}\pi R_E^3} = \frac{5.98 \times 10^{24}\ \mathrm{kg}}{\frac{4}{3}\pi(6.38 \times 10^6\ \mathrm{m})^3} = 5.50 \times 10^3\ \mathrm{kg/m^3}$$

Since this value is about twice the density of most rocks at the Earth's surface, we conclude that the inner core of the Earth has a density much higher than the average value.

Now consider a body of mass *m* a distance *h* above the Earth's surface, or a distance *r* from the Earth's center, where $r = R_E + h$. The magnitude of the gravitational force acting on this mass is

$$F_g = G\frac{M_E m}{r^2} = G\frac{M_E m}{(R_E + h)^2}$$

If the body is in free-fall, then $F_g = mg'$ and we see that g', the free-fall acceleration experienced by an object at the altitude h, is

$$g' = \frac{GM_E}{r^2} = \frac{GM_E}{(R_E + h)^2}$$

(14.5)   Variation of g with altitude

Thus, it follows that g' *decreases* with *increasing altitude*. Since the true weight of a body is mg', we see that as $r \rightarrow \infty$, the true weight approaches zero.

EXAMPLE 14.2   Variation of g with Altitude h

Determine the magnitude of the free-fall acceleration at an altitude of 500 km. By what percentage is the weight of a body reduced at this altitude?

Solution   Using Equation 14.5 with $h = 500$ km, $R_E = 6.38 \times 10^6$ m, and $M_E = 5.98 \times 10^{24}$ kg gives

$$g' = \frac{GM_E}{(R_E + h)^2}$$

$$= \frac{(6.67 \times 10^{-11} \text{ N} \cdot \text{m}^2/\text{kg}^2)(5.98 \times 10^{24} \text{ kg})}{(6.38 \times 10^6 + 0.500 \times 10^6)^2 \text{ m}^2}$$

$$= 8.43 \text{ m/s}^2$$

Since $g'/g = 8.43/9.8 = 0.86$, we conclude that the weight of a body is reduced by about 14% at an altitude of 500 km. Values of g' at other altitudes are listed in Table 14.1.

TABLE 14.1   Free-fall Acceleration, g', at Various Altitudes Above the Earth's Surface

Altitude h (km)	g' (m/s²)
1000	7.33
2000	5.68
3000	4.53
4000	3.70
5000	3.08
6000	2.60
7000	2.23
8000	1.93
9000	1.69
10 000	1.49
50 000	0.13
∞	0

14.4   KEPLER'S LAWS

The movements of the planets, stars, and other celestial bodies have been observed by people for thousands of years. In early history, scientists regarded the Earth as the center of the Universe. This so-called geocentric model was elaborated and formalized by the Greek astronomer Claudius Ptolemy (100–170) in the second century A.D. and was accepted for the next 1400 years. In 1543, the Polish astronomer Nicolaus Copernicus (1473–1543) suggested that the Earth and the other planets revolve in circular orbits about the Sun (the heliocentric model).

The Danish astronomer Tycho Brahe (1546–1601) made accurate astronomical measurements over a period of 20 years and provided a rigorous test of the alternate models of the Solar System. It is interesting to note that these precise observations, made on the planets and 777 stars visible to the naked eye, were carried out with a large sextant and compass without a telescope, which had not yet been invented.

The German astronomer Johannes Kepler, who was Brahe's assistant, acquired Brahe's astronomical data and spent about 16 years trying to deduce a mathemati-

Johannes Kepler (1571–1630), a German astronomer, is best known for developing the laws of planetary motion based on the careful observations of Tycho Brahe. Throughout his life, Kepler was sidetracked by mystic ideas dating back to the ancient Greeks. For example, he believed in the motion of the "music of the spheres" proposed by Pythagoras, in which each planet in its motion sounds out an exact musical note. After spending several years trying to work out a "regular-solid theory" of the planets, he concluded that the Copernican view of circular planetary orbits had to be abandoned for the view that the planetary orbits are ellipses with the Sun always at one of the foci. *(Art Resource)*

cal model for the motion of the planets. After many laborious calculations, he found that Brahe's precise data on the revolution of Mars about the Sun provided the answer. Such data are difficult to sort out because the Earth is also in motion about the Sun. Kepler's analysis first showed that the concept of circular orbits about the Sun had to be abandoned. He eventually discovered that the orbit of Mars could be accurately described by an ellipse with the Sun at one focal point. He then generalized this analysis to include the motion of all planets. The complete analysis is summarized in three statements, known as **Kepler's laws:**

1. All planets move in elliptical orbits with the Sun at one of the focal points.
2. The radius vector drawn from the Sun to a planet sweeps out equal areas in equal time intervals.
3. The square of the orbital period of any planet is proportional to the cube of the semimajor axis of the elliptical orbit.

Half a century later, Newton demonstrated that these laws are the consequence of a simple force that exists between any two masses. Newton's law of gravity, together with his development of the laws of motion, provides the basis for a full mathematical solution to the motion of planets and satellites. More important, Newton's law of gravity correctly describes the gravitational attractive force between *any* two masses.

14.5 THE LAW OF GRAVITY AND THE MOTION OF PLANETS

In formulating his law of gravity, Newton used the following observation, which supports the assumption that the gravitational force is proportional to the inverse square of the separation. Let us compare the acceleration of the Moon in its orbit with the acceleration of an object falling near the Earth's surface, such as the legendary apple (Fig. 14.4). Assume that both accelerations have the same cause, namely, the gravitational attraction of the Earth. From the inverse-square law, Newton found that the acceleration of the Moon toward the Earth (centripetal acceleration) should be proportional to $1/r_M^2$, where r_M is the Earth-Moon separation. Furthermore, the acceleration of the apple toward the Earth should vary as $1/R_E^2$, where R_E is the radius of the Earth. Using the values $r_M = 3.84 \times 10^8$ m and $R_E = 6.37 \times 10^6$ m, we predict the ratio of the Moon's acceleration, a_M, to the

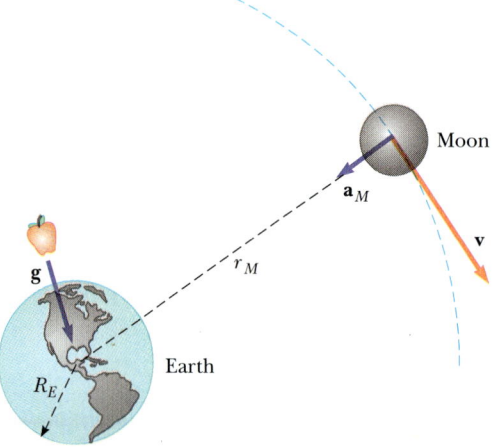

FIGURE 14.4 As it revolves about the Earth, the Moon experiences a centripetal acceleration $\mathbf{a}_M$ directed toward the Earth. An object near the Earth's surface, such as the apple shown here, experiences an acceleration g. (Dimensions are not to scale.)

apple's acceleration, g, to be

$$\frac{a_M}{g} = \frac{(1/r_M)^2}{(1/R_E)^2} = \left(\frac{R_E}{r_M}\right)^2 = \left(\frac{6.37 \times 10^6 \text{ m}}{3.84 \times 10^8 \text{ m}}\right)^2 = 2.75 \times 10^{-4}$$

Therefore

$$a_M = (2.75 \times 10^{-4})(9.80 \text{ m/s}^2) = 2.70 \times 10^{-3} \text{ m/s}^2$$

The centripetal acceleration of the Moon can also be calculated from a knowledge of its mean distance from the Earth and its orbital period, $T = 27.32$ days $= 2.36 \times 10^6$ s. In a time T, the Moon travels a distance $2\pi r_M$, which equals the circumference of its orbit. Therefore, its orbital speed is $2\pi r_M/T$, and its centripetal acceleration is

$$a_M = \frac{v^2}{r_M} = \frac{(2\pi r_M/T)^2}{r_M} = \frac{4\pi^2 r_M}{T^2} = \frac{4\pi^2(3.84 \times 10^8 \text{ m})}{(2.36 \times 10^6 \text{ s})^2} = 2.72 \times 10^{-3} \text{ m/s}^2$$

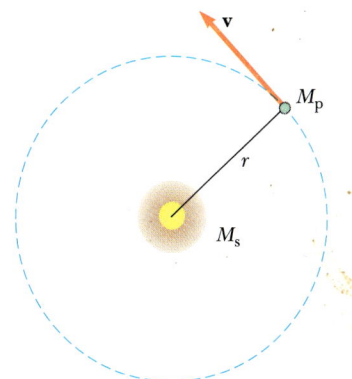

FIGURE 14.5   A planet of mass M_p moving in a circular orbit about the Sun. The orbits of all planets except Mars, Mercury, and Pluto are nearly circular.

The agreement between this value and the value obtained above using g provides strong evidence that the inverse-square law of force is correct.

Although these results must have been very encouraging to Newton, he was deeply troubled by an assumption made in the analysis. In order to evaluate the acceleration of an object at the Earth's surface, the Earth was treated as if its mass were all concentrated at its center. That is, Newton assumed that the Earth acts as a particle as far as its influence on an exterior object is concerned. Several years later in 1686, and based on his pioneering work in the development of the calculus, Newton proved this point.

Kepler's Third Law

It is informative to show that Kepler's third law can be predicted from the inverse-square law for circular orbits.[1] Consider a planet of mass M_p moving about the Sun of mass M_S in a circular orbit, as in Figure 14.5. Since the gravitational force exerted on the planet by the Sun is equal to the central force needed to keep the planet moving in a circle,

$$\frac{GM_SM_p}{r^2} = \frac{M_p v^2}{r}$$

But the orbital speed of the planet is simply $2\pi r/T$, where T is its period; therefore, the above expression becomes

$$\frac{GM_S}{r^2} = \frac{(2\pi r/T)^2}{r}$$

$$T^2 = \left(\frac{4\pi^2}{GM_S}\right)r^3 = K_S r^3 \qquad (14.6)$$

Kepler's third law

where K_S is a constant given by

$$K_S = \frac{4\pi^2}{GM_S} = 2.97 \times 10^{-19} \text{ s}^2/\text{m}^3$$

[1] The orbits of all planets except Mars, Mercury, and Pluto are very close to being circular. For example, the ratio of the semiminor to the semimajor axis for the Earth is $b/a = 0.999\,86$.

(Left) The Hubble Space Telescope's Faint Object Camera has obtained the clearest image ever of Pluto and its moon, Charon. Pluto is the bright object at the center of the frame; Charon is the fainter object in the lower left. Charon's orbit around Pluto is a circle seen nearly edge-on from Earth. *(Right)* Separate views of Jupiter and of Periodic Comet Shoemaker-Levy 9, both taken with the Hubble Space Telescope about two months before Jupiter and the comet collided in July 1994, were put together in a computer. Their relative sizes and distance were altered. The black spot on Jupiter is the shadow of its moon Io.   *(Both photos courtesy NASA)*

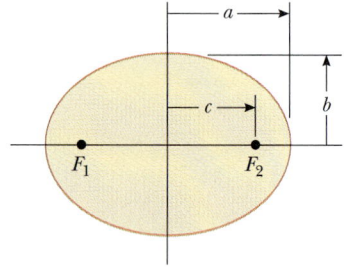

FIGURE 14.6   Plot of an ellipse. The semimajor axis has a length a, and the semiminor axis has a length b. The focal points are located at a distance c from the center, where $a^2 = b^2 + c^2$, and the eccentricity is defined as $e = c/a$.

Equation 14.6 is Kepler's third law. The law is also valid for elliptical orbits if we replace r by the length of the semimajor axis, a (Fig. 14.6). Note that the constant of proportionality, K_S, is independent of the mass of the planet. Therefore, Equation 14.6 is valid for *any* planet. If we were to consider the orbit of a satellite about the Earth, such as the Moon, then the constant would have a different value, with the Sun's mass replaced by the Earth's mass. In this case, the proportionality constant equals $4\pi^2/GM_E$.

A collection of useful planetary data is given in Table 14.2. The last column of this table verifies that T^2/r^3 is a constant whose value is $K_S = 4\pi^2/GM_S = 2.97 \times 10^{-19}$ s^2/m^3.

TABLE 14.2   Useful Planetary Data

Body	Mass (kg)	Mean Radius (m)	Period (s)	Distance from Sun (m)	$\dfrac{T^2}{r^3}\left(\dfrac{s^2}{m^3}\right)$
Mercury	3.18×10^{23}	2.43×10^6	7.60×10^6	5.79×10^{10}	2.97×10^{-19}
Venus	4.88×10^{24}	6.06×10^6	1.94×10^7	1.08×10^{11}	2.99×10^{-19}
Earth	5.98×10^{24}	6.37×10^6	3.156×10^7	1.496×10^{11}	2.97×10^{-19}
Mars	6.42×10^{23}	3.37×10^6	5.94×10^7	2.28×10^{11}	2.98×10^{-19}
Jupiter	1.90×10^{27}	6.99×10^7	3.74×10^8	7.78×10^{11}	2.97×10^{-19}
Saturn	5.68×10^{26}	5.85×10^7	9.35×10^8	1.43×10^{12}	2.99×10^{-19}
Uranus	8.68×10^{25}	2.33×10^7	2.64×10^9	2.87×10^{12}	2.95×10^{-19}
Neptune	1.03×10^{26}	2.21×10^7	5.22×10^9	4.50×10^{12}	2.99×10^{-19}
Pluto	$\approx 1.4 \times 10^{22}$	$\approx 1.5 \times 10^6$	7.82×10^9	5.91×10^{12}	2.96×10^{-19}
Moon	7.36×10^{22}	1.74×10^6	—	—	—
Sun	1.991×10^{30}	6.96×10^8	—	—	—

EXAMPLE 14.3 The Mass of the Sun

Calculate the mass of the Sun using the fact that the period of the Earth is 3.156×10^7 s and its distance from the Sun is 1.496×10^{11} m.

Solution   Using Equation 14.6, we get

$$M_S = \frac{4\pi^2 r^3}{GT^2}$$

$$= \frac{4\pi^2 (1.496 \times 10^{11} \text{ m})^3}{\left(6.67 \times 10^{-11} \dfrac{\text{N} \cdot \text{m}^2}{\text{kg}^2}\right)(3.156 \times 10^7 \text{ s})^2}$$

$$= \boxed{1.99 \times 10^{30} \text{ kg}}$$

Note that the Sun is 333 000 times as massive as the Earth!

Kepler's Second Law and Conservation of Angular Momentum

Consider a planet of mass M_p moving about the Sun in an elliptical orbit (Fig. 14.7). The gravitational force acting on the planet is always along the radius vector, directed toward the Sun. Such a force directed toward or away from a fixed point (that is, one that is a function of r only) is called a **central force**. The torque acting on the planet due to this central force is clearly zero since F is parallel to r. That is,

$$\boldsymbol{\tau} = \mathbf{r} \times \mathbf{F} = \mathbf{r} \times F(r)\hat{\mathbf{r}} = 0$$

But recall from Equation 11.19 that the torque equals the time rate of change of angular momentum, or $\boldsymbol{\tau} = d\mathbf{L}/dt$. Therefore,

> because $\boldsymbol{\tau} = 0$, the angular momentum L of the planet is a constant of the motion:
>
> $$\mathbf{L} = \mathbf{r} \times \mathbf{p} = M_p \mathbf{r} \times \mathbf{v} = \text{constant}$$

Since L is a constant of the motion, we see that the planet's motion at any instant is restricted to the plane formed by r and v.

We can relate this result to the following geometric consideration. The radius vector r in Figure 14.7b sweeps out an area dA in a time dt. This area equals one-half the area $|\mathbf{r} \times d\mathbf{r}|$ of the parallelogram formed by the vectors r and $d\mathbf{r}$. Since the displacement of the planet in a time dt is $d\mathbf{r} = \mathbf{v}\,dt$, we get

$$dA = \tfrac{1}{2}|\mathbf{r} \times d\mathbf{r}| = \tfrac{1}{2}|\mathbf{r} \times \mathbf{v}\,dt| = \frac{L}{2 M_p}\,dt$$

$$\frac{dA}{dt} = \frac{L}{2 M_p} = \text{constant} \qquad (14.7)$$

where L and M_p are both constants of the motion. Thus, we conclude that

> the radius vector from the Sun to a planet sweeps out equal areas in equal times.

It is important to recognize that this result, which is Kepler's second law, is a consequence of the fact that the force of gravity is a central force, which in turn

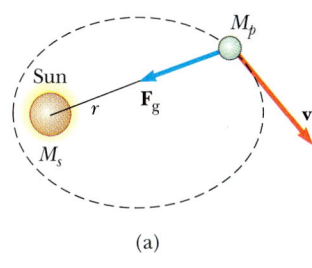

(a)

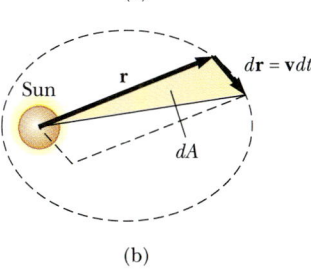

(b)

FIGURE 14.7   (a) The gravitational force acting on a planet is directed toward the Sun, along the radius vector. (b) As a planet orbits the Sun, the area swept out by the radius vector in a time dt is equal to one half the area of the parallelogram formed by the vectors r and $d\mathbf{r} = \mathbf{v}\,dt$.

Kepler's second law

implies that angular momentum remains constant. Therefore, the second law applies to *any* situation that involves a central force, whether inverse-square or not.

The inverse-square nature of the force of gravity is not revealed by Kepler's second law. Although we do not prove it here, Kepler's first law is a direct consequence of the fact that the gravitational force varies as $1/r^2$. That is, under an inverse-square force law, the orbits of the planets can be shown to be ellipses with the Sun at one focus.

EXAMPLE 14.4   Motion in an Elliptical Orbit

A satellite of mass m moves in an elliptical orbit about the Earth (Fig. 14.8). The minimum and maximum distances of the planet from the Earth are called the *perihelion* (indicated by p in Fig. 14.8) and *aphelion* (indicated by a), respectively. If the speed of the satellite at p is v_p, what is its speed at a?

Solution   The angular momentum of the planet relative to the Earth is $M_p\mathbf{r} \times \mathbf{v}$. At the points a and p, $\mathbf{v}$ is perpendicular to $\mathbf{r}$. Therefore, the magnitude of the angular momentum at these positions is $L_a = M_p v_a r_a$ and $L_p = M_p v_p r_p$. Because angular momentum is constant, we see that

$$M_p v_a r_a = M_p v_p r_p$$

$$v_a = \frac{r_p}{r_a} v_p$$

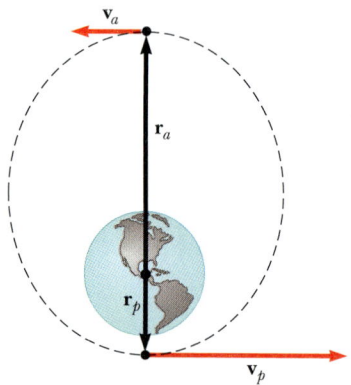

FIGURE 14.8   (Example 14.4) As a satellite moves about the Earth in an elliptical orbit, its angular momentum is constant. Therefore, $M_a v_a r_a = M_p v_p r_p$, where the subscripts a and p represent aphelion and perihelion, respectively.

14.6   THE GRAVITATIONAL FIELD

When Newton first published his theory of gravitation, his contemporaries found it difficult to accept the concept of a field force that could act through a distance. They asked how it was possible for two masses to interact even though they were not in contact with each other. Although Newton himself could not answer this question, his theory was considered a success because it satisfactorily explained the motion of the planets.

An alternative approach in describing the gravitational interaction, therefore, is to introduce the concept of a **gravitational field** that covers every point in space. When a particle of mass m is placed at a point where the field is the vector $\mathbf{g}$, the particle experiences a force $\mathbf{F}_g = m\mathbf{g}$. In other words, the field exerts a force on the particle. Hence, the gravitational field is defined by

Gravitational field

$$\mathbf{g} \equiv \frac{\mathbf{F}_g}{m} \tag{14.8}$$

That is, the gravitational field at a point in space equals the gravitational force experienced by a test mass placed at that point divided by the test mass. As an example, consider an object of mass m near the Earth's surface. The gravitational force on the object is directed toward the center of the Earth and has a magnitude mg. Since the gravitational force on the object has a magnitude $GM_E m/r^2$ (where

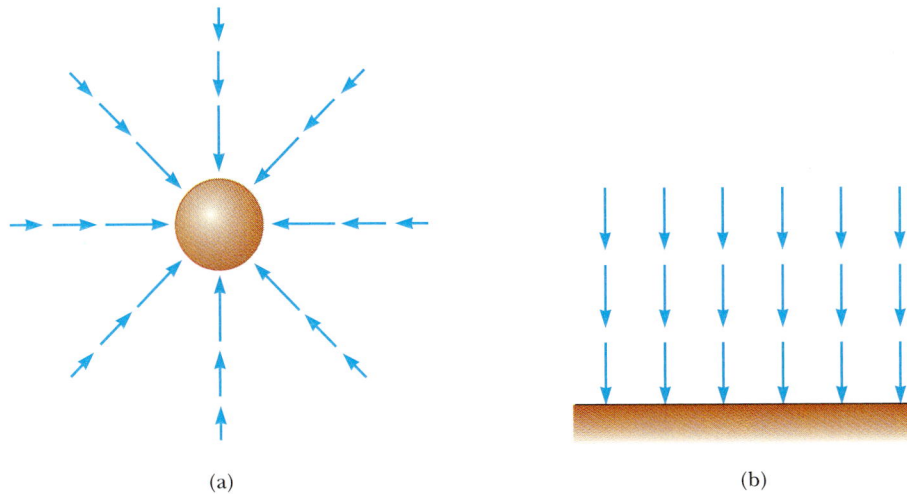

(a)   (b)

FIGURE 14.9 (a) The gravitational field vectors in the vicinity of a uniform spherical mass such as the Earth vary in both direction and magnitude. (b) The gravitational field vectors in a small region near the Earth's surface are uniform; that is, they match in both direction and magnitude.

M_E is the mass of the Earth), the field **g** at a distance r from the center of the Earth is

$$\mathbf{g} = \frac{\mathbf{F}_g}{m} = -\frac{GM_E}{r^2}\,\hat{\mathbf{r}} \tag{14.9}$$

where $\hat{\mathbf{r}}$ is a unit vector pointing radially outward from the Earth, and the minus sign indicates that the field points toward the center of the Earth, as in Figure 14.9a. Note that the field vectors at different points surrounding the Earth vary in both direction and magnitude. In a small region near the Earth's surface, the downward field g is approximately constant and uniform, as indicated in Figure 14.9b. Equation 14.9 is valid at all points *outside* the Earth's surface, assuming that the Earth is spherical. At the Earth's surface, where $r = R_E$, **g** has a magnitude of 9.80 N/kg.

CONCEPTUAL EXAMPLE 14.5

How would you explain the fact that Saturn and Jupiter have periods much greater than one year?

Reasoning Kepler's third law (Eq. 14.6), which applies to all the planets, tells us that the period of a planet is propor-

tional to $r^{3/2}$. Because Saturn and Jupiter are farther than Earth from the Sun, they have longer periods. The Sun's gravitational field (whose magnitude is GM_S/r^2) is much weaker at a distant Jovian planet. Thus, an outer planet experiences much smaller centripetal acceleration than Earth, and a correspondingly longer period.

14.7 GRAVITATIONAL POTENTIAL ENERGY

In Chapter 8 we introduced the concept of gravitational potential energy, that is, the energy associated with the position of a particle. We emphasized the fact that the gravitational potential energy function, $U = mgy$, is valid only when the particle is near the Earth's surface. Since the gravitational force between two particles

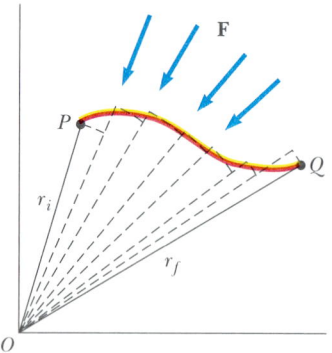

FIGURE 14.10 A particle moves from *P* to *Q* while under the action of a central force **F**, which is in the radial direction. The path is broken into a series of radial and circular segments. Since the work done along the circular segments is zero, the work done is independent of the path.

Work done by a central force

---

varies as $1/r^2$, we expect that the more general potential energy function — the one that is valid without the restriction of having to be near the Earth's surface — depends on the separation between the particles.

Before we calculate this general form for the gravitational potential energy function, we first verify that *the gravitational force is conservative*. In order to do this, we first note that the gravitational force is a central force. By definition, a central force is one that depends only on the polar coordinate *r*, and hence can be represented by $F(r)\hat{r}$, where $\hat{r}$ is a unit vector directed from the origin to the particle under consideration as in Figure 14.10. Such a force is directed parallel to the radius vector.

Consider a central force acting on a particle moving along the general path *P* to *Q* in Figure 14.10. The central force acts toward the point *O*. The path from *P* to *Q* can be approximated by a series of radial and circular segments. By definition, a central force is always directed along one of the radial segments; therefore, the work done by **F** along any *radial segment* is

$$dW = \mathbf{F} \cdot d\mathbf{r} = F(r)\ dr$$

You should recall that, by definition, the work done by a force that is perpendicular to the displacement is zero. Hence, the work done along any circular segment is zero because **F** is perpendicular to the displacement along these segments. Therefore, the total work done by **F** is the sum of the contributions along the radial segments:

$$W = \int_{r_i}^{r_f} F(r)\ dr$$

where the subscripts *i* and *f* refer to the initial and final positions. This result applies to *any* path from *P* to *Q*. Therefore, we conclude that *any central force is conservative*. We are now assured that a potential energy function can be obtained once the form of the central force is specified. You should recall from Chapter 8 that the change in the gravitational potential energy associated with a given displacement is defined as the negative of the work done by the gravitational force during that displacement, or

$$\Delta U = U_f - U_i = -\int_{r_i}^{r_f} F(r)\ dr \tag{14.10}$$

We can use this result to evaluate the gravitational potential energy function. Consider a particle of mass *m* moving between two points *P* and *Q* above the Earth's surface (Fig. 14.11). The particle is subject to the gravitational force given by Equation 14.1. We can express this force as

$$\mathbf{F}_g = -\frac{GM_E m}{r^2}\ \hat{r}$$

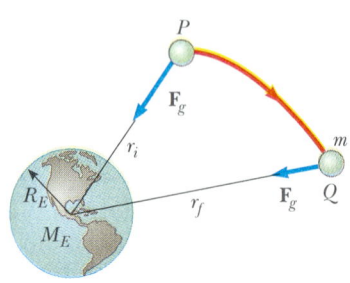

FIGURE 14.11 As a particle of mass *m* moves from *P* to *Q* above the Earth's surface, the gravitational potential energy changes according to Equation 14.10.

Change in gravitational potential energy

---

where $\hat{r}$ is a unit vector directed from the Earth to the particle and the negative sign indicates that the force is attractive. Substituting this expression for $\mathbf{F}_g$ into Equation 14.10, we can compute the change in the gravitational potential energy function:

$$U_f - U_i = GM_E m \int_{r_i}^{r_f} \frac{dr}{r^2} = GM_E m \left[-\frac{1}{r} \right]_{r_i}^{r_f}$$

$$U_f - U_i = - GM_E m \left(\frac{1}{r_f} - \frac{1}{r_i} \right) \tag{14.11}$$

As always, the choice of a reference point for the potential energy is completely arbitrary. It is customary to choose the reference point where the force is zero. Taking $U_i = 0$ at $r_i = \infty$, we obtain the important result

$$U(r) = -\frac{GM_E m}{r} \qquad (14.12)$$

This expression applies to the Earth-particle system where the two masses are separated by a distance r, provided that $r \geq R_E$. The result is not valid for particles moving inside the Earth, where $r < R_E$. (The situation where $r < R_E$ is treated in Section 14.10.) Because of our choice of U_i, the function $U(r)$ is always negative (Fig. 14.12).

Although Equation 14.12 was derived for the particle-Earth system, it can be applied to any two particles. That is, the gravitational potential energy associated with any pair of particles of masses m_1 and m_2 separated by a distance r is

$$U = -\frac{Gm_1 m_2}{r} \qquad (14.13)$$

This expression shows that the gravitational potential energy for any pair of particles varies as $1/r$, whereas the force between them varies as $1/r^2$. Furthermore, the potential energy is negative because the force is attractive and we have taken the potential energy as zero when the particle separation is infinity. Because the force between the particles is attractive, we know that an external agent must do positive work to increase the separation between them. The work done by the external agent produces an increase in the potential energy as the two particles are separated. That is, U becomes less negative as r increases.[2]

When two particles are separated by a distance r, an external agent has to supply an energy at least equal to $+Gm_1 m_2/r$ in order to separate the particles by an infinite distance. It is convenient to think of the absolute value of the potential energy as the *binding energy* of the system. If the external agent supplies an energy greater than the binding energy, $Gm_1 m_2/r$, the additional energy of the system will be in the form of kinetic energy when the particles are at an infinite separation.

We can extend this concept to three or more particles. In this case, the total potential energy of the system is the sum over all pairs of particles.[3] Each pair contributes a term of the form given by Equation 14.13. For example, if the system contains three particles as in Figure 14.13, we find that

$$U_{\text{total}} = U_{12} + U_{13} + U_{23} = -G\left(\frac{m_1 m_2}{r_{12}} + \frac{m_1 m_3}{r_{13}} + \frac{m_2 m_3}{r_{23}}\right) \qquad (14.14)$$

The absolute value of U_{total} represents the work needed to separate the particles by an infinite distance. If the system consists of four particles, there are six terms in the sum, corresponding to the six distinct pairs of interaction forces.

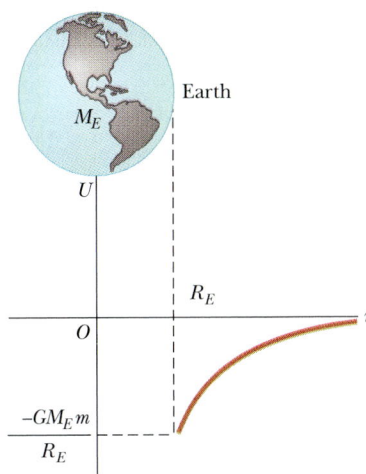

FIGURE 14.12   Graph of the gravitational potential energy, U, versus r for a particle above the Earth's surface. The potential energy goes to zero as r approaches ∞.

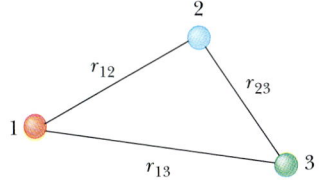

FIGURE 14.13   Diagram of three interacting particles.

---

[2] Note that part of the work done can also produce a change in kinetic energy of the system. That is, if the work done in separating the particles exceeds the increase in potential energy, the excess energy is accounted for by the increase in kinetic energy of the system.

[3] The fact that potential energy terms can be added for all pairs of particles stems from the experimental fact that gravitational forces obey the superposition principle. That is, if $\Sigma F = F_{12} + F_{13} + F_{23} + \cdots$ then there exists a potential energy term for each interaction F_{ij}.

EXAMPLE 14.6   The Change in Potential Energy

A particle of mass m is displaced through a small vertical distance Δy near the Earth's surface. Let us show that the general expression for the change in gravitational potential energy given by Equation 14.11 reduces to the familiar relationship $\Delta U = mg\,\Delta y$.

Solution   We can express Equation 14.11 in the form

$$\Delta U = -GM_E m\left(\frac{1}{r_f} - \frac{1}{r_i}\right) = GM_E m\left(\frac{r_f - r_i}{r_i r_f}\right)$$

If both the initial and the final position of the particle are close to the Earth's surface, then $r_f - r_i = \Delta y$ and $r_i r_f \approx R_E^2$. (Recall that r is measured from the center of the Earth.) Therefore, the change in potential energy becomes

$$\Delta U \approx \frac{GM_E m}{R_E^2}\,\Delta y = mg\,\Delta y$$

where we have used the fact that $g = GM_E/R_E^2$. Keep in mind that the reference point is arbitrary because it is the change in potential energy that is meaningful.

14.8   ENERGY CONSIDERATIONS IN PLANETARY AND SATELLITE MOTION

Consider a body of mass m moving with a speed v in the vicinity of a massive body of mass M, where $M \gg m$. The system might be a planet moving around the Sun or a satellite in orbit around the Earth. If we assume that M is at rest in an inertial reference frame, then the total energy E of the two-body system when the bodies are separated by a distance r is the sum of the kinetic energy of m and the potential energy of the system, given by Equation 14.13[4]:

$$E = K + U$$

$$E = \tfrac{1}{2}mv^2 - \frac{GMm}{r} \tag{14.15}$$

Furthermore, the total energy is constant if we assume the system is isolated. Therefore as m moves from P to Q in Figure 14.11, the total energy remains constant and Equation 14.15 gives

$$E = \tfrac{1}{2}mv_i^2 - \frac{GMm}{r_i} = \tfrac{1}{2}mv_f^2 - \frac{GMm}{r_f} \tag{14.16}$$

This result shows that E may be positive, negative, or zero, depending on the speed of m. However, for a bound system, such as the Earth and Sun, E is necessarily *less than zero*. We can easily establish that $E < 0$ for the system consisting of a mass m moving in a circular orbit about a body of mass M, where $M \gg m$ (Fig. 14.14). Newton's second law applied to m gives

$$\frac{GMm}{r^2} = \frac{mv^2}{r}$$

Multiplying both sides by r and dividing by 2 gives

$$\tfrac{1}{2}mv^2 = \frac{GMm}{2r} \tag{14.17}$$

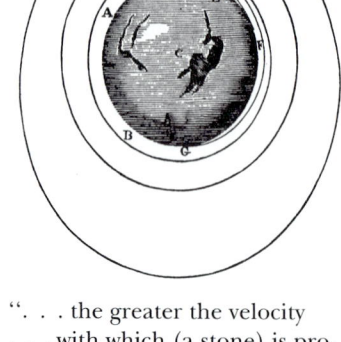

". . . the greater the velocity . . . with which (a stone) is projected, the farther it goes before it falls to the Earth. We may therefore suppose the velocity to be so increased, that it would describe an arc of 1, 2, 5, 10, 100, 1000 miles before it arrived at the Earth, till at last, exceeding the limits of the Earth, it should pass into space without touching."— Newton, *System of the World.*

[4] You might recognize that we have ignored the acceleration and kinetic energy of the larger mass. To see that this simplification is reasonable, consider an object of mass m falling toward the Earth. Since the center of mass of the object-Earth system is stationary, it follows that $mv = M_E v_E$. Thus, the Earth acquires a kinetic energy equal to

$$\tfrac{1}{2}M_E v_E^2 = \tfrac{1}{2}\frac{m^2}{M_E}v^2 = \frac{m}{M_E}K$$

where K is the kinetic energy of the object. Since $M_E \gg m$, the kinetic energy of the Earth is negligible.

Substituting this into Equation 14.15, we obtain

$$E = \frac{GMm}{2r} - \frac{GMm}{r}$$

$$E = -\frac{GMm}{2r} \qquad (14.18)$$

This clearly shows that *the total energy must be negative in the case of circular orbits*. Note that *the kinetic energy is positive and equal to one half the magnitude of the potential energy*. The absolute value of E is also equal to the binding energy of the system.

The total mechanical energy is also negative in the case of elliptical orbits. The expression for E for elliptical orbits is the same as Equation 14.18 with r replaced by the semimajor axis length, a.

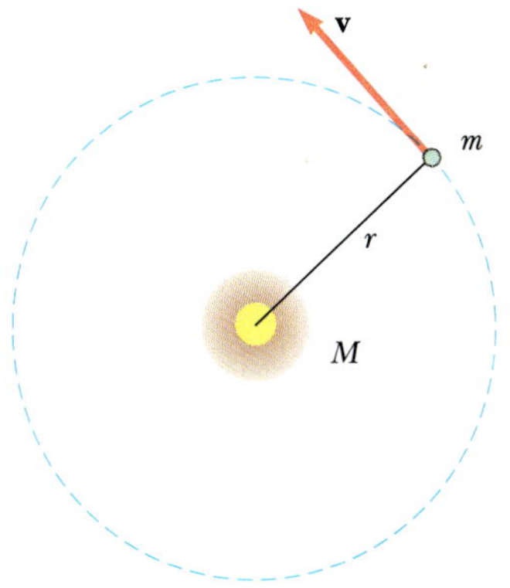

FIGURE 14.14   A body of mass m moving in a circular orbit about a much larger body of mass M.

> Both the total energy and the total angular momentum of a planet-Sun system are constants of the motion.

EXAMPLE 14.7   Changing the Orbit of a Satellite

Calculate the work required to move an Earth satellite of mass m from a circular orbit of radius $2R_E$ to one of radius $3R_E$.

Solution   Applying Equation 14.18, we get for the total initial and final energies

$$E_i = -\frac{GM_E m}{4R_E} \qquad E_f = -\frac{GM_E m}{6R_E}$$

Therefore, the work required to increase the energy of the system is

$$W = E_f - E_i = -\frac{GM_E m}{6R_E} - \left(-\frac{GM_E m}{4R_E}\right) = \boxed{\frac{GM_E m}{12R_E}}$$

For example, if we take $m = 10^3$ kg, we find that the work required is $W = 5.2 \times 10^9$ J, which is the energy equivalent of 39 gal of gasoline.

If we wish to determine how the energy is distributed after work is done on the system, we find from Equation 14.17 that the change in kinetic energy is $\Delta K = -GM_E m/12R_E$ (it decreases), while the corresponding change in potential energy is $\Delta U = GM_E m/6R_E$ (it increases). Thus, the work done on the system is $W = \Delta K + \Delta U = GM_E m/12R_E$, as we calculated above. In other words, part of the work done goes into increasing the potential energy and part goes into decreasing the kinetic energy.

Escape Speed

Suppose an object of mass m is projected vertically upward from the Earth's surface with an initial speed v_i, as in Figure 14.15. We can use energy considerations to find the minimum value of the initial speed such that the object will escape the Earth's gravitational field. Equation 14.16 gives the total energy of the object at any point when its speed and distance from the center of the Earth are known. At the surface of the Earth, $v_i = v$ and $r_i = R_E$. When the object reaches its maximum altitude, $v_f = 0$ and $r_f = r_{max}$. Because the total energy of the system is constant, substitution of these conditions into Equation 14.16 gives

$$\tfrac{1}{2}mv_i^2 - \frac{GM_E m}{R_E} = -\frac{GM_E m}{r_{max}}$$

Solving for v_i^2 gives

$$v_i^2 = 2GM_E\left(\frac{1}{R_E} - \frac{1}{r_{max}}\right) \qquad (14.19)$$

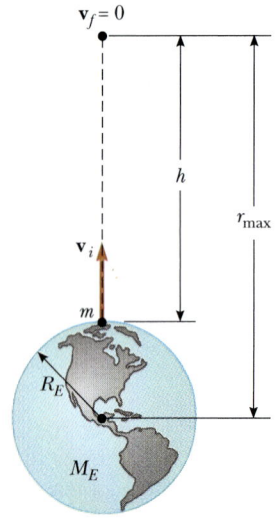

FIGURE 14.15 An object of mass m projected upward from the Earth's surface with an initial speed v_i reaches a maximum altitude h.

Therefore, if the initial speed is known, this expression can be used to calculate the maximum altitude h, because we know that $h = r_{max} - R_E$.

We are now in a position to calculate the minimum speed the object must have at the Earth's surface in order to escape from the influence of the Earth's gravitational field. Traveling at this minimum speed, the object can *just* reach infinity with a final speed of zero. Setting $r_{max} = \infty$ in Equation 14.19 and taking $v_i = v_{esc}$ (the escape speed), we get

$$v_{esc} = \sqrt{\frac{2\,GM_E}{R_E}} \qquad (14.20)$$

Note that this expression for v_{esc} is independent of the mass of the object. In other words, a spacecraft has the same escape speed as a molecule. Furthermore, the result is independent of the direction of the velocity, provided the trajectory does not intersect the Earth.

If the object is given an initial speed equal to v_{esc}, its total energy is equal to zero. This can be seen by noting that when $r = \infty$, the object's kinetic energy and its potential energy are both zero. If v_i is greater than v_{esc}, the total energy is greater than zero and the object has some residual kinetic energy at $r = \infty$.

EXAMPLE 14.8   Escape Speed of a Rocket

Calculate the escape speed from the Earth for a 5000-kg spacecraft, and determine the kinetic energy it must have at the Earth's surface in order to escape the Earth's gravitational field.

Solution   Using Equation 14.20 with $M_E = 5.98 \times 10^{24}$ kg and $R_E = 6.37 \times 10^6$ m gives

$$v_{esc} = \sqrt{\frac{2\,GM_E}{R_E}}$$

$$= \sqrt{\frac{2(6.67 \times 10^{-11}\ \text{N} \cdot \text{m}^2/\text{kg}^2)(5.98 \times 10^{24}\ \text{kg})}{6.37 \times 10^6\ \text{m}}}$$

$$= \boxed{1.12 \times 10^4\ \text{m/s}}$$

This corresponds to about 25 000 mi/h.

The kinetic energy of the spacecraft is

$$K = \tfrac{1}{2}mv_{esc}^2 = \tfrac{1}{2}(5.00 \times 10^3\ \text{kg})(1.12 \times 10^4\ \text{m/s})^2$$

$$= \boxed{3.14 \times 10^{11}\ \text{J}}$$

Finally, you should note that Equations 14.19 and 14.20 can be applied to objects projected from any planet. That is, in general, the escape speed from the surface of any planet of mass M and radius R is

Escape speed

$$v_{esc} = \sqrt{\frac{2\,GM}{R}}$$

A list of escape speeds for the planets, the Moon, and the Sun is given in Table 14.3. Note that the values vary from 1.1 km/s for Pluto to about 618 km/s for the Sun. These results, together with some ideas from the kinetic theory of gases (Chapter 21), explain why some planets have atmospheres and others do not. As we shall see later, a gas molecule has an average kinetic energy that depends on its temperature. Hence, lighter molecules, such as hydrogen and helium, have a higher average speed than the heavier species at the same temperature. When the speed of the lighter molecules is not much less than the escape speed, a significant fraction of them have a chance to escape from the planet.

This mechanism also explains why the Earth does not retain hydrogen and helium molecules in its atmosphere while much heavier molecules, such as oxygen and nitrogen, do not escape. On the other hand, the very large escape speed for Jupiter enables that planet to retain hydrogen, the primary constituent of its atmosphere. Similarly, the very large mass of the Sun allows it to retain hydrogen and helium, as well as heavier gases. In contrast, Mercury, which is small and hot, has no atmosphere.

*14.9   THE GRAVITATIONAL FORCE BETWEEN AN EXTENDED OBJECT AND A PARTICLE

We have emphasized that the law of universal gravitation given by Equation 14.3 is valid only if the interacting objects are considered as particles. In view of this, how can we calculate the force between a particle and an object having finite dimensions? This is accomplished by treating the extended object as a collection of particles and making use of integral calculus. We first evaluate the potential energy function, from which the force can be calculated.

The potential energy associated with a system consisting of a point mass m and an extended body of mass M is obtained by dividing the body into segments of mass ΔM_i (Fig. 14.16). The potential energy associated with this element and with the particle of mass m is $-Gm\,\Delta M_i/r_i$, where r_i is the distance from the particle to the element ΔM_i. The total potential energy of the system is obtained by taking the sum over all segments as $\Delta M_i \rightarrow 0$. In this limit, we can express U in integral form as

$$U = -Gm \int \frac{dM}{r} \tag{14.21}$$

Once U has been evaluated, the force exerted on m can be obtained by taking the negative derivative of this scalar function (see Section 8.6). If the extended body has spherical symmetry, the function U depends only on r and the force is given by $-dU/dr$. We treat this situation in Section 14.10. In principle, one can evaluate U for any geometry; however, the integration can be cumbersome.

An alternative approach to evaluating the gravitational force between a particle and an extended body is to perform a vector sum over all segments of the body. Using the procedure outlined in evaluating U and the law of gravity (Eq. 14.3), we obtain for the total force on the particle

$$\mathbf{F}_g = -Gm \int \frac{dM}{r^2} \hat{\mathbf{r}} \tag{14.22}$$

where $\hat{\mathbf{r}}$ is a unit vector directed from the element dM toward the particle (see Fig. 14.16). This procedure is not always recommended, because working with a vector function is more difficult than working with the scalar potential energy function. However, if the geometry is simple, as in the following example, the evaluation of $\mathbf{F}$ can be straightforward.

TABLE 14.3	Escape Speeds from the Surfaces of the Planets, the Moon, and the Sun
Planet	v_{esc} (km/s)
Mercury	4.3
Venus	10.3
Earth	11.2
Mars	5.0
Jupiter	60
Saturn	36
Uranus	22
Neptune	24
Pluto	1.1
Moon	2.3
Sun	618

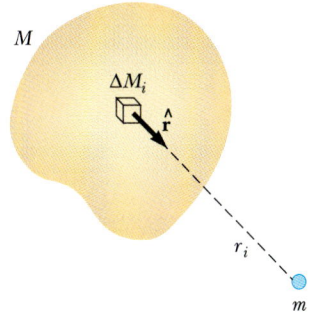

FIGURE 14.16   A particle of mass m interacting with an extended object of mass M. The total gravitational force exerted on the particle by the object can be obtained by taking a vector sum over all forces due to each segment of the object.

Force on a particle due to a spherical shell

EXAMPLE 14.9   Gravitational Force Between a Mass and a Bar

A homogeneous bar of length ℓ and mass M is at a distance h from a point mass m (Fig. 14.17). Calculate the total gravitational force exerted on m by the bar.

Solution   The segment of the bar that has a length dx has a mass dM. Since the mass per unit length is a constant, it then follows that the ratio of masses, dM/M, is equal to the ratio of

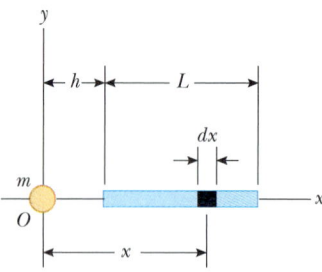

FIGURE 14.17   (Example 14.9) The gravitational force exerted on a particle at the origin by the bar is directed to the right. Note that the bar is *not* equivalent to a particle of mass M located at the center of mass of the bar.

lengths, dx/ℓ, and so $dM = (M/\ell)\,dx$. The variable r in Equation 14.22 is x in our case, and the force on m is to the

right; therefore, we get

$$\mathbf{F}_g = Gm \int_h^{\ell+h} \frac{M}{\ell} \frac{dx}{x^2} \mathbf{i}$$

$$\mathbf{F}_g = \frac{GmM}{\ell} \left[-\frac{1}{x} \right]_h^{\ell+h} \mathbf{i} = \frac{GmM}{h(\ell+h)} \mathbf{i}$$

We see that the force on m is in the positive x direction, as expected, since the gravitational force is attractive.

Note that in the limit $\ell \to 0$, the force varies as $1/h^2$, which is what is expected for the force between two point masses. Furthermore, if $h \gg \ell$, the force also varies as $1/h^2$. This can be seen by noting that the denominator of the expression for $\mathbf{F}_g$ can be expressed in the form $h^2 \left(1 + \dfrac{\ell}{h} \right)$, which is approximately equal to h^2. Thus, when bodies are separated by distances that are large compared with their characteristic dimensions, they behave like particles.

*14.10   GRAVITATIONAL FORCE BETWEEN A PARTICLE AND A SPHERICAL MASS

In this section we describe the gravitational force between a particle and a spherically symmetric mass distribution. We have already stated that a large sphere attracts a particle outside it as if the total mass of the sphere were concentrated at its center. Let us describe the nature of the force on a particle when the extended body is either a spherical shell or a solid sphere, and then apply these facts to some interesting systems.

Spherical Shell

Case 1.   If a particle of mass m is located outside a spherical shell of mass M (say, point P in Fig. 14.18), the shell attracts the particle as though the mass of the shell were concentrated at its center.

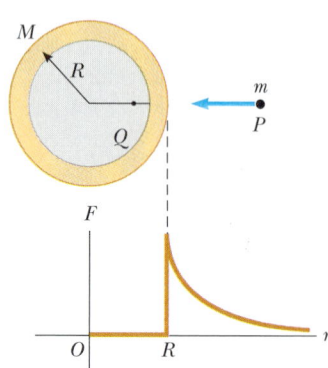

FIGURE 14.18   The gravitational force on a particle when it is outside the spherical shell is GMm/r^2 and acts toward the center. The force on the particle is zero everywhere inside the shell.

Case 2.   If the particle is located inside the shell (point Q in Fig. 14.18), the force on it is zero. We can express these two important results in the following way:

$$\mathbf{F}_g = -\frac{GMm}{r^2} \hat{\mathbf{r}} \qquad \text{for } r \geq R \tag{14.23a}$$

$$\mathbf{F}_g = 0 \qquad\qquad \text{for } r < R \tag{14.23b}$$

The force as a function of the distance r is plotted in Figure 14.18. Note that the shell does not act as a gravitational shield. Even when inside the shell, the particle may experience forces due to other masses outside the shell.

Solid Sphere

Case 1.   If a particle of mass m is located outside a homogeneous solid sphere of mass M (at point P in Fig. 14.19), the sphere attracts the particle as though the mass of the sphere were concentrated at its center. That is, Equation 14.23a applies in this situation. This follows from Case 1 above, since a solid sphere can be considered a collection of concentric spherical shells.

Case 2. If a particle of mass m is located inside a homogeneous solid sphere of mass M (at point Q in Fig. 14.19), the force on m is due *only* to the mass M' contained within the sphere of radius $r < R$, represented by the dashed circle in Figure 14.19. In other words,

$$\mathbf{F}_g = -\frac{GmM}{r^2}\,\hat{\mathbf{r}} \qquad \text{for } r \geq R \qquad (14.24a)$$

$$\mathbf{F}_g = -\frac{GmM'}{r^2}\,\hat{\mathbf{r}} \qquad \text{for } r < R \qquad (14.24b)$$

Because the sphere is assumed to have a uniform density, it follows that the ratio of masses M'/M is equal to the ratio of volumes V/V', where V is the total volume of the sphere and V' is the volume within the dotted surface. That is,

$$\frac{M'}{M} = \frac{V'}{V} = \frac{\frac{4}{3}\pi r^3}{\frac{4}{3}\pi R^3} = \frac{r^3}{R^3}$$

Solving this equation for M' and substituting the value obtained into Equation 14.24b, we get

$$\mathbf{F}_g = -\frac{GmM}{R^3}\,r\hat{\mathbf{r}} \qquad \text{for } r < R \qquad (14.25)$$

That is, the force goes to zero at the center of the sphere, as we would intuitively expect. The force as a function of r is plotted in Figure 14.19.

Case 3. If a particle is located inside a solid sphere having a density ρ that is spherically symmetric but not uniform, then M' in Equation 14.24b is given by an integral of the form $M' = \int \rho \, dV$, where the integration is taken over the volume contained within the dashed circle in Figure 14.19. This integral can be evaluated if the radial variation of ρ is given. The integral is easily evaluated if the mass distribution has spherical symmetry, that is, if ρ is a function of r only. In this case, we take the volume element dV as the volume of a spherical shell of radius r and thickness dr, so that $dV = 4\pi r^2 \, dr$. For example, if $\rho(r) = Ar$, where A is a constant, it is left as a problem (Problem 63) to show that $M' = \pi A r^4$. Hence we see from Equation 14.24b that F is proportional to r^2 in this case and is zero at the center.

Force on a particle due to a solid sphere

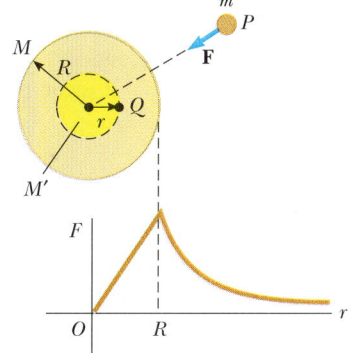

FIGURE 14.19 The gravitational force on a particle when it is outside a uniform solid sphere is GMm/r^2 and is directed toward the center. The force on the particle when it is inside such a sphere is proportional to r and goes to zero at the center.

CONCEPTUAL EXAMPLE 14.10

A particle is projected through a small hole into the interior of a large spherical shell. Describe the subsequent motion of the particle in the interior of the shell.

Reasoning The gravitational force (and field) is zero inside the spherical shell (Eq. 14.23b). Because the force on

the particle is zero once inside the shell, it moves with constant velocity in the direction of its original motion until it hits the inside wall of the shell. Its path thereafter depends on the nature of the collision of the object with the wall and the direction in which it was projected as it passed through the hole.

EXAMPLE 14.11   A Free Ride, Thanks to Gravity

An object moves in a smooth, straight tunnel dug between two points on the Earth's surface (Fig. 14.20). Show that the object moves with simple harmonic motion and find the period of its motion. Assume that the Earth's density is uniform throughout its volume.

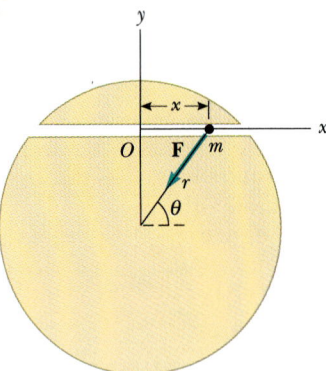

FIGURE 14.20   (Example 14.11) A particle moves along a tunnel dug through the Earth. The component of the gravitational force $\mathbf{F}_g$ along the x axis is the driving force for the motion. Note that this component always acts toward the origin O.

Solution   When the object is in the tunnel, the gravitational force exerted on the object acts toward the Earth's center and is given by Equation 14.25:

$$F_g = -\frac{GmM_E}{R_E{}^3}\, r$$

The y component of this force is balanced by the normal force exerted by the tunnel wall, and the x component is

$$F_x = -\frac{GmM_E}{R_E{}^3}\, r \cos\theta$$

Since the x coordinate of the object is $x = r\cos\theta$, we can write

$$F_x = -\frac{GmM_E}{R_E{}^3}\, x$$

Applying Newton's second law to the motion along x gives

$$F_x = -\frac{GmM_E}{R_E{}^3}\, x = ma$$

$$a = -\frac{GM_E}{R_E{}^3}\, x = -\omega^2 x$$

But this is the equation of simple harmonic motion with angular speed ω (Chapter 13), where

$$\omega = \sqrt{\frac{GM_E}{R_E{}^3}}$$

The period is calculated using the data in Table 14.2 and the above result:

$$T = \frac{2\pi}{\omega} = 2\pi\sqrt{\frac{R_E{}^3}{GM_E}}$$

$$= 2\pi\sqrt{\frac{(6.37\times10^6)^3}{(6.67\times10^{-11})(5.98\times10^{24})}}$$

$$= 5.06\times10^3 \text{ s} = \boxed{84.3 \text{ min}}$$

This period is the same as that of a satellite in a circular orbit just above the Earth's surface. Note that the result is independent of the length of the tunnel.

It has been proposed to operate a mass-transit system between any two cities using this principle. A one-way trip would take about 42 min. A more precise calculation of the motion must account for the fact that the Earth's density is not uniform as we have assumed. More important, there are many practical problems to consider. For instance, it would be impossible to achieve a frictionless tunnel, and so some auxiliary power source would be acquired. Can you think of other problems?

SUMMARY

Newton's law of gravity states that the gravitational force of attraction between any two particles of masses m_1 and m_2 separated by a distance r has the magnitude

$$F_g = G\frac{m_1 m_2}{r^2} \tag{14.1}$$

where G is the universal gravitational constant, which has the value $6.672 \times 10^{-11} \text{ N}\cdot\text{m}^2/\text{kg}^2$.

An object at a distance h above the Earth's surface experiences a gravitational force of magnitude mg', where g' is the **free-fall acceleration** at that elevation:

$$g' = \frac{GM_E}{r^2} = \frac{GM_E}{(R_E + h)^2} \tag{14.5}$$

In this expression, M_E is the mass of the Earth and R_E is the radius of the Earth. Thus, the weight of an object decreases as it moves away from the Earth's surface.

Kepler's laws of planetary motion state that

1. All planets move in elliptical orbits with the Sun at one of the focal points.
2. The radius vector drawn from the Sun to a planet sweeps out equal areas in equal time intervals.
3. The square of the orbital period of any planet is proportional to the cube of the semimajor axis for the elliptical orbit.

Kepler's second law is a consequence of the fact that the force of gravity is a *central force,* that is, one that is directed toward a fixed point. This implies that the angular momentum of the planet-Sun system is a constant of the motion.

Kepler's third law is consistent with the inverse-square nature of the law of universal gravitation. Newton's second law, together with the force law given by Equation 14.1, verifies that the period T and radius r of the orbit of a planet about the Sun are related by

$$T^2 = \left(\frac{4\pi^2}{GM_S}\right)r^3 \tag{14.6}$$

where M_S is the mass of the Sun.

Most planets have nearly circular orbits about the Sun. For elliptical orbits, Equation 14.6 is valid if r is replaced by the semimajor axis, a.

The gravitational force is conservative, and therefore a potential energy function can be defined. The **gravitational potential energy** associated with two particles separated by a distance r is

$$U = -\frac{Gm_1 m_2}{r} \tag{14.13}$$

where U is taken to be zero at $r = \infty$. The total potential energy for a system of particles is the sum of energies for all pairs of particles, with each pair represented by a term of the form given by Equation 14.13.

If an isolated system consists of a particle of mass m moving with a speed v in the vicinity of a massive body of mass M, the *total energy E* of the system is the sum of the kinetic and potential energies:

$$E = \tfrac{1}{2}mv^2 - \frac{GMm}{r} \tag{14.15}$$

The total energy is a constant of the motion.

If m moves in a circular orbit of radius r about M, where $M \gg m$, the **total energy of the system** is

$$E = -\frac{GMm}{2r} \tag{14.18}$$

The total energy is negative for any bound system, that is, one in which the orbit is closed, such as an elliptical orbit.

QUESTIONS

1. Estimate the gravitational force between you and a person 2 m away from you.
2. Use Kepler's second law to convince yourself that the Earth must move faster in its orbit during December, when it is closest to the Sun, than during June, when it is farthest from the Sun.
3. If a system consists of five particles, how many terms appear in the expression for the total potential energy?
4. Is it possible to calculate the potential energy function associated with a particle and an extended body without knowing the geometry or mass distribution of the extended body?
5. Does the escape speed of a rocket depend on its mass? Explain.
6. Compare the energies required to reach the Moon for a 10^5-kg spacecraft and a 10^3-kg satellite.
7. Explain why it takes more fuel for a spacecraft to travel from the Earth to the Moon than for the return trip. Estimate the difference.
8. Is the potential energy associated with the Earth-Moon system greater than, less than, or equal to the kinetic energy of the Moon relative to the Earth?
9. Explain why there is no work done on a planet as it moves in a circular orbit around the Sun, even though a gravitational force is acting on the planet. What is the net work done on a planet during each revolution as it moves around the Sun in an elliptical orbit?
10. Explain why the force exerted on a particle by a uniform sphere must be directed toward the center of the sphere.

Would this be the case if the mass distribution of the sphere were not spherically symmetric?
11. Neglecting the density variation of the Earth, what would be the period of a particle moving in a smooth hole dug through the Earth's center?
12. At what position in its elliptical orbit is the speed of a planet a maximum? At what position is the speed a minimum?
13. If you are given the mass and radius of planet X, how would you calculate the free-fall acceleration on the surface of this planet?
14. If a hole could be dug to the center of the Earth, do you think that the force on a mass m would still obey Equation 14.1 there? What do you think the force on m would be at the center of the Earth?
15. In his 1798 experiment, Cavendish was said to have "weighed the Earth." Explain this statement.
16. The *Voyager* spacecraft was accelerated toward escape speed from the Sun by the gravitational force exerted on the spacecraft by Jupiter. How is this possible?
17. How would you find the mass of the Moon?
18. The *Apollo 13* spaceship developed trouble in the oxygen system about halfway to the Moon. Why did the mission continue on around the Moon, and then return home, rather than immediately turn back to Earth?
19. By how much is the free-fall acceleration at the Earth's equator reduced because of the rotation of the Earth? How does this effect vary with latitude?

PROBLEMS

Review Problem

A satellite of mass m is in an orbit of radius R around a planet of mass M in the equatorial plane of the planet. The satellite remains above the same point on the planet at all times. If the free-fall acceleration on the surface of the planet is g, find (a) the speed of the satellite, (b) the period of the satellite, (c) the kinetic energy of the satellite, (d) the potential energy of the satellite, (e) the radius of the planet, (f) the minimum possible period for the satellite, (g) the maximum kinetic energy of the satellite, (h) the minimum potential energy of the satellite, and (i) the escape speed of the satellite from this orbit.

Section 14.1 through Section 14.3

1. On the way to the Moon the Apollo astronauts reach a point where the Moon's gravitational pull is stronger than that of Earth's. (a) Determine the distance of this point from the center of the Earth. (b) What is the acceleration due to the Earth's gravity at this point?
2. A 200-kg mass and a 500-kg mass are separated by 0.400 m. (a) Find the net gravitational force exerted by these masses on a 50.0-kg mass placed midway between them. (b) At what position (other than infinitely remote ones) does the 50.0-kg mass experience a net force of zero?
3. A student proposes to measure the gravitational con-

☐ indicates problems that have full solutions available in the Student Solutions Manual and Study Guide.

stant, *G*, by suspending two spherical masses from the ceiling of a tall cathedral and measuring the deflection from the vertical. If two 100.0-kg masses are suspended at the end of 45.00-m-long cables, and the cables are attached to the ceiling 1.000 m apart, what is the separation of the masses?

4. (a) Determine the change and fractional change in gravitational force that the Sun exerts on a 50.0-kg woman standing on the equator at noon and midnight. (*Hint:* Since Δr is so small, use differentials.) (b) By what percent does the weight of the 50.0-kg woman decrease during a total eclipse of the Sun (Fig. P14.4)?

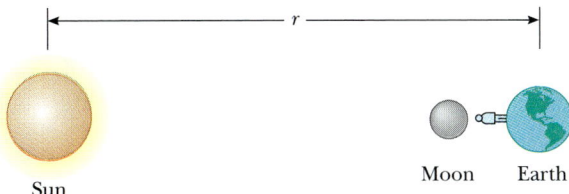

FIGURE P14.4

5. Three equal masses are located at three corners of a square of edge length ℓ as in Figure P14.5. Find the gravitational field **g** at the fourth corner due to these masses.

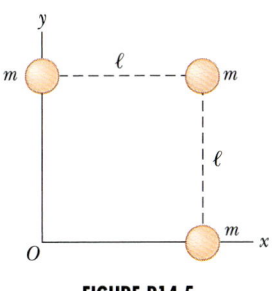

FIGURE P14.5

6. When a falling meteor is at a distance $d = 3R_E$ above the Earth's surface, what is its free-fall acceleration?

7. An astronaut weighs 140 N on the Moon's surface. When he is in a circular orbit about the Moon at an altitude $h = R_M$, what gravitational force does the Moon exert on him?

8. Two objects attract each other with a gravitational force of magnitude 1.0×10^{-8} N when separated by 20 cm. If the total mass of the two objects is 5.0 kg, what is the mass of each?

9. If the mass of Mars is $0.107M_E$ and its radius is $0.53R_E$, estimate the gravitational field *g* at the surface of Mars.

10. The free-fall acceleration on the surface of the Moon

is about one-sixth that on the surface of the Earth. If the radius of the Moon is about $0.25R_E$, find the ratio of their densities, $\rho_{\text{Moon}}/\rho_{\text{Earth}}$.

11. Plaskett's binary system consists of two stars that revolve in a circular orbit about a center of gravity midway between them. This means that the masses of the two stars are equal (Fig. P14.11). If the orbital speed of each star is 220 km/s and the orbital period of each is 14.4 days, find the mass *M* of each star. (For comparison, the mass of our Sun is 2×10^{30} kg.)

11A. Plaskett's binary system consists of two stars that revolve in a circular orbit about a center of gravity midway between them. This means that the masses of the two stars are equal (Fig. P14.11). If the orbital speed of each star is *v* and the orbital period of each is *T*, find the mass *M* of each star.

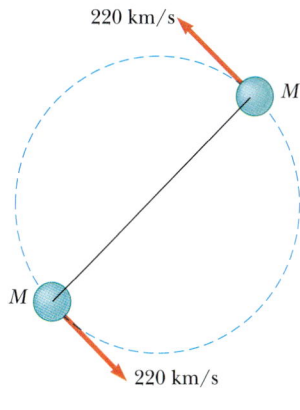

FIGURE P14.11

12. The Moon is 384 400 km distant from the Earth's center, and it completes an orbit in 27.3 days. (a) Determine the Moon's orbital speed. (b) How far does the Moon "fall" toward Earth in 1.00 s?

Section 14.4 Kepler's Laws
Section 14.5 The Law of Gravity and the Motion of Planets

13. During a solar eclipse, the Moon, Earth, and Sun all lie on the same line, with the Moon between the Earth and the Sun. (a) What force is exerted on the Moon by the Sun? (b) What force is exerted on the Moon by the Earth? (c) What force is exerted on the Earth by the Sun?

14. The *Explorer VIII* satellite, placed into orbit November 3, 1960, to investigate the ionosphere, had the following orbit parameters: perigee 459 km and apogee 2289 km (both distances above the Earth's surface); period 112.7 min. Find the ratio v_p/v_a.

15. Io, a small Moon of Jupiter, has an orbital period of 1.77 days and an orbital radius of 4.22×10^5 km. From these data, determine the mass of Jupiter.

16. A particle of mass m moves along a straight line with constant speed in the x direction a distance b from the x axis (Fig. P14.16). Show that Kepler's second law is satisfied by showing that the two shaded triangles in the figure have the same area when $t_4 - t_3 = t_2 - t_1$.

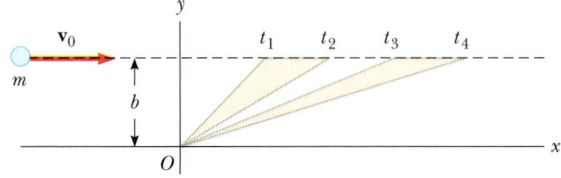

FIGURE P14.16

17. Two planets X and Y travel counterclockwise in circular orbits about a star as in Figure P14.17. The radii of their orbits are in the ratio $3 : 1$. At some time, they are aligned as in Figure P14.17a, making a straight line with the star. Five years later, planet X has rotated through $90°$ as in Figure P14.17b. Where is planet Y at this time?

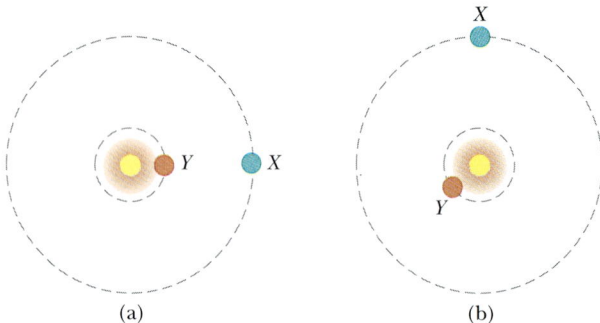

(a)   (b)

FIGURE P14.17

18. Geosynchronous satellites orbit the Earth 42 000 km from the Earth's center. Their angular speed at this height is the same as the rotational speed of the Earth, and so they appear stationary in the sky. What is the force acting on a 1000-kg satellite at this height?

19. A synchronous satellite, which always remains above the same point on a planet's equator, is put in orbit around Jupiter to study the famous red spot. Jupiter rotates once every 9.9 h. Use the data of Table 14.2 to find the altitude of the satellite.

20. Halley's comet approaches the Sun to within 0.57 A.U., and its orbital period is 75.6 years. (A.U. is the abbreviation for astronomical unit, where 1 A.U. = 1.50×10^8 km is the mean Earth-Sun distance.) How far from the Sun will Halley's comet travel before it starts its return journey? (Fig. P14.20.)

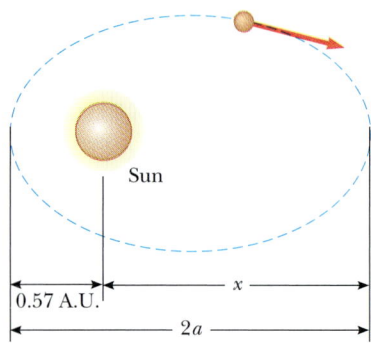

Sun

0.57 A.U.

$2a$

FIGURE P14.20

Section 14.6   The Gravitational Field

21. Compute the magnitude and direction of the gravitational field at a point P on the perpendicular bisector of two equal masses separated by a distance $2a$ as shown in Figure P14.21.

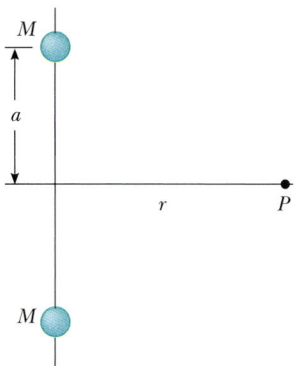

M

a

r P

M

FIGURE P14.21

22. Find the gravitational field at a distance r along the axis of a thin ring of mass M and radius a.

23. At what point along the line connecting the Earth and the Moon is there zero gravitational force on an object? (Ignore the presence of the Sun and the other planets.)

Section 14.7   Gravitational Potential Energy

(*Note:* Assume $U = 0$ at $r = \infty$.)

24. A satellite of the Earth has a mass of 100 kg and is at an altitude of 2.00×10^6 m. (a) What is the potential energy of the satellite-Earth system? (b) What is the magnitude of the gravitational force exerted by the Earth on the satellite?

25. How much work is done by the Moon's gravitational field as a 1000-kg meteor comes in from outer space and impacts on the Moon's surface?

26. How much energy is required to move a 1000-kg mass from the Earth's surface to an altitude $h = 2R_E$?

26A. How much energy is required to move a mass m from the Earth's surface to an altitude h?

27. After it exhausts its nuclear fuel, the ultimate fate of our Sun is possibly to collapse to a *white dwarf,* which is a star that has approximately the mass of the Sun, but the radius of the Earth. Calculate (a) the average density of the white dwarf, (b) the free-fall acceleration at its surface, and (c) the gravitational potential energy of a 1.00-kg object at its surface.

Section 14.8 Energy Considerations in Planetary and Satellite Motion

28. Determine the escape speed for a rocket on the far side of Ganymede, the largest of Jupiter's moons. The radius of Ganymede is 2.64×10^6 m, and its mass is 1.495×10^{23} kg. The mass of Jupiter is 1.90×10^{27} kg, and the distance between Jupiter and Ganymede is 1.071×10^9 m. Be sure to include the gravitational effect due to Jupiter, but you may ignore the motion of Jupiter and Ganymede as they revolve about their center of mass (Fig. P14.28).

Ganymede

Jupiter

FIGURE P14.28

29. A spaceship is fired from the Earth's surface with an initial speed of 2.00×10^4 m/s. What will its speed be when it is very far from the Earth? (Neglect friction.)

30. A 1000-kg satellite orbits the Earth at an altitude of 100 km. It is desired to increase the altitude of the orbit to 200 km. How much energy must be added to the system to effect this change in altitude?

30A. A satellite of mass m orbits the Earth at an altitude h_1. It is desired to increase the altitude of the orbit to h_2. How much energy must be added to the system to effect this change in altitude?

31. In Robert Heinlein's *The Moon Is a Harsh Mistress,* the colonial inhabitants of the Moon threaten to launch

rocks down onto the Earth if they are not given independence (or at least representation). Assuming that a rail gun could launch a rock of mass m at twice the lunar escape speed, calculate the speed of the rock as it enters the Earth's atmosphere.

32. A rocket is fired vertically, ejecting sufficient mass to move upward at a constant acceleration of $2g$. After 40.0 s, the rocket motors are turned off, and the rocket subsequently moves under the action of gravity alone, with negligible air resistance. Ignoring the variation of g with altitude, find (a) the maximum height the rocket reaches and (b) the total flight time from launch until the rocket returns to Earth. (c) Sketch a freehand (qualitative) graph of speed versus time for the flight.

33. A satellite moves in a circular orbit just above the surface of a planet. Show that the orbital speed v and escape speed of the satellite are related by the expression $v_{esc} = \sqrt{2}\,v$.

34. A satellite moves in an elliptical orbit about the Earth such that, at perigee and apogee positions, the distances from the Earth's center are, respectively, D and $4D$. Find the ratios (a) v_p/v_a and (b) E_p/E_a.

35. A 500-kg satellite is in a circular orbit at an altitude of 500 km above the Earth's surface. Because of air friction, the satellite eventually is brought to the Earth's surface, and it hits the Earth with a speed of 2.00 km/s. How much energy was absorbed by the atmosphere through friction?

35A. A satellite of mass m is in a circular orbit at an altitude h above the Earth's surface. Because of air friction, the satellite eventually is brought to the Earth's surface, and it hits the Earth with a speed v. How much energy was absorbed by the atmosphere through friction?

36. An artificial Earth satellite is "parked" in an equatorial circular orbit at an altitude of 1.00×10^3 km. What is the minimum additional speed that must be imparted to the satellite if it is to escape from Earth's gravitational attraction? How does this compare with the minimum escape speed for leaving from the Earth's surface?

37. (a) What is the minimum speed necessary for a spacecraft to escape the Solar System, starting at the Earth's orbit? (b) *Voyager 1* achieved a maximum speed of 125 000 km/h on its way to photograph Jupiter. Beyond what distance from the Sun is this speed sufficient to escape the Solar System?

*Section 14.9 The Gravitational Force Between an Extended Object and a Particle

38. A uniform rod of mass M is in the shape of a semicircle of radius R (Fig. P14.38). Calculate the force on a point mass m placed at the center of the semicircle.

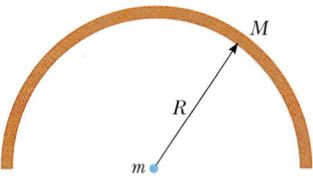

FIGURE P14.38

39. A spacecraft in the shape of a long cylinder has a length of 100 m and its mass with occupants is 1000 kg. It has strayed in too close to a 1.0-km radius black hole having a mass 100 times that of the Sun (Fig. P14.39). If the nose of the spacecraft points toward the center of the black hole, and if distance between the nose of the spaceship and the black hole's center is 10 km, (a) determine the average acceleration of the spaceship. (b) What is the difference in the acceleration experienced by the occupants in the nose of the ship and those in the rear of the ship farthest from the black hole?

39A. A spacecraft in the shape of a long cylinder has a length ℓ and its mass with occupants is m. It has strayed in too close to a black hole of radius R and mass M. If the nose of the spacecraft points toward the center of the black hole, and if distance between the nose of the spaceship and the black hole's center is d, (a) determine the average acceleration of the spaceship. (b) What is the difference in the acceleration experienced by the occupants in the nose of the ship and those in the rear of the ship farthest from the black hole?

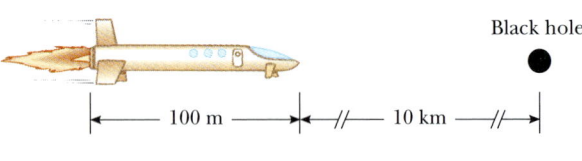

Black hole

|← 100 m →|←// 10 km //→|

FIGURE P14.39

Section 14.10   Gravitational Force Between a Particle and a Spherical Mass

40. (a) Show that the period calculated in Example 14.11 can be written as

$$T = 2\pi\sqrt{\frac{R_E}{g}}$$

where g is the free-fall acceleration. (b) What would this period be if tunnels were made through the Moon? (c) What practical problem regarding these tunnels on Earth would be removed if they were built on the Moon?

41. A 500-kg uniform solid sphere has a radius of 0.400 m. Find the magnitude of the gravitational force exerted by the sphere on a 50.0-g particle located (a) 1.50 m from the center of the sphere, (b) at

the surface of the sphere, and (c) 0.200 m from the center of the sphere.

42. A uniform solid sphere of mass m_1 and radius R_1 is inside and concentric with a spherical shell of mass m_2 and radius R_2 (Fig. P14.42). Find the gravitational force exerted by the sphere on a particle of mass m located at (a) $r = a$, (b) $r = b$, (c) $r = c$, where r is measured from the center of the spheres.

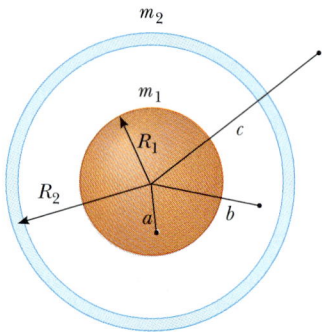

FIGURE P14.42

ADDITIONAL PROBLEMS

43. Let Δg_m represent the difference in the gravitational field produced by the Moon at the points on the Earth's surface nearest to and farthest from the Moon. Find the fraction $\Delta g_m/g$, where g is the free-fall acceleration due to the Earth's gravitational field. (This difference is responsible for the occurrence of the *lunar tides* on the Earth.)

44. *Voyagers 1* and *2* surveyed the surface of Jupiter's moon Io and photographed active volcanoes spewing liquid sulfur to heights of 70 km above the surface of this moon. Estimate the speed with which the liquid sulfur left the volcano. Io's mass is 8.9×10^{22} kg, and its radius is 1820 km.

45. A cylindrical habitat in space 6.0 km in diameter and 30 km long has been proposed (by G. K. O'Neill, 1974). Such a habitat would have cities, land, and lakes on the inside surface and air and clouds in the center. This would all be held in place by rotation of the cylinder about its long axis. How fast would the cylinder have to rotate to imitate the Earth's gravitational field at the walls of the cylinder?

45A. A cylindrical habitat in space having diameter d and length L has been proposed (by G. K. O'Neill, 1974). Such a habitat would have cities, land, and lakes on the inside surface and air and clouds in the center. This would all be held in place by rotation of the cylinder about its long axis. How fast would the cylinder have to rotate to imitate the Earth's gravitational field at the walls of the cylinder?

46. Two spheres having masses M and $2M$ and radii R and $3R$, respectively, are released from rest when the

distance between their centers is $12R$. How fast will each sphere be moving when they collide? Assume that the two spheres interact only with each other.

47. In introductory physics laboratories, a typical Cavendish balance for measuring the gravitational constant G uses lead spheres of masses 1.50 kg and 15.0 g whose centers are separated by 4.50 cm. Calculate the gravitational force between these spheres, treating each as a point mass located at the center of the sphere.

48. Consider two identical uniform rods of length L and mass m lying along the same line and having their closest points separated by a distance d (Fig. P14.48). Show that the mutual gravitational force between these rods has a magnitude

$$F = \frac{Gm^2}{L^2} \ln\left(\frac{(L+d)^2}{d(2L+d)}\right)$$

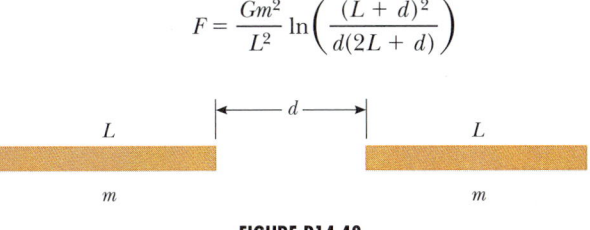

FIGURE P14.48

49. An object of mass m moves in a smooth straight tunnel of length L dug through a chord of the Earth as discussed in Example 14.11 (Fig. 14.20). (a) Determine the effective force constant of the harmonic motion and the amplitude of the motion. (b) Using energy considerations, find the maximum speed of the object. Where does this maximum speed occur? (c) Obtain a numerical value for the maximum speed if $L = 2500$ km.

50. For any body orbiting the Sun, Kepler's third law may be written $T^2 = kr^3$ where T is the orbital period and r is the semimajor axis of the orbit. (a) What is the value of k if T is measured in years and r is measured in A.U.? (See Problem 20.) (b) Use your value of k to find the orbital period of Jupiter if its mean radius from the Sun is 5.2 A.U.

51. Three point objects having masses m, $2m$, and $3m$ are fixed at the corners of a square of side length a such that the lighter object is at the upper left-hand corner, the heavier object is at the lower left-hand corner, and the remaining object is at the upper right-hand corner. Determine the magnitude and direction of the resulting gravitational field at the center of the square.

52. An airplane in a wide "outside" loop can create an apparent weight of zero inside the aircraft cabin. What must be the radius of curvature of the flight path for an aircraft moving at 480 km/h to create a condition of weightlessness inside the aircraft?

53. What angular speed (in revolutions per minute) is needed for a centrifuge to produce an acceleration of $1000g$ at a radius of 10.0 cm?

54. The Earth-Sun distance is 1.521×10^{11} m at the aphelion and 1.471×10^{11} m at the perihelion. If the Earth's orbital speed at the perihelion is 3.027×10^4 m/s, determine (a) its orbital speed at the aphelion, (b) the kinetic and potential energy at the perihelion, and (c) the kinetic and potential energy at the aphelion. Is the total energy constant? (Neglect the effect of the Moon and other planets.)

55. Two hypothetical planets of masses m_1 and m_2 and radii r_1 and r_2, respectively, are at rest when they are an infinite distance apart. Because of their gravitational attraction, they head toward each other on a collision course. (a) When their center-to-center separation is d, find the speed of each planet and their relative velocity. (b) Find the kinetic energy of each planet just before they collide if $m_1 = 2.0 \times 10^{24}$ kg, $m_2 = 8.0 \times 10^{24}$ kg, $r_1 = 3.0 \times 10^6$ m, and $r_2 = 5.0 \times 10^6$ m. (*Hint:* Both energy and momentum are conserved.)

56. After a supernova explosion, a star may undergo a gravitational collapse to an extremely dense state known as a neutron star, in which all the electrons and protons are squeezed together to form neutrons. A neutron star having a mass about equal to that of the Sun would have a radius of about 10 km. Find (a) the free-fall acceleration at its surface, (b) the weight of a 70-kg person at its surface, and (c) the energy required to remove a neutron of mass 1.67×10^{-27} kg from its surface to infinity.

57. When it orbited the Moon, the *Apollo 11* spacecraft's mass was 9.979×10^3 kg, its period was 119 min, and its mean distance from the Moon's center was 1.849×10^6 m. Assuming its orbit was circular and the Moon to be a uniform sphere, find (a) the mass of the Moon, (b) the orbital speed of the spacecraft, and (c) the minimum energy required for the craft to leave the orbit and escape the Moon's gravitational field.

58. Studies of the relationship of the Sun to its galaxy—the Milky Way—have revealed that the Sun is located near the outer edge of the galactic disc, about 30 000 light years from the center. Furthermore, it has been found that the Sun has an orbital speed of approximately 250 km/s around the galactic center. (a) What is the period of the Sun's galactic motion? (b) What is the approximate mass of the Milky Way galaxy? Using the fact that the Sun is a typical star, estimate the number of stars in the Milky Way.

59. X-ray pulses from Cygnus X-1, a celestial x-ray source, have been recorded during high-altitude rocket flights. The signals can be interpreted as originating when a blob of ionized matter orbits a black hole with a period of 5.0 ms. If the blob were in a circular orbit about a black hole whose mass is $20M_{Sun}$, what is the orbit radius?

60. *Vanguard I*, launched March 3, 1958, is the oldest man-made satellite still in orbit. Its initial orbit had

an apogee of 3970 km and a perigee of 650 km. Its maximum speed was 8.23 km/s and it had a mass of 1.60 kg. (a) Determine the period of the orbit. (Use the semimajor axis.) (b) Determine the speeds at apogee and perigee. (c) Find the total energy of the satellite.

61. A 200-kg satellite is placed in Earth orbit 200 km above the surface. (a) Assuming a circular orbit, how long does the satellite take to complete one orbit? (b) What is the satellite's speed? (c) What is the minimum energy necessary to place this satellite in orbit (assuming no air friction)?

61A. A satellite of mass m is placed in Earth orbit at an altitude h. (a) Assuming a circular orbit, how long does the satellite take to complete one orbit? (b) What is the satellite's speed? (c) What is the minimum energy necessary to place this satellite in orbit (assuming no air friction)?

62. In Larry Niven's science fiction novel *Ringworld*, a ring of material rotates about a star (Fig. P14.62). The rotational speed of the ring is 1.25×10^6 m/s and its radius is 1.53×10^{11} m. The inhabitants of this ringworld experience a normal contact force **n**. Acting alone, this normal force would produce an inward acceleration of 9.90 m/s². Additionally, the star at the center of the ring exerts a gravitational force on the ring and its inhabitants. (a) Show that the total centripetal acceleration of the inhabitants is 10.2 m/s². (b) The difference between the total acceleration and the acceleration provided by the normal force is due to the gravitational attraction of the central star. Show that the mass of the star is approximately 10^{32} kg.

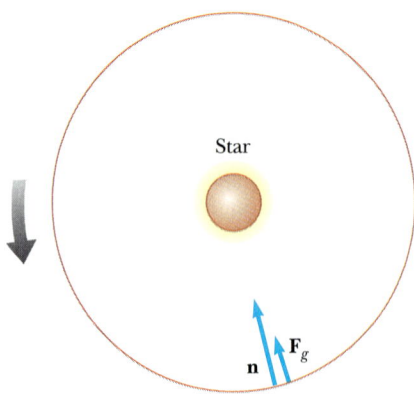

FIGURE P14.62

63. A sphere of mass M and radius R has a nonuniform density that varies with r, the distance from its center, according to the expression $\rho = Ar$, for $0 \le r \le R$. (a) What is the constant A in terms of M and R? (b) Determine the force on a particle of mass m placed outside the sphere. (c) Determine the force on the particle if it is inside the sphere. (*Hint:* See Section 14.10 and note that the distribution is spherically symmetric.)

64. Two stars of masses M and m, separated by a distance d, revolve in circular orbits about their center of mass (Fig. P14.64). Show that each star has a period given by

$$T^2 = \frac{4\pi^2}{G(M + m)} d^3$$

(*Hint:* Apply Newton's second law to each star, and note that the center-of-mass condition requires that $Mr_2 = mr_1$, where $r_1 + r_2 = d$.)

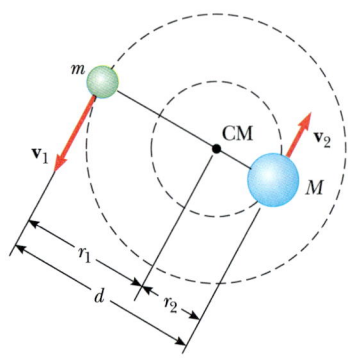

FIGURE P14.64

65. In 1978, astronomers at the U.S. Naval Observatory discovered that Pluto has a Moon, called Charon, that eclipses the planet every 6.4 days. Given that the center-to-center separation between Pluto and Charon is 19 700 km, find the total mass $(M + m)$ of the two bodies. (*Hint:* See Problem 64.)

66. In an effort to explain large meteor collisions with the Earth, scientists have postulated the existence of a companion star that is extremely dim and extremely far from the Sun. If this star (which some astronomers call Nemesis) has an orbital period of 3.0×10^7 years around the Sun-Nemesis center of mass, and a mass of $0.20M_{Sun}$, determine the average distance of this star from the Sun. $M_{Sun} = 2.0 \times 10^{30}$ kg.

67. A satellite is in a circular orbit about a planet of radius R. If the altitude of the satellite is h and its period is T, (a) show that the density of the planet is

$$\rho = \frac{3\pi}{GT^2} \left(1 + \frac{h}{R}\right)^3$$

(b) Calculate the average density of the planet if the period is 200 min and the satellite's orbit is close to the planet's surface.

68. It is claimed that a commercially available, portable gravity meter is sensitive enough to detect changes in

g to 1 part in 10^{11}. At the Earth's surface, what change in elevation would produce this variation? Assume the radius of the Earth is 6.0×10^6 m.

69. A particle of mass m is located inside a uniform solid sphere of radius R and mass M. If the particle is at a distance r from the center of the sphere, (a) show that the gravitational potential energy of the system is $U = (GmM/2R^3)r^2 - 3GmM/2R$. (b) How much work is done by the gravitational force in bringing the particle from the surface of the sphere to its center?

SPREADSHEET PROBLEMS

S1. Four point masses are fixed as shown in Figure PS14.1. The gravitational potential energy for a test particle of mass m moving along the x axis in the gravitational field of the fixed masses can be written as

$$U(x) = -\frac{GM_1 m}{r_1} - \frac{GM_2 m}{r_2} - \frac{GM_3 m}{r_3} - \frac{GM_4 m}{r_4}$$

where $r_1 = r_4 = [b^2 + (x + a^2)]^{1/2}$ and $r_2 = r_3 = [b^2 + (x - a)^2]^{1/2}$. Spreadsheet 14.1 calculates $U(x)$ and the force $F(x)$ exerted on the test particle $= -dU(x)/dx$. The quantities M_1, M_2, M_3, M_4, a, and b are input parameters. The mass of the test particle is 1.00 kg. Use $a = 0$, $b = 1.00$ m, $M_1 = M_3 = 0$, $M_2 = M_4 = 1.00$ kg. (This is the case of two particles on the y axis at $y = +1.00$ m and $y = -1.00$ m.) (a) Plot $U(x)$ and $F(x)$ versus x for $x = -4.00$ m to $+4.00$ m. (b) If the test particle has an energy of -7.50×10^{-11} J, describe the particle's motion. (c) How does the force on the particle vary as the particle moves along the x axis?

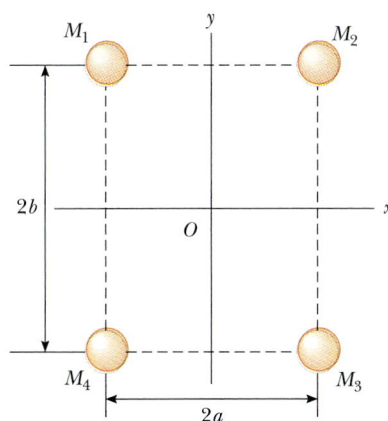

FIGURE PS14.1

S2. In Spreadsheet 14.1, let $a = 1.00$ m, $b = 0.50$ m, and $M_1 = M_2 = M_3 = M_4 = 1.00$ kg. The mass of the test

particle is 1.00 kg. (a) Plot $U(x)$ and $F(x)$ versus x. (b) If the test particle has an energy of -2.00×10^{-10} J, describe its motion. (c) If the particle has an energy of -3.00×10^{-10} J, what motion is possible?

S3. A projectile is fired straight up from the surface of the Earth with an initial speed v_0. If v_0 is less than the escape speed $v_{esc} = \sqrt{2gR_E} = 11.2$ km/s, the projectile will rise to a maximum height h above the Earth's surface and return. Neglecting air resistance and using conservation of energy, show that

$$h = \frac{R_E v_0^2}{2gR_E - v_0}$$

where R_E is the Earth's radius. Write your own spreadsheet to calculate h for $v_0 = 1, 2, 3, \ldots,$ 11 km/s. Plot h/R_E versus v_0. What happens as v_0 approaches the escape speed v_{esc}?

S4. The acceleration of an object moving in the gravitational field of the Earth is

$$a = -GM_E \frac{\mathbf{r}}{r^3}$$

where r is the position vector directed from the center of the Earth to the object. Choosing the origin at the center of the Earth, and assuming the object is moving in the xy-plane, the acceleration has rectangular components

$$a_x = -\frac{GM_E x}{(x^2 + y^2)^{3/2}}$$

$$a_y = -\frac{GM_E y}{(x^2 + y^2)^{3/2}}$$

Write your own spreadsheet (or computer program) to find the position of the object as a function of time. (Use the techniques developed in Spreadsheet 6.1 as a model for your program. Assume that the initial position of the object is at $x = 0$ and $y = 2R_E$, where R_E is the radius of the Earth, and give the object an initial velocity of 5 km/s in the x direction. The time increment should be made as small as practical. Try 5 s. Plot the x and y coordinates of the object as functions of time. Does the object hit the Earth? Vary the initial velocity until a circular orbit is found.

S5. Modify your spreadsheet (or program) from Problem S4 to calculate the kinetic, potential, and total energies as functions of time. How do they vary with time for the various cases you tried?

S6. Modify your spreadsheet (or program) from Problem S14.3 to calculate the angular momentum of the object. In rectangular coordinates $L = m(xv_y - yv_x)$. How does the angular momentum vary with time?

Kepler's second law of planetary motion implies that L remains constant. Is your calculated L constant?

S7. Imagine that you are an intrepid space traveler and you discover a Universe where your spaceship no longer behaves normally. You suspect that the law of gravity does not have the inverse square dependence that it should. Modify the spreadsheet (or program) in Problem S4 to allow for a gravitational law of any inverse power. Investigate the shape of the orbits for different inverse powers. Is energy constant for orbital motion in this strange Universe? Are Kepler's laws still valid?

B.C.   by John Hart

By permission of John Hart and Field Enterprises, Inc.

CHAPTER 15

Fluid Mechanics

A scuba diver plays with an octopus in Poor Knights Island, New Zealand. As the diver descends to greater depths, the water pressure increases above atmospheric pressure, and the internal pressure of the body also increases accordingly to maintain equilibrium. Scuba divers have been able to swim at depths of over 1000 ft.   *(Darryl Torckler/Tony Stone Images)*

Matter is normally classified as being in one of three states: solid, liquid, or gaseous. Everyday experience tells us that a solid has a definite volume and shape. A brick maintains its familiar shape and size day in and day out. We also know that a liquid has a definite volume but no definite shape. Finally, a gas has neither definite volume nor definite shape. These definitions help us to picture the states of matter, but they are somewhat artificial. For example, asphalt and plastics are normally considered solids, but over long periods of time they tend to flow like liquids. Likewise, most substances can be a solid, liquid, or gas (or combinations of these), depending on the temperature and pressure. In general, the time it takes a particular substance to change its shape in response to an external force determines whether we treat the substance as a solid, liquid, or gas.

A **fluid** is a collection of molecules that are randomly arranged and held together by weak cohesive forces and forces exerted by the walls of a container. Both liquids and gases are fluids.

In our treatment of the mechanics of fluids, we shall see that no new physical principles are needed to explain such effects as the buoyant force on a submerged object and the dynamic lift on an airplane wing. First, we consider a fluid at rest and derive an expression for the pressure exerted by the fluid as a function of its density and depth. We then treat fluids in motion, an area of study called fluid dynamics. A fluid in motion can be described by a model in which certain simplifying assumptions are made. We use this model to analyze some situations of practical importance. An underlying principle known as the *Bernoulli principle* enables us to determine relationships between the pressure, density, and velocity at every point in a fluid. We conclude the chapter with a brief discussion of internal friction in a fluid and turbulent motion.

15.1   PRESSURE

The study of fluid mechanics involves the density of a substance, defined as its mass per unit volume. For this reason, Table 15.1 lists the densities of various substances. These values vary slightly with temperature, since the volume of a substance is temperature dependent (as we shall see in Chapter 19). Note that under standard conditions (0°C and atmospheric pressure) the densities of gases are about 1/1000 the densities of solids and liquids. This difference implies that the average molecular spacing in a gas under these conditions is about ten times greater than in a solid or liquid.

Fluids do not sustain shearing stresses, and thus the only stress that can exist on an object submerged in a fluid is one that tends to compress the object. The force exerted by the fluid on the object is always perpendicular to the surfaces of the object, as shown in Figure 15.1.

The pressure at a specific point in a fluid can be measured with the device pictured in Figure 15.2. The device consists of an evacuated cylinder enclosing a light piston connected to a spring. As the device is submerged in a fluid, the fluid presses down on the top of the piston and compresses the spring until the inward force of the fluid is balanced by the outward force of the spring. The fluid pressure can be measured directly if the spring is calibrated in advance. This is accomplished by applying a known force to the spring to compress it a given distance.

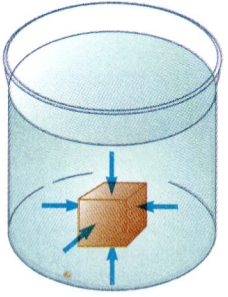

FIGURE 15.1   The force of the fluid on a submerged object at any point is perpendicular to the surface of the object. The force of the fluid on the walls of the container is perpendicular to the walls at all points.

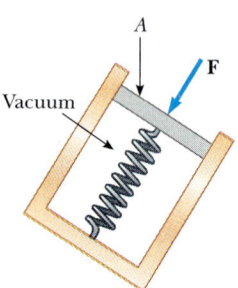

FIGURE 15.2   A simple device for measuring pressure in a fluid.

TABLE 15.1	**Densities of Some Common Substances**		
Substance	$\rho \, (kg/m^3)^a$	Substance	$\rho \, (kg/m^3)^a$
Ice	0.917×10^3	Water	1.00×10^3
Aluminum	2.70×10^3	Sea water	1.03×10^3
Iron	7.86×10^3	Ethyl alcohol	0.806×10^3
Copper	8.92×10^3	Benzene	0.879×10^3
Silver	10.5×10^3	Mercury	13.6×10^3
Lead	11.3×10^3	Air	1.29
Gold	19.3×10^3	Oxygen	1.43
Platinum	21.4×10^3	Hydrogen	8.99×10^{-2}
Glycerine	1.26×10^3	Helium	1.79×10^{-1}

a All values are at standard atmospheric pressure and temperature (STP), that is, atmospheric pressure and 0°C. To convert to grams per cubic centimeter, multiply by 10^{-3}.

If *F* is the magnitude of the normal force on the piston and *A* is the surface area of the piston, then the pressure, *P*, of the fluid at the level to which the device has been submerged is defined as the ratio of force to area:

$$P \equiv \frac{F}{A} \qquad \qquad (15.1)$$

Definition of pressure

To define the pressure at a specific point, consider a fluid enclosed as shown in Figure 15.2. If the normal force exerted by the fluid is *F* over a surface element of area δA that contains the point in question, then the pressure at that point is

$$P = \lim_{\delta A \to 0} \frac{F}{\delta A} = \frac{dF}{dA} \qquad \qquad (15.2)$$

As we see in the next section, the pressure in a fluid varies with depth. Therefore, to get the total force on a flat wall of a container, we have to integrate Equation 15.2 over the surface.

Since pressure is force per unit area, it has units of N/m² in the SI system. Another name for the SI unit of pressure is **pascal** (Pa).

$$1 \text{ Pa} \equiv 1 \text{ N/m}^2 \qquad \qquad (15.3)$$

CONCEPTUAL EXAMPLE 15.1

A woman wearing high-heeled shoes is invited into a home in which the kitchen has vinyl floor covering. Why should the homeowner be concerned?

Reasoning   She can exert enough pressure on the floor to dent or puncture the floor covering. The large pressure is caused by the fact that her weight is distributed over the very small cross-sectional area of her high heels. She should be asked to remove her high heels and put on some slippers.

CONCEPTUAL EXAMPLE 15.2

The daring physics professor, after a long lecture, stretches out for a nap on a bed of nails as in the photograph. How is this possible?

Reasoning   If you try to support your entire weight on a single nail, the pressure on your body is your weight divided by the very small area of the nail. This pressure is sufficiently large to penetrate the skin. However, if you distribute your weight over several hundred nails, as the professor is doing, the pressure is considerably reduced because the area that supports your weight is the total area of all nails in contact with your body. (Note that lying on a bed of nails is much more comfortable than sitting on the bed. Standing on the bed without shoes is not recommended.)

Conceptual Example 15.2.   *(Jim Lehman)*

Snowshoes prevent the person from sinking into the soft snow because the person's weight is spread over a larger area, which reduces the pressure on the snow's surface. *(Earl Young/FPG)*

15.2 VARIATION OF PRESSURE WITH DEPTH

As divers well know, the pressure in the sea or a lake increases as they dive to greater depths. Likewise, atmospheric pressure decreases with increasing altitude. For this reason, aircraft flying at high altitudes must have pressurized cabins.

We now show how the pressure in a liquid increases linearly with depth. Consider a liquid of density ρ at rest and open to the atmosphere as in Figure 15.3. Let us select a sample of the liquid contained within an imaginary cylinder of cross-sectional area A extending from the surface of the liquid to a depth h. The pressure exerted by the fluid on the bottom face is P, and the pressure on the top face of the cylinder is atmospheric pressure, P_0. Therefore, the upward force exerted by the liquid on the bottom of the cylinder is PA, and the downward force exerted by the atmosphere on the top is P_0A. Because the mass of liquid in the cylinder is $\rho V = \rho Ah$, the weight of the fluid in the cylinder is $w = \rho gV = \rho gAh$. Because the cylinder is in equilibrium, the upward force at the bottom must be greater than the downward force at the top of the sample to support its weight:

$$PA - P_0A = \rho hgA$$

Variation of pressure with depth

or

$$P = P_0 + \rho gh \qquad (15.4)$$

where we usually take atmospheric pressure to be $P_0 = 1.00$ atm $\approx 1.01 \times 10^5$ Pa. In other words,

> the absolute pressure P at a depth h below the surface of a liquid open to the atmosphere is *greater* than atmospheric pressure by an amount ρgh.

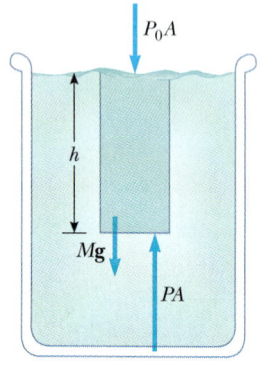

FIGURE 15.3 The variation of pressure with depth in a fluid. The net force on the volume of water within the darker region must be zero.

This result also verifies that the pressure is the same at all points having the same depth, independent of the shape of the container.

In view of the fact that the pressure in a liquid depends only upon depth, any increase in pressure at the surface must be transmitted to every point in the fluid. This was first recognized by the French scientist Blaise Pascal (1623–1662) and is called **Pascal's law:**

> A change in the pressure applied to an enclosed liquid is transmitted undiminished to every point of the liquid and to the walls of the container.

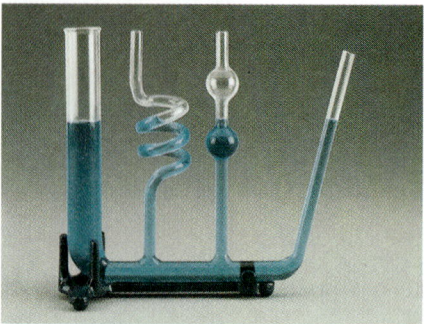

This photograph illustrates that the pressure in a liquid is the same at all points having the same elevation. Note that the shape of the vessel does not affect the pressure. *(Courtesy of Central Scientific Co.)*

(a)   (b)

FIGURE 15.4 (a) Diagram of a hydraulic press. Since the increase in pressure is the same at the left and right sides, a small force F_1 at the left produces a much larger force F_2 at the right. (b) A bus under repair is supported by a hydraulic lift in a garage. *(Superstock)*

An important application of Pascal's law is the hydraulic press illustrated by Figure 15.4. A force F_1 is applied to a small piston of area A_1. The pressure is transmitted through a liquid to a larger piston of area A_2. Since the pressure is the same on both sides, we see that $P = F_1/A_1 = F_2/A_2$. Therefore, the force F_2 is larger than F_1 by the multiplying factor A_2/A_1. Hydraulic brakes, car lifts, hydraulic jacks, and forklifts all make use of this principle.

CONCEPTUAL EXAMPLE 15.3

A typical silo on a farm has many bands wrapped around its perimeter as shown in the photograph. Why is the spacing between successive bands smaller at the lower portions of the silo?

Reasoning If you think of the grain stored in the silo as a fluid, the pressure the grain exerts on the walls of the silo increases with increasing depth just as water pressure in a lake increases with increasing depth. Thus, the spacing between bands is made smaller at the lower portions to overcome the larger outward forces on the walls in these regions.

Conceptual Example 15.3. *(Henry Leap)*

EXAMPLE 15.4 The Car Lift

In a car lift used in a service station, compressed air exerts a force on a small piston of circular cross-section having a radius of 5.00 cm. This pressure is transmitted by a liquid to a second piston of radius 15.0 cm. What force must the compressed air exert in order to lift a car weighing 13 300 N? What air pressure will produce this force?

Solution   Because the pressure exerted by the compressed air is transmitted undiminished throughout the fluid, we have

$$F_1 = \left(\frac{A_1}{A_2}\right)F_2 = \frac{\pi(5.00 \times 10^{-2}\ \mathrm{m})^2}{\pi(15.0 \times 10^{-2}\ \mathrm{m})^2}\,(1.33 \times 10^4\ \mathrm{N})$$

$$= \boxed{1.48 \times 10^3\ \mathrm{N}}$$

The air pressure that will produce this force is

$$P = \frac{F_1}{A_1} = \frac{1.48 \times 10^3\ \mathrm{N}}{\pi(5.00 \times 10^{-2}\ \mathrm{m})^2} = \boxed{1.88 \times 10^5\ \mathrm{Pa}}$$

This pressure is approximately twice atmospheric pressure.
   The input work (the work done by F_1) is equal to the output work (the work done by F_2), so that energy is conserved.

EXAMPLE 15.5   The Water Bed

A water bed is 2.00 m on a side and 30.0 cm deep. (a) Find its weight.

Solution   Since the density of water is 1000 kg/m³ (Table 15.1), the mass of the bed is

$$M = \rho V = (1000\ \mathrm{kg/m^3})(1.20\ \mathrm{m^3}) = 1.20 \times 10^3\ \mathrm{kg}$$

and its weight is

$$w = Mg = (1.20 \times 10^3\ \mathrm{kg})(9.80\ \mathrm{m/s^2}) = \boxed{1.18 \times 10^4\ \mathrm{N}}$$

This is equivalent to approximately 2640 lb (as compared with a regular bed that weighs approximately 300 lb). In order to support such a heavy load, you would be well advised to keep your water bed in the basement or on a sturdy, well-supported floor.

(b) Find the pressure exerted on the floor when the bed rests in its normal position. Assume that the entire lower surface of the bed makes contact with the floor.

Solution   The weight of the bed is 1.18×10^4 N. The cross-sectional area is 4.00 m² when the bed is in its normal position. This gives a pressure exerted on the floor of

$$P = \frac{1.18 \times 10^4\ \mathrm{N}}{4.00\ \mathrm{m^2}} = \boxed{2.95 \times 10^3\ \mathrm{Pa}}$$

Exercise   Calculate the pressure exerted on the floor when the bed rests on its side.

Answer   Since the area of its side is 0.600 m², the pressure is 1.96×10^4 Pa.

EXAMPLE 15.6   Pressure in the Ocean

Calculate the pressure at an ocean depth of 1000 m. Assume the density of sea water is 1.024×10^3 kg/m³ and take $P_0 = 1.01 \times 10^5$ Pa.

Solution

$$P = P_0 + \rho g h$$
$$= 1.01 \times 10^5\ \mathrm{Pa} + (1.024 \times 10^3\ \mathrm{kg/m^3})$$
$$\times\ (9.80\ \mathrm{m/s^2})(1.00 \times 10^3\ \mathrm{m})$$

$$P = \boxed{1.01 \times 10^7\ \mathrm{Pa}}$$

This is 100 times greater than atmospheric pressure! Obviously, the design and construction of vessels that will withstand such enormous pressures are not trivial matters.

Exercise   Calculate the total force exerted on the outside of a circular submarine window of diameter 30.0 cm at this depth.

Answer 7.00×10^5 N.

EXAMPLE 15.7   The Force on a Dam

Water is filled to a height H behind a dam of width w (Fig. 15.5). Determine the resultant force on the dam.

Reasoning   We cannot calculate the force on the dam by simply multiplying the area times the pressure, because the pressure varies with depth. The problem can be solved by finding the force dF on a narrow horizontal strip at depth h,

and then integrate the expression to find the total force on the dam.

Solution   The pressure at the depth h beneath the surface at the shaded portion is

$$P = \rho g h = \rho g(H - y)$$

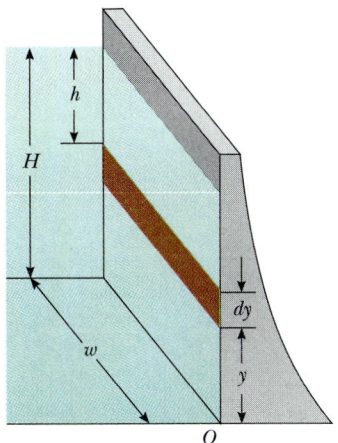

FIGURE 15.5 (Example 15.7) The total force on a dam must be obtained from the expression $F = \int P \, dA$, where dA is the area of the dark strip.

(We have left out atmospheric pressure because it acts on both sides of the dam.) Using Equation 15.2, we find the force on the shaded strip of area $dA = w \, dy$ to be

$$dF = P \, dA = \rho g (H - y) w \, dy$$

Therefore, the total force on the dam is

$$F = \int P \, dA = \int_0^H \rho g (H - y) w \, dy = \boxed{\tfrac{1}{2} \rho g w H^2}$$

Note that because the pressure increases with depth, the dam is designed such that its thickness increases with depth, as in Figure 15.5.

Exercise Use the fact that the pressure increases linearly with depth to find the average pressure on the dam and the total force on the dam.

15.3 PRESSURE MEASUREMENTS

One simple device for measuring pressure is the open-tube manometer illustrated in Figure 15.6a. One end of a U-shaped tube containing a liquid is open to the atmosphere, and the other end is connected to a system of unknown pressure P. The difference in pressure $P - P_0$ is equal to $\rho g h$. Therefore, we see that $P = P_0 + \rho g h$. The pressure P is called the **absolute pressure**, while the difference $P - P_0$ is called the **gauge pressure**. For example, the pressure you measure in your bicycle tire is gauge pressure.

Another instrument used to measure pressure is the common barometer, invented by Evangelista Torricelli (1608–1647). A long tube closed at one end is filled with mercury and then inverted into a dish of mercury (Fig. 15.6b). The closed end of the tube is nearly a vacuum, so its pressure can be taken as zero. Therefore, it follows that $P_0 = \rho_{Hg} g h$, where ρ_{Hg} is the density of the mercury and h is the height of the mercury column. One atmosphere (1 atm) of pressure is defined to be the pressure equivalent of a column of mercury that is exactly 0.7600 m in height at 0°C, with $g = 9.80665 \text{ m/s}^2$. At this temperature, mercury has a density of $13.595 \times 10^3 \text{ kg/m}^3$; therefore,

$$P_0 = \rho_{Hg} g h = (13.595 \times 10^3 \text{ kg/m}^3)(9.80665 \text{ m/s}^2)(0.7600 \text{ m})$$
$$= 1.013 \times 10^5 \text{ Pa}$$

15.4 BUOYANT FORCES AND ARCHIMEDES' PRINCIPLE

Archimedes' principle can be stated as follows:

> Any body completely or partially submerged in a fluid is buoyed up by a force equal to the weight of the fluid displaced by the body.

Everyone has experienced Archimedes' principle. As an example of a common experience, recall that it is relatively easy to lift someone if the person is in a

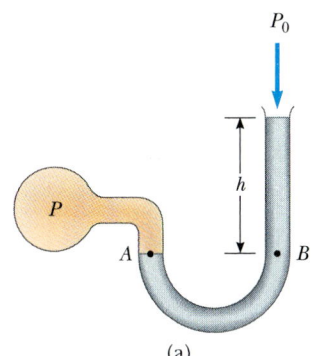

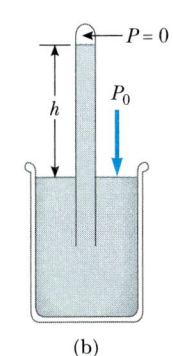

FIGURE 15.6 Two devices for measuring pressure: (a) an open-tube manometer and (b) a mercury barometer.

Archimedes, a Greek mathematician, physicist, and engineer, was perhaps the greatest scientist of antiquity. He was the first to compute accurately the ratio of a circle's circumference to its diameter and also showed how to calculate the volume and surface area of spheres, cylinders, and other geometric shapes. He is well known for discovering the nature of the buoyant force acting on objects and was also a gifted inventor. One of his practical inventions, still in use today, is the Archimedes' screw, an inclined rotating coiled tube used originally to lift water from the holds of ships. He also invented the catapult and devised systems of levers, pulleys, and weights for raising heavy loads. Such inventions were successfully used by the soldiers to defend his native city, Syracuse, during a two-year siege by the Romans.

According to legend, Archimedes was asked by King Hieron to determine whether the king's crown was made of pure gold or had been alloyed with some other metal. The task was to be performed without damaging the crown. Archimedes presumably arrived at a solution while taking a bath, noting a partial loss of weight after submerging his arms and legs in the water. As the story goes, he was so excited about his great discovery that he ran through the streets of Syracuse naked shouting "Eureka!," which is Greek for "I have found it."

Archimedes

| 2 8 7 – 2 1 2 B.C. |

swimming pool whereas lifting that same individual on dry land is much harder. Evidently, water provides partial support to any object placed in it. The upward force that the fluid exerts on an object submerged in it is called the **buoyant force.** According to Archimedes' principle,

the magnitude of the buoyant force always equals the weight of the fluid displaced by the object.

The buoyant force acts vertically upward through what was the center of gravity of the displaced fluid.

Archimedes' principle can be verified in the following manner. Suppose we focus our attention on the indicated cube of fluid in the container of Figure 15.7. This cube of fluid is in equilibrium under the action of the forces on it. One of these forces is its weight. What cancels this downward force? Apparently, the rest of the fluid inside the container is holding it in equilibrium. Thus, the buoyant force **B** on the cube of fluid is exactly equal in magnitude to the weight of the fluid inside the cube:

$$B = w$$

Now, imagine that the cube of fluid is replaced by a cube of steel of the same dimensions. What is the buoyant force on the steel? The fluid surrounding a cube behaves in the same way whether a cube of fluid or a cube of steel is being buoyed up; therefore, *the buoyant force acting on the steel is the same as the buoyant force acting on a cube of fluid of the same dimensions.* This result applies for a submerged object of any shape, size, or density.

Let us show explicitly that the buoyant force is equal in magnitude to the weight of the displaced fluid. The pressure at the bottom of the cube in Figure 15.7

FIGURE 15.7 The external forces on the cube of water are the force of gravity w and the buoyancy force **B**. Under equilibrium conditions, $B = w$.

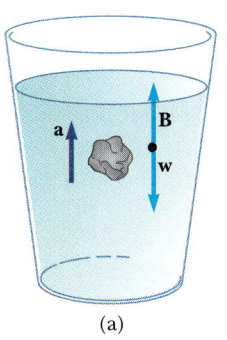
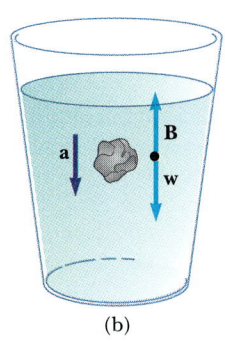

(a)   (b)

FIGURE 15.8 (a) A totally submerged object that is less dense than the fluid in which it is submerged will experience a net upward force. (b) A totally submerged object that is denser than the fluid sinks.

Hot-air balloons over Albuquerque, New Mexico. Since hot air is less dense than cold air, there is a net upward buoyant force on the balloons. *(William Moriarity/ Rainbow)*

is greater than the pressure at the top by an amount $\rho_f gh$, where ρ_f is the density of the fluid and h is the height of the cube. Since the pressure difference, ΔP, is equal to the buoyant force per unit area, that is, $\Delta P = B/A$, we see that $B = (\Delta P)A = (\rho_f gh)A = \rho_f gV$, where V is the volume of the cube. Since the mass of the fluid in the cube is $M = \rho_f V$, we see that

$B = w = \rho_f gV$ where w is the weight of the displaced fluid.

Before proceeding with a few examples, it is instructive to compare the forces acting on a totally submerged object with those acting on a floating object.

Case I. A Totally Submerged Object When an object is totally submerged in a fluid of density ρ_f, the upward buoyant force is given by $B = \rho_f V_0 g$, where V_0 is the volume of the object. If the object has a density ρ_0, its weight is equal to $w = Mg = \rho_0 V_0 g$, and the net force on it is $B - w = (\rho_f - \rho_0)V_0 g$. Hence, if the density of the object is less than the density of the fluid as in Figure 15.8a, the unsupported object will accelerate upward. If the density of the object is greater than the density of the fluid as in Figure 15.8b, the unsupported object will sink.

Case II. A Floating Object Now consider an object in static equilibrium floating on a fluid, that is, an object that is only partially submerged. In this case, the upward buoyant force is balanced by the downward weight of the object. If V is the volume of the fluid displaced by the object (which corresponds to that volume of the object beneath the fluid level), then the buoyant force has a magnitude $B = \rho_f Vg$. Since the weight of the object is $w = Mg = \rho_0 V_0 g$, and $w = B$, we see that $\rho_f Vg = \rho_0 V_0 g$, or

$$\frac{\rho_0}{\rho_f} = \frac{V}{V_0} \tag{15.6}$$

Under normal conditions, the average density of a fish is slightly greater than the density of water. This being the case, a fish would sink if it did not have some mechanism for adjusting its density. The fish accomplishes this by internally regulating the size of its swim bladder. In this manner, fish are able to swim to various depths.

CONCEPTUAL EXAMPLE 15.8

Steel is much denser than water. How, then, do ships made of steel float?

Reasoning The hull of the ship is full of air, and the density of air is about one thousandth the density of water. Hence, the total weight of the ship is less than the weight of an equal volume of water.

CONCEPTUAL EXAMPLE 15.9

A person in a boat floating in a small pond throws an anchor overboard. Does the level of the pond rise, fall, or remain the same?

Reasoning   The level of the pond falls. This is because the anchor displaces more water while in the boat. A floating object displaces a volume of water whose weight is equal to the weight of the object. A submerged object displaces a volume of water equal to the volume of the object. Because the density of the anchor is greater than that of water, a volume of water that weighs the same as the anchor will be greater than the volume of the anchor.

EXAMPLE 15.10   A Submerged Object

A piece of aluminum is suspended from a string and then completely immersed in a container of water (Fig. 15.9). The mass of the aluminum is 1.0 kg, and its density is 2.7×10^3 kg/m³. Calculate the tension in the string before and after the aluminum is immersed.

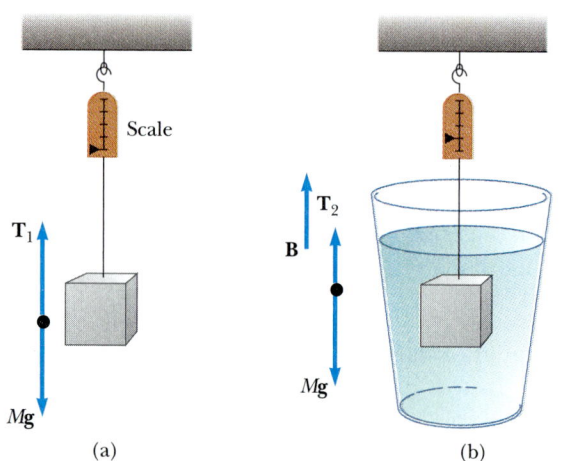

FIGURE 15.9   (Example 15.10) (a) When the aluminum is suspended in air, the scale reads the true weight, Mg (neglecting the buoyancy of air). (b) When the aluminum is immersed in water, the buoyant force **B** reduces the scale reading to $T_2 = Mg - B$.

Solution   When the aluminum is suspended in air, as in Figure 15.9a, the tension in the string, T_1 (the reading on the scale), is equal to the weight, Mg, of the aluminum, assuming that the buoyant force of air can be neglected:

$$T_1 = Mg = (1.0 \text{ kg})(9.80 \text{ m/s}^2) = \boxed{9.8 \text{ N}}$$

When immersed in water, the aluminum experiences an upward buoyant force **B**, as in Figure 15.9b, which reduces the tension in the string. Since the system is in equilibrium,

$$T_2 + B - Mg = 0$$
$$T_2 = Mg - B = 9.8 \text{ N} - B$$

In order to calculate B, we must first calculate the volume of the aluminum:

$$V_{Al} = \frac{M}{\rho_{Al}} = \frac{1.0 \text{ kg}}{2.7 \times 10^3 \text{ kg/m}^3} = 3.7 \times 10^{-4} \text{ m}^3$$

Since the buoyant force equals the weight of the water displaced, we have

$$B = M_w g = \rho_w V_{Al} g$$
$$= (1.0 \times 10^3 \text{ kg/m}^3)(3.7 \times 10^{-4} \text{ m}^3)(9.80 \text{ m/s}^2)$$
$$= 3.6 \text{ N}$$

Therefore,

$$T_2 = 9.8 \text{ N} - B = 9.8 \text{ N} - 3.6 \text{ N} = \boxed{6.2 \text{ N}}$$

EXAMPLE 15.11   The Floating Ice Cube

An ice cube floats in a glass of water as in Figure 15.10. What fraction of the cube lies above the water level?

Reasoning and Solution   This problem corresponds to Case II described in the text. The weight of the ice cube is $w = \rho_i V_i g$, where $\rho_i = 917$ kg/m³ and V_i is the volume of the whole ice cube. The upward buoyant force equals the weight of the displaced water; that is, $B = \rho_w V g$, where V is the volume of the ice cube beneath the water (the shaded region in

FIGURE 15.10   (Example 15.11).

Fig. 15.10) and ρ_w is the density of the water, $\rho_w = 1000$ kg/m³. Since $\rho_i V_i g = \rho_w V g$, the fraction of ice beneath water is $V/V_i = \rho_i/\rho_w$. Hence, the fraction of ice above the water level is

$$f = 1 - \frac{\rho_i}{\rho_f} = 1 - \frac{917 \text{ kg/m}^3}{1000 \text{ kg/m}^3} = \boxed{0.083} \quad \text{or} \quad 8.3\%$$

Exercise   What fraction of the volume of an iceberg lies below the level of the sea? The density of sea water is 1024 kg/m³.

Answer   0.896 or 89.6%.

15.5   FLUID DYNAMICS

Thus far, our study of fluids has been restricted to fluids at rest. We now turn our attention to fluid dynamics, that is, fluids in motion. Instead of trying to study the motion of each particle of the fluid as a function of time, we describe the properties of the fluid at each point as a function of time.

Flow Characteristics

When fluid is in motion, its flow can be characterized as being one of two main types. The flow is said to be **steady** or **laminar** if each particle of the fluid follows a smooth path, so that the paths of different particles never cross each other, as in Figure 15.11. Thus, in steady flow, the velocity of the fluid at any point remains constant in time.

Above a certain critical speed, fluid flow becomes **nonsteady** or **turbulent**. Turbulent flow is an irregular flow characterized by small whirlpool-like regions as in Figure 15.12. As an example, the flow of water in a stream becomes turbulent in regions where rocks and other obstructions are encountered, often forming "white water" rapids.

The term **viscosity** is commonly used in fluid flow to characterize the degree of internal friction in the fluid. This internal friction or viscous force is associated

FIGURE 15.11   A car undergoes an aerodynamic test in a wind tunnel. The streamlines in the airflow are made visible by smoke particles. What can you conclude about the speed of the airflow from this photograph?   *(Andy Sacks/Tony Stone Images)*

High-speed photograph of secondary drop formation after the impact of a falling drop on a pool of water. After cratering, the surface collapsed to form a column of water. At the top of the column, water surface tension effects create a secondary drop. This can be seen to contain a very high proportion of red dye, so little mixing has taken place. The water column itself is beginning to collapse.   *(Dr. David Gorham and Dr. Ian Hutchings/SPL/Photo Researchers)*

(a)

(b)

FIGURE 15.12   (a) Turbulent flow: the tip of a rotating blade (the dark region) forms a vortex in air that is being heated by an alcohol lamp (the wick is at the bottom). Note the air turbulence on both sides of the rotating blade.   *(© 1973 Kim Vandiver and Harold Edgerton, Palm Press, Inc.)* (b) Hot gases from a cigarette made visible by smoke particles. The smoke first moves in streamline flow at the bottom and then in turbulent flow above.   *(Werner Wolff/Black Star)*

with the resistance of two adjacent layers of the fluid to move relative to each other. Because of viscosity, part of the kinetic energy of a fluid is converted to thermal energy. This is similar to the mechanism by which an object sliding on a rough horizontal surface loses kinetic energy.

Because the motion of a real fluid is complex and not yet fully understood, we make some simplifying assumptions in our approach. As we shall see, many features of real fluids in motion can be understood by considering the behavior of an ideal fluid. In our model of an **ideal fluid,** we make four assumptions:

- **Nonviscous fluid.**   In a nonviscous fluid, internal friction is neglected. An object moving through the fluid experiences no viscous force.
- **Steady flow.**   In steady flow, we assume that the velocity of the fluid at each point remains constant in time.
- **Incompressible fluid.**   The density of an incompressible fluid is assumed to remain constant in time.
- **Irrotational flow.**   Fluid flow is irrotational if there is no angular momentum of the fluid about any point. If a small wheel placed anywhere in the fluid does not rotate about the wheel's center of mass, the flow is irrotational. (If the wheel were to rotate, as it would if turbulence were present, the flow would be rotational.)

15.6   STREAMLINES AND THE EQUATION OF CONTINUITY

The path taken by a fluid particle under steady flow is called a **streamline.** The velocity of the fluid particle is always tangent to the streamline, as shown in Figure 15.13. No two streamlines can cross each other, for if they did, a fluid particle could move either way at the crossover point and then the flow would not be steady. A set of streamlines as shown in Figure 15.13 forms what is called a *tube of flow.* Note that fluid particles cannot flow into or out of the sides of this tube, since if they did, the streamlines would be crossing each other.

Properties of an ideal fluid

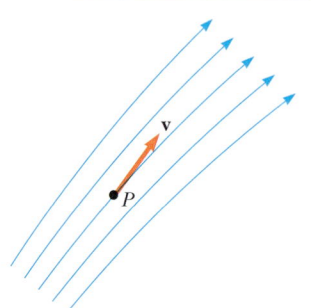

FIGURE 15.13   This diagram represents a set of streamlines (blue lines). A particle at *P* follows one of these streamlines, and its velocity is tangent to the streamline at each point along its path.

Consider an ideal fluid flowing through a pipe of nonuniform size as in Figure 15.14. The particles in the fluid move along the streamlines in steady flow. At all points the velocity of any particle is tangent to the streamline along which it moves.

In a small time interval Δt, the fluid at the bottom end of the pipe moves a distance $\Delta x_1 = v_1 \Delta t$. If A_1 is the cross-sectional area in this region, then the mass contained in the shaded region is $\Delta m_1 = \rho A_1 \Delta x_1 = \rho A_1 v_1 \Delta t$. Similarly, the fluid that moves through the upper end of the pipe in the time Δt has a mass $\Delta m_2 = \rho A_2 v_2 \Delta t$. However, since *mass is conserved* and because the flow is steady, the mass that crosses A_1 in a time Δt must equal the mass that crosses A_2 in a time Δt. That is, $\Delta m_1 = \Delta m_2$, or $\rho A_1 v_1 = \rho A_2 v_2$. If the density is common to both sides of this expression, we get

$$A_1 v_1 = A_2 v_2 = \text{constant} \tag{15.7}$$

This expression is called the **equation of continuity**. It says that

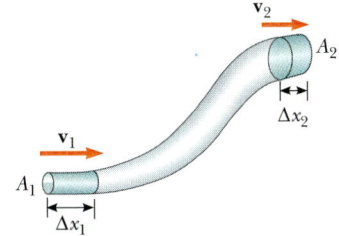

FIGURE 15.14 A fluid moving with streamline flow through a pipe of varying cross-sectional area. The volume of fluid flowing through A_1 in a time interval Δt must equal the volume flowing through A_2 in the same time interval. Therefore, $A_1 v_1 = A_2 v_2$.

> the product of the area and the fluid speed at all points along the pipe is a constant for an incompressible fluid.

Therefore, as one would expect, the speed is high where the tube is constricted and low where the tube is wide. The product Av, which has the dimensions of volume/time, is called the *volume flux*, or flow rate. The condition $Av = \text{constant}$ is equivalent to the fact that the amount of fluid that enters one end of the tube in a given time interval equals the amount leaving in the same time interval, assuming no leaks.

EXAMPLE 15.12 Filling a Water Bucket

A water hose 2.00 cm in diameter is used to fill a 20.0-liter bucket. If it takes 1.00 min to fill the bucket, what is the speed v at which the water leaves the hose? ($1 \text{ L} = 10^3 \text{ cm}^3$)

Solution The cross-sectional area of the hose is

$$A = \pi r^2 = \pi \frac{d^2}{4} = \pi \left(\frac{2.00^2}{4} \right) \text{cm}^2 = \pi \text{ cm}^2$$

According to the data given, the flow rate is equal to 20.0 liters/min. Equating this to the product Av gives

$$Av = 20.0 \, \frac{\text{L}}{\text{min}} = \frac{20.0 \times 10^3 \text{ cm}^3}{60.0 \text{ s}}$$

$$v = \frac{20.0 \times 10^3 \text{ cm}^3}{(\pi \text{ cm}^2)(60.0 \text{ s})} = 106 \text{ cm/s}$$

Exercise If the diameter of the hose is reduced to 1.00 cm, what will the speed of the water be as it leaves the hose, assuming the same flow rate?

Answer 424 cm/s.

15.7 BERNOULLI'S EQUATION

As a fluid moves through a pipe of varying cross-section and elevation, the pressure changes along the pipe. In 1738 the Swiss physicist Daniel Bernoulli first derived an expression that relates the pressure to fluid speed and elevation.

Consider the flow of an ideal fluid through a nonuniform pipe in a time Δt, as illustrated in Figure 15.15. The force on the lower end of the fluid is $P_1 A_1$, where P_1 is the pressure in section 1. The work done by this force is $W_1 = F_1 \Delta x_1 =$

Daniel Bernoulli

| 1 7 0 0 – 1 7 8 2 |

aniel Bernoulli was a Swiss physicist and mathematician who made important discoveries in hydrodynamics. Born into a family of mathematicians on February 8, 1700, he was the only member of the family to make a mark in physics. He was educated and received his doctorate in Basel, Switzerland.

Bernoulli's most famous work, *Hydrodynamica*, was published in 1738; it is both a theoretical and a practical study of equilibrium, pressure, and velocity of fluids. He showed that as the velocity of fluid flow increases, its pressure decreases. Referred to as "Bernoulli's principle," his work is used to produce a vacuum in chemical laboratories by connecting a vessel to a tube through which water is running rapidly.

Bernoulli's *Hydrodynamica* also attempted the first explanation of the behavior of gases with changing pressure and temperature; this was the beginning of the kinetic theory of gases.

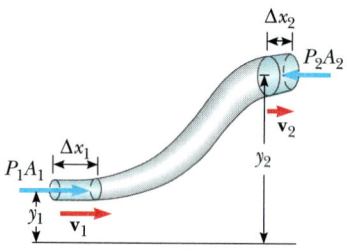

FIGURE 15.15   A fluid flowing through a constricted pipe with streamline flow. The fluid in the section of length Δx_1 moves to the section of length Δx_2. The volumes of fluid in the two sections are equal.

$P_1 A_1 \Delta x_1 = P_1 \Delta V$, where ΔV is the volume of section 1. In a similar manner, the work done on the fluid at the upper end in the time Δt is $W_2 = - P_2 A_2 \Delta x_2 = - P_2 \Delta V$. (The volume that passes through section 1 in a time Δt equals the volume that passes through section 2 in the same time interval.) This work is negative because the fluid force opposes the displacement. Thus the net work done by these forces in the time Δt is

$$W = (P_1 - P_2) \Delta V$$

Part of this work goes into changing the kinetic energy of the fluid, and part goes into changing the gravitational potential energy. If Δm is the mass passing through the pipe in the time Δt, then the change in its kinetic energy is

$$\Delta K = \tfrac{1}{2}(\Delta m) v_2{}^2 - \tfrac{1}{2}(\Delta m) v_1{}^2$$

The change in gravitational potential energy is

$$\Delta U = \Delta m g y_2 - \Delta m g y_1$$

We can apply the work-energy theorem in the form $W = \Delta K + \Delta U$ to this volume of fluid to give

$$(P_1 - P_2) \Delta V = \tfrac{1}{2}(\Delta m) v_2{}^2 - \tfrac{1}{2}(\Delta m) v_1{}^2 + \Delta m g y_2 - \Delta m g y_1$$

If we divide each term by ΔV and recall that $\rho = \Delta m / \Delta V$, the above expression reduces to

$$P_1 - P_2 = \tfrac{1}{2}\rho v_2{}^2 - \tfrac{1}{2}\rho v_1{}^2 + \rho g y_2 - \rho g y_1$$

Rearranging terms, we get

$$P_1 + \tfrac{1}{2}\rho v_1{}^2 + \rho g y_1 = P_2 + \tfrac{1}{2}\rho v_2{}^2 + \rho g y_2 \tag{15.8}$$

This is **Bernoulli's equation** as applied to an ideal fluid. It is often expressed as

Bernoulli's equation

$$P + \tfrac{1}{2}\rho v^2 + \rho g y = \text{constant} \tag{15.9}$$

Bernoulli's equation says that the sum of the pressure, (P), the kinetic energy per unit volume ($\tfrac{1}{2}\rho v^2$), and gravitational potential energy per unit volume ($\rho g y$) has the same value at all points along a streamline.

Note that Bernoulli's equation is *not* a sum of energy density terms, because
P is pressure, not energy density.

When the fluid is at rest, $v_1 = v_2 = 0$ and Equation 15.8 becomes

$$P_1 - P_2 = \rho g(y_2 - y_1) = \rho g h$$

which agrees with Equation 15.4.

EXAMPLE 15.13   The Venturi Tube

The horizontal constricted pipe illustrated in Figure 15.16, known as a *Venturi tube,* can be used to measure flow speeds in an incompressible fluid. Let us determine the flow speed at point 2 if the pressure difference $P_1 - P_2$ is known.

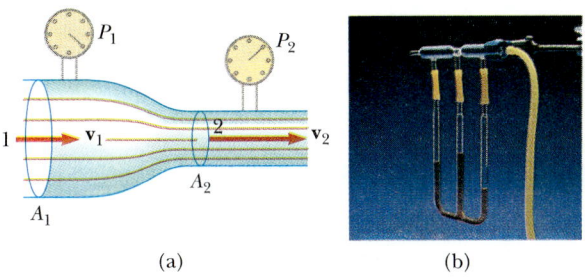

(a)   (b)

FIGURE 15.16   (a) (Example 15.13) The pressure P_1 is greater than the pressure P_2 since $v_1 < v_2$. This device can be used to measure the speed of fluid flow. (b) A Venturi tube.   *(Courtesy of Central Scientific Company)*

Solution   Since the pipe is horizontal, $y_1 = y_2$ and Equation 15.8 applied to points 1 and 2 gives

$$P_1 + \tfrac{1}{2}\rho v_1{}^2 = P_2 + \tfrac{1}{2}\rho v_2{}^2$$

From the equation of continuity (Eq. 15.7), we see that $A_1v_1 = A_2v_2$ or

$$v_1 = \frac{A_2}{A_1} v_2$$

Substituting this expression into the previous equation gives

$$P_1 + \tfrac{1}{2}\rho \left(\frac{A_2}{A_1}\right)^2 v_2{}^2 = P_2 + \tfrac{1}{2}\rho v_2{}^2$$

$$v_2 = A_1 \sqrt{\frac{2(P_1 - P_2)}{\rho(A_1{}^2 - A_2{}^2)}}$$

We can also obtain an expression for v_1 using this result and the continuity equation. Note that since $A_2 < A_1$, it follows that $P_1 > P_2$. In other words, the pressure is reduced in the constricted part of the pipe. This result is somewhat analogous to the following situation: Consider a very crowded room, where people are squeezed together. As soon as a door is opened and people begin to exit, the squeezing (pressure) is least near the door where the motion (flow) is greatest.

EXAMPLE 15.14   Torricelli's Law (Speed of Efflux)

A tank containing a liquid of density ρ has a hole in its side at a distance y_1 from the bottom (Fig. 15.17). The diameter of the hole is small compared to the diameter of the tank. The air above the liquid is maintained at a pressure P. Determine the speed at which the fluid leaves the hole when the liquid level is a distance h above the hole.

Solution   Because $A_2 \gg A_1$, the fluid is approximately at rest at the top, point 2. Applying Bernoulli's equation to points 1 and 2 and noting that at the hole $P_1 = P_0$, we get

$$P_0 + \tfrac{1}{2}\rho v_1{}^2 + \rho g y_1 = P + \rho g y_2$$

But $y_2 - y_1 = h$, and so this reduces to

$$v_1 = \sqrt{\frac{2(P - P_0)}{\rho} + 2gh}$$

The flow rate from the hole is A_1v_1. When P is large compared with atmospheric pressure P_0 (and therefore the term

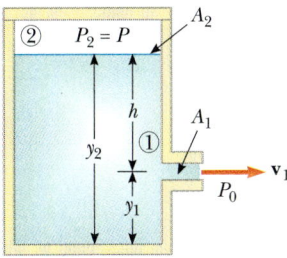

FIGURE 15.17   (Example 15.14) The water speed, v_1, from the hole in the side of the container is given by $v_1 = \sqrt{2gh}$.

$2gh$ can be neglected), the speed of efflux is mainly a function of P. Finally, if the tank is open to the atmosphere, then $P = P_0$ and $v_1 = \sqrt{2gh}$. In other words, the speed of efflux for an open tank is equal to that acquired by a body falling freely through a vertical distance h. This is known as **Torricelli's law.**

An instructor at the Massachusetts Institute of Technology tests his invention, a dimpled baseball bat, against a regulation bat in a wind tunnel. Just as dimples on jet aircraft and golf balls allow air to flow over them more efficiently, the dimpled bat also moved more efficiently and thus could strike a ball with more energy.   *(Bruce Dale © National Geographic Society)*

*15.8   OTHER APPLICATIONS OF BERNOULLI'S EQUATION

Consider the streamlines that flow around an airplane wing as shown in Figure 15.18. Let us assume that the airstream approaches the wing horizontally from the right with a velocity $\mathbf{v}_1$. The tilt of the wing causes the airstream to be deflected downward with a velocity $\mathbf{v}_2$. Because the airstream is deflected by the wing, the wing must exert a force on the airstream. According to Newton's third law, the airstream must exert an equal and opposite force $\mathbf{F}$ on the wing. This force has a vertical component called the **lift** (or aerodynamic lift) and a horizontal component called **drag**. The lift depends on several factors, such as the speed of the airplane, the area of the wing, its curvature, and the angle between the wing and the horizontal. As this angle increases, turbulent flow can set in above the wing to reduce the lift.

The lift on the wing is consistent with Bernoulli's equation. The speed of the airstream is greater above the wing, hence the air pressure above the wing is less than the pressure below the wing, resulting in a net upward force.

In general, an object experiences lift by any effect that causes the fluid to change its direction as it flows past the object. Some factors that influence the lift

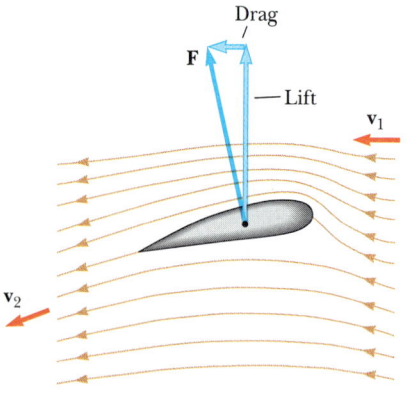

FIGURE 15.18 Streamline flow around a moving airplane wing. The air approaching from the right with a velocity $\mathbf{v}_1$ is deflected downward by the wing, leaving the trailing edge of the wing with a velocity $\mathbf{v}_2$. Because the airstream is deflected, it exerts a force $\mathbf{F}$ on the wing that has a vertical and horizontal component.

are the shape of the object, its orientation with respect to the fluid flow, spinning motion (for example, a spinning baseball), and the texture of the object's surface.

A number of devices operate in the manner described in Figure 15.19. A stream of air passing over an open tube reduces the pressure above the tube. This reduction in pressure causes the liquid to rise into the airstream. The liquid is then dispersed into a fine spray of droplets. You might recognize that this so-called atomizer is used in perfume bottles and paint sprayers. The same principle is used in the carburetor of a gasoline engine. In this case, the low-pressure region in the carburetor is produced by air drawn in by the piston through the air filter. The gasoline vaporizes, mixes with the air, and enters the cylinder of the engine for combustion.

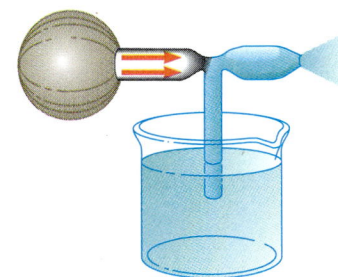

FIGURE 15.19 A stream of air passing over a tube dipped into a liquid will cause the liquid to rise in the tube as shown.

Bernoulli's equation explains one symptom called vascular flutter in a person with advanced arteriosclerosis. The artery is constricted as a result of an accumulation of plaque on its inner walls (Fig. 15.20). In order to maintain a constant flow rate through such a constricted artery, the driving pressure must increase. Such an increase in pressure requires a greater demand on the heart muscle. If the blood speed is sufficiently high in the constricted region, the artery may collapse under external pressure, causing a momentary interruption in blood flow. At this point, Bernoulli's equation does not apply, and the vessel reopens under arterial pressure. As the blood rushes through the constricted artery, the internal pressure drops and again the artery closes. Such variations in blood flow can be heard with a stethoscope. If the plaque becomes dislodged and ends up in a smaller vessel that delivers blood to the heart, the person can suffer a heart attack.

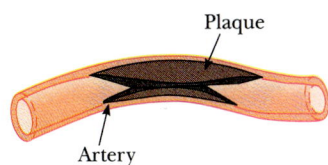

FIGURE 15.20   Blood must travel faster than normal through a constricted artery.

CONCEPTUAL EXAMPLE 15.15

A tornado or hurricane will often lift the roof of a house. Use Bernoulli's equation to explain how this occurs.

Reasoning   The rapidly moving air characteristic of a tornado or hurricane causes the external pressure to fall below atmospheric pressure. However, the stationary air inside the house remains at normal atmospheric pressure. The pressure difference between the inside and outside results in an upward force on the roof, which can be large enough to lift it off the house.

CONCEPTUAL EXAMPLE 15.16

When a fast-moving train passes a train at rest, the two tend to be drawn together. How does Bernoulli's equation explain this phenomenon?

Reasoning   As air is displaced by the moving train, the air passing between the trains has a higher relative speed than the air on the outside, which is free to expand. Thus, the air pressure is lower between the trains than on the outside, resulting in a net force that draws the trains closer to each other.

---

*15.9   ENERGY FROM THE WIND

Although the wind is a large potential source of energy (about 5 kW per acre in the United States), it has been harnessed only on a small scale. It has been estimated that, on a global scale, the winds account for a total available power of 2×10^{10} kW (about three times the current world power consumption). Therefore, if only a small percentage of the available power could be harnessed, wind power would represent a significant fraction of our energy needs. As with all indirect energy resources, wind power systems have some disadvantages, which in this case arise mainly from the variability of wind velocities.

We can use some of the ideas developed in this chapter to estimate wind power. Any wind-energy machine involves the conversion of the kinetic energy of moving air to the mechanical energy of an object, usually by means of a rotating shaft. The kinetic energy per unit volume of a moving column of air is

$$\frac{KE}{\text{volume}} = \tfrac{1}{2}\rho v^2$$

where ρ is the density of air and v is its speed. The rate of flow of air through a column of cross-sectional area A is Av (Fig. 15.21). This can be considered as the volume of air crossing a given surface area each second. In the working machine, A is the cross-sectional area of the wind-collecting system, such as a set of rotating propeller blades. Multiplying the kinetic energy per unit volume by the flow rate gives the rate at which energy is transferred, or, in other words, the power:

$$\text{Power} = \frac{KE}{\text{volume}} \times \frac{\text{volume}}{\text{time}} = (\tfrac{1}{2}\rho v^2)(Av) = \tfrac{1}{2}\rho v^3 A \qquad (15.10)$$

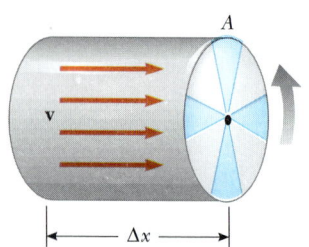

FIGURE 15.21   Wind moving through a cylindrical column of cross-sectional area A with a speed v.

Therefore, the available power per unit area is

$$\frac{\text{Power}}{A} = \tfrac{1}{2}\rho v^3 \qquad (15.11)$$

According to this result, if the moving air column could be brought to rest, a power of $\tfrac{1}{2}\rho v^3$ would be available for each square meter that was intercepted. For example, if we assume a moderate speed of 12 m/s (27 mi/h) and take $\rho = 1.3$ kg/m³, we find that

$$\frac{\text{Power}}{A} = \tfrac{1}{2}(1.3 \text{ kg/m}^3)(12 \text{ m/s})^3 \approx 1100 \text{ W/m}^2 = 1.1 \text{ kW/m}^2$$

Because the power per unit area varies as the cube of the speed, its value doubles if v increases by only 26%. Conversely, the power output is halved if the speed decreases by 21%.

This calculation is based on ideal conditions and assumes that all of the kinetic energy is available for power. In reality, the airstream emerges from the wind generator with some residual speed, and more refined calculations show that, at

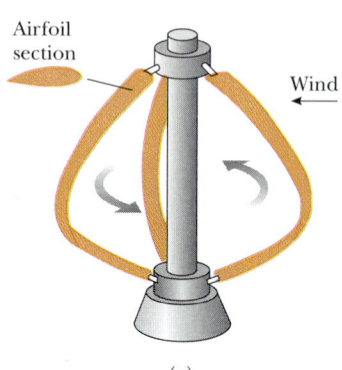

(a)

(b)

(c)

FIGURE 15.22   (a) A vertical-axis wind generator. (b) A horizontal-axis wind generator. (c) Photograph of a vertical-axis wind generator. *(Courtesy of U.S. Department of Energy)*

best, one can extract only 59.3% of this quantity. The expression for the maximum available power per unit area for the ideal wind generator is

$$\frac{\text{Maximum power}}{A} = \frac{8}{27}\rho v^3 \qquad (15.12)$$

In a real wind machine, further losses resulting from the nonideal nature of the propeller, gearing, and generator reduce the total available power to around 15% of the value predicted by Equation 15.11. Sketches of two types of wind turbines are shown in Figure 15.22.

EXAMPLE 15.17 **Power Output of a Windmill**

Calculate the power output of a wind generator having a blade diameter of 80 m, assuming a wind speed of 10 m/s and an overall efficiency of 15%.

Solution Since the radius of the blade is 40 m, the cross-sectional area of the propellers is

$$A = \pi r^2 = \pi(40 \text{ m})^2 = 5.0 \times 10^3 \text{ m}^2$$

If 100% of the available wind energy could be extracted, the maximum available power would be

$$\text{Maximum power} = \tfrac{1}{2}\rho A v^3$$
$$= \tfrac{1}{2}(1.2 \text{ kg/m}^3)(5.0 \times 10^3 \text{ m}^2)(10 \text{ m/s})^3$$

$$= 3.0 \times 10^6 \text{ W} = 3.0 \text{ MW}$$

Since the overall efficiency is 15%, the output power is

$$\text{Power} = 0.15(\text{maximum power}) = \boxed{0.45 \text{ MW}}$$

In comparison, a large steam-turbine plant has a power output of about 1 GW. Hence, 2200 such wind generators would be required to equal this output under these conditions. The large number of generators required for reasonable output power is clearly a major disadvantage of wind power. (See Problem 52.)

---

*15.10 VISCOSITY

A fluid does not support a shearing stress. However, fluids do offer some degree of resistance to shearing motion. This resistance to shearing motion is a form of internal friction called *viscosity*. The viscosity arises because of a frictional force between adjacent layers of the fluid as they slide past one another. The degree of viscosity of a fluid can be understood with the following example. If two plates of glass are separated by a layer of fluid such as oil, with one plate fixed in position, it is easy to slide one plate over the other (Fig. 15.23). However, if the fluid separating the plates is tar, the task of sliding one plate over the other becomes much more difficult. Thus, we would conclude that tar has a higher viscosity than oil. In Figure 15.23, note that the speed of successive layers of fluid increases linearly from 0 to v as one moves from a layer adjacent to the fixed plate to a layer adjacent to the moving plate.

Recall that in a solid a shearing stress gives rise to a relative displacement of adjacent layers (Section 12.4). In an analogous fashion, adjacent layers of a fluid under shear stress are set into relative motion. Again, consider two parallel layers, one fixed and one moving to the right under the action of an external force **F** as in Figure 15.23. Because of this motion, a portion of the fluid is distorted from its original shape, *ABCD*, at one instant to the shape *AEFD* after a short time interval. If you refer to Section 12.4, you will recognize that the fluid has undergone a shear strain. By definition, the shear stress on the fluid is equal to the ratio F/A, while the shear strain is defined by the ratio $\Delta x/\ell$:

$$\text{Shear stress} = \frac{F}{A} \qquad \text{Shear strain} = \frac{\Delta x}{\ell}$$

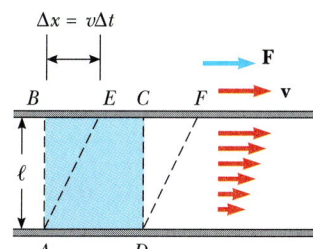

FIGURE 15.23 A layer of liquid between two solid surfaces in which the lower surface is fixed and the upper surface moves to the right with a velocity **v**.

TABLE 15.2 Coefficients of Viscosity of Various Fluids		
Fluid	T(°C)	η (N·s/m²)
Water	20	1.0×10^{-3}
Water	100	0.3×10^{-3}
Whole blood	37	2.7×10^{-3}
Glycerine	20	830×10^{-3}
Motor oil (SAE 10)	30	250×10^{-3}
Air	20	1.8×10^{-5}

The upper plate moves with a speed v, and the fluid adjacent to this plate has the same speed. Thus, in a time Δt, the fluid at the moving plate travels a distance $\Delta x = v \, \Delta t$, and we can express the shear strain per unit time as

$$\frac{\text{Shear strain}}{\Delta t} = \frac{\Delta x/\ell}{\Delta t} = \frac{v}{\ell}$$

This equation states that the rate of change of shearing strain is v/ℓ.

The **coefficient of viscosity**, η, for the fluid is defined as the ratio of the shearing stress to the rate of change of the shear strain:

Coefficient of viscosity

$$\eta \equiv \frac{F/A}{v/\ell} = \frac{F\ell}{Av} \tag{15.13}$$

The SI unit of the coefficient of viscosity is N·s/m². The coefficients of viscosity for some fluids are given in Table 15.2.

The expression for η given by Equation 15.13 is valid only if the fluid speed varies linearly with position. In this case, it is common to say that the speed gradient, v/ℓ, is uniform. If the speed gradient is *not* uniform, we must express η in the general form

$$\eta \equiv \frac{F/A}{dv/dy} \tag{15.14}$$

where the speed gradient dv/dy is the change in speed with position as measured perpendicular to the direction of velocity.

EXAMPLE 15.18   Measuring the Coefficient of Viscosity

A metal plate of surface area 0.15 m² is connected to an 8.0-g mass via a string that passes over an ideal pulley (massless and frictionless), as in Figure 15.24. A lubricant having a film thickness of 0.30 mm is placed between plate and surface. When released, the plate moves to the right with a constant speed of 0.085 m/s. Find the coefficient of viscosity of the lubricant.

Solution   Because the plate moves with constant speed, its acceleration is zero. It moves to the right under the action of the force T exerted by the string and the frictional force f

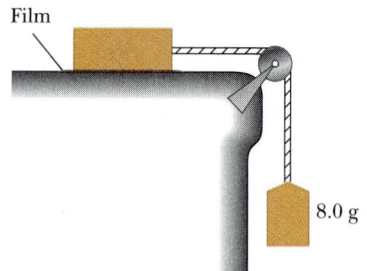

FIGURE 15.24   (Example 15.18).

associated with the viscous fluid. In this case, the tension is equal in magnitude to the suspended weight; therefore,

$$f = T = mg = (8.0 \times 10^{-3} \text{ kg})(9.80 \text{ m/s}^2) = 7.8 \times 10^{-2} \text{ N}$$

The lubricant in contact with the horizontal surface is at rest, while the layer in contact with the plate moves at the speed of the plate. Assuming the speed gradient is uniform, we have

$$\eta = \frac{F\ell}{Av} = \frac{(7.8 \times 10^{-2} \text{ N})(0.30 \times 10^{-3} \text{ m})}{(0.15 \text{ m}^2)(0.085 \text{ m/s})}$$

$$= 5.5 \times 10^{-3} \text{ N} \cdot \text{s/m}^2$$

SUMMARY

The **pressure,** P, in a fluid is the force per unit area that the fluid exerts on any surface:

$$P \equiv \frac{F}{A} \tag{15.1}$$

In the SI system, pressure has units of N/m², and 1 N/m² = 1 pascal (Pa).

The pressure in a fluid varies with depth h according to the expression

$$P = P_0 + \rho g h \tag{15.4}$$

where P_0 is atmospheric pressure ($= 1.013 \times 10^5$ N/m²) and ρ is the density of the fluid, assumed uniform.

Pascal's law states that when pressure is applied to an enclosed fluid, the pressure is transmitted undiminished to every point in the fluid and to every point on the walls of the container.

When an object is partially or fully submerged in a fluid, the fluid exerts an upward force on the object called the **buoyant force.** According to **Archimedes' principle,** the buoyant force is equal to the weight of the fluid displaced by the body.

Various aspects of fluid dynamics can be understood by assuming that the fluid is nonviscous and incompressible and that the fluid motion is a steady flow with no turbulence.

Using these assumptions, two important results regarding fluid flow through a pipe of nonuniform size can be obtained:

- The flow rate through the pipe is a constant, which is equivalent to stating that the product of the cross-sectional area, A, and the speed, v, at any point is a constant:

$$A_1 v_1 = A_2 v_2 = \text{constant} \tag{15.7}$$

- The sum of the pressure, kinetic energy per unit volume, and gravitational potential energy per unit volume has the same value at all points along a streamline:

$$P + \tfrac{1}{2}\rho v^2 + \rho g y = \text{constant} \tag{15.9}$$

QUESTIONS

1. Two drinking glasses having equal weights but different shapes and different cross-sectional areas are filled to the same level with water. According to the expression $P =$ $P_0 + \rho g h$, the pressure is the same at the bottom of both glasses. In view of this, why does one weigh more than the other?

2. If the top of your head has an area of 100 cm², what is the weight of the air above your head?

3. When you drink a liquid through a straw, you reduce the pressure in your mouth and let the atmosphere move the liquid. Explain how this works. Could you use a straw to sip a drink on the Moon?

4. Pascal used a barometer with water as the working fluid. Why is it impractical to use water for a typical barometer?

5. A helium-filled balloon rises until its density becomes the same as that of the air. If a sealed submarine begins to sink, will it go all the way to the bottom of the ocean or will it stop when its density becomes the same as that of the surrounding water?

6. A fish rests on the bottom of a bucket of water while the bucket is being weighed. When the fish begins to swim around, does the weight change?

7. Will a ship ride higher in the water of an inland lake or in the ocean? Why?

8. If 1 000 000 N of weight were placed on the deck of the World War II battleship North Carolina, the ship would sink only 2.5 cm lower in the water. What is the cross-sectional area of the ship at water level?

9. Lead has a greater density than iron, and both are denser than water. Is the buoyant force on a lead object greater than, less than, or equal to the buoyant force on an iron object of the same volume?

10. An ice cube is placed in a glass of water. What happens to the level of the water as the ice melts?

11. The water supply for a city is often provided from reservoirs built on high ground. Water flows from the reservoir, through pipes, and into your home when you turn the tap on your faucet. Why is the water flow more rapid out of a faucet on the first floor of a building than in an apartment on a higher floor?

12. Smoke rises in a chimney faster when a breeze is blowing. Use Bernoulli's equation to explain this phenomenon.

13. Why do many trailer trucks use wind deflectors on the top of their cabs? How do such devices reduce fuel consumption?

14. If you suddenly turn on your shower water at full speed, why is the shower curtain pushed inward?

15. If air from a hair dryer is blown over the top of a Ping-Pong ball, the ball can be suspended in air. Explain.

16. When ski-jumpers are airborne, why do they bend their bodies forward and keep their hands at their sides?

17. When an object is immersed in a liquid at rest, why is the net force on the object in the horizontal direction equal to zero?

18. Explain why a sealed bottle partially filled with a liquid can float.

19. When is the buoyant force on a swimmer greater—after the swimmer is exhaling or after inhaling?

20. A piece of unpainted wood is partially submerged in a container filled with water. If the container is sealed and pressurized above atmospheric pressure, does the wood rise, fall, or remain at the same level? (*Hint:* Wood is porous.)

21. A flat plate is immersed in a liquid at rest. For what orientation of the plate is the pressure on its flat surface uniform?

22. Because atmospheric pressure is about 10^5 N/m² and the area of a person's chest is about 0.13 m², the force of the atmosphere on one's chest is around 13 000 N. In view of this enormous force, why don't our bodies collapse?

23. How would you determine the density of an irregularly shaped rock?

24. Why do airplane pilots prefer to take off into the wind?

25. If you release a ball while inside a freely falling elevator, the ball remains in front of you rather than falling to the floor, because the ball, the elevator, and you all experience the same downward acceleration, **g**. What happens if you repeat this experiment with a helium-filled balloon? (This one is tricky.)

26. Two identical ships set out to sea. One is loaded with a cargo of Styrofoam, and the other is empty. Which ship is more submerged?

27. A small piece of steel is tied to a block of wood. When the wood is placed in a tub of water with the steel on top, half of the block is submerged. If the block is inverted so that the steel is under water, does the amount of the block submerged increase, decrease, or remain the same? What happens to the water level in the tub when the block is inverted?

28. Prairie dogs ventilate their burrows by building a mound over one entrance, which is open to a stream of air. A second entrance at ground level is open to almost stagnant air. How does this construction create an air flow through the burrow?

29. An unopened can of diet cola floats when placed in a tank of water, whereas a can of regular cola of the same brand sinks in the tank. What could explain this behavior?

30. The photograph (top left, p. 443) shows a glass cylinder containing four liquids of different densities. From top to bottom, the liquids are oil (orange), water (yellow), salt water (green), and mercury (silver). The cylinder also contains, from top to bottom, a Ping-Pong ball, a piece of wood, an egg, and a steel ball. (a) Which of these liquids has the lowest density, and which has the greatest? (b) What can you conclude about the density of each object?

31. In the photograph (top right, p. 443), an air stream moves from right to left through a tube that is constricted at the middle. Three Ping-Pong balls are levitated in

Question 16.   *(Galen Powell/Peter Arnold, Inc.)*

Question 30. *(Henry Leap and Jim Lehman)*

equilibrium above the vertical columns through which the air escapes. (a) Why is the ball at the right higher than the one in the middle? (b) Why is the ball at the left lower than the ball at the right even though the horizontal tube has the same dimensions at these two points?

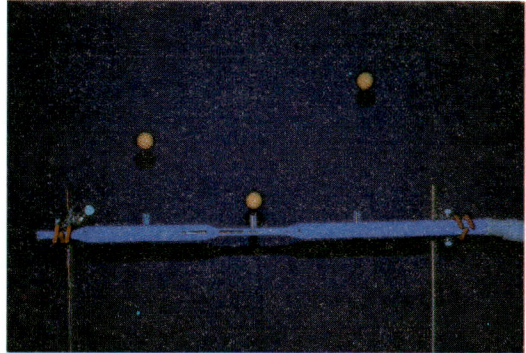

Question 31. *(Henry Leap and Jim Lehman)*

PROBLEMS

Section 15.1 Pressure

1. Calculate the mass of a solid iron sphere that has a diameter of 3.0 cm.

2. A small ingot of shiny grey metal has a volume of 25 cm^3 and a mass of 535 g. What is the metal? (See Table 15.1.)

3. Estimate the density of the *nucleus* of an atom. What does this result suggest concerning the structure of matter? (Use the fact that the mass of a proton is 1.67×10^{-27} kg and its radius is approximately 10^{-15} m.)

4. A king orders a gold crown having a mass of 0.5 kg. When it arrives from the metalsmith, the volume of the crown is found to be 185 cm^3. Is the crown made of solid gold?

5. A 50-kg woman balances on one heel of a pair of high-heeled shoes. If the heel is circular with radius 0.5 cm, what pressure does she exert on the floor?

6. What is the total mass of the Earth's atmosphere? (The radius of the Earth is 6.37×10^6 m, and atmospheric pressure at the surface is $1.01 \times 10^5 \text{ N/m}^2$.)

7. Estimate the density of a neutron star. Such an object is thought to have a radius of only 10 km and a mass equal to that of the Sun. ($M_{\text{Sun}} = 1.99 \times 10^{30}$ kg.)

Section 15.2 Variation of Pressure with Depth

8. Determine the absolute pressure at the bottom of a lake that is 30 m deep.

9. A cubic sealed vessel with edge L is placed on a cart, which is moving horizontally with an acceleration a as in Figure P15.9. The cube is filled with a fluid having density ρ. Determine the pressure P at the center of the cube.

FIGURE P15.9

10. The small piston of a hydraulic lift has a cross-sectional area of 3.00 cm^2, and the large piston has an area of 200 cm^2 (Fig. 15.4). What force must be applied to the small piston to raise a load of 15.0 kN? (In service stations this is usually accomplished with compressed air.)

11. The spring of the pressure gauge shown in Figure 15.2 has a force constant of 1000 N/m, and the piston has a diameter of 2.0 cm. Find the depth in water for which the spring compresses by 0.50 cm.

12. A swimming pool has dimensions 30 m $\times$ 10 m and a flat bottom. When the pool is filled to a depth of 2.0 m with fresh water, what is the total force due to

☐ indicates problems that have full solutions available in the Student Solutions Manual and Study Guide.

the water on the bottom? On each end? On each side?

13. What must be the contact area between a suction cup (completely exhausted) and a ceiling in order to support the weight of an 80.0-kg student?

14. The tank in Figure P15.14 is filled 2.0-m deep with water. At the bottom of the tank there is a rectangular hatch 1.0 m high and 2.0 m wide that is hinged at the top of the hatch. (a) Determine the net force on the hatch. (b) Find the torque exerted about the hinges.

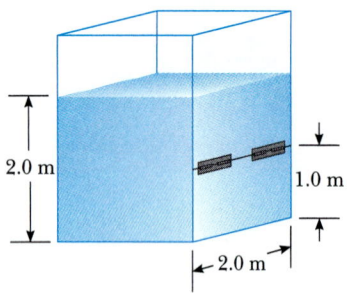

FIGURE P15.14

15. A solid copper ball with a diameter of 3.00 m at sea level is placed at the bottom of the ocean, at a depth of 10.000 km. If the density of seawater is 1030 kg/m^3, by how much (approximately) does the diameter of the ball decrease when it reaches bottom? The bulk modulus of copper is 14×10^{10} N/m^2.

16. What is the hydrostatic force on the back of Grand Coulee Dam if the water in the reservoir is 150 m deep and the width of the dam is 1200 m?

17. In some places, the Greenland ice sheet is 1.0 km thick. Estimate the pressure on the ground underneath the ice. ($\rho_{ice} = 920$ kg/m^3.)

Section 15.3   Pressure Measurements

18. Mercury is poured into a U-tube as in Figure P15.18a. The left arm of the tube has a cross-sectional area of $A_1 = 10.0$ cm^2, and the right arm has a cross-sectional area of $A_2 = 5.00$ cm^2. One hundred grams of water are then poured into the right arm as in Figure P15.18b. (a) Determine the length of the water column in the right arm of the U-tube. (b) Given that the density of mercury is 13.6 g/cm^3, what distance, h, does the mercury rise in the left arm?

19. A U-tube of constant cross-sectional area, open to the atmosphere, is partially filled with mercury. Water is then poured into both arms. If the equilibrium configuration of the tube is as shown in Figure P15.19, with $h_2 = 1.00$ cm, determine the value of h_1.

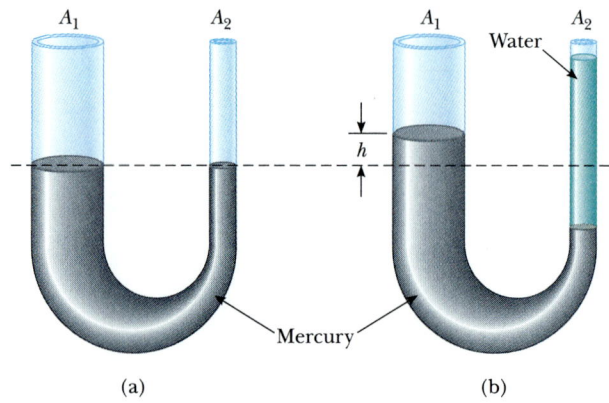

FIGURE P15.18

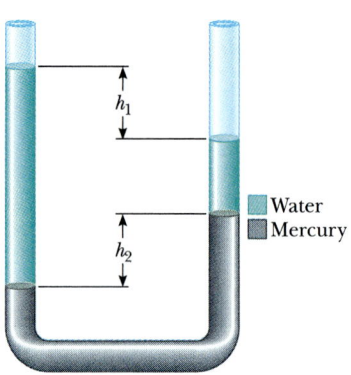

FIGURE P15.19

20. The open vertical tube in Figure P15.20 contains two fluids of densities ρ_1 and ρ_2, which do not mix. Show that the pressure at the depth $h_1 + h_2$ is given by the expression $P = P_0 + \rho_1 g h_1 + \rho_2 g h_2$.

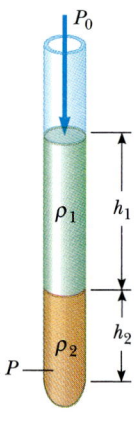

FIGURE P15.20

21. Blaise Pascal duplicated Torricelli's barometer using (as a Frenchman would) a red Bordeaux wine as the working liquid (Fig. P15.21). The density of the wine he used was 0.984×10^3 kg/m³. What was the height h of the wine column for normal atmospheric pressure? Would you expect the vacuum above the column to be as good as for mercury?

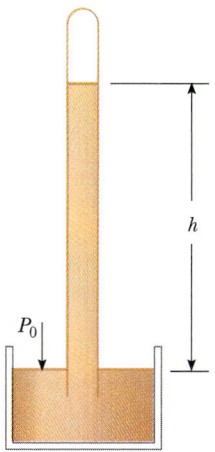

h

P_0

FIGURE P15.21

22. Normal atmospheric pressure is 1.013×10^5 Pa. The approach of a storm causes the height of a mercury barometer to drop by 20 mm from the normal height. What is the atmospheric pressure? (The density of mercury is 13.59 g/cm³.)

23. A simple U-tube that is open at both ends is partially filled with water (Fig. P15.23). Kerosene ($\rho_k = 0.82 \times 10^3$ kg/m³) is then poured into one arm of the tube, forming a column 6.0 cm in height, as shown in the diagram. What is the difference h in the heights of the two liquid surfaces?

23A. A simple U-tube that is open at both ends is partially filled with water (Fig. P15.23). Kerosene of density ρ_k is then poured into one arm of the tube, forming a column of height h_k, as shown in the diagram. What is the difference h in the heights of the two liquid surfaces?

Section 15.4   Buoyant Forces and Archimedes' Principle

24. Helium balloons having mass 5.0 g when deflated and radii 20.0 cm each are used by a 20.0-kg boy in order to lift himself off the ground. How many balloons are needed if the density of helium is 0.18 kg/m³, and the density of air is 1.29 kg/m³?

25. What is the true weight of one cubic meter of balsa wood that has a specific gravity of 0.15? Note the true weight of an object is its weight in vacuum.

26. A long cylindrical tube of radius r is weighted on one end so that it floats upright in a fluid having density ρ. It is pushed down a distance x from its equilibrium position and released. Show that it will execute simple harmonic motion if the resistive effects of the water are neglected, and determine the period of the oscillations.

27. A cube of wood 20 cm on a side and having a density of 0.65×10^3 kg/m³ floats on water. (a) What is the distance from the top face of the cube to the water level? (b) How much lead weight has to be placed on top of the cube so that its top is just level with the water? (Assume its top face remains parallel to the water's surface.)

28. A balloon is filled with 400 m³ of helium. How big a payload can the balloon lift? (The density of air is 1.29 kg/m³; the density of helium is 0.180 kg/m³.)

29. A plastic sphere floats in water with 50% of its volume submerged. This same sphere floats in oil with 40% of its volume submerged. Determine the densities of the oil and the sphere.

30. A 10-kg block of metal measuring 12 cm × 10 cm × 10 cm is suspended from a scale and immersed in water as in Figure 15.9b. The 12-cm dimension is vertical, and the top of the block is 5.0 cm from the surface of the water. (a) What are the forces on the top and bottom of the block? (Take $P_0 = 1.0130 \times 10^5$ N/m².) (b) What is the reading of the spring scale? (c) Show that the buoyant force equals the difference between the forces at the top and bottom of the block.

31. A frog in a hemispherical pod (Fig. P15.31) finds that he just floats without sinking in a blue-green sea

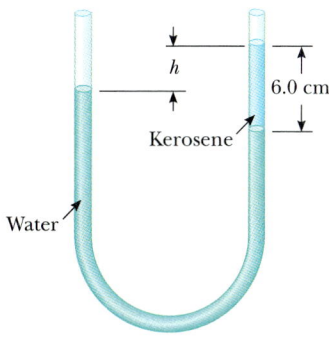

h

6.0 cm

Kerosene

Water

FIGURE P15.23

FIGURE P15.31

(density 1.35 g/cm³). If the pod has a radius of 6.0 cm and has negligible mass, what is the mass of the frog?

32. A light balloon filled with helium of density 0.180 kg/m³ is tied to a light string of length $L = 3.00$ m. The string is tied to the ground, forming an "inverted" simple pendulum as in Figure P15.32a. If the balloon is displaced slightly from equilibrium, as in Figure P15.32b, (a) show that the motion is simple harmonic and (b) determine the period of the motion. Take the density of air to be 1.29 kg/m³, and ignore any energy lost due to air friction.

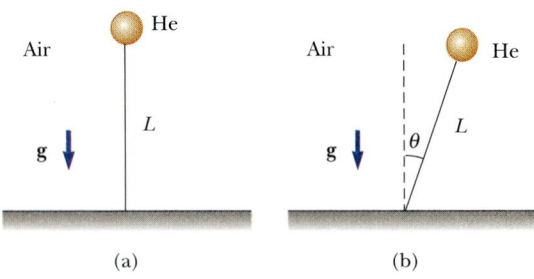

FIGURE P15.32

33. A Styrofoam slab has a thickness of 10 cm and a density of 300 kg/m³. What is the area of the slab if it floats just awash in fresh water when a 75-kg swimmer is aboard?

33A. A Styrofoam slab has a thickness h and a density ρ_s. What is the area of the slab if it floats just awash in fresh water when a swimmer of mass m is aboard?

34. A balloon is used to suspend a 0.020 m³ block of aluminum in water by filling it with air. (a) What volume of air is necessary to just suspend this with the top of the balloon at the water's surface? (b) If instead of being solid, the aluminum had a 0.0060 m³ hollow cavity in it, what fraction of the balloon would be above the water? Ignore the mass of the air in the balloon.

35. How many cubic meters of helium are required to lift a balloon with a 400-kg payload to a height of 8000 m? ($\rho_{He} = 0.18$ kg/m³.) Assume the balloon maintains a constant volume and that the density of air decreases with altitude z according to the expression $\rho_{air} = \rho_0 e^{-z/8000}$, where z is in meters, and ρ_0 ($= 1.29$ kg/m³) is the density of air at sea level.

Sections 15.5–15.7   Fluid Dynamics and Bernoulli's Equation

36. The rate of flow of water through a horizontal pipe is 2.00 m³/min. Determine the speed of flow at a point where the diameter of the pipe is (a) 10.0 cm, (b) 5.0 cm.

37. A large storage tank filled with water develops a small hole in its side at a point 16 m below the water level. If the rate of flow from the leak is 2.5×10^{-3} m³/min, determine (a) the speed at which the water leaves the hole and (b) the diameter of the hole.

37A. A large storage tank filled with water develops a small hole in its side at a point a distance h below the water level. If the rate of flow from the leak is R m³/min, determine (a) the speed at which the water leaves the hole and (b) the diameter of the hole.

38. The legendary Dutch boy who saved Holland by placing his finger in the hole of a dike had a finger 1.2 cm in diameter. Assuming the hole was 2.0 m below the surface of the sea (density 1030 kg/m³), (a) what was the force on his finger? (b) If he removed his finger from the hole, how long would it take the released water to fill one acre of land to a depth of one foot assuming the hole remained constant in size? (A typical U.S. family of four uses one acre-foot, 1234 m³, of water each year.)

39. A horizontal pipe 10.0 cm in diameter has a smooth reduction to a pipe 5.0 cm in diameter. If the pressure of the water in the larger pipe is 8.0×10^4 Pa and the pressure in the smaller pipe is 6.0×10^4 Pa, at what rate does water flow through the pipes?

40. Water is pumped from the Colorado River to Grand Canyon Village through a 15.0-cm-diameter pipe. The river is at 564 m elevation and the village is at 2096 m. (a) What is the minimum pressure the water must be pumped at to arrive at the village? (b) If 4500 m³ are pumped per day, what is the speed of the water in the pipe? (c) What additional pressure is necessary to deliver this flow? (*Note:* You may assume that the gravitational field strength and the density of air are constant over this range of elevations.)

41. Water flows through a fire hose of diameter 6.35 cm at a rate of 0.0120 m³/s. The fire hose ends in a nozzle of inner diameter 2.20 cm. What is the speed with which the water exits the nozzle?

42. The Garfield Thomas water tunnel at Pennsylvania State University has a circular cross-section that constricts from a diameter of 3.6 m to the test section, which is 1.2 m in diameter. If the speed of flow is 3.0 m/s in the larger-diameter pipe, determine the speed of flow in the test section.

43. Old Faithful Geyser in Yellowstone Park erupts at approximately 1-hour intervals, and the height of the fountain reaches 40 m. (a) With what speed does the water leave the ground? (b) What is the pressure (above atmospheric) in the heated underground chamber if its depth is 175 m?

*Section 15.8   Other Applications of Bernoulli's Equation

44. An airplane has a mass of 1.6×10^4 kg, and each wing has an area of 40.0 m². During level flight, the pressure on the lower wing surface is 7.0×10^4 Pa. Determine the pressure on the upper wing surface.

45. A pump is designed as a horizontal cylinder with a cross-sectional area of A and the open hole of cross-sectional area a with a very much smaller than A. A fluid having density ρ is forced out of the pump by a piston moving at a constant speed of v by an applied constant force F (Fig. P15.45). Determine the speed u of the jet of fluid.

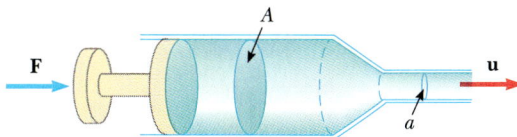

FIGURE P15.45

46. A Venturi tube may be used as a fluid flow meter (Fig. 15.16). If the difference in pressure $P_1 - P_2 = 21$ kPa, find the fluid flow rate in m³/s given that the radius of the outlet tube is 1.0 cm, the radius of the inlet tube is 2.0 cm, and the fluid is gasoline ($\rho = 700$ kg/m³).

47. A Pitot tube can be used to determine the velocity of air flow by measuring the difference between the total pressure and the static pressure (Fig. P15.47). If the fluid in the tube is mercury, density $\rho_{Hg} = 13\,600$ kg/m³, and $\Delta h = 5.00$ cm, find the speed of air flow. (Assume that the air is stagnant at point A and take $\rho_{air} = 1.25$ kg/m³.)

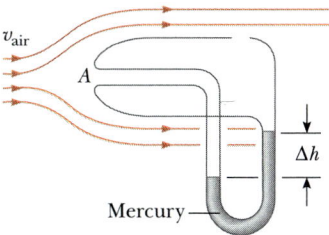

FIGURE P15.47

48. A siphon is used to drain water from a tank, as indicated in Figure P15.48. The siphon has a uniform diameter. Assume steady flow. (a) If the distance $h = 1.00$ m, find the speed of outflow at the end of the siphon. (b) What is the limitation on the height of the top of the siphon above the water surface? (In order to have a continuous flow of liquid, the pressure must not drop below the vapor pressure of the liquid.)

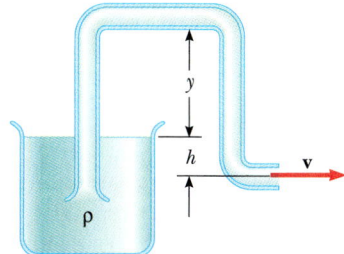

FIGURE P15.48

49. A large storage tank is filled to a height h_0. If the tank is punctured at a height h from the bottom of the tank (Fig. P15.49), how far from the tank will the stream land?

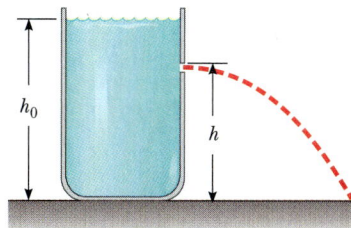

FIGURE P15.49

50. A hole is punched in the side of a 20-cm-tall container, full of water as shown in Figure P15.49. If the water is to shoot as far as possible horizontally, (a) how far from the bottom of the container should the hole be punched? (b) Neglecting friction losses, how far (initially) from the side of the container will the water land?

50A. A hole is punched in the side of a container of height h_0, full of water as shown in Figure P15.49. If the water is to shoot as far as possible horizontally, (a) how far from the bottom of the container should the hole be punched? (b) Neglecting friction losses, how far (initially) from the side of the container will the water land?

*Section 15.9   Energy from the Wind

51. Calculate the power output of a windmill having blades 10.0 m in diameter if the wind speed is 8.0 m/s. Assume that the efficiency of the system is 20%.

52. According to one rather ambitious plan, it would take 50 000 windmills, each 800 ft in diameter, to obtain an average output of 200 GW. These would be strategically located through the Great Plains, along the Aleutian Islands, and on floating platforms along the Atlantic and Gulf coasts and on the Great Lakes.

The annual energy consumption in the United States in 1980 was about 8.3×10^{19} J. What fraction of this could be supplied by the array of windmills?

53. Consider a windmill with blades of cross-sectional area A, as in Figure P15.53, and assume the mill is facing directly into the wind. If the wind speed is v, show that the kinetic energy of the air passing freely through an area A in a time Δt is $K = \frac{1}{2}\rho A v^3 \, \Delta t$. Why is it not possible for the blades to extract this amount of kinetic energy?

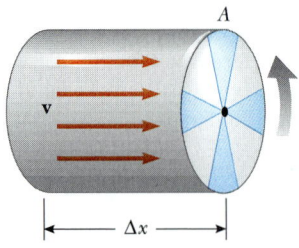

FIGURE P15.53

ADDITIONAL PROBLEMS

54. The water supply of a building is fed through a main 6.0-cm-diameter pipe. A 2.0-cm-diameter faucet tap located 2.0 m above the main pipe is observed to fill a 25-liter container in 30.0 s. (a) What is the speed at which the water leaves the faucet? (b) What is the *gauge pressure* in the 6.0-cm main pipe? (Assume the faucet is the only "leak" in the building.)

55. A Ping-Pong ball has a diameter of 3.8 cm and average density of 0.084 g/cm^3. What force is required to hold it completely submerged under water?

56. Figure P15.56 shows a water tank with a valve at the bottom. If this valve is opened, what is the maximum height attained by the water stream coming out of the right side of the tank? Assume that $h = 10$ m, $L = 2.0$ m, and $\theta = 30°$ and that the cross-sectional area at A is very large compared with that at B.

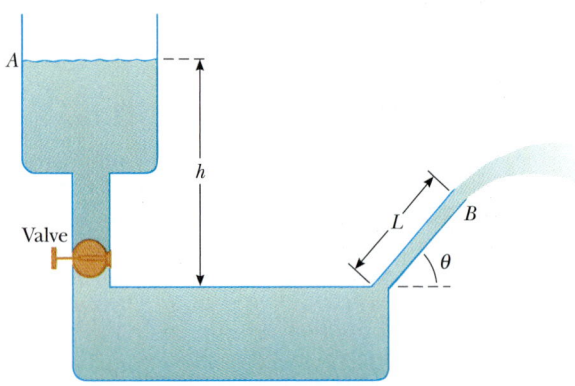

FIGURE P15.56

57. A helium-filled balloon is tied to a 2.0-m-long, 0.050-kg uniform string. The balloon is spherical with a radius of 0.40 m. When released, it lifts a length, h, of string and then remains in equilibrium, as in Figure P15.57. Determine the value of h. When deflated, the balloon has a mass of 0.25 kg.

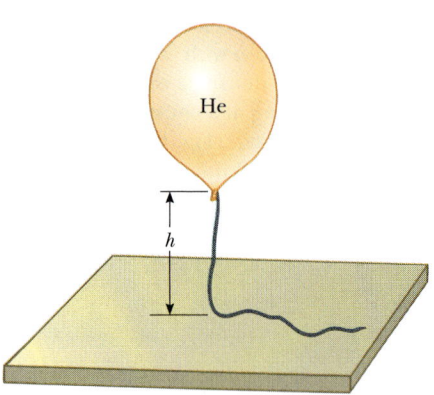

FIGURE P15.57

58. A tube of uniform cross-sectional area is open to the atmosphere and has the shape shown in Figure P15.58, with $\theta = 30.0°$. It is initially filled with water as in Figure P15.58a, and then oil (density = 750 kg/m^3) is poured into the left arm, forming a column 0.80 m long (slant height, s) as in Figure P15.58b. By what distance, h, does the water column rise in the right arm?

58A. A tube of uniform cross-sectional area is open to the atmosphere and has the shape shown in Figure P15.58. It is initially filled with water as in Figure P15.58a, and then oil of density ρ_0 is poured into the left arm, forming a column of slant height s, as in Figure P15.58b. By what distance, h, does the water column rise in the right arm?

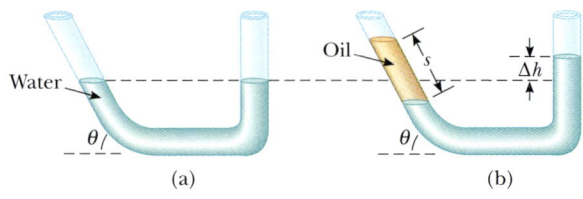

FIGURE P15.58

59. Water is forced out of a fire extinguisher by air pressure, as shown in Figure P15.59. How much gauge air pressure (above atmospheric) is required for a water jet to have a speed of 30 m/s when the water level is 0.50 m below the nozzle?

60. Torricelli was the first to realize that we live at the bottom of an ocean of air. He correctly surmised that

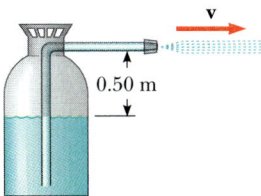

FIGURE P15.59

the pressure of our atmosphere is due to the weight of the air. The density of air at the Earth's surface is 1.29 kg/m³. The density decreases with increasing altitude as the atmosphere thins out. If we assume that the density is constant (1.29 kg/m³) up to some altitude h, and zero above, then h would represent the thickness of our atmosphere. Use this model to determine the value of h that gives a pressure of 1.0 atm at the surface of the Earth. Would the peak of Mt. Everest rise above the surface of such an atmosphere?

61. The true weight of a body is its weight when measured in a vacuum where there are no buoyant forces. A body of volume V is weighed in air on a balance using weights of density ρ. If the density of air is ρ_a and the balance reads w', show that the true weight w is

$$w = w' + \left(V - \frac{w'}{\rho g}\right)\rho_a g$$

62. A wooden dowel has a diameter of 1.20 cm. It floats in water with 0.40 cm of its diameter above water (Fig. P15.62). Determine the density of the dowel.

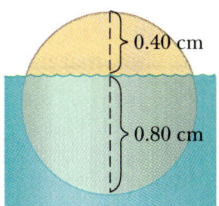

0.40 cm

0.80 cm

FIGURE P15.62

63. A pipe carrying water has a diameter of 2.5 cm. Estimate the maximum flow speed if the flow is to be laminar. Assume the temperature is 20°C.

64. A 1.0-kg beaker containing 2.0 kg of oil (density = 916.0 kg/m³) rests on a scale. A 2.0-kg block of iron is suspended from a spring scale and completely submerged in the oil as in Figure P15.64. Determine the equilibrium readings of both scales.

64A. A beaker of mass m_b containing oil of mass m_o (density = ρ_o) rests on a scale. A block of iron of

mass m_i is suspended from a spring scale and completely submerged in the oil as in Figure P15.64. Determine the equilibrium readings of both scales.

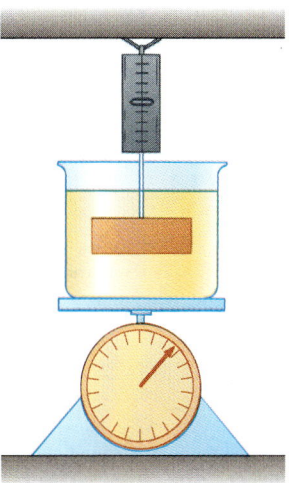

FIGURE P15.64

65. If a 1-megaton nuclear weapon is exploded at ground level, the peak overpressure (that is, the pressure increase above normal atmospheric pressure) will be 0.20 atm at a distance of 6.0 km. What force due to such an explosion will be exerted on the side of a house with dimensions 4.5 m × 22 m?

66. A light spring of constant $k = 90.0$ N/m rests vertically on a table (Fig. P15.66a). A 2.00-g balloon is filled with helium (density = 0.180 kg/m³) to a volume of 5.00 m³ and connected to the spring, causing it to expand as in Figure P15.66b. Determine the expansion length L when the balloon is in equilibrium.

66A. A light spring of constant k rests vertically on a table (Fig. P15.66a). A balloon of mass m is filled with helium (density = ρ_{He}) to a volume V and connected to the spring, causing it to expand as in Figure P15.66b. Determine the expansion length L when the balloon is in equilibrium.

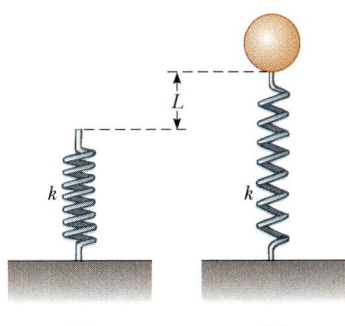

FIGURE P15.66   (a)   (b)

67. With reference to Figure 15.5, show that the total torque exerted by the water behind the dam about an axis through O is $\frac{1}{6}\rho g w H^3$. Show that the effective line of action of the total force exerted by the water is at a distance $\frac{1}{3}H$ above O.

68. In 1657 Otto von Guericke, inventor of the air pump, evacuated a sphere made of two brass hemispheres. Two teams of eight horses each could pull the hemispheres apart only on some trials, and then "with greatest difficulty" (Fig. P15.68). (a) Show that the force F required to pull the evacuated hemispheres apart is $\pi R^2 (P_0 - P)$, where R is the radius of the hemispheres and P is the pressure inside the hemispheres, which is much less than P_0. (b) Determine the force if $P = 0.10P_0$ and $R = 0.30$ m.

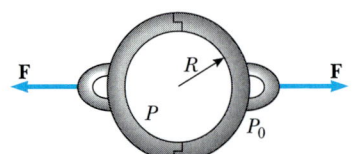

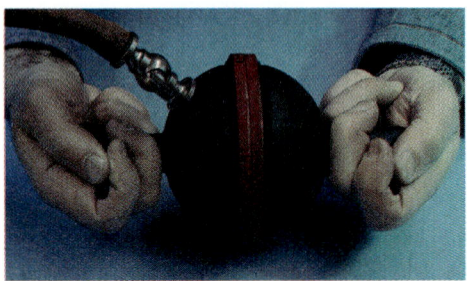

FIGURE P15.68   *(Henry Leap and Jim Lehman)*

69. In 1983, the United States began coining the cent piece out of copper-clad zinc rather than pure copper. If the mass of the old copper cent is 3.083 g while that of the new cent is 2.517 g, calculate the percent of zinc (by volume) in the new cent. The density of copper is 8.960 g/cm³ and that of zinc is 7.133 g/cm³. The new and old coins have the same volume.

70. How much air must be pushed downward at 40.0 m/s in order to keep an 800-kg helicopter aloft?

71. The flow rate of the Columbia River is approximately 3200 m³/s. What would be the maximum power output of the turbines in a dam if the water were to fall a vertical distance of 160 m?

72. A thin spherical shell of mass 4.0 kg and diameter 0.20 m is filled with helium (density = 0.180 kg/m³). It is then released from rest on the bottom of a pool of water that is 4.0 m deep. (a) Neglecting frictional effects, show that the shell rises with constant acceleration, and determine the value of that

acceleration. (b) How long will it take for the top of the shell to reach the water surface?

72A. A thin spherical shell of mass m and diameter d is filled with helium (density = ρ_{He}). It is then released from rest on the bottom of a pool of water of depth h. (a) Neglecting frictional effects, show that the shell rises with constant acceleration, and determine the value of that acceleration. (b) How long will it take for the top of the shell to reach the water surface?

73. An incompressible, nonviscous fluid initially rests in the vertical portion of the pipe shown in Figure P15.73a, where $L = 2.0$ m. When the valve is opened, the fluid flows into the horizontal section of the pipe. What is the speed of the fluid when it is entirely in the horizontal section as in Figure P15.73b? Assume the cross-sectional area of the entire pipe is constant.

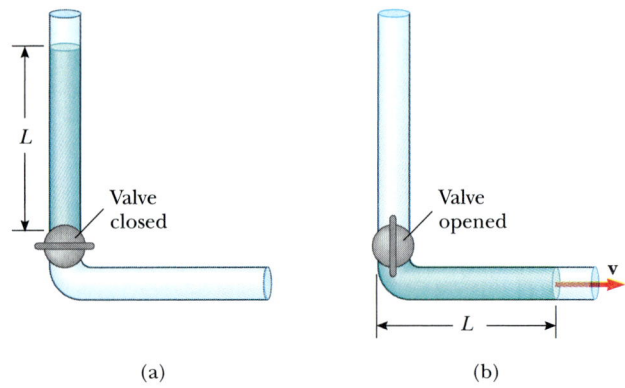

(a)   (b)

FIGURE P15.73

74. Water falls over a dam of height h meters at a rate of R kg/s. (a) Show that the power available from the water is

$$P = Rgh$$

where g is the free-fall acceleration. (b) Each hydroelectric unit at the Grand Coulee Dam discharges water at a rate of 8.5×10^5 kg/s from a height of 87 m. The power developed by the falling water is converted to electric power with an efficiency of 85%. How much electric power is produced by each hydroelectric unit?

75. A U-tube open at both ends is partially filled with water (Fig. P15.75a). Oil (density = 750 kg/m³) is then poured into the right arm and forms a column $L = 5.00$ cm high (Fig. P15.75b). (a) Determine the difference, h, in the heights of the two liquid surfaces. Assume that the density of air is 1.29 kg/m³, but be sure to include differences in the atmospheric pressure due to differences in altitude. (b) The right arm is then shielded from any air motion while air is

blown across the top of the left arm until the surfaces of the two liquids are at the same height (Fig. P15.75c). Determine the speed of the air being blown across the left arm.

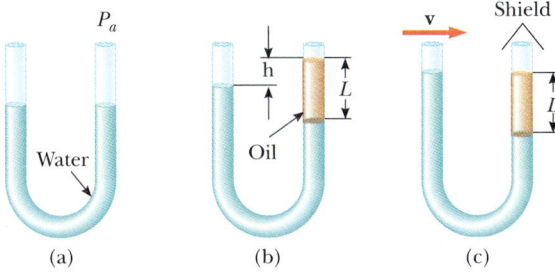

FIGURE P15.75

76. Show that the variation of atmospheric pressure with altitude is given by $P = P_0 e^{-\alpha h}$, where $\alpha = \rho_0 g/P_0$, P_0 is atmospheric pressure at some reference level, and ρ_0 is the atmospheric density at this level. Assume that the decrease in atmospheric pressure with increasing altitude is given by Equation 15.4 (so that $dP/dy = -\rho g$) and that the density of air is proportional to the pressure.

77. A cube of ice whose edge is 20 mm is floating in a glass of ice-cold water with one of its faces parallel to the water surface. (a) How far below the water surface is the bottom face of the block? (b) Ice-cold ethyl alcohol is gently poured onto the water surface to form a layer 5 mm thick above the water. When the ice cube attains hydrostatic equilibrium again, what will be the distance from the top of the water to the bottom face of the block? (c) Additional cold ethyl alcohol is poured onto the water surface until the top surface of the alcohol coincides with the top surface of the ice cube (in hydrostatic equilibrium). How thick is the required layer of ethyl alcohol?

78. The *spirit-in-glass thermometer*, invented in Florence, Italy, around 1654, consists of a tube of liquid (the spirit) containing a number of submerged glass spheres with slightly different masses (Fig. P15.78). At sufficiently low temperatures all the spheres float, but as the temperature rises, the spheres sink one after the other. The device is a crude but interesting

FIGURE P15.78   *(Courtesy Jeanne Maier)*

tool for measuring temperature. Suppose the tube is filled with ethyl alcohol, whose density is 0.78945 g/cm³ at 20.0°C and decreases to 0.78097 g/cm³ at 30.0°C. (a) If one of the spheres has a radius of 1.000 cm and is in equilibrium halfway up the tube at 20.0°C, determine its mass. (b) When the temperature increases to 30.0°C, what mass must a second sphere of the same radius have in order to be in equilibrium at the halfway point? (c) At 30.0°C the first sphere has fallen to the bottom of the tube. What upward force does the bottom of the tube exert on this sphere?

Aerial view of windsurfer riding on the crest of a wave. This dramatic photograph illustrates many physical principles that are described in the text. For example, the water wave carries energy and momentum as it travels from one location to another. The surfer and the surfboard move in a complex path under the action of several forces, including gravity, wind resistance, and the force of water on the surfboard. (© Darrell Wong/Tony Stone Images)

Mechanical Waves

The impetus is much quicker than the water, for it often happens that the wave flees the place of its creation, while the water does not; like the waves made in a field of grain by the wind, where we see the waves running across the field while the grain remains in place.

L E O N A R D O   D A   V I N C I

As we look around us, we find many examples of objects that vibrate: a pendulum, the strings of a guitar, an object suspended on a spring, the piston of an engine, the head of a drum, the reed of a saxophone. Most elastic objects vibrate when an impulse is applied to them. That is, once they are distorted, their shape tends to be restored to some equilibrium configuration. Even at the atomic level, the atoms in a solid vibrate about some position as if they were connected to their neighbors by imaginary springs.

Wave motion is closely related to the phenomenon of vibration.

Sound waves, earthquake waves, waves on stretched strings, and water waves are all produced by some source of vibration. As a sound wave travels through some medium, such as air, the molecules of the medium vibrate back and forth; as a water wave travels across a pond, the water molecules vibrate up and down and backward and forward. As waves travel through a medium, the particles of the medium move in repetitive cycles. Therefore, the motion of the particles bears a strong resemblance to the periodic motion of a vibrating pendulum or a mass attached to a spring.

There are many other phenomena in nature whose explanation requires us to understand the concepts of vibrations and waves. For instance, although many large structures, such as skyscrapers and bridges, appear to be rigid, they actually vibrate, a fact that must be taken into account by the architects and engineers who design and build them. To understand how radio and television work, we must understand the origin and nature of electromagnetic waves and how they propagate through space. Finally, much of what scientists have learned about atomic structure has come from information carried by waves. Therefore, we must first study waves and vibrations in order to understand the concepts and theories of atomic physics.

Wave Motion

Large waves sometimes travel great distances over the surface of the ocean, yet the water does not flow with the wave. The crests and troughs of the wave often form repetitive patterns.   *(Superstock)*

Most of us experienced waves as children when we dropped a pebble into a pond. At the point where the pebble hits the water surface, waves are created by the impact. These waves move outward from the creation point in expanding circles until they finally reach the shore. If you were to examine carefully the motion of a leaf floating on the disturbance, you would see that the leaf moves up, down, and sideways about its original position but does not undergo any net displacement away from or toward the point where the pebble hits the water. The water molecules just beneath the leaf, as well as all the other water molecules on the pond surface, behave in the same way. That is, the water wave moves from one place to another, and yet the water is not carried with it.

An excerpt from a book by Einstein and Infeld gives the following remarks concerning wave phenomena.[1]

---

[1] A. Einstein and L. Infeld, *The Evolution of Physics*, New York, Simon and Schuster, 1961. Excerpt from "What is a Wave?".

A bit of gossip starting in Washington reaches New York very quickly, even though not a single individual who takes part in spreading it travels between these two cities. There are two quite different motions involved, that of the rumor, Washington to New York, and that of the persons who spread the rumor. The wind, passing over a field of grain, sets up a wave which spreads out across the whole field. Here again we must distinguish between the motion of the wave and the motion of the separate plants, which undergo only small oscillations. . . . The particles constituting the medium perform only small vibrations, but the whole motion is that of a progressive wave. The essentially new thing here is that for the first time we consider the motion of something which is not matter, but energy propagated through matter.

Water waves and the waves across a grainfield are only two examples of physical phenomena that have wavelike characteristics. The world is full of waves, the two main types of which are mechanical waves and electromagnetic waves. We have already mentioned examples of mechanical waves: sound waves, water waves, and ''grain waves.'' In each case, there is some physical medium being disturbed—air molecules, water molecules, and stalks of grain in our three particular examples. Electromagnetic waves are a special class of waves that do not require a medium in order to propagate, some examples being visible light, radio waves, television signals, and x-rays. Here in Part II of this book, we shall study only mechanical waves.

The wave concept is abstract. When we observe what we call a water wave, what we see is a rearrangement of the water's surface. Without the water, there would be no wave. A wave traveling on a string would not exist without the string. Sound waves travel through air as a result of pressure variations from point to point. In such cases involving mechanical waves, what we interpret as a wave corresponds to the disturbance of a body or medium. Therefore, we can consider a wave to be the *motion of a disturbance.*

The mathematics used to describe wave phenomena is common to all waves. In general, we shall find that mechanical wave motion is described by specifying the positions of all points of the disturbed medium as a function of time.

16.1 INTRODUCTION

The mechanical waves discussed in this chapter require (1) some source of disturbance, (2) a medium that can be disturbed, and (3) some physical connection through which adjacent portions of the medium can influence each other. We shall find that all waves carry energy. The amount of energy transmitted through a medium and the mechanism responsible for that transport of energy differ from case to case. For instance, the power of ocean waves during a storm is much greater than the power of sound waves generated by a single human voice.

Three physical characteristics are important in characterizing waves: wavelength, frequency, and wave speed. One **wavelength** is the *minimum distance between any two points on a wave that behave identically,* as shown in Figure 16.1.

Most waves are periodic, and the **frequency** of such periodic waves is *the time rate at which the disturbance repeats itself.*

Waves travel with a specific speed, which depends on the properties of the medium being disturbed. For instance, sound waves travel through air at 20°C with a speed of about 344 m/s (781 mi/h), whereas the speed of sound in most solids is higher than 344 m/s. Electromagnetic waves travel very swiftly through a vacuum with a speed of approximately 3.00×10^8 m/s (186 000 mi/s).

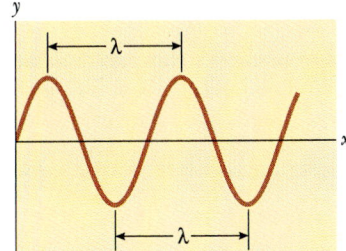

FIGURE 16.1 The wavelength λ of a wave is the distance between adjacent crests or adjacent troughs.

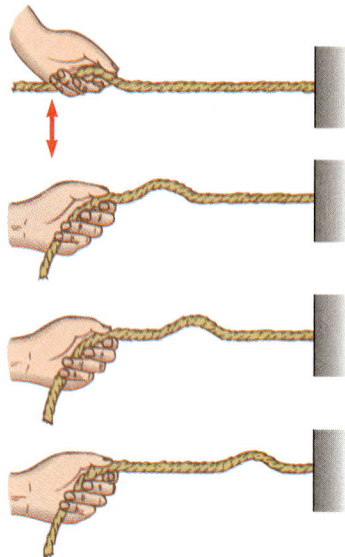

FIGURE 16.2   A wave pulse traveling down a stretched rope. The shape of the pulse is approximately unchanged as it travels along the rope.

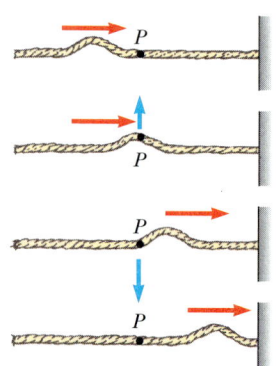

FIGURE 16.3   A pulse traveling on a stretched rope is a transverse wave. That is, any element *P* on the rope moves (blue arrows) in a direction perpendicular to the wave motion (red arrows).

16.2   TYPES OF WAVES

One way to demonstrate wave motion is to flick one end of a long rope that is under tension and has its opposite end fixed, as in Figure 16.2. In this manner, a single wave bump (called a wave pulse) is formed and travels (to the right in Fig. 16.2) with a definite speed. This type of disturbance is called a **traveling wave,** and Figure 16.2 represents four consecutive "snapshots" of the traveling wave. As we shall see later, the speed of the wave depends on the tension in the rope and on the properties of the rope. The rope is the medium through which the wave travels. The shape of the wave pulse changes very little as it travels along the rope.[2]

As the wave pulse travels, *each segment of the rope that is disturbed moves in a direction perpendicular to the wave motion.* Figure 16.3 illustrates this point for one particular segment, labeled *P.* Note that no part of the rope ever moves in the direction of the wave.

> A traveling wave that causes the particles of the disturbed medium to move perpendicular to the wave motion is called a **transverse wave.**

> A traveling wave that causes the particles of the medium to move parallel to the direction of wave motion is called a **longitudinal wave.**

Sound waves, which we discuss in Chapter 17, are one example of longitudinal waves. Sound waves in air are a series of high- and low-pressure regions, or disturbances, traveling in the same direction as the displacements. A longitudinal pulse can be easily produced in a stretched spring, as in Figure 16.4. The left end of the spring is given a sudden movement (consisting of a brief push to the right and equally brief pull to the left) along the length of the spring; this movement creates a sudden compression of the coils. The compressed region *C* (pulse) travels along the spring, and so we see that the disturbance is parallel to the wave motion. The compressed region is followed by a region where the coils are stretched.

Some waves in nature are neither transverse nor longitudinal, but a combination of the two. Surface water waves are a good example. When a water wave travels on the surface of deep water, water molecules at the surface move in nearly circular paths, as shown in Figure 16.5, where the water surface is drawn as a series of crests and troughs. Note that the disturbance has both transverse and longitudinal components. As the wave passes, water molecules at the crests move in the direction of

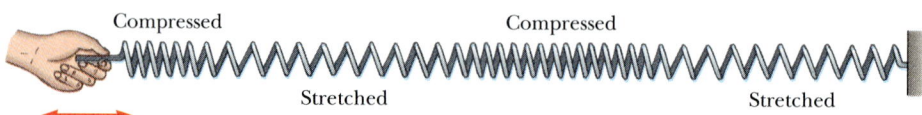

FIGURE 16.4   A longitudinal pulse along a stretched spring. The displacement of the coils is in the direction of the wave motion. For the starting motion described in the text, the compressed region is followed by a stretched region.

[2] Strictly speaking, the pulse will change its shape and gradually spread out during the motion. This effect is called *dispersion* and is common to many mechanical waves.

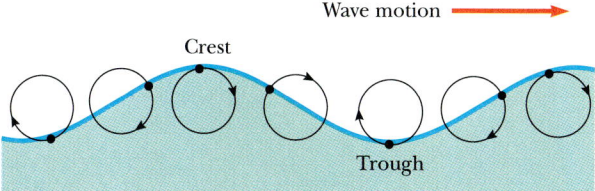

FIGURE 16.5 Wave motion on the surface of water. The molecules at the water's surface move in nearly circular paths. Each molecule is displaced horizontally and vertically from its equilibrium position, represented by circles.

the wave, and molecules at the troughs move in the opposite direction. Since the molecule at the crest in Figure 16.5 will soon be at a trough, its movement in the direction of the wave will soon be canceled by its movement in the opposite direction. Since this argument holds for every disturbed water molecule, we conclude that there is no net displacement of any water molecule.

16.3 ONE-DIMENSIONAL TRAVELING WAVES

Let us now give a mathematical description of a one-dimensional traveling wave. Consider again a wave pulse traveling to the right with constant speed v on a long taut string, as in Figure 16.6. The pulse moves along the x axis (the axis of the string), and the transverse displacement of the string (the medium) is measured with the coordinate y.

Figure 16.6a represents the shape and position of the pulse at time $t = 0$. At this time, the shape of the pulse, whatever it may be, can be represented as $y = f(x)$. That is, y is some definite function of x. The *maximum displacement of the string, A,* is called the **amplitude** of the wave. Since the speed of the wave pulse is v, it travels to the right a distance vt in a time t (Fig. 16.6b).

If the shape of the wave pulse doesn't change with time, we can represent the string displacement y for all later times measured in a stationary frame having the origin at 0 as

$$y = f(x - vt) \qquad (16.1)$$

Wave traveling to the right

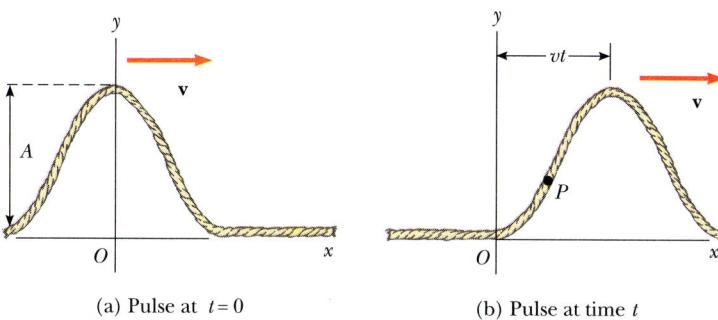

(a) Pulse at $t = 0$ (b) Pulse at time t

FIGURE 16.6 A one-dimensional wave pulse traveling to the right with a speed v. (a) At $t = 0$, the shape of the pulse is given by $y = f(x)$. (b) At some later time t, the shape remains unchanged and the vertical displacement of any point P of the medium is given by $y = f(x - vt)$.

If the wave pulse travels to the *left*, the string displacement is

$$y = f(x + vt) \qquad (16.2)$$

The displacement y, sometimes called the *wave function*, depends on the two variables x and t. For this reason, it is often written $y(x, t)$, which is read "y as a function of x and t."

It is important to understand the meaning of y. Consider a particular point P on the string, identified by a particular value of its coordinates. As the wave passes P, the y coordinate of this point increases, reaches a maximum, and then decreases to zero. Therefore, the **wave function** y *represents the y coordinate of any medium point P at any time t.* Furthermore, if t is fixed, then the wave function y as a function of x *defines a curve representing the shape of the pulse at this time.* This curve is equivalent to a "snapshot" of the wave at this time.

For a pulse that moves without changing shape, the speed of the pulse is the same as that of any feature along the pulse, such as the crest in Figure 16.6b. To find the speed of the pulse, we can calculate how far the crest moves in a short time and then divide this distance by the time interval. In order to follow the motion of the crest, some particular value, say x_0, must be substituted in Equation 16.1 for $x - vt$. Regardless of how x and t change individually, we must require that $x - vt = x_0$ in order to stay with the crest. This expression, therefore, represents the equation of motion of the crest. At $t = 0$, the crest is at $x = x_0$; at a time dt later, the crest is at $x = x_0 + v\,dt$. Therefore, in a time dt, the crest has moved a distance $dx = (x_0 + v\,dt) - x_0 = v\,dt$. Hence, the wave speed is

$$v = \frac{dx}{dt} \qquad (16.3)$$

As noted above, the wave velocity must not be confused with the transverse velocity (which is in the y direction) of a particle in the medium (nor with the longitudinal velocity for a longitudinal wave).

EXAMPLE 16.1   A Pulse Moving to the Right

A wave pulse moving to the right along the x axis is represented by the wave function

$$y(x, t) = \frac{2}{(x - 3.0t)^2 + 1}$$

where x and y are measured in centimeters and t is in seconds. Let us plot the waveform at $t = 0$, $t = 1.0$ s, and $t = 2.0$ s.

Solution   First, note that this function is of the form $y = f(x - vt)$. By inspection, we see that the speed of the wave is $v = 3.0$ cm/s. Furthermore, the wave amplitude (the maximum value of y) is given by $A = 2.0$ cm. At times $t = 0$, $t = 1.0$ s, and $t = 2.0$ s, the wave function expressions are

$$y(x, 0) = \frac{2}{x^2 + 1} \qquad \text{at } t = 0$$

$$y(x, 1.0) = \frac{2}{(x - 3.0)^2 + 1} \qquad \text{at } t = 1.0 \text{ s}$$

$$y(x, 2.0) = \frac{2}{(x - 6.0)^2 + 1} \qquad \text{at } t = 2.0 \text{ s}$$

We can now use these expressions to plot the wave function versus x at these times. For example, let us evaluate $y(x, 0)$ at $x = 0.50$ cm:

$$y(0.50, 0) = \frac{2}{(0.50)^2 + 1} = 1.6 \text{ cm}$$

Likewise, $y(1.0, 0) = 1.0$ cm, $y(2.0, 0) = 0.40$ cm, and so on. A continuation of this procedure for other values of x yields the waveform shown in Figure 16.7a. In a similar manner, one obtains the graphs of $y(x, 1.0)$ and $y(x, 2.0)$, shown in

Figures 16.7b and 16.7c, respectively. These snapshots show that the wave pulse moves to the right without changing its shape and has a constant speed of 3.0 cm/s.

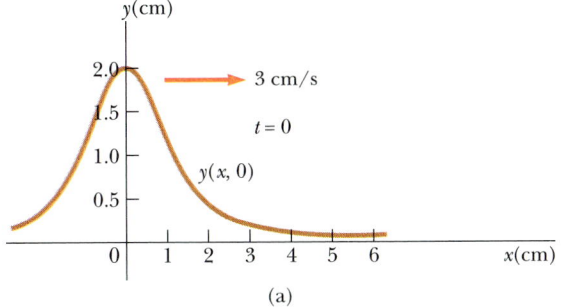

(a)

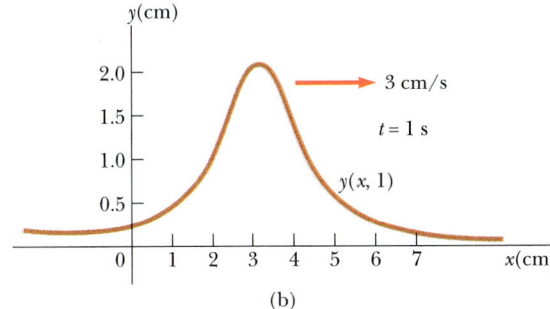

(b)

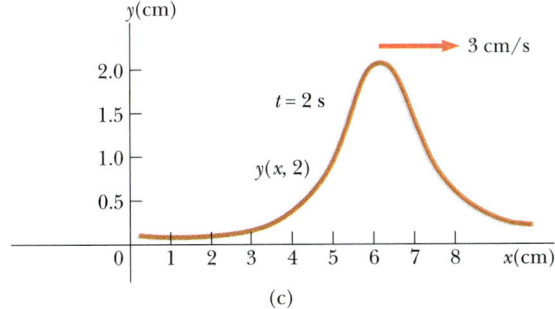

(c)

FIGURE 16.7 (Example 16.1) Graphs of the function $y(x, t) = 2/[(x - 3t)^2 + 1]$. (a) $t = 0$, (b) $t = 1$ s, and (c) $t = 2$ s.

16.4 SUPERPOSITION AND INTERFERENCE OF WAVES

Many interesting wave phenomena in nature cannot be described by a single moving pulse. Instead, one must analyze complex waveforms in terms of a combination of many traveling waves. To analyze such wave combinations, one can make use of the **superposition principle:**

> If two or more traveling waves are moving through a medium, the resultant wave function at any point is the algebraic sum of the wave functions of the individual waves.

Linear waves obey the superposition principle

Waves that obey this principle are called *linear waves,* and they are generally characterized by small wave amplitudes. Waves that violate the superposition principle are called *nonlinear waves* and are often characterized by large amplitudes. In this book, we deal only with linear waves.

One consequence of the superposition principle is that *two traveling waves can pass through each other without being destroyed or even altered.* For instance, when two pebbles are thrown into a pond and hit the surface at two places, the expanding circular surface waves do not destroy each other. In fact, they pass right through each other. The complex pattern that is observed can be viewed as two independent sets of expanding circles. Likewise, when sound waves from two sources move

Wave Motion

This simulator allows you to model wave motion involving one or two traveling waves. For the case of a single wave, you will be able to specify the velocity and shape of the wave by selecting values from a given list, or define your own wave shape and velocity. For a wave traveling on a string, you can specify whether an end of the string is fixed or free and examine how the wave is reflected at this end in each case. When studying two waves traveling on a string, you can investigate their superposition as they move through each other.

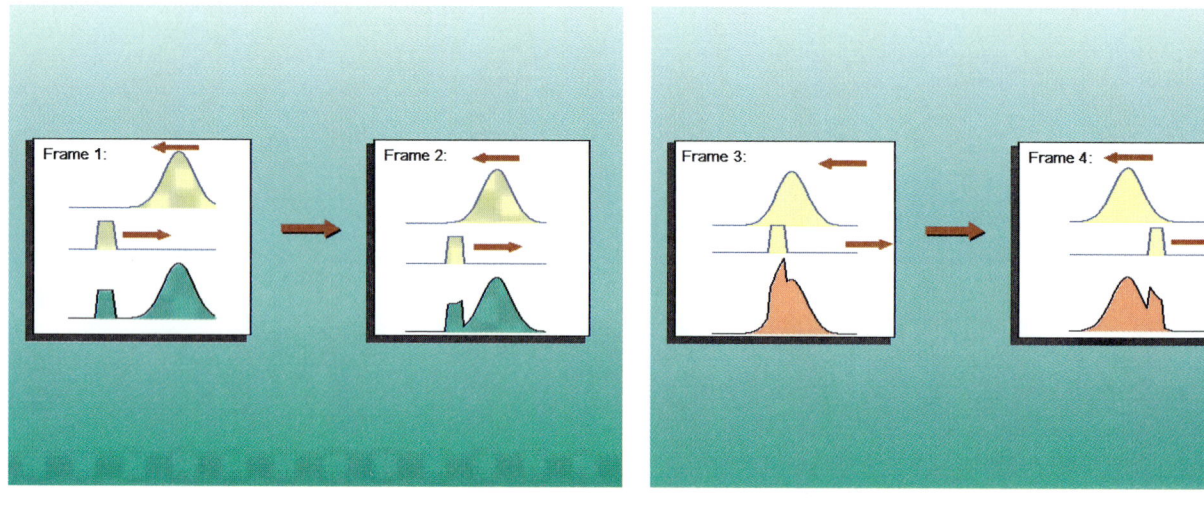

through air, they also pass through each other. The resulting sound one hears at a given point is the resultant of both disturbances.

A simple pictorial representation of the superposition principle is obtained by considering two pulses traveling in opposite directions on a taut string, as in Figure 16.8. The wave function for the pulse moving to the right is y_1, and the wave function for the pulse moving to the left is y_2. The pulses have the same speed but different shapes. Each pulse is assumed to be symmetric, and the displacement of the medium is in the positive y direction for both pulses. (Note that the superposition principle applies even if the two pulses are not symmetric and even when they travel at different speeds.) When the waves begin to overlap (Fig. 16.8b), the resulting complex waveform is given by $y_1 + y_2$. When the crests of the pulses coincide (Fig. 16.8c), the resulting waveform $y_1 + y_2$ is symmetric. The two pulses finally separate and continue moving in their original directions (Fig. 16.8d). Note that the final waveforms remain unchanged, as if the two pulses had never met!

The combination of separate waves in the same region of space to produce a resultant wave is called **interference**. For the two pulses shown in Figure 16.8, the displacement of the medium is in the positive y direction for both pulses, and the resultant waveform (when the pulses overlap) exhibits a displacement greater than those of the individual pulses. Since the displacements caused by the two pulses are in the same direction, we refer to their superposition as **constructive interference**.

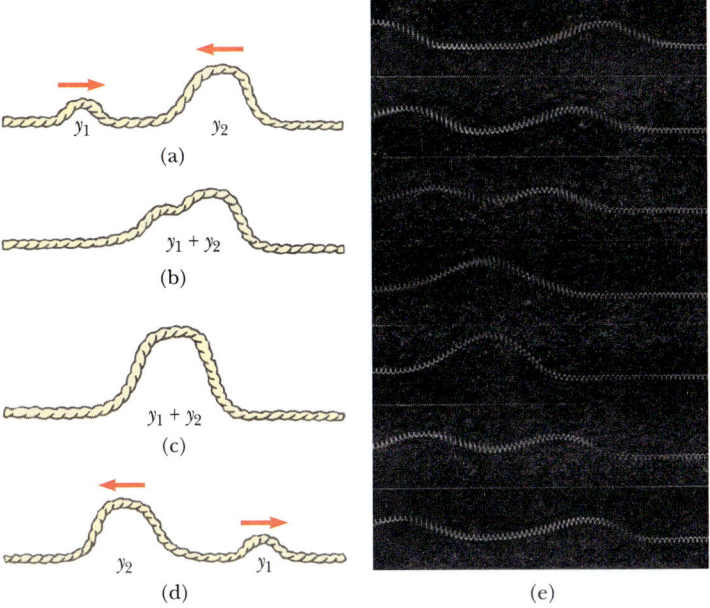

(a)

(b)

(c)

(d)

(e)

FIGURE 16.8 *(Left)* Two wave pulses traveling on a stretched string in opposite directions pass through each other. When the pulses overlap, as in (b) and (c), the net displacement of the string equals the sum of the displacements produced by each pulse. Since each pulse produces positive displacements of the string, we refer to their superposition as *constructive interference.* *(Right)* Photograph of superposition of two equal and symmetric pulses traveling in opposite directions on a stretched spring. *(Photo, Education Development Center, Newton, Mass.)*

Now consider two pulses traveling in opposite directions on a taut string, where now one is inverted relative to the other, as in Figure 16.9. In this case, when the pulses begin to overlap, the resultant waveform is given by $y_1 - y_2$. Again the two pulses pass through each other as indicated. Since the displacements caused by the two pulses are in opposite directions, we refer to their superposition as **destructive interference.**

Interference patterns produced by outward spreading waves from several drops of water falling into a pond. *(Martin Dohrn/SPL/Photo Researchers)*

Interference of water waves produced in a ripple tank. The sources of the waves are two objects that vibrate perpendicularly to the surface of the tank. *(Courtesy of Central Scientific CO.)*

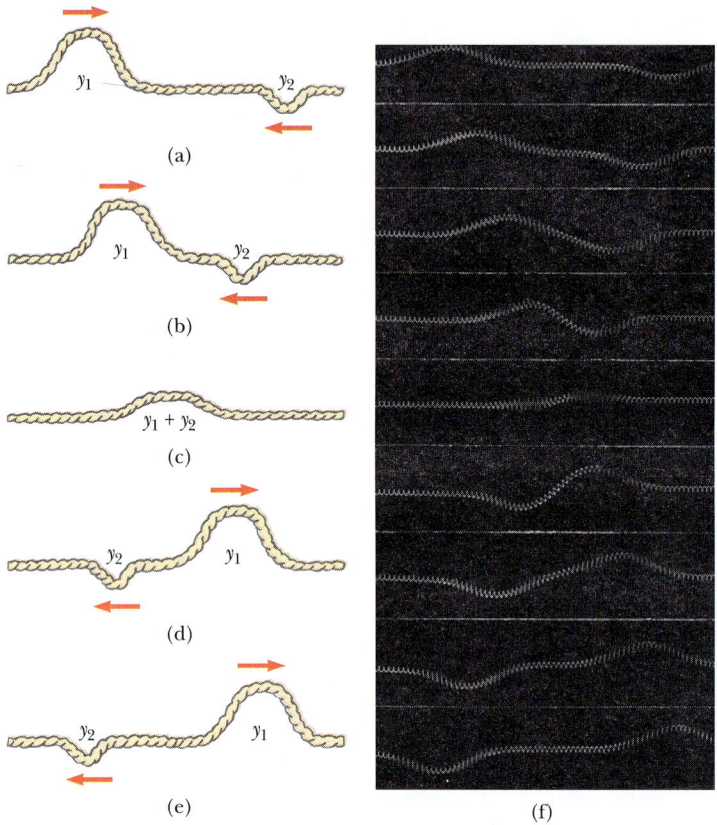

FIGURE 16.9   *(Left)* Two wave pulses traveling in opposite directions with displacements that are inverted relative to each other. When the two overlap as in (c), their displacements subtract from each other. *(Right)* Photograph of superposition of two symmetric pulses traveling in opposite directions, where one is inverted relative to the other.   *(Photo, Education Development Center, Newton, Mass.)*

16.5   THE SPEED OF WAVES ON STRINGS

The speed of linear mechanical waves depends only on the properties of the medium through which the wave travels. In this section, we focus on determining the speed of a transverse pulse traveling on a taut string. If the tension in the string is F and its mass per unit length is μ, then as we shall show, the wave speed is

Speed of a wave on a stretched string

$$v = \sqrt{\frac{F}{\mu}} \qquad (16.4)$$

First, let us verify that this expression is dimensionally correct. The dimensions of F are MLT^{-2}, and the dimensions of μ are ML^{-1}. Therefore, the dimensions of F/μ are L^2/T^2; hence the dimensions of $\sqrt{F/\mu}$ are L/T, which are indeed the dimensions of speed. No other combination of F and μ is dimensionally correct if we assume that they are the only variables relevant to the situation.

Now let us use a mechanical analysis to derive the above expression. Consider a pulse moving to the right with a uniform speed v, measured relative to a stationary frame of reference. Instead of staying in this frame, it is more convenient to choose as our reference frame one that moves along with the pulse with the same speed, so that the pulse is at rest in this frame, as in Figure 16.10a. This change of reference frame is permitted because Newton's laws are valid in either a stationary frame or one that moves with constant velocity.

A small segment of the string of length Δs forms an approximate arc of a circle of radius R, as shown in Figure 16.10a and magnified in Figure 16.10b. In the pulse's frame of reference (which is moving to the right along with the pulse), the shaded segment is moving with a speed v. This small segment has a centripetal acceleration equal to v^2/R, which is supplied by the force of tension $\mathbf{F}$ in the string. The force $\mathbf{F}$ acts on each side of the segment, tangent to the arc, as in Figure 16.10b. The horizontal components of $\mathbf{F}$ cancel, and each vertical component $F \sin \theta$ acts radially inward toward the center of the arc. Hence, the total radial force is $2F \sin \theta$. Since the segment is small, θ is small and we can use the familiar small-angle approximation $\sin \theta \approx \theta$. Therefore, the total radial force can be expressed as

$$F_r = 2F \sin \theta \approx 2F\theta$$

The small segment has a mass $m = \mu\,\Delta s$. Since the segment forms part of a circle and subtends an angle 2θ at the center, $\Delta s = R(2\theta)$, and hence

$$m = \mu\,\Delta s = 2\mu R\theta$$

If we apply Newton's second law to this segment, the radial component of motion gives

$$F_r = \frac{mv^2}{R} \quad \text{or} \quad 2F\theta = \frac{2\mu R\theta v^2}{R}$$

where F_r is the force that supplies the centripetal acceleration of the segment and maintains the curvature at this point. Solving for v gives Equation 16.4. Notice that this derivation is based on the assumption that the pulse height is small relative to the length of the string. Using this assumption, we were able to use the approximation that $\sin \theta \approx \theta$. Furthermore, the model assumes that the tension F is not affected by the presence of the pulse, so that F is the same at all points on the string. Finally, this proof does *not* assume any particular shape for the pulse. Therefore, we conclude that a pulse of *any shape* will travel on the string with speed $v = \sqrt{F/\mu}$ without any change in pulse shape.

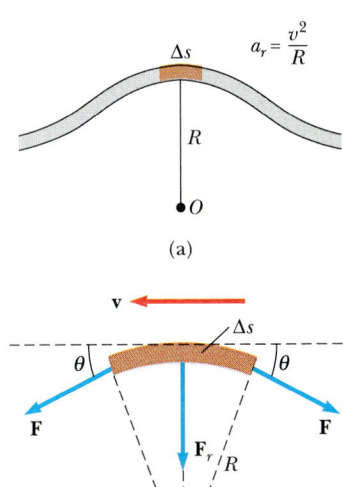

FIGURE 16.10   (a) To obtain the speed v of a wave on a stretched string, it is convenient to describe the motion of a small segment of the string in a moving frame of reference. (b) The net force on a small segment of length Δs is in the radial direction. The horizontal components of the tension force cancel.

EXAMPLE 16.2   The Speed of a Pulse on a Cord

A uniform cord has a mass of 0.300 kg and a length of 6.00 m (Fig. 16.11). Find the speed of a pulse on this cord.

Solution   The tension F in the cord is equal to the weight of the suspended 2.00-kg mass:

$$F = mg = (2.00 \text{ kg})(9.80 \text{ m/s}^2) = 19.6 \text{ N}$$

(This calculation of the tension neglects the small mass of the cord. Strictly speaking, the cord can never be exactly horizontal, and therefore the tension is not uniform.)

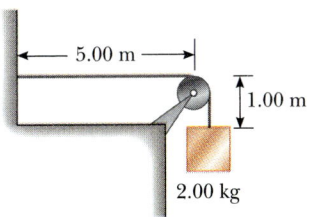

FIGURE 16.11   (Example 16.2) The tension F in the cord is maintained by the suspended mass. The wave speed is given by the expression $v = \sqrt{F/\mu}$.

The mass per unit length μ is

$$\mu = \frac{m}{\ell} = \frac{0.300 \text{ kg}}{6.00 \text{ m}} = 0.0500 \text{ kg/m}$$

Therefore, the wave speed is

$$v = \sqrt{\frac{F}{\mu}} = \sqrt{\frac{19.6 \text{ N}}{0.0500 \text{ kg/m}}} = \boxed{19.8 \text{ m/s}}$$

Exercise   Find the time it takes the pulse to travel from the wall to the pulley.

Answer   0.253 s.

16.6   REFLECTION AND TRANSMISSION OF WAVES

Whenever a traveling wave reaches a boundary, part or all of the wave is reflected. For example, consider a pulse traveling on a string fixed at one end (Fig. 16.12). When the pulse reaches the wall, it is reflected. Because the support attaching the string to the wall is rigid, the pulse does not transmit any part of the disturbance to the wall and its amplitude does not change.

Note that the reflected pulse is inverted. This can be explained as follows. When the pulse reaches the fixed end of the string, the string produces an upward force on the support. By Newton's third law, the support must then exert an equal and opposite (downward) reaction force on the string. This downward force causes the pulse to invert upon reflection.

Now consider another case, where this time the pulse arrives at the end of a string that is free to move vertically, as in Figure 16.13. The tension at the free end is maintained by tying the string to a ring of negligible mass that is free to slide vertically on a smooth post. Again the pulse is reflected, but this time it is not inverted. As the pulse reaches the post, it exerts a force on the free end of the string, causing the ring to accelerate upward. In the process, the ring would overshoot the height of the incoming pulse except that it is pulled back by the downward component of the tension force. This ring movement produces a reflected pulse that is not inverted, and whose amplitude is the same as that of the incoming pulse.

Finally, we may have a situation in which the boundary is intermediate between these two extreme cases, that is, one in which the boundary is neither rigid nor free. In this case, part of the incident pulse is transmitted and part is reflected. For instance, suppose a light string is attached to a heavier string as in Figure 16.14. When a pulse traveling on the light string reaches the boundary between the two,

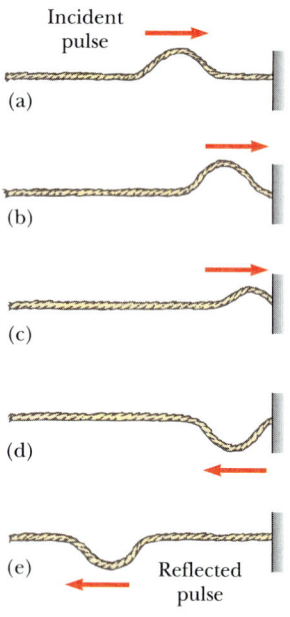

FIGURE 16.12   The reflection of a traveling wave pulse at the fixed end of a stretched string. The reflected pulse is inverted, but its shape remains the same.

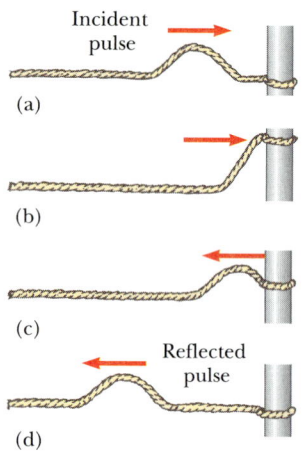

FIGURE 16.13   The reflection of a traveling wave pulse at the free end of a stretched string. The reflected pulse is not inverted.

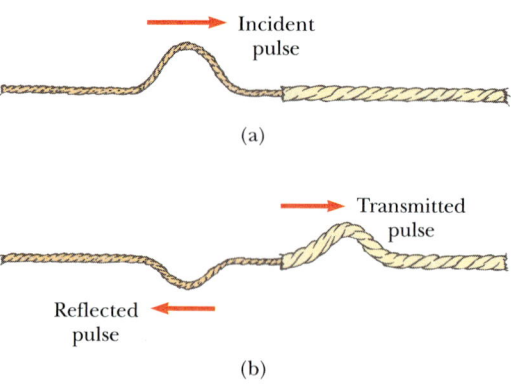

FIGURE 16.14   (a) A pulse traveling to the right on a light string attached to a heavier string. (b) Part of the incident pulse is reflected (and inverted), and part is transmitted to the heavier string. (Note that the change in pulse width is not shown.)

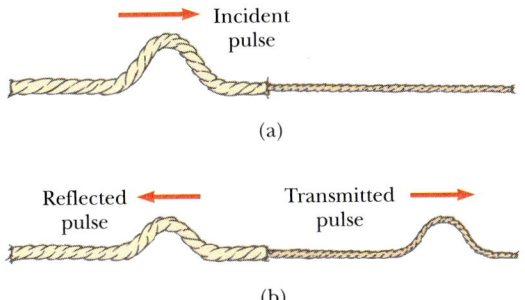

(a)

(b)

FIGURE 16.15 (a) A pulse traveling to the right on a heavy string attached to a lighter string. (b) The incident pulse is partially reflected and partially transmitted. In this case, the reflected pulse is not inverted. (Note that the change in pulse width is not shown.)

part of the pulse is reflected and inverted and part is transmitted to the heavier string. As one would expect, the reflected pulse has a smaller amplitude than the incident pulse, because part of the incident energy is transferred to the pulse in the heavier string. The reflected pulse is inverted for the same reasons described above for the fixed-support case.

When a pulse traveling on a heavy string strikes the boundary between that string and a lighter one, as in Figure 16.15, again part is reflected and part is transmitted. In this case, however, the reflected pulse is not inverted.

In either case, the relative heights of the reflected and transmitted pulses depend on the relative densities of the two strings. If the strings are identical, there is no discontinuity at the boundary, and hence no reflection takes place.

In the previous section, we found that the speed of a wave on a string increases as the mass per unit length of the string decreases. In other words, a pulse travels more slowly on a heavy string than on a light string if both are under the same tension. The following general rules apply to reflected waves: *When a wave pulse travels from medium A to medium B and $v_A > v_B$ (that is, when B is denser than A), the pulse is inverted upon reflection. When a wave pulse travels from medium A to medium B and $v_A < v_B$ (A is denser than B), the pulse is not inverted upon reflection.*

16.7 SINUSOIDAL WAVES

In this section, we introduce an important waveform known as a **sinusoidal wave** whose shape is shown in Figure 16.16. The red curve represents a snapshot of the traveling sinusoidal wave at $t = 0$, and the blue curve represents a snapshot of the wave at some later time t. At $t = 0$, the vertical displacement of the curve can be written

$$y = A \sin\left(\frac{2\pi}{\lambda} x\right) \tag{16.5}$$

where the constant A represents the wave amplitude and the constant λ is the symbol we use for the wavelength. Thus, we see that the vertical displacement repeats itself whenever x is increased by an integral multiple of λ. If the wave moves to the right with a speed v, the wave function at some later time t is

$$y = A \sin\left[\frac{2\pi}{\lambda}(x - vt)\right] \tag{16.6}$$

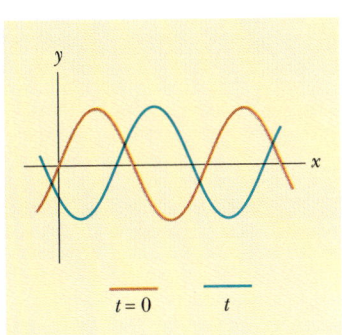

FIGURE 16.16 A one-dimensional sinusoidal wave traveling to the right with a speed v. The red curve represents a snapshot of the wave at $t = 0$, and the blue curve represents a snapshot at some later time t.

That is, the traveling sinusoidal wave moves to the right a distance vt in the time t, as in Figure 16.16. Note that the wave function has the form $f(x - vt)$ and represents a wave traveling to the right. If the wave were traveling to the left, the quantity $x - vt$ would be replaced by $x + vt$, as we learned when we developed Equations 16.1 and 16.2.

The time it takes the wave to travel a distance of one wavelength is called the **period,** T. Therefore, the wave speed, wavelength, and period are related by

$$v = \frac{\lambda}{T} \tag{16.7}$$

Substituting this into Equation 16.6, we find that

$$y = A \sin\left[2\pi\left(\frac{x}{\lambda} - \frac{t}{T}\right)\right] \tag{16.8}$$

This form of the wave function clearly shows the *periodic* nature of y. That is, at any given time t (a snapshot of the wave), y has the *same* value at the positions x, $x + \lambda$, $x + 2\lambda$, and so on. Furthermore, at any given position x, the value of y at times t, $t + T$, $t + 2T$, and so on is the same.

We can express the wave function in a convenient form by defining two other quantities, called the **angular wave number** k and the **angular frequency** ω:

Angular wave number

$$k \equiv \frac{2\pi}{\lambda} \tag{16.9}$$

Angular frequency

$$\omega \equiv \frac{2\pi}{T} \tag{16.10}$$

Using these definitions, we see that Equation 16.8 can be written in the more compact form

Wave function for a sinusoidal wave

$$y = A \sin(kx - \omega t) \tag{16.11}$$

The frequency of a sinusoidal wave, which we denote by the symbol f, is related to the period by the relationship

Frequency

$$f = \frac{1}{T} \tag{16.12}$$

The most common unit for frequency, as we learned in Chapter 13, is s^{-1}, or hertz (Hz). The corresponding unit for T is seconds.

Using Equations 16.9, 16.10, and 16.12, we can express the wave speed v in the alternative forms

Speed of a sinusoidal wave

$$v = \frac{\omega}{k} \tag{16.13}$$

$$v = \lambda f \tag{16.14}$$

The wave function given by Equation 16.11 assumes that the vertical displacement y is zero at $x = 0$ and $t = 0$. This need not be the case. If the vertical displacement is not zero at $x = 0$ and $t = 0$, we generally express the wave function in the form

$$y = A \sin(kx - \omega t - \phi) \qquad (16.15)$$

General relation for a sinusoidal wave

where ϕ is again called the **phase constant,** just as it was in our study of periodic motion in Chapter 13. This constant can be determined from the initial conditions.

EXAMPLE 16.3   A Traveling Sinusoidal Wave

A sinusoidal wave traveling in the positive x direction has an amplitude of 15.0 cm, a wavelength of 40.0 cm, and a frequency of 8.00 Hz. The vertical displacement of the medium at $t = 0$ and $x = 0$ is also 15.0 cm, as shown in Figure 16.17. (a) Find the angular wave number, period, angular frequency, and speed of the wave.

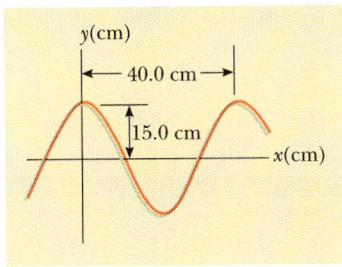

FIGURE 16.17   (Example 16.3) A sinusoidal wave of wavelength $\lambda = 40.0$ cm and amplitude $A = 15.0$ cm. The wave function can be written in the form $y = A \cos(kx - \omega t)$.

Solution   Using Equations 16.9, 16.10, 16.12, and 16.14, we find the following:

$$k = \frac{2\pi}{\lambda} = \frac{2\pi \text{ rad}}{40.0 \text{ cm}} = \boxed{0.157 \text{ rad/cm}}$$

$$T = \frac{1}{f} = \frac{1}{8.00 \text{ s}^{-1}} = \boxed{0.125 \text{ s}}$$

$$\omega = 2\pi f = 2\pi(8.00 \text{ s}^{-1}) = \boxed{50.3 \text{ rad/s}}$$

$$v = f\lambda = (8.00 \text{ s}^{-1})(40.0 \text{ cm}) = \boxed{320 \text{ cm/s}}$$

(b) Determine the phase constant ϕ, and write a general expression for the wave function.

Solution   Since $A = 15.0$ cm and since it is given that $y = 15.0$ cm at $x = 0$ and $t = 0$, substitution into Equation 16.15 gives

$$15 = 15 \sin(-\phi) \qquad \text{or} \qquad \sin(-\phi) = 1$$

Since $\sin(-\phi) = -\sin\phi$, we see that $\phi = -\pi/2$ rad (or $-90°$). Hence, the wave function is of the form

$$y = A \sin\left(kx - \omega t + \frac{\pi}{2}\right) = A \cos(kx - \omega t)$$

That the wave function must have this form can be seen by inspection, noting that the cosine argument is displaced by $90°$ from the sine function. Substituting the values for A, k, and ω into this expression gives

$$\boxed{y = (15.0 \text{ cm}) \cos(0.157x - 50.3t)}$$

Sinusoidal Waves on Strings

In Figure 16.2 we described how to create a wave pulse by jerking a taut string up and down once. To create a train of such pulses, normally referred to as a "wave train" or just plain "wave," we can replace the hand with a vibrating blade. Figure 16.18 represents snapshots of the wave created in this way at intervals of one quarter of a period. Note that because the blade vibrates in simple harmonic motion, *each particle of the string, such as P, also oscillates vertically with simple harmonic motion.* This must be the case because each particle follows the simple harmonic motion of the blade. Therefore, every segment of the string can be treated as a simple harmonic oscillator vibrating with a frequency equal to the frequency of

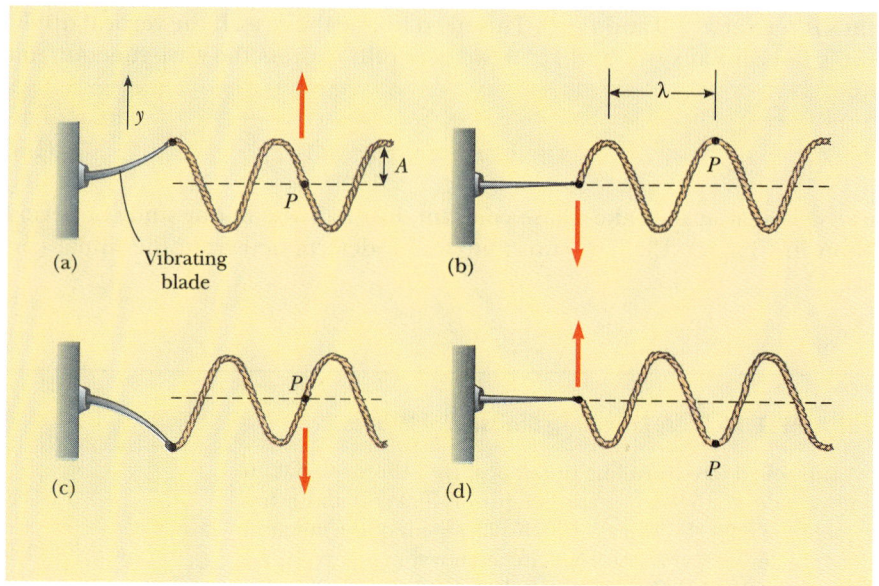

FIGURE 16.18   One method for producing a train of sinusoidal wave pulses on a continuous string. The left end of the string is connected to a blade that is set into vibration. Every segment of the string, such as the point P, oscillates with simple harmonic motion in the vertical direction.

vibration of the blade.[3] Note that although each segment oscillates in the y direction, the wave travels in the x direction with a speed v. Of course, this is the definition of a transverse wave. In this case, the energy carried by the traveling wave is supplied by the vibrating blade.

If the waveform at $t = 0$ is as described in Figure 16.18b, then the wave function can be written

$$y = A \sin(kx - \omega t)$$

We can use this expression to describe the motion of any point on the string. The point P (or any other point on the string) moves only vertically, and so *its x coordinate remains constant*. Therefore, the *transverse speed*, v_y (not to be confused with the wave speed v), and *transverse acceleration*, a_y, are

$$v_y = \frac{dy}{dt}\bigg]_{x=\text{constant}} = \frac{\partial y}{\partial t} = -\omega A \cos(kx - \omega t) \tag{16.16}$$

$$a_y = \frac{dv_y}{dt}\bigg]_{x=\text{constant}} = \frac{\partial v_y}{\partial t} = -\omega^2 A \sin(kx - \omega t) \tag{16.17}$$

The maximum values of these quantities are simply the absolute values of the coefficients of the cosine and sine functions:

$$(v_y)_{\text{max}} = \omega A \tag{16.18}$$

$$(a_y)_{\text{max}} = \omega^2 A \tag{16.19}$$

---

[3] In this arrangement, we are assuming that the mass always oscillates in a vertical line. The tension in the string would vary if the mass were allowed to move sideways. Such a motion would make the analysis very complex.

You should recognize that the transverse speed and transverse acceleration do not reach their maximum values simultaneously. In fact, the transverse speed reaches its maximum value (ωA) when $y = 0$, whereas the transverse acceleration reaches its maximum value ($\omega^2 A$) when $y = \pm A$. Finally, Equations 16.18 and 16.19 are identical to the corresponding equations for simple harmonic motion.

EXAMPLE 16.4   A Sinusoidally Driven String

The string shown in Figure 16.18 is driven at a frequency of 5.00 Hz. The amplitude of the motion is 12.0 cm, and the wave speed is 20.0 m/s. Determine the angular frequency and wave number for this wave, and write an expression for the wave function.

Solution   Using Equations 16.10, 16.12, and 16.13 gives

$$\omega = \frac{2\pi}{T} = 2\pi f = 2\pi(5.00\ \text{Hz}) = \boxed{31.4\ \text{rad/s}}$$

$$k = \frac{\omega}{v} = \frac{31.4\ \text{rad/s}}{20.0\ \text{m/s}} = \boxed{1.57\ \text{rad/m}}$$

Since $A = 12.0\ \text{cm} = 0.120\ \text{m}$, we have

$$y = A \sin(kx - \omega t) = \boxed{(0.120\ \text{m})\ \sin(1.57x - 31.4t)}$$

Exercise   Calculate the maximum values for the transverse speed and transverse acceleration of any point on the string.

Answer   3.77 m/s; 118 m/s².

16.8   ENERGY TRANSMITTED BY SINUSOIDAL WAVES ON STRINGS

As waves propagate through a medium, they transport energy. This is easily demonstrated by hanging a mass on a taut string and then sending a pulse down the string, as in Figure 16.19. When the pulse meets the suspended mass, the mass is momentarily displaced, as in Figure 16.19b. In the process, energy is transferred to the mass since work must be done in moving it upward.

In this section, we describe the rate at which energy is transported along a string by a one-dimensional sinusoidal wave. Later, we shall extend these ideas to three-dimensional waves.

Consider a sinusoidal wave traveling on a string (Fig. 16.20). The source of the energy is some external agent at the left end of the string, which does work in producing the train of wave pulses. Let us focus on an element of the string of length Δx and mass Δm. Each such segment moves vertically with simple harmonic motion. Furthermore, all segments have the same angular frequency, ω, and the same amplitude, A. As we found in Chapter 13, the total energy E associated with a particle moving with simple harmonic motion is $E = \frac{1}{2}kA^2 = \frac{1}{2}m\omega^2 A^2$, where k is the equivalent force constant of the restoring force. If we apply this to the element of length Δx, we see that the total energy of this element is

$$\Delta E = \tfrac{1}{2}(\Delta m)\omega^2 A^2$$

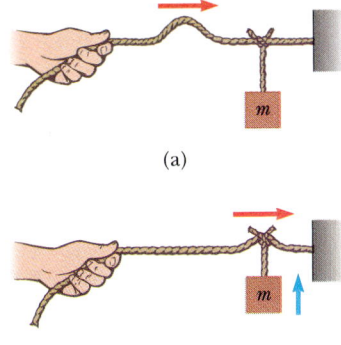

FIGURE 16.19   (a) A pulse traveling to the right on a stretched string on which a mass has been suspended. (b) Energy is transmitted to the suspended mass when the pulse arrives.

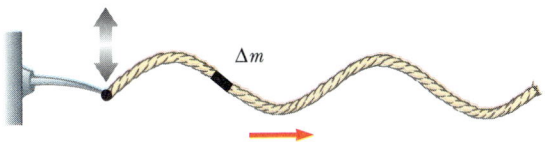

FIGURE 16.20   A sinusoidal wave traveling along the x axis on a stretched string. Every segment moves vertically, and each segment has the same total energy. The power transmitted by the wave equals the energy contained in one wavelength divided by the period of the wave.

If μ is the mass per unit length of the string, then the element of length Δx has a mass $\Delta m = \mu \, \Delta x$. Hence, we can express the energy ΔE as

$$\Delta E = \tfrac{1}{2}(\mu \, \Delta x)\,\omega^2 A^2 \tag{16.20}$$

If the wave travels from left to right as in Figure 16.20, the energy ΔE arises from the work done on the element Δm by the string element to the left of Δm. Similarly, the element Δm does work on the element to its right, and so we see that energy is transmitted to the right. The rate at which energy is transmitted along the string —in other words, the power—is dE/dt. If we let Δx approach 0, Equation 16.20 gives

$$\text{Power} = \frac{dE}{dt} = \tfrac{1}{2}\left(\mu \, \frac{dx}{dt}\right)\omega^2 A^2$$

Since dx/dt is equal to the wave speed, v, we have

Power

$$\text{Power} = \tfrac{1}{2}\mu\omega^2 A^2 v \tag{16.21}$$

This expression is generally valid, so that we can say *the power transmitted by any sinusoidal wave is proportional to the square of the frequency and to the square of the amplitude.*

Thus, we see that a wave traveling through a medium corresponds to energy transport through the medium, with no net transfer of matter. An oscillating source provides the energy and produces a sinusoidal disturbance in the medium. The disturbance is able to propagate through the medium as the result of the interaction between adjacent particles. In order to verify Equation 16.20 by direct experiment, one would have to design some device at the far end of the string to extract the energy of the wave without producing any reflections.

EXAMPLE 16.5   Power Supplied to a Vibrating String

A taut string having mass per unit length of $\mu = 5.00 \times 10^{-2}$ kg/m is under a tension of 80.0 N. How much power must be supplied to the rope to generate sinusoidal waves at a frequency of 60.0 Hz and an amplitude of 6.00 cm?

Solution   The wave speed on the string is

$$v = \sqrt{\frac{F}{\mu}} = \left(\frac{80.0 \text{ N}}{5.00 \times 10^{-2} \text{ kg/m}}\right)^{1/2} = 40.0 \text{ m/s}$$

Since $f = 60.0$ Hz, the angular frequency ω of the sinusoidal

waves on the string has the value

$$\omega = 2\pi f = 2\pi(60.0 \text{ Hz}) = 377 \text{ s}^{-1}$$

Using these values in Equation 16.21 for the power, with $A = 6.00 \times 10^{-2}$ m, gives

$$\begin{aligned}
\text{Power} &= \tfrac{1}{2}\mu\omega^2 A^2 v \\
&= \tfrac{1}{2}(5.00 \times 10^{-2} \text{ kg/m})(377 \text{ s}^{-1})^2 \\
&\quad \times (6.00 \times 10^{-2} \text{ m})^2(40.0 \text{ m/s}) \\
&= \boxed{512 \text{ W}}
\end{aligned}$$

---

*16.9   THE LINEAR WAVE EQUATION

In Section 16.3 we introduced the concept of the wave function to represent waves traveling on a string. All wave functions $y(x, t)$ represent solutions of an equation called the *linear wave equation*. This equation gives a complete description of the wave motion, and from it one can derive an expression for the wave speed. Furthermore, the wave equation is basic to many forms of wave motion. In this section, we derive the wave equation as applied to waves on strings.

Consider a small segment of a string of length Δx and tension F, on which a traveling wave is propagating (Fig. 16.21). Let us assume that the ends of the segment make small angles θ_A and θ_B with the x axis. The net force on the segment in the vertical direction is

$$\sum F_y = F \sin \theta_B - F \sin \theta_A = F(\sin \theta_B - \sin \theta_A)$$

Since we have assumed that the angles are small, we can use the small-angle approximation $\sin \theta \approx \tan \theta$ and express the net force as

$$\sum F_y \approx F(\tan \theta_B - \tan \theta_A)$$

However, the tangents of the angles at A and B are defined as the slope of the curve at these points. Since the slope of a curve is given by $\partial y / \partial x$, we have[4]

$$\sum F_y \approx F\left[\left(\frac{\partial y}{\partial x}\right)_B - \left(\frac{\partial y}{\partial x}\right)_A\right] \tag{16.22}$$

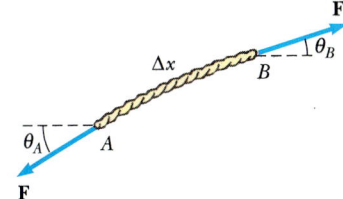

FIGURE 16.21 A segment of a string under tension F. Note that the slopes at points A and B are given by $\tan \theta_A$ and $\tan \theta_B$, respectively.

We now apply Newton's second law to the segment, with the mass of the segment given by $m = \mu \, \Delta x$, where μ is the mass per unit length of the string:

$$\sum F_y = ma_y = \mu \, \Delta x \left(\frac{\partial^2 y}{\partial t^2}\right) \tag{16.23}$$

Equating Equation 16.23 to Equation 16.22 gives

$$\mu \, \Delta x \left(\frac{\partial^2 y}{\partial t^2}\right) = F\left[\left(\frac{\partial y}{\partial x}\right)_B - \left(\frac{\partial y}{\partial x}\right)_A\right]$$

$$\frac{\mu}{F} \frac{\partial^2 y}{\partial t^2} = \frac{(\partial y / \partial x)_B - (\partial y / \partial x)_A}{\Delta x} \tag{16.24}$$

The right side of this equation can be expressed in a different form if we note that the partial derivative of any function is defined as

$$\frac{\partial f}{\partial x} \equiv \lim_{\Delta x \to 0} \frac{f(x + \Delta x) - f(x)}{\Delta x}$$

If we associate $f(x + \Delta x)$ with $(\partial y / \partial x)_B$ and $f(x)$ with $(\partial y / \partial x)_A$, we see that in the limit $\Delta x \to 0$, Equation 16.24 becomes

$$\frac{\mu}{F} \frac{\partial^2 y}{\partial t^2} = \frac{\partial^2 y}{\partial x^2} \tag{16.25}$$

Linear wave equation

This is the linear wave equation as it applies to waves on a string.

We now show that the sinusoidal wave function represents a solution of this wave equation. If we take the sinusoidal wave function to be of the form $y(x, t) = A \sin(kx - \omega t)$, the appropriate derivatives are

$$\frac{\partial^2 y}{\partial t^2} = -\omega^2 A \sin(kx - \omega t)$$

$$\frac{\partial^2 y}{\partial x^2} = -k^2 A \sin(kx - \omega t)$$

Substituting these expressions into Equation 16.25 gives

$$k^2 = (\mu / F) \omega^2$$

[4] It is necessary to use partial derivatives because y depends on both x and t.

Using the relationship $v = \omega/k$ in the above expression, we see that

$$v^2 = \frac{\omega^2}{k^2} = \frac{F}{\mu}$$

$$v = \sqrt{\frac{F}{\mu}}$$

which is Equation 16.4. This derivation represents another proof of the expression for the wave speed on a taut string.

The linear wave equation is often written in the form

Linear wave equation in general

$$\frac{\partial^2 y}{\partial x^2} = \frac{1}{v^2}\frac{\partial^2 y}{\partial t^2} \qquad (16.26)$$

This expression applies in general to various types of traveling waves. For waves on strings, y represents the vertical displacement of the string. For sound waves, y corresponds to variations in the pressure or density of a gas. In the case of electromagnetic waves, y corresponds to electric or magnetic field components.

We have shown that the sinusoidal wave function is one solution of the linear wave equation. Although we do not prove it here, the linear wave equation is satisfied by *any* wave function having the form $y = f(x \pm vt)$. Furthermore, we have seen that the wave equation is a direct consequence of Newton's second law applied to any segment of the string.

SUMMARY

A **transverse wave** is one in which the particles of the medium move in a direction *perpendicular* to the direction of the wave velocity. An example is a wave on a taut string.

A **longitudinal wave** is one in which the particles of the medium move in a direction *parallel* to the direction of the wave velocity. Sound waves in fluids are longitudinal.

Any one-dimensional wave traveling with a speed v in the x direction can be represented by a wave function of the form

$$y = f(x \pm vt) \qquad (16.1, 16.2)$$

where the $+$ sign applies to a wave traveling in the negative x direction and the $-$ sign applies to a wave traveling in the positive x direction. The shape of the wave at any instant (a snapshot of the wave) is obtained by holding t constant.

The **superposition principle** says that when two or more waves move through a medium, the resultant wave function equals the algebraic sum of the individual wave functions. Waves that obey this principle are said to be *linear*. When two waves combine in space, they interfere to produce a resultant wave. The **interference** may be **constructive** (when the individual displacements are in the same direction) or **destructive** (when the displacements are in opposite directions).

The **speed** of a wave traveling on a taut string of mass per unit length μ and tension F is

$$v = \sqrt{\frac{F}{\mu}} \qquad (16.4)$$

When a **pulse** traveling on a string meets a fixed end, the pulse is reflected and inverted. If the pulse reaches a free end, it is reflected but not inverted.

The **wave function** for a one-dimensional sinusoidal wave traveling to the right can be expressed as

$$y = A \sin\left[\frac{2\pi}{\lambda}(x - vt)\right] = A \sin(kx - \omega t) \qquad \text{(16.6, 16.11)}$$

where A is the amplitude, λ is the wavelength, k is the angular wave number, and ω is the angular frequency. If T is the period (the time it takes the wave to travel a distance equal to one wavelength) and f is the frequency, then v, k, and ω can be written

$$v = \frac{\lambda}{T} = \lambda f \qquad \text{(16.7, 16.14)}$$

$$k \equiv \frac{2\pi}{\lambda} \qquad \text{(16.9)}$$

$$\omega = \frac{2\pi}{T} = 2\pi f \qquad \text{(16.10, 16.12)}$$

The **power** transmitted by a sinusoidal wave on a stretched string is

$$\text{Power} = \tfrac{1}{2}\mu\omega^2 A^2 v \qquad \text{(16.21)}$$

QUESTIONS

1. Why is a wave pulse traveling on a string considered a transverse wave?

2. How would you set up a longitudinal wave in a stretched spring? Would it be possible to set up a transverse wave in a spring?

3. By what factor would you have to increase the tension in a taut string in order to double the wave speed?

4. When traveling on a taut string, does a wave pulse always invert upon reflection? Explain.

5. Can two pulses traveling in opposite directions on the same string reflect from each other? Explain.

6. Does the vertical speed of a segment of a horizontal taut string through which a wave is traveling depend on the wave speed?

7. If you were to shake one end of a taut rope periodically three times each second, what would be the period of the sinusoidal waves set up in the rope?

8. A vibrating source generates a sinusoidal wave on a string under constant tension. If the power delivered to the string is doubled, by what factor does the amplitude change? Does the wave speed change under these circumstances?

9. Consider a wave traveling on a taut rope. What is the difference, if any, between the speed of the wave and the speed of a small section of the rope?

10. If a long rope is hung from a ceiling and waves are sent up the rope from its lower end, they do not ascend with constant speed. Explain.

11. What happens to the wavelength of a wave on a string when the frequency is doubled? Assume the tension in the string remains the same.

12. What happens to the speed of a wave on a taut string when the frequency is doubled? Assume the tension in the string remains the same.

13. How do transverse waves differ from longitudinal waves?

14. When all the strings on a guitar are stretched to the same tension, will the speed of a wave along the more massive bass strings be faster or slower than the speed of a wave on the lighter strings?

15. If you stretch a rubber hose and pluck it, you can observe a pulse traveling up and down the hose. What happens to the speed if you stretch the hose tighter? If you fill the hose with water?

16. In a longitudinal wave in a spring, the coils move back and forth in the direction of wave motion. Does the speed of the wave depend on the maximum speed of each coil?

17. When two waves interfere, can the amplitude of the resultant wave be larger than either of the two original waves? Under what conditions?

18. A solid can transport both longitudinal waves and transverse waves, but a fluid can transport only longitudinal waves. Why?

19. In an earthquake both S (transverse) and P (longitudinal) waves are sent out. The S waves travel through the Earth more slowly than the P waves (4.5 km/s versus 7.8 km/s). By detecting the time of arrival of the waves, how can one determine how far away the epicenter of the quake was? How many detection centers are necessary to pinpoint the location of the epicenter?

PROBLEMS

Section 16.3   One-Dimensional Traveling Waves

1. At $t = 0$, a transverse wave pulse in a wire is described by the function

$$y = \frac{6}{x^2 + 3}$$

where x and y are in meters. Write the function $y(x, t)$ that describes this wave if it is traveling in the positive x direction with a speed of 4.5 m/s.

2. Two wave pulses **A** and **B** are moving in opposite directions along a taut string with a speed of 2 cm/s. The amplitude of **A** is twice the amplitude of **B**. The pulses are shown in Figure P16.2 at $t = 0$. Sketch the shape of the string at $t = 1, 1.5, 2, 2.5,$ and 3 s.

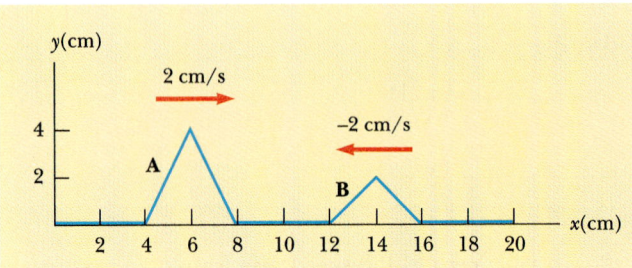

FIGURE P16.2

3. Two points, A and B, on the Earth are at the same longitude and 60.0° apart in latitude. An earthquake at point A sends two waves toward B. A transverse wave travels along the surface of the Earth at 4.50 km/s, and a longitudinal wave travels through the body of the Earth at 7.80 km/s. (a) Which wave arrives at B first? (b) What is the time difference between the arrivals of the two waves at B? Take the radius of the Earth to be 6370 km.

4. A wave moving along the x axis is described by

$$y(x, t) = 5.0 e^{-(x + 5.0t)^2}$$

where x is in meters and t is in seconds. Determine (a) the direction of the wave motion and (b) the speed of the wave.

5. Ocean waves with a crest-to-crest distance of 10 m can be described by

$$y(x, t) = (0.80 \text{ m}) \sin[0.63(x - vt)]$$

where $v = 1.2$ m/s. (a) Sketch $y(x, t)$ at $t = 0$. (b) Sketch $y(x, t)$ at $t = 2.0$ s. Note how the entire wave form has shifted 2.4 m in the positive x direction in this time interval.

Section 16.4   Superposition and Interference of Waves

6. Two waves in one string are described by the relationships

$$y_1 = 3.0 \cos(4.0x - 5.0t)$$
$$y_2 = 4.0 \sin(5.0x - 2.0t)$$

where y and x are in centimeters and t is in seconds. Find the superposition of the waves $y_1 + y_2$ at the points (a) $x = 1.0$, $t = 1.0$, (b) $x = 1.0$, $t = 0.50$, (c) $x = 0.50$, $t = 0$. (Remember that the arguments of the trigonometric functions are in radians.)

7. Two sinusoidal waves in a string are defined by the functions

$$y_1 = (2.0 \text{ cm}) \sin(20x - 30t)$$
$$y_2 = (2.0 \text{ cm}) \sin(25x - 40t)$$

where y and x are in centimeters and t is in seconds. (a) What is the phase difference between these two waves at the point $x = 5.0$ cm at $t = 2.0$ s? (b) What is the positive x value closest to the origin for which the two phases differ by $\pm \pi$ at $t = 2.0$ s? (This is where the two waves add to zero.)

8. Two waves are traveling in the same direction along a stretched string. Each has an amplitude of 4.0 cm, and they are 90° out of phase. Find the amplitude of the resultant wave.

8A. Two waves are traveling in the same direction along a stretched string. Each has an amplitude A, and they are out of phase by an angle ϕ. Find the amplitude of the resultant wave.

9. Two pulses traveling on the same string are described by

$$y_1 = \frac{5}{(3x - 4t)^2 + 2}$$

and

$$y_2 = \frac{-5}{(3x + 4t - 6)^2 + 2}$$

(a) In which direction does each pulse travel? (b) At what time do the two cancel? (c) At what point do the two waves always cancel?

Section 16.5   The Speed of Waves on Strings

10. Transverse waves with a speed of 50 m/s are to be produced in a taut string. A 5.0-m length of string with a total mass of 0.060 kg is used. What is the required tension?

□ indicates problems that have full solutions available in the Student Solutions Manual and Study Guide.

11. A piano string of mass per unit length 5.00×10^{-3} kg/m is under a tension of 1350 N. Find the speed with which a wave travels on this string.

12. An astronaut on the Moon wishes to measure the local value of g by timing pulses traveling down a wire that has a large mass suspended from it. Assume a wire of mass 4.00 g is 1.60 m long and has a 3.00-kg mass suspended from it. A pulse requires 36.1 ms to traverse the length of the wire. Calculate g from these data. (You may neglect the mass of the wire when calculating the tension in it.)

12A. An astronaut on the Moon wishes to measure the local value of g by timing pulses traveling down a wire that has a large mass suspended from it. Assume a wire of mass m and length L has a mass M suspended from it. A pulse requires a time t to traverse the length of the wire. Calculate g from these data. (You may neglect the mass of the wire when finding the tension in it.)

13. Transverse waves travel with a speed of 20.0 m/s in a string under a tension of 6.00 N. What tension is required for a wave speed of 30.0 m/s in the same string?

14. A simple pendulum consists of a ball of mass M hanging from a uniform string of mass m and length L, with $m \ll M$. If the period of oscillation for the pendulum is T, determine the speed of a transverse wave in the string when the pendulum hangs vertically.

15. The elastic limit of a piece of steel wire is 2.7×10^9 Pa. What is the maximum speed at which transverse wave pulses can propagate along this wire without exceeding this stress? (The density of steel is 7.86×10^3 kg/m^3.)

16. A light string of mass 10.0 g and length $L = 3.00$ m has its ends tied to two walls that are separated by the distance $D = 2.00$ m. Two masses, each of mass $M = 2.00$ kg, are suspended from the string as in Figure P16.16. If a wave pulse is sent from point A, how long does it take to travel to point B?

16A. A light string of mass m and length L has its ends tied to two walls that are separated by the distance D. Two masses, each of mass M, are suspended from the string as in Figure P16.16. If a wave pulse is sent from point A, how long does it take to travel to point B?

17. A 30.0-m steel wire and a 20.0-m copper wire, both with 1.00-mm diameters, are connected end to end and stretched to a tension of 150 N. How long does it take a transverse wave to travel the entire length of the two wires?

18. A light string of mass per unit length 8.00 g/m has its ends tied to two walls separated by a distance equal to three fourths the length of the string (Fig. P16.18). A mass m is suspended from the center of the string,

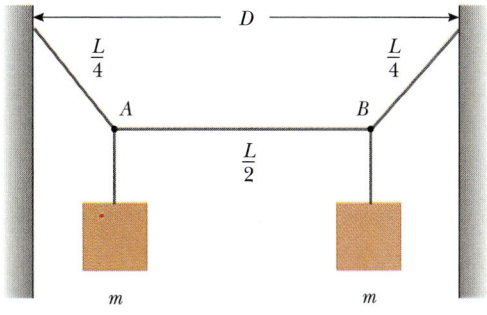

FIGURE P16.16

putting a tension in the string. (a) Find an expression for the transverse wave speed in the string as a function of the hanging mass. (b) How much mass should be suspended from the string to have a wave speed of 60.0 m/s?

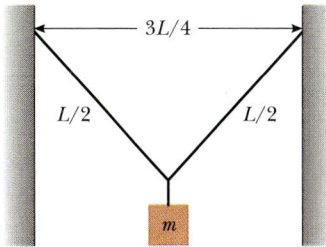

FIGURE P16.18

Section 16.7 Sinusoidal Waves

19. (a) Plot y versus t at $x = 0$ for a sinusoidal wave of the form $y = (15.0 \text{ cm}) \cos(0.157x - 50.3t)$, where x and y are in centimeters and t is in seconds. (b) Determine the period of vibration from this plot and compare your result with the value found in Example 16.3.

20. For a certain transverse wave, the distance between two successive maxima is 1.2 m and eight maxima pass a given point along the direction of travel every 12 s. Calculate the wave speed.

20A. For a certain transverse wave, the distance between two successive maxima is λ and N maxima pass a given point along the direction of travel every t s. Find an expression for the wave speed.

21. A sinusoidal wave is traveling along a rope. The oscillator that generates the wave completes 40.0 vibrations in 30.0 s. Also, a given maximum travels 425 cm along the rope in 10.0 s. What is the wavelength?

22. When a particular wire is vibrating with a frequency of 4.00 Hz, a transverse wave of wavelength 60.0 cm is produced. Determine the speed of wave pulses along the wire.

23. A sinusoidal wave traveling in the $-x$ direction (to the left) has an amplitude of 20.0 cm, a wavelength of 35.0 cm, and a frequency of 12.0 Hz. The displacement of the wave at $t = 0$, $x = 0$ is $y = -3.00$ cm, and the wave has a positive velocity here. (a) Sketch the wave at $t = 0$. (b) Find the angular wave number, period, angular frequency, and phase velocity of the wave. (c) Write an expression for the wave function $y(x, t)$.

24. A sinusoidal wave train is described by

$$y = (0.25 \text{ m}) \sin(0.30x - 40t)$$

where x and y are in meters and t is in seconds. Determine for this wave the (a) amplitude, (b) angular frequency, (c) angular wave number, (d) wavelength, (e) wave speed, and (f) direction of motion.

25. Two waves are described by

$$y_1(x, t) = 5.0 \sin(2.0x - 10t)$$

and

$$y_2(x, t) = 10 \cos(2.0x - 10t),$$

where x is in meters and t is in seconds. Show that the resulting wave is sinusoidal, and determine the amplitude and phase of this sinusoidal wave.

26. A bat can detect small objects such as an insect whose size is approximately equal to one wavelength of the sound the bat makes. If bats emit a chirp at a frequency of 60.0 kHz, and if the speed of sound in air is 340 m/s, what is the smallest insect a bat can detect?

27. (a) Write the expression for y as a function of x and t for a sinusoidal wave traveling along a rope in the *negative x* direction with the following characteristics: $A = 8.00$ cm, $\lambda = 80.0$ cm, $f = 3.00$ Hz, and $y(0, t) = 0$ at $t = 0$. (b) Write the expression for y as a function of x for the wave in part (a) assuming that $y(x, 0) = 0$ at the point $x = 10.0$ cm.

28. A transverse wave on a string is described by

$$y = (0.12 \text{ m}) \sin \pi(x/8 + 4t)$$

(a) Determine the transverse speed and acceleration of the string at $t = 0.20$ s for the point on the string located at $x = 1.6$ m. (b) What are the wavelength, period, and speed of propagation of this wave?

29. A transverse sinusoidal wave on a string has a period $T = 25.0$ ms and travels in the negative x direction with a speed of 30.0 m/s. At $t = 0$, a particle on the string at $x = 0$ has a displacement of 2.00 cm and travels to the left with a speed of 2.0 m/s. (a) What is the amplitude of the wave? (b) What is the initial phase angle? (c) What is the maximum transverse speed of the string? (d) Write the wave function for the wave.

30. A sinusoidal wave of wavelength 2.0 m and amplitude 0.10 m travels with a speed of 1.0 m/s on a string. Initially, the left end of the string is at the

origin and the wave moves from left to right. Find (a) the frequency and angular frequency, (b) the angular wave number, and (c) the wave function for this wave. Determine the equation of motion for (d) the left end of the string and (e) the point on the string at $x = 1.5$ m to the right of the left end. (f) What is the maximum speed of any point on the string?

31. A wave is described by $y = (2.0 \text{ cm}) \sin(kx - \omega t)$, where $k = 2.11$ rad/m, $\omega = 3.62$ rad/s, x is in meters, and t is in seconds. Determine the amplitude, wavelength, frequency, and speed of the wave.

32. A sinusoidal wave on a string is described by

$$y = (0.51 \text{ cm}) \sin(kx - \omega t)$$

where $k = 3.1$ rad/cm and $\omega = 9.3$ rad/s. How far does a wave crest move in 10 s? Does it move in the positive or negative x direction?

33. A transverse traveling wave on a taut wire has an amplitude of 0.200 mm and a frequency of 500 Hz and travels with a speed of 196 m/s. (a) Write an equation in SI units of the form $y = A \sin(kx - \omega t)$ for this wave. (b) The mass per unit length of this wire is 4.10 g/m. Find the tension in the wire.

34. A wave on a string is described by the wave function $y = (0.10 \text{ m}) \sin(0.50x - 20t)$. (a) Show that a particle in the string at $x = 2.0$ m executes harmonic motion. (b) Determine the frequency of oscillation of this particular point.

Section 16.8 Energy Transmitted by Sinusoidal Waves on Strings

35. A taut rope has a mass of 0.18 kg and a length of 3.6 m. What power must be supplied to the rope in order to generate sinusoidal waves having an amplitude of 0.10 m and a wavelength of 0.50 m and traveling with a speed of 30 m/s?

35A. A taut rope has a mass M and length L. What power must be supplied to the rope in order to generate sinusoidal waves having an amplitude A and wavelength λ and traveling with a speed v?

36. A two-dimensional water wave spreads in circular wavefronts. Show that the amplitude A at a distance r from the initial disturbance is proportional to $1/\sqrt{r}$. (*Hint:* Consider the energy concentrated in the outward-moving ripple.)

37. Transverse waves are being generated on a rope under constant tension. By what factor is the required power increased or decreased if (a) the length of the rope is doubled and the angular frequency remains constant, (b) the amplitude is doubled and the angular frequency is halved, (c) both the wavelength and the amplitude are doubled, and (d) both the length of the rope and the wavelength are halved?

38. Sinusoidal waves 5.00 cm in amplitude are to be transmitted along a string that has a linear density of 4.00×10^{-2} kg/m. If the maximum power delivered by the source is 300 W and the string is under a tension of 100 N, what is the highest vibrational frequency at which the source can operate?

39. A sinusoidal wave on a string is described by the equation

$$y = (0.15 \text{ m}) \sin(0.80x - 50t)$$

where x and y are in meters and t is in seconds. If the mass per unit length of this string is 12 g/m, determine (a) the speed of the wave, (b) the wavelength, (c) the frequency, and (d) the power transmitted to the wave.

40. A horizontal string can transmit a maximum power of P (without breaking) if a wave with amplitude A and angular frequency ω is traveling along it. In order to increase this maximum power, a student folds the string and uses this "double string" as a transmitter. Determine the maximum power that can be transmitted along the "double string."

*Section 16.9   The Linear Wave Equation

41. Show that the wave function $y = \ln[b(x - vt)]$ is a solution to Equation 16.26, where b is a constant.

42. Show that the wave function $y = e^{b(x - vt)}$ is a solution of the wave equation (Eq. 16.26), where b is a constant.

43. (a) Show that the function $y(x, t) = x^2 + v^2t^2$ is a solution to the wave equation. (b) Show that the function above can be written as $f(x + vt) + g(x - vt)$, and determine the functional forms for f and g. (c) Repeat parts (a) and (b) for the function, $y(x, t) = \sin(x) \cos(vt)$.

ADDITIONAL PROBLEMS

44. A traveling wave propagates according to the expression $y = (4.0 \text{ cm}) \sin(2.0x - 3.0t)$ where x is in centimeters and t is in seconds. Determine (a) the amplitude, (b) the wavelength, (c) the frequency, (d) the period, and (e) the direction of travel of the wave.

45. The wave function for a linearly polarized wave on a taut string is (in SI units)

$$y(x, t) = (0.35 \text{ m}) \sin(10\pi t - 3\pi x + \pi/4)$$

(a) What are the speed and direction of travel of the wave? (b) What is the vertical displacement of the string at $t = 0$, $x = 0.10$ m? (c) What are the wavelength and frequency of the wave? (d) What is the maximum magnitude of the transverse speed of the string?

46. A block of mass $M = 2.0$ kg, supported by a string, rests on an incline making an angle of $\theta = 45°$ with the horizontal (Fig. P16.46). The length of the string is $L = 0.5$ m and its mass is $m = 2.0$ g, and hence much less than M. Determine the time it takes a transverse wave to travel from one end of the string to the other.

46A. A block of mass M, supported by a string, rests on an incline making an angle of θ with the horizontal (Fig. P16.46). The length of the string is L and its mass is $m \ll M$. Determine the time it takes a transverse wave to travel from one end of the string to the other.

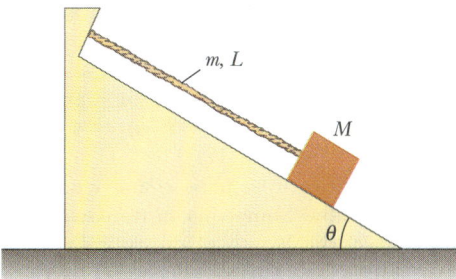

FIGURE P16.46

47. (a) Determine the speed of transverse waves on a string under a tension of 80.0 N if the string has a length of 2.00 m and a mass of 5.00 g. (b) Calculate the power required to generate these waves if they have a wavelength of 16.0 cm and an amplitude of 4.00 cm.

48. A sinusoidal wave in a rope is described by the wave function $y = (0.20 \text{ m}) \sin[\pi(0.75x + 18t)]$ where x and y are in meters and t is in seconds. The rope has a linear mass density of 0.25 kg/m. If the tension in the rope is provided by an arrangement like the one illustrated in Figure 16.11, what is the value of the suspended mass?

49. A 2.0-kg block hangs from a rubber string, being supported so that the string is not stretched. The unstretched length of the string is 0.5 m and its mass is 5.0 g. The "spring constant" for the string is 100.0 N/m. The block is released and stops at the lowest point. (a) Determine the tension in the string when the block is at this lowest point. (b) What is the length of the string in this "stretched" position? (c) Find the speed of a transverse wave in the string if the block is held in this lowest position.

49A. A block of mass M hangs from a rubber string, being supported so that the string is not stretched. The unstretched length of the string is L_0 and its mass is m. The "spring constant" for the string is k. The block is released and stops at the lowest point. (a) Determine the tension in the string when the block is at this lowest point. (b) What is the length of the string in this "stretched" position? (c) Find the speed of a transverse wave in the string if the block is held in this lowest position.

50. A wire of density ρ is tapered so that its cross-sectional area varies with x, according to

$$A = (1.0 \times 10^{-3}\, x + 0.010)\ \text{cm}^2$$

(a) If the wire is subject to a tension F, derive a relationship for the speed of a wave as a function of position. (b) If the wire is aluminum and is subject to a tension of 24 N, determine the speed at the origin and at $x = 10$ m.

51. Determine the speed and direction of propagation of each of the following sinusoidal waves, assuming that x is measured in meters and t in seconds.
    (a) $y = 0.60 \cos(3.0x - 15t + 2)$
    (b) $y = 0.40 \cos(3.0x + 15t - 2)$
    (c) $y = 1.2 \sin(15t + 2.0x)$
    (d) $y = 0.20 \sin(12t - x/2 + \pi)$

52. A rope of total mass m and length L is suspended vertically. Show that a transverse wave pulse will travel the length of the rope in a time $t = 2\sqrt{L/g}$. (*Hint:* First find an expression for the wave speed at any point a distance x from the lower end by considering the tension in the rope as resulting from the weight of the segment below that point.)

53. If mass M is suspended from the bottom of the rope in Problem 52, (a) show that the time for a transverse wave to travel the length of the rope is

$$t = 2\sqrt{\frac{L}{g}}\left(\frac{\sqrt{M + m} - \sqrt{M}}{\sqrt{m}}\right)$$

(b) Show that this reduces to the result of Problem 52 when $M = 0$. (c) Show that for $m \ll M$, the expression in part (a) reduces to

$$t = \sqrt{\frac{mL}{Mg}}$$

54. As a sound wave travels through the air, it produces pressure variations (above and below atmospheric pressure) given by $P = 1.27 \sin \pi(x - 340t)$ in SI units. Find (a) the amplitude of the pressure variations, (b) the frequency, (c) the wavelength in air, and (d) the speed of the sound wave.

55. An aluminum wire is clamped at each end under zero tension at room temperature (22°C). The tension in the wire is increased by reducing the temperature, which results in a decrease in the wire's equilibrium length. What strain ($\Delta L/L$) results in a transverse wave speed of 100 m/s? Take the cross-sectional area of the wire to be 5.0×10^{-6} m², the density to be 2.7×10^3 kg/m³, and Young's modulus to be 7.0×10^{10} N/m².

56. (a) Show that the speed of longitudinal waves along a spring of force constant k is $v = \sqrt{kL/\mu}$, where L is the unstretched length of the spring and μ is the mass per unit length. (b) A spring of mass 0.40 kg has an unstretched length of 2.0 m and a force constant of 100 N/m. Using the results to part (a), determine the speed of longitudinal waves along this spring.

57. It is stated in Problem 52 that a wave pulse travels from the bottom to the top of a rope of length L in a time $t = 2\sqrt{L/g}$. Use this result to answer the following questions. (It is *not* necessary to set up any new integrations.) (a) How long does it take for a wave pulse to travel halfway up the rope? (Give your answer as a fraction of the quantity $2\sqrt{L/g}$.) (b) A pulse starts traveling up the rope. How far has it traveled after a time $\sqrt{L/g}$?

58. A string of length L consists of two sections. The left half has mass per unit length $\mu = \mu_0/2$, while the right has a mass per unit length $\mu' = 3\mu = 3\mu_0/2$. Tension in the string is F_0. Notice from the data given that this string has the same total mass as a uniform string of length L and mass per unit length μ_0. (a) Find the speeds v and v' at which transverse wave pulses travel in the two sections. Express the speeds in terms of F_0 and μ_0, and also as multiples of the speed $v_0 = \sqrt{F_0/\mu_0}$. (b) Find the time required for a pulse to travel from one end of the string to the other. Give your result as a multiple of $T_0 = L/v_0$.

59. A wave pulse traveling along a string of linear mass density μ is described by the relationship

$$y = [A_0 e^{-bx}]\sin(kx - \omega t)$$

where the factors in brackets before the sine are said to be the amplitude. (a) What is the power $P(x)$ carried by this wave at a point x? (b) What is the power carried by this wave at the origin? (c) Compute the ratio $P(x)/P(0)$.

SPREADSHEET PROBLEM

S1. Two transverse wave pulses traveling in opposite directions along the x axis are represented by the following wave functions:

$$y_1(x, t) = \frac{6}{(x - 3t)^2} \qquad y_2(x, t) = -\frac{3}{(x + 3t)^2}$$

where x and y are measured in centimeters and t is in seconds. Write a spreadsheet or program to add the two pulses and obtain the shape of the composite waveform $y_{\text{tot}} = y_1 + y_2$ as a function of time. Plot y_{tot} versus x. Make separate plots for $t = 0, 0.5, 1, 1.5, 2, 2.5,$ and 3.0 s.

CHAPTER 17

Sound Waves

Fennec foxes are small animals with very large ears, about four inches long. The sensitivity of their auditory system is enhanced by these large ears, which enable them to hear very faint sounds as they collect a larger cross-section of sound waves. *(Tom McHugh/Photo Researchers)*

Sound waves are the most important example of longitudinal waves. They can travel through any material medium with a speed that depends on the properties of the medium. As the waves travel, the particles in the medium vibrate to produce density and pressure changes along the direction of motion of the wave. These changes result in a series of high- and low-pressure regions called *condensations* and *rarefactions,* respectively. If the source of the sound waves vibrates sinusoidally, the pressure variations are also sinusoidal. We shall find that the mathematical description of harmonic sound waves is identical to that of harmonic string waves discussed in the previous chapter.

There are three categories of mechanical waves that cover different ranges of frequency: (1) *Audible waves* (usually called **sound waves**) are waves that lie within the range of sensitivity of the human ear, typically, 20 Hz to 20 000 Hz. They can be generated in a variety of ways, such as by musical instruments, human vocal cords, and loudspeakers. (2) *Infrasonic waves* are waves having frequencies below the audible range. Earthquake waves are an example. (3) *Ultrasonic waves* are waves having frequencies above the audible range. For example, they can be generated by inducing vibrations in a quartz crystal with an applied alternating electric field. All may be longitudinal or transverse in solids but only longitudinal in fluids.

Any device that transforms one form of power into another is called a *transducer.* In addition to the loudspeaker (which transforms electric power to power in audible waves) and the quartz crystal (electric power to ultrasonic power), ceramic and magnetic phonograph pickups are common examples of sound transducers.

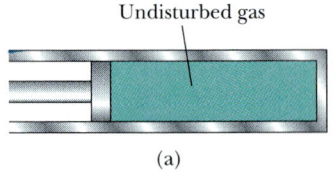

Undisturbed gas

(a)

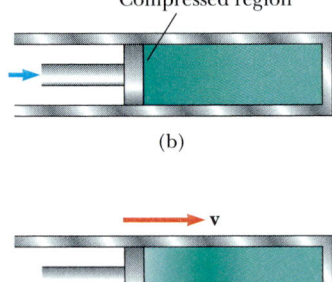

Compressed region

(b)

(c)

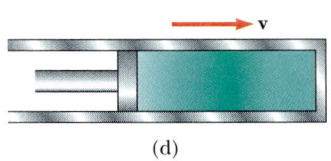

(d)

FIGURE 17.1   Motion of a longitudinal pulse through a compressible medium. The compression (darker region) is produced by the moving piston.

17.1   SPEED OF SOUND WAVES

The speed of sound waves depends on the compressibility and the inertia of the medium. If the medium has a bulk modulus B (Section 12.4) and an equilibrium density ρ, the speed of sound waves in that medium is

$$v = \sqrt{\frac{B}{\rho}} \qquad (17.1)$$

It is interesting to compare Equation 17.1 with the expression for the speed of transverse waves on a string, $v = \sqrt{F/\mu}$, discussed in the previous chapter. In both cases, the wave speed depends on an elastic property of the medium (B or F) and on an inertial property of the medium (ρ or μ). In fact, the speed of *all mechanical waves* follows an expression of the general form

$$v = \sqrt{\frac{\text{elastic property}}{\text{inertial property}}}$$

Let us describe pictorially the motion of a one-dimensional longitudinal pulse moving through a long tube containing a compressible gas or liquid (Fig. 17.1). A piston at the left end can be moved to the right to compress the fluid and create the pulse. Before the piston is moved, the medium is undisturbed and of uniform density, as represented by the uniformly shaded region in Figure 17.1a. When the piston is suddenly pushed to the right (Fig. 17.1b), the medium just in front of it is compressed (represented by the more heavily shaded region); the pressure and density in this region are higher than normal. When the piston comes to rest (Fig. 17.1c), the compressed region of the medium continues to move to the right, corresponding to a longitudinal pulse traveling down the tube with a speed v. Note that the piston speed does *not* equal v. Furthermore, the compressed region does not "stay with" the piston until the piston stops (in other words, once the pulse is created, $v_{\text{pulse}} > v_{\text{piston}}$).

EXAMPLE 17.1   Sound Waves in a Solid Bar

If a solid bar is struck at one end with a hammer, a longitudinal pulse propagates down the bar with a speed

$$v = \sqrt{\frac{Y}{\rho}}$$

where Y is the Young's modulus for the material (Section 12.4). Find the speed of sound in an aluminum bar.

Solution   From Table 12.1 we get $Y = 7.0 \times 10^{10}$ N/m² for aluminum and from Table 1.5 we get $\rho = 2.7 \times 10^3$ kg/m³. Therefore,

$$v_{\text{A1}} = \sqrt{\frac{Y}{\rho}} = \sqrt{\frac{7.0 \times 10^{10} \text{ N/m}^2}{2.7 \times 10^3 \text{ kg/m}^3}} \approx 5.1 \text{ km/s}$$

This is a typical value for the speed of sound in solids, much larger than the speed of sound in gases, as Table 17.1 shows. This difference in speeds makes sense because the molecules of a solid are bound together in a much more rigid structure than a gas.

TABLE 17.1   Speed of Sound in Various Media

Medium	v(m/s)
Gases	
Air (0°C)	331
Air (20°C)	343
Hydrogen (0°C)	1286
Oxygen (0°C)	317
Helium (0°C)	972
Liquids at 25°C	
Water	1493
Methyl alcohol	1143
Seawater	1533
Solids	
Aluminum	5100
Copper	3560
Iron	5130
Lead	1322
Vulcanized rubber	54

EXAMPLE 17.2 Speed of Sound in a Liquid

Find the speed of sound in water, which has a bulk modulus of about 2.1×10^9 N/m^2 and a density of 1.00×10^3 kg/m^3.

Solution Using Equation 17.1, we find that

$$v_{\text{water}} = \sqrt{\frac{B}{\rho}} \approx \sqrt{\frac{2.1 \times 10^9 \text{ N/m}^2}{1.00 \times 10^3 \text{ kg/m}^3}} = \boxed{1.5 \text{ km/s}}$$

In general, sound waves travel more slowly in liquids than in solids. This is because liquids are more compressible than solids and hence have a smaller bulk modulus.

17.2 PERIODIC SOUND WAVES

As just noted in the previous section, one can produce a one-dimensional periodic sound wave in a long, narrow tube containing a gas by means of a vibrating piston at one end, as in Figure 17.2. The darker regions in this figure represent regions where the gas is compressed, and so in these regions the density and pressure are above their equilibrium values.

A compressed region is formed whenever the piston is being pushed into the tube. This compressed region, called a **condensation**, moves down the tube as a pulse, continuously compressing the layers in front of it. When the piston is withdrawn from the tube, the gas in front of it expands and the pressure and density in this region fall below their equilibrium values (represented by the lighter regions in Figure 17.2). These low-pressure regions, called **rarefactions**, also propagate along the tube, following the condensations. Both regions move with a speed equal to the speed of sound in that medium (about 343 m/s in air at 20°C).

As the piston oscillates sinusoidally, regions of condensation and rarefaction are continuously set up. The distance between two successive condensations (or two successive rarefactions) equals the wavelength, λ. As these regions travel down the tube, any small volume of the medium moves with simple harmonic motion parallel to the direction of the wave. If $s(x, t)$ is the displacement of a small volume element measured from its equilibrium position, we can express this harmonic displacement function as

$$s(x, t) = s_{\text{max}} \cos(kx - \omega t) \tag{17.2}$$

where s_{max} is the *maximum displacement of the medium from equilibrium* (in other words, the **displacement amplitude**), k is the angular wave number, and ω is the angular frequency of the piston. Note that the displacement of the medium is along x, the direction of motion of the sound wave, which of course means we are describing a longitudinal wave.

The variation in the pressure of the gas, ΔP, measured from its equilibrium value is also periodic and given by

$$\Delta P = \Delta P_{\text{max}} \sin(kx - \omega t) \tag{17.3}$$

The derivation of this expression follows:

The **pressure amplitude** ΔP_{max} is the *maximum change in pressure from the equilibrium value.* As we show later, the pressure amplitude is proportional to the displacement amplitude, s_{max}:

$$\Delta P_{\text{max}} = \rho v \omega s_{\text{max}} \tag{17.4}$$

where ωs_{max} is the maximum longitudinal speed of the medium in front of the piston.

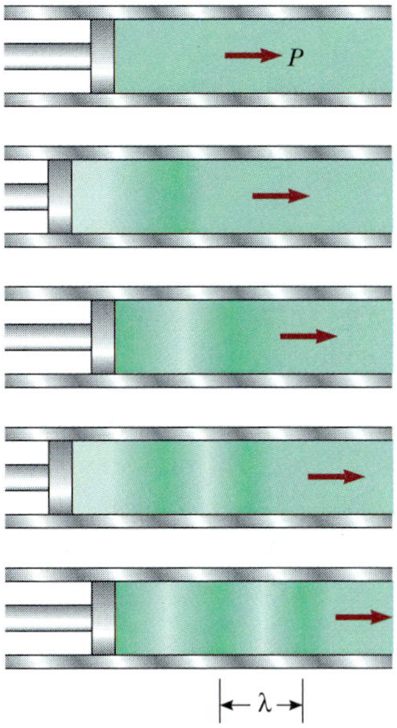

FIGURE 17.2 A sinusoidal longitudinal wave propagating down a tube filled with a compressible gas. The source of the wave is a vibrating piston at the left. The high- and low-pressure regions are dark and light, respectively.

Pressure amplitude

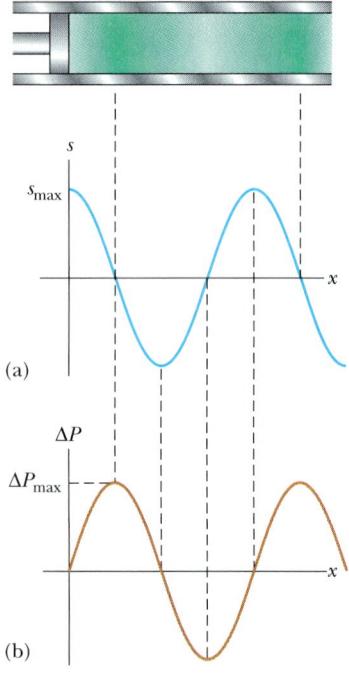

FIGURE 17.3   (a) Displacement amplitude versus position and (b) pressure amplitude versus position for a sinusoidal longitudinal wave. The displacement wave is 90° out of phase with the pressure wave.

Thus, we see that a sound wave may be considered as either a displacement wave or a pressure wave. A comparison of Equations 17.2 and 17.3 shows that *the pressure wave is 90° out of phase with the displacement wave*. Graphs of these functions are shown in Figure 17.3. Note that the pressure variation is a maximum when the displacement is zero, whereas the displacement is a maximum when the pressure variation is zero. Since pressure is proportional to density, the variation in density from the equilibrium value follows an expression similar to Equation 17.3.

We now derive Equations 17.3 and 17.4. From the definition of bulk modulus (Equation 12.8), we see that the pressure variation in a gas is

$$\Delta P = -B \frac{\Delta V}{V}$$

The volume of a medium segment that has a thickness Δx in the horizontal direction and a cross-sectional area A is $V = A \Delta x$. The change in the volume ΔV accompanying the pressure change is equal to $A \Delta s$, where Δs is the difference between the value of s at $x + \Delta x$ and the value of s at x. Hence, we can express ΔP as

$$\Delta P = -B \frac{\Delta V}{V} = -B \frac{A \Delta s}{A \Delta x} = -B \frac{\Delta s}{\Delta x}$$

As Δx approaches zero, the ratio $\Delta s/\Delta x$ becomes $\partial s/\partial x$. (The partial derivative is used here to indicate that we are interested in the variation of s with position at a *fixed* time.) Therefore,

$$\Delta P = -B \frac{\partial s}{\partial x}$$

If the displacement is the simple sinusoidal function given by Equation 17.2, we find that

$$\Delta P = -B \frac{\partial}{\partial x} [s_{max} \cos(kx - \omega t)] = B s_{max} k \sin(kx - \omega t)$$

Since the bulk modulus is given by $B = \rho v^2$ (Eq. 17.1), the pressure variation reduces to

$$\Delta P = \rho v^2 s_{max} k \ \sin(kx - \omega t)$$

Furthermore, from Equation 16.13, we can write $k = \omega/v$; hence ΔP can be expressed as

$$\Delta P = \rho \omega s_{max} v \ \sin(kx - \omega t)$$

Taking the maximum value of each side, we find

$$\Delta P_{max} = \rho v \omega s_{max}$$

which is Equation 17.4. Then, with this substitution, we arrive at Equation 17.3:

$$\Delta P = \Delta P_{max} \sin(kx - \omega t)$$

17.3   INTENSITY OF PERIODIC SOUND WAVES

In the previous chapter, we showed that a wave traveling on a taut string transports energy. The same concepts are now applied to sound waves. Consider a layer of air of mass Δm and width Δx in front of a piston oscillating with angular frequency ω,

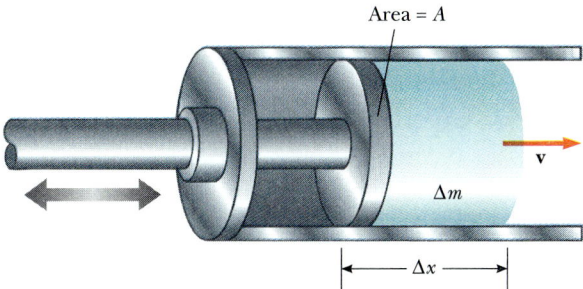

FIGURE 17.4   An oscillating piston transfers energy to the gas in the tube, causing the layer of width Δx and mass Δm to oscillate with an amplitude s_{max}.

as in Figure 17.4. The piston transmits energy to the layer of air.[1] Since the average kinetic energy equals the average potential energy in simple harmonic motion (Chapter 13), the average total energy of the mass Δm equals its maximum kinetic energy. Therefore, we can express the average energy of the moving layer of air as

$$\Delta E = \tfrac{1}{2}\Delta m(\omega s_{max})^2 = \tfrac{1}{2}(\rho A\,\Delta x)(\omega s_{max})^2$$

where $A\,\Delta x$ is the volume of the layer. The time rate at which energy is transferred to each layer—in other words the power—is

$$\text{Power} = \frac{\Delta E}{\Delta t} = \tfrac{1}{2}\rho A\left(\frac{\Delta x}{\Delta t}\right)(\omega s_{max})^2 = \tfrac{1}{2}\rho A v(\omega s_{max})^2$$

where $v = \Delta x/\Delta t$ is the speed of the disturbance to the right.

> We define the intensity I of a wave, or the power per unit area, to be the rate at which the energy being transported by the wave flows through a unit area A perpendicular to the direction of travel of the wave.

In our present case, therefore, the intensity is

$$I = \frac{\text{power}}{\text{area}} = \tfrac{1}{2}\rho(\omega s_{max})^2 v \tag{17.5}$$

Thus, we see that the intensity of a periodic sound wave is proportional to the square of the amplitude and to the square of the frequency (as in the case of a periodic string wave). This can also be written in terms of the pressure amplitude ΔP_{max}, using Equation 17.4, which gives

$$I = \frac{\Delta P_{max}^2}{2\rho v} \tag{17.6}$$

Intensity of a sound wave

---

[1] Although it is not proved here, the work done by the piston equals the energy carried away by the wave. For a detailed mathematical treatment of this concept, see Frank S. Crawford, Jr., *Waves*, New York, McGraw-Hill, 1968, Berkeley Physics Course, Volume 3, Chapter 4.

EXAMPLE 17.3   Hearing Limitations

The faintest sounds the human ear can detect at a frequency of 1000 Hz correspond to an intensity of about 1.00×10^{-12} W/m^2 (the so-called *threshold of hearing*). The loudest sounds that the ear can tolerate correspond to an intensity of about 1.00 W/m^2 *(the threshold of pain)*. Determine the pressure amplitudes and maximum displacements associated with these two limits.

Solution   First, consider the faintest sounds. Using Equation 17.6 and taking $v = 343$ m/s to be the speed of sound waves in air and the density of air to be $\rho = 1.29$ kg/m³, we get

$$\Delta P_{max} = \sqrt{2\rho v I}$$
$$= \sqrt{2(1.29 \text{ kg/m}^3)(343 \text{ m/s})(1.00 \times 10^{-12} \text{ W/m}^2)}$$
$$= 2.97 \times 10^{-5} \text{ N/m}^2$$

Since atmospheric pressure is about 10^5 N/m², this result

tells us that the ear can discern pressure fluctuations as small as 3 parts in 10^{10}!

The corresponding maximum displacement can be calculated using Equation 17.4, recalling that $\omega = 2\pi f$:

$$s_{max} = \frac{\Delta P_{max}}{\rho \omega v} = \frac{2.97 \times 10^{-5} \text{ N/m}^2}{(1.29 \text{ kg/m}^3)(2\pi \times 10^3 \text{ s}^{-1})(343 \text{ m/s})}$$
$$= 1.07 \times 10^{-11} \text{ m}$$

This is a remarkably small number! If we compare this result for s_{max} with the diameter of an atom (about 10^{-10} m), we see that the ear is an extremely sensitive detector of sound waves.

In a similar manner, one finds that the loudest sounds the human ear can tolerate correspond to a pressure amplitude of about 30 N/m² and a maximum displacement of about 1.1×10^{-5} m.

The pressure amplitudes, called acoustic pressure, correspond to fluctuations taking place above and below atmospheric pressure.

Sound Level in Decibels

The previous example illustrates the wide range of intensities the human ear can detect. Because this range is so wide, it is convenient to use a logarithmic scale, where the **sound level** β is defined by the equation

$$\beta \equiv 10 \log\left(\frac{I}{I_0}\right) \qquad (17.7)$$

Sound level in decibels

The constant I_0 is the *reference intensity*, taken to be at the threshold of hearing ($I_0 = 1.00 \times 10^{-12}$ W/m²), and I is the intensity in watts per square meter at the sound level β, where β is measured in decibels (dB).[2] On this scale, the threshold of pain ($I = 1.00$ W/m²) corresponds to a sound level of $\beta = 10 \log(1/10^{-12}) = 10 \log(10^{12}) = 120$ dB, and the threshold of hearing corresponds to a sound level $\beta = 10 \log(1/1) = 0$ dB.

Prolonged exposure to high sound levels may produce serious damage to the ear. Ear plugs are recommended whenever sound levels exceed 90 dB. Recent evidence also suggests that "noise pollution" may be a contributing factor to high blood pressure, anxiety, and nervousness. Table 17.2 gives some typical values of the sound levels of various sources.

TABLE 17.2   Sound Levels for Some Sources in Decibels

Source of Sound	β(dB)
Nearby jet airplane	150
Jackhammer; machine gun	130
Siren; rock concert	120
Subway; power mower	100
Busy traffic	80
Vacuum cleaner	70
Normal conversation	50
Mosquito buzzing	40
Whisper	30
Rustling leaves	10
Threshold of hearing	0

17.4   SPHERICAL AND PLANE WAVES

If a spherical body oscillates so that its radius varies sinusoidally with time, a spherical sound wave is produced (Fig. 17.5). The wave moves outward from the source at a constant speed if the medium is uniform.

[2] The "bel" is named after the inventor of the telephone, Alexander Graham Bell (1847–1922). The prefix *deci-* is the metric system scale factor that stands for 10^{-1}.

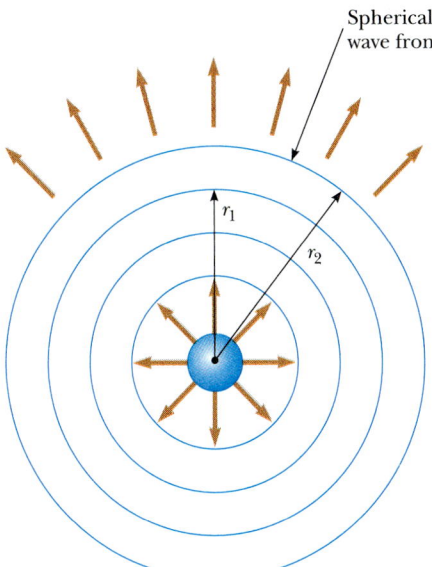

Spherical wave front

FIGURE 17.5 A spherical wave propagating radially outward from an oscillating spherical body. The intensity of the spherical wave varies as $1/r^2$.

Since all points on any given sphere behave in the same way, we conclude that the energy in a spherical wave propagates equally in all directions. That is, no one direction is preferred over any other. If P_{av} is the average power emitted by the source, then this power at any distance r from the source must be distributed over a spherical surface of area $4\pi r^2$. Hence, the wave intensity at a distance r from the source is

$$I = \frac{P_{av}}{A} = \frac{P_{av}}{4\pi r^2} \tag{17.8}$$

Since P_{av} is the same throughout any spherical surface centered at the source, we see that the intensities at distances r_1 and r_2 are

$$I_1 = \frac{P_{av}}{4\pi r_1{}^2} \quad \text{and} \quad I_2 = \frac{P_{av}}{4\pi r_2{}^2}$$

Therefore, the ratio of intensities on these two spherical surfaces is

$$\frac{I_1}{I_2} = \frac{r_2{}^2}{r_1{}^2}$$

In Equation 17.5 we found that the intensity is proportional to $s_{max}{}^2$, the square of the wave displacement amplitude. Comparing this result with Equation 17.8, we conclude that the displacement amplitude of a spherical wave must vary as $1/r$. Therefore, we can write the wave function ψ (Greek letter "psi") for an outgoing spherical wave in the form

$$\psi(r, t) = \frac{s_0}{r} \sin(kr - \omega t) \tag{17.9}$$

where s_0, the displacement amplitude at $t = 0$, is a constant.

It is useful to represent spherical waves by a series of circular arcs concentric with the source, as in Figure 17.6. Each arc represents a surface over which the phase of the wave is constant. We call such a surface of constant phase a **wave front**.

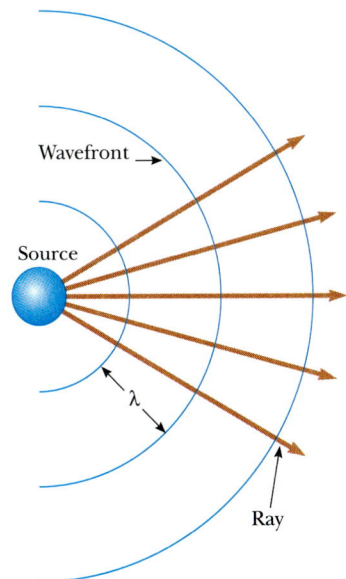

Wavefront

Source

λ

Ray

FIGURE 17.6 Spherical waves emitted by a point source. The circular arcs represent the spherical wave fronts concentric with the source. The rays are radial lines pointing outward from the source perpendicular to the wave fronts.

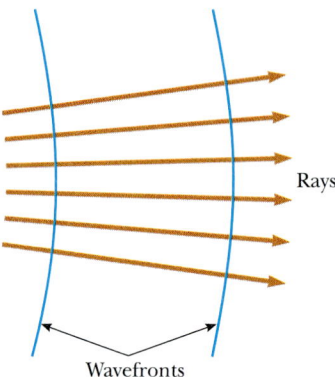

FIGURE 17.7 Far away from a point source, the wave fronts are nearly parallel planes and the rays are nearly parallel lines perpendicular to the planes. Hence, a small segment of a spherical wave front is approximately a plane wave.

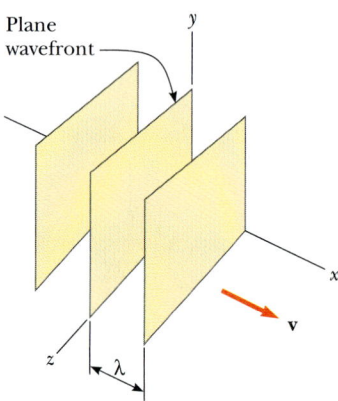

FIGURE 17.8 A representation of a plane wave moving in the positive *x* direction with a speed *v*. The wave fronts are planes parallel to the *yz* plane.

The distance between adjacent wave fronts equals the wavelength, λ. The radial lines pointing outward from the source are called **rays.**

Now consider a small portion of the wave fronts far from the source, as in Figure 17.7. In this case, the rays are nearly parallel to each other and the wave fronts are very close to being planar. Therefore, at distances from the source that are large compared with the wavelength, we can approximate the wave fronts by parallel planes. We call such a wave a **plane wave.** Any small portion of a spherical wave that is far from the source can be considered a plane wave.

Figure 17.8 illustrates a plane wave propagating along the *x* axis, which means the wave fronts are parallel to the *yz* plane. In this case, the wave function depends only on *x* and *t* and has the form

Plane wave representation

$$\psi(x, t) = A \sin(kx - \omega t) \tag{17.10}$$

That is, the wave function for a plane wave is identical in form to that of a one-dimensional traveling wave. The intensity is the same all over any given wave front of the plane wave.

EXAMPLE 17.4   Intensity Variations of a Point Source

A point source emits sound waves with an average power output of 80.0 W. (a) Find the intensity 3.00 m from the source.

Reasoning and Solution   A point source emits energy in the form of spherical waves (Fig. 17.5). At a distance *r* from the source, the power is distributed over the surface area of a sphere, $4\pi r^2$. Therefore, the intensity at a distance *r* from the source is given by Equation 17.8:

$$I = \frac{P_{av}}{4\pi r^2} = \frac{80.0 \text{ W}}{4\pi(3.00 \text{ m})^2} = \boxed{0.707 \text{ W/m}^2}$$

which is close to the threshold of pain.

(b) Find the distance at which the sound reduces to a level of 40 dB.

Solution   We can find the intensity at the 40-dB level by using Equation 17.7 with $I_0 = 1.00 \times 10^{-12}$ W/m²:

$$10 \log\left(\frac{I}{I_0}\right) = 40$$

$$I = 1.00 \times 10^4 \, I_0 = 1.00 \times 10^{-8} \text{ W/m}^2$$

Using this value for I in Equation 17.8 and solving for r, we get

$$r = \sqrt{\frac{P_{av}}{4\pi I}} = \sqrt{\frac{80.0 \text{ W}}{4\pi \times 1.00 \times 10^{-8} \text{ W/m}^2}}$$

$$= \boxed{2.52 \times 10^4 \text{ m}}$$

which equals about 16 miles!

*17.5   THE DOPPLER EFFECT

When a car or truck is moving while its horn is blowing, the frequency of the sound you hear is higher as the vehicle approaches you and lower as it moves away from you. This is one example of the **Doppler effect**.[3]

> In general, a Doppler effect is experienced whenever there is relative motion between source and observer. When the source and observer are moving toward each other, the frequency heard by the observer is *higher* than the frequency of the source. When the source and observer are moving away from each other, the frequency heard by the observer is *lower* than the source frequency.

"I love hearing that lonesome wail of the train whistle as the magnitude of the frequency of the wave changes due to the Doppler effect."

Although the Doppler effect is most commonly experienced with sound waves, it is a phenomenon common to all harmonic waves. For example, there is a shift in frequencies of light waves (electromagnetic waves) produced by the relative motion of source and observer. The Doppler effect is used in police radar systems to measure the speed of motor vehicles. Likewise, astronomers use the effect to determine the relative motion of stars, galaxies, and other celestial objects.

First, let us consider the case where the observer O is moving and the sound source S is stationary. For simplicity, we shall assume that the air is also stationary and that the observer moves directly toward the source. Figure 17.9 describes the situation when the observer moves with a speed v_O toward the source (considered as a point source), which is at rest ($v_S = 0$). In general, "at rest" means at rest with respect to the medium, air.

We shall take the frequency of the source to be f, the wavelength to be λ, and the speed of sound to be v. If the observer were also stationary, clearly he or she would detect f wave fronts per second. (That is, when $v_O = 0$ and $v_S = 0$, the observed frequency equals the source frequency.) When the observer moves toward the source, the speed of the waves relative to the observer is $v' = v + v_O$, but the wavelength λ is unchanged. Hence, the frequency heard by the observer is *increased* and given by

$$f' = \frac{v'}{\lambda} = \frac{v + v_O}{\lambda}$$

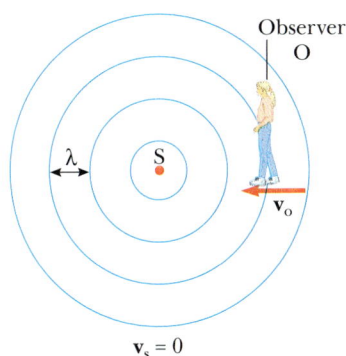

FIGURE 17.9   An observer O moving with a speed v_O toward a stationary point source S hears a frequency f' that is greater than the source frequency.

---

[3] Named after the Austrian physicist Christian Johann Doppler (1803–1853), who first suggested that the change in frequency observed for sound waves could also be applicable to light waves.

Since $\lambda = v/f$, we can express f' as

$$f' = f\left(1 + \frac{v_O}{v}\right) \qquad \text{(Observer moving toward the source)} \qquad \textbf{(17.11)}$$

Similarly, if the observer is moving *away* from the source, the speed of the wave relative to the observer is $v' = v - v_O$. The frequency heard by the observer in this case is *lowered* and is

$$f' = f\left(1 - \frac{v_O}{v}\right) \qquad \text{(Observer moving away from the source)} \qquad \textbf{(17.12)}$$

In general, when an observer moves with a speed v_O relative to a stationary source, the frequency heard by the observer is

Frequency heard with an
observer in motion

$$f' = f\left(1 \pm \frac{v_O}{v}\right) \qquad \textbf{(17.13)}$$

where the *positive* sign is used when the observer moves *toward* the source and the *negative* sign holds when the observer moves *away* from the source.

Now consider the situation in which the source is in motion and the observer is at rest. If the source moves directly toward observer A in Figure 17.10a, the wave fronts seen by the observer are closer together as a result of the motion of the source in the direction of the outgoing wave. As a result, the wavelength λ' measured by observer A is shorter than the wavelength λ of the source. During each vibration, which lasts for a time T (the period), the source moves a distance $v_S T = v_S/f$ and the wave length is *shortened* by this amount. Therefore, the observed wavelength λ' is

$$\lambda' = \lambda - \Delta\lambda = \lambda - \frac{v_S}{f}$$

Since $\lambda = v/f$, the frequency heard by observer A is

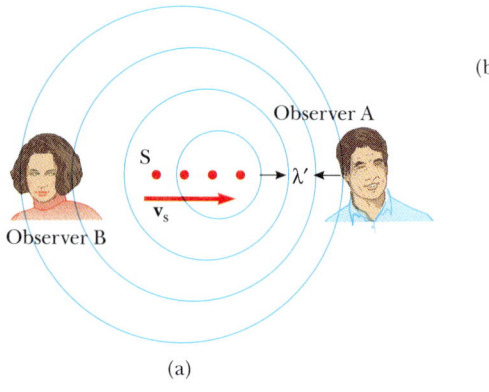

(a)

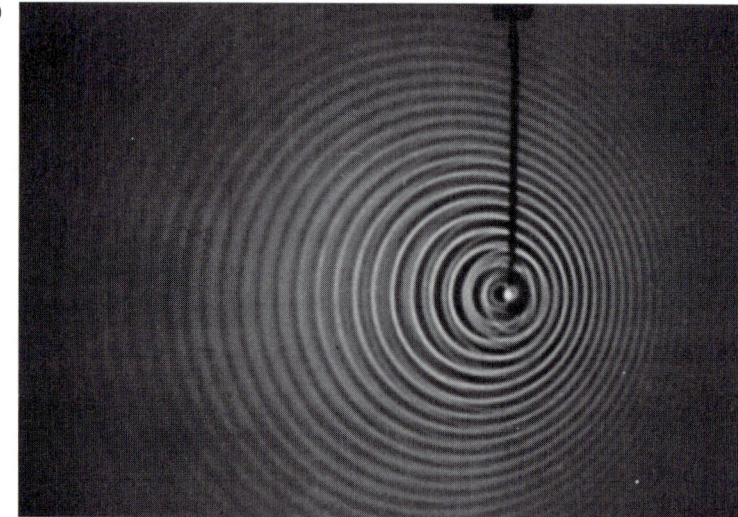

(b)

FIGURE 17.10   (a) A source S moving with a speed v_S toward a stationary observer A and away from a stationary observer B. Observer A hears an increased frequency, and observer B hears a decreased frequency. (b) The Doppler effect in water observed in a ripple tank.   *(Courtesy Educational Development Center, Newton, Mass.)*

$$f' = \frac{v}{\lambda'} = \frac{v}{\lambda - \dfrac{v_S}{f}} = \frac{v}{\dfrac{v}{f} - \dfrac{v_S}{f}}$$

$$f' = f\left(\frac{1}{1 - \dfrac{v_S}{v}}\right) \tag{17.14}$$

That is, the observed frequency is *increased* when the source moves toward the observer.

In a similar manner, when the source moves away from an observer B at rest (where observer B is to the left of the source, as in Fig. 17.10a), observer B measures a wavelength λ' that is *greater* than λ and hears a *decreased* frequency

$$f' = f\left(\frac{1}{1 + \dfrac{v_S}{v}}\right) \tag{17.15}$$

Combining Equations 17.14 and 17.15, we can express the general relationship for the observed frequency when the source is moving and the observer is at rest as

$$f' = f\left(\frac{1}{1 \mp \dfrac{v_S}{v}}\right) \tag{17.16}$$

Frequency heard with source in motion

Finally, if both the source and the observer are in motion, we find the following general relationship for the observed frequency:

$$f' = f\left(\frac{v \pm v_O}{v \mp v_S}\right) \tag{17.17}$$

Frequency heard with observer and source in motion

In this expression, the *upper* signs ($+ v_O$ and $- v_S$) refer to motion of one *toward* the other, and the lower signs ($- v_O$ and $+ v_S$) refer to motion of one *away from* the other.

A convenient rule to remember concerning signs when working with all Doppler effect problems is the following:

The word *toward* is associated with an *increase* in the observed frequency. The words *away from* are associated with a *decrease* in the observed frequency.

EXAMPLE 17.5   The Moving Train Whistle

A train moving at a speed of 40 m/s sounds its whistle, which has a frequency of 500 Hz. Determine the frequencies heard by a stationary observer as the train approaches and then recedes from the observer.

Solution   We can use Equation 17.14 to get the apparent frequency as the train approaches the observer. Taking $v =$ 343 m/s for the speed of sound in air gives

$$f' = f\left(\frac{1}{1 - \frac{v_S}{v}}\right) = (500 \text{ Hz})\left(\frac{1}{1 - \frac{40 \text{ m/s}}{343 \text{ m/s}}}\right) = \boxed{566 \text{ Hz}}$$

Likewise, Equation 17.15 can be used to obtain the frequency

heard as the train recedes from the observer:

$$f' = f\left(\frac{1}{1 + \frac{v_S}{v}}\right) = (500 \text{ Hz})\left(\frac{1}{1 + \frac{40 \text{ m/s}}{343 \text{ m/s}}}\right) = \boxed{448 \text{ Hz}}$$

EXAMPLE 17.6   The Noisy Siren

An ambulance travels down a highway at a speed of 33.5 m/s (75 mi/h). Its siren emits sound at a frequency of 400 Hz. What is the frequency heard by a passenger in a car traveling at 24.6 m/s (55 mi/h) in the opposite direction as the car approaches the ambulance and as the car moves away from the ambulance?

Solution   Let us take the speed of sound in air to be $v =$ 343 m/s. We can use Equation 17.17 in both cases. As the ambulance and car approach each other, the observed apparent frequency is

$$f' = f\left(\frac{v + v_O}{v - v_S}\right) = (400 \text{ Hz})\left(\frac{343 \text{ m/s} + 24.6 \text{ m/s}}{343 \text{ m/s} - 33.5 \text{ m/s}}\right)$$

$$= \boxed{475 \text{ Hz}}$$

Likewise, as they recede from each other, a passenger in the

car hears a frequency

$$f' = f\left(\frac{v - v_O}{v + v_S}\right) = (400 \text{ Hz})\left(\frac{343 \text{ m/s} - 24.6 \text{ m/s}}{343 \text{ m/s} + 33.5 \text{ m/s}}\right)$$

$$= \boxed{338 \text{ Hz}}$$

The *change* in frequency as detected by the passenger in the car is $475 - 338 = 137$ Hz, which is more than 30% of the actual frequency emitted.

Exercise   Suppose that the passenger car is parked on the side of the highway as the ambulance travels down the highway at the speed of 33.5 m/s. What frequency will the passenger in the car hear as the ambulance (a) approaches the parked car and (b) recedes from the parked car?

Answer   (a) 443 Hz; (b) 364 Hz.

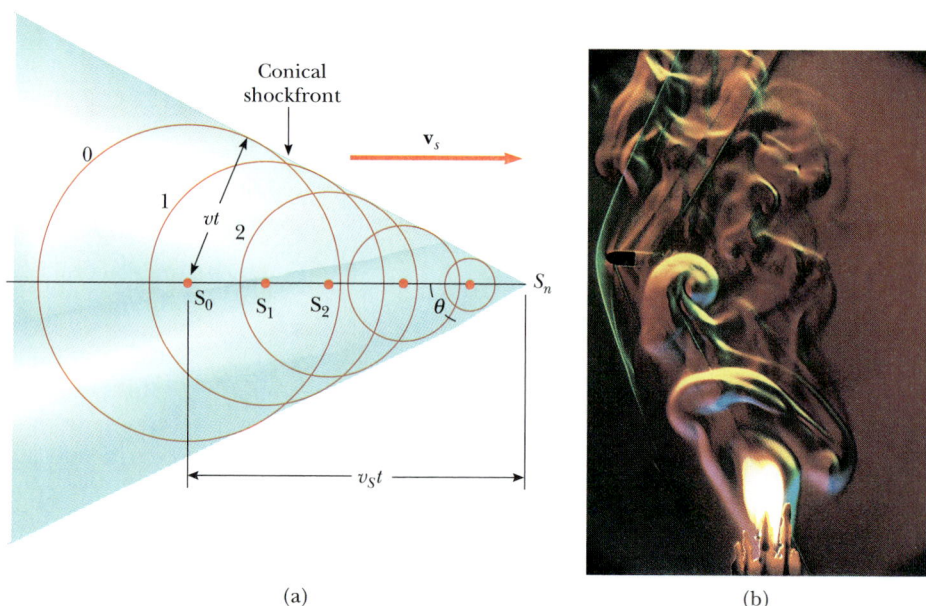

(a)   (b)

FIGURE 17.11   (a) A representation of a shock wave produced when a source moves from S_0 to S_n with a speed v_S, which is greater than the wave speed v in that medium. The envelope of the wavefronts forms a cone whose apex half-angle is given by $\sin\theta = v/v_S$. (b) A stroboscopic photograph of a bullet moving at supersonic speed through the hot air above a candle. Note the shock wave in the vicinity of the bullet.   (*© The Harold E. Edgerton 1992 Trust. Courtesy of Palm Press, Inc.*)

Shock Waves

Now let us consider what happens when the source speed v_S *exceeds* the wave speed v. This situation is described graphically in Figure 17.11. The circles represent spherical wave fronts emitted by the source at various times during its motion. At $t = 0$, the source is at S_0, and at some later time t, the source is at S_n. In the time t, the wave front centered at S_0 reaches a radius of vt. In this same interval, the source travels a distance $v_S t$ to S_n. At the instant the source is at S_n, waves are just beginning to be generated and so the wave front has zero radius at this point. The line drawn from S_n to the wave front centered on S_0 is tangent to all other wave fronts generated at intermediate times. Thus, we see that the envelope of these waves is a cone whose apex half-angle θ is

$$\sin\theta = \frac{v}{v_S}$$

The V-shaped wavefront that follows the duck occurs because the duck travels at a speed greater than the speed of water waves. This is analogous to shock waves produced by airplanes traveling at supersonic speeds. *(© Harry Engels)*

The ratio v_S/v is referred to as the *Mach number*. The conical wave front produced when $v_S > v$ (supersonic speeds) is known as a *shock wave*. An interesting analogy to shock waves is the V-shaped wave fronts produced by a duck (the bow wave) when the duck's speed exceeds the speed of the surface water waves.

Jet airplanes traveling at supersonic speeds produce shock waves, which are responsible for the loud explosion, or "sonic boom," one hears. The shock wave carries a great deal of energy concentrated on the surface of the cone, with correspondingly large pressure variations. Such shock waves are unpleasant to hear and can cause damage to buildings when aircraft fly supersonically at low altitudes. In fact, an airplane flying at supersonic speeds produces a double boom because two shock fronts are formed, one from the nose of the plane and one from the tail (Fig. 17.12).

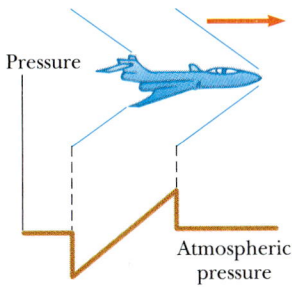

FIGURE 17.12 Two shock waves produced by the nose and tail of a jet airplane traveling at supersonic speeds.

SUMMARY

Sound waves are longitudinal and travel through a compressible medium with a speed that depends on the compressibility and inertia of that medium. The **speed of sound** in a medium having a bulk modulus B and density ρ is

$$v = \sqrt{\frac{B}{\rho}} \qquad (17.1)$$

In the case of sinusoidal sound waves, the **variation in pressure** from the equilibrium value is given by

$$\Delta P = \Delta P_{\max} \sin(kx - \omega t) \qquad (17.3)$$

where $\Delta P_{\max}$ is the **pressure amplitude**. The pressure wave is 90° out of phase with the displacement wave. If the displacement amplitude is $s_{\max}$, then $\Delta P_{\max}$ has the value

$$\Delta P_{\max} = \rho v \omega s_{\max} \qquad (17.4)$$

The **intensity of a harmonic sound wave**, which is the power per unit area, is

$$I = \tfrac{1}{2}\rho(\omega s_{\max})^2 v = \frac{\Delta P_{\max}^2}{2\rho v} \qquad (17.5, 17.6)$$

The **intensity of a spherical wave** produced by a point source is proportional to the average power emitted and inversely proportional to the square of the distance from the source.

The change in frequency heard by an observer whenever there is relative motion between the source and observer is called the **Doppler effect.** If the *observer moves* with a speed v_O and the source is at rest, the observed frequency f' is

$$f' = f\left(1 \pm \frac{v_O}{v}\right) \tag{17.13}$$

where the positive sign is used when the observer moves toward the source and the negative sign refers to motion away from the source.

If the *source moves* with a speed v_S and the observer is at rest, the observed frequency is

$$f' = f\left(\frac{1}{1 \mp \frac{v_S}{v}}\right) \tag{17.16}$$

where $- v_S$ refers to motion *toward* the observer and $+ v_S$ refers to motion *away* from the observer.

When the *observer and source are both moving,* the observed frequency is

$$f' = f\left(\frac{v \pm v_O}{v \mp v_S}\right) \tag{17.17}$$

QUESTIONS

1. Why are sound waves characterized as longitudinal?
2. As a result of a distant explosion, an observer senses a ground tremor and then hears the explosion. Explain.
3. Some sound waves are harmonic, whereas others are not. Give an example of each.
4. If the distance from a point source is tripled, by what factor does the intensity decrease?
5. Explain how the Doppler effect is used with microwaves to determine the speed of an automobile.
6. If you are in a moving vehicle, explain what happens to the frequency of your echo as you move *toward* a canyon wall. What happens to the frequency as you move *away* from the wall?
7. Suppose an observer and a source of sound are both at rest and a strong wind blows toward the observer. Describe the effect of the wind (if any) on (a) the observed wavelength, (b) the observed frequency, and (c) the wave velocity.
8. Of the following sounds, which is most likely to have an intensity level of 60 dB: a rock concert, the turning of a page in this text, normal conversation, a cheering crowd at a football game, or background noise at a church?
9. Estimate the decibel level of each of the sounds in Question 10.
10. A binary star system consists of two stars revolving about each other. If we observe the light reaching us from one of these stars as it makes one complete revolution about the other, what does the Doppler effect predict will happen to this light?
11. How could an object move with respect to an observer such that the sound from it is not shifted in frequency?
12. Why is it not possible to use sonar (sound waves) to determine the speed of an object traveling faster than the speed of sound in that medium?
13. Why is it so quiet after a snowfall?
14. Why is the intensity of an echo less than that of the original sound?
15. If the wavelength of a sound source is reduced by a factor of 2, what happens to its frequency? Its speed?
16. A sound wave travels in air at a frequency of 500 Hz. If part of the wave travels from the air into water, does its frequency change? Does its wavelength change? Justify your answers.
17. In a recent discovery, a nearby star was found to have a large planet orbiting about it, although the planet could not be seen. In terms of the concept of systems rotating about their center of mass and the Doppler shift for light (which is in many ways similar to that of sound), explain how an astronomer could determine the presence of the invisible planet.
18. Explain how the distance to a lightning bolt may be determined by counting the seconds between the flash and the sound of the thunder. Does the speed of the light signal have to be taken into account?

PROBLEMS

1. Suppose that you hear a thunder clap 16.2 s after seeing the associated lightning stroke. The speed of sound waves in air is 343 m/s and the speed of light in air is 3.0×10^8 m/s. How far are you from the lightning stroke?

2. A stone is dropped into a deep canyon and is heard to strike the bottom 10.2 s after release. The speed of sound waves in air is 343 m/s. How deep is the canyon? What would be the percentage error in the depth if the time required for the sound to reach the canyon rim were ignored?

2A. A stone is dropped into a deep canyon and is heard to strike the bottom t seconds after release. The speed of sound waves in air is v. How deep is the canyon? What would be the percentage error in the depth if the time required for the sound to reach the canyon rim were ignored?

3. Find the speed of sound in mercury, which has a bulk modulus of approximately 2.8×10^{10} N/m² and a density of 13 600 kg/m³.

4. A flower pot is knocked off a balcony 20.0 m above the sidewalk and is heading for a 1.75-m-tall man standing below. How high from the ground can the flower pot be after which it would be too late for a shouted warning to reach the man in time? Assume that the man below requires 0.300 s to respond to the warning.

5. The speed of sound in air is $v = \sqrt{\gamma P/\rho}$, where γ is a constant equal to 7/5, P is the air pressure, and ρ is the density of air. Calculate the speed of sound for $P = 1$ atm $= 1.013 \times 10^5$ Pa and $\rho = 1.29$ kg/m³.

Section 17.2 Periodic Sound Waves

(*Note:* In this section, use the following values as needed unless otherwise specified: the equilibrium density of air, $\rho = 1.20$ kg/m³; the speed of sound in air, $v = 343$ m/s. Also, pressure variations ΔP are measured relative to atmospheric pressure.)

6. The density of aluminum is 2.7×10^3 kg/m³. Use the value for the speed of sound in aluminum given in Table 17.1 to calculate Young's modulus for this material.

7. You are watching a pier being constructed on the far shore of a saltwater inlet when some blasting occurs. You hear the sound in the water 4.5 s before it reaches you through the air. How wide is the inlet? (*Hint:* See Table 17.1. Assume the air temperature is 20°C.)

8. A rescue plane flies horizontally at a constant speed searching for a disabled boat. When the plane is di-

rectly above the boat, the boat's crew blows a loud horn. By the time the plane's sound detector perceives the horn's sound, the plane has traveled a distance equal to one-half its altitude above the ocean. If it takes the sound 2.0 s to reach the plane, determine (a) the speed of the plane and (b) its altitude. Take the speed of sound to be 343 m/s.

9. The speed of sound in air (in m/s) depends on temperature according to the expression

$$v = 331.5 + 0.607 T_C$$

where T_C is the Celsius temperature. In dry air the temperature decreases about 1°C for every 150-m rise in altitude. (a) Assuming this change is constant up to an altitude of 9000 m, how long will it take the sound from an airplane flying at 9000 m to reach the ground on a day when the ground temperature is 30°C? (b) Compare this to the time it would take if the air were a constant 30°C. Which time is longer?

10. Calculate the pressure amplitude of a 2.0-kHz sound wave in air if the displacement amplitude is equal to 2.0×10^{-8} m.

11. A sound wave in air has a pressure amplitude equal to 4.0×10^{-3} Pa. Calculate the displacement amplitude of the wave at a frequency of 10.0 kHz.

12. A sound wave in a cylinder is described by Equations 17.2 through 17.4. Show that $\Delta P = \pm \rho v \omega \sqrt{s_{max}^2 - s^2}$.

13. An experimenter wishes to generate in air a sound wave that has a displacement amplitude equal to 5.5×10^{-6} m. The pressure amplitude is to be limited to 8.4×10^{-1} Pa. What is the minimum wavelength the sound wave can have?

14. A sound wave in air has a pressure amplitude of 4.0 Pa and a frequency of 5.0 kHz. $\Delta P = 0$ at the point $x = 0$ when $t = 0$. (a) What is ΔP at $x = 0$ when $t = 2.0 \times 10^{-4}$ s, and (b) what is ΔP at $x = 0.020$ m when $t = 0$?

15. A sinusoidal sound wave is described by the displacement

$$s(x, t) = (2.00 \ \mu m) \cos[(15.7 \ m^{-1})x - (858 \ s^{-1}) t]$$

(a) Find the amplitude, wavelength, and speed of this wave and state what material this sound wave is traveling through. (See Table 17.1.) (b) Determine the instantaneous displacement of the molecules at the position $x = 0.0500$ m at $t = 3.00$ ms. (c) Determine the maximum speed of the molecules' oscillatory motion.

16. The tensile stress in a copper rod is 99.5% of its elastic breaking point of 13×10^{10} N/m². If a 500-Hz sound wave is transmitted along the rod, (a) what displacement amplitude will cause the rod to break

$\square$ indicates problems that have full solutions available in the Student Solutions Manual and Study Guide.

and (b) what is the maximum speed of the particles at this moment?

17. Write an expression that describes the pressure variation as a function of position and time for a sinusoidal sound wave in air if $\lambda = 0.10$ m and $\Delta P_{max} = 0.20$ Pa.

18. Write the function that describes the displacement wave corresponding to the pressure wave in Problem 17.

Section 17.3   Intensity of Periodic Sound Waves

19. Calculate the sound level in dB of a sound wave that has an intensity of 4.0 μW/m².

20. A vacuum cleaner has a measured sound level of 70 dB. What is the intensity of this sound in W/m²?

21. Show that the difference in decibel levels, β_1 and β_2, of a sound source is related to the ratio of its distances, r_1 and r_2, from the receivers by

$$\beta_2 - \beta_1 = 20 \log\left(\frac{r_1}{r_2}\right)$$

22. The intensity of a sound wave at a fixed distance from a speaker vibrating at 1.00 kHz is 0.600 W/m². (a) Determine the intensity if the frequency is increased to 2.50 kHz while a constant displacement amplitude is maintained. (b) Calculate the intensity if the frequency is reduced to 0.500 kHz and the displacement amplitude is doubled.

22A. The intensity of a sound wave at a fixed distance from a speaker vibrating at a frequency f is I. (a) Determine the intensity if the frequency is increased to f' while a constant displacement amplitude is maintained. (b) Calculate the intensity if the frequency is reduced to $f/2$ and the displacement amplitude is doubled.

23. A speaker is placed between two observers who are 110 m apart, along the line connecting them. If one observer records an intensity level of 60 dB, and the other records an intensity level of 80 dB, how far is the speaker from each observer?

24. An explosive charge is detonated at a height of several kilometers in the atmosphere. At a distance of 400 m from the explosion the acoustic pressure reaches a maximum of 10 Pa. Assuming that the atmosphere is homogeneous over the distances considered, what will be the sound level (in dB) at 4 km from the explosion? (Sound waves in air are absorbed at a rate of approximately 7 dB/km.)

25. Two small speakers emit sound waves of different frequencies. Speaker A has an output of 1.0 mW and speaker B has an output of 1.5 mW. Determine the sound intensity level (in dB) at point C (Fig. P17.25) if (a) only speaker A emits sound, (b) only speaker B emits sound, (c) both speakers emit sound.

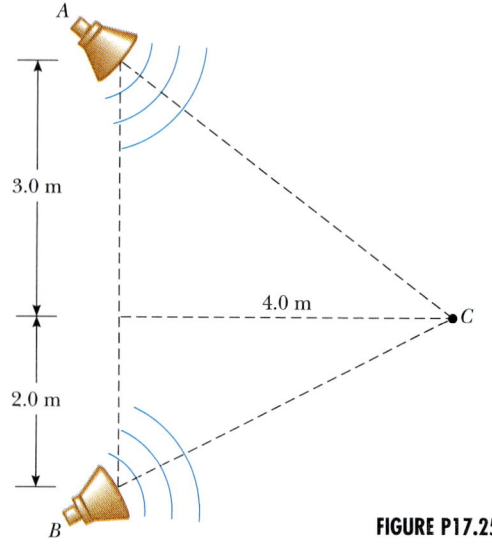

FIGURE P17.25

26. Two sources have sound levels of 75 dB and 80 dB. If they are sounding simultaneously, (a) what is the combined sound level? (b) What is their combined intensity in W/m²?

Section 17.4   Spherical and Plane Waves

27. An experiment requires a sound intensity of 1.2 W/m² at a distance of 4 m from a speaker. What power output is required?

28. A source of sound (1000 Hz) emits uniformly in all directions. An observer 3.0 m from the source measures a sound level of 40 dB. Calculate the average power output of the source.

29. The sound level at a distance of 3.0 m from a source is 120 dB. At what distance will the sound level be (a) 100 dB and (b) 10 dB?

30. A fireworks rocket explodes at a height of 100 m above the ground. An observer on the ground directly under the explosion experiences an average sound intensity of 7.0×10^{-2} W/m² for 0.20 s. (a) What is the total sound energy of the explosion? (b) What is the sound level in decibels heard by the observer?

30A. A fireworks rocket explodes at a height h above the ground. An observer on the ground directly under the explosion experiences an average sound intensity I for a time t. (a) What is the total sound energy of the explosion? (b) What is the sound level heard by the observer?

31. A rock group is playing in a studio. Sound emerging from an open door spreads uniformly in all directions. If the sound level of the music is 80.0 dB at a distance of 5.0 m from the door, at what distance is the music just barely audible to a person with a normal threshold of hearing (0 dB)? Disregard absorption.

32. A spherical wave is radiating from a point source and is described by the following:

$$y(r, t) = \left(\frac{25.0}{r}\right) \sin(1.25r - 1870t)$$

where y is in pascals, r in meters, and t in seconds. (a) What is the maximum pressure amplitude 4.00 m from the source? (b) Determine the speed of the wave and hence the material the wave is in. (c) Find the intensity of the wave in dB at a distance 4.00 m from the source. (d) Find the instantaneous pressure 5.00 m from the source at 0.0800 s.

*Section 17.5 The Doppler Effect

33. A bullet fired from a rifle travels at Mach 1.38 (that is, $v_S/v = 1.38$). What angle does the shock front make with the path of the bullet?

34. A block with a speaker bolted to it is connected to a spring having spring constant $k = 20.0$ N/m as in Figure P17.34. The total mass of the block and speaker is 5.00 kg, and the amplitude of this unit's motion is 0.500 m. (a) If the speaker emits sound waves of frequency 440 Hz, determine the range in frequencies heard by the person to the right of the speaker. (b) If the maximum intensity level heard by the person is 60 dB when he is closest to the speaker, 1.00 m away, what is the minimum intensity level heard by the observer? Assume that the speed of sound is 343 m/s.

34A. A block with a speaker bolted to it is connected to a spring having spring constant k as in Figure P17.34. The total mass of the block and speaker is m, and the amplitude of this unit's motion is A. (a) If the speaker emits sound waves of frequency f, determine the range in frequencies heard by the person to the right of the speaker. (b) If the maximum intensity level heard by the person is β when he is closest to the speaker, a distance d away, what is the minimum intensity level heard by the observer?

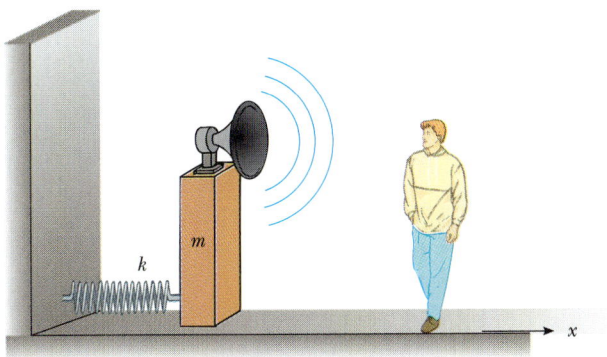

FIGURE P17.34

35. A jet fighter plane travels in horizontal flight at Mach 1.2 (that is, 1.2 times the speed of sound in air). At the instant an observer on the ground hears the shock wave, what is the angle her line of sight makes with the horizontal as she looks at the plane?

36. A helicopter drops a paratrooper carrying a buzzer that emits a 500-Hz signal. A sound receiver on the plane monitors the signal as the paratrooper falls. If the perceived frequency becomes constant at 450 Hz, what is the terminal speed of the paratrooper? Take the speed of sound in air to be 343 m/s and assume the paratrooper always remains below the helicopter.

37. Standing at a crosswalk, you hear a frequency of 560 Hz from the siren on an approaching police car. After the police car passes, the observed frequency of the siren is 480 Hz. Determine the car's speed from these observations.

38. A fire engine moving to the right at 40 m/s sounds its horn (frequency 500 Hz) at the two vehicles shown in Figure P17.38. The car is moving to the right at 30 m/s, while the van is at rest. (a) What frequency is heard by the passengers in the car? (b) What is the frequency as heard by the passengers in the van? (c) When the fire engine is 200 m from the car and 250 m from the van, the passengers in the car hear a sound intensity level of 90 dB. At that moment, what intensity level is heard by the passengers in the van?

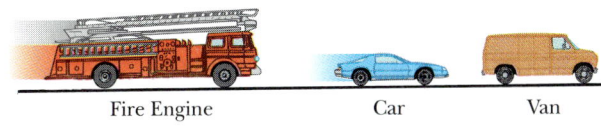

Fire Engine   Car   Van

FIGURE P17.38

39. A train is moving parallel to a highway at 20 m/s. A car is traveling in the same direction as the train at 40 m/s. The car horn sounds at 510 Hz and the train whistle sounds at 320 Hz. (a) When the car is behind the train what frequency does an occupant of the car observe for the train whistle? (b) When the car is in front of the train what frequency does a train passenger observe for the car horn just after passing?

40. A tuning fork vibrating at 512 Hz falls from rest and accelerates at 9.80 m/s². How far below the point of release is the tuning fork when waves of frequency 485 Hz reach the release point? Take the speed of sound in air to be 340 m/s.

41. When high-energy, charged particles move through a transparent medium with a speed greater than the speed of light in that medium, a shock wave, or bow wave, of light is produced. This phenomenon is called the *Cerenkov effect* and can be observed in the vicinity of the core of a swimming-pool reactor due to high-speed electrons moving through the water. In a particular case, the Cerenkov radiation produces a wavefront with an apex half-angle of 53°. Calculate the speed of the electrons in the water. (The speed of light in water is 2.25×10^8 m/s.)

42. A driver traveling northbound on a highway is driving at a speed of 25 m/s. A police car driving southbound at a speed of 40 m/s approaches with its siren sounding at a base frequency of 2500 Hz. (a) What frequency is observed by the driver as the police car approaches? (b) What frequency is detected by the driver after the police car passes him? (c) Repeat parts (a) and (b) for the case when the police car is traveling northbound.

43. A supersonic jet traveling at Mach 3 at an altitude of 20 000 m is directly overhead at time $t = 0$ as in Figure P17.43. (a) How long will it be before one encounters the shock wave? (b) Where will the plane be when it is finally heard? (Assume the speed of sound in air is uniform at 335 m/s.)

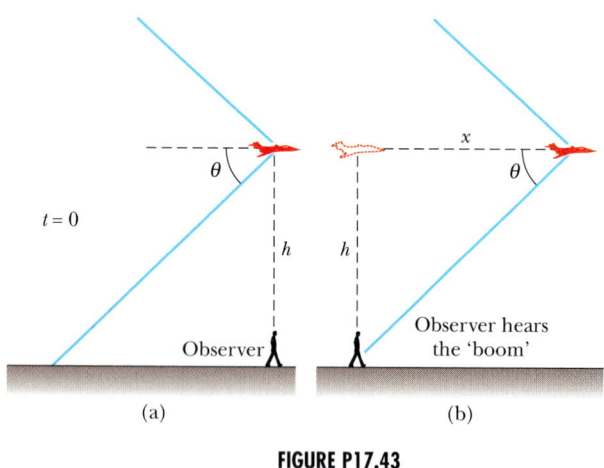

FIGURE P17.43

ADDITIONAL PROBLEMS

44. A copper rod is given a sharp compressional blow at one end. The sound of the blow, traveling through air at 0°C, reaches the opposite end of the rod 6.4 ms later than the sound transmitted through the rod. What is the length of the rod? (Refer to Table 17.1.)

45. An earthquake on the ocean floor in the Gulf of Alaska induces a *tsunami* (sometimes called a "tidal wave") that reaches Hilo, Hawaii, 4450 km distant, in a time of 9 h 30 min. Tsunamis have enormous wavelengths (100–200 km), and for such waves the propagation speed is $v \approx \sqrt{g\bar{d}}$, where $\bar{d}$ is the average depth of the water. From the information given, find the average wave speed and the average ocean depth between Alaska and Hawaii. (This method was used in 1856 to estimate the average depth of the Pacific Ocean long before soundings were made to give a direct determination.)

46. The power output of a certain stereo speaker is 6.0 W. (a) At what distance from the speaker would the sound be painful to the ear? (b) At what distance from the speaker would the sound be barely audible?

47. A jet flies toward higher altitude at a constant speed of 1963 m/s in a direction making an angle θ with the horizontal (Fig. P17.47). An observer on the ground hears the jet for the first time when it is directly overhead. Determine the value of θ if the speed of sound in air is 340.0 m/s.

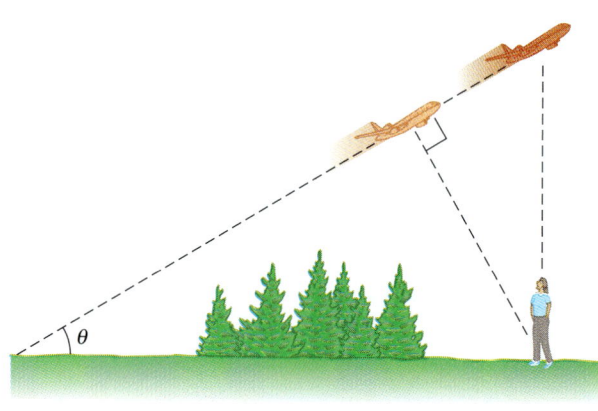

FIGURE P17.47

48. A microwave oven generates a sound level of 40.0 dB when consuming 1.00 kW of power. Estimate the fraction of this power that is converted into the energy of sound waves.

49. Two ships are moving along a line due east. The trailing vessel has a speed relative to a land-based observation point of 64 km/h, and the leading ship has a speed of 45 km/h relative to that point. The two ships are in a region of the ocean where the current is moving uniformly due west at 10 km/h. The trailing ship transmits a sonar signal at a frequency of 1200 Hz. What frequency is monitored by the leading ship? (Use 1520 m/s as the speed of sound in ocean water.)

49A. Two ships are moving along a line due east. The trailing vessel has a speed relative to a land-based observation point of v_1, and the leading ship has a speed $v_2 < v_1$ relative to that point. The two ships are in a region of the ocean where the current is moving uniformly due west at a speed v_c. The trailing ship transmits a sonar signal at a frequency f. What frequency is monitored by the leading ship? (Use v_S as the speed of sound in ocean water.)

50. Consider a longitudinal (compressional) wave of wavelength λ traveling with speed v along the x direction through a medium of density ρ. The *displacement* of the molecules of the medium from their equilibrium position is

$$s = s_{max} \sin(kx - \omega t)$$

Show that the pressure variation in the medium is

$$P = - \left(\frac{2\pi\rho v^2}{\lambda} s_{max} \right) \cos(kx - \omega t)$$

51. A meteoroid the size of a truck enters the Earth's atmosphere at a speed of 20 km/s and is not significantly slowed before entering the ocean. (a) What is the Mach angle of the shock wave from the meteoroid in the atmosphere? (Use 331 m/s as the sound speed.) (b) Assuming that the meteoroid survives the impact with the ocean surface, what is the (initial) Mach angle of the shock wave that the meteoroid produces in the water? (Use the wave speed for seawater given in Table 17.1.)

52. In the afternoon, the sound level of a busy freeway is 80 dB with 100 cars passing a given point every minute. Late at night, the traffic flow is only five cars per minute. What is the late-night sound level?

53. By proper excitation, it is possible to produce both longitudinal and transverse waves in a long metal rod. A particular metal rod is 150 cm long and has a radius of 0.20 cm and a mass of 50.9 g. Young's modulus for the material is 6.8×10^{10} N/m². What must the tension (or compression) in the rod be if the ratio of the speed of longitudinal waves to the speed of transverse waves is 8?

54. An earthquake emits both P waves and S waves that travel at different speeds through the Earth. A P wave travels at a speed of 9000 m/s and an S wave travels at 5000 m/s. If P waves are received at a seismic station 1 minute before an S wave arrives, how far away is the earthquake center?

55. A siren creates a sound level of 60.00 dB at 500.0 m from the speaker. The siren is powered by a battery that delivers a total energy of 1.00 kJ. Assuming that the efficiency of the siren is 30% (i.e., 30% of the supplied energy is transformed into sound energy), determine the total time the siren can sound.

55A. A siren creates a sound level β at a distance d from the speaker. The siren is powered by a battery that delivers a total energy E. Assuming that the efficiency of the siren is 30% (i.e., 30% of the supplied energy is transformed into sound energy), determine the total time the siren can sound.

56. The Doppler equation presented in the text is valid when the motion between the observer and the source occurs on a straight line, so that the source and observer are moving either directly toward or directly away from each other. If this restriction is relaxed, one must use the more general Doppler equation

$$f' = \left(\frac{v + v_O \cos\theta_O}{v - v_S \cos\theta_S} \right) f$$

where θ_O and θ_S are defined in Figure P17.56a. (a) If both observer and source are moving away from each other, show that the preceding equation reduces to Equation 17.17 with lower signs. (b) Use the preceding equation to solve the following problem. A train moves at a constant speed of 25.0 m/s toward the intersection shown in Figure P17.56b. A car is stopped near the intersection, 30.0 m from the tracks. If the train's horn emits a frequency of 500 Hz, what is the frequency heard by the passengers in the car when the train is 40.0 m from the intersection? Take the speed of sound to be 343 m/s.

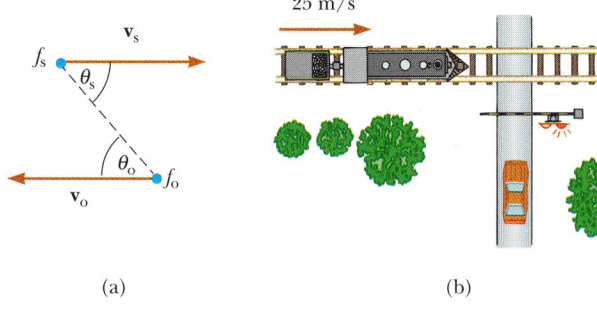

(a)   (b)

FIGURE P17.56

57. In order to be able to determine her speed, a skydiver carries a tone generator. A friend on the ground at the landing site has equipment for receiving and analyzing sound waves. While the skydiver is falling at terminal speed, her tone generator emits a steady tone of 1800 Hz. (Assume that the air is calm and that the sound speed is 343 m/s, independent of altitude.) (a) If her friend on the ground (directly beneath the skydiver) receives waves of frequency 2150 Hz, what is the skydiver's speed of descent? (b) If the skydiver were also carrying sound-receiving equipment sensitive enough to detect waves reflected from the ground, what frequency would she receive?

58. A train whistle ($f = 400$ Hz) sounds higher or lower in pitch depending on whether it approaches or re-

cedes. (a) Prove that the difference in frequency between the approaching and receding train whistle is

$$\Delta f = \frac{2f\left(\dfrac{u}{v}\right)}{1 - \dfrac{u^2}{v^2}} \qquad \begin{array}{l} u = \text{speed of train} \\ v = \text{speed of sound} \end{array}$$

(b) Calculate this difference for a train moving at a
speed of 130 km/h. Take the speed of sound in air
to be 340 m/s.

59. Three metal rods are located relative to each other as
shown in Figure P17.59, where $L_1 + L_2 = L_3$. Values
of density and Young's modulus for the three materi-
als are $\rho_1 = 2.7 \times 10^3$ kg/m^3, $Y_1 = 7.0 \times 10^{10}$ N/m^2;
$\rho_2 = 11.3 \times 10^3$ kg/m^3, $Y_2 = 1.6 \times 10^{10}$ N/m^2; and
$\rho_3 = 8.8 \times 10^3$ kg/m^3, $Y_3 = 11 \times 10^{10}$ N/m^2. (a) If
$L_3 = 1.5$ m, what must the ratio L_1/L_2 be if a sound
wave is to travel the length of rods 1 and 2 in the same
time for the wave to travel the length of rod 3? (b) If
the frequency of the source is 4.00 kHz, determine
the phase difference between the wave traveling
along rods 1 and 2 and the one traveling along rod 3.

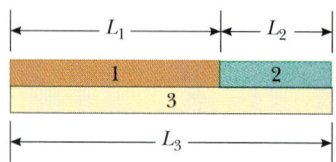

FIGURE P17.59

60. A bat, moving at 5.00 m/s, is chasing a flying insect.
If the bat emits a 40.0-kHz chirp and receives back an
echo at 40.4 kHz, at what speed is the insect moving
toward or away from the bat? (Take the speed of
sound in air to be $v = 340$ m/s.)

61. A supersonic aircraft is flying parallel to the ground.
When the aircraft is directly overhead, an observer
sees a rocket fired from the aircraft. Ten seconds
later the observer hears the sonic boom, followed
2.8 s later by the sound of the rocket engine. What is
the Mach number of the aircraft?

62. The volume knob on a radio has what is known as a
"logarithmic taper." The electrical device connected
to the knob (called a potentiometer) has a resistance
R whose logarithm is proportional to the angular position of the knob: that is, $\log R \propto \theta$. If the intensity of
the sound I (in W/m^2) produced by the speaker is
proportional to the resistance R, show that the sound
level β (in dB) is a linear function of θ.

COMPUTER PROBLEMS

63. The program DOPPLER plots the wavefronts of a
sound source moving with speed u. The only input
parameter is u/v, where v is the speed of sound. The
ratio u/v can be in the range -2.0 to 2.0. Select
$u/v = 0$, and measure the distance between adjacent
wavefronts on the computer screen. This is the wavelength λ_0 of the sound wave. Now select $u/v = 0.50$
and measure the wavelength ahead of and behind
the source. Calculate the change in the wavelength
$\Delta\lambda = \lambda - \lambda_0$ for both cases. Does $|\Delta\lambda/\lambda_0| = u/v$?
Enter *stop* to exit the program at the prompt $u/v =$.

64. Run the program DOPPLER for a series of values
of $u/v \geq 1.0$. A shock wave is produced at these
speeds. Note that as u/v increases, the angle of the
shock wave decreases. Verify that $\sin\theta = v/u$.

CHAPTER 18

Superposition and Standing Waves

Even when silent, this organ at the Mormon Tabernacle conveys a sense of the power of sound waves.

(Courtesy of Henry Leap)

An important aspect of waves is the combined effect of two or more of them traveling in the same medium. For instance, what happens to a string when a wave traveling toward its fixed end is reflected back on itself? What is the pressure variation in the air when the instruments of an orchestra sound together?

In a linear medium, that is, one in which the restoring force of the medium is proportional to the displacement of the medium, the principle of superposition can be applied to obtain the resultant disturbance. We discussed this principle in Chapter 16 as it applies to wave pulses. The term *interference* was used to describe the effect produced by combining two wave pulses moving simultaneously through a medium.

This chapter is concerned with the superposition principle as it applies to sinusoidal waves. If the sinusoidal waves that combine in a given medium have the same frequency and wavelength, one finds that a stationary pattern, called a *standing wave*, can be produced at certain frequencies under certain circumstances. For example, a taut string fixed at both ends has a discrete set of oscillation patterns, called *modes of vibration*, that depend upon the tension and mass per unit length of the string. These modes of vibration are found in stringed musical instruments. Other musical instruments, such as the organ and flute, make use of the natural frequencies of sound waves in hollow pipes. Such frequencies depend upon the length of the pipe and its shape and upon whether the pipe is open at both ends or open at one and closed at the other.

We also consider the superposition and interference of waves with different frequencies and wavelengths. When two sound waves with nearly the same frequency interfere, one hears variations in the loudness called *beats*. The beat frequency corresponds to the rate of alternation between constructive and destructive interference. Finally, we describe how any complex periodic wave can, in general, be described by a sum of sine and cosine functions.

18.1 SUPERPOSITION AND INTERFERENCE OF SINUSOIDAL WAVES

The superposition principle tells us that when two or more waves move in the same linear medium, the net displacement of the medium (the resultant wave) at any point equals the algebraic sum of the displacements caused by all the waves. Let us apply this principle to two sinusoidal waves traveling in the same direction in a medium. If the two waves are traveling to the right and have the same frequency, wavelength, and amplitude but differ in phase, we can express their individual wave functions as

$$y_1 = A_0 \sin(kx - \omega t) \qquad y_2 = A_0 \sin(kx - \omega t - \phi)$$

Hence, the resultant wave function y is

$$y = y_1 + y_2 = A_0[\sin(kx - \omega t) + \sin(kx - \omega t - \phi)]$$

To simplify this expression, it is convenient to make use of the trigonometric identity

$$\sin a + \sin b = 2 \cos\left(\frac{a - b}{2}\right) \sin\left(\frac{a + b}{2}\right)$$

If we let $a = kx - \omega t$ and $b = kx - \omega t - \phi$, we find that the resultant wave function y reduces to

Resultant of two traveling sinusoidal waves

$$y = 2A_0 \cos\left(\frac{\phi}{2}\right) \sin\left(kx - \omega t - \frac{\phi}{2}\right) \qquad (18.1)$$

There are several important features of this result. The resultant wave function y is also harmonic and has the same frequency and wavelength as the individual waves. The amplitude of the resultant wave is $2A_0 \cos(\phi/2)$, and its phase is equal to $\phi/2$. If the phase constant ϕ equals 0, then $\cos(\phi/2) = \cos 0 = 1$ and the amplitude of the resultant wave is $2A_0$. In other words, the amplitude of the resultant wave is twice the amplitude of either individual wave. In this case, the waves are said to be

Constructive interference

everywhere *in phase* and thus **interfere constructively**. That is, the crests and troughs of the individual waves (y_1 and y_2) occur at the same positions as the crests and troughs of the resultant wave (y), as shown by the blue curve in Figure 18.1a. In general, constructive interference occurs when $\cos(\phi/2) = \pm 1$, which means when $\phi = 0, 2\pi, 4\pi, \ldots$ rad. On the other hand, if ϕ is equal to π rad, or to any *odd* multiple of π, then $\cos(\phi/2) = \cos(\pi/2) = 0$ and the resultant wave has *zero*

Destructive interference

amplitude everywhere. In this case, the two waves **interfere destructively**. That is, the crest of one wave coincides with the trough of the second (Fig. 18.1b) and their displacements cancel at every point. Finally, when the phase constant has an arbitrary value between 0 and π rad, as in Figure 18.1c, the resultant wave has an amplitude whose value is somewhere between 0 and $2A_0$.

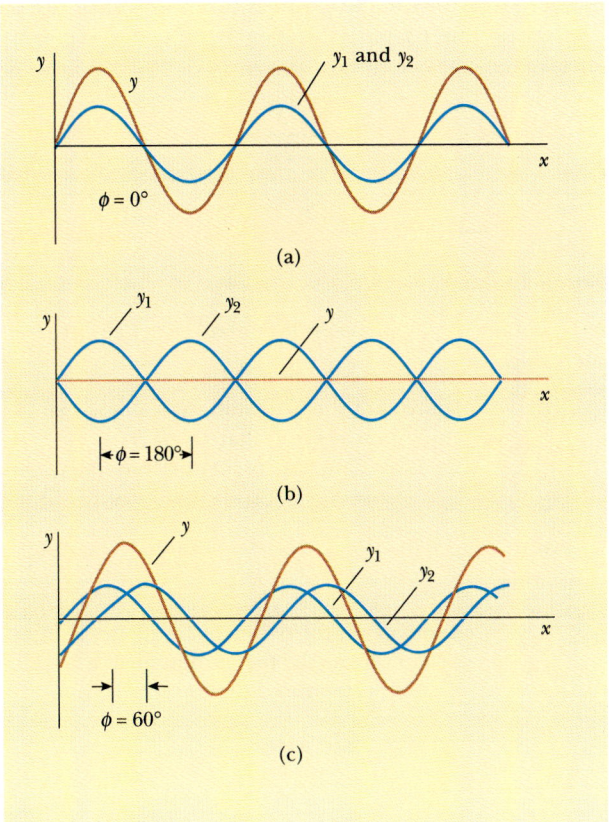

FIGURE 18.1   The superposition of two waves with amplitudes y_1 and y_2, where $y_1 = y_2$. (a) When the two waves are in phase, the result is constructive interference. (b) When the two waves are 180° out of phase, the result is destructive interference. (c) When the phase angle lies in the range $0 < \phi < 180°$, the resultant y falls somewhere between that shown in part (a) and that shown in part (b).

Interference of Sound Waves

One simple device for demonstrating interference of sound waves is illustrated in Figure 18.2. Sound from a loudspeaker S is sent into a tube at P, where there is a T-shaped junction. Half the sound power travels in one direction and half in the opposite direction. Thus, the sound waves that reach the receiver R at the other side can travel along two different paths. The distance along any path from speaker to receiver is called the *path length, r*. The lower path length r_1 is fixed but the upper path length r_2 can be varied by sliding the U-shaped tube, similar to that on a slide trombone. When the difference in the path lengths $\Delta r = |r_2 - r_1|$ is either zero or some integral multiple of the wavelength λ, the two waves reaching the receiver are in phase and interfere constructively, as in Figure 18.1a. For this case, a maximum in the sound intensity is detected at the receiver. If the path length r_2 is adjusted such that the path difference Δr is $\lambda/2, 3\lambda/2, \ldots, n\lambda/2$ (for n odd), the two waves are exactly 180° out of phase at the receiver and hence cancel each other. In this case of destructive interference, no sound is detected at the receiver. This simple experiment demonstrates that a phase difference may arise between two waves generated by the same source when they travel along paths of unequal lengths.

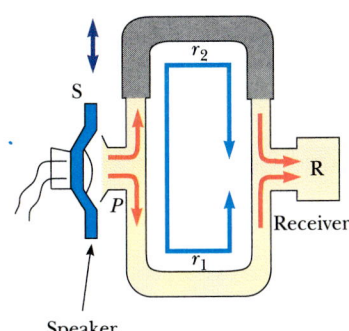

FIGURE 18.2   An acoustical system for demonstrating interference of sound waves. Sound from the speaker propagates into a tube and splits into two parts at P. The two waves, which superimpose at the opposite side, are detected at R. The upper path length r_2 can be varied by the sliding section.

It is often useful to express the path difference in terms of the phase difference ϕ between the two waves. Since a path difference of one wavelength corresponds to a phase difference of 2π rad, we obtain the ratio $\phi/2\pi = \Delta r/\lambda$, or

$$\Delta r = \frac{\lambda}{2\pi} \phi \qquad (18.2)$$

EXAMPLE 18.1   Two Speakers Driven by the Same Source

A pair of speakers placed 3.00 m apart are driven by the same oscillator (Fig. 18.3). A listener is originally at point O, which is located 8.00 m from the center of the line connecting the two speakers. The listener then walks to point P, which is a perpendicular distance 0.350 m from O before reaching the *first minimum* in sound intensity. What is the frequency of the oscillator?

Solution   The first minimum occurs when the two waves reaching the listener at P are 180° out of phase—in other words, when their path difference equals $\lambda/2$. In order to calculate the path difference, we must first find the path lengths r_1 and r_2. Making use of the two shaded triangles in Figure 18.3, we find the path lengths to be

$$r_1 = \sqrt{(8.00 \text{ m})^2 + (1.15 \text{ m})^2} = 8.08 \text{ m}$$

$$r_2 = \sqrt{(8.00 \text{ m})^2 + (1.85 \text{ m})^2} = 8.21 \text{ m}$$

Hence, the path difference is $r_2 - r_1 = 0.13$ m. Since we require that this path difference be equal to $\lambda/2$ for the first minimum, we find that $\lambda = 0.26$ m.

To obtain the oscillator frequency, we can use $v = \lambda f$, where v is the speed of sound in air, 343 m/s:

$$f = \frac{v}{\lambda} = \frac{343 \text{ m/s}}{0.26 \text{ m}} = \boxed{1.3 \text{ kHz}}$$

Exercise   If the oscillator frequency is adjusted such that the listener hears the first minimum at a distance of 0.75 m from O, what is the new frequency?

Answer   0.63 kHz.

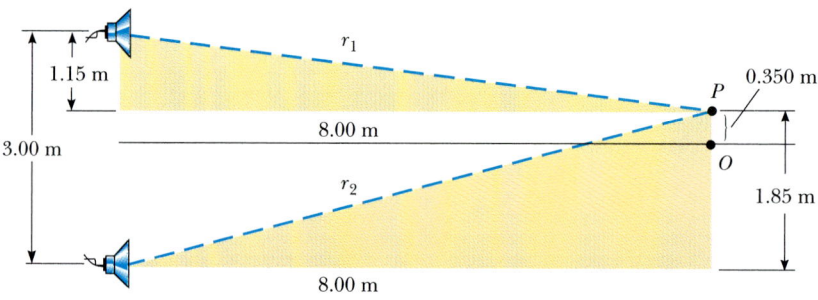

FIGURE 18.3   (Example 18.1).

18.2   STANDING WAVES

If a taut string is clamped at both ends, traveling waves are reflected from the fixed ends, creating waves traveling in both directions. The incident and reflected waves combine according to the superposition principle.

Consider two sinusoidal waves in the same medium with the same amplitude, frequency, and wavelength but traveling in opposite directions. Their wave functions can be written

$$y_1 = A_0 \sin(kx - \omega t) \qquad y_2 = A_0 \sin(kx + \omega t)$$

where y_1 represents a wave traveling to the right and y_2 represents a wave traveling to the left. Adding these two functions gives the resultant wave function y:

$$y = y_1 + y_2 = A_0 \sin(kx - \omega t) + A_0 \sin(kx + \omega t)$$

where $k = 2\pi/\lambda$ and $\omega = 2\pi f$, as usual. When we use the trigonometric identity $\sin(a \pm b) = \sin a \cos b \pm \cos a \sin b$, this expression reduces to

$$y = (2A_0 \sin kx) \cos \omega t \tag{18.3}$$

Wave function for a standing wave

which is the wave function of a **standing wave.** A standing wave is a stationary vibration pattern formed by the superposition of two waves of the same frequency traveling in opposite directions. From Equation 18.3, we see that a standing wave has an angular frequency ω and an amplitude $2A_0 \sin kx$. That is, every particle of the string vibrates in simple harmonic motion with the same frequency. However, the amplitude of motion of a given particle depends on x. This is in contrast to the situation involving a traveling sinusoidal wave, in which all particles oscillate with both the same amplitude and the same frequency.

Because the amplitude of the standing wave at any value of x is equal to $2A_0 \sin kx$, we see that the maximum amplitude has the value $2A_0$. This maximum occurs when the coordinate x satisfies the condition $\sin kx = \pm 1$, or when

$$kx = \frac{\pi}{2}, \frac{3\pi}{2}, \frac{5\pi}{2}, \cdots$$

Since $k = 2\pi/\lambda$, the positions of maximum amplitude, called **antinodes,** are

$$x = \frac{\lambda}{4}, \frac{3\lambda}{4}, \frac{5\lambda}{4}, \cdots = \frac{n\lambda}{4} \qquad n = 1, 3, 5, \ldots \tag{18.4}$$

Position of antinodes

Note that adjacent antinodes are separated by $\lambda/2$.

Similarly, the standing wave has a minimum amplitude of zero when x satisfies the condition $\sin kx = 0$, or when

$$kx = \pi, 2\pi, 3\pi, \ldots$$

giving

$$x = \frac{\lambda}{2}, \lambda, \frac{3\lambda}{2}, \cdots = \frac{n\lambda}{2} \qquad n = 0, 1, 2, 3, \ldots \tag{18.5}$$

Position of nodes

These points of zero amplitude, called **nodes,** are also spaced by $\lambda/2$. The distance between a node and an adjacent antinode is $\lambda/4$.

A graphical description of the standing wave patterns produced at various times by two waves traveling in opposite directions is shown in Figure 18.4. The top and middle waves in each part of the figure represent the individual traveling waves, and the bottom waves represent the standing wave patterns. The nodes of the standing wave are labeled N, and the antinodes are labeled A. At $t = 0$ (Fig. 18.4a), the two waves are identical spatially, giving a standing wave of maximum amplitude, $2A_0$. One quarter of a period later, at $t = T/4$ (Fig. 18.4b), the individual waves have moved one quarter of a wavelength (one to the right and the other to the left). At this time, the individual displacements are equal and opposite for all values of x, and hence the resultant wave has zero displacement everywhere. At $t = T/2$ (Fig. 18.4c), the individual waves are again identical spatially, producing a standing wave pattern that is inverted relative to the $t = 0$ pattern.

It is instructive to describe the energy associated with the motion of a standing wave. To illustrate this point, consider a standing wave formed on a taut string fixed at each end, as in Figure 18.5. Except for the nodes that are stationary, all points on the string oscillate vertically with the same frequency. Furthermore, the various points have different amplitudes of motion. Figure 18.5 represents snap-

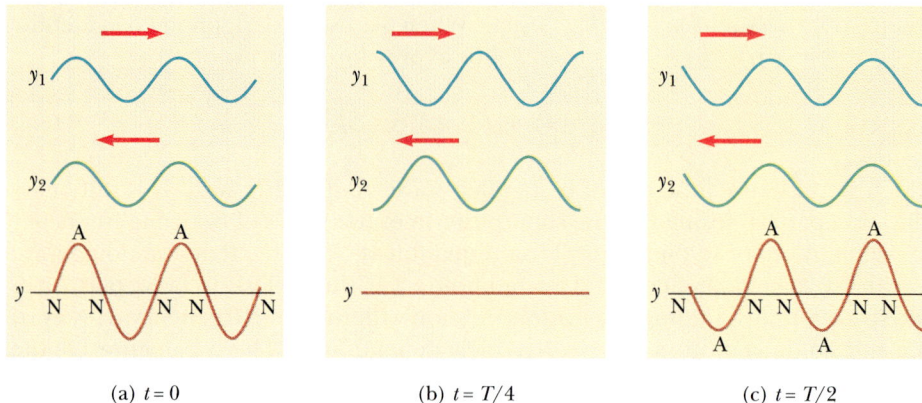

(a) $t = 0$ (b) $t = T/4$ (c) $t = T/2$

FIGURE 18.4   Standing wave patterns at various times produced by two waves of equal amplitude traveling in opposite directions. For the resultant wave y, the nodes (N) are points of zero displacement and the antinodes (A) are points of maximum displacement.

shots of the standing wave at various times over one half of a period. The nodal points do not move and may be grasped or clamped without affecting the standing wave. If dissipation of energy by the string is ignored, the oscillation is constant with time and may be called a **stationary wave;** no energy is transmitted along the wave. (In a real string, some energy is transmitted to maintain the oscillation.) Each point on the string executes simple harmonic motion in the vertical direction. That is, one can view the standing wave as a large number of oscillators vibrating parallel to each other. The energy of the vibrating string continuously alternates between elastic potential energy, at which time the string is momentarily stationary (Fig. 18.5a), and kinetic energy, at which time the string is horizontal and the particles have their maximum speed (Fig. 18.5c). At intermediate times (Figs. 18.5b and 18.5d), the string particles have both potential energy and kinetic energy.

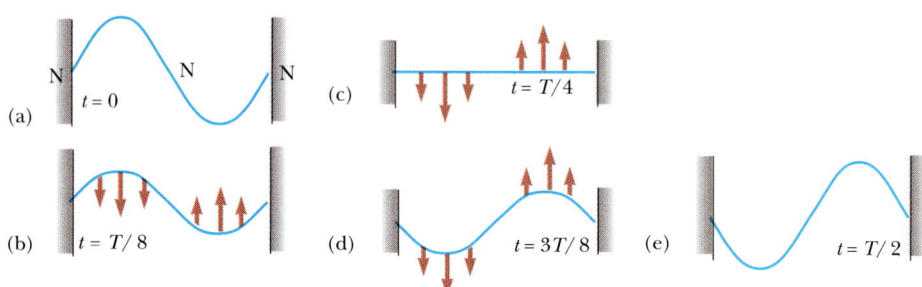

FIGURE 18.5   A standing wave pattern in a taut string showing snapshots during one-half cycle. (a) At $t = 0$, the string is momentarily at rest, and so $K = 0$ and all of the energy is potential energy U associated with the vertical displacements of the string segments. (b) At $t = T/8$, the string is in motion, and the energy is half kinetic and half potential. (c) At $t = T/4$, the string is horizontal (undeformed) and, therefore, $U = 0$; all of the energy is kinetic. The motion continues as indicated, and ultimately the initial configuration in part (a) is repeated.

EXAMPLE 18.2   Formation of a Standing Wave

Two waves traveling in opposite directions produce a standing wave. The individual wave functions are

$$y_1 = (4.0 \text{ cm}) \sin(3.0x - 2.0t)$$

$$y_2 = (4.0 \text{ cm}) \sin(3.0x + 2.0t)$$

where x and y are in centimeters. (a) Find the maximum displacement of the motion at $x = 2.3$ cm.

Solution   When the two waves are summed, the result is a standing wave whose function is given by Equation 18.3, with $A_0 = 4.0$ cm and $k = 3.0$ rad/cm:

$$y = (2A_0 \sin kx) \cos \omega t = [(8.0 \text{ cm}) \sin 3.0x] \cos \omega t$$

Thus, the maximum displacement of the motion at the position $x = 2.3$ cm is

$$y_{max} = (8.0 \text{ cm}) \sin 3.0x \big|_{x=2.3}$$

$$= (8.0 \text{ cm}) \sin(6.9 \text{ rad}) = \boxed{4.6 \text{ cm}}$$

(b) Find the positions of the nodes and antinodes.

Solution   Since $k = 2\pi/\lambda = 3$ rad/cm, we see that $\lambda = 2\pi/3$ cm. Therefore, from Equation 18.4 we find that the antinodes are located at

$$x = n\left(\frac{\pi}{6}\right) \text{ cm} \qquad (n = 1, 3, 5, \ldots)$$

and from Equation 18.5 we find that the nodes are located at

$$x = n\frac{\lambda}{2} = n\left(\frac{\pi}{3}\right) \text{ cm} \qquad (n = 1, 2, 3, \ldots)$$

18.3   STANDING WAVES IN A STRING FIXED AT BOTH ENDS

Consider a string of length L that is fixed at both ends, as in Figure 18.6. Standing waves are set up in the string by a continuous superposition of waves incident on and reflected from the ends. The string has a number of natural patterns of vibration, called **normal modes.** Each of these has a characteristic frequency that is easily calculated.

First, note that the ends of the string are nodes by definition because these points are *fixed.* In general, the motion of a vibrating string fixed at both ends is described by the superposition of several normal modes. The modes that are present depend on how the vibration is started. For example, when a guitar string

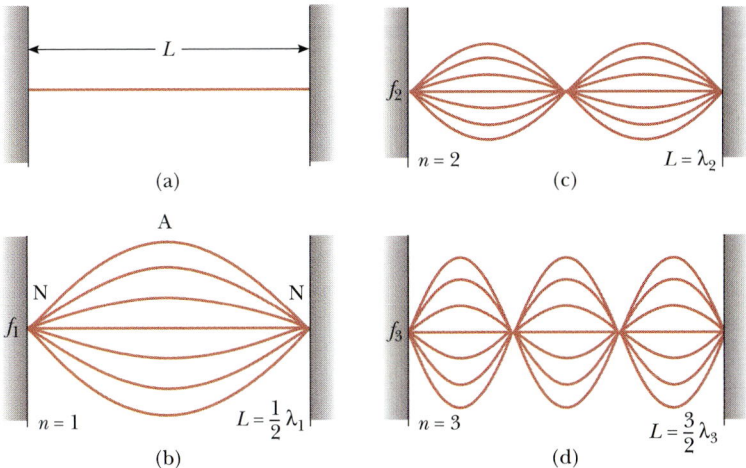

FIGURE 18.6   (a) A string of length L fixed at both ends. The normal modes of vibration form a harmonic series: (b) the fundamental frequency, or first harmonic; (c) the second harmonic; and (d) the third harmonic.

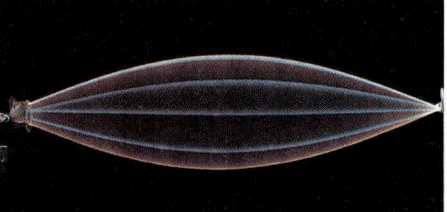 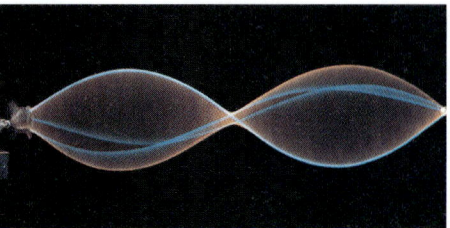 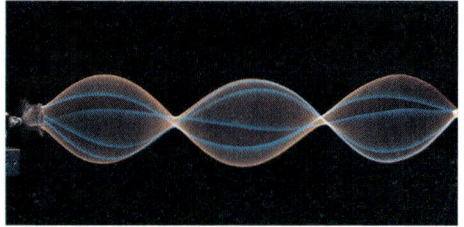

Multiflash photographs of standing wave patterns in a cord driven by a vibrator at the left end. The single loop pattern at the left represents the fundamental ($n = 1$), the two-loop pattern in the middle represents the second harmonic ($n = 2$), and the three-loop pattern at the right represents the third harmonic ($n = 3$). *(Richard Megna 1991, Fundamental Photographs)*

is plucked near the middle, the modes shown in Figure 18.6b and 18.6d having antinodes at the center are excited, as well as other modes not shown. The first normal mode, shown in Figure 18.6b, has nodes at its ends and one antinode in the middle. This normal mode occurs when the wavelength λ_1 equals twice the length of the string, that is, when $\lambda_1 = 2L$. The next normal mode, of wavelength λ_2 (Fig. 18.6c), occurs when the wavelength equals the length of the string, that is, when $\lambda_2 = L$. The third normal mode (Fig. 18.6d) corresponds to the case where the wavelength is two-thirds the length of the string, $\lambda_3 = 2L/3$. In general, the wavelengths of the various normal modes can be conveniently expressed as

| Wavelengths of normal modes |

$$\lambda_n = \frac{2L}{n} \qquad (n = 1, 2, 3, \ldots) \qquad (18.6)$$

where the index n refers to the nth normal mode of vibration. The natural frequencies associated with these modes are obtained from the relationship $f = v/\lambda$, where the *wave speed v is the same for all frequencies.* Using Equation 18.6, we find that the frequencies of the normal modes are

| Frequencies of normal modes as functions of wave speed and length of string |

$$f_n = \frac{v}{\lambda_n} = \frac{n}{2L} v \qquad (n = 1, 2, 3, \ldots) \qquad (18.7)$$

Because $v = \sqrt{F/\mu}$ (Eq. 16.4), where F is the tension in the string and μ is its mass per unit length, we can also express the natural frequencies of a taut string as

| Frequencies of normal modes as functions of string tension and linear mass density |

$$f_n = \frac{n}{2L} \sqrt{\frac{F}{\mu}} \qquad (n = 1, 2, 3, \ldots) \qquad (18.8)$$

The lowest frequency, corresponding to $n = 1$, is called the *fundamental* or the **fundamental frequency,** f_1, and is given by

| Fundamental frequency of a taut string |

$$f_1 = \frac{1}{2L} \sqrt{\frac{F}{\mu}} \qquad (18.9)$$

Clearly, the frequencies of the remaining normal modes (sometimes called *overtones*) are integral multiples of the fundamental frequency. These higher natural frequencies, together with the fundamental frequency, form a **harmonic series.** The fundamental, f_1, is the first harmonic; the frequency $f_2 = 2f_1$ is the second harmonic; the frequency f_n is the nth harmonic.

We can obtain the preceding results in an alternative manner. Since we require that the string be fixed at $x = 0$ and $x = L$, the wave function $y(x, t)$ given by

Equation 18.3 must be zero at these points for all times. That is, the boundary conditions require that $y(0, t) = 0$ and $y(L, t) = 0$ for all values of t. Since $y = (2A_0 \sin kx) \cos \omega t$, the first condition, $y(0, t) = 0$, is automatically satisfied because $\sin kx = 0$ at $x = 0$. To meet the second condition, $y(L, t) = 0$, we require that $\sin kL = 0$. This condition is satisfied when the angle kL equals an integral multiple of π (180°). Therefore, the allowed values of k are[1]

$$k_n L = n\pi \qquad (n = 1, 2, 3, \ldots) \qquad (18.10)$$

Since $k_n = 2\pi/\lambda_n$, we find that

$$\left(\frac{2\pi}{\lambda_n}\right) L = n\pi \qquad \text{or} \qquad \lambda_n = \frac{2L}{n}$$

which is identical to Equation 18.6.

When a taut string is distorted such that its distorted shape corresponds to any one of its harmonics, after being released it will vibrate at the frequency of that harmonic. However, if the string is struck or bowed such that its distorted shape is not just one single harmonic, the resulting vibration will include frequencies of various harmonics. In effect, the string "selects" the normal-mode frequencies when disturbed by a nonharmonic disturbance (which happens, for example, when a guitar string is plucked).

Figure 18.7 shows a taut string vibrating at its first and second harmonics simultaneously. In this figure, the combined vibration is the superposition of the vibrations shown in Figures 18.6b and 18.6c. The large loop corresponds to the fundamental frequency of vibration, f_1, and the smaller loops correspond to the second harmonic, f_2. In general, the resulting motion, or displacement, can be described by a superposition of the various harmonic wave functions, with different frequencies and amplitudes. Hence, the sound that one hears corresponds to a complex wave associated with these various modes of vibration. We shall return to this point in Section 18.8.

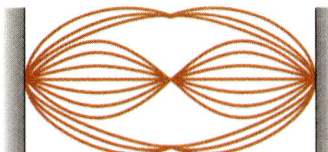

FIGURE 18.7   Multiple exposures of a string vibrating simultaneously in its first and second harmonics.

The frequency of a stringed instrument can be changed either by varying the tension F or by changing the length L. For example, the tension in guitar and violin strings is varied by a screw adjustment mechanism or by turning pegs located on the neck of the instrument. As the tension is increased, the frequency of the normal modes increases according to Equation 18.8. Once the instrument is "tuned," players vary the frequency by moving their fingers along the neck, thereby changing the length of the vibrating portion of the string. As the length is shortened, the frequency increases because the normal-mode frequencies are inversely proportional to string length.

EXAMPLE 18.3   Give Me a C Note

A middle C string of the C-major scale on a piano has a fundamental frequency of 264 Hz, and the A note has a fundamental frequency of 440 Hz. (a) Calculate the frequencies of the next two harmonics of the C string.

Solution   Since $f_1 = 264$ Hz, we can use Equations 18.8 and 18.9 to find the frequencies f_2 and f_3:

$$f_2 = 2f_1 = \boxed{528 \text{ Hz}}$$

$$f_3 = 3f_1 = \boxed{792 \text{ Hz}}$$

(b) If the strings for the A and C notes are assumed to have the same mass per unit length and the same length, determine the ratio of tensions in the two strings.

---

[1] We exclude $n = 0$ since this corresponds to the trivial case where no wave exists ($k = 0$).

Solution   Using Equation 18.8 for the two strings vibrating at their fundamental frequencies gives

$$f_{1A} = \frac{1}{2L} \sqrt{F_A/\mu} \quad \text{and} \quad f_{1C} = \frac{1}{2L} \sqrt{F_C/\mu}$$

$$f_{1A}/f_{1C} = \sqrt{F_A/F_C}$$

$$F_A/F_C = (f_{1A}/f_{1C})^2 = (440/264)^2 = \boxed{2.78}$$

(c) In a real piano, the assumption we made in part (b) is only half true. The string densities are equal, but the A string is 64% as long as the C string. What is the ratio of their tensions?

$$f_{1A}/f_{1C} = (L_C/L_A)\sqrt{F_A/F_C} = (100/64)\sqrt{F_A/F_C}$$

$$F_A/F_C = (0.64)^2(440/264)^2 = \boxed{1.14}$$

18.4   RESONANCE

We have seen that a system such as a taut string is capable of oscillating in one or more natural modes of vibration. *If a periodic force is applied to such a system, the resulting amplitude of motion of the system is larger when the frequency of the applied force is equal or nearly equal to one of the natural frequencies of the system* than when the driving force is applied at some other frequency. We have already discussed this phenomenon, known as *resonance,* for mechanical systems. The corresponding natural frequencies of oscillation of the system are often referred to as **resonant frequencies.**

Figure 18.8 shows the response of a vibrating system to various driving frequencies, where one of the resonant frequencies of the system is denoted by f_0. Note that the amplitude is largest when the frequency of the driving force equals the resonant frequency. The driving force continues to deliver energy to the oscillating system causing the amplitude to increase. The maximum amplitude of the motion is limited by friction in the system. Once maximum amplitude is reached, the work done by the periodic force is used only to overcome friction. A system is said to be *weakly damped* when the amount of friction to be overcome is small. Such a system has a large amplitude of motion when driven at one of its resonant frequencies and the oscillations persist for a long time after the driving force is removed. A system with considerable friction to be overcome, that is, one that is *strongly damped,* undergoes small amplitude oscillations that decrease rapidly with time once the driving force is removed.

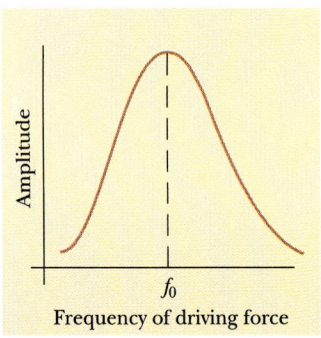

FIGURE 18.8   The amplitude (response) versus driving frequency for an oscillating system. The amplitude is a maximum at the resonance frequency, f_0.

Examples of Resonance

A playground swing is a pendulum with a natural frequency that depends on its length. Whenever we push a child in a swing with a series of regular impulses, the swing goes higher if the frequency of the periodic force equals the natural frequency of the swing. One can demonstrate a similar effect by suspending several pendula of different lengths from a horizontal support, as in Figure 18.9. If pendulum A is set into oscillation, the other pendula will soon begin to oscillate as a result of the longitudinal waves transmitted along the beam. However, you will find that a pendulum, such as C, whose length is close to the length of A oscillates with a much larger amplitude than those such as B and D whose lengths are much different from the length of A. This is because the natural frequency of C is nearly the same as the driving frequency associated with A.

Next, consider a taut string fixed at one end and connected at the opposite end to a vibrating blade as in Figure 18.10. The fixed end is a node, and the point that is near the end connected to the vibrating blade is very nearly a node, since the amplitude of the blade's motion is small compared with that of the string. As the blade oscillates, transverse waves sent down the string are reflected from the fixed

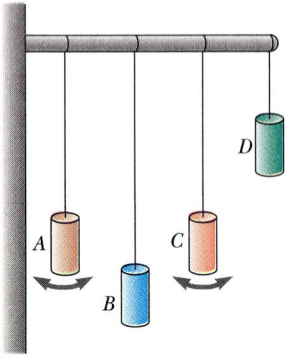

FIGURE 18.9   An example of resonance. If pendulum A is set into oscillation, only pendulum C, whose length matches that of A, will eventually oscillate with large amplitude, or resonate. The arrows indicate motion perpendicular to the page.

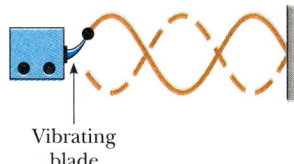

Vibrating
blade

FIGURE 18.10 Standing waves are set up in a string by connecting one end of the string to a vibrating blade. When the blade vibrates at one of the natural frequencies of the string, large-amplitude standing waves are created.

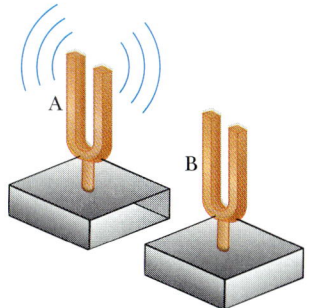

FIGURE 18.11 If tuning fork A is set into vibration, an identical tuning fork B eventually vibrates at the same frequency, or resonates.

end. As we found in Section 18.3, the string has natural frequencies of vibration that are determined by its length, tension, and mass per unit length (Eq. 18.8). When the frequency of the vibrating blade nearly equals one of the natural frequencies of the string, standing waves are produced and the string vibrates with a large amplitude. In this case, the wave being generated by the vibrating blade is in phase with the reflected wave, and so the string absorbs energy from the blade at resonance.

Once the amplitude of the standing-wave oscillations reaches a maximum, the energy delivered by the blade and absorbed by the system is lost because of the damping forces caused by friction in the system. If the applied frequency differs from one of the natural frequencies, however, energy is first transferred to the string from the blade, but later the phase of the wave becomes such that it forces the blade to receive energy from the string, thereby reducing the energy in the string.

As a final example of resonance, consider two identical tuning forks mounted on separate hollow boxes (Fig. 18.11). The hollow boxes augment the sound wave power generated by the vibrating forks. If fork A is set into vibration (say, by striking it), fork B is set into vibration as longitudinal sound waves are received from A. The frequencies of vibration of A and B are the same, assuming the forks are identical. The energy exchange, or resonance behavior, does not occur if the two have different natural frequencies of vibration.

CONCEPTUAL EXAMPLE 18.4

Some singers are able to shatter a wine glass by maintaining a certain pitch in their voice over a period of several seconds (with the help of an amplifier). What mechanism causes the glass to break?

Reasoning The wine glass has natural frequencies of vibration. If the singer's voice has a strong frequency component that corresponds to one of these natural frequencies, sound waves are coupled into the glass, setting up forced vibrations that produce large stresses in the glass causing it to shatter.

(Conceptual Example 18.4) A wine glass shattered by the amplified sound of a human voice. *(© Ben Rose 1992/The IMAGE Bank)*

CONCEPTUAL EXAMPLE 18.5

At certain speeds, an automobile driven on a washboard road will vibrate disastrously and lose traction and braking effectiveness. At other speeds, the vibration is more manageable. Explain. Why are "rumble strips" sometimes used just before stop signs?

Reasoning   At certain speeds the car crosses the ridges on the washboard road at a rate that matches one of the car's natural frequencies of vibration. This causes the car to go into a large amplitude vibration that can be dangerous. At a speed slightly slower or faster, the natural frequency of the car and the ridges on the road are out of step, and the car moves more smoothly. Rumble strips are used to get your attention at dangerous intersections.

18.5   STANDING WAVES IN AIR COLUMNS

Standing waves can be set up in a tube of air, such as an organ pipe, as the result of interference between longitudinal waves traveling in opposite directions. The phase relationship between the incident wave and the wave reflected from one end depends on whether that end is open or closed. This is analogous to the phase relationships between incident and reflected transverse waves at the ends of a string. *The closed end of an air column is a displacement node because the wall at this end does not allow molecular motion.* As a result, at a closed end of a tube of air, the reflected wave is 180° out of phase with the incident wave. Furthermore, since the pressure wave is 90° out of phase with the displacement wave (Section 17.2), *the closed end of an air column corresponds to a pressure antinode* (that is, a point of maximum pressure variation).

The open end of an air column is approximately a displacement antinode and a pressure node. The fact that there is no pressure variation at the open end can be understood by noting that the end of an air column in a tube is open to the atmosphere, and the pressure at this end must remain constant. If the pressure at this end were to change, air would flow into or out of the column. The wave reflected from an open end is nearly in phase with the incident wave when the tube's diameter is small relative to the wavelength of the sound.

Strictly speaking, the open end of an air column is not exactly an antinode. A condensation reaching an open end does not reflect until it passes beyond the end. For a thin-walled tube of circular cross-section, this end correction is approximately $0.6R$, where R is the tube's radius. Hence, the effective length of the tube is longer than the true length L. We ignore this end correction in what follows.

The first three normal modes of vibration of a pipe open at both ends are shown in Figure 18.12a. When air is directed against an edge at the left, longitudinal standing waves are formed and the pipe resonates at its natural frequencies. All modes of vibration are excited simultaneously (although not with the same amplitude). Note that the ends are displacement antinodes (approximately). In the fundamental mode, the wavelength is twice the length of the pipe, and hence the frequency of the fundamental, $f_1 = v/2L$. The frequencies of the higher harmonics are $2f_1, 3f_1, \ldots$. Thus,

in a pipe open at both ends, the natural frequencies of vibration form a harmonic series, that is, the higher harmonics are integral multiples of the fundamental frequency.

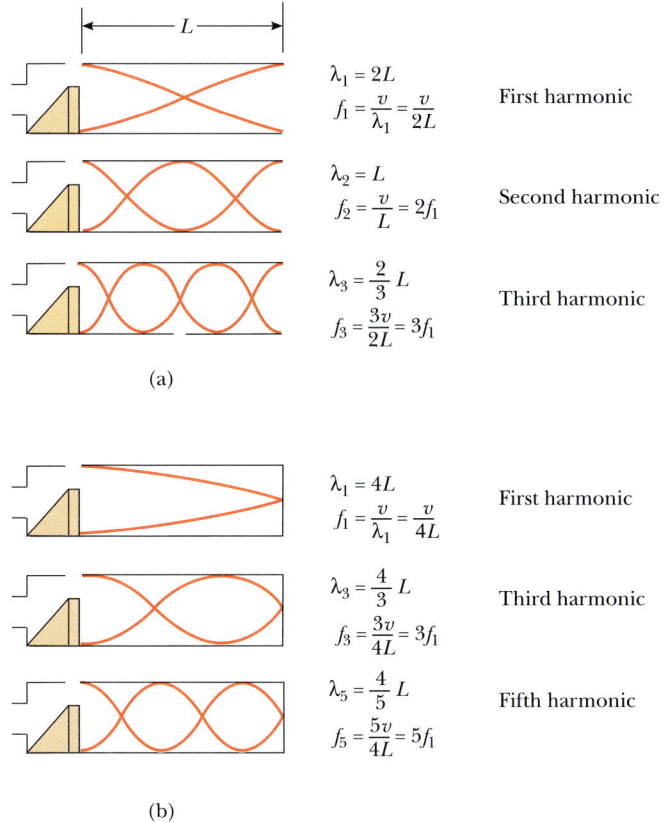

(a)

(b)

FIGURE 18.12 (a) Standing longitudinal waves in an organ pipe open at both ends. The natural frequencies that form a harmonic series are f_1, $2f_1$, $3f_1$, ... (b) Standing longitudinal waves in an organ pipe closed at one end. Only the *odd* harmonics are present, and so the natural frequencies are f_1, $3f_1$, $5f_1$, ...

Since all harmonics are present in a pipe open at both ends, we can express the natural frequencies of vibration as

$$f_n = n\frac{v}{2L} \qquad (n = 1, 2, 3, \ldots) \qquad (18.11)$$

Natural frequencies of a pipe open at both ends

where v is the speed of sound in air.

If a pipe is closed at one end and open at the other, the closed end is a displacement node (Fig. 18.12b). In this case, the wavelength for the fundamental mode is four times the length of the tube. Hence, the fundamental, f_1, is equal to $v/4L$, and the frequencies of the higher harmonics are equal to $3f_1$, $5f_1$, ... That is,

in a pipe closed at one end, only odd harmonics are present, and the natural frequencies are

$$f_n = n\frac{v}{4L} \qquad (n = 1, 3, 5, \ldots) \qquad (18.12)$$

Natural frequencies of a pipe closed at one end and open at the other

EXAMPLE 18.6 Resonance in a Pipe

A pipe has a length of 1.23 m. (a) Determine the frequencies of the first three harmonics if the pipe is open at each end. Take $v = 343$ m/s as the speed of sound in air.

Solution The first harmonic of a pipe open at both ends is

$$f_1 = \frac{v}{2L} = \frac{343 \text{ m/s}}{2(1.23 \text{ m})} = \boxed{139 \text{ Hz}}$$

Since all harmonics are present, the second and third harmonics are $f_2 = 2f_1 = 278$ Hz and $f_3 = 3f_1 = 417$ Hz.

(b) What are the three frequencies determined in part (a) if the pipe is closed at one end?

Solution The fundamental frequency of a pipe closed at one end is

$$f_1 = \frac{v}{4L} = \frac{343 \text{ m/s}}{4(1.23 \text{ m})} = \boxed{69.7 \text{ Hz}}$$

In this case, only odd harmonics are present, and so the next two resonances have frequencies $f_3 = 3f_1 = 209$ Hz and $f_5 = 5f_1 = 349$ Hz.

(c) For the pipe open at both ends, how many harmonics are present in the normal human hearing range (20 to 20 000 Hz)?

Solution Since all harmonics are present, $f_n = nf_1$. For $f_n = 20\,000$ Hz, we have $n = 20\,000/139 = 144$, so that 144 harmonics are present in the audible range. Actually, only the first few harmonics have sufficient amplitude to be heard.

EXAMPLE 18.7 Measuring the Frequency of a Tuning Fork

A simple apparatus for demonstrating resonance in a tube is described in Figure 18.13a. A long, vertical tube open at both ends is partially submerged in a beaker of water, and a vibrating tuning fork of unknown frequency is placed near the top. The length of the air column, L, is adjusted by moving the tube vertically. The sound waves generated by the fork are reinforced when the length of the air column corresponds to one of the resonant frequencies of the tube.

For a certain tube, the smallest value of L for which a peak occurs in the sound intensity is 9.00 cm. From this measurement, determine the frequency of the tuning fork and the value of L for the next two resonant modes.

Reasoning Although the tube is open at both ends to allow the water in, the water surface acts like a wall at one end of the length L. Therefore this setup represents a pipe closed at one end and the fundamental has a frequency of $v/4L$ (Fig. 18.13b).

Solution Taking $v = 343$ m/s for the speed of sound in air and $L = 0.0900$ m, we get

$$f_1 = \frac{v}{4L} = \frac{343 \text{ m/s}}{4(0.0900 \text{ m})} = \boxed{953 \text{ Hz}}$$

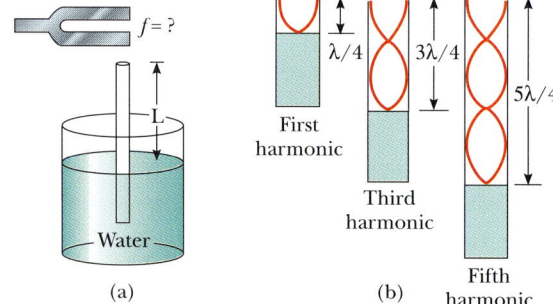

FIGURE 18.13 (Example 18.7) (a) Apparatus for demonstrating the resonance of sound waves in a tube closed at one end. The length L of the air column is varied by moving the tube vertically while it is partially submerged in water. (b) The first three normal modes of the system shown in (a).

From this information about the fundamental mode, we see that the wavelength is $\lambda = 4L = 0.360$ m. Since the frequency of the source is constant, the next two resonance modes (Fig. 18.13b) correspond to lengths of $3\lambda/4 = 0.270$ m and $5\lambda/4 = 0.450$ m.

*18.6 STANDING WAVES IN RODS AND PLATES

Standing waves can also be set up in rods and plates. If a rod is clamped in the middle and stroked at one end, it will undergo longitudinal vibrations as described in Figure 18.14a. Note that the broken lines in Figure 18.14 represent *longitudinal* displacements of various parts of the rod. The midpoint is a displacement node since it is fixed by the clamp, whereas the ends are displacement antinodes since they are free to vibrate. This setup is analogous to vibrations set up in a pipe open at each end. The broken lines in Figure 18.14a represent the fundamental mode,

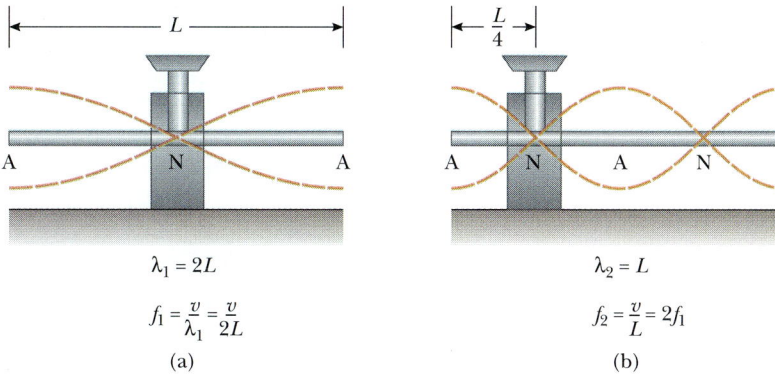

$$\lambda_1 = 2L$$

$$f_1 = \frac{v}{\lambda_1} = \frac{v}{2L}$$

(a)

$$\lambda_2 = L$$

$$f_2 = \frac{v}{L} = 2f_1$$

(b)

FIGURE 18.14   Normal-mode longitudinal vibrations of a rod of length L (a) clamped at the middle and (b) clamped at an approximate distance of $L/4$ from one end.

for which the wavelength is $2L$ and the frequency is $v/2L$, where v is the speed of longitudinal waves in the rod. Other modes may be excited by clamping the rod at different points. For example, the second harmonic (Fig. 18.14b) is excited by clamping the rod at a point that is a distance $L/4$ away from one end.

Two-dimensional vibrations can be set up in a flexible membrane stretched over a circular hoop, such as a drumhead. As the membrane is struck at some point, wave pulses that arrive at the fixed boundary are reflected many times. The resulting sound is not harmonic but rather explosive in nature. This is because the vibrating drumhead and the drum's hollow interior produce a disorganized set of waves that create a sound of indefinite pitch when they reach a listener's ear. This is in contrast to wind and stringed instruments, which produce sounds of definite pitch.

Some possible normal modes of oscillation of a vibrating, two-dimensional, circular membrane are shown in Figure 18.15. Note that the nodes are curves rather than points, which was the case for a vibrating string. The fixed circumference is one such nodal curve, and some other nodal curves are also indicated. The lowest mode of vibration with frequency f_1 (the fundamental) is a symmetric mode with one nodal curve, the circumference of the membrane. The other possible modes of vibration are *not* integral multiples of f_1; hence, the normal frequencies *do not* form a harmonic series. When a drum is struck, many of these modes are excited simultaneously. However, the higher-frequency modes damp out more rapidly. With this information, one can understand why the drum containing a membrane with fixed perimeters is a nonmelodious instrument. In contrast, a loudspeaker is designed to reproduce musical frequencies by unclamping the edge of the speaker diaphragm.

*18.7   BEATS: INTERFERENCE IN TIME

The interference phenomena we have been dealing with so far involve the superposition of two or more waves with the same frequency traveling in opposite directions. Since the resultant wave in this case depends on the coordinates of the disturbed medium, we can refer to the phenomenon as *spatial interference.* Standing waves in strings and pipes are common examples of spatial interference.

We now consider another type of interference, one that results from the superposition of two waves with slightly *different frequencies* traveling in the *same direction.* In this case, when the two waves are observed at a given point, they are periodically in and out of phase. That is, there is a temporal alternation between constructive

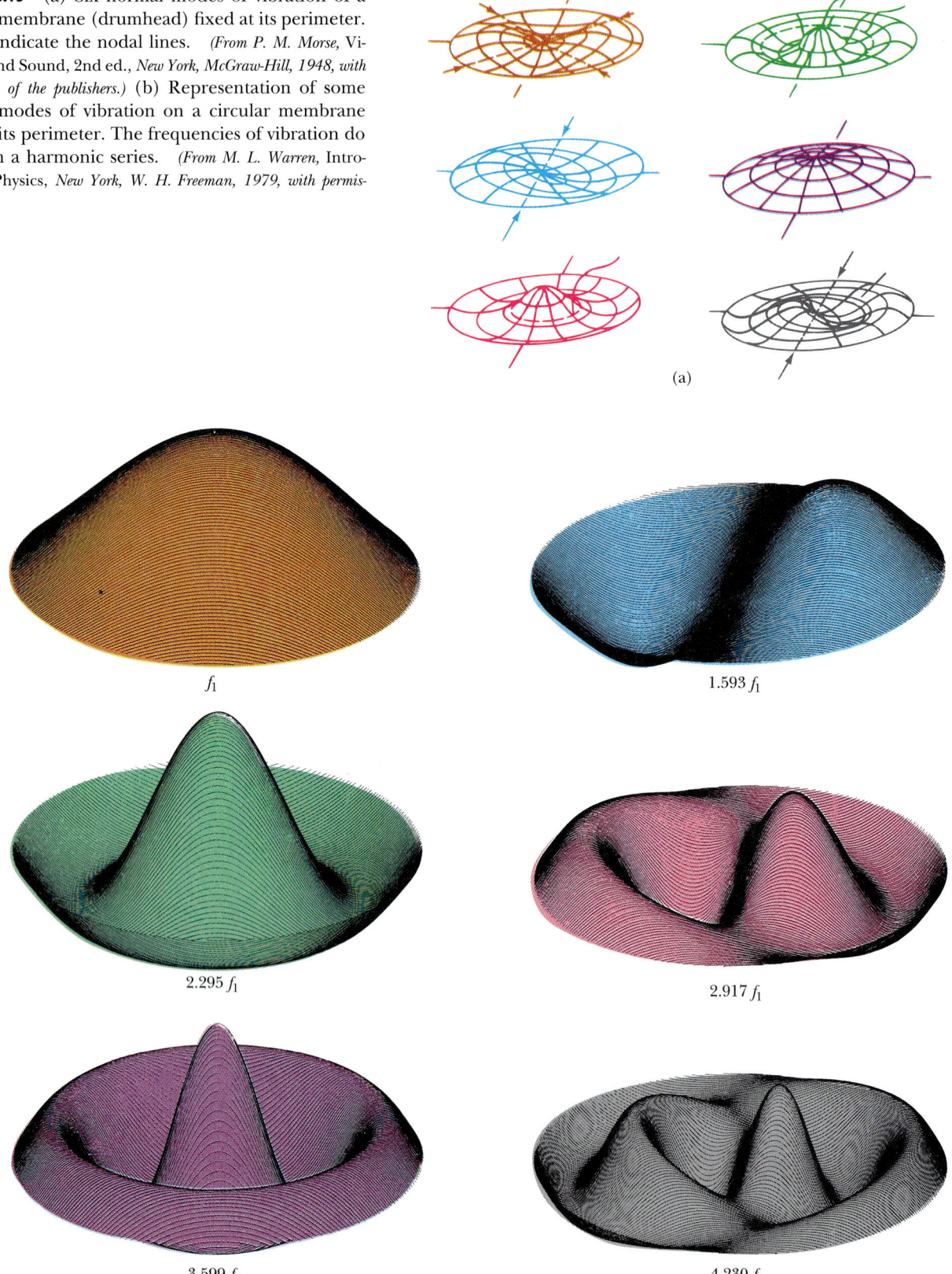

FIGURE 18.15 (a) Six normal modes of vibration of a circular membrane (drumhead) fixed at its perimeter. Arrows indicate the nodal lines. *(From P. M. Morse, Vibration and Sound, 2nd ed., New York, McGraw-Hill, 1948, with permission of the publishers.)* (b) Representation of some natural modes of vibration on a circular membrane fixed at its perimeter. The frequencies of vibration do not form a harmonic series. *(From M. L. Warren,* Introductory Physics, *New York, W. H. Freeman, 1979, with permission.)*

(a)

f_1

$1.593\,f_1$

$2.295\,f_1$

$2.917\,f_1$

$3.599\,f_1$

$4.230\,f_1$

(b)

and destructive interference. Thus, we refer to this phenomenon as *interference in time or temporal interference*. For example, if two tuning forks of slightly different frequencies are struck, one hears a sound of pulsating intensity, called a **beat:**

A beat is the periodic variation in intensity at a given point due to the superposition of two waves having slightly different frequencies.

Definition of beat

The number of beats one hears per second, or *beat frequency*, equals the difference in frequency between the two sources. The maximum beat frequency that the human ear can detect is about 20 beats/s. When the beat frequency exceeds this value, it blends indistinguishably with the compound sounds producing the beats.

One can use beats to tune a stringed instrument, such as a piano, by beating a note against a reference tone of known frequency. The string can then be adjusted to equal the frequency of the reference by tightening or loosening it until the beats become too infrequent to notice.

Consider two waves of equal amplitude traveling through a medium in the same direction but with slightly different frequencies, f_1 and f_2. We can represent the displacement each wave produces at a point as

$$y_1 = A_0 \cos 2\pi f_1 t \qquad y_2 = A_0 \cos 2\pi f_2 t$$

Using the superposition principle, we find that the resultant displacement at that point is

$$y = y_1 + y_2 = A_0 (\cos 2\pi f_1 t + \cos 2\pi f_2 t)$$

It is convenient to write this equation in a form that uses the trigonometric identity

$$\cos a + \cos b = 2 \cos\left(\frac{a - b}{2}\right) \cos\left(\frac{a + b}{2}\right)$$

Letting $a = 2\pi f_1 t$ and $b = 2\pi f_2 t$, we find that

$$y = 2 A_0 \cos 2\pi\left(\frac{f_1 - f_2}{2}\right) t \cos 2\pi\left(\frac{f_1 + f_2}{2}\right) t \qquad (18.13)$$

Resultant of two waves of different frequencies but equal amplitude

Graphs demonstrating the individual waves as well as the resultant wave are shown in Figure 18.16. From the factors in Equation 18.13, we see that the resultant

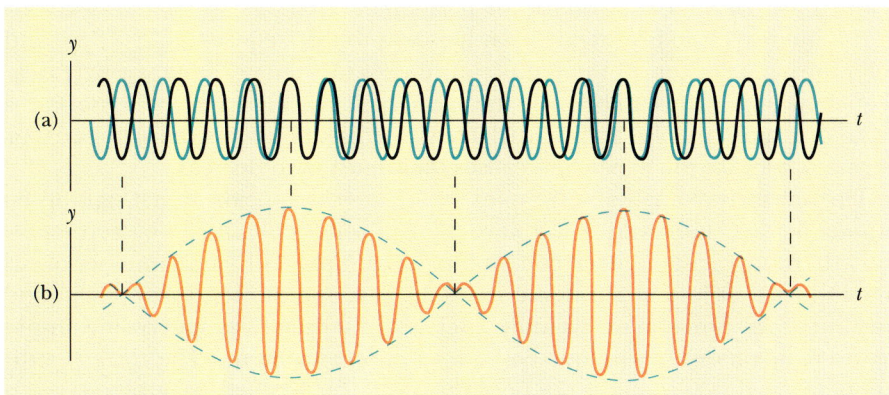

FIGURE 18.16   Beats are formed by the combination of two waves of slightly different frequencies traveling in the same direction. (a) The individual waves. (b) The combined wave has an amplitude (broken line) that oscillates in time.

vibration for a listener standing at any given point has an effective frequency equal to the average frequency, $(f_1 + f_2)/2$, and an amplitude

$$A = 2A_0 \cos 2\pi \left(\frac{f_1 - f_2}{2} \right) t \qquad (18.14)$$

That is, the *amplitude varies in time* and the frequency at which the amplitude maxima are heard is $(f_1 - f_2)/2$. When f_1 is close to f_2, this amplitude variation is slow, as illustrated by the envelope (broken line) of the resultant wave in Figure 18.16b.

Note that a beat, or a maximum in amplitude, is detected whenever

$$\cos 2\pi \left(\frac{f_1 - f_2}{2} \right) t = \pm 1$$

That is, there are *two* maxima in each cycle. Since the amplitude varies with frequency as $(f_1 - f_2)/2$, the number of beats per second, or the beat frequency f_b, is twice this value. That is,

Beat frequency

$$f_b = |f_1 - f_2| \qquad (18.15)$$

For instance, if one tuning fork vibrates at 438 Hz and a second tuning fork vibrates at 442 Hz, the resultant sound wave of the combination has a frequency of 440 Hz (the musical note A) and a beat frequency of 4 Hz. A listener would hear the 440-Hz sound wave go through an intensity maximum four times every second.

*18.8 COMPLEX WAVES

The sound wave patterns produced by most musical instruments are very complex. Characteristic patterns produced by a tuning fork, a harmonic flute, and a clarinet, each playing the same pitch, are shown in Figure 18.17. The term *pitch* refers to the human perception to the frequency of sound. Although each instrument has its own characteristic pattern, Figure 18.17 shows that each pattern is periodic. Furthermore, note that a struck tuning fork produces only one harmonic (the fundamental), whereas the flute and clarinet produce many frequencies, which include the fundamental and various harmonics. Thus, the complex wave patterns produced by a violin or clarinet, and the corresponding richness of musical tones, are the result of the superposition of various harmonics. This is in contrast to a drum, in which the overtones do not form a harmonic series.

It is interesting to investigate what happens to the frequencies of a flute and a violin during a concert as the temperature rises. A flute goes sharp (increases in frequency) as it warms up because the speed of sound increases in the warmer air inside the flute. A violin goes flat (decreases in frequency) as the strings expand thermally because the expansion causes their tension to decrease.

The problem of analyzing complex wave patterns appears at first sight to be a formidable task. However, if the wave pattern is periodic, it can be represented with arbitrary precision by the combination of a sufficiently large number of sinusoidal waves that form a harmonic series. In fact, one can represent any periodic function or any function over a finite interval as a series of sine and cosine terms by using a mathematical technique based on **Fourier's theorem.**[2] The corresponding sum of terms that represents the periodic waveform is called a **Fourier series.**

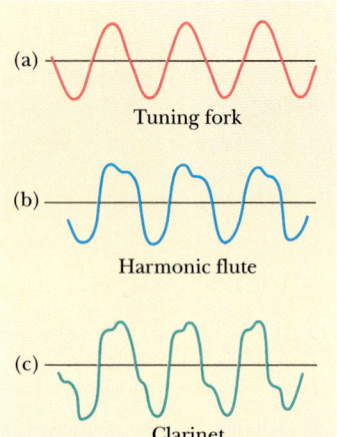

(a) Tuning fork
(b) Harmonic flute
(c) Clarinet

FIGURE 18.17 Waveform produced by (a) a tuning fork, (b) a harmonic flute, and (c) a clarinet, each at approximately the same frequency. *(Adapted from C. A. Culver, Musical Acoustics, 4th ed., New York, McGraw-Hill, 1956, p. 128.)*

[2] Developed by Jean Baptiste Joseph Fourier (1786–1830).

This simulator enables you to generate complex signals using the principle of superposition or Fourier synthesis. Up to ten sinusoidal waves of varying amplitudes, frequencies, and phase angles can be added together to generate complex waveforms that may represent actual phenomena. Depending upon the computer you are using, you may be able to play back and listen to the sound that your signal generates.

Complex Waves—The Fourier Synthesizer

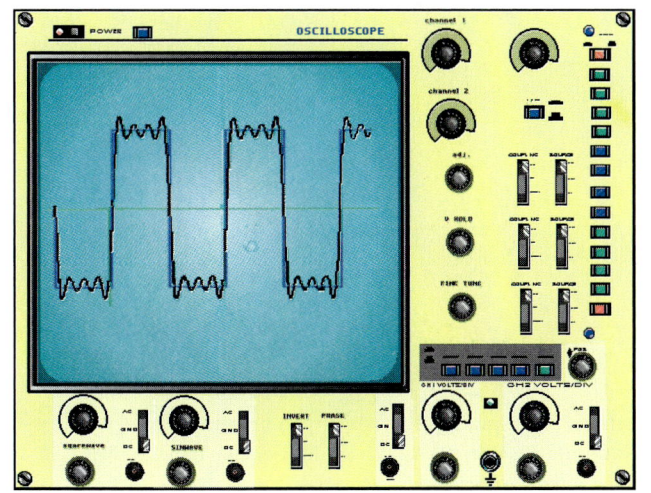
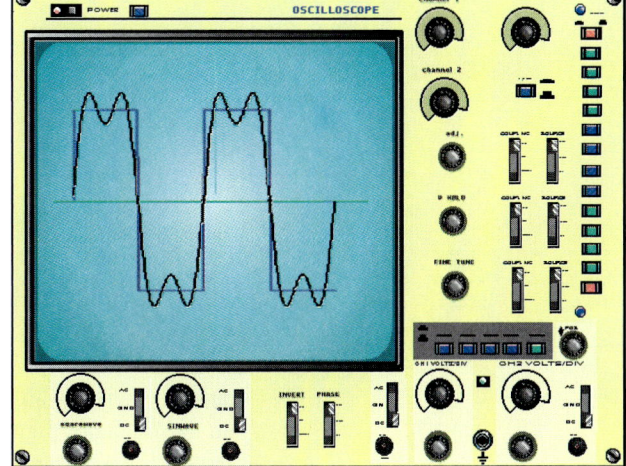

Let $y(t)$ be any function that is periodic in time with period T, such that $y(t + T) = y(t)$. Fourier's theorem states that this function can be written

$$y(t) = \sum_n (A_n \sin 2\pi f_n t + B_n \cos 2\pi f_n t) \tag{18.16}$$

where the lowest frequency is $f_1 = 1/T$.

Fourier's theorem

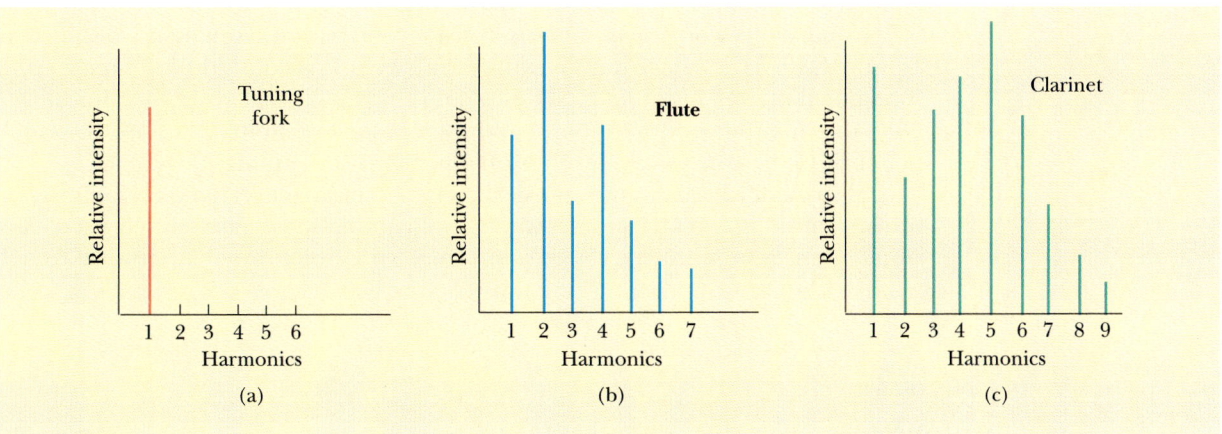

FIGURE 18.18   Harmonics of the waveforms shown in Figure 18.17. Note the variations in intensity of the various harmonics.   *(Adapted from C. A. Culver, Musical Acoustics, 4th ed., New York, McGraw-Hill, 1956.)*

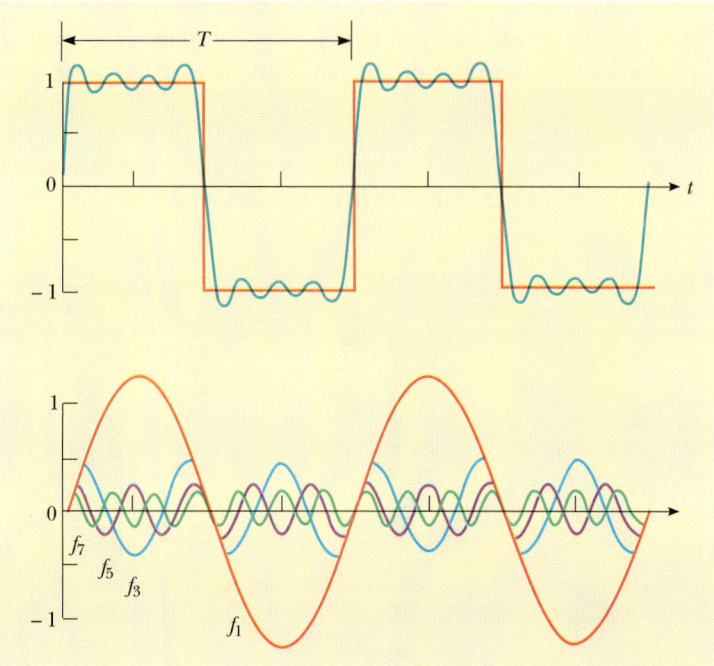

FIGURE 18.19   Harmonic synthesis of a square wave, which can be represented by the sum of odd harmonics of the fundamental.   *(From M. L. Warren,* Introductory Physics, *New York, W. H. Freeman, 1979, p. 178; by permission of the publisher.)*

The higher frequencies are integral multiples of the fundamental, so that $f_n = nf_1$. The coefficients A_n and B_n represent the amplitudes of the various waves. The amplitude of the nth harmonic is proportional to $\sqrt{A_n{}^2 + B_n{}^2}$, and its intensity is proportional to $A_n{}^2 + B_n{}^2$.

Figure 18.18 represents a harmonic analysis of the wave patterns shown in Figure 18.17. Note the variation in relative intensity of the various harmonics for the flute and clarinet. In general, any musical sound (of definite pitch) contains components that are members of a harmonic set with varying relative intensities.

As an example of *Fourier synthesis,* consider the periodic square wave shown in Figure 18.19. The square wave is synthesized by a series of odd harmonics of the fundamental. The series contains only sine functions (that is, $B_n = 0$ for all n). Only the first four odd harmonics and their respective amplitudes are shown. A better fit to the true wave pattern is obtained by adding more harmonics.

Using modern technology, musical sounds can be generated electronically by mixing any number of harmonics with varying amplitudes. These widely used electronic music synthesizers are able to produce an infinite variety of musical tones.

SUMMARY

When two waves with equal amplitudes and frequencies superimpose, the resultant wave has an amplitude that depends on the phase angle ϕ between the two waves. **Constructive interference** occurs when the two waves are in phase everywhere, corresponding to $\phi = 0, 2\pi, 4\pi, \ldots$ rad. **Destructive interference** occurs when the two waves are 180° out of phase everywhere, corresponding to $\phi = \pi$, $3\pi, 5\pi, \ldots$ rad.

Standing waves are formed from the superposition of two harmonic waves having the same frequency, amplitude, and wavelength but traveling in opposite directions. The resultant standing wave is described by the wave function

$$y = (2A_0 \sin kx) \cos \omega t \qquad (18.3)$$

Hence, its amplitude varies as $\sin kx$. The maximum amplitude points (called **antinodes**) occur at $x = n\pi/2k = n\lambda/4$ (for odd n). The points of zero amplitude (called **nodes**) occur at $x = n\pi/k = n\lambda/2$ (for integral values of n).

The natural frequencies of vibration of a taut string of length L, fixed at both ends are

$$f_n = \frac{n}{2L} \sqrt{\frac{F}{\mu}} \qquad (n = 1, 2, 3, \ldots) \qquad (18.8)$$

where F is the tension in the string and μ is its mass per unit length. The natural frequencies of vibration form a **harmonic series,** that is, $f_1, 2f_1, 3f_1, \ldots$.

A system capable of oscillating is said to be in **resonance** with some driving force whenever the frequency of the driving force matches one of the natural frequencies of the system. When the system is resonating, it responds by oscillating with a relatively large amplitude.

Standing waves can be produced in a tube of air. If the tube is open at both ends, all harmonics are present and the natural frequencies of vibration are

$$f_n = n \frac{v}{2L} \qquad (n = 1, 2, 3, \ldots) \qquad (18.11)$$

If the tube is open at one end and closed at the other, only the odd harmonics are present, and the natural frequencies of vibration are

$$f_n = n \frac{v}{4L} \qquad (n = 1, 3, 5, \ldots) \qquad (18.12)$$

QUESTIONS

1. For certain positions of the movable section in Figure 18.2, there is no sound detected at the receiver, corresponding to destructive interference. This suggests that perhaps energy is somehow lost! What happens to the energy transmitted by the speaker?

2. Does the phenomenon of wave interference apply only to sinusoidal waves?

3. When two waves interfere constructively or destructively, is there any gain or loss in energy? Explain.

4. A standing wave is set up on a string as in Figure 18.5. Explain why no energy is transmitted along the string.

5. What is common to *all* points (other than the nodes) on a string supporting a standing wave?

6. What limits the amplitude of motion of a real vibrating system that is driven at one of its resonant frequencies?

7. If the temperature of the air in an organ pipe increases, what happens to the resonance frequencies?

8. Explain why your voice seems to sound better than usual when you sing in the shower.

9. What is the purpose of the slide on a trombone or the valves on a trumpet?

10. Explain why all harmonics are present in an organ pipe open at both ends, but only the odd harmonics are present in a pipe closed at one end.

11. Explain how a musical instrument such as a piano may be tuned using the phenomenon of beats.

12. An airplane mechanic notices that the sound from a twin-engine aircraft rapidly varies in loudness when both engines are running. What could be causing this variation from loud to soft?

13. When the base of a vibrating tuning fork is placed against a chalk board, the sound becomes louder due to resonance. How does this affect the length of time for which the fork vibrates? Does this agree with conservation of energy?

14. Stereo speakers are supposed to be "phased" when set up. That is, the waves emitted from them should be in phase with each other. What would the sound be like along the center line of the speakers if one speaker were wired up backwards, that is, out of phase?

15. To keep animals away from their cars, some people mount short thin pipes on the fenders. The pipes give out a high-pitched wail when the cars are moving. How do they create the sound?

16. If you wet your fingers and lightly run them around the rim of a fine wine glass, a high-pitched sound is heard. Why? How could you produce various musical notes with a set of wine glasses?

17. When a bell is rung, standing waves are set up around the bell's circumference. What boundary conditions must be satisfied by the resonant wavelengths? How does a crack in the bell, such as in the Liberty Bell, affect the satisfying of the boundary conditions and the sound emanating from the bell?

PROBLEMS

Review Problem

For the arrangement shown in the diagram, $\theta = 30°$, the inclined plane is frictionless, the mass M rests at the bottom of the plane, and the string has a mass m that is small compared to M. When the system is in equilibrium, the vertical part of the string of length h is fixed so that the tension remains constant. If standing waves are set up in the vertical string, find (a) the tension in the string, (b) the length of the string, (c) the mass per unit length of the string, (d) the wave speed in the vertical part of the string, (e) the lowest frequency standing wave, (f) the period of the standing wave having three nodes, (g) the wavelength of the standing wave having three nodes, and (h) the frequency of the beats resulting from the interference of the sound wave of lowest frequency generated by the string with another sound wave having a frequency that is 2% greater.

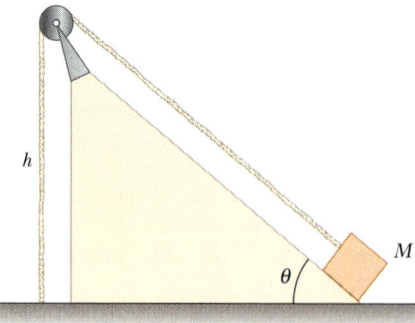

Section 18.1   Superposition and Interference of Sinusoidal Waves

1. Two harmonic waves are described by

$$y_1 = (5.0 \text{ m}) \sin[\pi(4.0x - 1200t)]$$

$$y_2 = (5.0 \text{ m}) \sin[\pi(4.0x - 1200t - 0.25)]$$

where x, y_1, and y_2 are in meters and t is in seconds.

(a) What is the amplitude of the resultant wave?
(b) What is the frequency of the resultant wave?

2. Two harmonic waves are described by

$$y_1 = (6.0 \text{ m}) \sin\left(\frac{\pi}{15} x - \frac{\pi}{0.0050} t\right)$$

$$y_2 = (6.0 \text{ m}) \sin\left(\frac{\pi}{15} x - \frac{\pi}{0.0050} t - \phi\right)$$

where x, y_1, and y_2 are in meters and t is in seconds. (a) What is the amplitude of the resultant wave when $\phi = (\pi/6)$ rad? (b) For what values of ϕ will the amplitude of the resultant wave have its maximum value?

3. Two identical harmonic waves with wavelengths of 3.0 m travel in the same direction at a speed of 2.0 m/s. The second wave originates from the same point as the first, but at a later time. Determine the minimum possible time interval between the starting moments of the two waves if the amplitude of the resultant wave is the same as that of the two initial waves.

3A. Two identical harmonic waves with wavelengths λ travel in the same direction at a speed v. The second wave originates from the same point as the first, but at a later time. Determine the minimum possible time interval between the starting moments of the two waves if the amplitude of the resultant wave is the same as that of the two initial waves.

4. Two speakers are driven by the same oscillator of frequency 200 Hz. They are located on a vertical pole a distance of 4.00 m from each other. A man walks toward one of the speakers in a direction perpendicular to the pole as shown in Figure P18.4. (a) How many times will he hear a minimum in sound intensity and (b) how far is he from the wall at these moments? Take the speed of sound to be 330 m/s and ignore any sound reflections coming off the floor.

□ indicates problems that have full solutions available in the Student Solutions Manual and Study Guide.

4A. Two speakers are driven by the same oscillator of frequency *f*. They are located on a vertical pole a distance *d* from each other. A man walks toward one of the speakers in a direction perpendicular to the pole as shown in Figure P18.4. (a) How many times will he hear a minimum in sound intensity and (b) how far is he from the wall at these moments? Take the speed of sound to be *v* and ignore any sound reflections coming off the floor.

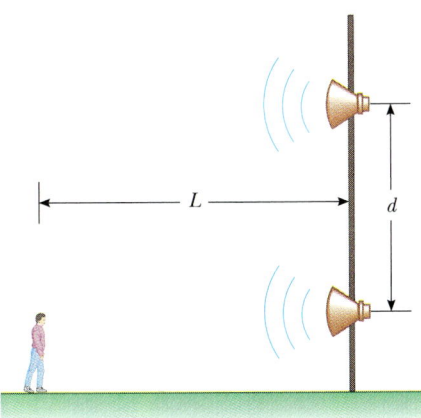

FIGURE P18.4

5. Two identical speakers separated by 10.0 m are driven by the same oscillator with a frequency of $f = 21.5$ Hz (Fig. P18.5). (a) Explain why a receiver at point *A* records a minimum in sound intensity from the two speakers. (b) If the receiver is moved in the horizontal plane of the speakers, what path should it take so that the intensity remains at a minimum? That is, determine the relationship between *x* and *y* (the coordinates of the receiver) that causes the receiver to record a minimum in sound intensity. Take the speed of sound to be 344 m/s.

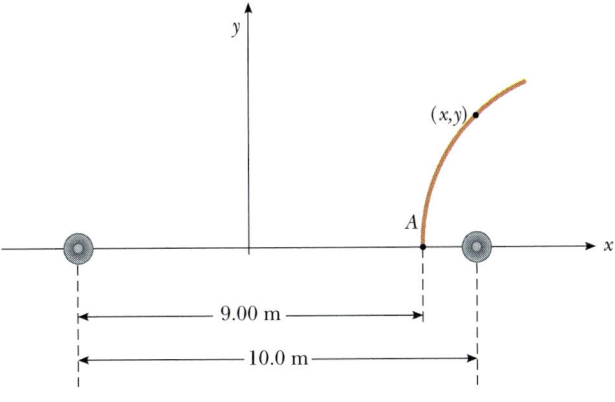

FIGURE P18.5

6. A tuning fork generates sound waves with a frequency of 246 Hz. The waves travel in opposite directions along a hallway, are reflected by walls, and return. What is the phase difference between the reflected waves when they meet? The corridor is 47 m long and the tuning fork is located 14 m from one end. The speed of sound in air is 343 m/s.

7. Two speakers are driven by a common oscillator at 800 Hz and face each other at a distance of 1.25 m. Locate the points along a line joining the two speakers where relative minima would be expected. (Use $v = 343$ m/s.)

8. For the arrangement shown in Figure 18.2, let the path length $r_1 = 1.20$ m and the path length $r_2 = 0.80$ m. (a) Calculate the three lowest speaker frequencies that will result in intensity maxima at the receiver. (b) What is the highest frequency within the audible range (20–20 000 Hz) that will result in a minimum at the receiver? (Use $v = 340$ m/s.)

Section 18.2 Standing Waves

9. Two harmonic waves are described by

$$y_1 = (3.0 \text{ cm}) \sin \pi(x + 0.60t)$$

$$y_2 = (3.0 \text{ cm}) \sin \pi(x - 0.60t)$$

where *x* is in centimeters and *t* is in seconds. Determine the *maximum* displacement of the motion at (a) $x = 0.25$ cm, (b) $x = 0.50$ cm, and (c) $x = 1.5$ cm. (d) Find the three smallest values of *x* corresponding to antinodes.

10. Use the trigonometric identity

$$\sin(a \pm b) = \sin a \cos b \pm \cos a \sin b$$

to show that the resultant of two wave functions each of amplitude A_0, angular frequency ω, and propagation number *k* and traveling in opposite directions can be written

$$y = (2A_0 \sin kx) \cos \omega t$$

11. The wave function for a standing wave in a string is

$$y = (0.30 \text{ m}) \sin(0.25x) \cos(120\pi t)$$

where *x* is in meters and *t* is in seconds. Determine the wavelength and frequency of the interfering traveling waves.

12. Two harmonic waves traveling in opposite directions interfere to produce a standing wave described by

$$y = (1.50 \text{ m}) \sin(0.400x) \cos(200t)$$

where *x* is in meters and *t* is in seconds. Determine the wavelength, frequency, and speed of the interfering waves.

13. A standing wave is formed by the interference of two traveling waves, each of which has an amplitude $A = \pi$ cm, angular wave number $k = (\pi/2)$ cm^{-1}, and

angular frequency $\omega = 10\pi$ rad/s. (a) Calculate the distance between the first two antinodes. (b) What is the amplitude of the standing wave at $x = 0.25$ cm?

14. Verify by direct substitution that the wave function for a standing wave given in Equation 18.3,

$$y = 2A_0 \sin kx \cos \omega t$$

is a solution of the general linear wave equation, Equation 16.26:

$$\frac{\partial^2 y}{\partial x^2} = \frac{1}{v^2} \frac{\partial^2 y}{\partial t^2}$$

15. Two waves given by $y_1(x, t) = A \sin(kx - \omega t)$, and $y_2(x, t) = A \sin(2kx + \omega t)$ interfere. (a) Determine all x values where there are stationary nodes. (b) Determine all x values where there are nodes that depend on the time t.

16. Two waves in a long string are given by

$$y_1 = (0.015 \text{ m}) \cos\left(\frac{x}{2} - 40t\right)$$

$$y_2 = (0.015 \text{ m}) \cos\left(\frac{x}{2} + 40t\right)$$

where the y's and x are in meters and t is in seconds. (a) Determine the positions of the nodes of the resulting standing wave. (b) What is the maximum displacement at the position $x = 0.40$ m?

Section 18.3   Standing Waves in a String Fixed at Both Ends

17. A standing wave is established in a 120-cm-long string fixed at both ends. The string vibrates in four segments when driven at 120 Hz. (a) Determine the wavelength. (b) What is the fundamental frequency?

18. A stretched string is 160 cm long and has a linear density of 0.015 g/cm. What tension in the string will result in a second harmonic of 460 Hz?

19. Consider a tuned guitar string of length L. At what point along the string (fraction of length from one end) should the string be plucked and at what point should the finger be held lightly against the string so that the second harmonic is the most prominent mode of vibration?

20. A string 50 cm long has a mass per unit length of 20×10^{-5} kg/m. To what tension should this string be stretched if its fundamental frequency is to be (a) 20 Hz and (b) 4500 Hz?

21. Find the fundamental frequency and the next three frequencies that could cause a standing wave pattern on a string that is 30 m long, has a mass per unit length 9.0×10^{-3} kg/m, and is stretched to a tension of 20 N.

22. A string of linear density 1.0×10^{-3} kg/m and length 3.0 m is stretched between two points. One end is vibrated transversely at 200 Hz. What tension

in the string will establish a standing-wave pattern with three loops along the string's length?

23. A sphere having mass $M = 2.0$ kg is supported by a string that passes over a light horizontal rod of length $L = 1.0$ m (Fig. P18.23). Given the angle is $\theta = 35°$ and the fundamental frequency of standing waves in the string is $f = 50.0$ Hz, determine the mass of the string.

23A. A sphere of mass M is supported by a string that passes over a light horizontal rod of length L (Fig. P18.23). Given the angle is θ and the fundamental frequency of standing waves in the string is f, determine the mass of the string above the horizontal rod.

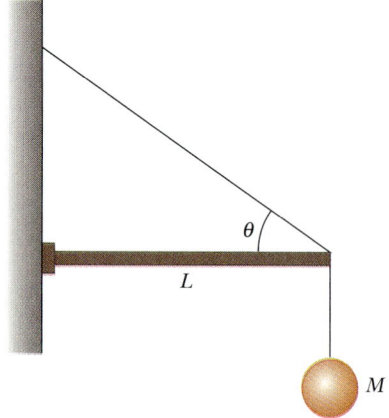

FIGURE P18.23

24. A 2.0-m-long wire having a mass of 0.10 kg is fixed at both ends. The tension in the wire is maintained at 20 N. A node is observed at a point 0.40 m from one end. What are the frequencies of the first three allowed modes of vibration?

25. A cello A-string vibrates in its fundamental mode with a frequency of 220 vibrations/s. The vibrating segment is 70 cm long and has a mass of 1.2 g. (a) Find the tension in the string. (b) Determine the frequency of the harmonic that causes the string to vibrate in three segments.

26. In the arrangement shown in Figure P18.26, a mass can be hung from a string (with linear mass density $\mu = 0.0020$ kg/m) around a light pulley. The string is connected to a vibrator (of constant frequency, f), and the length of the string between point P and the pulley is $L = 2.0$ m. When the mass m is either 16 kg or 25 kg, standing waves are observed, but no standing waves are observed for any mass between these values. (a) What is the frequency of the vibrator? (*Hint:* The greater the tension in the string, the smaller the number of nodes in the standing wave.) (b) What is the largest mass for which standing waves could be observed?

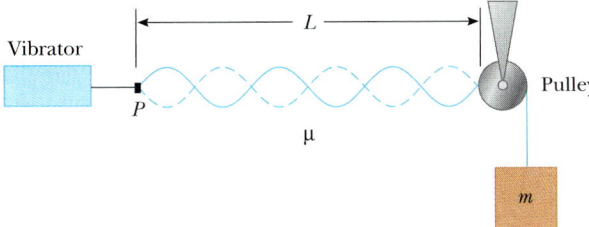

Vibrator

L

Pulley

P

μ

m

FIGURE P18.26

27. A 60-cm guitar string under a tension of 50 N has a mass per unit length of 0.10 g/cm. What is the highest resonant frequency that can be heard by a person capable of hearing frequencies up to 20 000 Hz?

28. A stretched wire vibrates in its fundamental mode at a frequency of 400 vibrations/s. What would be the fundamental frequency if the wire were half as long, with twice the diameter and with four times the tension?

29. A low C ($f = 65$ Hz) is sounded on a piano. If the length of the piano wire is 2.0 m and its linear mass density is 5.0 g/m, what is the tension in the wire?

30. A violin string has a length of 0.35 m and is tuned to concert G, $f_G = 392$ Hz. Where must the violinist place her finger to play concert A, $f_A = 440$ Hz? If this position is to remain correct to one half the width of a finger (i.e., to within 0.60 cm), by what fraction may the string tension be allowed to slip?

Section 18.5 Standing Waves in Air Columns

(In this section, unless otherwise indicated, assume that the speed of sound in air is 344 m/s.)

31. An open pipe 0.40 m in length is placed vertically in a cylindrical bucket having a bottom area of 0.10 m². Water is poured into the bucket until a sounding tuning fork of frequency 440 Hz, placed over the pipe, produces resonance. Find the mass of water in the bucket at this moment.

31A. An open pipe of length L is placed vertically in a cylindrical bucket having a bottom area A. Water is poured into the bucket until a sounding tuning fork of frequency f, placed over the pipe, produces resonance. Find the mass of water in the bucket at this moment.

32. If an organ pipe is to resonate at 20.0 Hz, what is its required length if it is (a) open at both ends and (b) closed at one end?

33. A tuning fork of frequency 512 Hz is placed near the top of the tube shown in Figure 18.13a. The water level is lowered so that the length L slowly increases from an initial value of 20.0 cm. Determine the next two values of L that correspond to resonant modes.

34. A student uses an audio oscillator of adjustable frequency to measure the depth of a water well. Two successive resonant frequencies are heard at 52.0 Hz and 60.0 Hz. What is the depth of the well?

34A. A student uses an audio oscillator of adjustable frequency to measure the depth of a water well. Two successive resonant frequencies are heard at f_1 and f_2. What is the depth of the well?

35. Calculate the length for a pipe that has a fundamental frequency of 240 Hz if the pipe is (a) closed at one end and (b) open at both ends.

36. A tunnel beneath a river is approximately 2.0 km long. At what frequencies can this tunnel resonate?

37. An organ pipe open at both ends is vibrating in its third harmonic with a frequency of 748 Hz. The length of the pipe is 0.70 m. Determine the speed of sound in air in the pipe.

38. The longest pipe on an organ that has pedal stops is often 4.88 m. What is the fundamental frequency (at 0.0°C) if the nondriven end of the pipe is (a) closed and (b) open? (c) What will be the frequencies at 20.0°C?

39. The overall length of a piccolo is 32 cm. The resonating air column vibrates as a pipe open at both ends. (a) Find the frequency of the lowest note a piccolo can play, assuming the speed of sound in air is 340 m/s. (b) Opening holes in the side effectively shortens the length of the resonant column. If the highest note a piccolo can sound is 4000 Hz, find the distance between adjacent nodes for this mode of vibration.

40. A piece of metal pipe is just the right length so that when it is cut into two pieces, their lowest resonant frequencies are 256 Hz for one and 440 Hz for the other. (a) What resonant frequency would have been produced by the original length of pipe and (b) how long was the original piece?

41. An air column 2.00 m in length is open at both ends. The frequency of a certain harmonic is 410 Hz, and the frequency of the next higher harmonic is 492 Hz. Determine the speed of sound in the air column.

42. A piece of cardboard tubing, closed at one end, is just the right length so that when it is cut into two pieces, the lowest resonant frequency is 256 Hz for the piece with the closed end and 440 Hz for the other. (a) What resonant frequency would have been produced by the original cardboard tubing and (b) how long was the original piece?

43. A glass tube is open at one end and closed at the other (by a movable piston). The tube is filled with 30.0°C air, and a 384-Hz tuning fork is held at the open end. Resonance is heard when the piston is 22.8 cm from the open end and again when it is 68.3 cm from the open end. (a) What speed of sound

is implied by these data? (b) Where would the piston be for the next resonance?

44. Water is pumped into a long cylinder at a rate of 18.0 cm³/s as in Figure P18.44. The radius of the cylinder is 4.0 cm, and at the top of the cylinder there is a tuning fork vibrating with a frequency of 200 Hz. As the column rises, how much time elapses between successive resonances?

44A. Water is pumped into a long cylinder at a rate R (cm³/s) as in Figure P18.44. The radius of the cylinder is r (cm), and at the top of the cylinder there is a tuning fork vibrating with a frequency f. As the column rises, how much time elapses between successive resonances?

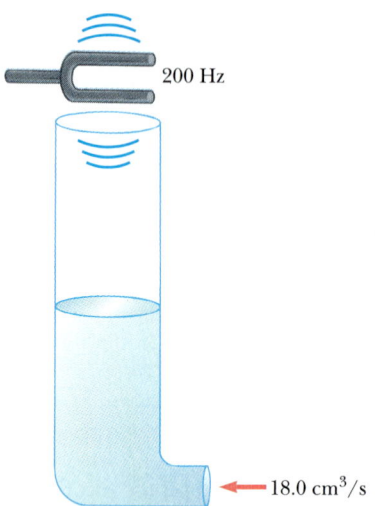

200 Hz

18.0 cm³/s

FIGURE P18.44

45. A shower stall measures 86 cm × 86 cm × 210 cm. When you sing in the shower, which frequencies will sound the richest (resonate), assuming the shower acts as a pipe closed at both ends (nodes at both sides)? Assume also that the human voice ranges from 130 Hz to 2000 Hz (not necessarily one person's voice, however). Let the speed of sound in the hot shower stall be 355 m/s.

*Section 18.6   Standing Waves in Rods and Plates

46. An aluminum rod is clamped at the one-quarter position and set into longitudinal vibration by a variable-frequency driving source. The lowest frequency that produces resonance is 4400 Hz. The speed of sound in aluminum is 5100 m/s. Determine the length of the rod.

47. A 60.0-cm metal bar that is clamped at one end is struck with a hammer. If the speed of longitudinal (compressional) waves in the bar is 4500 m/s, what is the lowest frequency with which the struck bar will resonate?

48. Longitudinal waves move with a speed v in a bar of length L. Write an expression for the frequencies of the longitudinal vibrations of a metal bar that is (a) clamped at its center, as shown in Figure 18.14a, and (b) clamped at one-fourth the length of the bar from one end, as shown in Figure 18.14b.

49. An aluminum rod 1.6 m long is held at its center. It is stroked with a rosin-coated cloth to set up longitudinal vibrations in the fundamental mode. (a) What is the frequency of the waves established in the rod? (b) What harmonics are set up in the rod held in this manner? (c) What would be the fundamental frequency if the rod were copper?

*Section 18.7   Beats: Interference in Time

50. In certain ranges of a piano keyboard, more than one string is tuned to the same note to provide extra loudness. For example, the note at 110 Hz has two strings at this pitch. If one string slips from its normal tension of 600 N to 540 N, what beat frequency will be heard when the two strings are struck simultaneously?

51. Two waves with equal amplitude but with slightly different frequencies are traveling in the same direction through a medium. At a given point the separate displacements are described by

$$y_1 = A_0 \cos \omega_1 t \qquad \text{and} \qquad y_2 = A_0 \cos \omega_2 t$$

Use the trigonometric identity

$$\cos a + \cos b = 2 \cos \left(\frac{a - b}{2} \right) \cos \left(\frac{a + b}{2} \right)$$

to show that the resultant displacement due to the two waves is given by

$$y = 2A_0 \left[\cos \left(\frac{\omega_1 - \omega_2}{2} \right) t \right] \left[\cos \left(\frac{\omega_1 + \omega_2}{2} \right) t \right]$$

52. While attempting to tune a C note at 523 Hz, a piano tuner hears 3 beats per second between the oscillator and the string. (a) What are the possible frequencies of the string? (b) By what percentage should the tension in the string be changed to bring the string into tune?

53. A student holds a tuning fork oscillating at 256 Hz. He walks toward a wall at a constant speed of 1.33 m/s. (a) What beat frequency does he observe between the tuning fork and its echo? (b) How fast must he walk away from the wall to observe a beat frequency of 5 Hz?

ADDITIONAL PROBLEMS

54. (a) What is the fundamental frequency of a 5.00 × 10⁻³-kg steel piano wire of length 1.00 m, under a tension of 1350 N? (b) What is the fundamental frequency of an organ pipe, 1.00 m in length, closed at the bottom and open at the top?

55. Two loudspeakers are placed on a wall 2.00 m apart. A listener stands directly in front of one of the speakers, 3.00 m from the wall. The speakers are being driven by a single oscillator at a frequency of 300 Hz. (a) What is the phase difference between the two waves when they reach the observer? (b) What is the frequency closest to 300 Hz to which the oscillator may be adjusted such that the observer will hear minimal sound?

56. On a marimba, the wooden bar that sounds a tone when struck vibrates as a transverse standing wave with three antinodes and two nodes. The lowest frequency note is 87 Hz, produced by a bar 40 cm long. (a) Find the speed of transverse waves on the bar. (b) The loudness and duration of the emitted sound are enhanced by a resonant pipe suspended vertically below the center of the bar. If the pipe is open at the top end only and the speed of sound in air is 340 m/s, what is the length of the pipe required to resonate with the bar in part (a)?

57. Jane waits on a railroad platform, while two trains approach from the same direction at equal speeds of 8.0 m/s. Both trains are blowing their whistles (which have the same frequency), and one train is some distance behind the other. After the first train passes Jane, but before the second train passes her, she hears beats of frequency 4.0 Hz. What is the frequency of the train whistles?

58. A 0.010-kg and 2.0-m-long wire is fixed at both ends and vibrates in its simplest mode under tension 200 N. When a tuning fork is placed near the wire, a beat frequency of 5.0 Hz is heard. (a) What is the frequency (frequencies) of the tuning fork? (b) What should the tension in the wire be to make the beat disappear?

59. If two adjacent natural frequencies of an organ pipe are determined to be 0.55 kHz and 0.65 kHz, calculate the fundamental frequency and length of this pipe. (Use $v = 340$ m/s.)

60. Two loudspeakers are mounted on the ends of a horizontal 2.0-m stick that is set in rotation about a vertical axis through its center such that the stick makes one complete rotation every 5.0 s. A listener is situated far from the rotation axis between the speakers as shown in Figure P18.60. If the speakers each produce a middle C tone of 262 Hz, what is the beat frequency heard by the listener? Take the speed of sound to be 343 m/s.

60A. Two loudspeakers are mounted on the ends of a horizontal stick of length L that is set in rotation about a vertical axis through its center such that the stick makes one complete rotation every t seconds. A listener is situated far from the rotation axis between the speakers as shown in Figure P18.60. If the speakers each produce a frequency f, what is the beat frequency heard by the listener? Take the speed of sound to be v.

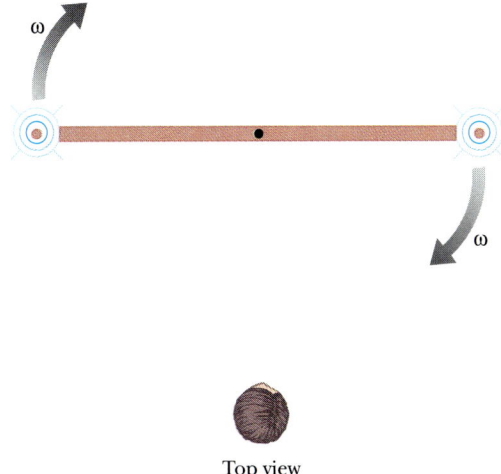

Top view

FIGURE P18.60

61. A 12-kg mass hangs in equilibrium from a string of total length $L = 5.0$ m and linear mass density $\mu = 0.0010$ kg/m. The string is wrapped around two light, frictionless pulleys that are separated by the distance $d = 2.0$ m (Fig. P18.61a). (a) Determine the tension in the string. (b) At what frequency must the string between the pulleys vibrate in order to form the standing wave pattern shown in Figure P18.61b?

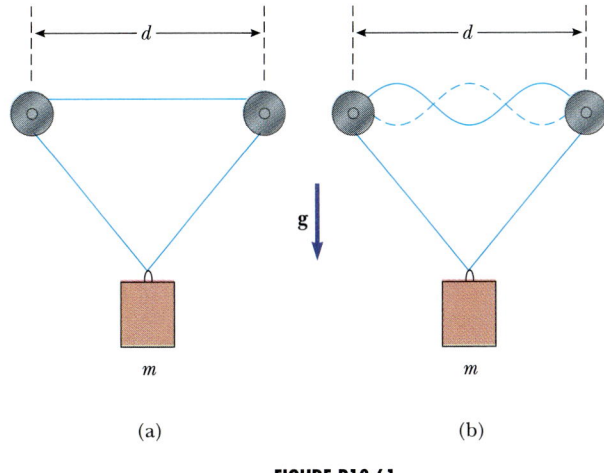

(a)   (b)

FIGURE P18.61

62. While waiting for Stan Speedy to arrive on a late passenger train, Kathy Kool notices beats occurring as a result of two trains blowing their whistles simultaneously. One train is at rest and the other is approaching her at a speed of 20 km/h. Assume that both whistles have the same frequency and that the speed of sound is 344 m/s. If Kathy hears 4 beats per second, what is the frequency of the whistles?

63. A string (mass = 4.8 g, length = 2.0 m, and tension = 48 N), fixed at both ends, vibrates in its second ($n = 2$) natural mode. What is the wavelength in air of the sound emitted by this vibrating string?

64. In an arrangement like the one shown in Figure 18.2, paths r_1 and r_2 are each 1.75 m in length. The top portion of the tube (corresponding to r_2) is filled with air at 0°C (273 K). Air in the lower portion is quickly heated to 200°C (473 K). What is the lowest speaker frequency that will produce an intensity maximum at the receiver? (You may determine the speed of sound in air in different temperatures by using the expression $v = 331(T/273)^{1/2}$ m/s, where T is in K.)

65. Two train whistles have identical frequencies of 180 Hz. When one train is at rest in the station sounding its whistle, a beat frequency of 2 Hz is heard from a moving train. What two possible speeds and directions can the moving train have?

66. In a major chord on the physical pitch musical scale, the frequencies are in the ratios 4:5:6:8. A set of pipes, closed at one end, are to be cut so that when sounded in their fundamental mode, they will sound out a major chord. (a) What is the ratio of the lengths of the pipes? (b) What length pipes are needed if the lowest frequency of the chord is 256 Hz? (c) What are the frequencies of this chord?

67. Two wires are welded together. The wires are of the same material, but one is twice the diameter of the other one. They are subjected to a tension of 4.6 N. The thin wire has a length of 40 cm and a linear mass density of 2.0 g/m. The combination is fixed at both ends and vibrated in such a way that two antinodes are present with the central node being right at the weld. (a) What is the frequency of vibration? (b) How long is the thick wire?

68. Two identical strings, each fixed at both ends, are arranged near each other. If string A starts oscillating in its fundamental mode, it is observed that string B will begin vibrating in its third ($n = 3$) natural mode. Determine the ratio of the tension of string B to the tension of string A.

69. A standing wave is set up in a string of variable length and tension by a vibrator of variable frequency. When the vibrator has a frequency f in a string of length L and tension F there are n antinodes set up in the string. (a) If the length of the string is doubled, by what factor should the frequency be changed to get the same number of antinodes? (b) If the frequency and length are held constant, what tension will produce $n + 1$ antinodes? (c) If the frequency is tripled and the length halved, by what factor should the tension be changed to get twice as many antinodes?

70. Radar detects the speed of a car, using the Doppler shift of microwaves that are reflected off the moving car, by beating the received wave with the transmitted wave and measuring the difference. The Doppler shift for light is

$$f = f_0 \sqrt{\frac{c + v}{c - v}}$$

where f_0 is the transmitted frequency, c is the speed of light (3×10^8 m/s), and v the relative speed of the two objects. (a) Show that the wave that reflects back to the source has a frequency

$$f = f_0 \frac{(c + v)}{(c - v)}$$

(b) Show that the expression for the beat frequency of the microwaves may be written as $f_b = 2 v/\lambda$. (Since the beat frequency is much smaller than the transmitted frequency, use the approximation $f + f_0 = 2f_0$.) (c) What beat frequency is measured for a speed of 30 m/s (67 mph) if the microwaves have a frequency of 10 GHz? (1 GHz = 10^9 Hz.) (d) If the beat frequency measurement is accurate to ± 5 Hz, how accurate is the velocity measurement?

SPREADSHEET PROBLEMS

S1. Spreadsheet 18.1 adds two traveling waves at some fixed time t. The resultant wave function is

$$y = y_1 + y_2 = A_1 \sin(k_1 x - \omega_1 t + \phi_1) + A_2 \sin(k_2 x - \omega_2 t + \phi_2)$$

Add two waves traveling in opposite directions having the same wavelengths, the same phases, and the same speeds. (a) Use $A_1 = A_2 = 0.10$ m, $\omega_1 = -\omega_2 = 3.0$ rad/s, $\phi_1 = \phi_2 = 0$, and $k_1 = k_2 = 2.0$ rad/m. View the associated graph. Choose different values of t to see the time evolution of the resultant wave function. Do you get standing waves? (b) Repeat part (a) using $A_1 = 0.10$ m, $A_2 = 0.20$ m.

S2. Use Spreadsheet 18.1 to add two traveling waves that differ only in phase. (a) Choose $A_1 = A_2 = 0.10$ m, $\omega_1 = \omega_2 = 2.5$ rad/s, $k_1 = k_2 = 1.0$ rad/m, $\phi_1 = 0$, and $\phi_2 = 0, \pi/8, \pi/4, \pi/2$, and π. View the associated graphs. For which values of ϕ_2 do you get constructive interference? destructive interference? (b) Repeat part (a) using $A_2 = 0.20$ m.

S3. Use Spreadsheet 18.1 to add two traveling waves with different wavelengths. (a) Choose $A_1 = A_2 = 0.10$ m, $\omega_1 = \omega_2 = 3.0$ rad/s, $k_1 = 2.0$ rad/m, $k_2 = 1.0$ rad/m, and $\phi_1 = \phi_2 = 0$. View the associated graph. (b) Repeat part (a) with $k_2 = 3, 4, 5, 6, 7, 8, 9$, and 10. Examine the graph for each case. Explain the appearance of the graphs.

S4. Three waves with the same frequency, wavelength, and amplitude are traveling in the same direction. Each differs in phase such that $(\phi_2 - \phi_1) = (\phi_3 - \phi_2) = \Delta\phi$. Modify Spreadsheet 18.1 to add the three waves at some fixed time t. Calculate the resultant wave function at $t = 0.0$, 1.0 s, and 2.0 s. Use $A_1 = A_2 = A_3 = 0.050$ m, $\omega_1 = \omega_2 = \omega_3 = 2.5$ rad/s, and $k_1 = k_2 = k_3 = 1.0$ rad/m. Choose $\Delta\phi = \pi/6$. View the associated graph. Repeat for different values of $\Delta\phi$.

S5. Write a spreadsheet or computer program to add two traveling waves of different frequencies:

$$y_1 = y_0 \cos 2\pi f_1 t \qquad y_2 = y_0 \cos 2\pi f_2 t$$

If the two frequencies are close to each other, beats will be produced. Does your program show the beats? From your numerical results, how does the beat frequency relate to the frequencies of the two waves?

S6. The Fourier theorem states that any periodic wave of frequency f, no matter how complicated, can be expressed as a sum of even or odd harmonic functions.

That is,

$$y(t) = \sum_{n=0}^{\infty} A_n \sin(n\omega t + \phi_n)$$

where $\omega = 2\pi f$ is the fundamental angular frequency. (a) Use $A_1 = 1$, $A_2 = 1/2$, $A_3 = 1/3$, and so on. All the ϕ_n's are zero. (b) Use $A_1 = 1$, $A_3 = 1/3$, $A_5 = 1/5$, and so on. All the even A_n's and all the ϕ_n's are zero. Write a spreadsheet or computer program that will calculate this sum. You should design your spreadsheet or program so that the number of terms in the sum can be easily specified. Start with just the first term ($n = 1$). Then rerun the calculation with more terms in the sum ($n = 2, 3, 4, \ldots$, 20). Plot the sum so that one or two periods are displayed. Note how the waveform builds up as more terms are added.

Water dripping from tree branches and mountain ledges forms ice crystals when the temperature falls below 0°C. As the crystals form, a phase transition from the liquid to the solid state occurs, and water molecules become arranged in a highly ordered structure. (©John Beatty/Tony Stone Images)

PART III

Thermodynamics

When dining, I had often observed that some particular dishes retained their Heat much longer than others; and that apple pies, and apples and almonds mixed (a dish in great repute in England) remained hot a surprising length of time. Much struck with this extraordinary quality of retaining Heat, which apples appeared to possess, it frequently occurred to my recollection; and I never burnt my mouth with them, or saw others meet with the same misfortune, without endeavouring, but in vain, to find out some way of accounting, in a satisfactory manner, for this surprising phenomenon.

BENJAMIN THOMPSON
(COUNT RUMFORD)

We now turn to the study of thermodynamics, which is concerned with the concepts of heat and temperature. As we shall see, thermodynamics is very successful in explaining the bulk properties of matter and the correlation between these properties and the mechanics of atoms and molecules.

Historically, the development of thermodynamics paralleled the development of the atomic theory of matter. By the 1820s, chemical experiments provided solid evidence for the existence of atoms. At that time, scientists recognized that there must be a connection between the theory of heat and temperature and the structure of matter. In 1827, the botanist Robert Brown reported that grains of pollen suspended in a liquid move erratically from one place to another, as if under constant agitation. In 1905, Albert Einstein explained this erratic motion, now called Brownian motion. Einstein explained this phenomenon by assuming that the grains of pollen are under constant bombardment by "invisible" molecules in the liquid, which themselves undergo an erratic motion. This important experiment and Einstein's insight gave scientists a means of discovering vital information concerning molecular motion. It also gave reality to the concept of the atomic constituents of matter.

Have you ever wondered how a refrigerator is able to cool its contents or what types of transformations occur in a power plant or in the engine of your automobile or what happens to the kinetic energy of an object when it falls to the ground and comes to rest? The laws of thermodynamics and the concepts of heat and temperature enable us to answer such practical questions. In general, thermodynamics is concerned with transformations of matter in all of its forms: solid, liquid, and gas.

Temperature

I n the study of mechanics, such concepts as mass, force, and kinetic energy were carefully defined in order to make the subject quantitative. Likewise, a quantitative description of thermal phenomena requires a careful definition of the concepts of temperature, heat, and internal energy. The laws of thermodynamics provide us with a relationship between heat flow, work, and the internal energy of a system.

The composition of a body is an important factor when dealing with thermal phenomena. For example, liquids and solids expand only slightly when heated, whereas gases expand appreciably when heated. If the gas is not free to expand, its pressure rises when heated. Certain substances may melt, boil, burn, or explode, depending on their composition and structure. Thus, the thermal behavior of a substance is closely related to its structure.

This chapter concludes with a study of ideal gases. We shall approach this study on two levels. The first will examine ideal gases on the macroscopic scale. Here we shall be concerned with the relationships among such quantities as pressure, volume, and temperature. On the second level, we shall examine gases on a microscopic scale, using a model that pictures the components of a gas as small particles.

The latter approach, called the kinetic theory of gases, will help us to understand what happens on the atomic level to affect such macroscopic properties as pressure and temperature.

19.1   TEMPERATURE AND THE ZEROTH LAW OF THERMODYNAMICS

We often associate the concept of **temperature** with how hot or cold an object feels when we touch it. Thus, our senses provide us with a qualitative indication of temperature. However, our senses are unreliable and often misleading. For example, if we remove a metal ice tray and a cardboard box of frozen vegetables from the freezer, the ice tray feels colder to the hand than the box even though both are at the same temperature. The two objects feel different because metal is a better heat conductor than cardboard. What we need, therefore, is a reliable and reproducible method for establishing the relative hotness or coldness of bodies. Scientists have developed a variety of thermometers for making such quantitative measurements.

We are all familiar with the fact that two objects at differential initial temperatures eventually reach some intermediate temperature when placed in contact with each other. For example, a piece of meat placed on a block of ice in a well-insulated container eventually reaches a temperature near 0°C. Likewise, if an ice cube is dropped into a cup of hot coffee, the ice eventually melts and the coffee's temperature decreases.

In order to understand the concept of temperature, it is useful to first define two often-used phrases, *thermal contact* and *thermal equilibrium*. To grasp the meaning of thermal contact, imagine two objects placed in an insulated container so that they interact with each other but not with the rest of the world. If the objects are at different temperatures, energy is exchanged between them. The energy exchanged between objects because of a temperature difference is called **heat**. We shall examine the concept of heat in more detail in Chapter 20. For purposes of the current discussion, we shall assume that two objects are in **thermal contact** with each other if heat can be exchanged between them. **Thermal equilibrium** is a situation in which two objects in thermal contact with each other cease to have any exchange of heat.

Now consider two objects, A and B, which are not in thermal contact, and a third object, C, which is our thermometer. We wish to determine whether or not A and B are in thermal equilibrium with each other. The thermometer (object C) is first placed in thermal contact with A until thermal equilibrium is reached. From that moment on, the thermometer's reading remains constant, and we record it. The thermometer is then removed from A and placed in thermal contact with B, and its reading is recorded after thermal equilibrium is reached. If the two readings are the same, then A and B are in thermal equilibrium with each other.

We can summarize these results in a statement known as the **zeroth law of thermodynamics** (the law of equilibrium):

Molten lava flowing down a mountain in Kilauea, Hawaii. In this case, the hot lava flows smoothly out of a central crater until it cools and solidifies to form the mountains. However, violent eruptions sometimes occur, as in the case of Mount St. Helens in 1980, which can cause both local and global (atmospheric) damage. *(Ken Sakomoto, Black Star)*

Zeroth law of thermodynamics

> If objects A and B are separately in thermal equilibrium with a third object, C, then A and B are in thermal equilibrium with each other if placed in thermal contact.

This statement may easily be proven experimentally and is very important because it can be used to define temperature. We can think of temperature as the property that determines whether or not an object is in thermal equilibrium with other objects. *Two objects in thermal equilibrium with each other are at the same temperature.* Conversely, if two objects have different temperatures, they are *not* in thermal equilibrium with each other. The zeroth law can be shown to be a consequence of the second law of thermodynamics, which we meet later, in Chapter 22.

19.2 THERMOMETERS AND TEMPERATURE SCALES

Thermometers are devices used to define and measure the temperature of a system. All thermometers make use of the change in some physical property with temperature. Some of these physical properties are (1) the change in volume of a liquid, (2) the change in length of a solid, (3) the change in pressure of a gas at constant volume, (4) the change in volume of a gas at constant pressure, (5) the change in electric resistance of a conductor, and (6) the change in color of some object. For a given substance and a given temperature range, a temperature scale can be established based on any one of these physical quantities.

The most common thermometer in everyday use consists of a mass of liquid—usually mercury or alcohol—that expands into a glass capillary tube when heated (Fig. 19.1). In this case the physical property is the change in volume of a liquid. Any temperature change can be defined to be proportional to the change in length of the liquid column. The thermometer can be calibrated by placing it in thermal contact with some natural systems that remain at constant temperature. One such system is a mixture of water and ice in thermal equilibrium at atmospheric pressure, which is defined to have a temperature of zero degrees Celsius, written 0°C; this temperature is called the ice point of water. Another commonly used system is a mixture of water and steam in thermal equilibrium at atmospheric pressure; its temperature is 100°C, the steam point of water. Once the liquid levels in the thermometer have been established at these two points, the column is divided into 100 equal segments, each denoting a change in temperature of one Celsius degree.

Thermometers calibrated in this way do present problems when extremely accurate readings are needed. For instance, an alcohol thermometer calibrated at the ice and steam points of water might agree with a mercury thermometer only at the calibration points. Because mercury and alcohol have different thermal expansion properties, when one thermometer reads a temperature of 50°C, for example, the other may indicate a slightly different value. The discrepancies between thermometers are especially large when the temperatures to be measured are far from the calibration points.[1]

An additional practical problem of any thermometer is its limited temperature range. A mercury thermometer, for example, cannot be used below the freezing point of mercury, which is − 39°C, and an alcohol thermometer is not useful for temperatures above 85°C. To surmount these problems, we need a universal thermometer whose readings are independent of the substance used. The gas thermometer approaches this requirement.

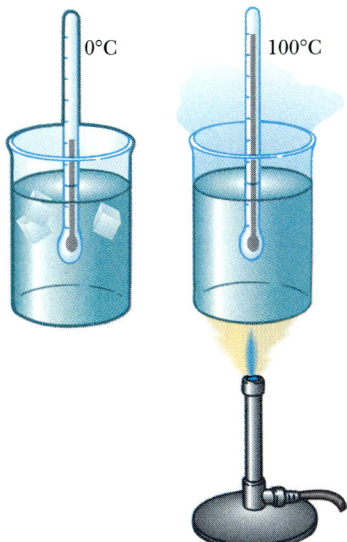

FIGURE 19.1 Schematic diagram of a mercury thermometer. As a result of thermal expansion, the level of the mercury rises as the mercury is heated from 0°C (the ice point) to 100°C (the steam point).

---

[1] Thermometers that use the same material may also give different readings. This is due in part to difficulties in constructing uniform-bore glass capillary tubes.

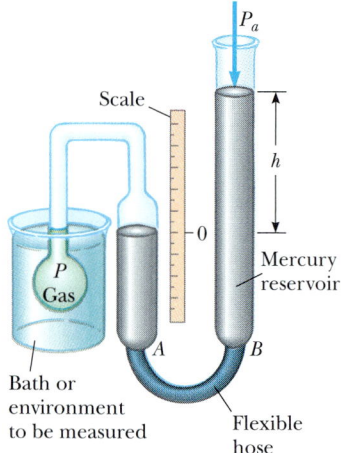

FIGURE 19.2 A constant-volume gas thermometer measures the pressure of the gas contained in the flask immersed in the bath. The volume of gas in the flask is kept constant by raising or lowering reservoir *B* such that the mercury level in column *A* remains constant.

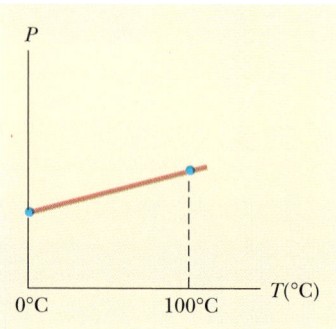

FIGURE 19.3 A typical graph of pressure versus temperature taken with a constant-volume gas thermometer. The dots represent known reference temperatures (the ice point and the steam point).

19.3 THE CONSTANT-VOLUME GAS THERMOMETER AND THE KELVIN SCALE

In a gas thermometer, the temperature readings are nearly independent of the substance used in the thermometer. One version is the constant-volume gas thermometer shown in Figure 19.2. The physical property being exploited in this device is how the pressure of a fixed volume of gas varies with temperature. When the constant-volume gas thermometer was developed, it was calibrated using the ice and steam points of water, as follows. (A different calibration procedure, to be discussed shortly, is now used.) The gas flask was inserted into an ice bath, and mercury reservoir *B* was raised or lowered until the volume of the confined gas was at some value, indicated by the zero point on the scale. The height *h*, the difference between the mercury levels in the reservoir and column *A*, indicated the pressure in the flask at 0°C. The flask was inserted into water at the steam point, and reservoir *B* was readjusted until the height in column *A* was again brought to zero on the scale, ensuring that the gas volume was the same as it had been in the ice bath (hence, the designation "constant volume"). This gave a value for pressure at 100°C. These pressure and temperature values were then plotted as in Figure 19.3. The line connecting the two points serves as a calibration curve for unknown temperatures. If we wanted to measure the temperature of a substance, we would place the gas flask in thermal contact with the substance and adjust the height of the mercury column until the gas occupied its specified volume. The height of the mercury column would tell us the pressure of the gas, and we could then find the temperature of the substance from the graph.

Now suppose that temperatures are measured with various gas thermometers containing different gases. Experiments show that the thermometer readings are nearly independent of the type of gas used, so long as the gas pressure is low and the temperature is well above the point at which the gas liquefies (Fig. 19.4). The agreement among thermometers using various gases improves as the pressure is reduced.

If you extend the curves in Figure 19.4 back toward negative temperatures, you find, in every case, that the pressure is zero when the temperature is $-273.15°C$. This significant temperature is used as the basis for the Kelvin temperature scale, which sets $-273.15°C$ as its zero point (0 K). The size of a degree on the Kelvin scale is identical to the size of a degree on the Celsius scale. Thus, the relationship that enables conversion between these temperatures is

$$T_C = T - 273.15 \qquad (19.1)$$

where T_C is the **Celsius temperature** and T is the **Kelvin temperature** (sometimes called the **absolute temperature**).

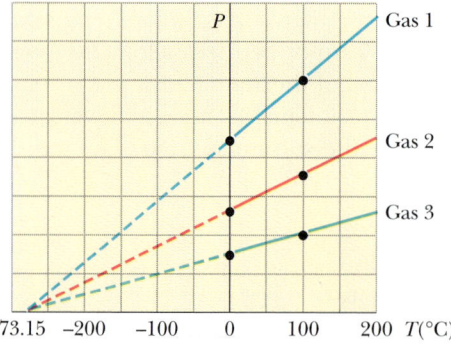

FIGURE 19.4 Pressure versus temperature for dilute gases. Note that, for all gases, the pressure extrapolates to zero at the unique temperature of $-273.15°C$.

TABLE 19.1 Fixed-Point Temperatures

Fixed Point	Temperature (°C)	Temperature (K)
Triple point of hydrogen	−259.34	13.81
Boiling point of hydrogen at 33.36 kPa pressure	−256.108	17.042
Boiling point of hydrogen	−252.87	20.28
Triple point of neon	−246.048	27.102
Triple point of oxygen	−218.789	54.361
Boiling point of oxygen	−182.962	90.188
Triple point of water	0.01	273.16
Boiling point of water	100.00	373.15
Freezing point of tin	231.9681	505.1181
Freezing point of zinc	419.58	692.73
Freezing point of silver	961.93	1235.08
Freezing point of gold	1064.43	1337.58

All values from National Bureau of Standards Special Publication 420; U.S. Department of Commerce, May 1975. All values at standard atmospheric pressure except as noted.

Early gas thermometers made use of the ice point and steam point according to the procedure just described. However, these points are experimentally difficult to duplicate. For this reason, a new temperature scale based on a single fixed point with b equal to zero was adopted in 1954 by the International Committee on Weights and Measures. The assigned temperatures of particular fixed points associated with various substances are given in Table 19.1. The *triple point of water,* which corresponds to the single temperature and pressure at which water, water vapor, and ice can coexist in equilibrium, was chosen as a convenient and reproducible reference temperature for this new scale. This triple point occurs at a temperature of approximately 0.01°C and a pressure of 4.58 mm of mercury. On the new scale, the temperature of water at the triple point was set at 273.16 kelvin, abbreviated 273.16 K.[2] This choice was made so that the old temperature scale based on the ice and steam points would agree closely with the new scale based on the triple point. This new scale is called the **thermodynamic temperature scale,** and the SI unit of thermodynamic temperature, the **kelvin,** *is defined as the fraction 1/273.16 of the temperature of the triple point of water.*

Figure 19.5 shows the Kelvin temperature for various physical processes and structures. The temperature 0 K is often referred to as **absolute zero,** and as Figure 19.5 shows, this temperature has never been achieved, although laboratory experiments have come close.

What would happen to a gas if its temperature could reach 0 K? As Figure 19.4 indicates, the pressure it exerted on the walls of its container would be zero. In Chapter 21 we show that the pressure of a gas is proportional to the kinetic energy of the molecules of that gas. Thus, according to classical physics, the kinetic energy of the gas would become zero, and molecular motion would cease; hence, the molecules would settle out on the bottom of the container. Quantum theory modifies this model and shows that there would be some residual energy, called the zero-point energy, at this low temperature.

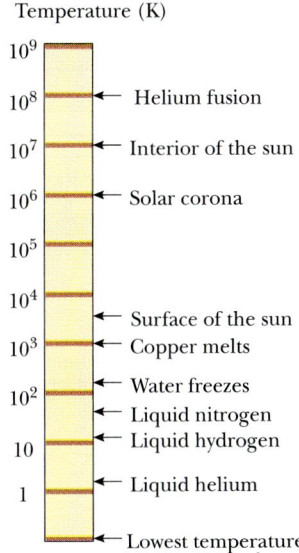

FIGURE 19.5 Absolute temperatures at which various selected physical processes occur. Note that the scale is logarithmic.

---

[2] We shall describe the meaning of this point in Chapter 22 when we discuss the second law of thermodynamics.

The Celsius, Fahrenheit, and Kelvin Temperature Scales[3]

Equation 19.1 shows that the Celsius temperature T_C is shifted from the absolute (Kelvin) temperature T by 273.15°. Because the size of a degree is the same on the two scales, a temperature difference of 5°C is equal to a temperature difference of 5 K. The two scales differ only in the choice of the zero point. Thus, the ice point (273.15 K) corresponds to 0.00°C, and the steam point (373.15 K) is equivalent to 100.00°C.

The most common temperature scale in everyday use in the United States is the **Fahrenheit scale.** This scale sets the temperature of the ice point at 32°F and the temperature of the steam point at 212°F. The relationship between the Celsius and Fahrenheit temperature scales is

$$T_F = \tfrac{9}{5}T_C + 32°\text{F} \tag{19.2}$$

Equation 19.2 can be used to find a relationship between changes in temperature on the Celsius and Fahrenheit scales.

$$\Delta T_C = \Delta T = \tfrac{5}{9}\Delta T_F \tag{19.3}$$

The change in temperature on the Celsius scale equals the change on the Kelvin scale. That is, $\Delta T = \Delta T_C$.

EXAMPLE 19.1   Converting Temperatures

On a day when the temperature reaches 50°F, what is the temperature in degrees Celius and in kelvin?

Solution   Substituting $T_F = 50°\text{F}$ into Equation 19.2, we get

$$T_C = \tfrac{5}{9}(T_F - 32) = \tfrac{5}{9}(50 - 32) = \boxed{10°\text{C}}$$

From Equation 19.1, we find that

$$T = T_C + 273.15 = \boxed{283 \text{ K}}$$

EXAMPLE 19.2   Heating a Pan of Water

A pan of water is heated from 25°C to 80°C. What is the change in its temperature on the Kelvin scale and on the Fahrenheit scale?

Solution   From Equation 19.1, we see that the change in temperature on the Celsius scale equals the change on the Kelvin scale. Therefore,

$$\Delta T = \Delta T_C = 80 - 25 = 55°\text{C} = \boxed{55 \text{ K}}$$

From Equation 19.3, we find

$$\Delta T_F = \tfrac{9}{5}\Delta T_C = \tfrac{9}{5}(80 - 25) = \boxed{99°\text{F}}$$

19.4   THERMAL EXPANSION OF SOLIDS AND LIQUIDS

Our discussion of the liquid thermometer made use of one of the best-known changes that occurs in a substance: As its temperature increases, its volume increases. (As we shall see shortly, in some substances the volume decreases when the temperature increases.) This phenomenon, known as **thermal expansion,** plays an important role in numerous engineering applications. For example, thermal

---

[3] Named after Anders Celsius (1701–1744), Gabriel Fahrenheit (1686–1736), and William Thomson, Lord Kelvin (1824–1907).

Thermal expansion joints are used to separate sections of roadways on bridges. Without these joints, the surfaces would buckle due to thermal expansion on very hot days or crack due to contraction on very cold days. *(Frank Siteman, Stock/Boston)*

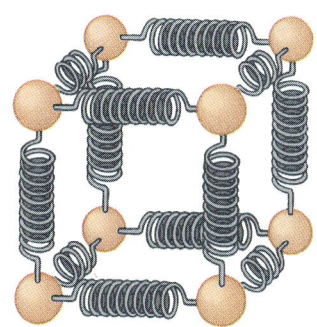

FIGURE 19.6 A mechanical model of a crystalline solid. The atoms (solid spheres) are imagined to be attached to each other by springs, which reflect the elastic nature of the interatomic forces.

expansion joints must be included in bridges and some other structures to compensate for changes in dimensions with temperature variations.

The overall thermal expansion of a body is a consequence of the change in the average separation between its constituent atoms or molecules. To understand this, consider how the atoms in a solid substance behave. Imagine that the atoms of the solid are connected by a set of stiff springs as in Figure 19.6. At ordinary temperatures, the atoms vibrate about their equilibrium positions with an amplitude of approximately 10^{-11} m and a frequency of approximately 10^{13} Hz. The average spacing between the atoms is approximately 10^{-10} m. As the temperature of the solid increases, the atoms vibrate with larger amplitudes and the average separation between them increases.[4] Consequently, the solid expands. If the thermal expansion of an object is sufficiently small compared with its initial dimensions, then the change in any dimension is, to a good approximation, dependent on the first power of the temperature change.

Suppose an object has an initial length L_0 along some direction at some temperature and that the length increases by an amount ΔL for the change in temperature ΔT. Experiments show that, when ΔT is small, ΔL is proportional to ΔT and to L_0:

$$\Delta L = \alpha L_0 \, \Delta T \tag{19.4}$$

or

$$L - L_0 = \alpha L_0 (T - T_0) \tag{19.5}$$

where L is the final length, T is the final temperature, and the proportionality constant α is called the **average coefficient of linear expansion** for a given material and has units of $(°C)^{-1}$.

It may be helpful to think of thermal expansion as an effective magnification or as a photographic enlargement of an object when it is heated. For example, as a metal washer is heated (Fig. 19.7) all dimensions, including the radius of the hole, increase according to Equation 19.4. Table 19.2 lists the average coefficient of linear expansion for various materials. Note that for these materials α is positive, indicating an increase in length with increasing temperature. This is not always the case. For example, some substances, such as calcite ($CaCO_3$), expand along one

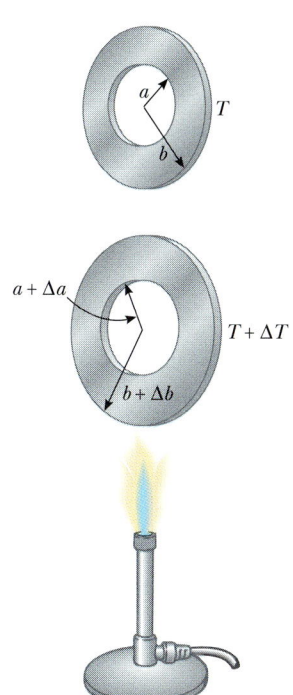

FIGURE 19.7 Thermal expansion of a homogeneous metal washer. Note that as the washer is heated, all dimensions increase. (The expansion is exaggerated in this figure.)

[4] Strictly speaking, thermal expansion arises from the *asymmetric* nature of the potential energy curve for the atoms in a solid. If the oscillators were truly harmonic, the average atomic separations would not change regardless of the amplitude of vibration.

TABLE 19.2   Expansion Coefficients for Some Materials Near Room Temperature

Material	Linear Expansion Coefficient $\alpha(°C)^{-1}$	Material	Volume Expansion Coefficient $\beta(°C)^{-1}$
Aluminum	24×10^{-6}	Alcohol, ethyl	1.12×10^{-4}
Brass and bronze	19×10^{-6}	Benzene	1.24×10^{-4}
Copper	17×10^{-6}	Acetone	1.5×10^{-4}
Glass (ordinary)	9×10^{-6}	Glycerine	4.85×10^{-4}
Glass (Pyrex)	3.2×10^{-6}	Mercury	1.82×10^{-4}
Lead	29×10^{-6}	Turpentine	9.0×10^{-4}
Steel	11×10^{-6}	Gasoline	9.6×10^{-4}
Invar (Ni–Fe alloy)	0.9×10^{-6}	Air at 0°C	3.67×10^{-3}
Concrete	12×10^{-6}	Helium at 0°C	3.665×10^{-3}

dimension (positive α) and contract along another (negative α) as their temperature is increased.

Because the linear dimensions of an object change with temperature, it follows that surface area and volume do so also. The change in volume at constant pressure is proportional to the original volume V and to the change in temperature according to the relationship

The change in volume of a solid at constant pressure is proportional to the change in temperature

$$\Delta V = \beta V \Delta T \qquad (19.6)$$

where β is the **average coefficient of volume expansion.** *For a solid, the coefficient of volume expansion is approximately three times the linear expansion coefficient, or $\beta = 3\alpha$.* (This assumes that the coefficient of linear expansion of the solid is the same in all directions.) Therefore, Equation 19.6 can be written

$$\Delta V = 3\alpha V \Delta T \qquad (19.7)$$

To show that $\beta = 3\alpha$ for a solid, consider an object in the shape of a box of dimensions ℓ, w, and h. Its volume at some temperature T is $V = \ell wh$. If the temperature changes to $T + \Delta T$, its volume changes to $V + \Delta V$, where each dimension changes according to Equation 19.6. Therefore,

$$\begin{aligned} V + \Delta V &= (\ell + \Delta \ell)(w + \Delta w)(h + \Delta h) \\ &= (\ell + \alpha \ell \Delta T)(w + \alpha w \Delta T)(h + \alpha h \Delta T) \\ &= \ell wh(1 + \alpha \Delta T)^3 \\ &= V[1 + 3\alpha \Delta T + 3(\alpha \Delta T)^2 + (\alpha \Delta T)^3] \end{aligned}$$

Hence, the fractional change in volume is

$$\frac{\Delta V}{V} = 3\alpha \Delta T + 3(\alpha \Delta T)^2 + (\alpha \Delta T)^3$$

Since the product $\alpha \Delta T$ is small compared with unity for typical values of ΔT (less than $\approx 100°C$), we can neglect the terms $3(\alpha \Delta T)^2$ and $(\alpha \Delta T)^3$. In this approximation, we see that

$$\beta = \frac{1}{V} \frac{\Delta V}{\Delta T} = 3\alpha$$

A sheet or flat plate can be described by its area. You should show (Problem 53) that the change in the area of a plate is

$$\Delta A = 2\alpha A \, \Delta T \tag{19.8}$$

As Table 19.2 indicates, each substance has its own characteristic coefficients of expansion. For example, when the temperature of a brass rod and a steel rod of equal length are raised by the same amount from some common initial value, the brass rod expands more than the steel rod because brass has a larger coefficient of expansion than steel. A simple device called a bimetallic strip that utilizes this principle is found in practical devices such as thermostats. As the temperature of the strip increases, the two metals expand by different amounts, and the strip bends as in Figure 19.8.

> The change in area of a plate is proportional to the change in temperature

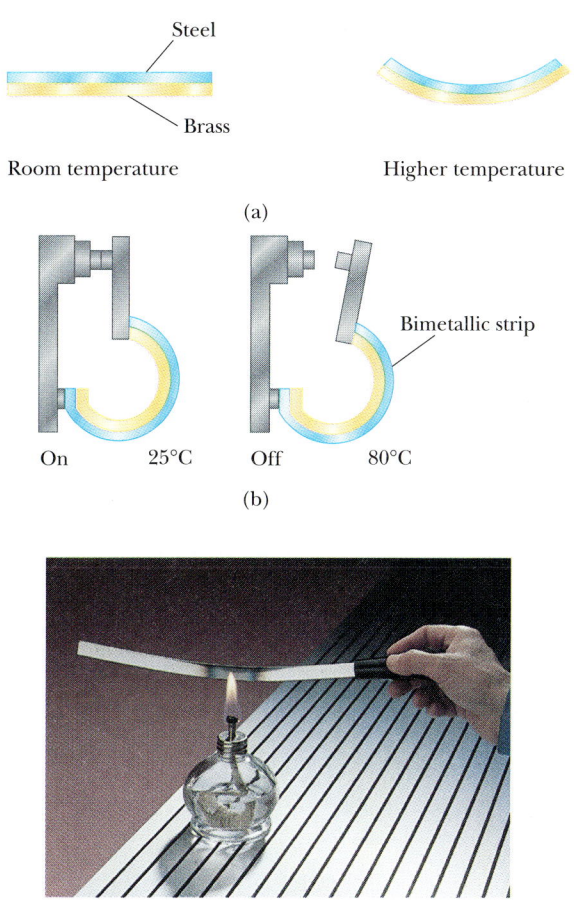

(a)

On   25°C   Off   80°C

(b)

(c)

FIGURE 19.8   (a) A bimetallic strip bends as the temperature changes because the two metals have different expansion coefficients. (b) A bimetallic strip used in a thermostat to break or make electrical contact. (c) The two metals that form this bimetallic strip are bonded along their longest dimension. The strip in this photograph was straight before being heated and bends when heated. Which way would it bend if it were cooled? *(Courtesy of Central Scientific Company)*

EXAMPLE 19.3   Expansion of a Railroad Track

A steel railroad track has a length of 30.0 m when the temperature is 0.0°C. (a) What is its length on a hot day when the temperature is 40.0°C?

Solution   Making use of Table 19.2 and noting that the change in temperature is 40.0°C, we find that the increase in length is

$$\Delta L = \alpha L \, \Delta T = [11 \times 10^{-6} (°C)^{-1}](30.0 \text{ m})(40.0°C)$$
$$= 0.013 \text{ m}$$

Therefore, the track's length at 40.0°C is 30.013 m.

(b) Suppose the ends of the rail are rigidly clamped at 0.0°C so as to prevent expansion. Calculate the thermal stress set up in the rail if its temperature is raised to 40.0°C.

Solution   From the definition of Young's modulus for a solid (Chapter 12), we have

$$\text{Tensile stress} = \frac{F}{A} = Y \frac{\Delta \ell}{\ell}$$

Since Y for steel is 20×10^{10} N/m² (Table 12.1), we have

$$\frac{F}{A} = \left(20 \times 10^{10} \, \frac{\text{N}}{\text{m}^2}\right)\left(\frac{0.013 \text{ m}}{30.0 \text{ m}}\right) = 8.7 \times 10^7 \text{ N/m}^2$$

Thermal expansion: The extreme heat of a July day in Asbury Park, New Jersey, caused these railroad tracks to buckle. *(Wide World Photos)*

Exercise   If the rail has a cross-sectional area of 30.0 cm², calculate the force of compression in the rail.

Answer 2.6×10^5 N or 58 000 lb!

The Unusual Behavior of Water

Liquids generally increase in volume with increasing temperature and have average volume-expansion coefficients about ten times greater than those of solids. Water is an exception to this rule, as we can see from its density-versus-temperature curve in Figure 19.9. As the temperature increases from 0°C to 4°C, water con-

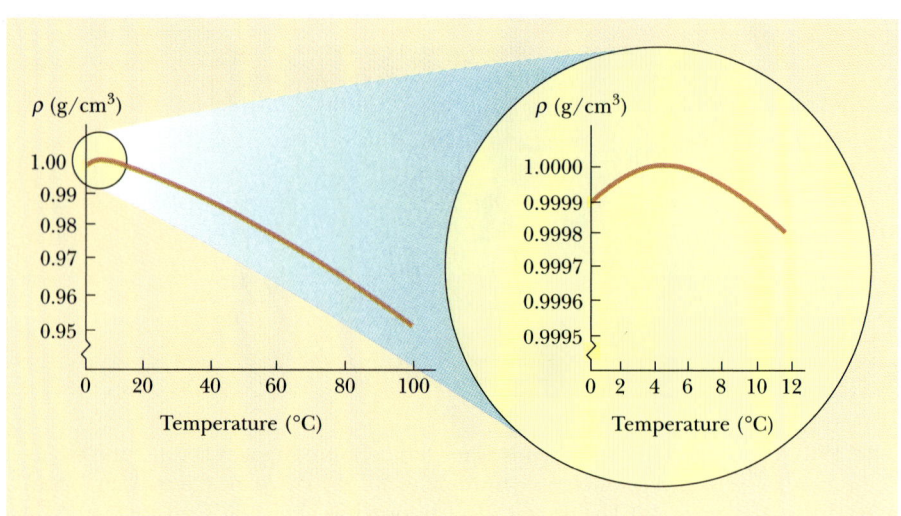

FIGURE 19.9   The variation of density with temperature for water at atmospheric pressure. The inset at the right shows that the maximum density of water occurs at 4°C.

tracts and thus its density increases. Above 4°C, water expands with increasing temperature. In other words, the density of water reaches a maximum value of 1000 kg/m³ at 4°C.

We can use this unusual thermal expansion behavior of water to explain why a pond freezes starting at the surface. When the atmospheric temperature drops from, say, 7°C to 6°C, the water at the surface of the pond also cools and consequently decreases in volume. This means that the surface water is denser than the water below it, which has not cooled and decreased in volume. As a result, the surface water sinks and warmer water from below is forced to the surface to be cooled. When the atmospheric temperature is between 4°C and 0°C, however, the surface water expands as it cools, becoming less dense than the water below it. The mixing process stops, and eventually the surface water freezes. As the water freezes, the ice remains on the surface because ice is less dense than water. The ice continues to build up at the surface, while water near the bottom remains at 4°C. If this did not happen, fish and other forms of marine life would not survive.

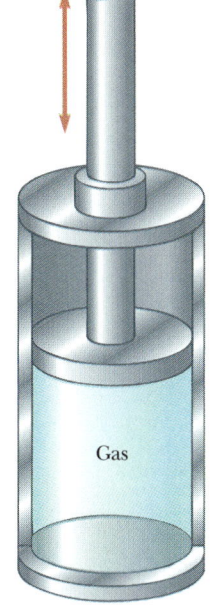

FIGURE 19.10   An ideal gas confined to a cylinder whose volume can be varied with a movable piston. The state of the gas is defined by any two properties, such as pressure, volume, and temperature.

19.5   MACROSCOPIC DESCRIPTION OF AN IDEAL GAS

In this section we are concerned with the properties of a gas of mass m confined to a container of volume V at a pressure P and temperature T. It is useful to know how these quantities are related. In general, the equation that interrelates these quantities, called the *equation of state,* is very complicated. However, if the gas is maintained at a very low pressure (or low density), the equation of state is found experimentally to be quite simple. Such a low-density gas is commonly referred to as an **ideal gas.** Most gases at room temperature and atmospheric pressure behave approximately as ideal gases.[5]

It is convenient to express the amount of gas in a given volume in terms of the number of moles, n. As we learned in Section 1.3, **one mole** of any substance is that mass of the substance that contains Avogadro's number, $N_A = 6.022 \times 10^{23}$, of molecules. The number of moles of a substance is related to its mass m through the expression

$$n = \frac{m}{M} \tag{19.9}$$

where M is a quantity called the **molar mass** of the substance, usually expressed in grams per mole. For example, the molar mass of oxygen, O_2, is 32.0 g/mol. Therefore, the mass of one mole of oxygen is 32.0 g.

Now suppose an ideal gas is confined to a cylindrical container whose volume can be varied by means of a movable piston, as in Figure 19.10. We assume that the cylinder does not leak, and so the mass (or the number of moles) of the gas remains constant. For such a system, experiments provide the following information. First, when the gas is kept at a constant temperature, its pressure is inversely proportional to the volume (Boyle's law). Second, when the pressure of the gas is kept constant, the volume is directly proportional to the temperature (the law of Charles and Gay-Lussac). These observations can be summarized by the **equation of state for an ideal gas:**

$$PV = nRT \tag{19.10}$$

Equation of state for an ideal gas

---

[5] To be more specific, the assumption here is that the temperature of the gas must not be too low (it must not condense into a liquid) nor too high, and the pressure must be low.

In this expression, called the **ideal gas law,** R is a universal constant that is the same for all gases, and T is the absolute temperature in kelvin. Experiments on several gases show that as the pressure approaches zero, the quantity PV/nT approaches the same value R for all gases. For this reason, R is called the **universal gas constant.** In SI units, where pressure is expressed in pascals and volume in cubic meters, the product PV has units of newton·meters, or joules, and R has the value

The universal gas constant

$$R = 8.31 \text{ J/mol·K} \qquad (19.11)$$

If the pressure is expressed in atmospheres and the volume in liters (1 L = 10^3 cm^3 = 10^{-3} m^3), then R has the value

$$R = 0.0821 \text{ L·atm/mol·K}$$

Using this value of R and Equation 19.10, we find that the volume occupied by 1 mole of any gas at atmospheric pressure and 0°C (273 K) is 22.4 L.

The ideal gas law is often expressed in terms of the total number of molecules, N. Since the total number of molecules equals the product of the number of moles and Avogadro's number, we can write Equation 19.10 as

$$PV = nRT = \frac{N}{N_\text{A}} RT$$

$$PV = Nk_\text{B} T \qquad (19.12)$$

where k_B is called **Boltzmann's constant,** which has the value

Boltzmann's constant

$$k_\text{B} = \frac{R}{N_\text{A}} = 1.38 \times 10^{-23} \text{ J/K} \qquad (19.13)$$

We have defined an ideal gas as one that obeys the equation of state, $PV = nRT$, under all conditions. In reality, an ideal gas does not exist. However, the concept of an ideal gas is very useful in view of the fact that real gases at low pressures behave as ideal gases. It is common to call quantities such as P, V, and T the **thermodynamic variables** of the system. If the equation of state is known, then one of the variables can always be expressed as some function of the other two. That is, given two of the variables, the third can be determined from the equation of state.

Other thermodynamic systems are often described with different thermodynamic variables. For example, for a wire under tension at constant pressure, the thermodynamic variables of the system are the length of the wire, the tension in it, and its temperature.

EXAMPLE 19.4   How Many Gas Molecules Are in a Container?

An ideal gas occupies a volume of 100 cm^3 at 20°C and a pressure of 100 Pa. Determine the number of moles of gas in the container.

Solution   The quantities given are volume, pressure, and temperature: $V = 100 \text{ cm}^3 = 1.00 \times 10^{-4}$ m^3, $P = 100$ Pa, and $T = 20°C = 293$ K. Using Equation 19.12, we get

$$n = \frac{PV}{RT} = \frac{(100 \text{ Pa})(1.00 \times 10^{-4} \text{ m}^3)}{(8.31 \text{ J/mol·K})(293 \text{ K})} = 4.11 \times 10^{-6} \text{ mol}$$

Note that you must express T as an absolute temperature (K) when using the ideal gas law.

Exercise   Calculate the number of molecules in the container, using the fact that Avogadro's number is 6.02×10^{23} molecules/mol.

Answer 2.47×10^{18} molecules.

EXAMPLE 19.5 Squeezing a Tank of Gas

Pure helium gas is admitted into a tank containing a movable piston. The initial volume, pressure, and temperature of the gas are 15×10^{-3} m³, 200 kPa, and 300 K. If the volume is decreased to 12×10^{-3} m³ and the pressure increased to 350 kPa, find the final temperature of the gas. (Assume that helium behaves like an ideal gas.)

Solution If no gas escapes from the tank, the number of moles remains constant; therefore, using $PV = nRT$ at the initial and final points gives

$$\frac{P_i V_i}{T_i} = \frac{P_f V_f}{T_f}$$

where i and f refer to the initial and final values. Solving for T_f, we get

$$T_f = \left(\frac{P_f V_f}{P_i V_i}\right) T_i = \frac{(350 \text{ kPa})(12 \times 10^{-3} \text{ m}^3)}{(200 \text{ kPa})(15 \times 10^{-3} \text{ m}^3)} (300 \text{ K})$$

$$= \boxed{420 \text{ K}}$$

EXAMPLE 19.6 Heating a Bottle of Air

A sealed glass bottle containing air at atmospheric pressure (101 kPa) and having a volume of 30 cm³ is at 27°C. It is then tossed into an open fire. When the temperature of the air in the bottle reaches 200°C, what is the pressure inside the bottle? Assume any volume changes of the bottle are negligible.

Solution This example is approached in the same fashion as that used in Example 19.5. We start with the expression

$$\frac{P_i V_i}{T_i} = \frac{P_f V_f}{T_f}$$

Since the initial and final volumes of the gas are assumed equal, this expression reduces to

$$\frac{P_i}{T_i} = \frac{P_f}{T_f}$$

This gives

$$P_f = \left(\frac{T_f}{T_i}\right) P_i = \left(\frac{473 \text{ K}}{300 \text{ K}}\right)(101 \text{ kPa}) = \boxed{159 \text{ kPa}}$$

Obviously, the higher the temperature, the higher the pressure exerted by the trapped air. Of course, if the pressure rises high enough, the bottle will shatter.

Exercise In this example, we neglected the change in volume of the bottle. If the coefficient of volume expansion for glass is 27×10^{-6} (°C)$^{-1}$, find the magnitude of this volume change.

Answer 0.14 cm³.

SUMMARY

Two bodies are in **thermal equilibrium** with each other if they have the same temperature.

The **zeroth law of thermodynamics** states that if bodies A and B are separately in thermal equilibrium with a third body, C, then A and B are in thermal equilibrium with each other.

The SI unit of thermodynamic temperature is the **kelvin,** which is defined to be the fraction 1/273.16 of the temperature of the triple point of water.

When the temperature of an object is changed by an amount ΔT, its length changes by an amount ΔL that is proportional to ΔT and to its initial length L_0:

$$\Delta L = \alpha L_0 \, \Delta T \tag{19.4}$$

where the constant α is the **average coefficient of linear expansion.** The **average volume expansion coefficient,** β, for a substance is equal to 3α.

An **ideal gas** is one that obeys the *equation of state,*

$$PV = nRT \qquad\qquad (19.10)$$

where n equals the number of moles of gas, V is its volume, R is the universal gas constant (8.31 J/mol·K), and T is the absolute temperature in kelvin. A real gas behaves approximately as an ideal gas if it is far from liquefaction. An ideal gas is used as the working substance in a constant-volume gas thermometer, which defines the absolute temperature scale in kelvin. This absolute temperature T is related to temperatures on the Celsius scale by $T = T_C + 273.15$.

QUESTIONS

1. Is it possible for two objects to be in thermal equilibrium if they are not in contact with each other? Explain.
2. A piece of copper is dropped into a beaker of water. If the water's temperature rises, what happens to the temperature of the copper? Under what conditions are the water and copper in thermal equilibrium?
3. In principle, any gas can be used in a constant-volume gas thermometer. Why is it not possible to use oxygen for temperatures as low as 15 K? What gas would you use? (Look at the data in Table 19.1.)
4. Rubber has a negative average coefficient of linear expansion. What happens to the size of a piece of rubber as it is warmed?
5. Why should the amalgam used in dental fillings have the same average coefficient of expansion as a tooth? What would occur if they were mismatched?
6. Explain why a column of mercury in a thermometer first descends slightly and then rises when placed in hot water.
7. Explain why the thermal expansion of a spherical shell made of a homogeneous solid is equivalent to that of a solid sphere of the same material.
8. A steel ring bearing has an inside diameter that is 1 mm smaller than an axle. How can it be made to fit onto the axle without removing any material?
9. Markings to indicate length are placed on a steel tape in a room that has a temperature of 22°C. Are measurements made with the tape on a day when the temperature is 27°C too long, too short, or accurate? Defend your answer.
10. What would happen if the glass of a thermometer expanded more upon heating than did the liquid inside?
11. Determine the number of grams in one mole of the following gases: (a) hydrogen, (b) helium, and (c) carbon monoxide.
12. Why is it necessary to use absolute temperature when doing calculations with the ideal gas law?
13. An inflated rubber balloon filled with air is immersed in a flask of liquid nitrogen that is at 77 K. Describe what happens to the balloon, assuming that it remains flexible while being cooled.
14. Two cylinders at the same temperature each contain the same kind and quantity of gas. If the volume of cylinder A is three times greater than the volume of cylinder B, what can you say about the relative pressures in the cylinders?
15. The suspension of a certain pendulum clock is made of brass. When the temperature increases, does the period of the clock increase, decrease, or remain the same? Explain.
16. An automobile radiator is filled to the brim with water while the engine is cool. What happens to the water when the engine is running and the water is heated? What do modern automobiles have in their cooling systems to prevent the loss of coolants?
17. Metal lids on glass jars can often be loosened by running hot water over them. How is this possible?
18. When the metal ring and metal sphere in Figure 19.11 are both at room temperature, the sphere can just be passed through the ring. After the sphere is heated, it cannot be passed through the ring. Explain.

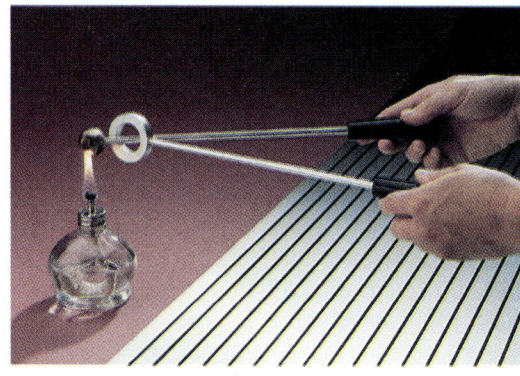

FIGURE 19.11   (Question 18).   *(Courtesy Central Scientific Co.)*

PROBLEMS

Section 19.3   The Constant-Volume Gas Thermometer and the Kelvin Scale

(*Note:* A pressure of 1.00 atm = 1.01×10^5 Pa = 101 kPa.)

1. A constant-volume gas thermometer is calibrated in dry ice (which is carbon dioxide in the solid state and has a temperature of $-80.0°C$) and in boiling ethyl alcohol ($78.0°C$). The two pressures are 0.900 atm and 1.635 atm. (a) What value of absolute zero does the calibration yield? What is the pressure at (b) the freezing point of water and (c) the boiling point of water?

2. Suppose the temperature (in units of kelvin) and pressure in an ideal gas thermometer are related by a *quadratic* equation, $T = aP^2 + bP$. If the temperature and pressure at the triple point of water are T_3 and P_3, respectively, and if the temperature and pressure at the boiling point of water are T_B and P_B, respectively, determine a and b in terms of T_3, P_3, T_B, and P_B.

3. A constant-volume gas thermometer registers a pressure of 0.062 atm when it is at a temperature of 450 K. (a) What is the pressure at the triple point of water? (b) What is the temperature when the pressure reads 0.015 atm?

4. In a constant-volume gas thermometer, the pressure at $20°C$ is 0.980 atm. (a) What is the pressure at $45°C$? (b) What is the temperature if the pressure is 0.500 atm?

5. A constant-volume gas thermometer is filled with helium. When immersed in boiling liquid nitrogen (77.34 K), the absolute pressure is 25.00 kPa. (a) What is the temperature in degrees Celsius and kelvin when the pressure is 45.00 kPa? (b) What is the pressure when the thermometer is immersed in boiling liquid hydrogen?

6. The melting point of gold is $1064°C$, and the boiling point is $2660°C$. (a) Express these temperatures in kelvin. (b) Compute the difference between these temperatures in Celsius degrees and kelvin.

7. Liquid nitrogen has a boiling point of $-195.81°C$ at atmospheric pressure. Express this temperature in (a) degrees Fahrenheit and (b) kelvin.

8. The highest recorded temperature on Earth is $136°F$, at Azizia, Libya, in 1922. The lowest recorded temperature is $-127°F$, at Vostok Station, Antarctica, in 1960. Express these temperature extremes in degrees Celsius.

9. Oxygen condenses to a liquid at approximately 90 K. What temperature, in degrees Fahrenheit, does this correspond to?

10. On a Strange temperature scale, the freezing point of water is $-15°S$ and the boiling point is $+60°S$. Develop a *linear* conversion equation between this temperature scale and the Celsius scale.

11. The temperature difference between the inside and the outside of an automobile engine is $450°C$. Express this temperature difference on the (a) Fahrenheit scale and (b) Kelvin scale.

12. The normal human body temperature is $98.6°F$. A person with a fever may record $102°F$. Express these temperatures in degrees Celsius.

13. A substance is heated from $-12°F$ to $150°F$. What is its change in temperature on (a) the Celsius scale and (b) the Kelvin scale?

14. Initially, an object has a temperature, which has the same numerical value in degrees Celsius and degrees Fahrenheit. Then, its temperature is changed so that the numerical value of the new temperature in degrees Celsius is one-third as large as small as that in kelvin. Find the change in the temperature in kelvin.

15. At what temperature are the readings from a Fahrenheit thermometer and Celsius thermometer the same?

Section 19.4   Thermal Expansion of Solids and Liquids

(*Note:* Use Table 19.2.)

16. An aluminum tube is 3.0000 m long at $20.0°C$. What is its length at (a) $100.0°C$ and (b) $0.0°C$?

17. A copper telephone wire has essentially no sag between two poles 35.0 m apart on a winter's day when the temperature is $-20°C$. How much longer is the wire on a summer's day when $T = 35°C$?

18. A concrete walk is poured on a day when the temperature is $20°C$ in such a way that the ends are unable to move. (a) What is the stress in the cement on a hot day of $50°C$? (b) Does the concrete fracture? Take Young's modulus for concrete to be 7.0×10^6 N/m^2 and the tensile strength to be 2.0×10^6 N/m^2.

19. A structural steel I-beam is 15.0 m long when installed at $20.0°C$. How much does its length change over the temperature extremes $-30.0°C$ to $50.0°C$?

20. The New River Gorge Bridge in West Virginia is a steel arch bridge 518 m in length. How much does its length change between temperature extremes of $-20.0°C$ and $35.0°C$?

21. The average volume coefficient of expansion for carbon tetrachloride is 5.81×10^{-4} (°C)$^{-1}$. If a 50.0-gal steel container is filled completely with carbon tetrachloride when the temperature is $10.0°C$, how much will spill over when the temperature rises to $30.0°C$?

☐ indicates problems that have full solutions available in the Student Solutions Manual and Study Guide.

22. A steel rod undergoes a stretching force of 500 N. Its cross-sectional area is 2.00 cm². Find the change in temperature that would elongate the rod by the same amount produced by the 500-N force. (*Hint:* Refer to Tables 12.1 and 19.2.)

23. A steel rod 4.0 cm in diameter is heated so that its temperature increases by 70°C and then is fastened between two rigid supports. The rod is allowed to cool to its original temperature. Assuming that Young's modulus for the steel is 20.6×10^{10} N/m² and that its average coefficient of linear expansion is 11×10^{-6} (°C)$^{-1}$, calculate the tension in the rod.

24. A brass ring of diameter 10.00 cm at 20.0°C is heated and slipped over an aluminum rod of diameter 10.01 cm at 20.0°C. Assuming the average coefficients of linear expansion are constant, (a) to what temperature must this combination be cooled to separate them? Is this attainable? (b) What if the aluminum rod were 10.02 cm in diameter?

25. The concrete sections of a certain superhighway are designed to have a length of 25.0 m. The sections are poured and cured at 10°C. What minimum spacing should the engineer leave between the sections to eliminate buckling if the concrete is to reach a temperature of 50°C?

26. A square hole 8.0 cm along each side is cut in a sheet of copper. Calculate the change in the area of this hole if the temperature of the sheet is increased by 50 K.

27. At 20.0°C, an aluminum ring has an inner diameter of 5.000 cm and a brass rod has a diameter of 5.050 cm. (a) To what temperature must the ring be heated so that it will just slip over the rod? (b) To what temperature must both be heated so that the ring just slips over the rod? Would the latter process work?

28. A pair of eyeglass frames is made of epoxy plastic. At room temperature (assume 20.0°C), the frames have circular lens holes 2.2 cm in radius. To what temperature must the frames be heated in order to insert lenses 2.21 cm in radius? The average coefficient of linear expansion for epoxy is 1.3×10^{-4} (°C)$^{-1}$.

29. A hollow aluminum cylinder 20.0 cm deep has an internal capacity of 2.000 L at 20.0°C. It is completely filled with turpentine, and then warmed to 80.0°C. (a) How much turpentine overflows? (b) If it is then cooled back to 20.0°C, how far below the surface of the cylinder's rim is the turpentine surface?

30. A copper rod and steel rod are heated. At 0°C the copper rod has a length of L_C, the steel one has a length L_S. When the rods are being heated or cooled, a difference of 5.0 cm is maintained between their lengths. Determine the values of L_C, and L_S.

30A. A copper rod and steel rod are heated. At T(°C) the copper rod has a length of L_C, the steel one has a length L_S. When the rods are being heated or cooled, a difference of ΔL is maintained between their lengths. Determine the values of L_C and L_S.

31. An automobile fuel tank is filled to the brim with 45 L of gasoline at 10°C. Immediately afterward, the vehicle is parked in the Sun where the temperature is 35°C. How much gasoline overflows from the tank as a result of expansion? (Neglect the expansion of the tank.)

32. A volumetric glass flask made of Pyrex is calibrated at 20.0°C. It is filled to the 100-mL mark with 35.0°C acetone. (a) What is the volume of the acetone when it cools to 20.0°C? (b) How significant is the change in volume of the flask?

33. The active element of a certain laser is made of a glass rod 30.0 cm long by 1.5 cm in diameter. If the temperature of the rod increases by 65°C, find the increase in (a) its length, (b) its diameter, and (c) its volume. (Take $\alpha = 9.0 \times 10^{-6}$ (°C)$^{-1}$.)

Section 19.5 Macroscopic Description of an Ideal Gas

34. An ideal gas is held in a container at constant volume. Initially, its temperature is 10.0°C and its pressure is 2.50 atm. What is its pressure when its temperature is 80.0°C?

35. A helium-filled balloon has a volume of 1.00 m³. As it rises in the Earth's atmosphere, its volume expands. What is its new volume (in cubic meters) if its original temperature and pressure are 20.0°C and 1.00 atm and its final temperature and pressure are −40.0°C and 0.10 atm?

36. An auditorium has dimensions 10.0 m × 20.0 m × 30.0 m. How many molecules of air are needed to fill the auditorium at 20.0°C and 101 kPa pressure?

37. A full tank of oxygen (O_2) contains 12.0 kg of oxygen under a gauge pressure of 40.0 atm. Determine the mass of oxygen that has been withdrawn from the tank when the pressure reading is 25.0 atm. Assume the temperature of the tank remains constant.

38. A car tire gauge is used to fill a tire to a gauge pressure of 32 lb/in.² on a cold morning when the temperature is −10°C. What would the tire gauge read when the tire has been heated up to 35°C?

39. The mass of a hot-air balloon and its cargo (not including the air inside) is 200 kg. The air outside is at 10.0°C and 101 kPa. The volume of the balloon is 400 m³. To what temperature must the air in the balloon be heated before the balloon will lift off? (Air density at 10.0°C is 1.25 kg/m³.)

40. A tank having a volume of 0.10 m³ contains helium gas at 150 atm. How many balloons can the tank blow up if each filled balloon is a sphere 0.30 m in diameter at an absolute pressure of 1.2 atm?

41. A room of volume 80.0 m³ contains air having an average molar mass of 29.0 g/mol. If the temperature of the room is raised from 18.0°C to 25.0°C, what mass of air (in kg) will leave the room? Assume that the air pressure in the room is maintained at 101 kPa.

41A. A room of volume V contains air having an average molar mass of M. If the temperature of the room is raised from T_1 to T_2, what mass of air will leave the room? Assume that the air pressure in the room is maintained at P_0.

42. At 25.0 m below the surface of the sea (density = 1025 kg/m³), where the temperature is 5.00°C, a diver exhales an air bubble having a volume of 1.00 cm³. If the surface temperature of the sea is 20.0°C, what is the volume of the bubble right before it breaks the surface?

42A. At a depth h below the surface of the sea (density = ρ), where the temperature is T_C, a diver exhales an air bubble having a volume V_0. If the surface temperature of the sea is T_h, what is the volume of the bubble right before it breaks the surface?

43. If 9.0 g of water is placed into a 2.0-L pressure cooker and heated to 500°C, what is the pressure inside the container?

44. In state-of-the-art vacuum systems, pressures as low as 1.00×10^{-9} Pa are being attained. Calculate the number of molecules in a 1.00-m³ vessel at this pressure if the temperature is 27°C.

45. The tire on a bicycle is filled with air to a gauge pressure of 550 kPa at 20°C. What is the gauge pressure in the tire after a ride on a hot day when the tire air temperature is 40°C? (Assume constant volume and a constant atmospheric pressure of 101 kPa.)

46. Show that 1.00 mol of any gas at atmospheric pressure (101 kPa) and standard temperature (273 K) occupies a volume of 22.4 L.

47. An automobile tire is inflated using air originally at 10°C and normal atmospheric pressure. During the process, the air is compressed to 28% of its original volume and the temperature is increased to 40°C. What is the tire pressure? After the car is driven at high speed, the tire air temperature rises to 85°C and the interior volume of the tire increases by 2%. What is the new tire pressure in pascals (absolute)?

48. A diving bell in the shape of a cylinder with a height of 2.50 m is closed at the upper end and open at the lower end. The bell is lowered from air into sea water (ρ = 1.025 g/cm³). The air in the bell is initially at 20.0°C. The bell is lowered to a depth (measured to the bottom of the bell) of 45.0 fathoms or 82.3 m. At this depth the water temperature is 4.0°C, and the bell is in thermal equilibrium with the water. (a) How high does sea water rise in the bell? (b) To what minimum pressure must the air in the bell be raised to expel the water that entered?

49. A bubble of marsh gas rises from the bottom of a freshwater lake at a depth of 4.2 m and a temperature of 5.0°C to the surface, where the water temperature is 12°C. What is the ratio of the bubble diameter at the two locations? (Assume that the bubble gas is in thermal equilibrium with the water at each location.)

50. An expandable cylinder has its top connected to a spring of constant 2.0×10^3 N/m (Fig. P19.50). The cylinder is filled with 5.0 L of gas with the spring relaxed at 1.0 atm and 20°C. (a) If the lid has a cross-sectional area of 0.010 m² and negligible mass, how high does the lid rise when the temperature is raised to 250°C? (b) What is the pressure of the gas at 250°C?

50A. An expandable cylinder has its top connected to a spring of constant k (Fig. P19.50). The cylinder is filled with V liters of gas with the spring relaxed at atmospheric pressure P_0 and temperature T_0. (a) If the lid has a cross-sectional area A and negligible mass, how high does the lid rise when the temperature is raised to T? (b) What is the pressure of the gas at this higher temperature?

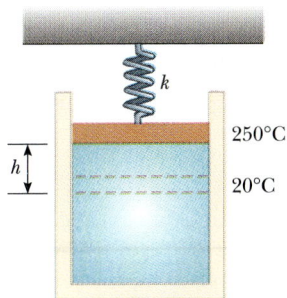

FIGURE P19.50

ADDITIONAL PROBLEMS

51. Precise temperature measurements are often made using the change in electrical resistance of a metal with temperature. The resistance varies according to the expression $R = R_0(1 + AT_C)$, where R_0 and A are constants. A certain element has a resistance of 50.0 ohms at 0°C and 71.5 ohms at the freezing point of tin (231.97°C). (a) Determine the constants A and R_0. (b) At what temperature is the resistance equal to 89.0 ohms?

52. A pendulum clock with a brass suspension system has a period of 1.000 s at 20.0°C. If the temperature increases to 30.0°C, (a) by how much does the period

change and (b) how much time does the clock gain or lose in one week?

53. The rectangular plate shown in Figure P19.53 has an area A equal to ℓw. If the temperature increases by ΔT, show that the increase in area is $\Delta A = 2\alpha A \, \Delta T$, where α is the average coefficient of linear expansion. What approximation does this expression assume? (*Hint:* Note that each dimension increases according to $\Delta\ell = \alpha\ell \, \Delta T$.)

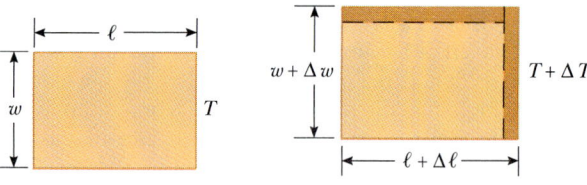

FIGURE P19.53

54. Consider an object with any one of the shapes displayed in Table 10.2. What is the percentage increase in the moment of inertia of the object when it is heated from 0°C to 100°C, if it is composed of (a) copper or (b) aluminum? (See Table 19.2. Assume that the average linear expansion coefficients do not vary between 0°C and 100°C.)

55. A mercury thermometer is constructed as in Figure P19.55. The capillary tube has a diameter of 0.0040 cm, and the bulb has a diameter of 0.25 cm. Neglecting the expansion of the glass, find the change in height of the mercury column for a temperature change of 30°C.

55A. A mercury thermometer is constructed as in Figure P19.55. The capillary tube has a diameter d_1, and the bulb has a diameter d_2. Neglecting the expansion of the glass, find the change in height of the mercury column for a temperature change ΔT.

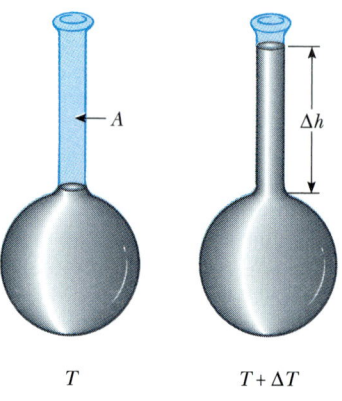

FIGURE P19.55

56. A liquid has a density ρ. (a) Show that the fractional change in density for a change in temperature ΔT is $\Delta\rho/\rho = -\beta \, \Delta T$. What does the negative sign signify? (b) Fresh water has a maximum density of 1.000 g/cm³ at 4.0°C. At 10.0°C, its density is 0.9997 g/cm³. What is β for water over this temperature interval?

57. A student measures the length of a brass rod with a steel tape at 20°C. The reading is 95.00 cm. What will the tape indicate for the length of the rod when the rod and the tape are at (a) -15°C and (b) 55°C?

58. (a) Derive an expression for the buoyant force on a spherical balloon as it is submerged in water as a function of the depth below the surface, the volume of the balloon at the surface, the pressure at the surface, and the density of the water. (Assume water temperature does not change with depth.) (b) Does the buoyant force increase or decrease as the balloon is submerged? (c) At what depth does the buoyant force decrease to one-half the surface value?

59. (a) Show that the density of an ideal gas occupying a volume V is given by $\rho = PM/RT$, where M is the molar mass. (b) Determine the density of oxygen gas at atmospheric pressure and 20.0°C.

60. Steel rails for an interurban rapid transit system form a continuous track that is held rigidly in place in concrete. (a) If the track was laid when the temperature was 0°C, what is the stress in the rails on a warm day when the temperature is 25°C? (b) What fraction of the yield strength of 52.2×10^7 N/m² does this stress represent?

61. Starting with Equation 19.12, show that the total pressure P in a container filled with a mixture of several ideal gases is $P = P_1 + P_2 + P_3 + \ldots$, where $P_1, P_2, \ldots$ are the pressures that each gas would exert if it alone filled the container (these individual pressures are called the *partial pressures* of the respective gases). This is known as *Dalton's law of partial pressures.*

62. A sample of air that has a mass of 100.00 g, collected at sea level, is analyzed and found to consist of the following gases:

$$\text{nitrogen } (N_2) = 75.52 \text{ g}$$
$$\text{oxygen } (O_2) = 23.15 \text{ g}$$
$$\text{argon } (Ar) = 1.28 \text{ g}$$
$$\text{carbon dioxide } (CO_2) = 0.05 \text{ g}$$

plus trace amounts of neon, helium, methane, and other gases. (a) Calculate the partial pressure (see Problem 61) of each gas when the pressure is 1.013×10^5 Pa. (b) Determine the volume occupied by the 100-g sample at a temperature of 15.00°C and a pressure of 1.013×10^5 Pa. What is the density of the air for these conditions? (c) What is the effective molar mass of the air sample?

63. Two concrete spans of a 250-m-long bridge are placed end to end so that there is no room for expansion (Fig. P19.63a). If a temperature increase of 20.0°C occurs, find the height, y, at which the spans buckle (Fig. P19.63b).

63A. Two concrete spans of a bridge of length L are placed end to end so that there is no room for expansion (Fig. P19.63a). If a temperature increase of ΔT occurs, find the height, y, at which the spans buckle (Fig. P19.63b).

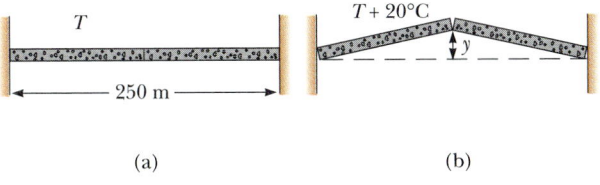

(a)　　　　　　　　(b)

FIGURE P19.63

64. A steel ball bearing is 4.000 cm in diameter at 20.0°C. A bronze plate has a hole in it that is 3.994 cm in diameter at 20.0°C. What common temperature must they have in order that the ball just squeeze through the hole?

65. A brass pendulum on a clock is adjusted to have a period of 1.000 s at 20°C. What is the temperature of a room in which the clock loses exactly 1 minute each week?

66. A vertical cylinder of cross-sectional area A is fitted with a tight-fitting, frictionless piston of mass m (Fig. P19.66). (a) If there are n mol of an ideal gas in the cylinder at a temperature T, determine the height h at which the piston is in equilibrium under its own weight. (b) What is the value for h if $n = 0.20$ mol, $T = 400$ K, $A = 0.0080$ m², and $m = 20.0$ kg?

67. An air bubble originating from a deep sea diver has a radius of 5.0 mm at some depth h. When the bubble reaches the surface of the water, it has a radius of 7.0 mm. Assuming the temperature of the air in the bubble remains constant, determine (a) the depth h of the diver and (b) the absolute pressure at this depth.

68. Figure P19.68 shows a circular steel casting with a gap. If the casting is heated, (a) does the width of the gap increase or decrease? (b) The gap width is 1.600 cm when the temperature is 30.0°C. Determine the gap width when the temperature is 190°C.

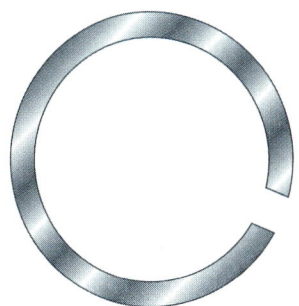

FIGURE P19.68

69. A cylinder that has a 40.0-cm radius and is 50.0 cm deep is filled with air at 20.0°C and 1.00 atm (Fig. P19.69a). A 20.0-kg piston is now lowered into the cylinder, compressing the air trapped inside (Fig. P19.69b). Finally, a 75.0-kg man stands on the piston, further compressing the air which remains at 20°C (Fig. P19.69c). (a) How far down (Δh) does the

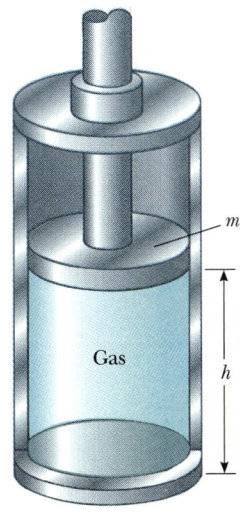

FIGURE P19.66

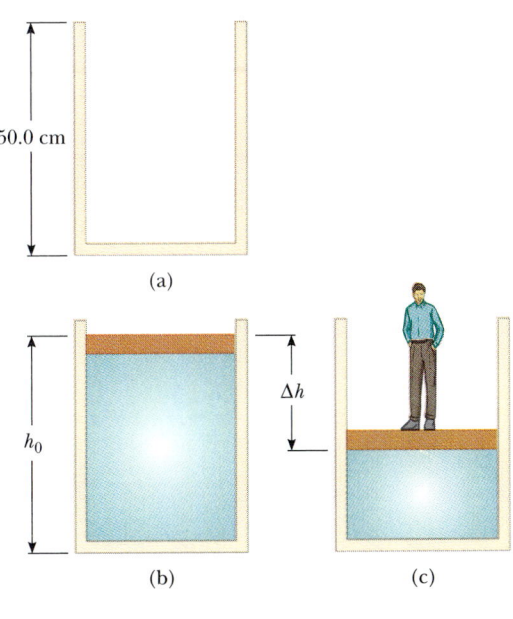

FIGURE P19.69

piston move when the man stands on it? (b) To what temperature should the gas be heated to raise the piston and man back to h_0?

70. An aluminum pot has the shape of a cylinder. The pot is initially at 4.0°C, at which temperature it has an inside diameter of 28.00 cm. The pot contains 3.000 gal of water at 4.0°C. (a) What is the depth of water in the pot? (1 gal = 3785 cm³.) (b) The pot and the water in it are heated to 90.0°C. Allowing for the expansion of the water but ignoring the expansion of the pot, what is the change in depth of the water? Express the change as a percentage of the original depth and also in millimeters. (The density of water is 1.000 g/cm³ at 4.0°C and 0.965 g/cm³ at 90.0°C.) (c) Modify your solution for part (b) to allow for the expansion of the pot. (Refer to Table 19.2.)

71. A sphere 20 cm in diameter contains an ideal gas at 1.00 atm and 20.0°C. As the sphere is heated to 100.0°C, gas is allowed to escape. The valve is closed and the sphere is placed in an ice-water bath.(a) How many moles of gas escape from the sphere as it warms? (b) What is the pressure in the sphere when it is in the ice water?

72. The relationship $L = L_0(1 + \alpha \Delta T)$ is an approximation that works when the average coefficient of expansion is small. If α is sizable, the relationship $dL/dT = \alpha L$ must be integrated to determine the final length. (a) Assuming the coefficient of linear expansion is constant, determine a general expression for the final length. (b) Given a rod of length 1.00 m and a temperature change of 100.0°C, determine the error caused by the approximation when $\alpha = 2.00 \times 10^{-5}$ (°C)$^{-1}$ (the normal value for common metals) and when $\alpha = 0.020$ (°C)$^{-1}$ (an unrealistically large value for comparison).

73. A steel guitar string with a diameter of 1.00 mm is stretched between supports 80.0 cm apart. The temperature is 0.0°C. (a) Find the mass per unit length of this string. (Use the value 7.86×10^3 kg/m³ for the density.) (b) The fundamental frequency of transverse oscillations of the string is 200 Hz. What is the tension in the string? (c) If the temperature is raised to 30.0°C, find the resulting values of the tension and the fundamental frequency. [Assume that both the Young's modulus (Table 12.1) and the average coefficient of expansion (Table 19.2) have constant values between 0.0°C and 30.0°C.]

74. A steel wire and a copper wire, each of diameter 2.000 mm, are joined end to end. At 40.0°C, each

has an unstretched length of 2.000 m; they are connected between two fixed supports 4.000 m apart on a tabletop, so that the steel wire extends from $x = -2.000$ m to $x = 0$, the copper wire extends from $x = 0$ to $x = 2.000$ m, and the tension is negligible. The temperature is then lowered to 20.0°C. At this lower temperature, find the tension in the wire and the x coordinate of the junction between the wires. (Refer to Tables 12.1 and 19.2.)

75. A bimetallic bar is made of two thin strips of dissimilar metals bonded together. As they are heated, the one with the larger average coefficient of expansion expands more than the other, forcing the bar into an arc, with the outer radius having a larger circumference (Fig. P19.75). (a) Derive an expression for the angle of bending θ as a function of the initial length of the strips, their average coefficients of linear expansion, the change in temperature, and the separation of the centers of the strips ($\Delta r = r_2 - r_1$). (b) Show that the angle of bending goes to zero when ΔT goes to zero or when the two coefficients of expansion become equal. (c) What happens if the bar is cooled?

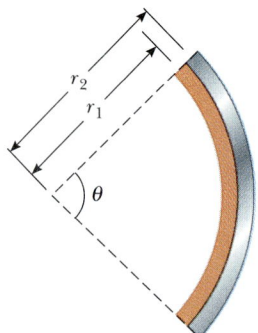

FIGURE P19.75

SPREADSHEET PROBLEM

S1. A 1.00-km steel railroad rail is fastened securely at both ends when the temperature is 20.0°C. As the temperature increases, the rail begins to buckle. If the shape of the buckle is the arc of a circle, find the height h of the center of the buckle when the temperature is 25.0°C. This problem requires you to solve a transcendental equation. Use a spreadsheet to solve the equation graphically.

CHAPTER 20

Heat and the First Law of Thermodynamics

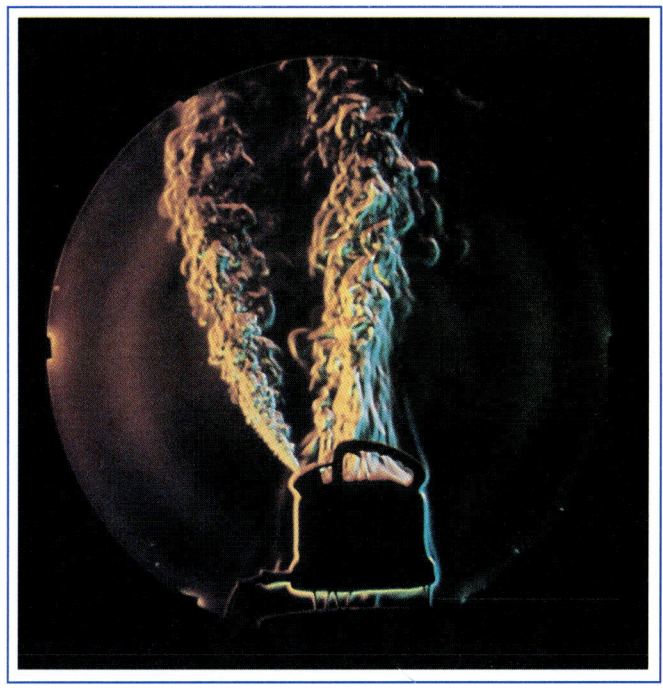

This Schlieren photograph of a steaming teakettle was produced by placing the teakettle between two spherical mirrors. Hotter areas appear white and cooler areas appear purple and black. *(Gary Settles/Science Source/Photo Researchers)*

Until about 1850, the fields of heat and mechanics were considered to be two distinct branches of science and the law of conservation of energy seemed to describe only certain kinds of mechanical systems. Mid-19th century experiments performed by the Englishman James Joule (1818–1889) and others showed that energy may be added to (or removed from) a system either as thermal energy or as work done on (or by) the system. Now thermal energy is treated as a form of energy that can be transformed into mechanical energy. Once the concept of energy was broadened to include thermal energy, the law of conservation of energy emerged as a universal law of nature.

This chapter focuses on the concept of heat, the first law of thermodynamics, processes by which thermal energy is transferred, and some important applications. The first law of thermodynamics is merely the law of conservation of energy. It tells us only that an increase in one form of energy must be accompanied by a decrease in some other form of energy. The first law places no restrictions on the types of energy conversions that can occur. Furthermore, it makes no distinction between heat and work. According to the first law, a system's internal energy can be increased either by transfer of thermal energy to the system or by work done on the system. An important difference between thermal energy and mechanical energy is not evident from the first law; it is possible to convert work completely to thermal energy but impossible to convert thermal energy completely to mechanical energy in a process at constant temperature.

20.1   HEAT AND THERMAL ENERGY

A major distinction must be made between internal energy, thermal energy, and heat. **Internal energy** is all of the energy belonging to a system while it is stationary (neither translating nor rotating), including nuclear energy, chemical energy, and strain energy (as for a compressed or stretched spring), as well as thermal energy. **Thermal energy** is that portion of the internal energy that changes when the temperature of the system changes. **Thermal energy transfer** is the transfer of thermal energy caused by a temperature difference between the system and its surroundings, *which may or may not change the amount of thermal energy in the system.*

Thermal energy

In practice, the term "**heat**" is used to mean both thermal energy and thermal energy transfer. Hence, one must always examine the context of the term *heat* to determine its intended meaning.

In the next chapter, we show that the thermal energy of a monatomic ideal gas is associated with the internal motion of its atoms. In this special case, the thermal energy is simply the kinetic energy on a microscopic scale: the higher the temperature of the gas, the greater the kinetic energy of the atoms and the greater the thermal energy of the gas. More generally, however, thermal energy includes other forms of molecular energy, such as rotational energy and vibrational kinetic and potential energy.

James Prescott Joule

| 1 8 1 8 – 1 8 8 9 |

(North Wind Picture Archives)

As an analogy, consider the distinction between work and energy that we discussed in Chapter 7. The work done on (or by) a system is a measure of energy transfer between the system and its surroundings, whereas the mechanical energy of the system (kinetic and/or potential) is a consequence of its motion and coordinates. Thus, when a person does work on a system, energy is transferred from the person to the system. It makes no sense to talk about the work of a system—one can refer only to the *work done on or by a system* when some process has occurred in which energy has been transferred to or from the system. Likewise, it makes no sense to use the term *heat* unless energy has been transferred as a result of a temperature difference.

It is also important to recognize that energy can be transferred between two systems even when no thermal energy transfer occurs. For example, when a gas is compressed by a piston, the gas is warmed and its thermal energy increases, but there is no transfer of thermal energy; if the gas then expands rapidly, it cools and its thermal energy decreases, but there is no transfer of thermal energy to the surroundings. In each case, energy is transferred to or from the system as work, but the energy appears within the system as an increase or decrease of thermal energy. The changes in internal energy in these examples are equal to changes in thermal energy and are measured by corresponding changes in temperature.

Units of Heat

Before it was understood that heat is a form of energy, scientists defined heat in terms of the temperature changes it produced in a body. Hence, the **calorie** (cal) was defined as *the amount of heat necessary to raise the temperature of 1 g of water from 14.5° C to 15.5° C*.[1] (Note that the "Calorie," with a capital C, used in describing the chemical energy content of foods, is actually a kilocalorie.) Likewise, the unit of heat in the British system is the **British thermal unit** (Btu), defined as *the heat required to raise the temperature of 1 lb of water from 63°F to 64°F*.

Since heat is now recognized as a form of energy, scientists are increasingly using the SI unit of energy, the *joule*, for heat. In this textbook, heat will usually be measured in joules.

The Mechanical Equivalent of Heat

When the concept of mechanical energy was introduced in Chapters 7 and 8, we found that whenever friction is present in a mechanical system, some mechanical energy is lost, or is not conserved. Various experiments show that this lost mechanical energy does not simply disappear but is transformed into thermal energy. Although this connection between mechanical and thermal energy was first suggested by Thompson's crude cannon-boring experiment, it was Joule (pronounced ''jewel'') who first established the equivalence of the two forms of energy.

A schematic diagram of Joule's most famous experiment is shown in Figure 20.1. The system of interest is the water in a thermally insulated container. Work is done on the water by a rotating paddle wheel, which is driven by weights falling at a constant speed. The water, which is stirred by the paddles, is warmed due to the friction between it and the paddles. If the energy lost in the bearings and through the walls is neglected, then the loss in potential energy of the weights equals the

Benjamin Thompson (1753–1814). "Being engaged, lately, in superintending the boring of cannon, in the workshops of the military arsenal at Munich, I was struck with the very considerable degree of Heat which a brass gun acquires, in a short time, in being bored; and with the still more intense Heat (much greater than that of boiling water, as I found by experiment) of the metallic chips separated from it by the borer." *(North Wind Picture Archives)*

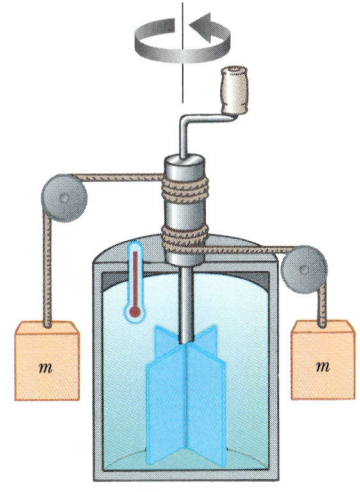

Thermal insulator

FIGURE 20.1   An illustration of Joule's experiment for determining the mechanical equivalent of heat. The falling weights rotate the paddles, causing the temperature of the water to increase.

---

[1] Originally, the calorie was defined as the heat necessary to raise the temperature of 1 g of water by 1°C. However, careful measurements showed that energy depends somewhat on temperature; hence, a more precise definition evolved.

work done by the paddle wheel on the water. If the two weights fall through a distance h, the loss in potential energy is $2mgh$, and it is this energy that is used to heat the water. By varying the conditions of the experiment, Joule found that the loss in mechanical energy, $2mgh$, is proportional to the increase in temperature of the water, ΔT. The proportionality constant was found to be equal to approximately 4.18 J/g · °C. Hence, 4.1858 J of mechanical energy raises the temperature of 1 g of water from 14.5°C to 15.5°C. We adopt this "15 degree calorie" value:

Mechanical equivalent of heat

$$1 \text{ cal} \equiv 4.186 \text{ J} \tag{20.1}$$

This is known, for purely historical reasons, as the **mechanical equivalent of heat.**

EXAMPLE 20.1   Losing Weight the Hard Way

A student eats a dinner rated at 2000 (food) Calories. He wishes to do an equivalent amount of work in the gymnasium by lifting a 50.0-kg mass. How many times must he raise the mass to expend this much energy? Assume that he raises it a distance of 2.00 m each time and that he regains no energy when it is dropped to the floor.

Solution   Since 1 Calorie $= 1.00 \times 10^3$ cal, the work required is 2.00×10^6 cal. Converting this to J, we have for the total work required

$$W = (2.00 \times 10^6 \text{ cal})(4.186 \text{ J/cal}) = 8.37 \times 10^6 \text{ J}$$

The work done in lifting the mass a distance h is equal to mgh, and the work done in lifting it n times is $nmgh$. We equate this to the total work required:

$$W = nmgh = 8.37 \times 10^6 \text{ J}$$

$$n = \frac{8.37 \times 10^6 \text{ J}}{(50.0 \text{ kg})(9.80 \text{ m/s}^2)(2.00 \text{ m})} = 8.54 \times 10^3 \text{ times}$$

If the student is in good shape and lifts the weight once every 5 s, it will take him about 12 h to perform this feat. Clearly, it is much easier to lose weight by dieting.

20.2   HEAT CAPACITY AND SPECIFIC HEAT

When heat is added to a substance (with no work done), its temperature usually rises. (An exception to this statement occurs when a substance undergoes a phase transition, say from a liquid to a gas or when a gas expands, which is discussed in the next section.) The quantity of heat energy required to raise the temperature of a given mass of a substance by some amount varies from one substance to another. For example, the heat required to raise the temperature of 1 kg of water by 1°C is 4186 J, but the heat required to raise the temperature of 1 kg of copper by 1°C is only 387 J. The **heat capacity,** C', of a particular sample of a substance is defined as the amount of heat needed to raise the temperature of that sample by one degree Celsius. From this definition, we see that if Q units of thermal energy when added to a substance produce a change in temperature of ΔT, then

Heat capacity

$$Q = C' \Delta T \tag{20.2}$$

The **specific heat** c of a substance is the heat capacity per unit mass. Thus, if Q units of thermal energy are transferred to m kg of a substance, thereby changing its temperature by ΔT, the specific heat of the substance is

Specific heat

$$c \equiv \frac{Q}{m \, \Delta T} \tag{20.3}$$

From this definition, we can express the thermal energy Q transferred between a substance of mass m and its surroundings for a temperature change ΔT as

$$Q = mc \, \Delta T \tag{20.4}$$

For example, the energy required to raise the temperature of 0.5 kg of water by 3°C is equal to $(0.5 \text{ kg})(4186 \text{ J}/\text{kg} \cdot °\text{C})(3°\text{C}) = 6280 \text{ J}$. Note that when the temperature increases, Q and ΔT are taken to be positive, corresponding to thermal energy flowing into the system. When the temperature decreases, Q and ΔT are negative and thermal energy flows out of the system.

The **molar specific heat** of a substance is defined as the heat capacity per mole. Hence, if the substance contains n mol, its molar specific heat is equal to C'/n. Table 20.1 also gives the specific heats and molar specific heats of various substances.

It is important to realize that specific heat varies with temperature. If the temperature intervals are not too great, the temperature variation can be ignored and c can be treated as a constant.[2] For example, the specific heat of water varies by only about 1% from 0°C to 100°C at atmospheric pressure. Unless stated otherwise, we shall neglect such variations.

When specific heats are measured, the values obtained are also found to depend on the conditions of the experiment. In general, measurements made at constant pressure are different from those made at constant volume. For solids and liquids, the difference between the two values is usually no more than a few percent and is often neglected. The values given in Table 20.1 were measured at atmospheric pressure and room temperature. As we shall see in Chapter 21, the

Molar specific heat

TABLE 20.1 Specific Heats of Some Substances at 25°C and Atmospheric Pressure

Substance	Specific Heat, c		Molar Specific Heats
	J/kg·°C	cal/g·°C	J/mol·°C
Elemental Solids			
Aluminum	900	0.215	24.3
Beryllium	1830	0.436	16.5
Cadmium	230	0.055	25.9
Copper	387	0.0924	24.5
Germanium	322	0.077	23.4
Gold	129	0.0308	25.4
Iron	448	0.107	25.0
Lead	128	0.0305	26.4
Silicon	703	0.168	19.8
Silver	234	0.056	25.4
Other Solids			
Brass	380	0.092	
Wood	1700	0.41	
Glass	837	0.200	
Ice (−5°C)	2090	0.50	
Marble	860	0.21	
Liquids			
Alcohol (ethyl)	2400	0.58	
Mercury	140	0.033	
Water (15°C)	4186	1.00	

[2] The definition given by Equation 20.4 assumes that the specific heat does not vary with temperature over the interval ΔT. In general, if c varies with temperature over the range T_i to T_f, the correct expression for Q is

$$Q = m \int_{T_i}^{T_f} c \, dT$$

specific heats for gases measured under constant pressure conditions are quite different from values measured under constant volume conditions.

It is interesting to note from Table 20.1 that water has the highest specific heat of common Earth materials. The high specific heat of water is responsible, in part, for the moderate temperatures found in regions near large bodies of water. As the temperature of a body of water decreases during the winter, heat is transferred from the water to the air, which in turn carries the heat landward when prevailing winds are favorable. For example, the prevailing winds of the western coast of the United States are toward the land (eastward). Hence the heat liberated by the Pacific Ocean as it cools keeps coastal areas much warmer than they would be otherwise. This explains why the western coastal states generally have more favorable winter weather than the eastern coastal states, where the prevailing winds do not tend to carry the heat toward land.

Conservation of Energy: Calorimetry

Situations in which mechanical energy is converted to thermal energy occur frequently. We shall see some in the examples following this section and in the problems at the end of the chapter, but most of our attention here will be directed toward a particular kind of conservation-of-energy situation. In problems using the procedure we shall describe, called *calorimetry* problems, only the thermal energy transfer between the system and its surroundings is considered.

One technique for measuring the specific heat of solids or liquids is simply to heat the substance to some known temperature, place it in a vessel containing water of known mass and temperature, and measure the temperature of the water after equilibrium is reached. Since a negligible amount of mechanical work is done in the process, the law of conservation of energy requires that the thermal energy that leaves the warmer substance (of unknown specific heat) equals the thermal energy that enters the water.[3] Devices in which this thermal energy transfer occurs are called **calorimeters.**

For example, suppose that m_x is the mass of a substance whose specific heat we wish to determine, c_x its specific heat, and T_x its initial temperature. Likewise, let m_w, c_w, and T_w represent corresponding values for the water. If T is the final equilibrium temperature after everything is mixed, then from Equation 20.3, we find that the thermal energy gained by the water is $m_w c_w (T - T_w)$, and the thermal energy lost by the substance of unknown c is $- m_x c_x (T - T_x)$. Assuming that the combined system (water + unknown) does not lose or gain any thermal energy, it follows that the thermal energy gained by the water must equal the thermal energy lost by the unknown (conservation of energy):

$$m_w c_w (T - T_w) = - m_x c_x (T - T_x)$$

Solving for c_x gives

$$c_x = \frac{m_w c_w (T - T_w)}{m_x (T_x - T)} \tag{20.5}$$

[3] For precise measurements, the container for the water should be included in our calculations, since it also exchanges heat. This would require a knowledge of its mass and composition. However, if the mass of the water is large compared with that of the container, we can neglect the heat gained by the container. Furthermore, precautions must be taken in such measurements to minimize heat transfer between the system and the surroundings.

EXAMPLE 20.2 Cooling a Hot Ingot

A 0.0500-kg ingot of metal is heated to 200.0°C and then dropped into a beaker containing 0.400 kg of water initially at 20.0°C. If the final equilibrium temperature of the mixed system is 22.4°C, find the specific heat of the metal.

Solution Because the thermal energy lost by the ingot equals the thermal energy gained by the water, we can write

$$m_x c_x (T_i - T_f) = m_w c_w (T_f - T_i)$$

$$(0.0500 \text{ kg})(c_x)(200.0°C - 22.4°C) = (0.400 \text{ kg})$$
$$(4186 \text{ J/kg} \cdot °C)(22.4°C - 20.0°C)$$

from which we find that

$$c_x = \boxed{453 \text{ J/kg} \cdot °C}$$

The ingot is most likely iron, as can be seen by comparing this result with the data in Table 20.1.

Exercise What is the total thermal energy transferred to the water as the ingot is cooled?

Answer 4020 J.

EXAMPLE 20.3 Fun Time for a Cowboy

A cowboy fires a silver bullet of mass 2.00 g with a muzzle speed of 200 m/s into the pine wall of a saloon. Assume that all the internal energy generated by the impact remains with the bullet. What is the temperature change of the bullet?

Solution The kinetic energy of the bullet is

$$\tfrac{1}{2}mv^2 = \tfrac{1}{2}(2.00 \times 10^{-3} \text{ kg})(200 \text{ m/s})^2 = 40.0 \text{ J}$$

Nothing in the environment is hotter than the bullet, so the bullet gains no thermal energy. Its temperature increases because the 40.0 J of kinetic energy becomes 40.0 J of extra internal energy. The temperature change would be the same as if 40.0 J of thermal energy were transferred from a stove to

the bullet, and we imagine this process to compute ΔT from

$$Q = mc\,\Delta T$$

Since the specific heat of silver is 234 J/kg·°C (Table 20.1), we get

$$\Delta T = \frac{Q}{mc} = \frac{40.0 \text{ J}}{(2.00 \times 10^{-3} \text{ kg})(234 \text{ J/kg} \cdot °C)} = \boxed{85.5°C}$$

Exercise Suppose the cowboy runs out of silver bullets and fires a lead bullet of the same mass and speed into the wall. What is the temperature change of the bullet?

Answer 157°C.

20.3 LATENT HEAT

A substance usually undergoes a change in temperature when thermal energy is transferred between the substance and its surroundings. There are situations, however, in which the transfer of thermal energy does not result in a change in temperature. This is the case whenever the physical characteristics of the substance change from one form to another, commonly referred to as a **phase change.** Some common phase changes are solid to liquid (melting), liquid to gas (boiling), and a change in crystalline structure of a solid. All such phase changes involve a change in internal energy.

The thermal energy required to change the phase of a given mass, *m*, of a pure substance is

$$Q = mL \tag{20.6}$$

where *L* is called the **latent heat** ("hidden" heat) of the substance[4] and depends on the nature of the phase change as well as on the properties of the substance. **Latent heat of fusion,** L_f, is the term used when the phase change is from solid to

---

[4] The word *latent* is from the Latin *latere*, meaning *to hide* or *to conceal.*

TABLE 20.2   Latent Heats of Fusion and Vaporization

Substance	Melting Point (°C)	Latent Heat of Fusion (J/kg)	Boiling Point (°C)	Latent Heat of Vaporization (J/kg)
Helium	−269.65	5.23×10^3	−268.93	2.09×10^4
Nitrogen	−209.97	2.55×10^4	−195.81	2.01×10^5
Oxygen	−218.79	1.38×10^4	−182.97	2.13×10^5
Ethyl alcohol	−114	1.04×10^5	78	8.54×10^5
Water	0.00	3.33×10^5	100.00	2.26×10^6
Sulfur	119	3.81×10^4	444.60	3.26×10^5
Lead	327.3	2.45×10^4	1750	8.70×10^5
Aluminum	660	3.97×10^5	2450	1.14×10^7
Silver	960.80	8.82×10^4	2193	2.33×10^6
Gold	1063.00	6.44×10^4	2660	1.58×10^6
Copper	1083	1.34×10^5	1187	5.06×10^6

liquid ("fuse" means to melt, to liquefy), and **latent heat of vaporization, L_v,** is used when the phase change is from liquid to gas (the liquid vaporizes).[5] For example, the latent heat of fusion for ice at atmospheric pressure is 3.33×10^5 J/kg, and the latent heat of vaporization of water is 2.26×10^6 J/kg. The latent heats of various substances vary considerably, as Table 20.2 shows.

Consider, for example, the thermal energy required to convert a 1.00-g block of ice at −30.0°C to steam (water vapor) at 120.0°C. Figure 20.2 indicates the experimental results obtained when thermal energy is gradually added to the ice. Let us examine each portion of the curve.

Part A   On this portion of the curve the temperature of the ice is changing from −30.0°C to 0.0°C. Since the specific heat of ice is 2090 J/kg·°C, we can calculate

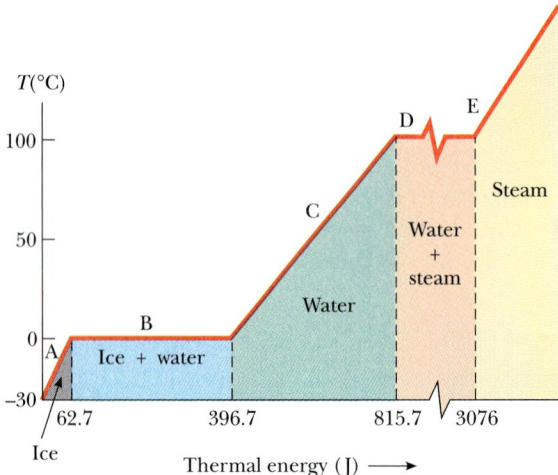

FIGURE 20.2 A plot of temperature versus thermal energy added when 1 g of ice initially at −30°C is converted to steam.

[5] When a gas cools, it eventually returns to the liquid phase, or *condenses.* The heat given up per unit mass is called the *heat of condensation,* which equals the negative of the heat of vaporization. Likewise, when a liquid cools it eventually solidifies, and the *heat of solidification* equals the negative of the heat of fusion.

the amount of thermal energy added from Equation 20.4:

$$Q = m_i c_i \, \Delta T = (1.00 \times 10^{-3} \text{ kg})(2090 \text{ J/kg} \cdot {}^{\circ}\text{C})(30.0{}^{\circ}\text{C}) = 62.7 \text{ J}$$

Part B When the ice reaches 0.0°C, the ice/water mixture remains at this temperature—even though thermal energy is being added—until all the ice melts. The thermal energy required to melt 1.00 g of ice at 0.0°C is, from Equation 20.6,

$$Q = mL_f = (1.00 \times 10^{-3} \text{ kg})(3.33 \times 10^5 \text{ J/kg}) = 333 \text{ J}$$

Part C Between 0.0°C and 100.0°C, nothing surprising happens. No phase change occurs in this region. The thermal energy added to the water is being used to increase its temperature. The amount of thermal energy necessary to increase the temperature from 0.0°C to 100.0°C is

$$Q = m_w c_w \, \Delta T = (1.00 \times 10^{-3} \text{ kg})(4.19 \times 10^3 \text{ J/kg} \cdot {}^{\circ}\text{C})(100.0{}^{\circ}\text{C}) = 419 \text{ J}$$

Part D At 100.0°C, another phase change occurs as the water changes from water at 100.0°C to steam at 100.0°C. Just as in Part B, the water/steam mixture remains at 100.0°C—even though thermal energy is being added—until all of the liquid has been converted to steam. The thermal energy required to convert 1.00 g of water to steam at 100.0°C is

$$Q = mL_v = (1.00 \times 10^{-3} \text{ kg})(2.26 \times 10^6 \text{ J/kg}) = 2.26 \times 10^3 \text{ J}$$

Part E On this portion of the curve, heat is being added to the steam with no phase change occurring. The thermal energy that must be added to raise the temperature of the steam to 120.0°C is

$$Q = m_s c_s \, \Delta T = (1.00 \times 10^{-3} \text{ kg})(2.01 \times 10^3 \text{ J/kg} \cdot {}^{\circ}\text{C})(20.0{}^{\circ}\text{C}) = 40.2 \text{ J}$$

The *total amount of thermal energy* that must be added to change one gram of ice at $-30.0{}^{\circ}\text{C}$ to steam at 120.0°C is approximately 3.11×10^3 J. Conversely, to cool one gram of steam at 120.0°C down to the point at which we have ice at $-30.0{}^{\circ}\text{C}$, we must remove 3.11×10^3 J of thermal energy.

Phase changes can be described in terms of a rearrangement of molecules when thermal energy is added or removed from a substance. Consider first the liquid-to-gas phase change. The molecules in a liquid are close together, and the forces between them are stronger than those between the more widely separated molecules of a gas. Therefore, work must be done on the liquid against these attractive molecular forces in order to separate the molecules. The latent heat of vaporization is the amount of energy that must be added to the liquid to accomplish this separation.

Similarly, at the melting point of a solid, we imagine that the amplitude of vibration of the atoms about their equilibrium position becomes large enough to allow the atoms to pass the barriers of adjacent atoms and move to their new positions. The new locations are, on the average, less symmetrical and therefore have higher energy. The latent heat of fusion is equal to the work required at the molecular level to transform the mass from the highly ordered solid phase to the less ordered liquid phase.

The average distance between atoms in the gas phase is much larger than in either the liquid or the solid phase. Each atom or molecule is removed from its neighbors, without the compensation of attractive forces to new neighbors. Therefore, it is not surprising that more work is required at the molecular level to vaporize a given mass of substance than to melt it; thus, the latent heat of vaporization is much larger than the latent heat of fusion for a given substance (Table 20.2).

> ### Problem Solving Strategy
> ### Calorimetry Problems
>
> If you are having difficulty with calorimetry problems, make the following considerations.
>
> - Be sure your units are consistent throughout. For instance, if you are using specific heats in cal/g · °C, be sure that masses are in grams and temperatures are in Celsius units throughout.
> - Losses and gains in thermal energy are found by using $Q = mc\,\Delta T$ only for those intervals in which no phase changes occur. Likewise, the equations $Q = mL_f$ and $Q = mL_v$ are to be used only when phase changes *are* taking place.
> - Often sign errors occur in heat loss = heat gain equations. One way to check your equation is to examine the signs of all ΔT's that appear in it.

EXAMPLE 20.4   Cooling the Steam

What mass of steam initially at 130°C is needed to warm 200 g of water in a 100-g glass container from 20.0°C to 50.0°C?

Solution   This is a heat-transfer problem in which we must equate the thermal energy lost by the steam to the thermal energy gained by the water and glass container. There are three stages as the steam loses thermal energy. In the first stage, the steam is cooled to 100°C. The thermal energy liberated in the process is

$$Q_1 = m_s c_s\,\Delta T = m_s(2.01 \times 10^3\,\text{J/kg}\cdot\text{°C})(30.0\text{°C})$$
$$= m_s(6.03 \times 10^4\,\text{J/kg})$$

In the second stage, the steam is converted to water. In this case, to find the thermal energy removed, we use the latent heat of vaporization and $Q = mL_v$:

$$Q_2 = m_s(2.26 \times 10^6\,\text{J/kg})$$

In the last stage, the temperature of the water is reduced to 50.0°C. This liberates an amount of thermal energy

$$Q_3 = m_s c_w\,\Delta T = m_s(4.19 \times 10^3\,\text{J/kg}\cdot\text{°C})(50.0\text{°C})$$
$$= m_s(2.09 \times 10^5\,\text{J/kg})$$

If we equate the thermal energy lost by the steam to the thermal energy gained by the water and glass and use the given information, we find

$$m_s(6.03 \times 10^4\,\text{J/kg}) + m_s(2.26 \times 10^6\,\text{J/kg})$$
$$+ m_s(2.09 \times 10^5\,\text{J/kg})$$
$$= (0.200\,\text{kg})(4.19 \times 10^3\,\text{J/kg}\cdot\text{°C})(30.0\text{°C})$$
$$+ (0.100\,\text{kg})(837\,\text{J/kg}\cdot\text{°C})(30.0\text{°C})$$

$$m_s = \boxed{1.09 \times 10^{-2}\,\text{kg} = 10.9\,\text{g}}$$

EXAMPLE 20.5   Boiling Liquid Helium

Liquid helium has a very low boiling point, 4.2 K, and a very low heat of vaporization, 2.09×10^4 J/kg (Table 20.2). A constant power of 10.0 W is transferred to a container of liquid helium from an immersed electric heater. At this rate, how long does it take to boil away 1.00 kg of liquid helium?

Reasoning and Solution   Since $L_v = 2.09 \times 10^4$ J/kg for liquid helium, we must supply 2.09×10^4 J of energy to boil away 1.00 kg. The power supplied to the helium is 10.0 W = 10.0 J/s. That is, in 1.00 s, 10.0 J of energy is transferred to

the helium. Therefore, the time it takes to transfer an energy of 2.09×10^4 J is

$$t = \frac{2.09 \times 10^4\,\text{J}}{10.0\,\text{J/s}} = 2.09 \times 10^3\,\text{s} \approx \boxed{35\,\text{min}}$$

Exercise   If 10.0 W of power is supplied to 1.00 kg of water at 100°C, how long will it take for the water to completely boil away?

Answer   62.8 h.

20.4   WORK AND HEAT IN THERMODYNAMIC PROCESSES

In the macroscopic approach to thermodynamics we describe the *state* of a system with such variables as pressure, volume, temperature, and internal energy. The number of macroscopic variables needed to characterize a system depends on the system's nature. For a homogeneous system, such as a gas containing only one type of molecule, usually only two variables are needed. However, it is important to note that a *macroscopic state* of an isolated system can be specified only if the system is in thermal equilibrium internally. In the case of a gas in a container, internal thermal equilibrium requires that every part of the container be at the same pressure and temperature.

Consider gas contained in a cylinder fitted with a movable piston (Fig. 20.3). In equilibrium, the gas occupies a volume V and exerts a uniform pressure P on the cylinder walls and piston. If the piston has a cross-sectional area A, the force exerted by the gas on the piston is $F = PA$. Now let us assume that the gas expands **quasi-statically,** that is, slowly enough to allow the system to remain essentially in thermodynamic equilibrium at all times. As the piston moves up a distance dy, the work done by the gas on the piston is

$$dW = F\, dy = PA\, dy$$

Since $A\, dy$ is the increase in volume of the gas dV, we can express the work done as

$$dW = P\, dV \qquad (20.7)$$

Since the gas expands, dV is positive and the work done by the gas is positive, whereas if the gas is compressed, dV is negative, indicating that the work done by the gas is negative.[6] (In the latter case, negative work can be interpreted as work

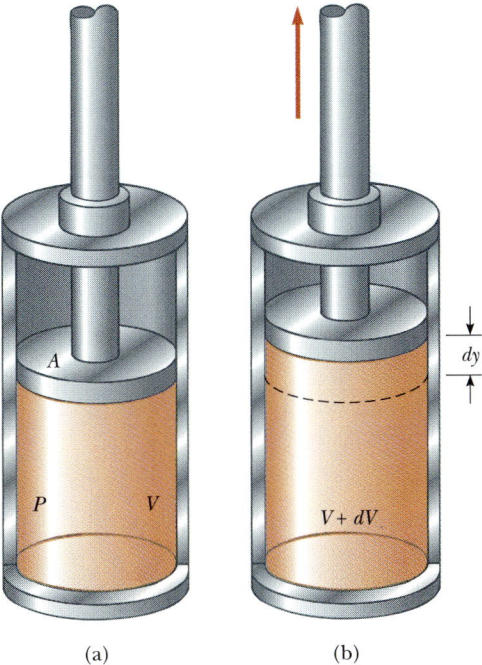

(a)   (b)

FIGURE 20.3   Gas contained in a cylinder at a pressure P does work on a moving piston as the system expands from a volume V to a volume $V + dV$.

---

[6] For historical reasons, we choose to let W represent the work done by the system here. In other parts of the text, W is the work done on the system. The change affects only the sign of W.

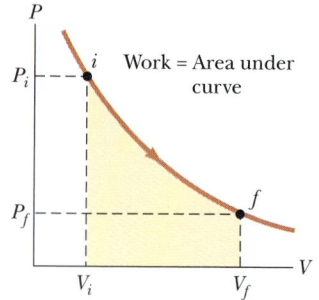

FIGURE 20.4   A gas expands reversibly (slowly) from state i to state f. The work done by the gas equals the area under the PV curve.

Work equals area under the curve in a PV diagram

being done *on* the gas.) Clearly, the work done by the gas is zero when the volume remains constant. The total work done by the gas as its volume changes from V_i to V_f is given by the integral of Equation 20.7:

$$W = \int_{V_i}^{V_f} P \, dV \tag{20.7}$$

To evaluate this integral, one must know how the pressure varies during the expansion process. (Note that a *process* is *not* specified merely by giving the initial and final states. Rather, a process is a *fully specified* change in state of a system.) In general, the pressure is not constant, but depends on the volume and temperature. If the pressure and volume are known at each step of the process, the states of the gas can then be represented as a curve on a PV diagram, as in Figure 20.4.

> The work done in the expansion from the initial state to the final state is the area under the curve in a PV diagram.

As Figure 20.4 shows, the work done in the expansion from the initial state, i, to the final state, f, depends on the path taken between these two states. To illustrate this important point, consider several different paths connecting i and f (Fig. 20.5). In the process depicted in Figure 20.5a, the pressure of the gas is first reduced from P_i to P_f by cooling at constant volume V_i, and the gas then expands from V_i to V_f at constant pressure P_f. The work done along this path is $P_f(V_f - V_i)$. In Figure 20.5b, the gas first expands from V_i to V_f at constant pressure P_i, and then its pressure is reduced to P_f at constant volume V_f. The work done along this path is $P_i(V_f - V_i)$, which is greater than that for the process described in Figure 20.5a. Finally, for the process described in Figure 20.5c, where both P and V change continuously, the work done has some value intermediate between the values obtained in the first two processes. To evaluate the work in this case, the shape of the PV curve must be known. Therefore, we see that the work done by a system depends on the process by which the system goes from the initial to the final state. In other words, *the work done depends on the initial, final, and intermediate states of the system.*

In a similar manner, the thermal energy transferred into or out of the system also depends on the process. This can be demonstrated by considering the situa-

This device, called Hero's engine, was invented around 150 B.C. by Hero in Alexandria. When water is boiled in the flask, which is suspended by a cord, steam exits through two tubes at the sides of the flask (in opposite directions), creating a torque that rotates the flask. *(Courtesy of Central Scientific Co.)*

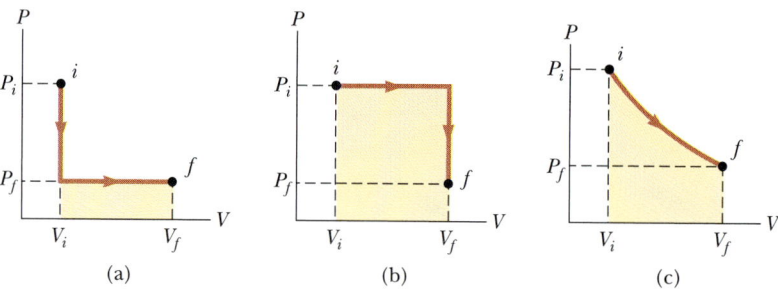

FIGURE 20.5   The work done by a gas as it is taken from an initial state to a final state depends on the path between these states.

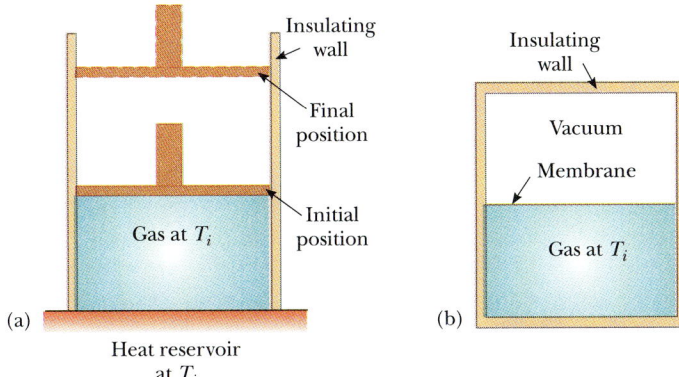

FIGURE 20.6   (a) A gas at temperature T_i expands slowly by absorbing thermal energy from a reservoir at the same temperature. (b) A gas expands rapidly into an evacuated region after a membrane is broken.

tions depicted in Figure 20.6. In each case, the gas has the same initial volume, temperature, and pressure and is assumed to be ideal. In Figure 20.6a, the gas is in thermal contact with a heat reservoir. If the pressure of the gas is infinitesimally greater than atmospheric pressure, the gas, because it absorbs thermal energy, expands and causes the piston to rise. During this expansion to some final volume V_f, just enough thermal energy to maintain a constant temperature T_i is transferred from the reservoir to the gas.

Now consider the thermally insulated system shown in Figure 20.6b. When the membrane is broken, the gas expands rapidly into the vacuum until it occupies a volume V_f and is at a pressure P_f. In this case, the gas does no work because there is no movable piston. Furthermore, no thermal energy is transferred through the insulating wall.

The initial and final states of the ideal gas in Figure 20.6a are identical to the initial and final states in Figure 20.6b, but the paths are different. In the first case, thermal energy is transferred slowly to the gas, and the gas does work on the piston. In the second case, no thermal energy is transferred and the work done is zero. Therefore, we conclude that *thermal energy transfer, like the work done, depends on the initial, final, and intermediate states of the system.* Furthermore, since thermal energy and work depend on the path, neither quantity is determined by the end points of a thermodynamic process.

Free expansion of a gas

20.5   THE FIRST LAW OF THERMODYNAMICS

When the law of conservation of energy was introduced in Chapter 8, it was stated that the mechanical energy of a system is constant in the absence of nonconservative forces, such as friction. That is, the changes in the internal energy of the system were not included in this mechanical model. The first law of thermodynamics, which we discuss in this section, is a generalization known as the law of conservation of energy and encompasses possible changes in internal energy. It is a universally valid law that can be applied to all kinds of processes. Furthermore, it provides us with a connection between the microscopic and macroscopic worlds.

We have seen that energy can be transferred between a system and its surroundings in two ways. One is work done by (or on) the system, which requires that there be a macroscopic displacement of the point of application of a force (or pressure). The other is thermal energy transfer, which occurs through random molecular collisions. Each of these represents a change of energy of the system and, therefore, usually results in measurable changes in the macroscopic variables of the system, such as pressure, temperature, and volume of a gas.

To put these ideas on a more quantitative basis, suppose a thermodynamic system undergoes a change from an initial state to a final state. During this change, positive Q is the thermal energy transferred *to* the system, and positive W is the work done *by* the system. As an example, suppose the system is a gas whose pressure and volume change from P_i, V_i to P_f, V_f. If the quantity $Q - W$ is measured for various paths connecting the initial and final equilibrium states (that is, for various *processes*), we find that $Q - W$ is the same for all paths connecting the initial and final states. We conclude that the quantity $Q - W$ is determined completely by the initial and final states of the system, and we call the quantity $Q - W$ the *change in the energy of the system*. Although Q and W both depend on the path, the quantity $Q - W$ *is independent of the path*. If we represent the energy function with the letter U, then the *change* in energy, $\Delta U = U_f - U_i$, can be expressed as

Change in energy	

First-law equation

$$\Delta U = U_f - U_i = Q - W \qquad (20.9)$$

where all quantities must have the same energy units. Equation 20.9 is known as the **first-law equation** and is a key equation to many applications. When it is used in this form, we use the convention that Q is positive when thermal energy enters the system and negative when thermal energy leaves the system. Likewise, W is positive when the system does work on the surroundings and negative if work is done on the system.

When a system undergoes an infinitesimal change in state, where a small amount of thermal energy dQ is transferred and a small amount of work dW is done, the energy also changes by a small amount dU. Thus, for infinitesimal processes we can express the first-law equation as[7]

First-law equation for infinitesimal changes

$$dU = dQ - dW$$

On a microscopic level, the internal energy of a system includes the kinetic and potential energies of the molecules making up the system. Part of the energy, U, is the internal energy of the system. In thermodynamics, we do not concern ourselves with the specific form of the internal energy. An analogy can be made between the internal energy of a system and the potential energy function associated with a body moving under the influence of gravity without friction. The potential energy function is independent of the path, and it is only changes in it that is of concern. Likewise, the change in internal energy of a thermodynamic system is what matters, since only differences are defined. Because absolute values are not defined, any reference state can be chosen for the internal energy.

The internal energy of an isolated system is constant

Let us look at some special cases in which the only changes in energy are changes in internal energy. First consider an *isolated system,* that is, one that does not interact with its surroundings. In this case, no thermal energy transfer takes place and the work done is zero; hence, the internal energy remains constant. That is, since $Q = W = 0$, $\Delta U = 0$, and so $U_i = U_f$. We conclude that *the internal energy of an isolated system remains constant.*

---

[7] Note that dQ and dW are not true differential quantities, although dU is a true differential. In fact, dQ and dW are *inexact differentials* and are often represented by $đQ$ and $đW$. For further details on this point, see an advanced text in thermodynamics, e.g., R. P. Bauman, *Modern Thermodynamics and Statistical Mechanics,* New York, Macmillan, 1992.

Next consider a process in which a system (one not isolated from its surroundings) is taken through a **cyclic process,** that is, one that originates and ends at the same state. In this case, the change in the internal energy must again be zero and, therefore, the thermal energy added to the system must equal the work done during the cycle. That is, in a cyclic process,

$$\Delta U = 0 \quad \text{and} \quad Q = W$$

Cyclic process

Note that *the net work done per cycle equals the area enclosed by the path representing the process on a PV diagram.*

If a process occurs in which the work done is zero, then the change in internal energy equals the thermal energy entering or leaving the system. If thermal energy enters the system, Q is positive and the internal energy increases. For a gas, we can associate this increase in internal energy with an increase in the kinetic energy of the molecules. On the other hand, if a process occurs in which the thermal energy transferred is zero and work is done by the system, then the magnitude of the change in internal energy equals the negative of the work done by the system. That is, the internal energy of the system decreases. For example, if a gas is compressed with no thermal energy transferred (by a moving piston, for example), the work done by the gas is negative and the internal energy again increases. This is because kinetic energy is transferred from the moving piston to the gas molecules.

No distinction exists between thermal energy transfer and work on a microscopic scale. Both can produce a change in the internal energy of a system. Although the macroscopic quantities Q and W are *not* properties of a system, they are related to the changes of the internal energy of a stationary system through the first-law equation. Once a process, or path, is defined, Q and W can be either calculated or measured, and the change in internal energy can be found from the first-law equation. One of the important consequences of the first law is that there is a quantity called internal energy, the value of which is determined by the state of the system. The internal energy function is therefore called a *state function.*

20.6   SOME APPLICATIONS OF THE FIRST LAW OF THERMODYNAMICS

In order to apply the first law of thermodynamics to specific systems, it is useful to first define some common thermodynamic processes. An **adiabatic process** is one during which no thermal energy enters or leaves the system, that is, $Q = 0$. An adiabatic process can be achieved either by thermally insulating the system from its surroundings (as in Fig. 20.6b) or by performing the process rapidly. Applying the first law of thermodynamics to an adiabatic process, we see that

Adiabatic process

$$\Delta U = - W \qquad (20.10)$$

First-law equation for an adiabatic process

From this result, we see that if a gas expands adiabatically, W is positive, so ΔU is negative and the temperature of the gas decreases. Conversely, the gas temperature rises when it is compressed adiabatically.

Adiabatic processes are very important in engineering practice. Some common examples include the expansion of hot gases in an internal combustion engine, the liquefaction of gases in a cooling system, and the compression stroke in a diesel engine.

The process described in Figure 20.6b, called an **adiabatic free expansion,** is an adiabatic process in which no work is done on or by the gas. Since $Q = 0$ and $W = 0$, we see from the first law that $\Delta U = 0$ for this process. That is, *the initial and final internal energies of a gas are equal in an adiabatic free expansion.* As we shall see in

Adiabatic free expansion

the next chapter, the internal energy of an ideal gas depends only on its temperature. Thus, we would expect no change in temperature during an adiabatic free expansion. This is in accord with experiments performed at low pressures. (Careful experiments at high pressures for real gases show a slight decrease or increases in temperature after the expansion.)

Isobaric process

A process that occurs at constant pressure is called an **isobaric process.** When such a process occurs, the thermal energy transferred and the work done are both usually nonzero. The work done is simply $P(V_f - V_i)$.

A process that takes place at constant volume is called an **isovolumetric process.** In such an expansion process, the work done is clearly zero. Hence from the first law we see that in an isovolumetric process with $W = 0$

First-law equation for a constant-volume process

$$\Delta U = Q \tag{20.11}$$

This tells us that *if thermal energy is added to a system kept at constant volume, all of the thermal energy goes into increasing the internal energy of the system.* When a mixture of gasoline vapor and air explodes in the cylinder of an engine, the temperature and pressure rise suddenly because the cylinder volume doesn't change appreciably during the short duration of the explosion.

Isothermal process

A process that occurs at constant temperature is called an **isothermal process,** and a plot of P versus V at constant temperature for an ideal gas yields a hyperbolic curve called an isotherm. The internal energy of an ideal gas is a function of temperature only. Hence, in an isothermal process of an ideal gas, $\Delta U = 0$.

Isothermal Expansion of an Ideal Gas

Suppose an ideal gas is allowed to expand quasi-statically at constant temperature as described by the PV diagram in Figure 20.7. The curve is a hyperbola, and the equation of this curve is $PV = $ constant. Let us calculate the work done by the gas in the expansion from state i to state f.

The isothermal expansion of the gas can be achieved by placing the gas in good thermal contact with a heat reservoir at the same temperature, as in Figure 20.6a.

The work done by the gas is given by Equation 20.7. Since the gas is ideal and the process is quasi-static, we can apply $PV = nRT$ for each point on the path. Therefore, we have

$$W = \int_{V_i}^{V_f} P \, dV = \int_{V_i}^{V_f} \frac{nRT}{V} \, dV$$

Since T is constant in this case, it can be removed from the integral along with n and R:

$$W = nRT \int_{V_i}^{V_f} \frac{dV}{V} = nRT \ln V \Big]_{V_i}^{V_f}$$

To evaluate the integral, we used $\int (dx/x) = \ln x$ (Table B.5 in Appendix B). Thus, we find

Work done in an isothermal process

$$W = nRT \ln \left(\frac{V_f}{V_i} \right) \tag{20.12}$$

Numerically, this work equals the shaded area under the PV curve in Figure 20.7. Because the gas expands isothermally, $V_f > V_i$ and we see that the work done by the gas is positive, as we would expect. If the gas is compressed isothermally, then

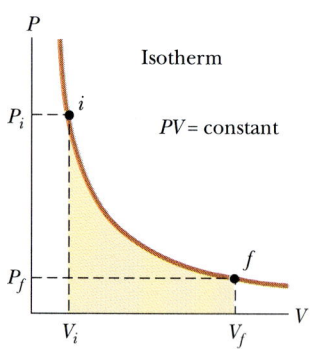

FIGURE 20.7 The PV diagram for an isothermal expansion of an ideal gas from an initial state to a final state. The curve is a hyperbola.

$V_f < V_i$ and the work done by the gas is negative. Hence, for an isothermal process $\Delta U = 0$, and from the first law we conclude that the thermal energy given up by the reservoir (and transferred to the gas) equals the work done by the gas, or $Q = W$.

EXAMPLE 20.6 Work Done During an Isothermal Expansion

Calculate the work done by 1.0 mol of an ideal gas that is kept at 0.0°C during an expansion from 3.0 liters to 10.0 liters.

Solution Substituting these values into Equation 20.12 gives

$$W = nRT \ln\left(\frac{V_f}{V_i}\right)$$

$$= (1.0\ \text{mol})(8.31\ \text{J/mol·K})(273\ \text{K}) \ln\left(\frac{10.0}{3.0}\right)$$

$$= 2.7 \times 10^3\ \text{J}$$

The thermal energy that must be supplied to the gas from the reservoir to keep T constant is also 2.7×10^3 J.

Boiling

Suppose that a liquid of mass m vaporizes at constant pressure P. Its volume in the liquid state is V_ℓ, and its volume in the vapor state is V_v. Let us find the work done in the expansion and the change in internal energy of the system.

Since the expansion takes place at constant pressure, the work done by the system is

$$W = \int_{V_\ell}^{V_v} P\,dV = P\int_{V_\ell}^{V_v} dV = P(V_v - V_\ell)$$

The thermal energy that must be transferred to the liquid to vaporize all of it is equal to $Q = mL_v$, where L_v is the latent heat of vaporization of the liquid. Using the first law and the result above, we get

$$\Delta U = Q - W = mL_v - P(V_v - V_\ell) \tag{20.13}$$

EXAMPLE 20.7 Boiling Water

One gram of water occupies a volume of 1.00 cm³ at atmospheric pressure. When this amount of water is boiled, it becomes 1671 cm³ of steam. Calculate the change in internal energy for this process.

Solution Since the latent heat of vaporization of water is 2.26×10^6 J/kg at atmospheric pressure, the heat required to boil 1.00 g is

$$Q = mL_v = (1.00 \times 10^{-3}\ \text{kg})(2.26 \times 10^6\ \text{J/kg}) = 2260\ \text{J}$$

The work done by the system is positive and equal to

$$W = P(V_v - V_\ell)$$
$$= (1.013 \times 10^5\ \text{N/m}^2)[(1671 - 1.00) \times 10^{-6}\ \text{m}^3]$$
$$= 169\ \text{J}$$

Hence, the change in internal energy is

$$\Delta U = Q - W = 2260\ \text{J} - 169\ \text{J} = 2.09\ \text{kJ}$$

The internal energy of the system increases because ΔU is positive. We see that most (93%) of the thermal energy transferred to the liquid goes into increasing the internal energy. Only 7% goes into external work.

EXAMPLE 20.8 Heat Transferred to a Solid

A 1.0-kg bar of copper is heated at atmospheric pressure. If its temperature increases from 20°C to 50°C, (a) find the work done by the copper.

Solution The change in volume of the copper can be calculated using Equation 19.6 and the volume expansion coefficient for copper taken from Table 19.2 (remembering that

$\beta = 3\alpha$):

$$\Delta V = \beta V \Delta T = [5.1 \times 10^{-5}(°C)^{-1}](50°C - 20°C)V$$
$$= 1.5 \times 10^{-3} V$$

But the volume is equal to m/ρ, and the density of copper is 8.92×10^3 kg/m³. Hence,

$$\Delta V = (1.5 \times 10^{-3})\left(\frac{1.0 \text{ kg}}{8.92 \times 10^3 \text{ kg/m}^3}\right) = 1.7 \times 10^{-7} \text{ m}^3$$

Since the expansion takes place at constant pressure, the work done is

$$W = P \Delta V = (1.013 \times 10^5 \text{ N/m}^2)(1.7 \times 10^{-7} \text{ m}^3)$$
$$= 1.7 \times 10^{-2} \text{ J}$$

(b) What quantity of thermal energy is transferred to the copper?

Solution   Taking the specific heat of copper from Table 20.1 and using Equation 20.4, we find that the thermal energy transferred is

$$Q = mc \Delta T = (1.0 \text{ kg})(387 \text{ J/kg} \cdot °C)(30°C)$$
$$= 1.2 \times 10^4 \text{ J}$$

(c) What is the increase in internal energy of the copper?

Solution   From the first law of thermodynamics, the increase in internal energy is

$$\Delta U = Q - W = 1.2 \times 10^4 \text{ J}$$

Note that almost all of the thermal energy transferred goes into increasing the internal energy. The fraction of thermal energy that is used to do work against the atmosphere is only about 10^{-6}! Hence, in the thermal expansion of a solid or a liquid, the small amount of work done is usually ignored.

20.7   HEAT TRANSFER

In practice, it is important to understand the rate at which thermal energy is transferred between a system and its surroundings and the mechanisms responsible for the transfer. You may have used a Thermos bottle or some other thermally insulated vessel to store hot coffee for a length of time. The vessel reduces thermal energy transfer between the outside air and the hot coffee. Ultimately, of course, the liquid will reach air temperature because the vessel is not a perfect insulator. There is no thermal energy transfer between a system and its surroundings when they are at the same temperature.

Heat Conduction

Melted snow pattern on a parking lot indicates the presence of underground hot water pipes used to aid snow removal. Heat from the water is conducted to the pavement from the pipes, causing the snow to melt. *(Courtesy of Dr. Albert A. Bartlett, University of Colorado, Boulder)*

The easiest thermal energy transfer process to describe quantitatively is called *conduction*. In this process, the thermal energy transfer can be viewed on an atomic scale as an exchange of kinetic energy between molecules, where the less energetic particles gain energy by colliding with the more energetic particles. For example, if you insert a metallic bar into a flame while holding one end, you will find that the temperature of the metal in your hand increases. The thermal energy reaches your hand through conduction. The manner in which thermal energy is transferred from the flame, through the bar, and to your hand can be understood by examining what is happening to the atoms and electrons of the metal. Initially, before the rod is inserted into the flame, the metal atoms and electrons are vibrating about their equilibrium positions. As the flame heats the rod, those metal atoms and electrons near the flame begin to vibrate with larger and larger amplitudes. These, in turn, collide with their neighbors and transfer some of their energy in the collisions. Slowly, metal atoms and electrons farther down the rod increase their amplitude of vibration, until the large-amplitude vibrations arrive at the end being held. The effect of this increased vibration is an increase in temperature of the metal, and possibly a burned hand.

Although the transfer of thermal energy through a metal can be partially explained by atomic vibrations and electron motion, the rate of conduction also depends on the properties of the substance being heated. For example, it is possible to hold a piece of asbestos in a flame indefinitely. This implies that very little thermal energy is being conducted through the asbestos. In general, metals are good conductors of thermal energy, and materials such as asbestos, cork, paper, and fiber glass are poor conductors. Gases also are poor conductors because of their dilute nature. Metals are good conductors of thermal energy because they contain large numbers of electrons that are relatively free to move through the metal and can transport energy from one region to another. Thus, in a good conductor, such as copper, conduction takes place via the vibration of atoms and via the motion of free electrons.

Conduction occurs only if there is a difference in temperature between two parts of the conducting medium. Consider a slab of material of thickness Δx and cross-sectional area A with its opposite faces at different temperatures T_1 and T_2, where $T_2 > T_1$ (Fig. 20.8). It is found from experiment that the thermal energy Q transferred in a time Δt flows from the hotter end to the colder end. The rate at which heat flows, $Q/\Delta t$, is found to be proportional to the cross-sectional area and the temperature difference, and inversely proportional to the thickness:

$$\frac{Q}{\Delta t} \propto A \frac{\Delta T}{\Delta x}$$

It is convenient to use the symbol H to represent the thermal energy transfer rate. That is, we take $H = Q/\Delta t$. Note that H has units of watts when Q is in joules and Δt is in seconds. For a slab of infinitesimal thickness dx and temperature difference dT, we can write the **law of heat conduction**

$$H = -kA \frac{dT}{dx} \qquad (20.14)$$

Law of heat conduction

where the proportionality constant k is called the **thermal conductivity** of the material, and dT/dx is the **temperature gradient** (the variation of temperature with position). The minus sign in Equation 20.14 denotes the fact that thermal energy flows in the direction of decreasing temperature.

Suppose a substance is in the shape of a long uniform rod of length L, as in Figure 20.9, and is insulated so that thermal energy cannot escape from its surface

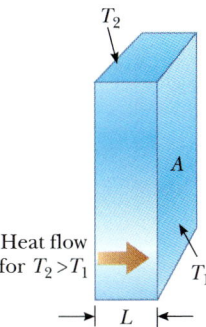

FIGURE 20.8 Heat transfer through a conducting slab of cross-sectional area A and thickness Δx. The opposite faces are at different temperatures, T_1 and T_2.

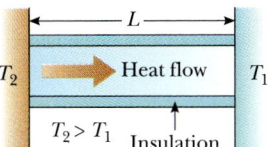

FIGURE 20.9 Conduction of heat through a uniform, insulated rod of length L. The opposite ends are in thermal contact with heat reservoirs at different temperatures.

Substance	Thermal Conductivity W/m·°C
Metals (at 25°C)	
Aluminum	238
Copper	397
Gold	314
Iron	79.5
Lead	34.7
Silver	427
Gases (at 20°C)	
Air	0.0234
Helium	0.138
Hydrogen	0.172
Nitrogen	0.0234
Oxygen	0.0238
Nonmetals (approximate values)	
Asbestos	0.08
Concrete	0.8
Diamond	2300
Glass	0.8
Ice	2
Rubber	0.2
Water	0.6
Wood	0.08

TABLE 20.3 **Thermal Conductivities of Some Substances**

except at the ends, which are in thermal contact with heat reservoirs having temperatures T_1 and T_2. When a steady state has been reached, the temperature at each point along the rod is constant in time. In this case, the temperature gradient is the same everywhere along the rod and is

$$\frac{dT}{dx} = \frac{T_1 - T_2}{L}$$

Thus the thermal energy transfer rate is

$$H = kA\frac{(T_2 - T_1)}{L} \qquad (20.15)$$

Substances that are good thermal conductors have large thermal conductivity values, whereas good thermal insulators have low thermal conductivity values. Table 20.3 lists thermal conductivities for various substances. We see that metals are generally better thermal conductors than nonmetals.

For a compound slab containing several materials of thicknesses $L_1, L_2, \ldots$ and thermal conductivities $k_1, k_2, \ldots$, the rate of thermal energy transfer through the slab at steady state is

$$H = \frac{A(T_2 - T_1)}{\Sigma_i (L_i / k_i)} \qquad (20.16)$$

where T_1 and T_2 are the temperatures of the outer extremities of the slab (which are held constant) and the summation is over all slabs. The following example is a proof of this result.

EXAMPLE 20.9 **Heat Transfer Through Two Slabs**

Two slabs of thickness L_1 and L_2 and thermal conductivities k_1 and k_2 are in thermal contact with each other as in Figure 20.10. The temperatures of their outer surfaces are T_1 and T_2, respectively, and $T_2 > T_1$. Determine the temperature at the interface and the rate of thermal energy transfer through the slabs in the steady-state condition.

Solution If T is the temperature at the interface, then the rate at which thermal energy is transferred through slab 1 is

$$(1) \qquad H_1 = \frac{k_1 A (T - T_1)}{L_1}$$

Likewise, the rate at which thermal energy is transferred through slab 2 is

$$(2) \qquad H_2 = \frac{k_2 A (T_2 - T)}{L_2}$$

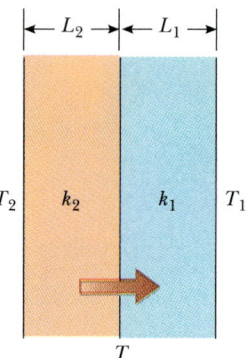

FIGURE 20.10 (Example 20.9) Heat transfer by conduction through two slabs in thermal contact with each other. At steady state, the rate of heat transfer through slab 1 equals the rate of heat transfer through slab 2.

When a steady state is reached, these two rates must be equal; hence,

$$\frac{k_1 A(T - T_1)}{L_1} = \frac{k_2 A(T_2 - T)}{L_2}$$

Solving for T gives

$$(3) \qquad T = \frac{k_1 L_2 T_1 + k_2 L_1 T_2}{k_1 L_2 + k_2 L_1}$$

Substituting (3) into either (1) or (2), we get

$$H = \frac{A(T_2 - T_1)}{(L_1/k_1) + (L_2/k_2)}$$

An extension of this model to several slabs of materials leads to Equation 20.16.

*Home Insulation

If you would like to do some calculating to determine whether or not to add insulation to a ceiling or to some other portion of a building, what you have just learned about conduction needs to be modified slightly because the insulating properties of materials used in buildings are usually expressed in engineering rather than SI units. For example, measurements stamped on a package of fiberglass insulating board will be in units such as British thermal units, feet, and degrees Fahrenheit.

In engineering practice, the term L/k for a particular substance is referred to as the R value of the material. Thus, Equation 20.16 reduces to

$$H = \frac{A(T_2 - T_1)}{\Sigma_i R_i} \qquad (20.17)$$

where $R_i = L_i/k_i$. The R values for a few common building materials are given in Table 20.4 (note the units).

Also, near any vertical surface there is a very thin, stagnant layer of air that must be considered when finding the total R value for a wall. The thickness of this stagnant layer on an outside wall depends on the speed of the wind. As a result, thermal energy loss from a house on a day when the wind is blowing hard is greater than loss on a day when the wind speed is zero. A representative R value for this stagnant layer of air is given in Table 20.4.

TABLE 20.4   *R* Values for Some Common Building Materials

Material	R Value (ft$^2 \cdot$ °F$\cdot$h/BTU)
Hardwood siding (1 in. thick)	0.91
Wood shingles (lapped)	0.87
Brick (4 in. thick)	4.00
Concrete block (filled cores)	1.93
Fiberglass batting (3.5 in. thick)	10.90
Fiberglass batting (6 in. thick)	18.80
Fiberglass board (1 in. thick)	4.35
Cellulose fiber (1 in. thick)	3.70
Flat glass (0.125 in. thick)	0.89
Insulating glass (0.25-in. space)	1.54
Vertical air space (3.5 in. thick)	1.01
Air film	0.17
Drywall (0.5 in. thick)	0.45
Sheathing (0.5 in. thick)	1.32

This thermogram of a home, made during cold weather, shows colors ranging from white and yellow (areas of greatest heat loss) to blue and purple (areas of least heat loss). *(Daedalus Enterprises, Inc./Peter Arnold, Inc.)*

EXAMPLE 20.10   The *R* Value of a Typical Wall

Calculate the total *R* value for a wall constructed as shown in Figure 20.11a. Starting outside the house (to the left in Fig. 20.11a) and moving inward, the wall consists of brick, 0.5 in. of sheathing, an air space 3.5 in. thick, and 0.5 in. of dry wall. Do not forget the air layers inside and outside the house.

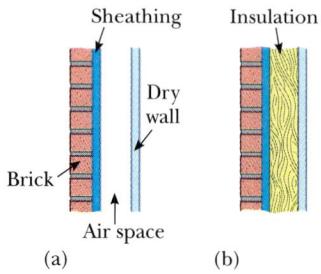

FIGURE 20.11   (Example 20.10) Cross-sectional view of an exterior wall containing (a) an air space and (b) insulation.

Solution   Referring to Table 20.4, we find the total *R* value for the wall:

R_1 (outside air film)	$= 0.17 \ \text{ft}^2 \cdot {}^{\circ}\text{F} \cdot \text{h}/\text{BTU}$
R_2 (brick)	$= 4.00$
R_3 (sheathing)	$= 1.32$
R_4 (air space)	$= 1.01$
R_5 (dry wall)	$= 0.45$
R_6 (inside air film)	$= 0.17$
R_{total}	$= 7.12 \ \text{ft}^2 \cdot {}^{\circ}\text{F} \cdot \text{h}/\text{BTU}$

Exercise   If a layer of fiber-glass insulation 3.5 in. thick is placed inside the wall to replace the air space as in Figure 20.11b, what is the new total *R* value? By what factor is the thermal energy loss reduced?

Answer $R = 17 \ \text{ft}^2 \cdot {}^{\circ}\text{F} \cdot \text{h}/\text{BTU}$; a factor of 2.4.

Convection

At one time or another you probably have warmed your hands by holding them over an open flame. In this situation, the air directly above the flame is heated and expands. As a result, the density of the air decreases and the air rises. This warmed mass of air heats your hands as it flows by. *Thermal energy transferred by the movement of a heated substance is said to have been transferred by* **convection.** When the movement results from differences in density, as in the example of air around a fire, it is referred to as *natural convection.* When the heated substance is forced to move by a fan or pump, as in some hot-air and hot-water heating systems, the process is called *forced convection.*

The circulating pattern of air flow at a beach is an example of convection. Likewise, the mixing that occurs as water is cooled and eventually freezes at its surface (Chapter 19) is an example of convection in nature. Recall that the mixing by convection currents ceases when the water temperature reaches 4°C. Since the water in the lake cannot be cooled by convection below 4°C, and because water is a relatively poor conductor of thermal energy, the water near the bottom remains near 4°C for a long time. As a result, fish have a comfortable temperature in which to live even in periods of prolonged cold weather.

If it were not for convection currents, it would be very difficult to boil water. As water is heated in a teakettle, the lower layers are warmed first. These heated regions expand and rise to the top because their density is lowered. At the same time, the denser cool water replaces the warm water at the bottom of the kettle so that it can be heated.

The same process occurs when a room is heated by a radiator. The hot radiator warms the air in the lower regions of the room. The warm air expands and rises to the ceiling because of its lower density. The denser regions of cooler air from above replace the warm air, setting up the continuous air current pattern shown in Figure 20.12.

FIGURE 20.12 Convection currents are set up in a room heated by a radiator.

Radiation

The third way of transferring thermal energy is through **radiation.** All objects radiate energy continuously in the form of electromagnetic waves, which we shall discuss in Chapter 34. The type of radiation associated with the transfer of thermal energy from one location to another is referred to as infrared radiation.

Through electromagnetic radiation, approximately 1340 J of thermal energy from the Sun strikes 1 m² of the top of the Earth's atmosphere every second. Some of this energy is reflected back into space and some is absorbed by the atmosphere, but enough arrives at the surface of the Earth each day to supply all of our energy needs on this planet hundreds of times over—if it could be captured and used efficiently. The growth in the number of solar houses in this country is one example of an attempt to make use of this free energy.

Radiant energy from the Sun affects our day-to-day existence in a number of ways. It influences the Earth's average temperature, ocean currents, agriculture, rain patterns, and so on. For example, consider what happens to the atmospheric temperature at night. If there is a cloud cover above the Earth, the water vapor in the clouds reflects back a part of the infrared radiation emitted by the Earth and consequently the temperature remains at moderate levels. In the absence of this cloud cover, however, there is nothing to prevent this radiation from escaping into space, and thus the temperature drops more on a clear night than when it is cloudy.

The rate at which an object emits radiant energy is proportional to the fourth power of its absolute temperature. This is known as **Stefan's law** and is expressed in equation form as

$$P = \sigma A e T^4 \qquad (20.18)$$

Stefan's law

where P is the power radiated by the body in watts, σ is a constant equal to 5.6696×10^{-8} W/m²·K⁴, A is the surface area of the object in square meters, e is a constant called the **emissivity,** and T is temperature in kelvin. The value of e can vary between zero and unity, depending on the properties of the surface.

An object radiates energy at a rate given by Equation 20.18. At the same time, the object also absorbs electromagnetic radiation. If the latter process did not occur, an object would eventually radiate all of its energy and its temperature would reach absolute zero. The energy that a body absorbs comes from its surroundings, which consist of other objects that radiate energy. If an object is at a temperature T and its surroundings are at a temperature T_0, the net energy gained or lost each second by the object as a result of radiation is

$$P_{net} = \sigma Ae(T^4 - T_0{}^4) \tag{20.19}$$

When an object is in equilibrium with its surroundings, it radiates and absorbs energy at the same rate, and so its temperature remains constant. When an object is hotter than its surroundings, it radiates more energy than it absorbs and so it cools. An **ideal absorber** is defined as an object that absorbs all of the energy incident on it. The emissivity of an ideal absorber is equal to unity. Such an object is often referred to as a **black body.** An ideal absorber is also an ideal radiator of energy. In contrast, an object with an emissivity equal to zero absorbs none of the energy incident on it. Such an object reflects all the incident energy and so is a perfect reflector.

The Dewar Flask

The Thermos bottle, called a *Dewar flask*[8] in the scientific community, is a practical example of a container designed to minimize thermal energy losses by conduction, convection, and radiation. Such a container is used to store either cold or hot liquids for long periods of time. The standard construction (Fig. 20.13) consists of a double-walled Pyrex vessel with silvered inner walls. The space between the walls is evacuated to minimize thermal energy transfer by conduction and convection. The silvered surfaces minimize thermal energy transfer by radiation by reflecting most of the radiant heat. Very little thermal energy is lost over the neck of the flask since glass is not a very good conductor, and the cross-section of the glass in the direction of conduction is small. A further reduction in thermal energy loss is obtained by reducing the size of the neck. Dewar flasks are commonly used to store liquid nitrogen (boiling point 77 K) and liquid oxygen (boiling point 90 K).

In order to confine liquid helium, which has a very low heat of vaporization (boiling point 4.2 K), it is often necessary to use a double Dewar system in which the Dewar flask containing the liquid is surrounded by a second Dewar flask. The space between the two flasks is filled with liquid nitrogen.

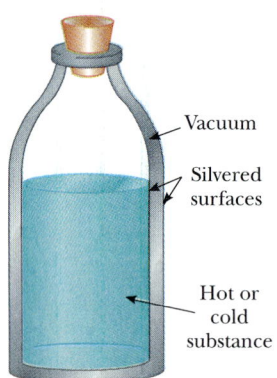

FIGURE 20.13   A cross-sectional view of a Dewar vessel, used to store hot or cold liquids or other substances.

Vacuum

Silvered surfaces

Hot or cold substance

EXAMPLE 20.11   Who Turned Down the Thermostat?

An unclothed student is in a 20°C room. If the skin temperature of the student is 37°C, how much heat is lost from his body in 10 min, assuming that the emissivity of skin is 0.90 and the surface area of the student is 1.5 m²?

Solution   Using Equation 20.19, the rate of thermal energy loss from the skin is

$$P_{net} = \sigma Ae(T^4 - T_0{}^4)$$

$$= (5.67 \times 10^{-8} \text{ W/m}^2 \cdot \text{K}^4)(1.5 \text{ m}^2)$$
$$\times (0.90)[(310 \text{ K})^4 - (293 \text{ K})^4]$$
$$= 140 \text{ J/s}$$

(Why is the temperature given in kelvin?) At this loss rate, the total thermal energy lost by the skin in 10 min is

$$Q = P_{net} \times \text{time} = (140 \text{ J/s})(600 \text{ s}) = \boxed{8.6 \times 10^4 \text{ J}}$$

---

[8] Invented by Sir James Dewar (1842–1923).

SUMMARY

Thermal energy transfer is a form of energy transfer that takes place as a consequence of a temperature difference. The **internal energy** of a substance is a function of its state and generally increases with increasing temperature.

The **calorie** is the amount of heat necessary to raise the temperature of 1 g of water from 14.5°C to 15.5°C. The **mechanical equivalent of heat** is 4.186 J/cal.

The **heat capacity** C of any substance is defined as the amount of thermal energy needed to raise the temperature of the substance by one degree Celsius. The thermal energy required to change the temperature of a substance by ΔT is

$$Q = mc \, \Delta T \tag{20.4}$$

where m is the mass of the substance and c is its **specific heat**.

The thermal energy required to change the phase of a pure substance of mass m is

$$Q = mL \tag{20.6}$$

The parameter L is called the **latent heat** of the substance and depends on the nature of the phase change and the properties of the substance.

The **work done** by a gas as its volume changes from some initial value V_i to some final value V_f is

$$W = \int_{V_i}^{V_f} P \, dV \tag{20.8}$$

where P is the pressure, which may vary during the process. In order to evaluate W, the nature of the process must be specified—that is, P and V must be known during each step. Since the work done depends on the initial, final, and intermediate states, it therefore depends on the path taken between the initial and final states.

The **first law of thermodynamics** states that when a system undergoes a change from one state to another, the change in its internal energy is

$$\Delta U = Q - W \tag{20.9}$$

where Q is the thermal energy transferred into (or out of) the system and W is the work done by (or on) the system. Although Q and W both depend on the path taken from the initial state to the final state, the quantity ΔU is path-independent.

In a **cyclic process** (one that originates and terminates at the same state), $\Delta U = 0$, and therefore $Q = W$. That is, the thermal energy transferred into the system equals the work done during the cycle.

An **adiabatic process** is one in which no thermal energy is transferred between the system and its surroundings ($Q = 0$). In this case, the first law gives $\Delta U = -W$. That is, the internal energy changes as a consequence of work being done by (or on) the system. In an **adiabatic free expansion** of a gas, $Q = 0$ and $W = 0$, and so $\Delta U = 0$. That is, the internal energy of the gas does not change in such a process.

An **isovolumetric process** is one that occurs at constant volume. No expansion work is done in such a process.

An **isobaric process** is one that occurs at constant pressure. The work done in such a process is $P \Delta V$.

An **isothermal process** is one that occurs at constant temperature. The work done by an ideal gas during a reversible isothermal process is

$$W = nRT \ln\left(\frac{V_f}{V_i}\right) \qquad (20.12)$$

Heat may be transferred by conduction, convection, and radiation. **Conduction** can be viewed as an exchange of kinetic energy between colliding molecules or electrons. The rate at which heat flows by conduction through a slab of area A is

$$H = -kA\frac{dT}{dx} \qquad (20.14)$$

where k is the **thermal conductivity** and dT/dx is the **temperature gradient.**

In **convection,** the heated substance moves from one place to another.

All bodies radiate and absorb energy in the form of electromagnetic waves. A body that is hotter than its surroundings radiates more energy than it absorbs, whereas a body that is cooler than its surroundings absorbs more energy than it radiates. A **black body** (or ideal absorber) is one that absorbs all energy incident on it; furthermore, a black body emits as much energy as is possible for a body of its size, shape, and temperature.

QUESTIONS

1. Ethyl alcohol has about one-half the specific heat of water. If equal masses of alcohol and water in separate beakers are supplied with the same amount of heat, compare the temperature increases of the two liquids.

2. Give one reason why coastal regions tend to have a more moderate climate than inland regions.

3. A small crucible is taken from a 200°C oven and immersed in a tub full of water at room temperature (this process is often referred to as *quenching*). What is the approximate final equilibrium temperature?

4. What is the major problem that arises in measuring specific heats if a sample with a temperature above 100°C is placed in water?

5. In a daring lecture demonstration, an instructor dips his wetted fingers into molten lead (327°C) and withdraws them quickly, without getting burned. How is this possible? (This is a dangerous experiment, which you should *not* attempt.)

6. The pioneers found that a large tub of water placed in a storage cellar would prevent their food from freezing on really cold nights. Explain why this is so.

7. What is wrong with the statement: "Given any two bodies, the one with the higher temperature contains more heat."?

8. Why is it possible to hold a lighted match, even when it is burned, to within a few millimeters of your fingertips?

9. Figure 20.14 shows a pattern formed by snow on the roof of a barn. What causes the alternating pattern of snow-covered and exposed roof?

10. Why is a person able to remove a piece of dry aluminum

FIGURE 20.14   (Question 9) Alternating pattern on a snow-covered roof.   *(Courtesy of Dr. Albert A. Bartlett, University of Colorado, Boulder)*

foil from a hot oven with bare fingers, while if there is moisture on the foil, a burn results?

11. A tile floor in a bathroom may feel uncomfortably cold to your bare feet, but a carpeted floor in an adjoining room at the same temperature will feel warm. Why?

12. Why can potatoes be baked more quickly when a metal skewer has been inserted through them?

13. Give reasons for the silvered walls and the vacuum jacket in a Thermos bottle.

14. A piece of paper is wrapped around a rod made half of wood and half of copper. When held over a flame, the paper in contact with the wood burns but the half in contact with the metal does not. Explain.

15. Why is it necessary to store liquid nitrogen or liquid oxygen in vessels equipped with either Styrofoam insulation or a double-evacuated wall?

16. Why do heavy draperies over the windows help keep a home warm in the winter and cool in the summer?

17. If you wish to cook a piece of meat thoroughly on an open fire, why should you not use a high flame? (Note that carbon is a good thermal insulator.)

18. When insulating a wood-frame house, is it better to place the insulation against the cooler outside wall or against the warmer inside wall? (In either case, there is an air barrier to consider.)

19. In an experimental house, Styrofoam beads were pumped into the air space between the double windows at night in the winter and pumped out to holding bins during the day. How would this assist in conserving heat energy in the house?

20. Pioneers stored fruits and vegetables in underground cellars. Discuss as fully as possible this choice for a storage site.

21. Why can you get a more severe burn from steam at 100°C than from water at 100°C?

22. Concrete has a higher specific heat than soil. Use this fact to explain (partially) why cities have a higher average night-time temperature than the surrounding countryside. If a city is hotter than the surrounding countryside, would you expect breezes to blow from city to country or from country to city? Explain.

23. When camping in a canyon on a still night, one notices that as soon as the Sun strikes the surrounding peaks, a breeze begins to stir. What causes the breeze?

24. Updrafts of air are familiar to all pilots. What causes these currents?

25. If water is a poor conductor of heat, why can it be heated quickly when placed over a flame?

26. The U.S. penny is now made of copper-coated zinc. Can a calorimetric experiment be devised to test for the metal content in a collection of pennies? If so, describe the procedure you would use.

27. If you hold water in a paper cup over a flame, you can bring the water to a boil without burning the cup. How is this possible?

28. When a sealed Thermos bottle full of hot coffee is shaken, what are the changes, if any, in (a) the temperature of the coffee and (b) the internal energy of the coffee?

29. Using the first law of thermodynamics, explain why the *total* energy of an isolated system is always constant.

30. Is it possible to convert internal energy to mechanical energy? Explain with examples.

31. Suppose you pour hot coffee for your guests, and one of them chooses to drink the coffee after it has been in the cup for several minutes. In order to have the warmest coffee, should the person add the cream just after the coffee is poured or just before drinking? Explain.

32. Two identical cups both at room temperature are filled with the same amount of hot coffee. One cup contains a metal spoon, while the other does not. If you wait for several minutes, which of the two will have the warmer coffee? Which heat transfer process explains your answer?

33. A warning sign often seen on highways just before a bridge is "Caution—Bridge surface freezes before road surface." Which of the three heat transfer processes is most important in causing a bridge surface to freeze before a road surface on very cold days?

PROBLEMS

Section 20.1   Heat and Thermal Energy

1. Consider Joule's apparatus described in Figure 20.1. The two masses are 1.50 kg each, and the tank is filled with 200 g of water. What is the increase in the temperature of the water after the masses fall through a distance of 3.00 m?

2. An 80-kg weight watcher wishes to climb a mountain to work off the equivalent of a large piece of chocolate cake rated at 700 (food) Calories. How high must the person climb?

3. Water at the top of Niagara Falls has a temperature of 10°C. If it falls through a distance of 50 m and all of its potential energy goes into heating the water, calculate the temperature of the water at the bottom of the falls.

Sections 20.2 and 20.3   Heat Capacity, Specific Heat, and Latent Heat

4. How many calories of heat are required to raise the temperature of 3.0 kg of aluminum from 20°C to 50°C?

5. The temperature of a silver bar rises by 10.0°C when

it absorbs 1.23 kJ of heat. The mass of the bar is 525 g. Determine the specific heat of silver.

6. If 100 g of water at 100°C is poured into a 20-g aluminum cup containing 50 g of water at 20°C, what is the equilibrium temperature of the system?

6A. If a mass m_h of water at T_h is poured into an aluminum cup of mass m_{Al} containing m_c of cold water at T_c, where $T_h > T_c$, what is the equilibrium temperature of the system?

7. What is the final equilibrium temperature when 10 g of milk at 10°C is added to 160 g of coffee at 90°C? (Assume the heat capacities of the two liquids are the same as that of water, and neglect the heat capacity of the container.)

8. (a) A calorimeter contains 500 ml of water at 30°C and 25 g of ice at 0°C. Determine the final temperature of the system. (b) Repeat part (a) if 250 g of ice is initially present at 0°C.

9. A 1.5-kg iron horseshoe initially at 600°C is dropped into a bucket containing 20 kg of water at 25°C. What is the final temperature? (Neglect the heat capacity of the container.)

10. The air temperature above coastal areas is profoundly influenced by the large specific heat of water. One reason is that the heat released when 1 cubic meter of water cools by 1.0°C will raise the temperature of an enormously larger volume of air by 1.0°C. Estimate this volume of air. The specific heat of air is approximately 1.0 kJ/kg·°C. Take the density of air to be 1.25 kg/m³.

11. If 200 g of water is contained in a 300-g aluminum vessel at 10°C and an additional 100 g of water at 100°C is poured into the container, what is the final equilibrium temperature of the system?

12. A student inhales 22°C air and exhales 37°C air. The average volume of air in one breath is 200 cm³. Neglecting evaporation of water into the air, estimate the amount of heat absorbed in one day by the air breathed by the student. The density of air is approximately 1.25 kg/m³, and the specific heat of air is 1000 J/kg·°C.

13. How much heat must be added to 20 g of aluminum at 20°C to melt it completely?

14. An insulated vessel contains a saturated vapor being cooled as cold water flows in a pipe that passes through the vessel. The temperature of the entering water is 273 K. When the flow speed is 3.0 m/s, the temperature of the exiting water is 303 K. Determine the temperature of the exiting water when the flow speed is decreased to 2.0 m/s. Assume that the rate of condensation remains unchanged.

14A. An insulated vessel contains a saturated vapor being cooled as cold water flows in a pipe that passes through the vessel. The temperature of the entering

water is T_1. When the flow speed is v_1, the temperature of the exiting water is T_{out}. Determine the temperature of the exiting water when the flow speed of the water is decreased to v_2. Assume that the rate of condensation remains unchanged.

15. A water heater is operated by solar power. If the solar collector has an area of 6.0 m² and the power delivered by sunlight is 550 W/m², how long does it take to increase the temperature of 1.0 m³ of water from 20°C to 60°C?

16. A 1.0-kg block of copper at 20°C is dropped into a large vessel of liquid nitrogen at 77 K. How many kilograms of nitrogen boil away by the time the copper reaches 77 K? (The specific heat of copper is 0.092 cal/g·°C. The latent heat of vaporization of nitrogen is 48 cal/g.)

17. How much heat is required to vaporize a 1.0-g ice cube initially at 0°C? The latent heat of fusion of ice is 80 cal/g and the latent heat of vaporization of water is 540 cal/g.

18. One liter of water at 30°C is used to make iced tea. How much ice at 0°C must be added to lower the temperature of the tea to 10°C?

19. When a driver brakes an automobile, the friction between the brake drums and brake shoes converts the car's kinetic energy to heat. If a 1500-kg automobile traveling at 30 m/s comes to a halt, how much does the temperature rise in each of the four 8.0-kg iron brake drums? (Neglect heat loss to the surroundings.)

20. If 90.0 g of molten lead at 327.3°C is poured into a 300.0-g casting made of iron initially at 20.0°C, what is the final temperature of the system? (Assume there are no heat losses.)

21. In an insulated vessel, 250 g of ice at 0°C is added to 600 g of water at 18°C. (a) What is the final temperature of the system? (b) How much ice remains when the system reaches equilibrium?

22. A 50.0-g ice cube at −20.0°C is dropped into a container of water at 0.0°C. How much water freezes onto the ice?

23. An iron nail is driven into a block of ice by a single blow of a hammer. The hammerhead has a mass of 0.50 kg and an initial speed of 2.0 m/s. Nail and hammer are at rest after the blow. How much ice melts? Assume the temperature of the nail is 0.0°C before and after.

24. Two speeding 5.0-g lead bullets, both at temperature 20°C, collide head-on when each is moving at 500 m/s. Assuming a perfectly inelastic collision and no loss of heat to the atmosphere, describe the final state of the two-bullet system.

25. A 3.0-g copper penny at 25°C drops 50 m to the ground. (a) If 60% of its initial potential energy goes into increasing the internal energy, determine its final temperature. (b) Does the result depend on the mass of the penny? Explain.

26. Lake Erie contains roughly 4.0×10^{11} m^3 of water. (a) How much heat is required to raise the temperature of that volume of water from 11°C to 12°C? (b) Approximately how many years would it take to supply this amount of heat by using the full output of a 1000-MW electric power plant?

27. A 3.0-g lead bullet is traveling at 240 m/s when it embeds in a block of ice at 0°C. If all the heat generated goes into melting ice, what quantity of ice is melted? (The latent heat of fusion for ice is 80 kcal/kg, and the specific heat of lead is 0.030 kcal/kg·°C.)

Section 20.4 Work and Heat in Thermodynamic Processes

28. One mole of an ideal gas is heated slowly so that it goes from the state (P_0, V_0) to the state $(3P_0, 3V_0)$. This change occurs in such a way that the gas pressure is directly proportional to the volume. (a) How much work is done in the process? (b) How is the temperature of the gas related to its volume during this process?

29. A gas expands from I to F along three possible paths as indicated by Figure P20.29. Calculate the work in joules done by the gas along the paths *IAF*, *IF*, and *IBF*.

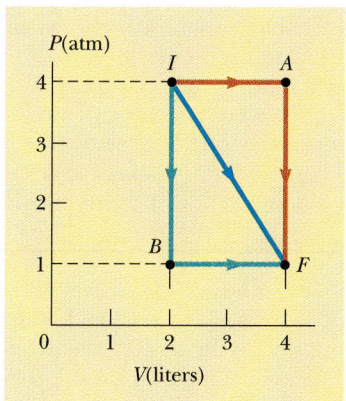

FIGURE P20.29

30. Gas in a container is at a pressure of 1.5 atm and a volume of 4.0 m^3. What is the work done by the gas if (a) it expands at constant pressure to twice its initial volume and (b) it is compressed at constant pressure to one-quarter its initial volume?

31. An ideal gas is enclosed in a cylinder that has a movable piston on top. The piston has a mass of 8000 g and an area of 5.0 cm^2 and is free to slide up and down, keeping the pressure of the gas constant. How much work is done as the temperature of 0.20 mol of the gas is raised from 20°C to 300°C?

31A. An ideal gas is enclosed in a cylinder that has a movable piston on top. The piston has a mass m and an area A and is free to slide up and down, keeping the pressure of the gas constant. How much work is done as the temperature of n mol of the gas is raised from T_1 to T_2?

32. An ideal gas undergoes a thermodynamic process that consists of two isobaric and two isothermal steps as shown in Figure P20.32. Show that the net work done during the four steps is

$$W_{\text{net}} = P_1 (V_2 - V_1) \ln\left(\frac{P_2}{P_1}\right)$$

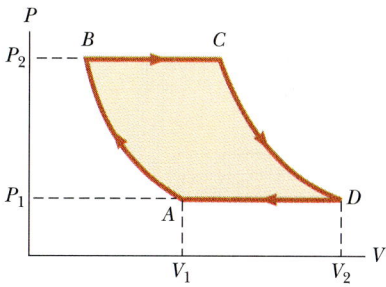

FIGURE P20.32

33. A sample of ideal gas is expanded to twice its original volume of 1.0 m^3 in a quasi-static process for which $P = \alpha V^2$, with $\alpha = 5.0$ atm/m^6, as shown in Figure P20.33. How much work was done by the expanding gas?

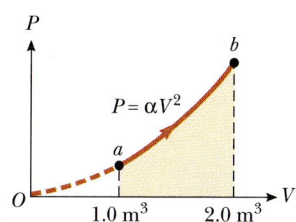

FIGURE P20.33

34. An ideal gas at STP (1 atm and 0°C) is taken through a process in which the volume is expanded from 25 L to 80 L. During this process the pressure varies inversely as the volume squared, $P = 0.5 \, aV^{-2}$. (a) Determine the constant a in standard SI units. (b) Find the final pressure and temperature. (c) Determine a general expression for the work done by the gas during this process. (d) Compute the actual work in joules done by the gas in this process.

35. One mole of an ideal gas does 3000 J of work on the surroundings as it expands isothermally to a final pressure of 1 atm and volume of 25 L. Determine

(a) the initial volume and (b) the temperature of the gas.

Section 20.5   The First Law of Thermodynamics

36. An ideal gas undergoes the cyclic process shown in Figure P20.36 from A to B to C and back to A. (a) Sketch a *PV* diagram for this cycle, and identify the steps during which heat is absorbed and those during which heat is evolved. (b) What is the overall result of the cycle in terms of U, Q, and W?

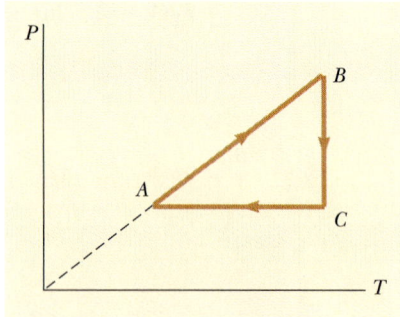

FIGURE P20.36

37. A gas is compressed at a constant pressure of 0.80 atm from 9.0 L to 2.0 L. In the process, 400 J of thermal energy leaves the gas. (a) What is the work done by the gas? (b) What is the change in its internal energy?

37A. A gas is compressed at a constant pressure P from V_1 to V_2. In the process, Q joules of thermal energy leaves the gas. (a) What is the work done by the gas? (b) What is the change in its internal energy?

38. A thermodynamic system undergoes a process in which its internal energy decreases by 500 J. If at the same time, 220 J of work is done on the system, find the thermal energy transferred to or from it.

39. A gas is taken through the cyclic process described in Figure P20.39. (a) Find the net thermal energy transferred to the system during one complete cycle.

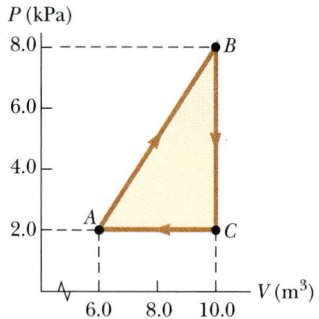

FIGURE P20.39

(b) If the cycle is reversed, that is, the process goes along *ACBA*, what is the net thermal energy transferred per cycle?

40. An ideal gas system goes through the process shown in Figure P20.40. From *A* to *B*, the process is adiabatic, and from *B* to *C*, it is isobaric with 100 kJ of heat flow into the system. From *C* to *D*, the process is isothermal, and from *D* to *A*, it is isobaric with 150 kJ of heat flow out of the system. Determine the difference in internal energy $U_B - U_A$.

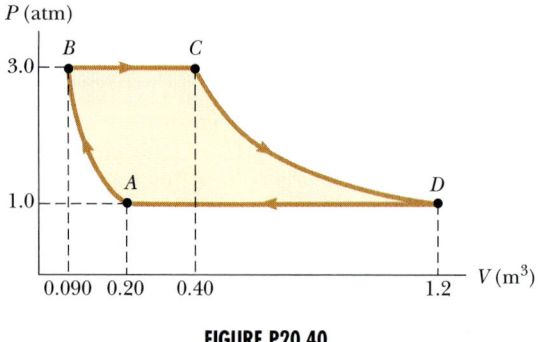

FIGURE P20.40

Section 20.6   Some Applications of the First Law of Thermodynamics

41. Five moles of an ideal gas expands isothermally at 127°C to four times its initial volume. Find (a) the work done by the gas and (b) the thermal energy transferred to the system, both in joules.

42. How much work is done by the steam when 1.0 mol of water at 100°C boils and becomes 1.0 mol of steam at 100°C at 1.0 atm pressure? Determine the change in internal energy of the steam as it vaporizes. Consider the steam to be an ideal gas.

43. Helium is heated at constant pressure from 273 K to 373 K. If the gas does 20.0 J of work during the process, what is the mass of the helium?

44. One mole of an ideal gas is heated at constant pressure so that its temperature triples. Then the gas is heated at constant temperature so that its volume triples. Find the ratio of the work done during the isothermal process to that done during the isobaric process.

45. An ideal gas initially at 300 K undergoes an isobaric expansion at 2.50 kPa. If the volume increases from 1.00 m³ to 3.00 m³ and 12.5 kJ of thermal energy is transferred to the gas, find (a) the change in its internal energy and (b) its final temperature.

46. Two moles of helium gas initially at 300 K and 0.40 atm are compressed isothermally to 1.2 atm. Find (a) the final volume of the gas, (b) the work done by the gas, and (c) the thermal energy transferred. Consider the helium to behave as an ideal gas.

47. One mole of water vapor at 373 K cools to 283 K. The heat given off by the cooling water vapor is absorbed by 10 mol of an ideal gas, and this heat absorption causes the gas to expand at a constant temperature of 273 K. If the final volume of the ideal gas is 20.0 L, determine its initial volume.

47A. One mole of water vapor at a temperature T_h cools to liquid at T_c. The heat, given off by the cooling water vapor, is absorbed by n mol of an ideal gas, and this heat absorption causes the gas to expand at a constant temperature of T_0. If the final volume of the ideal gas is V_f, determine its initial volume.

48. During a controlled expansion, the pressure of a gas is

$$P = 12e^{-bV} \text{ atm} \qquad b = \frac{1}{12 \text{ m}^3}$$

where the volume is in m^3 (Fig. P20.48). Determine the work performed when the gas expands from 12 m^3 to 36 m^3.

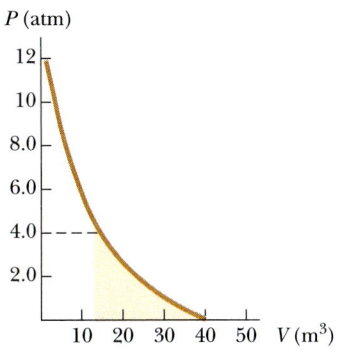

FIGURE P20.48

49. A 1.0-kg block of aluminum is heated at atmospheric pressure such that its temperature increases from 22°C to 40°C. Find (a) the work done by the aluminum, (b) the thermal energy added to it, and (c) the change in its internal energy.

50. In Figure P20.50, the change in internal energy of a gas that is taken from A to C is $+800$ J. The work done along path ABC is $+500$ J. (a) How much thermal energy has to be added to the system as it goes from A through B to C? (b) If the pressure at point A is five times that of point C, what is the work done by the system in going from C to D? (c) What is the thermal energy exchanged with the surroundings as the cycle goes from C to A? (d) If the change in internal energy in going from point D to point A is $+500$ J, how much thermal energy must be added to the system as it goes from point C to point D?

51. Helium with an initial volume of 1.00 liter and an initial pressure of 10.0 atm expands to a final volume of 1.00 m^3. The relationship between pressure and

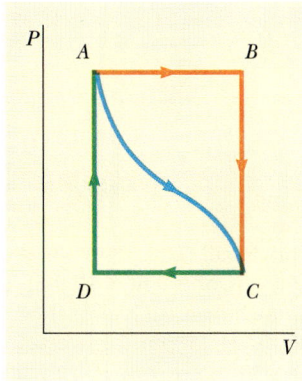

FIGURE P20.50

volume during the expansion is $PV = $ constant. Determine (a) the value of the constant, (b) the final pressure, and (c) the work done by the helium during the expansion.

Section 20.7 Heat Transfer

52. A glass window pane has an area of 3.0 m^2 and a thickness of 0.60 cm. If the temperature difference between its faces is 25°C, how much heat flows through the window per hour?

53. A Thermopane window of area 6.0 m^2 is constructed of two layers of glass, each 4.0 mm thick separated by an air space of 5.0 mm. If the inside is at 20°C and the outside is at -30°C, what is the heat loss through the window?

54. A silver bar of length 30.0 cm and cross-sectional area 1.00 cm^2 is used to transfer heat from a 100.0°C reservoir to a 0.0°C reservoir. How much heat is transferred per second?

55. A bar of gold is in thermal contact with a bar of silver of the same length and area (Fig. P20.55). One end of the compound bar is maintained at 80.0°C while the opposite end is at 30.0°C. When the heat flow reaches steady state, find the temperature at the junction.

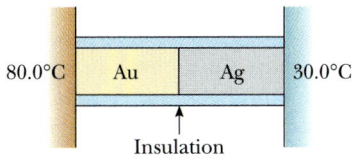

FIGURE P20.55

56. Two rods of the same length but of different materials and cross-sectional areas are placed side by side as in Figure P20.56. Determine the rate of heat flow in terms of the thermal conductivity and area of each rod. Generalize this to several rods.

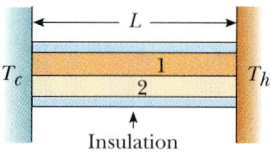

FIGURE P20.56

57. The brick wall ($k = 0.80$ W/m·°C) of a building has dimensions of 4.0 m × 10.0 m and is 15 cm thick. How much heat (in joules) flows through the wall in a 12-h period when the average inside and outside temperatures are, respectively, 20°C and 5.0°C?

58. A steam pipe is covered with 1.5-cm-thick insulating material of heat conductivity 0.200 cal/cm·°C·s. How much heat is lost every second when the steam is at 200°C and the surrounding air is at 20°C? The pipe has a circumference of 20 cm and a length of 50 m. Neglect losses through the ends of the pipe.

59. A box with a total surface area of 1.20 m² and a wall thickness of 4.00 cm is made of an insulating material. A 10.0-W electric heater inside the box maintains the inside temperature at 15.0°C above the outside temperature. Find the thermal conductivity k of the insulating material.

59A. A box with a total surface area A and a wall thickness t is made of an insulating material. An electric heater that delivers P watts inside the box maintains the inside temperature at a temperature T_0 above the outside temperature. Find the thermal conductivity k of the insulating material.

60. A Styrofoam box has a surface area of 0.80 m² and a wall thickness of 2.0 cm. The temperature inside is 5°C, and that outside is 25°C. If it takes 8.0 h for 5.0 kg of ice to melt in the container, determine the thermal conductivity of the Styrofoam.

61. The roof of a house built to absorb the solar radiation incident upon it has an area of 7.0 m × 10.0 m. The solar radiation at the Earth's surface is 840 W/m². On the average, the Sun's rays make an angle of 60° with the plane of the roof. (a) If 15% of the incident energy is converted to useful electrical power, how many kilowatt hours per day of useful energy does this source provide? Assume that the Sun shines for an average of 8.0 h/day. (b) If the average household user pays 6 cents/kWh, what is the monetary saving of this energy source per day?

62. The surface of the Sun has a temperature of about 5800 K. Taking the radius of the Sun to be 6.96 × 10⁸ m, calculate the total energy radiated by the Sun each day. (Assume $e = 1$.)

63. Calculate the R value of (a) a window made of a single pane of glass 1/8 in. thick and (b) a thermal pane window made of two single panes each 1/8 in. thick and separated by a 1/4-in. air space. (c) By what factor is the heat loss reduced by using the thermal window instead of the single pane window?

ADDITIONAL PROBLEMS

64. A cooking vessel on a slow burner contains 10.0 kg of water and an unknown mass of ice in equilibrium at 0°C at time $t = 0$. The temperature of the mixture is measured at various times, and the result is plotted in Figure P20.64. During the first 50 min, the mixture remains at 0°C. From 50 min to 60 min, the temperature increases to 2.0°C. Neglecting the heat capacity of the vessel, determine the initial mass of the ice.

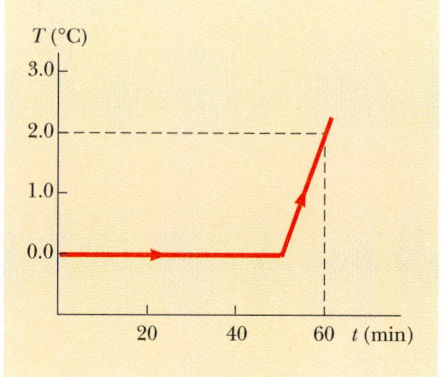

FIGURE P20.64

65. Around a crater formed by an iron meteorite, 75.0 kg of rock has melted under the impact of the meteorite. The rock has a specific heat of 0.800 kcal/kg·°C, a melting point of 500°C, and a latent heat of fusion of 48.0 kcal/kg. The original temperature of the ground was 0.0°C. If the meteorite hit the ground while moving at 600 m/s, what is the minimum mass of the meteorite? Assume no heat loss to the surrounding unmelted rock or the atmosphere during the impact. Disregard the heat capacity of the meteorite.

66. A *flow calorimeter* is an apparatus used to measure the specific heat of a liquid. The technique is to measure the temperature difference between the input and output points of a flowing stream of the liquid while adding heat at a known rate. In one particular experiment, a liquid of density 0.78 g/cm³ flows through the calorimeter at the rate of 4.0 cm³/s. At steady state, a temperature difference of 4.8°C is established between the input and output points when heat is supplied at the rate of 30 J/s. What is the specific heat of the liquid?

66A. A *flow calorimeter* is an apparatus used to measure the specific heat of a liquid. The technique is to measure the temperature difference between the input and output points of a flowing stream of the liquid while adding heat at a known rate. In one particular experiment, a liquid of density ρ flows through the calorimeter at the rate of R cm³/s. At steady state, a temperature difference ΔT is established between the input and output points when heat is supplied at the rate of P J/s. What is the specific heat of the liquid?

67. One mole of an ideal gas initially at 300 K is cooled at constant volume so that the final pressure is one-fourth the initial pressure. Then the gas expands at constant pressure until it reaches the initial temperature. Determine the work done by the gas.

68. An electric teakettle is boiling, and the electric power consumed by the water in the kettle is 1.0 kW. Assuming that the vapor pressure in the kettle equals atmospheric pressure, determine the speed at which water vapor exits from the spout, which has a cross-sectional area of 2.0 cm².

68A. An electric teakettle is boiling, and the electric power consumed by the water in the kettle is P. Assuming that the vapor pressure in the kettle equals atmospheric pressure, determine the speed at which water vapor exits from the spout, which has a cross-sectional area A.

69. An aluminum rod, 0.50 m in length and of cross-sectional area 2.5 cm², is inserted into a thermally insulated vessel containing liquid helium at 4.2 K. The rod is initially at 300 K. (a) If one half of the rod is inserted into the helium, how many liters of helium boil off by the time the inserted half cools to 4.2 K? (Assume the upper half does not cool.) (b) If the upper portion of the rod is maintained at 300 K, what is the approximate boil-off rate of liquid helium after the lower half has reached 4.2 K? (Note that aluminum has a thermal conductivity of 31 J/s·cm·K at 4.2 K, a specific heat of 0.21 cal/g·°C, and a density of 2.7 g/cm³. The density of liquid helium is 0.125 g/cm.³)

70. A heat pipe 0.025 m in diameter and 0.30 m long can transfer 3600 J of heat per second with a temperature difference across the ends of 10°C. How does this performance compare with the heat transfer of a solid silver bar of the same dimensions if the thermal conductivity of silver is $k = 427$ W/m·°C? (Silver is the best heat conductor of all metals.)

71. A 5.00-g lead bullet traveling at 300 m/s and at an initial temperature of 0°C strikes a flat steel plate and stops. If the collision is inelastic, will the bullet melt? Lead has a melting point of 327°C, a specific heat of 0.128 J/g·°C, and a latent heat of fusion of 24.5 J/g.

72. The average thermal conductivity of the walls (including windows) and roof of the house in Figure P20.72 is 0.48 W/m·°C, and their average thickness is 21.0 cm. The house is heated with natural gas having a heat of combustion (heat given off per cubic meter of gas burned) of 9300 kcal/m³. How many cubic meters of gas must be burned each day to maintain an inside temperature of 25.0°C if the outside temperature is 0.0°C? Disregard radiation and heat loss through the ground.

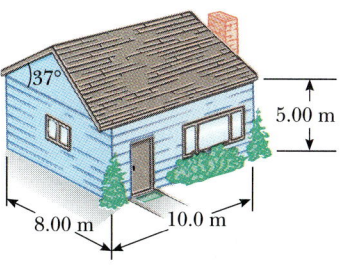

FIGURE P20.72

73. A class of 10 students taking an exam has a power output per student of about 200 W. Assume that the initial temperature of the room is 20°C and that its dimensions are 6.0 m by 15 m by 3.0 m. What is the temperature of the room at the end of 1.0 h if all the heat remains in the air in the room and none is added by an outside source? The specific heat of air is 837 J/kg·°C, and its density is about 1.3×10^{-3} g/cm³.

74. An ideal gas initially at P_0, V_0, and T_0 is taken through a cycle as described in Figure P20.74. (a) Find the net work done by the gas per cycle. (b) What is the net heat added to the system per cycle? (c) Obtain a numerical value for the net work done per cycle for 1 mol of gas initially at 0°C.

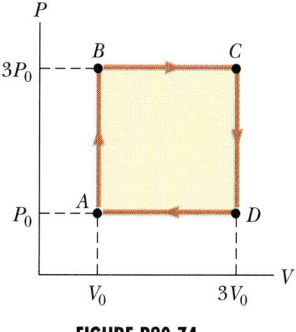

FIGURE P20.74

75. A one-person research submarine has a spherical iron hull 1.50 m in outer radius and 2.00 cm thick, lined with an equal thickness of rubber. If the submarine is used in Arctic waters (temperature 0°C) and the total rate of heat release within the sub (including the occupant's metabolic heat) is 1500 W, find the equilibrium temperature of the interior.

76. An iron plate is held against an iron wheel so that there is a sliding frictional force of 50.0 N acting between the two pieces of metal. The relative speed at which the two surfaces slide over each other is 40.0 m/s. (a) Calculate the rate at which mechanical energy is converted to thermal energy. (b) The plate and the wheel have a mass of 5.00 kg each, and each receives 50% of the thermal energy. If the system is run as described for 10.0 s and each object is then allowed to reach a uniform internal temperature, what is the resultant temperature increase?

77. A vessel in the shape of a spherical shell has an inner radius a and outer radius b. The wall has a thermal conductivity k. If the inside is maintained at a temperature T_1 and the outside is at a temperature T_2, show that the rate of heat flow between the surfaces is

$$\frac{dQ}{dt} = \left(\frac{4\pi kab}{b - a}\right)(T_1 - T_2)$$

78. The inside of a hollow cylinder is maintained at a temperature T_a while the outside is at a lower temperature, T_b (Fig. P20.78). The wall of the cylinder has a thermal conductivity k. Neglecting end effects, show that the rate of heat flow from the inner to the outer wall in the radial direction is

$$\frac{dQ}{dt} = 2\pi Lk\left[\frac{T_a - T_b}{\ln(b/a)}\right]$$

(*Hint:* The temperature gradient is dT/dr. Note that a radial heat current passes through a concentric cylinder of area $2\pi rL$.)

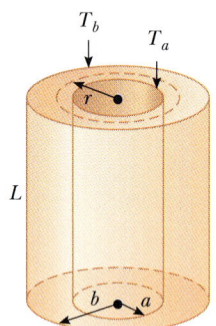

FIGURE P20.78

79. The passenger section of a jet airliner is in the shape of a cylindrical tube of length 35 m and inner radius 2.5 m. Its walls are lined with a 6.0-cm thickness of insulating material of thermal conductivity 4.0×10^{-5} cal/s·cm·°C. The inside is to be maintained at 25°C while the outside is at -35°C. What heating rate is required to maintain this temperature difference? (Use the result from Problem 78.)

80. A Thermos bottle in the shape of a cylinder has an inner radius of 4.0 cm, outer radius of 4.5 cm, and length of 30.0 cm. The insulating walls have a thermal conductivity equal to 2.0×10^{-5} cal/s·cm·°C. One liter of hot coffee at 90°C is poured into the bottle. If the outside wall remains at 20°C, how long does it take for the coffee to cool to 50°C? (Use the result from Problem 78 and assume that coffee has the same properties as water.)

81. A "solar cooker" consists of a curved reflecting mirror that focuses sunlight onto the object to be heated (Fig. P20.81). The solar power per unit area reaching the Earth at some location is 600 W/m², and the cooker has a diameter of 0.60 m. Assuming that 40% of the incident energy is converted into thermal energy, how long would it take to completely boil off 0.50 liter of water initially at 20°C? (Neglect the heat capacity of the container.)

FIGURE P20.81

82. A pond of water at 0°C is covered with a layer of ice 4.0 cm thick. If the air temperature stays constant at -10°C, how long will it be before the ice thickness is 8.0 cm? (*Hint:* To solve this problem, utilize Equation 20.14 in the form

$$\frac{dQ}{dt} = kA\frac{\Delta T}{x}$$

and note that the incremental heat dQ extracted from the water through the thickness x of ice is the amount required to freeze a thickness dx of ice. That is $dQ = L\rho A dx$, where ρ is the density of the ice, A is the area, and L is the latent heat of freezing.)

83. A student obtains the following data in a method-of-mixtures experiment designed to measure the specific heat of aluminum:

Initial temperature of water and calorimeter: 70°C
Mass of water: 0.400 kg

Mass of calorimeter: 0.040 kg
Specific heat of calorimeter: 0.63 kJ/kg·°C
Initial temperature of aluminum: 27°C
Mass of aluminum: 0.200 kg
Final temperature of mixture: 66.3°C

Use these data to determine the specific heat of aluminum. Your result should be within 15% of the value listed in Table 20.1.

SPREADSHEET PROBLEMS

S1. The rate at which an object with an initial temperature T_i cools when the surrounding temperature is T_0 is given by Newton's law of cooling:

$$\frac{dQ}{dt} = hA(T - T_0)$$

where A is the surface area of the object, and h is a constant that characterizes the rate of cooling, called the surface coefficient of heat transfer. The object's temperature T at any time t is

$$T = T_0 + (T_i - T_0)e^{-hAt/mc}$$

where m is the object's mass, and c is its specific heat. Spreadsheet 20.1 evaluates and plots the temperature of an object as a function of time. Consider the cooling of a cup of coffee whose initial temperature is 75°C. After 3 min, it cools to a drinkable 45°C. Room temperature is 22°C, the effective area of the coffee is 125 cm², and its mass is 200 g. Take the specific heat c to be that of water. Use Spreadsheet 19.1 to find h for this cup of coffee. That is, find the value of h for which $T = 45°C$ at $t = 180$ s.

S2. In the Einstein model of a crystalline solid the molar specific heat at constant volume is

$$C_V = 3R \left(\frac{T_E}{T}\right)^2 \frac{e^{T_E/T}}{(e^{T_E/T} - 1)^2}$$

where T_E is a characteristic temperature called the Einstein temperature and T is the temperature in Kelvin. (a) Verify that $C_V \approx 3R$, the Dulong-Petit law, for $T \gg T_E$, by evaluating $C_V - 3R$. (b) For diamond, T_E is approximately 1060 K. If 1 mol of diamond is heated from 300 K to 600 K, numerically integrate

$$\Delta U = \int C_V \, dT$$

to find the increase in the internal energy of the material. Use your spreadsheet to carry out a simple rectangular numeric integration. (You may also want to try to use the trapezoid and Simpson's methods to carry out the integration.)

The Kinetic Theory of Gases

The glass vessel contains dry ice (solid carbon dioxide). The white cloud is carbon dioxide vapor, which is denser than air and hence falls from the vessel. *(R. Folwell/Science Photo Library)*

I n Chapter 19 we discussed the properties of an ideal gas using such macroscopic variables as pressure, volume, and temperature. We shall now show that such large-scale properties can be described on a microscopic scale, where matter is treated as a collection of molecules. Newton's laws of motion applied in a statistical manner to a collection of particles provide a reasonable description of thermodynamic processes. In order to keep the mathematics relatively simple, we shall consider only the molecular behavior of gases, where the interactions between molecules are much weaker than in liquids or solids. In the established model of gas behavior, called the *kinetic theory* for historical reasons, gas molecules move about in a random fashion, colliding with the walls of their container and with each other. Perhaps the most important consequence of this theory is that it shows how the kinetic theory of molecular motion contributes to the internal energy of the system. Furthermore, the kinetic theory provides us with a physical basis upon which the concept of thermal energy can be understood.

In the simplest model of a gas, each molecule is considered to be a hard sphere that collides elastically with other molecules and with the container wall. The hard-sphere model assumes that the molecules do not interact with each other except during collisions and that they are not deformed by collisions. This description is adequate only for monatomic gases, where the energy is entirely translational kinetic energy. One must modify the theory for more complex molecules, such as O_2 and CO_2, to include the internal energy associated with rotations and vibrations of the molecules.

21.1 MOLECULAR MODEL OF AN IDEAL GAS

We begin this chapter by developing a microscopic model of an ideal gas. The model shows that the pressure that a gas exerts on the walls of its container is a consequence of the collisions of the gas molecules with the walls. As we shall see, the model is consistent with the macroscopic description of the preceding chapter. In developing this model, we make the following assumptions:

- *The number of molecules is large, and the average separation between them is large* compared with their dimensions. This means that the volume of the molecules is negligible when compared with the volume of the container.
- *The molecules obey Newton's laws of motion, but as a whole they move randomly.* By "randomly" we mean that any molecule can move in any direction at any speed. We also assume that the distribution of speeds does not change in time, despite the collisions between molecules. That is, at any given moment, a certain percentage of molecules moves at high speeds, a certain percentage moves at low speeds, and so on.

Assumptions of the molecular model of an ideal gas

This simulator is concerned with systems consisting of a large number of particles, rather than just the one or two objects dealt with in the previous simulators. After specifying the characteristics of a many-particle system, you will be able to observe particles colliding with each other and with the walls of a container. The results of the simulations should help you better understand the concepts of pressure, temperature, and related interesting phenomena.

Systems of Particles

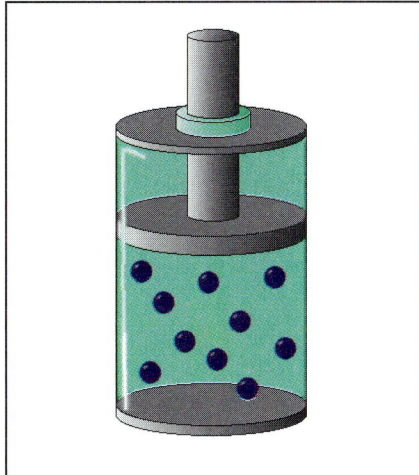
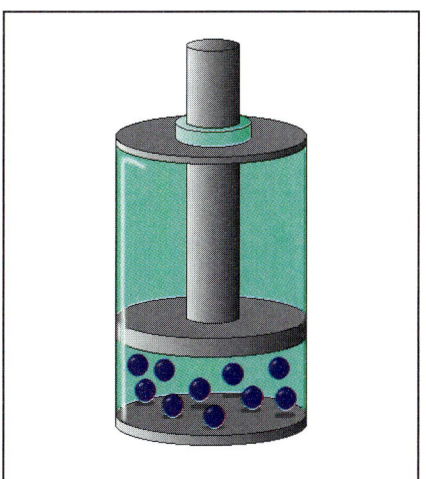

- *The molecules undergo elastic collisions with each other and with the walls of the container that are elastic on the average.* Thus, in the collisions both kinetic energy and momentum are constant.
- *The forces between molecules are negligible except during a collision.* The forces between molecules are short-range, so the molecules interact with each other only during collisions.
- *The gas under consideration is a pure substance.* That is, all molecules are identical.

Although we often picture an ideal gas as consisting of single atoms, molecular gases exhibit equally good approximations to ideal gas behavior at low pressures. Effects of molecular rotations or vibrations have no effect, on the average, on the motions considered here.

Now let us derive an expression for the pressure of an ideal gas consisting of N molecules in a container of volume V. The container is a cube with edges of length d (Fig. 21.1). Consider the collision of one molecule moving with a velocity $\mathbf{v}$ toward the right-hand face of the box. The molecule has velocity components v_x, v_y, and v_z. As it collides with the wall elastically, its x component of velocity is reversed, while its y and z components of velocity remain unaltered (Fig. 21.2). Since the x component of momentum of the molecule is mv_x before the collision and $-mv_x$ afterward, the change in momentum of the molecule is

$$\Delta p_x = -mv_x - (mv_x) = -2mv_x$$

Because the momentum of the system consisting of the wall and the molecule must be constant, we see that since the change in momentum of the molecule is $-2mv_x$, the change in momentum of the wall must be $2mv_x$. If F_1 is the magnitude of the average force exerted by a molecule on the wall in the time Δt, applying the impulse-momentum theorem to the wall gives

$$F_1\,\Delta t = \Delta p = 2mv_x$$

In order for the molecule to make two collisions with the same wall, it must travel a distance $2d$ along the x direction in a time Δt. Therefore, the time interval between two collisions with the same wall is $\Delta t = 2d/v_x$, and the force imparted to the wall by a single collision is

$$F_1 = \frac{2mv_x}{\Delta t} = \frac{2mv_x}{2d/v_x} = \frac{mv_x^2}{d} \tag{21.1}$$

The total force exerted on the wall by all the molecules is found by adding the forces exerted by the individual molecules:

$$F = \frac{m}{d}\left(v_{x1}^2 + v_{x2}^2 + \cdots\right)$$

In this equation, v_{x1} is the x component of velocity of molecule 1, v_{x2} is the x component of velocity of molecule 2, and so on. The summation terminates when we reach N molecules because there are N molecules in the container.

To proceed further, note that the average value of the square of the velocity in the x direction for N molecules is

$$\overline{v_x^2} = \frac{v_{x1}^2 + v_{x2}^2 + \cdots + v_{xN}^2}{N}$$

Thus, the total force on the wall can be written

$$F = \frac{Nm}{d}\,\overline{v_x^2}$$

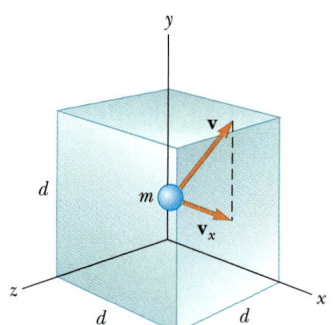

FIGURE 21.1 A cubical box with sides of length d containing an ideal gas. The molecule shown moves with velocity **v**.

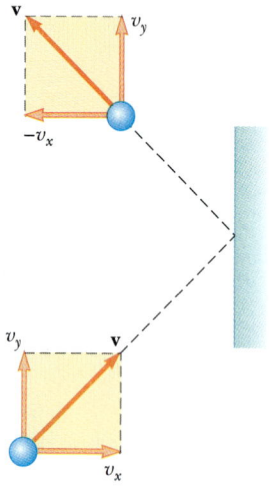

FIGURE 21.2 A molecule makes an elastic collision with the wall of the container. Its x component of momentum is reversed, thereby imparting momentum to the wall, while its y component remains unchanged. In this construction, we assume that the molecule moves in the xy plane.

Now let us focus on one molecule in the container whose velocity components are v_x, v_y, and v_z. The Pythagorean theorem relates the square of the speed to the square of these components:

$$v^2 = v_x^2 + v_y^2 + v_z^2$$

Hence, the average value of v^2 for all the molecules in the container is related to the average values of $\overline{v_x^2}$, $\overline{v_y^2}$, and $\overline{v_z^2}$ according to the expression

$$\overline{v^2} = \overline{v_x^2} + \overline{v_y^2} + \overline{v_z^2}$$

Because the motion is completely random, the average values $\overline{v_x^2}$, $\overline{v_y^2}$, and $\overline{v_z^2}$ are equal to each other. Using this fact and the above result, we find that

$$\overline{v^2} = 3\overline{v_x^2}$$

Thus, the total force on the wall is

$$F = \frac{N}{3}\left(\frac{m\overline{v^2}}{d}\right)$$

This expression allows us to find the total pressure exerted on the wall:

$$P = \frac{F}{A} = \frac{F}{d^2} = \tfrac{1}{3}\left(\frac{N}{d^3}\,m\overline{v^2}\right) = \tfrac{1}{3}\left(\frac{N}{V}\right)m\overline{v^2}$$

$$P = \tfrac{2}{3}\left(\frac{N}{V}\right)\left(\tfrac{1}{2}m\overline{v^2}\right) \qquad (21.2)$$

Relationship between pressure and molecular kinetic energy

This result shows that *the pressure is proportional to the number of molecules per unit volume and to the average translational kinetic energy of the molecule*, $\tfrac{1}{2}m\overline{v^2}$. With this simplified model of an ideal gas, we have arrived at an important result that relates the large-scale quantity of pressure to an atomic quantity, the average value of the square of the molecular speed. Thus, we have a key link between the atomic world and the large-scale world.

You should note that Equation 21.2 verifies some features of pressure that are probably familiar to you. One way to increase the pressure inside a container is to increase the number of molecules per unit volume in the container. You do this when you add air to a tire. The pressure in the tire can also be increased by increasing the average translational kinetic energy of the molecules in the tire. As we shall see shortly, this can be accomplished by increasing the temperature of the gas inside the tire. That is why the pressure inside a tire increases as the tire heats up during long trips. The continuous flexing of the tires as they move along the road surface generates heat that is transferred to the air inside the tires, increasing the air's temperature, which in turn produces an increase in pressure.

Molecular Interpretation of Temperature

We can obtain some insight into the meaning of temperature by first writing Equation 21.2 in the more familiar form

$$PV = \tfrac{2}{3}N(\tfrac{1}{2}m\overline{v^2})$$

Let us now compare this with the empirical equation of state for an ideal gas (Eq. 19.12):

$$PV = Nk_B T$$

These colorful balloons rise as a large gas burner heats the air inside them. Because warm air is less dense than cool air, the buoyant force upward can exceed the total force downward, causing the balloons to rise. The vertical motion can also be controlled by venting hot air at the top of the balloon.   *(Richard Megna/Fundamental Photographs)*

Recall that the equation of state is based on experimental facts concerning the macroscopic behavior of gases. Equating the right sides of these expressions, we find that

Temperature is proportional to average kinetic energy

$$T = \frac{2}{3k_B}\left(\tfrac{1}{2}m\overline{v^2}\right) \tag{21.3}$$

That is, *temperature is a direct measure of average molecular kinetic energy.*

By rearranging Equation 21.3, we can relate the translational molecular kinetic energy to the temperature:

Average kinetic energy per molecule

$$\tfrac{1}{2}m\overline{v^2} = \tfrac{3}{2}k_B T \tag{21.4}$$

That is, the average translational kinetic energy per molecule is $\tfrac{3}{2}k_B T$. Since $\overline{v_x^2} = \tfrac{1}{3}\overline{v^2}$, it follows that

$$\tfrac{1}{2}m\overline{v_x^2} = \tfrac{1}{2}k_B T \tag{21.5}$$

In a similar manner, for the y and z motions it follows that

$$\tfrac{1}{2}m\overline{v_y^2} = \tfrac{1}{2}k_B T \qquad \text{and} \qquad \tfrac{1}{2}m\overline{v_z^2} = \tfrac{1}{2}k_B T$$

Thus, each translational degree of freedom contributes an equal amount of energy to the gas, namely, $\tfrac{1}{2}k_B T$. (In general, "degrees of freedom" refers to the number of independent means by which a molecule can possess energy.) A generalization of this result, known as the **theorem of equipartition of energy,** says that, with certain restrictions,

Theorem of equipartition of energy

the energy of a system in thermal equilibrium is equally divided among all degrees of freedom.

The total translational kinetic energy of N molecules of gas is simply N times the average energy per molecule, which is given by Equation 21.4:

Total kinetic energy of N molecules

$$E = N\left(\tfrac{1}{2}m\overline{v^2}\right) = \tfrac{3}{2}Nk_B T = \tfrac{3}{2}nRT \tag{21.6}$$

where we have used $k_B = R/N_A$ for Boltzmann's constant and $n = N/N_A$ for the number of moles of gas. This result, together with Equation 21.2, implies that the pressure exerted by an ideal gas depends only on the number of molecules per unit volume and the temperature.

The square root of $\overline{v^2}$ is called the *root mean square* (rms) *speed* of the molecules. From Equation 21.4 we get, for the rms speed,

Root mean square speed

$$v_{\text{rms}} = \sqrt{\overline{v^2}} = \sqrt{\frac{3k_B T}{m}} = \sqrt{\frac{3RT}{M}} \tag{21.7}$$

where M is the molar mass in kg/mol. This expression shows that, at a given temperature, lighter molecules move faster, on the average, than heavier molecules. For example, at a given temperature, hydrogen, with a molar mass of 2×10^{-3} kg/mol, has an average speed of four times that of oxygen, whose molar mass is 32×10^{-3} kg/mol.

Table 21.1 lists the rms speeds for various molecules at 20°C.

TABLE 21.1   Some rms Speeds

Gas	Molecular Mass (g/mol)	v_{rms} at 20°C (m/s)
H_2	2.02	1902
He	4.0	1352
H_2O	18	637
Ne	20.1	603
N_2 or CO	28	511
NO	30	494
CO_2	44	408
SO_2	64	338

EXAMPLE 21.1   A Tank of Helium

A tank of volume 0.300 m³ contains 2.00 mol of helium gas at 20.0°C. Assuming the helium behaves like an ideal gas, (a) find the total thermal energy of the system.

Solution   Using Equation 21.6 with $n = 2.00$ and $T = 293$ K, we get

$$E = \tfrac{3}{2}nRT = \tfrac{3}{2}(2.00 \text{ mol})(8.31 \text{ J/mol} \cdot \text{K})(293 \text{ K})$$

$$= \boxed{7.30 \times 10^3 \text{ J}}$$

(b) What is the average kinetic energy per molecule?

Solution   From Equation 21.4, we see that the average kinetic energy per molecule is

$$\tfrac{1}{2}m\overline{v^2} = \tfrac{3}{2}k_B T = \tfrac{3}{2}(1.38 \times 10^{-23} \text{ J/K})(293 \text{ K})$$

$$= \boxed{6.07 \times 10^{-21} \text{ J}}$$

Exercise   Using the fact that the molar mass of helium is 4.00×10^{-3} kg/mol, determine the rms speed of the atoms at 20.0°C.

Answer 1.35×10^3 m/s.

21.2 SPECIFIC HEAT OF AN IDEAL GAS

The heat required to raise the temperature of n moles of gas from T_i to T_f depends on the path taken between the initial and final states. To understand this, consider an ideal gas undergoing several reversible processes such that the change in temperature is $\Delta T = T_f - T_i$ for all processes. The temperature change can be achieved by traveling along a variety of paths from one isotherm to another, as in Figure 21.3. Because ΔT is the same for each path, the change in internal energy ΔU is the same for all paths. However, the first law, $Q = \Delta U + W$, tells us that the heat Q required for each path will be different because W (the area under the curves) is different for each path. Thus the heat required to produce a given change in temperature does not have a unique value.

This difficulty is resolved by defining specific heats for only two processes that frequently occur: changes at constant volume and changes at constant pressure. Because the number of moles is a convenient measure of the amount of gas, we define the **molar specific heats** associated with these processes with the following equations:

$$Q = nC_V \, \Delta T \qquad \text{(constant volume)} \qquad (21.8)$$

$$Q = nC_P \, \Delta T \qquad \text{(constant pressure)} \qquad (21.9)$$

where C_v is the **molar specific heat at constant volume**, while C_p is the **molar specific heat at constant pressure**.

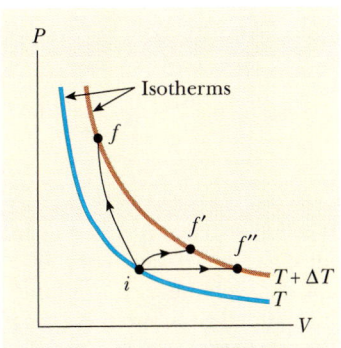

FIGURE 21.3   An ideal gas is taken from one isotherm at temperature T to another at temperature $T + \Delta T$ along three different paths.

In the previous section, we found that the temperature of a gas is a measure of the average translational kinetic energy of the gas molecules. This kinetic energy is associated with the motion of the center of mass of each molecule. It does not include the energy associated with the internal motion of the molecule, namely, vibrations and rotations about the center of mass.

In view of this, let us first consider the simplest case of an ideal monatomic gas, that is, a gas containing one atom per molecule, such as helium, neon, or argon. All of the kinetic energy of such molecules is associated with the motion of their centers of mass. When energy is added to a monatomic gas in a container of fixed volume (by heating, for example), all of the added energy goes into increasing the translational kinetic energy of the atoms. There is no other way to store the energy in a monatomic gas. Therefore, from Equation 21.6 we see that the total thermal energy U of N molecules (or n mol) of an ideal monatomic gas is

Internal energy of an ideal monatomic gas is proportional to its temperature

$$U = \tfrac{3}{2} N k_B T = \tfrac{3}{2} n R T \qquad (21.10)$$

It is important to note that for an ideal gas (only), U is a function of T only. If heat is transferred to the system at *constant volume,* the work done by the system is zero. That is, $W = \int P \, dV = 0$ for a constant volume process. Hence, from the first-law equation we see that

$$Q = \Delta U = \tfrac{3}{2} n R \, \Delta T \qquad (21.11)$$

In other words, all of the heat transferred goes into increasing the internal energy (and temperature) of the system. The constant-volume process from i to f is described in Figure 21.4, where ΔT is the temperature difference between the two isotherms. Substituting the value for Q given by Equation 21.8 into Equation 21.11, we get

$$n C_V \, \Delta T = \tfrac{3}{2} n R \, \Delta T$$

$$C_V = \tfrac{3}{2} R \qquad (21.12)$$

Note that this expression predicts a value of $\tfrac{3}{2} R = 12.5 \, \text{J/mol} \cdot \text{K}$ for all monatomic gases. This is in excellent agreement with measured values of molar specific heats for such gases as helium, neon, argon, xenon, etc. over a wide range of temperatures (Table 21.2).

The change in internal energy for an ideal gas can be expressed as

$$\Delta U = n C_V \, \Delta T \qquad (21.13)$$

In the limit of infinitesimal changes, we find the molar specific heat at constant volume to be

$$C_V = \frac{1}{n} \frac{dU}{dT} \qquad (21.14)$$

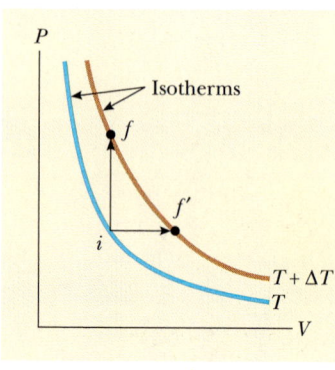

FIGURE 21.4 Heat is added to an ideal gas in two ways. For the constant-volume path *if*, all the heat added goes into increasing the internal energy of the gas since no work is done. Along the constant-pressure path *if'*, part of the heat added goes into work done by the gas. Note that the internal energy is constant along any isotherm.

Now suppose that the gas is taken along the constant-pressure path $i \rightarrow f'$ in Figure 21.4. Along this path, the temperature again increases by ΔT. The heat that must be transferred to the gas in this process is $Q = n C_P \, \Delta T$. Since the volume increases in this process, the work done by the gas is $W = P \, \Delta V$. Applying the first-law equation to this process gives

$$\Delta U = Q - W = n C_P \, \Delta T - P \, \Delta V \qquad (21.15)$$

In this case, the energy added to or removed from the gas is transferred in two forms. Part of it does work on the surroundings by moving a piston, and the

TABLE 21.2 Molar Specific Heats of Various Gases

	Molar Specific Heat (J/mol·K)			
	C_p	C_v	$C_p - C_v$	$\gamma = C_p/C_v$
Monatomic Gases				
He	20.8	12.5	8.33	1.67
Ar	20.8	12.5	8.33	1.67
Ne	20.8	12.7	8.12	1.64
Kr	20.8	12.3	8.49	1.69
Diatomic Gases				
H_2	28.8	20.4	8.33	1.41
N_2	29.1	20.8	8.33	1.40
O_2	29.4	21.1	8.33	1.40
CO	29.3	21.0	8.33	1.40
Cl_2	34.7	25.7	8.96	1.35
Polyatomic Gases				
CO_2	37.0	28.5	8.50	1.30
SO_2	40.4	31.4	9.00	1.29
H_2O	35.4	27.0	8.37	1.30
CH_4	35.5	27.1	8.41	1.31

Note: All values obtained at 300 K.

remainder is transferred as thermal energy to the gas. But the change in internal energy for the process $i \rightarrow f'$ is equal to the change for the process $i \rightarrow f$, because U depends only on temperature for an ideal gas and ΔT is the same for each process. In addition, since $PV = nRT$, we note that for a constant-pressure process, $P \Delta V = nR \Delta T$. Substituting this value for $P \Delta V$ into Equation 21.15 with $\Delta U = nC_V \Delta T$ (Eq. 21.13) gives

$$nC_V \Delta T = nC_P \Delta T - nR \Delta T$$

$$C_P - C_V = R \qquad (21.16)$$

This expression applies to *any* ideal gas. It shows that the molar specific heat of an ideal gas at constant pressure is greater than the molar specific heat at constant volume by an amount R, the universal gas constant (which has the value 8.31 J/mol·K). This result is in good agreement with real gases under standard conditions, that is, 0°C and atmospheric pressure (Table 21.2).

Since $C_V = \frac{3}{2}R$ for a monatomic ideal gas, Equation 21.16 predicts a value $C_P = \frac{5}{2}R = 20.8$ J/mol·K for the molar specific heat of a monatomic gas at constant pressure. The ratio of these heat capacities is a dimensionless quantity γ:

$$\gamma = \frac{C_P}{C_V} = \frac{\frac{5}{2}R}{\frac{3}{2}R} = \frac{5}{3} = 1.67 \qquad (21.17)$$

Ratio of molar specific heats for a monatomic ideal gas

The values of C_P and γ are in excellent agreement with experimental values for monatomic gases, but in serious disagreement with the values for the more complex gases (Table 21.2). This is not surprising because the value $C_V = \frac{3}{2}R$ was derived for a monatomic ideal gas, and we expect some additional contribution to the molar specific heat from the internal structure of the more complex molecules. In Section 21.4, we describe the effect of molecular structure on the specific heat of a gas. We shall find that the internal energy and hence the specific heat of a complex gas must include contributions from the rotational and vibrational motions of the molecule.

We have seen that the specific heats of gases at constant pressure are greater than the specific heats at constant volume. This difference is a consequence of the fact that in a constant-volume process, no work is done and all of the heat goes into increasing the internal energy (and temperature) of the gas, whereas in a constant-pressure process some of the thermal energy is transformed into work done by the gas. In the case of solids and liquids heated at constant pressure, very little work is done since the thermal expansion is small. Consequently C_P and C_V are approximately equal for solids and liquids.

EXAMPLE 21.2   Heating a Cylinder of Helium

A cylinder contains 3.00 mol of helium gas at a temperature of 300 K. (a) How much heat must be transferred to the gas to increase its temperature to 500 K if it is heated at constant volume?

Solution   For the constant-volume process, the work done is zero. Therefore from Equation 21.11, we get

$$Q_1 = \tfrac{3}{2}nR\,\Delta T = nC_V\,\Delta T$$

But $C_V = 12.5$ J/mol·K for He and $\Delta T = 200$ K; therefore,

$$Q_1 = (3.00 \text{ mol})(12.5 \text{ J/mol·K})(200 \text{ K}) = \boxed{7.50 \times 10^3 \text{ J}}$$

(b) How much thermal energy must be transferred to the gas at constant pressure to raise the temperature to 500 K?

Solution   Making use of Table 21.2, we get

$$Q_2 = nC_P\,\Delta T = (3.00 \text{ mol})(20.8 \text{ J/mol·K})(200 \text{ K})$$

$$= \boxed{12.5 \times 10^3 \text{ J}}$$

Exercise   What is the work done by the gas in this process?

Answer $W = Q_2 - Q_1 = 5.00 \times 10^3$ J.

21.3   ADIABATIC PROCESSES FOR AN IDEAL GAS

As you will recall, an adiabatic process is one in which there is no thermal energy transfer between a system and its surroundings. In reality, true adiabatic processes cannot occur because there is no such thing as a perfect thermal insulator. However, there are processes that are nearly adiabatic. For example, if a gas is compressed (or expanded) very rapidly, very little thermal energy flows into (or out of) the system, and so the process is nearly adiabatic. Such processes occur in the cycle of a gasoline engine, which we discuss in detail in the next chapter.

Another example of an adiabatic process is the very slow expansion of a gas that is thermally insulated from its surroundings. In general,

Definition of a reversible
adiabatic process

a **reversible adiabatic process** is one that is slow enough to allow the system to always be near equilibrium but fast compared with the time it takes the system to exchange thermal energy with its surroundings.

Suppose that an ideal gas undergoes a reversible adiabatic expansion. At any time during the process, we assume that the gas is in an equilibrium state, so that the equation of state, $PV = nRT$, is valid. The pressure and volume at any time during the process are related by the expression

Relationship between *P* and *V* for
a reversible adiabatic process
involving an ideal gas

$$PV^\gamma = \text{constant} \tag{21.18}$$

where $\gamma = C_P/C_V$ is assumed to be constant during the process. Thus, we see that all the thermodynamic variables—P, V, and T—change during a reversible adiabatic process.

Proof That $PV^{\gamma} = $ Constant for a Reversible Adiabatic Process

When a gas expands adiabatically in a thermally insulated cylinder, there is no thermal energy transferred between the gas and its surroundings, and so $Q = 0$. Let us take the infinitesimal change in volume to be dV and the infinitesimal change in temperature to be dT. The work done by the gas is $P\,dV$. Since the internal energy of an ideal gas depends only on temperature, the change in internal energy is $dU = nC_V\,dT$. Hence, the first-law equation, $\Delta U = Q - W$, becomes

$$dU = nC_V\,dT = -P\,dV$$

Taking the total differential of the equation of state of an ideal gas, $PV = nRT$, we see that

$$P\,dV + V\,dP = nR\,dT$$

Eliminating dT from these two equations, we find that

$$P\,dV + V\,dP = -\frac{R}{C_V}P\,dV$$

Substituting $R = C_P - C_V$ and dividing by PV, we get

$$\frac{dV}{V} + \frac{dP}{P} = -\left(\frac{C_P - C_V}{C_V}\right)\frac{dV}{V} = (1 - \gamma)\frac{dV}{V}$$

$$\frac{dP}{P} + \gamma\frac{dV}{V} = 0$$

Integrating this expression gives

$$\ln P + \gamma \ln V = \text{constant}$$

which is equivalent to Equation 21.18:

$$PV^{\gamma} = \text{constant}$$

The PV diagram for a reversible adiabatic expansion is shown in Figure 21.5. Because $\gamma > 1$, the PV curve is steeper than that for an isothermal expansion. As the gas expands adiabatically, no thermal energy is transferred in or out of the system. Hence, from the first law, we see that ΔU is negative and so ΔT is also negative. Thus, we see that the gas cools ($T_f < T_i$) during an adiabatic expansion. Conversely, the temperature increases if the gas is compressed adiabatically. Applying Equation 21.18 to the initial and final states, we see that

$$P_i V_i^{\gamma} = P_f V_f^{\gamma} \tag{21.19}$$

Using the ideal gas law, Equation 21.19 can also be expressed as

$$T_i V_i^{\gamma - 1} = T_f V_f^{\gamma - 1} \tag{21.20}$$

Note that the above analysis is valid only in a reversible adiabatic process.

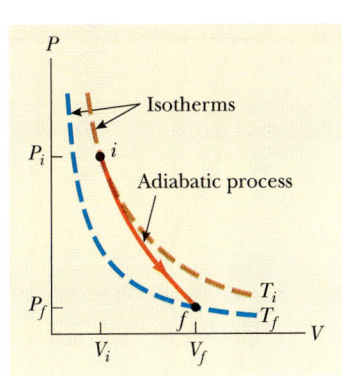

FIGURE 21.5 The PV diagram for a reversible adiabatic expansion. Note that $T_f < T_i$ in this process.

Reversible adiabatic process

EXAMPLE 21.3   A Diesel Engine Cylinder

Air in the cylinder of a diesel engine at 20.0°C is compressed from an initial pressure of 1.00 atm and volume of 800.0 cm³ to a volume of 60.0 cm³. Assuming that air behaves as an ideal gas ($\gamma = 1.40$) and that the compression is adiabatic and reversible, find the final pressure and temperature.

Solution   Using Equation 21.19, we find that

$$P_f = P_i\left(\frac{V_i}{V_f}\right)^\gamma = (1.00 \text{ atm})\left(\frac{800.0 \text{ cm}^3}{60.0 \text{ cm}^3}\right)^{1.4} = \boxed{37.6 \text{ atm}}$$

Since $PV = nRT$ is always valid during the process and since no gas escapes from the cylinder,

$$\frac{P_i V_i}{T_i} = \frac{P_f V_f}{T_f}$$

$$T_f = \frac{P_f V_f}{P_i V_i} T_i = \frac{(37.6 \text{ atm})(60.0 \text{ cm}^3)}{(1.00 \text{ atm})(800.0 \text{ cm}^3)} (293 \text{ K})$$

$$= \boxed{826 \text{ K} = 553°\text{C}}$$

21.4   THE EQUIPARTITION OF ENERGY

We have found that model predictions based on specific heat agree quite well with the behavior of monatomic gases but not with the behavior of complex gases (Table 21.2). Furthermore, the value predicted by the model for the quantity $C_P - C_V = R$ is the same for all gases. This is not surprising, because this difference is the result of the work done by the gas, which is independent of its molecular structure.

In order to explain the variations in C_V and C_P in going from monatomic gases to the more complex gases, let us explain the origin of the specific heat. So far, we have assumed that the sole contribution to the thermal energy of a gas is the translational kinetic energy of the molecules. However, the thermal energy of a gas actually includes contributions from the translational, vibrational, and rotational motion of the molecules. The rotational and vibrational motions of molecules can be activated by collisions and therefore are "coupled" to the translational motion of the molecules. The branch of physics known as *statistical mechanics* has shown that, for a large number of particles obeying the laws of Newtonian mechanics, the available energy is, on the average, shared equally by each independent degree of freedom. Recall from Section 21.1 that the equipartition theorem states that, at equilibrium, each degree of freedom contributes, on the average, $\frac{1}{2}k_B T$ of energy per molecule.

Let us consider a diatomic gas, in which the molecules have the shape of a dumbbell (Fig. 21.6). In this model, the center of mass of the molecule can translate in the x, y, and z directions (Fig. 21.6a). In addition, the molecule can rotate about three mutually perpendicular axes (Fig. 21.6b). We can neglect the rotation about the y axis for reasons that are discussed later. If the two atoms are taken to be point masses, then I_y is identically zero. Thus, there are five degrees of freedom: three associated with the translational motion and two associated with the rotational motion. Since *each degree of freedom contributes, on the average, $\frac{1}{2}k_B T$ of energy per molecule,* the total energy for N molecules is

$$U = 3N(\tfrac{1}{2}k_B T) + 2N(\tfrac{1}{2}k_B T) = \tfrac{5}{2}Nk_B T = \tfrac{5}{2}nRT$$

We can use this result and Equation 21.14 to get the molar specific heat at constant volume:

$$C_V = \frac{1}{n}\frac{dU}{dT} = \frac{1}{n}\frac{d}{dT}\left(\tfrac{5}{2}nRT\right) = \tfrac{5}{2}R$$

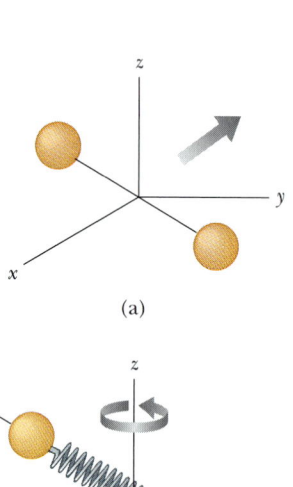

(a)

(b)

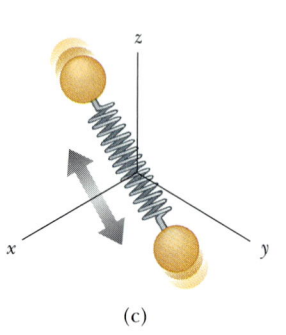

(c)

FIGURE 21.6   Possible motions of a diatomic molecule: (a) translational motion of the center of mass, (b) rotational motion about the various axes, and (c) vibrational motion along the molecular axis.

From Equations 21.16 and 21.17 we find that

$$C_P = C_V + R = \tfrac{7}{2}R$$

$$\gamma = \frac{C_P}{C_V} = \frac{\tfrac{7}{2}R}{\tfrac{5}{2}R} = \frac{7}{5} = 1.40$$

These results agree quite well with most of the data given in Table 21.2 for diatomic molecules. This is rather surprising because we have not yet accounted for the possible vibrations of the molecule. In the vibratory model, the two atoms are joined by an imaginary spring. The vibrational motion adds two more degrees of freedom, corresponding to the kinetic and potential energies associated with vibrations along the length of the molecule. Hence, the equipartition theorem predicts a thermal energy of $\tfrac{7}{2}nRT$ and a higher specific heat than is observed. Examination of the experimental data (Table 21.2) suggests that some diatomic molecules, such as H_2 and N_2, do not vibrate at room temperature, and others, such as Cl_2, do. For molecules with more than two atoms, the number of degrees of freedom is even larger and the vibrations are more complex. This results in an even higher predicted specific heat, which is in qualitative agreement with experiment.

We have seen that the equipartition theorem is successful in explaining some features of the specific heat of gas molecules with structure. However, the equipartition theorem does not explain the observed temperature variation in specific heats. As an example of such a temperature variation, C_V for the hydrogen molecule is $\tfrac{5}{2}R$ from about 250 K to 750 K and then increases steadily to about $\tfrac{7}{2}R$ well above 750 K (Fig. 21.7). This suggests that vibrations occur at very high temperatures. At temperatures well below 250 K, C_V has a value of about $\tfrac{3}{2}R$, suggesting that the molecule has only translational energy at low temperatures.

A Hint of Energy Quantization

The failure of the equipartition theorem to explain such phenomena is due to the inadequacy of classical mechanics when applied to molecular systems. For a more satisfactory description, it is necessary to use a quantum-mechanical model in

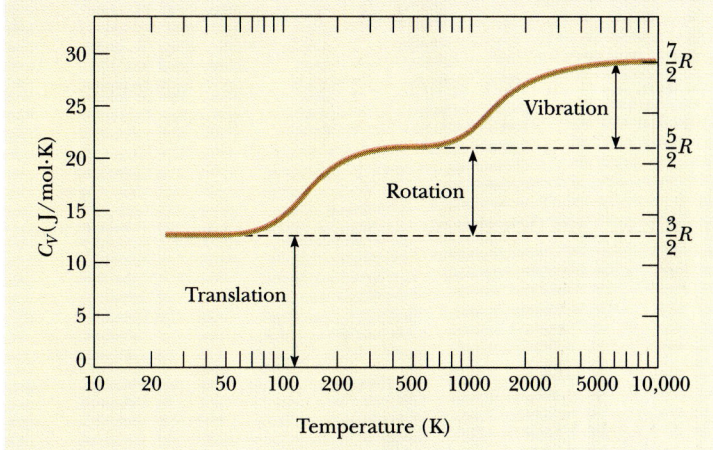

FIGURE 21.7 The molar specific heat of hydrogen as a function of temperature. The horizontal scale is logarithmic. Note that hydrogen liquefies at 20 K.

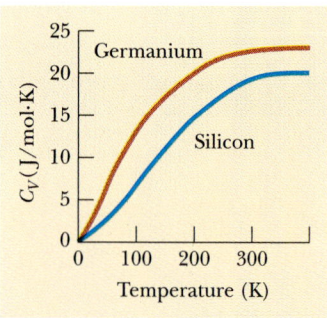

FIGURE 21.8   Molar specific heat of silicon and germanium. As T approaches zero, the specific heat also approaches zero.   *(From C. Kittel, Introduction to Solid State Physics, New York, John Wiley, 1971.)*

which the energy of an individual molecule is quantized. The energy separation between adjacent vibrational energy levels for a molecule such as H_2 is about ten times as great as the average kinetic energy of the molecule at room temperature. Consequently, collisions between molecules at low temperatures do not provide enough energy to change the vibrational state of the molecule. It is often stated that such degrees of freedom are "frozen out." This explains why the vibrational energy does not contribute to the specific heats of molecules at low temperatures.

The rotational energy levels are also quantized, but the spacing of most levels at ordinary temperatures is small compared with $k_B T$. The major exception is rotation of atoms or of linear molecules about their linear axes, for which the moments of inertia are very small and therefore the rotational spacings are very large. If the spacing between rotational levels is so small compared with $k_B T$, the system behaves classically. However, at sufficiently low temperatures (typically less than 50 K), where $k_B T$ is small compared with the spacing between rotational levels, intermolecular collisions may not be energetic enough to alter the rotational states. This explains why C_V reduces to $\frac{3}{2}R$ for H_2 in the range from 20 K to approximately 100 K.

The Specific Heat of Solids

Measurements of the specific heats of solids also show a marked temperature dependence. Solids have specific heats that generally decrease in a nonlinear manner with decreasing temperature and approach zero as the absolute temperature approaches zero. At high temperatures (usually above 300 K), the specific heats approach the value of about $3R \approx 25$ J/mol·K, a result known as the *DuLong-Petit law*. The typical data shown in Figure 21.8 demonstrate the temperature dependence of the molar specific heats for two semiconducting solids, silicon and germanium.

The specific heat of a solid at high temperatures can be explained using the equipartition theorem. For small displacements of an atom from its equilibrium position, each atom executes simple harmonic motion in the x, y, and z directions. The energy associated with vibrational motion in the x direction is

$$E_x = \tfrac{1}{2}mv_x^2 + \tfrac{1}{2}kx^2$$

There are analogous expressions for E_y and E_z. Therefore, each atom of the solid has six degrees of freedom. According to the equipartition theorem, this corresponds to an average vibrational energy of $6(\frac{1}{2}k_B T) = 3k_B T$ per atom. Therefore, the total thermal energy of a solid consisting of N atoms is

| Total thermal energy of a solid |

$$U = 3Nk_B T = 3nRT \tag{21.21}$$

From this result, we find that the molar specific heat of a solid at constant volume is

| Molar specific heat of a solid at constant volume |

$$C_V = \frac{1}{n}\frac{dU}{dT} = 3R \tag{21.22}$$

which agrees with the empirical law of DuLong and Petit. The discrepancies between this model and the experimental data at low temperatures are again due to the inadequacy of classical physics in the microscopic world. One can attribute the decrease in specific heat with decreasing temperature to a "freezing out" of various vibrational excitations.

*21.5   THE BOLTZMANN DISTRIBUTION LAW

Thus far we have neglected the fact that not all molecules in a gas have the same speed and energy. Their motion is extremely chaotic. Any individual molecule is colliding with others at the enormous rate of typically a billion times per second. Each collision results in a change in the speed and direction of motion of each of the participant molecules. From Equation 21.10, we see that average molecular speeds increase with increasing temperature. What we would like to know now is the distribution of molecular speeds. For example, how many molecules of a gas have a speed in the range from, say, 400 to 410 m/s? Intuitively, we expect that the speed distribution depends on temperature. Furthermore, we expect that the distribution peaks in the vicinity of v_{rms}. That is, few molecules are expected to have speeds much less than or much greater than v_{rms}, since these extreme speeds will result only from an unlikely chain of collisions.

As we examine the distribution of particles in space, we shall find that the particles distribute themselves among states of different energy in a specific way that depends exponentially on the energy, as first noted by Maxwell and extended by Boltzmann.

Ludwig Boltzmann (1844– 1906), an Austrian theoretical physicist, made many important contributions to the development of the kinetic theory of gases, electromagnetism, and thermodynamics. His pioneering work in the field of kinetic theory led to the branch of physics known as statistical mechanics.   *(Courtesy of AIP Niels Bohr Library, Lande Collection)*

The Exponential Atmosphere

We begin by considering the distribution of molecules in our atmosphere. Specifically, we determine how the number of molecules per unit volume varies with altitude. Our model assumes that the atmosphere is at a constant temperature T. (This assumption is not correct because the temperature of our atmosphere decreases by about 2°C per 300 m of altitude, but the model illustrates the basic features of the distribution.)

According to the ideal gas law that we studied in Chapter 19, a gas of N particles in thermal equilibrium obeys the relationship $PV = Nk_B T$. It is convenient to rewrite this equation in terms of the number of particles per unit volume of gas, $n_V = N/V$. This quantity is important because it can vary from one point to another. In fact, our goal is to determine how n_V changes in our atmosphere. We can express the ideal gas law in terms of n_V as $P = n_V k_B T$. Thus, if the number density n_V is known, we can find the pressure and vice versa. The pressure in the atmosphere decreases as the altitude increases because a given layer of air has to support the weight of all the atmosphere above it—the greater the altitude, the less the weight of the air above that layer, and the lower the pressure.

To determine the variation in pressure with altitude, consider an atmospheric layer of thickness dy and cross-sectional area A as in Figure 21.9. Because the air is in static equilibrium, the upward force on the bottom of this layer, PA, must exceed the downward force on the top of the layer, $(P + dP)A$, by an amount equal to the weight of gas in this thin layer. If the mass of a gas molecule in the layer is m, and there are a total of N molecules in the layer, then the weight of the layer is $w = mgN = mgn_V V = mgn_V A\, dy$. Thus, we see that

$$PA - (P + dP)A = mgn_V A\, dy$$

which reduces to

$$dP = -mgn_V\, dy$$

Because $P = n_V k_B T$, and T is assumed to remain constant, we see that $dP = k_B T\, dn_V$. Substituting this into the above expression for dP and rearranging gives

$$\frac{dn_V}{n_V} = -\frac{mg}{k_B T}\, dy$$

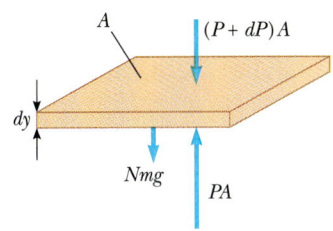

FIGURE 21.9   An atmospheric layer of gas in equilibrium.

Integrating this expression, we find

$$n(y) = n_0 e^{-mgy/k_B T} \qquad (21.23)$$

where the constant n_0 is the density at $y = 0$. This result is known as the **law of atmospheres**.

According to Equation 21.23, the density in thermal equilibrium decreases exponentially with increasing altitude. The density of our atmosphere at sea level is about $n_0 = 2.69 \times 10^{25}$ molecules/m^3. Because the pressure is $P = nk_B T$, we see from Equation 21.23 that the pressure varies with altitude as

$$P = P_0 e^{-mgy/k_B T} \qquad (21.24)$$

where $P_0 = n_0 k_B T$. A comparison of this model with the actual atmospheric pressure as a function of altitude shows that the exponential form is a reasonable approximation to the Earth's atmosphere.

Because our atmosphere contains different gases, each with a different molecular mass, one finds a higher concentration of heavier molecules at the lower altitudes, while the lighter molecules are more concentrated at the higher altitudes.

EXAMPLE 21.4 Distribution of Molecules in the Atmosphere

What is the density of air at an altitude of 12.0 km compared with the density at sea level?

Solution The density of our atmosphere decreases exponentially with altitude according to the law of atmospheres, Equation 21.23. We assume a temperature of 0°C ($T = 273$ K) and an average molecular mass of 28.8 u = 4.78×10^{-26} kg. Taking $y = 12.0$ km, the power of the exponential in Equation 21.23 is calculated to be

$$\frac{mgy}{k_B T} = \frac{(4.78 \times 10^{-26} \text{ kg})(9.80 \text{ m/s}^2)(12\ 000 \text{ m})}{(1.38 \times 10^{-23} \text{ J/K})(273 \text{ K})} = 1.49$$

Thus, Equation 21.23 gives

$$n = n_0 e^{-mgy/k_B T} = n_0 e^{-1.49} = \boxed{0.225 n_0}$$

That is, the density of air at an altitude of 12.0 km is only 22.5% of the density at sea level.

Computing Average Values

The exponential function $e^{-mgy/k_B T}$ that appears in Equation 21.23 can be interpreted as a probability distribution that gives the relative probability of finding a gas molecule at some height y. Thus, the probability distribution $p(y)$ is proportional to the density distribution $n(y)$. This concept allows us to determine many properties of the gas, such as the fraction of molecules below a certain height or the average potential energy of a molecule.

As an example, let us determine the average height $\bar{y}$ of a molecule in the atmosphere at temperature T. The expression for this average height is

$$\bar{y} = \frac{\int_0^\infty yn(y)\ dy}{\int_0^\infty n(y)\ dy} = \frac{\int_0^\infty ye^{-mgy/k_B T}\ dy}{\int_0^\infty e^{-mgy/k_B T}\ dy}$$

where the height of a molecule can range from 0 to ∞. The numerator in the preceding expression represents the sum of the heights of the particles times their number, while the denominator is the sum of the numbers of particles. The denominator is needed to give the correct average value. After performing the indicated integrations, we find

$$\bar{y} = \frac{(k_B T/mg)^2}{k_B T/mg} = \frac{k_B T}{mg}$$

This tells us that the average height of a molecule increases as T increases, as expected.

We can use a similar procedure to determine the average potential energy of a gas molecule. Because the gravitational potential energy of a molecule at height y is $U = mgy$, the average potential energy is equal to $mg\bar{y}$. Since $\bar{y} = k_B T/mg$, we see that $\bar{U} = mg(k_B T/mg) = k_B T$. This important result shows that the average gravitational potential energy of a molecule depends only on temperature, and not on m or g. Thus, the energy source that distributes the molecules in the atmosphere is thermal energy.

The Boltzmann Distribution

Because the gravitational potential energy of a molecule at height y is $U = mgy$, we can express the distribution law (Eq. 21.23) as

$$n = n_0 e^{-U/k_B T}$$

This means that gas molecules in thermal equilibrium are distributed in space with a probability that depends on gravitational potential energy according to the exponential factor $e^{-U/k_B T}$.

This can be extended to three dimensions, noting that the gravitational potential energy of a particle depends in general on three coordinates. That is, $U = U(x, y, z)$, hence the distribution of particles in space is

$$n(x, y, z) = n_0 e^{-U(x, y, z)/k_B T}$$

where n_0 is the number of particles where $U = 0$.

This kind of distribution applies to any energy the particles have, such as kinetic energy. In general, the relative number of particles having energy E is

$$n(E) = n_0 e^{-E/k_B T} \tag{21.25}$$

Boltzmann distribution law

This is called the **Boltzmann distribution law** and is important in describing the statistical mechanics of a large number of particles. It states that *the probability of finding the particles in a particular energy state varies exponentially as the negative of the energy divided by* $k_B T$. All the particles would fall into the lowest energy level, except that the thermal energy $k_B T$ tends to excite the particles to higher energy levels.

EXAMPLE 21.5 Thermal Excitation of Atomic Energy Levels

As we discussed briefly in Chapter 8, the electrons of an atom can occupy only certain discrete energy levels. Consider a gas at a temperature of 2500 K whose atoms can occupy only two energy levels separated by 1.5 eV (Fig. 21.10). Determine the ratio of the number of atoms in the higher energy level to the number in the lower energy level.

Solution Equation 21.25 gives the relative number of atoms in a given energy level. In our case, the atom has two possible energies, E_1 and E_2, where E_1 is the lower energy level. Hence, the ratio of the number of atoms in the higher

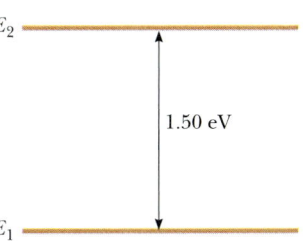

FIGURE 21.10 A gas whose atoms can occupy two energy levels.

energy level to the number in the lower level is

$$\frac{n(E_2)}{n(E_1)} = \frac{n_0 e^{-E_2/k_B T}}{n_0 e^{-E_1/k_B T}} = e^{-(E_2 - E_1)/k_B T}$$

In this problem, $E_2 - E_1 = 1.5$ eV and since 1 eV = 1.60×10^{-19} J,

$$k_B T = (1.38 \times 10^{-23} \text{ J/K})(2500 \text{ K})/1.60 \times 10^{-19} \text{ J/eV}$$
$$= 0.216 \text{ eV}$$

Therefore, the required ratio is

$$\frac{n(E_2)}{n(E_1)} = e^{-1.50 \text{ eV}/0.216 \text{ eV}} = e^{-6.94} = \boxed{9.64 \times 10^{-4}}$$

This result shows that at $T = 2500$ K, only a small fraction of the atoms are in the higher energy level. In fact, for every atom in the higher energy level, there are about 1000 atoms in the lower level. The number of atoms in the higher level increases at even higher temperatures, but the distribution law tells us that at equilibrium, there are always more atoms in the lower level than in the higher level.

*21.6   DISTRIBUTION OF MOLECULAR SPEEDS

In 1860 James Clerk Maxwell (1831–1879) derived an expression that describes the distribution of molecular speeds in a very definite manner. His work and developments by other scientists shortly thereafter were highly controversial, since experiments at that time were not capable of directly detecting molecules. However, about 60 years later, experiments were devised that confirmed Maxwell's predictions.

The observed speed distribution of gas molecules in thermal equilibrium is shown in Figure 21.11. The quantity N_v, called the **Maxwell-Boltzmann distribution function**, is defined as follows: If N is the total number of molecules, then the number of molecules with speeds between v and $v + dv$ is $dN = N_v \, dv$. This number is also equal to the area of the shaded rectangle in Figure 21.11. Furthermore, the fraction of molecules with speeds between v and $v + dv$ is $N_v \, dv/N$. This fraction is also equal to the probability that a molecule has a speed in the range from v to $v + dv$.

The fundamental expression that describes the most probable distribution of speeds of N gas molecules is

$$N_v = 4\pi N \left(\frac{m}{2\pi k_B T} \right)^{3/2} v^2 e^{-mv^2/2k_B T} \tag{21.26}$$

where m is the mass of a gas molecule, k_B is Boltzmann's constant, and T is the absolute temperature.[1]

As indicated in Figure 21.11, the average speed, $\bar{v}$, is somewhat lower than the rms speed. The most probable speed, v_{mp}, is the speed at which the distribution curve reaches a peak. Using Equation 21.26, one finds that

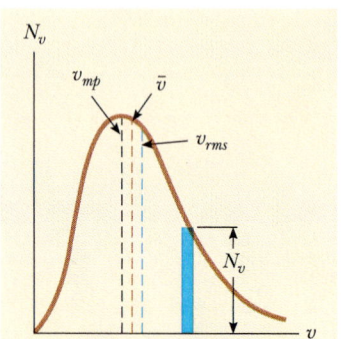

FIGURE 21.11 The speed distribution of gas molecules at some temperature. The number of molecules in the range Δv is equal to the area of the shaded rectangle, $N_v \Delta v$. The function N_v approaches zero as v approaches infinity.

rms speed

$$v_{rms} = \sqrt{\overline{v^2}} = \sqrt{3k_B T/m} = 1.73\sqrt{k_B T/m} \tag{21.27}$$

Average speed

$$\bar{v} = \sqrt{8k_B T/\pi m} = 1.60\sqrt{k_B T/m} \tag{21.28}$$

Most probable speed

$$v_{mp} = \sqrt{2k_B T/m} = 1.41\sqrt{k_B T/m} \tag{21.29}$$

[1] For the derivation of this expression, see any text on thermodynamics, such as that by R. P. Bauman, *Modern Thermodynamics and Statistical Mechanics,* New York, Macmillan, 1992.

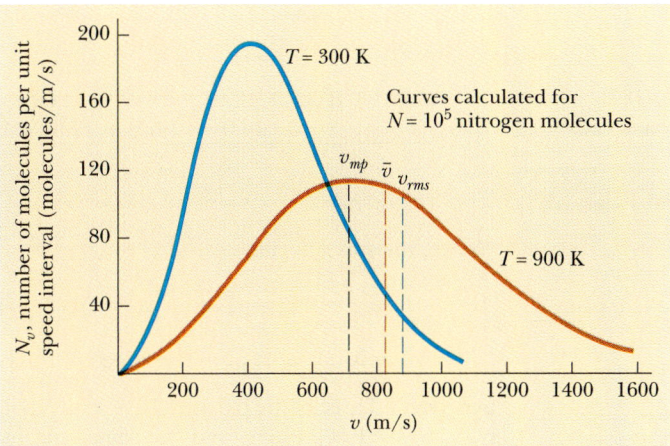

FIGURE 21.12   The speed distribution function for 10^5 nitrogen molecules at 300 K and 900 K. The total area under either curve is equal to the total number of molecules, which in this case equals 10^5. Note that $v_{\text{rms}} > \bar{v} > v_{\text{mp}}$.

The details of these calculations are left for the student (Problems 44 and 64), but from these equations we see that $v_{\text{rms}} > \bar{v} > v_{\text{mp}}$.

Figure 21.12 represents speed distribution curves for nitrogen molecules. The curves were obtained by using Equation 21.26 to evaluate the distribution function at various speeds and at two temperatures. Note that the curve shifts to the right as T increases, indicating that the average speed increases with increasing temperature, as expected. The asymmetric shape of the curves is due to the fact that the lowest speed possible is zero while the upper classical limit of the speed is infinity. Furthermore, as temperature increases the distribution curve broadens and the range of speeds also increases.

Equation 21.26 shows that the distribution of molecular speeds in a gas depends on mass as well as temperature. At a given temperature, the fraction of particles with speeds exceeding a fixed value increases as the mass decreases. This explains why lighter molecules, such as hydrogen and helium, escape more readily from the Earth's atmosphere than heavier molecules, such as nitrogen and oxygen. (See the discussion of escape speed in Chapter 14. Gas molecules escape even more readily from the Moon's surface because the escape speed on the Moon is lower.)

The speed distribution of molecules in a liquid is similar to that shown in Figure 21.12. The phenomenon of evaporation of a liquid can be understood from this distribution in speeds using the fact that some molecules in the liquid are more energetic than others. Some of the faster-moving molecules in the liquid penetrate the surface and leave the liquid even at temperatures well below the boiling point. The molecules that escape the liquid by evaporation are those that have sufficient energy to overcome the attractive forces of the molecules in the liquid phase. Consequently, the molecules left behind in the liquid phase have a lower average kinetic energy, causing the temperature of the liquid to decrease. Hence evaporation is a cooling process. For example, an alcohol-soaked cloth is often placed on a feverish head to cool and comfort the patient.

The evaporation process

EXAMPLE 21.6   A System of Nine Particles

Nine particles have speeds of 5.00, 8.00, 12.0, 12.0, 12.0, 14.0, 14.0, 17.0, and 20.0 m/s. (a) Find the average speed.

Solution   The average speed is the sum of the speeds divided by the total number of particles:

$$\bar{v} = \frac{(5.00 + 8.00 + 12.0 + 12.0 + 12.0 + 14.0 + 14.0 + 17.0 + 20.0)\ \text{m/s}}{9}$$

$$= \boxed{12.7\ \text{m/s}}$$

(b) What is the rms speed?

Solution   The average value of the square of the speed is

$$\bar{v^2} = \frac{\begin{array}{c}(5.00^2 + 8.00^2 + 12.0^2 + 12.0^2 + 12.0^2 \\ + 14.0^2 + 14.0^2 + 17.0^2 + 20.0^2)\ \text{m}^2/\text{s}^2\end{array}}{9}$$

$$= 178\ \text{m}^2/\text{s}^2$$

Hence, the rms speed is

$$v_{\text{rms}} = \sqrt{\bar{v^2}} = \sqrt{178\ \text{m}^2/\text{s}^2} = \boxed{13.3\ \text{m/s}}$$

(c) What is the most probable speed of the particles?

Solution   Three of the particles have a speed of 12 m/s, two have a speed of 14 m/s, and the remaining have different speeds. Hence, we see that the most probable speed, v_{mp}, is 12 m/s.

---

*21.7   MEAN FREE PATH

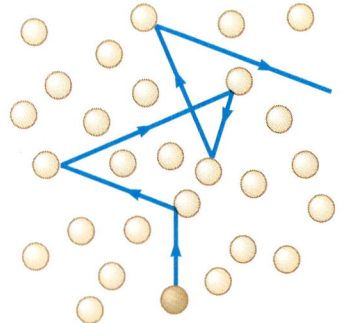

FIGURE 21.13   A molecule moving through a gas collides with other molecules in a random fashion. This behavior is sometimes referred to as a *random-walk process.* The mean free path increases as the number of molecules per unit volume decreases. Note that the motion is not limited to the plane of the paper.

Most of us are familiar with the fact that the strong odor associated with a gas such as ammonia may take a fraction of a minute to diffuse through a room. However, since average molecular speeds are typically several hundred meters per second at room temperature, we might expect a time much less than one second. To understand this apparent contradiction, we note that molecules collide with each other, because they are not geometrical points. Therefore, they do not travel from one side of a room to the other in a straight line. Between collisions, the molecules move with constant speed along straight lines. The average distance between collisions is called the **mean free path.** The path of individual molecules is random and resembles that shown in Figure 21.13. As we would expect from this description, the mean free path is related to the diameter of the molecules and the density of the gas.

We now describe how to estimate the mean free path for a gas molecule. For this calculation we assume that the molecules are spheres of diameter *d*. We see from Figure 21.14a that no two molecules collide unless their centers are less than a distance *d* apart as they approach each other. An equivalent description of the collisions is to imagine that one of the molecules has a diameter 2*d* and the rest are geometrical points (Fig. 21.14b). In a time *t*, the molecule having the speed that

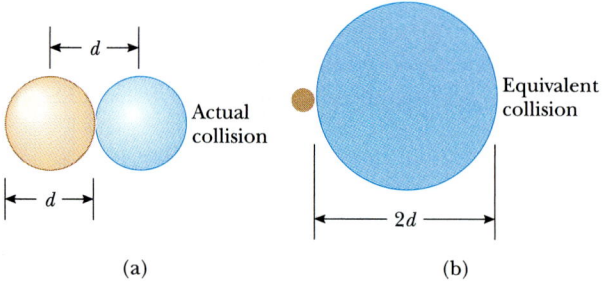

(a)   (b)

FIGURE 21.14   (a) Two spherical molecules, each of diameter *d*, collide if their centers are within a distance *d* of each other. (b) The collision between the two molecules is equivalent to a point mass colliding with a molecule having an effective diameter of 2*d*.

we take to be the average speed, $\bar{v}$, travels a distance $\bar{v}t$. In this same time interval, our molecule with equivalent diameter $2d$ sweeps out a cylinder having a cross-sectional area πd^2 and a length $\bar{v}t$ (Fig. 21.15). Hence, the volume of the cylinder is $\pi d^2 \bar{v}t$. If n_V is the number of molecules per unit volume, then the number of molecules in the cylinder is $(\pi d^2 \bar{v}t)\, n_V$. The molecule of equivalent diameter $2d$ collides with every molecule in this cylinder in the time t. Hence, the number of collisions in the time t is equal to the number of molecules in the cylinder, $(\pi d^2 \bar{v}t)\, n_V$.

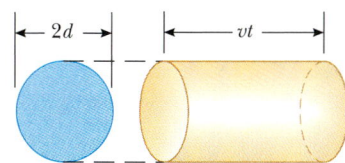

FIGURE 21.15   In a time t, a molecule of effective diameter $2d$ sweeps out a cylinder of length $\bar{v}t$, where $\bar{v}$ is its average speed. In this time, it collides with every molecule within this cylinder.

The **mean free path, ℓ**, which is the mean distance between collisions, equals the average distance $\bar{v}t$ traveled in a time t divided by the number of collisions that occurs in the time:

$$\ell = \frac{\bar{v}t}{(\pi d^2 \bar{v}t)\, n_V} = \frac{1}{\pi d^2 n_V}$$

Since the number of collisions in a time t is $(\pi d^2 \bar{v}t)\, n_V$, the number of collisions per unit time, or **collision frequency f**, is

$$f = \pi d^2 \bar{v} n_V$$

The inverse of the collision frequency is the average time between collisions, called the **mean free time.**

Our analysis has assumed that molecules in the cylinder are stationary. When the motion of these molecules is included in the calculation, the correct results are

$$\ell = \frac{1}{\sqrt{2}\, \pi d^2 n_V} \qquad (21.30) \qquad \text{Mean free path}$$

$$f = \sqrt{2}\, \pi d^2 \bar{v} n_V = \frac{\bar{v}}{\ell} \qquad (21.31) \qquad \text{Collision frequency}$$

EXAMPLE 21.7   A Collection of Nitrogen Molecules

Calculate the mean free path and collision frequency for nitrogen molecules at 20.0°C and 1.00 atm. Assume a molecular diameter of 2.00×10^{-10} m.

Solution   Assuming the gas is ideal, we can use the equation $PV = NkT$ to obtain the number of molecules per unit volume under these conditions:

$$n_V = \frac{N}{V} = \frac{P}{kT} = \frac{1.01 \times 10^5 \text{ N/m}^2}{(1.38 \times 10^{-23}\text{ J/K})(293\text{ K})}$$

$$= 2.50 \times 10^{25}\ \frac{\text{molecules}}{\text{m}^3}$$

Hence, the mean free path is

$$\ell = \frac{1}{\sqrt{2}\, \pi d^2 n_V}$$

$$= \frac{1}{\sqrt{2}\, \pi (2.00 \times 10^{-10}\text{ m})^2 \left(2.50 \times 10^{25}\ \dfrac{\text{molecules}}{\text{m}^3}\right)}$$

$$= 2.25 \times 10^{-7}\text{ m}$$

This is approximately 10^3 times greater than the molecular diameter. Since the average speed of a nitrogen molecule at 20.0°C is about 511 m/s (Table 21.1), the collision frequency is

$$f = \frac{\bar{v}}{\ell} = \frac{511 \text{ m/s}}{2.25 \times 10^{-7}\text{ m}} = 2.27 \times 10^9/\text{s}$$

The molecule collides with other molecules at the average rate of about two billion times each second!

The mean free path, ℓ, is *not* the same as the average separation between particles. In fact, the average separation, d, between particles is given approximately by $n_V^{-1/3}$. In this example, the average molecular separation is

$$d = \frac{1}{n_V^{1/3}} = \frac{1}{(2.5 \times 10^{25})^{1/3}} = 3.4 \times 10^{-9}\text{ m}$$

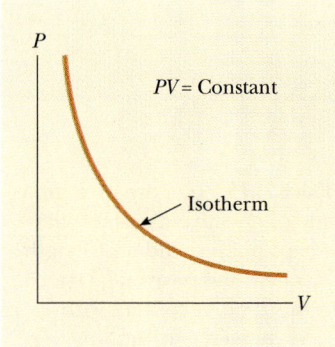

FIGURE 21.16 The *PV* diagram of an isothermal process for an ideal gas. In this case, the pressure and volume are related by *PV* = constant.

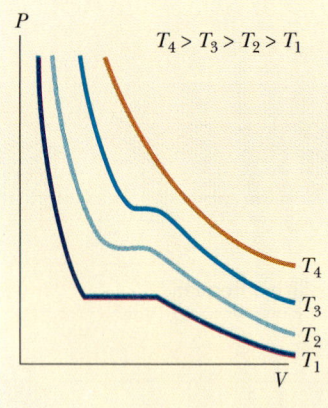

FIGURE 21.17 Isotherms for a real gas at various temperatures. At higher temperatures, such as T_4, the behavior is nearly ideal. The behavior is not ideal at the lower temperatures.

---

van der Waals' equation of state

*21.8 VAN DER WAALS' EQUATION OF STATE

Thus far we have assumed all gases to be ideal, that is, to obey the equation of state, $PV = nRT$. To a good approximation, real gases behave as ideal gases at ordinary temperatures and pressures. In the kinetic theory derivation of the ideal gas law, we neglected the volume occupied by the molecules and assumed that intermolecular forces were negligible. Now let us investigate the qualitative behavior of real gases and the conditions under which deviations from ideal gas behavior are expected.

Consider a gas contained in a cylinder fitted with a movable piston. As noted in Chapter 20, if the temperature is kept constant while the pressure is measured at various volumes, a plot of *P* versus *V* yields a hyperbolic curve (an *isotherm*) as predicted by the ideal gas law (Fig. 21.16).

Now let us describe what happens to a real gas. Figure 21.17 gives some typical experimental curves taken on a gas at various temperatures. At the higher temperatures, the curves are approximately hyperbolic and the gas behavior is close to ideal. However, as the temperature is lowered, the deviations from the hyperbolic shape are very pronounced.

There are two major reasons for this behavior. First, we must account for the volume occupied by the gas molecules. If *V* is the volume of the container and *b* is the volume occupied by the molecules, then $V - nb$ is the empty volume available to the gas, where *b* is a constant. As *V* decreases for a given quantity of gas, the fraction of the volume occupied by the molecules increases.

The second important effect concerns the intermolecular forces when the molecules are close together. At close separations, the molecules attract each other, as we might expect, since gases can condense to form liquids. When weak attractions between molecules are considered, the short-range forces alter the trajectories of molecules, introducing curvature and thereby increasing traversal time. This results in a decrease in the frequency of wall collisions and hence a decrease in pressure exerted on the walls. The magnitude of the effect depends on the number of collisions, and thus on the square of the number density. The average kinetic energy of the molecules is unchanged, so temperature is not affected. The net pressure is reduced by a factor proportional to the square of the density, which varies as $1/V^2$. Hence, the actual pressure, *P*, must be supplemented to give a corrected pressure, $P + n^2a/V^2$, that fits the form of the ideal gas law; *a* is (approximately) a constant.

The two effects just described can be incorporated into a modified equation of state proposed by J. D. van der Waals (1837–1923) in 1873. **Van der Waals' equation of state** is

$$\left(P + \frac{n^2a}{V^2} \right)\left(\frac{V}{n} - b \right) = RT \qquad (21.32)$$

The constants *a* and *b* are empirical and are chosen to provide the best fit to the experimental data for a particular gas.

The experimental curves in Figure 21.18 for CO_2 are described quite accurately by van der Waals' equation at the higher temperatures (T_3, T_4, and T_5) and outside the shaded regions. Within the yellow region there are major discrepancies. If the van der Waals' equation of state is used to predict the *PV* relationship at a temperature such as T_1, then a nonlinear curve is obtained that is unlike the observed flat portion of the curve in the figure.

The departure from the predictions of van der Waals' equation at the lower temperatures and higher densities is due to the onset of liquefaction. That is, the gas begins to liquefy at the pressure P_c, called the **critical pressure**. In the region within the dotted line below P_c, the gas is partially liquefied and the gas vapor and liquid coexist. However, liquefaction cannot occur (even at very high pressures) unless the temperature is below a critical value called the **critical temperature**. In the flat portions of the low-temperature isotherms, as the volume is decreased more gas liquefies and the pressure remains constant. At even lower volumes, the gas is completely liquefied. Any further decrease in volume leads to large increases in pressure because liquids are not easily compressed.

It is now realized that, because of the complex nature of the intermolecular forces, a real gas cannot be rigorously described by any simple equation of state, such as Equation 21.32. Nevertheless, the basic concepts involved in Equation 21.32 are correct. At very low temperatures, the low-energy molecules attract each other and the gas tends to liquefy. A further increase in pressure accelerates the rate of liquefaction. At the higher temperatures, the average kinetic energy is large enough to overcome the attractive intermolecular forces; hence, the molecules do not bind together at the higher temperatures and the gas phase is maintained.

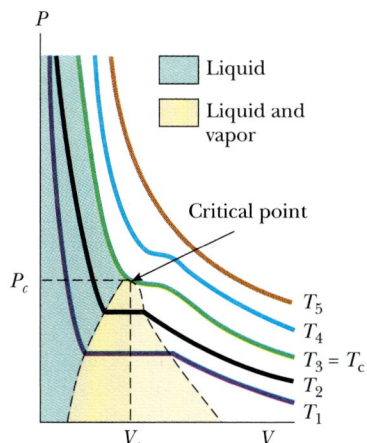

FIGURE 21.18 Isotherms for CO_2 at various temperatures. Below the critical temperature, T_c, the substance could be in the liquid state, the liquid-vapor equilibrium state, or the gaseous state, depending on the pressure and volume. *(Adapted from K. Mendelssohn, The Quest for Absolute Zero, New York, McGraw-Hill, World University Library, 1966.)*

SUMMARY

The pressure of N molecules of an ideal gas contained in a volume V is

$$P = \tfrac{2}{3} \frac{N}{V} \left(\tfrac{1}{2} m \overline{v^2} \right) \tag{21.2}$$

where $\tfrac{1}{2} m \overline{v^2}$ is the average kinetic energy per molecule.

The temperature of an ideal gas is related to the average kinetic energy per molecule through the expression

$$T = \frac{2}{3 k_B} \left(\tfrac{1}{2} m \overline{v^2} \right) \tag{21.3}$$

where k_B is Boltzmann's constant.

The average translational kinetic energy per molecule of a gas is

$$\tfrac{1}{2} m \overline{v^2} = \tfrac{3}{2} k_B T \tag{21.4}$$

Each translational degree of freedom (x, y, or z) has $\tfrac{1}{2} k_B T$ of energy associated with it.

The **equipartition of energy theorem** states that, with certain restrictions, the energy of a system in thermal equilibrium is equally divided among all degrees of freedom.

The total energy of N molecules (or n mol) of an ideal monatomic gas is

$$U = \tfrac{3}{2} N k_B T = \tfrac{3}{2} n R T \tag{21.10}$$

The change in internal energy for n mol of any ideal gas that undergoes a change in temperature ΔT is

$$\Delta U = n C_V \Delta T \tag{21.13}$$

where C_V is the molar specific heat at constant volume.

The molar specific heat of an ideal monatomic gas at constant volume is $C_V = \frac{3}{2}R$; the molar specific heat at constant pressure is $C_P = \frac{5}{2}R$. The ratio of specific heats is $\gamma = C_P/C_V = 5/3$.

If an ideal gas undergoes an adiabatic expansion or compression, the first law of thermodynamics together with the equation of state, shows that

$$PV^\gamma = \text{constant} \qquad (21.18)$$

QUESTIONS

1. Dalton's law of partial pressures states: *The total pressure of a mixture of gases is equal to the sum of the partial pressures of gases making up the mixture.* Give a convincing argument of this law based on the kinetic theory of gases.
2. One container is filled with helium gas and another with argon gas. If both containers are at the same temperature, which molecules have the higher rms speed?
3. If you wished to manufacture an after-shave lotion with a scent that is less "likely to get there before you do," would you use a high- or low-molecular-mass lotion?
4. A gas consists of a mixture of He and N_2 molecules. Do the lighter He molecules travel faster than the N_2 molecules? Explain.
5. Although the average speed of gas molecules in thermal equilibrium at some temperature is greater than zero, the average velocity is zero. Explain.
6. Why does a fan make you feel cooler on a hot day?
7. Alcohol taken internally makes you feel warmer. Yet when it is rubbed on your body, it lowers body temperature. Explain the latter effect.
8. A liquid partially fills a container. Explain why the temperature of the liquid decreases when the container is partially evacuated. (It is possible to freeze water using this technique.)
9. A vessel containing a fixed volume of gas is cooled. Does the mean free path of the gas molecules increase, decrease, or remain constant in the cooling process? What about the collision frequency?
10. A gas is compressed at a constant temperature. What hap-

pens to the mean free path of the molecules in this process?
11. If a helium-filled balloon initially at room temperature is placed in a freezer, will its volume increase, decrease, or remain the same?
12. What happens to a helium-filled balloon released into the air? Will it expand or contract? Will it stop rising at some height?
13. Why does a diatomic gas have a greater thermal energy content per mole than a monatomic gas at the same temperature?
14. What happens to the van der Waals' equation (Eq. 21.32) as the volume per mole, V, increases?
15. Ideal gas is contained in a vessel at 300 K. If the temperature is increased to 900 K, (a) by what factor does the rms speed of each molecule change? (b) By what factor does the pressure in the vessel change?
16. At room temperature, the average speed of an air molecule is several hundred meters per second. A molecule traveling at this speed should travel across a room in a small fraction of a second. In view of this, why does it take the odor of perfume (or other smells) several minutes to travel across the room?
17. A vessel is filled with gas at some equilibrium pressure and temperature. Can all gas molecules in the vessel have the same speed?
18. In our model of the kinetic theory of gases, molecules were viewed as hard spheres colliding elastically with the walls of the container. Is this model realistic?

PROBLEMS

Section 21.1   Molecular Model of an Ideal Gas

1. Find the rms speed of nitrogen molecules under standard conditions, 0.0°C and 1.00 atm pressure. Recall that 1 mol of any gas occupies a volume of 22.4 liters under standard conditions.
2. Two moles of oxygen gas is confined to a 5.00-liter vessel at a pressure of 8.00 atm. Find the average translational kinetic energy of an oxygen molecule under these conditions. (The mass of an O_2 molecule is 5.31×10^{-26} kg.)

3. A high-altitude research balloon contains helium gas. At its maximum altitude of 20.0 km, the outside temperature is $-50.0°C$ and the pressure has dropped to $\frac{1}{19}$ atm. The volume of the balloon at this location is 800 m³. Assuming the helium has the same temperature and pressure as the surrounding atmosphere, find the number of moles of helium in the balloon.
4. For the previous problem, find (a) the mass of the helium and (b) the volume of the balloon when it

---

☐ indicates problems that have full solutions available in the Student Solutions Manual and Study Guide.

was launched from the ground at standard pressure and temperature (1.00 atm, 0.0°C). (c) What volume tank at 27.0°C and 170 atm will supply this much helium?

5. A spherical balloon of volume 4000 cm³ contains helium at an (inside) pressure of 120 kPa. How many moles of helium are in the balloon, if each helium atom has an average kinetic energy of 3.6×10^{-22} J?

6. In a 30-s interval, 500 hailstones strike a glass window of area 0.60 m² at an angle of 45° to the window surface. Each hailstone has a mass of 5.0 g and a speed of 8.0 m/s. If the collisions are elastic, find the average force and pressure on the window.

6A. In a time t, N hailstones strike a glass window of area A at an angle θ to the window surface. Each hailstone has a mass m and a speed v. If the collisions are elastic, find the average force and pressure on the window.

7. A cylinder contains a mixture of helium and argon gas in equilibrium at 150°C. What is the average kinetic energy of each gas molecule?

8. Calculate the rms speed of an H_2 molecule at 250°C.

9. (a) Determine the temperature at which the rms speed of an He atom equals 500 m/s. (b) What is the rms speed of He on the surface of the Sun, where the temperature is 5800 K?

10. Gaseous helium is in thermal equilibrium with liquid helium at 4.20 K. Determine the most probable speed of a helium atom (mass = 6.65×10^{-27} kg).

11. If the rms speed of a helium atom at room temperature is 1350 m/s, what is the rms speed of an oxygen (O_2) molecule at this temperature? (The molar mass of O_2 is 32, and the molar mass of He is 4.)

12. A 5.00-liter vessel contains nitrogen gas at 27.0°C and 3.00 atm. Find (a) the total translational kinetic energy of the gas molecules and (b) the average kinetic energy per molecule.

13. One mole of xenon gas at 20.0°C occupies 0.0224 m³. What is the pressure exerted by the Xe atoms on the walls of a container?

14. (a) How many atoms of helium gas are required to fill a balloon to diameter 30.0 cm at 20.0°C and 1.00 atm? (b) What is the average kinetic energy of each helium atom? (c) What is the rms speed of each helium atom?

Section 21.2   Specific Heat of an Ideal Gas

(*Note:* Use data in Table 21.2.)

15. Calculate the change in internal energy of 3.0 mol of helium gas when its temperature is increased by 2.0 K.

16. One mole of a diatomic gas has pressure P and volume V. By heating the gas, its pressure is tripled and

its volume is doubled. If this heating process includes two steps, one at constant pressure and the other at constant volume, determine the amount of heat transferred to the gas.

17. One mole of an ideal monatomic gas is at an initial temperature of 300 K. The gas undergoes an isovolumetric process acquiring 500 J of heat. It then undergoes an isobaric process losing this same amount of heat. Determine (a) the new temperature of the gas and (b) the work done on the gas.

17A. One mole of an ideal monatomic gas is at an initial temperature T_0. The gas undergoes an isovolumetric process acquiring heat Q. It then undergoes an isobaric process losing this same amount of heat. Determine (a) the new temperature of the gas and (b) the work done on the gas.

18. One mol of air ($C_v = 5R/2$) at 300 K confined in a cylinder under a heavy piston occupies a volume of 5.0 liters. Determine the new volume of the gas if 4.4 kJ of heat is transferred to the air.

19. One mole of hydrogen gas is heated at constant pressure from 300 K to 420 K. Calculate (a) the heat transferred to the gas, (b) the increase in its internal energy, and (c) the work done by the gas.

20. In a constant-volume process, 209 J of heat is transferred to 1 mol of an ideal monatomic gas initially at 300 K. Find (a) the increase in internal energy of the gas, (b) the work it does, and (c) its final temperature.

21. What is the thermal energy of 100 g of He gas at 77 K? How much more energy must be supplied to heat this gas to 24°C?

22. A container has a mixture of two gases: n_1 mol of gas 1 having molar specific heat C_1 and n_2 mol of gas 2 of molar specific heat C_2. (a) Find the molar specific heat of the mixture. (b) What is the molar specific heat if the mixture has m gases having moles, n_1, n_2, n_3, . . . , n_m, and molar specific heats C_1, C_2, C_3, . . . , C_m, respectively?

23. A room of a well-insulated house has a volume of 100 m³ and is filled with air at 300 K. (a) Estimate the energy required to increase the temperature of this volume of air by 1.0°C. (b) If this energy could be used to lift an object of mass m to a height of 2.0 m, calculate the value of m.

24. How much thermal energy is in the air in a 20.0 m³ room at (a) 0.0°C and (b) 20.0°C? Assume that the pressure remains at 1.00 atm.

Section 21.3   Adiabatic Processes for an Ideal Gas

25. Two moles of an ideal gas ($\gamma = 1.40$) expands slowly and adiabatically from a pressure of 5.00 atm and a volume of 12.0 liters to a final volume of 30.0 liters. (a) What is the final pressure of the gas? (b) What are the initial and final temperatures?

26. Four liters of a diatomic ideal gas ($\gamma = 1.40$) confined to a cylinder are put through a closed cycle. The gas is initially at 1.0 atm and at 300 K. First, its pressure is tripled under constant volume. It then expands adiabatically to its original pressure and finally is compressed isobarically to its original volume. (a) Draw a *PV* diagram of this cycle. (b) Determine the volume at the end of the adiabatic expansion. Find (c) the temperature of the gas at the start of the adiabatic expansion and (d) the temperature at the end of this process. (e) What was the net work done for this cycle?

26A. A diatomic ideal gas ($\gamma = 1.40$) confined to a cylinder is put through a closed cycle. Initially the gas is at P_0, V_0, and T_0. First, its pressure is tripled under constant volume. It then expands adiabatically to its original pressure and finally is compressed isobarically to its original volume. (a) Draw a *PV* diagram of this cycle. (b) Determine the volume at the end of the adiabatic expansion. Find (c) the temperature of the gas at the start of the adiabatic expansion and (d) the temperature at the end of this process. (e) What was the net work done for this cycle?

27. Air ($\gamma = 1.4$) at 27°C and atmospheric pressure is drawn into a bicycle pump that has a cylinder with an inner diameter of 2.5 cm and length 50.0 cm. The down stroke adiabatically compresses the air, which reaches a gauge pressure of 800 kPa before entering the tire. Determine (a) the volume of the compressed air and (b) the temperature of the compressed air. (c) The pump is made of steel and has an inner wall which is 2.0 mm thick. Assume that 4.0 cm of the cylinder's length is allowed to come to thermal equilibrium with the air. What will be the increase in wall temperature?

28. During the power stroke in a four-stroke automobile engine, the piston is forced down as the mixture of gas and air undergoes a reversible adiabatic expansion. Find the average power generated during the expansion by assuming (a) the engine is running at 2500 rpm, (b) the gauge pressure right before the expansion is 20 atm, (c) the volumes of the mixture right before and after the expansion are 50 and 400 cm³, respectively (Fig. P21.28), (d) the time involved in the expansion is one-fourth that of the total cycle, and (e) the mixture behaves like an ideal diatomic gas.

29. During the compression stroke of a certain gasoline engine, the pressure increases from 1.00 atm to 20.0 atm. Assuming that the process is adiabatic and reversible and the gas is ideal with $\gamma = 1.40$, (a) by what factor does the volume change and (b) by what factor does the temperature change?

30. Helium gas at 20.0°C is compressed reversibly without heat loss to one-fifth its initial volume. (a) What

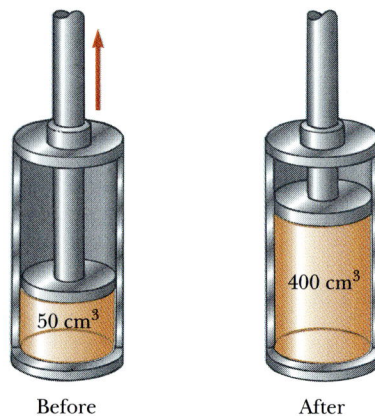

Before   After

FIGURE P21.28

is its temperature after compression? (b) What if the gas is dry air (77% N_2, 23% O_2)?

31. Air in a thundercloud expands as it rises. If its initial temperature was 300 K, and no heat is lost on expansion, what is its temperature when the initial volume is doubled?

32. How much work is required to compress 5.00 mol of air at 20.0°C and 1.00 atm to 1/10 of the original volume by (a) an isothermal process and (b) a reversible adiabatic process? (c) What are the final pressures for the two cases?

33. One mole of an ideal diatomic gas occupies a volume of one liter at a pressure of 0.10 atm. The gas undergoes a process in which the pressure is proportional to the volume, and at the end of the process, it is found that the speed of sound in the gas has doubled from its initial value. Determine the amount of heat transferred to the gas.

33A. One mole of an ideal diatomic gas occupies a volume V_0 at a pressure P_0. The gas undergoes a process in which the pressure is proportional to the volume, and at the end of the process, it is found that the speed of sound in the gas has doubled from its initial value. Determine the amount of heat transferred to the gas.

Section 21.4   The Equipartition of Energy

34. If a molecule has *f* degrees of freedom, show that a gas consisting of such molecules has the following properties: (1) its total thermal energy is $fnRT/2$; (2) its molar specific heat at constant volume is $fR/2$; (3) its molar specific heat at constant pressure is $(f + 2)R/2$; and (4) the ratio $\gamma = C_P/C_V = (f + 2)/f$.

35. A 5.00-liter vessel contains 0.125 mol of an ideal gas at 1.50 atm. What is the average translational kinetic energy of a single molecule?

36. Inspecting the magnitudes of C_V and C_P for the diatomic and polyatomic gases in Table 21.2, we find that the values increase with increasing molecular mass. Give a qualitative explanation of this observation.

37. In a crude model (Fig. P21.37) of a rotating diatomic molecule of chlorine (Cl_2), the two Cl atoms are 2.0×10^{-10} m apart and rotate about their center-of-mass with angular speed $\omega = 2.0 \times 10^{12}$ rad/s. What is the rotational kinetic energy of one molecule of Cl_2, which has a molar mass of 70?

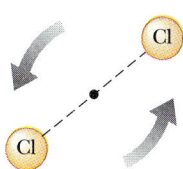

FIGURE P21.37

38. Consider 2 mol of an ideal diatomic gas. Find the total heat capacity at constant volume and at constant pressure if (a) the molecules rotate but do not vibrate and (b) the molecules rotate and vibrate.

*Section 21.5   The Boltzmann Distribution Law
*Section 21.6   Distribution of Molecular Speeds

39. The latent heat of vaporization for water at room temperature is 2430 J/g. (a) How much kinetic energy does each water molecule that evaporates possess before evaporating? (b) Find the average speed before evaporating of a water molecule that is evaporating. (c) What is the effective temperature of these molecules? Why don't these molecules burn you?

40. Fifteen identical particles have the following speeds: one has speed 2.0 m/s; two have speed 3.0 m/s; three have speed 5.0 m/s; four have speed 7.0 m/s; three have speed 9.0 m/s; two have speed 12.0 m/s. Find (a) the average speed, (b) the rms speed, and (c) the most probable speed of these particles.

41. It is reported that there is only one particle per cubic meter in deep space. Using the average temperature of 3.0 K and assuming the particle is H_2 with a diameter of 0.20 nm, (a) determine the mean free path of the particle and the average time between collisions. (b) Repeat part (a) assuming only one particle per cubic centimeter.

42. The chemical composition of the atmosphere changes slightly with altitude because the various molecules have different masses. Use the law of the atmospheres to determine how the ratio of oxygen to nitrogen molecules changes between sea level and 10 km. Assume a temperature of 300 K and take the masses to be 32 u for oxygen (O_2) and 28 u for nitrogen (N_2).

43. (a) Find the ratio of speeds for the two isotopes of chlorine, ^{35}Cl and ^{37}Cl, as they diffuse through air. (b) Which isotope moves faster?

44. Show that the most probable speed of a gas molecule is given by Equation 21.29. Note that the most probable speed corresponds to the point where the slope of the speed distribution curve, dN_v/dv, is zero.

45. At what temperature would the average speed of helium atoms equal (a) the escape speed from Earth, 1.12×10^4 m/s and (b) the escape speed from the Moon, 2.37×10^3 m/s? (See Chapter 14 for a discussion of escape speed, and note that the mass of a helium atom is 6.65×10^{-27} kg.)

46. A gas is at 0°C. To what temperature must it be heated to double the rms speed of its molecules?

*Section 21.7   Mean Free Path

47. In an ultrahigh vacuum system, the pressure is measured to be 1.00×10^{-10} torr (where 1 torr = 133 Pa). If the gas molecules have a molecular diameter of 3.00×10^{-10} m and the temperature is 300 K, find (a) the number of molecules in a volume of 1.00 m^3, (b) the mean free path of the molecules, and (c) the collision frequency, assuming an average speed of 500 m/s.

48. Show that the mean free path for the molecules of an ideal gas is

$$\ell = \frac{k_B T}{\sqrt{2}\pi d^2 P}$$

where d is the molecular diameter.

49. In a tank full of oxygen, how many molecular diameters d (on average) will an oxygen molecule travel (at 1.00 atm and 20.0°C) before colliding with another O_2 molecule? (The diameter of the O_2 molecule is approximately 3.60×10^{-10} m.)

50. Argon gas at atmospheric pressure and 20.0°C is confined in a 1.00-m^3 vessel. The effective hard-sphere diameter of the argon atom is 3.10×10^{-10} m. (a) Determine the mean free path ℓ. (b) Find the pressure when $\ell = 1.00$ m. (c) Find the pressure when $\ell = 3.10 \times 10^{-10}$ m.

*Section 21.8   Van der Waals' Equation of State

51. The constant b that appears in van der Waals' equation of state for oxygen is measured to be 31.8 cm^3/mol. Assuming a spherical shape, estimate the diameter of the molecule.

52. Show that the work done in expanding 1 mol of a van

der Waals' gas from an initial volume V_i to final volume V_f at constant temperature is

$$W = RT \ln\left(\frac{V_f - b}{V_i - b}\right) + a(V_f^{-1} - V_i^{-1})$$

ADDITIONAL PROBLEMS

53. A mixture of two gases will diffuse through a filter at rates proportional to their rms speeds. If the molecules of the two gases have masses m_1 and m_2, show that the ratio of their rms speeds (or the ratio of diffusion rates) is

$$\frac{(v_1)_{rms}}{(v_2)_{rms}} = \sqrt{\frac{m_2}{m_1}}$$

54. A cylinder containing n mol of an ideal gas undergoes a reversible adiabatic process. (a) Starting with the expression $W = \int P\,dV$ and using $PV^\gamma =$ constant, show that the work done is

$$W = \left(\frac{1}{\gamma - 1}\right)(P_i V_i - P_f V_f)$$

   (b) Starting with the first-law equation in differential form, prove that the work done is also equal to $nC_V(T_i - T_f)$. Show that this result is consistent with the equation in part (a).

55. Twenty particles, each of mass m and confined to a volume v, have the following speeds: two have speed v; three have speed $2v$; five have speed $3v$; four have speed $4v$; three have speed $5v$; two have speed $6v$; one has speed $7v$. Find (a) the average speed, (b) the rms speed, (c) the most probable speed, (d) the pressure they exert on the walls of the vessel, and (e) the average kinetic energy per particle.

56. A vessel contains 1.00×10^4 oxygen molecules at 500 K. (a) Make an accurate graph of the Maxwell speed distribution function versus speed with points at speed intervals of 100 m/s. (b) Determine the most probable speed from this graph. (c) Calculate the average and rms speeds for the molecules and label these points on your graph. (d) From the graph, estimate the fraction of molecules with speeds in the range 300 m/s to 600 m/s.

57. One cubic meter of atomic hydrogen at 0°C at atmospheric pressure contains approximately 2.7×10^{25} atoms. The first excited state of the hydrogen atom has an energy of 10.2 eV above the lowest energy level called the ground state. Use the Boltzmann factor to find the number of atoms in the first excited state at 0°C and at 10 000°C.

58. On a day when the atmospheric pressure is 1.00 atm and the temperature is 20.0°C, a diving bell in the shape of a cylinder 4.0 m tall, closed at the upper end, is lowered into water to aid in the construction of an underground foundation for a bridge tower.

The water inside the diving bell rises to within 1.5 m of the top, and the temperature drops to 8.0°C. (a) Find the air pressure inside the bell. (b) How far below the surface of the water is the bell located? (In actual use, additional air is pumped in, forcing the water out to provide working space for the construction workers.)

58A. On a day when the atmospheric pressure is P_0 and the temperature is T_0, a diving bell in the shape of a cylinder of height h, closed at the upper end, is lowered into water to aid in the construction of an underground foundation for a bridge tower. The water inside the diving bell rises to within a distance $d < h$ of the top, and the temperature drops to T. (a) Find the air pressure inside the bell. (b) How far below the surface of the water is the bell located?

59. Oxygen at pressures much above 1 atm becomes toxic to lung cells. What ratio, by weight, of helium gas (He) to oxygen (O_2) must be used by a scuba diver who is to descend to an ocean depth of 50.0 m?

60. The compressibility, κ, of a substance is defined as the fractional change in volume of that substance for a given change in pressure:

$$\kappa = -\frac{1}{V}\frac{dV}{dP}$$

   (a) Explain why the negative sign in this expression ensures that κ is always positive. (b) Show that if an ideal gas is compressed isothermally, its compressibility is given by $\kappa_1 = 1/P$. (c) Show that if an ideal gas is compressed adiabatically, its compressibility is given by $\kappa_2 = 1/\gamma P$. (d) Determine values for κ_1 and κ_2 for a monatomic ideal gas at a pressure of 2.00 atm.

61. One mole of a gas obeying van der Waals' equation of state is compressed isothermally. At some critical temperature, T_c, the isotherm has a point of zero slope, as in Figure 21.17. That is, at $T = T_c$,

$$\frac{\partial P}{\partial V} = 0 \quad \text{and} \quad \frac{\partial^2 P}{\partial V^2} = 0$$

   Using Equation 21.32 and these conditions, show that at the critical point, $P_c = a/27b^2$, $V_c = 3b$, and $T_c = 8a/27Rb$.

62. Consider the particles in a gas centrifuge, a device used to separate particles of different mass by whirling them in a circular path of radius r at angular speed ω. The central force acting on a particle is $mr\omega^2$. (a) Discuss how a gas centrifuge can be used to separate particles of different mass. (b) Show that the density of the particles as a function of r is

$$n(r) = n_0 e^{-mr^2\omega^2/2k_B T}$$

63. Consider a system of 1.00×10^4 oxygen molecules at a temperature T. Write a program that will enable

you to calculate the Maxwell distribution function N_v as a function of the speed of the molecules and the temperature. Use your program to evaluate N_v for speeds ranging from $v = 0$ to $v = 2000$ m/s (in intervals of 100 m/s) at temperatures of (a) 300 K and (b) 1000 K. (c) Make graphs of your results (N_v versus v) and use the graph at $T = 1000$ K to calculate the number of molecules having speeds between 800 m/s and 1000 m/s at $T = 1000$ K.

64. Verify Equations 21.27 and 21.28 for the rms and average speeds of the molecules of a gas at a temperature T. Note that the average value of v^n is

$$\overline{v^n} = \frac{1}{N}\int_0^\infty v^n N_v\, dv$$

and make use of the integrals

$$\int_0^\infty x^3 e^{-ax^2}\, dx = \frac{1}{2a^2} \qquad \int_0^\infty x^4 e^{-ax^2}\, dx = \frac{3}{8a^2}\sqrt{\frac{\pi}{a}}$$

65. (a) Show that the fraction of particles below an altitude h in the atmosphere is

$$f = 1 - e^{-mgh/k_B T}$$

(b) Use this result to show that half the particles are below the altitude $h' = k_B T \ln(2)/mg$. What is the value of h' for Earth? (Assume a uniform temperature of 270 K and note that the average molecular mass for air is 29.0 u.)

66. By volume, air is composed of approximately 78% nitrogen (N_2), 21% oxygen (O_2), and 1% other gases. Ignoring the 1% other gases, (a) use these facts to find the mass of a cubic meter of air at standard conditions (1.00 atm, 0.0°C). (b) Given this result, calculate the lifting force on a helium-filled balloon with a volume of 1.00 m³ at a pressure of 1.00 atm. (c) Show that a helium-filled balloon has 92.6% the lifting force of a similar hydrogen-filled balloon.

67. There are roughly 10^{59} neutrons and protons in an average star and about 10^{11} stars in a typical galaxy. Galaxies tend to form in clusters of (on the average) about 10^3 galaxies, and there are about 10^9 clusters in the known part of the Universe. (a) Approximately how many neutrons and protons are there in the known Universe? (b) Suppose all this matter were compressed into a sphere of nuclear matter such that each nuclear particle occupied a volume of 10^{-45} m³ (about the "volume" of a neutron or proton). What would be the radius of this sphere of nuclear matter? (c) How many moles of nuclear particles are there in the observable Universe?

68. (a) If it has enough kinetic energy, a molecule at the surface of the Earth can escape the Earth's gravitation. Using energy conservation, show that the minimum kinetic energy needed to escape is mgR, where

m is the mass of the molecule, g is the free-fall acceleration at the surface, and R is the radius of the Earth. (b) Calculate the temperature for which the minimum escape kinetic energy equals ten times the average kinetic energy of an oxygen molecule.

69. Using multiple laser beams, physicists have been able to cool and trap sodium atoms in a small region. In one experiment the temperature of the atoms was reduced to 2.4×10^{-4} K. (a) Determine the rms speed of the sodium atoms at this temperature. The atoms can be trapped for about 1.0 s. The trap has a linear dimension of roughly 1.0 cm. (b) Approximately how long would it take an atom to wander out of the trap region if there were no trapping action?

70. For a Maxwellian gas, use a computer or programmable calculator to find the numerical value of the ratio $\{N_v(v)/N_v(v_{mp})\}$ for the following values of v: $v = (v_{mp}/50)$, $(v_{mp}/10)$, $(v_{mp}/2)$, $2v_{mp}$, $10v_{mp}$, $50v_{mp}$. Give your results to three significant figures.

SPREADSHEET PROBLEMS

S1. For a gas consisting of N molecules, the number of molecules that have speeds in the range between v and $v + dv$ is $dN = N_v\, dv$, where N_v is the Maxwell-Boltzmann distribution function given by Equation 21.26. If the interval dv is small enough, N_v/N can be considered to be constant over the interval. Spreadsheet 21.1 calculates the fraction of the total number of molecules $dN/N = N_v\, dv/N$ having a particular speed in a narrow range of speeds dv and plots dN/N versus v. The spreadsheet also calculates the most probable speed and the rms speed of the molecules. (a) Use Spreadsheet 21.1 to find the fraction of molecules of hydrogen gas with speeds between 1000 m/s and 1100 m/s and between 3000 m/s and 3100 m/s at 100 K. (b) Repeat part (a) for $T = 273$ K. (c) Repeat part (a) for $T = 1000$ K.

S2. Repeat Problem S1. for oxygen molecules.

S3. Spreadsheet 21.2 calculates and plots three isotherms for the van der Waals' equation of state:

$$\left(P + \frac{an^2}{v^2}\right)(V - bn) = nRT$$

It also calculates the critical temperature, critical pressure, and critical volume for the gas. The constants a and b and the experimental values for the critical temperature and pressure of several real gases shown in the table below. For a van der Waals gas, it can be shown that the critical temperature is $8a/27Rb$, the critical pressure is $a/27b^2$, and the critical volume is $3bn$. Use Spreadsheet 21.2 to plot the isotherms for each of the gases in the table. Compare the calculated and experimental values for the critical temperatures and pressures for each gas.

Gas	$a(L^2 \cdot atm/mol^2)$	$b(L/mol)$	$T_c(K)$	$P_c(atm)$
N_2	1.390	0.03913	126	33.5
O_2	1.360	0.03183	155	50.1
NO_2	5.284	0.04424	431	100
He	0.03412	0.02370	5.25	2.26
C_6H_6	18.00	0.11540	562	48.6

S4. Modify Spreadsheet 21.1 so that it plots dN/N versus v for three temperatures on the same graph. Plot the distributions for hydrogen molecules at $T = 273$ K, 293 K, and 393 K. Compare the fraction of molecules having speeds between 3000 m/s and 3050 m/s for the three temperatures.

B.C.   **By John Hart**

By permission of John Hart and Field Enterprises, Inc.

CHAPTER 22

Heat Engines, Entropy, and the Second Law of Thermodynamics

This steam-driven locomotive runs from Durango to Silverton, Colorado. Early steam-driven locomotives obtained their energy by burning wood or coal. The generated heat produces the steam, which powers the locomotive. Modern trains use electricity or diesel fuel to power their locomotives. All heat engines extract heat from a burning fuel and convert only a fraction of this energy to mechanical energy. (© Lois Moulton, Tony Stone Worldwide, Ltd.)

The first law of thermodynamics, studied in Chapter 20, is a statement of conservation of energy, generalized to include heat as a form of energy transfer. This law tells us only that an increase in one form of energy must be accompanied by a decrease in some other form of energy. It places no restrictions on the types of energy conversions that can occur. Furthermore, it makes no distinction between heat and work. According to the first law, the internal (thermal) energy of a body may be increased either by adding heat to it or by doing work on it. An important distinction exists between heat and work, however, that is not evident from the first law. One manifestation of this difference is the fact that it is impossible to convert thermal energy to mechanical energy in an isothermal process.

Contrary to what the first law implies, only certain types of energy conversions can take place. The second law of thermodynamics establishes which processes in nature can and which cannot occur. The following are examples of processes that are consistent with the first law of thermodynamics but proceed in an order governed by the second law.

- When two objects at different temperatures are placed in thermal contact with each other, thermal energy always flows from the warmer to the cooler object, never from the cooler to the warmer.
- A rubber ball that is dropped to the ground bounces several times and eventually comes to rest, but a ball lying on the ground never begins bouncing on its own.
- An oscillating pendulum eventually comes to rest because of collisions with air molecules and friction at the point of suspension. The initial mechanical energy is converted to thermal energy; the reverse conversion of energy never occurs.

These are all *irreversible* processes, that is, processes that occur naturally in only one direction. No irreversible process ever runs backward, because if it did, it would violate the second law of thermodynamics.[1]

From an engineering viewpoint, perhaps the most important application of the second law of thermodynamics is the limited efficiency of heat engines. The second law says that a machine capable of continuously converting thermal energy completely to other forms of energy in a cyclic process cannot be constructed.

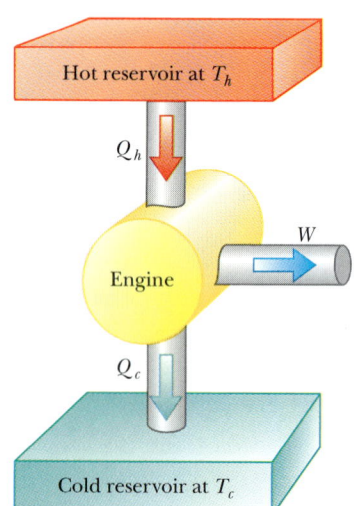

FIGURE 22.1   Schematic representation of a heat engine. The engine absorbs thermal energy Q_h from the hot reservoir, expels thermal energy Q_c to the cold reservoir, and does work W.

22.1   HEAT ENGINES AND THE SECOND LAW OF THERMODYNAMICS

As we learned in Chapter 20, a heat engine is a device that converts thermal energy to other useful forms, such as mechanical and electrical energy. In a typical process for producing electricity in a power plant, for instance, coal or some other fuel is burned and the thermal energy produced is used to convert water to steam. This steam is directed at the blades of a turbine, setting it into rotation. Finally, the mechanical energy associated with this rotation is used to drive an electric generator. Another heat engine, the internal combustion engine in your automobile, extracts thermal energy from a burning fuel and converts a fraction of this energy to mechanical energy.

A heat engine carries some working substance through a cyclic process during which (1) thermal energy is absorbed from a source at a high temperature, (2) work is done by the engine, and (3) thermal energy is expelled by the engine to a source at a lower temperature. As an example, consider the operation of a steam engine in which the working substance is water. The water is carried through a cycle in which it first evaporates to steam in a boiler and then expands against a piston. After the steam is condensed with cooling water, the liquid water produced is returned to the boiler and the process is repeated. It is useful to represent a heat engine schematically as in Figure 22.1. The engine absorbs a quantity of heat Q_h from the hot reservoir, does work W, and then gives up heat Q_c to the cold reservoir. Because the working substance goes through a cycle, its initial and final internal energies are equal, so $\Delta U = 0$. Hence, from the first law of thermodynamics we see that *the net work W done by a heat engine equals the net heat flowing into it.*

A model steam engine equipped with a built-in horizontal boiler. The water is heated electrically, which generates steam that is used to power the electric generator at the right.   *(Courtesy of Central Scientific Co.)*

---

[1] To be more precise, we should say that the set of events in the time-reversed sense is highly improbable. From this viewpoint, events occur with a vastly higher probability in one direction than in the opposite direction.

As we can see from Figure 22.1, $Q_{net} = Q_h - Q_c$; therefore,

$$W = Q_h - Q_c \tag{22.1}$$

where Q_h and Q_c are taken to be positive quantities.

The net work done for a cyclic process is the area enclosed by the curve representing the process on a PV diagram. This is shown for an arbitrary cyclic process in Figure 22.2.

> The **thermal efficiency**, *e*, of a heat engine is defined as the ratio of the net work done to the thermal energy absorbed at the higher temperature during one cycle:
>
> $$e = \frac{W}{Q_h} = \frac{Q_h - Q_c}{Q_h} = 1 - \frac{Q_c}{Q_h} \tag{22.2}$$

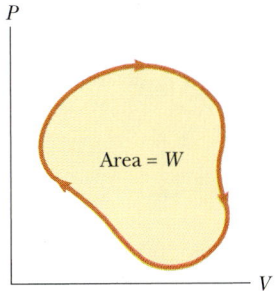

FIGURE 22.2 The *PV* diagram for an arbitrary cyclic process. The net work done equals the area enclosed by the curve.

We can think of the efficiency as the ratio of what you get (mechanical work) to what you give (thermal energy at the higher temperature). Equation 22.2 shows that a heat engine has 100% efficiency ($e = 1$) only if $Q_c = 0$—that is, if no thermal energy is expelled to the cold reservoir. In other words, a heat engine with perfect efficiency would have to convert all of the absorbed thermal energy to mechanical work. One of the consequences of the second law of thermodynamics is that this is impossible.

In practice, it is found that all heat engines convert only a fraction of the absorbed thermal energy to mechanical work. For example, a good automobile engine has an efficiency of about 20% and diesel engines have efficiencies ranging from 35% to 40%. On the basis of this fact, the **Kelvin-Planck** form of the **second law of thermodynamics** states the following:

> It is impossible to construct a heat engine that, operating in a cycle, produces no other effect than the absorption of thermal energy from a reservoir and the performance of an equal amount of work.

This form of the second law is useful in understanding the operation of heat engines. With reference to Equation 22.2, the second law says that, during the operation of a heat engine, *W* can never be equal to Q_h or, alternatively, that some thermal energy Q_c must be rejected to the environment. As a result, it is theoretically impossible to construct an engine that works with 100% efficiency. Figure 22.3 is a schematic diagram of the impossible "perfect" heat engine.

Our assessment of the first two laws of thermodynamics can be summed up as follows: The first law says that *we cannot get more energy out of a cyclic process than the amount of thermal energy we put in,* and the second law says that *we cannot break even because we must put more thermal energy in, at the higher temperature, than the net amount of work output.*

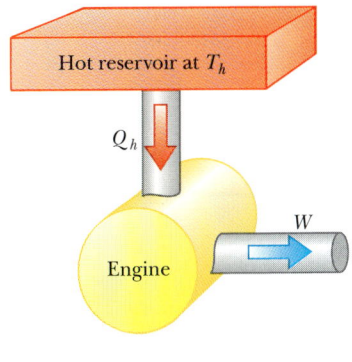

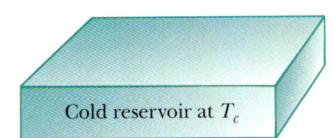

The impossible engine

FIGURE 22.3 Schematic diagram of a heat engine that absorbs thermal energy Q_h from a hot reservoir and does an equivalent amount of work. It is impossible to construct such a perfect engine.

Refrigerators and Heat Pumps

Refrigerators and heat pumps are heat engines running in reverse (Fig. 22.4). The engine absorbs thermal energy Q_c from the cold reservoir and expels thermal energy Q_h to the hot reservoir. This can be accomplished only if work is done *on*

Lord Kelvin, British physicist and mathematician (1824–1907). Born William Thomson in Belfast, Kelvin was the first to propose the use of an absolute scale of temperature. The kelvin scale, named in his honor, is discussed in Section 19.3. Kelvin's study of Carnot's theory led to the idea that heat cannot pass spontaneously from a colder body to a hotter body.

(J. L. Charmet/SPL/Photo Researchers)

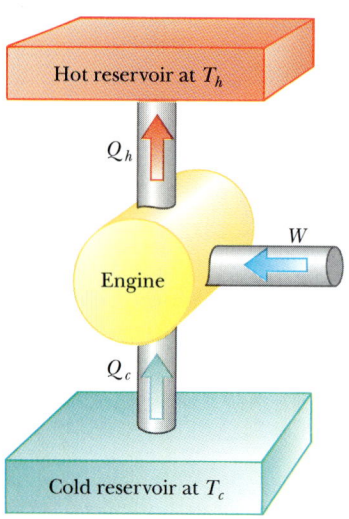

FIGURE 22.4 Schematic diagram for a refrigerator, which absorbs thermal energy Q_c from the cold reservoir and expels thermal energy Q_h to the hot reservoir. Work W is done *on* the refrigerator.

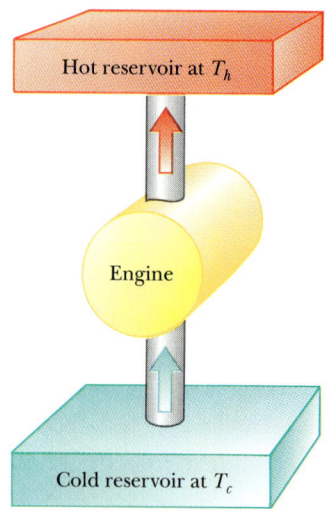

FIGURE 22.5 Schematic diagram of the impossible refrigerator, that is, one that absorbs thermal energy Q_c from a cold reservoir and expels an equivalent amount of thermal energy to the hot reservoir with $W = 0$.

the refrigerator. From the first law, we see that the thermal energy given up to the hot reservoir must equal the sum of the work done and the thermal energy absorbed from the cold reservoir. Therefore, we see that the refrigerator transfers thermal energy from a colder body (the contents of the refrigerator) to a hotter body (the room). In practice, it is desirable to carry out this process with a minimum of work. If it could be accomplished without doing any work, we would have a "perfect" refrigerator (Fig. 22.5). Again, this is in violation of the second law of thermodynamics, which in the form of the **Clausius statement**[2] says the following:

> It is impossible to construct a machine operating in a cycle that produces no other effect than to transfer thermal energy continuously from one object to another object at a higher temperature.

In simpler terms, *thermal energy does not flow spontaneously from a cold object to a hot object.* For example, homes are cooled in summer by pumping thermal energy out; the work done on the air conditioner is supplied by the power company.

The Clausius and Kelvin-Planck statements of the second law appear, at first sight, to be unrelated. They are, in fact, equivalent in all respects. Although we do not prove it here, it can be shown that if either statement is false, so is the other.[3]

---

[2] First expressed by Rudolf Clausius (1822–1888).

[3] See, for example, R. P. Bauman, *Modern Thermodynamics and Statistical Mechanics*, New York, Macmillan, 1992.

EXAMPLE 22.1   The Efficiency of an Engine

Find the efficiency of an engine that introduces 2000 J of heat during the combustion phase and loses 1500 J at exhaust.

Solution   The efficiency of the engine is given by Equation 22.2:

$$e = 1 - \frac{Q_c}{Q_h} = 1 - \frac{1500 \text{ J}}{2000 \text{ J}} = 0.25, \text{ or } \boxed{25\%}$$

22.2   REVERSIBLE AND IRREVERSIBLE PROCESSES

In the next section we shall discuss a theoretical heat engine that is the most efficient engine possible. In order to understand its nature, we must first examine the meaning of reversible and irreversible processes. A **reversible process** is one that can be performed so that, at its conclusion, both the system and its surroundings have been returned to their exact initial conditions. A process that does not satisfy these requirements is **irreversible.**

All natural processes are known to be irreversible. From the endless number of examples that can be selected, let us examine the free expansion of a gas, already discussed in Section 20.6, and show that it cannot be reversible. The gas is in an insulated container, as in Figure 22.6, with a membrane separating the gas from a vacuum. If the membrane is punctured, the gas expands freely into the vacuum. Because the gas does not exert a force through a distance on the surroundings, it does no work as it expands. In addition, no thermal energy is transferred to or from the gas because the container is insulated from its surroundings. Thus, in this adiabatic process, the system has changed but the surroundings have not. Now imagine that we try to reverse the process by first compressing the gas to its original volume. Let's say an engine is being used to force the piston inward. This action is changing both the system and its surroundings. The surroundings are changing because work is being done by an outside agent on the system, and the system is changing because the compression is increasing the temperature of the gas. We can lower the temperature of the gas by allowing it to come into contact with an external heat reservoir. Although this second procedure returns the gas to its original state, the surroundings are again affected because thermal energy is being added to the surroundings. If this thermal energy could somehow be used to drive the engine and compress the gas, the system and its surroundings could be returned to their initial states. However, our statement of the second law says that this extracted thermal energy cannot be completely converted to mechanical energy isothermally. We must conclude that a reversible process has not occurred.

Although real processes are always irreversible, some are *almost* reversible. If a real process occurs very slowly so that the system is virtually always in equilibrium, the process can be considered reversible. For example, imagine compressing a gas very slowly by dropping some grains of sand onto a frictionless piston as in Figure 22.7. The process is made isothermal by placing the gas in thermal contact with a heat reservoir, and just enough thermal energy is transferred from the gas to the reservoir during the process to keep the temperature constant. The pressure, volume, and temperature of the gas are well defined during the isothermal compression. Each time a grain of sand is added to the piston, the volume decreases slightly while the pressure increases slightly. Each added grain represents a change to a

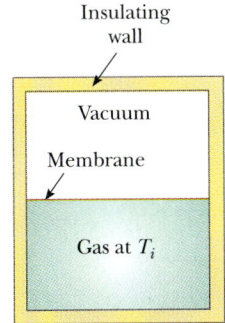

FIGURE 22.6   Free expansion of a gas.

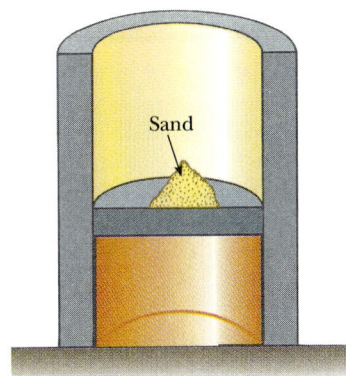

FIGURE 22.7   A gas in thermal contact with a heat reservoir is compressed slowly by dropping grains of sand onto the piston. The compression is isothermal and reversible.

new equilibrium state. The process can be reversed by slowly removing grains from the piston. Although adding and removing grains of sand is an irreversible process for the surroundings, changes in the system have occurred as if the entire process were reversible.

A general characteristic of a reversible process is that there can be no dissipative effects present, such as turbulence or friction, that convert mechanical energy to thermal energy. In reality, such effects are impossible to eliminate completely, and hence it is not surprising that real processes in nature are irreversible.

22.3 THE CARNOT ENGINE

In 1824 a French engineer named Sadi Carnot (1796–1832) described a theoretical engine, now called a **Carnot engine,** that is of great importance from both practical and theoretical viewpoints. He showed that a heat engine operating in an ideal, reversible cycle—called a **Carnot cycle**—between two heat reservoirs is the most efficient engine possible. Such an ideal engine establishes an upper limit on the efficiencies of all engines. That is, the net work done by a working substance taken through the Carnot cycle is the greatest amount of work possible for a given amount of thermal energy supplied to the substance at the upper temperature. **Carnot's theorem** can be stated as follows:

> No real heat engine operating between two heat reservoirs can be more efficient than a Carnot engine operating between the same two reservoirs.

Let us briefly describe some aspects of this theorem. First, we assume that the second law is valid. Next, imagine two heat engines operating between the same heat reservoirs, one which is a Carnot engine with efficiency e_c, and the other whose efficiency e is greater than e_c. If the more efficient engine is used to drive the Carnot engine as a refrigerator, the net result is a transfer of heat from the cold to the hot reservoir. According to the second law, this is impossible. Hence, the assumption that $e > e_c$ must be false.

Sadi Carnot, a French physicist, was the first to show the quantitative relationship between work and heat. Carnot was born in Paris on June 1, 1796, and was educated at the École Polytechnique in Paris and at the École Genie in Metz. His interests included mathematics, tax reform, industrial development, and the fine arts.

In 1824 he published his only work —*Reflections on the Motive Power of Heat*—which reviewed the industrial, political, and economic importance of the steam engine. In it he defined work as "weight lifted through a height." Carnot began to study the

Sadi Carnot

| 1 7 9 6 – 1 8 3 2 | (FPG)

physical properties of gases in 1831, particularly the relationship between temperature and pressure.

On August 24, 1832, he died suddenly of cholera. In accordance with the custom of his time, all of his personal effects were burned, but some of his notes fortunately escaped destruction. Carnot's notes led Lord Kelvin to confirm and extend the science of thermodynamics in 1850.

To describe the Carnot cycle, we assume that the substance working between temperatures T_c and T_h is an ideal gas contained in a cylinder with a movable piston at one end. The cylinder walls and the piston are thermally nonconducting. Four stages of the Carnot cycle are shown in Figure 22.8, and the PV diagram for the cycle is shown in Figure 22.9. The Carnot cycle consists of two adiabatic and two isothermal processes, all reversible.

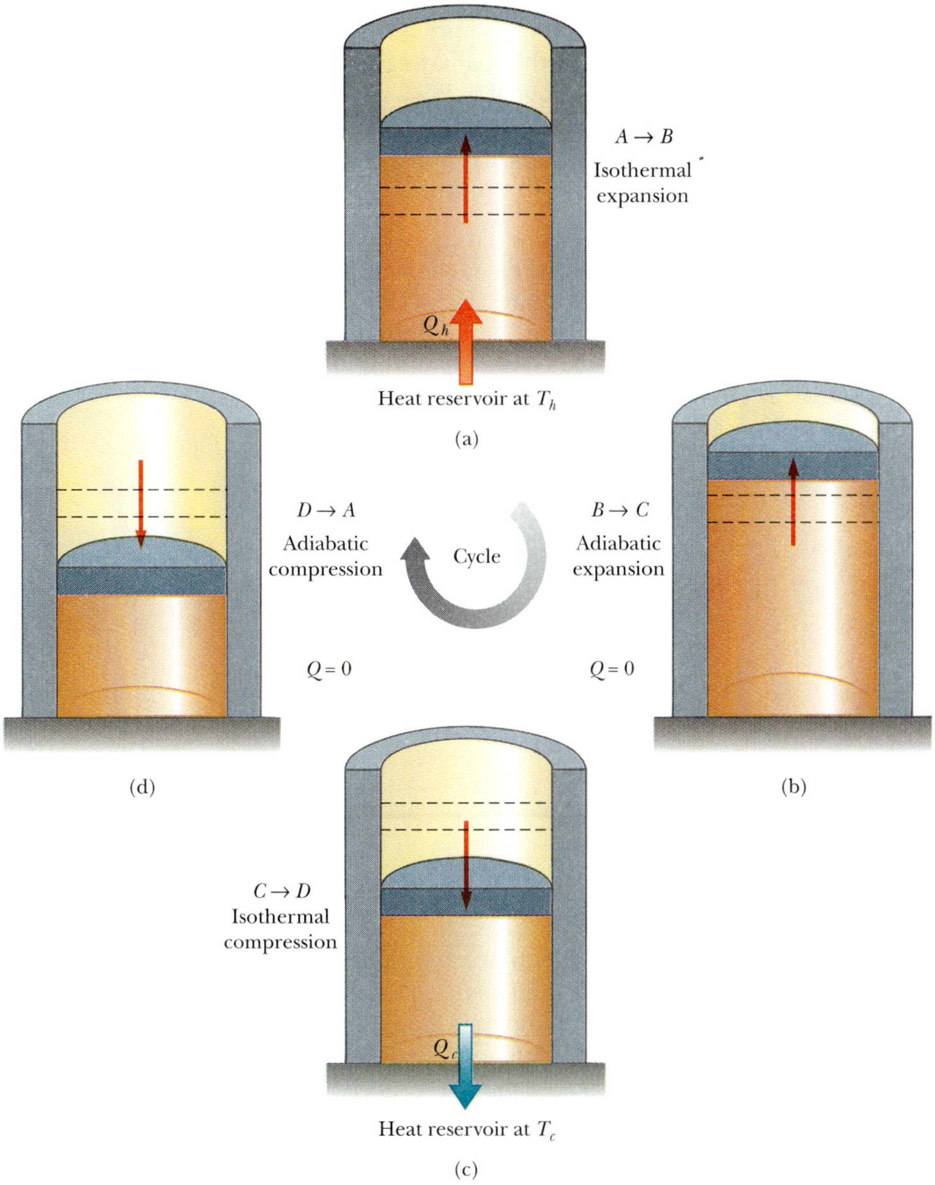

$A \rightarrow B$
Isothermal
expansion

Q_h

Heat reservoir at T_h

(a)

$D \rightarrow A$
Adiabatic
compression

Cycle

$B \rightarrow C$
Adiabatic
expansion

$Q = 0$

$Q = 0$

(d)

(b)

$C \rightarrow D$
Isothermal
compression

Q_c

Heat reservoir at T_c

(c)

FIGURE 22.8   The Carnot cycle. In process $A \rightarrow B$, the gas expands isothermally while in contact with a reservoir at T_h. In process $B \rightarrow C$, the gas expands adiabatically ($Q = 0$). In process $C \rightarrow D$, the gas is compressed isothermally while in contact with a reservoir at $T_c < T_h$. In process $D \rightarrow A$, the gas is compressed adiabatically. The upward arrows on the piston indicate weights being removed during the expansions, and the downward arrows indicate the addition of weights during the compressions.

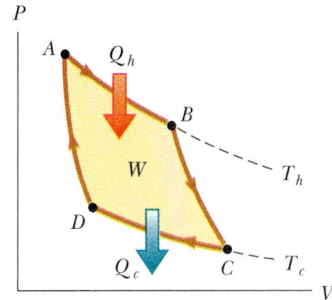

FIGURE 22.9   The PV diagram for the Carnot cycle. The net work done, W, equals the net heat received in one cycle, $Q_h - Q_c$. Note that $\Delta U = 0$ for the cycle.

- The process $A \rightarrow B$ is an isothermal expansion at temperature T_h, in which the gas is placed in thermal contact with a heat reservoir at temperature T_h (Fig. 22.8a). During the expansion, the gas absorbs thermal energy Q_h from the reservoir through the base of the cylinder and does work W_{AB} in raising the piston.
- In the process $B \rightarrow C$, the base of the cylinder is replaced by a thermally nonconducting wall and the gas expands adiabatically; that is, no thermal energy enters or leaves the system (Fig. 22.8b). During the expansion, the temperature falls from T_h to T_c and the gas does work W_{BC} in raising the piston.
- In the process $C \rightarrow D$, the gas is placed in thermal contact with a heat reservoir at temperature T_c (Fig. 22.8c) and is compressed isothermally at temperature T_c. During this time, the gas expels thermal energy Q_c to the reservoir and the work done on the gas by an external agent is W_{CD}.
- In the final stage, $D \rightarrow A$, the base of the cylinder is replaced by a nonconducting wall (Fig. 22.8d) and the gas is compressed adiabatically. The temperature of the gas increases to T_h and the work done on the gas by an external agent is W_{DA}.

The net work done in this reversible, cyclic process is equal to the area enclosed by the path *ABCDA* of the *PV* diagram (Fig. 22.9). As we showed in Section 22.1, the net work done in one cycle equals the net heat transferred into the system, $Q_h - Q_c$, since the change in internal energy is zero. Hence, the thermal efficiency of the engine is given by Equation 22.2:

$$e = \frac{W}{Q_h} = 1 - \frac{Q_c}{Q_h}$$

In Example 22.2, we will show that for a Carnot cycle,

Ratio of heats for a Carnot cycle

$$\frac{Q_c}{Q_h} = \frac{T_c}{T_h} \tag{22.3}$$

Hence, the thermal efficiency of a Carnot engine is

Efficiency of a Carnot engine

$$e_c = 1 - \frac{T_c}{T_h} \tag{22.4}$$

According to this result,

all Carnot engines operating between the same two temperatures in a reversible manner have the same efficiency. From Carnot's theorem, the efficiency of any reversible engine operating in a cycle between two temperatures is greater than the efficiency of any irreversible (real) engine operating between the same two temperatures.

Equation 22.4 can be applied to any working substance operating in a Carnot cycle between two heat reservoirs. According to this result, the efficiency is zero if $T_c = T_h$, as one would expect. The efficiency increases as T_c is lowered and as T_h increases. However, the efficiency can be unity (100%) only if $T_c = 0$ K. Such reservoirs are not available, and so the maximum efficiency is always less than 100%. In most practical cases, the cold reservoir is near room temperature, about

300 K. Therefore, one usually strives to increase the efficiency by raising the temperature of the hot reservoir. *All real engines are less efficient than the Carnot engine since they are subject to such practical difficulties as friction and heat losses by conduction.*

EXAMPLE 22.2   Efficiency of the Carnot Engine

Show that the efficiency of a heat engine operating in a Carnot cycle using an ideal gas is given by Equation 22.4.

Solution   During the isothermal expansion, $A \rightarrow B$ (Fig. 22.8a), the temperature does not change and so the internal energy remains constant. The work done by the gas is given by Equation 20.12. According to the first law, the thermal energy absorbed, Q_h, equals the work done, so that

$$Q_h = W_{AB} = nRT_h \ln \frac{V_B}{V_A}$$

In a similar manner, the thermal energy rejected to the cold reservoir during the isothermal compression $C \rightarrow D$ is

$$Q_c = |W_{CD}| = nRT_c \ln \frac{V_C}{V_D}$$

Dividing these expressions, we find that

$$(1) \qquad \frac{Q_c}{Q_h} = \frac{T_c \ln(V_C/V_D)}{T_h \ln(V_B/V_A)}$$

We now show that the ratio of the logarithmic quantities is unity by obtaining a relation between the ratio of volumes.

   For any quasi-static, adiabatic process, the pressure and volume are related by Equation 21.18:

$$PV^\gamma = \text{constant}$$

During any reversible, quasi-static process, the ideal gas must also obey the equation of state, $PV = nRT$. Substituting this into the above expression to eliminate the pressure, we find that

$$TV^{\gamma-1} = \text{constant}$$

Applying this result to the adiabatic processes $B \rightarrow C$ and $D \rightarrow A$, we find that

$$T_h V_B{}^{\gamma-1} = T_c V_C{}^{\gamma-1}$$
$$T_h V_A{}^{\gamma-1} = T_c V_D{}^{\gamma-1}$$

Dividing these equations, we obtain

$$(V_B/V_A)^{\gamma-1} = (V_C/V_D)^{\gamma-1}$$

$$(2) \qquad \frac{V_B}{V_A} = \frac{V_C}{V_D}$$

Substituting (2) into (1), we see that the logarithmic terms cancel and we obtain the relationship

$$\frac{Q_c}{Q_h} = \frac{T_c}{T_h}$$

Using this result and Equation 22.2, we see that the thermal efficiency of the Carnot engine is

$$e_c = 1 - \frac{Q_c}{Q_h} = 1 - \frac{T_c}{T_h} = \frac{T_h - T_c}{T_h}$$

EXAMPLE 22.3   The Steam Engine

A steam engine has a boiler that operates at 500 K. The heat changes water to steam, and this steam then drives the piston. The exhaust temperature is that of the outside air, approximately 300 K. What is the maximum thermal efficiency of this steam engine?

Solution   From the expression for the efficiency of a Carnot engine, we find the maximum thermal efficiency for any engine operating between these temperatures:

$$e_c = 1 - \frac{T_c}{T_h} = 1 - \frac{300 \text{ K}}{500 \text{ K}} = 0.4, \text{ or } \boxed{40\%}$$

You should note that this is the highest theoretical efficiency of the engine. In practice, the efficiency is considerably lower.

Exercise   Determine the maximum work the engine can perform in each cycle of operation if it absorbs 200 J of heat from the hot reservoir during each cycle.

Answer   80 J.

EXAMPLE 22.4   The Carnot Efficiency

The highest theoretical efficiency of a gasoline engine, based on the Carnot cycle, is 30%. If this engine expels its gases into the atmosphere, which has a temperature of 300 K, what is the temperature in the cylinder immediately after combustion?

Solution   The Carnot efficiency is used to find T_h:

$$e_c = 1 - \frac{T_c}{T_h}$$

$$T_h = \frac{T_c}{1 - e_c} = \frac{300 \text{ K}}{1 - 0.3} = \boxed{429 \text{ K}}$$

Exercise   If the heat engine absorbs 837 J of heat from the hot reservoir during each cycle, how much work can it perform in each cycle?

Answer   251 J.

22.4   THE ABSOLUTE TEMPERATURE SCALE

In Chapter 19 we defined temperature scales in terms of how certain physical properties of materials change as the temperature changes. It is desirable to define a temperature scale that is independent of material properties. The Carnot cycle provides us with the basis for such a temperature scale. Equation 22.3 tells us that the ratio Q_c / Q_h depends *only* on the temperatures of the two heat reservoirs. The ratio T_c / T_h can be obtained by operating a reversible heat engine in a Carnot cycle between these two temperatures and carefully measuring Q_c and Q_h. A temperature scale can be determined with reference to some fixed-point temperatures. The **absolute,** or **kelvin, temperature scale** is defined by choosing 273.16 K as the temperature of the triple point of water.

The temperature of any substance can be obtained in the following manner: (1) take the substance through a Carnot cycle; (2) measure the thermal energy Q absorbed or expelled by the system at some temperature T; and (3) measure the thermal energy Q_3 absorbed or expelled by the system when it is at the temperature of the triple point of water. From Equation 22.3 and this procedure, we find that the unknown temperature is

$$T = (273.16 \text{ K}) \frac{Q}{Q_3}$$

The absolute temperature scale is identical to the ideal gas temperature scale and is independent of the properties of the working substance. Therefore it can be applied even at very low temperatures.

In the previous section, we found that the thermal efficiency of any Carnot engine is given by $e_c = 1 - (T_c / T_h)$. This result shows that a 100% efficient engine is possible only if a temperature of absolute zero is maintained for T_c. If this were possible, any Carnot engine operating between T_h and $T_c = 0$ K would convert all of the absorbed thermal energy to work.[4] Using this idea, Lord Kelvin defined absolute zero as follows: *Absolute zero is the temperature of a reservoir at which a Carnot engine will expel no heat.*

---

[4] Experimentally, it is not possible to reach absolute zero. Temperatures as low as about 10^{-5} K have been achieved with enormous difficulties using a technique called *nuclear demagnetization*. Even lower temperatures have recently been obtained using lasers to stop the motion of atoms of a gas. The fact that absolute zero may be approached but never reached is a consequence of a law of nature known as the *third law of thermodynamics*.

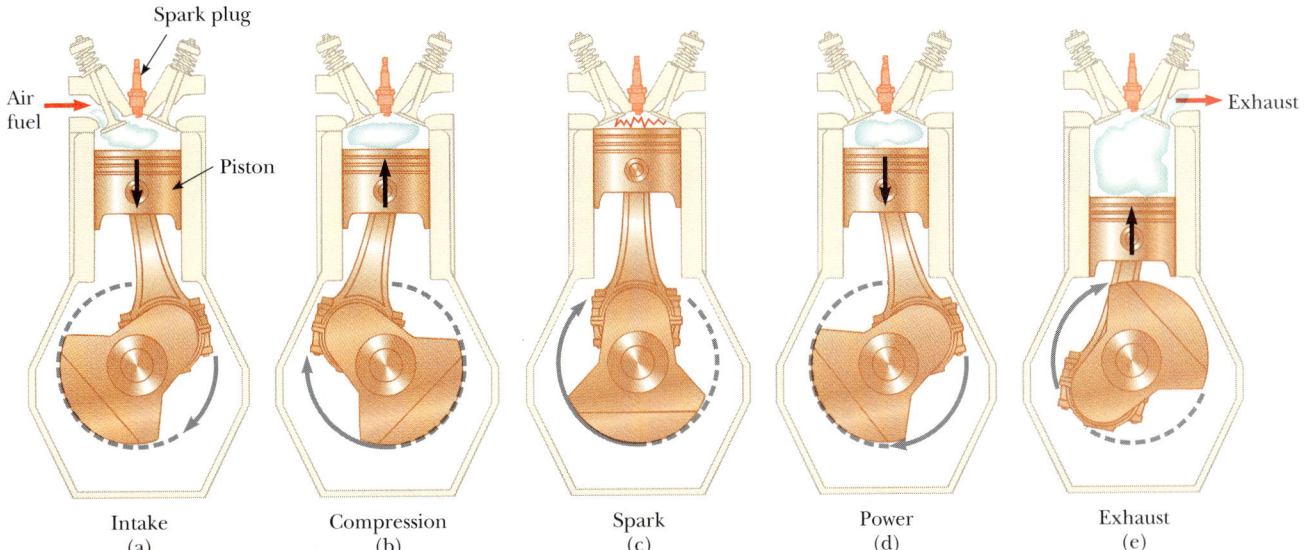

Intake   Compression   Spark   Power   Exhaust
(a)   (b)   (c)   (d)   (e)

FIGURE 22.10   The four-stroke cycle of a conventional gasoline engine. (a) In the intake stroke, air is mixed with fuel. (b) The intake valve is then closed, and the air-fuel mixture is compressed by the piston. (c) The mixture is ignited by the spark plug raising it to a higher temperature. (d) In the power stroke, the gas expands against the piston. (e) Finally, the residual gases are expelled and the cycle repeats.

22.5   THE GASOLINE ENGINE

In this section we discuss the efficiency of the common gasoline engine. Five successive processes occur in each cycle, as illustrated in Figure 22.10. These processes can be approximated by the **Otto cycle,** a PV diagram of which is illustrated in Figure 22.11:

- During the intake stroke $O \rightarrow A$, air is drawn into the cylinder at atmospheric pressure and the volume increases from V_2 to V_1.
- In the process $A \rightarrow B$ (compression stroke), the air-fuel mixture is compressed adiabatically from volume V_1 to volume V_2, and the temperature increases from T_A to T_B. The work done on the gas is the area under the curve AB.
- In the process $B \rightarrow C$, combustion occurs and thermal energy Q_h is added to the gas. This is not an inflow of thermal energy but rather a release of thermal energy from the combustion process. During this time the pressure and temperature rise rapidly, but the volume remains approximately constant. No work is done on the gas.
- In the process $C \rightarrow D$ (power stroke), the gas expands adiabatically from V_2 to V_1, causing the temperature to drop from T_C to T_D. The work done by the gas equals the area under the curve CD.
- In the process $D \rightarrow A$, thermal energy Q_c is extracted from the gas as its pressure decreases at constant volume when an exhaust valve is opened. No work is done during this process.
- In the final process of the exhaust stroke $A \rightarrow O$, the residual gases are exhausted at atmospheric pressure, and the volume decreases from V_1 to V_2. The cycle then repeats itself.

If the air-fuel mixture is assumed to be an ideal gas, the efficiency of the Otto cycle that is shown in Example 22.7 will be

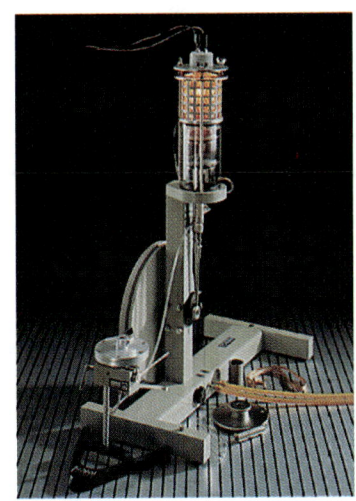

Model of a Stirling engine. This device uses a "regenerator," which recycles some of the heat that would normally be lost back into the system; for this reason, the Stirling cycle has an efficiency equal to that of the Carnot cycle.   *(Courtesy of Central Scientific Co.)*

$$e = 1 - \frac{1}{(V_1/V_2)^{\gamma-1}} \qquad (22.5)$$

where γ is the ratio of the molar specific heats C_p/C_v and V_1/V_2 is called the **compression ratio.** This expression shows that the efficiency increases with increasing compression ratios. For a typical compression ratio of 8 and $\gamma = 1.4$, a theoretical efficiency of 56% is predicted for an engine operating in the idealized Otto cycle. This is much higher than what is achieved in real engines (15% to 20%) because of such effects as friction, heat loss to the cylinder walls, and incomplete combustion of the air-fuel mixture. Diesel engines have higher efficiencies than gasoline engines because of their higher compression ratios and therefore higher combustion temperatures.

CONCEPTUAL EXAMPLE 22.5

A diesel engine operates with a compression ratio of about 16. A gasoline engine has a compression ratio of about 8. Which engine runs hotter? Which engine (at least theoretically) can operate more efficiently?

Reasoning   An internal-combustion engine takes in fuel and air at the environmental temperature and then compresses it. During the rapid (adiabatic) compression, work is delivered to the fuel-air mixture (the system). Heat has no time to leave the system, so the internal energy and temperature rise. The increase in temperature depends only on the nature of the molecules and the decrease in volume, or the so-called compression ratio. Because a diesel engine has a larger compression ratio, its fuel and air mixture will be hotter before and after combustion, compared with a gasoline engine with spark plugs. Its Carnot efficiency limit is set by the higher temperature at which it can take in heat from the chemical energy in the fuel, so the diesel engine can have higher efficiency.

CONCEPTUAL EXAMPLE 22.6

Electrical energy can be converted to thermal energy with an efficiency of 100%. Why is this number misleading with regard to heating a home? That is, what other factors must be considered in comparing the cost of electric heating with the cost of hot-air or hot-water heating?

Reasoning   Most electric power plants produce their electricity by burning fossil fuels, with an efficiency of approximately 30%. When you pay your electric bill, you are indirectly paying for the oil or coal used to generate the electricity, in addition to the other services provided. In conventional hot-air or hot-water heating systems, oil or gas is used to directly heat the air or water. You must compare the lower cost of this direct heating with the cost of maintenance, initial installation, and so forth.

EXAMPLE 22.7   Efficiency of the Otto Cycle

Show that the thermal efficiency of an engine operating in an idealized Otto cycle (Fig. 22.11) is given by Equation 22.5. Treat the working substance as an ideal gas.

Solution   First, let us calculate the work done by the gas during each cycle. No work is done during the processes $B \rightarrow C$ and $D \rightarrow A$. Work is done on the gas during the adiabatic compression $A \rightarrow B$, and work is done by the gas during the adiabatic expansion $C \rightarrow D$. The net work done equals the area bounded by the closed curve in Figure 22.11. Because the change in internal energy is zero for one cycle, we see from the first law that the net work done for each cycle equals the net thermal energy into the system:

$$W = Q_h - Q_c$$

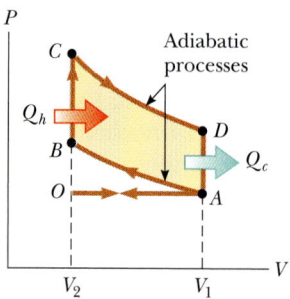

FIGURE 22.11   The *PV* diagram for the Otto cycle, which approximately represents the processes in the internal combustion engine.

Because the processes $B \rightarrow C$ and $D \rightarrow A$ take place at constant volume and since the gas is ideal, we find from the definition of specific heat that

$$Q_h = nC_v(T_C - T_B) \quad \text{and} \quad Q_c = nC_v(T_D - T_A)$$

Using these expressions together with Equation 22.2, we obtain for the thermal efficiency

(1) $\quad e = \dfrac{W}{Q_h} = 1 - \dfrac{Q_c}{Q_h} = 1 - \dfrac{T_D - T_A}{T_C - T_B}$

We can simplify this expression by noting that the processes $A \rightarrow B$ and $C \rightarrow D$ are adiabatic and hence obey the relationship $TV^{\gamma-1} = $ constant. Using this condition and the facts that $V_A = V_D = V_1$ and $V_B = V_C = V_2$, we find

(2) $\quad \dfrac{T_D - T_A}{T_C - T_B} = \left[\dfrac{V_2}{V_1} \right]^{\gamma - 1}$

Substituting (2) into (1) gives for the thermal efficiency

(3) $\quad e = 1 - \dfrac{1}{(V_1/V_2)^{\gamma-1}}$

This can also be expressed in terms of a ratio of temperatures by noting that since $T_A V_1^{\gamma-1} = T_B V_2^{\gamma-1}$, it follows that

$$\left[\dfrac{V_2}{V_1} \right]^{\gamma-1} = \dfrac{T_A}{T_B} = \dfrac{T_D}{T_C}$$

Therefore, (3) becomes

(4) $\quad e = 1 - \dfrac{T_A}{T_B} = 1 - \dfrac{T_D}{T_C}$

During this cycle, the lowest temperature is T_A and the highest temperature is T_C. Therefore the efficiency of a Carnot engine operating between reservoirs at these two temperatures, given by $e_c = 1 - (T_A/T_C)$, would be *greater* than the efficiency of the Otto cycle, given by (4).

22.6   HEAT PUMPS AND REFRIGERATORS

A **heat pump** is a mechanical device that moves thermal energy from a region at lower temperature to a region at higher temperature. Heat pumps have long been popular for cooling and are now becoming increasingly popular for heating purposes as well. In the heating mode, a circulating fluid absorbs thermal energy from the outside and releases it to the interior of the building. The fluid is usually in the form of a low-pressure vapor when in the coils of a unit outside the building, where it absorbs heat from either the air or the ground. This gas is then compressed and enters the building as a hot, high-pressure vapor and enters the interior part of the unit, where it condenses to a liquid and releases its stored thermal energy. An air conditioner is simply a heat pump installed backward, with "exterior" and "interior" interchanged.

Figure 22.12 is a schematic representation of a heat pump. The outside temperature is T_c, the inside temperature is T_h, and the thermal energy absorbed by the circulating fluid is Q_c. The heat pump does work W on the fluid, and the thermal energy transferred from the pump into the building is Q_h.

The effectiveness of a heat pump, in its heating mode, is described in terms of a number called the **coefficient of performance**, COP. This is defined as the ratio of the heat transferred into the hot reservoir and the work required to transfer that heat:

$$\text{COP (heat pump)} \equiv \frac{\text{heat transferred}}{\text{work done by pump}} = \frac{Q_h}{W} \qquad (22.6)$$

If the outside temperature is 25°F or higher, the COP for a heat pump is about 4. That is, the heat transferred into the house is about four times greater than the work done by the motor in the heat pump. However, as the outside temperature decreases, it becomes more difficult for the heat pump to extract sufficient heat from the air and the COP drops. In fact, the COP can fall below unity for temperatures below the midteens.

A Carnot-cycle heat engine run in reverse constitutes a heat pump—in fact, the heat pump with the highest possible coefficient of performance for

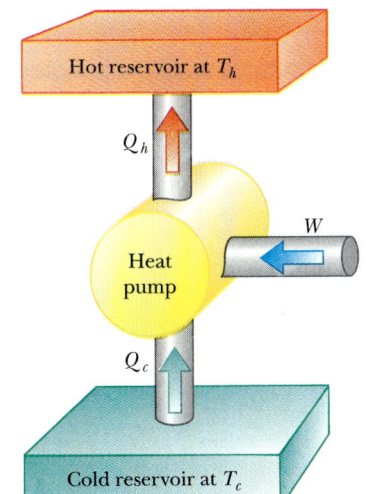

FIGURE 22.12 Schematic diagram of a heat pump, which absorbs thermal energy Q_c from the cold reservoir and expels thermal energy Q_h to the hot reservoir.

Rudolph Clausius (1822–1888). "I propose . . . to call S the entropy of a body, after the Greek word 'transformation.' I have designedly coined the word 'entropy' to be similar to energy, for these two quantities are analogous in their physical significance, that an analogy of denominations seems to be helpful." *(SPL/Photo Researchers)*

the temperatures between which it operates. The maximum coefficient of performance is

$$\text{COP}_c \text{ (heat pump)} = \frac{T_h}{T_h - T_c}$$

The refrigerator works much like a heat pump; it cools its interior by pumping thermal energy from the food storage compartments into the warmer air outside. During its operation, a refrigerator removes a quantity of thermal energy Q_c from the interior of the refrigerator, and in the process (like the heat pump) its motor does work W. The coefficient of performance of a refrigerator or of a heat pump is defined in terms of Q_c:

$$\text{COP (refrigerator)} = \frac{Q_c}{W} \qquad (22.7)$$

An efficient refrigerator is one that removes the greatest amount of thermal energy from the cold reservoir for the least amount of work. Thus, a good refrigerator should have a high coefficient of performance, typically 5 or 6.

The highest possible coefficient of performance is again that of a refrigerator whose working substance is carried through a Carnot heat-engine cycle in reverse:

$$\text{COP}_c \text{ (refrigerator)} = \frac{T_c}{T_h - T_c}$$

As the difference between temperatures of the two reservoirs approaches zero, the theoretical coefficient of performance of a Carnot heat pump approaches infinity. In practice, the low temperature of the cooling coils and the high temperature at the compressor limit the COP to values below 10.

22.7   ENTROPY

The zeroth law of thermodynamics involves the concept of temperature, and the first law involves the concept of internal energy. Temperature and internal energy are both state functions; that is, they can be used to describe the thermodynamic state of a system. Another state function, this one related to the second law of thermodynamics, is the **entropy function,** *S.* In this section we define entropy on a macroscopic scale as it was first expressed by Rudolph Clausius in 1865.

Consider any infinitesimal process for a system between two equilibrium states. If dQ_r is the amount of thermal energy that would be transferred *if the system had followed a reversible path,* then the change in entropy dS, *regardless of the actual path followed,* is equal to this amount of thermal energy transferred along the reversible path divided by the absolute temperature of the system:

Clausius definition of change in entropy

$$dS = \frac{dQ_r}{T} \qquad (22.8)$$

The subscript r on the term dQ_r, is a reminder that the thermal energy transfer is to be measured along a reversible path, even though the system may actually have followed some irreversible path. When thermal energy is absorbed by the system, dQ_r is positive and hence the entropy increases. When thermal energy is expelled by the system, dQ_r is negative and the entropy decreases. Note that Equation 22.8

defines not entropy, but rather the *change* in entropy. This is consistent with the fact that a change in state always accompanies heat transfer. Hence, the meaningful quantity in describing a process is the *change* in entropy.

Entropy originally found its place in thermodynamics, but its importance grew tremendously as the field of statistical mechanics developed because this method of analysis provided an alternative way of interpreting entropy. In statistical mechanics, the behavior of a substance is described in terms of the statistical behavior of the atoms and molecules contained in the substance. One of the main results of this treatment is that

isolated systems tend toward disorder, and entropy is a measure of this disorder.

For example, consider the molecules of a gas in the air in your room. If all the gas molecules moved together like soldiers marching in step, this would be a very ordered state. It is also an unlikely state. If you could see the molecules, you would see that they moved haphazardly in all directions, bumping into one another, changing speed upon collision, some going fast, some slowly. This is a highly disordered state. It is also highly possible that the actual state of the system is such a highly disordered state.

The cause of this perpetual drive toward disorder is easily seen. For any given energy of the system, only certain states are possible, or accessible. Among those states, it is assumed that all are equally probable. However, when such possible states are examined, it is found that far more of them are disordered states than ordered states. Because each of the states is equally probable, it is highly probable that the actual state will be one of the highly disordered states, or more precisely, that the actual state will be one in which the system moves between states of equivalent amounts of disorder. In the following discussion, therefore, the term *state* indicates a collection of states of equivalent amounts of disorder.

All physical processes tend toward more probable states for the system and its surroundings. The more probable state is always one of higher disorder. Because entropy is a measure of disorder, an alternative way of saying this is that

the entropy of the Universe increases in all processes.

This statement is yet another way of stating the second law of thermodynamics.

In judging the relative order or disorder of two states, it is important to examine the spatial order. For example, a perfect crystal has the highest possible spatial order. But one must also consider the order of velocities. At 0 K, all particles have the same speed (in classical physics), but as the temperature increases, the variation in speeds increases, corresponding to an increase in disorder and therefore in entropy. An isolated supercooled liquid may convert to a crystal (corresponding to a decrease in spatial entropy), but only if the temperature rises (corresponding to an increase in velocity entropy), for a net increase in entropy of the system.

To calculate the change in entropy for a finite process, we must recognize that T is generally not constant. If dQ_r is the thermal energy transferred when the system is at a temperature T, then the change in entropy in an arbitrary reversible process between an initial state and a final state is

$$\Delta S = \int_i^f dS = \int_i^f \frac{dQ_r}{T} \qquad \text{(reversible path)} \qquad (22.9)$$

Because entropy is a state function, the change in entropy of a system in going from one state to another has the same value for *all* paths connecting the two states. That is, *the change in entropy of a system depends only on the properties of the initial and final equilibrium state.*

In the case of a *reversible, adiabatic* process, no thermal energy is transferred between the system and its surroundings, and therefore $\Delta S = 0$ in this case. Since there is no change in entropy, such a process is often referred to as an **isentropic process.**

Consider the changes in entropy that occur in a Carnot heat engine operating between the temperatures T_c and T_h. In one cycle, the engine absorbs thermal energy Q_h from the hot reservoir and rejects thermal energy Q_c to the cold reservoir. Thus, the total change in entropy for one cycle is

$$\Delta S = \frac{Q_h}{T_h} - \frac{Q_c}{T_c}$$

where the negative sign represents the fact that thermal energy Q_c is expelled by the system. In Example 22.2 we showed that for a Carnot cycle,

$$\frac{Q_c}{Q_h} = \frac{T_c}{T_h}$$

Using this result in the previous expression for ΔS, we find that the total change in entropy for a Carnot engine operating in a cycle is *zero:*

$$\Delta S = 0$$

Now consider a system taken through an arbitrary cycle. Since the entropy function is a state function and hence depends only on the properties of a given equilibrium state, we conclude that $\Delta S = 0$ for *any* cycle. In general, we can write this condition in the mathematical form

$$\oint \frac{dQ_r}{T} = 0 \qquad (22.10)$$

where the symbol $\oint$ indicates that the integration is over a closed path.

Another important property of entropy is the fact that the entropy of the Universe is not changed by a reversible process. This can be understood by noting that two bodies A and B that interact with each other reversibly must always be in thermal equilibrium with each other. That is, their temperatures must always be equal. Therefore, when a small amount of thermal energy dQ is transferred from A to B, the increase in entropy of B is dQ/T, while the corresponding change in entropy of A is $-dQ/T$. Thus the total change in entropy of the system (A + B) is zero, and the entropy of the Universe is unaffected by the reversible process.

Quasi-Static, Reversible Process for an Ideal Gas

An ideal gas undergoes a quasi-static, reversible process from an initial state T_i, V_i to a final state T_f, V_f. Let us calculate the change in entropy for this process.

According to the first law, $dQ_r = dU + dW$, where $dW = P\,dV$. For an ideal gas, recall that $dU = nC_V\,dT$ and from the ideal gas law, we have $P = nRT/V$. Therefore, we can express the thermal energy transferred as

$$dQ_r = dU + P\, dV = nC_V\, dT + nRT\frac{dV}{V}$$

We cannot integrate this expression as it stands because the last term contains two variables, T and V. However, if we divide each term by T, we can integrate both terms on the right-hand side:

$$\frac{dQ_r}{T} = nC_V\frac{dT}{T} + nR\frac{dV}{V} \tag{22.11}$$

Assuming that C_V is constant over the interval in question, and integrating Equation 22.11 from T_i, V_i to T_f, V_f, we get

$$\Delta S = \int_i^f \frac{dQ_r}{T} = nC_V \ln\frac{T_f}{T_i} + nR\ln\frac{V_f}{V_i} \tag{22.12}$$

This expression shows that ΔS *depends only on the initial and final states and is independent of the reversible path.* Furthermore, ΔS can be positive or negative depending on whether the gas absorbs or expels thermal energy during the process. Finally, for a cyclic process ($T_i = T_f$ and $V_i = V_f$), we see that $\Delta S = 0$.

EXAMPLE 22.8 Change in Entropy—Melting Process

A solid substance that has a latent heat of fusion L_f melts at a temperature T_m. Calculate the change in entropy that occurs when m grams of this substance is melted.

$$\Delta S = \int \frac{dQ_r}{T} = \frac{1}{T_m}\int dQ = \frac{Q}{T_m} = \frac{mL_f}{T_m}$$

Note that we were able to remove T_m from the integral because the process is isothermal.

Solution Let us assume that the melting occurs so slowly that it can be considered a reversible process. In that case the temperature can be regarded as constant and equal to T_m. Making use of Equations 22.9 and the fact that the latent heat of fusion $Q = mL_f$, (Eq. 20.6), we find that

Exercise Calculate the change in entropy when 0.30 kg of lead melts at 327°C. Lead has a latent heat of fusion equal to 24.5 kJ/kg.

Answer $\Delta S = 12.3\,\text{J/K}$.

22.8 ENTROPY CHANGES IN IRREVERSIBLE PROCESSES

By definition, calculation of the change in entropy requires information about a reversible path connecting the initial and final equilibrium states. In order to calculate changes in entropy for real (irreversible) processes, we must first recognize that the entropy function (like internal energy) depends only on the *state* of the system. That is, entropy is a state function. Hence, the change in entropy when a system moves between any two equilibrium states depends only on the initial and final states. Experimentally it is found that the entropy change is the same for all processes that can occur between a given set of initial and final states. It is possible to show that if this were not the case, the second law of thermodynamics would be violated.

We now calculate entropy changes for irreversible processes between two equilibrium states by devising a reversible process (or series of reversible processes) between the same two states and computing $\int dQ_r/T$ for the reversible process. The entropy change for the irreversible process is the same as that of the reversible process between the same two equilibrium states. In irreversible processes, it is critically important to distinguish between Q, the actual thermal energy transfer in the process, and Q_r, the thermal energy that would have been transferred along a reversible path. It is only the second process that gives the correct value for the entropy change.

An illustration from Flammarion's novel *La Fin du Monde*, depicting the "heat death" of the Universe.

As we shall see in the following examples, the change in entropy for the system plus its surroundings is always positive for an irreversible process. In general, the total entropy (and disorder) always increases in an irreversible process. From these considerations, the second law of thermodynamics can be stated as follows: *The total entropy of an isolated system that undergoes a change cannot decrease.* Furthermore, if the process is *irreversible,* the total entropy of an isolated system always *increases.* On the other hand, in a reversible process, the total entropy of an isolated system remains constant.

When dealing with interacting objects that are not isolated from the environment, remember that the increase of entropy applies to the system *and* its surroundings. When two objects interact in an irreversible process, the increase in entropy of one part of the system is greater than the decrease in entropy of the other part. Hence, we conclude that the change in entropy of the Universe must be greater than zero for an irreversible process and equal to zero for a reversible process. Ultimately, the entropy of the Universe should reach a maximum value. At this point the Universe will be in a state of uniform temperature and density. All physical, chemical, and biological processes will cease, since a state of perfect disorder implies that no energy is available for doing work. This gloomy state of affairs is sometimes referred to as the heat death of the Universe.

CONCEPTUAL EXAMPLE 22.9

A living system, such as a tree, combines unorganized molecules (CO_2, H_2O), using sunlight, to produce leaves and branches. Is this reduction of entropy in the tree a violation of the second law of thermodynamics?

Reasoning   Living things survive on the flow of solar radiation from Sun to Earth. A snake basking in the Sun diverts thermal energy temporarily, before the energy is finally radiated into outer space. A tree builds organized cellulose molecules and humans construct buildings, all out of a stream of energy from the Sun. The second law is not violated, because the local reductions in entropy are created at the expense of increase in the total entropy of the Universe.

Heat Conduction

Consider the reversible transfer of thermal energy Q from a hot reservoir at temperature T_h to a cold reservoir at temperature T_c. Since the cold reservoir absorbs thermal energy Q, its entropy increases by Q/T_c. At the same time, the hot reservoir loses thermal energy Q and its entropy decreases by Q/T_h. The increase in entropy of the cold reservoir is greater than the decrease in entropy of the hot reservoir since T_c is less than T_h. Therefore, the total change in entropy of the system (and of the Universe) is greater than zero:

$$\Delta S_u = \frac{Q}{T_c} - \frac{Q}{T_h} > 0$$

EXAMPLE 22.10   Which Way Does the Heat Flow?

A large cold object is at 273 K, and a large hot object is at 373 K. Show that it is impossible for a small amount of thermal energy, say 8.00 J, to be transferred from the cold object to the hot one without decreasing the entropy of the Universe.

Reasoning   We assume that, during the heat transfer, the two objects do not undergo a temperature change. This is not a necessary assumption; it is used to avoid reliance on the techniques of integral calculus. The process as described is irreversible, so we must find an equivalent reversible process. It is sufficient to assume that the hot and cold objects are connected by a poor thermal conductor whose temperature spans the range from 273 K to 373 K, which transfers thermal energy but whose state does not change during the process. Then, the thermal energy transfer to or from each object is reversible, and we may set $Q = Q_r$.

Solution   The entropy change of the hot object is

$$\Delta S_h = \frac{Q}{T_h} = \frac{8.00 \text{ J}}{373 \text{ K}} = 0.0214 \text{ J/K}$$

The cold object loses thermal energy, and its entropy change is

$$\Delta S_c = \frac{Q}{T_c} = \frac{-8.00 \text{ J}}{273 \text{ K}} = -0.0293 \text{ J/K}$$

The net entropy change of the Universe is

$$\Delta S_u = \Delta S_c + \Delta S_h = -0.0079 \text{ J/K}$$

This is in violation of the concept that the entropy of the Universe always increases in natural processes. That is, *the spontaneous transfer of thermal energy from a cold to a hot object cannot occur.*

Exercise   In the preceding example, suppose that 8.00 J of thermal energy is transferred from the hot to the cold object. What is the net entropy change of the Universe?

Answer $+0.0079$ J/K.

Free Expansion

An ideal gas in an insulated container initially occupies a volume V_i (Fig. 22.13). A partition separating the gas from an evacuated region is suddenly broken so that the gas expands (irreversibly) to a volume V_f. Let us find the change in entropy of the gas and the Universe.

The process is clearly neither reversible nor quasi-static. The work done by the gas against the vacuum is zero, and since the walls are insulating, no thermal energy is transferred during the expansion. That is, $W = 0$ and $Q = 0$. Using the first law, we see that the change in internal energy is zero, therefore $U_i = U_f$. Since the gas is ideal, U depends on temperature only, so we conclude that $T_i = T_f$.

To apply Equation 22.9, we must find Q_r; that is, we must find an equivalent reversible path that shares the same initial and final states. A simple choice is an isothermal, reversible expansion in which the gas pushes slowly against a piston. Since T is constant in this process, Equation 22.9 gives

$$\Delta S = \int \frac{dQ_r}{T} = \frac{1}{T} \int_i^f dQ_r$$

But $\int dQ_r$ is simply the work done by the gas during the isothermal expansion from V_i to V_f, which is given by Equation 20.12. Using this result, we find that

$$\Delta S = nR \ln \frac{V_f}{V_i} \qquad (22.13)$$

Since $V_f > V_i$, we conclude that ΔS is positive, and this positive result tells us that both the entropy and the disorder of the gas (and the Universe) increase as a result of the irreversible, adiabatic expansion.

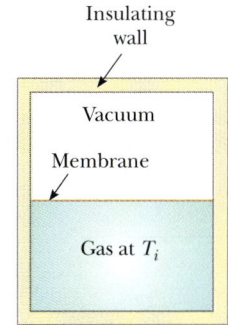

FIGURE 22.13   Free expansion of a gas. When the partition separating the gas from the evacuated region is ruptured, the gas expands freely and irreversibly so that it occupies a greater final volume. The container is thermally insulated from its surroundings, so $Q = 0$.

EXAMPLE 22.11   Free Expansion of a Gas

Calculate the change in entropy of 2.00 mol of an ideal gas that undergoes a free expansion to three times its initial volume.

Solution   Using Equation 22.13 with $n = 2.00$ and $V_f = 3V_i$, we find that

$$\Delta S = nR \ln \frac{V_f}{V_i} = (2.00 \text{ mol})(8.31 \text{ J/mol·K}) \ln 3$$

$$= 18.3 \text{ J/K}$$

Irreversible Heat Transfer

A substance of mass m_1, specific heat c_1, and initial temperature T_1 is placed in thermal contact with a second substance of mass m_2, specific heat c_2, and initial temperature T_2, where $T_2 > T_1$. The two substances are contained in an insulated box so no heat is lost to the surroundings. The system is allowed to reach thermal equilibrium. What is the total entropy change for the system?

First, let us calculate the final equilibrium temperature, T_f. Energy conservation requires that the heat lost by one substance equals the heat gained by the other. Since by definition, $Q = mc\,\Delta T$ for each substance, we get $Q_1 = -Q_2$, or

$$m_1 c_1 \,\Delta T_1 = - m_2 c_2 \,\Delta T_2$$

$$m_1 c_1 \,(T_f - T_1) = - m_2 c_2 \,(T_f - T_2)$$

Solving for T_f gives

$$T_f = \frac{m_1 c_1 T_1 + m_2 c_2 T_2}{m_1 c_1 + m_2 c_2} \tag{22.14}$$

Note that $T_1 < T_f < T_2$, as expected.

The process is irreversible because the system goes through a series of non-equilibrium states. During such a transformation, the temperature at any time is not well defined. However, we can imagine that the hot substance at the initial temperature T_i is slowly cooled to the temperature T_f by placing it in contact with a series of reservoirs differing infinitesimally in temperature, the first reservoir being at T_i and the last at T_f. Such a series of very small changes in temperature would approximate a reversible process. Applying Equation 22.9 and noting that $dQ = mc\,dT$ for an infinitesimal change, we get

$$\Delta S = \int_1 \frac{dQ_1}{T} + \int_2 \frac{dQ_2}{T} = m_1 c_1 \int_{T_1}^{T_f} \frac{dT}{T} + m_2 c_2 \int_{T_2}^{T_f} \frac{dT}{T}$$

where we have assumed that the specific heats remain constant. Integrating, we find

Change in entropy for a heat transfer process

$$\Delta S = m_1 c_1 \ln \frac{T_f}{T_1} + m_2 c_2 \ln \frac{T_f}{T_2} \tag{22.15}$$

where T_f is given by Equation 22.14. If Equation 22.14 is substituted into Equation 22.15, we can show that one of the terms in Equation 22.15 is always positive and the other always negative. (You may want to verify this for yourself.) However, the positive term is always larger than the negative term, resulting in a positive value for ΔS. Thus, we conclude that the entropy of the Universe increases in this irreversible process.

Finally, you should note that Equation 22.15 is valid only when no mixing occurs. If the substances are liquids or gases and mixing occurs, the result applies only if the two fluids are identical, as in the following example.

EXAMPLE 22.12 Calculating ΔS for a Mixing Process

Suppose 1.00 kg of water at 0°C is mixed with an equal mass of water at 100°C. After equilibrium is reached, the mixture has a uniform temperature of 50°C. What is the change in entropy of the system?

Solution The change in entropy can be calculated from Equation 22.15 using the values $m_1 = m_2 = 1.00$ kg, $c_1 =$

$c_2 = 4186$ J/kg·K, $T_1 = 0$°C   ($= 273$ K), $T_2 = 100$°C ($= 373$ K), and $T_f = 50$°C ($= 323$ K).

$$\Delta S = m_1 c_1 \ln \frac{T_f}{T_1} + m_2 c_2 \ln \frac{T_f}{T_2}$$

$$= (1.00 \text{ kg})(4186 \text{ J/kg·K}) \ln\left(\frac{323 \text{ K}}{273 \text{ K}}\right)$$

$$= + (1.00 \text{ kg})(4186 \text{ J/kg·K}) \ln\left(\frac{323 \text{ K}}{373 \text{ K}}\right)$$

$$= 704 \text{ J/K} - 602 \text{ J/K} = \boxed{102 \text{ J/K}}$$

That is, as a result of this irreversible process, the increase in entropy of the cold water is greater than the decrease in entropy of the warm water. Consequently, the increase in entropy of the system is 102 J/K.

*22.9   ENTROPY ON A MICROSCOPIC SCALE[5]

As we have seen, entropy can be approached via macroscopic concepts, using parameters such as pressure and temperature. Entropy can also be treated from a microscopic viewpoint through statistical analysis of molecular motions. We shall use a microscopic model to investigate the free expansion of an ideal gas discussed in the preceding section.

In the kinetic theory of gases, gas molecules are taken to be particles moving in a random fashion. Suppose the gas is confined to the volume V_i (Fig. 22.14a). When the partition separating V_i from the larger container is removed, the molecules eventually are distributed in some fashion throughout the greater volume, V_f (Fig. 22.14b). The exact nature of the distribution is a matter of probability.

Such a probability can be determined by first finding the probabilities for the variety of molecular locations involved in the free expansion process. The instant after the partition is removed (and before the molecules have had a chance to rush into the other half of the container), all the molecules are in the initial volume. Let us estimate the probability of the molecules arriving at a particular configuration through natural random motions into some larger volume V. Assume each molecule occupies some microscopic volume V_m. Then the total number of possible locations of that molecule, in a macroscopic initial volume V_i, is the ratio $W_i = (V_i/V_m)$, which is a huge number. We assume each of these locations is equally probable. (W_i is not to be confused with work.)

As more molecules are added to the system, the number of possible states multiply together. Neglecting the very small probability of two molecules trying to occupy the same location, each molecule may go into any of the (V_i/V_m) locations, so the number of possibilities is $W_i^N = (V_i/V_m)^N$. Although a large number of possible states would be considered unusual, the total number of states is so enormous that these unusual states have a negligible probability of occurring. Hence, the number of equivalent states is proportional to $(V_i/V_m)^N$. Similarly, if the volume is increased to V_f, the number of equivalent states increases to $W_f^N = (V_f/V_m)^N$. Hence, the probabilities are $P_i = cW_i^N$ and $P_f = cW_f^N$, where the constant, c, has been left undetermined. The ratio of these probabilities is

$$\frac{W_f^N}{W_i^N} = \frac{(V_f/V_m)^N}{(V_i/V_m)^N} = \left(\frac{V_f}{V_i}\right)^N$$

If we now take the natural logarithm of this equation and multiply by Boltzmann's constant, we find

$$N k_B \ln\left(\frac{W_f}{W_i}\right) = n N_A k_B \ln\left(\frac{V_f}{V_i}\right)$$

where we write the number of molecules, N, as nN_A, the number of moles times Avogadro's number (Section 19.5). Furthermore, we know that $N_A k_B$ is the uni-

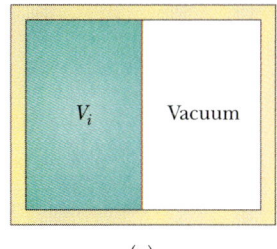

(a)

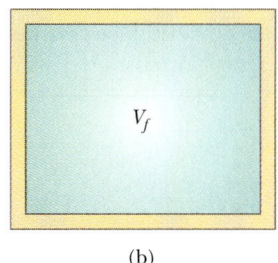

(b)

FIGURE 22.14   In a free expansion, the gas is allowed to expand into a larger volume that was previously a vacuum.

A full house is a very good hand in the game of poker. Can you calculate the probability of being dealt this hand in a standard deck of 52 cards?   (*Tom Mareschel, The IMAGE Bank*)

[5] This section was adapted from A. Hudson and R. Nelson, *University Physics*, Philadelphia, Saunders College Publishing, 1990, used with permission of the publisher.

versal gas constant, R, so this equation may be written as

$$Nk_{\mathrm{B}} \ln W_f - Nk_{\mathrm{B}} \ln W_i = nR \ln\left(\frac{V_f}{V_i}\right) \tag{22.16}$$

From thermodynamic considerations (specifically, Eq. 22.13) we know that when n mol of a gas undergoes a free expansion from V_i to V_f, the change in entropy is

$$S_f - S_i = nR \ln\left(\frac{V_f}{V_i}\right) \tag{22.17}$$

Note that the right-hand sides of Equations 22.16 and 22.17 are identical. We thus make the following important connection between *entropy* and *probability*.

Entropy (microscopic definition)

$$S \equiv Nk_{\mathrm{B}} \ln W \tag{22.18}$$

In real processes, the disorder in the system increases

Although our discussion used the specific example of the free expansion of an ideal gas, a more rigorous development of the statistical interpretation of entropy leads to the same conclusion. *Entropy is a measure of microscopic disorder.*

Imagine that the container of gas in Figure 22.15a has molecules with speeds above the mean value on the left side, and molecules with lower than the mean value on the right side (an ordered arrangement). Compare this with an even mixture of fast- and slow-moving molecules as in Figure 22.15b (a disordered arrangement). You might expect the ordered arrangement to be very unlikely because random motions tend to mix the slow- and fast-moving molecules uniformly. Yet each of these arrangements, *individually*, is equally probable. However, there are far more disordered arrangements than ordered arrangements, so *collectively*, the ordered arrangements are much less probable.

Let us estimate this probability for 100 molecules. The chance of any one molecule being in the left part of the container as a result of random motion is $1/2$. If the molecules move independently, the probability of 50 faster molecules being found in the left part at any instant is $(1/2)^{50}$. Likewise, the probability of the remaining 50 slower molecules being found in the right part at any instant is $(1/2)^{50}$. Therefore, the probability of finding this fast-slow separation through random motion is the product $(1/2)^{50}(1/2)^{50} = (1/2)^{100}$, which corresponds to about 1 chance in 10^{30}. When this calculation is extrapolated from 100 molecules to 1 mol of gas (6.02×10^{23} molecules), the ordered arrangement is found to be *extremely* improbable!

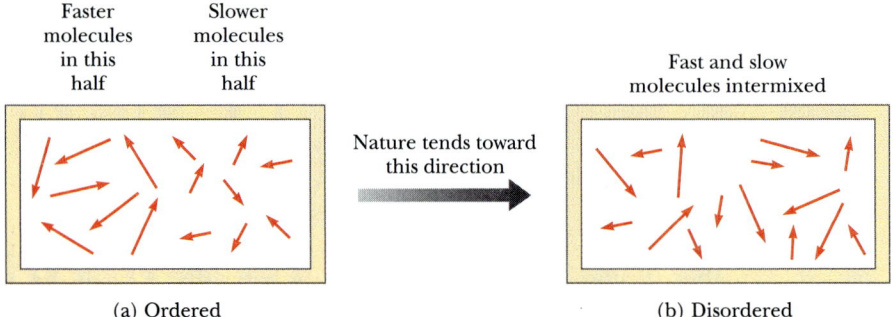

Faster molecules in this half   Slower molecules in this half

Nature tends toward this direction

Fast and slow molecules intermixed

(a) Ordered   (b) Disordered

FIGURE 22.15 A box of gas in two equally probable states of molecular motion that may exist. (a) An ordered arrangement; one of a few, and therefore a collectively unlikely set. (b) A disordered arrangement; one of many, and therefore a collectively likely set.

EXAMPLE 22.13 Free Expansion of an Ideal Gas—Revisited

Again consider the free expansion of an ideal gas. Let us verify that the macroscopic and microscopic approaches lead to the same conclusion. Suppose that 1 mol of gas undergoes a free expansion to four times its initial volume. The initial and final temperatures are, of course, the same. (a) Using a macroscopic approach, calculate the entropy change. (b) Find the probability, P_i, that all molecules will, through random motions, be found simultaneously in the original volume. (c) Using the probability considerations of part (b), calculate the change in entropy for the free expansion and show that it agrees with part (a).

Solution (a) From Equation 22.13, we obtain

$$\Delta S = nR \ln\left(\frac{V_f}{V_i}\right) = (1) R \ln\left(\frac{4V_i}{V_i}\right) = \boxed{R \ln 4}$$

(b) The number of states available to a single molecule in the initial volume V_i is $W_i = (V_i / V_m)$. For one mole (N_A molecules), the number of available states is

$$P_i = cW_i^{N_A} = c\left(\frac{V_i}{V_m}\right)^{N_A}$$

(c) The number of states for all N_A molecules in the volume $V_f = 4V_i$ is

$$P_f = cW_f^{N_A} = c\left(\frac{V_f}{V_m}\right)^{N_A} = c\left(\frac{4V_i}{V_m}\right)^{N_A}$$

From Equation 22.18 we obtain

$$\Delta S = k_B \ln W_f^{N_A} - k_B \ln W_i^{N_A} = k_B \ln\left(\frac{W_f^{N_A}}{W_i^{N_A}}\right) = k_B \ln(4^{N_A})$$

$$= N_A k_B \ln 4 = \boxed{R \ln 4}$$

The answer is the same as that in part (a), which dealt with microscopic parameters.

CONCEPTUAL EXAMPLE 22.14

Suppose you have a bag of 100 marbles, of which 50 are red and 50 are green. You are allowed to draw four marbles from the bag according to the following rules: Draw one marble, record its color, return it to the bag, and draw another. Continue this process until four marbles have been drawn. What are the possible outcomes for this set of events?

Reasoning Because each marble is returned to the bag before the next one is drawn, the probability of drawing a red marble is always the same as that of drawing a green one. All the possible outcomes are shown in Table 22.1. As this table indicates, there is only one way to draw four red marbles. However, there are four possible sequences that could produce one green and three red marbles, six sequences that could produce two green and two red, four sequences that could produce three green and one red, and one sequence that could produce all green. The most likely outcome—two red and two green—corresponds to the most disordered

TABLE 22.1 Possible Results of Drawing Four Marbles from a Bag

End Result	Possible Draws	Total Number of Same Results
All R	RRRR	1
1G, 3R	RRRG, RRGR, RGRR, GRRR	4
2G, 2R	RRGG, RGRG, GRRG, RGGR, GRGR, GGRR	6
3G, 1R	GGGR, GGRG, GRGG, RGGG	4
All G	GGGG	1

state. The probability that you would draw four red or four green marbles, these being the most ordered states, is much lower.

SUMMARY

A **heat engine** is a device that converts thermal energy to other useful forms of energy. The net work done by a heat engine in carrying a substance through a cyclic process ($\Delta U = 0$) is

$$W = Q_h - Q_c \tag{22.1}$$

where Q_h is the thermal energy absorbed from a hot reservoir and Q_c is the thermal energy rejected to a cold reservoir.

The **thermal efficiency, e,** of a heat engine is

$$e = \frac{W}{Q_h} = 1 - \frac{Q_c}{Q_h} \tag{22.2}$$

The **second law of thermodynamics** can be stated in many ways:

- It is impossible to construct a heat engine that, operating in a cycle, produces no effect other than the absorption of thermal energy from a reservoir and the performance of an equal amount of work (Kelvin-Planck statement).
- It is impossible to construct a cyclical machine whose sole effect is to transfer heat continuously from one body to another body at a higher temperature (Clausius statement).

A **reversible cyclic process** is one that can be performed so that, at its conclusion, both the system and its surroundings have been returned to their exact initial conditions. A process that does not satisfy these requirements is **irreversible.**

Carnot's theorem states that no real heat engine operating (irreversibly) between the temperatures T_c and T_h can be more efficient than an engine operating reversibly in a Carnot cycle between the same two temperatures.

The *efficiency of a heat engine* operating in the **Carnot cycle** is

$$e_c = 1 - \frac{T_c}{T_h} \tag{22.4}$$

The second law of thermodynamics states that when real (irreversible) processes occur, the degree of disorder in the system plus the surroundings increases. When a process occurs in an isolated system, an ordered state is converted into a more disordered state. The measure of disorder in a system is called **entropy, S.**

The **change in entropy,** dS, of a system moving between two equilibrium states is

$$dS = \frac{dQ_r}{T} \tag{22.8}$$

The change in entropy of a system in an arbitrary reversible process moving between an initial state and a final state is

$$\Delta S = \int_i^f \frac{dQ_r}{T} \tag{22.9}$$

The value of ΔS is the same for all paths connecting the initial and final states. The change in entropy for any reversible, cyclic process is zero, and when such a process occurs, the entropy of the Universe remains constant.

Entropy is a state function; that is, it depends on the state of the system. The change in entropy for a system undergoing a real (irreversible) process between two equilibrium states is the same as that of a reversible process between the same states.

In an irreversible process, the total entropy of an isolated system always increases. In general, the total entropy (and disorder) always increases in any irreversible process. Furthermore, the change in entropy of the Universe is greater than zero for an irreversible process.

From a microscopic viewpoint, **entropy,** S, is defined as

$$S \equiv Nk_B \ln W \tag{22.18}$$

where k_B is Boltzmann's constant and W is the number of (microscopic) states available to the system in its (macroscopic) state. Because of the statistical tendency of systems to proceed toward states of greater probability and greater disorder, all natural processes are irreversible and entropy increases. Thus, *entropy is a measure of microscopic disorder.*

QUESTIONS

1. Distinguish clearly among temperature, heat (Q) and internal energy.
2. When a sealed Thermos bottle full of hot coffee is shaken, what are the changes, if any, in (a) the temperature of the coffee and (b) its internal energy?
3. Use the first law of thermodynamics to explain why the total energy of an isolated system is always constant.
4. Is it possible to convert internal energy to mechanical energy?
5. What are some factors that affect the efficiency of automobile engines?
6. In practical heat engines, which do we have more control of, the temperature of the hot reservoir or the temperature of the cold reservoir? Explain.
7. A steam-driven turbine is one major component of an electric power plant. Why is it advantageous to have the temperature of the steam as high as possible?
8. Is it possible to construct a heat engine that creates no thermal pollution?
9. Discuss three common examples of natural processes that involve an increase in entropy. Be sure to account for all parts of each system under consideration.
10. Discuss the change in entropy of a gas that expands (a) at constant temperature and (b) adiabatically.
11. In solar ponds constructed in Israel, the Sun's energy is concentrated near the bottom of a salty pond. With the proper layering of salt in the water, convection is prevented, and temperatures of 100°C may be reached. Can you guess the maximum efficiency with which useful energy can be extracted from the pond?
12. The vortex tube (Fig. 22.16) is a T-shaped device that takes in compressed air at 20 atm and 20°C and produces cold air at −20°C out one flared end and 60°C hot air out the other flared end. Does the operation of this device violate the second law of thermodynamics?

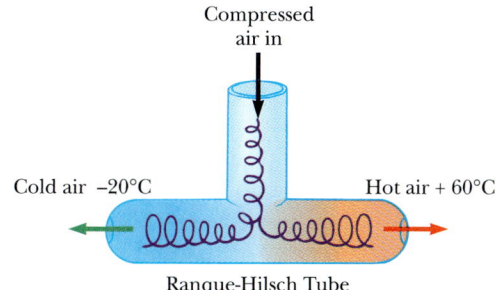

Compressed air in

Cold air −20°C   Hot air + 60°C

Ranque-Hilsch Tube

FIGURE 22.16   (Question 12).

13. Why does your automobile burn more gas in winter than in summer?
14. Can a heat pump have a coefficient of performance less than unity? Explain.
15. Give some examples of irreversible processes that occur in nature.

16. Give an example of a process in nature that is nearly reversible.
17. A thermodynamic process occurs in which the entropy of a system changes by −8.0 J/K. According to the second law of thermodynamics, what can you conclude about the entropy change of the environment?
18. If a supersaturated sugar solution is allowed to slowly evaporate, sugar crystals form in the container. Hence, sugar molecules go from a disordered form (in solution) to a highly ordered crystalline form. Does this process violate the second law of thermodynamics? Explain.
19. How could you increase the entropy of 1 mol of a metal that is at room temperature? How could you decrease its entropy?
20. A heat pump is to be installed in a region where the average outdoor temperature in the winter months is −20°C. In view of this, why would it be advisable to place the outdoor compressor unit deep in the ground? Why are heat pumps not commonly used for heating in cold climates?
21. Suppose your roommate is "Mr. Clean" and tidies up your messy room after a big party. Since more order is being created by your roommate, does this represent a violation of the second law of thermodynamics?
22. Discuss the entropy changes that occur when you (a) bake a loaf of bread and (b) consume the bread.
23. The device in Figure 22.17, called a thermoelectric converter, uses a series of semiconductor cells to convert thermal energy to electrical energy. In the photograph at the left, both legs of the device are at the same temperature, and no electrical energy is produced. However, when one leg is at a higher temperature than the other, as in the photograph on the right, electrical energy is produced as the device extracts energy from the hot reservoir and drives a small electric motor. (a) Why does the temperature differential produce electrical energy in this demonstration? (b) In what sense does this intriguing experiment demonstrate the second law of thermodynamics?

FIGURE 22.17   (Question 23). *(Courtesy of PASCO Scientific Co.)*

PROBLEMS

Review Problem

One mole of a monatomic ideal gas is at an initial pressure P and an initial volume V. The gas is taken through the cycle described in the *PV* diagram. Find (a) the heat absorbed during the cycle, (b) the heat rejected during the cycle, (c) the work done by the gas during the cycle, (d) the efficiency of the cycle, (e) the minimum temperature of the gas during the cycle, (f) the maximum temperature of the gas during the cycle, (g) the efficiency of a Carnot engine operating between the minimum and maximum temperatures, (h) the coefficient of performance of a Carnot refrigerator operating between the minimum and maximum temperatures, (i) the maximum work that can be done by a Carnot engine that absorbs the heat found in (a), and (j) the change in entropy during the process 1–2.

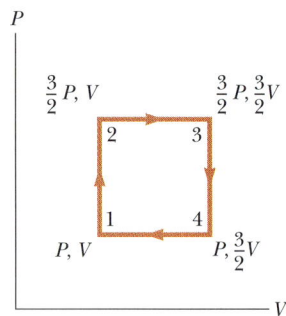

Section 22.1   Heat Engines and the Second Law of Thermodynamics

1. A heat engine absorbs 360 J of thermal energy and performs 25 J of work in each cycle. Find (a) the efficiency of the engine and (b) the thermal energy expelled in each cycle.

2. A heat engine performs 200 J of work in each cycle and has an efficiency of 30%. For each cycle, how much thermal energy is (a) absorbed and (b) expelled?

3. The heat absorbed by an engine is three times greater than the work it performs. (a) What is its thermal efficiency? (b) What fraction of the heat absorbed is expelled to the cold reservoir?

4. Determine the change in the internal energy of a system that (a) absorbs 500 cal of thermal energy while doing 800 J of external work, (b) absorbs 500 cal of thermal energy while 500 J of external work is done on the system, and (c) is maintained at a

constant volume while 1000 cal is removed from the system.

5. An ideal gas is compressed to half its original volume while its temperature is held constant. (a) If 1000 J of energy is removed from the gas during the compression, how much work is done on the gas? (b) What is the change in the internal energy of the gas during the compression?

6. A particular engine has a power output of 5.0 kW and an efficiency of 25%. If the engine expels 8000 J of thermal energy in each cycle, find (a) the heat absorbed in each cycle and (b) the time for each cycle.

7. An engine absorbs 1600 J from a hot reservoir and expels 1000 J to a cold reservoir in each cycle. (a) What is the efficiency of the engine? (b) How much work is done in each cycle? (c) What is the power output of the engine if each cycle lasts for 0.30 s?

Section 22.3   The Carnot Engine

8. A heat engine operates between two reservoirs at 20°C and 300°C. What is the maximum efficiency possible for this engine?

9. A power plant operates at a 32% efficiency during the summer when the sea water for cooling is at 20°C. The plant uses 350°C steam to drive turbines. Assuming that the plant's efficiency changes in the same proportion as the ideal efficiency, what is the plant's efficiency in the winter when the sea water is at 10°C?

10. A Carnot engine has a power output of 150 kW. The engine operates between two reservoirs at 20°C and 500°C. (a) How much thermal energy is absorbed per hour? (b) How much thermal energy is lost per hour?

11. A power plant has been proposed that would make use of the temperature gradient in the ocean. The system is to operate between 20°C (surface water temperature) and 5°C (water temperature at a depth of about 1 km). (a) What is the maximum efficiency of such a system? (b) If the power output of the plant is 75 MW, how much thermal energy is absorbed per hour? (c) What compensating factor made this proposal of interest despite the value calculated in part (a)?

12. A heat engine operates in a Carnot cycle between 80°C and 350°C. It absorbs 2.0×10^4 J of thermal energy per cycle from the hot reservoir. The duration of each cycle is 1.0 s. (a) What is the maximum power

☐ indicates problems that have full solutions available in the Student Solutions Manual and Study Guide.

output of this engine? (b) How much thermal energy does it expel in each cycle?

13. One of the most efficient engines ever built (42%) operates between 430°C and 1870°C. (a) What is its maximum theoretical efficiency? (b) How much power does the engine deliver if it absorbs 1.4×10^5 J of thermal energy each second?

14. Steam enters a turbine at 800°C and is exhausted at 120°C. What is the maximum efficiency of this turbine?

15. The efficiency of a 1000-MW nuclear power plant is 33%; that is, 2000 MW of heat is rejected to the environment for every 1000 MW of electrical energy produced. If a river of flow rate 10^6 kg/s were used to transport the excess thermal energy away, what would be the average temperature increase of the river?

16. At point A in a Carnot cycle, 2.34 mol of a monatomic gas has a pressure of 1400 kPa, a volume of 10.0 liters, and a temperature of 720 K. It expands isothermally to point B, and then expands adiabatically to point C, where its volume is 24.0 liters. An isothermal compression brings it to point D, where its new volume is 15.0 liters. An adiabatic process returns the gas to point A. (a) Determine all the unknown pressures, volumes, and temperatures by filling in the following table.

	P	V	T
A	1400 kPa	10.0 liters	720 K
B			
C		24.0 liters	
D		15.0 liters	

(b) Find the heat added, work done, and the change in internal energy for each of the steps, AB, BC, CD, and DA. (c) Show that $W_{net}/Q_{in} = 1 - T_C/T_A$, the Carnot efficiency.

17. A steam engine is operated in a cold climate where the exhaust temperature is 0°C. (a) Calculate the theoretical maximum efficiency of the engine using an intake steam temperature of 100°C. (b) If, instead, superheated steam at 200°C is used, find the maximum possible efficiency.

18. A Carnot engine has an efficiency of 25% when the hot reservoir temperature is 500°C. If we want to improve the efficiency to 30%, what should be the temperature of the hot reservoir, assuming everything else remains unchanged?

19. A 20%-efficient engine is used to speed up a train from rest to 5.0 m/s. It is known that an ideal (Carnot) engine using the same cold and hot reservoirs would accelerate the same train from rest to a speed of 6.5 m/s using the same amount of fuel. If the engines use air at 300 K as a cold reservoir, find the temperature of the steam serving as the hot reservoir.

Section 22.5   The Gasoline Engine

20. A gasoline engine has a compression ratio of 6 and uses a gas for which $\gamma = 1.4$. (a) What is the efficiency of the engine if it operates in an idealized Otto cycle? (b) If the actual efficiency is 15%, what fraction of the fuel is wasted as a result of friction and unavoidable heat losses? (Assume complete combustion of the air-fuel mixture.)

21. A 1.6-liter gasoline engine with a compression ratio of 6.2 has a power output of 102 hp. If the engine operates in an idealized Otto cycle, find the heat absorbed and exhausted by the engine each second. Assume the fuel-air mixture behaves like an ideal diatomic gas.

22. In a cylinder of an automobile engine, just after combustion, the gas is confined to a volume of 50 cm³ and has an initial pressure of 3.0×10^6 Pa. The piston moves outward to a final volume of 300 cm³ and the gas expands without heat loss. If $\gamma = 1.40$ for the gas, what is the final pressure?

23. How much work is done by the gas in Problem 22 in expanding from $V_1 = 50$ cm³ to $V_2 = 300$ cm³?

Section 22.6   Heat Pumps and Refrigerators

24. An ideal refrigerator or ideal heat pump is equivalent to a Carnot engine running in reverse. That is, heat Q_c is absorbed from a cold reservoir and heat Q_h is rejected to a hot reservoir. (a) Show that the work that must be supplied to run the refrigerator or pump is

$$W = \frac{T_h - T_c}{T_c} Q_c$$

(b) Show that the coefficient of performance of the ideal refrigerator is

$$COP = \frac{T_c}{T_h - T_c}$$

25. A refrigerator has a coefficient of performance equal to 5. If the refrigerator absorbs 120 J of thermal energy from a cold reservoir in each cycle, find (a) the work done in each cycle and (b) the thermal energy expelled to the hot reservoir.

26. What is the maximum coefficient of performance of a heat pump that brings heat from outdoors at −3°C into a 22°C house? (*Hint:* The heat pump does work W, which is also available to heat the house.)

27. How much work is required, using an ideal Carnot refrigerator, to remove 1.0 J of thermal energy from helium gas at 4.0 K and reject this thermal energy to a room-temperature (293 K) environment?

27A. How much work is required, using an ideal Carnot refrigerator, to remove Q joules of thermal energy from helium gas at T_c and reject this thermal energy to a room-temperature environment at T_h?

Section 22.7   Entropy

28. What is the change in entropy when 1 mol of silver (108 g) is melted at 961°C?

29. What is the entropy decrease in 1 mol of helium gas which is cooled at 1 atm from room temperature 293 K to a final temperature 4 K? (C_p of helium = 21 J/mol·K.)

30. An airtight freezer contains air at 25°C and 1.0 atm. The air is then cooled to −18°C. (a) What is the change in entropy of the air if the volume is held constant? (b) What would the change be if the pressure were maintained at 1 atm during the cooling?

31. Calculate the change in entropy of 250 g of water heated slowly from 20°C to 80°C. (*Hint:* Note that $dQ = mc \, dT$.)

32. An ice tray contains 500 g of water at 0°C. Calculate the change in entropy of the water as it freezes completely and slowly at 0°C.

33. A 1.0-kg iron horseshoe is taken from a furnace at 900°C and dropped into 4.0 kg of water at 10°C. If no heat is lost to the surroundings, determine the total entropy change.

34. At a pressure of 1 atm, liquid helium boils at 4.2 K. The latent heat of vaporization is 20.5 kJ/kg. Determine the entropy change (per kilogram) resulting from vaporization.

Section 22.8   Entropy Changes in Irreversible Processes

35. The surface of the Sun is approximately 5700 K, and the temperature of the Earth's surface is approximately 290 K. What entropy change occurs when 1000 J of thermal energy is transferred from the Sun to the Earth?

36. A 100 000-kg iceberg at −5°C breaks away from the polar ice shelf and floats away into the ocean at 5°C. What is the final change in the entropy of the system when the iceberg has completely melted? (The specific heat of ice is 2090 J/kg·°C.)

37. One mole of H_2 gas is contained in the left-hand side of the container shown in Figure P22.37, which has equal volumes left and right. The right-hand side is evacuated. When the valve is opened, the gas streams into the right side. What is the final entropy change? Does the temperature of the gas change?

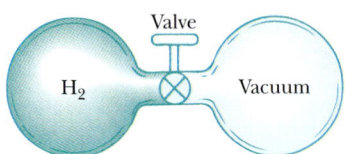

FIGURE P22.37

38. A 2.0-liter container has a center partition that divides it into two equal parts, as shown in Figure

P22.38. The left side contains H_2 gas and the right side contains O_2 gas. Both gases are at room temperature and at atmospheric pressure. The partition is removed and the gases are allowed to mix. What is the entropy increase?

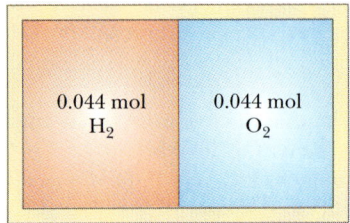

FIGURE P22.38

39. One mole of an ideal monatomic gas, initially at a pressure of 1.000 atm and a volume of 0.025 m³, is heated to a final state where the pressure is 2.000 atm and the volume is 0.040 m³. Determine the change in entropy for this process.

40. One mole of a diatomic ideal gas, initially having pressure P and volume V, expands to having pressure $2P$ and volume $2V$. If, during the expansion, the pressure remains directly proportional to the volume, determine the entropy change in the process.

ADDITIONAL PROBLEMS

41. An 18-g ice cube at 0.0°C is heated until it vaporizes as steam. (a) How much does the entropy increase? (See Table 20.2.) (b) How much energy was required to vaporize the ice cube?

42. If a 35%-efficient Carnot heat engine (Fig. 22.1) is run in reverse so as to form a refrigerator (Fig. 22.4), what would be this refrigerator's coefficient of performance?

43. A house loses thermal energy through the exterior walls and roof at a rate of 5000 J/s = 5 kW when the interior temperature is 22°C and the outside temperature is −5°C. Calculate the electric power required to maintain the interior temperature at 22°C for the following two cases: (a) The electric power is used in electric resistance heaters (which convert all of the electricity supplied to thermal energy). (b) The electric power is used to operate the compressor of a heat pump (which has a coefficient of performance equal to 60% of the Carnot cycle value).

44. Calculate the increase in entropy of the Universe when you add 20 g of 5°C cream to 200 g of 60°C coffee. The specific heat of cream and coffee is 4.2 J/g·°C.

45. One mole of an ideal monatomic gas is taken through the cycle shown in Figure P22.45. The process AB is a reversible isothermal expansion. Calcu-

late (a) the net work done by the gas, (b) the thermal energy added to the gas, (c) the thermal energy expelled by the gas, and (d) the efficiency of the cycle.

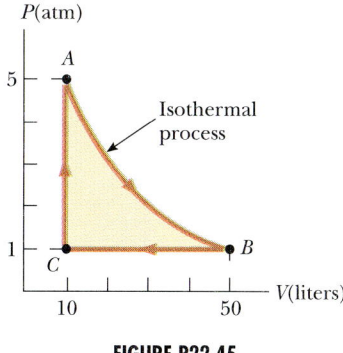

FIGURE P22.45

46. Using an ideal Carnot refrigerator, how much work is required to change 0.50 kg of tap water at $10°C$ into ice at $-20°C$? Assume the freezer compartment is held at $-20°C$ and the refrigerator exhausts heat into a room at $20°C$.

47. Figure P22.47 represents n mol of an ideal monatomic gas being taken through a reversible cycle consisting of two isothermal processes at temperatures $3T_0$ and T_0 and two constant-volume processes. For each cycle, determine in terms of n, R, and T_0 (a) the net thermal energy transferred to the gas and (b) the efficiency of an engine operating in this cycle.

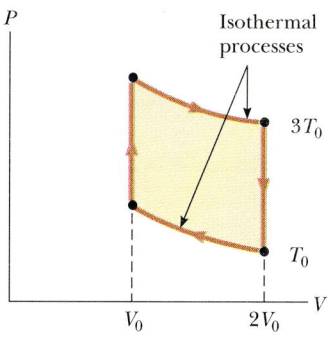

FIGURE P22.47

48. An ideal (Carnot) freezer has a constant temperature of 260 K, while the external air has a constant temperature of 300 K. Suppose the insulation for the freezer is not perfect so that some heat flows into the freezer at a rate of 0.15 W. Determine the average power of the freezer's motor that is needed to maintain the constant temperature in the freezer.

49. One mole of a monatomic ideal gas is taken through the reversible cycle shown in Figure P22.49. At point A, the pressure, volume, and temperature are P_0, V_0, and T_0, respectively. In terms of R and T_0, find

(a) the total heat entering the system per cycle, (b) the total heat leaving the system per cycle, (c) the efficiency of an engine operating in this reversible cycle, and (d) the efficiency of an engine operating in a Carnot cycle between the same temperature extremes.

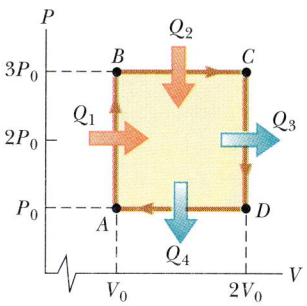

FIGURE P22.49

50. One mole of an ideal gas expands isothermally. (a) If the gas doubles its volume, show that the work of expansion is $W = RT \ln 2$. (b) Since the internal energy U of an ideal gas depends solely on its temperature, there is no change in U during the expansion. It follows from the first law that the thermal energy absorbed by the gas during the expansion is converted completely to work. Why does this *not* violate the second law?

51. A system consisting of n mol of an ideal gas undergoes a reversible, *isobaric* process from a volume V_0 to a volume $3V_0$. Calculate the change in entropy of the gas. (*Hint:* Imagine that the system goes from the initial state to the final state first along an isotherm and then along an adiabatic path—there is no change in entropy along the adiabatic path.)

52. An electrical power plant has an overall efficiency of 15%. The plant is to deliver 150 MW of power to a city, and its turbines use coal as the fuel. The burning coal produces steam which drives the turbines. This steam is then condensed to water at $25°C$ by passing it through cooling coils in contact with river water. (a) How many metric tons of coal does the plant consume each day (1 metric ton = 10^3 kg)? (b) What is the total cost of the fuel per year if the delivered price is $8/metric ton? (c) If the river water is delivered at $20°C$, at what minimum rate must it flow over the cooling coils in order that its temperature not exceed $25°C$? (*Note:* The heat of combustion of coal is 33 kJ/g.)

53. A power plant, having a Carnot efficiency, produces 1000 MW of electrical power from turbines that take in steam at 500 K and reject water at 300 K into a flowing river. If the water downstream is 6 K warmer

due to the output of the power plant, determine the flow rate of the river.

53A. A power plant, having a Carnot efficiency, produces electrical power P (in MW) from turbines that take in steam at T_h and reject water at T_c into a flowing river. If the water downstream is ΔT warmer due to the output of the power plant, determine the flow rate of the river.

54. Suppose you are working in a patent office, and an inventor comes to you with the claim that her heat engine, which employs water as a working substance, has a thermodynamic efficiency of 0.61. She explains that it operates between heat reservoirs at 4°C and 0°C. It is a very complicated device, with many pistons, gears, and pulleys, and the cycle involves freezing and melting. Does her claim that $e = 0.61$ warrant serious consideration? Explain.

55. An idealized Diesel engine operates in a cycle known as the *air-standard Diesel cycle*, shown in Figure P22.55. Fuel is sprayed into the cylinder at the point of maximum compression, B. Combustion occurs during the expansion $B \rightarrow C$, which is approximated as an isobaric process. The rest of the cycle is the same as in the gasoline engine, described in Figure 22.11. Show that the efficiency of an engine operating in this idealized Diesel cycle is

$$e = 1 - \frac{1}{\gamma}\left(\frac{T_D - T_A}{T_C - T_B}\right)$$

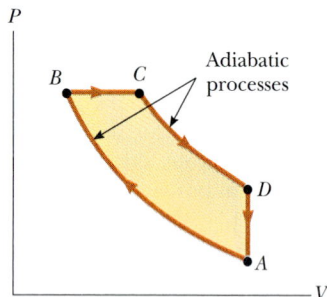

FIGURE P22.55

56. One mole of an ideal gas $(\gamma = 1.4)$ is carried through the Carnot cycle described in Figure 22.9. At point A, the pressure is 25 atm and the temperature is 600 K. At point C, the pressure is 1 atm and the temperature is 400 K. (a) Determine the pressures and volumes at points A, B, C, and D. (b) Calculate the net work done per cycle. (c) Determine the efficiency of an engine operating in this cycle.

57. A typical human has a mass of 70 kg and produces about 2000 kcal $(2.0 \times 10^6$ cal) of metabolic heat per day. (a) Find the rate of heat production in watts and in calories per hour. (b) If none of the metabolic heat were lost, and assuming that the specific heat of the human body is 1.0 cal/g·°C, find the rate at which body temperature would rise. Give your answer in °C per hour and in °F per hour.

58. Suppose 1.00 kg of water at 10.0°C is mixed with 1.00 kg of water at 30.0°C at constant pressure. When the mixture has reached equilibrium, (a) what is the final temperature? (b) Take $C_p = 4.19$ kJ/kg·K for water and show that the entropy of the system increases by

$$\Delta S = 4.19 \ln\left[\left(\frac{293}{283}\right)\left(\frac{293}{303}\right)\right] \text{kJ/K}$$

(c) Verify numerically that $\Delta S > 0$. (d) Is the mixing an irreversible process?

59. The Stirling engine described in Figure P22.59 operates between the isotherms T_1 and T_2, where $T_2 > T_1$. Assuming that the operating gas is an ideal monatomic gas, calculate the efficiency of an engine whose constant-volume processes occur at the volumes V_1 and V_2.

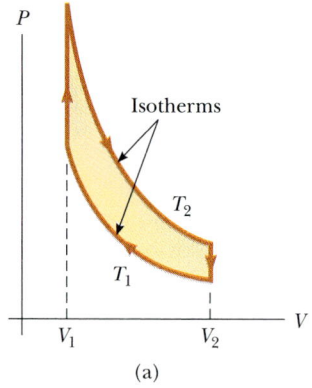

(a)

FIGURE P22.59

60. Suppose a heat engine is connected to two heat reservoirs, one a pool of molten aluminum (660°C) and the other a block of solid mercury (-38.9°C). The engine runs by freezing 1.00 g of aluminum and melting 15.0 g of mercury during each cycle. The latent heat of fusion of aluminum is 3.97×10^5 J/kg, and that of mercury is 1.18×10^4 J/kg. (a) What is the efficiency of this engine? (b) How does the efficiency compare with that of a Carnot engine?

61. A gas is taken through the cyclic process described by Figure P22.61. (a) If Q is negative for the process BC and ΔU is negative for the process CA, determine the signs of Q, W, and ΔU associated with each process. (b) Find the net heat transferred to the system during one complete cycle. (c) If the cycle is reversed—that is, the process follows the path $ACBA$—what is the net heat transferred per cycle?

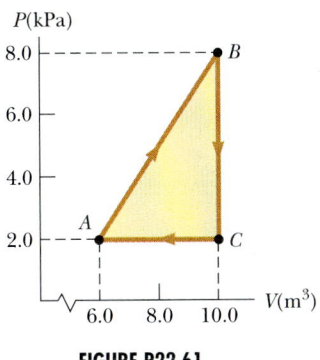

FIGURE P22.61

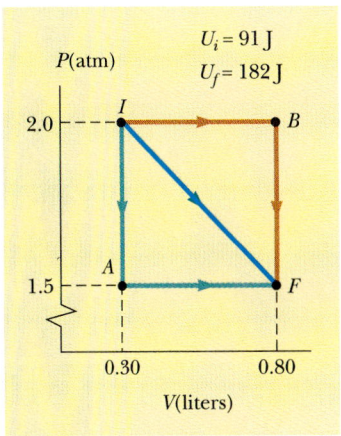

FIGURE P22.62

62. One mole of gas is initially at a pressure of 2.0 atm and a volume of 0.30 L and has an internal energy equal to 91 J. In its final state the gas is at a pressure of 1.5 atm and a volume of 0.80 L, and its internal energy equals 180 J. For the paths *IAF*, *IBF*, and *IF* in Figure P22.62, calculate (a) the work done by the gas and (b) the net heat-transferred to the gas in the process.

63. Powdered steel is placed in a container filled with oxygen and fitted with a piston that moves to keep the pressure in the container constant at one atmo-sphere. A chemical reaction occurs that produces heat. To keep the contents at a constant temperature of 22°C, it is necessary to remove 8.3×10^5 J of heat from the container as it contracts. During the chemical reaction, it is found that 1.5 mol of oxygen is consumed. Find the internal energy change for the system of iron and oxygen.

This dramatic photograph captures multiple lightning bolts near some rural homes. The most intense bolt strikes a tree adjacent to one of the homes.

(© Johnny Autery)

PART IV

Electricity and Magnetism

For the sake of persons of . . . different types, scientific truth should be presented in different forms, and should be regarded as equally scientific, whether it appears in the robust form and the vivid coloring of a physical illustration, or in the tenuity and paleness of a symbolic expression.

JAMES CLERK MAXWELL

We now begin the study of that branch of physics that is concerned with electric and magnetic phenomena. The laws of electricity and magnetism play a central role in the operation of various devices such as radios, televisions, electric motors, computers, high-energy accelerators, and a host of electronic devices used in medicine. However, more fundamentally, we now know that the interatomic and intermolecular forces that are responsible for the formation of solids and liquids are electric in origin. Furthermore, such forces as the pushes and pulls between objects and the elastic force in a spring arise from electric forces at the atomic level.

Evidence in Chinese documents suggests that magnetism was known as early as around 2000 B.C. The ancient Greeks observed electric and magnetic phenomena possibly as early as 700 B.C. They found that a piece of amber, when rubbed, becomes electrified and attracts pieces of straw or feathers. The existence of magnetic forces was known from observations that pieces of a naturally occurring stone called *magnetite* (Fe_3O_4) are attracted to iron. (The word *electric* comes from the Greek word for amber, *elecktron.* The word *magnetic* comes from the name of *Magnesia,* on the coast of Turkey, where magnetite was found.

In 1600, William Gilbert discovered that electrification was not limited to amber but is a general phenomenon. Scientists went on to electrify a variety of objects, including chickens and people! Experiments by Charles Coulomb in 1785 confirmed the inverse-square law for electric forces.

It was not until the early part of the 19th century that scientists established that electricity and magnetism are, in fact, related phenomena. In 1820, Hans Oersted discovered that a compass needle is deflected when placed near a circuit carrying an electric current. In 1831, Joseph Henry in America, and almost simultaneously Michael Faraday in Britain, showed that when a wire is moved near a magnet (or, equivalently, when a magnet is moved near a wire), an electric current is established in the wire. In 1873, James Clerk Maxwell used these observations and other experimental facts as a basis for formulating the laws of electromagnetism as we know them today. (*Electromagnetism* is a name given to the combined fields of electricity and magnetism.) Shortly thereafter (around 1888), Heinrich Hertz verified Maxwell's predictions by producing electromagnetic waves in the laboratory. This led to such practical developments as radio and television.

Maxwell's contributions to the field of electromagnetism were especially significant because the laws he formulated are basic to *all* forms of electromagnetic phenomena. His work is as important as Newton's work on the laws of motion and the theory of gravitation.

CHAPTER 23

Electric Fields

This metallic sphere is charged by a generator to a very high voltage. The high concentration of charge formed on the sphere creates a strong electric field around the sphere. The charges then leak through the gas, producing the pink glow. (E. R. Degginger/ H. Armstrong Roberts)

The electromagnetic force between charged particles is one of the fundamental forces of nature. In this chapter, we begin by describing some of the basic properties of electric forces. We then discuss Coulomb's law, which is the fundamental law of force between any two charged particles. The concept of an electric field associated with a charge distribution is then introduced, and its effect on other charged particles is described. The method of using Coulomb's law to calculate electric fields of a given charge distribution is discussed, and several examples are given. Then the motion of a charged particle in a uniform electric field is discussed. We conclude the chapter with a brief discussion of the oscilloscope.

23.1 PROPERTIES OF ELECTRIC CHARGES

A number of simple experiments demonstrate the existence of electric forces and charges. For example, after running a comb through your hair on a dry day, you will find that the comb attracts bits of paper. The attractive force is often strong enough to suspend the pieces of paper. The same effect occurs when materials such as glass or rubber are rubbed with silk or fur.

Another simple experiment is to rub an inflated balloon with wool. The balloon then adheres to the wall or the ceiling of a room, often for hours. When materials behave in this way, they are said to be *electrified,* or to have become **electrically charged.** You can easily electrify your body by vigorously rubbing your shoes on a wool rug. The charge on your body can be sensed and removed by lightly touching (and startling) a friend. Under the right conditions, a visible spark is seen when you touch one another, and a slight tingle will be felt by both parties. (Experiments such as these work best on a dry day, since an excessive amount of moisture can lead to a leakage of charge from the electrified body to the Earth by various conducting paths.)

In a systematic series of simple experiments, it is found that there are two kinds of electric charges, which were given the names **positive** and **negative** by Benjamin Franklin (1706–1790). To demonstrate this fact, consider a hard rubber rod that has been rubbed with fur and then suspended by a nonmetallic thread as in Figure 23.1. When a glass rod that has been rubbed with silk is brought near the rubber rod, the rubber rod is attracted toward the glass rod. On the other hand, if two charged rubber rods (or two charged glass rods) are brought near each other, as in Figure 23.1b, the force between them is repulsive. This observation shows that the rubber and glass are in two different states of electrification. On the basis of these observations, we conclude that *like charges repel one another and unlike charges attract one another.* Using the convention suggested by Franklin, the electric charge on the glass rod is called *positive* and that on the rubber rod is called *negative.* Therefore, any charged body that is attracted to a charged rubber rod (or repelled by a charged glass rod) must have a positive charge. Conversely, any charged body that is repelled by a charged rubber rod (or attracted to a charged glass rod) has a negative charge on it.

Another important aspect of Franklin's model of electricity is the implication that *electric charge is always conserved.* That is, when one body is rubbed against another, charge is not created in the process. The electrified state is due to a *transfer* of charge from one body to the other. Therefore, one body gains some

Charge is conserved

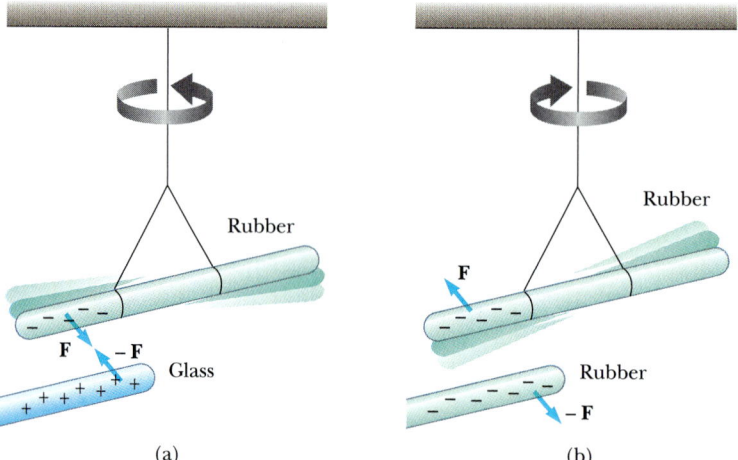

FIGURE 23.1   (a) A negatively charged rubber rod, suspended by a thread, is attracted to a positively charged glass rod. (b) A negatively charged rubber rod is repelled by another negatively charged rubber rod.

amount of negative charge while the other gains an equal amount of positive charge. For example, when a glass rod is rubbed with silk, the silk obtains a negative charge that is equal in magnitude to the positive charge on the glass rod. We now know from our understanding of atomic structure that *it is the negatively charged electrons that are transferred* from the glass to the silk in the rubbing process. Likewise, when rubber is rubbed with fur, electrons are transferred from the fur to the rubber, giving the rubber a net negative charge and the fur a net positive charge. This is consistent with the fact that neutral, uncharged matter contains as many positive charges (protons within atomic nuclei) as negative charges (electrons).

In 1909, Robert Millikan (1868–1953) confirmed that electric charge always occurs as some integral multiple of some fundamental unit of charge, *e*. In modern terms, the charge *q* is said to be **quantized,** where *q* is the standard symbol used for charge. That is, electric charge exists as discrete "packets." Thus, we can write $q = Ne$, where *N* is some integer. Other experiments in the same period showed that the electron has a charge $-e$ and the proton has an equal and opposite charge, $+e$. Some elementary particles, such as the neutron, have no charge. A neutral atom must contain as many protons as electrons.

Electric forces between charged objects were measured quantitatively by Coulomb using the torsion balance, which he invented (Fig. 23.2). Using this apparatus, Coulomb confirmed that the electric force between two small charged spheres is proportional to the inverse square of their separation, that is, $F \propto 1/r^2$. The operating principle of the torsion balance is the same as that of the apparatus used by Cavendish to measure the gravitational constant (Section 14.2), with masses replaced by charged spheres. The electric force between the charged spheres produces a twist in the suspended fiber. Since the restoring torque of the twisted fiber is proportional to the angle through which it rotates, a measurement of this angle provides a quantitative measure of the electric force of attraction or repulsion. If the spheres are charged by rubbing, the electrical force between the spheres is very large compared with the gravitational attraction; hence, the gravitational force can be neglected.

From our discussion thus far, we conclude that electric charge has the following important properties:

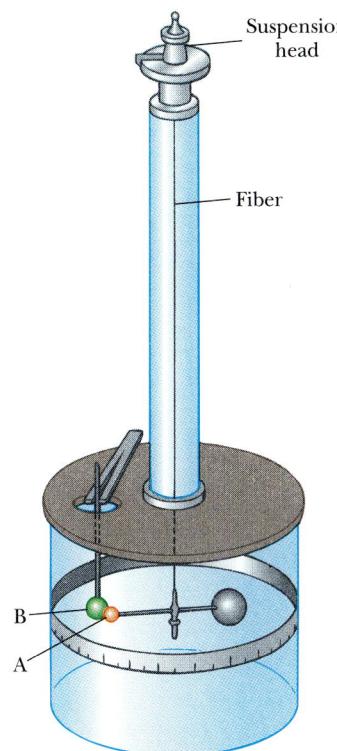

FIGURE 23.2 Coulomb's torsion balance, which was used to establish the inverse-square law for the electrostatic force between two charges.

- There are two kinds of charges in nature, with the property that unlike charges attract one another and like charges repel one another.
- The force between charges varies as the inverse square of their separation.
- Charge is conserved.
- Charge is quantized.

Properties of electric charge

23.2 INSULATORS AND CONDUCTORS

It is convenient to classify substances in terms of their ability to conduct electrical charge.

Conductors are materials in which electric charges move quite freely, whereas **insulators** are materials that do not readily transport charge.

Materials such as glass, rubber, and lucite fall into the category of insulators. When such materials are charged by rubbing, only the area that is rubbed becomes charged and the charge is unable to move to other regions of the material.

In contrast, materials such as copper, aluminum, and silver are good conductors. When such materials are charged in some small region, the charge readily distributes itself over the entire surface of the conductor. If you hold a copper rod in your hand and rub it with wool or fur, it will not attract a small piece of paper. This might suggest that a metal cannot be charged. On the other hand, if you hold the copper rod by a lucite handle and then rub, the rod will remain charged and attract the piece of paper. This is explained by noting that in the first case, the electric charges produced by rubbing will readily move from copper through your body and finally to Earth. In the second case, the insulating lucite handle prevents the flow of charge to Earth.

Semiconductors are a third class of materials, and their electrical properties are somewhere between those of insulators and those of conductors. Silicon and germanium are well-known examples of semiconductors commonly used in the fabrication of a variety of electronic devices. The electrical properties of semiconductors can be changed over many orders of magnitude by adding controlled amounts of certain foreign atoms to the materials.

When a conductor is connected to Earth by means of a conducting wire or pipe, it is said to be **grounded**. The Earth can then be considered an infinite "sink" to which electrons can easily migrate. With this in mind, we can understand how to charge a conductor by a process known as **induction.**

To understand induction, consider a negatively charged rubber rod brought near a neutral (uncharged) conducting sphere insulated from ground. That is, there is no conducting path to ground (Fig. 23.3a). The region of the sphere nearest the rod obtains an excess of positive charge, while the region of the sphere farthest from the rod obtains an equal excess of negative charge. (That is, electrons in the part of the sphere nearest the rod migrate to the opposite side of the sphere.) If the same experiment is performed with a conducting wire connected from the sphere to ground (Fig. 23.3b), some of the electrons in the conductor are so strongly repelled by the presence of the negative charge that they move out of the sphere through the ground wire and into the Earth. If the wire to ground is then removed (Fig. 23.3c), the conducting sphere contains an excess of *induced* positive charge. Finally, when the rubber rod is removed from the vicinity of the sphere (Fig. 23.3d), the induced positive charge remains on the ungrounded sphere. Note that the charge remaining on the sphere is uniformly distributed over its surface because of the repulsive forces among the like charges. In the process, the electrified rubber rod loses none of its negative charge.

Thus, we see that charging an object by induction requires no contact with the body inducing the charge. This is in contrast to charging an object by rubbing (that is, charging by *conduction*), which does require contact between the two objects.

A process very similar to charging by induction in conductors also takes place in insulators. In most neutral atoms and molecules, the center of positive charge coincides with the center of negative charge. However, in the presence of a charged object, these centers may shift slightly, resulting in more positive charge on one side of the molecule than on the other. This effect, known as polarization, is discussed more completely in Chapter 26. This realignment of charge within individual molecules produces an induced charge on the surface of the insulator, as shown in Figure 23.4. With these ideas, you should be able to explain why a comb that has been rubbed through hair attracts bits of neutral paper, or why a balloon that has been rubbed against your clothing is able to stick to a neutral wall.

Metals are good conductors

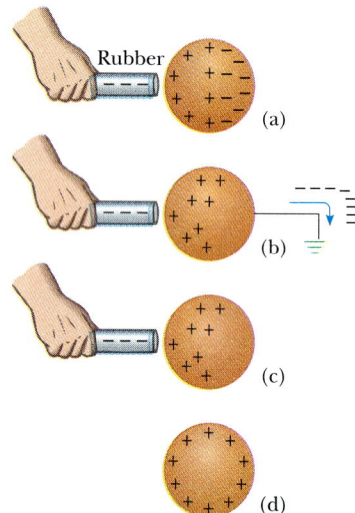

FIGURE 23.3   Charging a metallic object by induction. (a) The charge on a neutral metallic sphere is redistributed when a charged rubber rod is placed near the sphere. (b) The sphere is grounded, and some of the electrons leave. (c) The ground connection is removed, and the sphere has a net nonuniform positive charge. (d) When the rubber rod is removed, the excess positive charge becomes uniformly distributed over the surface of the sphere.

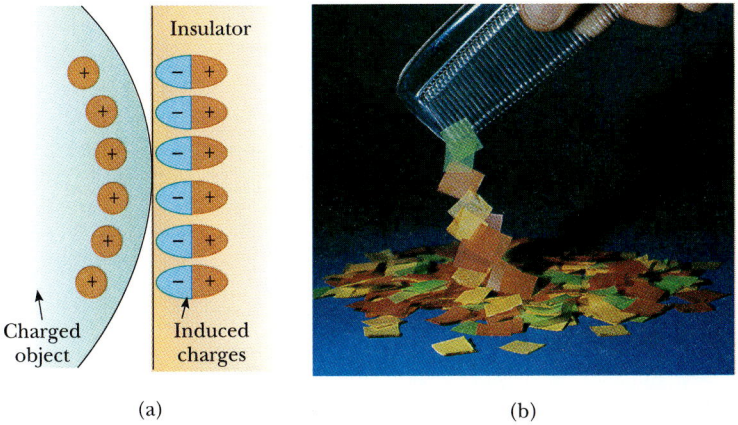

(a)   (b)

FIGURE 23.4   (a) The charged object induces charges on the surface of an insulator. (b) A charged comb attracts bits of paper because charges are displaced in the paper. The paper is neutral but polarized.   *(© 1968 Fundamental Photographs)*

CONCEPTUAL EXAMPLE 23.1

If a suspended object A is attracted to object B, which is charged, can we conclude that object A is charged?

Reasoning   No. Object A might have a charge opposite in sign to that of B, but it also might be neutral. In the latter case, object B causes A to be polarized, pulling charge of one sign to the near face of A and pushing an equal amount of charge of the opposite sign to the far face, as in Figure 23.5. Then the force of attraction exerted on B by the induced charge on the near side of A is slightly larger than the force of repulsion exerted on B by the induced charge on the far side of A. Therefore, the net force on A is toward B.

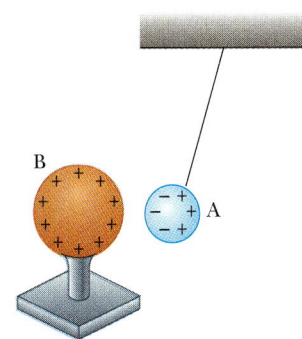

FIGURE 23.5   (Conceptual Example 23.1) Electrostatic attraction between a charged sphere B and a neutral conductor A.

23.3   COULOMB'S LAW

In 1785, Coulomb established the fundamental law of electric force between two stationary, charged particles. Experiments show that an **electric force** has the following properties:

- The force is inversely proportional to the square of the separation, r, between the two particles and directed along the line joining the particles.
- The force is proportional to the product of the charges q_1 and q_2 on the two particles.
- The force is attractive if the charges are of opposite sign and repulsive if the charges have the same sign.

From these observations, we can express the magnitude of the electric force between the two charges as

$$F = k_e \frac{|q_1||q_2|}{r^2}$$

(23.1)   Coulomb's law

Charles Coulomb

| 1 7 3 6 – 1 8 0 6 |

Charles Coulomb, the great French physicist after whom the unit of electric charge called the *coulomb* was named, was born in Angoulême in 1736. He was educated at the École du Génie in Mézieres, graduating in 1761 as a military engineer with a rank of First Lieutenant. Coulomb served in the West Indies for nine years, where he supervised the building of fortifications in Martinique.

In 1774, Coulomb became a correspondent to the Paris Academy of Science. There he shared the Academy's first prize for his paper on magnetic compasses and also received first prize for his classic work on friction, a study that was unsurpassed for 150 years. During the next 25 years, he presented 25 papers to the Academy on electricity, magnetism, torsion, and applications to the torsion balance, as well as several hundred committee reports on engineering and civil projects.

Coulomb took full advantage of the various positions he held during his lifetime. For example, his experience as an engineer led him to investigate the strengths of materials and determine the forces that affect objects on beams, thereby contributing to the field of structural mechanics. He also contributed to the field of ergonom-ics. His research provided a fundamental understanding of the ways in which people and animals can best do work and greatly influenced the subsequent research of Gaspard Coriolis (1792–1843).

Coulomb's major contribution to science was in the field of electrostatics and magnetism, in which he made use of the torsion balance he developed (see Fig. 23.2). The paper describing this invention also contained a design for a compass using the principle of torsion suspension. His next paper gave proof of the inverse square law for the electrostatic force between two charges.

Coulomb died in 1806, five years after becoming president of the Institut de France (formerly the Paris Academy of Science). His research on electricity and magnetism brought this area of physics out of traditional natural philosophy and made it an exact science.

(Photograph courtesy of AIP Niels Bohr Library, E. Scott Barr Collection)

where k_e is a constant called the **Coulomb constant**. In his experiments, Coulomb was able to show that the value of the exponent of r was 2 to within an uncertainty of a few percent. Modern experiments have shown that the exponent is 2 to a precision of a few parts in 10^{16}.

The Coulomb constant has a value that depends on the choice of units. The unit of charge in SI units is the coulomb (C). The coulomb is defined in terms of a unit current called the *ampere* (A), where current equals the rate of flow of charge. (The ampere is defined in Chapter 27.) When the current in a wire is 1 A, the amount of charge that flows past a given point in the wire in 1 s is 1 C. The Coulomb constant k_e in SI units has the value

Coulomb constant

$$k_e = 8.9875 \times 10^9 \text{ N} \cdot \text{m}^2/\text{C}^2$$

To simplify our calculations, we shall use the approximate value

$$k_e \cong 8.99 \times 10^9 \text{ N} \cdot \text{m}^2/\text{C}^2$$

The constant k_e is also written

$$k_e = \frac{1}{4\pi\epsilon_0}$$

where the constant ϵ_0 is known as the *permittivity of free space* and has the value

$$\epsilon_0 = 8.8542 \times 10^{-12} \text{ C}^2/\text{N}\cdot\text{m}^2$$

The smallest unit of charge known in nature is the charge on an electron or proton.[1] The charge of an electron or proton has a magnitude

$$|e| = 1.60219 \times 10^{-19} \text{ C}$$

Charge on an electron or proton

Therefore, 1 C of charge is equal to the charge of 6.3×10^{18} electrons. This number can be compared with the number of free electrons[2] in 1 cm^3 of copper, which is of the order of 10^{23}. Note that 1 C is a substantial amount of charge. In typical electrostatic experiments, where a rubber or glass rod is charged by friction, a net charge of the order of 10^{-6} C is obtained. In other words, only a very small fraction of the total available charge is transferred between the rod and the rubbing material.

The charges and masses of the electron, proton, and neutron are given in Table 23.1.

When dealing with Coulomb's force law, you must remember that force is a vector quantity and must be treated accordingly. Furthermore, note that *Coulomb's law applies exactly only to point charges or particles*. The electric force exerted on q_2 due to a charge q_1, written $\mathbf{F}_{21}$, can be expressed in vector form as

$$\mathbf{F}_{21} = k_e \frac{q_1 q_2}{r^2} \hat{\mathbf{r}} \tag{23.2}$$

where $\hat{\mathbf{r}}$ is a unit vector directed from q_1 to q_2 as in Figure 23.6a. Since Coulomb's law obeys Newton's third law, the electric force exerted on q_1 by q_2 is equal in magnitude to the force exerted on q_2 by q_1 and in the opposite direction, that is, $\mathbf{F}_{12} = -\mathbf{F}_{21}$. Finally, from Equation 23.2 we see that if q_1 and q_2 have the same sign, the product $q_1 q_2$ is positive and the force is repulsive, as in Figure 23.6a. If q_1 and q_2 are of opposite sign, as in Figure 23.6b, the product $q_1 q_2$ is negative and the force is attractive.

When more than two charges are present, the force between any pair of them is given by Equation 23.2. Therefore, the resultant force on any one of them equals the vector sum of the forces exerted by the various individual charges. For exam-

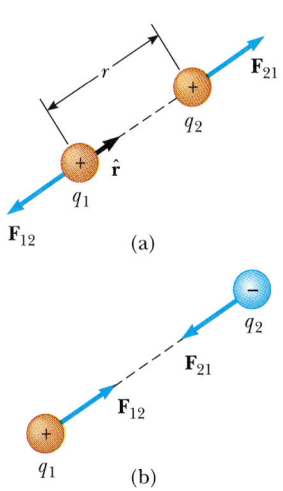

FIGURE 23.6   Two point charges separated by a distance r exert a force on each other given by Coulomb's law. Note that the force on q_1 is equal to and opposite the force on q_2. (a) When the charges are of the same sign, the force is repulsive. (b) When the charges are of the opposite sign, the force is attractive.

TABLE 23.1   **Charge and Mass of the Electron, Proton, and Neutron**

Particle	Charge (C)	Mass (kg)
Electron (e)	$-1.6021917 \times 10^{-19}$	9.1095×10^{-31}
Proton (p)	$+1.6021917 \times 10^{-19}$	1.67261×10^{-27}
Neutron (n)	0	1.67492×10^{-27}

[1] No unit of charge smaller than e has been detected as a free charge; however, some recent theories have proposed the existence of particles called *quarks* having charges $e/3$ and $2e/3$. Although there is experimental evidence for such particles inside nuclear matter, *free* quarks have never been detected. We discuss other properties of quarks in Chapter 47 of the extended version of this text.

[2] A metal atom, such as copper, contains one or more outer electrons, which are weakly bound to the nucleus. When many atoms combine to form a metal, the so-called free electrons are these outer electrons, which are not bound to any one atom. These electrons move about the metal in a manner similar to gas molecules moving in a container.

ple, if there are four charges, then the resultant force on particle 1 due to particles 2, 3, and 4 is

$$\mathbf{F}_1 = \mathbf{F}_{12} + \mathbf{F}_{13} + \mathbf{F}_{14}$$

This principle of superposition as applied to electrostatic forces is an experimentally observed fact.

EXAMPLE 23.2   Find the Resultant Force

Consider three point charges located at the corners of a triangle, as in Figure 23.7, where $q_1 = q_3 = 5.0\ \mu C$, $q_2 = -2.0\ \mu C$, and $a = 0.10$ m. Find the resultant force on q_3.

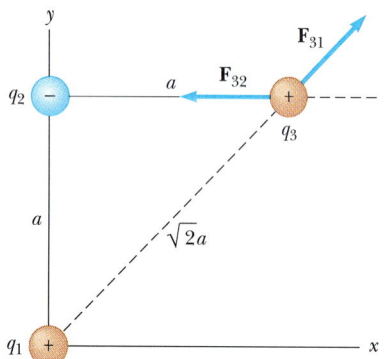

FIGURE 23.7   (Example 23.2) The force exerted on q_3 by q_1 is $\mathbf{F}_{31}$. The force exerted on q_3 by q_2 is $\mathbf{F}_{32}$. The resultant force $\mathbf{F}_3$ exerted on q_3 is the vector sum $\mathbf{F}_{31} + \mathbf{F}_{32}$.

Solution   First, note the direction of the individual forces exerted on q_3 by q_1 and q_2. The force exerted on q_3 by q_2 is attractive because q_2 and q_3 have opposite signs. The force exerted on q_3 by q_1 is repulsive because they are both positive.

Now let us calculate the magnitude of the forces on q_3. The magnitude of $\mathbf{F}_{32}$ is

$$F_{32} = k_e \frac{|q_3||q_2|}{a^2}$$

$$= \left(8.99 \times 10^9\ \frac{\mathrm{N \cdot m^2}}{\mathrm{C^2}}\right) \frac{(5.0 \times 10^{-6}\ \mathrm{C})(2.0 \times 10^{-6}\ \mathrm{C})}{(0.10\ \mathrm{m})^2}$$

$$= 9.0\ \mathrm{N}$$

Note that since q_3 and q_2 have opposite signs, $\mathbf{F}_{32}$ is to the left, as shown in Figure 23.7.

The magnitude of the force exerted on q_3 by q_1 is

$$F_{31} = k_e \frac{|q_3||q_1|}{(\sqrt{2}a)^2}$$

$$= \left(8.99 \times 10^9\ \frac{\mathrm{N \cdot m^2}}{\mathrm{C^2}}\right) \frac{(5.0 \times 10^{-6}\ \mathrm{C})(5.0 \times 10^{-6}\ \mathrm{C})}{2(0.10\ \mathrm{m})^2}$$

$$= 11\ \mathrm{N}$$

The force $\mathbf{F}_{31}$ is repulsive and makes an angle of $45°$ with the x axis. Therefore, the x and y components of $\mathbf{F}_{31}$ are equal, with magnitude given by $F_{31} \cos 45° = 7.9\ \mathrm{N}$. The force $\mathbf{F}_{32}$ is in the negative x direction. Hence, the x and y components of the resultant force on q_3 are

$$F_x = F_{31x} + F_{32} = 7.9\ \mathrm{N} - 9.0\ \mathrm{N} = \boxed{-1.1\ \mathrm{N}}$$

$$F_y = F_{31y} = \boxed{7.9\ \mathrm{N}}$$

We can also express the resultant force on q_3 in unit-vector form as $\mathbf{F}_3 = (-1.1\mathbf{i} + 7.9\mathbf{j})\ \mathrm{N}$.

Exercise   Find the magnitude and direction of the resultant force on q_3.

Answer   8.0 N at an angle of $98°$ with the x axis.

EXAMPLE 23.3   Where is the Resultant Force Zero?

Three charges lie along the x axis as in Figure 23.8. The positive charge $q_1 = 15.0\ \mu C$ is at $x = 2.00$ m, and the positive charge $q_2 = 6.00\ \mu C$ is at the origin. Where must a negative charge q_3 be placed on the x axis such that the resultant force on it is zero?

Solution   Since q_3 is negative and both q_1 and q_2 are positive, the forces $\mathbf{F}_{31}$ and $\mathbf{F}_{32}$ are both attractive, as indicated in Figure 23.8. If we let x be the coordinate of q_3, then the forces $\mathbf{F}_{31}$ and $\mathbf{F}_{32}$ have magnitudes

$$F_{31} = k_e \frac{|q_3||q_1|}{(2.00 - x)^2} \quad \text{and} \quad F_{32} = k_e \frac{|q_3||q_2|}{x^2}$$

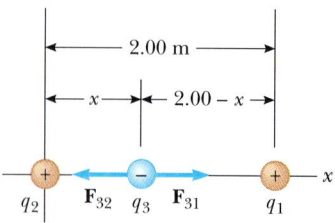

FIGURE 23.8   (Example 23.3) Three point charges are placed along the x axis. The charge q_3 is negative, whereas q_1 and q_2 are positive. If the net force on q_3 is zero, then the force on q_3 due to q_1 must be equal and opposite to the force on q_3 due to q_2.

In order for the resultant force on q_3 to be zero, $\mathbf{F}_{32}$ must be equal to and opposite $\mathbf{F}_{31}$, or

$$k_e \frac{|q_3||q_2|}{x^2} = k_e \frac{|q_3||q_1|}{(2.00 - x)^2}$$

Since k_e and q_3 are common to both sides, we solve for x and

find that

$$(2.00 - x)^2 |q_2| = x^2 |q_1|$$

$$(4.00 - 4.00x + x^2)(6.00 \times 10^{-6} \text{ C}) = x^2 (15.0 \times 10^{-6} \text{ C})$$

Solving this quadratic equation for x, we find that $x = 0.775$ m. Why is the negative root not acceptable?

EXAMPLE 23.4   The Hydrogen Atom

The electron and proton of a hydrogen atom are separated (on the average) by a distance of approximately 5.3×10^{-11} m. Find the magnitude of the electric force and the gravitational force between the two particles.

Solution   From Coulomb's law, we find that the attractive electric force has the magnitude

$$F_e = k_e \frac{|e|^2}{r^2} = 8.99 \times 10^9 \frac{\text{N} \cdot \text{m}^2}{\text{C}^2} \frac{(1.60 \times 10^{-19} \text{ C})^2}{(5.3 \times 10^{-11} \text{ m})^2}$$

$$= 8.2 \times 10^{-8} \text{ N}$$

Using Newton's law of gravity and Table 23.1 for the parti-

cle masses, we find that the gravitational force has the magnitude

$$F_g = G \frac{m_e m_p}{r^2} = \left(6.7 \times 10^{-11} \frac{\text{N} \cdot \text{m}^2}{\text{kg}^2} \right)$$

$$\times \frac{(9.11 \times 10^{-31} \text{ kg})(1.67 \times 10^{-27} \text{ kg})}{(5.3 \times 10^{-11} \text{ m})^2}$$

$$= 3.6 \times 10^{-47} \text{ N}$$

The ratio $F_e/F_g \approx 2 \times 10^{39}$. Thus the gravitational force between charged atomic particles is negligible compared with the electric force.

EXAMPLE 23.5   Find the Charge on the Spheres

Two identical small charged spheres, each having a mass of 3.0×10^{-2} kg, hang in equilibrium as shown in Figure 23.9a. If the length of each string is 0.15 m and the angle $\theta = 5.0°$, find the magnitude of the charge on each sphere.

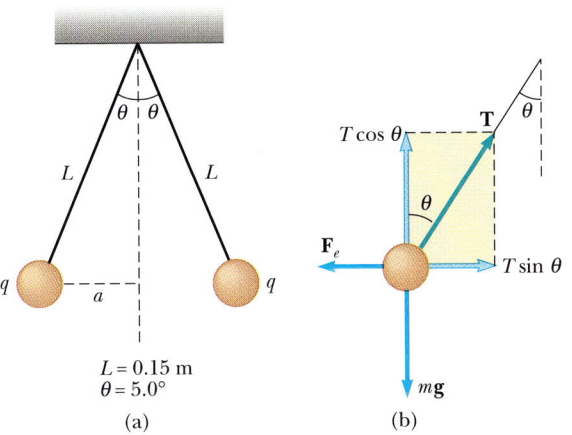

FIGURE 23.9   (Example 23.5) (a) Two identical spheres, each with the same charge q, suspended in equilibrium by strings. (b) The free-body diagram for the charged spheres on the left side.

Solution   From the right triangle in Figure 23.9a, we see that $\sin \theta = a/L$. Therefore,

$$a = L \sin \theta = (0.15 \text{ m}) \sin 5.0° = 0.013 \text{ m}$$

The separation of the spheres is $2a = 0.026$ m.

The forces acting on one of the spheres are shown in Figure 23.9b. Because the sphere is in equilibrium, the resultants of the forces in the horizontal and vertical directions must separately add up to zero:

$$(1) \qquad \sum F_x = T \sin \theta - F_e = 0$$

$$(2) \qquad \sum F_y = T \cos \theta - mg = 0$$

From (2), we see that $T = mg/\cos \theta$, and so T can be eliminated from (1) if we make this substitution. This gives a value for the electric force, F_e:

$$(3) \qquad F_e = mg \tan \theta$$

$$= (3.0 \times 10^{-2} \text{ kg})(9.80 \text{ m/s}^2) \tan(5.0°)$$

$$= 2.6 \times 10^{-2} \text{ N}$$

From Coulomb's law (Eq. 23.1), the electric force between the charges has magnitude

$$F_e = k_e \frac{|q|^2}{r^2}$$

where $r = 2a = 0.026$ m and $|q|$ is the magnitude of the charge on each sphere. (Note that the term $|q|^2$ arises here because the charge is the same on both spheres.) This equation can be solved for $|q|^2$ to give

$$|q|^2 = \frac{F_e r^2}{k_e} = \frac{(2.6 \times 10^{-2}\,\text{N})(0.026\,\text{m})^2}{8.99 \times 10^9\,\text{N} \cdot \text{m}^2/\text{C}^2}$$

$$|q| = 4.4 \times 10^{-8}\,\text{C}$$

Exercise　If the charge on the spheres is negative, how many electrons had to be added to them to give a net charge of -4.4×10^{-8} C?

Answer　2.7×10^{11} electrons.

23.4　THE ELECTRIC FIELD

The gravitational field **g** at a point in space was defined in Chapter 14 to be equal to the gravitational force **F** acting on a test mass m_0 divided by the test mass: $\mathbf{g} \equiv \mathbf{F}/m_0$. In a similar manner, an electric field at a point in space can be defined in terms of the electric force acting on a test charge q_0 placed at that point. To be more precise,

Definition of electric field

> **the electric field vector E at a point in space is defined as the electric force F acting on a positive test charge placed at that point divided by the magnitude of the test charge q_0:**
>
> $$\mathbf{E} \equiv \frac{\mathbf{F}}{q_0} \qquad (23.3)$$

Note that **E** is the field produced by some charge *external* to the test charge—not the field produced by the test charge. This is analogous to the gravitational field set up by some body such as the Earth. The vector **E** has the SI units of newtons per coulomb (N/C). The direction of **E** is in the direction of **F** because we have assumed that **F** acts on a positive test charge. Thus, we can say that *an electric field exists at a point if a test charge at rest placed at that point experiences an electric force.* Once the electric field is known at some point, the force on *any* charged particle placed at that point can be calculated from Equation 23.3. Furthermore, the electric field is said to exist at some point (even empty space) regardless of whether or not a test charge is located at that point.

When Equation 23.3 is applied, we must assume that the test charge q_0 is small enough that it does not disturb the charge distribution responsible for the electric field. For instance, if a vanishingly small test charge q_0 is placed near a uniformly charged metallic sphere as in Figure 23.10a, the charge on the metallic sphere,

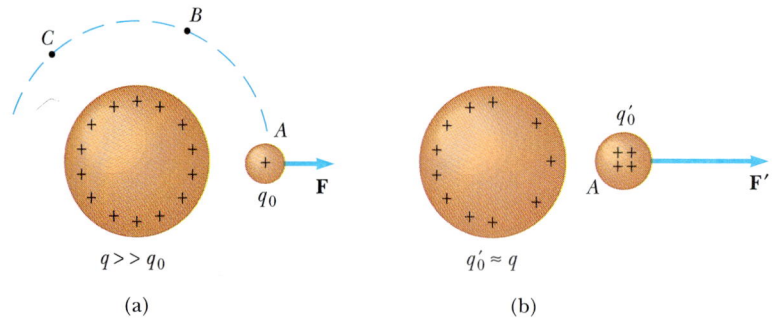

FIGURE 23.10　(a) When a small test charge q_0 is placed near a conducting sphere of charge q (where $q \gg q_0$), the charge on the conducting sphere remains uniform. (b) If the test charge q_0' is of the order of the charge on the sphere, the charge on the sphere is nonuniform.

which produces the electric field, remains uniformly distributed. Furthermore, the force **F** on the test charge has the same magnitude at *A, B,* and *C*, which are equidistant from the sphere. If the test charge is large enough ($q_0' \gg q_0$) as in Figure 23.10b, the charge on the metallic sphere is redistributed and the ratio of the force to the test charge at *A* is different: ($F'/q_0' \neq F/q_0$). That is, because of this redistribution of charge on the metallic sphere, the electric field at *A* set up by the sphere in Figure 23.10b must be different from that of the field at *A* in Figure 23.10a. Furthermore, the distribution of charge on the sphere changes as q_0' is moved from *A* to *B* or *C*.

Consider a point charge *q* located a distance *r* from a test charge q_0. According to Coulomb's law, the force exerted on the test charge by *q* is

$$\mathbf{F} = k_e \frac{qq_0}{r^2} \, \hat{\mathbf{r}}$$

Since the electric field at the position of the test charge is defined by $\mathbf{E} = \mathbf{F}/q_0$, we find that, at the position of q_0, the electric field created by *q* is

$$\mathbf{E} = k_e \frac{q}{r^2} \, \hat{\mathbf{r}} \tag{23.4}$$

where $\hat{\mathbf{r}}$ is a unit vector directed from *q* toward q_0 (Fig. 23.11). If *q* is positive, as in Figure 23.11a, the electric field is directed radially outward from it. If *q* is negative, as in Figure 23.11b, the field is directed toward it.

In order to calculate the electric field at a point *P* due to a group of point charges, we first calculate the electric field vectors at *P* individually using Equation 23.4 and then add them vectorially. In other words,

> the total electric field due to a group of charges equals the vector sum of the electric fields of all the charges.

This superposition principle applied to fields follows directly from the superposition property of electric forces. Thus, the electric field of a group of charges (excluding the test charge q_0) can be expressed as

$$\mathbf{E} = k_e \sum_i \frac{q_i}{r_i^2} \, \hat{\mathbf{r}}_i \tag{23.5}$$

where r_i is the distance from the *i*th charge, q_i, to the point *P* (the location of the test charge) and $\hat{\mathbf{r}}_i$ is a unit vector directed from q_i toward *P*.

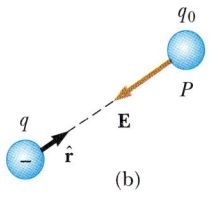

FIGURE 23.11   A test charge q_0 at point *P* is a distance *r* from a point charge *q*. (a) If *q* is positive, the electric field at *P* points radially outward from *q*. (b) If *q* is negative, the electric field at *P* points radially inward toward *q*.

CONCEPTUAL EXAMPLE 23.6

An uncharged, metallic coated Styrofoam ball is suspended in the region between two vertical metal plates as in Figure 23.12. If the two plates are charged, one positive and one negative, describe the motion of the ball after it is brought into contact with one of the plates.

Reasoning   The two charged plates create a region of uniform electric field between them, directed from the positive toward the negative plate. Once the ball is disturbed so as to touch one plate, say the negative one, some negative charge will be transferred to the ball and it experiences an electric force that accelerates it to the positive plate. Once the charge touches the positive plate, it releases its negative charge, acquires a positive charge, and accelerates back to the negative plate. The ball continues to move back and forth between the plates until it has transferred all its net charge, thereby making both plates neutral.

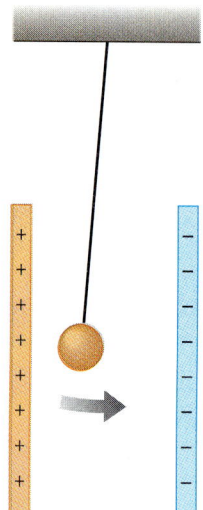

FIGURE 23.12 (Conceptual Example 23.6) An edge view of a sphere suspended between two oppositely charged plates.

EXAMPLE 23.7   Electric Force on a Proton

Find the electric force on a proton placed in an electric field of 2.0×10^4 N/C directed along the positive x axis.

Solution   Since the charge on a proton is $+e = 1.6 \times 10^{-19}$ C, the electric force on it is

$$\mathbf{F} = e\mathbf{E} = (1.6 \times 10^{-19} \text{ C})(2.0 \times 10^4 \mathbf{i} \text{ N/C})$$

$$= 3.2 \times 10^{-15} \mathbf{i} \text{ N}$$

where $\mathbf{i}$ is a unit vector in the positive x direction.

The weight of the proton is $mg = (1.67 \times 10^{-27}$ kg$)$ $(9.8 \text{ m/s}^2) = 1.6 \times 10^{-26}$ N. Hence, we see that the magnitude of the gravitational force in this case is negligible compared with the electric force.

EXAMPLE 23.8   Electric Field Due to Two Charges

A charge $q_1 = 7.0$ μC is located at the origin, and a second charge $q_2 = -5.0$ μC is located on the x axis 0.30 m from the origin (Figure 23.13). Find the electric field at the point P, which has coordinates (0, 0.40) m.

Solution   First, let us find the magnitude of the electric field due to each charge. The fields $\mathbf{E}_1$ due to the 7.0-μC charge and $\mathbf{E}_2$ due to the -5.0-μC charge are shown in Figure 23.13. Their magnitudes are

$$E_1 = k_e \frac{|q_1|}{r_1{}^2} = \left(8.99 \times 10^9 \frac{\text{N} \cdot \text{m}^2}{\text{C}^2}\right) \frac{(7.0 \times 10^{-6} \text{ C})}{(0.40 \text{ m})^2}$$

$$= 3.9 \times 10^5 \text{ N/C}$$

$$E_2 = k_e \frac{|q_2|}{r_2{}^2} = \left(8.99 \times 10^9 \frac{\text{N} \cdot \text{m}^2}{\text{C}^2}\right) \frac{(5.0 \times 10^{-6} \text{ C}}{(0.50 \text{ m})^2}$$

$$= 1.8 \times 10^5 \text{ N/C}$$

The vector $\mathbf{E}_1$ has only a y component. The vector $\mathbf{E}_2$ has an x component given by $E_2 \cos \theta = \frac{3}{5} E_2$ and a negative y component given by $-E_2 \sin \theta = -\frac{4}{5} E_2$. Hence, we can express the

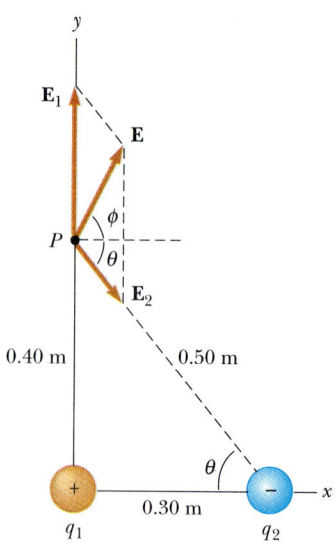

FIGURE 23.13 (Example 23.8) The total electric field $\mathbf{E}$ at P equals the vector sum $\mathbf{E}_1 + \mathbf{E}_2$, where $\mathbf{E}_1$ is the field due to the positive charge q_1 and $\mathbf{E}_2$ is the field due to the negative charge q_2.

vectors as

$$E_1 = 3.9 \times 10^5 j \text{ N/C}$$

$$E_2 = (1.1 \times 10^5 i - 1.4 \times 10^5 j) \text{ N/C}$$

The resultant field **E** at *P* is the superposition of E_1 and E_2:

$$E = E_1 + E_2 = \boxed{(1.1 \times 10^5 i + 2.5 \times 10^5 j) \text{ N/C}}$$

From this result, we find that **E** has a magnitude of 2.7×10^5 N/C and makes an angle ϕ of 66° with the positive *x* axis.

Exercise   Find the electric force on a positive test charge of 2×10^{-8} C placed at *P*.

Answer 5.4×10^{-3} N in the same direction as **E**.

EXAMPLE 23.9   Electric Field of a Dipole

An **electric dipole** consists of a positive charge *q* and a negative charge $-q$ separated by a distance 2*a*, as in Figure 23.14. Find the electric field **E** due to these charges along the *y* axis at the point *P*, which is a distance *y* from the origin. Assume that $y \gg a$.

Solution   At *P*, the fields E_1 and E_2 due to the two charges are equal in magnitude, since *P* is equidistant from the two equal and opposite charges. The total field $E = E_1 + E_2$, where

$$E_1 = E_2 = k_e \frac{q}{r^2} = k_e \frac{q}{y^2 + a^2}$$

The *y* components of E_1 and E_2 cancel each other. The *x* components are equal since they are both along the *x* axis. Therefore, **E** is parallel to the *x* axis and has a magnitude equal to $2E_1 \cos \theta$. From Figure 23.14 we see that $\cos \theta = a/r = a/(y^2 + a^2)^{1/2}$. Therefore,

$$E = 2E_1 \cos \theta = 2k_e \frac{q}{(y^2 + a^2)} \frac{a}{(y^2 + a^2)^{1/2}}$$

$$= k_e \frac{2qa}{(y^2 + a^2)^{3/2}}$$

Using the approximation $y \gg a$, we can neglect a^2 in the denominator and write

$$\boxed{E \approx k_e \frac{2qa}{y^3}}$$

Thus we see that along the *y* axis the field of a dipole at a distant point varies as $1/r^3$, whereas the more slowly varying field of a point charge goes as $1/r^2$. This is because at distant points, the fields of the two equal and opposite charges almost cancel each other. The $1/r^3$ variation in *E* for the dipole is also obtained for a distant point along the *x* axis (Problem

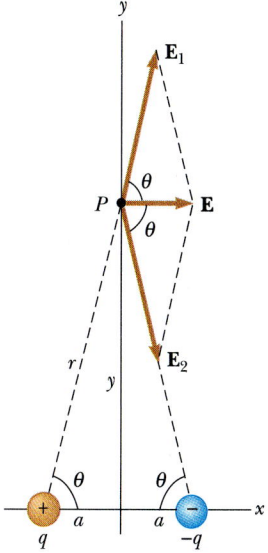

FIGURE 23.14   (Example 23.9) The total electric field **E** at *P* due to two equal and opposite charges (an electric dipole) equals the vector sum $E_1 + E_2$. The field E_1 is due to the positive charge *q*, and E_2 is the field due to the negative charge $-q$.

61) and for a general distant point. The dipole is a good model of many molecules, such as HCl.

As we shall see in later chapters, neutral atoms and molecules behave as dipoles when placed in an external electric field. Furthermore, many molecules, such as HCl, are permanent dipoles. (HCl is partially described as an H^+ ion combined with a Cl^- ion.) The effect of such dipoles on the behavior of materials subjected to electric fields is discussed in Chapter 26.

23.5   ELECTRIC FIELD OF A CONTINUOUS CHARGE DISTRIBUTION

Very often a group of charges are located very close together compared with their distances to points of interest (for example, a point where the electric field is to be calculated). In such situations, the system of charges can be considered to be *continuous*. That is, we imagine that the system of closely spaced charges is equiva-

A continuous charge distribution

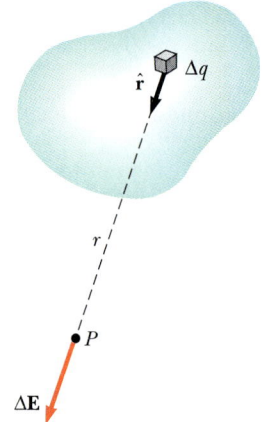

FIGURE 23.15   The electric field at *P* due to a continuous charge distribution is the vector sum of the fields due to all the elements Δq of the charge distribution.

lent to a total charge that is continuously distributed along a line, over some surface, or throughout a volume.

To evaluate the electric field of a continuous charge distribution, the following procedure is used. First, we divide the charge distribution into small elements each of which contains a small charge Δq, as in Figure 23.15. Next, we use Coulomb's law to calculate the electric field due to one of these elements at a point *P*. Finally, we evaluate the total field at *P* due to the charge distribution by summing the contributions of all the charge elements (that is, by applying the superposition principle).

The electric field at *P* due to one element of charge Δq is

$$\Delta \mathbf{E} = k_e \frac{\Delta q}{r^2} \hat{\mathbf{r}}$$

where *r* is the distance from the element to point *P* and $\hat{\mathbf{r}}$ is a unit vector directed from the charge element toward *P*. The total electric field at *P* due to all elements in the charge distribution is approximately

$$\mathbf{E} \approx k_e \sum_i \frac{\Delta q_i}{r_i^2} \hat{\mathbf{r}}_i$$

where the index *i* refers to the *i*th element in the distribution. If the separation between elements in the charge distribution is small compared with the distance to *P*, the charge distribution can be approximated to be continuous. Therefore, the total field at *P* in the limit $\Delta q_i \to 0$ becomes

Electric field of a continuous charge distribution

$$\mathbf{E} = k_e \lim_{\Delta q_1 \to 0} \sum_i \frac{\Delta q_i}{r_i^2} \hat{\mathbf{r}}_i = k_e \int \frac{dq}{r^2} \hat{\mathbf{r}} \qquad (23.6)$$

where the integration is a vector operation and must be treated with caution. We illustrate this type of calculation with several examples. In these examples, we assume that the charge is uniformly distributed on a line or a surface or throughout some volume. When performing such calculations, it is convenient to use the concept of a charge density along with the following notations:

- If a charge *Q* is uniformly distributed throughout a volume *V*, the *charge per unit volume*, ρ, is defined by

Volume charge density

$$\rho \equiv \frac{Q}{V}$$

where ρ has units of C/m^3.

- If a charge *Q* is uniformly distributed on a surface of area *A*, the *surface charge density*, σ, is defined by

Surface charge density

$$\sigma \equiv \frac{Q}{A}$$

where σ has units of C/m^2.

- Finally, if a charge *Q* is uniformly distributed along a line of length ℓ, the *linear charge density*, λ, is defined by

Linear charge density

$$\lambda \equiv \frac{Q}{\ell}$$

where λ has units of C/m.

If the charge is nonuniformly distributed over a volume, surface, or line, we have to express the charge densities as

$$\rho = \frac{dQ}{dV} \qquad \sigma = \frac{dQ}{dA} \qquad \lambda = \frac{dQ}{d\ell}$$

where dQ is the amount of charge in a small volume, surface, or length element.

EXAMPLE 23.10   The Electric Field Due to a Charged Rod

A rod of length ℓ has a uniform positive charge per unit length λ and a total charge Q. Calculate the electric field at a point P along the axis of the rod, a distance d from one end (Fig. 23.16).

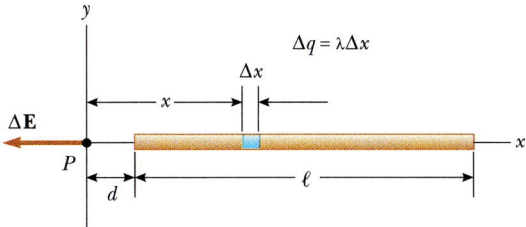

FIGURE 23.16   (Example 23.10) The electric field at P due to a uniformly charged rod lying along the x axis. The field at P due to the segment of charge Δq is $k_e\,\Delta q/x^2$. The total field at P is the vector sum over all segments of the rod.

Reasoning and Solution   For this calculation, the rod is taken to be along the x axis. Let us use Δx to represent the length of one small segment of the rod and let Δq be the charge on the segment. The ratio of Δq to Δx is equal to the ratio of the total charge to the total length of the rod. That is, $\Delta q/\Delta x = Q/\ell = \lambda$. Therefore, the charge Δq on the small segment is $\Delta q = \lambda\,\Delta x$.

The field $\Delta \mathbf{E}$ due to this segment at the point P is in the negative x direction, and its magnitude is[3]

$$\Delta E = k_e \frac{\Delta q}{x^2} = k_e \frac{\lambda\,\Delta x}{x^2}$$

Note that each element produces a field in the negative x direction, and so the problem of summing their contributions is particularly simple in this case. The total field at P due to all segments of the rod, which are at different dis-

tances from P, is given by Equation 23.6, which in this case becomes

$$E = \int_d^{\ell + d} k_e \lambda \frac{dx}{x^2}$$

where the limits on the integral extend from one end of the rod ($x = d$) to the other ($x = \ell + d$). Since k_e and λ are constants, they can be removed from the integral. Thus, we find that

$$E = k_e \lambda \int_d^{\ell + d} \frac{dx}{x^2} = k_e \lambda \left[-\frac{1}{x} \right]_d^{\ell + d}$$

$$= k_e \lambda \left(\frac{1}{d} - \frac{1}{\ell + d} \right) = \boxed{\frac{k_e Q}{d(\ell + d)}}$$

where we have used the fact that the total charge $Q = \lambda \ell$.

From this result we see that if the point P is far from the rod ($d \gg \ell$), then ℓ in the denominator can be neglected, and $E \approx k_e Q/d^2$. This is just the form you would expect for a point charge. Therefore, at large values of d/ℓ the charge distribution appears to be a point charge of magnitude Q. The use of the limiting technique ($d/\ell \rightarrow \infty$) is often a good method for checking a theoretical formula.

---

[3] It is important that you understand the procedure being used to carry out integrations such as this. First, choose an element whose parts are all equidistant from the point at which the field is being calculated. Next, express the charge element Δq in terms of the other variables within the integral (in this example, there is one variable, x). The integral must be over scalar quantities, and therefore must be expressed in terms of components. Then reduce the form to an integral over a single variable (or multiple integrals, each over a single variable). In examples that have spherical or cylindrical symmetry, the variables will be radial coordinates.

EXAMPLE 23.11   The Electric Field of a Uniform Ring of Charge

A ring of radius a has a uniform positive charge per unit length, with a total charge Q. Calculate the electric field along the axis of the ring at a point P lying a distance x from the center of the ring (Fig. 23.17a).

Reasoning and Solution   The magnitude of the electric field at P due to the segment of charge dq is

$$dE = k_e \frac{dq}{r^2}$$

This field has an x component $dE_x = dE \cos \theta$ along the axis of the ring and a component $dE_\perp$ perpendicular to the axis. But as we see in Figure 23.17b, the resultant field at P must lie along the x axis because the perpendicular components sum to zero. That is, the perpendicular component of any element is canceled by the perpendicular component of an element on the opposite side of the ring. Since $r = (x^2 + a^2)^{1/2}$ and $\cos \theta = x/r$, we find that

$$dE_x = dE \cos \theta = \left(k_e \frac{dq}{r^2} \right) \frac{x}{r} = \frac{k_e x}{(x^2 + a^2)^{3/2}} \, dq$$

In this case, all segments of the ring give the same contribution to the field at P since they are all equidistant from this point. Thus, we can integrate the expression above to get the total field at P:

$$E_x = \int \frac{k_e x}{(x^2 + a^2)^{3/2}} \, dq = \frac{k_e x}{(x^2 + a^2)^{3/2}} \int dq$$

$$= \frac{k_e x}{(x^2 + a^2)^{3/2}} \, Q$$

This result shows that the field is zero at $x = 0$. Does this surprise you?

Exercise Show that at large distances from the ring $(x \gg a)$ the electric field along the axis approaches that of a point charge of magnitude Q.

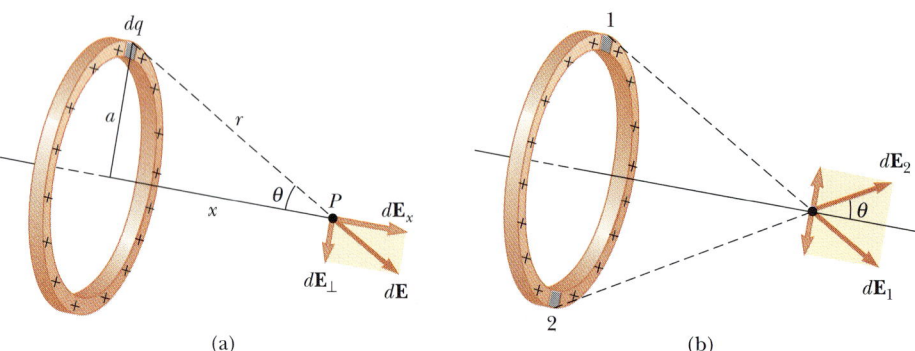

FIGURE 23.17 (Example 23.11) A uniformly charged ring of radius a. (a) The field at P on the x axis due to an element of charge dq. (b) The total electric field at P is along the x axis. Note that the perpendicular component of the electric field at P due to segment 1 is canceled by the perpendicular component due to segment 2, which is located on the ring opposite segment 1.

EXAMPLE 23.12   The Electric Field of a Uniformly Charged Disk

A disk of radius R has a uniform charge per unit area σ. Calculate the electric field at a point P that lies along the central axis of the disk and a distance x from its center (Fig. 23.18).

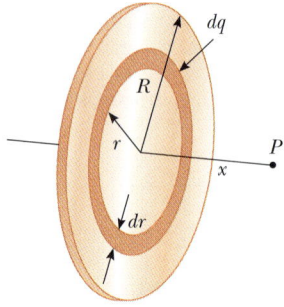

FIGURE 23.18 (Example 23.12) A uniformly charged disk of radius R. The electric field at an axial point P is directed along this axis, perpendicular to the plane of the disk.

Reasoning The solution to this problem is straightforward if we consider the disk as a set of concentric rings. We can then make use of Example 23.11, which gives the field of a ring of radius r, and sum up contributions of all rings making up the disk. By symmetry, the field on an axial point must be parallel to this axis.

Solution The ring of radius r and width dr has an area equal to $2\pi r \, dr$ (Fig. 23.18). The charge dq on this ring is equal to the area of the ring multiplied by the charge per unit area, or $dq = 2\pi\sigma r \, dr$. Using this result in the equation given for E_x in Example 23.11 (with a replaced by r) gives for the field due to the ring the expression

$$dE = \frac{k_e x}{(x^2 + r^2)^{3/2}} \, (2\pi\sigma r \, dr)$$

To get the total field at P, we integrate this expression over the limits $r = 0$ to $r = R$, noting that x is a constant. This gives

$$E = k_e x \pi \sigma \int_0^R \frac{2r \, dr}{(x^2 + r^2)^{3/2}}$$

$$= k_e x \pi \sigma \left[\frac{(x^2 + r^2)^{-1/2}}{-1/2} \right]_0^R$$

$$= 2\pi k_e \sigma \left(\frac{x}{|x|} - \frac{x}{(x^2 + R^2)^{1/2}} \right) \qquad (1)$$

The result is valid for all values of x. The field close to the

disk along an axial point can also be obtained from (1) by assuming $R \gg x$:

$$E = 2\pi k_e \sigma = \boxed{\frac{\sigma}{2\epsilon_0}} \qquad (2)$$

where ϵ_0 is the permittivity of free space. As we shall find in the next chapter, the same result is obtained for the field of a uniformly charged infinite sheet.

23.6   ELECTRIC FIELD LINES

A convenient aid for visualizing electric field patterns is to draw lines pointing in the same direction as the electric field vector at any point. These lines, called **electric field lines,** are related to the electric field in any region of space in the following manner:

- The electric field vector **E** is tangent to the electric field line at each point.
- The number of lines per unit area through a surface perpendicular to the lines is proportional to the strength of the electric field in that region. Thus E is large when the field lines are close together and small when they are far apart.

These properties are illustrated in Figure 23.19. The density of lines through surface A is greater than the density of lines through surface B. Therefore, the electric field is more intense on surface A than on surface B. Furthermore, the fact that the lines at different locations point in different directions tells us that the field is nonuniform.

Some representative electric field lines for a single positive point charge are shown in Figure 23.20a. Note that in this two-dimensional drawing we show only the field lines that lie in the plane containing the point charge. The lines are actually directed radially outward from the charge in all directions, somewhat like the needles protruding from a curled-up porcupine. Since a positive test charge placed in this field would be repelled by the positive point charge, the lines are directed radially away from the positive point charge. Similarly, the electric field lines for a single negative point charge are directed toward the charge (Fig.

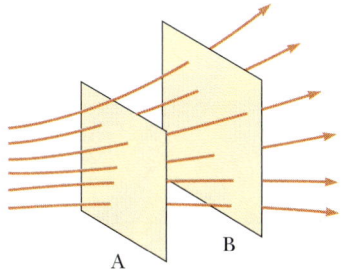

FIGURE 23.19   Electric field lines penetrating two surfaces. The magnitude of the field is greater on surface A than on surface B.

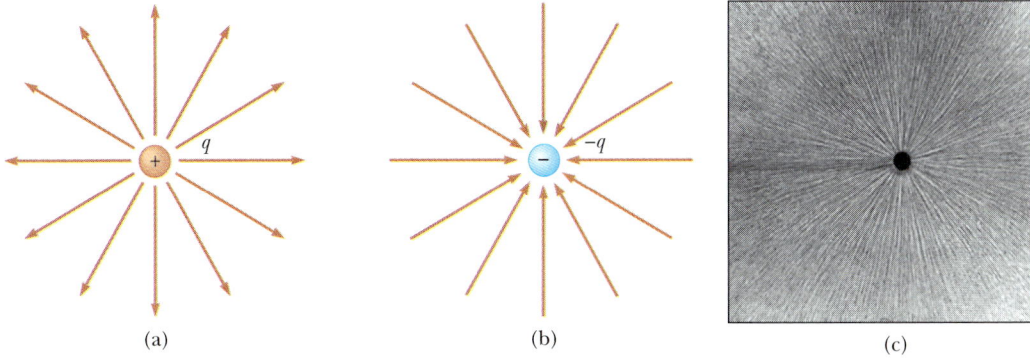

FIGURE 23.20   The electric field lines for a point charge. (a) For a positive point charge, the lines are radially outward. (b) For a negative point charge, the lines are radially inward. Note that the figures show only those field lines that lie in the plane containing the charge. (c) The dark areas are small pieces of thread suspended in oil, which align with the electric field produced by a small charged conductor at the center.   *(Photo courtesy of Harold M. Waage, Princeton University)*

23.20b). In either case, the lines are along the radial direction and extend all the way to infinity. Note that the lines are closer together as they get near the charge, indicating that the strength of the field is increasing.

The rules for drawing electric field lines for any charge distribution are as follows:

Rules for drawing electric field lines

- The lines must begin on positive charges and terminate on negative charges, although if the net charge is not zero, lines may begin or terminate at infinity.
- The number of lines drawn leaving a positive charge or approaching a negative charge is proportional to the magnitude of the charge.
- No two field lines can cross or touch.

Is this visualization of the electric field in terms of field lines consistent with Coulomb's law? To answer this question, consider an imaginary spherical surface of radius r concentric with the charge. From symmetry, we see that the magnitude of the electric field is the same everywhere on the surface of the sphere. The number of lines, N, that emerge from the charge is equal to the number that penetrate the spherical surface. Hence, the number of lines per unit area on the sphere is $N/4\pi r^2$ (where the surface area of the sphere is $4\pi r^2$). Since E is proportional to the number of lines per unit area, we see that E varies as $1/r^2$. This is consistent with the result obtained from Coulomb's law, that is, $E = k_e q/r^2$.

It is important to note that electric field lines are not material objects. They are used only to provide us with a qualitative description of the electric field. One problem with this model is the fact that we always draw a finite number of lines from each charge, which makes it appear as if the field were quantized and acted only along certain lines. The field, in fact, is continuous—existing at every point. Another problem with this model is the danger of getting the wrong impression from a two-dimensional drawing of field lines being used to describe a three-dimensional situation.

Since charge is quantized, the number of lines leaving any material object must be 0, $\pm C'e$, $\pm 2C'e$, . . . , where C' is an arbitrary (but fixed) proportionality constant. Once C' is chosen, the number of lines is fixed. For example, if object 1 has charge Q_1 and object 2 has charge Q_2, then the ratio of number of lines is $N_2/N_1 = Q_2/Q_1$.

The electric field lines for two point charges of equal magnitude but opposite signs (the electric dipole) are shown in Figure 23.21. In this case, the number of lines that begin at the positive charge must equal the number that terminate at the negative charge. At points very near the charges, the lines are nearly radial. The high density of lines between the charges indicates a region of strong electric field.

Figure 23.22 shows the electric field lines in the vicinity of two equal positive point charges. Again, the lines are nearly radial at points close to either charge. The same number of lines emerge from each charge since the charges are equal in magnitude. At large distances from the charges, the field is approximately equal to that of a single point charge of magnitude $2q$.

Finally, in Figure 23.23 we sketch the electric field lines associated with a positive charge $+2q$ and a negative charge $-q$. In this case, the number of lines leaving $+2q$ is twice the number entering $-q$. Hence only half of the lines that leave the positive charge enter the negative charge. The remaining half terminate on a negative charge we assume to be at infinity. At distances that are large compared

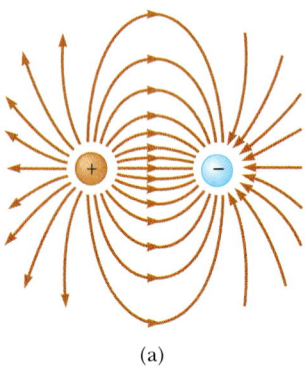

(a)

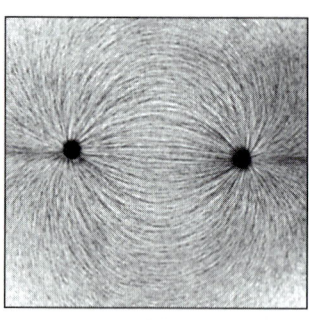

(b)

FIGURE 23.21   (a) The electric field lines for two equal and opposite point charges (an electric dipole). Note that the number of lines leaving the positive charge equals the number terminating at the negative charge. (b) The photograph was taken using small pieces of thread suspended in oil, which align with the electric field.   *(Photo courtesy of Harold M. Waage, Princeton University)*

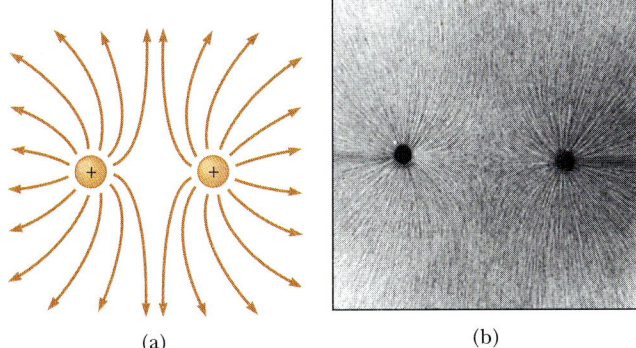

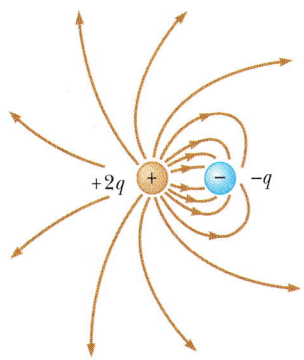

FIGURE 23.22   (a) The electric field lines for two positive point charges. (b) The photograph was taken using small pieces of thread suspended in oil, which align with the electric field.   *(Photo courtesy of Harold M. Waage, Princeton University)*

FIGURE 23.23   The electric field lines for a point charge $+2q$ and a second point charge $-q$. Note that two lines leave the charge $+2q$ for every one that terminates on $-q$.

with the charge separation, the electric field lines are equivalent to those of a single charge $+q$.

23.7   MOTION OF CHARGED PARTICLES IN A UNIFORM ELECTRIC FIELD

In this section we describe the motion of a charged particle in a uniform electric field. As we shall see, the motion is equivalent to that of a projectile moving in a uniform gravitational field. When a particle of charge q and mass m is placed in an electric field $\mathbf{E}$, the electric force on the charge is $q\mathbf{E}$. If this is the only force exerted on the charge, then Newton's second law applied to the charge gives

$$\mathbf{F} = q\mathbf{E} = m\mathbf{a}$$

The acceleration of the particle is therefore

$$\mathbf{a} = \frac{q\mathbf{E}}{m} \tag{23.7}$$

This simulator enables you to investigate the influence of electric fields on the motion of charged particles. You will be able to specify the charge on the particle and its velocity and and also have control over the magnitude and direction of an applied electric field. By varying these parameters, you can observe the motion of such particles as electrons and protons in the presence (or absence) of an electric field as well as the behavior of a system of charged particles.

Motion in an Electric Field

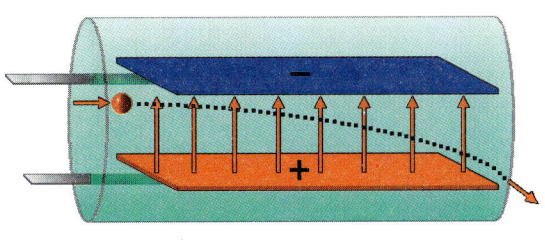

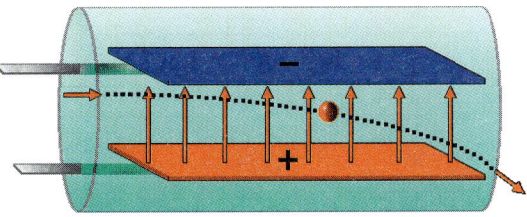

If **E** is uniform (that is, constant in magnitude and direction), we see that the acceleration is a constant of the motion. If the charge is positive, the acceleration is in the direction of the electric field. If the charge is negative, the acceleration is in the direction opposite the electric field.

EXAMPLE 23.13   An Accelerating Positive Charge

A positive point charge q of mass m is released from rest in a uniform electric field **E** directed along the x axis as in Figure 23.24. Describe its motion.

Reasoning and Solution   The acceleration of the charge is constant and given by qE/m. The motion is simple linear motion along the x axis. Therefore, we can apply the equations of kinematics in one dimension (from Chapter 2):

$$x - x_0 = v_0 t + \tfrac{1}{2}at^2 \qquad v = v_0 + at$$

$$v^2 = v_0{}^2 + 2a(x - x_0)$$

Taking $x_0 = 0$ and $v_0 = 0$ gives

$$x = \tfrac{1}{2}at^2 = \frac{qE}{2m}\, t^2$$

$$v = at = \frac{qE}{m}\, t$$

$$v^2 = 2\,ax = \left(\frac{2qE}{m}\right)x$$

The kinetic energy of the charge after it has moved a distance x is

$$K = \tfrac{1}{2}mv^2 = \tfrac{1}{2}m\left(\frac{2qE}{m}\right)x = qEx$$

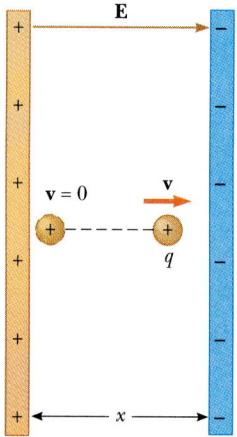

FIGURE 23.24   (Example 23.13) A positive point charge q in a uniform electric field E undergoes constant acceleration in the direction of the field.

This result can also be obtained from the work-energy theorem, since the work done by the electric force is $F_e x = qEx$ and $W = \Delta K$.

The electric field in the region between two oppositely charged flat metal plates is approximately uniform (Fig. 23.25). Suppose an electron of charge $-e$ is projected horizontally into this field with an initial velocity $v_0\mathbf{i}$. Since the electric field **E** is in the positive y direction, the acceleration of the electron is in the negative y direction. That is,

$$\mathbf{a} = -\frac{eE}{m}\,\mathbf{j} \tag{23.8}$$

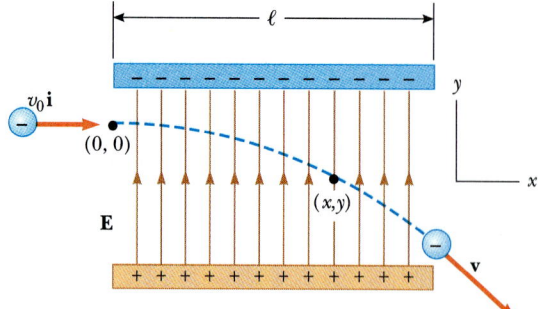

FIGURE 23.25   An electron is projected horizontally into a uniform electric field produced by two charged plates. The electron undergoes a downward acceleration (opposite **E**), and its motion is parabolic.

Because the acceleration is constant, we can apply the equations of kinematics in two dimensions (from Chapter 4) with $v_{x0} = v_0$ and $v_{y0} = 0$. After it has been in the electric field for a time t, the components of velocity of the electron are

$$v_x = v_0 = \text{constant} \tag{23.9}$$

$$v_y = at = -\frac{eE}{m}t \tag{23.10}$$

Likewise, the coordinates of the electron after a time t in the electric field are

$$x = v_0 t \tag{23.11}$$

$$y = \tfrac{1}{2}at^2 = -\tfrac{1}{2}\frac{eE}{m}t^2 \tag{23.12}$$

Substituting the value $t = x/v_0$ from Equation 23.11 into Equation 23.12, we see that y is proportional to x^2. Hence, the trajectory is a parabola. After the electron leaves the region of uniform electric field, it continues to move in a straight line with a speed $v > v_0$.

Note that we have neglected the gravitational force on the electron. This is a good approximation when dealing with atomic particles. For an electric field of 10^4 N/C, the ratio of the electric force, eE, to the gravitational force, mg, for the electron is of the order of 10^{14}. The corresponding ratio for a proton is of the order of 10^{11}.

EXAMPLE 23.14   An Accelerated Electron

An electron enters the region of a uniform electric field as in Figure 23.25, with $v_0 = 3.00 \times 10^6$ m/s and $E = 200$ N/C. The width of the plates is $\ell = 0.100$ m. (a) Find the acceleration of the electron while in the electric field.

Solution   Since the charge on the electron has a magnitude of 1.60×10^{-19} C and $m = 9.11 \times 10^{-31}$ kg, Equation 23.8 gives

$$\mathbf{a} = -\frac{eE}{m}\mathbf{j} = -\frac{(1.60 \times 10^{-19}\text{ C})(200\text{ N/C})}{9.11 \times 10^{-31}\text{ kg}}\mathbf{j}$$

$$= -3.51 \times 10^{13}\,\mathbf{j}\text{ m/s}^2$$

(b) Find the time it takes the electron to travel through the region of the electric field.

Solution   The horizontal distance traveled by the electron while in the electric field is $\ell = 0.100$ m. Using Equation 23.11 with $x = \ell$, we find that the time spent in the electric field is

$$t = \frac{\ell}{v_0} = \frac{0.100\text{ m}}{3.00 \times 10^6\text{ m/s}} = \boxed{3.33 \times 10^{-8}\text{ s}}$$

(c) What is the vertical displacement y of the electron while it is in the electric field?

Solution   Using Equation 23.12 and the results from (a) and (b), we find that

$$y = \tfrac{1}{2}at^2 = -\tfrac{1}{2}(3.51 \times 10^{13}\text{ m/s}^2)(3.33 \times 10^{-8}\text{ s})^2$$

$$= -0.0195\text{ m} = -1.95\text{ cm}$$

If the separation between the plates is smaller than this, the electron will strike the positive plate.

Exercise   Find the speed of the electron as it emerges from the electric field.

Answer 3.22×10^6 m/s.

---

*23.8   THE OSCILLOSCOPE

The oscilloscope is an electronic instrument widely used in making electrical measurements. Its main component is the cathode ray tube (CRT), shown in Figure 23.26. This tube is commonly used to obtain a visual display of electronic information for other applications, including radar systems, television receivers, and computers. The CRT is a vacuum tube in which electrons are accelerated and deflected under the influence of electric fields.

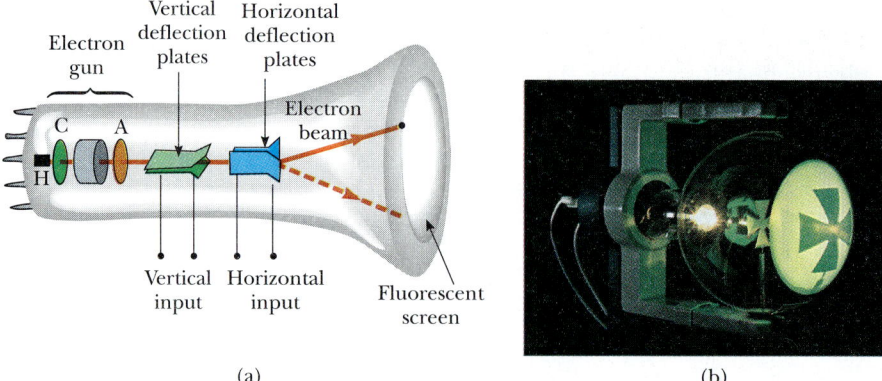

(a)   (b)

FIGURE 23.26   (a) Schematic diagram of a cathode ray tube. Electrons leaving the hot cathode C are accelerated to the anode A. The electron gun is also used to focus the beam, and the plates deflect the beam. (b) Photograph of a ''Maltese Cross'' tube showing the shadow of a beam of cathode rays falling on the tube's screen. The hot filament also produces a beam of light and a second shadow of the cross.   *(Courtesy of Central Scientific Co.)*

The electron beam is produced by an assembly called an *electron gun,* located in the neck of the tube. The assembly shown in Figure 23.26 consists of a heater (H), a negatively charged cathode (C), and a positively charged anode (A). An electric current maintained in the heater causes its temperature to rise, which in turn heats the cathode. The cathode reaches temperatures high enough to cause electrons to be ''boiled off.'' Although they are not shown in the figure, the electron gun also includes an element that focuses the electron beam and one that controls the number of electrons reaching the anode (that is, a brightness control). The anode has a hole in its center that allows the electrons to pass through without striking the anode. These electrons, if left undisturbed, travel in a straight-line path until they strike the front of the CRT, ''the screen,'' which is coated with a material that emits visible light when bombarded with electrons. This emission results in a visible spot of light on the screen.

The electrons are deflected in various directions by two sets of plates placed at right angles to each other in the neck of the tube. In order to understand how the deflection plates operate, first consider the horizontal deflection plates in Figure 23.26a. An external electric circuit is used to control and change the amount of charge present on these plates, with positive charge being placed on one plate and negative on the other. (In Chapter 25 we shall see that this can be accomplished by applying a voltage across the plates.) This increasing charge creates an increasing electric field between the plates, which causes the electron beam to be deflected from its straight-line path. The screen is slightly phosphorescent and therefore glows briefly after the electron beam moves from one point to another on it. Slowly increasing the charge on the horizontal plates causes the electron beam to move gradually from the center toward the side of the screen. Because of the phosphorescence, however, one sees a horizontal line extending across the screen instead of the simple movement of the dot. The horizontal line can be maintained on the screen by rapid, repetitive tracing.

The vertical deflection plates act in exactly the same way as the horizontal plates, except that changing the charge on them causes a vertical line on the screen. In practice, the horizontal and vertical deflection plates are used simultaneously.

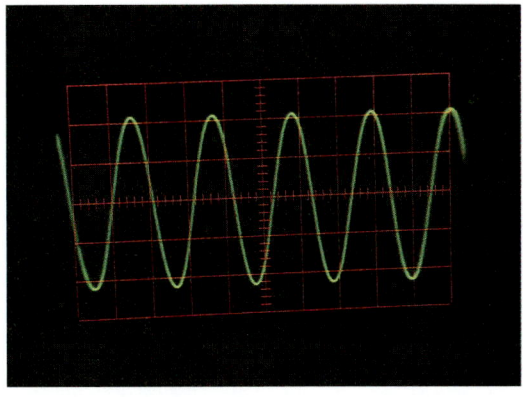

FIGURE 23.27 A sinusoidal wave produced by a vibrating tuning fork and displayed on an oscilloscope screen. *(Henry Leap and Jim Lehman)*

To see how the oscilloscope can display visual information, let us examine how we could observe the sound wave from a tuning fork on the screen. For this purpose, the charge on the horizontal plates changes in such a manner that the beam sweeps across the face of the tube at a constant rate. The tuning fork is then sounded into a microphone, which changes the sound signal to an electric signal that is applied to the vertical plates. The combined effect of the horizontal and vertical plates causes the beam to sweep the tube horizontally and up and down at the same time, with the vertical motion corresponding to the tuning fork signal. A pattern such as that shown in Figure 23.27 is seen on the screen.

SUMMARY

Electric charges have the following important properties:

- Unlike charges attract one another and like charges repel one another.
- Electric charge is always conserved.
- Charge is quantized, that is, it exists in discrete packets that are some integral multiple of the electronic charge.
- The force between charged particles varies as the inverse square of their separation.

Conductors are materials in which charges move freely. **Insulators** are materials that do not readily transport charge.

Coulomb's law states that the electrostatic force between two stationary, charged particles separated by a distance r has magnitude

$$F = k_e \frac{|q_1||q_2|}{r^2} \tag{23.1}$$

where the constant k_e, called the Coulomb constant, has the value

$$k_e = 8.9875 \times 10^9 \text{ N} \cdot \text{m}^2/\text{C}^2$$

The smallest unit of charge known to exist in nature is the charge on an electron or proton:

$$|e| = 1.60219 \times 10^{-19} \text{ C}$$

The electric field **E** at some point in space is defined as the electric force **F** that acts on a small positive test charge placed at that point divided by the magnitude of the test charge q_0:

$$\mathbf{E} \equiv \frac{\mathbf{F}}{q_0} \tag{23.3}$$

The electric field due to a point charge q at a distance r from the charge is

$$\mathbf{E} = k_e \frac{q}{r^2} \hat{\mathbf{r}} \tag{23.4}$$

where $\hat{\mathbf{r}}$ is a unit vector directed from the charge to the point in question. The electric field is directed radially outward from a positive charge and directed toward a negative charge.

The electric field due to a group of charges can be obtained using the superposition principle. That is, the total electric field equals the vector sum of the electric fields of all the charges at some point:

$$\mathbf{E} = k_e \sum_i \frac{q_i}{r_i^2} \hat{\mathbf{r}}_i \tag{23.5}$$

The electric field of a continuous charge distribution at some point is

$$\mathbf{E} = k_e \int \frac{dq}{r^2} \hat{\mathbf{r}} \tag{23.6}$$

where dq is the charge on one element of the charge distribution and r is the distance from the element to the point in question.

Electric field lines are useful for describing the electric field in any region of space. The electric field vector $\mathbf{E}$ is always tangent to the electric field lines at every point. The number of lines per unit area through a surface perpendicular to the lines is proportional to the magnitude of $\mathbf{E}$ in that region.

A charged particle of mass m and charge q moving in an electric field $\mathbf{E}$ has an acceleration

$$\mathbf{a} = \frac{q\mathbf{E}}{m} \tag{23.7}$$

If the electric field is uniform, the acceleration is constant and the motion of the charge is similar to that of a projectile moving in a uniform gravitational field.

Problem-Solving Strategy and Hints
Finding the Electric Field

- **Units:** When performing calculations that involve the Coulomb constant k_e $(= 1/4\pi\epsilon_0)$, charges must be in coulombs and distances in meters. If they appear in other units, you must convert them.
- **Applying Coulomb's law to point charges:** Use the superposition principle properly when dealing with a collection of interacting charges. When several charges are present, the resultant force on any one of them is the vector sum of the forces exerted by the individual charges. You must be very careful in the algebraic manipulation of vector quantities. It may be useful to review the material on vector addition in Chapter 3.
- **Calculating the electric field of point charges:** The superposition principle can be applied to electric fields. To find the total electric field at a given point, first calculate the electric field at the point due to each individual

charge. The resultant field at the point is the vector sum of the fields due to the individual charges.

- **Continuous charge distributions:** When you are confronted with problems that involve a continuous distribution of charge, the vector sums for evaluating the total electric field at some point must be replaced by vector integrals. The charge distribution is divided into infinitesimal pieces, and the vector sum is carried out by integrating over the entire charge distribution. You should review Examples 23.10–23.12, which demonstrate such procedures.
- **Symmetry:** Whenever dealing with either a distribution of point charges or a continuous charge distribution, you should take advantage of any symmetry in the system to simplify your calculations.

QUESTIONS

1. Sparks are often observed (or heard) on a dry day when clothes are removed in the dark. Explain.
2. Explain from an atomic viewpoint why charge is usually transferred by electrons.
3. A balloon is negatively charged by rubbing and then clings to a wall. Does this mean that the wall is positively charged? Why does the balloon eventually fall?
4. A light, uncharged metal sphere suspended from a thread is attracted to a charged rubber rod. After touching the rod, the sphere is repelled by the rod. Explain.
5. Explain what we mean by a neutral atom.
6. Why do some clothes cling together and to your body after being removed from a dryer?
7. A large metal sphere insulated from ground is charged with an electrostatic generator while a person standing on an insulating stool holds the sphere. Why is it safe to do this? Why wouldn't it be safe for another person to touch the sphere after it has been charged?
8. What is the difference between charging an object by induction and charging by conduction?
9. What are the similarities and differences between Newton's law of gravity, $F = Gm_1 m_2 / r^2$, and Coulomb's law, $F = k_e q_1 q_2 / r^2$?
10. Assume that someone proposes a theory that says people are bound to the Earth by electric forces rather than by gravity. How could you prove this theory wrong?
11. Would life be different if the electron were positively charged and the proton were negatively charged? Does the choice of signs have any bearing on physical and chemical interactions? Explain.
12. When defining the electric field, why is it necessary to specify that the magnitude of the test charge be very small (that is, take the limit of F/q as $q \rightarrow 0$)?
13. Two charged conducting spheres, each of radius a, are separated by a distance $r > 2a$. Is the force on either

sphere given by Coulomb's law? Explain. (*Hint:* Refer to Chapter 14 on gravitation and Fig. 23.10.)
14. When is it valid to approximate a charge distribution by a point charge?
15. Is it possible for an electric field to exist in empty space? Explain.
16. Explain why electric field lines do not form closed loops.
17. Explain why electric field lines never cross. (*Hint:* **E** must have a unique direction at all points.)
18. A free electron and free proton are placed in an identical electric field. Compare the electric forces on each particle. Compare their accelerations.
19. Explain what happens to the magnitude of the electric field of a point charge as r approaches zero.
20. A negative charge is placed in a region of space where the electric field is directed vertically upward. What is the direction of the electric force experienced by this charge?
21. A charge $4q$ is a distance r from a charge $-q$. Compare the number of electric field lines leaving the charge $4q$ with the number entering the charge $-q$.
22. In Figure 23.23, where do the extra lines leaving the charge $+2q$ end?
23. Consider two equal point charges separated by some distance d. At what point (other than ∞) would a third test charge experience no net force?
24. A negative point charge $-q$ is placed at the point P near the positively charged ring shown in Figure 23.17. If $x \ll a$, describe the motion of the point charge if it is released from rest.
25. Explain the differences between linear, surface, and volume charge densities, and give examples of when each would be used.
26. If the electron in Figure 23.25 is projected into the electric field with an arbitrary velocity v_0 (at an angle to **E**), will its trajectory still be parabolic? Explain.

27. If a metal object receives a positive charge, does the mass of the object increase, decrease, or stay the same? What happens to its mass if the object is given a negative charge?

28. It has been reported that in some instances people near where a lightning bolt strikes the Earth have had their clothes thrown off. Explain why this might happen.

29. Why should a ground wire be connected to the metal support rod for a television antenna?

30. A light piece of aluminum foil is draped over a wooden rod. When a rod carrying a positive charge is brought close to the foil, the two parts of the foil stand apart. Why? What kind of charge is on the foil?

31. Why is it more difficult to charge an object by friction on a humid day than on a dry day?

32. How would you experimentally distinguish an electric field from a gravitational field?

PROBLEMS

Section 23.3 Coulomb's Law

1. Suppose that 1.00 g of hydrogen is separated into electrons and protons. Suppose also that the protons are placed at the Earth's North Pole and the electrons are placed at the South Pole. What is the resulting compressional force on the Earth?

2. (a) Calculate the number of electrons in a small silver pin, electrically neutral, that has a mass of 10.0 g. Silver has 47 electrons per atom, and its atomic mass is 107.87. (b) Electrons are added to the pin until the net charge is 1.00 mC. How many electrons are added for every 10^9 electrons already present?

3. Two protons in a molecule are separated by 3.80×10^{-10} m. (a) Find the electrostatic force exerted by one proton on the other. (b) How does the magnitude of this force compare with the magnitude of the gravitational force between the two protons? (c) What must be the charge-to-mass ratio of a particle if the magnitude of the gravitational force between it and a particle equals the magnitude of electrostatic force between them?

4. Can an electron remain suspended between a neutral insulating horizontal surface and a fixed positive charge, q, 7.62 m from the electron? Is this observation possible? Explain.

5. (a) What magnitude of charge must be placed equally on the Earth and Moon to make the magnitude of the electrical force between these two bodies equal the gravitational force between them? (b) What would be the electric field on the Moon due to the Earth's charge?

6. In fission, a nucleus of uranium-238, which contains 92 protons, divides into two smaller spheres, each having 46 protons and a radius of 5.9×10^{-15} m. What is the magnitude of the repulsive electric force pushing the two spheres apart?

7. Three point charges are located at the corners of an equilateral triangle as in Figure P23.7. Calculate the net electric force on the 7.0-μC charge.

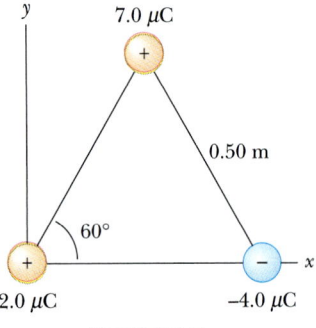

FIGURE P23.7

8. Two identical point charges $+q$ are fixed in space and separated by a distance d. A third point charge $-Q$ of mass m is free to move and lies initially at rest on a perpendicular bisector of line connecting the two fixed charges a distance x from the line (Fig. P23.8). (a) Show that if x is small relative with d, the motion of $-Q$ is simple harmonic along the bisector, and determine the period of that motion. (b) How

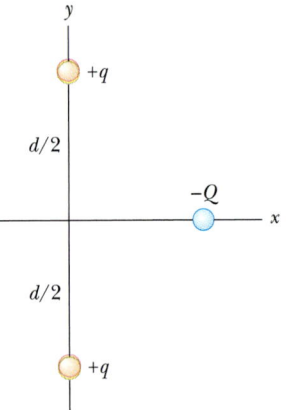

FIGURE P23.8

☐ indicates problems that have full solutions available in the Student Solutions Manual and Study Guide.

fast is $-Q$ moving when it is at the midpoint between the two fixed charges?

9. Four identical point charges ($q = +10.0$ μC) are located on the corners of a rectangle as shown in Figure P23.9. The dimensions of the rectangle are $L = 60.0$ cm and $W = 15.0$ cm. Calculate the magnitude and direction of the net electric force exerted on the charge at the lower left corner by the other three charges.

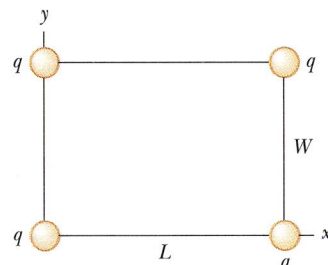

FIGURE P23.9

10. Three identical point charges, each of mass $m = 0.100$ kg and charge q, hang from three strings, as in Figure P23.10. If the lengths of the left and right strings are $L = 30.0$ cm and angle $\theta = 45.0°$, determine the value of q.

10A. Three identical point charges, each of mass m and charge q, hang from three strings, as in Figure P23.10. Determine the value of q in terms of m, L, and θ.

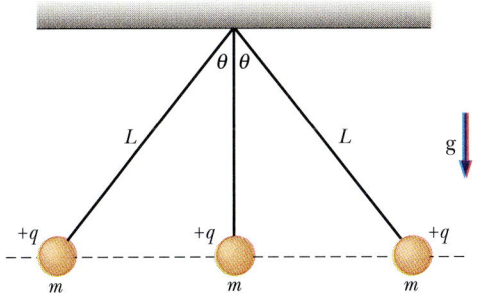

FIGURE P23.10

11. Two small silver spheres, each with a mass of 100 g, are separated by 1.0 m. Calculate the fraction of the electrons in one sphere that must be transferred to the other in order to produce an attractive force of 1.0×10^4 N (about a ton) between the spheres. (The number of electrons per atom of silver is 47, and the number of atoms per gram is Avogadro's number divided by the molar mass of silver, 107.87.)

12. Richard Feynman once said that if two persons stood at arm's length from each other and each person had 1% more electrons than protons, the force of repulsion between them would be enough to lift a "weight" equal to that of the entire Earth. Carry out an order-of-magnitude calculation to substantiate this assertion.

13. In a thundercloud there may be an electric charge of $+40$ C near the top and -40 C near the bottom. These charges are separated by approximately 2.0 km. What is the electric force between them?

Section 23.4   The Electric Field

14. An airplane is flying through a thundercloud at a height of 2000 m. (This is a very dangerous thing to do because of updrafts, turbulence, and the possibility of electric discharge.) If there is a charge concentration of $+40$ C at height 3000 m within the cloud and -40 C at height 1000 m, what is the electric field E at the aircraft?

15. What are the magnitude and direction of the electric field that will balance the weight of (a) an electron and (b) a proton? (Use the data in Table 23.1.)

16. An object having a net charge of 24 μC is placed in a uniform electric field of 610 N/C directed vertically. What is the mass of this object if it "floats" in the field?

16A. An object having a net charge Q is placed in a uniform electric field of magnitude E directed vertically. What is the mass of this object if it "floats" in the field?

17. On a dry winter day, if you scuff your feet across a carpet, you build up a charge and get a shock when you touch a metal doorknob. In a dark room you can actually see a spark about 2.0 cm long. Air breaks down at a field strength of 3.0×10^6 N/C. Assume that just before the spark occurs, all the charge is in your finger, drawn there by induction due to the proximity of the doorknob. Approximate your fingertip as a sphere of diameter 1.5 cm, and assume that there is an equal amount of charge on the doorknob 2.0 cm away. (a) How much charge have you built up? (b) How many electrons does this correspond to?

18. A point having charge q is located at (x_0, y_0) in the xy plane. Show that the x and y components of the electric field at (x, y) due to this charge are

$$E_x = \frac{k_e q(x - x_0)}{[(x - x_0)^2 + (y - y_0)^2]^{3/2}}$$

$$E_y = \frac{k_e q(y - y_0)}{[(x - x_0)^2 + (y - y_0)^2]^{3/2}}$$

19. Two 2.0-μC point charges are located on the x axis. One is at $x = 1.0$ m, and the other is at $x = -1.0$ m. (a) Determine the electric field on the y axis at $y = 0.50$ m. (b) Calculate the electric force on a -3.0-μC charge placed on the y axis at $y = 0.50$ m.

20. Determine the magnitude of the electric field at the surface of a lead-208 nucleus, which contains 82 pro-

tons and 126 neutrons. Assume the lead nucleus has a volume 208 times that of one proton, and consider a proton to be a hard sphere of radius 1.2×10^{-15} m.

21. Consider n equal positive point charges each of magnitude q/n placed symmetrically around a circle of radius R. (a) Calculate the magnitude of the electric field at a point a distance x from the plane on the line passing through the center of the circle and perpendicular to the plane of the circle. (b) Explain why this result is identical to the calculation done in Example 23.11.

22. Three point charges, q, $2q$, and $3q$, are arranged on the vertices of an equilateral triangle. Determine the magnitude of electric field at the geometric center of the triangle.

23. Three equal positive charges q are at the corners of an equilateral triangle of sides a as in Figure P23.23. (a) At what point in the plane of the charges (other than ∞) is the electric field zero? (b) What are the magnitude and direction of the electric field at P due to the two charges at the base?

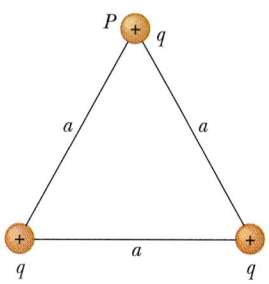

FIGURE P23.23

24. Three charges are at the corners of an equilateral triangle as in Figure P23.7. Calculate the electric field intensity at the position of the 2.0-μC charge due to the 7.0-μC and -4.0-μC charges.

25. Four point charges are at the corners of a square of side a as in Figure P23.25. (a) Determine the magnitude and direction of the electric field at the location of charge q. (b) What is the resultant force on q?

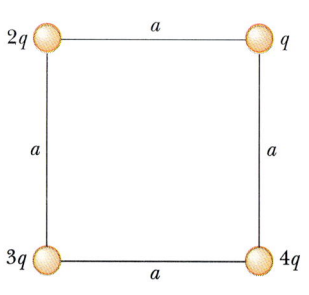

FIGURE P23.25

26. A charge of -4.0 μC is located at the origin, and a charge of -5.0 μC is located along the y axis at $y = 2.0$ m. At what point along the y axis is the electric field zero?

26A. A charge $-q_1$ is located at the origin, and a charge $-q_2$ is located along the y axis at $y = d$. At what point along the y axis is the electric field zero?

27. Consider an infinite number of identical charges (each of charge q) placed along the x axis at distances, a, $2a$, $3a$, $4a$, . . . , from the origin. What is the electric field at the origin due to this distribution? *Hint:* Use the fact that

$$1 + \frac{1}{2^2} + \frac{1}{3^2} + \frac{1}{4^2} + \cdots = \frac{\pi^2}{6}$$

Section 23.5 Electric Field of a Continuous Charge Distribution

28. A rod 14 cm long is uniformly charged and has a total charge of -22 μC. Determine the magnitude and direction of the electric field along the axis of the rod at a point 36 cm from its center.

29. A continuous line of charge lies along the x axis, extending from $x = +x_0$ to positive infinity. The line carries a uniform linear charge density λ_0. What are the magnitude and direction of the electric field at the origin?

30. A line of charge starts at $x = +x_0$ and extends to positive infinity. If the linear charge density is $\lambda = \lambda_0 x_0 / x$, determine the electric field at the origin.

31. A uniformly charged ring of radius 10 cm has a total charge of 75 μC. Find the electric field on the axis of the ring at (a) 1.0 cm, (b) 5.0 cm, (c) 30 cm, and (d) 100 cm from the center of the ring.

32. Show that the maximum field strength $E_{\max}$ along the axis of a uniformly charged ring occurs at $x = a/\sqrt{2}$ (see Fig. 23.17) and has the value $Q/(6\sqrt{3}\pi\epsilon_0 a^2)$.

33. (a) Consider a uniformly charged right circular cylindrical shell having total charge Q, radius R, and height h. Determine the electric field at a point a distance d from the right side of the cylinder as in Figure P23.33. (*Hint:* Use the result of Example 23.11 and treat the cylinder as a collection of ring charges.) (b) Use the result of Example 23.12 to

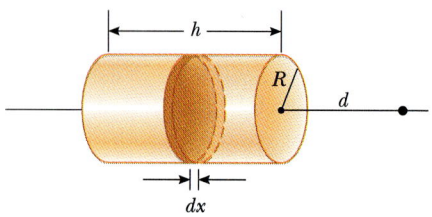

FIGURE P23.33

solve the same problem, but this time assume the cylinder is solid.

34. A uniformly charged disk of radius 35 cm carries a charge density of 7.9×10^{-3} C/m^2. Calculate the electric field on the axis of the disk at (a) 5.0 cm, (b) 10 cm, (c) 50 cm, and (d) 200 cm from the center of the disk.

35. Example 23.12 derives the exact expression for the electric field at a point on the axis of a uniformly charged disk. Consider a disk of radius $R = 3.0$ cm, having a uniformly distributed charge of $+5.2$ μC. (a) Using the result of Example 23.12, compute the electric field at a point on the axis and 3.0 mm from the center. Compare this answer with the field computed from the near-field approximation. (b) Using the result of Example 23.12, compute the electric field at a point on the axis and 30 cm from the center of the disk. Compare this with the electric field obtained by treating the disk as a $+5.2$-μC point charge at a distance of 30 cm.

36. The electric field along the axis of a uniformly charged disk of radius R and total charge Q was calculated in Example 23.12. Show that the electric field at distances x that are large compared with R approaches that of a point charge $Q = \sigma \pi R^2$. (*Hint:* First show that $x/(x^2 + R^2)^{1/2} = (1 + R^2/x^2)^{-1/2}$ and use the binomial expansion $(1 + \delta)^n \approx 1 + n\delta$ when $\delta \ll 1$.)

37. A uniformly charged ring and a uniformly charged disk each have a charge of $+25$ μC and a radius of 3.0 cm. For each of these charged objects, determine the electric field at a point along the axis 4.0 cm from the center of the object.

38. A 10.0-g piece of Styrofoam carries a net charge of -0.700 μC and floats above the center of a very large horizontal sheet of plastic that has a uniform charge density on its surface. What is the charge per unit area on the plastic sheet?

38A. A piece of Styrofoam having a mass m carries a net charge of $-q$ and floats above the center of a very large horizontal sheet of plastic that has a uniform charge density on its surface. What is the charge per unit area on the plastic sheet?

39. A uniformly charged insulating rod of length 14 cm is bent into the shape of a semicircle as in Figure P23.39. If the rod has a total charge of -7.5 μC, find the magnitude and direction of the electric field at O, the center of the semicircle.

Section 23.6 Electric Field Lines

40. A positively charged disk has a uniform charge per unit area as described in Example 23.12. Sketch the electric field lines in a plane perpendicular to the plane of the disk passing through its center.

41. A negatively charged rod of finite length has a uni-

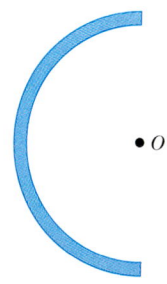

FIGURE P23.39

form charge per unit length. Sketch the electric field lines in a plane containing the rod.

42. A positive point charge is at a distance $R/2$ from the center of an uncharged thin conducting spherical shell of radius R. Sketch the electric field lines both inside and outside the shell.

43. Figure P23.43 shows the electric field lines for two point charges separated by a small distance. (a) Determine the ratio q_1/q_2. (b) What are the signs of q_1 and q_2?

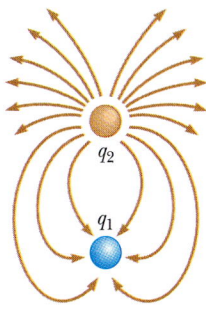

FIGURE P23.43

Section 23.7 Motion of Charged Particles in a Uniform Electric Field

44. An electron and a proton are each placed at rest in an electric field of 520 N/C. Calculate the speed of each particle 48 ns after being released.

45. A proton accelerates from rest in a uniform electric field of 640 N/C. At some later time, its speed is 1.20×10^6 m/s (nonrelativistic since v is much less than the speed of light). (a) Find the acceleration of the proton. (b) How long does it take the proton to reach this speed? (c) How far has it moved in this time? (d) What is its kinetic energy at this time?

46. An electron moves at 3×10^6 m/s into a uniform electric field of magnitude 1000 N/C. The field is parallel to the electron's velocity and acts to decelerate the electron. How far does the electron travel before it is brought to rest?

47. The electrons in a particle beam each have a kinetic energy of 1.60×10^{-17} J. What are the magnitude and direction of the electric field that will stop these electrons in a distance of 10.0 cm?

47A. The electrons in a particle beam each have a kinetic energy K. What are the magnitude and direction of the electric field that will stop these electrons in a distance d?

48. An electron traveling with an initial velocity equal to $8.6 \times 10^5 \mathbf{i}$ m/s enters a region of a uniform electric field given by $\mathbf{E} = 4.1 \times 10^3$ N/C. (a) Find the acceleration of the electron. (b) Determine the time it takes for the electron to come to rest after it enters the field. (c) How far does the electron move in the electric field before coming to rest?

49. A proton is projected in the positive x direction into a region of a uniform electric field $\mathbf{E} = -6.00 \times 10^5 \mathbf{i}$ N/C. The proton travels 7.00 cm before coming to rest. Determine (a) the acceleration of the proton, (b) its initial speed, and (c) the time it takes the proton to come to rest.

50. A positively charged 1.00-g bead initially at rest in a vacuum falls 5.00 m through a uniform vertical electric field of magnitude 1.00×10^4 N/C. The bead hits the ground at 21.0 m/s. Determine (a) the direction of the electric field (up or down) and (b) the charge on the bead.

50A. A positively charged bead having a mass m falls from rest in a vacuum at a height h in a uniform vertical electric field of magnitude E. The bead hits the ground at a speed $v > \sqrt{2gh}$. Determine (a) the direction of the electric field (up or down) and (b) the charge on the bead.

51. A proton moves at 4.50×10^5 m/s in the horizontal direction. It enters a uniform electric field of 9.60×10^3 N/C directed vertically downward. Ignore any gravitational effects and find (a) the time it takes the proton to travel 5.00 cm horizontally, (b) its vertical displacement after it has traveled 5.00 cm horizontally, and (c) the horizontal and vertical components of its velocity after it has traveled 5.00 cm horizontally.

52. An electron is projected at an angle of 30° above the horizontal at a speed of 8.2×10^5 m/s, in a region where the electric field is $\mathbf{E} = 390\mathbf{j}$ N/C. Neglect gravity and find (a) the time it takes the electron to return to its initial height, (b) the maximum height it reaches, and (c) its horizontal displacement when it reaches its maximum height.

52A. An electron is projected at an angle θ above the horizontal at a speed v, in a region where the electric field is $\mathbf{E} = E_0\mathbf{j}$ N/C. Neglect gravity and find (a) the time it takes the electron to return to its initial

height, (b) the maximum height it reaches, and (c) its horizontal displacement when it reaches its maximum height.

53. Protons are projected with an initial speed $v_0 = 9.55 \times 10^3$ m/s into a region where a uniform electric field $\mathbf{E} = (-720\mathbf{j})$ N/C is present, as in Figure P23.53. The protons are to hit a target that lies at a horizontal distance of 1.27 mm from the point where the protons are launched. Find (a) the two projection angles θ that will result in a hit and (b) the total time of flight for each trajectory.

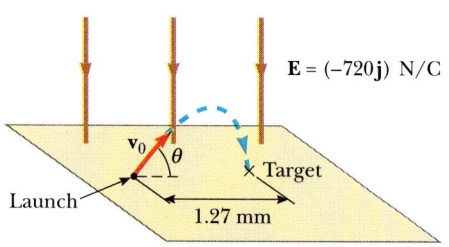

FIGURE P23.53

ADDITIONAL PROBLEMS

54. A small 2.00-g plastic ball is suspended by a 20.0-cm-long string in a uniform electric field as in Figure P23.54. If the ball is in equilibrium when the string makes a 15.0° angle with the vertical, what is the net charge on the ball?

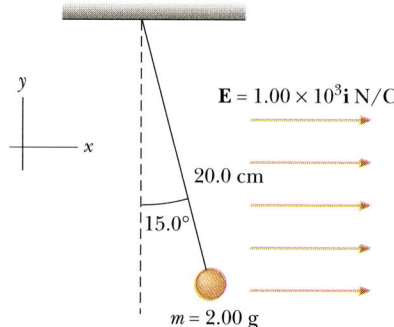

FIGURE P23.54

55. A charged cork ball of mass 1.00 g is suspended on a light string in the presence of a uniform electric field as in Figure P23.55. When $\mathbf{E} = (3.00\mathbf{i} + 5.00\mathbf{j}) \times$

10^5 N/C, the ball is in equilibrium at $\theta = 37.0°$. Find (a) the charge on the ball and (b) the tension in the string.

55A. A charged cork ball of mass m is suspended on a light string in the presence of a uniform electric field as in Figure P23.55. When $\mathbf{E} = (E_x\mathbf{i} + E_y\mathbf{j})$ N/C, the ball is in equilibrium at the angle θ. Find (a) the charge on the ball and (b) the tension in the string.

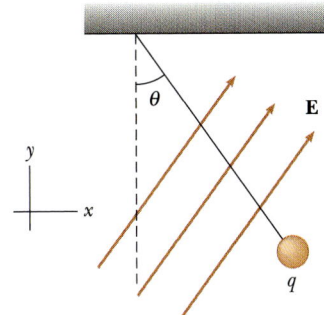

FIGURE P23.55

56. Two small spheres each of mass 2.00 g are suspended by light strings 10.0 cm in length (Fig. P23.56). A uniform electric field is applied in the x direction. If the spheres have charges equal to -5.00×10^{-8} C and $+5.00 \times 10^{-8}$ C, determine the electric field that enables the spheres to be in equilibrium at an angle of $\theta = 10.0°$.

56A. Two small spheres each of mass m are suspended by light strings of length L (Fig. P23.56). A uniform electric field is applied in the x direction. If the spheres have charges equal to $-q$ and $+q$, determine the electric field that enables the spheres to be in equilibrium at an angle θ.

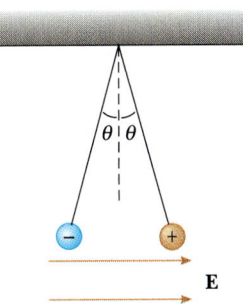

FIGURE P23.56

57. Two small spheres of mass m are suspended from strings of length ℓ that are connected at a common point. One sphere has charge Q; the other has charge $2Q$. Assume the angles θ_1 and θ_2 that the strings make with the vertical are small. (a) How are θ_1 and θ_2 related? (b) Show that the distance r between the spheres is

$$r \cong \left(\frac{4k_eQ^2\ell}{mg}\right)^{1/3}$$

58. Three charges of equal magnitude q are fixed in position at the vertices of an equilateral triangle (Fig. P23.58). A fourth charge Q is free to move along the positive x axis under the influence of the forces exerted by the three fixed charges. Find a value for s for which Q is in equilibrium.

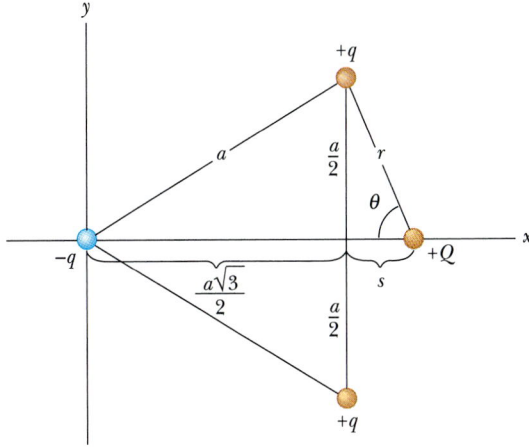

FIGURE P23.58

59. Three identical small Styrofoam balls ($m = 2.00$ g) are suspended from a fixed point by three nonconducting threads, each with a length of 50.0 cm and with negligible mass. At equilibrium the three balls form an equilateral triangle with sides of 30.0 cm. What is the common charge q carried by each ball?

60. A uniform electric field of magnitude 640 N/C exists between two parallel plates that are 4.00 cm apart. A proton is released from the positive plate at the same instant that an electron is released from the negative plate. (a) Determine the distance from the positive plate that the two pass each other. (Ignore the electrostatic attraction between the proton and electron.) (b) Repeat part (a) for a sodium ion (Na^+) and a chlorine ion (Cl^-).

61. Consider the electric dipole shown in Figure P23.61. Show that the electric field at a *distant* point along the x axis is $E_x \cong 4k_eqa/x^3$.

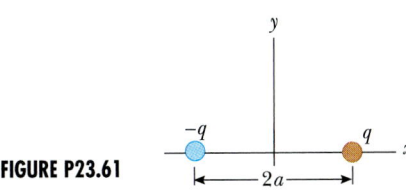

FIGURE P23.61

62. Four identical point charges each having charge $+q$ are fixed at the corners of a square of side d. A fifth point charge $-Q$ lies a distance z from the square, along the line that is perpendicular to the plane of the square and passes through its center (Fig. P23.62). (a) Show that the force exerted on $-Q$ by the other four charges is

$$\mathbf{F} = -\frac{4k_e qQz}{\left(z^2 + \dfrac{d^2}{2}\right)^{3/2}}\mathbf{k}$$

Note that this force is directed toward the center of the square whether z is positive ($-Q$ above the square) or negative ($-Q$ below the square). (b) If $z \ll d$, the above expression reduces to $\mathbf{F} \approx -(\text{const})z\,\mathbf{k}$. Why does this result imply that the motion of $-Q$ is simple harmonic, and what would be the period of this motion if the mass of $-Q$ is m?

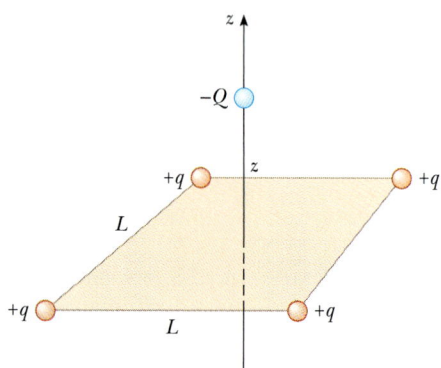

FIGURE P23.62

63. Three charges of equal magnitude q reside at the corners of an equilateral triangle of side length a (Fig. P23.63). (a) Find the magnitude and direction of the electric field at point P, midway between the negative charges, in terms of k_e, q, and a. (b) Where must a $-4q$ charge be placed so that any charge located at P will experience no net electric force? In part (b) let the distance between the $+q$ charge and P be 1.00 m.

64. Two identical beads each have a mass $m = 0.300$ kg and charge q. When placed in a spherical bowl with frictionless, nonconducting walls, the beads move until at equilibrium they are $R = 0.750$ m apart (Fig. P23.64). If the radius of the bowl is also $R = 0.750$ m, determine the charge on each bead.

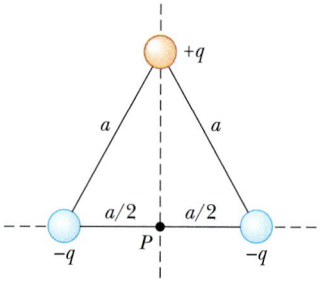

FIGURE P23.63

64A. Two identical beads each have a mass m and charge q. When placed in a hemispherical bowl of radius R with frictionless, nonconducting walls, the beads move until at equilibrium they are a distance R apart (Fig. P23.64). Determine the charge on each bead.

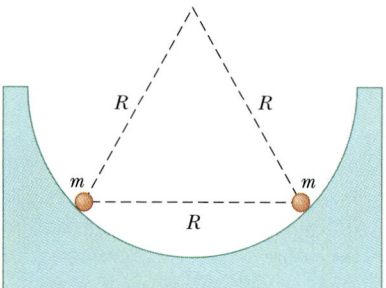

FIGURE P23.64

65. A 1.00-g cork ball carrying a charge of 2.00 μC is suspended vertically on a light string in a uniform, downward-directed electric field of magnitude $E = 1.00 \times 10^5$ N/C. If the ball is displaced slightly from the vertical, it oscillates like a simple pendulum. (a) Determine the period of this oscillation if the string is 0.500 m long. (b) Should gravity be included in the calculation for part (a)? Explain.

65A. A cork ball having mass m and charge q is suspended vertically on a light string of length L in a uniform, downward-directed electric field of magnitude E. If the ball is displaced slightly from the vertical, it oscillates like a simple pendulum. (a) Determine the period of this oscillation. (b) Should gravity be included in the calculation for part (a)? Explain.

66. A line of positive charge is formed into a semicircle of radius $R = 60.0$ cm as shown in Figure P23.66. The charge per unit length along the semicircle is described by the expression $\lambda = \lambda_0 \cos\theta$. The total charge on the semicircle is 12.0 μC. Calculate the total force on a charge of 3.00 μC placed at the center of curvature.

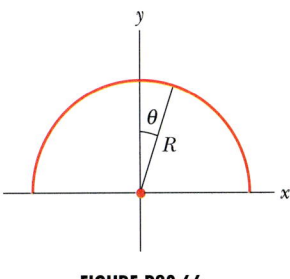

FIGURE P23.66

67. Air breaks down (loses its insulating quality) and sparking results if the electric field strength gets above 3.0×10^6 N/C. What acceleration does an electron experience in such a field? If the electron starts from rest, in what distance does it acquire a speed equal to 10% of the speed of light?

68. A line charge of length ℓ and oriented along the x axis as in Figure 23.16 has a charge per unit length λ, which varies with x as $\lambda = \lambda_0 (x - d)/d$, where d is the distance of the line from the origin (point P in the figure) and λ_0 is a constant. Find the electric field at the origin. (*Hint:* An infinitesimal element has a charge $dq = \lambda \, dx$, but note that λ is not constant.)

69. A thin rod of length ℓ and uniform charge per unit length λ lies along the x axis as shown in Figure P23.69. (a) Show that the electric field at P, a distance y from the rod, along the perpendicular bisector has no x component and is given by $E = 2k_e\lambda \sin \theta_0/y$. (b) Using your result to part (a), show that the field of a rod of infinite length is $E = 2k_e\lambda/y$. (*Hint:* First calculate the field at P due to an element of length dx, which has a charge $\lambda \, dx$. Then change variables from x to θ using the facts that $x = y \tan \theta$ and $dx = y \sec^2 \theta \, d\theta$ and integrate over θ.)

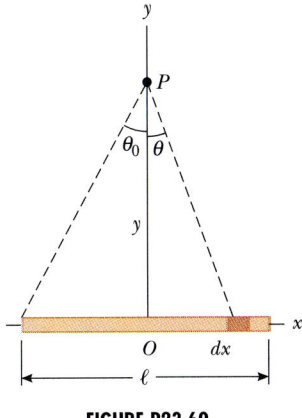

FIGURE P23.69

70. A charge q of mass M is free to move along the x axis. It is in equilibrium at the origin, midway between a pair of identical point charges, q, located on the x

axis at $x = +a$ and $x = -a$. The charge at the origin is displaced a small distance $x \ll a$ and released. Show that it can undergo simple harmonic motion with an angular frequency

$$\omega = \left(\frac{4k_e q^2}{Ma^3} \right)^{1/2}$$

71. Eight point charges, each of magnitude q, are located on the corners of a cube of side s, as in Figure P23.71. (a) Determine the x, y, and z components of the resultant force exerted on the charge located at point A by the other charges. (b) What are the magnitude and direction of this resultant force?

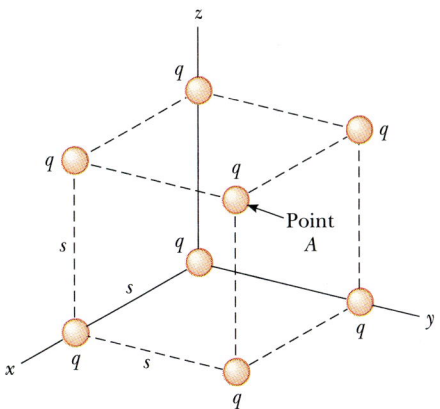

FIGURE P23.71

72. Consider the charge distribution shown in Figure P23.71. (a) Show that the magnitude of the electric field at the center of any face of the cube has a value of $2.18k_e q/s^2$. (b) What is the direction of the electric field at the center of the top face of the cube?

73. Three point charges q, $-2q$, and q are located along the x axis as in Figure P23.73. Show that the electric field at P ($y \gg a$) along the y axis is

$$\mathbf{E} = -k_e \frac{3qa^2}{y^4} \mathbf{j}$$

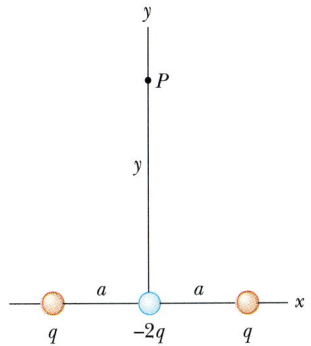

FIGURE P23.73

This charge distribution, which is essentially that of two electric dipoles, is called an *electric quadrupole*. Note that **E** varies as r^{-4} for the quadrupole, compared with variations of r^{-3} for the dipole and r^{-2} for the monopole (a single charge).

74. An electric dipole in a uniform electric field is displaced slightly from its equilibrium position, as in Figure P23.74, where θ is small. The moment of inertia of the dipole is I. If the dipole is released from this position, show that it exhibits simple harmonic motion with a frequency

$$f = \frac{1}{2\pi}\sqrt{\frac{2qaE}{I}}$$

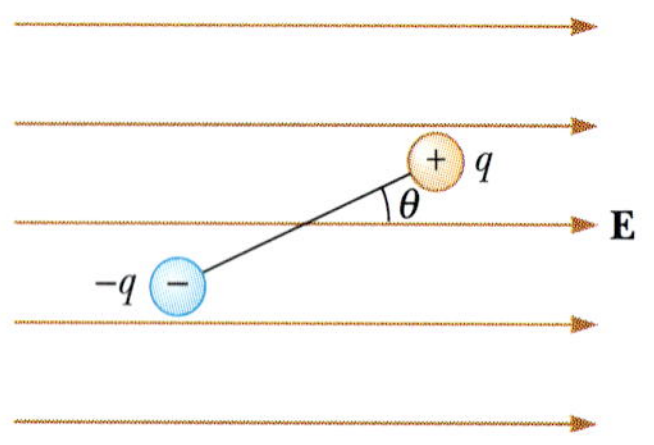

FIGURE P23.74

75. A negatively charged particle $-q$ is placed at the center of a uniformly charged ring, where the ring has a total positive charge Q as in Example 23.11. The particle, confined to move along the x axis, is displaced a *small* distance x along the axis (where $x \ll a$) and released. Show that the particle oscillates with simple harmonic motion along the x axis with a frequency given by

$$f = \frac{1}{2\pi}\left(\frac{k_e qQ}{ma^3}\right)^{1/2}$$

SPREADSHEET PROBLEMS

S1. Spreadsheet 23.1 calculates the x and y components of the electric field **E** along the x axis due to any two point charges Q_1 and Q_2. Enter the values of the charges and their x and y coordinates. Choose $Q_1 = 4\ \mu C$ at the origin, and set $Q_2 = 0$. Plot the electric field along the x axis. *Note:* You will have to choose a step size and adjust the scaling for the graph because the electric field at the position of the point charge is infinite.

S2. (a) Using Spreadsheet 23.1, place $Q_1 = 4\ \mu C$ at the origin and $Q_2 = 4\ \mu C$ at $x = 0.08$ m, $y = 0$. Plot the components of the electric field **E** as a function of x. (b) Change Q_2 to $-4\ \mu C$, and plot the components of the electric field as a function of x. (This is a dipole.) You will need to adjust the scales of your graphs.

S3. Using Spreadsheet 23.1, take $Q_1 = 6$ nC at $x = 0$, $y = 0.03$ m and $Q_2 = 6$ nC at $x = 0$, $y = -0.03$ m. Plot the components of the electric field as a function of x. (b) Change Q_2 to -6 nC and repeat part (a).

S4. The electric field on the perpendicular bisector of a uniformly charged wire of length L is

$$E_y = \frac{2\ k_e \lambda}{y}\sin\theta_0$$

(Problem 69), where

$$\sin\theta_0 = \frac{L/2}{\sqrt{(L/2)^2 + x^2}},$$

y is the distance from the wire, and λ is the charge per unit length. At sufficiently large distances from the wire, the field looks like that of a point charge. At very close distances, the field approaches that of an infinite line charge. Spreadsheet 23.2 calculates the fields at a distance y from (1) a point charge Q, (2) a finite wire of length L and total charge Q, and (3) an infinite charged wire with a linear charge density the same as that of the finite wire. Plot the three fields as a function of y. Choose $Q = 3\ \mu C$ and $L = 0.05$ m. (a) At what distance from the finite wire is its electric field within 2 percent of that due to an infinite wire? (b) At what distance from the finite wire is its electric field within 2 percent of that due to a point charge? (c) Change Q to $10\ \mu C$ and repeat parts (a) and (b). Do your answers to the questions change? Why? (d) Change L to 0.10 m, keeping $Q = 3\ \mu C$. Repeat parts (a) and (b). Do your answers change? (e) Spreadsheet 23.2 also includes a column giving the values of y/L. Answer the questions in parts (a) and (b) in terms of y/L for each case considered. Do your answers change? Why?

S5. Consider an electric dipole in which a charge $+q$ is on the x axis at $x = a$ and a charge $-q$ is on the x axis at $x = -a$. Take $q = 6$ nC and $a = 0.02$ m. Use Spreadsheet 23.1 to plot the components of the electric field along the x axis for values of x greater than 4 cm. Modify Spreadsheet 23.1 to add a plot of $2k_e p/x^3$ (the E field in the dipole approximation $x \gg a$), where $p = 2aq$ is the dipole moment of the charge distribution. For what range of values of x is the dipole approximation within 20 percent of the actual value? Within 5 percent?

S6. A quadrupole consists of two dipoles placed next to each other as in Figure PS23.6. The effective charge at the origin is $-2q$, and the other charges on the y axis at $y = a$ and $y = -a$ are each $+q$. Modify Spreadsheet 23.1 to calculate and plot the components of the electric field along the x axis, taking $q = 1\ \mu C$ and $a = 1.5$ cm.

S7. The electric field on the symmetry axis of a uniformly charged disk with surface charge density σ and radius

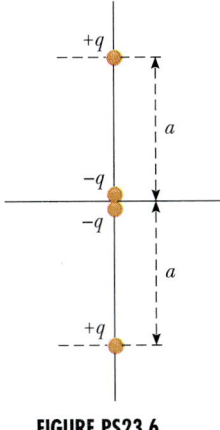

FIGURE PS23.6

R at a distance x from the disk is derived in Example 23.12:

$$E_{disk} = 2\pi k_e \sigma \left(1 - \frac{x}{\sqrt{x^2 + R^2}} \right)$$

For points sufficiently close to the disk, the field approaches that of an infinite sheet of charge; $E_{sheet} = 2\pi k_e \sigma$. For points sufficiently far away, the field approaches that of a point charge $Q = \sigma \pi R^2$. To examine these limits, note that the equation for E_{disk} can

be written in the dimensionless form

$$\frac{E_{disk}}{E_{sheet}} = 1 - \frac{x/R}{\sqrt{1 + (x/R)^2}}$$

The ratio E_{disk}/E_{sheet} approaches 1 as x/R approaches 0. The field of the point charge Q defined above can be written as

$$E_{point} = \frac{k_e Q}{x^2} = \frac{E_{sheet}}{2(x/R)^2}$$

So,

$$\frac{E_{disk}}{E_{point}} = 2(x/R)^2 \frac{E_{disk}}{E_{sheet}}$$

$$= 2(x/R)^2 \left(1 - \frac{x/R}{\sqrt{1 + (x/R)^2}} \right)$$

This ratio E_{disk}/E_{point} approaches 1 as x/R gets large. (a) Using a spreadsheet, plot the quantity E_{disk}/E_{sheet} as a function of x/R. From your graph, for what values of x/R can you approximate the field of a disk by that of a sheet of charge to within 99%? (That is, $E_{disk}/E_{sheet} \le 0.99$.) 90%? 50%? (b) Plot E_{disk}/E_{point} as a function of x/R. From your graph, for what values of x/R is the field of a disk approximated by that of a point charge to within 99%? 90%? 50%? (*Hint:* Choose the scales for each graph in such a way as to express your results clearly.)

Gauss's Law

This beautiful closed surface generated by a computer is one example of a mathematical construct in hyperbolic space. In this chapter, we use simpler mathematical constructs to calculate the electric fields due to charge distributions that have spherical, cylindrical, or planar symmetry. *(Courtesy of Wolfram Research, Inc.)*

I n the preceding chapter we showed how to calculate the electric field generated by a given charge distribution from Coulomb's law. In this chapter we describe an alternative procedure for calculating electric fields. This procedure is known as *Gauss's law*. This formulation is based on the fact that the fundamental electrostatic force between point charges is an inverse-square law. Although Gauss's law is a consequence of Coulomb's law, Gauss's law is much more convenient for calculating the electric field of highly symmetric charge distributions. Furthermore, Gauss's law serves as a guide for understanding more complicated problems.

24.1   ELECTRIC FLUX

The concept of electric field lines is described qualitatively in the previous chapter. We now use the concept of electric flux to put this idea on a quantitative basis. *Electric flux is represented by the number of electric field lines penetrating some surface.* When the surface being penetrated encloses some net charge, the net number of lines that go through the surface is proportional to the net charge within the surface. The number of lines counted is independent of the shape of the surface enclosing the charge. This is essentially a statement of Gauss's law, which we describe in the next section.

First consider an electric field that is uniform in both magnitude and direction, as in Figure 24.1. The electric field lines penetrate a rectangular surface of area A, which is perpendicular to the field. Recall that the number of lines per unit area is proportional to the magnitude of the electric field. Therefore, the number of lines penetrating the surface is proportional to the product EA. The product of the electric field strength E and a surface area A perpendicular to the field is called the **electric flux, Φ:**

$$\Phi = EA \qquad (24.1)$$

From the SI units of E and A, we see that electric flux has the units of $\mathrm{N \cdot m^2/C}$.

If the surface under consideration is not perpendicular to the field, the number of lines (in other words, the flux) through it must be less than that given by Equation 24.1. This can be understood by considering Figure 24.2, where the normal to the surface of area A is at an angle θ to the uniform electric field. Note that the number of lines that cross this area is equal to the number that cross the projected area A', which is perpendicular to the field. From Figure 24.2 we see that the two areas are related by $A' = A \cos \theta$. Since the flux through the area A equals the flux through A', we conclude that the flux through A is

$$\Phi = EA \cos \theta \qquad (24.2)$$

From this result, we see that the flux through a surface of fixed area has the maximum value, EA, when the surface is perpendicular to the field (in other words, when the normal to the surface is parallel to the field, that is, $\theta = 0°$); the flux is zero when the surface is parallel to the field (when the normal to the surface is perpendicular to the field, that is, $\theta = 90°$).

In more general situations, the electric field may vary over the surface in question. Therefore, our definition of flux given by Equation 24.2 has meaning only over a small element of area. Consider a general surface divided up into a large number of small elements, each of area ΔA. The variation in the electric field over the element can be neglected if the element is small enough. It is convenient to define a vector $\Delta \mathbf{A}_i$ whose magnitude represents the area of the ith element and whose direction is *defined to be perpendicular* to the surface, as in Figure 24.3. The electric flux $\Delta \Phi_i$ through this small element is

$$\Delta \Phi_i = E_i \, \Delta A_i \cos \theta = \mathbf{E}_i \cdot \Delta \mathbf{A}_i$$

where we have used the definition of the scalar product of two vectors ($\mathbf{A} \cdot \mathbf{B} = AB \cos \theta$). By summing the contributions of all elements, we obtain the total flux

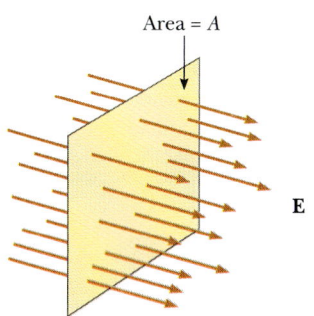

FIGURE 24.1   Field lines of a uniform electric field penetrating a plane of area A perpendicular to the field. The electric flux, Φ, through this area is equal to EA.

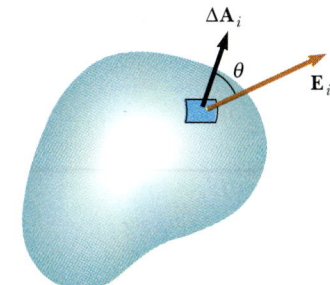

FIGURE 24.3   A small element of a surface of area ΔA_i. The electric field makes an angle θ with the normal to the surface (the direction of ΔA_i), and the flux through the element is equal to $E_i \, \Delta A_i \cos \theta$.

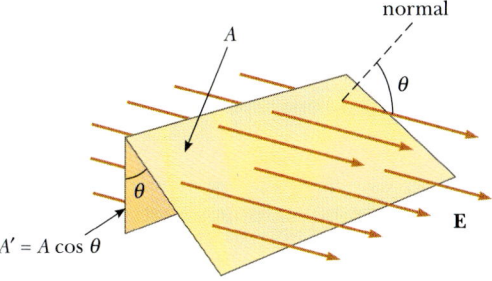

FIGURE 24.2   Field lines for a uniform electric field through an area A that is at an angle θ to the field. Since the number of lines that go through the shaded area A' is the same as the number that go through A, we conclude that the flux through A' is equal to the flux through A and is given by $\Phi = EA \cos \theta$.

through the surface.[1] If we let the area of each element approach zero, then the number of elements approaches infinity and the sum is replaced by an integral. Therefore *the general definition of electric flux is*

$$\Phi \equiv \lim_{\Delta A_i \to 0} \sum \mathbf{E}_i \cdot \Delta \mathbf{A}_i = \int_{\text{surface}} \mathbf{E} \cdot d\mathbf{A} \tag{24.3}$$

Equation 24.3 is a surface integral, which must be evaluated over the hypothetical surface in question. In general, the value of Φ depends both on the field pattern and on the specified surface.

We are usually interested in evaluating the flux through a *closed surface*. (A **closed surface** is defined as one that divides space into an inside and an outside region, so that one cannot move from one region to the other without crossing the surface. The surface of a sphere, for example, is a closed surface.) Consider the closed surface in Figure 24.4. Note that the vectors $\Delta \mathbf{A}_i$ point in different directions for the various surface elements. At each point, these vectors are normal to the surface and, by convention, always point outward. At the elements labeled ① and ②, **E** is outward and $\theta < 90°$; hence, the flux $\Delta \Phi = \mathbf{E} \cdot \Delta \mathbf{A}$ through these elements is positive. For elements such as ③, where the field lines are directed into the surface, $\theta > 90°$ and the flux becomes negative because $\cos \theta$ is negative. The net flux through the surface is proportional to the net number of lines leaving the surface (where the net number means *the number leaving the surface minus the number entering the surface*). If there are more lines leaving than entering, the net flux is positive. If more lines enter than leave, the net flux is negative. Using the symbol $\oint$ to represent an *integral over a closed surface*, we can write the net flux, Φ_c, through a closed surface

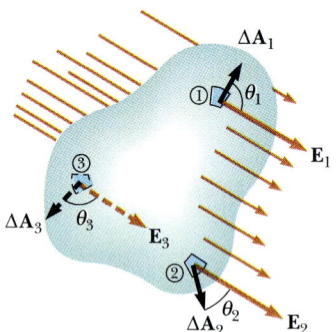

FIGURE 24.4 A closed surface in an electric field. The area vectors $\Delta \mathbf{A}_i$ are, by convention, normal to the surface and point outward. The flux through an area element can be positive (elements ① and ②) or negative (element ③).

$$\Phi_c = \oint \mathbf{E} \cdot d\mathbf{A} = \oint E_n \, dA \tag{24.4}$$

where E_n represents the component of the electric field normal to the surface and the subscript c denotes a closed surface. Evaluating the net flux through a closed surface could be very cumbersome. However, if the field is normal to the surface at each point and constant in magnitude, the calculation is straightforward. The following example illustrates this point.

EXAMPLE 24.1   Flux Through a Cube

Consider a uniform electric field **E** oriented in the x direction. Find the net electric flux through the surface of a cube of edges ℓ oriented as shown in Figure 24.5.

Solution   The net flux can be evaluated by summing up the fluxes through each face of the cube. First, note that the flux through four of the faces is zero, because **E** is perpendicular to $d\mathbf{A}$ on these faces. In particular, the orientation of $d\mathbf{A}$ is

perpendicular to **E** for the faces labeled ③ and ④ in Figure 24.5. Therefore, $\theta = 90°$, so that $\mathbf{E} \cdot d\mathbf{A} = E \, dA \cos 90° = 0$. The flux through each of the planes parallel to the yx plane is also zero for the same reason.

Now consider the faces labeled ① and ②. The net flux through these faces is

$$\Phi_c = \int_1 \mathbf{E} \cdot d\mathbf{A} + \int_2 \mathbf{E} \cdot d\mathbf{A}$$

---

[1] It is important to note that drawings with field lines have their inaccuracies, since a small area (depending on its location) may happen to have too many or too few penetrating lines, and the density of lines in three-dimensional space is not well represented in a two-dimensional cross-section. At any rate, it is stressed that the basic definition of electric flux is $\int \mathbf{E} \cdot d\mathbf{A}$. The use of lines is only an aid for visualizing the concept.

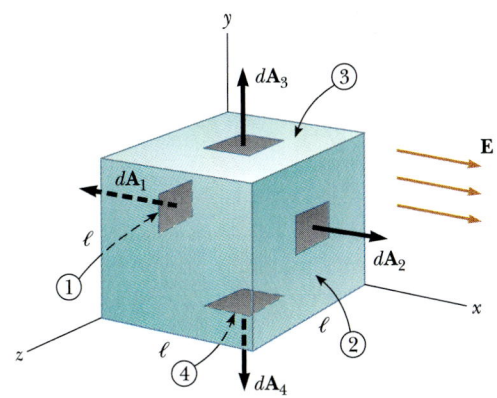

FIGURE 24.5 (Example 24.1) A hypothetical surface in the shape of a cube in a uniform electric field parallel to the x axis. The net flux through the surface is zero.

For face ①, **E** is constant and inward while $d\mathbf{A}$ is outward ($\theta = 180°$), so that we find that the flux through this face is

$$\int_1 \mathbf{E} \cdot d\mathbf{A} = \int_1 E \cdot dA \cos 180° = -E \int_1 dA = -EA = -E\ell^2$$

since the area of each face is $A = \ell^2$.

Likewise, for ②, **E** is constant and outward and in the same direction as $d\mathbf{A}$ ($\theta = 0°$), so that the flux through this face is

$$\int_2 \mathbf{E} \cdot d\mathbf{A} = \int_2 E \cdot dA \cos 0° = E \int_2 dA = +EA = E\ell^2$$

Hence, the net flux over all faces is zero, because

$$\Phi_c = -E\ell^2 + E\ell^2 = 0$$

24.2 GAUSS'S LAW

In this section we describe a general relationship between the net electric flux through a closed surface (often called a *gaussian surface*) and the charge enclosed by the surface. This relationship, known as *Gauss's law,* is of fundamental importance in the study of electric fields.

First, let us consider a positive point charge q located at the center of a sphere of radius r as in Figure 24.6. From Coulomb's law we know that the magnitude of the electric field everywhere on the surface of the sphere is $E = k_e q/r^2$. Furthermore, the field lines are radial outward and hence are perpendicular to the surface at each point. That is, at each point, **E** is parallel to the vector $\Delta\mathbf{A}_i$ representing the local element of area ΔA_i. Therefore,

$$\mathbf{E} \cdot \Delta\mathbf{A}_i = E_n \, \Delta A_i = E \, \Delta A_i$$

and from Equation 24.4 we find that the net flux through the gaussian surface is

$$\Phi_c = \oint E_n \, dA = \oint E \, dA = E \oint dA$$

Karl Friedrich Gauss (1777–1855).

since by symmetry E is constant over the surface and given by $E = k_e q/r^2$. Furthermore, for a spherical gaussian surface, $\oint dA = A = 4\pi r^2$ (the surface area of a sphere). Hence, the net flux through the gaussian surface is

$$\Phi_c = \frac{k_e q}{r^2} (4\pi r^2) = 4\pi k_e q$$

Recalling from Section 23.3 that $k_e = 1/4\pi\epsilon_0$, we can write the above expression in the form

$$\Phi_c = \frac{q}{\epsilon_0} \tag{24.5}$$

Note that this result, which is independent of r, says that the net flux through a spherical gaussian surface is proportional to the charge q *inside* the surface. The fact that the flux is independent of the radius is a consequence of the inverse-square dependence of the electric field given by Coulomb's law. That is, E varies as

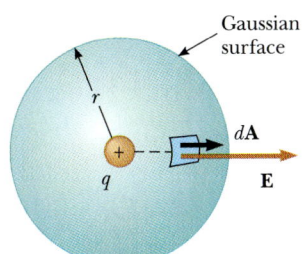

FIGURE 24.6 A spherical gaussian surface of radius r surrounding a point charge q. When the charge is at the center of the sphere, the electric field is normal to the surface and constant in magnitude everywhere on the surface.

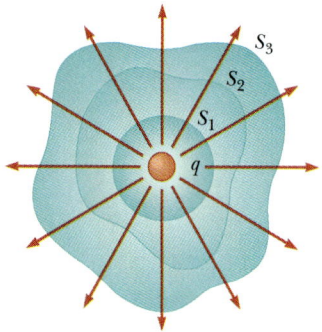

FIGURE 24.7   Closed surfaces of various shapes surrounding a charge q. Note that the net electric flux through each surface is the same.

The net flux through a closed surface is zero if there is no charge inside

---

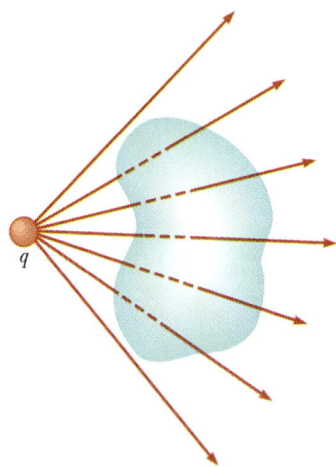

FIGURE 24.8   A point charge located *outside* a closed surface. In this case, note that the number of lines entering the surface equals the number leaving the surface.

$1/r^2$, but the area of the sphere varies as r^2. Their combined effect produces a flux that is independent of r.

Now consider several closed surfaces surrounding a charge q as in Figure 24.7. Surface S_1 is spherical, whereas surfaces S_2 and S_3 are nonspherical. The flux that passes through surface S_1 has the value q/ϵ_0. As we discussed in the previous section, the flux is proportional to the number of electric field lines passing through that surface. The construction in Figure 24.7 shows that the number of electric field lines through the spherical surface S_1 is equal to the number of electric field lines through the nonspherical surfaces S_2 and S_3. Therefore, it is reasonable to conclude that the net flux through any closed surface is independent of the shape of that surface. (One can prove that this is the case if $E \propto 1/r^2$.) In fact, *the net flux through any closed surface surrounding a point charge q is given by q/ϵ_0.*

Now consider a point charge located *outside* a closed surface of arbitrary shape, as in Figure 24.8. As you can see from this construction, some electric field lines enter the surface, and others leave the surface. However, *the number of electric field lines entering the surface equals the number leaving the surface.* Therefore, we conclude that *the net electric flux through a closed surface that surrounds no charge is zero.* If we apply this result to Example 24.1, we can easily see that the net flux through the cube is zero, since there is no charge inside the cube.

Let us extend these arguments to the generalized case of either many point charges or a continuous distribution of charge. We once again use the superposition principle, which says that *the electric field due to many charges is the vector sum of the electric fields produced by the individual charges.* That is, we can express the flux through any closed surface as

$$\oint \mathbf{E} \cdot d\mathbf{A} = \oint (\mathbf{E}_1 + \mathbf{E}_2 + \mathbf{E}_3) \cdot d\mathbf{A}$$

where $\mathbf{E}$ is the total electric field at any point on the surface and $\mathbf{E}_1$, $\mathbf{E}_2$, and $\mathbf{E}_3$ are the fields produced by the individual charges at that point. Consider the system of charges shown in Figure 24.9. The surface S surrounds only one charge, q_1; hence, the net flux through S is q_1/ϵ_0. The flux through S due to the charges outside it is zero because each electric field line that enters S at one point leaves it at another. The surface S' surrounds charges q_2 and q_3; hence, the net flux through S' is $(q_2 + q_3)/\epsilon_0$. Finally, the net flux through surface S'' is zero because there is no charge inside this surface. That is, *all* the electric field lines that enter S'' at one point leave S'' at another.

Gauss's law, which is a generalization of the above discussion, states that the net flux through *any* closed surface is

$$\Phi_c = \oint \mathbf{E} \cdot d\mathbf{A} = \frac{q_{\text{in}}}{\epsilon_0} \qquad (24.6)$$

where q_{in} represents the *net charge inside* the surface and $\mathbf{E}$ represents the electric field at any point on the surface. In words, **Gauss's law** states that

---

Gauss's law

the net electric flux through any closed surface is equal to the net charge inside the surface divided by ϵ_0.

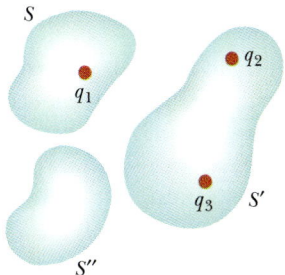

FIGURE 24.9 The net electric flux through any closed surface depends only on the charge *inside* that surface. The net flux through surface S is q_1/ϵ_0, the net flux through surface S' is $(q_2 + q_3)/\epsilon_0$, and the net flux through surface S'' is zero.

A formal proof of Gauss's law is presented in Section 24.6. When using Equation 24.6, you should note that although the charge q_{in} is the net charge inside the gaussian surface, the **E** that appears in Gauss's law represents the *total electric field,* which includes contributions from charges both inside and outside the gaussian surface.

In principle, Gauss's law can always be used to calculate the electric field of a system of charges or a continuous distribution of charge. However, in practice, *the technique is useful only in a limited number of situations where there is a high degree of symmetry.* As we see in the next section, *Gauss's law can be used to evaluate the electric field for charge distributions that have spherical, cylindrical, or planar symmetry.* If one carefully chooses the gaussian surface surrounding the charge distribution, the integral in Equation 24.6 is easy to evaluate. You should also note that a gaussian surface is a mathematical surface and need not coincide with any real physical surface.

> Gauss's law is useful for evaluating E when the charge distribution has symmetry

---

CONCEPTUAL EXAMPLE 24.2

If the net flux through a gaussian surface is zero, which of the following statements are true? (a) There are no charges inside the surface. (b) The net charge inside the surface is zero. (c) The electric field is zero everywhere on the surface. (d) The number of electric field lines entering the surface equals the number leaving the surface.

Reasoning Statements (b) and (d) are true and follow from Gauss's law. Statement (a) is not necessarily true be-

cause Gauss's law says that the net flux through any closed surface equals the net charge inside the surface divided by ϵ_0. For example, an electric dipole (whose net charge is zero) might be inside the surface. Statement (c) is not necessarily true. Although the net flux through the surface is zero, the electric field in that region may not be zero (Fig. 24.8).

CONCEPTUAL EXAMPLE 24.3

A spherical gaussian surface surrounds a point charge q. Describe what happens to the total flux through the surface if (a) the charge is tripled, (b) the volume of the sphere is doubled, (c) the surface is changed to a cube, and (d) the charge is moved to another location *inside* the surface.

Reasoning (a) If the charge is tripled, the flux through the surface is also tripled, because the net flux is proportional to

the charge inside the surface. (b) The flux remains constant when the volume changes, because the surface surrounds the same amount of charge, regardless of its volume. (c) The total flux does not change when the shape of the closed surface changes. (d) The total flux through the closed surface remains unchanged as the charge inside the surface is moved to another location inside that surface. All of these conclusions are arrived at through an understanding of Gauss's law.

24.3 APPLICATION OF GAUSS'S LAW TO CHARGED INSULATORS

Gauss's law is useful when there is a high degree of symmetry in the charge distribution, as in the case of uniformly charged spheres, long cylinders, and flat sheets. In such cases, it is possible to find a simple gaussian surface over which the surface integral given by Equation 24.6 is easily evaluated. The surface should always be chosen to take advantage of the symmetry of the charge distribution.

EXAMPLE 24.4   The Electric Field Due to a Point Charge

Starting with Gauss's law, calculate the electric field due to an isolated point charge q and show that Coulomb's law follows from this result.

Solution   For this situation we choose a spherical gaussian surface of radius r and centered on the point charge, as in Figure 24.10. The electric field of a positive point charge is radial outward by symmetry and is therefore normal to the surface at every point. That is, **E** is parallel to $d\mathbf{A}$ at each

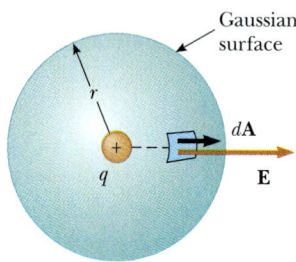

Gaussian surface

$d\mathbf{A}$

E

q

FIGURE 24.10   (Example 24.4) The point charge q is at the center of the spherical gaussian surface, and **E** is parallel to $d\mathbf{A}$ at every point on the surface.

point, and so $\mathbf{E} \cdot d\mathbf{A} = E\, dA$ and Gauss's law gives

$$\Phi_c = \oint \mathbf{E} \cdot d\mathbf{A} = \oint E\, dA = \frac{q}{\epsilon_0}$$

By symmetry, E is constant everywhere on the surface, and so it can be removed from the integral. Therefore,

$$\oint E\, dA = E \oint dA = E(4\pi r^2) = \frac{q}{\epsilon_0}$$

where we have used the fact that the surface area of a sphere is $4\pi r^2$. Hence, the magnitude of the field a distance r from q is

$$E = \frac{q}{4\pi\epsilon_0 r^2} = k_e \frac{q}{r^2}$$

If a second point charge q_0 is placed at a point where the field is E, the electric force on this charge has a magnitude

$$F = q_0 E = k_e \frac{qq_0}{r^2}$$

Previously we obtained Gauss's law from Coulomb's law. Here we show that Coulomb's law follows from Gauss's law. They are equivalent.

EXAMPLE 24.5   A Spherically Symmetric Charge Distribution

An insulating sphere of radius a has a uniform charge density ρ and a total positive charge Q (Fig. 24.11). (a) Calculate the magnitude of the electric field at a point outside the sphere.

Solution   Since the charge distribution is spherically symmetric, we again select a spherical gaussian surface of radius r, concentric with the sphere, as in Figure 24.11a. Following the line of reasoning given in Example 24.4, we find that

$$E = k_e \frac{Q}{r^2} \qquad \text{(for } r > a\text{)}$$

Note that this result is identical to that obtained for a point charge. Therefore, we conclude that, for a uniformly charged sphere, the field in the region external to the sphere

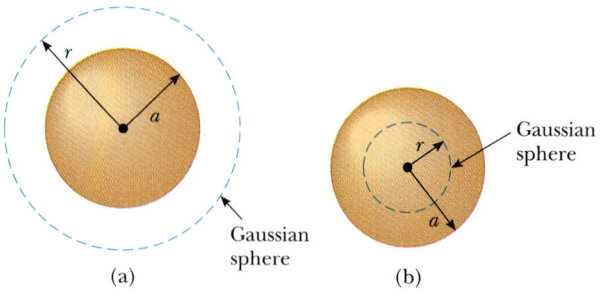

FIGURE 24.11   (Example 24.5) A uniformly charged insulating sphere of radius a and total charge Q. (a) The field at a point exterior to the sphere is $k_e Q/r^2$. (b) The field inside the sphere is due only to the charge *within* the gaussian surface and is given by $(k_e Q/a^3)r$.

is *equivalent* to that of a point charge located at the center of the sphere.

(b) Find the magnitude of the electric field at a point inside the sphere.

Reasoning and Solution   In this case we select a spherical gaussian surface with radius $r < a$, concentric with the charge distribution (Fig. 24.11b). Let us denote the volume of this smaller sphere by V'. To apply Gauss's law in this situation, it is important to recognize that the charge q_{in} within the gaussian surface of volume V' is a quantity less than the total charge Q. To calculate the charge q_{in}, we use the fact that $q_{in} = \rho V'$, where ρ is the charge per unit volume and V' is the volume enclosed by the gaussian surface, given by $V' = \frac{4}{3}\pi r^3$ for a sphere. Therefore,

$$q_{in} = \rho V' = \rho(\tfrac{4}{3}\pi r^3)$$

As in Example 24.4, the magnitude of the electric field is constant everywhere on the spherical gaussian surface and is normal to the surface at each point. Therefore, Gauss's law in the region $r < a$ gives

$$\oint E\,dA = E\oint dA = E(4\pi r^2) = \frac{q_{in}}{\epsilon_0}$$

Solving for E gives

$$E = \frac{q_{in}}{4\pi\epsilon_0 r^2} = \frac{\rho\frac{4}{3}\pi r^3}{4\pi\epsilon_0 r^2} = \frac{\rho}{3\epsilon_0}r$$

Since by definition $\rho = Q/\frac{4}{3}\pi a^3$, this can be written

$$E = \frac{Qr}{4\pi\epsilon_0 a^3} = \frac{k_e Q}{a^3}r \qquad \text{(for } r < a)$$

Note that this result for E differs from that obtained in part (a). It shows that $E \to 0$ as $r \to 0$, as you might have guessed based on the spherical symmetry of the charge distribution. Therefore, the result fortunately eliminates the singularity that would exist at $r = 0$ if E varied as $1/r^2$ inside the sphere. That is, if $E \propto 1/r^2$, the field would be infinite at $r = 0$, which is clearly a physically impossible situation. A plot of E versus r is shown in Figure 24.12.

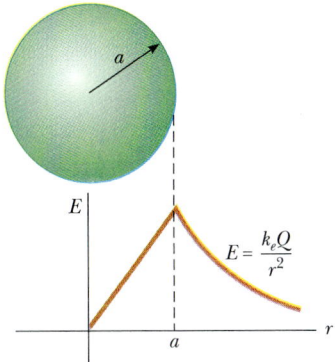

FIGURE 24.12   (Example 24.5) A plot of E versus r for a uniformly charged insulating sphere. The field inside the sphere ($r < a$) varies linearly with r. The field outside the sphere ($r > a$) is the same as that of a point charge Q located at the origin.

EXAMPLE 24.6   The Electric Field Due to a Thin Spherical Shell

A thin spherical shell of radius a has a total charge Q distributed uniformly over its surface (Fig. 24.13). Find the electric field at points inside and outside the shell.

Reasoning and Solution   The calculation of the field outside the shell is identical to that already carried out for the solid sphere in Example 24.5a. If we construct a spherical gaussian surface of radius $r > a$, concentric with the shell, then the charge inside this surface is Q. Therefore, the field at a point outside the shell is equivalent to that of a point charge Q at the center:

$$E = k_e \frac{Q}{r^2} \qquad \text{(for } r > a)$$

The electric field inside the spherical shell is zero. This also follows from Gauss's law applied to a spherical surface of

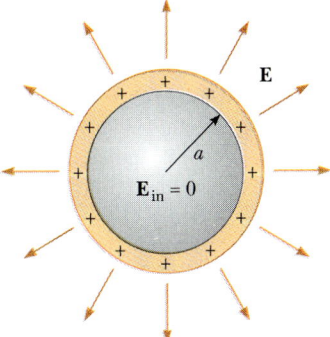

FIGURE 24.13   (Example 24.6) The electric field inside a uniformly charged spherical shell is *zero*. The field outside is the same as that of a point charge having a total charge Q located at the center of the shell.

radius $r < a$. Since the net charge inside the surface is zero, and because of the spherical symmetry of the charge distribution, application of Gauss's law shows that $E = 0$ in the region $r < a$.

The same results can be obtained using Coulomb's law and integrating over the charge distribution. This calculation is rather complicated and will be omitted.

EXAMPLE 24.7 A Cylindrically Symmetric Charge Distribution

Find the electric field a distance r from a uniform positive line charge of infinite length whose charge per unit length is λ = constant (Fig. 24.14).

Reasoning The symmetry of the charge distribution shows that E must be perpendicular to the line charge and directed outward as in Figure 24.14a. The end view of the line charge shown in Figure 24.14b should help visualize the directions of the electric field lines. In this situation, we select a cylindrical gaussian surface of radius r and length ℓ that is coaxial with the line charge. For the curved part of this surface, E is constant in magnitude and perpendicular to the surface at each point. Furthermore, the flux through the ends of the gaussian cylinder is zero because E is parallel to these surfaces.

Solution The total charge inside our gaussian surface is $\lambda\ell$. Applying Gauss's law and noting that E is parallel to dA everywhere on the cylindrical surface, we find that

$$\Phi_c = \oint \mathbf{E} \cdot d\mathbf{A} = E \oint dA = \frac{q_{in}}{\epsilon_0} = \frac{\lambda\ell}{\epsilon_0}$$

But the area of the curved surface is $A = 2\pi r\ell$; therefore,

$$E(2\pi r\ell) = \frac{\lambda\ell}{\epsilon_0}$$

$$E = \frac{\lambda}{2\pi\epsilon_0 r} = 2k_e\frac{\lambda}{r} \qquad (24.7)$$

Thus, we see that the field of a cylindrically symmetric charge distribution varies as $1/r$, whereas the field external to a spherically symmetric charge distribution varies as $1/r^2$. Equation 24.7 can also be obtained using Coulomb's law and integration; however, the mathematical techniques necessary for this calculation are more cumbersome.

If the line charge has a finite length, the result for E is not that given by Equation 24.7. For points close to the line charge and far from the ends, Equation 24.7 gives a good approximation of the value of the field. It turns out that Gauss's law is not useful for calculating E for a finite line charge. This is because the magnitude of the electric field is no longer constant over the surface of the gaussian cylinder. Furthermore, E is not perpendicular to the cylindrical sur-

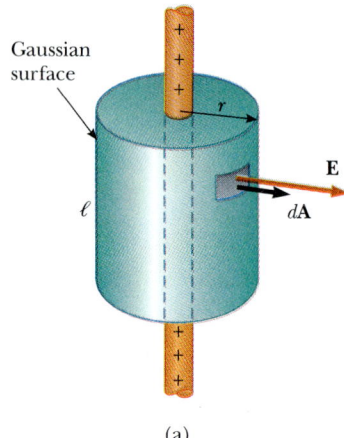

Gaussian surface

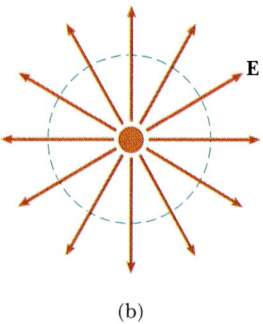

FIGURE 24.14 (Example 24.7) (a) An infinite line of charge surrounded by a cylindrical gaussian surface concentric with the line charge. (b) An end view shows that the field on the cylindrical surface is constant in magnitude and perpendicular to the surface.

face at all points. When there is little symmetry in the charge distribution, as in this situation, it is necessary to calculate E using Coulomb's law.

It is left as a problem (Problem 35) to show that the E field inside a uniformly charged rod of finite thickness is proportional to r.

EXAMPLE 24.8 A Nonconducting Plane Sheet of Charge

Find the electric field due to a nonconducting, infinite plane with uniform charge per unit area σ.

Reasoning and Solution The symmetry of the situation shows that **E** must be perpendicular to the plane and that the direction of **E** on one side of the plane must be opposite its direction on the other side, as in Figure 24.15. It is conve-

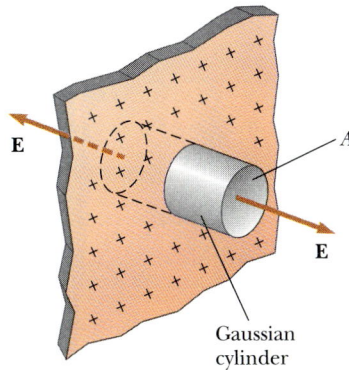

E

A

E

Gaussian
cylinder

FIGURE 24.15 (Example 24.8) A cylindrical gaussian surface penetrating an infinite sheet of charge. The flux through each end of the gaussian surface is *EA*. There is no flux through the cylinder's curved surface.

nient to choose for our gaussian surface a small cylinder whose axis is perpendicular to the plane and whose ends each have an area *A* and are equidistant from the plane. Here we see that since **E** is parallel to the cylindrical surface, there is no flux through this surface. The flux out of each end of the cylinder is *EA* (since **E** is perpendicular to the ends); hence, the total flux through our gaussian surface is 2*EA*.

Solution Noting that the total charge inside the surface is σA, we use Gauss's law to get

$$\Phi_c = 2EA = \frac{q_{in}}{\epsilon_0} = \frac{\sigma A}{\epsilon_0}$$

$$E = \frac{\sigma}{2\epsilon_0} \qquad (24.8)$$

Since the distance of the surfaces from the plane does not appear in Equation 24.8, we conclude that $E = \sigma/2\epsilon_0$ at any distance from the plane. That is, the field is uniform everywhere.

An important configuration related to this example is the case of two parallel planes of charge, with charge densities σ and $-\sigma$, respectively (Problem 58). In this situation, the electric field is σ/ϵ_0 between the planes and approximately zero elsewhere.

CONCEPTUAL EXAMPLE 24.9

Explain why Gauss's law cannot be used to calculate the electric field of (a) an electric dipole, (b) a charged disk, and (c) three point charges at the corners of a triangle.

Reasoning The electric field patterns of each of these three configurations do not have sufficient symmetry to make the calculations practical. (Gauss's law is only useful for

calculating the electric field of highly symmetric charge distributions, such as uniformly charged spheres, cylinders, and sheets.) In order to apply Gauss's law, you must be able to find a closed surface surrounding the charge distribution, which can be subdivided so that the field over the separate regions of the surface is constant. Such a surface cannot be found for these cases.

24.4 CONDUCTORS IN ELECTROSTATIC EQUILIBRIUM

As we learned in Section 23.2, a good electrical conductor, such as copper, contains charges (electrons) that are not bound to any atom and are free to move about within the material. When there is no net motion of charge within the conductor, the conductor is in **electrostatic equilibrium.** As we shall see, a conductor in electrostatic equilibrium has the following properties:

- The electric field is zero everywhere inside the conductor.
- Any charge on an isolated conductor resides on its surface.
- The electric field just outside a charged conductor is perpendicular to the

*Properties of a conductor in
electrostatic equilibrium*

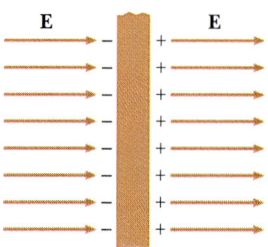

FIGURE 24.16 A conducting slab in an external electric field **E**. The charges induced on the surfaces of the slab produce an electric field that opposes the external field, giving a resultant field of zero inside the conductor.

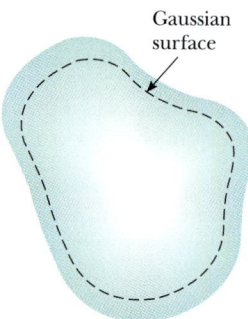

FIGURE 24.17 An insulated conductor of arbitrary shape. The broken line represents a gaussian surface just inside the conductor.

surface of the conductor and has a magnitude σ/ϵ_0, where σ is the charge per unit area at that point.

- On an irregularly shaped conductor, charge tends to accumulate at locations where the radius of curvature of the surface is the smallest, that is, at sharp points.

The first property can be understood by considering a conducting slab placed in an external field **E** (Fig. 24.16). In electrostatic equilibrium, the electric field inside the conductor must be zero. If this were not the case, the free charges would accelerate under the action of the field. Before the external field is applied, the electrons are uniformly distributed throughout the conductor. When the external field is applied, the free electrons accelerate to the left, causing a buildup of negative charge on the left surface (excess electrons) and of positive charge on the right (where electrons have been removed). These charges create their own internal electric field, which opposes the external field. The surface charge density increases until the magnitude of the internal electric field equals that of the external field, giving a net field of zero inside the conductor. In a good conductor, the time it takes the conductor to reach equilibrium is of the order of 10^{-16} s, which for most purposes can be considered instantaneous.

We can use Gauss's law to verify the second and third properties of a conductor in electrostatic equilibrium. Figure 24.17 shows an arbitrarily shaped insulated conductor. A gaussian surface is drawn inside the conductor and can be as close to the surface as we wish. As we have just shown, the electric field everywhere inside the conductor is zero when it is in electrostatic equilibrium. Therefore, the electric field must also be zero at every point on the gaussian surface, so that the net flux through this surface is zero. From this result and Gauss's law, we conclude that the net charge inside the gaussian surface is zero. Since there can be no net charge inside the gaussian surface (which is arbitrarily close to the conductor's surface), *any net charge on the conductor must reside on its surface.* Gauss's law does not tell us how this excess charge is distributed on the surface. In Section 25.6 we prove the fourth property of a conductor in electrostatic equilibrium.

We can use Gauss's law to relate the electric field just outside the surface of a charged conductor in equilibrium to the charge distribution on the conductor. To do this, it is convenient to draw a gaussian surface in the shape of a small cylinder

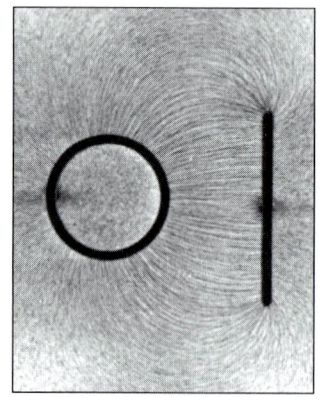

Electric field pattern of a charged conducting plate near an oppositely charged conducting cylinder. Small pieces of thread suspended in oil align with the electric field lines. Note that (1) the electric field lines are perpendicular to the conductors and (2) there are no lines inside the cylinder $(E = 0)$. *(Courtesy of Harold M. Waage, Princeton University)*

with end faces parallel to the surface (Fig. 24.18). Part of the cylinder is just outside the conductor, and part is inside. There is no flux through the face on the inside of the cylinder because $E = 0$ inside the conductor. Furthermore, the field is normal to the surface. If **E** had a tangential component, the free charges would move along the surface creating surface currents, and the conductor would not be in equilibrium. There is no flux through the cylindrical part of the gaussian surface because **E** is tangent to this part. Hence, the net flux through the gaussian surface is $E_n A$, where E_n is the electric field just outside the conductor. Applying Gauss's law to this surface gives

$$\Phi_c = \oint E_n \, dA = E_n A = \frac{q_{in}}{\epsilon_0} = \frac{\sigma A}{\epsilon_0}$$

We have used the fact that the charge inside the gaussian surface is $q_{in} = \sigma A$, where A is the area of the cylinder's face and σ is the (local) charge per unit area. Solving for E_n gives

$$E_n = \frac{\sigma}{\epsilon_0} \qquad (24.9)$$

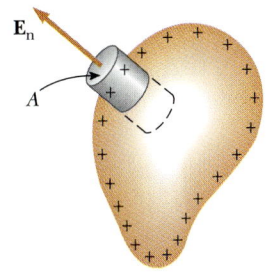

FIGURE 24.18   A gaussian surface in the shape of a small cylinder is used to calculate the electric field just outside a charged conductor. The flux through the gaussian surface is $E_n A$. Note that **E** is zero inside the conductor.

EXAMPLE 24.10   A Sphere Inside a Spherical Shell

A solid conducting sphere of radius a has a net positive charge $2Q$ (Fig. 24.19). A conducting spherical shell of inner radius b and outer radius c is concentric with the solid sphere and has a *net* charge $-Q$. Using Gauss's law, find the electric field in the regions labeled ①, ②, ③, and ④ and the charge distribution on the spherical shell.

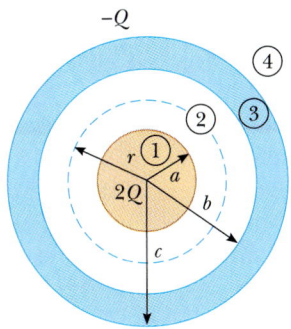

FIGURE 24.19   (Example 24.10) A solid conducting sphere of radius a and charge $2Q$ surrounded by a conducting spherical shell of charge $-Q$.

Reasoning and Solution   First note that the charge distribution on both spheres has spherical symmetry, since they are concentric. To determine the electric field at various distances r from the center, we construct spherical gaussian surfaces of radius r.

To find E inside the solid sphere of radius a (region ①), consider a gaussian surface of radius $r < a$. Since there can be no charge inside a conductor in electrostatic equilibrium, we see that $q_{in} = 0$, and so from Gauss's law and symmetry,

$E_1 = 0$ for $r < a$. Thus, we conclude that the net charge $2Q$ on the solid sphere is distributed on its outer surface.

In region ② between the spheres, where $a < r < b$, we construct a spherical gaussian surface of radius r and note that the charge inside this surface is $+2Q$ (the charge on the inner sphere). Because of the spherical symmetry, the electric field lines must be radial outward and constant in magnitude on the gaussian surface. Following Example 24.4 and using Gauss's law, we find that

$$E_2 A = E_2(4\pi r^2) = \frac{q_{in}}{\epsilon_0} = \frac{2Q}{\epsilon_0}$$

$$E_2 = \frac{2Q}{4\pi\epsilon_0 r^2} = \frac{2k_e Q}{r^2} \qquad \text{(for } a < r < b\text{)}$$

In region ④, where $r > c$, the spherical gaussian surface surrounds a total charge of $q_{in} = 2Q + (-Q) = Q$. Therefore, Gauss's law applied to this surface gives

$$E_4 = \frac{k_e Q}{r^2} \qquad \text{(for } r > c\text{)}$$

Finally, consider region ③, where $b < r < c$. The electric field must be zero in this region because the spherical shell is also a conductor in equilibrium. If we construct a gaussian surface of this radius, we see that q_{in} must be zero since $E_3 = 0$. From this argument, we conclude that the charge on the inner surface of the spherical shell must be $-2Q$ to cancel the charge $+2Q$ on the solid sphere. (The charge $-2Q$ is induced by the charge $+2Q$.) Furthermore, since the net charge on the shell is $-Q$, we conclude that the outer surface of the shell must have a charge equal to $+Q$.

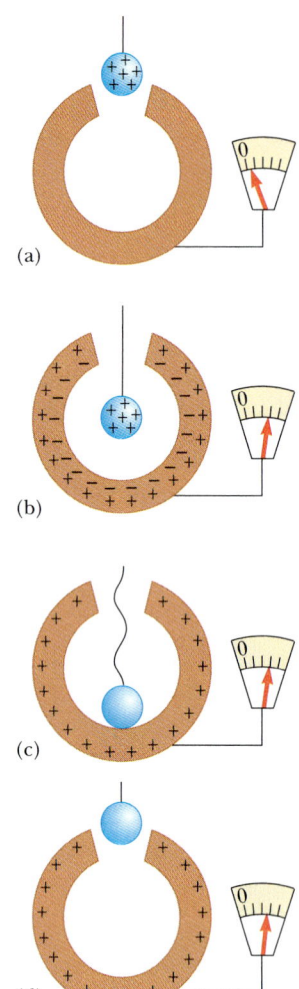

(a)

(b)

(c)

(d)

FIGURE 24.20 An experiment showing that any charge transferred to a conductor resides on its surface in electrostatic equilibrium. The hollow conductor is insulated from ground, and the small metal ball is supported by an insulating thread.

*24.5 EXPERIMENTAL PROOF OF GAUSS'S LAW AND COULOMB'S LAW

When a net charge is placed on a conductor, the charge distributes itself on the surface in such a way that the electric field inside is zero. Since $E = 0$ inside a conductor in electrostatic equilibrium, Gauss's law shows that there can be no net charge inside the conductor. We have seen that Gauss's law is a consequence of Coulomb's law (Example 24.4). Hence, it is possible to test the validity of the inverse-square law of force by attempting to detect a net charge inside a conductor. If a net charge is detected anywhere but on the conductor's surface, Gauss's law, and hence Coulomb's law, are invalid. Many experiments, including early work by Faraday, Cavendish, and Maxwell, have been performed to show that the net charge on a conductor resides on its surface. In all reported cases, no electric field could be detected in a closed conductor. Experiments by Williams, Faller, and Hill in 1971 showed that the exponent of r in Coulomb's law is $(2 + \delta)$, where $\delta = (2.7 \pm 3.1) \times 10^{-16}$!

The following experiment can be performed to verify that the net charge on a conductor resides on its surface. A positively charged metal ball at the end of a silk thread is lowered into an uncharged, hollow conductor through a small opening[2] (Fig. 24.20a). The hollow conductor is insulated from ground. The positively charged ball induces a negative charge on the inner wall of the hollow conductor, leaving an equal positive charge on the outer wall (Fig. 24.20b). The presence of positive charge on the outer wall is indicated by the deflection of an electrometer (a device used to measure charge). The deflection of the electrometer remains unchanged when the ball touches the inner surface of the hollow conductor (Fig. 24.20c). When the ball is removed, the electrometer reading remains the same and the ball is found to be uncharged (Fig. 24.20d). This experiment shows that charge is transferred from the ball to the hollow conductor. Furthermore, *the charge on the hollow conductor resides on its outer surface.* A small charged metal ball now lowered into the center of the charged hollow conductor is not attracted or repelled by the hollow conductor. This shows that $\mathbf{E} = 0$ at the center of the hollow conductor. A small positively charged ball placed near the outside of the conductor is repelled by the conductor, showing that $\mathbf{E} \neq 0$ outside the conductor.

*24.6 DERIVATION OF GAUSS'S LAW

One way of deriving Gauss's law involves the concept of the *solid angle.* Consider a spherical surface of radius r containing an area element ΔA. The solid angle $\Delta \Omega$ subtended by this element at the center of the sphere is defined to be

$$\Delta \Omega \equiv \frac{\Delta A}{r^2}$$

From this expression, we see that $\Delta \Omega$ has no dimensions, since ΔA and r^2 both have the dimension of L^2. The dimensionless unit of a solid angle is the **steradian.** Since the total surface area of a sphere is $4\pi r^2$, the total solid angle subtended by the sphere at the center is

$$\Omega = \frac{4\pi r^2}{r^2} = 4\pi \text{ steradians}$$

---

[2] The experiment is often referred to as *Faraday's ice-pail experiment,* since it was first performed by Faraday using an ice pail for the hollow conductor.

Now consider a point charge q surrounded by a closed surface of arbitrary shape (Fig. 24.21). The total flux through this surface can be obtained by evaluating $\mathbf{E} \cdot \Delta\mathbf{A}$ for each element of area and summing over all elements of the surface. The flux through the element of area ΔA is

$$\Delta\Phi = \mathbf{E} \cdot \Delta\mathbf{A} = E \cos\theta \, \Delta A = k_e q \frac{\Delta A \cos\theta}{r^2}$$

where we have used the fact that $E = k_e Q/r^2$ for a point charge. But the quantity $\Delta A \cos\theta / r^2$ is equal to the solid angle $\Delta\Omega$ subtended at the charge q by the surface element ΔA. From Figure 24.22 we see that $\Delta\Omega$ is equal to the solid angle subtended by the element of a spherical surface of radius r. Since the total solid angle at a point is 4π steradians, we see that the total flux through the closed surface is

$$\Phi_c = k_e q \oint \frac{dA \cos\theta}{r^2} = k_e q \oint d\Omega = 4\pi k_e q = \frac{q}{\epsilon_0}$$

Thus we have derived Gauss's law, Equation 24.6. Note that this result is independent of the shape of the closed surface and independent of the position of the charge within the surface.

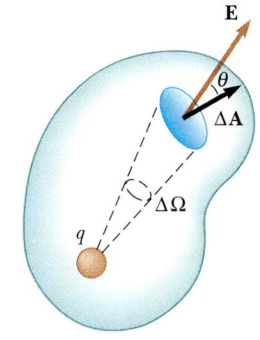

FIGURE 24.21 A closed surface of arbitrary shape surrounds a point charge q. The net flux through the surface is independent of the shape of the surface.

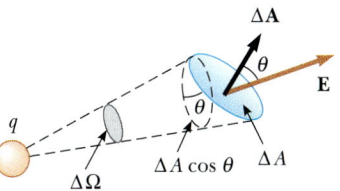

FIGURE 24.22 The area element ΔA subtends a solid angle $\Delta\Omega = (\Delta A \cos\theta)/r^2$ at the charge q.

SUMMARY

Electrix flux is represented by the number of electric field lines that penetrate a surface. If the electric field is uniform and makes an angle θ with the normal to the surface, the electric flux through the surface is

$$\Phi = EA \cos\theta \qquad (24.2)$$

TABLE 24.1 Typical Electric Field Calculations Using Gauss's Law		
Charge Distribution	**Electric Field**	**Location**
Insulating sphere of radius R, uniform charge density, and total charge Q	$k_e \dfrac{Q}{r^2}$	$r > R$
	$k_e \dfrac{Q}{R^3} r$	$r < R$
Thin spherical shell of radius R and total charge Q	$k_e \dfrac{Q}{r^2}$	$r > R$
	0	$r < R$
Line charge of infinite length and charge per unit length λ	$2k_e \dfrac{\lambda}{r}$	Outside the line charge
Nonconducting, infinite charged plane with charge per unit area σ	$\dfrac{\sigma}{2\epsilon_0}$	Everywhere outside the plane
Conductor of surface charge per unit area σ	$\dfrac{\sigma}{\epsilon_0}$	Just outside the conductor
	0	Inside the conductor

In general, the electric flux through a surface is

$$\Phi = \int\limits_{\text{surface}} \mathbf{E} \cdot dA \tag{24.3}$$

Gauss's law says that the net electric flux, Φ_c, through any closed gaussian surface is equal to the *net* charge inside the surface divided by ϵ_0:

$$\Phi_c = \oint \mathbf{E} \cdot d\mathbf{A} = \frac{q_{\text{in}}}{\epsilon_0} \tag{24.6}$$

Using Gauss's law, one can calculate the electric field due to various symmetric charge distributions. Table 24.1 lists some typical results.

A conductor in electrostatic equilibrium has the following properties:

- The electric field is zero everywhere inside it.
- Any excess charge on it resides entirely on its surface.
- The electric field just outside it is perpendicular to its surface and has a magnitude σ/ϵ_0, where σ is the charge per unit area at that point.
- On an irregularly shaped conductor, charge tends to accumulate where the radius of curvature of the surface is the smallest, that is, at sharp points.

Problem-Solving Strategy and Hints
Applying Gauss's Law

Gauss's law may seem mysterious to you, and it is usually one of the most difficult concepts to understand in introductory physics. However, as we have seen, it is very powerful in solving problems having a high degree of symmetry. In this chapter, you will encounter only problems with three kinds of symmetry: planar, cylindrical, and spherical. It is important to review Examples 24.4 through 24.10 and to use the following procedure:

- First, select a gaussian surface that *has a symmetry to match the charge distribution.* For point charges or spherically symmetric charge distributions, the gaussian surface should be a sphere centered on the charge as in Examples 24.4, 24.5, 24.6, and 24.10. For uniform line charges or uniformly charged cylinders, your choice of a gaussian surface should be a cylindrical surface that is coaxial with the line charge or cylinder as in Example 24.7. For sheets of charge having plane symmetry, the gaussian surface should be a cylinder that straddles the sheet as in Example 24.8. Note that in all cases, the gaussian surface is selected such that the electric field has the same magnitude everywhere on the surface and is directed perpendicularly to the surface. This enables you to easily evaluate the surface integral that appears on the left side of Gauss's law, which represents the total electric flux through that surface.
- Now evaluate the right side of Gauss's law, which amounts to calculating the total electric charge, q_{in}, inside the gaussian surface. If the charge density is uniform, as is usually the case (that is, if λ, σ, or ρ is constant), simply multiply that charge density by the length, area, or volume enclosed by the gaussian surface. However, if the charge distribution is *nonuniform,* you must integrate the charge density over the region enclosed by the gaussian surface. For example, if the charge is distributed along a line, you would integrate the expression $dq = \lambda \, dx$, where dq is the charge on an

infinitesimal element dx and λ is the charge per unit length. For a plane of charge, you would integrate $dq = \sigma\, dA$, where σ is the charge per unit area and dA is an infinitesimal element of area. Finally, for a volume of charge you would integrate $dq = \rho\, dV$, where ρ is the charge per unit volume and dV is an infinitesimal element of volume.

- Once the left and right sides of Gauss's law have been evaluated, you can calculate the electric field on the gaussian surface assuming the charge distribution is given in the problem. Conversely, if the electric field is known, you can calculate the charge distribution that produces the field.

QUESTIONS

1. If the electric field in a region of space is zero, can you conclude there are no electric charges in that region? Explain.

2. If there are more electric field lines leaving a gaussian surface than entering, what can you conclude about the net charge enclosed by that surface?

3. A uniform electric field exists in a region of space in which there are no charges. What can you conclude about the net electric flux through a gaussian surface placed in this region of space?

4. If the total charge inside a closed surface is known but the distribution of the charge is unspecified, can you use Gauss's law to find the electric field? Explain.

5. Explain why the electric flux through a closed surface with a given enclosed charge is independent of the size or shape of the surface.

6. Consider the electric field due to a nonconducting infinite plane having a uniform charge density. Explain why the electric field does not depend on the distance from the plane in terms of the spacing of the electric field lines.

7. Use Gauss's law to explain why electric field lines must begin and end on electric charges. (*Hint:* Change the size of the gaussian surface.)

8. A point charge is placed at the center of an uncharged metallic spherical shell insulated from ground. As the point charge is moved off center, describe what happens to (a) the total induced charge on the shell and (b) the distribution of charge on the interior and exterior surfaces of the shell.

9. Explain why excess charge on an isolated conductor must reside on its surface, using the repulsive nature of the force between like charges and the freedom of motion of charge within the conductor.

10. A person is placed in a large hollow metallic sphere that is insulated from ground. If a large charge is placed on the sphere, will the person be harmed upon touching the inside of the sphere? Explain what will happen if the person also has an initial charge whose sign is opposite that of the charge on the sphere.

11. How would the observations described in Figure 24.20 differ if the hollow conductor were grounded? How would they differ if the small charged ball were an insulator rather than a conductor?

12. What other experiment might be performed on the ball in Figure 24.20 to show that its charge was transferred to the hollow conductor?

13. What would happen to the electrometer reading if the charged ball in Figure 24.20 touched the inner wall of the conductor? the outer wall?

14. Two solid spheres, both of radius R, carry identical total charges, Q. One sphere is a good conductor while the other is an insulator. If the charge on the insulating sphere is uniformly distributed throughout its interior volume, how do the electric fields outside these two spheres compare? Are the fields identical inside the two spheres?

PROBLEMS

Review Problem

A solid insulating sphere of radius a has a net positive charge $3Q$. Concentric with this sphere is a conducting spherical shell of inner radius b and outer radius c, and having a net negative charge $-Q$ as in the figure. (a) Construct a spherical gaussian surface of radius $r > c$ and find the net charge enclosed by this surface. (b) What is the direction of the electric field at $r > c$? (c) Find the electric field at $r \geq c$. (d) Construct a spherical gaussian sur-

□ indicates problems that have full solutions available in the Student Solutions Manual and Study Guide.

face of radius r, where $b < r < c$, and find the net charge enclosed by this surface. (e) Find the electric field in the region $b < r < c$. (f) Construct a spherical gaussian surface of radius r, where $a < r < b$. What is the net charge enclosed by this surface? (g) Use your result to part (f) and Gauss's law to find the electric field in the region $a < r < b$. (h) Construct a spherical gaussian surface of radius $r < a$ and find an expression for the net charge inside that surface as a function of r. Note that the charge inside this surface is less than $3Q$. (i) Use your result to part (h) and Gauss's law to find the electric field in the region $r < a$. (j) From what you have learned in the previous parts, determine the net charge on the inner surface of the conducting shell and the net charge on the outer surface of the conducting shell. (k) Make a plot of the magnitude of the electric field versus r in the interior and exterior regions of the spherical shell.

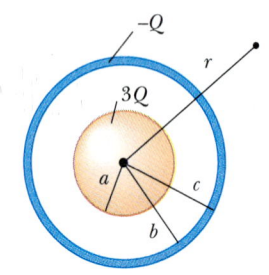

Section 24.1 Electric Flux

1. A spherical shell is placed in a uniform electric field. Determine the total electric flux through the shell.
2. An electric field of magnitude 3.5×10^3 N/C is applied along the x axis. Calculate the electric flux through a rectangular plane 0.35 m wide and 0.70 m long if the plane (a) is parallel to the yz plane, (b) is parallel to the xy plane, and (c) contains the y axis and its normal makes an angle of $40°$ with the x axis.
3. A uniform electric field $a\mathbf{i} + b\mathbf{j}$ intersects a surface of area A. What is the flux through this area if the surface lies (a) in the yz plane? (b) in the xz plane? (c) in the xy plane?
4. Consider a closed triangular box resting within a horizontal electric field of magnitude $E = 7.8 \times 10^4$

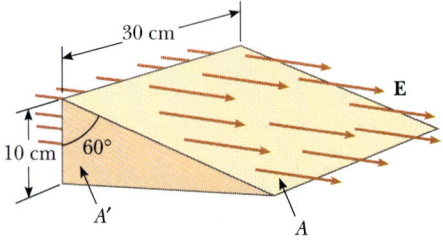

FIGURE P24.4

N/C as in Figure P24.4. Calculate the electric flux through (a) the vertical surface, (b) the slanted surface, and (c) the entire surface of the box.

5. A 40-cm-diameter loop is rotated in a uniform electric field until the position of maximum electric flux is found. The flux in this position is measured to be 5.2×10^5 N·m²/C. What is the electric field strength?

5A. A loop of diameter d is rotated in a uniform electric field until the position of maximum electric flux is found. The flux in this position is measured to be Φ. What is the electric field strength?

6. A point charge q is located at the center of a uniform ring having linear charge density λ and radius a. Determine the total electric flux through a sphere centered at the point charge and having radius R, where $R < a$ (Fig. P24.6).

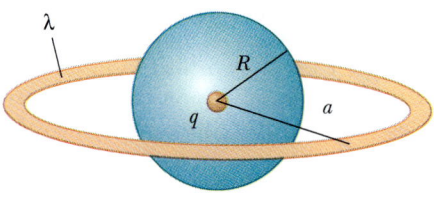

FIGURE P24.6

7. A cone of base radius R and height h is located on a horizontal table, and a horizontal uniform electric field E penetrates the cone, as in Figure P24.7. Determine the electric flux entering the cone.

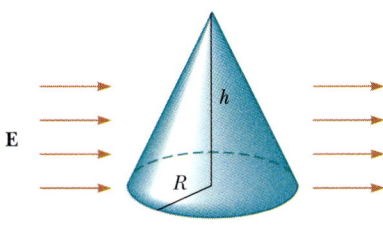

FIGURE P24.7

8. An electric field of magnitude 2.0×10^4 N/C and directed perpendicular to the Earth's surface exists on a day when a thunderstorm is brewing. A car that can be approximated as a rectangle 6.0 m by 3.0 m is traveling along a road that is inclined $10°$ relative to the ground. Determine the electric flux through the bottom of the car.

9. A pyramid with a 6.0-m-square base and height of 4.0 m is placed in a vertical electric field of 52 N/C. Calculate the total electric flux through the pyramid's four slanted surfaces.

Section 24.2 Gauss's Law

10. Four closed surfaces, S_1 through S_4, together with the charges $-2Q$, Q, and $-Q$ are sketched in Figure P24.10. Find the electric flux through each surface.

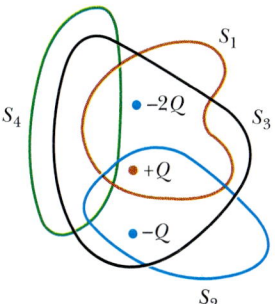

FIGURE P24.10

11. (a) A point charge q is located a distance d from an infinite plane. Determine the electric flux through the plane due to the point charge. (b) A point charge q is located a *very small* distance from the center of a *very large* square on the line perpendicular to the square and going through its center. Determine the approximate electric flux through the square due to the point charge. (c) Explain why the answers to parts (a) and (b) are identical.

12. If the constant electric field in Figure P24.12 has a magnitude E_0, calculate the total electric flux through the paraboloidal surface.

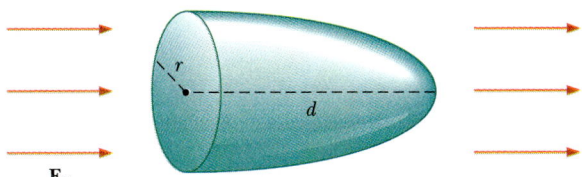

FIGURE P24.12

13. A point charge of 12 μC is placed at the center of a spherical shell of radius 22 cm. What is the total electric flux through (a) the surface of the shell and (b) any hemispherical surface of the shell? (c) Do the results depend on the radius? Explain.

13A. A point charge Q is placed at the center of a spherical shell of radius R. What is the total electric flux through (a) the surface of the shell and (b) any hemispherical surface of the shell? (c) Do the results depend on the radius? Explain.

14. A charge of 12 μC is at the geometric center of a cube. What is the electric flux through one of the faces?

15. The following charges are located inside a submarine: 5.0 μC, -9.0 μC, 27 μC, and -84 μC. Calculate the net electric flux through the submarine. Compare the number of electric field lines leaving the submarine with the number entering it.

16. A point charge of 0.0462 μC is inside a pyramid. Determine the total electric flux through the surface of the pyramid.

17. An infinitely long line of charge having a uniform charge per unit length λ lies a distance d from a point O, as in Figure P24.17. Determine the total electric flux through the surface of a sphere of radius R centered at O. (*Hint*: Consider both $R < d$ and $R > d$.)

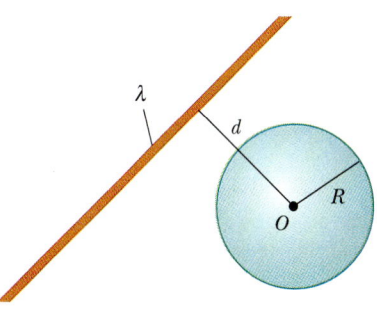

FIGURE P24.17

18. A point charge $Q = 5.0$ μC is located at the center of a cube of side $L = 0.10$ m. Six other point charges, each carrying a charge $q = -1.0$ μC, are positioned symmetrically around Q as in Figure P24.18. Determine the electric flux through one face of the cube.

18A. A point charge Q is located at the center of a cube of side L. Six other point charges, each carrying a charge $-q$, are positioned symmetrically around Q as in Figure P24.18. Determine the electric flux through one face of the cube.

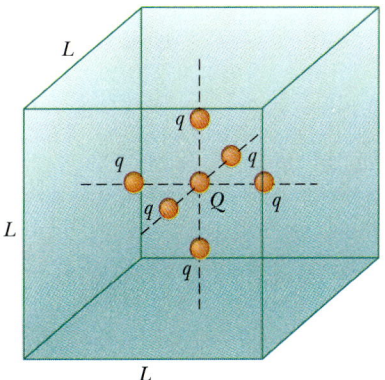

FIGURE P24.18

19. A 10.0-μC charge located at the origin of a cartesian coordinate system is surrounded by a nonconducting hollow sphere of radius 10.0 cm. A drill with a radius of 1.00 mm is aligned along the z axis, and a hole is drilled in the sphere. Calculate the electric flux through the hole.

20. A charge of 170 μC is at the center of a cube of side 80.0 cm. (a) Find the total flux through each face of the cube. (b) Find the flux through the whole surface of the cube. (c) Would your answers to parts (a) or (b) change if the charge were not at the center? Explain.

20A. A charge Q is at the center of a cube of side L. (a) Find the total flux through each face of the cube. (b) Find the flux through the whole surface of the cube. (c) Would your answers to parts (a) or (b) change if the charge were not at the center? Explain.

21. The total electric flux through a closed surface in the shape of a cylinder is 8.60×10^4 N·m²/C. (a) What is the net charge within the cylinder? (b) From the information given, what can you say about the charge within the cylinder? (c) How would your answers to parts (a) and (b) change if the net flux were -8.60×10^4 N·m²/C?

22. The line *ag* in Figure P24.22 is a diagonal of the cube, and a point charge q is located very close to vertex a (on the extension of *ag*) as in Figure P24.22. Determine the electric flux through each side of the cube which contains the point a.

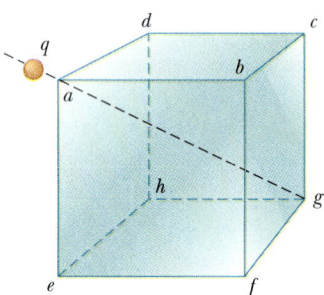

FIGURE P24.22

23. A point charge Q is located just above the center of the flat face of a hemisphere of radius R as in Figure P24.23. What is the electric flux (a) through the curved surface and (b) through the flat face?

Section 24.3   Application of Gauss's Law to Charged Insulators

24. On a clear, sunny day, there is a vertical electrical field of about 130 N/C pointing down over flat ground or water. What is the surface charge density on the ground for these conditions?

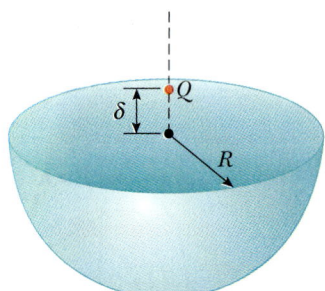

FIGURE P24.23

25. Consider a thin spherical shell of radius 14.0 cm with a total charge of 32.0 μC distributed uniformly on its surface. Find the electric field (a) 10.0 cm and (b) 20.0 cm from the center of the charge distribution.

26. An inflated balloon in the shape of a sphere of radius 12.0 cm has a total charge of 7.00 μC uniformly distributed on its surface. Calculate the magnitude of the electric field (a) 10.0 cm, (b) 12.5 cm, and (c) 30.0 cm from the center of the balloon.

27. An insulating sphere is 8.00 cm in diameter and carries a 5.70-μC charge uniformly distributed throughout its interior volume. Calculate the charge enclosed by a concentric spherical surface with radius (a) $r = 2.00$ cm and (b) $r = 6.00$ cm.

28. Consider an infinitely long line of charge having uniform charge per unit length λ. Determine the total electric flux through a closed right circular cylinder of length L and radius R that is parallel to the line, if the distance between the axis of the cylinder and the line is d. (*Hint:* Consider both when $R < d$ and $R > d$.)

29. A solid sphere of radius 40.0 cm has a total positive charge of 26.0 μC uniformly distributed throughout its volume. Calculate the magnitude of the electric field (a) 0 cm, (b) 10.0 cm, (c) 40.0 cm, and (d) 60.0 cm from the center of the sphere.

30. A sphere of radius a carries a volume charge density $\rho = \rho_0 (r/a)^2$ for $r < a$. Calculate the electric field inside and outside the sphere.

31. A spherically symmetric charge distribution has a charge density given by $\rho = a/r$, where a is constant. Find the electric field as a function of r. (See the note in Problem 54.)

32. The charge per unit length on a long, straight filament is -90.0 μC/m. Find the electric field (a) 10.0 cm, (b) 20.0 cm, and (c) 100 cm from the filament, where distances are measured perpendicular to the length of the filament.

33. A uniformly charged, straight filament 7.00 m in length has a total positive charge of 2.00 μC. An uncharged cardboard cylinder 2.00 cm in length and 10.0 cm in radius surrounds the filament at its

center, with the filament as the axis of the cylinder. Using any reasonable approximations, find (a) the electric field at the surface of the cylinder and (b) the total electric flux through the cylinder.

34. An infinitely long cylindrical insulating shell of inner radius a and outer radius b has a uniform volume charge density ρ (C/m^3). A line of charge density λ (C/m) is placed along the axis of the shell. Determine the electric field intensity everywhere.

35. Consider a long cylindrical charge distribution of radius R with a uniform charge density ρ. Find the electric field at distance r from the axis where $r < R$.

36. A nonconducting wall carries a uniform charge density of 8.6 $\mu C/cm^2$. What is the electric field 7.0 cm in front of the wall? Does your result change as the distance from the wall is varied?

37. A large flat sheet of charge has a charge per unit area of 9.0 $\mu C/m^2$. Find the electric field intensity just above the surface of the sheet, measured from its midpoint.

Section 24.4 Conductors in Electrostatic Equilibrium

38. A conducting spherical shell of radius 15 cm carries a net charge of -6.4 μC uniformly distributed on its surface. Find the electric field at points (a) just outside the shell and (b) inside the shell.

39. A long, straight metal rod has a radius of 5.00 cm and a charge per unit length of 30.0 nC/m. Find the electric field (a) 3.00 cm, (b) 10.0 cm, and (c) 100 cm from the axis of the rod, where distances are measured perpendicular to the rod.

40. A charge q is placed on a conducting sheet that has a finite but very small thickness and area A. What is the approximate magnitude of the electric field at an external point that is a very small distance above the center point?

41. A thin conducting plate 50.0 cm on a side lies in the xy plane. If a total charge of 4.00×10^{-8} C is placed on the plate, find (a) the charge density on the plate, (b) the electric field just above the plate, and (c) the electric field just below the plate.

42. Two identical conducting spheres each having a radius of 0.500 cm are connected by a light 2.00-m-long conducting wire. Determine the tension in the wire if 60.0 μC is placed on one of the conductors. (*Hint:* Assume that the surface distribution of charge on each sphere is uniform.)

42A. Two identical conducting spheres each having a radius R are connected by a light conducting wire of length L, where $L \gg R$. Determine the tension in the wire if a charge Q is placed on one of the conductors. (*Hint:* Assume that the surface distribution of charge on each sphere is uniform.)

43. A conducting spherical shell having an inner radius of 4.0 cm and outer radius of 5.0 cm carries a net charge of $+10$ μC. If a $+2.0$-μC point charge is placed at the center of this shell, determine the surface charge density on (a) the inner surface and (b) the outer surface.

43A. A conducting spherical shell having an inner radius a and outer radius b carries a net charge Q. If a point charge q is placed at the center of this shell, determine the surface charge density on (a) the inner surface and (b) the outer surface.

44. A hollow conducting sphere is surrounded by a larger concentric, spherical, conducting shell. The inner sphere has a charge $-Q$, and the outer sphere has a charge $3Q$. The charges are in electrostatic equilibrium. Using Gauss's law, find the charges and the electric fields everywhere.

45. A solid conducting sphere of radius 2.0 cm has a charge 8.0 μC. A conducting spherical shell of inner radius 4.0 cm and outer radius 5.0 cm is concentric with the solid sphere and has a charge -4.0 μC. Find the electric field at (a) $r = 1.0$ cm, (b) $r = 3.0$ cm, (c) $r = 4.5$ cm, and (d) $r = 7.0$ cm from the center of this charge configuration.

46. Two identical metal blocks resting on a frictionless horizontal surface are connected by a light metal spring for which the spring constant is $k = 100$ N/m and the unstretched length is 0.30 m as in Figure P24.46a. A charge Q is slowly placed on the system causing the spring to stretch to an equilibrium length of 0.40 m as in Figure P24.46b. Determine the value of Q, assuming that all the charge resides on the blocks and that the blocks can be treated as point charges.

46A. Two identical metal blocks resting on a frictionless horizontal surface are connected by a light metal spring having constant k and unstretched length L_0 as in Figure P24.46a. A charge Q is slowly placed on

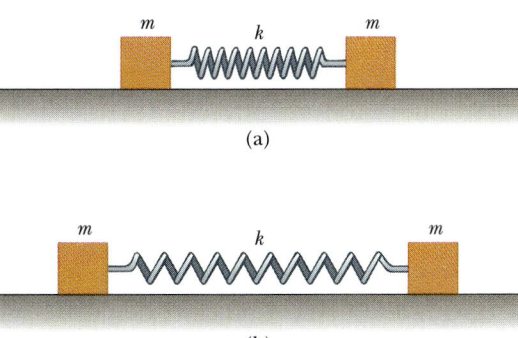

FIGURE P24.46

the system, causing the spring to stretch to an equilibrium length *L* as in Figure P24.46b. Determine the value of *Q*, assuming that all the charge resides on the blocks and that the blocks can be treated as point charges.

47. A long, straight wire is surrounded by a hollow metal cylinder whose axis coincides with that of the wire. The wire has a charge per unit length of λ, and the cylinder has a net charge per unit length of 2λ. From this information, use Gauss's law to find (a) the charge per unit length on the inner and outer surfaces of the cylinder and (b) the electric field outside the cylinder, a distance *r* from the axis.

48. Consider two identical conducting spheres whose surfaces are separated by a small distance. One sphere is given a large net positive charge while the other is given a small net positive charge. It is found that the force between them is attractive even though both spheres have net charges of the same sign. Explain how this is possible.

Section 24.6   Derivation of Gauss's Law

49. A sphere of radius *R* surrounds a point charge *Q*, located at its center. (a) Show that the electric flux through a circular cap of half-angle θ (Fig. P24.49) is

$$\Phi = \frac{Q}{2\epsilon_0} (1 - \cos \theta)$$

What is the flux for (b) $\theta = 90°$ and (c) $\theta = 180°$?

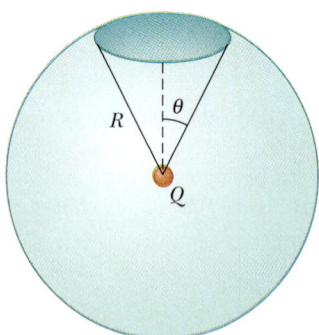

FIGURE P24.49

ADDITIONAL PROBLEMS

50. For the configuration shown in Figure P24.50, suppose that *a* = 5.0 cm, *b* = 20 cm, and *c* = 25 cm. Furthermore, suppose that the electric field at a point 10 cm from the center is measured to be 3.6 × 10^3 N/C radially inward while the electric field at a point 50 cm from the center is 2.0 × 10^2 N/C radially outward. From this information, find (a) the

charge on the insulating sphere, (b) the net charge on the hollow conducting sphere, and (c) the total charge on the inner and outer surfaces of the hollow conducting sphere.

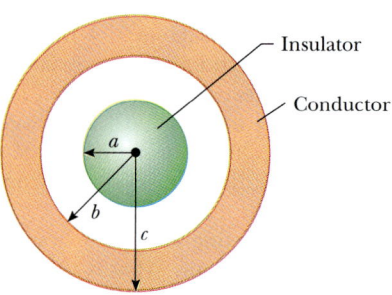

FIGURE P24.50

51. A solid insulating sphere of radius *a* has a uniform charge density ρ and a total charge *Q*. Concentric with this sphere is an uncharged, conducting hollow sphere whose inner and outer radii are *b* and *c*, as in Figure P24.50. (a) Find the magnitude of the electric field in the regions $r < a$, $a < r < b$, $b < r < c$, and $r > c$. (b) Determine the induced charge per unit area on the inner and outer surfaces of the hollow sphere.

52. Consider an insulating sphere of radius *R* and having a uniform volume charge density ρ. Plot the magnitude of the electric field *E* as a function of the distance from the center of the sphere, *r*. Let *r* range over the interval $0 < r < 3R$ and plot *E* in units of $\rho R / \epsilon_0$.

53. An early (incorrect) model of the hydrogen atom, suggested by J. J. Thomson, proposed that a positive cloud of charge $+e$ was uniformly distributed throughout the volume of a sphere of radius *R*, with the electron an equal-magnitude negative point charge $-e$ at the center. (a) Using Gauss's law, show that the electron would be in equilibrium at the center and, if displaced from the center a distance $r < R$, would experience a restoring force of the form $F = -Kr$, where *K* is a constant. (b) Show that $K = k_e e^2 / R^3$. (c) Find an expression for the frequency *f* of simple harmonic oscillations that an electron would undergo if displaced a short distance ($<R$) from the center and released. (d) Calculate a numerical value for *R* that would result in a frequency of 2.47 × 10^{15} Hz, the most intense line in the hydrogen spectrum.

54. Consider a solid insulating sphere of radius *b* with nonuniform charge density $\rho = Cr$ for $0 < r \leq b$. Find the charge contained within the radius when (a) $r < b$ and (b) $r > b$. (*Note:* The volume element *dV* for a spherical shell of radius *r* and thickness *dr* is equal to $4\pi r^2 \, dr$.)

55. A solid insulating sphere of radius R has a nonuniform charge density that varies with r according to the expression $\rho = Ar^2$, where A is a constant and $r < R$ is measured from the center of the sphere. (a) Show that the electric field outside $(r > R)$ the sphere is $E = AR^5/5\epsilon_0 r^2$. (b) Show that the electric field inside $(r < R)$ the sphere is $E = Ar^3/5\epsilon_0$. (*Hint:* Note that the total charge Q on the sphere is equal to the integral of $\rho\, dV$, where r extends from 0 to R; also note that the charge q within a radius $r < R$ is less than Q. To evaluate the integrals, note that the volume element dV for a spherical shell of radius r and thickness dr is equal to $4\pi r^2\, dr$.)

56. An infinitely long insulating cylinder of radius R has a volume charge density that varies with the radius as

$$\rho = \rho_0\left(a - \frac{r}{b}\right)$$

where ρ_0, a, and b are positive constants and r is the distance from the axis of the cylinder. Use Gauss's law to determine the magnitude of the electric field at radial distances (a) $r < R$ and (b) $r > R$.

57. (a) Using the fact that Newton's law of gravitation is mathematically similar to Coulomb's law, show that Gauss's law for gravitation can be written

$$\oint \mathbf{g} \cdot d\mathbf{A} = -4\pi G m_{\text{in}}$$

where m_{in} is the net mass inside the gaussian surface and $\mathbf{g} = \mathbf{F}_g / m$ is the gravitational field at any point on the surface. (b) Determine the gravitational field at a point a distance r from the center of the Earth, where $r < R_E$, assuming that the Earth's mass density is uniform.

58. Two infinite, nonconducting sheets of charge are parallel to each other as in Figure P24.58. The sheet on the left has a uniform surface charge density σ, and the one on the right has a uniform charge density $-\sigma$. Calculate the value of the electric field at points (a) to the left of, (b) in between, and (c) to the right of the two sheets. (*Hint:* See Example 24.8.)

59. Repeat the calculations for Problem 58 when both sheets have positive uniform charge densities σ.

60. A closed surface with dimensions $a = b = 0.40$ m and $c = 0.60$ m is located as in Figure P24.60. The electric field throughout the region is nonuniform and given by $\mathbf{E} = (3.0 + 2.0x^2)\mathbf{i}$ N/C, where x is in meters. Calculate the net electric flux leaving the closed surface. What net charge is enclosed by the surface?

60A. A closed surface with dimensions a, b, and c is located as in Figure P24.60. The electric field throughout the region is nonuniform and given by $\mathbf{E} = (A + Bx^2)\mathbf{i}$ N/C, where A and B are constants and x is in meters. Calculate the net electric flux leaving the closed surface. What net charge is enclosed by the surface?

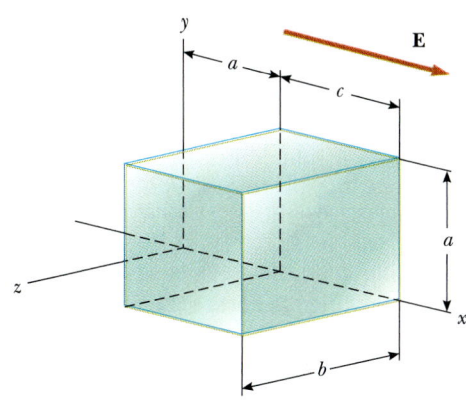

FIGURE P24.60

61. A slab of insulating material (infinite in two of its three dimensions) has a uniform positive charge density ρ. An edge view of the slab is shown in Figure P24.61. (a) Show that the electric field a distance x from its center and inside the slab is $E = \rho x/\epsilon_0$. (b) Suppose an electron of charge $-e$ and mass m is placed inside the slab. If it is released from rest at a distance x from the center, show that the electron exhibits simple harmonic motion with a frequency

$$f = \frac{1}{2\pi}\sqrt{\frac{\rho e}{m\epsilon_0}}$$

62. A slab of insulating material has a nonuniform positive charge density $\rho = Cx^2$, where x is measured from the center of the slab as in Figure P24.61, and C is a constant. The slab is infinite in the y and z directions. Derive expressions for the electric field in (a) the exterior regions and (b) the interior region of the slab $(-d/2 < x < d/2)$.

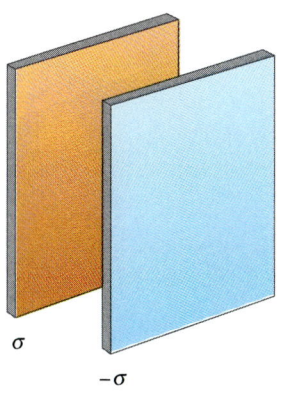

σ

$-\sigma$

FIGURE P24.58

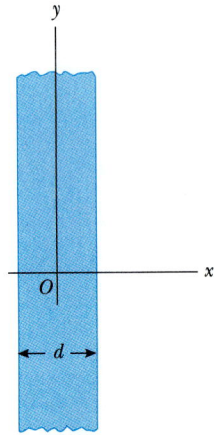

FIGURE P24.61

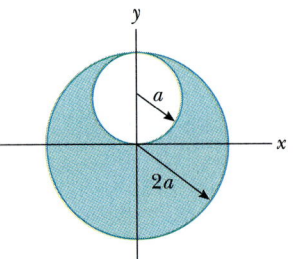

FIGURE P24.63

64. A point charge Q is located on the axis of a disk of radius R at a distance b from the plane of the disk (Fig. P24.64). Show that if one fourth of the electric flux from the charge passes through the disk, then $R = \sqrt{3}\,b$.

63. A sphere of radius $2a$ is made of a nonconducting material that has a uniform volume charge density ρ. (Assume that the material does not affect the electric field.) A spherical cavity of radius a is now removed from the sphere as shown in Figure P24.63. Show that the electric field within the cavity is uniform and is given by $E_x = 0$ and $E_y = \rho a/3\epsilon_0$. (*Hint:* The field within the cavity is the superposition of the field due to the original uncut sphere, plus the field due to a sphere the size of the cavity with a uniform negative charge density $-\rho$.)

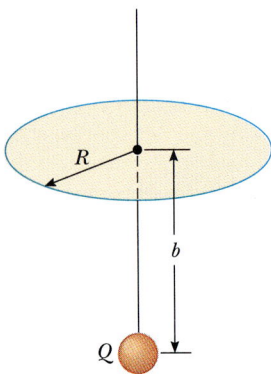

FIGURE P24.64

CHAPTER 25

Electric Potential

Jennifer is holding on to a charged sphere that reaches a potential of about 100 000 volts. The device that generates this high potential is called a Van de Graaff generator. Why do you suppose Jennifer's hair stands on end like the needles of a porcupine? Why is it important that she stand on a pedestal insulated from ground?

(Courtesy of Henry Leap and Jim Lehman)

The concept of potential energy was first introduced in Chapter 8 in connection with such conservative forces as the force of gravity and the elastic force of a spring. By using the law of energy conservation, we were often able to avoid working directly with forces when solving various mechanical problems. In this chapter we see that the energy concept is also of great value in the study of electricity. Since the electrostatic force given by Coulomb's law is conservative, electrostatic phenomena can be described conveniently in terms of an electrical potential energy. This idea enables us to define a scalar quantity called *electric potential*. Because the potential is a scalar function of position, it offers a simpler way of describing electrostatic phenomena than does the electric field. In later chapters we shall see that the concept of the electric potential is of great practical value. In fact, the measured voltage between any two points in an electrical circuit is simply the difference in electric potential between the points.

25.1 POTENTIAL DIFFERENCE AND ELECTRIC POTENTIAL

When a test charge q_0 is placed in an electric field $\mathbf{E}$, the electric force on the test charge is $q_0\mathbf{E}$. This force is the vector sum of the individual forces exerted on q_0 by the various charges producing the field $\mathbf{E}$. It follows that the force $q_0\mathbf{E}$ is conservative, because the individual forces governed by Coulomb's law are conservative. When a charge is moved within an electric field, the work done on q_0 by the electric field is equal to the negative of the work done by the external agent causing the displacement. For an infinitesimal displacement $d\mathbf{s}$, the work done by the electric field is $\mathbf{F}\cdot d\mathbf{s} = q_0\mathbf{E}\cdot d\mathbf{s}$. This decreases the potential energy of the electric field by an amount $dU = -q_0\mathbf{E}\cdot d\mathbf{s}$. For a finite displacement of the test charge between points A and B, the change in potential energy $\Delta U = U_B - U_A$ is

| Change in potential energy |

$$\Delta U = -q_0 \int_A^B \mathbf{E}\cdot d\mathbf{s} \tag{25.1}$$

The integration is performed along the path by which q_0 moves from A to B, and the integral is called either a *path integral* or a *line integral*. Since the force $q_0\mathbf{E}$ is conservative, *this line integral does not depend on the path taken between A and B.*

The potential energy per unit charge, U/q_0, is independent of the value of q_0 and has a unique value at every point in an electric field. The quantity U/q_0 is called the **electric potential** (or simply the **potential**), V. Thus, the electric potential at any point in an electric field is

$$V = \frac{U}{q_0} \tag{25.2}$$

Because potential energy is a scalar quantity, electric potential is also a scalar quantity.

The **potential difference,** $\Delta V = V_B - V_A$, between the points A and B is defined as the change in potential energy divided by the test charge q_0:

| Potential difference |

$$\Delta V = \frac{\Delta U}{q_0} = -\int_A^B \mathbf{E}\cdot d\mathbf{s} \tag{25.3}$$

Potential difference should not be confused with difference in potential energy. The potential difference is proportional to the change in potential energy, and we see from Equation 25.3 that the two are related by $\Delta U = q_0\,\Delta V$. The potential is a scalar characteristic of the field, independent of the charges that may be placed in the field. The potential energy resides in the field, also. However, because we are interested in the potential at the location of a charge, and in the potential energy caused by the interaction of the charge with the field, we follow the common convention of speaking of the potential energy as if it belonged to the charge, except when we specifically require information about transfer of energy between the field and the charge.

Because, as already noted, the change in potential energy of the charge is the negative of the work done by the electric force, the potential difference ΔV equals the work per unit charge that an external agent must perform to move a test charge from A to B without a change in kinetic energy of the test charge.

Equation 25.3 defines a potential difference only. That is, only *differences* in V are meaningful. The electric potential function is often taken to be zero at some convenient point. We arbitrarily set the potential to be zero for a point that is infinitely remote from the charges producing the electric field. With this choice, we can say that the *electric potential at an arbitrary point equals the work required per unit charge to bring a positive test charge from infinity to that point.* Thus, if we take $V_A = 0$ at infinity in Equation 25.3, then the potential at any point P is

$$V_P = -\int_{\infty}^{P} \mathbf{E} \cdot d\mathbf{s} \tag{25.4}$$

In reality, V_P represents the potential difference between the point P and a point at infinity. (Equation 25.4 is a special case of Eq. 25.3.)

Since potential difference is a measure of energy per unit charge, the SI unit of potential is joules per coulomb, defined to be equal to a unit called the **volt (V)**:

$$1 \text{ V} \equiv 1 \text{ J}/1 \text{ C}$$

Definition of a volt

That is, 1 J of work must be done to take a 1-C charge through a potential difference of 1 V. Equation 25.3 shows that the potential difference also has units of electric field times distance. From this, it follows that the SI unit of electric field (N/C) can also be expressed as volts per meter:

$$\frac{\text{N}}{\text{C}} = \frac{\text{V}}{\text{m}}$$

A unit of energy commonly used in atomic and nuclear physics is the **electron volt (eV)**, which is defined as *the energy that an electron (or proton) gains or loses by moving through a potential difference of 1 V.* Since $1 \text{ V} = 1 \text{ J}/1 \text{ C}$ and since the fundamental charge is equal to 1.60×10^{-19} C, we see that the electron volt (eV) is related to the joule as follows:

The electron volt

$$1 \text{ eV} = 1.60 \times 10^{-19} \text{ C} \cdot \text{V} = 1.60 \times 10^{-19} \text{ J} \tag{25.5}$$

For instance, an electron in the beam of a typical television picture tube (or cathode ray tube) has a speed of 5.0×10^7 m/s. This corresponds to a kinetic energy of 1.1×10^{-15} J, which is equivalent to 7.1×10^3 eV. Such an electron has to be accelerated from rest through a potential difference of 7.1 kV to reach this speed.

25.2   POTENTIAL DIFFERENCES IN A UNIFORM ELECTRIC FIELD

In this section, we describe the potential difference between any two points in a uniform electric field. The potential difference is independent of the path between these two points; that is, the work done in taking a test charge from point A to point B is the same along all paths. This confirms that a static, uniform electric field is conservative.

First, consider a uniform electric field directed along the negative y axis, as in Figure 25.1a. Let us calculate the potential difference between two points, A and B, separated by a distance d, where d is measured parallel to the field lines. If we apply Equation 25.3 to this situation, we get

$$V_B - V_A = \Delta V = -\int_A^B \mathbf{E} \cdot d\mathbf{s} = -\int_A^B E \cos 0° \, ds = -\int_A^B E \, ds$$

Since E is constant, it can be removed from the integral sign, giving

$$\Delta V = -E \int_A^B d\mathbf{s} = -Ed \tag{25.6}$$

Potential difference in a uniform **E** field

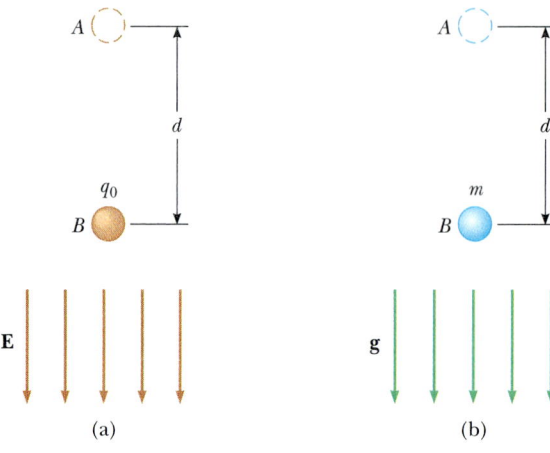

FIGURE 25.1 (a) When the electric field **E** is directed downward, point *B* is at a lower electric potential than point *A*. A positive test charge that moves from *A* to *B* loses electric potential energy. (b) A mass *m* moving downward in the direction of the gravitational field **g** loses gravitational potential energy.

The minus sign results from the fact that point *B* is at a lower potential than point *A*; that is, $V_B < V_A$. *Electric field lines always point in the direction of decreasing electric potential*, as in Figure 25.1a.

Now suppose that a test charge q_0 moves from *A* to *B*. The change in its potential energy can be found from Equations 25.3 and 25.6:

$$\Delta U = q_0 \, \Delta V = - q_0 E d \qquad (25.7)$$

From this result, we see that if q_0 is positive, ΔU is negative. This means that *an electric field does work on a positive charge when the positive charge moves in the direction of the electric field*. (This is analogous to the work done by the gravitational field on a falling mass, as in Fig. 25.1b.) If a positive test charge is released from rest in this electric field, it experiences an electric force $q_0 E$ in the direction of **E** (downward in Fig. 25.1a). Therefore, it accelerates downward, gaining kinetic energy. *As the charged particle gains kinetic energy, the field loses an equal amount of potential energy.*

If the test charge q_0 is negative, then ΔU is positive and the situation is reversed. *A negative charge gains electric potential energy when it moves in the direction of the electric field.* If a negative charge is released from rest in the field **E**, it accelerates in a direction opposite the electric field.

Now consider the more general case of a charged particle moving between any two points in a uniform electric field directed along the *x* axis, as in Figure 25.2. If **s** represents the displacement vector between points *A* and *B*, Equation 25.3 gives

$$\Delta V = - \int_A^B \mathbf{E} \cdot d\mathbf{s} = - \mathbf{E} \cdot \int_A^B d\mathbf{s} = - \mathbf{E} \cdot \mathbf{s} \qquad (25.8)$$

where again we are able to remove **E** from the integral since it is constant. Further, the change in potential energy of the charge is

$$\Delta U = q_0 \, \Delta V = - q_0 \mathbf{E} \cdot \mathbf{s} \qquad (25.9)$$

Finally, our results show that all points in a plane perpendicular to a uniform electric field are at the same potential. This can be seen in Figure 25.2, where the potential difference $V_B - V_A$ is equal to $V_C - V_A$. Therefore, $V_B = V_C$.

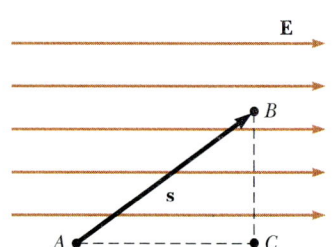

FIGURE 25.2 A uniform electric field directed along the positive *x* axis. Point *B* is at a lower potential than point *A*. Points *B* and *C* are at the *same* potential.

An equipotential surface

The name **equipotential surface** is given to any surface consisting of a continuous distribution of points having the same electric potential.

The line *CB* in Figure 25.2 is actually the cross-section of an equipotential surface, which in this case is a plane perpendicular to the electric field. Note that since $\Delta U = q_0 \, \Delta V$, no work is done in moving a test charge between any two points on an equipotential surface. The equipotential surfaces of a uniform electric field consist of a family of planes all perpendicular to the field. Equipotential surfaces for fields with other symmetries are described in later sections.

CONCEPTUAL EXAMPLE 25.1

If a proton is released from rest in a uniform electric field, does the electric potential increase or decrease? What about its electric potential energy?

Reasoning The proton moves in the direction of the electric field to a position of lower potential and lower potential energy. The decrease in electric potential energy is accompanied by an equal increase in kinetic energy as required by the principle of conservation of energy. An electron moves in the direction opposite to the electric field so its electric potential increases, but the electric potential energy decreases.

EXAMPLE 25.2   The Electric Field Between Two Parallel Plates of Opposite Charge

A 12-V battery is connected between two parallel plates as in Figure 25.3. The separation between the plates is 0.30 cm, and the electric field is assumed to be uniform. (This assumption is reasonable if the plate separation is small relative to the plate size and if we do not consider points near the edges of the plates.) Find the magnitude of the electric field between the plates.

Solution The electric field is directed from the positive plate toward the negative plate. We see that the positive plate is at a higher potential than the negative plate. Note that the potential difference between plates must equal the potential difference between the battery terminals. This can be understood by noting that all points on a conductor in equilibrium are at the same potential,[1] and hence there is no potential difference between a terminal of the battery and any portion of the plate to which it is connected. Therefore, the magnitude of the electric field between the plates is

$$E = \frac{|V_B - V_A|}{d} = \frac{12 \text{ V}}{0.30 \times 10^{-2} \text{ m}} = 4.0 \times 10^3 \text{ V/m}$$

This configuration, which is called a *parallel-plate capacitor*, is examined in more detail in the next chapter.

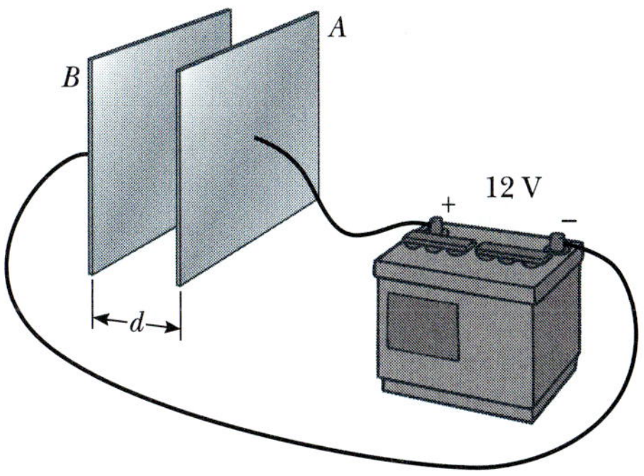

FIGURE 25.3 (Example 25.2) A 12-V battery connected to two parallel plates. The electric field between the plates has a magnitude given by the potential difference divided by the plane separation *d*.

[1] The electric field vanishes within a conductor in electrostatic equilibrium, and so the path integral $\int \mathbf{E} \cdot d\mathbf{s}$ between any two points within the conductor must be zero. A fuller discussion of this point is given in Section 25.6.

EXAMPLE 25.3   Motion of a Proton in a Uniform Electric Field

A proton is released from rest in a uniform electric field of magnitude 8.0×10^4 V/m directed along the positive *x* axis (Fig. 25.4). The proton undergoes a displacement of 0.50 m in the direction of **E**. (a) Find the change in the electric potential between the points *A* and *B*.

Solution From Equation 25.6 we have

$$\Delta V = -Ex = -\left(8.0 \times 10^4 \, \frac{\text{V}}{\text{m}}\right)(0.50 \text{ m})$$

$$= -4.0 \times 10^4 \text{ V}$$

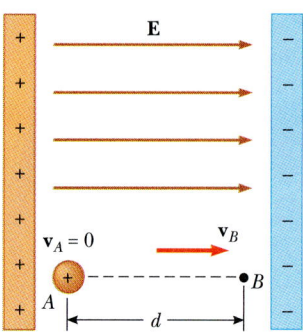

FIGURE 25.4 (Example 25.3) A proton accelerates from A to B in the direction of the electric field.

This negative result tells us that the electric potential of the proton decreases as it moves from A to B.

(b) Find the change in potential energy of the proton for this displacement.

Solution

$$\Delta U = q_0 \, \Delta V = e \, \Delta V$$
$$= (1.6 \times 10^{-19} \, \text{C})(-4.0 \times 10^4 \, \text{V})$$
$$= -6.4 \times 10^{-15} \, \text{J}$$

The negative sign here means that the potential energy of the proton decreases as it moves in the direction of the electric field. This makes sense since as the proton accelerates in the direction of the field, it gains kinetic energy and at the same time (the field) loses electrical potential energy (energy is conserved).

Exercise Apply the principle of energy conservation to find the speed of the proton after it has moved 0.50 m, starting from rest.

Answer 2.77×10^6 m/s.

25.3 ELECTRIC POTENTIAL AND POTENTIAL ENERGY DUE TO POINT CHARGES

Consider an isolated positive point charge q (Fig. 25.5). Recall that such a charge produces an electric field that is radially outward from the charge. In order to find the electric potential at a field point located a distance r from the charge, we begin with the general expression for the potential difference:

$$V_B - V_A = -\int_A^B \mathbf{E} \cdot d\mathbf{s}$$

Since the electric field due to the point charge is given by $\mathbf{E} = k_e q \hat{\mathbf{r}} / r^2$, where $\hat{\mathbf{r}}$ is a unit vector directed from the charge to the field point, the quantity $\mathbf{E} \cdot d\mathbf{s}$ can be expressed as

$$\mathbf{E} \cdot d\mathbf{s} = k_e \frac{q}{r^2} \hat{\mathbf{r}} \cdot d\mathbf{s}$$

The dot product $\hat{\mathbf{r}} \cdot d\mathbf{s} = ds \cos \theta$, where θ is the angle between $\hat{\mathbf{r}}$ and $d\mathbf{s}$ as in Figure 25.5. Furthermore, note that $ds \cos \theta$ is the projection of $d\mathbf{s}$ onto r, so that $ds \cos \theta = dr$. That is, any displacement of the charge $d\mathbf{s}$ produces a change dr in the magnitude of $\mathbf{r}$. With these substitutions, we find that $\mathbf{E} \cdot d\mathbf{s} = (k_e q/r^2) dr$, so that the expression for the potential difference becomes

$$V_B - V_A = -\int E_r \, dr = -k_e q \int_{r_A}^{r_B} \frac{dr}{r^2} = k_e q \left[\frac{1}{r} \right]_{r_A}^{r_B}$$

$$V_B - V_A = k_e q \left[\frac{1}{r_B} - \frac{1}{r_A} \right] \tag{25.10}$$

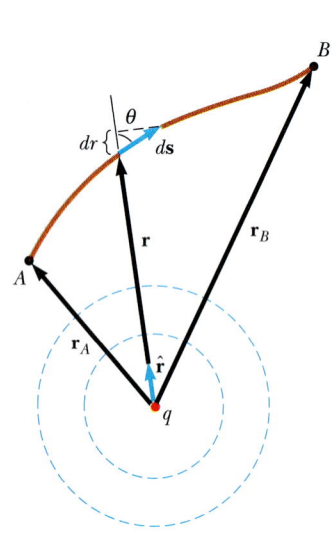

FIGURE 25.5 The potential difference between points A and B due to a point charge q depends *only* on the initial and final radial coordinates, r_A and r_B, respectively. The two dashed circles represent cross-sections of spherical equipotential surfaces.

The integral of $\mathbf{E} \cdot d\mathbf{s}$ is *independent* of the path between A and B—as it must be, because the electric field of a point charge is a conservative field. Furthermore, Equation 25.10 expresses the important result that the potential difference between any two points A and B depends only on the radial coordinates r_A and r_B. It is customary to choose the reference of potential to be zero at $r_A = \infty$. (This is quite natural because then $V \propto 1/r_A$ and as $r_A \to \infty$, $V \to 0$.) With this choice, the electric potential due to a point charge at any distance r from the charge is

$$V = k_e \frac{q}{r} \qquad (25.11)$$

Electric potential of a point charge

From this we see that V is constant on any spherical surface when the point charge is at its center. Hence, we conclude that *the equipotential surfaces (surfaces on which V remains constant) for an isolated point charge consist of a family of spheres concentric with the charge*, as shown in Figure 25.5. Note that the blue dashed lines in Figure 25.5 are cross-sections of spherical equipotential surfaces that are perpendicular to the lines of electric force, as was the case for a uniform electric field.

The electric potential of two or more point charges is obtained by applying the superposition principle. That is, the total potential at some point P due to several point charges is the sum of the potentials due to the individual charges. For a group of charges, we can write the total potential at P in the form

$$V = k_e \sum_i \frac{q_i}{r_i} \qquad (25.12)$$

The electric potential of several point charges

where the potential is again taken to be zero at infinity and r_i is the distance from the point P to the charge q_i. Note that the sum in Equation 25.12 is an algebraic sum of scalars rather than a vector sum (which is used to calculate the electric field of a group of charges). Thus, it is much easier to evaluate V than to evaluate **E**.

We now consider the potential energy of interaction of a system of charged particles. If V_1 is the electric potential at a point P due to charge q_1, then the work required to bring a second charge q_2 from infinity to the point P without acceleration is $q_2 V_1$. By definition, this work equals the potential energy U of the two-particle system when the particles are separated by a distance r_{12} (Fig. 25.6). Therefore, we can express the potential energy as

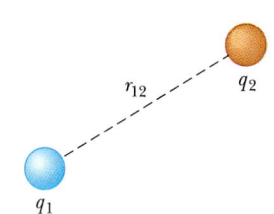

$$U = q_2 V_1 = k_e \frac{q_1 q_2}{r_{12}} \qquad (25.13)$$

FIGURE 25.6 If two point charges are separated by a distance r_{12}, the potential energy of the pair of charges is given by $kq_1 q_2 / r_{12}$.

Note that if the charges are of the same sign, U is positive.[2] This is consistent with the fact that like charges repel, and so positive work must be done on the system to bring the two charges near one another. If the charges are of opposite sign, the force is attractive and U is negative. This means that negative work must be done to bring the unlike charges near one another.

If there are more than two charged particles in the system, the total potential energy can be obtained by calculating U for every pair of charges and summing the terms algebraically. As an example, the total potential energy of the three charges shown in Figure 25.7 is

$$U = k_e \left(\frac{q_1 q_2}{r_{12}} + \frac{q_1 q_3}{r_{13}} + \frac{q_2 q_3}{r_{23}} \right) \qquad (25.14)$$

Physically, we can interpret this as follows: Imagine that q_1 is fixed at the position shown in Figure 25.7, but q_2 and q_3 are at infinity. The work required to bring q_2

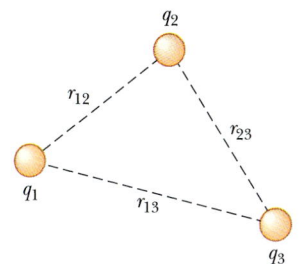

FIGURE 25.7 Three point charges are fixed at the positions shown. The potential energy of this system of charges is given by Equation 25.14.

---

[2] The expression for the electric potential energy for two point charges, Equation 25.13, is of the *same* form as the gravitational potential energy of two point masses given by $Gm_1 m_2 /r$ (Chapter 14). The similarity is not surprising in view of the fact that both are derived from an inverse-square force law.

from infinity to its position near q_1 is $k_e q_1 q_2 / r_{12}$, which is the first term in Equation 25.14. The last two terms in Equation 25.14 represent the work required to bring q_3 from infinity to its position near q_1 and q_2. (The result is independent of the order in which the charges are transported.)

CONCEPTUAL EXAMPLE 25.4

If the electric potential at some point is zero, can you conclude that there are no charges in the vicinity of that point?

Reasoning   No. Suppose there are several charges in the vicinity of the point in question. If some charges are positive and some negative, their contributions to the potential at the point may cancel. For example, the electric potential at the midpoint of two equal but opposite charges is zero.

EXAMPLE 25.5   The Potential Due to Two Point Charges

A 2.00-μC point charge is located at the origin, and a second point charge of -6.00 μC is located on the y axis at the position (0, 3.00) m, as in Figure 25.8a. (a) Find the total electric potential due to these charges at the point P, whose coordinates are (4.00, 0) m.

Solution   For two charges, the sum in Equation 25.12 gives

$$V_P = k_e \left(\frac{q_1}{r_1} + \frac{q_2}{r_2} \right)$$

In this example, $q_1 = 2.00$ μC, $r_1 = 4.00$ m, $q_2 = -6.00$ μC, and $r_2 = 5.00$ m. Therefore, V_P reduces to

$$V_P = 8.99 \times 10^9 \ \frac{\text{N} \cdot \text{m}^2}{\text{C}^2} \left(\frac{2.00 \times 10^{-6} \ \text{C}}{4.00 \ \text{m}} - \frac{6.00 \times 10^{-6} \ \text{C}}{5.00 \ \text{m}} \right)$$

$$= -6.29 \times 10^3 \ \text{V}$$

(b) How much work is required to bring a 3.00-μC point charge from infinity to the point P?

Solution

$$W = q_3 V_P = (3.00 \times 10^{-6} \ \text{C})(-6.29 \times 10^3 \ \text{V})$$

$$= -18.9 \times 10^{-3} \ \text{J}$$

The negative sign means that work is done by the field on the charge as it is displaced from infinity to P. Therefore, positive work would have to be done by an external agent to remove the charge from P back to infinity.

Exercise   Find the total potential energy of the system of three charges in the configuration shown in Figure 25.8b.

Answer -5.48×10^{-2} J.

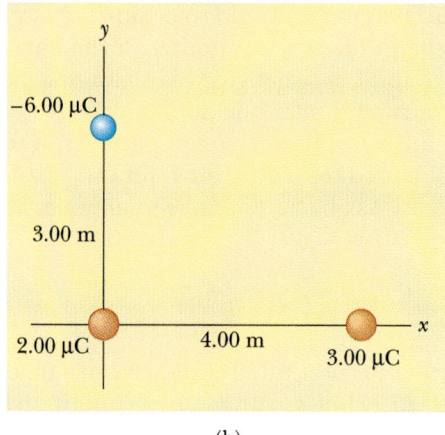

(a)   (b)

FIGURE 25.8   (Example 25.5) (a) The electric potential at the point P due to the two point charges q_1 and q_2 is the algebraic sum of the potentials due to the individual charges. (b) What is the potential energy of the system of three charges?

25.4   OBTAINING E FROM THE ELECTRIC POTENTIAL

The electric field **E** and the potential V are related as shown in Equation 25.3. Both quantities are determined by a specific charge distribution. We now show how to calculate the electric field if the electric potential is known in a certain region.

From Equation 25.3 we can express the potential difference dV between two points a distance ds apart as

$$dV = -\mathbf{E} \cdot d\mathbf{s} \qquad (25.15)$$

If the electric field has only one component, E_x, then $\mathbf{E} \cdot d\mathbf{s} = E_x \, dx$. Therefore, Equation 25.15 becomes $dV = -E_x \, dx$, or

$$E_x = -\frac{dV}{dx} \qquad (25.16)$$

That is, the electric field is equal to the negative of the derivative of the potential with respect to some coordinate. Note that the potential change is zero for any displacement perpendicular to the electric field. This is consistent with the notion of equipotential surfaces being perpendicular to the field, as in Figure 25.9a.

If the charge distribution has spherical symmetry, where the charge density depends only on the radial distance r, then the electric field is radial. In this case, $\mathbf{E} \cdot d\mathbf{s} = E_r \, dr$, and so we can express dV in the form $dV = -E_r \, dr$. Therefore,

$$E_r = -\frac{dV}{dr} \qquad (25.17)$$

For example, the potential of a point charge is $V = k_e q/r$. Since V is a function of r only, the potential function has spherical symmetry. Applying Equation 25.17, we find that the electric field due to the point charge is $E_r = k_e q/r^2$, a familiar result. Note that the potential changes only in the radial direction, not in a direction

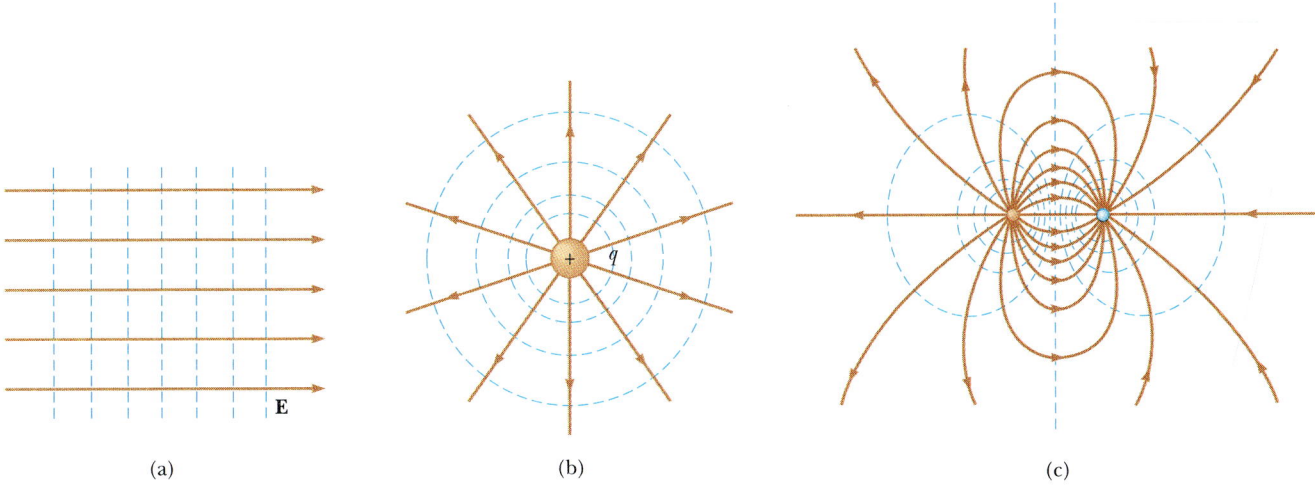

(a)   (b)   (c)

FIGURE 25.9   Equipotential surfaces (dashed blue lines) and electric field lines (red lines) for (a) a uniform electric field produced by an infinite sheet of charge, (b) a point charge, and (c) an electric dipole. In all cases, the equipotential surfaces are *perpendicular* to the electric field lines at every point. A more accurate computer-generated plot is shown on page 717.

Equipotential surfaces are always perpendicular to the electric field lines

perpendicular to r. Thus V (like E_r) is a function only of r. Again, this is consistent with the idea that *equipotential surfaces are perpendicular to field lines*. In this case the equipotential surfaces are a family of spheres concentric with the spherically symmetric charge distribution (Figure 25.9b).

The equipotential surfaces for an electric dipole are sketched in Figure 25.9c. When a test charge is displaced by a vector $d\mathbf{s}$ that lies within any equipotential surface, then by definition $dV = -\mathbf{E} \cdot d\mathbf{s} = 0$. This shows that the equipotential surfaces must *always be perpendicular* to the electric field lines.

In general, the electric potential is a function of all three spatial coordinates. If $V(r)$ is given in terms of the rectangular coordinates, the electric field components E_x, E_y, and E_z can readily be found from $V(x, y, z)$:

$$E_x = -\frac{\partial V}{\partial x} \qquad E_y = -\frac{\partial V}{\partial y} \qquad E_z = -\frac{\partial V}{\partial z}$$

In these expressions, the derivatives are called *partial derivatives*. In the operation $\partial V/\partial x$, we take a derivative with respect to x while y and z are held constant.[3] For example, if $V = 3x^2y + y^2 + yz$, then

$$\frac{\partial V}{\partial x} = \frac{\partial}{\partial x}(3x^2y + y^2 + yz) = \frac{\partial}{\partial x}(3x^2y) = 3y\frac{d}{dx}(x^2) = 6xy$$

---

[3] In vector notation, $\mathbf{E}$ is often written

$$\mathbf{E} = -\nabla V = -\left(\mathbf{i}\frac{\partial}{\partial x} + \mathbf{j}\frac{\partial}{\partial y} + \mathbf{k}\frac{\partial}{\partial z}\right)V$$ where ∇ is called the *gradient operator*.

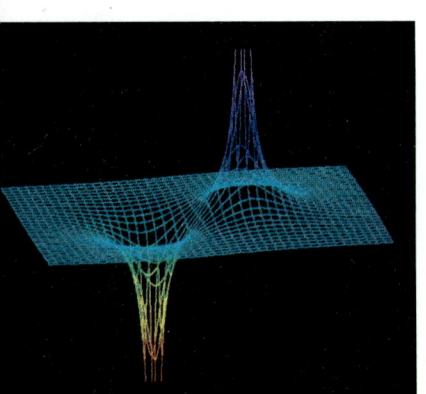

Computer-generated plot of the electric potential associated with an electric dipole. The charges lie in the horizontal plane, at the centers of the potential spikes. The contour lines help visualize the size of the potential, whose values are plotted vertically. *(Richard Megna 1990, Fundamental Photographs)*

CONCEPTUAL EXAMPLE 25.6

If the electric potential is constant in some region, what can you conclude about the electric field in that region? If the electric field is zero in some region, what can you say about the electric potential in that region?

Reasoning   If V is constant in some region, the electric field must be zero in that region. In a one-dimensional situation, $E = -dV/dx$, so if $V = $ constant, $E = 0$. Likewise, if $E = 0$ in some region, one can only conclude that V is a constant in that region.

EXAMPLE 25.7   The Electric Potential of a Dipole

An electric dipole consists of two equal and opposite charges separated by a distance $2a$, as in Figure 25.10. The dipole is along the x axis and is centered at the origin. Calculate the electric potential and the electric field at P.

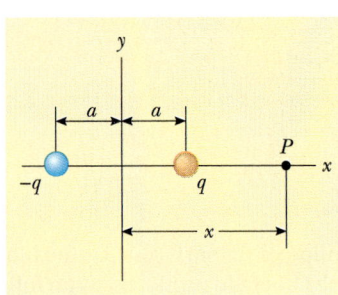

FIGURE 25.10   (Example 25.7) An electric dipole located on the x axis.

Solution

$$V = k_e \sum \frac{q_i}{r_i} = k_e \left(\frac{q}{x - a} - \frac{q}{x + a} \right) = \frac{2 k_e q a}{x^2 - a^2}$$

If P is far from the dipole, so that $x \gg a$, then a^2 can be neglected in the term $x^2 - a^2$ and V becomes

$$V \approx \frac{2 k_e q a}{x^2} \qquad (x \gg a)$$

Using Equation 25.16 and this result, we calculate the electric field at P:

$$E = -\frac{dV}{dx} = \frac{4 k_e q a}{x^3} \qquad \text{for } x \gg a$$

25.5 ELECTRIC POTENTIAL DUE TO CONTINUOUS CHARGE DISTRIBUTIONS

The electric potential due to a continuous charge distribution can be calculated in two ways. If the charge distribution is known, we can start with Equation 25.11 for the potential of a point charge. We then consider the potential due to a small charge element dq, treating this element as a point charge (Fig. 25.11). The potential dV at some point P due to the charge element dq is

$$dV = k_e \frac{dq}{r} \qquad (25.18)$$

This simulator can be used to map the electric field lines and electric potential for an arbitrary collection of fixed point charges. After arranging these point charges in an appropriate manner, you will be able to map the electric fields and lines of equal potential for configurations such as line charges, charges on a ring, and geometries that you can create.

Mapping the Electric Field

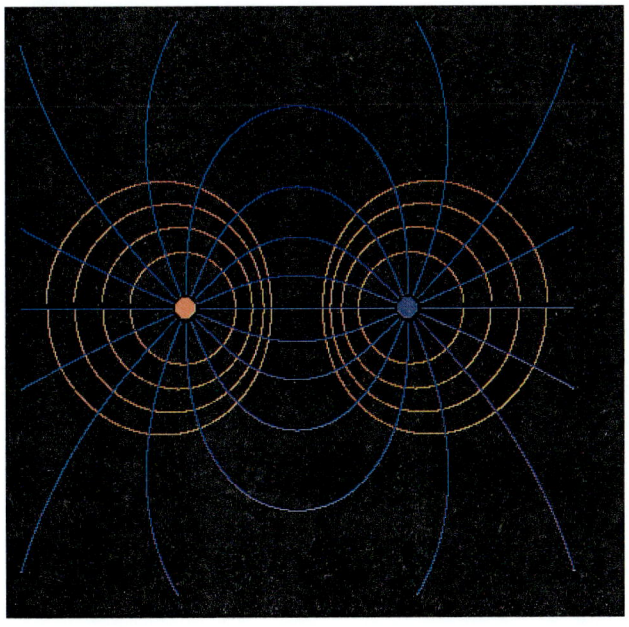

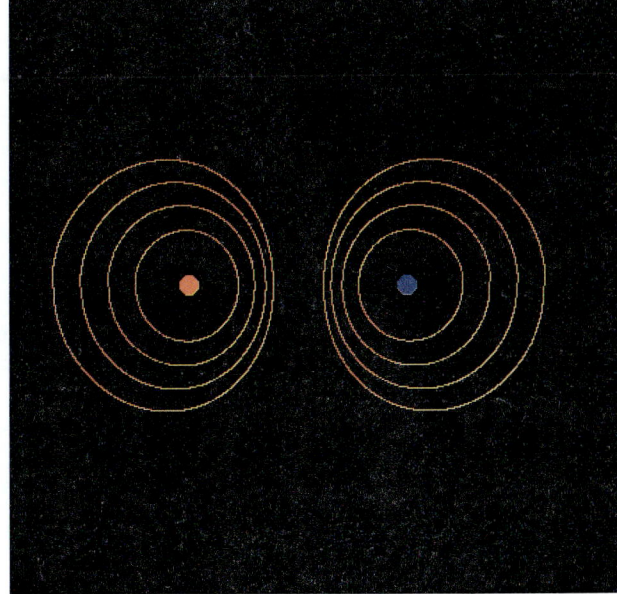

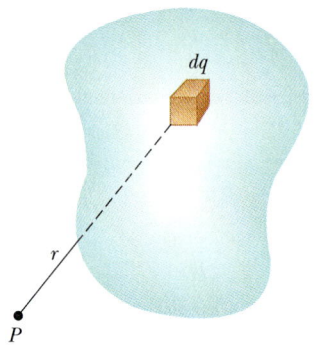

FIGURE 25.11   The electric potential at the point *P* due to a continuous charge distribution can be calculated by dividing the charged body into segments of charge *dq* and summing the potential contributions over all segments.

where *r* is the distance from the charge element to the point *P*. To get the total potential at *P*, we integrate Equation 25.18 to include contributions from all elements of the charge distribution. Since each element is, in general, at a different distance from *P* and since k_e is a constant, we can express *V* as

$$V = k_e \int \frac{dq}{r} \qquad (25.19)$$

In effect, we have replaced the sum in Equation 25.12 with an integral. Note that this expression for *V* uses a particular reference: The potential is taken to be zero when *P* is infinitely far from the charge distribution.

The second method for calculating the potential of a continuous charge distribution makes use of Equation 25.3. This procedure is useful when the electric field is already known from other considerations, such as Gauss's law. If the charge distribution is highly symmetric, we first evaluate **E** at any point using Gauss's law and then substitute the value obtained into Equation 25.3 to determine the potential difference between any two points. We then choose *V* to be zero at some convenient point. Let us illustrate both methods with several examples.

EXAMPLE 25.8   Potential Due to a Uniformly Charged Ring

Find the electric potential at a point *P* located on the axis of a uniformly charged ring of radius *a* and total charge *Q*. The plane of the ring is chosen perpendicular to the *x* axis (Fig. 25.12).

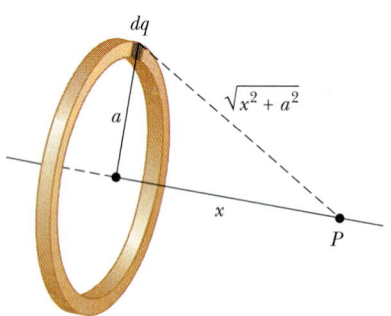

FIGURE 25.12   (Example 25.8) A uniformly charged ring of radius *a*, whose plane is perpendicular to the *x* axis. All segments of the ring are at the same distance from any axial point *P*.

Reasoning and Solution   Let us take *P* to be at a distance *x* from the center of the ring, as in Figure 25.12. The charge element *dq* is at a distance equal to $\sqrt{x^2 + a^2}$ from the point *P*. Hence, we can express *V* as

$$V = k_e \int \frac{dq}{r} = k_e \int \frac{dq}{\sqrt{x^2 + a^2}}$$

In this case, each element *dq* is at the same distance from *P*. Therefore, the term $\sqrt{x^2 + a^2}$ can be removed from the integral, and *V* reduces to

$$V = \frac{k_e}{\sqrt{x^2 + a^2}} \int dq = \frac{k_e Q}{\sqrt{x^2 + a^2}} \qquad (25.20)$$

The only variable in this expression for *V* is *x*. This is not surprising, since our calculation is valid only for points along the *x* axis, where *y* and *z* are both zero. From the symmetry, we see that along the *x* axis **E** can have only an *x* component. Therefore, we can use Equation 25.16 to find the electric field at *P*:

$$E_x = -\frac{dV}{dx} = -k_e Q \frac{d}{dx} (x^2 + a^2)^{-1/2}$$

$$= -k_e Q (-\tfrac{1}{2})(x^2 + a^2)^{-3/2}(2x)$$

$$= \frac{k_e Q x}{(x^2 + a^2)^{3/2}} \qquad (25.21)$$

This result agrees with that obtained by direct integration (see Example 23.11). Note that $E_x = 0$ at $x = 0$ (the center of the ring). Could you have guessed this from Coulomb's law?

Exercise   What is the electric potential at the center of the uniformly charged ring? What does the field at the center imply about this result?

Answer $V = k_e Q/a$ at $x = 0$. Because $E = 0$, *V* must have a maximum or minimum value; it is in fact a maximum.

EXAMPLE 25.9   Potential of a Uniformly Charged Disk

Find the electric potential along the axis of a uniformly charged disk of radius a and charge per unit area σ (Fig. 25.13).

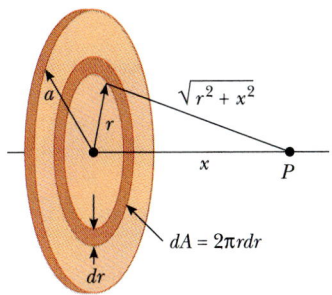

FIGURE 25.13   (Example 25.9) A uniformly charged disk of radius a, whose plane is perpendicular to the x axis. The calculation of the potential at an axial point P is simplified by dividing the disk into rings of area $2\pi r\,dr$.

Reasoning and Solution   Again we choose the point P to be at a distance x from the center of the disk and take the plane of the disk perpendicular to the x axis. The problem is simplified by dividing the disk into a series of charged rings. The potential of each ring is given by Equation 25.20 in Example 25.8. Consider one such ring of radius r and width dr,

as indicated in Figure 25.13. The area of the ring is $dA = 2\pi r\,dr$ (the circumference multiplied by the width), and the charge on the ring is $dq = \sigma\,dA = \sigma 2\pi r\,dr$. Hence, the potential at the point P due to this ring is

$$dV = \frac{k_e\,dq}{\sqrt{r^2 + x^2}} = \frac{k_e\sigma 2\pi r\,dr}{\sqrt{r^2 + x^2}}$$

To find the *total* potential at P, we sum over all rings making up the disk. That is, we integrate dV from $r = 0$ to $r = a$:

$$V = \pi k_e\sigma \int_0^a \frac{2r\,dr}{\sqrt{r^2 + x^2}} = \pi k_e\sigma \int_0^a (r^2 + x^2)^{-1/2}2r\,dr$$

This integral is of the form $u^n\,du$ and has the value $u^{n+1}/(n + 1)$, where $n = -\frac{1}{2}$ and $u = r^2 + x^2$. This gives the result

$$V = 2\pi k_e\sigma[(x^2 + a^2)^{1/2} - x] \qquad (25.22)$$

As in Example 25.8, we can find the electric field at any axial point by taking the negative of the derivative of V with respect to x:

$$E_x = -\frac{dV}{dx} = 2\pi k_e\sigma\left(1 - \frac{x}{\sqrt{x^2 + a^2}}\right) \qquad (25.23)$$

The calculation of V and E for an arbitrary point off the axis is more difficult to perform.

EXAMPLE 25.10   Potential of a Finite Line of Charge

A rod of length ℓ located along the x axis has a uniform charge per unit length and a total charge Q. Find the electric potential at a point P along the y axis a distance d from the origin (Figure 25.14).

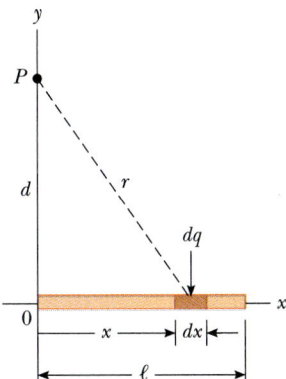

FIGURE 25.14   (Example 25.10) A uniform line charge of length ℓ located along the x axis. To calculate the potential at P, the line charge is divided into segments each of length dx, having a charge $dq = \lambda\,dx$.

Solution   The element of length dx has a charge $dq = \lambda\,dx$, where λ is the charge per unit length, Q/ℓ. Since this element is a distance $r = \sqrt{x^2 + d^2}$ from P, we can express the potential at P due to this element as

$$dV = k_e\frac{dq}{r} = k_e\frac{\lambda\,dx}{\sqrt{x^2 + d^2}}$$

To get the total potential at P, we integrate this expression over the limits $x = 0$ to $x = \ell$. Noting that k_e, λ, and d are constants, we find that

$$V = k_e\lambda \int_0^\ell \frac{dx}{\sqrt{x^2 + d^2}} = k_e\frac{Q}{\ell}\int_0^\ell \frac{dx}{\sqrt{x^2 + d^2}}$$

This integral, found in most integral tables, has the value

$$\int \frac{dx}{\sqrt{x^2 + d^2}} = \ln(x + \sqrt{x^2 + d^2})$$

Evaluating V, we find that

$$V = \frac{k_e Q}{\ell}\ln\left(\frac{\ell + \sqrt{\ell^2 + d^2}}{d}\right) \qquad (25.24)$$

EXAMPLE 25.11   **Potential of a Uniformly Charged Sphere**

An insulating solid sphere of radius R has a uniform positive charge density with total charge Q (Fig. 25.15). (a) Find the electric potential at a point outside the sphere, that is, for $r > R$. Take the potential to be zero at $r = \infty$.

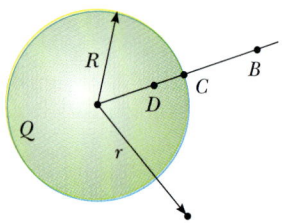

FIGURE 25.15   (Example 25.11) A uniformly charged insulating sphere of radius R and total charge Q. The electric potentials at points B and C are equivalent to those produced by a point charge Q located at the center of the sphere.

Solution   In Example 24.5, we found from Gauss's law that the magnitude of the electric field outside a uniformly charged sphere is

$$E_r = k_e \frac{Q}{r^2} \quad \text{(for } r > R\text{)}$$

where the field is directed radially outward when Q is positive. To obtain the potential at an exterior point, such as B in Figure 25.15, we substitute this expression for E into Equation 25.4. Since $\mathbf{E} \cdot d\mathbf{s} = E_r\, dr$ in this case, we get

$$V_B = -\int_\infty^r E_r\, dr = -k_e Q \int_\infty^r \frac{dr}{r^2}$$

$$V_B = k_e \frac{Q}{r} \quad \text{(for } r > R\text{)}$$

Note that the result is identical to that for the electric potential due to a point charge. Since the potential must be continuous at $r = R$, we can use this expression to obtain the potential at the surface of the sphere. That is, the potential at a point such as C in Figure 25.15 is

$$V_C = k_e \frac{Q}{R} \quad \text{(for } r = R\text{)}$$

(b) Find the potential at a point inside the charged sphere, that is, for $r < R$.

Solution   In Example 24.5 we found that the electric field inside a uniformly charged sphere is

$$E_r = \frac{k_e Q}{R^3} r \quad \text{(for } r < R\text{)}$$

We can use this result and Equation 25.3 to evaluate the potential difference $V_D - V_C$, where D is an interior point:

$$V_D - V_C = -\int_R^r E_r\, dr = -\frac{k_e Q}{R^3} \int_R^r r\, dr = \frac{k_e Q}{2R^3}(R^2 - r^2)$$

Substituting $V_C = k_e Q/R$ into this expression and solving for V_D, we get

$$V_D = \frac{k_e Q}{2R}\left(3 - \frac{r^2}{R^2}\right) \quad \text{(for } r < R\text{)} \quad (25.25)$$

At $r = R$, this expression gives a result for the potential that agrees with that for the potential at the surface, that is, V_C. A plot of V versus r for this charge distribution is given in Figure 25.16.

Exercise   What are the electric field and the electric potential at the center of a uniformly charged sphere?

Answer $E = 0$ and $V_0 = 3k_e Q/2R$.

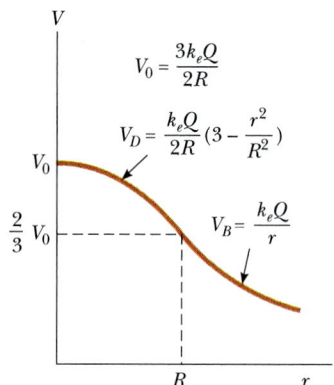

FIGURE 25.16   (Example 25.11) A plot of the electric potential V versus the distance r from the center of a uniformly charged, insulating sphere of radius R. The curve for V_D inside the sphere is parabolic and joins smoothly with the curve for V_B outside the sphere, which is a hyperbola. The potential has a maximum value V_0 at the center of the sphere.

25.6   POTENTIAL OF A CHARGED CONDUCTOR

In the previous chapter we found that when a solid conductor in equilibrium carries a net charge, the charge resides on the outer surface of the conductor. Furthermore, we showed that the electric field just outside the surface of a con-

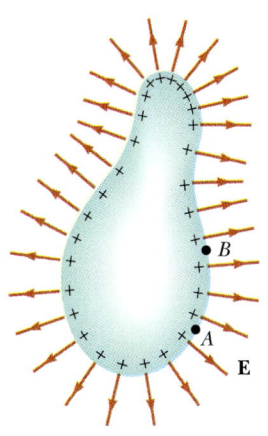

FIGURE 25.17 An arbitrarily shaped conductor with an excess positive charge. When the conductor is in electrostatic equilibrium, all of the charge resides at the surface, E = 0 inside the conductor, and the electric field just outside the conductor is perpendicular to the surface. The potential is constant inside the conductor and is equal to the potential at the surface. The surface charge density is nonuniform.

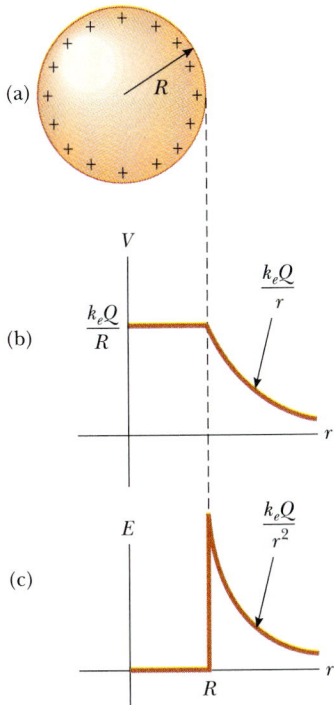

FIGURE 25.18 (a) The excess charge on a conducting sphere of radius R is uniformly distributed on its surface. (b) The electric potential versus the distance r from the center of the charged conducting sphere. (c) The electric field intensity versus the distance r from the center of the charged conducting sphere.

ductor in equilibrium is perpendicular to the surface while the field inside the conductor is zero. If the electric field had a component parallel to the surface, this would cause surface charges to move, creating a current and a nonequilibrium situation.

We now show that *every point on the surface of a charged conductor in equilibrium is at the same potential.* Consider two points A and B on the surface of a charged conductor, as in Figure 25.17. Along a surface path connecting these points, E is always perpendicular to the displacement $d\mathbf{s}$; therefore, $\mathbf{E} \cdot d\mathbf{s} = 0$. Using this result and Equation 25.3, we conclude that the potential difference between A and B is necessarily zero:

$$V_B - V_A = -\int_A^B \mathbf{E} \cdot d\mathbf{s} = 0$$

This result applies to any two points on the surface. Therefore, V is constant everywhere on the surface of a charged conductor in equilibrium. That is,

> the surface of any charged conductor in equilibrium is an equipotential surface. Furthermore, since the electric field is zero inside the conductor, we conclude that the potential is constant everywhere inside the conductor and equal to its value at the surface.

Therefore, no work is required to move a test charge from the interior of a charged conductor to its surface. (Note that the potential is not zero inside the conductor even though the electric field is zero.)

For example, consider a solid metal sphere of radius R and total positive charge Q, as shown in Figure 25.18a. The electric field outside the charged sphere is $k_e Q/r^2$ and points radially outward. Following Example 25.11, we see that the potential at the interior and surface of the sphere must be $k_e Q/R$ relative to infinity. The potential outside the sphere is $k_e Q/r$. Figure 25.18b is a plot of the potential as a function of r, and Figure 25.18c shows the variations of the electric field with r.

When a net charge is placed on a spherical conductor, the surface charge density is uniform, as indicated in Figure 25.18a. However, if the conductor is nonspherical, as in Figure 25.17, the surface charge density is high where the radius of curvature is small and convex and low where the radius of curvature is

The surface of a charged conductor is an equipotential surface

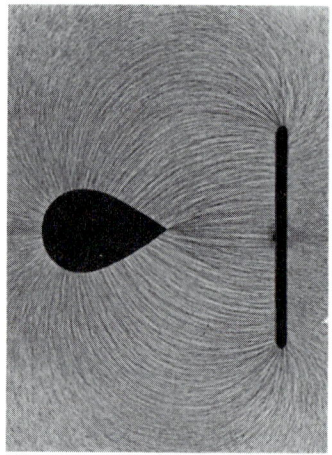

Electric field pattern of a charged conducting plate near an oppositely charged pointed conductor. Small pieces of thread suspended in oil align with the electric field lines. Note that the electric field is most intense near the pointed part of the conductor and at other points where the radius of curvature is small.   *(Courtesy of Harold M. Waage, Princeton University)*

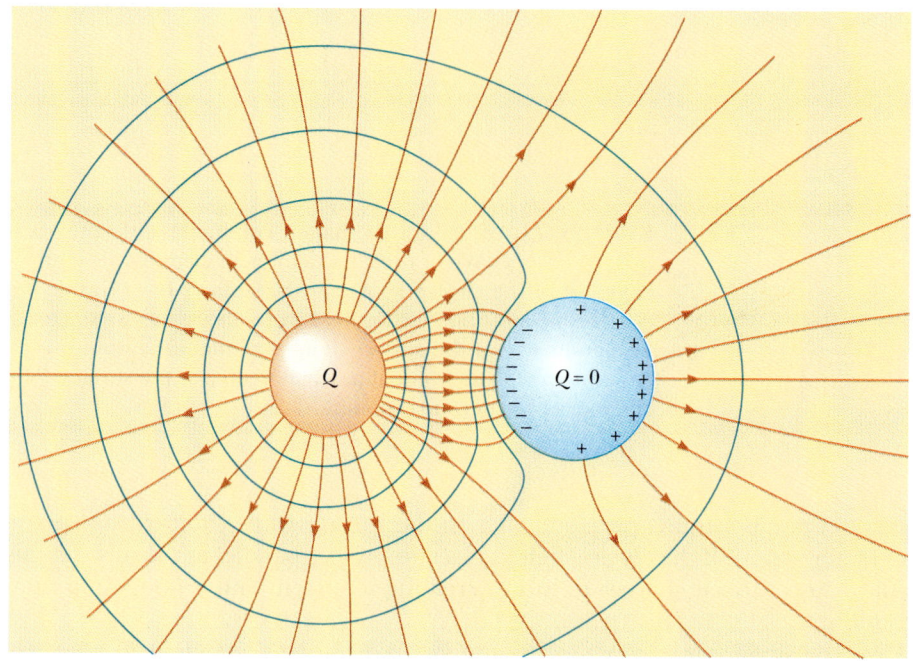

FIGURE 25.19   The electric field lines (in red) around two spherical conductors. The smaller sphere on the left has a net charge Q, and the sphere on the right has zero net charge. The blue lines represent the edges of the equipotential surfaces.   *(From E. Purcell, Electricity and Magnetism, New York, McGraw-Hill, 1965, with permission of the Education Development Center, Inc.)*

small and concave. Since the electric field just outside a charged conductor is proportional to the surface charge density, σ, we see that *the electric field is large near points having small convex radii of curvature and reaches very high values at sharp points.*

Figure 25.19 shows the electric field lines around two spherical conductors, one with a net charge Q and a larger one with zero net charge. In this case, the surface charge density is not uniform on either conductor. The sphere having zero net charge has negative charges induced on its side that faces the charged sphere and positive charge on its side opposite the charged sphere. The blue lines in Figure 25.19 represent the boundaries of the equipotential surfaces for this charge configuration. Again, at all points the field lines are perpendicular to the conducting surfaces. Furthermore, the equipotential surfaces are perpendicular to the field lines at the boundaries of the conductor and everywhere else in space.

A Cavity Within a Conductor

Now consider a conductor of arbitrary shape containing a cavity as in Figure 25.20. Let us assume there are no charges inside the cavity. *The electric field inside the cavity must be zero,* regardless of the charge distribution on the outside surface of the conductor. Furthermore, the field in the cavity is zero even if an electric field exists outside the conductor.

In order to prove this point, we use the fact that every point on the conductor is at the same potential, and therefore any two points A and B on the surface of the cavity must be at the same potential. Now imagine that a field $\mathbf{E}$ exists in the cavity, and evaluate the potential difference $V_B - V_A$ defined by the expression

$$V_B - V_A = -\int_A^B \mathbf{E} \cdot d\mathbf{s}$$

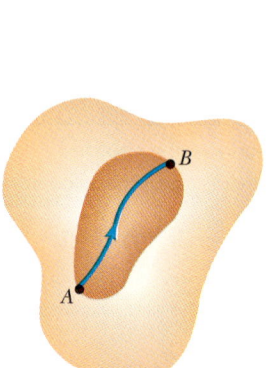

FIGURE 25.20   A conductor in electrostatic equilibrium containing an empty cavity. The electric field in the cavity is *zero*, regardless of the charge on the conductor.

If **E** is nonzero, we can always find a path between A and B for which $\mathbf{E} \cdot d\mathbf{s}$ is always a positive number, and so the integral must be positive. However, since $V_B - V_A = 0$, the integral must also be zero. This contradiction can be reconciled only if $\mathbf{E} = 0$ inside the cavity. Thus, we conclude that a cavity surrounded by conducting walls is a field-free region as long as there are no charges inside the cavity.

This result has some interesting applications. For example, it is possible to shield an electronic circuit or even an entire laboratory from external fields by surrounding it with conducting walls. Shielding is often necessary when performing highly sensitive electrical measurements.

Corona Discharge

A phenomenon known as **corona discharge** is often observed near sharp points of a conductor raised to a high potential. The discharge appears as a greenish glow visible to the naked eye. In this process, air becomes a conductor as a result of the ionization of air molecules in regions of high electric fields. At standard temperature and pressure, this discharge occurs at electric field strengths equal to or greater than approximately 3×10^6 V/m. Because air contains a small number of ions (produced, for example, by cosmic rays), a charged conductor attracts ions of the opposite sign from the air. Near sharp points, where the field is very high, the ions in the air are accelerated to high speeds. These energetic ions collide with other air molecules, producing more ions and an increase in conductivity of the air. The discharge of the conductor is often accompanied by a visible glow surrounding the sharp points.

EXAMPLE 25.12 Two Connected Charged Spheres

Two spherical conductors of radii r_1 and r_2 are separated by a distance much larger than the radius of either sphere. The spheres are connected by a conducting wire as in Figure 25.21. If the charges on the spheres in equilibrium are q_1 and q_2, respectively, find the ratio of the field strengths at the surfaces of the spheres.

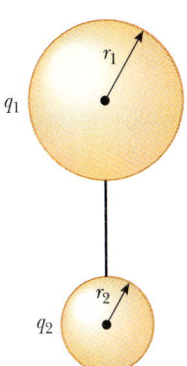

FIGURE 25.21 (Example 25.12) Two charged spherical conductors connected by a conducting wire. The spheres are at the *same* potential, *V*.

Solution Since the spheres are connected by a conducting wire, they must both be at the same potential

$$V = k_e \frac{q_1}{r_1} = k_e \frac{q_2}{r_2}$$

Therefore, the ratio of charges is

$$(1) \qquad \frac{q_1}{q_2} = \frac{r_1}{r_2}$$

Since the spheres are very far apart, their surfaces are uniformly charged, and we can express the magnitude of the electric fields at their surfaces as

$$E_1 = k_e \frac{q_1}{r_1^2} \qquad \text{and} \qquad E_2 = k_e \frac{q_2}{r_2^2}$$

Taking the ratio of these two fields and making use of (1), we find that

$$(2) \qquad \frac{E_1}{E_2} = \frac{r_2}{r_1}$$

Hence, the field is more intense in the vicinity of the smaller sphere.

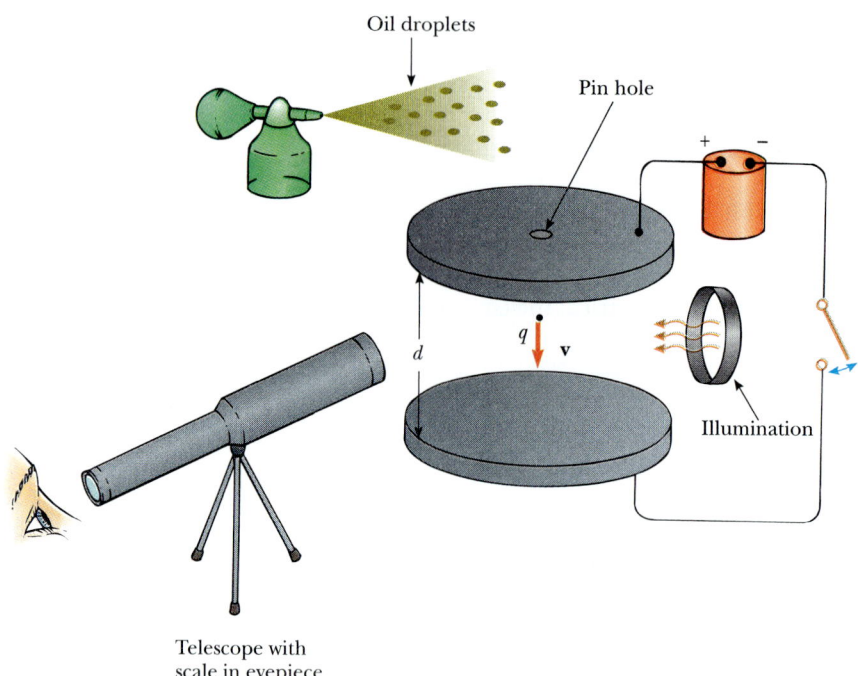

FIGURE 25.22   A schematic view of the Millikan oil-drop apparatus.

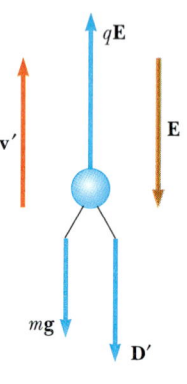

(a) Field off

(b) Field on

FIGURE 25.23   The forces on a charged oil droplet in the Millikan experiment.

*25.7 THE MILLIKAN OIL-DROP EXPERIMENT

During the period 1909 to 1913, Robert A. Millikan (1868–1953) performed a brilliant set of experiments in which he measured the elementary charge on an electron, *e*, and demonstrated the quantized nature of the electronic charge. The apparatus used by Millikan, diagrammed in Figure 25.22, contains two parallel metal plates. Charged oil droplets from an atomizer are allowed to pass through a small hole in the upper plate. A light beam directed horizontally is used to illuminate the oil droplets, which are viewed by a telescope whose axis is at right angles to the light beam. When the droplets are viewed in this manner, they appear as shining stars against a dark background, and the rate of fall of individual drops may be determined.[4]

Let us assume a single drop having a mass *m* and carrying a charge *q* is being viewed, and that its charge is negative. If there is no electric field present between the plates, the two forces acting on the charge are the force of gravity, *mg*, acting downward, and an upward viscous drag force **D**, as indicated in Figure 25.23a. The drag force is proportional to the speed of the drop. When the drop reaches its terminal speed *v*, the two forces balance each other ($mg = D$).

Now suppose that an electric field is set up between the plates by connecting a battery such that the upper plate is at the higher potential. In this case, a third force $q\mathbf{E}$ acts on the charged drop. Since *q* is negative and **E** is downward, this electric force is directed upward as in Figure 25.23b. If this force is large enough, the drop moves upward and the drag force D' acts downward. When the upward

---

[4] At one time, the oil droplets were termed "Millikan's Shining Stars." Perhaps this description has lost its popularity because of the generations of physics students who have experienced hallucinations, near blindness, migraine headaches, and so forth while repeating this experiment!

electric force qE balances the sum of the force of gravity and the drag force, both acting downward, the drop reaches a new terminal speed v'.

With the field turned on, a drop moves slowly upward, typically at rates of hundredths of a centimeter per second. The rate of fall in the absence of a field is comparable. Hence, a single droplet with constant mass and radius may be followed for hours, alternately rising and falling, by simply turning the electric field on and off.

After making measurements on thousands of droplets, Millikan and his co-workers found that all droplets, to within about 1% precision, had a charge equal to some integer multiple of the elementary charge e:

$$q = ne \qquad n = 0, -1, -2, -3, \ldots$$

where $e = 1.60 \times 10^{-19}$ C. Millikan's experiment is conclusive evidence that charge is quantized. He was awarded the Nobel Prize in Physics in 1923 for this work.

*25.8 APPLICATIONS OF ELECTROSTATICS

The principles of electrostatics have been used in various applications, a few of which we briefly discuss in this section. Some of the more practical applications include electrostatic precipitators, used to reduce atmospheric pollution from coal-burning power plants, and xerography, the process that has revolutionized imaging process technology. Scientific applications of electrostatic principles include electrostatic generators for accelerating elementary charged particles and the field-ion microscope, which is used to image atoms on the surface of metallic samples.

The Van de Graaff Generator

In the previous chapter we described an experiment that demonstrates a method for transferring charge to a hollow conductor (the Faraday ice-pail experiment). When a charged conductor is placed in contact with the inside of a hollow conductor, all of the charge of the first conductor is transferred to the hollow conductor. In principle, the charge on the hollow conductor and its potential can be increased without limit by repeating the process.

In 1929, Robert J. Van de Graaff used this principle to design and build an electrostatic generator. This type of generator is used extensively in nuclear physics research. The basic idea is described in Figure 25.24. Charge is delivered continuously to a high-voltage electrode on a moving belt of insulating material. The high-voltage electrode is a hollow conductor mounted on an insulating column. The belt is charged at A by means of a corona discharge between comb-like metallic needles and a grounded grid. The needles are maintained at a positive potential of typically 10^4 V. The positive charge on the moving belt is transferred to the high-voltage electrode by a second comb of needles at B. Since the electric field inside the hollow conductor is negligible, the positive charge on the belt easily transfers to the high-voltage electrode, regardless of its potential. In practice, it is possible to increase the potential of the high-voltage electrode until electrical discharge occurs through the air. Since the "breakdown" electric field in air is equal to about 3×10^6 V/m, a sphere 1 m in radius can be raised to a maximum potential of 3×10^6 V. The potential can be increased further by increasing the radius of the hollow conductor and by placing the entire system in a container filled with high-pressure gas.

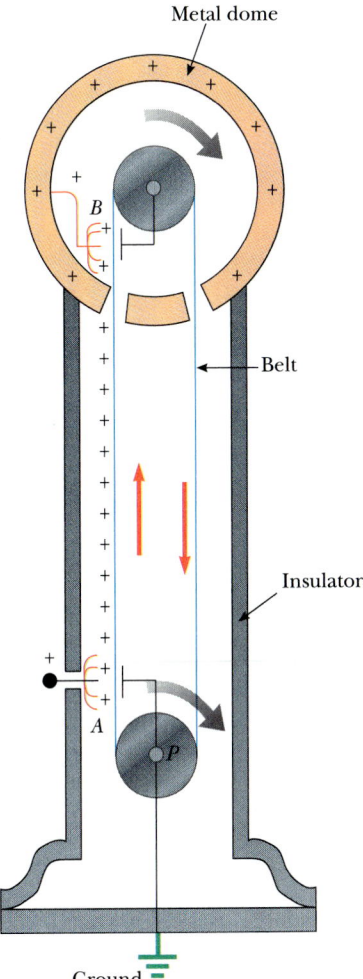

FIGURE 25.24   Schematic diagram of a Van de Graaff generator. Charge is transferred to the hollow conductor at the top by means of a moving belt. The charge is deposited on the belt at point A and is transferred to the hollow conductor at point B.

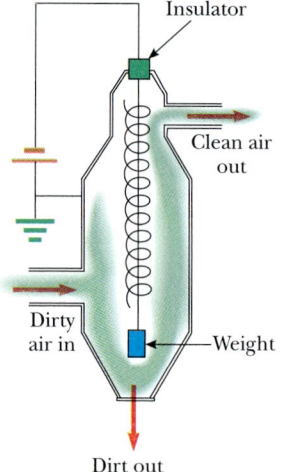

FIGURE 25.25 Schematic diagram of an electrostatic precipitator. The high negative voltage maintained on the central wire creates an electrical discharge in the vicinity of the wire.

Van de Graaff generators can produce potential differences as high as 20 million volts. Protons accelerated through such potential differences receive enough energy to initiate nuclear reactions between themselves and various target nuclei.

The Electrostatic Precipitator

One important application of electrical discharge in gases is a device called an *electrostatic precipitator*. This device is used to remove particulate matter from combustion gases, thereby reducing air pollution. They are especially useful in coal-burning power plants and in industrial operations that generate large quantities of smoke. Current systems are able to eliminate more than 99% of the ash and dust (by weight) from the smoke.

Figure 25.25 shows the basic idea of the electrostatic precipitator. A high voltage (typically 40 kV to 100 kV) is maintained between a wire running down the center of a duct and the outer wall, which is grounded. The wire is maintained at a negative potential with respect to the walls, and so the electric field is directed toward the wire. The electric field near the wire reaches high enough values to cause a corona discharge around the wire and the formation of positive ions, electrons, and negative ions, such as O_2^-. As the electrons and negative ions are accelerated toward the outer wall by the nonuniform electric field, the dirt particles in the streaming gas become charged by collisions and ion capture. Since most of the charged dirt particles are negative, they are also drawn to the outer wall by the electric field. By periodically shaking the duct, the particles fall loose and are collected at the bottom.

In addition to reducing the level of particulate matter in the atmosphere, the electrostatic precipitator recovers valuable materials from the stack in the form of metal oxides.

Xerography

The process of xerography is widely used for making photocopies of letters, documents, and other printed materials. The basic idea for the process was developed by Chester Carlson, for which he was granted a patent in 1940. The one idea that makes the process unique is the use of a photoconductive material to form an image. (A photoconductor is a material that is a poor conductor in the dark but becomes a good electrical conductor when exposed to light.)

The process is illustrated in Figure 25.26. First, the surface of a plate or drum is coated with a thin film of the photoconductive material (usually selenium or some compound of selenium), and the photoconductive surface is given a positive electrostatic charge in the dark. The page to be copied is then projected onto the charged surface. The photoconducting surface becomes conducting only in areas where light strikes. In these areas, the light produces charge carriers in the photoconductor, which neutralize the positively charged surface. However, the charges remain on those areas of the photoconductor not exposed to light, leaving a latent (hidden) image of the object in the form of a positive surface charge distribution.

Next, a negatively charged powder called a *toner* is dusted onto the photoconducting surface. The charged powder adheres only to those areas of the surface that contain the positively charged image. At this point, the image becomes visible. The image is then transferred to the surface of a sheet of positively charged paper.

Finally, the toner material is "fixed" to the surface of the paper through the application of heat. This results in a permanent copy of the original.

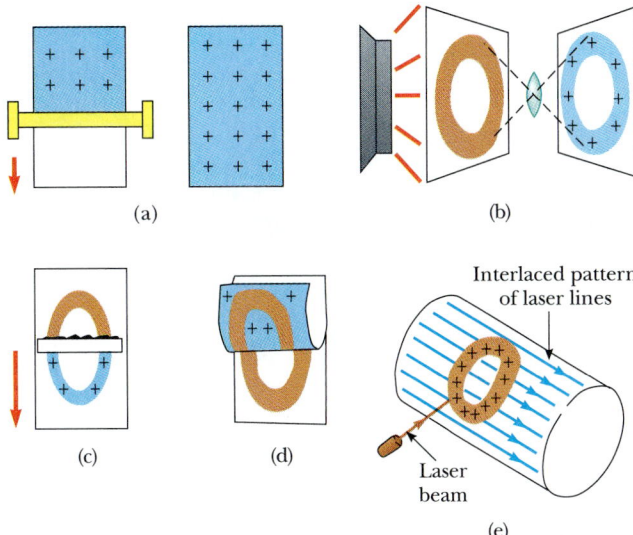

(a)

(b)

(c)

(d)

Interlaced pattern
of laser lines

Laser
beam

(e)

FIGURE 25.26 The xerographic process: (a) The photoconductive surface is positively charged. (b) Through the use of a light source and lens, an image is formed on the surface in the form of hidden positive charges. (c) The surface containing the image is covered with a charged powder, which adheres only to the image area. (d) A piece of paper is placed over the surface and given a charge. This transfers the visible image to the paper, which is then heat-treated to ''fix'' the powder to the paper. (e) A laser printer operates similarly except the image is produced by turning a laser beam on and off as it sweeps across the selenium-coated drum.

The Field-Ion Microscope

In Section 25.6 we pointed out that the electric field intensity can be very high in the vicinity of a sharp point on a charged conductor. A device that makes use of this intense field is the *field-ion microscope,* invented in 1956 by E. W. Mueller of the Pennsylvania State University. The purpose of this device is to look at individual atoms on the surface of a very small specimen.

The basic construction is shown in Figure 25.27. A specimen to be studied is fabricated from a fine wire, and a sharp needle-shaped tip is formed, usually by etching the wire in an acid. Typically, the diameter of the tip is about 0.1 μm. The specimen is placed at the center of an evacuated glass tube containing a fluorescent screen. Next, a small amount of helium is introduced into the vessel. A very high voltage (potential difference) is applied between the specimen and the screen, producing a very intense electric field near the tip of the specimen. The helium atoms in the vicinity of this high-field region are ionized, stripped of an electron. This leaves the helium positively charged. The positively charged He$^+$ ions then accelerate to the negatively charged fluorescent screen. This results in a pattern on the screen that represents an image of the tip of the specimen.

It is important to cool the tip of the specimen to at least the temperature of liquid nitrogen to slow down the atoms and obtain pictures. Under the proper conditions (low specimen temperature and high vacuum), the images of the individual atoms on the surface of the sample are visible, and the atomic arrangement on the surface can be studied. Unfortunately, the high electric fields also set up large mechanical stresses near the tip of the specimen, which limits the application of the technique to strong metallic elements, such as tungsten and rhenium. Figure 25.28 represents a typical field-ion microscope pattern of a platinum crystal.

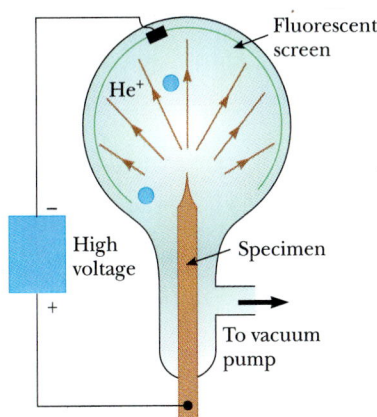

FIGURE 25.27 Schematic diagram of a field-ion microscope. The electric field is very intense at the tip of the needle-shaped specimen.

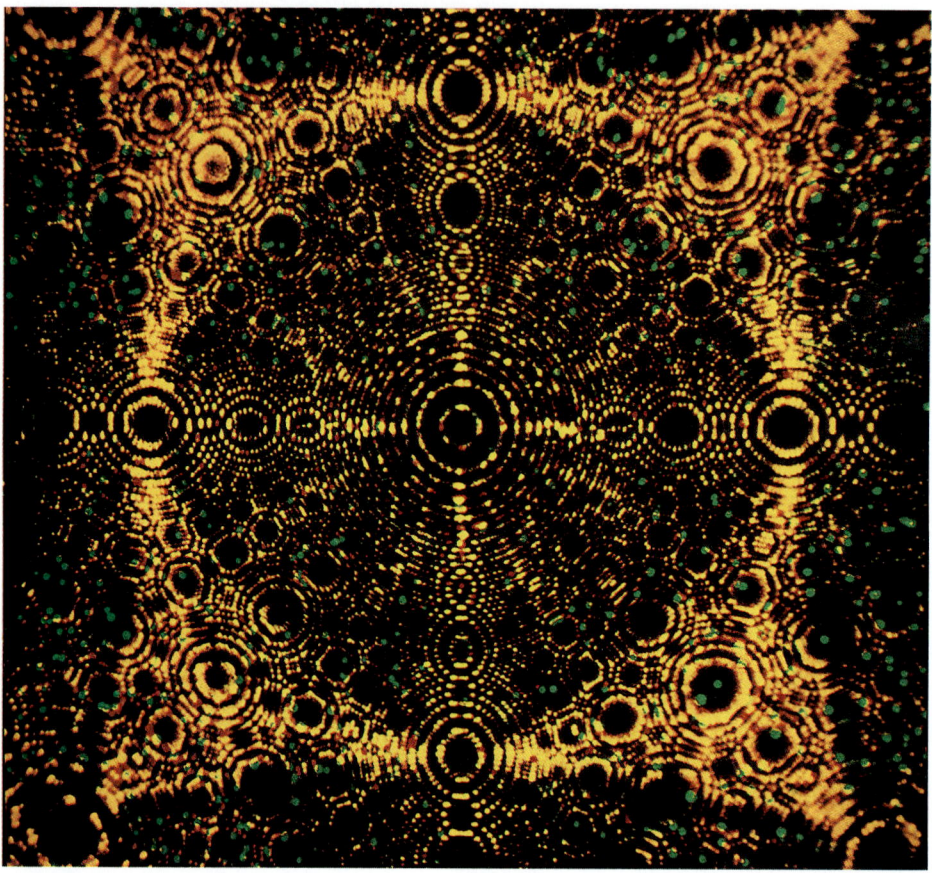

FIGURE 25.28   Field-ion microscope image of the surface of a platinum crystal with a magnification of 1 000 000×. Individual atoms can be seen on surface layers using this technique.   *(Manfred Kage/Peter Arnold, Inc.)*

SUMMARY

When a positive test charge q_0 is moved between points A and B in an electric field **E**, the **change in the potential energy** is

$$\Delta U = - q_0 \int_A^B \mathbf{E} \cdot d\mathbf{s} \tag{25.1}$$

The **potential difference** ΔV between points A and B in an electric field **E** is defined as

$$\Delta V = \frac{\Delta U}{q_0} = - \int_A^B \mathbf{E} \cdot d\mathbf{s} \tag{25.3}$$

where the electric potential V is a scalar and has the units of J/C, where $1\,\text{J/C} \equiv 1\,\text{V}$.

The potential difference between two points A and B in a uniform electric field **E** is

$$\Delta V = - Ed \tag{25.6}$$

where d is the displacement in the direction parallel to **E**.

Equipotential surfaces are surfaces on which the electric potential remains constant. Equipotential surfaces are perpendicular to electric field lines. The potential due to a point charge q at any distance r from the charge is

$$V = k_e \frac{q}{r} \qquad (25.11)$$

The potential due to a group of point charges is obtained by summing the potentials due to the individual charges. Since V is a scalar, the sum is a simple algebraic operation.

The **potential energy of a pair of point charges** separated by a distance r_{12} is

$$U = k_e \frac{q_1 q_2}{r_{12}} \qquad (25.13)$$

This energy represents the work required to bring the charges from an infinite separation to the separation r_{12}. The potential energy of a distribution of point charges is obtained by summing terms like Equation 25.13 over all pairs of particles.

If the electric potential is known as a function of coordinates x, y, z, the components of the electric field can be obtained by taking the negative derivative of the potential with respect to the coordinates. For example, the x component of the electric field is

$$E_x = -\frac{dV}{dx} \qquad (25.16)$$

The **electric potential due to a continuous charge distribution** is

$$V = k_e \int \frac{dq}{r} \qquad (25.19)$$

Every point on the surface of a charged conductor in electrostatic equilibrium is at the same potential. Furthermore, the potential is constant everywhere inside the conductor and equal to its value at the surface. Table 25.1 lists potentials due to several charge distributions.

TABLE 25.1 Potentials Due to Various Charge Distributions

Charge Distribution	Electrical Potential	Location
Uniformly charged ring of radius a	$V = k_e \dfrac{Q}{\sqrt{x^2 + a^2}}$	Along the axis of the ring, a distance x from its center
Uniformly charged disk of radius a	$V = 2\pi k_e \sigma[(x^2 + a^2)^{1/2} - x]$	Along the axis of the disk, a distance x from its center
Uniformly charged, *insulating* solid sphere of radius R and total charge Q	$\begin{cases} V = k_e \dfrac{Q}{r} \\[2mm] V = \dfrac{k_e Q}{2R}\left(3 - \dfrac{r^2}{R^2}\right) \end{cases}$	$r \geq R$ $r < R$

Problem-Solving Strategy and Hints
Calculating the Electric Potential

- When working problems involving electric potential, remember that potential is a scalar quantity and so there are no components to worry about. Therefore, when using the superposition principle to evaluate the electric potential at a point due to a system of point charges, you simply take the algebraic sum of the potentials due to each charge. However, you must keep track of signs. The potential for each positive charge ($V = k_e q/r$) is positive, while the potential for each negative charge is negative.
- Just as in mechanics, only *changes* in potential are significant; hence, the point where you choose the potential to be zero is arbitrary. When dealing with point charges or a finite-sized charge distribution, we usually define $V = 0$ to be at a point infinitely far from the charges. However, if the charge distribution itself extends to infinity, or if the problem concerns the potential difference between two charge distributions, some other nearby point must be selected as the reference point.
- The electric potential at some point P due to a continuous distribution of charge can be evaluated by dividing the charge distribution into infinitesimal elements of charge dq located at a distance r from the point P. You then treat this element as a point charge, so that the potential at P due to the element is $dV = k_e\, dq/r$. The total potential at P is obtained by integrating dV over the entire charge distribution. In performing the integration for most problems, it is necessary to express dq and r in terms of a single variable. In order to simplify the integration, it is important to give careful consideration of the geometry involved in the problem. You should review Examples 25.9 through 25.11 as guides for using this method.
- Another method that can be used to obtain the potential due to a finite continuous charge distribution is to start with the definition of the potential difference given by Equation 25.3. If **E** is known or can be obtained easily (say from Gauss's law), then the line integral of $\mathbf{E} \cdot d\mathbf{s}$ can be evaluated. An example of this method is given in Example 25.11.
- Once you know the electric potential at a point, it is possible to obtain the electric field at that point by remembering that the electric field is equal to the negative of the derivative of the potential with respect to some coordinate. Example 25.8 illustrates how to use this procedure.

QUESTIONS

1. Distinguish between electric potential and electrical potential energy.
2. A negative charge moves in the direction of a uniform electric field. Does the potential energy of the charge increase or decrease? Does it move to a position of higher or lower potential?
3. Give a physical explanation of the fact that the potential energy of a pair of like charges is positive whereas the potential energy of a pair of unlike charges is negative.
4. A uniform electric field is parallel to the x axis. In what direction can a charge be displaced in this field without any external work being done on the charge?
5. Explain why equipotential surfaces are always perpendicular to electric field lines.
6. Describe the equipotential surfaces for (a) an infinite line of charge and (b) a uniformly charged sphere.
7. Explain why, under static conditions, all points in a conductor must be at the same electric potential.

8. The electric field inside a hollow, uniformly charged sphere is zero. Does this imply that the potential is zero inside the sphere? Explain.

9. The potential of a point charge is defined to be zero at an infinite distance. Why can we not define the potential of an infinite line of charge to be zero at $r = \infty$?

10. Two charged conducting spheres of different radii are connected by a conducting wire as in Figure 25.21. Which sphere has the greater charge density?

11. What determines the maximum potential to which the dome of a Van de Graaff generator can be raised?

12. In what type of weather would a car battery be more likely to discharge and why?

13. Explain the origin of the glow sometimes observed around the cables of a high-voltage power line.

14. Why is it important to avoid sharp edges or points on conductors used in high-voltage equipment?

15. How would you shield an electronic circuit or laboratory from stray electric fields? Why does this work?

16. Why is it relatively safe to stay in an automobile with a metal body during a severe thunderstorm?

17. Walking across a carpet and then touching someone can result in a shock. Explain why this occurs.

PROBLEMS

Review Problem

A uniformly charged insulating sphere of radius R has a total positive charge Q. (a) Use Gauss's law to find the magnitude of the electric field at a point outside the sphere, that is, for $r > R$. (b) Use Gauss's law to find the magnitude of the electric field at a point inside the sphere, that is, for $r < R$. (c) Use the result to part (a) and find an expression for the electric potential at a point outside the sphere, taking the potential to be zero at $r = \infty$. (d) Use the results to (b) and (c) to find the electric potential at a point inside the sphere. (e) Sketch graphs of the electric field versus r and the electric potential versus r. (f) If a negative point charge $-q$ having mass m is located at an exterior point, what is the electric force on the charge? (g) When the negative point charge is at an exterior point, what is the potential energy of the system? (h) If the negative point charge is released from an exterior point, find its speed when it reaches the surface of the sphere.

Section 25.1   Potential Difference and Electric Potential

1. The gap between electrodes in a spark plug is 0.060 cm. To produce an electric spark in a gasoline-air mixture, an electric field of 3.0×10^6 V/m must be achieved. When starting the car, what minimum voltage must be supplied by the ignition circuit?

2. What change in potential energy does a 12-μC charge experience when it is moved between two points for which the potential difference is 65 V? Express the answer in electron volts.

3. (a) Calculate the speed of a proton that is accelerated from rest through a potential difference of 120 V. (b) Calculate the speed of an electron that is accelerated through the same potential difference.

4. Through what potential difference would an electron need to be accelerated for it to achieve a speed of 40% of the speed of light ($c = 3.0 \times 10^8$ m/s), starting from rest?

5. A deuteron (a nucleus that consists of one proton and one neutron) is accelerated through a 2.7-kV potential difference. (a) How much energy does it gain? (b) How fast is it going if it starts from rest?

6. What potential difference is needed to stop an electron having an initial speed of 4.2×10^5 m/s?

7. An ion accelerated through a potential difference of 115 V experiences an increase in kinetic energy of 7.37×10^{-17} J. Calculate the charge on the ion.

8. In a Van de Graaff accelerator a proton is accelerated through a potential difference of 14×10^6 V. Assuming that the proton starts from rest, calculate its (a) final kinetic energy in joules, (b) final kinetic energy in electron volts, and (c) final speed.

9. A positron has the same mass as an electron. When a positron is accelerated from rest between two points at a fixed potential difference, it acquires a speed that is 30% of the speed of light. What speed is achieved by a proton accelerated from rest between the same two points?

Section 25.2   Potential Differences in a Uniform Electric Field

10. Consider two points in an electric field. The potential at P_1 is $V_1 = -30$ V, and the potential at P_2 is $V_2 = +150$ V. How much work is done by an external force in moving a charge $q = -4.7$ μC from P_2 to P_1?

11. How much work is done (by a battery, generator, or other source of electrical energy) in moving Avoga-

□ indicates problems that have full solutions available in the Student Solutions Manual and Study Guide.

dro's number of electrons from an initial point where the electric potential is 9 V to a point where the potential is -5 V? (The potential in each case is measured relative to a common reference point.)

12. Two parallel plates are separated by 0.30 mm. If a 20-V potential difference is maintained between those plates, calculate the electric field strength in the region between them.

13. The magnitude of the electric field between two charged parallel plates separated by 1.8 cm is 2.4×10^4 N/C. Find the potential difference between the two plates. How much kinetic energy is gained by a deuteron in accelerating from the positive to the negative plate?

14. On planet Tehar, the gravitational field strength is the same as that on Earth but on Tehar there is also a strong downward electric field that is uniform close to the planet's surface. A 2.00-kg ball carrying a charge of 5.00 μC is thrown upward at 20.1 m/s and hits the ground after 4.10 s. What is the potential difference between the starting point and the top point of the trajectory?

14A. On planet Tehar, the gravitational field strength is the same as that on Earth but on Tehar there is also a strong downward electric field that is uniform close to the planet's surface. A ball of mass m carrying a charge q is thrown upward at a speed v and hits the ground after an interval t. What is the potential difference between the starting point and the top point of the trajectory?

15. An electron moving parallel to the x axis has an initial speed of 3.7×10^6 m/s at the origin. Its speed is reduced to 1.4×10^5 m/s at the point $x = 2.0$ cm. Calculate the potential difference between the origin and this point. Which point is at the higher potential?

16. A positron has the same charge as a proton but the same mass as an electron. Suppose a positron moves 5.2 cm in the direction of a uniform 480-V/m electric field. (a) How much potential energy does it gain or lose? (b) How much kinetic energy?

17. An electron in the beam of a typical television picture tube is accelerated through a potential difference of 20 kV before striking the face of the tube. (a) What is the energy of this electron, in electron volts, and what is its speed when it strikes the screen? (b) How much momentum is imparted to the screen by the electron?

18. A 4.00-kg block carrying a charge $Q = 50.0$ μC is connected to a spring for which $k = 100$ N/m. The block lies on a frictionless horizontal track, the system is immersed in a uniform electric field of magnitude $E = 5.00 \times 10^5$ V/m and is directed as in Figure P25.18. If the block is released from rest when the spring is unstretched (at $x = 0$), (a) by what max-

mum amount does the spring expand? (b) What will be the equilibrium position of the block? (c) Show that the block's motion is simple harmonic, and determine its period. (d) Repeat part (a) if the coefficient of kinetic friction between block and surface is 0.200.

18A. A block having mass m and charge Q is connected to a spring having constant k. The block lies on a frictionless horizontal track, the system is immersed in a uniform electric field of magnitude E and is directed as in Figure P25.18. If the block is released from rest when the spring is unstretched (at $x = 0$), (a) by what maximum amount does the spring expand? (b) What is the equilibrium position of the block? (c) Show that the block's motion is simple harmonic, and determine its period. (d) Repeat part (a) if the coefficient of kinetic friction between block and surface is μ.

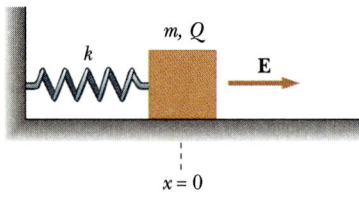

FIGURE P25.18

19. An insulating rod having linear charge density $\lambda = 40.0$ μC/m and linear mass density $\mu = 0.100$ kg/m is released from rest in a uniform electric field $E = 100$ V/m and directed perpendicular to the rod (Fig. P25.19). (a) Determine the speed of the rod after it has traveled 2.00 m. (b) How does your answer to part (a) change if the electric field is not perpendicular to the rod? Explain.

19A. An insulating rod having linear charge density λ and linear mass density μ is released from rest in a uniform electric field E and directed perpendicular to the rod (Fig. P25.19). (a) Determine the speed of the rod after it has traveled a distance d. (b) How does your answer to part (a) change if the electric field is not perpendicular to the rod? Explain.

20. A particle having charge $q = +2.0$ μC and mass $m = 0.01$ kg is connected to a string that is $L = 1.5$ m long and is tied to the pivot point P in Figure P25.20. The particle, string, and pivot point all lie on a horizontal table. The particle is released from rest when the string makes an angle $\theta = 60°$ with a uniform electric field of magnitude $E = 300$ V/m. Determine the speed of the particle when the string is parallel to the electric field (point a in Fig. P25.20).

20A. A particle having charge q and mass m in Figure P25.20 is connected to a string of length L and tied to

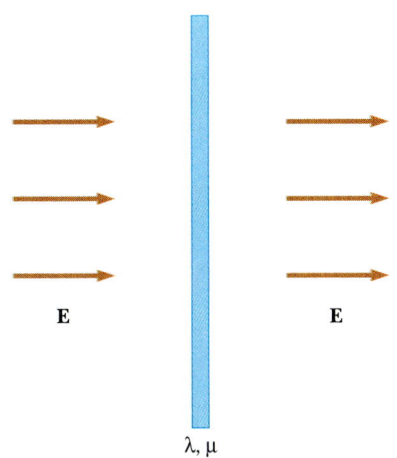

FIGURE P25.19

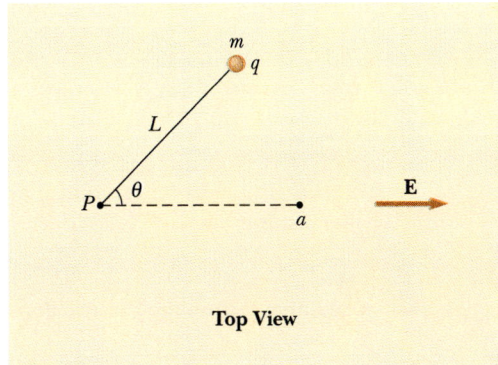

Top View

FIGURE P25.20

the pivot point P. The particle, string, and pivot point all lie on a horizontal table. The particle is released from rest when the string makes an angle θ with a uniform electric field of magnitude E. Determine the speed of the particle when the string is parallel to the electric field (point a in Fig. P25.20).

Section 25.3   Electric Potential and Potential Energy Due to Point Charges

(Note: Unless stated otherwise, assume a reference $V = 0$ at $r = \infty$.)

21. At what distance from a point charge of 8.0 μC does the electric potential equal 3.6×10^4 V?

22. A small spherical object carries a charge of 8.0 nC. At what distance from the center of the object is the potential equal to 100 V? 50 V? 25 V? Is the spacing of the equipotentials proportional to the change in V?

23. At a distance r away from a point charge q, the elec-

tric potential is $V = 400$ V and the magnitude of the electric field is $E = 150$ N/C. Determine the values of q and r.

24. Given two 2.00-μC charges, as in Figure P25.24, and a positive test charge $q = 1.28 \times 10^{-18}$ C at the origin, (a) what is the net force exerted on q by the two 2.00-μC charges? (b) What is the electric field at the origin due to the two 2.00-μC charges? (c) What is the electric potential at the origin due to the two 2.00-μC charges?

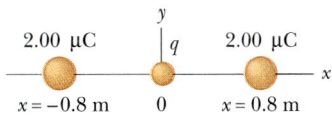

FIGURE P25.24

25. The electrostatic potential due to a set of point charges on a Cartesian grid is

$$V = \frac{36}{\sqrt{(x+1)^2 + y^2}} - \frac{45}{\sqrt{x^2 + (y-2)^2}}$$

where V is in volts. Determine the position and magnitude of all charges in this distribution.

26. A charge $+q$ is at the origin. A charge $-2q$ is at $x = 2.0$ m on the x axis. For what finite value(s) of x is (a) the electric field zero? (b) the electric potential zero?

27. The three charges in Figure P25.27 are at the vertices of an isosceles triangle. Calculate the electric potential at the midpoint of the base, taking $q = 7.00$ μC.

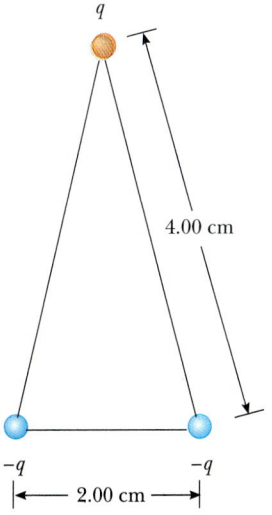

FIGURE P25.27

28. In Rutherford's famous scattering experiments that led to the planetary model of the atom, alpha parti-

cles (charge $+2e$, mass $= 6.6 \times 10^{-27}$ kg) were fired at a gold nucleus (charge $+79e$). An alpha particle initially very far from the gold nucleus is fired at 2.0×10^7 m/s directly toward the center of the nucleus. How close does the alpha particle get to this center before turning around?

29. Two point charges, $Q_1 = +5.00$ nC and $Q_2 = -3.00$ nC, are separated by 35.0 cm. (a) What is the potential energy of the pair? What is the significance of the algebraic sign of your answer? (b) What is the electric potential at a point midway between the charges?

30. An electron starts from rest 3.00 cm from the center of a uniformly charged insulating sphere of radius 2.00 cm and total charge of 1.00 nC. What is the speed of the electron when it reaches the surface of the sphere?

30A. An electron starts from rest a distance d from the center of a uniformly charged insulating sphere of radius R and total charge Q. If $d > R$, what is the speed of the electron when it reaches the surface of the sphere?

31. The Bohr model of the hydrogen atom states that the electron can exist only in certain allowed orbits. The radius of each Bohr orbit is $r = n^2 (0.0529$ nm$)$ where $n = 1, 2, 3, \ldots$. Calculate the electric potential energy of a hydrogen atom when the electron is in the (a) first allowed orbit, $n = 1$, (b) second allowed orbit, $n = 2$, and (c) when the electron has escaped from the atom, $r = \infty$. Express your answers in electron volts.

32. Calculate the energy required to assemble the array of charges shown in Figure P25.32, where $a = 0.20$ m, $b = 0.40$ m, and $q = 6.0$ μC.

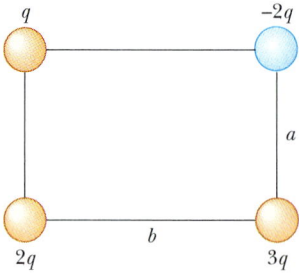

FIGURE P25.32

33. Show that the amount of work required to assemble four identical point charges of magnitude Q at the corners of a square of side s is $5.41 k_e Q^2 / s$.

34. Two insulating spheres having radii 0.30 cm and 0.50 cm, masses 0.10 kg and 0.70 kg, and charges -2.0 μC and 3.0 μC are released from rest when their centers are separated by 1.0 m. (a) How fast is

each moving when they collide? (*Hint:* Consider conservation of energy and conservation of linear momentum.) (b) If the spheres are conductors, will the speeds calculated in part (a) be larger or smaller? Explain.

34A. Two insulating spheres having radii r_1 and r_2, masses m_1 and m_2, and charges $-q_1$ and q_2 are released from rest when their centers are separated by a distance d. (a) How fast is each moving when they collide? (*Hint:* Consider conservation of energy and conservation of linear momentum.) (b) If the spheres are conductors, will the speeds calculated in part (a) be larger or smaller? Explain.

35. Four identical particles each have charge $q = 0.50$ μC and mass 0.010 kg. They are released from rest at the vertices of a square of side 0.10 m. How fast is each charge moving when their distance from the center of the square doubles?

35A. Four identical particles each have charge q and mass m. They are released from rest at the vertices of a square of side L. How fast is each charge moving when their distance from the center of the square doubles?

36. How much work is required to assemble eight identical point charges, each of magnitude q, at the corners of a cube of side s?

Section 25.4   Obtaining E from the Electric Potential

37. An infinite sheet of charge that has a surface charge density of 25.0 nC/m² lies in the yz plane, passes through the origin, and is at a potential of 1.0 kV at the point $y = 0$, $z = 0$. A long wire having a linear charge density of 80.0 nC/m lies parallel to the y axis and intersects the x axis at $x = 3.0$ m. (a) Determine, as a function of x, the potential along the x axis between wire and sheet. (b) What is the potential energy of a 2.0-nC charge placed at $x = 0.8$ m?

37A. An infinite sheet of charge with a surface charge density σ lies in the yz plane, passes through the origin, and is at a potential V_0 at the point $y = 0$, $z = 0$. A long wire having a linear charge density λ lies parallel to the y axis and intersects the x axis at $x = d$. (a) Determine the potential along the x axis between the wire and the sheet as a function of x. (b) What is the potential energy of a charge q placed at $x = d/4$?

38. The electric potential in a certain region is $V = 4xz - 5y + 3z^2$ V. Find the magnitude of the electric field at $(+2, -1, +3)$, where all distances are in meters.

39. Over a certain region of space, the electric potential is $V = 5x - 3x^2y + 2yz^2$. Find the expressions for the x, y, and z components of the electric field over

this region. What is the magnitude of the field at the point P, which has coordinates $(1, 0, -2)$ m?

40. The electric potential in a certain region is $V = ax^2 + bx + c$, where $a = 12$ V/m^2, $b = -10$ V/m, and $c = 62$ V. Determine (a) the magnitude and direction of the electric field at $x = +2.0$ m and (b) the position where the electric field is zero.

41. The potential in a region between $x = 0$ and $x = 6.0$ m is $V = a + bx$ where $a = 10$ V and $b = -7.0$ V/m. Determine (a) the potential at $x = 0$, 3.0 m, and 6.0 m and (b) the magnitude and direction of the electric field at $x = 0$, 3.0 m, and 6.0 m.

42. The electric potential inside a charged spherical conductor of radius R is given by $V = k_e Q/R$ and outside the potential is given by $V = k_e Q/r$. Using $E_r = -dV/dr$, derive the electric field (a) inside and (b) outside this charge distribution.

43. When an uncharged conducting sphere of radius a is placed at the origin of an xyz coordinate system that lies in an initially uniform electric field $\mathbf{E} = E_0 \mathbf{k}$, the resulting electrostatic potential is $V(x, y, z) = V_0$ for points inside the sphere, and

$$V(x, y, z) = V_0 - E_0 z + \frac{E_0 a^3 z}{(x^2 + y^2 + z^2)^{3/2}}$$

for points outside the sphere, where V_0 is the (constant) electrostatic potential on the conductor. Use this equation to determine the x, y, and z components of the resulting electric field.

44. An uncharged, infinitely long conducting cylinder of radius a is placed in an initially uniform electric field $\mathbf{E} = E_0 \mathbf{i}$, such that the cylinder's axis lies along the z axis. The resulting electrostatic potential is $V(x, y, z) = V_0$ for points inside the cylinder, and

$$V(x, y, z) = V_0 - E_0 x + \frac{E_0 a^2 x}{x^2 + y^2}$$

for points outside the cylinder, where V_0 is the (constant) electrostatic potential on the conductor. Use this equation to determine the x, y, and z components of the resulting electric field.

Section 25.5 Electric Potential Due to Continuous Charge Distributions

45. Consider a ring of radius R with total charge Q spread uniformly over its perimeter. What is the potential difference between the point at the center of the ring and a point on its axis a distance $2R$ from the center?

46. Consider two coaxial rings of 30.0-cm radius and separated by 30.0 cm. (a) Calculate the electric potential at a point on their common axis midway between the two rings, assuming that each ring carries a uniformly distributed charge of 5.00 μC. (b) What is the potential at this point if the two rings carry equal and opposite charges?

47. A rod of length L (Fig. P25.47) lies along the x axis with its left end at the origin and has a nonuniform charge density $\lambda = \alpha x$ (where α is a positive constant). (a) What are the units of α? (b) Calculate the electric potential at A.

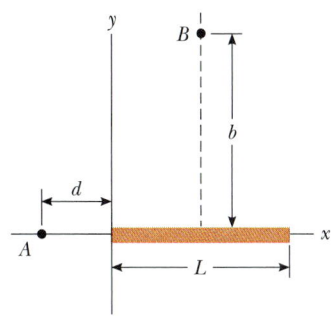

FIGURE P25.47

48. For the arrangement described in the previous problem, calculate the electric potential at point B that lies on the perpendicular bisector of the rod a distance b above the x axis.

49. Calculate the electric potential at point P on the axis of the annulus shown in Figure P25.49, which has a uniform charge density σ.

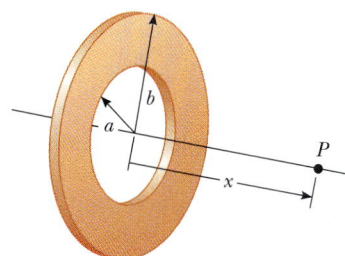

FIGURE P25.49

50. A wire of finite length that has a uniform linear charge density λ is bent into the shape shown in Figure P25.50. Find the electrical potential at point O.

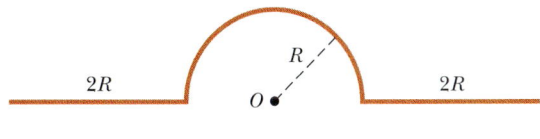

FIGURE P25.50

Section 25.6 Potential of a Charged Conductor

51. How many electrons should be removed from an initially uncharged spherical conductor of radius 0.300 m to produce a potential of 7.50 kV at the surface?

52. Calculate the surface charge density, σ (in C/m^2), for a solid spherical conductor of radius $R = 0.250$ m if the potential 0.500 m from the center of the sphere is 1.30 kV.

53. A spherical conductor has a radius of 14.0 cm and charge of 26.0 μC. Calculate the electric field and the electric potential (a) $r = 10.0$ cm, (b) $r = 20.0$ cm, and (c) $r = 14.0$ cm from the center.

54. Two concentric spherical conducting shells of radii $a = 0.400$ m and $b = 0.500$ m are connected by a thin wire as in Figure P25.54. If a total charge $Q = 10.0$ μC is placed on the system, how much charge settles on each sphere?

54A. Two concentric spherical conducting shells of radii a and b are connected by a thin wire as in Figure P25.54. If a total charge Q is placed on the system, how much charge settles on each sphere?

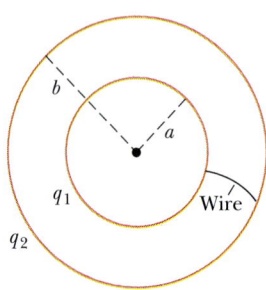

FIGURE P25.54

55. Two charged spherical conductors are connected by a long conducting wire, and a charge of 20.0 μC is placed on the combination. (a) If one sphere has a radius of 4.00 cm and the other has a radius of 6.00 cm, what is the electric field near the surface of each sphere? (b) What is the electrical potential of each sphere?

56. An egg-shaped conductor has a charge of 43 nC placed on its surface. It has a total surface area of 38 cm². What are (a) the average surface charge density, (b) the electric field inside the conductor, and (c) the (average) electric field just outside the conductor?

*Section 25.8   Applications of Electrostatics

57. Consider a Van de Graaff generator with a 30.0-cm-diameter dome operating in dry air. (a) What is the maximum potential of the dome? (b) What is the maximum charge on the dome?

58. (a) Calculate the largest amount of charge possible on the surface of a Van de Graaff generator that has a 40.0-cm-diameter dome surrounded by air. (b) What is the potential of this dome when it carries this charge?

59. What charge must be placed on the surface of a Van de Graaff generator, whose dome has a radius of 15 cm, to produce a spark across a 10-cm air gap?

ADDITIONAL PROBLEMS

60. Three point charges having magnitudes of 8.00 μC, -3.00 μC, and 5.00 μC are located at corners of a triangle whose sides are each 9.00 cm long. Calculate the electric potential at the center of this triangle.

61. At a certain distance from a point charge, the magnitude of the electric field is 500 V/m and the electric potential is -3.00 kV. (a) What is the distance to the charge? (b) What is the magnitude of the charge?

62. Two parallel plates having equal but opposite charge are separated by 12.0 cm. Each plate has a surface charge density of 36.0 nC/m². A proton is released from rest at the positive plate. Determine (a) the potential difference between the plates, (b) the energy of the proton when it reaches the negative plate, (c) the speed of the proton just before it strikes the negative plate, (d) the acceleration of the proton, and (e) the force on the proton. (f) From the force, find the electric field intensity and show that it is equal to the electric field intensity found from the charge densities on the plates.

63. A Van de Graaff generator is operated until the spherical dome has a measured potential of 6.0×10^5 V and an electric field of maximum value for a dome surrounded by air (3.0×10^6 V/m). Determine (a) the charge on the dome and (b) the radius of the dome.

64. (a) Consider a uniformly charged cylindrical shell having total charge Q, radius R, and height h. Determine the electrostatic potential at a point a distance d from the right side of the cylinder as in Figure P25.64. (*Hint:* Use the result of Example 25.8 by treating the cylinder as a collection of ring charges.) (b) Use the result of Example 25.9 to solve the same problem for a solid cylinder.

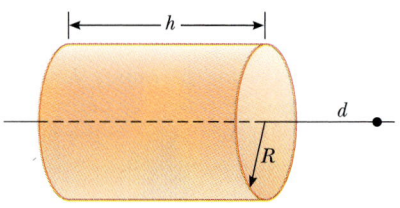

FIGURE P25.64

65. Equal charges ($q = 2.0$ μC) are placed at 30° intervals around the equator of a sphere that has a radius of 1.2 m. What is the electric potential (a) at the center of the sphere and (b) at its north pole?

66. The charge distribution shown in Figure P25.66 is referred to as a linear quadrupole. (a) Show that the potential at a point on the *x* axis where $x > d$ is

$$V = \frac{2k_eQd^2}{x^3 - xd^2}$$

(b) Show that the expression obtained in (a) when $x \gg d$ reduces to

$$V = \frac{2k_eQd^2}{x^3}$$

66A. Use the exact result from Problem 66a to evaluate the potential for the linear quadrupole at $x = 3d$ if $d = 2.00$ mm and $Q = 3.00$ μC. Compare this answer with what you obtain when you use the approximate result (Problem 66b) valid when $x \gg d$.

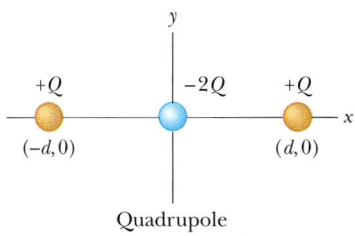

Quadrupole

FIGURE P25.66

67. (a) Use the exact result from Problem 66 to find the electric field at any point along the axis of the linear quadrupole for $x > d$. (b) Evaluate E at $x = 3d$ if $d = 2.00$ mm and $Q = 3.00$ μC.

68. A ring of radius 0.20 m carries a uniformly distributed positive charge, as in Figure P25.68. The linear charge density of the ring is 0.10 μC/m, and an electron is located 0.10 above the plane of the ring on the central perpendicular. If this electron is released from rest, what is its speed when it reaches the center of the ring?

68A. A ring of radius R carries a uniformly distributed positive charge, as in Figure P25.68. The linear charge density of the ring is λ, and an electron is located a distance d above the plane of the ring on the central perpendicular axis. If this electron is released from rest, what is its speed when it reaches the center of the ring?

69. Two point charges of equal magnitude are located along the *y* axis equal distances above and below the *x* axis, as in Figure P25.69. (a) Plot a graph of the potential at points along the *x* axis over the interval $-3a < x < 3a$. You should plot the potential in units of k_eQ/a. (b) Let the charge located at $-a$ be negative and plot the potential along the *y* axis over the interval $-4a < y < 4a$.

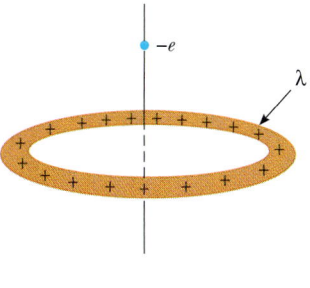

FIGURE P25.68

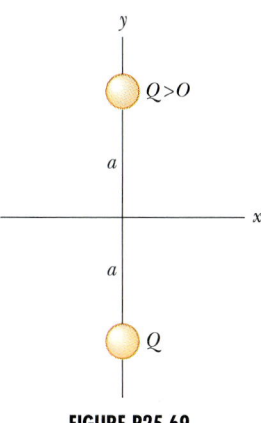

FIGURE P25.69

70. The liquid-drop model of the nucleus suggests that high-energy oscillations of certain nuclei can split the nucleus into two unequal fragments plus a few neutrons. The fragments acquire kinetic energy from their mutual Coulomb repulsion. Calculate the electric potential energy (in electron volts) of two spherical fragments from a uranium nucleus having the following charges and radii: $38e$ and 5.5×10^{-15} m; $54e$ and 6.2×10^{-15} m. Assume that the charge is distributed uniformly throughout the volume of each spherical fragment and that their surfaces are initially in contact at rest. (The electrons surrounding the nucleus can be neglected.)

71. Two identical raindrops, each carrying surplus electrons on its surface to give a net charge $-q$ on each, collide and form a single drop of larger size. Before the collision, the characteristics of each drop are as follows: (a) surface charge density σ_0, (b) electric field E_0 at the surface, (c) electric potential V_0 at the surface (where $V = 0$ at $r = \infty$). For the combined drop, find these three quantities in terms of their original values.

72. Use the results to Example 25.11 and $E_r = -dV/dr$ to derive the electric field (a) inside and (b) outside a uniformly charged insulating sphere.

73. Calculate the work that must be done to charge a spherical shell of radius R to a total charge Q.

74. A point charge q is located at $x = -R$, and a point charge $-2q$ is located at the origin. Prove that the equipotential surface that has zero potential is a sphere centered at $(-4R/3, 0, 0)$ and having a radius $r = 2R/3$.

75. From Gauss's law, the electric field set up by a uniform line of charge is

$$\mathbf{E} = \left(\frac{\lambda}{2\pi\epsilon_0 r}\right)\hat{\mathbf{r}}$$

where $\hat{\mathbf{r}}$ is a unit vector pointing radially away from the line and λ is the charge per meter along the line. Derive an expression for the potential difference between $r = r_1$ and $r = r_2$.

76. Consider two thin, conducting, spherical shells as in Figure P25.76. The inner shell has a radius $r_1 = 15$ cm and a charge of 10 nC. The outer shell has a radius $r_2 = 30$ cm and a charge of -15 nC. Find (a) the electric field $\mathbf{E}$ and (b) the electric potential V in regions A, B, and C, with $V = 0$ at $r = \infty$.

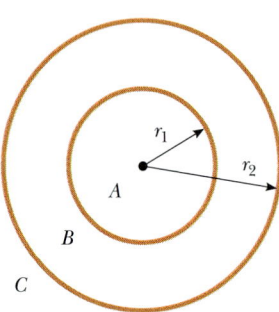

FIGURE P25.76

77. The x axis is the symmetry axis of a uniformly charged ring of radius R and charge Q (Fig. P25.77). A point charge Q of mass M is located at the center of the ring. When it is displaced slightly, the point charge accelerates along the x axis to infinity. Show that the ultimate speed of the point charge is

$$v = \left(\frac{2k_e Q^2}{MR}\right)^{1/2}$$

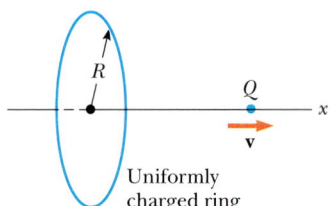

FIGURE P25.77

78. The thin, uniformly charged rod shown in Figure P25.78 has a linear charge density λ. Find an expression for the electric potential at P.

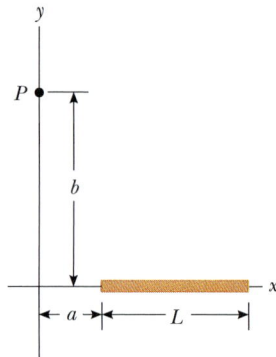

FIGURE P25.78

79. It is shown in Example 25.10 that the potential at a point P a distance d above one end of a uniformly charged rod of length ℓ lying along the x axis is

$$V = \frac{k_e Q}{\ell}\ln\left(\frac{\ell + \sqrt{\ell^2 + d^2}}{d}\right)$$

Use this result to derive an expression for the y component of the electric field at P. (*Hint:* Replace d with y.)

80. Figure P25.80 shows several equipotential lines each labeled by its potential in volts. The distance between the lines of the square grid represents 1 cm. (a) Is the magnitude of the field bigger at A or at B? Why? (b) What is $\mathbf{E}$ at B? (c) Represent what the field looks like by drawing at least eight field lines.

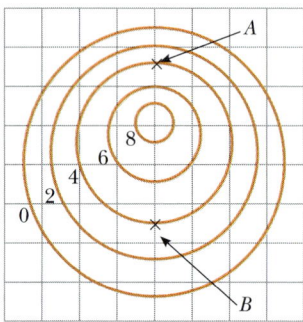

FIGURE P25.80

81. A dipole is located along the y axis as in Figure P25.81. (a) At a point P, which is far from the dipole ($r \gg a$), the electric potential is

$$V = k_e \frac{p\cos\theta}{r^2}$$

where $p = 2qa$. Calculate the radial component of the associated electric field, E_r, and the perpendicular component, E_θ. Note that $E_\theta = -\dfrac{1}{r}\left(\dfrac{\partial V}{\partial \theta}\right)$. Do these results seem reasonable for $\theta = 90°$ and $0°$? for $r = 0$? (b) For the dipole arrangement shown, express V in terms of rectangular coordinates using $r = (x^2 + y^2)^{1/2}$ and

$$\cos\theta = \frac{y}{(x^2 + y^2)^{1/2}}$$

Using these results and taking $r \gg a$, calculate the field components E_x and E_y.

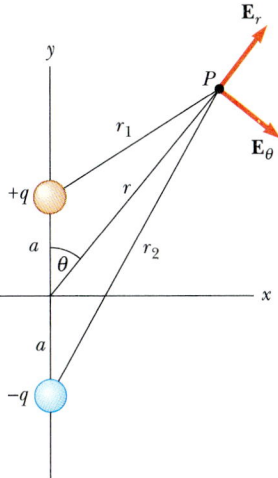

FIGURE P25.81

82. A disk of radius R has a nonuniform surface charge density $\sigma = Cr$, where C is a constant and r is measured from the center of the disk (Fig. P25.82). Find (by direct integration) the potential at P.

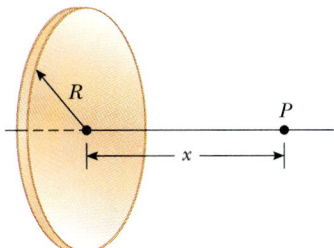

FIGURE P25.82

83. A solid sphere of radius R has a uniform charge density ρ and total charge Q. Derive an expression for its total electric potential energy. (*Hint:* Imagine that the sphere is constructed by adding successive layers

of concentric shells of charge $dq = (4\pi r^2\, dr)\rho$ and use $dU = V\, dq$.)

84. A Geiger-Müller counter is a radiation detector that essentially consists of a hollow cylinder (the cathode) of inner radius r_a and a coaxial cylindrical wire (the anode) of radius r_b (Fig. P25.84). The charge per unit length on the anode is λ, while the charge per unit length on the cathode is $-\lambda$. (a) Show that the magnitude of the potential difference between the wire and the cylinder in the sensitive region of the detector is

$$V = 2k_e\lambda \ln\left(\frac{r_a}{r_b}\right)$$

(b) Show that the magnitude of the electric field over that region is given by

$$E = \frac{V}{\ln(r_a/r_b)}\left(\frac{1}{r}\right)$$

where r is the distance from the center of the anode to the point where the field is to be calculated.

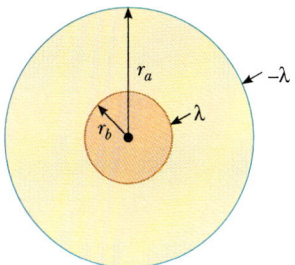

FIGURE P25.84

85. Three identical charges lie at the vertices of an equilateral triangle having sides 2.000 m long (Fig. P25.85). Locate positions of electrostatic equilibrium within the triangle.

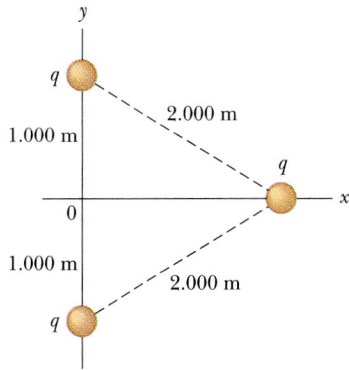

FIGURE P25.85

SPREADSHEET PROBLEMS

S1. Spreadsheet 25.1 calculates the electrical potential in the two-dimensional region around two charges. The region is divided into a 10 × 10 grid. The upper left corner of the grid corresponds to $x = 1.0$ m, $y = 1.0$ m, and the lower-right corner corresponds to $x = 10$ m, $y = 10$ m. The two charges can be placed anywhere in this grid. The first two tables calculate the distances from each charge to each grid point. The spreadsheet then calculates the electric potential at each grid point. Place charges of $+2.0$ μC at $x = 3.0$ m, $y = 5.0$ m and $x = 7.0$ m, $y = 5.0$ m. Print out the electric potential matrix. The location of the point charges will appear on the grid in the cells denoted by *ERR* (in Lotus 1-2-3) or *#DIV/0* (in Excel) since the potential is infinite at the location of the point charges. Sketch by hand the equipotential lines for $V = 20$ kV, 15 kV, and 10 kV. Sketch in a representative set of electric-field lines. (*Hint:* By adjusting the cell width on the spreadsheet and the printer line spacing, you can get the grid to print out approximately as a square.)

S2. Repeat Problem S1 with one charge replaced by -2.0 μC. Print out the electric potential matrix. Sketch the equipotential lines for $V = \pm20$ kV, ±15 kV, and ±10 kV. Sketch in a representative set of electric-field lines.

S3. Place two charges anywhere in the grid in Problem S1. Choose any values for the charges. Print out the electric potential matrix. Sketch representative equipotentials and electric-field lines.

S4. Modify Spreadsheet 25.1 to include additional charges. Choose locations and values for each charge. Sketch the equipotentials and the electric-field lines. Place a row of charges together. Are the field lines for this case uniform? If not, why not?

S5. Spreadsheet 14.1 calculates the gravitational potential energy of a mass moving along the x axis in the gravitational field of four fixed point masses. Rework this spreadsheet to calculate the electrical potential along the x axis for four fixed point charges at the same locations as the four point masses. Find the electric field on the x axis as well. (*Hint:* $GM_1 m$ must be replaced by $k_e Q_1$, and so forth.) Investigate various choices for the charges and their locations.

CHAPTER 26

Capacitance and Dielectrics

Electrical discharge visible in a section of the Particle Beam Fusion Accelerator II, the nation's most powerful x-ray source, at Sandia National Laboratories. The discharges are due to air breakdown at the surface of the water that covers this section. At the center of the machine, high-energy electrons are converted to x-rays, which are used to determine their effect on weapons systems and other components. *(Courtesy of Sandia National Laboratories. Photo by Walter Dickenman)*

T his chapter is concerned with the properties of capacitors, which are devices that store charge. Capacitors are commonly used in a variety of electrical circuits. For instance, they are used (1) to tune the frequency of radio receivers, (2) as filters in power supplies, (3) to eliminate sparking in automobile ignition systems, and (4) as energy-storing devices in electronic flash units.

A capacitor basically consists of two conductors separated by an insulator. We shall see that the capacitance of a given device depends on its geometry and on the material separating the charged conductors, called a *dielectric*.

26.1 DEFINITION OF CAPACITANCE

Consider two conductors having a potential difference V between them. Let us assume that the conductors have equal and opposite charges as in Figure 26.1. Such a combination of two conductors is called a **capacitor**. The potential differ-

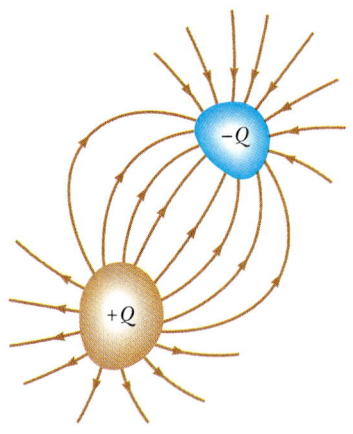

FIGURE 26.1   A capacitor consists of two conductors electrically isolated from each other and their surroundings. Once the capacitor is charged, the two conductors carry equal but opposite charges.

ence V is proportional to the magnitude of the charge Q on the capacitor.[1]

> The **capacitance,** C, of a capacitor is defined as the ratio of the magnitude of the charge on either conductor to the magnitude of the potential difference between them:
>
> $$C \equiv \frac{Q}{V} \qquad (26.1)$$

Note that by definition *capacitance is always a positive quantity.* Furthermore, since the potential difference increases as the stored charge increases, the ratio Q/V is constant for a given capacitor. Therefore, the capacitance of a device is a measure of its ability to store charge and electrical potential energy.

From Equation 26.1, we see that capacitance has the SI unit coulomb per volt. The SI unit of capacitance is the **farad** (F), in honor of Michael Faraday:

$$[\text{Capacitance}] = 1 \text{ F} = \frac{1 \text{ C}}{1 \text{ V}}$$

The farad is a very large unit of capacitance. In practice, typical devices have capacitances ranging from microfarads to picofarads. As a practical note, capacitors are often labeled mF for microfarads and mmF for micromicrofarads (picofarads).

As we show in the next section, the capacitance of a device depends, among other things, on the geometrical arrangement of the conductors. To illustrate this point, let us calculate the capacitance of an isolated spherical conductor of radius R and charge Q. (The second conductor can be taken as a concentric hollow conducting sphere of infinite radius.) Since the potential of the sphere is simply k_eQ/R (where $V = 0$ at infinity), its capacitance is

$$C = \frac{Q}{V} = \frac{Q}{k_eQ/R} = \frac{R}{k_e} = 4\pi\epsilon_0 R \qquad (26.2)$$

This shows that the capacitance of an isolated charged sphere is proportional to its radius and is independent of both the charge and the potential difference. For example, an isolated metallic sphere of radius 0.15 m has a capacitance of

$$C = 4\pi\epsilon_0 R = 4\pi(8.85 \times 10^{-12} \text{ C}^2/\text{N}\cdot\text{m}^2)(0.15 \text{ m}) = 17 \text{ pF}$$

26.2   CALCULATION OF CAPACITANCE

The capacitance of a pair of oppositely charged conductors can be calculated in the following manner. A charge of magnitude Q is assumed, and the potential difference is calculated using the techniques described in the previous chapter. One then simply uses $C = Q/V$ to evaluate the capacitance. As you might expect, the calculation is relatively easy to perform if the geometry of the capacitor is simple.

Let us illustrate this with three geometries that we are familiar with, namely, two parallel plates, two coaxial cylinders, and two concentric spheres. In these

---

[1] The proportionality between the potential difference and charge on the conductors can be proved from Coulomb's law or by experiment.

examples, we assume that the charged conductors are separated by a vacuum. The effect of a dielectric material placed between the conductors is treated in Section 26.5.

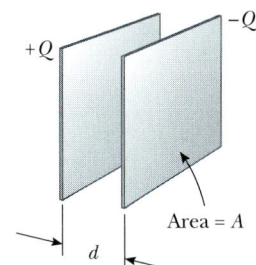

The Parallel-Plate Capacitor

Two parallel plates of equal area A are separated by a distance d as in Figure 26.2. One plate has a charge Q, the other, charge $-Q$. The charge per unit area on either plate is $\sigma = Q/A$. If the plates are very close together (compared with their length and width), we can neglect edge effects and assume that the electric field is uniform between the plates and zero elsewhere. According to Example 24.8, the electric field between the plates is

FIGURE 26.2   A parallel-plate capacitor consists of two parallel plates each of area A, separated by a distance d. When the capacitor is charged, the plates carry equal charges of opposite sign.

$$E = \frac{\sigma}{\epsilon_0} = \frac{Q}{\epsilon_0 A}$$

where ϵ_0 is the permittivity of free space. The potential difference between the plates equals Ed; therefore,

$$V = Ed = \frac{Qd}{\epsilon_0 A}$$

Substituting this result into Equation 26.1, we find that the capacitance is

$$C = \frac{Q}{V} = \frac{Q}{Qd/\epsilon_0 A}$$

$$C = \frac{\epsilon_0 A}{d} \qquad\qquad (26.3)$$

That is, *the capacitance of a parallel-plate capacitor is proportional to the area of its plates and inversely proportional to the plate separation.*

EXAMPLE 26.1   Parallel-Plate Capacitor

A parallel-plate capacitor has an area $A = 2.00 \times 10^{-4}$ m² and a plate separation $d = 1.00$ mm. Find its capacitance.

Solution   From Equation 26.3, we find

$$C = \epsilon_0 \frac{A}{d} = \left(8.85 \times 10^{-12} \frac{C^2}{N \cdot m^2}\right)\left(\frac{2.00 \times 10^{-4} \text{ m}^2}{1.00 \times 10^{-3} \text{ m}}\right)$$

$$= 1.77 \times 10^{-12} \text{ F} = \boxed{1.77 \text{ pF}}$$

Exercise   If the plate separation is increased to 3.00 mm, find the capacitance.

Answer   0.590 pF.

As you can see from the definition of capacitance, $C = Q/V$, the amount of charge a given capacitor is able to store for a given potential difference across its plates increases as the capacitance increases. Therefore, it seems reasonable that a capacitor constructed from plates having a large area should be able to store a large charge. The amount of charge needed to produce a given potential difference increases with decreasing plate separation.

A careful inspection of the electric field lines for a parallel-plate capacitor

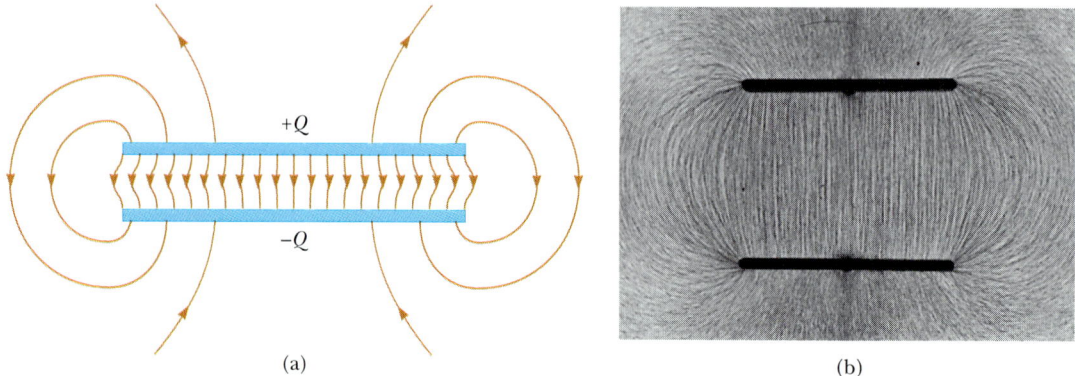

(a)   (b)

FIGURE 26.3 (a) The electric field between the plates of a parallel-plate capacitor is uniform near its center, but is nonuniform near its edges. (b) Electric field pattern of two oppositely charged conducting parallel plates. Small pieces of thread on an oil surface align with the electric field. Note the nonuniform nature of the electric field at the ends of the plates. Such end effects can be neglected if the plate separation is small compared to the length of the plates. *(Courtesy of Harold M. Waage, Princeton University)*

reveals that the field is uniform in the central region between the plates as shown in Figure 26.3a. However, the field is nonuniform at the edges of the plates. Figure 26.3b is a photograph of the electric field pattern of a parallel-plate capacitor showing the nonuniform field lines at its edges.

EXAMPLE 26.2 The Cylindrical Capacitor

A cylindrical conductor of radius a and charge Q is coaxial with a larger cylindrical shell of radius b and charge $-Q$ (Fig. 26.4a). Find the capacitance of this cylindrical capacitor if its length is ℓ.

Reasoning and Solution If we assume that ℓ is long compared with a and b, we can neglect end effects. In this case, the field is perpendicular to the axis of the cylinders and is confined to the region between them (Fig. 26.4b). We must first calculate the potential difference between the two cylinders, which is given in general by

$$V_b - V_a = -\int_a^b \mathbf{E} \cdot d\mathbf{s}$$

where $\mathbf{E}$ is the electric field in the region $a < r < b$. In Chapter 24, we showed using Gauss's law that the electric field of a cylinder of charge per unit length λ is $E = 2k_e\lambda/r$. The same result applies here, since the outer cylinder does not contribute to the electric field inside it. Using this result and noting that $\mathbf{E}$ is along r in Figure 26.4b, we find that

$$V_b - V_a = -\int_a^b E_r\, dr = -2k_e\lambda \int_a^b \frac{dr}{r} = -2k_e\lambda \ln\!\left(\frac{b}{a}\right)$$

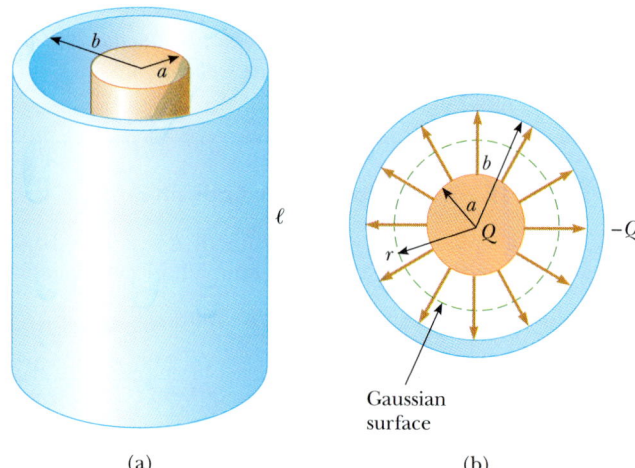

(a)   (b)

FIGURE 26.4 (Example 26.2) (a) A cylindrical capacitor consists of a cylindrical conductor of radius a and length ℓ surrounded by a coaxial cylindrical shell of radius b. (b) The end view of a cylindrical capacitor. The dashed line represents the end of the cylindrical gaussian surface of radius r and length ℓ.

Substituting this into Equation 26.1 and using the fact that $\lambda = Q/\ell$, we get

$$C = \frac{Q}{V} = \frac{Q}{\dfrac{2k_eQ}{\ell}\ln\left(\dfrac{b}{a}\right)} = \frac{\ell}{2k_e\ln\left(\dfrac{b}{a}\right)} \qquad (26.4)$$

where V is the magnitude of the potential difference, given by $2k_e\lambda \ln (b/a)$, a positive quantity. That is, $V = V_a - V_b$ is positive because the inner cylinder is at the higher potential. Our result for C makes sense because it shows that the capacitance is proportional to the length of the cylinders. As you

might expect, the capacitance also depends on the radii of the two cylindrical conductors. As an example, a coaxial cable consists of two concentric cylindrical conductors of radii a and b separated by an insulator. The cable carries currents in opposite directions in the inner and outer conductors. Such a geometry is especially useful for shielding an electrical signal from external influences. From Equation 26.4, we see that the capacitance per unit length of a coaxial cable is

$$\frac{C}{\ell} = \frac{1}{2k_e\ln\left(\dfrac{b}{a}\right)}$$

EXAMPLE 26.3 The Spherical Capacitor

A spherical capacitor consists of a spherical conducting shell of radius b and charge $-Q$ that is concentric with a smaller conducting sphere of radius a and charge Q (Fig. 26.5). Find its capacitance.

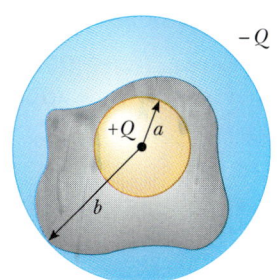

FIGURE 26.5 (Example 26.3) A spherical capacitor consists of an inner sphere of radius a surrounded by a concentric spherical shell of radius b. The electric field between the spheres is radial outward if the inner sphere is positively charged.

Reasoning and Solution As we showed in Chapter 24, the field outside a spherically symmetric charge distribution is

radial and given by k_eQ/r^2. In this case, this corresponds to the field between the spheres ($a < r < b$). (The field is zero elsewhere.) From Gauss's law we see that only the inner sphere contributes to this field. Thus, the potential difference between the spheres is given by

$$V_b - V_a = -\int_a^b E_r\,dr = -k_eQ\int_a^b \frac{dr}{r^2} = k_eQ\left[\frac{1}{r}\right]_a^b$$
$$= k_eQ\left(\frac{1}{b} - \frac{1}{a}\right)$$

The magnitude of the potential difference is

$$V = V_a - V_b = k_eQ\frac{(b-a)}{ab}$$

Substituting this into Equation 26.1, we get

$$C = \frac{Q}{V} = \frac{ab}{k_e(b-a)} \qquad (26.5)$$

Exercise Show that as the radius b of the outer sphere approaches infinity, the capacitance approaches the value $a/k_e = 4\pi\epsilon_0 a$. This is consistent with Equation 26.2.

26.3 COMBINATIONS OF CAPACITORS

Two or more capacitors are often combined in circuits in several ways. The equivalent capacitance of certain combinations can be calculated using methods described in this section. The circuit symbols for capacitors and batteries, together with their color codes, are given in Figure 26.6. The positive terminal of the battery is at the higher potential and is represented by the longer vertical line in the battery symbol.

Parallel Combination

Two capacitors connected as shown in Figure 26.7a are known as a *parallel combination* of capacitors. The left plates of the capacitors are connected by a conducting wire to the positive terminal of the battery and are therefore both at the same

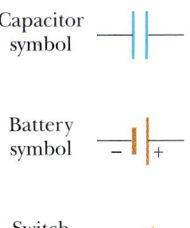

FIGURE 26.6 Circuit symbols for capacitors, batteries, and switches. Note that capacitors are in blue, and batteries and switches are in red.

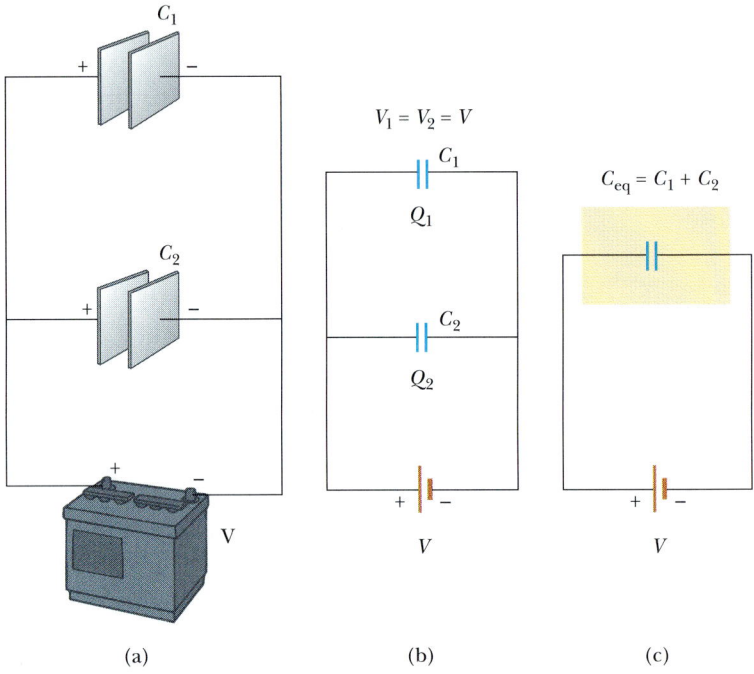

FIGURE 26.7 (a) A parallel combination of two capacitors. (b) The circuit diagram for the parallel combination. (c) The potential difference is the same across each capacitor, and the equivalent capacitance is $C_{eq} = C_1 + C_2$.

potential as the positive terminal. Likewise, the right plates are connected to the negative terminal of the battery and are therefore both at the same potential as the negative terminal. When the capacitors are first connected in the circuit, electrons are transferred through the battery from the left plates to the right plates, leaving the left plates positively charged and the right plates negatively charged. The energy source for this charge transfer is the internal chemical energy stored in the battery, which is converted to electrical energy. The flow of charge ceases when the voltage across the capacitors is equal to that of the battery. The capacitors reach their maximum charge when the flow of charge ceases. Let us call the maximum charges on the two capacitors Q_1 and Q_2. Then the *total charge, Q*, stored by the two capacitors is

$$Q = Q_1 + Q_2 \qquad (26.6)$$

Suppose we wish to replace these two capacitors by one equivalent capacitor having a capacitance C_{eq}. This equivalent capacitor must have exactly the same external effect on the circuit as the original two. That is, it must store Q units of charge. We see from Figure 26.7b that

the potential difference across each capacitor in the parallel circuit is the same and is equal to the voltage of the battery, V.

From Figure 26.7c, we see that the voltage across the equivalent capacitor is also V.

Thus, we have

$$Q_1 = C_1 V \qquad Q_2 = C_2 V$$

and, for the equivalent capacitor,

$$Q = C_{eq} V$$

Substituting these relationships into Equation 26.6 gives

$$C_{eq} V = C_1 V + C_2 V$$

or

$$C_{eq} = C_1 + C_2 \qquad \left(\begin{array}{c} \text{parallel} \\ \text{combination} \end{array} \right)$$

If we extend this treatment to three or more capacitors connected in parallel, the equivalent capacitance is found to be

$$C_{eq} = C_1 + C_2 + C_3 + \cdots \qquad \left(\begin{array}{c} \text{parallel} \\ \text{combination} \end{array} \right) \qquad (26.7)$$

Thus we see that *the equivalent capacitance of a parallel combination of capacitors is larger than any of the individual capacitances.*

Series Combination

Now consider two capacitors connected in *series,* as illustrated in Figure 26.8a.

> For this series combination of capacitors, the magnitude of the charge must be the same on all the plates.

To see why this must be true, let us consider the charge transfer process in some detail. We start with uncharged capacitors and follow what happens just after a

When the key on a computer keyboard is depressed, the spacing between the plates of a capacitor beneath the key changes, causing a change in capacitance. An electrical signal derived from this capacitance change is used to register the keystroke. *(Ray Serway)*

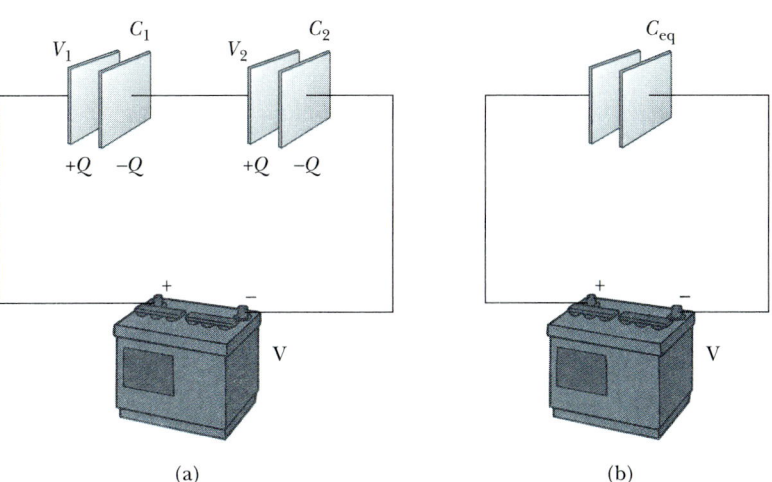

(a)   (b)

FIGURE 26.8 A series combination of two capacitors. The charge on each capacitor is the same, and the equivalent capacitance can be calculated from the relationship

$$1/C_{eq} = 1/C_1 + 1/C_2$$

battery is connected to the circuit. When the battery is connected, electrons are transferred from the left plate of C_1 to the right plate of C_2 through the battery. As this negative charge accumulates on the right plate of C_2, an equivalent amount of negative charge is forced off the left plate of C_2, leaving it with an excess positive charge. The negative charge leaving the left plate of C_2 accumulates on the right plate of C_1, where again an equivalent amount of negative charge leaves the left plate. The result of this is that *all of the right plates gain a charge of* $-Q$ *while all of the left plates have a charge of* $+Q$.

Suppose an equivalent capacitor performs the same function as the series combination. After it is fully charged, *the equivalent capacitor must have a charge of* $-Q$ *on its right plate and* $+Q$ *on its left plate*. By applying the definition of capacitance to the circuit shown in Figure 26.8b, we have

$$V = \frac{Q}{C_{eq}}$$

where V is the potential difference between the terminals of the battery and C_{eq} is the equivalent capacitance. From Figure 26.8a, we see that

$$V = V_1 + V_2 \qquad (26.8)$$

where V_1 and V_2 are the potential differences across capacitors C_1 and C_2. In general, the potential difference across any number of capacitors in series is equal to the sum of the potential differences across the individual capacitors. Since $Q = CV$ can be applied to each capacitor, the potential difference across each is

$$V_1 = \frac{Q}{C_1} \qquad V_2 = \frac{Q}{C_2}$$

Substituting these expressions into Equation 26.8, and noting that $V = Q/C_{eq}$, we have

$$\frac{Q}{C_{eq}} = \frac{Q}{C_1} + \frac{Q}{C_2}$$

Cancelling Q, we arrive at the relationship

$$\frac{1}{C_{eq}} = \frac{1}{C_1} + \frac{1}{C_2} \qquad \left(\begin{array}{l}\text{series} \\ \text{combination}\end{array}\right)$$

If this analysis is applied to three or more capacitors connected in series, the equivalent capacitance is found to be

$$\frac{1}{C_{eq}} = \frac{1}{C_1} + \frac{1}{C_2} + \frac{1}{C_3} + \cdots \qquad \left(\begin{array}{l}\text{series} \\ \text{combination}\end{array}\right) \qquad (26.9)$$

This shows that *the equivalent capacitance of a series combination is always less than any individual capacitance in the combination.*

EXAMPLE 26.4   Equivalent Capacitance

Find the equivalent capacitance between a and b for the combination of capacitors shown in Figure 26.9a. All capacitances are in microfarads.

Solution   Using Equations 26.7 and 26.9, we reduce the combination step by step as indicated in the figure. The 1.0-μF and 3.0-μF capacitors are in parallel and combine according to $C_{eq} = C_1 + C_2$. Their equivalent capacitance is 4.0 μF. Likewise, the 2.0-μF and 6.0-μF capacitors are also in parallel and have an equivalent capacitance of 8.0 μF. The upper branch in Figure 26.9b now consists of two 4.0-μF capacitors in series, which combine according to

$$\frac{1}{C_{eq}} = \frac{1}{C_1} + \frac{1}{C_2} = \frac{1}{4.0 \ \mu F} + \frac{1}{4.0 \ \mu F} = \frac{1}{2.0 \ \mu F}$$

$$C_{eq} = 2.0 \ \mu F$$

Likewise, the lower branch in Figure 26.9b consists of two

8.0-μF capacitors in *series*, which give an equivalent of 4.0 μF. Finally, the 2.0-μF and 4.0-μF capacitors in Figure 26.9c are in parallel and have an equivalent capacitance of 6.0 μF. Hence, the equivalent capacitance of the circuit is 6.0 μF, as shown in Figure 26.9d.

Exercise Consider three capacitors having capacitances of 3.0 μF, 6.0 μF, and 12 μF. Find their equivalent capacitance if they are connected (a) in parallel and (b) in series.

Answer (a) 21 μF, (b) 1.7 μF.

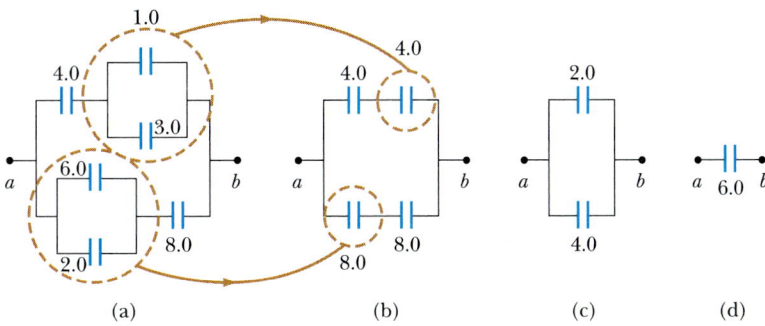

FIGURE 26.9 (Example 26.4) To find the equivalent combination of the capacitors in (a), the various combinations are reduced in steps as indicated in (b), (c), and (d), using the series and parallel rules described in the text.

26.4 ENERGY STORED IN A CHARGED CAPACITOR

If the plates of a charged capacitor are connected together by a conductor, such as a wire, charge moves from one plate to the other until the two are uncharged. The discharge can often be observed as a visible spark. If you should accidentally touch the opposite plates of a charged capacitor, your fingers act as a pathway by which the capacitor could discharge, and the result is an electric shock. The degree of shock you receive depends on the capacitance and voltage applied to the capacitor. Such a shock could be fatal where high voltages are present, such as in the power supply of a television set.

Consider a parallel-plate capacitor that is initially uncharged, so that the initial potential difference across the plates is zero. Now imagine that the capacitor is connected to a battery which supplies it with charge Q. We assume that the capacitor is charged slowly so that the problem can be considered as an electrostatic system. The final potential difference across the capacitor is $V = Q/C$. Because the initial potential difference is zero, the average potential difference during the charging process is $V/2 = Q/2C$. From this we might conclude that the work needed to charge the capacitor is $W = QV/2 = Q^2/2C$. Although this result is correct, a more detailed proof is desirable and is now given.

Suppose that q is the charge on the capacitor at some instant during the charging process. At the same instant, the potential difference across the capacitor is $V = q/C$. The work necessary[2] to transfer an increment of charge dq from the plate

---

[2] One mechanical analog of this process is the work required to raise a mass through some vertical distance in the presence of gravity.

of charge $-q$ to the plate of charge q (which is at the higher potential) is

$$dW = V \, dq = \frac{q}{C} \, dq$$

Thus, the total work required to charge the capacitor from $q = 0$ to some final charge $q = Q$ is

$$W = \int_0^Q \frac{q}{C} \, dq = \frac{Q^2}{2C}$$

But the work done in charging the capacitor can be considered as potential energy U stored in the capacitor. Using $Q = CV$, we can express the electrostatic potential energy stored in a charged capacitor in the following alternative forms:

Energy stored in a charged capacitor

$$U = \frac{Q^2}{2C} = \tfrac{1}{2}QV = \tfrac{1}{2}CV^2 \qquad (26.10)$$

This result applies to any capacitor, regardless of its geometry. We see that for a given capacitance, the stored energy increases as the charge increases and as the potential difference increases. In practice, there is a limit to the maximum energy (or charge) that can be stored. This is because electrical discharge ultimately occurs between the plates of the capacitor at a sufficiently large value of V. For this reason, capacitors are usually labeled with a maximum operating voltage.

The energy stored in a capacitor can be considered as being stored in the electric field created between the plates as the capacitor is charged. This description is reasonable in view of the fact that the electric field is proportional to the charge on the capacitor. For a parallel-plate capacitor, the potential difference is related to the electric field through the relationship $V = Ed$. Furthermore, its capacitance is $C = \epsilon_0 A/d$. Substituting these expressions into Equation 26.10 gives

Energy stored in a parallel-plate capacitor

$$U = \tfrac{1}{2} \frac{\epsilon_0 A}{d} (E^2 d^2) = \tfrac{1}{2}(\epsilon_0 Ad)E^2 \qquad (26.11)$$

Since the volume occupied by the electric field is Ad, the *energy per unit volume* $u_E = U/Ad$, called the *energy density*, is

Energy density in an electric field

$$u_E = \tfrac{1}{2}\epsilon_0 E^2 \qquad (26.12)$$

Although Equation 26.12 was derived for a parallel-plate capacitor, the expression is generally valid. That is, the *energy density in any electric field is proportional to the square of the electric field in the unit volume.*

EXAMPLE 26.5   Rewiring Two Charged Capacitors

Two capacitors C_1 and C_2 (where $C_1 > C_2$) are charged to the same potential difference V_0, but with opposite polarity. The charged capacitors are removed from the battery, and their plates are connected as shown in Figure 26.10a. The switches S_1 and S_2 are then closed as in Figure 26.10b. (a) Find the final potential difference between a and b after the switches are closed.

Solution The charges on the left-hand plates of the capacitors before the switches are closed are

$$Q_1 = C_1 V_0 \qquad \text{and} \qquad Q_2 = -C_2 V_0$$

The negative sign for Q_2 is necessary since this capacitor's polarity is opposite that of capacitor C_1. After the switches are closed, the charges on the plates redistribute until the total charge Q shared by the capacitors is

$$Q = Q_1 + Q_2 = (C_1 - C_2)V_0$$

The two capacitors are now in parallel, and so the final po-

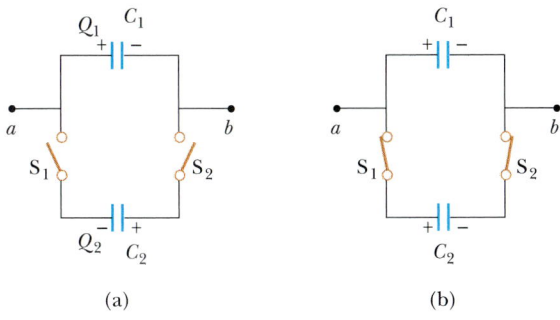

FIGURE 26.10   (Example 26.5).

tential difference across each is the same:

$$V = \frac{Q}{C_1 + C_2} = \left(\frac{C_1 - C_2}{C_1 + C_2}\right) V_0$$

(b) Find the total energy stored in the capacitors before and after the switches are closed.

Solution   Before the switches are closed, the total energy stored in the capacitors is

$$U_i = \tfrac{1}{2}C_1 V_0^2 + \tfrac{1}{2}C_2 V_0^2 = \tfrac{1}{2}(C_1 + C_2)V_0^2$$

After the switches are closed and the capacitors have reached an equilibrium charge, the total energy stored in them is

$$U_f = \tfrac{1}{2}C_1 V^2 + \tfrac{1}{2}C_2 V^2 = \tfrac{1}{2}(C_1 + C_2)V^2$$

$$= \tfrac{1}{2}(C_1 + C_2)\left(\frac{C_1 - C_2}{C_1 + C_2}\right)^2 V_0^2 = \left(\frac{C_1 - C_2}{C_1 + C_2}\right)^2 U_i$$

Therefore, the ratio of the final to the initial energy stored is

$$\frac{U_f}{U_i} = \left(\frac{C_1 - C_2}{C_1 + C_2}\right)^2$$

This shows that the final energy is less than the initial energy. At first, you might think that energy conservation has been violated, but this is not the case. Part of the missing energy appears as thermal energy in the connecting wires, and part is radiated away in the form of electromagnetic waves (Chapter 34).

26.5   CAPACITORS WITH DIELECTRICS

A **dielectric** is a nonconducting material, such as rubber, glass, or waxed paper. When a dielectric material is inserted between the plates of a capacitor, the capacitance increases. If the dielectric completely fills the space between the plates, the capacitance increases by a dimensionless factor κ, called the **dielectric constant.**

The following experiment can be performed to illustrate the effect of a dielectric in a capacitor. Consider a parallel-plate capacitor of charge Q_0 and capacitance C_0 in the absence of a dielectric. The potential difference across the capacitor as measured by a voltmeter is $V_0 = Q_0/C_0$ (Fig. 26.11a). Notice that the capacitor

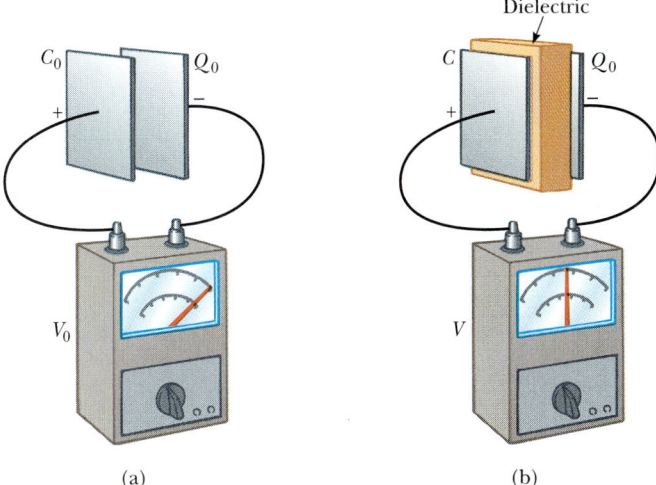

FIGURE 26.11   When a dielectric is inserted between the plates of a charged capacitor, the charge on the plates remains unchanged, but the potential difference as recorded by an electrostatic voltmeter is reduced from V_0 to $V = V_0/\kappa$. Thus, the capacitance *increases* in the process by the factor κ.

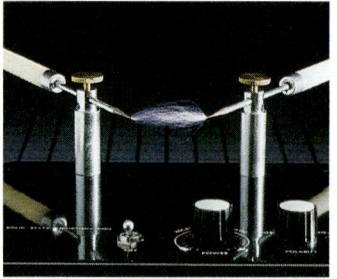

Sparks leap between two brass spheres connected to an electrostatic generator. The potential difference between the spheres is approximately 4000 V. *(Courtesy of Central Scientific Co.)*

circuit is open, that is, the plates of the capacitor are not connected to a battery and charge cannot flow through an ideal voltmeter. (We discuss the voltmeter further in Chapter 28.) Hence, there is no path by which charge can flow and alter the charge on the capacitor. If a dielectric is now inserted between the plates as in Figure 26.11b, it is found that the voltmeter reading decreases by a factor κ to a value V, where

$$V = \frac{V_0}{\kappa}$$

Since $V < V_0$, we see that $\kappa > 1$.

Since the charge Q_0 on the capacitor *does not change,* we conclude that the capacitance must change to the value

$$C = \frac{Q_0}{V} = \frac{Q_0}{V_0/\kappa} = \kappa \frac{Q_0}{V_0}$$

$$C = \kappa C_0 \qquad (26.13)$$

where C_0 is the capacitance in the absence of the dielectric. That is, the capacitance *increases* by the factor κ when the dielectric completely fills the region between the plates.[3] For a parallel-plate capacitor, where $C_0 = \epsilon_0 A/d$ (Eq. 26.3), we can express the capacitance when the capacitor is filled with a dielectric as

$$C = \kappa \frac{\epsilon_0 A}{d} \qquad (26.14)$$

The capacitance of a filled capacitor is greater than that of an empty one by a factor κ.

TABLE 26.1   Dielectric Constants and Dielectric Strengths of Various Materials at Room Temperature

Material	Dielectric Constant κ	Dielectric Strength[a] (V/m)
Vacuum	1.00000	—
Air (dry)	1.00059	3×10^6
Bakelite	4.9	24×10^6
Fused quartz	3.78	8×10^6
Pyrex glass	5.6	14×10^6
Polystyrene	2.56	24×10^6
Teflon	2.1	60×10^6
Neoprene rubber	6.7	12×10^6
Nylon	3.4	14×10^6
Paper	3.7	16×10^6
Strontium titanate	233	8×10^6
Water	80	—
Silicone oil	2.5	15×10^6

[a] The dielectric strength equals the maximum electric field that can exist in a dielectric without electrical breakdown.

[3] If another experiment is performed in which the dielectric is introduced while the potential difference remains constant by means of a battery, the charge increases to a value $Q = \kappa Q_0$. The additional charge is supplied by the battery and the capacitance still increases by the factor κ.

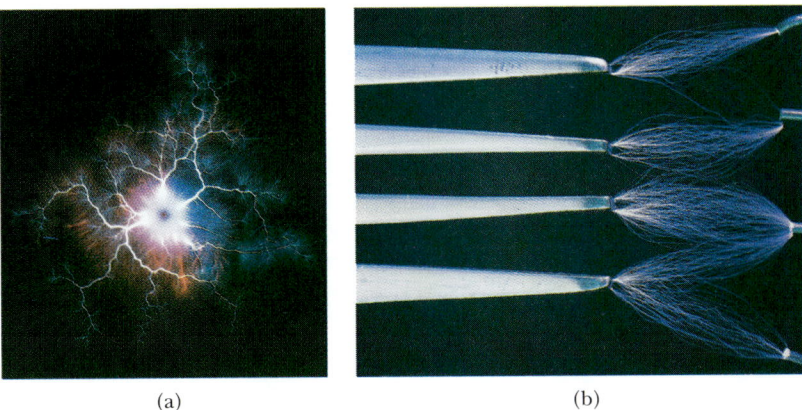

(a)   (b)

(a) Kirlian photograph created by dropping a steel ball into a high-energy electric field. This technique is also known as electrophotography. *(Henry Dakin/Science Photo Library)* (b) Sparks from static electricity discharge between a fork and four electrodes. Many sparks were used to make this image, because only one spark will form for a given discharge. Each spark follows the line of least resistance through the air at the time. Note that the bottom prong of the fork forms discharges to both electrodes at bottom right. The light of each spark is created by the excitations of gas atoms along its path. *(Adam Hart-Davis/Science Photo Library)*

From Equations 26.3 and 26.14, it would appear that the capacitance could be made very large by decreasing d, the distance between the plates. In practice, the lowest value of d is limited by the electrical discharge that could occur through the dielectric medium separating the plates. For any given separation d, the maximum voltage that can be applied to a capacitor without causing a discharge depends on the *dielectric strength* (maximum electric field strength) of the dielectric, which for air is equal to 3×10^6 V/m. If the field strength in the medium exceeds the dielectric strength, the insulating properties break down and the medium begins to conduct. Most insulating materials have dielectric strengths and dielectric constants greater than that of air, as Table 26.1 indicates. Thus, we see that a dielectric provides the following advantages:

- Increases the capacitance of a capacitor
- Increases the maximum operating voltage of a capacitor
- May provide mechanical support between the conducting plates

Types of Capacitors

Commercial capacitors are often made using metal foil interlaced with thin sheets of paraffin-impregnated paper or Mylar, which serves as the dielectric material. These alternate layers of metal foil and dielectric are then rolled into a cylinder to form a small package (Fig. 26.12a). High-voltage capacitors commonly consist of a number of interwoven metal plates immersed in silicone oil (Fig. 26.12b). Small capacitors are often constructed from ceramic materials. Variable capacitors (typically 10 to 500 pF) usually consist of two interwoven sets of metal plates, one fixed and the other movable, with air as the dielectric.

An *electrolytic capacitor* is often used to store large amounts of charge at relatively low voltages. This device, shown in Figure 26.12c, consists of a metal foil in contact

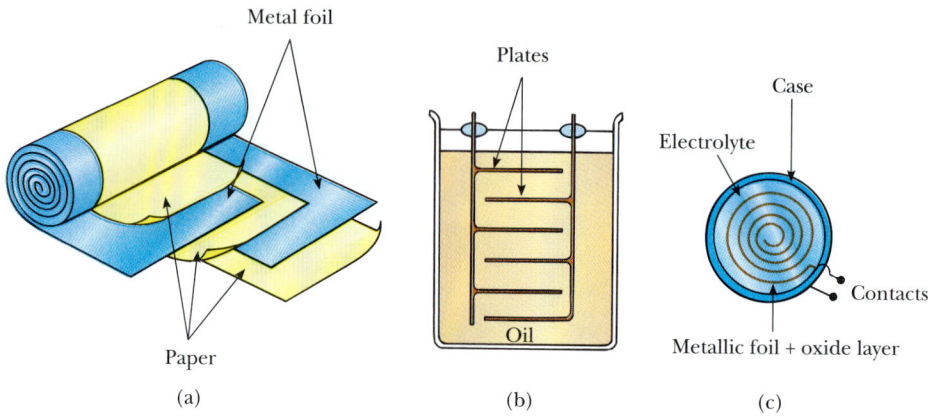

FIGURE 26.12   Three commercial capacitor designs. (a) A tubular capacitor whose plates are separated by paper and then rolled into a cylinder. (b) A high-voltage capacitor consists of many parallel plates separated by insulating oil. (c) An electrolytic capacitor.

with an electrolyte—a solution that conducts electricity by virtue of the motion of ions contained in the solution. When a voltage is applied between the foil and the electrolyte, a thin layer of metal oxide (an insulator) is formed on the foil, and this layer serves as the dielectric. Very large values of capacitance can be obtained because the dielectric layer is very thin and thus the plate separation is very small.

When electrolytic capacitors are used in circuits, the polarity (the plus and minus signs on the device) must be installed properly. If the polarity of the applied voltage is opposite what is intended, the oxide layer is removed and the capacitor conducts electricity rather than stores charge.

EXAMPLE 26.6   A Paper-Filled Capacitor

A parallel-plate capacitor has plates of dimensions 2.0 cm × 3.0 cm separated by a 1.0-mm thickness of paper. (a) Find the capacitance of this device.

Solution   Since $\kappa = 3.7$ for paper (Table 26.1), we get

$$C = \kappa \frac{\epsilon_0 A}{d} = 3.7 \left(8.85 \times 10^{-12} \, \frac{C^2}{N \cdot m^2} \right) \left(\frac{6.0 \times 10^{-4} \, m^2}{1.0 \times 10^{-3} \, m} \right)$$

$$= 20 \times 10^{-12} \, F = \boxed{20 \text{ pF}}$$

(b) What is the maximum charge that can be placed on the capacitor?

Solution   From Table 26.1 we see that the dielectric strength of paper is 16×10^6 V/m. Since the thickness of the

paper is 1.0 mm, the maximum voltage that can be applied before breakdown is

$$V_{max} = E_{max} d = \left(16 \times 10^6 \, \frac{V}{m} \right) (1.0 \times 10^{-3} \, m)$$

$$= 16 \times 10^3 \, V$$

Hence, the maximum charge is

$$Q_{max} = CV_{max} = (20 \times 10^{-12} \, F)(16 \times 10^3 \, V) = \boxed{0.32 \, \mu C}$$

Exercise   What is the maximum energy that can be stored in the capacitor?

Answer 2.5×10^{-3} J.

EXAMPLE 26.7   Energy Stored Before and After

A parallel-plate capacitor is charged with a battery to a charge Q_0, as in Figure 26.13a. The battery is then removed, and a slab of material that has a dielectric constant κ is in-

serted between the plates, as in Figure 26.13b. Find the energy stored in the capacitor before and after the dielectric is inserted.

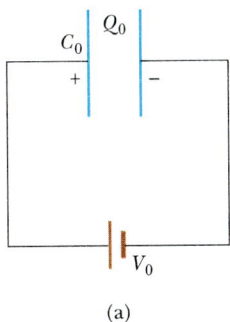

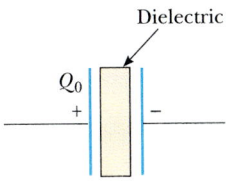

FIGURE 26.13   (Example 26.7).

Solution   The energy stored in the capacitor in the absence of the dielectric is

$$U_0 = \tfrac{1}{2} C_0 V_0^2$$

Since $V_0 = Q_0/C_0$, this can be expressed as

$$U_0 = \frac{Q_0^2}{2C_0}$$

After the battery is removed and the dielectric is inserted between the plates, the *charge on the capacitor remains the same.* Hence, the energy stored in the presence of the dielectric is

$$U = \frac{Q_0^2}{2C}$$

But the capacitance in the presence of the dielectric is $C = \kappa C_0$, and so U becomes

$$U = \frac{Q_0^2}{2\kappa C_0} = \frac{U_0}{\kappa}$$

Since $\kappa > 1$, we see that the final energy is less than the initial energy by the factor $1/\kappa$. This missing energy can be accounted for by noting that when the dielectric is inserted into the capacitor, it gets pulled into the device. An external agent must do negative work to keep the slab from accelerating. This work is simply the difference $U - U_0$. (Alternatively, the positive work done by the system on the external agent is $U_0 - U$.)

Exercise   Suppose that the capacitance in the absence of a dielectric is 8.50 pF, and the capacitor is charged to a potential difference of 12.0 V. If the battery is disconnected, and a slab of polystyrene ($\kappa = 2.56$) is inserted between the plates, calculate the energy difference $U - U_0$.

Answer   373 pJ.

As we have seen, the energy of a capacitor is lowered when a dielectric is inserted between the plates, which means that work is done on the dielectric. This, in turn, implies that a force must be acting on the dielectric that draws it into the capacitor. This force originates from the nonuniform nature of the electric field of the capacitor near its edges, as indicated in Figure 26.14. The horizontal compo-

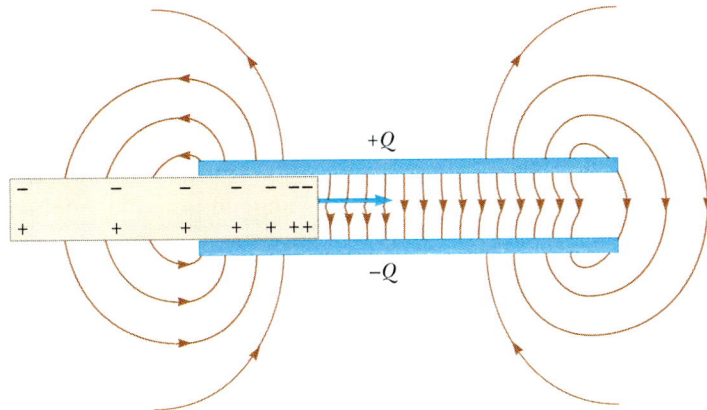

FIGURE 26.14   The nonuniform electric field near the edges of a parallel-plate capacitor causes a dielectric to be pulled into the capacitor. Note that the field acts on the induced surface charges on the dielectric that are nonuniformly distributed.

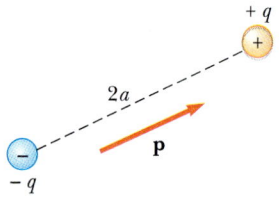

FIGURE 26.15 An electric dipole consists of two equal and opposite charges separated by a distance $2a$. The electric dipole moment p is directed from $-q$ to $+q$.

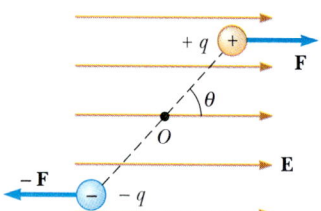

FIGURE 26.16 An electric dipole in a uniform electric field. The dipole moment p is at an angle θ with the field, and the dipole experiences a torque.

Torque on an electric dipole in an external electric field

---

nent of this fringe field acts on the induced charges on the surface of the dielectric, producing a net horizontal force directed into the capacitor.

*26.6 ELECTRIC DIPOLE IN AN EXTERNAL ELECTRIC FIELD

The electric dipole, discussed briefly in Example 23.9, consists of two equal and opposite charges separated by a distance $2a$, as in Figure 26.15. Let us define the **electric dipole moment** of this configuration as the vector **p** that is directed from $-q$ to $+q$ along the line joining the charges, and whose magnitude is $2aq$:

$$p \equiv 2aq \tag{26.15}$$

Now suppose an electric dipole is placed in a uniform electric field **E** as in Figure 26.16, where the dipole moment makes an angle θ with the field. The forces on the two charges are equal and opposite as shown, each having a magnitude

$$F = qE$$

Thus, we see that the net force on the dipole is zero. However, the two forces produce a net torque on the dipole, and the dipole tends to rotate such that its axis is aligned with the field. The torque due to the force on the positive charge about an axis through O is $Fa \sin\theta$, where $a \sin\theta$ is the moment arm of F about O. In Figure 26.16, this force tends to produce a clockwise rotation. The torque about O on the negative charge is also $Fa \sin\theta$, and so the net torque about O is

$$\tau = 2Fa \sin\theta$$

Because $F = qE$ and $p = 2aq$, we can express τ as

$$\tau = 2aqE \sin\theta = pE \sin\theta \tag{26.16}$$

It is convenient to express the torque in vector form as the cross product of the vectors **p** and **E**:

$$\boldsymbol{\tau} = \mathbf{p} \times \mathbf{E} \tag{26.17}$$

We can also determine the potential energy of an electric dipole as a function of its orientation with respect to the external electric field. In order to do this, you should recognize that work must be done by an external agent to rotate the dipole through a given angle in the field. The work done increases the potential energy in the system, which consists of the dipole and the external field. The work dW required to rotate the dipole through an angle $d\theta$ is $dW = \tau\, d\theta$ (Chapter 10). Because $\tau = pE \sin\theta$, and because the work is transformed to potential energy U, we find that for a rotation from θ_0 to θ, the change in potential energy is

$$U - U_0 = \int_{\theta_0}^{\theta} \tau\, d\theta = \int_{\theta_0}^{\theta} pE \sin\theta\, d\theta = pE \int_{\theta_0}^{\theta} \sin\theta\, d\theta$$

$$U - U_0 = pE[-\cos\theta]_{\theta_0}^{\theta} = pE(\cos\theta_0 - \cos\theta)$$

The term involving $\cos\theta_0$ is a constant that depends on the initial orientation of the dipole. It is convenient to choose $\theta_0 = 90°$, so that $\cos\theta_0 = \cos 90° = 0$. Furthermore, let us choose $U_0 = 0$ at $\theta_0 = 90°$ as our reference of potential energy. Hence, we can express U as

$$U = -pE \cos\theta \tag{26.18}$$

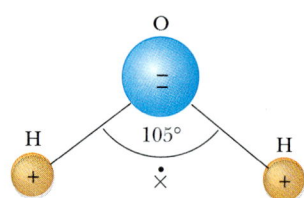

FIGURE 26.17 The water molecule, H_2O, has a permanent polarization resulting from its bent geometry. The center of positive charge is at the point x.

This can be written as the dot product of the vectors **p** and **E**:

$$U = -\mathbf{p} \cdot \mathbf{E} \qquad (26.19)$$

Potential energy of a dipole in an electric field

Molecules are said to be polarized when there is a separation between the "center of gravity" of the negative charges and that of the positive charges on the molecule. In some molecules, such as water, this condition is always present. This can be understood by inspecting the geometry of the water molecule. The molecule is arranged so that the oxygen atom is bonded to the hydrogen atoms with an angle of 105° between the two bonds (Fig. 26.17). The center of negative charge is near the oxygen atom, and the center of positive charge lies at a point midway along the line joining the hydrogen atoms (point *x* in the diagram). Materials composed of molecules that are permanently polarized in this fashion have large dielectric constants. For example, the dielectric constant of water is quite large ($\kappa = 80$).

A symmetrical molecule (Fig. 26.18a) might have no permanent polarization, but a polarization can be induced by an external electric field. A field directed to the left, as in Figure 26.18b, would cause the center of positive charge to shift to the right from its initial position and the center of negative charge to shift to the left. This *induced polarization* is the effect that predominates in most materials used as dielectrics in capacitors.

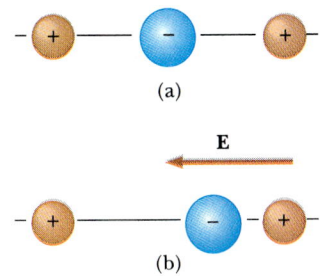

FIGURE 26.18   (a) A symmetric molecule has no permanent polarization. (b) An external electric field induces a polarization in the molecule.

EXAMPLE 26.8   The H₂O Molecule

The H_2O molecule has a dipole moment of 6.3×10^{-30} C·m. A sample contains 10^{21} such molecules, whose dipole moments are all oriented in the direction of an electric field of 2.5×10^5 N/C. How much work is required to rotate the dipoles from this orientation ($\theta = 0°$) to one in which all of the moments are perpendicular to the field ($\theta = 90°$)?

Solution   The work required to rotate one molecule by 90° is equal to the difference in potential energy between the 90° orientation and the 0° orientation. Using Equation 26.18

gives

$$W = U_{90} - U_0 = (-pE \cos 90°) - (-pE \cos 0°)$$
$$= pE = (6.3 \times 10^{-30} \text{ C·m})(2.5 \times 10^5 \text{ N/C})$$
$$= 1.6 \times 10^{-24} \text{ J}$$

Since there are 10^{21} molecules in the sample, the *total* work required is

$$W_{\text{total}} = (10^{21})(1.6 \times 10^{-24} \text{ J}) = \boxed{1.6 \times 10^{-3} \text{ J}}$$

---

*26.7   AN ATOMIC DESCRIPTION OF DIELECTRICS

In Section 26.5 we found that the potential difference between the plates of a capacitor is reduced by the factor κ when a dielectric is introduced. Since the potential difference between the plates equals the product of the electric field and the separation d, the electric field is also reduced by the factor κ. Thus, if E_0 is the electric field without the dielectric, the field in the presence of a dielectric is

$$\mathbf{E} = \frac{\mathbf{E}_0}{\kappa} \qquad (26.20)$$

This relationship can be understood by noting that a dielectric can be polarized. At the atomic level, a polarized material is one in which the positive and negative charges are slightly separated. If the molecules of the dielectric possess permanent electric dipole moments in the absence of an electric field, they are called **polar molecules** (water is an example). The dipoles are randomly oriented in the ab-

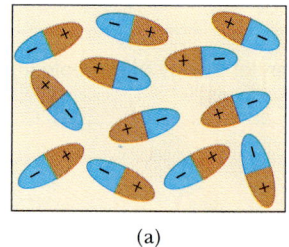

(a)

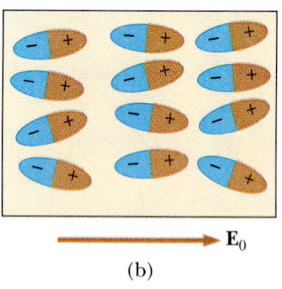

$\mathbf{E}_0$

(b)

FIGURE 26.19 (a) Molecules with a permanent dipole moment are randomly oriented in the absence of an external electric field. (b) When an external field is applied, the dipoles are partially aligned with the field.

sence of an electric field, as shown in Figure 26.19a. When an external field $\mathbf{E}_0$ is applied, a torque is exerted on the dipoles, causing them to be partially aligned with the field, as in Figure 26.19b. The degree of alignment depends on temperature and on the magnitude of the applied field. In general, the alignment increases with decreasing temperature and with increasing electric field strength. The partially aligned dipoles produce an internal electric field that opposes the external field, thereby causing a reduction in the net field within the dielectric.

If the molecules of the dielectric do not possess a permanent dipole moment, they are called **nonpolar molecules.** In this case, an external electric field produces some charge separation and an induced dipole moment. These induced dipole moments tend to align with the external field, causing a reduction in the internal electric field.

With these ideas in mind, consider a slab of dielectric material in a uniform electric field $\mathbf{E}_0$ as in Figure 26.20a. The external electric field in the capacitor is directed toward the right and exerts forces on the molecules of the dielectric material. Under the influence of these forces, electrons in the dielectric are shifted from their equilibrium positions toward the left. Hence, the applied electric field polarizes the dielectric. The net effect on the dielectric is the formation of an "induced" positive surface charge density σ_i on the right face and an equal negative surface charge density on the left face, as in Figure 26.20b. These induced surface charges on the dielectric give rise to an induced internal electric field $\mathbf{E}_i$ that opposes the external field $\mathbf{E}_0$. Therefore, the net electric field $\mathbf{E}$ in the dielectric has a magnitude given by

$$E = E_0 - E_i \tag{26.21}$$

In the parallel-plate capacitor shown in Figure 26.21, the external field E_0 is related to the free charge density σ on the plates through the relationship $E_0 = \sigma/\epsilon_0$. The induced electric field in the dielectric is related to the induced charge density σ_i through the relationship $E_i = \sigma_i/\epsilon_0$. Since $E = E_0/\kappa = \sigma/\kappa\epsilon_0$, substitution into Equation 26.21 gives

$$\frac{\sigma}{\kappa\epsilon_0} = \frac{\sigma}{\epsilon_0} - \frac{\sigma_i}{\epsilon_0}$$

$$\sigma_i = \left(\frac{\kappa - 1}{\kappa}\right)\sigma \tag{26.22}$$

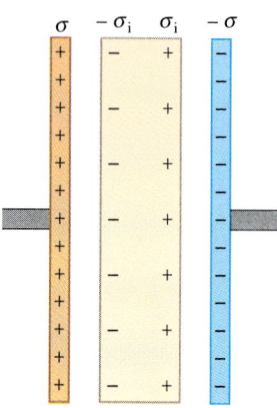

$\sigma \quad -\sigma_i \quad \sigma_i \quad -\sigma$

FIGURE 26.21 Induced charge on a dielectric placed between the plates of a charged capacitor. Note that the induced charge density on the dielectric is *less* than the free charge density on the plates.

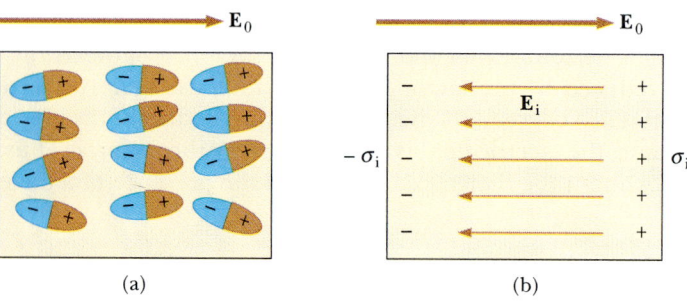

(a)   (b)

FIGURE 26.20 (a) When a dielectric is polarized, the molecular dipole moments in the dielectric are partially aligned with the external field $\mathbf{E}_0$. (b) This polarization causes an induced negative surface charge on one side of the dielectric and an equal positive surface charge on the opposite side. This results in a reduction in the net electric field within the dielectric.

Because $\kappa > 1$, this expression shows that the charge density σ_i induced on the dielectric is less than the free charge density σ on the plates. For instance, if $\kappa = 3$, we see that the induced charge density on the dielectric is two-thirds the free charge density on the plates. If there is no dielectric present, $\kappa = 1$ and $\sigma_i = 0$ as expected. However, if the dielectric is replaced by an electrical conductor, for which $E = 0$, then Equation 26.21 shows that $E_0 = E_i$, corresponding to $\sigma_i = \sigma$. That is, the surface charge induced on the conductor is equal to and opposite that on the plates, resulting in a net field of zero in the conductor.

EXAMPLE 26.9   A Partially Filled Capacitor

A parallel-plate capacitor has a capacitance C_0 in the absence of a dielectric. A slab of dielectric material of dielectric constant κ and thickness $\frac{1}{3}d$ is inserted between the plates (Fig. 26.22a). What is the new capacitance when the dielectric is present?

Reasoning   This capacitor is equivalent to two parallel-plate capacitors of the same area A connected in series, one with a plate separation $d/3$ (dielectric filled) and the other with a plate separation $2d/3$ and air between the plates (Fig. 26.22b). (This breaking-into-two step is permissible because there is no potential difference between the lower plate of C_1 and the upper plate of C_2.)[4]

From Equations 26.3 and 26.13, the two capacitances are

$$C_1 = \frac{\kappa \epsilon_0 A}{d/3} \quad \text{and} \quad C_2 = \frac{\epsilon_0 A}{2d/3}$$

Solution   Using Equation 26.9 for two capacitors combined in series, we get

$$\frac{1}{C} = \frac{1}{C_1} + \frac{1}{C_2} = \frac{d/3}{\kappa \epsilon_0 A} + \frac{2d/3}{\epsilon_0 A}$$

$$\frac{1}{C} = \frac{d}{3\epsilon_0 A}\left(\frac{1}{\kappa} + 2\right) = \frac{d}{3\epsilon_0 A}\left(\frac{1 + 2\kappa}{\kappa}\right)$$

$$C = \left(\frac{3\kappa}{2\kappa + 1}\right)\frac{\epsilon_0 A}{d}$$

Since the capacitance without the dielectric is $C_0 = \epsilon_0 A/d$, we see that

$$C = \left(\frac{3\kappa}{2\kappa + 1}\right)C_0$$

[4] You could also imagine placing two thin metallic plates (with a coiled-up conducting wire between them) at the lower surface of the dielectric in Figure 26.22a and then pulling the assembly out until it becomes like Figure 26.22b.

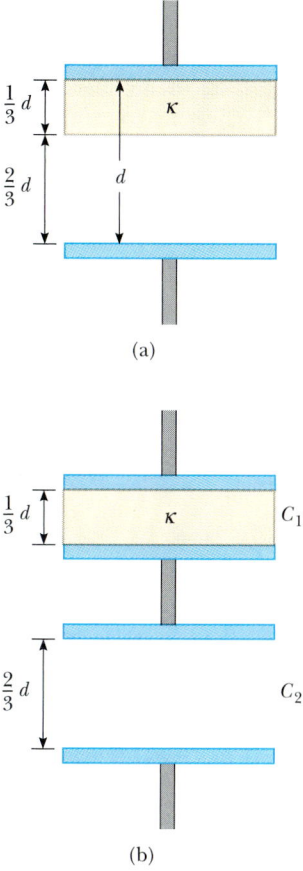

FIGURE 26.22   (Example 26.9) (a) A parallel-plate capacitor of plate separation d partially filled with a dielectric of thickness $d/3$. (b) The equivalent circuit of the capacitor consists of two capacitors connected in series.

EXAMPLE 26.10   Effect of a Metal Slab

A parallel-plate capacitor has a plate separation d and plate area A. An uncharged metal slab of thickness a is inserted midway between the plates, as shown in Figure 26.23a. Find the capacitance of the device.

Reasoning   This problem can be solved by noting that any charge that appears on one plate of the capacitor must induce an equal and opposite charge on the metal slab, as in Figure 26.23a. Consequently, the net charge on the metal

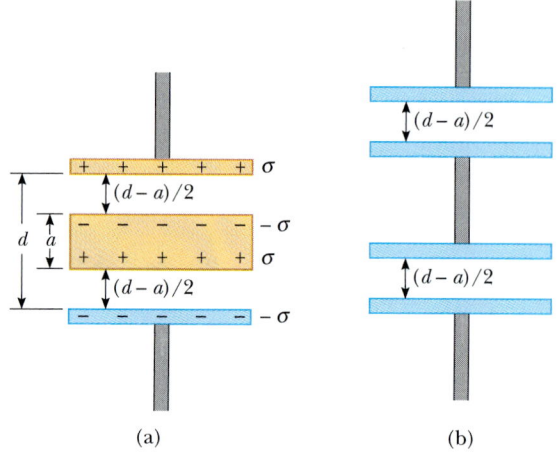

(a)   (b)

FIGURE 26.23   (Example 26.10) (a) A parallel-plate capacitor of plate separation d partially filled with a metal slab of thickness a. (b) The equivalent circuit of the device in (a) consists of two capacitors in series, each with a plate separation $(d - a)/2$.

slab remains zero, and the field inside the slab is zero. Hence, the capacitor is equivalent to two capacitors in series, each having a plate separation $(d - a)/2$, as in Figure 26.23b.

Solution   Using the rule for adding two capacitors in series we get

$$\frac{1}{C} = \frac{1}{C_1} + \frac{1}{C_2} = \frac{1}{\dfrac{\epsilon_0 A}{(d - a)/2}} + \frac{1}{\dfrac{\epsilon_0 A}{(d - a)/2}}$$

$$C = \frac{\epsilon_0 A}{d - a}$$

Note that C approaches infinity as a approaches d. Why?

SUMMARY

A *capacitor* consists of two equal and oppositely charged conductors separated by a distance that is very small compared with the dimensions of the plates. The **capacitance** C of any capacitor is defined to be the ratio of the charge Q on either conductor to the potential difference V between them:

$$C \equiv \frac{Q}{V} \tag{26.1}$$

The SI unit of capacitance is coulomb per volt, or farad (F), and 1 F = 1 C/V.

The capacitance of several capacitors is summarized in Table 26.2. The formulas apply when the charged conductors are separated by a vacuum.

TABLE 26.2 **Capacitance and Geometry**		
Geometry	Capacitance	Equation
Isolated charged sphere of radius R	$C = 4\pi\epsilon_0 R$	(26.2)
Parallel-plate capacitor of plate area A and plate separation d	$C = \epsilon_0 \dfrac{A}{d}$	(26.3)
Cylindrical capacitor of length ℓ and inner and outer radii a and b, respectively	$C = \dfrac{\ell}{2k_e \ln\left(\dfrac{b}{a}\right)}$	(26.4)
Spherical capacitor with inner and outer radii a and b, respectively	$C = \dfrac{ab}{k_e(b - a)}$	(26.5)

If two or more capacitors are connected in parallel, the potential difference is the same across all of them. The equivalent capacitance of a parallel combination of capacitors is

$$C_{eq} = C_1 + C_2 + C_3 + \cdots \qquad (26.7)$$

If two or more capacitors are connected in series, the charge is the same on all of them, and the equivalent capacitance of the series combination is

$$\frac{1}{C_{eq}} = \frac{1}{C_1} + \frac{1}{C_2} + \frac{1}{C_3} + \cdots \qquad (26.9)$$

Work is required to charge a capacitor, since the charging process consists of transferring charges from one conductor at a lower potential to another conductor at a higher potential. The work done in charging the capacitor to a charge Q equals the electrostatic potential energy U stored in the capacitor, where

$$U = \frac{Q^2}{2C} = \tfrac{1}{2} QV = \tfrac{1}{2} CV^2 \qquad (26.10)$$

When a dielectric material is inserted between the plates of a capacitor, the capacitance generally increases by a dimensionless factor κ called the **dielectric constant**:

$$C = \kappa C_0 \qquad (26.13)$$

where C_0 is the capacitance in the absence of the dielectric. The increase in capacitance is due to a decrease in the electric field in the presence of the dielectric and to a corresponding decrease in the potential difference between the plates — assuming the charging battery is removed from the circuit before the dielectric is inserted. The decrease in E arises from an internal electric field produced by aligned dipoles in the dielectric. This internal field produced by the dipoles opposes the original applied field, and the result is a reduction in the net electric field.

An *electric dipole* consists of two equal and opposite charges separated by a distance $2a$. The **electric dipole moment p** of this configuration has a magnitude

$$p \equiv 2aq \qquad (26.15)$$

The **torque** acting on an electric dipole in a uniform electric field E is

$$\boldsymbol{\tau} = \mathbf{p} \times \mathbf{E} \qquad (26.17)$$

The **potential energy** of an electric dipole in a uniform external electric field E is

$$U = -\mathbf{p} \cdot \mathbf{E} \qquad (26.19)$$

Problem-Solving Strategy and Hints
Capacitors

- Be careful with your choice of units. To calculate capacitance in farads, make sure that distances are in meters and use the SI value of ϵ_0. When checking consistency of units, remember that the units for electric fields can be either N/C or V/m.

- When two or more unequal capacitors are connected in series, they carry the same charge, but the potential differences are not the same. Their capacitances add as reciprocals, and the equivalent capacitance of the combination is always less than the smallest individual capacitor.
- When two or more capacitors are connected in parallel, the potential difference across each is the same. The charge on each capacitor is proportional to its capacitance; hence, the capacitances add directly to give the equivalent capacitance of the parallel combination.
- The effect of a dielectric is to increase the capacitance of a capacitor by a factor κ (the dielectric constant) over its empty capacitance. The reason for this is that induced surface charges on the dielectric reduce the electric field inside the material from E to E/κ.
- Be careful about problems in which you may be connecting or disconnecting a battery to a capacitor. It is important to note whether modifications to the capacitor are being made while the capacitor is connected to the battery or after it has been disconnected. If the capacitor remains connected to the battery, the voltage across the capacitor necessarily remains the same (equal to the battery voltage), and the charge is proportional to the capacitance *however it may be modified* (say, by inserting a dielectric). If you disconnect the capacitor from the battery before making any modifications to the capacitor, then its charge remains the same. In this case, as you vary the capacitance, the voltage across the plates changes as $V = Q/C$.

QUESTIONS

1. What happens to the charge on a capacitor if the potential difference between the conductors is doubled?

2. The plates of a capacitor are connected to a battery. What happens to the charge on the plates if the connecting wires are removed from the battery? What happens to the charge if the wires are removed from the battery and connected to each other?

3. A farad is a very large unit of capacitance. Calculate the length of one side of a square, air-filled capacitor that has a plate separation of 1 m. Assume it has a capacitance of 1 F.

4. A pair of capacitors are connected in parallel while an identical pair are connected in series. Which pair would be more dangerous to handle after being connected to the same voltage source? Explain.

5. If you are given three different capacitors C_1, C_2, C_3, how many different combinations of capacitance can you produce?

6. What advantage might there be in using two identical capacitors in parallel connected in series with another identical parallel pair, rather than using a single capacitor?

7. Is it always possible to reduce a combination of capacitors to one equivalent capacitor with the rules we have developed? Explain.

8. Since the net charge in a capacitor is always zero, what does a capacitor store?

9. Since the charges on the plates of a parallel-plate capacitor are equal and opposite, they attract each other. Hence, it would take positive work to increase the plate separation. What happens to the external work done in this process?

10. Explain why the work needed to move a charge Q through a potential difference V is $W = QV$ whereas the energy stored in a charged capacitor is $U = \frac{1}{2}QV$. Where does the $\frac{1}{2}$ factor come from?

11. If the potential difference across a capacitor is doubled, by what factor does the energy stored change?

12. Why is it dangerous to touch the terminals of a high-voltage capacitor even after the applied voltage has been turned off? What can be done to make the capacitor safe to handle after the voltage source has been removed?

13. If you want to increase the maximum operating voltage of a parallel-plate capacitor, describe how you can do this for a fixed plate separation.

14. An air-filled capacitor is charged, then disconnected from the power supply, and finally connected to a voltmeter. Explain how and why the voltage reading changes when a dielectric is inserted between the plates of the capacitor.

15. Using the polar molecule description of a dielectric, explain how a dielectric affects the electric field inside a capacitor.

16. Explain why a dielectric increases the maximum operating voltage of a capacitor although the physical size of the capacitor does not change.
17. What is the difference between dielectric strength and the dielectric constant?
18. Explain why a water molecule is permanently polarized. What type of molecule has no permanent polarization?
19. If a dielectric-filled capacitor is heated, how will its capac-

itance change? (Neglect thermal expansion and assume that the dipole orientations are temperature dependent.)
20. In terms of induced charges, explain why a charged comb attracts small bits of paper.
21. If you were asked to design a capacitor where small size and large capacitance were required, what factors would be important in your design?

PROBLEMS

Section 26.1   Definition of Capacitance

1. The excess charge on each conductor of a parallel-plate capacitor is 53.0 μC. What is the potential difference between the conductors if the capacitance of the system is 4.00×10^{-3} μF?
2. If a drop of liquid had a capacitance of 1.0 pF in air, (a) what is its radius? (b) If its radius is 2.0 mm, what is its capacitance? (c) What is the charge on the smaller drop if its potential is 100 V?
3. Two conductors having net charges of $+10.0$ μC and -10.0 μC have a potential difference of 10.0 V. Determine (a) the capacitance of the system and (b) the potential difference between the two conductors if the charges on each are increased to $+100.0$ μC and -100.0 μC.
4. Two conductors insulated from each other are charged by transferring electrons from one conductor to the other. After 1.6×10^{12} electrons have been transferred, the potential difference between the conductors is 14 V. What is the capacitance of the system?
5. A parallel-plate capacitor has a capacitance of 19.0 μF. What charge on each plate produces a potential difference of 36.0 V between the plates?
6. By what factor does the capacitance of a metal sphere increase if its volume is tripled?
7. An isolated charged conducting sphere of radius 12.0 cm creates an electric field of 4.90×10^4 N/C at a distance 21.0 cm from its center. (a) What is its surface charge density? (b) What is its capacitance?
8. Two conducting spheres with diameters of 0.40 m and 1.0 m are separated by a distance that is large compared with the diameters. The spheres are connected by a thin wire and are charged to 7.0 μC. (a) How is this total charge shared between the spheres? (Neglect any charge on the wire.) (b) What is the potential of the system of spheres when the reference potential is taken to be $V = 0$ at $r = \infty$?
9. Two spherical conductors with radii $R_1 = 0.15$ cm and $R_2 = 0.23$ cm are separated by a distance large

enough to make induction effects negligible. The spheres are connected by a thin conducting wire and are brought to the same potential of 775 V relative to $V = 0$ at $r = \infty$. (a) Determine the capacitance of the system. (b) What is the charge ratio Q_1/Q_2?
9A. Two spherical conductors with radii R_1 and R_2 are separated by a distance large enough to make induction effects negligible. The spheres are connected by a thin conducting wire and are brought to the same potential V relative to $V = 0$ at $r = \infty$. (a) Determine the capacitance C of the system, where $C = (Q_1 + Q_2)/V$. (b) What is the charge ratio Q_1/Q_2?

Section 26.2   Calculation of Capacitance

10. An air-filled parallel-plate capacitor is to have a capacitance of 1.00 F. If the distance between the plates is 1.00 mm, calculate the required surface area of each plate. Convert your answer to square miles.
11. A parallel-plate capacitor has a plate area of 12.0 cm^2 and a capacitance of 7.00 pF. What is the plate separation?
12. The plates of a parallel-plate capacitor are separated by 0.20 mm. If the space between the plates is air, what plate area is required to provide a capacitance of 9.0 pF?
13. When a potential difference of 150 V is applied to the plates of a parallel-plate capacitor, the plates carry a surface charge density of 30 nC/cm^2. What is the spacing between the plates?

14. A small object with a mass of 350 mg carries a charge of 30 nC and is suspended by a thread between the vertical plates of a parallel-plate capacitor. The plates are separated by 4.0 cm. If the thread makes an angle of 15° with the vertical, what is the potential difference between the plates?
14A. A small object of mass m carries a charge q and is suspended by a thread between the vertical plates of

---

☐ indicates problems that have full solutions available in the Student Solutions Manual and Study Guide.

a parallel-plate capacitor. The plate separation is d. If the thread makes an angle θ with the vertical, what is the potential difference between the plates?

15. An air-filled capacitor consists of two parallel plates, each with an area of 7.60 cm², separated by a distance of 1.80 mm. If a 20.0-V potential difference is applied to these plates, calculate (a) the electric field between the plates, (b) the surface charge density, (c) the capacitance, and (d) the charge on each plate.

16. A 1-megabit computer memory chip contains many 60-fF capacitors. Each capacitor has a plate area of 21×10^{-12} m². Determine the plate separation of such a capacitor (assume a parallel-plate configuration). The characteristic atomic diameter is 10^{-10} m = 1 Å. Express the plate separation in Å.

17. A variable air capacitor used in tuning circuits is made of 10 semicircular plates each of radius 2.5 cm and positioned 0.80 cm from each other. A second identical set that is free to rotate is enmeshed with the first set of plates (Fig. P26.17). Determine the capacitance as a function of the angle of rotation θ, where $\theta = 0$ corresponds to the maximum capacitance.

17A. A variable air capacitor used in tuning circuits is made of N semicircular plates each of radius R and positioned a distance d from each other. A second identical set of plates that is free to rotate is enmeshed with the first set (Fig. P26.17). Determine the capacitance as a function of the angle of rotation θ, where $\theta = 0$ corresponds to the maximum capacitance.

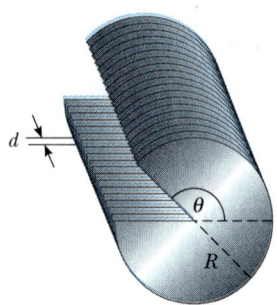

FIGURE P26.17

18. A capacitor is constructed of interlocking plates as shown in Figure P26.18 (a cross-sectional view). The separation between adjacent plates is 0.80 mm, and the total effective area of all plates combined is 7.0 cm². Ignoring side effects, calculate the capacitance of the unit.

FIGURE P26.18

19. A 20.0-μF spherical capacitor is composed of two metal spheres, one having a radius twice as large as the other. If the region between the spheres is a vacuum, determine the volume of this region.

19A. A spherical capacitor having a capacitance C is composed of two metal spheres, one having a radius twice as large as the other. If the region between the spheres is a vacuum, determine the volume of this region.

20. An air-filled cylindrical capacitor has a capacitance of 10 pF and is 6.0 cm in length. If the radius of the outside conductor is 1.5 cm, what is the required radius of the inner conductor?

21. A 50.0-m length of coaxial cable has an inner conductor that has a diameter of 2.58 mm and carries a charge of 8.10 μC. The surrounding conductor has an inner diameter of 7.27 mm and a charge of -8.10 μC. (a) What is the capacitance of this cable? (b) What is the potential difference between the two conductors? Assume the region between the conductors is air.

22. A cylindrical capacitor has outer and inner conductors whose radii are in the ratio of $b/a = 4/1$. The inner conductor is to be replaced by a wire whose radius is one-half the radius of the original inner conductor. By what factor should the length be increased in order to obtain a capacitance equal to that of the original capacitor?

23. An air-filled spherical capacitor is constructed with inner and outer shell radii of 7.00 and 14.0 cm, respectively. (a) Calculate the capacitance of the device. (b) What potential difference between the spheres results in a charge of 4.00 μC on the capacitor?

24. Find the capacitance of the Earth. (*Hint:* The outer conductor of the "spherical capacitor" may be considered as a conducting sphere at infinity where $V \equiv 0$.)

25. A spherical capacitor consists of a conducting ball of radius 10.0 cm that is centered inside a grounded conducting spherical shell of inner radius 12.0 cm. What capacitor charge is required to achieve a potential of 1000 V on the ball?

26. Estimate the maximum voltage to which a smooth, metallic sphere 10 cm in diameter can be charged without exceeding the dielectric strength of the dry air around the sphere.

Section 26.3 Combinations of Capacitors

27. (a) Two capacitors, $C_1 = 2.00 \ \mu\text{F}$ and $C_2 = 16.0 \ \mu\text{F}$, are connected in parallel. What is the equivalent capacitance of the combination? (b) Calculate the equivalent capacitance of the two capacitors in part (a) if they are connected in series.

28. Two capacitors when connected in parallel give an equivalent capacitance of 9.0 pF and an equivalent capacitance of 2.0 pF when connected in series. What is the capacitance of each capacitor?

28A. Two capacitors give an equivalent capacitance C_p and an equivalent capacitance C_s when connected in series. What is the capacitance of each capacitor?

29. (a) Determine the equivalent capacitance for the capacitor network shown in Figure P26.29. (b) If the network is connected to a 12-V battery, calculate the potential difference across each capacitor and the charge on each capacitor.

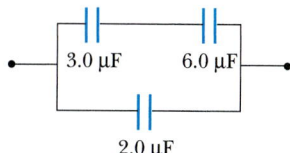

FIGURE P26.29

30. Evaluate the equivalent capacitance of the configuration shown in Figure P26.30. All the capacitors are identical, and each has capacitance C.

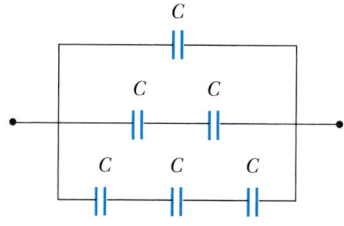

FIGURE P26.30

31. Four capacitors are connected as shown in Figure P26.31. (a) Find the equivalent capacitance between points a and b. (b) Calculate the charge on each capacitor if $V_{ab} = 15$ V.

32. For three $2.0\text{-}\mu\text{F}$ capacitors, sketch the arrangement that gives (a) the largest equivalent capacitance,

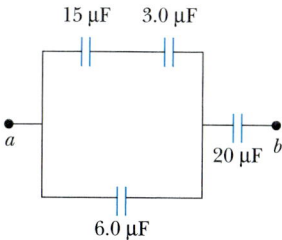

FIGURE P26.31

(b) the smallest equivalent capacitance, (c) an equivalent capacitance of $3.0 \ \mu\text{F}$.

33. Consider the circuit shown in Figure P26.33, where $C_1 = 6.00 \ \mu\text{F}$, $C_2 = 3.00 \ \mu\text{F}$, and $V = 20.0$ V. Capacitor C_1 is first charged by the closing of switch S_1. Switch S_1 is then opened, and the charged capacitor is connected to the uncharged capacitor by the closing of S_2. Calculate the initial charge acquired by C_1 and the final charge on each.

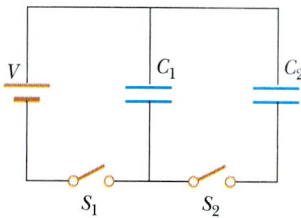

FIGURE P26.33

34. The circuit in Figure P26.34 consists of two identical parallel metal plates connected by identical metal springs to a 100-V battery. With the switch open, the plates are uncharged, are separated by a distance $d = 8.0$ mm, and have a capacitance $C = 2.0 \ \mu\text{F}$. When the switch is closed, the distance between the

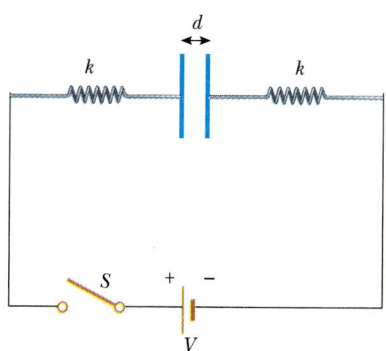

FIGURE P26.34

plates decreases by a factor of 0.5. (a) How much charge collects on each plate and (b) what is the spring constant for each spring? (*Hint:* Use the result of Problem 51.)

34A. The circuit in Figure P26.34 consists of two identical parallel metal plates connected by identical metal springs to a battery of voltage *V*. With the switch open, the plates are uncharged, are separated by a distance *d*, and have a capacitance *C*. When the switch is closed, the distance between the plates decreases by a factor of 0.5. (a) How much charge collects on each plate and (b) what is the spring constant for each spring? (*Hint:* Use the result of Problem 51.)

35. Figure P26.35 shows six concentric conducting spheres, A, B, C, D, E, and F, having radii R, $2R$, $3R$, $4R$, $5R$, and $6R$, respectively. Spheres B and C are connected by a conducting wire as are spheres D and E. Determine the equivalent capacitance of this system.

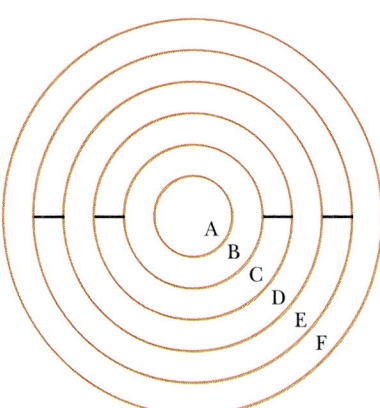

FIGURE P26.35

36. How many 0.25-pF capacitors must be connected in parallel in order to store 1.2 μC of charge when connected to a battery providing a potential difference of 10 V?

37. A group of identical capacitors is connected first in series and then in parallel. The combined capacitance in parallel is 100 times larger than for the series connection. How many capacitors are in the group?

38. Find the equivalent capacitance between points a and b for the group of capacitors connected as shown in Figure P26.38 if $C_1 = 5.00\ \mu$F, $C_2 = 10.0\ \mu$F, and $C_3 = 2.00\ \mu$F.

39. For the network described in the previous problem if the potential between points a and b is 60.0 V, what charge is stored on C_3?

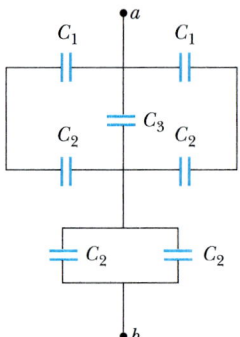

FIGURE P26.38

40. Determine the equivalent capacitance between points a and b in Figure P26.40.

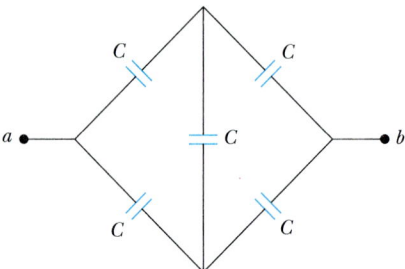

FIGURE P26.40

41. Find the equivalent capacitance between points a and b in the combination of capacitors shown in Figure P26.41.

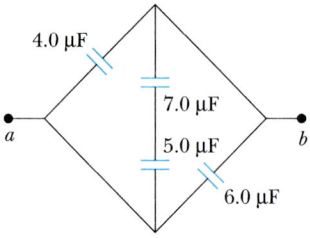

FIGURE P26.41

42. How should four 2.0-μF capacitors be connected to have a total capacitance of (a) 8.0 μF, (b) 2.0 μF, (c) 1.5 μF, and (d) 0.50 μF?

43. A conducting slab of thickness d and area A is inserted into the space between the plates of a parallel-plate capacitor with spacing s and surface area A, as in Figure P26.43. What is the capacitance of the system?

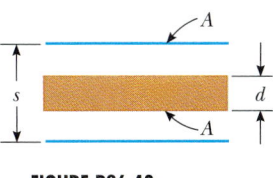

FIGURE P26.43

Section 26.4 Energy Stored in a Charged Capacitor

44. A certain storm cloud has a potential difference of 1.00×10^8 V relative to a tree. If, during a lightning storm, 50.0 C of charge is transferred through this potential difference and 1 percent of the energy is absorbed by the tree, how much water (sap in the tree) initially at 30°C can be boiled away? Water has a specific heat of 4186 J/kg·°C, a boiling point of 100°C, and a heat of vaporization of 2.26×10^6 J/kg.

45. Calculate the energy stored in an 18.0-μF capacitor when it is charged to a potential of 100 V.

46. The energy stored in a particular capacitor is increased fourfold. What is the accompanying change in (a) the charge and (b) the potential difference across the capacitor?

47. The energy stored in a 12.0-μF capacitor is 130 μJ. Determine (a) the charge on the capacitor and (b) the potential difference across it.

48. Einstein said that energy is associated with mass according to the famous relationship, $E = mc^2$. Estimate the radius of an electron, assuming that its charge is distributed uniformly over the surface of a sphere of radius R and that the mass-energy of the electron is equal to the total energy stored in the resulting nonzero electric field between R and infinity. (See Problem 53.)

49. A 16.0-pF parallel-plate capacitor is charged by a 10.0-V battery. If each plate of the capacitor has an area of 5.00 cm², what is the energy stored in the capacitor? What is the energy density (energy per unit volume) in the electric field of the capacitor if the plates are separated by air?

50. The energy stored in a 52.0-μF capacitor is used to melt a 6.00-mg sample of lead. To what voltage must the capacitor be initially charged, assuming that the initial temperature of the lead is 20.0°C? Lead has a specific heat of 128 J/kg·°C, a melting point of 327.3°C, and a latent heat of fusion of 24.5 kJ/kg.

51. A parallel-plate capacitor has a charge Q and plates of area A. Show that the force exerted on each plate by the other is $F = Q^2/2\epsilon_0 A$. (*Hint:* Let $C = \epsilon_0 A/x$ for an arbitrary plate separation x; then, require that the work done in separating the two charged plates be $W = \int F\,dx$.)

52. A uniform electric field $E = 3000$ V/m exists within a certain region. What volume of space contains an energy equal to 1.00×10^{-7} J? Express your answer in cubic meters and in liters.

53. Show that the energy associated with a conducting sphere of radius R and charge Q surrounded by a vacuum is $U = k_e Q^2/2R$.

54. Plate a of a parallel-plate, air-filled capacitor is connected to a spring having force constant k, and plate b is fixed. They rest on a table top as shown (top view) in Figure P26.54. If a charge $+Q$ is placed on plate a and a charge $-Q$ is placed on plate b, by how much does the spring expand?

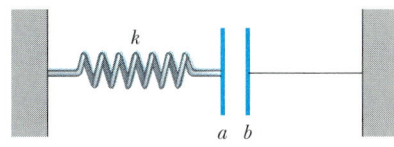

FIGURE P26.54

Section 26.5 Capacitors with Dielectrics
*Section 26.7 An Atomic Description of Dielectrics

55. A parallel-plate capacitor in air has a plate separation of 1.50 cm and a plate area of 25.0 cm². The plates are charged to a potential difference of 250 V and disconnected from the source. The capacitor is then immersed in distilled water. Determine (a) the charge on the plates before and after immersion, (b) the capacitance and voltage after immersion, and (c) the change in energy of the capacitor. Neglect the conductance of the water.

55A. A parallel-plate capacitor in air has a plate separation d and a plate area A. The plates are charged to a potential difference V and disconnected from the source. The capacitor is then immersed in a liquid of dielectric constant κ. Determine (a) the charge on the plates before and after immersion, (b) the capacitance and voltage after immersion and (c) the change in energy of the capacitor.

56. A parallel-plate capacitor is to be constructed using paper as a dielectric. If a maximum voltage before breakdown of 2500 V is desired, what thickness of dielectric is needed? (See Table 26.1 for other dielectric properties.)

57. A parallel-plate capacitor has a plate area of 0.64 cm². When the plates are in a vacuum, the capacitance of the device is 4.9 pF. (a) Calculate the value of the capacitance if the space between the plates is filled with nylon. (b) What is the maximum potential difference that can be applied to the plates without causing dielectric breakdown?

58. A capacitor is constructed from two square metal plates of side length L and separated by a distance d (Fig. P26.58). One half of the space between the plates (top to bottom) is filled with polystyrene (κ = 2.56), and the other half is filled with neoprene rubber (κ = 6.7). Calculate the capacitance of the device, taking L = 2.0 cm and d = 0.75 mm. (*Hint:* The capacitor can be considered as two capacitors connected in parallel.)

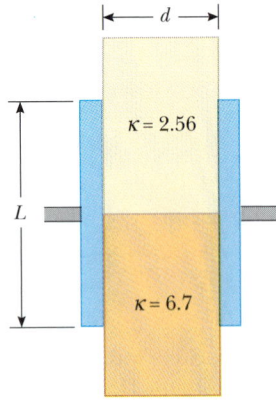

FIGURE P26.58

59. A commercial capacitor is constructed as in Figure 26.12a. This particular capacitor is rolled from two strips of aluminum separated by two strips of paraffin-coated paper. Each strip of foil and paper is 7.0 cm wide. The foil is 0.0040 mm thick, and the paper is 0.025 mm thick and has a dielectric constant of 3.7. What length should the strips be if a capacitance of 9.5×10^{-8} F is desired? (Use the parallel-plate formula.)

60. A detector of radiation called a Geiger tube consists of a closed, hollow, conducting cylinder with a fine wire along its axis. Suppose that the internal diameter of the cylinder is 2.5 cm and that the wire along the axis has a diameter of 0.20 mm. If the dielectric strength of the gas between the central wire and the cylinder is 1.2×10^{6} V/m, calculate the maximum voltage that can be applied between the wire and the cylinder before breakdown occurs in the gas.

61. The plates of an isolated, charged capacitor are 1.0 mm apart, and the potential difference across them is V_0. The plates are now separated to 4.0 mm (while the charge on them is preserved), and a slab of dielectric material is inserted, filling the space between the plates. The potential difference across the capacitor is now $V_0/2$. Find the dielectric constant of the material.

62. (a) What is the capacitance of a square parallel-plate capacitor measuring 5.0 cm on a side and having a

0.20-mm gap between the plates if this gap is filled with Teflon? (b) What maximum voltage can this capacitor withstand? (c) What maximum energy can this capacitor store?

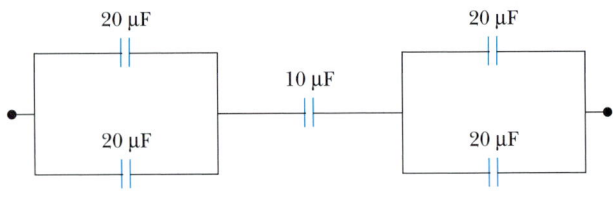

FIGURE P26.63

64. A conducting spherical shell has inner radius a and outer radius c. The space between these two surfaces is filled with a dielectric for which the dielectric constant is κ_1 between a and b and κ_2 between b and c (Fig. P26.64). Determine the capacitance of this system.

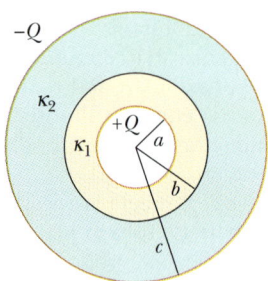

FIGURE P26.64

65. A sheet of 0.10-mm-thick paper is inserted between the plates of a 340-pF air-filled capacitor that has a plate separation of 0.40 mm. Calculate the new capacitance.

66. A wafer of titanium dioxide (κ = 173) has an area of 1.0 cm^2 and a thickness of 0.10 mm. Aluminum is evaporated on the parallel faces to form a parallel-plate capacitor. (a) Calculate the capacitance. (b) When the capacitor is charged with a 12-V battery, what is the magnitude of charge delivered to each plate? (c) For the situation in part (b), what are the free and induced surface charge densities? (d) What is the electric field strength E?

ADDITIONAL PROBLEMS

67. When two capacitors are connected in parallel, the equivalent capacitance is 4.00 μF. If the same capacitors are reconnected in series, the equivalent capacitance is one-fourth the capacitance of one of the two capacitors. Determine the two capacitances.

68. For the system of capacitors shown in Figure P26.68, find (a) the equivalent capacitance of the system, (b) the potential across each capacitor, (c) the charge on each capacitor, and (d) the total energy stored by the group.

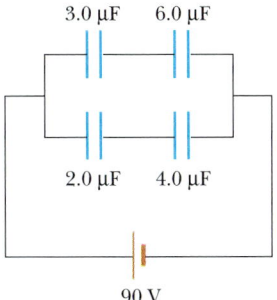

3.0 μF   6.0 μF

2.0 μF   4.0 μF

90 V

FIGURE P26.68

69. (a) Two spheres have radii a and b and their centers are a distance d apart. Show that the capacitance of this system is

$$C \approx \frac{4\pi\epsilon_0}{\dfrac{1}{a} + \dfrac{1}{b} - \dfrac{2}{d}}$$

provided that d is large compared with a and b. (*Hint:* Since the spheres are very far apart, assume that the potential on each equals the sum of the potentials due to each sphere, and when calculating those potentials, assume that $V = k_e Q/r$ applies.) (b) Show that as d approaches $+\infty$, the above result reduces to that of two spherical capacitors in series.

70. A parallel-plate capacitor with air between its plates has a capacitance C_0. A slab of dielectric material with a dielectric constant κ and a thickness equal to a fraction f of the separation of the plates is inserted between the plates in contact with one plate. Find the capacitance C in terms of f, κ, and C_0. Check your result by first letting f approach zero and then letting it approach unity.

71. When a certain air-filled parallel-plate capacitor is connected across a battery, it acquires a charge (on each plate) of 150 μC. While the battery connection is maintained, a dielectric slab is inserted into and fills the region between the plates. This results in the accumulation of an *additional* charge of 200 μC on each plate. What is the dielectric constant of the slab?

71A. When a certain air-filled parallel-plate capacitor is connected across a battery, it acquires a charge (on each plate) of q_0. While the battery connection is maintained, a dielectric slab is inserted into and fills the region between the plates. This results in the accumulation of an *additional* charge q on each plate. What is the dielectric constant of the slab?

72. A capacitor is constructed from two square plates of sides ℓ and separation d. A material of dielectric constant κ is inserted a distance x into the capacitor, as in Figure P26.72. (a) Find the equivalent capacitance of the device. (b) Calculate the energy stored in the capacitor if the potential difference is V. (c) Find the direction and magnitude of the force exerted on the dielectric, assuming a constant potential difference V. Neglect friction. (d) Obtain a numerical value for the force assuming that $\ell = 5.0$ cm, $V = 2000$ V, $d = 2.0$ mm, and the dielectric is glass ($\kappa = 4.5$). (*Hint:* The system can be considered as two capacitors connected in *parallel*.)

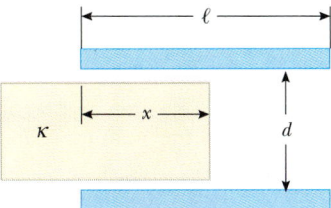

ℓ

x

κ

d

FIGURE P26.72

73. Three capacitors—8.0 μF, 10.0 μF, and 14 μF—are connected to the terminals of a 12-V battery. How much energy does the battery supply if the capacitors are connected (a) in series and (b) in parallel?

74. When considering the energy supply for an automobile, the energy per unit mass of the energy source is an important parameter. Using the following data, compare the energy per unit mass (J/kg) for gasoline, lead-acid batteries, and capacitors.

- *Gasoline:* 126 000 Btu/gal; density = 670 kg/m^3
- *Lead-acid battery:* 12 V; 100 A·h; mass = 16 kg
- *Capacitor:* potential difference at full charge = 12 V; capacitance = 0.10 F; mass = 0.10 kg

75. An isolated capacitor of unknown capacitance has been charged to a potential difference of 100 V. When the charged capacitor is then connected in parallel to an uncharged 10-μF capacitor, the voltage across the combination is 30 V. Calculate the unknown capacitance.

75A. An isolated capacitor of unknown capacitance has been charged to a potential difference of V_0. When the charged capacitor is then connected in parallel to an uncharged capacitor C, the voltage across the combination is $V < V_0$. Calculate the unknown capacitance.

76. A certain electronic circuit calls for a capacitor having a capacitance of 1.2 pF and a breakdown potential of 1000 V. If you have a supply of 6.0-pF capacitors, each having a breakdown potential of 200 V, how could you meet this circuit requirement?

77. A 2.0-μF capacitor and a 3.0-μF capacitor have the same maximum voltage rating V_{max}. Due to this voltage limitation, the maximum potential difference that can be applied to a series combination of these capacitors is 800 V. Calculate the maximum voltage rating of the individual capacitors.

78. A 2.0-nF parallel-plate capacitor is charged to an initial potential difference $V_i = 100$ V and then isolated. The dielectric material between the plates is mica ($\kappa = 5.0$). (a) How much work is required to withdraw the mica sheet? (b) What is the potential difference of the capacitor after the mica is withdrawn?

79. A parallel-plate capacitor is constructed using a dielectric material whose dielectric constant is 3.0 and whose dielectric strength is 2.0×10^8 V/m. The desired capacitance is 0.25 μF, and the capacitor must withstand a maximum potential difference of 4000 V. Find the minimum area of the capacitor plates.

80. A parallel-plate capacitor is constructed using three dielectric materials, as in Figure P26.80. (a) Find an expression for the capacitance of the device in terms of the plate area A and d, κ_1, κ_2, and κ_3. (b) Calculate the capacitance using the values $A = 1.0$ cm^2, $d = 2.00$ mm, $\kappa_1 = 4.9$, $\kappa_2 = 5.6$, and $\kappa_3 = 2.1$.

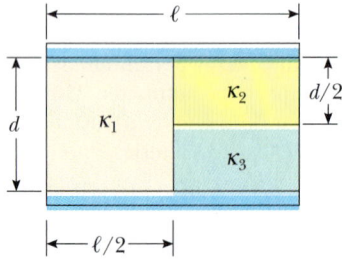

FIGURE P26.80

81. In the arrangement shown in Figure P26.81, a potential V is applied, and C_1 is adjusted so that the voltmeter between points b and d reads zero. This "balance" occurs when $C_1 = 4.00$ μF. If $C_3 = 9.00$ μF and $C_4 = 12.0$ μF, calculate the value of C_2.

82. It is possible to obtain large potential differences by first charging a group of capacitors connected in parallel and then activating a switch arrangement that in

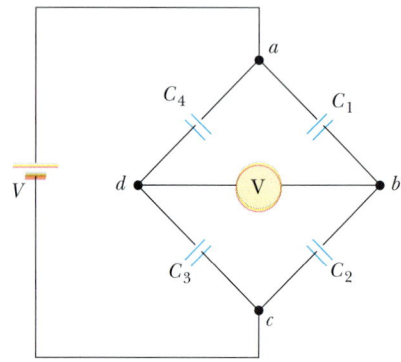

FIGURE P26.81

effect disconnects the capacitors from the charging source and from each other and reconnects them in a series arrangement. The group of charged capacitors is then discharged in series. What is the maximum potential difference that can be obtained in this manner by using ten capacitors each of 500 μF and a charging source of 800 V?

83. A parallel-plate capacitor of plate separation d is charged to a potential difference V_0. A dielectric slab of thickness d and dielectric constant κ is introduced between the plates *while the battery remains connected to the plates*. (a) Show that the ratio of energy stored after the dielectric is introduced to the energy stored in the empty capacitor is $U/U_0 = \kappa$. Give a physical explanation for this increase in stored energy. (b) What happens to the charge on the capacitor? (Note that this situation is not the same as Example 26.7, in which the battery was removed from the circuit before the dielectric was introduced.)

84. A parallel-plate capacitor with plates of area A and plate separation d has the region between the plates filled with two dielectric materials as in Figure P26.84. Assume $d \ll L$ and $d \ll W$. (a) Determine the capacitance, and (b) show that when $\kappa_1 = \kappa_2 = \kappa$, your result becomes the same as that for a capacitor containing a single dielectric: $C = \kappa \epsilon_0 A/d$.

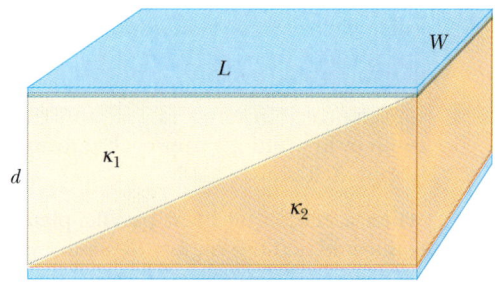

FIGURE P26.84

85. A vertical parallel-plate capacitor is half filled with a dielectric for which the dielectric constant is 2.0 (Fig.

P26.85a). When this capacitor is positioned horizontally, what fraction of it should be filled with the same dielectric (Fig. P26.85b) in order for the two capacitors to have equal capacitance?

(a)   (b)

FIGURE P26.85

86. Capacitors $C_1 = 6.00$ μF and $C_2 = 2.00$ μF are charged as a parallel combination across a 250-V battery. The capacitors are disconnected from the battery and from each other. They are then connected positive plate to negative plate and negative plate to positive plate. Calculate the resulting charge on each capacitor.

87. A stack of N plates has alternate plates connected to form a capacitor similar to that shown in Figure P26.18. Adjacent plates are separated by a dielectric of thickness d. The dielectric constant is κ, and the area of overlap of adjacent plates is A. Show that the capacitance of this stack of plates is $C = \dfrac{\kappa \epsilon_0 A}{d} (N - 1)$.

88. A coulomb balance is constructed of two parallel plates, each 10.0 cm square. The upper plate is movable. A 25.0-mg mass is placed on the upper plate and the plate is observed to lower; the mass is then removed. When a potential difference is applied to the plates, it is found that the applied voltage must be 375 V to cause the upper plate to lower the same amount as it lowered when the mass was on it. If the force exerted on each plate by the other is $F = Q^2/2\epsilon_0 A$, calculate the following, assuming an applied voltage of 375 V: (a) the charge on the plates, (b) the electric field between them, (c) their separation distance, and (d) the capacitance of this capacitor.

89. The inner conductor of a coaxial cable has a radius of 0.80 mm, and the outer conductor's inside radius is 3.0 mm. The space between the conductors is filled with polyethylene, which has a dielectric constant of

2.3 and a dielectric strength of 18×10^6 V/m. What is the maximum potential difference that this cable can withstand?

90. You are optimizing coaxial cable design for a major manufacturer. Show that for a given outer conductor radius b, maximum potential difference capability is attained when the radius of the inner conductor is $a = b/e$ where e is the base of natural logarithms.

91. Calculate the equivalent capacitance between the points a and b in Figure P26.91. Note that this is not a simple series or parallel combination. (*Hint:* Assume a potential difference V between points a and b. Write expressions for V_{ab} in terms of the charges and capacitances for the various possible pathways from a to b, and require conservation of charge for those capacitor plates that are connected to each other.)

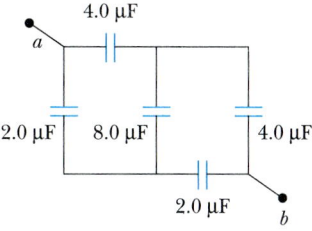

FIGURE P26.91

92. Determine the effective capacitance of the combination shown in Figure P26.92. (*Hint:* Consider the symmetry involved!)

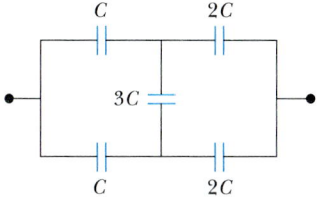

FIGURE P26.92

93. Consider two *long*, parallel, and oppositely charged wires of radius d with their centers separated by a distance D. Assuming the charge is distributed uniformly on the surface of each wire, show that the capacitance per unit length of this pair of wires is

$$\frac{C}{\ell} = \frac{\pi \epsilon_0}{\ln\left(\dfrac{D - d}{d}\right)}$$

CHAPTER 27

Current and Resistance

Photograph of a carbon filament incandescent lamp. The resistance of such a lamp is typically 10 Ω, but its value changes with temperature. Most modern lightbulbs use tungsten filaments, whose resistance increases with increasing temperature. *(Courtesy of Central Scientific Co.)*

Thus far our discussion of electrical phenomena has been confined to charges at rest, or electrostatics. We now consider situations involving electric charges in motion. The term *electric current,* or simply *current,* is used to describe the rate of flow of charge through some region of space. Most practical applications of electricity deal with electric currents. For example, the battery of a flashlight supplies current to the filament of the bulb when the switch is turned on. A variety of home appliances operate on alternating current. In these common situations, the flow of charge takes place in a conductor, such as a copper wire. It is also possible for currents to exist outside of a conductor. For instance, a beam of electrons in a TV picture tube constitutes a current.

In this chapter we first define current and current density. A microscopic description of current is given, and some of the factors that contribute to the resistance to the flow of charge in conductors are discussed. Mechanisms responsible for the electrical resistance of various materials depend on the composition of the material and on temperature. A classical model is used to describe electrical conduction in metals, and some of the limitations of this model are pointed out.

27.1   ELECTRIC CURRENT

Whenever electric charges of like sign move, a *current* is established. To define current more precisely, suppose the charges are moving perpendicular to a surface of area A as in Figure 27.1. (This area could be the cross-sectional area of a wire, for example.) The **current** is *the rate at which charge flows through this surface.* If ΔQ is the amount of charge that passes through this area in a time interval Δt, the **average current**, I_{av}, is equal to the charge that passes through A per unit time:

$$I_{av} = \frac{\Delta Q}{\Delta t} \qquad (27.1)$$

If the rate at which charge flows varies in time, the current also varies in time and we define the **instantaneous current** I as the differential limit of Equation 27.1:

$$I \equiv \frac{dQ}{dt} \qquad (27.2)$$

As we learned in Section 23.3, the SI unit of current is the **ampere** (A):

$$1\text{A} = \frac{1\ C}{1\ s} \qquad (27.3)$$

That is, 1 A of current is equivalent to 1 C of charge passing through the surface area in 1 s.

The charges passing through the surface in Figure 27.1 can be positive, negative, or both. *It is conventional to give the current the same direction as the flow of positive charge.* In a conductor such as copper, the current is due to the motion of negatively charged electrons. Therefore, when we speak of current in an ordinary conductor, such as a copper wire, *the direction of the current is opposite the direction of flow of electrons.* On the other hand, if a beam of positively charged protons in an accelerator is considered, the current is in the direction of motion of the protons. In some cases—gases and electrolytes, for instance—the current is the result of the flow of both positive and negative charges. It is common to refer to a moving charge (whether it is positive or negative) as a mobile *charge carrier.* For example, the charge carriers in a metal are electrons.

It is instructive to relate current to the motion of the charged particles. To illustrate this point, consider the current in a conductor of cross-sectional area A (Fig. 27.2). The volume of an element of the conductor of length Δx (the shaded region in Fig. 27.2) is $A\,\Delta x$. If n represents the number of mobile charge carriers per unit volume, then the number of mobile charge carriers in the volume element is $nA\,\Delta x$. Therefore, the charge ΔQ in this element is

$$\Delta Q = \text{number of charges} \times \text{charge per particle} = (nA\,\Delta x)\,q$$

where q is the charge on each particle. If the charge carriers move with a speed v_d, the distance they move in a time Δt is $\Delta x = v_d\,\Delta t$. Therefore, we can write ΔQ in the form

$$\Delta Q = (nAv_d\,\Delta t)\,q$$

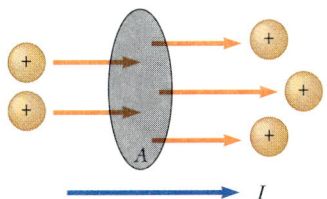

FIGURE 27.1   Charges in motion through an area A. The time rate of flow of charge through the area is defined as the current I. The direction of the current is the direction in which positive charge would flow if free to do so.

Electric current

The direction of the current

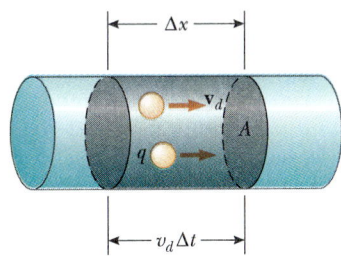

FIGURE 27.2   A section of a uniform conductor of cross-sectional area A. The charge carriers move with a speed v_d, and the distance they travel in a time Δt is given by $\Delta x = v_d\,\Delta t$. The number of mobile charge carriers in the section of length Δx is given by $nAv_d\,\Delta t$, where n is the number of mobile carriers per unit volume.

If we divide both sides of this equation by Δt, we see that the current in the conductor is given by

Current in a conductor

$$I = \frac{\Delta Q}{\Delta t} = nqv_d A \qquad (27.4)$$

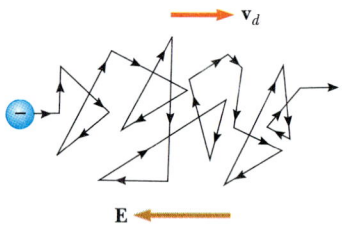

FIGURE 27.3 A schematic representation of the zigzag motion of a charge carrier in a conductor. The changes in direction are due to collisions with atoms in the conductor. Note that the net motion of electrons is opposite the direction of the electric field. The zigzag paths are actually parabolic segments.

The speed of the charge carriers, v_d, is actually an average speed and is called the **drift speed**. To understand its meaning, consider a conductor in which the charge carriers are free electrons. If the conductor is isolated, these electrons undergo random motion similar to that of gas molecules. When a potential difference is applied across the conductor (say, by means of a battery), an electric field is set up in the conductor, and this field creates an electric force on the electrons and hence a current. In reality, the electrons do not move in straight lines along the conductor. Instead, they undergo repeated collisions with the metal atoms, and the result is a complicated zigzag motion (Fig. 27.3). The energy transferred from the electrons to the metal atoms during collision causes an increase in the vibrational energy of the atoms and a corresponding increase in the temperature of the conductor. However, despite the collisions, the electrons move slowly along the conductor (in a direction opposite **E**) at the drift velocity, $\mathbf{v}_d$. The work done by the field on the electrons exceeds the average loss in energy due to collisions, and this provides a steady current. One can think of the collisions of the electrons within a conductor as being an effective internal friction (or drag force), similar to that experienced by the molecules of a liquid flowing through a pipe stuffed with steel wool.

EXAMPLE 27.1   Drift Speed in a Copper Wire

A copper wire of cross-sectional area 3.00×10^{-6} m² carries a current of 10.0 A. Find the drift speed of the electrons in this wire. The density of copper is 8.95 g/cm³.

Solution   From the periodic table of the elements in Appendix C, we find that the atomic mass of copper is 63.5 g/mol. Recall that one atomic mass of any substance contains Avogadro's number of atoms, 6.02×10^{23} atoms. Knowing the density of copper enables us to calculate the volume occupied by 63.5 g of copper:

$$V = \frac{m}{\rho} = \frac{63.5 \text{ g}}{8.95 \text{ g/cm}^3} = 7.09 \text{ cm}^3$$

If we now assume that each copper atom contributes one free electron to the body of the material, we have

$$n = \frac{6.02 \times 10^{23} \text{ electrons}}{7.09 \text{ cm}^3}$$

$$= 8.48 \times 10^{22} \text{ electrons/cm}^3$$

$$= \left(8.48 \times 10^{22} \frac{\text{electrons}}{\text{cm}^3} \right) \left(10^6 \frac{\text{cm}^3}{\text{m}^3} \right)$$

$$= 8.48 \times 10^{28} \text{ electrons/m}^3$$

From Equation 27.4, we find that the drift speed is

$$v_d = \frac{I}{nqA}$$

$$= \frac{10.0 \text{ C/s}}{(8.48 \times 10^{28} \text{ m}^{-3})(1.60 \times 10^{-19} \text{ C})(3.00 \times 10^{-6} \text{ m}^2)}$$

$$= 2.46 \times 10^{-4} \text{ m/s}$$

Example 27.1 shows that typical drift speeds are very small. In fact, the drift speed is much smaller than the average speed between collisions. For instance, electrons traveling with this drift speed would take approximately 68 min to travel 1 m! In view of this low speed, you might wonder why a light turns on almost instantaneously when a switch is thrown. This can be explained by considering the flow of water through a pipe. If a drop of water is forced in one end of a pipe that is already filled with water, a drop must be pushed out the other end. While it may take individual drops of water a long time to make it through the pipe, a flow initiated at one end produces a similar flow at the other end very quickly. In a conductor, the electric field that drives the free electrons travels through the conductor with a speed close to that of light. Thus, when you flip a light switch, the message for the electrons to start moving through the wire (the electric field) reaches them at a speed on the order of 10^8 m/s.

27.2 RESISTANCE AND OHM'S LAW

Earlier, we found that there can be no electric field inside a conductor. However, this statement is true *only* if the conductor is in static equilibrium. The purpose of this section is to describe what happens when the charges are allowed to move in the conductor.

Charges move in a conductor to produce a current under the action of an electric field inside the conductor. An electric field can exist in the conductor in this case because we are dealing with charges in motion, a *nonelectrostatic* situation.

Consider a conductor of cross-sectional area A carrying a current I. The **current density** J in the conductor is defined to be the current per unit area. Since the current $I = nqv_dA$, the current density is

$$J \equiv \frac{I}{A} = nqv_d \qquad (27.5)$$

where J has SI units of A/m^2. This expression is valid only if the current density is uniform and only if the surface of cross-sectional area A is perpendicular to the direction of the current. In general, the current density is a *vector quantity:*

$$\mathbf{J} = nq\mathbf{v}_d \qquad (27.6)$$

Current density

From this definition, we see once again that the current density, like the current, is in the direction of charge motion for positive charge carriers and opposite the direction of motion for negative charge carriers.

A current density $\mathbf{J}$ *and an electric field* $\mathbf{E}$ *are established in a conductor when a potential difference is maintained across the conductor.* If the potential difference is constant, the current is also constant. Very often, the current density is proportional to the electric field:

$$\mathbf{J} = \sigma\mathbf{E} \qquad (27.7)$$

Ohm's law

where the constant of proportionality σ is called the **conductivity** of the conductor.[1] Materials that obey Equation 27.7 are said to follow Ohm's law, named after Georg Simon Ohm (1787–1854). More specifically, **Ohm's law** states that

for many materials (including most metals), the ratio of the current density to the electric field is a constant, σ, that is independent of the electric field producing the current.

Materials that obey Ohm's law and hence demonstrate this linear behavior between $\mathbf{E}$ and $\mathbf{J}$ are said to be *ohmic*. The electrical behavior of most materials is quite

[1] Do not confuse the conductivity σ with the surface charge density, for which the same symbol is used.

linear for very small changes in the current. Experimentally, it is found that not all materials have this property, however. Materials that do not obey Ohm's law are said to be *nonohmic*. Ohm's law is not a fundamental law of nature but rather an empirical relationship valid only for certain materials.

A form of Ohm's law useful in practical applications can be obtained by considering a segment of a straight wire of cross-sectional area A and length ℓ, as in Figure 27.4. A potential difference $V = V_b - V_a$ is maintained across the wire, creating an electric field in the wire and a current. If the electric field in the wire is assumed to be uniform, the potential difference is related to the electric field through the relationship[2]

$$V = E\ell$$

Therefore, we can express the magnitude of the current density in the wire as

$$J = \sigma E = \sigma \frac{V}{\ell}$$

Since $J = I/A$, the potential difference can be written

$$V = \frac{\ell}{\sigma} J = \left(\frac{\ell}{\sigma A}\right) I$$

The quantity $\ell/\sigma A$ is called the **resistance** R of the conductor. From the last expression, we can define the resistance as the ratio of the potential difference across the conductor to the current:

Resistance of a conductor

$$R = \frac{\ell}{\sigma A} \equiv \frac{V}{I} \tag{27.8}$$

From this result we see that resistance has SI units of volts per ampere. One volt per ampere is defined to be one ohm (Ω):

$$1\ \Omega \equiv \frac{1\ \text{V}}{1\ \text{A}} \tag{27.9}$$

That is, if a potential difference of 1 V across a conductor causes a current of 1 A, the resistance of the conductor is 1 Ω. For example, if an electrical appliance connected to a 120-V source carries a current of 6 A, its resistance is 20 Ω.

The inverse of conductivity is **resistivity** ρ:

Resistivity

$$\rho \equiv \frac{1}{\sigma} \tag{27.10}$$

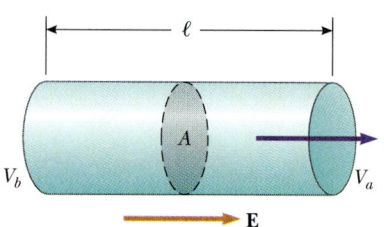

FIGURE 27.4   A uniform conductor of length ℓ and cross-sectional area A. A potential difference $V_b - V_a$ maintained across the conductor sets up an electric field $\mathbf{E}$ in the conductor, and this field produces a current I that is proportional to the potential difference.

---

[2] This result follows from the definition of potential difference:

$$V_b - V_a = -\int_a^b \mathbf{E} \cdot d\mathbf{s} = E \int_0^\ell dx = E\ell$$

Using this definition and Equation 27.8, we can express the resistance as

$$R = \rho \frac{\ell}{A} \qquad (27.11)$$

where ρ has the units ohm-meters ($\Omega \cdot$m). (The symbol ρ for resistivity should not be confused with the same symbol used earlier in the text for mass density or charge density.) Every ohmic material has a characteristic resistivity that depends on the properties of the material and on temperature. On the other hand, as you can see from Equation 27.11, the resistance of a substance depends on simple geometry as well as on resistivity. Good electrical conductors have very low resistivity (or high conductivity), and good insulators have very high resistivity (low conductivity). Table 27.1 gives the resistivities of a variety of materials at 20°C. Note the enormous range in resistivities, from very low values for good conductors, such as copper and silver, to very high values for good insulators, such as glass and rubber. An ideal, or "perfect," conductor would have zero resistivity, and an ideal insulator would have infinite resistivity.

Equation 27.11 shows that the resistance of a given cylindrical conductor is proportional to its length and inversely proportional to its cross-sectional area. If the length of a wire is doubled, its resistance doubles. If its cross-sectional area is doubled, its resistance drops by one half. The situation is analogous to the flow of a liquid through a pipe. As the length of the pipe is increased, the resistance to flow increases. As its cross-sectional area is increased, the pipe can more readily transport liquid.

An assortment of resistors used for various applications in electronic circuits.   *(Henry Leap and Jim Lehman)*

TABLE 27.1 Resistivities and Temperature Coefficients of Resistivity for Various Materials

Material	Resistivity[a] ($\Omega \cdot$m)	Temperature Coefficient $\alpha[(°C)^{-1}]$
Silver	1.59×10^{-8}	3.8×10^{-3}
Copper	1.7×10^{-8}	3.9×10^{-3}
Gold	2.44×10^{-8}	3.4×10^{-3}
Aluminum	2.82×10^{-8}	3.9×10^{-3}
Tungsten	5.6×10^{-8}	4.5×10^{-3}
Iron	10×10^{-8}	5.0×10^{-3}
Platinum	11×10^{-8}	3.92×10^{-3}
Lead	22×10^{-8}	3.9×10^{-3}
Nichrome[b]	1.50×10^{-6}	0.4×10^{-3}
Carbon	3.5×10^{-5}	-0.5×10^{-3}
Germanium	0.46	-48×10^{-3}
Silicon	640	-75×10^{-3}
Glass	$10^{10}-10^{14}$	
Hard rubber	$\approx 10^{13}$	
Sulfur	10^{15}	
Quartz (fused)	75×10^{16}	

[a] All values at 20°C.

[b] A nickel-chromium alloy commonly used in heating elements.

TABLE 27.2	**Color Code for Resistors**		
Color	Number	Multiplier	Tolerance (%)
Black	0	1	
Brown	1	10^1	
Red	2	10^2	
Orange	3	10^3	
Yellow	4	10^4	
Green	5	10^5	
Blue	6	10^6	
Violet	7	10^7	
Gray	8	10^8	
White	9	10^9	
Gold		10^{-1}	5%
Silver		10^{-2}	10%
Colorless			20%

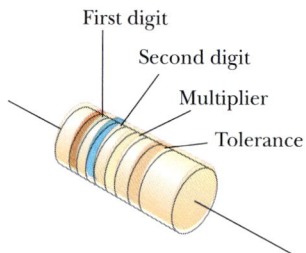

First digit
Second digit
Multiplier
Tolerance

FIGURE 27.5 The colored bands on a resistor represent a code for determining the value of its resistance. The first two colors give the first two digits in the resistance value. The third color represents the power of ten for the multiplier of the resistance value. The last color is the tolerance of the resistance value. As an example, if the four colors are orange, blue, yellow, and gold, the resistance value is $36 \times 10^4 \ \Omega$ or 360 kΩ, with a tolerance value of 18 kΩ (5%).

Most electric circuits make use of devices called **resistors** to control the current level in the various parts of the circuit. Two common types of resistors are the composition resistor containing carbon, which is a semiconductor, and the wire-wound resistor, which consists of a coil of wire. Resistors are normally color coded to give their values in ohms, as shown in Figure 27.5 and Table 27.2.

Ohmic materials have a linear current-voltage relationship over a large range of applied voltage (Fig. 27.6a). The slope of the I versus V curve in the linear region yields a value for R. Nonohmic materials have a nonlinear current-voltage relationship. One common semiconducting device that has nonlinear I versus V characteristics is the diode (Fig. 27.6b). The resistance of this device (inversely proportional to the slope of its I versus V curve) is small for currents in one direction (positive V) and large for currents in the reverse direction (negative V). In fact, most modern electronic devices, such as transistors, have nonlinear current-voltage relationships; their proper operation depends on the particular way in which they violate Ohm's law.

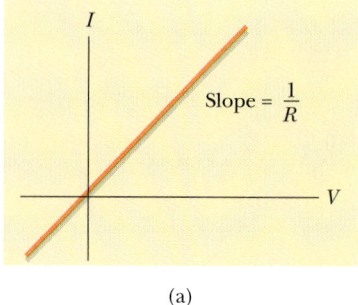

I

Slope = $\dfrac{1}{R}$

V

(a)

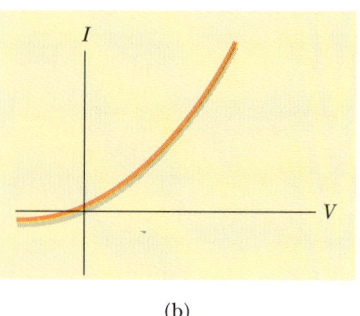

I

V

(b)

FIGURE 27.6 (a) The current-voltage curve for an ohmic material. The curve is linear, and the slope gives the resistance of the conductor. (b) A nonlinear current-voltage curve for a semiconducting diode. This device does not obey Ohm's law.

CONCEPTUAL EXAMPLE 27.2   Is There a Paradox?

We have seen that an electric field must exist inside a conductor that carries a current. How is this possible in view of the fact that in electrostatics we concluded that the electric field is zero inside a conductor?

Reasoning In the electrostatic case where charges are stationary (Chapters 23 and 24), the internal electric field must

be zero because a nonzero field would produce a current (by interacting with the free electrons in the conductor), which would violate the condition of static equilibrium. In this chapter we deal with conductors that carry current, a non-electrostatic situation. The current arises because of a potential difference applied between the ends of the conductor, which produces an internal electric field. So there is no paradox.

EXAMPLE 27.3   The Resistance of a Conductor

Calculate the resistance of an aluminum cylinder that is 10.0 cm long and has a cross-sectional area of 2.00×10^{-4} m². Repeat the calculation for a glass cylinder of resistivity 3.0×10^{10} Ω·m.

Solution From Equation 27.11 and Table 27.1, we can calculate the resistance of the aluminum cylinder:

$$R = \rho \frac{\ell}{A} = (2.82 \times 10^{-8}\ \Omega\cdot m)\left(\frac{0.100\ m}{2.00 \times 10^{-4}\ m^2}\right)$$

$$= \ 1.41 \times 10^{-5}\ \Omega$$

Similarly, for glass we find

$$R = \rho \frac{\ell}{A} = (3.0 \times 10^{10}\ \Omega\cdot m)\left(\frac{0.100\ m}{2.00 \times 10^{-4}\ m^2}\right)$$

$$= \ 1.5 \times 10^{13}\ \Omega$$

As you might expect, aluminum has a much lower resistance than glass. This is why aluminum is considered a good electrical conductor and glass is a poor conductor.

EXAMPLE 27.4   The Resistance of Nichrome Wire

(a) Calculate the resistance per unit length of a 22-gauge Nichrome wire, which has a radius of 0.321 mm.

Solution The cross-sectional area of this wire is

$$A = \pi r^2 = \pi (0.321 \times 10^{-3}\ m)^2 = 3.24 \times 10^{-7}\ m^2$$

The resistivity of Nichrome is 1.5×10^{-6} Ω·m (Table 27.1). Thus, we can use Equation 27.11 to find the resistance per unit length:

$$\frac{R}{\ell} = \frac{\rho}{A} = \frac{1.5 \times 10^{-6}\ \Omega\cdot m}{3.24 \times 10^{-7}\ m^2} = \ 4.6\ \Omega/m$$

(b) If a potential difference of 10 V is maintained across a 1.0-m length of the nichrome wire, what is the current in the wire?

Solution Since a 1.0-m length of this wire has a resistance of 4.6 Ω, Ohm's law gives

$$I = \frac{V}{R} = \frac{10\ V}{4.6\ \Omega} = \ 2.2\ A$$

Note from Table 27.1 that the resistivity of nichrome wire is about 100 times that of copper. Therefore, a copper wire of the same radius would have a resistance per unit length of only 0.052 Ω/m. A 1.0-m length of copper wire of the same radius would carry the same current (2.2 A) with an applied voltage of only 0.11 V.

Because of its high resistivity and its resistance to oxidation, Nichrome is often used for heating elements in toasters, irons, and electric heaters.

Exercise What is the resistance of a 6.0-m length of 22-gauge Nichrome wire? How much current does it carry when connected to a 120-V source?

Answer 28 Ω, 4.3 A.

Exercise Calculate the current density and electric field in the wire assuming that it carries a current of 2.2 A.

Answer 6.7×10^6 A/m²; 10 N/C.

EXAMPLE 27.5   The Resistance of a Coaxial Cable

A coaxial cable consists of two cylindrical conductors. The gap between the conductors is completely filled with silicon as in Figure 27.7a. The radius of the inner conductor is $a = 0.500$ cm, the radius of the outer one is $b = 1.75$ cm, and their length is $L = 15.0$ cm. Calculate the total resistance of the silicon when measured between the inner and outer conductors.

Reasoning   In this type of problem, we must divide the object whose resistance we are calculating into elements of infinitesimal thickness over which the area may be considered constant. We start by using the differential form of Equation 27.11, which is $dR = \rho \, d\ell / A$, where dR is the resistance of a section of silicon of thickness $d\ell$ and area A. In this example, we take as our element a hollow cylinder of thickness dr and length L as in Figure 27.7b. Any current that passes between the inner and outer conductors must pass radially through such elements, and the area through which this current passes is $A = 2\pi rL$. (This is the surface area of our hollow cylinder neglecting the area of its ends.) Hence, we can write

the resistance of our hollow cylinder as

$$dR = \frac{\rho}{2\pi rL} \, dr$$

Solution   Since we wish to know the total resistance of the silicon, we must integrate this expression over dr from $r = a$ to $r = b$:

$$R = \int_a^b dR = \frac{\rho}{2\pi L} \int_a^b \frac{dr}{r} = \frac{\rho}{2\pi L} \ln\left(\frac{b}{a}\right)$$

Substituting in the values given, and using $\rho = 640$ $\Omega \cdot$m for silicon, we get

$$R = \frac{640 \ \Omega \cdot \text{m}}{2\pi(0.150 \ \text{m})} \ln\left(\frac{1.75 \ \text{cm}}{0.500 \ \text{cm}}\right) = \boxed{851 \ \Omega}$$

Exercise   If a potential difference of 12.0 V is applied between the inner and outer conductors, calculate the total current that passes between them.

Answer   14.1 mA.

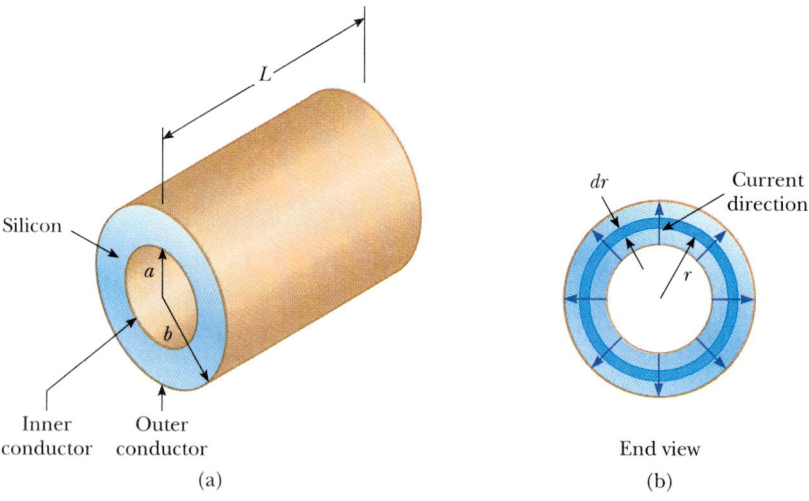

Silicon · Inner conductor · Outer conductor · a · b · L

(a)

dr · Current direction · r · End view

(b)

FIGURE 27.7   (Example 27.5).

27.3   RESISTANCE AND TEMPERATURE

Resistivity depends on a number of factors, one of which is temperature. For all metals, resistivity increases with increasing temperature. The resistivity of a conductor varies approximately linearly with temperature over a limited temperature range according to the expression

Variation of ρ with temperature

$$\rho = \rho_0[1 + \alpha(T - T_0)] \tag{27.12}$$

where ρ is the resistivity at some temperature T (in °C), ρ_0 is the resistivity at some reference temperature T_0 (usually taken to be 20°C), and α is called the **temperature coefficient of resistivity**. From Equation 27.12, we see that the temperature coefficient of resistivity can also be expressed as

$$\alpha = \frac{1}{\rho_0} \frac{\Delta\rho}{\Delta T} \qquad (27.13)$$

where $\Delta\rho = \rho - \rho_0$ is the change in resistivity in the temperature interval $\Delta T = T - T_0$. (The temperature coefficients for various materials are given in Table 27.1.)

Since resistance is proportional to resistivity according to Equation 27.11, the variation of resistance with temperature can be written as

$$R = R_0 [1 + \alpha (T - T_0)] \qquad (27.14)$$

Precise temperature measurements are often made using this property, as shown in the following example.

EXAMPLE 27.6   A Platinum Resistance Thermometer

A resistance thermometer, which measures temperature by measuring the change in resistance of a conductor, is made from platinum and has a resistance of 50.0 Ω at 20.0°C. When immersed in a vessel containing melting indium, its resistance increases to 76.8 Ω. What is the melting point of indium?

Solution   Solving Equation 27.14 for ΔT, and getting α from Table 27.1, we get

$$\Delta T = \frac{R - R_0}{\alpha R_0} = \frac{76.8\ \Omega - 50.0\ \Omega}{[3.92 \times 10^{-3}\ (°C)^{-1}](50.0\ \Omega)} = 137°C$$

Since $T_0 = 20.0°C$, we find that $T = 157°C$.

For several metals, resistivity is nearly proportional to temperature, as shown in Figure 27.8. In reality, however, there is always a nonlinear region at very low temperatures, and the resistivity usually approaches some finite value near absolute zero (see magnified insert in Fig. 27.8). This residual resistivity near absolute zero is due primarily to collisions of electrons with impurities and imperfections in the metal. In contrast, the high-temperature resistivity (the linear region) is dominated by collisions of electrons with the metal atoms. We describe this process in more detail in Section 27.5.

The resistivity of semiconductors decreases with increasing temperature, corresponding to a negative temperature coefficient of resistivity (Fig. 27.9). This

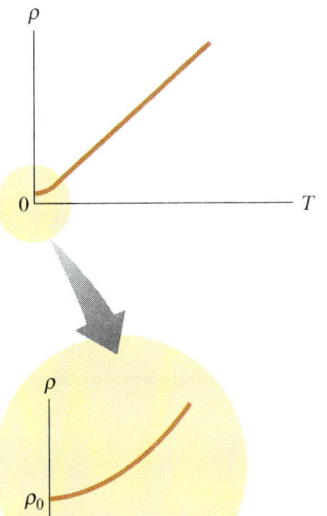

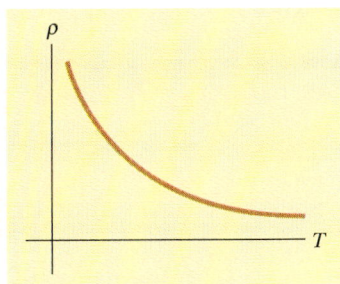

FIGURE 27.8   Resistivity versus temperature for a normal metal, such as copper. The curve is linear over a wide range of temperatures, and ρ increases with increasing temperature. As T approaches absolute zero (insert), the resistivity approaches a finite value ρ_0.

FIGURE 27.9   Resistivity versus temperature for a pure semiconductor, such as silicon or germanium.

behavior is due to the increase in the density of charge carriers at the higher temperatures. Since the charge carriers in a semiconductor are often associated with impurity atoms, the resistivity is very sensitive to the type and concentration of such impurities. The **thermistor** is a semiconducting thermometer that makes use of the large changes in its resistivity with temperature. We shall return to the study of semiconductors in the extended version of this text, Chapter 43.

27.4   SUPERCONDUCTORS

There is a class of metals and compounds whose resistance goes to zero below a certain temperature, T_c, called the *critical temperature*. These materials are known as **superconductors.** The resistance-temperature graph for a superconductor follows that of a normal metal at temperatures above T_c (Fig. 27.10). When the temperature is at or below T_c, the resistivity drops suddenly to zero. This phenomenon was discovered in 1911 by the Dutch physicist Heike Kamerlingh-Onnes as he worked with mercury, which is a superconductor below 4.2 K. Recent measurements have shown that the resistivities of superconductors below T_c are less than $4 \times 10^{-25} \, \Omega \cdot m$, which is around 10^{17} times smaller than the resistivity of copper and considered to be zero in practice.

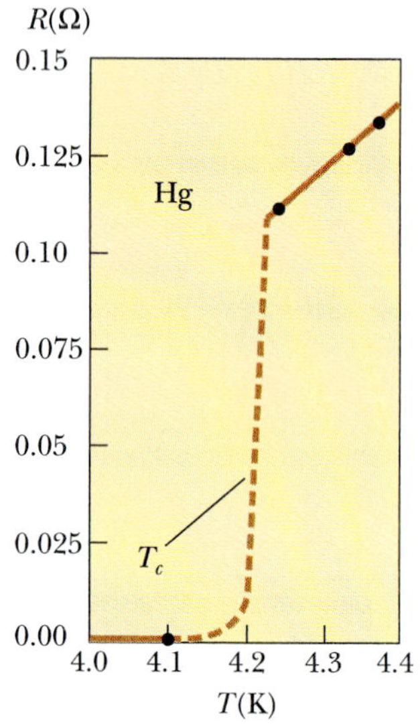

Today there are thousands of known superconductors, and Table 27.3 lists the critical temperatures of several. The value of T_c is sensitive to chemical composition, pressure, and crystalline structure. It is interesting to note that copper, silver, and gold, which are excellent conductors, do not exhibit superconductivity.

One of the truly remarkable features of superconductors is that once a current is set up in them, it persists *without any applied voltage* (since $R = 0$). In fact, steady currents have been observed to persist in superconducting loops for several years with no apparent decay!

One of the most important recent developments in physics that has created much excitement in the scientific community has been the discovery of high-temperature copper-oxide superconductors. The excitement began with a 1986 publication by George Bednorz and K. Alex Müller, two scientists working at the IBM Zurich Research Laboratory, who reported evidence for superconductivity near 30 K in an oxide of barium, lanthanum, and copper. Bednorz and Müller were awarded the Nobel Prize in 1987 for their remarkable and important discovery. Shortly thereafter, a new family of compounds was open for investigation, and research activity in the field of superconductivity proceeded vigorously. In early 1987, groups at the University of Alabama at Huntsville and the University of Houston announced the discovery of superconductivity at about 92 K in an oxide of yttrium, barium, and copper ($YBa_2Cu_3O_7$). Late in 1987, teams of scientists from Japan and the United States reported superconductivity at 105 K in an oxide of bismuth, strontium, calcium, and copper. Most recently, scientists have reported superconductivity as high as 134 K in the compound. At this point, the possibility of room temperature superconductivity cannot be ruled out, and the search for novel superconducting materials continues. These developments are very exciting and important both for scientific reasons and because practical applications become more probable and widespread as the critical temperature is raised.

An important and useful application of superconductivity has been the construction of superconducting magnets in which the magnetic field strengths are about ten times greater than those of the best normal electromagnets. Such superconducting magnets are being considered as a means of storing energy. The idea of using superconducting power lines for transmitting power efficiently is also

FIGURE 27.10   Resistance versus temperature for mercury. The graph follows that of a normal metal above the critical temperature, T_c. The resistance drops to zero at the critical temperature, which is 4.2 K for mercury.

TABLE 27.3   Critical Temperatures for Various Superconductors

Material	T_c(K)
$YBa_2Cu_3O_{7-\delta}$	92
Bi-Sr-Ca-Cu-O	105
Tl-Ba-Ca-Cu-O	125
$HgBa_2Ca_2Cu_3O_8$	134
Nb_3Ge	23.2
Nb_3Sn	18.05
Nb	9.46
Pb	7.18
Hg	4.15
Sn	3.72
Al	1.19
Zn	0.88

receiving some consideration. Modern superconducting electronic devices consisting of two thin-film superconductors separated by a thin insulator have been constructed. They include magnetometers (a magnetic-field measuring device) and various microwave devices.

27.5   A MODEL FOR ELECTRICAL CONDUCTION

In this section we describe a classical model of electrical conduction in metals. This model leads to Ohm's law and shows that resistivity can be related to the motion of electrons in metals.

Consider a conductor as a regular array of atoms containing free electrons (sometimes called *conduction* electrons). These electrons are free to move through the conductor and are approximately equal in number to the number of atoms. In the absence of an electric field, the free electrons move in random directions through the conductor with average speeds of the order of 10^6 m/s. The situation is similar to the motion of gas molecules confined in a vessel. In fact, some scientists refer to conduction electrons in a metal as an *electron gas*. The conduction electrons are not totally free because they are confined to the interior of the conductor and undergo frequent collisions with the atoms. These electron-atom collisions are the predominant mechanism for the resistivity of a metal at normal temperatures. Note that there is no current through a conductor in the absence of an electric field because the drift velocity of the free electrons is zero. That is, on the average, just as many electrons move in one direction as in the opposite direction, and so there is no net flow of charge.

The situation is modified when an electric field is applied to the conductor. In addition to the random motion just described, the free electrons drift slowly in a direction opposite that of the electric field, with an average drift speed v_d, that is much smaller (typically 10^{-4} m/s) than the average speed between collisions (typically 10^6 m/s). Figure 27.11 provides a crude description of the motion of free electrons in a conductor. In the absence of an electric field, there is no net displacement after many collisions (Fig. 27.11a). An electric field **E** modifies the random motion and causes the electrons to drift in a direction opposite that of **E** (Fig. 27.11b). The slight curvature in the paths in Figure 27.11b results from the acceleration of the electrons between collisions, caused by the applied field. One

Photograph of a small permanent magnet levitated above a disk of the superconductor $YBa_2Cu_3O_{7-\delta}$, which is at 77 K.
(Tony Stone/Worldwide)

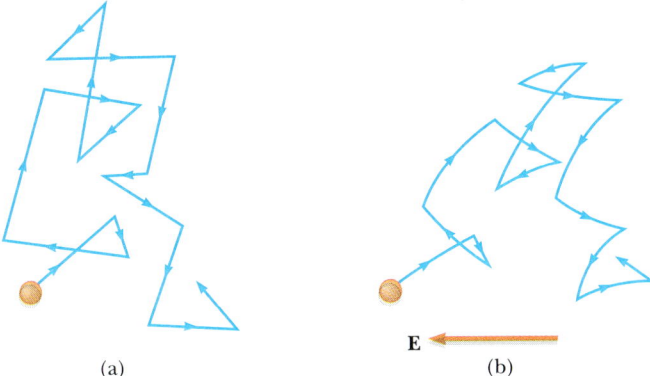

(a)   (b)

FIGURE 27.11   (a) A schematic diagram of the random motion of a charge carrier in a conductor in the absence of an electric field. The drift velocity is zero. (b) The motion of a charge carrier in a conductor in the presence of an electric field. Note that the random motion is modified by the field, and the charge carrier has a drift velocity.

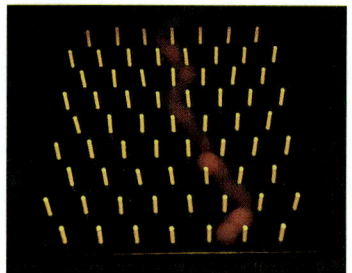

FIGURE 27.12   A mechanical system somewhat analogous to the motion of charge carriers in the presence of an electric field. The collisions of the ball with the pegs represent the resistance to the ball's motion down the incline. *(Jim Lehman)*

mechanical system somewhat analogous to this situation is a ball rolling down a slightly inclined plane through an array of closely spaced, fixed pegs (Fig. 27.12). The ball represents a conduction electron, the pegs represent defects in the crystal lattice, and the component of the gravitational force along the incline represents the electric force $e\mathbf{E}$.

In our model, we assume that the excess energy acquired by the electrons in the electric field is lost to the conductor in the collision process. The energy given up to the atoms in the collisions increases the vibrational energy of the atoms, causing the conductor to heat up. The model also assumes that the motion of an electron after a collision is independent of its motion before the collision.

We are now in a position to obtain an expression for the drift velocity. When a free electron of mass m and charge q is subjected to an electric field $\mathbf{E}$, it experiences a force $q\mathbf{E}$. Since $\mathbf{F} = m\mathbf{a}$, we conclude that the acceleration of the electron is

$$\mathbf{a} = \frac{q\mathbf{E}}{m} \tag{27.15}$$

This acceleration, which occurs for only a short time between collisions, enables the electron to acquire a small drift velocity. If t is the time since the last collision and $\mathbf{v}_0$ is the initial velocity, then the velocity of the electron after a time t is

$$\mathbf{v} = \mathbf{v}_0 + \mathbf{a}t = \mathbf{v}_0 + \frac{q\mathbf{E}}{m} t \tag{27.16}$$

We now take the average value of $\mathbf{v}$ over all possible times t and all possible values of $\mathbf{v}_0$. If the initial velocities are assumed to be randomly distributed in space, we see that the average value of $\mathbf{v}_0$ is zero. The term $(q\mathbf{E}/m)t$ is the velocity added by the field during one trip between atoms. If the electron starts with zero velocity, the average value of the second term of Equation 27.16 is $(q\mathbf{E}/m)\tau$, where τ is the *average time interval between successive collisions*. Because the average of $\mathbf{v}$ is equal to the drift velocity,[3] we have

Drift velocity

$$\mathbf{v}_d = \frac{q\mathbf{E}}{m} \tau \tag{27.17}$$

Substituting this result into Equation 27.6, we find that the magnitude of the current density is

Current density

$$J = nq v_d = \frac{nq^2 E}{m} \tau \tag{27.18}$$

Comparing this expression with Ohm's law, $J = \sigma E$, we obtain the following relationships for conductivity and resistivity:

Conductivity

$$\sigma = \frac{nq^2 \tau}{m} \tag{27.19}$$

Resistivity

$$\rho = \frac{1}{\sigma} = \frac{m}{nq^2 \tau} \tag{27.20}$$

[3] Since the collision process is random, each collision event is *independent* of what happened earlier. This is analogous to the random process of throwing a die. The probability of rolling a particular number on one throw is independent of the result of the previous throw. On the average, it would take six throws to come up with that number, starting at any arbitrary time.

According to this classical model, conductivity and resistivity do not depend on the electric field. This feature is characteristic of a conductor obeying Ohm's law. The average time between collisions is related to the average distance between collisions ℓ (the mean free path, Section 21.7) and the average speed $\bar{v}$ through the expression

$$\tau = \frac{\ell}{\bar{v}} \tag{27.21}$$

EXAMPLE 27.7   Electron Collisions in Copper

(a) Using the data and results from Example 27.1 and the classical model of electron conduction, estimate the average time between collisions for electrons in copper at 20°C.

Solution   From Equation 27.20 we see that

$$\tau = \frac{m}{nq^2\rho}$$

where $\rho = 1.7 \times 10^{-8} \ \Omega \cdot m$ for copper and the carrier density $n = 8.48 \times 10^{28}$ electrons/m^3 for the wire described in Example 27.1. Substitution of these values into the expression above gives

$$\tau = \frac{(9.11 \times 10^{-31} \ \text{kg})}{(8.48 \times 10^{28} \ \text{m}^{-3})(1.6 \times 10^{-19} \ \text{C})^2(1.7 \times 10^{-8} \ \Omega \cdot m)}$$

$$= 2.5 \times 10^{-14} \ \text{s}$$

(b) Assuming the average speed for free electrons in copper to be 1.6×10^6 m/s and using the result from part (a), calculate the mean free path for electrons in copper.

Solution

$$\ell = \bar{v}\tau = (1.6 \times 10^6 \ \text{m/s})(2.5 \times 10^{-14} \ \text{s})$$

$$= 4.0 \times 10^{-8} \ \text{m}$$

which is equivalent to 40 nm (compared with atomic spacings of about 0.2 nm). Thus, although the time between collisions is very short, the electrons travel about 200 atomic distances before colliding with an atom.

Although this classical model of conduction is consistent with Ohm's law, it is not satisfactory for explaining some important phenomena. For example, classical calculations for $\bar{v}$ using the ideal-gas model are about a factor of 10 smaller than the true values. Furthermore, according to Equations 27.20 and 27.21, the temperature variation of the resistivity is predicted to vary as $\bar{v}$, which according to an ideal-gas model (Chapter 21) is proportional to $\sqrt{T}$. This is in disagreement with the linear dependence of resistivity with temperature for pure metals (Fig. 27.8). It is possible to account for such observations only by using a quantum mechanical model, which we shall describe briefly.

According to quantum mechanics, electrons have wave-like properties. If the array of atoms is regularly spaced (that is, periodic), the wave-like character of the electrons makes it possible for them to move freely through the conductor, and a collision with an atom is unlikely. For an idealized conductor, there would be no collisions, the mean free path would be infinite, and the resistivity would be zero. Electron waves are scattered only if the atomic arrangement is irregular (not periodic) as a result of, for example, structural defects or impurities. At low temperatures, the resistivity of metals is dominated by scattering caused by collisions between the electrons and impurities. At high temperatures, the resistivity is dominated by scattering caused by collisions between the electrons and the atoms of the conductor, which are continuously displaced as a result of thermal agitation. The thermal motion of the atoms causes the structure to be irregular (compared with an atomic array at rest), thereby reducing the electron's mean free path.

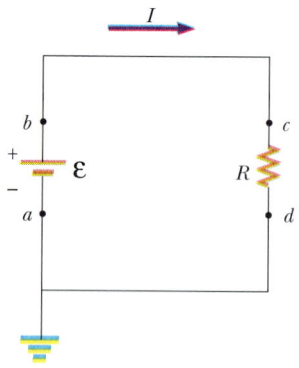

FIGURE 27.13   A circuit consisting of a battery of emf $\mathcal{E}$ and resistance R. Positive charge flows in the clockwise direction, from the negative to the positive terminal of the battery. Points a and d are grounded.

27.6   ELECTRICAL ENERGY AND POWER

If a battery is used to establish an electric current in a conductor, there is a continuous transformation of chemical energy stored in the battery to kinetic energy of the charge carriers. This kinetic energy is quickly lost as a result of collisions between the charge carriers and the atoms making up the conductor, resulting in an increase in the temperature of the conductor. Therefore, we see that the chemical energy stored in the battery is continuously transformed to thermal energy.

Consider a simple circuit consisting of a battery whose terminals are connected to a resistor R, as shown in Figure 27.13. (Resistors are designated by the symbol $-\mathsf{WWv}-$.) Now imagine following a positive quantity of charge ΔQ moving around the circuit from point a through the battery and resistor and back to a. Point a is a reference point that is grounded (ground symbol $\overset{\perp}{=}$), and its potential is taken to be zero. As the charge moves from a to b through the battery, its electrical potential energy *increases* by an amount $V \Delta Q$ (where V is the potential at b) while the chemical potential energy in the battery *decreases* by the same amount. (Recall from Eq. 25.9 that $\Delta U = q \Delta V$.) However, as the charge moves from c to d through the resistor, it *loses* this electrical potential energy as it undergoes collisions with atoms in the resistor, thereby producing thermal energy. If we neglect the resistance of the interconnecting wires, there is no loss in energy for paths bc and da. When the charge returns to point a, it must have the same potential energy (zero) as it had at the start.[4]

The rate at which the charge ΔQ loses potential energy in going through the resistor is

$$\frac{\Delta U}{\Delta t} = \frac{\Delta Q}{\Delta t} V = IV$$

where I is the current in the circuit. Of course, the charge regains this energy when it passes through the battery. Since the rate at which the charge loses energy equals the power P dissipated in the resistor, we have

Power

$$P = IV \qquad (27.22)$$

In this case, the power is supplied to a resistor by a battery. However, Equation 27.22 can be used to determine the power transferred to *any* device carrying a current I and having a potential difference V between its terminals.

Using Equation 27.22 and the fact that $V = IR$ for a resistor, we can express the power dissipated by the resistor in the alternative forms

Power loss in a conductor

$$P = I^2 R = \frac{V^2}{R} \qquad (27.23)$$

When I is in amperes, V in volts, and R in ohms, the SI unit of power is the watt.

---

[4] Note that when the current reaches its steady-state value, there is *no* change with time in the kinetic energy associated with the current.

The power lost as heat in a conductor of resistance R is called *joule heat*[5]; however, it is often referred to as an I^2R loss.

A battery or any other device that provides electrical energy is called a source of *electromotive force*, usually referred to as an *emf*. The concept of emf is discussed in more detail in Chapter 28. (The phrase *electromotive force* is an unfortunate one, because it describes not a force but rather a potential difference in volts.) *Neglecting the internal resistance of the battery, the potential difference between points a and b in Figure 27.13 is equal to the emf ε of the battery.* That is, $V = V_b - V_a = \varepsilon$, and the current in the circuit is $I = V/R = \varepsilon/R$. Since $V = \varepsilon$, the power supplied by the emf can be expressed as $P = I\varepsilon$, which, of course, equals the power dissipated in the resistor, I^2R.

CONCEPTUAL EXAMPLE 27.8

Two lightbulbs A and B are connected across the same potential difference as in Figure 27.14. The resistance of A is twice that of B. Which lightbulb dissipates more power? Which carries the greater current?

Reasoning   Because the voltage across each lightbulb is the same, and the power dissipated by a conductor is $P = V^2/R$, the conductor with the lower resistance will dissipate more power. In this case, the power dissipated by B is twice that of A and provides twice as much illumination. Furthermore, because $P = IV$, we see that the current carried by B is twice that of A.

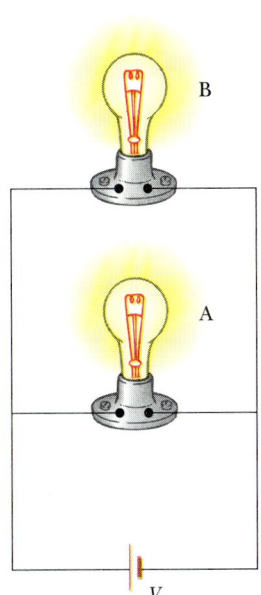

FIGURE 27.14   (Conceptual Example 27.8) Two bulbs connected across the same potential difference.

EXAMPLE 27.9   **Power in an Electric Heater**

An electric heater is constructed by applying a potential difference of 110 V to a Nichrome wire of total resistance 8.00 Ω. Find the current carried by the wire and the power rating of the heater.

Solution   Since $V = IR$, we have

$$I = \frac{V}{R} = \frac{110 \text{ V}}{8.00 \ \Omega} = \boxed{13.8 \text{ A}}$$

We can find the power rating using $P = I^2R$:

$$P = I^2R = (13.8 \text{ A})^2(8.00 \ \Omega) = \boxed{1.52 \text{ kW}}$$

If we doubled the applied voltage, the current would double but the power would quadruple.

---

[5] It is called *joule heat* even though its dimensions are *energy per unit time,* which are dimensions of power.

EXAMPLE 27.10   Electrical Rating of a Lightbulb

A lightbulb is rated at 120 V/75 W, which means its operating voltage is 120 V and it has a power rating of 75.0 W. The bulb is powered by a 120-V direct-current power supply. Find the current in the bulb and its resistance.

Solution   Since the power rating of the bulb is 75.0 W and the operating voltage is 120 V, we can use $P = IV$ to find the current:

$$I = \frac{P}{V} = \frac{75.0 \text{ W}}{120 \text{ V}} = \boxed{0.625 \text{ A}}$$

Using Ohm's law, $V = IR$, the resistance is calculated to be

$$R = \frac{V}{I} = \frac{120 \text{ V}}{0.625 \text{ A}} = \boxed{192 \ \Omega}$$

Exercise   What would the resistance be in a lamp rated at 120 V and 100 W?

Answer   144 Ω.

EXAMPLE 27.11   The Cost of Operating a Lightbulb

How much does it cost to burn a 100-W lightbulb for 24 h if electricity costs eight cents per kilowatt hour?

Solution   Since the energy consumed equals power × time, the amount of energy you must pay for, expressed in kWh, is

$$\text{Energy} = (0.10 \text{ kW})(24 \text{ h}) = 2.4 \text{ kWh}$$

If energy is purchased at eight cents per kWh, the cost is

$$\text{Cost} = (2.4 \text{ kWh})(\$0.080/\text{kWh}) = \boxed{\$0.19}$$

That is, it costs 19 cents to operate the lightbulb for one day. This is a small amount, but when larger and more complex devices are being used, the costs go up rapidly.

Demands on our energy supplies have made it necessary to be aware of the energy requirements of our electric devices. This is true not only because they are becoming more expensive to operate but also because, with the dwindling of the coal and oil resources that ultimately supply us with electrical energy, increased awareness of conservation becomes necessary. On every electric appliance is a label that contains the information you need to calculate the power requirements of the appliance. The power consumption in watts is often stated directly, as on a lightbulb. In other cases, the amount of current used by the device and the voltage at which it operates are given. This information and Equation 27.22 are sufficient to calculate the operating cost of any electric device.

Exercise   If electricity costs eight cents per kilowatt hour, what does it cost to operate an electric oven, which operates at 20.0 A and 220 V, for 5.00 h?

Answer $1.76.

EXAMPLE 27.12   Current in an Electron Beam

In a certain accelerator, electrons emerge with energies of 40.0 MeV (1 MeV = 1.60×10^{-13} J). The electrons do not emerge in a steady stream, but in pulses that repeat 250 times per second. This corresponds to a time between each pulse of 4.00 ms in Figure 27.15. Each pulse lasts for 200 ns and the electrons in the pulse constitute a current of 250 mA. The current is zero between pulses. (a) How many electrons are delivered by the accelerator per pulse?

Solution   We can use Equation 27.2 in the form $dQ = I \, dt$, and integrate to find the charge per pulse. While the pulse is on, the current is constant, therefore

$$Q_{\text{pulse}} = I \int dt = It = (250 \times 10^{-3} \text{ A})(200 \times 10^{-9} \text{ s})$$

$$= 5.00 \times 10^{-8} \text{ C}$$

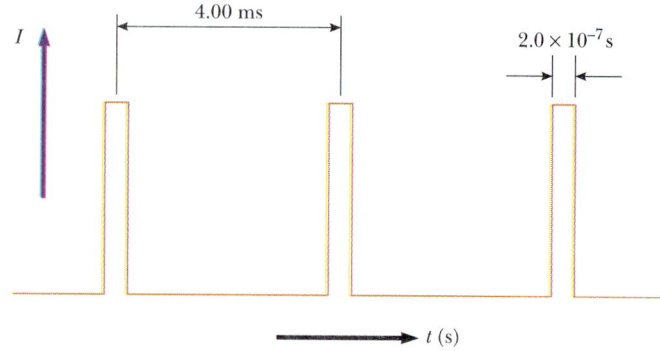

FIGURE 27.15   (Example 27.12) Current versus time for a pulsed beam of electrons.

This quantity of charge per pulse divided by the electronic charge gives the number of electrons per pulse:

$$\text{No. of electrons per pulse} = \frac{5.00 \times 10^{-8} \text{ C/pulse}}{1.60 \times 10^{-19} \text{ C/electron}}$$

$$= 3.13 \times 10^{11} \text{ electrons/pulse}$$

(b) What is the average current delivered by the accelerator?

Solution The average current is given by Equation 27.1, $I = \Delta Q/\Delta t$. Since the duration of one pulse is 4.00 ms, and the charge per pulse is known from part (a), we get

$$I_{av} = \frac{Q_{pulse}}{\Delta t} = \frac{5.00 \times 10^{-8} \text{ C}}{4.00 \times 10^{-3} \text{ s}} = 12.5 \ \mu\text{A}$$

This represents only 0.0005% of the peak current.

(c) What is the maximum power delivered by the electron beam?

Solution By definition, power is the energy delivered per unit time. Thus, the maximum power is equal to the energy delivered by the beam during the pulse period:

$$P = \frac{E}{\Delta t} = \frac{(3.13 \times 10^{11} \text{ electrons/pulse})(40.0 \text{ MeV/electron})}{2.00 \times 10^{-7} \text{ s/pulse}}$$

$$= (6.26 \times 10^{19} \text{ MeV/s})(1.60 \times 10^{-13} \text{ J/MeV})$$

$$= 1.00 \times 10^7 \text{ W} = 10.0 \text{ MW}$$

SUMMARY

The **electric current** I in a conductor is defined as

$$I \equiv \frac{dQ}{dt} \tag{27.2}$$

where dQ is the charge that passes through a cross-section of the conductor in a time dt. The SI unit of current is the ampere (A), where 1 A = 1 C/s.

The current in a conductor is related to the motion of the charge carriers through the relationship

$$I = nqv_d A \tag{27.4}$$

where n is the density of charge carriers, q is their charge, v_d is the drift speed, and A is the cross-sectional area of the conductor.

The **current density J** in a conductor is defined as the current per unit area:

$$\mathbf{J} = nq\mathbf{v}_d \tag{27.6}$$

The current density in a conductor is proportional to the electric field according to the expression

$$\mathbf{J} = \sigma \mathbf{E} \tag{27.7}$$

The constant σ is called the **conductivity** of the material. The inverse of σ is called the **resistivity**, ρ. That is, $\rho = 1/\sigma$.

A material is said to obey Ohm's law if its conductivity is independent of the applied field.

The **resistance** R of a conductor is defined as the ratio of the potential difference across the conductor to the current:

$$R \equiv \frac{V}{I} \tag{27.8}$$

If the resistance is independent of the applied voltage, the conductor obeys Ohm's law.

If the conductor has a uniform cross-sectional area A and a length ℓ, its resistance is

$$R = \frac{\ell}{\sigma A} = \rho \frac{\ell}{A} \tag{27.11}$$

The SI unit of resistance is volt per ampere, which is defined to be 1 ohm (Ω). That is, $1\ \Omega = 1\ \text{V/A}$.

The resistivity of a conductor varies with temperature in an approximately linear fashion, that is

$$\rho = \rho_0 [1 + \alpha (T - T_0)] \tag{27.12}$$

where α is the temperature coefficient of resistivity and ρ_0 is the resistivity at some reference temperature T_0.

In a classical model of electronic conduction in a metal, the electrons are treated as molecules of a gas. In the absence of an electric field, the average velocity of the electrons is zero. When an electric field is applied, the electrons move (on the average) with a **drift velocity** $\mathbf{v}_d$, which is opposite the electric field and given by

$$\mathbf{v}_d = \frac{q\mathbf{E}}{m} \tau \tag{27.17}$$

where τ is the average time between collisions with the atoms of the metal. The resistivity of the material according to this model is

$$\rho = \frac{m}{nq^2 \tau} \tag{27.20}$$

where n is the number of free electrons per unit volume.

If a potential difference V is maintained across a resistor, the **power,** or rate at which energy is supplied to the resistor, is

$$P = IV \tag{27.22}$$

Since the potential difference across a resistor is given by $V = IR$, we can express the power dissipated in a resistor in the form

$$P = I^2 R = \frac{V^2}{R} \tag{27.23}$$

The electrical energy supplied to a resistor appears in the form of internal energy (thermal energy) in the resistor.

QUESTIONS

1. In an analogy between traffic flow and electrical current, what would correspond to the charge Q? What would correspond to the current I?

2. What factors affect the resistance of a conductor?

3. What is the difference between resistance and resistivity?

4. Two wires A and B of circular cross-section are made of the same metal and have equal lengths, but the resistance of wire A is three times greater than that of wire B. What is the ratio of their cross-sectional areas? How do their radii compare?

5. What is required in order to maintain a steady current in a conductor?

6. Do all conductors obey Ohm's law? Give examples to justify your answer.

7. When the voltage across a certain conductor is doubled, the current is observed to increase by a factor of 3. What can you conclude about the conductor?

8. In the water analogy of an electric circuit, what corresponds to the power supply, resistor, charge, and potential difference?

9. Why might a "good" electrical conductor also be a "good" thermal conductor?

10. Use the atomic theory of matter to explain why the resistance of a material should increase as its temperature increases.

11. How does the resistance change with temperature for copper and silicon? Why are they different?

12. Explain how a current can persist in a superconductor without any applied voltage.

13. What single experimental requirement makes superconducting devices expensive to operate? In principle, can this limitation be overcome?

14. What would happen to the drift velocity of the electrons in a wire and to the current in the wire if the electrons could move freely without resistance through the wire?

15. If charges flow very slowly through a metal, why does it not require several hours for a light to come on when you throw a switch?

16. In a conductor, the electric field that drives the electrons through the conductor propagates with a speed close to the speed of light, although the drift velocity of the electrons is very small. Explain how these can both be true.

Does the same electron move from one end of the conductor to the other?

17. Two conductors of the same length and radius are connected across the same potential difference. One conductor has twice the resistance of the other. Which conductor will dissipate more power?

18. When incandescent lamps burn out, they usually do so just after they are switched on. Why?

19. If you were to design an electric heater using nichrome wire as the heating element, what parameters of the wire could you vary to meet a specific power output, such as 1000 W?

20. Two lightbulbs both operate from 110 V, but one has a power rating of 25 W and the other of 100 W. Which bulb has the higher resistance? Which bulb carries the greater current?

21. A typical monthly utility rate structure might go something like this: $1.60 for the first 16 kWh, 7.05 cents/kWh for the next 34 kWh used, 5.02 cents/kWh for the next 50 kWh, 3.25 cents/kWh for the next 100 kWh, 2.95 cents/kWh for the next 200 kWh, 2.35 cents/kWh for all in excess of 400 kWh. Based on these rates, what would be the charge for 327 kWh?

PROBLEMS

Section 27.1   Electric Current

1. In the Bohr model of the hydrogen atom, an electron in the lowest energy state follows a circular path, 5.29×10^{-11} m from the proton. (a) Show that the speed of the electron is 2.19×10^{6} m/s. (b) What is the effective current associated with this orbiting electron?

2. In a particular cathode ray tube, the measured beam current is 30 μA. How many electrons strike the tube screen every 40 s?

3. A small sphere that carries a charge of 8.00 nC is whirled in a circle at the end of an insulating string. The angular frequency of rotation is 100π rad/s. What average current does this rotating charge represent?

3A. A small sphere that carries a charge q is whirled in a circle at the end of an insulating string. The angular frequency of rotation is ω. What average current does this rotating charge represent?

4. The quantity of charge q (in C) passing through a surface of area 2.0 cm^2 varies with time as $q = 4t^3 + 5t + 6$, where t is in seconds. (a) What is the instantaneous current through the surface at $t = 1.0$ s? (b) What is the value of the current density?

5. An electric current is given by $I(t) = 100.0$

$\sin(120\pi t)$, where I is in amperes and t is in seconds. What is the total charge carried by the current from $t = 0$ to $t = 1/240$ s?

6. Suppose that the current through a conductor decreases exponentially with time according to

$$I(t) = I_0 e^{-t/\tau}$$

where I_0 is the initial current (at $t = 0$), and τ is a constant having dimensions of time. Consider a fixed observation point within the conductor. (a) How much charge passes this point between $t = 0$ and $t = \tau$? (b) How much charge passes this point between $t = 0$ and $t = 10\tau$? (c) How much charge passes this point between $t = 0$ and $t = \infty$?

7. A Van de Graaff generator produces a beam of 2.0-MeV deuterons, which are heavy hydrogen nuclei containing a proton and a neutron. (a) If the beam current is 10.0 μA, how far apart are the deuterons in the beam? (b) Is their electrostatic repulsion a factor of beam stability? Explain.

8. Calculate the average drift speed of electrons traveling through a copper wire with a cross-sectional area of 1.00 mm^2 when carrying a current of 1.00 A (values similar to those for the electric wire to your study lamp). It is known that about one electron per atom of copper contributes to the current. The

$\square$ indicates problems that have full solutions available in the Student Solutions Manual and Study Guide.

atomic weight of copper is 63.54, and its density is 8.92 g/cm³.

9. A copper bus bar has a cross-section of 5.0 cm × 15.0 cm and carries a current with a density of 2000 A/cm². (a) What is the total current in the bus bar? (b) How much charge passes a given point in the bar per hour?

10. Figure P27.10 represents a section of a circular conductor of nonuniform diameter carrying a current of 5.0 A. The radius of cross-section A_1 is 0.40 cm. (a) What is the magnitude of the current density across A_1? (b) If the current density across A_2 is one-fourth the value across A_1, what is the radius of the conductor at A_2?

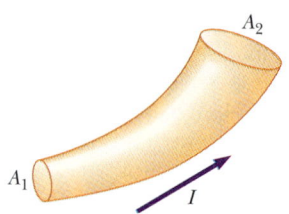

FIGURE P27.10

11. A coaxial conductor with a length of 20 m consists of an inner cylinder with a radius of 3.0 mm and a concentric outer cylindrical tube with an inside radius of 9.0 mm. A uniformly distributed leakage current of 10 μA flows between the two conductors. Determine the leakage current density (in A/m²) through a cylindrical surface (concentric with the conductors) that has a radius of 6.0 mm.

Section 27.2   Resistance and Ohm's Law

12. A conductor of uniform radius 1.2 cm carries a current of 3.0 A produced by an electric field of 120 V/m. What is the resistivity of the material?

13. An electric field of 2100 V/m is applied to a section of silver of uniform cross-section. Calculate the resulting current density if the specimen is at a temperature of 20°C.

14. A solid cube of silver (specific gravity = 10.50) has a mass of 90 g. (a) What is the resistance between opposite faces of the cube? (b) If there is one conduction electron for each silver atom, find the average drift speed of electrons when a potential difference of 1.0×10^{-5} V is applied to opposite faces. The atomic number of silver is 47, and its atomic mass is 107.87.

15. Calculate the resistance at 20°C of a 40-m length of silver wire having a cross-sectional area of 0.40 mm².

16. Eighteen-gauge wire has a diameter of 1.024 mm. Calculate the resistance of 15.0 m of 18-gauge copper wire at 20.0°C.

17. While traveling through Death Valley on a day when the temperature is 58°C, Bill Hiker finds that a certain voltage applied to a copper wire produces a current of 1.000 A. Bill then travels to Antarctica and applies the same voltage to the same wire. What current does he register there if the temperature is −88°C? Assume no change in the wire's shape and size.

18. A circular disk of radius R and thickness d is made of material with resistivity ρ. Show that the resistance between points a and b (Fig. P27.18) is independent of the radius and is given by $R = \pi\rho/2d$.

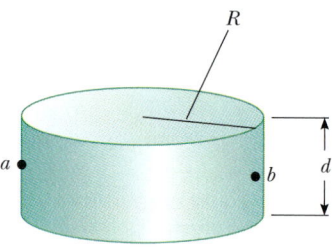

FIGURE P27.18

19. A 12.0-Ω metal wire is cut into three equal pieces that are then connected side by side to form a new wire the length of which is equal to one-third the original length. What is the resistance of this new wire?

19A. A metal wire of resistance R is cut into three equal pieces that are then connected side by side to form a new wire the length of which is equal to one-third the original length. What is the resistance of this new wire?

20. The temperature coefficients of resistivity in Table 27.1 are at 20°C. What would they be at 0°C? (*Hint:* The temperature coefficient of resistivity at 20°C satisfies $\rho = \rho_0[1 + \alpha(T - T_0)]$, where ρ_0 is the resistivity of the material at $T_0 = 20$°C. The temperature coefficient of resistivity, α', at 0°C must satisfy $\rho = \rho_0'[1 + \alpha' T]$, where ρ_0' is the resistivity of the material at 0°C.)

21. A wire with a resistance R is lengthened to 1.25 times its original length by pulling it through a small hole. Find the resistance of the wire after it is stretched.

22. Aluminum and copper wires of equal length are found to have the same resistance. What is the ratio of their radii?

23. Suppose that you wish to fabricate a uniform wire out of 1.0 g of copper. If the wire is to have a resistance of $R = 0.50$ Ω, and all of the copper is to be used, what will be (a) the length and (b) the diameter of this wire?

24. A platinum resistance thermometer, when placed in thermal equilibrium with an object, uses the measured resistance of the platinum to determine the

temperature of the object. If such a thermometer has a resistance of 200.0 Ω when placed in a 0°C ice bath and of 253.8 Ω when immersed in a crucible containing melting potassium, what is the melting point of potassium? (*Hint:* First determine the measured resistance of the thermometer at room temperature, 20°C.)

25. A 0.90-V potential difference is maintained across a 1.5-m length of tungsten wire that has a cross-sectional area of 0.60 mm². What is the current in the wire?

26. The electron beam emerging from a certain high-energy electron accelerator has a circular cross-section of radius 1.00 mm. (a) If the beam current is 8.00 μA, find the current density in the beam, assuming that it is uniform throughout. (b) The speed of the electrons is so close to the speed of light that their speed can be taken as $c = 3.00 \times 10^8$ m/s with negligible error. Find the electron density in the beam. (c) How long does it take for an Avogadro's number of electrons to emerge from the accelerator?

27. A resistor is constructed of a carbon rod that has a uniform cross-sectional area of 5.0 mm². When a potential difference of 15 V is applied across the ends of the rod, there is a current of 4.0×10^{-3} A in the rod. Find (a) the resistance of the rod and (b) the rod's length.

28. A current density of 6.0×10^{-13} A/m² exists in the atmosphere where the electric field (due to charged thunderclouds in the vicinity) is 100 V/m. Calculate the electrical conductivity of the Earth's atmosphere in this region.

Section 27.3   Resistance and Temperature

29. An aluminum wire with a diameter of 0.10 mm has a uniform electric field of 0.20 V/m imposed along its entire length. The temperature of the wire is 50°C. Assume one free electron per atom. (a) Use the information in Table 27.1 and determine the resistivity. (b) What is the current density in the wire? (c) What is the total current in the wire? (d) What is the drift speed of the conduction electrons? (e) What potential difference must exist between the ends of a 2.0-m length of the wire to produce the stated electric field strength?

30. The rod in Figure P27.30 is made of two materials. Both have a square cross-section 3.0 mm on a side. The first material has a resistivity of 4.00×10^{-3} $\Omega \cdot$m and is 25.0 cm long, while the second material has a resistivity of 6.00×10^{-3} $\Omega \cdot$m and is 40.0 cm long. What is the resistance between the ends of the rod?

30A. The rod in Figure P27.30 is made of two materials. Both have a square cross-section with side d. The first

material has resistivity ρ_1 and length L_1, while the second material has resistivity ρ_2 and length L_2. What is the resistance between the ends of the rod?

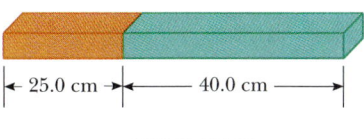

FIGURE P27.30

31. What is the fractional change in the resistance of an iron filament when its temperature changes from 25°C to 50°C?

32. The resistance of a platinum wire is to be calibrated for low-temperature measurements. A platinum wire with resistance 1.00 Ω at 20.0°C is immersed in liquid nitrogen at 77 K (−196°C). If the temperature response of the platinum wire is linear, what is the expected resistance of the platinum wire at −196°C? ($\alpha_{\text{platinum}} = 3.92 \times 10^{-3}$/°C)

33. If a copper wire has a resistance of 18 Ω at 20°C, what resistance will it have at 60°C? (Neglect any change in length or cross-sectional area due to the change in temperature.)

34. A carbon wire and a Nichrome wire are connected in series. If the combination has a resistance of 10.0 kΩ at 0°C, what is the resistance of each wire at 0°C so that the resistance of the combination does not change with temperature?

34A. A carbon wire and a Nichrome wire are connected in series. If the combination has a resistance R at temperature T_0, what is the resistance of each wire at T_0 so that the resistance of the combination does not change with temperature?

35. At what temperature will tungsten have a resistivity four times that of copper? (Assume that the copper is at 20°C.)

36. A segment of Nichrome wire is initially at 20°C. Using the data from Table 27.1, calculate the temperature to which the wire must be heated to double its resistance.

37. At 45.0°C, the resistance of a segment of gold wire is 85.0 Ω. When the wire is placed in a liquid bath, the resistance decreases to 80.0 Ω. What is the temperature of the bath?

38. A 500-W heating coil designed to operate from 110 V is made of Nichrome wire 0.50 mm in diameter. (a) Assuming that the resistivity of the Nichrome remains constant at its 20°C value, find the length of wire used. (b) Now consider the variation of resistivity with temperature. What power will the coil of part (a) actually deliver when it is heated to 1200°C?

Section 27.5   A Model for Electrical Conduction

39. Calculate the current density in a gold wire in which an electric field of 0.74 V/m exists.

40. If the drift velocity of free electrons in a copper wire is 7.84×10^{-4} m/s, calculate the electric field in the conductor.

41. Use data from Example 27.1 to calculate the collision mean free path of electrons in copper if the average thermal speed of conduction electrons is 8.6×10^5 m/s.

42. If the current carried by a conductor is doubled, what happens to the (a) charge carrier density? (b) current density? (c) electron drift velocity? (d) average time between collisions?

Section 27.6   Electrical Energy and Power

43. A 10-V battery is connected to a 120-Ω resistor. Neglecting the internal resistance of the battery, calculate the power dissipated in the resistor.

44. A coil of Nichrome wire is 25.0 m long. The wire has a diameter of 0.40 mm and is at 20.0°C. If it carries a current of 0.50 A, what are (a) the electric field intensity in the wire and (b) the power dissipated in it? (c) If the temperature is increased to 340°C and the voltage across the wire remains constant, what is the power dissipated?

45. Suppose that a voltage surge produces 140 V for a moment. By what percentage will the output of a 120-V, 100-W lightbulb increase, assuming its resistance does not change?

46. A particular type of automobile storage battery is characterized as "360-ampere-hour, 12 V." What total energy can the battery deliver?

47. Batteries are rated in terms of ampere hours (A·h), where a battery rated at 1.0 A·h can produce a current of 1.0 A for 1.0 h. (a) What is the total energy, in kilowatt hours, stored in a 12.0-V battery rated at 55.0 A·h? (b) At $0.06 per kilowatt hour, what is the value of the electricity produced by this battery?

48. In a hydroelectric installation, a turbine delivers 1500 hp to a generator, which in turn converts 80% of the mechanical energy into electrical energy. Under these conditions, what current will the generator deliver at a terminal potential difference of 2000 V?

49. Suppose that you want to install a heating coil that will convert electric energy to heat at a rate of 300 W for a current of 1.5 A. (a) Determine the resistance of the coil. (b) The resistivity of the coil wire is 1.0×10^{-6} $\Omega \cdot$m, and its diameter is 0.30 mm. Determine its length.

50. It is estimated that in the United States (population 250 million) there is one electric clock per person, with each clock using energy at a rate of 2.5 W. To supply this energy, about how many metric tons of coal are burned per hour in coal-fired electric generating plants that are, on average, 25% efficient. The heat of combustion for coal is 33.0 MJ/kg.

51. What is the required resistance of an immersion heater that will increase the temperature of 1.5 kg of water from 10°C to 50°C in 10 min while operating at 110 V?

51A. What is the required resistance of an immersion heater that will increase the temperature of m kg of water from T_1 to T_2 in a time t while operating at a voltage V?

52. The heating element of a coffee maker operates at 120 V and carries a current of 2.0 A. Assuming that all of the heat generated is absorbed by the water, how long does it take to heat 0.50 kg of water from room temperature (23°C) to the boiling point?

53. Compute the cost per day of operating a lamp that draws 1.7 A from a 110-V line if the cost of electrical energy is $0.06/kWh.

54. It requires about 10.0 W of electric power per square foot to heat a room having ceilings 7.5 ft high. At a cost of $0.080/kWh, how much does it cost per day to use electric heat to heat a room 10.0 ft × 15.0 ft?

55. A certain toaster has a heating element made of Nichrome resistance wire. When first connected to a 120-V voltage source (and the wire is at a temperature of 20.0°C) the initial current is 1.80 A but begins to decrease as the resistive element heats up. When the toaster has reached its final operating temperature, the current has dropped to 1.53 A. (a) Find the power the toaster consumes when it is at its operating temperature. (b) What is the final temperature of the heating element?

ADDITIONAL PROBLEMS

56. An electric utility company supplies a customer's house from the main power lines (120 V) with two copper wires, each 50.0 m long and having a resistance of 0.108 Ω per 300 m. (a) Find the voltage at the customer's house for a load current of 110 A. For this load current, find (b) the power the customer is receiving and (c) the power dissipated in the copper wires.

57. The potential difference across the filament of a lamp is maintained at a constant level while equilibrium temperature is being reached. It is observed that the steady-state current in the lamp is only one tenth of the current drawn by the lamp when it is first turned on. If the temperature coefficient of resistivity for the lamp at 20°C is 0.0045 (°C)$^{-1}$, and if the resistance increases linearly with increasing temperature, what is the final operating temperature of the filament?

58. The current in a resistor decreases by 3.0 A when the

voltage applied across the resistor decreases from 12.0 V to 6.0 V. Find the resistance of the resistor.

59. An electric car is designed to run off a bank of 12-V batteries with total energy storage of 2.0×10^7 J. (a) If the electric motor draws 8.0 kW, what is the current delivered to the motor? (b) If the electric motor draws 8.0 kW as the car moves at a steady speed of 20 m/s, how far will the car travel before it is "out of juice"?

60. When a straight wire is heated, its resistance changes according to Equation 27.14, where α is the temperature coefficient of resistivity. (a) Show that a more precise result, one that includes the fact that the length and area of the wire do change when heated, is

$$R = \frac{R_0[1 + \alpha(T - T_0)][1 + \alpha'(T - T_0)]}{[1 - 2\alpha'(T - T_0)]}$$

where α' is the coefficient of linear expansion (Chapter 19). (b) Compare these two results for a 2.00-m-long copper wire of radius 0.100 mm, first initially at 20.0°C and then heated to 100.0°C.

61. A Wheatstone bridge can be used to measure the strain ($\Delta L/L_0$) of a wire (see Section 12.4), where L_0 is the length before stretching, L is the length after stretching, and $\Delta L = L - L_0$. Let $\alpha = \Delta L/L_0$. Show that the resistance is $R = R_0(1 + 2\alpha + \alpha^2)$ for any length where $R_0 = \rho L_0/A_0$. Assume that the resistivity and volume of the wire stay constant.

62. The current in a wire decreases with time according to the relationship $I = 2.5e^{-at}$ mA where $a = 0.833$ s^{-1}. Determine the total charge that passes through the wire by the time the current has diminished to zero.

63. A resistor is constructed by forming a material of resistivity ρ into the shape of a hollow cylinder of length L and inner and outer radii r_a and r_b, respectively (Fig. P27.63). In use, a potential difference is applied between the ends of the cylinder, producing a current parallel to the axis. (a) Find a general expression for the resistance of such a device in terms of L, ρ, r_a, and r_b. (b) Obtain a numerical value for R when $L = 4.0$ cm, $r_a = 0.50$ cm, $r_b = 1.2$ cm, and the resistivity $\rho = 3.5 \times 10^5$ $\Omega \cdot$m. (c) Suppose now that

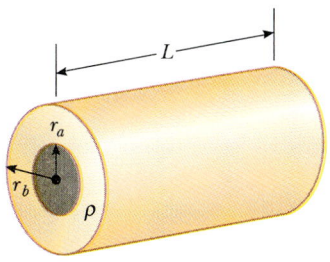

FIGURE P27.63

the potential difference is applied between the inner and outer surfaces so that the resulting current flows radially outward. (a) Find a general expression for the resistance of the device in terms of L, ρ, r_a, and r_b. (d) Calculate the value of R using the parameter values given in part (b).

64. In a certain stereo system, each speaker has a resistance of 4.00 Ω. The system is rated at 60.0 W in each channel, and each speaker circuit includes a fuse rated at 4.00 A. Is this system adequately protected against overload? Explain.

65. A more general definition of the temperature coefficient of resistivity is

$$\alpha = \frac{1}{\rho} \frac{d\rho}{dT}$$

where ρ is the resistivity at temperature T. (a) Assuming that α is constant, show that

$$\rho = \rho_0 e^{\alpha(T - T_0)}$$

where ρ_0 is the resistivity at temperature T_0. (b) Using the series expansion ($e^x \approx 1 + x$; $x \ll 1$), show that the resistivity is given approximately by the expression $\rho = \rho_0[1 + \alpha(T - T_0)]$ for $\alpha(T - T_0) \ll 1$.

66. There is a close analogy between the flow of heat because of a temperature difference (Section 20.7) and the flow of electrical charge because of a potential difference. The thermal energy dQ and the electrical charge dq are both transported by free electrons in the conducting material. Consequently, a good electrical conductor is usually a good heat conductor as well. Consider a thin conducting slab of thickness dx, area A, and electrical conductivity σ, with a potential difference dV between opposite faces. Show that the current $I = dq/dt$ is

Charge conduction	Analogous heat conduction
	(Eq. 20.14)
$\dfrac{dq}{dt} = -\sigma A \dfrac{dV}{dx}$	$\dfrac{dQ}{dt} = -kA \dfrac{dT}{dx}$

In the analogous heat conduction equation, the rate of heat flow dQ/dt (in SI units of joules per second) is due to a temperature gradient dT/dx, in a material of thermal conductivity k. What is the origin of the minus sign in the charge conduction equation?

67. Material with uniform resistivity ρ is formed into a wedge as shown in Figure P27.67. Show that the resistance between face A and face B of this wedge is

$$R = \rho \frac{L}{w(y_2 - y_1)} \ln\left(\frac{y_2}{y_1}\right)$$

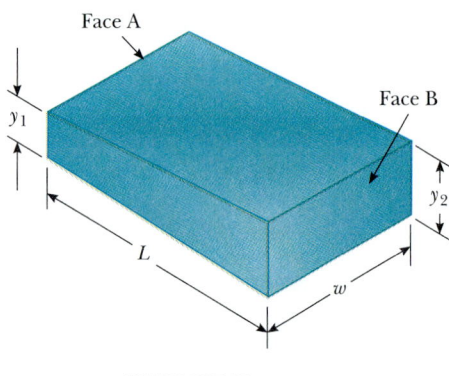

FIGURE P27.67

68. A material of resistivity ρ is formed into the shape of a truncated cone of altitude h as in Figure P27.68. The bottom end has a radius b and the top end has a radius a. Assuming a uniform current density through any circular cross-section of the cone, show that the resistance between the two ends is

$$R = \frac{\rho}{\pi}\left(\frac{h}{ab}\right)$$

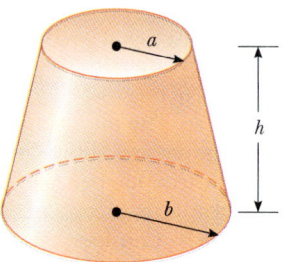

FIGURE P27.68

69. An experiment is conducted to measure the electrical resistivity of nichrome in the form of wires with different lengths and cross-sectional areas. For one set of measurements, a student uses #30-gauge wire, which has a cross-sectional area of 7.3×10^{-8} m². The voltage across the wire and the current in the wire are measured with a voltmeter and ammeter, respectively. For each of the measurements given in the table below taken on wires of three different lengths, calculate the resistance of the wires and the corresponding values of the resistivity. What is the average value of the resistivity, and how does it compare with the value given in Table 27.1?

L (m)	V (V)	I (A)	R (Ω)	ρ ($\Omega \cdot$m)
0.54	5.22	0.500		
1.028	5.82	0.276		
1.543	5.94	0.187		

SPREADSHEET PROBLEMS

S1. Spreadsheet 27.1 calculates the average annual lighting cost per bulb for fluorescent and incandescent bulbs and the average yearly savings realized with fluorescent bulbs. It also graphs the average annual lighting cost per bulb versus the cost of electrical energy. (a) Suppose that a fluorescent bulb costs $5, lasts for 5000 h, and consumes 40 W of power, but provides the light intensity of a 100-W incandescent bulb. Assume that a 100-W incandescent bulb costs $0.65 and lasts for 750 h. If the average house has six 100-W incandescent bulbs on at all times and if energy costs 8.3 cents per kilowatt hour, how much does a consumer save each year by switching to fluorescent bulbs? (b) Check with your local electric company for their current rates, and find the cost of bulbs in your area. Would it pay you to switch to fluorescent bulbs? (c) Vary the parameters for bulbs of different wattages and reexamine the annual savings.

S2. The current-voltage characteristic curve for a semiconductor diode as a function of temperature T is given by

$$I = I_0 (e^{eV/k_B T} - 1)$$

where e is the charge on the electron, k_B is Boltzmann's constant, and T is the absolute temperature. Set up a spreadsheet to calculate I and $R = V/I$ for $V = 0.40$ V to 0.60 V in increments of 0.01 V. Assume $I_0 = 1.0$ nA. Plot R versus V for $T = 280$ K, 300 K, and 320 K.

CHAPTER 28

Direct Current Circuits

Electrical discharges in tubes filled with neon and other gases are used to produce light of various colors. The characteristic color of each tube depends on the type of gas contained in the tube. As examples, neon gives off reddish light, argon produces violet light, and sodium emits yellow light.

(Dan McCoy/Rainbow)

This chapter is concerned with the analysis of some simple circuits whose elements include batteries, resistors, and capacitors in various combinations. The analysis of these circuits is simplified by the use of two rules known as *Kirchhoff's rules,* both of which follow from the laws of conservation of energy and conservation of charge. Most of the circuits analyzed are assumed to be in *steady state,* which means that the currents are constant in magnitude and direction. In Section 28.4, however, we discuss circuits for which the current varies with time. Finally, a number of common electrical devices and techniques are described for measuring current, potential differences, resistance, and emfs.

28.1 ELECTROMOTIVE FORCE

In the previous chapter we found that a constant current can be maintained in a closed circuit through the use of a source of energy, an *emf* (ee-em-eff), from the historical, but inaccurate, term electromotive force. A source of emf is any device (such as a battery or generator) that produces an electric field and thus may cause charges to move around a circuit. One can think of a source of emf as a "charge

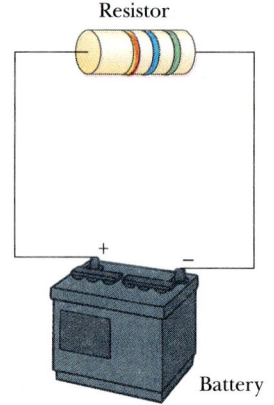

FIGURE 28.1   A circuit consisting of a resistor connected to the terminals of a battery.

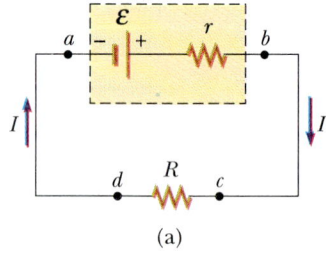

(a)

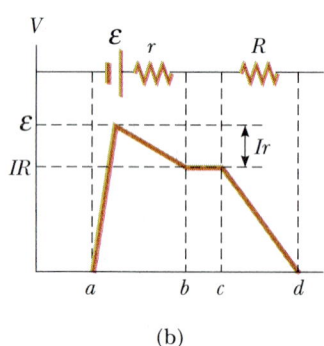

(b)

FIGURE 28.2   (a) Circuit diagram of an emf $\mathcal{E}$ of internal resistance r connected to an external resistor R. (b) Graphical representation showing how the potential changes as the series circuit in part (a) is traversed clockwise.

pump." When a potential can be defined, the source moves charges "uphill" to a higher potential. The emf, $\mathcal{E}$, describes the work done per unit charge, and hence the SI unit of emf is the volt.

Consider the circuit shown in Figure 28.1, consisting of a battery connected to a resistor. We assume that the connecting wires have no resistance. The positive terminal of the battery is at a higher potential than the negative terminal. If we neglect the internal resistance of the battery, then the potential difference across it (the terminal voltage) equals its emf. However, because a real battery always has some internal resistance r, the terminal voltage is not equal to the emf. The circuit shown in Figure 28.1 can be described by the circuit diagram in Figure 28.2a. The battery within the dashed rectangle is represented by an emf $\mathcal{E}$ in series with the internal resistance r. Now imagine a positive charge moving from a to b in Figure 28.2a. As the charge passes from the negative to the positive terminal, its potential *increases* by $\mathcal{E}$. However, as it moves through the resistance r, its potential *decreases* by an amount Ir, where I is the current in the circuit. Thus, the terminal voltage of the battery, $V = V_b - V_a$, is[1]

$$V = \mathcal{E} - Ir \qquad (28.1)$$

From this expression, note that $\mathcal{E}$ is equivalent to the **open-circuit voltage,** that is, the *terminal voltage when the current is zero.* Figure 28.2b is a graphical representation of the changes in potential as the circuit is traversed in the clockwise direction. By inspecting Figure 28.2a we see that the terminal voltage V must also equal the potential difference across the external resistance R, often called the **load resistance.** That is, $V = IR$. Combining this expression with Equation 28.1, we see that

$$\mathcal{E} = IR + Ir \qquad (28.2)$$

Solving for the current gives

$$I = \frac{\mathcal{E}}{R + r} \qquad (28.3)$$

This equation shows that the current in this simple circuit depends on both the resistance R external to the battery and the internal resistance r. If R is much greater than r, we can neglect r.

If we multiply Equation 28.2 by the current I, we get

$$I\mathcal{E} = I^2R + I^2r \qquad (28.4)$$

This equation tells us that, because power $P = IV$ (Eq. 27.22), the total power output of the device emf, $I\mathcal{E}$, is converted to power dissipated as joule heat in the load resistance, I^2R, plus power dissipated in the internal resistance, I^2r. Again, if $r \ll R$, then most of the power delivered by the battery is transferred to the load resistance.

---

[1] The terminal voltage in this case is less than the emf by an amount Ir. In some situations, the terminal voltage may *exceed* the emf by an amount Ir. This happens when the direction of the current is *opposite* that of the emf, as in the case of charging a battery with another source of emf.

EXAMPLE 28.1   Terminal Voltage of a Battery

A battery has an emf of 12.0 V and an internal resistance of 0.05 Ω. Its terminals are connected to a load resistance of 3.00 Ω. (a) Find the current in the circuit and the terminal voltage of the battery.

Solution   Using first Equations 28.3 and then 28.1, we get

$$I = \frac{\mathcal{E}}{R + r} = \frac{12.0 \text{ V}}{3.05 \text{ Ω}} = \boxed{3.93\text{A}}$$

$$V = \mathcal{E} - Ir = 12.0 \text{ V} - (3.93 \text{ A})(0.05 \text{ Ω}) = \boxed{11.8 \text{ V}}$$

As a check of this result, we can calculate the voltage drop across the load resistance R:

$$V = IR = (3.93 \text{ A})(3.00 \text{ Ω}) = 11.8 \text{ V}$$

(b) Calculate the power dissipated in the load resistor, the power dissipated by the internal resistance of the battery, and the power delivered by the battery.

Solution   The power dissipated by the load resistor is

$$P_R = I^2R = (3.93 \text{ A})^2(3.00 \text{ Ω}) = \boxed{46.3 \text{ W}}$$

The power dissipated by the internal resistance is

$$P_r = I^2r = (3.93 \text{ A})^2(0.05 \text{ Ω}) = \boxed{0.772 \text{ W}}$$

Hence, the power delivered by the battery is the sum of these quantities, or 47.1 W. This can be checked using the expression $P = I\mathcal{E}$.

EXAMPLE 28.2   Matching the Load

Show that the maximum power lost in the load resistance R in Figure 28.2a occurs when $R = r$, that is, when the load resistance matches the internal resistance.

Solution   The power dissipated in the load resistance is equal to I^2R, where I is given by Equation 28.3:

$$P = I^2R = \frac{\mathcal{E}^2R}{(R + r)^2}$$

When P is plotted versus R as in Figure 28.3, we find that P reaches a maximum value of $\mathcal{E}^2/4r$ at $R = r$. This can also be proved by differentiating P with respect to R, setting the result equal to zero, and solving for R. The details are left as a problem (Problem 79).

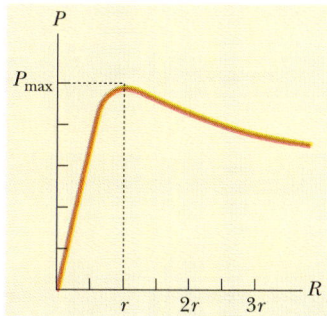

FIGURE 28.3   (Example 28.2) Graph of the power P delivered to a load resistor as a function of R. The power delivered to R is a maximum when the load resistance of the circuit equals the internal resistance of the battery.

28.2   RESISTORS IN SERIES AND IN PARALLEL

When two or more resistors are connected together so that they have only one common point per pair, they are said to be in *series*. Figure 28.4 shows two resistors connected in series. Note that

the current is the same through each resistor because any charge that flows through R_1 must also flow through R_2.

> For a series connection of resistors, the current is the same in each resistor

Since the potential drop from a to b in Figure 28.4b equals IR_1 and the potential drop from b to c equals IR_2, the potential drop from a to c is

$$V = IR_1 + IR_2 = I(R_1 + R_2)$$

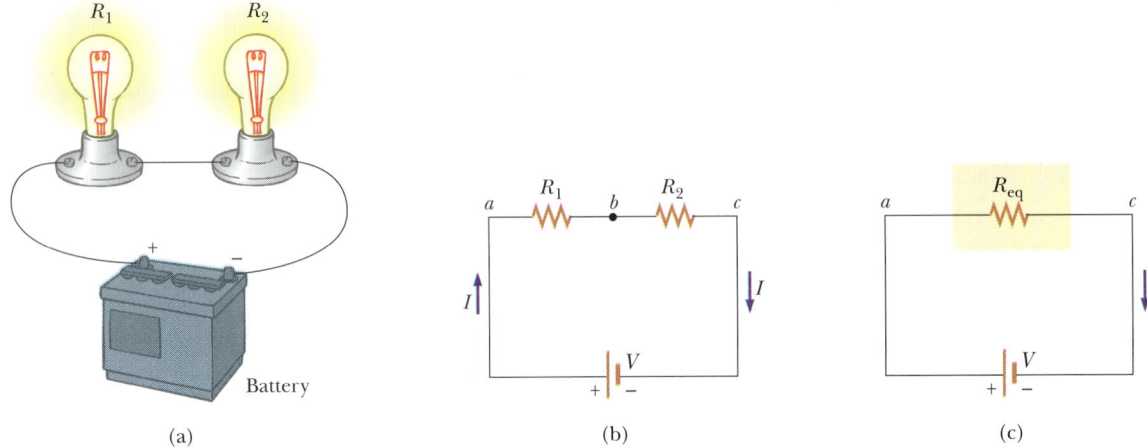

FIGURE 28.4   Series connection of two resistors, R_1 and R_2. The current is the same in each resistor.

Therefore, we can replace the two resistors in series by a single *equivalent resistance* R_{eq} whose value is the *sum* of the individual resistances:

$$R_{eq} = R_1 + R_2 \qquad (28.5)$$

The resistance R_{eq} is equivalent to the series combination $R_1 + R_2$ in the sense that the circuit current is unchanged when R_{eq} replaces $R_1 + R_2$. The equivalent resistance of three or more resistors connected in series is simply

$$R_{eq} = R_1 + R_2 + R_3 + \cdots \qquad (28.6)$$

Therefore, *the equivalent resistance of a series connection of resistors is always greater than any individual resistance.*

Note that if the filament of one lightbulb in Figure 28.4 were to break, or "burn out," the circuit would no longer be complete (an open-circuit condition) and the second bulb would also go out. Some Christmas tree light sets (especially older ones) are connected in this way, and the experience of determining which bulb is burned out was once a familiar one. Frustrating experiences such as this illustrate how inconvenient it would be to have all appliances in a house connected in series. In many circuits, fuses are used in series with other circuit elements for safety purposes. The conductor in the fuse is designed to melt and open the circuit at some maximum current, the value of which depends on the nature of the circuit. If a fuse is not used, excessive currents could damage circuit elements, overheat wires, and perhaps cause a fire. In modern home construction, circuit breakers are used in place of fuses. When the current in a circuit exceeds some value (typically 15 A), the circuit breaker acts as a switch and opens the circuit.

Now consider two resistors connected in *parallel* as shown in Figure 28.5.

In this case, there is an equal potential difference across each resistor.

However, the current in each resistor is in general not the same. When the current I reaches point a, called a *junction*, it splits into two parts, I_1 going through R_1 and

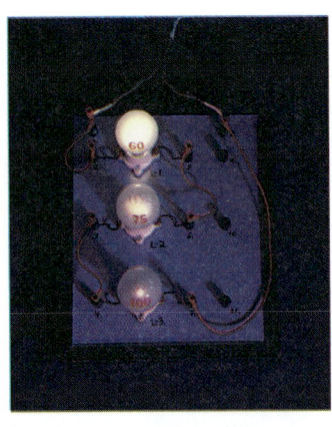

A series connection of three lamps, all rated at 120 V, with power ratings of 60 W, 75 W, and 200 W. Why are the intensities of the lamps different? Which lamp has the greatest resistance? How would their relative intensities differ if they were connected in parallel?   *(Henry Leap and Jim Lehman)*

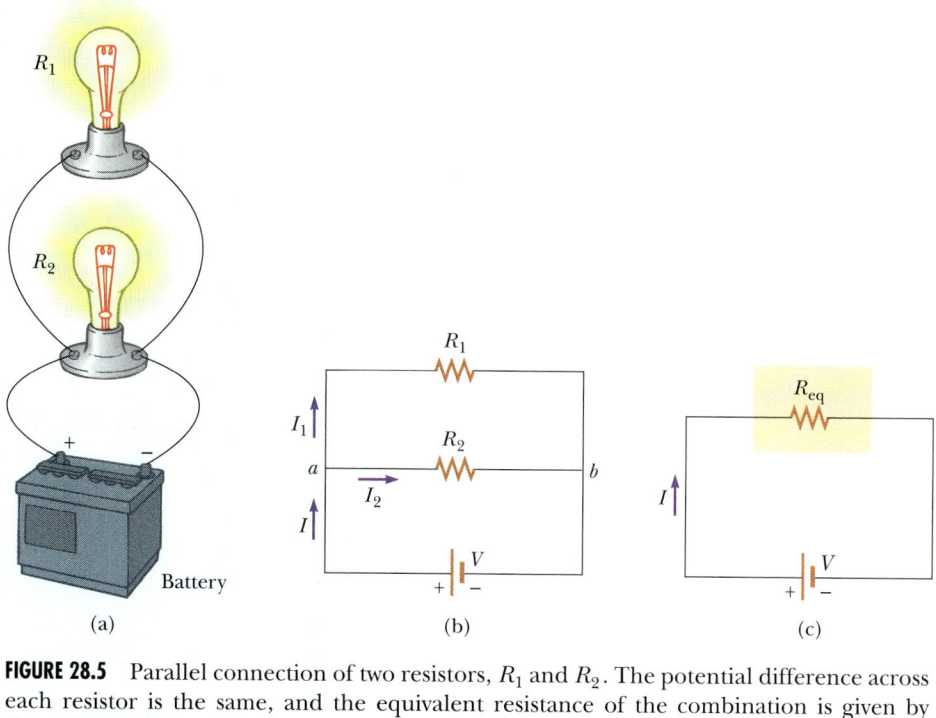

FIGURE 28.5 Parallel connection of two resistors, R_1 and R_2. The potential difference across each resistor is the same, and the equivalent resistance of the combination is given by $R_{eq} = R_1 R_2 / (R_1 + R_2)$.

Three incandescent lamps with power ratings of 25 W, 75 W, and 150 W connected in parallel to a voltage source of about 100 V. All lamps are rated at the same voltage. Why do the intensities of the lamps differ? Which lamp draws the most current? Which has the least resistance? *(Henry Leap and Jim Lehman)*

I_2 going through R_2. (A **junction** is any point in a circuit where a current can split.) If R_1 is greater than R_2, then I_1 will be less than I_2. That is, the moving charge tends to take the path of least resistance. Clearly, since charge must be conserved, the current I that enters point a must equal the total current leaving this point:

$$I = I_1 + I_2$$

Since the potential drop across each resistor must be the same, Ohm's law gives

$$I = I_1 + I_2 = \frac{V}{R_1} + \frac{V}{R_2} = V\left(\frac{1}{R_1} + \frac{1}{R_2}\right) = \frac{V}{R_{eq}}$$

From this result, we see that the equivalent resistance of two resistors in parallel is

$$\frac{1}{R_{eq}} = \frac{1}{R_1} + \frac{1}{R_2} \tag{28.7}$$

$$R_{eq} = \frac{R_1 R_2}{R_1 + R_2}$$

An extension of this analysis to three or more resistors in parallel gives

$$\frac{1}{R_{eq}} = \frac{1}{R_1} + \frac{1}{R_2} + \frac{1}{R_3} + \cdots \tag{28.8}$$

It can be seen from this expression that the equivalent resistance of two or more resistors connected in parallel is always less than the smallest resistance in the group.

This versatile circuit enables the experimenter to examine the properties of circuit elements such as capacitors and resistors and their effect on circuit behavior. *(Courtesy of Central Scientific Company)*

Household circuits are always wired such that the lightbulbs (or appliances, etc.) are connected in parallel. Connected this way, each device operates independently of the others, so that if one is switched off, the others remain on. Equally important, each device operates on the same voltage.

EXAMPLE 28.3 Find the Equivalent Resistance

Four resistors are connected as shown in Figure 28.6a.
(a) Find the equivalent resistance between a and c.

Solution   The circuit can be reduced in steps as shown in Figure 28.6. The 8.0-Ω and 4.0-Ω resistors are in series, and so the equivalent resistance between a and b is 12 Ω (Eq. 28.5). The 6.0-Ω and 3.0-Ω resistors are in parallel, and so from Equation 28.7 we find that the equivalent resistance from b to c is 2.0 Ω. Hence, the equivalent resistance from a to c is 14 Ω.

(b) What is the current in each resistor if a potential difference of 42 V is maintained between a and c?

Solution   The current in the 8.0-Ω and 4.0-Ω resistors is the same because they are in series. Using Ohm's law and the results from part (a), we get

$$I = \frac{V_{ac}}{R_{eq}} = \frac{42\ V}{14\ \Omega} = 3.0\ A$$

When this current enters the junction at b, it splits, part passing through the 6.0-Ω resistor (I_1) and part through the 3.0-Ω resistor (I_2). Since the potential difference across these resistors, V_{bc}, is the same (they are in parallel), we see that $6I_1 = 3I_2$, or $I_2 = 2I_1$. Using this result and the fact that $I_1 + I_2 = 3.0$ A, we find that $I_1 = 1.0$ A and $I_2 = 2.0$ A. We could have guessed this from the start by noting that the current through the 3.0-Ω resistor has to be twice the current through the 6.0-Ω resistor in view of their relative resistances and the fact that the same voltage is applied to each of them.

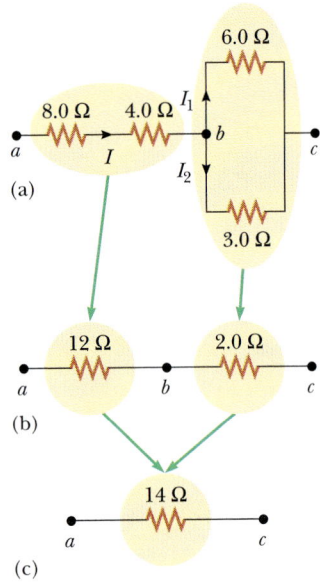

FIGURE 28.6   (Example 28.3) The resistances of the four resistors shown in (a) can be reduced in steps to an equivalent 14-Ω resistor.

As a final check, note that $V_{bc} = 6I_1 = 3I_2 = 6.0$ V and $V_{ab} = 12I = 36$ V; therefore, $V_{ac} = V_{ab} + V_{bc} = 42$ V, as it must.

EXAMPLE 28.4 Three Resistors in Parallel

Three resistors are connected in parallel as in Figure 28.7. A potential difference of 18 V is maintained between points a and b. (a) Find the current in each resistor.

Solution   The resistors are in parallel, and the potential difference across each is 18 V. Applying $V = IR$ to each resistor gives

$$I_1 = \frac{V}{R_1} = \frac{18\ V}{3.0\ \Omega} = 6.0\ A$$

$$I_2 = \frac{V}{R_2} = \frac{18\ V}{6.0\ \Omega} = 3.0\ A$$

$$I_3 = \frac{V}{R_3} = \frac{18\ V}{9.0\ \Omega} = 2.0\ A$$

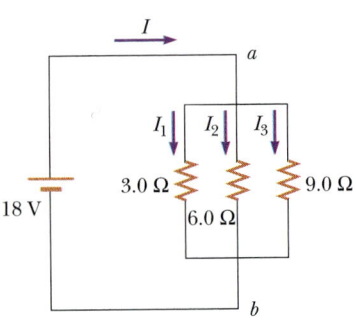

FIGURE 28.7   (Example 28.4) Three resistors connected in parallel. The voltage across each resistor is 18 V.

(b) Calculate the power dissipated by each resistor and the total power dissipated by the three resistors.

Solution   Applying $P = I^2R$ to each resistor gives

3.0-Ω: $P_1 = I_1^2R_1 = (6.0 \text{ A})^2(3.0 \text{ }\Omega) = \boxed{110 \text{ W}}$

6.0-Ω: $P_2 = I_2^2R_2 = (3.0 \text{ A})^2(6.0 \text{ }\Omega) = \boxed{54 \text{ W}}$

9.0-Ω: $P_3 = I_3^2R_3 = (2.0 \text{ A})^2(9.0 \text{ }\Omega) = \boxed{36 \text{ W}}$

This shows that the smallest resistor dissipates the most power since it carries the most current. (Note that you can also use $P = V^2/R$ to find the power dissipated by each resis-

tor.) Summing the three quantities gives a total power of 200 W.

(c) Calculate the equivalent resistance of the three resistors. We can use Equation 28.7 to find R_{eq}:

Solution

$$\frac{1}{R_{eq}} = \frac{1}{3.0} + \frac{1}{6.0} + \frac{1}{9.0}$$

$$R_{eq} = \frac{18}{11} \text{ }\Omega = \boxed{1.6 \text{ }\Omega}$$

Exercise   Use R_{eq} to calculate the total power dissipated in the circuit.

Answer   200 W.

EXAMPLE 28.5   Finding R_{eq} by Symmetry Arguments

Consider the five resistors connected as shown in Figure 28.8a. Find the equivalent resistance between points a and b.

Solution   In this type of problem, it is convenient to assume a current entering junction a and then apply symmetry arguments. Because of the symmetry in the circuit (all 1-Ω resistors in the outside loop), the currents in branches ac and ad

must be equal; hence, the potentials at points c and d must be equal. Since $V_c = V_d$, points c and d may be connected together without affecting the circuit, as in Figure 28.8b. Thus, the 5-Ω resistor may be moved from the circuit, and the circuit may be reduced as shown in Figures 28.8c and 28.8d. From this reduction, we see that the equivalent resistance of the combination is 1 Ω. Note that the result is 1 Ω regardless of what resistor is connected between c and d.

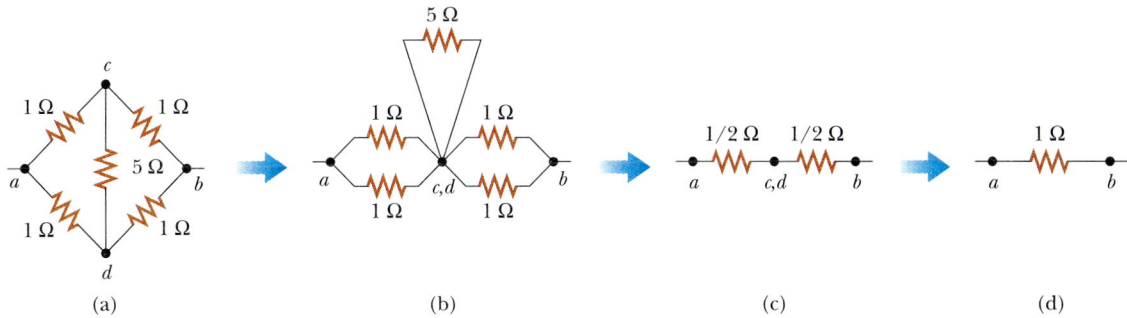

FIGURE 28.8   (Example 28.5)   Because of the symmetry in this circuit, the 5-Ω resistor does not contribute to the resistance between points a and b and can be disregarded.

CONCEPTUAL EXAMPLE 28.6   Operation of a Three-Way Lightbulb

Figure 28.9 illustrates how a three-way lightbulb is constructed to provide three levels of light intensity. The socket of the lamp is equipped with a three-way switch for selecting different light intensities. The bulb contains two filaments. Why are the filaments connected in parallel? Explain how the two filaments are used to provide three different light intensities.

Reasoning   If the filaments were connected in series, and one of them were to burn out, no current could pass through the bulb, and the bulb would give no illumination, regardless of the switch position. However, when the filaments are connected in parallel, and one of them (say the 75-W filament) burns out, the bulb would still operate in one of the switch positions as current passes through the other (100 W) fila-

ment. The three light intensities are made possible by selecting one of three values of filament resistance, using a single
value of 120 V for the applied voltage. The 75-W filament
offers one value of resistance, the 100-W filament offers a
second value, and the third resistance is obtained by combining the two filaments in parallel. When switch 1 is closed and
switch 2 is opened, current passes only through the 75-W
filament. When switch 1 is open and switch 2 is closed,
current passes only through the 100-W filament. When both
switches are closed, current passes through both filaments,
and a total illumination of 175 W is obtained.

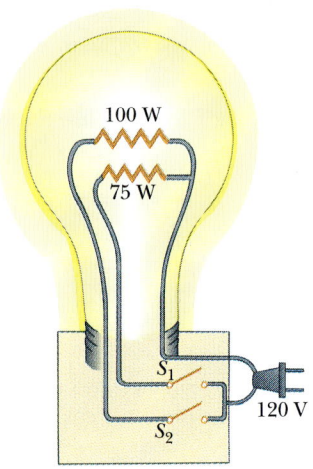

FIGURE 28.9   (Conceptual Example 28.6) Three-way lightbulb.

Georg Simon Ohm (1787–
1854).   *(Courtesy of North Wind
Picture Archives)*

28.3   KIRCHHOFF'S RULES

As we saw in the previous section, simple circuits can be analyzed using Ohm's law
and the rules for series and parallel combinations of resistors. Very often it is not
possible to reduce a circuit to a single loop, however. The procedure for analyzing
more complex circuits is greatly simplified by the use of two simple rules called
Kirchhoff's rules:

> • The sum of the currents entering any junction must equal the sum of the
>   currents leaving that junction.
> • The algebraic sum of the changes in potential across all of the elements
>   around any closed circuit loop must be zero.

The first rule is a statement of conservation of charge. All current that enters a
given point in a circuit must leave that point, because charge cannot build up at a
point. If we apply this rule to the junction shown in Figure 28.10a, we get

$$I_1 = I_2 + I_3$$

Figure 28.10b represents a mechanical analog to this situation, in which water
flows through a branched pipe with no leaks. The flow rate into the pipe equals the
total flow rate out of the two branches.

The second rule follows from conservation of energy. A charge that moves
around any closed loop in a circuit (the charge starts and ends at the same point)
must gain as much energy as it loses if a potential is defined for each point in the
circuit. Its energy may decrease in the form of a potential drop, $-IR$, across a
resistor or as the result of having the charge move in the reverse direction through
an emf. In a practical application of the latter case, electrical energy is converted to
chemical energy when a battery is charged; similarly, electrical energy may be
converted to mechanical energy for operating a motor. However, Kirchhoff's second rule applies only for circuits in which a potential is defined at each point,
which may not be satisfied if there are changing electromagnetic fields.

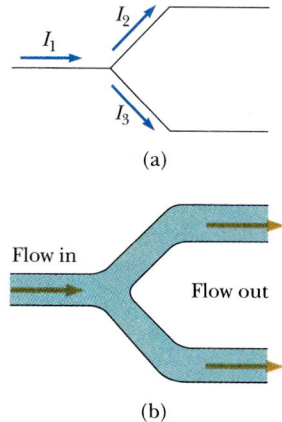

(a)

Flow in

Flow out

(b)

FIGURE 28.10 (a) A schematic diagram illustrating Kirchhoff's junction rule. Conservation of charge requires that all current entering a junction must leave that junction. Therefore, in this case, $I_1 = I_2 + I_3$. (b) A mechanical analog of the junction rule: The flow out must equal the flow in.

(a)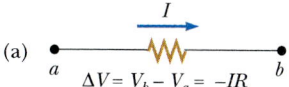

$\Delta V = V_b - V_a = -IR$

(b)

$\Delta V = V_b - V_a = +IR$

(c)

$\mathcal{E}$

$\Delta V = V_b - V_a = +\mathcal{E}$

(d)

$\mathcal{E}$

$\Delta V = V_b - V_a = -\mathcal{E}$

FIGURE 28.11 Rules for determining the potential changes across a resistor and a battery, assuming the battery has no internal resistance.

As an aid in applying the second rule, the following rules should be noted:

- If a resistor is traversed in the direction of the current, the change in potential across the resistor is $-IR$ (Fig. 28.11a).
- If a resistor is traversed in the direction *opposite* the current, the change in potential across the resistor is $+IR$ (Fig. 28.11b).
- If an emf is traversed in the direction of the emf (from $-$ to $+$ on the terminals), the change in potential is $+\mathcal{E}$ (Fig. 28.11c).
- If an emf is traversed in the direction opposite the emf (from $+$ to $-$ on the terminals), the change in potential is $-\mathcal{E}$ (Fig. 28.11d).

There are limitations on the number of times you can use the junction rule and the loop rule. The junction rule can be used as often as needed so long as each time you write an equation, you include in it a current that has not been used in a previous junction rule equation. In general, the number of times the junction rule must be used is one fewer than the number of junction points in the circuit. The loop rule can be used as often as needed so long as a new circuit element (resistor or battery) or a new current appears in each new equation. In general, *the number of independent equations you need equals the number of unknowns in order to solve a particular circuit problem.*

Complex networks with many loops and junctions generate large numbers of independent linear equations and a corresponding large number of unknowns. Such situations can be handled formally using matrix algebra. Computer programs can also be written to solve for the unknowns.

Gustav Robert Kirchhoff (1824–1887) *(Courtesy of North Wind Picture Archives)*

The following examples illustrate the use of Kirchhoff's rules in analyzing circuits. In all cases, it is assumed that the circuits have reached steady-state conditions, that is, the currents in the various branches are constant. If a capacitor is included, as an example, in one of the branches, *it acts as an open circuit,* that is, the current in the branch containing the capacitor is zero under steady-state conditions.

Problem-Solving Strategy and Hints
Kirchhoff's Rules

- Draw the circuit diagram and label all the known and unknown quantities. You must assign a *direction* to the currents in each part of the circuit. Do not be alarmed if you guess the direction of a current incorrectly; your result will be negative, but *its magnitude will be correct.* Although the assignment of current directions is arbitrary, you must adhere *rigorously* to the assigned directions when applying Kirchhoff's rule.
- Apply the junction rule (Kirchhoff's first rule) to any junction in the circuit that provides a relationship between the various currents.
- Apply Kirchhoff's second rule to as many loops in the circuit as are needed to solve for the unknowns. In order to apply this rule, you must correctly identify the change in potential as you cross each element in traversing the closed loop (either clockwise or counterclockwise). Watch out for signs!
- Solve the equations simultaneously for the unknown quantities.

EXAMPLE 28.7   A Single-Loop Circuit

A single-loop circuit contains two external resistors and two batteries as in Figure 28.12. (Neglect the internal resistances of the batteries.) (a) Find the current in the circuit.

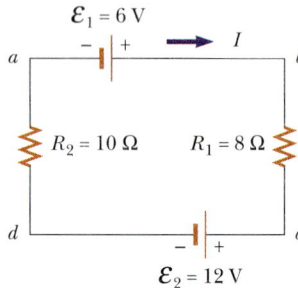

FIGURE 28.12   (Example 28.7) A series circuit containing two batteries and two resistors, where the polarities of the batteries are in opposition to each other.

Reasoning   There are no junctions in this single-loop circuit, and so the current is the same in all elements. Let us assume that the current is in the clockwise direction as shown

in Figure 28.12. Traversing the circuit in this direction, starting at a, we see that $a \rightarrow b$ represents a potential increase of $+ \mathcal{E}_1$, $b \rightarrow c$ represents a potential decrease of $- IR_1$, $c \rightarrow d$ represents a potential decrease of $- \mathcal{E}_2$, and $d \rightarrow a$ represents a potential decrease of $- IR_2$. Applying Kirchhoff's second rule gives

$$\sum_i \Delta V_i = 0$$

$$\mathcal{E}_1 - IR_1 - \mathcal{E}_2 - IR_2 = 0$$

Solution   Solving for I and using the values given in Figure 28.12, we get

$$I = \frac{\mathcal{E}_1 - \mathcal{E}_2}{R_1 + R_2} = \frac{6 \text{ V} - 12 \text{ V}}{8 \ \Omega + 10 \ \Omega} = -\frac{1}{3} \text{ A}$$

The negative sign for I indicates that the direction of the current is opposite the assumed direction.

(b) What is the power lost in each resistor?

Solution

$$P_1 = I^2 R_1 = (\tfrac{1}{3}\,\text{A})^2 (8\,\Omega) = \frac{8}{9}\,\text{W}$$

$$P_2 = I^2 R_2 = (\tfrac{1}{3}\,\text{A})^2 (10\,\Omega) = \frac{10}{9}\,\text{W}$$

Hence, the total power lost is $P_1 + P_2 = 2$ W. Note that the 12-V battery delivers $I\mathcal{E}_2 = 4$ W. Half of this power is delivered to the external resistors. The other half is delivered to the 6-V battery, which is being charged by the 12-V battery. If we had included the internal resistances of the batteries, some of the power would be dissipated as heat in the batteries, so that less power would be delivered to the 6-V battery.

EXAMPLE 28.8   Applying Kirchhoff's Rules

Find the currents I_1, I_2, and I_3 in the circuit shown in Figure 28.13.

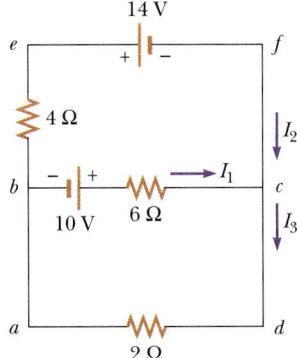

FIGURE 28.13   (Example 28.8) A circuit containing three loops.

Reasoning   We choose the directions of the currents as in Figure 28.13. Applying Kirchhoff's first rule to junction c gives

(1) $I_1 + I_2 = I_3$

There are three loops in the circuit, *abcda, befcb,* and *aefda* (the outer loop). Therefore, we need only two loop equations to determine the unknown currents. The third loop equation would give no new information. Applying Kirchhoff's second rule to loops *abcda* and *befcb* and traversing these loops in the clockwise direction, we obtain the expressions

(2)   Loop *abcda:* $10\,\text{V} - (6\,\Omega)I_1 - (2\,\Omega)I_3 = 0$

(3)   Loop *befcb:* $-14\,\text{V} - 10\,\text{V} + (6\,\Omega)I_1 - (4\,\Omega)I_2 = 0$

Note that in loop *befcb,* a positive sign is obtained when traversing the 6-Ω resistor because the direction of the path is opposite the direction of I_1. A third loop equation for *aefda* gives $-14 = 2I_3 + 4I_2$, which is just the sum of (2) and (3).

Solution   Expressions (1), (2), and (3) represent three independent equations with three unknowns. We can solve the problem as follows: Substituting (1) into (2) gives

$$10 - 6I_1 - 2(I_1 + I_2) = 0$$

(4) $10 = 8I_1 + 2I_2$

Dividing each term in (3) by 2 and rearranging the equation gives

(5) $-12 = -3I_1 + 2I_2$

Subtracting (5) from (4) eliminates I_2, giving

$$22 = 11I_1$$

$$I_1 = 2\,\text{A}$$

Using this value of I_1 in (5) gives a value for I_2:

$$2I_2 = 3I_1 - 12 = 3(2) - 12 = -6$$

$$I_2 = -3\,\text{A}$$

Finally, $I_3 = I_1 + I_2 = -1$ A. Hence, the currents have the values

$$I_1 = \boxed{2\,\text{A}} \qquad I_2 = \boxed{-3\,\text{A}} \qquad I_3 = \boxed{-1\,\text{A}}$$

The fact that I_2 and I_3 are both negative indicates only that we chose the wrong direction for these currents. However, the numerical values are correct.

Exercise   Find the potential difference between points b and c.

Answer $V_b - V_c = 2$ V.

EXAMPLE 28.9   A Multiloop Circuit

(a) Under steady-state conditions, find the unknown currents in the multiloop circuit shown in Figure 28.14.

Reasoning   First note that *the capacitor represents an open circuit, and hence there is no current along path ghab under steady-state*

conditions. Therefore, $I_{gf} = I_1$. Labeling the currents as shown in Figure 28.14 and applying Kirchhoff's first rule to junction c, we get

(1) $I_1 + I_2 = I_3$

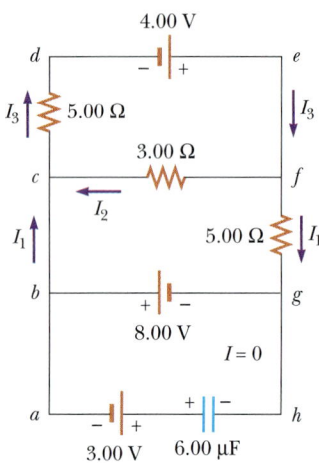

FIGURE 28.14   (Example 28.9) A multiloop circuit. Note that Kirchhoff's loop equation can be applied to *any* closed loop, including one containing the capacitor.

Kirchhoff's second rule applied to loops *defcd* and *cfgbc* gives

(2)   Loop *defcd:* $4.00 \text{ V} - (3.00 \text{ }\Omega)I_2 - (5.00 \text{ }\Omega)I_3 = 0$

(3)   Loop *cfgbc:* $8.00 \text{ V} - (5.00 \text{ }\Omega)I_1 + (3.00 \text{ }\Omega)I_2 = 0$

Solution   From (1) we see that $I_1 = I_3 - I_2$, which when substituted into (3) gives

(4) $8.00 \text{ V} - (5.00 \text{ }\Omega)I_3 + (8.00 \text{ }\Omega)I_2 = 0$

Subtracting (4) from (2), we eliminate I_3 and find

$$I_2 = -\tfrac{4}{11}\text{ A} = \boxed{-0.364 \text{ A}}$$

Since I_2 is negative, we conclude that I_2 is from *c* to *f* through the 3.00-Ω resistor. Using this value of I_2 in (3) and (1) gives the following values for I_1 and I_3:

$$I_1 = \boxed{1.38 \text{ A}} \qquad I_3 = \boxed{1.02 \text{ A}}$$

Under state-steady conditions, the capacitor represents an *open* circuit, and so there is no current in the branch *ghab*.

(b) What is the charge on the capacitor?

Solution   We can apply Kirchhoff's second rule to loop *abgha* (or any other loop that contains the capacitor) to find the potential difference V_c across the capacitor:

$$-8.00 \text{ V} + V_c - 3.00 \text{ V} = 0$$

$$V_c = 11.0 \text{ V}$$

Since $Q = CV_c$, the charge on the capacitor is

$$Q = (6.00 \text{ }\mu\text{F})(11.0 \text{ V}) = \boxed{66.0 \text{ }\mu\text{C}}$$

Why is the left side of the capacitor positively charged?

Exercise   Find the voltage across the capacitor by traversing any other loop.

Answer   11.0 V.

28.4   RC CIRCUITS

So far we have been concerned with circuits with constant currents, or so-called *steady-state circuits.* We now consider circuits containing capacitors, in which the currents may vary in time.

Charging a Capacitor

Consider the series circuit shown in Figure 28.15. Let us assume that the capacitor is initially uncharged. There is no current when the switch S is open (Fig. 28.15b). If the switch is closed at $t = 0$, charges begin to flow, setting up a current in the circuit, and the capacitor begins to charge (Fig. 28.15c). Note that during the charging process, charges do not jump across the plates of the capacitor because the gap between the plates represents an open circuit. Instead, charge is transferred from one plate to the other through the resistor, switch, and battery until the capacitor is fully charged. The value of the maximum charge depends on the voltage of the battery. Once the maximum charge is reached, the current in the circuit is zero.

To put this discussion on a quantitative basis, let us apply Kirchhoff's second rule to the circuit after the switch is closed. Doing so gives

$$\mathcal{E} - IR - \frac{q}{C} = 0 \qquad (28.9)$$

where IR is the potential drop across the resistor and q/C is the potential drop

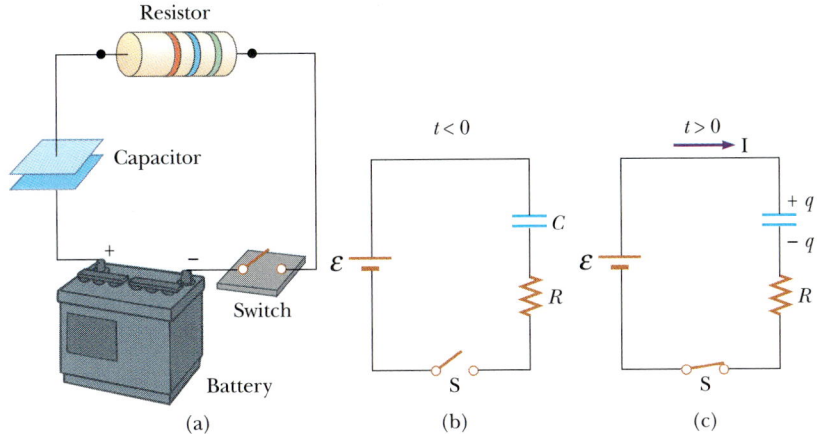

FIGURE 28.15 (a) A capacitor in series with a resistor, battery, and switch. (b) Circuit diagram representing this system before the switch is closed, $t < 0$. (c) Circuit diagram after the switch is closed, $t > 0$.

across the capacitor. Note that q and I are instantaneous values of the charge and current, respectively, as the capacitor is being charged.

We can use Equation 28.9 to find the initial current in the circuit and the maximum charge on the capacitor. At the instant the switch is closed ($t = 0$), the charge on the capacitor is zero, and from Equation 28.9 we find that the initial current in the circuit I_0 is a maximum and equal to

$$I_0 = \frac{\mathcal{E}}{R} \qquad \text{(current at } t = 0\text{)} \qquad (28.10)$$ **Maximum current**

At this time, *the potential drop is entirely across the resistor.* Later, when the capacitor is charged to its maximum value Q, charges cease to flow, the current in the circuit is zero, and *the potential drop is entirely across the capacitor.* Substituting $I = 0$ into Equation 28.9 gives

$$Q = C\mathcal{E} \qquad \text{(maximum charge)} \qquad (28.11)$$ **Maximum charge on the capacitor**

To determine analytical expressions for the time dependence of the charge and current, we must solve Equation 28.9, a single equation containing two variables, q and I. In order to do this, let us substitute $I = dq/dt$ and rearrange the equation:

$$\frac{dq}{dt} = \frac{\mathcal{E}}{R} - \frac{q}{RC}$$

An expression for q may be found in the following manner. Rearrange the equation by placing terms involving q on the left side and those involving t on the right side. Then integrate both sides:

$$\frac{dq}{(q - C\mathcal{E})} = -\frac{1}{RC}\, dt$$

$$\int_0^q \frac{dq}{(q - C\mathcal{E})} = -\frac{1}{RC} \int_0^t dt$$

$$\ln\left(\frac{q - C\mathcal{E}}{-C\mathcal{E}}\right) = -\frac{t}{RC}$$

From the definition of the natural logarithm, we can write this expression as

$$q(t) = C\mathcal{E}[1 - e^{-t/RC}] = Q[1 - e^{-t/RC}] \qquad (28.12)$$

where e is the base of the natural logarithm.

An expression for the charging current may be found by differentiating Equation 28.12 with respect to time. Using $I = dq/dt$, we find

$$I(t) = \frac{\mathcal{E}}{R} e^{-t/RC} \qquad (28.13)$$

where $I_0 = \mathcal{E}/R$ is the initial current in the circuit.

Plots of charge and current versus time are shown in Figure 28.16. Note that the charge is zero at $t = 0$ and approaches the maximum value of $C\mathcal{E}$ as $t \to \infty$ (Fig. 28.16a). Furthermore, the current has its maximum value $I_0 = \mathcal{E}/R$ at $t = 0$ and decays exponentially to zero as $t \to \infty$ (Fig. 28.16b). The quantity RC, which appears in the exponents of Equations 28.12 and 28.13, is called the **time constant**, τ, of the circuit. It represents the time it takes the current to decrease to $1/e$ of its initial value; that is, in a time τ, $I = e^{-1}I_0 = 0.368I_0$. In a time 2τ, $I = e^{-2}I_0 = 0.135I_0$, and so forth. Likewise, in a time τ the charge increases from zero to $C\mathcal{E}[1 - e^{-1}] = 0.632 C\mathcal{E}$.

The following dimensional analysis shows that τ has the unit of time:

$$[\tau] = [RC] = \left[\frac{V}{I} \times \frac{Q}{V} \right] = \left[\frac{Q}{Q/T} \right] = [T]$$

The work done by the battery during the charging process is $Q\mathcal{E} = C\mathcal{E}^2$. After the capacitor is fully charged, the energy stored in the capacitor is $\frac{1}{2}Q\mathcal{E} = \frac{1}{2}C\mathcal{E}^2$, which is just half the work done by the battery. It is left as a problem to show that the remaining half of the energy supplied by the battery goes into joule heat in the resistor (Problem 82).

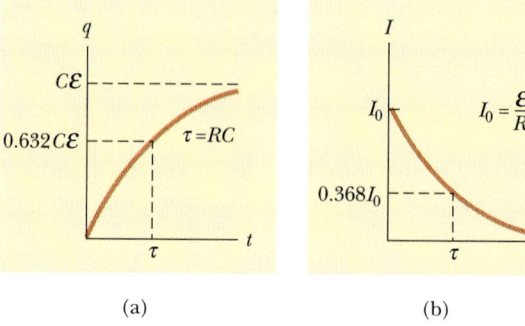

(a)   (b)

FIGURE 28.16   (a) Plot of capacitor charge versus time for the circuit shown in Figure 28.15. After one time constant, τ, the charge is 63.2% of the maximum value, $C\mathcal{E}$. The charge approaches its maximum value as t approaches infinity. (b) Plot of current versus time for the RC circuit shown in Figure 28.15. The current has its maximum value, $I_0 = \mathcal{E}/R$, at $t = 0$ and decays to zero exponentially as t approaches infinity. After one time constant, τ, the current decreases to 36.8% of its initial value.

Discharging a Capacitor

Now consider the circuit in Figure 28.17, consisting of a capacitor with an initial charge Q, a resistor, and a switch. When the switch is open (Fig. 28.17a), there is a potential difference of Q/C across the capacitor and zero potential difference across the resistor since $I = 0$. If the switch is closed at $t = 0$, the capacitor begins to discharge through the resistor. At some time during the discharge, the current in the circuit is I and the charge on the capacitor is q (Fig. 28.17b). From Kirchhoff's second rule, we see that the potential drop across the resistor, IR, must equal the potential difference across the capacitor, q/C:

$$IR = \frac{q}{C} \tag{28.14}$$

However, the current in the circuit must equal the rate of *decrease* of charge on the capacitor. That is, $I = -dq/dt$, and so Equation 28.14 becomes

$$-R\frac{dq}{dt} = \frac{q}{C}$$

$$\frac{dq}{q} = -\frac{1}{RC}dt$$

Integrating this expression using the fact that $q = Q$ at $t = 0$ gives

$$\int_Q^q \frac{dq}{q} = -\frac{1}{RC}\int_0^t dt$$

$$\ln\left(\frac{q}{Q}\right) = -\frac{t}{RC}$$

$$q(t) = Qe^{-t/RC} \tag{28.15}$$

Differentiating Equation 28.15 with respect to time gives the current as a function of time:

$$I(t) = -\frac{dq}{dt} = I_0 e^{-t/RC} \tag{28.16}$$

where the initial current $I_0 = Q/RC$. Therefore, we see that both the charge on the capacitor and the current decay exponentially at a rate characterized by the time constant $\tau = RC$.

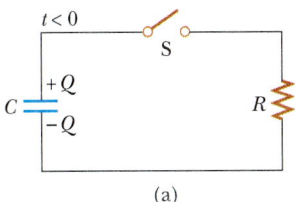

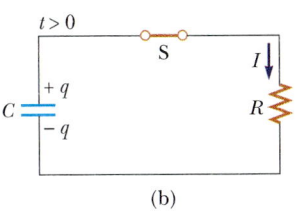

FIGURE 28.17   (a) A charged capacitor connected to a resistor and a switch, which is open at $t < 0$. (b) After the switch is closed, a nonsteady current is set up in the direction shown and the charge on the capacitor decreases exponentially with time.

Charge versus time for a discharging capacitor

Current versus time for a discharging capacitor

CONCEPTUAL EXAMPLE 28.10   Intermittent Windshield Wipers

Many automobiles are equipped with windshield wipers that can be used intermittently during a light rainfall. How does the operation of this feature depend on the charging and discharging of a capacitor?

Reasoning   The wipers are part of an RC circuit whose time constant can be varied by selecting different values of R through a multi-positioned switch. The brief time that the wipers remain on, and the time they are off, is determined by the value of the time constant of the circuit.

EXAMPLE 28.11   Charging a Capacitor in an *RC* Circuit

An uncharged capacitor and a resistor are connected in series to a battery as in Figure 28.18. If $\mathcal{E} = 12.0$ V, $C = 5.00$ μF, and $R = 8.00 \times 10^5$ Ω, find the time constant of the circuit, the maximum charge on the capacitor, the maximum current in the circuit, and the charge and current as a function of time.

Exercise   Calculate the charge on the capacitor and the current in the circuit after one time constant has elapsed.

Answer   37.9 μC, 5.52 μA.

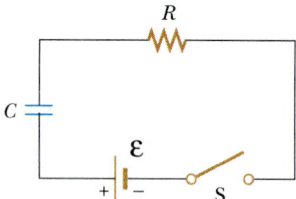

FIGURE 28.18   (Example 28.11) The switch of this series *RC* circuit is closed at $t = 0$.

Solution   The time constant of the circuit is $\tau = RC = (8.00 \times 10^5$ Ω$)(5.00 \times 10^{-6}$ F$) = 4.00$ s. The maximum charge on the capacitor is $Q = C\mathcal{E} = (5.00 \times 10^{-6}$ F$)(12.0$ V$) = 60.0$ μC. The maximum current in the circuit is $I_0 = \mathcal{E}/R = (12.0$ V$)/(8.00 \times 10^5$ Ω$) = 15.0$ μA. Using these values and Equations 28.12 and 28.13, we find that

$$q(t) = \boxed{60.0[1 - e^{-t/4}]\,\mu C}$$

$$I(t) = \boxed{15.0e^{-t/4}\,\mu A}$$

Graphs of these functions are given in Figure 28.19.

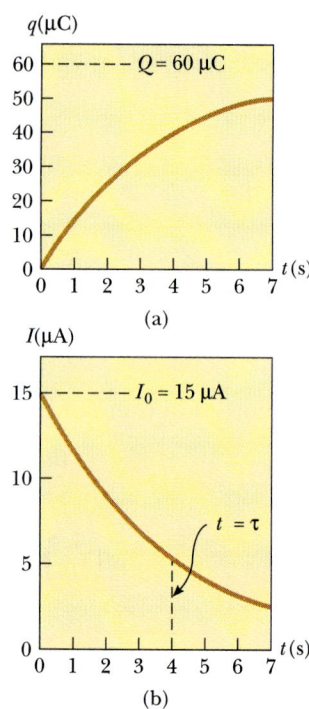

FIGURE 28.19   (Example 28.11) Plots of (a) charge versus time and (b) current versus time for the *RC* circuit shown in Figure 28.18, with $\mathcal{E} = 12.0$ V, $R = 8.00 \times 10^5$ Ω, and $C = 5.00$ μF.

EXAMPLE 28.12   Discharging a Capacitor in an *RC* Circuit

Consider a capacitor *C* being discharged through a resistor *R* as in Figure 28.17. (a) After how many time constants is the charge on the capacitor one fourth of its initial value?

Solution   The charge on the capacitor varies with time according to Equation 28.15, $q(t) = Qe^{-t/RC}$. To find the time it takes the charge q to drop to one fourth of its initial value, we substitute $q(t) = Q/4$ into this expression and solve for t:

$$\tfrac{1}{4}Q = Qe^{-t/RC}$$

$$\tfrac{1}{4} = e^{-t/RC}$$

Taking logarithms of both sides, we find

$$-\ln 4 = -\frac{t}{RC}$$

$$t = RC\ln 4 = \boxed{1.39\,RC}$$

(b) The energy stored in the capacitor decreases with time as it discharges. After how many time constants is this stored energy one fourth of its initial value?

Solution   Using Equations 26.10 and 28.15, we can express the energy stored in the capacitor at any time t as

$$U = \frac{q^2}{2C} = \frac{Q^2}{2C}\, e^{-2t/RC} = U_0 e^{-2t/RC}$$

where U_0 is the initial energy stored in the capacitor. As in part (a), we now set $U = U_0/4$ and solve for t:

$$\tfrac{1}{4} U_0 = U_0 e^{-2t/RC}$$

$$\tfrac{1}{4} = e^{-2t/RC}$$

Again, taking logarithms of both sides and solving for t gives

$$t = \tfrac{1}{2} RC \ln 4 = \boxed{0.693 RC}$$

Exercise   After how many time constants is the current in the RC circuit one half of its initial value?

Answer $0.693 RC$.

EXAMPLE 28.13   Energy Loss in a Resistor

A 5.00-μF capacitor is charged to a potential difference of 800 V and is then discharged through a 25.0-kΩ resistor. How much energy is lost as joule heating in the time it takes to fully discharge the capacitor?

Solution   We shall solve this problem in two ways. The first method is to note that the initial energy in the system equals the energy stored in the capacitor, $C\mathcal{E}^2/2$. Once the capacitor is fully discharged, the energy stored in it is zero. Since energy is conserved, the initial energy stored in the capacitor is transformed to thermal energy dissipated in the resistor. Using the given values of C and $\mathcal{E}$, we find

$$\text{Energy} = \tfrac{1}{2} C\mathcal{E}^2 = \tfrac{1}{2}(5.00 \times 10^{-6}\ \text{F})(800\ \text{V})^2 = \boxed{1.60\ \text{J}}$$

The second method, which is more difficult but perhaps more instructive, is to note that as the capacitor discharges through the resistor, the rate at which heat is generated in the resistor (or the power loss) is given by RI^2, where I is the instantaneous current given by Equation 28.16. Since power is defined as the rate of change of energy, we conclude that

the energy lost in the resistor in the form of heat must equal the time integral of $RI^2\, dt$:

$$\text{Energy} = \int_0^{\infty} RI^2\, dt = \int_0^{\infty} R(I_0 e^{-t/RC})^2\, dt$$

To evaluate this integral, we note that the initial current $I_0 = \mathcal{E}/R$ and all parameters are constants except for t. Thus, we find

$$\text{Energy} = \frac{\mathcal{E}^2}{R} \int_0^{\infty} e^{-2t/RC}\, dt$$

This integral has a value of $RC/2$, and so we find

$$\text{Energy} = \tfrac{1}{2} C\mathcal{E}^2$$

which agrees with the simpler approach, as it must. Note that this second approach can be used to find the energy lost as heat at *any* time after the switch is closed by simply replacing the upper limit in the integral by that specific value of t.

Exercise   Show that the integral given in this example has the value of $RC/2$.

---

*28.5   ELECTRICAL INSTRUMENTS

The Ammeter

Current is one of the most important quantities that one would like to measure in an electric circuit. A device that measures current is called an **ammeter**. The current to be measured must pass directly through the ammeter, because the ammeter must be connected in the current, as in Figure 28.20. The wires usually must be cut to make connections to the ammeter. When using an ammeter to measure direct currents, you must be sure to connect it so that current enters the positive terminal of the instrument and exits at the negative terminal. **Ideally, an ammeter should have zero resistance so as not to alter the current being measured.** In the circuit shown in Figure 28.20, this condition requires that the ammeter's resistance be small compared with $R_1 + R_2$. Since any ammeter always has some

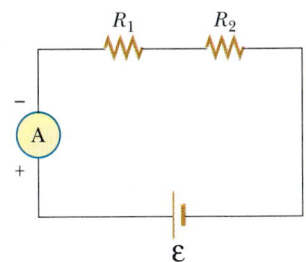

FIGURE 28.20   The current in a circuit can be measured with an ammeter connected in series with the resistor and battery. An ideal ammeter has zero resistance.

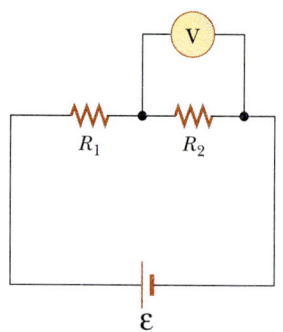

FIGURE 28.21 The potential difference across a resistor can be measured with a voltmeter connected in parallel with the resistor. An ideal voltmeter has infinite resistance and does not affect the circuit.

resistance, its presence in the circuit slightly reduces the current from its value when the ammeter is not present.

The Voltmeter

A device that measures potential differences is called a **voltmeter.** The potential difference between any two points in the circuit can be measured by simply attaching the terminals of the voltmeter between these points without breaking the circuit, as in Figure 28.21. The potential difference across resistor R_2 is measured by connecting the voltmeter in parallel with R_2. Again, it is necessary to observe the polarity of the instrument. The positive terminal of the voltmeter must be connected to the end of the resistor at the higher potential, and the negative terminal to the low-potential end of the resistor. **An ideal voltmeter has infinite resistance so that no current passes through it.** In Figure 28.21, this condition requires that the voltmeter have a resistance that is very large relative to R_2. In practice, if this condition is not met, corrections should be made for the known resistance of the voltmeter.

The Galvanometer

The **galvanometer** is the main component used in the construction of ammeters and voltmeters. The essential features of a common type, called the *D'Arsonval galvanometer,* are shown in Figure 28.22. It consists of a coil of wire mounted so that it is free to rotate on a pivot in a magnetic field provided by a permanent magnet. The basic operation of the galvanometer makes use of the fact that a torque acts on a current loop in the presence of a magnetic field. (The reason for this is discussed in detail in Chapter 29.) The torque experienced by the coil is proportional to the current through it. This means that the larger the current, the larger the torque and the more the coil rotates before the spring tightens enough to stop the rotation. Hence, the amount of deflection is proportional to the current. Once the instrument is properly calibrated, it can be used in conjunction with other circuit elements to measure either currents or potential differences.

A typical off-the-shelf galvanometer is often not suitable for use as an ammeter, mainly because a typical galvanometer has a resistance of about 60 Ω. An ammeter resistance this large considerably alters the current in the circuit in which it is placed. This can be understood by considering the following example. Suppose you construct a simple series circuit containing a 3-V battery and a 3-Ω resistor. The current in such a circuit is 1 A. However, if you insert a 60-Ω galvanometer in the circuit to measure the current, the total resistance of the circuit is 63 Ω and the current is reduced to 0.048 A.

A second factor that limits the use of a galvanometer as an ammeter is the fact that a typical galvanometer gives a full-scale deflection for very low currents, of the order of 1 mA or less. Consequently, such a galvanometer cannot be used directly to measure currents greater than this. However, a galvanometer can be converted to an ammeter by placing a resistor R_p in parallel with the galvanometer as in Figure 28.23a. The value of R_p, sometimes called the *shunt resistor,* must be very small relative to the resistance of the galvanometer so that most of the current to be measured passes through the shunt resistor.

A galvanometer can also be used as a voltmeter by adding an external resistor R_s in series with it, as in Figure 28.23b. In this case, the external resistor must have a value that is very large relative to the resistance of the galvanometer. This ensures that the galvanometer does not significantly alter the voltage to be measured.

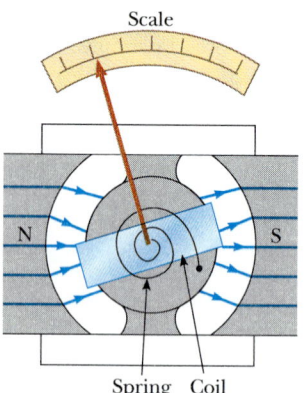

FIGURE 28.22 The principal components of a D'Arsonval galvanometer. When current passes through the coil, situated in a magnetic field, the magnetic torque causes the coil to twist. The angle through which the coil rotates is proportional to the current through it because of the spring's torque.

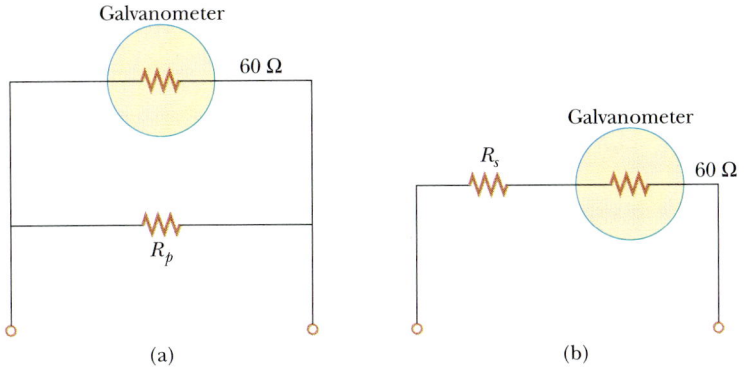

(a)   (b)

FIGURE 28.23 (a) When a galvanometer is to be used as an ammeter, a resistor, R_p, is connected in parallel with the galvanometer. (b) When the galvanometer is used as a voltmeter, a resistor, R_s, is connected in series with the galvanometer.

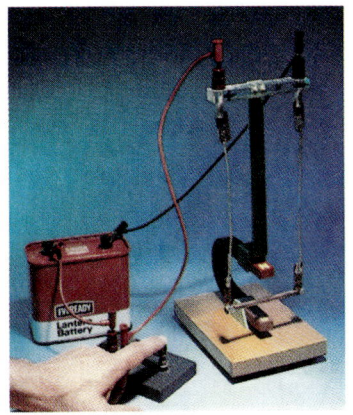

Large-scale model of a galvanometer movement. Why does the coil rotate about the vertical axis after the switch is closed? *(Henry Leap and Jim Lehman)*

The Wheatstone Bridge

Unknown resistances can be accurately measured using a circuit known as a **Wheatstone bridge** (Fig. 28.24). This circuit consists of the unknown resistance, R_x, three known resistors, R_1, R_2, and R_3 (where R_1 is a calibrated variable resistor), a galvanometer, and a battery. The known resistor R_1 is varied until the galvanometer reading is zero, that is, until there is no current from a to b. Under this condition the bridge is said to be balanced. Since the potential at point a must equal the potential at point b when the bridge is balanced, the potential difference across R_1 must equal the potential difference across R_2. Likewise, the potential difference across R_3 must equal the potential difference across R_x. From these considerations, we see that

$$(1) \qquad I_1 R_1 = I_2 R_2$$

$$(2) \qquad I_1 R_3 = I_2 R_x$$

Dividing (1) by (2) eliminates the currents, and solving for R_x we find

$$R_x = \frac{R_2 R_3}{R_1} \qquad (28.17)$$

Since R_1, R_2, and R_3 are known quantities, R_x can be calculated. There are a number of similar devices that use the null measurement (when the galvanometer reads zero), such as a capacitance bridge used to measure unknown capacitances. These devices do not require the use of calibrated meters and can be used with any voltage source.

When very high resistances are to be measured (above $10^5 \, \Omega$), the Wheatstone bridge method becomes difficult for technical reasons. As a result of recent advances in the technology of such solid state devices as the field-effect transistor, modern electronic instruments can measure resistances as high as $10^{12} \, \Omega$. Such instruments have an extremely high resistance between their input terminals. For example, input resistances of $10^{10} \, \Omega$ are common in most digital multimeters. (A multimeter is a device that is used to measure voltage, current, and resistance.)

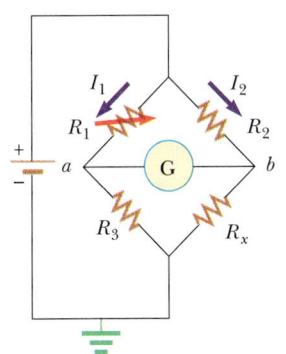

FIGURE 28.24 Circuit diagram for a Wheatstone bridge. This circuit is often used to measure an unknown resistance R_x in terms of known resistances R_1, R_2, and R_3. When the bridge is balanced, there is no current in the galvanometer.

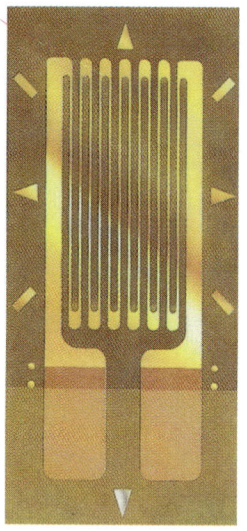

The strain gauge, a device used for experimental stress analysis, consists of a thin coiled wire bonded to a flexible plastic backing. Stresses are measured by detecting changes in resistance of the coil as the strip bends. Resistance measurements are made with the gauge as one element of a Wheatstone bridge. These devices are commonly used in modern electronic balances to measure the mass of an object.

The Potentiometer

A **potentiometer** is a circuit used to measure an unknown emf $\mathcal{E}_x$ by comparison with a known emf. Figure 28.25 shows the essential components of the potentiometer. Point d represents a sliding contact used to vary the resistance (and hence the potential difference) between points a and d. In a common version of the potentiometer, called a **slide-wire potentiometer,** the variable resistor is a wire with the contact point d at some position on the wire. The other required components in this circuit are a galvanometer, a battery with emf $\mathcal{E}$, and the unknown emf $\mathcal{E}_x$.

With the currents in the directions shown in Figure 28.25, we see from Kirchhoff's first rule that the current through the resistor R_x is $I - I_x$, where I is the current in the left branch (through the battery of emf $\mathcal{E}$) and I_x is the current in the right branch. Kirchhoff's second rule applied to loop *abcda* gives

$$- \mathcal{E}_x + (I - I_x) R_x = 0$$

where R_x is the resistance between points a and d. The sliding contact at d is now adjusted until the galvanometer reads zero (a balanced circuit). Under this condition, the current in the galvanometer and in the unknown cell is zero, and the potential difference between a and d equals the unknown emf $\mathcal{E}_x$. That is,

$$\mathcal{E}_x = IR_x$$

Next, the battery of unknown emf is replaced by a standard battery of known emf $\mathcal{E}_s$ and the above procedure is repeated. That is, the moving contact at d is varied until a balance is obtained. If R_s is the resistance between a and d when balance is achieved, then

$$\mathcal{E}_s = IR_s$$

where it is assumed that I remains the same.

Combining this expression with the previous equation, we see that

$$\mathcal{E}_x = \frac{R_x}{R_s} \mathcal{E}_s \tag{28.18}$$

This result shows that the unknown emf can be determined from a knowledge of the standard-battery emf and the ratio of the two resistances.

Voltages, currents, and resistances are frequently measured by digital multimeters like the one shown in this photograph. *(Henry Leap and Jim Lehman)*

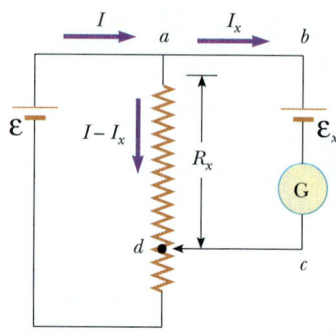

FIGURE 28.25   Circuit diagram for a potentiometer. The circuit is used to measure an unknown emf $\mathcal{E}_x$.

If the resistor is a wire of resistivity ρ, its resistance can be varied using sliding contacts to vary the length of the circuit. With the substitutions $R_s = \rho L_s / A$ and $R_x = \rho L_x / A$, Equation 28.18 reduces to

$$\mathcal{E}_x = \frac{L_x}{L_s} \mathcal{E}_s \qquad (28.19)$$

According to this result, the unknown emf can be obtained from a measurement of the two wire lengths and the magnitude of the standard emf.

*28.6   HOUSEHOLD WIRING AND ELECTRICAL SAFETY

Household circuits represent a practical application of some of the ideas we have presented in this chapter. In our world of electrical appliances, it is useful to understand the power requirements and limitations of conventional electrical systems and the safety measures that should be practiced to prevent accidents.

In a conventional installation, the utilities company distributes electrical power to individual homes with a pair of power lines. Each user is connected in parallel to these lines, as in Figure 28.26. The potential difference between these wires is about 120 V. The voltage alternates in time, with one of the wires connected to ground, and the potential of the other, "live," wire oscillates relative to ground.[2] For the present discussion, we assume a constant voltage (direct current). (Alternating voltages and currents are discussed in Chapter 33.)

A meter and circuit breaker (or in older installations, a fuse) are connected in series with one of the wires entering the house. The wire and circuit breaker are carefully selected to meet the current demands for that circuit. If a circuit is to carry currents as large as 30 A, a heavy wire and appropriate circuit breaker must be selected to handle this current. Other individual household circuits normally used to power lamps and small appliances often require only 15 A. Each circuit has its own circuit breaker to accomodate various load conditions.

As an example, consider a circuit in which a toaster, a microwave oven, and a heater are in the same circuit (corresponding to R_1, R_2, . . . in Figure 28.26). We can calculate the current drawn by each appliance using the expression $P = IV$. The toaster, rated at 1000 W, draws a current of $1000/120 = 8.33$ A. The microwave oven, rated at 800 W, draws 6.67 A, and the electric heater, rated at 1300 W, draws 10.8 A. If the three appliances are operated simultaneously, they draw a total current of 25.8 A. Therefore, the circuit should be wired to handle at least this much current. In order to accomodate a small additional load, such as a 100-W lamp, a 30-A circuit should be installed. Alternatively, the toaster and microwave oven could be operated on one 20-A circuit and the heater on a separate 20-A circuit.

Many heavy-duty appliances, such as electric ranges and clothes dryers, require 240 V for their operation. The power company supplies this voltage by providing a third wire that is 120 V below ground potential (Fig. 28.27). The potential difference between this wire and the other live wire (which is 120 V above ground potential) is 240 V. An appliance that operates from a 240-V line requires half the current of one operating from a 120-V line; therefore, smaller wires can be used in the higher-voltage circuit without overheating becoming a problem.

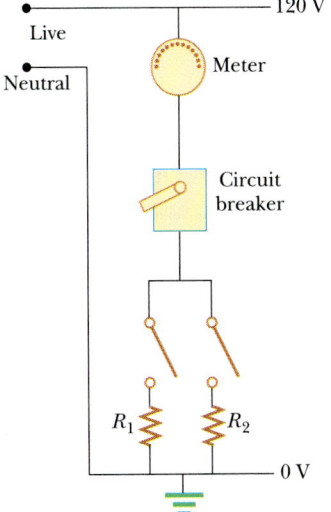

FIGURE 28.26   Wiring diagram for a household circuit. The resistances R_1 and R_2 represent appliances or other electric devices that operate with an applied voltage of 120 V.

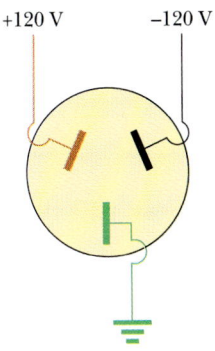

FIGURE 28.27   Power connections for a 240-V appliance.

---

[2] The phrase *live wire* is common jargon for a conductor whose potential is above or below ground.

Electrical Safety

When the live wire of an electrical outlet is connected directly to ground, the circuit is completed and a short-circuit condition exists. When this happens accidentally, a properly operating circuit breaker opens the circuit. On the other hand, a person in contact with ground can be electrocuted by touching the live wire of a frayed cord or other exposed conductors. An exceptionally good ground contact is made either by the person's touching a water pipe (normally at ground potential) or by standing on ground with wet feet, since water is a conductor. Such situations should be avoided at all costs.

Electrical shock can result in fatal burns, or it can cause the muscles of vital organs, such as the heart, to malfunction. The degree of damage to the body depends on the magnitude of the current, the length of time it acts, the location of the contact, and the part of the body through which the current passes. Currents of 5 mA or less can cause a sensation of shock, but ordinarily do little or no damage. If the current is larger than about 10 mA, the muscles contract and the person may be unable to release the live wire. If a current of about 100 mA passes through the body for only a few seconds, the result can be fatal. Such large currents paralyze the respiratory muscles and prevent breathing. In some cases, currents of about 1 A through the body can produce serious (and sometimes fatal) burns. In practice, no contact with live wires (voltages above 24 V) is regarded as safe.

Many 120-V outlets are designed to accept a three-pronged power cord. (This feature is required in all new electrical installations.) One of these prongs is the live wire, a second called the "neutral," carries current to ground, and the third, round prong is a safety ground wire that carries no current. The additional grounded wire is an important safety feature with the safety ground wire connected directly to the casing of the appliance. If the live wire is accidentally shorted to the casing (which often occurs when the wire insulation wears off), the current takes the low-resistance path through the appliance to ground. In contrast, if the casing of the appliance is not properly grounded and a short occurs, anyone in contact with the appliance experiences an electric shock because the body provides a low-resistance path to ground.

Special power outlets called ground-fault interrupters (GFIs) are now being used in kitchens, bathrooms, basements, exterior outlets, and other hazardous areas of new homes. These devices are designed to protect persons from electrical shock by sensing small currents (≈ 5 mA) leaking to ground. (The principle of their operation is described in Chapter 31.) When an excessive leakage current is detected, the current is shut off in less than 1 ms.

SUMMARY

The **emf** of a battery is equal to the voltage across its terminals when the current is zero. That is, the emf is equivalent to the open-circuit voltage of the battery.

The **equivalent resistance** of a set of resistors connected in **series** is

$$R_{eq} = R_1 + R_2 + R_3 + \cdots \tag{28.6}$$

The **equivalent resistance** of a set of resistors connected in **parallel** is

$$\frac{1}{R_{eq}} = \frac{1}{R_1} + \frac{1}{R_2} + \frac{1}{R_3} + \cdots \tag{28.8}$$

Complex circuits involving more than one loop are conveniently analyzed using **Kirchhoff's rules:**

- The sum of the currents entering any junction must equal the sum of the currents leaving that junction.
- The sum of the potential differences across each element around any closed-circuit loop must be zero.

The first rule is a statement of **conservation of charge.** The second rule is equivalent to a statement of **conservation of energy.**

When a resistor is traversed in the direction of the current, the change in potential, ΔV, across the resistor is $- IR$. If a resistor is traversed in the direction opposite the current, $\Delta V = + IR$. If a source of emf is traversed in the direction of the emf (negative to positive) the change in potential is $+ \mathcal{E}$. If it is traversed opposite the emf (positive to negative), the change in potential is $- \mathcal{E}$.

If a capacitor is charged with a battery through a resistance R, the charge on the capacitor and the current in the circuit vary in time according to the expressions

$$q(t) = Q[1 - e^{-t/RC}] \qquad (28.12)$$

$$I(t) = \frac{\mathcal{E}}{R} e^{-t/RC} \qquad (28.13)$$

where $Q = C\mathcal{E}$ is the maximum charge on the capacitor. The product RC is called the **time constant** of the circuit.

If a charged capacitor is discharged through a resistance R, the charge and current decrease exponentially in time according to the expressions

$$q(t) = Qe^{-t/RC} \qquad (28.15)$$

$$I(t) = I_0 e^{-t/RC} \qquad (28.16)$$

where $I_0 = Q/RC$ is the initial current and Q is the initial charge on the capacitor.

QUESTIONS

1. Explain the difference between load resistance and internal resistance for a battery.
2. Under what condition does the potential difference across the terminals of a battery equal its emf? Can the terminal voltage ever exceed the emf? Explain.
3. Is the direction of current through a battery always from negative to positive on the terminals? Explain.
4. Two sets of Christmas-tree lights are available. For set A, when one bulb is removed (or burns out), the remaining bulbs remain illuminated. For set B, when one bulb is removed, the remaining bulbs do not operate. Explain the difference in wiring for the two sets.
5. How would you connect resistors so that the equivalent resistance is larger than the individual resistances? Give an example involving two or three resistors.
6. How would you connect resistors so that the equivalent resistance is smaller than the individual resistances? Give an example involving two or three resistors.
7. Given three lightbulbs and a battery, sketch as many different electric circuits as you can.
8. When resistors are connected in series, which of the following would be the same for each resistor: potential difference, current, power?
9. When resistors are connected in parallel, which of the following would be the same for each resistor: potential difference, current, power?
10. What advantage might there be in using two identical resistors in parallel connected in series with another identical parallel pair, rather than just using a single resistor?
11. Are the two headlights on a car wired in series or in parallel? How can you tell?
12. An incandescent lamp connected to a 120-V source with a short extension cord provides more illumination than the same lamp connected to the same source with a very long extension cord. Explain.
13. Embodied in Kirchhoff's rules are two conservation laws, what are they?
14. When can the potential difference across a resistor be positive?
15. In Figure 28.14, suppose the wire between points g and

h is replaced by a 10-Ω resistor. Explain why this change does not affect the currents calculated in Example 28.9.

16. In Figure 28.28, describe what happens to the lightbulb after the switch is closed. Assume the capacitor has a large capacitance and is initially uncharged, and assume that the light illuminates when connected directly across the battery terminals.

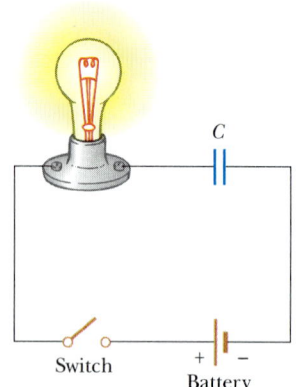

FIGURE 28.28   (Question 16).

17. What is the internal resistance of an ideal ammeter and voltmeter? Do real meters attain this ideal case?
18. Although the internal resistance of the unknown and known emfs was neglected in the treatment of the potentiometer (Section 28.5), it is really not necessary to make this assumption. Explain why the internal resistances play no role in this measurement.
19. Why is it dangerous to turn on a light when you are in the bathtub?
20. Why is it possible for a bird to sit on a high-voltage wire without being electrocuted?
21. Suppose you fall from a building and on the way down grab a high-voltage wire. Assuming that the wire holds you, will you be electrocuted? If the wire then breaks,

should you continue to hold onto an end of the wire as you fall?
22. Would a fuse work successfully if it were placed in parallel with the device it is supposed to protect?
23. What advantage does 120-V operation offer over 240 V? What disadvantages?
24. When electricians work with potentially live wires, they often use the backs of their hands or fingers to move wires. Why do you suppose they use this technique?
25. What procedure would you use to try to save a person who is "frozen" to a live high-voltage wire without endangering your own life?
26. If it is the current flowing through the body that determines how serious a shock will be, why do we see warnings of high voltage rather than high current near electric equipment?
27. Suppose you are flying a kite when it strikes a high-voltage wire. What factors determine how great a shock you receive?
28. A series circuit consists of three identical lamps connected to a battery as in Figure 28.29. When the switch S is closed, what happens (a) to the intensities of lamps A and B; (b) to the intensity of lamp C; (c) to the current in the circuit; and (d) to the voltage drop across the three lamps? (e) Does the power dissipated in the circuit increase, decrease, or remain the same?

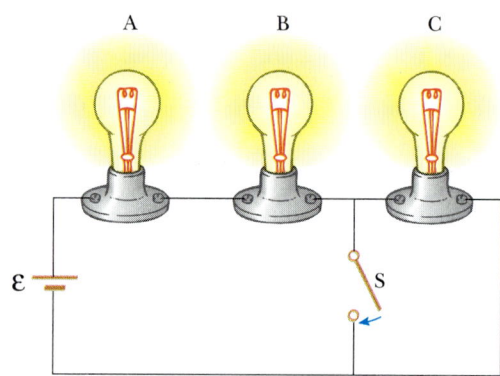

FIGURE 28.29   (Question 28).

PROBLEMS

Review Problem

In the circuit shown, the values of *I* and *R* are known, while the emf and internal resistance of the battery are unknown. When the switch S is closed, the ammeter reads $20I$ before burning out. When the switch is open, find (a) the total exter-

nal resistance, (b) the potential difference between *a* and *b*, (c) the emf of the battery, (d) the internal resistance *r* of the battery, (e) the current in the resistor marked *, and (f) the power dissipated by the resistor marked *. If points 1 and 2 are connected by a wire, find (g) the total resistance of the external circuit, (h) the current in the battery, and (i) the power dissipated in the circuit. (j) In the original circuit, suppose an uncharged capacitor C is inserted between b and d as indicated by the dashed circuit symbol. How long does it take the capacitor to acquire a charge $q = CIR$?

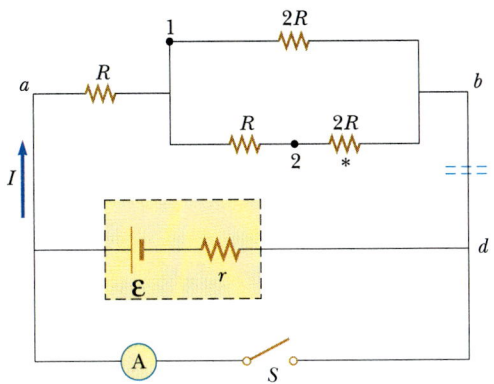

Section 28.1 Electromotive Force

1. A battery with an emf of 12 V and internal resistance of 0.90 Ω is connected across a load resistor R. (a) If the current in the circuit is 1.4 A, what is the value of R? (b) What power is dissipated in the internal resistance of the battery?

2. A 9.00-V battery delivers 117 mA when connected to a 72.0-Ω load. Determine the internal resistance of the battery.

3. (a) What is the current in a 5.6-Ω resistor connected to a battery that has a 0.2-Ω internal resistance if the terminal voltage of the battery is 10 V? (b) What is the emf of the battery?

4. If the emf of a battery is 15 V and a current of 60 A is measured when the battery is shorted, what is the internal resistance of the battery?

5. The current in a loop circuit that has a resistance of R_1 is 2 A. The current is reduced to 1.6 A when an additional resistor $R_2 = 3$ Ω is added in series with R_1. What is the value of R_1?

6. A typical fresh AA dry cell has an emf of 1.50 V and an internal resistance of 0.311 Ω. (a) Find the terminal voltage of the battery when it supplies 58 mA to a circuit. (b) What is the resistance R of the external circuit?

7. A battery has an emf of 15.0 V. The terminal voltage of the battery is 11.6 V when it is delivering 20.0 W of power to an external load resistor R. (a) What is the value of R? (b) What is the internal resistance of the battery?

8. What potential difference is measured across an 18-Ω load resistor when it is connected across a battery of emf 5.0 V and internal resistance 0.45 Ω?

9. Two 1.50-V batteries—with their positive terminals in the same direction—are inserted in series into the barrel of a flashlight. One battery has an internal resistance of 0.255 Ω, the other an internal resistance of 0.153 Ω. When the switch is closed, a current of 600 mA occurs in the lamp. (a) What is the lamp's resistance? (b) What fraction of the power dissipated is dissipated in the batteries?

Section 28.2 Resistors in Series and in Parallel

10. Two circuit elements with fixed resistances R_1 and R_2 are connected in series with a 6.0-V battery and a switch. The battery has an internal resistance of 5.0 Ω, $R_1 = 132$ Ω, and $R_2 = 56$ Ω. (a) What is the current through R_1 when the switch is closed? (b) What is the voltage across R_2 when the switch is closed?

11. Using only three resistors—2.0 Ω, 3.0 Ω, and 4.0 Ω—find 17 resistance values that may be obtained by various combinations of one or more resistors. Tabulate the combinations in order of increasing resistance.

12. (a) You need a 45-Ω resistor, but the stockroom has only 20-Ω and 50-Ω resistors. How can the desired resistance be achieved under these circumstances? (b) What can you do if you need a 35-Ω resistor?

13. The current in a circuit is tripled by connecting a 500-Ω resistor in parallel with the resistance of the circuit. Determine the resistance of the circuit in the absence of the 500-Ω resistor.

14. (a) Find the equivalent resistance between points a and b in Figure P28.14. (b) If a potential difference of 34 V is applied between points a and b, calculate the current in each resistor.

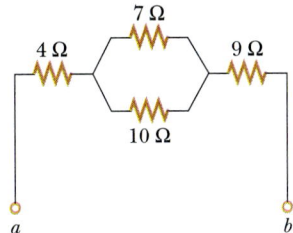

FIGURE P28.14

15. The resistance between terminals a and b in Figure P28.15 is 75 Ω. If the resistors labeled R have the same value, determine R.

16. Four copper wires of equal length are connected in series. Their cross-sectional areas are 1.0 cm^2, 2.0 cm^2, 3.0 cm^2, and 5.0 cm^2. If a voltage of 120 V is

---

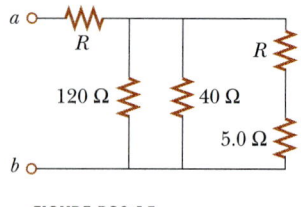

FIGURE P28.15

applied to the arrangement, determine the voltage across the 2.0-cm² wire.

17. The power dissipated in the top part of the circuit in Figure P28.17 does not depend on whether the switch is open or closed. If $R = 1.0 \; \Omega$, determine R'. Neglect the internal resistance of the voltage source.

17A. The power dissipated in the top part of the circuit in Figure P28.17 does not depend on whether the switch is open or closed. Determine R' in terms of R. Neglect the internal resistance of the voltage source.

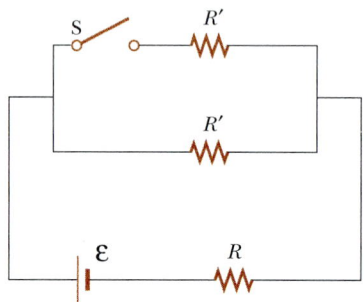

FIGURE P28.17

18. In Figures 28.4 and 28.5, $R_1 = 11 \; \Omega$, $R_2 = 22 \; \Omega$, and the battery has a terminal voltage of 33 V. (a) In the parallel circuit shown in Figure 28.5, which resistor uses more power? (b) Verify that the sum of the power (I^2R) used by each resistor equals the power supplied by the battery (IV). (c) In the series circuit, which resistor uses more power? (d) Verify that the sum of the power (I^2R) used by each resistor equals the power supplied by the battery ($P = IV$). (e) Which circuit configuration uses more power?

19. Calculate the power dissipated in each resistor in the circuit of Figure P28.19.

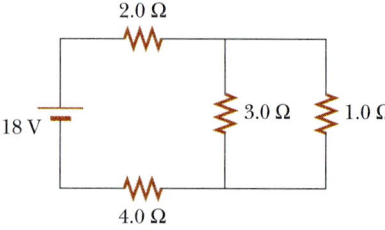

FIGURE P28.19

20. Determine the equivalent resistance between the terminals a and b for the network illustrated in Figure P28.20.

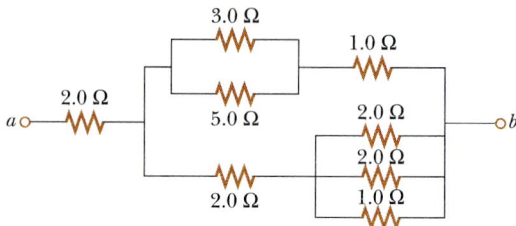

FIGURE P28.20

21. In Figure P28.21, find (a) the current in the 20-Ω resistor and (b) the potential difference between points a and b.

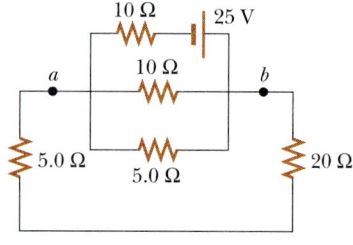

FIGURE P28.21

22. The resistance between points a and b in Figure P28.22 drops to one-half its original value when switch S is closed. Determine the value of R.

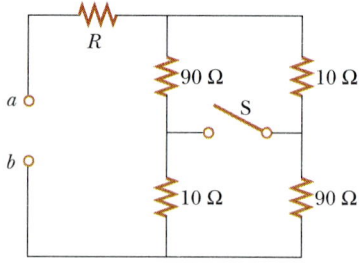

FIGURE P28.22

23. Two resistors connected in series have an equivalent resistance of 690 Ω. When they are connected in parallel, their equivalent resistance is 150 Ω. Find the resistance of each resistor.

23A. Two resistors connected in series have an equivalent resistance of R_s. When they are connected in parallel, their equivalent resistance is R_p. Find the resistance of each resistor.

24. Three 100-Ω resistors are connected as shown in Figure P28.24. The maximum power dissipated in any

one resistor is 25 W. (a) What is the maximum voltage that can be applied to the terminals *a* and *b*? (b) For the voltage determined in part (a), what is the power dissipation in each resistor? What is the total power dissipation?

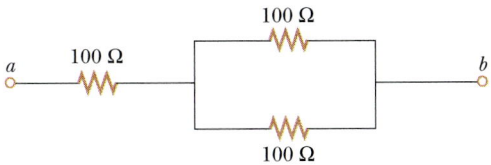

100 Ω

100 Ω

100 Ω

FIGURE P28.24

Section 28.3 Kirchhoff's Rules

(*Note:* The currents are not necessarily in the direction shown for some circuits.)

25. (a) Find the potential difference between points *a* and *b* in Figure P28.25. (b) Find the currents I_1, I_2, and I_3 in Figure P28.25.

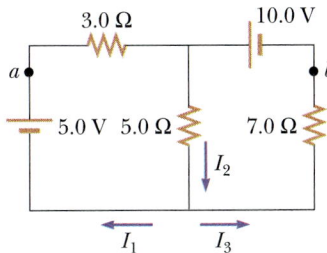

3.0 Ω   10.0 V

a   *b*

5.0 V   5.0 Ω   7.0 Ω

I_2

I_1 I_3

FIGURE P28.25

26. If $R = 1.0$ kΩ and $\mathcal{E} = 250$ V in Figure P28.26, determine the direction and magnitude of the current in the horizontal wire between *a* and *e*.

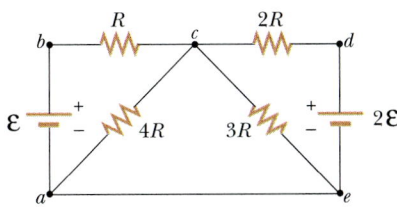

R   *c*   *2R*   *d*

$\mathcal{E}$ *4R*   *3R*   *2ε*

a   *e*

FIGURE P28.26

27. Determine the current in each branch in Figure P28.27.

28. In the circuit of Figure P28.28, determine the current in each resistor and the voltage across the 200-Ω resistor.

29. A dead battery is charged by connecting it to the live battery of another car (Fig. P28.29). Determine the current in the starter and in the dead battery.

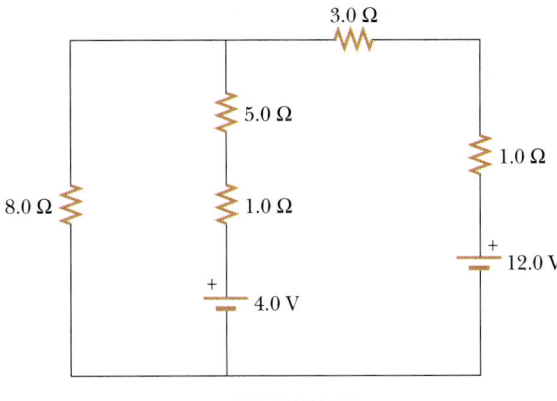

3.0 Ω

5.0 Ω

8.0 Ω   1.0 Ω   1.0 Ω

4.0 V   12.0 V

FIGURE P28.27

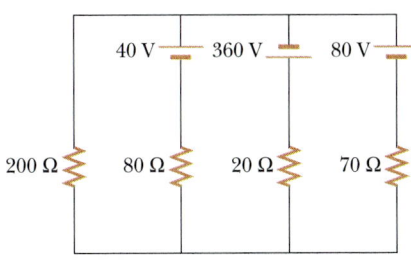

40 V   360 V   80 V

200 Ω   80 Ω   20 Ω   70 Ω

FIGURE P28.28

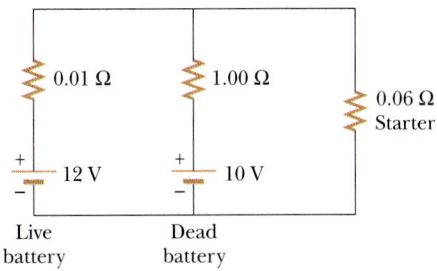

0.01 Ω   1.00 Ω

0.06 Ω
Starter

12 V   10 V

Live
battery

Dead
battery

FIGURE P28.29

30. For the network shown in Figure P28.30, show that the resistance $R_{ab} = (27/17)\,\Omega$.

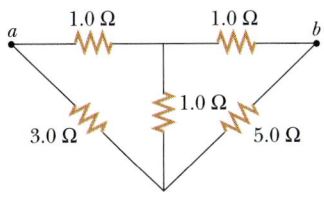

1.0 Ω   1.0 Ω
a   *b*

1.0 Ω

3.0 Ω   5.0 Ω

FIGURE P28.30

31. Calculate I_1, I_2, and I_3 in Figure P28.31.

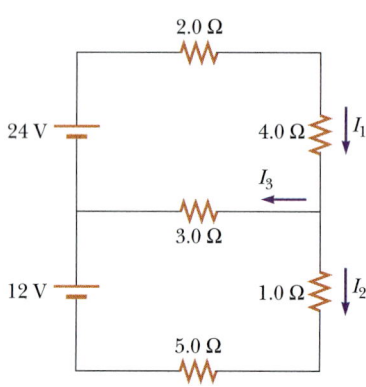

FIGURE P28.31

32. The ammeter in Figure P28.32 reads 2.0 A. Find I_1, I_2, and $\mathcal{E}$.

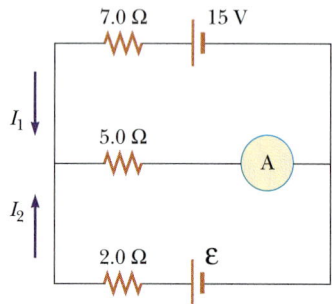

FIGURE P28.32

33. Using Kirchhoff's rules, (a) find the current in each resistor in Figure P28.33. (b) Find the potential difference between points c and f. Which point is at the higher potential?

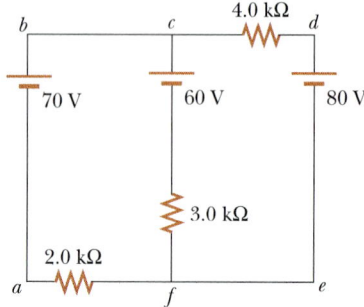

FIGURE P28.33

34. Find I_1, I_2, and I_3 in Figure P28.34.
35. An automobile battery has an emf of 12.6 V and an internal resistance of 0.080 Ω. The headlights have total resistance 5.00 Ω (assumed constant). What is the potential difference across the headlight bulbs

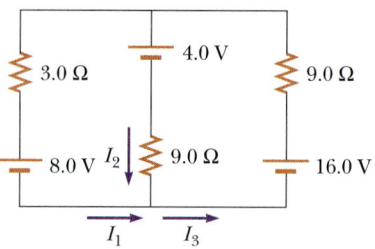

FIGURE P28.34

(a) when they are the only load on the battery and (b) when the starter motor is operated, taking an additional 35.0 A from the battery?

36. In Figure P28.36, calculate (a) the equivalent resistance of the network outside the battery, (b) the current through the battery, and (c) the current in the 6.0-Ω resistor.

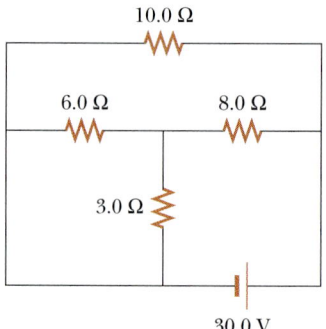

FIGURE P28.36

37. For the circuit shown in Figure P28.37, calculate (a) the current in the 2.0-Ω resistor and (b) the potential difference between points a and b.

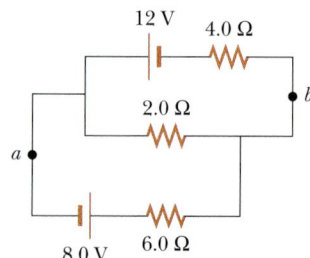

FIGURE P28.37

38. The resistor R in Figure P28.38 dissipates 20 W of power. Determine the value of R.
39. Calculate the power dissipated in each resistor in Figure P28.39.
40. Calculate the power dissipated in each resistor in Figure P28.40.

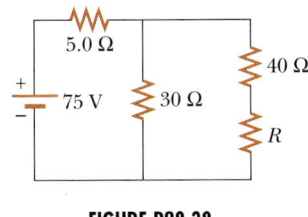

FIGURE P28.38

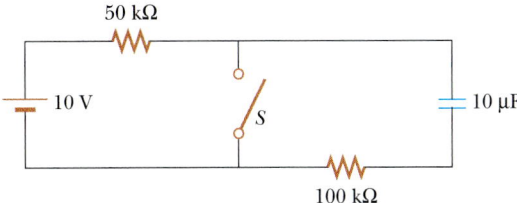

FIGURE P28.43

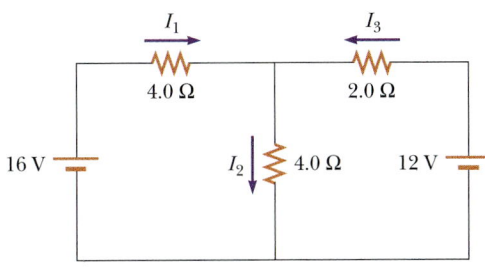

FIGURE P28.39

44. At $t = 0$, an uncharged capacitor of capacitance C is connected through a resistance R to a battery of constant emf $\mathcal{E}$ (Fig. P28.44). How long does it take the capacitor to (a) reach one-half its final charge and (b) to become fully charged?

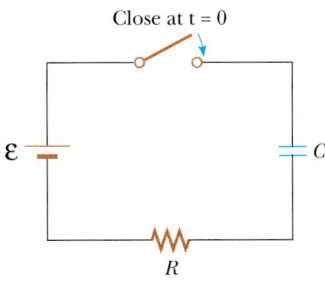

FIGURE P28.44

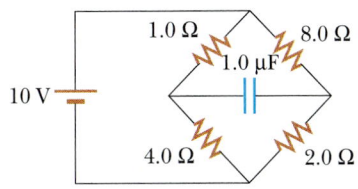

FIGURE P28.40

Section 28.4 *RC* Circuits

41. A fully charged capacitor stores 12 J of energy. How much energy remains when its charge has decreased to half its original value?

41A. A fully charged capacitor stores energy U_0. How much energy remains when its charge has decreased to half its original value?

42. Consider a series RC circuit (Fig. 28.15) for which $R = 1.00$ MΩ, $C = 5.00$ μF, and $\mathcal{E} = 30.0$ V. Find (a) the time constant of the circuit and (b) the maximum charge on the capacitor after the switch is closed. (c) If the switch is closed at $t = 0$, find the current in the resistor 10.0 s later.

43. In the circuit of Figure P28.43, the switch S has been open for a long time. It is then suddenly closed. Determine the time constant (a) before the switch is closed and (b) after the switch is closed. (c) If the switch is closed at $t = 0$ s, determine the current through it as a function of time.

45. A 4.00-MΩ resistor and a 3.00-μF capacitor are connected in series with a 12.0-V power supply. (a) What is the time constant for the circuit? (b) Express the current in the circuit and the charge on the capacitor as functions of time.

46. A 750-pF capacitor has an initial charge of 6.00 μC. It is then connected to a 150-MΩ resistor and allowed to discharge through the resistor. (a) What is the time constant for the circuit? (b) Express the current in the circuit and the charge on the capacitor as functions of time.

47. The circuit in Figure P28.47 has been connected for a long time. (a) What is the voltage across the capacitor? (b) If the battery is disconnected, how long does it take the capacitor to discharge to 1/10 of its initial voltage?

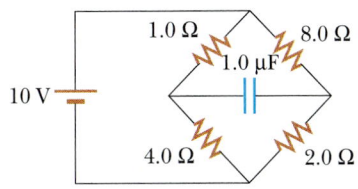

FIGURE P28.47

48. A 2.0×10^{-3}-μF capacitor with an initial charge of 5.1 μC is discharged through a 1.3-kΩ resistor. (a) Calculate the current through the resistor 9.0 μs after the resistor is connected across the terminals of the capacitor. (b) What charge remains on the capacitor after 8.0 μs? (c) What is the maximum current in the resistor?

49. Find the current through the ammeter 9.5 μs after the switch in Figure P28.49 is thrown from position a to position b.

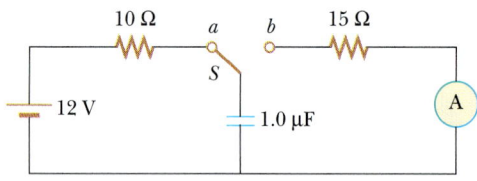

FIGURE P28.49

50. A capacitor in an RC circuit is charged to 60% of its maximum value in 0.90 s. What is the time constant of the circuit?

51. Dielectric materials used in the manufacture of capacitors are characterized by conductivities that are small but not zero. Therefore, a charged capacitor slowly loses its charge by "leaking" across the dielectric. If a certain 3.60-μF capacitor leaks charge such that the potential difference decreases to half its initial value in 4.00 s, what is the equivalent resistance of the dielectric?

51A. Dielectric materials used in the manufacture of capacitors are characterized by conductivities that are small but not zero. Therefore, a charged capacitor slowly loses its charge by "leaking" across the dielectric. If a capacitor having capacitance C leaks charge such that the potential difference decreases to half its initial value in a time t, what is the equivalent resistance of the dielectric?

*Section 28.5   Electrical Instruments

52. A typical galvanometer, which requires a current of 1.50 mA for full-scale deflection and has a resistance of 75.0 Ω, may be used to measure currents of much larger values. To enable an operator to measure large currents without damage to the galvanometer, a relatively small shunt resistor is wired in parallel with the galvanometer similar to Figure 28.23a. Most of the current then flows through the shunt resistor. Calculate the value of the shunt resistor that enables the galvanometer to be used to measure a current of 1.00 A at full-scale deflection. (*Hint:* Use Kirchhoff's laws.)

53. The same galvanometer described in the previous problem may be used to measure voltages. In this

case a large resistor is wired in series with the galvanometer similar to Figure 28.22b, which in effect limits the current that flows through the galvanometer when large voltages are applied. Most of the potential drop occurs across the resistor placed in series. Calculate the value of the resistor that enables the galvanometer to measure an applied voltage of 25.0 V at full-scale deflection.

54. For each voltage setting, a galvanometer having an internal resistance of 100 Ω deflects full scale when the current is 1.0 mA. For the multiscale voltmeter in Figure P28.54, what are the values of R_1, R_2, and R_3?

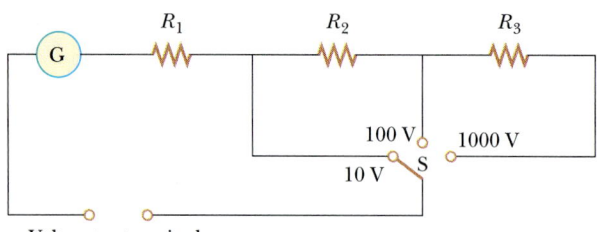

Voltmeter terminals

FIGURE P28.54

55. Assume that a galvanometer has an internal resistance of 60.0 Ω and requires a current of 0.500 mA to produce full-scale deflection. What resistance must be connected in parallel with the galvanometer if the combination is to serve as an ammeter that has a full-scale deflection for a current of 0.100 A?

56. An ammeter is constructed using a galvanometer that requires a potential difference of 50.0 mV across the galvanometer and a current of 1.00 mA through the galvanometer to cause a full-scale deflection. Find the shunt resistance R that produces a full-scale deflection when a current of 5.00 A enters the ammeter.

57. A galvanometer having a full-scale sensitivity of 1.00 mA requires a 900-Ω series resistor to make a voltmeter reading full scale when 1.00 V is measured across the terminals. What series resistor is required to make the same galvanometer into a 50.0-V (full-scale) voltmeter?

58. A current of 2.50 mA causes a given galvanometer to deflect full scale. The resistance of the galvanometer is 200 Ω. (a) Show by means of a circuit diagram, using two resistors and three external jacks, how the galvanometer may be made into a dual-range voltmeter. (b) Determine the values of the resistors needed to make the high range 0–200 V and the low range 0–20.0 V. Indicate these values on the diagram.

59. For the same galvanometer as in the previous problem, (a) show by means of a circuit diagram, using two resistors and three external jacks, how the galvanometer may be made into a dual-range ammeter.

(b) Determine the values of the resistors needed to make the high range 0–10.0 A and the low range 0–1.00 A. Indicate these values on the diagram.

60. A Wheatstone bridge of the type shown in Figure 28.24 is used to make a precise measurement of the resistance of a wire connector. If $R_3 = 1.00$ kΩ and the bridge is balanced by adjusting R_1 such that $R_1 = 2.50 R_2$, what is R_x?

61. Consider the case when the Wheatstone bridge shown in Figure 28.24 is unbalanced. Calculate the current through the galvanometer when $R_x = R_3 = 7.00$ Ω, $R_2 = 21.0$ Ω, and $R_1 = 14.0$ Ω. Assume the voltage across the bridge is 70.0 V, and neglect the galvanometer's resistance.

62. When the Wheatstone bridge shown in Figure 28.24 is balanced, the voltage drop across R_x is 3.20 V and $I_1 = 200$ μA. If the total current drawn from the power supply is 500 μA, what is R_x?

63. The Wheatstone bridge in Figure 28.24 is balanced when $R_1 = 10.0$ Ω, $R_2 = 20.0$ Ω, and $R_3 = 30.0$ Ω. Calculate R_x.

64. Consider the potentiometer circuit shown in Figure 28.25. When a standard battery of emf 1.0186 V is used in the circuit and the resistance between a and d is 36.0 Ω, the galvanometer reads zero. When the standard battery is replaced by an unknown emf, the galvanometer reads zero when the resistance is adjusted to 48.0 Ω. What is the value of the unknown emf?

Section 28.6   Household Wiring and Electrical Safety

65. An electric heater is rated at 1500 W, a toaster at 750 W, and an electric grill at 1000 W. The three appliances are connected to a common 120-V circuit. (a) How much current does each draw? (b) Is a 25-A circuit sufficient in this situation? Explain.

66. A 1000-W toaster, 800-W microwave oven, and 500-W coffee pot are all plugged into the same 120-V outlet. If the circuit is protected by a 20-A fuse, will the fuse blow if all these appliances are used at once?

67. An 8-foot extension cord has two 18-gauge copper wires, each having a diameter of 1.024 mm. How much power does this cord dissipate when carrying a current of (a) 1.0 A and (b) 10 A?

68. Sometimes aluminum wiring is used instead of copper for economic reasons. According to the National Electrical Code, the maximum allowable current for 12-gauge copper wire with rubber insulation is 20 A. What should be the maximum allowable current in a 12-gauge aluminum wire if it is to dissipate the same power per unit length as the copper wire?

69. A 4.0-kW heater is wired for 240-V operation with Nichrome wire having a total mass M. (a) How much current does the heater require? (b) How much current does a 120-V, 4.0-kW heater require? (c) If a 240-V, 4.0-kW heater and a 120-V, 4.0-kW heater

have the same length of wires in them, how does the mass of the wire in the 120-V heater compare with the mass of the wire in the 240-V heater?

ADDITIONAL PROBLEMS

70. Calculate the potential difference between points a and b in Figure P28.70 and identify which point is at the higher potential.

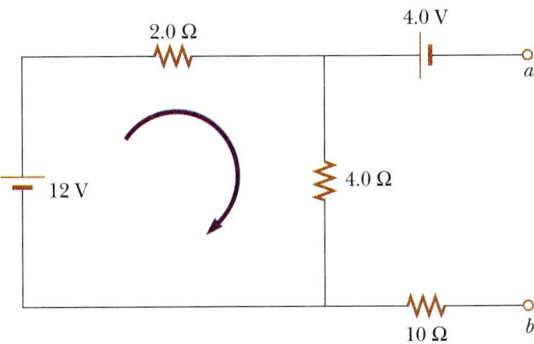

FIGURE P28.70

71. When two unknown resistors are connected in series with a battery, 225 W is dissipated with a total current of 5.00 A. For the same total current, 50.0 W is dissipated when the resistors are connected in parallel. Determine the values of the two resistors.

71A. When two unknown resistors are connected in series with a battery, a total power P_s is dissipated with a total current of I. For the same total current, a total power P_p is dissipated when the resistors are connected in parallel. Determine the values of the two resistors.

72. Before the switch is closed in Figure P28.72, there is no charge stored by the capacitor. Determine the currents in R_1, R_2, and C (a) at the instant the switch is closed (that is, at $t = 0$) and (b) after the switch is closed for a long time (that is, as $t \rightarrow \infty$).

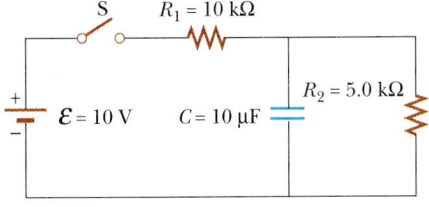

FIGURE P28.72

73. Three resistors, each of value 3.0 Ω, are connected in two arrangements as in Figure P28.73. If the maximum allowable power for each resistor is 48 W, cal-

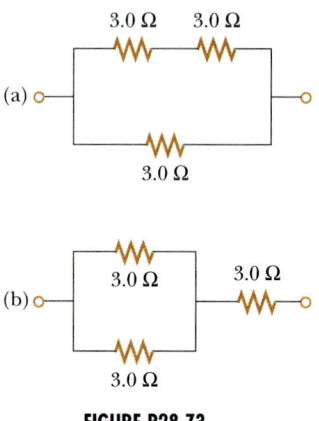

(a)

(b)

FIGURE P28.73

culate the maximum power that can be dissipated by
(a) the circuit in Figure P28.73a and (b) the circuit
in Figure P28.73b.

74. Arrange nine 100-Ω resistors in a series-parallel network so that the total resistance of the network is also
100 Ω. All nine resistors must be used.

75. Three 60-W, 120-V lightbulbs are connected across a
120-V power source, as shown in Figure P28.75. Find
(a) the total power dissipated in the three bulbs and
(b) the voltage across each. Assume that the resistance of each bulb conforms to Ohm's law (even
though in reality the resistance increases markedly
with current).

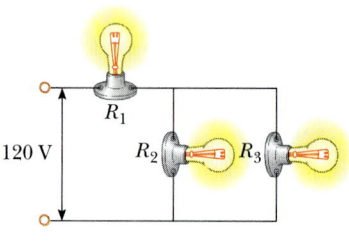

FIGURE P28.75

76. A series combination of a 12-kΩ resistor and an unknown capacitor is connected to a 12-V battery. One
second after the circuit is completed, the voltage
across the capacitor is 10 V. Determine the capacitance.

77. The value of a resistor *R* is to be determined using
the ammeter-voltmeter setup shown in Figure
P28.77. The ammeter has a resistance of 0.50 Ω, and
the voltmeter has a resistance of 20 000.0 Ω. Within
what range of actual values of *R* will the measured
values be correct to within 5% if the measurement is
made using the circuit shown in (a) Figure P28.77a
and (b) Figure P28.77b?

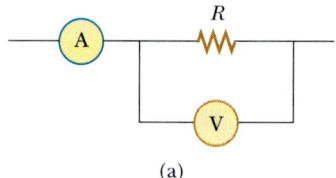

(a)

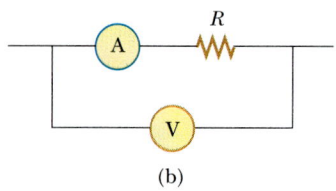

(b)

FIGURE P28.77

78. A power supply has an open-circuit voltage of 40.0 V
and an internal resistance of 2.0 Ω. It is used to
charge two storage batteries connected in series,
each having an emf of 6.0 V and internal resistance
of 0.30 Ω. If the charging current is to be 4.0 A,
(a) what additional resistance should be added in
series? (b) Find the power lost in the supply, the batteries, and the added series resistance. (c) How
much power is converted to chemical energy in the
batteries?

79. A battery has an emf $\mathcal{E}$ and internal resistance *r*. A
variable resistor *R* is connected across the terminals
of the battery. Find the value of *R* such that (a) the
potential difference across the terminals is a maximum, (b) the current in the circuit is a maximum,
(c) the power delivered to the resistor is a maximum.

80. Consider the circuit shown in Figure P28.80. (a) Calculate the current in the 5.0-Ω resistor. (b) What
power is dissipated by the circuit? (c) Determine the
potential difference between points *a* and *b*. Which
point is at the higher potential?

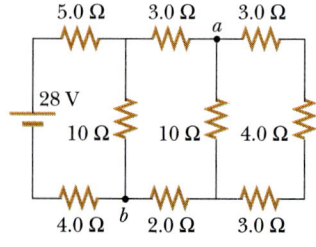

FIGURE P28.80

81. The values of the components in a simple series *RC*
circuit containing a switch (Fig. 28.15) are *C* =

$1.00 \ \mu F$, $R = 2.00 \times 10^6 \ \Omega$, and $\mathcal{E} = 10.0$ V. At the instant 10.0 s after the switch is closed, calculate (a) the charge on the capacitor, (b) the current in the resistor, (c) the rate at which energy is being stored in the capacitor, and (d) the rate at which energy is being delivered by the battery.

82. A battery is used to charge a capacitor through a resistor, as in Figure 28.15. Show that half the energy supplied by the battery is dissipated as heat in the resistor and half is stored in the capacitor.

83. The switch in Figure P28.83a closes when $V_c \geq 2V/3$ and opens when $V_c \leq V/3$. The voltmeter reads a voltage as plotted in Figure P28.83b. What is the period, T, of the waveform in terms of R_A, R_B, and C?

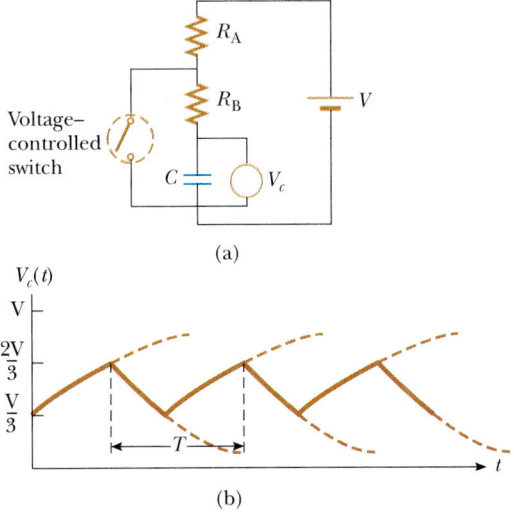

(a)

(b)

FIGURE P28.83

84. Design a multirange voltmeter capable of full-scale deflection for (a) 20.0 V, (b) 50.0 V, and (c) 100.0 V. Assume a meter that has a resistance of $60.0 \ \Omega$ and gives a full-scale deflection for a current of 1.00 mA.

85. Design a multirange ammeter capable of full-scale deflection for (a) 25.0 mA, (b) 50.0 mA, and (c) 100.0 mA. Assume a meter that has a resistance of $25.0 \ \Omega$ and gives a full-scale deflection for 1.00 mA.

86. A particular galvanometer serves as a 2.00-V full-scale voltmeter when a 2500-Ω resistor is connected in series with it. It serves as a 0.500-A full-scale ammeter when a 0.220-Ω resistor is connected in parallel with it. Determine the internal resistance of the galvanometer and the current required to produce full-scale deflection.

87. In Figure P28.87, suppose the switch has been closed sufficiently long for the capacitor to become fully charged. Find (a) the steady-state current through

each resistor and (b) the charge Q on the capacitor. (c) The switch is now opened at $t = 0$. Write an equation for the current i_{R_2} through R_2 as a function of time and (d) find the time that it takes for the charge on the capacitor to fall to one-fifth its initial value.

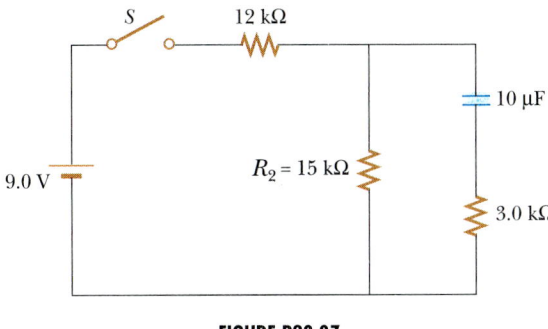

FIGURE P28.87

88. A 10.0-μF capacitor is charged by a 10.0-V battery through a resistance R. The capacitor reaches a potential difference of 4.00 V in a time 3.00 s after charging begins. Find R.

89. (a) Determine the charge on the capacitor in Figure P28.89 when $R = 10 \ \Omega$. (b) For what value of R is the charge on the capacitor zero?

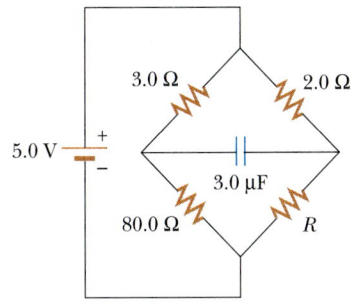

FIGURE P28.89

90. (a) Using symmetry arguments, show that the current through any resistor in the configuration of Figure P28.90 is either $I/3$ or $I/6$. All resistors have the same resistance r. (b) Show that the equivalent resistance between points a and b is $(5/6)r$.

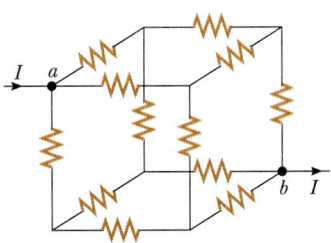

FIGURE P28.90

91. The circuit shown in Figure P28.91 is set up in the laboratory to measure an unknown capacitance C using a voltmeter of resistance $R = 10.0$ MΩ and a battery whose emf is 6.19 V. The data given in the table below are the measured voltages across the capacitor as a function of time, where $t = 0$ represents the time the switch is open. (a) Construct a graph of $\ln(\mathcal{E}/V)$ versus t, and do a linear least-squares fit on the data. (b) From the slope of your graph, obtain a value for the time constant of the circuit and a value for the capacitance.

V(V)	t(s)	$\ln(\mathcal{E}/V)$
6.19	0	
5.55	4.87	
4.93	11.1	
4.34	19.4	
3.72	30.8	
3.09	46.6	
2.47	67.3	
1.83	102.2	

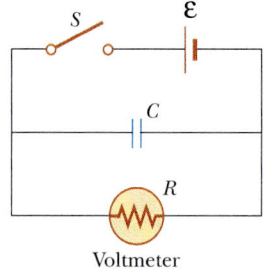

FIGURE P28.91

92. The student engineer of a campus radio station wishes to verify the effectiveness of the lightning rod on the antenna mast (Fig. P28.92). The unknown resistance R_x is between points C and E. Point E is a true ground but is inaccessible for direct measurement because it is several meters below the Earth's surface. Two identical rods are driven into the ground at A and B, introducing an unknown resistance, R_y. The procedure for determining the effectiveness of the rod is as follows: Measure resistance R_1 between A and B, then connect A and B with a heavy conducting wire and measure resistance R_2 between A and C. (a) Derive a formula for R_x in terms of the observable resistances, R_1 and R_2. (b) A satisfactory ground resistance would be $R_x < 2.0$ Ω. Is the grounding of the station adequate if measurements give $R_1 = 13$ Ω and $R_2 = 6.0$ Ω?

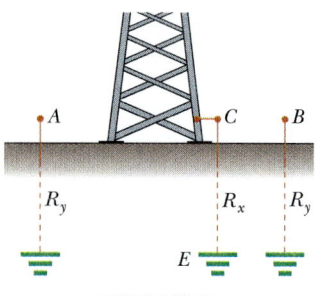

FIGURE P28.92

93. Three 2.0-Ω resistors are connected as in Figure P28.93. Each can dissipate a maximum power of 32 W without being excessively heated. Determine the maximum power the network can dissipate.

93A. Three resistors each having resistance R are connected as in Figure P28.93. Each can dissipate a maximum power P without being excessively heated. Determine the maximum power the network can dissipate.

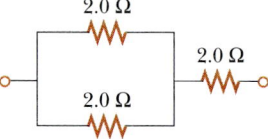

FIGURE P28.93

94. In Figure P28.94, $R_1 = 2.0$ kΩ, $R_2 = 3.0$ kΩ, $C_1 = 2.0$ μF, $C_2 = 3.0$ μF, and $\mathcal{E} = 120$ V. If there are no charges on the capacitors before switch S is closed, determine the charges q_1 and q_2 on capacitors C_1 and C_2, after the switch is closed. (*Hint:* First reconstruct the circuit so that it becomes a simple RC circuit containing a single resistor and single capacitor in series, in series with the battery, and then determine the total charge q stored in the circuit.)

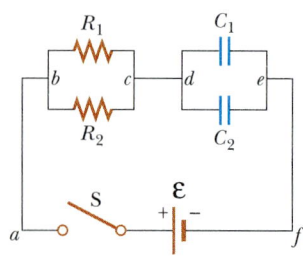

FIGURE P28.94

SPREADSHEET PROBLEM

S1. The application of Kirchhoff's rules to a dc circuit leads to a set of n linear equations in n unknowns. It is very tedious to solve these algebraically if $n > 3$. The purpose of this problem is to solve for the currents in a moderately complex circuit using matrix operations on a spreadsheet. You can solve equations very easily this way, and you can also readily explore the consequences of changing the values of the circuit parameters. (a) Consider the circuit in Figure S28.1. Assume the four unknown currents are in the directions shown.

- Apply Kirchhoff's rules to get four independent equations for the four unknown currents I_i, $i = 1$, 2, 3, and 4.
- Write these equations in matrix form $\mathbf{AI} = \mathbf{B}$, that is,

$$\sum_{j=1}^{4} A_{ij} I_j = B_i, \; i = 1, 2, 3, 4$$

The solution is $\mathbf{I} = \mathbf{A}^{-1}\mathbf{B}$, where $\mathbf{A}^{-1}$ is the inverse matrix of $\mathbf{A}$.

- Set $R_1 = 2 \, \Omega$, $R_2 = 4 \, \Omega$, $R_3 = 6 \, \Omega$, $R_4 = 8 \, \Omega$, $\mathcal{E}_1 = 3 \, \text{V}$, $\mathcal{E}_2 = 9 \, \text{V}$, and $\mathcal{E}_3 = 12 \, \text{V}$.
- Enter the matrix $\mathbf{A}$ into your spreadsheet, one value per cell. Use the matrix inversion operation of the spreadsheet to calculate $\mathbf{A}^{-1}$.

- Find the currents by using the matrix multiplication operation of the spreadsheet to calculate $\mathbf{I} = \mathbf{A}^{-1}\mathbf{B}$.

(b) Change the sign of $\mathcal{E}_3$, and repeat the calculations in part (a). This is equivalent to changing the polarity of $\mathcal{E}_3$. (c) Set $\mathcal{E}_1 = \mathcal{E}_2 = 0$ and repeat the calculations in part (a). For these values, the circuit can be solved using simple series-parallel rules. Compare your results using both methods. (d) Investigate any other cases of interest. For example, see how the currents change if you vary R_4.

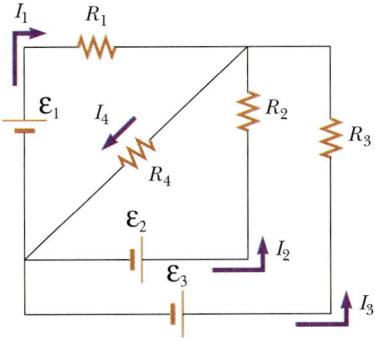

FIGURE S28.1

Magnetic Fields

Aurora Borealis, the Northern Lights, photographed near Fairbanks, Alaska. Auroras occur when cosmic rays — electrically charged particles originating mainly from the Sun — become trapped in the Earth's atmosphere over Earth's magnetic poles and collide with other atoms, resulting in the emission of visible light. *(Jack Finch/SPL/Photo Researchers)*

Many historians of science believe that the compass, which uses a magnetic needle, was used in China as early as the 13th century B.C., its invention being of Arabic or Indian origin. The early Greeks knew about magnetism as early as 800 B.C. They discovered that certain stones, now called magnetite (Fe_3O_4), attract pieces of iron. Legend ascribes the name *magnetite* to the shepherd Magnes, the nails of whose shoes and the tip of whose staff stuck fast in a magnetic field while he pastured his flocks. In 1269 Pierre de Maricourt mapped out the directions taken by a needle when it was placed at various points on the surface of a spherical natural magnet. He found that the directions formed lines that encircled the sphere and passed through two points diametrically opposite each other, which he called the *poles* of the magnet. Subsequent experiments showed that every magnet, regardless of its shape, has two poles, called *north* and *south poles,* which exhibit forces on each other in a manner analogous to electric charges. That is, like poles repel each other and unlike poles attract each other. The poles received their names because of the behavior of a magnet in the presence of the Earth's magnetic field. If a bar magnet is suspended from its midpoint by a piece of string so that it can swing freely in a horizontal plane, it will rotate until its "north" pole points to the north of the Earth and its "south" pole points to the south. (The same idea is used to construct a simple compass.)

In 1600 William Gilbert extended these experiments to a variety of materials. Using the fact that a compass needle orients in preferred directions, he suggested that the Earth itself is a large permanent magnet. In 1750 John Michell used a torsion balance to show that magnetic poles exert attractive or repulsive forces on each other and that these forces vary as the inverse square of their separation. Although the force between two magnetic poles is similar to the force between two electric charges, there is an important difference. Electric charges can be isolated (witness the electron or proton), whereas *magnetic poles cannot be isolated*. That is, *magnetic poles are always found in pairs*. All attempts thus far to detect an isolated magnetic monopole have been unsuccessful. No matter how many times a permanent magnet is cut, each piece will always have a north and a south pole. (There is some theoretical basis for speculating that magnetic monopoles—isolated north or south poles—may exist in nature, and attempts to detect them currently make up an active experimental field of investigation. However, none of these attempts has proven successful.)

The relationship between magnetism and electricity was discovered in 1819 when, during a lecture demonstration, the Danish scientist Hans Christian Oersted (1777–1851) found that an electric current in a wire deflected a nearby compass needle.[1] Shortly thereafter, André Ampère (1775–1836) formulated quantitative laws for calculating the magnetic force between current-carrying conductors. He also suggested that electric current loops of molecular size are responsible for *all* magnetic phenomena.

In the 1820s, further connections between electricity and magnetism were demonstrated by Faraday and independently by Joseph Henry (1797–1878). They showed that an electric current can be produced in a circuit either by moving a magnet near the circuit or by changing the current in another nearby circuit. These observations demonstrate that a changing magnetic field produces an electric field. Years later, theoretical work by Maxwell showed that the reverse is also true: A changing electric field gives rise to a magnetic field.

There is a similarity between electric and magnetic effects that has provided methods of making permanent magnets. In Chapter 23 we learned that when rubber and wool are rubbed together, both become charged, one positively and the other negatively. In an analogous fashion, an unmagnetized piece of iron can be magnetized by stroking it with a magnet. Magnetism can also be induced in iron (and other materials) by other means. For example, if a piece of unmagnetized iron is placed near a strong magnet, the piece of iron eventually becomes magnetized. The process of magnetizing the piece of iron in the presence of a strong external field can be accelerated either by heating and cooling the iron or by hammering.

This chapter examines forces on moving charges and on current-carrying wires in the presence of a magnetic field. The source of the magnetic field itself is described in Chapter 30.

An electromagnet is used to move tons of scrap metal. *(Ray Pfortner/Peter Arnold, Inc.)*

Hans Christian Oersted (1777–1851), Danish physicist. *(The Bettmann Archive)*

29.1 THE MAGNETIC FIELD

In earlier chapters we found it convenient to describe the interaction between charged objects in terms of electric fields. Recall that an electric field surrounds any electric charge. The region of space surrounding a *moving* charge includes a

---

[1] It is interesting to note that the same discovery was reported in 1802 by an Italian jurist, Gian Dominico Romognosi, but was overlooked, probably because it was published in a newspaper, *Gazetta de Trentino*, rather than in a scholarly journal.

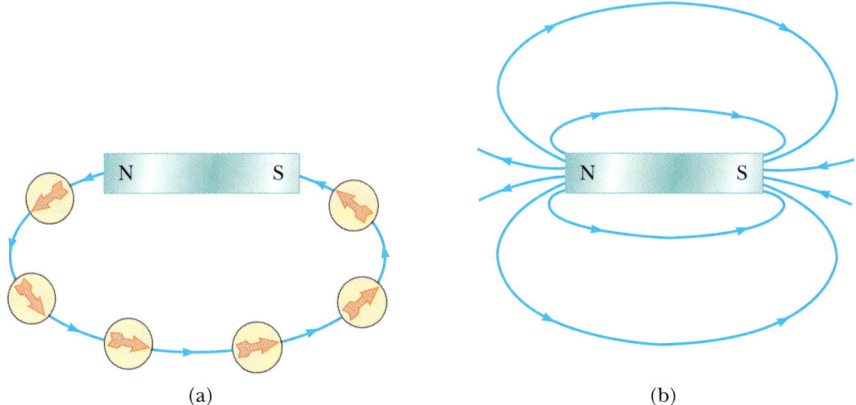

FIGURE 29.1   (a) Tracing the magnetic field lines of a bar magnet. (b) Several magnetic field lines of a bar magnet.

magnetic field in addition to the electric field. A magnetic field also surrounds any magnetic substance.

In order to describe any type of field, we must define its magnitude, or strength, and its direction. The direction of the magnetic field **B** at any location is in the direction in which the north pole of a compass needle points at that location. Figure 29.1a shows how the magnetic field of a bar magnet can be traced with the aid of a compass. Several magnetic field lines of a bar magnet traced out in this manner are shown in Figure 29.1b. Magnetic field patterns can be displayed by small iron filings, as in Figure 29.2.

We can define a magnetic field **B** at some point in space in terms of the magnetic force exerted on an appropriate test object. Our test object is a charged particle moving with a velocity **v**. For the time being, let us assume there are no electric or gravitational fields present in the region of the charge. Experiments on

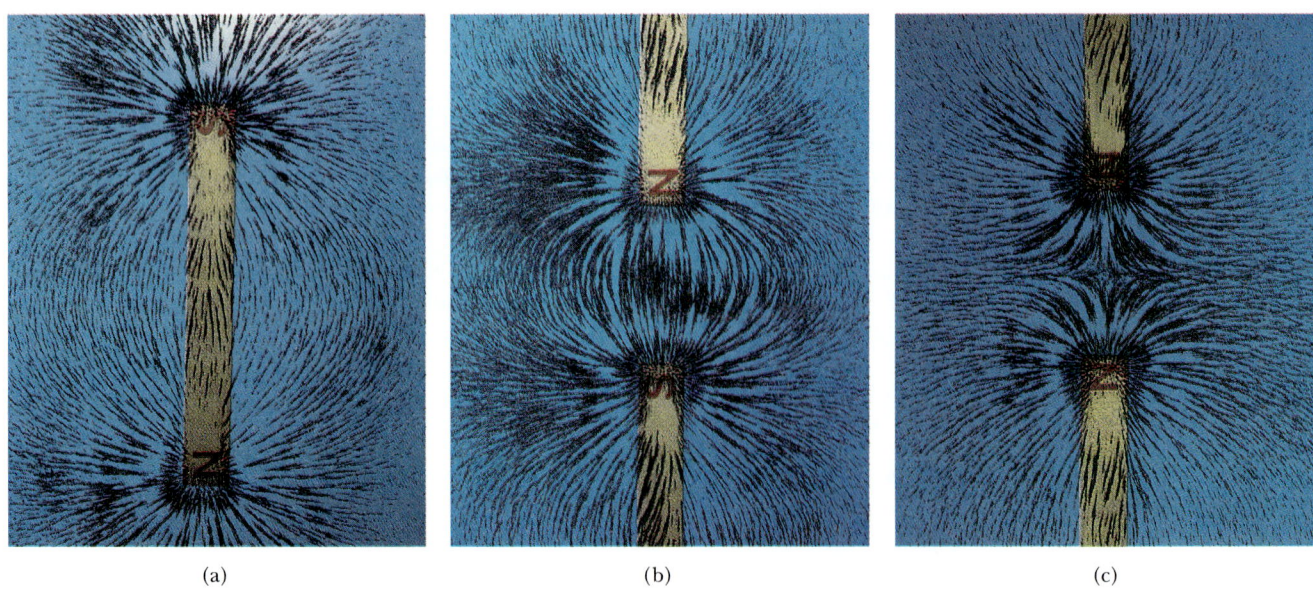

| (a) | (b) | (c) |

FIGURE 29.2   (a) Magnetic field patterns surrounding a bar magnet as displayed with iron filings. (b) Magnetic field patterns between *unlike* poles of two bar magnets. (c) Magnetic field pattern between *like* poles of two bar magnets.   *(Courtesy of Henry Leap and Jim Lehman)*

the motion of various charged particles moving in a magnetic field give the following results:

- The magnitude of the magnetic force is proportional to the charge q and speed v of the particle.
- The magnitude and direction of the magnetic force depend on the velocity of the particle and on the magnitude and direction of the magnetic field.
- When a charged particle moves parallel to the magnetic field vector, the magnetic force on the charge is zero.
- When the velocity vector makes an angle θ with the magnetic field, the magnetic force acts in a direction perpendicular to both **v** and **B**; that is, **F** is perpendicular to the plane formed by **v** and **B** (Fig. 29.3a).
- The magnetic force on a positive charge is in the direction opposite the direction of the force on a negative charge moving in the same direction (Fig. 29.3b).
- If the velocity vector makes an angle θ with the magnetic field, the magnitude of the magnetic force is proportional to $\sin \theta$.

> Properties of the magnetic force on a charge moving in a **B** field

These observations can be summarized by writing the magnetic force in the form

$$\mathbf{F} = q\mathbf{v} \times \mathbf{B} \tag{29.1}$$

where the direction of the magnetic force is in the direction of **v** × **B** if Q is positive, which, by definition of the cross product, is perpendicular to both **v** and **B**. We can regard this equation as an operational definition of the magnetic field at some point in space. That is, the magnetic field is defined in terms of a sideways force acting on a moving charged particle.

The white arc in this photograph indicates the circular path followed by an electron beam moving in a magnetic field. The vessel contains gas at very low pressure, and the beam is made visible as the electrons collide with the gas atoms, which in turn emit visible light. The magnetic field is produced by two coils (not shown). The apparatus can be used to measure the charge/mass ratio for the electron.   (*Courtesy of Central Scientific Company*)

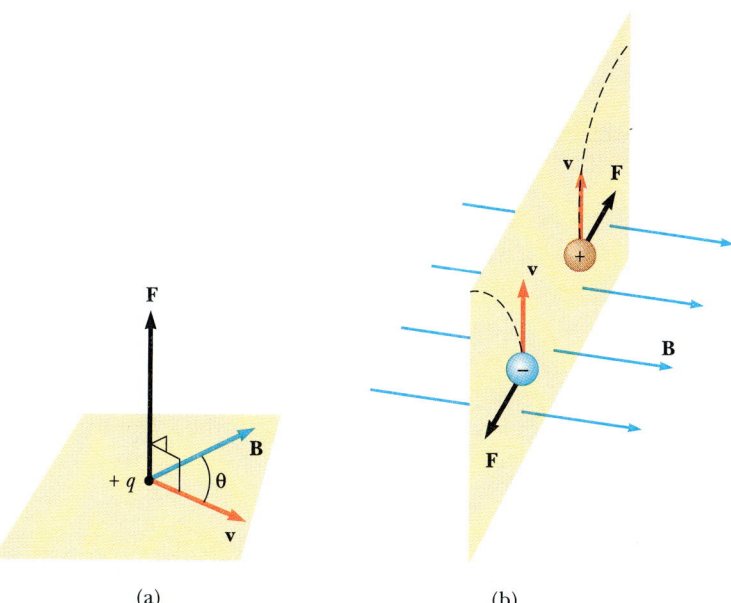

(a)   (b)

FIGURE 29.3   The direction of the magnetic force on a charged particle moving with a velocity **v** in the presence of a magnetic field **B**. (a) When **v** is at an angle θ to **B**, the magnetic force is perpendicular to both **v** and **B**. (b) In the presence of a magnetic field, the moving charged particles are deflected as indicated by the dotted lines.

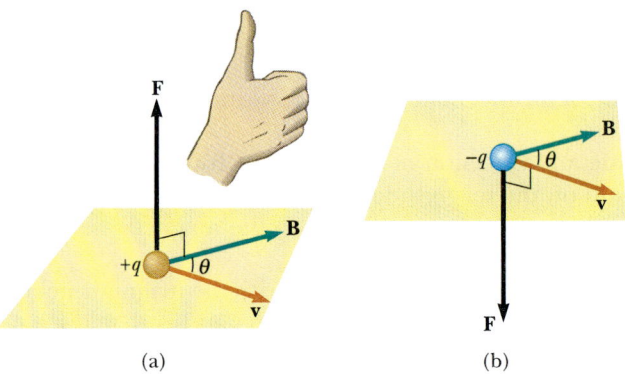

FIGURE 29.4 The right-hand rule for determining the direction of the magnetic force **F** acting on a charge *q* moving with a velocity **v** in a magnetic field **B**. If *q* is positive, **F** is upward in the direction of the thumb. If *q* is negative, **F** is downward.

Figure 29.4 reviews the right-hand rule for determining the direction of the cross product **v** × **B**. You point the four fingers of your right hand along the direction of **v**, and then turn them until they point along the direction of **B**. The thumb then points in the direction of **v** × **B**. Since **F** = *q***v** × **B**, **F** is in the direction of **v** × **B** if *q* is positive (Fig. 29.4a) and opposite the direction of **v** × **B** if *q* is negative (Fig. 29.4b). The magnitude of the magnetic force has the value

Magnetic force on a charged particle moving in a magnetic field

$$\mathbf{F} = qvB \sin\theta \tag{29.2}$$

where θ is the smaller angle between **v** and **B**. From this expression, we see that *F* is zero when **v** is parallel to **B** ($\theta = 0$ or $180°$) and maximum ($F_{max} = qvB$) when **v** is perpendicular to **B** ($\theta = 90°$).

There are several important differences between electric and magnetic forces:

Differences between electric and magnetic fields

- The electric force is always in the direction of the electric field, whereas the magnetic force is perpendicular to the magnetic field.
- The electric force acts on a charged particle independent of the particle's velocity, whereas the magnetic force acts on a charged particle only when the particle is in motion.
- The electric force does work in displacing a charged particle, whereas the magnetic force associated with a steady magnetic field does no work when a particle is displaced.

This last statement is a consequence of the fact that when a charge moves in a steady magnetic field, the magnetic force is always *perpendicular* to the displacement. That is, $\mathbf{F} \cdot d\mathbf{s} = (\mathbf{F} \cdot \mathbf{v}) \, dt = 0$, because the magnetic force is a vector perpendicular to **v**. From this property and the work-energy theorem, we conclude that the kinetic energy of a charged particle cannot be altered by a magnetic field alone. In other words,

A steady magnetic field cannot change the speed of a particle

when a charge moves with a velocity **v**, an applied magnetic field can alter the direction of the velocity vector, but it cannot change the speed of the particle. That is, a static magnetic field changes the direction of the velocity but does not affect the speed or kinetic energy of the charged particle.

The SI unit of the magnetic field is the **weber per square meter** (Wb/m^2), also called the **tesla** (T). This unit can be related to the fundamental units by using Equation 29.1: A 1-C charge moving through a field of 1 T with a velocity of 1 m/s perpendicular to the field experiences a force of 1 N:

$$[B] = T = \frac{Wb}{m^2} = \frac{N}{C \cdot m/s} = \frac{N}{A \cdot m}$$

Another non-SI unit in common use, called the **gauss** (G), is related to the tesla through the conversion 1 T = 10^4 G.

Conventional laboratory magnets can produce magnetic fields as large as 2.5 T. Superconducting magnets that can generate magnetic fields as high as 25 T have been constructed. This value can be compared with the Earth's magnetic field near its surface, which is about 0.5 × 10^{-4} T.

EXAMPLE 29.1 A Proton Moving in a Magnetic Field

A proton moves with a speed of 8.0 × 10^6 m/s along the x axis. It enters a region where there is a magnetic field of magnitude 2.5 T, directed at an angle of 60° to the x axis and lying in the xy plane (Fig. 29.5). Calculate the initial magnetic force on and acceleration of the proton.

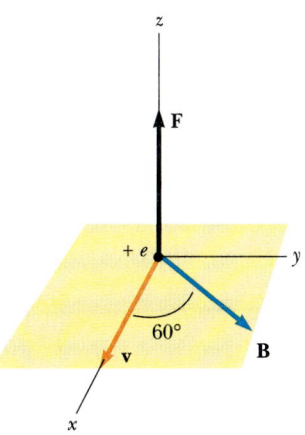

FIGURE 29.5 (Example 29.1) The magnetic force F on a proton is in the positive z direction when v and B lie in the xy plane.

Solution From Equation 29.2, we get

$$F = qvB \sin \theta$$
$$= (1.60 \times 10^{-19} \text{ C})(8.0 \times 10^6 \text{ m/s})(2.5 \text{ T})(\sin 60°)$$
$$= 2.8 \times 10^{-12} \text{ N}$$

Because v × B is in the positive z direction (the right-hand rule), and the charge is positive, F is in the positive z direction.

The mass of the proton is 1.67 × 10^{-27} kg, and so its initial acceleration is

$$a = \frac{F}{m} = \frac{2.8 \times 10^{-12} \text{ N}}{1.67 \times 10^{-27} \text{ kg}} = 1.7 \times 10^{15} \text{ m/s}^2$$

in the positive z direction.

Exercise Calculate the acceleration of an electron that moves through the same magnetic field at the same speed as the proton.

Answer 3.0 × 10^{18} m/s^2.

29.2 MAGNETIC FORCE ON A CURRENT-CARRYING CONDUCTOR

If a magnetic force is exerted on a single charged particle when it moves through a magnetic field, it should not surprise you that a current-carrying wire also experiences a force when placed in a magnetic field. This follows from the fact that the current represents a collection of many charged particles in motion; hence, the resultant force on the wire is due to the sum of the individual forces exerted on the charged particles. The force on the particles is transmitted to the "bulk" of the wire through collisions with the atoms making up the wire.

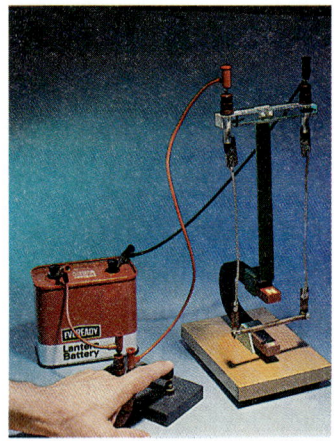

This apparatus demonstrates the force on a current-carrying conductor in an external magnetic field. Why does the bar swing *into* the magnet after the switch is closed? *(Henry Leap and Jim Lehman)*

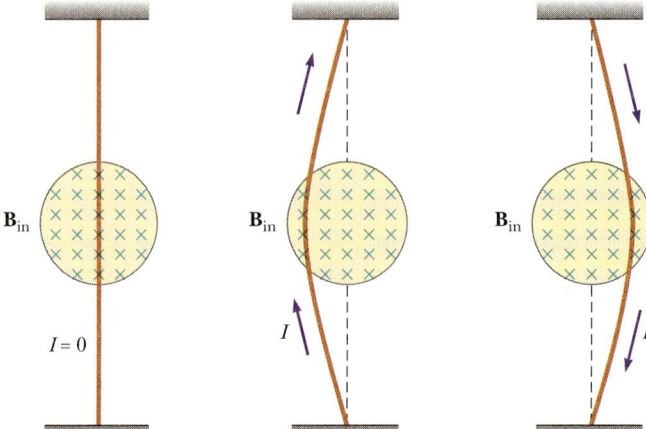

FIGURE 29.6   A segment of a flexible vertical wire is stretched between the poles of a magnet with the magnetic field (blue crosses) directed into the paper. (a) When there is no current in the wire, it remains vertical. (b) When the current is upward, the wire deflects to the left. (c) When the current is downward, the wire deflects to the right.

Before we continue our discussion, some explanation is in order concerning notation in many of our figures. To indicate the direction of **B**, we use the following convention. If **B** is directed into the page as in Figure 29.6a, we use a series of blue crosses, which represent the tails of arrows. If **B** is directed out of the page, we use a series of blue dots, which represent the tips of arrows. If **B** lies in the plane of the page, we use a series of blue field lines with arrowheads.

The force on a current-carrying conductor can be demonstrated by hanging a wire between the poles of a magnet as in Figure 29.6. In this figure (which shows only the farther pole face), the magnetic field is directed into the page and covers the region within the shaded circles. When the current in the wire is zero, the wire remains vertical as in Figure 29.6a. However, when a current is set up in the wire directed upwards as in Figure 29.6b, the wire deflects to the left. If we reverse the current, as in Figure 29.6c, the wire deflects to the right.

Let us quantify this discussion by considering a straight segment of wire of length L and cross-sectional area A, carrying a current I in a uniform magnetic field **B** as in Figure 29.7. The magnetic force on a charge q moving with a drift velocity $\mathbf{v}_d$ is $q\mathbf{v}_d \times \mathbf{B}$. To find the total force on the wire, we multiply the force on one charge, $q\mathbf{v}_d \times \mathbf{B}$, by the number of charges in the segment. Since the volume of the segment is AL the number of charges in the segment is nAL, where n is the number of charges per unit volume. Hence, the total magnetic force on the wire of length L is

$$\mathbf{F} = (q\mathbf{v}_d \times \mathbf{B})\, nAL$$

This can be written in a more convenient form by noting that, from Equation 27.4, the current in the wire is $I = nqv_d A$. Therefore,

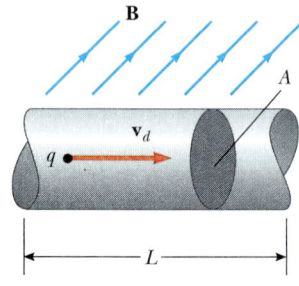

FIGURE 29.7   A section of a wire containing moving charges in a magnetic field **B**. The magnetic force on each charge is $q\mathbf{v}_d \times \mathbf{B}$, and the net force on the segment of length L is $I\mathbf{L} \times \mathbf{B}$.

$$\mathbf{F} = I\mathbf{L} \times \mathbf{B} \tag{29.3}$$

where **L** is a vector in the direction of the current I; the magnitude of **L** equals the length L of the segment. Note that this expression applies only to a straight segment of wire in a uniform magnetic field.

Now consider an arbitrarily shaped wire segment of uniform cross-section in a magnetic field, as in Figure 29.8. It follows from Equation 29.3 that the magnetic force on a very small segment $d\mathbf{s}$ in the presence of a field $\mathbf{B}$ is

$$d\mathbf{F} = I\,d\mathbf{s} \times \mathbf{B} \qquad (29.4)$$

where $d\mathbf{F}$ is directed out of the page for the directions assumed in Figure 29.8. We can consider Equation 29.4 as an alternative definition of $\mathbf{B}$. That is, the field $\mathbf{B}$ can be defined in terms of a measurable force on a current element, where the force is a maximum when $\mathbf{B}$ is perpendicular to the element and zero when $\mathbf{B}$ is parallel to the element.

To get the total force $\mathbf{F}$ on the wire, we integrate Equation 29.4 over the length of the wire:

$$\mathbf{F} = I\int_{a}^{b} d\mathbf{s} \times \mathbf{B} \qquad (29.5)$$

where a and b represent the end points of the wire. When this integration is carried out, the magnitude of the magnetic field and the direction the field makes with the vector $d\mathbf{s}$ (that is, the element orientation) may differ at different points.

Now let us consider two special cases involving the application of Equation 29.5. In both cases, the magnetic field is taken to be constant in magnitude and direction.

Case I

Consider a curved wire carrying a current I; the wire is located in a uniform magnetic field $\mathbf{B}$ as in Figure 29.9a. Since the field is uniform, $\mathbf{B}$ can be taken outside the integral in Equation 29.5, and we get

$$\mathbf{F} = I\left(\int_{a}^{b} d\mathbf{s}\right) \times \mathbf{B} \qquad (29.6)$$

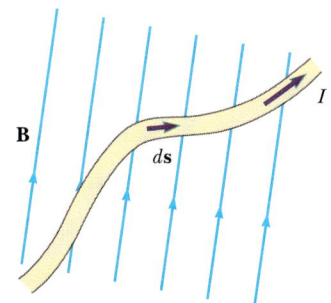

FIGURE 29.8   A wire segment of arbitrary shape carrying a current I in a magnetic field $\mathbf{B}$ experiences a magnetic force. The force on any segment $d\mathbf{s}$ is $I\,d\mathbf{s} \times \mathbf{B}$ and is directed *out* of the page. You should use the right-hand rule to confirm this direction.

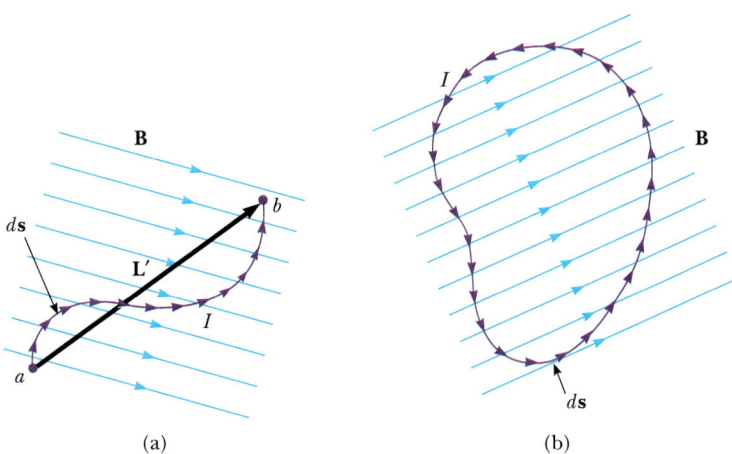

(a)   (b)

FIGURE 29.9   (a) A curved conductor carrying a current I in a uniform magnetic field. The magnetic force on the conductor is equivalent to the force on a straight segment of length L' running between the ends of the wire, a and b. (b) A current-carrying loop of arbitrary shape in a uniform magnetic field. The net magnetic force on the loop is 0.

But the quantity $\int_a^b d\mathbf{s}$ represents the *vector sum* of all the displacement elements from *a* to *b*. From the law of vector addition (Chapter 3), the sum equals the vector **L'**, which is directed from *a* to *b*. Therefore, Equation 29.6 reduces to

$$\mathbf{F} = I\mathbf{L'} \times \mathbf{B} \qquad (29.7)$$

Case II

An arbitrarily shaped, closed loop carrying a current *I* is placed in a uniform magnetic field as in Figure 29.9b. Again, we can express the force in the form of Equation 29.6. In this case, the vector sum of the displacement vectors must be taken over the closed loop. That is,

$$\mathbf{F} = I\left(\oint d\mathbf{s}\right) \times \mathbf{B}$$

Since the set of displacement vectors forms a closed polygon (Fig. 29.9b), the vector sum must be zero. This follows from the graphical procedure of adding vectors by the polygon method (Chapter 3). Since $\oint d\mathbf{s} = 0$, we conclude that **F** = 0. That is,

> the total magnetic force on any closed current loop in a uniform magnetic field is zero.

EXAMPLE 29.2   Force on a Semicircular Conductor

A wire bent into the shape of a semicircle of radius *R* forms a closed circuit and carries a current *I*. The circuit lies in the *xy* plane, and a uniform magnetic field is present along the positive *y* axis as in Figure 29.10. Find the magnetic force on the straight portion of the wire and on the curved portion.

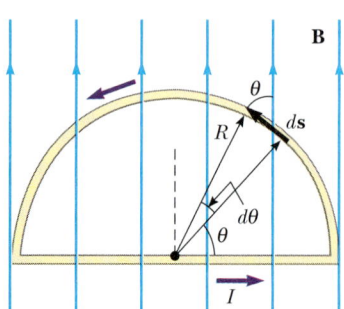

FIGURE 29.10   (Example 29.2) The net force on a closed-current loop in a uniform magnetic field is zero. In this case, the force on the straight portion is $2IRB$ and outward, while the force on the curved portion is also $2IRB$ and inward.

Reasoning and Solution   The force on the straight portion of the wire has a magnitude $F_1 = ILB = 2IRB$, since $L = 2R$ and the wire is perpendicular to **B**. The direction of $\mathbf{F}_1$ is out of the paper since $\mathbf{L} \times \mathbf{B}$ is outward. (That is, **L** is to the right, in the direction of the current, and so by the rule of cross products, $\mathbf{L} \times \mathbf{B}$ is outward.)

To find the force on the curved part, we must first write an expression for the force $d\mathbf{F}_2$ on the element *d***s**. If θ is the angle between **B** and *d***s** in Figure 29.10, then the magnitude of $d\mathbf{F}_2$ is

$$dF_2 = I|\, d\mathbf{s} \times \mathbf{B}\,| = IB \sin\theta\, ds$$

where *ds* is the length of the small element measured along the circular arc. In order to integrate this expression, we must express *ds* in terms of θ. Since $s = R\theta$, $ds = R\, d\theta$, and the expression for dF_2 can be written

$$dF_2 = IRB \sin\theta\, d\theta$$

To get the total force F_2 on the curved portion, we can integrate this expression to account for contributions from all elements. Note that the direction of the force on every element is the same: into the paper (since $ds \times \mathbf{B}$ is inward). Therefore, the resultant force $\mathbf{F}_2$ on the curved wire must also be into the paper. Integrating dF_2 over the limits $\theta = 0$ to $\theta = \pi$ (that is, the entire semicircle) gives

$$F_2 = IRB \int_0^\pi \sin \theta \, d\theta = IRB[-\cos \theta]_0^\pi$$

$$= -IRB(\cos \pi - \cos 0) = -IRB(-1 - 1) = \boxed{2IRB}$$

Since $F_2 = 2IRB$ and is directed into the paper while the force on the straight wire $F_1 = 2IRB$ is out of the paper, we see that the net force on the closed loop is zero. This result is consistent with Case II as described above.

29.3 TORQUE ON A CURRENT LOOP IN A UNIFORM MAGNETIC FIELD

In the previous section we showed how a force is exerted on a current-carrying conductor when the conductor is placed in a magnetic field. With this as a starting point, we now show that a torque is exerted on a current loop placed in a magnetic field. The results of this analysis will be of great practical value when we discuss generators in a future chapter.

Consider a rectangular loop carrying a current I in the presence of a uniform magnetic field directed parallel to the plane of the loop, as in Figure 29.11a. The forces on the sides of length a are zero because these wires are parallel to the field; hence, $ds \times \mathbf{B} = 0$ for these sides. The magnitude of the forces on the sides of length b, however, is

$$F_1 = F_2 = IbB$$

The direction of $\mathbf{F}_1$, the force on the left side of the loop, is out of the paper and that of $\mathbf{F}_2$, the force on the right side of the loop, is into the paper. If we were to view the loop by standing underneath the bottom and looking up, we would see the view shown in Figure 29.11b, and the forces would be as shown there. If we assume that the loop is pivoted so that it can rotate about point O, we see that these two forces produce a torque about O that rotates the loop clockwise. The magnitude of this torque, $\tau_{\max}$, is

$$\tau_{\max} = F_1 \frac{a}{2} + F_2 \frac{a}{2} = (IbB) \frac{a}{2} + (IbB) \frac{a}{2} = IabB$$

where the moment arm about O is $a/2$ for each force. Since the area of the loop is $A = ab$, the maximum torque can be expressed as

$$\tau_{\max} = IAB \qquad (29.8)$$

Remember that this result is valid only when the magnetic field is parallel to the plane of the loop. The sense of the rotation is clockwise when viewed from the bottom end, as indicated in Figure 29.11b. If the current were reversed, the forces would reverse and the rotational tendency would be counterclockwise.

Now suppose the uniform magnetic field makes an angle θ with a line perpendicular to the plane of the loop, as in Figure 29.12a. For convenience, we assume that $\mathbf{B}$ is perpendicular to the sides of length b. (The end view of these sides is shown in Fig. 29.12b.) In this case, the magnetic forces $\mathbf{F}_3$ and $\mathbf{F}_4$ on the sides of length a cancel each other and produce no torque because they pass through a common origin. However, the forces acting on the sides of length b, $\mathbf{F}_1$ and $\mathbf{F}_2$, form a couple and hence produce a torque about *any point*. Referring to the end view in Figure 29.12b, we note that the moment arm of $\mathbf{F}_1$ about the point O is equal to $(a/2) \sin \theta$. Likewise, the moment arm of $\mathbf{F}_2$ about O is also $(a/2) \sin \theta$.

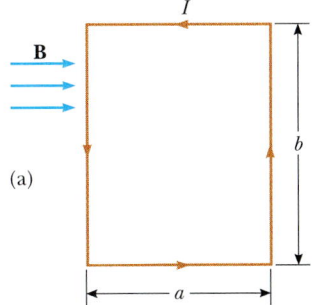

(a)

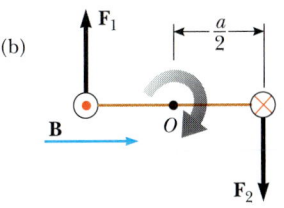

(b)

FIGURE 29.11 (a) Front view of a rectangular loop in a uniform magnetic field. There are no forces acting on the sides of width a parallel to $\mathbf{B}$, but there are forces acting on the sides of length b. (b) Bottom view of the rectangular loop shows that the forces $\mathbf{F}_1$ and $\mathbf{F}_2$ on the sides of length b create a torque that tends to twist the loop clockwise as shown.

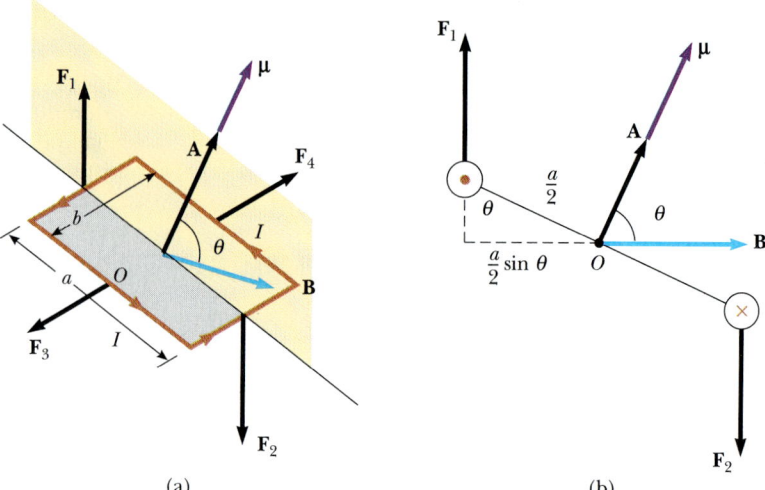

FIGURE 29.12   (a) A rectangular current loop whose normal makes an angle θ with a uniform magnetic field. The forces on the sides of length a cancel while the forces on the sides of width b create a torque on the loop. (b) An end view of the loop. The magnetic moment $\boldsymbol{\mu}$ is in the direction normal to the plane of the loop.

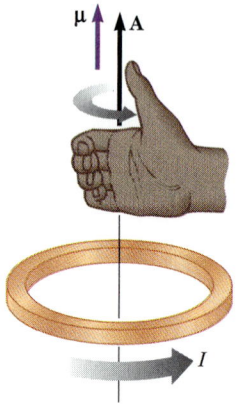

FIGURE 29.13   A right-hand rule for determining the direction of the vector **A**. The magnetic moment $\boldsymbol{\mu}$ is also in the direction of **A**.

Torque on a current loop

Magnetic moment of a current loop

Since $F_1 = F_2 = IbB$, the net torque about O has the magnitude

$$\tau = F_1 \frac{a}{2} \sin \theta + F_2 \frac{a}{2} \sin \theta$$

$$= IbB \left(\frac{a}{2} \sin \theta \right) + IbB \left(\frac{a}{2} \sin \theta \right) = IabB \sin \theta$$

$$= IAB \sin \theta$$

where $A = ab$ is the area of the loop. This result shows that the torque has its maximum value IAB when the field is parallel to the plane of the loop ($\theta = 90°$), as we saw when discussing Figure 29.11, and is zero when the field is perpendicular to the plane of the loop ($\theta = 0$). As we see in Figure 29.12, the loop tends to rotate in the direction of decreasing values of θ (that is, so that the normal to the plane of the loop rotates toward the direction of the magnetic field).

A convenient expression for the torque is

$$\boldsymbol{\tau} = I\mathbf{A} \times \mathbf{B} \tag{29.9}$$

where **A**, a vector perpendicular to the plane of the loop, has a magnitude equal to the area of the loop. The sense of **A** is determined by the right-hand rule as described in Figure 29.13. By rotating the four fingers of the right hand in the direction of the current in the loop, the thumb points in the direction of **A**. The product $I\mathbf{A}$ is defined to be the **magnetic moment** $\boldsymbol{\mu}$ of the loop. That is,

$$\boldsymbol{\mu} = I\mathbf{A} \tag{29.10}$$

The SI unit of magnetic moment is ampere-meter^2 (A·m^2). Using this definition, the torque can be expressed as

$$\boldsymbol{\tau} = \boldsymbol{\mu} \times \mathbf{B} \tag{29.11}$$

Note that this result is analogous to the torque acting on an electric dipole in the presence of an external electric field E, where $\tau = \mathbf{p} \times \mathbf{E}$, where p is the dipole moment defined by Equation 26.15.

Although we obtained the torque for a particular orientation of B with respect to the loop, the equation $\tau = \boldsymbol{\mu} \times \mathbf{B}$ is valid for any orientation. Furthermore, although we derived the torque expression for a rectangular loop, the result is valid for a loop of any shape.

If a coil consists of N turns of wire, each carrying the same current and having the same area, the total magnetic moment of the coil is the product of the number of turns and the magnetic moment for one turn. The torque on an N-turn coil is N times greater than that on a one-turn coil. That is, $\tau = N\boldsymbol{\mu} \times \mathbf{B}$.

EXAMPLE 29.3   The Magnetic Moment of a Coil

A rectangular coil of dimensions 5.40 cm × 8.50 cm consists of 25 turns of wire. The coil carries a current of 15.0 mA. (a) Calculate the magnitude of its magnetic moment.

Solution   The magnitude of the magnetic moment of a current loop is $\mu = IA$ (see Eq. 29.10). In this case, $A = (0.0540 \text{ m})(0.0850 \text{ m}) = 4.59 \times 10^{-3} \text{ m}^2$. Since the coil has 25 turns, and assuming that each turn has the same area A, we have

$$\mu_{coil} = NIA = (25)(15.0 \times 10^{-3} \text{ A})(4.59 \times 10^{-3} \text{ m}^2)$$

$$= 1.72 \times 10^{-3} \text{ A} \cdot \text{m}^2$$

(b) Suppose a magnetic field of magnitude 0.350 T is applied parallel to the plane of the loop. What is the magnitude of the torque acting on the loop?

Solution   The torque is given by Equation 29.11, $\tau = \boldsymbol{\mu} \times \mathbf{B}$. In this case, B is *perpendicular* to $\boldsymbol{\mu}_{coil}$, so that

$$\tau = \mu_{coil}B = (1.72 \times 10^{-3} \text{ A} \cdot \text{m}^2)(0.350 \text{ T})$$

$$= 6.02 \times 10^{-4} \text{ N} \cdot \text{m}$$

Note that this is the basic principle behind the operation of a galvanometer coil discussed in Chapter 28.

Exercise   Show that the units $\text{A} \cdot \text{m}^2 \cdot \text{T}$ reduce to $\text{N} \cdot \text{m}$.

Exercise   Calculate the magnitude of the torque on the coil when the 0.350-T magnetic field makes angles of (a) 60° and (b) 0° with $\boldsymbol{\mu}$.

Answer   (a) $5.21 \times 10^{-4} \text{ N} \cdot \text{m}$; (b) zero.

29.4   MOTION OF A CHARGED PARTICLE IN A MAGNETIC FIELD

In Section 29.1 we found that the magnetic force acting on a charged particle moving in a magnetic field is perpendicular to the velocity of the particle and that consequently the work done on the particle by the magnetic force is zero. Consider now the special case of a positively charged particle moving in a uniform magnetic field with its initial velocity vector perpendicular to the field. Let us assume that the magnetic field is into the page. Figure 29.14 shows that

the charged particle moves in a circle whose plane is perpendicular to the magnetic field.

The particle moves in this way because the magnetic force F is at right angles to v and B and has a constant magnitude equal to qvB. As the force deflects the particle, the directions of v and F change continuously, as in Figure 29.14. Therefore F acts like a central force, which changes only the direction of v and not its magnitude. The sense of the rotation, as shown in Figure 29.14, is counterclockwise for a positive charge. If q were negative, the sense of the rotation would be clockwise.

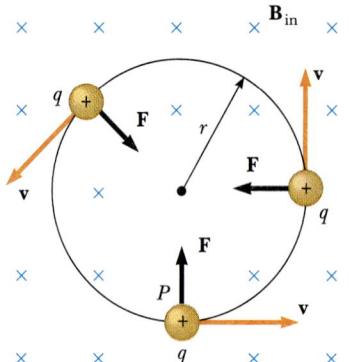

FIGURE 29.14 When the velocity of a charged particle is perpendicular to a uniform magnetic field, the particle moves in a circular path whose plane is perpendicular to **B**. In this case, **B** is directed into the page and the charge is positive. The magnetic force **F** on the charge is always directed toward the center of the circle.

Since **F** (which is in the radial direction) has a magnitude qvB, we can equate this to the required central force, which is the mass m multiplied by the centripetal acceleration v^2/r. From Newton's second law, we find that

$$F = qvB = \frac{mv^2}{r}$$

$$r = \frac{mv}{qB} \tag{29.12}$$

That is, the radius of the path is proportional to the momentum mv of the particle and inversely proportional to the magnitude of the magnetic field. The angular frequency of the rotating charged particle is

$$\omega = \frac{v}{r} = \frac{qB}{m} \tag{29.13}$$

The period of its motion (the time for one revolution) is equal to the circumference of the circle divided by the speed of the particle:

$$T = \frac{2\pi r}{v} = \frac{2\pi}{\omega} = \frac{2\pi m}{qB} \tag{29.14}$$

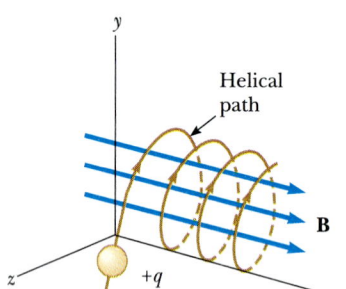

FIGURE 29.15 A charged particle having a velocity vector that has a component parallel to a uniform magnetic field moves in a helical path.

These results show that the angular frequency and period of the circular motion do not depend on the speed of the particle or the radius of the orbit. The frequency, $\omega/2\pi$, is often referred to as the **cyclotron frequency** because charged particles circulate at this frequency in one type of accelerator called a *cyclotron,* discussed in Section 29.5.

If a charged particle moves in a uniform magnetic field with its velocity at some arbitrary angle to **B**, its path is a helix. For example, if the field is in the *x* direction as in Figure 29.15, there is no component of force in the *x* direction, and hence $a_x = 0$ and the *x* component of velocity remains constant. On the other hand, the magnetic force $q\mathbf{v} \times \mathbf{B}$ causes the components v_y and v_z to change in time, and the resulting motion is a helix having its axis parallel to the **B** field. The projection of the path onto the *yz* plane (viewed along the *x* axis) is a circle. (The projections of the path onto the *xy* and *xz* planes are sinusoids!) Equations 29.12 to 29.14 still apply, provided that v is replaced by $v_\perp = \sqrt{v_y^2 + v_z^2}$.

EXAMPLE 29.4   A Proton Moving Perpendicular to a Uniform Magnetic Field

A proton is moving in a circular orbit of radius 14 cm in a uniform magnetic field of magnitude 0.35 T directed perpendicular to the velocity of the proton. Find the orbital speed of the proton.

Solution   From Equation 29.12, we get

$$v = \frac{qBr}{m} = \frac{(1.60 \times 10^{-19}\ \text{C})(0.35\ \text{T})(14 \times 10^{-2}\ \text{m})}{1.67 \times 10^{-27}\ \text{kg}}$$

$$= 4.7 \times 10^6\ \text{m/s}$$

Exercise   If an electron moves perpendicular to the same magnetic field with this speed, what is the radius of its circular orbit?

Answer 7.6×10^{-5} m.

EXAMPLE 29.5 The Bending of an Electron Beam

In an experiment designed to measure the strength of a uniform magnetic field produced by a set of coils, the electrons are accelerated from rest through a potential difference of 350 V, and the beam associated with the electrons is measured to have a radius of 7.5 cm, shown in Figure 29.16. Assuming the magnetic field is perpendicular to the beam, (a) what is the magnitude of the magnetic field?

FIGURE 29.16 (Example 29.5) The bending of an electron beam in a magnetic field. The tube contains gas at very low pressure, and the beam is made visible as the electrons collide with the gas atoms, which in turn emit visible light. The apparatus used to take this photograph is part of a system designed to measure the ratio e/m. *(Henry Leap and Jim Lehman)*

Reasoning First, we must calculate the speed of the electrons using the fact that the increase in their kinetic energy must equal the change in their potential energy, $|e|V$ (because of conservation of energy). Then we use Equation 29.12 to find the strength of the magnetic field.

Solution Since $K_i = 0$ and $K_f = mv^2/2$, we have

$$\tfrac{1}{2}mv^2 = |e|V$$

$$v = \sqrt{\frac{2|e|V}{m}} = \sqrt{\frac{2(1.60 \times 10^{-19}\,\text{C})(350\,\text{V})}{9.11 \times 10^{-31}\,\text{kg}}}$$

$$= 1.11 \times 10^7\,\text{m/s}$$

We now use Equation 29.12 to find B:

$$B = \frac{mv}{|e|r} = \frac{(9.11 \times 10^{-31}\,\text{kg})(1.11 \times 10^7\,\text{m/s})}{(1.60 \times 10^{-19}\,\text{C})(0.075\,\text{m})}$$

$$= \boxed{8.4 \times 10^{-4}\,\text{T}}$$

(b) What is the angular frequency of the electrons?

Solution Using Equation 29.13, we find

$$\omega = \frac{v}{r} = \frac{1.11 \times 10^7\,\text{m/s}}{0.075\,\text{m}} = \boxed{1.5 \times 10^8\,\text{rad/s}}$$

Exercise What is the period of revolution of the electrons?

Answer $T = 43$ ns.

When charged particles move in a nonuniform magnetic field, the motion is complex. For example, in a magnetic field that is strong at the ends and weak in the middle, as in Figure 29.17, the particles can oscillate back and forth between the end points. Such a field can be produced by two current loops as in Figure 29.17. In this case, a charged particle starting at one end spirals along the field lines until it reaches the other end, where it reverses its path and spirals back. This configuration is known as a *magnetic bottle* because charged particles can be trapped in it. This setup has been used to confine plasmas, and such a plasma-confinement scheme could play a crucial role in controlling nuclear fusion, a process that could supply us with an almost endless source of energy. Unfortunately, the magnetic bottle has its problems. If a large number of particles are trapped, collisions between the particles cause them to eventually leak from the system.

The Van Allen radiation belts, discovered in 1958 by a team of researchers under the direction of James Van Allen, consist of charged particles (mostly electrons and protons) surrounding the Earth in doughnut-shaped regions (Fig. 29.18). The particles, trapped by the Earth's nonuniform magnetic field, spiral around the field lines from pole to pole. These particles originate mainly from the Sun, but some come from stars and other heavenly objects. For this reason, the particles are called *cosmic rays*. Most cosmic rays are deflected by the Earth's magnetic field and never reach the Earth. However, some become trapped, and these are the ones that make up the Van Allen belts. When the particles are in the Earth's atmosphere over the poles, they frequently collide with atoms, causing

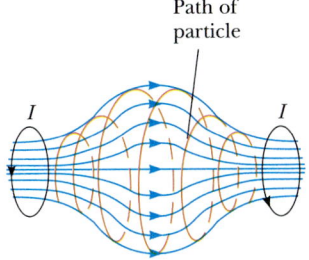

FIGURE 29.17 A charged particle moving in a nonuniform magnetic field represented by blue lines (a magnetic bottle) spirals about the field (red path) and oscillates between the end points.

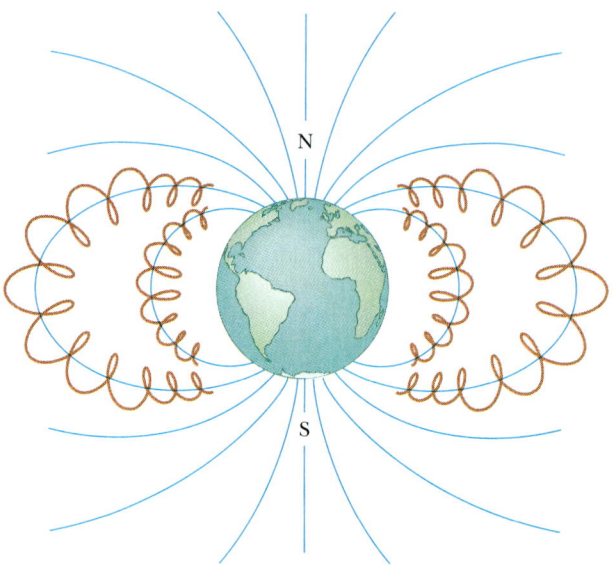

FIGURE 29.18   The Van Allen belts are made up of charged particles (electrons and protons) trapped by the Earth's nonuniform magnetic field. The magnetic field lines are in blue and the particle paths in red.

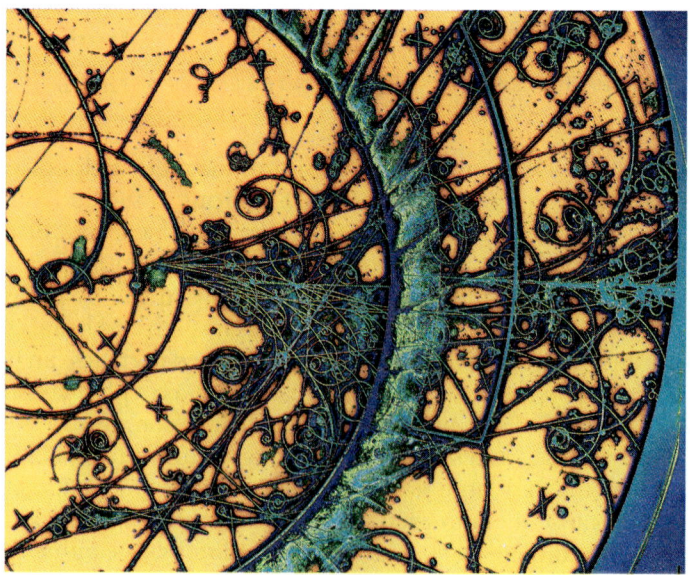

This color-enhanced photograph, taken at CERN, the particle physics laboratory outside Geneva, Switzerland, shows a collection of tracks left by subatomic particles in a bubble chamber. A bubble chamber is a container filled with liquid hydrogen which is super-heated, that is, momentarily raised above its normal boiling point by a sudden drop in pressure in the container. Any charged particle passing through the liquid in this state leaves behind a trail of tiny bubbles as the liquid boils in its wake. These bubbles are seen as fine tracks, showing the characteristic paths of different types of particles. The paths are curved due to an intense applied magnetic field. The tightly wound spiral tracks are due to electrons and positrons.   *(Patrice Loiez, CERN/SPL/Photo Researchers)*

them to emit visible light. This is the origin of the beautiful Aurora Borealis, or Northern Lights. A similar phenomenon seen in the southern hemisphere is called the Aurora Australis.

*29.5   APPLICATIONS OF THE MOTION OF CHARGED PARTICLES IN A MAGNETIC FIELD

In this section we describe some important devices that involve the motion of charged particles in uniform magnetic fields. For many situations, the charge moves with a velocity **v** in the presence of both an electric field **E** and a magnetic field **B**. Therefore, the charge experiences both an electric force $q\mathbf{E}$ and a magnetic force $q\mathbf{v} \times \mathbf{B}$, and so the total force, called the **Lorentz force**, is

$$\mathbf{F} = q\mathbf{E} + q\mathbf{v} \times \mathbf{B} \qquad (29.15)$$

Lorentz force

Velocity Selector

In many experiments involving the motion of charged particles, it is important to have particles that move with essentially the same velocity. This can be achieved by applying a combination of an electric field and a magnetic field oriented as shown in Figure 29.19. A uniform electric field vertically downward is provided by a pair of charged parallel plates, while a uniform magnetic field is applied perpendicular to the page. For q positive, the magnetic force $q\mathbf{v} \times \mathbf{B}$ is upward and the electric force $q\mathbf{E}$ is downward. If the fields are chosen so that the electric force balances the

The technology underlying many appliances such as television sets and computer monitors requires an understanding of the motion of charged particles in electric and/or magnetic fields. Some of the most fundamental properties of electrons and other subatomic particles have been determined in controlled experiments involving electric and magnetic fields. Using this simulator, you will be able to reproduce these famous experiments and observe the motion of a particle as you modify the magnetic field, the electric field, the charge on the particle, and its mass and velocity.

Motion of Charged Particles in Electric and Magnetic Fields

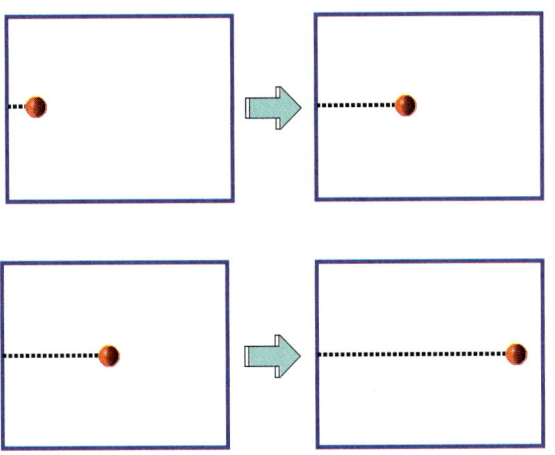

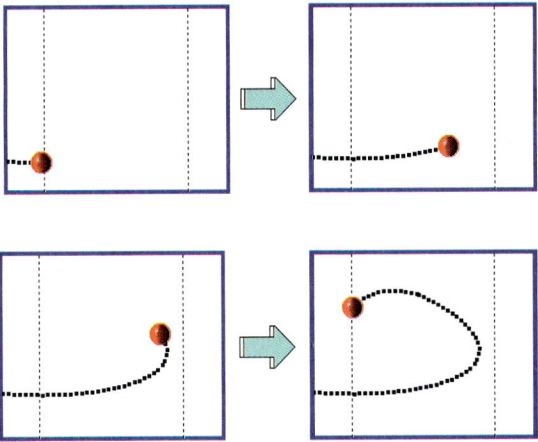

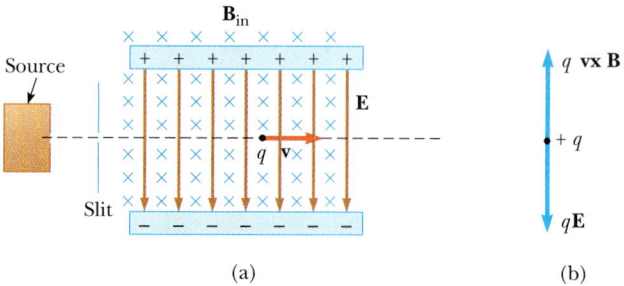

FIGURE 29.19   (a) A velocity selector. When a positively charged particle is in the presence of both an inward magnetic field and a downward electric field, it experiences both an electric force $q\mathbf{E}$ downward and a magnetic force $q\mathbf{v} \times \mathbf{B}$ upward. (b) When these forces balance each other, the particle moves in a horizontal line through the fields.

magnetic force, the particle moves in a straight horizontal line through the region of the fields. If we adjust the magnitudes of the two fields so that $qvB = qE$, we get

$$v = \frac{E}{B} \tag{29.16}$$

Only those particles having this speed pass undeflected through the perpendicular electric and magnetic fields. The magnetic force acting on particles with speeds greater than this is stronger than the electric force, and these particles are deflected upward. Those having speeds less than this are deflected downward.

The Mass Spectrometer

The **mass spectrometer** separates ions according to their mass-to-charge ratio. In one version, known as the *Bainbridge mass spectrometer,* a beam of ions first passes through a velocity selector and then enters a second uniform magnetic field $\mathbf{B}_0$ directed into the paper (Fig. 29.20). Upon entering the second magnetic field, the ions move in a semicircle of radius r before striking a photographic plate at P. From Equation 29.12, we can express the ratio m/q as

$$\frac{m}{q} = \frac{rB_0}{v}$$

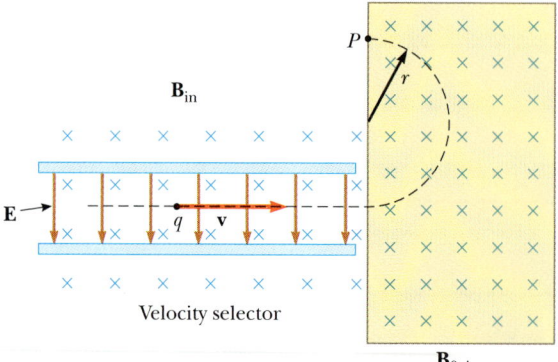

FIGURE 29.20   A mass spectrometer. Charged particles are first sent through a velocity selector. They then enter a region where the magnetic field $\mathbf{B}_0$ (inward) causes positive ions to move in a semicircular path and strike a photographic film at P.

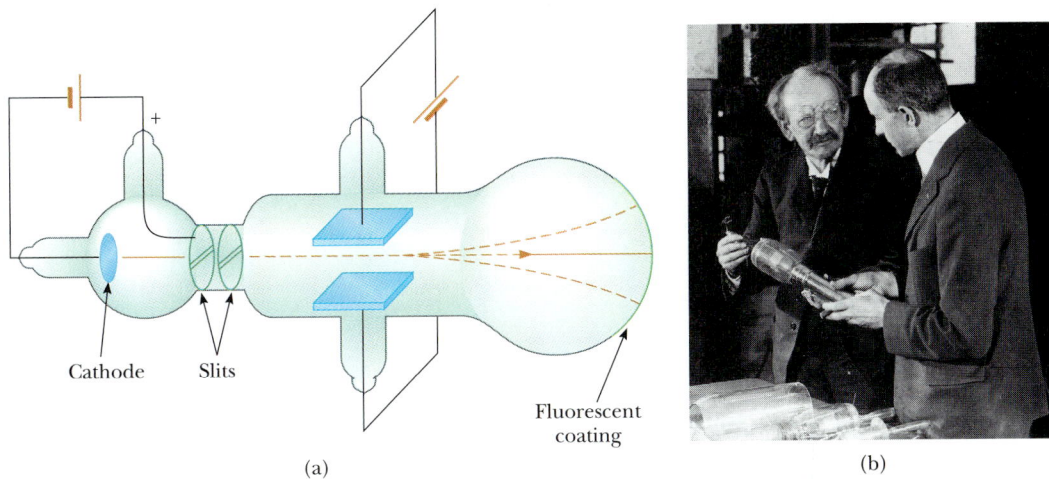

FIGURE 29.21   (a) Thomson's apparatus for measuring e/m. Electrons are accelerated from the cathode, pass through two slits, and are deflected by both an electric field and a magnetic field (not shown, but directed perpendicular to the electric field). The deflected beam then strikes a phosphorescent screen. (b) J. J. Thomson (*left*) with Frank Baldwin Jewett at Western Electric Company in 1923.   *(Bell Telephone Labs/Courtesy Emilio Segrè Visual Archives)*

Using Equation 29.16, we find that

$$\frac{m}{q} = \frac{rB_0 B}{E} \tag{29.17}$$

Therefore, m/q can be determined by measuring the radius of curvature and knowing the fields B, B_0, and E. In practice, one usually measures the masses of various isotopes of a given ion with the same charge q. Hence, the mass ratios can be determined even if q is unknown.

A variation of this technique was used by J. J. Thomson (1856–1940) in 1897 to measure the ratio e/m for electrons. Figure 29.21a shows the basic apparatus he used. Electrons are accelerated from the cathode to the anodes, collimated by slits in the anodes, and then allowed to drift into a region of crossed (perpendicular) electric and magnetic fields. The simultaneously applied **E** and **B** fields are first adjusted to produce an undeflected beam. If the **B** field is then turned off, the **E** field alone produces a measurable beam deflection on the phosphorescent screen. From the size of the deflection and the measured values of E and B, the charge to mass ratio may be determined. The results of this crucial experiment represent the discovery of the electron as a fundamental particle of nature.

The Cyclotron

The **cyclotron,** invented in 1934 by E. O. Lawrence and M. S. Livingston, can accelerate charged particles to very high speeds. Both electric and magnetic forces play a key role. The energetic particles produced are used to bombard other nuclei and thereby produce nuclear reactions of interest to researchers. A number of hospitals use cyclotron facilities to produce radioactive substances for diagnosis and treatment.

A schematic drawing of a cyclotron is shown in Figure 29.22. The charges move in two semicircular containers, D_1 and D_2, referred to as *dees*. The dees are evacuated in order to minimize energy losses resulting from collisions between the ions and air molecules. A high-frequency alternating voltage is applied to the dees, and

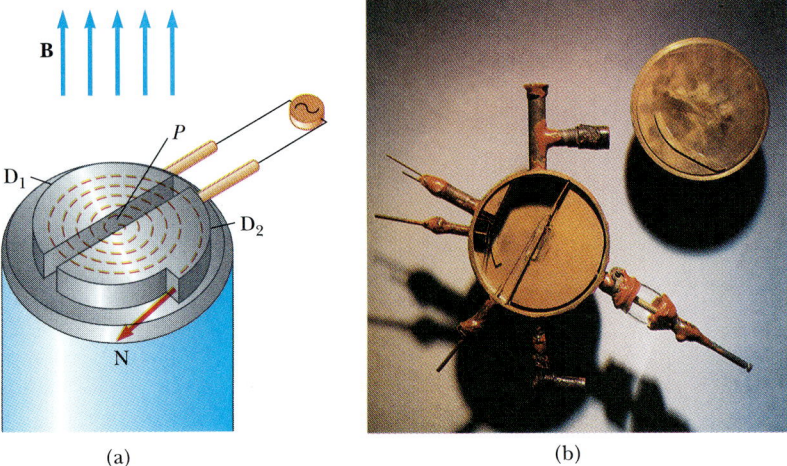

FIGURE 29.22   (a) The cyclotron consists of an ion source, two dees across which an alternating voltage is applied, and a uniform magnetic field provided by an electromagnet. (The south pole of the magnet is not shown.) (b) The first cyclotron, invented by E. O. Lawrence and M. S. Livingston in 1934.   *(Courtesy of Lawrence Berkeley Laboratory, University of California)*

a uniform magnetic field provided by an electromagnet is directed perpendicular to them. Positive ions released at *P* near the center of the magnet move in a semicircular path and arrive back at the gap in a time *T*/2, where *T* is the period of revolution, given by Equation 29.14. The frequency of the applied voltage is adjusted so that the polarity of the dees is reversed in the same time it takes the ions to complete one half of a revolution. If the phase of the applied voltage is adjusted such that D_2 is at a lower potential than D_1 by an amount *V*, the ion accelerates across the gap to D_2 and its kinetic energy increases by an amount *qV*. The ion then continues to move in D_2 in a semicircular path of larger radius (since its speed has increased). After a time *T*/2, it again arrives at the gap. By this time, the potential across the dees is reversed (so that D_1 is now negative) and the ion is given another "kick" across the gap. The motion continues so that for each half revolution, the ion gains additional kinetic energy equal to *qV*. When the radius of its orbit is nearly that of the dees, the energetic ion leaves the system through the exit slit.

It is important to note that the operation of the cyclotron is based on the fact that the time for one revolution is independent of the speed of the ion or the radius of its orbit.

We can obtain an expression for the kinetic energy of the ion when it exits from the cyclotron in terms of the radius *R* of the dees. From Equation 29.12 we know that $v = qBR/m$. Hence, the kinetic energy is

$$K = \tfrac{1}{2}mv^2 = \frac{q^2B^2R^2}{2m} \tag{29.18}$$

When the energy of the ions exceeds about 20 MeV, relativistic effects come into play and the ion masses are no longer constant. (Such effects are discussed in Chapter 39.) For this reason, the period of the orbit increases and the rotating ions do not remain in phase with the applied voltage. Some accelerators solve this problem by modifying the period of the applied voltage so that it remains in phase with the rotating ion. In 1977, protons were accelerated to 400 GeV (1 GeV = 10^9 eV) in an accelerator in Batavia, Illinois. The system incorporates 954 magnets and has a circumference of 6.3 km (4.1 miles)!

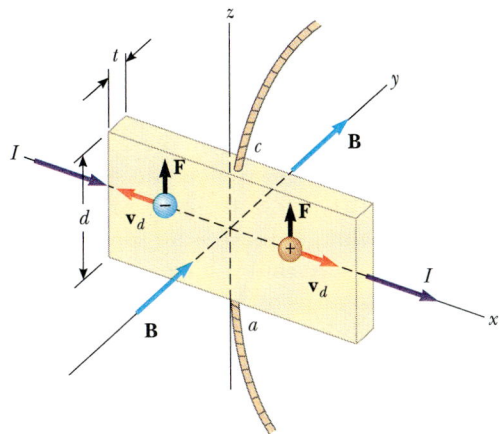

FIGURE 29.23   To observe the Hall effect, a magnetic field is applied to a current-carrying conductor. When I is in the x direction and **B** in the y direction as shown, both positive and negative charge carriers are deflected upward in the magnetic field. The Hall voltage is measured between points a and c.

*29.6   THE HALL EFFECT

In 1879 Edwin Hall discovered that, when a current-carrying conductor is placed in a magnetic field, a voltage is generated in a direction perpendicular to both the current and the magnetic field. This observation, known as the *Hall effect*, arises from the deflection of charge carriers to one side of the conductor as a result of the magnetic force they experience. It gives information regarding the sign of the charge carriers and their density and also provides a convenient technique for measuring magnetic fields.

The arrangement for observing the Hall effect consists of a conductor in the form of a flat strip carrying a current I in the x direction as in Figure 29.23. A uniform magnetic field **B** is applied in the y direction. If the charge carriers are electrons moving in the negative x direction with a velocity v_d, they experience an upward magnetic force **F**, are deflected upward, and accumulate at the upper edge leaving an excess positive charge at the lower edge (Fig. 29.24a). This accumulation of charge at the edges increases until the electrostatic field set up by the charge separation balances the magnetic force on the carriers. When this equilibrium condition is reached, the electrons are no longer deflected upward. A sensi-

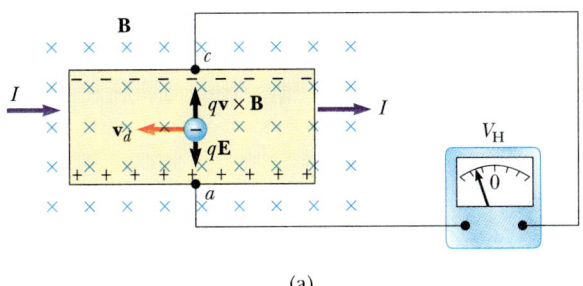

(a)

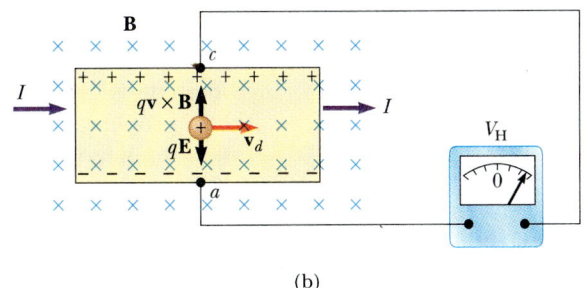

(b)

FIGURE 29.24   (a) When the charge carriers are negative, the upper edge becomes negatively charged, and c is at a lower potential than a. (b) When the charge carriers are positive, the upper edge becomes positively charged and c is at a higher potential than a. In either case, the charge carriers are no longer deflected when the edges become fully charged, that is, when there is a balance between the electrostatic force trying to combine the charges and the magnetic deflection force.

tive voltmeter or potentiometer connected across the sample as in Figure 29.24 can be used to measure the potential difference generated across the conductor, known as the **Hall voltage** V_H. If the charge carriers are positive and hence move in the positive x direction as in Figure 29.24b, they also experience an upward magnetic force $q\mathbf{v}_d \times \mathbf{B}$. This produces a buildup of positive charge on the upper edge and leaves an excess of negative charge on the lower edge. Hence, the sign of the Hall voltage generated in the sample is opposite the sign of the voltage resulting from the deflection of electrons. The sign of the charge carriers can therefore be determined from a measurement of the polarity of the Hall voltage.

To find an expression for the Hall voltage, first note that the magnetic force on the charge carriers has a magnitude $q v_d B$. In equilibrium, this force is balanced by the electrostatic force $q E_H$, where E_H is the electric field due to the charge separation (sometimes referred to as the *Hall field*). Therefore,

$$q v_d B = q E_H$$

$$E_H = v_d B$$

If d is the width of the conductor, the Hall voltage V_H is equal to $E_H d$, or

$$V_H = E_H d = v_d B d \qquad (29.19)$$

Thus, we see that the measured Hall voltage gives a value for the drift speed of the charge carriers if d and B are known.

The number of charge carriers per unit volume (or charge density), n, can be obtained by measuring the current in the sample. From Equation 27.4, the drift speed can be expressed as

$$v_d = \frac{I}{nqA} \qquad (29.20)$$

where A is the cross-sectional area of the conductor. Substituting Equation 29.20 into Equation 29.19, we obtain

$$V_H = \frac{IBd}{nqA} \qquad (29.21)$$

Since $A = td$, where t is the thickness of the conductor, we can also express Equation 29.21 as

$$V_H = \frac{IB}{nqt} \qquad (29.22)$$

The quantity $1/nq$ is referred to as the **Hall coefficient** R_H. Equation 29.22 shows that a properly calibrated conductor can be used to measure the strength of an unknown magnetic field.

Since all quantities appearing in Equation 29.22 other than nq can be measured, a value for the Hall coefficient is readily obtained. The sign and magnitude of R_H give the sign of the charge carriers and their density. In metals, the charge carriers are electrons and the charge density determined from Hall-effect measurements is in good agreement with calculated values for monovalent metals, such as Li, Na, Cu, and Ag, where n is approximately equal to the number of valence electrons per unit volume. However, this classical model is not valid for metals such as Fe, Bi, and Cd or for semiconductors. These discrepancies can be explained only by using a model based on the quantum nature of solids.

EXAMPLE 29.6 The Hall Effect for Copper

A rectangular copper strip 1.5 cm wide and 0.10 cm thick carries a current of 5.0 A. A 1.2-T magnetic field is applied perpendicular to the strip as in Figure 29.24. Find the resulting Hall voltage.

Solution If we assume there is one electron per atom available for conduction, then we can take the charge density to be $n = 8.48 \times 10^{28}$ electrons/m³ (Example 27.1). Substituting this value and the given data into Equation 29.22 gives

$$V_H = \frac{IB}{nqt}$$

$$= \frac{(5.0 \text{ A})(1.2 \text{ T})}{(8.48 \times 10^{28} \text{ m}^{-3})(1.6 \times 10^{-19} \text{ C})(0.10 \times 10^{-2} \text{ m})}$$

$$V_H = \boxed{0.44 \ \mu V}$$

Hence, the Hall voltage is quite small in good conductors. Note that the width of this sample is not needed in this calculation.

In semiconductors, where n is much smaller than in monovalent metals, one finds a larger Hall voltage since V_H varies as the inverse of n. Current levels of the order of 1 mA are generally used for such materials. Consider a piece of silicon with the same dimensions as the copper strip, with $n = 1.0 \times 10^{20}$ electrons/m³. Taking $B = 1.2$ T and $I = 0.10$ mA, we find that $V_H = 7.5$ mV. Such a voltage is readily measured with a potentiometer.

---

*29.7 THE QUANTUM HALL EFFECT

Although the Hall effect was discovered over 100 years ago, it continues to be one of the most valuable techniques for helping scientists understand the electronic properties of metals and semiconductors. For example, in 1980, scientists reported that at low temperatures and in very strong magnetic fields, a two-dimensional system of electrons in a semiconductor has a conductivity $\sigma = i(e^2/h)$, where i is a small integer, e is the electronic charge, and h is Planck's constant. This behavior manifests itself as a series of plateaus in the Hall voltage as the applied magnetic field is varied. The quantized nature of this two-dimensional conductivity was totally unanticipated. As its discoverer, Klaus von Klitzing stated, "It is quite astonishing that it is the total macroscopic conductance of the Hall device which is quantized rather than some idealized microscopic conductivity."

One of the important consequences of the quantum Hall effect is the ability to measure the ratio e^2/h to an accuracy of at least one part in 10^5. This provides a very accurate measure of the dimensionless fine-structure constant, given by $\alpha = e^2/hc \approx 1/137$, since c is an exactly defined quantity (the speed of light). In addition, the quantum Hall effect provides scientists with a new and convenient standard of resistance. The 1985 Nobel Prize in Physics was awarded to von Klitzing for this fundamental discovery.

Another great surprise occurred in 1982 when scientists announced that in some nearly ideal samples at very low temperatures, the Hall conductivity could take on both integer and fractional values of e^2/h. Undoubtedly, future discoveries in this and related areas of science will continue to improve our understanding of the nature of matter.

SUMMARY

The magnetic force that acts on a charge q moving with a velocity **v** in a magnetic field **B** is

$$\mathbf{F} = q\mathbf{v} \times \mathbf{B} \tag{29.1}$$

This magnetic force is in a direction perpendicular both to the velocity of the particle and to the field. The magnitude of this force is

$$F = qvB \sin \theta \tag{29.2}$$

where θ is the smaller angle between **v** and **B**. The SI unit of **B** is the **weber per square meter** (Wb/m^2), also called the **tesla** (T), where $[B] = \mathrm{T} = \mathrm{Wb/m^2} = \mathrm{N/A \cdot m}$.

When a charged particle moves in a magnetic field, the work done by the magnetic force on the particle is zero because the displacement is always perpendicular to the direction of the magnetic force. The magnetic field can alter the direction of the velocity vector, but it cannot change the speed of the particle.

If a straight conductor of length L carries a current I, the force on that conductor when placed in a uniform magnetic field **B** is

$$\mathbf{F} = I\mathbf{L} \times \mathbf{B} \tag{29.3}$$

where the direction of **L** is in the direction of the current and $|\mathbf{L}| = L$.

If an arbitrarily shaped wire carrying a current I is placed in a magnetic field, the magnetic force on a very small segment $d\mathbf{s}$ is

$$d\mathbf{F} = I\,d\mathbf{s} \times \mathbf{B} \tag{29.4}$$

To determine the total magnetic force on the wire, one has to integrate Equation 29.4, keeping in mind that both **B** and $d\mathbf{s}$ may vary at each point.

The force on a current-carrying conductor of arbitrary shape in a uniform magnetic field is

$$\mathbf{F} = I\mathbf{L}' \times \mathbf{B} \tag{29.7}$$

where **L**$'$ is a vector directed from one end of the conductor to the opposite end.

The net magnetic force on any closed loop carrying a current in a uniform magnetic field is zero.

The **magnetic moment** $\boldsymbol{\mu}$ of a current loop carrying a current I is

$$\boldsymbol{\mu} = I\mathbf{A} \tag{29.10}$$

where **A** is perpendicular to the plane of the loop and $|\mathbf{A}|$ is equal to the area of the loop. The SI unit of $\boldsymbol{\mu}$ is $\mathrm{A \cdot m^2}$.

The torque $\boldsymbol{\tau}$ on a current loop when the loop is placed in a uniform *external* magnetic field **B** is

$$\boldsymbol{\tau} = \boldsymbol{\mu} \times \mathbf{B} \tag{29.11}$$

If a charged particle moves in a uniform magnetic field so that its initial velocity is perpendicular to the field, the particle moves in a circle whose plane is perpendicular to the magnetic field. The radius r of the circular path is

$$r = \frac{mv}{qB} \tag{29.12}$$

where m is the mass of the particle and q is its charge. The angular frequency of the rotating charged particle is

$$\omega = \frac{qB}{m} \tag{29.13}$$

QUESTIONS

1. At a given instant, a proton moves in the positive x direction in a region where there is a magnetic field in the negative z direction. What is the direction of the magnetic force? Does the proton continue to move in the positive x direction? Explain.

2. Two charged particles are projected into a region where there is a magnetic field perpendicular to their velocities. If the charges are deflected in opposite directions, what can you say about them?

3. If a charged particle moves in a straight line through some region of space, can you say that the magnetic field in that region is zero?

4. Suppose an electron is chasing a proton up this page when suddenly a magnetic field is formed perpendicular to the page. What happens to the particles?

5. Why does the picture on a TV screen become distorted when a magnet is brought near the screen?

6. How can the motion of a moving charged particle be used to distinguish between a magnetic field and an electric field? Give a specific example to justify your argument.

7. List several similarities and differences in electric and magnetic forces.

8. Justify the following statement: "It is impossible for a constant (in other words, a time independent) magnetic field to alter the speed of a charged particle."

9. In view of the above statement, what is the role of a magnetic field in a cyclotron?

10. A current-carrying conductor experiences no magnetic force when placed in a certain manner in a uniform magnetic field. Explain.

11. Is it possible to orient a current loop in a uniform magnetic field so that the loop does not tend to rotate? Explain.

12. How can a current loop be used to determine the presence of a magnetic field in a given region of space?

13. What is the net force on a compass needle in a uniform magnetic field?

14. What type of magnetic field is required to exert a resultant force on a magnetic dipole? What is the direction of the resultant force?

15. A proton moving horizontally enters a region where there is a uniform magnetic field perpendicular to the proton's velocity, as shown in Figure 29.25. Describe the subsequent motion of the proton. How would an electron behave under the same circumstances?

16. In a magnetic bottle, what reverses the direction of the velocity of the confined charged particles? (*Hint:* Find the direction of the magnetic force on these particles in a region where the field becomes stronger and the field lines converge.)

17. In the cyclotron, why do particles of different velocities take the same amount of time to complete one half of a revolution?

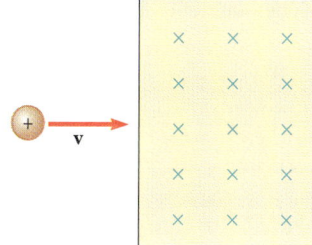

FIGURE 29.25   (Question 15).

18. The *bubble chamber* is a device used for observing tracks of particles that pass through the chamber, which is immersed in a magnetic field. If some of the tracks are spirals and others are straight lines, what can you say about the particles?

19. Can a magnetic field set a resting electron into motion? If so, how?

20. You are designing a magnetic probe that uses the Hall effect to measure magnetic fields. Assume that you are restricted to using a given material and that you have already made the probe as thin as possible. What, if anything, can be done to increase the Hall voltage produced for a given magnetic field strength?

21. The electron beam in Figure 29.26 is projected to the right. The beam deflects downward in the presence of a magnetic field produced by a pair of current-carrying coils. (a) What is the direction of the magnetic field? (b) What would happen to the beam if the current in the coils were reversed?

FIGURE 29.26   (Question 21).   (*Courtesy of Central Scientific Company*)

PROBLEMS

Section 29.1   The Magnetic Field

1. Consider an electron near the equator. In which direction does it tend to deflect if its velocity is directed (a) downward, (b) northward, (c) westward, or (d) southeastward?

2. An electron moving along the positive x axis perpendicular to a magnetic field experiences a magnetic deflection in the negative y direction. What is the direction of the magnetic field?

3. An electron in a uniform electric and magnetic field has a velocity of 1.2×10^4 m/s in the positive x direction and an acceleration of 2.0×10^{12} m/s^2 in the positive z direction. If the electric field has strength of 20 N/C (in the positive z direction), what is the magnetic field in the region?

4. What magnetic force is experienced by a proton moving north to south at 4.8×10^6 m/s at a location where the vertical component of the Earth's magnetic field is 75 μT directed downward? In what direction is the proton deflected?

5. A proton moving at 4.0×10^6 m/s through a magnetic field of 1.7 T experiences a magnetic force of magnitude 8.2×10^{-13} N. What is the angle between the proton's velocity and the field?

6. A metal ball having net charge $Q = 5.0$ μC is thrown out of a window horizontally at a speed $v = 20$ m/s. The window is at a height $h = 20$ m above the ground. A uniform horizontal field of magnitude $B = 0.010$ T is perpendicular to the plane of the ball's trajectory. Find the magnetic force acting on the ball just before it hits the ground.

6A. A metal ball having net charge Q is thrown out of a window horizontally at a speed v. The window is at a height h above the ground. A uniform horizontal magnetic field of magnitude B is perpendicular to the plane of the ball's trajectory. Find the magnetic force acting on the ball just before it hits the ground.

7. A duck flying due north at 15 m/s passes over Atlanta, where the Earth's magnetic field is 5.0×10^{-5} T in a direction 60° below the horizontal line running north and south. If the duck has a net positive charge of 0.040 μC, what is the magnetic force acting on it?

8. An electron is projected into a uniform magnetic field $\mathbf{B} = (1.4\mathbf{i} + 2.1\mathbf{j})$ T. Find the vector expression for the force on the electron when its velocity is $\mathbf{v} = 3.7 \times 10^5 \mathbf{j}$ m/s.

9. A proton moves with a velocity of $\mathbf{v} = (2\mathbf{i} - 4\mathbf{j} + \mathbf{k})$ m/s in a region in which the magnetic field is $\mathbf{B} =$ $(\mathbf{i} + 2\mathbf{j} - 3\mathbf{k})$ T. What is the magnitude of the magnetic force this charge experiences?

10. A proton moves perpendicular to a uniform magnetic field $\mathbf{B}$ at 1.00×10^7 m/s and experiences an acceleration of 2.00×10^{13} m/s^2 in the $+x$ direction when its velocity is in the $+z$ direction. Determine the magnitude and direction of the field.

11. Show that the work done by the magnetic force on a charged particle moving in a magnetic field is zero for any displacement of the particle.

Section 29.2   Magnetic Force on a Current-Carrying Conductor

12. A wire 40 cm long carries a current of 20 A. It is bent into a loop and placed with its plane perpendicular to a magnetic field having a flux density of 0.52 T. What is the torque on the loop if it is bent into (a) an equilateral triangle, (b) a square, (c) a circle. (d) Which torque is greatest?

13. A wire carries a steady current of 2.4 A. A straight section of the wire is 0.75 m long and lies along the x axis within a uniform magnetic field, $\mathbf{B} = (1.6\mathbf{k})$ T. If the current is in the $+x$ direction, what is the magnetic force on the section of wire?

14. A conductor suspended by two flexible wires as in Figure P29.14 has a mass per unit length of 0.040 kg/m. What current must exist in the conductor in order for the tension in the supporting wires to be zero when the magnetic field is 3.6 T into the page? What is the required direction for the current?

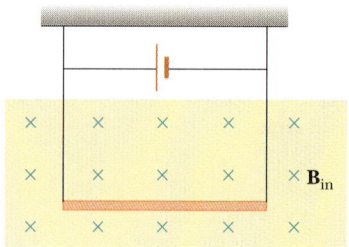

FIGURE P29.14

15. A wire having a mass per unit length of 0.50 g/cm carries a 2.0-A current horizontally to the south. What are the direction and magnitude of the minimum magnetic field needed to lift this wire vertically upward?

16. A rectangular loop with dimensions 10.0 cm × 20.0 cm is suspended by a string, and the lower hori-

zontal section of the loop is immersed in a magnetic field confined to a circular region (Fig. P29.16). If a current of 3.00 A is maintained in the loop in the direction shown, what are the direction and magnitude of the magnetic field required to produce a tension of 4.00×10^{-2} N in the supporting string? (Neglect the mass of the loop.)

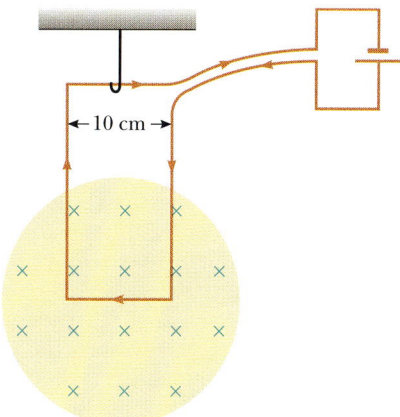

FIGURE P29.16

17. A wire 2.8 m in length carries a current of 5.0 A in a region where a uniform magnetic field has a magnitude of 0.39 T. Calculate the magnitude of the magnetic force on the wire if the angle between the magnetic field and the current is (a) 60°, (b) 90°, (c) 120°.

18. In Figure P29.18, the cube is 40.0 cm on each edge. Four straight segments of wire — ab, bc, cd, and da — form a closed loop that carries a current $I = 5.0$ A in the direction shown. A uniform magnetic field $\mathbf{B} = 0.020$ T is in the positive direction. Determine the magnitude and direction of the magnetic force on each segment.

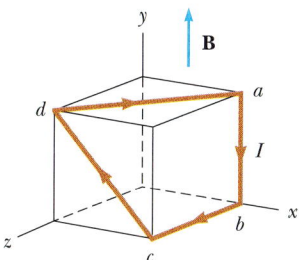

FIGURE P29.18

19. Imagine a very long, uniform wire that has a linear mass density of 1.0 g/m and that encircles the Earth at its magnetic equator. What are the magnitude and direction of the current in the wire that keep it levitated just above the ground?

20. A rod of mass 0.72 kg and radius 6.0 cm rests on two parallel rails (Fig. P29.20) that are $d = 12$ cm apart and $L = 45$ cm long. The rod carries a current $I = 48$ A in the direction shown and rolls along the rails without slipping. If the rod starts from rest, what is its speed as it leaves the rails if there is a uniform 0.24-T magnetic field directed perpendicular to the rod and the rails?

20A. A rod of mass m and radius R rests on two parallel rails (Fig. P29.20) that are a distance d apart and have a length L. The rod carries a current I in the direction shown and rolls along the rails without slipping. If the rod starts from rest, what is its speed as it leaves the rails if there is a uniform magnetic field $\mathbf{B}$ directed perpendicular to the rod and the rails?

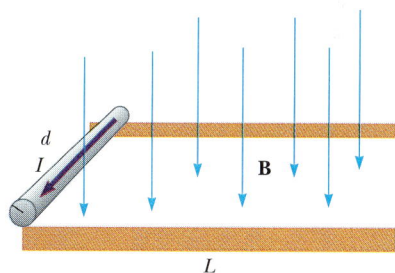

FIGURE P29.20

21. A strong magnet is placed under a horizontal conducting ring of radius r that carries a current I as in Figure P29.21. If the magnetic field $\mathbf{B}$ makes an angle θ with the vertical at the ring's location, what are the magnitude and direction of the resultant force on the ring?

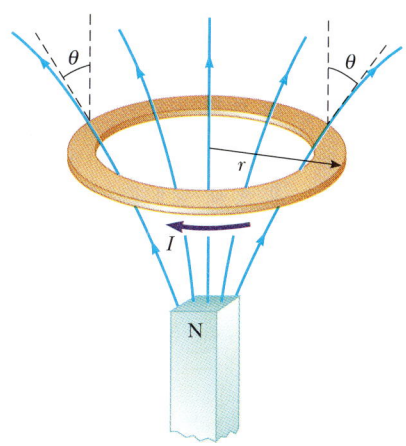

FIGURE P29.21

Section 29.3   Torque on a Current Loop in a Uniform Magnetic Field

22. A current of 17.0 mA is maintained in a single circular loop of 2.00 m circumference. A magnetic field of 0.800 T is directed parallel to the plane of the loop. (a) Calculate the magnetic moment of the loop. (b) What is the magnitude of the torque exerted on the loop by the magnetic field?

23. A rectangular loop consists of $N = 100$ closely wrapped turns and has dimensions $a = 0.40$ m and $b = 0.30$ m. The loop is hinged along the y axis, and its plane makes an angle $\theta = 30°$ with the x axis (Fig. P29.23). What is the magnitude of the torque exerted on the loop by a uniform magnetic field $B = 0.80$ T directed along the x axis when the current is $I = 1.2$ A in the direction shown? What is the expected direction of rotation of the loop?

23A. A rectangular loop consists of N closely wrapped turns and has dimensions a and b. The loop is hinged along the y axis, and its plane makes an angle θ with the x axis (Fig. P29.23). What is the magnitude of the torque exerted on the loop by a uniform magnetic field B directed along the x axis when the current is I in the direction shown? What is the expected direction of rotation of the loop?

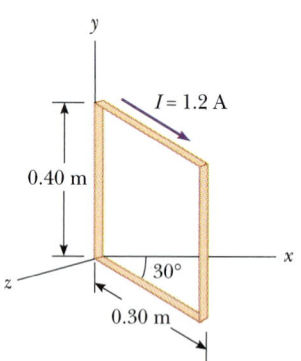

FIGURE P29.23

24. A square loop, hinged along one side, is made of a wire that has a mass per unit length of 0.10 kg/m and carries a current of 5.0 A. A 0.010-T uniform magnetic field directed perpendicular to the hinged side exists in the region. Determine the maximum linear acceleration of the side of the loop opposite the hinged side.

25. A circular coil of 225 turns and area 0.45 m² is in a uniform magnetic field of 0.21 T. The maximum torque exerted on the coil by the field is 8.0×10^{-3} N·m. (a) Calculate the current in the coil. (b) Would this value be different if the 225 turns of wire were used to form a single-turn coil with the same shape but much larger area? Explain.

26. A wire is formed into a circle having a diameter of 10.0 cm and placed in a uniform magnetic field of 3.00×10^{-3} T. A current of 5.00 A passes through the wire. Find (a) the maximum torque on the wire and (b) the range of potential energy the wire possesses for different orientations.

27. A long piece of wire of mass 0.10 kg and length 4.0 m is used to make a square coil 0.10 m on a side. The coil is hinged along a horizontal side, carries a 3.4-A current, and is placed in a vertical magnetic field of magnitude 0.010 T. (a) Determine the angle that the plane of the coil makes with the vertical when the coil is in equilibrium. (b) Find the torque acting on the coil due to the magnetic force at equilibrium.

27A. A long piece of wire of mass m and length L is used to make a square coil having side d. The coil is hinged along a horizontal side, carries a current I, and is placed in a vertical magnetic field of magnitude B. (a) Determine the angle that the plane of the coil makes with the vertical when the coil is in equilibrium. (b) Find the torque acting on the coil due to the magnetic force at equilibrium.

Section 29.4   Motion of a Charged Particle in a Magnetic Field

28. The magnetic field of the Earth at a certain location is directed vertically downward and has a magnitude of 0.5×10^{-4} T. A proton is moving horizontally towards the west in this field with a speed of 6.2×10^6 m/s. (a) What are the direction and magnitude of the magnetic force the field exerts on this charge? (b) What is the radius of the circular arc followed by this proton?

29. A singly charged positive ion has a mass of 3.20×10^{-26} kg. After being accelerated from rest through a potential difference of 833 V, the ion enters a magnetic field of 0.920 T along a direction perpendicular to the direction of the field. Calculate the radius of the path of the ion in the field.

30. One electron collides with a second electron initially at rest. After the collision, the radii of their trajectories are 1.0 cm and 2.4 cm. The trajectories are perpendicular to a uniform magnetic field of magnitude 0.044 T. Determine the energy (in keV) of the incident electron.

30A. One electron collides with a second electron initially at rest. After the collision, the radii of their trajectories are R_1 and R_2. The trajectories are perpendicular to a uniform magnetic field of magnitude B. Determine the energy (in keV) of the incident electron.

31. A proton moving in a circular path perpendicular to a constant magnetic field takes 1.00 μs to complete one revolution. Determine the magnitude of the field.

32. An electron moves in a circular path perpendicular to a constant magnetic field of magnitude 1.00 mT. If the angular momentum of the electron about the center of the circle is 4.00×10^{-25} J·s, determine (a) the radius of the path and (b) the speed of the electron.

33. A proton (charge $+e$, mass m_p), a deuteron (charge $+e$, mass $2m_p$), and an alpha particle, (charge $+2e$, mass $4m_p$) are accelerated through a common potential difference, V. The particles enter a uniform magnetic field, $\mathbf{B}$, in a direction perpendicular to $\mathbf{B}$. The proton moves in a circular path of radius r_p. Determine the values of the radii of the circular orbits for the deuteron, r_d, and the alpha particle, r_α, in terms of r_p.

34. Calculate the cyclotron frequency of a proton in a magnetic field of magnitude 5.2 T.

35. A cosmic-ray proton in interstellar space has an energy of 10 MeV and executes a circular orbit having a radius equal to that of Mercury's orbit around the Sun (5.8×10^{10} m). What is the magnetic field in that region of space?

36. A singly charged ion of mass m is accelerated from rest by a potential difference V. It is then deflected by a uniform magnetic field (perpendicular to the ion's velocity) into a semicircle of radius R. Now a doubly charged ion of mass m' is accelerated through the same potential difference and deflected by the same magnetic field into a semicircle of radius $R' = 2R$. What is the ratio of the ions' masses?

37. A singly charged positive ion moving at 4.60×10^5 m/s leaves a circular track of radius 7.94 mm along a direction perpendicular to the 1.80-T magnetic field of a bubble chamber. Compute the mass (in atomic mass units) of this ion, and, from that value, identify it.

38. The accelerating voltage that is applied to an electron gun is 15 kV, and the horizontal distance from the gun to a viewing screen is 35 cm. What is the deflection caused by the vertical component of the Earth's magnetic field (4.0×10^{-5} T), assuming that any change in the horizontal component of the beam velocity is negligible.

*Section 29.5 Applications of the Motion of Charged Particles in a Magnetic Field

39. A crossed-field velocity selector has a magnetic field of magnitude 1.00×10^{-2} T. What electric field strength is required if 10.0-keV electrons are to pass through undeflected?

40. A velocity selector consists of magnetic and electric fields described by $\mathbf{E} = E\mathbf{k}$ and $\mathbf{B} = B\mathbf{j}$. If $B = 0.015$ T, find the value of E such that a 750-eV electron moving along the positive x axis is undeflected.

41. At the equator, near the surface of the Earth, the magnetic field is approximately 50 μT northward and the electric field is about 100 N/C downward (in other words, toward the ground). Find the gravitational, electric, and magnetic forces on a 100-eV electron moving eastward in a straight line in this environment.

42. (a) Singly charged uranium-238 ions are accelerated through a potential difference of 2.00 kV and enter a uniform magnetic field of 1.20 T directed perpendicular to their velocities. Determine the radius of their circular path. (b) Repeat for uranium-235 ions. How does the ratio of these path radii depend on the accelerating voltage and the magnetic field strength?

43. Consider the mass spectrometer shown schematically in Figure 29.20. The electric field between the plates of the velocity selector is 2500 V/m, and the magnetic field in both the velocity selector and the deflection chamber has a magnitude of 0.0350 T. Calculate the radius of the path for a singly charged ion having a mass $m = 2.18 \times 10^{-26}$ kg.

44. What is the required radius of a cyclotron designed to accelerate protons to energies of 34 MeV using a magnetic field of 5.2 T?

45. What is the minimum size of a cyclotron designed to accelerate protons to an energy of 18.0 MeV with a cyclotron frequency of 30.0 MHz?

46. At the Fermilab accelerator in Batavia, Illinois, protons having momentum 4.8×10^{-16} kg·m/s are held in a circular orbit of radius 1.0 km by an upward magnetic field. What is the magnitude of this field?

47. A cyclotron designed to accelerate protons has a magnetic field of magnitude 0.45 T over a region of radius 1.2 m. What are (a) the cyclotron frequency and (b) the maximum speed acquired by the protons?

48. Deuterium ions are accelerated through a potential difference of 45 kV. The ions enter a velocity selector in which the electric field intensity is 2.5 kV/m. They then continue into a uniform magnetic field that has the same flux density and direction as the magnetic field in the velocity selector. What are (a) the radius of the deuterons' orbit and (b) their speed? (c) What is the strength of the magnetic field?

49. The picture tube in a television uses magnetic deflection coils rather than electric deflection plates. Suppose an electron beam is accelerated through a 50.0-kV potential difference and then travels through a region of uniform magnetic field 1.00 cm wide. The screen is located 10.0 cm from the center of the coils and is 50.0 cm wide. When the field is turned off, the electron beam hits the center of the screen. What field strength is necessary to deflect the beam to the side of the screen?

*Section 29.6 The Hall Effect

50. A flat ribbon of silver having a thickness $t = 0.20$ mm is used in a Hall-effect measurement of a uniform

magnetic field perpendicular to the ribbon, as in Figure P29.50. The Hall coefficient for silver is $R_H = 0.84 \times 10^{-10}$ m³/C. (a) What is the density of charge carriers in silver? (b) If a current $I = 20$ A produces a Hall voltage $V_H = 15$ μV, what is the magnitude of the applied magnetic field?

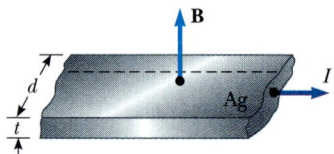

FIGURE P29.50

51. A section of conductor 0.40 cm thick is used in a Hall-effect measurement. If a Hall voltage of 35 μV is measured for a current of 21 A in a magnetic field of 1.8 T, calculate the Hall coefficient for the conductor.

52. The thickness of a thin film of copper is to be determined using the Hall effect. In the experiment, a Hall voltage of 27.0 μV is measured for a current of 17.0 A in a 2.10-T magnetic field. Calculate the thickness of the film.

53. In an experiment designed to measure the Earth's magnetic field using the Hall effect, a copper bar 0.50 cm thick is positioned along an east-west direction. If a current of 8.0 A in the conductor results in a Hall voltage of 5.1×10^{-12} V, what is the magnitude of the Earth's magnetic field? (Assume that $n = 8.48 \times 10^{28}$ electrons/m³ and that the plane of the bar is rotated to be perpendicular to the direction of **B**.)

54. A flat copper ribbon 0.33 mm thick carries a steady current of 50 A and is located in a uniform 1.3-T magnetic field directed perpendicular to the plane of the ribbon. If a Hall voltage of 9.6 μV is measured across the ribbon, what is the charge density of the free electrons? What effective number of free electrons per atom does this result indicate?

54A. A flat copper ribbon of thickness t carries a steady current I and is located in a uniform magnetic field B directed perpendicular to the plane of the ribbon. If a Hall voltage V_H is measured across the ribbon, what is the charge density of the free electrons? What effective number of free electrons per atom does this result indicate?

55. The Hall effect can be used to measure n, the number of conduction electrons per unit volume for an unknown sample. The sample is 15 mm thick and when placed in a 1.8-T magnetic field produces a Hall voltage of 0.122 μV while carrying a 12-A current. What is the value of n?

56. A Hall-effect probe operates with a 120-mA current. When the probe is placed in a uniform magnetic field of magnitude 0.080 T, it produces a Hall voltage of 0.70 μV. (a) When it is measuring an unknown magnetic field, the Hall voltage is 0.33 μV. What is the unknown field strength? (b) If the thickness of the probe in the direction of **B** is 2.0 mm, find the charge-carrier density (each of charge e).

ADDITIONAL PROBLEMS

57. A wire having a linear mass density of 1.0 g/cm is placed on a horizontal surface that has a coefficient of friction of 0.20. The wire carries a current of 1.5 A toward the east and moves horizontally to the north. What are the magnitude and direction of the smallest magnetic field that enables the wire to move in this fashion?

58. Indicate the initial direction of the deflection of the charged particles as they enter the magnetic fields shown in Figure P29.58.

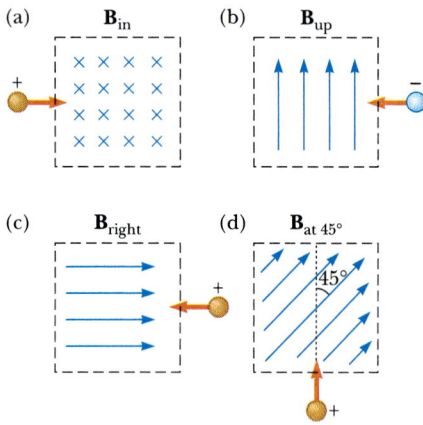

FIGURE P29.58

59. A positive charge $q = 3.2 \times 10^{-19}$ C moves with a velocity $\mathbf{v} = (2\mathbf{i} + 3\mathbf{j} - \mathbf{k})$ m/s through a region where both a uniform magnetic field and a uniform electric field exist. (a) Calculate the total force on the moving charge (in unit-vector notation) if $\mathbf{B} = (2\mathbf{i} + 4\mathbf{j} + \mathbf{k})$ T and $\mathbf{E} = (4\mathbf{i} - \mathbf{j} - 2\mathbf{k})$ V/m. (b) What angle does the force vector make with the positive x axis?

60. A cosmic-ray proton traveling at half the speed of light is heading directly toward the center of the Earth in the plane of the Earth's equator. Will it hit the Earth? Assume that the magnitude of the Earth's magnetic field is 5.0×10^{-5} T and extends out one Earth diameter, or 1.3×10^7 m. Calculate the radius of curvature of the proton in this magnetic field.

61. The circuit in Figure P29.61 consists of wires at the top and bottom and identical metal springs as the left and right sides. The wire at the bottom has a mass of 10 g and is 5.0 cm long. The springs stretch 0.50 cm under the weight of the wire and the circuit has a total resistance of 12 Ω. When a magnetic field is turned on, directed out of the page, the springs stretch an additional 0.30 cm. What is the strength of the magnetic field? (The upper portion of the circuit is fixed.)

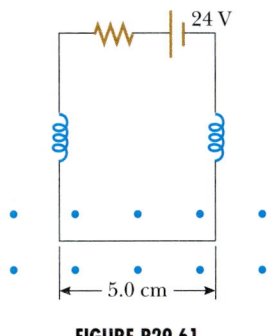

24 V

5.0 cm

FIGURE P29.61

62. An electron enters a region traveling perpendicular to the linear boundary of a 0.10-T magnetic field. The direction of the field is perpendicular to the velocity of the electron. (a) Determine the time it takes for the electron to leave the "field-filled" region, given that it travels a semicircular path. (b) Find the kinetic energy of the electron if the maximum depth of penetration in the field is 2.0 cm.

63. Sodium melts at 99°C. Liquid sodium, an excellent thermal conductor, is used in some nuclear reactors to remove thermal energy from the reactor core. The liquid sodium is moved through pipes by pumps that exploit the force on a moving charge in a magnetic field. The principle is as follows: Imagine the liquid metal to be in a pipe having a rectangular cross-section of width w and height h. A uniform magnetic field perpendicular to the pipe affects a section of length L (Fig. P29.63). An electric current directed perpendicular to the pipe and to the magnetic field produces a current density J. (a) Explain why this

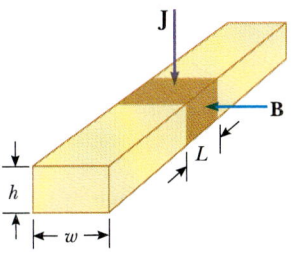

FIGURE P29.63

arrangement produces on the liquid a force that is directed along the length of the pipe. (b) Show that the section of liquid in the magnetic field experiences a pressure increase JLB.

64. A metal rod having a mass per unit length of 0.010 kg/m carries a current of $I = 5.0$ A. The rod hangs from two vertical wires in a uniform vertical magnetic field as in Figure P29.64. If the wires make an angle $\theta = 45°$ with the vertical when in equilibrium, determine the strength of the magnetic field.

64A. A metal rod having a mass per unit length μ carries a current I. The rod hangs from two vertical wires in a uniform vertical magnetic field as in Figure P29.64. If the wires make an angle θ with the vertical when in equilibrium, determine the strength of the magnetic field.

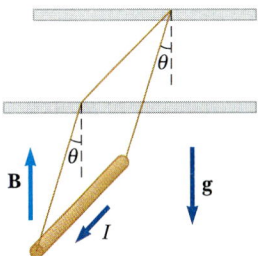

FIGURE P29.64

65. Cyclotrons are sometimes used for carbon-dating by accelerating C^{14} and C^{12} ions from a sample of material. If the cyclotron has a 2.4-T magnetic field, what is the difference in cyclotron frequencies for the two ions?

66. A 0.20-kg metal rod carrying a current of 10 A glides on two horizontal rails 0.50 m apart. What vertical magnetic field is required to keep the rod moving at a constant speed if the coefficient of kinetic friction between rod and rails is 0.10?

66A. A metal rod of mass m and carrying a current I glides on two horizontal rails separated by a distance d. What vertical magnetic field is required to keep the rod moving at a constant speed if the coefficient of kinetic friction between rod and rails is μ?

67. A singly charged ion completes five revolutions in a uniform magnetic field of magnitude 5.00×10^{-2} T in 1.50 ms. Calculate the approximate mass of the ion in kilograms.

68. A uniform magnetic field of magnitude 0.15 T is directed along the positive x axis. A positron moving at 5.0×10^6 m/s enters the field along a direction that makes an angle of 85° with the x axis (Fig. P29.68). The motion of the particle is expected to be a helix, as described in Section 29.4. Calculate (a) the pitch p and (b) the radius r of the trajectory.

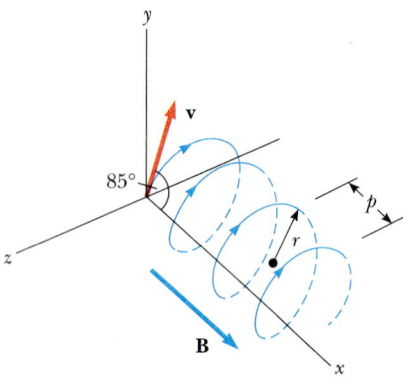

FIGURE P29.68

69. Consider an electron orbiting a proton and maintained in a fixed circular path of radius $R = 5.29 \times 10^{-11}$ m by the Coulomb force. Treating the orbiting charge as a current loop, calculate the resulting torque when the system is in a magnetic field of 0.400 T directed perpendicular to the magnetic moment of the electron.

70. A proton moving in the plane of the page has a kinetic energy of 6.0 MeV. It enters a magnetic field of magnitude $B = 1.0$ T directed into the page at an angle $\theta = 45°$ to the linear boundary of the field as shown in Figure P29.70. (a) Find x, the distance from the point of entry to where the proton will leave the field. (b) Determine θ', the angle between the boundary and the proton's velocity vector as it leaves the field.

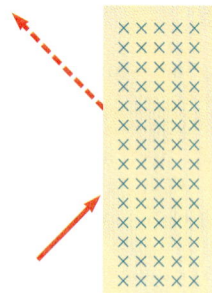

FIGURE P29.70

71. Protons having a kinetic energy of 5.00 MeV are moving in the positive x direction and enter a magnetic field $\mathbf{B} = (0.0500 \text{ T})\mathbf{k}$ directed out of the plane of the page and extending from $x = 0$ to $x = 1.00$ m as in Figure P29.71. (a) Calculate the y component of the protons' momentum as they leave the magnetic field. (b) Find the angle α between the initial velocity vector of the proton beam and the velocity vector after the beam emerges from the field. (*Hint:* Ne-

glect relativistic effects and note that 1 eV = 1.60×10^{-19} J.)

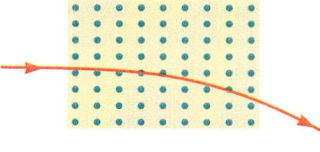

FIGURE P29.71

72. Table 29.1 shows measurements of a Hall voltage and corresponding magnetic field for a probe used to measure magnetic fields. (a) Plot these data, and deduce a relationship between the two variables. (b) If the measurements were taken with a current of 0.20 A and the sample is made from a material having a charge-carrier density of $1.0 \times 10^{26}/\text{m}^3$, what is the thickness of the sample?

TABLE 29.1

V_H (μV)	B (T)
0	0.00
11	0.10
19	0.20
28	0.30
42	0.40
50	0.50
61	0.60
68	0.70
79	0.80
90	0.90
102	1.00

SPREADSHEET PROBLEMS

S1. A galvanometer consists of a coil of wire suspended in a radial magnetic field by a thin, flexible fiber. When a current I passes through the coil, a torque is produced that causes the coil to rotate. The fiber, in turn, supplies a restoring torque τ that is proportional to the angle θ through which it has been twisted. That is, $\tau = \kappa\theta$, where κ is the torsion constant of the supporting fiber. The current I through the coil is

$$I = \frac{\kappa\theta}{NAB}$$

where N is the number of turns in the coil, A is the coil's area, and B is magnitude of the magnetic field.

A student passes small currents through a coil and measures its angular deflections for each current. The following data are obtained for a coil of area 2.0 cm² with 100 turns of wire and $B = 0.015$ T:

$I(\mu A)$	$\theta(\text{deg})$
0.10	3
0.15	6
0.20	7
0.30	11
0.50	18
0.75	27

Using a spreadsheet, plot these data points on a graph of current versus angular deflection. (Plot the data as *points*, not as a connected line.) Use the spreadsheet's regression (least-squares) routine to determine the straight line that best fits the data. Add the best-fit line to the plot, and obtain the torsion constant κ from the slope of this line.

S2. An electron has an initial velocity v_0 at the origin of a coordinate system at $t = 0$. A uniform electric field $\mathbf{E} = (1.00 \times 10^3 \text{ V/m})\mathbf{i}$ and a uniform magnetic field $\mathbf{B} = (0.500 \times 10^{-4} \text{ T})\mathbf{j}$ are turned on at $t = 0$. Write a spreadsheet or computer program to integrate the equations of motion. Since v_0 can be in any direction, you must consider the motion along x, y, and z directions. Describe the general motion of the electron for various values of v_0.

Sources of the Magnetic Field

Oxygen, a paramagnetic substance, is attracted to a magnetic field. The liquid oxygen in this photograph is suspended between the poles of a permanent magnet. Paramagnetic substances contain atoms (or ions) that have permanent magnetic dipole moments. These dipoles interact weakly with each other and are randomly oriented in the absence of an external magnetic field. When the substance is placed in an external magnetic field, its atomic dipoles tend to line up with the field. *(Courtesy of Leon Lewandowski)*

The preceding chapter treated a class of problems involving the magnetic force on a charged particle moving in a magnetic field. To complete the description of the magnetic interaction, this chapter deals with the origin of the magnetic field, namely, moving charges or electric currents. We begin by showing how to use the law of Biot and Savart to calculate the magnetic field produced at a point by a current element. Using this formalism and the superposition principle, we then calculate the total magnetic field due to a distribution of currents for several geometries. Next, we show how to determine the force between two current-carrying conductors, a calculation that leads to the definition of the ampere. We also introduce Ampère's law, which is very useful for calculating the magnetic field of highly symmetric configurations carrying steady currents. We apply Ampère's law to determine the magnetic field for several current configurations.

This chapter is also concerned with some aspects of the complex processes that occur in magnetic materials. All magnetic effects in matter can be explained on the basis of magnetic dipole moments similar to those associated with current loops. These atomic magnetic moments arise both from the orbital motion of the electrons and from an intrinsic property of the electrons known as spin. Our description of magnetism in matter is based in part on the experimental fact that the presence of bulk matter generally modifies the magnetic field produced by currents. For example, when a material is placed inside a current-carrying solenoid, the material sets up its own magnetic field, which adds vectorially to the field previously present.

30.1   THE BIOT-SAVART LAW

Shortly after Oersted's discovery in 1819 that a compass needle is deflected by a current-carrying conductor, Jean Baptiste Biot and Felix Savart reported that a conductor carrying a steady current exerts a force on a magnet. From their experimental results, Biot and Savart arrived at an expression that gives the magnetic field at some point in space in terms of the current that produces the field. The Biot-Savart law says that if a wire carries a steady current I, the magnetic field $d\mathbf{B}$ at a point P associated with an element of the wire $d\mathbf{s}$ (Fig. 30.1) has the following properties:

- The vector $d\mathbf{B}$ is perpendicular both to $d\mathbf{s}$ (which is a vector having units of length and in the direction of the current) and to the unit vector $\hat{\mathbf{r}}$ directed from the element to P.
- The magnitude of $d\mathbf{B}$ is inversely proportional to r^2, where r is the distance from the element to P.
- The magnitude of $d\mathbf{B}$ is proportional to the current and to the length ds of the element.
- The magnitude of $d\mathbf{B}$ is proportional to $\sin\theta$, where θ is the angle between the vectors $d\mathbf{s}$ and $\hat{\mathbf{r}}$.

Properties of the magnetic field created by an electric current

The **Biot-Savart law** can be summarized

$$d\mathbf{B} = k_m \frac{I\,d\mathbf{s} \times \hat{\mathbf{r}}}{r^2} \tag{30.1}$$

Biot-Savart law

where k_m is a constant that in SI units is exactly $10^{-7}\ \text{T}\cdot\text{m/A}$. This constant is usually written $\mu_0/4\pi$, where μ_0 is another constant, called the **permeability of free space**:

$$\frac{\mu_0}{4\pi} = k_m = 10^{-7}\ \text{T}\cdot\text{m/A} \tag{30.2}$$

$$\mu_0 = 4\pi k_m = 4\pi \times 10^{-7}\ \text{T}\cdot\text{m/A} \tag{30.3}$$

Permeability of free space

Hence, Equation 30.1 can also be written

$$d\mathbf{B} = \frac{\mu_0}{4\pi} \frac{I\,d\mathbf{s} \times \hat{\mathbf{r}}}{r^2} \tag{30.4}$$

Biot-Savart law

It is important to note that the Biot-Savart law gives the magnetic field at a point only for a small element of the conductor. To find the total magnetic field **B** created at some point by a conductor of finite size, we must sum up contributions from all current elements making up the conductor. That is, we must evaluate **B** by integrating Equation 30.4:

$$\mathbf{B} = \frac{\mu_0 I}{4\pi} \int \frac{d\mathbf{s} \times \hat{\mathbf{r}}}{r^2} \tag{30.5}$$

where the integral is taken over the entire conductor. This expression must be handled with special care because the integrand is a vector quantity.

There are interesting similarities between the Biot-Savart law of magnetism and Coulomb's law of electrostatics. The current element produces a magnetic field, whereas a point charge produces an electric field. Furthermore, *the magnitude of the magnetic field varies as the inverse square of the distance from the current element,* as does the electric field due to a point charge. However, the directions of the two fields are quite different. The electric field created by a point charge is radial. In the case of a positive point charge, E is directed from the charge to any point. The magnetic field created by a current element is perpendicular to both the element and the radius vector. Hence, if the conductor lies in the plane of the paper, as in Figure 30.1, *d*B points out of the paper at *P* and into the paper at *P'*.

The examples that follow illustrate how to use the Biot-Savart law for calculating the magnetic field of several important geometric arrangements. It is important that you recognize that the magnetic field described in these calculations is *the field created by a current-carrying conductor.* This is not to be confused with any external field that may be applied to the conductor.

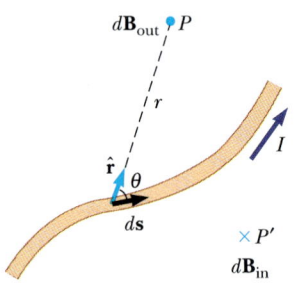

FIGURE 30.1 The magnetic field *d*B at a point *P* due to a current element *d*s is given by the Biot-Savart law. The field is out of the page at *P* and into the page at *P'*.

EXAMPLE 30.1   Magnetic Field Surrounding a Thin, Straight Conductor

Consider a thin, straight wire carrying a constant current *I* and placed along the *x* axis as in Figure 30.2. Calculate the total magnetic field at *P*.

Solution   An element *d*s is a distance *r* from *P*. The direction of the field at *P* due to this element is out of the paper because *d*s × r̂ is out of the paper. In fact, *all* elements produce a magnetic field directed out of the paper at *P*. Therefore, we have only to determine the magnitude of the field at *P*. Taking the origin at *O* and letting *P* be along the positive *y* axis, with **k** being a unit vector pointing out of the paper, we see that

$$d\mathbf{s} \times \hat{\mathbf{r}} = \mathbf{k} \,|\, d\mathbf{s} \times \hat{\mathbf{r}} \,| = \mathbf{k}(dx \sin \theta)$$

Substitution into Equation 30.4 gives *d*B = **k** *dB*, with

$$(1) \qquad dB = \frac{\mu_0 I}{4\pi} \frac{dx \sin \theta}{r^2}$$

In order to integrate this expression, we must relate the variables θ, *x*, and *r*. One approach is to express *x* and *r* in terms of θ. From the geometry in Figure 30.2a and some simple differentiation, we obtain the following relationship:

$$(2) \qquad r = \frac{a}{\sin \theta} = a \csc \theta$$

Since tan θ = −*a*/*x* from the right triangle in Figure 30.2a, we have

$$x = -a \cot \theta$$

$$(3) \qquad dx = a \csc^2 \theta \, d\theta$$

Substitution of (2) and (3) into (1) gives

$$(4) \qquad dB = \frac{\mu_0 I}{4\pi} \frac{a \csc^2 \theta \sin \theta \, d\theta}{a^2 \csc^2 \theta}$$

$$= \frac{\mu_0 I}{4\pi a} \sin \theta \, d\theta$$

Thus, we have reduced the expression to one involving only the variable θ. We can now obtain the total field at *P* by integrating (4) over all elements subtending angles ranging from θ₁ to θ₂ as defined in Figure 30.2b:

$$B = \frac{\mu_0 I}{4\pi a} \int_{\theta_1}^{\theta_2} \sin \theta \, d\theta = \frac{\mu_0 I}{4\pi a} (\cos \theta_1 - \cos \theta_2) \tag{30.6}$$

We can apply this result to find the magnetic field of any straight wire if we know the geometry and hence the angles θ₁ and θ₂.

Consider the special case of an infinitely long, straight wire. In this case, θ₁ = 0 and θ₂ = π, as can be seen from

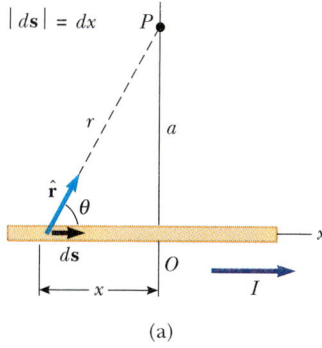

$|d\mathbf{s}| = dx$

(a)

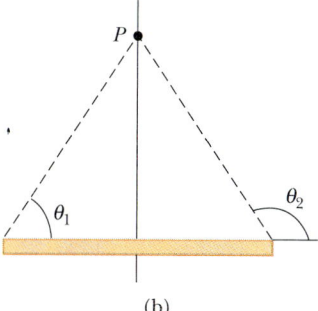

(b)

FIGURE 30.2 (Example 30.1) (a) A straight wire segment carrying a current *I*. The magnetic field at *P* due to each element *d***s** is out of the paper, and so the net field is also out of the paper. (b) The limiting angles θ_1 and θ_2 for this geometry.

Figure 30.2b, for segments ranging from $x = -\infty$ to $x = +\infty$. Since $(\cos\theta_1 - \cos\theta_2) = (\cos 0 - \cos\pi) = 2$, Equation 30.6 becomes

$$B = \frac{\mu_0 I}{2\pi a} \qquad (30.7)$$

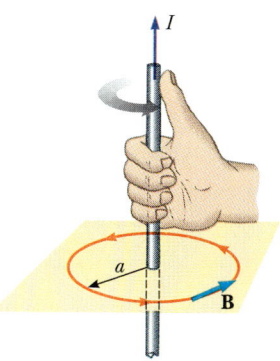

FIGURE 30.3 The right-hand rule for determining the direction of the magnetic field surrounding a long, straight wire carrying a current. Note that the magnetic field lines form circles around the wire.

A three-dimensional view of the direction of **B** for a long, straight wire is shown in Figure 30.3. The field lines are circles concentric with the wire and are in a plane perpendicular to the wire. The magnitude of **B** is constant on any circle of radius *a* and is given by Equation 30.7. A convenient rule for determining the direction of **B** is to grasp the wire with the right hand, with the thumb along the direction of the current. The four fingers wrap in the direction of the magnetic field.

Our result shows that the magnitude of the magnetic field is proportional to the current and decreases as the distance from the wire increases, as one might intuitively expect. Notice that Equation 30.7 has the same mathematical form as the expression for the magnitude of the electric field due to a long charged wire (Eq. 24.7).

Exercise Calculate the magnitude of the magnetic field 4.0 cm from a long, straight wire carrying a current of 5.0 A.

Answer 2.5×10^{-5} T.

EXAMPLE 30.2   Magnetic Field Due to a Wire Segment

Calculate the magnetic field at the point *O* for the wire segment shown in Figure 30.4. The wire consists of two straight portions and a circular arc of radius *R*, which subtends an angle θ.

Reasoning First, note that the magnetic field at *O* due to the straight segments *AA′* and *CC′* is identically zero, because *d***s** is parallel to $\hat{\mathbf{r}}$ along these paths so that $d\mathbf{s} \times \hat{\mathbf{r}} = 0$.

Note that each element along the path *AC* is at the same distance *R* from *O*, and each gives a contribution *d***B**, which is directed into the paper at *O*. Furthermore, at

every point on the path *AC*, *d***s** is perpendicular to $\hat{\mathbf{r}}$, so that $|d\mathbf{s} \times \hat{\mathbf{r}}| = ds$.

Solution Using this information and Equation 30.4, we get for the field at *O* due to the segment *d***s**

$$dB = \frac{\mu_0 I}{4\pi} \frac{ds}{R^2}$$

Since *I* and *R* are constants, we can easily integrate this expression:

$$B = \frac{\mu_0 I}{4\pi R^2} \int ds = \frac{\mu_0 I}{4\pi R^2} s = \frac{\mu_0 I}{4\pi R} \theta \qquad (30.8)$$

where we have used the fact that $s = R\theta$, where θ is measured in radians. The direction of **B** is into the paper at O because $d\mathbf{s} \times \hat{\mathbf{r}}$ is into the paper for every segment.

Exercise   A loop in the form of a full circle of radius R carries a current I. What is the magnitude of the magnetic field at its center?

Answer $\mu_0 I/2R$.

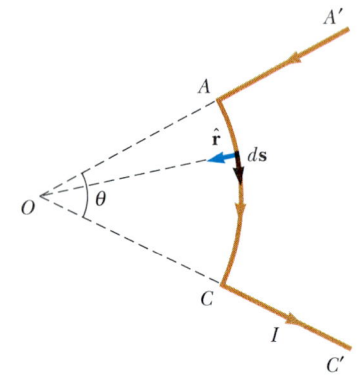

FIGURE 30.4   (Example 30.2) The magnetic field at O due to the curved segment AC is into the paper. The contribution to the field at O due to the straight segments is zero.

EXAMPLE 30.3   Magnetic Field on the Axis of a Circular Current Loop

Consider a circular loop of wire of radius R located in the yz plane and carrying a steady current I, as in Figure 30.5. Calculate the magnetic field at an axial point P a distance x from the center of the loop.

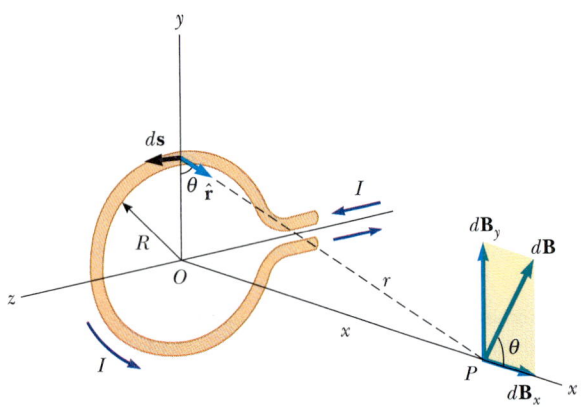

FIGURE 30.5   (Example 30.3) The geometry for calculating the magnetic field at an axial point P for a current loop. Note that by symmetry the total field **B** is along the x axis.

Reasoning   In this situation, note that any element $d\mathbf{s}$ is perpendicular to $\hat{\mathbf{r}}$. Furthermore, all elements around the loop are at the same distance r from P, where $r^2 = x^2 + R^2$. Hence, the magnitude of $d\mathbf{B}$ due to the element $d\mathbf{s}$ is

$$dB = \frac{\mu_0 I}{4\pi} \frac{|d\mathbf{s} \times \hat{\mathbf{r}}|}{r^2} = \frac{\mu_0 I}{4\pi} \frac{ds}{(x^2 + R^2)}$$

The direction of the magnetic field $d\mathbf{B}$ due to the element $d\mathbf{s}$ is perpendicular to the plane formed by $\hat{\mathbf{r}}$ and $d\mathbf{s}$, as in Figure 30.5. The vector $d\mathbf{B}$ can be resolved into a component dB_x, along the x axis, and a component dB_y, which is perpen-

dicular to the x axis. When the components perpendicular to the x axis are summed over the whole loop, the result is zero. That is, by symmetry any element on one side of the loop sets up a perpendicular component that cancels the component set up by an element diametrically opposite it.

Solution   For the reasons given above, *the resultant field at P must be along the x axis* and can be found by integrating the components $dB_x = dB \cos\theta$, where this expression is obtained from resolving the vector $d\mathbf{B}$ into its components as shown in Figure 30.5. That is, $\mathbf{B} = \mathbf{i}B_x$, where

$$B_x = \oint dB \cos\theta = \frac{\mu_0 I}{4\pi} \oint \frac{ds \cos\theta}{x^2 + R^2}$$

and the integral must be taken over the entire loop. Because θ, x, and R are constants for all elements of the loop and since $\cos\theta = R/(x^2 + R^2)^{1/2}$, we get

$$B_x = \frac{\mu_0 IR}{4\pi(x^2 + R^2)^{3/2}} \oint ds = \frac{\mu_0 IR^2}{2(x^2 + R^2)^{3/2}} \qquad (30.9)$$

where we have used the fact that $\oint ds = 2\pi R$ (the circumference of the loop).

To find the magnetic field at the center of the loop, we set $x = 0$ in Equation 30.9. At this special point, this gives

$$B = \frac{\mu_0 I}{2R} \qquad \text{(at } x = 0\text{)} \qquad (30.10)$$

It is also interesting to determine the behavior of the magnetic field far from the loop, that is, when x is large compared with R. In this case, we can neglect the term R^2 in the denominator of Equation 30.9 and get

$$B \approx \frac{\mu_0 IR^2}{2x^3} \qquad \text{(for } x \gg R\text{)} \qquad (30.11)$$

Since the magnitude of the magnetic dipole moment μ of the loop is defined as the product of the current and the area (Eq. 29.10), $\mu = I(\pi R^2)$ and we can express Equation 30.11 in the form

$$B = \frac{\mu_0}{2\pi} \frac{\mu}{x^3} \qquad (30.12)$$

This result is similar in form to the expression for the electric field due to an electric dipole, $E = k_e p / y^3$ (Example 23.9), where p is the electric dipole moment. The pattern of the magnetic field lines for a circular loop is shown in Figure 30.6. For clarity, the lines are drawn only for one plane that contains the axis of the loop. The field pattern is axially symmetric.

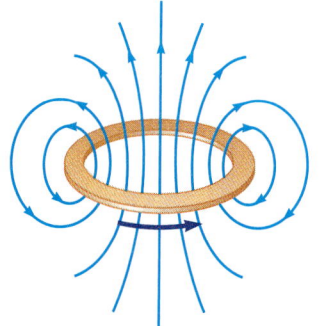

FIGURE 30.6 Magnetic field lines for a current loop. Far from the loop, the field lines are identical in form to those of an electric dipole.

30.2 THE MAGNETIC FORCE BETWEEN TWO PARALLEL CONDUCTORS

In the previous chapter we described the magnetic force that acts on a current-carrying conductor when the conductor is placed in an external magnetic field. Since a current in a conductor sets up its own magnetic field, it is easy to understand that two current-carrying conductors exert magnetic forces on each other. As we shall see, such forces can be used as the basis for defining the ampere and the coulomb.

Consider two long, straight, parallel wires separated by a distance a and carrying currents I_1 and I_2 in the same direction, as in Figure 30.7. We can easily determine the force on one wire due to a magnetic field set up by the other wire. Wire 2, which carries a current I_2, creates a magnetic field $\mathbf{B}_2$ at the position of wire 1. The direction of $\mathbf{B}_2$ is perpendicular to wire 1, as shown in Figure 30.7. According to Equation 29.3, the magnetic force on a length ℓ of wire 1 is $\mathbf{F}_1 = I_1 \boldsymbol{\ell} \times \mathbf{B}_2$. Since $\boldsymbol{\ell}$ is perpendicular to $\mathbf{B}_2$, the magnitude of $\mathbf{F}_1$ is $F_1 = I_1 \ell B_2$. Since the field created by wire 2 is given by Equation 30.7, we see that

$$F_1 = I_1 \ell B_2 = I_1 \ell \left(\frac{\mu_0 I_2}{2\pi a} \right) = \frac{\mu_0 I_1 I_2}{2\pi a} \ell$$

We can rewrite this in terms of the force per unit length as

$$\frac{F_1}{\ell} = \frac{\mu_0 I_1 I_2}{2\pi a} \qquad (30.13)$$

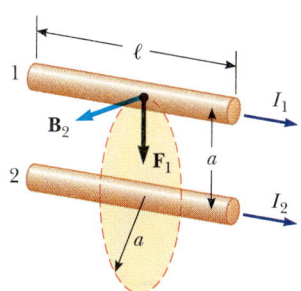

FIGURE 30.7 Two parallel wires each carrying a steady current exert a force on each other. The field $\mathbf{B}_2$ at wire 1 due to the current in wire 2 produces a force on wire 1 given by $F_1 = I_1 \ell B_2$. The force is attractive if the currents are parallel as shown and repulsive if the currents are antiparallel.

The direction of $\mathbf{F}_1$ is downward, toward wire 2, because $\boldsymbol{\ell} \times \mathbf{B}_2$ is downward. If the field set up at wire 2 by wire 1 is considered, the force $\mathbf{F}_2$ on wire 2 is found to be equal to and opposite $\mathbf{F}_1$. This is what would be expected, if Newton's third law of action-reaction is to be obeyed.[1]

---

[1] Although the total force on wire 1 is equal to and opposite the total force on wire 2, Newton's third law does not apply when two small elements of the wires that are not opposite each other are considered in isolation. Resolution of this discrepancy for interacting current loops is described in more advanced treatments on electricity and magnetism.

When the currents are in opposite directions, the forces are reversed and the wires repel each other. Hence, we find that

> parallel conductors carrying currents in the same direction attract each other, whereas parallel conductors carrying currents in opposite directions repel each other.

The force between two parallel wires each carrying a current is used to define the **ampere** as follows:

Definition of the ampere

> If two long, parallel wires 1 m apart carry the same current and the force per unit length on each wire is 2×10^{-7} N/m, then the current is defined to be 1 A.

The numerical value of 2×10^{-7} N/m is obtained from Equation 30.13, with $I_1 = I_2 = 1$ A and $a = 1$ m. Therefore, a mechanical measurement can be used to standardize the ampere. For instance, the National Institute of Standards and Technology uses an instrument called a current balance for primary current measurements. These results are then used to standardize other, more conventional instruments, such as ammeters.

The SI unit of charge, the **coulomb**, is defined in terms of the ampere as follows:

Definition of the coulomb

> If a conductor carries a steady current of 1 A, then the quantity of charge that flows through a cross-section of the conductor in 1 s is 1 C.

30.3 AMPÈRE'S LAW

An experiment first carried out by Oersted in 1820 demonstrates that a current-carrying conductor produces a magnetic field. Several compass needles are placed in a horizontal plane near a long vertical wire, as in Figure 30.8a. When there is no

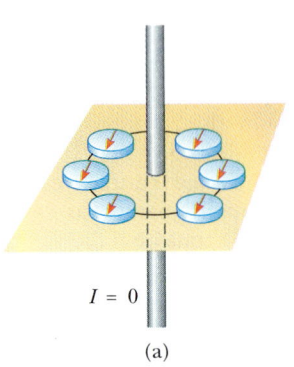

(a)

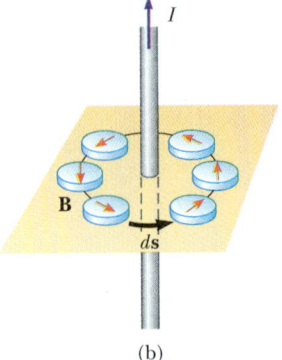

(b)

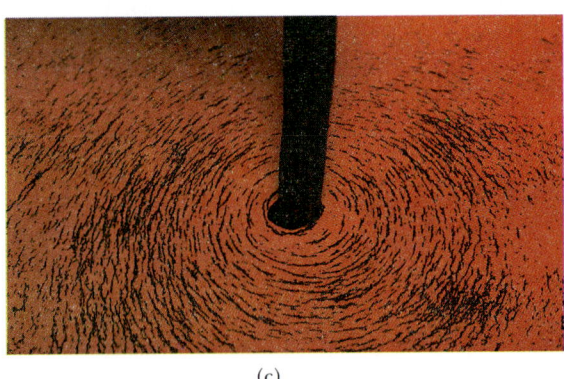

(c)

FIGURE 30.8   (a) When there is no current in the vertical wire, all compass needles point in the same direction. (b) When the wire carries a strong current, the compass needles deflect in a direction tangent to the circle, which is the direction of **B** due to the current. (c) Circular magnetic field lines surrounding a current-carrying conductor as displayed with iron filings. The photograph was taken using 30 parallel wires each carrying a current of 0.50 A.   *(Henry Leap and Jim Lehman)*

André-Marie Ampère was a French mathematician, chemist, and philosopher who founded the science of electrodynamics. The unit of measure for electric current was named in his honor.

Ampère's genius, particularly in mathematics, became evident early in his life: He had mastered advanced mathematics by the age of 12. In his first publication, *Considerations on the Mathematical Theory of Games,* an early contribution to the theory of probability, he proposed the inevitability of a player losing a game of chance to a player with greater financial resources.

Ampère is credited with the discovery of electromagnetism—the relationship between electric current and magnetic fields. His work in this field was influenced by the findings of Danish physicist Hans Christian Oersted. Ampère presented a series of papers expounding the theory and basic laws of electromagnetism, which

André-Marie Ampère

| 1 7 7 5 – 1 8 3 6 |

he called electrodynamics, to differentiate it from the study of stationary electric forces, which he called electrostatics.

The culmination of Ampère's studies came in 1827 when he published his *Mathematical Theory of Electrodynamic Phenomena Deduced Solely from Experiment,* in which he derived precise mathematical formulations of electromagnetism, notably Ampère's law.

Many stories are told of Ampère's absentmindedness, a trait he shared with Newton. In one instance, he forgot to honor an invitation to dine with the Emperor Napoleon.

Ampère's personal life was filled with tragedy. His father, a wealthy city official, was guillotined during the French Revolution, and his wife's death in 1803 was a major blow. Ampère died at the age of 61 of pneumonia. His judgment of his life is clear from the epitaph he chose for his gravestone: *Tandem felix* (Happy at last).

(Photo courtesy of AIP Niels Bohr Library)

current in the wire, all the needles point in the same direction (that of the Earth's magnetic field), as would be expected. When the wire carries a strong, steady current, the needles all deflect in a direction tangent to the circle, as in Figure 30.8b. These observations show that the direction of **B** is consistent with the right-hand rule described in Figure 30.3.

> If the wire is grasped in the right hand with the thumb in the direction of the current, the fingers curl in the direction of **B**.

When the current is reversed, the needles in Figure 30.8b also reverse.

Because the compass needles point in the direction of **B**, we conclude that the lines of **B** form circles around the wire, as we discussed in the previous section. By symmetry, the magnitude of **B** is the same everywhere on a circular path centered on the wire and lying in a plane that is perpendicular to the wire. By varying the current and distance r from the wire, it is found that B is proportional to the current and inversely proportional to distance from the wire.

Now let us evaluate the product $\mathbf{B} \cdot d\mathbf{s}$ and sum these products over the closed circular path centered on the wire. Along this path, the vectors $d\mathbf{s}$ and **B** are parallel at each point (Fig. 30.8b), so that $\mathbf{B} \cdot d\mathbf{s} = B\,ds$. Furthermore, **B** is constant in magnitude on this circle and given by Equation 30.7. Therefore, the sum of the products $B\,ds$ over the closed path, which is equivalent to the line integral of

$\mathbf{B} \cdot d\mathbf{s}$, is

$$\oint \mathbf{B} \cdot d\mathbf{s} = B \oint ds = \frac{\mu_0 I}{2\pi r} (2\pi r) = \mu_0 I \qquad (30.14)$$

where $\oint ds = 2\pi r$ is the circumference of the circle.

This result was calculated for the special case of a circular path surrounding a wire. However, it holds when an arbitrary closed path is threaded by a *steady current.* This general case, known as **Ampère's law,** can be stated as follows:

> The line integral of $\mathbf{B} \cdot d\mathbf{s}$ around any closed path equals $\mu_0 I$, where I is the total steady current passing through any surface bounded by the closed path.

Ampère's law

$$\oint \mathbf{B} \cdot d\mathbf{s} = \mu_0 I \qquad (30.15)$$

Ampère's law is valid only for steady currents and is useful only for calculating the magnetic field of current configurations having a high degree of symmetry.

EXAMPLE 30.4   The Magnetic Field Created by a Long, Current-Carrying Wire

A long, straight wire of radius R carries a steady current I_0 that is uniformly distributed through the cross-section of the wire (Fig. 30.9). Calculate the magnetic field a distance r from the center of the wire in the regions $r \geq R$ and $r < R$.

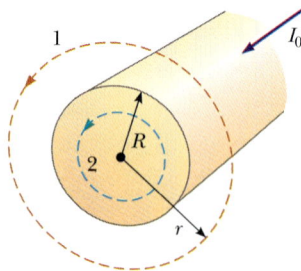

FIGURE 30.9   (Example 30.4) A long, straight wire of radius R carrying a steady current I_0 uniformly distributed across the wire. The magnetic field at any point can be calculated from Ampère's law using a circular path of radius r, concentric with the wire.

Solution   In region 1, where $r \geq R$, let us choose for our path of integration a circle of radius r centered at the wire. From symmetry, we see that **B** must be constant in magnitude and parallel to $d\mathbf{s}$ at every point on this circle. Since the total

current passing through the plane of the circle is I_0, Ampère's law applied to the circle gives

$$\oint \mathbf{B} \cdot d\mathbf{s} = B \oint ds = B(2\pi r) = \mu_0 I_0$$

$$B = \frac{\mu_0 I_0}{2\pi r} \qquad \text{(for } r \geq R\text{)} \qquad (30.16)$$

which is identical in meaning to Equation 30.7.

Now consider the interior of the wire, that is, region 2, where $r < R$. Here the current I passing through the plane of the circle of radius $r < R$ is less than the total current I_0. Since the current is uniform over the cross-section of the wire, the fraction of the current enclosed by the circle of radius $r < R$ must equal the ratio of the area πr^2 enclosed by circle 2 and the cross-sectional area πR^2 of the wire [2]:

$$\frac{I}{I_0} = \frac{\pi r^2}{\pi R^2}$$

$$I = \frac{r^2}{R^2} I_0$$

---

[2] Alternatively, the current linked by path 2 must equal the product of the current density, $J = I_0 / \pi R^2$, and the area πr^2 enclosed by path 2.

Following the same procedure as for circle 1, we apply Ampère's law to circle 2:

$$\oint \mathbf{B} \cdot d\mathbf{s} = B(2\pi r) = \mu_0 I = \mu_0 \left(\frac{r^2}{R^2} I_0 \right)$$

$$B = \left(\frac{\mu_0 I_0}{2\pi R^2} \right) r \qquad \text{(for } r < R) \qquad (30.17)$$

The magnetic field strength versus r for this configuration is sketched in Figure 30.10. Note that inside the wire, $B \to 0$ as $r \to 0$. This result is similar in form to that of the electric field inside a uniformly charged rod.

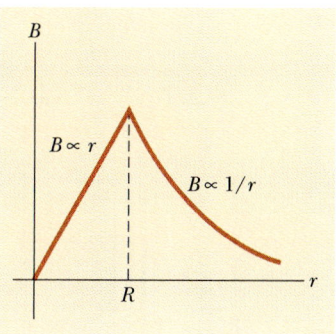

FIGURE 30.10   A sketch of the magnetic field versus r for the wire described in Example 30.4. The field is proportional to r inside the wire and varies as $1/r$ outside the wire.

EXAMPLE 30.5   The Magnetic Field Created by a Toroid

A toroid consists of N turns of wire wrapped around a ring-shaped structure as in Figure 30.11. You can think of a toroid as a solenoid bent into the shape of a doughnut. Assuming that the turns are closely spaced, calculate the magnetic field inside the toroid, a distance r from the center.

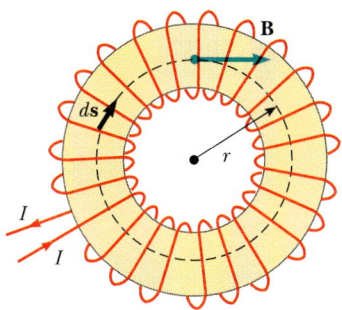

FIGURE 30.11   (Example 30.5) A toroid consists of many turns of wire wrapped around a doughnut-shaped structure (called a torus). If the coils are closely spaced, the field in the interior of the toroid is tangent to the dashed circle and varies as $1/r$, and the exterior field is zero.

Reasoning   To calculate the field inside the toroid, we evaluate the line integral of $\mathbf{B} \cdot d\mathbf{s}$ over a circle of radius r. By

symmetry, we see that the magnetic field is constant in magnitude on this circle and tangent to it, so that $\mathbf{B} \cdot d\mathbf{s} = B \, ds$. Furthermore, note that the closed path threads N loops of wire, each of which carries a current I. Therefore, the right side of Equation 30.15 is $\mu_0 NI$ in this case.

Solution   Ampère's law applied to the circle gives

$$\oint \mathbf{B} \cdot d\mathbf{s} = B \oint ds = B(2\pi r) = \mu_0 NI$$

$$B = \frac{\mu_0 NI}{2\pi r} \qquad (30.18)$$

This result shows that B varies as $1/r$ and hence is nonuniform within the coil. However, if r is large compared with the cross-sectional radius of the toroid, then the field is approximately uniform inside the coil.

Furthermore, for an ideal toroid, where the turns are closely spaced, the external magnitude field is zero. This can be seen by noting that the net current threaded by any circular path lying outside the toroid is zero (including the region of the "hole in the doughnut"). Therefore, from Ampère's law we find that $B = 0$ in the regions exterior to the coil. In reality, the turns of a toroid form a helix rather than circular loops. As a result, there is always a small field external to the coil.

EXAMPLE 30.6   Magnetic Field Created by an Infinite Current Sheet

An infinite sheet lying in the yz plane carries a surface current of density $\mathbf{J}_s$. The current is in the y direction, and J_s represents the current per unit length measured along the z axis. Find the magnetic field near the sheet.

Reasoning and Solution   To evaluate the line integral in Ampère's law, let us take a rectangular path through the sheet as in Figure 30.12. The rectangle has dimensions ℓ and w, where the sides of length ℓ are parallel to the surface of the

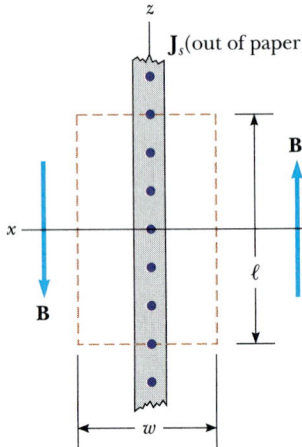

FIGURE 30.12   (Example 30.6) A top view of an infinite current sheet lying in the *yz* plane, where the current is in the *y* direction (out of the paper). This view shows the direction of **B** on both sides of the sheet.

sheet. The net current passing through the plane of the rectangle is $J_s\ell$. Hence, applying Ampère's law over the rectangle and noting that the two sides of length *w* do not contribute to the line integral (because the component of **B** along the direction of these paths is zero), we get

$$\oint \mathbf{B} \cdot d\mathbf{s} = \mu_0 I = \mu_0 J_s \ell$$

$$2B\ell = \mu_0 J_s \ell$$

$$B = \mu_0 \frac{J_s}{2} \qquad (30.19)$$

The result shows that *the magnetic field is independent of distance from the current sheet.* In fact, the magnetic field is uniform and is everywhere parallel to the plane of the sheet. This is reasonable since we are dealing with an infinite sheet of current. The result is analogous to the uniform electric field associated with an infinite sheet of charge. (See Example 24.8.)

EXAMPLE 30.7   The Magnetic Force on a Current Segment

Wire 1 in Figure 30.13 is oriented along the *y* axis and carries a steady current I_1. A rectangular circuit located to the right of the wire carries a current I_2. Find the magnetic force exerted on the top horizontal wire (wire 2) of the rectangular circuit.

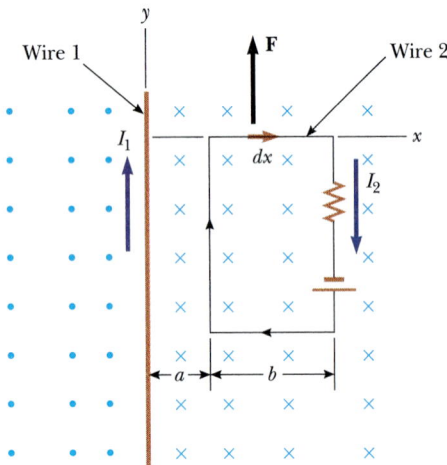

FIGURE 30.13   (Example 30.7).

Reasoning and Solution   In this problem, you may be tempted to use Equation 30.13 to obtain the force exerted on a small segment of the horizontal wire. However, this result applies only to two parallel wires, and cannot be used here.

The correct approach is to consider the force on a small segment *ds* of the conductor. This force is given by $d\mathbf{F} = I\,d\mathbf{s} \times \mathbf{B}$ (Eq. 29.4), where $I = I_2$ and **B** is the magnetic field created by the current in wire 1 at the position of *ds*. From Ampère's law, the field at a distance *x* from wire 1 is

$$\mathbf{B} = \frac{\mu_0 I_1}{2\pi x}(-\mathbf{k})$$

where the field points into the page as indicated by the unit vector notation $(-\mathbf{k})$. Because wire 2 is along the *x* axis, $d\mathbf{s} = dx\,\mathbf{i}$, and we find

$$d\mathbf{F} = \frac{\mu_0 I_1 I_2}{2\pi x}[\mathbf{i} \times (-\mathbf{k})]\,dx = \frac{\mu_0 I_1 I_2}{2\pi}\frac{dx}{x}\mathbf{j}$$

Integrating this equation over the limits $x = a$ to $x = a + b$ gives

$$\mathbf{F} = \frac{\mu_0 I_1 I_2}{2\pi}\ln x \Big]_a^{a+b}\mathbf{j} = \frac{\mu_0 I_1 I_2}{2\pi}\ln\left(1 + \frac{b}{a}\right)\mathbf{j}$$

The force points upward as indicated by the notation **j**, and as shown in Figure 30.13.

Exercise   What is the force on the wire of length *b* at the bottom of the rectangular circuit?

Answer   The force has the same magnitude as the force on wire 2, but is directed downward.

30.4   THE MAGNETIC FIELD OF A SOLENOID

A **solenoid** is a long wire wound in the form of a helix. With this configuration, a reasonably uniform magnetic field can be produced in the space surrounded by the turns of wire. When the turns are closely spaced, each can be regarded as a circular loop, and the net magnetic field is the vector sum of the fields due to all the turns.

Figure 30.14 shows the magnetic field lines of a loosely wound solenoid. Note that the field lines in the space surrounded by the coil are nearly parallel, uniformly distributed, and close together, indicating that the field in this space is uniform. The field lines between the turns tend to cancel each other. The field at exterior points, such as *P*, is weak because the field due to current elements on the upper portions tends to cancel the field due to current elements on the lower portions.

If the turns are closely spaced and the solenoid is of finite length, the field lines are as shown in Figure 30.15. In this case, the field lines diverge from one end and converge at the opposite end. An inspection of this field distribution shows a similarity with the field of a bar magnet. Hence, one end of the solenoid behaves like the north pole of a magnet while the opposite end behaves like the south pole. As the length of the solenoid increases, the field in the space enclosed by the coils becomes more and more uniform. The case of an *ideal solenoid* is approached when the turns are closely spaced and the length is long compared with the radius. In this case, the field outside the solenoid is weak compared with the field in the space enclosed by the coils and the field there is uniform over a large volume.

We can use Ampère's law to obtain an expression for the magnetic field in the space surrounded by an ideal solenoid. A longitudinal cross-section of part of our ideal solenoid (Fig. 30.16) carries a current *I*. For the ideal solenoid, **B** in the interior space (blue area) is uniform and parallel to the axis and **B** in the space

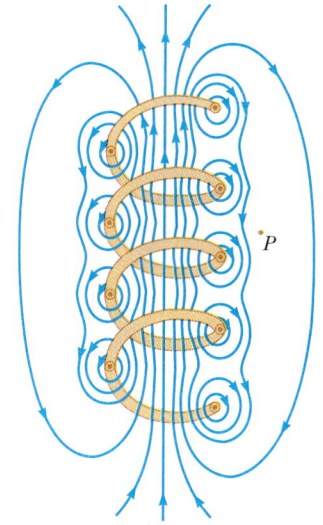

FIGURE 30.14   The magnetic field lines for a loosely wound solenoid. *(Adapted from D. Halliday and R. Resnick, Physics, New York, Wiley, 1978)*

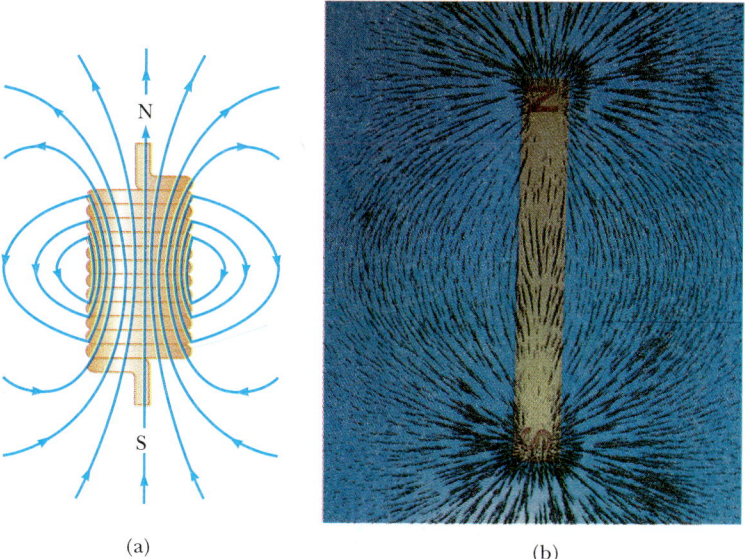

(a)   (b)

FIGURE 30.15   (a) Magnetic field lines for a tightly wound solenoid of finite length carrying a steady current. The field in the space enclosed by the solenoid is nearly uniform and strong. Note that the field lines resemble those of a bar magnet, so that the solenoid effectively has north and south poles. (b) Magnetic field pattern of a bar magnet, as displayed by small iron filings on a sheet of paper.   *(Henry Leap and Jim Lehman)*

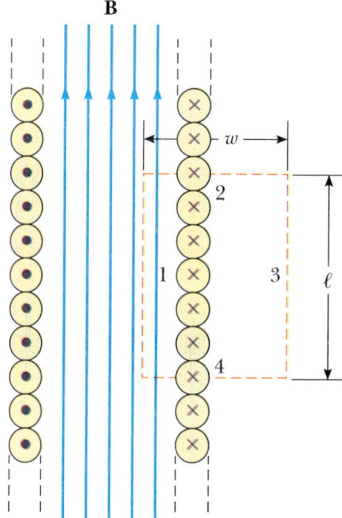

FIGURE 30.16   A cross-sectional view of a tightly wound solenoid. If the solenoid is long relative to its radius, we can assume that the magnetic field inside is uniform and the field outside is zero. Ampère's law applied to the red dashed rectangular path can then be used to calculate the field inside the solenoid.

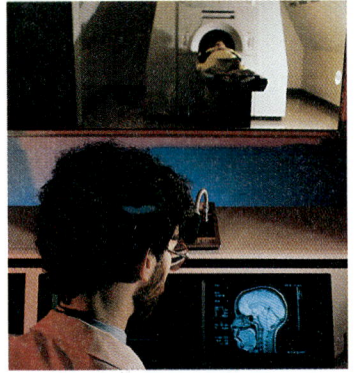

A technician studies the scan of a head. The scan was obtained using a medical diagnostic technique known as magnetic resonance imaging (MRI). This instrument makes use of strong magnetic fields produced by superconducting solenoids. *(Hank Morgan, Science Source)*

surrounding the coil (yellow area) is zero. Consider a rectangular path of length ℓ and width w as shown in Figure 30.16. We can apply Ampère's law to this path by evaluating the integral of $\mathbf{B} \cdot d\mathbf{s}$ over each side of the rectangle. The contribution along side 3 is clearly zero, because $\mathbf{B} = 0$ in this region. The contributions from sides 2 and 4 are both zero because $\mathbf{B}$ is perpendicular to $d\mathbf{s}$ along these paths. Side 1 gives a contribution $B\ell$ to the integral because $\mathbf{B}$ is uniform and parallel to $d\mathbf{s}$ along this path. Therefore, the integral over the closed rectangular path is

$$\oint \mathbf{B} \cdot d\mathbf{s} = \int_{\text{path 1}} \mathbf{B} \cdot d\mathbf{s} = B \int_{\text{path 1}} ds = B\ell$$

The right side of Ampère's law involves the total current that passes through the area bound by the path of integration. In our case, the total current through the rectangular path equals the current through each turn multiplied by the number of turns. If N is the number of turns in the length ℓ, then the total current through the rectangle equals NI. Therefore, Ampère's law applied to this path gives

$$\oint \mathbf{B} \cdot d\mathbf{s} = B\ell = \mu_0 NI$$

$$B = \mu_0 \frac{N}{\ell} I = \mu_0 nI \qquad (30.20)$$

where $n = N/\ell$ is the number of turns per unit length.

We also could obtain this result in a simpler manner by reconsidering the magnetic field of a toroid (Example 30.5). If the radius r of the toroid containing N turns is large compared with its cross-sectional radius a, then a short section of the toroid approximates a solenoid with $n = N/2\pi r$. In this limit, we see that Equation 30.18 derived for the toroid agrees with Equation 30.20.

Equation 30.20 is valid only for points near the center (that is, far from the ends) of a very long solenoid. As you might expect, the field near each end is smaller than the value given by Equation 30.20. At the very end of a long solenoid, the magnitude of the field is about one-half that of the field at the center. The field at arbitrary axial points of the solenoid is derived in Section 30.5.

*30.5   THE MAGNETIC FIELD ALONG THE AXIS OF A SOLENOID

Consider a solenoid of length ℓ and radius R containing N closely spaced turns and carrying a steady current I. Let us determine an expression for the magnetic field at an axial point P lying in the space enclosed by the solenoid, as indicated in Figure 30.17.

Perhaps the simplest way to obtain the desired result is to consider the solenoid as a distribution of current loops. The field of any one loop along the axis is given by Equation 30.9. Hence, the net field in the solenoid is the superposition of fields from all loops. The number of turns in a length dx of the solenoid is $(N/\ell)\,dx$; therefore the total current in a width dx is given by $I(N/\ell)\,dx$. Then, using Equation 30.9, we find that the field at P due to the section dx is

$$dB = \frac{\mu_0 R^2}{2(x^2 + R^2)^{3/2}} I \left(\frac{N}{\ell} \right) dx \qquad (30.21)$$

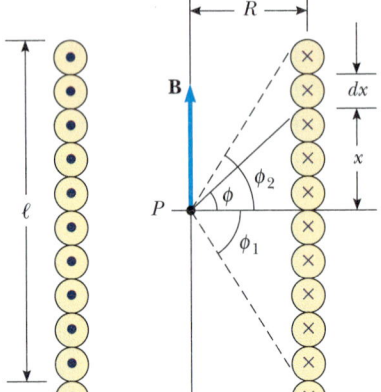

FIGURE 30.17   The geometry for calculating the magnetic field at an axial point P lying in the space enclosed by a tightly wound solenoid.

This expression contains the variable x, which can be expressed in terms of the variable ϕ, defined in Figure 30.17. That is, $x = R \tan \phi$, so that we have $dx = R \sec^2 \phi \, d\phi$. Substituting these expressions into Equation 30.21 and integrating from ϕ_1 to ϕ_2, we get

$$B = \frac{\mu_0 NI}{2\ell} \int_{\phi_1}^{\phi_2} \cos \phi \, d\phi = \frac{\mu_0 NI}{2\ell}(\sin \phi_2 - \sin \phi_1) \qquad (30.22)$$

If P is at the midpoint of the solenoid and if we assume that the solenoid is long compared with R, then $\phi_2 \approx 90°$ and $\phi_1 \approx -90°$; therefore,

$$B \approx \frac{\mu_0 NI}{2\ell}(1 + 1) = \frac{\mu_0 NI}{\ell} = \mu_0 nI \qquad \text{(at the center)}$$

which is in agreement with our previous result, Equation 30.20.

If P is a point at the end of a long solenoid (say, the bottom), then $\phi_1 \approx 0°$, $\phi_2 \approx 90°$, and

$$B \approx \frac{\mu_0 NI}{2\ell}(1 + 0) = \tfrac{1}{2}\mu_0 nI \qquad \text{(at the ends)}$$

This shows that the field at each end of a solenoid approaches one-half the value at the solenoid's center as the length ℓ approaches infinity.

A sketch of the field at axial points versus x for a solenoid is shown in Figure 30.18. If the length ℓ is large compared with R, the axial field is uniform over most of the solenoid and the curve is flat except at points near the ends. If ℓ is comparable to R, then the field has a value somewhat less than $\mu_0 nI$ at the middle and is uniform only over a small region.

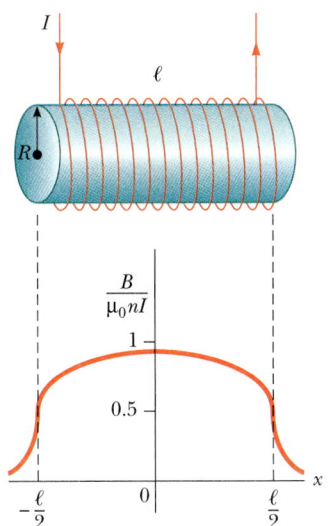

FIGURE 30.18   A sketch of the magnetic field along the axis versus x for a tightly wound solenoid. Note that the magnitude of the field at the ends is about one-half the value of the center.

30.6   MAGNETIC FLUX

The flux associated with a magnetic field is defined in a manner similar to that used to define the electric flux. Consider an element of area dA on an arbitrarily shaped surface, as in Figure 30.19. If the magnetic field at this element is $\mathbf{B}$, then the magnetic flux through the element is $\mathbf{B} \cdot d\mathbf{A}$, where $d\mathbf{A}$ is a vector perpendicular to the surface whose magnitude equals the area dA. Hence, the total magnetic flux Φ_B through the surface is

$$\Phi_B = \int \mathbf{B} \cdot d\mathbf{A} \qquad (30.23)$$

Consider the special case of a plane of area A and a uniform field $\mathbf{B}$ that makes an angle θ with the vector $d\mathbf{A}$. The magnetic flux through the plane in this case is

$$\Phi_B = BA \cos \theta \qquad (30.24)$$

If the magnetic field lies in the plane as in Figure 30.20a, then $\theta = 90°$ and the flux is zero. If the field is perpendicular to the plane as in Figure 30.20b, then $\theta = 0°$ and the flux is BA (the maximum value).

The unit of flux is the weber (Wb), where 1 Wb = 1 T·m².

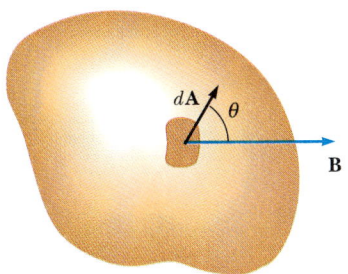

FIGURE 30.19   The magnetic flux through an area element dA is $\mathbf{B} \cdot d\mathbf{A} = BdA \cos \theta$. Note that $d\mathbf{A}$ is perpendicular to the surface.

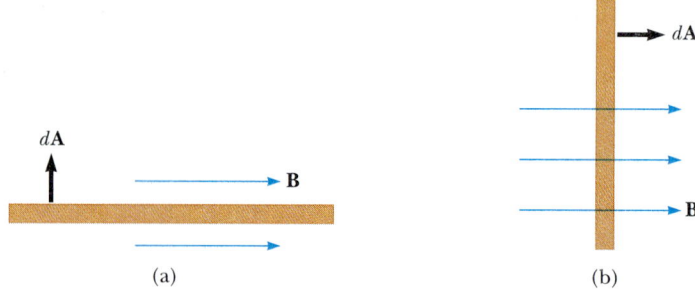

FIGURE 30.20 Edge view of a plane lying in a magnetic field. (a) The flux through the plane is zero when the magnetic field is parallel to the surface of the plane. (b) The flux through the plane is a maximum when the magnetic field is perpendicular to the plane.

EXAMPLE 30.8 Magnetic Flux Through a Rectangular Loop

A rectangular loop of width a and length b is located a distance c from a long wire carrying a current I (Fig. 30.21). The wire is parallel to the long side of the loop. Find the total magnetic flux through the loop.

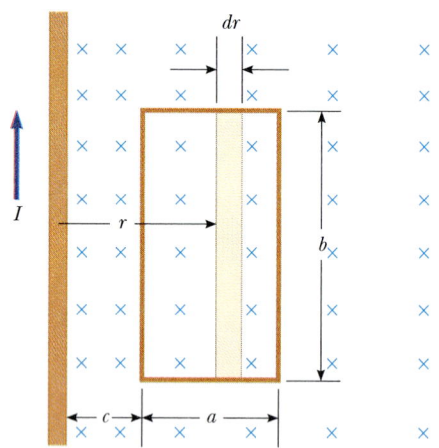

FIGURE 30.21 (Example 30.8) The magnetic field due to the wire carrying a current I is not uniform over the rectangular loop.

Reasoning From Ampère's law, we know that the strength of the magnetic field created by the long current-carrying wire at a distance r from the wire is

$$B = \frac{\mu_0 I}{2\pi r}$$

That is, the field varies over the loop and is directed into the page as shown in Figure 30.21. Since $\mathbf{B}$ is parallel to $d\mathbf{A}$, we can express the magnetic flux through an area element dA as

$$\Phi_B = \int B \, dA = \int \frac{\mu_0 I}{2\pi r} \, dA$$

Note that because $\mathbf{B}$ is not uniform but rather depends on r, it cannot be removed from the integral.

Solution In order to integrate, we first express the area element (the blue region in Fig. 30.21) as $dA = b \, dr$. Since r is the only variable that now appears in the integral, the expression for Φ_B becomes

$$\Phi_B = \frac{\mu_0 I}{2\pi} b \int_c^{a+c} \frac{dr}{r} = \frac{\mu_0 Ib}{2\pi} \ln r \Big]_c^{a+c}$$

$$= \frac{\mu_0 Ib}{2\pi} \ln\left(\frac{a+c}{c}\right)$$

30.7 GAUSS'S LAW IN MAGNETISM

In Chapter 24 we found that the flux of the electric field through a closed surface surrounding a net charge is proportional to that charge (Gauss's law). In other words, the number of electric field lines leaving the surface depends only on the net charge within it. This property is based in part on the fact that electric field lines originate on electric charges.

The situation is quite different for magnetic fields, which are continuous and form closed loops. Magnetic field lines created by currents do not begin or end at any point. The magnetic field lines of the bar magnet in Figure 30.22 illustrate this point. Note that for any closed surface, the number of lines entering that surface equals the number leaving that surface, and so the net magnetic flux is zero. This is in contrast to the case of a surface surrounding one charge of an electric dipole (Fig. 30.23), where the net electric flux is not zero.

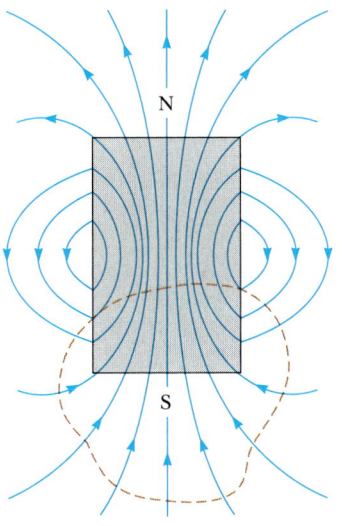

> **Gauss's law for magnetism** states that the net magnetic flux through any closed surface is always zero:
>
> $$\oint \mathbf{B} \cdot d\mathbf{A} = 0 \qquad (30.25)$$

This statement is based on the experimental fact that *isolated magnetic poles (or monopoles) have never been detected and perhaps do not even exist.*

30.8   DISPLACEMENT CURRENT AND THE GENERALIZED AMPÈRE'S LAW

We have seen that charges in motion, or currents, produce magnetic fields. When a current-carrying conductor has high symmetry, we can calculate the magnetic field using Ampère's law, given by Equation 30.15:

$$\oint \mathbf{B} \cdot d\mathbf{s} = \mu_0 I$$

where the line integral is over *any closed path through which the conduction current passes*, and the conduction current is defined by $I = dQ/dt$. (In this section, we use

FIGURE 30.22   The magnetic field lines of a bar magnet form closed loops. Note that the net flux through the closed surface surrounding one of the poles (or any other closed surface) is zero.

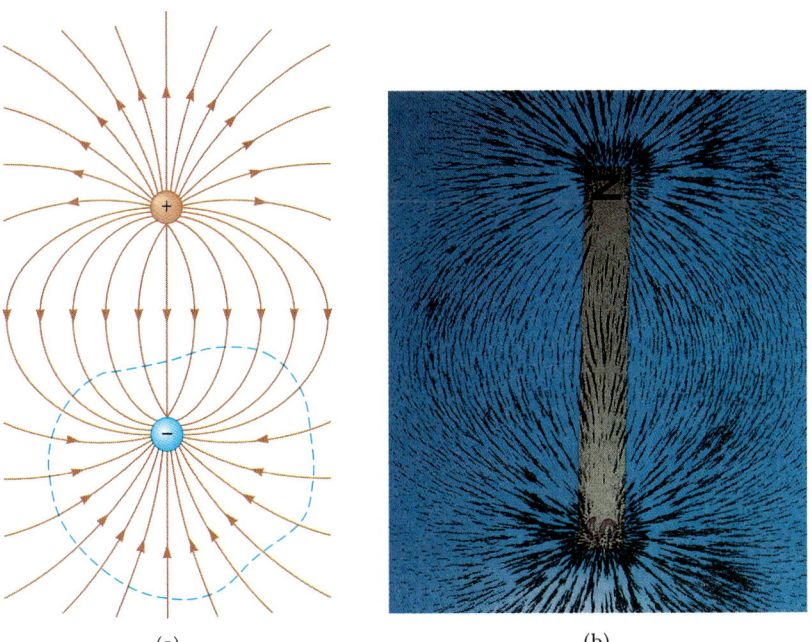

(a)   (b)

FIGURE 30.23   (a) The electric field lines of an electric dipole begin on the positive charge and terminate on the negative charge. The electric flux through a closed surface surrounding one of the charges is not zero. (b) Magnetic field pattern of a bar magnet.   *(Henry Leap and Jim Lehman)*

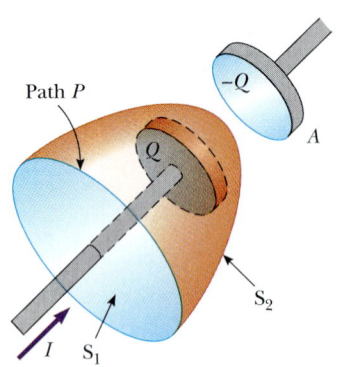

FIGURE 30.24   The surfaces S_1 and S_2 are bounded by the same path P. The conduction current in the wire passes only through S_1. This leads to a contradiction in Ampère's law that is resolved only if one postulates a displacement current through S_2.

the term *conduction current* to refer to the current carried by the wire.) We now show that *Ampère's law in this form is valid only if the conduction current is constant in time.* Maxwell recognized this limitation and modified Ampère's law to include all possible situations.

We can understand this problem by considering a capacitor being charged as in Figure 30.24. When the conduction current changes with time, the charge on the plate changes, but *no conduction current passes between the plates.* Now consider the two surfaces S_1 and S_2 in Figure 30.24 bounded by the same path P. Ampère's law says that the line integral of $\mathbf{B} \cdot d\mathbf{s}$ around this path must equal $\mu_0 I$, where I is the total current through any surface bounded by the path P.

When the path P is considered as bounding S_1, the result of the integral is $\mu_0 I$ because the current passes through S_1. When the path bounds S_2, however, the result is zero since no current passes through S_2. Thus, we have a contradictory situation that arises from the discontinuity of the current! Maxwell solved this problem by postulating an additional term on the right side of Equation 30.15, called the **displacement current,** I_d, defined as

$$I_d \equiv \epsilon_0 \frac{d\Phi_E}{dt} \qquad (30.26)$$

Recall that Φ_E is the flux of the electric field, defined as $\Phi_E = \int \mathbf{E} \cdot d\mathbf{A}$.

As the capacitor is being charged (or discharged), the changing electric field between the plates may be thought of as a sort of current that bridges the discontinuity in the conduction current in the wire. When the expression for the displacement current given by Equation 30.26 is added to the right side of Ampère's law, the difficulty represented by Figure 30.24 is resolved. No matter what surface bounded by the path P is chosen, some combination of conduction and displacement current will pass through it. With this new term I_d, we can express the generalized form of Ampère's law (sometimes called the **Ampère-Maxwell law**) as[3]

Ampère-Maxwell law

$$\oint \mathbf{B} \cdot d\mathbf{s} = \mu_0 (I + I_d) = \mu_0 I + \mu_0 \epsilon_0 \frac{d\Phi_E}{dt} \qquad (30.27)$$

The meaning of this expression can be understood by referring to Figure 30.25. The electric flux through S_2 is $\Phi_E = \int \mathbf{E} \cdot d\mathbf{A} = EA$, where A is the area of the capacitor plates and E is the uniform electric field strength between the plates. If Q is the charge on the plates at any instant, then $E = Q/\epsilon_0 A$ (Section 26.2). Therefore, the electric flux through S_2 is simply

$$\Phi_E = EA = \frac{Q}{\epsilon_0}$$

Hence, the displacement current I_d through S_2 is

$$I_d = \epsilon_0 \frac{d\Phi_E}{dt} = \frac{dQ}{dt} \qquad (30.28)$$

That is, the displacement current is precisely equal to the conduction current I through S_1!

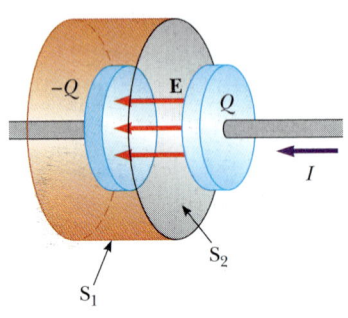

FIGURE 30.25   The conduction current $I = dQ/dt$ passes through S_1. The displacement current $I_d = \epsilon_0 d\Phi_E/dt$ passes through S_2. The two currents must be equal for continuity. In general, the total current through any surface bounded by some path is $I + I_d$.

---

[3] Strictly speaking, this expression is valid only in a vacuum. If a magnetic material is present, a magnetizing current I_m must also be included on the right side of Equation 30.27 to make Ampère's law fully general.

The central point of this formalism is the fact that

> magnetic fields are produced both by conduction currents and by changing electric fields.

EXAMPLE 30.9   Displacement Current in a Capacitor

A sinusoidal voltage is applied directly across an 8.00-μF capacitor. The frequency of the source is 3.00 kHz, and the voltage amplitude is 30.0 V. Find the displacement current between the plates of the capacitor.

Solution   The angular frequency of the source is given by $\omega = 2\pi f = 2\pi(3.00 \times 10^3 \text{ Hz}) = 6\pi \times 10^3 \text{ s}^{-1}$. Hence, the voltage across the capacitor in terms of t is

$$V = V_{max} \sin \omega t = (30.0 \text{ V}) \sin(6\pi \times 10^3 t)$$

We can use Equation 30.28 and the fact that the charge on the capacitor is $Q = CV$ to find the displacement current:

$$I_d = \frac{dQ}{dt} = \frac{d}{dt}(CV) = C\frac{dV}{dt}$$

$$= (8.00 \times 10^{-6})\frac{d}{dt}[30.0 \sin(6\pi \times 10^3 t)]$$

$$= \boxed{(4.52 \text{ A}) \cos(6\pi \times 10^3 t)}$$

Hence, the displacement current varies sinusoidally with time and has a maximum value of 4.52 A.

---

*30.9   MAGNETISM IN MATTER

The magnetic field produced by a current in a coil gives us a hint as to what might cause certain materials to exhibit strong magnetic properties. Earlier we found that a coil like that shown in Figure 30.15 has a north and a south pole. In general, any current loop has a magnetic field and a corresponding magnetic moment. Similarly, the magnetic moments in a magnetized substance may be described as arising from internal currents on the atomic level. For electrons moving about the nucleus, this is consistent with the Bohr model (after modifying quantum numbers). There is also an intrinsic magnetic moment for electrons, protons, neutrons, and other particles that can only be roughly modeled as arising from rotating charges.

We begin with a brief discussion of the magnetic moments due to electrons. The mutual forces between these magnetic dipole moments and their interaction with an external magnetic field are of fundamental importance in understanding the behavior of magnetic materials. We shall describe three categories of materials as paramagnetic, ferromagnetic, and diamagnetic. **Paramagnetic** and **ferromagnetic** materials are those that have atoms with permanent magnetic dipole moments. **Diamagnetic** materials are those whose atoms have no permanent magnetic dipole moments.

The Magnetic Moments of Atoms

It is instructive to begin our discussion with a classical model of the atom in which electrons move in circular orbits around the much more massive nucleus. In this model, an orbiting electron constitutes a tiny current loop (because it is a moving charge) and the atomic magnetic moment is associated with this orbital motion. Although this model has many deficiencies, its predictions are in good agreement with the correct theory from quantum physics.

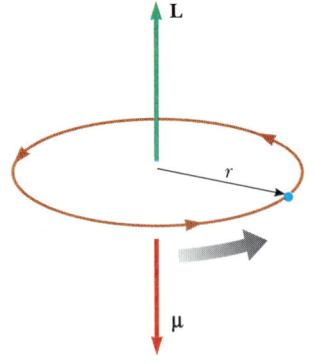

FIGURE 30.26 An electron moving in a circular orbit of radius r has an angular momentum L in one direction and a magnetic moment $\boldsymbol{\mu}$ in the opposite direction.

Consider an electron moving with constant speed v in a circular orbit of radius r about the nucleus, as in Figure 30.26. Because the electron travels a distance of $2\pi r$ (the circumference of the circle) in a time T, where T is the time for one revolution, its orbital speed is $v = 2\pi r/T$. The effective current associated with this orbiting electron equals its charge divided by the time for one revolution. Using $T = 2\pi/\omega$ and $\omega = v/r$, we have

$$I = \frac{e}{T} = \frac{e\omega}{2\pi} = \frac{ev}{2\pi r}$$

The magnetic moment associated with this effective current loop is $\mu = IA$, where $A = \pi r^2$ is the area of the orbit. Therefore,

$$\mu = IA = \left(\frac{ev}{2\pi r}\right)\pi r^2 = \tfrac{1}{2}evr \tag{30.29}$$

Since the magnitude of the orbital angular momentum of the electron is $L = mvr$, the magnetic moment can be written as

Orbital magnetic moment

$$\mu = \left(\frac{e}{2m}\right)L \tag{30.30}$$

This result says that *the magnetic moment of the electron is proportional to its orbital angular momentum.* Note that because the electron is negatively charged, the vectors $\boldsymbol{\mu}$ and L point in opposite directions. Both vectors are perpendicular to the plane of the orbit, as indicated in Figure 30.26.

A fundamental outcome of quantum physics is that orbital angular momentum is quantized, and always some integer multiple of $\hbar = h/2\pi = 1.06 \times 10^{-34}\,\text{J}\cdot\text{s}$, where h is Planck's constant. That is

Angular momentum is quantized

$$L = 0,\ \hbar,\ 2\hbar,\ 3\hbar,\ \ldots$$

Hence, the smallest nonzero value of the magnetic moment is

$$\mu = \frac{e}{2m}\hbar \tag{30.31}$$

Since all substances contain electrons, you may wonder why all substances are not magnetic. The main reason is that in most substances, the magnetic moment of one electron in an atom is canceled by the magnetic moment of another electron in the atom orbiting in the opposite direction. The net result is that, for most materials, *the magnetic effect produced by the orbital motion of the electrons is either zero or very small.*

So far we considered only the contribution an orbiting electron makes to the magnetic moment of an atom. However, an electron has another intrinsic property, called *spin*, that also contributes to the magnetic moment. In this regard, the electron can be viewed as a charged sphere spinning about its axis as it orbits the nucleus, as in Figure 30.27, but this classical description should not be taken literally. The property of spin arises from relativistic dynamics, which may be incorporated into quantum mechanics. The magnitude of this so-called spin magnetic moment is of the same order of magnitude as the magnetic moment due to the first effect current loop, the orbital motion. The magnitude of the spin angular momentum predicted by quantum theory is

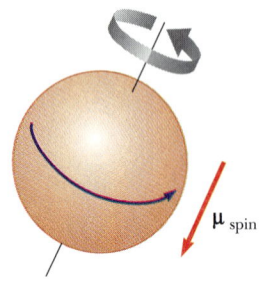

FIGURE 30.27 Classical model of a spinning electron. This model gives an incorrect magnitude, incorrect quantum numbers, and too many degrees of freedom.

Spin angular momentum

$$S = \frac{\hbar}{2} = 5.2729 \times 10^{-35}\,\text{J}\cdot\text{s}$$

The intrinsic magnetic moment associated with the spin of an electron has the value

$$\mu_B = \frac{e}{2m}\hbar = 9.27 \times 10^{-24} \text{ J/T} \qquad (30.32)$$

which is called the **Bohr magneton.**

In atoms or ions containing many electrons, the electrons usually pair up with their spins opposite each other, a situation that results in a cancellation of the spin magnetic moments. However, atoms with an odd number of electrons must have at least one "unpaired" electron and a corresponding spin magnetic moment. The total magnetic moment of an atom is the vector sum of the orbital and spin magnetic moments, and a few examples are given in Table 30.1. Note that helium and neon have zero moments because their individual moments cancel.

The nucleus of an atom also has a magnetic moment associated with its constituent protons and neutrons. However, the magnetic moment of a proton or neutron is small compared with the magnetic moment of an electron and can usually be neglected. This can be understood by inspecting Equation 30.32. Since the masses of the proton and neutron are much greater than that of the electron, their magnetic moments are smaller than that of the electron by a factor of approximately 10^3.

TABLE 30.1	**Magnetic Moments of Some Atoms and Ions**
Atom (or Ion)	Magnetic Moment (10^{-24} J/T)
H	9.27
He	0
Ne	0
Ce³⁺	19.8
Yb³⁺	37.1

Magnetization and Magnetic Field Strength

The magnetic state of a substance is described by a quantity called the **magnetization vector, M.** *The magnitude of the magnetization vector is equal to the magnetic moment per unit volume of the substance.* As you might expect, the total magnetic field in a substance depends on both the applied (external) field and the magnetization of the substance.

Consider a region where there exists a magnetic field $\mathbf{B}_0$ produced by a current-carrying conductor. If we now fill that region with a magnetic substance, the total magnetic field $\mathbf{B}$ in that region is $\mathbf{B} = \mathbf{B}_0 + \mathbf{B}_m$ where $\mathbf{B}_m$ is the field produced by the magnetic substance. This contribution can be expressed in terms of the magnetization vector as $\mathbf{B}_m = \mu_0\mathbf{M}$; hence, the total magnetic field in the region becomes

$$\mathbf{B} = \mathbf{B}_0 + \mu_0\mathbf{M} \qquad (30.33)$$

It is convenient to introduce a field quantity **H**, called the **magnetic field strength.** This vector quantity is defined by the relationship $\mathbf{H} = (\mathbf{B}/\mu_0) - \mathbf{M}$, or

$$\mathbf{B} = \mu_0(\mathbf{H} + \mathbf{M}) \qquad (30.34)$$

In SI units, the dimensions of both **H** and **M** are amperes per meter.

To better understand these expressions, consider the region inside the space enclosed by a toroid that carries a current I. (We call this space the *core* of the toroid.) If this space is a vacuum, then $\mathbf{M} = 0$ and $\mathbf{B} = \mathbf{B}_0 = \mu_0\mathbf{H}$. Since $B_0 = \mu_0 nI$ in the core, where n is the number of turns per unit length of the toroid, then $H = B_0/\mu_0 = \mu_0 nI/\mu_0$ or

$$H = nI \qquad (30.35)$$

That is, the magnetic field strength in the core of the toroid is due to the current in its windings.

If the toroid core is now filled with some substance and the current I is kept constant, then **H** inside the substance remains unchanged and has magnitude nI.

This is because the magnetic field strength **H** is due solely to the current in the toroid. The total field **B**, however, changes when the substance is introduced. From Equation 30.34, we see that part of **B** arises from the term $\mu_0\mathbf{H}$ associated with the current in the toroid; the second contribution to **B** is the term $\mu_0\mathbf{M}$ due to the magnetization of the substance filling the core.

Classification of Magnetic Substances

For a large class of substances, specifically paramagnetic and diamagnetic substances, the magnetization vector **M** is proportional to the magnetic field strength **H**. In these substances, we can write

$$\mathbf{M} = \chi\mathbf{H} \tag{30.36}$$

Magnetic susceptibility

where χ (Greek letter chi) is a dimensionless factor called the **magnetic susceptibility**. If the sample is paramagnetic, χ is positive, in which case **M** is in the same direction as **H**. If the substance is diamagnetic, χ is negative, and **M** is opposite **H**. It is important to note that this linear relationship between **M** and **H** does not apply to ferromagnetic substances. The susceptibilities of some substances are given in Table 30.2. Substituting Equation 30.36 for **M** into Equation 30.34 gives

$$\mathbf{B} = \mu_0(\mathbf{H} + \mathbf{M}) = \mu_0(\mathbf{H} + \chi\mathbf{H}) = \mu_0(1 + \chi)\mathbf{H}$$

or

$$\mathbf{B} = \mu_m\mathbf{H} \tag{30.37}$$

where the constant μ_m is called the **magnetic permeability** of the substance and has the value

Magnetic permeability

$$\mu_m = \mu_0(1 + \chi) \tag{30.38}$$

Substances may also be classified in terms of how their magnetic permeability μ_m compares with μ_0 (the permeability of free space) as follows:

$$\text{Ferromagnetic} \qquad \mu_m \gg \mu_0$$

$$\text{Paramagnetic} \qquad \mu_m > \mu_0$$

$$\text{Diamagnetic} \qquad \mu_m < \mu_0$$

Since χ is very small for paramagnetic and diamagnetic substances (Table 30.2), μ_m is nearly equal to μ_0 in these cases. For ferromagnetic substances, however, μ_m

TABLE 30.2   Magnetic Susceptibilities of Some Paramagnetic and Diamagnetic Substances at 300 K

Paramagnetic Substance	χ	Diamagnetic Substance	χ
Aluminum	2.3×10^{-5}	Bismuth	-1.66×10^{-5}
Calcium	1.9×10^{-5}	Copper	-9.8×10^{-6}
Chromium	2.7×10^{-4}	Diamond	-2.2×10^{-5}
Lithium	2.1×10^{-5}	Gold	-3.6×10^{-5}
Magnesium	1.2×10^{-5}	Lead	-1.7×10^{-5}
Niobium	2.6×10^{-4}	Mercury	-2.9×10^{-5}
Oxygen (STP)	2.1×10^{-6}	Nitrogen (STP)	-5.0×10^{-9}
Platinum	2.9×10^{-4}	Silver	-2.6×10^{-5}
Tungsten	6.8×10^{-5}	Silicon	-4.2×10^{-6}

is typically several thousand times larger than μ_0. Although Equation 30.37 provides a simple relationship between **B** and **H**, it must be interpreted with care when dealing with ferromagnetic substances. As mentioned earlier, **M** is not a linear function of **H** for ferromagnetic substances. This is because the value of μ_m *is not a characteristic of the substance,* but rather depends on the previous state and treatment of the sample.

EXAMPLE 30.10 **An Iron-Filled Toroid**

A toroid wound with 60.0 turns/m of wire carries 5.00 A. The core is iron, which has a magnetic permeability of $5000\mu_0$ under the given conditions. Find H and B inside the iron.

Solution Using Equations 30.35 and 30.37, we get

$$H = nI = \left(60.0\ \frac{\text{turns}}{\text{m}}\right)(5.00\ \text{A}) = \boxed{300\ \frac{\text{A} \cdot \text{turns}}{\text{m}}}$$

$$B = \mu_m H = 5000\ \mu_0 H$$

$$= 5000\left(4\pi \times 10^{-7}\ \frac{\text{T} \cdot \text{m}}{\text{A}}\right)\left(300\ \frac{\text{A} \cdot \text{turns}}{\text{m}}\right) = \boxed{1.88\ \text{T}}$$

This value of B is 5000 times larger than the value in the absence of iron!

Exercise Determine the magnitude and direction of the magnetization inside the iron core.

Answer $M = 1.5 \times 10^6\ \text{A/m}$; **M** is in the direction of **H**.

Ferromagnetism

Certain crystalline substances, whose atomic constituents have permanent magnetic dipole moments, exhibit strong magnetic effects called **ferromagnetism**. Examples of ferromagnetic substances include iron, cobalt, nickel, gadolinium, and dysprosium. Such substances contain atomic magnetic moments that tend to align parallel to each other even in a weak external magnetic field. Once the moments are aligned, the substance remains magnetized after the external field is removed. This permanent alignment is due to a strong coupling between neighboring moments, which can be understood only in quantum mechanical terms.

All ferromagnetic materials contain microscopic regions called **domains**, within which all magnetic moments are aligned. These domains have volumes of about 10^{-12} to $10^{-8}\ \text{m}^3$ and contain 10^{17} to 10^{21} atoms. The boundaries between the various domains having different orientations are called **domain walls.** In an unmagnetized sample, the domains are randomly oriented so that the net magnetic moment is zero as shown in Figure 30.28a. When the sample is placed in an external magnetic field, the magnetic moments of some domains may tend to align with the field, which results in a magnetized sample, as in Figure 30.28b. Observations show that domains initially oriented along the external field will grow in size at the expense of the less favorably oriented domains. When the external field is removed, the sample may retain a net magnetization in the direction of the original field. At ordinary temperatures, thermal agitation is not sufficiently high to disrupt this preferred orientation of magnetic moments.

A typical experimental arrangement used to measure the magnetic properties of a ferromagnetic material consists of a toroid-shaped sample wound with N turns of wire, as in Figure 30.29. This configuration is sometimes referred to as the **Rowland ring.** A secondary coil connected to a galvanometer is used to measure the magnetic flux. The magnetic field **B** within the core of the toroid is measured by increasing the current in the toroid coil from zero to I. As the current changes, the magnetic flux through the secondary coil changes by BA, where A is the cross-sectional area of the toroid. Because of this changing flux, an emf is induced in the secondary coil that is proportional to the rate of change in magnetic flux. If the

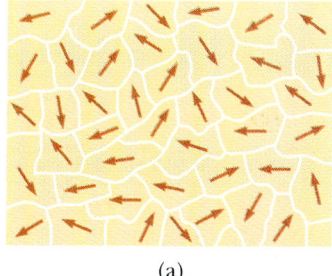

(a)

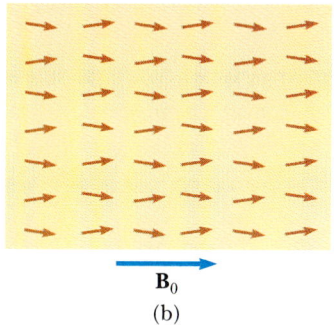

$\mathbf{B}_0$

(b)

FIGURE 30.28 (a) Random orientation of atomic magnetic dipoles in an unmagnetized substance. (b) When an external field $\mathbf{B}_{\text{ext}}$ is applied, the atomic magnetic dipoles tend to align with the field, giving the sample a net magnetization **M**.

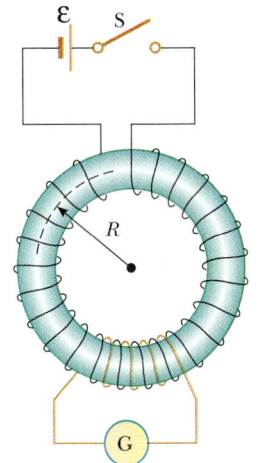

FIGURE 30.29   A toroidal winding arrangement used to measure the magnetic properties of a substance. The material under study fills the core of the toroid, and the circuit containing the galvanometer measures the magnetic flux.

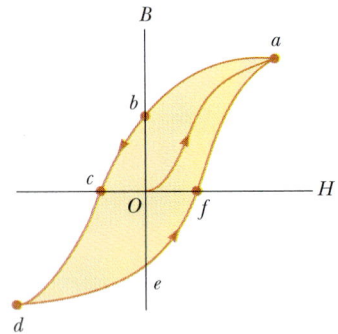

FIGURE 30.30   Hysteresis curve for a ferromagnetic material.

galvanometer in the secondary circuit is properly calibrated, a value for **B** corresponding to any value of the current in the toroid can be obtained. The magnetic field **B** is measured first in the empty coil and then with the same coil filled with the magnetic substance. The magnetic properties of the substance are then obtained from a comparison of the two measurements.

Now consider a toroid whose core consists of unmagnetized iron. If the current in the windings is increased from zero to some value I, the field intensity H increases linearly with I according to the expression $H = nI$. Furthermore, the total field B also increases with increasing current as shown in Figure 30.30. At point O, the domains are randomly oriented, corresponding to $B_m = 0$. As the external field increases, the domains become more aligned until all are nearly aligned at point a. At this point, the iron core is approaching saturation. (The condition of saturation corresponds to the case where all domains are aligned in the same direction.) Next, suppose the current is reduced to zero, thereby eliminating the external field. The B versus H curve, called a **magnetization curve,** now follows the path ab shown in Figure 30.30. Note that at point b, the field **B** is not zero, although the external field is $B_0 = 0$. This is explained by the fact that the iron core is now magnetized due to the alignment of a large number of domains (that is, $B = B_m$). At this point, the iron is said to have a remanent magnetization. If the external field is reversed in direction and increased in strength by reversing the current, the domains reorient until the sample is again unmagnetized at point c, where $B = 0$. A further increase in the reverse current causes the iron to be magnetized in the opposite direction, approaching saturation at point d. A similar sequence of events occurs as the current is reduced to zero and then increased in the original (positive) direction. In this case, the magnetization curve follows the path def. If the current is increased sufficiently, the magnetization curve returns to point a, where the sample again has its maximum magnetization.

The effect just described, called **magnetic hysteresis,** shows that the magnetization of a ferromagnetic substance depends on the history of the substance as well as the strength of the applied field. (The word hysteresis literally means to "lag behind.") It is often said that a ferromagnetic substance has a "memory" since it remains magnetized after the external field is removed. The closed loop in Figure 30.30 is referred to as a hysteresis loop. Its shape and size depend on the properties of the ferromagnetic substance and on the strength of the maximum applied field. The hysteresis loop for "hard" ferromagnetic materials (used in permanent magnets) is characteristically wide as in Figure 30.31a, corresponding to a large remanent magnetization. Such materials cannot be easily demagnetized by an external field. This is in contrast to "soft" ferromagnetic materials, such as iron, that have a

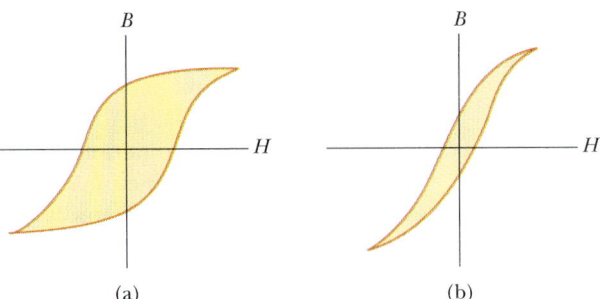

(a)   (b)

FIGURE 30.31   Hysteresis curves for (a) a hard ferromagnetic material and (b) a soft ferromagnetic material.

very narrow hysteresis loop and a small remanent magnetization (Fig. 30.31b). Such materials are easily magnetized and demagnetized. An ideal soft ferromagnet would exhibit no hysteresis and, hence, would have no remanent magnetization. A ferromagnetic substance can be demagnetized by carrying the substance through successive hysteresis loops, gradually decreasing the applied field as in Figure 30.32.

The magnetization curve is useful for another reason. *The area enclosed by the magnetization curve represents the work required to take the material through the hysteresis cycle.* The energy acquired by the sample in the magnetization process originates from the source of the external field, that is, the emf in the circuit of the toroidal coil. When the magnetization cycle is repeated, dissipative processes within the material due to realignment of the domains result in a transformation of magnetic energy into internal thermal energy, which raises the temperature of the substance. For this reason, devices subjected to alternating fields (such as transformers) use cores made of soft ferromagnetic substances, which have narrow hysteresis loops and a correspondingly small energy loss per cycle.

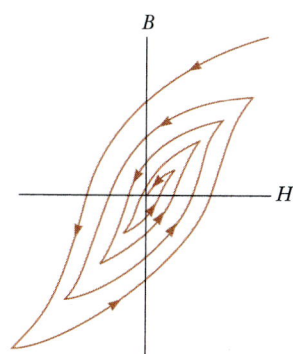

FIGURE 30.32   Demagnetizing a ferromagnetic material by carrying it through successive hysteresis loops.

Paramagnetism

Paramagnetic substances have a positive but small susceptibility $(0 < \chi \ll 1)$, which is due to the presence of atoms (or ions) with permanent magnetic dipole moments. These dipoles interact only weakly with each other and are randomly oriented in the absence of an external magnetic field. When the substance is placed in an external magnetic field, its atomic dipoles tend to line up with the field. However, this alignment process must compete with the effects of thermal motion, which tends to randomize the dipole orientations.

Experimentally, it is found that the magnetization of a paramagnetic substance is proportional to the applied magnetic field and inversely proportional to the absolute temperature under a wide range of conditions. That is,

$$M = C\frac{B}{T} \tag{30.39}$$

This is known as **Curie's law** after its discoverer Pierre Curie (1859–1906), and the constant C is called **Curie's constant.** This shows that the magnetization increases with increasing applied field and with decreasing temperature. When $B = 0$, the magnetization is zero, corresponding to a random orientation of dipoles. At very high fields or very low temperatures, the magnetization approaches its maximum, or saturation, value corresponding to a complete alignment of its dipoles and Equation 30.39 is no longer valid.

It is interesting to note when the temperature of a ferromagnetic substance reaches or exceeds a critical temperature, called the **Curie temperature,** the substance loses its spontaneous magnetization and becomes paramagnetic (see Fig. 30.33). Below the Curie temperature, the magnetic moments are aligned and the substance is ferromagnetic. Above the Curie temperature, the thermal energy is large enough to cause a random orientation of dipoles, hence the substance becomes paramagnetic. For example, the Curie temperature for iron is 1043 K. A list of Curie temperatures for several ferromagnetic substances is given in Table 30.3.

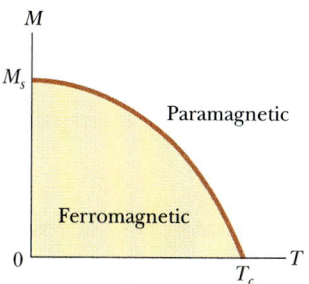

FIGURE 30.33   Plot of the magnetization versus absolute temperature for a ferromagnetic substance. The magnetic moments are aligned (ordered) below the Curie temperature T_c, where the substance is ferromagnetic. The substance becomes paramagnetic (disordered) above T_c.

TABLE 30.3	Curie Temperature for Several Ferromagnetic Substances
Substance	T_c **(K)**
Iron	1043
Cobalt	1394
Nickel	631
Gadolinium	317
Fe_2O_3	893

Diamagnetism

A diamagnetic substance is one whose atoms have no permanent magnetic dipole moment. When an external magnetic field is applied to a diamagnetic substance such as bismuth or silver, a weak magnetic dipole moment is induced in the direc-

A small permanent magnet floats freely above a ceramic disk of the superconductor $YBa_2Cu_3O_7$ cooled by liquid nitrogen at 77 K. The superconductor has zero electric resistance at temperatures below 92 K and expels any applied magnetic field. *(D.O.E./Science Source/Photo Researchers)*

tion opposite the applied field. Although the effect of diamagnetism is present in all matter, it is weak compared to paramagnetism or ferromagnetism.

We can obtain some understanding of diamagnetism by considering two electrons of an atom orbiting the nucleus in opposite directions but with the same speed. The electrons remain in these circular orbits because of the attractive electrostatic force (the centripetal force) of the positively charged nucleus. Because the magnetic moments of the two electrons are equal in magnitude and opposite in direction, they cancel each other and the dipole moment of the atom is zero. When an external magnetic field is applied, the electrons experience an additional force $q\mathbf{v} \times \mathbf{B}$. This added force modifies the centripetal force so as to increase the orbital speed of the electron whose magnetic moment is antiparallel to the field and decreases the speed of the electron whose magnetic moment is parallel to the field. As a result, the magnetic moments of the electrons no longer cancel, and the substance acquires a net dipole moment that opposes the applied field.

As you recall from Chapter 27, superconductors are substances whose dc resistance is zero below some critical temperature characteristic of the substance. Certain types of superconductors also exhibit perfect diamagnetism in the superconducting state. As a result, an applied magnetic field is expelled by the superconductor so that the field is zero in its interior. This phenomenon of flux expulsion is known as the **Meissner effect.** If a permanent magnet is brought near a superconductor, the two substances will repel each other. This is illustrated in the photograph, which shows a small permanent magnet levitated above a superconductor maintained at 77 K. A more detailed description of the unusual properties of superconductors is presented in Chapter 44 of the extended version of this text.

EXAMPLE 30.11   Saturation Magnetization

Estimate the maximum magnetization in a long cylinder of iron, assuming there is one unpaired electron spin per atom.

Solution   The maximum magnetization, called the saturation magnetization, is obtained when all the magnetic moments in the sample are aligned. If the sample contains n atoms per unit volume, then the saturation magnetization M_s has the value

$$M_s = n\mu$$

where μ is the magnetic moment per atom. Since the molecular weight of iron is 55 g/mol and its density is 7.9 g/cm³,

the value of n is 8.5×10^{28} atoms/m³. Assuming each atom contributes one Bohr magneton (due to one unpaired spin) to the magnetic moment, we get

$$M_s = \left(8.5 \times 10^{28}\, \frac{\text{atoms}}{\text{m}^3} \right) \left(9.27 \times 10^{-24}\, \frac{\text{A} \cdot \text{m}^2}{\text{atom}} \right)$$

$$= \boxed{7.9 \times 10^5\,\text{A/m}}$$

This is about one-half the experimentally determined saturation magnetization for annealed iron, which indicates that there are actually two unpaired electron spins per atom.

---

*30.10   MAGNETIC FIELD OF THE EARTH

When we speak of a small bar magnet's having a north and a south pole, we should more properly say that it has a "north-seeking" and a "south-seeking" pole. By this we mean that if such a magnet is used as a compass, one end will seek, or point to, the north geographic pole of the Earth. Thus, we conclude that *a north magnetic pole is located near the south geographic pole, and a south magnetic pole is located near the north geographic pole.* In fact, the configuration of the Earth's magnetic field, pictured in

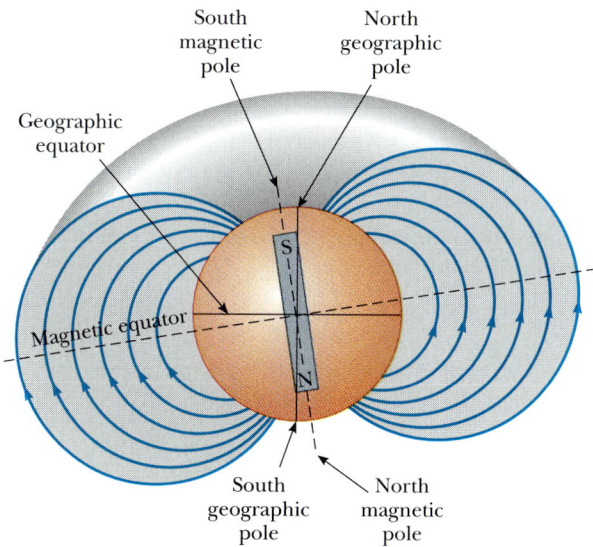

FIGURE 30.34   The Earth's magnetic field lines. Note that a south magnetic pole is at the north geographic pole and a north magnetic pole is at the south geographic pole.

Figure 30.34, is very much like that which would be achieved by burying a bar magnet deep in the interior of the Earth.

   If a compass needle is suspended in bearings that allow it to rotate in the vertical plane as well as in the horizontal plane, the needle is horizontal with respect to the Earth's surface only near the equator. As the device is moved northward, the needle rotates so that it points more and more toward the surface of the Earth. Finally, at a point just north of Hudson Bay in Canada, the north pole of the needle would point directly downward. This location, first found in 1832, is considered to be the location of the south-seeking magnetic pole of the Earth. This site is approximately 1300 mi from the Earth's geographic North Pole and varies with time. Similarly, the north-seeking magnetic pole of the Earth is about 1200 miles away from the Earth's geographic South Pole. Thus, it is only approximately correct to say that a compass needle points north. The difference between true north, defined as the geographic North Pole, and north indicated by a compass varies from point to point on the Earth, and the difference is referred to as magnetic declination. For example, along a line through Florida and the Great Lakes, a compass indicates true north, whereas in Washington state, it aligns 25° east of true north.

   Although the magnetic field pattern of the Earth is similar to that which would be set up by a bar magnet deep within the Earth, it is easy to understand why the source of the Earth's magnetic field cannot be large masses of permanently magnetized material. The Earth does have large deposits of iron ore deep beneath its surface, but the high temperatures in the Earth's core prevent the iron from retaining any permanent magnetization. It is considered more likely that the true source is charge-carrying convection currents in the Earth's core. Charged ions or electrons circling in the liquid interior could produce a magnetic field, just as a current in a loop of wire produces a magnetic field. There is also strong evidence to indicate that the strength of a planet's field is related to the planet's rate of rotation. For example, Jupiter rotates faster than the Earth, and recent space probes indicate that Jupiter's magnetic field is stronger than ours. Venus, on the other hand, rotates more slowly than the Earth, and its magnetic field is found to be weaker. Investigation into the cause of the Earth's magnetism remains open.

There is an interesting sidelight concerning the Earth's magnetic field. It has been found that the direction of the field has been reversed several times during the last million years. Evidence for this is provided by basalt (a type of rock that contains iron) that is spewed forth by volcanic activity on the ocean floor. As the lava cools, it solidifies and retains a picture of the Earth's magnetic field direction. The rocks can be dated by other means to provide the evidence for these periodic reversals of the magnetic field.

SUMMARY

The **Biot-Savart law** says that the magnetic field $d\mathbf{B}$ at a point P due to a current element $d\mathbf{s}$ carrying a steady current I is

$$d\mathbf{B} = k_m \frac{I\, d\mathbf{s} \times \hat{\mathbf{r}}}{r^2} \tag{30.1}$$

where $k_m = 10^{-7}\ \text{Wb/A} \cdot \text{m}$ and r is the distance from the element to the point P. To find the total field at P due to a current-carrying conductor, we must integrate this vector expression over the entire conductor.

The **magnetic field** at a distance a from a long, straight wire carrying a current I is given by

$$B = \frac{\mu_0 I}{2\pi a} \tag{30.7}$$

where $\mu_0 = 4\pi \times 10^{-7}\ \text{Wb/A} \cdot \text{m}$ is the **permeability of free space**. The field lines are circles concentric with the wire.

The force per unit length between two parallel wires separated by a distance a and carrying currents I_1 and I_2 has a magnitude given by

$$\frac{F}{\ell} = \frac{\mu_0 I_1 I_2}{2\pi a} \tag{30.13}$$

The force is attractive if the currents are in the same direction and repulsive if they are in opposite directions.

Ampère's law says that the line integral of $\mathbf{B} \cdot d\mathbf{s}$ around any closed path equals $\mu_0 I$, where I is the total steady current passing through any surface bounded by the closed path. That is,

$$\oint \mathbf{B} \cdot d\mathbf{s} = \mu_0 I \tag{30.15}$$

Using Ampère's law, it is found that the fields inside a toroid and solenoid are

$$B = \frac{\mu_0 N I}{2\pi r} \quad \text{(toroid)} \tag{30.18}$$

$$B = \mu_0 \frac{N}{\ell} I = \mu_0 n I \quad \text{(solenoid)} \tag{30.20}$$

where N is the total number of turns.

The **magnetic flux** Φ_B through a surface is defined by the surface integral

$$\Phi_B = \int \mathbf{B} \cdot d\mathbf{A} \qquad (30.23)$$

Gauss's law of magnetism states that the net magnetic flux through any closed surface is zero. That is, isolated magnetic poles (or magnetic monopoles) do not exist.

A **displacement current** I_d arises from a time-varying electric flux Φ_E and is defined by

$$I_d \equiv \epsilon_0 \frac{d\Phi_E}{dt} \qquad (30.26)$$

The **generalized form of Ampère's law**, which includes the displacement current, is

$$\oint \mathbf{B} \cdot d\mathbf{s} = \mu_0 I + \mu_0 \epsilon_0 \frac{d\Phi_E}{dt} \qquad (30.27)$$

This law describes the fact that magnetic fields are produced both by conduction currents and by changing electric fields.

QUESTIONS

1. Is the magnetic field created by a current loop uniform? Explain.
2. A current in a conductor produces a magnetic field that can be calculated using the Biot-Savart law. Since current is defined as the rate of flow of charge, what can you conclude about the magnetic field produced by stationary charges? What about moving charges?
3. Two parallel wires carry currents in opposite directions. Describe the nature of the resultant magnetic field created by the two wires at points (a) between the wires and (b) outside the wires in a plane containing them.
4. Explain why two parallel wires carrying currents in opposite directions repel each other.
5. When an electric circuit is being assembled, a common practice is to twist together two wires carrying equal and opposite currents. Why does this technique reduce stray magnetic fields?
6. Is Ampère's law valid for all closed paths surrounding a conductor? Why is it not useful for calculating **B** for all such paths?
7. Compare Ampère's law with the Biot-Savart law. Which is the more general method for calculating **B** for a current-carrying conductor?
8. Is the magnetic field inside a toroid uniform? Explain.
9. Describe the similarities between Ampère's law in magnetism and Gauss's law in electrostatics.
10. A hollow copper tube carries a current. Why is **B** = 0 inside the tube? Is **B** nonzero outside the tube?
11. Why is **B** nonzero outside a solenoid? Why is **B** = 0 outside a toroid? (The lines of **B** must form closed paths.)

12. Describe the change in the magnetic field in the space enclosed by a solenoid carrying a steady current I if (a) the length of the solenoid is doubled but the number of turns remains the same and (b) the number of turns is doubled but the length remains the same.
13. A flat conducting loop is located in a uniform magnetic field directed along the x axis. For what orientation of the loop is the flux through it a maximum? A minimum?
14. What new concept did Maxwell's generalized form of Ampère's circuital law include?
15. A magnet attracts a piece of iron. The iron can then attract another piece of iron. On the basis of domain alignment, explain what happens in each piece of iron.
16. You are an astronaut stranded on a planet with no test equipment for minerals around. The planet does not even have a magnetic field. You have two bars of iron in your possession: one is magnetized, one is not. How could you determine which is which?
17. Why does hitting a magnet with a hammer cause the magnetism to be reduced?
18. Will a nail be attracted to either pole of a magnet? Explain what is happening inside the nail when placed near the magnet.
19. The north-seeking pole of a magnet is attracted toward the geographic North Pole of the Earth. Yet, like poles repel. What is the way out of this dilemma?
20. A Hindu ruler once suggested that he be entombed in a magnetic coffin with the polarity arranged so that he would be forever suspended between Heaven and Earth. Is such magnetic levitation possible? Discuss.

21. Why is **M** = 0 in a vacuum? What is the relationship between **B** and **H** in a vacuum?

22. Explain why some atoms have permanent magnetic dipole moments and others do not.

23. What factors contribute to the total magnetic dipole moment of an atom?

24. Why is the susceptibility of a diamagnetic substance negative?

25. Why can the effect of diamagnetism be neglected in a paramagnetic substance?

26. Explain the significance of the Curie temperature for a ferromagnetic substance.

27. Discuss the difference between ferromagnetic, paramagnetic, and diamagnetic substances.

28. What is the difference between hard and soft ferromagnetic materials?

29. Should the surface of a computer disc be made from a hard or a soft ferromagnetic substance?

30. Explain why it is desirable to use hard ferromagnetic materials to make permanent magnets.

31. Why is an unmagnetized steel nail attracted to a permanent magnet?

32. Would you expect the tape from a tape recorder to be attracted to a magnet? (Try it, but not with a recording you wish to save.)

33. Given only a strong magnet and a screwdriver, how would you first magnetize and then demagnetize the screwdriver?

34. Figure 30.35 shows two permanent magnets, each having a hole through its center. Note that the upper magnet is levitated above the lower one. (a) How does this occur? (b) What purpose does the pencil serve? (c) What can you say about the poles of the magnets from this observation? (d) If the upper magnet were inverted, what do you suppose would happen?

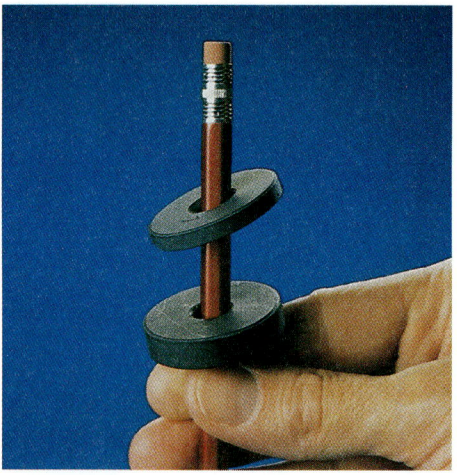

FIGURE 30.35   (Question 34) Magnetic levitation using two ceramic magnets. *(Courtesy of Central Scientific Co.)*

PROBLEMS

Review Problem

The square loop *abcd* shown in the figure has a total resistance $4R$. Each side of the loop has a length d and the wire has a radius r. If a battery of voltage V is connected as shown, find

(a) the currents in each branch of the loop,

(b) the magnetic field at the center of the loop,

(c) the magnetic force between wires *ab* and *cd*,

(d) the magnetic force between wires *bc* and *ad*, and

(e) the equilibrium position (in the plane of the page) of a current-carrying wire that is parallel to wires *bc* and *ad*.

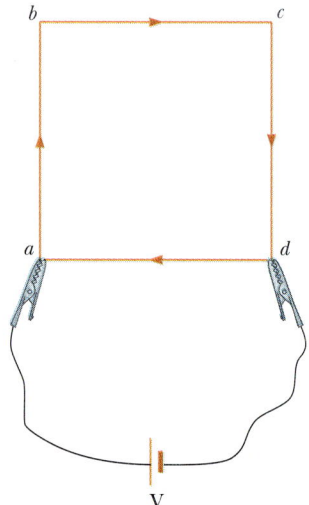

FIGURE RP30

☐ indicates problems that have full solutions available in the Student Solutions Manual and Study Guide.

Section 30.1   The Biot-Savart Law

1. Calculate the magnitude of the magnetic field at a point 100 cm from a long, thin conductor carrying a current of 1.0 A.

2. A long, thin conductor carries a current of 10.0 A. At what distance from the conductor is the magnitude of the resulting magnetic field 1.00×10^{-4} T?

3. A wire in which there is a current of 5.00 A is to be formed into a circular loop of one turn. If the required value of the magnetic field at the center of the loop is 10.0 μT, what is the required radius?

4. In Niels Bohr's 1913 model of the hydrogen atom, an electron circles the proton at a distance of 5.3×10^{-11} m with a speed of 2.2×10^6 m/s. Compute the magnetic field strength that this motion produces at the location of the proton.

5. (a) A conductor in the shape of a square of edge length $\ell = 0.4$ m carries a current $I = 10$ A (Fig. P30.5). Calculate the magnitude and direction of the magnetic field at the center of the square. (b) If this conductor is formed into a single circular turn and carries the same current, what is the value of the magnetic field at the center?

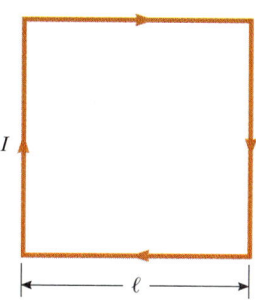

FIGURE P30.5

6. A 12-cm × 16-cm rectangular loop of superconducting wire carries a current of 30 A. What is the magnetic field at the center of the loop?

7. A long, straight wire lies on a horizontal table and carries a current of 1.2 μA. A proton moves parallel to the wire (opposite the current) with a constant speed of 2.3×10^4 m/s at a distance d above the wire. Determine the value of d. You may ignore the magnetic field due to the Earth.

8. A conductor consists of a circular loop of radius R = 0.100 m and two straight, long sections, as in Figure P30.8. The wire lies in the plane of the paper and carries a current of $I = 7.00$ A. Determine the magnitude and direction of the magnetic field at the center of the loop.

8A. A conductor consists of a circular loop of radius R and two straight, long sections, as in Figure P30.8.

The wire lies in the plane of the paper and carries a current I. Determine the magnitude and direction of the magnetic field at the center of the loop.

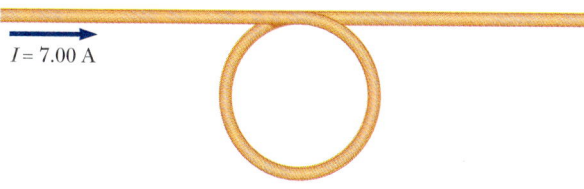

$I = 7.00$ A

FIGURE P30.8

9. Determine the magnetic field at a point P located a distance x from the corner of an infinitely long wire bent at a right angle, as in Figure P30.9. The wire carries a steady current I.

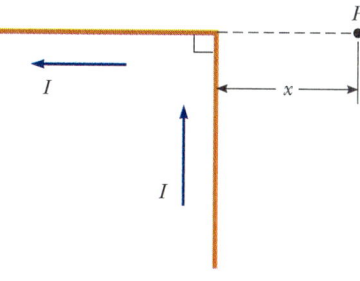

FIGURE P30.9

10. A segment of wire of total length $4r$ is formed into the shape shown in Figure P30.10 and carries a current $I = 6.00$ A. Find the magnitude and direction of the magnetic field at P when $r = 2\pi$ cm.

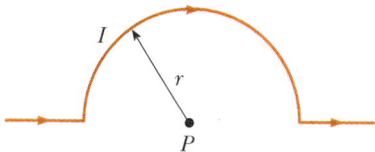

FIGURE P30.10

11. The segment of wire in Figure P30.11 carries a current of $I = 5.00$ A, and the radius of the circular arc is $R = 3.00$ cm. Determine the magnitude and direction of the magnetic field at the origin.

11A. The segment of wire in Figure P30.11 carries a current I, and the radius of the circular arc is R. Determine the magnitude and direction of the magnetic field at the origin.

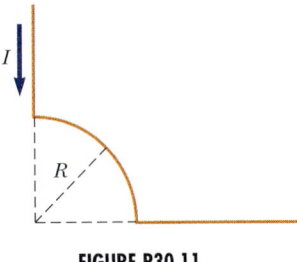

FIGURE P30.11

12. Consider the current-carrying loop shown in Figure P30.12, formed of radial lines and segments of circles whose centers are at point P. Find the magnitude and direction of **B** at P.

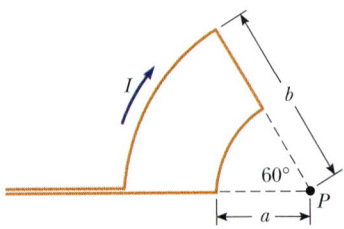

FIGURE P30.12

13. Determine the magnetic field (in terms of I, a, and d) at the origin due to the current loop in Figure P30.13.

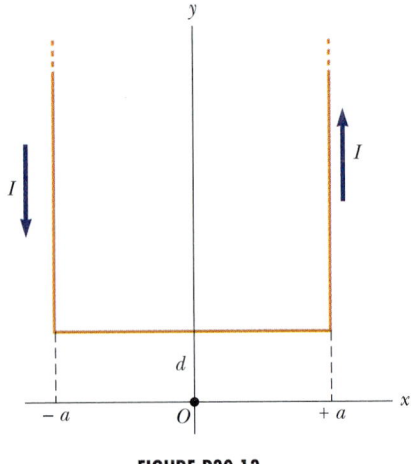

FIGURE P30.13

14. The loop in Figure P30.14 carries a current $I = 2.50$ A. The semicircular portion has a radius $R = 5.00$ cm and $L = 10.0$ cm. Determine the magnitude and direction of the magnetic field at point A.

14A. The loop in Figure P30.14 carries a current I. Determine the magnetic field at point A in terms of I, R, and L.

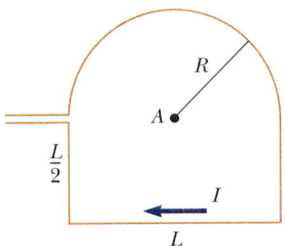

FIGURE P30.14

Section 30.2   The Magnetic Force Between Two Parallel Conductors

15. Two long, parallel conductors separated by 10.0 cm carry currents in the same direction. If $I_1 = 5.00$ A and $I_2 = 8.00$ A, what is the force per unit length exerted on each conductor by the other?

16. Two long, parallel wires, each having a mass per unit length of 40 g/m, are supported in a horizontal plane by strings 6.0 cm long, as in Figure P30.16. Each wire carries the same current I, causing the wires to repel each other so that the angle θ between the supporting strings is 16°. (a) Are the currents in the same or opposite directions? (b) Find the magnitude of each current.

16A. Two long, parallel wires, each having a mass per unit length μ, are supported in a horizontal plane by strings of length L, as in Figure P30.16. Each wire carries the same current I, causing the wires to repel each other so that the angle between the supporting strings is θ. (a) Are the currents in the same or opposite directions? (b) Find the magnitude of each current.

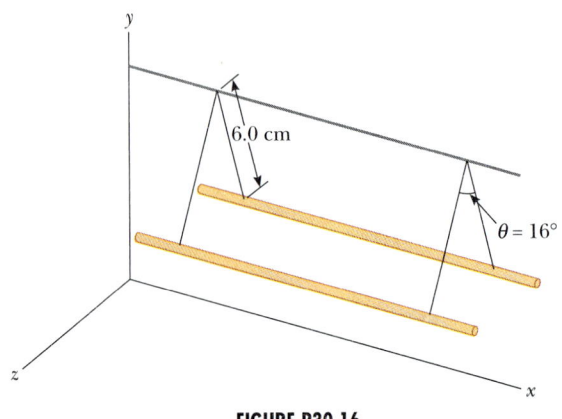

FIGURE P30.16

17. Two parallel copper rods are 1.00 cm apart. Lightning sends a 10.0-kA pulse of current along each conductor. Calculate the force per unit length on one conductor. Is the force attractive or repulsive?

18. Compute the magnetic force per unit length between two adjacent windings of a solenoid if each carries a current $I = 100$ A and the center-to-center distance between the wires is 4 mm.

19. In Figure P30.19, the current in the long, straight wire is $I_1 = 5.00$ A and the wire lies in the plane of the rectangular loop, which carries 10.0 A. The dimensions are $c = 0.100$ m, $a = 0.150$ m, and $\ell = 0.450$ m. Find the magnitude and direction of the net force exerted on the loop by the magnetic field created by the wire.

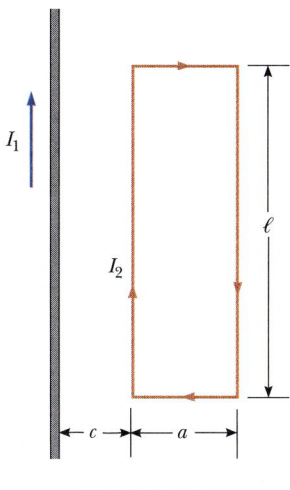

FIGURE P30.19

20. Three long, parallel conductors each carry a 2.0 A-current. Figure P30.20 is an end view of the conductors, with each current coming out of the page. If $a = 1.0$ cm, determine the magnitude and direction of the magnetic field at points *A*, *B*, and *C*.

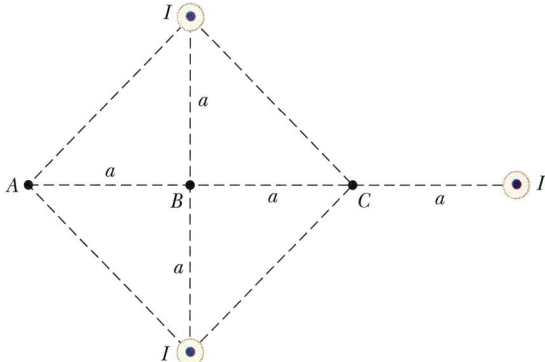

FIGURE P30.20

21. Two long, parallel conductors carry currents $I_1 = 3.00$ A and $I_2 = 3.00$ A, both directed into the page in Figure P30.21. Determine the magnitude and direction of the resultant magnetic field at *P*.

21A. Two long, parallel conductors carry currents I_1 and I_2, both directed into the page in Figure P30.21. Determine the magnitude and direction of the resultant magnetic field at *P*.

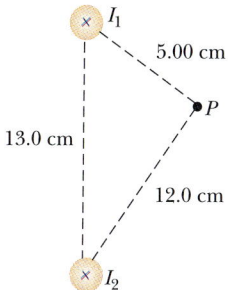

FIGURE P30.21

Section 30.3 Ampère's Law and
Section 30.4 The Magnetic Field of a Solenoid

22. A closely wound, long solenoid of overall length 30.0 cm has a magnetic field of 5.00×10^{-4} T at its center produced by a current of 1.00 A through its windings. How many turns of wire are on the solenoid?

23. A superconducting solenoid is to generate a magnetic field of 10.0 T. (a) If the solenoid winding has 2000 turns/m, what is the required current? (b) What force per unit length is exerted on the windings by this magnetic field?

24. What current is required in the windings of a long solenoid that has 1000 turns uniformly distributed over a length of 0.40 m in order to produce at the center of the solenoid a magnetic field of magnitude 1.0×10^{-4} T?

25. Some superconducting alloys at very low temperatures can carry very high currents. For example, Nb_3Sn wire at 10 K can carry 1.0×10^3 A and maintain its superconductivity. Determine the maximum value of *B* achievable in a solenoid of length 25 cm if 1000 turns of Nb_3Sn wire are wrapped on the outside surface.

26. Imagine a long cylindrical wire of radius *R* that has a current density $J(r) = J_0(1 - r^2/R^2)$ for $r \leq R$ and $J(r) = 0$, for $r > R$, where *r* is the distance from a point of interest to the central axis running along the length of the wire. (a) Find the resulting magnetic field inside ($r \leq R$) and outside ($r > R$) the wire. (b) Plot the magnitude of the magnetic field as a function of *r*. (c) Find the location where the magnetic field strength is a maximum, and the value of that maximum field.

27. The magnetic coils of a tokamak fusion reactor are in the shape of a toroid having an inner radius of 0.70 m and outer radius of 1.30 m. If the toroid has 900 turns of large-diameter wire, each of which carries a current of 14 kA, find the magnetic field strength along (a) the inner radius and (b) the outer radius.

28. A charged-particle beam that is shot horizontally moves into a region where there is a constant magnetic field of magnitude 2.45×10^{-3} T that points straight down. The particles then move in a circular path of radius 2.00 cm. If they accelerated through a potential difference of 211 V, determine their charge-to-mass ratio.

29. You are given a certain volume of copper from which you can make copper wires. The wires are then used to produce a tightly wound solenoid of maximum magnetic field at the center. (a) Should you make the wires long and thin or short and thick? (b) Should you make the solenoid radius small or large? Explain.

30. Niobium metal becomes a superconductor when cooled below 9 K. If superconductivity is destroyed when the surface magnetic field exceeds 0.10 T, determine the maximum current a 2.0-mm-diameter niobium wire can carry and remain superconducting.

31. A packed bundle of 100 long, straight, insulated wires forms a cylinder of radius $R = 0.50$ cm. (a) If each wire carries 2.0 A, what are the magnitude and direction of the magnetic force per unit length acting on a wire located 0.2 cm from the center of the bundle? (b) Would a wire on the outer edge of the bundle experience a force greater or smaller than the value calculated in part (a)?

32. In Figure P30.32, both currents are in the negative *x* direction. (a) Sketch the magnetic field pattern in the *yz* plane. (b) At what distance *d* along the *z* axis is the magnetic field a maximum?

*Section 30.5   The Magnetic Field Along the Axis of a Solenoid

33. A solenoid of radius $R = 5.00$ cm is made of a long piece of wire of radius $r = 2.00$ mm, length $\ell = 10.0$ m $(\ell \gg R)$, and resistivity $\rho = 1.7 \times 10^{-8}$ $\Omega \cdot$m. Find the magnetic field at the center of the solenoid if the wire is connected to a battery having an emf $\mathcal{E} = 20.0$ V.

33A. A solenoid of radius *R* is made of a long piece of wire of radius *r*, length ℓ $(\ell \gg R)$, and resistivity ρ. Find the magnetic field at the center of the solenoid if the wire is connected to a battery having an emf $\mathcal{E}$.

34. A solenoid 80 cm long has 900 turns and a radius of 2.5 cm. If it carries a current of 3.0 A, calculate the magnetic field along its axis at (a) its center and (b) a point near one end.

35. A solenoid has 500 turns, a length of 50.0 cm, and a radius of 5.0 cm. If it carries 4.0 A, calculate the magnetic field at an axial point located 15 cm from the center (that is, 10 cm from one end).

Section 30.6   Magnetic Flux

36. A toroid is constructed from *N* rectangular turns of wire. Each turn has height *h*. The toroid has an inner radius *a* and outer radius *b*. (a) If the toroid carries a current *I*, show that the total magnetic flux through the turns of the toroid is proportional to $\ln(b/a)$. (b) Evaluate this flux if $N = 200$ turns, $h = 1.5$ cm, $a = 2.0$ cm, $b = 5.0$ cm, and $I = 2.0$ A.

37. A cube of edge length $\ell = 2.5$ cm is positioned as shown in Figure P30.37. There is throughout the region a uniform magnetic field given by $\mathbf{B} = (5.0\mathbf{i} + 4.0\mathbf{j} + 3.0\mathbf{k})$ T. (a) Calculate the flux through the shaded face. (b) What is the total flux through the six faces?

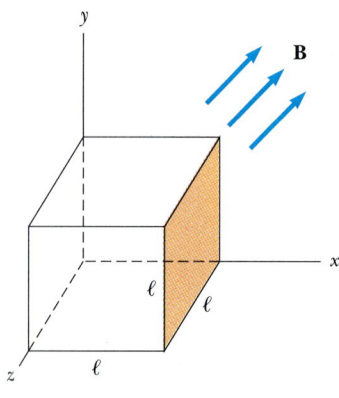

FIGURE P30.37

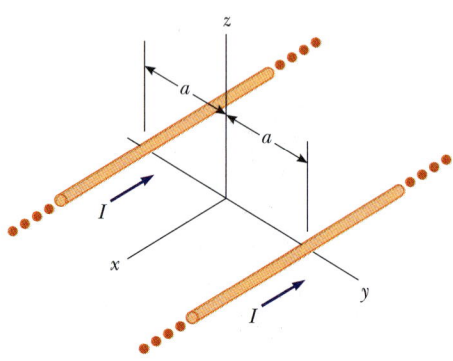

FIGURE P30.32

38. A solenoid 2.5 cm in diameter and 30 cm long has 300 turns and carries 12 A. (a) Calculate the flux through the surface of a disk of radius 5.0 cm that is positioned perpendicular to and centered on the axis of the solenoid, as in Figure P30.38a. (b) Figure P30.38b shows an enlarged end view of the solenoid described above. Calculate the flux through the blue area, which is defined by an annulus that has an inner radius of 0.40 cm and outer radius of 0.80 cm.

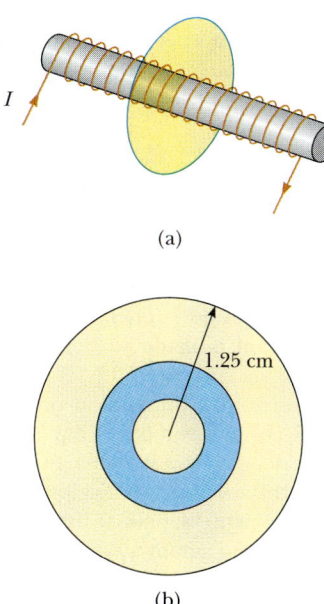

(a)

(b)

FIGURE P30.38

39. Consider the cube in Figure P30.39 having side L. A uniform magnetic field **B** is directed perpendicular to face *abfe*. Find the magnetic flux through the imaginary planar loops (a) *dfhd* and (b) *acfa*.

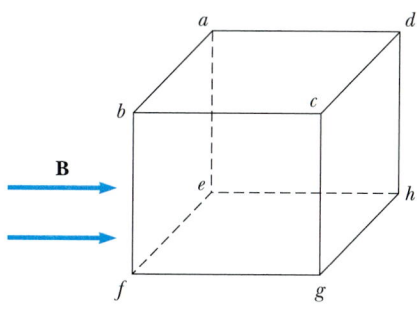

FIGURE P30.39

40. Consider the hemispherical closed surface in Figure P30.40. If the hemisphere is in a uniform magnetic field that makes an angle θ with the vertical, calculate the magnetic flux through (a) the flat surface S_1 and (b) the hemispherical surface S_2.

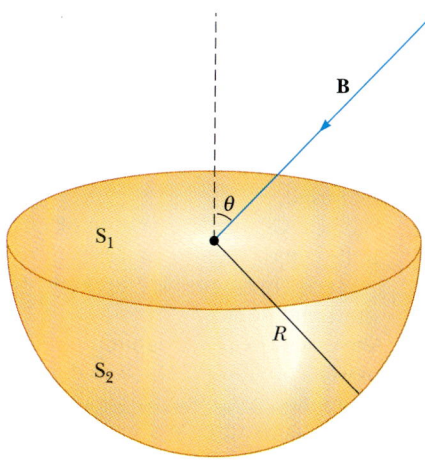

FIGURE P30.40

Section 30.8   Displacement Current and the Generalized Ampère's Law

41. The applied voltage across the plates of a 4.00-μF capacitor varies in time according to the expression

$$V_{app} = (8.00 \text{ V})(1 - e^{-t/4})$$

where t is in seconds. Calculate (a) the displacement current as a function of time and (b) the value of the current at $t = 4.00$ s.

42. A capacitor of capacitance C has a charge Q at $t = 0$. At that time, a resistor of resistance R is connected to the capacitor plates. (a) Find the displacement current in the dielectric between the plates as a function of time. (b) Evaluate this displacement current at time $t = 0.10$ s if $C = 2.0$ μF, $Q = 20$ μC, and $R = 500$ kΩ. (c) At what rate is the electric flux between the plates changing at $t = 0.10$ s?

43. A 0.10 A current is charging a capacitor that has square plates, 5.0 cm on a side. If the plate separation is 4.0 mm, find (a) the time rate of change of electric flux between the plates and (b) the displacement current between the plates.

44. A 0.20-A current is charging a capacitor that has circular plates, 10 cm in radius. If the plate separation is 4.0 mm, (a) what is the time rate of increase of electric field between the plates? (b) What is the magnetic field between the plates 5.0 cm from the center?

*Section 30.9   Magnetism in Matter

45. What is the relative permeability of a material that has a magnetic susceptibility of 10^{-4}?

46. An iron-core toroid is wrapped with 250 turns of wire per meter of its length. The current in the winding is 8.00 A. Taking the magnetic permeability of iron to

be $\mu_m = 5000\mu_0$, calculate (a) the magnetic field strength, H, and (b) the field, B.

47. A toroid with a mean radius of 20 cm and 630 turns (Fig. 30.29) is filled with powdered steel whose magnetic susceptibility χ is 100. If the current in the windings is 3.00 A, find B (assumed uniform) inside the toroid.

48. A toroid has an average radius of 9.0 cm. The current in the coil is 0.50 A. How many turns are required to produce a magnetic field strength of 700 A · turns/m within the toroid?

49. A magnetic field of magnitude 1.3 T is to be set up in an iron-core toroid. The toroid has a mean radius of 10 cm, and magnetic permeability of $5000\mu_0$. What current is required if there are 470 turns of wire in the winding?

50. A toroidal solenoid has an average radius of 10 cm and a cross-sectional area of 1.0 cm². There are 400 turns of wire on the soft iron core, which has a permeability of $800\mu_0$. Calculate the current necessary to produce a magnetic flux of 5.0×10^{-4} Wb through a cross-section of the core.

51. A coil of 500 turns is wound on an iron ring ($\mu_m = 750\mu_0$) of 20-cm mean radius and 8.0-cm² cross-sectional area. Calculate the magnetic flux Φ in this Rowland ring when the current in the coil is 0.50 A.

52. A uniform ring of radius 2.0 cm and total charge 6.0 μC rotates with a constant angular speed of 4.0 rad/s around an axis perpendicular to the plane of the ring and passing through its center. What is the magnetic moment of the rotating ring?

52A. A uniform ring of radius R and total charge Q rotates with constant angular speed ω around an axis perpendicular to the plane of the ring and passing through its center. What is the magnetic moment of the rotating ring?

53. In the text, we found that an alternative description for magnetic field B in terms of magnetic field strength H and magnetization M is $\mathbf{B} = \mu_0\mathbf{H} + \mu_0\mathbf{M}$. Relate the magnetic susceptibility χ to $|\mathbf{H}|$ and $|\mathbf{M}|$ for paramagnetic or diamagnetic materials.

54. Calculate the magnetic field strength H of a magnetized substance characterized by a magnetization of 0.88×10^6 A · turns/m and a magnetic field of magnitude 4.4 T. (*Hint:* See Problem 53.)

55. A magnetized cylinder of iron has a magnetic field $B = 0.040$ T in its interior. The magnet is 3.0 cm in diameter and 20 cm long. If the same magnetic field is to be produced by a 5.0-A current carried by an air-core solenoid having the same dimensions as the cylindrical magnet, how many turns of wire must be on the solenoid?

56. In Bohr's 1913 model of the hydrogen atom, the electron is in a circular orbit of radius 5.3×10^{-11} m

and its speed is 2.2×10^6 m/s. (a) What is the magnitude of the magnetic moment due to the electron's motion? (b) If the electron orbits counterclockwise in a horizontal circle, what is the direction of this magnetic moment vector?

57. At saturation, the alignment of spins in iron can contribute as much as 2.0 T to the total magnetic field B. If each electron contributes a magnetic moment of 9.27×10^{-24} A · m² (one Bohr magneton), how many electrons per atom contribute to the saturated field of iron? (*Hint:* There are 8.5×10^{28} iron atoms/m³.)

*Section 30.10 Magnetic Field of the Earth

58. A circular coil of five turns and a diameter of 30 cm is oriented in a vertical plane with its axis perpendicular to the horizontal component of the Earth's magnetic field. A horizontal compass placed at the center of the coil is made to deflect 45° from magnetic north by a current of 0.60 A in the coil. (a) What is the horizontal component of the Earth's magnetic field? (b) If a compass "dip" needle oriented in a vertical north-south plane makes an angle of 13° from the vertical, what is the total strength of the Earth's magnetic field at this location?

59. The magnetic moment of the Earth is approximately 8.0×10^{22} A · m². (a) If this were caused by the complete magnetization of a huge iron deposit, how many unpaired electrons would this correspond to? (b) At two unpaired electrons per iron atom, how many kilograms of iron would this correspond to? (The density of iron is 7900 kg/m³, and there are approximately 8.5×10^{28} iron atoms/m³.)

ADDITIONAL PROBLEMS

60. A lightning bolt may carry a current of 1.0×10^4 A for a short period of time. What is the resulting magnetic field 100 m from the bolt?

61. Measurements of the magnetic field of a large tornado were made at the Geophysical Observatory in Tulsa, Oklahoma, in 1962. If the tornado's field was $B = 1.5 \times 10^{-8}$ T pointing north when the tornado was 9.0 km east of the observatory, what current was carried up/down the funnel of the tornado?

62. Figure P30.62 is a cross-sectional view of a coaxial cable. The center conductor is surrounded by a rubber layer, which is surrounded by an outer conductor, which is surrounded by another rubber layer. The current in the inner conductor is 1.00 A out of the page and the current in the outer conductor is 3.00 A into the page. Determine the magnitude and direction of the magnetic field at points a and b.

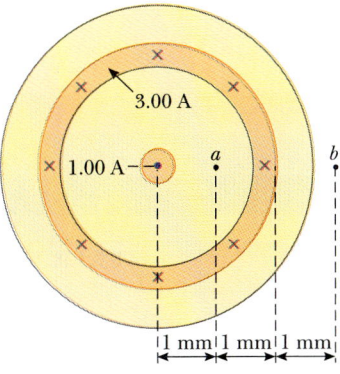

FIGURE P30.62

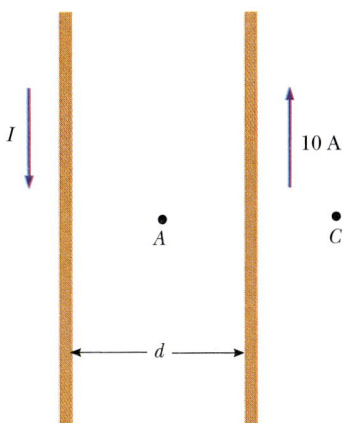

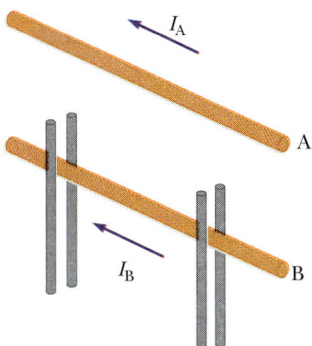

FIGURE P30.63

(a) the value of the current I and (b) the value of the magnetic field at A.

FIGURE P30.64

63. Two long, parallel conductors carry currents in the same direction as in Figure P30.63. Conductor A carries a current of 150 A and is held firmly in position. Conductor B carries a current I_B and is allowed to slide freely up and down (parallel to A) between a set of nonconducting guides. If the mass per unit length of conductor B is 0.10 g/cm, what value of current I_B will result in equilibrium when the distance between the two conductors is 2.5 cm?

63A. Two long, parallel conductors carry currents in the same direction as in Figure P30.63. Conductor A carries a current of I_A and is held firmly in position. Conductor B carries a current I_B and is allowed to slide freely up and down (parallel to A) between a set of nonconducting guides. If the mass per unit length of conductor B is μ, what value of current I_B will result in equilibrium when the distance between the two conductors is d?

64. Two parallel conductors carry current in opposite directions as shown in Figure P30.64. One conductor carries a current of 10 A. Point A is at the midpoint between the wires and point C is a distance $d/2$ to the right of the 10-A current. If $d = 18$ cm and I is adjusted so that the magnetic field at C is zero, find

65. A very long, thin strip of metal of width w carries a current I along its length as in Figure P30.65. Find the magnetic field in the plane of the strip (at an external point P) a distance b from one edge.

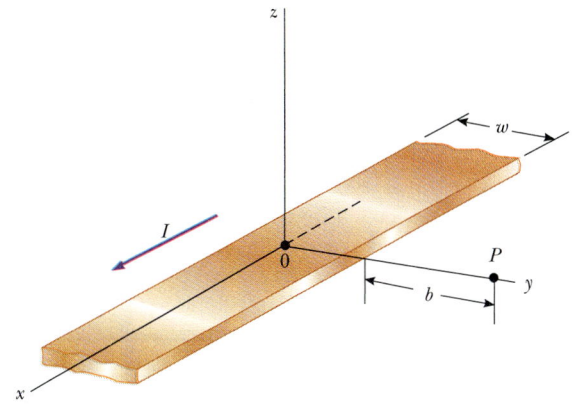

FIGURE P30.65

66. A large nonconducting belt with a uniform surface charge density σ moves with a speed v on a set of rollers as shown in Figure P30.66. Consider a point

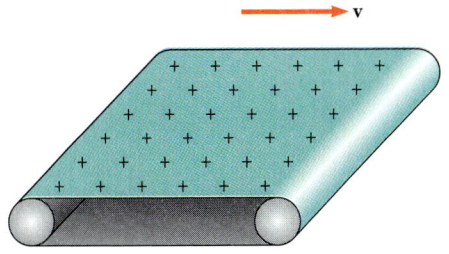

FIGURE P30.66

just above the surface of the moving belt. (a) Find an expression for the magnitude of the magnetic field **B** at this point. (b) If the belt is positively charged, what is the direction of **B**? (Note that the belt may be considered as an infinite sheet.)

67. A straight wire located at the equator is oriented parallel to the Earth along the east-west direction. The Earth's magnetic field at this point is horizontal and has a magnitude of 3.3×10^{-5} T. If the mass per unit length of the wire is 2.0×10^{-3} kg/m, what current must the wire carry in order that the magnetic force balance the weight of the wire?

68. The magnitude of the Earth's magnetic field at either pole is approximately 7.0×10^{-5} T. Using a model in which you assume that this field is produced by a current loop around the equator, determine the current that would generate such a field. (The radius of the Earth is $R_E = 6.37 \times 10^6$ m.)

69. A nonconducting ring of radius R is uniformly charged with a total positive charge q. The ring rotates at a constant angular speed ω about an axis through its center, perpendicular to the plane of the ring. If $R = 0.10$ m, $q = 10$ μC, and $\omega = 20$ rad/s, what is the magnitude of the magnetic field on the axis of the ring 0.050 m from its center?

69A. A nonconducting ring of radius R is uniformly charged with a total positive charge q. The ring rotates at a constant angular speed ω about an axis through its center, perpendicular to the plane of the ring. What is the magnitude of the magnetic field on the axis of the ring a distance $R/2$ from its center?

70. Consider a thin disk of radius R mounted to rotate about the x axis in the yz plane. The disk has a positive uniform surface charge density σ and angular velocity ω. Show that the magnetic field at the center of the disk is $B = \frac{1}{2}\mu_0\sigma\omega R$.

71. Two circular coils of radius R are each perpendicular to a common axis. The coil centers are a distance R apart and a steady current I flows in the same direction around each coil as shown in Figure P30.71. (a) Show that the magnetic field on the axis at a distance x from the center of one coil is

$$B = \frac{\mu_0 I R^2}{2} \left[\frac{1}{(R^2 + x^2)^{3/2}} \right.$$

$$\left. + \frac{1}{(2R^2 + x^2 - 2Rx)^{3/2}} \right]$$

(b) Show that $\dfrac{dB}{dx}$ and $\dfrac{d^2B}{dx^2}$ are both zero at a point midway between the coils. This means the magnetic field in the region midway between the coils is uniform. Coils in this configuration are called **Helmholtz coils**.

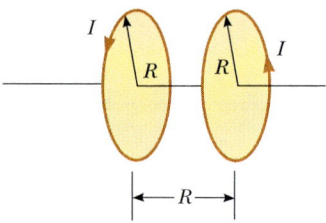

FIGURE P30.71

72. Two identical, flat, circular coils of wire each have 100 turns and a radius of 0.50 m. If these coils are arranged as a set of Helmholtz coils and each coil carries a current of 10 A, determine the magnitude of the magnetic field at a point halfway between the coils and on the axis of the coils. (See Problem 71.)

73. A long cylindrical conductor of radius R carries a current I as in Figure P30.73. The current density J, however, is not uniform over the cross-section of the conductor but is a function of the radius according to $J = br$, where b is a constant. Find an expression for the magnetic field B (a) at a distance $r_1 < R$ and (b) at a distance $r_2 > R$, measured from the axis.

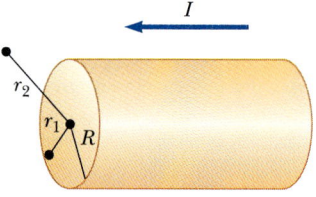

FIGURE P30.73

74. A bar magnet (mass = 39.4 g, magnetic moment = 7.65 J/T, length = 10 cm) is connected to the ceiling by a string. A uniform external magnetic field is applied horizontally, as shown in Figure P30.74. The magnet is in equilibrium, making an angle θ with the horizontal. If $\theta = 5.0°$, determine the strength of the applied magnetic field.

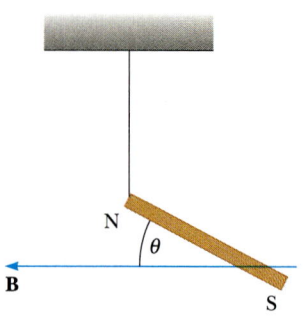

FIGURE P30.74

75. Two circular loops are parallel, coaxial, and almost in contact, 1.0 mm apart (Fig. P30.75). Each loop is 10 cm in radius. The top loop carries a clockwise current of 140 A. The bottom loop carries a counterclockwise current of 140 A. (a) Calculate the magnetic force that the bottom loop exerts on the top loop. (b) The upper loop has a mass of 0.021 kg. Calculate its acceleration, assuming that the only forces acting on it are the force in part (a) and its weight.

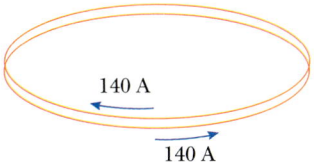

140 A

140 A

FIGURE P30.75

76. For a research project, a student needs a solenoid that produces an interior magnetic field of 0.030 T. She decides to use a current of 1.0 A and a wire 0.50 mm in diameter. She winds the solenoid as layers on an insulating form 1.0 cm in diameter and 10.0 cm long. Determine the number of layers of wire needed and the total length of the wire.

77. A toroid filled with a magnetic substance carries a steady current of 2.00 A. The coil contains 1505 turns, has an average radius of 4.00 cm, and its core has a cross-sectional area of 1.21 cm². The total magnetic flux through a cross-section of the toroid is 3.00×10^{-5} Wb. Assume the flux density is constant. (a) What is the magnetic field strength H within the core? (b) Determine the permeability of the core material if its susceptibility is 3.38×10^{-4}.

78. A paramagnetic substance achieves 10% of its saturation magnetization when placed in a magnetic field of 5.0 T at a temperature of 4.0 K. The density of magnetic atoms in the sample is 8.0×10^{27} atoms/ m³, and the magnetic moment per atom is 5.0 Bohr magnetons. Calculate the Curie constant for this substance.

79. An infinitely long, straight wire carrying a current $I_1 = 50.0$ A is partially surrounded by a loop as in Figure P30.79. The loop has a length $L = 35.0$ cm, and a radius $R = 15.0$ cm, and carries a current $I_2 = 20.0$ A. The central perpendicular axis of the loop coincides with the wire. Calculate the force exerted on the loop.

79A. An infinitely long, straight wire carrying a current I_1 is partially surrounded by a loop as in Figure P30.79. The loop has length L, and radius R, and carries a current I_2. The axis of the loop coincides with the wire. Calculate the force exerted on the loop.

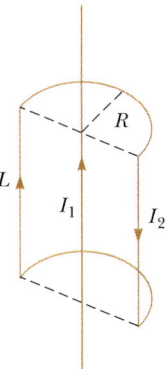

R

L

I_1 I_2

FIGURE P30.79

80. The force on a magnetic dipole M aligned with a nonuniform magnetic field in the x direction is given by $F_x = M \, dB/dx$. Suppose that two flat loops of wire each have radius R and carry current I. (a) If the loops are arranged coaxially and separated by a large variable distance x, show that the magnetic force between them varies as $1/x^4$. (b) Evaluate the magnitude of this force if $I = 10$ A, $R = 0.5$ cm, and $x = 5.0$ cm.

81. A wire is formed into the shape of a square of edge length L (Fig. P30.81). Show that when the current in the loop is I, the magnetic field at point P a distance x from the center of the square along its axis is

$$B = \frac{\mu_0 I L^2}{2\pi \left(x^2 + \frac{L^2}{4}\right)\sqrt{x^2 + \frac{L^2}{2}}}$$

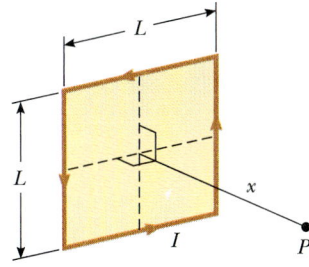

L

L

x

I

P

FIGURE P30.81

82. A wire is bent into the shape shown in Figure P30.82a, and the magnetic field is measured at P_1 when the current in the wire is I. The same wire is then formed into the shape shown in Figure P30.82b, and the magnetic field measured at point P_2 when the current is again I. If the total length of wire is the same in each case, what is the ratio of B_1/B_2?

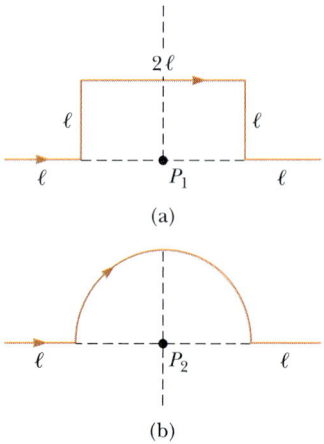

FIGURE P30.82

83. A wire carrying a current I is bent into the shape of an exponential spiral, $r = e^\theta$, from $\theta = 0$ to $\theta = 2\pi$ as in Figure P30.83. To complete a loop, the ends of the spiral are connected by a straight wire along the x axis. Find the magnitude and direction of **B** at the origin. *Hints:* Use the Biot-Savart law. The angle β between a radial line and its tangent line at any point on the curve $r = f(\theta)$ is related to the function in the following way:

$$\tan \beta = \frac{r}{dr/d\theta}$$

Thus in this case $r = e^\theta$, $\tan \beta = 1$, and $\beta = \pi/4$. Therefore, the angle between ds and $\hat{r}$ is $\pi - \beta = 3\pi/4$. Also

$$ds = \frac{dr}{\sin \pi/4} = \sqrt{2}\, dr$$

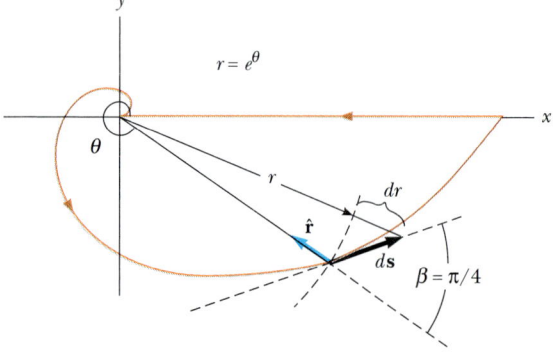

FIGURE P30.83

84. A long cylindrical conductor of radius a has two cylindrical cavities of diameter a through its entire length, as shown in Figure P30.84. A current, I, is directed out of the page and is uniform through a cross-section of the conductor. Find the magnitude and direction of the magnetic field at (a) point P_1 and (b) point P_2 in terms of μ_0, I, r, and a.

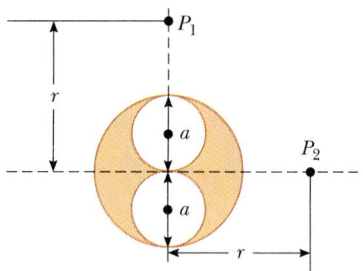

FIGURE P30.84

85. Four long, parallel conductors all carry 5.00 A. An end view of the conductors is shown in Figure P30.85. The current direction is out of the page at points A and B (indicated by the dots) and into the page at points C and D (indicated by the crosses). Calculate the magnitude and direction of the magnetic field at point P, located at the center of the square.

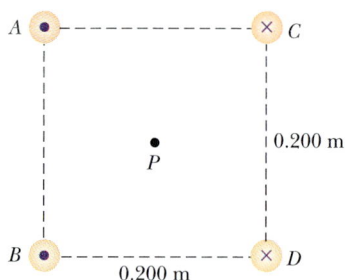

FIGURE P30.85

86. Consider a flat circular current loop of radius R carrying current I. Choose the x axis to be along the axis of the loop with the origin at the center of the loop. Plot a graph of the ratio of the magnitude of the magnetic field at coordinate x to that at the origin for $x = 0$ to $x = 5R$. It may be useful to use a programmable calculator or small computer to solve this problem.

87. A sphere of radius R has a constant volume charge density ρ. Determine the magnetic field at the center of the sphere when it rotates as a rigid body with angular velocity ω about an axis through its center (Fig. P30.87).

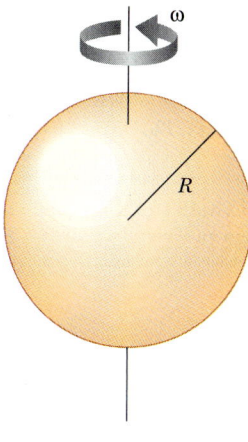

FIGURE P30.87

88. Table 30.4 is data taken on a ferromagnetic material. (a) Construct a magnetization curve from the data. Remember that $\mathbf{B} = \mathbf{B}_0 + \mu_0\mathbf{M}$. (b) Determine the ratio B/B_0 for each pair of values of B and B_0, and construct a graph of B/B_0 versus B_0. (B/B_0 is called the relative permeability, and it is a measure of the induced magnetic field.)

TABLE 30.4

B (T)	B_0 (T)
0.2	4.8×10^{-5}
0.4	7.0×10^{-5}
0.6	8.8×10^{-5}
0.8	1.2×10^{-4}
1.0	1.8×10^{-4}
1.2	3.1×10^{-4}
1.4	8.7×10^{-4}
1.6	3.4×10^{-3}
1.8	1.2×10^{-1}

89. A sphere of radius R has a constant volume charge density ρ. Determine the magnetic dipole moment of the sphere when it rotates as a rigid body with angular velocity ω about an axis through its center (Fig. P30.87).

SPREADSHEET PROBLEMS

S1. Spreadsheet 30.1 calculates the magnetic field along the x axis of a solenoid of length l, radius R, and n turns per unit length carrying a current I. The solenoid is centered at the origin with its axis along the x axis. The field at any point x is

$$B = \frac{1}{2}\mu_0 nI[\,f_1(x) - f_2(x)\,]$$

where

$$f_1(x) = \frac{x + \frac{1}{2}l}{\left[R^2 + \left(x + \frac{1}{2}l\right)^2\right]^{1/2}}$$

$$f_2(x) = \frac{x - \frac{1}{2}l}{\left[R^2 + \left(x - \frac{1}{2}l\right)^2\right]^{1/2}}$$

(a) Let $l = 10$ cm, $R = 0.50$ cm, $I = 4.0$ A, and $n = 1500$ turns/m. Plot B versus x for positive values of x ranging from $x = 0$ to $x = 20$ cm. (b) Change R to the following values, and observe the effects on the graph in each case: $R = 0.10$ cm, 1.0 cm, 2.0 cm, 5.0 cm, and 10.0 cm. For small values of R (small compared to l), the field inside the solenoid should look like that of an ideal solenoid; for large values of R, the field should look like that of a loop. Does it?

S2. If $x \gg l$ (where x is measured from the center of the solenoid), the magnetic field of a finite solenoid along the x axis approaches

$$\frac{\mu_0}{4\pi}\frac{2\mu}{x^3}$$

where $\mu = NI(\pi R^2)$ is the magnetic dipole moment of the solenoid. Modify Spreadsheet 30.1 to calculate and plot $x^3 B(x)$ as a function of x. Verify that $x^3 B(x)$ approaches a constant for $x \gg l$.

S3. A circular loop of wire of radius R carries a current I. Using the Biot-Savart law, the magnetic field in the plane of the loop at a distance r from its center for $r < R$ is found to be

$$B = \frac{\mu_0 I}{4\pi R}F(r/R)$$

where

$$F(y) = \int_0^{2\pi}\frac{(1 - y\sin\theta)\,d\theta}{(1 + y^2 - 2y\sin\theta)^{3/2}}$$

Numerically integrate this expression to find F as a function of $y = r/R$. Use a trapezoidal or Simpson's method integration scheme to carry out the numerical integration. Choose any values for I and R; plot B as a function of r/R to show how the field varies inside the loop. What happens as r approaches R?

S4. Consider a multi-layered solenoid of length l centered at the origin whose axis is along the x axis. Construct a computer spreadsheet or write your own computer program to calculate the magnetic field **B** along the x axis. **B** is given by

$$B(x) = \frac{\mu_0 nIN}{2(R_0 - R_i)} \left[(x + a) \ln \frac{R_0 + \sqrt{(x + a)^2 + R_0^2}}{R_i + \sqrt{(x + a)^2 + R_i^2}} \right.$$

$$\left. - (x - a) \ln \frac{R_0 + \sqrt{(x - a)^2 + R_0^2}}{R_i + \sqrt{(x - a)^2 + R_i^2}} \right]$$

where $a = l/2$, N is the total number of layers, n is the number of turns per unit length in each layer, R_1 is the inside radius, and R_0 is the outside radius of the solenoid.[4] Use any reasonable values for the parameters. Plot your results.

[4] For a derivation of this equation, see L. Golden, J. Klein, and L. Tongson, *American Journal of Physics* 56:846–848, 1988.

CHAPTER 31

Faraday's Law

Induced emf. When a strong magnet is moved toward or away from the coil attached to a galvanometer, an electric current is induced, indicated by the momentary deflection of the galvanometer during the movement of the magnet. *(Richard Megna/Fundamental Photographs)*

O ur studies so far have been concerned with electric fields due to stationary charges and magnetic fields produced by moving charges. This chapter deals with electric fields that originate from changing magnetic fields.

Experiments conducted by Michael Faraday in England in 1831 and independently by Joseph Henry in the United States that same year showed that an electric current could be induced in a circuit by a changing magnetic field. The results of these experiments led to a very basic and important law of electromagnetism known as Faraday's law of induction. This law says that the magnitude of the emf induced in a circuit equals the time rate of change of the magnetic flux through the circuit.

As we shall see, an induced emf can be produced in many ways. For instance, an induced emf and an induced current can be produced in a closed loop of wire when the wire moves into a magnetic field. We shall describe such experiments along with a number of important applications that make use of the phenomenon of electromagnetic induction.

With the treatment of Faraday's law, we complete our introduction to the fundamental laws of electromagnetism. These laws can be summarized in a set of four equations called Maxwell's equations. Together with the Lorentz force law, which we shall discuss briefly, they represent a complete theory for describing the interaction of charged objects. Maxwell's equations relate electric and magnetic fields to each other and to their ultimate source, namely, electric charges.

31.1 FARADAY'S LAW OF INDUCTION

We begin by describing two simple experiments that demonstrate that a current can be produced by a changing magnetic field. First, consider a loop of wire connected to a galvanometer as in Figure 31.1. If a magnet is moved toward the loop, the galvanometer needle will deflect in one direction, as in Figure 31.1a. If the magnet is moved away from the loop, the galvanometer needle will deflect in the opposite direction, as in Figure 31.1b. If the magnet is held stationary relative to the loop, no deflection is observed. Finally, if the magnet is held stationary and the coil is moved either toward or away from the magnet, the needle will also deflect. From these observations, it can be concluded that *a current is set up in the circuit as long as there is relative motion between the magnet and the coil.*[1]

These results are quite remarkable in view of the fact that *a current is set up in the circuit even though there are no batteries in the circuit!* We call such a current an induced current, which is produced by an induced emf.

Now let us describe an experiment, first conducted by Faraday, that is illustrated in Figure 31.2. Part of the apparatus consists of a coil connected to a switch

---

[1] The exact magnitude of the current depends on the particular resistance of the circuit, but the existence of the current (or the algebraic sign) does not.

Joseph Henry, an American physicist who carried out early experiments in electrical induction, was born in Albany, New York, in 1797. The son of a laborer, Henry had little schooling and was forced to go to work at a very young age. After working his way through Albany Academy to study medicine, then engineering, Henry became professor of mathematics and physics in 1826. He later became professor of natural philosophy at New Jersey College (now Princeton University).

In 1848, Henry became the first director of the Smithsonian Institute, where he introduced a weather-forecasting system based on meteorological information received by the electric telegraph. He was also the first president of the Academy of Natural Science, a position he held until his death in 1878.

Many of Henry's early experiments were with electromagnetism. He improved the electromagnet of William Sturgeon and made one of the first electromagnetic motors. By 1830, Henry had made powerful electromagnets by using many turns of fine insulated wire wound around iron cores. He discovered the phenomenon of self-induction but failed to publish his findings; as a result, credit was given to Michael Faraday.

Henry's contribution to science was ultimately recognized: In 1893 the unit of inductance was named the henry.

Joseph Henry

| 1 7 9 7 – 1 8 7 8 |

(North Wind Picture Archives)

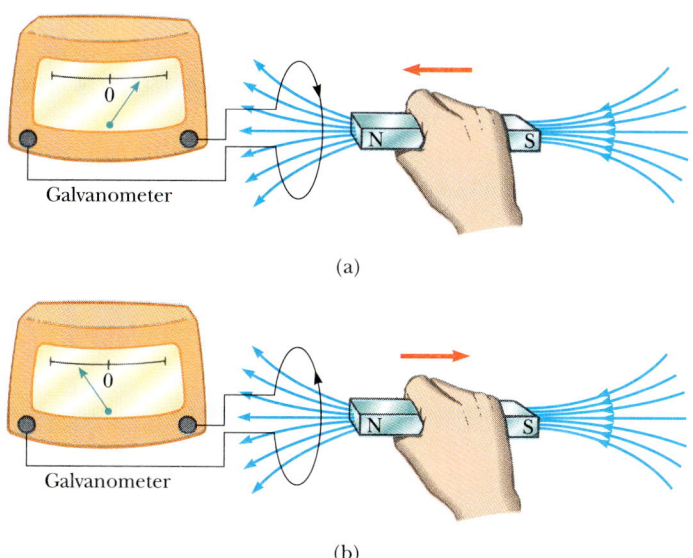

(a)

(b)

FIGURE 31.1 (a) When a magnet is moved toward a loop of wire connected to a galvanometer, the galvanometer deflects as shown. This shows that a current is induced in the loop. (b) When the magnet is moved away from the loop, the galvanometer deflects in the opposite direction, indicating that the induced current is opposite that shown in part (a).

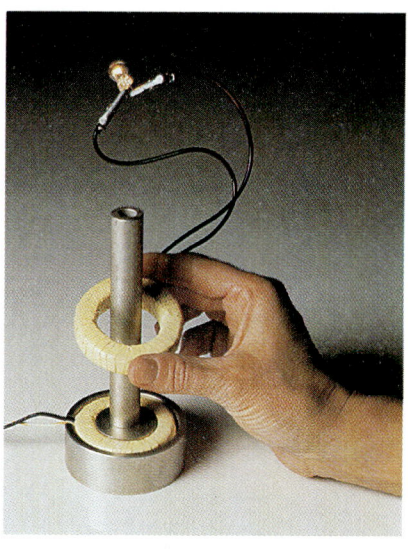

A demonstration of electromagnetic induction. An ac voltage is applied to the lower coil. A voltage is induced in the upper coil as indicated by the illuminated lamp connected to the upper coil. What do you think happens to the lamp's intensity as the upper coil is moved over the vertical tube? *(Courtesy of Central Scientific Co.)*

and a battery. We shall refer to this coil as the primary coil and to the corresponding circuit as the primary circuit. The coil is wrapped around an iron ring to intensify the magnetic field produced by the current through the coil. A second coil, at the right, is also wrapped around the iron ring and is connected to a galvanometer. We shall refer to this as the secondary coil and to the corresponding circuit as the secondary circuit. There is no battery in the secondary circuit and the secondary coil is not connected to the primary coil. The only purpose of this circuit is to show that a current is produced by a change in the magnetic field.

At first sight, you might guess that no current would ever be detected in the secondary circuit. However, something quite amazing happens when the switch in the primary circuit is suddenly closed or opened. At the instant the switch in the primary circuit is closed, the galvanometer in the secondary circuit deflects in one direction and then returns to zero. When the switch is opened, the galvanometer

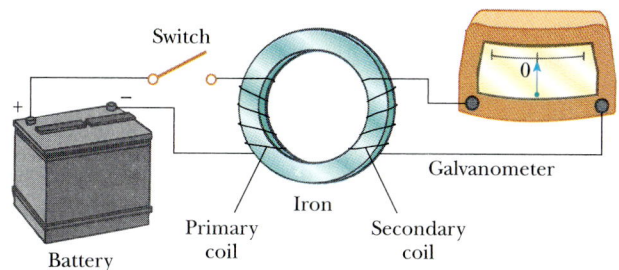

FIGURE 31.2 Faraday's experiment. When the switch in the primary circuit at the left is closed, the galvanometer in the secondary circuit at the right deflects momentarily. The emf induced in the secondary circuit is caused by the changing magnetic field through the coil in this circuit.

deflects in the opposite direction and again returns to zero. Finally, the galvanometer reads zero when there is a steady current in the primary circuit.

As a result of these observations, Faraday concluded that *an electric current can be produced by a changing magnetic field*. A current cannot be produced by a steady magnetic field. The current that is produced in the secondary circuit occurs for only an instant while the magnetic field through the secondary coil is changing. In effect, the secondary circuit behaves as though there were a source of emf connected to it for a short instant. It is customary to say that

> an induced emf is produced in the secondary circuit by the changing magnetic field.

These two experiments have one thing in common. In both cases, an emf is induced in a circuit when the magnetic flux through the circuit changes with time. In fact, a general statement that summarizes such experiments involving induced currents and emfs is as follows:

> The emf induced in a circuit is directly proportional to the time rate of change of magnetic flux through the circuit.

This statement, known as **Faraday's law of induction,** can be written

Faraday's law

$$\varepsilon = -\frac{d\Phi_B}{dt} \qquad (31.1)$$

where Φ_B is the magnetic flux threading the circuit (Section 30.6), which can be expressed as

$$\Phi_B = \int \mathbf{B} \cdot d\mathbf{A} \qquad (31.2)$$

The integral given by Equation 31.2 is taken over the area bounded by the circuit. The meaning of the negative sign in Equation 31.1 is a consequence of Lenz's law and will be discussed in Section 31.3. If the circuit is a coil consisting of *N* loops all of the same area and if the flux threads all loops, the induced emf is

$$\varepsilon = -N\frac{d\Phi_B}{dt} \qquad (31.3)$$

Suppose the magnetic field is uniform over a loop of area *A* lying in a plane as in Figure 31.3. In this case, the flux through the loop is equal to *BA* cos θ; hence, the induced emf can be expressed as

$$\varepsilon = -\frac{d}{dt}\,(BA \cos \theta) \qquad (31.4)$$

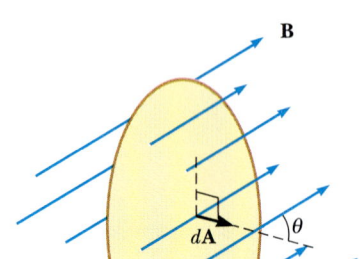

FIGURE 31.3 A conducting loop of area *A* in the presence of a uniform magnetic field **B**, which is at an angle θ with the normal to the loop.

From this expression, we see that an emf can be induced in the circuit in several ways:

- The magnitude of **B** can vary with time.
- The area of the circuit can change with time.
- The angle θ between **B** and the normal to the plane can change with time.
- Any combination of the above can occur.

Some Applications of Faraday's Law

The ground fault interrupter (GFI) is an interesting safety device that protects users of electrical power against electric shock when they touch appliances. Its operation makes use of Faraday's law. Figure 31.4 shows its essential parts. Wire 1 leads from the wall outlet to the appliance to be protected, and wire 2 leads from the appliance back to the wall outlet. An iron ring surrounds the two wires so as to confine the magnetic field set up by each wire. A sensing coil, which can activate a circuit breaker when changes in magnetic flux occur, is wrapped around part of the iron ring. Because the currents in the wires are in opposite directions, the net magnetic field through the sensing coil due to the currents is zero. However, if an insulation fault occurs that connects either the hot or neutral wire to the case, and thus to the safety ground or to the user holding the case, the net magnetic flux through the sensing coil is no longer zero. Because the current is alternating, the magnetic flux through the sensing coil changes with time, producing an induced voltage in the coil. This induced voltage is used to trigger a circuit breaker, stopping the current before it might be harmful to the person using the appliance.

Another interesting application of Faraday's law is the production of sound in an electric guitar. A vibrating string induces an emf in a coil (Fig. 31.5). The pickup coil is placed near the vibrating guitar string, which is made of a metal that can be magnetized. The permanent magnet inside the coil magnetizes the portion of the string nearest the coil. When the string vibrates at some frequency, its magnetized segment produces a changing magnetic flux through the pickup coil. The changing flux induces a voltage in the coil, and the voltage is fed to an

This modern electric range cooks food using the principle of induction. An oscillating current is passed through a coil placed below the cooking surface made of a special glass. The current produces an oscillating magnetic field, which induces a current in the cooking utensil. Since the cooking utensil has some electrical resistance, the electrical energy associated with the induced current transforms into thermal energy, causing the utensil and its contents to heat up.   *(Corning Inc.)*

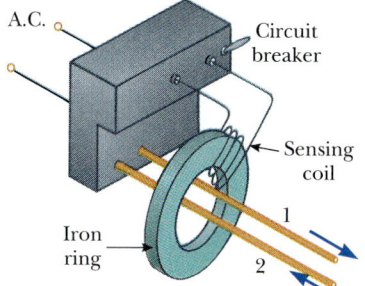

FIGURE 31.4 Essential components of a ground fault interrupter.

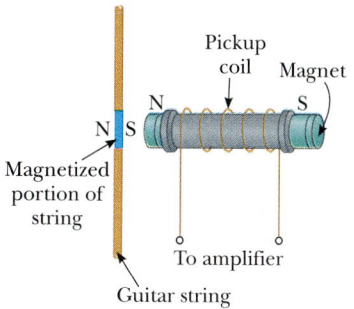

FIGURE 31.5 In an electric guitar, a vibrating string induces a voltage in a pickup coil.

amplifier. The output of the amplifier is sent to the loudspeakers, producing the sound waves that we hear.

EXAMPLE 31.1   One Way to Induce an Emf in a Coil

A coil is wrapped with 200 turns of wire on the perimeter of a square frame of sides 18 cm. Each turn has the same area, equal to that of the frame, and the total resistance of the coil is 2.0 Ω. A uniform magnetic field is turned on perpendicular to the plane of the coil. If the field changes linearly from 0 to 0.50 Wb/m² in a time of 0.80 s, find the magnitude of the induced emf in the coil while the field is changing.

Solution   The area of the loop is $(0.18 \text{ m})^2 = 0.0324 \text{ m}^2$. The magnetic flux through the loop at $t = 0$ is zero since $B = 0$. At $t = 0.80$ s, the magnetic flux through the loop is

$\Phi_B = BA = (0.50 \text{ Wb/m}^2)(0.0324 \text{ m}^2) = 0.0162 \text{ Wb}$. Therefore, the magnitude of the induced emf is

$$|\varepsilon| = \frac{N \Delta\Phi_B}{\Delta t} = \frac{200(0.0162 \text{ Wb} - 0 \text{ Wb})}{0.80 \text{ s}} = \boxed{4.1 \text{ V}}$$

(Note that 1 Wb = 1 V · s.)

Exercise   What is the magnitude of the induced current in the coil while the field is changing?

Answer   2.0 A.

Michael Faraday

| 1 7 9 1 – 1 8 6 7 |

Michael Faraday was a British physicist and chemist who is often regarded as the greatest experimental scientist of the 1800s. His many contributions to the study of electricity include the invention of the electric motor, electric generator, and transformer, as well as the discovery of electromagnetic induction, the laws of electrolysis, the discovery of benzene, and the theory that the plane of polarization of light is rotated in an electric field.

Faraday was born in 1791 in rural England, but his family moved to London shortly thereafter. One of ten children and the son of a blacksmith, Faraday received a minimal education and became apprenticed to a bookbinder at age 14. He was fascinated by articles on electricity and chemistry and was fortunate to have an employer who allowed him to read books and attend scientific lectures. He received some education in science from the City Philosophical Society.

When Faraday finished his apprenticeship in 1812, he expected to devote himself to bookbinding rather than to science. That same year, Faraday attended a lecture by Humphry Davy, who made many contributions in the field of heat and thermodynamics. Faraday sent 386 pages of notes, bound in leather, to Davy; Davy was impressed and appointed Faraday his permanent assistant at the Royal Institution. Faraday toured France and Italy from 1813 to 1815 with Davy, visiting leading scientists of the time such as Volta and Vauquelin.

Despite his limited mathematical ability, Faraday succeeded in making the basic discoveries on which virtually all our uses of electricity depend. He conceived the fundamental nature of magnetism and, to a degree, that of electricity and light.

A modest man who was content to serve science as best he could, Faraday declined a knighthood and an offer to become president of the Royal Society. He was also a moral man; he refused to take part in the preparation of poison gas for use in the Crimean War.

Faraday died in 1867. His many achievements are recognized by the use of his name. The Faraday constant is the quantity of electricity required to deliver a standard amount of substance in electrolysis, and the SI unit of capacitance is the farad.

(By kind permission of the President and Council of the Royal Society)

EXAMPLE 31.2   An Exponentially Decaying B Field

A plane loop of wire of area A is placed in a region where the magnetic field is perpendicular to the plane. The magnitude of $\mathbf{B}$ varies in time according to the expression $B = B_0 e^{-at}$. That is, at $t = 0$ the field is B_0, and for $t > 0$, the field decreases exponentially in time (Fig. 31.6). Find the induced emf in the loop as a function of time.

Solution   Since $\mathbf{B}$ is perpendicular to the plane of the loop, the magnetic flux through the loop at time $t > 0$ is

$$\Phi_B = BA = AB_0 e^{-at}$$

Also, since the coefficient AB_0 and the parameter a are constants, the induced emf can be calculated from Equation 31.1:

$$\mathcal{E} = -\frac{d\Phi_B}{dt} = -AB_0 \frac{d}{dt} e^{-at} = \boxed{aAB_0 e^{-at}}$$

That is, the induced emf decays exponentially in time. Note that the maximum emf occurs at $t = 0$, where $\mathcal{E}_{max} = aAB_0$. Why is this true? The plot of $\mathcal{E}$ versus t is similar to the B versus t curve shown in Figure 31.6.

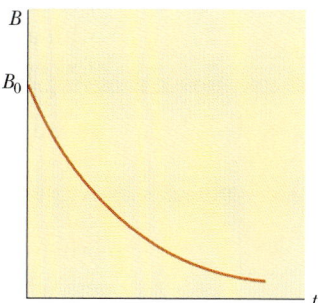

FIGURE 31.6   (Example 31.2) Exponential decrease of the magnetic field with time. The induced emf and induced current have similar time variations.

31.2   MOTIONAL EMF

In Examples 31.1 and 31.2, we considered cases in which an emf is produced in a circuit when the magnetic field changes with time. In this section we describe the so-called **motional emf**, which is the emf induced in a conductor moving through a magnetic field.

First, consider a straight conductor of length ℓ moving with constant velocity through a uniform magnetic field directed into the paper as in Figure 31.7. For simplicity, we shall assume that the conductor is moving perpendicularly to the field. The electrons in the conductor will experience a force along the conductor given by $\mathbf{F} = q\mathbf{v} \times \mathbf{B}$. Under the influence of this force, the electrons will move to the lower end and accumulate there, leaving a net positive charge at the upper end. An electric field is therefore produced within the conductor as a result of this charge separation. The charge at the ends builds up until the magnetic force qvB is balanced by the electric force qE. At this point, charge stops flowing and the condition for equilibrium requires that

$$qE = qvB \qquad \text{or} \qquad E = vB$$

Since the electric field is constant, the electric field produced in the conductor is related to the potential difference across the ends according to the relationship $V = E\ell$. Thus,

$$V = E\ell = B\ell v$$

where the upper end is at a higher potential than the lower end. Thus, *a potential difference is maintained between the ends of the conductor as long as there is motion through the field. If the motion is reversed, the polarity of V is also reversed.*

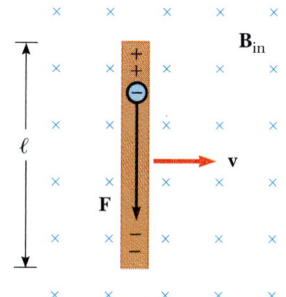

FIGURE 31.7   A straight conducting bar of length ℓ moving with a velocity $\mathbf{v}$ through a uniform magnetic field $\mathbf{B}$ directed perpendicular to $\mathbf{v}$. An emf equal to $B\ell v$ is induced between the ends of the bar.

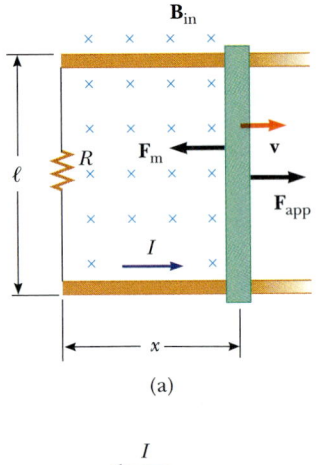

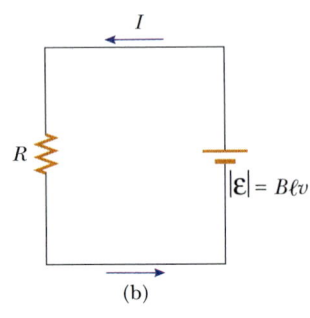

FIGURE 31.8 (a) A conducting bar sliding with a velocity **v** along two conducting rails under the action of an applied force $\mathbf{F}_{app}$. The magnetic force $\mathbf{F}_m$ opposes the motion, and a counterclockwise current is induced in the loop. (b) The equivalent circuit of (a).

A more interesting situation occurs if the moving conductor is part of a closed conducting path. This situation is particularly useful for illustrating how a changing magnetic flux can cause an induced current in a closed circuit. Consider a circuit consisting of a conducting bar of length ℓ sliding along two fixed parallel conducting rails as in Figure 31.8a. For simplicity, we assume that the moving bar has zero resistance and that the stationary part of the circuit has a resistance R. A uniform and constant magnetic field **B** is applied perpendicularly to the plane of the circuit. As the bar is pulled to the right with a velocity **v**, under the influence of an applied force $\mathbf{F}_{app}$, free charges in the bar experience a magnetic force along the length of the bar. This force, in turn, sets up an induced current since the charges are free to move in a closed conducting path. In this case, the rate of change of magnetic flux through the loop and the corresponding induced emf across the moving bar are proportional to the change in area of the loop as the bar moves through the magnetic field. As we shall see, if the bar is pulled to the right with a constant velocity, the work done by the applied force is dissipated in the form of joule heating in the circuit's resistive element.

Since the area of the circuit at any instant is ℓx, the external magnetic flux through the circuit is

$$\Phi_B = B\ell x$$

where x is the width of the circuit, which changes with time. Using Faraday's law, we find that the induced emf is

$$\mathcal{E} = -\frac{d\Phi_B}{dt} = -\frac{d}{dt}(B\ell x) = -B\ell\frac{dx}{dt}$$

$$\mathcal{E} = -B\ell v \tag{31.5}$$

If the resistance of the circuit is R, the magnitude of the induced current is

$$I = \frac{|\mathcal{E}|}{R} = \frac{B\ell v}{R} \tag{31.6}$$

The equivalent circuit diagram for this example is shown in Figure 31.8b.

Let us examine the system using energy considerations. Since there is no battery in the circuit, one might wonder about the origin of the induced current and the electrical energy in the system. We can understand this by noting that the external force does work on the conductor, thereby moving charges through a magnetic field. This causes the charges to move along the conductor with some average drift velocity, and hence a current is established. From the viewpoint of energy conservation, the total work done by the applied force during some time interval should equal the electrical energy that the induced emf supplied in that same period. Furthermore, if the bar moves with constant speed, the work done must equal the energy dissipated as heat in the resistor in this time interval.

As the conductor of length ℓ moves through the uniform magnetic field **B**, it experiences a magnetic force $\mathbf{F}_m$ of magnitude $I\ell B$ (Section 29.2). The direction of this force is opposite the motion of the bar, or to the left in Figure 31.8a.

If the bar is to move with a constant velocity, the applied force must be equal to and opposite the magnetic force, or to the right in Figure 31.8a. If the magnetic force acted in the direction of motion, it would cause the bar to accelerate once it was in motion, thereby increasing its velocity. This state of affairs would represent a violation of the principle of energy conservation. Using Equation 31.6 and the fact that $F_{app} = I\ell B$, we find that the power delivered by the applied force is

$$P = F_{app}v = (I\ell B)v = \frac{B^2\ell^2 v^2}{R} = \frac{V^2}{R} \qquad (31.7)$$

This power is equal to the rate at which energy is dissipated in the resistor, $I^2 R$, as we would expect. It is also equal to the power $I\mathcal{E}$ supplied by the induced emf. This example is a clear demonstration of the conversion of mechanical energy into electrical energy and finally into thermal energy (joule heating).

CONCEPTUAL EXAMPLE 31.3

A circular loop of wire is located in a uniform and constant magnetic field. Describe how an emf can be induced in the loop.

Reasoning   According to Faraday's law, an emf is induced in a wire loop if the magnetic flux through the loop changes with time. In this situation, an emf can be induced by either rotating the loop around an arbitrary axis or by changing the shape of the loop.

CONCEPTUAL EXAMPLE 31.4

A spacecraft orbiting the Earth has a coil of wire in it. An astronaut measures a small current in the coil although there is no battery connected to it and there are no magnets on the spacecraft. What is causing the current?

Reasoning   As the spacecraft moves through space, it is apparently moving from a region of one magnetic field strength to a region of a different magnetic field strength. The changing magnetic field through the coil induces an emf and a corresponding induced current in the coil.

EXAMPLE 31.5   Emf Induced in a Rotating Bar

A conducting bar of length ℓ rotates with a constant angular speed ω about a pivot at one end. A uniform magnetic field **B** is directed perpendicularly to the plane of rotation, as in Figure 31.9. Find the emf induced between the ends of the bar.

Solution   Consider a segment of the bar of length dr, whose velocity is **v**. According to Equation 31.5, the emf induced in a conductor of this length moving perpendicularly to a field **B** is

$$(1) \qquad d\mathcal{E} = Bv\, dr$$

Each segment of the bar is moving perpendicularly to **B**, so there is an emf generated across each segment; the value of this emf is given by (1). Summing up the emfs induced across

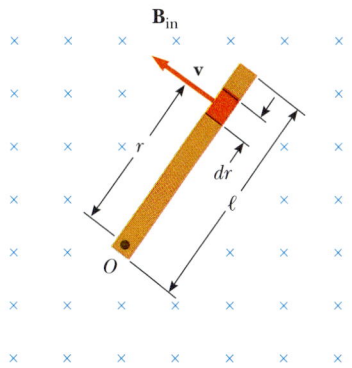

FIGURE 31.9 (Example 31.5) A conducting bar rotating around a pivot at one end in a uniform magnetic field that is perpendicular to the plane of rotation. An emf is induced across the ends of the bar.

all elements, which are in series, gives the total emf between the ends of the bar. That is,

$$\mathcal{E} = \int Bv \, dr$$

In order to integrate this expression, note that the linear speed of an element is related to the angular speed ω through the relationship $v = r\omega$. Therefore, since B and ω are constants, we find that

$$\mathcal{E} = B \int v \, dr = B\omega \int_0^\ell r \, dr = \tfrac{1}{2} B\omega \ell^2$$

EXAMPLE 31.6 Magnetic Force on a Sliding Bar

A bar of mass m and length ℓ moves on two frictionless parallel rails in the presence of a uniform magnetic field directed into the paper (Fig. 31.10). The bar is given an initial velocity v_0 to the right and is released. Find the velocity of the bar as a function of time.

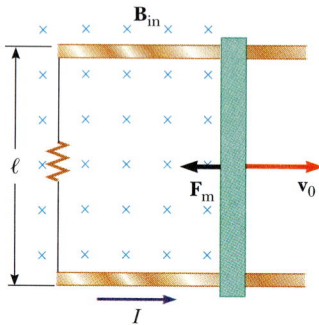

FIGURE 31.10   (Example 31.6) A conducting bar of length ℓ sliding on two fixed conducting rails is given an initial velocity v_0 to the right.

Solution   First note that the induced current is counterclockwise and the magnetic force is $F_m = -I\ell B$, where the negative sign denotes that the force is to the left and retards the motion. This is the only horizontal force acting on the bar, and hence Newton's second law applied to motion in the horizontal direction gives

$$F_x = ma = m\frac{dv}{dt} = -I\ell B$$

Since the induced current is given by Equation 31.6, $I = B\ell v/R$, we can write this expression as

$$m\frac{dv}{dt} = -\frac{B^2 \ell^2}{R} v$$

$$\frac{dv}{v} = -\left(\frac{B^2 \ell^2}{mR}\right) dt$$

Integrating this last equation using the initial condition that $v = v_0$ at $t = 0$, we find that

$$\int_{v_0}^{v} \frac{dv}{v} = \frac{-B^2 \ell^2}{mR} \int_0^t dt$$

$$\ln\left(\frac{v}{v_0}\right) = -\left(\frac{B^2 \ell^2}{mR}\right) t = -\frac{t}{\tau}$$

where the constant $\tau = mR/B^2 \ell^2$. From this, we see that the velocity can be expressed in the exponential form

$$v = v_0 e^{-t/\tau}$$

Therefore, the velocity of the bar decreases exponentially with time under the action of the magnetic retarding force. Furthermore, if we substitute this result into Equations 31.5 and 31.6, we find that the induced emf and induced current also decrease exponentially with time. That is,

$$I = \frac{B\ell v}{R} = \frac{B\ell v_0}{R} e^{-t/\tau}$$

$$\mathcal{E} = IR = B\ell v_0 e^{-t/\tau}$$

31.3 LENZ'S LAW

The direction of the induced emf and induced current can be found from **Lenz's law**,[2] which can be stated as follows:

> The polarity of the induced emf is such that it tends to produce a current that will create a magnetic flux to oppose the change in magnetic flux through the loop.

---

[2] Developed by the German physicist Heinrich Lenz (1804–1865).

That is, the induced current tends to keep the original flux through the circuit from changing. The interpretation of this statement depends on the circumstances. As we shall see, this law is a consequence of the law of conservation of energy.

In order to obtain a better understanding of Lenz's law, let us return to the example of a bar moving to the right on two parallel rails in the presence of a uniform magnetic field directed into the paper (Fig. 31.11a). As the bar moves to the right, the magnetic flux through the circuit increases with time since the area of the loop increases. Lenz's law says that the induced current must be in a direction so that the flux it produces opposes the change in the external magnetic flux. Since the flux due to the external field is increasing into the paper, the induced current, if it is to oppose the change, must produce a flux out of the paper. Hence, the induced current must be counterclockwise when the bar moves to the right to give a counteracting flux out of the paper in the region inside the loop. (Use the right-hand rule to verify this direction.) On the other hand, if the bar is moving to the left, as in Figure 31.11b, the magnetic flux through the loop decreases with time. Since the flux is into the paper, the induced current has to be clockwise to produce a flux into the paper inside the loop. In either case, the induced current tends to maintain the original flux through the circuit.

Let us look at this situation from the viewpoint of energy considerations. Suppose that the bar is given a slight push to the right. In the above analysis, we found that this motion leads to a counterclockwise current in the loop. Let us see what happens if we assume that the current is clockwise. For a clockwise current I, the direction of the magnetic force on the sliding bar would be to the right. This force would accelerate the rod and increase its velocity. This, in turn, would cause the area of the loop to increase more rapidly, thus increasing the induced current, which would increase the force, which would increase the current, which would . . . . In effect, the system would acquire energy with no additional input energy. This is clearly inconsistent with all experience and with the law of conservation of energy. Thus, we are forced to conclude that the current must be counterclockwise.

Consider another situation, one in which a bar magnet is moved to the right toward a stationary loop of wire, as in Figure 31.12a. As the magnet moves to the right toward the loop, the magnetic flux through the loop increases with time. To counteract this increase in flux to the right, the induced current produces a flux to the left, as in Figure 31.12b; hence, the induced current is in the direction shown.

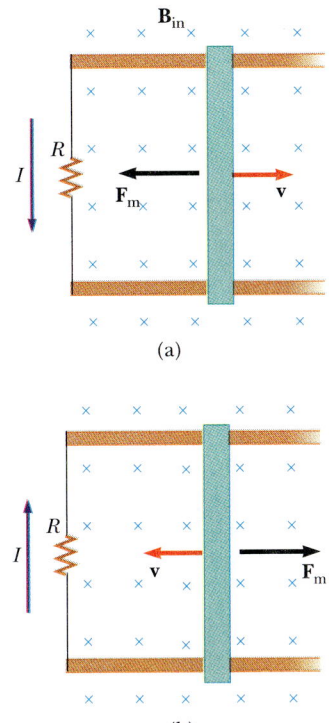

FIGURE 31.11   (a) As the conducting bar slides on the two fixed conducting rails, the magnetic flux through the loop increases in time. By Lenz's law, the induced current must be counterclockwise so as to produce a counteracting flux out of the paper. (b) When the bar moves to the left, the induced current must be clockwise. Why?

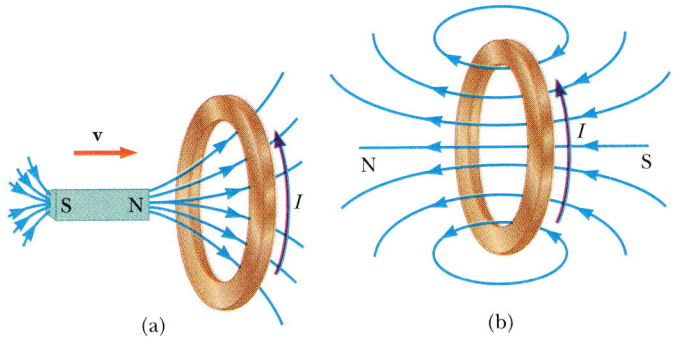

FIGURE 31.12   (a) When the magnet is moved toward the stationary conducting loop, a current is induced in the direction shown. (b) This induced current produces its own flux to the left to counteract the increasing external flux to the right.

Note that the magnetic field lines associated with the induced current oppose the motion of the magnet. Therefore, the left face of the current loop is a north pole and the right face is a south pole.

On the other hand, if the magnet were moving to the left, its flux through the loop, which is toward the right, would decrease in time. Under these circumstances, the induced current in the loop would be in a direction so as to set up a field through the loop directed from left to right in an effort to maintain a constant number of flux lines. Hence, the induced current in the loop would be opposite that shown in Figure 31.12b. In this case, the left face of the loop would be a south pole and the right face would be a north pole.

CONCEPTIONAL EXAMPLE 31.7   Application of Lenz's Law

A coil of wire is placed near an electromagnet as shown in Figure 31.13a. Find the direction of the induced current in the coil (a) at the instant the switch is closed, (b) after the switch has been closed for several seconds, and (c) when the switch is opened.

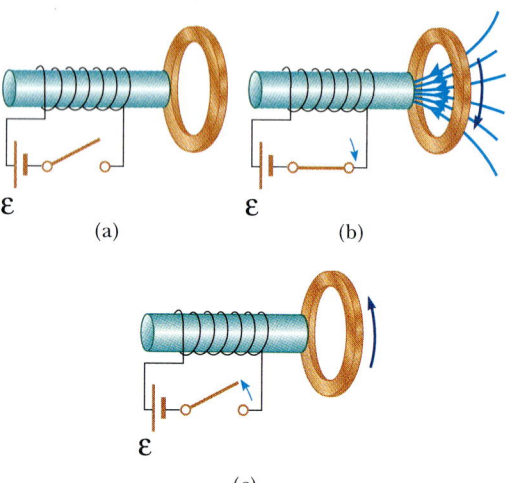

FIGURE 31.13   (Conceptual Example 31.7).

Reasoning   (a) When the switch is closed, the situation changes from a condition in which no lines of flux pass through the coil to one in which lines of flux pass through in the direction shown in Figure 31.13b. To counteract this change in the number of lines, the coil must set up a field from left to right in the figure. This requires a current directed as shown in Figure 31.13b.

(b) After the switch has been closed for several seconds, there is no change in the number of lines through the loop; hence, the induced current is zero.

(c) Opening the switch causes the magnetic field to change from a condition in which flux lines thread through the coil from right to left to a condition of zero flux. The induced current must then be as shown in Figure 31.13c, so as to set up its own field from right to left.

EXAMPLE 31.8   A Loop Moving Through a B Field

A rectangular loop of dimensions ℓ and w and resistance R moves with constant speed v to the right, as in Figure 31.14a. It continues to move with this speed through a region containing a uniform magnetic field **B** directed into the paper and extending a distance $3w$. Plot (a) the flux, (b) the induced emf, and (c) the external force acting on the loop as a function of the position of the loop in the field.

Reasoning and Solution   (a) Figure 31.14b shows the flux through the loop as a function of loop position. Before the loop enters the field, the flux is zero. As it enters the field, the flux increases linearly with position. Finally, the flux decreases linearly to zero as the loop leaves the field.

(b) Before the loop enters the field, there is no induced emf since there is no field present (Fig. 31.14c). As the right

side of the loop enters the field, the flux inward begins to increase. Hence, according to Lenz's law, the induced current is counterclockwise and the induced emf is given by $-B\ell v$. This motional emf arises from the magnetic force experienced by charges in the right side of the loop. When the loop is entirely in the field, the change in flux is zero, and hence the induced emf vanishes.

From another point of view, the right and left sides of the loop experience magnetic forces that tend to set up currents that cancel one another. As the right side of the loop leaves the field, the flux inward begins to decrease, a clockwise current is induced, and the induced emf is $B\ell v$. As soon as the left side leaves the field, the emf drops to zero.

(c) The external force that must act on the loop to maintain this motion is plotted in Figure 31.14d. When the loop is not in the field, there is no magnetic force on it; hence, the external force on it must be zero if v is constant. When the right side of the loop enters the field, the external force necessary to maintain constant speed must be equal to and opposite the magnetic force on that side, given by $F_m = -I\ell B = -B^2\ell^2 v/R$. When the loop is entirely in the field, the flux through the loop is not changing with time. Hence, the net emf induced in the loop is zero, and the current is also zero. Therefore, no external force is needed to maintain the motion of the loop. (From another point of view, the right and left sides of the loop experience equal and opposite forces; hence, the net force is zero.) Finally, as the right side leaves the field, the external force must be equal to and opposite the magnetic force on the left side of the loop. From this analysis, we conclude that power is supplied only when the loop is either entering or leaving the field. Furthermore, this example shows that the induced emf in the loop can be zero even when there is motion through the field! Again, it is emphasized that an emf is induced in the loop only when the magnetic flux through the loop changes in time.

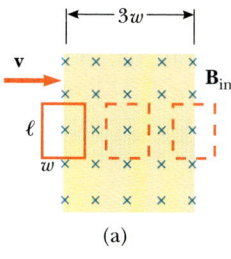
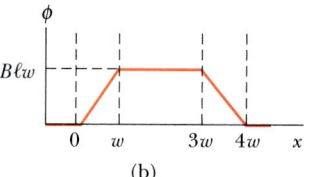
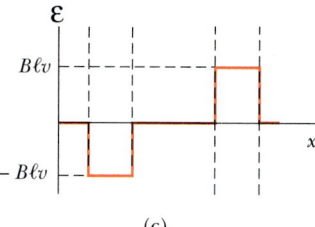
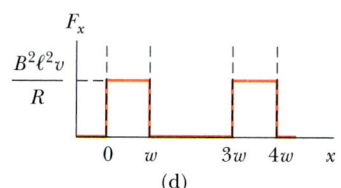

FIGURE 31.14 (Example 31.8) (a) A conducting rectangular loop of width w and length ℓ moving with a velocity **v** through a uniform magnetic field extending a distance $3w$. (b) A plot of the flux as a function of the position of the loop. (c) A plot of the induced emf versus the position of the leading edge. (d) A plot of the force versus position such that the velocity of the loop remains constant.

31.4 INDUCED EMFS AND ELECTRIC FIELDS

We have seen that a changing magnetic flux induces an emf and a current in a conducting loop. We therefore must conclude that *an electric field is created in the conductor as a result of the changing magnetic flux.* In fact, the law of electromagnetic induction shows that *an electric field is always generated by a changing magnetic flux, even in free space where no charges are present.* However, this induced electric field has properties that are quite different from those of an electrostatic field *produced by stationary charges.*

We can illustrate this point by considering a conducting loop of radius r situated in a uniform magnetic field that is perpendicular to the plane of the loop, as in Figure 31.15. If the magnetic field changes with time, then Faraday's law tells us that an emf given by $\mathcal{E} = -d\Phi_B/dt$ is induced in the loop. The induced current that is produced implies the presence of an induced electric field **E**, which must be tangent to the loop since all points on the loop are equivalent. The work done in moving a test charge q once around the loop is equal to $q\mathcal{E}$. Since the electric force on the charge is q**E**, the work done by this force in moving the charge once around

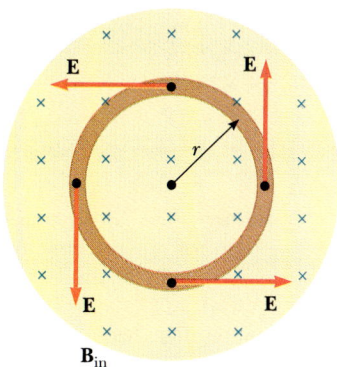

FIGURE 31.15 A loop of radius r in a uniform magnetic field perpendicular to the plane of the loop. If **B** changes in time, an electric field is induced in a direction tangent to the loop.

the loop is given by $qE(2\pi r)$, where $2\pi r$ is the circumference of the loop. These two expressions for the work must be equal; therefore, we see that

$$q\mathcal{E} = qE(2\pi r)$$

$$E = \frac{\mathcal{E}}{2\pi r}$$

Using this result, Faraday's law, and the fact that $\Phi_B = BA = \pi r^2 B$ for a circular loop, we find that the induced electric field can be expressed as

$$E = -\frac{1}{2\pi r}\frac{d\Phi_B}{dt} = -\frac{r}{2}\frac{dB}{dt} \qquad (31.8)$$

If the time variation of the magnetic field is specified, the induced electric field can easily be calculated from Equation 31.8. The negative sign indicates that the induced electric field **E** opposes the change in the magnetic field. It is important to understand that *this result is also valid in the absence of a conductor.* That is, a free charge placed in a changing magnetic field will also experience the same electric field.

The emf for any closed path can be expressed as the line integral of $\mathbf{E} \cdot d\mathbf{s}$ over that path. In more general cases, E may not be constant, and the path may not be a circle. Hence, Faraday's law of induction, $\mathcal{E} = -d\Phi_B/dt$, can be written as

Faraday's law in general form

$$\oint \mathbf{E} \cdot d\mathbf{s} = -\frac{d\Phi_B}{dt} \qquad (31.9)$$

It is important to recognize that *the induced electric field* **E** *that appears in Equation 31.9 is a nonconservative, time-varying field that is generated by a changing magnetic field.* The field **E** that satisfies Equation 31.9 could not possibly be an electrostatic field for the following reason. If the field were electrostatic, and hence conservative, the line integral of $\mathbf{E} \cdot d\mathbf{s}$ over a closed loop would be zero, contrary to Equation 31.9.

EXAMPLE 31.9   Electric Field Due to a Solenoid

A long solenoid of radius R has n turns per unit length and carries a time-varying current that varies sinusoidally as $I = I_0 \cos \omega t$, where I_0 is the maximum current and ω is the angular frequency of the current source (Fig. 31.16). (a) Determine the electric field outside the solenoid, a distance r from its axis.

Solution   First, let us consider an external point and take the path for our line integral to be a circle centered on the solenoid, as in Figure 31.16. By symmetry we see that the magnitude of **E** is constant on this path and tangent to it. The magnetic flux through this path is given by $BA = B(\pi R^2)$, and hence Equation 31.9 gives

$$\oint \mathbf{E} \cdot d\mathbf{s} = -\frac{d}{dt}[B(\pi R^2)] = -\pi R^2 \frac{dB}{dt}$$

$$E(2\pi r) = -\pi R^2 \frac{dB}{dt}$$

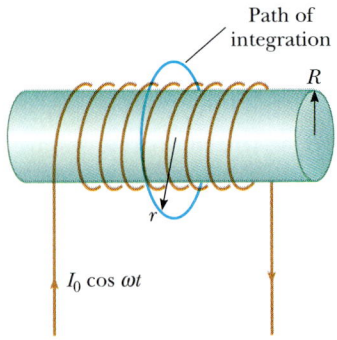

FIGURE 31.16   (Example 31.9) A long solenoid carrying a time-varying current given by $I = I_0 \cos \omega t$. An electric field is induced both inside and outside the solenoid.

Since the magnetic field inside a long solenoid is given by Equation 30.20, $B = \mu_0 nI$, and $I = I_0 \cos \omega t$, we find that

$$E(2\pi r) = -\pi R^2 \mu_0 nI_0 \frac{d}{dt}(\cos \omega t) = \pi R^2 \mu_0 nI_0 \omega \sin \omega t$$

$$E = \frac{\mu_0 nI_0 \omega R^2}{2r} \sin \omega t \qquad \text{(for } r > R)$$

Hence, the electric field varies sinusoidally with time, and its amplitude falls off as $1/r$ outside the solenoid.

   (b) What is the electric field inside the solenoid, a distance r from its axis?

Solution   For an interior point ($r < R$), the flux threading an integration loop is given by $B(\pi r^2)$. Using the same procedure as in part (a), we find that

$$E(2\pi r) = -\pi r^2 \frac{dB}{dt} = \pi r^2 \mu_0 nI_0 \omega \sin \omega t$$

$$E = \frac{\mu_0 nI_0 \omega}{2} r \sin \omega t \qquad \text{(for } r < R)$$

This shows that the amplitude of the electric field inside the solenoid increases linearly with r and varies sinusoidally with time.

*31.5   GENERATORS AND MOTORS

Generators and motors are important devices that operate on the principle of electromagnetic induction. First, let us consider the **alternating current generator** (or ac generator), a device that converts mechanical energy to electrical energy. In its simplest form, the ac generator consists of a loop of wire rotated by some external means in a magnetic field (Fig. 31.17a). In commercial power plants, the energy required to rotate the loop can be derived from a variety of sources. For example, in a hydroelectric plant, falling water directed against the blades of a turbine produces the rotary motion; in a coal-fired plant, the heat produced by burning coal is used to convert water to steam and this steam is directed against the turbine blades. As the loop rotates, the magnetic flux through it changes with time, inducing an emf and a current in an external circuit. The ends of the loop are connected to slip rings that rotate with the loop. Connections to the external circuit are made by stationary brushes in contact with the slip rings.

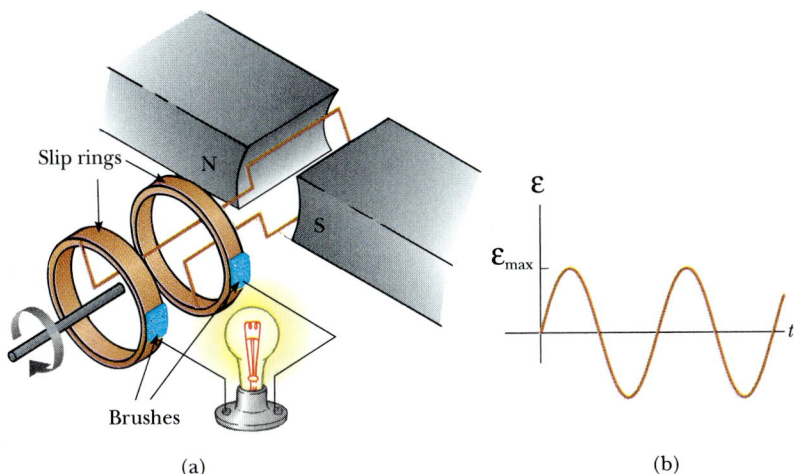

(a)

(b)

FIGURE 31.17   (a) Schematic diagram of an ac generator. An emf is induced in a coil that rotates by some external means in a magnetic field. (b) The alternating emf induced in the loop plotted versus time.

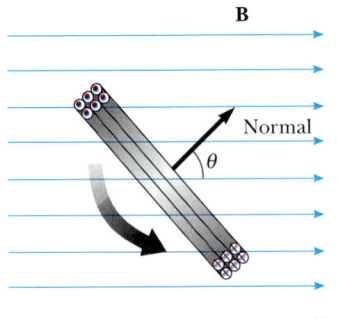

FIGURE 31.18   A loop of area A containing N turns, rotating with constant angular speed ω in the presence of a magnetic field. The emf induced in the loop varies sinusoidally in time.

To put our discussion of the generator on a quantitative basis, suppose that the loop has N turns (a more practical situation), all of the same area A, and suppose that the loop rotates with a constant angular speed ω. If θ is the angle between the magnetic field and the normal to the plane of the loop as in Figure 31.18, then the magnetic flux through the loop at any time t is

$$\Phi_B = BA \cos \theta = BA \cos \omega t$$

where we have used the relationship between angular displacement and angular speed, $\theta = \omega t$. (We have set the clock so that $t = 0$ when $\theta = 0$.) Hence, the induced emf in the coil is

$$\mathcal{E} = -N\frac{d\Phi_B}{dt} = -NAB\frac{d}{dt}(\cos \omega t) = NAB\omega \sin \omega t \qquad (31.10)$$

This result shows that the emf varies sinusoidally with time, as plotted in Figure 31.17b. From Equation 31.10 we see that the maximum emf has the value

$$\mathcal{E}_{\text{max}} = NAB\omega \qquad (31.11)$$

which occurs when $\omega t = 90°$ or $270°$. In other words, $\mathcal{E} = \mathcal{E}_{\text{max}}$ when the magnetic field is in the plane of the coil, and the time rate of change of flux is a maximum. Furthermore, the emf is zero when $\omega t = 0$ or $180°$, that is, when **B** is perpendicular to the plane of the coil, and the time rate of change of flux is zero. The frequency for commercial generators in the United States and Canada is 60 Hz, whereas in some European countries, 50 Hz is used. (Recall that $\omega = 2\pi f$, where f is the frequency in hertz.)

EXAMPLE 31.10   Emf Induced in a Generator

An ac generator consists of eight turns of wire each of area $A = 0.0900$ m^2 and total resistance 12.0 Ω. The loop rotates in a magnetic field $B = 0.500$ T at a constant frequency of 60.0 Hz. (a) Find the maximum induced emf.

Solution   First note that

$$\omega = 2\pi f = 2\pi(60.0 \text{ Hz}) = 377 \text{ s}^{-1}.$$

Using Equation 31.11 with the appropriate numerical values gives

$$\mathcal{E}_{\text{max}} = NAB\omega = 8(0.0900 \text{ m}^2)(0.500 \text{ T})(377 \text{ s}^{-1})$$

$$= \boxed{136 \text{ V}}$$

(b) What is the maximum induced current?

Solution   From Ohm's law and the results to part (a), we find that the maximum induced current is

$$I_{\text{max}} = \frac{\mathcal{E}_{\text{max}}}{R} = \frac{136 \text{ V}}{12.0 \text{ }\Omega} = \boxed{11.3 \text{ A}}$$

Exercise   Determine the time variation of the induced emf and induced current when the output terminals are connected by a low-resistance conductor.

Answers

$$\mathcal{E} = \mathcal{E}_{\text{max}} \sin \omega t = (136 \text{ V}) \sin 377t$$

$$I = I_{\text{max}} \sin \omega t = (11.3 \text{ A}) \sin 377t$$

The **direct current (dc) generator** is illustrated in Figure 31.19a. Such generators are used, for instance, to charge storage batteries in older style cars. The components are essentially the same as those of the ac generator, except that the contacts to the rotating loop are made using a split ring, or commutator.

In this configuration, the output voltage always has the same polarity and the current is a pulsating direct current as in Figure 31.19b. The reason for this can be

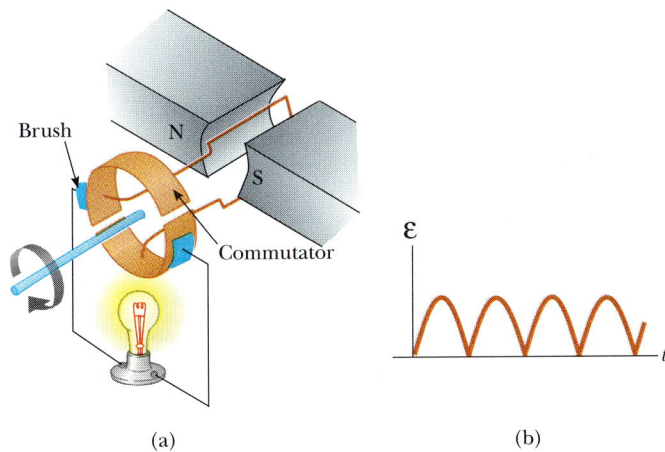

(a)   (b)

FIGURE 31.19   (a) Schematic diagram of a dc generator. (b) The emf versus time fluctuates in magnitude but always has the same polarity.

understood by noting that the contacts to the split ring reverse their roles every half cycle. At the same time, the polarity of the induced emf reverses; hence, the polarity of the split ring (which is the same as the polarity of the output voltage) remains the same.

A pulsating dc current is not suitable for most applications. To obtain a more steady dc current, commercial dc generators use many armature coils and commutators distributed so that the sinusoidal pulses from the various coils are out of phase. When these pulses are superimposed, the dc output is almost free of fluctuations.

Motors are devices that convert electrical energy into mechanical energy. Essentially, *a motor is a generator operating in reverse.* Instead of generating a current by rotating a loop, a current is supplied to the loop by a battery and the torque acting on the current-carrying loop causes it to rotate.

Useful mechanical work can be done by attaching the rotating armature to some external device. However, as the loop rotates, the changing magnetic flux induces an emf in the loop; this induced emf always acts to reduce the current in the loop. If this were not the case, Lenz's law would be violated. The back emf increases in magnitude as the rotational speed of the armature increases. (The phrase *back emf* is used to indicate an emf that tends to reduce the supplied current.) Since the voltage available to supply current equals the difference between the supply voltage and the back emf, the current through the armature coil is limited by the back emf.

When a motor is first turned on, there is initially no back emf and the current is very large because it is limited only by the resistance of the coil. As the coils begin to rotate, the induced back emf opposes the applied voltage and the current in the coils is reduced. If the mechanical load increases, the motor will slow down, which causes the back emf to decrease. This reduction in the back emf increases the current in the coils and therefore also increases the power needed from the external voltage source. For this reason, the power requirements are greater for starting a motor and for running it under heavy loads. If the motor is allowed to run under no mechanical load, the back emf reduces the current to a value just large enough to overcome energy losses due to heat and friction.

EXAMPLE 31.11   The Induced Current in a Motor

Assume that a motor having coils with a resistance of 10 Ω is supplied by a voltage of 120 V. When the motor is running at its maximum speed, the back emf is 70 V. Find the current in the coils (a) when the motor is first turned on and (b) when the motor has reached maximum speed.

Solution   (a) When the motor is first turned on, the back emf is zero. (The coils are motionless.) Thus the current in the coils is a maximum and equal to

$$I = \frac{\mathcal{E}}{R} = \frac{120 \text{ V}}{10 \text{ }\Omega} = \boxed{12 \text{ A}}$$

(b) At the maximum speed, the back emf has its maximum value. Thus, the effective supply voltage is now that of the external source minus the back emf. Hence, the current is reduced to

$$I = \frac{\mathcal{E} - \mathcal{E}_{\text{back}}}{R} = \frac{120 \text{ V} - 70 \text{ V}}{10 \text{ }\Omega} = \frac{50 \text{ V}}{10 \text{ }\Omega} = \boxed{5.0 \text{ A}}$$

Exercise   If the current in the motor is 8.0 A at some instant, what is the back emf at this time?

Answer   40 V.

*31.6   EDDY CURRENTS

As we have seen, an emf and a current are induced in a circuit by a changing magnetic flux. In the same manner, circulating currents called **eddy currents** are set up in bulk pieces of metal moving through a magnetic field. This can easily be demonstrated by allowing a flat metal plate at the end of a rigid bar to swing as a pendulum through a magnetic field (Fig. 31.20). The metal should be a material such as aluminum or copper. As the plate enters the field, the changing flux creates an induced emf in the plate, which in turn causes the free electrons in the metal to move, producing the swirling eddy currents. According to Lenz's law, the direction of the eddy currents must oppose the change that causes them. For this reason, the eddy currents must produce effective magnetic poles on the plate, which are repelled by the poles of the magnet, thus giving rise to a repulsive force that opposes the motion of the pendulum. (If the opposite were true, the pendulum would accelerate and its energy would increase after each swing, in violation of the law of energy conservation.) Alternatively, the retarding force can be "felt" by pulling a metal sheet through the field of a strong magnet.

As indicated in Figure 31.21, with **B** into the paper, the eddy current is counterclockwise as the swinging plate enters the field in position 1. This is because the external flux into the paper is increasing, and hence by Lenz's law the induced current must provide a flux out of the paper. The opposite is true as the plate leaves the field in position 2, where the current is clockwise. Since the induced eddy current always produces a retarding force **F** when the plate enters or leaves the field, the swinging plate eventually comes to rest.

If slots are cut in the metal plate as in Figure 31.22, the eddy currents and the corresponding retarding force are greatly reduced. This can be understood since the cuts in the plate result in open circuits for any large current loops that might otherwise be formed.

The braking systems on many subway and rapid transit cars make use of electromagnetic induction and eddy currents. An electromagnet, which can be energized with a current, is positioned near the steel rails. The braking action occurs when a large current is passed through the electromagnet. The relative motion of the magnet and rails induces eddy currents in the rails, and the direction of these currents produces a drag force on the moving vehicle. The loss in mechanical energy of the vehicle is transformed into joule heat. Since the eddy currents decrease steadily in magnitude as the vehicle slows down, the braking effect is quite

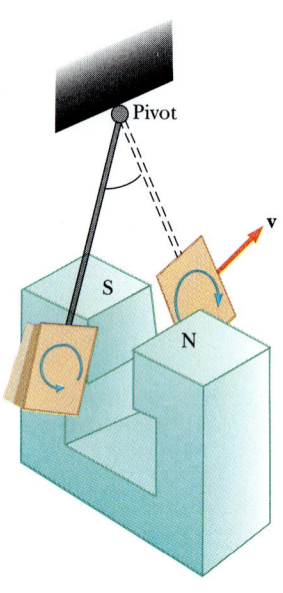

FIGURE 31.20   An apparatus that demonstrates the formation of eddy currents in a conductor moving through a magnetic field. As the plate enters or leaves the field, the changing flux sets up an induced emf, which causes the eddy currents in the plate.

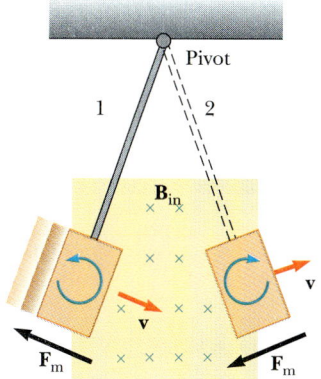

FIGURE 31.21 As the conducting plate enters the field in position 1, the eddy currents are counter-clockwise. However, in position 2, the currents are clockwise. In either case, the plate is repelled by the magnet and eventually comes to rest.

FIGURE 31.22 When slots are cut in the conducting plate, the eddy currents are reduced and the plate swings more freely through the magnetic field.

smooth. Eddy current brakes are also used in some mechanical balances and in various machines.

Eddy currents are often undesirable since they dissipate energy in the form of heat. To reduce this energy loss, the moving conducting parts are often laminated, that is, built up in thin layers separated by a nonconducting material such as lacquer or a metal oxide. This layered structure increases the resistance of the possible paths of the eddy currents and effectively confines the currents to individual layers. Such a laminated structure is used in the cores of transformers and motors to minimize eddy currents and thereby increase the efficiency of these devices.

31.7 MAXWELL'S WONDERFUL EQUATIONS

We conclude this chapter by presenting four equations that can be regarded as the basis of all electrical and magnetic phenomena. These equations, known as Maxwell's equations, after James Clerk Maxwell, are as fundamental to electromagnetic phenomena as Newton's laws are to the study of mechanical phenomena. In fact, the theory developed by Maxwell was more far-reaching than even he imagined at the time, because it turned out to be in agreement with the special theory of relativity, as Einstein showed in 1905. As we shall see, Maxwell's equations represent laws of electricity and magnetism that have already been discussed. However, the equations have additional important consequences. In Chapter 34 we shall show that these equations predict the existence of electromagnetic waves (traveling patterns of electric and magnetic fields), which travel with a speed $c = 1/\sqrt{\mu_0 \epsilon_0} \approx 3 \times 10^8$ m/s, the speed of light. Furthermore, the theory shows that such waves are radiated by accelerating charges.

For simplicity, we present **Maxwell's equations** as applied to free space, that is, in the absence of any dielectric or magnetic material. These are the four equations:

Gauss's law

Gauss's law in magnetism

Faraday's law

Ampère-Maxwell law

$$\oint \mathbf{E} \cdot d\mathbf{A} = \frac{Q}{\epsilon_0} \tag{31.12}$$

$$\oint \mathbf{B} \cdot d\mathbf{A} = 0 \tag{31.13}$$

$$\oint \mathbf{E} \cdot d\mathbf{s} = -\frac{d\Phi_B}{dt} \tag{31.14}$$

$$\oint \mathbf{B} \cdot d\mathbf{s} = \mu_0 I + \epsilon_0 \mu_0 \frac{d\Phi_E}{dt} \tag{31.15}$$

Let us discuss these equations one at a time. Equation 31.12 is Gauss's law, which states that *the total electric flux through any closed surface equals the net charge inside that surface divided by* ϵ_0. This law relates the electric field to the charge distribution, where electric field lines originate on positive charges and terminate on negative charges.

Equation 31.13, which can be considered *Gauss's law in magnetism,* says that *the net magnetic flux through a closed surface is zero.* That is, the number of magnetic field lines that enter a closed volume must equal the number that leave that volume. This implies that magnetic field lines cannot begin or end at any point. If they did, this would mean that isolated magnetic monopoles existed at those points. The fact that isolated magnetic monopoles have not been observed in nature can be taken as a confirmation of Equation 31.13.

Equation 31.14 is *Faraday's law of induction,* which describes the relationship between an electric field and a changing magnetic flux. This law states that *the line integral of the electric field around any closed path (which equals the emf) equals the rate of change of magnetic flux through any surface area bounded by that path.* One consequence of Faraday's law is the current induced in a conducting loop placed in a time-varying magnetic field.

Equation 31.15 is the generalized form of Ampère's law, which describes a relationship between magnetic and electric fields and electric currents. That is, *the line integral of the magnetic field around any closed path is determined by the sum of the net conduction current through that path and the rate of change of electric flux through any surface bounded by that path.*

Once the electric and magnetic fields are known at some point in space, the force on a particle of charge q can be calculated from the expression

The Lorentz force

$$\mathbf{F} = q\mathbf{E} + q\mathbf{v} \times \mathbf{B} \tag{31.16}$$

This is called the **Lorentz force.** Maxwell's equations, together with this force law, give a complete description of all classical electromagnetic interactions.

It is interesting to note the symmetry of Maxwell's equations. Equations 31.12 and 31.13 are symmetric, apart from the absence of a magnetic monopole term in Equation 31.13. Furthermore, Equations 31.14 and 31.15 are symmetric in that the line integrals of **E** and **B** around a closed path are related to the rate of change of magnetic flux and electric flux, respectively. "Maxwell's wonderful equations," as they were called by John R. Pierce,[3] are of fundamental importance not only to electronics but to all of science. Heinrich Hertz once wrote, "One cannot escape the feeling that these mathematical formulas have an independent existence and

---

[3] John R. Pierce, *Electrons and Waves,* New York, Doubleday Science Study Series, 1964. Chapter 6 of this interesting book is recommended as supplemental reading.

an intelligence of their own, that they are wiser than we are, wiser even than their discoverers, that we get more out of them than we put into them.''

SUMMARY

Faraday's law of induction states that the emf induced in a circuit is directly proportional to the time rate of change of magnetic flux through the circuit. That is,

$$\mathcal{E} = -\frac{d\Phi_B}{dt} \tag{31.1}$$

where Φ_B is the magnetic flux:

$$\Phi_B = \int \mathbf{B} \cdot d\mathbf{A} \tag{31.2}$$

When a conducting bar of length ℓ moves through a magnetic field $\mathbf{B}$ with a speed v such that $\mathbf{B}$ is perpendicular to the bar, the emf induced in the bar (the so-called **motional emf**) is

$$\mathcal{E} = -B\ell v \tag{31.5}$$

Lenz's law states that the induced current and induced emf in a conductor are in such a direction as to oppose the change that produced them.

A general form of **Faraday's law of induction** is

$$\mathcal{E} = \oint \mathbf{E} \cdot d\mathbf{s} = -\frac{d\Phi_B}{dt} \tag{31.9}$$

where $\mathbf{E}$ is a nonconservative, time-varying electric field that is produced by the changing magnetic flux.

When used with the Lorentz force law, $\mathbf{F} = q\mathbf{E} + q\mathbf{v} \times \mathbf{B}$, **Maxwell's equations,** given below in integral form, describe all electromagnetic phenomena:

$$\oint \mathbf{E} \cdot d\mathbf{A} = \frac{Q}{\epsilon_0} \tag{31.12}$$

$$\oint \mathbf{B} \cdot d\mathbf{A} = 0 \tag{31.13}$$

$$\oint \mathbf{E} \cdot d\mathbf{s} = -\frac{d\Phi_B}{dt} \tag{31.14}$$

$$\oint \mathbf{B} \cdot d\mathbf{s} = \mu_0 I + \epsilon_0 \mu_0 \frac{d\Phi_E}{dt} \tag{31.15}$$

The last two equations are of particular importance for the material discussed in this chapter. Faraday's law describes how an electric field can be induced by a changing magnetic flux. Similarly, the Ampère-Maxwell law describes how a magnetic field can be produced by both a conduction current and a changing electric flux.

QUESTIONS

1. What is the difference between magnetic flux and magnetic field?

2. A loop of wire is placed in a uniform magnetic field. For what orientation of the loop is the magnetic flux a maximum? For what orientation is the flux zero?

3. As the conducting bar in Figure 31.23 moves to the right, an electric field is set up directed downward. If the bar were moving to the left, explain why the electric field would be upward.

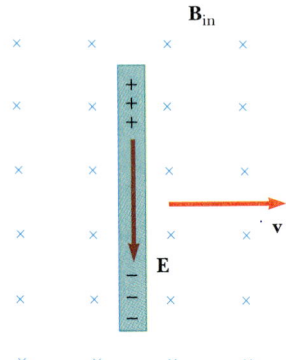

FIGURE 31.23   (Questions 3 and 4).

4. As the bar in Figure 31.23 moves perpendicular to the field, is an external force required to keep it moving with constant speed?

5. The bar in Figure 31.24 moves on rails to the right with a velocity **v**, and the uniform, constant magnetic field is directed out of the page. Why is the induced current clockwise? If the bar were moving to the left, what would be the direction of the induced current?

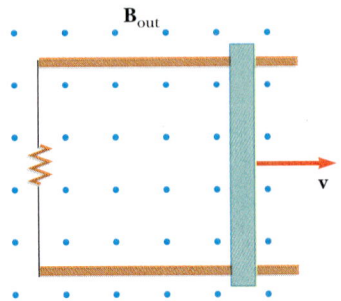

FIGURE 31.24   (Questions 5 and 6).

6. Explain why an external force is necessary to keep the bar in Figure 31.24 moving with a constant speed.

7. A large circular loop of wire lies in the horizontal plane. A bar magnet is dropped through the loop. If the axis of the magnet remains horizontal as it falls, describe the emf induced in the loop. How is the situation altered if the axis of the magnet remains vertical as it falls?

8. When a small magnet is moved toward a solenoid, an emf is induced in the coil. However, if the magnet is moved around inside a toroid, there is no induced emf. Explain.

9. Will dropping a magnet down a long copper tube produce a current in the tube? Explain.

10. How is electrical energy produced in dams (that is, how is the energy of motion of the water converted to ac electricity)?

11. In a beam-balance scale, an aluminum plate is sometimes used to slow the oscillations of the beam near equilibrium. The plate is mounted at the end of the beam and moves between the poles of a small horseshoe magnet attached to the frame. Why are the oscillations of the beam strongly damped near equilibrium?

12. What happens when the speed at which the coil of a generator is rotated is increased?

13. Could a current be induced in a coil by rotating a magnet inside the coil? If so, how?

14. When the switch in Figure 31.25a is closed, a current is set up in the coil and the metal ring springs upward (Fig. 31.25b). Explain this behavior.

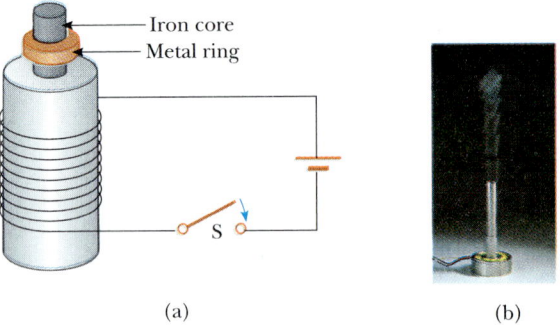

(a)

(b)

FIGURE 31.25   (Questions 14 and 15).   *(Photo courtesy of Central Scientific Co.)*

15. Assume that the battery in Figure 31.25a is replaced by an ac source and the switch is held closed. If held down, the metal ring on top of the solenoid becomes hot. Why?

16. Do Maxwell's equations allow for the existence of magnetic monopoles?

PROBLEMS

Section 31.1   Faraday's Law of Induction

1. A 50-turn rectangular coil of dimensions 5.0 cm × 10.0 cm is dropped from a position where $B = 0$ to a new position where $B = 0.50$ T and is directed perpendicularly to the plane of the coil. Calculate the resulting average emf induced in the coil if the displacement occurs in 0.25 s.

2. A flat loop of wire consisting of a single turn of cross-sectional area 8.0 cm² is perpendicular to a magnetic field that increases uniformly in magnitude from 0.50 T to 2.50 T in 1.0 s. What is the resulting induced current if the loop has a resistance of 2.0 Ω?

3. A powerful electromagnet has a field of 1.6 T and a cross-sectional area of 0.20 m². If we place a coil having 200 turns and a total resistance of 20 Ω around the electromagnet and then turn off the power to the electromagnet in 20 ms, what is the current induced in the coil?

4. In Figure P31.4 find the current through section *PQ*, which has a length $a = 65.0$ cm. The circuit is located in a magnetic field whose magnitude varies with time according to the expression $B = (1.00 \times 10^{-3} \text{ T/s})\, t$. Assume the resistance per unit length of the wire is 0.100 Ω/m.

4A. In Figure P31.4 find the current through section *PQ*, which has resistance R and length a. The circuit is located in a magnetic field whose magnitude varies with time according to the expression $B = At$, where A is a constant having units of teslas/second.

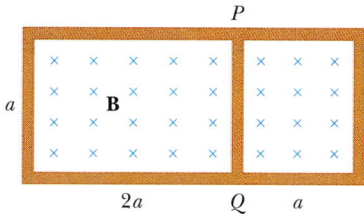

FIGURE P31.4

5. The 10.0-Ω square loop in Figure P31.5, is placed in a uniform 0.10-T magnetic field directed perpendicular to the plane of the loop. The loop, which is hinged at each vertex, is pulled as shown until the separation between points A and B is 3.0 m. If the process takes 0.10 s, what is the average current generated in the loop?

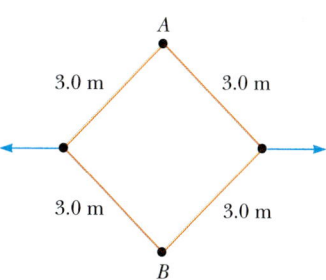

FIGURE P31.5

6. A tightly wound circular coil has 50 turns, each of radius 0.10 m. A uniform magnetic field is turned on along a direction perpendicular to the plane of the coil. If the field increases linearly from 0 to 0.60 T in 0.20 s, what emf is induced in the coil?

7. A 30-turn circular coil of radius 4 cm and resistance 1 Ω is placed in a magnetic field directed perpendicularly to the plane of the coil. The magnitude of the magnetic field varies in time according to the expression $B = 0.010t + 0.040t^2$, where t is in seconds and B is in tesla. Calculate the induced emf in the coil at $t = 5.0$ s.

8. A circular wire loop of radius 0.50 m lies in a plane perpendicular to a uniform magnetic field of magnitude 0.40 T. If in 0.10 s the wire is reshaped into a square but remains in the same plane, what is the magnitude of the average induced emf in the wire during this time?

9. A flat loop of wire of area 14 cm² and two turns is perpendicular to a magnetic field whose magnitude decays in time according to $B = (0.50 \text{ T}) e^{-t/7}$. What is the induced emf as a function of time?

10. A rectangular loop of area A is placed in a region where the magnetic field is perpendicular to the plane of the loop. The magnitude of the field is allowed to vary in time according to $B = B_0 e^{-t/\tau}$, where B_0 and τ are constants. The field has a value of B_0 at $t \le 0$. (a) Use Faraday's law to show that the emf induced in the loop is

$$\mathcal{E} = \frac{AB_0}{\tau} e^{-t/\tau}$$

(b) Obtain a numerical value for $\mathcal{E}$ at $t = 4.0$ s when $A = 0.16$ m², $B_0 = 0.35$ T, and $\tau = 2.0$ s. (c) For the values of A, B_0, and τ given in part (b), what is the maximum value of $\mathcal{E}$?

□ indicates problems that have full solutions available in the Student Solutions Manual and Study Guide.

11. A long solenoid has 400 turns per meter and carries a current $I = (30 \text{ A})(1 - e^{-1.6t})$. Inside the solenoid and coaxial with it is a loop that has a radius of 6.0 cm and consists of a total of 250 turns of fine wire. What emf is induced in the loop by the changing current? (See Fig. P31.11.)

11A. A long solenoid has n turns per meter and carries a current $I = I_0(1 - e^{-\alpha t})$. Inside the solenoid and coaxial with it is a loop that has a radius R and consists of a total of N turns of fine wire. What emf is induced in the loop by the changing current? (See Fig. P31.11.)

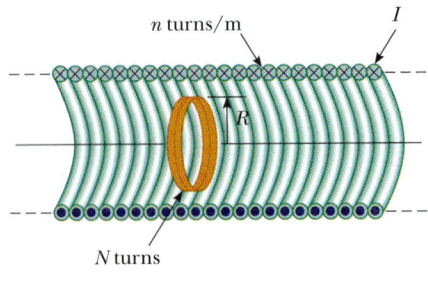

FIGURE P31.11

12. A magnetic field of 0.20 T exists in the region enclosed by a solenoid that has 500 turns and a diameter of 10 cm. Within what period of time must the field be reduced to zero if the average magnitude of the induced emf within the coil during this time interval is to be 10 kV?

13. A coil formed by wrapping 50 turns of wire in the shape of a square is positioned in a magnetic field so that the normal to the plane of the coil makes an angle of 30° with the direction of the field. When the magnitude of the magnetic field is increased uniformly from 200 μT to 600 μT in 0.40 s, an emf of 80 mV is induced in the coil. What is the total length of the wire?

14. A long, straight wire carries a current $I = I_0 \sin(\omega t + \phi)$ and lies in the plane of a rectangular loop of N turns of wire, as shown in Figure P31.14. The quantities I_0, ω, and ϕ are all constants. Determine the emf induced in the loop by the magnetic field created by the current in the straight wire. Assume $I_0 = 50$ A, $\omega = 200\pi$ s^{-1}, $N = 100$, $a = b = 5.0$ cm, and $\ell = 20$ cm.

15. A two-turn circular wire loop of radius 0.500 m lies in a plane perpendicular to a uniform magnetic field of magnitude 0.40 T. If the wire is reshaped from a two-turn circle to a one-turn circle in 0.10 s (while remaining in the same plane), what is the magnitude of the average induced emf in the wire during this time?

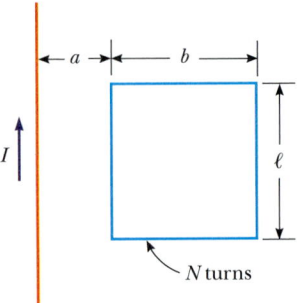

FIGURE P31.14

15A. A two-turn circular wire loop of radius R lies in a plane perpendicular to a uniform magnetic field of magnitude B. If the wire is reshaped from a two-turn circle to a one-turn circle in a time t (while remaining in the same plane), what is the magnitude of the average induced emf in the wire during this time?

16. A toroid having a rectangular cross-section ($a = 2.0$ cm by $b = 3.0$ cm) and inner radius $R = 4.0$ cm consists of 500 turns of wire that carries a current $I = I_0 \sin \omega t$, with $I_0 = 50$ A and a frequency $f = \omega/2\pi = 60$ Hz. A loop that consists of 20 turns of wire links the toroid, as in Figure P31.16. Determine the emf induced in the loop by the changing current I.

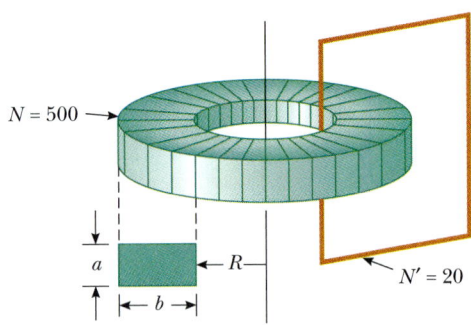

FIGURE P31.16

Section 31.2   Motional Emf
Section 31.3   Lenz's Law

17. A Boeing-747 jet with a wing span of 60.0 m is flying horizontally at a speed of 300 m/s over Phoenix, where the direction of the Earth's magnetic field is 58° below the horizontal. If the magnitude of the magnetic field is 50.0 μT, what is the voltage generated between the wing tips?

18. Consider the arrangement shown in Figure P31.18. Assume that $R = 6.0$ Ω, $\ell = 1.2$ m, and a uniform 2.5-T magnetic field is directed into the page. At

what speed should the bar be moved to produce a current of 0.50 A in the resistor?

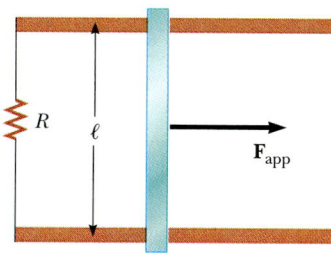

FIGURE P31.18

19. In the arrangement shown in Figure P31.18, the resistor is 6.0 Ω and a 2.5-T magnetic field is directed into the paper. Let $\ell = 1.2$ m and neglect the mass of the bar. (a) Calculate the applied force required to move the bar to the right at a constant speed of 2.0 m/s. (b) At what rate is energy dissipated in the resistor?

20. In Figure P31.20, the bar magnet is moved toward the loop. Is $V_a - V_b$ positive, negative, or zero? Explain.

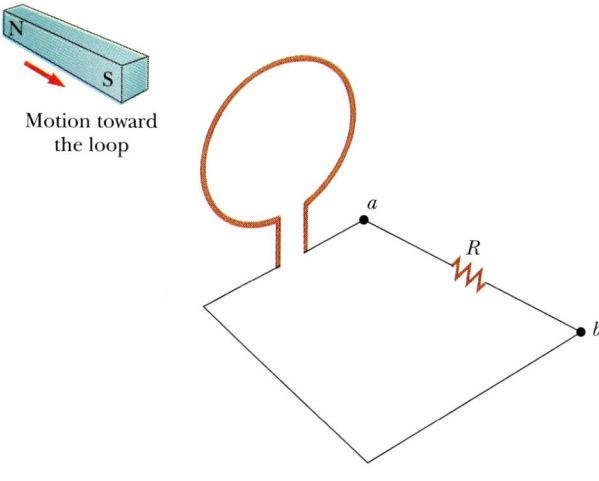

FIGURE P31.20

21. Over a region where the vertical component of the Earth's magnetic field is 40.0 μT directed downward, a 5.00-m length of wire is held along an east-west direction and moved horizontally to the north at 10.0 m/s. Calculate the potential difference between the ends of the wire and determine which end is positive.

22. A metal rod is sliding on a metal ring of radius R, shown in Figure P31.22. There is a uniform magnetic field inside the ring, and the rod moves at constant speed v. (a) Find the induced emf in the rod when it

is at a distance d from the center of the ring. (b) Plot the emf as a function of d.

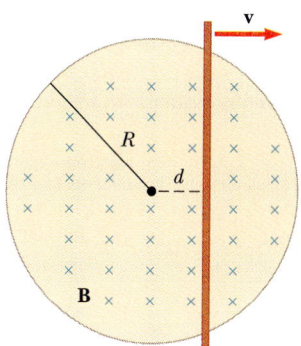

FIGURE P31.22

23. A helicopter has blades of length 3.0 m, rotating at 2.0 rev/s around a central hub. If the vertical component of the Earth's magnetic field is 0.50×10^{-4} T, what is the emf induced between the blade tip and the center hub?

24. A metal bar spins at a constant rate in the magnetic field of the Earth as in Figure 31.9. The rotation occurs in a region where the component of the Earth's magnetic field perpendicular to the plane of rotation is 3.3×10^{-5} T. If the bar is 1.0 m in length and its angular speed is 5π rad/s, what potential difference is developed between its ends?

25. Two parallel rails having negligible resistance are 10.0 cm apart and are connected by a 5.00-Ω resistor. The circuit also contains 10.0-Ω and 15.0-Ω metal rods sliding along the rails and moving away from the resistor at the velocities shown in Figure P31.25. A uniform 0.01-T magnetic field is applied perpendicular to the plane of the rails. Determine the current in the 5.00-Ω resistor.

25A. Two parallel rails having negligible resistance are a distance d apart and are connected by resistor R_1. The circuit also contains two metal rods having resistances R_2 and R_3 sliding along the rails (Fig. P31.25). The rods move away from the resistor at constant speeds v_2 and v_3, respectively. A uniform magnetic field of magnitude B is applied perpendic-

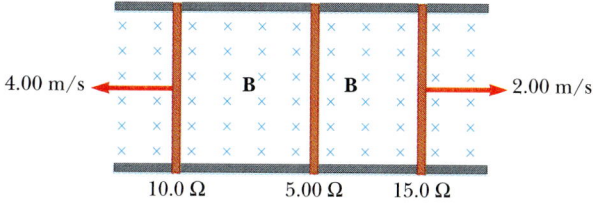

FIGURE P31.25

ularly to the plane of the rails. Determine the current in the resistor R_1.

26. Find the power dissipated in the 12.0-Ω resistor in Figure P31.26. The 0.675-T uniform magnetic field is directed into the plane of the circuit and the 50.0-cm-long conductor moves at a speed $v = 4.20$ m/s.

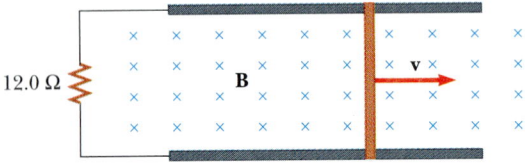

FIGURE P31.26

27. Use Lenz's law to answer the following questions concerning the direction of induced currents. (a) What is the direction of the induced current in resistor R in Figure P31.27a when the bar magnet is moved to the left? (b) What is the direction of the current induced in the resistor R right after the switch S in Figure P31.27b is closed? (c) What is the direction of the induced current in R when the current I in Figure P31.27c decreases rapidly to zero? (d) A copper bar is moved to the right while its axis is maintained perpendicularly to a magnetic field, as in Figure P31.27d. If the top of the bar becomes positive relative to the bottom, what is the direction of the magnetic field?

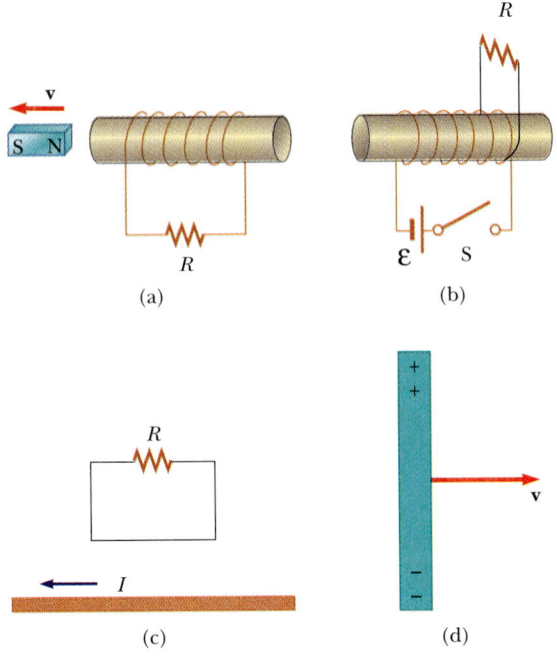

FIGURE P31.27

28. A conducting rectangular loop of mass M, resistance R, and dimensions w by ℓ falls from rest into a magnetic field B as in Figure P31.28. The loop accelerates until it reaches a terminal speed v_t. (a) Show that

$$v_t = \frac{MgR}{B^2 w^2}$$

(b) Why is v_t proportional to R? (c) Why is it inversely proportional to B^2?

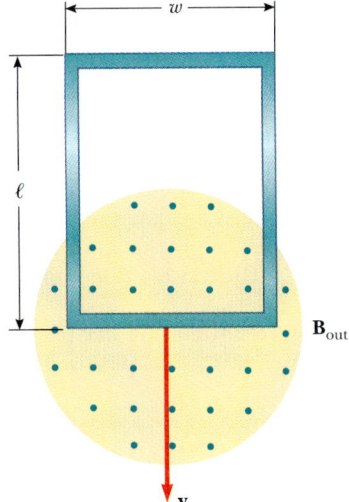

FIGURE P31.28

29. A 0.15-kg wire in the shape of a closed rectangle 1.0 m wide and 1.5 m long has a total resistance of 0.75 Ω. The rectangle is allowed to fall through a magnetic field directed perpendicularly to the direction of motion of the rectangle (Fig. P31.28). The rectangle accelerates downward until it acquires a constant speed of 2.0 m/s with its top not yet in that region of the field. Calculate the magnitude of **B**.

Section 31.4   Induced Emfs and Electric Fields

30. The current in a solenoid is increasing at a rate of 10 A/s. The cross-sectional area of the solenoid is π cm², and there are 300 turns on its 15-cm length. What is the induced emf opposing the increasing current?

31. A single-turn, circular loop of radius R is coaxial with a long solenoid of radius 0.030 m and length 0.75 m and having 1500 turns (Fig. P31.31). The variable resistor is changed so that the solenoid current decreases linearly from 7.2 A to 2.4 A in 0.30 s. Calculate the induced emf in the loop.

31A. A single-turn, circular loop of radius R is coaxial with a long solenoid of radius r and length ℓ and having N

turns (Fig. P31.31). The variable resistor is changed so that the solenoid current decreases linearly from I_1 to I_2 in an interval Δt. Find the induced emf in the loop.

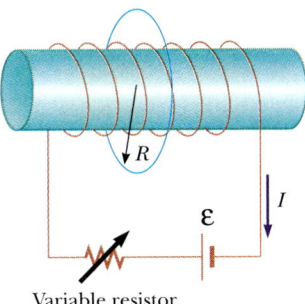

Variable resistor

FIGURE P31.31

32. A coil of 15 turns and radius 10 cm surrounds a long solenoid of radius 2.0 cm and 1.0×10^3 turns/meter (Fig. P31.32). If the current in the solenoid changes as $I = (5.0 \text{ A}) \sin(120t)$, what is the induced emf in the 15-turn coil?

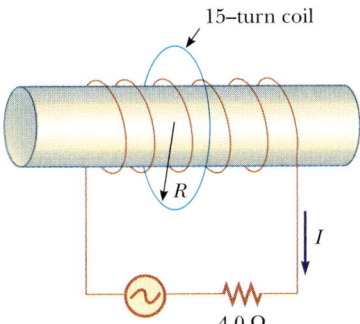

15–turn coil

R

I

4.0 Ω

FIGURE P31.32

33. A magnetic field directed into the page changes with time according to $B = (0.030t^2 + 1.4)$ T, where t is in seconds. The field has a circular cross-section of radius $R = 2.5$ cm (Fig. P31.33). What are the magnitude and direction of the electric field at point P_1 when $t = 3.0$ s and $r_1 = 0.020$ m?

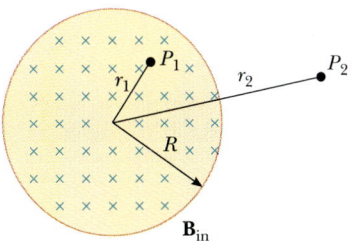

FIGURE P31.33

34. For the situation described in Figure P31.33, the magnetic field changes with time according to $B = (2.0t^3 - 4.0t^2 + 0.80)$ T, and $r_2 = 2R = 5.0$ cm. (a) Calculate the magnitude and direction of the force exerted on an electron located at point P_2 when $t = 2.0$ s. (b) At what time is this force equal to zero?

35. A long solenoid with 1000 turns/meter and radius 2.0 cm carries an oscillating current $I = (5.0 \text{ A}) \sin(100\pi t)$. What is the electric field induced at a radius $r = 1.0$ cm from the axis of the solenoid? What is the direction of this electric field when the current is increasing counterclockwise in the coil?

36. In 1832 Faraday proposed that the apparatus shown in Figure P31.36 could be used to generate electric current from the flowing water in the Thames River. Two conducting planes of lengths a and widths b are placed facing one another on opposite sides of the river, a distance w apart. The flow velocity of the river is v and the vertical component of the Earth's magnetic field is B. (a) Show that the current in the load resistor R is

$$I = \frac{abvB}{\rho + abR/w}$$

where ρ is the electrical resistivity of the water. (b) Calculate the short-circuit current ($R = 0$) if $a = 100$ m, $b = 5.00$ m, $v = 3.00$ m/s, $B = 0.500 \ \mu\text{T}$, and $\rho = 100 \ \Omega \cdot \text{m}$.

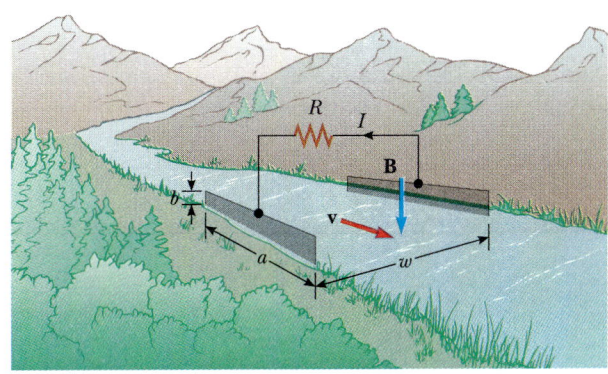

FIGURE P31.36

37. An aluminum ring of radius 5.0 cm and resistance $3.0 \times 10^{-4} \ \Omega$ is placed on top of a long air-core solenoid with 1000 turns per meter and radius 3.0 cm as in Figure P31.37. At the location of the ring, the magnetic field due to the current in the solenoid is one-half that at the center of the solenoid. If the current in the solenoid is increasing at a rate of 270 A/s, (a) what is the induced current in the ring? (b) At the center of the ring, what is the magnetic field produced by the induced current in the ring? (c) What is the direction of this field?

37A. An aluminum ring of radius r_1 and resistance R is placed on top of a long air-core solenoid with n turns per meter and radius r_2 as in Figure P31.37. At the location of the ring, the magnetic field due to the current in the solenoid is one-half that at the center of the solenoid. If the current in the solenoid is increasing at a rate $\Delta I/\Delta t$, (a) what is the induced current in the ring? (b) At the center of the ring, what is the magnetic field produced by the induced current in the ring? (c) What is the direction of this field?

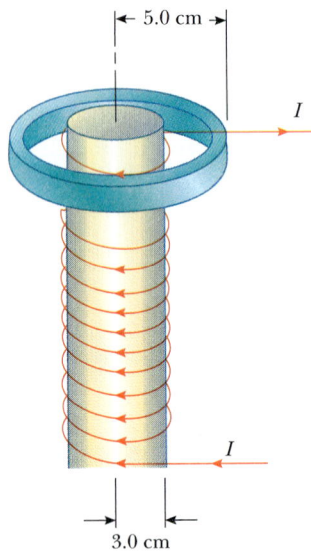

← 5.0 cm →

I

I

3.0 cm

FIGURE P31.37

38. A circular coil enclosing an area of 100 cm² is made of 200 turns of copper wire as shown in Figure P31.38. Initially, a 1.10-T uniform magnetic field points perpendicularly upward through the plane of the coil. The direction of the field then reverses. During the time the field is changing its direction, how much charge flows through the coil if $R = 5.0\ \Omega$?

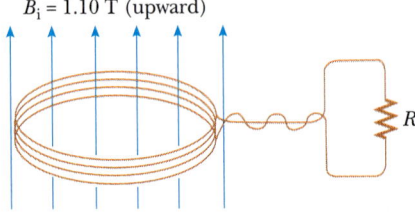

$B_i = 1.10$ T (upward)

R

FIGURE P31.38

*Section 31.5   Generators and Motors

39. A square coil (20 cm × 20 cm) that consists of 100 turns of wire rotates about a vertical axis at 1500 rev/min, as indicated in Figure P31.39. The horizontal component of the Earth's magnetic field at the location of the coil is 2.0×10^{-5} T. Calculate the maximum emf induced in the coil by this field.

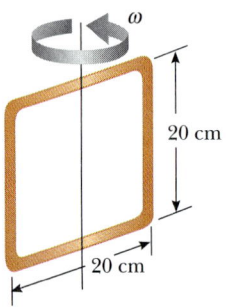

ω

20 cm

20 cm

FIGURE P31.39

40. A 25-turn circular loop of wire has a diameter of 1.0 m. In 0.20 s it is flipped 180° at a location where the magnitude of the Earth's magnetic field is 50 μT. What is the emf generated in the loop?

41. A loop of area 0.10 m² is rotating at 60 rev/s with the axis of rotation perpendicular to a 0.20-T magnetic field. (a) If there are 1000 turns on the loop, what is the maximum voltage induced in it? (b) When the maximum induced voltage occurs, what is the orientation of the loop with respect to the magnetic field?

42. In a 250-turn automotive alternator, the magnetic flux in each turn is $\Phi = (2.5 \times 10^{-4}$ Wb$) \cos(\omega t)$, where ω is the angular frequency of the alternator. The alternator rotates three times for each engine rotation. When the engine is running at 1000 rpm, determine (a) the induced emf in the alternator as a function of time and (b) the maximum emf in the alternator.

43. A long solenoid, the axis of which coincides with the x axis, consists of 200 turns/m of wire that carries a steady current of 15 A. A coil is formed by wrapping 30 turns of thin wire around a frame that has a radius of 8.0 cm. The coil is placed inside the solenoid and mounted on an axis that is a diameter of the coil and coincides with the y axis. The coil is then rotated with an angular speed of 4π rad/s. (The plane of the coil is in the yz plane at $t = 0$.) Determine the emf developed in the coil.

44. The rotating loop in an ac generator is a square 10 cm on a side. It is rotated at 60 Hz in a uniform field of 0.80 T. Calculate (a) the flux through the loop as a function of time, (b) the emf induced in the loop, (c) the current induced in the loop for a loop resistance of 1.0 Ω, (d) the power dissipated in the

loop, and (e) the torque that must be exerted to rotate the loop.

45. (a) What is the maximum torque delivered by an electric motor if it has 80 turns of wire wrapped on a rectangular coil, of dimensions 2.5 cm by 4.0 cm? Assume that the motor uses 10 A of current and that a uniform 0.80-T magnetic field exists within the motor. (b) If the motor rotates at 3600 rev/min, what is the peak power produced by the motor?

46. A semicircular conductor of radius $R = 0.25$ m is rotated about the axis AC at a constant rate of 120 rev/min (Fig. P31.46). A uniform magnetic field in all of the lower half of the figure is directed out of the plane of rotation and has a magnitude of 1.3 T. (a) Calculate the maximum value of the emf induced in the conductor. (b) What is the value of the average induced emf for each complete rotation? (c) How would the answers to (a) and (b) change if **B** were allowed to extend a distance R above the axis of rotation? Sketch the emf versus time (d) when the field is as drawn in Figure P31.46 and (e) when the field is extended as described in (c).

46A. A semicircular conductor of radius R rotates about the axis AC at a constant angular speed ω (Fig. P31.46). A uniform magnetic field in all of the lower half of the figure is directed out of the plane of rotation and has a magnitude B. (a) Find the maximum emf induced in the conductor. (b) What is the value of the average induced emf for each complete rotation?

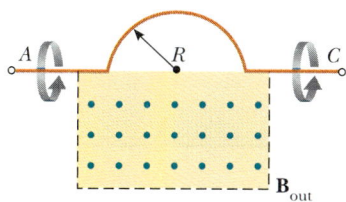

FIGURE P31.46

47. A small rectangular coil composed of 50 turns of wire has an area of 30 cm² and carries a current of 1.5 A. When the plane of the coil makes an angle of 30° with a uniform magnetic field, the torque on the coil is 0.10 N·m. What is the magnitude of the magnetic field?

48. A bar magnet is spun at constant angular speed ω around an axis as shown in Figure P31.48. A flat rectangular conducting loop surrounds the magnet, and at $t = 0$ the magnet is oriented as shown. Sketch the induced current in the loop as a function of time, plotting counterclockwise currents as positive and clockwise currents as negative.

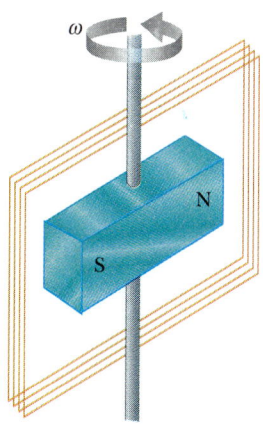

FIGURE P31.48

*Section 31.6   Eddy Currents

49. A rectangular loop with resistance R has N turns, each of length ℓ and width w as shown in Figure P31.49. The loop moves into a uniform magnetic field **B** with velocity **v**. What are the magnitude and direction of the resultant force on the loop (a) as it enters the magnetic field, (b) as it moves within the field, (c) as it leaves the field?

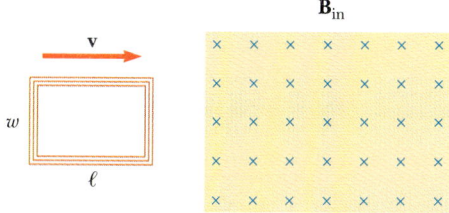

FIGURE P31.49

50. A dime is suspended from a thread and hung between the poles of a strong horseshoe magnet as shown in Figure P31.50. The dime rotates at constant angular speed ω about a vertical axis. Letting θ be the angle between the direction of **B** and the normal to the face of the dime, sketch a graph of the torque due to induced currents as a function of θ for $0 \leq \theta \leq 2\pi$.

Section 31.7   Maxwell's Wonderful Equations

51. A proton moves through a uniform electric field $\mathbf{E} = 50\mathbf{j}$ V/m and a uniform magnetic field $\mathbf{B} = (0.20\mathbf{i} + 0.30\mathbf{j} + 0.40\mathbf{k})$ T. Determine the acceleration of the proton when it has a velocity $\mathbf{v} = 200\mathbf{i}$ m/s.

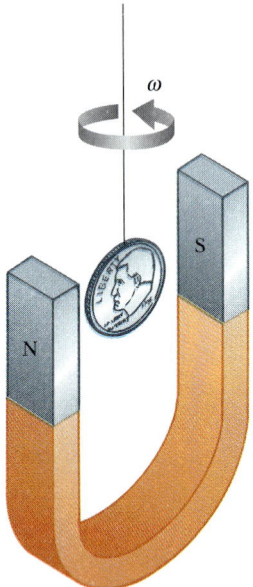

FIGURE P31.50

52. An electron moves through a uniform electric field $\mathbf{E} = (2.5\mathbf{i} + 5.0\mathbf{j})$ V/m and a uniform magnetic field $\mathbf{B} = 0.40\mathbf{k}$ T. Determine the acceleration of the electron when it has a velocity $\mathbf{v} = 10\mathbf{i}$ m/s.

ADDITIONAL PROBLEMS

53. A conducting rod moves with a constant velocity v perpendicular to a long, straight wire carrying a current *I* as in Figure P31.53. Show that the emf generated between the ends of the rod is

$$|\boldsymbol{\mathcal{E}}| = \frac{\mu_0 v I}{2\pi r}\ell$$

In this case, note that the emf decreases with increasing *r*, as you might expect.

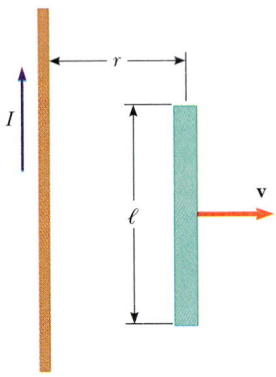

FIGURE P31.53

54. A circular loop of wire 5.0 cm in radius is in a uniform magnetic field, with the plane of the loop perpendicular to the direction of the field (Fig. P31.54). The magnetic field varies with time according to $B(t) = a + bt$, where $a = 0.20$ T and $b = 0.32$ T/s. (a) Calculate the magnetic flux through the loop at $t = 0$. (b) Calculate the emf induced in the loop. (c) If the resistance of the loop is 1.2 Ω, what is the induced current? (d) At what rate is electric energy being dissipated in the loop?

54A. A circular loop of wire of radius *r* is in a uniform magnetic field, with the plane of the loop perpendicular to the direction of the field (Fig. P31.54). The magnetic field varies with time according to $B(t) = a + bt$, where *a* and *b* are constants. (a) Calculate the magnetic flux through the loop at $t = 0$. (b) Calculate the emf induced in the loop. (c) If the resistance of the loop is *R*, what is the induced current? (d) At what rate is electric energy being dissipated in the loop?

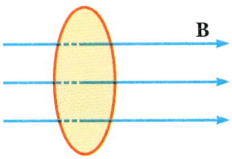

FIGURE P31.54

55. Consider a long solenoid of length ℓ containing a core of permeability μ. The core material is magnetized by increasing the current in the coil, so as to produce a changing field dB/dt. (a) Show that the rate at which work is done against the induced emf in the coil is

$$\frac{dW}{dt} = I\boldsymbol{\mathcal{E}} = HA\ell\frac{dB}{dt}$$

where *A* is the cross-sectional area of the solenoid. (*Hint:* Use Faraday's law to find $\boldsymbol{\mathcal{E}}$ and use Equation 30.34.) (b) Use the result of part (a) to show that the total work done in a complete hysteresis cycle equals the area enclosed by the *B* versus *H* curve (Fig. 30.31).

56. In Figure P31.56, a uniform magnetic field decreases at a constant rate $dB/dt = -K$, where *K* is a positive constant. A circular loop of wire of radius *a* containing a resistance *R* and a capacitance *C* is placed with its plane normal to the field. (a) Find the charge *Q* on the capacitor when it is fully charged. (b) Which plate is at the higher potential? (c) Discuss the force that causes the separation of charges.

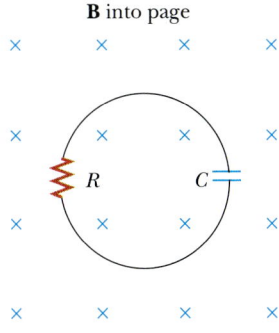

B into page

FIGURE P31.56

57. A rectangular coil of 60 turns, dimensions 0.10 m by 0.20 m, and total resistance 10 Ω rotates with angular speed 30 rad/s about the y axis in a region where a 1.0-T magnetic field is directed along the x axis. The rotation is initiated so that the plane of the coil is perpendicular to the direction of **B** at $t = 0$. Calculate (a) the maximum induced emf in the coil, (b) the maximum rate of change of magnetic flux through the coil, (c) the induced emf at $t = 0.050$ s, and (d) the torque exerted on the loop by the magnetic field at the instant when the emf is a maximum.

58. A bar of mass m, length d, and resistance R slides without friction on parallel rails as shown in Figure P31.58. A battery that maintains a constant emf $\mathcal{E}$ is connected between the rails, and a constant magnetic field **B** is directed perpendicularly to the plane of the page. If the bar starts from rest, show that at time t it moves with a speed

$$v = \frac{\mathcal{E}}{Bd}(1 - e^{-B^2 d^2 t / mR})$$

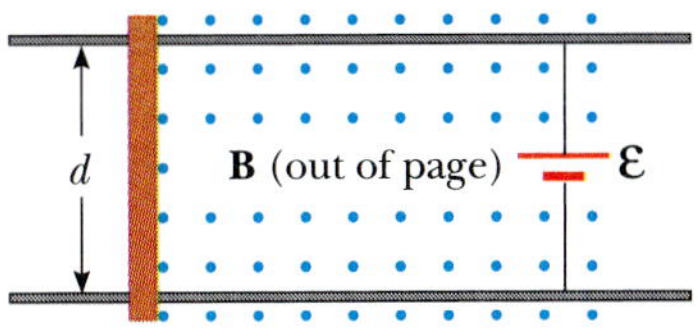

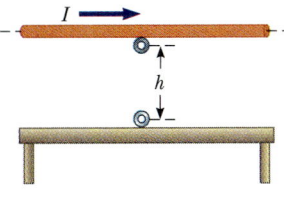

I →

h

FIGURE P31.59

60. To monitor the breathing of a hospital patient, a thin belt is girded around the patient's chest. The belt is a 200-turn coil. When the patient inhales, the area encircled by the coil increases by 39 cm². The magnitude of the Earth's magnetic field is 50 μT and makes an angle of 28° with the plane of the coil. If a patient takes 1.8 s to inhale, find the average induced emf in the coil during this time.

61. An automobile has a vertical radio antenna 1.2 m long. The automobile travels at 65 km/h on a horizontal road where the Earth's magnetic field is 50 μT directed downward (toward the north) at an angle of 65° below the horizontal. (a) Specify the direction that the automobile should move in order to generate the maximum motional emf in the antenna, with the top of the antenna positive relative to the bottom. (b) Calculate the magnitude of this induced emf.

62. A long, straight wire is parallel to one edge and in the plane of a single-turn rectangular loop as in Figure P31.62. (a) If the current in the long wire varies in time as $I = I_0 e^{-t/\tau}$, show that the induced emf in the loop is

$$\mathcal{E} = \frac{\mu_0 bI}{2\pi\tau} \ln\left(1 + \frac{a}{d}\right)$$

(b) Calculate the induced emf at $t = 5.0$ s taking $I_0 = 10$ A, $d = 3.0$ cm, $a = 6.0$ cm, $b = 15$ cm, and $\tau = 5.0$ s.

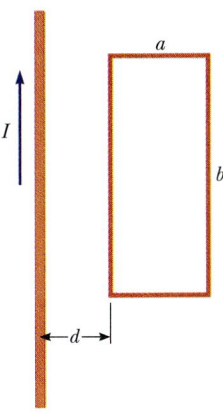

a

I

b

d

FIGURE P31.62

FIGURE P31.58 (d, **B** (out of page), $\mathcal{E}$)

59. A small circular copper washer of radius 0.500 cm is held directly below a long, straight wire carrying a current of 10.0 A and 0.500 m above the top of the table in Figure P31.59. (a) If the washer is dropped from rest, what is the magnitude of the average emf induced in it from the time it is released to the moment it hits the tabletop? Assume that the magnetic field through the washer is the same as the magnetic field at its center. (b) What is the direction of the induced current in the washer?

63. A conducting rod of length ℓ moves with velocity v parallel to a long wire carrying a steady current I. The axis of the rod is maintained perpendicular to the wire with the near end a distance r away, as shown in Figure P31.63. Show that the emf induced in the rod is

$$|\mathcal{E}| = \frac{\mu_0 I}{2\pi} v \ln\left(1 + \frac{\ell}{r}\right)$$

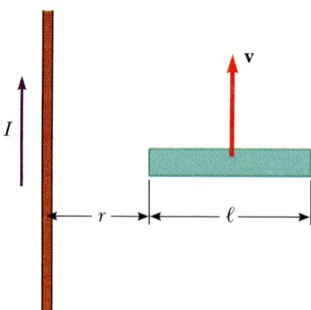

FIGURE P31.63

64. A rectangular loop of dimensions ℓ and w moves with a constant velocity v away from a long wire that carries a current I in the plane of the loop (Fig. P31.64). The total resistance of the loop is R. Derive an expression that gives the current in the loop at the instant the near side is a distance r from the wire.

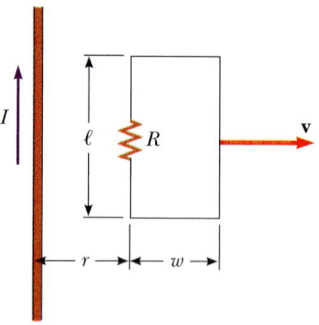

FIGURE P31.64

65. A plane of a square loop of wire with edge length $a = 0.20$ m is perpendicular to the Earth's magnetic field at a point where $B = 15$ μT, as in Figure P31.65. The total resistance of the loop and the wires connecting it to the galvanometer is 0.50 Ω. If the loop is suddenly collapsed by horizontal forces as shown, what total charge passes through the galvanometer?

65A. A plane of a square loop of wire with edge length a is perpendicular to the Earth's magnetic field at a point where the magnitude of the magnetic field is B,

as in Figure P31.65. The total resistance of the loop and the wires connecting it to the galvanometer is R. If the loop is suddenly collapsed by horizontal forces as shown, what total charge passes through the galvanometer?

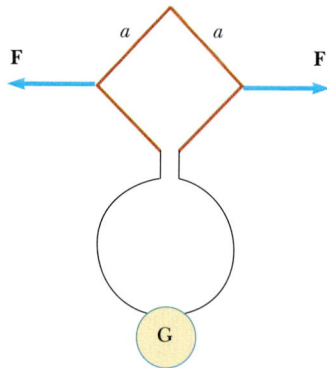

FIGURE P31.65

66. A horizontal wire is free to slide on the vertical rails of a conducting frame, as in Figure P31.66. The wire has mass m and length ℓ, and the resistance of the circuit is R. If a uniform magnetic field is directed perpendicularly to the frame, what is the terminal speed of the wire as it falls under the influence of gravity?

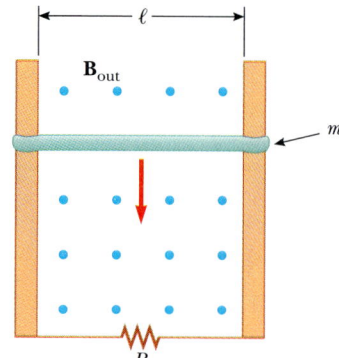

FIGURE P31.66

67. The magnetic flux threading a metal ring varies with time t according to $\Phi_B = 3(at^3 - bt^2)$ T·m², with $a = 2.0$ s^{-3} and $b = 6.0$ s^{-2}. The resistance of the ring is 3.0 Ω. Determine the maximum current induced in the ring during the interval from $t = 0$ to $t = 2.0$ s.

67A. The magnetic flux threading a metal ring varies with time t according to $\Phi_B = 3(at^3 - bt^2)$ T·m², where a and b are constants. The resistance of the ring is R.

Determine the maximum current induced in the ring during the interval from $t = 0$ to some later time t.

68. The bar of mass m in Figure P31.68 is pulled horizontally across parallel rails by a massless string that passes over an ideal pulley and is attached to a suspended mass M. The uniform magnetic field has a magnitude B, and the distance between the rails is ℓ. The rails are connected at one end by a load resistor R. Derive an expression that gives the horizontal speed of the bar as a function of time, assuming that the suspended mass is released with the bar at rest at $t = 0$. Assume no friction between rails and bar.

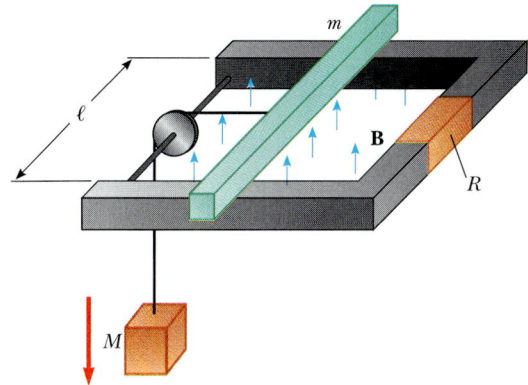

FIGURE P31.68

69. In Figure P31.69, the rolling axle, 1.5 m long, is pushed along horizontal rails at a constant speed $v = 3.0$ m/s. A resistor $R = 0.40\ \Omega$ is connected to the rails at points a and b, directly opposite each other. (The wheels make good electrical contact with the rails, and so the axle, rails, and R form a closed-loop circuit. The only significant resistance in the circuit is R.) There is a uniform magnetic field $B = 0.08$ T vertically downward. (a) Find the induced current I in the resistor. (b) What horizontal force F is required to keep the axle rolling at constant speed? (c) Which end of the resistor, a or b, is at the higher electric potential? (d) After the axle rolls past the resistor, does the current in R reverse direction?

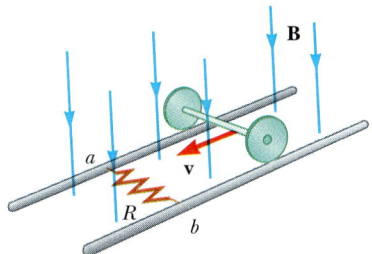

FIGURE P31.69

70. Two infinitely long solenoids (seen in cross-section) thread a circuit as in Figure P31.70. The magnitude of **B** inside each is the same and is increasing at the rate of 100 T/s. What is the current in each resistor?

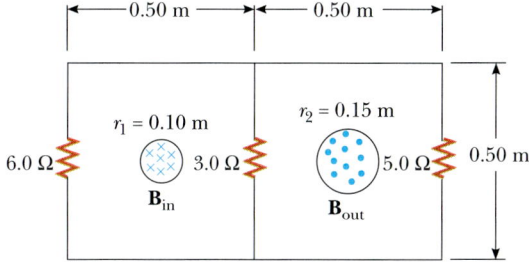

FIGURE P31.70

71. Figure P31.71 shows a circular loop of radius r that has a resistance R spread uniformly throughout its length. The loop's plane is normal to the magnetic field **B** that decreases at a constant rate $dB/dt = -K$, where K is a positive constant. What are (a) the direction and (b) the value of the induced current? (c) Which point, a or b, is at the higher potential? Explain. (d) Discuss what force causes the current in the loop.

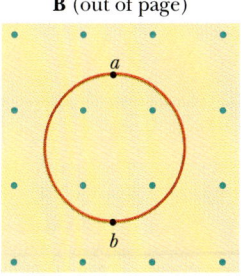

FIGURE P31.71

72. A wire 30.0 cm long is held parallel to and 80.0 cm above a long wire carrying 200 A and resting on the floor (Fig. P31.72). The 30.0-cm wire is released and falls, remaining parallel with the current-carrying wire as it falls. Assume that the falling wire accelerates at 9.80 m/s² and derive an equation for the emf induced in it. Express your result as a function of the time t after the wire is dropped. What is the induced emf 0.30 s after the wire is released?

72A. A wire of length L is held parallel to and a distance h above a long wire carrying a current I and resting on the floor (Fig. P31.72). The wire of length L is released and falls, remaining parallel with the current-carrying wire as it falls. Assuming that the falling wire is in free fall with an acceleration g, derive an equation for the emf induced in it. Express your result as a function of the time t after the wire is dropped.

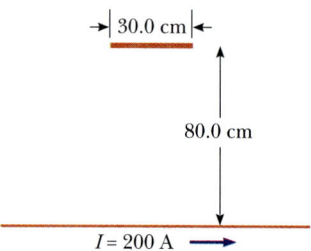

FIGURE P31.72

73. (a) A loop of wire in the shape of a rectangle of width w and length L and a long, straight wire carrying a current I lie on a tabletop as shown in Figure P31.73. (a) Determine the magnetic flux through the loop. (b) Suppose the current is changing with time according to $I = a + bt$, where a and b are constant. Determine the induced emf in the loop if $b = 10.0$ A/s, $h = 1.00$ cm, $w = 10.0$ cm, and $L = 100$ cm.

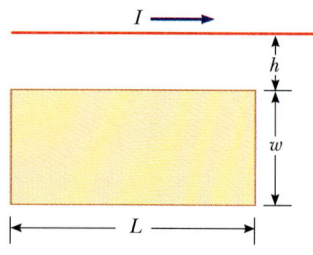

FIGURE P31.73

74. The wire shown in Figure P31.74 is bent in the shape of a tent, with $\theta = 60°$ and $L = 1.5$ m, and is placed in a 0.30-T magnetic field directed perpendicular to the tabletop. The wire is hinged at points a and b. If

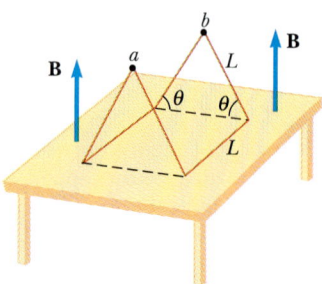

FIGURE P31.74

the tent is flattened out on the table in 0.10 s, what is the average emf induced in the wire during this time?

SPREADSHEET PROBLEMS

S1. The size and orientation of a flat surface of area A can be described by a vector $\mathbf{A} = A\hat{\mathbf{n}}$, where $\hat{\mathbf{n}}$ is a unit vector perpendicular to the surface. Suppose that a magnetic field $\mathbf{B}$ exists in the region of this surface. If the field is constant over the area A, then the magentic flux, ϕ, through the surface is $\phi = \mathbf{B} \cdot \mathbf{A} = BA \cos \theta = B_x A_x + B_y A_y + B_z A_z$, where $\mathbf{A} = A_x \mathbf{i} + A_y \mathbf{j} + A_z \mathbf{k}$ and $\mathbf{B} = B_x \mathbf{i} + B_y \mathbf{j} + B_z \mathbf{k}$. (Fig. PS31.1). Spreadsheet 31.1 will calculate the flux as a function of time. It will also numerically differentiate the flux to find the induced emf, $\mathcal{E} = -\Delta\phi/\Delta t$. (a) Set $A_x = 0$, $A_y = 0$, $A_z = 0.2$ m², $B_x = 0$, $B_y = 0.5$ T, and $B_z = 0.6$ T at $t = 0$. Copy these values down through $t = 4.5$ s. What is the induced emf? (b) Modify the spreadsheet so that B_z increases with time, $B_z = 0.2t$, where t is in seconds and B_z is in teslas. What is the induced emf?

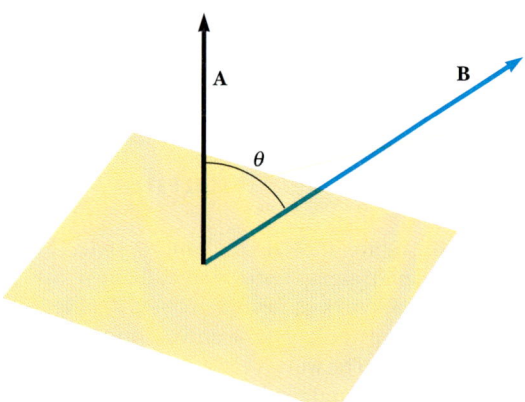

FIGURE PS31.1   Faraday's Law—Induced emf.

S2. Modify Spreadsheet 31.1 so that $\mathbf{A} = 0.02 \sin(\omega t)\mathbf{i} + 0.02 \cos(\omega t)\mathbf{k}$, where A is in square meters. Set $\mathbf{B} = 0.5$ T $\mathbf{k}$. (a) What physical situation does this variation in $\mathbf{A}$ correspond to? (b) Let $\omega = 1$ rad/s. View the included graph, and describe the induced emf. (c) Increase ω to 2 rad/s. How does the induced emf change from part (b)?

S3. (a) With $\mathbf{A}$ as given in Problem S2, modify Spreadsheet 31.1 so that $B_z = 0.2t$, where t is in seconds and B_z is in teslas. View the included graph, and explain your numerical results. (b) Choose any other time dependence for $\mathbf{B}$. Explore the consequences of your choice.

CHAPTER 32

Inductance

This photograph of water-driven generators was taken at the Bonneville Dam in Oregon. Hydroelectric power is generated when water from a dam passes through the generators under the influence of gravity, which causes turbines in the generator to rotate. The mechanical energy of the rotating turbines is transformed into electrical energy using the principle of electromagnetic induction, which you will study in this chapter. *(David Weintraub/Photo Researchers)*

I n the previous chapter, we saw that currents and emfs are induced in a circuit when the magnetic flux through the circuit changes with time. This electromagnetic induction has some practical consequences, which we describe in this chapter. First, we describe an effect known as *self-induction*, in which a time-varying current in a conductor induces in the conductor an emf that opposes the external emf that set up the current. Self-induction is the basis of the *inductor*, an electrical element that plays an important role in circuits that use time-varying currents. We discuss the energy stored in the magnetic field of an inductor and the energy density associated with a magnetic field.

Next, we study how an emf is induced in a circuit as a result of a changing magnetic flux produced by an external circuit, which is the basic principle of *mutual induction*. Finally, we examine the characteristics of circuits containing inductors, resistors, and capacitors in various combinations.

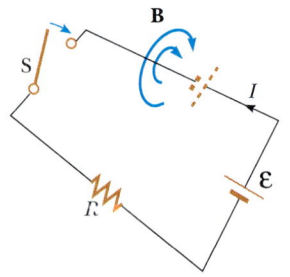

FIGURE 32.1   After the switch is closed, the current produces a magnetic flux through the loop. As the current increases toward its equilibrium value, the flux changes in time and induces an emf in the loop. The battery drawn with dashed lines is a symbol for the self-induced emf.

32.1   SELF-INDUCTANCE

Consider a circuit consisting of a switch, resistor, and source of emf, as in Figure 32.1. At the moment the switch is thrown to its closed position, the current doesn't immediately jump from zero to its maximum value, $\mathcal{E}/R$. The law of electromagnetic induction (Faraday's law) prevents this from occurring. What happens is the following: As the current increases with time, the magnetic flux through the loop due to this current also increases with time. This increasing flux induces in the circuit an emf that opposes the change in the net magnetic flux through the loop. By Lenz's law, the direction of the electric field induced in the wires must be opposite the direction of the current, and this opposing emf results in a gradual increase in the current. This effect is called *self-induction* because the changing flux through the circuit arises from the circuit itself. The emf $\mathcal{E}_L$ setup in this case is called a **self-induced emf**.

To obtain a quantitative description of self-induction, we recall from Faraday's law that the induced emf is equal to the negative time rate of change of the magnetic flux. The magnetic flux is proportional to the magnetic field, which in turn is proportional to the current in the circuit. Therefore, *the self-induced emf is always proportional to the time rate of change of the current.* For a closely spaced coil of N turns (a toroid or an ideal solenoid), we find that

Self-induced emf

$$\mathcal{E}_L = -N\frac{d\Phi_B}{dt} = -L\frac{dI}{dt} \tag{32.1}$$

where L is a proportionality constant, called the **inductance** of the coil, that depends on the geometry of the circuit and other physical characteristics. From this expression, we see that the inductance of a coil containing N turns is

Inductance of an N-turn coil

$$L = \frac{N\Phi_B}{I} \tag{32.2}$$

where it is assumed that the same flux passes through each turn. Later we shall use this equation to calculate the inductance of some special current geometries.

From Equation 32.1, we can also write the inductance as the ratio

Inductance

$$L = -\frac{\mathcal{E}_L}{dI/dt} \tag{32.3}$$

This is usually taken to be the defining equation for the inductance of any coil, regardless of its shape, size, or material characteristics. Just as resistance is a measure of the opposition to current, inductance is a measure of the opposition to any change in current.

The SI unit of inductance is the **henry** (H), which, from Equation 32.3, is seen to be equal to 1 volt-second per ampere:

$$1 \text{ H} = 1\frac{\text{V} \cdot \text{s}}{\text{A}}$$

As we shall see, *the inductance of a device depends on its geometry.* Induction calculations can be quite difficult for complicated geometries, but the following examples involve simple situations for which inductances are easily evaluated.

EXAMPLE 32.1   Inductance of a Solenoid

Find the inductance of a uniformly wound solenoid having N turns and length ℓ. Assume that ℓ is long compared with the radius and that the core of the solenoid is air.

Solution   In this case, we can take the interior magnetic field to be uniform and given by Equation 30.20:

$$B = \mu_0 n I = \mu_0 \frac{N}{\ell} I$$

where n is the number of turns per unit length, N/ℓ. The magnetic flux through each turn is

$$\Phi_B = BA = \mu_0 \frac{NA}{\ell} I$$

where A is the cross-sectional area of the solenoid. Using this expression and Equation 32.2, we find that

$$L = \frac{N\Phi_B}{I} = \frac{\mu_0 N^2 A}{\ell} \qquad (32.4)$$

This shows that L depends on geometry and is proportional to the square of the number of turns. Because $N = n\ell$, we can also express the result in the form

$$L = \mu_0 \frac{(n\ell)^2}{\ell} A = \mu_0 n^2 A \ell = \mu_0 n^2 \text{ (volume)} \qquad (32.5)$$

where $A\ell$ is the volume of the solenoid.

EXAMPLE 32.2   Calculating Inductance and Emf

(a) Calculate the inductance of a solenoid containing 300 turns if the length of the solenoid is 25.0 cm and its cross-sectional area is $4.00 \text{ cm}^2 = 4.00 \times 10^{-4} \text{ m}^2$.

Solution   Using Equation 32.4 we get

$$L = \frac{\mu_0 N^2 A}{\ell} = (4\pi \times 10^{-7} \text{ Wb/A} \cdot \text{m}) \frac{(300)^2 (4.00 \times 10^{-4} \text{ m}^2)}{25.0 \times 10^{-2} \text{ m}}$$

$$= 1.81 \times 10^{-4} \text{ Wb/A} = \boxed{0.181 \text{ mH}}$$

(b) Calculate the self-induced emf in the solenoid described in part (a) if the current through it is decreasing at the rate of 50.0 A/s.

Solution   Using Equation 32.1 and given that $dI/dt = -50.0$ A/s, we get

$$\mathcal{E}_L = -L\frac{dI}{dt} = -(1.81 \times 10^{-4} \text{ H})(-50.0 \text{ A/s})$$

$$= \boxed{9.05 \text{ mV}}$$

32.2   *RL* CIRCUITS

Any circuit that contains a coil, such as a solenoid, has a self-inductance that prevents the current from increasing or decreasing instantaneously. A circuit element that has a large inductance is called an **inductor,** symbol ⎍⎍⎍. We always assume that the self-inductance of the remainder of a circuit is negligible compared with that of the inductor.

Consider the circuit shown in Figure 32.2, where the battery has negligible internal resistance. Suppose the switch S is closed at $t = 0$. The current begins to increase, and because of the increasing current, the inductor produces a back emf that opposes the increasing current. In other words, the inductor acts like a battery whose polarity is opposite that of the real battery in the circuit. The back emf is

$$\mathcal{E}_L = -L\frac{dI}{dt}$$

Since the current is increasing, dI/dt is positive; therefore, $\mathcal{E}_L$ is negative. This negative value corresponds to the fact that there is a potential drop in going from a to b across the inductor. For this reason, point a is at a higher potential than point b, as illustrated in Figure 32.2.

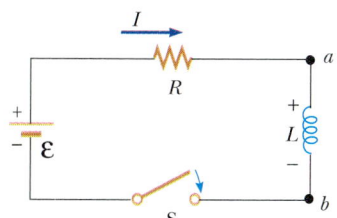

FIGURE 32.2   A series *RL* circuit. As the current increases toward its maximum value, the inductor produces an emf that opposes the increasing current.

With this in mind, we can apply Kirchhoff's loop equation to this circuit:

$$\mathcal{E} - IR - L\frac{dI}{dt} = 0 \tag{32.6}$$

where IR is the voltage drop across the resistor. We must now look for a solution to this differential equation, which is similar to that of the RC circuit (Section 28.4).

To obtain a mathematical solution of Equation 32.6, it is convenient to change variables by letting $x = \dfrac{\mathcal{E}}{R} - I$, so that $dx = -dI$. With these substitutions, Equation 32.6 can be written

$$x + \frac{L}{R}\frac{dx}{dt} = 0$$

$$\frac{dx}{x} = -\frac{R}{L}dt$$

Integrating this last expression gives

$$\ln\frac{x}{x_0} = -\frac{R}{L}t$$

where the integrating constant is taken to be $-\ln x_0$. Taking the antilog of this result gives

$$x = x_0 e^{-Rt/L}$$

Since at $t = 0$, $I = 0$, we note that $x_0 = \mathcal{E}/R$. Hence, the last expression is equivalent to

$$\frac{\mathcal{E}}{R} - I = \frac{\mathcal{E}}{R}e^{-Rt/L}$$

$$I = \frac{\mathcal{E}}{R}(1 - e^{-Rt/L})$$

which represents the solution of Equation 32.6.

This mathematical solution of Equation 32.6, which represents the current as a function of time, can also be written:

$$I(t) = \frac{\mathcal{E}}{R}(1 - e^{-t/\tau}) \tag{32.7}$$

where the constant τ is the time constant of the RL circuit:

$$\tau = L/R \tag{32.8}$$

Physically, τ is the time it takes the current to reach $(1 - e^{-1}) = 0.632$ of its final value, $\mathcal{E}/R$.

Figure 32.3 graphs the current versus time, where $I = 0$ at $t = 0$. Note that the equilibrium value of the current, which occurs at $t = \infty$, is $\mathcal{E}/R$. This can be seen by setting dI/dt equal to zero in Equation 32.6 (at equilibrium, the change in the

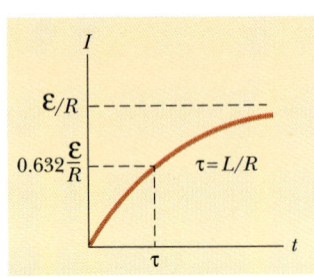

FIGURE 32.3   Plot of the current versus time for the RL circuit shown in Figure 32.2. The switch is closed at $t = 0$, and the current increases toward its maximum value, $\mathcal{E}/R$. The time constant τ is the time it takes I to reach 63% of its maximum value.

current is zero) and solving for the current. Thus, we see that the current rises very fast initially and then gradually approaches the equilibrium value $\mathcal{E}/R$ as $t \to \infty$.

We can show that Equation 32.7 is a solution of Equation 32.6 by computing dI/dt and noting that $I = 0$ at $t = 0$. Taking the first time derivative of Equation 32.7, we get

$$\frac{dI}{dt} = \frac{\mathcal{E}}{L}\, e^{-t/\tau} \qquad (32.9)$$

Substitution of this result for the dI/dt term in Equation 32.6 together with the value of I given by Equation 32.7 will indeed verify that our solution satisfies Equation 32.6. That is,

$$\mathcal{E} - IR - L\frac{dI}{dt} = 0$$

$$\mathcal{E} - \frac{\mathcal{E}}{R}(1 - e^{-t/\tau})R - L\left(\frac{\mathcal{E}}{L}\, e^{-t/\tau}\right) = 0$$

$$\mathcal{E} - \mathcal{E} + \mathcal{E}e^{-t/\tau} - \mathcal{E}e^{-t/\tau} = 0$$

From Equation 32.9 we see that the time rate of increase of current is a maximum (equal to $\mathcal{E}/L$) at $t = 0$ and falls off exponentially to zero as $t \to \infty$ (Fig. 32.4).

Now consider the *RL* circuit arranged as shown in Figure 32.5. The circuit contains two switches that operate so that when one is closed, the other is opened. Suppose that S_1 is closed for a long enough time to allow the current to reach its equilibrium value, $\mathcal{E}/R$. If we call $t = 0$ the instant at which S_1 is thrown open and S_2 is simultaneously thrown closed, we have a circuit with no battery ($\mathcal{E} = 0$). If we

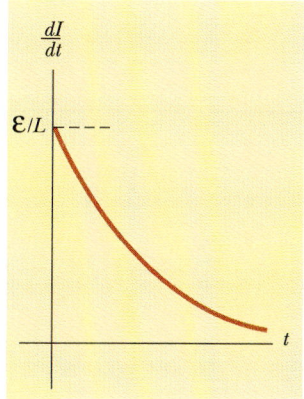

FIGURE 32.4   Plot of dI/dt versus time for the *RL* circuit shown in Figure 32.2. The time rate of change of current is a maximum at $t = 0$ when the switch is closed. The rate decreases exponentially with time as *I* increases toward its maximum value.

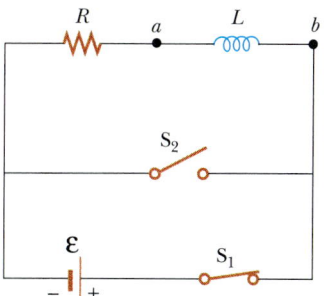

FIGURE 32.5   An *RL* circuit containing two switches. When S_1 is closed and S_2 is open as shown, the battery is in the circuit. At the instant S_2 is closed, S_1 is opened and the battery is removed from the circuit.

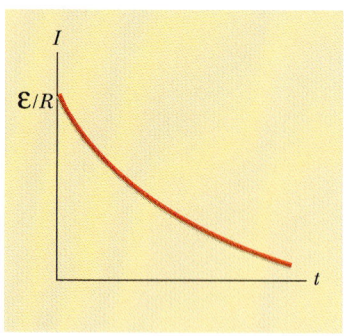

FIGURE 32.6 Current versus time for the circuit shown in Figure 32.5. At $t < 0$, S_1 is closed and S_2 is open. At $t = 0$, S_2 is closed, S_1 is open, and the current has its maximum value $\mathcal{E}/R$.

apply Kirchhoff's circuit law to the upper loop containing the resistor and inductor, we obtain

$$IR + L\frac{dI}{dt} = 0$$

It is left as a problem (Problem 18) to show that the solution of this differential equation is

$$I(t) = \frac{\mathcal{E}}{R} e^{-t/\tau} = I_0 e^{-t/\tau} \tag{32.10}$$

where the current at $t = 0$ is $I_0 = \mathcal{E}/R$ and $\tau = L/R$.

The graph of the current versus time (Fig. 32.6) shows that the current is continuously decreasing with time, as would be expected. Furthermore, note that the slope, dI/dt, is always negative and has its maximum value at $t = 0$. The negative slope signifies that $\mathcal{E}_L = -L(dI/dt)$ is now positive; that is, point a is at a lower potential than point b in Figure 32.5.

EXAMPLE 32.3   Time Constant of an *RL* Circuit

The switch in Figure 32.7a is closed at $t = 0$. (a) Find the time constant of the circuit.

Solution   The time constant is given by Equation 32.8

$$\tau = \frac{L}{R} = \frac{30.0 \times 10^{-3}\ \text{H}}{6.00\ \Omega} = 5.00\ \text{ms}$$

(b) Calculate the current in the circuit at $t = 2.00$ ms.

Solution   Using Equation 32.7 for the current as a function of time (with t and τ in milliseconds), we find that

at $t = 2.00$ ms

$$I = \frac{\mathcal{E}}{R}(1 - e^{-t/\tau}) = \frac{12.0\ \text{V}}{6.00\ \Omega}(1 - e^{-0.400}) = 0.659\ \text{A}$$

A plot of Equation 32.7 for this circuit is given in Figure 32.7b.

Exercise   Calculate the current in the circuit and the voltage across the resistor after one time constant has elapsed.

Answer   1.26 A, 7.56 V.

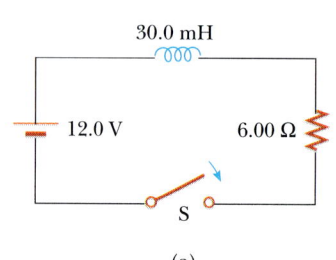

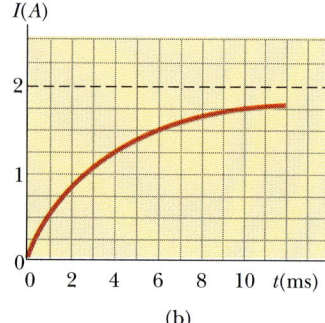

FIGURE 32.7   (Example 32.3) (a) The switch in this *RL* circuit is closed at $t = 0$. (b) A graph of the current versus time for the circuit in part (a).

32.3   ENERGY IN A MAGNETIC FIELD

Because the emf induced by an inductor prevents a battery from establishing an instantaneous current, the battery has to do work against the inductor to create a current. Part of the energy supplied by the battery goes into joule heat dissipated in the resistor, while the remaining energy is stored in the inductor. If we multiply each term in Equation 32.6 by I and rearrange the expression, we get

$$I\mathcal{E} = I^2 R + LI\frac{dI}{dt} \qquad (32.11)$$

This expression tells us that the rate at which energy is supplied by the battery, $I\mathcal{E}$, equals the sum of the rate at which joule heat is dissipated in the resistor, $I^2 R$, and the rate at which energy is stored in the inductor, $LI(dI/dt)$. Thus, Equation 32.11 is simply an expression of energy conservation. If we let U_B denote the energy stored in the inductor at any time (where the subscript B denotes energy stored in the magnetic field of the inductor), then the rate dU_B/dt at which energy is stored can be written

$$\frac{dU_B}{dt} = LI\frac{dI}{dt}$$

To find the total energy stored in the inductor, we can rewrite this expression as $dU_B = LI\,dI$ and integrate:

$$U_B = \int_0^{U_B} dU_B = \int_0^I LI\,dI$$

$$U_B = \tfrac{1}{2}LI^2 \qquad (32.12)$$

Energy stored in an inductor

where L is constant and can be removed from the integral. Equation 32.12 represents the energy stored in the magnetic field of the inductor when the current is I. Note that this equation is similar in form to the equation for the energy stored in the electric field of a capacitor, $Q^2/2C$ (Eq. 26.10). In either case, we see that it takes work to establish a field.

We can also determine the energy per unit volume, or energy density, stored in a magnetic field. For simplicity, consider a solenoid whose inductance is given by Equation 32.5:

$$L = \mu_0 n^2 A\ell$$

The magnetic field of a solenoid is given by Equation 30.20:

$$B = \mu_0 nI$$

Substituting the expression for L and $I = B/\mu_0 n$ into Equation 32.12 gives

$$U_B = \tfrac{1}{2}LI^2 = \tfrac{1}{2}\mu_0 n^2 A\ell\left(\frac{B}{\mu_0 n}\right)^2 = \frac{B^2}{2\mu_0}(A\ell) \qquad (32.13)$$

Because $A\ell$ is the volume of the solenoid, the energy stored per unit volume in a magnetic field is

$$u_B = \frac{U_B}{A\ell} = \frac{B^2}{2\mu_0} \qquad (32.14)$$

Magnetic energy density

Although Equation 32.14 was derived for the special case of a solenoid, it is valid for any region of space in which a magnetic field exists. Note that Equation 32.14 is similar in form to the equation for the energy per unit volume stored in an electric field, given by $\tfrac{1}{2}\epsilon_0 E^2$ (Eq. 26.12). In both cases, the energy density is proportional to the square of the field strength.

EXAMPLE 32.4   What Happens to the Energy in the Inductor?

Consider once again the *RL* circuit shown in Figure 32.5, in which switch S_2 is closed at the instant S_1 is opened (at $t = 0$). Recall that the current in the upper loop decays exponentially with time according to the expression $I = I_0 e^{-t/\tau}$, where $I_0 = \mathcal{E}/R$ is the initial current in the circuit and $\tau = L/R$ is the time constant. Let us show explicitly that all the energy stored in the magnetic field of the inductor gets dissipated as heat in the resistor.

Solution   That rate at which energy is dissipated in the resistor, dU/dt, (or the power) is equal to $I^2 R$, where I is the instantaneous current:

$$\frac{dU}{dt} = I^2 R = (I_0 e^{-Rt/L})^2 R = I_0^2 R e^{-2Rt/L}$$

To find the total energy dissipated in the resistor, we inte-grate this expression over the limits $t = 0$ to $t = \infty$ (∞ because it takes an infinite time for the current to reach zero):

$$U = \int_0^\infty I_0^2 R e^{-2Rt/L}\, dt = I_0^2 R \int_0^\infty e^{-2Rt/L}\, dt \quad (1)$$

The value of the definite integral is $L/2R$, and so U becomes

$$U = I_0^2 R \left(\frac{L}{2R} \right) = \frac{1}{2} L I_0^2$$

Note that this is equal to the initial energy stored in the magnetic field of the inductor, given by Equation 32.13, as we set out to prove.

Exercise   Show that the integral on the right-hand side of Equation (1) has the value $L/2R$.

EXAMPLE 32.5   The Coaxial Cable

A long coaxial cable consists of two concentric cylindrical conductors of radii a and b and length ℓ, as in Figure 32.8. The inner conductor is assumed to be a thin cylindrical shell. Each conductor carries a current I (the outer one being a return path). (a) Calculate the self-inductance L of this cable.

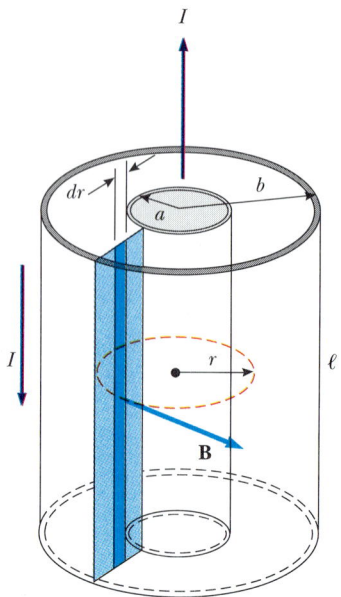

FIGURE 32.8   (Example 32.5) Section of a long coaxial cable. The inner and outer conductors carry equal and opposite currents.

Solution   To obtain L, we must know the magnetic flux through any cross-section between the two conductors. From Ampère's law (Section 30.3), it is easy to see that the magnetic field between the conductors is $B = \mu_0 I/2\pi r$. The field is zero outside the conductors ($r > b$) because the net current through a circular path surrounding both wires is zero, and hence from Ampère's law, $\oint \mathbf{B} \cdot d\mathbf{s} = 0$. The field is zero inside the inner conductor because it is hollow and there is no current within a radius $r < a$.

The magnetic field is perpendicular to the shaded rectangular strip of length ℓ and width $(b - a)$, the cross-section of interest. Dividing this rectangle into strips of width dr, we see that the area of each strip is $\ell\, dr$ and the flux through each strip is $B\, dA = B\ell\, dr$. Hence, the total flux through any cross-section is

$$\Phi_B = \int B\, dA = \int_a^b \frac{\mu_0 I}{2\pi r} \ell\, dr = \frac{\mu_0 I \ell}{2\pi} \int_a^b \frac{dr}{r} = \frac{\mu_0 I \ell}{2\pi} \ln\!\left(\frac{b}{a} \right)$$

Using this result, we find that the self-inductance of the cable is

$$L = \frac{\Phi_B}{I} = \frac{\mu_0 \ell}{2\pi} \ln\!\left(\frac{b}{a} \right)$$

(b) Calculate the total energy stored in the magnetic field of the cable.

Solution   Using Equation 32.13 and the results to part (a) gives

$$U_B = \tfrac{1}{2} L I^2 = \frac{\mu_0 \ell I^2}{4\pi} \ln\!\left(\frac{b}{a} \right)$$

*32.4   MUTUAL INDUCTANCE

Very often the magnetic flux through a circuit varies with time because of varying currents in nearby circuits. This condition induces an emf through a process known as mutual induction, so-called because it depends on the interaction of two circuits.

Consider two closely wound coils as shown in the cross-sectional view of Figure 32.9. The current I_1 in coil 1, which has N_1 turns, creates magnetic field lines, some of which pass through coil 2, which has N_2 turns. The corresponding flux through coil 2 produced by coil 1 is represented by Φ_{21}. We define the **mutual inductance** M_{21} of coil 2 with respect to coil 1 as the ratio of $N_2\Phi_{21}$ and the current I_1:

$$M_{21} \equiv \frac{N_2\Phi_{21}}{I_1} \qquad (32.15)$$

$$\Phi_{21} = \frac{M_{21}}{N_2}I_1$$

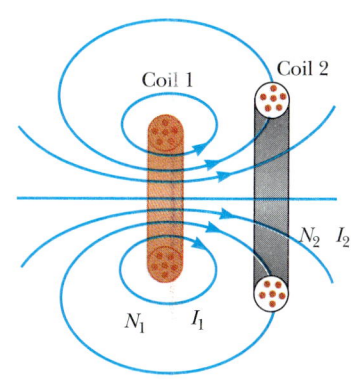

FIGURE 32.9   A cross-sectional view of two adjacent coils. A current in coil 1 sets up a magnetic flux, part of which passes through coil 2.

The mutual inductance depends on the geometry of both circuits and on their orientation with respect to one another. Clearly, as the circuit separation increases, the mutual inductance decreases because the flux linking the circuits decreases.

If the current I_1 varies with time, we see from Faraday's law and Equation 32.15 that the emf induced in coil 2 by coil 1 is

$$\mathcal{E}_2 = -N_2\frac{d\Phi_{21}}{dt} = -M_{21}\frac{dI_1}{dt} \qquad (32.16)$$

Similarly, if the current I_2 varies with time, the emf induced in coil 1 by coil 2 is

$$\mathcal{E}_1 = -M_{12}\frac{dI_2}{dt} \qquad (32.17)$$

These results are similar in form to Equation 32.1 for the self-induced emf $\mathcal{E} = -L(dI/dt)$. *The emf induced by mutual induction in one coil is always proportional to the rate of current change in the other coil.* If the rates at which the currents change with time are equal (that is, if $dI_1/dt = dI_2/dt$), then $\mathcal{E}_1 = \mathcal{E}_2$. Although the proportionality constants M_{12} and M_{21} appear to be different, it can be shown that they are equal. Thus, taking $M_{12} = M_{21} = M$, Equations 32.16 and 32.17 become

$$\mathcal{E}_2 = -M\frac{dI_1}{dt} \qquad \text{and} \qquad \mathcal{E}_1 = -M\frac{dI_2}{dt}$$

The unit of mutual inductance is also the henry.

EXAMPLE 32.6   Mutual Inductance of Two Solenoids

A long solenoid of length ℓ has N_1 turns, carries a current I, and has a cross-sectional area A. A second coil containing N_2 turns is wound around the center of the first coil, as in Figure 32.10. Find the mutual inductance of the system.

Solution   If the solenoid carries a current I_1, the magnetic field at its center is

$$B = \frac{\mu_0 N_1 I_1}{\ell}$$

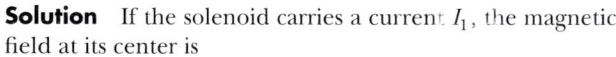

Since the flux Φ_{21} through coil 2 due to coil 1 is BA, the mutual inductance is

$$M = \frac{N_2 \Phi_{21}}{I_1} = \frac{N_2 BA}{I_1} = \mu_0 \frac{N_1 N_2 A}{\ell}$$

Exercise   Calculate the mutual inductance of two solenoids for which $N_1 = 500$ turns, $A = 3 \times 10^{-3}$ m^2, $\ell = 0.5$ m, and $N_2 = 8$ turns.

Answer   30.2 μH.

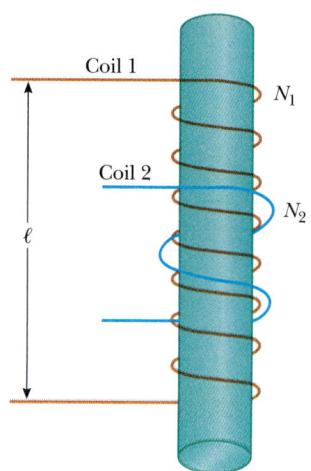

FIGURE 32.10   (Example 32.6) A small coil of N_2 turns wrapped around the center of a long solenoid of N_1 turns.

32.5   OSCILLATIONS IN AN *LC* CIRCUIT

When a charged capacitor is connected to an inductor as in Figure 32.11 and the switch is then closed, both the current and the charge on the capacitor oscillate. If the resistance of the circuit is zero, no energy is dissipated as joule heat and the oscillations persist. In this section we neglect the resistance in the circuit.

In the following analysis, let us assume that the capacitor has an initial charge Q_{max} (the maximum charge) and that the switch is thrown closed at $t = 0$. It is convenient to describe what happens from an energy viewpoint.

When the capacitor is fully charged, the total energy U in the circuit is stored in the electric field of the capacitor and is equal to $Q_{max}^2/2C$ (Eq. 26.10). At this time, the current is zero and so there is no energy stored in the inductor. As the capacitor begins to discharge, the energy stored in its electric field decreases. At the same time, the current increases and some energy is now stored in the magnetic field of the inductor. Thus, we see that energy is transferred from the electric field of the capacitor to the magnetic field of the inductor. When the capacitor is fully discharged, it stores no energy. At this time, the current reaches its maximum value and all of the energy is stored in the inductor. The process then repeats in the reverse direction. The energy continues to oscillate between the inductor and capacitor indefinitely.

A graphical description of this energy transfer is shown in Figure 32.12. The circuit behavior is analogous to the oscillating mass-spring system studied in Chapter 13. The potential energy stored in a stretched spring, $\frac{1}{2}kx^2$, is analogous to the potential energy stored in the capacitor, $Q_{max}^2/2C$. The kinetic energy of the moving mass, $\frac{1}{2}mv^2$, is analogous to the energy stored in the inductor, $\frac{1}{2}LI^2$, which requires the presence of moving charges. In Figure 32.12a, all of the energy is stored as potential energy in the capacitor at $t = 0$ (since $I = 0$). In Figure 32.12b, all of the energy is stored as "kinetic" energy in the inductor, $\frac{1}{2}LI_{max}^2$, where I_{max} is the maximum current. At intermediate points, part of the energy is potential energy and part is kinetic energy.

Consider some arbitrary time t after the switch is closed, so that the capacitor has a charge Q and the current is I. At this time, both elements store energy, but the sum of the two energies must equal the total initial energy U stored in the fully charged capacitor at $t = 0$:

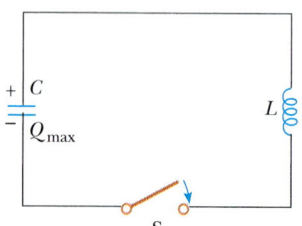

FIGURE 32.11   A simple *LC* circuit. The capacitor has an initial charge Q_{max}, and the switch is thrown closed at $t = 0$.

Total energy stored in the *LC* circuit

$$U = U_C + U_L = \frac{Q^2}{2C} + \frac{1}{2}LI^2 \tag{32.18}$$

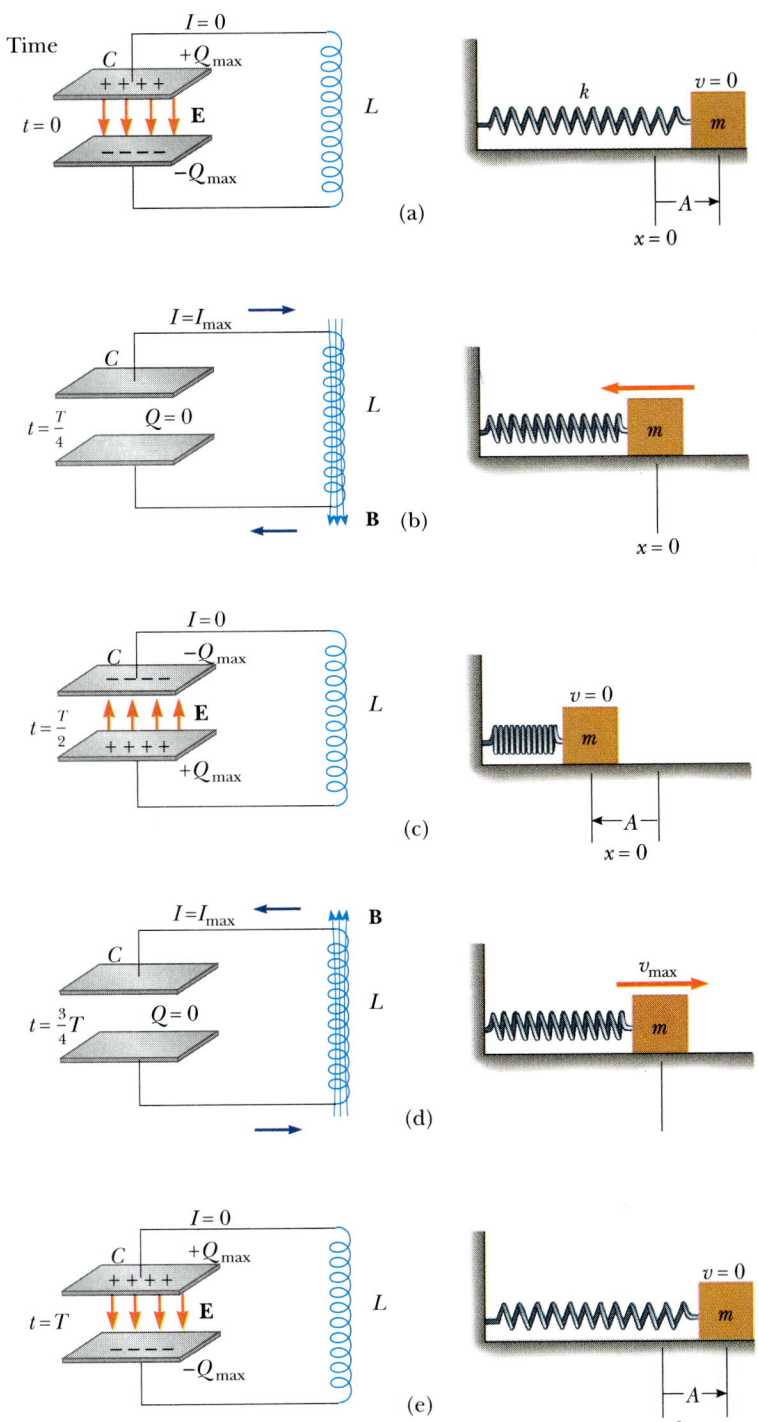

FIGURE 32.12   Energy transfer in a resistanceless *LC* circuit. The capacitor has a charge Q_{max} at $t = 0$, the instant at which the switch is thrown closed. The mechanical analog of this circuit is a mass–spring system.

Since we have assumed the circuit resistance to be zero, no energy is dissipated as joule heat and hence *the total energy must remain constant in time.* This means that $dU/dt = 0$. Therefore, by differentiating Equation 32.18 with respect to time while noting that Q and I vary with time, we get

$$\frac{dU}{dt} = \frac{d}{dt}\left(\frac{Q^2}{2C} + \tfrac{1}{2}LI^2\right) = \frac{Q}{C}\frac{dQ}{dt} + LI\frac{dI}{dt} = 0 \qquad (32.19)$$

We can reduce this to a differential equation in one variable by using the relationship $I = dQ/dt$. From this, it follows that $dI/dt = d^2Q/dt^2$. Substitution of these relationships into Equation 32.19 gives

$$L\frac{d^2Q}{dt^2} + \frac{Q}{C} = 0$$

$$\frac{d^2Q}{dt^2} = -\frac{1}{LC}Q \qquad (32.20)$$

We can solve for Q by noting that Equation 32.20 is of the same form as the analogous Equation 13.14 for a mass-spring system:

$$\frac{d^2x}{dt^2} = -\frac{k}{m}x = -\omega^2 x$$

where k is the spring constant, m is the mass, and $\omega = \sqrt{k/m}$. The solution of this equation has the general form

$$x = A\cos(\omega t + \phi)$$

where ω is the angular frequency of the simple harmonic motion, A is the amplitude of motion (the maximum value of x), and ϕ is the phase constant; the values of A and ϕ depend on the initial conditions. Since it is of the same form as the differential equation of the simple harmonic oscillator, Equation 32.20 has the solution

$$Q = Q_{max}\cos(\omega t + \phi) \qquad (32.21)$$

where Q_{max} is the maximum charge of the capacitor and the angular frequency ω is

$$\omega = \frac{1}{\sqrt{LC}} \qquad (32.22)$$

Note that *the angular frequency of the oscillations depends solely on the inductance and capacitance of the circuit.*

Because Q varies harmonically, the current also varies harmonically. This is easily shown by differentiating Equation 32.21 with respect to time:

$$I = \frac{dQ}{dt} = -\omega Q_{max}\sin(\omega t + \phi) \qquad (32.23)$$

To determine the value of the phase angle ϕ, we examine the initial conditions, which in our situation require that at $t = 0$, $I = 0$, and $Q = Q_{max}$. Setting $I = 0$ at $t = 0$ in Equation 32.23 gives

$$0 = -\omega Q_{max} \sin \phi$$

which shows that $\phi = 0$. This value for ϕ is also consistent with Equation 32.21 and with the condition that $Q = Q_{max}$ at $t = 0$. Therefore, in our case, the time variation of Q and that of I are

$$Q = Q_{max} \cos \omega t \tag{32.24}$$

$$I = -\omega Q_{max} \sin \omega t = -I_{max} \sin \omega t \tag{32.25}$$

where $I_{max} = \omega Q_{max}$ is the maximum current in the circuit.

Graphs of Q versus t and I versus t are shown in Figure 32.13. Note that the charge on the capacitor oscillates between the extreme values Q_{max} and $-Q_{max}$, and the current oscillates between I_{max} and $-I_{max}$. Furthermore, the current is 90° out of phase with the charge. That is, when the charge reaches an extreme value, the current is zero, and when the charge is zero, the current has an extreme value.

Let us return to the energy of the *LC* circuit. Substituting Equations 32.24 and 32.25 in Equation 32.18, we find that the total energy is

$$U = U_C + U_L = \frac{Q_{max}^2}{2C} \cos^2 \omega t + \frac{LI_{max}^2}{2} \sin^2 \omega t \tag{32.26}$$

This expression contains all of the features described qualitatively at the beginning of this section. It shows that the energy of the system continuously oscillates between energy stored in the electric field of the capacitor and energy stored in the magnetic field of the inductor. When the energy stored in the capacitor has its maximum value, $Q_{max}^2/2C$, the energy stored in the inductor is zero. When the energy stored in the inductor has its maximum value, $\frac{1}{2}LI_{max}^2$, the energy stored in the capacitor is zero.

Plots of the time variations of U_C and U_L are shown in Figure 32.14. Note that the sum $U_C + U_L$ is a constant and equal to the total energy, $Q_{max}^2/2C$. An analytical proof of this is straightforward. The maximum energy stored in the capacitor (when $I = 0$) must equal the maximum energy stored in the inductor (when $Q = 0$), so

$$\frac{Q_{max}^2}{2C} = \frac{1}{2}LI_{max}^2$$

Using this expression in Equation 32.26 for the total energy gives

$$U = \frac{Q_{max}^2}{2C} (\cos^2 \omega t + \sin^2 \omega t) = \frac{Q_{max}^2}{2C} \tag{32.27}$$

because $\cos^2 \omega t + \sin^2 \omega t = 1$.

In our idealized situation, the oscillations in the circuit persist indefinitely, but you should note that the total energy U remains constant only if energy losses are neglected. In actual circuits, there is always some resistance and so energy will be lost in the form of heat. (In fact, even when the energy losses due to wire resistance are neglected, energy is also lost in the form of electromagnetic waves radiated by the circuit.)

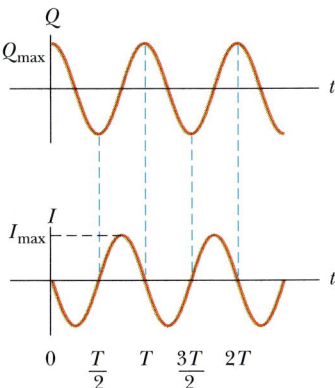

FIGURE 32.13 Graphs of charge versus time and current versus time for a resistanceless *LC* circuit. Note that Q and I are 90° out of phase with each other.

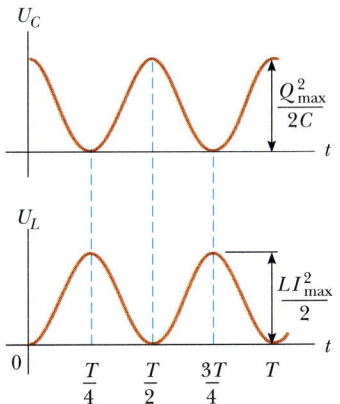

FIGURE 32.14 Plots of U_C versus t and U_L versus t for a resistanceless *LC* circuit. The sum of the two curves is a constant and equal to the total energy stored in the circuit.

EXAMPLE 32.7   An Oscillatory *LC* Circuit

In Figure 32.15, the capacitor is initially charged when S_1 is open and S_2 is closed. Then S_1 is thrown closed at the same instant that S_2 is opened so that the capacitor is shorted across the inductor. (a) Find the frequency of oscillation.

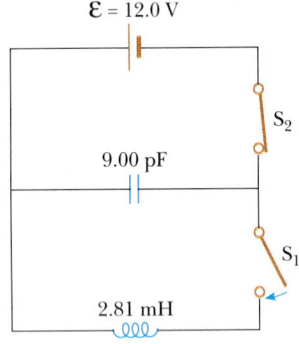

$\mathcal{E} = 12.0$ V

S_2

9.00 pF

S_1

2.81 mH

FIGURE 32.15   (Example 32.7) First the capacitor is fully charged with the switch S_1 open and S_2 closed. Then, S_1 is thrown closed at the same time that S_2 is thrown open.

Solution   Using Equation 32.22 gives for the frequency

$$f = \frac{\omega}{2\pi} = \frac{1}{2\pi\sqrt{LC}}$$

$$= \frac{1}{2\pi[(2.81 \times 10^{-3} \text{ H})(9.00 \times 10^{-12} \text{ F})]^{1/2}}$$

$$= \boxed{1.00 \times 10^6 \text{ Hz}}$$

(b) What are the maximum values of charge on the capacitor and current in the circuit?

Solution   The initial charge on the capacitor equals the maximum charge, and since $C = Q/V$, we get

$$Q_{max} = CV = (9.00 \times 10^{-12} \text{ F})(12.0 \text{ V}) = \boxed{1.08 \times 10^{-10} \text{ C}}$$

From Equation 32.25, we see that the maximum current is related to the maximum charge:

$$I_{max} = \omega Q_{max} = 2\pi f Q_{max}$$
$$= (2\pi \times 10^6 \text{ s}^{-1})(1.08 \times 10^{-10} \text{ C})$$

$$= \boxed{6.79 \times 10^{-4} \text{ A}}$$

(c) Determine the charge and current as functions of time.

Solution   Equations 32.24 and 32.25 give the following expressions for the time variation of Q and I:

$$Q = Q_{max} \cos \omega t = \boxed{(1.08 \times 10^{-10} \text{ C}) \cos \omega t}$$

$$I = -I_{max} \sin \omega t = \boxed{(-6.79 \times 10^{-4} \text{ A}) \sin \omega t}$$

where

$$\omega = 2\pi f = 2\pi \times 10^6 \text{ rad/s}$$

Exercise   What is the total energy stored in the circuit?

Answer 6.48×10^{-10} J.

*32.6   THE *RLC* CIRCUIT

We now turn our attention to a more realistic circuit consisting of an inductor, a capacitor, and a resistor connected in series, as in Figure 32.16. We assume that the capacitor has an initial charge Q_{max} before the switch is closed. Once the switch is thrown closed and a current is established, the total energy stored in the circuit at any time is given, as before, by Equation 32.18. That is, the energy stored in the capacitor is $Q^2/2C$, and the energy stored in the inductor is $\frac{1}{2}LI^2$. However, the total energy is no longer constant, as it was in the *LC* circuit, because of the presence of a resistor, which dissipates energy as heat. Since the rate of energy dissipation through a resistor is I^2R, we have

$$\frac{dU}{dt} = -I^2R$$

where the negative sign signifies that U is decreasing in time. Substituting this result into Equation 32.19 gives

$$L\frac{dI}{dt} + \frac{Q}{C}\frac{dQ}{dt} = -I^2R \tag{32.28}$$

Using the fact that $I = dQ/dt$ and $dI/dt = d^2Q/dt^2$, and dividing Equation 32.28 by I, we get

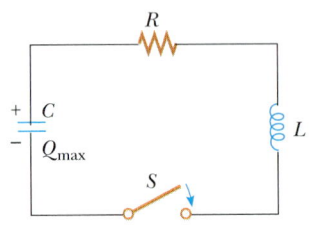

R

C

Q_{max}

L

S

FIGURE 32.16   A series *RLC* circuit. The capacitor has a charge Q_{max} at $t = 0$, the instant at which the switch is thrown closed.

$$L \frac{d^2Q}{dt^2} + R \frac{dQ}{dt} + \frac{Q}{C} = 0 \qquad (32.29)$$

Note that the *RLC* circuit is analogous to the damped harmonic oscillator discussed in Section 13.6 and illustrated in Figure 32.17. The equation of motion for this mechanical system is

$$m \frac{d^2x}{dt^2} + b \frac{dx}{dt} + kx = 0 \qquad (32.30)$$

Comparing Equations 32.29 and 32.30, we see that Q corresponds to x, L to m, R to the damping constant b, and C to $1/k$, where k is the force constant of the spring.

Because the analytical solution of Equation 32.29 is cumbersome, we give only a qualitative description of the circuit behavior.

In the simplest case, when $R = 0$, Equation 32.29 reduces to that of a simple *LC* circuit, as expected, and the charge and the current oscillate sinusoidally in time.

Next consider the situation where R is reasonably small. In this case, the solution of Equation 32.29 is

$$Q = Q_{max} e^{-Rt/2L} \cos \omega_d t \qquad (32.31)$$

where

$$\omega_d = \left[\frac{1}{LC} - \left(\frac{R}{2L} \right)^2 \right]^{1/2} \qquad (32.32)$$

That is, the charge oscillates with damped harmonic motion in analogy with a mass-spring system moving in a viscous medium. From Equation 32.32, we see that, when $R \ll \sqrt{4L/C}$, the frequency ω_d of the damped oscillator is close to that of the undamped oscillator, $1/\sqrt{LC}$. Because $I = dQ/dt$, it follows that the current also undergoes damped harmonic motion. A plot of the charge versus time for the damped oscillator is shown in Figure 32.18. Note that the maximum value of Q decreases after each oscillation, just as the amplitude of a damped harmonic oscillator decreases in time.

When we consider larger values of R, we find that the oscillations damp out more rapidly; in fact, there exists a critical resistance value R_c above which no oscillations occur. The critical value is given by $R_c = \sqrt{4L/C}$. A system with $R = R_c$ is said to be critically damped. When R exceeds R_c, the system is said to be overdamped (Fig. 32.19).

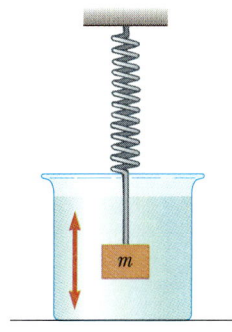

FIGURE 32.17   A mass-spring system moving in a viscous medium with damped harmonic motion is analogous to an *RLC* circuit.

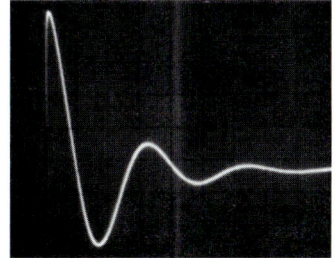

Oscilloscope pattern showing the decay in the oscillations of an *RLC* circuit. The parameters used were $R = 75\ \Omega$, $L = 10$ mH, $C = 0.19\ \mu$F, and $f = 300$ Hz.   *(Courtesy of J. Rudmin)*

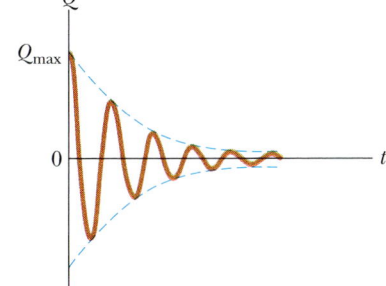

FIGURE 32.18   Charge versus time for a damped *RLC* circuit. This occurs for $R \ll \sqrt{4L/C}$. The Q versus t curve represents a plot of Equation 32.31.

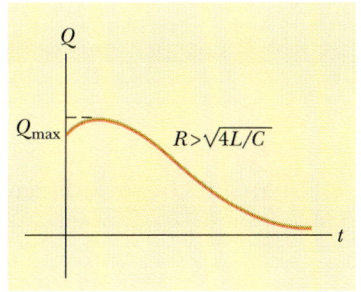

FIGURE 32.19   Plot of Q versus t for an overdamped *RLC* circuit, which occurs for values of $R > \sqrt{4L/C}$.

SUMMARY

When the current in a coil changes with time, an emf is induced in the coil according to Faraday's law. The **self-induced emf** is

$$\mathcal{E}_L = -L\frac{dI}{dt} \tag{32.1}$$

where L is the **inductance** of the coil. Inductance is a measure of the opposition of a device to a change in current. Inductance has the SI unit of **henry** (H), where $1\ \text{H} = 1\ \text{V}\cdot\text{s/A}$.

The inductance of any coil is

$$L = \frac{N\Phi_B}{I} \tag{32.2}$$

where Φ_B is the magnetic flux through the coil and N is the total number of turns. The inductance of a device depends on its geometry. For example, the inductance of an air-core solenoid is

$$L = \frac{\mu_0 N^2 A}{\ell} \tag{32.4}$$

where A is the cross-sectional area and ℓ is the length of the solenoid.

If a resistor and inductor are connected in series to a battery of emf $\mathcal{E}$, and a switch in the circuit is thrown closed at $t = 0$, the current in the circuit varies in time according to the expression

$$I(t) = \frac{\mathcal{E}}{R}\left(1 - e^{-t/\tau}\right) \tag{32.7}$$

where $\tau = L/R$ is the time constant of the RL circuit. That is, the current rises to an equilibrium value of $\mathcal{E}/R$ after a time that is long compared to τ. If the battery is removed from the circuit, the current decays exponentially with time according to the expression

$$I(t) = \frac{\mathcal{E}}{R}e^{-t/\tau} \tag{32.10}$$

where $\mathcal{E}/R$ is the initial current in the circuit.

The energy stored in the magnetic field of an inductor carrying a current I is

$$U_B = \tfrac{1}{2}LI^2 \tag{32.12}$$

The energy per unit volume at a point where the magnetic field is B is

$$u_B = \frac{B^2}{2\mu_0} \tag{32.14}$$

In an LC circuit with zero resistance, the charge on the capacitor and the current in the circuit vary in time according to the expressions

$$Q = Q_{\max}\cos(\omega t + \phi) \tag{32.21}$$

$$I = \frac{dQ}{dt} = -\omega Q_{\max}\sin(\omega t + \phi) \tag{32.23}$$

where Q_{max} is the maximum charge on the capacitor, ϕ is a phase constant, and ω is the angular frequency of oscillation:

$$\omega = \frac{1}{\sqrt{LC}}$$

(32.22)

The energy in an *LC* circuit continuously transfers between energy stored in the capacitor and energy stored in the inductor. The total energy of the *LC* circuit at any time *t* is

$$U = U_C + U_L = \frac{Q_{max}^2}{2C} \cos^2 \omega t + \frac{LI_{max}^2}{2} \sin^2 \omega t$$

(32.26)

At $t = 0$, all of the energy is stored in the electric field of the capacitor ($U = Q_{max}^2/2C$). Eventually, all of this energy is transferred to the inductor ($U = LI_{max}^2/2$). However, the total energy remains constant because the energy losses are neglected in the ideal *LC* circuit.

The charge and current in an *RLC* circuit exhibit a damped harmonic behavior for small values of *R*. This is analogous to the damped harmonic motion of a mass-spring system in which friction is present.

QUESTIONS

1. Why is the induced emf that appears in an inductor called a "counter" or "back" emf?

2. The current in a circuit containing a coil, resistor, and battery has reached a constant value. Does the coil have an inductance? Does the coil affect the value of the current?

3. Does the inductance of a coil depend on the current in the coil? What other parameters affect the inductance of a coil?

4. How can a long piece of wire be wound on a spool so that the wire has a negligible self-inductance?

5. A long, fine wire is wound as a solenoid with a self-inductance *L*. If it is connected across the terminals of a battery, how does the maximum current depend on *L*?

6. For the series *RL* circuit shown in Figure 32.20, can the back emf ever be greater than the battery emf? Explain.

7. Suppose the switch in Figure 32.20 has been closed for a long time and is suddenly opened. Does the current instantaneously drop to zero? Why does a spark appear at the switch contacts at the instant the switch is thrown open?

8. If the current in an inductor is doubled, by what factor does the stored energy change?

9. Discuss the similarities between the energy stored in the electric field of a charged capacitor and the energy stored in the magnetic field of a current-carrying coil.

10. What is the inductance of two inductors connected in series?

11. Discuss how mutual inductance arises between the primary and secondary coils in a transformer.

12. The centers of two circular loops are separated by a fixed distance. For what relative orientation of the loops is their mutual inductance a maximum? a minimum?

13. Two solenoids are connected in series so that each carries the same current at any instant. Is mutual induction present? Explain.

14. In the *LC* circuit shown in Figure 32.12, the charge on the capacitor is sometimes zero, even though there is current in the circuit. How is this possible?

15. If the resistance of the wires in an *LC* circuit were not zero, would the oscillations persist? Explain.

16. How can you tell whether an *RLC* circuit is overdamped or underdamped?

17. What is the significance of critical damping in an *RLC* circuit?

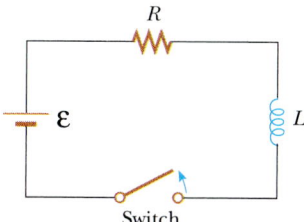

FIGURE 32.20 (Questions 6 and 7).

PROBLEMS

Section 32.1   Self-Inductance

1. A 2.00-H inductor carries a steady current of 0.500 A. When the switch in the circuit is thrown open, the current disappears in 10.0 ms. What is the average induced emf in the inductor during this time?

2. A spring has a radius of 4.00 cm and an inductance of 125 μH when extended to a length of 2.00 m. Find an approximate value for the total number of turns in the spring?

3. A coiled telephone cord has 70 turns, a cross-sectional diameter of 1.3 cm, and an unstretched length of 60 cm. Determine an approximate value for the self-inductance of the unstretched cord.

4. A small air-core solenoid has a length of 4.0 cm and a radius of 0.25 cm. If the inductance is to be 0.060 mH, how many turns per centimeter are required?

5. An emf induced in a solenoid of inductance L changes in time as $\mathcal{E} = \mathcal{E}_0 e^{-kt}$. Find the total charge that passes through the solenoid.

6. Calculate the magnetic flux through a 300-turn, 7.20-mH coil when the current in the coil is 10.0 mA.

6A. Calculate the magnetic flux through an N-turn coil having inductance L when the coil carries a current I.

7. A 40.0-mA current is carried by a uniformly wound air-core solenoid with 450 turns, a 15.0-mm diameter, and 12.0-cm length. Compute (a) the magnetic field inside the solenoid, (b) the magnetic flux through each turn, and (c) the inductance of the solenoid. (d) Which of these quantities depends on the current?

8. A 0.388-mH inductor has a length that is four times its diameter. If it is wound with 22 turns per centimeter, what is its length?

9. An emf of 36 mV is induced in a 400-turn coil at an instant when the current has a value of 2.8 A and is changing at a rate of 12 A/s. What is the total magnetic flux through the coil?

10. The current in a 90-mH inductor changes with time as $I = t^2 - 6t$ (in SI units). Find the magnitude of the induced emf at (a) $t = 1.0$ s and (b) $t = 4.0$ s. (c) At what time is the emf zero?

11. A 10.0-mH inductor carries a current $I = I_{max} \sin \omega t$, with $I_{max} = 5.00$ A and $\omega/2\pi = 60.0$ Hz. What is the back emf as a function of time?

12. A solenoid inductor is 20.0 cm long and has a cross-sectional area of 5.00 cm^2. When the current through the solenoid decreases at a rate of 0.625 A/s, the induced emf is 200 μV. Find the number of turns/unit length of the solenoid.

13. Two coils, A and B, are wound using equal lengths of wire. Each coil has the same number of turns per unit length, but coil A has twice as many turns as coil B. What is the ratio of the self-inductance of A to the self-inductance of B? (*Note:* The radii of the two coils are not equal.)

14. A toroid has a major radius R and a minor radius r, and is tightly wound with N turns of wire, as shown in Figure P32.14. If $R \gg r$, the magnetic field within the region of the toroid of cross-sectional area $A = \pi r^2$ is essentially that of a long solenoid that has been bent into a large circle of radius R. Using the uniform field of a long solenoid, show that the self-inductance of such a toroid is approximately

$$L \cong \frac{\mu_0 N^2 A}{2 \pi R}$$

(An exact expression for the inductance of a toroid with a rectangular cross-section is derived in Problem 78.)

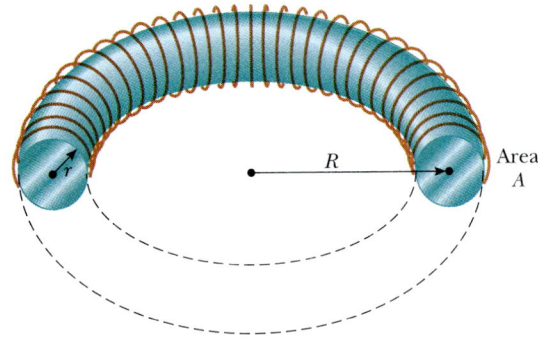

FIGURE P32.14

15. A solenoid has 120 turns uniformly wrapped around a wooden core, which has a diameter of 10.0 mm and a length of 9.00 cm. (a) Calculate the inductance of the solenoid. (b) The wooden core is replaced with a soft iron rod that has the same dimensions, but a magnetic permeability $\mu_m = 800\mu_0$. What is the new inductance?

16. An inductor in the form of a solenoid contains 420 turns, is 16.0 cm in length, and has a cross-sectional area of 3.00 cm^2. What uniform rate of decrease of current through the inductor induces an emf of 175 μV?

□ indicates problems that have full solutions available in the Student Solutions Manual and Study Guide.

16A. An inductor in the form of a solenoid contains N turns, has a length ℓ, and has cross-sectional area A. What uniform rate of decrease of current through the inductor induces an emf $\mathcal{E}$?

Section 32.2   *RL* Circuits

17. The switch in Figure P32.17 is closed at time $t = 0$. Find the current in the inductor and the current through the switch as functions of time.

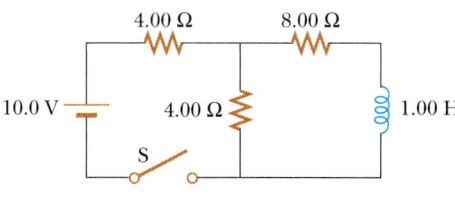

FIGURE P32.17

18. Show that $I = I_0 e^{-t/\tau}$ is a solution of the differential equation

$$IR + L\frac{dI}{dt} = 0$$

where $\tau = L/R$ and $I_0 = \mathcal{E}/R$ is the current at $t = 0$.

19. Calculate the resistance in an *RL* circuit in which $L = 2.5$ H and the current increases to 90% of its final value in 3.0 s.

20. In an *RL* series circuit (Fig. 32.2), the maximum current is 0.500 A. After the switch is thrown, a current of 0.250 A is reached in 0.150 s, and then the voltage across the resistor is 20.0 V. Calculate the values of R, L, and $\mathcal{E}$ for this circuit.

21. A series *RL* circuit with $L = 3.00$ H and a series *RC* circuit with $C = 3.00$ μF have the same time constant. If the two circuits have the same resistance R, (a) what is the value of R and (b) what is the time constant?

21A. A series *RL* circuit and a series *RC* circuit have the same time constant. If the two circuits have the same resistance R, (a) what is the value of R in terms of L and C, and (b) what is the time constant in terms of L and C?

22. An inductor that has an inductance of 15.0 H and a resistance of 30.0 Ω is connected across a 100-V battery. What is the rate of increase of the current (a) at $t = 0$ and (b) at $t = 1.50$ s?

23. A 12-V battery is about to be connected to a series circuit containing a 10-Ω resistor and a 2.0-H inductor. How long will it take the current to reach (a) 50% and (b) 90% of its final value?

24. A current pulse is fed to the partial circuit shown in Figure P32.24. The current begins at zero, then becomes 10.0 A for 200 μs, and then is zero once again.

Determine the voltage across the inductor as a function of time.

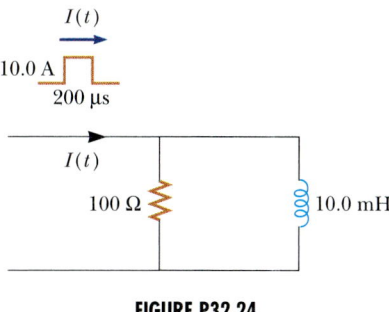

FIGURE P32.24

25. A 140-mH inductor and a 4.9-Ω resistor are connected with a switch to a 6.0-V battery as shown in Figure P32.25. (a) If the switch is thrown to the left (connecting the battery), how much time elapses before the current reaches 220 mA? (b) What is the current in the inductor 10.0 s after the switch is closed? (c) Now the switch is quickly thrown from A to B. How much time elapses before the current falls to 160 mA?

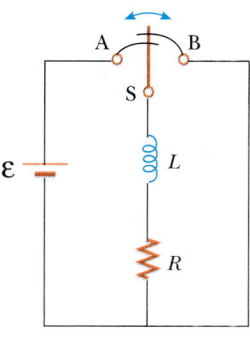

FIGURE P32.25

26. When the switch in Figure P32.26 is thrown closed, the current takes 3.00 ms to reach 98% of its final value. If $R = 10.0$ Ω, what is the inductance?

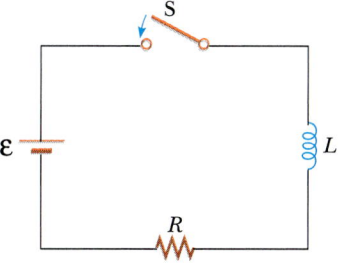

FIGURE P32.26

27. For $\mathcal{E} = 6.00$ V, $L = 24.0$ mH, and $R = 10.0\ \Omega$ in Figure P32.26, find (a) the current 0.500 ms after S is thrown closed and (b) the maximum current.

28. Consider two ideal inductors, L_1 and L_2, that have *zero* internal resistance and are far apart so that their mutual inductance is zero. (a) If these inductors are connected in series, show that they are equivalent to a single ideal inductor having $L_{eq} = L_1 + L_2$. (b) If these same inductors are connected in parallel, show that they are equivalent to a single ideal inductor having $1/L_{eq} = 1/L_1 + 1/L_2$. (c) Now consider two inductors, L_1 and L_2, that have *nonzero* internal resistance R_1 and R_2, respectively, but are still far apart so that their mutual inductance is zero. If these inductors are connected in series show that they are equivalent to a single inductor having $L_{eq} = L_1 + L_2$ and $R_{eq} = R_1 + R_2$. (d) If these same inductors are now connected in parallel, is it necessarily true that they are equivalent to a single ideal inductor having $1/L_{eq} = 1/L_1 + 1/L_2$ and $1/R_{eq} = 1/R_1 + 1/R_2$? Explain.

29. Let $L = 3.00$ H, $R = 8.00\ \Omega$, and $\mathcal{E} = 36.0$ V in Figure P32.26. (a) Calculate the ratio of the potential difference across the resistor to that across the inductor when $I = 2.00$ A. (b) Calculate the voltage across the inductor when $I = 4.50$ A.

30. In the circuit shown in Figure P32.30, switch S_1 is closed first, and after a certain period of time the current through L_1 is 0.60 A. Then switch S_2 is closed. Find the currents through the coils the moment switch S_2 is closed. Assume that the resistances of the coils are negligible, and take $\mathcal{E} = 12.0$ V, $r = 10.0\ \Omega$, $L_1 = 2.00$ H, and $L_2 = 6.00$ H.

30A. In the circuit shown in Figure P32.30, switch S_1 is closed first, and after a certain period of time the current through L_1 is I_1. Then switch S_2 is closed. Find the currents through the coils the moment switch S_2 is closed. Assume that the resistances of the coils are negligible.

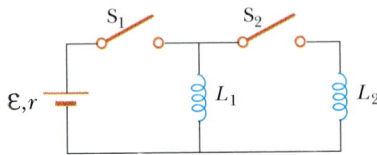

FIGURE P32.30

31. One application of an *RL* circuit is the generation of time-varying high-voltage from a low-voltage source, as shown in Figure P32.31. (a) What is the current in the circuit a long time after the switch has been in position A? (b) Now the switch is thrown quickly from A to B. Compute the initial voltage across each resistor and the inductor. (c) How much time elapses before the voltage across the inductor drops to 12 V?

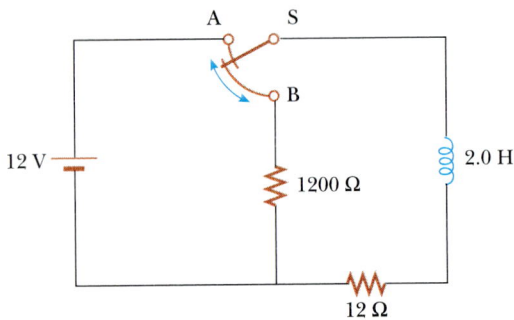

FIGURE P32.31

Section 32.3   Energy in a Magnetic Field

32. Calculate the energy associated with the magnetic field of a 200-turn solenoid in which a current of 1.75 A produces a flux of 3.70×10^{-4} Wb in each turn.

33. An air-core solenoid with 68 turns is 8.0 cm long and has a diameter of 1.2 cm. How much energy is stored in its magnetic field when it carries a current of 0.77 A?

34. Consider the circuit shown in Figure P32.34. What energy is stored in the inductor when the current reaches its final equilibrium value after the switch is closed?

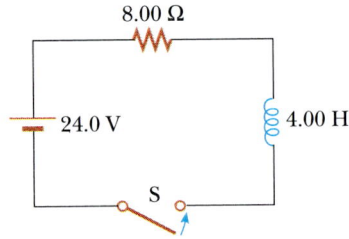

FIGURE P32.34

35. A 10.0-V battery, a 5.00-Ω resistor, and a 10.0-H inductor are connected in series. After the current in the circuit has reached its maximum value, calculate (a) the power supplied by the battery, (b) the power dissipated in the resistor, (c) the power dissipated in the inductor, and (d) the energy stored in the magnetic field of the inductor.

36. At $t = 0$, an emf of 500 V is applied to a coil that has an inductance of 0.80 H and a resistance of 30 Ω. (a) Find the energy stored in the magnetic field when the current reaches half its maximum value. (b) After the emf is connected, how long does it take the current to reach this value?

37. The magnetic field inside a superconducting solenoid is 4.5 T. The solenoid has an inner diameter of 6.2 cm and a length of 26 cm. Determine (a) the

magnetic energy density in the field and (b) the energy stored in the magnetic field within the solenoid.

38. A uniform electric field of magnitude 6.80×10^5 V/m throughout a cylindrical volume results in a total energy of $3.40 \ \mu J$. What magnetic field over this same region stores the same total energy?

39. On a clear day, there is a 100-V/m vertical electric field near the Earth's surface. At the same time, the Earth's magnetic field has a magnitude of 0.500×10^{-4} T. Compute the energy density of the two fields.

40. A 15.0-V battery is connected to an RL circuit for which $L = 0.600$ H and $R = 7.00 \ \Omega$. When the current has reached one half of its final value, what is the total magnetic energy stored in the inductor?

41. Two inductors, $L_1 = 85 \ \mu H$ and $L_2 = 200 \ \mu H$, are connected in series with an 850-mA dc power supply. Calculate the energy stored in each inductor.

42. In the circuit of Figure P32.42, $\mathcal{E} = 50.0$ V, $R = 250 \ \Omega$, and $C = 0.50 \ \mu F$. The switch S is closed for a long time and no voltage is measured across the capacitor. After the switch is opened, the voltage across the capacitor reaches a maximum value of 150 V. What is the inductance, L?

42A. In the circuit of Figure P32.42, the switch S is closed for a long time, and no voltage is measured across the capacitor. After the switch is opened, the voltage across the capacitor reaches a maximum value of V. Find an expression for the inductance, L.

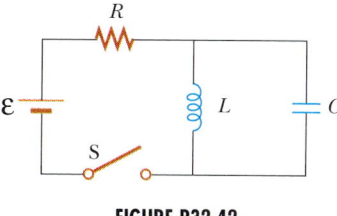

FIGURE P32.42

43. An inductor has a self-inductance of 20.0 H and a resistance of $10.0 \ \Omega$. At $t = 0.10$ s after this inductor is connected to a 12.0-V battery, calculate (a) the stored magnetic power, (b) the power dissipated in the resistor, and (c) the power delivered by the battery.

44. The magnitude of the magnetic field outside a sphere of radius R is $B = B_0 (R/r)^2$, where B_0 is a constant. Determine the total energy stored in the magnetic field outside the sphere and evaluate your result for $B_0 = 5.0 \times 10^{-5}$ T and $R = 6.0 \times 10^6$ m, values appropriate for the Earth's magnetic field.

*Section 32.4 Mutual Inductance

45. Two coils are close to each other. The first coil carries a time-varying current given by $I(t) = (5.0 \text{ A}) e^{-0.025t}$

$\sin(377t)$. At $t = 0.80$ s, the voltage measured across the second coil is -3.2 V. What is the mutual inductance of the coils?

46. Two coils, held in fixed positions, have a mutual inductance of $100 \ \mu H$. What is the maximum voltage in one when the second coil carries a sinusoidal current given by $I(t) = (10.0 \text{ A}) \sin(1000t)$?

47. An emf of 96.0 mV is induced in the windings of a coil when the current in a nearby coil is increasing at the rate of 1.20 A/s. What is the mutual inductance of the two coils?

48. Two inductors having self-inductances $L_1 = 10.0$ H and $L_2 = 5.00$ H are connected in parallel as in Figure P32.48a. The mutual inductance between the two inductors is $M = 6.50$ H. Determine the equivalent self-inductance, L_{eq}, for the system (Fig. P32.48b).

48A. Two inductors having self-inductances L_1 and L_2 are connected in parallel as in Figure P32.48a. The mutual inductance between the two inductors is M. Determine the equivalent self-inductance, L_{eq}, for the system (Fig. P32.48b).

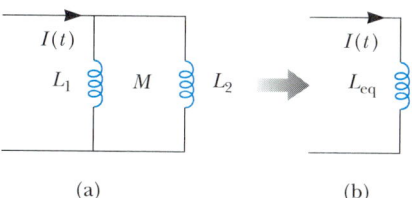

(a)   (b)

FIGURE P32.48

49. A coil of 50 turns is wound on a long solenoid as shown in Figure 32.10. The solenoid has a cross-sectional area of 8.8×10^{-3} m² and is wrapped uniformly with 1000 turns per meter of length. Calculate the mutual inductance of the two windings.

50. A 70-turn solenoid is 5.0 cm long and 1.0 cm in diameter and carries a 2.0-A current. A single loop of wire, 3.0 cm in diameter, is held so that the plane of the loop is perpendicular to the long axis of the solenoid. What is the mutual inductance of the two if the plane of the loop passes through the solenoid 2.5 cm from one end?

51. Two solenoids A and B spaced closely to each other and sharing the same cylindrical axis have 400 and 700 turns, respectively. A current of 3.5 A in coil A produces a flux of 300 μWb at the center of A and a flux of 90 μWb at the center of B. (a) Calculate the mutual inductance of the two solenoids. (b) What is the self-inductance of A? (c) What emf is induced in B when the current in A increases at the rate of 0.50 A/s?

52. Two single-turn circular loops of wire have radii R

and r, with $R \gg r$. The loops lie in the same plane and are concentric. (a) Show that the mutual inductance of the pair is $M = \mu_0 \pi r^2 / 2R$. (*Hint:* Assume that the larger loop carries a current I and compute the resulting flux through the smaller loop.) (b) Evaluate M for $r = 2.00$ cm and $R = 20.0$ cm.

Section 32.5   Oscillations in an *LC* Circuit

53. A 1.00-μF capacitor is charged by a 40.0-V power supply. The fully charged capacitor is then discharged through a 10.0-mH inductor. Find the maximum current in the resulting oscillations.

54. An *LC* circuit consists of a 20-mH inductor and a 0.50-μF capacitor. If the maximum instantaneous current is 0.10 A, what is the greatest potential difference across the capacitor?

55. A 1.00-mH inductor and a 1.00-μF capacitor are connected in series. The current in the circuit is described by $I = 20.0t$, where t is in seconds and I is in amperes. The capacitor initially has no charge. Determine (a) the voltage across the inductor as a function of time, (b) the voltage across the capacitor as a function of time, (c) the time when the energy stored in the capacitor first exceeds that in the inductor.

55A. An inductor having inductance L and a capacitor having capacitance C are connected in series. The current in the circuit is described by $I = Kt$, where t is in seconds and I is in amperes. The capacitor initially has no charge. Determine (a) the voltage across the inductor as a function of time, (b) the voltage across the capacitor as a function of time, and (c) the time when the energy stored in the capacitor first exceeds that in the inductor.

56. Calculate the inductance of an *LC* circuit that oscillates at 120 Hz when the capacitance is 8.00 μF.

57. A fixed inductance $L = 1.05$ μH is used in series with a variable capacitor in the tuning section of a radio. What capacitance tunes the circuit into the signal from a station broadcasting at 6.30 MHz?

58. An *LC* circuit like the one in Figure 32.11 contains an 82-mH inductor and a 17-μF capacitor that initially carries a 180-μC charge. The switch is thrown closed at $t = 0$. (a) Find the frequency (in hertz) of the resulting oscillations. At $t = 1.0$ ms, find (b) the charge on the capacitor and (c) the current in the circuit.

59. (a) What capacitance must be combined with a 45.0-mH inductor in order to achieve a resonant frequency of 125 Hz? (b) What time interval elapses between accumulations of maximum charge of the same sign on a given plate of the capacitor?

60. The switch in Figure P32.60 is connected to point a for a long time. After the switch is thrown to point b,

find (a) the frequency of oscillation in the *LC* circuit, (b) the maximum charge that builds up on the capacitor, (c) the maximum current in the inductor, and (d) the total energy stored in the circuit at $t = 3.0$ s.

60A. The switch in Figure P32.60 is connected to point a for a long time. After the switch is thrown to point b, find (in terms of $\mathcal{E}$, R, L, and C) (a) the frequency of oscillation in the *LC* circuit, (b) the maximum charge on the capacitor, (c) the maximum current in the inductor, and (d) the total energy stored in the circuit at time t.

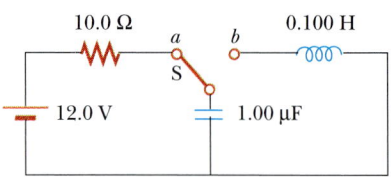

FIGURE P32.60

61. An *LC* circuit like that in Figure 32.11 consists of a 3.30-H inductor and an 840-pF capacitor, initially carrying a 105-μC charge. At $t = 0$ the switch is thrown closed. Compute the following quantities at $t = 2.00$ ms: (a) the energy stored in the capacitor, (b) the energy stored in the inductor, and (c) the total energy in the circuit.

62. A 6.0-V battery is used to charge a 50-μF capacitor. The capacitor is then discharged through a 0.34-mH inductor. Find (a) the maximum charge on the capacitor, (b) the maximum current in the circuit, and (c) the maximum energy stored in each component.

*Section 32.6   The *RLC* Circuit

63. In Figure 32.16, let $R = 7.60$ Ω, $L = 2.20$ mH, and $C = 1.80$ μF. (a) Calculate the frequency of the damped oscillation of the circuit. (b) What is the critical resistance?

64. Consider a series *LC* circuit in which $L = 2.18$ H and $C = 6.00$ nF. What is the maximum value of a resistor that, inserted in series with L and C, allows the circuit to continue to oscillate?

65. Consider an *LC* circuit in which $L = 500$ mH and $C = 0.100$ μF. (a) What is the resonant frequency (ω_0)? (b) If a resistance of 1.00 kΩ is introduced into this circuit, what is the frequency of the (damped) oscillations? (c) What is the percent difference between the two frequencies?

66. Electrical oscillations are initiated in a series circuit containing a capacitance C, inductance L, and resistance R. (a) If $R \ll \sqrt{4L/C}$ (weak damping), how much time elapses before the current falls off to 50% of its initial value? (b) How long does it take the energy to decrease to 50% of its initial value?

67. Show that Equation 32.29 in the text is consistent with Kirchhoff's loop law as applied to Figure 32.16.

68. Consider an *RLC* series circuit consisting of a charged 500-μF capacitor connected to a 32-mH inductor and a resistor *R*. Calculate the frequency of the oscillations (in Hertz) for (a) $R = 0$ (no damping); (b) $R = 16 \, \Omega$ (critical damping: $R = \sqrt{4L/C}$); (c) $R = 4.0 \, \Omega$ (underdamped: $R < \sqrt{4L/C}$); and (d) $R = 64 \, \Omega$ (overdamped: $R > \sqrt{4L/C}$).

ADDITIONAL PROBLEMS

69. An inductor that has a resistance of 0.50 Ω is connected to a 5.0-V battery. One second after the switch is closed, the current in the circuit is 4.0 A. Calculate the inductance.

70. A soft iron rod ($\mu = 800\mu_0$) is used as the core of a solenoid. The rod has a diameter of 24 mm and is 10 cm long. A 10-m piece of 22-gauge copper wire (diameter = 0.644 mm) is wrapped around the rod in a single uniform layer, except for a 10-cm length at each end to be used for connections. (a) How many turns of this wire can be wrapped around the rod? (*Hint:* The radius of the wire adds to the diameter of the rod in determining the circumference of each turn. Also, the wire spirals diagonally along the surface of the rod.) (b) What is the resistance of this inductor? (c) What is its inductance?

71. An 820-turn wire coil of resistance 24.0 Ω is placed on top of a 12 500-turn, 7.00-cm-long solenoid, as in Figure P32.71. Both coil and solenoid have cross-sectional areas of $1.00 \times 10^{-4} \, \text{m}^2$. (a) How long does it take the solenoid current to reach 63.2 percent its maximum value? Determine (b) the average back emf caused by the self-inductance of the solenoid

during this interval, (c) the average rate of change in magnetic flux through the coil during this interval, and (d) the magnitude of the average induced current in the coil.

72. A capacitor in a series *LC* circuit has an initial charge *Q* and is being discharged. Find, in terms of *L* and *C*, the flux through the coil when the charge on the capacitor is $Q/2$.

73. The inductor in Figure P32.73 has negligible resistance. When the switch is thrown open after having been closed for a long time, the current in the inductor drops to 0.25 A in 0.15 s. What is the inductance of the inductor?

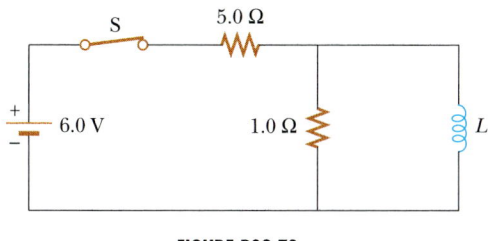

FIGURE P32.73

74. A platinum wire 2.5 mm in diameter is connected in series to a 100-μF capacitor and a $1.2 \times 10^{-3} \, \mu$H inductor to form an *RLC* circuit. The resistivity of platinum is $11 \times 10^{-8} \, \Omega \cdot \text{m}$. Calculate the maximum length of wire for which the current oscillates.

75. Assume that the switch in Figure P32.75 is initially in position 1. Show that if the switch is thrown from position 1 to position 2, all the energy stored in the magnetic field of the inductor is dissipated as thermal energy in the resistor.

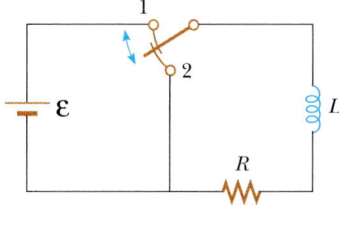

FIGURE P32.75

76. The lead-in wires from a TV antenna are often constructed in the form of two parallel wires (Fig. P32.76). (a) Why does this configuration of conductors have an inductance? (b) What constitutes the flux loop for this configuration? (c) Neglecting any magnetic flux inside the wires, show that the inductance of a length *x* of this type of lead-in is

$$L = \frac{\mu_0 x}{\pi} \ln\left(\frac{w - a}{a}\right)$$

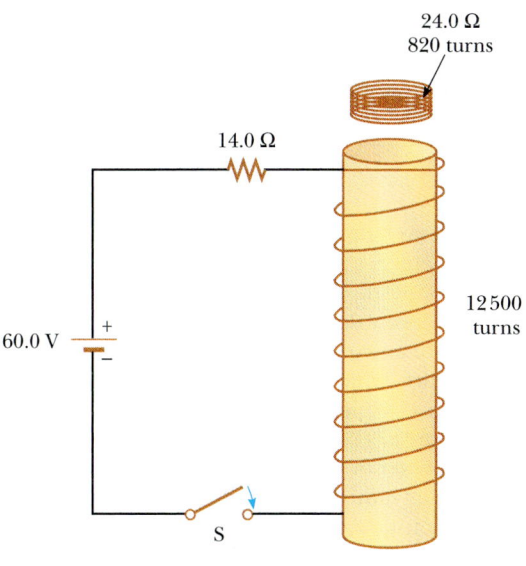

FIGURE P32.71

where a is the radius of the wires and w is their center-to-center separation.

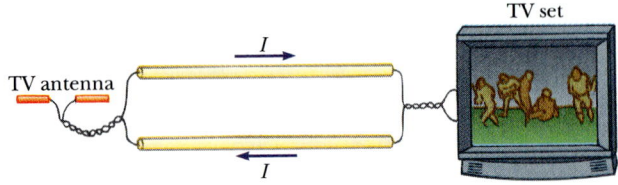

FIGURE P32.76

77. At $t = 0$, the switch in Figure P32.77 is thrown closed. By using Kirchhoff's laws for the instantaneous currents and voltages in this two-loop circuit, show that the current in the inductor is

$$I(t) = \frac{\mathcal{E}}{R_1}[1 - e^{-(R'/L)t}]$$

where $R' = R_1 R_2 / (R_1 + R_2)$.

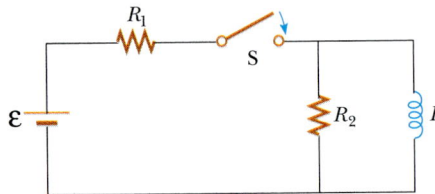

FIGURE P32.77

78. The toroid in Figure P32.78 consists of N turns and has a rectangular cross-section. Its inner and outer radii are a and b, respectively. (a) Show that

$$L = \frac{\mu_0 N^2 h}{2\pi} \ln\left(\frac{b}{a}\right)$$

(b) Using this result, compute the self-inductance of a 500-turn toroid for which $a = 10.0$ cm, $b = 12.0$ cm, and $h = 1.00$ cm. (c) In Problem 14, an approximate formula for the inductance of a toroid with $R \gg r$ was derived. To get a feel for the accuracy of this result, use the expression in Problem 14 to compute the (approximate) inductance of the toroid described in part (b).

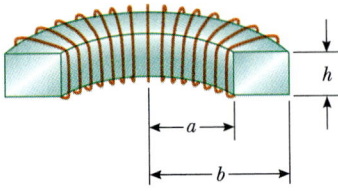

FIGURE P32.78

79. In Figure P32.79, the switch is closed at $t < 0$, and steady-state conditions are established. The switch is now thrown open at $t = 0$. (a) Find the initial voltage $\mathcal{E}_0$ across L just after $t = 0$. Which end of the coil is at the higher potential: a or b? (b) Make freehand graphs of the currents in R_1 and in R_2 as a function of time, treating the steady-state directions as positive. Show values before and after $t = 0$. (c) How long after $t = 0$ is the current in R_2 2.0 mA?

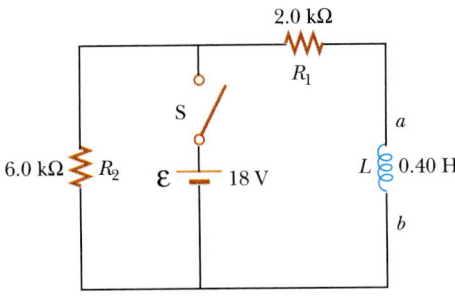

FIGURE P32.79

80. The switch in Figure P32.80 is thrown closed at $t = 0$. Before the switch is closed, the capacitor is uncharged and all currents are zero. Determine the currents in L, C, and R and the potential differences across L, C, and R (a) the instant after the switch is closed and (b) long after it is closed.

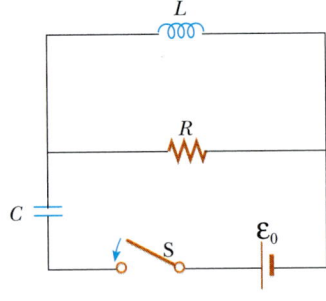

FIGURE P32.80

81. Two long parallel wires, each of radius a, have their centers a distance d apart and carry equal currents in opposite directions. Neglecting the flux within the wires, calculate the inductance per unit length.

82. An air-core solenoid 0.50 m in length contains 1000 turns and has a cross-sectional area of 1.0 cm². (a) Neglecting end effects, what is the self-inductance? (b) A secondary winding wrapped around the center of the solenoid has 100 turns. What is the mutual inductance? (c) The secondary winding carries a constant current of 1.0 A, and the solenoid is connected to a load of 1.0 kΩ. The constant current is suddenly stopped. How much charge flows through the load resistor?

83. To prevent damage from arcing in an electric motor, a discharge resistor is sometimes placed in parallel with the armature. If the motor is suddenly unplugged while running, this resistor limits the voltage that appears across the armature coils. Consider a 12-V dc motor that has an armature that has a resistance of 7.5 Ω and an inductance of 450 mH. Assume the back emf in the armature coils is 10 V when the motor is running at normal speed. (The equivalent circuit for the armature is shown in Fig. P32.83.) Calculate the maximum resistance R that limits the voltage across the armature to 80 V when the motor is unplugged.

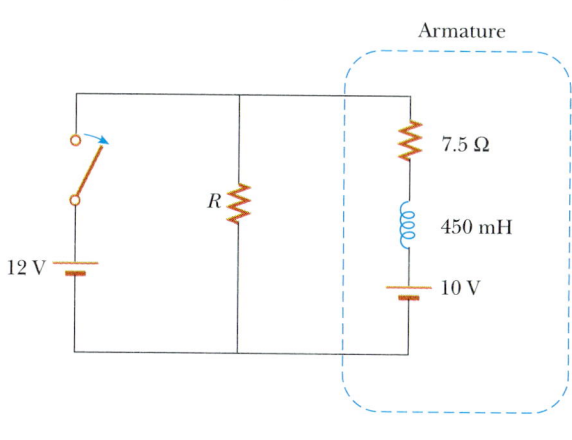

FIGURE P32.83

84. A battery is in series with a switch and a 2.0-H inductor whose windings have a resistance R. After the switch is thrown closed, the current rises to 80% of its final value in 0.40 s. Find R.

85. Initially, the capacitor in a series LC circuit is charged. A switch is closed, allowing the capacitor to discharge, and 0.50 μs later the energy stored in the capacitor is one-fourth its initial value. Determine L if $C = 5.0$ pF.

85A. Initially, the capacitor in a series LC circuit is charged. A switch is closed, allowing the capacitor to discharge, and t seconds later the energy stored in the capacitor is one-fourth its initial value. Determine L if C is known.

86. A toroid has two sets of windings, each spread uniformly around the toroid, with total turns N_1 and N_2, respectively. The toroid has a circumferential length ℓ and a cross-sectional area A. (a) Write expressions for the self-inductances L_1 and L_2 when each coil is used alone. (b) Derive an expression for the mutual inductance M of the two coils. (c) Show that $M^2 = L_1 L_2$. (This expression is true only when all the flux through one coil also passes through the other coil.)

SPREADSHEET PROBLEMS

S1. Spreadsheet 32.1 calculates the current and the energy stored in the magnetic field of an RL circuit when the circuit is charging and when it is discharging. Use $R = 1000$ Ω, $L = 0.35$ H, and $\mathcal{E} = 10$ V. (a) What is the time constant of the circuit? From the graph of the current versus time when the circuit is charging, how much time has elapsed when the current is 50% of its maximum value? 90%? 99%? Give your answers both in seconds and in multiples of the time constant. (b) Repeat for the case when the circuit is discharging. (c) How much time has elapsed when the energy is 50% of its maximum value for the two cases? 90%? 99%?

S2. A coil of self-inductance L carries a current given by $I = I_{max} \sin 2\pi ft$. The self-induced emf in the coil is $\mathcal{E}_L = -L \, dI/dt$. Develop a spreadsheet to calculate I as a function of time. Numerically differentiate the current and calculate $\mathcal{E}$. Choose $I_{max} = 2.00$ A, $f = 60.0$ Hz, and $L = 10.0$ mH. Plot $\mathcal{E}$ versus t.

CHAPTER 33

Alternating Current Circuits

Line repair crews working on transmission lines.   *(Tom Pantages)*

I n this chapter, we describe alternating current (ac) circuits. We investigate the characteristics of circuits containing familiar elements and driven by a sinusoidal voltage. Our discussion is limited to simple series circuits containing resistors, inductors, and capacitors, and we find that the ac current in each element is proportional to the instantaneous ac voltage across the element. We also find that when the applied voltage is sinusoidal, the current in each element is also sinusoidal but not necessarily in phase with the applied voltage. We conclude the chapter with two sections concerning the characteristics of *RC* filters, transformers, and power transmission.

33.1 AC SOURCES AND PHASORS

An ac circuit consists of circuit elements and a generator that provides the alternating current. The basic principle of the ac generator is a direct consequence of Faraday's law of induction. When a coil is rotated in a magnetic field at constant angular frequency ω, a sinusoidal voltage (emf) is induced in the coil. This instantaneous voltage v is

$$v = V_{max} \sin \omega t$$

where V_{max} is the maximum output voltage of the ac generator, or the **voltage amplitude.** The angular frequency is given by

$$\omega = 2\pi f = \frac{2\pi}{T}$$

where f is the frequency of the source and T is the period. Commercial electric-power plants in the United States use a frequency of 60 Hz, which corresponds to an angular frequency of 377 rad/s.

The primary aim of this chapter can be summarized as follows: Consider an ac generator connected to a series circuit containing R, L, and C elements. If the voltage amplitude and frequency of the generator are given, together with the values of R, L, and C, find the amplitude and phase constant of the current. In order to simplify our analysis of circuits containing two or more elements, we use graphical constructions called *phasor diagrams*. In these constructions, alternating quantities, such as current and voltage, are represented by rotating vectors called **phasors.** The length of the phasor represents the amplitude (maximum value) of the quantity, while the projection of the phasor onto the vertical axis represents the instantaneous value of that quantity. As we shall see, the method of combining several sinusoidally varying currents or voltages with different phases is greatly simplified using this procedure.

33.2 RESISTORS IN AN AC CIRCUIT

Consider a simple ac circuit consisting of a resistor and an ac generator (—◯—), as in Figure 33.1. At any instant, the algebraic sum of the potential increases and decreases around a closed loop in a circuit must be zero (Kirchhoff's loop equation). Therefore, $v - v_R = 0$, or

$$v = v_R = V_{max} \sin \omega t \tag{33.1}$$

where v_R is the *instantaneous voltage drop across the resistor.* Therefore, the instantaneous current in the resistor is

$$i_R = \frac{v}{R} = \frac{V_{max}}{R} \sin \omega t = I_{max} \sin \omega t \tag{33.2}$$

where I_{max} is the maximum current:

$$I_{max} = \frac{V_{max}}{R}$$

From Equations 33.1 and 33.2, we see that the instantaneous voltage drop across the resistor is

$$v_R = I_{max} R \sin \omega t \tag{33.3}$$

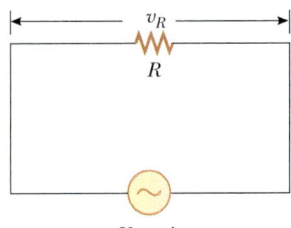

FIGURE 33.1 A circuit consisting of a resistor R connected to an ac generator, designated by the symbol —◯—.

$v = V_{max} \sin \omega t$

Maximum current in a resistor

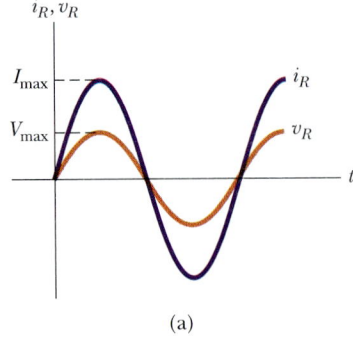

(a)

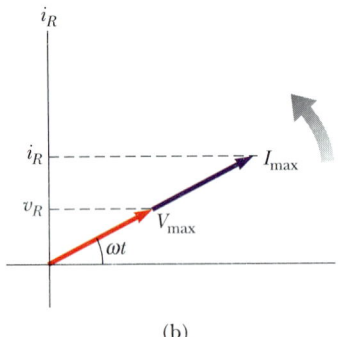

(b)

FIGURE 33.2 (a) Plots of the current and voltage across a resistor as functions of time. The current is in phase with the voltage, which means that the voltage is zero when the current is zero, maximum when the current is maximum, and minimum when the current is minimum. (b) A phasor diagram for the resistive circuit, showing that the current is in phase with the voltage.

Because i_R and v_R both vary as $\sin \omega t$ and reach their maximum values at the same time as in Figure 33.2, they are said to be in phase. A phasor diagram is used to represent phase relationships. The lengths of the arrows correspond to V_{max} and I_{max}. The projections of the arrows onto the vertical axis give v_R and i_R. In the case of the single-loop resistive circuit, the current and voltage phasors lie along the same line, as in Figure 33.2b, because i_R and v_R are in phase.

Note that *the average value of the current over one cycle is zero*. That is, the current is maintained in the positive direction for the same amount of time and at the same magnitude as it is maintained in the negative direction. However, the direction of the current has no effect on the behavior of the resistor. This can be understood by realizing that collisions between electrons and the fixed atoms of the resistor result in an increase in the temperature of the resistor. Although this temperature increase depends on the magnitude of the current, it is independent of the direction of the current.

This discussion can be made quantitative by recalling that the rate at which electrical energy is converted to heat in a resistor is the power $P = i^2R$, where i is the instantaneous current in the resistor. Since the heating effect of a current is proportional to the square of the current, it makes no difference whether the current is direct or alternating, that is, whether the sign associated with the current is positive or negative. However, the heating effect produced by an alternating current having a maximum value I_{max} is not the same as that produced by a direct current of the same value. This is because the alternating current is at this maximum value for only a very brief instant of time during each cycle. What is of importance in an ac circuit is an average value of current referred to as the rms current. The notation **rms** refers to root mean square, which means the square root of the average value of the square of the current. Because I^2 varies as $\sin^2 \omega t$, and because the average value of i^2 is $\frac{1}{2}I_{max}^2$ (Fig. 33.3), the rms current is[1]

$$I_{rms} = \frac{I_{max}}{\sqrt{2}} = 0.707 I_{max} \qquad (33.4)$$

This equation says that an alternating current whose maximum value is 2.00 A produces in a resistor the same heating effect as a direct current of $(0.707)(2.00) = 1.41$ A. Thus, we can say that the average power dissipated in a resistor that carries an alternating current is $P_{av} = I_{rms}^2 R$.

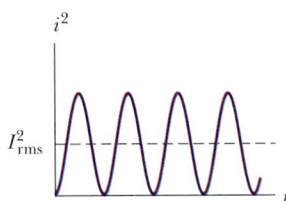

FIGURE 33.3 Plot of the square of the current in a resistor versus time. The rms current is the square root of the average of the square of the current.

[1] The fact that the square root of the average value of the square of the current is equal to $I_{max}/\sqrt{2}$ can be shown as follows. The current in the circuit varies with time according to the expression $i = I_{max} \sin \omega t$, so that $i^2 = I_{max}^2 \sin^2 \omega t$. Therefore we can find the average value of i^2 by calculating the average value of $\sin^2 \omega t$. Note that a graph of $\cos^2 \omega t$ versus time is identical to a graph of $\sin^2 \omega t$ versus time, except that the points are shifted on the time axis. Thus, the time average of $\sin^2 \omega t$ is equal to the time average of $\cos^2 \omega t$ when taken over one or more complete cycles. That is,

$$(\sin^2 \omega t)_{av} = (\cos^2 \omega t)_{av}$$

With this fact and the trigonometric identity $\sin^2 \theta + \cos^2 \theta = 1$, we get

$$(\sin^2 \omega t)_{av} + (\cos^2 \omega t)_{av} = 2(\sin^2 \omega t)_{av} = 1$$

$$(\sin^2 \omega t)_{av} = \tfrac{1}{2}$$

When this result is substituted in the expression $i^2 = I_{max}^2 \sin^2 \omega t$, we get $(i^2)_{av} = I_{rms}^2 = I_{max}^2/2$, or $I_{rms} = I_{max}/\sqrt{2}$, where I_{rms} is the rms current. The factor of $1/\sqrt{2}$ is only valid for sinusoidally varying currents. Other waveforms such as sawtooth variations have different factors.

TABLE 33.1	**Notation Used in This Chapter**	
	Voltage	Current
Instantaneous value	v	i
Maximum value	V_{max}	I_{max}
rms value	V_{rms}	I_{rms}

Alternating voltages are also best discussed in terms of rms voltages, and the relationship here is identical to the above; that is, the rms voltage is

$$V_{rms} = \frac{V_{max}}{\sqrt{2}} = 0.707\,V_{max} \qquad (33.5)$$

rms voltage

When speaking of measuring a 120-V ac voltage from an electric outlet, we are really referring to an rms voltage of 120 V. A quick calculation using Equation 33.5 shows that such an ac voltage actually has a maximum value of about 170 V. In this chapter we use rms values when discussing alternating currents and voltages. One reason for this is that ac ammeters and voltmeters are designed to read rms values. Furthermore, with rms values, many of the equations we use have the same form as their direct-current counterparts. Table 33.1 summarizes the notation used in this chapter.

EXAMPLE 33.1   What Is the rms Current?

The output of a generator is given by $v = 200 \sin \omega t$. Find the rms current in the circuit when this generator is connected to a 100-Ω resistor.

Solution   Comparing this expression for the voltage output with the general form, $v = V_{max} \sin \omega t$, we see that $V_{max} = 200$ V. Thus, the rms voltage is

$$V_{rms} = \frac{V_{max}}{\sqrt{2}} = \frac{200 \text{ V}}{\sqrt{2}} = 141 \text{ V}$$

The calculated rms voltage can be used with Ohm's law to find the rms current in the circuit:

$$I_{rms} = \frac{V_{rms}}{R} = \frac{141 \text{ V}}{100 \ \Omega} = \boxed{1.41 \text{ A}}$$

Exercise   Find the maximum current in the circuit.

Answer   2.00 A.

33.3   INDUCTORS IN AN AC CIRCUIT

Now consider an ac circuit consisting only of an inductor connected to the terminals of an ac generator, as in Figure 33.4. Because the induced emf in the inductor is $L\,dI/dt$, Kirchhoff's loop rule applied to the circuit gives

$$v - L\frac{di}{dt} = 0$$

When we rearrange this equation and substitute $V_{max} \sin \omega t$ for v, we get

$$L\frac{di}{dt} = V_{max} \sin \omega t \qquad (33.6)$$

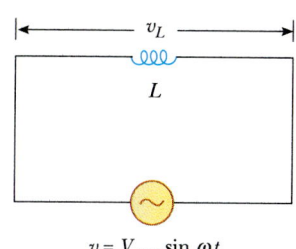

FIGURE 33.4   A circuit consisting of an inductor L connected to an ac generator.

Integrating this expression[2] gives the current as a function of time:

$$i_L = \frac{V_{max}}{L} \int \sin \omega t \, dt = -\frac{V_{max}}{\omega L} \cos \omega t \tag{33.7}$$

When we use the trigonometric identity $\cos \omega t = -\sin(\omega t - \pi/2)$, we can express Equation 33.7 as

$$i_L = \frac{V_{max}}{\omega L} \sin\left(\omega t - \frac{\pi}{2}\right) \tag{33.8}$$

Comparing this result with Equation 33.6 clearly shows that the current and voltage are out of phase with each other by $\pi/2$ rad, or 90°. A plot of voltage and current versus time is given in Figure 33.5a. The voltage reaches its maximum value one quarter of an oscillation period before the current reaches its maximum value. The corresponding phasor diagram for this circuit is shown in Figure 33.5b. Thus, we see that

The current in an inductor lags the voltage by 90°

> for a sinusoidal applied voltage, the current in an inductor always lags behind the voltage across the inductor by 90°.

This lag can be understood by noting that because the voltage across the inductor is proportional to di/dt, the value of v_L is largest when the current is changing most rapidly. Since i versus t is a sinusoidal curve, di/dt (the slope) is maximum when the curve goes through zero. This shows that v_L reaches its maximum value when the current is zero.

From Equation 33.7 we see that the current reaches its maximum values when $\cos \omega t = 1$:

Maximum current in an inductor

$$I_{max} = \frac{V_{max}}{\omega L} = \frac{V_{max}}{X_L} \tag{33.9}$$

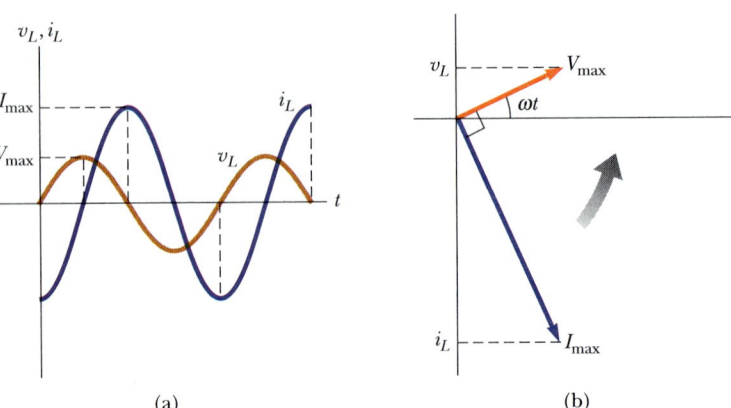

(a)   (b)

FIGURE 33.5   (a) Plots of the current and voltage across an inductor as functions of time. The current lags the voltage by 90°. (b) The phasor diagram for the inductive circuit. The angle between the current phasor and voltage phasor is 90°.

---

[2] The constant of integration is neglected here since it depends on the initial conditions, which are not important for this situation.

where the quantity X_L, called the **inductive reactance**, is

$$X_L = \omega L \qquad (33.10)$$

Inductive reactance

The expression for the rms current is similar to Equation 33.9, with V_{max} replaced by V_{rms}.

Inductive reactance, like resistance, has units of ohms. The maximum current decreases as the inductive reactance increases for a given applied voltage. However, unlike resistance, reactance depends on frequency as well as the characteristics of the inductor. Note that the reactance of an inductor increases as the frequency of the current increases. This is because at higher frequencies, the instantaneous current must change more rapidly than it does at the lower frequencies, which in turn causes an increase in the induced emf associated with a given peak current.

Using Equations 33.6 and 33.9, we find that the instantaneous voltage drop across the inductor is

$$v_L = L\frac{di}{dt} = V_{max} \sin \omega t = I_{max} X_L \sin \omega t \qquad (33.11)$$

We can think of Equation 33.11 as Ohm's law for an inductive circuit. As an exercise, you should show that X_L has the SI unit of ohm.

EXAMPLE 33.2　A Purely Inductive ac Circuit

In a purely inductive ac circuit (Fig. 33.4), $L = 25.0$ mH and the rms voltage is 150 V. Find the inductive reactance and rms current in the circuit if the frequency is 60.0 Hz.

Solution　First, recall from Equation 13.4 that $\omega = 2\pi f = 2\pi(60.0) = 377$ s^{-1}. Equation 33.10 then gives

$$X_L = \omega L = (377 \text{ s}^{-1})(25.0 \times 10^{-3} \text{ H}) = \boxed{9.43 \ \Omega}$$

The rms current is

$$I_{rms} = \frac{V_L}{X_L} = \frac{150 \text{ V}}{9.43 \ \Omega} = \boxed{15.9 \text{ A}}$$

Exercise　Calculate the inductive reactance and rms current in the circuit if the frequency is 6.00 kHz.

Answers　943 Ω, 0.159 A.

33.4　CAPACITORS IN AN AC CIRCUIT

Figure 33.6 shows an ac circuit consisting of a capacitor connected across the terminals of an ac generator. Kirchhoff's loop rule applied to this circuit gives $v - v_C = 0$, or

$$v = v_C = V_{max} \sin \omega t \qquad (33.12)$$

where v_C is the *instantaneous voltage drop across the capacitor*. From the definition of capacitance, $v_C = Q/C$, and this value for v_C substituted into Equation 33.12 gives

$$Q = CV_{max} \sin \omega t \qquad (33.13)$$

Since $i = dQ/dt$, differentiating Equation 33.13 gives the instantaneous current in the circuit:

$$i_C = \frac{dQ}{dt} = \omega CV_{max} \cos \omega t \qquad (33.14)$$

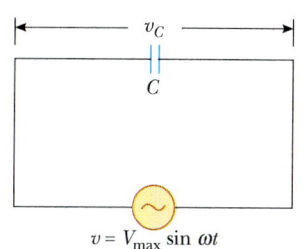

FIGURE 33.6　A circuit consisting of a capacitor C connected to an ac generator.

Here again we see that the current is not in phase with the voltage drop across the capacitor, given by Equation 33.12. Using the trigonometric identity $\cos \omega t = \sin(\omega t + \frac{\pi}{2})$, we can express Equation 33.14 in the alternative form

$$i_C = \omega C V_{max} \sin\left(\omega t + \frac{\pi}{2}\right) \tag{33.15}$$

Comparing this expression with Equation 33.12, we see that the current is $\pi/2$ rad $= 90°$ out of phase with the voltage across the capacitor. A plot of current and voltage versus time (Fig. 33.7a) shows that the current reaches its maximum value one quarter of a cycle sooner than the voltage reaches its maximum value. The corresponding phasor diagram in Figure 33.7b also shows that

The current leads the voltage across a capacitor by 90°

for a sinusoidally applied emf, the current always leads the voltage across a capacitor by 90°.

From Equation 33.14, we see that the current in the circuit reaches its maximum value when $\cos \omega t = 1$:

$$I_{max} = \omega C V_{max} = \frac{V_{max}}{X_C} \tag{33.16}$$

where X_C is called the **capacitive reactance**:

Capacitive reactance

$$X_C = \frac{1}{\omega C} \tag{33.17}$$

The SI unit of X_C is also the ohm. The rms current is given by an expression similar to Equation 33.16, with V_{max} replaced by V_{rms}.

Combining Equations 33.12 and 33.16, we can express the instantaneous voltage drop across the capacitor as

$$v_C = V_{max} \sin \omega t = I_{max} X_C \sin \omega t \tag{33.18}$$

As the frequency of the circuit increases, the maximum current increases but the

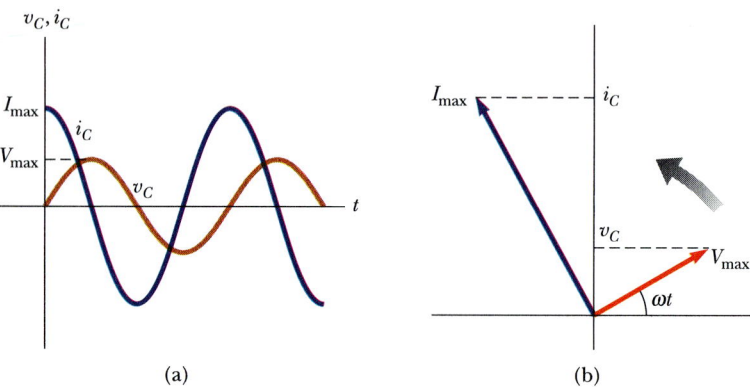

(a)   (b)

FIGURE 33.7   (a) Plots of the current and voltage across the capacitor as functions of time. The voltage lags behind the current by 90°. (b) Phasor diagram for the purely capacitive circuit. Projections of the phasors onto the vertical axis gives the instantaneous values v_C and i_C.

reactance decreases. For a given maximum applied voltage V_{max}, the current increases as the frequency increases. As the frequency approaches zero, the capacitive reactance approaches infinity and so the current approaches zero. This makes sense because the circuit approaches dc conditions as $\omega \to 0$. Of course, no current passes through a capacitor under steady-state dc conditions.

CONCEPTUAL EXAMPLE 33.3

Explain why the reactance of a capacitor decreases with increasing frequency, while the reactance of an inductor increases with increasing frequency.

Reasoning   As the frequency of a capacitive circuit increases, the polarities of the charged plates must change more rapidly with time, corresponding to a larger current. The capacitive reactance varies as the inverse of the frequency, and hence approaches zero as f approaches infinity.

The current is zero in a dc capacitive circuit, which corresponds to zero frequency and infinite reactance. The inductive reactance is proportional to the frequency and, therefore, increases with increasing frequency. At the higher frequencies, the current changes more rapidly, which according to Faraday's law results in an increase in the back emf associated with an inductor and a corresponding decrease in current.

EXAMPLE 33.4   A Purely Capacitive ac Circuit

An 8.00-μF capacitor is connected to the terminals of a 60.0-Hz ac generator whose rms voltage is 150 V. Find the capacitive reactance and the rms current in the circuit.

Solution   Using the Equation 33.17 and the fact that $\omega = 2\pi f = 377$ s^{-1} (Eq. 13.4) gives

$$X_C = \frac{1}{\omega C} = \frac{1}{(377\ \text{s}^{-1})(8.00 \times 10^{-6}\ \text{F})} = \boxed{332\ \Omega}$$

Hence, the rms current is

$$I_{rms} = \frac{V_{rms}}{X_C} = \frac{150\ \text{V}}{332\ \Omega} = \boxed{0.452\ \text{A}}$$

Exercise   If the frequency is doubled, what happens to the capacitive reactance and the current?

Answer X_C halved, I_{max} doubled.

33.5   THE *RLC* SERIES CIRCUIT

Figure 33.8a shows a circuit containing a resistor, an inductor, and a capacitor connected in series across an ac-voltage source. As before, we assume that the applied voltage varies sinusoidally with time. It is convenient to assume that the applied voltage is given by

$$v = V_{max} \sin \omega t$$

while the current varies as

$$i = I_{max} \sin(\omega t - \phi)$$

where ϕ is the **phase angle** between the current and the applied voltage. Our aim is to determine ϕ and I_{max}. Figure 33.8b shows the voltage versus time across each element in the circuit and their phase relationships.

In order to solve this problem, we must analyze the phasor diagram for this circuit. First, note that because the elements are in series, the current everywhere in the circuit must be the same at any instant. That is, *the current at all points in a series ac circuit has the same amplitude and phase*. Therefore, as we found in the previous sections, the voltage across each element has different amplitudes and

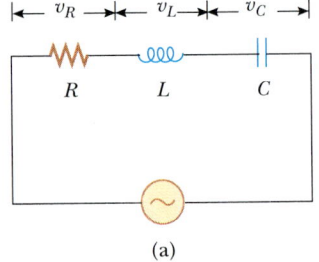

(a)

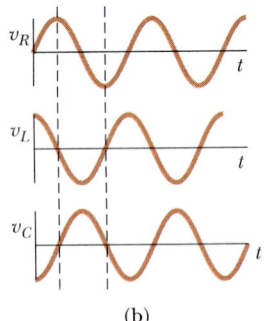

(b)

FIGURE 33.8 (a) A series circuit consisting of a resistor, an inductor, and a capacitor connected to an ac generator. (b) Phase relationships in the series *RLC* circuit.

phases, as summarized in Figure 33.9. In particular, the voltage across the resistor is in phase with the current, the voltage across the inductor leads the current by 90°, and the voltage across the capacitor lags behind the current by 90°. Using these phase relationships, we can express the instantaneous voltage drops across the three elements as

$$v_R = I_{max} R \sin \omega t = V_R \sin \omega t \tag{33.19}$$

$$v_L = I_{max} X_L \sin\left(\omega t + \frac{\pi}{2}\right) = V_L \cos \omega t \tag{33.20}$$

$$v_C = I_{max} X_C \sin\left(\omega t - \frac{\pi}{2}\right) = -V_C \cos \omega t \tag{33.21}$$

where V_R, V_L, and V_C are the voltage amplitudes across each element:

$$V_R = I_{max} R \qquad V_L = I_{max} X_L \qquad V_C = I_{max} X_C$$

At this point, we could proceed by noting that the instantaneous voltage v across the three elements equals the sum

$$v = v_R + v_L + v_C$$

Although this analytical approach is correct, it is simpler to obtain the sum by examining the phasor diagram.

Because the current in each element is the same at any instant, we can obtain the resulting phasor diagram by combining the three phasor pairs shown in Figure 33.9 to obtain Figure 33.10a, where a single phasor I_{max} is used to represent the current in each element. To obtain the vector sum of these voltages, it is convenient to redraw the phasor diagram as in Figure 33.10b. From this diagram, we see that the vector sum of the voltage amplitudes V_R, V_L, and V_C equals a phasor whose length is the maximum applied voltage, V_{max}, where the phasor V_{max} makes an angle ϕ with the current phasor, I_{max}. Note that the voltage phasors V_L and V_C are in opposite directions along the same line, and hence we are able to construct the difference phasor $V_L - V_C$, which is perpendicular to the phasor V_R. From the right triangle in Figure 33.10b, we see that

$$V_{max} = \sqrt{V_R^2 + (V_L - V_C)^2} = \sqrt{(I_{max} R)^2 + (I_{max} X_L - I_{max} X_C)^2}$$

$$V_{max} = I_{max} \sqrt{R^2 + (X_L - X_C)^2} \tag{33.22}$$

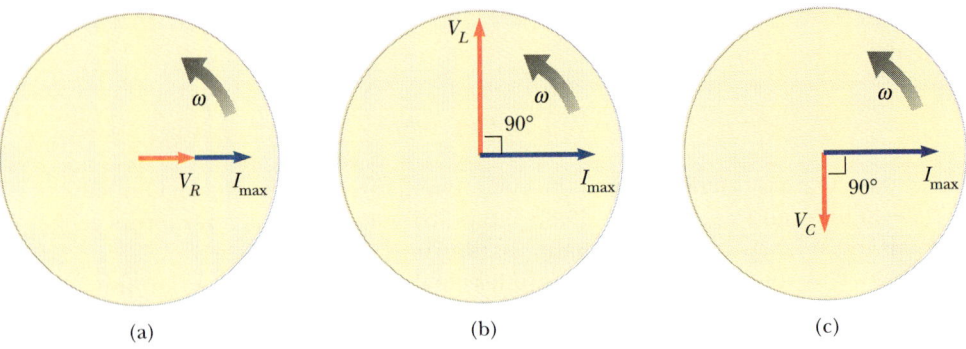

(a)   (b)   (c)

FIGURE 33.9 Phase relationships between the peak voltage and current phasors for (a) a resistor, (b) an inductor, and (c) a capacitor.

Therefore, we can express the maximum current as

$$I_{\max} = \frac{V_{\max}}{\sqrt{R^2 + (X_L - X_C)^2}}$$

The **impedance** Z of the circuit is defined to be

$$Z \equiv \sqrt{R^2 + (X_L - X_C)^2} \qquad (33.23)$$

where impedance also has the SI unit of ohm. Therefore, we can write Equation 33.22 in the form

$$V_{\max} = I_{\max} Z \qquad (33.24)$$

We can regard Equation 33.24 as a generalized Ohm's law applied to an ac circuit. Note that the current in the circuit depends upon the resistance, the inductance, the capacitance, and the frequency since the reactances are frequency dependent.

By removing the common factor $I_{\max}$ from each phasor in Figure 33.10, we can also construct an impedance triangle, shown in Figure 33.11. From this phasor diagram, we find that the phase angle ϕ between the current and voltage is

$$\tan \phi = \frac{X_L - X_C}{R} \qquad (33.25)$$

For example, when $X_L > X_C$ (which occurs at high frequencies), the phase angle is positive, signifying that the current lags behind the applied voltage, as in Figure 33.10. On the other hand, if $X_L < X_C$, the phase angle is negative, signifying that the current leads the applied voltage. Finally, when $X_L = X_C$, the phase angle is zero. In this case, the ac impedance equals the resistance and the current has its maximum value, given by $V_{\max}/R$. The frequency at which this occurs is called the *resonance frequency,* and is described further in Section 33.7.

Figure 33.12 gives impedance values and phase angles for various series circuits containing different combinations of circuit elements.

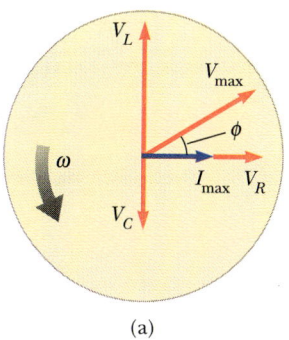

(a)

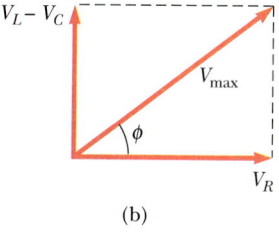

(b)

FIGURE 33.10   (a) The phasor diagram for the series *RLC* circuit shown in Figure 33.8. Note that the phasor V_R is in phase with the current phasor $I_{\max}$, the phasor V_L leads $I_{\max}$ by 90°, and the phasor V_C lags $I_{\max}$ by 90°. The total voltage $V_{\max}$ makes an angle ϕ with $I_{\max}$. (b) Simplified version of the phasor diagram shown in part (a).

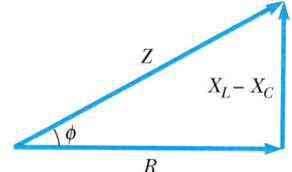

FIGURE 33.11   The impedance triangle for a series *RLC* circuit gives the relationship $Z = \sqrt{R^2 + (X_L - X_C)^2}$.

Circuit Elements	Impedance, Z	Phase angle, ϕ
R ⌇⌇⌇	R	$0°$
C ⊣⊢	X_C	$-90°$
L ⌿⌿⌿	X_L	$+90°$
R C	$\sqrt{R^2 + X_C^2}$	Negative, between $-90°$ and $0°$
R L	$\sqrt{R^2 + X_L^2}$	Positive, between $0°$ and $90°$
R L C	$\sqrt{R^2 + (X_L - X_C)^2}$	Negative if $X_C > X_L$ Positive if $X_C < X_L$

FIGURE 33.12   The impedance values and phase angles for various circuit element combinations. In each case, an ac voltage (not shown) is applied across the combination of elements (that is, across the dots).

Oscilloscope Simulator

The oscilloscope is a principal laboratory instrument for measuring and viewing electrical phenomena. It is especially important for electrical engineering students or anyone who intends to work in areas involving electronics. The oscilloscope simulator duplicates the behavior and appearance of an oscilloscope. The simulator offers two modes of operation. In the tutorial mode, you can click on a component to view a message describing how that component operates. In the second mode of operation, the experimental mode, you can select different input signals and observe the signal as it would appear on an oscilloscope screen.

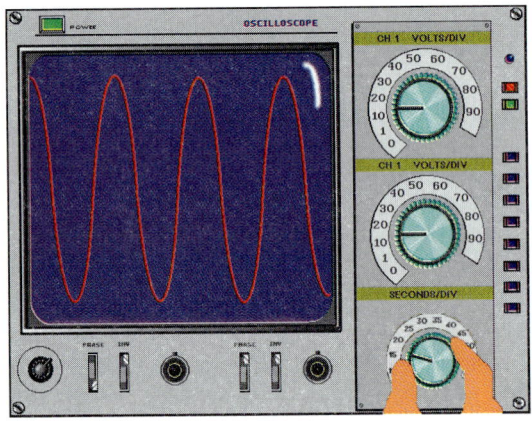

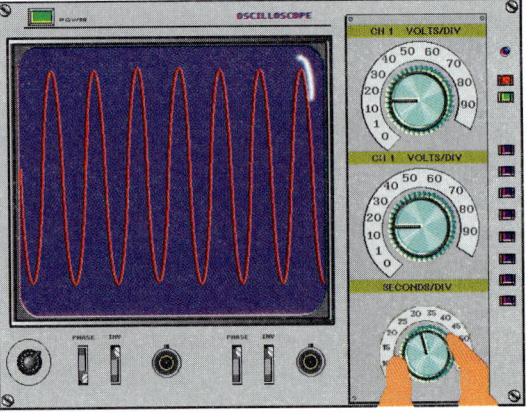

EXAMPLE 33.5 Analyzing a Series *RLC* Circuit

Analyze a series *RLC* ac circuit for which $R = 250\ \Omega$, $L = 0.600$ H, $C = 3.50\ \mu$F, $\omega = 377$ s^{-1}, and $V_{max} = 150$ V.

Solution The reactances are $X_L = \omega L = 226\ \Omega$ and $X_C = 1/\omega C = 758\ \Omega$. Therefore, the impedance is

$$Z = \sqrt{R^2 + (X_L - X_C)^2}$$
$$= \sqrt{(250\ \Omega)^2 + (226\ \Omega - 758\ \Omega)^2} = 588\ \Omega$$

The maximum current is

$$I_{max} = \frac{V_{max}}{Z} = \frac{150\ \text{V}}{588\ \Omega} = 0.255\ \text{A}$$

The phase angle between the current and voltage is

$$\phi = \tan^{-1}\left(\frac{X_L - X_C}{R}\right) = \tan^{-1}\left(\frac{226 - 758}{250}\right) = -64.8°$$

Since the circuit is more capacitive than inductive, ϕ is negative and the current leads the applied voltage.

The maximum voltages across each element are given by

$$V_R = I_{max}R = (0.255\ \text{A})(250\ \Omega) = 63.8\ \text{V}$$

$$V_L = I_{max}X_L = (0.255\ \text{A})(226\ \Omega) = 57.6\ \text{V}$$

$$V_C = I_{max}X_C = (0.255\ \text{A})(758\ \Omega) = 193\ \text{V}$$

Using Equations 33.19, 33.20, and 33.21, we find that the instantaneous voltages across the three elements can be written

$$v_R = (63.8\ \text{V}) \sin 377t$$

$$v_L = (57.6\ \text{V}) \cos 377t$$

$$v_C = (-193\ \text{V}) \cos 377t$$

and the applied voltage is $v = 150 \sin(\omega t - 64.8°)$. The sum of the maximum voltages across each element is $V_R + V_L + V_C = 314$ V, which is much larger than the maximum voltage of the generator, 150 V. The former is a meaningless quantity. This is because when harmonically varying quantities are added, *both their amplitudes and their phases* must be taken into account and we know that the peak voltages across the different circuit elements occur at different times. That is, the voltages must be added in a way that takes account of the different phases. When this is done, Equation 33.22 is satisfied. You should verify this result.

EXAMPLE 33.6   **Finding *L* from a Phasor Diagram**

In a series *RLC* circuit, the applied voltage has a maximum value of 120 V and oscillates at a frequency of 60.0 Hz. The circuit contains an inductor whose inductance can be varied, $R = 800\ \Omega$, and $C = 4.00\ \mu\text{F}$. Determine the value of *L* such that the voltage across the capacitor is out of phase with the applied voltage by 30.0°, with V_{max} leading V_C.

Solution   The phase relationships for the voltage drops across the elements in the circuit are shown in Figure 33.13. From the figure, we see that the phase angle is 60.0°. This is because the phasors representing I_{max} and V_R are in the same direction (they are in phase). From Equation 33.25, we find that

$$X_L = X_C + R \tan \phi \qquad (1)$$

Substituting Equations 33.10 and 33.17 into (1) gives

$$2\pi f L = \frac{1}{2\pi fC} + R \tan \phi$$

or

$$L = \frac{1}{2\pi f}\left[\frac{1}{2\pi fC} + R \tan \phi\right] \qquad (2)$$

Substituting the given values into (2) gives $L = 5.44$ H.

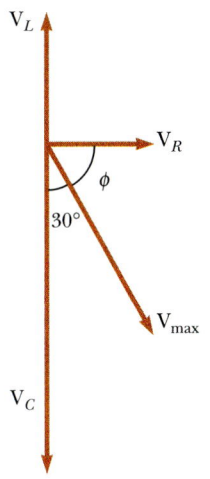

FIGURE 33.13   (Example 33.6).

33.6   POWER IN AN AC CIRCUIT

As we see in this section, *there are no power losses associated with pure capacitors and pure inductors in an ac circuit.* (A pure inductor is defined as one having no resistance or capacitance.) First, let us analyze the power dissipated in an ac circuit containing only a generator and a capacitor.

When the current begins to increase in one direction in an ac circuit, charge begins to accumulate on the capacitor and a voltage drop appears across it. When this voltage drop reaches its peak value, the energy stored in the capacitor is $\frac{1}{2}CV_{\text{max}}^2$. However, this energy storage is only momentary. The capacitor is charged and discharged twice during each cycle. In this process, charge is delivered to the capacitor during two quarters of the cycle and is returned to the voltage source during the remaining two quarters. Therefore, *the average power supplied by the source is zero.* In other words, *a capacitor in an ac circuit does not dissipate energy.*

Similarly, the voltage source must do work against the back emf of the inductor, which carries a current. When the current reaches its peak value, the energy stored in the inductor is a maximum and is given by $\frac{1}{2}LI_{\text{max}}^2$. When the current begins to decrease in the circuit, this stored energy is returned to the source as the inductor attempts to maintain the current in the circuit.

When we studied dc circuits in Chapter 28, we found that the power delivered by a battery to a circuit is equal to the product of the current and the emf of the battery. Likewise, the instantaneous power delivered by an ac generator to any circuit is the product of the generator current and the applied voltage. For the *RLC* circuit shown in Figure 33.8, we can express the instantaneous power *P* as

$$P = iv = I_{\text{max}} \sin(\omega t - \phi)\, V_{\text{max}} \sin \omega t$$
$$= I_{\text{max}} V_{\text{max}} \sin \omega t \sin(\omega t - \phi) \qquad (33.26)$$

Clearly this result is a complicated function of time and therefore not very useful from a practical viewpoint. What is generally of interest is the average power over one or more cycles. Such an average can be computed by first using the trigonometric identity $\sin(\omega t - \phi) = \sin \omega t \cos \phi - \cos \omega t \sin \phi$. Substituting this into Equation 33.26 gives

$$P = I_{max} V_{max} \sin^2 \omega t \cos \phi - I_{max} V_{max} \sin \omega t \cos \omega t \sin \phi \qquad (33.27)$$

We now take the time average of P over one or more cycles, noting that I_{max}, V_{max}, ϕ, and ω are all constants. The time average of the first term on the right of Equation 33.27 involves the average value of $\sin^2 \omega t$, which is $\frac{1}{2}$, as shown in footnote 1. The time average of the second term on the right is identically zero because $\sin \omega t \cos \omega t = \frac{1}{2} \sin 2\omega t$, and the average value of $\sin 2\omega t$ is zero. Therefore, we can express the **average power** P_{av} as

$$P_{av} = \tfrac{1}{2} I_{max} V_{max} \cos \phi \qquad (33.28)$$

It is convenient to express the average power in terms of the rms current and rms voltage defined by Equations 33.4 and 33.5. Using these defined quantities, the average power becomes

$$P_{av} = I_{rms} V_{rms} \cos \phi \qquad (33.29)$$

where the quantity $\cos \phi$ is called the **power factor.** By inspecting Figure 33.10, we see that the maximum voltage drop across the resistor is given by $V_R = V_{max} \cos \phi = I_{max} R$. Using Equation 33.3 and the fact that $\cos \phi = I_{max} R / V_{max}$, we find that P_{av} can be expressed as

$$P_{av} = I_{rms} V_{rms} \cos \phi = I_{rms} \left(\frac{V_{max}}{\sqrt{2}} \right) \frac{I_{max} R}{V_{max}} = I_{rms} \frac{I_{max} R}{\sqrt{2}}$$

$$P_{av} = I_{rms}^2 R \qquad (33.30)$$

In other words, the *average power delivered by the generator is dissipated as heat in the resistor,* just as in the case of a dc circuit. *There is no power loss in an ideal inductor or capacitor.* When the load is purely resistive, then $\phi = 0$, $\cos \phi = 1$, and from Equation 33.29 we see that $P_{av} = I_{rms} V_{rms}$.

EXAMPLE 33.7   Average Power in a *RLC* Series Circuit

Calculate the average power delivered to the series *RLC* circuit described in Example 33.5.

Solution   First, let us calculate the rms voltage and rms current:

$$V_{rms} = \frac{V_{max}}{\sqrt{2}} = \frac{150 \text{ V}}{\sqrt{2}} = 106 \text{ V}$$

$$I_{rms} = \frac{I_{max}}{\sqrt{2}} = \frac{V_{max}/Z}{\sqrt{2}} = \frac{0.255 \text{ A}}{\sqrt{2}} = 0.180 \text{ A}$$

Since $\phi = -64.8°$, the power factor, $\cos \phi$, is 0.426, and hence the average power is

$$P_{av} = I_{rms} V_{rms} \cos \phi = (0.180 \text{ A})(106 \text{ V})(0.426)$$

$$= 8.13 \text{ W}$$

The same result can be obtained using Equation 33.30.

33.7   RESONANCE IN A SERIES *RLC* CIRCUIT

A series *RLC* circuit is said to be in resonance when the current has its maximum value. In general, the rms current can be written

$$I_{rms} = \frac{V_{rms}}{Z} \tag{33.31}$$

where *Z* is the impedance. Substituting Equation 33.23 into 33.31 gives

$$I_{rms} = \frac{V_{rms}}{\sqrt{R^2 + (X_L - X_C)^2}} \tag{33.32}$$

Because the impedance depends on the frequency of the source, we see that the current in the *RLC* circuit also depends on the frequency. Note that the current reaches its peak when $X_L = X_C$, corresponding to $Z = R$. The frequency ω_0 at which this occurs is called the **resonance frequency** of the circuit. To find ω_0, we use the condition $X_L = X_C$, from which we get $\omega_0 L = 1/\omega_0 C$, or

$$\omega_0 = \frac{1}{\sqrt{LC}} \tag{33.33}$$

Resonance frequency

Note that this frequency also corresponds to the natural frequency of oscillation of an *LC* circuit (Section 32.5). Therefore, the current in a series *RLC* circuit reaches its maximum value when the frequency of the applied voltage matches the natural oscillator frequency, which depends only on *L* and *C*. Furthermore, at this frequency the current is in phase with the applied voltage.

A plot of the rms current versus frequency for a series *RLC* circuit is shown in Figure 33.14a. The data that are plotted assume a constant rms voltage of 5.0 mV,

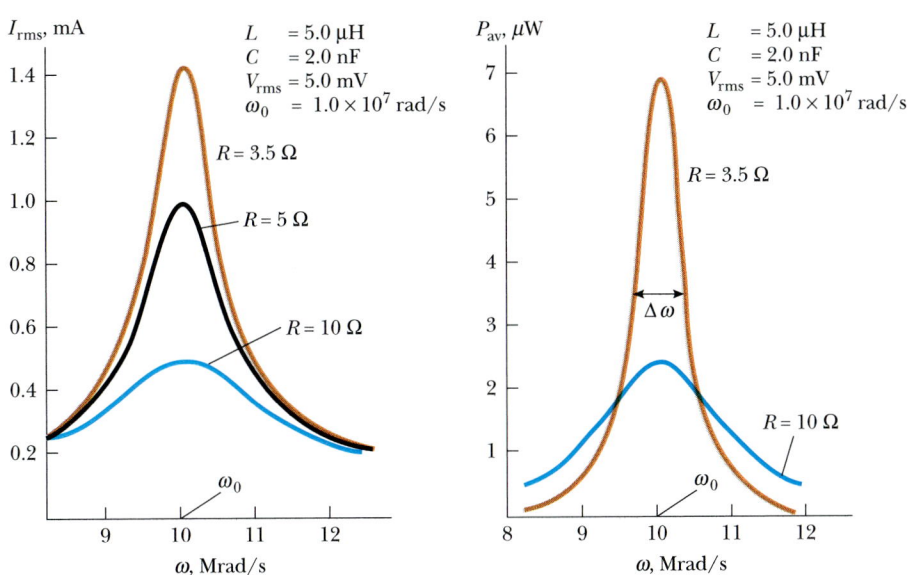

FIGURE 33.14   (a) Plots of the rms current versus frequency for a series *RLC* circuit for three values of *R*. Note that the current reaches its peak value at the resonance frequency ω_0. (b) Plots of the average power versus frequency for the series *RLC* circuit for two values of *R*.

$L = 5.0 \ \mu\text{H}$, and $C = 2.0$ nF. The three curves correspond to three values of R. Note that in each case, the current reaches its maximum value at the resonance frequency, ω_0. Furthermore, the curves become narrower and taller as the resistance decreases.

By inspecting Equation 33.32, it must be concluded that, when $R = 0$, the current would become infinite at resonance. Although the equation predicts this, real circuits always have some resistance, which limits the value of the current.

Mechanical systems can also exhibit resonances. For example, when an undamped mass-spring system is driven at its natural frequency of oscillation, its amplitude increases with time, as we discussed in Chapter 13. Large-amplitude mechanical vibrations can be disastrous, as in the case of the Tacoma Narrows Bridge collapse.

It is also interesting to calculate the average power as a function of frequency for a series *RLC* circuit. Using Equations 33.30 and 33.31, we find that

$$P_{av} = I_{rms}^2 R = \frac{V_{rms}^2}{Z^2} R = \frac{V_{rms}^2 R}{R^2 + (X_L - X_C)^2} \tag{33.34}$$

Since $X_L = \omega L$, $X_C = 1/\omega C$, and $\omega_0^2 = 1/LC$, the factor $(X_L - X_C)^2$ can be expressed as

$$(X_L - X_C)^2 = \left(\omega L - \frac{1}{\omega C} \right)^2 = \frac{L^2}{\omega^2}(\omega^2 - \omega_0^2)^2$$

Using this result in Equation 33.34 gives

Power in an *RLC* circuit

$$P_{av} = \frac{V_{rms}^2 R \omega^2}{R^2 \omega^2 + L^2(\omega^2 - \omega_0^2)^2} \tag{33.35}$$

This expression shows that at resonance, when $\omega = \omega_0$, the *average power is a maximum* and has the value V_{rms}^2 / R. A plot of average power versus frequency is shown in Figure 33.14b for the series *RLC* circuit described in Figure 33.14a, taking $R = 3.5 \ \Omega$ and $R = 10 \ \Omega$. As the resistance is made smaller, the curve becomes sharper in the vicinity of the resonance frequency. The sharpness of the curve is usually described by a dimensionless parameter known as the **quality factor**, denoted by Q_0 (not to be confused with the symbol for charge)[3]:

$$Q_0 = \frac{\omega_0}{\Delta \omega}$$

where $\Delta \omega$ is the width of the curve measured between the two values of ω for which P_{av} has half its maximum value (half-power points, see Fig. 33.14b). It is left as a problem (Problem 87) to show that the width at the half-power points has the value $\Delta \omega = R/L$, so that

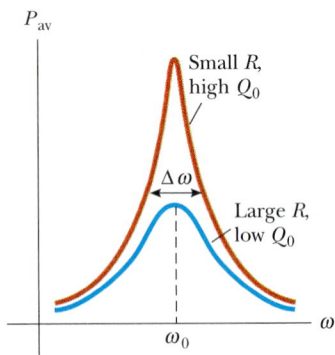

$$Q_0 = \frac{\omega_0 L}{R} \tag{33.36}$$

FIGURE 33.15 Plots of the average power versus frequency for a series *RLC* circuit (see Eq. 33.35). The upper, narrow curve is for a small value of *B*, and the lower, broad curve is for a large value of *R*. The width $\Delta \omega$ of each curve is measured between points where the power is half its maximum value. The power is a maximum at the resonance frequency, ω_0.

That is, Q_0 is equal to the ratio of the inductive reactance to the resistance evaluated at the resonance frequency, ω_0.

The curves plotted in Figure 33.15 show that a high-Q_0 circuit responds to a very narrow range of frequencies, whereas a low-Q_0 circuit responds to a much

---

[3] The quality factor is also defined as the ratio $2\pi E/\Delta E$, where E is the energy stored in the oscillating system and ΔE is the energy lost per cycle of oscillation. The quality factor for a mechanical system such as a damped oscillator can also be defined.

broader range of frequencies. Typical values of Q_0 in electronic circuits range from 10 to 100.

The receiving circuit of a radio is an important application of a resonant circuit. The radio is tuned to a particular station (which transmits a specific radio-frequency signal) by varying a capacitor, which changes the resonant frequency of the receiving circuit. When the resonance frequency of the circuit matches that of the incoming radio wave, the current in the receiving circuit increases. This signal is then amplified and fed to a speaker. Since many signals are often present over a range of frequencies, it is important to design a high-Q_0 circuit in order to eliminate unwanted signals. In this manner, stations whose frequencies are near but not at the resonance frequency give negligibly small signals at the receiver relative to the one that matches the resonance frequency.

EXAMPLE 33.8   A Resonating Series *RLC* Circuit

Consider a series *RLC* circuit for which $R = 150\ \Omega$, $L = 20.0$ mH, $V_{rms} = 20.0$ V, and $\omega = 5000\ \text{s}^{-1}$. Determine the value of the capacitance for which the current has its peak value.

$$\omega_0 = 5.00 \times 10^3\ \text{s}^{-1} = \frac{1}{\sqrt{LC}}$$

$$C = \frac{1}{\omega_0^2 L} = \frac{1}{(25.0 \times 10^6\ \text{s}^{-2})(2.00 \times 10^{-3}\ \text{H})} = \boxed{2.00\ \mu\text{F}}$$

Solution   The current has its peak value at the resonance frequency ω_0, which should be made to match the "driving" frequency of $5000\ \text{s}^{-1}$ in this problem:

Exercise   Calculate the maximum value of the rms current in the circuit.

Answer   0.133 A.

---

*33.8   FILTER CIRCUITS

A filter circuit is used to smooth out or eliminate a time-varying signal. For example, radios are usually powered by a 60-Hz ac voltage, which is converted to dc using a rectifier circuit. After rectification, however, the voltage still contains a small ac component at 60 Hz (sometimes called ripple), which must be filtered. By "filter," we mean that the 60-Hz ripple must be reduced to a value much smaller than the audio signal to be amplified, because without filtering, the resulting audio signal includes an annoying hum at 60 Hz.

First, consider the simple series *RC* circuit shown in Figure 33.16a. The input

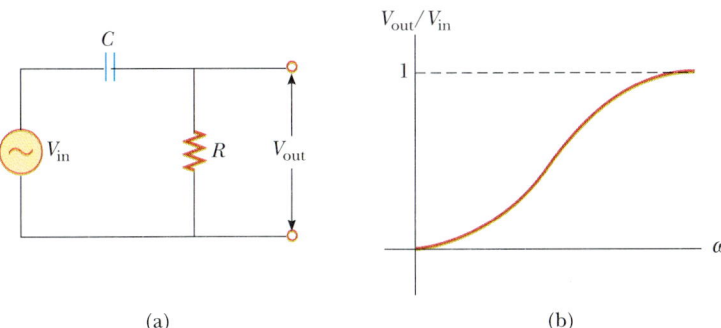

(a)   (b)

FIGURE 33.16   (a) A simple *RC* high-pass filter. (b) Ratio of the output voltage to the input voltage for an *RC* high-pass filter.

voltage is across the two elements and is represented by $V_{max} \sin \omega t$. Since we are interested only in maximum values, we can use Equation 33.24, which shows that the maximum input voltage is related to the maximum current by

$$V_{in} = I_{max} Z = I_{max} \sqrt{R^2 + \left(\frac{1}{\omega C}\right)^2}$$

If the voltage across the resistor is considered to be the output voltage, V_{out}, then from Ohm's law the maximum output voltage is

$$V_{out} = I_{max} R$$

Therefore, the ratio of the output voltage to the input voltage is

High-pass filter

$$\frac{V_{out}}{V_{in}} = \frac{R}{\sqrt{R^2 + \left(\frac{1}{\omega C}\right)^2}} \qquad (33.37)$$

A plot of Equation 33.37, given in Figure 33.16b, shows that at low frequencies, V_{out} is small compared with V_{in}, whereas at high frequencies the two voltages are equal. Since the circuit preferentially passes signals of higher frequency while low frequencies are filtered (or attenuated), the circuit is called an RC high-pass filter. Physically, a high-pass filter is a result of the "blocking action" of a capacitor to direct current or low frequencies.

Now consider the RC series circuit shown in Figure 33.17a, where the output voltage is taken across the capacitor. In this case, the maximum voltage equals the voltage across the capacitor. Because the impedance across the capacitor is given by $X_C = 1/\omega C$,

$$V_{out} = I_{max} X_C = \frac{I_{max}}{\omega C}$$

Therefore, the ratio of the output voltage to the input voltage is

Low-pass filter

$$\frac{V_{out}}{V_{in}} = \frac{1/\omega C}{\sqrt{R^2 + \left(\frac{1}{\omega C}\right)^2}} \qquad (33.38)$$

This ratio, plotted in Figure 33.17b, shows that in this case the circuit preferentially passes signals of low frequency. Hence, the circuit is called an RC low-pass filter. Filters can be designed to block, or pass, a narrow band of frequencies.

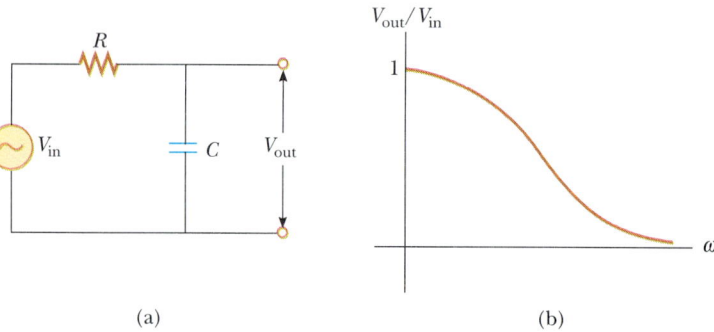

(a)   (b)

FIGURE 33.17   (a) A simple RC low-pass filter. (b) Ratio of the output voltage to the input voltage for an RC low-pass filter.

*33.9   THE TRANSFORMER AND POWER TRANSMISSION

When electrical power is transmitted over large distances, it is economical to use a high voltage and low current to minimize the I^2R heating loss in the transmission lines. For this reason, 350-kV lines are common, and in many areas even higher-voltage (765 kV) lines are under construction. Such high-voltage transmission systems have met with considerable public resistance because of the potential safety and environmental problems they pose. At the receiving end of such lines, the consumer requires power at a low voltage and high current (for safety and efficiency in design) to operate such things as appliances and motor-driven machines. Therefore, a device is required that can increase (or decrease) the ac voltage and current without causing appreciable changes in the power delivered. The ac transformer is the device used for this purpose.

In its simplest form, the ac transformer consists of two coils of wire wound around a core of soft iron as in Figure 33.18. The coil on the left, which is connected to the input ac voltage source and has N_1 turns, is called the primary winding (or primary). The coil on the right, consisting of N_2 turns and connected to a load resistor R, is called the secondary. The purpose of the common iron core is to increase the magnetic flux and to provide a medium in which nearly all the flux through one coil passes through the other coil. Eddy current losses are reduced by using a laminated iron core. Soft iron is used as the core material to reduce hysteresis losses. Joule heat losses caused by the finite resistance of the coil wires are usually quite small. Typical transformers have power efficiencies ranging from 90% to 99%. In what follows, we assume an *ideal transformer,* one in which the energy losses in the windings and core are zero.

First, let us consider what happens in the primary circuit when the switch in the secondary circuit of Figure 33.18 is open. If we assume that the resistance of the primary coil is negligible relative to its inductive reactance, then the primary circuit is equivalent to a simple circuit consisting of an inductor connected to an ac generator (described in Section 33.3). Since the current is 90° out of phase with the voltage, the power factor, $\cos \phi$, is zero, and hence the average power delivered from the generator to the primary circuit is zero. Faraday's law tells us that the voltage V_1 across the primary coil is

$$V_1 = - N_1 \frac{d\Phi_B}{dt} \tag{33.39}$$

where Φ_B is the magnetic flux through each turn. If we assume that no flux leaks out of the iron core, then the flux through each turn of the primary equals the flux through each turn of the secondary. Hence, the voltage across the secondary coil is

$$V_2 = - N_2 \frac{d\Phi_B}{dt} \tag{33.40}$$

Since $d\Phi_B/dt$ is common to Equations 33.39 and 33.40, we find that

$$V_2 = \frac{N_2}{N_1} V_1 \tag{33.41}$$

When $N_2 > N_1$, the output voltage V_2 exceeds the input voltage V_1. This setup is referred to as a step-up transformer. When $N_2 < N_1$, the output voltage is less than the input voltage, and we speak of a step-down transformer.

When the switch in the secondary circuit is closed, a current I_2 is induced in the secondary. If the load in the secondary circuit is a pure resistance, the induced current is in phase with the induced voltage. The power supplied to the secondary

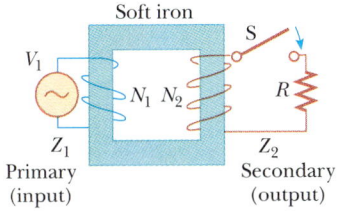

FIGURE 33.18   An ideal transformer consists of two coils wound on the same soft iron core. An ac voltage V_1 is applied to the primary coil, and the output voltage V_2 is across the load resistance R.

Nikola Tesla (1856–1943) was born in Croatia but spent most of his professional life as an inventor in the United States. He was a key figure in the development of alternating-current electricity, high-voltage transformers, and the transport of electrical power using ac transmission lines. Tesla's viewpoint was at odds with the ideas of Edison, who committed himself to the use of direct current in power transmission. Tesla's ac approach won out. *(UPI/Bettmann)*

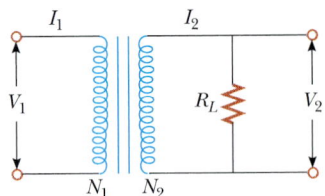

FIGURE 33.19 Conventional circuit diagram for a transformer.

circuit must be provided by the ac generator connected to the primary circuit, as in Figure 33.19. In an ideal transformer, the power supplied by the generator, $I_1 V_1$, is equal to the power in the secondary circuit, $I_2 V_2$. That is,

$$I_1 V_1 = I_2 V_2 \tag{33.42}$$

Clearly, the value of the load resistance R determines the value of the secondary current, since $I_2 = V_2/R$. Furthermore, the current in the primary is $I_1 = V_1/R_{eq}$, where

$$R_{eq} = \left(\frac{N_1}{N_2}\right)^2 R \tag{33.43}$$

is the equivalent resistance of the load resistance when viewed from the primary side. From this analysis, we see that a transformer may be used to match resistances between the primary circuit and the load. In this manner, maximum power transfer can be achieved between a given power source and the load resistance.

We can now understand why transformers are useful for transmitting power over long distances. Because the generator voltage is stepped up, the current in the transmission line is reduced, thereby reducing $I^2 R$ losses. In practice, the voltage is stepped up to around 230 000 V at the generating station, then stepped down to around 20 000 V at a distributing station, and finally stepped down to 120–220 V at the customer's utility poles. The power is supplied by a three-wire cable. In the United States, two of these wires are "hot," with voltages of 120 V with respect to a common ground wire. Home appliances operating on 120 V are connected in parallel between one of the hot wires and ground. Larger appliances, such as electric stoves and clothes dryers, require 220 V. This is obtained across the two hot wires, which are 180° out of phase so that the voltage difference between them is 220 V.

There is a practical upper limit to the voltages that can be used in transmission lines. Excessive voltages could ionize the air surrounding the transmission lines, which could result in a conducting path to ground or to other objects in the vicinity. This, of course, would present a serious hazard to any living creatures. For this reason, a long string of insulators is used to keep high-voltage wires away from their supporting metal towers. Other insulators are used to maintain separation between wires.

EXAMPLE 33.9 A Step-Up Transformer

A generator produces 10 A (rms) of current at 400 V. The voltage is stepped up to 4500 V by an ideal transformer and transmitted a long distance through a power line of total resistance 30 Ω. (a) Determine the percentage of power lost when the voltage is stepped up.

Solution Using Equation 33.42 for an ideal transformer, we find that the current in the transmission line is

$$I_2 = \frac{I_1 V_1}{V_2} = \frac{(10 \text{ A})(400 \text{ V})}{4500 \text{ V}} = 0.89 \text{ A}$$

Hence, the power lost in the transmission line is

$$P_{lost} = I_2^2 R = (0.89 \text{ A})^2 (30 \text{ Ω}) = 24 \text{ W}$$

Since the output power of the generator is $P = IV = (10 \text{ A})(400 \text{ V}) = 4000$ W, the percentage of power lost is

$$\% \text{ power lost} = \left(\frac{24}{4000}\right) \times 100 = \boxed{0.60\%}$$

(b) What percentage of the original power would be lost in the transmission line if the voltage were not stepped up?

Solution If the voltage were not stepped up, the current in the transmission line would be 10 A and the power lost in the line would be $I^2 R = (10 \text{ A})^2 (30 \text{ } \Omega) = 3000$ W. Hence, the percentage of power lost would be

$$\% \text{ power lost} = \left(\frac{3000}{4000} \right) \times 100 = \boxed{75\%}$$

This example illustrates the advantage of high-voltage transmission lines.

Exercise If the transmission line is cooled so that the resistance is reduced to 5.0 Ω, how much power will be lost in the line if it carries a current of 0.89 A?

Answer 4.0 W.

SUMMARY

If an ac circuit consists of a generator and a resistor, the current in the circuit is in phase with the voltage. That is, the current and voltage reach their peak values at the same time.

The **rms current** and **rms voltage** in an ac circuit in which the voltages and current vary sinusoidally are given by

$$I_{rms} = \frac{I_{max}}{\sqrt{2}} = 0.707 I_{max} \qquad (33.4)$$

$$V_{rms} = \frac{V_{max}}{\sqrt{2}} = 0.707 V_{max} \qquad (33.5)$$

where I_{max} and V_{max} are the maximum values.

If an ac circuit consists of a generator and an inductor, the current lags behind the voltage by 90°. That is, the voltage reaches its maximum value one quarter of a period before the current reaches its maximum value.

If an ac circuit consists of a generator and a capacitor, the current leads the voltage by 90°. That is, the current reaches its maximum value one quarter of a period before the voltage reaches its maximum value.

In ac circuits that contain inductors and capacitors, it is useful to define the **inductive reactance** X_L and **capacitive reactance** X_C as

$$X_L = \omega L \qquad (33.10)$$

$$X_C = \frac{1}{\omega C} \qquad (33.17)$$

where ω is the angular frequency of the ac generator. The SI unit of reactance is the ohm.

The **impedance** Z of an RLC series ac circuit, which also has the unit of ohm, is

$$Z \equiv \sqrt{R^2 + (X_L - X_C)^2} \qquad (33.23)$$

The applied voltage and current are out of phase, where the **phase angle** ϕ between the current and voltage is

$$\tan \phi = \frac{X_L - X_C}{R} \qquad (33.25)$$

The sign of ϕ can be positive or negative, depending on whether X_L is greater or less than X_C. The phase angle is zero when $X_L = X_C$.

The **average power** delivered by the generator in an RLC ac circuit is

$$P_{av} = I_{rms} V_{rms} \cos \phi \qquad (33.29)$$

An equivalent expression for the average power is

$$P_{av} = I_{rms}^2 R \qquad (33.30)$$

The average power delivered by the generator is dissipated as heat in the resistor. There is no power loss in an ideal inductor or capacitor.

The rms current in a series *RLC* circuit is

$$I_{rms} = \frac{V_{rms}}{\sqrt{R^2 + (X_L - X_C)^2}} \qquad (33.32)$$

where V_{rms} is the rms value of the applied voltage.

A series *RLC* circuit is in resonance when the inductive reactance equals the capacitive reactance. When this condition is met, the current given by Equation 33.32 reaches its maximum value. Setting $X_L = X_C$, the **resonance frequency** ω_0 of the circuit has the value

$$\omega_0 = \frac{1}{\sqrt{LC}} \qquad (33.33)$$

The current in a series *RLC* circuit reaches its maximum value when the frequency of the generator equals ω_0, that is, when the "driving" frequency matches the resonance frequency.

QUESTIONS

1. What is meant by the statement "the voltage across an inductor leads the current by 90°"?

2. A night watchman is fired by his boss for being wasteful and keeping all the lights on in the building. The night watchman defends himself by claiming that the building is electrically heated, so his boss's claim is unfounded. Who should win the argument if this were to end up in a court of law?

3. Why does a capacitor act as a short circuit at high frequencies? Why does it act as an open circuit at low frequencies?

4. Explain how the acronym "ELI the ICE man" can be used to recall whether current leads voltage or voltage leads current in *RLC* circuits.

5. Why is the sum of the maximum voltages across each of the elements in a series *RLC* circuit usually greater than the maximum applied voltage? Doesn't this violate Kirchhoff's voltage law?

6. Does the phase angle depend on frequency? What is the phase angle when the inductive reactance equals the capacitive reactance?

7. In a series *RLC* circuit, what is the possible range of values for the phase angle?

8. If the frequency is doubled in a series *RLC* circuit, what happens to the resistance, the inductive reactance, and the capacitive reactance?

9. Energy is delivered to a series *RLC* circuit by a generator. This energy is dissipated as heat in the resistor. What is the source of this energy?

10. Explain why the average power delivered to an *RLC* circuit by the generator depends on the phase between the current and applied voltage.

11. A particular experiment requires a beam of light of very stable intensity. Why would an ac voltage be unsuitable for powering the light source?

12. What is the impedance of an *RLC* circuit at the resonance frequency?

13. Consider a series *RLC* circuit in which *R* is an incandescent lamp, *C* is some fixed capacitor, and *L* is a variable inductance. The source is 120 V ac. Explain why the lamp glows brightly for some values of *L* and does not glow at all for other values.

14. What is the advantage of transmitting power at high voltages?

15. What determines the maximum voltage that can be used on a transmission line?

16. Why do power lines carry electrical energy at several thousand volts potential, but it is always stepped down to 240 V as it enters your home?

17. Will a transformer operate if a battery is used for the input voltage across the primary? Explain.

18. How can the average value of a current be zero and yet

the square root of the average squared current not be zero?

19. What is the time average of a sinusoidal potential with amplitude V_{max}? What is its rms voltage?

20. What is the time average of the "square-wave" potential shown in Figure 33.20? What is its rms voltage?

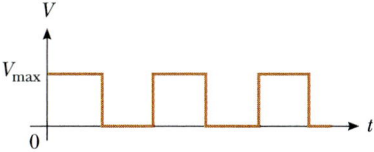

FIGURE 33.20 (Question 20).

21. Do ac ammeters and voltmeters read maximum, rms, or average values?

22. Is the voltage applied to a circuit always in phase with the current in a resistor in the circuit?

23. Would an inductor and a capacitor used together in an ac circuit dissipate any power?

24. Show that the peak current in an *RLC* circuit occurs when the circuit is in resonance.

25. Explain how the quality factor is related to the response characteristics of a receiver. Which variable most strongly determines the quality factor?

26. List some applications for a filter circuit.

27. The approximate efficiency of an incandescent lamp for converting electrical energy to heat is (a) 30%, (b) 60%, (c) 100%, or (d) 10%.

28. Why are the primary and secondary coils of a transformer wrapped on an iron core that passes through both coils?

29. With reference to Figure 33.21, explain why the capacitor prevents a dc signal from passing between A and B, yet allows an ac signal to pass from A to B. (The circuits are said to be capacitively coupled.)

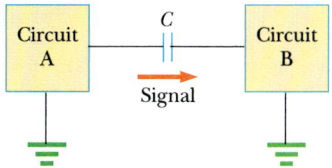

FIGURE 33.21 (Question 29).

30. With reference to Figure 33.22, if *C* is made sufficiently large, an ac signal passes from A to ground rather than from A to B. Hence, the capacitor acts as a filter. Explain.

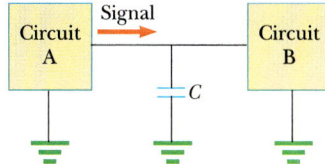

FIGURE 33.22 (Question 30).

PROBLEMS

Review Problem

In the circuit shown below, all parameters except for *C* are given. Find (a) the current as a function of time, (b) the power dissipated in the circuit, (c) the current as a function of

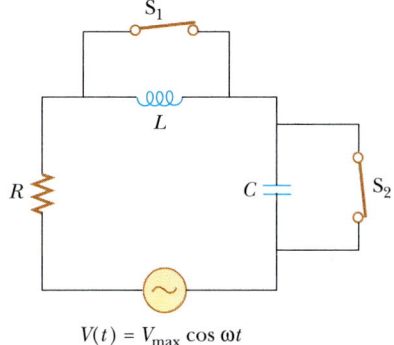

$V(t) = V_{max} \cos \omega t$

time after *only* switch 1 is opened, (d) the capacitance *C* if the current and voltage are in phase after switch 2 is *also* opened, (e) the impedance of the circuit when both switches are open, (f) the maximum energy stored in the capacitor during oscillations, (g) the maximum energy stored in the inductor during oscillations, (h) the phase change between the current and voltage if the frequency of the voltage source is doubled, and (i) the frequency that makes the inductive reactance one-half the capacitive reactance.

Assume all ac voltages and currents are sinusoidal, unless stated otherwise.

Section 33.2 Resistors in an ac Circuit

1. Show that the rms value for the sawtooth voltage shown in Figure P33.1 is $V_{max}/\sqrt{3}$.

2. (a) What is the resistance of a lightbulb that uses an average power of 75 W when connected to a 60-Hz

□ indicates problems that have full solutions available in the Student Solutions Manual and Study Guide.

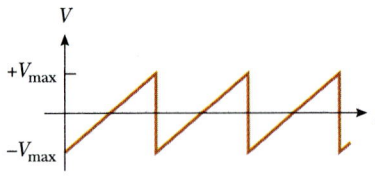

FIGURE P33.1

power source having a maximum voltage of 170 V?
(b) What is the resistance of a 100-W bulb?

3. An ac power supply produces a maximum voltage
$V_{max} = 100$ V. This power supply is connected to a
24-Ω resistor, and the current and resistor voltage
are measured with an ideal ac ammeter and voltmeter, as in Figure P33.3. What does each meter
read?

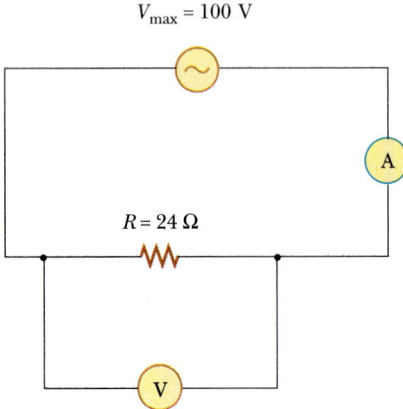

FIGURE P33.3

4. Figure P33.4 shows three lamps connected to a 120-V
ac (rms) household supply voltage. Lamps 1 and 2
have 150-W bulbs and lamp 3 has a 100-W bulb. Find
the rms current and resistance of each bulb.

4A. Figure P33.4 shows three lamps connected to an ac
(rms) voltage V. Lamps 1 and 2 have bulbs that each
dissipate power P_1 and lamp 3 has a bulb that dissipates power P_2. Find the rms current and resistance
of each bulb.

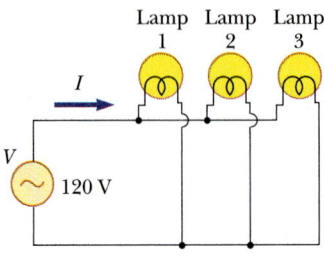

FIGURE P33.4

5. An audio amplifier, represented by the ac source and
resistor in Figure P33.5, delivers to the speaker alternating voltage at audio frequencies. If the output
voltage has an amplitude of 15.0 V, $R = 8.20$ Ω, and
the speaker is equivalent to a resistance of 10.4 Ω,
what is the time-averaged power input to it?

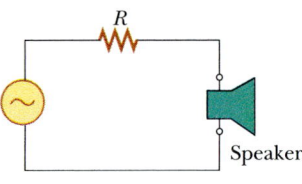

FIGURE P33.5

6. In the simple ac circuit shown in Figure 33.1, $R =
70$ Ω, and $v = V_{max} \sin \omega t$. (a) If $V_R = 0.25 V_{max}$ at
$t = 0.010$ s, what is the angular frequency of the gen-
erator? (b) What is the next value of t for which
$V_R = 0.25 V_{max}$?

7. The current in the circuit shown in Figure 33.1
equals 60% of the maximum current at $t = 7.0$ ms,
and $v = V_{max} \sin \omega t$. What is the smallest frequency
of the generator that gives this current?

Section 33.3   Inductors in an ac Circuit

8. Determine the maximum magnetic flux through an
inductor connected to a standard outlet ($V_{rms} =
120$ V, $f = 60$ Hz).

9. In a purely inductive ac circuit, as in Figure 33.4,
$V_{max} = 100$ V. (a) If the maximum current is 7.5 A at
50 Hz, calculate the inductance L. (b) At what angular frequency ω is the maximum current 2.5 A?

10. When a particular inductor is connected to a sinusoidal voltage with a 120-V amplitude, a peak current of
3.0 A appears in the inductor. (a) What is the maximum current if the frequency of the applied voltage
is doubled? (b) What is the inductive reactance at
these two frequencies?

11. An inductor is connected to a 20.0-Hz power supply
that produces a 50.0-V rms voltage. What inductance
is needed to keep the instantaneous current in the
circuit below 80.0 mA?

12. An inductor has a 54.0-Ω reactance at 60.0 Hz. What
is the maximum current if this inductor is connected
to a 50.0-Hz source that produces a 100-V rms voltage?

13. For the circuit shown in Figure 33.4, $V_{max} = 80.0$ V,
$\omega = 65\pi$ rad/s, and $L = 70.0$ mH. Calculate the
current in the inductor at $t = 15.5$ ms.

14. (a) If $L = 310$ mH and $V_{max} = 130$ V in Figure 33.4,
at what frequency is the inductive reactance 40.0 Ω?
(b) Calculate the maximum current at this frequency.

15. A 20.0-mH inductor is connected to a standard outlet ($V_{\text{rms}} = 120$ V, $f = 60.0$ Hz). Determine the energy stored in the inductor at $t = (1/180)$ s, assuming that this energy is zero at $t = 0$.

Section 33.4 Capacitors in an ac Circuit

16. A 1.0-mF capacitor is connected to a standard outlet ($V_{\text{rms}} = 120$ V, $f = 60.0$ Hz). Determine the current in the capacitor at $t = (1/180)$ s, assuming that at $t = 0$, the energy stored in the capacitor is zero.

17. (a) For what linear frequencies does a 22.0-μF capacitor have a reactance below 175 Ω? (b) Over this same frequency range, what is the reactance of a 44.0-μF capacitor?

18. What maximum current is delivered by a 2.2-μF capacitor when connected across (a) a North American outlet having $V_{\text{rms}} = 120$ V, $f = 60$ Hz and (b) a European outlet having $V_{\text{rms}} = 240$ V, $f = 50$ Hz?

19. A 98.0-pF capacitor is connected to a 60.0-Hz power supply that produces a 20.0-V rms voltage. What is the maximum charge that appears on either of the capacitor plates?

19A. A capacitor C is connected to a power supply that operates at a frequency f and produces an rms voltage V. What is the maximum charge that appears on either of the capacitor plates?

20. A sinusoidal voltage $v(t) = V_{\text{max}} \cos \omega t$ is applied to a capacitor as in Figure P33.20. (a) Write an expression for the instantaneous charge on the capacitor in terms of V_{max}, C, t, and ω. (b) What is the instantaneous current in the circuit?

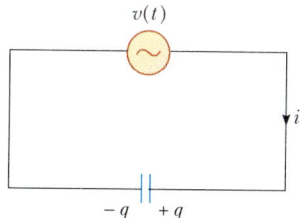

$v(t)$

i

$-q \quad +q$

FIGURE P33.20

21. What maximum current is delivered by an ac generator with $V_{\text{max}} = 48$ V and $f = 90$ Hz when connected across a 3.7-μF capacitor?

22. A variable-frequency ac generator with $V_{\text{max}} = 18$ V is connected across a 9.4×10^{-8}-F capacitor. At what frequency should the generator be operated to provide a maximum current of 5.0 A?

23. The generator in a purely capacitive ac circuit (Fig. 33.6) has an angular frequency of 100π rad/s and $V_{\text{max}} = 220$ V. If $C = 20.0$ μF, what is the current in the circuit at $t = 4.00$ ms?

Section 33.5 The *RLC* Series Circuit

24. At what frequency does the inductive reactance of a 57-μH inductor equal the capacitive reactance of a 57-μF capacitor?

25. A series ac circuit contains the following components: $R = 150$ Ω, $L = 250$ mH, $C = 2.00$ μF, and a generator with $V_{\text{max}} = 210$ V operating at 50.0 Hz. Calculate the (a) inductive reactance, (b) capacitive reactance, (c) impedance, (d) maximum current, and (e) phase angle.

26. A sinusoidal voltage $v(t) = (40.0$ V$) \sin(100t)$ is applied to a series *RLC* circuit with $L = 160$ mH, $C = 99.0$ μF, and $R = 68.0$ Ω. (a) What is the impedance of the circuit? (b) What is the maximum current? (c) Determine the numerical values for I_{max}, ω, and ϕ in the equation $i(t) = I_{\text{max}} \sin(\omega t - \phi)$.

27. An *RLC* circuit consists of a 150-Ω resistor, a 21-μF capacitor, and a 460-mH inductor, connected in series with a 120-V, 60-Hz power supply. (a) What is the phase angle between the current and the applied voltage? (b) Which reaches its maximum earlier, the current or the voltage?

28. A resistor ($R = 900$ Ω), a capacitor ($C = 0.25$ μF), and an inductor ($L = 2.5$ H) are connected in series across a 240-Hz ac source for which $V_{\text{max}} = 140$ V. Calculate the (a) impedance of the circuit, (b) peak current delivered by the source, and (c) phase angle between the current and voltage. (d) Is the current leading or lagging behind the voltage?

29. A person is working near the secondary of a transformer, as shown in Figure P33.29. The primary voltage is 120 V at 60.0 Hz. The capacitance C_s, which is the capacitance between hand and secondary winding, is 20.0 pF. Assuming the person has a body resistance to ground $R_b = 50.0$ kΩ, determine the rms voltage across the body. (*Hint:* Redraw the circuit with the secondary of the transformer as a simple ac source.)

29A. A person is working near the secondary of a transformer, as in Figure P33.29. The primary rms voltage

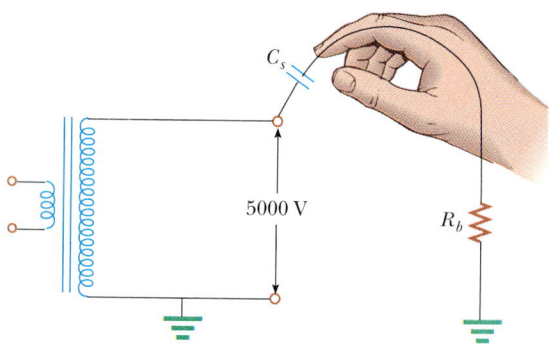

C_s

5000 V

R_b

FIGURE P33.29

is *V* at a frequency *f*. The stray capacitance between hand and secondary winding is C_s. Assuming the person has a body resistance to ground R_b, determine the rms voltage across the body. (*Hint:* Redraw the circuit with the secondary of the transformer as a simple ac source.)

30. The voltage source in Figure P33.30 has an output $V_{rms} = (100 \text{ V}) \cos(1000t)$. Determine (a) the current in the circuit and (b) the power supplied by the source. (c) Show that the power dissipated in the resistor is equal to the power supplied by the source.

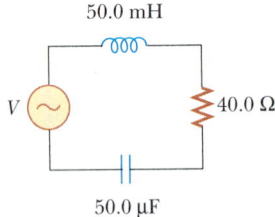

50.0 mH

V 40.0 Ω

50.0 μF

FIGURE P33.30

31. An ac source with $V_{max} = 150$ V and $f = 50.0$ Hz is connected between points *a* and *d* in Figure P33.31. Calculate the maximum voltages between points (a) *a* and *b*, (b) *b* and *c*, (c) *c* and *d*, and (d) *b* and *d*.

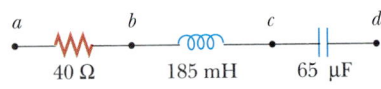

| a | b | c | d |
| 40 Ω | 185 mH | 65 μF | |

FIGURE P33.31

32. Draw to scale a phasor diagram showing *Z*, X_L, X_C, and ϕ for an ac series circuit for which $R = 300$ Ω, $C = 11$ μF, $L = 0.20$ H, and $f = 500/\pi$ Hz.

33. An inductor ($L = 400$ mH), a capacitor ($C = 4.43$ μF), and a resistor ($R = 500$ Ω) are connected in series. A 50.0-Hz ac generator produces a peak current of 250 mA in the circuit. (a) Calculate the required peak voltage V_{max}. (b) Determine the angle by which the current leads or lags the applied voltage.

34. A series *RLC* circuit in which $R = 1500$ Ω and $C = 15.0$ nF is connected to an ac generator whose frequency can be varied. When the frequency is adjusted to 50.5 kHz, the rms current in the circuit reaches a maximum at 0.140 A. Determine (a) the inductance and (b) the rms value of the generator voltage.

Section 33.6   Power in an ac Circuit

35. If 100 MW of power at 50.0-kV is to be transmitted over 100 km with only 1.00 percent loss, copper wire

of what diameter should be used? Assume uniform current density in the conductors.

35A. If power *P* is to be transmitted over a distance *d* at a voltage *V* with only 1.00 percent loss, copper wire of what diameter should be used? Assume uniform current density in the conductors.

36. A diode is a device that allows current to pass in only one direction (indicated by the arrowhead). Find, in terms of *V* and *R*, the average power dissipated in the diode circuit shown in Figure P33.36.

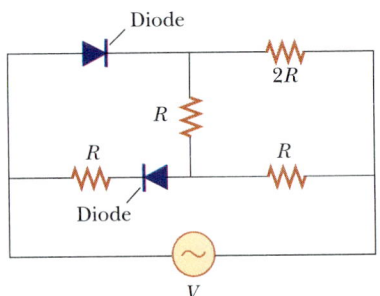

Diode

2R

R

R   R

Diode

V

FIGURE P33.36

37. An ac voltage of the form $v = (100 \text{ V}) \sin(1000t)$ is applied to a series *RLC* circuit. If $R = 400$ Ω, $C = 5.0$ μF, and $L = 0.50$ H, find the average power dissipated in the circuit.

38. An ac voltage with an amplitude of 100 V is applied to a series combination of a 200-μF capacitor, a 100-mH inductor, and a 20.0-Ω resistor. Calculate the power dissipated and the power factor for a frequency of (a) 60.0 Hz and (b) 50.0 Hz.

39. The rms output voltage of an ac generator is 200 V and the operating frequency is 100 Hz. Write the equation giving the output voltage as a function of time.

40. The average power in a circuit for which the rms current is 5.00 A is 450 W. Calculate the resistance of the circuit.

41. In a certain series *RLC* circuit, $I_{rms} = 9.0$ A, $V_{rms} = 180$ V, and the current leads the voltage by 37°. (a) What is the total resistance of the circuit? (b) Calculate the magnitude of the reactance of the circuit ($X_L - X_C$).

42. A series *RLC* circuit has a resistance of 45 Ω and an impedance of 75 Ω. What average power is delivered to this circuit when $V_{rms} = 210$ V?

Section 33.7   Resonance in a Series *RLC* Circuit

43. Calculate the resonance frequency of a series *RLC* circuit for which $C = 8.40$ μF and $L = 120$ mH.

44. Show that the *Q* value in a series *RLC* circuit is

$$Q_0 = \frac{1}{R}\sqrt{\frac{L}{C}}$$

45. An *RLC* circuit is used in a radio to tune into an FM station broadcasting at 99.7 MHz. The resistance in the circuit is 12.0 Ω, and the inductance is 1.40 μH. What capacitance should be used?

46. The tuning circuit of an AM radio is a parallel *LC* combination that has 1.00-Ω resistance. The inductance is 0.200 mH, and the capacitor is variable, so that the circuit can resonate between 550 kHz and 1650 kHz. Find the range of values for *C*.

47. A coil of resistance 35.0 Ω and inductance 20.5 H is in series with a capacitor and a 200-V (rms), 100-Hz source. The rms current in the circuit is 4.00 A. (a) Calculate the capacitance in the circuit. (b) What is V_{rms} across the coil?

48. A series *RLC* circuit has the following values: $L =$ 20.0 mH, $C =$ 100 nF, $R =$ 20.0 Ω, and $V_{max} =$ 100 V, with $v = V_{max} \sin \omega t$. Find (a) the resonant frequency, (b) the amplitude of the current at the resonant frequency, (c) the Q of the circuit, and (d) the amplitude of the voltage across the inductor at resonance.

49. A 10.0-Ω resistor, 10.0-mH inductor, and 100-μF capacitor are connected in series to a 50.0-V (rms) source having variable frequency. Find the heat dissipated in the circuit during one period if the operating frequency is twice the resonance frequency.

49A. A resistor *R*, inductor *L*, and capacitor *C* are connected in series to an ac source of rms voltage *V* and variable frequency. Find the heat dissipated in the circuit during one period if the operating frequency is twice the resonance frequency.

*Section 33.8   Filter Circuits

50. Consider the circuit shown in Figure 33.16, with $R =$ 800 Ω and $C =$ 0.090 μF. Calculate the ratio V_{out}/V_{in} for (a) $\omega =$ 300 s^{-1} and (b) $\omega = 7.0 \times 10^5$ s^{-1}.

51. The *RC* high-pass filter shown in Figure 33.16 has a resistance $R =$ 0.50 Ω. (a) What capacitance gives an output signal that has one-half the amplitude of a 300-Hz input signal? (b) What is the gain (V_{out}/V_{in}) for a 600-Hz signal?

52. The *RC* low-pass filter shown in Figure 33.17 has a resistance $R =$ 90.0 Ω and a capacitance $C =$ 8.00 nF. Calculate the gain (V_{out}/V_{in}) for an input frequency of (a) 600 Hz and (b) 600 kHz.

53. The circuit shown in Figure P33.53 represents a high-pass filter in which the inductor has internal resistance. Determine the source frequency if the output voltage V_2 is one-half the input voltage.

54. (a) For the circuit shown in Figure P33.54, show that the maximum possible value of the ratio V_{out}/V_{in} is unity. (b) At what frequency (expressed in terms of *R*, *L*, and *C*) does this maximum value occur?

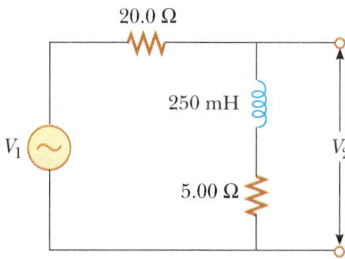

FIGURE P33.53

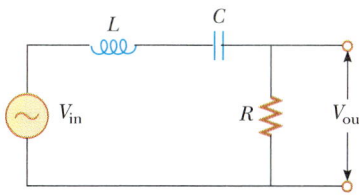

FIGURE P33.54

55. The circuit shown in Figure P33.54 can be used as a filter to pass signals that lie in a certain frequency band. (a) Show that the gain (V_{out}/V_{in}) for an input voltage of frequency ω is

$$\frac{V_{out}}{V_{in}} = \frac{1}{\sqrt{1 + \left[\dfrac{(\omega^2/\omega_0^2) - 1}{\omega RC}\right]^2}}$$

(b) Let $R =$ 100 Ω, $C =$ 0.0500 μF, and $L =$ 0.127 H. Compute the gain of this circuit for input frequencies $f_1 =$ 1.50 kHz, $f_2 =$ 2.00 kHz, and $f_3 =$ 2.50 kHz.

56. Show that two successive high-pass filters having the same values of *R* and *C* give a combined gain

$$\frac{V_{out}}{V_{in}} = \frac{1}{1 + (1/\omega RC)^2}$$

57. Consider a low-pass filter followed by a high-pass filter, as shown in Figure P33.57. If $R =$ 1000 Ω and $C =$ 0.050 μF, determine V_{out}/V_{in} for a 2.0-kHz input frequency.

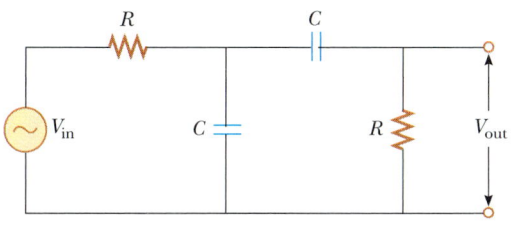

FIGURE P33.57

*Section 33.9   The Transformer and Power Transmission

58. The primary winding of an electric-train transformer has 400 turns, and the secondary has 50. If the input voltage is 120 V (rms), what is the output voltage?

59. A transformer has $N_1 = 350$ turns and $N_2 = 2000$ turns. If the input voltage is $v(t) = (170\ \text{V})\cos \omega t$, what rms voltage is developed across the secondary coil?

60. In an LR circuit, a 120-V (rms), 60-Hz source is in series with a 25-mH inductor and a 20-Ω resistor. What are (a) the rms current and (b) the power factor? (c) What capacitor must be added in series to make the power factor 1?

61. A step-down transformer is used for recharging the batteries of portable devices such as tape players. The turns ratio inside the transformer is 13 : 1 and is used with 120-V (rms) household service. If a particular tape player draws 0.35 A from the house outlet, what are (a) the voltage and (b) the current supplied from the transformer? (c) How much power is delivered?

62. A step-up transformer is designed to have an output voltage of 2200 V (rms) when the primary is connected across a 110-V (rms) source. (a) If there are 80 turns on the primary winding, how many turns are required on the secondary? (b) If a load resistor across the secondary draws a current of 1.5 A, what is the current in the primary, assuming ideal conditions? (c) If the transformer has an efficiency of 95 percent, what is the current in the primary when the secondary current is 1.2 A?

63. In the transformer shown in Figure P33.63, the load resistor is 50.0 Ω. The turn ratio $N_1 : N_2$ is 5 : 2, and the source voltage is 80.0 V (rms). If a voltmeter across the load measures 25.0 V (rms), what is the source resistance R_s?

63A. In the transformer shown in Figure P33.63, the load resistor is R_L and the source resistor is R_s. The turn ratio is $N_1 : N_2$, and the rms source voltage is V_s. If a voltmeter across the load measures an rms voltage V_2, what is the source resistance?

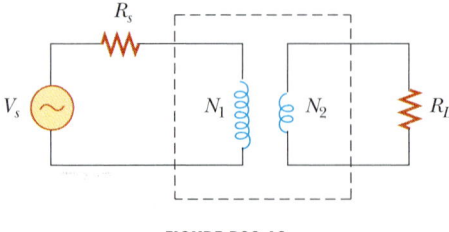

FIGURE P33.63

64. The secondary voltage of an ignition transformer used in a furnace is 10.0 kV. When the primary operates at an rms voltage of 120 V, the primary imped-

ance is 24.0 Ω and the transformer is 90 percent efficient. (a) What turn ratio is required? What are (b) the current and (c) the impedance in the secondary?

ADDITIONAL PROBLEMS

65. A series RLC circuit consists of an 8.00-Ω resistor, a 5.00-μF capacitor, and a 50.0-mH inductor. A variable frequency source of amplitude 400 V (rms) is applied across the combination. Determine the power delivered to the circuit when the frequency is equal to one half the resonance frequency.

66. A series RLC circuit has $R = 10.0\ \Omega$, $L = 2.00$ mH, and $C = 4.00\ \mu$F. Determine (a) the impedance at 60.0 Hz, (b) the resonant frequency in hertz, (c) the impedance at resonance, and (d) the impedance at a frequency equal to one-half the resonant frequency.

67. In a series RLC ac circuit, $R = 21.0\ \Omega$, $L = 25.0$ mH, $C = 17.0\ \mu$F, $V_{\max} = 150$ V, and $\omega = (2000/\pi)\ \text{s}^{-1}$. (a) Calculate the maximum current in the circuit. (b) Determine the maximum voltage across each element. (c) What is the power factor for the circuit? (d) Show X_L, X_C, R, and ϕ in a phasor diagram.

68. An RL series combination consisting of a 1.50-Ω resistor and a 2.50-mH inductor is connected to a 12.5-V (rms), 400-Hz generator. Determine (a) the impedance of the circuit, (b) the rms current, (c) the rms voltage across the resistor, and (d) the rms voltage across the inductor.

69. As a way of determining the inductance of a coil used in a research project, a student first connects the coil to a 12-V battery and measures a current of 0.63 A. The student then connects the coil to a 24-V (rms), 60-Hz generator and measures an rms current of 0.57 A. What is the inductance?

70. In Figure P33.70, find the current delivered by the 45-V (rms) power supply when (a) the frequency is very large and (b) the frequency is very small.

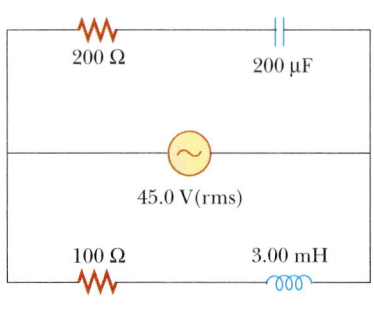

200 Ω 200 μF

45.0 V(rms)

100 Ω 3.00 mH

FIGURE P33.70

71. A transmission line that has a resistance per unit length of $4.50 \times 10^{-4}\ \Omega$/m is to be used to transmit 5.00 MW over 400 miles (6.44×10^5 m). The output voltage of the generator is 4.50 kV. (a) What is the

line loss if a transformer is used to step up the voltage to 500 kV? (b) What fraction of the input power is lost to the line under these circumstances? (c) What difficulties would be encountered on attempting to transmit the 5.00 MW at the generator voltage of 4.50 kV?

72. A transformer operating from 120 V (rms) supplies a 12-V lighting system for a garden. Eight lights, each rated 40 W, are installed in parallel. (a) Find the equivalent resistance of the system. (b) What is the current in the secondary circuit? (c) What single resistance, connected across the 120-V supply, would consume the same power as when the transformer is used? Show that this resistance equals the answer to part (a) times the square of the turns ratio.

73. *LC* filters are used as both high- and low-pass filters as were the *RC* filters in Section 33.8. However, all real inductors have resistance, as indicated in Figure P33.73, and this resistance must be taken into account. (a) Determine which circuit in Figure P33.73 is the high-pass filter and which is the low-pass filter. (b) Derive the output/input formulas for each circuit following the procedure used for the *RC* filters in Section 33.8.

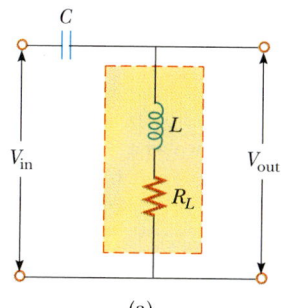

(a)

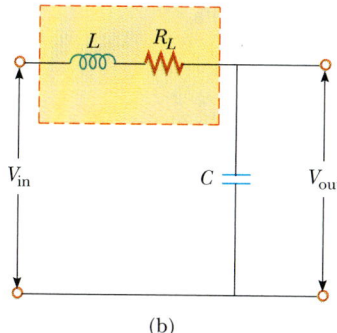

(b)

FIGURE P33.73

74. An 80.0-Ω resistor and a 200-mH inductor are connected in parallel across a 100-V (rms), 60.0-Hz source. (a) What is the rms current in the resistor?

(b) By what angle does the total current lead or lag behind the voltage?

75. A power station sends 1.50 MW to a town 10.0 km from the station. The total resistance in the connecting wires is 1.10 Ω. Find the power lost as heat in the wires if the power is sent at (a) 120 V, (b) 12.0 kV, and (c) 120 kV.

76. The average power delivered to a series *RLC* circuit at frequency ω (Section 33.7) is given by Equation 33.35. (a) Show that the maximum current can be written

$$I_{max} = \omega V_{max}[L^2(\omega_0^2 - \omega^2)^2 + (\omega R)^2]^{-1/2}$$

where ω is the operating frequency of the circuit and ω_0 is the resonance frequency. (b) Show that the phase angle can be expressed as

$$\phi = \tan^{-1}\left[\frac{L}{R}\left(\frac{\omega_0^2 - \omega^2}{\omega}\right)\right]$$

77. Consider a series *RLC* circuit having the following circuit parameters: $R = 200$ Ω, $L = 663$ mH, and $C = 26.5$ μF. The applied voltage has an amplitude of 50.0 V and a frequency of 60.0 Hz. Find the following amplitudes: (a) The current i, including its phase constant ϕ relative to the applied voltage v; (b) the voltage V_R across the resistor and its phase relative to the current; (c) the voltage V_C across the capacitor and its phase relative to the current; and (d) the voltage V_L across the inductor and its phase relative to the current.

78. A voltage $v = (100 \text{ V}) \sin \omega t$ (in SI units) is applied across a series combination of a 2.00-H inductor, a 10.0-μF capacitor, and a 10.0-Ω resistor. (a) Determine the angular frequency ω_0 at which the power dissipated in the resistor is a maximum. (b) Calculate the power dissipated at that frequency. (c) Determine the two angular frequencies ω_1 and ω_2 at which the power dissipated is one-half the maximum value. [The Q of the circuit is approximately $\omega_0/(\omega_2 - \omega_1)$.]

79. *Impedance matching:* A transformer may be used to provide maximum power transfer between two ac circuits that have different impedances. (a) Show that the ratio of turns N_1/N_2 needed to meet this condition is

$$\frac{N_1}{N_2} = \sqrt{\frac{Z_1}{Z_2}}$$

(b) Suppose you want to use a transformer as an impedance-matching device between an audio amplifier that has an output impedance of 8.00 kΩ and a speaker that has an input impedance of 8.00 Ω. What should your N_1/N_2 ratio be?

80. An ac source has an internal resistance of 3.20 kΩ. In order for the maximum power to be transferred to an 8.00-Ω resistive load R_2, a transformer is used be-

tween the source and the load. Assuming an ideal transformer, (a) find the appropriate turns ratio of the transformer. If the output voltage of the source is 80.0 V (rms), determine (b) the rms voltage across the load resistor and (c) the rms current in the load resistor. (d) Calculate the power dissipated in the load. (e) Verify that the ratio of currents is inversely proportional to the turns ratio.

81. Figure P33.81a shows a parallel *RLC* circuit, and the corresponding phasor diagram is given in Figure P33.81b. The instantaneous voltage (and rms voltage) across each of the three circuit elements is the same, and each is in phase with the current through the resistor. The currents in *C* and *L* lead (or lag behind) the current in the resistor, as shown in Figure P33.81b. (a) Show that the rms current delivered by the source is

$$I_{rms} = V_{rms}\left[\frac{1}{R^2} + \left(\omega C - \frac{1}{\omega L}\right)^2\right]^{1/2}$$

(b) Show that the phase angle ϕ between V_{rms} and I_{rms} is

$$\tan\phi = R\left(\frac{1}{X_C} - \frac{1}{X_L}\right)$$

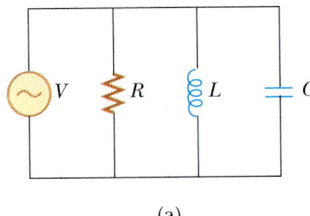

(a)

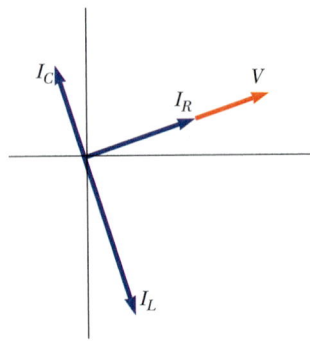

(b)

FIGURE P33.81

82. An 80.0-Ω resistor, a 200-mH inductor, and a 0.150-μF capacitor are connected in parallel across a 120-V (rms) source operating at 374 rad/s. (a) What is the resonant frequency of the circuit? (b) Calculate the rms current in the resistor, inductor, and capacitor.

(c) What is the rms current delivered by the source?
(d) Is the current leading or lagging behind the voltage? By what angle?

83. Consider the phase-shifter circuit shown in Figure P33.83. The input voltage is described by the expression $v = (10\text{ V})\sin 200t$ (in SI units). Assuming that $L = 500$ mH, find (a) the value of R such that the output voltage v_{out} lags behind the input voltage by 30° and (b) the amplitude of the output voltage.

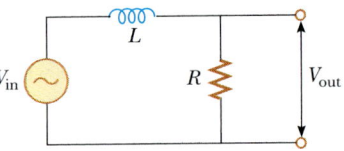

FIGURE P33.83

84. A series *RLC* circuit is operating at 2000 Hz. At this frequency, $X_L = X_C = 1884$ Ω. The resistance of the circuit is 40 Ω. (a) Prepare a table showing the values of X_L, X_C, and Z for $f = 300, 600, 800, 1000, 1500, 2000, 3000, 4000, 6000$, and 10 000 Hz. (b) Plot on the same set of axes X_L, X_C, and Z as a function of ln f.

85. Suppose the high-pass filter shown in Figure 33.16 has $R = 1000$ Ω and $C = 0.050$ μF. (a) At what frequency does $V_{out}/V_{in} = \frac{1}{2}$? (b) Plot $\log_{10}(V_{out}/V_{in})$ versus $\log_{10}(f)$ over the frequency range from 1 Hz to 1 MHz. (This log–log plot of gain versus frequency is known as a **Bode plot**.)

86. Suppose the low-pass filter shown in Figure 33.17 has $R = 1000$ Ω and $C = 0.050$ μF. (a) At what frequency does $V_{out}/V_{in} = \frac{1}{2}$? (b) Plot $\log_{10}(V_{out}/V_{in})$ versus $\log_{10}(f)$ over the frequency range from 1 Hz to 1 MHz.

87. A series *RLC* circuit in which $R = 1.00$ Ω, $L = 1.00$ mH, and $C = 1.00$ nF is connected to an ac generator delivering 1.00 V (rms). Make a careful plot of the power delivered to the circuit as a function of the frequency and verify that the half-width of the resonance peak is $R/2\pi L$.

SPREADSHEET PROBLEMS

S1. Spreadsheet 33.1 calculates the impedances (Z, X_L, X_C) and the voltage amplitudes (V_R, V_L, V_C) across each circuit element in an ac series *RLC* circuit as a function of frequency ω. In addition, the resonance frequency ω_0 is calculated. Use $R = 100$ Ω, $L = 0.50$ H, and $C = 1.0$ μF. The amplitude V_{max} of the generator is 5.0 V. (a) Plot the voltage amplitudes versus ω and note their relative magnitudes. (b) At the resonance frequency ω_0, what are the magnitudes of V_L and V_C? (c) For a very low frequency ($\omega \ll \omega_0$), how does V_L compare to V_C and V_R?

(d) Repeat part (c) for a frequency much higher than the resonance frequency. (e) The tuner of your radio is really an *RLC* circuit. Vary *R*, *L*, and *C* such that a very sharp resonance is obtained at the frequency of your favorite radio station.

S2. Spreadsheet 33.2 calculates the voltages across the resistor, capacitor, and inductor of an ac series *RLC* circuit as functions of time. Input parameters are *R*, *L*, *C*, and V_{max} and the generator angular frequency is ω. The spreadsheet also calculates X_L, X_C, Z, ω_0, and the phase angle ϕ. (a) Use $R = 1000\ \Omega$, $L = 0.37\ H$, $C = 1.0\ \mu F$, and $V_{max} = 5.0\ V$. Initially choose $\omega = 1000\ rad/s$. Plot the voltages versus time and note the magnitudes and phase differences of the various voltages. (b) Increase ω to 2200 rad/s in steps of 200 rad/s. Note how the relative phase differences change. (c) Choose $\omega = \omega_0$. What is the phase difference between V_L and V_C? What are their relative magnitudes? What is the sum of V_L and V_C? (d) Choose other values for *R*, *L*, and *C* and repeat this investigation.

S3. A series *RLC* circuit contains a switch but no battery. The capacitor has an initial charge Q_0 with the switch open, and then the switch is thrown closed at $t = 0$. The current in this *RLC* circuit for $t > 0$ is given by

$$I = -\frac{I_0}{\cos\phi}\sin(\omega_0 t + \phi)e^{-Rt/2L}$$

where $I_0 = \omega' Q_0$, $\omega' = \sqrt{(1/LC) - (R/2L)^2}$ and $\tan\phi = R/2L\omega'$.

Construct a spreadsheet to calculate and plot the current versus time for this circuit. Input parameters should be *R*, *L*, *C*, and Q_0. Use $R = 10\ \Omega$, $L = 10\ mH$, $C = 4.0\ \mu F$, and $Q_0 = 2.0\ \mu C$. Verify that the ratio of successive amplitude peaks is a constant. Why is this the case? Vary *R* from $0\ \Omega$ to $100\ \Omega$ in steps of $10\ \Omega$ and investigate what happens to the damping.

Electromagnetic Waves

Satellite receiver-transmitter dish at night. The photograph was taken at the Land Earth Station at Goonhilly in Cornwall, United Kingdom. This Earth Station serves the INMARSAT (International Maritime Satellite) organization. It provides telephone, telex, data, and facsimile operations to the shipping, aviation, offshore, and land mobile industries. *(Photo Researchers, Inc.)*

The waves described in Chapters 16, 17, and 18 are mechanical waves. By definition, mechanical disturbances, such as sound waves, water waves, and waves on a string, require the presence of a medium. This chapter is concerned with the properties of electromagnetic waves that (unlike mechanical waves) can propagate through empty space.

In Section 31.7 we gave a brief description of Maxwell's equations, which form the theoretical basis of all electromagnetic phenomena.[1] The consequences of Maxwell's equations are far reaching and very dramatic for the history of physics. One of them, the Ampère-Maxwell law, predicts that a time-varying electric field produces a magnetic field just as a time-varying magnetic field produces an electric field (Faraday's law). From this generalization, Maxwell introduced the concept of displacement current, a new source of a magnetic field. Thus, Maxwell's theory provided the final important link between electric and magnetic fields.

Astonishingly, Maxwell's formalism also predicts the existence of electromagnetic waves that propagate through space with the speed of light. This prediction was confirmed experimentally by Heinrich Hertz, who generated and detected

[1] The reader should review Section 31.7 as a background for the material in this chapter.

James Clerk Maxwell is generally regarded as the greatest theoretical physicist of the 19th century. Born in Edinburgh to a well-known Scottish family, he entered the University of Edinburgh at age 15, around the time that he discovered an original method for drawing a perfect oval. Maxwell was appointed to his first professorship in 1856 at Aberdeen. This was the beginning of a career during which he would develop the electromagnetic theory of light and explanations of the nature of Saturn's rings, and contribute to the kinetic theory of gases.

Maxwell's development of the electromagnetic theory of light took many years and began with the paper "On Faraday's Lines of Force," in which Maxwell expanded upon Faraday's theory that electric and magnetic effects result from force fields surrounding conductors and magnets. His next publication, "On Physi-

James Clerk Maxwell

| 1 8 3 1 – 1 8 7 9 |

cal Lines of Force," included a series of papers explaining the known effects and the nature of electromagnetism.

Maxwell's other important contributions to theoretical physics were made in the area of the kinetic theory

of gases. Here, he furthered the work of Rudolf Clausius, who in 1858 had shown that a gas must consist of molecules in constant motion colliding with one another and with the walls of the container. This resulted in Maxwell's distribution of molecular speeds in addition to important applications of the theory to viscosity, conduction of heat, and diffusion of gases.

Maxwell's successful interpretation of Faraday's concept of the electromagnetic field resulted in the field equation bearing Maxwell's name. Formidable mathematical ability combined with great insight enabled Maxwell to lead the way in the study of the two most important areas of physics at that time. Maxwell died of cancer before he was 50.

(North Wind Picture Archives)

electromagnetic waves. This discovery has led to many practical communication systems, including radio, television, and radar. On a conceptual level, Maxwell unified the subjects of light and electromagnetism by developing the idea that light is a form of electromagnetic radiation.

Electromagnetic waves are generated by oscillating electric charges. The radiated waves consist of oscillating electric and magnetic fields, which are *at right angles to each other* and also *at right angles to the direction of wave propagation*. Thus, electromagnetic waves are transverse in nature. Maxwell's theory shows that the electric and magnetic field amplitudes in an electromagnetic wave are related by $E = cB$. At large distances from the source of the waves, these amplitudes diminish with distance, in proportion to $1/r$. The radiated waves can be detected at great distances from the oscillating charges. Furthermore, electromagnetic waves carry energy and momentum and hence exert pressure on a surface.

Electromagnetic waves cover a wide range of frequencies. For example, radio waves (frequencies of about 10^7 Hz) are electromagnetic waves produced by oscillating currents in a radio tower's transmitting antenna. Light waves are a high-frequency form of electromagnetic radiation (about 10^{14} Hz) produced by electrons within atomic systems.

34.1 MAXWELL'S EQUATIONS AND HERTZ'S DISCOVERIES

In his unified theory of electromagnetism, Maxwell showed that electromagnetic waves are a natural consequence of the fundamental laws expressed in four equations:

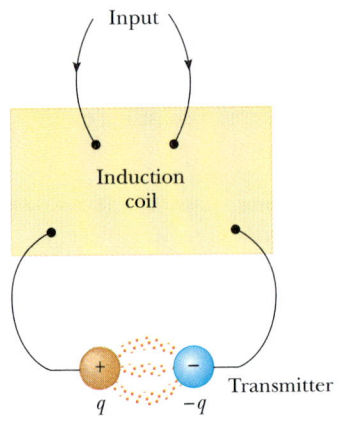

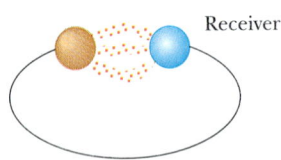

FIGURE 34.1 Schematic diagram of Hertz's apparatus for generating and detecting electromagnetic waves. The transmitter consists of two spherical electrodes connected to an induction coil, which provides short voltage surges to the spheres, setting up oscillations in the discharge. The receiver is a nearby loop containing a second spark gap.

$$\oint \mathbf{E} \cdot d\mathbf{A} = \frac{Q}{\epsilon_0} \tag{34.1}$$

$$\oint \mathbf{B} \cdot d\mathbf{A} = 0 \tag{34.2}$$

$$\oint \mathbf{E} \cdot d\mathbf{s} = -\frac{d\Phi_B}{dt} \tag{34.3}$$

$$\oint \mathbf{B} \cdot d\mathbf{s} = \mu_0 I + \mu_0 \epsilon_0 \frac{d\Phi_E}{dt} \tag{34.4}$$

As we shall see in the next section, Equations 34.3 and 34.4 can be combined to obtain a wave equation for both the electric and the magnetic fields. In empty space ($Q = 0$, $I = 0$), the solution to these two equations shows that the wave speed $(\mu_0 \epsilon_0)^{-1/2}$ equals the measured speed of light. This result led Maxwell to the prediction that light waves are a form of electromagnetic radiation.

Electromagnetic waves were generated and detected by Hertz in 1887, using electrical sources. His experimental apparatus is shown schematically in Figure 34.1. An induction coil is connected to two spherical electrodes having a narrow gap between them (the transmitter). The coil provides short voltage surges to the spheres, making one positive, the other negative. A spark is generated between the spheres when the voltage between them reaches the breakdown voltage for air. As the air in the gap is ionized, it conducts more readily and the discharge between the spheres becomes oscillatory. From an electrical circuit viewpoint, this is equivalent to an LC circuit, where the inductance is that of the loop and the capacitance is due to the spherical electrodes.

Since L and C are quite small, the frequency of oscillation is very high, ≈ 100 MHz. (Recall from Eq. 32.22 that $\omega = 1/\sqrt{LC}$ for an LC circuit.) Electromagnetic waves are radiated at this frequency as a result of the oscillation (and hence acceleration) of free charges in the loop. Hertz was able to detect these waves using a single loop of wire with its own spark gap (the receiver). This loop, placed several

Heinrich Rudolf Hertz was born in 1857 in Hamburg, Germany. He studied physics under Helmholtz and Kirchhoff at the University of Berlin. In 1885, Hertz accepted the position of Professor of Physics at Karlsruhe; it was here that he demonstrated radio waves in 1887, his most important accomplishment.

In 1889 Hertz succeeded Rudolf Clausius as Professor of Physics at the University of Bonn. Hertz's subsequent experiments involving metal penetration by cathode rays led him to the conclusion that cathode rays are waves rather than particles.

Exploring radio waves, demon-

Heinrich Rudolf Hertz

| 1 8 5 7 – 1 8 9 4 |

strating their generation, and determining their speed are among Hertz's many achievements. After finding that the speed of a radio wave was the same as that of light, Hertz showed that radio waves, like light waves, could be reflected, refracted, and diffracted.

Hertz died of blood poisoning at the age of 36. During his short life, he made many contributions to science. The hertz, equal to one complete vibration or cycle per second, is named after him.

(The Bettmann Archive)

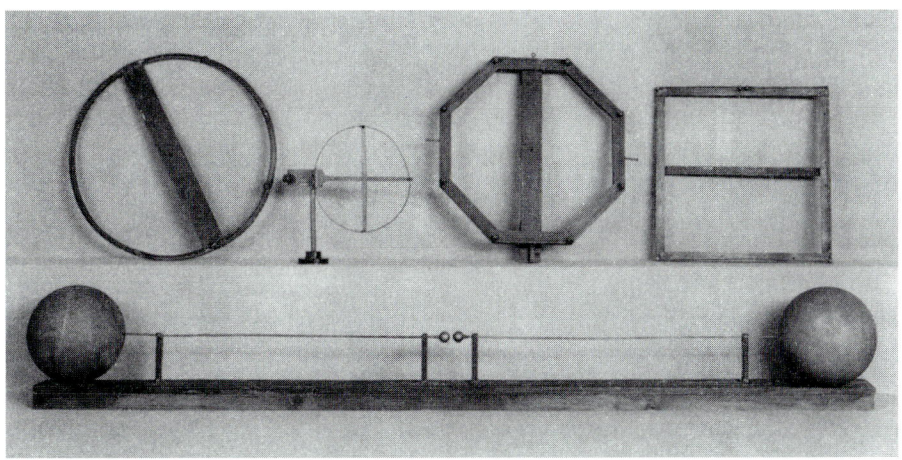

Large oscillator (bottom) as well as circular, octagonal, and square resonators used by Heinrich Hertz. *(Photo Deutsches Museum Munich)*

meters from the transmitter, has its own effective inductance, capacitance, and natural frequency of oscillation. Sparks were induced across the gap of the receiving electrodes when the frequency of the receiver was adjusted to match that of the transmitter. Thus, Hertz demonstrated that the oscillating current induced in the receiver was produced by electromagnetic waves radiated by the transmitter. His experiment is analogous to the mechanical phenomenon in which a tuning fork picks up the vibrations from another, identical oscillating tuning fork.

In a series of experiments, Hertz also showed that the radiation generated by his spark-gap device exhibited the wave properties of interference, diffraction, reflection, refraction, and polarization, all of which are properties exhibited by light. Thus, it became evident that the radio-frequency waves had properties similar to light waves and differed only in frequency and wavelength. Perhaps his most convincing experiment was the measurement of the speed of this radiation. Radio-frequency waves of known frequency were reflected from a metal sheet and created an interference pattern whose nodal points (where E was zero) could be detected. The measured distance between the nodal points allowed determination of the wavelength λ. Using the relation $v = \lambda f$, Hertz found that v was close to 3×10^8 m/s, the known speed of visible light.

34.2 PLANE ELECTROMAGNETIC WAVES

The properties of electromagnetic waves can be deduced from Maxwell's equations. One approach to deriving such properties is to solve the second-order differential equation obtained from Maxwell's third and fourth equations. A rigorous mathematical treatment of this sort is beyond the scope of this text. To circumvent this problem, we assume that the electric and magnetic vectors have a specific space-time behavior that is consistent with Maxwell's equations.

First, we assume that the electromagnetic wave is a plane wave, that is, one that travels in one direction. The plane wave we are describing has the following properties. It travels in the x direction (the direction of propagation), the electric field **E** is in the y direction, and the magnetic field **B** is in the z direction, as in Figure 34.2. Waves in which the electric and magnetic fields are restricted to being paral-

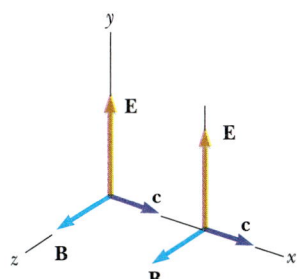

FIGURE 34.2 A plane-polarized electromagnetic wave traveling in the positive x direction. The electric field is along the y direction, and the magnetic field is along the z direction. These fields depend only on x and t.

lel to certain lines in the *yz* plane are said to be **linearly polarized waves.**[2] Furthermore, we assume that *E* and *B* at any point *P* depend only upon *x* and *t* and not upon the *y* or *z* coordinates of *P*.

We can relate *E* and *B* to each other by Equations 34.3 and 34.4. In empty space, where $Q = 0$ and $I = 0$, Equation 34.3 remains unchanged and Equation 34.4 becomes

$$\oint \mathbf{B} \cdot d\mathbf{s} = \epsilon_0 \mu_0 \frac{d\Phi_E}{dt} \tag{34.5}$$

Using Equations 34.3 and 34.5 and the plane wave assumption, the following differential equations relating *E* and *B* are obtained. (We do this more formally later in this section.) For simplicity of notation, we drop the subscripts on the components E_y and B_z:

$$\frac{\partial E}{\partial x} = -\frac{\partial B}{\partial t} \tag{34.6}$$

$$\frac{\partial B}{\partial x} = -\mu_0 \epsilon_0 \frac{\partial E}{\partial t} \tag{34.7}$$

Note that the derivatives here are partial derivatives. For example, when $\partial E/\partial x$ is evaluated, we assume that *t* is constant. Likewise, when evaluating $\partial B/\partial t$, *x* is held constant. Taking the derivative of Equation 34.6 and combining this with Equation 34.7 we get

$$\frac{\partial^2 E}{\partial x^2} = -\frac{\partial}{\partial x}\left(\frac{\partial B}{\partial t}\right) = -\frac{\partial}{\partial t}\left(\frac{\partial B}{\partial x}\right) = -\frac{\partial}{\partial t}\left(-\mu_0 \epsilon_0 \frac{\partial E}{\partial t}\right)$$

Wave equations for electromagnetic waves in free space

$$\frac{\partial^2 E}{\partial x^2} = \mu_0 \epsilon_0 \frac{\partial^2 E}{\partial t^2} \tag{34.8}$$

In the same manner, taking a derivative of Equation 34.7 and combining it with Equation 34.8, we get

$$\frac{\partial^2 B}{\partial x^2} = \mu_0 \epsilon_0 \frac{\partial^2 B}{\partial t^2} \tag{34.9}$$

Equations 34.8 and 34.9 both have the form of the general wave equation,[3] with the wave speed *v* replaced by *c*, where

$$c = \frac{1}{\sqrt{\mu_0 \epsilon_0}} \tag{34.10}$$

---

[2] Waves with other particular patterns of vibrations of **E** and **B** include circularly polarized waves. The most general polarization pattern is elliptical.

[3] The general wave equation is of the form $(\partial^2 f/\partial x^2) = (1/v^2)(\partial^2 f/\partial t^2)$, where *v* is the speed of the wave and *f* is the wave amplitude. The wave equation was first introduced in Chapter 16, and it would be useful for the reader to review this material.

Taking $\mu_0 = 4\pi \times 10^{-7}$ Wb/A·m and $\epsilon_0 = 8.85418 \times 10^{-12}$ C²/N·m² in Equation 34.10, we find that $c = 2.99792 \times 10^8$ m/s. Since this speed is precisely the same as the speed of light in empty space, we are led to believe (correctly) that light is an electromagnetic wave.

The simplest plane wave solution is a sinusoidal wave, for which the field amplitudes E and B vary with x and t according to the expressions

$$E = E_{max} \cos(kx - \omega t) \tag{34.11}$$

$$B = B_{max} \cos(kx - \omega t) \tag{34.12}$$

where E_{max} and B_{max} are the maximum values of the fields. The constant $k = 2\pi/\lambda$, where λ is the wavelength, and the angular frequency $\omega = 2\pi f$, where f is the number of cycles per second. The ratio ω/k equals the speed c, since

$$\frac{\omega}{k} = \frac{2\pi f}{2\pi/\lambda} = \lambda f = c$$

Figure 34.3 is a pictorial representation at one instant of a sinusoidal, linearly polarized plane wave moving in the positive x direction.

Taking partial derivatives of Equations 34.11 and 34.12, we find that

$$\frac{\partial E}{\partial x} = -kE_{max} \sin(kx - \omega t)$$

$$-\frac{\partial B}{\partial t} = -\omega B_{max} \sin(kx - \omega t)$$

Since these must be equal, according to Equation 34.6, we find that at any instant

$$kE_{max} = \omega B_{max}$$

$$\frac{E_{max}}{B_{max}} = \frac{\omega}{k} = c$$

Using these results together with Equations 34.11 and 34.12, we see that

$$\frac{E_{max}}{B_{max}} = \frac{E}{B} = c \tag{34.13}$$

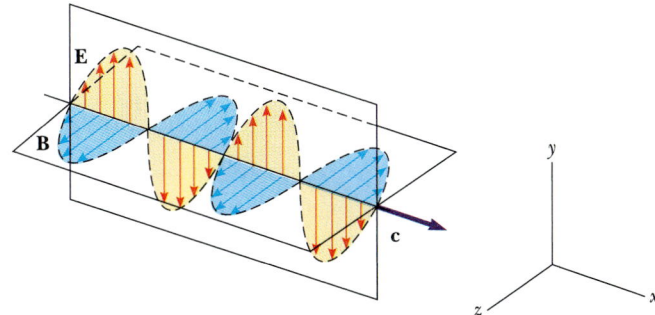

FIGURE 34.3 Representation of a sinusoidal, plane-polarized electromagnetic wave moving in the positive x direction with a speed c. The drawing represents a snapshot, that is, the wave at some instant. Note the sinusoidal variations of E and B with x.

That is, *at every instant the ratio of the electric field to the magnetic field of an electromagnetic wave equals the speed of light.*

Finally, it should be noted that electromagnetic waves obey the superposition principle, because the differential equations involving E and B are linear equations. For example, two waves traveling in opposite directions with the same frequency could be added by simply adding the wave fields algebraically. Furthermore, we now have a theoretical value for c, given by the relationship $c = 1/\sqrt{\mu_0 \epsilon_0}$.

Let us summarize the properties of electromagnetic waves as we have described them:

Properties of electromagnetic waves

- The solutions of Maxwell's third and fourth equations are wavelike, where both E and B satisfy the same wave equation.
- Electromagnetic waves travel through empty space with the speed of light, $c = 1/\sqrt{\epsilon_0 \mu_0}$.
- The electric and magnetic field components of plane electromagnetic waves are perpendicular to each other and also perpendicular to the direction of wave propagation. The latter property can be summarized by saying that electromagnetic waves are transverse waves.
- The relative magnitudes of **E** and **B** in empty space are related by $E/B = c$.
- Electromagnetic waves obey the principle of superposition.

EXAMPLE 34.1   An Electromagnetic Wave

A plane electromagnetic sinusoidal wave of frequency 40.0 MHz travels in free space in the x direction, as in Figure 34.4. At some point and at some instant, the electric field has its maximum value of 750 N/C and is along the y axis. (a) Determine the wavelength and period of the wave.

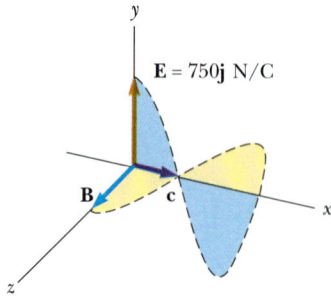

E = 750**j** N/C

FIGURE 34.4   (Example 34.1) At some instant, a plane electromagnetic wave moving in the x direction has a maximum electric field of 750 N/C in the positive y direction. The corresponding magnetic field at that point has a magnitude E/c and is in the z direction.

Solution   Because $c = \lambda f$ and $f = 40.0$ MHz $= 4.00 \times 10^7$ s^{-1}, we get

$$\lambda = \frac{c}{f} = \frac{3.00 \times 10^8 \text{ m/s}}{4.00 \times 10^7 \text{ s}^{-1}} = 7.50 \text{ m}$$

The period of the wave T equals the inverse of the frequency, and so

$$T = \frac{1}{f} = \frac{1}{4.00 \times 10^7 \text{ s}^{-1}} = 2.50 \times 10^{-8} \text{ s}$$

(b) Calculate the magnitude and direction of the magnetic field when $\mathbf{E} = 750\mathbf{j}$ N/C.

Solution   From Equation 34.13 we see that

$$B_{max} = \frac{E_{max}}{c} = \frac{750 \text{ N/C}}{3.00 \times 10^8 \text{ m/s}} = 2.50 \times 10^{-6} \text{ T}$$

Since **E** and **B** must be perpendicular to each other and both must be perpendicular to the direction of wave propagation (x in this case), we conclude that **B** is in the z direction.

(c) Write expressions for the space-time variation of the electric and magnetic field components for this wave.

Solution   We can apply Equations 34.11 and 34.12 directly:

$$E = E_{max} \cos(kx - \omega t) = (750 \text{ N/C}) \cos(kx - \omega t)$$

$$B = B_{max} \cos(kx - \omega t) = (2.50 \times 10^{-6} \text{ T}) \cos(kx - \omega t)$$

where

$$\omega = 2\pi f = 2\pi(4.00 \times 10^7 \text{ s}^{-1}) = 8\pi \times 10^7 \text{ rad/s}$$

$$k = \frac{2\pi}{\lambda} = \frac{2\pi}{7.50 \text{ m}} = 0.838 \text{ rad/m}$$

*Derivation of Equations 34.6 and 34.7

In this optional section, we derive Equations 34.6 and 34.7. To derive Equation 34.6, we start with Faraday's law, that is, Equation 34.3:

$$\oint \mathbf{E} \cdot d\mathbf{s} = -\frac{d\Phi_B}{dt}$$

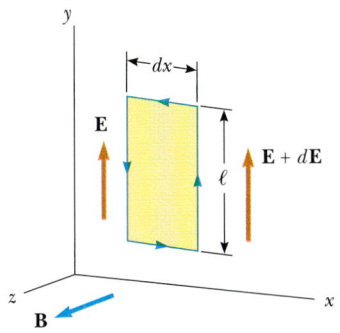

Again, let us assume that the electromagnetic wave is a plane wave traveling in the x direction, with the electric field $\mathbf{E}$ in the positive y direction and the magnetic field $\mathbf{B}$ in the positive z direction.

Consider a thin rectangle of width dx and height ℓ lying in the xy plane, as in Figure 34.5. To apply Equation 34.3, we must first evaluate the line integral of $\mathbf{E} \cdot d\mathbf{s}$ around this rectangle. The contributions from the top and bottom of the rectangle are zero because $\mathbf{E}$ is perpendicular to $d\mathbf{s}$ for these paths. We can express the electric field on the right side of the rectangle as

$$E(x + dx, t) \approx E(x, t) + \frac{dE}{dx}\bigg]_{t\text{ constant}} dx = E(x, t) + \frac{\partial E}{\partial x} dx$$

while the field on the left side is simply $E(x, t)$. Therefore, the line integral over this rectangle becomes approximately[4]

$$\oint \mathbf{E} \cdot d\mathbf{s} = E(x + dx, t) \cdot \ell - E(x, t) \cdot \ell \approx (\partial E / \partial x) \, dx \cdot \ell \qquad (34.14)$$

Because the magnetic field is in the z direction, the magnetic flux through the rectangle of area $\ell \, dx$ is approximately

$$\Phi_B = B\ell \, dx$$

(This assumes that dx is small compared with the wavelength of the wave.) Taking the time derivative of the flux gives

$$\frac{d\Phi_B}{dt} = \ell \, dx \frac{dB}{dt}\bigg]_{x\text{ constant}} = \ell \, dx \frac{\partial B}{\partial t} \qquad (34.15)$$

Substituting Equations 34.14 and 34.15 into Equation 34.3 gives

$$\left(\frac{\partial E}{\partial x}\right) dx \cdot \ell = -\ell \, dx \frac{\partial B}{\partial t}$$

$$\frac{\partial E}{\partial x} = -\frac{\partial B}{\partial t}$$

This expression is equivalent to Equation 34.6.

In a similar manner, we can verify Equation 34.7 by starting with Maxwell's fourth equation in empty space (Eq. 34.5). In this case, we evaluate the line integral of $\mathbf{B} \cdot d\mathbf{s}$ around a rectangle lying in the xz plane and having width dx and length ℓ, as in Figure 34.6. Using the sense of the integration shown and noting that the magnetic field changes from $B(x, t)$ to $B(x + dx, t)$ over the width dx, we get

$$\oint \mathbf{B} \cdot d\mathbf{s} = B(x, t) \cdot \ell - B(x + dx, t) \cdot \ell = -(\partial B / \partial x) \, dx \cdot \ell \qquad (34.16)$$

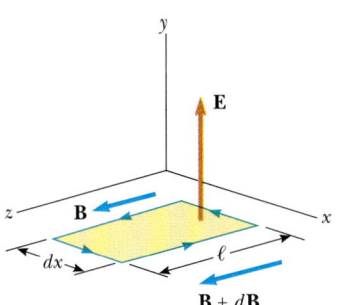

FIGURE 34.5 As a plane wave passes through a rectangular path of width dx lying in the xy plane, the electric field in the y direction varies from $\mathbf{E}$ to $\mathbf{E} + d\mathbf{E}$. This spatial variation in $\mathbf{E}$ gives rise to a time-varying magnetic field along the z direction, according to Equation 34.6.

FIGURE 34.6 As a plane wave passes through a rectangular curve of width dx lying in the xz plane, the magnetic field along z varies from $\mathbf{B}$ to $\mathbf{B} + d\mathbf{B}$. This spatial variation in $\mathbf{B}$ gives rise to a time-varying electric field along the y direction, according to Equation 34.7.

---

[4] Since dE/dx means the change in E with x at a given instant t, dE/dx is equivalent to the partial derivative $\partial E/\partial x$. Likewise, dB/dt means the change in B with time at a particular position x, and so we can replace dB/dt by $\partial B/\partial t$.

The electric flux through the rectangle is

$$\Phi_E = E\ell \, dx$$

which when differentiated with respect to time gives

$$\frac{\partial \Phi_E}{\partial t} = \ell \, dx \frac{\partial E}{\partial t} \tag{34.17}$$

Substituting Equations 34.16 and 34.17 into Equation 34.5 gives

$$-(\partial B/\partial x) \, dx \cdot \ell = \mu_0 \epsilon_0 \ell \, dx (\partial E/\partial t)$$

$$\frac{\partial B}{\partial x} = -\mu_0 \epsilon_0 \frac{\partial E}{\partial t}$$

which is equivalent to Equation 34.7.

34.3   ENERGY CARRIED BY ELECTROMAGNETIC WAVES

Electromagnetic waves carry energy, and as they propagate through space they can transfer energy to objects placed in their path. The rate of flow of energy in an electromagnetic wave is described by a vector **S**, called the **Poynting vector**, defined by the expression

Poynting vector

$$\mathbf{S} \equiv \frac{1}{\mu_0} \mathbf{E} \times \mathbf{B} \tag{34.18}$$

The magnitude of the Poynting vector represents the rate at which energy flows through a unit surface area perpendicular to the flow and its direction is along the direction of wave propagation (Fig. 34.7). The SI units of the Poynting vector are $J/s \cdot m^2 = W/m^2$. (These are the units **S** must have because it represents the power per unit area, where the unit area is oriented at right angles to the direction of wave propagation.)

As an example, let us evaluate the magnitude of **S** for a plane electromagnetic wave where $|\mathbf{E} \times \mathbf{B}| = EB$. In this case

$$S = \frac{EB}{\mu_0} \tag{34.19}$$

Because $B = E/c$, we can also express this as

$$S = \frac{E^2}{\mu_0 c} = \frac{c}{\mu_0} B^2$$

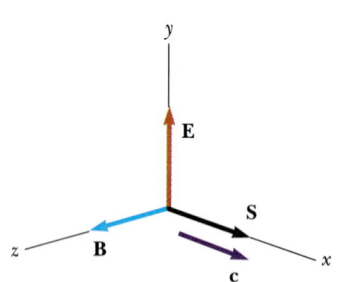

FIGURE 34.7   The Poynting vector **S** for a plane electromagnetic wave is along the direction of propagation.

These equations for S apply at any instant of time.

What is of more interest for a sinusoidal plane electromagnetic wave is the time average of S over one or more cycles, called the wave intensity, I. When this average is taken, we obtain an expression involving the time average of $\cos^2(kx - \omega t)$, which equals $\frac{1}{2}$. Hence, the average value of S (or the intensity of the wave) is

Wave intensity

$$I = S_{av} = \frac{E_{max} B_{max}}{2\mu_0} = \frac{E_{max}^2}{2\mu_0 c} = \frac{c}{2\mu_0} B_{max}^2 \tag{34.20}$$

The constant $\mu_0 c$, called the **impedance of free space,** has the SI unit of ohms and the value

$$\mu_0 c = \sqrt{\frac{\mu_0}{\epsilon_0}} = 120\pi \ \Omega \approx 377 \ \Omega$$

Impedance of free space

Recall that the energy per unit volume u_E, which is the instantaneous energy density associated with an electric field (Section 26.4), is given by Equation 26.12:

$$u_E = \tfrac{1}{2}\epsilon_0 E^2$$

and that the instantaneous energy density u_B associated with a magnetic field (Section 32.3) is given by Equation 32.14:

$$u_B = \frac{B^2}{2\mu_0}$$

Because E and B vary with time for an electromagnetic wave, the energy densities also vary with time. Using the relationships $B = E/c$ and $c = 1/\sqrt{\epsilon_0\mu_0}$, Equation 32.15 becomes

$$u_B = \frac{(E/c)^2}{2\mu_0} = \frac{\epsilon_0\mu_0}{2\mu_0} E^2 = \tfrac{1}{2}\epsilon_0 E^2$$

Comparing this result with the expression for u_E we see that

$$u_B = u_E = \tfrac{1}{2}\epsilon_0 E^2 = \frac{B^2}{2\mu_0}$$

Total energy density

That is, *for an electromagnetic wave, the instantaneous energy density associated with the magnetic field equals the instantaneous energy density associated with the electric field.* Hence, in a given volume the energy is equally shared by the two fields.

The **total instantaneous energy density** u is equal to the sum of the energy densities associated with the electric and magnetic fields:

$$u = u_E + u_B = \epsilon_0 E^2 = \frac{B^2}{\mu_0}$$

When this is averaged over one or more cycles of an electromagnetic wave, we again get a factor of $\frac{1}{2}$. Hence, the total average energy per unit volume of an electromagnetic wave is

$$u_{av} = \epsilon_0 (E^2)_{av} = \tfrac{1}{2}\epsilon_0 E_{max}^2 = \frac{B_{max}^2}{2\mu_0} \tag{34.21}$$

Average energy density of an electromagnetic wave

Comparing this result with Equation 34.20 for the average value of S, we see that

$$I = S_{av} = cu_{av} \tag{34.22}$$

In other words, *the intensity of an electromagnetic wave equals the average energy density multiplied by the speed of light.*

EXAMPLE 34.2   Fields Due to a Point Source

A point source of electromagnetic radiation has an average power output of 800 W. Calculate the maximum values of the electric and magnetic fields at a point 3.50 m from the source.

Solution   Recall from Chapter 17 that the wave intensity, I, a distance r from a point source is

$$I = \frac{P_{av}}{4\pi r^2}$$

where P_{av} is the average power output of the source and $4\pi r^2$ is the area of a sphere of radius r centered on the source. Because the intensity of an electromagnetic wave is also given by Equation 34.20, we have

$$I = \frac{P_{av}}{4\pi r^2} = \frac{E^2_{max}}{2\mu_0 c}$$

Solving for E_{max} gives

$$E_{max} = \sqrt{\frac{\mu_0 c P_{av}}{2\pi r^2}}$$

$$= \sqrt{\frac{(4\pi \times 10^{-7}\ \text{N/A}^2)(3.00 \times 10^8\ \text{m/s})(800\ \text{W})}{2\pi(3.50\ \text{m})^2}}$$

$$= 62.6\ \text{V/m}$$

We calculate the maximum value of the magnetic field using this result and the relationship $B_{max} = E_{max}/c$ (Eq. 34.13):

$$B_{max} = \frac{E_{max}}{c} = \frac{62.6\ \text{V/m}}{3.00 \times 10^8\ \text{m/s}} = 2.09 \times 10^{-7}\ \text{T}$$

Exercise   Calculate the energy density 3.50 m from the point source.

Answer $1.73 \times 10^{-8}\ \text{J/m}^3$.

34.4   MOMENTUM AND RADIATION PRESSURE

Electromagnetic waves transport linear momentum as well as energy. Hence, it follows that pressure is exerted on a surface when an electromagnetic wave impinges on it. In what follows, we assume that the electromagnetic wave transports a total energy U to a surface in a time t. If the surface absorbs all the incident energy U in this time, Maxwell showed that the total momentum **p** delivered to this surface has a magnitude

Momentum delivered to an absorbing surface

$$p = \frac{U}{c} \qquad \text{(complete absorption)} \tag{34.23}$$

Furthermore, if the Poynting vector of the wave is **S**, the radiation pressure P (force per unit area) exerted on the perfect absorbing surface is

Radiation pressure exerted on a perfect absorbing surface

$$P = \frac{S}{c} \tag{34.24}$$

A black body is such a perfectly absorbing surface (Chapter 20), for which all of the incident energy is absorbed (none is reflected).

On the other hand, if the surface is a perfect reflector, then the momentum delivered in a time t for normal incidence is twice that given by Equation 34.24. That is, a momentum U/c is delivered first by the incident wave and then again by the reflected wave, in analogy with a ball colliding elastically with a wall. Therefore,

$$p = \frac{2U}{c} \qquad \text{(complete reflection)} \tag{34.25}$$

The momentum delivered to a surface having a reflectivity somewhere between these two extremes has a value between U/c and $2U/c$, depending on the properties of the surface. Finally, the radiation pressure exerted on a perfect reflecting

A solar home in Oregon. (*John Neal/Photo Researchers, Inc.*)

surface for normal incidence of the wave is given by[5]

$$P = \frac{2S}{c} \qquad (34.26)$$

Although radiation pressures are very small (about 5×10^{-6} N/m² for direct sunlight), they have been measured using torsion balances such as the one shown in Figure 34.8. Light is allowed to strike either a mirror or a black disk, both of which are suspended from a fine fiber. Light striking the black disk is completely absorbed, and so all of its momentum is transferred to the disk. Light striking the mirror (normal incidence) is totally reflected, hence the momentum transfer is twice as great as that transferred to the disk. The radiation pressure is determined by measuring the angle through which the horizontal connecting rod rotates. The apparatus must be placed in a high vacuum to eliminate the effects of air currents.

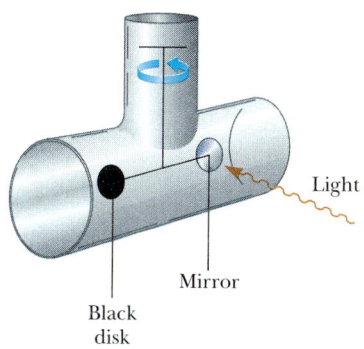

Light

Mirror

Black disk

FIGURE 34.8 An apparatus for measuring the pressure exerted by light. In practice, the system is contained in a high vacuum.

EXAMPLE 34.3 Solar Energy

The Sun delivers about 1000 W/m² of electromagnetic flux to the Earth's surface. (a) Calculate the total power incident on a roof of dimensions 8.00 m × 20.0 m.

Solution The Poynting vector has a magnitude $S = 1000$ W/m², which represents the power per unit area, or the light intensity. Assuming the radiation is incident normal to the roof (Sun directly overhead), we get

$$\text{Power} = SA = (1000 \text{ W/m}^2)(8.00 \times 20.0 \text{ m}^2)$$

$$= 1.60 \times 10^5 \text{ W}$$

If this power could all be converted to electrical energy, it would provide more than enough power for the average home. However, solar energy is not easily harnessed, and the prospects for large-scale conversion are not as bright as they may appear from this simple calculation. For example, the conversion efficiency from solar to electrical energy is typically 10% for photovoltaic cells. Roof systems for converting

solar energy to thermal energy are approximately 50% efficient; however, there are other practical problems with solar energy that must be considered, such as initial cost, overcast days, geographic location, and energy storage.

(b) Determine the radiation pressure and radiation force on the roof assuming the roof covering is a perfect absorber.

Solution Using Equation 34.24 with $S = 1000$ W/m², we find that the radiation pressure is

$$P = \frac{S}{c} = \frac{1000 \text{ W/m}^2}{3.00 \times 10^8 \text{ m/s}} = 3.33 \times 10^{-6} \text{ N/m}^2$$

Because pressure equals force per unit area, this corresponds to a radiation force of

$$F = PA = (3.33 \times 10^{-6} \text{ N/m}^2)(160 \text{ m}^2) = 5.33 \times 10^{-4} \text{ N}$$

Exercise How much solar energy (in joules) is incident on the roof in 1 h?

Answer 5.76×10^8 J.

---

[5] For oblique incidence, the momentum transferred is $2U \cos \theta / c$ and the pressure is given by $P = 2S \cos^2 \theta / c$, where θ is the angle between the normal to the surface and the direction of propagation.

EXAMPLE 34.4   Poynting Vector for a Wire

A long, straight wire of resistance R, radius a, and length ℓ carries a constant current I as in Figure 34.9. Calculate the Poynting vector for this wire.

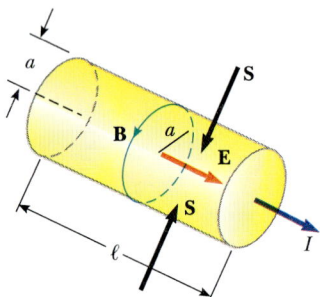

FIGURE 34.9   (Example 34.4) A wire of length ℓ, resistance R, and radius a carrying a current I. The Poynting vector S is directed radially inward.

Solution   First, let us find the electric field E along the wire. If V is the potential difference across its ends, then $V = IR$ and

$$E = \frac{V}{\ell} = \frac{IR}{\ell}$$

Recall that the magnetic field at the surface of the wire (Example 30.4) is

$$B = \frac{\mu_0 I}{2\pi a}$$

The vectors E and B are mutually perpendicular, as shown in Figure 34.9, and therefore $|\mathbf{E} \times \mathbf{B}| = EB$. Hence, the Poynting vector S is directed radially inward and has a magnitude

$$S = \frac{EB}{\mu} = \frac{1}{\mu} \frac{IR}{\ell} \frac{\mu_0 I}{2\pi a} = \frac{I^2 R}{2\pi a \ell} = \boxed{\frac{I^2 R}{A}}$$

where $A = 2\pi a\ell$ is the surface area of the wire, and the total area through which S passes. From this result, we see that

$$SA = I^2 R$$

where SA has units of power (J/s = W). That is, *the rate at which electromagnetic energy flows into the wire, SA, equals the rate of energy (or power) dissipated as joule heat, $I^2 R$.*

Exercise   A heater wire of radius 0.30 mm, length 1.0 m, and resistance 5.0 Ω carries a current of 2.0 A. Determine the magnitude and direction of the Poynting vector for this wire.

Answer 1.1×10^4 W/m^2 directed radially inward.

*34.5   RADIATION FROM AN INFINITE CURRENT SHEET

In this section, we describe the fields radiated by a conductor carrying a time-varying current. In the symmetric plane geometry we treat, the mathematics is less complex than that required in lower-symmetry situations.

Consider an infinite conducting sheet lying in the yz plane and carrying a surface current per unit length $\mathbf{J}_S$ in the y direction, as in Figure 34.10. Let us assume that J_S varies sinusoidally with time as

$$J_S = J_0 \cos \omega t$$

A similar problem for the case of a steady current was treated in Example 30.6, where we found that the magnetic field outside the sheet is everywhere parallel to the sheet and lies along the z axis. The magnetic field was found to have a magnitude

$$B_z = -\mu_0 \frac{J_S}{2}$$

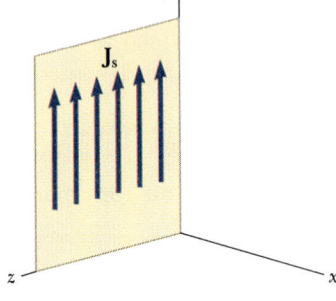

FIGURE 34.10   An infinite current sheet lying in the yz plane. The current density is sinusoidal and given by $J_s = J_0 \cos \omega t$. The magnetic field is everywhere parallel to the sheet and lies along z.

In the present situation, where J_S varies with time, this equation for B_z is valid only for distances close to the sheet. That is,

$$B_z = -\frac{\mu_0}{2} J_0 \cos \omega t \qquad \text{(for small values of } x\text{)}$$

To obtain the expression for B_z for arbitrary values of x, we can investigate the following solution[6]:

$$B_z = -\frac{\mu_0 J_0}{2} \cos(kx - \omega t) \qquad (34.27)$$

Radiated magnetic field

There are two things to note about this solution, which is unique to the geometry under consideration. First, it agrees with our original solution for small values of x. Second, it satisfies the wave equation as it is expressed in Equation 34.9. Hence, we conclude that the magnetic field lies along the z axis and is characterized by a transverse traveling wave having an angular frequency ω, angular wave number $k = 2\pi/\lambda$, and wave speed c.

We can obtain the radiated electric field that accompanies this varying magnetic field by using Equation 34.13:

$$E_y = cB_z = -\frac{\mu_0 J_0 c}{2} \cos(kx - \omega t) \qquad (34.28)$$

Radiated electric field

That is, the electric field is in the y direction, perpendicular to **B**, and has the same space and time dependencies.

These expressions for B_z and E_y show that the radiation field of an infinite current sheet carrying a sinusoidal current is a plane electromagnetic wave propagating with a speed c along the x axis, as shown in Figure 34.11.

We can calculate the Poynting vector for this wave by using Equation 34.19 together with Equations 34.27 and 34.28:

$$S = \frac{EB}{\mu_0} = \frac{\mu_0 J_0^2 c}{4} \cos^2(kx - \omega t) \qquad (34.29)$$

The intensity of the wave, which equals the average value of S, is

$$S_{av} = \frac{\mu_0 J_0^2 c}{8} \qquad (34.30)$$

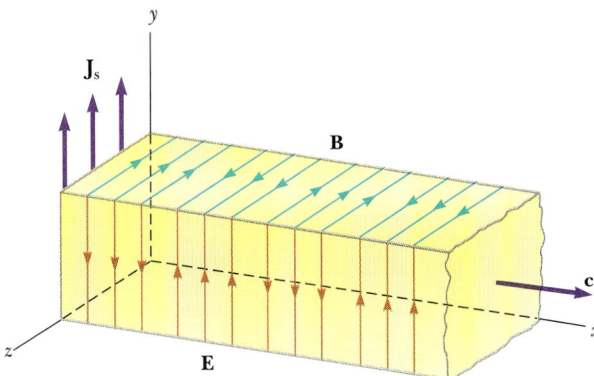

FIGURE 34.11 Representation of the plane electromagnetic wave radiated by an infinite current sheet lying in the yz plane. Note that **B** is in the z direction, **E** is in the y direction, and the direction of wave motion is along x. Both vectors have a $\cos(kx - \omega t)$ behavior.

---

[6] Note that the solution could also be written in the form $\cos(\omega t - kx)$, which is equivalent to $\cos(kx - \omega t)$. That is, $\cos\theta$ is an even function, which means that $\cos(-\theta) = \cos\theta$.

The intensity given by Equation 34.30 represents the average intensity of the outgoing wave on each side of the sheet. The total rate of energy emitted per unit area of the conductor is $2S_{av} = \mu_0 J_0^2 c/4$.

EXAMPLE 34.5   An Infinite Sheet Carrying a Sinusoidal Current

An infinite current sheet lying in the yz plane carries a sinusoidal current the density of which has a maximum value of 5.00 A/m. (a) Find the maximum values of the radiated magnetic field and electric field.

Solution   From Equations 34.28 and 34.29, we see that the maximum values of B_z and E_y are

$$B_{max} = \frac{\mu_0 J_0}{2} \quad \text{and} \quad E_{max} = \frac{\mu_0 J_0 c}{2}$$

Using the values $\mu_0 = 4\pi \times 10^{-7}$ Wb/A·m, $J_0 = 5.00$ A/m, and $c = 3.00 \times 10^8$ m/s, we get

$$B_{max} = \frac{(4\pi \times 10^{-7}\ \text{Wb/A·m})(5.00\ \text{A/m})}{2}$$

$$= 3.14 \times 10^{-16}\ \text{T}$$

$$E_{max} = \frac{(4\pi \times 10^{-7}\ \text{Wb/A·m})(5.00\ \text{A/m})(3.00 \times 10^8\ \text{m/s})}{2}$$

$$= 942\ \text{V/m}$$

(b) What is the average power incident on a plane surface that is parallel to the sheet and has an area of 3.00 m²? (The length and width of this surface are both much larger than the wavelength of the light.)

Solution   The power per unit area (the average value of the Poynting vector) radiated in each direction by the current sheet is given by Equation 34.30. Multiplying this by the area of the plane in question gives the incident power:

$$P = \left(\frac{\mu_0 J_0^2 c}{8}\right) A$$

$$= \frac{(4\pi \times 10^{-7}\ \text{Wb/A·m})}{8}(5.00\ \text{A/m})^2(3.00 \times 10^8\ \text{m/s})}{8}(3.00\ \text{m}^2)$$

$$= 3.54 \times 10^3\ \text{W}$$

The result is independent of the distance from the current sheet because we are dealing with a plane wave.

---

*34.6   THE PRODUCTION OF ELECTROMAGNETIC WAVES BY AN ANTENNA

Electromagnetic waves arise as a consequence of two effects:

- A changing magnetic field produces an electric field;
- A changing electric field produces a magnetic field.

Therefore, it is clear that neither stationary charges nor steady currents can produce electromagnetic waves. Whenever the current through a wire changes with time, however, the wire emits electromagnetic radiation.

Accelerating charges produce EM radiation.

> The fundamental mechanism responsible for this radiation is the acceleration of a charged particle. Whenever a charged particle accelerates, it must radiate energy.

An alternating voltage applied to the wires of an antenna forces an electric charge in the antenna to oscillate. This is a common technique for accelerating charged particles and is the source of the radio waves emitted by the antenna of a radio station. Figure 34.12 shows how this is done. Two metal rods are connected to an ac generator, which causes charges to oscillate between the two rods. The output voltage of the generator is sinusoidal. At $t = 0$, the upper rod is given a maximum positive charge and the bottom rod an equal negative charge, as in Figure 34.12a. The electric field near the antenna at this instant is also shown in

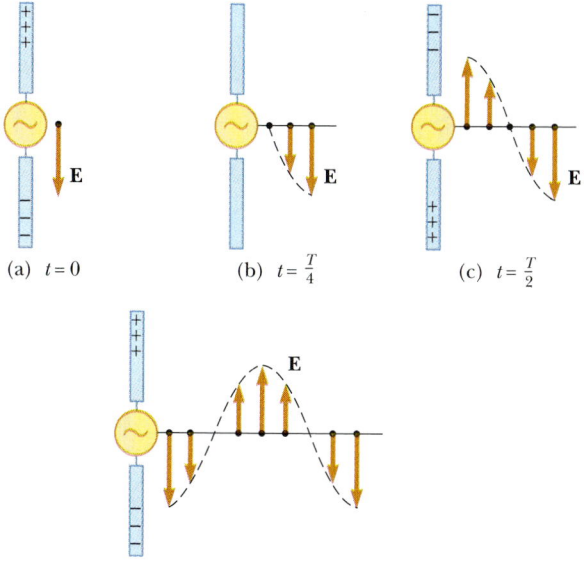

(a) $t = 0$

(b) $t = \frac{T}{4}$

(c) $t = \frac{T}{2}$

(d) $t = T$

FIGURE 34.12   The electric field set up by charges oscillating in an antenna. The field moves away from the antenna with the speed of light.

Figure 34.12a. As the charges oscillate, the rods become less charged, the field near the rods decreases in strength, and the downward-directed maximum electric field produced at $t = 0$ moves away from the rod. When the charges are neutralized, as in Figure 34.12b, the electric field at the rod has dropped to zero. This occurs at a time equal to one quarter of the period of oscillation.

Continuing in this fashion, the upper rod soon obtains a maximum negative charge and the lower rod becomes positive, as in Figure 34.12c, resulting in an electric field near the rod directed upward. This occurs after a time equal to one-half the period of oscillation. The oscillations continue as indicated in Figure 34.12d. (A magnetic field oscillating perpendicular to the plane of the diagram in Figure 34.12 accompanies the oscillating electric field, but it is not shown for clarity.) The electric field near the antenna oscillates in phase with the charge distribution. That is, the field points down when the upper rod is positive and up when the upper rod is negative. Furthermore, the magnitude of the field at any instant depends on the amount of charge on the rods at that instant.

As the charges continue to oscillate (and accelerate) between the rods, the electric field they set up moves away from the antenna at the speed of light. As you can see from Figure 34.12 one cycle of charge oscillation produces one wavelength in the electric field pattern.

Next, consider what happens when two conducting rods are connected to the opposite ends of a battery (Fig. 34.13). Before the switch is closed, the current is zero, and so there are no fields present (Fig. 34.13a). Just after the switch is closed, charge of opposite signs begins to build up on the rods (Fig. 34.13b), which corresponds to a time-varying current. The changing charge causes the electric field to change, which in turn produces a magnetic field around the rods.[7] Finally, when the rods are fully charged, the current is zero and there is no magnetic field (Fig. 34.13c).

---

[7] We have neglected the field caused by the wires leading to the rods. This is a good approximation if the circuit dimensions are small relative to the length of the rods.

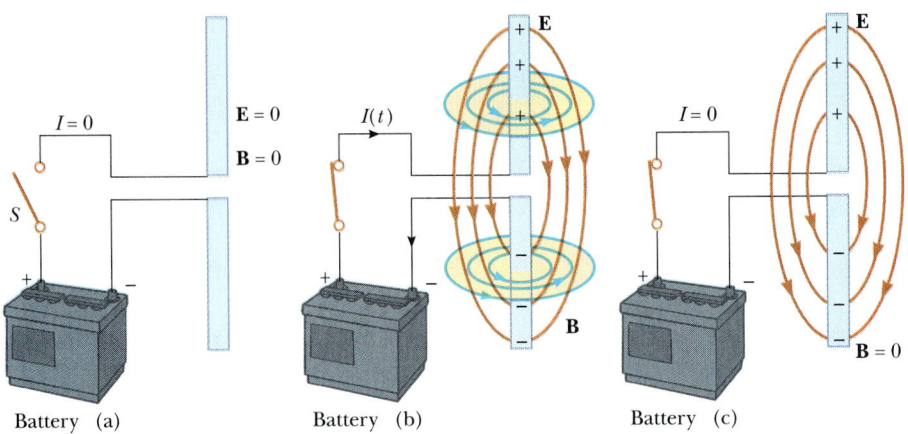

FIGURE 34.13   A pair of metal rods connected to a battery. (a) When the switch is open and there is no current, the electric and magnetic fields are both zero. (b) After the switch is closed and the rods are being charged (so that a current exists), the rods generate changing electric and magnetic fields. (c) When the rods are fully charged, the current is zero, the electric field is a maximum, and the magnetic field is zero.

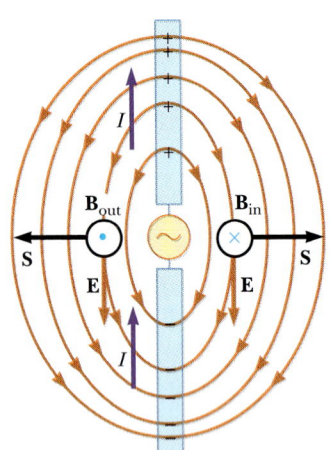

FIGURE 34.14   A half-wave (dipole) antenna consists of two metal rods connected to an alternating voltage source. The diagram shows **E** and **B** at an instant when the current is upward. Note that the electric field lines resemble those of a dipole.

Now let us consider the production of electromagnetic waves by a half-wave antenna. In this arrangement, two conducting rods, each one quarter of a wavelength long, are connected to a source of alternating emf (such as an *LC* oscillator), as in Figure 34.14. The oscillator forces charges to accelerate back and forth between the two rods. Figure 34.14 shows the field configuration at some instant when the current is upward. The electric field lines resemble those of an electric dipole, that is, two equal and opposite charges. Since these charges are continuously oscillating between the two rods, the antenna can be approximated by an oscillating electric dipole. The magnetic field lines form concentric circles around the antenna and are perpendicular to the electric field lines at all points. The magnetic field is zero at all points along the axis of the antenna. Furthermore, **E** and **B** are 90° out of phase in time, that is, **E** at some point reaches its maximum value when **B** is zero and vice versa. This is because when the charges at the ends of the rods are at a maximum, the current is zero.

At the two points where the magnetic field is shown in Figure 34.14, the Poynting vector **S** is radially outward. This indicates that energy is flowing away from the antenna at this instant. At later times, the fields and the Poynting vector change direction as the current alternates. Since **E** and **B** are 90° out of phase at points near the dipole, the net energy flow is zero. From this, we might conclude (incorrectly) that no energy is radiated by the dipole.

Since the dipole fields fall off as $1/r^3$ (as in the case of a static dipole discussed in Chapter 23), they are not important at large distances from the antenna. However, at these large distances, another effect produces the radiation field. The source of this radiation is the continuous induction of an electric field by a time-varying magnetic field and the induction of a magnetic field by a time-varying electric field, predicted by Equations 34.3 and 34.4. The electric and magnetic fields produced in this manner are in phase with each other and vary as $1/r$. The result is an outward flow of energy at all times.

The electric field lines produced by an oscillating dipole at some instant are shown in Figure 34.15. Note that the intensity (and the power radiated) is a maximum in a plane that is perpendicular to the antenna and passing through its midpoint. Furthermore, the power radiated is zero along the antenna's axis. A

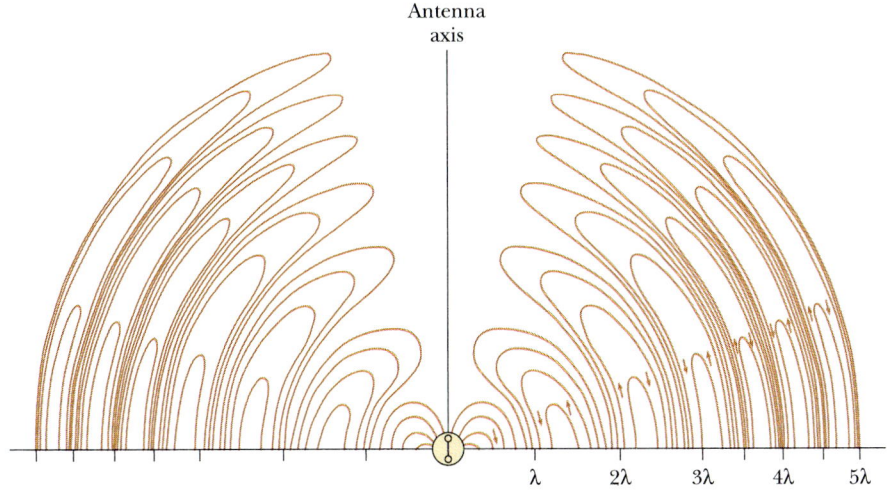

Antenna
axis

λ 2λ 3λ 4λ 5λ

FIGURE 34.15   Electric field lines surrounding an oscillating dipole at a given instant. The radiation fields propagate outward from the dipole with a speed c.

mathematical solution to Maxwell's equations for the oscillating dipole shows that the intensity of the radiation field varies as $\sin^2 \theta / r^2$, where θ is measured from the axis of the antenna. The angular dependence of the radiation intensity (power per unit area) is sketched in Figure 34.16.

Electromagnetic waves can also induce currents in a receiving antenna. The response of a dipole-receiving antenna at a given position is a maximum when the antenna axis is parallel to the electric field at that point and zero when the axis is perpendicular to the electric field.

34.7   THE SPECTRUM OF ELECTROMAGNETIC WAVES

Since all electromagnetic waves travel through vacuum with a speed c, their frequency f and wavelength λ are related by the important expression

$$c = f\lambda \qquad (34.31)$$

For instance, a radio wave of frequency 5.00 MHz (a typical value) has a wavelength

$$\lambda = \frac{c}{f} = \frac{3.00 \times 10^8 \text{ m/s}}{5.00 \times 10^6 \text{ s}^{-1}} = 60.0 \text{ m}$$

The various types of electromagnetic waves are listed in Figure 34.17. Note the wide range of frequencies and wavelengths. There is no sharp dividing point between one kind of wave and the next. It should be noted that all forms of radiation are produced (classically) by accelerating charges.

Radio waves, as we just saw, are the result of charges accelerating through conducting wires. They are generated by such electronic devices as LC oscillators and are used in radio and television communication systems.

Microwaves have wavelengths ranging between approximately 1 mm and 30 cm and are also generated by electronic devices. Because of their short wavelength, they are well suited for the radar systems used in aircraft navigation and for study-

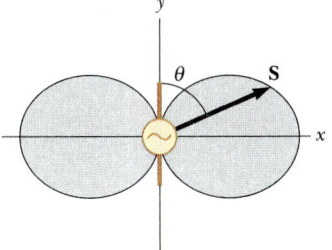

FIGURE 34.16   Angular dependence of the intensity of radiation produced by an oscillating electric dipole.

Most widely used in rural locations, satellite-dish television antennas receive television-station signals from satellites in orbit around the Earth. *(© Hank Delespinasse/The IMAGE Bank)*

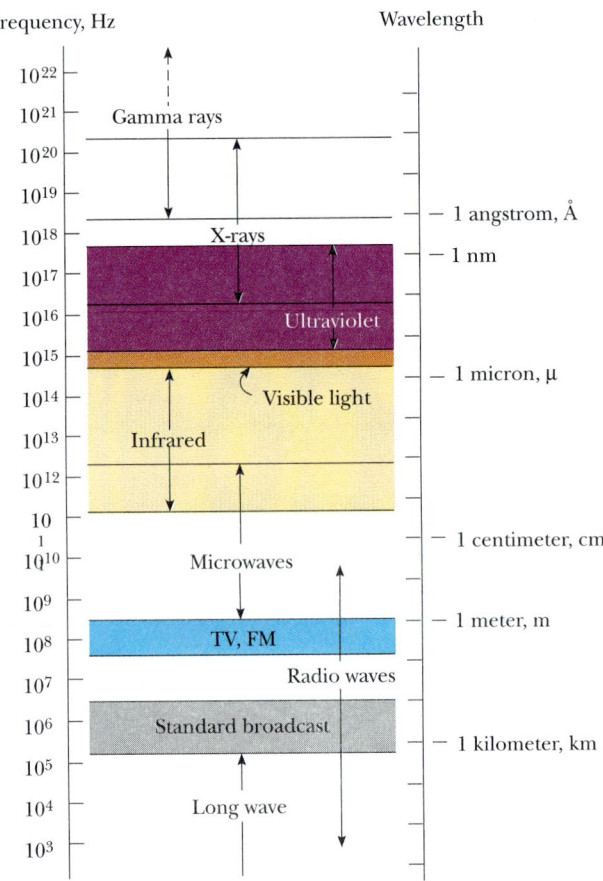

FIGURE 34.17   The electromagnetic spectrum. Note the overlap between adjacent wave types.

ing the atomic and molecular properties of matter. Microwave ovens represent an interesting domestic application of these waves. It has been suggested that solar energy could be harnessed by beaming microwaves down to Earth from a solar collector in space.[8]

Infrared waves

Infrared waves (sometimes called heat waves) have wavelengths ranging from approximately 1 mm to the longest wavelength of visible light, 7×10^{-7} m. These waves, produced by hot bodies and molecules, are readily absorbed by most materials. The infrared energy absorbed by a substance appears as heat because the energy agitates the atoms of the body, increasing their vibrational and rotational motion, which results in a temperature rise. Infrared radiation has many practical and scientific applications, including physical therapy, infrared photography, and vibrational spectroscopy.

Visible waves

Visible light, the most familiar form of electromagnetic waves, is that part of the electromagnetic spectrum that the human eye can detect. Light is produced by the rearrangement of electrons in atoms and molecules. The various wavelengths of visible light are classified with colors ranging from violet ($\lambda \approx 4 \times 10^{-7}$ m) to red ($\lambda \approx 7 \times 10^{-7}$ m). The eye's sensitivity is a function of wavelength, the sensitivity being a maximum at a wavelength of about 5.6×10^{-7} m (yellow-green).

Ultraviolet waves

Ultraviolet light covers wavelengths ranging from approximately 3.8×10^{-7} m (380 nm) down to 6×10^{-10} m (0.6 nm). The Sun is an important source of ultravi-

[8] P. Glaser, ''Solar Power from Satellites,'' *Physics Today,* February, 1977, p. 30.

olet light, which is the main cause of suntans. Most of the ultraviolet light from the Sun is absorbed by atoms in the upper atmosphere, or stratosphere. This is fortunate since uv light in large quantities produces harmful effects on humans. One important constituent of the stratosphere is ozone (O_3), which results from reactions of oxygen with ultraviolet radiation. This ozone shield converts lethal high-energy ultraviolet radiation to heat, which in turn warms the stratosphere. Recently, there has been a great deal of controversy concerning the possible depletion of the protective ozone layer as a result of the use of the freons used in aerosol spray cans and as refrigerants.

X-rays are electromagnetic waves with wavelengths in the range of approximately 10^{-8} m (10 nm) down to 10^{-13} m (10^{-4} nm). The most common source of x-rays is the deceleration of high-energy electrons bombarding a metal target. X-rays are used as a diagnostic tool in medicine and as a treatment for certain forms of cancer. Since x-rays damage or destroy living tissues and organisms, care must be taken to avoid unnecessary exposure or overexposure. X-rays are also used in the study of crystal structure, since x-ray wavelengths are comparable to the atomic separation distances (≈ 0.1 nm) in solids.

X-rays

Gamma rays are electromagnetic waves emitted by radioactive nuclei (such as ^{60}Co and ^{137}Cs) and during certain nuclear reactions. They have wavelengths ranging from approximately 10^{-10} m to less than 10^{-14} m. They are highly penetrating and produce serious damage when absorbed by living tissues. Consequently, those working near such dangerous radiation must be protected with heavily absorbing materials, such as thick layers of lead.

Gamma rays

SUMMARY

Electromagnetic waves, which are predicted by Maxwell's equations, have the following properties:

- The electric and magnetic fields satisfy the following wave equations, which can be obtained from Maxwell's third and fourth equations:

$$\frac{\partial^2 E}{\partial x^2} = \mu_0 \epsilon_0 \frac{\partial^2 E}{\partial t^2} \tag{34.8}$$

$$\frac{\partial^2 B}{\partial x^2} = \mu_0 \epsilon_0 \frac{\partial^2 B}{\partial t^2} \tag{34.9}$$

- Electromagnetic waves travel through a vacuum with the speed of light c, where

$$c = \frac{1}{\sqrt{\mu_0 \epsilon_0}} = 3.00 \times 10^8 \text{ m/s} \tag{34.10}$$

- The electric and magnetic fields of an electromagnetic wave are perpendicular to each other and perpendicular to the direction of wave propagation. (Hence, electromagnetic waves are transverse waves.)
- The instantaneous magnitudes of $|\mathbf{E}|$ and $|\mathbf{B}|$ in an electromagnetic wave are related by the expression

$$\frac{E}{B} = c \tag{34.13}$$

- Electromagnetic waves carry energy. The rate of flow of energy crossing a unit area is described by the Poynting vector **S**, where

$$\mathbf{S} \equiv \frac{1}{\mu_0} \mathbf{E} \times \mathbf{B} \qquad (34.18)$$

- Electromagnetic waves carry momentum and hence can exert pressure on surfaces. If an electromagnetic wave whose Poynting vector is **S** is completely absorbed by a surface upon which it is normally incident, the radiation pressure on that surface is

$$P = \frac{S}{c} \qquad \text{(complete absorption)} \qquad (34.24)$$

If the surface totally reflects a normally incident wave, the pressure is doubled.

The electric and magnetic fields of a sinusoidal plane electromagnetic wave propagating in the positive x direction can be written

$$E = E_{max} \cos(kx - \omega t) \qquad (34.11)$$

$$B = B_{max} \cos(kx - \omega t) \qquad (34.12)$$

where ω is the angular frequency of the wave and k is the angular wave number. These equations represent special solutions to the wave equations for E and B. Since $\omega = 2\pi f$ and $k = 2\pi/\lambda$, where f and λ are the frequency and wavelength, respectively, it is found that

$$\frac{\omega}{k} = \lambda f = c$$

The average value of the Poynting vector for a plane electromagnetic wave has a magnitude

$$S_{av} = \frac{E_{max} B_{max}}{2\mu_0} = \frac{E_{max}^2}{2\mu_0 c} = \frac{c}{2\mu_0} B_{max}^2 \qquad (34.20)$$

The intensity of a sinusoidal plane electromagnetic wave equals the average value of the Poynting vector taken over one or more cycles.

The electromagnetic spectrum includes waves covering a broad range of frequencies and wavelengths. The frequency f and wavelength λ of a given wave are related by

$$c = f\lambda \qquad (34.31)$$

QUESTIONS

1. For a given incident energy of an electromagnetic wave, why is the radiation pressure on a perfect reflecting surface twice as large as that on a perfect absorbing surface?
2. Describe the physical significance of the Poynting vector.
3. Do all current-carrying conductors emit electromagnetic waves? Explain.
4. What is the fundamental source of electromagnetic radiation?
5. Electrical engineers often speak of the radiation resistance of an antenna. What do you suppose they mean by this phrase?

6. If a high-frequency current is passed through a solenoid containing a metallic core, the core heats up by induction. Explain why the materials heat up in these situations.
7. Certain orientations of the receiving antenna on a television set give better reception than others. Furthermore, the best orientation varies from station to station. Explain.
8. Does a wire connected to a battery emit an electromagnetic wave? Explain.
9. If you charge a comb by running it through your hair and

then hold the comb next to a bar magnet, do the electric and magnetic fields produced constitute an electromagnetic wave?

10. An empty plastic or glass dish removed from a microwave oven is cool to the touch right after it is removed from the oven. How can this be possible? (You can assume that your electric bill has been paid.)

11. Often when you touch the indoor antenna on a radio or television receiver, the reception instantly improves. Why?

12. Explain how the (dipole) VHF antenna of a television set works.

13. Explain how the UHF (loop) antenna of a television set works.

14. Explain why the voltage induced in a UHF (loop) antenna depends on the frequency of the signal, while the voltage in a VHF (dipole) antenna does not.

15. List as many similarities and differences as you can between sound waves and light waves.

16. What does a radio wave do to the charges in the receiving antenna to provide a signal for your car radio?

17. What determines the height of an AM radio station's broadcast antenna?

18. Some radio transmitters use a "phased array" of antennas. What is their purpose?

19. What happens to the radio reception in an airplane as it flies over the (vertical) dipole antenna of the control tower?

20. When light (or other electromagnetic radiation) travels across a given region, what is it that moves?

21. Why should an infrared photograph of a person look different from a photograph taken with visible light?

PROBLEMS

Section 34.2 Plane Electromagnetic Waves

1. If the North Star, Polaris, were to burn out today, in what year would it disappear from our vision? The distance from the Earth to Polaris is approximately 6.44×10^{18} m.

2. An electromagnetic wave in vacuum has an electric field amplitude of 220 V/m. Calculate the amplitude of the corresponding magnetic field.

3. (a) Use the relationship $B = \mu_0 H$ (Eq. 30.37) for free space where $\mu_m = \mu_0$ together with the properties of E and B described in Section 34.2 to show that $E/H = \sqrt{\mu_0/\epsilon_0}$. Recall that H is the magnetic intensity. (b) Calculate the numerical value of this ratio and show that it has SI units of ohms. (Because $E/H = \sqrt{\mu_0/\epsilon_0} = \mu_0 c$, we see that ratio E/H is equal to the impedance of free space discussed in Section 34.3.)

4. Show that $E = f(x - ct) + g(x + ct)$ satisfies Equation 34.8, where f and g are any functions.

5. The magnetic field amplitude of an electromagnetic wave is 5.4×10^{-7} T. Calculate the electric field amplitude if the wave is traveling (a) in free space and (b) in a medium in which the speed of the wave is $0.8 c$.

6. The speed of an electromagnetic wave traveling in a transparent substance is $v = 1/\sqrt{\kappa \mu_0 \epsilon_0}$, where κ is the dielectric constant of the substance. Determine the speed of light in water, which has a dielectric constant at optical frequencies of 1.78.

7. Calculate the maximum value of the magnetic field in a medium where the speed of light is two-thirds that of the speed of light in vacuum and the electric field amplitude is 7.6 mV/m.

8. Write down expressions for the electric and magnetic fields of a sinusoidal plane electromagnetic wave having a frequency of 3.00 GHz and traveling in the positive x direction. The amplitude of the electric field is 300 V/m.

9. Figure 34.3 shows a plane electromagnetic sinusoidal wave propagating in the x direction. The wavelength is 50.0 m, and the electric field vibrates in the xy plane with an amplitude of 22.0 V/m. Calculate (a) the sinusoidal frequency and (b) the magnitude and direction of **B** when the electric field has its maximum value in the negative y direction. (c) Write an expression for B in the form

$$B = B_{max} \cos(kx - \omega t)$$

with numerical values for B_{max}, k, and ω.

10. Verify that the following equations are solutions to Equations 34.8 and 34.9, respectively:

$$E = E_{max} \cos(kx - \omega t) \qquad B = B_{max} \cos(kx - \omega t)$$

11. In SI units, the electric field in an electromagnetic wave is described by

$$E_y = 100 \sin(1.00 \times 10^7 x - \omega t)$$

Find (a) the amplitude of the corresponding magnetic wave, (b) the wavelength λ, and (c) the frequency f.

Section 34.3 Energy Carried by Electromagnetic Waves

12. How much electromagnetic energy is contained per cubic meter near the Earth's surface if the intensity of sunlight under clear skies is 1000 W/m²?

□ indicates problems that have full solutions available in the Student Solutions Manual and Study Guide.

13. At what distance from a 100-W electromagnetic wave point source does $E_{max} = 15$ V/m?

14. A 10-mW laser has a beam diameter of 1.6 mm. (a) What is the intensity of the light, assuming it is uniform across the circular beam? (b) What is the average energy density of the beam?

15. What is the average magnitude of the Poynting vector 5.0 miles from a radio transmitter broadcasting isotropically with an average power of 250 kW?

16. The Sun radiates electromagnetic energy at the rate of $P_{Sun} = 3.85 \times 10^{26}$ W. (a) At what distance from the Sun does the intensity of its radiation fall to 1000 W/m²? (Compare this distance to the radius of the Earth's orbit.) (b) At the distance you just found, what is the average energy density of the Sun's radiation?

17. The amplitude of the electric field is 0.20 V/m 10 km from a radio transmitter. What is the total power emitted by the transmitter?

18. In a region of free space the electric field intensity at an instant of time is $\mathbf{E} = (80\mathbf{i} + 32\mathbf{j} - 64\mathbf{k})$ N/C and the magnetic field intensity is $\mathbf{B} = (0.20\mathbf{i} + 0.080\mathbf{j} + 0.29\mathbf{k})$ μT. (a) Show that the two fields are perpendicular to each other. (b) Determine the Poynting vector for these fields.

19. The filament of an incandescent lamp has a 150-Ω resistance and carries a direct current of 1.0 A. The filament is 8.0 cm long and 0.90 mm in radius. (a) Calculate the Poynting vector at the surface of the filament. (b) Find the magnitude of the electric and magnetic fields at the surface of the filament.

20. A monochromatic light source emits 100 W of electromagnetic power uniformly in all directions. (a) Calculate the average electric-field energy density 1 m from the source. (b) Calculate the average magnetic-field energy density at the same distance from the source. (c) Find the wave intensity at this location.

21. A helium-neon laser intended for instructional use operates at 5.0 mW. (a) Determine the maximum value of the electric field at a point where the cross-sectional area of the beam is 4.0 mm². (b) Calculate the electromagnetic energy in a 1.0-m length of the beam.

22. A lightbulb's filament has a resistance of 110 Ω. The bulb is plugged into a standard 110-V (rms) outlet and emits 1.0% of the joule heat as electromagnetic radiation of frequency f. Assuming that the bulb is covered with a filter, absorbing all other frequencies, find the amplitude of the magnetic field 1.0 m from the bulb.

23. At one location on the Earth, the rms value of the magnetic field caused by solar radiation is 1.8 μT. From this value, calculate (a) the average electric field due to solar radiation, (b) the average energy density of the solar component of electromagnetic radiation at this location, and (c) the magnitude of the Poynting vector (S_{av}) for the Sun's radiation. (d) Compare the value found in part (c) to the value of the solar flux given in Example 34.3.

24. High-power lasers in factories are used to cut through cloth and metal. One such laser has a beam diameter of 1.00 mm and generates an electric field at the target having an amplitude 0.70 MV/m. Find (a) the amplitude of the magnetic field produced, (b) the intensity of the laser, and (c) the power dissipated.

Section 34.4   Momentum and Radiation Pressure

25. A radio wave transmits 25 W/m² of power per unit area. A flat surface of area A is perpendicular to the direction of propagation of the wave. Calculate the radiation pressure on it if the surface is a perfect absorber.

26. A 100-mW laser beam is reflected back upon itself by a mirror. Calculate the force on the mirror.

27. A 15-mW helium-neon laser ($\lambda = 632.8$ nm) emits a beam of circular cross-section whose diameter is 2.0 mm. (a) Find the maximum electric field in the beam. (b) What total energy is contained in a 1.0-m length of the beam? (c) Find the momentum carried by a 1.0-m length of the beam.

28. Given that the solar radiation incident on the upper atmosphere of the Earth is 1340 W/m², determine (a) the solar radiation incident on Mars, (b) the total power incident on Mars, and (c) the total force acting on Mars. (d) Compare this force with the gravitational attraction between Mars and the Sun. (See Table 14.2.)

29. A plane electromagnetic wave has an energy flux of 750 W/m². A flat, rectangular surface of dimensions 50 cm × 100 cm is placed perpendicularly to the direction of the wave. If the surface absorbs half of the energy and reflects half, calculate (a) the total energy absorbed by the surface in 1.0 min and (b) the momentum absorbed in this time.

30. Lasers have been used to suspend spherical glass beads in the Earth's gravitational field. (a) If a bead has a mass of 1.00 μg and a density of 0.20 g/cm³, determine the radiation intensity needed to support the bead. (b) If the beam has a radius of 0.20 cm, what is the power required for this laser?

30A. Lasers have been used to suspend spherical glass beads in the Earth's gravitational field. (a) If a bead has a mass m and a density ρ, determine the radiation intensity needed to support the bead. (b) If the beam has a radius r, what is the power required for this laser?

*Section 34.5   Radiation from an Infinite Current Sheet

31. A rectangular surface of dimensions 120 cm × 40 cm is parallel to and 4.4 m from a much larger

conducting sheet in which there is a sinusoidally varying surface current that has a maximum value of 10 A/m. (a) Calculate the average power incident on the smaller sheet. (b) What power per unit area is radiated by the larger sheet?

32. A large current-carrying sheet is expected to radiate in each direction (normal to the plane of the sheet) at a rate equal to 570 W/m². What maximum value of sinusoidal current density is required?

*Section 34.6 The Production of Electromagnetic Waves by an Antenna

33. What is the length of a half-wave antenna designed to broadcast 20-MHz radio waves?

34. An AM radio station broadcasts isotropically with an average power of 4.0 kW. A dipole-receiving antenna 65 cm long is located 4.0 miles from the transmitter. Compute the emf induced by this signal between the ends of the receiving antenna.

35. A television set uses a dipole-receiving antenna for VHF channels and a loop antenna for UHF channels. The UHF antenna produces a voltage from the changing magnetic flux through the loop. (a) Using Faraday's law, derive an expression for the amplitude of the voltage that appears in a single-turn circular loop antenna with a radius r. The TV station broadcasts a signal with a frequency f, and the signal has an electric field amplitude E_{max} and magnetic field amplitude B_{max} at the receiving antenna's location. (b) If the electric field in the signal points vertically, what should be the orientation of the loop for best reception?

36. Figure 34.14 shows a Hertz antenna (also known as a half-wave antenna, since its length is $\lambda/2$). The antenna is located far enough from the ground that reflections do not significantly affect its radiation pattern. Most AM radio stations, however, use a Marconi antenna, which consists of the top half of a Hertz antenna. The lower end of this (quarter-wave) antenna is connected to Earth ground, and the ground itself serves as the missing lower half. What is the antenna length for a radio station broadcasting at (a) 560 kHz and (b) 1600 kHz?

37. Accelerating charges can radiate electromagnetic waves. Calculate the wavelength of radiation produced by a proton in a cyclotron of radius 0.50 m and magnetic field strength 0.35 T.

37A. Accelerating charges can radiate electromagnetic waves. Calculate the wavelength of radiation produced by a proton in a cyclotron of radius R and magnetic field strength B.

38. Two radio-transmitting antennas are separated by half the broadcast wavelength and are driven in phase with each other. In which directions are (a) the strongest and (b) the weakest signals radiated?

Section 34.7 The Spectrum of Electromagnetic Waves

39. What is the wavelength of an electromagnetic wave in free space that has a frequency of (a) 5.00×10^{19} Hz and (b) 4.00×10^9 Hz?

40. Suppose you are located 180 m from a radio transmitter. How many wavelengths are you from the transmitter if the station calls itself (a) 1150 AM (the AM band frequencies are in kilohertz) and (b) 98.1 FM (the FM band frequencies are in megahertz)?

41. An important news announcement is transmitted by radio waves to people who are 100 km from the station, listening to their radio, and by sound waves to people sitting across the newsroom, 3.0 m from the newscaster. Who receives the news first? Explain. Take the speed of sound in air to be 343 m/s.

42. What are the wavelength ranges in (a) the AM radio band (540–1600 kHz) and (b) the FM radio band (88–108 MHz)?

43. There are 12 VHF television channels (Channels 2–13) that lie in a frequency range from 54 MHz to 216 MHz. Each channel has a width of 6 MHz, with the two ranges 72–76 MHz and 88–174 MHz reserved for non-TV purposes. (Channel 2, for example, lies between 54 and 60 MHz.) Calculate the wavelength range for (a) Channel 4, (b) Channel 6, and (c) Channel 8.

ADDITIONAL PROBLEMS

44. Assume that the solar radiation incident on the Earth is 1340 W/m². (a) Calculate the total power radiated by the Sun, taking the average Earth-Sun separation to be 1.49×10^{11} m. (b) Determine the maximum values of the electric and magnetic fields at the Earth's surface due to solar radiation.

45. A community plans to build a facility to convert solar radiation to electrical power. They require 1.00 MW of power, and the system to be installed has an efficiency of 30% (that is, 30% of the solar energy incident on the surface is converted to electrical energy). What must be the effective area of a perfectly absorbing surface used in such an installation, assuming a constant energy flux of 1000 W/m²?

46. A microwave source produces pulses of 20-GHz radiation, with each pulse lasting 1.0 ns. A parabolic reflector ($R = 6.0$ cm) is used to focus these into a parallel beam of radiation, as in Figure P34.46. The average power during each pulse is 25 kW. (a) What is the wavelength of these microwaves? (b) What is the total energy contained in each pulse? (c) Com-

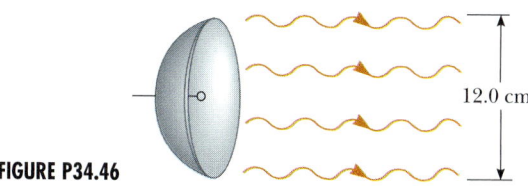

FIGURE P34.46

12.0 cm

pute the average energy density inside each pulse. (d) Determine the amplitude of the electric and magnetic fields in these microwaves. (e) If this pulsed beam strikes an absorbing surface, compute the force exerted on the surface during the 1.0-ns duration of each pulse.

47. A dish antenna having a diameter of 20 m receives (at normal incidence) a radio signal from a distant source, as shown in Figure P34.47. The radio signal is a continuous sinusoidal wave with amplitude $E_{max} = 0.20 \ \mu V/m$. Assume the antenna absorbs all the radiation that falls on the dish. (a) What is the amplitude of the magnetic field in this wave? (b) What is the intensity of the radiation received by this antenna? (c) What is the power received by the antenna? (d) What force is exerted on the antenna by the radio waves?

FIGURE P34.47

48. The electromagnetic power radiated by a nonrelativistic moving point charge q having an acceleration a is

$$P = \frac{q^2 a^2}{6 \pi \epsilon_0 c^3}$$

where ϵ_0 is the permittivity of vacuum and c is the speed of light in vacuum. (a) Show that the right side of this equation is in watts. (b) If an electron is placed in a constant electric field of 100 N/C, determine its acceleration and the electromagnetic power it radiates. (c) If a proton is placed in a cyclotron of radius 0.50 m and magnetic field of magnitude 0.35 T, what is the electromagnetic power radiated by this proton?

49. A thin tungsten filament of length 1.0 m radiates 60 W of power in the form of electromagnetic waves. A perfectly absorbing surface in the form of a hollow cylinder of radius 5.0 cm and length 1.0 m is placed concentrically with the filament. Calculate the radiation pressure acting on the cylinder. (Assume that the radiation is emitted in the radial direction, and neglect end effects.)

50. A group of astronauts plans to propel a spaceship by using a "sail" to reflect solar radiation. The sail is totally reflecting, oriented with its plane perpendicular to the direction of the Sun, and 1.00 km × 1.50 km in size. What is the maximum acceleration that can be expected for a spaceship of 4.00 metric tons (4000 kg)? (Use the solar radiation data from Problem 44 and neglect gravitational forces.)

51. In 1965, Penzias and Wilson discovered the cosmic microwave radiation left over from the Big Bang expansion of the Universe. The energy density of this radiation is $4.0 \times 10^{-14} \ J/m^3$. Determine the corresponding electric field amplitude.

52. A possible means of space flight is to place a perfectly reflecting aluminized sheet into Earth's orbit and use the light from the Sun to push this solar sail. Suppose a sail of area $6.00 \times 10^4 \ m^2$ and mass 6000 kg is placed in orbit facing the Sun. (a) What force is exerted on the sail? (b) What is the sail's acceleration? (c) How long does it take the sail to reach the Moon, 3.84×10^8 m away? Ignore all gravitational effects, assume that the acceleration calculated in part (b) remains constant, and assume a solar intensity of 1380 W/m^2. (*Hint:* The radiation pressure exerted by a reflected wave is given as twice the average power per area divided by the speed of light.)

53. An astronaut in a spacecraft moving with constant velocity wishes to increase the speed of the craft by using a laser beam attached to the spaceship. The laser beam emits 100 J of electromagnetic energy per pulse, and the laser is pulsed at the rate of 0.2 pulse/s. If the mass of the spaceship plus its contents is 5000 kg, for how long a time must the beam be on in order to increase the speed of the vehicle by 1 m/s in the direction of its initial motion? In what direction should the beam be pointed to achieve this?

54. Consider a small, spherical particle of radius r located in space a distance R from the Sun. (a) Show that the ratio $F_{rad}/F_{grav} \propto 1/r$, where F_{rad} = the force exerted by solar radiation and F_{grav} = the force of gravitational attraction. (b) The result of part (a) means that, for a sufficiently small value of r, the force exerted on the particle by solar radiation exceeds the force of gravitational attraction. Calculate the value of r for which the particle is in equilibrium under the two forces. (Assume that the particle has a perfectly absorbing surface and a mass density of 1.50 g/cm^3. Let the particle be located 3.75×10^{11} m from the Sun and use 214 W/m^2 as the value of the solar flux at that point.)

55. The torsion balance shown in Figure 34.8 is used in an experiment to measure radiation pressure. The torque constant (elastic restoring torque) of the suspension fiber is 1.0×10^{-11} N·m/deg, and the length of the horizontal rod is 6.0 cm. The beam from a 3.0-mW helium-neon laser is incident on the black disk, and the mirror disk is completely

shielded. Calculate the angle between the equilibrium positions of the horizontal bar when the beam is switched from "off" to "on."

56. The Earth reflects approximately 38% of the incident sunlight by reflection from its clouds and oceans. (a) Given that the intensity of solar radiation is 1340 W/m², find the radiation pressure at the location on Earth where the Sun is straight overhead. (b) Compare this to normal atmospheric pressure, which is 1.01×10^5 N/m² at the Earth's surface.

57. A linearly polarized microwave of wavelength 1.50 cm is directed along the positive x axis. The electric field vector has a maximum value of 175 V/m and vibrates in the xy plane. (a) Assume that the magnetic field component of the wave can be written in the form $B = B_{max} \sin(kx - \omega t)$ and give values for B_{max}, k, and ω. Also, determine in which plane the magnetic field vector vibrates. (b) Calculate the magnitude of the Poynting vector for this wave. (c) What maximum radiation pressure would this wave exert if directed at normal incidence onto a perfectly reflecting sheet? (d) What maximum acceleration would be imparted to a 500-g sheet (perfectly reflecting and at normal incidence) of dimensions 1.0 m × 0.75 m?

58. A police radar unit transmits at 10.525 GHz, better known as the X-band. (a) Show that when this radar wave is reflected from the front of a moving car, the reflected wave is shifted in frequency by the amount $\Delta f \cong 2fv/c$, where v is the speed of the car. (*Hint:* Treat the reflection of a wave as two separate processes: First, the motion of the car produces a "moving-observer" Doppler shift in the frequency of the waves striking the car, and then these Doppler-shifted waves are re-emitted by a moving source. Also, automobile speeds are low enough that you can use the binomial expansion $(1 + x)^n \cong 1 + nx$ in the acoustic Doppler formulas derived in Chapter 17.) (b) The unit is usually calibrated before and after an arrest for speeds of 35 mph and 80 mph. Calculate the frequency shift produced by these two speeds.

59. An astronaut, stranded in space "at rest" 10 m from his spacecraft, has a mass (including equipment) of 110 kg. Having a 100-W light source that forms a directed beam, he decides to use the beam as a photon rocket to propel himself continuously toward the spacecraft. (a) Calculate how long it takes him to reach the spacecraft by this method. (b) Suppose, instead, he decides to throw the light source away in a direction opposite the spacecraft. If the mass of the light source is 3.0 kg and, after being thrown, moves at 12 m/s *relative to the recoiling astronaut,* how long does he take to reach the spacecraft?

60. A "laser cannon" on a spacecraft has a beam of cross-sectional area A. The maximum electric field in the beam is E. At what rate a will an asteroid acceler-

ate away from the spacecraft if the laser beam strikes the asteroid perpendicularly to its surface, and the surface is nonreflecting? The mass of the asteroid is m. Neglect the acceleration of the spacecraft.

61. (a) For a parallel-plate capacitor having a small plate separation compared with the length or width of a plate, show that the displacement current is

$$I_d = C \frac{dV}{dt}$$

(b) Calculate the value of dV/dt required to produce a displacement current of 1.00 A in a 1.00-μF capacitor.

62. Consider the situation shown in Figure P34.62. An electric field of 300 V/m is confined to a circular area 10 cm in diameter and directed outward from the plane of the figure. If the field is increasing at a rate of 20 V/m·s, what are the direction and magnitude of the magnetic field at the point P, 15 cm from the center of the circle?

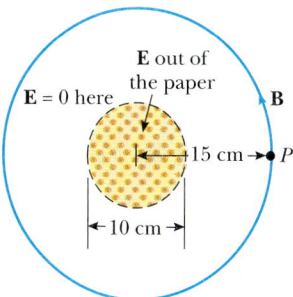

FIGURE P34.62

63. A plane electromagnetic wave varies sinusoidally at 90.0 MHz as it travels along the $+x$ direction. The peak value of the electric field is 2.00 mV/m, and it is directed along the $\pm y$ direction. (a) Find the wavelength, the period, and the maximum value of the magnetic field. (b) Write expressions in SI units for the space and time variations of the electric field and of the magnetic field. Include numerical values as well as subscripts to indicate coordinate directions. (c) Find the average power per unit area that this wave propagates through space. (d) Find the average energy density in the radiation (in joules per cubic meter). (e) What radiation pressure would this wave exert upon a perfectly reflecting surface at normal incidence?

64. A 1.00-m diameter mirror focuses the Sun's rays onto an absorbing plate 2.00 cm in radius, which holds a can containing 1.00 liter of water at 20°C. (a) If the solar intensity is 1.00 kW/m², what is the intensity on the absorbing plate? (b) What are the maximum **E** and **B** field intensities? (c) If 40% of the energy is absorbed, how long would it take to bring the water to its boiling point?

Photo of the image of an old building in Venice with gondolas in the foreground. The distorted image was formed by light reflected from a wavy water surface. *(John McDermott/Tony Stone Images)*

Light and Optics

"I procured me a Triangular glass-Prisme to try therewith the celebrated Pheaenomena of Colours. . . . I placed my Prisme at his entrance (the sunlight), that it might thereby be refracted to the opposite wall. It was a very pleasing divertisement to view the vivid and intense colours produced thereby; . . . I have often with admiration beheld that all the colours of the Prisme being made to converge, and thereby to be again mixed, as they were in the light before it was incident upon the Prisme, reproduced light, entirely and perfectly white, and not at all sensibly differing from the direct light of the sun. . . ."

ISAAC NEWTON

Scientists have long been intrigued by the nature of light, and philosophers have had endless arguments concerning the proper definition and perception of light. It is important to understand the nature of light because it is one of the basic ingredients of life on Earth. Plants convert light energy from the Sun to chemical energy through photosynthesis. Light is the principal means by which we are able to transmit and receive information from objects around us and throughout the Universe.

The nature and properties of light have been a subject of great interest and speculation since ancient times. The Greeks believed that light consisted of tiny particles (corpuscles) that were emitted by a light source and then stimulated the perception of vision upon striking the observer's eye. Newton used this corpuscular theory to explain the reflection and refraction of light. In 1670, one of Newton's contemporaries, the Dutch scientist Christian Huygens, was able to explain many properties of light by proposing that light is made up of waves. In 1801, Thomas Young showed that light beams can interfere with one another, giving strong support to the wave theory. In 1865, Maxwell developed a brilliant theory where he demonstrated that all electromagnetic waves travel at the same speed, c. (Chapter 34). By this time, the wave theory of light seemed to be on firm ground.

However, at the beginning of the 20th century, Albert Einstein returned to the corpuscular theory of light in order to explain the emission of electrons from a metal surface exposed to light (the photoelectric effect). We discuss these and other topics in the last part of this text, which is concerned with modern physics.

Today, scientists view light as having a dual nature. In some experiments, light displays particle-like behavior and in other experiments it exhibits wave properties.

In this part of the book, we concentrate on those aspects of light best understood through the wave model. First, we discuss the reflection of light at the boundary between two media and the refraction of light as it travels from one medium into another. Next we use these ideas to study the refraction of light as it passes through lenses and the reflection of light from surfaces. Then we describe how lenses and mirrors are used to view objects through such instruments as cameras, telescopes, and microscopes. Finally, we shall discuss the phenomena of diffraction, polarization, and interference as they apply to light.

CHAPTER 35

The Nature of Light and the Laws of Geometric Optics

This photograph taken in Salamanca, Spain, shows the reflection of the New Cathedral in the Tormes River. Are you able to distinguish the cathedral from its image? *(David Parker / Photo Researchers)*

35.1 THE NATURE OF LIGHT

Before the beginning of the 19th century, light was considered to be a stream of particles that were emitted by a light source and then stimulated the sense of sight upon entering the eye. The chief architect of this particle theory of light was, once again, Isaac Newton, who was able to explain on the basis of the particle theory some known experimental facts concerning the nature of light, namely, reflection and refraction.

Most scientists accepted Newton's particle theory of light. However, during his lifetime another theory was proposed—one that argued that light is some sort of wave motion. In 1678, a Dutch physicist and astronomer, Christian Huygens (1629–1695), showed that a wave theory of light could also explain the laws of reflection and refraction. The wave theory did not receive immediate acceptance for several reasons. First, all waves known at the time (sound, water, and so forth) traveled through some sort of medium. Because light could travel to us from the Sun through the vacuum of space, it could not possibly be a wave because wave travel requires a medium. Furthermore, it was argued that, if light were some form of wave, it would bend around obstacles; hence, we should be able to see around corners. It is now known that light does indeed bend around the edges of objects. This phenomenon, known as diffraction, is not easy to observe because light waves have such short wavelengths. Thus, although experimental evidence for the diffraction of light was discovered by Francesco Grimaldi (1618–1663) in approximately 1660, most scientists rejected the wave theory and adhered to Newton's particle theory for more than a century. This was, for the most part, due to Newton's great reputation as a scientist.

The first clear demonstration of the wave nature of light was provided in 1801 by Thomas Young (1773–1829), who showed that, under appropriate conditions, light rays interfere with each other. At that time, such behavior could not be explained by a particle theory because there is no conceivable way by which two or more particles could come together and cancel one another. Several years later, a French physicist, Augustin Fresnel (1788–1829), performed a number of experiments dealing with interference and diffraction. In 1850, Jean Foucault (1791–1868) provided further evidence of the inadequacy of the particle theory by showing that the speed of light in glasses and liquids is less than in air. According to the particle model, the speed of light would be higher in glasses and liquids than in air.

Additional developments during the 19th century led to the general acceptance of the wave theory of light, the most important being the work of Maxwell, who in 1873 asserted that light was a form of high-frequency electromagnetic wave (Chapter 34). As discussed in Chapter 34, Hertz provided experimental confirmation of Maxwell's theory by producing and detecting electromagnetic waves. Furthermore, Hertz and other investigators showed that *these waves exhibited reflection, refraction, and all the other characteristic properties of waves.*

Although the classical theory of electricity and magnetism was able to explain most known properties of light via the wave model, some subsequent experiments could not be explained by it. The most striking of these is the photoelectric effect, also discovered by Hertz: When light strikes a metal surface, electrons are sometimes ejected from the surface. As one example of the difficulties that arose, experiments showed that the kinetic energy of an ejected electron is independent of the light intensity. This was in contradiction of the wave theory, which held that a more intense beam of light should add more energy to the electron. An explanation of the photoelectric effect was proposed by Einstein in 1905 in a theory that used the concept of quantization developed by Max Planck (1858–1947) in 1900. The quantization model assumes that the energy of a light wave is present in bundles of energy called photons; hence, the energy is said to be quantized. (Any quantity that appears in discrete bundles is said to be quantized, just as charge and other properties are quantized.) According to Planck's theory, the energy of a photon is proportional to the frequency of the electromagnetic wave:

Energy of a photon

$$E = hf \tag{35.1}$$

where the constant of proportionality $h = 6.63 \times 10^{-34}$ J·s is Planck's constant. It is important to note that this theory retains some features of both the wave theory and the particle theory of light. As we discuss later, the photoelectric effect is the result of energy transfer from a single photon to an electron in the metal. That is, the electron interacts with one photon of light as if the electron had been struck by a particle. Yet this photon has wave-like characteristics because its energy is determined by the frequency (a wave-like quantity).

In view of these developments, light must be regarded as having a dual nature. That is, *in some cases light acts like a wave and in others it acts like a particle.* Classical electromagnetic wave theory provides an adequate explanation of light propagation and of the effects of interference, whereas the photoelectric effect and other experiments involving the interaction of light with matter are best explained by assuming that light is a particle. Light is light, to be sure. However, the question, "Is light a wave or a particle?" is an inappropriate one. In some experiments we measure its wave properties; in other experiments we measure its particle properties. In the next few chapters, we investigate the wave nature of light.

35.2 MEASUREMENTS OF THE SPEED OF LIGHT

Light travels at such a high speed ($c \approx 3 \times 10^8$ m/s) that early attempts to measure its speed were unsuccessful. Galileo attempted to measure the speed of light by positioning two observers in towers separated by approximately 5 mi. Each observer carried a shuttered lantern. One observer would open his lantern first, and then the other would open his lantern at the moment he saw the light from the first lantern. The speed could then be obtained, in principle, knowing the transit time of the light beams between lanterns. The results were inconclusive. Today, we realize (and as Galileo himself concluded) that it is impossible to measure the speed of light in this manner because the transit time is very small compared with the reaction time of the observers.

We now describe two methods for determining the speed of light.

Roemer's Method

The first successful estimate of the speed of light was made in 1675 by the Danish astronomer Ole Roemer (1644–1710). His technique involved astronomical observations of one of the moons of Jupiter, Io, which has a period of approximately 42.5 h; this was measured by observing the eclipse of Io as it passed behind Jupiter. The period of Jupiter is about 12 years, and so as the Earth moves through 180° around the Sun, Jupiter revolves through only 15° (Fig. 35.1).

Using the orbital motion of Io as a clock, its orbit around Jupiter would be expected to have a constant period over long time intervals. However, Roemer observed a systematic variation in Io's period during a year's time. He found that the periods were getting larger when the Earth was receding from Jupiter and smaller when the Earth was approaching Jupiter. For example, if Io had a constant period, Roemer should have seen an eclipse occurring at a particular instant and been able to predict when an eclipse should begin at a later time in the year. However, when he checked the second eclipse, he found that if the Earth was receding from Jupiter, the eclipse was late. In fact, if the interval between observations was three months, the delay was approximately 600 s. Roemer attributed this variation in period to the fact that the distance between the Earth and Jupiter was changing from one observation to the next. In three months (one quarter of the period of the Earth), the light from Jupiter has to travel an additional distance equal to the radius of the Earth's orbit.

Using Roemer's data, Huygens estimated the lower limit for the speed of light

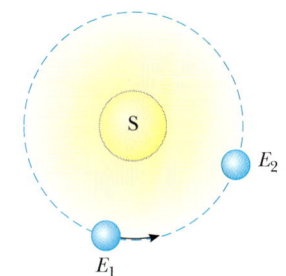

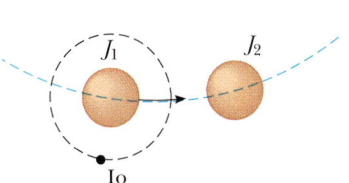

FIGURE 35.1 Roemer's method for measuring the speed of light. In the time it takes the Earth to travel 180° around the Sun (6 months), Jupiter travels only 15°.

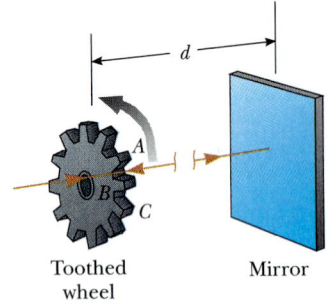

FIGURE 35.2   Fizeau's method for measuring the speed of light using a rotating toothed wheel.

to be approximately 2.3×10^8 m/s. This experiment is important historically because it demonstrated that light does have a finite speed and gave an estimate of this speed.

Fizeau's Technique

The first successful method for measuring the speed of light using purely terrestrial techniques was developed in 1849 by Armand H. L. Fizeau (1819–1896). Figure 35.2 represents a simplified diagram of his apparatus. The basic idea is to measure the total time it takes light to travel from some point to a distant mirror and back. If d is the distance between source and mirror and if the transit time for one round trip is t, then the speed of light is $c = 2d/t$. To measure the transit time, Fizeau used a rotating toothed wheel, which converts an otherwise continuous beam of light into a series of light pulses. Additionally, the rotation of the wheel controls what an observer at the light source sees. For example, if the light passing the opening at point A in Figure 35.2 should return at the instant that tooth B had rotated into position to cover the return path, the light would not reach the observer. At a faster rate of rotation, the opening at point C could move into position to allow the reflected beam to reach the observer. Knowing the distance d, the number of teeth in the wheel, and the angular speed of the wheel, Fizeau arrived at a value of $c = 3.1 \times 10^8$ m/s. Similar measurements made by subsequent investigators yielded more precise values for c, approximately 2.9977×10^8 m/s.

EXAMPLE 35.1   Measuring the Speed of Light with Fizeau's Toothed Wheel

Assume Fizeau's wheel has 360 teeth and is rotating at 27.5 rev/s when a burst of light passing through A in Figure 35.2 is blocked by tooth B on return. If the distance to the mirror is 7500 m, find the speed of light.

Solution   Because the wheel has 360 teeth, it turns through an angle of 1/720 rev in the time that passes while the light makes its round trip. From the definition of angular speed,

we see that the time is

$$t = \frac{\theta}{\omega} = \frac{(1/720) \text{ rev}}{27.5 \text{ rev/s}} = 5.05 \times 10^{-5} \text{ s}$$

Hence, the speed of light is

$$c = \frac{2d}{t} = \frac{2(7500 \text{ m})}{5.05 \times 10^{-5} \text{ s}} = \boxed{2.97 \times 10^8 \text{ m/s}}$$

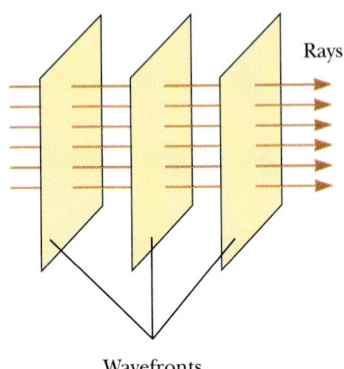

FIGURE 35.3   A plane wave propagating to the right. Note that the rays, which always point in the direction of wave motion, are straight lines perpendicular to the wave fronts.

35.3   THE RAY APPROXIMATION IN GEOMETRIC OPTICS

In studying geometric optics here and in Chapter 36, we use what is called the **ray approximation**. To understand this approximation, first note that the rays of a given wave are straight lines perpendicular to the wave fronts, as illustrated in Figure 35.3 for a plane wave. In the ray approximation, we assume that a wave moving through a medium travels in a straight line in the direction of its rays.

If the wave meets a barrier in which there is a circular opening whose diameter is large relative to the wavelength, as in Figure 35.4a, the wave emerging from the opening continues to move in a straight line (apart from some small edge effects); hence, the ray approximation continues to be valid. If, however, the diameter of the opening is of the order of the wavelength, as in Figure 35.4b, the waves spread out from the opening in all directions. We say that the outgoing wave is noticeably diffracted. Finally, if the opening is small relative to the wavelength, the opening can be approximated as a point source of waves (Fig. 34.4c). Thus, the effect of diffraction is more pronounced as the ratio d/λ approaches zero. Similar effects

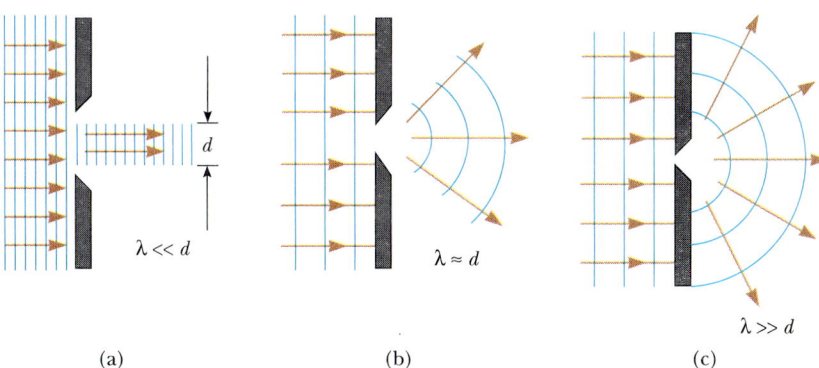

FIGURE 35.4 A plane wave of wavelength λ is incident on a barrier in which there is an opening of diameter d. (a) When $\lambda \ll d$, there is almost no observable diffraction and the ray approximation remains valid. (b) When $\lambda \approx d$, diffraction becomes significant. (c) When $\lambda \gg d$, the opening behaves as a point source emitting spherical waves.

are seen when waves encounter an opaque circular object of diameter d. In this case, when $\lambda \ll d$, the object casts a sharp shadow.

The ray approximation and the assumption that $\lambda \ll d$ are used here and in Chapter 36, both dealing with geometric optics. This approximation is very good for the study of mirrors, lenses, prisms, and associated optical instruments, such as telescopes, cameras, and eyeglasses. We return to the subject of diffraction (where $\lambda \approx d$) in Chapter 38.

35.4 REFLECTION AND REFRACTION

Reflection of Light

When a light ray traveling in a medium encounters a boundary leading into a second medium, part or all of the incident ray is reflected back into the first medium. Figure 35.5a shows several rays of a beam of light incident on a smooth, mirror-like, reflecting surface. The reflected rays are parallel to each other, as indicated in the figure. Reflection of light from such a smooth surface is called **specular reflection.** If the reflecting surface is rough, as in Figure 35.5b, the surface reflects the rays not as a parallel set but in various directions. Reflection from any rough surface is known as **diffuse reflection.** A surface behaves as a smooth surface as long as the surface variations are small compared with the wavelength of the incident light. Photographs of specular reflection and diffuse reflection using laser light are shown in Figures 35.5c and 35.5d. In this book, we concern ourselves only with specular reflection and use the term *reflection* to mean specular reflection.

Consider a light ray traveling in air and incident at an angle on a flat, smooth surface, as in Figure 35.6. The incident and reflected rays make angles θ_1 and θ_1', respectively, with a line drawn perpendicular to the surface at the point where the incident ray strikes. We call this line the normal to the surface. Experiments show that *the angle of reflection equals the angle of incidence:*

$$\theta_1' = \theta_1 \tag{35.2}$$

By convention, the angles of incidence and reflection are measured from the normal to the surface rather than from the surface itself.

The pencil partially immersed in water appears bent because light is refracted as it travels across the boundary between water and air. *(Jim Lehman)*

Law of reflection

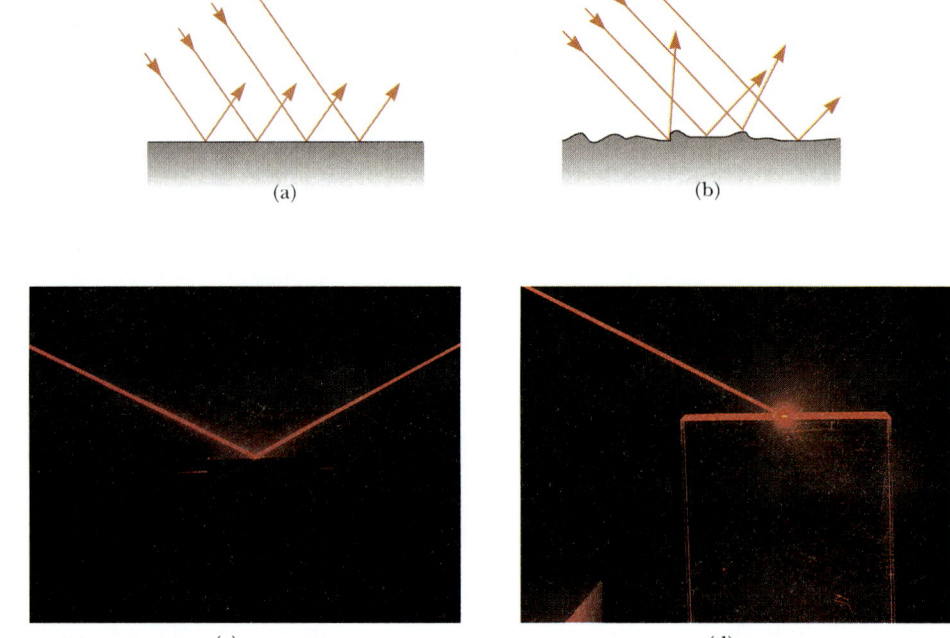

(a)

(b)

FIGURE 35.6 According to the law of reflection, $\theta_1 = \theta_1'$. The incident ray, the reflected ray, and the normal all lie in the same plane.

(c)

(d)

FIGURE 35.5 Schematic representation of (a) specular reflection, where the reflected rays are all parallel to each other, and (b) diffuse reflection, where the reflected rays travel in random directions. (c) and (d) Photographs of specular and diffuse reflection using laser light. *(Henry Leap and Jim Lehman)*

EXAMPLE 35.2   The Double-Reflected Light Ray

Two mirrors make an angle of 120° with each other, as in Figure 35.7. A ray is incident on mirror M_1 at an angle of 65° to the normal. Find the direction of the ray after it is reflected from mirror M_2.

Reasoning and Solution   From the law of reflection, we know that the first reflected ray also makes an angle of 65° with the normal. Thus, this ray makes an angle of $90° - 65°$, or 25°, with the horizontal. From the triangle made by the first reflected ray and the two mirrors, we see that the first reflected ray makes an angle of 35° with M_2 (since the sum of the interior angles of any triangle is 180°). This means that this ray makes an angle of 55° with the normal to M_2. Hence, from the law of reflection, the second reflected ray makes an angle of 55° with the normal to M_2. By comparing the direction of the ray incident on M_1 with its direction after reflecting from M_2, we see that the ray is reflected through 120°, which happens to correspond to the angle between the mirrors.

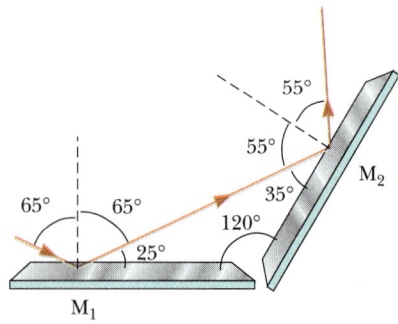

FIGURE 35.7 (Example 35.2) Mirrors M_1 and M_2 make an angle of 120° with each other.

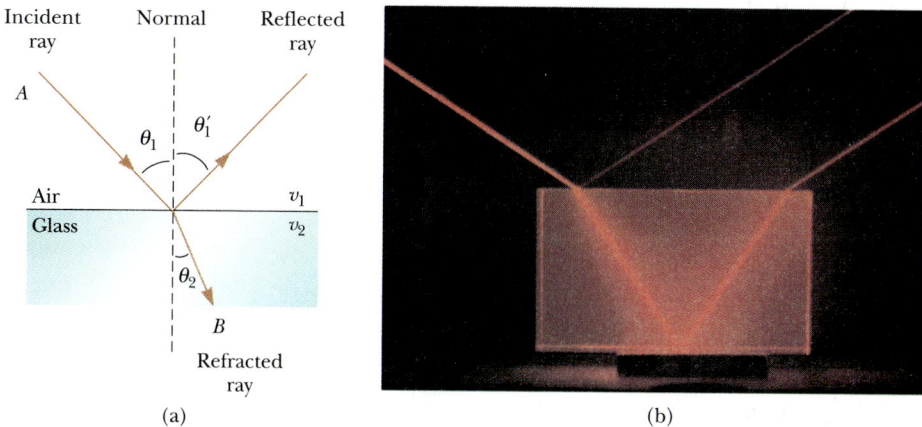

FIGURE 35.8   (a) A ray obliquely incident on an air-glass interface. The refracted ray is bent toward the normal because $v_2 < v_1$. All rays and the normal lie in the same plane. (b) Light incident on the Lucite block bends both when it enters the block and when it leaves the block.   *(Henry Leap and Jim Lehman)*

Refraction of Light

When a ray of light traveling through a transparent medium encounters a boundary leading into another transparent medium, as in Figure 35.8, part of the ray is reflected and part enters the second medium. The ray that enters the second medium is bent at the boundary and is said to be refracted. *The incident ray, the reflected ray, the normal, and the refracted ray all lie in the same plane.* The angle of refraction, θ_2 in Figure 35.8, depends on the properties of the two media and on the angle of incidence through the relationship[1]

$$\frac{\sin \theta_2}{\sin \theta_1} = \frac{v_2}{v_1} = \text{constant} \qquad (35.3)$$

where v_1 is the speed of light in medium 1 and v_2 is the speed of light in medium 2.

   The path of a light ray through a refracting surface is reversible. For example, the ray in Figure 35.8 travels from point A to point B. If the ray originated at B, it would follow the same path to reach point A. In the latter case, however, the reflected ray would be in the glass.

   When light moves from a material in which its speed is high to a material in which its speed is lower, the angle of refraction, θ_2, is less than the angle of incidence, as shown in Figure 35.9a. If the ray moves from a material in which it moves slowly to a material in which it moves more rapidly, it is bent away from the normal, as shown in Figure 35.9b.

   The behavior of light as it passes from air into another substance and then re-emerges into air is often a source of confusion to students. Let us take a look at what happens and see why this behavior is so different from other occurrences in our daily lives. When light travels in air, its speed is equal to 3×10^8 m/s, but this speed is reduced to approximately 2×10^8 m/s when the light enters a block of

---

[1] This relationship was also deduced from the particle theory of light by René Descartes (1596–1650) and hence is known as *Descartes' law* in France.

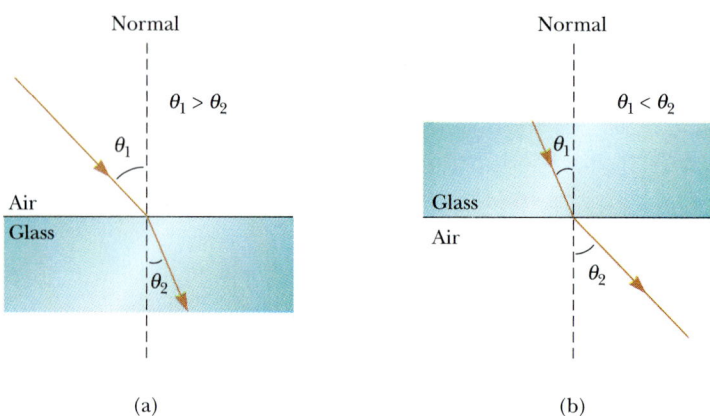

(a)   (b)

FIGURE 35.9   (a) When the light beam moves from air into glass, its path is bent toward the normal. (b) When the beam moves from glass into air, its path is bent away from the normal.

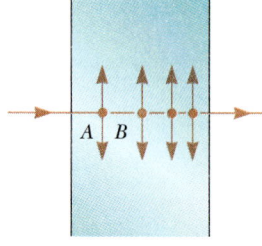

FIGURE 35.10   Light passing from one atom to another in a medium. The dots are electrons, and the vertical arrows represent their oscillations.

glass. When the light re-emerges into air, its speed instantaneously increases to its original value of 3×10^8 m/s. This is far different from what happens, for example, when a bullet is fired through a block of wood. In this case, the speed of the bullet is reduced as it moves through the wood because some of its original energy is used to tear apart the fibers of the wood. When the bullet enters the air once again, it emerges at the speed it had just before leaving the block of wood.

To see why light behaves as it does, consider Figure 35.10, which represents a beam of light entering a piece of glass from the left. Once inside the glass, the light may encounter an electron bound to an atom, indicated as point A in the figure. Let us assume that light is absorbed by the atom, which causes the electron to oscillate (a detail represented by the double-headed vertical arrows in the drawing). The oscillating electron then acts as an antenna and radiates the beam of light toward an atom at B, where the light is again absorbed by an atom. The details of these absorptions and radiations are best explained in terms of quantum mechanics, a subject we study in the extended version of this text. For now, it is sufficient to think of the process as one in which the light passes from one atom to another through the glass. (The situation is somewhat analogous to a relay race in which a baton is passed between runners on the same team.) Although light travels from one glass atom to another at 3×10^8 m/s, the absorption and radiation processes that take place in the glass cause the light speed to slow to 2×10^8 m/s. Once the light emerges into the air, the absorptions and radiations cease and its speed returns to the original value.

Index of Refraction

The speed of light in any material is less than the speed in vacuum except near very strong absorption bands.[2]

Index of refraction

$$n = \frac{\text{speed of light in vacuum}}{\text{speed of light in a medium}} = \frac{c}{v} \qquad (35.4)$$

---

[2] There are two speeds for a wave, which may be the same or different. The group speed, or signal speed, for light cannot be greater than the speed of light in vacuum. The speed considered here is the phase speed, which may be greater than c, without violating special relativity.

From this definition, we see that the index of refraction is a dimensionless number usually greater than unity because v is usually less than c. Furthermore, n is equal to unity for vacuum. The indices of refraction for various substances are listed in Table 35.1.

As light travels from one medium to another, *its frequency does not change but its wavelength does.* To see why this is so, consider Figure 35.11. Wave fronts pass an observer at point A in medium 1 with a certain frequency and are incident on the boundary between medium 1 and medium 2. The frequency with which the wave fronts pass an observer at point B in medium 2 must equal the frequency at which they arrive at point A in medium 1. If this were not the case, either wave fronts would be piling up at the boundary or they would be destroyed or created at the boundary. Since there is no mechanism for this to occur, the frequency must be a constant as a light ray passes from one medium into another.

Therefore, because the relationship $v = f\lambda$ (Eq. 16.14) must be valid in both media and because $f_1 = f_2 = f$, we see that

$$v_1 = f\lambda_1 \quad \text{and} \quad v_2 = f\lambda_2 \tag{35.5}$$

Because $v_1 \neq v_2$, it follows that $\lambda_1 \neq \lambda_2$. A relationship between index of refraction and wavelength can be obtained by dividing the first of Equations 35.5 by the other and making use of the definition of the index of refraction given by Equation 35.4:

$$\frac{\lambda_1}{\lambda_2} = \frac{v_1}{v_2} = \frac{c/n_1}{c/n_2} = \frac{n_2}{n_1} \tag{35.6}$$

which gives

$$\lambda_1 n_1 = \lambda_2 n_2$$

If medium 1 is vacuum, or for all practical purposes air, then $n_1 = 1$. Hence, it follows from Equation 35.6 that the index of refraction of any medium can be

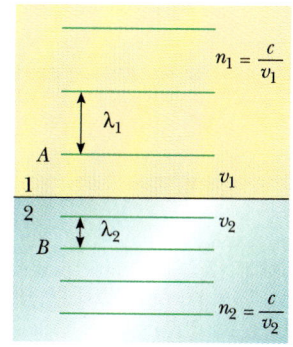

FIGURE 35.11   As a wave front moves from medium 1 to medium 2, its wavelength changes but its frequency remains constant.

TABLE 35.1	Index of Refraction for Various Substances Measured with Light of Vacuum Wavelength $\lambda_0 = 589$ nm		
Substance	**Index of Refraction**	**Substance**	**Index of Refraction**
Solids at 20°C		**Liquids at 20°C**	
Diamond (C)	2.419	Benzene	1.501
Fluorite (CaF_2)	1.434	Carbon disulfide	1.628
Silica (SiO_2)	1.458	Carbon tetrachloride	1.461
Glass, crown	1.52	Ethyl alcohol	1.361
Glass, flint	1.66	Glycerine	1.473
Ice (H_2O)	1.309	Water	1.333
Polystyrene	1.49		
Sodium chloride (NaCl)	1.544	**Gases at 0°C, 1 atm**	
Zircon	1.923	Air	1.000293
		Carbon dioxide	1.00045

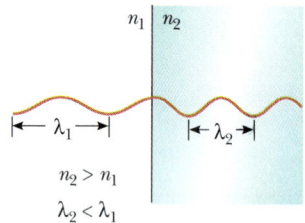

FIGURE 35.12 Schematic diagram of the reduction in wavelength when light travels from a medium of low index of refraction to one of higher index of refraction.

expressed as the ratio

$$n = \frac{\lambda_0}{\lambda_n} \tag{35.7}$$

where λ_0 is the wavelength of light in vacuum and λ_n is the wavelength in the medium whose index of refraction is n. A schematic representation of this reduction in wavelength is shown in Figure 35.12.

We are now in a position to express Snell's law of refraction (Eq. 35.3) in an alternative form. If we substitute Equation 35.6 into Equation 35.3, we get

$$n_1 \sin \theta_1 = n_2 \sin \theta_2 \tag{35.8}$$

The experimental discovery of this relationship is usually credited to Willebrord Snell (1591–1627) and is therefore known as **Snell's law.**

EXAMPLE 35.3   An Index of Refraction Measurement

A beam of light of wavelength 550 nm traveling in air is incident on a slab of transparent material. The incident beam makes an angle of 40.0° with the normal, and the refracted beam makes an angle of 26.0° with the normal. Find the index of refraction of the material.

Solution   Snell's law of refraction (Eq. 35.8) with these data gives

$$n_1 \sin \theta_1 = n_2 \sin \theta_2$$

$$n_2 = \frac{n_1 \sin \theta_1}{\sin \theta_2} = (1.00)\frac{\sin 40.0°}{\sin 26.0°} = \frac{0.643}{0.438} = \boxed{1.47}$$

If we compare this value with the data in Table 35.1, we see that the material may be silica.

Exercise   What is the wavelength of light in the material?

Answer   374 nm.

EXAMPLE 35.4   Angle of Refraction for Glass

A light ray of wavelength 589 nm traveling through air is incident on a smooth, flat slab of crown glass at an angle of 30.0° to the normal, as sketched in Figure 35.13. Find the angle of refraction.

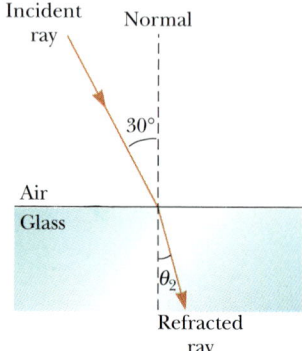

FIGURE 35.13   (Example 35.4) Refraction of light by glass.

Solution   Snell's law given by Equation 35.8 can be rearranged as

$$\sin \theta_2 = \frac{n_1}{n_2} \sin \theta_1$$

From Table 35.1, we find that $n_1 = 1.00$ for air and $n_2 = 1.52$ for crown glass. Therefore,

$$\sin \theta_2 = \left(\frac{1.00}{1.52}\right)(\sin 30.0°) = 0.329$$

$$\theta_2 = \sin^{-1}(0.329) = \boxed{19.2°}$$

The ray is bent toward the normal, as expected.

Exercise   If the light ray moves from inside the glass toward the glass-air interface at an angle of 30.0° to the normal, determine the angle of refraction.

Answer   49.5° away from the normal.

EXAMPLE 35.5   The Speed of Light in Silica

Light of wavelength 589 nm in vacuum passes through a piece of silica ($n = 1.458$). (a) Find the speed of light in silica.

Solution   The speed of light in silica can be easily obtained from Equation 35.4:

$$v = \frac{c}{n} = \frac{3.00 \times 10^8 \text{ m/s}}{1.458} = 2.06 \times 10^8 \text{ m/s}$$

(b) What is the wavelength of this light in silica?

Solution   We use Equation 35.7 to calculate the wavelength in silica, noting that we are given the wavelength in vacuum to be $\lambda_0 = 589$ nm:

$$\lambda_n = \frac{\lambda_0}{n} = \frac{589 \text{ nm}}{1.458} = 404 \text{ nm}$$

Exercise   Find the frequency of the light.

Answer 5.09×10^{14} Hz.

EXAMPLE 35.6   Light Passing Through a Slab

A light beam passes from medium 1 to medium 2 with the latter being a thick slab of material whose index of refraction is n_2 (Fig. 35.14). Show that the emerging beam is parallel to the incident beam.

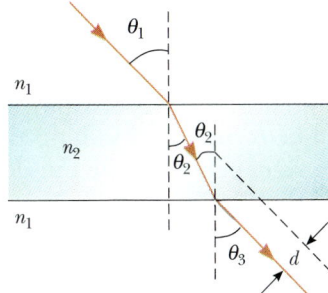

FIGURE 35.14   (Example 35.6) When light passes through a flat slab of material, the emerging beam is parallel to the incident beam, and therefore $\theta_1 = \theta_3$. The green dashed line represents the path the light would take if the slab were not there.

Solution   First, let us apply Snell's law to the upper surface:

$$(1) \qquad \sin \theta_2 = \frac{n_1}{n_2} \sin \theta_1$$

Applying Snell's law to the lower surface gives

$$(2) \qquad \sin \theta_3 = \frac{n_2}{n_1} \sin \theta_2$$

Substituting (1) into (2) gives

$$\sin \theta_3 = \frac{n_2}{n_1}\left(\frac{n_1}{n_2} \sin \theta_1\right) = \sin \theta_1$$

That is, $\theta_3 = \theta_1$, and so the layer does not alter the direction of the beam as shown by the green dashed line. It does, however, produce a displacement of the beam. The same result is obtained when light passes through multiple layers of materials.

*35.5   DISPERSION AND PRISMS

An important property of the index of refraction is that, for a given material, the index varies with the wavelength of light passing through the material as Figure 35.15 shows. Since n is a function of wavelength, Snell's law indicates that light of different wavelengths is bent at different angles when incident on a refracting material. As we see from Figure 35.15, the index of refraction decreases with increasing wavelength. This means that blue light bends more than red light when passing into a refracting material. The various wavelengths are refracted at different angles. This phenomenon is known as **dispersion.**

To understand the effects that dispersion can have on light, let us consider

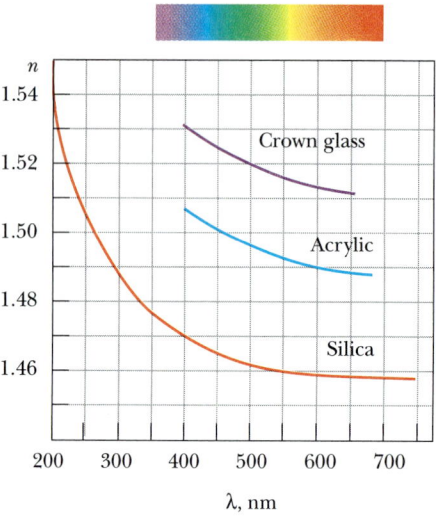

FIGURE 35.15   Variation of index of refraction with vacuum wavelength for three materials.

what happens when light strikes a prism, as in Figure 35.16. A ray of single-wavelength light incident on the prism from the left emerges bent away from its original direction of travel by an angle δ, called the **angle of deviation.** Now suppose a beam of white light (a combination of all visible wavelengths) is incident on a prism, as in Figure 35.17. The rays that emerge from the second face spread out in a series of colors known as a **spectrum.** These colors, in order of decreasing wavelength, are red, orange, yellow, green, blue, indigo, and violet. Newton showed that each color has a particular angle of deviation, that the spectrum cannot be further broken down, and that the colors can be recombined to form the original white light. Clearly, the angle of deviation, δ, depends on the wavelength of a given

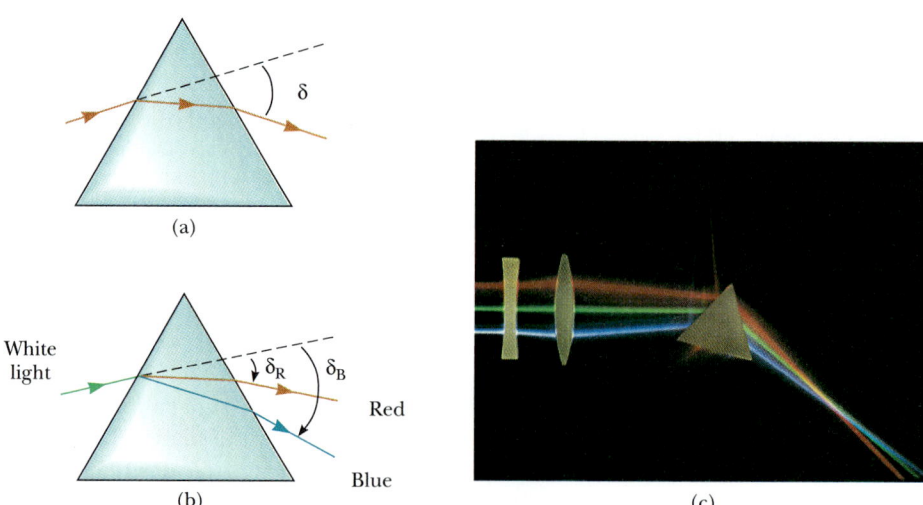

FIGURE 35.16   (a) A prism refracts single-wavelength light and deviates the light through an angle δ. (b) When light is incident on a prism, the blue light is bent more than the red. (c) Light of different colors passes through a prism and two lenses. As the light passes through the prism, different wavelengths are refracted at different angles. *(David Parker, SPL/Photo Researchers)*

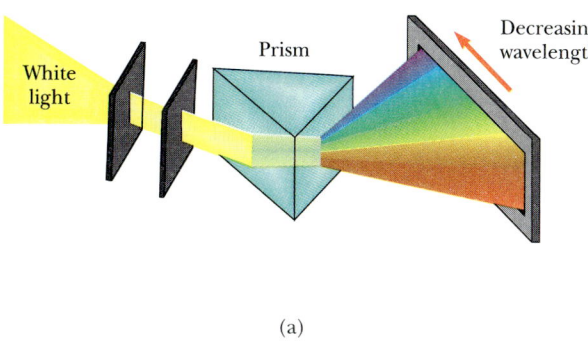

White light

Prism

Decreasing wavelength

(a)

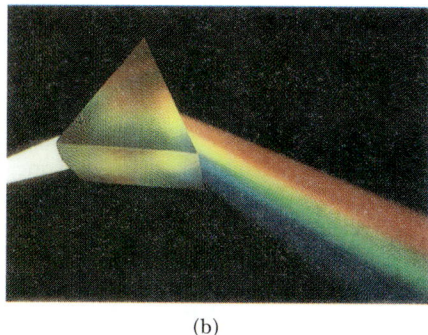

(b)

FIGURE 35.17 (a) Dispersion of white light by a prism. Because *n* varies with wavelength, the prism disperses the white light into its various spectral components. (b) Different colors are refracted at different angles because the index of refraction of the glass depends on wavelength. Violet light deviates the most; red light deviates the least. *(Photograph courtesy of Bausch and Lomb)*

color. Violet light deviates the most, red the least, and the remaining colors in the visible spectrum fall between these extremes.

A prism is often used in an instrument known as a **prism spectroscope,** the essential elements of which are shown in Figure 35.18. The instrument is commonly used to measure the wavelengths emitted by a light source. Light from the source is sent through a narrow, adjustable slit to produce a collimated beam. The light then passes through the prism and is dispersed into a spectrum. The refracted light is observed through a telescope. The experimenter sees an image of the slit through the eyepiece of the telescope. The telescope can be moved or the prism rotated in order to view the various images formed by different wavelengths at different angles of deviation.

All hot, low-pressure gases emit their own characteristic spectra. Thus, one use of a prism spectroscope is to identify gases. For example, sodium emits two wavelengths in the visible spectrum, which appear as two closely spaced yellow lines. Thus, a gas emitting these colors can be identified as having sodium as one of its constituents. Likewise, mercury vapor has its own characteristic spectrum, consisting of four prominent wavelengths—orange, green, blue, and violet lines—along with some wavelengths of lower intensity. The particular wavelengths emitted by a gas serve as "fingerprints" of that gas.

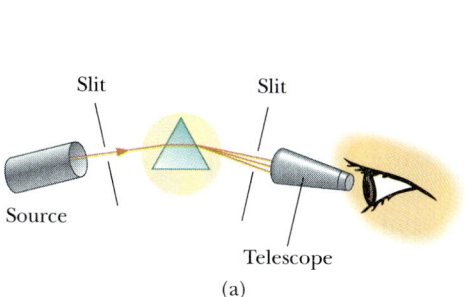

Slit

Slit

Source

Telescope

(a)

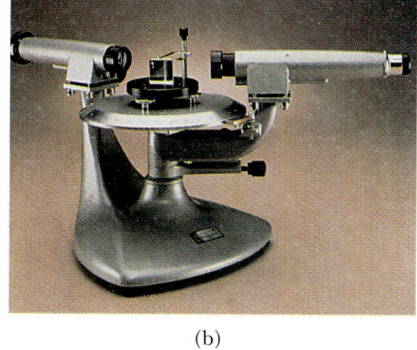

(b)

FIGURE 35.18 (a) Diagram of a prism spectrometer. The various colors in the spectrum are viewed through a telescope. (b) Photograph of a prism spectrometer. *(Courtesy of Central Scientific Co.)*

EXAMPLE 35.7   Measuring *n* Using a Prism

A prism is often used to measure the index of refraction of a transparent solid. Although we do not prove it here, the minimum angle of deviation, δ_{min}, occurs at the angle of incidence θ_1 where the refracted ray inside the prism makes the same angle α with the normal to the two prism faces,[3] as in Figure 35.19. Let us obtain an expression for the index of refraction of the prism material.

Using the geometry shown, we find that $\theta_2 = \Phi/2$, where Φ is the apex angle and

$$\theta_1 = \theta_2 + \alpha = \frac{\Phi}{2} + \frac{\delta_{min}}{2} = \frac{\Phi + \delta_{min}}{2}$$

From Snell's law of refraction,

$$\sin \theta_1 = n \sin \theta_2$$

$$\sin\left(\frac{\Phi + \delta_{min}}{2}\right) = n \sin(\Phi/2)$$

$$n = \frac{\sin\left(\dfrac{\Phi + \delta_{min}}{2}\right)}{\sin(\Phi/2)} \qquad (35.9)$$

[3] For details, see F. A. Jenkins and H. E. White, *Fundamentals of Optics,* New York, McGraw-Hill, 1976, Chapter 2.

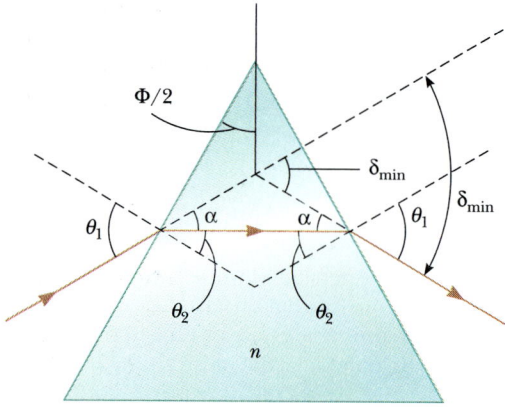

FIGURE 35.19   (Example 35.7) A light ray passing through a prism at the minimum angle of deviation, δ_{min}.

Hence, knowing the apex angle Φ of the prism and measuring δ_{min}, we can calculate the index of refraction of the prism material. Furthermore, we can use a hollow prism to determine the values of n for various liquids.

35.6   HUYGENS' PRINCIPLE

In this section, we develop the laws of reflection and refraction by using a method proposed by Huygens in 1678. As noted in Section 35.1, Huygens assumed that light is some form of wave motion rather than a stream of particles. He had no knowledge of the nature of light or of its electromagnetic character. Nevertheless, his simplified wave model is adequate for understanding many practical aspects of the propagation of light.

Huygens' principle is a construction for using knowledge of an earlier wave front to determine the position of a new wave front at some instant. In Huygens' construction,

Huygens' principle

> all points on a given wave front are taken as point sources for the production of spherical secondary waves, called wavelets, which propagate outward with speeds characteristic of waves in that medium. After some time has elapsed, the new position of the wave front is the surface tangent to the wavelets.

Figure 35.20 illustrates two simple examples of Huygens' construction. First, consider a plane wave moving through free space, as in Figure 35.20a. At $t = 0$, the wave front is indicated by the plane labeled AA'. In Huygens' construction, each

C hristian Huygens was a Dutch physicist and astronomer who is best known for his contributions to the fields of dynamics and optics. As a physicist, his accomplishments include invention of the pendulum clock and the first exposition of a wave theory of light. As an astronomer, he was the first to recognize the rings around Saturn and to discover Titan, a satellite of Saturn.

Huygens was born in 1629 into a prominent family in The Hague. He was the son of Constantin Huygens, one of the most important figures of the Renaissance in Holland. Educated at the University of Leyden, Christian became a close friend of René Descartes, a frequent guest at the Huygens' home. Huygens published his first paper in 1651 on the subject of the quadrature of various curves.

Huygens' reputation in optics and dynamics spread throughout Europe, and in 1663 he was elected a charter member of the Royal Society. Louis XIV lured Huygens to France in 1666, in accordance with the king's policy of collecting scholars for the glory of his regime. While in France, Huygens became one of the founders of the French Academy of Science.

Christian Huygens

| 1 6 2 9 – 1 6 9 5 |

In 1673, in Paris, Huygens published *Horologium Oscillatorium*. In this work he described a solution to the problem of the compound pendulum, for which he calculated the equivalent simple pendulum length. In the same publication he also derived a formula for computing the period of oscillation of a simple pendulum and explained his laws of centrifugal force for uniform motion in a circle.

Huygens returned to Holland in 1681, constructed some lenses of large focal lengths, and invented the achromatic eyepiece for telescopes. Shortly after returning from a visit to England, where he met Newton, Huygens published his treatise on the wave theory of light. To Huygens, light was a vibratory motion in the ether, spreading out and producing the sensation of light when impinging on the eye. On the basis of this theory, he was able to deduce the laws of reflection and refraction and to explain the phenomenon of double refraction.

Huygens, second only to Newton among the greatest scientists in the second half of the 17th century, was the first to proceed in the field of dynamics beyond the point reached by Galileo and Descartes. It was Huygens who essentially solved the problem of centrifugal force. A solitary man, Huygens did not attract students or disciples and was very slow in publishing his findings. He died in 1695 after a long illness.

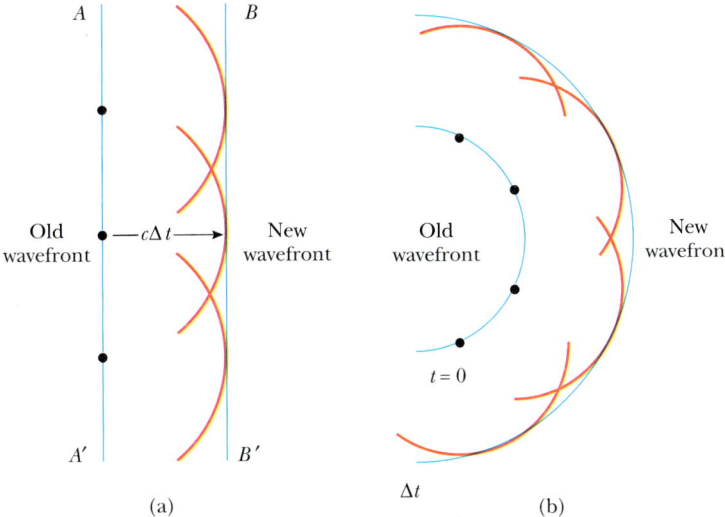

FIGURE 35.20   Huygens' construction for (a) a plane wave propagating to the right and (b) a spherical wave.

FIGURE 35.21 Water waves in a ripple tank are used to demonstrate Huygens' wavelets. A plane wave is incident on a barrier with two small openings. Each opening acts as a source of circular wavelets. *(Erich Schrempp/Photo Researchers)*

point on this wave front is considered a point source for generating other waves. For clarity, only a few points on AA' are shown. With these points as sources for the wavelets, we draw circles each of radius $c\,\Delta t$, where c is the speed of light in free space and Δt is the time of propagation from one wave front to the next. The surface drawn tangent to these wavelets is the plane BB', which is parallel to AA'. In a similar manner, Figure 35.20b shows Huygens' construction for an outgoing spherical wave.

A convincing demonstration of Huygens' principle is obtained with water waves in a shallow tank (called a ripple tank), as in Figure 35.21. Plane waves produced below the slits emerge above the slits as two-dimensional circular waves propagating outward.

Huygens' Principle Applied to Reflection and Refraction

The laws of reflection and refraction were stated earlier in this chapter without proof. We now derive these laws using Huygens' principle.

Consider a plane wave incident on a reflecting surface as in Figure 35.22. At $t = 0$, suppose the wave front labeled I_1 is in contact with the surface at A. Point A is a source of Huygens' waves, and in a time interval Δt, these waves expand radially outward a distance $c\,\Delta t$ from A. In the same time interval, the incident wave travels a distance $c\,\Delta t$ (corresponding to the new wave front I_2), striking the surface at C.

From Huygens' construction, the reflected wave front R travels in the direction perpendicular to the wave front. From the construction in Figure 35.22, we see that $BC = AD = c\,\Delta t$, and the hypotenuse AC is common to both right triangles ADC and ABC. Hence, the two triangles are congruent, and $\theta_1 = \theta_1'$. Because the incident ray is perpendicular to AB, and the reflected ray is perpendicular to DC, the

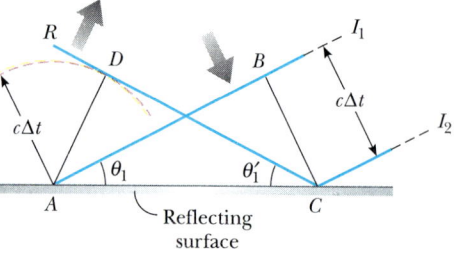

FIGURE 35.22 Huygens' construction for proving the law of reflection. Triangles ABC and ADC are congruent.

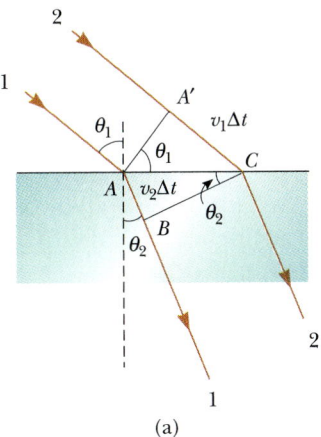

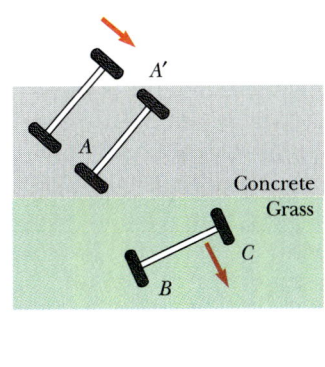

FIGURE 35.23   (a) Huygens' construction for proving the law of refraction. (b) A mechanical analog of refraction.

angles that these rays make with the normal to the surface are also the angles of incidence and reflection. Hence, we have proved the law of reflection.

Now let us use Huygens' principle and Figure 35.23a to derive Snell's law of refraction. Note that in the time interval Δt, ray 1 moves from A to B and ray 2 moves from A' to C. The radius of the outgoing spherical wavelet centered at A is equal to $v_2 \Delta t$. The distance $A'C$ is equal to $v_1 \Delta t$. Geometric considerations show that angle $A'AC$ equals θ_1 and angle ACB equals θ_2. From triangles $AA'C$ and ACB, we find that

$$\sin \theta_1 = \frac{v_1 \Delta t}{AC} \quad \text{and} \quad \sin \theta_2 = \frac{v_2 \Delta t}{AC}$$

Dividing these two equations, we get

$$\frac{\sin \theta_1}{\sin \theta_2} = \frac{v_1}{v_2}$$

But from Equation 35.4 we know that $v_1 = c/n_1$ and $v_2 = c/n_2$. Therefore,

$$\frac{\sin \theta_1}{\sin \theta_2} = \frac{c/n_1}{c/n_2} = \frac{n_2}{n_1}$$

$$n_1 \sin \theta_1 = n_2 \sin \theta_2$$

which is Snell's law of refraction.

A mechanical analog of refraction is shown in Figure 35.23b. The wheels on a device such as a wagon change their direction as they move from a concrete surface to a grass surface.

35.7   TOTAL INTERNAL REFLECTION

An interesting effect called total internal reflection can occur when light attempts to move from a medium having a given index of refraction to one having a lower index of refraction. Consider a light beam traveling in medium 1 and meeting the boundary between medium 1 and medium 2, where n_1 is greater than n_2 (Fig. 35.24). Various possible directions of the beam are indicated by rays 1 through 5. The refracted rays are bent away from the normal because n_1 is greater than n_2.

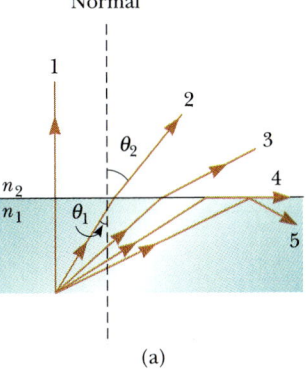

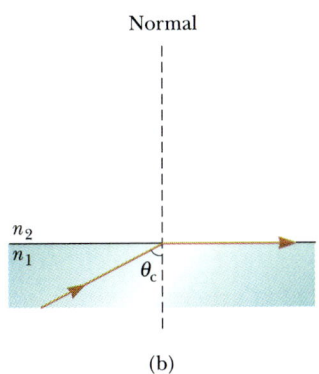

FIGURE 35.24   (a) Rays from a medium of index of refraction n_1 travel into a medium of index of refraction n_2, where $n_2 < n_1$. As the angle of incidence increases, the angle of refraction θ_2 increases until θ_2 is 90° (ray 4). For this and larger angles of incidence, total internal reflection occurs (ray 5). (b) The angle of incidence producing an angle of refraction equal to 90° is the critical angle, θ_c.

This photograph shows nonparallel light rays entering a glass prism. The bottom two rays undergo total internal reflection at the longest side of the prism. The top three rays are refracted at the longest side as they leave the prism. *(Henry Leap and Jim Lehman)*

(Remember that when light refracts at the interface between the two media, it is also partially reflected. For simplicity, we ignore these reflected rays here.) At some particular angle of incidence, θ_c, called the **critical angle,** the refracted light ray would move parallel to the boundary so that $\theta_2 = 90°$ (Fig. 35.24b). For angles of incidence greater than or equal to θ_c, the beam is entirely reflected at the boundary as shown by ray 5 in Figure 35.24a. This ray is reflected at the boundary as though it had struck a perfectly reflecting surface. This ray, and all those like it, obey the law of reflection; that is, the angle of incidence equals the angle of reflection.

We can use Snell's law of refraction to find the critical angle. When $\theta_1 = \theta_c$, $\theta_2 = 90°$ and Snell's law (Eq. 35.8) gives

$$n_1 \sin \theta_c = n_2 \sin 90° = n_2$$

$$\sin \theta_c = \frac{n_2}{n_1} \qquad (\text{for } n_1 > n_2) \qquad (35.10)$$

In this equation n_1 is always greater than n_2. That is,

total internal reflection occurs only when light attempts to move from a medium of given index of refraction to a medium of lower index of refraction.

If n_1 were less than n_2, Equation 35.10 would give $\sin \theta_c > 1$, which is meaningless because the sine of an angle can never be greater than unity.

The critical angle for total internal reflection will be small when n_1 is considerably larger than n_2. Examples of this combination are diamond ($n = 2.42$ and $\theta_c = 24°$) and crown glass ($n = 1.52$ and $\theta_c = 41°$). This property combined with proper faceting causes diamonds and crystal glass to sparkle.

A prism and the phenomenon of total internal reflection can be used to alter the direction of travel of a propagation beam. Two such possibilities are illustrated in Figure 35.25. In one case the light beam is deflected through 90° (Fig. 35.25a), and in the second case the path of the beam is reversed (Fig. 35.25b). A common

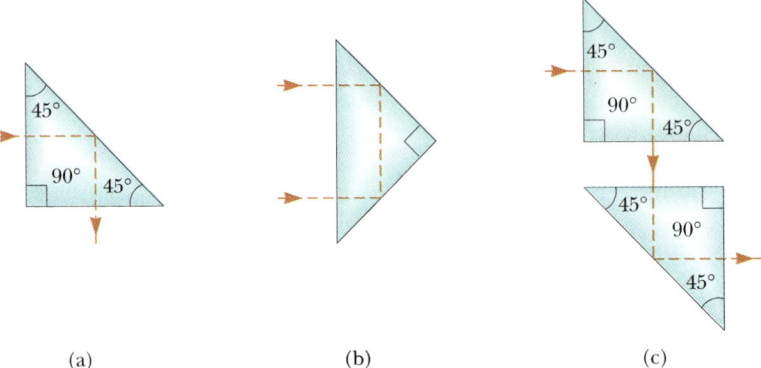

(a)   (b)   (c)

FIGURE 35.25 Internal reflection in a prism. (a) The ray is deviated by 90°. (b) The direction of the ray is reversed. (c) Two prisms used as a periscope.

application of total internal reflection is in a submarine periscope. In this device, two prisms are arranged as in Figure 35.25c so that an incident beam of light follows the path shown and one is able to "see around corners."

EXAMPLE 35.8 A View from the Fish's Eye

(a) Find the critical angle for a water-air boundary if the index of refraction of water is 1.33.

Solution Applying Equation 35.10, we find the critical angle to be

$$\sin \theta_c = \frac{n_2}{n_1} = \frac{1}{1.33} = 0.752$$

$$\theta_c = \boxed{48.8°}$$

(b) Use the results of part (a) to predict what a fish sees when it looks upward toward the water surface at an angle of 40°, 49°, and 60° (Fig. 35.26).

Reasoning Examine Figure 35.24a. Note that because the path of a light ray is reversible, light traveling from medium 2 into medium 1 follows the paths shown in Figure 35.24a, but in the *opposite* direction. This situation is illustrated in Figure 35.26 for the case of a fish looking upward toward the water surface. A fish can see out of the water if it looks toward the surface at an angle less than the critical angle. Thus, at 40°, the fish can see into the air above the water. At an angle of 49°, the critical angle for water, the light that reaches the fish

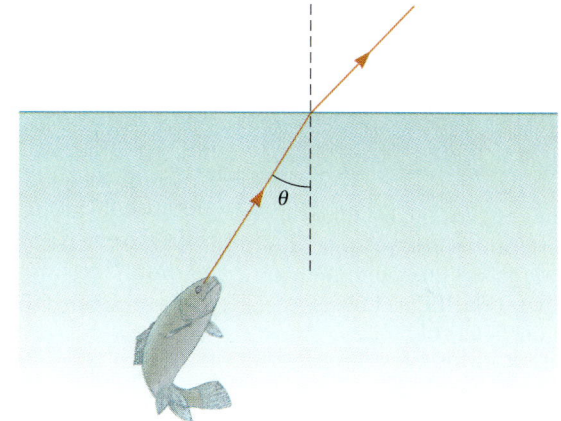

FIGURE 35.26 (Example 35.8).

has to skim along the water surface before being refracted to the fish's eye. At angles greater than the critical angle, the light reaching the fish comes via internal reflection at the surface. Thus, at 60°, the fish sees a reflection of some object on the bottom of the pool.

Fiber Optics

Another interesting application of total internal reflection is the use of glass or transparent plastic rods to "pipe" light from one place to another. As indicated in Figure 35.27, light is confined to traveling within the rods, even around gentle curves, as the result of successive internal reflections. Such a "light pipe" is flexible if thin fibers are used rather than thick rods. If a bundle of parallel fibers is used to construct an optical transmission line, images can be transferred from one point to another.

This technique is used in a sizable industry known as fiber optics. There is very little light intensity lost in these fibers as a result of reflections on the sides. Any loss in intensity is due essentially to reflections from the two ends and absorption by the fiber material. These devices are particularly useful when an image of an object that is at an inaccessible location is to be viewed. For example, physicians often use this technique to examine internal organs of the body. The field of fiber optics is finding increasing use in telecommunications, since the fibers can carry a much higher volume of telephone calls or other forms of communication than electrical wires.

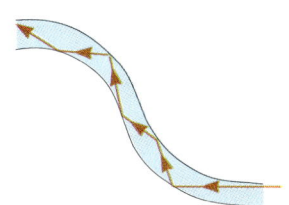

FIGURE 35.27 Light travels in a curved transparent rod by multiple internal reflections.

Strands of glass optical fibers are used to carry voice, video, and data signals in telecommunication networks. Typical fibers have diameters of 60 μm. (© *Richard Megna 1983, Fundamental Photographs*)

*35.8 FERMAT'S PRINCIPLE

A general principle that can be used to determine light paths was developed by Pierre de Fermat (1601–1665). **Fermat's principle** states that

> when a light ray travels between any two points *P* and *Q*, its path is the one that requires the least time, or constant time.[4]

An obvious consequence of this principle is that when the rays travel in a single, homogeneous medium, the paths are straight lines because a straight line is the shortest distance between two points. Let us illustrate how to use Fermat's principle to derive the law of refraction.

Suppose a light ray is to travel from *P* to *Q*, where *P* is in medium 1 and *Q* is in medium 2 (Fig. 35.28). The points *P* and *Q* are at perpendicular distances *a* and *b*, respectively, from the interface. The speed of light is c/n_1 in medium 1 and c/n_2 in medium 2. Using the geometry of Figure 35.28, we see that the time it takes the ray to travel from *P* to *Q* is

$$t = \frac{r_1}{v_1} + \frac{r_2}{v_2} = \frac{\sqrt{a^2 + x^2}}{c/n_1} + \frac{\sqrt{b^2 + (d - x)^2}}{c/n_2}$$

We obtain the least time, or the minimum value of *t*, by taking the derivative of *t* with respect to *x* (the variable) and setting the derivative equal to zero:

$$\frac{dt}{dx} = \frac{n_1}{c} \frac{d}{dx} \sqrt{a^2 + x^2} + \frac{n_2}{c} \frac{d}{dx} \sqrt{b^2 + (d - x)^2}$$

$$= \frac{n_1}{c} \left(\frac{1}{2}\right) \frac{2x}{(a^2 + x^2)^{1/2}} + \frac{n_2}{c} \left(\frac{1}{2}\right) \frac{2(d - x)(-1)}{[b^2 + (d - x)^2]^{1/2}}$$

$$= \frac{n_1 x}{c(a^2 + x^2)^{1/2}} - \frac{n_2(d - x)}{c[b^2 + (d - x)^2]^{1/2}} = 0$$

From Figure 35.28 and recognizing that in this equation,

$$\sin \theta_1 = \frac{x}{(a^2 + x^2)^{1/2}} \qquad \sin \theta_2 = \frac{d - x}{[b^2 + (d - x)^2]^{1/2}}$$

we find that

$$n_1 \sin \theta_1 = n_2 \sin \theta_2$$

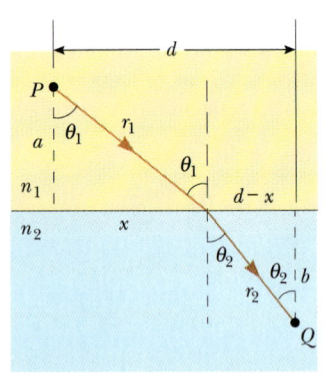

FIGURE 35.28 Geometry for deriving the law of refraction using Fermat's principle.

which is Snell's law of refraction.

It is a simple matter to use a similar procedure to derive the law of reflection. The calculation is left for you to carry out (Problem 47).

SUMMARY

In geometric optics, we use the **ray approximation,** which assumes that a wave travels through a medium in straight lines in the direction of the rays.

The basic laws of geometric optics are the laws of reflection and refraction for

---

[4] More generally, Fermat's principle requires only that the time be an extremum with respect to small variations in path.

light rays. The **law of reflection** states that the angle of reflection, θ_1', equals the angle of incidence, θ_1. The **law of refraction,** or **Snell's law,** states that

$$\frac{\sin \theta_2}{\sin \theta_1} = \frac{v_2}{v_1} = \text{constant} \qquad (35.3)$$

where θ_2 is the angle of refraction.

The **index of refraction** of a medium, n, is defined by the ratio

$$n \equiv \frac{c}{v} \qquad (35.4)$$

where c is the speed of light in a vacuum and v is the speed of light in the medium. In general, n varies with wavelength and is given by

$$n = \frac{\lambda_0}{\lambda_n} \qquad (35.7)$$

where λ_0 is the vacuum wavelength and λ_n is the wavelength in the medium.

An alternate form of Snell's law of refraction is

$$n_1 \sin \theta_1 = n_2 \sin \theta_2 \qquad (35.8)$$

where n_1 and n_2 are the indices of refraction in the two media. The incident ray, the reflected ray, the refracted ray, and the normal to the surface all lie in the same plane.

Huygens' principle states that all points on a wave front can be taken as point sources for the production of secondary wavelets. At some later time, the new position of the wave front is the surface tangent to these wavelets.

Total internal reflection can occur when light travels from a medium of high index of refraction to one of lower index of refraction. The minimum angle of incidence, θ_c, for which total reflection occurs at an interface is given by

$$\sin \theta_c = \frac{n_2}{n_1} \qquad \text{(where } n_1 > n_2\text{)} \qquad (35.10)$$

Fermat's principle states that when a light ray travels between two points, its path is the one that requires the least time (or more generally an extremum, which may be a maximum).

QUESTIONS

1. Light of wavelength λ is incident on a slit of width d. Under what conditions is the ray approximation valid? Under what circumstances does the slit produce enough diffraction to make the ray approximation invalid?

2. Sound waves have much in common with light waves, including the properties of reflection and refraction. Give examples of these phenomena for sound waves.

3. Does a light ray traveling from one medium into another always bend toward the normal as in Figure 35.8? Explain.

4. As light travels from one medium to another, does the wavelength of the light change? Does the frequency change? Does the speed change? Explain.

5. A laser beam passing through a nonhomogeneous sugar solution follows a curved path. Explain.

6. A laser beam ($\lambda = 632.8$ nm) is incident on a piece of Lucite as in Figure 35.29. Part of the beam is reflected and part is refracted. What information can you get from this photograph?

7. Suppose blue light were used instead of red light in the experiment shown in Figure 35.29. Would the refracted beam be bent at a larger or smaller angle?

8. The level of water in a clear, colorless glass is easily observed with the naked eye. The level of liquid helium in a clear glass vessel is extremely difficult to see with the naked eye. Explain.

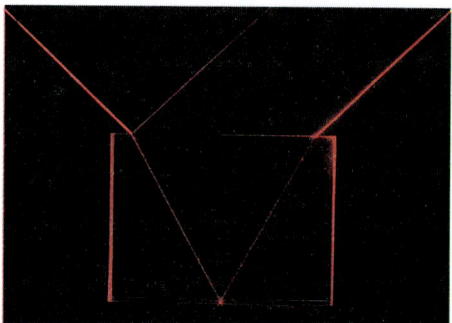

FIGURE 35.29 (Questions 6 and 7) Light from a helium-neon laser beam ($\lambda = 632.8$ nm, red light) is incident on a block of Lucite. The photograph shows both reflected and refracted rays. Can you identify the incident, reflected, and refracted rays? From this photograph, estimate the index of refraction of Lucite at this wavelength. *(Henry Leap and Jim Lehman)*

9. Describe an experiment in which internal reflection is used to determine the index of refraction of a medium.
10. Why does a diamond show flashes of color when observed under white light?
11. Explain why a diamond sparkles more than a glass crystal of the same shape and size.
12. Explain why an oar in the water appears bent.
13. Redesign the periscope of Figure 35.25c so that it can show you where you have been rather than where you are going.
14. Under certain circumstances, sound can be heard over extremely long distances. This frequently happens over a body of water, where the air near the water surface is cooler than the air higher up. Explain how the refraction of sound waves in such a situation could increase the distance over which the sound can be heard.
15. Why do astronomers looking at distant galaxies talk about looking backward in time?
16. A solar eclipse occurs when the Moon gets between the Earth and the Sun. Use a diagram to show why some areas of the Earth see a total eclipse, other areas see a partial eclipse, and most areas see no eclipse.
17. Some department stores have their windows slanted slightly inward at the bottom. This is to decrease the glare from streetlights or the Sun, which would make it difficult

for shoppers to see the display inside. Sketch a light ray reflecting off such a window to show how this technique works.
18. When two colors of light (X and Y) are sent through a glass prism, X is bent more than Y. Which color travels more slowly in the prism?
19. Figure 35.30 represents sunlight striking a drop of water in the atmosphere. Use the laws of refraction and reflection and the fact that sunlight consists of a wide range of wavelengths to discuss the formation of rainbows.

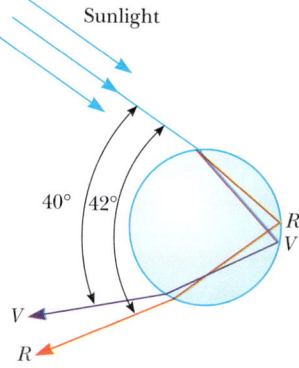

FIGURE 35.30 (Question 19) Refraction of sunlight by a spherical raindrop.

20. Why does the arc of a rainbow appear with red on top and violet on the bottom?
21. How is it possible that a complete circle of rainbow can sometimes be seen from an airplane?
22. You can make a corner reflector by placing three flat mirrors in the corner of a room where the ceiling meets the walls. Show that no matter where you are in the room, you can see yourself reflected in the mirrors—upside down.
23. Several corner reflectors were left on the Moon's Sea of Tranquility by the astronauts of Apollo 11. How can scientists utilize a laser beam sent from Earth even today to determine the precise distance from the Earth to the Moon?
24. Under what conditions is a mirage formed? On a hot day, what are we seeing when we observe "water on the road"?

PROBLEMS

Section 35.2   Measurements of the Speed of Light

1. During the Apollo XI Moon landing, a highly reflecting screen was erected on the Moon's surface. The speed of light is found by measuring the time it takes a laser beam to travel from Earth, reflect from the screen, and return to Earth. If this interval is measured to be 2.51 s, what is the measured speed of

light? Take the center-to-center distance from Earth to Moon to be 3.84×10^8 m, and do not neglect the sizes of the Earth and Moon.

2. As a result of his observations, Roemer concluded that the time interval between eclipses of Io by Jupiter increased by 22 min during a 6-month period as the Earth moved from a point in its orbit where its motion is toward Jupiter to a diametrically opposite point where it moves away from Jupiter. Using 1.5×10^8 km as the average radius of the Earth's orbit around the Sun, calculate the speed of light from these data.

3. Figure P35.3 shows an apparatus used to measure the speed distribution of gas molecules. It consists of two slotted rotating disks separated by a distance s, with the slots displaced by the angle θ. Suppose the speed of light is measured by sending a light beam toward the right disk of this apparatus. (a) Show that a light beam will be seen in the detector (that is, will make it through both slots) only if its speed is given by $c = \omega s / \theta$, where ω is the angular speed of the disks and θ is measured in radians. (b) What is the measured speed of light if the distance between the two slotted rotating disks is 2.5 m, the slot in the second disk is displaced 0.017° from the slot in the first disk, and the disks are rotating at 5555 rev/s?

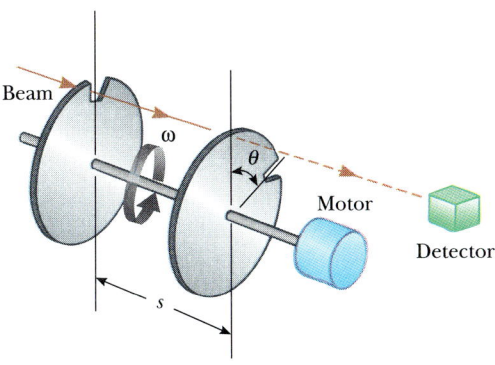

Beam

ω

θ

Motor

Detector

s

FIGURE P35.3

4. Albert Michelson used an improved version of the technique developed by Fizeau to measure the speed of light. In one of Michelson's experiments, the toothed wheel was replaced by a wheel with 32 identical mirrors mounted on its perimeter, with the plane of each mirror perpendicular to a radius of the wheel. The total light path was 8 mi (obtained by multiple reflections of a light beam within an evacuated tube 1 mi long). For what minimum angular speed of the mirror would Michelson have calculated the speed of light to be 2.998×10^8 m/s?

5. In an experiment to measure the speed of light using the apparatus of Fizeau (Fig. 35.2), the distance between light source and mirror was 11.45 km and the wheel had 720 notches. The experimentally determined value of c was 2.998×10^8 m/s. Calculate the minimum angular speed of the wheel for this experiment.

6. If the Fizeau experiment is performed such that the round-trip distance for the light is 40 m, find the two lowest speeds of rotation that allow the light to pass through the notches. Assume that the wheel has 360 teeth and that the speed of light is 3×10^8 m/s. Repeat for a round-trip distance of 4000 m.

7. Use Roemer's value of 22 min discussed in Problem 2 and the presently accepted value of the speed of light in vacuum to find an average value for the distance between the Earth and the Sun.

Section 35.4 Reflection and Refraction

(*Note:* In this section if an index of refraction value is not given, refer to Table 35.1.)

8. A coin is on the bottom of a swimming pool 1.00 m deep. What is the apparent depth of the coin, seen from above the water surface?

9. The wavelength of red helium-neon laser light in air is 632.8 nm. (a) What is its frequency? (b) What is its wavelength in glass that has an index of refraction of 1.50? (c) What is its speed in the glass?

10. A narrow beam of sodium yellow light is incident from air onto a smooth water surface at an angle $\theta_1 = 35.0°$. Determine the angle of refraction θ_2 and the wavelength of the light in water.

11. An underwater scuba diver sees the Sun at an apparent angle of 45° from the vertical. Where is the Sun?

12. A light ray in air is incident on a water surface at an angle of 30.0° with respect to the normal to the surface. What is the angle of the refracted ray relative to this normal?

13. A ray of light in air is incident on a planar surface of silica. The refracted ray makes an angle of 37.0° with the normal. Calculate the angle of incidence.

14. A light ray initially in water enters a transparent substance at an angle of incidence of 37.0°, and the transmitted ray is refracted at an angle of 25.0°. Calculate the speed of light in the transparent substance.

15. A ray of light strikes a flat block of glass ($n = 1.50$) of thickness 2.00 cm at an angle of 30.0° with the normal. Trace the light beam through the glass, and find the angles of incidence and refraction at each surface.

16. Find the speed of light in (a) flint glass, (b) water, and (c) zircon.

17. Light of wavelength 436 nm in air enters a fishbowl filled with water, then exits through the crown glass wall of the container. What is the wavelength of the light (a) in the water and (b) in the glass?

18. The reflecting surfaces of two intersecting plane

mirrors are at an angle of θ $(0° < \theta < 90°)$, as in Figure P35.18. If a light ray strikes the horizontal mirror, show that the emerging ray will intersect the incident ray at an angle of $\beta = 180° - 2\theta$.

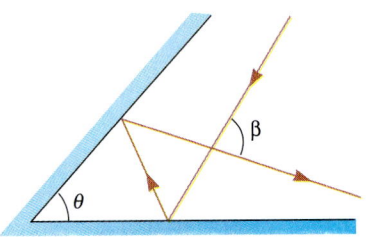

FIGURE P35.18

19. A 1.00-cm-thick by 4.00-cm-long glass plate is made up of two fused prisms. The top prism has an index of refraction of 1.486 for blue light and 1.472 for red light. The bottom prism has an index of refraction of 1.878 for blue light and 1.862 for red light. A ray consisting of red and blue light is incident at 50.0° on the top face as shown in Figure P35.19. Determine the exit angles for both rays that pass through the prisms and the angle between them.

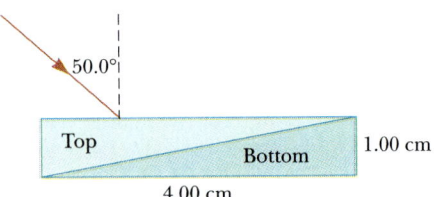

FIGURE P35.19

20. A glass block having $n = 1.52$ and surrounded by air measures 10.0 cm × 10.0 cm. For an angle of incidence of 45.0°, what is the maximum distance x

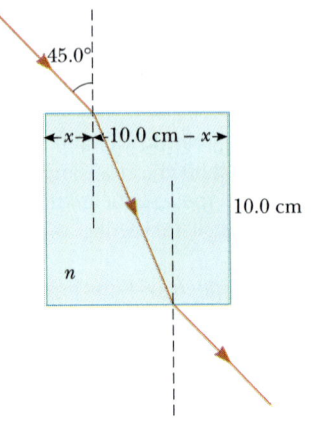

FIGURE P35.20

shown in Figure P35.20 so that the ray will emerge from the opposite side?

***Section 35.5 Dispersion and Prisms**

21. A ray of light strikes the midpoint of one face of an equiangular glass prism ($n = 1.50$) at an angle of incidence of 30.0°. Trace the path of the light ray through the glass and find the angles of incidence and refraction at each surface.

22. A light ray enters the atmosphere of a planet along a radius and then descends to the surface 20.0 km below. The index of refraction where the light enters the atmosphere is 1.000, and it increases linearly to the surface, where it has a value of 1.005. (a) How long does it take the ray to traverse this path? (b) Compare this with the time needed to cover the same distance in a vacuum.

22A. A light ray enters the atmosphere of a planet along a radius and then descends to the surface a distance h below. The index of refraction where the light enters the atmosphere is 1.000, and it increases linearly to the surface, where it has a value of n. (a) How long does it take the ray to traverse this path? (b) Compare this with the time needed to cover the same distance in a vacuum.

23. A narrow white light ray is incident on a block of silica at an angle of 30.0°. Find the angular width of the light ray inside the silica.

23A. A narrow white light ray is incident on a block of fused silica at an angle θ. Find the angular width of the light ray inside the silica.

24. A certain kind of glass has an index of refraction of 1.6500 for blue light (430 nm) and an index of 1.6150 for red light (680 nm). If a beam containing these two colors is incident at an angle of 30° on a piece of this glass, what is the angle between the two beams inside the glass?

25. Show that if the apex angle Φ of a prism is small, an approximate value for the angle of minimum deviation is $\delta_{min} = (n - 1)\Phi$.

26. The index of refraction for red light in water is 1.331, and that for blue light is 1.340. If a ray of white light enters the water at an angle of incidence of 83.00°, what are the underwater angles of refraction for the blue and red components of the light?

27. An experimental apparatus includes a prism made of sodium chloride. The angle of minimum deviation for light of wavelength 589 nm is to be 10.0°. What is the required apex angle of the prism?

28. Light of wavelength 700 nm is incident on the face of a silica prism at an angle of 75° (with respect to the normal to the surface). The apex angle of the prism is 60°. Use the value of n from Figure 35.15 and calculate the angle (a) of refraction at this first surface,

(b) of incidence at the second surface, (c) of refraction at the second surface, and (d) between the incident and emerging rays.

29. A prism that has an apex angle of 50.0° is made of cubic zirconia, with $n = 2.20$. What is its angle of minimum deviation?

30. A triangular glass prism with apex angle 60.0° has an index of refraction $n = 1.50$. (a) What is the smallest angle of incidence θ_1 for which a light ray can emerge from the other side? (See Figure 35.19.) (b) For what angle of incidence θ_1 does the light ray leave at the same angle θ_1?

31. The index of refraction for violet light in silica flint glass is 1.66, and that for red light is 1.62. What is the average angular dispersion of visible light passing through a prism of apex angle 60.0° if the angle of incidence is 50.0° (Fig. P35.31)?

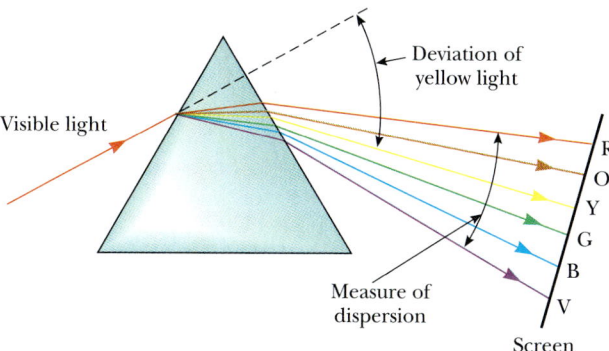

FIGURE P35.31

Section 35.7   Total Internal Reflection

32. A large Lucite cube ($n = 1.59$) has a small air bubble (a defect in the casting process) below one surface. When a penny (diameter 1.90 cm) is placed directly over the bubble on the outside of the cube, the bubble cannot be seen by looking down into the cube at any angle. However, when a dime (radius 1.75 cm) is placed directly over it, the bubble can be seen by looking down into the cube. What is the range of the possible depths of the air bubble beneath the surface?

33. A fiber optic cable ($n = 1.50$) is submerged in water ($n = 1.33$). What is the critical angle for light to stay inside the cable?

34. A glass cube is placed on a newspaper, which rests on a table. A person reads the news through the vertical side of the cube. Determine the maximum index of refraction of the cube.

35. For 589-nm light, calculate the critical angle for the following materials surrounded by air: (a) diamond, (b) flint glass, and (c) ice.

36. Repeat Problem 35 when the materials are surrounded by water.

37. Consider a common mirage formed by super-heated air just above the roadway. If an observer viewing from 2.00 m above the road (where $n = 1.0003$) sees water up the road at $\theta_1 = 88.8°$, find the index of refraction of the air just above the road surface. (*Hint:* Treat this as a problem in total internal reflection.)

38. Traveling inside a diamond, a light ray is incident on the interface between diamond and air. What is the critical angle for total internal reflection? Use Table 35.1. (The smallness of θ_c for diamond means that light is easily "trapped" within a diamond and eventually emerges from the many cut faces; this makes a diamond more brilliant than stones with smaller n and larger θ_c.)

39. An optical fiber is made of a clear plastic for which the index of refraction is 1.50. For what angles with the surface does light remain contained within the fiber?

ADDITIONAL PROBLEMS

40. A 4.00-m-long pole stands vertically in a river having a depth of 2.00 m. When the Sun is 40.0° above the horizontal, determine the length of the pole's shadow on the bottom of the river. Take the index of refraction for water to be 1.33.

40A. A pole of length L stands vertically in a river having a depth d. When the Sun is at an angle θ above the horizontal, determine the length of the pole's shadow on the bottom of the river. Take the index of refraction for water to be n.

41. A specimen of glass has an index of refraction of 1.61 for the wavelength corresponding to the prominent bright line in the sodium spectrum. If an equiangular prism is made from this glass, what angle of incidence results in minimum deviation of the sodium line?

42. A coin is at the bottom of a 6.00-cm-deep beaker. The beaker is filled to the top with 3.50 cm of water (index of refraction = 1.33) covered by 2.50 cm of ether (index of refraction = 1.36). How deep does the coin appear to be when seen from the top of the beaker?

43. A small underwater pool light is 1.00 m below the surface. The light emerging from the water forms a circle on the water surface. What is the diameter of this circle?

44. When the Sun is directly overhead, a narrow shaft of light enters a cathedral through a small hole in the ceiling and forms a spot on the floor 10.0 m below. (a) At what speed (in centimeters per minute) does the spot move across the (flat) floor? (b) If a mirror is placed on the floor to intercept the light, at what speed does the reflected spot move across the ceiling?

45. A drinking glass is 4.00 cm wide at the bottom, as shown in Figure P35.45. When an observer's eye is placed as shown, the observer sees the edge of the bottom of the glass. When this glass is filled with water, the observer sees the center of the bottom of the glass. Find the height of the glass.

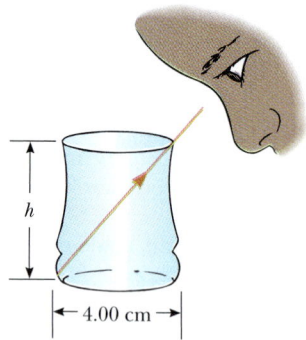

FIGURE P35.45

46. A material having an index of refraction $n = 2.0$ is surrounded by a vacuum and is in the shape of a quarter circle of radius $R = 10$ cm (Fig. P35.46). A light ray parallel to the base of the material is incident from the left at a distance of $L = 5.0$ cm above the base and emerges out of the material at the angle θ. Determine the value of θ.

46A. A material having an index of refraction n is surrounded by a vacuum and is in the shape of a quarter circle of radius R (Fig. P35.46). A light ray parallel to the base of the material is incident from the left at a distance of L above the base and emerges out of the material at the angle θ. Determine an expression for θ in terms of L, n, and R.

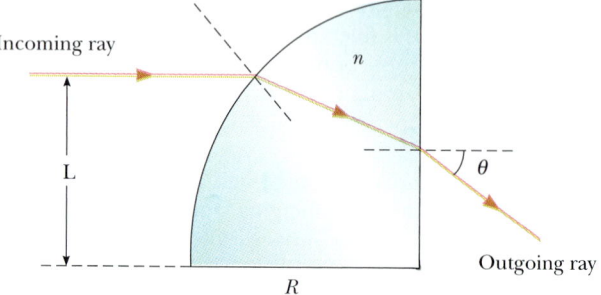

FIGURE P35.46

47. Derive the law of reflection (Eq. 35.2) from Fermat's principle of least time. (See the procedure outlined in Section 35.8 for the derivation of the law of refraction from Fermat's principle.)

48. The angle between the two mirrors in Figure P35.48 is a right angle. The beam of light in the vertical plane *P* strikes mirror 1 as shown. (a) Determine the distance the reflected light beam travels before striking mirror 2. (b) In what direction does the light beam travel after being reflected from mirror 2?

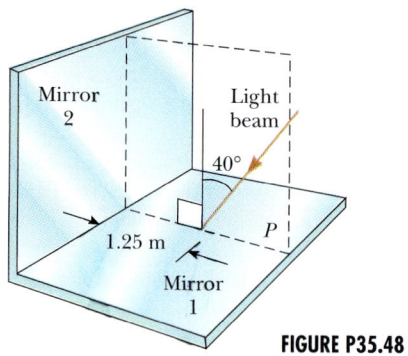

FIGURE P35.48

49. A light ray of wavelength 589 nm is incident at an angle θ on the top surface of a block of polystyrene, as shown in Figure P35.49. (a) Find the maximum value of θ for which the refracted ray undergoes total internal reflection at the left vertical face of the block. Repeat the calculation for the case in which the polystyrene block is immersed in (b) water and (c) carbon disulfide.

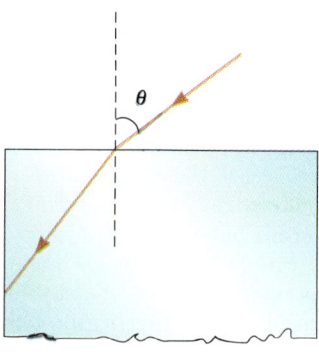

FIGURE P35.49

50. A cylindrical material of radius $R = 2.00$ m has a mirrored surface on its right half, as in Figure P35.50. A light ray traveling in air is incident on the left side of the cylinder. If the incident light ray and exiting light ray are parallel and $d = 2.00$ m, determine the index of refraction of the material.

51. A hiker stands on a mountain peak near sunset and observes a rainbow caused by water droplets in the air about 8 km away. The valley is 2 km below the mountain peak and entirely flat. What fraction of the complete circular arc of the rainbow is visible to the hiker? See Question 19.

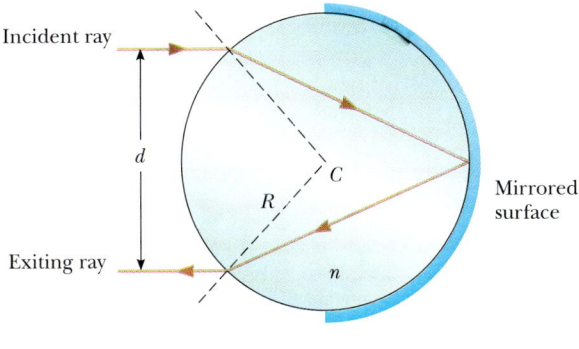

FIGURE P35.50

52. A fish is at a depth d under water. Show that when viewed from an angle of incidence θ_1, the apparent depth z of the fish is

$$z = \frac{3d \cos \theta_1}{\sqrt{7 + 9 \cos^2 \theta_1}}$$

53. A light ray is incident on a prism and refracted at the first surface as shown in Figure P35.53. Let Φ represent the apex angle of the prism and n its index of refraction. Find in terms of n and Φ the smallest allowed value of the angle of incidence at the first surface for which the refracted ray does not undergo internal reflection at the second surface.

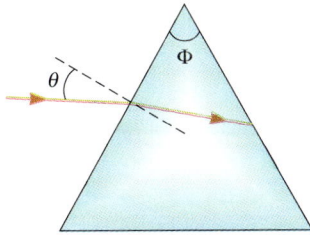

FIGURE P35.53

54. The prism shown in Figure P35.54 has an index of refraction of 1.55. Light is incident at an angle of

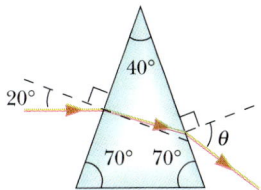

FIGURE P35.54

20°. Determine the angle θ at which the light emerges.

55. A laser beam strikes one end of a slab of material, as shown in Figure P35.55. The index of refraction of the slab is 1.48. Determine the number of internal reflections of the beam before it emerges from the opposite end of the slab.

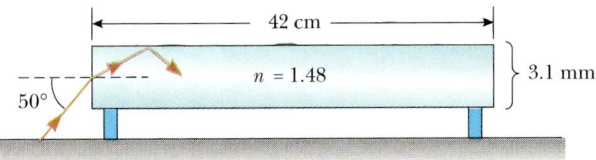

FIGURE P35.55

56. A. H. Pfund's method for measuring the index of refraction of glass is illustrated in Figure P35.56. One face of a slab of thickness t is painted white, and a small hole scraped clear at point P serves as a source of diverging rays when the slab is illuminated from below. Ray PBB' strikes the clear surface at the critical angle and is totally reflected, as are rays such as PCC'. Rays such as PAA' emerge from the clear surface. On the painted surface there appears a dark circle of diameter d, surrounded by an illuminated region, or halo. (a) Derive a formula for n in terms of the measured quantities d and t. (b) What is the diameter of the dark circle if $n = 1.52$ for a slab 0.600 cm thick? (c) If white light is used, the critical angle depends on color caused by dispersion. Is the inner edge of the white halo tinged with red light or violet light? Explain.

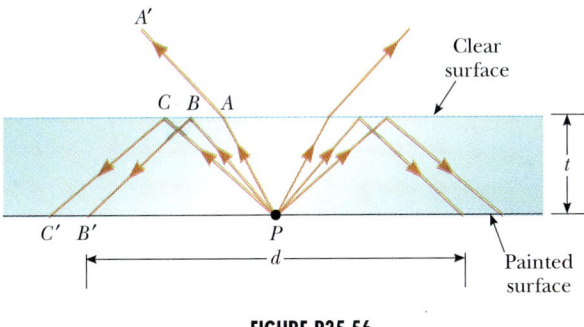

FIGURE P35.56

57. The light beam in Figure P35.57 strikes surface 2 at the critical angle. Determine the angle of incidence θ_1.

58. Students allow a narrow beam of laser light to strike a water surface. They arrange to measure the angle of refraction for selected angles of incidence and

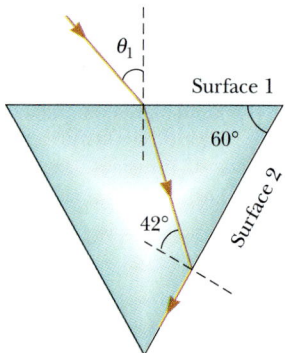

FIGURE P35.57

record the data shown in the accompanying table. Use the data to verify Snell's law of refraction by plotting the sine of the angle of incidence versus the sine of the angle of refraction. Use the resulting plot to deduce the index of refraction of water.

Angle of Incidence (degrees)	Angle of Refraction (degrees)
10.0	7.5
20.0	15.1
30.0	22.3
40.0	28.7
50.0	35.2
60.0	40.3
70.0	45.3
80.0	47.7

59. A light ray traveling in air is incident on one face of a right-angle prism of index of refraction $n = 1.5$ as in Figure P35.59, and the ray follows the path shown in the figure. If $\theta = 60°$ and the base of the prism is mirrored, determine the angle ϕ made by the outgoing ray with the normal to the right face of the prism.

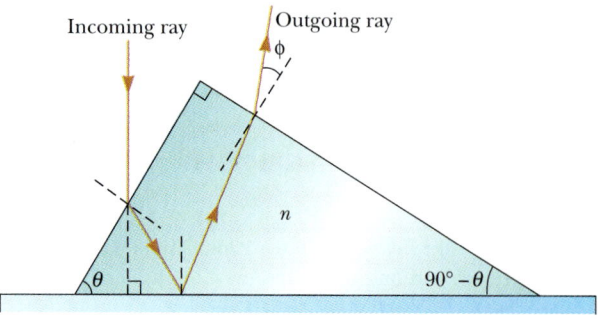

FIGURE P35.59

59A. A light ray traveling in air is incident normally on the base of a right-angle prism of index of refraction n. The ray follows the path shown in Figure P35.59. If the base of the prism is mirrored, determine the angle ϕ made by the outgoing ray with the normal to the right face of the prism in terms of n and the angle θ.

60. A piece of wire is bent through an angle θ. The bent wire is partially submerged in benzene (index of refraction = 1.50) so that looking along the "dry" part of the wire, it appears to be straight and makes a 30.0° angle with the horizontal. Determine the value of θ.

61. A light ray enters a rectangular block of plastic at an angle of $\theta_1 = 45°$ and emerges at an angle of $\theta_2 = 76°$, as in Figure P35.61. (a) Determine the index of refraction for the plastic. (b) If the light ray enters the plastic at a point $L = 50$ cm from the bottom edge, how long does it take the light ray to travel through the plastic?

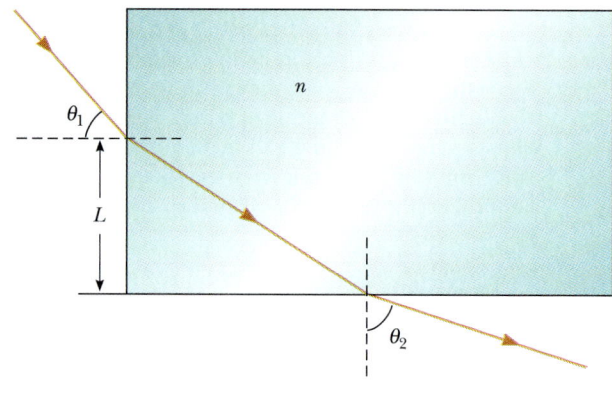

FIGURE P35.61

SPREADSHEET PROBLEM

S1. Spreadsheet 35.1 calculates the angle of refraction and angle of deviation as light beams of different incidence angles travel through a prism of apex angle Φ. Input parameters are the index of refraction n and the apex angle Φ (Fig. PS35.1). As the angle of incidence θ_1 increases from zero, the angle of deviation δ decreases, reaches a minimum value, and then increases. At minimum deviation

$$n \sin \frac{\Phi}{2} = \sin \frac{\Phi + \delta_{min}}{2} = \sin \theta_1$$

This equation is often used to experimentally find n of any transparent prism. (a) For crown glass ($n = 1.52$ and $\Phi = 60°$), find the minimum angle of de-

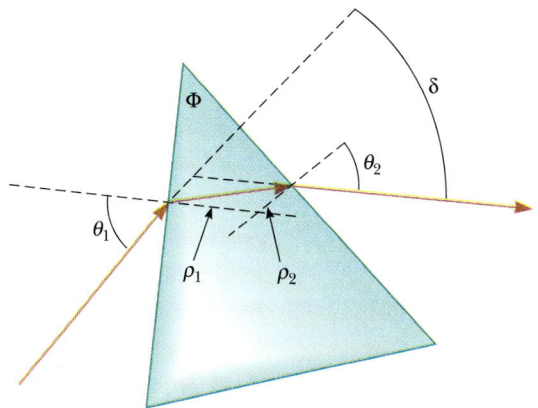

FIGURE PS35.1

iation from the graph. Using the spreadsheet, verify these equations. At the angle of minimum deviation, how is angle ρ_1 related to ρ_2? (b) Because the index of refraction depends slightly on wavelength, vary n between 1.50 and 1.55, and note how the minimum angle of deviation changes with wavelength. (c) Choose several different values for Φ, and note the minimum angles of deviation for each.

''Most mirrors reverse left and right. This one reverses top and bottom.''

Geometric Optics

Nearly parallel light rays are incident on a diverging (concave) lens at the left. The red and blue rays refracted by the diverging lens are bent away from the horizontal, while the same rays are bent toward the horizontal as they pass through the converging lens at the right. The net effect is to produce nearly parallel rays at the right. *(Richard Megna/ Fundamental Photographs)*

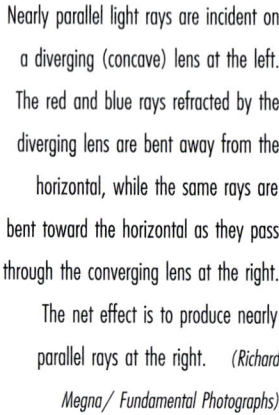

This chapter is concerned with the images formed when spherical waves fall on flat and spherical surfaces. We find that images can be formed by reflection or by refraction and that mirrors and lenses work because of this reflection and refraction. Such devices, commonly used in optical instruments and systems, are described in detail. In this chapter, we continue to use the ray approximation and to assume that light travels in straight lines, both steps valid because here we are studying the field called *geometric optics*. The field of wave optics is discussed in subsequent chapters.

36.1   IMAGES FORMED BY FLAT MIRRORS

In this chapter we discuss the manner in which optical instruments such as lenses and mirrors form images. We begin this investigation by considering the simplest possible mirror, the flat mirror.

Consider a point source of light placed at *O* in Figure 36.1, a distance *p* in front of a flat mirror. The distance *p* is called the **object distance.** Light rays leave the source and are reflected from the mirror. After reflection, the rays diverge (spread apart), but they appear to the viewer to come from a point *I* located behind the mirror. Point *I* is called the **image** of the object at *O*. Regardless of the system under study, images are always formed in the same way. *Images are formed either at the point where rays of light actually intersect or at the point from which they appear to originate.* Since the rays in Figure 36.1 appear to originate at *I*, which is a distance *q* behind the mirror, this is the location of the image. The distance *q* is called the **image distance.**

Images are classified as real or virtual. A **real image** is *one in which rays converge at the image point;* a **virtual image** is *one in which the light rays do not converge to the image point but appear to emanate from that point.* The image formed by the mirror in Figure 36.1 is virtual. The images seen in flat mirrors *are always virtual.* Real images can usually be displayed on a screen (as at a movie), but virtual images cannot be displayed on a screen.

We examine the properties of the images formed by flat mirrors by using the simple geometric techniques shown in Figure 36.2. In order to find out where an image is formed, it is always necessary to follow at least two rays of light as they reflect from the mirror. One of those rays starts at *P*, follows a horizontal path to the mirror, and reflects back on itself. The second ray follows the oblique path *PR* and reflects as shown. An observer to the left of the mirror would trace the two reflected rays back to the point from which they appear to have originated, that is, point *P'*. A continuation of this process for points on the object other than *P* would result in a virtual image (drawn as a yellow arrow) to the right of the mirror. Since triangles *PQR* and *P'QR* are congruent, *PQ = P'Q*. Hence, we conclude that *the image formed by an object placed in front of a flat mirror is as far behind the mirror as the object is in front of the mirror.*

Geometry also shows that the object height, *h*, equals the image height, *h'*. Let us define **lateral magnification,** *M*, as follows:

$$M \equiv \frac{\text{image height}}{\text{object height}} = \frac{h'}{h} \tag{36.1}$$

This is a general definition of the lateral magnification of any type of mirror. *M* = 1 for a flat mirror because *h' = h* in this case.

The image formed by a flat mirror has one more important property: it reverses forward and back, but it does not reverse left and right or top and bottom. If you stand in front of the mirror and raise your right hand, the image raises the hand on your right, which you interpret to be the left hand of the image because the image is facing the opposite direction from you.

Thus, we conclude that the image formed by a flat mirror has the following properties:

- The image is as far behind the mirror as the object is in front.
- The image is unmagnified, virtual, and upright. (By upright we mean that, if the object arrow points upward as in Figure 36.2, so does the image arrow.)
- The image has forward-back reversal.

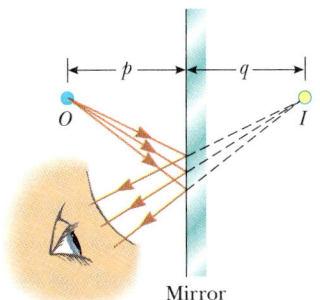

FIGURE 36.1   An image formed by reflection from a flat mirror. The image point, *I*, is located behind the mirror at a distance *q*, which is equal to the object distance, *p*.

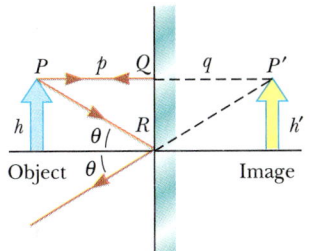

FIGURE 36.2   Geometric construction used to locate the image of an object placed in front of a flat mirror. Because the triangles *PQR* and *P'QR* are congruent, *p = q* and *h = h'*.

CONCEPTUAL EXAMPLE 36.1 Multiple Images Formed by Two Mirrors

Two flat mirrors are at right angles to each other, as in Figure 36.3, and an object is placed at point *O*. In this situation, multiple images are formed. Locate the positions of these images.

Reasoning The image of the object is at I_1 in mirror 1 and at I_2 in mirror 2. In addition, a third image is formed at I_3, which is the image of I_1 in mirror 2 or, equivalently, the image of I_2 in mirror 1. That is, the image at I_1 (or I_2) serves as the object for I_3. Note that in order to form this image at I_3, the rays reflect twice after leaving the object at *O*.

Exercise Sketch the rays corresponding to viewing the images at I_1 and I_2 and show that the light is reflected only once in these cases.

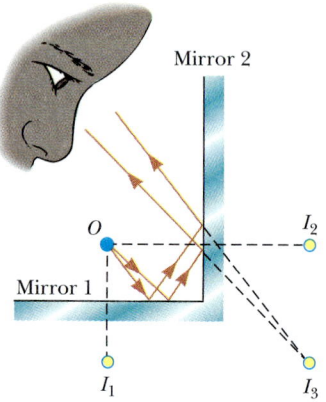

FIGURE 36.3 (Conceptual Example 36.1) When an object is placed in front of two mutually perpendicular mirrors as shown, three images are formed.

CONCEPTUAL EXAMPLE 36.2 The Levitated Professor

The professor in the box shown in Figure 36.4 appears to be balancing himself on a few finger's with both of his feet elevated from the floor. The professor can maintain this position for a long time, and he appears to defy gravity. How do you suppose this illusion was created?

Reasoning This is one example of an optical illusion, used by magicians, that makes use of a mirror. The box that the professor is standing in is a cubical frame that contains a flat vertical mirror through a diagonal plane. The professor straddles the mirror so that one foot is in front of the mirror which you see, and one foot is behind the mirror which you cannot see. When he raises the foot that you see in front of the mirror, the reflection of this foot also rises, so he appears to float in air.

FIGURE 36.4 (Conceptual Example 36.2) An optical illusion. *(Henry Leap and Jim Lehman)*

CONCEPTUAL EXAMPLE 36.3 The Tilting Rearview Mirror

Most rearview mirrors on cars have a day setting and a night setting. The night setting greatly diminishes the intensity of the image so that lights from trailing vehicles do not blind the driver. How does such a mirror work?

Reasoning Consider Figure 36.5, which represents a cross-sectional view of the mirror for the two settings. The mirror is a wedge of glass with a reflecting mirror on the back side. When the mirror is in the day setting, as in Figure 36.5a, the light from an object behind the car strikes the mirror at point 1. Most of the light enters the wedge, is refracted, and reflects from the back of the mirror to return to the front surface, where it is refracted again as it re-enters the air as ray *B* (for *bright*). In addition, a small portion of the light is reflected at the front surface, as indicated by ray *D* (for *dim*). This dim reflected light is responsible for the image observed when the mirror is in the night setting, as in Figure 36.5b. In this case, the wedge is rotated so that the path followed by the bright light (ray *B*) does not lead to the eye. Instead, the dim light reflected from the front surface travels to the eye, and the brightness of trailing headlights does not become a hazard.

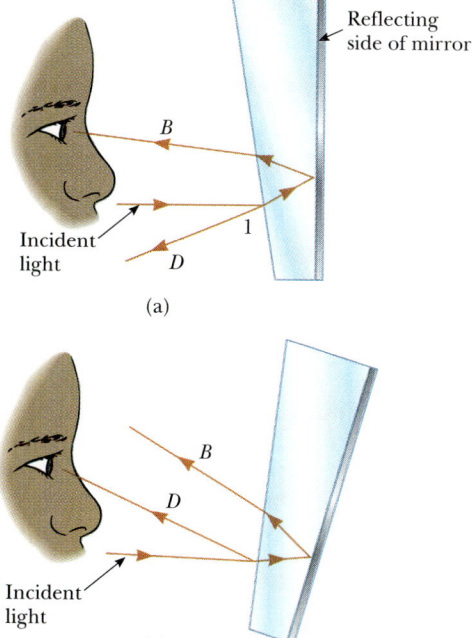

FIGURE 36.5   (Conceptual Example 36.3) A cross-sectional view of a rearview mirror. (a) The day setting forms a bright image, *B*. (b) The night setting forms a dim image, *D*.

36.2   IMAGES FORMED BY SPHERICAL MIRRORS

Concave Mirrors

A **spherical mirror,** as its name implies, has the shape of a segment of a sphere. Figure 36.6a shows the cross-section of a spherical mirror with its surface represented by the solid curved black line. (The blue band represents the area behind the mirror.) Such a mirror, in which light is reflected from the inner, concave surface, is called a **concave mirror.** The mirror has a radius of curvature *R*, and its center of curvature is located at point *C*. Point *V* is the center of the spherical

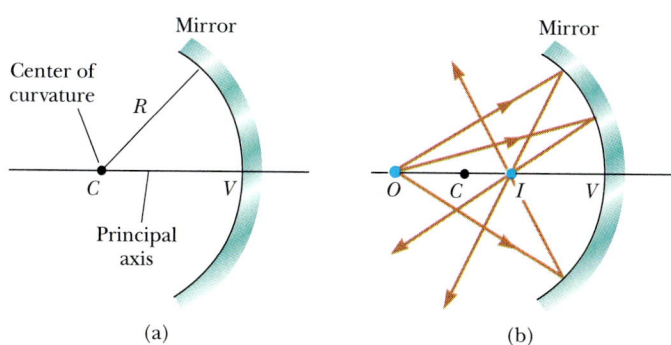

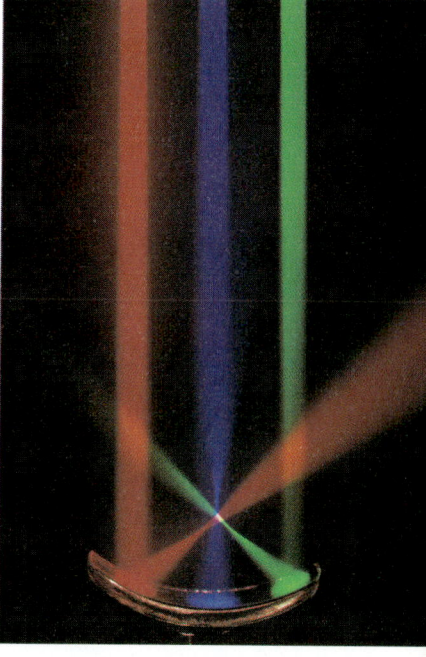

FIGURE 36.6   (a) A concave mirror of radius *R* whose center of curvature is at *C* located on the principal axis. (b) A point object placed at *O* in front of a concave spherical mirror of radius *R*, where *O* is any point on the principal axis that is farther than *R* from the mirror surface, forms a real image at *I*. If the rays diverge from *O* at small angles, they all reflect through the same image point.

Red, blue, and green light rays are reflected by a parabolic mirror. Note that the focal point where the three colors meet is white light.   (*Ken Kay/Fundamental Photographs*)

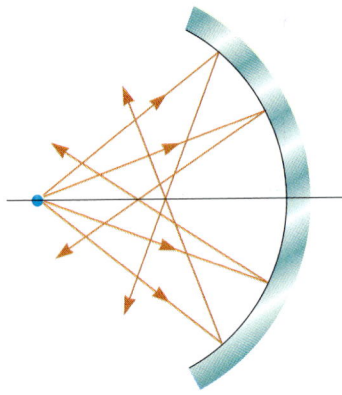

FIGURE 36.7 Rays at large angles from the horizontal axis reflect from a spherical concave mirror to intersect the principal axis at different points, resulting in a blurred image. This is called *spherical aberration.*

segment, and a line drawn from C to V is called the **principal axis** of the optical system.

Now consider a point source of light placed at point O in Figure 36.6b, located on the principal axis to the left of point C. Several diverging rays originating at O are shown. After reflecting from the mirror, these rays converge (come together) at the image point I. The rays then continue to diverge from I as if there were an object there. As a result, a real image is formed.

In what follows, we assume that all rays that diverge from the object make a small angle with the principal axis. Such rays are called **paraxial rays.** All such rays reflect through the image point, as in Figure 36.6b. Rays that are far from the principal axis, as in Figure 36.7, converge to other points on the principal axis, producing a blurred image. This effect, called **spherical aberration,** is present to some extent for any spherical mirror and is discussed in Section 36.5.

We can use the geometry shown in Figure 36.8 to calculate the image distance q from a knowledge of the object distance p and the mirror radius of curvature, R. By convention, these distances are measured from point V. Figure 36.8 shows two rays of light leaving the tip of the object. One of these rays passes through the center of curvature, C, of the mirror, falls normal to the surface of the mirror, and reflects back on itself. The second ray strikes the mirror at the center, point V, and reflects as shown, obeying the law of reflection. The image of the tip of the arrow is located at the point where these two rays intersect. From the largest right triangle in Figure 36.8 whose base is OV, we see that $\tan \theta = h/p$, while the blue-shaded right triangle gives $\tan \theta = -h'/q$. The negative sign is introduced because the image is inverted, and so h' is taken to be negative. Thus, from Equation 36.1 and these results, we find that the magnification of the mirror is

$$M = \frac{h'}{h} = -\frac{q}{p} \tag{36.2}$$

We also note from two other triangles in Figure 36.8 that

$$\tan \alpha = \frac{h}{p - R} \quad \text{and} \quad \tan \alpha = -\frac{h'}{R - q}$$

from which we find that

$$\frac{h'}{h} = -\frac{R - q}{p - R} \tag{36.3}$$

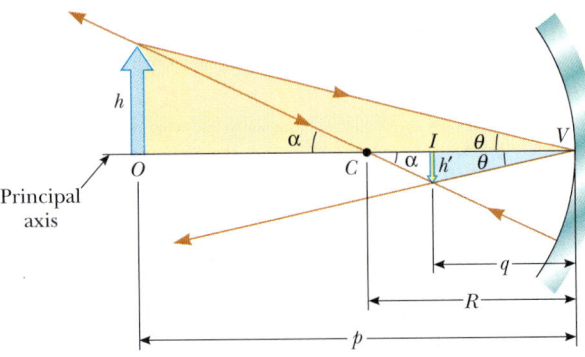

FIGURE 36.8 The image formed by a spherical concave mirror where the object O lies outside the center of curvature, C.

If we compare Equations 36.2 and 36.3, we see that

$$\frac{R-q}{p-R} = \frac{q}{p}$$

Simple algebra reduces this to

$$\frac{1}{p} + \frac{1}{q} = \frac{2}{R} \qquad (36.4)$$

This expression is called the **mirror equation.** It is applicable only to paraxial rays.

If the object is very far from the mirror, that is, if the object distance, p, is large enough compared with R that p can be said to approach infinity, then $1/p \approx 0$, and we see from Equation 36.4 that $q \approx R/2$. That is, when the object is very far from the mirror, *the image point is halfway between the center of curvature and the center of the mirror,* as in Figure 36.9a. The rays are essentially parallel in this figure because the source is assumed to be very far from the mirror. We call the image point in this special case the **focal point,** F, and the image distance the **focal length,** f, where

$$f = \frac{R}{2} \qquad (36.5)$$

Focal length

The mirror equation can therefore be expressed in terms of the focal length:

$$\frac{1}{p} + \frac{1}{q} = \frac{1}{f} \qquad (36.6)$$

Mirror equation

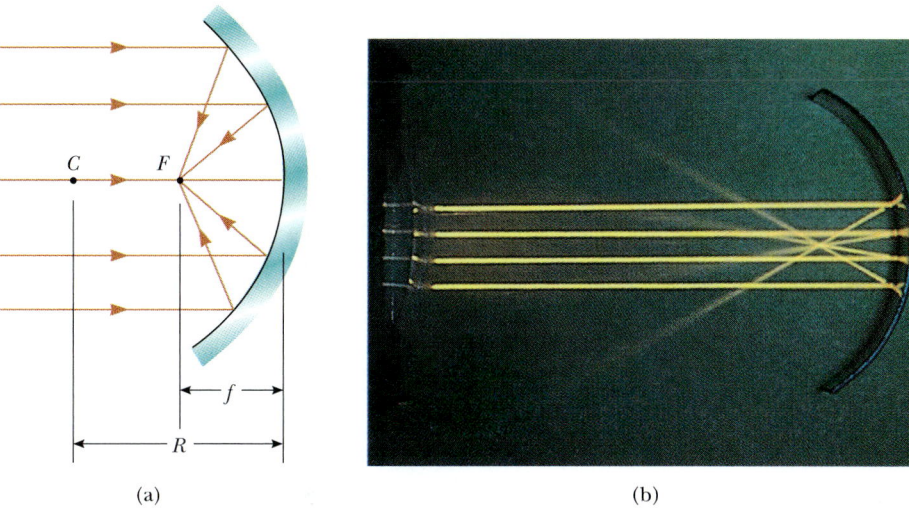

(a) (b)

FIGURE 36.9 (a) Light rays from a distant object ($p \approx \infty$) reflect from a concave mirror through the focal point, F. In this case, the image distance $q = R/2 = f$, where f is the focal length of the mirror. (b) Photograph of the reflection of parallel rays from a concave mirror. *(Henry Leap and Jim Lehman)*

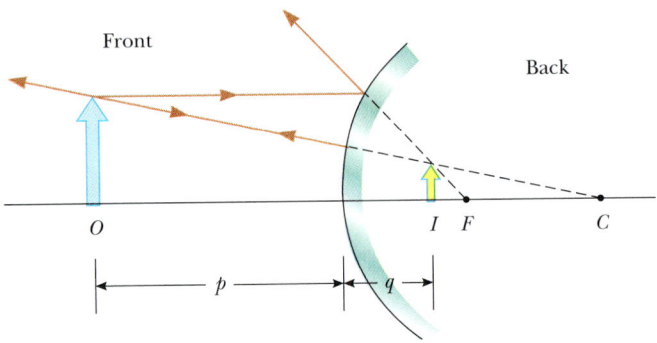

Front

Back

O I F C

p q

FIGURE 36.10   Formation of an image by a spherical convex mirror. The image formed by the real object is virtual and upright.

Convex cylindrical mirror: reflection of parallel lines. The image of any object in front of the mirror is virtual, erect, and diminished in size.   *(© Richard Megna 1990, Fundamental Photographs)*

Convex Mirrors

Figure 36.10 shows the formation of an image by a **convex mirror,** that is, one silvered so that light is reflected from the outer, convex surface. This is sometimes called a **diverging mirror** because the rays from any point on a real object diverge after reflection as though they were coming from some point behind the mirror. The image in Figure 36.10 is virtual because the reflected rays only appear to originate at the image point. Furthermore, the image is always upright and smaller than the object, as shown in the figure.

We do not derive any equations for convex spherical mirrors because we can use Equations 36.2, 36.4, and 36.6 for either concave or convex mirrors if we adhere to the following procedure. Let us refer to the region in which light rays move as the *front side* of the mirror and the other side, where virtual images are formed, as the *back side*. For example, in Figures 36.7 and 36.9, the side to the left of the mirrors is the front side and the side to the right of the mirrors is the back side. Figure 36.11 is helpful for understanding the rules for object and image distances, and Table 36.1 summarizes the sign conventions for all the necessary quantities.

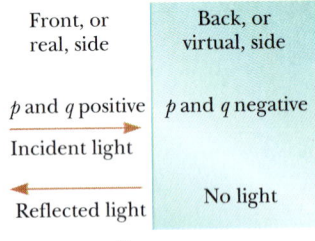

Front, or real side	Back, or virtual side
p and q positive	p and q negative
Incident light	
Reflected light	No light

Convex or concave mirror

FIGURE 36.11   A diagram for describing the signs of p and q for convex and concave mirrors.

Ray Diagrams for Mirrors

The position and size of images formed by mirrors can be conveniently determined by using *ray diagrams*. These graphical constructions tell us the total nature of the image and can be used to check results calculated from the mirror and

---

TABLE 36.1   Sign Convention for Mirrors

p is + if the object is in front of the mirror (real object).
p is − if the object is in back of the mirror (virtual object).

q is + if the image is in front of the mirror (real image).
q is − if the image is in back of the mirror (virtual image).

Both f and R are + if the center of curvature is in front of the mirror (concave mirror).
Both f and R are − if the center of curvature is in back of the mirror (convex mirror).

If M is positive, the image is upright.
If M is negative, the image is inverted.

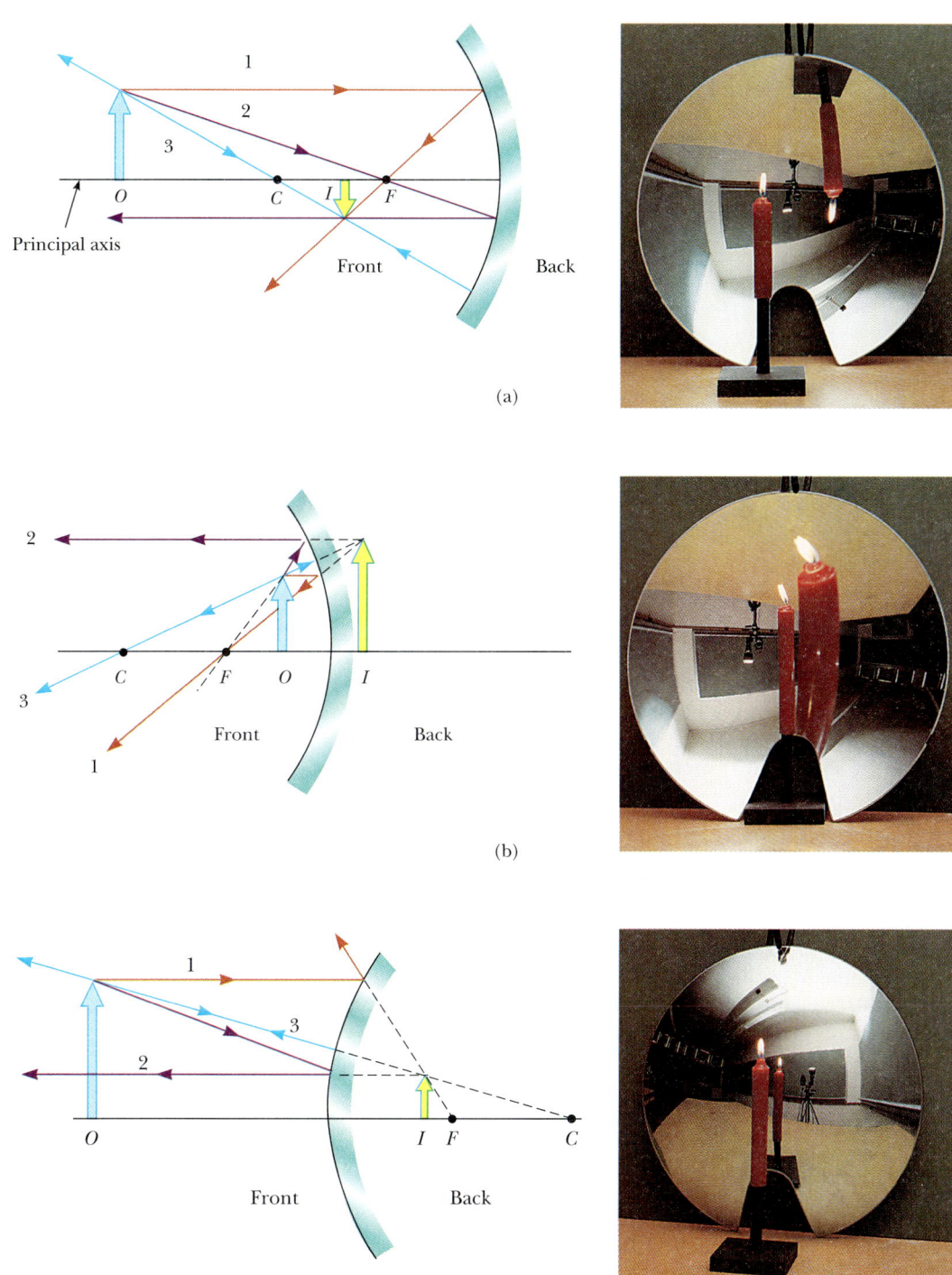

Principal axis

Front   Back

(a)

2
3
C   F   O   I

Front   Back

(b)

1
3
2
O   Front   Back   I F   C

(c)

FIGURE 36.12 Ray diagrams for spherical mirrors, and corresponding photographs of the images of candles. (a) When the object is located so that the center of curvature lies between the object and a concave mirror surface, the image is real, inverted, and reduced in size. (b) When the object is located between the focal point and a concave mirror surface, the image is virtual, upright, and enlarged. (c) When the object is in front of a spherical convex mirror, the image is virtual, upright, and reduced in size.

magnification equations. In these diagrams, we need to know the position of the object and the location of the center of curvature. In order to locate the image, three rays are then constructed, as shown by the various examples in Figure 36.12. These rays all start from any object point (although in our examples we always choose the tip of the arrow for simplicity) and are drawn as follows:

- Ray 1 is drawn from the top of the object parallel to the optical axis and is reflected back through the focal point, F.
- Ray 2 is drawn from the top of the object through the focal point. Thus, it is reflected parallel to the optic axis.
- Ray 3 is drawn from the top of the object through the center of curvature, C, and is reflected back on itself.

The intersection of any two of these rays locates the image. The third ray serves as a check on your construction. The image point obtained in this fashion must always agree with the value of q calculated from the mirror equation.

With concave mirrors, note what happens as the object is moved closer to the mirror. The real, inverted image in Figure 36.12a moves to the left as the object approaches the focal point. When the object is at the focal point, the image is infinitely far to the left. However, when the object lies between the focal point and the mirror surface, as in Figure 36.12b, the image is virtual and upright. Finally, for the convex mirror shown in Figure 36.12c, the image of a real object is always virtual and upright. In this case, as the object distance increases, the virtual image decreases in size and approaches the focal point as p approaches infinity. You should construct other diagrams to verify how the image position varies with object position.

EXAMPLE 36.4 The Image for a Concave Mirror

Assume that a certain concave spherical mirror has a focal length of 10.0 cm. Find the location of the image for object distances of (a) 25.0 cm, (b) 10.0 cm, and (c) 5.00 cm. Describe the image in each case.

Solution (a) For an object distance of 25.0 cm, we find the image distance using the mirror equation:

$$\frac{1}{p} + \frac{1}{q} = \frac{1}{f}$$

$$\frac{1}{25.0 \text{ cm}} + \frac{1}{q} = \frac{1}{10.0 \text{ cm}}$$

$$q = \boxed{16.7 \text{ cm}}$$

The magnification is given by Equation 36.2:

$$M = -\frac{q}{p} = -\frac{16.7 \text{ cm}}{25.0 \text{ cm}} = -0.668$$

This value of M means that the image is smaller than the object. The negative sign means that the image is inverted. Finally, because q is positive, the image is located on the front side of the mirror and is real. This situation is pictured in Figure 36.12a.

(b) When the object distance is 10.0 cm, the object is located at the focal point. Substituting the values $p = 10.0$ cm and $f = 10.0$ cm into the mirror equation, we find

$$\frac{1}{10.0 \text{ cm}} + \frac{1}{q} = \frac{1}{10.0 \text{ cm}}$$

$$q = \boxed{\infty}$$

Thus, we see that rays of light originating from an object located at the focal point of a mirror are reflected so that the image is formed at an infinite distance from the mirror; that is, the rays travel parallel to one another after reflection.

(c) When the object is at the position $p = 5.00$ cm, it lies between the focal point and the mirror surface. In this case, the mirror equation gives

$$\frac{1}{5.00 \text{ cm}} + \frac{1}{q} = \frac{1}{10.0 \text{ cm}}$$

$$q = \boxed{-10.0 \text{ cm}}$$

That is, the image is virtual because it is located behind the mirror. The magnification is

$$M = -\frac{q}{p} = -\left(\frac{-10.0 \text{ cm}}{5.00 \text{ cm}}\right) = 2.00$$

From this, we see that the image is twice as large as the object and the positive sign for M indicates that the image is upright (Fig. 36.12b). The negative value of q means that the image is behind the mirror and is virtual.

Note the characteristics of the images formed by a concave spherical mirror. When the focal point lies between the object and mirror surface, the image is inverted and real; with the object at the focal point, the image is formed at infinity; with the object between the focal point and mirror surface, the image is upright and virtual.

Exercise If the object distance is 20.0 cm, find the image distance and the magnification of the mirror.

Answer $q = 20.0$ cm, $M = -1.00$.

EXAMPLE 36.5 The Image for a Convex Mirror

An object 3.00 cm high is placed 20.0 cm from a convex mirror having a focal length of 8.00 cm. Find (a) the position of the final image and (b) the magnification.

Solution (a) Because the mirror is convex, its focal length is negative. To find the image position, we use the mirror equation:

$$\frac{1}{p} + \frac{1}{q} = \frac{1}{f} = -\frac{1}{8.00 \text{ cm}}$$

$$\frac{1}{q} = -\frac{1}{8.00 \text{ cm}} - \frac{1}{20.0 \text{ cm}}$$

$$q = \boxed{-5.71 \text{ cm}}$$

The negative value of q indicates that the image is virtual, or behind the mirror, as in Figure 36.12c.

(b) The magnification is

$$M = -\frac{q}{p} = -\left(\frac{-5.71 \text{ cm}}{20.0 \text{ cm}}\right) = \boxed{0.286}$$

The image is about 30% of the size of the object and upright because M is positive.

Exercise Find the height of the image.

Answer 0.857 cm.

36.3 IMAGES FORMED BY REFRACTION

In this section, we describe how images are formed by the refraction of rays at a spherical surface of a transparent material. Consider two transparent media with indices of refraction n_1 and n_2, where the boundary between the two media is a spherical surface of radius R (Fig. 36.13). We assume that the object at point O is in the medium whose index of refraction is n_1. Furthermore, of all the paraxial rays leaving O, let us consider only those that make a small angle with the axis and with each other. As we shall see, all such rays originating at the object point are refracted at the spherical surface and focus at a single point I, the image point.

Let us proceed by considering the geometric construction in Figure 36.14,

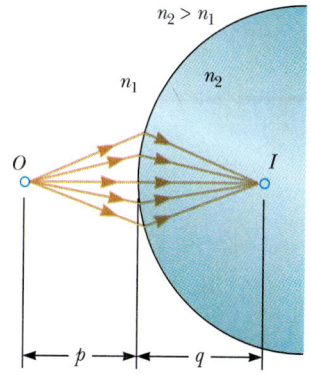

FIGURE 36.13 An image formed by refraction at a spherical surface. Rays making small angles with the optic axis diverge from a point object at O and pass through the image point, I.

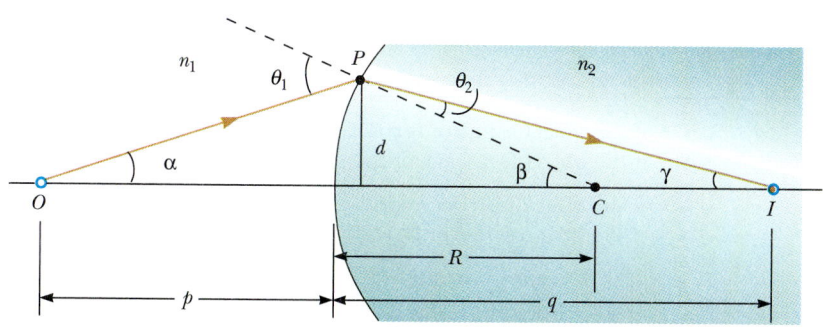

FIGURE 36.14 Geometry used to derive Equation 36.8.

which shows a single ray leaving point O and focusing at point I. Snell's law applied to this refracted ray gives

$$n_1 \sin \theta_1 = n_2 \sin \theta_2$$

Because the angles θ_1 and θ_2 are assumed small, we can use the small-angle approximation $\sin \theta \approx \theta$ (angles in radians). Therefore, Snell's law becomes

$$n_1 \theta_1 = n_2 \theta_2$$

Now we use the fact that an exterior angle of any triangle equals the sum of the two opposite interior angles. Applying this to the triangles OPC and PIC in Figure 36.14 gives

$$\theta_1 = \alpha + \beta$$

$$\beta = \theta_2 + \gamma$$

If we combine the last three expressions, and eliminate θ_1 and θ_2, we find

$$n_1 \alpha + n_2 \gamma = (n_2 - n_1)\beta \tag{36.7}$$

Again, in the small angle approximation, $\tan \theta \approx \theta$, and so we can write the approximate relationships

$$\alpha \cong \frac{d}{p} \qquad \beta \cong \frac{d}{R} \qquad \gamma \cong \frac{d}{q}$$

where d is the distance shown in Figure 36.14. We substitute these expressions into Equation 36.7 and divide through by d to give

$$\frac{n_1}{p} + \frac{n_2}{q} = \frac{n_2 - n_1}{R} \tag{36.8}$$

For a fixed object distance p, the image distance q is independent of the angle that the ray makes with the axis. This result tells us that all paraxial rays focus at the same point I.

As with mirrors, we must use a sign convention if we are to apply this equation to a variety of circumstances. First, note that real images are formed on the side of the surface that is opposite the side from which the light comes, in contrast to mirrors, where real images are formed on the same side of the reflecting surface. Therefore, *the sign convention for spherical refracting surfaces is similar to the convention for mirrors, recognizing the change in sides of the surface for real and virtual images.* For example, in Figure 36.14, p, q, and R are all positive.

The sign convention for spherical refracting surfaces is summarized in Table 36.2. (The same sign convention is used for thin lenses, which we discuss in the

---

TABLE 36.2   **Sign Convention for Refracting Surfaces**

p is $+$ if the object is in front of the surface (real object).
p is $-$ if the object is in back of the surface (virtual object).

q is $+$ if the image is in back of the surface (real image).
q is $-$ if the image is in front of the surface (virtual image).

R is $+$ if the center of curvature is in back of the surface.
R is $-$ if the center of curvature is in front of the surface.

next section.) As with mirrors, we assume that the front of the refracting surface is the side from which the light approaches the surface.

Flat Refracting Surfaces

If the refracting surface is flat, then R approaches infinity and Equation 36.8 reduces to

$$\frac{n_1}{p} = -\frac{n_2}{q}$$

$$q = -\frac{n_2}{n_1} p \tag{36.9}$$

From Equation 36.9 we see that the sign of q is opposite that of p. Thus, *the image formed by a flat refracting surface is on the same side of the surface as the object*. This is illustrated in Figure 36.15 for the situation in which n_1 is greater than n_2, where a virtual image is formed between the object and the surface. If n_1 is less than n_2, the image is still virtual but is formed to the left of the object.

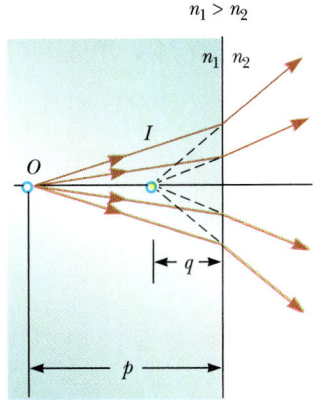

FIGURE 36.15 The image formed by a flat refracting surface is virtual; that is, it forms to the left of the refracting surface. All rays are assumed to be paraxial.

CONCEPTUAL EXAMPLE 36.6   Let's Go Scuba Diving

It is well known that objects viewed underwater with the naked eye appear blurred and out of focus. However, a scuba diver using goggles has a clear view of underwater objects. Explain how this works, using the fact that the indices of refraction of the cornea, water, and air are 1.376, 1.333, and 1.00029, respectively.

Reasoning   Because the cornea and water have almost identical indices of refraction, very little refraction occurs

when underwater objects are viewed with the naked eye. In this case, light from the object focuses behind the retina and causes a blurred image. When goggles are used, the air space between the eye and goggle surface provides the normal amount of refraction at the eye-air interface, and the light from the object is focused on the retina.

EXAMPLE 36.7   Gaze into the Crystal Ball

A coin 2.00 cm in diameter is embedded in a solid glass ball of radius 30.0 cm (Fig. 36.16). The index of refraction of the ball is $n_1 = 1.5$, and the coin is 20.0 cm from the surface. Find the position of the image.

Solution   Because $n_1 > n_2$, where $n_2 = 1.00$ is the index of refraction for air, the rays originating from the object are refracted away from the normal at the surface and diverge outward. Hence, the image is formed in the glass and is virtual. Applying Equation 36.8, we get

$$\frac{n_1}{p} + \frac{n_2}{q} = \frac{n_2 - n_1}{R}$$

$$\frac{1.50}{20.0 \text{ cm}} + \frac{1}{q} = \frac{1.00 - 1.50}{-30.0 \text{ cm}}$$

$$q = -17.1 \text{ cm}$$

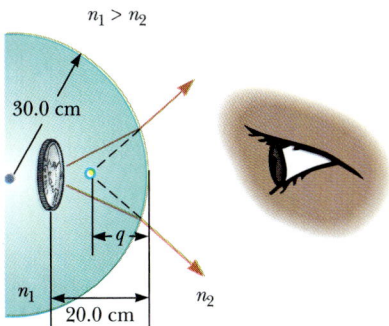

FIGURE 36.16   (Example 36.7) A coin embedded in a glass ball forms a virtual image between the coin and the glass surface. All rays are assumed to be paraxial.

The negative sign indicates that the image is in the same medium as the object (the side of incident light), in agreement with our ray diagram. Being in the same medium as the object, the image must be virtual.

EXAMPLE 36.8 **The One That Got Away**

A small fish is swimming at a depth d below the surface of a pond (Fig. 36.17). What is the apparent depth of the fish as viewed from directly overhead?

Solution   In this example, the refracting surface is flat, and so R is infinite. Hence, we can use Equation 36.9 to determine the location of the image. Using the facts that $n_1 = 1.33$ for water and $p = d$ gives

$$q = -\frac{n_2}{n_1}p = -\frac{1}{1.33}d = -0.750d$$

Again, since q is negative, the image is virtual, as indicated in Figure 36.17. The apparent depth is three-fourths the actual depth.

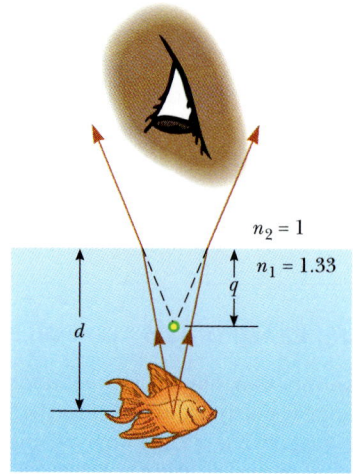

FIGURE 36.17   (Example 36.8) The apparent depth, q, of the fish is less than the true depth, d. All rays are assumed to be paraxial.

36.4   THIN LENSES

Lenses are commonly used to form images by refraction in optical instruments, such as cameras, telescopes, and microscopes. The essential idea in locating the final image of a lens is to *use the image formed by one refracting surface as the object for the second surface.*

Consider a lens having an index of refraction n and two spherical surfaces of radii of curvature R_1 and R_2, as in Figure 36.18. An object is placed at point O at a distance p_1 in front of surface 1. For this example, p_1 has been chosen so as to produce a virtual image I_1 to the left of the lens. This image is then used as the object for surface 2, which results in a real image I_2.

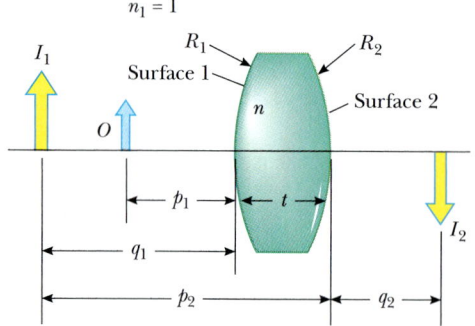

FIGURE 36.18   To locate the image of a lens, the image at I_1 formed by the first surface is used as the object for the second surface. The final image is at I_2.

Using Equation 36.8 and assuming $n_1 = 1$ because the lens is surrounded by air, we find that the image formed by surface 1 satisfies the equation

$$(1) \qquad \frac{1}{p_1} + \frac{n}{q_1} = \frac{n-1}{R_1}$$

Now we apply Equation 36.8 to surface 2, taking $n_1 = n$ and $n_2 = 1$. That is, light approaches surface 2 as if it had come from I_1. Taking p_2 as the object distance and q_2 as the image distance for surface 2 gives

$$(2) \qquad \frac{n}{p_2} + \frac{1}{q_2} = \frac{1-n}{R_2}$$

But $p_2 = -q_1 + t$, where t is the thickness of the lens. (Remember q_1 is a negative number and p_2 must be positive by our sign convention.) For a thin lens, we can neglect t. In this approximation and from Figure 36.18, we see that $p_2 = -q_1$. Hence, (2) becomes

$$(3) \qquad -\frac{n}{q_1} + \frac{1}{q_2} = \frac{1-n}{R_2}$$

Adding (1) and (3), we find that

$$(4) \qquad \frac{1}{p_1} + \frac{1}{q_2} = (n-1)\left(\frac{1}{R_1} - \frac{1}{R_2}\right)$$

For the thin lens, we can omit the subscripts on p_1 and q_2 in (4) and call the object distance p and the image distance q, as in Figure 36.19. Hence, we can write (4) in the form

$$\frac{1}{p} + \frac{1}{q} = (n-1)\left(\frac{1}{R_1} - \frac{1}{R_2}\right) \qquad (36.10)$$

This expression relates the image distance q of the image formed by a thin lens to the object distance p and to the thin lens properties (index of refraction and radii of curvature). It is valid only for paraxial rays and only when the lens thickness is small relative to R_1 and R_2.

We now define the focal length f of a thin lens as the image distance that corresponds to an infinite object distance, as we did with mirrors. According to this definition and from Equation 36.10, we see that as $p \to \infty$, $q \to f$; therefore, the inverse of the focal length for a thin lens is

$$\frac{1}{f} = (n-1)\left(\frac{1}{R_1} - \frac{1}{R_2}\right) \qquad (36.11)$$

Lens makers' equation

Equation 36.11 is called the **lens makers' equation** because it enables f to be calculated from the known properties of the lens. It can also be used to determine the values of R_1 and R_2 needed for a given index of refraction and desired focal length.

Using Equation 36.11, we can write Equation 36.10 in an alternate form identical to Equation 36.6 for mirrors:

$$\frac{1}{p} + \frac{1}{q} = \frac{1}{f} \qquad (36.12)$$

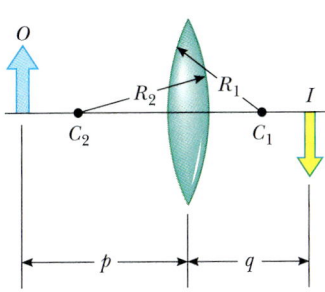

FIGURE 36.19 The biconvex lens.

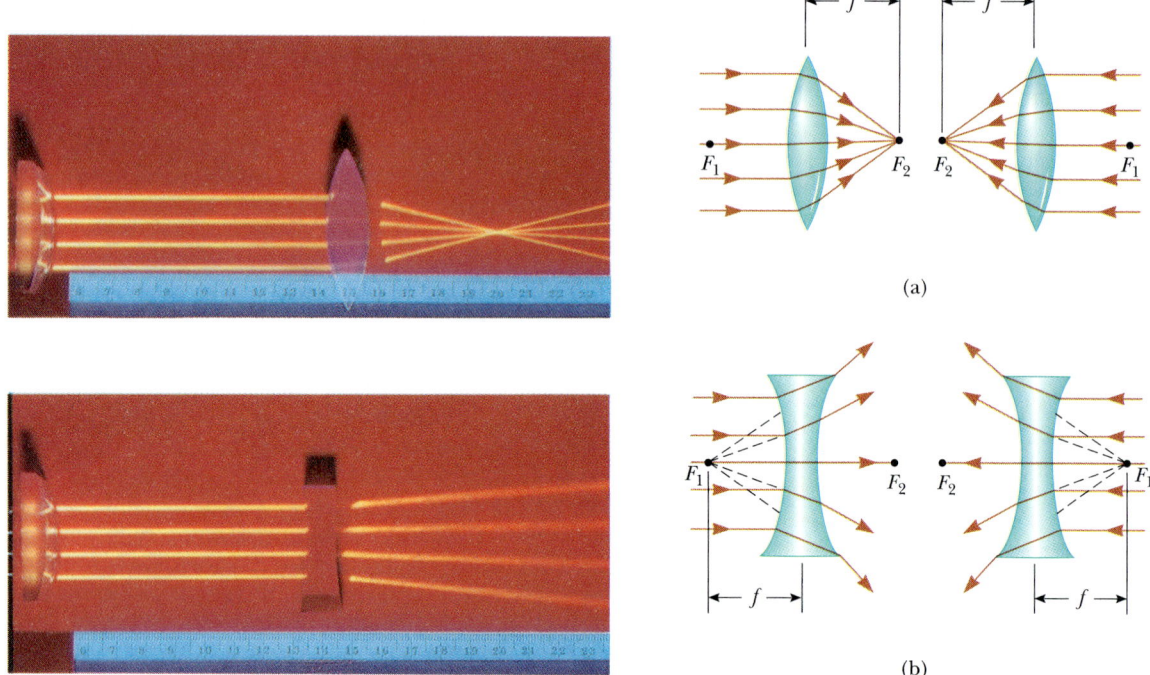

FIGURE 36.20 *(Left)* Photographs of the effect of converging and diverging lenses on parallel rays. *(Henry Leap and Jim Lehman) (Right)* The object and image focal points of (a) the biconvex lens and (b) the biconcave lens.

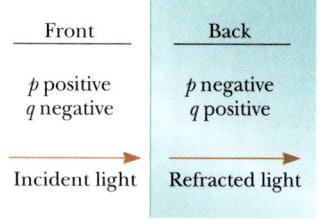

FIGURE 36.21 A diagram for obtaining the signs of p and q for a thin lens or a refracting surface.

A thin lens has two focal points, corresponding to incident parallel light rays traveling from the left or right. This is illustrated in Figure 36.20 for a biconvex lens (converging, positive f) and a biconcave lens (diverging, negative f). Focal point F_1 is sometimes called the *object focal point*, and F_2 is called the *image focal point*.

Figure 36.21 is useful for obtaining the signs of p and q, and Table 36.3 gives the complete sign conventions for lenses. Note that the sign conventions for thin lenses are the same as for refracting surfaces. Applying these rules to a converging lens, we see that when $p > f$, the quantities p, q, and R_1 are positive and R_2 is negative. Therefore, when a converging lens forms a real image from a real object, p, q, and f are all positive. For a diverging lens, p and R_2 are positive, q and R_1 are negative, and so f is negative for a diverging lens.

Sketches of various lens shapes are shown in Figure 36.22. In general, note that a converging lens is thicker at the center than at the edge, whereas a diverging lens is thinner at the center than at the edge.

TABLE 36.3   Sign Convention for Thin Lenses

p is + if the object is in front of the lens.
p is − if the object is in back of the lens.

q is + if the image is in back of the lens.
q is − if the image is in front of the lens.

R_1 and R_2 are + if the center of curvature is in back of the lens.
R_1 and R_2 are − if the center of curvature is in front of the lens.

Consider a single thin lens illuminated by a real object (that the object is real means that $p > 0$). As with mirrors, the lateral magnification of a thin lens is defined as the ratio of the image height h' to the object height h:

$$M = \frac{h'}{h} = -\frac{q}{p}$$

From this expression, it follows that when M is positive, the image is upright and on the same side of the lens as the object. When M is negative, the image is inverted and on the side of the lens opposite the object.

Ray Diagrams for Thin Lenses

Ray diagrams are very convenient for locating the image formed by a thin lens or a system of lenses. Such constructions also help clarify the sign conventions that have been discussed. Figure 36.23 illustrates this method for three single-lens situations. To locate the image of a converging lens (Figs. 36.23a and 36.23b), the following three rays are drawn from the top of the object:

- Ray 1 is drawn parallel to the optic axis. After being refracted by the lens, this ray passes through (or appears to come from) one of the focal points.
- Ray 2 is drawn through the center of the lens. This ray continues in a straight line.
- Ray 3 is drawn through the object focal point and emerges from the lens parallel to the optic axis.

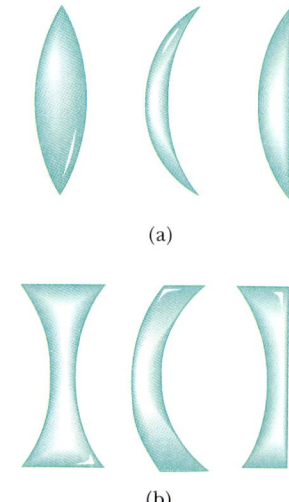

FIGURE 36.22   Various lens shapes: (a) Converging lenses have a positive focal length and are thickest at the middle. From left to right are biconvex, convex-concave, and plano-convex lenses. (b) Diverging lenses have a negative focal length and are thickest at the edges. From left to right are biconcave, convex-concave, and plano-concave lenses.

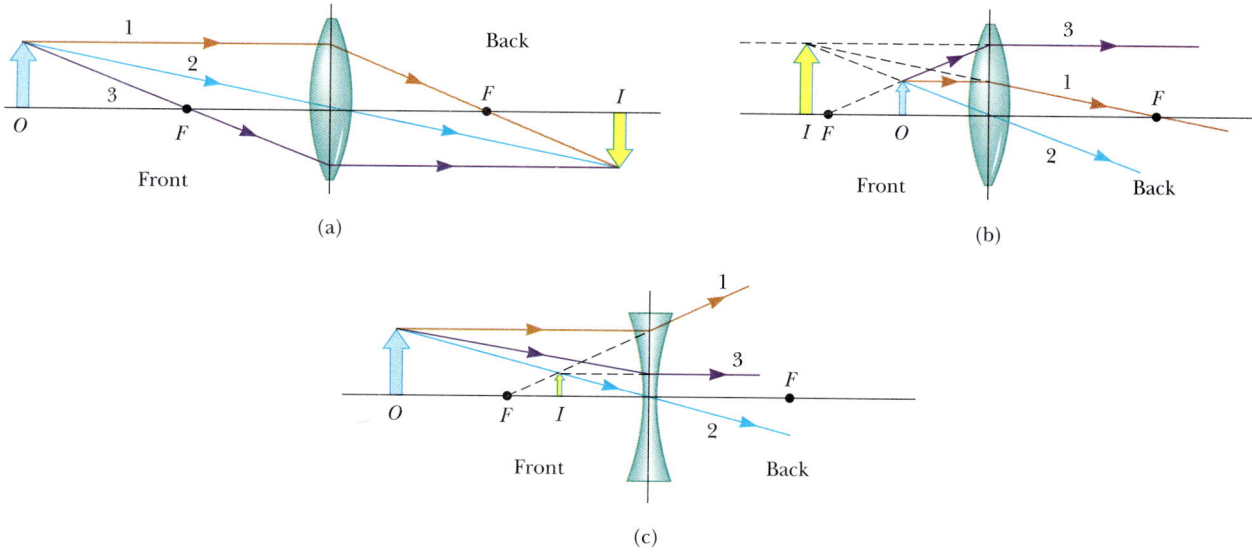

FIGURE 36.23   Ray diagrams for locating the image formed by a thin lens. (a) The object is located to the left of the object focal point of a converging lens. (b) The object is located between the object focal point and a converging lens. (c) The object is located to the left of the object focal point of a diverging lens.

A similar construction is used to locate the image of a diverging lens, as shown in Figure 36.23c.

For the converging lens in Figure 36.23a, where the object is to the left of the object focal point ($p > f$), the image is real and inverted. When the real object is between the object focal point and the lens ($p < f$), as in Figure 36.23b, the image is virtual and erect. Finally, for a diverging lens (Fig. 36.23c), the image is always virtual and upright. These geometric constructions are reasonably accurate only if the distance between the rays and the principal axis is small compared with the radii of the lens surfaces.

EXAMPLE 36.9   An Image Formed by a Diverging Lens

A diverging lens has a focal length of -20.0 cm. An object 2.00 cm in height is placed 30.0 cm in front of the lens. Locate the position of the image.

Solution   Using the thin lens equation (Eq. 36.12) with $p = 30.0$ cm and $f = -20.0$ cm, we get

$$\frac{1}{p} + \frac{1}{q} = \frac{1}{f}$$

$$\frac{1}{30.0 \text{ cm}} + \frac{1}{q} = -\frac{1}{20.0 \text{ cm}}$$

$$q = \boxed{-12.0 \text{ cm}}$$

The negative sign tells us that the image is virtual, as indicated in Figure 36.23c.

Exercise   Find the magnification and the height of the image.

Answer $M = 0.400$, $h' = 0.800$ cm.

EXAMPLE 36.10   An Image Formed by a Converging Lens

A converging lens of focal length 10.0 cm forms an image of an object placed (a) 30.0 cm, (b) 10.0 cm, and (c) 5.00 cm from the lens. Find the image distance and describe the image in each case.

Solution   (a) The thin lens equation, Equation 36.12, can be used to find the image distance:

$$\frac{1}{p} + \frac{1}{q} = \frac{1}{f}$$

$$\frac{1}{30.0 \text{ cm}} + \frac{1}{q} = \frac{1}{10.0 \text{ cm}}$$

$$q = \boxed{15.0 \text{ cm}}$$

The positive sign tells us that the image is real. The magnification is

$$M = -\frac{q}{p} = -\frac{15.0 \text{ cm}}{30.0 \text{ cm}} = \boxed{-0.500}$$

Thus, the image is reduced in size by one half, and the negative sign for M tells us that the image is inverted. The situation is like that pictured in Figure 36.23a.

(b) No calculation is necessary for this case because we know that, when the object is placed at the focal point, the image is formed at infinity. This is readily verified by substituting $p = 10.0$ cm into the lens equation.

(c) We now move inside the focal point, to an object distance of 5.00 cm. In this case, the thin lens equation gives

$$\frac{1}{5.00 \text{ cm}} + \frac{1}{q} = \frac{1}{10.0 \text{ cm}}$$

$$q = \boxed{-10.0 \text{ cm}}$$

$$M = -\frac{q}{p} = -\left(\frac{-10.0 \text{ cm}}{5.00 \text{ cm}}\right) = \boxed{2.00}$$

The negative image distance tells us that the image is virtual. The image is enlarged, and the positive sign for M tells us that the image is upright, as in Figure 36.23b.

There are two general cases for a converging lens. When the object distance is greater than the focal length ($p > f$), the image is real and inverted. When the object is between the object focal point and lens ($p < f$), the image is virtual, upright, and enlarged.

EXAMPLE 36.11   A Lens Under Water

A converging glass lens ($n = 1.52$) has a focal length of 40.0 cm in air. Find its focal length when it is immersed in water, which has an index of refraction of 1.33.

Solution   We can use the lens makers' formula (Eq. 36.11) in both cases, noting that R_1 and R_2 remain the same in air and water:

$$\frac{1}{f_{air}} = (n - 1)\left(\frac{1}{R_1} - \frac{1}{R_2}\right)$$

$$\frac{1}{f_{water}} = (n' - 1)\left(\frac{1}{R_1} - \frac{1}{R_2}\right)$$

where n' is the index of refraction of glass relative to water. That is, $n' = 1.52/1.33 = 1.14$. Dividing the two equations gives

$$\frac{f_{water}}{f_{air}} = \frac{n - 1}{n' - 1} = \frac{1.52 - 1}{1.14 - 1} = 3.71$$

Since $f_{air} = 40.0$ cm, we find that

$$f_{water} = 3.71 f_{air} = 3.71(40.0 \text{ cm}) = \boxed{148 \text{ cm}}$$

In fact, the focal length of any glass lens is increased by the factor $(n - 1)/(n' - 1)$ when immersed in water.

Combination of Thin Lenses

If two thin lenses are used to form an image, the system can be treated in the following manner. First, the image of the first lens is located as if the second lens were not present. The light then approaches the second lens as if it had come from the image formed by the first lens. In other words, the image of the first lens is treated as the object of the second lens. The image of the second lens is the final image of the system. If the image formed by the first lens lies to the right of the second lens, then that image is treated as a virtual object for the second lens (that is, p negative). The same procedure can be extended to a system of three or more lenses. The overall magnification of a system of thin lenses equals the product of the magnification of the separate lenses.

Now suppose two thin lenses of focal lengths f_1 and f_2 are placed in contact with each other. If p is the object distance for the combination, then application of the thin lens equation to the first lens gives

$$\frac{1}{p} + \frac{1}{q_1} = \frac{1}{f_1}$$

where q_1 is the image distance for the first lens. Treating this image as the object for the second lens, we see that the object distance for the second lens must be $-q_1$. Therefore, for the second lens

$$-\frac{1}{q_1} + \frac{1}{q} = \frac{1}{f_2}$$

where q is the final image distance from the second lens. Adding these equations eliminates q_1 and gives

$$\frac{1}{p} + \frac{1}{q} = \frac{1}{f_1} + \frac{1}{f_2}$$

$$\frac{1}{f} = \frac{1}{f_1} + \frac{1}{f_2} \qquad (36.13)$$

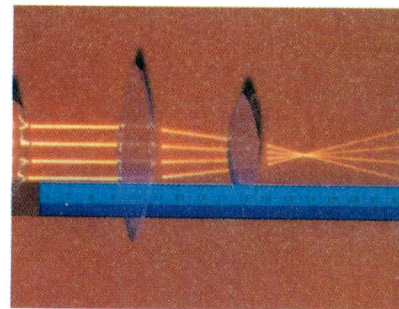

Light from a distant object brought into focus by two converging lenses. Can you estimate the overall focal length of this combination from the photograph?   *(Henry Leap and Jim Lehman)*

Focal length of two thin lenses in contact

If the two thin lenses are in contact with one another, then q is also the distance of the final image from the first lens. Therefore, *two thin lenses in contact are equivalent to a single thin lens whose focal length is given by Equation 36.13.*

EXAMPLE 36.12　Where Is the Final Image?

Two thin converging lenses of focal lengths 10.0 cm and 20.0 cm are separated by 20.0 cm, as in Figure 36.24. An object is placed 15.0 cm to the left of the first lens. Find the position of the final image and the magnification of the system.

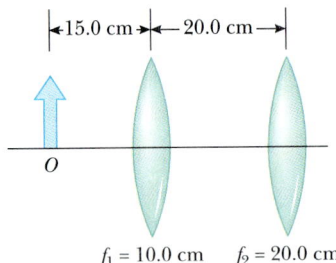

FIGURE 36.24　(Example 36.12) A combination of two converging lenses.

$f_1 = 10.0$ cm　　$f_2 = 20.0$ cm

Solution　First, we find the image position for the first lens while neglecting the second lens:

$$\frac{1}{p_1} + \frac{1}{q_1} = \frac{1}{15.0 \text{ cm}} + \frac{1}{q_1} = \frac{1}{10.0 \text{ cm}}$$

$$q_1 = 30.0 \text{ cm}$$

where q_1 is measured from the first lens.

Because q_1 is greater than the separation between the two lenses, the image of the first lens lies 10.0 cm to the right of the second lens. We take this as the object distance for the second lens. That is, we apply the thin lens equation to the second lens with $p_2 = -10.0$ cm, where distances are now measured from the second lens, whose focal length is 20.0 cm:

$$\frac{1}{p_2} + \frac{1}{q_2} = \frac{1}{f_2}$$

$$\frac{1}{-10.0 \text{ cm}} + \frac{1}{q_2} = \frac{1}{20.0 \text{ cm}}$$

$$q_2 = \boxed{6.67 \text{ cm}}$$

That is, the final image lies 6.67 cm to the right of the second lens.

The magnification of each lens separately is given by

$$M_1 = \frac{-q_1}{p_1} = -\frac{30.0 \text{ cm}}{15.0 \text{ cm}} = -2.00$$

$$M_2 = \frac{-q_2}{p_2} = -\frac{6.67 \text{ cm}}{-10.0 \text{ cm}} = 0.667$$

The total magnification M of the two lenses is the product $M_1 M_2 = (-2.00)(0.667) = -1.33$. Hence, the final image is real, inverted, and enlarged.

*36.5　LENS ABERRATIONS

One of the basic problems of lenses and lens systems is imperfect images, largely the result of defects in the shape and form of the lenses. The simple theory of mirrors and lenses assumes that rays make small angles with the optic axis. In this simple model, all rays leaving a point source focus at a single point, producing a sharp image. Clearly, this is not always true. When the approximations used in this theory do not hold, imperfect images are formed.

If one wishes to perform a precise analysis of image formation, it is necessary to trace each ray using Snell's law at each refracting surface. This procedure shows that the rays from a point object do not focus at a single point. That is, there is no single point image; instead, the image is blurred. The departures of real (imperfect) images from the ideal image predicted by the simple theory are called **aberrations**.

Spherical Aberrations

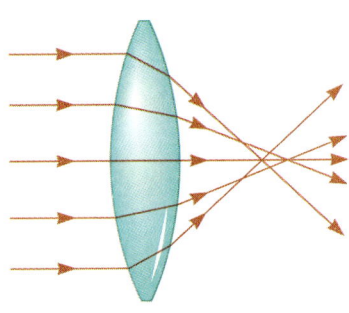

FIGURE 36.25　Spherical aberration caused by a converging lens. Does a diverging lens cause spherical aberration?

Spherical aberrations result from the fact that the focal points of light rays far from the optic axis of a spherical lens (or mirror) are different from the focal points of rays of the same wavelength passing near the center. Figure 36.25 illustrates spherical aberration for parallel rays passing through a converging lens. Rays near the middle of the lens are imaged farther from the lens than rays at the edges. Hence, there is no single focal length for a lens.

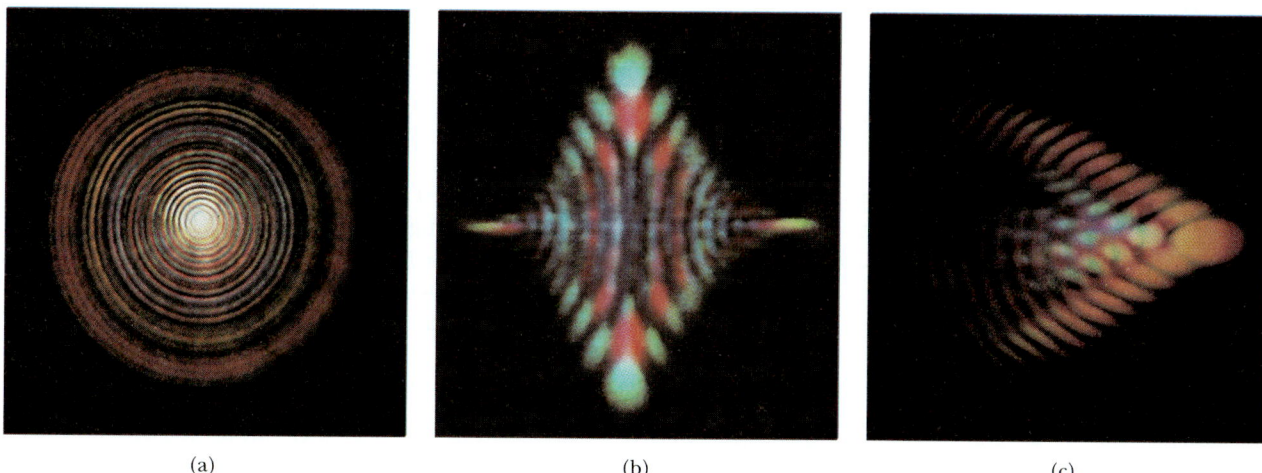

(a)	(b)	(c)

Lenses can produce various forms of aberrations, as shown by the blurred images of a point source in these photos. (a) Spherical aberration occurs when light passing through the lens at different distances from the optical axis is focused at different points. (b) Astigmatism is an aberration that occurs for objects not located on the optical axis of the lens. (c) Coma. This aberration occurs as light passing through the lens far from the optical axis and light passing near the center of the lens focus at different parts of the focal plane. *(Photographs by Norman Goldberg)*

Many cameras are equipped with a variable aperture to control the light intensity and reduce spherical aberration when possible. (A variable aperture is used to control the amount of light transmitted through the lens.) Sharper images are produced as the aperture size is reduced, because for small apertures only the central portion of the lens is exposed to the incident light. At the same time, however, less light is imaged. To compensate for this lower light intensity, a longer exposure time is used on the photographic film. A good example is the sharp image produced by a "pin-hole" camera, whose aperture size is approximately 1 mm.

In the case of mirrors used for very distant objects, spherical aberrations can be eliminated, or at least minimized, by using a parabolic surface rather than a spherical surface. Parabolic surfaces are not often used, however, because those with high-quality optics are very expensive to make. Parallel light rays incident on such a surface focus at a common point. Parabolic reflecting surfaces are used in many astronomical telescopes in order to enhance the image quality. They are also used in flashlights, where a nearby parallel light beam is produced from a small lamp placed at the focus of the surface.

Chromatic Aberrations

The fact that different wavelengths of light refracted by a lens focus at different points gives rise to chromatic aberrations. In Chapter 35, we described how the index of refraction of a material varies with wavelength. When white light passes through a lens, it is found, for example, that violet light rays are refracted more than red light rays (Fig. 36.26). From this we see that the focal length is larger for red light than for violet light. Other wavelengths (not shown in Fig. 36.26) have intermediate focal points. The chromatic aberration for a diverging lens is opposite that for a converging lens. Chromatic aberration can be greatly reduced by

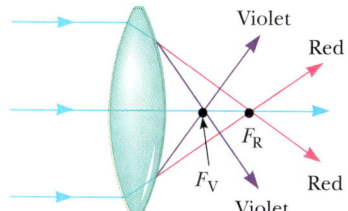

FIGURE 36.26 Chromatic aberration caused by a converging lens. Rays of different wavelengths focus at different points.

using a combination of a converging and diverging lens made from two different types of glass.

*36.6   THE CAMERA

The photographic **camera** is a simple optical instrument whose essential features are shown in Figure 36.27. It consists of a light-tight box, a converging lens that produces a real image, and a film behind the lens to receive the image. Focusing is accomplished by varying the distance between lens and film with an adjustable bellows in older style cameras or some other mechanical arrangement in modern cameras. For proper focusing, or sharp images, the lens-to-film distance will depend on the object distance as well as on the focal length of the lens. The shutter, located behind the lens, is a mechanical device that is opened for selected time intervals. With this arrangement, one can photograph moving objects by using short exposure times or dark scenes (low light levels) by using long exposure times. If this arrangement were not available, it would be impossible to take stop-action photographs. For example, a rapidly moving vehicle could move far enough in the time that the shutter was open to produce a blurred image. Another major cause of blurred image is the movement of the camera while the shutter is open. For this reason, short exposure times or a tripod should be used, even for stationary objects. Typical shutter speeds are 1/30, 1/60, 1/125, and 1/250 s. A stationary object is normally shot with a shutter speed of 1/60 s.

More expensive cameras also have an aperture of adjustable diameter either behind or between the lenses to further control the intensity of the light reaching the film. When an aperture of small diameter is used, only light from the central portion of the lens reaches the film and so the aberration is reduced somewhat.

The brightness of the image depends on the focal length and diameter D of the lens. Clearly, the light intensity I is proportional to the area of the lens. Since the area is proportional to D^2, we conclude that $I \propto D^2$. Furthermore, the intensity is a measure of the energy received by the film per unit area of the image. Since the area of the image is proportional to $(q)^2$, and $q \approx f$ (for objects with $p \gg f$), we conclude that the intensity is also proportional to $1/f^2$, so that $I \propto D^2/f^2$. The ratio f/D is defined to be the **f-number** of a lens:

$$f\text{-number} \equiv \frac{f}{D} \tag{36.14}$$

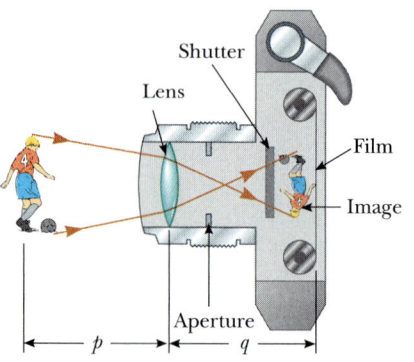

FIGURE 36.27   Cross-sectional view of a simple camera. Note that in reality, $p \gg q$.

Hence, the intensity of light incident on the film can be expressed as

$$I \propto \frac{1}{(f/D)^2} \propto \frac{1}{(f\text{-number})^2} \qquad (36.15)$$

The *f*-number is a measure of the "light-concentrating" power and determines the "speed" of the lens. A "fast" lens has a small *f*-number. Fast lenses, with an *f*-number as low as approximately 1.4, are more expensive because it is more difficult to keep aberrations acceptably small. Camera lenses are often marked with various *f*-numbers such as $f/2.8$, $f/4$, $f/5.6$, $f/8$, $f/11$, $f/16$. These settings are obtained by adjusting the aperture, which effectively changes D. The smallest *f*-number corresponds to the aperture wide open and the full lens area in use. Simple cameras for routine snapshots usually have a fixed focal length and fixed aperture size, with an *f*-number of about $f/11$.

EXAMPLE 36.13   Finding the Correct Exposure Time

The lens of a certain 35-mm camera (where 35 mm is the width of the film strip) has a focal length of 55 mm and a speed of $f/1.8$. The correct exposure time for this speed under certain conditions is known to be $(1/500)$ s. (a) Determine the diameter of the lens.

Solution   From Equation 36.14, we find that

$$D = \frac{f}{f\text{-number}} = \frac{55 \text{ mm}}{1.8} = \boxed{31 \text{ mm}}$$

(b) Calculate the correct exposure time if the *f*-number is changed to $f/4$ under the same lighting conditions.

Reasoning and Solution   The total light energy received by each part of the image is proportional to the product of the flux and the exposure time. If I is the light intensity reaching the film, then in a time t, the energy received by the film is It. Comparing the two situations, we require that $I_1 t_1 = I_2 t_2$, where t_1 is the correct exposure time for $f/1.8$ and t_2 is the correct exposure time for some other *f*-number. Using this result, together with Equation 36.15, we find that

$$\frac{t_1}{(f_1\text{-number})^2} = \frac{t_2}{(f_2\text{-number})^2}$$

$$t_2 = \left(\frac{f_2\text{-number}}{f_1\text{-number}} \right)^2 t_1 = \left(\frac{4}{1.8} \right)^2 \left(\frac{1}{500} \text{ s} \right) \approx \boxed{\frac{1}{100} \text{ s}}$$

That is, as the aperture is reduced in size, the exposure time must increase.

---

*36.7   THE EYE

The eye is an extremely complex part of the body, and because of its complexity, certain defects often arise that can cause impaired vision. In this section we describe the parts of the eye, their purpose, and some of the corrections that can be made when the eye does not function properly.

Like a camera, a normal eye focuses light and produces a sharp image. However, the mechanisms by which the eye controls the amount of light admitted and adjusts to produce correctly focused images are far more complex, intricate, and effective than those in even the most sophisticated camera. In all respects, the eye is an architectural wonder.

Figure 36.28 shows the essential parts of the eye. The front is covered by a transparent membrane called the *cornea*, behind which are a clear liquid region (the *aqueous humor*), a variable aperture (the *iris* and *pupil*), and the *crystalline lens*. Most of the refraction occurs in the cornea because the liquid medium surrounding the lens has an average index of refraction close to that of the lens. The iris, which is the colored portion of the eye, is a muscular diaphragm that controls pupil size. The iris regulates the amount of light entering the eye by dilating the

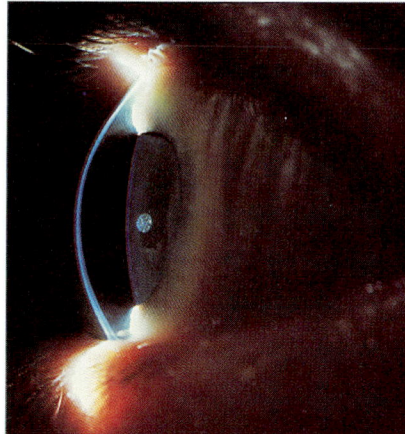

Close-up photograph of the cornea of the human eye. (*Lennart Nilsson*, Behold Man, *Little Brown and Company*)

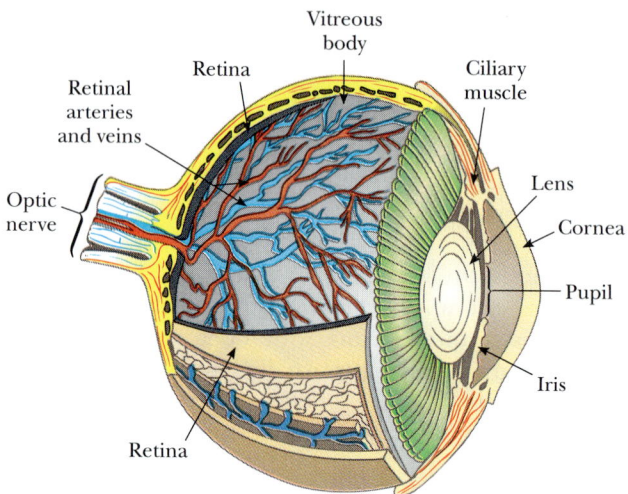

FIGURE 36.28 Essential parts of the eye. Note the similarity between the eye and the simple camera. Can you correlate the parts of the eye with those of the camera?

pupil in light of low intensity and contracting the pupil in high-intensity light. The *f*-number range of the eye is from about $f/2.8$ to $f/16$.

Light entering the eye is focused by the cornea-lens system onto the back surface of the eye, called the *retina*, which consists of millions of sensitive structures called *rods* and *cones*. When stimulated by light, these receptors send impulses via the optic nerve to the brain, where an image is perceived. By this process, a distinct image of an object is observed when the image falls on the retina.

The eye focuses on a given object by varying the shape of the pliable crystalline lens through an amazing process called **accommodation.** An important component in accommodation is the ciliary muscle, which is attached to the lens. When the eye is focused on distant objects, the ciliary muscle is relaxed. For an object distance of infinity, the focal length of the eye (the distance between the lens and the retina) is about 1.7 cm. The eye focuses on nearby objects by tensing the ciliary muscle. This action effectively decreases the focal length by slightly decreasing the radius of curvature of the lens, which allows the image to be focused on the retina. This lens adjustment takes place so swiftly that we are not even aware of the change. Again in this respect, even the finest electronic camera is a toy compared with the eye. It is evident that there is a limit to accommodation because objects that are very close to the eye produce blurred images. The **near point** represents the closest distance for which the lens of the relaxed eye can focus light on the retina. This distance usually increases with age and has an average value of 25 cm. Typically, at age ten the near point of the eye is about 18 cm. This increases to about 25 cm at age 20, to 50 cm at age 40, and to 500 cm or greater at age 60. The **far point** of the eye represents the largest distance for which the lens of the relaxed eye can focus light on the retina. A person with normal vision is able to see very distant objects, such as the Moon, and thus has a far point near infinity.

Conditions of the Eye

The eye may have several abnormalities, which can often be corrected with eyeglasses, contact lenses, or surgery.

When the relaxed (unaccommodated) eye produces an image of a distant object behind the retina, as in Figure 36.29a, the condition is known as **farsighted-**

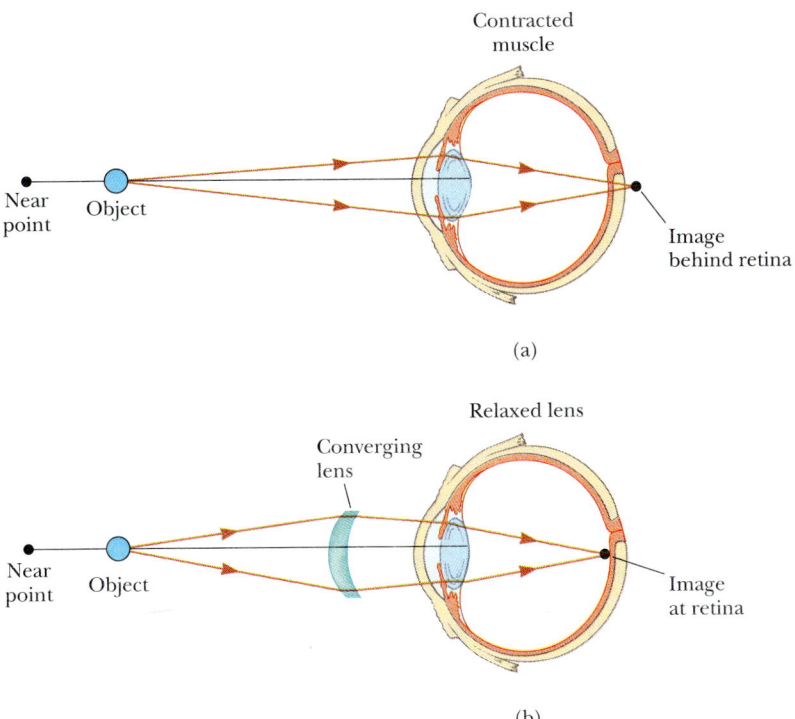

Contracted
muscle

Near
point   Object

Image
behind retina

(a)

Relaxed lens

Converging
lens

Near
point   Object

Image
at retina

(b)

FIGURE 36.29   (a) When a farsighted person looks at an object inside the near point, the image is formed behind the retina, resulting in blurred vision. The eye muscle contracts to try to bring the object into focus. (b) The condition of farsightedness can be corrected with a converging lens.

ness (or *hyperopia*). A farsighted person can usually see faraway objects clearly but cannot focus on nearby objects. Although the near point of a normal eye is approximately 25 cm, the near point of a farsighted person is much farther than this. The eye of a farsighted person tries to focus on objects closer than the near point by accommodation, that is, by shortening its focal length. However, because the focal length of the eye is longer than normal, the light from a nearby object forms a sharp image behind the retina, causing a blurred image. The condition can be corrected by placing a converging lens in front of the eye, as in Figure 36.29b. The lens refracts the incoming rays more toward the principal axis before entering the eye, allowing them to converge and focus on the retina.

Nearsightedness (or *myopia*) is a condition in which a person is able to focus on nearby objects but cannot see faraway objects clearly. In most cases, nearsightedness is due to an eye whose lens is too far from the retina. The far point of the nearsightedness is not at infinity, and may be as close as a few meters. The maximum focal length of the nearsighted eye is insufficient to produce a clearly formed image on the retina. In this case, rays from a distant object are focused in front of the retina, causing blurred vision (Fig. 36.30a). Nearsightedness can be corrected with a diverging lens, as in Figure 36.30b. The lens refracts the rays away from the principal axis before entering the eye, allowing them to focus on the retina.

Beginning with middle age, most people lose some of their accommodation power as the ciliary muscle weakens and the lens hardens. This causes an individual to become farsighted. Fortunately, the condition can be corrected with converging lenses.

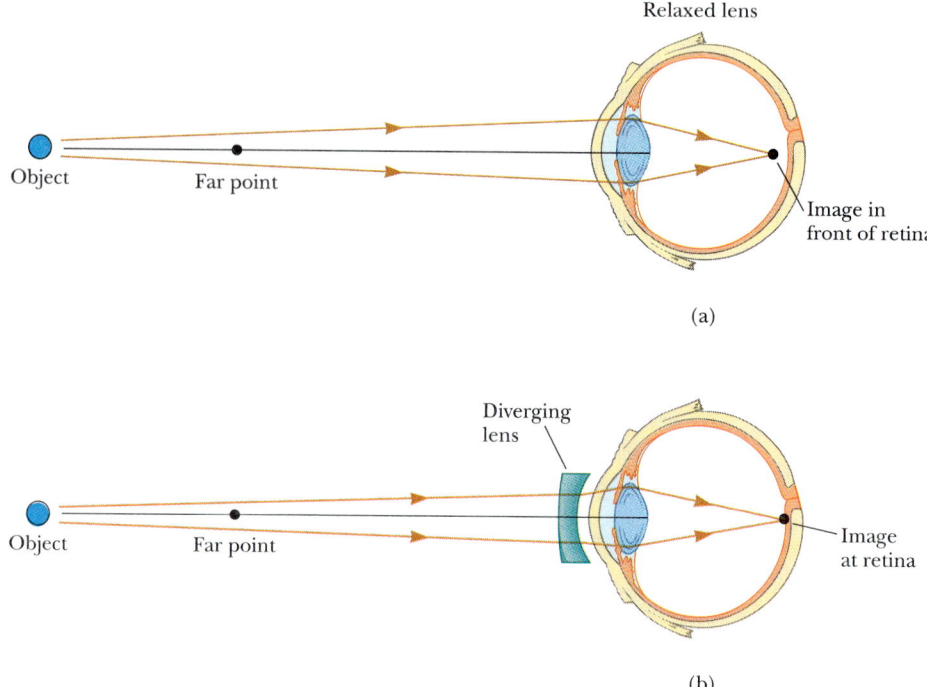

FIGURE 36.30 (a) When a nearsighted person looks at a distant object that lies beyond the far point, the image is formed in front of the retina, resulting in blurred vision. (b) The condition of nearsightedness can be corrected with a diverging lens.

A person may also have an eye defect known as **astigmatism,** in which light from a point source produces a line image on the retina. This condition arises either when the cornea or the lens or both are not perfectly symmetric. Astigmatism can be corrected with lenses having different curvatures in two mutually perpendicular directions.

Optometrists and ophthalmologists usually prescribe lenses measured in **diopters.**

> The **power,** P, of a lens in diopters equals the inverse of the focal length in meters, that is, $P = 1/f$.

For example, a converging lens whose focal length is $+20$ cm has a power of $+5.0$ diopters, and a diverging lens whose focal length is -40 cm has a power of -2.5 diopters.

EXAMPLE 36.14   A Case of Nearsightedness

A particular nearsighted person is unable to see objects clearly when they are beyond 2.5 m (the far point of the eye). What should the focal length of the lens prescribed to correct this problem be?

Solution The purpose of the lens in this instance is to "move" an object from infinity to a distance where it can be seen clearly. This is accomplished by having the lens produce an image at the far point of the eye. From the thin lens equation, we have

$$\frac{1}{p} + \frac{1}{q} = \frac{1}{\infty} - \frac{1}{2.5 \text{ m}} = \frac{1}{f}$$

$$f = -2.5 \text{ m}$$

Why did we use a negative sign for the image distance? As you should have suspected, the lens must be a diverging lens (negative focal length) to correct nearsightedness.

Exercise What is the power of this lens?

Answer -0.40 diopters.

*36.8 THE SIMPLE MAGNIFIER

The simple magnifier consists of a single converging lens. As the name implies, this device is used to increase the apparent size of an object. Suppose an object is viewed at some distance p from the eye, as in Figure 36.31. Clearly, the size of the image formed at the retina depends on the angle θ subtended by the object at the eye. As the object moves closer to the eye, θ increases and a larger image is observed.[1] However, an average normal eye is unable to focus on an object closer than about 25 cm, the near point (Fig. 36.32a). Try it! Therefore, θ is maximum at the near point.

To further increase the apparent angular size of an object, a converging lens can be placed in front of the eye with the object located at point O, just inside the focal point of the lens, as in Figure 36.32b. At this location, the lens forms a virtual, upright, and enlarged image, as shown. Clearly, the lens increases the angular size of the object. We define the **angular magnification**, m, as the ratio of the angle subtended by an object with a lens in use (angle θ in Fig. 36.32b) to that subtended by the object placed at the near point with no lens (angle θ_0 in Fig. 36.32a):

$$m \equiv \frac{\theta}{\theta_0} \qquad (36.16)$$

The angular magnification is a maximum when the image is at the near point of the eye, that is, when $q = -25$ cm. The object distance corresponding to this image distance can be calculated from the thin lens formula:

$$\frac{1}{p} + \frac{1}{-25 \text{ cm}} = \frac{1}{f}$$

$$p = \frac{25f}{25 + f}$$

where f is the focal length of the magnifier in centimeters. If we make the small angle approximations

$$\theta_0 \approx \frac{h}{25} \qquad \text{and} \qquad \theta \approx \frac{h}{p} \qquad (36.17)$$

Equation 36.16 becomes

$$m_{\text{max}} = \frac{\theta}{\theta_0} = \frac{h/p}{h/25} = \frac{25}{p} = \frac{25}{25f/(25 + f)}$$

$$m_{\text{max}} = 1 + \frac{25 \text{ cm}}{f} \qquad (36.18)$$

A magnifying glass is useful when viewing fine details on a map. *(Jim Lehman)*

Angular magnification with the object at the near point

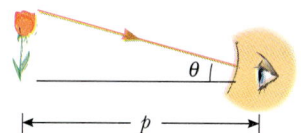

FIGURE 36.31 The size of the image formed on the retina depends on the angle θ subtended at the eye.

[1] Regular eyeglasses give some magnification because the lenses are not located at the lens of the eye. On the other hand, contact lenses minimize this effect because of their close proximity to the lens of the eye.

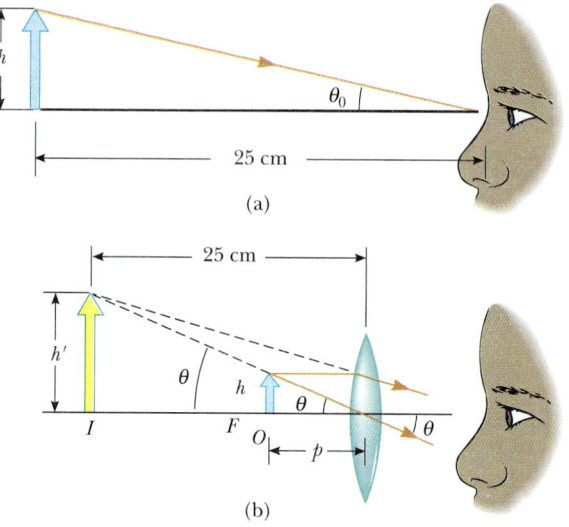

FIGURE 36.32 (a) An object placed at the near point of the eye ($p = 25$ cm) subtends an angle θ_0 at the eye, where $\theta_0 \approx h/25$. (b) An object placed near the focal point of a converging lens produces a magnified image, which subtends an angle $\theta \approx h'/25$ at the eye.

Although the eye can focus on an image formed anywhere between the near point and infinity, it is most relaxed when the image is at infinity (Section 36.7). For the image formed by the magnifying lens to appear at infinity, the object has to be at the focal point of the lens. In this case, Equations in 36.17 become

$$\theta_0 \approx \frac{h}{25} \qquad \text{and} \qquad \theta \approx \frac{h}{f}$$

and the magnification is

$$m_{min} = \frac{\theta}{\theta_0} = \frac{25 \text{ cm}}{f} \tag{36.19}$$

With a single lens, it is possible to obtain angular magnifications up to about 4 without serious aberrations. Magnifications up to about 20 can be achieved by using one or two additional lenses to correct for aberrations.

EXAMPLE 36.15 Maximum Magnification of a Lens

What is the maximum magnification of a lens having a focal length of 10 cm, and what is the magnification of this lens when the eye is relaxed?

Solution The maximum magnification occurs when the image is located at the near point of the eye. Under these circumstances, Equation 36.18 gives

$$m_{max} = 1 + \frac{25 \text{ cm}}{f} = 1 + \frac{25 \text{ cm}}{10 \text{ cm}} = \boxed{3.5}$$

When the eye is relaxed, the image is at infinity. In this case, we use Equation 36.19:

$$m_{min} = \frac{25 \text{ cm}}{f} = \frac{25 \text{ cm}}{10 \text{ cm}} = \boxed{2.5}$$

---

*36.9 THE COMPOUND MICROSCOPE

A simple magnifier provides only limited assistance in inspecting the minute details of an object. Greater magnification can be achieved by combining two lenses in a device called a compound microscope, a schematic diagram of which is shown in Figure 36.33a. It consists of an objective lens that has a very short focal length $f_0 < 1$ cm, and an eyepiece lens having a focal length, f_e, of a few centimeters. The two lenses are separated by a distance L, where L is much greater than either f_0 or f_e. The object, which is placed just to the left of the focal point of the objective,

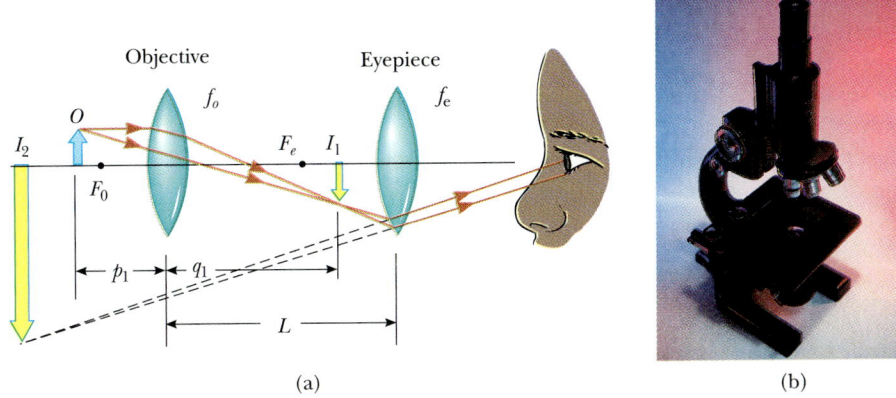

FIGURE 36.33   (a) Diagram of a compound microscope, which consists of an objective lens and an eyepiece lens. (b) A compound microscope. The three-objective turret allows the user to switch to several different powers of magnification. Combinations of eyepieces with different focal lengths and different objectives can produce a wide range of magnifications.   *(Henry Leap and Jim Lehman)*

forms a real, inverted image at I_1, which is at or close to the focal point of the eyepiece. The eyepiece, which serves as a simple magnifier, produces at I_2 an image of the image at I_1, and this image at I_2 is virtual and inverted. The lateral magnification, M_1, of the first image is $- q_1 / p_1$. Note from Figure 36.33a that q_1 is approximately equal to L, and recall that the object is very close to the focal point of the objective; thus, $p_1 \approx f_0$. This gives a magnification for the objective of

$$M_1 \approx - \frac{L}{f_0}$$

The angular magnification of the eyepiece for an object (corresponding to the image at I_1) placed at the focal point of the eyepiece is found from Equation 36.19 to be

$$m_e = \frac{25 \text{ cm}}{f_e}$$

The overall magnification of the compound microscope is defined as the product of the lateral and angular magnifications:

$$M = M_1 m_e = - \frac{L}{f_0} \left(\frac{25 \text{ cm}}{f_e} \right) \tag{36.20}$$

The negative sign indicates that the image is inverted.

The microscope has extended our vision to the previously unknown details of incredibly small objects. The capabilities of this instrument have steadily increased with improved techniques in the precision grinding of lenses. A question that is often asked about microscopes is, "If you were extremely patient and careful, would it be possible to construct a microscope that would enable you to see an atom?" The answer to this question is no, as long as light is used to illuminate the object. The reason is that, in order to be seen, the object under a microscope must be at least as large as a wavelength of light. An atom is many times smaller than the wavelengths of visible light, and so its mysteries have to be probed using other types of "microscopes."

The wavelength dependence of the "seeing" ability of a wave can be illustrated by water waves in a bathtub. Suppose you vibrate your hand in the water until waves having a wavelength of about 15 cm are moving along the surface. If you fix a small object, such as a toothpick, in the path of the waves, the waves are not disturbed appreciably by the toothpick but instead continue along their path, oblivious of the small object. Now suppose you fix a larger object, such as a toy sailboat, in the path of the waves. In this case, the waves are considerably "disturbed" by the object. In the first case, the toothpick is smaller than the wavelength of the waves, and as a result the waves do not "see" the toothpick. (The intensity of the scattered waves is low.) In the second case, the toy sailboat is about the same size as the wavelength of the waves and, hence, creates a disturbance. That is, the object acts as the source of scattered waves that appear to come from it. Light waves behave in this same general way. The ability of an optical microscope to view an object depends on the size of the object relative to the wavelength of the light used to observe it. Hence, we will never be able to observe atoms or molecules with a light microscope, because their dimensions are small (≈ 0.1 nm) relative to the wavelength of the light (≈ 500 nm).

*36.10   THE TELESCOPE

There are two fundamentally different types of **telescopes**, both designed to aid in viewing distant objects, such as the planets in our solar system. The **refracting telescope** uses a combination of lenses to form an image, and the **reflecting telescope** uses a curved mirror and a lens.

The telescope sketched in Figure 36.34a is a refracting telescope. The two lenses are arranged so that the objective forms a real, inverted image of the distant object very near the focal point of the eyepiece. Furthermore, the image at I_1 is formed at the focal point of the objective because the object is essentially at infinity. Hence, the two lenses are separated by a distance $f_0 + f_e$, which corresponds to the length of the telescope's tube. The eyepiece finally forms, at I_2, an enlarged, inverted image of the image at I_1.

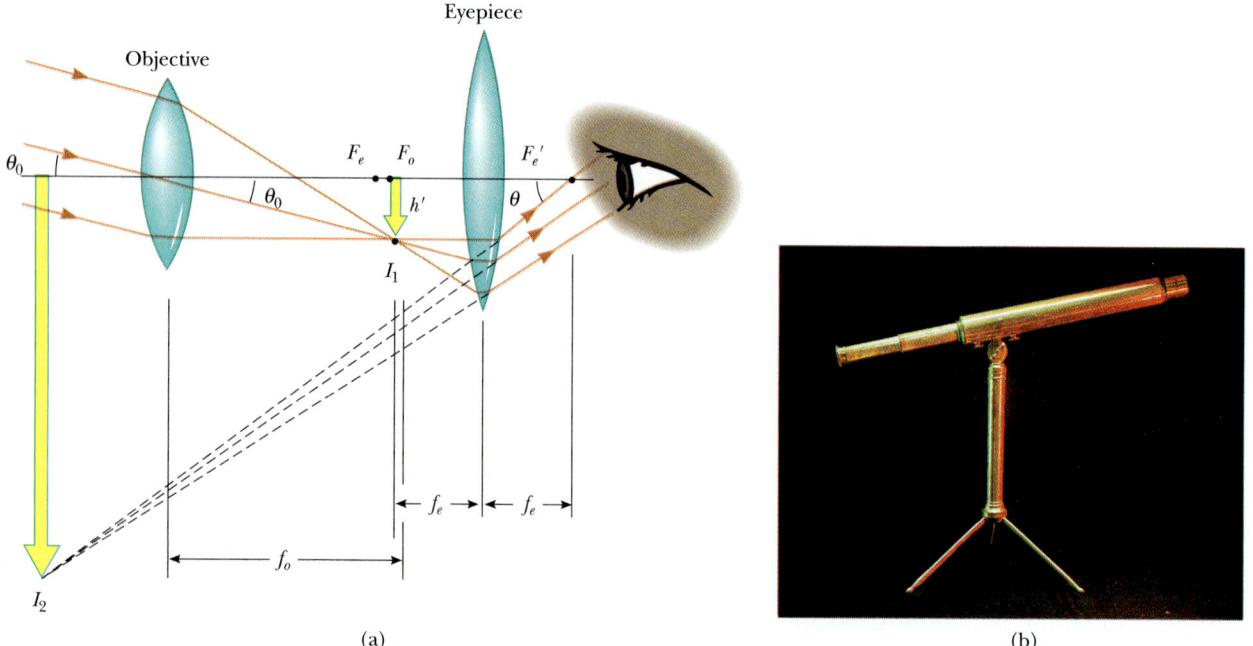

(a)

(b)

FIGURE 36.34   (a) Diagram of a refracting telescope, with the object at infinity. (b) Photograph of a refracting telescope.   *(Henry Leap and Jim Lehman)*

Eyeglasses, mirrors, microscopes, and telescopes are examples of optical instruments without which our lives would be dramatically different. This simulator enables you to construct a multitude of optical instruments and investigate how their performance is affected when you modify parameters such as focal length, wavelength, and index of refraction. The simulations you create can include lenses, prisms, mirrors, and light sources of various wavelengths. Furthermore, the model you create will automatically show a new image as you move or modify any optical component. The adventuresome student might even invent a new optical instrument.

Optical Instruments

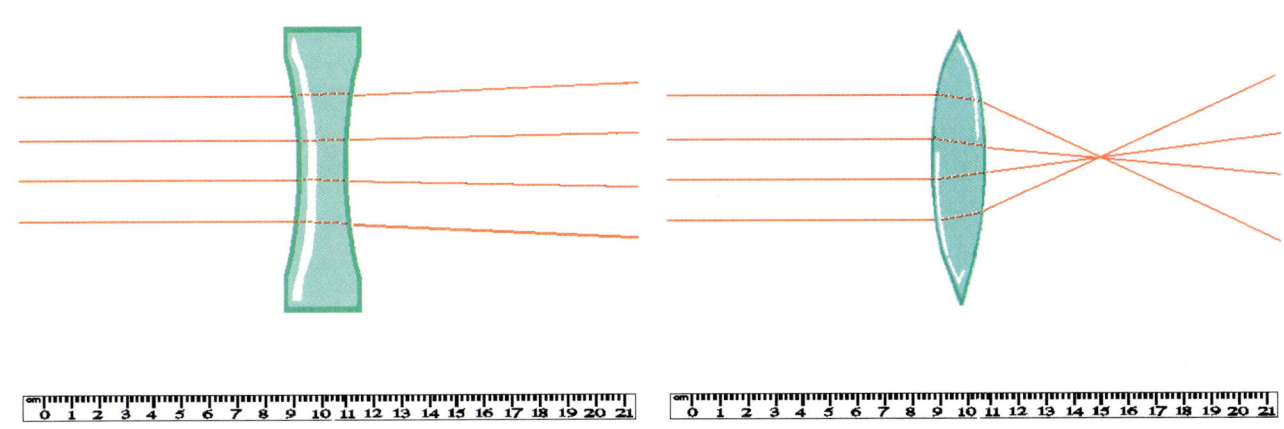

The angular magnification of the telescope is given by θ/θ_0, where θ_0 is the angle subtended by the object at the objective and θ is the angle subtended by the final image at the viewer's eye. From the triangles in Figure 36.34a, and for small angles, we have

$$\theta \approx -\frac{h'}{f_e} \quad \text{and} \quad \theta_0 \approx \frac{h'}{f_0}$$

Hence, the angular magnification of the telescope can be expressed as

$$m = \frac{\theta}{\theta_0} = \frac{-h'/f_e}{h'/f_0} = -\frac{f_0}{f_e} \tag{36.21}$$

The minus sign indicates that the image is inverted. This expression says that the angular magnification of a telescope equals the ratio of the objective focal length to the eyepiece focal length. Here again, the magnification is the ratio of the angular size seen with the telescope to the angular size seen with the unaided eye.

In some applications, such as observing nearby objects like the Sun, Moon, or planets, magnification is important. However, stars are so far away that they always appear as small points of light regardless of how much magnification is used. Large research telescopes used to study very distant objects must have a large diameter in order to gather as much light as possible. It is difficult and expensive to manufacture large lenses for refracting telescopes. Another difficulty with large lenses is that their large weight leads to sagging, which is an additional source of aberration. These problems can be partially overcome by replacing the objective lens with a reflecting, concave mirror. Figure 36.35 shows the design for a typical reflecting

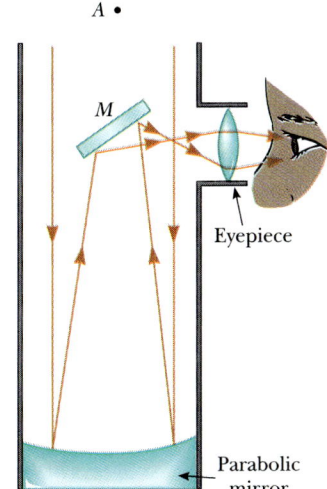

FIGURE 36.35 A reflecting telescope with a Newtonian focus.

telescope. Incoming light rays pass down the barrel of the telescope and are reflected by a parabolic mirror at the base. These rays converge toward point *A* in the figure, where an image would be formed. However, before this image is formed, a small flat mirror *M* reflects the light toward an opening in the side of the tube that passes into an eyepiece. This particular design is said to have a Newtonian focus because it was Newton who developed it. Note that the light never passes through glass in the reflecting telescope (except through the small eyepiece). As a result, problems associated with chromatic aberration are virtually eliminated.

The largest telescope in the world is the 6-m-diameter reflecting telescope on Mount Pastukhov in the Caucasus, Soviet Union. The largest reflecting telescope in the United States is the 5-m-diameter instrument on Mount Palomar in California. In contrast, the largest refracting telescope in the world, which is located at the Yerkes Observatory in Williams Bay, Wisconsin, has a diameter of only 1 m.

SUMMARY

The **lateral magnification** *M* of a mirror or lens is defined as the ratio of the image height h' to the object height h:

$$M = \frac{h'}{h} \tag{36.1}$$

In the paraxial ray approximation, the object distance p and image distance q for a spherical mirror of radius R are related by the **mirror equation**

$$\frac{1}{p} + \frac{1}{q} = \frac{2}{R} = \frac{1}{f} \tag{36.4, 36.6}$$

where $f = R/2$ is the **focal length** of the mirror.

An image can be formed by refraction from a spherical surface of radius R. The object and image distances for refraction from such a surface are related by

$$\frac{n_1}{p} + \frac{n_2}{q} = \frac{n_2 - n_1}{R} \tag{36.8}$$

where the light is incident in the medium of index of refraction n_1 and is refracted in the medium whose index of refraction is n_2.

The inverse of the **focal length** f of a thin lens in air is

$$\frac{1}{f} = (n - 1)\left(\frac{1}{R_1} - \frac{1}{R_2}\right) \tag{36.11}$$

Converging lenses have positive focal lengths, and **diverging lenses** have negative focal lengths.

For a thin lens, and in the paraxial ray approximation, the object and image distances are related by

$$\frac{1}{p} + \frac{1}{q} = \frac{1}{f} \tag{36.12}$$

QUESTIONS

1. When you look in a mirror, your image is reversed left to right but not top to bottom. Explain.

2. Using a simple ray diagram, as in Figure 36.2, show that a flat mirror whose top is at eye level need not be as long as your height for you to see your entire body.

3. Consider a concave spherical mirror with a real object. Is the image always inverted? Is the image always real? Give conditions for your answers.

4. Repeat the previous question for a convex spherical mirror.

5. Why does a clear stream always appear to be shallower than it actually is?

6. Consider the image formed by a thin converging lens. Under what conditions is the image (a) inverted, (b) upright, (c) real, (d) virtual, (e) larger than the object, and (f) smaller than the object?

7. Repeat Question 6 for a thin diverging lens.

8. If a cylinder of solid glass or clear plastic is placed above the words LEAD OXIDE and viewed from the side as shown in Figure 36.36, the LEAD appears inverted but the OXIDE does not. Explain.

FIGURE 36.36 (Question 8) *(Henry Leap and Jim Lehman)*

9. Describe two types of aberration common in a spherical lens.

10. Explain why a mirror cannot give rise to chromatic aberration.

11. What is the magnification of a flat mirror? What is its focal length?

12. Why do some emergency vehicles have the symbol AMBULANCE written on the front?

13. Explain why a fish in a spherical goldfish bowl appears larger than it really is.

14. Lenses used in eyeglasses, whether converging or diverging, are always designed such that the middle of the lens curves away from the eye, like the center lenses of Figure 36.22a and 36.22b. Why?

15. A mirage is formed when the air gets gradually cooler as the height above the ground increases. What might happen if the air grows gradually warmer as the height is increased? This often happens over bodies of water or snow-covered ground: The effect is called *looming.*

16. Consider a spherical concave mirror, with the object located to the left of the mirror beyond the focal point. Using ray diagrams, show that the image moves to the left as the object approaches the focal point.

17. In a Jules Verne novel, a piece of ice is shaped to form a magnifying lens to focus sunlight to start a fire. Is this possible?

18. The *f*-number of a camera is the focal length of the lens divided by its aperture (or diameter). How can the *f*-number of the lens be changed? How does changing this number affect the required exposure time?

19. A solar furnace can be constructed by using a concave mirror to reflect and focus sunlight into a furnace enclosure. What factors in the design of the reflecting mirror would guarantee very high temperatures?

20. One method for determining the position of an image, either real or virtual, is by means of *parallax.* If a finger or other object is placed at the position of the image, as in Figure 36.37, and the finger and image are viewed simultaneously (the image through the lens if it is virtual), the finger and image have the same parallax; that is, if it is viewed from different positions, the image will appear to move along with the finger. Use this method to locate the image formed by a lens. Explain why the method works.

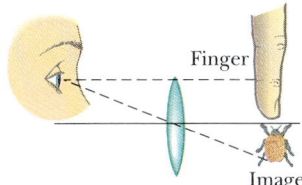

FIGURE 36.37 (Question 20)

FIGURE 36.38 (Question 21) *(M. C. Escher/Cordon Art-Baarn-Holland. All rights reserved.)*

21. Figure 36.38 shows a lithograph by M. C. Escher titled *Hand with Reflection Sphere (Self-Portrait in Spherical Mirror)*. Escher had this to say about the work: "The picture shows a spherical mirror, resting on a left hand. But as a print is the reverse of the original drawing on stone, it was my right hand that you see depicted. (Being left-handed, I needed my left hand to make the drawing.) Such a globe reflection collects almost one's whole surroundings in one disk-shaped image. The whole room, four walls, the floor, and the ceiling, everything, albeit distorted, is compressed into that one small circle. Your own head, or more exactly the point between your eyes, is the absolute center. No matter how you turn or twist yourself, you can't get out of that central point. You are immovably the focus, the unshakable core, of your world." Comment on the accuracy of Escher's description.

PROBLEMS

Review Problem

One of the surfaces of a lens is flat; the other one is convex and has a radius of curvature $R = 10.0$ cm. The speed of light in the lens material is $v = 1.50 \times 10^8$ m/s. Find (a) the index of refraction of the lens material, (b) the angle of total internal reflection, (c) the power of the lens, (d) the focal length of the lens, (e) the object distance if the magnification $M = 2.00$, (f) the image distance if an object is 1.00 m from the screen and an enlarged image appears on the screen, (g) the magnification M' in case (f), and (h) the angular magnification of the lens.

Section 36.1   Images Formed by Flat Mirrors

1. In a physics laboratory experiment, a torque is applied to a small-diameter wire suspended vertically under tensile stress. The small angle through which the wire turns as a consequence of the net torque is measured by attaching a small mirror to the wire and reflecting a beam of light off the mirror and onto a circular scale. Such an arrangement is known as an optical lever and is shown from a top view in Figure P36.1. Show that when the mirror turns through an angle θ, the reflected beam is rotated by an angle 2θ.

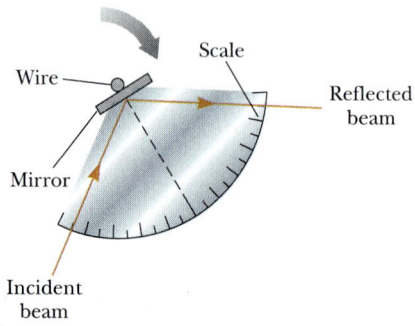

FIGURE P36.1

2. A light ray enters a small hole in a rectangular box that has three mirrored sides, as in Figure P36.2. (a) For what angle θ does the light ray emerge from the hole, assuming that it hits each mirror only once? (b) Show that your answer does not depend on the location of the hole.

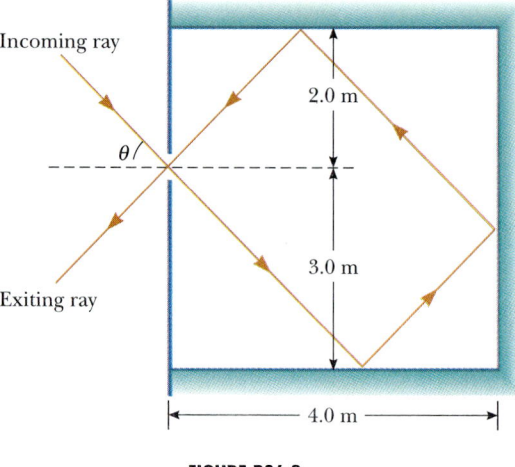

FIGURE P36.2

3. Determine the minimum height of a vertical flat mirror in which a person 5'10" in height can see his or her full image. (A ray diagram would be helpful.)

3A. Determine the minimum height of a vertical flat mirror in which a person of height h can see his or her full image. (A ray diagram would be helpful.)

4. Two flat mirrors have their reflecting surfaces facing one another, with the edge of one mirror in contact with an edge of the other, so that the angle between the mirrors is α. When an object is placed between the mirrors, a number of images are formed. In general, if the angle α is such that $n\alpha = 360°$, where n is an integer, the number of images formed is $n - 1$. Graphically, find all the image positions for

the case $n = 6$ when a point object is between the mirrors (but not on the angle bisector).

Section 36.2   Images Formed by Spherical Mirrors

5. A concave mirror has a focal length of 40.0 cm. Determine the object position for which the resulting image is upright and four times the size of the object.

6. A rectangle 10.0 cm × 20.0 cm is placed so that its right edge is 40.0 cm to the left of a concave spherical mirror, as in Figure P36.6. The radius of curvature of the mirror is 20.0 cm. (a) Draw the image seen through this mirror. (b) What is the area of the image?

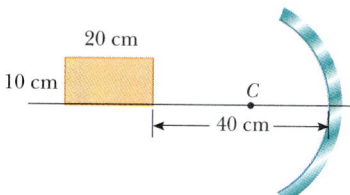

FIGURE P36.6

7. A concave mirror has a radius of curvature of 60 cm. Calculate the image position and magnification of an object placed in front of the mirror at distances of (a) 90 cm and (b) 20 cm. (c) Draw ray diagrams to obtain the image in each case.

8. The height of the real image formed by a concave mirror is four times larger than the object height when the object is 30.0 cm in front of the mirror. (a) What is the radius of curvature of the mirror? (b) Use a ray diagram to locate this image.

9. A dedicated sports car enthusiast polishes the inside and outside surfaces of a hubcap that is a section of a sphere. When she looks into one side of the hubcap, she sees an image of her face 30.0 cm in back of the hubcap. She then turns the hubcap over and sees another image of her face 10.0 cm in back of the hubcap. (a) How far is her face from the hubcap? (b) What is the radius of curvature of the hubcap?

9A. A dedicated sports car enthusiast polishes the inside and outside surfaces of a hubcap that is a section of a sphere. When she looks into one side of the hubcap, she sees an image of her face a distance d in back of the hubcap. She then turns the hubcap over and sees another image of her face a distance $d/3$ in back of the hubcap. (a) How far is her face from the hubcap? (b) What is the radius of curvature of the hubcap?

10. A candle is 49 cm in front of a convex spherical mirror that has a radius of curvature of 70 cm. (a) Where is the image? (b) What is the magnification?

11. An object 2.0 cm high is placed 10 cm in front of a mirror. What type of mirror and what radius of curvature are needed for an upright image that is 4.0 cm high?

12. A ball is dropped from rest 3.00 m directly above the vertex of a concave mirror having a radius of 1.00 m and lying in a horizontal plane. (a) Describe the motion of the ball's image in the mirror. (b) At what time do the ball and its image coincide?

13. A spherical mirror is to be used to form, on a screen located 5.0 m from the object, an image five times the size of the object. (a) Describe the type of mirror required. (b) Where should the mirror be positioned relative to the object?

14. (a) A concave mirror forms an inverted image four times larger than the object. Find the focal length of the mirror if the distance between object and image is 0.60 m. (b) A convex mirror forms a virtual image half the size of the object. If the distance between image and object is 20.0 cm, determine the radius of curvature of the mirror.

15. A spherical convex mirror has a radius of 40.0 cm. Determine the position of the virtual image and magnification for object distances of (a) 30.0 cm and (b) 60.0 cm. (c) Are the images upright or inverted?

16. An object is 15 cm from the surface of a reflective spherical Christmas-tree ornament 6.0 cm in diameter. What are the magnification and position of the image?

Section 36.3   Images Formed by Refraction

17. A smooth block of ice ($n = 1.309$) rests on the floor with one face parallel to the floor. The block has a vertical thickness of 50.0 cm. Find the location of the image of a pattern in the floor covering as formed by rays that are nearly perpendicular to the block.

18. One end of a long glass rod ($n = 1.50$) is formed into a convex surface of radius 6.0 cm. An object is located in air along the axis of the rod. Find the image positions corresponding to object distances of (a) 20.0 cm, (b) 10.0 cm, and (c) 3.0 cm from the end of the rod.

19. A goldfish is swimming at 2.00 cm/s toward the right side of a rectangular tank. What is the apparent speed of the fish measured by an observer looking in from outside the right side of the tank? The index of refraction for water is 1.33.

19A. A goldfish is swimming at a speed v toward the right side of a rectangular tank. What is the apparent speed of the goldfish as measured by an observer looking in from outside the right side of the tank? The index of refraction for water is n.

20. A lens made with a material of refractive index n has a focal length f in air. When immersed in a liquid having a refractive index n_1, the lens has a focal

length f'. Derive an expression for f' in terms of f, n, and n_1.

21. A glass sphere ($n = 1.50$) of radius 15 cm has a tiny air bubble located 5.0 cm from the center. The sphere is viewed along a direction parallel to the radius containing the bubble. What is the apparent depth of the bubble below the surface of the sphere?

22. A flint glass plate ($n = 1.66$) rests on the bottom of an aquarium tank. The plate is 8.0 cm thick (vertical dimension) and covered with water ($n = 1.33$) to a depth of 12 cm. Calculate the apparent thickness of the plate as viewed from above the water. (Assume nearly normal incidence.)

23. A glass hemisphere is used as a paperweight with its flat face resting on a stack of papers. The radius of the circular cross-section is 4.0 cm, and the index of refraction of the glass is 1.55. The center of the hemisphere is directly over a letter "O" that is 2.5 mm in diameter. What is the diameter of the image of the letter as seen looking along a vertical radius?

24. A goldfish is swimming in water inside a spherical plastic bowl of index of refraction 1.33. If the goldfish is 10 cm from the wall of the 15-cm-radius bowl, where does the goldfish appear to an observer outside the bowl?

25. A transparent sphere of unknown composition is observed to form an image of the Sun on the surface of the sphere opposite the Sun. What is the refractive index of the sphere material?

Section 36.4   Thin Lenses

26. An object located 32 cm in front of a lens forms an image on a screen 8.0 cm behind the lens. (a) Find the focal length of the lens. (b) Determine the magnification. (c) Is the lens converging or diverging?

27. The left face of a biconvex lens has a radius of curvature of 12 cm, and the right face has a radius of curvature of 18 cm. The index of refraction of the glass is 1.44. (a) Calculate the focal length of the lens. (b) Calculate the focal length if the radii of curvature of the two faces are interchanged.

28. A microscope slide is placed in front of a converging lens that has a focal length of 2.44 cm. The lens forms an image of the slide 12.9 cm from the slide. How far is the lens from the slide if the image is (a) real and (b) virtual?

29. What is the image distance of an object 1.0 m in front of a converging lens of focal length 20 cm? What is the magnification of the object? (See Figure 36.23a for a ray diagram.)

30. A person looks at a gem with a jeweler's microscope —a converging lens that has a focal length of 12.5 cm. The microscope forms a virtual image 30.0 cm from the lens. (a) Determine the magnification. Is the image upright or inverted? (b) Construct a ray diagram for this arrangement.

31. Figure P36.31 shows a thin glass ($n = 1.50$) converging lens for which the radii of curvature are $R_1 = 15.0$ cm and $R_2 = -12.0$ cm. To the left of the lens is a cube having a face area of 100.0 cm². The base of the cube is on the axis of the lens, and the right face is 20.0 cm to the left of the lens. (a) Determine the focal length of the lens. (b) Draw the image of the square face formed by the lens. What type of geometric figure is this? (c) Determine the area of the image.

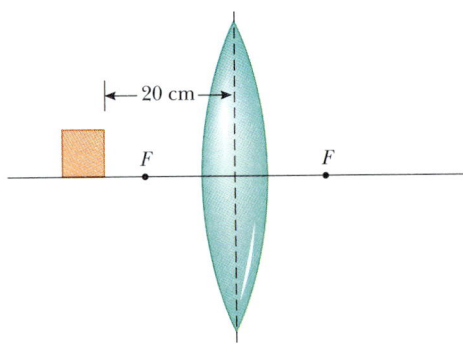

FIGURE P36.31

32. A converging lens has a focal length of 40 cm. Calculate the size of the real image of an object 4.0 cm in height for the following object distances: (a) 50 cm, (b) 60 cm, (c) 80 cm, (d) 100 cm, (e) 200 cm, (f) ∞.

33. An object is located 20 cm to the left of a diverging lens having a focal length $f = -32$ cm. Determine (a) the location and (b) the magnification of the image. (c) Construct a ray diagram for this arrangement.

34. An object is 5.00 m to the left of a flat screen. A converging lens for which the focal length is $f = 0.800$ m is placed between object and screen. (a) Show that there are two lens positions that form an image on the screen, and determine how far these positions are from the object. (b) How do the two images differ from each other?

34A. An object is at a distance d to the left of a flat screen. A converging lens having focal length $f < d$ is placed between object and screen. (a) Show that there are two lens positions that form an image on the screen, and determine how far these positions are from the object. (b) How do the two images differ from each other?

35. A slide 24.0 mm high is to be projected so that its image fills a screen 1.80 m high. The slide-screen distance is 3.00 m, as shown in Figure P36.35. (a) Determine the focal length of the lens in the projector. (b) How far from the slide should the lens of the projector be in order to form the image on the screen?

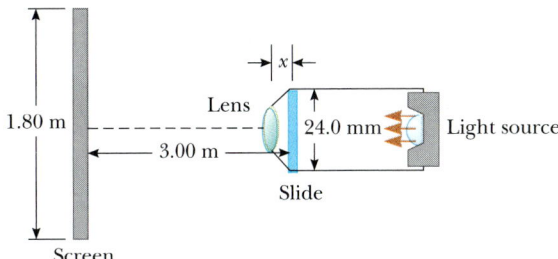

FIGURE P36.35

36. A diverging lens is used to form a virtual image of a real object. The object is positioned 80.0 cm to the left of the lens, and the image is located 40.0 cm to the left of the lens. (a) Determine the focal length of the lens. (b) If the surfaces of the lens have radii of curvature of $R_1 = -40.0$ cm and $R_2 = -50.0$ cm, what is its index of refraction?

37. The nickel's image in Figure P36.37 has twice the diameter of the nickel and is 2.84 cm from the lens. Determine the focal length of the lens.

FIGURE P36.37

*Section 36.6   The Camera and *Section 36.7   The Eye

38. A camera gives proper film exposure when set at $f/16$ and the shutter is open for $(1/32)$ s. Determine the correct exposure time if a setting of $f/8$ is used. (Assume the lighting conditions are unchanged.)

39. A camera is being used with correct exposure at $f/4$ and a shutter speed of $(1/16)$ s. In order to photograph a fast-moving subject, the shutter speed is changed to $(1/128)$ s. Find the new f-number setting needed to maintain satisfactory exposure.

40. A 1.7-m-tall woman stands 5.0 m in front of a camera equipped with a 50.0 mm focal length lens. What is the size of the image formed on the film?

41. A nearsighted woman cannot see objects clearly beyond 25 cm (the far point). If she has no astigmatism and contact lenses are prescribed, what are the power and type of lens required to correct her vision?

42. A camera that works properly with a 50.0-mm lens is to be reconfigured for underwater use. If the present double convex lens ($n = 1.50$, $R_1 = 10.0$ cm, $R_2 = -3.33$ cm) is to be retained, (a) what is its focal length under water and (b) at what distance from the lens should the film be placed when the camera is used under water?

43. If the aqueous humor of the eye has an index of refraction of 1.34 and the distance from the front of the cornea to the retina is 2.2 cm, what is the radius of curvature of the cornea for which distant objects are focused on the retina? (Assume all refraction occurs in the aqueous humor.)

44. A person sees clearly wearing eyeglasses that have a power of -4.0 diopters and sit 2.0 cm in front of the eyes. If the person wants to switch to contact lenses, which are placed directly on the eyes, what lens power should be prescribed?

45. A runner passes a spectator at 10.0 m/s. The spectator photographs the runner using a camera having a focal length of 5.00 cm and an exposure time of 20 ms. The maximum image blurring allowed in a good photograph is 0.500 mm. Find the minimum distance between runner and camera in order to obtain a good photograph.

45A. A runner passes a spectator at speed v. The spectator photographs the runner using a camera having a focal length f and an exposure time of Δt. The maximum image blurring allowed for a good photograph is Δh. Find the minimum distance between the runner and camera in order to obtain a good photograph.

46. The accommodation limits for nearsighted Nick's eyes are 18.0 cm and 80.0 cm. When he wears his glasses, he is able to see faraway objects clearly. At what minimum distance is he able to see objects clearly?

*Section 36.8   The Simple Magnifier, *Section 36.9   The Compound Microscope, and *Section 36.10   The Telescope

47. A philatelist examines the printing detail on a stamp using a convex lens of focal length 10.0 cm as a simple magnifier. The lens is held close to the eye, and the lens-to-object distance is adjusted so that the virtual image is formed at the normal near point (25 cm). Calculate the magnification.

48. A lens that has a focal length of 5.0 cm is used as a magnifying glass. (a) To obtain maximum magnification, where should the object be placed? (b) What is the magnification?

49. The Yerkes refracting telescope has a 1.0-m diameter objective lens of focal length 20 m and an eyepiece of focal length 2.5 cm. (a) Determine the magnification of the planet Mars as seen through this tele-

scope. (b) Are the martian polar caps right side up or upside down?

50. The Palomar reflecting telescope has a parabolic mirror that has an 80-m focal length. Determine the magnification achieved when an eyepiece of 2.5-cm focal length is used.

51. An astronomical telescope has an objective of focal length 75 cm and an eyepiece of focal length 4.0 cm. What is the magnifying power?

52. The distance between eyepiece and objective lens in a certain compound microscope is 23 cm. The focal length of the eyepiece is 2.5 cm, and that of the objective is 0.40 cm. What is the overall magnification of the microscope?

53. The desired overall magnification of a compound microscope is 140×. The objective alone produces a lateral magnification of 12×. Determine the required focal length of the eyepiece.

ADDITIONAL PROBLEMS

54. An object placed 10.0 cm from a concave spherical mirror produces a real image 8.0 cm from the mirror. If the object is moved to a new position 20.0 cm from the mirror, what is the position of the image? Is the latter image real or virtual?

55. Figure P36.55 shows a thin converging lens for which the radii are $R_1 = 9.00$ cm and $R_2 = -11.0$ cm. The lens is in front of a concave spherical mirror of radius $R = 8.00$ cm. (a) If its focal points F_1 and F_2 are 5.00 cm from the vertex of the lens, determine its index of refraction. (b) If the lens and mirror are 20.0 cm apart, and an object is placed 8.00 cm to the left of the lens, determine the position of the final image and its magnification as seen by the eye in the figure. (c) Is the final image inverted or upright? Explain.

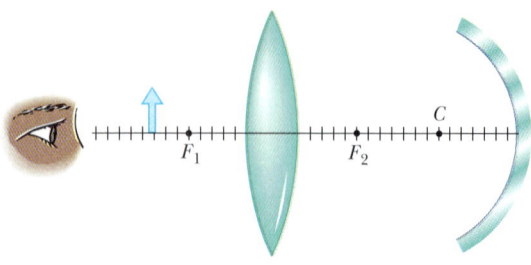

FIGURE P36.55

56. A compound microscope has an objective of focal length 0.300 cm and an eyepiece of focal length 2.50 cm. If an object is 3.40 mm from the objective, what is the magnification? (*Hint:* Use the lens equation for the objective.)

57. An object 2.00 cm tall is placed 40.0 cm to the left of

a converging lens having a focal length of 30.0 cm. A diverging lens having a focal length of −20.0 cm is placed 110 cm to the right of the converging lens. (a) Determine the final position and magnification of the final image. (b) Is the image upright or inverted? (c) Repeat (a) and (b) for the case where the second lens is a converging lens having a focal length of +20.0 cm.

58. An object is placed 15.0 cm to the left of a converging lens ($f_1 = 10.0$ cm). A diverging lens ($f_2 = -20.0$ cm) is placed 15.0 cm to the right of the converging lens. Locate and describe the final image formed by the two lenses.

59. A parallel beam of light enters a glass hemisphere perpendicular to the flat face, as shown in Figure P36.59. The radius is $R = 6.00$ cm, and the index of refraction is $n = 1.560$. Determine the point at which the beam is focused. (Assume paraxial rays.)

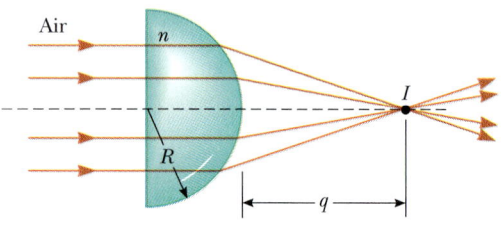

FIGURE P36.59

60. A thin lens having refractive index n is immersed in a liquid having index n'. Show that the focal length f of the lens is

$$\frac{1}{f} = \left(\frac{n}{n'} - 1\right)\left(\frac{1}{R_1} - \frac{1}{R_2}\right)$$

61. A thin lens of focal length 20.0 cm lies on a horizontal front-surfaced mirror. How far above the lens should an object be held if its image is to coincide with the object?

61A. A thin lens of focal length f lies on a horizontal front-surfaced mirror. How far above the lens should an object be held if its image is to coincide with the object?

62. Figure P36.62 shows a triangular enclosure whose inner walls are mirrors. A ray of light enters a small hole at the center of the short side. For each of the following, make a sketch showing the light path and find the angle θ for a ray that meets the stated conditions. (a) A light ray that is reflected once by each of the side mirrors and then exits through the hole. (b) A light ray that reflects only once and then exits. (c) Is there a path that reflects three times and then exits? If so, sketch the path and find θ. (d) A ray that reflects four times and then exits.

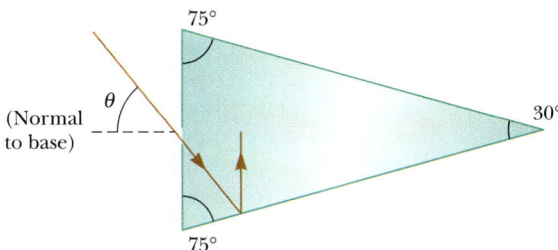

FIGURE P36.62

63. A cataract-impaired lens in an eye may be surgically removed and replaced by a manufactured lens. The focal length of the new lens is determined by the lens-to-retina distance, which is measured by a sonar-like device, and by the requirement that the implant provide for correct distant vision. (a) If the distance from lens to retina is 22.4 mm, calculate the power of the implanted lens in diopters. (b) Since there is no accommodation and the implant allows for correct distant vision, a corrective lens for close work or reading must be used. Assume a reading distance of 33 cm and calculate the power of the lens in the reading glasses.

64. The distance between an object and its upright image is 20.0 cm. If the magnification is 0.500, what is the focal length of the lens being used to form the image?

64A. The distance between an object and its upright image is d. If the magnification is M, what is the focal length of the lens being used to form the image?

65. An object is placed 12.0 cm to the left of a diverging lens of focal length -6.0 cm. A converging lens of focal length 12.0 cm is placed a distance d to the right of the diverging lens. Find the distance d so that the final image is at infinity. Draw a ray diagram for this case.

66. A converging lens has a focal length of 20.0 cm. Find the position of the image for object distances of (a) 50.0 cm, (b) 30.0 cm, and (c) 10.0 cm. (d) Determine the magnification of the lens for these object distances and whether the image is upright or inverted. (e) Draw ray diagrams for these object distances.

67. Two rays traveling parallel to the principal axis strike a plano-convex lens having a refractive index of 1.60 (Fig. P36.67). If the convex face is spherical, a ray near the edge does not strike the focal point (spherical aberration). If this face has a radius of curvature of 20.0 cm and the two rays are $h_1 = 0.500$ cm and $h_2 = 12.0$ cm from the principal axis, find the difference in the positions where each strikes the principal axis.

67A. Two rays traveling parallel to the principal axis strike a plano-convex lens having a refractive index n. If the convex face is spherical, a ray near the edge does not strike the focal point (spherical aberration). If this convex face has a radius of curvature R and the two rays are h_1 and h_2 from the principal axis, find the difference in the positions where each strikes the principal axis.

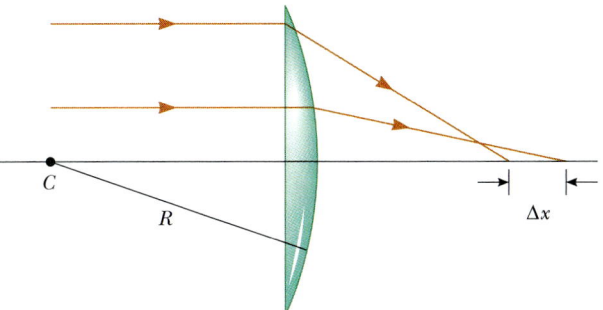

FIGURE P36.67

68. Consider light rays parallel to the principal axis approaching a concave spherical mirror. According to the mirror equation (Eq. 36.4), rays close to the principal axis focus at F, a distance $R/2$ from the mirror. What about a ray farther from the principal axis? Will it be reflected to the axis at a point closer or farther than F from the mirror? Illustrate with an accurately drawn light ray that obeys the law of reflection.

69. A colored marble is dropped into a large tank filled with benzene ($n = 1.50$). (a) What is the depth of the tank if the apparent depth of the marble when viewed from directly above the tank is 35 cm? (b) If the marble has a diameter of 1.5 cm, what is its apparent diameter when viewed from directly above the tank?

70. An object 1.0 cm in height is placed 4.0 cm to the left of a converging lens of focal length 8.0 cm. A diverging lens of focal length -16 cm is located 6.0 cm to the right of the converging lens. Find the position and size of the final image. Is it inverted or upright? Real or virtual?

71. The disk of the Sun subtends an angle of 0.50° at the Earth. What are the position and diameter of the solar image formed by a concave spherical mirror of radius of curvature 3.0 m?

72. A reflecting telescope with a 2.0-m focal length objective and a 10-cm focal length eyepiece is used to view the Moon. Calculate the size of the image formed at the viewer's near point, 25 cm from the eye. (The diameter of the Moon = 3.5×10^3 km; the Earth-Moon distance = 3.84×10^5 km.)

73. The cornea of an eye has a radius of curvature of 0.80 cm. (a) What is the focal length of the reflecting surface of the eye? (b) If a $20 gold piece 3.4 cm in diameter is held 25 cm from the cornea, what are the size and location of the reflected image?

74. Two converging lenses having focal lengths of 10.0 cm and 20.0 cm are located 50.0 cm apart as in Figure P36.74. The final image is to be located between the lenses at the position indicated. (a) How far to the left of the first lens should the object be? (b) What is the overall magnification? (c) Is the final image upright or inverted?

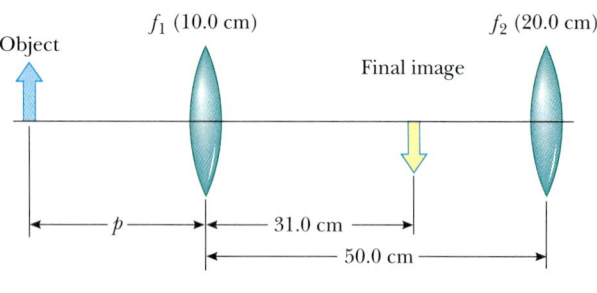

FIGURE P36.74

75. In a darkened room, a burning candle is placed 1.5 m from a white wall. A lens is placed between candle and wall at a location that causes a larger, inverted image to form on the wall. When the lens is moved 90 cm toward the wall, another image of the candle is formed. Find (a) the two object distances that produce the images stated above and (b) the focal length of the lens. (c) Characterize the second image.

76. A lens and mirror have focal lengths of + 80 cm and − 50 cm, respectively, and the lens is located 1.0 m to the left of the mirror. An object is placed 1.0 m to the left of the lens. Locate the final image. State whether the image is upright or inverted and determine the overall magnification.

77. A floating coin illusion consists of two parabolic mirrors, each having a focal length 7.5 cm, facing each other so that their centers are 7.5 cm apart (Fig. P36.77). If a few coins are placed on the lower mirror, an image of the coins is formed at the small opening at the center of the top mirror. Show that the final image is formed at that location and describe its characteristics. (*Note:* A very startling effect is to shine a flashlight beam on these images. Even at a glancing angle, the incoming light beam is seemingly reflected off the images! Do you understand why?)

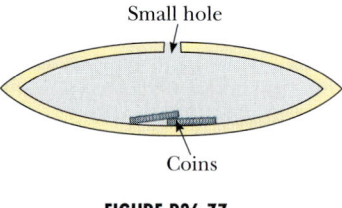

FIGURE P36.77

SPREADSHEET PROBLEM

S1. The equation for refraction by a spherical surface separating two media (Fig. 36.14) having indices of refraction n_1 and n_2 is

$$\frac{n_1}{p} + \frac{n_2}{q} = \frac{n_2 - n_1}{R}$$

This equation was derived using small-angle approximations. Spreadsheet 36.1 tests the range of validity of these approximations. The spreadsheet calculates the image distance for a series of incident angles α without making small-angle approximations. Input parameters are the indices of refraction, the radius of curvature of the surface, and the object distance. (a) Take $n_1 = 1.00$, $n_2 = 1.52$, $R = 10.0$ cm, and $p = 30.0$ cm. If the maximum acceptable error in the image distance is 3.00 percent, for what maximum angle α can the refraction equation be used? (b) Repeat part (a) using $p = 50.0$ cm. (c) Choose other values for the input parameters and repeat (a). (d) Is the error linearly proportional to α?

CHAPTER 37

Interference of Light Waves

A layer of bubbles on water, produced by soap film. The colors, produced just before the bubbles burst, are due to interference between light rays reflected from the front and back of the thin film of water making the bubble. The colors depend on the thickness of the film, ranging from black where the film is at its thinnest to red as the film gets thicker. *(Dr. Jeremy Burgess/Science Photo Library)*

I n the previous chapter on geometric optics, we used light rays to examine what happens when light passes through a lens or reflects from a mirror. The next two chapters are concerned with wave optics, which deals with the interference, diffraction, and polarization of light. These phenomena cannot be adequately explained with the ray optics of Chapter 36, but we describe how treating light as waves rather than as rays leads to a satisfying description of such phenomena.

37.1 CONDITIONS FOR INTERFERENCE

In Chapter 18, we found that two waves can add together constructively or destructively. In constructive interference, the amplitude of the resultant wave is greater than that of either of the individual waves. Light waves also interfere with each other. Fundamentally, all interference associated with light waves arises when the electromagnetic fields that constitute the individual waves combine.

To observe sustained interference in light waves, the following conditions must be met:

- The sources must be **coherent;** that is, they must maintain a constant phase with respect to each other.
- The sources must be **monochromatic,** that is, of a single wavelength.
- The superposition principle must apply.[1]

We now describe the characteristics of coherent sources. As we saw when we studied mechanical waves, two sources (producing two traveling waves) are needed to create interference. In order to produce a stable interference pattern, *the individual waves must maintain a constant phase relationship with one another.* When this is the case, the sources are said to be coherent. As an example, the sound waves emitted by two side-by-side loudspeakers driven by a single amplifier can interfere with each other because the two speakers are coherent—they respond to the amplifier in the same way at the same time.

If two light sources are placed side by side, no interference effects are observed because the light waves from one source are emitted independently of the other source; hence, the emissions from the two sources do not maintain a constant phase relationship with each other over the time of observation. Light from an ordinary light source undergoes such random changes about once every 10^{-8} s. Therefore, the conditions for constructive interference, destructive interference, or some intermediate state last for times of the order of 10^{-8} s. The result is that no interference effects are observed since the eye cannot follow such short-term changes. Such light sources are said to be **incoherent.**

A common method for producing two coherent light sources is to use one monochromatic source to illuminate a screen containing two small openings (usually in the shape of slits). The light emerging from the two slits is coherent because a single source produces the original light beam and the two slits serve only to separate the original beam into two parts (which, after all, is what was done to the sound signal from the side-by-side loudspeakers). Any random change in the light emitted by the source will occur in both beams at the same time, and as a result interference effects can be observed.

37.2   YOUNG'S DOUBLE-SLIT EXPERIMENT

Interference in light waves from two sources was first demonstrated by Thomas Young in 1801. A schematic diagram of the apparatus used in this experiment is shown in Figure 37.1a. Light is incident on a screen, in which there is a narrow slit S_0. The waves emerging from this slit arrive at a second screen, which contains two narrow, parallel slits, S_1 and S_2. These two slits serve as a pair of coherent light sources because waves emerging from them originate from the same wave front and therefore maintain a constant phase relationship. The light from the two slits produces on screen C a visible pattern of bright and dark parallel bands called **fringes** (Fig. 37.1b). When the light from S_1 and that from S_2 both arrive at a point on screen C such that constructive interference occurs at that location, a bright line appears. When the light from the two slits combines destructively at any loca-

---

[1] Later we will find that light consists of photons, and it would violate conservation of energy for one photon to annihilate another. The conditions specified above reflect the more fundamental condition that a photon can only interfere with itself.

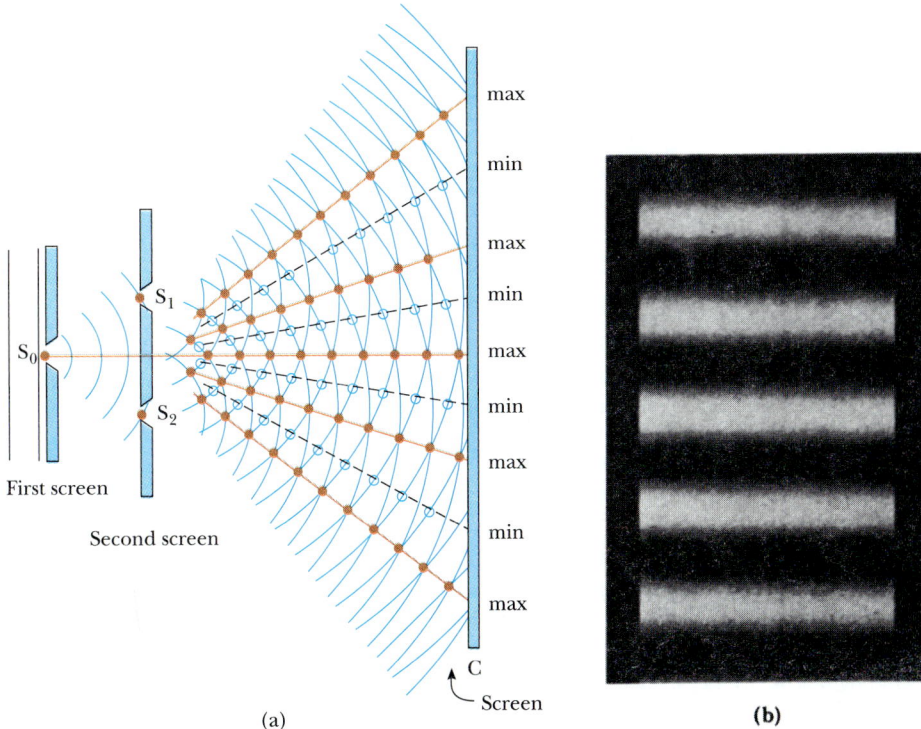

(a)

(b)

FIGURE 37.1 (a) Schematic diagram of Young's double-slit experiment. The narrow slits act as wave sources. Slits S_1 and S_2 behave as coherent sources that produce an interference pattern on screen C. (Note that this drawing is not to scale.) (b) The fringe pattern formed on screen C could look like this.

tion on the screen, a dark line results. Figure 37.2 is a photograph of an interference pattern produced by two coherent vibrating sources in a water tank.

Figure 37.3 is a schematic diagram of some of the ways the two waves can combine at the screen. In Figure 37.3a, the two waves, which leave the two slits in phase, strike the screen at the central point P. Since these waves travel an equal distance, they arrive at P in phase, and as a result constructive interference occurs at this location and a bright area is observed. In Figure 37.3b, the two light waves again start in phase, but now the upper wave has to travel one wavelength farther than the lower wave to reach point Q on the screen. Since the upper wave falls

FIGURE 37.2 An interference pattern involving water waves is produced by two vibrating sources at the water's surface. The pattern is analogous to that observed in Young's double-slit experiment. Note the regions of constructive and destructive interference. *(Richard Megna, Fundamental Photographs)*

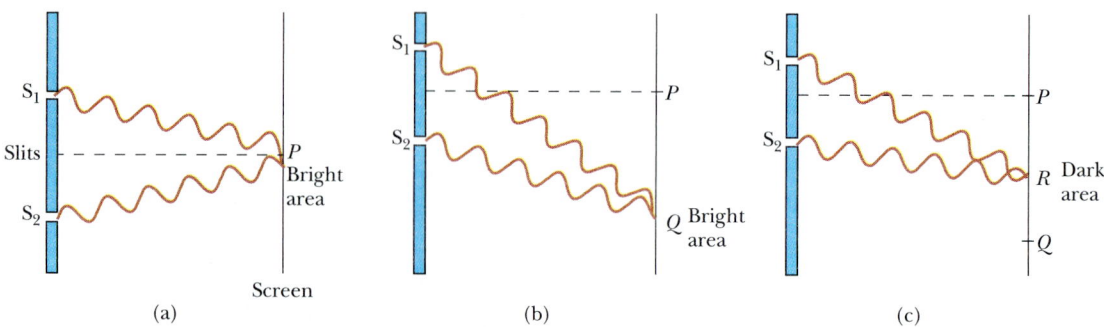

FIGURE 37.3 (a) Constructive interference occurs at *P* when the waves combine. (b) Constructive interference also occurs at *Q*. (c) Destructive interference occurs at *R* because the wave from the upper slit falls half a wavelength behind the wave from the lower slit. (Note that these figures are not drawn to scale.)

behind the lower one by exactly one wavelength, they still arrive in phase at *Q*, and so a second bright light appears at this location. Now consider point *R*, midway between *P* and *Q* in Figure 37.3c. At this location, the upper wave has fallen half a wavelength behind the lower wave. This means that the trough from the bottom wave overlaps the crest from the upper wave, giving rise to destructive interference at *R*. For this reason, a dark region is observed at this location.

We can describe Young's experiment quantitatively with the help of Figure 37.4. The screen is located a perpendicular distance *L* from the screen containing slits S_1 and S_2, which are separated by a distance *d* and the source is monochromatic. Under these conditions, the waves emerging from S_1 and S_2 have the same

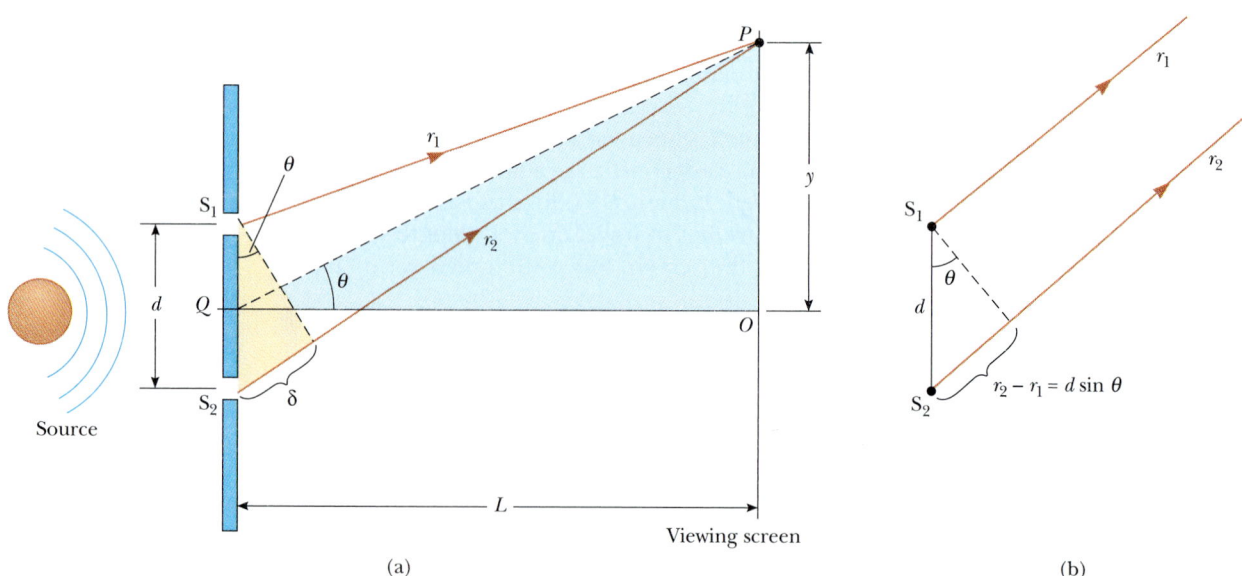

FIGURE 37.4 (a) Geometric construction for describing Young's double-slit experiment. (Note that this figure is not drawn to scale.) (b) When we use the approximation r_1 parallel to r_2, the path difference between the two rays is $r_2 - r_1 = d \sin \theta$. For approximation to be valid, it is essential that $L \gg d$.

frequency and amplitude and are in phase. The light intensity on the screen at any arbitrary point P is the resultant of the light coming from both slits. Note that, in order to reach P, a wave from the lower slit travels farther than a wave from the upper slit by a distance equal to $d \sin \theta$. This distance is called the **path difference,** δ, where

$$\delta = r_2 - r_1 = d \sin \theta \qquad (37.1)$$

Path difference

This equation assumes that r_1 and r_2 are parallel, which is approximately true because L is much greater than d. The value of this path difference determines whether or not the two waves are in phase when they arrive at P. If the path difference is either zero or some integral multiple of the wavelength, the two waves are in phase at P and constructive interference results. Therefore, the condition for bright fringes, or **constructive interference,** at P is

$$\delta = d \sin \theta = m\lambda \qquad (m = 0, \pm 1, \pm 2, \ldots) \qquad (37.2)$$

Conditions for constructive interference

The number m is called the **order number.** The central bright fringe at $\theta = 0$ ($m = 0$) is called the zeroth-order maximum. The first maximum on either side, when $m = \pm 1$, is called the first-order maximum, and so forth.

When the path difference is an odd multiple of $\lambda/2$, the two waves arriving at P are 180° out of phase and will give rise to destructive interference. Therefore, the condition for dark fringes, or **destructive interference,** at P is

$$\delta = d \sin \theta = (m + \tfrac{1}{2})\lambda \qquad (m = 0, \pm 1, \pm 2, \ldots) \qquad (37.3)$$

Conditions for destructive interference

It is useful to obtain expressions for the positions of the bright and dark fringes measured vertically from O to P. In addition to our assumption that $L \gg d$, we assume $d \gg \lambda$. This situation prevails in practice because L is often of the order of 1 m while d is a fraction of a millimeter and λ is a fraction of a micrometer for visible light. Under these conditions, θ is small, and so we can use the approximation $\sin \theta \approx \tan \theta$. From the triangle OPQ in Figure 37.4, we see that

$$\sin \theta \approx \tan \theta = \frac{y}{L} \qquad (37.4)$$

Using this result together with Equation 37.2, we see that the positions of the bright fringes measured from O are given by

$$y_{\text{bright}} = \frac{\lambda L}{d} m \qquad (37.5)$$

Similarly, using Equations 37.3 and 37.4, we find that the dark fringes are located at

$$y_{\text{dark}} = \frac{\lambda L}{d}(m + \tfrac{1}{2}) \qquad (37.6)$$

As we demonstrate in Example 37.1, Young's double-slit experiment provides a method for measuring the wavelength of light. In fact, Young used this technique to do just that. Additionally, the experiment gave the wave model of light a great deal of credibility.

EXAMPLE 37.1   Measuring the Wavelength of a Light Source

A viewing screen is separated from a double-slit source by 1.2 m. The distance between the two slits is 0.030 mm. The second-order bright fringe ($m = 2$) is 4.5 cm from the center line. (a) Determine the wavelength of the light.

Solution We can use Equation 37.5, with $m = 2$, $y_2 = 4.5 \times 10^{-2}$ m, $L = 1.2$ m, and $d = 3.0 \times 10^{-5}$ m:

$$\lambda = \frac{dy_2}{mL} = \frac{(3.0 \times 10^{-5}\ \text{m})(4.5 \times 10^{-2}\ \text{m})}{2 \times 1.2\ \text{m}}$$

$$= 5.6 \times 10^{-7}\ \text{m} = \boxed{560\ \text{nm}}$$

(b) Calculate the distance between adjacent bright fringes.

Solution From Equation 37.5 and the results to part (a), we get

$$y_{m+1} - y_m = \frac{\lambda L (m+1)}{d} - \frac{\lambda L m}{d}$$

$$= \frac{\lambda L}{d} = \frac{(5.6 \times 10^{-7}\ \text{m})(1.2\ \text{m})}{3.0 \times 10^{-5}\ \text{m}}$$

$$= 2.2 \times 10^{-2}\ \text{m} = \boxed{2.2\ \text{cm}}$$

EXAMPLE 37.2   The Distance Between Bright Fringes

A light source emits visible light of two wavelengths: $\lambda = 430$ nm and $\lambda' = 510$ nm. The source is used in a double-slit interference experiment in which $L = 1.5$ m and $d = 0.025$ mm. Find the separation between the third-order bright fringes.

Solution Using Equation 37.5 with $m = 3$, we find that the values of the fringe positions corresponding to these two wavelengths are

$$y_3 = \frac{\lambda L}{d}\, m = 3\, \frac{\lambda L}{d} = 7.74 \times 10^{-2}\ \text{m}$$

$$y_3' = \frac{\lambda' L}{d}\, m = 3\, \frac{\lambda' L}{d} = 9.18 \times 10^{-2}\ \text{m}$$

Hence, the separation between the two fringes is

$$\Delta y = y_3' - y_3 = \frac{3(\lambda' - \lambda)}{d}\, L$$

$$= 1.4 \times 10^{-2}\ \text{m} = \boxed{1.4\ \text{cm}}$$

37.3   INTENSITY DISTRIBUTION OF THE DOUBLE-SLIT INTERFERENCE PATTERN

We now calculate the distribution of light intensity associated with the double-slit interference pattern. Again, suppose the two slits represent coherent sources of sinusoidal waves. Hence, the waves have the same angular frequency ω and a constant phase difference ϕ. The total electric field intensity at point P on the screen in Figure 37.5 is the vector superposition of the two waves. Assuming the two waves have the same amplitude E_0, we can write the electric field intensities at P due to each wave separately as

$$E_1 = E_0 \sin \omega t \quad \text{and} \quad E_2 = E_0 \sin(\omega t + \phi) \tag{37.7}$$

Although the waves are in phase at the slits, *their phase difference ϕ at P depends on the path difference* $\delta = r_2 - r_1 = d \sin \theta$. Because a path difference of λ (constructive interference) corresponds to a phase difference of 2π rad, while a path difference of $\lambda/2$ (destructive interference) corresponds to a phase difference of π rad, we obtain the ratio

$$\frac{\delta}{\lambda} = \frac{\phi}{2\pi}$$

$$\phi = \frac{2\pi}{\lambda}\, \delta = \frac{2\pi}{\lambda}\, d \sin \theta \tag{37.8}$$

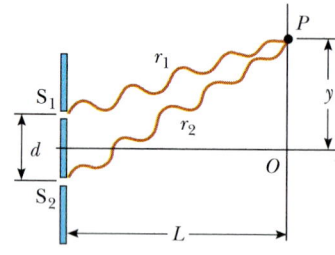

FIGURE 37.5 Construction for analyzing the double-slit interference pattern. A bright region, or intensity maximum, is observed at *O*.

This equation tells us precisely how the phase difference ϕ depends on the angle θ in Figure 37.4.

Using the superposition principle and Equation 37.7, we can obtain the resultant electric field at P:

$$E_P = E_1 + E_2 = E_0[\sin \omega t + \sin(\omega t + \phi)] \tag{37.9}$$

To simplify this expression, we use the trigonometric identity

$$\sin A + \sin B = 2 \sin\left(\frac{A + B}{2}\right) \cos\left(\frac{A - B}{2}\right)$$

Taking $A = \omega t + \phi$ and $B = \omega t$, we can write Equation 37.9 in the form

$$E_P = 2E_0 \cos\left(\frac{\phi}{2}\right) \sin\left(\omega t + \frac{\phi}{2}\right) \tag{37.10}$$

Hence, the electric field at P has the same frequency ω as the light at the slits, but its amplitude is multiplied by the factor $2 \cos(\phi/2)$. To check the consistency of this result, note that if $\phi = 0, 2\pi, 4\pi, \ldots$, the amplitude at P is $2E_0$, corresponding to the condition for constructive interference. Referring to Equation 37.8, we find that our result is consistent with Equation 37.2. Likewise, if $\phi = \pi, 3\pi, 5\pi, \ldots$, the amplitude at P is zero, which is consistent with Equation 37.3 for destructive interference.

Finally, to obtain an expression for the light intensity at P, recall that *the intensity of a wave is proportional to the square of the resultant electric field at that point* (Section 34.3). Using Equation 37.10, we can therefore express the intensity at P as

$$I \propto E_P^2 = 4E_0^2 \cos^2(\phi/2) \sin^2\left(\omega t + \frac{\phi}{2}\right)$$

Since most light-detecting instruments measure the time average light intensity and the time average value of $\sin^2(\omega t + \phi/2)$ over one cycle is $1/2$, we can write the average intensity at P as

$$I_{av} = I_0 \cos^2(\phi/2) \tag{37.11}$$

where I_0 is the maximum possible time average light intensity. [You should note that $I_0 \propto (E_0 + E_0)^2 = (2E_0)^2 = 4E_0^2$.] Substituting Equation 37.8 into Equation 37.11, we find that

$$I_{av} = I_0 \cos^2\left(\frac{\pi d \sin \theta}{\lambda}\right) \tag{37.12}$$

Alternatively, since $\sin \theta \approx y/L$ for small values of θ, we can write Equation 37.12 in the form

$$I_{av} = I_0 \cos^2\left(\frac{\pi d}{\lambda L} y\right) \tag{37.13}$$

Constructive interference, which produces intensity maxima, occurs when the quantity $\pi y d/\lambda L$ is an integral multiple of π, corresponding to $y = (\lambda L/d) m$. This is consistent with Equation 37.5.

A plot of intensity distribution versus $d \sin \theta$ is given in Figure 37.6. Note that the interference pattern consists of equally spaced fringes of equal intensity. However, the result is valid only if the slit-to-screen distance L is large relative to the slit separation, and only for small values of θ.

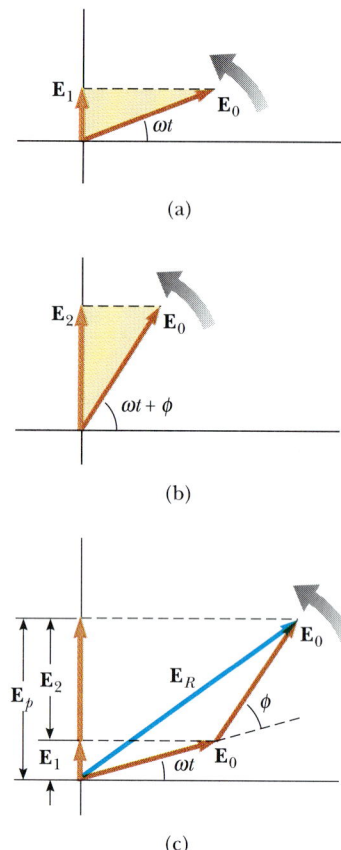

FIGURE 37.7 (a) Phasor diagram for the wave disturbance $E_1 = E_0 \sin \omega t$. The phasor is a vector of length E_0 rotating counterclockwise. (b) Phasor diagram for the wave $E_2 = E_0 \sin(\omega t + \phi)$. (c) E_R is the resultant phasor formed from the individual phasors shown in parts (a) and (b).

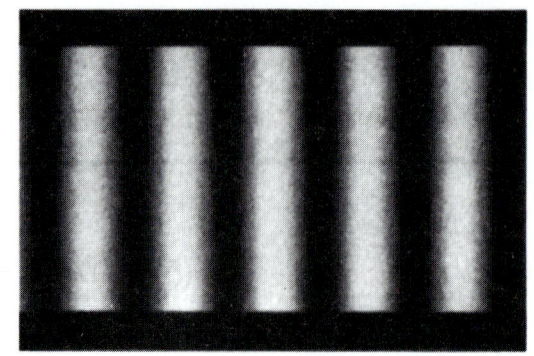

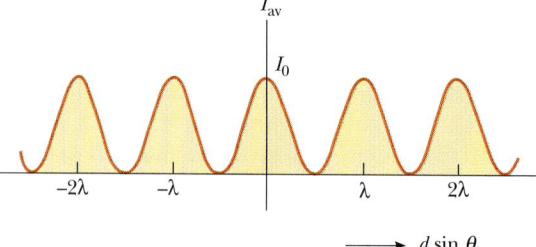

FIGURE 37.6 Intensity distribution versus $d \sin \theta$ or the double-slit pattern when the screen is far from the two slits $(L \gg d)$. *(Photo from M. Cagnet, M. Francon, and J. C. Thierr,* Atlas of Optical Phenomena, *Berlin, Springer-Verlag, 1962)*

We have seen that the interference phenomena arising from two sources depend on the relative phase of the waves at a given point. Furthermore, the phase difference at a given point depends on the path difference between the two waves. The *resultant intensity at a point is proportional to the square of the resultant amplitude.* That is, the intensity is proportional to $(E_1 + E_2)^2$. It would be incorrect to calculate the resultant intensity by adding the intensities of the individual waves. This procedure would give $E_1{}^2 + E_2{}^2$, which of course is not the same thing as $(E_1 + E_2)^2$. Finally, $(E_1 + E_2)^2$ has the same average value as $E_1{}^2 + E_2{}^2$ when the time average is taken over all values of the phase difference between E_1 and E_2. Hence, there is no violation of energy conservation.

37.4 PHASOR ADDITION OF WAVES

In the previous section we combined two waves algebraically to obtain the resultant wave amplitude at some point on a screen. Unfortunately, this analytical procedure becomes cumbersome when several wave amplitudes have to be added. Since we shall eventually be interested in combining a large number of waves, we now describe a graphical procedure for this purpose.

Again, consider a sinusoidal wave whose electric field component is given by

$$E_1 = E_0 \sin \omega t$$

where E_0 is the wave amplitude and ω is the angular frequency. This wave can be represented graphically by a vector of magnitude E_0, rotating about the origin counterclockwise with an angular frequency ω, as in Figure 37.7a. Note that the phasor makes an angle ωt with the horizontal axis. The projection of the phasor on the vertical axis represents E_1, the magnitude of the wave disturbance at some time t. Hence, as the phasor rotates in a circle, the projection E_1 oscillates along the vertical axis about the origin.

Now consider a second sinusoidal wave whose electric field is given by

$$E_2 = E_0 \sin(\omega t + \phi)$$

That is, this wave has the same amplitude and frequency as E_1, but its phase is ϕ with respect to E_1. The phasor representing the wave E_2 is shown in Figure 37.7b. The resultant wave, which is the sum of E_1 and E_2, can be obtained graphically by redrawing the phasors end to end, as in Figure 37.7c, where the tail of the second phasor is placed at the tip of the first. As with vector addition, the resultant phasor $\mathbf{E}_R$ runs from the tail of the first phasor to the tip of the second. Furthermore, $\mathbf{E}_R$ rotates along with the two individual phasors at the same angular frequency ω. The projection of $\mathbf{E}_R$ along the vertical axis equals the sum of the projections of the two phasors: $E_P = E_1 + E_2$.

It is convenient to construct the phasors at $t = 0$ as in Figure 37.8. From the geometry of the triangle, we see that

$$E_R = E_0 \cos \alpha + E_0 \cos \alpha = 2E_0 \cos \alpha$$

Because the sum of the two opposite interior angles equals the exterior angle ϕ, we see that $\alpha = \phi/2$, so that

$$E_R = 2E_0 \cos(\phi/2)$$

Hence, the projection of the phasor $\mathbf{E}_R$ along the vertical axis at any time t is

$$E_P = E_R \sin\left(\omega t + \frac{\phi}{2}\right) = 2E_0 \cos\left(\frac{\phi}{2}\right)\sin\left(\omega t + \frac{\phi}{2}\right)$$

This is consistent with the result obtained algebraically, Equation 37.10. The resultant phasor has an amplitude $2E_0 \cos(\phi/2)$ and makes an angle $\phi/2$ with the first phasor. Furthermore, the average intensity at P, which varies as E_P^2, is proportional to $\cos^2(\phi/2)$, as described in Equation 37.11.

We can now describe how to obtain the resultant of several waves that have the same frequency:

- Draw the phasors representing each wave end to end, as in Figure 37.9, remembering to maintain the proper phase relationship between waves.
- The resultant represented by the phasor $\mathbf{E}_R$ is the vector sum of the individual phasors. At each instant, the projection of $\mathbf{E}_R$ along the vertical axis represents the time variation of the resultant wave. The phase angle α of the resultant wave is the angle between $\mathbf{E}_R$ and the first phasor. From the construction in Figure 37.9, drawn for four phasors, we see that the phasor of the resultant wave is given by $E_P = E_R \sin(\omega t + \alpha)$.

Phasor Diagrams for Two Coherent Sources

As an example of the phasor method, consider the interference pattern produced by two coherent sources. Figure 37.10 represents the phasor diagrams for various values of the phase difference ϕ and the corresponding values of the path difference δ, which are obtained using Equation 37.8. The intensity at a point is a maximum when $\mathbf{E}_R$ is a maximum. This occurs at $\phi = 0, 2\pi, 4\pi, \ldots$. Likewise, the intensity at some observation point is zero when $\mathbf{E}_R$ is zero. The first zero-intensity point occurs at $\phi = 180°$, corresponding to $\delta = \lambda/2$, while the other zero points (not shown) occur at $\delta = 3\lambda/2, 5\lambda/2, \ldots$. These results are in complete agreement with the analytical procedure described in the previous section.

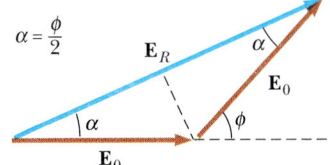

FIGURE 37.8 A reconstruction of the resultant phasor $\mathbf{E}_R$. From the geometry, note that $\alpha = \phi/2$.

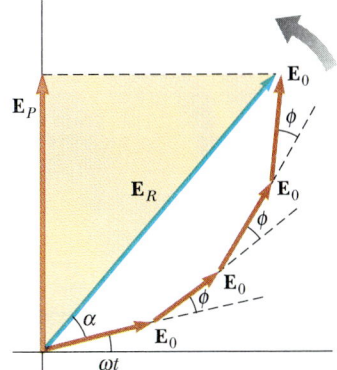

FIGURE 37.9 The phasor $\mathbf{E}_R$ is the resultant of four phasors of equal amplitude E_0. The phase of $\mathbf{E}_R$ with respect to the first phasor is α.

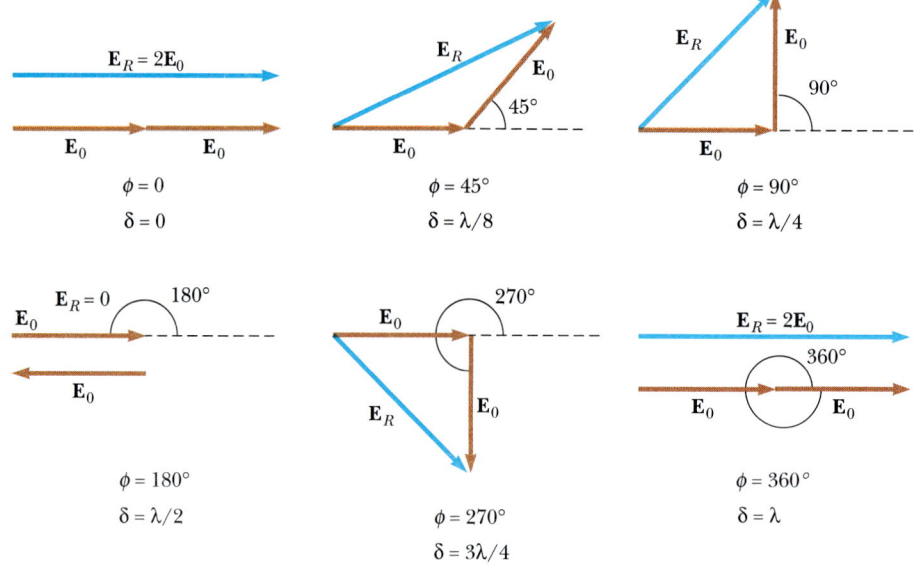

FIGURE 37.10 Phasor diagrams for the double-slit interference pattern. The resultant phasor E_R is a maximum when $\phi = 0, 2\pi, 4\pi, \ldots$ and is zero when $\phi = \pi, 3\pi, 5\pi, \ldots$.

Three-Slit Interference Pattern

Using phasor diagrams, let us analyze the interference pattern caused by three equally spaced slits. The electric fields at a point P on the screen caused by waves from the individual slits can be expressed as

$$E_1 = E_0 \sin \omega t$$

$$E_2 = E_0 \sin(\omega t + \phi)$$

$$E_3 = E_0 \sin(\omega t + 2\phi)$$

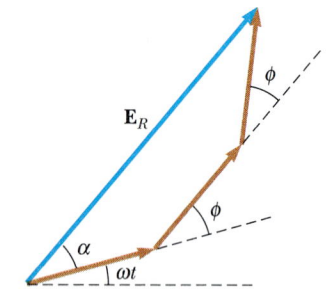

FIGURE 37.11 Phasor diagram for three equally spaced slits.

where ϕ is the phase difference between waves from adjacent slits. Hence, the resultant field at P can be obtained by using the phasor diagram shown in Figure 37.11.

The phasor diagrams for various values of ϕ are shown in Figure 37.12. Note that the resultant amplitude at P has a maximum value of $3E_0$ (called the primary

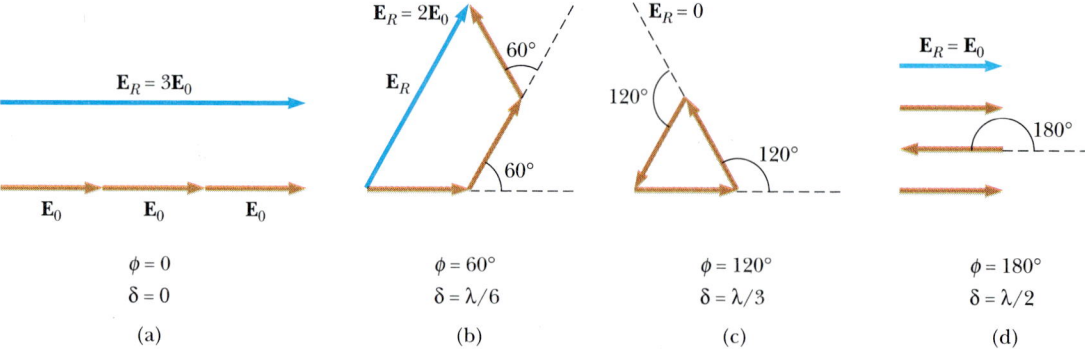

FIGURE 37.12 Phasor diagrams for three equally spaced slits at various values of ϕ. Note that there are primary maxima of amplitude $3E_0$ and secondary maxima of amplitude E_0.

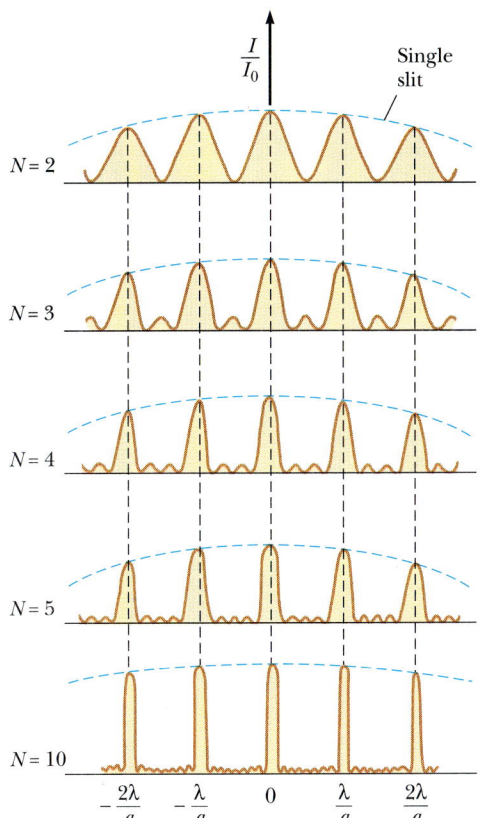

FIGURE 37.13 Multiple-slit interference patterns. The parameter a is the width of each slit. As the number of slits is increased, the primary maxima (the most intense bands) become narrower but remain fixed in position and the number of secondary maxima increases. For any value of N, the decrease in intensity in maxima to the left and right of the central maximum, indicated by the dashed lines, is due to diffraction, which is discussed in Chapter 38.

maximum) when $\phi = 0, \pm 2\pi, \pm 4\pi, \ldots$. This corresponds to the case where the three phasors are aligned as in Figure 37.12a. However, we also find that secondary maxima of amplitude E_0 occur between the primary maxima when $\phi = \pm \pi, \pm 3\pi$, $\ldots$. For these points, the wave from one slit exactly cancels that from another slit (Fig. 37.12d), which results in a total amplitude of E_0. Total destructive interference occurs whenever the three phasors form a closed triangle as in Figure 37.12c. These points where $E_0 = 0$ correspond to $\phi = \pm 2\pi/3, \pm 4\pi/3, \ldots$. You should be able to construct other phasor diagrams for values of ϕ greater than π.

Figure 37.13 shows multiple-slit interference patterns for a number of configurations. These patterns represent plots of the intensity for the various primary and secondary maxima. For three slits, note that the primary maxima are nine times more intense than the secondary maxima. This is because the intensity varies as E_R^2. Figure 37.13 also shows that as the number of slits increases, the number of secondary maxima also increases. In fact, the number of secondary maxima is always equal to $N - 2$, where N is the number of slits. Finally, as the number of slits increases, the primary maxima increase in intensity and become narrower, while the secondary maxima decrease in intensity relative to the primary maxima.

37.5 CHANGE OF PHASE DUE TO REFLECTION

Young's method for producing two coherent light sources involves illuminating a pair of slits with a single source. Another simple, yet ingenious, arrangement for producing an interference pattern with a single light source is known as Lloyd's mirror (Fig. 37.14). A light source is placed at S close to a mirror and a viewing

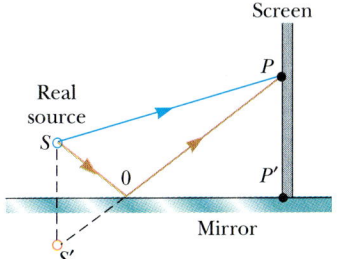

FIGURE 37.14 Lloyd's mirror. An interference pattern is produced on a screen at P as a result of the combination of the direct ray (blue) and the reflected ray (red). The reflected ray undergoes a phase change of 180°.

screen is placed at right angles to the mirror. Waves can reach P on the screen either by the direct path SP or by the path involving reflection from the mirror. The reflected ray can be treated as a ray originating from a virtual source at S'. Hence, at observation points far from the source, we would expect an interference pattern due to waves from S and S' just as is observed for two real coherent sources. An interference pattern is indeed observed. However, the positions of the dark and bright fringes are reversed relative to the pattern of two real coherent sources (Young's experiment). This is because the coherent sources at S and S' differ in phase by 180°, a phase change produced by reflection.

To illustrate this further, consider the point P', where mirror meets screen. This point is equidistant from S and S'. If path difference alone were responsible for the phase difference, we would expect to see a bright fringe at P' (since the path difference is zero for this point), corresponding to the central fringe of the two-slit interference pattern. Instead, we observe a dark fringe at P' because of the 180° phase change produced by reflection. In general,

an electromagnetic wave undergoes a phase change of 180° upon reflection from a medium of higher index of refraction than the one in which the wave is traveling.

It is useful to draw an analogy between reflected light waves and the reflections of a transverse wave on a stretched string when the wave meets a boundary (Section

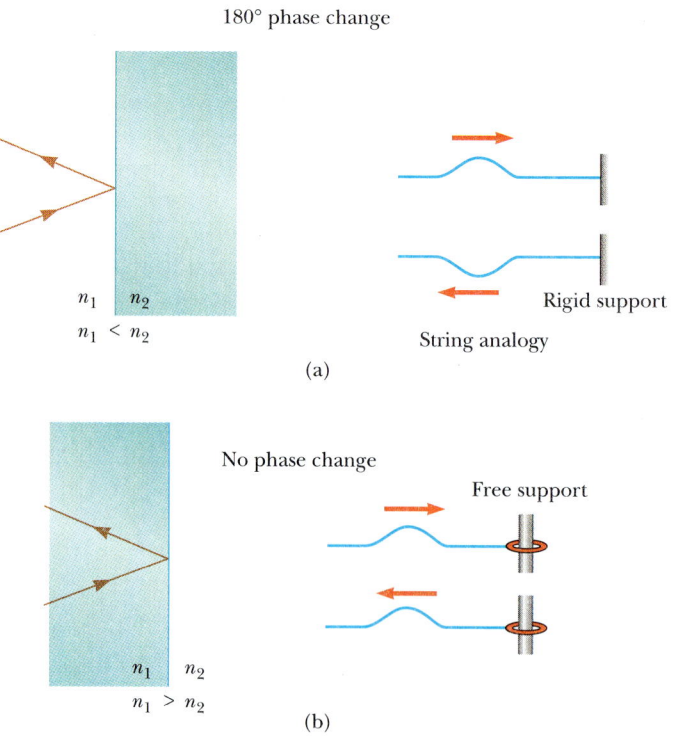

FIGURE 37.15 (a) A ray traveling in medium 1 reflecting from the surface of medium 2 undergoes a 180° phase change. The right side shows the analogy with a reflected pulse on a string fixed at one end. (b) A ray traveling in medium 1 reflecting from the surface of medium 2 with $n_1 > n_2$ undergoes no phase change. The right side shows the analogy with a reflected pulse on a string whose end is free.

16.6). The reflected pulse on a string undergoes a phase change of 180° when reflected from the boundary of a medium of higher index of refraction, but no phase change when reflected from the boundary of a medium of lower index of refraction. Similarly, an electromagnetic wave undergoes a 180° phase change when reflected from a boundary leading to a medium of higher index of refraction, but no phase change when reflected from a boundary leading to a medium of lower index of refraction. In either case, the part of the wave that crosses the boundary undergoes no phase change. These rules, summarized in Figure 37.15, can be deduced from Maxwell's equations, but the treatment is beyond the scope of this text.

37.6 INTERFERENCE IN THIN FILMS

Interference effects are commonly observed in thin films, such as thin layers of oil on water and soap bubbles. The varied colors observed when white light is incident on such films result from the interference of waves reflected from the two surfaces of the film.

Consider a film of uniform thickness t and index of refraction n, as in Figure 37.16. Let us assume that the light rays traveling in air are nearly normal to the two surfaces of the film. To determine whether the reflected rays interfere constructively or destructively, first note the following facts:

- A wave traveling from a medium of index of refraction n_1 toward a medium of index of refraction n_2 undergoes a 180° phase change upon reflection when $n_2 > n_1$. There is no phase change in the reflected wave if $n_2 < n_1$.
- The wavelength of light λ_n in a medium whose refraction index is n (Section 35.4) is

$$\lambda_n = \frac{\lambda}{n} \tag{37.14}$$

where λ is the wavelength of light in free space.

Let us apply these rules to the film of Figure 37.16. Ray 1, which is reflected from the upper surface (A), undergoes a phase change of 180° with respect to the incident wave, and ray 2, which is reflected from the lower surface (B), undergoes no phase change. Therefore, ray 1 is 180° out of phase with ray 2, which is equivalent to a path difference of $\lambda_n/2$. However, we must also consider that ray 2 travels an extra distance $2t$ before the waves recombine. For example, if $2t = \lambda_n/2$, rays 1 and 2 recombine in phase and the result is constructive interference. In general, the condition for constructive interference is

$$2t = (m + \tfrac{1}{2})\lambda_n \qquad (m = 0, 1, 2, \ldots) \tag{37.15}$$

This condition takes into account two factors: (a) the difference in optical path length for the two rays (the term $m\lambda_n$) and (b) the 180° phase change upon reflection (the term $\lambda_n/2$). Because $\lambda_n = \lambda/n$, we can write Equation 37.15 as

$$2nt = (m + \tfrac{1}{2})\lambda \qquad (m = 0, 1, 2, \ldots) \tag{37.16}$$

If the extra distance $2t$ traveled by ray 2 corresponds to a multiple of λ_n, the two waves combine out of phase and the result is destructive interference. The general

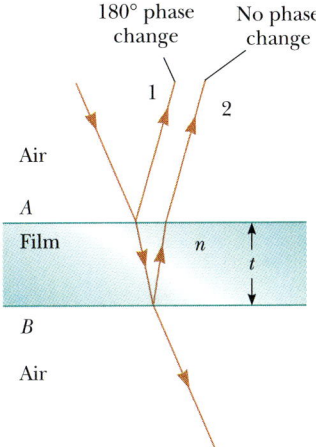

FIGURE 37.16 Interference in light reflected from a thin film is due to a combination of rays reflected from the upper and lower surfaces of the film.

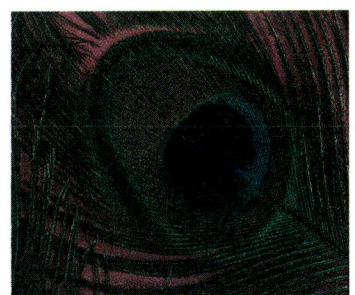

The brilliant colors in a peacock's feathers are due to interference, rather than absorption and reflection. The multilayer structure of the feathers causes constructive interference for certain colors such as blues and greens. The colors change as you view a peacock's feather from different angles. Other creatures such as butterflies exhibit similar interference effects. *(Werner H. Muller/Peter Arnold, Inc.)*

(Left) A thin film of oil on water displays interference, as shown by the pattern of colors when white light is incident on the film. The film thickness varies, thereby producing the interesting color pattern. *(Peter Aprahamian/Science Photo Library)*. *(Right)* Interference in a vertical soap film of variable thickness. The top of the film appears darkest where the film is thinnest. *(© 1983 Larry Mulvehill, Photo Researchers)*

equation for destructive interference is

Conditions for destructive interference in thin films

$$2nt = m\lambda \qquad (m = 0, 1, 2, \ldots) \qquad (37.17)$$

The foregoing conditions for constructive and destructive interference are valid only when the medium above the top surface of the film is the same as the medium below the bottom surface. The surrounding medium may have a refractive index less than or greater than that of the film. In either case, the rays reflected from the two surfaces are out of phase by 180°. If the film is placed between two different media, one with $n < n_{\text{film}}$ and the other with $n > n_{\text{film}}$, the conditions for constructive and destructive interference are reversed. In this case, either there is a phase change of 180° for both ray 1 reflecting from surface A and ray 2 reflecting from surface B or there is no phase change for either ray; hence, the net change in relative phase due to the reflections is zero.

Newton's Rings

Another method for observing interference of light waves is to place a plano-convex lens (one having one flat side and one convex side) on top of a flat glass surface as in Figure 37.17a. With this arrangement, the air film between the glass surfaces varies in thickness from zero at the point of contact to some value t at P. If the radius of curvature of the lens R is very large compared with the distance r, and if the system is viewed from above using light of wavelength λ, a pattern of light and dark rings is observed. A photograph of such a pattern is shown in Figure 37.17b. These circular fringes, discovered by Newton, are called **Newton's rings.** Newton's particle model of light explained the origin of the rings by assuming intermittent behavior of the particle-medium interaction.

The interference effect is due to the combination of ray 1, reflected from the flat plate, with ray 2, reflected from the lower part of the lens. Ray 1 undergoes a phase change of 180° upon reflection, because it is reflected from a medium of higher refractive index, whereas ray 2 undergoes no phase change. Hence, the conditions for constructive and destructive interference are given by Equations 37.16 and 37.17, respectively, with $n = 1$ because the film is air. Point O is dark, as seen in Figure 37.17b, because ray 1, reflected from the flat surface, undergoes a 180° phase change with respect to ray 2. Using the geometry shown in Figure

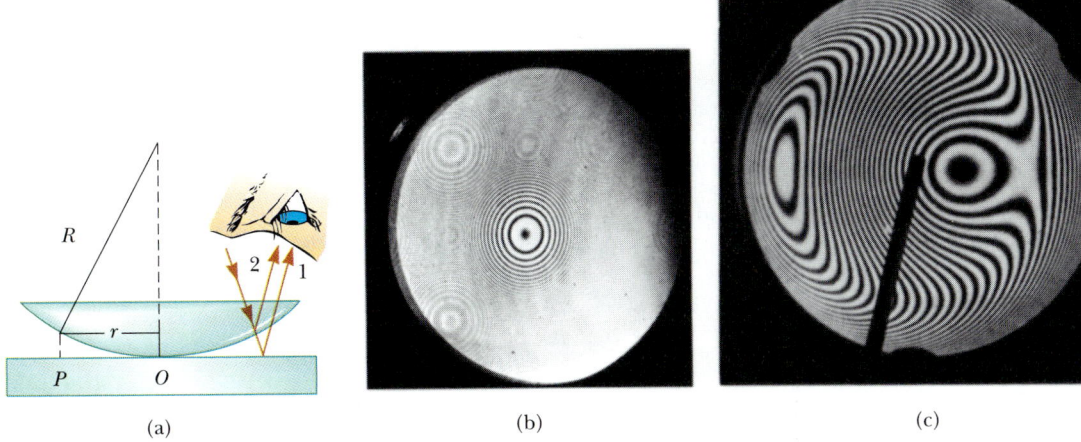

FIGURE 37.17   (a) The combination of rays reflected from the flat plate and the curved lens surface gives rise to an interference pattern known as Newton's rings. (b) Photograph of Newton's rings.   *(Courtesy of Bausch and Lomb Optical Co.)*   (c) This asymmetrical interference pattern indicates imperfections in the lens.   *(From Physical Science Study Committee,* College Physics, *Lexington, Mass., Heath, 1968)*

37.17a, expressions for the radii of the bright and dark bands can be obtained in terms of the radius of curvature R and wavelength λ. For example, the dark rings have radii given by $r \approx \sqrt{m\lambda R / n}$. The details are left as a problem for the reader (Problem 67). By measuring the radii of the rings, the wavelength can be obtained, provided R is known. Conversely, if the wavelength is accurately known, this effect can be used to obtain R.

One of the important uses of Newton's rings is in the testing of optical lenses. A circular pattern like that pictured in Figure 37.17b is obtained only when the lens is ground to a perfectly spherical curvature. Variations from such symmetry produce a pattern like that in Figure 37.17c. These variations indicate how the lens must be ground and polished to remove the imperfections.

Problem-Solving Strategy
Thin-Film Interference

The following ideas should be kept in mind when you work thin-film interference problems:

- Identify the thin film causing the interference.
- The type of interference that occurs is determined by the phase relationship between the portion of the wave reflected at the upper surface of the film and the portion reflected at the lower surface.
- Phase differences between the two portions of the wave have two causes: (1) differences in the distances traveled by the two portions and (2) phase changes that may occur upon reflection.
- When distance traveled and phase changes upon reflection are both taken into account, the interference is constructive if the path difference between the two waves is an integral multiple of λ, and destructive if the path difference is $\lambda/2$, $3\lambda/2$, $5\lambda/2$, and so forth.

EXAMPLE 37.3   Interference in a Soap Film

Calculate the minimum thickness of a soap bubble film ($n = 1.33$) that results in constructive interference in the reflected light if the film is illuminated with light whose wavelength in free space is 600 nm.

Solution   The minimum film thickness for constructive interference in the reflected light corresponds to $m = 0$ in Equation 37.16. This gives $2nt = \lambda/2$, or

$$t = \frac{\lambda}{4n} = \frac{600 \text{ nm}}{4(1.33)} = \boxed{113 \text{ nm}}$$

Exercise   What other film thicknesses produce constructive interference?

Answer   338 nm, 564 nm, 789 nm, and so on.

EXAMPLE 37.4   Nonreflecting Coatings for Solar Cells

Semiconductors such as silicon are used to fabricate solar cells—devices that generate electricity when exposed to sunlight. Solar cells are often coated with a transparent thin film, such as silicon monoxide (SiO, $n = 1.45$), in order to minimize reflective losses from the surface. A silicon solar cell ($n = 3.5$) is coated with a thin film of silicon monoxide for this purpose (Fig. 37.18). Determine the minimum film thickness that produces the least reflection at a wavelength of 552 nm, which is the center of the visible spectrum.

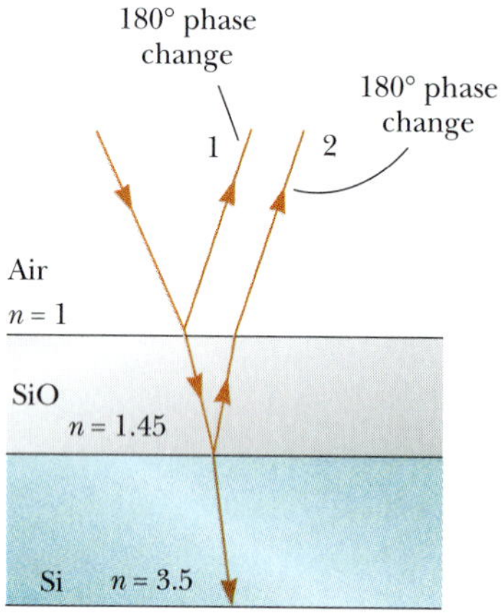

Reasoning   The reflected light is a minimum when rays 1 and 2 in Figure 37.18 meet the condition of destructive interference. Note that both rays undergo a 180° phase change upon reflection in this case, one from the upper and one from the lower surface. Hence, the net change in phase is zero due to reflection, and the condition for a reflection minimum requires a path difference of $\lambda_n/2$; hence, $2t = \lambda/2n$.

Solution   Since $2t = \lambda/2n$, the required thickness is

$$t = \frac{\lambda}{4n} = \frac{550 \text{ nm}}{4(1.45)} = \boxed{94.8 \text{ nm}}$$

Typically, such antireflecting coatings reduce the reflective loss from 30% (with no coating) to 10% (with coating), thereby increasing the cell's efficiency, since more light is available to create charge carriers in the cell. In reality, the coating is never perfectly nonreflecting because the required thickness is wavelength-dependent and the incident light covers a wide range of wavelengths.

Glass lenses used in cameras and other optical instruments are usually coated with a transparent thin film, such as magnesium fluoride (MgF_2), to reduce or eliminate unwanted reflection. More important, such coatings enhance the transmission of light through the lenses.

180° phase change

180° phase change

1   2

Air
$n = 1$

SiO
$n = 1.45$

Si $n = 3.5$

FIGURE 37.18   (Example 37.4) Reflective losses from a silicon solar cell are minimized by coating it with a thin film of silicon monoxide.

EXAMPLE 37.5   Interference in a Wedge-Shaped Film

A thin, wedge-shaped film of refractive index n is illuminated with monochromatic light of wavelength λ, as illustrated in Figure 37.19. Describe the interference pattern observed for this case.

Reasoning and Solution   The interference pattern is that of a thin film of variable thickness surrounded by air. Hence, the pattern is a series of alternating bright and dark parallel bands. A dark band corresponding to destructive interference appears at point O, the apex, because the upper reflected ray undergoes a 180° phase change while the lower one does not. According to Equation 37.17, other dark bands appear when $2nt = m\lambda$, so that $t_1 = \lambda/2n$, $t_2 = \lambda/n$, $t_3 = 3\lambda/2n$, and so on. Similarly, bright bands are observed when

the thickness satisfies the condition $2nt = (m + \frac{1}{2})\lambda$, corresponding to thicknesses of $\lambda/4n$, $3\lambda/4n$, $5\lambda/4n$, and so on. If white light is used, bands of different colors are observed at different points, corresponding to the different wavelengths of light.

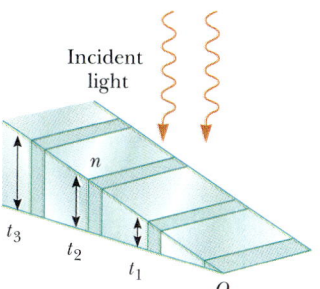

FIGURE 37.19 (Example 37.5) Interference bands in reflected light can be observed by illuminating a wedge-shaped film with monochromatic light. The dark blue areas correspond to positions of destructive interference.

*37.7 THE MICHELSON INTERFEROMETER

The **interferometer,** invented by the American physicist A. A. Michelson (1852–1931), splits a light beam into two parts and then recombines them to form an interference pattern. The device can be used to measure wavelengths or other lengths accurately.

A schematic diagram of the interferometer is shown in Figure 37.20. A beam of

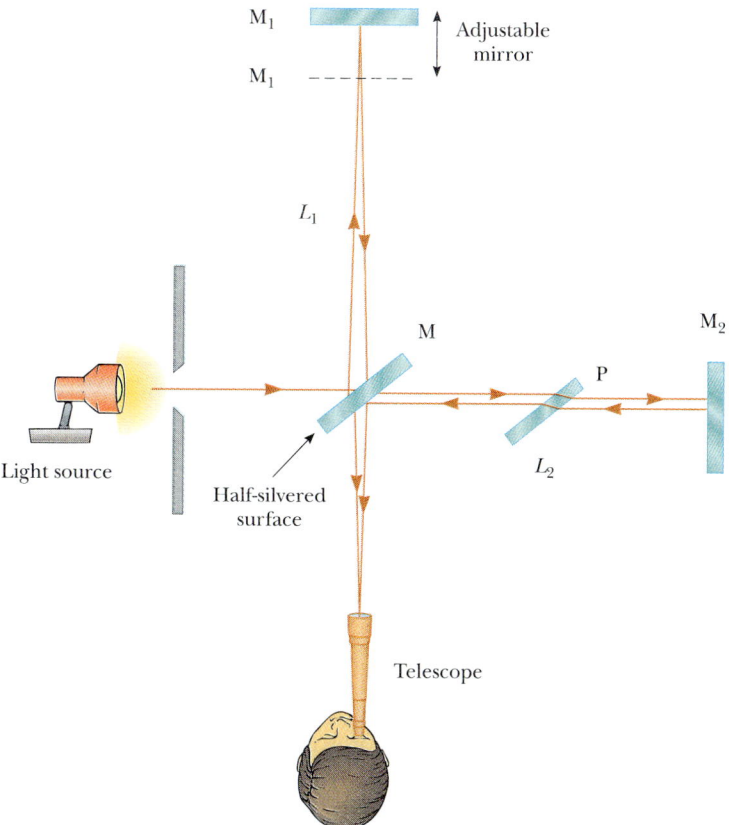

FIGURE 37.20 Diagram of the Michelson interferometer. A single beam is split into two rays by the partially silvered mirror M. The path difference between the two rays is varied with the adjustable mirror M_1.

light provided by a monochromatic source is split into two rays by a partially silvered mirror M inclined at 45° to the incident light beam. One ray is reflected vertically upward toward mirror M_1, while the second ray is transmitted horizontally through M toward mirror M_2. Hence, the two rays travel separate paths L_1 and L_2. After reflecting from M_1 and M_2, the two rays eventually recombine to produce an interference pattern, which can be viewed through a telescope. The glass plate P, equal in thickness to mirror M, is placed in the path of the horizontal ray in order to ensure that the two rays travel the same distance through glass.

The interference condition for the two rays is determined by the difference in their optical path lengths. When the two rays are viewed as shown, the image of M_2 is at M_2' parallel to M_1. Hence, M_2' and M_1 form the equivalent of an air film. The effective thickness of the air film is varied by moving mirror M_1 along the direction of the light beam with a finely threaded screw. Under these conditions, the interference pattern is a series of bright and dark circular rings. If a dark circle appears at the center of the pattern, the two rays interfere destructively. If M_1 is then moved a distance $\lambda/4$, the path difference changes by $\lambda/2$ (twice the separation between M_1 and M_2'). The two rays now interfere constructively, giving a bright circle in the middle. As M_1 is moved an additional distance $\lambda/4$, another dark circle appears. Thus, we see that successive dark or bright circles are formed each time M_1 is moved a distance of $\lambda/4$. The wavelength of light is then measured by counting the number of fringe shifts for a given displacement of M_1. Conversely, if the wavelength is accurately known (as with a laser beam), mirror displacements can be measured to within a fraction of the wavelength.

Since the interferometer can accurately measure displacement, it is often used to make highly precise measurements of the length of mechanical components.

SUMMARY

Interference in light waves occurs whenever two or more waves overlap at a given point. A sustained interference pattern is observed if (1) the sources are coherent (that is, they maintain a constant phase relationship with one another), (2) the sources have identical wavelengths, and (3) the linear superposition principle is applicable.

In Young's double-slit experiment, two slits separated by a distance d are illuminated by a single-wavelength light source. An interference pattern consisting of bright and dark fringes is observed on a viewing screen. The condition for bright fringes (**constructive interference**) is

$$d \sin \theta = m\lambda \qquad (m = 0, \pm 1, \pm 2, \ldots) \tag{37.2}$$

The condition for dark fringes (**destructive interference**) is

$$d \sin \theta = (m + \tfrac{1}{2})\lambda \qquad (m = 0, \pm 1, \pm 2, \ldots) \tag{37.3}$$

The number m is called the **order number** of the fringe.

The **average intensity** of the double-slit interference pattern is

$$I_{av} = I_0 \cos^2\left(\frac{\pi d \sin \theta}{\lambda}\right) \tag{37.12}$$

where I_0 is the maximum intensity on the screen.

When a series of N slits is illuminated, the diffraction pattern produced can be viewed as interference arising from the superposition of a large number of waves.

It is convenient to use phasor diagrams to simplify the analysis of interference from three or more equally spaced slits.

A wave traveling from a medium of index of refraction n_1 toward a medium of index of refraction n_2 undergoes a 180° phase change upon reflection when $n_2 > n_1$. There is no phase change in the reflected wave if $n_2 < n_1$.

The condition for constructive interference in a film of thickness t and refractive index n with a common medium on both sides of the film is

$$2nt = (m + \tfrac{1}{2})\lambda \qquad (m = 0, 1, 2, \ldots) \qquad (37.16)$$

Similarly, the condition for destructive interference in thin films is

$$2nt = m\lambda \qquad (m = 0, 1, 2, \ldots) \qquad (37.17)$$

QUESTIONS

1. What is the necessary condition on the path-length difference between two waves that interfere (a) constructively and (b) destructively?
2. Explain why two flashlights held close together do not produce an interference pattern on a distant screen.
3. If Young's double-slit experiment were performed underwater, how would the observed interference pattern be affected?
4. What is the difference between interference and diffraction?
5. In Young's double-slit experiment, why do we use monochromatic light? If white light is used, how would the pattern change?
6. As a soap bubble evaporates, it appears black just before it breaks. Explain this phenomenon in terms of the phase changes that occur upon reflection from the two surfaces of the soap film.
7. An oil film on water appears brightest at the outer regions, where it is thinnest. From this information, what can you say about the index of refraction of oil relative to that of water?
8. A soap film on a wire loop and held in air appears black in the thinnest regions when observed by reflected light and shows a variety of colors in thicker regions, as in Figure 37.21. Explain.
9. A simple way of observing an interference pattern is to look at a distant light source through a stretched handkerchief or an opened umbrella. Explain how this works.

FIGURE 37.21   (Question 8).

10. In order to observe interference in a thin film, why must the film not be very thick (on the order of a few wavelengths)?
11. A lens with outer radius of curvature R and index of refraction n rests on a flat glass plate and the combination is illuminated with white light from above. Is there a dark spot or a light spot at the center of the lens? What does it mean if the observed rings are noncircular?
12. Why is the lens on a good-quality camera coated with a thin film?
13. Would it be possible to place a nonreflective coating on an airplane to cancel radar waves of wavelength 3 cm?
14. Why is it so much easier to perform interference experiments with a laser than with an ordinary light source?

PROBLEMS

Section 37.2   Young's Double-Slit Experiment

1. A pair of narrow, parallel slits separated by 0.25 mm is illuminated by green light ($\lambda = 546.1$ nm). The interference pattern is observed on a screen 1.2 m away from the plane of the slits. Calculate the distance (a) from the central maximum to the first bright region on either side of the central maximum and (b) between the first and second dark bands.

□ indicates problems that have full solutions available in the Student Solutions Manual and Study Guide.

2. A laser beam (λ = 632.8 nm) is incident on two slits 0.20 mm apart. Approximately how far apart are the bright interference lines on a screen 5.0 m away from the double slits?

3. A Young's interference experiment is performed with monochromatic light. The separation between the slits is 0.50 mm, and the interference pattern on a screen 3.3 m away shows the first maximum 3.4 mm from the center of the pattern. What is the wavelength?

4. Light (λ = 442 nm) passes through a double-slit system that has a slit separation d = 0.40 mm. Determine how far away a screen must be placed so that dark fringes appear directly opposite both slits.

5. On a day when the speed of sound is 354 m/s, a 2000-Hz sound wave impinges on two slits 30.0 cm apart. (a) At what angle is the first maximum located? (b) If the sound wave is replaced by 3.00-cm microwaves, what slit separation gives the same angle for the first maximum? (c) If the slit separation is 1.00 μm, light of what frequency gives the same first-maximum angle?

6. A double slit with a spacing of 0.083 mm between the slits is 2.5 m from a screen. (a) If yellow light of wavelength 570 nm strikes the double slit, what is the separation between the zeroth- and first-order maxima on the screen? (b) If blue light of wavelength 410 nm strikes the double slit, what is the separation between the second- and fourth-order maxima? (c) Repeat parts (a) and (b) for the minima.

7. A riverside warehouse has two open doors as in Figure P37.7. A boat on the river sounds its horn. To person A the sound is loud and clear. To person B, the sound is barely audible. The principal wavelength of the sound waves is 3.00 m. Assuming B is at the position of the first minimum, determine the distance between the doors, center to center.

(b) the position (y) of the third-order maximum, and (c) the angular position (θ) of the $m = 1$ minimum.

9. A radio transmitter A, operating at 60.0 MHz is 10.0 m from a similar transmitter B that is 180° out of phase with A. How far must an observer move from A toward B along the line connecting the two in order to move between two points where the two beams are in phase?

9A. A radio transmitter A, operating at a frequency f, is a distance d from a similar transmitter B that is 180° out of phase with A. How far must an observer move from A toward B along the line connecting the two in order to move between two points where the two beams are in phase?

10. Two radio antennas separated by 300 m as in Figure P37.10 simultaneously transmit identical signals at the same wavelength. A radio in a car traveling due north receives the signals. (a) If the car is at the position of the second maximum, what is the wavelength of the signals? (b) How much farther must the car travel to encounter the next minimum in reception? (*Caution:* Do not use small-angle approximations in this problem.)

10A. Two radio antennas separated by a distance d as in Figure P37.10 simultaneously transmit identical signals at the same wavelength. A radio in a car traveling due north receives the signals. (a) If the car is at the position of the second maximum, what is the wavelength of the signals? (b) How much farther must the car travel to encounter the next minimum in reception? (*Caution:* Do not use small-angle approximations in this problem.)

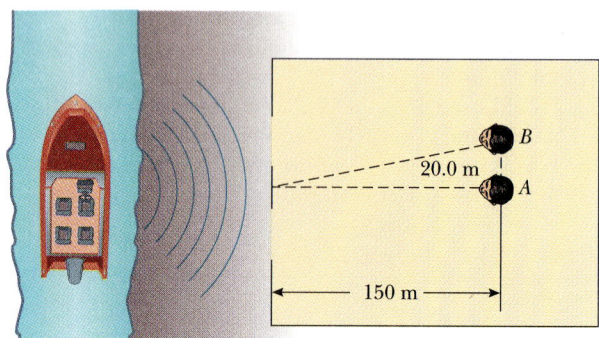

FIGURE P37.7

8. Monochromatic light illuminates a double-slit system with a slit separation d = 0.30 mm. The second-order maximum occurs at y = 4.0 mm on a screen 1.0 m from the slits. Determine (a) the wavelength,

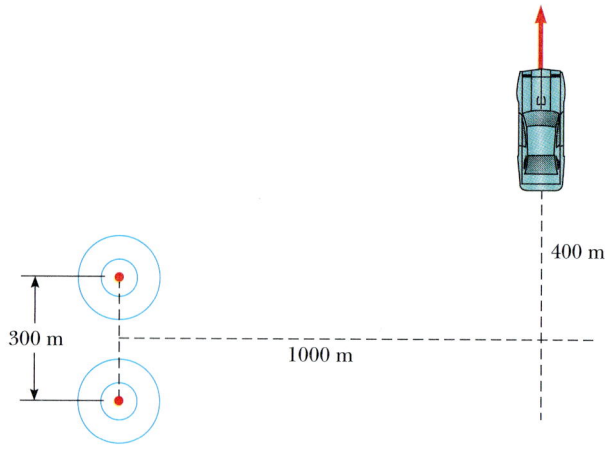

FIGURE P37.10

11. Two slits are separated by a distance d. Coherent light rays of wavelength λ strike the slits at an angle θ_1 as in Figure P37.11. If an interference maximum is

formed at an angle θ_2 far from the slits, show that $d(\sin \theta_2 - \sin \theta_1) = m\lambda$, where m is an integer.

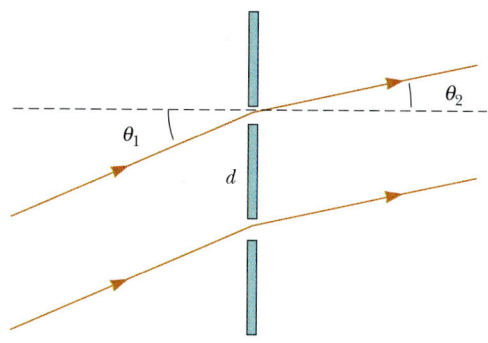

FIGURE P37.11

12. In a double-slit arrangement of Figure 37.4, $d = 0.15$ mm, $L = 140$ cm, $\lambda = 643$ nm, and $y = 1.8$ cm. (a) What is the path difference δ for the two rays from the two slits arriving at P? (b) Express this path difference in terms of λ. (c) Does P correspond to a maximum, a minimum, or an intermediate condition?

Section 37.3   Intensity Distribution of the Double-Slit Interference Pattern

13. In Figure 37.4, let $L = 120$ cm and $d = 0.25$ cm. The slits are illuminated with coherent 600-nm light. Calculate the distance y above the central maximum for which the average intensity on the screen is 75% of the maximum.

14. Two slits are separated by 0.18 mm. An interference pattern is formed on a screen 80 cm away by 656.3-nm light. Calculate the fraction of the maximum intensity 0.60 cm above the central maximum.

15. In Figure 37.4, $d = 0.20$ mm, $L = 160$ cm, and $y = 1.0$ mm. What wavelength results in an average intensity at P that is 36% of the maximum?

16. Monochromatic coherent light of amplitude E_0 and angular frequency ω passes through three parallel slits each separated by a distance d from its neighbor. (a) Show that the time-averaged intensity as a function of the angle θ is

$$I(\theta) = I_0 \left[1 + 2 \cos\left(\frac{2\pi d \sin \theta}{\lambda}\right) \right]^2$$

(b) Determine the ratio of the intensities of the two maxima.

17. In a Young's double-slit experiment using 350 nm light, a thin piece of Plexiglas ($n = 1.51$) covers one of the slits. If the center point on the screen is a dark spot, what is the minimum thickness of the Plexiglas?

17A. In a Young's double-slit experiment using light of wavelength λ, a thin piece of Plexiglas having index of refraction n covers one of the slits. If the center point on the screen is a dark spot, what is the minimum thickness of the Plexiglas?

18. In Figure 37.4 let $L = 1.2$ m and $d = 0.12$ mm and assume that the slit system is illuminated with monochromatic 500-nm light. Calculate the phase difference between the two wave fronts arriving at P when (a) $\theta = 0.50°$ and (b) $y = 5.0$ mm. (c) What is the value of θ for which (1) the phase difference is 0.333 rad and (2) the path difference is $\lambda/4$?

19. Two slits are separated by 0.32 mm. A beam of 500-nm light strikes them, producing an interference pattern. Determine the number of maxima observed in the angular range $-30° < \theta < 30°$.

20. The intensity on the screen at a certain point in a double-slit interference pattern is 64% of the maximum value. (a) What minimum phase difference (in radians) between sources produces this result? (b) Express this phase difference as a path difference for 486.1-nm light.

21. At a particular location in a Young's interference pattern, the intensity on the screen is 6.4% of maximum. (a) What minimum phase difference (in radians) between sources produces this result? (b) Determine the path difference for 587.5-nm light.

22. Monochromatic light ($\lambda = 632.8$ nm) is incident on two fine parallel slits separated by 0.20 mm. What is the distance to the first maximum and its intensity (relative to the central maximum) on a screen 2.0 m beyond the slits?

23. Two narrow parallel slits separated by 0.85 mm are illuminated by 600-nm light, and the viewing screen is 2.80 m away from the slits. (a) What is the phase difference between the two interfering waves on a screen at a point 2.50 mm from the central bright fringe? (b) What is the ratio of the intensity at this point to the intensity at the center of a bright fringe?

Section 37.4   Phasor Addition of Waves

24. The electric fields from three coherent sources are described by $E_1 = E_0 \sin \omega t$, $E_2 = E_0 \sin(\omega t + \phi)$, and $E_3 = E_0 \sin(\omega t + 2\phi)$. Let the resultant field be represented by $E_P = E_R \sin(\omega t + \alpha)$. Use phasors to find E_R and α when (a) $\phi = 20°$, (b) $\phi = 60°$, (c) $\phi = 120°$. (d) Repeat when $\phi = (3\pi/2)$ rad.

25. Suppose the slit openings in a Young's double-slit experiment have different sizes so that the electric field and intensity from one slit are different from those of the other slit. If $E_1 = E_{10} \sin(\omega t)$ and $E_2 = E_{20} \sin(\omega t + \phi)$, show that the resultant electric field is $E = E_0 \sin(\omega t + \theta)$, where

$$E_0 = \sqrt{E_{10}^2 + E_{20}^2 + 2E_{10}E_{20} \cos \phi}$$

and

$$\sin \theta = \frac{E_{20} \sin \phi}{E_0}$$

26. Use phasors to find the resultant (magnitude and phase angle) of two fields represented by $E_1 = 12 \sin \omega t$ and $E_2 = 18 \sin(\omega t + 60°)$. (Note that in this case the amplitudes of the two fields are unequal.)

27. Determine the resultant of the two waves $E_1 = 6.0 \sin(100 \pi t)$ and $E_2 = 8.0 \sin(100 \pi t + \pi/2)$.

28. Two coherent waves are described by

$$E_1 = E_0 \sin\left(\frac{2\pi x_1}{\lambda} - 2\pi f t + \frac{\pi}{6}\right)$$

$$E_2 = E_0 \sin\left(\frac{2\pi x_2}{\lambda} - 2\pi f t + \frac{\pi}{8}\right)$$

Determine the relationship between x_1 and x_2 that produces constructive interference when the two waves are superposed.

29. When illuminated, four equally spaced parallel slits act as multiple coherent sources, each differing in phase from the adjacent one by an angle ϕ. Use a phasor diagram to determine the smallest value of ϕ for which the resultant of the four waves (assumed to be of equal amplitude) is zero.

30. Sketch a phasor diagram to illustrate the resultant of $E_1 = E_{01} \sin \omega t$ and $E_2 = E_{02} \sin(\omega t + \phi)$, where $E_{02} = 1.5 E_{01}$ and $\pi/6 \leq \phi \leq \pi/3$. Use the sketch and the law of cosines to show that, for two coherent waves, the resultant intensity can be written in the form $I_R = I_1 + I_2 + 2\sqrt{I_1 I_2} \cos \phi$.

31. Consider N coherent sources described by $E_1 = E_0 \sin(\omega t + \phi)$, $E_2 = E_0 \sin(\omega t + 2\phi)$, $E_3 = E_0 \sin(\omega t + 3\phi)$, . . . , $E_N = E_0 \sin(\omega t + N\phi)$. Find the minimum value of ϕ for which $E_R = E_1 + E_2 + E_3 + \cdots + E_N$ is zero.

Section 37.6   Interference in Thin Films

32. A material having an index of refraction of 1.30 is used to coat a piece of glass ($n = 1.50$). What should be the minimum thickness of this film to minimize reflection of 500-nm light?

33. A film of MgF_2 ($n = 1.38$) having thickness 1.00×10^{-5} cm is used to coat a camera lens. Are any wavelengths in the visible spectrum intensified in the reflected light?

34. A soap bubble of index of refraction 1.33 strongly reflects both the red and the green components of white light. What film thickness allows this to happen? (In air, $\lambda_{red} = 700$ nm, $\lambda_{green} = 500$ nm.)

35. A thin film of oil ($n = 1.25$) covers a smooth wet pavement. When viewed perpendicular to the pavement, the film appears to be predominantly red (640 nm) and has no blue (512 nm). How thick is it?

36. A thin layer of oil ($n = 1.25$) is floating on water.

How thick is the oil in the region that reflects green light ($\lambda = 525$ nm)?

37. A thin layer of liquid methylene iodide ($n = 1.756$) is sandwiched between two flat parallel plates of glass. What must be the thickness of the liquid layer if normally incident 600-nm light is to be strongly reflected?

38. A beam of 580-nm light passes through two closely spaced glass plates, as shown in Figure P37.38. For what minimum nonzero value of the plate separation, d, is the transmitted light bright?

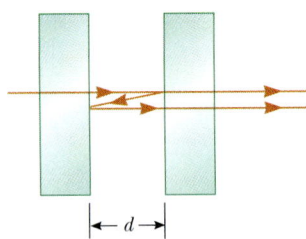

FIGURE P37.38

39. An oil film ($n = 1.45$) floating on water is illuminated by white light at normal incidence. The film is 280 nm thick. Find (a) the dominant observed color in the reflected light and (b) the dominant color in the transmitted light. Explain your reasoning.

40. A possible means for making an airplane invisible to radar is to coat the plane with an antireflective polymer. If radar waves have a wavelength of 3.00 cm and the index of refraction of the polymer is $n = 1.50$, how thick would you make the coating?

41. A thin film of cryolite ($n = 1.35$) is applied to a camera lens ($n = 1.50$). The coating is designed to reflect wavelengths at the blue end of the spectrum and transmit wavelengths in the near infrared. What minimum thickness gives high reflectivity at 450 nm and high transmission at 900 nm?

42. Two rectangular flat glass plates ($n = 1.52$) are in contact along one end and separated along the other end by a sheet of paper 4.0×10^{-3} cm thick (Fig. P37.42). The top plate is illuminated by monochromatic light ($\lambda = 546.1$ nm). Calculate the number of dark parallel bands crossing the top plate (include the dark band at zero thickness along the edge of contact between the two plates).

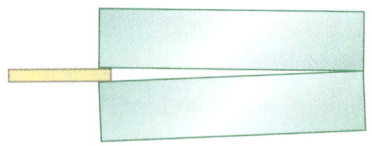

FIGURE P37.42

43. An air wedge is formed between two glass plates separated at one edge by a very fine wire as in Figure P37.42. When the wedge is illuminated from above by 600-nm light, 30 dark fringes are observed. Calculate the radius of the wire.

44. When a liquid is introduced into the air space between the lens and the plate in a Newton's-rings apparatus, the diameter of the tenth ring changes from 1.50 to 1.31 cm. Find the index of refraction of the liquid.

44A. When a liquid is introduced into the air space between the lens and the plate in a Newton's-rings apparatus, the diameter of the tenth ring changes from d_1 to d_2. Find the index of refraction of the liquid.

*Section 37.7 The Michelson Interferometer

45. Mirror M_1 in Figure 37.20 is displaced a distance ΔL. During this displacement, 250 fringes (formation of successive dark or bright bands) are counted. The light being used has a wavelength of 632.8 nm. Calculate the displacement ΔL.

46. Monochromatic light is beamed into a Michelson interferometer. The movable mirror is displaced 0.382 mm, causing the interferometer pattern to reproduce itself 1700 times. Determine the wavelength of the light. What color is it?

47. Light of wavelength 550.5 nm is used to calibrate a Michelson interferometer and mirror M_1 is moved 0.18 mm. How many dark fringes are counted?

48. One leg of a Michelson interferometer contains an evacuated cylinder that is 3.00 cm long and has a glass plate on each end. A gas is slowly leaked into the cylinder until a pressure of 1.00 atm is reached. If 35 bright fringes pass on the screen when 633-nm light is used, what is the index of refraction of the gas?

48A. One leg of a Michelson interferometer contains an evacuated cylinder that is of length L and has a glass plate on each end. A gas is slowly leaked into the cylinder until a pressure of P is reached. If N bright fringes pass on the screen when light of wavelength λ is used, what is the index of refraction of the gas?

ADDITIONAL PROBLEMS

49. Figure P37.49 shows a radio-wave transmitter and a receiver separated by $d = 600$ m and both $h = 30.0$ m high. The receiver can receive both direct signals from the transmitter and indirect ones reflected off the ground. Assuming that the ground is level between the two towers and that a $\lambda/2$ phase shift occurs upon reflection, determine the longest wavelengths that interfere (a) constructively and (b) destructively.

49A. Figure P37.49 shows a radio-wave transmitter and a receiver separated by a distance d and both a height h. The receiver can receive both direct signals from the transmitter and indirect ones reflected off the ground. Assuming that the ground is level between the towers and that a $\lambda/2$ phase shift occurs upon reflection, determine the longest wavelengths that interfere (a) constructively and (b) destructively.

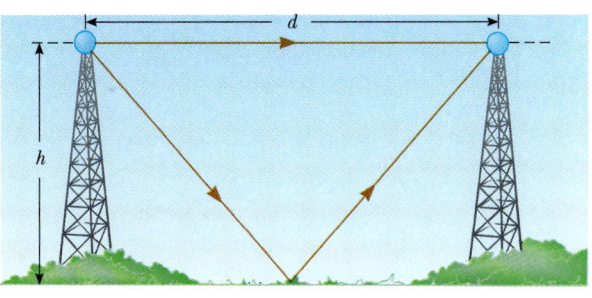

Transmitter   Receiver

FIGURE P37.49

50. Interference effects are produced at point P on a screen as a result of direct rays from a 500-nm source and reflected rays off the mirror, as in Figure 37.14. If the source is 100 m to the left of the screen, and 1.00 cm above the mirror, find the distance y (in millimeters) to the first dark band above the mirror.

51. Astronomers observed a 60-MHz radio source both directly and by reflection from the sea. If the receiving dish is 20 m above sea level, what is the angle of the radio source above the horizon at first maximum?

52. The waves from a radio station can reach a home receiver by two paths. One is a straight-line path from transmitter to home, a distance of 30 km. The second path is by reflection from the ionosphere (a layer of ionized air molecules near the top of the atmosphere). Assume this reflection takes place at a point midway between receiver and transmitter. If the wavelength broadcast by the radio station is 350 m, find the minimum height of the ionospheric layer that produces destructive interference between the direct and reflected beams. (Assume no phase changes on reflection.)

53. Measurements are made of the intensity distribution in a Young's interference pattern (Fig. 37.6). At a particular value of y, it is found that $I/I_0 = 0.81$ when 600-nm light is used. What wavelength of light should be used to reduce the relative intensity at the same location to 64%?

54. Waves broadcast by a 1500-kHz radio station arrive at a home receiver by two paths. One is a direct path,

and the second is from reflection off an airplane directly above the receiver. The airplane is approximately 100 m above the receiver, and the direct distance from station to home is 20.0 km. What is the exact height of the airplane if destructive interference is occurring? (Assume no phase change on reflection.)

54A. Waves broadcast by a radio station at frequency f arrive at a home receiver by two paths. One is a direct path, and the second is from reflection off an airplane directly above the receiver. The airplane is at an approximate height h above the receiver, and the direct distance from station to home is d. What is the exact height of the airplane if destructive interference is occurring? (Assume no phase change on reflection.)

55. In a Young's interference experiment, the two slits are separated by 0.15 mm, and the incident light includes light of wavelengths $\lambda_1 = 540$ nm and $\lambda_2 = 450$ nm. The overlapping interference patterns are formed on a screen 1.40 m from the slits. Calculate the minimum distance from the center of the screen to the point where a bright line of the λ_1 light coincides with a bright line of the λ_2 light.

56. In a Newton's-ring experiment, a plano-convex glass ($n = 1.52$) lens of diameter 10.0 cm is placed on a flat plate as in Figure 37.17a. When 650-nm light is incident normally, 55 bright rings are observed, with the last one right on the edge of the lens. (a) What is the radius of curvature of the convex surface of the lens? (b) What is the focal length of the lens?

57. Young's double-slit experiment is performed with 589-nm light and a slits-to-screen distance of 2.00 m. The tenth interference minimum is observed 7.26 mm from the central maximum. Determine the spacing of the slits.

58. A soap film 500 nm thick has an index of refraction of 1.35 and is illuminated with white light. (a) If the film is part of a bubble in air, what color is the reflected light? (b) If the film is on a flat glass plate, what color is the reflected light?

59. A hair is placed at one edge between two flat glass plates 8.00 cm long. When this arrangement is illuminated with 600-nm light, 121 dark bands are counted, starting at the point of contact of the two plates. How thick is the hair?

60. A glass plate ($n = 1.61$) is covered with a thin uniform layer of oil ($n = 1.20$). A nonmonochromatic light beam in air is incident normally on the oil surface. Observation of the reflected beam shows destructive interference at 500 nm and constructive interference at 750 nm with no intervening maxima or minima. Calculate the thickness of the oil layer.

61. A piece of transparent material having an index of refraction n is cut into the shape of a wedge, as shown in Figure P37.61. The angle of the wedge is small,

and monochromatic light of wavelength λ is normally incident from above. If the height of the wedge is h and the width is ℓ, show that bright fringes occur at the positions $x = \lambda\ell(m + \frac{1}{2})/2hn$ and dark fringes occur at the positions $x = \lambda\ell m/2hn$, where $m = 0$, 1, 2, . . . and x is measured as shown.

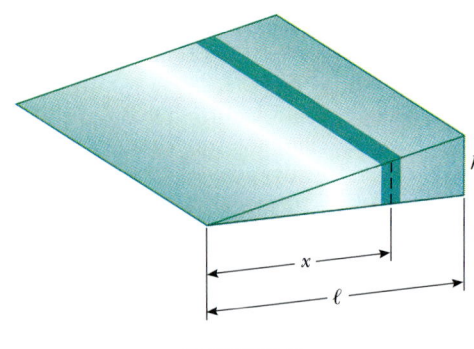

FIGURE P37.61

62. An air wedge is formed between two glass plates in contact along one edge and slightly separated at the opposite edge. When the plates are illuminated with monochromatic light from above, the reflected light has 85 dark fringes. Calculate the number of dark fringes that would appear if water ($n = 1.33$) were to replace the air between the plates.

63. The condition for constructive interference by reflection from a thin film in air as developed in Section 37.6 assumes nearly normal incidence. (a) Show that if the light is incident on the film at an angle $\phi_1 \gg 0$ (relative to the normal), then the condition for constructive interference is $2nt \cos \theta_2 = (m + \frac{1}{2})\lambda$, where θ_2 is the angle of refraction. (b) Calculate the minimum thickness for constructive interference if 590-nm light is incident at an angle of 30° on a film that has an index of refraction of 1.38.

64. Use phasor addition to find the resultant amplitude and phase constant when the following three harmonic functions are combined: $E_1 = \sin(\omega t + \pi/6)$, $E_2 = 3.0 \sin(\omega t + 7\pi/2)$, $E_3 = 6.0 \sin(\omega t + 4\pi/3)$.

65. A planoconvex lens having a radius of curvature $r = 4.00$ m is placed on a concave reflecting surface having a radius of curvature $R = 12.0$ m, as in Figure P37.65. Determine the radius of the 100th bright ring if 500-nm light is incident normal to the flat surface of the lens.

66. A soap film ($n = 1.33$) is contained within a rectangular wire frame. The frame is held vertically so that the film drains downward and becomes thicker at the bottom than at the top, where the thickness is essentially zero. The film is viewed in white light with near-normal incidence, and the first violet ($\lambda = 420$ nm) interference band is observed 3.0 cm from the top

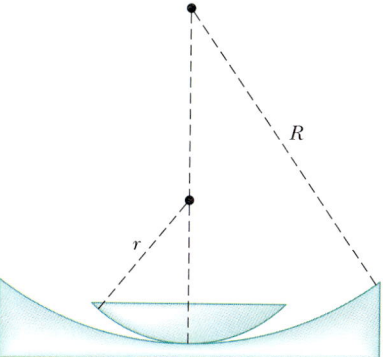

FIGURE P37.65

edge of the film. (a) Locate the first red ($\lambda =$ 680 nm) interference band. (b) Determine the film thickness at the positions of the violet and red bands. (c) What is the wedge angle of the film?

67. (a) A lens is made with its lower surface described by the function $y = f(x)$ rotated about the y axis (Fig. P37.67). The lens is placed on a flat glass plate. Light of wavelength λ is incident straight down. Show that dark rings form with radii

$$r = f^{-1}\left(\frac{m\lambda}{2n}\right)$$

where n is the index of refraction of the medium surrounding the lens, m is a non-negative integer, and f^{-1} is the inverse of the function f. (b) Show that the dark rings correspond to the Newton's rings described in the text, where

$$f(x) = R - \sqrt{R^2 - x^2}$$

for $0 < x < R$, R is the radius of curvature of the convex lens, and y is small compared to R.

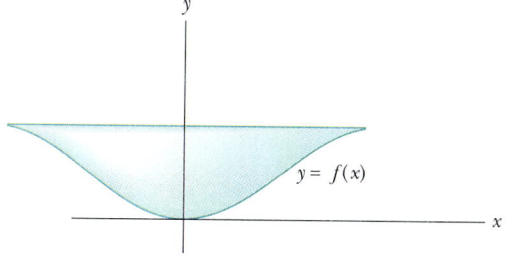

FIGURE P37.67

68. Interference fringes are produced using Lloyd's mirror and a 606-nm source as in Figure 37.14. If fringes 1.2 mm apart are formed on a screen 2.0 m from the real source S, find the vertical distance h of the source above the reflecting surface.

69. (a) Both sides of a uniform film that has index of refraction n and thickness d are in contact with air. For normal incidence of light, an intensity minimum is observed in the reflected light at λ_2 and an intensity maximum is observed at λ_1, where $\lambda_1 > \lambda_2$. If there are no intensity minima observed between λ_1 and λ_2, show that the integer m in Equations 37.16 and 37.17 is given by $m = \lambda_1/2(\lambda_1 - \lambda_2)$. (b) Determine the thickness of the film if $n = 1.40$, $\lambda_1 = 500$ nm, and $\lambda_2 = 370$ nm.

70. Consider the double-slit arrangement shown in Figure P37.70, where the separation d is 0.30 mm and the distance L is 1.00 m. A sheet of transparent plastic ($n = 1.50$) 0.050 mm thick (about the thickness of this page) is placed over the upper slit. As a result, the central maximum of the interference pattern moves upward a distance y'. Find this distance.

70A. Consider the double-slit arrangement shown in Figure P37.70, where the slit separation is d and the slit-to-screen distance is L. A sheet of transparent plastic having an index of refraction n and thickness t is placed over the upper slit. As a result, the central maximum of the interference pattern moves upward a distance y'. Find y'.

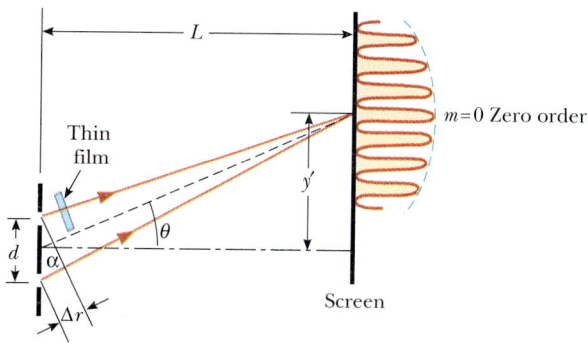

FIGURE P37.70

71. Slit 1 of a double slit is wider than slit 2 so that the light from slit 1 has an amplitude three times that of the light from slit 2. Show that for this situation, Equation 37.11 has the form $I = (4I_0/9)(1 + 3 \cos^2 \phi/2)$.

SPREADSHEET PROBLEMS

S1. To calculate the intensity distribution of the interference pattern for equally spaced sources, it is necessary to add a series of terms such as: $E_1 = A_0 \sin \alpha$, $E_2 = A_0 \sin(\alpha + \phi)$, $E_3 = A_0 \sin(\alpha + 2\phi)$, where ϕ is the phase difference caused by different path lengths. For N sources the time-averaged relative intensity as a function of the phase angle ϕ is

$$I_{rel} = f_N^2(\phi) + g_N^2(\phi)$$

where

$$f_N(\phi) = \sum_{n=0}^{N-1} \cos(n\phi) \qquad g_N(\phi) = \sum_{n=0}^{N-1} \sin(n\phi)$$

and I_{rel} is the ratio of the average intensity for N sources to that of one source. Spreadsheet 37.1 calculates and plots I_{rel} versus phase angle ϕ for up to a maximum of six sources. The coefficients a, b, c, d, e, and f used in the spreadsheet determine the number of sources. For two sources, for example, set $a = 1$, $b = 1$, and $c = d = e = f = 0$. For three sources set $a = b = c = 1$, $d = e = f = 0$, and so on. (a) For three sources, what is the ratio of the intensity of the principal maxima to that of a single source? (b) What is the ratio of the intensity of the secondary maxima to that of the principal maximum?

S2. Use Spreadsheet 37.1 to calculate the intensity pattern for four equally spaced sources. (a) What is the ratio of the intensity of the principal maxima to that of a single source? (b) What is the ratio of the intensity of the secondary maxima to that of the principal maxima? (c) Repeat parts (a) and (b) for five and six sources. (d) Are all the secondary maxima of equal intensity?

Diffraction and Polarization

These glass objects, called Prince Rupert drops, are made by dripping molten glass into water. The photograph was made by placing the objects between two crossed polarizers. The patterns that are observed represent the strain distribution in the glass. Studies of such patterns led to the development of tempered glass. *(James L. Amos/Peter Arnold, Inc.)*

W hen light waves pass through a small aperture, an interference pattern is observed rather than a sharp spot of light, showing that light spreads beyond the aperture into regions where a shadow would be expected if light traveled in straight lines. Other waves, such as sound waves and water waves, also have this property of being able to bend around corners. This phenomenon, known as diffraction, can be regarded as interference from a great number of coherent wave sources. In other words, diffraction and interference are basically equivalent.

In Chapter 34, we learned that electromagnetic waves are transverse. That is, the electric and magnetic field vectors are perpendicular to the direction of propagation. In this chapter, we see that under certain conditions, light waves can be polarized in various ways, such as by passing light through polarizing sheets.

38.1   INTRODUCTION TO DIFFRACTION

In Section 37.2 we learned that when two slits are illuminated by a single-wavelength light source, an interference pattern is formed on a viewing screen. If the light truly traveled in straight-line paths after passing through the slits, as in Figure

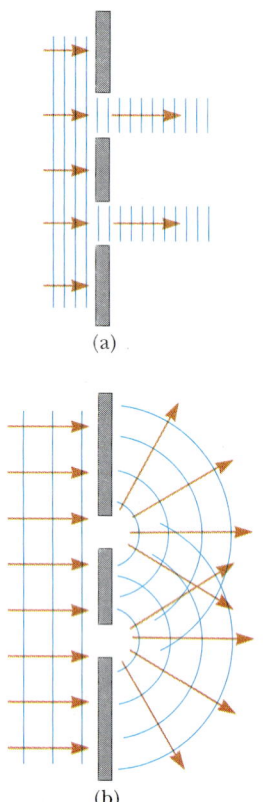

(a)

(b)

FIGURE 38.1 (a) If light waves did not spread out after passing through the slits, no interference would occur. (b) The light waves from the two slits overlap as they spread out, filling the expected shadowed regions with light and producing interference fringes.

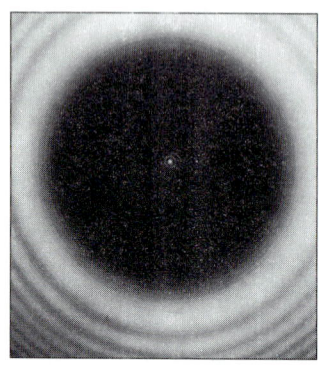

FIGURE 38.3 Diffraction pattern of a penny, taken with the penny midway between screen and source. *(Courtesy of P. M. Rinard, from Am. J. Phys. 44:70, 1976)*

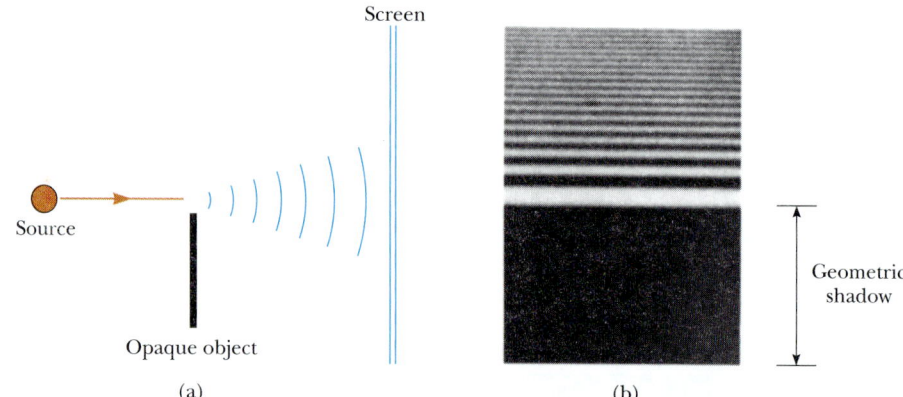

(a)

(b)

FIGURE 38.2 (a) We expect the shaded region to be completely shielded from the light by the opaque object. Instead, the light bends around the object and enters what "should be" a shadowed region. (b) Diffraction pattern of a straight edge.

38.1a, the waves would not overlap and no interference pattern would be seen. Instead, Huygens' principle requires that the waves spread out from the slits as shown in Figure 38.1b. In other words, the light deviates from a straight-line path and enters the region that would otherwise be shadowed. This divergence of light from its initial line of travel is called **diffraction.**

In general, diffraction occurs when waves pass through small openings, around obstacles, or past sharp edges. As an example of diffraction, consider the following. When an opaque object is placed between a point source of light and a screen, as in Figure 38.2a, the boundary between the shadowed and illuminated regions on the screen is not sharp. A careful inspection of the boundary shows that a small amount of light bends into the shadowed region. The region outside the shadow contains alternating light and dark bands, as in Figure 38.2b. The intensity in the first bright band is greater than the intensity in the region of uniform illumination.

Figure 38.3 shows the diffraction pattern and shadow of a penny. There are a bright spot at the center, circular fringes near the shadow's edge, and another set of fringes outside the shadow. The center bright spot can be explained only through the use of the wave theory of light, which predicts constructive interference at this point. From the viewpoint of geometric optics (light as a collection of particles), we expect the center of the shadow to be dark because that part of the viewing screen is completely shielded by the penny.

It is interesting to point out an historical incident that occurred shortly before the central bright spot was first observed. One of the supporters of geometric optics, Simeon Poisson, argued that if Augustin Fresnel's wave theory of light were valid, then a central bright spot should be observed in any shadow. Because the spot was observed shortly thereafter, Poisson's prediction reinforced the wave theory rather than disproved it.

Diffraction phenomena are usually classified into two types. **Fraunhofer diffraction** occurs when the rays reaching a viewing screen are approximately parallel. This can be achieved experimentally either by placing the screen far from the opening as in Figure 38.4a or by using a converging lens to focus the parallel rays on the screen. A bright fringe is observed along the axis at $\theta = 0$, with alternating dark and bright fringes on either side of the central bright one. Figure 38.4b is a photograph of a single-slit Fraunhofer diffraction pattern.

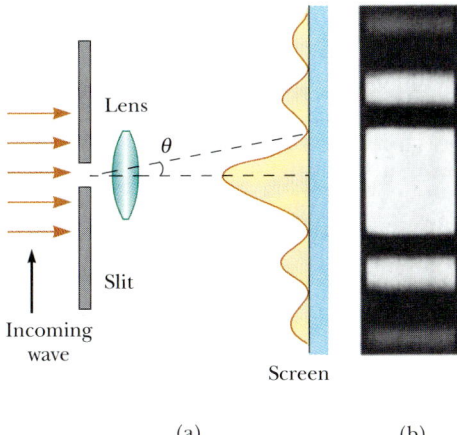

(a)   (b)

FIGURE 38.4   (a) Fraunhofer diffraction pattern of a single slit. The pattern consists of a central bright region flanked by much weaker maxima alternating with dark bands. (Note that this is not to scale.) (b) Photograph of a single-slit Fraunhofer diffraction pattern. *(From M. Cagnet, M. Francon, and J. C. Thierr,* Atlas of Optical Phenomena, *Berlin, Springer-Verlag, 1962, plate 18)*

When the observing screen is placed a finite distance from the slit and no lens is used to focus parallel rays, the observed pattern is called a **Fresnel diffraction** pattern (Fig. 38.5). The diffraction patterns shown in Figures 38.2b and 38.3 are examples of Fresnel diffraction. Because Fresnel diffraction is difficult to treat quantitatively, the following discussion is restricted to Fraunhofer diffraction.

38.2   SINGLE-SLIT DIFFRACTION

Up until now, we have assumed that slits are point sources of light. In this section, we abandon this assumption and determine how the finite width of slits is the basis for understanding Fraunhofer diffraction.

We can deduce some important features of this problem by examining waves coming from various portions of the slit, as shown in Figure 38.6. According to Huygens' principle, *each portion of the slit acts as a source of waves.* Hence, light from one portion of the slit can interfere with light from another portion, and the resultant intensity on the screen depends on the direction θ.

To analyze the diffraction pattern, it is convenient to divide the slit in two halves, as in Figure 38.6. All the waves that originate from the slit are in phase. Consider waves 1 and 3, which originate from a segment just above the bottom and just above the center of the slit, respectively. Wave 1 travels farther than wave 3 by an amount equal to the path difference $(a/2) \sin \theta$, where a is the width of the slit. Similarly, the path difference between waves 2 and 4 is also $(a/2) \sin \theta$. If this path difference is exactly one-half wavelength (corresponding to a phase difference of 180°), the two waves cancel each other and destructive interference results. This is true, in fact, for any two waves that originate at points separated by half the slit width because the phase difference between two such points is 180°. Therefore, waves from the upper half of the slit interfere destructively with waves from the lower half of the slit when

$$\frac{a}{2} \sin \theta = \frac{\lambda}{2}$$

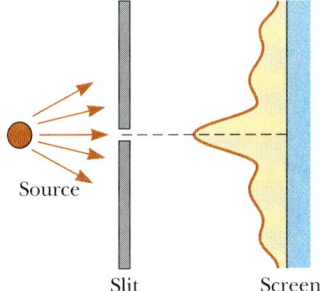

FIGURE 38.5   A Fresnel diffraction pattern of a single slit is observed when the rays are not parallel and the observing screen is a finite distance from the slit. (Note that this is not to scale.)

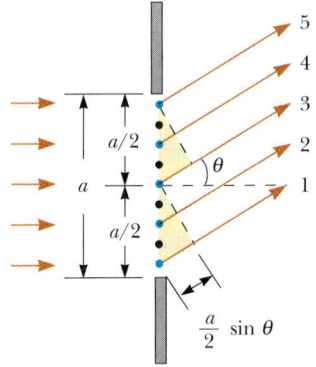

FIGURE 38.6   Diffraction of light by a narrow slit of width a. Each portion of the slit acts as a point source of waves. The path difference between rays 1 and 3 or between rays 2 and 4 is $(a/2) \sin \theta$. (Note that this is not to scale.)

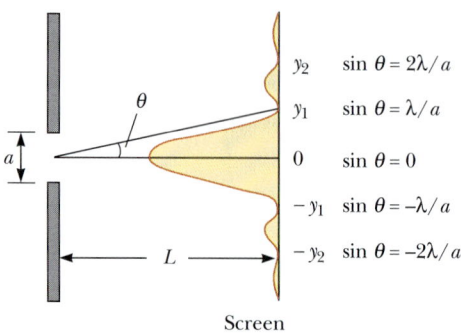

FIGURE 38.7 Positions of the minima for the Fraunhofer diffraction pattern of a single slit of width *a*. The pattern is obtained only if $L \gg a$. (Note that this is not to scale.)

or when

$$\sin \theta = \frac{\lambda}{a}$$

If we divide the slit into four parts rather than two and use similar reasoning, we find that the screen is also dark when

$$\sin \theta = \frac{2\lambda}{a}$$

Likewise, we can divide the slits into six parts and show that darkness occurs on the screen when

$$\sin \theta = \frac{3\lambda}{a}$$

Therefore, the general condition for destructive interference is

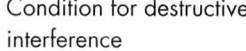
Condition for destructive interference

$$\sin \theta = m\frac{\lambda}{a} \qquad (m = \pm 1, \pm 2, \pm 3, \ldots) \tag{38.1}$$

Equation 38.1 gives the values of θ for which the diffraction pattern has zero intensity — that is, a dark fringe is formed. However, Equation 38.1 tells us nothing about the variation in intensity along the screen. The general features of the intensity distribution are shown in Figure 38.7. A broad central bright fringe is observed, flanked by much weaker, bright fringes alternating with dark fringes. The various dark fringes (points of zero intensity) occur at the values of θ that satisfy Equation 38.1. The position of the points of constructive interference lies approximately halfway between the dark fringes. Note that the central bright fringe is twice as wide as the weaker maxima.

EXAMPLE 38.1   Where Are the Dark Fringes?

Light of wavelength 580 nm is incident on a slit of width 0.300 mm. The observing screen is 2.00 m from the slit. Find the positions of the first dark fringes and the width of the central bright fringe.

Solution   The first dark fringes that flank the central bright fringe correspond to $m = \pm 1$ in Equation 38.1. Hence, we

find that

$$\sin \theta = \pm \frac{\lambda}{a} = \pm \frac{5.80 \times 10^{-7} \text{ m}}{0.300 \times 10^{-3} \text{ m}} = \pm 1.93 \times 10^{-3}$$

From the triangle in Figure 38.7, note that $\tan \theta = y_1/L$. Since θ is very small, we can use the approximation $\sin \theta \approx \tan \theta$, so that $\sin \theta \approx y_1/L$. Therefore, the positions of the first minima measured from the central axis are given by

$$y_1 \approx L \sin \theta = \pm L \frac{\lambda}{a} = \pm 3.87 \times 10^{-3} \text{ m}$$

The positive and negative signs correspond to the dark fringes on either side of the central bright fringe. Hence, the width of the central bright fringe is equal to $2|y_1| = 7.73 \times 10^{-3}$ m = 7.73 mm. Note that this value is much larger than the width of the slit. However, as the width of the slit is increased, the diffraction pattern narrows, corresponding to smaller values of θ. In fact, for large values of a, the various maxima and minima are so closely spaced that the only thing observed is a large central bright area, which resembles the geometric image of the slit. This matter is of great importance in the design of lenses used in telescopes, microscopes, and other optical instruments.

Exercise Determine the width of the first-order bright fringe.

Answer 3.87 mm.

Intensity of the Single-Slit Diffraction Pattern

We can use phasors to determine the intensity distribution for a single-slit diffraction pattern. Imagine a slit divided into a large number of small zones, each of width Δy as in Figure 38.8. Each zone acts as a source of coherent radiation, and each contributes an incremental electric field amplitude ΔE at some point P on the screen. The total electric field amplitude E at P is obtained by summing the contributions from all the zones. The light intensity at P is the square of the amplitude.

The incremental electric field amplitudes between adjacent zones are out of phase with one another by an amount $\Delta\beta$. The phase difference $\Delta\beta$ is related to the path difference $\Delta y \sin \theta$ between adjacent zones by

$$\text{Phase difference} = \left(\frac{2\pi}{\lambda}\right) \text{path difference}$$

$$\Delta\beta = \frac{2\pi}{\lambda} \Delta y \sin \theta \tag{38.2}$$

To find the total electric field amplitude on the screen at any angle θ, we sum the incremental amplitudes ΔE due to each zone. For small values of θ, we can assume that all the ΔE values are the same. It is convenient to use phasor diagrams for various angles as in Figure 38.9. When $\theta = 0$, all phasors are aligned as in Figure 38.9a, because all the waves from the various zones are in phase. In this case, the total amplitude of the center of the screen is $E_0 = N \Delta E$, where N is the number of zones. The amplitude E_θ at some small angle θ is shown in Figure 38.9b, where each phasor differs in phase from an adjacent one by an amount $\Delta\beta$. In this case, E_θ is the vector sum of the incremental amplitudes, and hence is given by the length of the chord. Therefore, $E_\theta < E_0$. The total phase difference β between

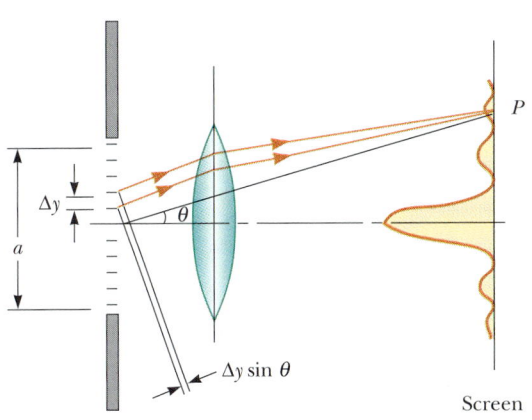

FIGURE 38.8 Fraunhofer diffraction by a single slit. The intensity at the point P is the resultant of all the incremental fields from zones of width Δy leaving the slit.

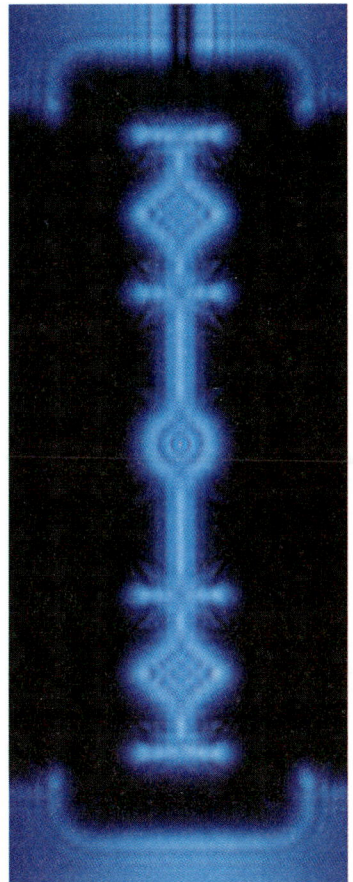

The diffraction pattern of a razor blade observed under monochromatic light. (© Ken Kay, 1987/Fundamental Photographs)

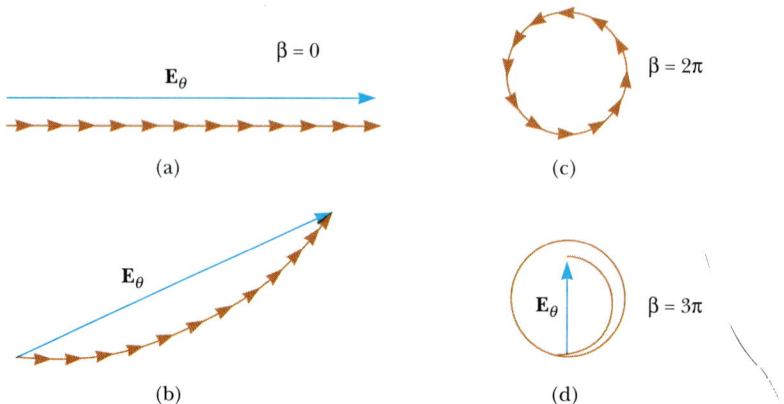

FIGURE 38.9   Phasor diagrams for obtaining the various maxima and minima of a single-slit diffraction pattern.

waves from the top and bottom portions of the slit is

$$\beta = N\Delta\beta = \frac{2\pi}{\lambda} N\Delta y \sin\theta = \frac{2\pi}{\lambda} a \sin\theta \qquad (38.3)$$

where $a = N\Delta y$ is the width of the slit.

As θ increases, the chain of phasors eventually forms a closed path as in Figure 38.9c. At this point, the vector sum is zero, and so $E_\theta = 0$, corresponding to the first minimum on the screen. Noting that $\beta = N\Delta\beta = 2\pi$ in this situation, we see from Equation 38.3 that

$$2\pi = \frac{2\pi}{\lambda} a \sin\theta$$

$$\sin\theta = \frac{\lambda}{a}$$

That is, the first minimum in the diffraction pattern occurs when $\sin\theta = \lambda/a$, which agrees with Equation 38.1.

At larger values of θ, the spiral chain of phasors continues. For example, Figure 38.9d represents the situation corresponding to the second maximum, which occurs when $\beta \cong 360° + 180° = 540°$ (3π rad). The second minimum (two complete spirals, not shown) corresponds to $\beta = 720°$ (4π rad), which satisfies the condition $\sin\theta = 2\lambda/a$.

The total amplitude and intensity at any point on the screen can be obtained by considering the limiting case where Δy becomes infinitesimal (dy) and $N \to \infty$. In this limit, the phasor chains in Figure 38.9 become smooth curves, as in Figure 38.10. From this figure, we see that at some angle θ, the wave amplitude on the screen, E_θ, is equal to the chord length, while E_0 is the arc length. From the triangle whose angle is $\beta/2$, we see that

$$\sin\frac{\beta}{2} = \frac{E_\theta/2}{R}$$

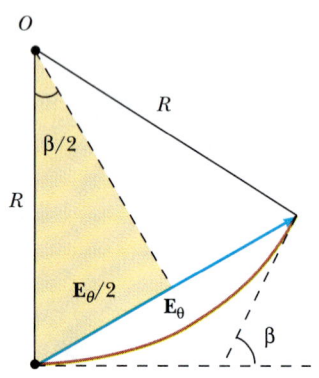

FIGURE 38.10   Phasor diagram for a large number of coherent sources. Note that all the ends of the phasors lie on a circular arc of radius R. The resultant amplitude E_θ equals the length of the chord.

where R is the radius of curvature. But the arc length E_0 is equal to the product $R\beta$, where β is in radians. Combining this with the expression above gives

$$E_\theta = 2R\sin\frac{\beta}{2} = 2\left(\frac{E_0}{\beta}\right)\sin\frac{\beta}{2} = E_0\left[\frac{\sin(\beta/2)}{\beta/2}\right]$$

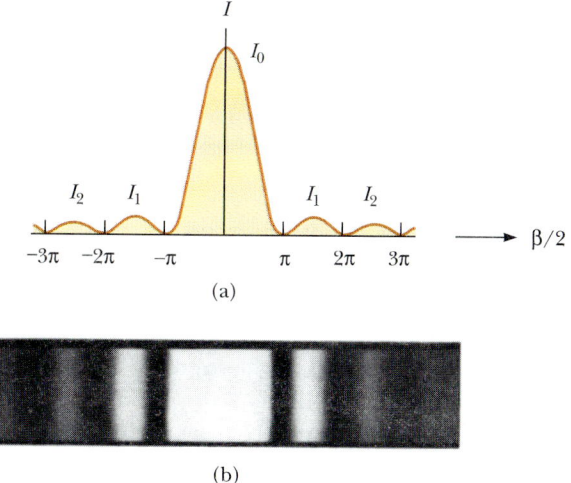

(a)

(b)

FIGURE 38.11 (a) A plot of the intensity I versus $\beta/2$ for the single-slit Fraunhofer diffraction pattern. (b) Photograph of a single-slit Fraunhofer diffraction pattern. *(From M. Cagnet, M. Francon, and J. C. Thierr,* Atlas of Optical Phenomena, *Berlin, Springer-Verlag, 1962, plate 18)*

Because the resultant intensity I_θ at P is proportional to the square of the amplitude E_θ, we find

$$I_\theta = I_0 \left[\frac{\sin(\beta/2)}{\beta/2} \right]^2 \qquad (38.4)$$

Intensity of a single-slit Fraunhofer diffraction pattern

where I_0 is the intensity at $\theta = 0$ (the central maximum) and $\beta = 2\pi a \sin \theta / \lambda$. Substitution of this expression for β into Equation 38.4 gives

$$I_\theta = I_0 \left[\frac{\sin(\pi a \sin \theta / \lambda)}{\pi a \sin \theta / \lambda} \right]^2 \qquad (38.5)$$

From this result, we see that minima occur when

$$\frac{\pi a \sin \theta}{\lambda} = m\pi$$

$$\sin \theta = m \frac{\lambda}{a}$$

Condition for intensity minima

where $m = \pm 1, \pm 2, \pm 3, \ldots$. This is in agreement with our earlier result, given by Equation 38.1.

Figure 38.11a represents a plot of Equation 38.4, and a photograph of a single-slit Fraunhofer diffraction pattern is shown in Figure 38.11b. Most of the light intensity is concentrated in the central bright fringe.

EXAMPLE 38.2 **Relative Intensities of the Maxima**

Find the ratio of intensities of the secondary maxima to the intensity of the central maximum for the single-slit Fraunhofer diffraction pattern.

Solution To a good approximation, the secondary maxima lie midway between the zero points. From Figure 38.11a, we see that this corresponds to $\beta/2$ values of $3\pi/2$, $5\pi/2$,

$7\pi/2, \ldots$ Substituting these into Equation 38.4 gives for the first two ratios

$$\frac{I_1}{I_0} = \left[\frac{\sin(3\pi/2)}{(3\pi/2)}\right]^2 = \frac{1}{9\pi^2/4} = 0.045$$

$$\frac{I_2}{I_0} = \left[\frac{\sin(5\pi/2)}{5\pi/2}\right]^2 = \frac{1}{25\pi^2/4} = 0.016$$

That is, the secondary maximum (the one adjacent to the central maximum) has an intensity of 4.5% that of the central bright fringe, and the next second-order maximum has an intensity of 1.6% that of the central bright fringe.

Exercise   Determine the intensity of the secondary maximum corresponding to $m = 3$ relative to the central maximum.

Answer   0.0083.

38.3   RESOLUTION OF SINGLE-SLIT AND CIRCULAR APERTURES

The ability of optical systems to distinguish between closely spaced objects is limited because of the wave nature of light. To understand this difficulty, consider Figure 38.12, which shows two light sources far from a narrow slit of width a. The sources can be considered as two noncoherent point sources, S_1 and S_2. For example, they could be two distant stars. If no diffraction occurred, two distinct bright spots (or images) would be observed on the viewing screen. However, because of diffraction, each source is imaged as a bright central region flanked by weaker bright and dark bands. What is observed on the screen is the sum of two diffraction patterns, one from S_1 and the other from S_2.

If the two sources are separated enough to keep their central maxima from overlapping, as in Figure 38.12a, their images can be distinguished and are said to be resolved. If the sources are close together, however, as in Figure 38.12b, the two central maxima overlap and the images are not resolved. To decide when two images are resolved, the following criterion is used:

When the central maximum of one image falls on the first minimum of another image, the images are said to be just resolved. This limiting condition of resolution is known as **Rayleigh's criterion.**

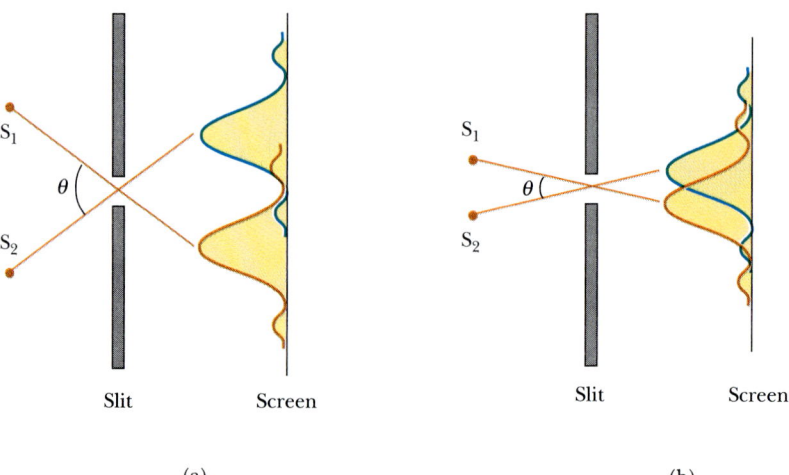

(a)   (b)

FIGURE 38.12   Two point sources far from a small aperture each produce a diffraction pattern. (a) The angle subtended by the sources at the aperture is large enough so that the diffraction patterns are distinguishable. (b) The angle subtended by the sources is so small that their diffraction patterns overlap and the images are not well resolved. (Note that the angles are greatly exaggerated.)

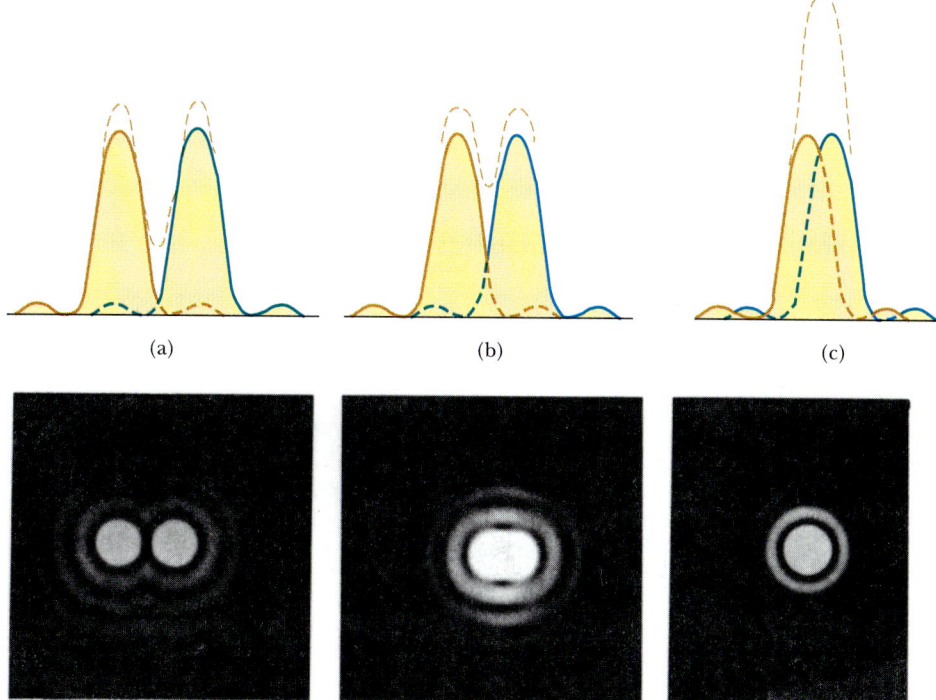

(a) (b) (c)

FIGURE 38.13 The diffraction patterns of two point sources (solid curves) and the resultant pattern (dashed curves) for various angular separations of the sources. In each case, the dashed curve is the sum of the two solid curves. (a) The sources are far apart, and the patterns are well resolved. (b) The sources are closer together, and the patterns are just resolved. (c) The sources are so close together that the patterns are not resolved. *(From M. Cagnet, M. Francon, and J. C. Thierr,* Atlas of Optical Phenomena, *Berlin, Springer-Verlag, 1962, plate 16)*

Figure 38.13 shows diffraction patterns for three situations. When the objects are far apart, their images are well resolved (Fig. 38.13a). The images are just resolved when the angular separation of the objects satisfies Rayleigh's criterion (Fig. 38.13b). Finally, the images are not resolved in Figure 38.13c.

From Rayleigh's criterion, we can determine the minimum angular separation, θ_{min}, subtended by the sources at the slit so that their images are just resolved. In Section 38.2, we found that the first minimum in a single-slit diffraction pattern occurs at the angle for which

$$\sin \theta = \frac{\lambda}{a}$$

where a is the width of the slit. According to Rayleigh's criterion, this expression gives the smallest angular separation for which the two images are resolved. Because $\lambda \ll a$ in most situations, $\sin \theta$ is small and we can use the approximation $\sin \theta \approx \theta$. Therefore, the limiting angle of resolution for a slit of width a is

$$\theta_{min} = \frac{\lambda}{a} \tag{38.6}$$

where θ_{min} is expressed in radians. Hence, the angle subtended by the two sources at the slit must be greater than λ / a if the images are to be resolved.

Many optical systems use circular apertures rather than rectangular slits. The diffraction pattern of a circular aperture, illustrated in Figure 38.14, consists of a

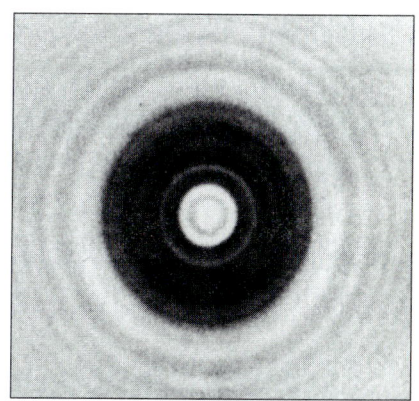

FIGURE 38.14 The diffraction pattern of a circular aperture consists of a central bright disk surrounded by concentric bright and dark rings. *(From M. Cagnet, M. Francon, and J. C. Thierr,* Atlas of Optical Phenomena, *Berlin, Springer-Verlag, 1962, plate 34)*

central circular bright disk surrounded by progressively fainter rings. The limiting angle of resolution of the circular aperture is

$$\theta_{min} = 1.22 \frac{\lambda}{D} \qquad (38.7)$$

where D is the diameter of the aperture. Note that Equation 38.7 is similar to Equation 38.6 except for the factor of 1.22, which arises from a complex mathematical analysis of diffraction from the circular aperture.

CONCEPTUAL EXAMPLE 38.3 Can You See an Atom?

Explain why it is theoretically impossible to see an object as small as an atom regardless of the quality of the light microscope being used.

Reasoning   In order to "see" an object, the wavelength of the light in the microscope must be comparable to the size of the object. An atom is much smaller than the wavelength of light in the visible region of the spectrum, so an atom can never be seen using visible light.

EXAMPLE 38.4 Limiting Resolution of a Microscope

Light of wavelength 589 nm is used to view an object under a microscope. If the aperture of the objective has a diameter of 0.900 cm, (a) find the limiting angle of resolution.

Solution   (a) From Equation 38.7, we find the limiting angle of resolution to be

$$\theta_{min} = 1.22 \left(\frac{589 \times 10^{-9} \text{ m}}{0.900 \times 10^{-2} \text{ m}} \right) = 7.98 \times 10^{-5} \text{ rad}$$

This means that any two points on the object subtending an angle less than approximately 8×10^{-5} rad at the objective cannot be distinguished in the image.

(b) Using visible light of any wavelength you desire, what is the maximum limit of resolution for this microscope?

Solution   To obtain the smallest angle corresponding to the maximum limit of resolution, we have to use the shortest wavelength available in the visible spectrum. Violet light (400 nm) gives us a limiting angle of resolution of

$$\theta_{min} = 1.22 \left(\frac{400 \times 10^{-9} \text{ m}}{0.900 \times 10^{-2} \text{ m}} \right) = 5.42 \times 10^{-5} \text{ rad}$$

(c) Suppose water ($n = 1.33$) fills the space between object and objective. What effect would this have on resolving power?

Solution   In this case, the wavelength of the sodium light in the water is found by $\lambda_w = \lambda_a / n$ (Eq. 35.7). Thus, we have

$$\lambda_w = \frac{\lambda_a}{n} = \frac{589 \text{ nm}}{1.33} = 443 \text{ nm}$$

The limiting angle of resolution at this wavelength is

$$\theta_m = 1.22 \left(\frac{443 \times 10^{-9} \text{ m}}{0.900 \times 10^{-2} \text{ m}} \right) = 6.00 \times 10^{-5} \text{ rad}$$

EXAMPLE 38.5 Resolution of a Telescope

The Hale telescope at Mount Palomar has a diameter of 200 in. What is its limiting angle of resolution for 600-nm light?

Solution   Because $D = 200$ in. $= 5.08$ m and $\lambda = 6.00 \times 10^{-7}$ m, Equation 38.7 gives

$$\theta_{min} = 1.22 \frac{\lambda}{D} = 1.22 \left(\frac{6.00 \times 10^{-7} \text{ m}}{5.08 \text{ m}} \right)$$

$$= 1.44 \times 10^{-7} \text{ rad} \cong 0.03 \text{ s of arc}$$

Therefore, any two stars that subtend an angle greater than or equal to this value are resolved (assuming ideal atmospheric conditions).

The Hale telescope can never reach its diffraction limit. Instead, its limiting angle of resolution is always set by atmospheric blurring. This seeing limit is usually approximately 1 s of arc and is never smaller than approximately 0.1 s of arc. (This is one of the reasons for the current interest in a large space telescope.)

Exercise The large radio telescope at Arecibo, Puerto Rico, has a diameter of 305 m and is designed to detect 0.75-m radio waves. Calculate the minimum angle of resolution for this telescope and compare your answer with that of the Hale telescope.

Answer 3.0×10^{-3} rad (10 min of arc), more than 10 000 times larger than the Hale minimum.

EXAMPLE 38.6 Resolution of the Eye

Calculate the limiting angle of resolution for the eye assuming its resolution is limited only by diffraction. Take the wavelength to be 500 nm and assume a pupil diameter of 2.00 mm.

Solution We use Equation 38.7, taking $\lambda = 500$ nm and $D = 2.00$ mm. This gives

$$\theta_{min} = 1.22 \frac{\lambda}{D} = 1.22 \left(\frac{5.00 \times 10^{-7}\ \text{m}}{2.00 \times 10^{-3}\ \text{m}} \right)$$

$$= 3.05 \times 10^{-4}\ \text{rad} = 0.0175°$$

We can use this result to calculate the minimum separation d between two point sources that the eye can distinguish if they are a distance L from the observer (Fig. 38.15). Since θ_{min} is small, we see that

$$\sin \theta_{min} \approx \theta_{min} \approx \frac{d}{L}$$

$$d = L\theta_{min}$$

For example, if the objects are 25 cm from the eye (the near

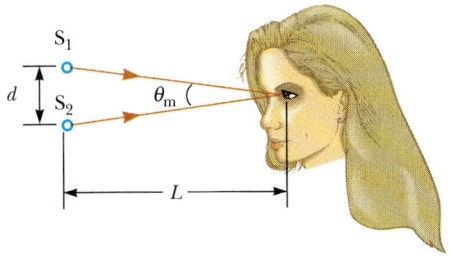

FIGURE 38.15 (Example 38.6) Two point sources separated by a distance d as observed by the eye.

point), then

$$d = (25\ \text{cm})(3.05 \times 10^{-4}\ \text{rad}) = 7.6 \times 10^{-3}\ \text{cm}$$

This is approximately equal to the thickness of a human hair.

Exercise Suppose the eye is dilated to a diameter of 5.0 mm and is looking at two point sources 3.0 cm away. How far apart must the sources be if this eye is to resolve them?

Answer 0.037 cm.

38.4 THE DIFFRACTION GRATING

The diffraction grating, a useful device for analyzing light sources, consists of a large number of equally spaced parallel slits. A grating can be made by cutting parallel lines on a glass plate with a precision ruling machine. In a transmission grating, the space between any two lines is transparent to the light and hence acts as a separate slit. Gratings with many lines very close to each other can have very small slit spacings. For example, a grating ruled with 5000 lines/cm has a slit spacing $d = (1/5000)$ cm $= 2.00 \times 10^{-4}$ cm.

A section of a diffraction grating is illustrated in Figure 38.16. A plane wave is incident from the left, normal to the plane of the grating. A converging lens brings the rays together at the point P. The pattern observed on the screen is the result of the combined effects of interference and diffraction. Each slit produces diffraction, and the diffracted beams interfere with each other to produce the final pattern. Moreover, each slit acts as a source of waves, where all waves start at the slits in phase. However, for some arbitrary direction θ measured from the horizontal, the waves must travel different path lengths before reaching P. From Figure 38.16, note that the path difference between waves from any two adjacent slits is

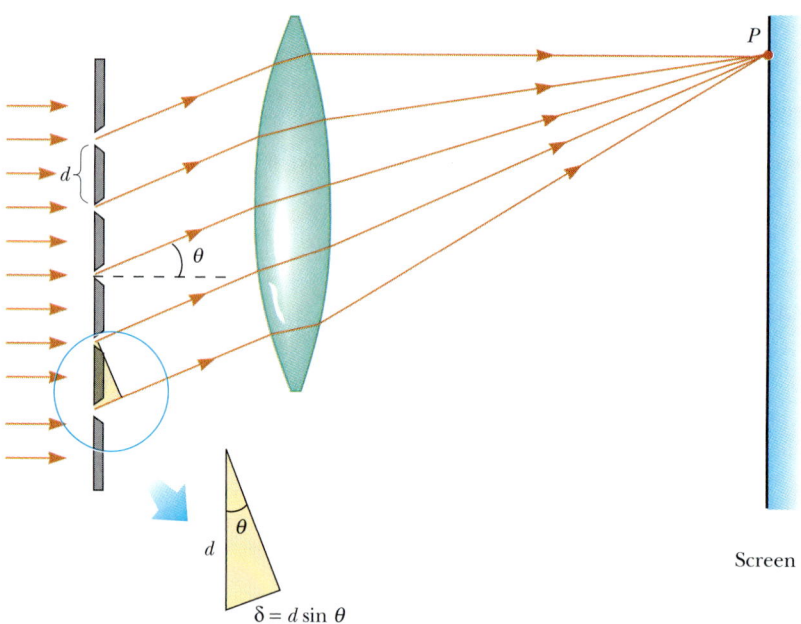

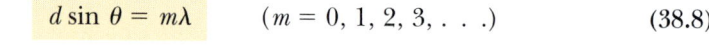

FIGURE 38.16   Side view of a diffraction grating. The slit separation is *d*, and the path difference between adjacent slits is *d* sin θ.

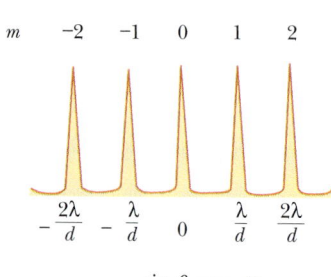

FIGURE 38.17   Intensity versus sin θ for a diffraction grating. The zeroth-, first-, and second-order maxima are shown.

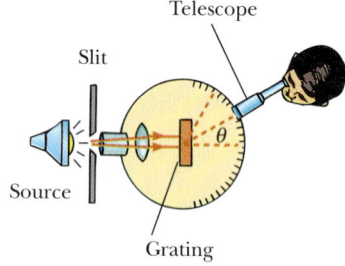

FIGURE 38.18   Diagram of a diffraction grating spectrometer. The collimated beam incident on the grating is diffracted into the various orders at the angles θ that satisfy the equation *d* sin θ = *m*λ, where *m* = 0, 1, 2,. . . . (In most spectrometers, the angle of incidence is approximately equal to the angle of diffraction, so *m*λ = *d*(sin θ + sin θ′), with θ ≈ ± θ′.)

equal to *d* sin θ. If this path difference equals one wavelength or some integral multiple of a wavelength, waves from all slits are in phase at *P* and a bright line is observed. Therefore, the condition for maxima in the interference pattern at the angle θ is

$$d \sin \theta = m\lambda \qquad (m = 0, 1, 2, 3, \ldots) \qquad (38.8)$$

This expression can be used to calculate the wavelength from a knowledge of the grating spacing and the angle of deviation θ. If the incident radiation contains several wavelengths, the *m*th order maximum for each wavelength occurs at a specific angle. All wavelengths are seen at θ = 0, corresponding to *m* = 0, the zeroth-order maximum. The first-order maximum (*m* = 1) is observed at an angle that satisfies the relationship sin θ = λ/*d*; the second-order maximum (*m* = 2) is observed at a larger angle θ, and so on.

The intensity distribution for a diffraction grating using a monochromatic source is shown in Figure 38.17. Note the sharpness of the principal maxima and the broadness of dark areas. This is in contrast to the broad bright fringes characteristic of the two-slit interference pattern (Fig. 37.1).

A simple arrangement used to measure orders of a diffraction pattern is shown in Figure 38.18. This is a form of a diffraction grating spectrometer. The light to be analyzed passes through a slit, and a parallel beam of light exits from the collimator, which is perpendicular to the grating. The diffracted light leaves the grating at angles that satisfy Equation 38.8, and a telescope is used to view the image of the slit. The wavelength can be determined by measuring the precise angles at which the images of the slit appear for the various orders.

CONCEPTUAL EXAMPLE 38.7   The Compact Disc Is a Diffraction Grating

Light reflected from the surface of a compact disc has a multicolored appearance as shown in Figure 38.19. Furthermore, the observation depends on the orientation of the disc relative to the eye and the position of the light source. Explain how this works.

Reasoning   The surface of a compact disc has a spiral grooved track (with a spacing of approximately 1 μm) that acts as a reflection grating. The light scattered by these closely spaced grooves interferes constructively only in certain directions that depend on the wavelength and on the direction of the incident light. Any one section of the disc serves as a diffraction grating for white light, sending different colors in different directions. The different colors you see when viewing one section of the disc change as the light source, the disc, or you move to change the angles of incidence or diffraction.

FIGURE 38.19   (Conceptual Example 38.7) A compact disc observed under white light. The colors that are observed in the reflected light and their intensities depend on the orientation of the disc relative to the eye and to the light source. Can you explain how this works?   *(Kristen Brochmann/Fundamental Photographs)*

EXAMPLE 38.8   The Orders of a Diffraction Grating

Monochromatic light from a helium-neon laser ($\lambda = 632.8$ nm) is incident normally on a diffraction grating containing 6000 lines/cm. Find the angles at which the first-order, second-order, and third-order maxima can be observed.

Solution   First, we must calculate the slit separation, which is equal to the inverse of the number of lines per cm:

$$d = (1/6000) \text{ cm} = 1.667 \times 10^{-4} \text{ cm} = 1667 \text{ nm}$$

For the first-order maximum ($m = 1$), we get

$$\sin \theta_1 = \frac{\lambda}{d} = \frac{632.8 \text{ nm}}{1667 \text{ nm}} = 0.3797$$

$$\theta_1 = \boxed{22.31°}$$

For $m = 2$, we find

$$\sin \theta_2 = \frac{2\lambda}{d} = \frac{2(632.8 \text{ nm})}{1667 \text{ nm}} = 0.7592$$

$$\theta_2 = \boxed{49.41°}$$

For $m = 3$, we find $\sin \theta_3 = 1.139$. Since $\sin \theta$ cannot exceed unity, this does not represent a realistic solution. Hence, only zeroth-, first-, and second-order maxima are observed for this situation.

Resolving Power of the Diffraction Grating

The diffraction grating is useful for measuring wavelengths accurately. Like the prism, the diffraction grating can be used to disperse a spectrum into its components. Of the two devices, the grating may be more precise if one wants to distinguish between two closely spaced wavelengths.

If λ_1 and λ_2 are the two nearly equal wavelengths between which the spectrom-

Resolving power

eter can just barely distinguish, the **resolving power** R is defined as

$$R \equiv \frac{\lambda}{\lambda_2 - \lambda_1} = \frac{\lambda}{\Delta\lambda} \tag{38.9}$$

where $\lambda = (\lambda_1 + \lambda_2)/2$ and $\Delta\lambda = \lambda_2 - \lambda_1$. Thus, a grating with a high resolving power can distinguish small differences in wavelength. Furthermore, if N lines of the grating are illuminated, it can be shown that the resolving power in the mth-order diffraction equals the product Nm:

Resolving power of a grating

$$R = Nm \tag{38.10}$$

Thus, resolving power increases with increasing order number. Furthermore, R is large for a grating that has a large number of illuminated slits. Note that for $m = 0$, $R = 0$, which signifies that *all wavelengths are indistinguishable* for the zeroth-order maximum. However, consider the second-order diffraction pattern ($m = 2$) of a grating that has 5000 rulings illuminated by the light source. The resolving power of such a grating in second order is $R = 5000 \times 2 = 10\,000$. Therefore, the minimum wavelength separation between two spectral lines that can be just resolved, assuming a mean wavelength of 600 nm, is $\Delta\lambda = \lambda/R = 6.00 \times 10^{-2}$ nm. For the third-order principal maximum, $R = 15\,000$ and $\Delta\lambda = 4.00 \times 10^{-2}$ nm, and so on.

EXAMPLE 38.9   Resolving the Sodium Spectral Lines

Two strong lines in the spectrum of sodium have wavelengths of 589.00 nm and 589.59 nm. (a) What must the resolving power of a grating be in order to distinguish these wavelengths?

Solution

$$R = \frac{\lambda}{\Delta\lambda} = \frac{589.30 \text{ nm}}{589.59 \text{ nm} - 589.00 \text{ nm}} = \frac{589.30}{0.59} = \boxed{999}$$

(b) In order to resolve these lines in the second-order spectrum ($m = 2$), how many lines of the grating must be illuminated?

Solution   From Equation 38.10 and the results to part (a), we find

$$N = \frac{R}{m} = \frac{999}{2} = \boxed{500 \text{ lines}}$$

*38.5   DIFFRACTION OF X-RAYS BY CRYSTALS

In principle, the wavelength of any electromagnetic wave can be determined if a grating of the proper spacing (of the order of λ) is available. X-rays, discovered by W. Roentgen (1845–1923) in 1895, are electromagnetic waves of very short wavelength (of the order of $= 0.1$ nm). Obviously, it would be impossible to construct a grating having such a small spacing. However, the atomic spacing in a solid is known to be about 10^{-10} m. In 1913, Max von Laue (1879–1960) suggested that the regular array of atoms in a crystal could act as a three-dimensional diffraction grating for x-rays. Subsequent experiments confirmed this prediction. The diffraction patterns that are observed are complicated because of the three-dimensional nature of the crystal. Nevertheless, x-ray diffraction has proved to be an invaluable technique for elucidating crystalline structures and for understanding the structure of matter.[1]

---

[1] For more details on this subject, see Sir Lawrence Bragg, "X-Ray Crystallography," *Scientific American*, July 1968.

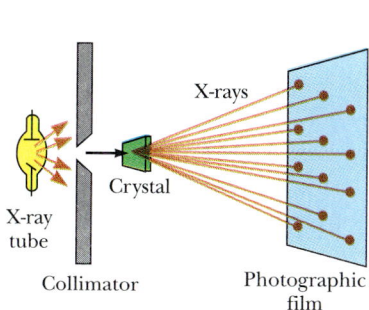

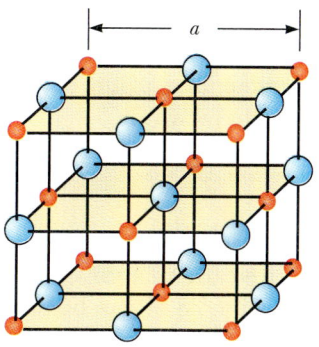

FIGURE 38.20 Schematic diagram of the technique used to observe the diffraction of x-rays by a crystal. The array of spots formed on the film is called a Laue pattern.

FIGURE 38.21 A model of the crystalline structure of sodium chloride. The blue spheres represent the Cl^- ions, and the red spheres represent the Na^+ ions. The length of the cube edge is $a = 0.562737$ nm.

Figure 38.20 is one experimental arrangement for observing x-ray diffraction from a crystal. A collimated beam of x-rays containing a continuous range of wavelengths is incident on a crystal. The diffracted beams are very intense in certain directions, corresponding to constructive interference from waves reflected from layers of atoms in the crystal. The diffracted beams can be detected by a counter or photographic film, forming an array of spots known as a *Laue pattern*. The crystalline structure is deduced by analyzing the positions and intensities of the various spots in the pattern.

The arrangement of atoms in a crystal of NaCl is shown in Figure 38.21. The smaller, red spheres represent Na^+ ions and the larger spheres represent Cl^- ions. Each unit cell (the set of atoms that repeats through the crystal) contains four Na^+ and four Cl^- ions. The unit cell is a cube whose edge length is a.

A careful examination of the NaCl structure shows that the ions lie in discrete planes, the shaded areas in Figure 38.21. Now suppose an incident x-ray beam makes an angle θ with one of the planes, as in Figure 38.22. The beam can be reflected from both the upper plane and the lower one. However, the geometric construction in Figure 38.22 shows that the beam reflected from the lower plane travels farther than the beam reflected from the upper one. The effective path difference between the two beams is $2d \sin \theta$. The two beams reinforce each other

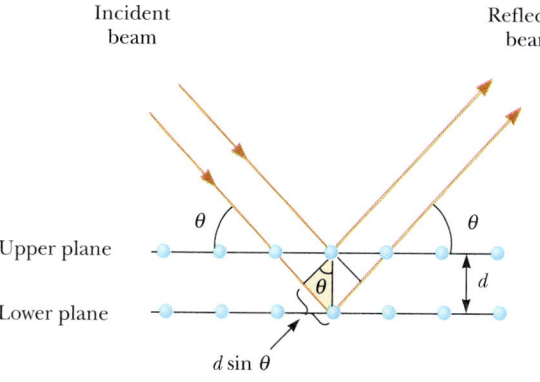

FIGURE 38.22 A two-dimensional description of the reflection of an x-ray beam from two parallel crystalline planes separated by a distance d. The beam reflected from the lower plane travels farther than the one reflected from the upper plane by a distance equal to $2d \sin \theta$.

(constructive interference) when this path difference equals some integral multiple of the wavelength λ. The same is true for reflection from the entire family of parallel planes. Hence, the condition for constructive interference (maxima in the reflected wave) is

Bragg's law

$$2d \sin \theta = m\lambda \qquad (m = 1, 2, 3, \ldots) \qquad (38.11)$$

This condition is known as **Bragg's law** after W. L. Bragg (1890–1971), who first derived the relationship. If the wavelength and diffraction angle are measured, Equation 38.11 can be used to calculate the spacing between atomic planes.

38.6 POLARIZATION OF LIGHT WAVES

In Chapter 34 we described the transverse nature of light and all other electromagnetic waves. Figure 38.23 shows that the electric and magnetic vectors associated with an electromagnetic wave are at right angles to each other and also to the direction of wave propagation. Polarization is firm evidence of the transverse nature of electromagnetic waves.

An ordinary beam of light consists of a large number of waves emitted by the atoms or molecules of the light source. Each atom produces a wave with its own orientation of E, as in Figure 38.23, corresponding to the direction of atomic vibration. The direction of polarization of the electromagnetic wave is defined to be the direction in which **E** is vibrating. However, because all directions of vibration are possible, the resultant electromagnetic wave is a superposition of waves produced by the individual atomic sources. The result is an **unpolarized** light wave, described in Figure 38.24a. The direction of wave propagation in this figure is perpendicular to the page. Note that *all* directions of the electric field vector, lying in a plane perpendicular to the direction of propagation, are equally probable. At any given point and at some instant of time, there is only one resultant electric field, hence you should not be misled by the meaning of Figure 38.24a.

A wave is said to be **linearly polarized** if **E** vibrates in the same direction *at all times* at a particular point, as in Figure 38.24b. (Sometimes such a wave is described as *plane-polarized,* or simply *polarized.*) The wave described in Figure 38.23 is an example of a wave linearly polarized in the *y* direction. As the wave propagates in the *x* direction, **E** is always in the *y* direction. The plane formed by **E** and the direction of propagation is called the *plane of polarization* of the wave. In Figure 38.23, the plane of polarization is the *xy* plane. It is possible to obtain a linearly polarized beam from an unpolarized beam by removing all waves from the beam except those whose electric field vectors oscillate in a single plane. We now discuss four processes for producing polarized light from unpolarized light.

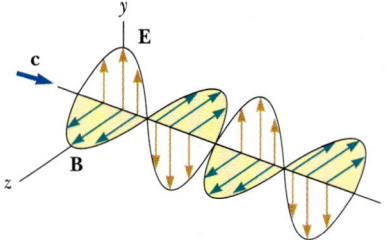

FIGURE 38.23 Schematic diagram of an electromagnetic wave propagating in the *x* direction. The electric field vector **E** vibrates in the *xy* plane, and the magnetic field vector **B** vibrates in the *xz* plane.

Polarization by Selective Absorption

The most common technique for polarizing light is to use a material that transmits waves whose electric field vectors vibrate in a given plane and absorbs waves whose electric field vectors vibrate in other directions.

In 1938, E. H. Land discovered a material, which he called **Polaroid,** that polarizes light through selective absorption by oriented molecules. This material is fabricated in thin sheets of long-chain hydrocarbons. The sheets are manufactured such that the molecules are aligned in long chains. After a sheet is dipped into a solution containing iodine, the molecules become good electrical conductors. However, the conduction takes place primarily along the hydrocarbon chains since the electrons of the molecules can move easily only along the chains. As a result, the molecules readily absorb light whose electric field vector is parallel to their length and transmit light whose electric field vector is perpendicular to their length. It is common to refer to the direction perpendicular to the molecular chains as the **transmission axis.** In an ideal polarizer, all light with **E** parallel to the transmission axis is transmitted, and all light with **E** perpendicular to the transmission axis is absorbed.

Figure 38.25 represents an unpolarized light beam incident on a first polarizing sheet, called the **polarizer,** where the transmission axis is indicated by the straight lines on the polarizer. The light passing through this sheet is polarized vertically as shown, where the transmitted electric field vector is E_0. A second polarizing sheet, called the **analyzer,** intercepts this beam because the analyzer transmission axis is set at an angle θ to the polarizer axis. The component of E_0 perpendicular to the axis of the analyzer is completely absorbed, and the component of E_0 parallel to the axis of the analyzer is $E_0 \cos \theta$. Note that when light emerges from a polarizer, it loses all information on its original plane of polarization and has the polarization of the most recent polarizer. Because the transmitted intensity varies as the square of the transmitted amplitude, we conclude that the intensity of the transmitted (polarized) light varies as

$$I = I_0 \cos^2 \theta \tag{38.12}$$

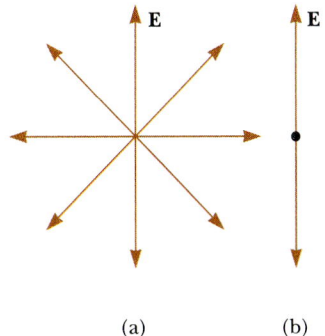

FIGURE 38.24 (a) An unpolarized light beam viewed along the direction of propagation (perpendicular to the page). The transverse electric field vector can vibrate in any direction with equal probability. (b) A linearly polarized light beam with the electric field vector vibrating in the vertical direction.

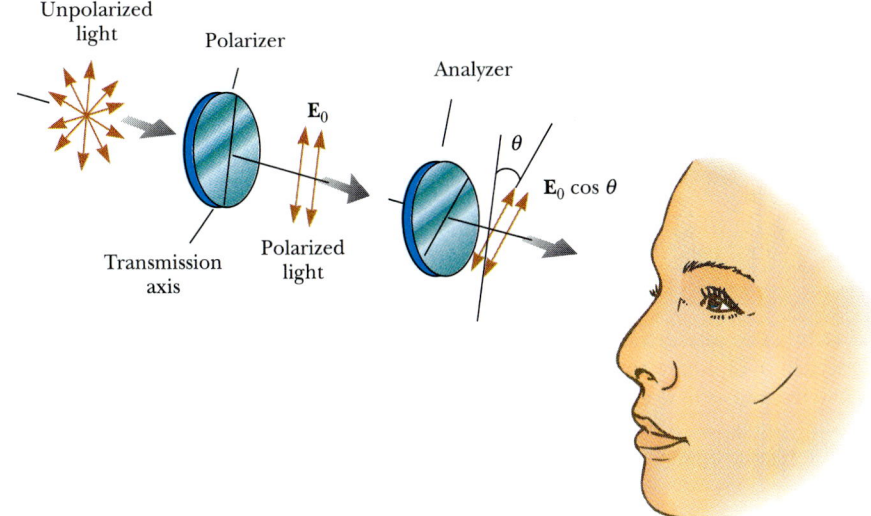

FIGURE 38.25 Two polarizing sheets whose transmission axes make an angle θ with each other. Only a fraction of the polarized light incident on the analyzer is transmitted.

where I_0 is the intensity of the polarized wave incident on the analyzer. This expression, known as **Malus's law,**[2] applies to any two polarizing materials whose transmission axes are at an angle θ to each other.

From this expression, note that the transmitted intensity is maximum when the transmission axes are parallel ($\theta = 0$ or $180°$) and zero (complete absorption by the analyzer) when the transmission axes are perpendicular to each other. This variation in transmitted intensity through a pair of polarizing sheets is illustrated in Figure 38.26.

Polarization by Reflection

When an unpolarized light beam is reflected from a surface, the reflected light is completely polarized, partially polarized, or unpolarized, depending on the angle of incidence. If the angle of incidence is either 0 or $90°$, the reflected beam is unpolarized. For intermediate angles of incidence, however, the reflected light is polarized to some extent, and for one particular angle of incidence, the reflected light is completely polarized. Let us now investigate reflection at that special angle.

Suppose an unpolarized light beam is incident on a surface as in Figure 38.27a. The beam can be described by two electric field components, one parallel to the surface (represented by the dots) and the other perpendicular both to the first component (represented by the red arrows) and to the direction of propagation. It is found that the parallel component reflects more strongly than the perpendicular component, and this results in a partially polarized reflected beam. Furthermore, the refracted beam is also partially polarized.

Now suppose the angle of incidence, θ_1, is varied until the angle between the reflected and refracted beams is $90°$ (Fig. 38.27b). At this particular angle of incidence, the reflected beam is completely polarized, with its electric field vector parallel to the surface, while the refracted beam is partially polarized. The angle of incidence at which this occurs is called the **polarizing angle,** θ_p.

The polarizing angle

(a)

(b)

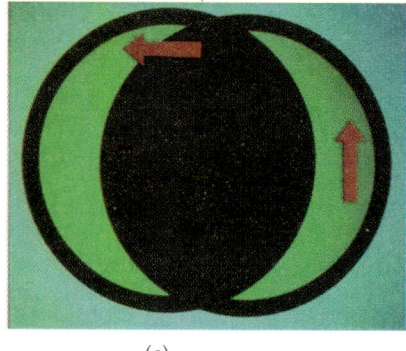
(c)

FIGURE 38.26   The intensity of light transmitted through two polarizers depends on the relative orientation of their transmission axes. (a) The transmitted light has maximum intensity when the transmission axes are aligned with each other. (b) The transmitted light intensity diminishes when the transmission axes are at an angle of $45°$ with each other. (c) The transmitted light intensity is a minimum when the transmission axes are at right angles to each other.   *(Henry Leap and Jim Lehman)*

---

[2] Named after its discoverer, E. L. Malus (1775–1812). Actually, Malus first discovered that reflected light was polarized by viewing it through a calcite ($CaCO_3$) crystal.

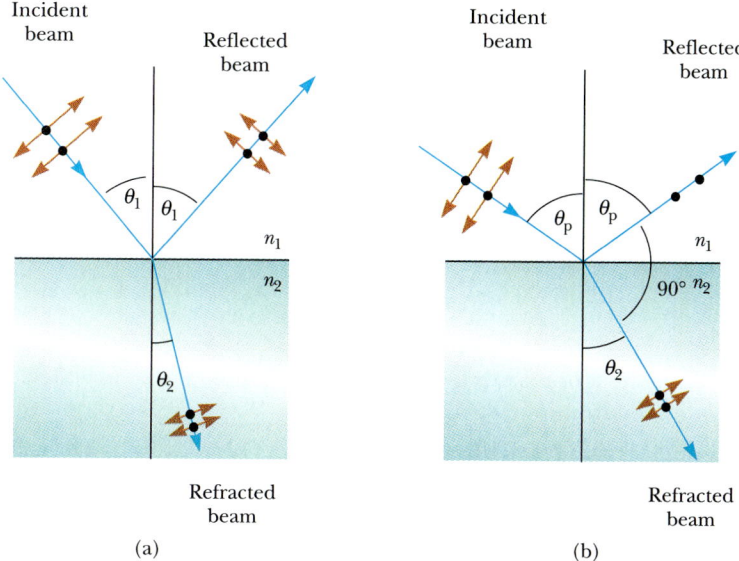

FIGURE 38.27 (a) When unpolarized light is incident on a reflecting surface, the reflected and refracted beams are partially polarized. (b) The reflected beam is completely polarized when the angle of incidence equals the polarizing angle, θ_p, which satisfies the equation $n = \tan \theta_p$.

An expression relating the polarizing angle to the index of refraction of the reflecting substance can be obtained from Figure 38.27b. From this figure, we see that $\theta_p + 90° + \theta_2 = 180°$, so that $\theta_2 = 90° - \theta_p$. Using Snell's law and taking $n_1 = 1.00$ and $n_2 = n$, we have

$$n = \frac{\sin \theta_1}{\sin \theta_2} = \frac{\sin \theta_p}{\sin \theta_2}$$

Because $\sin \theta_2 = \sin(90° - \theta_p) = \cos \theta_p$, the expression for n can be written as $n = \sin \theta_p / \cos \theta_p$, or

$$n = \tan \theta_p \qquad (38.13)$$

Brewster's law

This expression is called **Brewster's law,** and the polarizing angle θ_p is sometimes called **Brewster's angle,** after its discoverer, Sir David Brewster (1781–1868). For example, the Brewster's angle for crown glass ($n = 1.52$) is $\theta_p = \tan^{-1}(1.52) = 56.7°$. Because n varies with wavelength for a given substance, the Brewster's angle is also a function of wavelength.

Polarization by reflection is a common phenomenon. Sunlight reflected from water, glass, and snow is partially polarized. If the surface is horizontal, the electric field vector of the reflected light has a strong horizontal component. Sunglasses made of polarizing material reduce the glare of reflected light. The transmission axes of the lenses are oriented vertically so as to absorb the strong horizontal component of the reflected light.

Polarization by Double Refraction

Solids can be classified on the basis of internal structure. Those in which the atoms are arranged in a specific order are called *crystalline;* the sodium chloride structure

Unpolarized
light

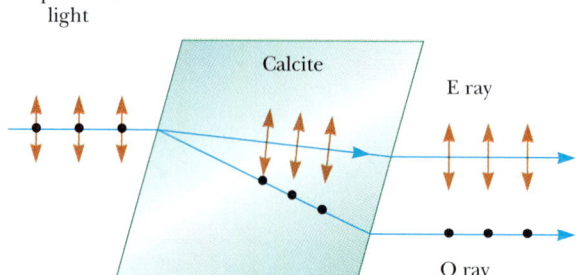

FIGURE 38.28 Unpolarized light incident on a calcite crystal splits into an ordinary (O) ray and an extraordinary (E) ray. These two rays are polarized in mutually perpendicular directions. (Note that this is not to scale.)

of Figure 38.21 is just one example of crystalline solids. Those in which the atoms are distributed randomly are called *amorphous*. When light travels through an amorphous material, such as glass, it travels with a speed that is the same in all directions. That is, glass has a single index of refraction. In certain crystalline materials, however, such as calcite and quartz, the speed of light is not the same in all directions. Such materials are characterized by two indices of refraction. Hence, they are often referred to as **double-refracting** or **birefringent** materials.

When unpolarized light enters a calcite crystal, it splits into two plane-polarized rays that travel with different speeds, corresponding to two angles of refraction, as in Figure 38.28. The two rays are polarized in two mutually perpendicular directions, as indicated by the dots and arrows. One ray, called the **ordinary (O) ray,** is characterized by an index of refraction, n_O, that is the same in all directions. This means that if one could place a point source of light inside the crystal, as in Figure 38.29, the ordinary waves would spread out from the source as spheres.

The second plane-polarized ray, called the **extraordinary (E) ray,** travels with different speeds in different directions and hence is characterized by an index of refraction, n_E, that varies with the direction of propagation. A point source of light inside such a crystal would send out an extraordinary wave having wave fronts that are elliptical in cross-section (Fig. 38.29). Note from Figure 38.29 that there is one direction, called the **optic axis,** along which the ordinary and extraordinary rays have the same speed, corresponding to the direction for which $n_O = n_E$. The difference in speed for the two rays is a maximum in the direction perpendicular to the optic axis. For example, in calcite, $n_O = 1.658$ at a wavelength of 589.3 nm and n_E varies from 1.658 along the optic axis to 1.486 perpendicular to the optic axis. Values for n_O and n_E for various double-refracting crystals are given in Table 38.1.

If a piece of calcite is placed on a sheet of paper and then we look through the crystal at any writing on the paper, two images are seen, as shown in Figure 38.30.

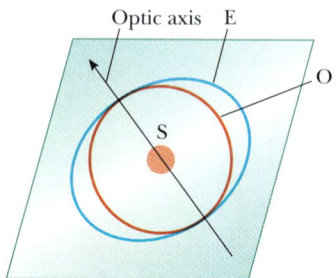

FIGURE 38.29 A point source, S, inside a double-refracting crystal produces a spherical wave front corresponding to the ordinary ray and an elliptical wavefront corresponding to the extraordinary ray. The two waves propagate with the same velocity along the optic axis.

FIGURE 38.30 A calcite crystal produces a double image because it is a birefringent (double-refracting) material. *(Henry Leap and Jim Lehman)*

TABLE 38.1	Indices of Refraction for Some Double-Refracting Crystals at a Wavelength of 589.3 nm		
Crystal	n_O	n_E	n_O/n_E
Calcite ($CaCO_3$)	1.658	1.486	1.116
Quartz (SiO_2)	1.544	1.553	0.994
Sodium nitrate ($NaNO_3$)	1.587	1.336	1.188
Sodium sulfite ($NaSO_3$)	1.565	1.515	1.033
Zinc chloride ($ZnCl_2$)	1.687	1.713	0.985
Zinc sulfide (ZnS)	2.356	2.378	0.991

As can be seen from Figure 38.28, these two images correspond to one formed by the ordinary ray and the second formed by the extraordinary ray. If the two images are viewed through a sheet of rotating polarizing glass, they alternately appear and disappear because the ordinary and extraordinary rays are plane-polarized along mutually perpendicular directions.

Polarization by Scattering

When light is incident on any material, the electrons in the material can absorb and reradiate part of the light. Such absorption and reradiation of light by electrons in the gas molecules that make up air are what causes sunlight reaching an observer on the Earth to be partially polarized. You can observe this effect—called **scattering**—by looking directly up through a pair of sunglasses whose lenses are made of polarizing material. Less light passes through at certain orientations of the lenses than at others.

Figure 38.31 illustrates how sunlight becomes partially polarized. An incident unpolarized beam of sunlight is traveling in the horizontal direction and on the verge of striking a molecule of one of the gases that make up air. This beam striking the gas molecule sets the electrons of the molecule into vibration. These vibrating charges act like the vibrating charges in an antenna, except that these charges are vibrating in a complicated pattern. The horizontal part of the electric field vector in the incident wave causes the charges to vibrate horizontally, and the vertical part of the vector simultaneously causes them to vibrate vertically. A horizontally polarized wave is emitted by the electrons as a result of their horizontal motion, and a vertically polarized wave is emitted parallel to the Earth as a result of their vertical motion.

Some phenomena involving the scattering of light in the atmosphere can be understood as follows: When light of various wavelengths is incident on gas molecules of diameter d, where $d \ll \lambda$, the relative intensity of the scattered light varies as $1/\lambda^4$. This condition is satisfied for scattering of sunlight in the Earth's atmosphere. Hence shorter wavelengths (blue light) are scattered more efficiently than longer wavelengths (red light). When sunlight is scattered by gas molecules in the air, therefore, the shorter wavelength radiation (blue part) is scattered more intensely than the longer wavelength radiation (red part), and as a result the sky appears blue.

Optical Activity

Many important applications of polarized light involve materials that display **optical activity**. A substance is said to be optically active if it rotates the plane of polarization of transmitted light. The angle through which the light is rotated by a specific material depends on the length of the sample and on the concentration if the substance is in solution. One optically active material is a solution of the common sugar dextrose. A standard method for determining the concentration of sugar solutions is to measure the rotation produced by a fixed length of the solution.

A material is optically active because of an asymmetry in the shape of its constituent molecules. For example, some proteins are optically active because of their spiral shape. Other materials, such as glass and plastic, become optically active when stressed. Suppose an unstressed piece of plastic is placed between a polarizer and an analyzer so that light passes from polarizer to plastic to analyzer. When the analyzer axis is perpendicular to the polarizer axis, none of the polarized light passes through the analyzer. In other words, the unstressed plastic has no

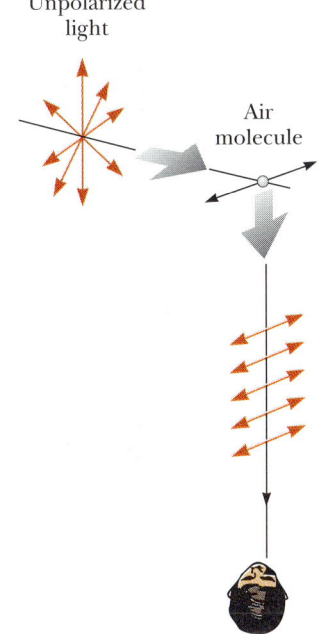

FIGURE 38.31   The scattering of unpolarized sunlight by air molecules. The light observed at right angles is plane-polarized because the vibrating molecule has a horizontal component of vibration.

Unpolarized light

Air molecule

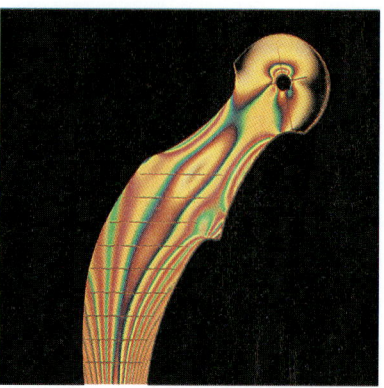 

FIGURE 38.32   (a) Photograph showing strain distribution in a plastic model of a hip replacement used in a medical research laboratory. The pattern is produced when the plastic model is viewed between a polarizer and an analyzer oriented perpendicularly to the polarizer. *(Sepp Seitz, 1981)* (b) A plastic model of an arch structure under load conditions observed between two crossed polarizers. Such patterns are useful in the optimum design of architectural components.   *(Peter Aprahamian/Science Photo Library)*

effect on the light passing through it. If the plastic is stressed, however, the regions of greatest stress rotate the polarized light through the largest angles. Hence, a series of bright and dark bands is observed in the transmitted light, with the bright bands corresponding to regions of greatest stress.

Engineers often use this technique, called *optical stress analysis,* to assist in designing structures ranging from bridges to small tools. A plastic model is built and analyzed under different load conditions to determine regions of potential weakness and failure under stress. Some examples of plastic models under stress are shown in Figure 38.32.

SUMMARY

Diffraction arises from interference in a very large number of coherent sources. Diffraction accounts for the deviation of light from a straight-line path when the light passes through an aperture or around obstacles.

The **Fraunhofer diffraction pattern** produced by a single slit of width a on a distant screen consists of a central bright maximum and alternating bright and dark regions of much lower intensities. The angles θ at which the diffraction pattern has zero intensity, corresponding to destructive interference, are given by

$$\sin \theta = m \frac{\lambda}{a} \qquad (m = \pm 1, \pm 2, \pm 3, \ldots) \qquad (38.1)$$

How the intensity I of a single-slit diffraction pattern varies with angle θ is given by

$$I_\theta = I_0 \left[\frac{\sin (\beta/2)}{\beta/2} \right]^2 \qquad (38.4)$$

where $\beta = 2\pi a \sin \theta / \lambda$ and I_0 is the intensity at $\theta = 0$.

Rayleigh's criterion, which is a limiting condition of resolution, says that two images formed by an aperture are just distinguishable if the central maximum of the diffraction pattern for one image falls on the first minimum of the diffraction pattern for the other image. The limiting angle of resolution for a slit of width a is $\theta_{min} = \lambda / a$, and the limiting angle of resolution for a circular aperture of diameter D is given by $\theta_{min} = 1.22\lambda / D$.

A **diffraction grating** consists of a large number of equally spaced, identical slits. The condition for intensity maxima in the interference pattern of a diffraction grating for normal incidence is

$$d \sin \theta = m\lambda \qquad (m = 0, 1, 2, 3, \ldots) \qquad (38.8)$$

where d is the spacing between adjacent slits and m is the order number of the diffraction pattern. The resolving power of a diffraction grating in the mth order of the diffraction pattern is

$$R = Nm \qquad (38.10)$$

where N is the number of rulings in the grating.

When polarized light of intensity I_0 is incident on a polarizing film, the light transmitted through the film has an intensity equal to $I_0 \cos^2 \theta$, where θ is the angle between the transmission axis of the polarizer and the electric field vector of the incident light.

In general, reflected light is partially polarized. However, the reflected light is completely polarized when the angle of incidence is such that the angle between the reflected and refracted beams is $90°$. This angle of incidence, called the **polarizing angle** θ_p, satisfies **Brewster's law**:

$$n = \tan \theta_p \qquad (38.13)$$

where n is the index of refraction of the reflecting medium.

QUESTIONS

1. What is the difference between Fraunhofer and Fresnel diffraction?

2. Although we can hear around corners, we cannot see around corners. How can you explain this in view of the fact that sound and light are both waves?

3. Describe the change in width of the central maximum of the single-slit diffraction pattern as the width of the slit is made narrower.

4. Assuming that the headlights of a car are point sources, estimate the maximum distance from an observer to the car at which the headlights are distinguishable from each other.

5. A laser beam is incident at a shallow angle on a machinist's ruler that has a finely calibrated scale. The rulings on the scale give rise to a diffraction pattern on a screen. Discuss how you can use this technique to obtain a measure of the wavelength of the laser light.

6. Certain sunglasses use a polarizing material to reduce the intensity of light reflected from shiny surfaces. What orientation of polarization should the material have to be most effective?

7. The path of a light beam can be made visible by placing dust in the air (perhaps by shaking a blackboard eraser in the path of the light beam). Explain why the beam can be seen under these circumstances.

8. Is light from the sky polarized? Why is it that clouds seen through Polaroid glasses stand out in bold contrast to the sky?

9. If a coin is glued to a glass sheet and this arrangement is held in front of a laser beam, the projected shadow has diffraction rings around its edge and a bright spot in the center. How is this possible?

10. If a fine wire is stretched across the path of a laser beam, is it possible to produce a diffraction pattern?

11. How could the index of refraction of a flat piece of dark obsidian glass be determined?

PROBLEMS

Review Problem

A wide screen is placed 1.00 m from a single slit of width 0.0200 mm. A beam of white light ($\lambda = 400 - 750$ nm is incident on the slit. Find (a) the number of spectra formed on the screen (*Hint*: The order number is a maximum for $\theta = 90°$), (b) the location of the second-order maximum at 750 nm, (c) the width of the second-order maximum at 400 nm, (d) the fractional light intensity (I/I_0) at 5.00 mm from the center when a 500-nm transmission filter is placed between the beam and the slit, (e) the resolution of the slit for 600-nm light, (f) the minimum separation between two candles (1.00 km from the slit) in order to be resolved by the slit, (g) the maximum resolving power (at 750 nm) of a diffraction grating that is 4.00 cm wide and has a line spacing of 0.0200 mm.

Section 38.2   Single-Slit Diffraction

1. Light from a He-Ne laser ($\lambda = 632.8$ nm) is incident on a single slit. What is the minimum width for which no diffraction minima are observed?

2. A Fraunhofer diffraction pattern is produced on a screen 140 cm from a single slit. The distance from the center of the central maximum to the first-order maximum is $1.00 \times 10^4 \lambda$. Calculate the slit width.

3. The second-order bright fringe in a single-slit diffraction pattern is 1.4 mm from the center of the central maximum. The screen is 80 cm from a slit of width 0.80 mm. Assuming monochromatic incident light, calculate the wavelength.

4. Helium-neon laser light ($\lambda = 632.8$ nm) is sent through a 0.300-mm-wide single slit. What is the width of the central maximum on a screen 1.00 m from the slit?

5. The pupil of a cat's eye narrows to a slit of width 0.50 mm in daylight. What is the angular resolution? (Use 500-nm light in your calculation.)

6. Light of wavelength 587.5 nm illuminates a single slit 0.75 mm in width. (a) At what distance from the slit should a screen be located if the first minimum in the diffraction pattern is to be 0.85 mm from the center of the screen? (b) What is the width of the central maximum?

7. A screen is placed 50 cm from a single slit, which is illuminated with 690-nm light. If the distance between the first and third minima in the diffraction pattern is 3.0 mm, what is the width of the slit?

8. In Equation 38.4, let $\beta/2 \equiv \phi$ and show that $I = 0.5I_0$ when $\sin \phi = \phi/\sqrt{2}$.

9. The equation $\sin \phi = \phi/\sqrt{2}$ found in Problem 8 is

known as a *transcendental equation,* which can be solved graphically. To illustrate this, let $\phi = \beta/2$, $y_1 = \sin \phi$, and $y_2 = \phi/\sqrt{2}$. Plot y_1 and y_2 on the same set of axes over a range from $\phi = 1$ rad to $\phi = \pi/2$ rad. Determine ϕ from the point of intersection of the two curves.

10. A beam of green light is diffracted by a slit of width 0.55 mm. The diffraction pattern forms on a wall 2.06 m beyond the slit. The distance between the positions of zero intensity ($m = \pm 1$) is 4.1 mm. Estimate the wavelength of the laser light.

11. A diffraction pattern is formed on a screen 120 cm away from a 0.40-mm-wide slit. Monochromatic 546.1-nm light is used. Calculate the fractional intensity I/I_0 at a point on the screen 4.1 mm from the center of the principal maximum.

12. If the light in Figure 38.6 strikes the single slit at an angle of β from the perpendicular direction, show that Equation 38.1, the condition for destructive interference, must be modified to read

$$\sin \theta = m\left(\frac{\lambda}{a}\right) - \sin \beta$$

Section 38.3   Resolution of Single-Slit and Circular Apertures

13. A helium-neon laser emits light that has a wavelength of 632.8 nm. The circular aperture through which the beam emerges has a diameter of 0.50 cm. Estimate the diameter of the beam 10.0 km from the laser.

14. The Moon is approximately 400 000 km from the Earth. Can two lunar craters 50 km apart be resolved by a telescope on the Earth if the telescope mirror has a diameter of 15 cm? Can craters 1.0 km apart be resolved? Take the wavelength to be 700 nm, and justify your answers with approximate calculations.

15. On the night of April 18, 1775, a signal was to be sent from the Old North Church steeple to Paul Revere, who was 1.8 miles away: "One if by land, two if by sea." At what minimum separation did the sexton have to set the lanterns so that Paul Revere could receive the correct message? Assume that Paul Revere's pupils had a diameter of 4.00 mm at night, and that the candlelight had a predominant wavelength of 580 nm.

16. If we were to send a ruby laser beam ($\lambda = 694.3$ nm) outward from the barrel of a 2.7-m-diameter telescope, what would be the diameter of the big red spot when the beam hit the Moon 384 000 km away? (Neglect atmospheric dispersion.)

☐ indicates problems that have full solutions available in the Student Solutions Manual and Study Guide.

17. Suppose you are standing on a straight highway and watching a car moving away from you at 20.0 m/s. The air is perfectly clear, and after 2.00 min you see only one taillight. If the diameter of your pupil is 7.00 mm, estimate the width of the car.

17A. Suppose you are standing on a straight highway and watching a car moving away from you at a speed v. The air is perfectly clear, and after a time t you see only one taillight. If the diameter of your pupil is d, estimate the width of the car.

18. Find the radius of a star image formed on the retina of the eye if the aperture diameter (the pupil) at night is 0.70 cm, and the length of the eye is 3.0 cm. Assume the wavelength of starlight in the eye is 500 nm.

19. At what distance could one theoretically distinguish two automobile headlights separated by 1.4 m? Assume a pupil diameter of 6.0 mm and yellow headlights ($\lambda = 580$ nm).

20. A binary star system in the constellation Orion has an angular separation between the two stars of 1.0×10^{-5} rad. If $\lambda = 500$ nm, what is the smallest diameter the telescope can have and just resolve the two stars?

21. The Impressionist painter Georges Seurat created paintings with an enormous number of dots of pure pigment approximately 2.0 mm in diameter. The idea was to have colors such as red and green next to each other to form a scintillating canvas. Outside what distance would one be unable to discern individual dots on the canvas? (Assume $\lambda = 500$ nm and a pupil diameter of 4.0 mm.)

22. The angular resolution of a radio telescope is to be 0.10° when the incident waves have a wavelength of 3.0 mm. Approximately what minimum diameter is required for the telescope's receiving dish?

23. A circular radar antenna on a navy ship has a diameter of 2.1 m and radiates at a frequency of 15 GHz. Two small boats are located 9.0 km away from the ship. How close together could the boats be and still be detected as two objects?

24. When Mars is nearest the Earth, the distance separating the two planets is 88.6×10^6 km. Mars is viewed through a telescope whose mirror has a diameter of 30 cm. (a) If the wavelength of the light is 590 nm, what is the angular resolution of the telescope? (b) What is the smallest distance that can be resolved between two points on Mars?

Section 38.4   The Diffraction Grating

25. Show that, whenever a continuous visible spectrum is passed through a diffraction grating of any spacing size, the third-order fringe of the violet end of the

visible spectrum always overlaps the second-order fringe of the red light at the other end.

26. A beam of 541-nm light is incident on a diffraction grating that has 400 lines/mm. (a) Determine the angle of the second-order ray. (b) If the entire apparatus is immersed in water, determine the new second-order angle of diffraction. (c) Show that the two diffracted rays of parts (a) and (b) are related through the law of refraction.

27. A grating with 250 lines/mm is used with an incandescent light source. Assume the visible spectrum to range in wavelength from 400 to 700 nm. In how many orders can one see (a) the entire visible spectrum and (b) the short-wavelength region?

28. Two wavelengths λ and $\lambda + \Delta\lambda$ ($\Delta\lambda \ll \lambda$) are incident on a diffraction grating. Show that the angular separation between the mth-order spectra is

$$\Delta\theta = \frac{\Delta\lambda}{\sqrt{(d/m)^2 - \lambda^2}}$$

where d is the grating constant and m is the order number.

29. The full width of a 3.00-cm-wide grating is illuminated by a sodium discharge tube. The lines in the grating are uniformly spaced at 775 nm. Calculate the angular separation in the first-order spectrum between the two wavelengths forming the sodium doublet ($\lambda_1 = 589.0$ nm and $\lambda_2 = 589.6$ nm).

30. Light from an argon laser strikes a diffraction grating that has 5310 lines per centimeter. The central and first-order principal maxima are separated by 0.488 m on a wall 1.72 m from the grating. Determine the wavelength of the laser light.

31. A source emits 531.62-nm and 531.81-nm light. (a) What minimum number of lines is required for a grating that resolves the two wavelengths in the first-order spectrum? (b) Determine the slit spacing for a grating 1.32 cm wide that has the required minimum number of lines.

32. A diffraction grating has 800 rulings per millimeter. A beam of light containing wavelengths from 500 to 700 nm hits the grating. Do the spectra of different orders overlap? Explain.

33. A diffraction grating has 4200 rulings per centimeter. On a screen 2.0 m from the grating, it is found that, for a particular order m, the maxima corresponding to two closely spaced wavelengths of sodium (589.0 and 589.6 nm) are separated by 1.5 mm. Determine the value of m.

34. A helium-neon laser ($\lambda = 632.8$ nm) is used to calibrate a diffraction grating. If the first-order maximum occurs at 20.5°, what is the line spacing, d?

35. White light is spread out into its spectral components by a diffraction grating. If the grating has 2000 lines

per centimeter, at what angle does red light ($\lambda =$ 640 nm) appear in first order?

36. Two spectral lines in a mixture of hydrogen (H_2) and deuterium (D_2) gas have wavelengths of 656.30 nm and 656.48 nm, respectively. What is the minimum number of lines a diffraction grating must have to resolve these two wavelengths in first order?

37. Monochromatic 632.8-nm light is incident on a diffraction grating containing 4000 lines per centimeter. Determine the angle of the first-order maximum.

*Section 38.5   Diffraction of X-Rays by Crystals

38. Potassium iodide (KI) has the same crystalline structure as NaCl, with $d = 0.353$ nm. A monochromatic x-ray beam shows a diffraction maximum when the grazing angle is 7.6°. Calculate the x-ray wavelength. (Assume first order.)

39. Monochromatic x-rays of the K_α line of potassium from a nickel target ($\lambda = 0.166$ nm) are incident on a KCl crystal surface. The interplanar distance in KCl is 0.314 nm. At what angle (relative to the surface) should the beam be directed so that a second-order maximum is observed?

40. A monochromatic x-ray beam is incident on a NaCl crystal surface that has an interplanar spacing of 0.281 nm. The second-order maximum in the reflected beam is found when the angle between incident beam and surface is 20.5°. Determine the wavelength of the x-rays.

41. Certain cubic crystals (such as NaCl) do not diffract the odd orders ($m = 1, 3, 5, \ldots$) because of destructive interference from the unit cell planes. For such a crystal having interplanar spacing of 0.2810 nm and illuminated with 0.0956-nm x-rays, find (a) the angles at which constructive interference occurs and (b) the radii of the concentric circles such patterns make on a screen 10.0 cm from the crystal.

42. The first-order diffraction is observed at $+ 12.6°$ for a crystal in which the interplanar spacing is 0.240 nm. How many other orders can be observed?

43. X-rays of wavelength 0.140 nm are reflected from a NaCl crystal, and the first-order maximum occurs at an angle of 14.4°. What value does this give for the interplanar spacing of NaCl?

44. A wavelength of 0.129 nm characterizes K_β x-rays from zinc. When a beam of these x-rays is incident on the surface of a crystal whose structure is similar to that of NaCl, a first-order maximum is observed at 8.15°. Calculate the interplanar spacing based on this information.

45. If the interplanar spacing of NaCl is 0.281 nm, what is the predicted angle at which 0.140-nm x-rays are diffracted in a first-order maximum?

46. In an x-ray diffraction experiment using x-rays of $\lambda = 0.500 \times 10^{-10}$ m, a first-order maximum occurred at 5.0°. Find the crystal plane spacing.

Section 38.6   Polarization of Light Waves

47. The angle of incidence of a light beam onto a reflecting surface is continuously variable. The reflected ray is found to be completely polarized when the angle of incidence is 48°. What is the index of refraction of the reflecting material?

48. Light is reflected from a smooth ice surface, and the reflected ray is completely polarized. Determine the angle of incidence. ($n = 1.309$ for ice)

49. A light beam is incident on heavy flint glass ($n = 1.65$) at the polarizing angle. Calculate the angle of refraction for the transmitted ray.

50. How far above the horizon is the Moon when its image reflected in calm water is completely polarized? ($n_{\text{water}} = 1.33$)

51. Consider a polarizer-analyzer arrangement as shown in Figure 38.25. At what angle is the axis of the analyzer to the axis of the polarizer if the original beam intensity is reduced by (a) 10, (b) 50, and (c) 90 percent after passing through both sheets?

52. An unpolarized light beam in water reflects off a glass plate ($n = 1.570$). At what angle should the beam be incident on the plate to be totally polarized upon reflection?

53. Plane-polarized light is incident on a single polarizing disk with the direction of E_0 parallel to the direction of the transmission axis. Through what angle should the disk be rotated so that the intensity in the transmitted beam is reduced by a factor of (a) 3, (b) 5, (c) 10?

54. Three polarizing disks whose planes are parallel are centered on a common axis. The direction of the transmission axis in each case is shown in Figure P38.54 relative to the common vertical direction. A plane-polarized beam of light with E_0 parallel to the vertical reference direction is incident from the left on the first disk with intensity $I_i = 10$ units (arbitrary). Calculate the transmitted intensity I_f when (a) $\theta_1 = 20°$, $\theta_2 = 40°$, and $\theta_3 = 60°$; (b) $\theta_1 = 0°$, $\theta_2 = 30°$, and $\theta_3 = 60°$.

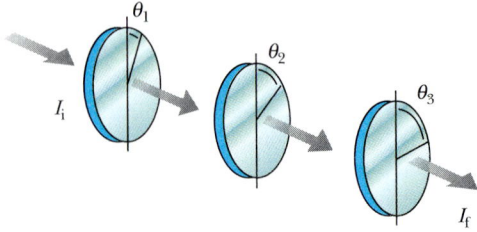

FIGURE P38.54

55. The critical angle for sapphire surrounded by air is 34.4°. Calculate the polarizing angle for sapphire.

56. For a particular transparent medium surrounded by air, show that the critical angle for internal reflection and the polarizing angle are related by $\cot \theta_p = \sin \theta_c$.

57. If the polarizing angle for cubic zirconia (ZrO₂) is 65.6°, what is the index of refraction for this material?

ADDITIONAL PROBLEMS

58. A solar eclipse is projected through a pinhole of diameter 0.50 mm and strikes a screen 2.0 meters away. (a) What is the diameter of the image? (b) What is the radius of the first diffraction minimum? Both the Sun and Moon have angular diameters of very nearly 0.50°. (Assume $\lambda = 550$ nm.)

59. The hydrogen spectrum has a red line at 656 nm and a blue line at 434 nm. What is the angular separation between two spectral lines obtained with a diffraction grating that has 4500 lines/cm?

60. What are the approximate dimensions of the smallest object on Earth that astronauts can resolve by eye when they are orbiting 250 km above the Earth? Assume $\lambda = 500$ nm light in the eye and a pupil diameter of 5.0 mm.

61. Grote Reber was a pioneer in radio astronomy. He constructed a radio telescope with a 10-m diameter receiving dish. What was the telescope's angular resolution for 2.0-m radio waves?

62. An unpolarized beam of light is reflected from a glass surface, and the reflected beam is linearly polarized when the light is incident from air at 58.6°. What is the refractive index of the glass?

63. A 750-nm light beam hits the flat surface of a certain liquid, and the beam is split into a reflected ray and a refracted ray. If the reflected rays are completely polarized at grazing angle 36°, what is the wavelength of the refracted ray?

64. Light strikes a water surface at the polarizing angle. The part of the beam refracted into the water strikes a submerged glass slab (index of refraction = 1.50) as in Figure P38.64. If the light reflected from the upper surface of the slab is completely polarized, find the angle between water surface and glass slab.

65. A diffraction grating of length 4.00 cm contains 6000 rulings over a width of 2.00 cm. (a) What is the resolving power of this grating in the first three orders? (b) If two monochromatic waves incident on this grating have a mean wavelength of 400 nm, what is their wavelength separation if they are just resolved in the third order?

66. An American standard television picture is composed of approximately 485 horizontal lines of varying light intensity. Assume that your ability to resolve the lines is limited only by the Rayleigh criterion and that the pupils of your eyes are 5.0 mm in diameter. Calculate the ratio of minimum viewing distance to the vertical dimension of the picture such that you will not be able to resolve the lines. Assume that the average wavelength of the light coming from the screen is 550 nm.

67. Light of wavelength 500 nm is incident normally on a diffraction grating. If the third-order maximum of the diffraction pattern is observed at 32°, (a) what is the number of rulings per centimeter for the grating? (b) Determine the total number of primary maxima that can be observed in this situation.

68. (a) If light traveling in a medium for which the index of refraction is n_1 is incident at an angle θ on the surface of a medium of index n_2 so that the angle between the reflected and refracted rays is β, show that

$$\tan \theta = \frac{n_2 \sin \beta}{n_1 - n_2 \cos \beta}$$

(*Hint:* Use the identity $\sin(A + B) = \sin A \cos B + \cos A \sin B$.) (b) Show that this expression for $\tan \theta$ reduces to Brewster's law when $\beta = 90°$, $n_1 = 1$, and $n_2 = n$.

69. Two polarizing sheets are placed together with their transmission axes crossed so that no light is transmitted. A third sheet is inserted between them with its transmission axis at an angle of 45° with respect to each of the other axes. Find the fraction of incident unpolarized light intensity transmitted by the three-sheet combination. (Assume each polarizing sheet is ideal.)

70. Figure P38.70a is a three-dimensional sketch of a birefringent crystal. The dotted lines illustrate how a thin parallel-faced slab of material could be cut from the larger specimen with the optic axis of the crystal parallel to the faces of the plate. A section cut from the crystal in this manner is known as a *retardation plate*. When a beam of light is incident on the plate perpendicular to the direction of the optic axis, as shown in Figure P38.70b, the O ray and the E ray travel along a straight line, but with different speeds. (a) Let the thickness of the plate be d and show that

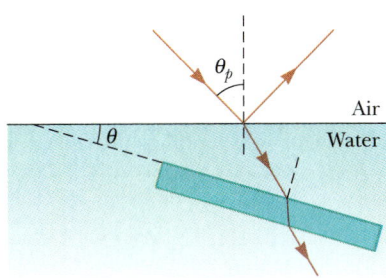

FIGURE P38.64

the phase difference between the O ray and the E ray is

$$\theta = \frac{2\pi d}{\lambda_0} \, |n_O - n_E|$$

where λ_0 is the wavelength in air. (Recall that the optical path length in a material is the product of geometric path and index of refraction.) (b) If in a particular case the incident light has a wavelength of 550 nm, find the minimum value of d for a quartz plate for which $\theta = \pi/2$. Such a plate is called a quarter-wave plate. (Use values of n_O and n_E from Table 38.1.)

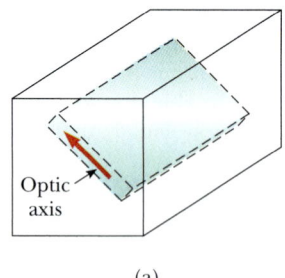

(a)

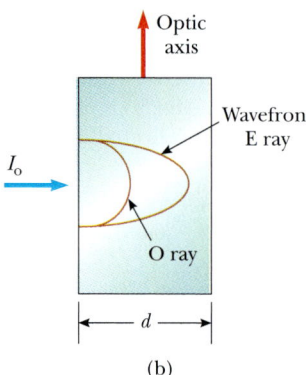

(b)

FIGURE P38.70

71. You want to rotate the plane of polarization of a polarized light beam by 45° for a maximum intensity reduction of 10 percent. (a) How many sheets of perfect polarizers do you need to use to achieve your goal? (b) What is the angle between the adjacent polarizers?

72. In Figure P38.54, suppose that the left and right polarizing disks have their transmission axes perpendicular to each other. Also, let the center disk be rotated on the common axis with an angular speed ω. Show that if unpolarized light is incident on the left disk with an intensity I_0, the intensity of the beam emerging from the right disk is

$$I = \frac{1}{16} I_0 (1 - \cos 4\omega t)$$

This means that the intensity of the emerging beam

is modulated at a rate that is four times the rate of rotation of the center disk. [*Hint:* Use the trigonometric identities $\cos^2 \theta = (1 + \cos 2\theta)/2$ and $\sin^2 \theta = (1 - \cos 2\theta)/2$, and recall that $\theta = \omega t$.]

73. Suppose that the single slit in Figure 38.7 is 6.0 cm wide and in front of a microwave source operating at 7.5 GHz. (a) Calculate the angle subtended by the first minimum in the diffraction pattern. (b) What is the relative intensity I/I_0 at $\theta = 15°$? (c) Consider the case when there are two such sources, separated laterally by 20 cm, behind the slit. What must the maximum distance between the plane of the sources and the slit be if the diffraction patterns are to be resolved? (In this case, the approximation $\sin \theta \approx \tan \theta$ is not valid because of the relatively small value of a/λ.)

74. Light of wavelength 632.8 nm illuminates a single slit, and a diffraction pattern is formed on a screen 1.00 m from the slit. Using the following data, plot relative intensity versus distance. Choose an appropriate value for the slit width a and, on the same graph used for the experimental data, plot the theoretical expression for the relative intensity

$$\frac{I_\theta}{I_0} = \frac{\sin^2(\beta/2)}{(\beta/2)^2}$$

What value of a gives the best fit of theory and experiment?

Relative Intensity	Distance from Center of Central Maximum (mm)
0.95	0.8
0.80	1.6
0.60	2.4
0.39	3.2
0.21	4.0
0.079	4.8
0.014	5.6
0.003	6.5
0.015	7.3
0.036	8.1
0.047	8.9
0.043	9.7
0.029	10.5
0.013	11.3
0.002	12.1
0.0003	12.9
0.005	13.7
0.012	14.5
0.016	15.3
0.015	16.1
0.010	16.9
0.0044	17.7
0.0006	18.5
0.0003	19.3
0.003	20.2

75. From Equation 38.4 show that, in the Fraunhofer diffraction pattern of a single slit, the angular width of the central maximum at the point where $I = 0.5I_0$ is $\Delta\theta = 0.886\lambda/a$. (*Hint:* In Equation 38.4, let $\beta/2 = \phi$ and solve the resulting transcendental equation graphically; see Problem 9.)

76. Another method to solve the equation $\phi = \sqrt{2}\sin\phi$ in Problem 9 is to use a calculator, guess a first value of ϕ, see if it fits, and continue to update your estimate until the equation balances. How many steps (iterations) did this take?

SPREADSHEET PROBLEM

S1. Figure 38.11 shows the relative intensity of a single-slit Fraunhofer diffraction pattern as a function of the parameter $\beta/2 = \pi a \sin\theta/\lambda$. Spreadsheet 38.1 plots the relative intensity I/I_0 as a function of θ, where θ is defined in Figure 38.8. Examine the cases where $\lambda = a$, $\lambda = 0.5a$, $\lambda = 0.1a$, and $\lambda = 0.05a$, and explain your results.

Computer simulation of a two-dimensional array of spheres as they would appear as they move past an observer moving at a relativistic speed. *(Mel Pruitt)*

Modern Physics

"Imagination is more important than knowledge."

ALBERT EINSTEIN

At the end of the 19th century, scientists believed that they had learned most of what there was to know about physics. Newton's laws of motion and his universal theory of gravitation, Maxwell's theoretical work in unifying electricity and magnetism, and the laws of thermodynamics and kinetic theory were highly successful in explaining a wide variety of phenomena.

As the century turned, however, a major revolution shook the world of physics. In 1900 Planck provided the basic ideas that led to the formulation of the quantum theory, and in 1905 Albert Einstein formulated his brilliant special theory of relativity. The excitement of the times is captured in Einstein's own words: ''It was a marvelous time to be alive.'' Both ideas were to have a profound effect on our understanding of nature. Within a few decades, these theories inspired new developments and theories in the fields of atomic physics, nuclear physics, and condensed-matter physics.

In Chapter 39 we introduce the special theory of relativity. Although the concepts underlying this theory seem to contradict our common sense, the theory provides us with a new and deeper view of physical laws.

You should keep in mind that, although modern physics has been developed during this century and has led to a multitude of important technological achievements, the story is still incomplete. Discoveries will continue to evolve during our lifetime, and many of these discoveries will deepen or refine our understanding of nature and the world around us. It is still a ''marvelous time to be alive.''

CHAPTER 39

Relativity

Albert Einstein (1879–1955), one of the greatest physicists of all times, is best known for developing the theory of relativity. He is shown here in a playful mood riding a bicycle. The photograph was taken in 1933 in Santa Barbara, California. *(From the California Institute of Technology archives)*

Most of our everyday experiences and observations have to do with objects that move at speeds much less than that of light. Newtonian mechanics and early ideas on space and time were formulated to describe the motion of such objects. This formalism is very successful in describing a wide range of phenomena that occur at low speeds. It fails, however, when applied to particles whose speeds approach that of light. Experimentally, the predictions of Newtonian theory can be tested at high speeds by accelerating electrons or other charged particles through a large electric potential difference. For example, it is possible to accelerate an electron to a speed of $0.99\,c$ (where c is the speed of light) by using a potential difference of several million volts. According to Newtonian mechanics, if the potential difference (as well as the corresponding energy) is increased by a factor of 4, the electron speed should jump to $1.98\,c$. However, experiments show that the speed of the electron — as well as the speeds of all other particles in the Universe — always remains less than the speed of light, regardless of the size of the accelerating voltage. Because it places no upper limit on speed, Newtonian mechanics is contrary to modern experimental results and is clearly a limited theory.

In 1905, at the age of only 26, Einstein published his special theory of relativity. Regarding the theory, Einstein wrote:

> The relativity theory arose from necessity, from serious and deep contradictions in the old theory from which there seemed no escape. The strength of the new theory lies in the consistency and simplicity with which it solves all these difficulties, using only a few very convincing assumptions. . . .[1]

Although Einstein made many important contributions to science, the theory of relativity alone represents one of the greatest intellectual achievements of the 20th century. With this theory, experimental observations can be correctly predicted over the range of speeds from $v = 0$ to speeds approaching the speed of light. Newtonian mechanics, which was accepted for over 200 years, is in fact a special case of Einstein's theory. This chapter gives an introduction to the special theory of relativity, with emphasis on some of its consequences.

Special relativity covers phenomena such as the slowing down of clocks and the contraction of lengths in moving reference frames as measured by a laboratory observer. We also discuss the relativistic forms of momentum and energy, and some consequences of the famous mass-energy formula, $E = mc^2$.

In addition to its well-known and essential role in theoretical physics, the theory of relativity has practical applications, including the design of accelerators and other devices that utilize high-speed particles. These devices will not work if designed according to nonrelativistic principles. We shall have occasion to use relativity in some subsequent chapters of this text, most often presenting only the outcome of relativistic effects.

39.1   THE PRINCIPLE OF NEWTONIAN RELATIVITY

To describe a physical event, it is necessary to establish a frame of reference. You should recall from Chapter 5 that Newton's laws are valid in all inertial frames of reference. Because an inertial frame is defined as one in which Newton's first law is valid, it can be said that *an inertial system is one in which a free body experiences no acceleration.* Furthermore, any system moving with constant velocity with respect to an inertial system must also be an inertial system.

Inertial frame of reference

There is no preferred frame. This means that the results of an experiment performed in a vehicle moving with uniform velocity will be identical to the results of the same experiment performed in a stationary vehicle. The formal statement of this result is called the **principle of Newtonian relativity**:

The laws of mechanics must be the same in all inertial frames of reference.

Let us consider an observation that illustrates the equivalence of the laws of mechanics in different inertial frames. Consider a pickup truck moving with a constant velocity as in Figure 39.1a. If a passenger in the truck throws a ball straight up in the air, the passenger observes that the ball moves in a vertical path. The

---

[1] A. Einstein and L. Infeld, *The Evolution of Physics*, New York, Simon and Schuster, 1961.

FIGURE 39.1   (a) The observer in the truck sees the ball move in a vertical path when thrown upward. (b) The Earth observer views the path of the ball to be a parabola.

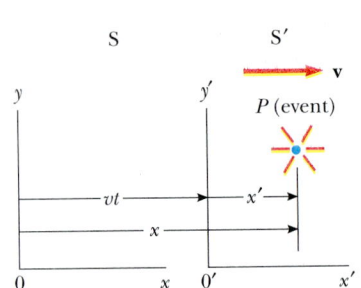

FIGURE 39.2   An event occurs at a point *P*. The event is observed by two observers in inertial frames S and S′, where S′ moves with a velocity *v* relative to S.

motion of the ball appears to be precisely the same as if the ball were thrown by a person at rest on Earth. The law of gravity and the equations of motion under constant acceleration are obeyed whether the truck is at rest or in uniform motion. Now consider the same experiment viewed by an observer at rest on Earth. This stationary observer sees the path of the ball as a parabola, as in Figure 39.1b. Furthermore, according to this observer, the ball has a horizontal component of velocity equal to the velocity of the truck. Although the two observers disagree on certain aspects of the experiment, they agree on the validity of Newton's laws and classical principles like conservation of energy and conservation of momentum. This agreement implies that no mechanical experiment can detect any difference between the two inertial frames. The only thing that can be detected is the relative motion of one frame with respect to the other. That is, the notion of absolute motion through space is meaningless, as is the notion of a preferred reference frame.

Suppose that some physical phenomenon, which we call an event, occurs in an inertial frame. The event's location and time of occurrence can be specified by the coordinates (x, y, z, t). We would like to be able to transform these coordinates from one inertial frame to another moving with uniform relative velocity. This is accomplished by using what is called a *Galilean transformation,* which owes its origin to Galileo.

Consider two inertial frames S and S′ (Fig. 39.2). The system S′ moves with a constant velocity **v** along the *xx′* axes, where **v** is measured relative to S. We assume that an event occurs at the point *P* and that the origins of S and S′ coincide at $t = 0$. An observer in S describes the event with space-time coordinates (x, y, z, t), while an observer in S′ uses (x', y', z', t') to describe the same event. As we can see from Figure 39.2, these coordinates are related by the equations

Galilean transformation of coordinates

$$x' = x - vt$$
$$y' = y$$
$$z' = z$$
$$t' = t$$

(39.1)

These equations constitute what is known as a **Galilean transformation of coordi-**

nates. Note that time is assumed to be the same in both inertial systems. That is, within the framework of classical mechanics, all clocks run at the same rate, regardless of their velocity, so that the time at which an event occurs for an observer in S is the same as the time for the same event in S'. Consequently, the time interval between two successive events should be the same for both observers. Although this assumption may seem obvious, it turns out to be incorrect when treating situations in which v is comparable to the speed of light. This point of equal time intervals represents one of the profound differences between Newtonian concepts and Einstein's theory of relativity.

Now suppose two events are separated by a distance dx and a time interval dt as measured by an observer in S. It follows from Equation 39.1 that the corresponding distance dx' measured by an observer in S' is $dx' = dx - v\,dt$, where dx is the distance between the two events measured by an observer in S. Since $dt = dt'$, we find that

$$\frac{dx'}{dt} = \frac{dx}{dt} - v$$

$$u'_x = u_x - v \tag{39.2}$$

Galilean addition law for velocities

where u_x and u'_x are the instantaneous velocities of the object relative to S and S', respectively. This result, which is called the **Galilean addition law for velocities** (or Galilean velocity transformation), is used in everyday observations and is consistent with our intuitive notion of time and space. As we will soon see, however, it leads to serious contradictions when applied to electromagnetic waves.

The Speed of Light

It is quite natural to ask whether the principle of Newtonian relativity in mechanics also applies to electricity, magnetism, and optics, and the answer is no. Recall from Chapter 34 that Maxwell in the 1860s showed that the speed of light in free space was given by $c = 3.00 \times 10^8$ m/s. Physicists of the late 1800s thought that light waves required a definite medium to move in called the *ether,* and that the speed of light was only c in a special, absolute frame at rest with respect to the ether. The Galilean addition law of velocities was expected to hold in any other frame moving at speed v relative to the absolute ether frame.

Since the existence of a preferred, absolute ether frame would show that light was similar to other classical waves, and that Newtonian ideas of an absolute frame were true, considerable importance was attached to establishing the existence of the ether frame. Because the speed of light is enormous, experiments involving light traveling in media moving at then attainable laboratory speeds had not been capable of detecting small changes of the size of $c \pm v$ prior to the late 1800s. Scientists of the period, realizing that the Earth moved around the Sun at 30 km/s, decided to use the Earth as the moving frame in an attempt to improve their chances of detecting these small changes in light speed.

From our point of view of observers fixed on Earth, we may say that we are stationary and that the absolute ether frame containing the medium for light propagation moves past us with speed v. Determining the speed of light under these circumstances is just like determining the speed of an aircraft by measuring the speed of the air next to it, and consequently we speak of an "ether wind" blowing through our apparatus fixed to the Earth.

A direct method for detecting an ether wind would be to measure its influence

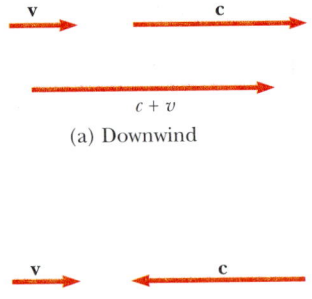

(a) Downwind

(b) Upwind

(c) Across wind

FIGURE 39.3 If the velocity of the ether wind relative to the Earth is **v**, and **c** is the velocity of light relative to the ether, the speed of light relative to the Earth is (a) $c + v$ in the downwind direction, (b) $c - v$ in the upwind direction, and (c) $(c^2 - v^2)^{1/2}$ in the direction perpendicular to the wind.

on the speed of light using an apparatus fixed in a frame of reference on Earth. If v is the speed of the ether relative to Earth, then the speed of light should have its maximum value, $c + v$, when propagating downwind as in Figure 39.3a. Likewise, the speed of light should have its minimum value, $c - v$, when propagating upwind as in Figure 39.3b, and some intermediate value, $(c^2 - v^2)^{1/2}$, in the direction perpendicular to the ether wind as in Figure 39.3c. If the Sun is assumed to be at rest in the ether, then the velocity of the ether wind would be equal to the orbital velocity of the Earth around the Sun, which has a magnitude of approximately 3×10^4 m/s. Since $c = 3 \times 10^8$ m/s, it should be possible to detect a change in speed of about 1 part in 10^4 for measurements in the upwind or downwind directions. However, as we shall see in the next section, all attempts to detect such changes and establish the existence of the ether (and hence the absolute frame) proved futile!

If it is assumed that the laws of electricity and magnetism are the same in all inertial frames, a paradox concerning the speed of light immediately arises. This can be understood by recalling that according to Maxwell's equations, the speed of light always has the fixed value $(\mu_0\epsilon_0)^{-1/2} \approx 3.00 \times 10^8$ m/s, a result in direct contradiction to what would be expected based on the Galilean law. According to Galileo, the speed of light should not be the same in all inertial frames.

For example, suppose a light pulse is sent out by an observer S′ standing in a boxcar moving with a velocity **v** (Fig. 39.4). The light pulse has a speed c relative to S′. According to Newtonian relativity, the speed of the pulse relative to the stationary observer S outside the boxcar should be $c + v$. This is in obvious contradiction to Einstein's theory, which, we shall see, postulates that the speed of the light pulse is the same for all observers.

In order to resolve this paradox, it must be concluded that either (1) the laws of electricity and magnetism are not the same for all inertial frames or (2) the Galilean law of addition for velocities is incorrect. If we assume the first alternative, we are forced to abandon the seemingly obvious notions of absolute time and absolute length that form the basis for the Galilean transformations. If we assume that the second alternative is true, then there must exist a preferred reference frame in which the speed of light has the value c and the speed must be greater or less than this value in any other reference frame, in accordance with the Galilean addition law for velocities. It is useful to draw an analogy with sound waves propagating through air. The speed of sound in air is about 330 m/s when measured in a

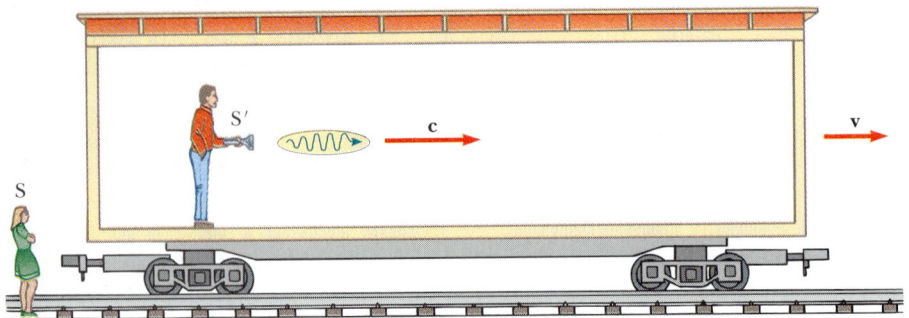

FIGURE 39.4 A pulse of light is sent out by a person in a moving boxcar. According to Newtonian relativity, the speed of the pulse should be $c + v$ relative to a stationary observer.

1. A. Piccard
2. E. Henriot
3. P. Ehrenfest
4. E. Herzen
5. Th. de Donder
6. E. Schroedinger
7. E. Verschaffelt
8. W. Pauli
9. W. Heisenberg
10. R.H. Fowler
11. L. Brillouin
12. P. Debye
13. M. Knudsen
14. W.L. Bragg
15. H.A. Kramers
16. P.A.M. Dirac
17. A.H. Compton
18. L.V. de Broglie
19. M. Born
20. N. Bohr
21. I. Langmuir
22. M. Planck
23. M. Curie
24. H.A. Lorentz
25. A. Einstein
26. P. Langevin
27. C.E. Guye
28. C.T.R. Wilson
29. O.W. Richardson

The "architects" of modern physics. This unique photograph shows many eminent scientists who participated in the fifth international congress of physics held in 1927 by the Solvay Institute in Brussels. At this and similar conferences, held regularly from 1911 on, scientists were able to discuss and share the many dramatic developments in atomic and nuclear physics. This elite company of scientists includes fifteen Nobel Prize winners in physics and three in chemistry. *(Photograph courtesy of AIP Niels Bohr Library)*

reference frame in which the air is stationary. However, the speed is greater or less than this value when measured from a reference frame that is moving relative to the sound source.

39.2   THE MICHELSON-MORLEY EXPERIMENT

The most famous experiment designed to detect small changes in the speed of light was performed in 1881 by Albert A. Michelson (1852–1931) and repeated under various conditions in 1887 with Edward W. Morley (1838–1923). We state at the outset that the outcome of the experiment was null, thus contradicting the ether hypothesis. The experiment was designed to determine the velocity of the Earth relative to the hypothetical ether. The experimental tool used was the Michelson interferometer, shown in Figure 39.5. One of the arms of the interferometer is aligned along the direction of the motion of Earth through space. The Earth moving through the ether is equivalent to the ether flowing past the Earth in the opposite direction with speed v as in Figure 39.5. This ether wind blowing in the direction opposite the direction of Earth's motion should cause the speed of light measured in the Earth's frame of reference to be $c - v$ as the light approaches mirror M_2 in Figure 39.5 and $c + v$ after reflection. The speed v is the speed of the Earth through space, and hence the speed of the ether wind, while c is the speed of light in the ether frame. The two beams of light reflected from M_1 and M_2 recombine, and an interference pattern consisting of alternating dark and bright fringes is formed.

During the experiment, the interference pattern was observed while the interferometer was rotated through an angle of 90°. This rotation changed the speed of the ether wind along the direction of the arms of the interferometer. The effect of this rotation should have been to cause the fringe pattern to shift slightly but measurably. Measurements failed to show any change in the interference pattern! The Michelson-Morley experiment was repeated at different times of the year when the ether wind was expected to change direction and magnitude, but the results were always the same: *No fringe shift of the magnitude required was ever observed.*[2]

The null results of the Michelson-Morley experiment not only contradicted the ether hypothesis, it also meant that it was impossible to measure the absolute velocity of the Earth with respect to the ether frame. However, as we see in the next section, Einstein offered a postulate for his theory of relativity that places quite a different interpretation on these null results. In later years, when more was known about the nature of light, the idea of an ether that permeates all of space was relegated to the ash heap of worn-out concepts. Light is now understood to be *an electromagnetic wave, which requires no medium for its propagation.* As a result, the idea of having an ether in which these waves could travel became unnecessary.

Details of the Michelson-Morley Experiment

To understand the outcome of the Michelson-Morley experiment, let us assume that the two arms of the interferometer in Figure 39.5 are of equal length L. First consider the light beam traveling along arm 1 parallel to the direction of the ether wind. As the light beam moves to the right, its speed is reduced by the wind

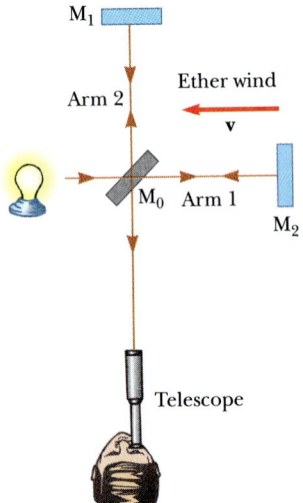

FIGURE 39.5   According to the ether wind theory, the speed of light should be $c - v$ as the beam approaches mirror M_2 and $c + v$ after reflection.

---

[2] From an Earth observer's point of view, changes in the Earth's speed and direction of motion in the course of a year are viewed as ether wind shifts. Even if the speed of the Earth with respect to the ether were zero at some time, six months later the speed of the Earth would be 60 km/s with respect to the ether and a clear fringe shift should be found. None has ever been observed, however.

blowing in the opposite direction and its speed relative to the Earth is $c - v$. On its return journey, as the light beam moves to the left along arm 1, its direction is the same as that of the ether wind and so its speed relative to the Earth is $c + v$. Thus, the time of travel to the right is $L/(c - v)$ and the time of travel to the left is $L/(c + v)$. The total time of travel for the round trip along the horizontal path is

$$t_1 = \frac{L}{c + v} + \frac{L}{c - v} = \frac{2Lc}{c^2 - v^2} = \frac{2L}{c}\left(1 - \frac{v^2}{c^2}\right)^{-1}$$

Now consider the light beam traveling along arm 2, perpendicular to the ether wind. Because the speed of the beam relative to the Earth is $(c^2 - v^2)^{1/2}$ in this case (see Fig. 39.5), then the time of travel for each half of this trip is given by $L/(c^2 - v^2)^{1/2}$, and the total time of travel for the round trip is

$$t_2 = \frac{2L}{(c^2 - v^2)^{1/2}} = \frac{2L}{c}\left(1 - \frac{v^2}{c^2}\right)^{-1/2}$$

Thus, the time difference between the light beam traveling horizontally (arm 1) and the beam traveling vertically (arm 2) is

$$\Delta t = t_1 - t_2 = \frac{2L}{c}\left[\left(1 - \frac{v^2}{c^2}\right)^{-1} - \left(1 - \frac{v^2}{c^2}\right)^{-1/2}\right]$$

Because $v^2/c^2 \ll 1$, this expression can be simplified by using the following binomial expansion after dropping all terms higher than second order:

$$(1 - x)^n \approx 1 - nx \qquad (\text{for } x \ll 1)$$

In our case, $x = v^2/c^2$, and we find

$$\Delta t = t_1 - t_2 \approx \frac{Lv^2}{c^3} \tag{39.3}$$

The two light beams in Figure 39.5 start out in phase and return to form an interference pattern. Let us assume that the interferometer is adjusted for parallel fringes and that a telescope is focused on one of these fringes. The time difference between the two instants at which the light beams arrive at the telescope gives rise to a phase difference between the beams, producing an interference pattern when they combine at the position of the telescope. A difference in the pattern should be detected by rotating the interferometer through 90° in a horizontal plane, so that the two arms of the interferometer exchange positions. This results in a net time difference of twice that given by Equation 39.3. Thus, the path difference that corresponds to this time difference is

$$\Delta d = c(2\,\Delta t) = \frac{2Lv^2}{c^2}$$

The corresponding fringe shift is equal to this path difference divided by the wavelength of light, since a change in path length of one wavelength corresponds to a shift one fringe:

$$\text{Shift} = \frac{2Lv^2}{\lambda c^2} \tag{39.4}$$

In the experiments by Michelson and Morley, each light beam was reflected by mirrors many times to give an effective path length L of approximately 11 m. Using this value, and taking v to be equal to 3.0×10^4 m/s, the speed of the Earth

Albert A. Michelson (1852–1931). A German-American physicist, Michelson invented the interferometer and spent much of his life making accurate measurements of the speed of light. In 1907 he was the first American to be awarded the Nobel prize, which he received for his work in optics. His most famous experiment, conducted with Edward Morley in 1887, implied that it was impossible to measure the absolute velocity of the Earth with respect to the ether. Subsequent work by Einstein in his special theory of relativity eliminated the ether concept by assuming that the speed of light has the same value in all inertial reference frames. *(AIP Emilio Segré Visual Archives, Michelson Collection)*

around the Sun, we get a path difference of

$$\Delta d = \frac{2(11 \text{ m})(3.0 \times 10^4 \text{ m/s})^2}{(3.0 \times 10^8 \text{ m/s})^2} = 2.2 \times 10^{-7} \text{ m}$$

This extra travel distance should produce a noticeable shift in the fringe pattern. Specifically, using light of wavelength 500 nm, we would expect a fringe shift for rotation through 90° of

$$\text{Shift} = \frac{\Delta d}{\lambda} = \frac{2.2 \times 10^{-7}}{5.0 \times 10^{-7}} \cong 0.40$$

The instrument used by Michelson and Morley had the capability of detecting shifts as small as 0.01 fringe. However, they *reported a shift that is less than one fourth of a fringe*. Since then, the experiment has been repeated many times by various scientists under various conditions, and no fringe shift has ever been detected. Thus, it was concluded that the motion of the Earth with respect to the ether cannot be detected.

Many efforts were made to explain the null results of the Michelson-Morley experiment and to save the ether frame concept and the Galilean addition law for the velocity of light. Since all of these proposals have been shown to be wrong, they will not be considered here. In the 1890s, G. F. Fitzgerald and Hendrik A. Lorentz independently tried to explain the null results by making the following ad hoc assumption. They proposed that the length of an object moving at speed v would contract along the direction of motion by a factor of $\sqrt{1 - v^2/c^2}$. The net result of this contraction would be a change in length of one of the arms of the interferometer such that no path difference would occur as the apparatus was rotated. Such a physical contraction would fully explain the original Michelson-Morley experiment but would be inconsistent with the same experiment when the two arms of the interferometer have different lengths.

No experiment in the history of physics has received such valiant efforts to explain the absence of an expected result as did the Michelson-Morley experiment. The stage was set for the brilliant Albert Einstein, who solved the problem in 1905 with his special theory of relativity.

39.3 EINSTEIN'S PRINCIPLE OF RELATIVITY

In the previous section we noted the impossibility of measuring the speed of the ether with respect to the Earth and the failure of the Galilean addition law for velocities in the case of light. Albert Einstein proposed a theory that boldly removed these difficulties and at the same time completely altered our notion of space and time.[3] Einstein based his special theory of relativity on two postulates:

The postulates of the special theory of relativity

1. **The Principle of Relativity:** All the laws of physics are the same in all inertial reference frames.
2. **The Constancy of the Speed of Light:** The speed of light in vacuum has the same value, $c = 3.00 \times 10^8$ m/s, in all inertial frames, regardless of the velocity of the observer or the velocity of the source emitting the light.

[3] A. Einstein, "On the Electrodynamics of Moving Bodies," *Ann. Physik* **17**:891 (1905). For an English translation of this article and other publications by Einstein, see the book by H. Lorentz, A. Einstein, H. Minkowski, and H. Weyl, *The Principle of Relativity,* Dover, 1958.

The first postulate asserts that *all* the laws of physics, those dealing with mechanics, electricity and magnetism, optics, thermodynamics, and so on, are the same in all reference frames moving with constant velocity relative to each other. This postulate is a sweeping generalization of the principle of Newtonian relativity that only refers to the laws of mechanics. From an experimental point of view, Einstein's principle of relativity means that any kind of experiment (measuring the speed of light for example), performed in a laboratory at rest, must give the same result when performed in a laboratory moving a constant velocity past the first one. Hence, no preferred inertial reference frame exists and it is impossible to detect absolute motion.

Note that postulate 2, the principle of the constancy of the speed of light, is required by postulate 1: If the speed of light were not the same in all inertial frames, it would be possible to distinguish between inertial frames and a preferred, absolute frame could be identified, in contradiction to postulate 1. Postulate 2 also eliminates the problem of measuring the speed of the ether by denying the existence of the ether and boldly asserting that light always moves with speed c relative to all inertial observers.

Although the Michelson-Morley experiment was performed before Einstein published his work on relativity, it is not clear whether or not Einstein was aware of the details of the experiment. Nonetheless, the null result of the experiment can be readily understood within the framework of Einstein's theory. According to his principle of relativity, the premises of the Michelson-Morley experiment were incorrect. In the process of trying to explain the expected results, we stated that when light traveled against the ether wind its speed was $c - v$, in accordance with the Galilean addition law for velocities. However, if the state of motion of the observer or of the source has no influence on the value found for the speed of light, one will always measure the value to be c. Likewise, the light makes the return trip after reflection from the mirror at speed c, not at speed $c + v$. Thus, the motion of the Earth does not influence the fringe pattern observed in the Michelson-Morley experiment, and a null result should be expected.

If we accept Einstein's theory of relativity, we must conclude that relative motion is unimportant when measuring the speed of light. At the same time, we must alter our common-sense notion of space and time and be prepared for some rather bizarre consequences.

39.4 CONSEQUENCES OF SPECIAL RELATIVITY

Before we discuss the consequences of special relativity, we must first understand how an observer located in an inertial reference frame describes an event. As mentioned earlier, an event is an occurrence describable by three space coordinates and one time coordinate. Different observers in different inertial frames usually describe the same event with different space-time coordinates.

The reference frame used to describe an event consists of a coordinate grid and a set of synchronized clocks located at the grid intersections as shown in Figure 39.6 in two dimensions. The clocks can be synchronized in many ways with the help of light signals. For example, suppose the observer is located at the origin with his master clock and sends out a pulse of light at $t = 0$. The light pulse takes a time r/c to reach a second clock located a distance r from the origin. Hence, the second clock is synchronized with the clock at the origin if the second clock reads a time r/c at the instant the pulse reaches it. This procedure of synchronization assumes that the speed of light has the same value in all directions and in all inertial frames.

FIGURE 39.6 In relativity, we use a reference frame consisting of a coordinate grid and a set of synchronized clocks.

Furthermore, the procedure concerns an event recorded by an observer in a specific inertial reference frame. An observer in some other inertial frame would assign different space-time coordinates to events being observed by using another coordinate grid and another array of clocks.

Almost everyone who has dabbled even superficially with science is aware of some of the startling predictions that arise because of Einstein's approach to relative motion. As we examine some of the consequences of relativity in the following three sections, we see that they conflict with our basic notions of space and time. We restrict our discussion to the concepts of length, time, and simultaneity, which are quite different in relativistic mechanics than in Newtonian mechanics. For example, *the distance between two points and the time interval between two events depend on the frame of reference in which they are measured.* That is, *there is no such thing as absolute length or absolute time in relativity.* Furthermore, *events at different locations that occur simultaneously in one frame are not simultaneous in another frame moving uniformly past the first.*

Simultaneity and the Relativity of Time

A basic premise of Newtonian mechanics is that a universal time scale exists that is the same for all observers. In fact, Newton wrote that "Absolute, true, and mathematical time, of itself, and from its own nature, flows equably without relation to anything external." Thus, Newton and his followers simply took simultaneity for granted. In his special theory of relativity, Einstein abandoned this assumption. According to Einstein, *a time interval measurement depends on the reference frame in which the measurement is made.*

Einstein devised the following thought experiment to illustrate this point. A boxcar moves with uniform velocity, and two lightning bolts strike its ends, as in Figure 39.7a, leaving marks on the boxcar and on the ground. The marks on the boxcar are labeled A' and B', while those on the ground are labeled A and B. An observer at O' moving with the boxcar is midway between A' and B', while a ground observer at O is midway between A and B. The events recorded by the observers are the light signals from the lightning bolts.

The two light signals reach the observer at O at the same time, as indicated in Figure 39.7b. This observer realizes that the light signals have traveled at the same speed over equal distances, and so rightly concludes that the events at A and B occurred simultaneously. Now consider the same events as viewed by the observer on the boxcar at O'. By the time the light has reached observer O, observer O' has moved as indicated in Figure 39.7b. Thus, the light signal from B' has already

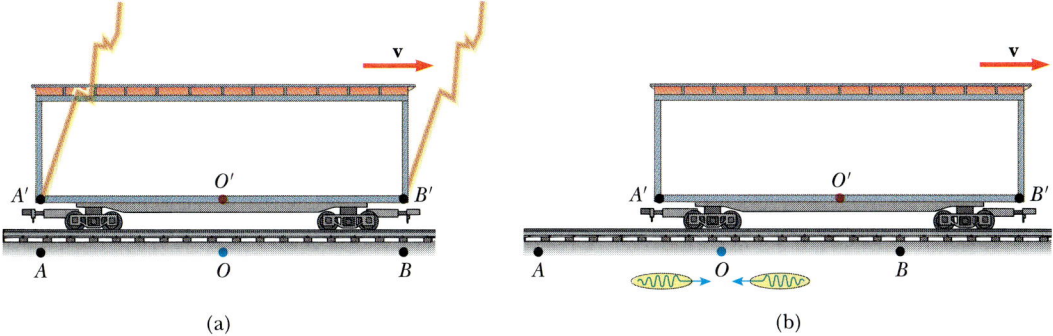

FIGURE 39.7   Two lightning bolts strike the end of a moving boxcar. (a) The events appear to be simultaneous to the stationary observer at O, who is midway between A and B. (b) The events do not appear to be simultaneous to the observer at O', who claims that the front of the train is struck before the rear.

swept past O', while the light from A' has not yet reached O'. According to Einstein, *observer O' must find that light travels at the same speed as that measured by observer O.* Therefore, the observer O' concludes that the lightning struck the front of the boxcar before it struck the back. This thought experiment clearly demonstrates that the two events, which appear to be simultaneous to observer O, do not appear to be simultaneous to observer O'. In other words,

> two events that are simultaneous in one reference frame are in general not simultaneous in a second frame moving relative to the first. That is, simultaneity is not an absolute concept but one that depends upon the state of motion of the observer.

At this point, you might wonder which observer is right concerning the two events. The answer is that *both are correct,* because the principle of relativity states that *there is no preferred inertial frame of reference.* Although the two observers reach different conclusions, both are correct in their own reference frame because the concept of simultaneity is not absolute. This, in fact, is the central point of relativity—any uniformly moving frame of reference can be used to describe events and do physics. However, observers in different inertial frames of reference always measure different time intervals with their clocks and different distances with their meter sticks. Nevertheless, all observers agree on the forms of the laws of physics in their respective frames, because these laws must be the same for all observers in uniform motion. It is the alteration of time and space that allows the laws of physics (including Maxwell's equations) to be the same for all observers in uniform motion.

Time Dilation

The fact that observers in different inertial frames always measure different time intervals between a pair of events can be illustrated by considering a vehicle moving to the right with a speed v as in Figure 39.8a. A mirror is fixed to the ceiling of the vehicle and observer O' at rest in this system holds a laser a distance d below the mirror. At some instant, the laser emits a pulse of light directed towards the mirror (event 1), and at some later time after reflecting from the mirror, the pulse

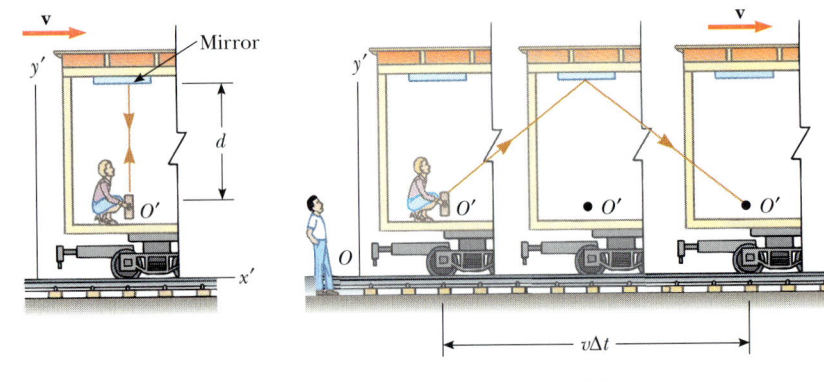

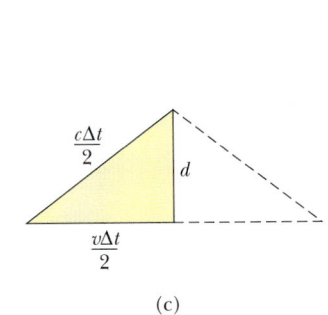

(a)   (b)   (c)

FIGURE 39.8   (a) A mirror is fixed to a moving vehicle, and a light pulse leaves O' at rest in the vehicle. (b) Relative to a stationary observer on Earth, the mirror and O' move with a speed v. Note that the distance the pulse travels is greater than $2d$ as measured by the stationary observer. (c) The right triangle for calculating the relationship between Δt and $\Delta t'$.

arrives back at the laser (event 2). Observer O' carries a clock C' that she uses to measure the time interval $\Delta t'$ between these two events. Because the light pulse has a speed c, the time it takes the pulse to travel from O' to the mirror and back to O' can be found from the definition of speed:

$$\Delta t' = \frac{\text{distance traveled}}{\text{speed}} = \frac{2d}{c} \tag{39.5}$$

This time interval $\Delta t'$ measured by O', who is at rest in the moving vehicle, requires only a single clock C' located at the same place in this frame.

Now consider the same pair of events as viewed by observer O in a second frame as in Figure 39.8b. According to this observer, the mirror and laser are moving to the right with a speed v. The sequence of events appears entirely different as viewed by this observer. By the time the light from the laser reaches the mirror, the mirror has moved a distance $v\,\Delta t/2$, where Δt is the time it takes the light to travel from O' to the mirror and back to O' as measured by observer O. In other words, the second observer concludes that, because of the motion of the vehicle, if the light is to hit the mirror, it must leave the laser at an angle with respect to the vertical direction. Comparing Figures 39.8a and 39.8b, we see that the light must travel farther in the second frame than in the first frame.

According to the second postulate of special relativity, both observers must measure c for the speed of light. Because the light travels farther in the second frame, it follows that the time interval Δt measured by the observer in the second frame is longer than the time interval $\Delta t'$ measured by the observer in the first frame. To obtain a relationship between these two time intervals, it is convenient to use the right triangle shown in Figure 39.8c. The Pythagorean theorem gives

$$\left(\frac{c\,\Delta t}{2}\right)^2 = \left(\frac{v\,\Delta t}{2}\right)^2 + d^2$$

Solving for Δt gives

$$\Delta t = \frac{2d}{\sqrt{c^2 - v^2}} = \frac{2d}{c\sqrt{1 - \dfrac{v^2}{c^2}}} \tag{39.6}$$

Albert Einstein, one of the greatest physicists of all times, was born in Ulm, Germany. He left the highly disciplined German school system after one teacher stated, "You will never amount to anything, Einstein." Following a vacation in Italy, he completed his education at the Swiss Federal Polytechnic School in 1901. Although Einstein attended very few lectures, he was able to pass the courses with the help of excellent lecture notes taken by a friend. Unable to find an academic position, Einstein accepted a position as a junior official in the Swiss Patent Office in Berne. In this setting, and during his "spare time," he continued his independent studies in theoretical physics. In 1905, at the age of 26, he published four scientific papers that revolutionized physics. (In that same year, he earned his Ph.D.) One of these papers, for which he was awarded the 1921 Nobel prize in physics, dealt with the photoelectric effect. Another was concerned with Brownian motion, the irregular motion of small particles suspended in a liquid. The remaining two papers were concerned with what is now considered his most important contribution of all, the special theory of relativity. In 1916, Einstein published his work on the general theory of relativity, which relates gravity to the structure of space and time. The most dramatic prediction of this theory is the degree to which light is deflected by a gravitational field. Measurements made by astronomers on bright stars in the vicinity of the eclipsed Sun in 1919 confirmed Einstein's prediction, and Einstein suddenly became a world celebrity.

In 1913, following academic appointments in Switzerland and Czechoslovakia, Einstein accepted a special position created for him at the Kaiser Wilhelm Institute in Berlin. This made it possible for him to devote all of his time to research free of financial troubles and routine duties. Einstein left Germany in 1933, which was then under Hitler's power, thereby escaping the fate of millions of other European Jews. In the same year he accepted a special position at the Institute for Advanced Study in Princeton where he remained for the rest of his life. He became an American citizen in 1940. Although he was a pacifist, Einstein was persuaded by Leo Szilard to write a letter to President Franklin D. Roosevelt urging him to initiate a program to develop a nuclear bomb. The result was the successful six-year Manhattan project and two nuclear explosions in Japan that ended World War II in 1945.

Einstein made many important contributions to the development of modern physics, including the concept of the light quantum and the idea of stimulated emission of radiation, which led to the invention of the laser 40 years later. He was deeply disturbed by the development of quantum mechanics in the 1920s despite his own role as a scientific revolutionary. In particular, he could never accept the probabilistic view of events in nature that is a central feature of the Copenhagen interpretation of quantum theory. He once said, "God does not play dice with nature." The last few decades of his life were devoted to an unsuccessful search for a unified theory that would combine gravitation and electromagnetism into one picture.

Albert Einstein

| 1 8 7 9 – 1 9 5 5 |

Photo credit: (Photograph courtesy of AIP Niels Bohr Library)

Because $\Delta t' = 2d/c$, we can express Equation 39.6 as

$$\Delta t = \frac{\Delta t'}{\sqrt{1 - \dfrac{v^2}{c^2}}} = \gamma \, \Delta t'$$

(39.7)   Time dilation

where $\gamma = (1 - v^2/c^2)^{-1/2}$. This result says that *the time interval Δt measured by an observer moving with respect to the clock is longer than the time interval $\Delta t'$ measured by the observer at rest with respect to the clock* because γ is always greater than unity. That is, $\Delta t > \Delta t'$. This effect is known as **time dilation**.

The time interval $\Delta t'$ in Equation 39.7 is called the **proper time.** In general, proper time is defined as *the time interval between two events measured by an observer who sees the events occur at the same point in space.* In our case, observer O' measures the proper time. That is, *proper time is always the time measured with a single clock at rest in the frame in which the event takes place.*

Because the time between ticks of a moving clock, $\gamma(2d/c)$, is observed to be longer than the time between ticks of an identical clock at rest, $2d/c$, it is often said "*A moving clock runs slower than a clock at rest by a factor γ.*" This is true for ordinary mechanical clocks as well as for the light clock just described. In fact, we can generalize these results by stating that *all physical, chemical, and biological processes slow down relative to a stationary clock when those processes occur in a moving frame.* For example, the heartbeat of an astronaut moving through space would keep time with a clock inside the spaceship. Both the astronaut's clock and heartbeat are slowed down relative to a stationary clock. The astronaut would not have any sensation of life slowing down in the spaceship. For the astronaut, it is the clock on Earth and the companions at Mission Control that are moving and therefore keep a slow time.

Time dilation is a verifiable phenomenon. For example, muons are unstable elementary particles that have a charge equal to that of the electron and a mass 207 times that of the electron. Muons can be produced by the collision of cosmic radiation with atoms high in the atmosphere. Muons have a lifetime of only 2.2 μs when measured in a reference frame at rest relative to them. If we take 2.2 μs as the average lifetime of a muon and assume that its speed is close to the speed of light, we find that these particles can travel a distance of only approximately 600 m before they decay (Fig. 39.9a). Hence, they cannot reach the Earth from the upper atmosphere where they are produced. However, experiments show that a large number of muons do reach the Earth. The phenomenon of time dilation explains this effect. Relative to an observer on Earth, the muons have a lifetime equal to $\gamma\tau$, where $\tau = 2.2$ μs is the lifetime in a frame of reference traveling with the muons. For example, for $v = 0.99c$, $\gamma \approx 7.1$ and $\gamma\tau \approx 16$ μs. Hence, the average distance traveled as measured by an observer on Earth is $\gamma v\tau \approx 4800$ m, as indicated in Figure 39.9b.

In 1976, at the laboratory of the European Council for Nuclear Research (CERN) in Geneva, muons injected into a large storage ring reached speeds of approximately $0.9994c$. Electrons produced by the decaying muons were detected by counters around the ring, enabling scientists to measure decay rate and, hence, the muon lifetime. The lifetime of the moving muons was measured to be approximately 30 times as long as that of the stationary muon (Fig. 39.10), in agreement with the prediction of relativity to within two parts in a thousand.

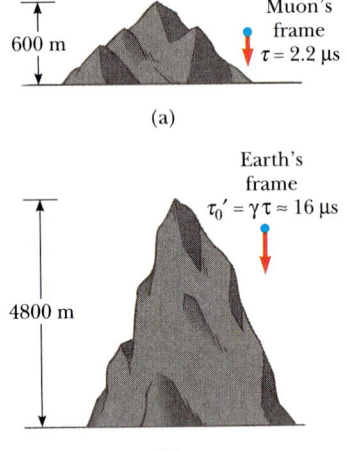

FIGURE 39.9 (a) Muons traveling with a speed of $0.99c$ travel only approximately 600 m as measured in the muons' reference frame, where their lifetime is about 2.2 μs. (b) The muons travel approximately 4800 m as measured by an observer on Earth. Because of time dilation, the muons' lifetime is longer as measured by the Earth observer.

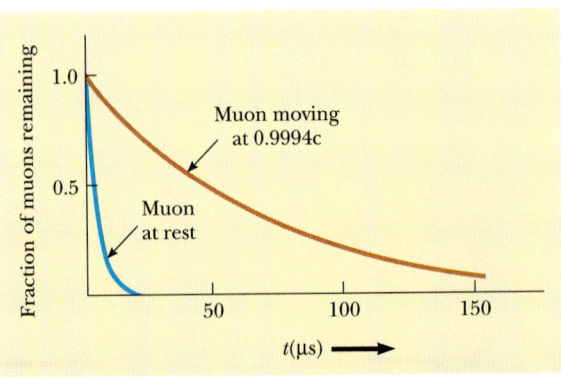

FIGURE 39.10 Decay curves for muons traveling at a speed of $0.9994c$ and for muons at rest.

EXAMPLE 39.1   What Is the Period of the Pendulum?

The period of a pendulum is measured to be 3.0 s in the rest frame of the pendulum. What is the period when measured by an observer moving at a speed of 0.95c relative to the pendulum?

Reasoning   Instead of the observer moving at 0.95c, we can take the equivalent point of view that the observer is at rest and that the pendulum is moving at 0.95c past the stationary observer. Hence the pendulum is an example of a moving clock.

Solution   The proper time is 3.0 s. Because a moving clock runs slower than a stationary clock by γ, Equation 39.7 gives

$$T = \gamma T' = \frac{1}{\sqrt{1 - \dfrac{(0.95\,c)^2}{c^2}}}\, T' = (3.2)(3.0\text{ s}) = \boxed{9.6\text{ s}}$$

That is, a moving pendulum slows down or takes longer to complete a period compared with one at rest.

Length Contraction

The measured distance between two points also depends on the frame of reference. The **proper length** of an object is defined as *the length of the object measured by someone who is at rest relative to the object.* The length of an object measured by someone in a reference frame that is moving with respect to the object is always less than the proper length. This effect is known as **length contraction.**

Consider a spaceship traveling with a speed v from one star to another. There are two observers, one on Earth and the other in the spaceship. The observer at rest on Earth (and also assumed to be at rest with respect to the two stars) measures the distance between the stars to be L_p, the proper length. According to this observer, the time it takes the spaceship to complete the voyage is $\Delta t = L_p/v$. What does an observer in the moving spaceship measure for the distance between the stars? Because of time dilation, the space traveler measures a smaller time of travel: $\Delta t' = \Delta t/\gamma$. The space traveler claims to be at rest and sees the destination star moving toward the spaceship with speed v. Because the space traveler reaches the star in the time $\Delta t'$, he concludes that the distance, L, between the stars is shorter than L_p. This distance measured by the space traveler is

$$L = v\,\Delta t' = v\,\frac{\Delta t}{\gamma}$$

Since $L_p = v\,\Delta t$, we see that $L = L_p/\gamma$ or

$$L = L_p\left(1 - \frac{v^2}{c^2}\right)^{1/2} \tag{39.8}$$

where $(1 - v^2/c^2)^{1/2}$ is a factor less than one. This result may be interpreted as follows:

If an object has a length L_p when it is at rest, then when it moves with speed v in a direction parallel to its length, it contracts to the length L, where $L = L_p(1 - v^2/c^2)^{1/2}$.

Note that *length contraction takes place only along the direction of motion.* For example, suppose a stick moves past a stationary Earth observer with speed v as in Figure 39.11. The length of the stick as measured by an observer in a frame attached to

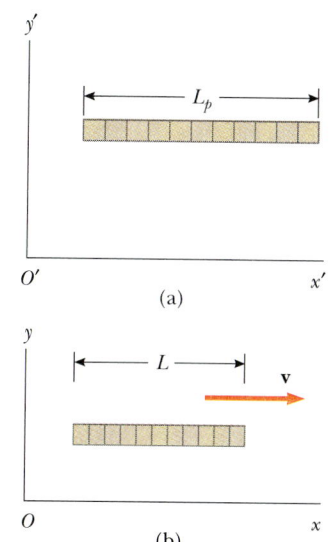

FIGURE 39.11   (a) A stick as viewed by an observer in a frame attached to the stick (i.e., both have the same velocity). (b) The stick as seen by an observer in a frame in which the stick has a velocity **v** relative to the frame. The length is shorter than the proper length, L_p, by a factor $(1 - v^2/c^2)^{1/2}$.

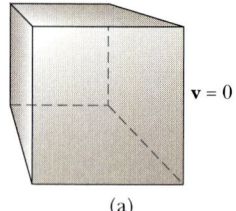

(a)

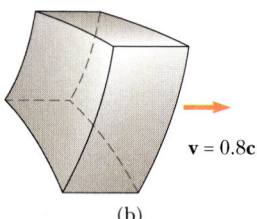

v = 0.8**c**

(b)

FIGURE 39.12 Computer-simulated photographs of a box (a) at rest relative to the camera and (b) moving at a velocity v = 0.8 *c* relative to the camera.

the stick is the proper length, L_p, as in Figure 39.11a. The length of the stick, L, measured by the Earth observer is shorter than L_p by the factor $(1 - v^2/c^2)^{1/2}$. Furthermore, length contraction is a symmetrical effect: If the stick is at rest on Earth, an observer in the moving frame would also measure its length to be shorter by the same factor $(1 - v^2/c^2)^{1/2}$.

It is important to emphasize that proper length and proper time are measured in different reference frames. As an example of this point, let us return to the decaying muons moving at speeds close to the speed of light. An observer in the muon's reference frame would measure the proper lifetime, while an Earth-based observer would measure the proper height of the mountain in Figure 39.9. In the muon's reference frame, there is no time dilation but the distance of travel is observed to be shorter when measured in this frame. Likewise, in the Earth observer's reference frame, there is time dilation, but the distance of travel is measured to be the actual height of the mountain. Thus, when calculations on the muon are performed in both frames, the effect of "offsetting penalties" is seen and the outcome of the experiment in one frame is the same as the outcome in the other frame!

If a box passing by at a speed close to *c* could be photographed, its image would show length contraction, but its shape would also be distorted. This is illustrated in Figure 39.12 for a box moving past a camera with a speed v = 0.8 *c*. When the shutter of the camera is opened, it records the shape of the box at a given instant of time. Since light from different parts of the box must arrive at the camera at the same instant (the instant at which the photograph is taken), light from the more distant parts must start its journey earlier than light from closer parts. Hence, the photograph records different parts of the box at different times. This results in a highly distorted image, which shows horizontal length contraction, vertical curvature, and image rotation.

EXAMPLE 39.2   The Contraction of a Spaceship

A spaceship is measured to be 120 m long while at rest relative to an observer. If this spaceship now flies by the observer with a speed 0.99 *c*, what length does the observer measure?

Solution   From Equation 39.8, the length measured by the observer is

$$L = L_p\sqrt{1 - \frac{v^2}{c^2}} = (120 \text{ m})\sqrt{1 - \frac{(0.99\,c)^2}{c^2}} = \boxed{17 \text{ m}}$$

Exercise   If the ship moves past the observer with a speed of 0.10000 *c*, what length will the observer measure?

Answer   119.40 m.

EXAMPLE 39.3   The Triangular Spaceship

A spaceship in the form of a triangle flies by an observer with a speed of 0.95 *c*. When the ship is at rest relative to the observer (Fig. 39.13a), the distances *x* and *y* are measured to be

52 m and 25 m, respectively. What is the shape of the ship as seen by an observer at rest when the ship is in motion along the direction shown in Figure 39.13b?

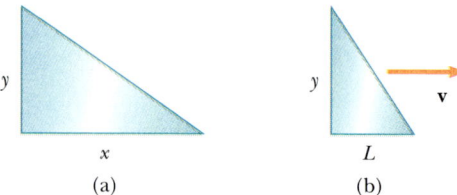

(a)   (b)

FIGURE 39.13   (Example 39.4) (a) When the spaceshift is at rest, its shape is as shown. (b) The spaceship appears to look like this when it moves to the right with a speed *v*. Note that only its *x* dimension is contracted in this case.

Solution   The observer sees the horizontal length of the ship to be contracted to a length

$$L = L_p\sqrt{1 - \frac{v^2}{c^2}} = (52 \text{ m})\sqrt{1 - \frac{(0.95\,c)^2}{c^2}} = 16 \text{ m}$$

The 25-m vertical height is unchanged because it is perpendicular to the direction of relative motion between observer and spaceship. Figure 39.13b represents the size of the spaceship as seen by the observer at rest.

CONCEPTUAL EXAMPLE 39.4   A Voyage to Sirius

An astronaut takes a trip to Sirius, located 8 lightyears from Earth. The astronaut measures the time of the one-way journey to be 6 y. If the spaceship moved at a constant speed of 0.8 c, how can the 8-lightyear distance be reconciled with the 6-y duration measured by the astronaut?

Reasoning   The 8 lightyears (ly) represents the proper length (distance) from Earth to Sirius measured by an observer seeing both nearly at rest. The astronaut sees Sirius

approaching her at 0.8 c, but she also sees the distance contracted to

$$\frac{8 \text{ ly}}{\gamma} = (8 \text{ ly})\sqrt{1 - \frac{v^2}{c^2}} = (8 \text{ ly})\sqrt{1 - \frac{(0.8\,c)^2}{c^2}} = 5 \text{ ly}$$

So the travel time measured on her clocks is

$$t = \frac{d}{v} = \frac{5 \text{ ly}}{0.8\,c} = 6 \text{ y}$$

---

*The Twins Paradox

An intriguing consequence of time dilation is the so-called twins paradox. Consider an experiment involving a set of twins named Speedo and Goslo who are, say, 21 years old. The twins carry with them identical clocks that have been synchronized (Fig. 39.14). Speedo, the more adventuresome of the two, sets out on an epic journey to Planet X, located 10 lightyears from Earth. Furthermore, his spaceship is capable of reaching a speed of 0.500 c relative to the inertial frame of his twin brother on Earth. After reaching Planet X, Speedo becomes homesick and immediately returns to Earth at the same high speed he had attained on the outbound journey. Upon his return, Speedo is shocked to discover that many things have changed in his absence. To Speedo, the most significant change to have occurred is that his twin brother Goslo has aged 40 years and is now 61 years old. Speedo, on the other hand, has aged by only 34.6 years.

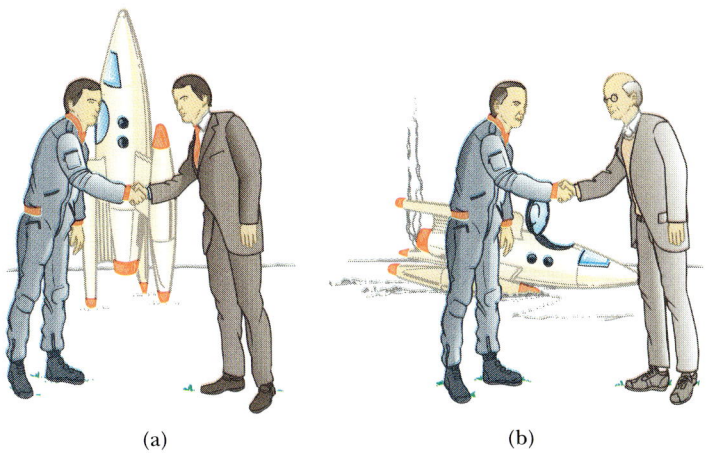

(a)   (b)

FIGURE 39.14   (a) As the twins depart, they are the same age. (b) When Speedo returns from his journey to Planet X, he is younger than his twin Goslo, who remained on Earth.

At this point, it is fair to raise the following question—which twin is the traveler and which twin is really younger as a result of this experiment? From Goslo's frame of reference, he was at rest while his brother traveled at a high speed. From Speedo's perspective, it is he who is at rest while Goslo is on the high-speed space journey. According to Speedo, it is Goslo and the Earth that have raced away on a 17.3-year journey and then headed back for another 17.3 years. This leads to an apparent contradiction. Which twin has developed signs of excess aging?

To resolve this apparent paradox, recall that the special theory of relativity deals with inertial frames of reference moving relative to each other at uniform speed. However, the trip situation is not symmetrical. Speedo, the space traveler, must experience a series of accelerations during his journey. As a result, his speed is not always uniform and consequently Speedo is not in an inertial frame. He cannot be regarded as always being at rest and Goslo to be in uniform motion, because to do so would be an incorrect application of the special theory of relativity. Therefore there is no paradox.

The conclusion that Speedo is in a noninertial frame is inescapable. The time required to accelerate and decelerate Speedo's spaceship may be made very small using large rockets, so that Speedo can claim that he spends most of his time traveling to Planet X at $0.500c$ in an inertial frame. However, Speedo must slow down, reverse his motion, and return to Earth in an altogether different inertial frame. At the very best, Speedo is in two different inertial frames during his journey. Only Goslo, who is in a single inertial frame, can apply the simple time dilation formula to Speedo's trip. Thus, Goslo finds that instead of aging 40 years, Speedo ages only $(1 - v^2/c^2)^{1/2}(40 \text{ years}) = 34.6$ years. On the other hand, Speedo spends 17.3 years traveling to Planet X and 17.3 years returning, for a total travel time of 34.6 years, in agreement with our earlier statement.

CONCEPTUAL EXAMPLE 39.5

Suppose astronauts were paid according to the time spent traveling in space. After a long voyage traveling at a speed near that of light, a crew of astronauts return to Earth and open their pay envelopes. What will their reaction be?

Reasoning   Assuming that their on-duty time was kept on Earth, they will be pleasantly surprised with a large paycheck. Less time will have passed for the astronauts in their frame of reference than for their employer back on Earth.

An experiment reported by Hafele and Keating provided direct evidence of time dilation.[4] Time intervals measured with four cesium beam atomic clocks in jet flight were compared with time intervals measured by Earth-based reference atomic clocks. In order to compare these results with the theory, many factors had to be considered, including periods of acceleration and deceleration relative to the Earth, variations in direction of travel, and the weaker gravitational field experienced by the flying clocks compared with the Earth-based clock. Their results were in good agreement with the predictions of the special theory of relativity and can be explained in terms of the relative motion between the Earth's rotation and the jet aircraft. In their paper, Hafele and Keating say: "Relative to the atomic time scale of the U.S. Naval Observatory, the flying clocks lost 59 ± 10 ns during the

---

[4] J. C. Hafele and R. E. Keating, "Around the World Atomic Clocks: Relativistic Time Gains Observed," *Science*, July 14, 1972, p. 168.

eastward trip and gained 273 ± 7 ns during the westward trip. . . . These results provide an unambiguous empirical resolution of the famous clock paradox with macroscopic clocks.''

39.5 THE LORENTZ TRANSFORMATION EQUATIONS

We have seen that the Galilean transformation is not valid when v approaches the speed of light. In this section, we state the correct transformation equations that apply for all speeds in the range $0 \leqslant v < c$.

Suppose an event that occurs at some point P is reported by two observers, one at rest in a frame S and another in a frame S′ that is moving to the right with speed v as in Figure 39.15. The observer in S reports the event with space-time coordinates (x, y, z, t), while the observer in S′ reports the same event using the coordinates (x', y', z', t'). We would like to find a relationship between these coordinates that is valid for all speeds. In Section 39.1, we found that the Galilean transformation of coordinates, given by Equation 39.1, does not agree with experiment at speeds comparable to the speed of light.

The equations that are valid from $v = 0$ to $v = c$ and enable us to transform coordinates from S to S′ are given by the **Lorentz transformation equations:**

$$x' = \gamma(x - vt)$$
$$y' = y$$
$$z' = z$$
$$t' = \gamma\left(t - \frac{v}{c^2}\,x\right)$$

(39.9)   Lorentz transformation for S → S′

This transformation, known as the Lorentz transformation, was developed by Hendrik A. Lorentz (1853–1928) in 1890 in connection with electromagnetism. However, it was Einstein who recognized their physical significance and took the bold step of interpreting them within the framework of the theory of relativity.

We see that the value for t' assigned to an event by an observer standing at O' depends both on the time t and on the coordinate x as measured by an observer at O. This is consistent with the notion that an event is characterized by four space-time coordinates (x, y, z, t). In other words, in relativity, space and time are not separate concepts but rather are closely interwoven with each other. This is unlike the case of the Galilean transformation in which $t = t'$.

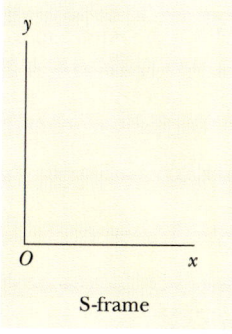

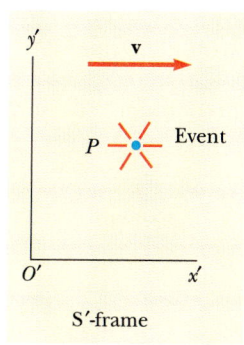

FIGURE 39.15 Representation of an event that occurs at some point P as observed by an observer at rest in the S frame and another in the S′ frame, which is moving to the right with a speed v.

If we wish to transform coordinates in the S' frame to coordinates in the S frame, we simply replace v by $-v$ and interchange the primed and unprimed coordinates in Equation 39.9:

Inverse Lorentz transformation for
S' → S

$$x = \gamma(x' + vt')$$

$$y = y'$$

$$z = z'$$

$$t = \gamma\left(t' + \frac{v}{c^2}x'\right)$$

(39.10)

When $v \ll c$, the Lorentz transformation should reduce to the Galilean transformation. To check this, note that as $v \to 0$, $v/c^2 \ll 1$ and $v^2/c^2 \ll 1$, so that $\gamma = 1$ and Equation 39.9 reduces in this limit to the Galilean coordinate transformation equations

$$x' = x - vt \qquad y' = y \qquad z' = z \qquad t' = t$$

In many situations, we would like to know the difference in coordinates between two events or the time interval between two events as seen by observers at O and O'. This can be accomplished by writing the Lorentz equations in a form suitable for describing pairs of events. From Equations 39.9 and 39.10, we can express the differences between the four variables x, x', t, and t' in the form

$$\Delta x' = \gamma(\Delta x - v\,\Delta t)$$
$$\Delta t' = \gamma\left(\Delta t - \frac{v}{c^2}\,\Delta x\right)$$
$$\left.\right\}S \to S'$$

(39.11)

$$\Delta x = \gamma(\Delta x' + v\,\Delta t')$$
$$\Delta t = \gamma\left(\Delta t' + \frac{v}{c^2}\,\Delta x'\right)$$
$$\left.\right\}S' \to S$$

(39.12)

where $\Delta x' = x'_2 - x'_1$ and $\Delta t' = t'_2 - t'_1$ are the differences measured by the observer at O', while $\Delta x = x_2 - x_1$ and $\Delta t = t_2 - t_1$ are the differences measured by the observer at O. We have not included the expressions for relating the y and z coordinates because they are unaffected by motion along the x direction.[5]

EXAMPLE 39.6   Simultaneity and Time Dilation Revisited

Use the Lorentz transformation equations in difference form to show that (a) simultaneity is not an absolute concept and (b) moving clocks run slower than stationary clocks.

Solution   (a) Suppose that two events are simultaneous according to a moving observer at O', so that $\Delta t' = 0$. From the expression for Δt given in Equation 39.12, we see that in this case, $\Delta t = \gamma v\,\Delta x'/c^2$. That is, the time interval for the same two events as measured by an observer at O is nonzero, and so they do not appear to be simultaneous in O.

(b) Suppose that an observer at O' finds that two events occur at the same place ($\Delta x' = 0$), but at different times ($\Delta t' \neq 0$). In this situation, the expression for Δt given in Equation 39.12 becomes $\Delta t = \gamma\,\Delta t'$. This is the equation for time dilation found earlier, Equation 39.7, where $\Delta t' = \Delta t_p$ is the proper time measured by the single clock located at O'.

Exercise   Use the Lorentz transformation equations in difference form to confirm that $L = L_p/\gamma$.

---

[5] Although motion along x does not change y and z coordinates, it does change velocity components along y and z.

Lorentz Velocity Transformation

Let us now derive the Lorentz velocity transformation, which is the relativistic counterpart of the Galilean velocity transformation. Once again S is our stationary frame of reference and S′ is our frame of reference that moves at a speed v relative to S. Suppose that an object is observed in the S′ frame with an instantaneous speed u_x' measured in S′ given by

$$u_x' = \frac{dx'}{dt'} \tag{39.13}$$

Using Equations 39.9, we have

$$dx' = \gamma(dx - v\,dt)$$

$$dt' = \gamma\left(dt - \frac{v}{c^2}\,dx\right)$$

Substituting these values into Equation 39.13 gives

$$u_x' = \frac{dx'}{dt'} = \frac{dx - v\,dt}{dt - \frac{v}{c^2}\,dx} = \frac{\frac{dx}{dt} - v}{1 - \frac{v}{c^2}\frac{dx}{dt}}$$

But dx/dt is just the velocity component u_x of the object measured in S, and so this expression becomes

$$u_x' = \frac{u_x - v}{1 - \frac{u_x v}{c^2}} \tag{39.14}$$

Lorentz velocity transformation for S → S′

Similarly, if the object has velocity components along y and z, the components in S′ are

$$u_y' = \frac{u_y}{\gamma\left(1 - \frac{u_x v}{c^2}\right)} \quad \text{and} \quad u_z' = \frac{u_z}{\gamma\left(1 - \frac{u_x v}{c^2}\right)} \tag{39.15}$$

When u_x and v are both much smaller than c (the nonrelativistic case), the denominator of Equation 39.14 approaches unity and so $u_x' \approx u_x - v$. This corresponds to the Galilean velocity transformations. In the other extreme, when $u_x = c$, Equation 39.14 becomes

$$u_x' = \frac{c - v}{1 - \frac{cv}{c^2}} = \frac{c\left(1 - \frac{v}{c}\right)}{1 - \frac{v}{c}} = c$$

From this result, we see that an object moving with a speed c relative to an observer in S also has a speed c relative to an observer in S′—independent of the relative motion of S and S′. Note that this conclusion is consistent with Einstein's second postulate, namely, that the speed of light must be c relative to all inertial frames of reference. Furthermore, the speed of an object can never exceed c. That is, the speed of light is the ultimate speed. We return to this point later when we consider the energy of a particle.

To obtain u_x in terms of u'_x, we replace v by $-v$ in Equation 39.15 and interchange the roles of u_x and u'_x:

$$u_x = \frac{u'_x + v}{1 + \dfrac{u'_x v}{c^2}}$$

(39.16)

EXAMPLE 39.7   Relative Velocity of Spaceships

Two spaceships A and B are moving in opposite directions, as in Figure 39.16. An observer on Earth measures the speed of A to be $0.750c$ and the speed of B to be $0.850c$. Find the velocity of B with respect to A.

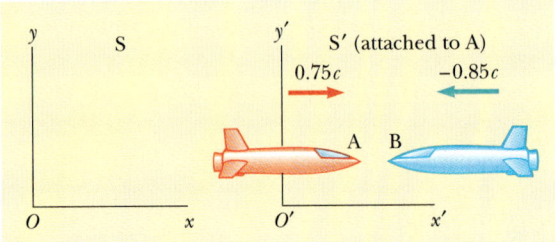

FIGURE 39.16   (Example 39.7) Two spaceships A and B move in opposite directions. The speed of B relative to A is *less* than c and is obtained by using the relativistic velocity transformation.

Solution   This problem can be solved by taking the S' frame as being attached to A, so that $v = 0.750c$ relative to the Earth observer (the S frame). Spaceship B can be considered as an object moving with a velocity $u_x = -0.850c$ relative to the Earth observer. Hence, the velocity of B with respect to A can be obtained using Equation 39.14:

$$u'_x = \frac{u_x - v}{1 - \dfrac{u_x v}{c^2}} = \frac{-0.850c - 0.750c}{1 - \dfrac{(-0.850c)(0.750c)}{c^2}} = \boxed{-0.980c}$$

The negative sign indicates that spaceship B is moving in the negative x direction as observed by A. Note that the result is less than c. That is, a body whose speed is less than c in one frame of reference must have a speed less than c in any other frame. (If the Galilean velocity transformation were used in this example, we would find that $u'_x = u_x - v = -0.850c - 0.750c = -1.60c$, which is greater than c. The Galilean transformation does not work in relativistic situations.)

EXAMPLE 39.8   The Speeding Motorcycle

Imagine a motorcycle rider moving with a speed $0.80c$ past a stationary observer, as shown in Figure 39.17. If the rider tosses a ball in the forward direction with a speed of $0.70c$ relative to himself, what is the speed of the ball as seen by the stationary observer?

Solution   In this situation, the velocity of the motorcycle with respect to the stationary observer is $v = 0.80c$. The velocity of the ball in the frame of reference of the motorcyclist is $0.70c$. Therefore, the velocity, u_x, of the ball relative to the stationary observer is

$$u_x = \frac{u'_x + v}{1 + \dfrac{u'_x v}{c^2}} = \frac{0.70c + 0.80c}{1 + \dfrac{(0.70c)(0.80c)}{c^2}} = \boxed{0.96c}$$

Exercise   Suppose that the motorcyclist turns on a beam of light that moves away from him with a speed c in the forward direction. What does the stationary observer measure for the speed of the light?

Answer c.

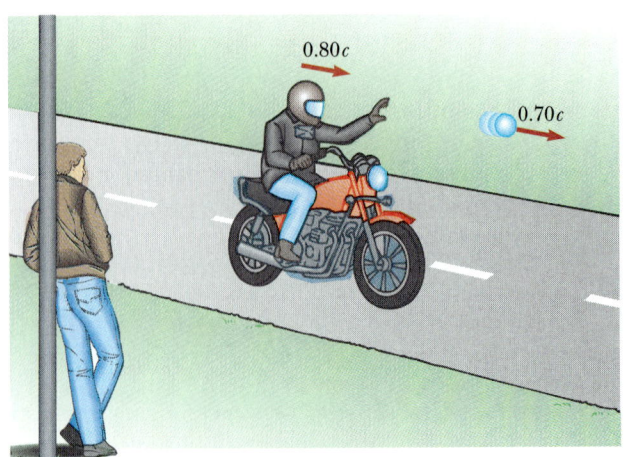

FIGURE 39.17   (Example 39.8) A motorcyclist moves past a stationary observer with a speed of $0.80c$ and throws a ball in the direction of motion with a speed of $0.70c$ relative to himself.

EXAMPLE 39.9   Relativistic Leaders of the Pack

Two motorcycle pack leaders named David and Emily are racing at relativistic speeds along perpendicular paths as in Figure 39.18. How fast does Emily recede over David's right shoulder as seen by David?

Solution   Figure 39.18 represents the situation as seen by a police person at rest in frame S, who observes the following:

$$\text{David:} \quad u_x = 0.75\,c \quad u_y = 0$$

$$\text{Emily:} \quad u_x = 0 \quad u_y = -0.90\,c$$

To get Emily's speed of recession as seen by David, we take S' to move along with David and we calculate u'_x and u'_y for Emily using Equations 39.14 – 39.15:

$$u'_x = \frac{u_x - v}{1 - \dfrac{u_x v}{c^2}} = \frac{0 - 0.75\,c}{1 - \dfrac{(0)(0.75\,c)}{c^2}} = -0.75\,c$$

$$u'_y = \frac{u_y}{\gamma\left(1 - \dfrac{u_x v}{c^2}\right)} = \frac{\sqrt{1 - \dfrac{(0.75\,c)^2}{c^2}}\,(-0.90\,c)}{\left(1 - \dfrac{(0)(0.75\,c)}{c^2}\right)}$$

$$= -0.60\,c$$

Thus, the speed of Emily as observed by David is

$$u' = \sqrt{(u'_x)^2 + (u'_y)^2} = \sqrt{(-0.75\,c)^2 + (-0.60\,c)^2} = \boxed{0.96\,c}$$

Note that this speed is less than c as required by special relativity.

Exercise   Calculate the classical speed of recession for Emily as observed by David using a Galilean transformation.

Answer $1.2\,c$.

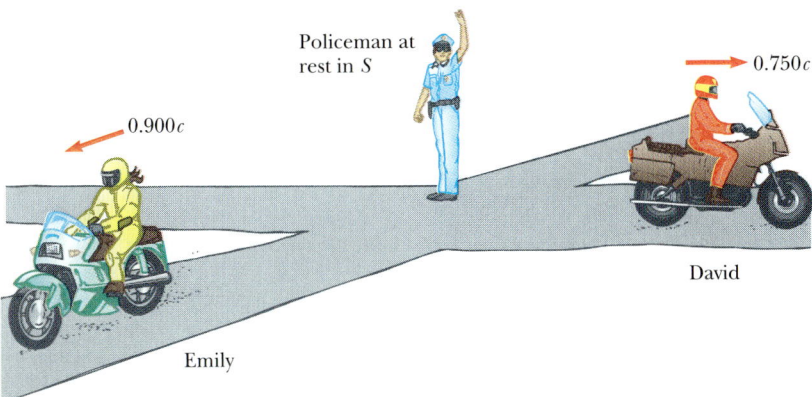

FIGURE 39.18   (Example 39.9) David moves to the east with a speed $0.750\,c$ relative to the policeman, while Emily travels south at a speed $0.900\,c$.

39.6   RELATIVISTIC MOMENTUM AND THE RELATIVISTIC FORM OF NEWTON'S LAWS

We have seen that to properly describe the motion of particles within the framework of special relativity, the Galilean transformation must be replaced by the Lorentz transformation. Because the laws of physics must remain unchanged under the Lorentz transformation, we must generalize Newton's laws and the definitions of momentum and energy to conform to the Lorentz transform and the principle of relativity. These generalized definitions should reduce to the classical (nonrelativistic) definitions for $v \ll c$.

First, recall that the law of conservation of momentum states that when two bodies collide, the total momentum remains constant, assuming the bodies are isolated. Suppose the collision is described in a reference frame S in which momentum is conserved. If the velocities in a second reference frame S' are calculated using the Lorentz transformation and the classical definition of momentum, $\mathbf{p} = m\mathbf{u}$ (where $\mathbf{u}$ is the velocity of the particle), it is found that momentum is *not* conserved in the second reference frame. However, because the laws of physics are the same in all inertial frames, the momentum must be conserved in all systems. In view of this condition and assuming the Lorentz transformation is correct, we must modify the definition of momentum to satisfy the following conditions:

- **p** must be conserved in all collisions
- **p** must approach the classical value $m\mathbf{u}$ as $\mathbf{u} \rightarrow 0$

The correct relativistic equation for momentum that satisfies these conditions is

Definition of relativistic momentum

$$\mathbf{p} \equiv \frac{m\mathbf{u}}{\sqrt{1 - \dfrac{u^2}{c^2}}} \qquad (39.17)$$

where **u** is the velocity of the particle. (We use the symbol **u** for particle velocity rather than **v**, which is used for the relative velocity of two reference frames.) When u is much less than c, the denominator of Equation 39.17 approaches unity, so that **p** approaches $m\mathbf{u}$. Therefore, the relativistic equation for **p** reduces to the classical expression when u is small compared with c. Because it is simpler, Equation 39.17 is often written as

$$\mathbf{p} = \gamma m\mathbf{u} \qquad (39.18)$$

where $\gamma = (1 - u^2/c^2)^{-1/2}$. Note that γ has the same functional form as the γ in the Lorentz transformation. The transformation is from that of the particle to the frame of the observer moving at speed u relative to the particle.

The relativistic force **F** on a particle whose momentum is **p** is defined as

$$\mathbf{F} \equiv \frac{d\mathbf{p}}{dt} \qquad (39.19)$$

where **p** is given by Equation 39.17. This expression is reasonable because it preserves classical mechanics in the limit of low velocities and requires conservation of momentum for an isolated system ($\mathbf{F} = 0$) both relativistically and classically.

It is left as an end-of-chapter problem (Problem 55) to show that the acceleration **a** of a particle decreases under the action of a constant force, in which case $a \propto (1 - u^2/c^2)^{3/2}$. From this formula note that as the particle's speed approaches

---

EXAMPLE 39.10   Momentum of an Electron

An electron, which has a mass of 9.11×10^{-31} kg, moves with a speed of $0.750c$. Find its relativistic momentum and compare this with the momentum calculated from the classical expression.

Solution   Using Equation 39.17 with $u = 0.75c$, we have

$$p = \frac{m u}{\sqrt{1 - \dfrac{u^2}{c^2}}}$$

$$p = \frac{(9.11 \times 10^{-31} \text{ kg})(0.750 \times 3.00 \times 10^8 \text{ m/s})}{\sqrt{1 - \dfrac{(0.750\,c)^2}{c^2}}}$$

$$= 3.10 \times 10^{-22} \text{ kg} \cdot \text{m/s}$$

The incorrect classical expression gives

$$\text{Momentum} = mu = 2.05 \times 10^{-22} \text{ kg} \cdot \text{m/s}$$

Hence, the correct relativistic result is 50% greater than the classical result!

c, the acceleration caused by any finite force approaches zero. Hence, it is impossible to accelerate a particle from rest to a speed $u \geq c$.

39.7   RELATIVISTIC ENERGY

We have seen that the definition of momentum and the laws of motion require generalization to make them compatible with the principle of relativity. This implies that the definition of kinetic energy must also be modified.

   To derive the relativistic form of the work–energy theorem, let us start with the definition of the work done on a particle by a force F and use the definition of relativistic force, Equation 39.19:

$$W = \int_{x_1}^{x_2} F \, dx = \int_{x_1}^{x_2} \frac{dp}{dt} \, dx \qquad (39.20)$$

for force and motion both along the *x* axis. In order to perform this integration, and find the work done on a particle, or the relativistic kinetic energy as a function of *u*, we first evaluate *dp/dt*:

$$\frac{dp}{dt} = \frac{d}{dt} \frac{mu}{\sqrt{1 - \dfrac{u^2}{c^2}}} = \frac{m(du/dt)}{\left(1 - \dfrac{u^2}{c^2}\right)^{3/2}}$$

Substituting this expression for *dp/dt* and *dx = u dt* into Equation 39.20 gives

$$W = \int_{x_1}^{x_2} \frac{m(du/dt)\, u \, dt}{\left(1 - \dfrac{u^2}{c^2}\right)^{3/2}} = m \int_0^u \frac{u}{\left(1 - \dfrac{u^2}{c^2}\right)^{3/2}} \, du$$

where we have assumed that the particle is accelerated from rest to some final speed *u*. Evaluating the integral, we find that

$$W = \frac{mc^2}{\sqrt{1 - \dfrac{u^2}{c^2}}} - mc^2 \qquad (39.21)$$

Recall from Chapter 7 that the work done by a force acting on a particle equals the change in kinetic energy of the particle. Because the initial kinetic energy is zero, we conclude that the work *W* is equivalent to the relativistic kinetic energy *K*:

$$K = \frac{mc^2}{\sqrt{1 - \dfrac{u^2}{c^2}}} - mc^2 = \gamma mc^2 - mc^2 \qquad (39.22)$$

Relativistic kinetic energy

This equation is routinely confirmed by experiments using high-energy particle accelerators.

   At low speeds, where $u/c \ll 1$, Equation 39.22 should reduce to the classical expression $K = \frac{1}{2}mu^2$. We can check this by using the binomial expansion $(1 - x^2)^{-1/2} \approx 1 + \frac{1}{2}x^2 + \cdots$ for $x \ll 1$, where the higher-order powers of *x* are neglected in the expansion. In our case, $x = u/c$, so that

$$\frac{1}{\sqrt{1 - \frac{u^2}{c^2}}} = \left(1 - \frac{u^2}{c^2}\right)^{-1/2} \approx 1 + \frac{1}{2}\frac{u^2}{c^2} + \cdots$$

Substituting this into Equation 39.22 gives

$$K \approx mc^2\left(1 + \frac{1}{2}\frac{u^2}{c^2} + \cdots\right) - mc^2 = \frac{1}{2}mu^2$$

which agrees with the classical result. A graph comparing the relativistic and nonrelativistic expressions is given in Figure 39.19. In the relativistic case, the particle speed never exceeds c, regardless of the kinetic energy. The two curves are in good agreement when $u \ll c$.

The constant term mc^2 in Equation 39.22, which is independent of the speed, is called the **rest energy** of the free particle E_R. The term γmc^2, which depends on the particle speed, is therefore the sum of the kinetic and rest energies. We define γmc^2 to be the **total energy** E, that is,

Total energy = kinetic energy + rest energy

$$E = \gamma mc^2 = K + mc^2 \qquad (39.23)$$

or, when γ is replaced by its equivalent,

$$E = \frac{mc^2}{\sqrt{1 - \frac{u^2}{c^2}}} \qquad (39.24)$$

This, of course, is Einstein's famous mass-energy equivalence equation. The relation $E = \gamma mc^2 = \gamma E_R$ shows that *mass is a property of energy*. Furthermore, this result shows that a small mass corresponds to an enormous amount of energy. This concept is fundamental to much of the field of nuclear physics.

In many situations, the momentum or energy of a particle is measured rather than its speed. It is therefore useful to have an expression relating the total energy E to the relativistic momentum p. This is accomplished by using the expressions $E = \gamma mc^2$ and $p = \gamma mu$. By squaring these equations and subtracting, we can eliminate u (Problem 23). The result, after some algebra, is

$$E^2 = p^2c^2 + (mc^2)^2 \qquad (39.25)$$

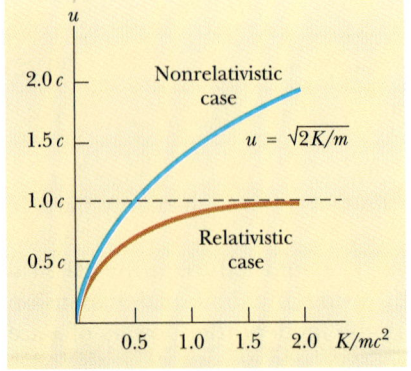

FIGURE 39.19   A graph comparing relativistic and nonrelativistic kinetic energy. The energies are plotted versus speed. In the relativistic case, u is always less than c.

When the particle is at rest, $p = 0$, and so $E = E_R = mc^2$. That is, the total energy equals the rest energy. For the case of particles that have zero mass, such as photons (massless, chargeless particles of light) and neutrinos (massless, chargeless particles associated with beta decay of a nucleus), we set $m = 0$ in Equation 39.25, and we see that

$$E = pc \qquad\qquad (39.26)$$

This equation is an exact expression relating energy and momentum for photons and neutrinos, which always travel at the speed of light.

Finally, note that since the mass m of a particle is independent of its motion, m must have the same value in all reference frames. For this reason, m is often called the *invariant mass*. On the other hand, the total energy and momentum of a particle depend on the reference frame in which they are measured, since they both depend on velocity. Since m is a constant, then according to Equation 39.26 the quantity $E^2 - p^2c^2$ must have the same value in all reference frames. That is, $E^2 - p^2c^2$ is invariant under a Lorentz transformation. These equations do not yet make provision for potential energy.

When dealing with subatomic particles, it is convenient to express their energy in electron volts (eV), because the particles are usually given this energy by acceleration through a potential difference. The conversion factor is

$$1 \text{ eV} = 1.60 \times 10^{-19} \text{ J}$$

For example, the mass of an electron is 9.11×10^{-31} kg. Hence, the rest energy of the electron is

$$mc^2 = (9.11 \times 10^{-31} \text{ kg})(3.00 \times 10^8 \text{ m/s})^2 = 8.20 \times 10^{-14} \text{ J}$$

Converting this to eV, we have

$$mc^2 = (8.20 \times 10^{-14} \text{ J})(1 \text{ eV}/1.60 \times 10^{-19} \text{ J}) = 0.511 \text{ MeV}$$

EXAMPLE 39.11 The Energy of a Speedy Electron

An electron moves with a speed $u = 0.850c$. Find its total energy and kinetic energy in electron volts.

Solution Using the fact that the rest energy of the electron is 0.511 MeV together with Equation 39.24 gives

$$E = \frac{mc^2}{\sqrt{1 - \dfrac{u^2}{c^2}}} = \frac{0.511 \text{ MeV}}{\sqrt{1 - \dfrac{(0.850\,c)^2}{c^2}}}$$

$$= 1.90(0.511 \text{ MeV}) = \boxed{0.970 \text{ MeV}}$$

The kinetic energy is obtained by subtracting the rest energy from the total energy:

$$K = E - mc^2 = 0.970 \text{ MeV} - 0.511 \text{ MeV} = \boxed{0.459 \text{ MeV}}$$

EXAMPLE 39.12 The Energy of a Speedy Proton

The total energy of a proton is three times its rest energy.
(a) Find the proton's rest energy in electron volts.

Solution

Rest energy $= mc^2 = (1.67 \times 10^{-27} \text{ kg})(3.00 \times 10^8 \text{ m/s})^2$

$$= (1.50 \times 10^{-10} \text{ J})(1.00 \text{ eV}/1.60 \times 10^{-19} \text{ J})$$

$$= \boxed{938 \text{ MeV}}$$

(b) With what speed is the proton moving?

Solution   Since the total energy E is three times the rest energy, $E = \gamma mc^2$ (Eq. 39.24) gives

$$E = 3\,mc^2 = \frac{mc^2}{\sqrt{1 - \dfrac{u^2}{c^2}}}$$

$$3 = \frac{1}{\sqrt{1 - \dfrac{u^2}{c^2}}}$$

Solving for u gives

$$\left(1 - \frac{u^2}{c^2}\right) = \frac{1}{9} \quad \text{or} \quad \frac{u^2}{c^2} = \frac{8}{9}$$

$$u = \frac{\sqrt{8}}{3}\,c = \boxed{2.83 \times 10^8 \text{ m/s}}$$

(c) Determine the kinetic energy of the proton in electron volts.

Solution

$$K = E - mc^2 = 3\,mc^2 - mc^2 = 2\,mc^2$$

Since $mc^2 = 938$ MeV

$$K = \boxed{1876 \text{ MeV}}$$

(d) What is the proton's momentum?

Solution   We can use Equation 39.25 to calculate the momentum with $E = 3\,mc^2$:

$$E^2 = p^2c^2 + (mc^2)^2 = (3\,mc^2)^2$$
$$p^2c^2 = 9(mc^2)^2 - (mc^2)^2 = 8(mc^2)^2$$

$$p = \sqrt{8}\,\frac{mc^2}{c} = \sqrt{8}\,\frac{(938 \text{ MeV})}{c} = \boxed{2650\,\frac{\text{MeV}}{c}}$$

The unit of momentum is written MeV/c for convenience.

39.8   EQUIVALENCE OF MASS AND ENERGY

To understand the equivalence of mass and energy, consider the following "thought experiment" proposed by Einstein in developing his famous equation $E = mc^2$. Imagine a box of mass M and length L initially at rest as in Figure 39.20a. Suppose that a pulse of light is emitted from the left side of the box as in Figure 39.20b. From Equation 39.26, we know that the light of energy E carries momentum $p = E/c$. Hence, the box must recoil to the left with a speed v to conserve momentum. Assuming the box is very massive, the recoil speed is small compared with the speed of light, and conservation of momentum gives $Mv = E/c$, or

$$v = \frac{E}{Mc}$$

The time it takes the light to move the length of the box is approximately $\Delta t = L/c$ (where, again, we assume that $v \ll c$). In this time interval, the box moves a small distance Δx to the left, where

$$\Delta x = v\,\Delta t = \frac{EL}{Mc^2}$$

The light then strikes the right end of the box, transfers its momentum to the box, causing the box to stop. With the box in its new position, it appears as if its center of mass has moved to the left. However, its center of mass cannot move because the box is an isolated system. Einstein resolved this perplexing situation by assuming that in addition to energy and momentum, light also carries mass. If m is the equivalent mass carried by the pulse of light, and the center of mass of the box is to remain fixed, then

$$mL = M\,\Delta x$$

Solving for m, and using the previous expression for Δx, we get

$$m = \frac{M\,\Delta x}{L} = \frac{M}{L}\frac{EL}{Mc^2} = \frac{E}{c^2}$$

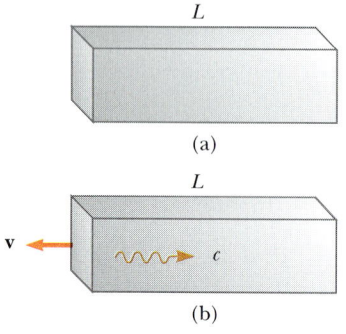

FIGURE 39.20   (a) A box of length L at rest. (b) When a light pulse is emitted at the left end of the box, the box recoils to the left as the pulse strikes the right end.

or

$$E = mc^2$$

Thus, Einstein reached the profound conclusion "If a body gives off the energy E in the form of radiation, its mass diminishes by E/c^2, ... The mass of a body is a measure of its energy content."

It follows that mass varies with speed (relative to the observer). We must therefore distinguish between the **rest mass**, m_o, which is the mass measured by an observer at rest relative to the particle (and at the same location), and the mass measured in real experiments. For a free particle, $m = \gamma m_o$ provides an adequate description. For a larger object, whose center of mass is at rest relative to the observer, $m = \Sigma_i \gamma m_{oi}$ (if we neglect the energy of particle interactions). In every case, the real mass is provided by the total energy, E, divided by the square of the speed of light.

Although we derived the relationship $E = mc^2$ for light energy, the equivalence of mass and energy is universal. Equation 39.24, $E = \gamma m_o c^2$, which represents the total energy of any particle, suggests that even when a particle is at rest ($\gamma = 1$) it still possesses enormous energy through its mass. Probably the clearest experimental proof of the equivalence of mass and energy occurs in nuclear and elementary particle interactions in which there are large amounts of energy released, accompanied by the release of mass. Because energy and mass are related, we see that the laws of conservation of energy and conservation of mass are one and the same. The left side of $E = mc^2$ is constant for an isolated system; therefore the right side must be constant, and thus m (but not m_o) is also constant.

Simply put, this law states that

the energy, and therefore the mass, of a system of particles before interaction must equal the energy, and therefore the mass, of the system after interaction, where the mass is defined as $m = E/c^2$. For free particles, $m = \gamma m_o$, so

$$E_i = \frac{m_{oi} c^2}{\sqrt{1 - \dfrac{u_i^2}{c^2}}}$$

Conservation of mass-energy

The release of enormous energy, accompanied by the change in masses of particles after they have lost their excess energy as they are brought to rest, is the basis of atomic and hydrogen bombs. In fact, whenever energy is released, as in chemical reactions, the residual mass (and energy) are decreased. In a conventional nuclear reactor, the uranium nucleus undergoes fission, a reaction that results in several lighter fragments having considerable kinetic energy. In the case of ^{235}U (the parent nucleus), which undergoes spontaneous fission, the fragments are two lighter nuclei and two neutrons. The total rest mass of the fragments is less than that of the parent nucleus by an amount Δm. The corresponding energy Δmc^2 associated with this mass difference is exactly equal to the total kinetic energy of the fragments. This kinetic energy is then used to produce heat and steam for the generation of electrical power.

Next, consider the basic fusion reaction in which two deuterium atoms combine to form one helium atom. This reaction is of major importance in current research and development of controlled-fusion reactors. The decrease in rest mass that results from the creation of one helium atom from two deuterium atoms is $\Delta m = 4.25 \times 10^{-29}$ kg. Hence, the corresponding excess energy that results from one fusion reaction is $\Delta mc^2 = 3.83 \times 10^{-12}$ J $= 23.9$ MeV. To appreciate the magnitude of this result, if 1 g of deuterium is converted to helium, the energy released is about 10^{12} J! At the 1995 cost of electrical energy, this would be worth about $60 000.

CONCEPTUAL EXAMPLE 39.13

Because mass is a measure of energy, can we conclude that a compressed spring has more mass than the same spring when it is not compressed?

Reasoning   Recall that when a spring of force constant k is compressed (or stretched) from its equilibrium position by a distance x, it stores elastic potential energy $U = kx^2/2$. According to the theory of special relativity, any change in the total energy of a system is equivalent to a change in mass of the system. Therefore, the mass of a compressed (or stretched) spring is greater than the mass of the spring in its equilibrium position by an amount U/c^2.

EXAMPLE 39.14   Binding Energy of the Deuteron

The mass of the deuteron, which is the nucleus of "heavy hydrogen," is not equal to the sum of the masses of its constituents, which are the proton and neutron. Calculate this mass difference and determine its energy equivalence.

Solution   Using atomic mass units (u), we have

$$m_p = \text{mass of proton} = 1.007276 \text{ u}$$

$$m_n = \text{mass of neutron} = 1.008665 \text{ u}$$

$$m_p + m_n = 2.015941 \text{ u}$$

Since the mass of the deuteron is 2.013553 u (Appendix A), we see that the mass difference Δm is 0.002388 u. By defini-

tion, 1 u $= 1.66 \times 10^{-27}$ kg, and therefore

$$\Delta m = 0.002388 \text{ u} = \boxed{3.96 \times 10^{-30} \text{ kg}}$$

Using $E = \Delta mc^2$, we find that

$$E = \Delta mc^2 = (3.96 \times 10^{-30} \text{ kg})(3.00 \times 10^8 \text{ m/s})^2$$

$$= 3.56 \times 10^{-13} \text{ J} = \boxed{2.23 \text{ MeV}}$$

Therefore, the minimum energy required to separate the proton from the neutron of the deuterium nucleus (the binding energy) is 2.23 MeV.

39.9   RELATIVITY AND ELECTROMAGNETISM

Relativity requires that the laws of physics must be the same in all inertial frames. The intimate connection between relativity and electromagnetism was recognized in his first paper on relativity entitled "On the Electrodynamics of Moving Bodies." As we have seen, a light wave in one frame must be a light wave in any other frame, and its speed must be c relative to any observer. This is consistent with Maxwell's theory of electromagnetism. That is, Maxwell's equations require no modification in relativity.

   To understand the relation between relativity and electromagnetism, we now describe a situation that shows how an electric field in one frame of reference is viewed as a magnetic field in another frame of reference. Consider a wire carrying a current, and suppose that a positive test charge q is moving parallel to the wire with velocity $\mathbf{u}$ as in Figure 39.21a. We assume that the net charge on the wire is

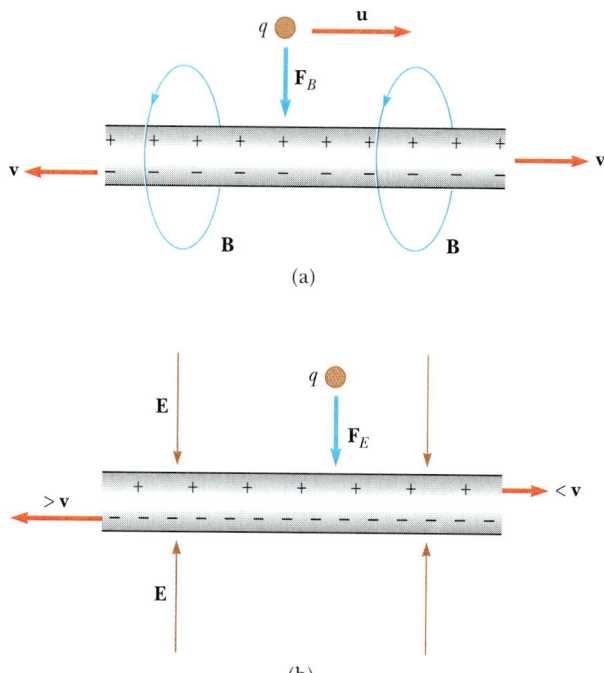

FIGURE 39.21 (a) A positive test charge moving to the right with a velocity **u** near a wire carrying a current. In the frame of the wire, positive and negative charges of equal densities move in opposite directions, the net charge is zero, and **E** = 0. A magnetic field **B** surrounds the wire, and the charge experiences a magnetic force toward the wire. (b) In the rest frame of the test charge, the negative charges in the wire are contracted more than the positive charges, so the wire has a net negative charge, creating an electric field **E** directed toward the wire. Hence, the test charge experiences an electric force toward the wire.

zero. The current in the wire produces a magnetic field that forms circles around the wire and is directed out of the page at the moving test charge. This results in a magnetic force $\mathbf{F}_B = q\mathbf{u} \times \mathbf{B}$ on the test charge acting toward the wire, but no electric force because the net charge on the wire is zero when viewed in this frame.

Now consider the same situation as viewed from the frame of the test charge as in Figure 39.21b. In this frame, the positive charges in the wire move slower relative to the test charge than the negative charges in the wire. Because of length contraction, distances between positive charges in the wire are smaller than distances between negative charges. Hence, there is a net negative charge on the wire when viewed in this frame. The net negative charge produces an electric field pointing toward the wire, and our positive test charge experiences an electric force toward the wire. Thus, what was viewed as a magnetic field in the frame of the wire transforms into an electric field in the frame of the test charge.

*39.10 GENERAL RELATIVITY

Up to this point, we have sidestepped a curious puzzle. Mass has two seemingly different properties: a *gravitational attraction* for other masses and an *inertial* property that resists acceleration. To designate these two attributes, we use the subscripts g and i and write

$$\text{Gravitational property} \qquad F_g = m_g g$$

$$\text{Inertial property} \qquad F = m_i a$$

The value for the gravitational constant G was chosen to make the magnitudes of m_g and m_i numerically equal. Regardless of how G is chosen, however, the strict equality of m_g and m_i has been measured to an extremely high degree: a few parts in 10^{12}. Thus, it appears that gravitational mass and inertial mass may indeed be exactly equal.

But why? They seem to involve two entirely different concepts: a force of mutual gravitational attraction between two masses and the resistance of a single mass to being accelerated. This question, which puzzled Newton and many other physicists over the years, was answered when Einstein published his theory of gravitation, known as *general relativity,* in 1916. Because it is a mathematically complex theory, we merely offer a hint of its elegance and insight.

In Einstein's view, the remarkable coincidence that m_g and m_i seemed to be exactly proportional was evidence for a very intimate and basic connection between the two concepts. He pointed out that no mechanical experiment (such as dropping a mass) could distinguish between the two situations illustrated in Figures 39.22a and 39.22b. In each case, a mass released by the observer undergoes a downward acceleration of g relative to the floor.

Einstein carried this idea further and proposed that *no* experiment, mechanical or otherwise, could distinguish between the two cases. This extension to include all phenomena (not just mechanical ones) has interesting consequences. For example, suppose that a light pulse is sent horizontally across the box, as in Figure 39.22c. The trajectory of the light pulse bends downward as the box accelerates upward to meet it. Therefore, Einstein proposed that a beam of light should

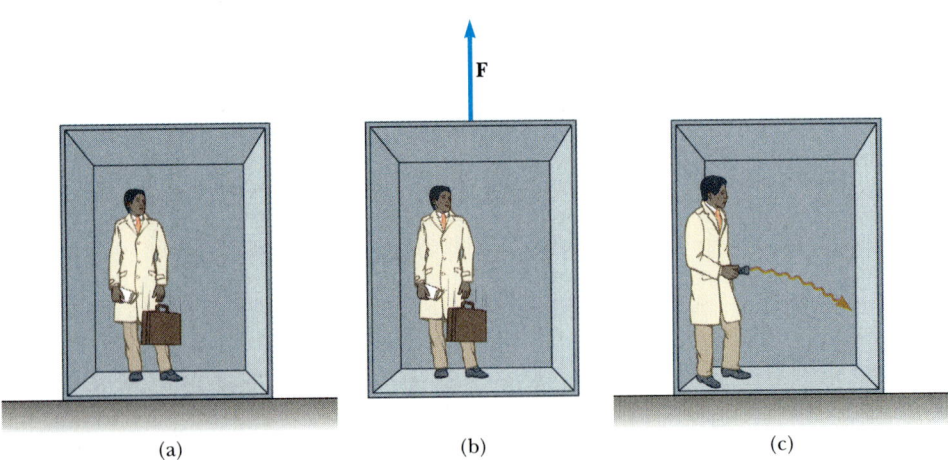

(a)   (b)   (c)

FIGURE 39.22   (a) The observer is at rest in a uniform gravitational field **g**. (b) The observer is in a region where gravity is negligible, but the frame of reference is accelerated by an external force **F** that produces an acceleration **g**. According to Einstein, the frames of reference in parts (a) and (b) are equivalent in every way. No local experiment could distinguish any difference between the two frames. (c) If parts (a) and (b) are truly equivalent, as Einstein proposed, then a ray of light would bend in a gravitational field.

also be bent downward by a gravitational field. (No such bending is predicted in Newton's theory of gravitation.)

The two postulates of Einstein's **general relativity** are as follows:

- All the laws of nature have the same form for observers in any frame of reference, whether accelerated or not.
- In the vicinity of any given point, a gravitational field is equivalent to an accelerated frame of reference in the absence of gravitational effects. (This is the *principle of equivalence*.)

The second postulate implies that gravitational mass and inertial mass are completely equivalent, not just proportional. What were thought to be two different types of mass are actually identical.

One interesting effect predicted by general relativity is that time scales are altered by gravity. A clock in the presence of gravity runs more slowly than one where gravity is negligible. Consequently, the frequencies of radiation emitted by atoms in the presence of a strong gravitational field are *red-shifted* to lower frequencies when compared with the same emissions in a weak field. This gravitational red shift has been detected in spectral lines emitted by atoms in massive stars. It has also been verified on the Earth by comparing the frequencies of gamma rays (a high energy form of electromagnetic radiation) emitted from nuclei separated vertically by about 20 m.

The second postulate suggests that a gravitational field may be "transformed away" at any point if we choose an appropriate accelerated frame of reference—a freely falling one. Einstein developed an ingenious method of describing the acceleration necessary to make the gravitational field "disappear." He specified a certain quantity, the *curvature of space-time,* that describes the gravitational effect at every point. In fact, the curvature of space-time completely replaces Newton's gravitational theory. According to Einstein, there is no such thing as a gravitational force. Rather, the presence of a mass causes a curvature of space-time in the vicinity of the mass, and this curvature dictates the space-time path that all freely moving objects must follow. As one physicist says: "Mass tells space-time how to curve; curved space-time tells mass how to move." One important test of general relativity is the prediction that a light ray passing near the Sun should be deflected by some angle. This prediction was confirmed by astronomers as bending of starlight during a total solar eclipse shortly following World War I (Fig. 39.23).

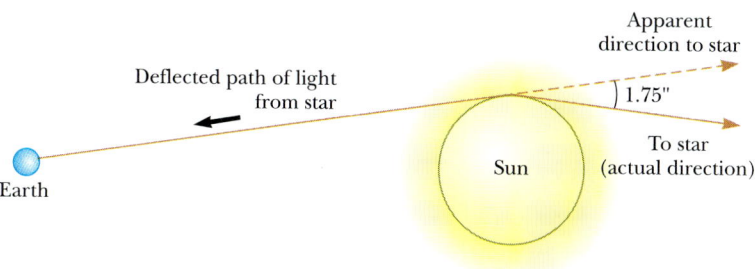

FIGURE 39.23   Deflection of starlight passing near the Sun. Because of this effect, the Sun and other remote objects can act as a *gravitational lens.* In his general theory of relativity, Einstein calculated that starlight just grazing the Sun's surface should be deflected by an angle of 1.75″.

If the concentration of mass becomes very great, as is believed to occur when a large star exhausts its nuclear fuel and collapses to a very small volume, a **black hole** may form. Here the curvature of space-time is so extreme that, within a certain distance from the center of the black hole, all matter and light become trapped.

SUMMARY

The two basic postulates of the special theory of relativity are

- All the laws of physics are the same in all inertial reference frames.
- The speed of light in vacuum has the same value, $c = 3.00 \times 10^8$ m/s, in all inertial frames, regardless of the velocity of the observer or the velocity of the source emitting the light.

Three consequences of the special theory of relativity are as follows:

- Events that are simultaneous for one observer are not simultaneous for another observer who is in motion relative to the first.
- Clocks in motion relative to an observer appear to be slowed down by a factor γ. This is known as **time dilation.**
- Lengths of objects in motion appear to be contracted in the direction of motion.

To satisfy the postulates of special relativity, the Galilean transformations must be replaced by the **Lorentz transformations:**

$$x' = \gamma(x - vt)$$
$$y' = y$$
$$z' = z \tag{39.9}$$
$$t' = \gamma\left(t - \frac{v}{c^2}x\right)$$

where $\gamma = (1 - v^2/c^2)^{-1/2}$.

The relativistic form of the **velocity transformation** is

$$u'_x = \frac{u_x - v}{1 - \frac{u_x v}{c^2}} \tag{39.14}$$

where u_x is the speed of an object as measured in the S frame and u'_x is its speed measured in the S′ frame.

The relativistic expression for the **momentum** of a particle moving with a velocity **u** is

$$\mathbf{p} \equiv \frac{m\mathbf{u}}{\sqrt{1 - \frac{u^2}{c^2}}} = \gamma m\mathbf{u} \tag{39.18}$$

The relativistic expression for the **kinetic energy** of a particle is

$$K = \gamma mc^2 - mc^2 \tag{39.22}$$

where mc^2 is called the **rest energy** of the particle.

The total energy E of a particle is related to the **mass** through the famous **energy-mass** equivalence expression:

$$E = \gamma mc^2 = \frac{mc^2}{\sqrt{1 - \dfrac{u^2}{c^2}}} \qquad (39.24)$$

The relativistic momentum is related to the total energy through the equation

$$E^2 = p^2 c^2 + (mc^2)^2 \qquad (39.25)$$

QUESTIONS

1. What two speed measurements do two observers in relative motion always agree on?
2. A spaceship in the shape of a sphere moves past an observer on Earth with a speed $0.5\,c$. What shape does the observer see as the spaceship moves past?
3. An astronaut moves away from the Earth at a speed close to the speed of light. If an observer on Earth measures the astronaut's size and pulse rate, what changes (if any) would the observer measure? Would the astronaut measure any changes?
4. Two identical clocks are synchronized. One is put in orbit directed eastward around the Earth while the other remains on Earth. Which clock runs slower? When the moving clock returns to Earth, are the two still synchronized?
5. Two lasers situated on a moving spacecraft are triggered simultaneously. An observer on the spacecraft claims to see the pulses of light simultaneously. What condition is necessary so that a secondary observer agrees?
6. When we say that a moving clock runs slower than a stationary one, does this imply that there is something physically unusual about the moving clock?
7. List some ways our day-to-day lives would change if the speed of light were only 50 m/s.
8. Give a physical argument that shows that it is impossible to accelerate an object of mass m to the speed of light, even with a continuous force acting on it.
9. It is said that Einstein, in his teenage years, asked the question, "What would I see in a mirror if I carried it in my hands and ran at the speed of light?" How would you answer this question?
10. What happens to the density of an object as its speed increases? Note that relativistic density is $m/V = E/c^2 V$.
11. Some of the distant stars, called quasars, are receding from us at half the speed of light (or greater). What is the speed of the light we receive from these quasars?
12. How is it possible that photons of light, which have zero rest mass, have momentum?
13. With regard to reference frames, how does general relativity differ from special relativity?

PROBLEMS

Section 39.1   The Principle of Newtonian Relativity

1. In a laboratory frame of reference, an observer notes that Newton's second law is valid. Show that it is also valid for an observer moving at a constant speed relative to the laboratory frame.
2. Show that Newton's second law is not valid in a reference frame moving past the laboratory frame of Problem 1 with a constant acceleration.
3. A 2000-kg car moving at 20 m/s collides with and sticks to a 1500-kg car at rest at a stop sign. Show that momentum is conserved in a reference frame moving at 10 m/s in the direction of the moving car.
4. A billiard ball of mass of 0.30 kg moving at 5.0 m/s collides elastically with a ball of mass 0.20 kg moving in the opposite direction at 3.0 m/s. Show that momentum is conserved in a frame of reference moving with a speed of 2.0 m/s in the direction of the second ball.

5. A ball is thrown at 20 m/s inside a boxcar moving along the tracks at 40 m/s. What is the speed of the ball relative to the ground if the ball is thrown (a) forward, (b) backward, (c) out the side door?
5A. A ball is thrown at a speed v_b inside a boxcar moving along the tracks at a speed v. What is the speed of the ball relative to the ground if the ball is thrown (a) forward, (b) backward, (c) out the side door?

□ indicates problems that have full solutions available in the Student Solutions Manual and Study Guide.

Section 39.4   Consequences of Special Relativity

6. At what speed does a clock have to move in order to run at a rate that is one-half the rate of a clock at rest?

7. In 1962, when Scott Carpenter orbited Earth 22 times, the press stated that for each orbit he aged 2.0×10^{-6} s less than he would have had he remained on Earth. (a) Assuming he was 160 km above Earth in a circular orbit, determine the time difference between someone on Earth and Carpenter for the 22 orbits. (*Hint:* Use the approximation $\sqrt{1-x} \approx 1 - x/2$ for small x.) (b) Was the press information accurate? Explain.

8. The proper length of one spaceship is three times that of another. The two ships are traveling in the same direction and, while both are passing overhead, an Earth observer measures them to have the same length. If the slower ship is moving at $0.35c$, determine the speed of the faster one.

8A. The proper length of one spaceship is N times that of another. The two ships are traveling in the same direction and, while both are passing overhead, an Earth observer measures them to have the same length. If the slower ship is moving with speed v, determine the speed of the faster spaceship.

9. A spaceship of proper length 300 m takes 0.75 μs to pass an Earth observer. Determine its speed as measured by the Earth observer.

9A. A spaceship of proper length L_p takes t seconds to pass an Earth observer. Determine its speed as measured by the Earth observer.

10. Muons move in circular orbits at a speed of $0.9994c$ in a storage ring of radius 500 m. If a muon at rest decays into other particles after $T = 2.20$ μs, how many trips around the storage ring do we expect the muons to make before they decay?

11. A spacecraft moves at $0.90c$. If its length is L_0 when measured from inside the spacecraft, what is its length measured by a ground observer?

12. The cosmic rays of highest energy are protons, having kinetic energy of 10^{13} MeV. (a) How long would it take a proton of this energy to travel across the Milky Way galaxy, of diameter 10^5 lightyears, as measured in the proton's frame? (b) From the point of view of the proton, how many kilometers across is the galaxy?

13. The pion has an average lifetime of 26.0 ns when at rest. In order for it to travel 10.0 m, how fast must it move?

14. If astronauts could travel at $v = 0.95c$, we on Earth would say it takes $(4.2/0.95) = 4.4$ years to reach Alpha Centauri, 4.2 lightyears away. The astronauts disagree. (a) How much time passes on the astronauts' clocks? (b) What distance to Alpha Centauri do the astronauts measure?

Section 39.5   The Lorentz Transformation Equations

15. A spaceship travels at $0.75c$ relative to Earth. If the spaceship fires a small rocket in the forward direction, what initial speed (relative to the ship) must the rocket have in order for it to travel at $0.95c$ relative to Earth?

16. A certain quasar recedes from the Earth at $v = 0.87c$. A jet of material ejected from the quasar toward the Earth moves at $0.55c$ relative to the quasar. Find the speed of the ejected material relative to the Earth.

17. Two jets of material from the center of a radio galaxy fly away in opposite directions. Both jets move at $0.75c$ relative to the galaxy. Determine the speed of one jet relative to the other.

18. A Klingon spaceship moves away from the Earth at a speed of $0.80c$ (Fig. P39.18). The Starship Enterprise pursues at a speed of $0.90c$ relative to the Earth. Observers on Earth see the Enterprise overtaking the Klingon ship at a relative speed of $0.10c$. With what speed is the Enterprise overtaking the Klingon ship as seen by the crew of the Enterprise?

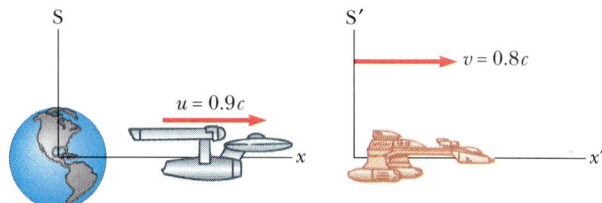

FIGURE P39.18   The Earth is frame S. The Klingon ship is frame S'. The Enterprise is the object whose motion is followed from S and S'.

19. A cube of steel has a volume of 1.0 cm^3 and a mass of 8.0 g when at rest on the Earth. If this cube is now given a speed $v = 0.90c$, what is its density as measured by a stationary observer? Note that relativistic density is $m/V = E/c^2V$.

19A. A cube of steel has a volume V and a mass m when at rest on the Earth. If this cube is now given a speed v, what is its density as measured by a stationary observer? Note that relativistic density is $m/V = E/c^2V$.

Section 39.6   Relativistic Momentum and Relativistic Form of Newton's Laws

20. Calculate the momentum of a proton moving at (a) $0.0100c$, (b) $0.500c$, (c) $0.900c$.

21. Find the momentum of a proton in MeV/c units if its total energy is twice its rest energy.

22. Show that the speed of an object having momentum p and mass m is

$$v = \frac{c}{\sqrt{1 + (mc/p)^2}}$$

Section 39.7 Relativistic Energy

23. Show that the energy-momentum relationship $E^2 = p^2c^2 + (mc^2)^2$ follows from the expressions $E = \gamma mc^2$ and $p = \gamma mu$.

24. Cherenkov radiation is given off when an electron travels faster than the speed of light in a medium, the relativistic equivalent of a sonic boom. Consider an electron traveling 10 percent faster than light in water. Determine (a) the electron's total energy and kinetic energy in electron volts and (b) its momentum in MeV/c.

25. A proton moves at $0.95\,c$. Calculate its (a) rest energy, (b) total energy, and (c) kinetic energy.

26. The total volume of water in the oceans is approximately 1.4×10^9 km^3. The density of sea water is 1030 kg/m^3, and its heat capacity is 4200 J/kg · °C. Estimate the increase in mass of sea water corresponding to an increase in temperature of 10°C.

27. Find the speed of a particle whose total energy is twice its rest energy.

28. A proton in a high-energy accelerator is given a kinetic energy of 50 GeV. Determine (a) its momentum and (b) its speed.

29. Determine the energy required to accelerate an electron from (a) $0.50\,c$ to $0.90\,c$ and (b) $0.90\,c$ to $0.99\,c$.

30. In a typical color television tube, the electrons are accelerated through a potential difference of 25 000 V. (a) What speed do the electrons have when they strike the screen? (b) What is their kinetic energy in joules?

31. Electrons are accelerated to an energy of 20 GeV in the 3.0-km-long Stanford Linear Accelerator. (a) What is the γ factor for the electrons? (b) What is their speed? (c) How long does the accelerator appear to them?

32. A spaceship of mass 1.0×10^6 kg is to be accelerated to $0.60\,c$. (a) How much energy does this require? (b) How many kilograms of mass (apart from its fuel) does the spaceship gain from burning its fuel?

33. A pion at rest $(m_\pi = 270\,m_e)$ decays to a muon $(m_\mu = 206\,m_e)$ and an antineutrino $(m_\nu = 0)$: $\pi^- \rightarrow \mu^- + \bar{\nu}$. Find the kinetic energy of the muon and the antineutrino in electron volts. (*Hint:* Relativistic momentum is conserved.)

Section 39.8 Equivalence of Mass and Energy

34. In a nuclear power plant, the fuel rods last three years. If a 1.0-GW plant operates at 80 percent capacity for the three years, what is the loss of mass (to the steam) of the fuel?

35. Consider the decay $^{55}_{24}\text{Cr} \rightarrow \,^{55}_{25}\text{Mn} + \text{e}$, where e is an electron. The ^{55}Cr nucleus has a mass of 54.9279 u, and the ^{55}Mn nucleus has a mass of 54.9244 u. (a) Calculate the mass difference between the two

nuclei in electron volts. (b) What is the maximum kinetic energy of the emitted electron?

36. A ^{57}Fe nucleus at rest emits a 14-keV photon. Use the conservation of energy and momentum to deduce the kinetic energy of the recoiling nucleus in electron volts. (Use $Mc^2 = 8.6 \times 10^{-9}$ J for the final state of the ^{57}Fe nucleus.)

37. The power output of the Sun is 3.8×10^{26} W. How much rest mass is converted to kinetic energy in the Sun each second?

38. A gamma ray (a high-energy photon of light) can produce an electron (e$^-$) and a positron (e$^+$) when it enters the electric field of a heavy nucleus: $(\gamma \rightarrow \text{e}^+ + \text{e}^-)$. What minimum γ-ray energy is required to accomplish this task? (*Hint:* The masses of the electron and the positron are equal.)

ADDITIONAL PROBLEMS

39. A spaceship moves away from Earth at $0.50\,c$ and fires a shuttle craft that then moves in the forward direction at $0.50\,c$ relative to the ship. The pilot of the shuttle launches a probe forward at speed $0.50\,c$ relative to the shuttle. Determine (a) the speed of the shuttle relative to Earth and (b) the speed of the probe relative to Earth.

39A. A spaceship moves away from Earth at a speed v and fires a shuttle craft that then moves in the forward direction at a speed v relative to the ship. The pilot of the shuttle launches a probe at speed v relative to the shuttle. Determine (a) the speed of the shuttle relative to Earth and (b) the speed of the probe relative to Earth.

40. An astronaut wishes to visit the Andromeda galaxy (2 million lightyears away) in a one-way trip that will take 30 years in the spaceship's frame of reference. Assuming that his speed is constant, how fast must he travel relative to the Earth?

41. The net nuclear reaction inside the Sun is $4\text{p} \rightarrow \,^4\text{He} + \Delta E$. If the rest mass of each proton is 938.2 MeV and the rest mass of the ^{4}He nucleus is 3727 MeV, calculate the percentage of the starting mass that is released as energy.

42. The annual energy requirement for the United States is on the order of 10^{20} J. How many kilograms of matter would have to be released as energy to meet this requirement?

43. A rocket moves toward a mirror at $0.80\,c$ relative to reference frame S in Figure P39.43. The mirror is stationary relative to S. A light pulse emitted by the rocket travels to the mirror and is reflected back to the rocket. The front of the rocket is 1.8×10^{12} m from the mirror (as measured by observers in S) at the moment the light pulse leaves the rocket. What is the total travel time of the pulse as measured by observers in (a) the S frame and (b) the front of the rocket?

43A. A rocket moves toward a mirror at speed v relative to the reference frame labeled by S in Figure P39.43. The mirror is stationary with respect to S. A light pulse emitted by the rocket travels toward the mirror and is reflected back to the rocket. The front of the rocket is a distance d from the mirror (as measured by observers in S) at the moment the light pulse leaves the rocket. What is the total travel time of the pulse as measured by observers in (a) the S frame and (b) the front of the rocket?

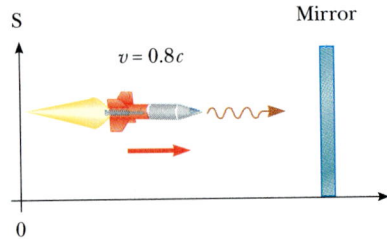

FIGURE P39.43

44. How fast would a motorist have to be going to make a red light appear green? ($\lambda_{red} = 650$ nm, $\lambda_{green} = 550$ nm.) In computing this, use the correct relativistic formula for Doppler shift:

$$\frac{\Delta\lambda}{\lambda} + 1 = \sqrt{\frac{c - v}{c + v}}$$

where v is the approach speed and λ is the source wavelength.

45. A physics professor on Earth gives an exam to students who are on a rocket ship traveling at speed v relative to Earth. The moment the ship passes the professor, she signals the start of the exam. If she wishes her students to have time T_0 (rocket time) to complete the exam, show that she should wait an Earth time

$$T = T_0\sqrt{\frac{1 - v/c}{1 + v/c}}$$

before sending a light signal telling them to stop. (*Hint:* Remember that it takes some time for the second light signal to travel from the professor to the students.)

46. Ted and Mary are playing catch in frame S, which is moving at $0.60c$ relative to frame S, while Jim in frame S watches the action (Fig. P39.46). Ted throws the ball to Mary at $0.80c$ (according to Ted) and their separation (measured in S') is 1.8×10^{12} m. (a) According to Mary, how fast is the ball moving? (b) According to Mary, how long does it take the ball to

reach her? (c) According to Jim, how far apart are Ted and Mary, and how fast is the ball moving? (d) According to Jim, how long does it take the ball to reach Mary?

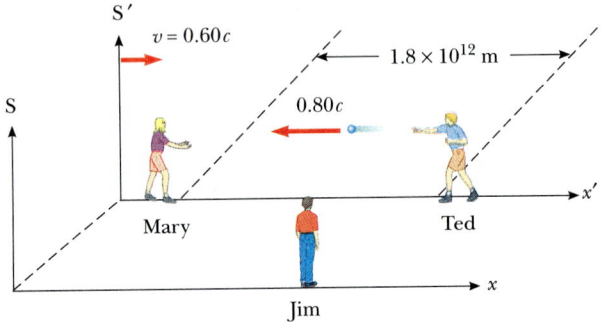

FIGURE P39.46

47. Spaceship I, which contains students taking a physics exam, approaches Earth with a speed of $0.60c$ (relative to Earth), while spaceship II, which contains professors proctoring the exam, moves at $0.28c$ (relative to Earth) directly toward the students. If the professors stop the exam after 50 min have passed on their clock, how long does the exam last as measured by (a) the students and (b) an observer on Earth?

47A. Spaceship I, which contains students taking a physics exam, approaches Earth with a speed v_I (relative to Earth), while spaceship II, which contains professors proctoring the exam, moves at a speed v_{II} (relative to Earth) directly toward the students. If the professors stop the exam after t_{II} minutes have passed on their clock, how long does the exam last as measured by (a) the students and (b) an observer on Earth?

48. A rod of length L_0 moving with a speed v along the horizontal direction makes an angle of θ_0 with respect to the x' axis. (a) Show that the length of the rod as measured by a stationary observer is $L = L_0[1 - (v^2/c^2)\cos^2\theta_0]^{1/2}$. (b) Show that the angle that the rod makes with the x axis is $\tan\theta = \gamma\tan\theta_0$. These results show that the rod is both contracted and rotated. (Take the lower end of the rod to be at the origin of the primed coordinate system.)

49. Imagine a spacecraft that starts from Earth moving at constant speed to the yet-to-be-discovered planet Retah, which is 20 lighthours away from Earth. It takes 25 h (according to an Earth observer) for the spacecraft to reach this planet. Assuming that clocks on Earth and in the spacecraft are synchronized at the beginning of the journey, compare the time elapsed in the spacecraft's frame for the one-way journey with the time elapsed in Earth's frame.

50. If the number of muons at $t = 0$ is N_0, the number at time t is $N = N_0 e^{-t/\tau}$ where τ is the mean lifetime,

equal to 2.2 μs. Suppose muons move at $0.95c$ and there are 5.0×10^4 of them at $t = 0$. (a) What is the observed lifetime of the muons? (b) How many remain after traveling 3.0 km?

51. Consider two inertial reference frames S and S′, where S′ is moving to the right with a constant speed of $0.60c$ relative to S. A stick of proper length 1.0 m moves to the left toward the origins of both S and S′, and the length of the stick is 50 cm as measured by an observer in S′. (a) Determine the speed of the stick as measured by observers in S and S′. (b) What is the length of the stick as measured by an observer in S?

51A. Consider two inertial reference frames S and S′, where S′ is moving to the right with a constant speed v relative to S. A stick of proper length L_p moves to the left toward the origins of both S and S′, and the length of the stick is $L′$ as measured by an observer in S′. (a) Determine the speed of the stick as measured by observers in S and S′. (b) What is the length of the stick as measured by an observer in S?

52. Suppose our Sun is about to explode. In an effort to escape, we depart in a spaceship at $v = 0.80c$ and head toward the star Tau Ceti, 12 lightyears away. When we reach the midpoint of our journey from the Earth, we see our Sun explode and, unfortunately, at the same instant we see Tau Ceti explode as well. (a) In the spaceship's frame of reference, should we conclude that the two explosions occurred simultaneously? If not, which occurred first? (b) In a frame of reference in which the Sun and Tau Ceti are at rest, did they explode simultaneously? If not, which exploded first?

53. Two rockets are on a collision course. They are moving at $0.800c$ and $0.600c$ and are initially 2.52×10^{12} m apart as measured by Liz, the Earth observer in Figure P39.53. Both rockets are 50.0 m in length as measured by Liz. (a) What are their respective proper lengths? (b) What is the length of each rocket as measured by an observer in the other rocket? (c) According to Liz, how long before the rockets

collide? (d) According to rocket 1, how long before they collide? (e) According to rocket 2, how long before they collide? (f) If both rocket crews are capable of total evacuation within 90 min (their own time), will there be any casualties?

54. *The red shift.* A light source recedes from an observer with a speed v_s, which is small compared with c. (a) Show that the fractional shift in the measured wavelength is given by the approximate expression

$$\frac{\Delta \lambda}{\lambda} \approx \frac{v_s}{c}$$

This result is known as the red shift, because the visible light is shifted toward the red. (b) Spectroscopic measurements of light at $\lambda = 397$ nm coming from a galaxy in Ursa Major reveal a red shift of 20 nm. What is the recessional speed of the galaxy?

55. A particle having charge q moves at speed v along a straight line in a uniform electric field E. If the motion and the electric field are both in the x direction, (a) show that the acceleration of the particle in the x direction is

$$a = \frac{dv}{dt} = \frac{qE}{m} \left(1 - \frac{v^2}{c^2} \right)^{3/2}$$

(b) Discuss the significance of the fact that the acceleration depends on the speed. (c) If the particle starts from rest at $x = 0$ at $t = 0$, how would you find its speed and position after a time t has elapsed?

56. Consider two inertial reference frames S and S′, where S′ is moving to the right with constant speed $0.60c$ relative to S. Jennifer is located 1.8×10^{11} m to the right of the origin of S and is fixed in S (as measured by an observer in S), while Matt is fixed in S′ at the origin (as measured by an observer in S′). At the instant their origins coincide, Matt throws a ball toward Jennifer at $0.80c$ as measured by Matt (Fig. P39.56). (a) What is the speed of the ball as measured by Jennifer? How long before Jennifer catches the ball, as measured by (b) Jennifer, (c) the ball, and (d) Matt?

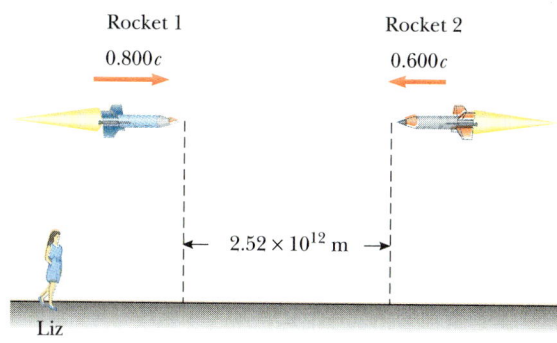

FIGURE P39.53

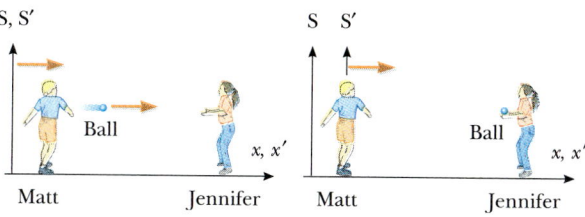

Matt throws the ball   Jennifer catches the ball

(a)   (b)

FIGURE P39.56

57. As measured by observers in a reference frame S, a particle having charge q moves with velocity **v** in a magnetic field **B** and an electric field **E**. The resulting force on the particle is $\mathbf{F} = q(\mathbf{E} + \mathbf{v} \times \mathbf{B})$. Another observer moves along with the particle and measures its charge to be q also but its electric field to be $\mathbf{E}'$. If both observers are to measure the same force **F**, show that $\mathbf{E}' = \mathbf{E} + \mathbf{v} \times \mathbf{B}$.

SPREADSHEET PROBLEMS

S1. Astronomers use the Doppler shift in the Balmer series of the hydrogen spectrum to determine the radial speed of a galaxy. The fractional change in the wavelength of the spectral line is given by

$$Z = \frac{\Delta\lambda}{\lambda_0} = \frac{\lambda - \lambda_0}{\lambda_0} = \sqrt{\frac{1 + v/c}{1 - v/c}} - 1$$

Once this quantity is measured for a particular receding galaxy, the speed of recession can be found by solving for v/c in terms of Z. Spreadsheet 39.1 calculates the speed of recession for a range of Z values. (a) What is the speed for galaxies having $Z = 0.2$, 0.5, 1.0, and 2.0? (b) The largest Z values, $Z \approx 3.8$, have been measured for several quasars (quasi-stellar radio sources). How fast are these quasars moving away from us?

S2. Astronauts in a starship traveling at a speed v relative to a starbase are given instructions from mission control to call back in 1 h as measured by the starship clocks. Spreadsheet 39.2 calculates how long mission control has to wait for the call for different starship speeds. How long does mission control have to wait if the ship is traveling at $v = 0.1\,c$, $0.2\,c$, $0.4\,c$, $0.6\,c$, $0.8\,c$, $0.9\,c$, $0.95\,c$, $0.995\,c$, $0.9995\,c$?

S3. Most astronomers believe that the Universe began at some instant with an explosion called the Big Bang and that the observed recession of the galaxies is a direct result of this explosion. If the galaxies recede from each other at a constant rate, then we expect that the galaxies moving fastest are now farthest away from Earth. This result, called Hubble's law after Edwin Hubble, can be written $v = Hr$, where H can be determined from observation, v is the speed of recession of the galaxy, and r is its distance from Earth. Current estimates of H range from 15 to 30 km/s/Mly, where Mly is the distance light travels in one million years, and a conservative estimate is $H = 20$ km/s/Mly. Use Spreadsheet 39.1 to calculate the distance from Earth for each galaxy in Problem S1.

S4. Design a spreadsheet program to calculate and plot the relativistic kinetic energy (Eq. 39.22) and the classical kinetic energy ($\frac{1}{2}mu^2$) of a macroscopic object. Plot the relativistic and classical energies versus speed on the same graph. (a) For an object of rest mass $m_0 = 3$ kg, at what speed does the classical kinetic energy underestimate the relativistic value by 1 percent? 5 percent? 50 percent? What is the relativistic kinetic energy at these speeds? Repeat part (a) for (b) an electron and (c) a proton.

CHAPTER 40

Introduction to Quantum Physics

In the drift-tube section of the Clinton P. Anderson Meson Physics Facility accelerator, particles are accelerated to an energy of 100 million eV. This accelerator provides beams of protons, pions, muons, neutrons, and neutrinos for scientists and students from more than 25 countries. *(Courtesy of Los Alamos National Laboratory)*

In the previous chapter, we discussed the fact that Newtonian mechanics must be replaced by Einstein's special theory of relativity when we are dealing with particle speeds comparable with the speed of light. Although many problems were indeed resolved by the theory of relativity in the early part of the 20th century, many experimental and theoretical problems remained unanswered. Attempts to apply the laws of classical physics to explain the behavior of matter on the atomic scale were consistently unsuccessful. Various phenomena, such as blackbody radiation, the photoelectric effect, and the emission of sharp spectral lines by atoms in a gas discharge could not be explained within the framework of classical physics.

As physicists sought new ways to solve these puzzles, another revolution took place in physics between 1900 and 1930. This new theory called *quantum mechanics* was highly successful in explaining the behavior of atoms, molecules, and nuclei. Like relativity, the quantum theory requires a modification of our ideas concerning the physical world.

The basic ideas of quantum theory were first introduced by Max Planck, but most of the subsequent mathematical developments and interpretations were made by a number of distinguished physicists, including Einstein, Bohr, Schrödinger, de Broglie, Heisenberg, Born, and Dirac. Despite the great success of the quantum theory, Einstein frequently played the role of critic, especially with regard to the manner in which the theory was interpreted. In particular, Einstein did not accept Heisenberg's interpretation of the uncertainty principle, which says that it is impossible to obtain a precise simultaneous measurement of the position and the velocity of a particle. According to this principle, it is possible to predict only the probability of the future of a system, contrary to the deterministic view held by Einstein.[1]

An extensive study of quantum theory is certainly beyond the scope of this book, and therefore this chapter is simply an introduction to its underlying ideas. We also discuss two simple applications of quantum theory: the photoelectric effect and the Compton effect.

40.1 BLACKBODY RADIATION AND PLANCK'S HYPOTHESIS

An object at any temperature emits radiation sometimes referred to as **thermal radiation.** The characteristics of this radiation depend on the temperature and properties of the object. At low temperatures, the wavelengths of the thermal radiation are mainly in the infrared region and, hence, not observed by the eye. As the temperature of the object increases, the object eventually begins to glow red —in other words, the thermal radiation shifts to the visible part of the spectrum. At sufficiently high temperatures, it appears to be white, as in the glow of the hot tungsten filament of a lightbulb. Careful study shows that, as the temperature of an object increases, the thermal radiation it emits consists of a continuous distribution of wavelengths from the infrared, visible, and ultraviolet portions of the spectrum.

From a classical viewpoint, thermal radiation originates from accelerated charges near the surface of the object; those charges emit radiation much as small antennas do. The thermally agitated charges can have a distribution of accelerations, which accounts for the continuous spectrum of radiation emitted by the object. By the end of the 19th century, however, it had become apparent that the classical theory of thermal radiation was inadequate. The basic problem was in understanding the observed distribution of wavelengths in the radiation emitted by a black body. As we saw in Section 20.7, a black body is an ideal system that absorbs all radiation incident on it. A good approximation to a black body is the inside of a hollow object, as shown in Figure 40.1. The nature of the radiation emitted through a small hole leading to the cavity depends only on the temperature of the cavity walls.

Experimental data for the distribution of energy for blackbody radiation at three temperatures are shown in Figure 40.2. The radiated energy varies with wavelength and temperature. As the temperature of the black body increases, the total amount of energy it emits increases. Also, with increasing temperatures, the

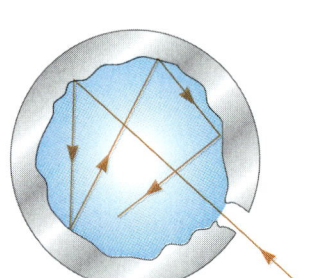

FIGURE 40.1 The opening to the cavity inside a body is a good approximation of a black body. Light entering the small opening strikes the far wall, where some of it is absorbed but some is reflected at some random angle. The light continues to be reflected, and at each reflection a portion of the light is absorbed by the cavity walls. After many reflections, essentially all of the incident energy is absorbed.

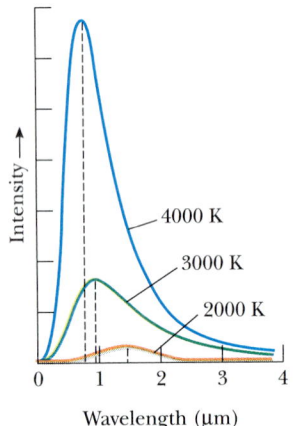

FIGURE 40.2 Intensity of blackbody radiation versus wavelength at three temperatures. Note that the amount of radiation emitted (the area under a curve) increases with increasing temperature.

---

[1] Einstein's views on the probabilistic nature of quantum theory are brought out in his statement, "God does not play dice with the Universe."

Max Planck was born in Kiel, Germany, and received his college education in Munich and Berlin. He joined the faculty at Munich in 1880 and five years later received a professorship at Kiel University. In 1889 he replaced Kirchhoff at the University of Berlin and remained there until 1926. He introduced the concept of a "quantum of action" (Planck's constant h) in an attempt to explain the spectral distribution of blackbody radiation, and this radically new concept laid the foundations for quantum theory. Planck was awarded the Nobel prize in 1918 for this discovery of the quantized nature of energy. The work leading to the blackbody radiation formula was described by Planck in his Nobel prize acceptance speech: "But even if the radiation formula proved to be perfectly correct, it would after all have been only an interpolation formula found by lucky guesswork and thus, would have left us rather unsatisfied. I

Max Planck

| 1 8 5 8 – 1 9 4 7 |

therefore strived from the day of its discovery, to give it a real physical interpretation and this led me to consider the relations between entropy and probability according to Boltzmann's ideas."

Planck's life was filled with per-

sonal tragedies. One of his sons was killed in action in World War I, and two daughters died during childbirth in the same period. His house was destroyed by bombs in World War II, and his son Erwin was executed by the Nazis in 1944 after being accused of plotting to assassinate Hitler.

Planck became president of the Kaiser Wilhelm Institute of Berlin in 1930. The institute was renamed the Max Planck Institute in his honor after World War II. Although Planck remained in Germany during the Hitler regime, he openly protested the Nazi treatment of his Jewish colleagues and consequently was forced to resign his presidency in 1937. Following World War II, he was renamed the president of the Max Planck Institute. He spent the last two years of his life in Göttingen as an honored and respected scientist and humanitarian.

(Courtesy of AIP Niels Bohr Library, W. F. Meggers Collection)

peak of the distribution shifts to shorter wavelengths. This shift was found to obey the following relationship, called **Wien's displacement law:**

$$\lambda_{\max} T = 0.2898 \times 10^{-2} \text{ m} \cdot \text{K} \qquad (40.1)$$

where $\lambda_{\max}$ is the wavelength at which the curve peaks and T is the absolute temperature of the object emitting the radiation. Early attempts to explain the shapes of the curves in Figure 40.2 based on classical theories failed.

To describe the radiation spectrum, it is useful to define $I(\lambda, T) \, d\lambda$ to be the power per unit area emitted in the wavelength interval $d\lambda$. The result of a calculation based on a classical model of blackbody radiation known as the **Rayleigh-Jeans law** is

$$I(\lambda, T) = \frac{2\pi c \, k_B T}{\lambda^4} \qquad (40.2)$$

where k_B is Boltzmann's constant. In this classical model of blackbody radiation, the atoms in the cavity walls are treated as a set of oscillators that emit electromagnetic waves at all wavelengths. This model leads to an average energy per oscillator that is proportional to T.

An experimental plot of the blackbody radiation spectrum is shown in Figure 40.3, together with the theoretical prediction of the Rayleigh-Jeans law. At long wavelengths, the Rayleigh-Jeans law is in reasonable agreement with experimental data. However, at short wavelengths there is major disagreement. This can be seen by noting that as λ approaches zero, the function $I(\lambda, T)$ given by Equation 40.2

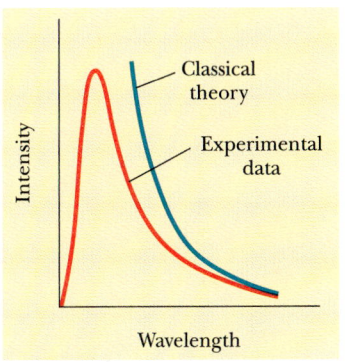

FIGURE 40.3 Comparison of the experimental results with the curve predicted by the Rayleigh-Jeans classical model for the distribution of blackbody radiation.

approaches infinity. Hence, not only should short-wavelength radiation predominate, but the energy density of the radiation emitted by *any* black body should become infinite in the limit of high frequencies or short wavelengths. This was called the ultraviolet catastrophe.

By contrast, the experimental data plotted in Figure 40.3 show that as λ approaches zero, $I(\lambda, T)$ also approaches zero. Not only does $I(\lambda, T)$ remain finite, experimentally, but the total power per unit area, given by $\int_0^\infty I(\lambda, T)\, d\lambda$, remains finite.[2]

In 1900 Planck discovered a formula for blackbody radiation that was in complete agreement with experiment at all wavelengths. Planck's analysis led to the curve in red shown in Figure 40.3. The function proposed by Planck is

$$I(\lambda, T) = \frac{2\pi hc^2}{\lambda^5 (e^{hc/\lambda k_B T} - 1)} \tag{40.3}$$

where h, Planck's constant, can be adjusted to fit the data. As we learned in Section 35.1, the value of h is

$$h = 6.626 \times 10^{-34}\ \text{J} \cdot \text{s} \tag{40.4}$$

You should show that at long wavelengths, Planck's expression, Equation 40.3, reduces to the Rayleigh-Jeans expression, Equation 40.2. Furthermore, at short wavelengths, Planck's law predicts an exponential decrease in $I(\lambda, T)$ with decreasing wavelength, in agreement with experimental results.

In his theory, Planck made two bold and controversial assumptions concerning the nature of the oscillating molecules at the surface of the black body:

- The molecules can have only *discrete* units of energy E_n,

$$E_n = nhf \tag{40.5}$$

where n is a positive integer called a **quantum number** and f is the frequency of vibration of the molecules. Because the energy of a molecule can have only discrete values given by Equation 40.5, we say the energy is *quantized*. Each discrete energy value represents a different *quantum state*, with each value of n representing a specific quantum state. When the molecule is in the $n = 1$ quantum state, its energy is hf; when it is in the $n = 2$ quantum state, its energy is $2hf$; and so on.

- The molecules emit or absorb energy in discrete packets called **photons**. A suggestive image of several photons, not to be taken too literally, is shown in Figure 40.4. The molecules emit or absorb these photons by "jumping" from one

---

[2] When $I(\lambda, T)$ is given by the Rayleigh-Jeans law, Eq. 40.2, the total power per unit area $I = \int_0^\infty I(\lambda, T)\, d\lambda$ diverges to ∞ when all wavelengths are allowed.

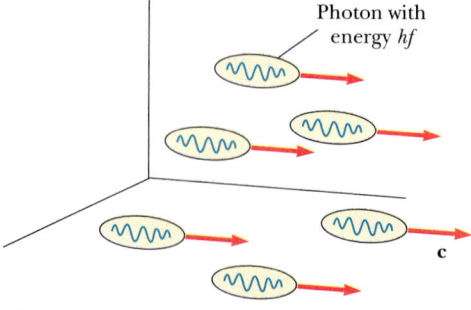

Photon with energy hf

FIGURE 40.4   A representation of photons. Each photon has a discrete energy, hf.

quantum state to another. If the jump is from one state to an adjacent state—say from the $n = 3$ state to the $n = 2$ state—Equation 40.5 shows that the amount of energy radiated by the molecule equals hf. Hence, the energy of one photon corresponding to the energy difference between two adjacent quantum states is

$$E = hf \qquad (40.6)$$

A molecule radiates or absorbs energy only when it changes quantum states. If it remains in one quantum state, no energy is absorbed or emitted. Figure 40.5 shows the quantized energy levels and allowed transitions proposed by Planck.

The key point in Planck's theory is the radical assumption of quantized energy states. This development marked the birth of the quantum theory. When Planck presented his theory, most scientists (including Planck!) did not consider the quantum concept to be realistic. Hence, Planck and others continued to search for a more rational explanation of blackbody radiation. However, subsequent developments showed that a theory based on the quantum concept (rather than on classical concepts) had to be used to explain many other phenomena at the atomic level.

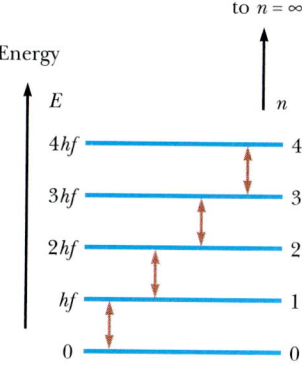

FIGURE 40.5   Allowed energy levels, as predicted by the old quantum theory, for a molecule that oscillates at its natural frequency f. Allowed transitions are indicated by the vertical arrows.

EXAMPLE 40.1   **Thermal Radiation from the Human Body**

The temperature of human skin is approximately 35°C. What is the peak wavelength in the radiation it emits?

Solution   From Wien's displacement law (Eq. 40.1), we have

$$\lambda_{max} T = 0.2898 \times 10^{-2}\ \text{m} \cdot \text{K}$$

Solving for λ_{max}, noting that 35°C corresponds to an absolute temperature of 308 K, we have

$$\lambda_{max} = \frac{0.2898 \times 10^{-2}\ \text{m} \cdot \text{K}}{308\ \text{K}} = 9.4\ \mu\text{m}$$

This radiation is in the infrared region of the spectrum.

EXAMPLE 40.2   **The Quantized Oscillator**

A 2.0-kg mass is attached to a massless spring of force constant $k = 25$ N/m. The spring is stretched 0.40 m from its equilibrium position and released. (a) Find the total energy and frequency of oscillation according to classical calculations.

Solution   The total energy of a simple harmonic oscillator having an amplitude A is $\frac{1}{2}kA^2$ (Eq. 13.19). Therefore,

$$E = \tfrac{1}{2}kA^2 = \tfrac{1}{2}(25\ \text{N/m})(0.40\ \text{m})^2 = \boxed{2.0\ \text{J}}$$

The frequency of oscillation is (Eq. 13.16)

$$f = \frac{1}{2\pi}\sqrt{\frac{k}{m}} = \frac{1}{2\pi}\sqrt{\frac{25\ \text{N/m}}{2.0\ \text{kg}}} = \boxed{0.56\ \text{Hz}}$$

(b) Assume that the energy is quantized and find the quantum number, n, for the system.

Solution   If the energy is quantized, we have $E_n = nhf$, and from the result of part (a) we get

$$E_n = nhf = n(6.626 \times 10^{-34}\ \text{J} \cdot \text{s})(0.56\ \text{Hz}) = 2.0\ \text{J}$$

$$n = \boxed{5.4 \times 10^{33}}$$

(c) How much energy is carried away in a one-quantum change?

Solution   The energy carried away in a one-quantum change is

$$E = hf = (6.63 \times 10^{-34}\ \text{J} \cdot \text{s})(0.56\ \text{Hz}) = \boxed{3.7 \times 10^{-34}\ \text{J}}$$

The energy carried away by a one-quantum change in energy is such a small fraction of the total energy of the oscillator that we could not expect to see such a small change in the system. Thus, even though the decrease in energy of a spring-mass system is quantized and does decrease by small quantum jumps, our senses perceive the decrease as continuous. Quantum effects become important and measurable only on the submicroscopic level of atoms and molecules.

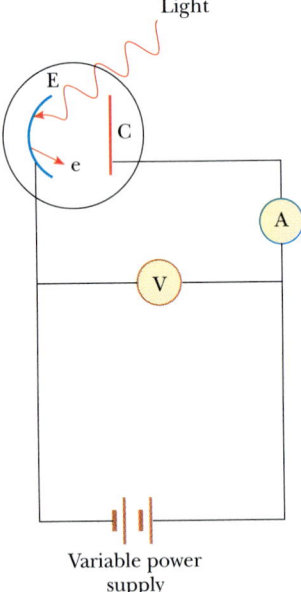

FIGURE 40.6   A circuit diagram for observing the photoelectric effect. When light strikes the plate E (the emitter), photoelectrons are ejected from the plate. Electrons collected at C (the collector) constitute a current in the circuit.

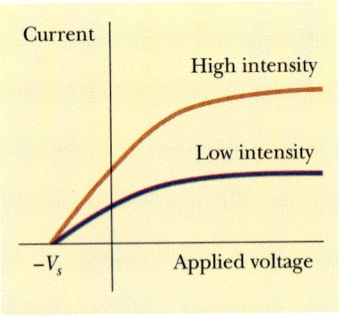

FIGURE 40.7   Photoelectric current versus applied voltage for two light intensities. The current increases with intensity but reaches a saturation level for large values of V. At voltages equal to or less than $-V_s$, the stopping potential, the current is zero.

40.2   THE PHOTOELECTRIC EFFECT

In the latter part of the 19th century, experiments showed that light incident on certain metallic surfaces caused electrons to be emitted from the surfaces. This phenomenon, which we first met in Section 35.1, is known as the **photoelectric effect,** and the emitted electrons are called **photoelectrons.**

Figure 40.6 is a diagram of an apparatus in which the photoelectric effect can occur. An evacuated glass or quartz tube contains a metal plate, E, connected to the negative terminal of a battery. Another metal plate, C, is maintained at a positive potential by the battery. When the tube is kept in the dark, the ammeter reads zero, indicating no current in the circuit. However, when monochromatic light of the appropriate wavelength shines on plate E, a current is detected by the ammeter, indicating a flow of charges across the gap between E and C. The current associated with this process arises from electrons emitted from the negative plate (the emitter) and collected at the positive plate (the collector).

Figure 40.7 is a plot of the photoelectric current versus the potential difference, V, between E and C for two light intensities. Note that for large values of V, the current reaches a maximum value. In addition, the current increases as the incident light intensity increases, as you might expect. Finally, when V is negative —that is, when the battery in the circuit is reversed to make E positive and C negative— the current drops to a very low value because most of the emitted photoelectrons are repelled by the negative plate, C. Only those electrons having a kinetic energy greater than eV will reach C, where e is the charge on the electron. When V is less than or equal to V_s, the **stopping potential,** no electrons reach C and the current is zero. The stopping potential is independent of the radiation intensity. The maximum kinetic energy of the photoelectrons is related to the stopping potential through the relationship

$$K_{max} = eV_s \qquad (40.7)$$

Several features of the photoelectric effect could not be explained with classical physics or with the wave theory of light:

- No electrons are emitted if the frequency of the incident light falls below some **cutoff frequency,** f_c, which is characteristic of the material being illuminated. This is inconsistent with the wave theory, which predicts that the photoelectric effect should occur at any frequency, provided the light intensity is high enough.
- The maximum kinetic energy of the photoelectrons is independent of light intensity.
- The maximum kinetic energy of the photoelectrons increases with increasing light frequency.
- Electrons are emitted from the surface almost instantaneously (less than 10^{-9} s after the surface is illuminated), even at low light intensities. Classically, we would expect the electrons to require some time to absorb the incident radiation before they acquire enough kinetic energy to escape from the metal.

A successful explanation of the photoelectric effect was given by Einstein in 1905, the same year he published his special theory of relativity. As part of a general paper on electromagnetic radiation, for which he received the Nobel prize in 1921, Einstein extended Planck's concept of quantization to electromagnetic waves. He assumed that light (or any electromagnetic wave) of frequency f can be considered a stream of photons. Each photon has an energy E, given by Equation 40.6, $E = hf$.

Einstein's picture was that a photon was so localized that it gave *all* its energy, hf, to a single electron in the metal. According to Einstein, the maximum kinetic energy for these liberated electrons is

$$K_{max} = hf - \phi \qquad (40.8)$$

Photoelectric effect equation

where ϕ is called the **work function** of the metal. The work function represents the minimum energy with which an electron is bound in the metal and is on the order of a few electron volts. Table 40.1 lists work functions measured for different metals.

With the photon theory of light, we can explain the above-mentioned features of the photoelectric effect that cannot be understood using classical concepts:

- That the effect is not observed below a certain cutoff frequency follows from the fact that the energy of the photon must be $\geq \phi$. If the energy of the incoming photon is not $\geq \phi$, the electrons are never ejected from the surface, regardless of the light intensity.
- That K_{max} is independent of light intensity can be understood with the following argument. If the light intensity is doubled, the number of photons is doubled, which doubles the number of photoelectrons emitted. However, their kinetic energy, which equals $hf - \phi$, depends only on the light frequency and the work function, not on the light intensity.
- That K_{max} increases with increasing frequency is easily understood with Equation 40.8.
- That the electrons are emitted almost instantaneously is consistent with the particle theory of light, in which the incident energy appears in small packets and there is a one-to-one interaction between photons and electrons. This is in contrast to having the energy of the photons distributed uniformly over a large area.

Experimental observation of a linear relationship between f and K_{max} would be a final confirmation of Einstein's theory. Indeed, such a linear relationship is observed, as sketched in Figure 40.8. The slope of this curve is h, and the intercept on the horizontal axis gives the cutoff frequency, which is related to the work function through the relationship $f_c = \phi / h$. The cutoff frequency corresponds to a **cutoff wavelength** of

$$\lambda_c = \frac{c}{f_c} = \frac{c}{\phi / h} = \frac{hc}{\phi} \qquad (40.9)$$

where c is the speed of light. Wavelengths greater than λ_c incident on a material having a work function ϕ do not result in the emission of photoelectrons.

TABLE 40.1 Work Functions of Selected Metals

Metal	ϕ (eV)
Na	2.46
Al	4.08
Cu	4.70
Zn	4.31
Ag	4.73
Pt	6.35
Pb	4.14
Fe	4.50

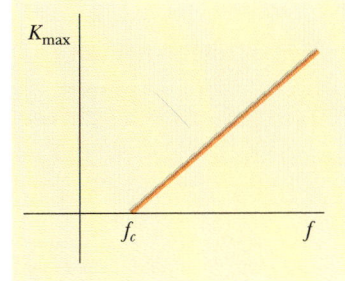

FIGURE 40.8 A sketch of K_{max} versus frequency of incident light for photoelectrons in a typical photoelectric effect experiment. Photons with frequency less than f_c do not have sufficient energy to eject an electron from the metal.

EXAMPLE 40.3 The Photoelectric Effect for Sodium

A sodium surface is illuminated with light of wavelength 300 nm. The work function for sodium metal is 2.46 eV. Find (a) the kinetic energy of the ejected photoelectrons and (b) the cutoff wavelength for sodium.

Solution (a) The energy of the illuminating light beam is

$$E = hf = \frac{hc}{\lambda} = \frac{(6.626 \times 10^{-34} \, \text{J} \cdot \text{s})(3.00 \times 10^{8} \, \text{m/s})}{300 \times 10^{-9} \, \text{m}}$$

$$= 6.63 \times 10^{-19} \, \text{J} = \frac{6.63 \times 10^{-19} \, \text{J}}{1.60 \times 10^{-19} \, \text{J/eV}} = 4.14 \, \text{eV}$$

Using Equation 40.8 gives

$$K_{max} = hf - \phi = 4.14 \, \text{eV} - 2.46 \, \text{eV} = \boxed{1.68 \, \text{eV}}$$

(b) The cutoff wavelength can be calculated from Equation 40.9 after we convert ϕ from electron volts to joules:

$$\phi = 2.46 \text{ eV} = (2.46 \text{ eV})(1.60 \times 10^{-19} \text{ J/eV})$$
$$= 3.94 \times 10^{-19} \text{ J}$$

Hence,

$$\lambda_c = \frac{hc}{\phi} = \frac{(6.626 \times 10^{-34} \text{ J} \cdot \text{s})(3.00 \times 10^8 \text{ m/s})}{3.94 \times 10^{-19} \text{ J}}$$

$$= 5.05 \times 10^{-7} \text{ m} = \boxed{505 \text{ nm}}$$

This wavelength is in the green region of the visible spectrum.

Exercise Calculate the maximum speed of the photoelectrons under the conditions described in this example.

Answer 7.68×10^5 m/s.

40.3 APPLICATIONS OF THE PHOTOELECTRIC EFFECT

The phototube, which operates on the basis of the photoelectric effect, acts much like a switch in an electric circuit in that it produces a current in an external circuit when light of sufficiently high frequency falls on a metal plate but produces no current in the dark. Many practical devices depend on the photoelectric effect. One of the first practical uses was as the detector in the light meter of a camera. Light reflected from the object to be photographed struck a photoelectric surface, causing it to emit electrons, which then passed through a sensitive ammeter. The magnitude of the current depended on the light intensity. Modern solid-state devices have now replaced light meters that used the photoelectric effect.

A second example of the photoelectric effect in action is the burglar alarm. This device often uses ultraviolet rather than visible light to make the beam invisible to the eye. A beam of light passes from the source to a photosensitive surface; the current produced is then amplified and used to energize an electromagnet that attracts a metal rod, as in Figure 40.9a. If an intruder breaks the light beam, the electromagnet switches off and the spring pulls the iron rod to the right (Fig. 40.9b). In this position, a completed circuit allows current to pass and activate the alarm system.

Figure 40.10 shows how the photoelectric effect is used to produce sound on a movie film. The soundtrack is located along the side of the film in the form of an optical pattern of light and dark lines. A beam of light in the projector is directed through the soundtrack toward a phototube. The variation in shading on the

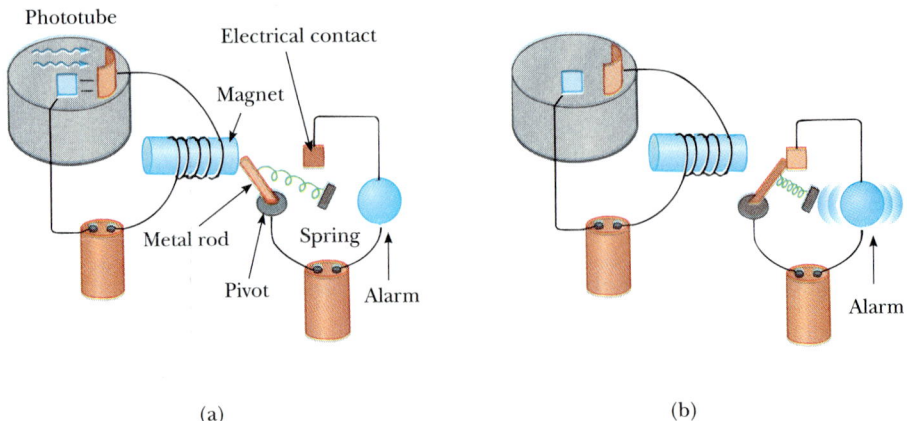

(a)

(b)

FIGURE 40.9 (a) When light strikes the phototube, the current in the circuit on the left energizes the magnet, breaking the burglar alarm circuit on the right. (b) When the light beam is intercepted (so that no light strikes the phototube), the alarm circuit is closed, and the alarm sounds.

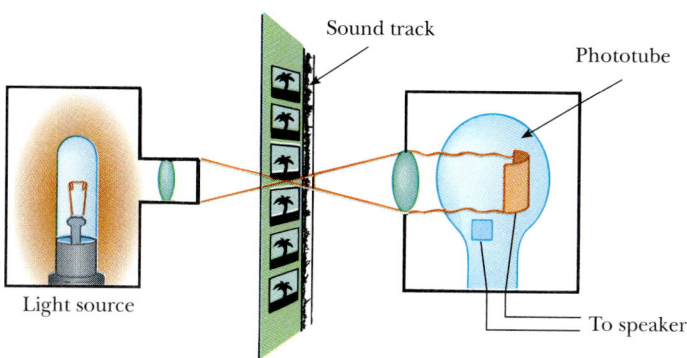

FIGURE 40.10   The shading on the soundtrack of a movie film varies the light intensity reaching the phototube and hence the current to the speaker.

soundtrack varies the light intensity falling on the plate of the phototube, thus changing the current in the circuit. This changing current electrically simulates the original sound wave and reproduces it in the speaker.

40.4   THE COMPTON EFFECT

In 1906, Einstein concluded that a photon of energy E travels in a single direction (unlike a spherical wave) and carries a momentum equal to $E/c = hf/c$. In his own words, "If a bundle of radiation causes a molecule to emit or absorb an energy packet hf, then momentum of quantity hf/c is transferred to the molecule, directed along the line of the bundle for absorption and opposite the bundle for emission." In 1923, Arthur Holly Compton (1892–1962) and Peter Debye (1884–1966) independently carried Einstein's idea of photon momentum farther. They realized that the scattering of x-ray photons from electrons could be explained by treating photons as point-like particles having energy hf and momentum hf/c, and by assuming that the energy and momentum of the photon-electron pair are conserved in a collision.

Prior to 1922, Compton and his coworkers had accumulated evidence that classical wave theory failed to explain the scattering of x-rays from electrons. According to classical theory, incident electromagnetic waves of frequency f_0 should accelerate electrons, forcing them to oscillate and reradiate at a frequency $f < f_0$ as in Figure 40.11a. Furthermore, the frequency or wavelength of the scattered radiation should depend on how long the sample was exposed to the incident radiation and on the intensity of the radiation.

Imagine the surprise when Compton showed, experimentally, that the wavelength shift of x-rays scattered at a given angle is absolutely independent of the intensity of radiation and length of exposure, and depends only on the scattering angle. Figure 40.11b shows the quantum picture of the transfer of momentum and energy between an x-ray photon and an electron. The quantum picture explains the lower scattered frequency f, because the incident photon transfers some of its energy to the recoiling electron.

Figure 40.12a is a schematic diagram of the apparatus used by Compton. In his original experiment Compton measured how scattered x-ray intensity depends on wavelength at three scattering angles. The wavelength was measured for x-rays scattered from a graphite target, analyzed with a rotating crystal spectrometer, and the intensity was measured with an ionization chamber that generated a current

Arthur Holly Compton (1892–1962), American physicist.   *(Courtesy of AIP Emilio Segrè Visual Archives)*

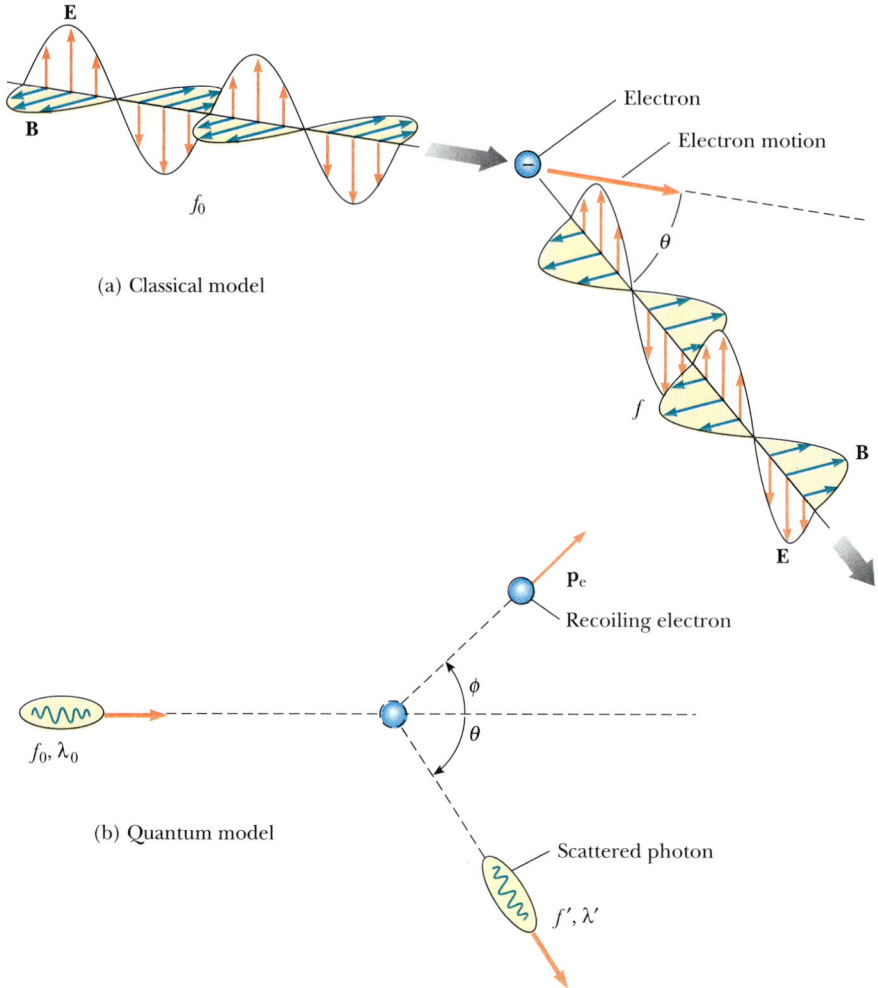

FIGURE 40.11   X-ray scattering from an electron: (a) the classical model and (b) the quantum model.

proportional to the x-ray intensity. The incident beam consisted of monochromatic x-rays of wavelength $\lambda_0 = 0.071$ nm. The experimental intensity versus wavelength plots observed by Compton for four scattering angles are shown in Figure 40.12b. They show two peaks, one at λ_0 and one at a longer wavelength λ'. The shifted peak at λ' is caused by the scattering of x-rays from free electrons, and it was predicted by Compton to depend on scattering angle as

Compton shift equation

$$\lambda' - \lambda_0 = \frac{h}{mc}\,(1 - \cos\theta) \qquad (40.10)$$

In this expression, known as the **Compton shift equation**, m is the mass of the electron; h/mc is called the **Compton wavelength** λ_C of the electron and has a currently accepted value of

Compton wavelength

$$\lambda_C = \frac{h}{mc} = 0.00243 \text{ nm}$$

The peak at λ_0 is caused by x-rays scattered from electrons tightly bound to the target atoms that have an effective mass equal to that of the carbon atom. This

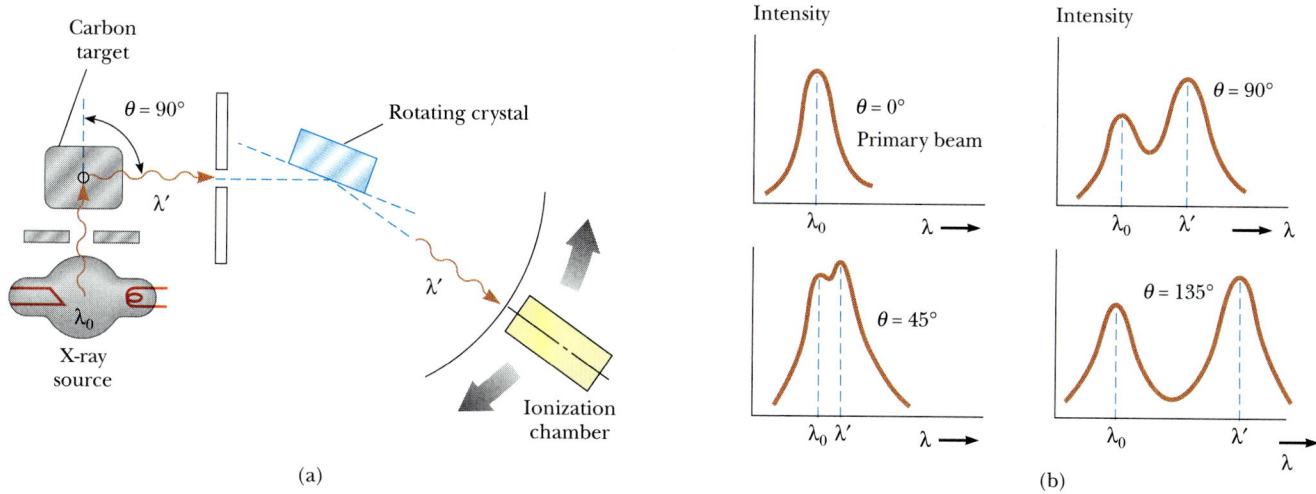

FIGURE 40.12   (a) Schematic diagram of Compton's apparatus. The wavelength was measured with a rotating crystal spectrometer using graphite (carbon) as the target. The intensity was determined by a movable ionization chamber that generated a current proportional to the x-ray intensity. (b) Scattered x-ray intensity versus wavelength for Compton scattering at $\theta = 0°$, $45°$, $90°$, and $135°$.

unshifted peak is also predicted by Equation 40.10 if the electron mass is replaced with the mass of carbon, which is about 23 000 times the mass of the electron. Compton's measurements were in excellent agreement with the predictions of Equation 40.10. It is fair to say that these results were the first to really convince most physicists of the fundamental validity of the quantum theory!

Derivation of the Compton Shift Equation

We can derive the Compton shift equation by assuming that the photon behaves like a particle and collides elastically with a free electron initially at rest, as in Figure 40.13. In this model, the photon is treated as a particle of energy $E = hf = hc/\lambda$ and mass zero. In the scattering process, the total energy and total momen-

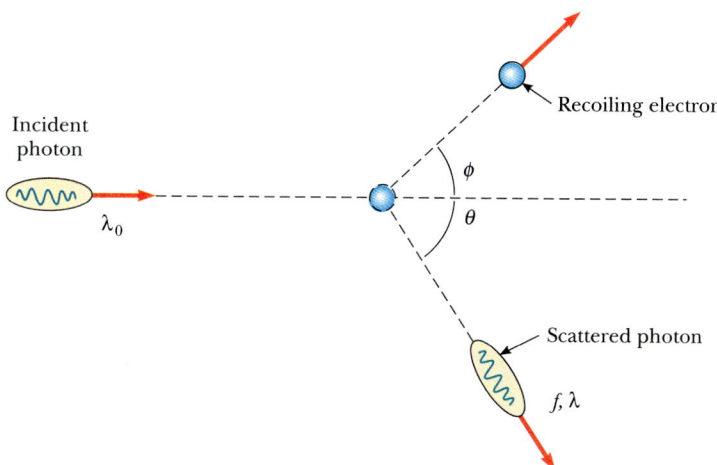

FIGURE 40.13   Diagram representing Compton scattering of a photon by an electron. The scattered photon has less energy (or longer wavelength) than the incident photon.

tum of the system must be conserved. Applying conservation of energy to this process gives

$$\frac{hc}{\lambda_0} = \frac{hc}{\lambda'} + K_e$$

where hc/λ_0 is the energy of the incident photon, hc/λ' is the energy of the scattered photon, and K_e is the kinetic energy of the recoiling electron. Because the electron may recoil at speeds comparable with the speed of light, we must use the relativistic expression $K_e = \gamma mc^2 - mc^2$ (Equation 39.23). Therefore,

$$\frac{hc}{\lambda_0} = \frac{hc}{\lambda'} + \gamma mc^2 - mc^2 \qquad (40.11)$$

where $\gamma = 1/\sqrt{1 - v^2/c^2}$.

Next, we apply the law of conservation of momentum to this collision, noting that *both the x and y components of momentum are conserved*. Because the momentum of a photon has a magnitude $p = E/c = h/\lambda$, and because the relativistic expression for the momentum of the recoiling electron is $p_e = \gamma mv$ (Eq. 39.18), we obtain the following expressions for the *x* and *y* components of linear momentum:

$$x \text{ component:} \qquad \frac{h}{\lambda_0} = \frac{h}{\lambda'} \cos\theta + \gamma mv \cos\phi \qquad (40.12)$$

$$y \text{ component:} \qquad 0 = \frac{h}{\lambda'} \sin\theta - \gamma mv \sin\phi \qquad (40.13)$$

By eliminating v and ϕ from Equations 40.11 to 40.13, we obtain a single expression that relates the remaining three variables (λ', λ_0, and θ). After some algebra (Problem 68), the Compton shift equation is obtained:

$$\Delta\lambda = \lambda' - \lambda_0 = \frac{h}{mc} (1 - \cos\theta)$$

EXAMPLE 40.4   Compton Scattering at 45°

X-rays of wavelength $\lambda_0 = 0.200$ nm are scattered from a block of material. The scattered x-rays are observed at an angle of 45.0° to the incident beam. Calculate their wavelength.

Solution The shift in wavelength of the scattered x-rays is given by Equation 40.10:

$$\Delta\lambda = \frac{h}{mc}(1 - \cos\theta)$$

$$\Delta\lambda = \frac{6.626 \times 10^{-34} \, \text{J} \cdot \text{s}}{(9.11 \times 10^{-31} \, \text{kg})(3.00 \times 10^8 \, \text{m/s})}(1 - \cos 45.0°)$$

$$= 7.10 \times 10^{-13} \, \text{m} = 0.000 \, 710 \, \text{nm}$$

Hence, the wavelength of the scattered x-ray at this angle is

$$\lambda' = \Delta\lambda + \lambda_0 = \boxed{0.200 \, 710 \, \text{nm}}$$

Exercise Find the fraction of energy lost by the photon in this collision.

Answer Fraction $= \Delta E/E = 0.003 \, 54$.

40.5 ATOMIC SPECTRA

As already pointed out in Section 40.1, all substances at a given temperature emit thermal radiation characterized by a continuous distribution of wavelengths. The shape of the distribution depends on the temperature and properties of the substance. In sharp contrast to this continuous distribution spectrum is the discrete **line spectrum** emitted by a low-pressure gaseous element subject to electric discharge. (Electric discharge occurs when the applied voltage exceeds the breakdown voltage of the gas.) When the light from such a low-pressure gas discharge is examined with a spectroscope, it is found to consist of a few bright lines of pure color on a generally dark background. This contrasts sharply with the continuous rainbow of colors seen when a glowing solid is viewed through a spectroscope. Furthermore, as you can see from Figure 40.14a, the wavelengths contained in a given line spectrum are characteristic of the element emitting the light. The simplest line spectrum is that for atomic hydrogen, and we describe this spectrum in detail. Other atoms emit completely different line spectra. Because no two elements emit the same line spectrum, this phenomenon represents a practical and sensitive technique for identifying the elements present in unknown samples.

Another form of spectroscopy very useful in analyzing substances is **absorption spectroscopy.** An absorption spectrum is obtained by passing light from a continuous source through a gas or dilute solution of the element being analyzed. The absorption spectrum consists of a series of dark lines superimposed on the otherwise continuous spectrum of that same element, as shown in Figure 40.14b for atomic hydrogen. In general, not all of the emission lines are present in the absorption spectrum.

The absorption spectrum of an element has many practical applications. For example, the continuous spectrum of radiation emitted by the Sun must pass through the cooler gases of the solar atmosphere and through the Earth's atmosphere. The various absorption lines observed in the solar spectrum have been used to identify elements in the solar atmosphere. In early studies of the solar spectrum, experimenters found some lines that did not correspond to any known

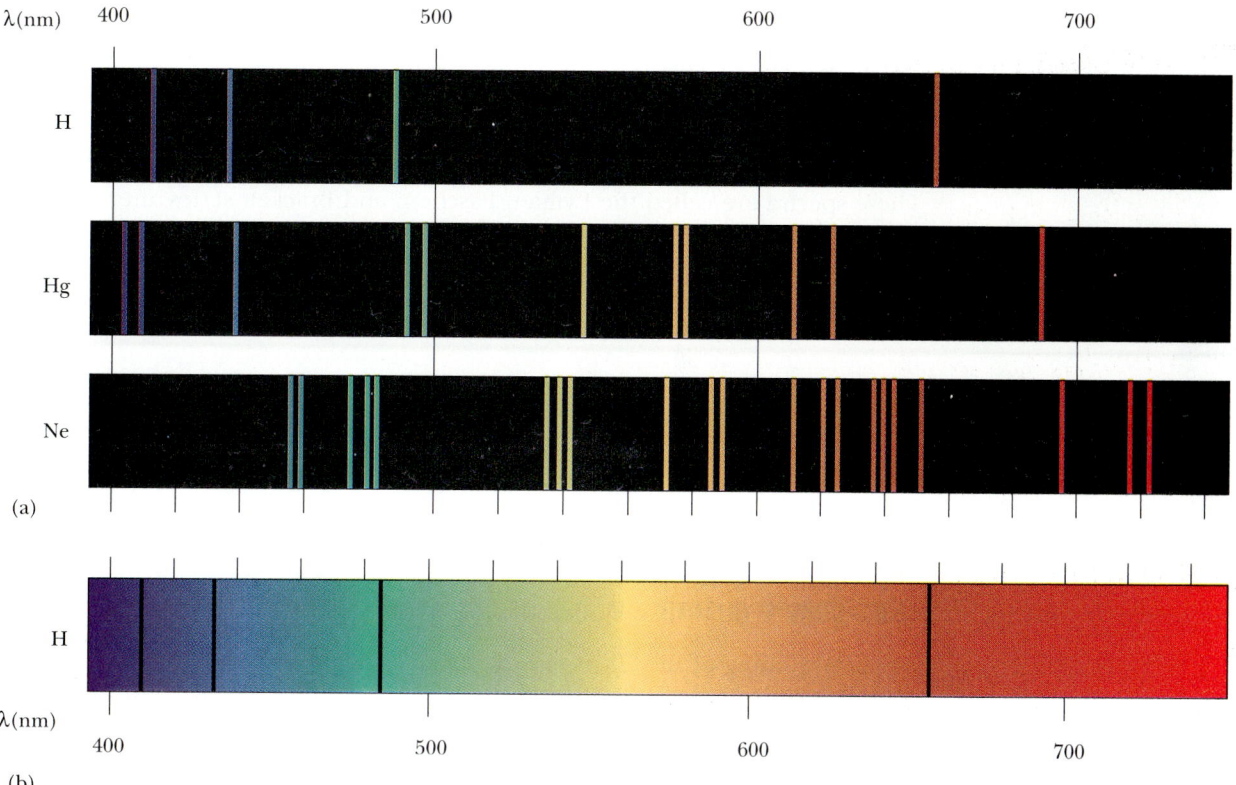

FIGURE 40.14 Visible spectra. (a) Line spectra produced by emission in the visible range for hydrogen, mercury, and neon. (b) The absorption spectrum for hydrogen. The dark absorption lines occur at the same wavelengths as the hydrogen emission lines in (a). (*K. W. Whitten, K. D. Gailey, and R. E. Davis,* General Chemistry, *3rd ed.,* Saunders College Publishing, *1992.*)

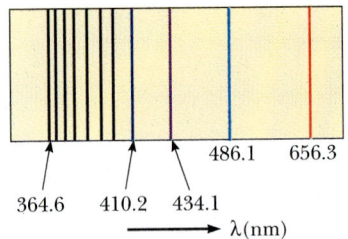

FIGURE 40.15   A series of spectral lines for atomic hydrogen. The prominent lines labeled are part of the Balmer series.

element. A new element had been discovered! The new element was named helium, after the Greek word for Sun, *helios*. Scientists are able to examine the light from stars other than our Sun in this fashion, but elements other than those present on Earth have never been detected. Atomic absorption spectroscopy has also been a useful technique in analyzing heavy-metal contamination of the food chain. For example, the first determination of high levels of mercury in tuna fish was made with atomic absorption.

From 1860 to 1885, scientists accumulated a great deal of data using spectroscopic measurements. In 1885, a Swiss schoolteacher, Johann Jacob Balmer (1825–1898) found a formula that correctly predicted the wavelengths of four visible emission lines of hydrogen: H_α (red), H_β (green), H_γ (blue), and H_δ (violet). Figure 40.15 shows these and other lines in the emission spectrum of hydrogen. The four visible lines occur at the wavelengths 656.3 nm, 486.1 nm, 434.1 nm, and 410.2 nm. The wavelengths of these lines (called the Balmer series) can be described by the empirical equation

> **Balmer series**

$$\frac{1}{\lambda} = R_H\left(\frac{1}{2^2} - \frac{1}{n^2}\right) \tag{40.14}$$

where n may have integral values of 3, 4, 5, . . . and R_H is a constant now called the **Rydberg constant**. If the wavelength is in meters, R_H has the value

> **Rydberg constant**

$$R_H = 1.0973732 \times 10^7 \text{ m}^{-1}$$

The first line in the Balmer series at 656.3 nm corresponds to $n = 3$ in Equation 40.14; the second line at 486.1 nm corresponds to $n = 4$, and so on. Much to Balmer's delight, the measured spectral lines agreed with his empirical formula to within 0.1%!

Other line spectra for hydrogen were found following Balmer's discovery. These spectra are called the Lyman, Paschen, and Brackett series after their discoverers. The wavelengths of the lines in these series can be calculated by the following empirical formulas:

> **Lyman series**

$$\frac{1}{\lambda} = R_H\left(1 - \frac{1}{n^2}\right) \qquad n = 2, 3, 4, . . . \tag{40.15}$$

> **Paschen series**

$$\frac{1}{\lambda} = R_H\left(\frac{1}{3^2} - \frac{1}{n^2}\right) \qquad n = 4, 5, 6, . . . \tag{40.16}$$

> **Brackett series**

$$\frac{1}{\lambda} = R_H\left(\frac{1}{4^2} - \frac{1}{n^2}\right) \qquad n = 5, 6, 7, . . . \tag{40.17}$$

40.6   BOHR'S QUANTUM MODEL OF THE ATOM

At the beginning of the 20th century, scientists were perplexed by the failure of classical physics in explaining the characteristics of atomic spectra. Why did atoms of a given element emit only certain spectral lines? Furthermore, why did the atoms absorb only those wavelengths they emitted? In 1913, Niels Bohr provided an explanation of atomic spectra that includes some features of the currently accepted theory. Bohr's theory contained a combination of ideas from Planck's original quantum theory, Einstein's photon theory of light, and Rutherford's model of the atom. Using the simplest atom, hydrogen, Bohr described a model of what he thought must be the atom's structure. His model of the hydrogen atom contains some classical features as well as some revolutionary postulates that could

Niels Bohr

| 1 8 8 5 – 1 9 6 2 |

Niels Bohr, a Danish physicist, proposed the first quantum model of the atom. He was an active participant in the early development of quantum mechanics and provided much of its philosophical framework. He made many other important contributions to theoretical nuclear physics, including the development of the liquid-drop model of the nucleus and work in nuclear fission. He was awarded the 1922 Nobel prize in physics for his investigation of the structure of atoms and of the radiation emanating from them.*

Bohr spent most of his life in Copenhagen and received his doctorate at the University of Copenhagen in 1911. The following year he traveled to England, where he worked under J. J. Thomson at Cambridge, and then went to Manchester where he worked under Ernest Rutherford. He was married in 1912 and returned to the University of Copenhagen in 1916 as professor of physics. During the 1920s and 1930s, Bohr headed the Institute for Advanced Studies in Copenhagen under the support of the Carlsberg

* For several interesting articles concerning Bohr, read the special Niels Bohr centennial issue of *Physics Today*, October 1985.

brewery. (This was certainly the greatest benefit provided by beer to the field of theoretical physics.) The institute was a magnet for many of the world's best physicists and provided a forum for the exchange of ideas. Bohr was always the initiator of probing questions, reflections, and discussions with his guests.

When Bohr visited the United States in 1939 to attend a scientific conference, he brought news that the

fission of uranium had been discovered by Hahn and Strassman in Berlin. The results, confirmed by other scientists shortly thereafter, were the foundations of the atomic bomb developed in the United States during World War II. He returned to Denmark and was there during the German occupation in 1940. He escaped to Sweden in 1943 to avoid imprisonment and helped to arrange the escape of many other imperiled Danish citizens.

Although Bohr worked on the Manhattan Project at Los Alamos until 1945, he strongly felt that openness between nations concerning nuclear weapons should be the first step in establishing their control. After the war, he committed himself to many human issues, including the development of peaceful uses of atomic energy. He was awarded the Atoms for Peace award in 1957. As John Archibald Wheeler summarizes, "Bohr was a great scientist. He was a great citizen of Denmark, of the World. He was a great human being."

(Photograph courtesy of AIP Niels Bohr Library, Margarethe Bohr Collection)

---

not be justified within the framework of classical physics. The basic ideas of the Bohr theory as it applies to the hydrogen atom are as follows:

- The electron moves in circular orbits around the proton under the influence of the Coulomb force of attraction, as in Figure 40.16. In this case, the Coulomb force is the centripetal force.
- Only certain electron orbits are stable. These stable orbits are ones in which the electron does not emit energy in the form of radiation. Hence, the total energy of the atom remains constant, and classical mechanics can be used to describe the electron's motion.
- Radiation is emitted by the atom when the electron "jumps" from a more energetic initial orbit to a lower orbit. This "jump" cannot be visualized or treated classically. In particular, the frequency f of the photon emitted in the jump is related to the change in the atom's energy and *is not equal to the frequency of the electron's orbital motion*. The frequency of the emitted radiation is

$$E_i - E_f = hf \qquad (40.18)$$

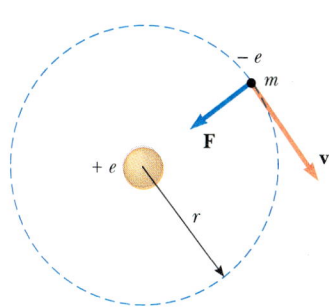

FIGURE 40.16   Diagram representing Bohr's model of the hydrogen atom, in which the orbiting electron is allowed to be only in specific orbits of discrete radii.

where E_i is the energy of the initial state, E_f is the energy of the final state, and $E_i > E_f$.

- The size of the allowed electron orbits is determined by a condition imposed on the electron's orbital angular momentum: The allowed orbits are those for which the electron's orbital angular momentum about the nucleus is an integral multiple of $\hbar = h/2\pi$,

$$mvr = n\hbar \qquad n = 1, 2, 3, \ldots \qquad (40.19)$$

Using these four assumptions, we can calculate the allowed energy levels and emission wavelengths of the hydrogen atom. Recall that the electrical potential energy of the system shown in Figure 40.16 is given by Equation 25.13, $U = qV = -k_e e^2/r$, where k_e is the Coulomb constant. Thus, the total energy of the atom, which contains both kinetic and potential energy terms, is

$$E = K + U = \tfrac{1}{2}mv^2 - k_e \frac{e^2}{r} \qquad (40.20)$$

Applying Newton's second law to this system, we see that the Coulomb attractive force on the electron, $k_e e^2/r^2$, must equal the mass times the centripetal acceleration ($a = v^2/r$) of the electron:

$$\frac{k_e e^2}{r^2} = \frac{mv^2}{r}$$

From this expression, we see that the kinetic energy of the electron is

$$K = \frac{mv^2}{2} = \frac{k_e e^2}{2r} \qquad (40.21)$$

Substituting this value of K into Equation 40.20, we find that the total energy of the atom is

$$E = -\frac{k_e e^2}{2r} \qquad (40.22)$$

Note that the total energy is negative, indicating a bound electron-proton system. This means that energy in the amount of $k_e e^2/2r$ must be added to the atom to just remove the electron and make the total energy zero.

An expression for r, the radius of the allowed orbits, may be obtained by solving Equations 40.19 and 40.21 for v and equating the results:

$$v^2 = \frac{n^2 \hbar^2}{m^2 r^2} = \frac{k_e e^2}{mr}$$

Radii of Bohr orbits in hydrogen

$$r_n = \frac{n^2 \hbar^2}{mk_e e^2} \qquad n = 1, 2, 3, \ldots \qquad (40.23)$$

This equation shows that the radii have discrete values, or are quantized. The result is based on the *assumption* that the electron can exist only in certain allowed orbits determined by the integer n.

The orbit with the smallest radius, called the **Bohr radius,** a_0, corresponds to $n = 1$ and has the value

$$a_0 = \frac{\hbar^2}{mk_e e^2} = 0.0529 \text{ nm} \qquad (40.24)$$

A general expression for the radius of any orbit in the hydrogen atom is obtained by substituting Equation 40.24 into Equation 40.23:

$$r_n = n^2 a_0 = n^2 (0.0529 \text{ nm}) \tag{40.25}$$

The fact that Bohr's theory gave a precise value for the radius of hydrogen from first principles without any empirical calibration of orbit size was considered a striking triumph for this theory. The first three Bohr orbits are shown to scale in Figure 40.17.

The quantization of the orbit radii immediately leads to energy quantization. This can be seen by substituting $r_n = n^2 a_0$ into Equation 40.22, giving for the allowed energy levels

$$E_n = -\frac{k_e e^2}{2a_0}\left(\frac{1}{n^2}\right) \qquad n = 1, 2, 3, \ldots \tag{40.26}$$

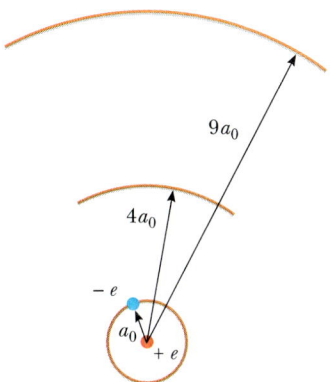

FIGURE 40.17   The first three circular orbits predicted by the Bohr model of the hydrogen atom.

Inserting numerical values into Equation 40.26 gives

$$E_n = -\frac{13.6}{n^2} \text{ eV} \qquad n = 1, 2, 3, \ldots \tag{40.27}$$

The lowest allowed energy level, called the **ground state**, has $n = 1$ and energy $E_1 = -13.6$ eV. The next energy level, the **first excited state**, has $n = 2$ and energy $E_2 = E_1/2^2 = -3.4$ eV. Figure 40.18 is an energy level diagram showing the energies of these discrete energy states and the corresponding quantum numbers. The uppermost level, corresponding to $n = \infty$ (or $r = \infty$) and $E = 0$, represents the state for which the electron is removed from the atom. The minimum energy required to ionize the atom (that is, to completely remove an electron in the ground state from the proton's influence) is called the **ionization energy.** As can be seen from Figure 40.18, the ionization energy for hydrogen based on Bohr's calculation is 13.6 eV. This constituted another major achievement for the Bohr theory because the ionization energy for hydrogen had already been measured to be precisely 13.6 eV.

Equation 40.26 together with Bohr's third postulate can be used to calculate the frequency of the photon emitted when the electron jumps from an outer orbit to an inner orbit:

$$f = \frac{E_i - E_f}{h} = \frac{k_e e^2}{2a_0 h}\left(\frac{1}{n_f^2} - \frac{1}{n_i^2}\right) \tag{40.28}$$

Since the quantity measured is wavelength, it is convenient to use $c = f\lambda$ to convert frequency to wavelength:

$$\frac{1}{\lambda} = \frac{f}{c} = \frac{k_e e^2}{2a_0 hc}\left(\frac{1}{n_f^2} - \frac{1}{n_i^2}\right) \tag{40.29}$$

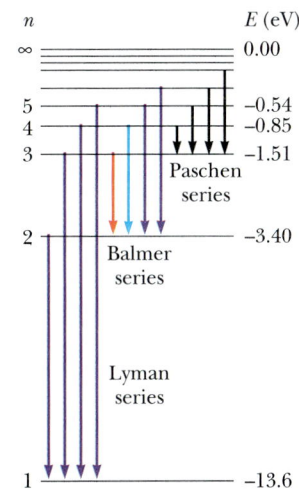

FIGURE 40.18   An energy level diagram for hydrogen. The discrete allowed energies are plotted on the vertical axis. Nothing is plotted on the horizontal axis, but the horizontal extent of the diagram is made large enough to show allowed transitions. Quantum numbers are given on the left and energies (in eV) are on the right.

The remarkable fact is that this expression, which is purely theoretical, is identical to a generalized form of the empirical relations discovered by Balmer and others and given by Equations 40.14 to 40.17:

$$\frac{1}{\lambda} = R_H\left(\frac{1}{n_f^2} - \frac{1}{n_i^2}\right) \tag{40.30}$$

provided the constant $k_e e^2/2 a_0 hc$ is equal to the experimentally determined Rydberg constant, $R_H = 1.0973732 \times 10^7$ m^{-1}. After Bohr demonstrated the agreement of these two quantities to a precision of approximately 1%, it was soon recognized as the crowning achievement of his new theory of quantum mechanics. Furthermore, Bohr showed that all of the spectral series for hydrogen have a natural interpretation in his theory. Figure 40.18 shows these spectral series as transitions between energy levels.

Bohr immediately extended his model for hydrogen to other elements in which all but one electron had been removed. Ionized elements such as He$^+$, Li^{++}, and Be^{+++} were suspected to exist in hot stellar atmospheres, where atomic collisions frequently have enough energy to completely remove one or more atomic electrons. Bohr showed that many mysterious lines observed in the Sun and several other stars could not be due to hydrogen but were correctly predicted by his theory if attributed to singly ionized helium. In general, to describe a single electron orbiting a fixed nucleus of charge $+ Ze$ (where Z is the *atomic number* of the element, defined as the number of protons in the nucleus), Bohr's theory gives

$$r_n = (n^2)\frac{a_0}{Z} \tag{40.31}$$

$$E_n = -\frac{k_e e^2}{2 a_0}\left(\frac{Z^2}{n^2}\right) \qquad n = 1, 2, 3, \ldots \tag{40.32}$$

EXAMPLE 40.5 Spectral Lines from the Star ξ-Puppis

Some mysterious lines observed in 1896 in the emission spectrum of the star ξ-Puppis fit the empirical formula

$$\frac{1}{\lambda} = R_H\left(\frac{1}{(n_f/2)^2} - \frac{1}{(n_i/2)^2}\right)$$

Show that these lines can be explained by the Bohr theory as originating from He$^+$.

Solution He$^+$ has $Z = 2$. Thus, the allowed energy levels are given by Equation 40.32 as

$$E_n = -\frac{k_e e^2}{2 a_0}\left(\frac{4}{n^2}\right)$$

Using $hf = E_i - E_f$, we find

$$f = \frac{E_i - E_f}{h} = \frac{k_e e^2}{2 a_0 h}\left(\frac{4}{n_f^2} - \frac{4}{n_i^2}\right)$$

$$= \frac{k_e e^2}{2 a_0 h}\left(\frac{1}{(n_f/2)^2} - \frac{1}{(n_i/2)^2}\right)$$

$$\frac{1}{\lambda} = \frac{f}{c} = \frac{k_e e^2}{2 a_0 hc}\left(\frac{1}{(n_f/2)^2} - \frac{1}{(n_i/2)^2}\right)$$

This is the desired solution, because $R_H \equiv k_e e^2/2 a_0 hc$.

EXAMPLE 40.6 An Electronic Transition in Hydrogen

The electron in the hydrogen atom makes a transition from the $n = 2$ energy state to the ground state (corresponding to $n = 1$). Find the wavelength and frequency of the emitted photon.

Solution We can use Equation 40.30 directly to obtain λ, with $n_i = 2$ and $n_f = 1$:

$$\frac{1}{\lambda} = R_H\left(\frac{1}{n_f^2} - \frac{1}{n_i^2}\right)$$

$$\frac{1}{\lambda} = R_H\left(\frac{1}{1^2} - \frac{1}{2^2}\right) = \frac{3R_H}{4}$$

$$\lambda = \frac{4}{3R_H} = \frac{4}{3(1.097 \times 10^7 \text{ m}^{-1})}$$

$$= 1.215 \times 10^{-7} \text{ m} = \boxed{121.5 \text{ nm}} \qquad \text{(ultraviolet)}$$

Since $c = f\lambda$, the frequency of the photon is

$$f = \frac{c}{\lambda} = \frac{3.00 \times 10^8 \text{ m/s}}{1.215 \times 10^{-7} \text{ m}} = \boxed{2.47 \times 10^{15} \text{ Hz}}$$

Exercise   What is the wavelength of the photon emitted by hydrogen when the electron makes a transition from the $n = 3$ state to the $n = 1$ state?

Answer $\dfrac{9}{8R_H} = 102.6$ nm.

EXAMPLE 40.7   The Balmer Series for Hydrogen

The Balmer series for the hydrogen atom corresponds to electronic transitions that terminate in the state of quantum number $n = 2$, as shown in Figure 40.19. (a) Find the longest-wavelength photon emitted in this series and determine its energy.

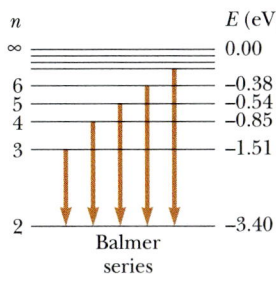

n	E (eV)
∞	0.00
6	−0.38
5	−0.54
4	−0.85
3	−1.51
2	−3.40

Balmer series

FIGURE 40.19   (Example 40.8) Transitions responsible for Balmer series for the hydrogen atom. All transitions in this series terminate at the $n = 2$ level.

Solution   The longest-wavelength photon in the Balmer series results from the transition from $n = 3$ to $n = 2$. Using Equation 40.30 gives

$$\frac{1}{\lambda} = R_H\left(\frac{1}{n_f^2} - \frac{1}{n_i^2}\right)$$

$$\frac{1}{\lambda_{\text{max}}} = R_H\left(\frac{1}{2^2} - \frac{1}{3^2}\right) = \frac{5}{36}R_H$$

$$\lambda_{\text{max}} = \frac{36}{5R_H} = \frac{36}{5(1.097 \times 10^7 \text{ m}^{-1})}$$

$$= \boxed{656.3 \text{ nm}} \qquad \text{(red)}$$

The energy of this photon is

$$E_{\text{photon}} = hf = \frac{hc}{\lambda_{\text{max}}}$$

$$= \frac{(6.626 \times 10^{-34} \text{ J} \cdot \text{s})(3.00 \times 10^8 \text{ m/s})}{656.3 \times 10^{-9} \text{ m}}$$

$$= 3.03 \times 10^{-19} \text{ J} = \boxed{1.89 \text{ eV}}$$

We could also obtain the energy by using the expression $hf = E_3 - E_2$, where E_2 and E_3 can be calculated from Equation 40.26. Note that this is the lowest-energy photon in this series because it involves the smallest energy change.

(b) Find the shortest-wavelength photon emitted in the Balmer series.

Solution   The shortest-wavelength photon in the Balmer series is emitted when the electron makes a transition from $n = \infty$ to $n = 2$. Therefore,

$$\frac{1}{\lambda_{\text{min}}} = R_H\left(\frac{1}{2^2} - \frac{1}{\infty}\right) = \frac{R_H}{4}$$

$$\lambda_{\text{min}} = \frac{4}{R_H} = \frac{4}{1.097 \times 10^7 \text{ m}^{-1}} = \boxed{364.6 \text{ nm}}$$

This wavelength is in the ultraviolet region and corresponds to the series limit.

Exercise   Find the energy of the shortest-wavelength photon emitted in the Balmer series for hydrogen.

Answer   3.40 eV.

Bohr's Correspondence Principle

In our study of relativity, we found that newtonian mechanics is a special case of relativistic mechanics and is usable only when v is much less than c. Similarly, *quantum mechanics is in agreement with classical physics where the difference between quantized levels vanishes.* This principle, first set forth by Bohr, is called the **correspondence principle.**

For example, consider an electron orbiting the hydrogen atom with $n > 10\ 000$. For such large values of n, the energy differences between adjacent levels approach zero and therefore the levels are nearly continuous. Consequently, the classical model is reasonably accurate in describing the system for large values of n. According to the classical picture, the frequency of the light emitted by the atom is equal to the frequency of revolution of the electron in its orbit about the nucleus. Calculations show that for $n > 10\,000$, this frequency is different from that predicted by quantum mechanics by less than 0.015%.

SUMMARY

The characteristics of blackbody radiation cannot be explained using classical concepts. Planck first introduced the quantum concept when he assumed that the atomic oscillators responsible for this radiation existed only in discrete states.

The **photoelectric effect** is a process whereby electrons are ejected from a metallic surface when light is incident on that surface. Einstein provided a successful explanation of this effect by extending Planck's quantum hypothesis to the electromagnetic field. In this model, light is viewed as a stream of photons, each with energy $E = hf$, where f is the frequency and h is Planck's constant. The maximum kinetic energy of the ejected photoelectron is

$$K_{\max} = hf - \phi \tag{40.8}$$

where ϕ is the **work function** of the metal.

X-rays from an incident beam are scattered at various angles by electrons in a target. In such a scattering event, a shift in wavelength is observed for the scattered x-rays, and the phenomenon is known as the **Compton effect.** Classical physics does not explain this effect. If the x-ray is treated as a photon, conservation of energy and momentum applied to the photon-electron collisions yields for the Compton shift:

$$\lambda' - \lambda_0 = \frac{h}{mc}(1 - \cos\theta) \tag{40.10}$$

where m is the mass of the electron, c is the speed of light, and θ is the scattering angle.

The Bohr model of the atom is successful in describing the spectra of atomic hydrogen and hydrogen-like ions. One of the basic assumptions of the model is that the electron can exist only in discrete orbits such that the angular momentum mvr is an integral multiple of $h/2\pi = \hbar$. Assuming circular orbits and a simple coulombic attraction between the electron and proton, the energies of the quantum states for hydrogen are calculated to be

$$E_n = -\frac{k_e e^2}{2a_0}\left(\frac{1}{n^2}\right) \qquad n = 1, 2, 3, \ldots \tag{40.26}$$

where k_e is the Coulomb constant, e is electronic charge, n is an integer called the **quantum number,** and $a_0 = 0.0529$ nm is the **Bohr radius.**

If the electron in the hydrogen atom makes a transition from an orbit whose quantum number is n_i to one whose quantum number is n_f, where $n_f < n_i$, a photon is emitted by the atom and the frequency of this photon is

$$f = \frac{k_e e^2}{2a_0 h}\left(\frac{1}{n_f^2} - \frac{1}{n_i^2}\right) \tag{40.28}$$

QUESTIONS

1. What assumptions were made by Planck in dealing with the problem of blackbody radiation? Discuss the consequences of these assumptions.
2. The classical model of blackbody radiation given by the Rayleigh-Jeans law has a major flaw. Identify it and explain how Planck's law deals with it.
3. If the photoelectric effect is observed for one metal, can you conclude that the effect will also be observed for another metal under the same conditions? Explain.
4. In the photoelectric effect, explain why the stopping potential depends on the frequency of light but not on the intensity.
5. Suppose the photoelectric effect occurs in a gaseous target rather than a solid plate. Will photoelectrons be produced at all frequencies of the incident photon? Explain.
6. How does the Compton effect differ from the photoelectric effect?
7. What assumptions were made by Compton in dealing with the scattering of a photon from an electron?
8. The Bohr theory of the hydrogen atom is based upon several assumptions. Discuss these assumptions and their significance. Do any of them contradict classical physics?
9. Suppose that the electron in the hydrogen atom obeyed classical mechanics rather than quantum mechanics. Why

should such a "hypothetical" atom emit a continuous spectrum rather than the observed line spectrum?
10. Can the electron in the ground state of hydrogen absorb a photon of energy (a) less than 13.6 eV and (b) greater than 13.6 eV?
11. Why would the spectral lines of the diatomic hydrogen be different from those of monatomic hydrogen?
12. Explain why, in the Bohr model, the total energy of the atom is negative.
13. An x-ray photon is scattered by an electron. What happens to the frequency of the scattered photon relative to that of the incident photon?
14. Why does the existence of a cutoff frequency in the photoelectric effect favor a particle theory for light rather than a wave theory?
15. Using Wien's displacement law, calculate the wavelength of highest intensity given off by a human body. Using this information, explain why an infrared detector would be a useful alarm for security work.
16. All objects radiate energy, but we are not able to see all objects in a dark room. Explain.
17. Which has more energy, a photon of ultraviolet radiation or a photon of yellow light?
18. Some stars look red, and some look blue. Which have the higher surface temperature? Explain.

PROBLEMS

Section 40.1 Blackbody Radiation and Planck's Hypothesis

1. Calculate the energy, in electron volts, of a photon whose frequency is (a) 6.2×10^{14} Hz, (b) 3.1 GHz, (c) 46 MHz. (d) Determine the corresponding wavelengths for these photons.
2. (a) Assuming that the tungsten filament of a lightbulb is a black body, determine its peak wavelength if its temperature is 2900 K. (b) Why does your answer to part (a) suggest that more energy from a lightbulb goes into heat than into light?
3. An FM radio transmitter has a power output of 150 kW and operates at a frequency of 99.7 MHz. How many photons per second does the transmitter emit?
4. The average power generated by the Sun is equal to 3.74×10^{26} W. Assuming the average wavelength of the Sun's radiation to be 500 nm, find the number of photons emitted by the Sun in 1 s.

5. A black body at 7500 K has a 0.0500-mm-diameter hole in it. Estimate the number of photons per second escaping the hole having wavelengths between 500 nm and 501 nm.
5A. A black body at temperature T has a hole of diameter d in it. Estimate the number of photons per second escaping the hole having wavelengths between λ_1 and λ_2.

6. A sodium-vapor lamp has a power output of 10 W. Using 589.3 nm as the average wavelength of this source, calculate the number of photons emitted per second.
7. Using Wien's displacement law, calculate the surface temperature of a red giant star that radiates with a peak wavelength of 650 nm.
8. The radius of our Sun is 6.96×10^8 m, and its total power output is 3.77×10^{26} W. (a) Assuming that

□ indicates problems that have full solutions available in the Student Solutions Manual and Study Guide.

the Sun's surface emits as an ideal black body, calculate its surface temperature. (b) Using the result of part (a), find λ_{max} from the Sun.

9. What is the peak wavelength emitted by the human body? Assume a body temperature of 98.6°F and use the Wien displacement law. In what part of the electromagnetic spectrum does this wavelength lie?

10. A tungsten filament is heated to 800°C. What is the wavelength of the most intense radiation?

11. The human eye is most sensitive to 560-nm light. What temperature black body would radiate most intensely at this wavelength?

12. A star moving away from the Earth at $0.280c$ emits radiation having a maximum intensity at a wavelength of 500 nm. Determine the surface temperature of this star. (*Hint:* See Problem S1 in Chapter 39.)

12A. A star moving away from the Earth at a speed v emits radiation having a maximum intensity at a wavelength λ. Determine the surface temperature of this star. (*Hint:* See Problem S1 in Chapter 39.)

13. Show that at short wavelengths or low temperatures, Planck's radiation law (Eq. 40.3) predicts an exponential decrease in $I(\lambda, T)$ given by Wien's radiation law:

$$I(\lambda, T) = \frac{2\pi hc^2}{\lambda^5} e^{-hc/\lambda kT}$$

Section 40.2 The Photoelectric Effect
Section 40.3 Applications of the Photoelectric Effect

14. In an experiment on the photoelectric effect, the photocurrent is stopped by a retarding potential of 0.54 V for 750-nm radiation. Find the work function for the material.

15. The work function for potassium is 2.24 eV. If potassium metal is illuminated with 480-nm light, find (a) the maximum kinetic energy of the photoelectrons and (b) the cutoff wavelength.

16. Molybdenum has a work function of 4.2 eV. (a) Find the cutoff wavelength and threshold frequency for the photoelectric effect. (b) Calculate the stopping potential if the incident light has a wavelength of 180 nm.

17. A student studying the photoelectric effect from two different metals records the following information: (i) the stopping potential for photoelectrons released from metal 1 is 1.48 V larger than that for metal 2, and (ii) the threshold frequency for metal 1 is 40% smaller than that for metal 2. Determine the work function for each metal.

18. When 445-nm light strikes a certain metal surface, the stopping potential is 70.0 percent of that which results when 410-nm light strikes the same metal surface. Based on this information and the following

table of work functions, identify the metal involved in the experiment.

Metal	Work Function (eV)
Cesium	1.90
Potassium	2.24
Silver	4.73
Tungsten	4.58

19. Two light sources are used in a photoelectric experiment to determine the work function for a particular metal surface. When green light from a mercury lamp ($\lambda = 546.1$ nm) is used, a retarding potential of 1.70 V reduces the photocurrent to zero. (a) Based on this measurement, what is the work function for this metal? (b) What stopping potential would be observed when using the yellow light from a helium discharge tube ($\lambda = 587.5$ nm)?

20. When 625-nm light shines on a certain metal surface, the photoelectrons have speeds up to 4.6×10^5 m/s. What are (a) the work function and (b) the cutoff frequency for this metal?

21. Lithium, beryllium, and mercury have work functions of 2.3 eV, 3.9 eV, and 4.5 eV, respectively. If 400-nm light is incident on each of these metals, determine (a) which metals exhibit the photoelectric effect and (b) the maximum kinetic energy for the photoelectron in each case.

22. Light of wavelength 300 nm is incident on a metallic surface. If the stopping potential for the photoelectric effect is 1.2 V, find (a) the maximum energy of the emitted electrons, (b) the work function, and (c) the cutoff wavelength.

23. A light source emitting radiation at 7.0×10^{14} Hz is incapable of ejecting photoelectrons from a certain metal. In an attempt to use this source to eject photoelectrons from the metal, the source is given a velocity toward the metal. (a) Explain how this procedure produces photoelectrons. (b) When the speed of the light source is equal to $0.28c$, photoelectrons just begin to be ejected from the metal. What is the work function of the metal? (c) When the speed of the light source is increased to $0.90c$, determine the maximum kinetic energy of the photoelectrons.

24. If a photodiode is exposed to green light (500 nm), it acquires the voltage of 1.4 V. Determine the voltage that will be caused by exposure of the same photodiode to violet light (400 nm).

Section 40.4 The Compton Effect

25. A 0.70-MeV photon scatters off a free electron such that the scattering angle of the photon is twice the scattering angle of the electron (Fig. P40.25). Determine (a) the scattering angle for the electron and (b) the final speed of the electron.

25A. A photon of energy E_0 scatters off a free electron such that the scattering angle of the photon is twice the scattering angle of the electron (Fig. P40.25). Determine (a) the scattering angle for the electron and (b) the final speed of the electron.

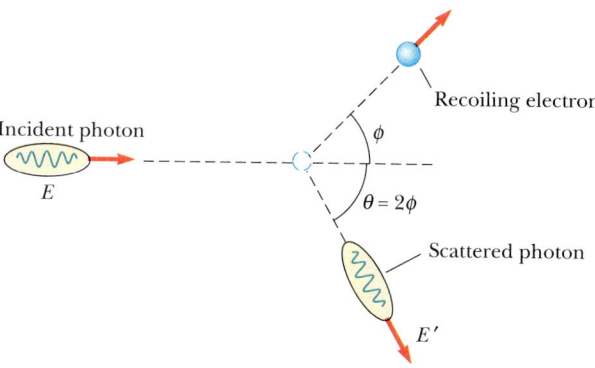

FIGURE P40.25

26. X-rays of wavelength 0.200 nm are scattered from a block of carbon. If the scattered radiation is detected at 60° to the incident beam, find (a) the Compton shift and (b) the kinetic energy imparted to the recoiling electron.

27. A photon having wavelength λ scatters off a free electron at A (Fig. P40.27) producing a second photon having wavelength λ'. This photon then scatters off another free electron at B producing a third photon having wavelength λ'', and moving in a direction directly opposite the original photon as in Figure P40.27. Determine the numerical value of $\Delta\lambda = \lambda'' - \lambda$.

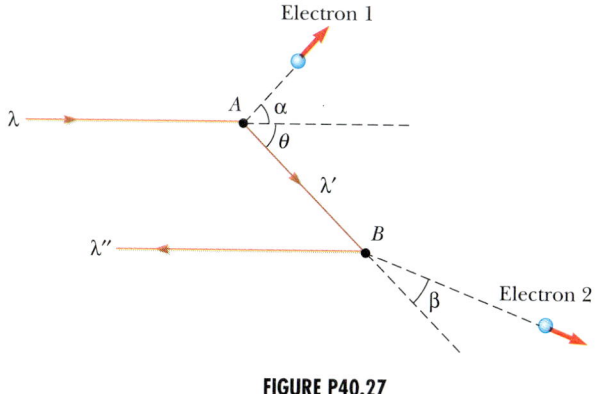

FIGURE P40.27

28. In a Compton scattering experiment, a photon is scattered through an angle of 90° and the electron is scattered through an angle of 20°. Determine the wavelength of the scattered photon.

29. A 0.667-MeV gamma ray scatters off an electron that is bound to a nucleus with an energy of 150 keV. If the photon scatters at an angle of 180°, (a) determine the energy and momentum of the recoiling electron after it has been freed from the atom. (b) Where does the missing momentum appear?

30. X-rays having an energy of 300 keV undergo Compton scattering from a target. If the scattered rays are detected at 37° relative to the incident rays, find (a) the Compton shift at this angle, (b) the energy of the scattered x-ray, and (c) the energy of the recoiling electron.

31. After a 0.80-nm x-ray photon scatters from a free electron, the electron recoils at 1.4×10^6 m/s. (a) What was the Compton shift in the photon's wavelength? (b) Through what angle was the photon scattered?

32. A 0.110-nm photon collides with a stationary electron. After the collision, the electron moves forward and the photon recoils backward. Find the momentum and kinetic energy of the electron.

33. A 0.88-MeV photon is scattered by a free electron initially at rest such that the scattering angle of the scattered electron equals that of the scattered photon ($\phi = \theta$ in Figure 40.11). Determine (a) the angles ϕ and θ, (b) the energy and momentum of the scattered photon, and (c) the kinetic energy and momentum of the scattered electron.

33A. A photon having energy E_0 is scattered by a free electron initially at rest such that the scattering angle of the scattered electron equals that of the scattered photon ($\phi = \theta$ in Fig. 40.11). Determine (a) the angles ϕ and θ, (b) the energy and momentum of the scattered photon, and (c) the kinetic energy and momentum of the scattered electron.

34. A 0.500-nm x-ray photon is deflected through 134° in a Compton scattering event. At what angle (with respect to the incident beam) is the recoiling electron found?

35. A 0.0016-nm photon scatters from a free electron. For what (photon) scattering angle do the recoiling electron and the scattered photon have the same kinetic energy?

Section 40.5 Atomic Spectra

36. Show that the wavelengths for the Balmer series satisfy the equation

$$\lambda = \frac{364.5\, n^2}{n^2 - 4}\ \text{nm}$$

where $n = 3, 4, 5, \ldots$.

37. (a) Suppose that the Rydberg constant were given by $R_H = 2.0 \times 10^7$ m^{-1}, what part of the electromagnetic spectrum would the Balmer series correspond to? (b) Repeat for $R_H = 0.5 \times 10^7$ m^{-1}.

38. (a) Compute the shortest wavelength in each of these hydrogen spectral series: Lyman, Balmer, Paschen, and Brackett. (b) Compute the energy (in electron volts) of the highest-energy photon produced in each series.

39. (a) What value of n is associated with the 94.96-nm line in the Lyman hydrogen series? (b) Could this wavelength be associated with the Paschen or Brackett series?

40. Liquid oxygen has a bluish color, meaning that it preferentially absorbs light toward the red end of the visible spectrum. Although the oxygen molecule (O_2) does not strongly absorb visible radiation, it does absorb strongly at 1269 nm, which is in the infrared region of the spectrum. Research has shown that it is possible for two colliding O_2 molecules to absorb a single photon, sharing its energy equally. The transition that both molecules undergo is the same transition that results when they absorb 1269-nm radiation. What is the wavelength of the single photon that causes this double transition? What is the color of this radiation?

Section 40.6   Bohr's Quantum Model of the Atom

41. Use Equation 40.23 to calculate the radius of the first, second, and third Bohr orbits of hydrogen.

42. For a hydrogen atom in its ground state, use the Bohr model to compute (a) the orbital speed of the electron, (b) its kinetic energy (in electron volts), and (c) the electrical potential energy (in electron volts) of the atom.

43. (a) Construct an energy level diagram for the He^+ ion, for which $Z = 2$. (b) What is the ionization energy for He^+?

44. A monochromatic beam of light is absorbed by a collection of ground-state hydrogen atoms in such a way that six different wavelengths are observed when the hydrogen relaxes back to the ground state. What is the wavelength of the incident beam?

45. What is the radius of the first Bohr orbit in (a) He^+, (b) Li^{2+}, and (c) Be^{3+}?

46. Two hydrogen atoms collide head-on and end up with zero kinetic energy. Each then emits a 121.6-nm photon ($n = 2$ to $n = 1$ transition). At what speed were the atoms moving before the collision?

46A. Two hydrogen atoms collide head-on and end up with zero kinetic energy. Each then emits a photon of wavelength λ. At what speed were the atoms moving before the collision?

47. A photon is emitted as a hydrogen atom undergoes a transition from the $n = 6$ state to the $n = 2$ state. Calculate (a) the energy, (b) the wavelength, and (c) the frequency of the emitted photon.

48. A particle of charge q and mass m, moving with a constant speed, v, perpendicular to a constant mag-

netic field, B, follows a circular path. If the angular momentum around the center of this circle is quantized so that $mvr = n\hbar$, show that the allowed radii for the particle are

$$r_n = \sqrt{\frac{n\hbar}{qB}}$$

for $n = 1, 2, 3, \ldots$.

49. Four possible transitions for a hydrogen atom are listed below.

(A)	$n_i = 2$; $n_f = 5$
(B)	$n_i = 5$; $n_f = 3$
(C)	$n_i = 7$; $n_f = 4$
(D)	$n_i = 4$; $n_f = 7$

(a) Which transition emits the photons having the shortest wavelength? (b) For which transition does the atom gain the most energy? (c) For which transition(s) does the atom lose energy?

50. An electron is in the nth Bohr orbit of the hydrogen atom. (a) Show that the period of the electron is $T = \tau_0 n^3$, and determine the numerical value of τ_0. (b) On the average, an electron remains in the $n = 2$ orbit for about 10 μs before it jumps down to the $n = 1$ (ground-state) orbit. How many revolutions does the electron make before it jumps to the ground state? (c) If one revolution of the electron is defined as an "electron year" (analogous to an Earth year being one revolution of the Earth around the Sun), does the electron in the $n = 2$ orbit "live" very long? Explain. (d) How does the above calculation support the "electron cloud" concept?

51. Find the potential energy and kinetic energy of the electron in the first excited state of the hydrogen atom.

ADDITIONAL PROBLEMS

52. A hydrogen atom is in its first excited state ($n = 2$). Using the Bohr theory of the atom, calculate (a) the radius of the orbit, (b) the linear momentum of the electron, (c) the angular momentum of the electron, (d) the kinetic energy, (e) the potential energy, and (f) the total energy.

53. Positronium is a hydrogen-like atom consisting of a positron (a positively charged electron) and an electron revolving around each other. Using the Bohr model, find the allowed radii (relative to the center of mass of the two particles) and the allowed energies of the system.

54. Figure P40.54 shows the stopping potential versus incident photon frequency for the photoelectric effect for sodium. Use these data points to find (a) the work function, (b) the ratio h/e, and (c) the cutoff

wavelength. (Data taken from R. A. Millikan, *Phys. Rev.* 7:362 [1916].)

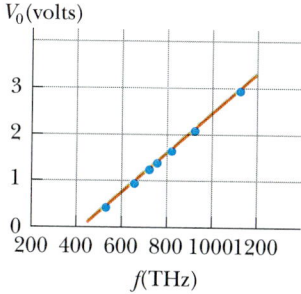

V_0(volts)

f(THz)

FIGURE P40.54

55. An electron in chromium moves from the $n = 2$ state to the $n = 1$ state without emitting a photon. Instead, the excess energy is transferred to an outer electron (one in the $n = 4$ state), which is then ejected by the atom. (This is called an Auger [pronounced 'ohjay'] process, and the ejected electron is referred to as an Auger electron.) Use the Bohr theory to estimate the kinetic energy of the Auger electron.

56. Photons of wavelength 450 nm are incident on a metal. The most energetic electrons ejected from the metal are bent into a circular arc of radius 20 cm by a magnetic field the strength of which is 2.0×10^{-5} T. What is the work function of the metal?

56A. Photons of wavelength λ are incident on a metal. The most energetic electrons ejected from the metal are bent into a circular arc of radius R by a magnetic field having a magnitude B. What is the work function of the metal?

57. Gamma rays (high-energy photons) of energy 1.02 MeV are scattered from electrons that are initially at rest. If the scattering is symmetric, that is, if $\theta = \phi$, find (a) the scattering angle θ and (b) the energy of the scattered photons.

58. A 200-MeV photon is scattered at 40° by a free proton initially at rest. (a) Find the energy (in MeV) of the scattered photon. (b) What kinetic energy (in MeV) does the proton acquire?

59. As we learned in Section 39.4, a muon has a charge of $-e$ and a mass equal to 207 times the mass of an electron. Muonic lead is formed when a lead nucleus captures a muon. According to the Bohr theory, what are the radius and energy of the ground state of muonic lead?

60. Use Bohr's model of the hydrogen atom to show that when the electron moves from the state n to the state $n - 1$, the frequency of the emitted light is

$$f = \frac{2\pi^2 m k_e^2 e^4}{h^3} \left[\frac{2n - 1}{(n - 1)^2 n^2} \right]$$

Show that as $n \to \infty$, this expression varies as $1/n^3$ and reduces to the classical frequency the atom is expected to emit. (*Hint:* To calculate the classical frequency, note that the frequency of revolution is $v/2\pi r$, where r is given by Eq. 40.23.)

61. A muon (Problem 59) is captured by a deuteron to form a muonic atom. (a) Find the energy of the ground state and the first excited state. (b) What is the wavelength of the photon emitted when the atom makes a transition from the first excited state to the ground state?

62. Show that a photon cannot transfer all of its energy to a free electron. (*Hint:* Note that energy and momentum must be conserved.)

63. A photon with wavelength λ moves toward a free electron that is moving with speed u in the same direction as the photon (Fig. P40.63a). If the photon scatters at an angle θ (Fig. P40.63b), show that the wavelength of the scattered photon is

$$\lambda' = \lambda \left(\frac{1 - (u/c)\cos\theta}{1 - u/c} \right)$$
$$+ \frac{h}{mc}\sqrt{\frac{1 + u/c}{1 - u/c}}\,(1 - \cos\theta)$$

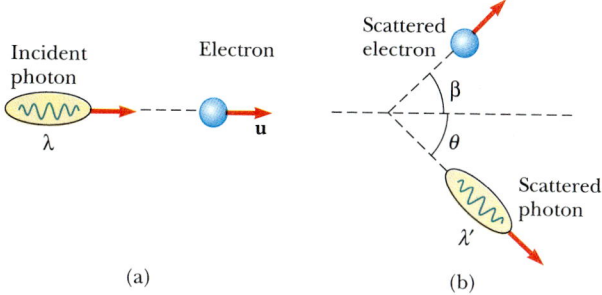

(a)   (b)

FIGURE P40.63

64. The total power per unit area radiated by a black body at a temperature T is the area under the $I(\lambda, T)$ versus λ curve, as in Figure 40.2. (a) Show that this power per unit area is

$$\int_0^\infty I(\lambda, T)\, d\lambda = \sigma T^4$$

where $I(\lambda, T)$ is given by Planck's radiation law and σ is a constant independent of T. This result is known as the Stefan-Boltzmann law. To carry out the integration, you should make the change of variable $x = hc/\lambda k_B T$ and use the fact that

$$\int_0^\infty \frac{x^3\, dx}{e^x - 1} = \frac{\pi^4}{15}$$

(b) Show that the Stefan-Boltzmann constant σ has

the value

$$\sigma = \frac{2\pi^5 k_B^4}{15 c^2 h^3} = 5.7 \times 10^{-8} \text{ W/m}^2 \cdot \text{K}^4$$

65. The Lyman series for a (new?) one-electron atom is observed in a distant galaxy. The wavelengths of the first four lines and the short-wavelength limit of this series are given by the energy-level diagram in Figure P40.65. Based on this information, calculate (a) the energies of the ground state and first four excited states for this one-electron atom and (b) the wavelengths of the first three lines and the short-wavelength limit in the Balmer series for this atom. (c) Show that the wavelengths of the first four lines and the short wavelength limit of the Lyman series for the hydrogen atom are all exactly 60% of the wavelengths for the Lyman series in the one-electron atom described in part (b). (d) Based on this observation, explain why this atom could be hydrogen. (*Hint:* See Problem S1 in Chapter 39.)

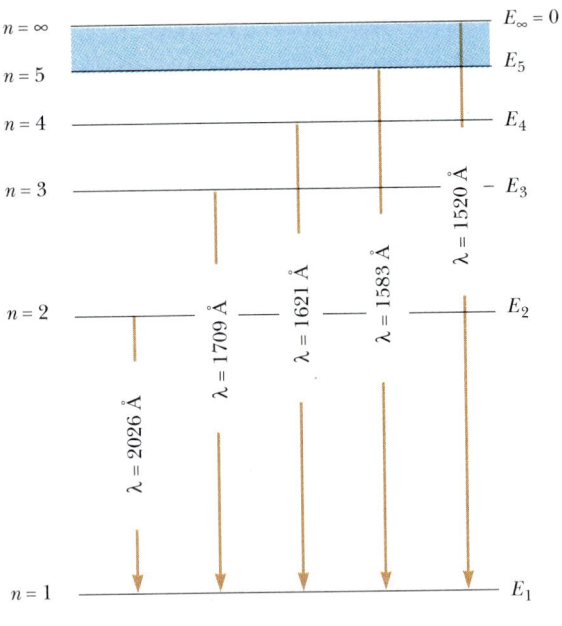

FIGURE P40.65

66. A photon of initial energy E_0 undergoes Compton scattering at an angle θ from a free electron (mass m) initially at rest. Using relativistic equations for energy and momentum conservation, derive the following relationship for the final energy E of the scattered photon:

$$E = E_0[1 - (E_0/mc^2)(1 - \cos\theta)]^{-1}$$

67. Show that the ratio of the Compton wavelength λ_C to the de Broglie wavelength $\lambda = h/p$ for a relativistic

electron is

$$\frac{\lambda_C}{\lambda} = \left[\left(\frac{E}{mc^2} \right)^2 - 1 \right]^{1/2}$$

where E is the total energy of the electron and m is its mass.

68. Derive the formula for the Compton shift (Eq. 40.10) from Equations 40.11, 40.12, and 40.13.

69. An electron initially at rest recoils from a head-on collision with a photon. Show that the kinetic energy acquired by the electron is $2hf\alpha/(1 + 2\alpha)$, where α is the ratio of the photon's initial energy to the rest energy of the electron.

70. The table below shows data obtained in a photoelectric experiment. (a) Using these data, make a graph similar to Figure 40.8 that plots as a straight line. From the graph, determine (b) an experimental value for Planck's constant (in joule-seconds) and (c) the work function (in electron volts) for the surface. (Two significant figures for each answer are sufficient.)

Wavelength (nm)	Maximum Kinetic Energy of Photoelectrons (eV)
588	0.67
505	0.98
445	1.35
399	1.63

71. The spectral distribution function $I(\lambda, T)$ for an ideal black body at absolute temperature T is shown in Figure P40.71. (a) Show that the percentage of the total power radiated per unit area in the range $0 \le \lambda \le \lambda_{\max}$ is

$$\frac{A}{A + B} = 1 - \frac{15}{\pi^4} \int_0^{4.965} \frac{x^3}{e^x - 1} \, dx$$

independent of the value of T. (b) Using numerical integration, show that this ratio is approximately $1/4$.

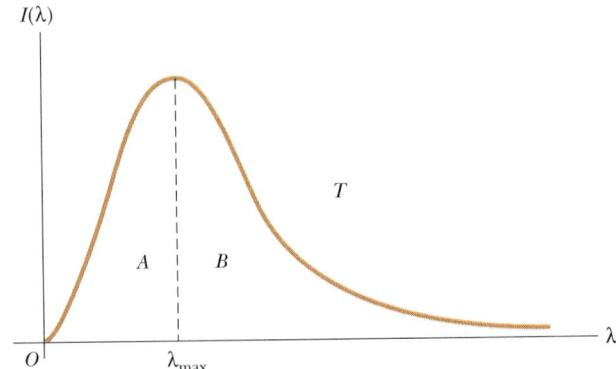

FIGURE P40.71

72. A tungsten filament of a lightbulb is equivalent to a black body at 2900 K. (a) Determine the percentage of the emitted radiant power per unit area that is in the form of visible light. This is equivalent to finding the ratio of areas shown in Figures P40.72a and P40.72b, where $\lambda_1 = 400$ nm and $\lambda_2 = 700$ nm. (*Hint:* You may find it useful to examine the function

$$F(x) = \int_0^x \frac{t^3}{e^t - 1} \, dt$$

which is the area under the curve in Figure P40.72c and is tabulated for various values of x in the table below.) (b) Based on your answer, would it be more appropriate to call it a "heat bulb" instead of a "lightbulb"? Explain.

x	$F(x)$	x	$F(x)$	x	$F(x)$
7.0	6.003	9.0	6.367	11.0	6.464
7.1	6.034	9.1	6.375	11.1	6.467
7.2	6.062	9.2	6.383	11.2	6.469
7.3	6.089	9.3	6.391	11.3	6.470
7.4	6.115	9.4	6.398	11.4	6.472
7.5	6.139	9.5	6.405	11.5	6.474
7.6	6.162	9.6	6.411	11.6	6.475
7.7	6.183	9.7	6.417	11.7	6.477
7.8	6.203	9.8	6.422	11.8	6.478
7.9	6.222	9.9	6.427	11.9	6.479
8.0	6.240	10.0	6.432	12.0	6.480
8.1	6.256	10.1	6.436	12.1	6.481
8.2	6.272	10.2	6.440	12.2	6.482
8.3	6.287	10.3	6.444	12.3	6.483
8.4	6.300	10.4	6.448	12.4	6.484
8.5	6.313	10.5	6.451	12.5	6.485
8.6	6.325	10.6	6.454	12.6	6.486
8.7	6.337	10.7	6.457	12.7	6.487
8.8	6.347	10.8	6.460	12.8	6.488
8.9	6.357	10.9	6.462	12.9	6.489

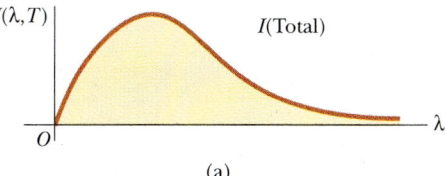

(a)

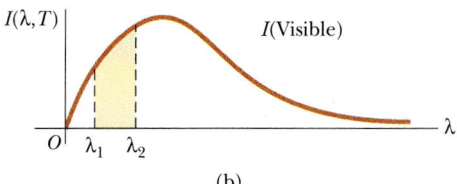

(b)

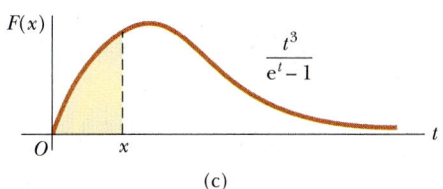

(c)

FIGURE P40.72

SPREADSHEET PROBLEM

S1. Compton scattering involves the scattering of an incident photon of wavelength λ_0 from a free electron. The scattered photon has a decreased wavelength λ' that depends on its scattering angle θ. The electron recoils with kinetic energy K_e at a scattering angle ϕ relative to the incident direction. Spreadsheet 40.1 calculates and plots both K_e and ϕ as functions of θ for a given λ. (a) For $\lambda = 1 \times 10^{-10}$ m, examine the values of K_e and ϕ at $\theta = 30°$, $90°$, $120°$, and $180°$. Do your results make sense? (Consider conservation of energy and momentum.) (b) Repeat part (a) for $\lambda = 1 \times 10^{-9}$ m, 1×10^{-11} m, and 1×10^{-12} m. Explain the overall variation with incident wavelength.

Quantum Mechanics

This photograph is a demonstration of a "quantum-corral" consisting of a ring of 48 iron atoms located on a copper surface. The diameter of the ring is 143 nm, and the photograph was obtained using a low-temperature scanning tunneling microscope. Corrals and other structures are able to confine surface-electron waves. The study of such structures will play an important role in determining the future of small electronic devices. *(IBM Corporation Research Division)*

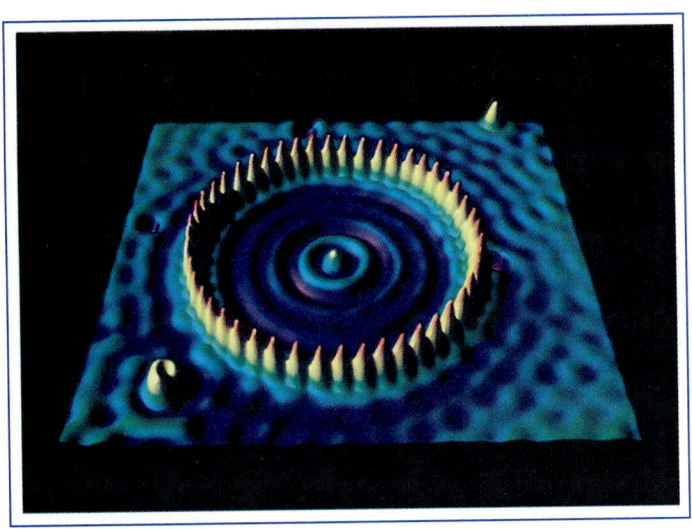

In Chapter 40, we pointed out that the Bohr model of the hydrogen atom has severe limitations. For example, it does not explain the spectra of complex atoms nor does it predict such details as variations in spectral line intensities and the splittings observed in certain spectral lines under controlled laboratory conditions. Finally, it does not enable us to understand how atoms interact with each other and how such interactions affect the observed physical and chemical properties of matter.

In this chapter, we introduce quantum mechanics, a theory for explaining atomic structure. This scheme, developed from 1925 to 1926 by Schrödinger, Heisenberg, and others, addresses the limitations of the Bohr model and lets us understand a host of phenomena involving atoms, molecules, nuclei, and solids. We begin by describing wave-particle duality, wherein a particle can be viewed as having wave-like properties, and its wavelength can be calculated if its momentum is known. Next, we describe some of the basic features of the formalism of quantum mechanics and their application to simple, one-dimensional systems. For example, we shall treat the problem of a particle confined to a potential well having infinitely high barriers.

41.1 PHOTONS AND ELECTROMAGNETIC WAVES

Phenomena such as the photoelectric effect and the Compton effect offer ironclad evidence that when light and matter interact, the light behaves as if it were composed of particles having energy hf and momentum h/λ. An obvious question at this point is, "How can light be considered a photon (in other words, a particle) when we know it is a wave?" On the one hand, we describe light in terms of photons having energy and momentum. On the other hand, we recognize that light and other electromagnetic waves exhibit interference and diffraction effects, which are consistent only with a wave interpretation.

Which model is correct? Is light a wave or a particle? The answer depends on the phenomenon being observed. Some experiments can be explained better, or solely, with the photon concept, whereas others are best described, or can be only described, with a wave model. The end result is that *we must accept both models and admit that the true nature of light is not describable in terms of any single classical picture.* However, you should recognize that the same light beam that can eject photoelectrons from a metal (meaning that the beam consists of photons) can also be diffracted by a grating (meaning that the beam is a wave). In other words, *the photon theory and the wave theory of light complement each other.*

The success of the particle model of light in explaining the photoelectric effect and the Compton effect raises many other questions. If light is a particle, what is the meaning of its "frequency" and "wavelength," and which of these two properties determines its energy and momentum? Is light simultaneously a wave and a particle? Although photons have no rest mass (a nonobservable quantity), is there a simple expression for the effective mass of a moving photon? If a photon has mass, do photons experience gravitational attraction? What is the spatial extent of a photon, and how does an electron absorb or scatter one photon? Although some of these questions are answerable, others demand a view of atomic processes that is too pictorial and literal. Furthermore, many of these questions stem from classical analogies such as colliding billiard balls and water waves breaking on a shore. Quantum mechanics gives light a more fluid and flexible nature by incorporating both the particle model and wave model of light as necessary and complementary. Hence,

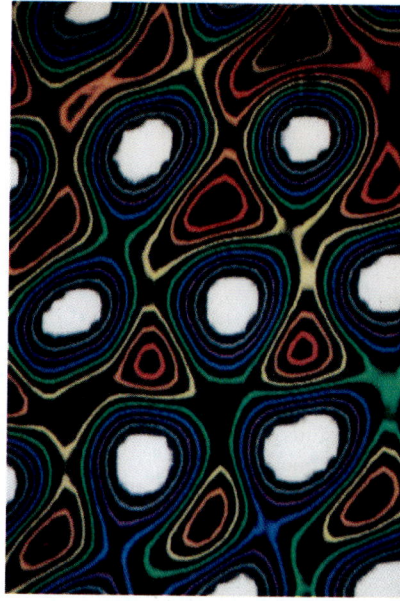

Viewing crystal surfaces. The surface of $TaSe_2$, "viewed" with a scanning tunneling microscope. The photograph is actually a charge density wave contour of the surface, where the various colors indicate regions of different charge densities. *(Courtesy of Prof. R. V. Coleman, University of Virginia)*

light has a dual nature: it exhibits both wave and particle characteristics.

The dual nature of light

To understand why photons are compatible with electromagnetic waves, consider 2.5-MHz radio waves as an example. The energy of a photon having this frequency is only about 10^{-8} eV, too small to be detected as a single photon. A sensitive radio receiver might require as many as 10^{10} of these photons to produce a detectable signal. Such a large number of photons would appear, on the average, as a continuous wave. With so many photons reaching the detector every second, it is unlikely that any graininess would appear in the detected signal. That is, with 2.5-MHz waves, one would not be able to detect the individual photons striking the antenna.

Now consider what happens as we go to shorter wavelengths. In the visible region, it is possible to observe both the photon and the wave characteristics of light. As we mentioned earlier, a beam of visible light shows interference phenomena (thus, it is a wave) and at the same time can produce photoelectrons (thus, it is a particle). At even shorter wavelengths, the momentum and energy of the photon increase. Consequently, the photon nature of light becomes more evident than its wave nature. For example, absorption of an x-ray photon is easily detected as a single event. However, as the wavelength decreases, wave effects become more difficult to observe.

Louis de Broglie (1892–1987), a French physicist, was awarded the Nobel prize in 1929 for his discovery of the wave nature of electrons. "It would seem that the basic idea of quantum theory is the impossibility of imaging an isolated quantity of energy without associating with it a certain frequency." *(AIP Niels Bohr Library)*

41.2   THE WAVE PROPERTIES OF PARTICLES

Students introduced to the dual nature of light often find the concept difficult to accept. In the world around us, we are accustomed to regarding such things as baseballs solely as particles and such things as sound waves solely as forms of wave motion. Every large-scale observation can be interpreted by considering either a wave explanation or a particle explanation, but in the world of photons and electrons, such distinctions are not as sharply drawn. Even more disconcerting is the fact that, under certain conditions, the things we unambiguously call "particles" exhibit wave characteristics!

In 1923, in his doctoral dissertation, Louis Victor de Broglie postulated that *because photons have wave and particle characteristics, perhaps all forms of matter have wave as well as particle properties.* This was a highly revolutionary idea with no experimental confirmation at that time. According to de Broglie, electrons have a dual particle-wave nature. An electron in motion exhibits wave character or wave properties. De Broglie explained the source of this assertion in his 1929 Nobel prize acceptance speech:

> On the one hand the quantum theory of light cannot be considered satisfactory since it defines the energy of a light corpuscle in terms of frequency *f.* Now a purely corpuscular theory contains nothing that enables us to define a frequency; for this reason alone, therefore, we are compelled, in the case of light, to introduce the idea of a corpuscle and that of periodicity simultaneously. On the other hand, determination of the stable motion of electrons in the atom introduces integers, and up to this point the only phenomena involving integers in physics were those of interference and of normal modes of vibration. This fact suggested to me the idea that electrons too could not be considered simply as corpuscles, but that periodicity must be assigned to them also.

In Chapter 39, we found that the relationship between the energy and momentum of a photon, which has a rest mass of zero, is $p = E/c$. We also know that the energy of a photon is $E = hf = hc/\lambda$. Thus, the momentum of a photon can be expressed as

Momentum of a photon

$$p = \frac{E}{c} = \frac{hc}{c\lambda} = \frac{h}{\lambda}$$

From this equation we see that the photon wavelength can be specified by its momentum, or $\lambda = h/p$. De Broglie suggested that material particles of momentum p have a characteristic wavelength $\lambda = h/p$. Because the momentum of a particle of mass m and speed v is $p = mv$, the **de Broglie wavelength** of that particle is

De Broglie wavelength

$$\lambda = \frac{h}{p} = \frac{h}{mv} \tag{41.1}$$

Furthermore, in analogy with photons, de Broglie postulated that the frequencies of matter waves (that is, waves associated with particles of nonzero rest mass) obey the Planck relationship $E = hf$, so that

Frequency of matter waves

$$f = \frac{E}{h} \tag{41.2}$$

The dual nature of matter is apparent in these two equations because each contains both particle concepts (mv and E) and wave concepts (λ and f). The fact that these relationships are established experimentally for photons makes the de Broglie hypothesis that much easier to accept.

Quantization of Angular Momentum in the Bohr Model

Bohr's model of the atom, discussed in Chapter 40, has many shortcomings and problems. For example, as the electrons revolve around the nucleus, how do we visualize the fact that only certain electronic energies are allowed? Also, why do all atoms of a given element have precisely the same physical properties regardless of the infinite variety of starting speeds and positions of the electrons in each atom?

De Broglie's great insight was to recognize that wave theories of matter use the concept of interference to explain these observations. Recall from Chapter 18 that a plucked guitar string, while initially subjected to a wide range of wavelengths, supports only standing wave patterns that have nodes at its ends. Any free vibration of the string consists of a superposition of varied amounts of many standing waves. This same reasoning can be applied to electron matter waves bent into a circle around the nucleus. All of the possible states of the electron are standing-wave states, each with its own wavelength, speed, and energy. Figure 41.1a illustrates this point when three complete wavelengths are contained in the circumference of one orbit. Similar patterns can be drawn for orbits containing two wavelengths, four wavelengths, and so forth. This situation is analogous to that of standing waves on a string that has preferred (resonant) frequencies of vibration. Figure 41.1b shows a standing wave pattern containing three wavelengths for a string that is fixed at each end. Now imagine the vibrating string is removed from its supports at *A* and *B* and bent into a circular shape so that points *A* and *B* are connected. The end result is a pattern like that shown in Figure 41.1a.

Another aspect of the Bohr theory that is easier to visualize physically by using de Broglie's hypothesis is the quantization of angular momentum. We simply need to assume that the allowed Bohr orbits arise because the electron matter waves form standing waves when an integral number of wavelengths exactly fits into the circumference of a circular orbit. Thus,

$$n\lambda = 2\pi r$$

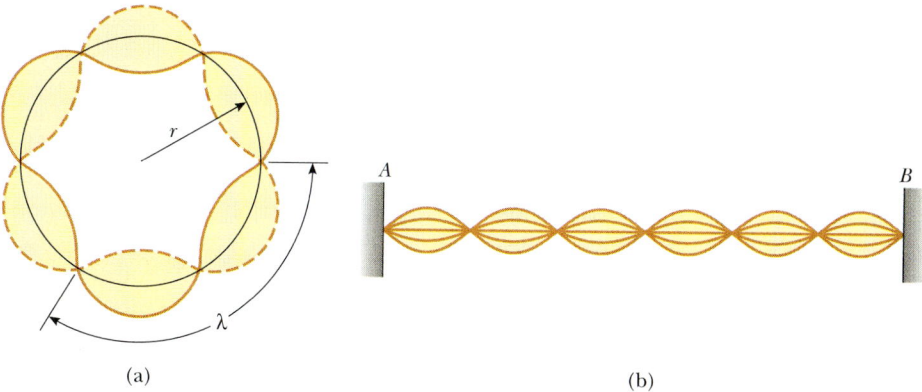

(a)   (b)

FIGURE 41.1   Standing wave patterns. (a) Three full wavelengths associated with an electron moving in a stable atomic orbit. (b) Three full wavelengths on a stretched string.

where *r* is the radius of the orbit and $n = 1, 2, 3, 4, \ldots$. Because the de Broglie wavelength is $\lambda = h/mv$, and because $\hbar = h/2\pi$, we can write the preceding condition as $n(h/mv) = 2\pi r$, or

$$mvr = n\hbar$$

Note that this is precisely the Bohr condition of quantization of angular momentum (Eq. 40.19).[1] Thus, electron waves that fit the orbits are standing waves because of the boundary conditions imposed. These standing waves have discrete frequencies, corresponding to the allowed wavelengths. If $n\lambda \neq 2\pi r$, a standing-wave pattern can never form a closed circular orbit as in Figure 41.1a because the waves interfere destructively.

The Davisson–Germer Experiment

De Broglie's proposal in 1923 that any kind of particle exhibits both wave and particle properties was regarded as pure speculation. If particles such as electrons had wave-like properties, then under the correct conditions they should exhibit diffraction effects. Only three years later, C. J. Davisson and L. H. Germer of the United States succeeded in measuring the wavelength of electrons. Their important discovery provided the first experimental confirmation of the matter waves proposed by de Broglie.

Interestingly, the intent of the initial Davisson–Germer experiment was not to confirm the de Broglie hypothesis. In fact, their discovery was made by accident (as is often the case). The experiment involved the scattering of low-energy electrons (about 54 eV) from a nickel target in a vacuum. During one experiment, the nickel surface was badly oxidized because of an accidental break in the vacuum system. After the target was heated in a flowing stream of hydrogen to remove the oxide coating, electrons scattered by it exhibited intensity maxima at specific angles. The experimenters finally realized that the nickel had formed large crystalline regions upon heating and that the regularly spaced planes of atoms in these regions served as a diffraction grating for electron matter waves.

Shortly thereafter, Davisson and Germer performed more extensive diffraction measurements on electrons scattered from single-crystal targets. Their results showed conclusively the wave nature of electrons and confirmed the de Broglie relationship $p = h/\lambda$, and a year later in 1928, G. P. Thomson of Scotland observed electron diffraction patterns by passing electrons through very thin gold foils. Diffraction patterns have since been observed for beams of helium atoms, hydrogen atoms, and neutrons. Hence, the universal nature of matter waves has been established in various ways.

The problem of understanding the dual nature of matter and radiation is conceptually difficult because the two models seem to contradict each other. This problem as it applies to light was discussed earlier. Bohr helped to resolve this problem in his **principle of complementarity**, which states that *the wave and particle models of either matter or radiation complement each other*. Neither model can be used exclusively to adequately describe matter or radiation. A complete understanding is obtained only when the two models are combined in a complementary manner.

The principle of complementarity

---

[1] De Broglie's analysis failed to explain the fact that states having an orbital angular momentum quantum number of zero do exist.

EXAMPLE 41.1   **The Wavelength of an Electron**

Calculate the de Broglie wavelength for an electron ($m = 9.11 \times 10^{-31}$ kg) moving at 1.00×10^7 m/s.

Solution   Equation 41.1 gives

$$\lambda = \frac{h}{mv} = \frac{6.63 \times 10^{-34}\,\text{J}\cdot\text{s}}{(9.11 \times 10^{-31}\,\text{kg})(1.00 \times 10^7\,\text{m/s})}$$

$$= 7.28 \times 10^{-11}\,\text{m}$$

This is the wavelength of an x-ray.

Exercise   Find the de Broglie wavelength of a proton moving at 1.00×10^7 m/s.

Answer 3.97×10^{-14} m.

EXAMPLE 41.2   **The Wavelength of a Rock**

A rock of mass 50 g is thrown with a speed of 40 m/s. What is its de Broglie wavelength?

Solution   From Equation 41.1, we have

$$\lambda = \frac{h}{mv} = \frac{6.63 \times 10^{-34}\,\text{J}\cdot\text{s}}{(50 \times 10^{-3}\,\text{kg})(40\,\text{m/s})} = 3.3 \times 10^{-34}\,\text{m}$$

This wavelength is much smaller than the size of any possible aperture through which the rock could pass. This means that we could not observe diffraction effects, and as a result we realize that the wave properties of large-scale objects cannot be observed.

EXAMPLE 41.3   **An Accelerated Charge**

A particle of charge q and mass m is accelerated from rest through a potential difference V. Find its de Broglie wavelength.

Solution   When a charge is accelerated from rest through a potential difference V, its gain in kinetic energy $\frac{1}{2}mv^2$ must equal its loss in potential energy qV:

$$\tfrac{1}{2}mv^2 = qV$$

Since $p = mv$, we can express this in the form

$$\frac{p^2}{2m} = qV \quad \text{or} \quad p = \sqrt{2mqV}$$

Substituting this expression for p into Equation 41.1 gives

$$\lambda = \frac{h}{p} = \frac{h}{\sqrt{2mqV}}$$

Exercise   Calculate the de Broglie wavelength of the electron accelerated through a potential difference of 50 V.

Answer   0.174 nm

41.3   THE DOUBLE-SLIT EXPERIMENT REVISITED

One way to crystallize our ideas about wave-particle duality is to consider the diffraction of electrons passing through a double slit. This experiment shows the impossibility of *simultaneously* measuring both wave and particle properties and illustrates the use of the wave function in determining interference effects.

Consider a beam of monoenergetic electrons incident on a double slit as in Figure 41.2. Because the slit openings are much smaller than the wavelength of the electrons, no structure is expected from single-slit diffraction. An electron detector is positioned far from the slits at a distance much greater than D. *If the detector collects electrons at different positions for a long enough time, one finds an interference pattern representing the number of electrons arriving at any position along the detector line.* Such an interference pattern would not be expected if the electrons behaved as classical particles.

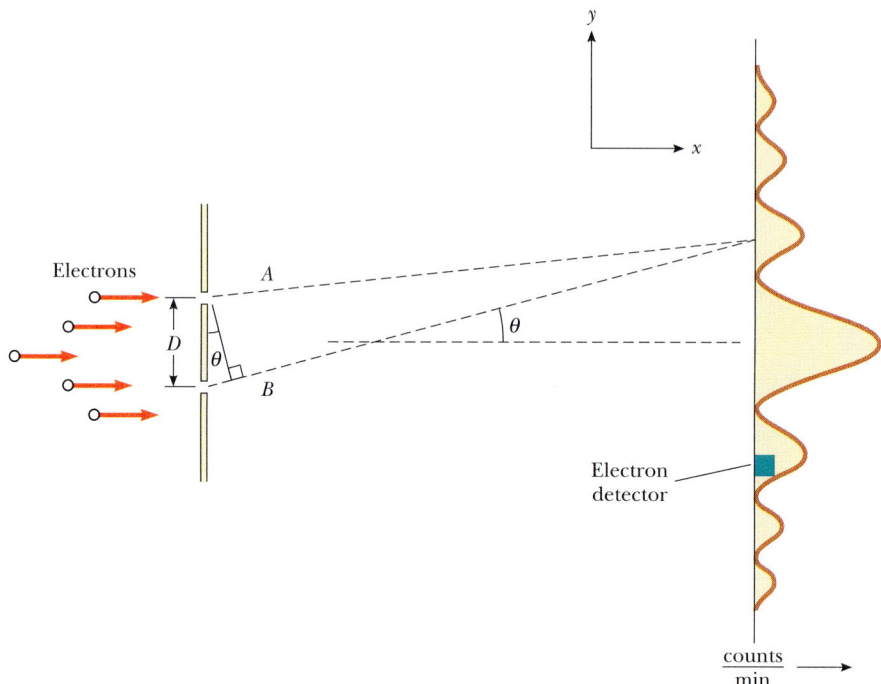

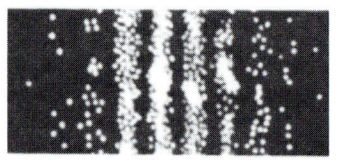

(a) After 28 electrons

FIGURE 41.2 Electron diffraction. The slit separation D is much greater than the individual slit widths and much smaller than the distance from the slits to the screen.

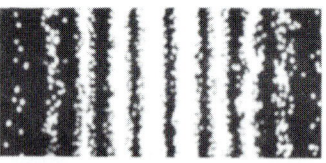

(b) After 1000 electrons

If the experiment is carried out at lower beam intensities, the interference pattern is still observed if the time of the measurement is sufficiently long. This is illustrated in the computer-simulated patterns in Figure 41.3. Note that the interference pattern becomes clearer as the number of electrons reaching the screen increases.

If a single electron producing in-phase "wavelets" as it reaches one of the slits is imagined, standard wave theory can be used to find the angular separation, θ, between the central probability maximum and its neighboring minimum. The minimum occurs when the path length difference between A and B is half a wavelength, or

(c) After 10,000 electrons

$$D \sin \theta = \frac{\lambda}{2}$$

As the electron's wavelength is given by $\lambda = h/p$, we see that, for small θ,

$$\sin \theta \approx \theta = \frac{h}{2p_x D}$$

(d) Two-slit electron pattern

FIGURE 41.3 (a), (b), (c) Computer-simulated interference patterns for a beam of electrons incident on a double slit. *(From E. R. Huggins, Physics I, New York, W. A. Benjamin, 1968)* (d) Photograph of a double-slit interference pattern produced by electrons. *(From C. Jönsson, Zeitschrift für Physik 161:454, 1961, used with permission)*

Thus the dual nature of the electron is clearly shown in this experiment: *The electrons are detected as particles at a localized spot at some instant of time, but the probability of arrival at that spot is determined by finding the intensity of two interfering matter waves.*

And there is more. If one slit is covered during the experiment, and the slit width is small compared to the wavelength of the electrons, a symmetric curve peaked around the center of the open slit, much like the pattern formed by bullets shot through a hole in armor plate, is obtained. Plots of the counts per minute or probability of arrival of electrons with the lower or upper slit closed are shown as blue curves in the central part of Figure 41.4. These curves are expressed as the square of the absolute value of some wave function, $|\psi_1|^2 = \psi_1^* \psi_1$ or

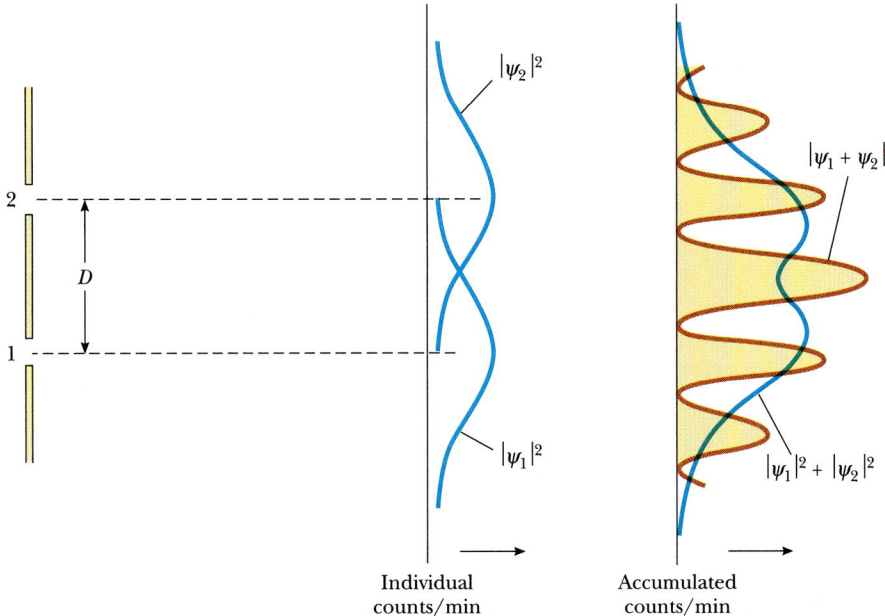

FIGURE 41.4 The blue curve on the right represents the accumulated pattern of counts per minute when each slit is closed half the time. The red curve represents the diffraction pattern with both slits open at the same time.

$|\psi_2|^2 = \psi_2^* \psi_2$, where ψ_1 and ψ_2 represent the electron passing through slit 1 and slit 2, respectively.

If an experiment is performed with slit 2 blocked half of the time and then slit 1 blocked during the remaining time, the accumulated pattern of counts/minute shown by the blue curve on the right side of Figure 41.4 is completely different from the case with both slits open (red curve). There is no longer a maximum probability of arrival at $\theta = 0$. In fact, *the interference pattern has been lost and the accumulated result is simply the sum of the individual results.* Because the electron must pass either through slit 1 or through slit 2, it has the same localizability and indivisibility at the slits as we measure at the detector. Thus the pattern must represent the sum of those electrons that come through slit 1, $|\psi_1|^2$, and those that come through slit 2, $|\psi_2|^2$.

When both slits are open, it is tempting to assume that the electron goes through either slit 1 or slit 2 and that the counts/minute are again given by $|\psi_1|^2 + |\psi_2|^2$. We know, however, that the experimental results indicated by the red interference pattern in Figure 41.4 contradict this assumption. Thus our assumption that the electron is localized and goes through only one slit when both slits are open must be wrong (a painful conclusion!). Because it exhibits interference, we are forced to conclude that — somehow — *the electron must be simultaneously present at both slits.* To find the probability of detecting the electron at a particular point on the screen when both slits are open, we may say that the electron is in a *superposition state* given by

$$\psi = \psi_1 + \psi_2$$

Thus the probability of detecting the electron at the screen is $|\psi_1 + \psi_2|^2$ and not $|\psi_1|^2 + |\psi_2|^2$. Because matter waves that start out in phase at the slits in general travel different distances to the screen, ψ_1 and ψ_2 possess a relative phase differ-

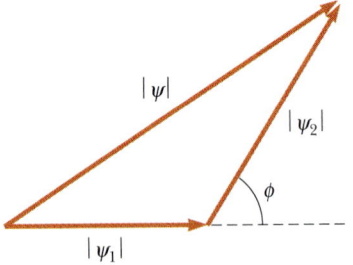

FIGURE 41.5   Phasor diagram to represent the addition of two complex quantities, ψ_1 and ψ_2.

ence ϕ at the screen. Using a phasor diagram (Fig. 41.5) to find $|\psi_1 + \psi_2|^2$ immediately yields

$$|\psi|^2 = |\psi_1 + \psi_2|^2 = |\psi_1|^2 + |\psi_2|^2 + 2|\psi_1||\psi_2|\cos\phi$$

where $|\psi_1|^2$ is the probability of detection if slit 1 is open and slit 2 is closed and $|\psi_2|^2$ is the probability of detection if slit 2 is open and slit 1 is closed. The term $2|\psi_1||\psi_2|\cos\phi$ in this expression is the interference term, which arises from the relative phase ϕ of the waves in analogy with the phasor addition used in wave optics (Chapter 37).

In order to interpret these results, we are forced to conclude that *an electron interacts with both slits simultaneously*. If we attempt to determine experimentally which slit the electron goes through, the act of measuring destroys the interference pattern. It is impossible to determine which slit the electron goes through. In effect, *we can say only that the electron passes through both slits!* The same arguments apply to photons.

The Electron Microscope

A practical device that relies on the wave characteristics of electrons is the **electron microscope** (Fig. 41.6), which is in many respects similar to an ordinary compound microscope. One important difference between the two is that the electron microscope has a much greater resolving power because electrons can be accelerated to very high kinetic energies, giving them very short wavelengths. Any microscope is capable of detecting details that are comparable in size to the wavelength of the radiation used to illuminate the object. Typically, the wavelengths of electrons are about 100 times shorter than those of the visible light used in optical microscopes. As a result, electron microscopes are able to distinguish details about 100 times smaller.

In operation, a beam of electrons falls on a thin slice of the material to be examined. The section to be examined must be very thin, typically only a few micrometers thick, in order to minimize undesirable effects, such as absorption or scattering of the electrons. The electron beam is controlled by electrostatic or magnetic deflection, which acts on the charges to focus the beam to an image. Rather than examining the image through an eyepiece as in an ordinary microscope, a magnetic lens forms an image on a fluorescent screen. (The viewing screen must be fluorescent because otherwise the image produced would not be visible.) A photograph taken by an electron microscope is shown in Figure 41.7.

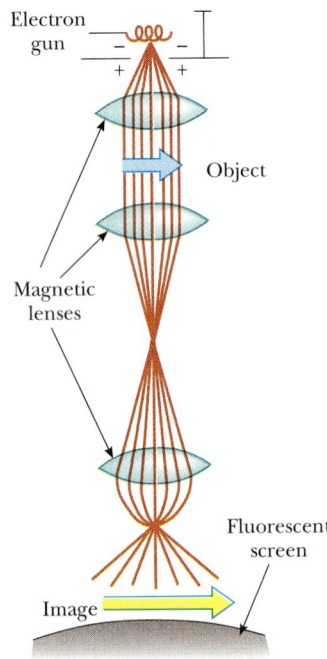

FIGURE 41.6   Diagram of an electron microscope. The "lenses" that control the electron beam are magnetic deflection coils.

41.4   THE UNCERTAINTY PRINCIPLE

If you were to measure the position and speed of a particle at any instant, you would always be faced with experimental uncertainties in your measurements. According to classical mechanics, there is no fundamental barrier to an ultimate refinement of the apparatus or experimental procedures. In other words, it is possible, in principle, to make such measurements with arbitrarily small uncertainty. Quantum theory predicts, however, that *it is fundamentally impossible to make simultaneous measurements of a particle's position and speed with infinite accuracy.*

In 1927, Werner Heisenberg (1901–1976) derived this notion, which is now known as the **Heisenberg uncertainty principle:**

If a measurement of position is made with precision Δx and a simultaneous measurement of momentum components, p_x, is made with precision Δp_x, then the product of the two uncertainties can never be smaller than $\hbar/2$. That is,

$$\Delta x \, \Delta p_x \geq \frac{\hbar}{2} \qquad (41.3)$$

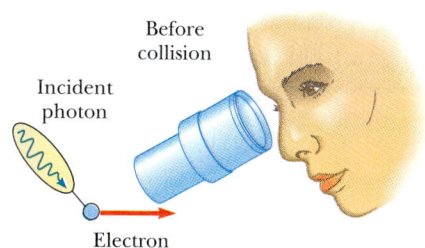

FIGURE 41.7 Scanning electron micrograph of an unidentified species of dust mite. Magnification ×200.   *(Andrew Syred/SPL/ Photo Researchers, Inc.)*

That is, *it is physically impossible to simultaneously measure the exact position and exact momentum of a particle.* If Δx is very small, then Δp_x is large, and vice versa. Heisenberg was careful to point out that the inescapable uncertainties Δx and Δp_x do not arise from imperfections in measuring instruments. Rather, they arise from the quantum structure of matter: from effects such as the unpredictable recoil of an electron when struck by an indivisible photon or the diffraction of light or electrons passing through a small opening.

To understand the uncertainty principle, consider the following thought experiment introduced by Heisenberg. Suppose you wish to measure the position and momentum of an electron as accurately as possible. You might be able to do this by viewing the electron with a powerful light microscope. In order to see the electron, and thus determine its location, at least one photon of light must bounce off the electron, as shown in Figure 41.8a, and then it must pass through the microscope into your eye as shown in Figure 41.8b. When it strikes the electron, however, the photon transfers some unknown amount of its momentum to the electron. Thus, in the process of locating the electron very accurately (that is, making Δx very small), the very light that allows you to succeed changes the electron's momentum to some undeterminable extent (making Δp_x large).

Let us analyze the collision by first noting that the incoming photon has momentum h/λ. As a result of the collision, the photon transfers part or all of its

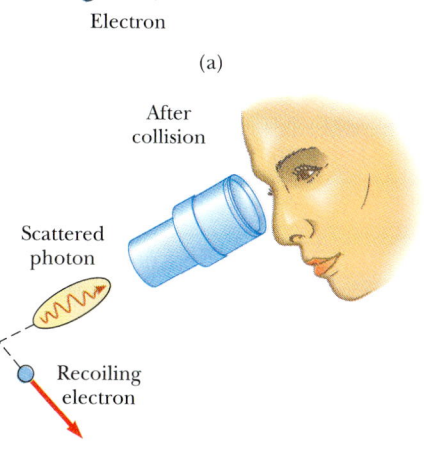

(a)

(b)

FIGURE 41.8 A thought experiment for viewing an electron with a powerful microscope. (a) The electron is moving to the right before colliding with the photon. (b) The electron recoils (its momentum changes) as the result of the collision with the photon.

Werner Heisenberg (1901–1976), a German theoretical physicist, obtained his PhD in 1923 at the University of Munich, where he studied under Arnold Sommerfeld and became an enthusiastic mountain climber and skier. While physicists such as de Broglie and Schrödinger tried to develop physical models of the atom, Heisenberg developed an abstract mathematical model called matrix mechanics to explain the wavelengths of spectral lines. The more successful wave mechanics by Schrödinger announced a few months later was shown to be equivalent to Heisenberg's approach. Heisenberg made many other significant contributions to physics, including his famous uncertainty principle, for which he received the Nobel Prize in 1932; the prediction of two forms of molecular hydrogen; and theoretical models of the nucleus.

momentum along the *x* axis to the electron. Thus, the uncertainty in the electron's momentum after the collision is as large as the momentum of the incoming photon. That is, $\Delta p_x = h/\lambda$. Furthermore, since light also has wave properties, we would expect to be able to determine the position of the electron to within one wavelength of the light being used to view it, so that $\Delta x = \lambda$. Multiplying these two uncertainties gives

$$\Delta x \, \Delta p_x = \lambda \left(\frac{h}{\lambda} \right) = h$$

This represents the minimum in the products of the uncertainties. Since the uncertainty can always be greater than this minimum, we have

$$\Delta x \, \Delta p_x \geqslant h$$

This agrees with Equation 41.3 (apart from the numerical factor of $1/4\pi$ introduced by Heisenberg's more precise analysis).

Heisenberg's uncertainty principle enables us to better understand the dual wave-particle nature of light and matter. We have seen that the wave description is quite different from the particle description. Therefore, if an experiment (such as the photoelectric effect) is designed to reveal the particle character of an electron, its wave character becomes less apparent. If the experiment (such as diffraction from a crystal) is designed to accurately measure the electron's wave properties, its particle character becomes less apparent.

Another uncertainty relationship sets a limit on the accuracy with which the energy of a system, ΔE, can be measured if a finite time interval, Δt, is allowed for the measurement. This energy-time uncertainty principle may be stated as

$$\Delta E \, \Delta t \geqslant \frac{\hbar}{2} \tag{41.4}$$

This relationship is plausible if a frequency measurement of any wave is considered. For example, consider the measurement of a 1000-Hz electrical wave. If our frequency-measuring device has a fixed phase sensitivity of ± 1 cycle, in 1 s we measure a frequency of (1000 ± 1) cycles/1 s, but in 2 s we measure a frequency of (2000 ± 1) cycles/2 s. Thus the uncertainty in frequency, Δf, is inversely proportional to Δt, the time during which the measurement is made. This may be stated as

$$\Delta f \, \Delta t \approx 1$$

Since all quantum systems are wave-like and can be described by the relationship $E = hf$, we may substitute $\Delta f = \Delta E/h$ into the expression above to obtain

$$\Delta E \, \Delta t \approx h$$

in basic agreement with Equation 41.4, apart from the factor of $1/4\pi$.

We conclude this section with two examples of the types of calculations that can be done with the uncertainty principle. In the spirit of Fermi or Heisenberg, these "back-of-the-envelope calculations" are surprising for their simplicity and for their essential description of quantum systems of which the details are unknown.

CONCEPTUAL EXAMPLE 41.4 Is the Bohr Model Realistic?

According to the Bohr model of the hydrogen atom, the electron in the ground state moves in a circular orbit of radius 0.529×10^{-10} m, and the speed of the electron in this state is 2.2×10^6 m/s. In view of the Heisenberg uncertainty principle, is this model of the atom realistic?

Reasoning Because the Bohr model assumes that the electron is located exactly at $r = 0.529$ nm, the uncertainty Δr in its radial position in this model is zero. But according to the uncertainty principle, the product $\Delta p_r \Delta r \geqslant \hbar/2$, where Δp_r is the uncertainty in the momentum of the electron in the radial direction. Because the momentum of the electron is mv, we can assume that the uncertainty in its momentum is

less than this value. That is,

$$\Delta p_r < mv = (9.11 \times 10^{-31} \text{ kg})(2.2 \times 10^6 \text{ m/s})$$
$$= 2.0 \times 10^{-24} \text{ kg} \cdot \text{m/s}$$

From the uncertainty principle, we can now estimate the minimum uncertainty in the radial position:

$$\Delta r_{\min} = \frac{\hbar}{2\Delta p_r} = \frac{1.05 \times 10^{-34} \text{ J} \cdot \text{s}}{2 \times 2 \times 10^{-24} \text{ kg} \cdot \text{m/s}} = 0.26 \times 10^{-10} \text{ m}$$

Because this uncertainty in position is approximately equal to the size of the Bohr radius, we must conclude that the Bohr model is an incorrect description of the atom.

EXAMPLE 41.5 Locating an Electron

The speed of an electron is measured to be 5.00×10^3 m/s to an accuracy of 0.0030%. Find the uncertainty in determining the position of this electron.

Solution The momentum of the electron is

$$p_x = mv = (9.11 \times 10^{-31} \text{ kg})(5.00 \times 10^3 \text{ m/s})$$
$$= 4.56 \times 10^{-27} \text{ kg} \cdot \text{m/s}$$

The uncertainty in p_x is 0.0030% of this value:

$$\Delta p_x = 0.000030 p_x = (0.000030)(4.56 \times 10^{-27} \text{ kg} \cdot \text{m/s})$$

$$= 1.37 \times 10^{-31} \text{ kg} \cdot \text{m/s}$$

The uncertainty in position can now be calculated by using this value of Δp_x and Equation 41.3:

$$\Delta x \, \Delta p_x \geqslant \frac{\hbar}{2}$$

$$\Delta x \geqslant \frac{\hbar}{2 \, \Delta p} = \frac{1.05 \times 10^{-34} \text{ J} \cdot \text{s}}{2(1.37 \times 10^{-31} \text{ kg} \cdot \text{m/s})}$$

$$= 0.38 \times 10^{-3} \text{ m} = \boxed{0.38 \text{ mm}}$$

EXAMPLE 41.6 The Width of Spectral Lines

Although an excited atom can radiate at any time from $t = 0$ to $t = \infty$, the average time after excitation at which a group of atoms radiates is called the **lifetime**, τ. (a) If $\tau = 1.0 \times 10^{-8}$ s, use the uncertainty principle to compute the line width Δf produced by this finite lifetime.

Solution We use $\Delta E \, \Delta t \geqslant \hbar/2$, where $\Delta E = h \, \Delta f$ and $\Delta t = 1.0 \times 10^{-8}$ s is the average time available to measure the excited state. Thus,

$$\Delta f = \frac{1}{4\pi \times 10^{-8} \text{ s}} = \boxed{8.0 \times 10^6 \text{ Hz}}$$

Note that ΔE is the uncertainty in energy of the excited state. It is also the uncertainty in the energy of the photon emitted by an atom in this state.

(b) If the wavelength of the spectral line involved in this process is 500 nm, find the fractional broadening $\Delta f/f$.

Solution First, we find the center frequency of this line:

$$f_0 = \frac{c}{\lambda} = \frac{3.00 \times 10^8 \text{ m/s}}{500 \times 10^{-9} \text{ m}} = 6.00 \times 10^{14} \text{ Hz}$$

Hence,

$$\frac{\Delta f}{f_0} = \frac{8.0 \times 10^6 \text{ Hz}}{6.00 \times 10^{14} \text{ Hz}} = \boxed{1.3 \times 10^{-8}}$$

This narrow natural line width could be seen with a sensitive interferometer. Usually, however, temperature and pressure effects overshadow the natural line width and broaden the line through mechanisms associated with the Doppler effect and collisions.

41.5   INTRODUCTION TO QUANTUM MECHANICS

Just as the wave theory of light gives only the probability of finding a photon at a given point at some instant, the wave theory of matter gives only the probability of finding a particle of matter at a given point at some instant. Matter waves are described by a complex-valued wave function (usually denoted by ψ) whose absolute square $|\psi|^2$ is proportional to the probability of finding the particle at a given point at some instant. The wave function contains within it all the information that can be known about the particle.

This interpretation of matter waves was first suggested by Max Born (1882–1970) in 1928. In 1926, Erwin Schrödinger (1887–1961) proposed a wave equation that describes how matter waves change in space and time. The Schrödinger wave equation represents a key element in the theory of quantum mechanics.

The concepts of quantum mechanics, strange as they sometimes seem, developed from classical ideas. In fact, when the techniques of quantum mechanics are applied to macroscopic systems, the results are essentially identical to those of classical physics. This blending of the two theories occurs when the de Broglie wavelength is small compared with the dimensions of the system. The situation is similar to the agreement between relativistic mechanics and classical mechanics when $v \ll c$.

A number of experiments show that matter has both a wave nature and a particle nature. A question that arises quite naturally in this regard is the following: If we are describing a particle, how do we view what is waving? In the cases of waves on strings, water waves, and sound waves, the wave is represented by some quantity that varies with time and position. In a similar manner, matter waves (de Broglie waves) are represented by a wave function ψ. In general, ψ depends on both the position of all the particles in a system and on time, and therefore is often written $\psi(x, y, z, t)$. If ψ is known for a particle, then the particular properties of that particle can be described. In fact, the fundamental problem of quantum mechanics is this: Given the wave function at some instant, find the wave function at some later time t.

In Section 41.2, we found that the de Broglie equation relates the momentum of a particle to its wavelength through the relationship $p = h/\lambda$. If a free particle has a precisely known momentum, its wave function is a sinusoidal wave of wavelength $\lambda = h/p$, and the particle has equal probability of being at any point along the x axis. The wave function for such a free particle moving along the x axis can be written as

$$\psi(x) = A \sin\left(\frac{2\pi x}{\lambda}\right) = A \sin(kx) \tag{41.5}$$

where $k = 2\pi/\lambda$ is the angular wave number and A is a constant amplitude. As we mentioned earlier, the wave function is generally a function of both position and time. Equation 41.5 represents that part of the wave function that depends on position only. For this reason, we can view $\psi(x)$ as a snapshot of the wave function at a given instant, as shown in Figure 41.9a. The wave function for a particle whose wavelength is not precisely defined is shown in Figure 41.9b. Since the wavelength is not precisely defined, it follows that the momentum is only approximately known. That is, if the momentum of the particle were measured, the result would have any value over some range, determined by the spread in wavelength.

Although we cannot measure ψ, we can measure $|\psi|^2$, a quantity that can be interpreted as follows. If ψ represents a single particle, then $|\psi|^2$—called the

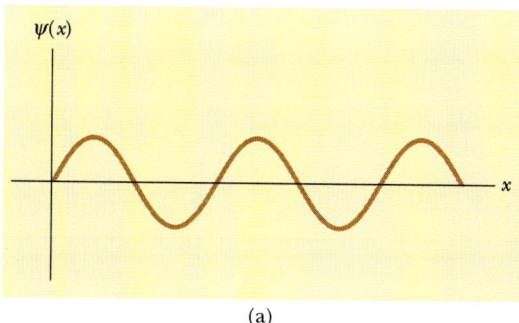

(a)

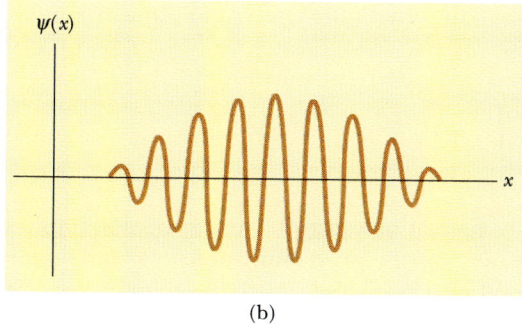

(b)

FIGURE 41.9   (a) Wave function segment for a particle whose wavelength is precisely known. (b) Wave function for a particle whose wavelength is not precisely known and hence whose momentum is known only over some range of values.

probability density—is the relative probability per unit volume that the particle will be found at any given point in the volume. This interpretation, first suggested by Born in 1928, can also be stated in the following manner. If dV is a small volume element surrounding some point, then the probability of finding the particle in that volume element is $|\psi|^2 \, dV$. In this chapter we deal only with one-dimensional systems, where the particle must be located along the x axis; thus, we replace dV by dx. In this case, the probability, $P(x) \, dx$, that the particle will be found in the infinitesimal interval dx around the point x is

$$P(x) \, dx = |\psi|^2 \, dx$$

> $|\psi|^2$ equals the probability density

Because the particle must be somewhere along the x axis, the sum of the probabilities over all values of x must be 1:

$$\int_{-\infty}^{\infty} |\psi|^2 \, dx = 1 \qquad (41.6)$$

> Normalization condition on ψ

Any wave function satisfying Equation 41.6 is said to be normalized. **Normalization** is simply a statement that the particle exists at some point at all times. If the probability were zero, the particle would not exist. Therefore, although it is not possible to specify the position of a particle with complete certainty, it is possible, through $|\psi|^2$, to specify the probability of observing it. Furthermore, *the probability of finding the particle in the interval $a \leq x \leq b$ is*

$$P_{ab} = \int_{a}^{b} |\psi|^2 \, dx \qquad (41.7)$$

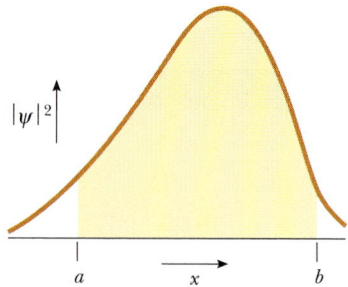

FIGURE 41.10 The probability of a particle being in the interval $a \leq x \leq b$ is the area under the curve from a to b.

The probability P_{ab} is the area under the curve of probability density versus x between the points $x = a$ and $x = b$ as in Figure 41.10.

Experimentally, there is always some probability of finding a particle at some point and at some instant, and so the value of the probability must lie between the limits 0 and 1. For example, if the probability is 0.3, there is a 30% chance of finding the particle.

The wave function ψ satisfies a wave equation, just as the electric field associated with an electromagnetic wave satisfies a wave equation that follows from Maxwell's equations. The wave equation satisfied by ψ, is the Schrödinger equation, and ψ can be computed from it. Although ψ is not a measurable quantity, all measurable quantities of a particle, such as its energy and momentum, can be derived from a knowledge of ψ. For example, once the wave function for a particle is known, it is possible to calculate the average position x of the particle, after many experimental trials. This average position is called the **expectation value** of x and is defined by the equation

Expectation value of x

$$\langle x \rangle \equiv \int_{-\infty}^{\infty} x \, |\psi|^2 \, dx \tag{41.8}$$

(Brackets, $\langle \ldots \rangle$, are used to denote expectation values.) This expression implies that the particle is in a definite state, so that the probability density is time-independent. Note that the expectation value is equivalent to the average value of x that would be obtained when dealing with a large number of particles in the same state. Furthermore, the expectation value of any function $f(x)$ can be found by using Equation 41.8 with x replaced by $f(x)$.

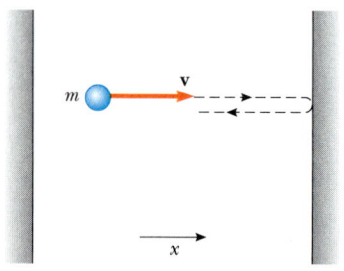

FIGURE 41.11 A particle of mass m and velocity **v** confined to moving parallel to the x axis and bouncing between two impenetrable walls.

41.6 A PARTICLE IN A BOX

From a classical viewpoint, if a particle is confined to moving along the x axis and to bouncing back and forth between two impenetrable walls (Fig. 41.11), its motion is easy to describe. If the speed of the particle is v, then the magnitude of its momentum (mv) remains constant, as does its kinetic energy. Furthermore, classical physics places no restrictions on the values of its momentum and energy. The quantum mechanics approach to this problem is quite different and requires that we find the appropriate wave function consistent with the conditions of the situation.

Before we address this problem, it is instructive to review the classical situation of standing waves on a stretched string (Sections 18.2 and 18.3). If a string of length L is fixed at each end, the standing waves set up in the string must have

Experiments in Modern Physics

This simulator enables you to investigate a number of important experiments in the field of modern physics. These experiments include the Millikan oil drop experiment, the photoelectric effect, Compton scattering, the measurement of e/m, Rutherford scattering, radioactive decay, tunneling phenomena, and atomic spectroscopy. You should take the opportunity to run these experiments under different conditions and check the outcome with theoretically expected results as described in the textbook.

nodes at the ends, as in Figure 41.12, because the wave function must vanish at the boundaries. Resonance is achieved only when the length is some integral multiple of half-wavelengths. That is, we require that

$$L = n\frac{\lambda}{2}$$

or

$$\lambda = \frac{2L}{n} \qquad n = 1, 2, 3, \ldots$$

This result shows that *the wavelength for a standing wave on a string is quantized.*

As we saw in Section 18.2, each point on a standing wave oscillates with simple harmonic motion. Furthermore, all points oscillate with the same frequency, but the amplitude of oscillation, y, differs from one point to the next and depends on how far a given point is from one end. We found that the position-dependent part of the wave function for a standing wave is

$$y(x) = A \sin(kx) \tag{41.9}$$

where A is the amplitude and $k = 2\pi/\lambda$. Since $\lambda = 2L/n$, we see that

$$k = \frac{2\pi}{\lambda} = \frac{2\pi}{2L/n} = n\frac{\pi}{L}$$

Substituting this into Equation 41.9 gives

$$y(x) = A \sin\left(\frac{n\pi x}{L}\right)$$

From this expression, we see that the wave function for a standing wave on a string meets the required boundary conditions, namely, that for all values of n, $y = 0$ at $x = 0$ and at $x = L$. The wave functions for $n = 1$, 2, and 3 are plotted in Figure 41.12.

Now let us return to the quantum mechanical description of a particle in a box. Because the walls are impenetrable, the wave function $\psi(x) = 0$ for $x \le 0$ and for $x \ge L$, where L is now the distance between the two walls. Only those wave functions that satisfy this condition are allowed. In analogy with standing waves on a string, the allowed wave functions for the particle in the box are sinusoidal and are given by

$$\psi(x) = A \sin\left(\frac{n\pi x}{L}\right) \qquad n = 1, 2, 3, \ldots \tag{41.10}$$

Allowed wave functions for a particle in a box

where A is the maximum value of the wave function. This expression shows that, for a particle confined to a box and having a well-defined de Broglie wavelength, ψ is represented by a sinusoidal wave. The allowed wavelengths are those for which the $L = n\lambda/2$. These allowed states of the system are called **stationary states** because they are constant with time.

Figure 41.13a and b are graphs of ψ versus x and $|\psi|^2$ versus x for $n = 1, 2$, and 3. As we shall soon see, these correspond to the three lowest allowed energies for the particle. Note that although ψ can be either positive or negative, $|\psi|^2$ is always positive. From any viewpoint, a negative value for $|\psi|^2$ is meaningless.

Further inspection of Figure 41.13b shows that $|\psi|^2$ is always zero at the boundaries, indicating that it is impossible to find the particle at these points. In

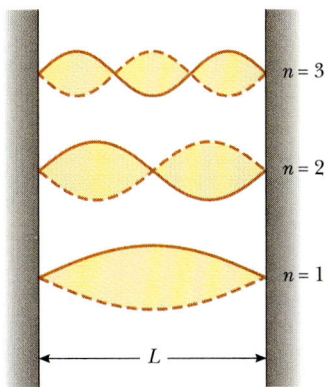

FIGURE 41.12   Standing waves set up in a stretched string of length L.

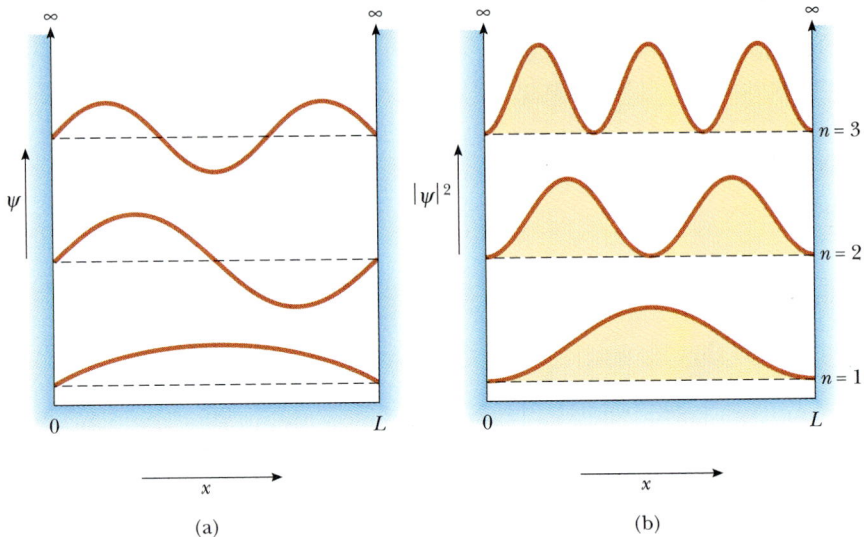

FIGURE 41.13   The first three allowed stationary states for a particle confined to a one-dimensional box. (a) The wave functions for $n = 1$, 2, and 3. (b) The probability distributions for $n = 1$, 2, and 3.

addition, $|\psi|^2$ is zero at other points, depending on the value of n. For $n = 2$, $|\psi|^2 = 0$ at $x = L/2$; for $n = 3$, $|\psi|^2 = 0$ at $x = L/3$ and at $x = 2L/3$. For $n = 1$, however, the probability of finding the particle is a maximum at $x = L/2$. For $n = 4$, $|\psi|^2$ has maxima also at $x = L/4$ and at $x = 3L/4$, and so on.

Because the wavelengths of the particle are restricted by the condition $\lambda = 2L/n$, the magnitude of the momentum is also restricted to specific values:

$$p = \frac{h}{\lambda} = \frac{h}{2L/n} = \frac{nh}{2L}$$

Using $p = mv$, we find that the allowed values of the kinetic energy are

$$E_n = \tfrac{1}{2}mv^2 = \frac{p^2}{2m} = \frac{(nh/2L)^2}{2m}$$

$$E_n = \left(\frac{h^2}{8mL^2}\right)n^2 \qquad n = 1, 2, 3, \ldots \qquad (41.11)$$

As we see from this expression, *the energy of the particle is quantized*, as we would expect. The lowest allowed energy corresponds to $n = 1$, for which $E_1 = h^2/8mL^2$. Since $E_n = n^2E_1$, the excited states corresponding to $n = 2, 3, 4, \ldots$ have energies given by $4E_1, 9E_1, 16E_1, \ldots$. Figure 41.14 is an energy-level diagram describing the positions of the allowed states. Note that the state $n = 0$ is not allowed. This means that, according to quantum mechanics, the particle can never be at rest. The least energy the particle can have, corresponding to $n = 1$, is called the zero-point energy. This result is clearly contradictory to the classical viewpoint, in which $E = 0$ is an acceptable state, as are all positive values of E. In this situation, only positive values of E are considered, because the total energy E equals the kinetic energy, and the potential energy is zero.

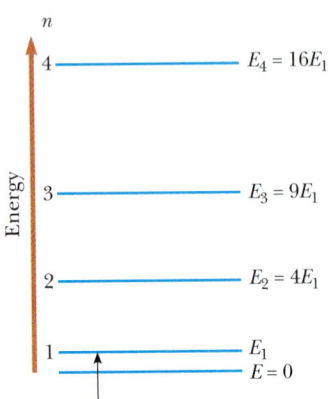

FIGURE 41.14   Energy-level diagram for a particle confined to a one-dimensional box of width L. The lowest allowed energy $E_1 = h^2/8mL^2$.

The energy levels are of special importance for the following reason. If the particle is electrically charged, it can emit a photon when it drops from an excited state, such as E_3, to a lower-lying state, such as E_2. It can also absorb a photon whose energy matches the difference in energy between two allowed states. For example, if the photon frequency is f, the particle jumps from state E_1 to state E_2 if $hf = E_2 - E_1$. The processes of photon emission or absorption can be observed by spectroscopy, in which spectral wavelengths are a direct measurement of such energy differences.

EXAMPLE 41.7   A Bound Electron

An electron is confined between two impenetrable walls 0.200 nm apart. Determine the energy levels for the states $n = 1, 2,$ and 3.

Solution   We can apply Equation 41.11, using $m = 9.11 \times 10^{-31}$ kg and $L = 0.200$ nm $= 2.00 \times 10^{-10}$ m. For the state $n = 1$, we get

$$E_1 = \frac{h^2}{8mL^2} = \frac{(6.63 \times 10^{-34}\,\text{J·s})^2}{8(9.11 \times 10^{-31}\,\text{kg})(2.00 \times 10^{-10}\,\text{m})^2}$$

$$= 1.51 \times 10^{-18}\,\text{J} = \boxed{9.42\ \text{eV}}$$

For $n = 2$ and $n = 3$, we find that $E_2 = 4E_1 = 37.7$ eV and $E_3 = 9E_1 = 84.8$ eV. Although this is a rather primitive model, it can be used to describe an electron trapped in a vacant crystal site.

EXAMPLE 41.8   Energy Quantization for a Macroscopic Object

A 1.00-mg object is confined to moving between two rigid walls separated by 1.00 cm. (a) Calculate the minimum speed of the object.

Solution   The minimum speed corresponds to the state for which $n = 1$. Using Equation 41.11 with $n = 1$ gives the zero-point energy:

$$E_1 = \frac{h^2}{8mL^2} = \frac{(6.63 \times 10^{-34}\,\text{J·s})^2}{8(1.00 \times 10^{-6}\,\text{kg})(1.00 \times 10^{-2}\,\text{m})^2}$$

$$= 5.49 \times 10^{-58}\,\text{J}$$

Since $E = \frac{1}{2}mv^2$, we can find v as follows:

$$\tfrac{1}{2}mv^2 = 5.49 \times 10^{-58}\,\text{J}$$

$$v = \left[\frac{2(5.49 \times 10^{-58}\,\text{J})}{1.00 \times 10^{-6}\,\text{kg}}\right]^{1/2} = \boxed{3.31 \times 10^{-26}\ \text{m/s}}$$

This speed is so small that the object can be considered to be at rest, which is what one would expect for a macroscopic object.

(b) If the speed of the object is 3.00×10^{-2} m/s, find the corresponding value of n.

Solution   The kinetic energy is

$$E = \tfrac{1}{2}mv^2 = \tfrac{1}{2}(1.00 \times 10^{-6}\,\text{kg})(3.00 \times 10^{-2}\,\text{m/s})^2$$

$$= 4.50 \times 10^{-10}\,\text{J}$$

Since $E_n = n^2 E_1$ and $E_1 = 5.49 \times 10^{-58}$ J, we find that

$$n^2 E_1 = 4.50 \times 10^{-10}\,\text{J}$$

$$n = \left(\frac{4.50 \times 10^{-10}\,\text{J}}{E_1}\right)^{1/2} = \left(\frac{4.50 \times 10^{-10}\,\text{J}}{5.49 \times 10^{-58}\,\text{J}}\right)^{1/2}$$

$$= \boxed{9.05 \times 10^{23}}$$

This value of n is so large that we would never be able to distinguish the quantized nature of the energy levels. That is, the difference in energy between the two adjacent states corresponding to the quantum numbers $n_1 = 9.05 \times 10^{23}$ and $n_2 = (9.05 \times 10^{23}) + 1$ is too small to be detected experimentally. This is another example that illustrates the working of the correspondence principle, that is, as m or L becomes large, the quantum description must agree with the classical result. In reality, the speed of the particle in this state cannot be measured because its position in the box cannot be specified.

EXAMPLE 41.9   Model of an Atom

An atom can be viewed as several electrons moving around a positively charged nucleus, where the electrons are subject mainly to the coulomb attraction of the nucleus (which is actually partially "screened" by the inner-core electrons). Figure 41.15 represents the potential energy of the electron as a function of r. Use the model of a particle in a box to

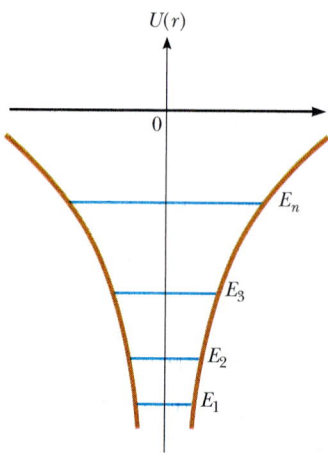

FIGURE 41.15 (Example 41.9) Model of the potential energy versus *r* for an atom.

estimate the energy (in eV) required to raise an electron from the state $n = 1$ to the state $n = 2$, assuming the atom has a radius of 0.1 nm.

Solution Using Equation 41.11 and taking the length *L* of the box to be 0.200 nm (the diameter of the atom) and *m* =

9.11 × 10⁻³¹ kg, we find that

$$E_n = \left(\frac{h^2}{8mL^2} \right) n^2$$

$$= \frac{(6.63 \times 10^{-34}\,\text{J}\cdot\text{s})^2}{8(9.11 \times 10^{-31}\,\text{kg})(2.00 \times 10^{-10}\,\text{m})^2} n^2$$

$$= (1.51 \times 10^{-18})\, n^2\,\text{J} = 9.42 n^2\,\text{eV}$$

Hence, the energy difference between the states $n = 1$ and $n = 2$ is

$$\Delta E = E_2 - E_1 = 9.42(2)^2 - 9.42(1)^2 = \boxed{28.3\ \text{eV}}$$

We could also calculate the wavelength of the photon that would cause this transition, using the fact that $\Delta E = hc/\lambda$:

$$\lambda = \frac{hc}{\Delta E} = \frac{(6.63 \times 10^{-34}\,\text{J}\cdot\text{s})(3.00 \times 10^8\,\text{m/s})}{(28.3\ \text{eV} \times 1.60 \times 10^{-19}\,\text{J/eV})}$$

$$= 4.40 \times 10^{-8}\ \text{m} = 44.0\ \text{nm}$$

This wavelength is in the far ultraviolet region, and it is interesting to note that the result is roughly correct. Although this oversimplified model gives a good estimate for transitions between lowest-lying levels of the atom, the estimate gets progressively worse for higher-energy transitions.

41.7   THE SCHRÖDINGER EQUATION

As we mentioned earlier, the wave function for de Broglie waves must satisfy an equation developed by Schrödinger. One of the methods of quantum mechanics is to determine a solution to this equation, which in turn yields the allowed wave functions and energy levels of the system under consideration. Proper manipulation of the wave functions allows calculation of all measurable features of the system.

In Chapter 16, we discussed the general form of the wave equation for waves traveling along the *x* axis:

$$\frac{\partial^2 \psi}{\partial x^2} = \frac{1}{v^2} \frac{\partial^2 \psi}{\partial t^2} \tag{41.12}$$

where *v* is the wave speed and where the wave function ψ depends on *x* and *t*. (We use ψ here instead of *y* because now we are dealing with de Broglie waves.)

In describing de Broglie waves, let us confine our discussion to bound systems whose total energy *E* remains constant. Since $E = hf$, the frequency of the de Broglie wave associated with the particle also remains constant. In this case, we can express the wave function $\psi(x, t)$ as the product of a term that depends only on *x* and a term that depends only on *t*:

$$\psi(x,\ t) = \psi(x)\ \cos(\omega t) \tag{41.13}$$

This is analogous to the case of standing waves on a string, where the wave function is represented by $y(x, t) = y(x) \cos \omega t$. The frequency-dependent part of the wave function is sinusoidal because the frequency is precisely known. Substituting

Equation 41.13 into Equation 41.12 gives

$$\cos(\omega t)\frac{\partial^2 \psi}{\partial x^2} = -\left(\frac{\omega^2}{v^2}\right)\psi\cos(\omega t)$$

$$\frac{\partial^2 \psi}{\partial x^2} = -\left(\frac{\omega^2}{v^2}\right)\psi \qquad (41.14)$$

Recall that $\omega = 2\pi f = 2\pi v/\lambda$ and, for de Broglie waves, $p = h/\lambda$. Therefore,

$$\frac{\omega^2}{v^2} = \left(\frac{2\pi}{\lambda}\right)^2 = \frac{4\pi^2}{h^2}\,p^2 = \frac{p^2}{\hbar^2}$$

Furthermore, we can express the total energy E as the sum of the kinetic energy and the potential energy:

$$E = K + U = \frac{p^2}{2m} + U$$

so that

$$p^2 = 2m(E - U)$$

and

$$\frac{\omega^2}{v^2} = \frac{p^2}{\hbar^2} = \frac{2m}{\hbar^2}(E - U)$$

Substituting this result into Equation 41.14 gives

$$\frac{\partial^2 \psi}{\partial x^2} = -\frac{2m}{\hbar^2}(E - U)\psi \qquad (41.15)$$

This is the famous **Schrödinger equation** as it applies to a particle confined to moving along the x axis. Because this equation is independent of time, it is commonly referred to as the *time-independent Schrödinger equation*. (We shall not discuss the time-dependent Schrödinger equation in this text.)

In principle, if the potential energy $U(x)$ is known for the system, we can solve Equation 41.15 and obtain the wave functions and energies for the allowed states. Since U may vary with position, it may be necessary to solve the equation in pieces. In the process, the wave functions for the different regions must join smoothly at the boundaries. In the language of mathematics, we require that $\psi(x)$ be *continuous*. Furthermore, so that $\psi(x)$ obeys the normalization condition, we require that $\psi(x)$ approaches zero as x approaches $\pm \infty$. Finally, $\psi(x)$ must be *single-valued* and $d\psi/dx$ must also be continuous for finite values of $U(x)$.

It is important to recognize that the steps leading to Equation 41.15 do not represent a derivation of the Schrödinger equation. Rather, the procedure represents a plausibility argument based upon an analogy with other wave phenomena that are already familiar to us.

The task of solving the Schrödinger equation may be very difficult, depending on the form of the potential energy function. As it turns out, the Schrödinger equation has been extremely successful in explaining the behavior of atomic and nuclear systems, whereas classical physics has failed to do so. Furthermore, when wave mechanics is applied to macroscopic objects, the results agree with classical physics, as required by the correspondence principle.

Erwin Schrödinger (1887–1961) was an Austrian theoretical physicist best known as the creator of wave mechanics. He also produced important papers in the fields of statistical mechanics, color vision, and general relativity. Schrödinger did much to hasten the universal acceptance of quantum theory by demonstrating the mathematical equivalence between his wave mechanics and the more abstract matrix mechanics developed by Heisenberg. In 1927 Schrödinger accepted the chair of theoretical physics at the University of Berlin, where he formed a close friendship with Max Planck. In 1933 he left Germany and eventually settled at the Dublin Institute of Advanced Study, where he spent 17 happy, creative years working on problems in general relativity, cosmology, and the application of quantum physics to biology. In 1956 he returned home to Austria and to his beloved Tirolean mountains, where he died in 1961.

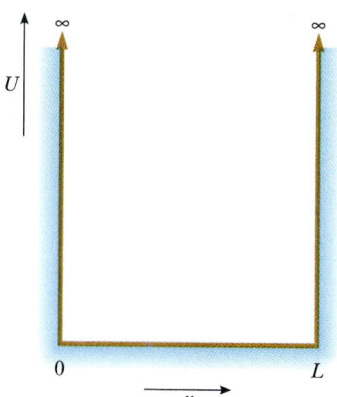

FIGURE 41.16   Diagram of a one-dimensional box of width L and infinitely high walls.

The Particle in a Box Revisited

Let us solve the Schrödinger equation for our particle in a one-dimensional box of width L (Fig. 41.16). The walls are infinitely high, corresponding to $U(x) = \infty$ for $x = 0$ and $x = L$. The potential energy is constant within the box, and it is convenient to set it equal to zero. Hence, in the region $0 < x < L$, we can express the Schrödinger equation in the form

$$\frac{d^2\psi}{dx^2} = -\frac{2mE}{\hbar^2}\psi = -k^2\psi \tag{41.16}$$

$$k = \frac{\sqrt{2mE}}{\hbar}$$

Because the walls are infinitely high, the particle cannot exist outside the box. Consequently, $\psi(x)$ must be zero outside the box and at the walls. The solution of Equation 41.16 that meets the boundary conditions $\psi(x) = 0$ at $x = 0$ and $x = L$ is

$$\psi(x) = A\sin(kx) \tag{41.17}$$

This can easily be verified by substitution into Equation 41.16. Note that the first boundary condition, $\psi(0) = 0$, is satisfied by Equation 41.17 because $\sin 0 = 0$. The second boundary condition, $\psi(L) = 0$, is satisfied only if kL is an integral multiple of π, that is, if $kL = n\pi$, where n is an integer. Since $k = \sqrt{2mE}/\hbar$, we get

$$kL = \frac{\sqrt{2mE}}{\hbar}L = n\pi$$

Solving for the allowed energies E_n gives

$$E_n = \left(\frac{h^2}{8mL^2}\right)n^2$$

Likewise, the allowed wave functions are given by

$$\psi_n(x) = A\sin\left(\frac{n\pi x}{L}\right)$$

These results agree with those obtained in the previous section. Normalizing this relationship shows that $A = (2/L)^{1/2}$. (See Problem 37.)

*41.8   A PARTICLE IN A WELL OF FINITE HEIGHT

Consider a particle whose potential energy is zero in the region $0 < x < L$, which we call the well, and has a finite value U outside this region as in Figure 41.17. If the energy E of the particle is less than U, classically the particle is permanently bound in the well. However, according to quantum mechanics, there is a finite probability that the particle can be found outside the well. That is, the wave function is generally not zero outside the well, in regions I and III in Figure 41.17, and so the probability density is also not zero in these regions. This demonstrates, again, the importance of what is known and measurable about the system, as contrasted with what we would expect to know about a classical system.

In region II, where $U = 0$, the allowed wave functions are again sinusoidal because they represent solutions of Equation 41.16. However, the boundary conditions no longer require that ψ be zero at the walls, as was the case with infinitely high walls.

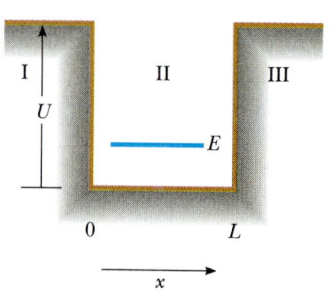

FIGURE 41.17   Potential energy diagram of a well of finite height U and width L. The energy E of the particle is less than U.

The Schrödinger equation for regions I and III may be written

$$\frac{d^2\psi}{dx^2} = \frac{2m(U-E)}{\hbar^2}\,\psi \qquad (41.18)$$

Because $U > E$, the coefficient on the right-hand side is necessarily positive. Therefore, we can express Equation 41.18 in the form

$$\frac{d^2\psi}{dx^2} = C^2\psi \qquad (41.19)$$

where $C^2 = 2m(U-E)/\hbar^2$ is a positive constant in regions I and III. As you can verify by substitution, the general solution of Equation 41.19 is

$$\psi = Ae^{Cx} + Be^{-Cx}$$

where A and B are constants.

We can use this solution as a starting point for determining the appropriate form of the solutions for regions I and III. The function we choose for our solution must remain finite over the entire region under consideration. In region I, where $x < 0$, we must rule out the term Be^{-Cx}. In other words, we must require that $B = 0$ in region I in order to avoid an infinite value for ψ for large values of x measured in the negative direction. Likewise, in region III, where $x > L$, we must rule out the term Ae^{Cx}; this is accomplished by setting $A = 0$ in this region. This choice avoids an infinite value for ψ for positive values of x. Hence, the solutions in regions I and III are

$$\psi_{\mathrm{I}} = Ae^{Cx} \qquad \text{for } x < 0$$

$$\psi_{\mathrm{III}} = Be^{-Cx} \qquad \text{for } x > L$$

In region II the wave function is sinusoidal and has the general form

$$\psi_{\mathrm{II}}(x) = F\sin(kx) + G\cos(kx)$$

where F and G are constants.

These results show that the wave functions in the exterior regions decay exponentially with distance. At large negative x values, ψ_{I} approaches zero exponentially; at large positive x values, ψ_{III} approaches zero exponentially. These functions, together with the sinusoidal solution in region II, are shown in Figure 41.18a for the first three energy states. In evaluating the complete wave function, we require that

$$\psi_{\mathrm{I}} = \psi_{\mathrm{II}} \quad \text{and} \quad \frac{d\psi_{\mathrm{I}}}{dx} = \frac{d\psi_{\mathrm{II}}}{dx} \qquad \text{at } x = 0$$

$$\psi_{\mathrm{II}} = \psi_{\mathrm{III}} \quad \text{and} \quad \frac{d\psi_{\mathrm{II}}}{dx} = \frac{d\psi_{\mathrm{III}}}{dx} \qquad \text{at } x = L$$

Figure 41.18b plots the probability densities for these states. Note that in each case the wave functions join smoothly at the boundaries of the potential well. These boundary conditions and plots follow from the Schrödinger equation. Further inspection of Figure 41.18a shows that the wave functions are not equal to zero at the walls of the potential well and in the exterior regions. Therefore, the probability density is nonzero at these points. The fact that ψ is nonzero at the walls increases the de Broglie wavelength in region II (compare the case of a particle in a potential well of infinite depth), and this in turn lowers the energy and momentum of the particle.

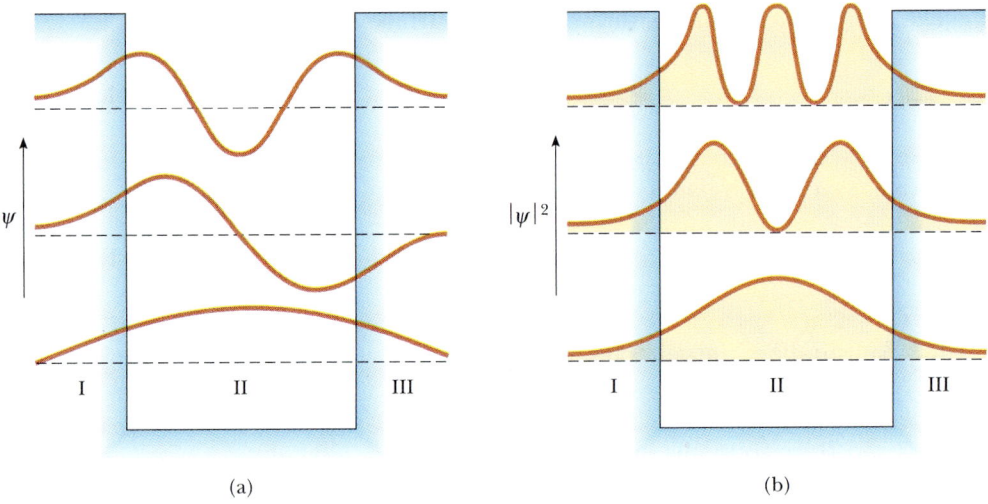

FIGURE 41.18   (a) Wave functions and (b) probability densities for the lowest three energy states for a particle in a potential well of finite height.

*41.9   TUNNELING THROUGH A BARRIER

A very interesting and peculiar phenomenon occurs when a particle strikes a barrier of finite height and width. Consider a particle of energy E incident on a rectangular barrier of height U and width L, where $E < U$ (Fig. 41.19). Classically, the particle is reflected because it does not have sufficient energy to cross or even penetrate the barrier. Thus, regions II and III are classically *forbidden* to the particle.

According to quantum mechanics, however, *all regions are accessible to the particle, regardless of its energy,* because the amplitude of the matter wave associated with the particle is nonzero everywhere (except at certain points). A typical waveform for this case, illustrated in Figure 41.19, shows the wave penetrating into the barrier and beyond. The wave functions are sinusoidal to the left (region I) and right (region III) of the barrier and join smoothly with an exponentially decaying function within the barrier (region II). Because the probability of locating the particle is proportional to $|\psi|^2$, we conclude that the chance of finding the particle beyond the barrier in region III is nonzero. This barrier penetration is in complete disagreement with classical physics. The possibility of having the particle penetrate the barrier is called **tunneling** or **barrier penetration**. Any attempt to observe the particle inside the barrier and confirm the value of its energy is frustrated by the uncertainty principle. If tunneling is to take place, the barrier must be sufficiently narrow and low, to allow a measurement of the particle's position and momentum consistent with expectations. This surprise result arises from our classical view of the barrier as being continuous. In practice, barriers are typically formed by particles of uncertain position, much like wandering sentinels outside camp. Occasionally, intruders are able to penetrate the barriers of sentinels and enter the camp.

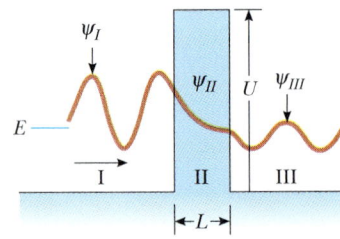

FIGURE 41.19   Wave function for a particle incident from the left on a barrier of height U. The wave function is sinusoidal in regions I and III but exponentially decaying in region II.

The probability of tunneling can be described with a transmission coefficient, T, and a reflection coefficient, R. The transmission coefficient measures the probability that the particle penetrates the barrier, and the reflection coefficient is the probability that the particle is reflected by the barrier. Because the incident particle is either reflected or transmitted, we must require that $T + R = 1$. An approximate expression for the transmission coefficient when $T \ll 1$ (a very high or wide barrier) is

$$T \cong e^{-2KL} \tag{41.20}$$

where K is given by

$$K = \frac{\sqrt{2m(U - E)}}{\hbar} \tag{41.21}$$

EXAMPLE 41.10   Transmission Coefficient for an Electron

A 30-eV electron is incident on a square barrier of height 40 eV. What is the probability that the electron will tunnel through the barrier if its width is (a) 1.0 nm and (b) 0.10 nm?

Solution   (a) In this situation, the quantity $U - E$ has the value

$$U - E = (40 \text{ eV} - 30 \text{ eV}) = 10 \text{ eV} = 1.6 \times 10^{-18} \text{ J}$$

Using Equation 41.21, and given that $L = 1.0$ nm, we find for the quantity $2KL$

$$2KL = 2 \frac{\sqrt{2(9.11 \times 10^{-31} \text{ kg})(1.6 \times 10^{-18} \text{ J})}}{1.054 \times 10^{-34} \text{ J} \cdot \text{s}} (1.0 \times 10^{-9} \text{ m})$$

$$= 32.4$$

Thus, the probability of tunneling through the barrier is

$$T \cong e^{-2KL} = e^{-32.4} = \boxed{8.5 \times 10^{-15}}$$

That is, the electron has only about 1 chance in 10^{14} to tunnel through the 1.0-nm-wide barrier.

(b) For $L = 0.10$ nm, $2KL = 3.24$, and

$$T \cong e^{-2KL} = e^{-3.24} = \boxed{0.039}$$

The electron has a high probability (4 percent chance) of penetrating the 0.10-nm barrier. Thus, reducing the width of the barrier by only one order of magnitude has increased the probability of tunneling by about 12 orders of magnitude!

Some Applications of Tunneling

As we have seen, tunneling is a quantum phenomenon, a manifestation of the wave nature of matter. There are many examples in nature on the atomic and nuclear scales which may be understood only on the basis of tunneling.

- **Tunnel diode**   The tunnel diode is a semiconductor device consisting of two oppositely charged regions separated by a very narrow neutral region. The electric current, or rate of tunneling, can be controlled over a wide range by varying the bias voltage, which changes the energy of the tunneling electrons.
- **Josephson junction**   A Josephson junction consists of two superconductors separated by a thin insulating oxide layer, 1 to 2 nm thick. Under appropriate conditions, electrons in the superconductors travel as pairs and tunnel from one superconductor to the other through the oxide layer. Several effects have been observed in this type of junction. For example, a direct current is observed across the junction *in the absence of electric or magnetic fields*. The current is proportional to sin ϕ, where ϕ is the phase difference between the wave functions in the two superconductors. When a bias voltage V is applied across the junction, one observes the current oscillating with a frequency $f = 2eV/h$, where e is the charge on the electron.
- **Alpha decay**   One form of radioactive decay is the emission of alpha particles (the nuclei of helium atoms) by unstable, heavy nuclei. In order to escape from the nucleus, an alpha particle must penetrate a barrier several times larger than its energy. The barrier is due to a combination of the attractive nuclear force and the Coulomb repulsion between the alpha particle and the rest of the nucleus. Occasionally, an alpha particle tunnels through the barrier, which explains the basic mechanism for this type of decay and the large variations in the mean lifetimes of various radioactive nuclei.

*41.10   THE SCANNING TUNNELING MICROSCOPE[3]

One of the basic phenomena of quantum mechanics—tunneling—is at the heart of a very practical device, the *scanning tunneling microscope,* or *STM,* which enables us to get highly detailed images of surfaces with resolution comparable to the size of a *single atom.*

Figure 41.20, an image of the surface of a piece of graphite, shows what the STM can do. Note the high quality of the image and the recognizable rings of carbon atoms. What makes this image so remarkable is that its *resolution*—the size of the smallest detail that can be discerned—is about 0.2 nm. For an ordinary microscope, the resolution is limited by the wavelength of the waves used to make the image. Thus, an optical microscope has a resolution no better than 200 nm, about half the wavelength of visible light, and so could never show the detail displayed in Figure 41.20. Electron microscopes can have a resolution of 0.2 nm by using electron waves of this wavelength, given by the de Broglie formula $\lambda = h/p$. The electron momentum p required to give this wavelength is 10 000 eV/c, corresponding to an electron speed of 2% of the speed of light. Electrons traveling at this speed would penetrate into the interior of the piece of graphite in Figure 41.20 and so couldn't give us information about individual surface atoms.

The STM achieves its very fine resolution by using the basic idea shown in Figure 41.21. A conducting probe with a very sharp tip is brought near the surface to be studied. Because it is attracted to the positive ions in the surface, an electron in the surface has a lower total energy than an electron in the empty space between surface and tip. The same thing is true for an electron in the probe tip, which is attracted to the positive ions in the tip. In Newtonian mechanics, this means that electrons cannot move between surface and tip because they lack the energy to escape either material. Because the electrons obey quantum mechanics, however, they can "tunnel" across the barrier of empty space. By applying a voltage between surface and tip, the electrons can be made to tunnel preferentially from surface to tip. In this way the tip samples the distribution of electrons just above the surface.

Because of the nature of tunneling, the STM is very sensitive to the distance z

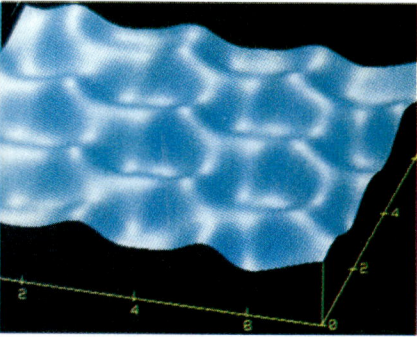

FIGURE 41.20 The surface of graphite as "viewed" with a scanning microscope. This technique enables scientists to see small details on surfaces with a lateral resolution of about 0.2 nm and a vertical resolution of 0.001 nm. The contours seen here represent the arrangement of individual carbon atoms on the crystal surface.

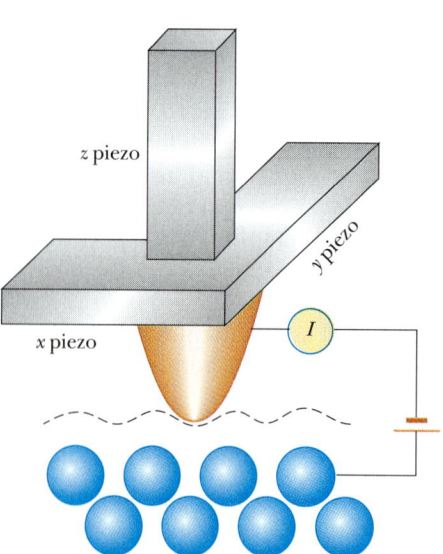

FIGURE 41.21 A schematic view of an STM. The tip, shown as a rounded cone, is mounted on a piezoelectric *x, y, z* scanner. A scan of the tip over the sample can reveal contours of the surface down to the atomic level. An STM image is composed of a series of scans displaced laterally from each other. *(Based on a drawing from P. K. Hansma, V. B. Elings, O. Marti, and C. Bracker,* Science *242:209, 1988, Copyright 1988 by the AAAS.)*

[3] This section was written by Roger A. Freedman and Paul K. Hansma, University of California, Santa Barbara.

from tip to surface. The reason is that in the empty space between tip and surface, the electron wave function falls off exponentially with a decay length of order 0.1 nm, that is, the wave function decreases by $1/e$ over that distance. For distances z greater than 1 nm (that is, beyond a few atomic diameters), essentially no tunneling takes place. This exponential behavior causes the current of electrons tunneling from surface to tip to depend very strongly on z. This sensitivity is the basis of the operation of the STM: By monitoring the tunneling current as the tip is scanned over the surface, scientists obtain a sensitive measure of the topography of the electron distribution on the surface. The result of this scan is used to make images like that in Figure 41.20. In this way the STM can measure the height of surface features to within 0.001 nm, approximately 1/100 of an atomic diameter!

You can see just how sensitive the STM is by examining Figure 41.20. Of the six carbon atoms in each ring, three *appear* lower than the other three. In fact all six atoms are at the same level, but they all have slightly different electron distributions. The three atoms that appear lower are bonded to other carbon atoms directly beneath them in the underlying atomic layer, and so their electron distributions—which are responsible for the bonding—extend downward beneath the surface. The atoms in the surface layer that appear higher do not lie directly over subsurface atoms and hence are not bonded to carbon atoms beneath them. For these higher-appearing atoms, the electron distribution extends upward into the space above the surface. This extra electron density is what makes these electrons appear higher in Figure 41.20, since what the STM maps is the topography of the electron distribution.

The STM has, however, one serious limitation: it depends on electrical conductivity of the sample and the tip. Unfortunately, most materials are not electrically conductive at their surface. Even metals such as aluminum are covered with nonconductive oxides. A newer microscope, the atomic force microscope, or AFM, overcomes this limitation. It measures the force between a tip and the sample rather than an electrical current. This force, which is typically a result of the exclusion principle, depends very strongly on the tip–sample separation just as the electron tunneling current does for the STM. Thus the AFM has comparable sensitivity for measuring topography and has become widely used for technological applications.

Perhaps the most remarkable thing about the STM is that its operation is based on a quantum-mechanical phenomenon—tunneling—that was well understood in the 1920s, even though the first STM was not built until the 1980s. What other applications of quantum mechanics may yet be waiting to be discovered?

*41.11   THE SIMPLE HARMONIC OSCILLATOR

Finally, let us consider the problem of a particle subject to a linear restoring force $F = -kx$, where x is the displacement of the particle from equilibrium ($x = 0$) and k is the force constant. The classical motion of a particle subject to such a force is simple harmonic motion, which was discussed in Chapter 13. The potential energy of the system is

$$U = \tfrac{1}{2}kx^2 = \tfrac{1}{2}m\omega^2 x^2$$

where the angular frequency of vibration is $\omega = \sqrt{k/m}$. Classically, if the particle is displaced from its equilibrium position and released, it oscillates between the points $x = -A$ and $x = A$, where A is the amplitude of motion. Furthermore, its total energy E is

$$E = K + U = \tfrac{1}{2}kA^2 = \tfrac{1}{2}m\omega^2 A^2$$

In the classical model, any value of E is allowed and the total energy may be zero if the particle is at rest at $x = 0$.

The Schrödinger equation for this problem is obtained by substituting $U = \frac{1}{2}m\omega^2 x^2$ into Equation 41.15:

$$\frac{d^2\psi}{dx^2} = -\left[\left(\frac{2mE}{\hbar^2}\right) - \left(\frac{m\omega}{\hbar}\right)^2 x^2\right]\psi \qquad (41.22)$$

The mathematical technique for solving this equation is beyond the level of this text. However, it is instructive to guess at a solution. We take as our guess the following wave function:

$$\psi = Be^{-Cx^2} \qquad (41.23)$$

Substituting this function into Equation 41.22, we find that Equation 41.23 is a satisfactory solution to the Schrödinger equation provided that

$$C = \frac{m\omega}{2\hbar} \qquad \text{and} \qquad E = \tfrac{1}{2}\hbar\omega$$

It turns out that the solution we have guessed corresponds to the ground state of the system, the state that has the lowest energy, $\frac{1}{2}\hbar\omega$, which is the zero-point energy of the system. Since $C = m\omega/2\hbar$, it follows from Equation 41.23 that the wave function for this state is

Wave function for the ground state of a simple harmonic oscillator

$$\psi = Be^{-(m\omega/2\hbar)x^2} \qquad (41.24)$$

This is only one solution to Equation 41.22. The remaining solutions, which describe the excited states, are more complicated, but all have the form of an exponential factor, e^{-Cx^2}, multiplied by a polynomial in x.

The energy levels of a harmonic oscillator are quantized, as we would expect. The energy of the state for which the quantum number is n is

Allowed energies for a simple harmonic oscillator

$$E_n = (n + \tfrac{1}{2})\hbar\omega \qquad n = 0, 1, 2, \ldots$$

The state $n = 0$ corresponds to the ground state, where $E_0 = \frac{1}{2}\hbar\omega$; the state $n = 1$ corresponds to the first excited state, where $E_1 = \frac{3}{2}\hbar\omega$; and so on. The energy-level diagram for this system is shown in Figure 41.22. Note that the separations between adjacent levels are equal and given by

$$\Delta E = \hbar\omega \qquad (41.25)$$

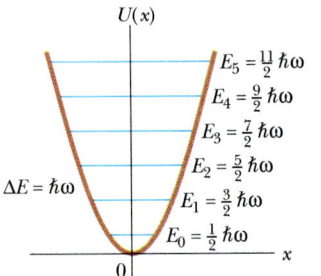

FIGURE 41.22 Energy-level diagram for a simple harmonic oscillator. The levels are equally spaced, with separation $\hbar\omega$. The zero-point energy is E_0.

The probability densities for the first three states of a harmonic oscillator are indicated by the red curves in Figure 41.23. The blue lines represent the classical probability densities corresponding to the same energy, provided for comparison. Note that as n increases, the agreement between the classical and quantum mechanics probabilities improves, as expected.

The quantum-mechanics solution to the simple harmonic oscillator problem predicts a series of equally spaced energy levels with separations equal to $\hbar\omega$. This result represents a justification of Planck's quantum hypothesis, made 25 years before the Schrödinger equation was developed. Furthermore, the solution presented is useful for describing more complicated problems, such as molecular vibrations, which can be approximated by the idealized model of simple harmonic motion.

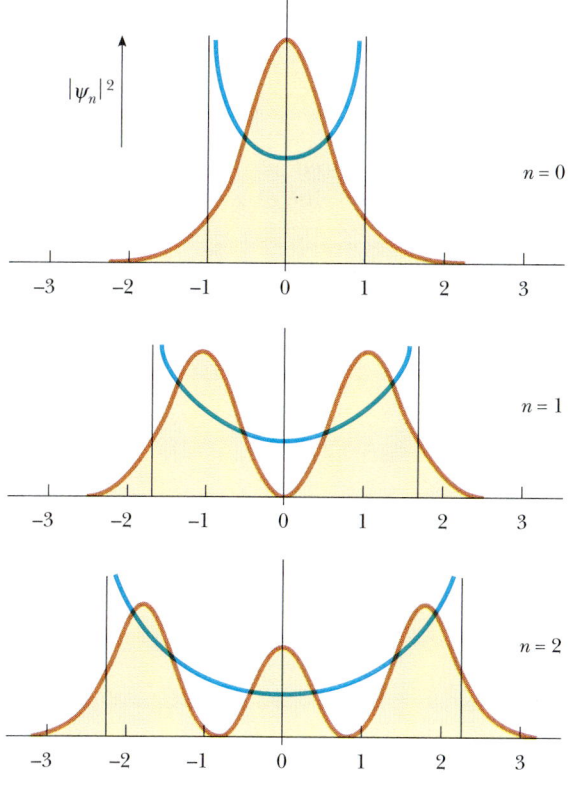

FIGURE 41.23   Red curves: probability densities for a few states of a harmonic oscillator represented. Blue curves: classical probabilities corresponding to the same energies.   *(From C. W. Sherwin, Introduction to Quantum Mechanics, New York, Holt, Rinehart and Winston, 1959; used with permission)*

SUMMARY

Light has a dual nature in that it has both wave and particle characteristics. Some experiments can be explained better, or solely, using the particle description, whereas others are best described, or can only be described, with a wave model.

Every object of mass m and momentum p has wave-like properties, with a wavelength given by the de Broglie relation

$$\lambda = \frac{h}{p} \tag{41.2}$$

By applying this wave theory of matter to electrons in atoms, de Broglie was able to explain quantization in the Bohr model of hydrogen as a standing-wave phenomenon.

The **uncertainty principle** states that if a measurement of position is made with precision Δx and a simultaneous measurement of momentum is made with precision Δp_x, the product of the two uncertainties can never be smaller than $\hbar/2$.

$$\Delta x \, \Delta p_x \geqslant \frac{\hbar}{2} \tag{41.3}$$

In quantum mechanics, matter waves (also called de Broglie waves) are represented by a wave function $\psi(x, y, z, t)$. The probability per unit volume (or probability density) that a particle will be found at a point is $|\psi|^2$. If the particle is confined to moving along the x axis, then the probability that it will be located in an interval dx is $|\psi|^2\, dx$. Furthermore, the sum of all these probabilities over all values of x must be 1:

$$\int_{-\infty}^{\infty} |\psi|^2\, dx = 1 \qquad (41.6)$$

This is called the **normalization condition**. The measured position x of the particle, averaged over many trials, is called the **expectation value** of x and is defined by

$$\langle x \rangle \equiv \int_{-\infty}^{\infty} x\,|\psi|^2\, dx \qquad (41.8)$$

If a particle of mass m is confined to moving in a one-dimensional box of width L whose walls are perfectly rigid, we require that ψ be zero at the walls and outside the box. The **allowed wave functions** for the particle are given by

$$\psi(x) = A \sin\left(\frac{n\pi x}{L}\right) \qquad n = 1, 2, 3, \ldots \qquad (41.10)$$

where A is the maximum value of ψ. The particle has a well-defined wavelength λ the values of which are such that $L = n\lambda/2$. These allowed states are called **stationary states** of the system. The energies of a particle in a box are quantized and are given by

$$E_n = \left(\frac{h^2}{8mL^2}\right)n^2 \qquad n = 1, 2, 3, \ldots \qquad (41.11)$$

The wave function must satisfy the **Schrödinger equation**. The time-independent Schrödinger equation for a particle confined to moving along the x axis is

$$\frac{\partial^2 \psi}{\partial x^2} = -\frac{2m}{\hbar^2}(E - U)\psi \qquad (41.15)$$

where E is the total energy of the system and U is the potential energy.

The approach of quantum mechanics is to solve Equation 41.15 for ψ and E, given the potential energy $U(x)$ for the system. In doing so, we must place special restrictions on $\psi(x)$. We require (1) that $\psi(x)$ be continuous, (2) that $\psi(x)$ approach zero as x approaches $\pm\infty$, (3) that $\psi(x)$ be single-valued, and (4) that $d\psi/dx$ be continuous for all finite values of $U(x)$.

QUESTIONS

1. Is light a wave or a particle? Support your answer by citing specific experimental evidence.

2. Is an electron a particle or a wave? Support your answer by citing some experimental results.

3. An electron and a proton are accelerated from rest through the same potential difference. Which particle has the longer wavelength?

4. If matter has a wave nature, why is this wave-like characteristic not observable in our daily experiences?

5. In what ways does Bohr's model of the hydrogen atom violate the uncertainty principle?

6. Why is it impossible to simultaneously measure with infinite accuracy the position and speed of a particle?

7. Suppose a beam of electrons is incident on three or more slits. How would that influence the interference pattern? Would the state of an electron depend on the number of slits? Explain.

8. In describing the passage of electrons through a slit and arriving at a screen, the physicist Richard Feynman said that "electrons arrive in lumps, like particles, but the probability of arrival of these lumps is determined as the intensity of the waves would be. It is in this sense that the electron behaves sometimes like a particle and sometimes like a wave." Elaborate on this point in your own words. (For a further discussion of this point, see R. Feynman, *The Character of Physical Law,* Cambridge, Mass., MIT Press, 1980, Chapter 6.)

9. For a particle in a box, the probability density at certain

points is zero, as seen in Figure 41.13b. Does this imply that the particle cannot move across these points? Explain.

10. Discuss the relationship between zero-point energy and the uncertainty principle.

11. As a particle of energy E is reflected from a potential barrier of height U, where $E < U$, how does the amplitude of the reflected wave change as the barrier height is reduced?

12. A philosopher once said that "it is necessary for the very existence of science that the same conditions always produce the same results." In view of what has been discussed in this section, present an argument showing that this statement is false. How might the statement be reworded to make it true?

13. In wave mechanics it is possible for the energy E of a parti-

cle to be less than the potential energy, but classically this is not possible. Explain.

14. Consider two square wells of the same width, one with finite walls and the other with infinite walls. Compare the energy and momentum of a particle trapped in the finite well with the energy and momentum of an identical particle in the infinite well.

15. Why cannot the lowest energy state of a harmonic oscillator be zero?

16. Why is an electron microscope more suitable than an optical microscope for "seeing" objects of an atomic size?

17. What is the Schrödinger equation? How is it useful in describing atomic phenomena?

18. Why was the Davisson-Germer diffraction of electrons an important experiment?

19. What is the significance of the wave function ψ?

PROBLEMS

Section 41.1 Photons and Electromagnetic Waves and
Section 41.2 The Wave Properties of Particles

1. The Sun's light reaches the Earth at an average intensity of 1350 W/m². Estimate the number of photons that reach the Earth's surface per second if the temperature of the Sun is 6000 K.

2. Calculate the de Broglie wavelength for an electron that has kinetic energy (a) 50 eV and (b) 50 keV.

3. After learning from de Broglie's hypothesis that particles of momentum p have wave characteristics with wavelength $\lambda = h/p$, an 80-kg student has grown concerned about being diffracted when passing through a 75-cm-wide doorway. Assuming that significant diffraction occurs when the width of the diffraction aperture is less than 10 times the wavelength of the wave being diffracted, (a) determine the maximum speed at which the student can pass through the doorway in order to be significantly diffracted. (b) With that speed, how long will it take the student to pass through the doorway if it is 15 cm thick? Compare your result with the currently accepted age of the Universe, which is 4×10^{17} s. (c) Should this student worry about being diffracted?

4. The seeing ability, or resolution, of radiation is determined by its wavelength. If the size of an atom is of the order of 0.10 nm, how fast must an electron travel to have a wavelength small enough to "see" an atom?

5. A constant force of 20.0 N is applied to a 3.00-g particle initially at rest. (a) After what time interval (in hours) will the particle's de Broglie wavelength equal its Compton wavelength, $\lambda_C = h/mc$? (b) How fast will the particle be moving at this time?

6. Calculate the de Broglie wavelength of a proton accelerated through a potential difference of 10 MV.

7. (a) Show that the frequency, f, and wavelength, λ, of a particle are related by the expression

$$\left(\frac{f}{c}\right)^2 = \frac{1}{\lambda^2} + \frac{1}{\lambda_C^2}$$

where $\lambda_C = h/mc$ is the Compton wavelength of the particle. (b) Is it ever possible for a photon and a particle (having nonzero mass) to have the same wavelength *and* frequency? Explain.

8. The distance between adjacent atoms in crystals is of the order of 0.10 nm. The use of electrons in diffraction studies of crystals requires that the de Broglie wavelength of the electrons be of the order of the distance between atoms of the crystals. What must be the minimum energy (in eV) of electrons to be used for this purpose?

9. What is the speed of an electron if its de Broglie wavelength equals its Compton wavelength? (*Hint:* If you get an answer of c, see Problem 55.)

10. For an electron to be confined to a nucleus, its de Broglie wavelength would have to be less than 10^{-14} m. (a) What would be the kinetic energy of an electron confined to this region? (b) On the basis of this result, would you expect to find an electron in a nucleus? Explain.

11. In the Davisson–Germer experiment, 54-eV electrons were diffracted from a nickel lattice. If the first maximum in the diffraction pattern was observed at $\phi = 50°$ (Fig. P41.11), what was the lattice spacing a?

12. Robert Hofstadter won the 1961 Nobel prize in physics for his pioneering work in scattering 20-GeV elec-

---

☐ indicates problems that have full solutions available in the Student Solutions Manual and Study Guide.

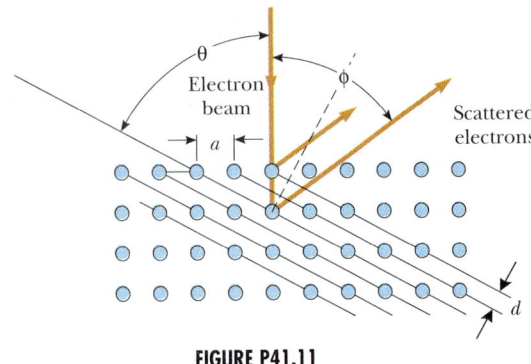

FIGURE P41.11

trons from nuclei. (a) What is the γ-factor for a 20-GeV electron, where $\gamma = (1 - v^2/c^2)^{-1/2}$? What is the momentum of the electron in kg·m/s? (b) What is the wavelength of a 20-GeV electron and how does it compare with the size of a nucleus?

13. Electrons are accelerated through 40 000 V in an electron microscope. What, theoretically, is the smallest observable distance between objects?

Section 41.3   The Double-Slit Experiment Revisited

14. A modified oscilloscope is used to perform an electron interference experiment. Electrons are incident on a pair of narrow slits 0.060 μm apart. The bright bands in the interference pattern are separated by 0.40 mm on a screen 20 cm from the slits. Determine the potential difference that the electrons were accelerated through to give this pattern.

15. A beam of electrons with a kinetic energy of 1.00 MeV strikes normally at an array of atoms separated by 0.25 nm. In what direction can we expect the electrons in the fifth order?

16. Neutrons traveling at 0.40 m/s are directed through a double slit having a 1.0-mm separation. An array of detectors is placed 10 m from the slit. (a) What is the de Broglie wavelength of the neutrons? (b) How far off axis is the first zero-intensity point on the detector array? (c) Can we say which slit the neutron passed through? Explain.

17. The resolving power of a microscope is proportional to the wavelength used. If one wished to use a microscope to "see" an atom, a resolution of approximately 10^{-11} m would have to be obtained. (a) If electrons are used (electron microscope), what minimum kinetic energy is required for the electrons? (b) If photons are used what minimum photon energy is needed to obtain the required resolution?

18. An air rifle is used to shoot 1.0-g particles at 100 m/s through a hole of diameter 2.0 mm. How far from the rifle must an observer be to see the beam spread by 1.0 cm because of the uncertainty principle? Compare this answer with the diameter of the Universe (2×10^{26} m).

Section 41.4   The Uncertainty Principle

19. A light source is used to determine the location of an electron in an atom to a precision of 0.05 nm. What is the uncertainty in the speed of the electron?

20. Suppose Fuzzy, a quantum-mechanical duck, lives in a world in which $h = 2\pi$ J·s. Fuzzy has a mass of 2.0 kg and is initially known to be within a region 1.0 m wide. (a) What is the minimum uncertainty in his speed? (b) Assuming this uncertainty in speed to prevail for 5.0 s, determine the uncertainty in position after this time.

21. (a) Suppose an electron is confined inside a nucleus of diameter 5.0×10^{-15} m. Use the uncertainty principle to determine if this electron is relativistic or nonrelativistic. (b) If this nucleus contains only protons and neutrons, are any of these particles relativistic? Explain.

22. An electron ($m = 9.11 \times 10^{-31}$ kg) and a bullet ($m = 0.020$ kg) each have a speed of 500 m/s, accurate to within 0.010%. Within what limits could we determine the positions of the objects?

23. A woman on a ladder drops small pellets toward a spot on the floor. (a) Show that, according to the uncertainty principle, the miss distance must be at least

$$\Delta x = \left(\frac{\hbar}{2m} \right)^{1/2} \left(\frac{H}{2g} \right)^{1/4}$$

where H is the initial height of each pellet above the floor and m is the mass of each pellet. (b) If $H = 2.0$ m and $m = 0.50$ g, what is Δx?

Section 41.5   Introduction to Quantum Mechanics

24. A free electron has a wave function

$$\psi(x) = A \sin(5.00 \times 10^{10} x)$$

where x is in meters. Find (a) the de Broglie wavelength, (b) the momentum, and (c) the energy in electron volts.

25. The wave function of an electron is

$$\psi(x) = \sqrt{\frac{2}{L}} \sin\left(\frac{2\pi x}{L} \right)$$

Find the probability of finding the electron between $x = 0$ and $x = \dfrac{L}{4}$.

Section 41.6   A Particle in a Box

26. An electron is confined to a one-dimensional region in which its ground-state ($n = 1$) energy is 2.00 eV. (a) What is the width of the region? (b) How much energy is required to promote the electron to its first excited state?

27. Use the particle-in-a-box model to calculate the first

three energy levels of a neutron trapped in a nucleus of diameter 2.00×10^{-5} nm. Are the energy-level differences realistic?

28. A particle in an infinite square well has a wave function given by

$$\psi_1(x) = \sqrt{\frac{2}{L}} \sin\left(\frac{\pi x}{L}\right)$$

for $0 \leqslant x \leqslant L$ and zero otherwise. (a) Determine the probability of finding the particle between $x = 0$ and $x = L/3$. (b) Use the results of this calculation and symmetry arguments to find the probability of finding the particle between $x = L/3$ and $x = 2L/3$. Do not reevaluate the integral.

29. A particle in an infinite square well has a wave function given by

$$\psi_2(x) = \sqrt{\frac{2}{L}} \sin\left(\frac{2\pi x}{L}\right)$$

for $0 \leqslant x \leqslant L$ and zero otherwise. Determine (a) the expectation value of x; (b) the probability of finding the particle near $L/2$, by calculating the probability the particle lies in the range, $0.49L \leqslant x \leqslant 0.51L$; and (c) the probability of finding the particle near $L/4$, by calculating the probability that the particle lies in the range, $0.24L \leqslant x \leqslant 0.26L$. (d) Reconcile these probabilities with the result for the average value of x found in part (a).

30. An electron that has an energy of approximately 6.0 eV moves between rigid walls 1.00 nm apart. Find (a) the quantum number n for the energy state that the electron occupies and (b) the energy of the electron.

31. An alpha particle in a nucleus can be modeled as a particle moving in a box of width 1.0×10^{-14} m (the approximate diameter of a nucleus). Using this model, estimate the energy and momentum of an alpha particle in its lowest energy state. ($m_\alpha = 4 \times 1.66 \times 10^{-27}$ kg)

32. An electron in an infinite square well has a wave function given by

$$\psi_2(x) = \sqrt{\frac{2}{L}} \sin\left(\frac{2\pi x}{L}\right)$$

for $0 \leqslant x \leqslant L$ and zero otherwise. What are the most probable positions of the electron?

33. An electron is contained in a one-dimensional box of width 0.100 nm. (a) Draw an energy-level diagram for the electron for levels up to $n = 4$. (b) Find the wavelengths of all photons that can be emitted by the electron in making transitions that will eventually get it from the $n = 4$ state to the $n = 1$ state.

34. Consider a particle moving in a one-dimensional box for which the walls are at $x = -L/2$ and $x = L/2$.

(a) Write the wave functions and probability densities for $n = 1$, $n = 2$, and $n = 3$. (b) Sketch the wave functions and probability densities. (*Hint:* Make an analogy to the case of a particle in a box for which the walls are at $x = 0$ and $x = L$.)

35. A ruby laser emits 694.3-nm light. If this light is due to transitions of an electron in a box from the $n = 2$ state to the $n = 1$ state, find the width of the box.

35A. A laser emits light of wavelength λ. If this light is due to transitions of an electron in a box from the $n = 2$ state to the $n = 1$ state, find the width of the box.

36. A proton is confined to moving in a one-dimensional box of width 0.2 nm. (a) Find the lowest possible energy of the proton. (b) What is the lowest possible energy of an electron confined to the same box? (c) How do you account for the large difference in your results for parts (a) and (b)?

Section 41.7   The Schrödinger Equation

37. The wave function for a particle confined to moving in a one-dimensional box is

$$\psi(x) = A \sin\left(\frac{n\pi x}{L}\right)$$

Use the normalization condition on ψ to show that

$$A = \sqrt{\frac{2}{L}}$$

Hint: Because the box width is L, the normalization condition (Eq. 41.6) is

$$\int_0^L |\psi|^2 \, dx = 1$$

38. The wave function for a particle is

$$\psi(x) = \sqrt{\frac{a}{\pi(x^2 + a^2)}}$$

for $a > 0$ and $-\infty < x < +\infty$. Determine the probability that the particle is located somewhere between $x = -a$ and $x = +a$.

39. A particle of mass m moves in a potential well of width $2L$ (from $x = -L$ to $x = +L$), and in this well is a potential given by

$$U(x) = \frac{-\hbar^2 x^2}{mL^2(L^2 - x^2)}$$

In addition, the particle is in a stationary state described by the wave function, $\psi(x) = A(1 - x^2/L^2)$ for $-L < x < +L$, and $\psi(x) = 0$ elsewhere. (a) Determine the energy of the particle in terms of $\hbar$, m and L. (*Hint:* Use the Schrödinger equation, Eq. 41.15.) (b) Show that $A = (15/16L)^{1/2}$. (c) Determine the probability that the particle is located between $x = -L/3$ and $x = +L/3$.

40. In a region of space, a particle with zero energy has a wave function

$$\psi(x) = Axe^{-x^2/L^2}$$

(a) Find the potential energy U as a function of x.
(b) Make a sketch of $U(x)$ versus x.

41. Show that the time-dependent wave function

$$\psi = Ae^{i(kx - \omega t)}$$

is a solution to the Schrödinger equation (Eq. 41.15). $(k = 2\pi/\lambda)$

*Section 41.8   A Particle in a Well of Finite Height

42. Sketch the wave function $\psi(x)$ and the probability density $|\psi(x)|^2$ for the $n = 4$ state of a particle in a finite potential well. (See Fig. 41.18.)

43. Suppose a particle is trapped in its ground state in a box that has infinitely high walls (Fig. 41.13a). Now suppose the left-hand wall is suddenly lowered to a finite height. (a) Qualitatively sketch the wave function for the particle a short time later. (b) If the box has a width L, what is the wavelength of the wave that penetrates the barrier?

44. A particle that has 7.00 eV of kinetic energy moves from a region where the potential energy is zero into one in which $U = 5.00$ eV (Fig. P41.44). Classically, one would expect the particle to continue on, although with less kinetic energy. According to quantum mechanics, the particle has a probability of being transmitted and a probability of being reflected. What are these probabilities?

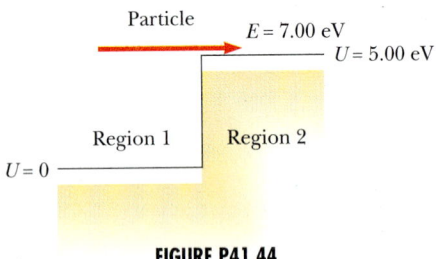

FIGURE P41.44

*Section 41.9   Tunneling Through a Barrier and
*Section 41.11   The Simple Harmonic Oscillator

45. A 5.0-eV electron is incident on a barrier 0.20 nm thick and 10 eV high (Fig. P41.45). What is the probability that the electron (a) will tunnel through the barrier and (b) will be reflected?

45A. An electron having energy E is incident on a barrier of width L and height U (Fig. P41.45). What is the probability that the electron (a) will tunnel through the barrier and (b) will be reflected?

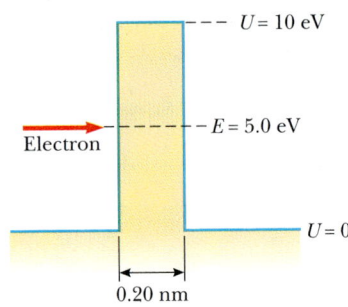

FIGURE P41.45

46. A one-dimensional harmonic oscillator wave function is

$$\psi = Axe^{-bx^2}$$

(a) Show that ψ satisfies Equation 41.22. (b) Find b and the total energy E. (c) Is this a ground state or a first excited state?

47. A simple pendulum has a length of 1.0 m and a mass of 1.0 kg. If the amplitude of oscillations of the pendulum is 3.0 cm, estimate the quantum number for the pendulum.

48. The total energy of a particle moving with simple harmonic motion along the x axis is

$$E = \frac{p_x^2}{2m} + \frac{Kx^2}{2}$$

where p_x is the momentum of the particle and K is the spring constant. (a) Using the uncertainty principle, show that this expression can also be written

$$E = \frac{p_x^2}{2m} + \frac{K\hbar^2}{2p_x^2}$$

(b) Show that the minimum kinetic energy of the harmonic oscillator is

$$K_{min} = \frac{p_x^2}{2m} = \frac{1}{2}\hbar\sqrt{\frac{K}{m}} = \frac{1}{2}\hbar\omega$$

ADDITIONAL PROBLEMS

49. The nuclear potential energy that binds protons and neutrons in a nucleus is often approximated by a square well. Imagine a proton confined in an infinitely high square well of width 1.0×10^{-5} nm, a typical nuclear diameter. Calculate the wavelength and energy associated with the photon emitted when the proton moves from the $n = 2$ state to the ground state. In what region of the electromagnetic spectrum does this wavelength belong?

50. Use the uncertainty relationship to show that an electron confined to a nucleus having a diameter of 2.0×10^{-15} m must be analyzed using Einsteinian

relativity, but a proton confined to the same nucleus can be analyzed using Newtonian relativity.

51. Johnny Jumper's favorite trick is to step out of his 16th-story window and fall 50 m into a pool. A news representative uses an exposure time of 5.0 ms to take a picture of 75-kg Johnny just before he makes a splash. Find (a) Johnny's de Broglie wavelength at this moment, (b) the uncertainty of his kinetic energy measurement during such a period of time, and (c) the percent error caused by such an uncertainty.

52. An atom in an excited state 1.8 eV above the ground state remains in that excited state 2.0 μs before moving to the ground state. Find (a) the frequency of the emitted photon, (b) its wavelength, and (c) its approximate uncertainty in energy.

52A. An atom in an excited state E above the ground state remains in that excited state for a time T before moving to the ground state. Find (a) the frequency of the emitted photon, (b) its wavelength, and (c) its approximate uncertainty in energy.

53. A π^0 meson is an unstable particle produced in high-energy particle collisions. It has a mass-energy equivalent of about 135 MeV, and it exists for an average lifetime of only 8.7×10^{-17} s before decaying into two gamma ways. Using the uncertainty principle, estimate the fractional uncertainty $\Delta m / m$ in its mass determination.

54. The neutron has a mass of 1.67×10^{-27} kg. Neutrons emitted in nuclear reactions can be slowed down via collisions with matter. They are referred to as thermal neutrons once they come into thermal equilibrium with their surroundings. The average kinetic energy $(3 k_B T / 2)$ of a thermal neutron is approximately 0.040 eV. Calculate the de Broglie wavelength of a neutron having this kinetic energy. Compare this wavelength with the characteristic atomic spacing in a crystal. Would you expect thermal neutrons to exhibit diffraction effects when scattered by a crystal?

55. Show that the speed of a particle having de Broglie wavelength λ and Compton wavelength $\lambda_C = h / (mc)$ is

$$v = \frac{c}{\sqrt{1 + (\lambda / \lambda_C)^2}}$$

56. A particle is described by the wave function

$$\psi(x) = \begin{cases} A \cos\left(\dfrac{2\pi x}{L}\right) & \text{for } -\dfrac{L}{4} \le x \le \dfrac{L}{4} \\ 0 & \text{for other values of } x \end{cases}$$

(a) Determine the normalization constant A. (b) What is the probability that the particle will be found between $x = 0$ and $x = L/8$ if its position is measured? (*Hint:* Use Eq. 41.7.)

57. For a particle described by a wave function $\psi(x)$, the expectation value of a physical quantity $f(x)$ associated with the particle is defined by

$$\langle f(x) \rangle \equiv \int_{-\infty}^{\infty} f(x) \mid \psi \mid^2 dx$$

For a particle in a one-dimensional box extending from $x = 0$ to $x = L$, show that

$$\langle x^2 \rangle = \frac{L^2}{3} - \frac{L^2}{2 n^2 \pi^2}$$

58. (a) Show that the wavelength of a neutron is

$$\lambda = \frac{2.86 \times 10^{-11}}{\sqrt{K_n}} \text{ m}$$

where K_n is the kinetic energy of the neutron in electron volts. (b) What is the wavelength of a 1.00-keV neutron?

59. Particles incident from the left are confronted with a step potential energy shown in Figure P41.59. The step has a height U_0, and the particles have energy $E > U_0$. Classically, all the particles would pass into the region of higher potential energy at the right. However, according to quantum mechanics, a fraction of the particles are reflected at the barrier. The reflection coefficient R for this case is

$$R = \frac{(k_1 - k_2)^2}{(k_1 + k_2)^2}$$

where $k_1 = 2\pi / \lambda_1$ and $k_2 = 2\pi / \lambda_2$ are the angular wave numbers for the incident and transmitted particles, respectively. If $E = 2U_0$, what fraction of the incident particles are reflected? (This situation is analogous to the partial reflection and transmission of light striking an interface between two different media.)

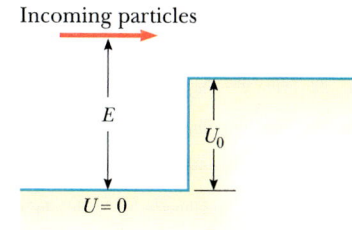

Incoming particles

FIGURE P41.59

60. A particle has a wave function

$$\psi(x) = \begin{cases} \sqrt{\dfrac{2}{a}} \, e^{-x/a} & \text{for } x > 0 \\ 0 & \text{for } x < 0 \end{cases}$$

(a) Find and sketch the probability density. (b) Find the probability that the particle will be at any point where $x < 0$. (c) Show that ψ is normalized and then find the probability that the particle will be found between $x = 0$ and $x = a$.

61. An electron of momentum p is at a distance r from a stationary proton. The electron has kinetic energy $K = p^2/2m$, potential energy $U = -k_e e^2/r$, and total energy $E = K + U$. If the electron is bound to the proton to form a hydrogen atom, its average position is at the proton, but the uncertainty in its position is approximately equal to the radius r of its orbit. The electron's average momentum is zero, but the uncertainty in its momentum is given by the uncertainty principle. Treating the atom as a one-dimensional system, (a) estimate the uncertainty in the electron's momentum in terms of r. (b) Estimate the electron's kinetic, potential, and total energies in terms of r. (c) The actual value of r is the one that *minimizes the total energy*, resulting in a stable atom. Find the value of r and the resulting total energy. Compare your answer with the predictions of the Bohr theory.

62. A certain electron microscope accelerates electrons to an energy of 65 keV. (a) Find the wavelength of these electrons. (b) If two points separated by at least 50 wavelengths can be resolved, what is the smallest separation (or minimum sized object) that can be resolved with this microscope?

63. A particle of mass m is placed in a one-dimensional box of width L. The box is so small that the particle's motion is *relativistic*, so that $E = p^2/2m$ is not valid. (a) Derive an expression for the energy levels of the particle. (b) If the particle is an electron in a box of width $L = 1.00 \times 10^{-12}$ m, find its lowest possible kinetic energy. By what percent is the nonrelativistic formula in error? (*Hint:* See Eq. 39.25.)

64. Consider a "crystal" consisting of two nuclei and two electrons as shown in Figure P41.64. (a) Taking into account all the pairs of interactions, find the potential energy of the system as a function of d. (b) Assuming the electrons to be restricted to a one-dimensional box of width $3d$, find the minimum kinetic energy of the two electrons. (c) Find the value of d for which the total energy is a minimum. (d) Compare this value of d with the spacing of atoms in lithium, which has a density of 0.53 g/cm^3 and an atomic mass of 7 u. (This type of calculation can be used to estimate the densities of crystals and certain stars.)

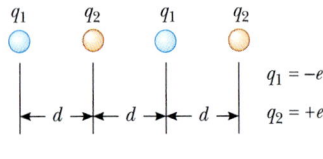

FIGURE P41.64

65. An electron is represented by the time-independent wave function

$$\psi(x) = \begin{cases} Ae^{-\alpha x} & \text{for } x > 0 \\ Ae^{+\alpha x} & \text{for } x < 0 \end{cases}$$

(a) Sketch the wave function as a function of x. (b) Sketch the probability that the electron is found between x and $x + dx$. (c) Why do you suppose this is a physically reasonable wave function? (d) Normalize the wave function. (e) Determine the probability of finding the electron somewhere in the range

$$x_1 = -\frac{1}{2\alpha} \quad \text{to} \quad x_2 = \frac{1}{2\alpha}$$

66. The normalized ground-state wave function for the electron in the hydrogen atom is

$$\psi(r, \theta, \phi) = \frac{2}{\sqrt{4\pi}} \left(\frac{1}{a_0}\right)^{3/2} e^{-r/a_0}$$

where r is the radial coordinate of the electron and a_0 is the Bohr radius. (a) Sketch the wave function versus r. (b) Show that the probability of finding the electron between r and $r + dr$ is $|\psi(r)|^2 \, 4\pi r^2 \, dr$. (c) Sketch the probability versus r and from your sketch find the radius at which the electron is most likely to be found. (d) Show that the wave function as given is normalized. (e) Find the probability of locating the electron between $x_1 = a_0/2$ and $x_2 = 3a_0/2$.

67. An electron having total energy $E = 4.5$ eV approaches a barrier where $U = 5.0$ eV and $L = 950$ pm as in Figure P41.67. Classically, the electron could not pass through the barrier because $E < U$. However, quantum-mechanically there is a finite probability of tunneling. (a) Use the transmission coefficient to calculate this probability. (b) By how much would the width L of the potential barrier have to be increased so that the chance of an incident 4.5-eV electron tunneling through the barrier is one in a million?

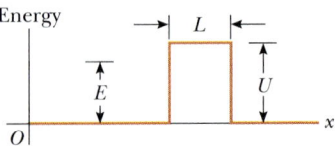

FIGURE P41.67

68. An electron is trapped at a defect in a crystal. The defect may be modeled as a one-dimensional, rigid-walled box of width 1.0 nm. (a) Sketch the wave functions and probability densities for the $n = 1$ and $n = 2$ states. (b) For the $n = 1$ state, calculate the probability of finding the electron between $x_1 = 0.15$ nm and $x_2 = 0.35$ nm, where $x = 0$ is the left side of the

box. (c) Repeat part (b) for the $n = 2$ state. (d) Calculate the energies in electron volts of the $n = 1$ and $n = 2$ states. *Hint:* For parts (b) and (c), use Equation 41.7 and note that

$$\int \sin^2 ax \, dx = \tfrac{1}{2}x - \frac{1}{4a} \sin 2ax$$

69. *The simple harmonic oscillator:* (a) Show that Equation 41.24 is a solution of Equation 41.22 with energy $E = \tfrac{1}{2}\hbar\omega$. (b) The wave function

$$\psi(x) = Cxe^{-(m\omega/2\hbar)x^2}$$

is also a solution to the simple harmonic oscillator problem. Find the energy corresponding to this state. Can you identify this state?

70. *Normalization of wave functions:* (a) Find the normalization constant A for a wave function made up of the two lowest states of a particle in a box:

$$\psi(x) = A\left[\sin\left(\frac{\pi x}{L}\right) + 4\sin\left(\frac{2\pi x}{L}\right)\right]$$

(b) A particle is described in the space $-a \leqslant x \leqslant a$ by the wave function

$$\psi(x) = A\cos\left(\frac{\pi x}{2a}\right) + B\sin\left(\frac{\pi x}{a}\right)$$

Determine values for A and B. (*Hint:* Use the identity $\sin 2\theta = 2\sin\theta\cos\theta$.)

CHAPTER 42

Atomic Physics

One of the many applications of laser technology is shown in this photo of a robot carrying laser scissors — a device used to cut fabric for the clothing industry. *(Philippe Plailly/SPL/Photo Researchers)*

I n Chapter 41, we introduced some of the basic concepts and techniques used in quantum mechanics, along with their applications to various simple systems. This chapter deals with the application of quantum mechanics to the real world of atomic structure.

A large portion of this chapter is an application of quantum mechanics to the study of the hydrogen atom. Understanding the hydrogen atom, the simplest atomic system, is especially important for several reasons:

- Much of what is learned about the hydrogen atom with its single electron can be extended to such single-electron ions as He^+ and Li^{2+}.
- The hydrogen atom is an ideal system for performing precise tests of theory against experiment and for improving our overall understanding of atomic structure.
- The quantum numbers used to characterize the allowed states of hydrogen can be used to describe the allowed states of more complex atoms. This enables us to understand the periodic table of the elements, one of the greatest triumphs of quantum mechanics.

- The basic ideas about atomic structure must be well understood before we attempt to deal with the complexities of molecular structures and the electronic structure of solids.

The full mathematical solution of the Schrödinger equation applied to the hydrogen atom gives a complete and beautiful description of its various properties. However, the mathematical procedures involved are beyond the scope of this text, and so the details are omitted. The solutions for some states of hydrogen are discussed, together with the quantum numbers used to characterize various allowed stationary states. We also discuss the physical significance of the quantum numbers and the effect of a magnetic field on certain quantum states.

A new physical idea, the *exclusion principle,* is also presented in this chapter. This physical principle is extremely important in understanding the properties of multielectron atoms and the arrangement of elements in the periodic table. In fact, the implications of the exclusion principle are almost as far-reaching as those of the Schrödinger equation. Finally, we apply our knowledge of atomic structure to describe the mechanisms involved in the production of x-rays and in the operation of a laser.

42.1   EARLY MODELS OF THE ATOM

The model of the atom in the days of Newton was a tiny, hard, indestructible sphere. Although this model provided a good basis for the kinetic theory of gases, new models had to be devised when later experiments revealed the electrical nature of atoms. J. J. Thomson suggested a model that describes the atom as a volume of positive charge with electrons embedded throughout the volume, much like the seeds in a watermelon (Fig. 42.1).

In 1911, Ernest Rutherford and his students Hans Geiger and Ernst Marsden performed a critical experiment that showed that Thomson's model could not be correct. In this experiment, a beam of positively charged **alpha particles,** now known to be the nuclei of helium atoms, was projected into a thin metal foil, as in Figure 42.2a. The results of the experiment were astounding: Most of the particles

Sir Joseph John Thomson (1856–1940), an English physicist and the recipient of the Nobel Prize in 1906. Thomson, usually considered the discoverer of the electron, opened up the field of subatomic particle physics with his extensive work on the deflection of cathode rays (electrons) in an electric field.   *(Stock Montage, Inc.)*

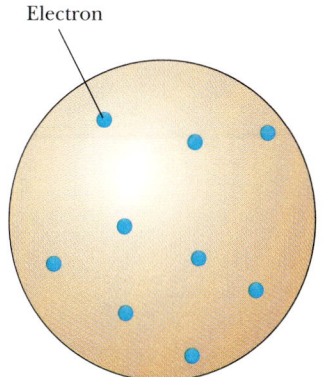

FIGURE 42.1   Thomson's model of the atom.

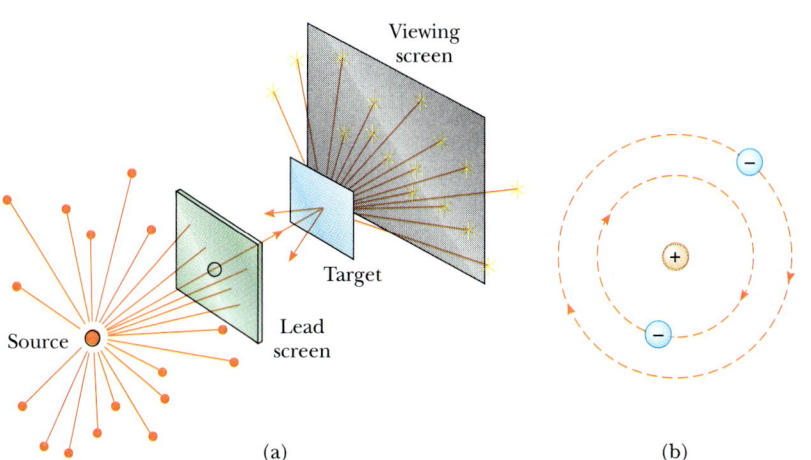

FIGURE 42.2   (a) Rutherford's technique for observing the scattering of alpha particles from a thin foil target. The source is a naturally occurring radioactive substance, such as radium. (b) Rutherford's planetary model of the atom.

passed through the foil as if it were empty space. But many of the particles deflected from their original direction of travel were scattered through large angles. Some particles were even deflected backward reversing their direction of travel. When Geiger informed Rutherford that some alpha particles were scattered backward, Rutherford wrote, "It was quite the most incredible event that has ever happened to me in my life. It was almost as incredible as if you fired a 15-inch shell at a piece of tissue paper and it came back and hit you."

Such large deflections were not expected on the basis of Thomson's model. According to this model, a positively charged alpha particle would never come close enough to a large enough charge to cause any large-angle deflections. Rutherford explained his astounding results by assuming that the positive charge in the atom was concentrated in a region that was small relative to the size of the atom. He called this concentration of positive charge the **nucleus** of the atom. Any electrons belonging to the atom were assumed to be in the relatively large volume outside the nucleus. To explain why these electrons were not pulled into the nucleus, the electrons were viewed as moving in orbits around the positively charged nucleus in the same manner as the planets orbit the Sun, as in Figure 42.2b.

There are two basic difficulties with Rutherford's planetary model. As we saw in Chapter 40, an atom emits certain characteristic frequencies of electromagnetic radiation and no others; the Rutherford model is unable to explain this phenomenon. A second difficulty is that Rutherford's electrons are undergoing a centripetal acceleration. According to Maxwell's theory of electromagnetism, centripetally accelerated charges revolving with frequency f should radiate electromagnetic waves of frequency f. Unfortunately, this classical model leads to disaster when applied to the atom. As the electron radiates energy, the radius of its orbit steadily decreases and its frequency of revolution increases. This leads to an ever-increasing frequency of emitted radiation and an ultimate collapse of the atom as the electron plunges into the nucleus (Fig. 42.3).

Now the stage was set for Bohr! In order to circumvent the erroneous deductions of electrons falling into the nucleus and a continuous emission spectrum from elements, Bohr postulated that classical radiation theory did not hold for atomic-sized systems. He overcame the problem of a classical electron that continuously loses energy by applying Planck's ideas of quantized energy levels to orbiting atomic electrons. Thus, Bohr postulated that electrons in atoms are generally confined to stable, nonradiating energy levels and orbits called stationary states. Furthermore, he applied Einstein's concept of the photon to arrive at an expression for the frequency of light emitted when the electron jumps from one stationary state to another.

One of the first indications the Bohr theory needed modifying arose when improved spectroscopic techniques were used to examine the spectral lines of hydrogen. It was found that many of the lines in the Balmer and other series were not single lines at all. Instead, each was a group of lines spaced very close together. An additional difficulty arose when it was observed that, in some situations, certain single spectral lines were split into three closely spaced lines when the atoms were placed in a strong magnetic field.

Efforts to explain these deviations from the Bohr model led to improvements in the theory. One of the changes introduced into the original theory was the postulate that the electron could spin on its axis. Also, Arnold Sommerfeld improved the Bohr theory by introducing the theory of relativity into the analysis of the electron's motion. An electron in an elliptical orbit has a continuously changing speed, with an average speed that depends on the eccentricity of the orbit.

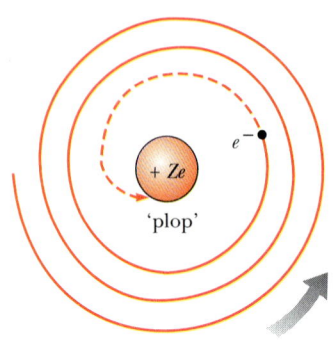

FIGURE 42.3   The classical model of the nuclear atom.

Different orbits have different average speeds and slightly different relativistic energies.

42.2   THE HYDROGEN ATOM REVISITED

The hydrogen atom consists of one electron and one proton. In Chapter 40, we described how the Bohr model views the electron as a particle orbiting around the nucleus in nonradiating, quantized energy levels. The de Broglie model gave electrons a wave nature by viewing them as standing waves in allowed orbits. This standing-wave description removed the objections to Bohr's postulates, which were quite arbitrary. However, the de Broglie model created other problems: The exact nature of the de Broglie wave was unspecified, and the model implied unlikely electron densities at large distances from the nucleus. Fortunately, these difficulties are removed when the methods of quantum mechanics are used to describe atoms.

The potential energy function for the hydrogen atom is

$$U(r) = -k_e \frac{e^2}{r} \tag{42.1}$$

where k_e is the Coulomb constant and r is the radial distance from the proton (situated at $r = 0$) to the electron. Figure 42.4 is a plot of this function versus r/a_0 where a_0 is the Bohr radius, 0.0529 nm.

The formal procedure for solving the problem of the hydrogen atom would be to substitute $U(r)$ into the Schrödinger equation and find appropriate solutions to the equation, as we did in Chapter 41. The present problem is more complicated, however, because it is three-dimensional and because U depends on the radial coordinate r. We do not attempt to carry out these solutions. Rather, we simply describe the properties of these solutions and some of their implications with regard to atomic structure.

According to quantum mechanics, the energies of the allowed states for the hydrogen atom are

$$E_n = -\left(\frac{k_e e^2}{2a_0}\right)\frac{1}{n^2} = -\frac{13.6}{n^2}\ \text{eV} \qquad n = 1, 2, 3, \ldots \tag{42.2}$$

Allowed energies for the hydrogen atom

This result is in exact agreement with that obtained in the Bohr theory. Note that the allowed energies depend only on the quantum number n.

In one-dimensional problems, only one quantum number is needed to characterize a stationary state. In the three-dimensional hydrogen atom, three quantum numbers are needed for each stationary state, corresponding to three independent degrees of freedom for the electron. The three quantum numbers that emerge from the theory are represented by the symbols n, ℓ, and m_ℓ. The quantum number n is called the **principal quantum number,** ℓ is called the **orbital quantum number,** and m_ℓ is called the **orbital magnetic quantum number.**

There are certain important relationships between these quantum numbers, as well as certain restrictions on their values:

- The values of n can range from 1 to ∞.
- The values of ℓ can range from 0 to $n - 1$.
- The values of m_ℓ can range from $-\ell$ to ℓ.

Restrictions on the values of quantum numbers

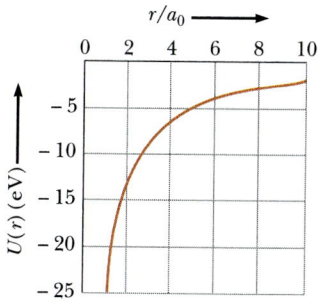

FIGURE 42.4   The potential energy $U(r)$ versus the ratio r/a_0 for the hydrogen atom. The constant a_0 is the Bohr radius, and r is the electron-proton separation.

TABLE 42.1   Three Quantum Numbers for the Hydrogen Atom

Quantum Number	Name	Allowed Values	Number of Allowed States
n	Principal quantum number	1, 2, 3, . . .	Any number
ℓ	Orbital quantum number	0, 1, 2, . . . , $n - 1$	n
m_ℓ	Orbital magnetic quantum number	$-\ell, -\ell + 1, \ldots,$ $0, \ldots, \ell - 1, \ell$	$2\ell + 1$

TABLE 42.2   Atomic Shell and Subshell Notations

n	Shell Symbol	ℓ	Subshell Symbol
1	K	0	s
2	L	1	p
3	M	2	d
4	N	3	f
5	O	4	g
6	P	5	h
. . .		. . .	

For example, if $n = 1$, only $\ell = 0$ and $m_\ell = 0$ are permitted. If $n = 2$, ℓ may be 0 or 1; if $\ell = 0$, then $m_\ell = 0$, but if $\ell = 1$, then m_ℓ may be 1, 0, or -1. Table 42.1 summarizes the rules for determining the allowed values of ℓ and m_ℓ for a given n.

For historical reasons, *all states having the same principal quantum number are said to form a* **shell**. Shells are identified by the letters K, L, M, . . . , which designate the states for which $n = 1, 2, 3, \ldots$. Likewise, *the states having the same values of n and ℓ are said to form a* **subshell**. The letters $s, p, d, f, g, h, \ldots$ are used to designate the states for which $\ell = 0, 1, 2, 3, \ldots$. For example, the state designated by $3p$ has the quantum numbers $n = 3$ and $\ell = 1$; the $2s$ state has the quantum numbers $n = 2$ and $\ell = 0$. These notations are summarized in Table 42.2.

States that violate the rules given in Table 42.1 cannot exist. For instance, the $2d$ state, which would have $n = 2$ and $\ell = 2$, cannot exist because the highest allowed value of ℓ is $n - 1$, or 1 in this case. Thus, for $n = 2$, $2s$ and $2p$ are allowed states but $2d$, $2f, \ldots$ are not. For $n = 3$, the allowed states are $3s$, $3p$, and $3d$.

EXAMPLE 42.1   The $n = 2$ Level of Hydrogen

For a hydrogen atom, determine the number of orbital states corresponding to the principal quantum number $n = 2$, and calculate the energies of these states.

Solution   When $n = 2$, ℓ can be 0 or 1. If $\ell = 0$, m_ℓ can only be 0; for $\ell = 1$, m_ℓ can be -1, 0, or 1. Hence, we have a state designated as the $2s$ state associated with the quantum numbers $n = 2$, $\ell = 0$, and $m_\ell = 0$, and three orbital states designated as $2p$ states for which the quantum numbers are $n = 2$, $\ell = 1$, $m_\ell = -1$; $n = 2$, $\ell = 1$, $m_\ell = 0$; and $n = 2$, $\ell = 1$, $m_\ell = 1$.

Because all of these orbital states have the same principal quantum number, they also have the same energy, which can be calculated using Equation 42.2, taking $n = 2$:

$$E_2 = -\frac{13.6}{2^2} \text{ eV} = -3.40 \text{ eV}$$

Exercise   How many possible states are there for the $n = 3$ level of hydrogen? For the $n = 4$ level?

Answers   9 and 16.

42.3   THE SPIN MAGNETIC QUANTUM NUMBER

In Example 42.1, we found four quantum states corresponding to $n = 2$. In reality, however, there are eight states rather than the four we found. These extra states can be explained by requiring a fourth quantum number for each state: the **spin magnetic quantum number**, m_s.

The need for this new quantum number came about because of an unusual feature in the spectra of certain gases, such as sodium vapor. Close examination of one of the prominent lines of sodium shows that this line is, in fact, two very closely spaced lines called a doublet. The wavelengths of these lines occur in the yellow region at 589.0 nm and 589.6 nm. In 1925, when this doublet was first observed, atomic theory could not explain it. To resolve this dilemma, Samuel Goudsmidt and George Uhlenbeck, following a suggestion by the Austrian physicist Wolfgang Pauli, proposed a new quantum number, called the spin quantum number.

In order to describe the spin quantum number, it is convenient (but incorrect) to think of the electron as spinning on its axis as it orbits the nucleus, just as the Earth spins about its axis as it orbits the Sun. There are only two directions that the electron can spin as it orbits the nucleus, as shown in Figure 42.5. If the direction of spin is as shown in Figure 42.5a, the electron is said to have "spin up." If the direction of spin is reversed, as in Figure 42.5b, the electron is said to have "spin down." In the presence of any other magnetic field, the energy of the electron is slightly different for the two spin directions, and this energy difference accounts for the sodium doublet. The quantum numbers associated with the spin of the electron are $m_s = \frac{1}{2}$ for the spin-up state and $m_s = -\frac{1}{2}$ for the spin-down state.

This classical description of electron spin is incorrect, because quantum mechanics tells us that a rotational degree of freedom would require too many quantum numbers, and more recent theory indicates that the electron is a point particle, without spatial extent. The electron cannot be considered to be spinning as pictured in Figure 42.5. In spite of this conceptual difficulty, all experimental evidence supports the fact that an electron does have some intrinsic property that can be described by the spin magnetic quantum number. The origin of this fourth quantum number was shown by Sommerfeld and Dirac to lie in the relativistic properties of the electron, which requires four quantum numbers to describe its location in four-dimensional space-time.

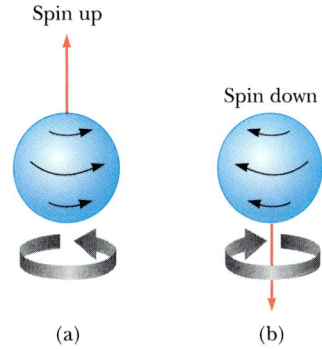

FIGURE 42.5 The spin of an electron can be either (a) up or (b) down.

EXAMPLE 42.2 Adding Some Spin on Hydrogen

For a hydrogen atom, determine the quantum numbers associated with the possible states that correspond to the principal quantum number $n = 2$.

Solution With the addition of the spin quantum number, we have the possibilities given in the table at the right.

Exercise Show that for $n = 3$, there are 18 possible states. (This follows from the restrictions that the maximum number of electrons in the $3s$ state is 2, the maximum number in the $3p$ state is 6, and the maximum number in the $3d$ state is 10.)

n	ℓ	m_ℓ	m_s	Subshell	Shell	Number of Electrons in Subshell
2	0	0	$\frac{1}{2}$			
2	0	0	$-\frac{1}{2}$	$2s$	L	2
2	1	1	$\frac{1}{2}$			
2	1	1	$-\frac{1}{2}$			
2	1	0	$\frac{1}{2}$			
2	1	0	$-\frac{1}{2}$	$2p$	L	6
2	1	-1	$\frac{1}{2}$			
2	1	-1	$-\frac{1}{2}$			

42.4 THE WAVE FUNCTIONS FOR HYDROGEN

Neglecting electron spin for the present, the potential energy of the hydrogen atom depends only on the radial distance r between nucleus and electron. We therefore expect that some of the allowed states for this atom can be represented by wave functions that depend only on r. This indeed is the case. The simplest wave function for hydrogen is the one that describes the $1s$ state and is designated $\psi_{1s}(r)$:

$$\psi_{1s}(r) = \frac{1}{\sqrt{\pi a_0^3}} e^{-r/a_0} \tag{42.3}$$

Wave function for hydrogen in its ground state

where a_0 is the Bohr radius:

Bohr radius

$$a_0 = \frac{\hbar^2}{mk_e e^2} = 0.0529 \text{ nm} \tag{42.4}$$

Note that ψ_{1s} satisfies the condition that it approach zero as r approaches ∞ and is normalized as presented. Furthermore, because ψ_{1s} depends only on r, it is *spherically symmetric*. This, in fact, is true for all s states.

Recall that the probability of finding the electron in any region is equal to an integral of $|\psi|^2$ over the region. The probability density for the $1s$ state is

$$|\psi_{1s}|^2 = \left(\frac{1}{\pi a_0^3}\right) e^{-2r/a_0} \tag{42.5}$$

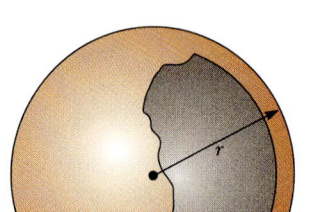

FIGURE 42.6   A spherical shell of radius r and thickness dr has a volume equal to $4\pi r^2 \, dr$.

and the actual probability of finding the electron in a volume element dV is $|\psi|^2 \, dV$. It is convenient to define the radial probability density function $P(r)$ as the probability of finding the electron in a spherical shell of radius r and thickness dr. The volume of such a shell equals its surface area, $4\pi r^2$, multiplied by the shell thickness, dr (Fig. 42.6), so that we get

$$P(r) \, dr = |\psi|^2 \, dV = |\psi|^2 \, 4\pi r^2 \, dr$$

$$P(r) = 4\pi r^2 |\psi|^2 \tag{42.6}$$

Substituting Equation 42.5 into Equation 42.6 gives the radial probability density function for the hydrogen atom in its ground state:

Radial probability density for the $1s$ state of hydrogen

$$P_{1s}(r) = \left(\frac{4r^2}{a_0^3}\right) e^{-2r/a_0} \tag{42.7}$$

A plot of the function $P_{1s}(r)$ versus r is presented in Figure 42.7a. The peak of the curve corresponds to the most probable value of r for this particular state. The spherical symmetry of the distribution function is shown in Figure 42.7b.

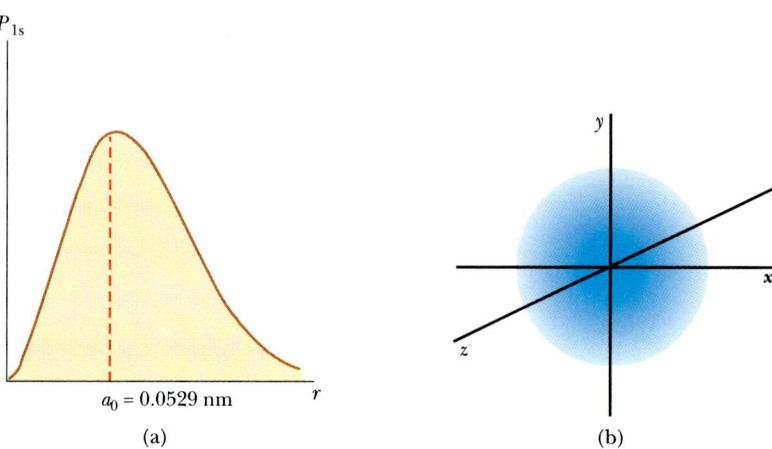

(a)   (b)

FIGURE 42.7   (a) The probability of finding the electron as a function of distance from the nucleus for the hydrogen atom in the $1s$ (ground) state. Note that the probability has its maximum value when r equals the first Bohr radius, a_0. (b) The spherical electron charge distribution for the hydrogen atom in its $1s$ state.

EXAMPLE 42.3   **The Ground State of Hydrogen**

Calculate the most probable value of r for an electron in the ground state of the hydrogen atom.

Solution   The most probable value of r corresponds to the peak of the plot of $P(r)$ versus r. The slope of the curve at this point is zero, and so we can evaluate the most probable value of r by setting $dP/dr = 0$ and solving for r. Using Equation 42.7, we get

$$\frac{dP}{dr} = \frac{d}{dr}\left[\left(\frac{4r^2}{a_0^{\,3}}\right)e^{-2r/a_0}\right] = 0$$

Carrying out the derivative operation and simplifying the expression, we get

$$e^{-2r/a_0}\frac{d}{dr}(r^2) + r^2\frac{d}{dr}(e^{-2r/a_0}) = 0$$

$$2re^{-2r/a_0} + r^2(-2/a_0)e^{-2r/a_0} = 0$$

$$2r[1 - (r/a_0)]e^{-2r/a_0} = 0$$

This expression is satisfied if

$$1 - \frac{r}{a_0} = 0$$

$$r = \boxed{a_0}$$

EXAMPLE 42.4   **Probabilities for the Electron in Hydrogen**

Calculate the probability that the electron in the ground state of hydrogen will be found outside the first Bohr radius.

Solution   The probability is found by integrating the radial probability density for this state, $P_{1s}(r)$, from the Bohr radius a_0 to ∞. Using Equation 42.7, we get

$$P = \int_{a_0}^{\infty} P_{1s}(r)\ dr = \frac{4}{a_0^{\,3}}\int_{a_0}^{\infty} r^2 e^{-2r/a_0}\ dr$$

We can put the integral in dimensionless form by changing variables from r to $z = 2r/a_0$. Noting that $z = 2$ when $r = a_0$ and that $dr = (a_0/2)\ dz$, we get

$$P = \frac{1}{2}\int_{2}^{\infty} z^2 e^{-z}\ dz = -\frac{1}{2}\{z^2 + 2z + 2\}e^{-z}\ \Big|_{2}^{\infty}$$

$$P = 5e^{-2} = \boxed{0.677}\quad\text{or}\quad\boxed{67.7\%}$$

Example 42.3 shows that, for the ground state of hydrogen, the most probable value of r equals the first Bohr radius, a_0. It turns out that the average value of r for the ground state of hydrogen is $\frac{3}{2}a_0$, which is 50% larger than the most probable value (see Problem 40). The reason for this is the large asymmetry in the radial distribution function shown in Figure 42.7a, which has a greater area to the right of the peak. According to quantum mechanics, there is no sharply defined boundary to the atom. Therefore, the probability distribution for the electron can be viewed as being a diffuse region of space, commonly referred to as an "electron cloud."

The next-simplest wave function for the hydrogen atom is the one corresponding to the $2s$ state ($n = 2$, $\ell = 0$). The normalized wave function for this state is

$$\psi_{2s}(r) = \frac{1}{4\sqrt{2\pi}}\left(\frac{1}{a_0}\right)^{3/2}\left[2 - \frac{r}{a_0}\right]e^{-r/2a_0} \qquad (42.8)$$

Wave function for hydrogen in the $2s$ state

Again, we see that ψ_{2s} depends only on r and is spherically symmetric. The energy corresponding to this state is $E_2 = -(13.6/4)$ eV $= -3.4$ eV. This energy level represents the first excited state of hydrogen. Plots of the radial distribution function for this state and several others are shown in Figure 42.8. The plot for the $2s$ state, which applies to a single electron, has two peaks. In this case, the most probable value corresponds to that value of r that has the highest value of P ($\approx 5a_0$). An electron in the $2s$ state would be much farther from the nucleus (on the average) than an electron in the $1s$ state. The average value of r is even greater for the $3d$, $3p$, and $4d$ states.

As we have mentioned, all s states have spherically symmetric wave functions. The other states are not spherically symmetric. For example, the three wave func-

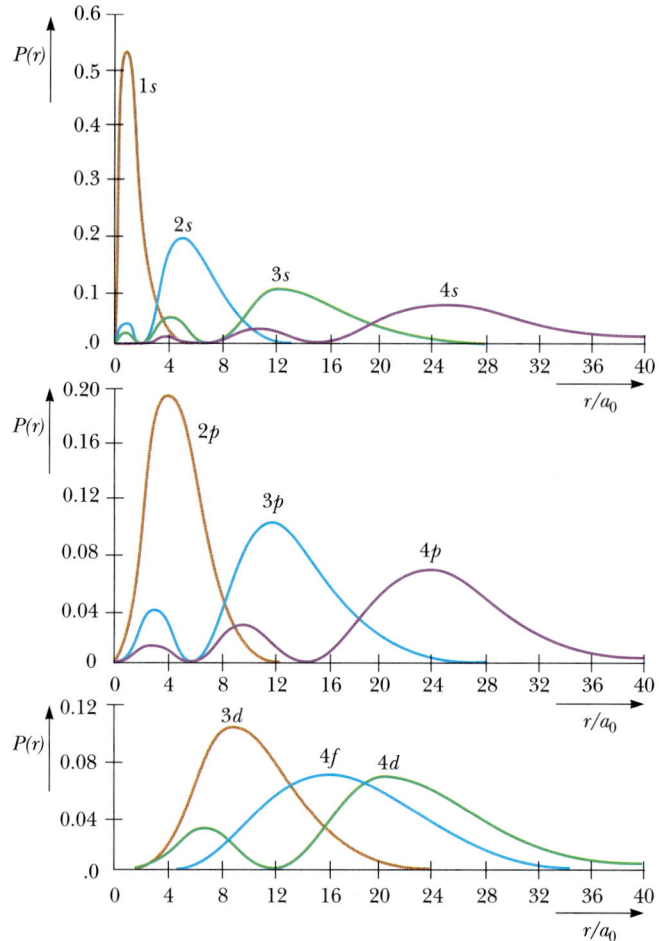

FIGURE 42.8   The radial probability density function versus r/a_0 for several states of the hydrogen atom.   *(From E. U. Condon and G. H. Shortley,* The Theory of Atomic Spectra, *Cambridge, Cambridge University Press, 1953; used with permission.)*

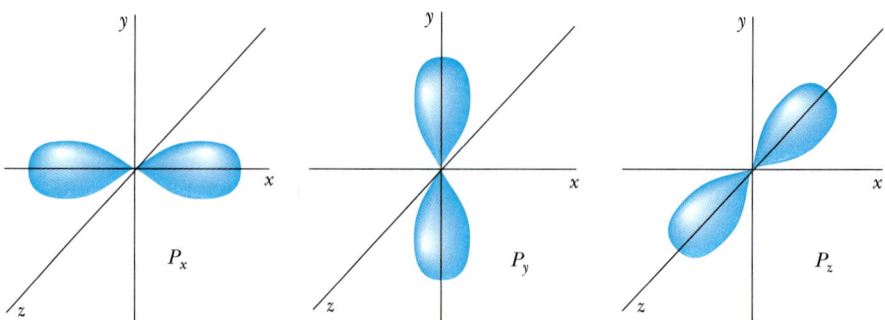

FIGURE 42.9   Electron charge distribution for an electron in a p state. The three charge distributions p_x, p_y, and p_z have the same structure and differ only in their orientation in space.

tions corresponding to the states for which $n = 2$, $\ell = 1$ ($m_\ell = 1, 0,$ or -1) can be expressed as appropriate linear combinations of the three p states. Although quantum mechanics limits our knowledge of angular momentum to the projection along any one axis at a time, these p states may be described mathematically as linear combinations of mutually perpendicular functions p_x, p_y, and p_z, as represented in Figure 42.9, where only the angular dependence of these functions is shown. Note that the three clouds have identical structure but differ in their orientation with respect to the x, y, and z axes. The nonspherical wave functions for these states are

$$\psi_{2p_x} = xF(r)$$
$$\psi_{2p_y} = yF(r) \qquad (42.9)$$
$$\psi_{2p_z} = zF(r)$$

Wave functions for the 2p state

where $F(r)$ is some exponential function of r. Wave functions that have a highly directional character, such as these, are convenient for descriptions of chemical bonding, the formation of molecules, and chemical properties.

42.5   THE "OTHER" QUANTUM NUMBERS

The energy of a particular state depends primarily on the principal quantum number. Now let us see what the other three quantum numbers contribute to our atomic model.

The Orbital Quantum Number

If a particle moves in a circle of radius r, the magnitude of its angular momentum relative to the center of the circle is $L = mvr$. The direction of L is perpendicular to the plane of the circle, and its sense is given by a right-hand rule.[1] According to classical physics, L can have any value. However, the Bohr model of hydrogen postulates that the angular momentum of the electron is restricted to multiples of $\hbar$; that is, $mvr = n\hbar$. This model must be modified because it predicts (incorrectly) that the ground state of hydrogen ($n = 1$) has one unit of angular momentum. Furthermore, if L is taken to be zero in the Bohr model, we are forced to accept a picture of the electron as a particle oscillating along a straight line through the nucleus, a physically unacceptable situation.

These difficulties are resolved with the quantum mechanical model of the atom. According to quantum mechanics, an atom in a state whose principal quantum number is n can take on the following *discrete* values of orbital angular momentum:

$$L = \sqrt{\ell(\ell + 1)}\,\hbar \qquad \ell = 0, 1, 2, 3, \ldots, n - 1 \qquad (42.10)$$

Allowed value of L

Because ℓ is restricted to the values $\ell = 0, 1, 2, \ldots, n - 1$, we see that $L = 0$ (corresponding to $\ell = 0$) is an acceptable value of the angular momentum. The fact that L can be zero in this model serves to point out the inherent difficulties in any attempt to describe results based on quantum mechanics in terms of a purely particle-like model. In the quantum mechanical interpretation, the electron cloud for the $L = 0$ state is spherically symmetric and has no fundamental axis of revolution.

---

[1] See Sections 11.2 and 11.3 for details on angular momentum if you have forgotten this material.

EXAMPLE 42.5   Calculating *L* for a *p* State

Calculate the orbital angular momentum of an electron in a *p* state of hydrogen.

Solution   Because we know that $\hbar = 1.054 \times 10^{-34}$ J·s, we can use Equation 42.10 to calculate *L*. With $\ell = 1$ for a *p* state, we have

$$L = \sqrt{1(1 + 1)}\hbar = \sqrt{2}\hbar = 1.49 \times 10^{-34} \text{ J·s}$$

This number is extremely small relative to, say, the orbital angular momentum of the Earth orbiting the Sun, which is approximately 2.7×10^{40} J·s. The quantum number that describes *L* for macroscopic objects, such as the Earth, is so large that the separation between adjacent states cannot be measured. Once again, the correspondence principle is upheld.

The Magnetic Orbital Quantum Number

Because angular momentum is a vector, its direction must also be specified. Recall from Chapter 30 that an orbiting electron can be considered an effective current loop with a corresponding magnetic moment. Such a moment placed in a magnetic field **B** interacts with the field. Suppose a weak magnetic field applied along the *z* axis defines a direction in space. According to quantum mechanics, L^2 and L_z, the projection of **L** along the *z* axis, can have discrete values. The magnetic orbital quantum m_ℓ specifies the allowed values of L_z according to the expression

Allowed values of L_z

$$L_z = m_\ell\hbar \tag{42.11}$$

Space quantization

The fact that the direction of **L** is quantized with respect to an external magnetic field is often referred to as **space quantization.**

Let us look at the possible orientations of **L** for a given value of ℓ. Recall that m_ℓ can have values ranging from $-\ell$ to ℓ. If $\ell = 0$, then $m_\ell = 0$ and $L_z = 0$. For L_z to be zero, **L** must be perpendicular to **B**. If $\ell = 1$, then the possible values of m_ℓ are $-1, 0,$ and 1, so that L_z may be $-\hbar, 0,$ or $\hbar$. If $\ell = 2$, m_ℓ can be $-2, -1, 0, 1,$ or 2, corresponding to L_z values of $-2\hbar, -\hbar, 0, \hbar,$ or $2\hbar$, and so on.

A vector model describing space quantization for $\ell = 2$ is shown in Figure 42.10a. Note that **L** can never be aligned parallel or antiparallel to **B** because L_z must be smaller than the total angular momentum *L*. From a three-dimensional viewpoint, **L** must lie on the surface of a cone that makes an angle θ with the *z* axis, as shown in Figure 42.10b. From the figure, we see that θ is also quantized and that

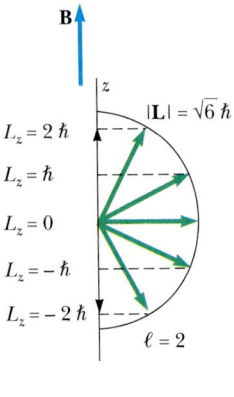

(a)

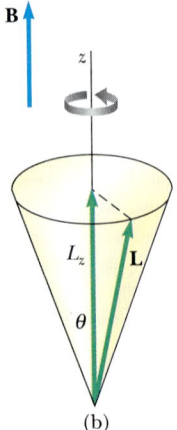

(b)

FIGURE 42.10   (a) The allowed projections of the orbital angular momentum for the case $\ell = 2$. (b) The orbital angular momentum vector lies on the surface of a cone and precesses about the *z* axis when a magnetic field **B** is applied in this direction.

its values are specified through the relationship

$$\cos \theta = \frac{L_z}{|\mathbf{L}|} = \frac{m_\ell}{\sqrt{\ell(\ell + 1)}} \qquad (42.12)$$

θ is quantized

Note that m_ℓ is never greater than ℓ, and therefore θ can never be zero. (Classically, θ can have any value.)

Because of the uncertainty principle, L does not point in a specific direction but rather traces out a cone in space. If L had a definite value, then all three components L_x, L_y, and L_z would be exactly specified. For the moment, let us assume that this is the case and let us suppose that the electron moves in the xy plane, so that L is in the z direction and $p_z = 0$. This means that p_z is precisely known, which is in violation of the uncertainty principle, $\Delta p_z \, \Delta z \geq \hbar//2$. In reality, only the magnitude of L and one component (say L_z) can have definite values. In other words, quantum mechanics allows us to specify L and L_z but not L_x and L_y. Because the direction of L is constantly changing as it precesses about the z axis, the average values of L_x and L_y are zero and L_z maintains a fixed value of $m_\ell \hbar$.

EXAMPLE 42.6 **Space Quantization for Hydrogen**

Consider the hydrogen atom in the $\ell = 3$ state. Calculate the magnitude of L and the allowed values of L_z and θ.

Solution Since $\ell = 3$, we can calculate the total angular momentum using Equation 42.10:

$$L = \sqrt{\ell(\ell + 1)}\hbar = \sqrt{3(3 + 1)}\hbar = 2\sqrt{3}\hbar$$

The allowed values of L_z can be calculated using $L_z = m_\ell \hbar$, with $m_\ell = -3, -2, -1, 0, 1, 2,$ and 3:

$$L_z = -3\hbar, -2\hbar, -\hbar, 0, \hbar, 2\hbar, \text{ and } 3\hbar$$

Finally, we calculate the allowed values of θ using Equation 42.12. Since $\ell = 3$, $\sqrt{\ell(\ell + 1)} = 2\sqrt{3}$ and we have

$$\cos \theta = \frac{m_\ell}{2\sqrt{3}}$$

Substituting the allowed values of m_ℓ gives

$$\cos \theta = \pm 0.866, \pm 0.577, \pm 0.289, \text{ and } 0$$

$$\theta = 30.0°, 54.8°, 73.2°, 90.0°, 107°, 125°, \text{ and } 150°$$

Electron Spin

In 1921, Stern and Gerlach performed an experiment that demonstrated the phenomenon of space quantization. However, their results were not in quantitative agreement with the theory that existed at that time. In their experiment, a beam of neutral silver atoms sent through a nonuniform magnetic field was split into two components (Fig. 42.11). The experiment was repeated using other atoms, and in each case the beam split into two or more components. The classical argument is as follows: If the z direction is chosen to be the direction of the maximum inhomogeneity of B, the net magnetic force on the atom is along the z axis and is proportional to the magnetic moment in the direction of $\boldsymbol{\mu}_z$. Classically, $\boldsymbol{\mu}_z$ can have any orientation, and so the deflected beam should be spread out continuously. According to quantum mechanics, however, the deflected beam has several components and the number of components determines the various possible values of μ_z. Hence, because the Stern-Gerlach experiment showed split beams, space quantization was at least qualitatively verified.

For the moment, let us assume that μ_z is due to the orbital angular momentum. Because μ_z is proportional to m_ℓ, the number of possible values of μ_z is $2\ell + 1$.

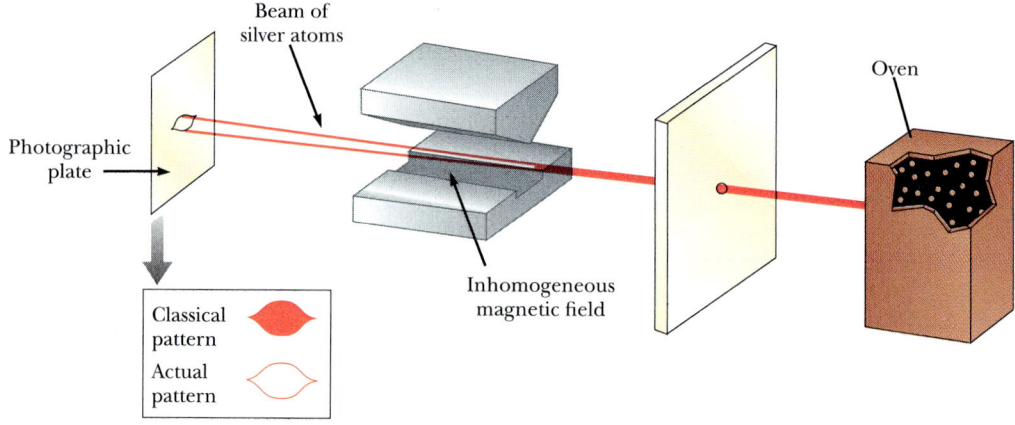

FIGURE 42.11   The apparatus used by Stern and Gerlach to verify space quantization. A beam of neutral silver atoms is split into two components by a nonuniform magnetic field.

Furthermore, because ℓ is an integer, the number of values of μ_z is always odd. This prediction is clearly not consistent with the observations of Stern and Gerlach, who observed only two components in the deflected beam of silver atoms. Hence, we are forced to conclude that either quantum mechanics is incorrect or the model is in need of refinement.

In 1927, Phipps and Taylor repeated the Stern-Gerlach experiment using a beam of hydrogen atoms. This experiment is important because it deals with an atom with a single electron in its ground state, for which the theory makes reliable predictions. Recall that $\ell = 0$ for hydrogen in its ground state, and so $m_\ell = 0$. Hence, we would not expect the beam to be deflected by the field because μ_z is zero. However, the beam in the Phipps-Taylor experiment is again split into two components. On the basis of this result, we can come to only one conclusion: Something other than the orbital motion is contributing to the magnetic moment.

As we learned earlier, Goudsmidt and Uhlenbeck had proposed that the electron has an intrinsic angular momentum apart from its orbital angular momentum. From a classical viewpoint, this intrinsic angular momentum is attributed to the charged electron spinning about its own axis and, hence, is called electron spin.[2] In other words, the total angular momentum of the electron in a particular electronic state contains both an orbital contribution L and a spin contribution S. The Phipps-Taylor result confirmed this hypothesis of Goudsmidt and Uhlenbeck.

In 1929, Dirac solved the relativistic wave equation for the electron in a potential well using the relativistic form of the total energy. Dirac's analysis confirmed the fundamental nature of electron spin. Furthermore, the theory showed that electron spin can be described by a single quantum number s, whose value could

---

[2] Physicists often use the word *spin* when referring to spin angular momentum. For example, it is common to use the phrase *the electron has a spin* of $\frac{1}{2}$. The spin angular momentum of the electron *never changes*. This notion contradicts classical laws, where a rotating charge slows down in the presence of an applied magnetic field because of the Faraday emf that accompanies the changing field. Furthermore, if the electron is viewed as a spinning ball of charge subject to classical laws, parts of it near its surface would be rotating with speeds exceeding the speed of light. Thus, the classical picture must not be pressed too far; ultimately, the spinning electron is a quantum entity defying any simple classical description.

be only $\frac{1}{2}$. The magnitude of the **spin angular momentum S** for the electron is

$$S = \sqrt{s(s + 1)}\hbar = \frac{\sqrt{3}}{2}\hbar \qquad (42.13)$$

Spin angular momentum of an electron

Like orbital angular momentum, spin angular momentum is quantized in space, as described in Figure 42.12. It can have two orientations, specified by the spin magnetic quantum number m_s, where $m_s = \pm\frac{1}{2}$. The z component of spin angular momentum is

$$S_z = m_s\hbar = \pm\tfrac{1}{2}\hbar \qquad (42.14)$$

The two values $\pm\hbar/2$ for S_z correspond to the two possible orientations for **S** shown in Figure 42.12. The value $m_s = +1/2$ refers to the up spin case, while the value $m_s = -1/2$ refers to the down spin case.

The spin magnetic moment of the electron, $\boldsymbol{\mu}_s$, is related to its spin angular momentum **S** by the expression

$$\boldsymbol{\mu}_s = -\frac{e}{m}\mathbf{S} \qquad (42.15)$$

Since $S_z = \pm\frac{1}{2}\hbar$, the z component of the spin magnetic moment can have the values

$$\mu_{sz} = \pm\frac{e\hbar}{2m} \qquad (42.16)$$

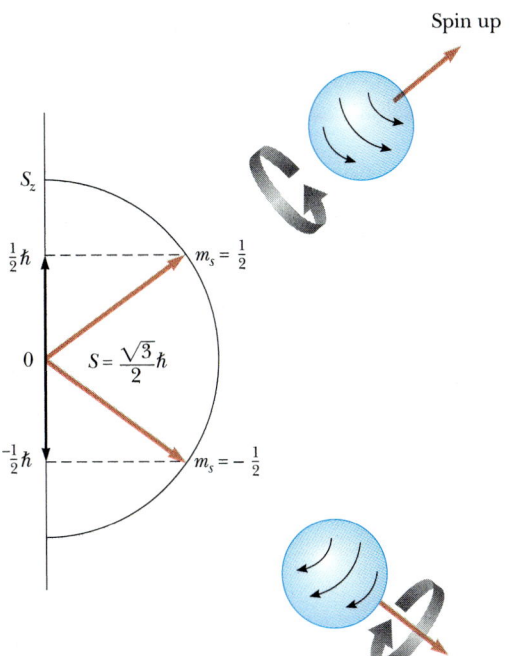

Spin up

Spin down

FIGURE 42.12 The spin angular momentum also exhibits space quantization. This figure shows the two allowed orientations of the spin vector **S** for a spin $\frac{1}{2}$ particle, such as the electron.

As we learned in Section 30.9, the quantity $e\hbar/2m$ is the Bohr magneton μ_B, and has the numerical value $9.27 \times 10^{-24}\,\text{J}/\text{T}$. Note that the spin contribution to the angular momentum is twice the contribution of the orbital motion. The factor of 2 is explained in a relativistic treatment of the system first carried out by Dirac.

Today physicists explain the Stern-Gerlach experiment as follows. The observed moments for both silver and hydrogen are due to spin angular momentum and not to orbital angular momentum. A single-electron atom such as hydrogen has its electron quantized in the magnetic field in such a way that its z component of spin angular momentum is either $\frac{1}{2}\hbar$ or $-\frac{1}{2}\hbar$, corresponding to $m_s = \pm\frac{1}{2}$. Electrons with spin $+\frac{1}{2}$ are deflected downward, and those with spin $-\frac{1}{2}$ are deflected upward. The Stern-Gerlach experiment provided two important results. First, it verified the concept of space quantization. Second, it showed that spin angular momentum exists even though this property was not recognized until four years after the experiments were performed.

42.6   THE EXCLUSION PRINCIPLE AND THE PERIODIC TABLE

Earlier we found that the state of a hydrogen atom is specified by four quantum numbers: n, ℓ, m_ℓ, and m_s. For example, an electron in the ground state of hydrogen might have quantum numbers of $n = 1$, $\ell = 0$, $m_\ell = 0$, $m_s = \frac{1}{2}$. As it turns out, the state of an electron in any other atom may also be specified by this same set of quantum numbers. In fact, these four quantum numbers can be used to describe all the electronic states of an atom regardless of the number of electrons in its structure.

An obvious question that arises here is, "How many electrons can have a particular set of quantum numbers?" This important question was answered by Pauli in 1925 in a powerful statement known as the **exclusion principle:**

Wolfgang Pauli and Niels Bohr watch a spinning top.   *(Courtesy of AIP Niels Bohr Library, Margarethe Bohr Collection)*

Exclusion principle

> No two electrons can ever be in the same quantum state; therefore, no two electrons in the same atom can have the same set of quantum numbers.

If this principle were not valid, every electron would end up in the lowest possible energy state of the atom and the chemical behavior of the elements would be grossly modified. Nature as we know it would not exist! In reality, we can view the electronic structure of complex atoms as a succession of filled levels increasing in energy.

As a general rule, the order of filling of an atom's subshells with electrons is as follows. Once one subshell is filled, the next electron goes into the lowest-energy vacant subshell. This principle can be understood by recognizing that if the atom were not in the lowest energy state available to it, it would radiate energy until it reached this state.

Before we discuss the electronic configuration of various elements, it is convenient to define an *orbital* as the state of an electron characterized by the quantum numbers n, ℓ, and m_ℓ. From the exclusion principle, we see that *there can be only two electrons in any orbital*. One of these electrons has a spin quantum number $m_s = +\frac{1}{2}$, and the other has $m_s = -\frac{1}{2}$. Because each orbital is limited to two electrons, the number of electrons that can occupy the various levels is also limited.

TABLE 42.3   Allowed Quantum Numbers for an Atom for Which $n = 3$

n	1	2				3								
ℓ	0	0	1			0	1			2				
m_ℓ	0	0	1	0	-1	0	1	0	-1	2	1	0	-1	-2
m_s	⇅	⇅	⇅	⇅	⇅	⇅	⇅	⇅	⇅	⇅	⇅	⇅	⇅	⇅

Table 42.3 shows the number of allowed quantum states for an atom for which $n = 3$. The arrows pointing upward indicate $m_s = \frac{1}{2}$, and those pointing downward indicate $m_s = -\frac{1}{2}$. The $n = 1$ shell can accommodate only two electrons because there is only one allowed orbital with $m_\ell = 0$. The $n = 2$ shell has two subshells, with $\ell = 0$ and $\ell = 1$. The $\ell = 0$ subshell is limited to only two electrons because $m_\ell = 0$. The $\ell = 1$ subshell has three allowed orbitals, corresponding to $m_\ell = 1, 0,$

Wolfgang Pauli was an extremely talented Austrian theoretical physicist who made important contributions in many areas of modern physics. At the age of 21, Pauli gained public recognition with a masterful review article on relativity, which is still considered to be one of the finest and most comprehensive introductions to the subject. Other major contributions were the discovery of the exclusion principle, the explanation of the connection between particle spin and statistics, theories of relativistic quantum electrodynamics, the neutrino hypothesis, and the hypothesis of nuclear spin. An article entitled "The Fundamental Principles of Quantum Mechanics," written by Pauli in 1933 for the *Handbuch der Physik*, is widely acknowledged to be one of the best treatments of quantum physics ever written. Pauli was a forceful and colorful character, well known for his witty and often caustic remarks directed at those who presented new theories in a less than perfectly clear manner. Pauli exerted great influence on his students and colleagues by forcing them with his sharp criticism to a deeper and clearer understanding. Victor Weisskopf, one of Pauli's famous students, has aptly described him as "the conscience of theoretical physics." Pauli's sharp sense of humor was also nicely captured by Weisskopf in the following anecdote:

"In a few weeks, Pauli asked me to come to Zurich. I came to the big door of his office, I knocked, and no answer. I knocked again and no answer. After about five minutes he said, rather roughly, 'Who is it? Come in!' I opened the door, and here was Pauli —it was a very big office—at the other side of the room, at his desk, writing and writing. He said, 'Who is this? First I must finish calculating.' Again he let me wait for about five minutes and then: 'Who is that?' 'I am Weisskopf.' 'Uhh, Weisskopf, ja, you are my new assistant.' Then he looked at me and said, 'Now, you see I wanted to take Bethe, but Bethe works now on the solid state. Solid state I don't like, although I started it. This is why I took you.' Then I said, 'What can I do for you sir?' and he said, 'I shall give you right away a problem.' He gave me a problem, some calculation, and then he said, 'Go and work.' So I went, and after 10 days or so, he came and said, 'Well, show me what you have done.' And I showed him. He looked at it and exclaimed: 'I should have taken Bethe!' "[1]

Wolfgang Pauli

| 1 9 0 0 – 1 9 5 8 |

[1] From *Physics in the Twentieth Century: Selected Essays, My Life as a Physicist*, Victor F. Weisskopf, The MIT Press, 1972, p. 10.

and -1. Because each orbital can accommodate two electrons, the $\ell = 1$ subshell can hold six electrons. The $n = 3$ shell has three subshells and nine orbitals and can accommodate up to 18 electrons. In general, each shell can accommodate up to $2n^2$ electrons.

The exclusion principle can be illustrated by an examination of the electronic arrangement in a few of the lighter atoms.

Hydrogen has only one electron, which, in its ground state, can be described by either of two sets of quantum numbers: $1, 0, 0, \frac{1}{2}$ or $1, 0, 0, -\frac{1}{2}$. The electronic configuration of this atom is often designated as $1s^1$. The notation $1s$ refers to a state for which $n = 1$ and $\ell = 0$, and the superscript indicates that one electron is present in the s subshell.

Neutral *helium* has two electrons. In the ground state, the quantum numbers for these two electrons are $1, 0, 0, \frac{1}{2}$ and $1, 0, 0, -\frac{1}{2}$. There are no other possible combinations of quantum numbers for this level, and we say that the K shell is filled. Helium is designated by $1s^2$.

Neutral *lithium* has three electrons. In the ground state, two of these are in the $1s$ subshell and the third is in the $2s$ subshell, because this subshell is slightly lower

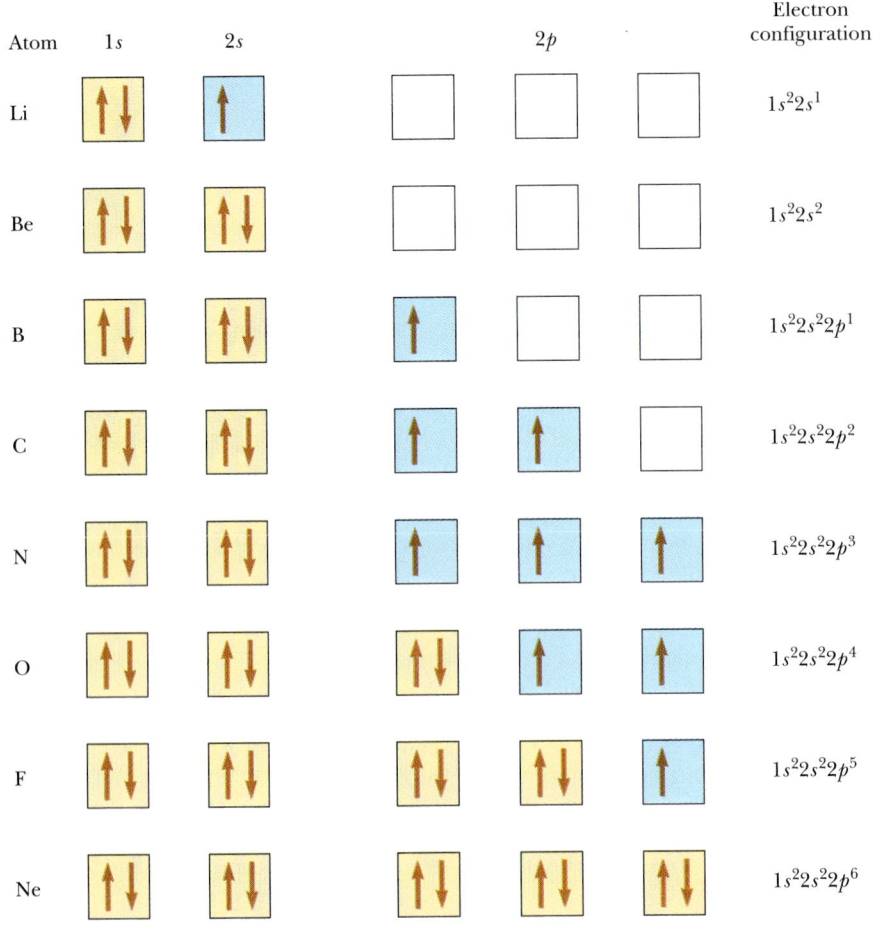

FIGURE 42.13 The filling of electronic states must obey the Pauli exclusion principle and Hund's rule. The electronic configurations for these elements are given at the right.

in energy than the $2p$ subshell. Hence, the electronic configuration for lithium is $1s^2 2s^1$.

The electronic configurations of some successive elements are given in Figure 42.13. Note that the electronic configuration of *beryllium*, with its four electrons, is $1s^2 2s^2$, and *boron* has a configuration of $1s^2 2s^2 2p^1$. The $2p$ electron in boron may be described by six sets of quantum numbers, corresponding to six states of equal energy.

Carbon has six electrons, and a question arises concerning how to assign the two $2p$ electrons. Do they go into the same orbital with paired spins (↑↓), or do they occupy different orbitals with unpaired spins (↑↑)? Experimental data show that the most stable configuration (that is, the one that is energetically preferred) is the latter, where the spins are unpaired. Hence, the two $2p$ electrons in carbon and the three $2p$ electrons in nitrogen have unpaired spins (Fig. 42.13). The general rule that governs such situations, called **Hund's rule,** states that

> when an atom has orbitals of equal energy, the order in which they are filled by electrons is such that a maximum number of electrons have unpaired spins.

Hund's rule

Some exceptions to this rule occur in elements having sublevels close to being filled or half-filled.

A complete list of electronic configurations is provided in Table 42.4. An early, attempt at finding some order among the elements was made by a Russian chemist, Dmitri Mendeleev, in 1871. He arranged the atoms in a table similar to that shown in Appendix C according to their atomic masses and chemical similarities. The atomic number Z is the number of protons in a nucleus. The first table Mendeleev proposed contained many blank spaces, and he boldly stated that the gaps were there only because the elements had not yet been discovered. By noting the column in which some missing elements should be located, he was able to make rough predictions about their chemical properties. Within 20 years of this announcement, most of these elements were indeed discovered.

The elements in the periodic table are arranged so that all those in a column have similar chemical properties. For example, consider the elements in the last column: He (helium), Ne (neon), Ar (argon), Kr (krypton), Xe (xenon), and Rn (radon). The outstanding characteristic of these elements is that they do not normally take part in chemical reactions—that is, they do not join with other atoms to form molecules—and they are therefore classified as being inert. Because of this aloofness, they are referred to as the *noble gases.*

We can partially understand this behavior by looking at the electronic configurations in Table 42.4. The electronic configuration for helium is $1s^2$—in other words, one shell is filled. Additionally, it is found that the energy of the electrons in this filled shell is considerably lower than the energy of the next available level, the $2s$ level. Next, look at the electronic configuration for neon, $1s^2 2s^2 2p^6$. Again, the outermost shell is filled and there is a wide gap in energy between the $2p$ level and the $3s$ level. Argon has the configuration $1s^2 2s^2 2p^6 3s^2 3p^6$. Here, the $3p$ subshell is filled and there is a wide gap in energy between the $3p$ subshell and the $3d$ subshell. We could continue this procedure through all the noble gases; the pattern remains the same. A noble gas is formed when either a shell or a subshell is filled and there is a large gap in energy before the next possible level is encountered.

TABLE 42.4 **Electronic Configuration of the Elements**

Z	Symbol	Ground-State Configuration		Ionization Energy (eV)
1	H		$1s^1$	13.595
2	He		$1s^2$	24.581
3	Li	[He]	$2s^1$	5.39
4	Be		$2s^2$	9.320
5	B		$2s^2 2p^1$	8.296
6	C		$2s^2 2p^2$	11.256
7	N		$2s^2 2p^3$	14.545
8	O		$2s^2 2p^4$	13.614
9	F		$2s^2 2p^5$	17.418
10	Ne		$2s^2 2p^6$	21.559
11	Na	[Ne]	$3s^1$	5.138
12	Mg		$3s^2$	7.644
13	Al		$3s^2 3p^1$	5.984
14	Si		$3s^2 3p^2$	8.149
15	P		$3s^2 3p^3$	10.484
16	S		$3s^2 3p^4$	10.357
17	Cl		$3s^2 3p^5$	13.01
18	Ar		$3s^2 3p^6$	15.755
19	K	[Ar]	$4s^1$	4.339
20	Ca		$4s^2$	6.111
21	Sc		$3d^1 4s^2$	6.54
22	Ti		$3d^2 4s^2$	6.83
23	V		$3d^3 4s^2$	6.74
24	Cr		$3d^5 4s^1$	6.76
25	Mn		$3d^5 4s^2$	7.432
26	Fe		$3d^6 4s^2$	7.87
27	Co		$3d^7 4s^2$	7.86
28	Ni		$3d^8 4s^2$	7.633
29	Cu		$3d^{10} 4s^1$	7.724
30	Zn		$3d^{10} 4s^2$	9.391
31	Ga		$3d^{10} 4s^2 4p^1$	6.00
32	Ge		$3d^{10} 4s^2 4p^2$	7.88
33	As		$3d^{10} 4s^2 4p^3$	9.81
34	Se		$3d^{10} 4s^2 4p^4$	9.75
35	Br		$3d^{10} 4s^2 4p^5$	11.84
36	Kr		$3d^{10} 4s^2 4p^6$	13.996
37	Rb	[Kr]	$5s^1$	4.176
38	Sr		$5s^2$	5.692
39	Y		$4d^1 5s^2$	6.377
40	Zr		$4d^2 5s^2$	
41	Nb		$4d^4 5s^1$	6.881
42	Mo		$4d^5 5s^1$	7.10
43	Tc		$4d^5 5s^2$	7.228
44	Ru		$4d^7 5s^1$	7.365
45	Rh		$4d^8 5s^1$	7.461
46	Pd		$4d^{10}$	8.33
47	Ag		$4d^{10} 5s^1$	7.574
48	Cd		$4d^{10} 5s^2$	8.991

TABLE 42.4   *(Continued)*

Z	Symbol	Ground-State Configuration		Ionization Energy (eV)
49	In	[Kr]	$4d^{10}5s^25p^1$	$5p^1$
50	Sn		$4d^{10}5s^25p^2$	7.342
51	Sb		$4d^{10}5s^25p^3$	8.639
52	Te		$4d^{10}5s^25p^4$	9.01
53	I		$4d^{10}5s^25p^5$	10.454
54	Xe		$4d^{10}5s^25p^6$	12.127
55	Cs	[Xe]	$6s^1$	3.893
56	Ba		$6s^2$	5.210
57	La		$5d^16s^2$	5.61
58	Ce		$4f^15d^16s^2$	6.54
59	Pr		$4f^36s^2$	5.48
60	Nd		$4f^46s^2$	5.51
61	Pm		$4f^56s^2$	
62	Fm		$4f^66s^2$	5.6
63	Eu		$4f^76s^2$	5.67
64	Gd		$4f^75d^16s^2$	6.16
65	Tb		$4f^96s^2$	6.74
66	Dy		$4f^{10}6s^2$	
67	Ho		$4f^{11}6s^2$	
68	Er		$4f^{12}6s^2$	
69	Tm		$4f^{13}6s^2$	
70	Yb		$4f^{14}6s^2$	6.22
71	Lu		$4f^{14}5d^16s^2$	6.15
72	Hf		$4f^{14}5d^26s^2$	7.0
73	Ta		$4f^{14}5d^36s^2$	7.88
74	W		$4f^{14}5d^46s^2$	7.98
75	Re		$4f^{14}5d^56s^2$	7.87
76	Os		$4f^{14}5d^66s^2$	8.7
77	Ir		$4f^{14}5d^76s^2$	9.2
78	Pt		$4f^{14}5d^86s^2$	8.88
79	Au	$[Xe, 4f^{14}5d^{10}]$	$6s^1$	9.22
80	Hg		$6s^2$	10.434
81	Tl		$6s^26p^1$	6.106
82	Pb		$6s^26p^2$	7.415
83	Bi		$6s^26p^3$	7.287
84	Po		$6s^26p^4$	8.43
85	At		$6s^26p^5$	
86	Rn		$6s^26p^6$	10.745
87	Fr	[Rn]	$7s^1$	
88	Ra		$7s^2$	5.277
89	Ac		$6d^17s^2$	6.9
90	Th		$6d^27s^2$	
91	Pa		$5f^26d^17s^2$	
92	U		$5f^36d^17s^2$	4.0
93	Np		$5f^46d^17s^2$	
94	Pu		$5f^67s^2$	
95	Am		$5f^77s^2$	
96	Cm		$5f^76d^17s^2$	

(Continued)

TABLE 42.4	(Continued)		
Z	Symbol	Ground-State Configuration	Ionization Energy (eV)
97	Bk	$5f^86d^17s^2$	
98	Cf	$5f^{10}7s^2$	
99	Es	$5f^{11}7s^2$	
100	Fm	$5f^{12}7s^2$	
101	Mv	$5f^{13}7s^2$	
102	No	$5f^{14}7s^2$	
103	Lw	$5f^{14}6d^17s^2$	
104	Ku	$5f^{14}6d^27s^2$	

Note: The bracket notation is used as a shorthand method to avoid repetition in indicating inner-shell electrons. Thus, [He] represents $1s^2$, [Ne] represents $1s^22s^22p^6$, [Ar] represents $1s^22s^22p^63s^23p^6$, and so on.

42.7 ATOMIC SPECTRA: VISIBLE AND X-RAY

In Chapter 40 we briefly discussed the origin of the spectral lines for hydrogen and hydrogen-like ions. Recall that an atom emits electromagnetic radiation if an electron in an excited state makes a transition to a lower energy state. The set of wavelengths observed when a specific species undergoes such processes is called the **emission spectrum** for that species. Likewise, atoms with electrons in the ground-state configuration can also absorb electromagnetic radiation at specific wavelengths, giving rise to an **absorption spectrum.** Such spectra can be used to identify elements.

The energy-level diagram for hydrogen is shown in Figure 42.14. The various diagonal lines drawn represent allowed transitions between stationary states. Whenever an electron makes a transition from a higher energy state to a lower one, a photon of light is emitted. The frequency of this photon is $f = \Delta E/h$, where ΔE is the energy difference between the two levels and h is Planck's constant. The **selection rules** for the *allowed transitions* are

Selection rules for allowed atomic transitions

$$\Delta \ell = \pm 1 \tag{42.17}$$

$$\Delta m_\ell = 0 \text{ or } \pm 1$$

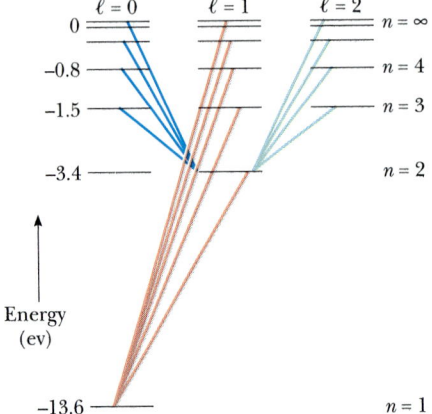

FIGURE 42.14 Some allowed electronic transitions for hydrogen, represented by the colored lines. These transitions must obey the selection rule $\Delta \ell = \pm 1$.

Transitions which do not obey the above selection rules are said to be *forbidden*. (Such transitions can occur, but their probability is negligible relative to the probability of allowed transitions.)

Because the orbital angular momentum of an atom changes when a photon is emitted or absorbed (that is, as a result of a transition) and because angular momentum must be conserved, we conclude that *the photon involved in the process must carry angular momentum*. In fact, the photon has an angular momentum equivalent to that of a particle having a spin of 1. Also, the angular momentum of the photon is consistent with the classical description of electromagnetic radiation. Hence, a photon has energy, linear momentum, and angular momentum.

It is interesting to plot the ionization energy versus atomic number Z, as in Figure 42.15. Note the pattern of $\Delta Z = 2, 8, 8, 18, 18, 32$ for the various peaks. This pattern follows from the Pauli exclusion principle and helps explain why the elements repeat their chemical properties in groups. For example, the peaks at $Z = 2, 10, 18$, and 36 correspond to the elements He, Ne, Ar, and Kr, which have filled shells. These elements have relatively high energies and similar chemical behavior.

Recall from Chapter 40 that the allowed energies for one-electron atoms, such as hydrogen or He^+, are

$$E_n = -\frac{13.6\,Z^2}{n^2}\ \text{eV} \qquad (42.18)$$

For multielectron atoms, the nuclear charge Ze is largely canceled or shielded by the negative charge of the inner-core electrons. Hence, the outer electrons interact with a net charge of the order of the electronic charge. The expression for the allowed energies for multielectron atoms has the same form as Equation 42.18

> The photon carries angular momentum.

> Allowed energies for one-electron atoms

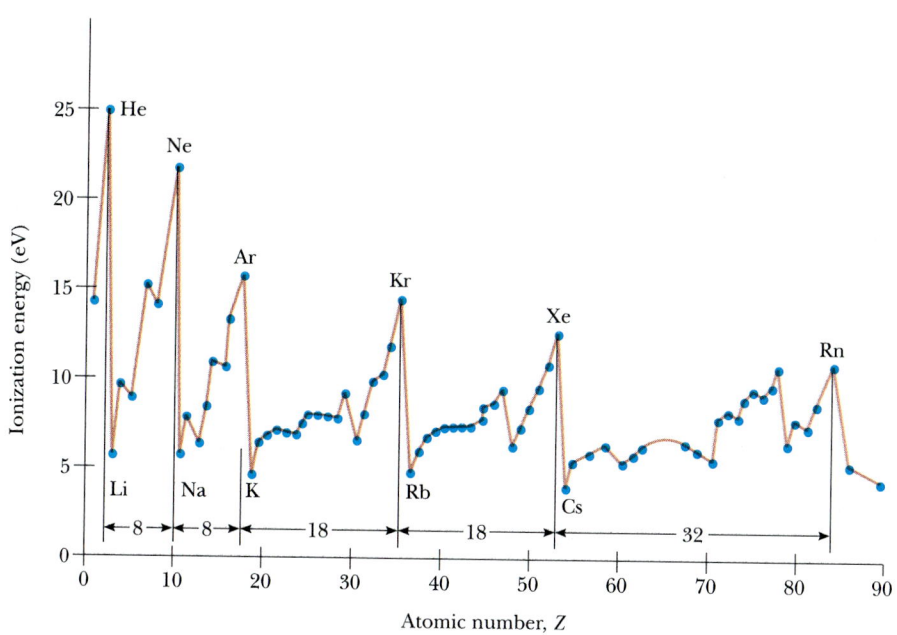

FIGURE 42.15   (a) Ionization energy of the elements versus atomic number Z. (b) Atomic volume of the elements versus atomic number.   (*Adapted from J. Orear, Physics, New York, Macmillan, 1979.*)

Allowed energies for
multielectron atoms

with Z replaced by an effective atomic number Z_{eff}. That is,

$$E_n = -\frac{13.6\,Z_{\text{eff}}^2}{n^2}\text{ eV} \qquad (42.19)$$

where Z_{eff} depends on n and ℓ. For the higher energy states, this reduction in charge increases and $Z_{\text{eff}} \to 1$.

X-Ray Spectra

X-rays are emitted when a metal target is bombarded by high-energy electrons or any other charged particle. The x-ray spectrum typically consists of a broad continuous band and a series of sharp lines that depend on the type of material used for the target, as shown in Figure 42.16. These lines are known as **characteristic lines**, as they are characteristic of the target material. They were discovered in 1908, but their origin remained unexplained until the details of atomic structure, particularly the shell structure of the atom, were developed.

The first step in the production of characteristic x-rays occurs when a bombarding electron that collides with an electron in an inner shell of a target atom has sufficient energy to remove the electron from the atom. The vacancy created in the shell is filled when an electron from a higher level drops down into the vacancy. The time it takes for this to happen is a function of Z and the level of the vacancy, but is typically less than 10^{-9} s. This transition is accompanied by the emission of a photon whose energy equals the difference in energy between the two levels. Typically, the energy of such transitions is greater than 10 000 eV, and the emitted x-ray photons have wavelengths in the range of 0.001 nm to 0.1 nm.

Let us assume that the incoming electron has dislodged an atomic electron from the innermost shell, the K shell. If the vacancy is filled by an electron dropping from the next higher shell, the L shell, the photon emitted in the process has an energy corresponding to the K_α line on the curve of Figure 42.16. If the vacancy is filled by an electron dropping from the M shell, the line produced is called the K_β line.

Other characteristic x-ray lines are formed when electrons drop from upper levels to vacancies other than those in the K shell. For example, L lines are produced when vacancies in the L shell are filled by electrons dropping from higher shells. An L_α line is produced as an electron drops from the M shell to the L shell, and an L_β line is produced by a transition from the N shell to the L shell.

We can estimate the energy of the emitted x-rays as follows. Consider two electrons in the K shell of an atom whose atomic number is Z. Each electron partially shields the other from the charge of the nucleus, Ze, and so each electron is subject to an effective nuclear charge $Z_{\text{eff}} = (Z - 1)e$. We can now use Equation 42.19 to estimate the energy of either electron in the K shell (with $n = 1$). This gives

$$E_K = -(Z - 1)^2(13.6\text{ eV}) \qquad (42.20)$$

As we show in the following example, the energy of an electron in an L or M shell can be estimated in a similar fashion. Taking the energy difference between these two levels, the energy and wavelength of the emitted photon can then be calculated.

In 1914, Henry G. J. Moseley plotted the Z values for a number of elements versus $\sqrt{1/\lambda}$, where λ is the wavelength of the K_α line of each element. He found that the plot is a straight line, as in Figure 42.17. This is consistent with rough

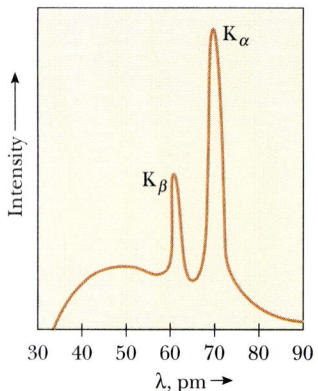

FIGURE 42.16 The x-ray spectrum of a metal target consists of a broad continuous spectrum plus a number of sharp lines, which are due to *characteristic x-rays*. The data shown were obtained when 35-keV electrons bombarded a molybdenum target. Note that 1 pm = 10^{-12} m = 10^{-3} nm.

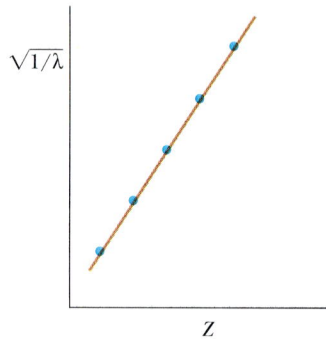

FIGURE 42.17 A Moseley plot is a plot of $\sqrt{1/\lambda}$, versus Z where λ is the wavelength of the K_α x-ray lines of the elements.

calculations of the energy levels given by Equation 42.20. From this plot, Moseley was able to determine the Z values of some missing elements, which provided a periodic chart in excellent agreement with the known chemical properties of the elements.

EXAMPLE 42.7   Estimating the Energy of an X-Ray

Estimate the energy of the characteristic x-ray emitted from a tungsten target when an electron drops from an M shell ($n = 3$ state) to a vacancy in the K shell ($n = 1$ state).

Solution   The atomic number for tungsten is $Z = 74$. Using Equation 42.20, we see that the energy of the electron in the K shell state is approximately

$$E_K = -(74 - 1)^2(13.6 \text{ eV}) = -72\,500 \text{ eV}$$

The electron in the M shell is subject to an effective nuclear charge that depends on the number of electrons in the $n = 1$ and $n = 2$ states, which shield M electron from the nucleus. Because there are eight electrons in the $n = 2$ state and one in the $n = 1$ state, roughly nine electrons shield M electron from the nucleus, and so $Z_{\text{eff}} = Z - 9$. Hence, the energy of an electron in the M shell, following Equation

42.19, is

$$E_M = -Z_{\text{eff}}^2 E_3 = -(Z - 9)^2 \frac{E_0}{3^2}$$

$$= -(74 - 9)^2 \frac{(13.6 \text{ eV})}{9} = -6380 \text{ eV}$$

where E_3 is the energy of an electron in the M shell of the hydrogen atom and E_0 is the ground state energy. Therefore, the emitted x-ray has an energy equal to $E_M - E_K = -6380$ eV $- (-72\,500 \text{ eV}) = 66\,100$ eV. Note that this energy difference is also equal to $hf = hc/\lambda$, where λ is the wavelength of the emitted x-ray.

Exercise   Calculate the wavelength of the emitted x-ray for this transition.

Answer   0.0188 nm.

42.8   ATOMIC TRANSITIONS

We have seen that an atom absorbs and emits radiation only at frequencies that correspond to the energy separation between the allowed states. Let us now look at the details of these processes. Consider an atom with many allowed energy states, labeled $E_1, E_2, E_3, \ldots$, as in Figure 42.18. When light is incident on the atom, only those photons whose energy hf matches the energy separation ΔE between two levels can be absorbed by the atom. Figure 42.19 is a schematic diagram

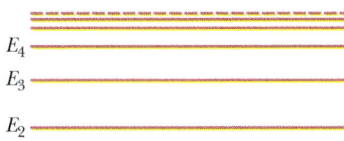

E_4 ——————————————

E_3 ——————————————

E_2 ——————————————

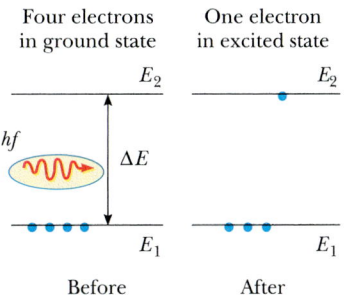

Four electrons in ground state ── One electron in excited state

hf ΔE

E_2 E_2

E_1 E_1

Before   After

FIGURE 42.19   Stimulated absorption of a photon. The dots represent electrons. One electron is transferred from the ground state to the excited state when the atom absorbs a photon whose energy $hf = E_2 - E_1$.

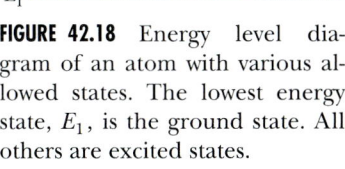

E_1 ——————————————

FIGURE 42.18   Energy level diagram of an atom with various allowed states. The lowest energy state, E_1, is the ground state. All others are excited states.

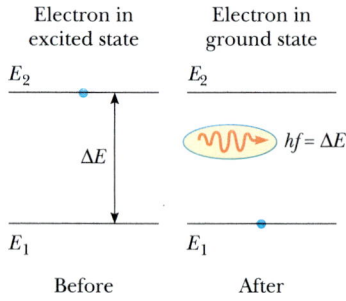

Electron in excited state   Electron in ground state

E_2 ——————•—————— E_2 ——————————————

ΔE $\left(\text{$\sim\!\!\sim\!\!\sim$}\right)$ $hf = \Delta E$

E_1 —————————————— E_1 ——————•——————

Before   After

FIGURE 42.20   Spontaneous emission of a photon by an atom initially in the excited state E_2. When the electron falls to the ground state, the atom emits a photon whose energy $hf = E_2 - E_1$.

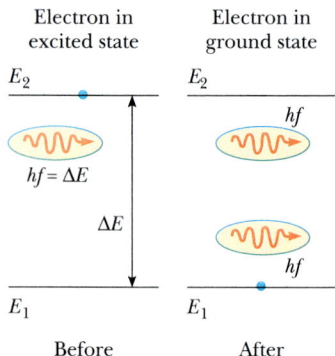

Electron in excited state   Electron in ground state

E_2 ——————•—————— E_2 ——————————————
                                              hf
$\left(\text{$\sim\!\!\sim\!\!\sim$}\right)$ $\left(\text{$\sim\!\!\sim\!\!\sim$}\right)$
$hf = \Delta E$

                    ΔE
                                              $\left(\text{$\sim\!\!\sim\!\!\sim$}\right)$
                                                    hf
E_1 —————————————— E_1 ——————•——————

Before   After

FIGURE 42.21   Stimulated emission of a photon by an incoming photon of energy hf. Initially, the atom is in the excited state. The incoming photon stimulates the atom to emit a second photon of energy $hf = E_2 - E_1$.

representing this **stimulated absorption.** At ordinary temperatures, most of the atoms in a sample are in the ground state. If a vessel containing many atoms of a gaseous element is illuminated with a light beam containing all possible photon frequencies (that is, a continuous spectrum), only those photons of energies $E_2 - E_1$, $E_3 - E_1$, $E_4 - E_1$, $E_3 - E_2$, $E_4 - E_2$, and so on, are absorbed by the atoms. As a result of this absorption, some of the atoms in the vessel are raised to allowed higher energy levels called **excited states.**

Once an atom is in an excited state, there is a certain probability that the excited electron will jump back to a lower level and emit a photon in the process, as in Figure 42.20. This process is known as **spontaneous emission.** Typically, an atom remains in an excited state for only about 10^{-8} s.

Finally, there is a third process, **stimulated emission,** that is of importance in lasers. Suppose an electron in an atom is in an excited state E_2, as in Figure 42.21, and a photon of energy $hf = E_2 - E_1$ is incident on it. The excited state must be a metastable state having a lifetime much longer than 10^{-8} s. The incoming photon increases the probability that the electron will return to the ground state and thereby emit a second photon having energy hf. Note that the incident photon is not absorbed, so after the stimulated emission, there are two nearly identical photons—the incident photon and the emitted photon; the emitted photon is in phase with the incident photon. These photons can, in turn, stimulate other atoms to emit photons in a chain of similar processes. The many photons produced in such a stimulated emission are the source of the intense, coherent light in a laser.

*42.9   LASERS AND HOLOGRAPHY

We have described how an incident photon can cause atomic transitions either upward (stimulated absorption) or downward (stimulated emission). The two processes are equally probable. When light is incident on a system of atoms, there is usually a net absorption of energy because, when the system is in thermal equilibrium, there are many more atoms in the ground state than in excited states. However, if the situation can be inverted so that there are more atoms in an excited state than in the ground state, a net emission of photons can result. Such a condition is called **population inversion.**

This, in fact, is the fundamental principle involved in the operation of a **laser,** an acronym for *light amplification by stimulated emission of radiation.* The amplification corresponds to a buildup of photons in the system as the result of a chain reaction of events. The following three conditions must be satisfied in order to achieve laser action:

- The system must be in a state of population inversion (more atoms in an excited state than in the ground state).
- The excited state of the system must be a *metastable state,* which means its lifetime must be long compared with the usually short lifetimes of excited states. When such is the case, stimulated emission occurs before spontaneous emission.
- The emitted photons must be confined in the system long enough to allow them to stimulate further emission from other excited atoms. This is achieved by using reflecting mirrors at the ends of the system. One end is made totally reflecting, and the other is slightly transparent to allow the laser beam to escape (Fig. 42.22a).

One device that exhibits stimulated emission of radiation is the helium-neon gas laser. Figure 42.23 is an energy-level diagram for the neon atom in this system.

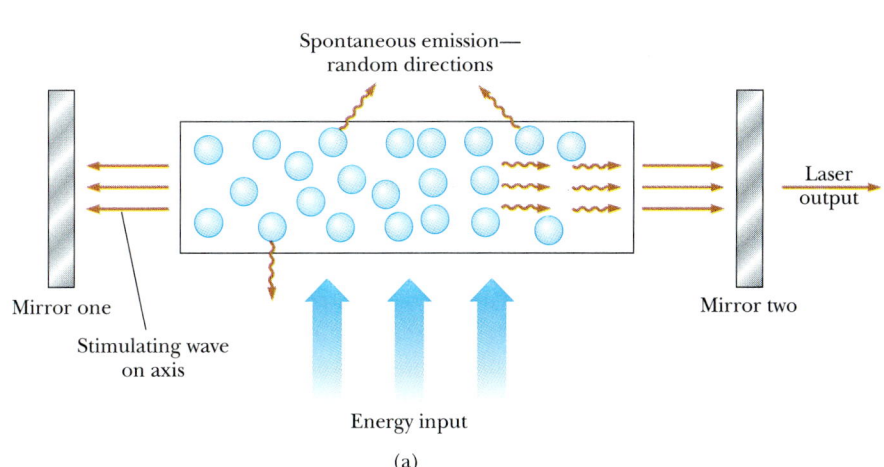

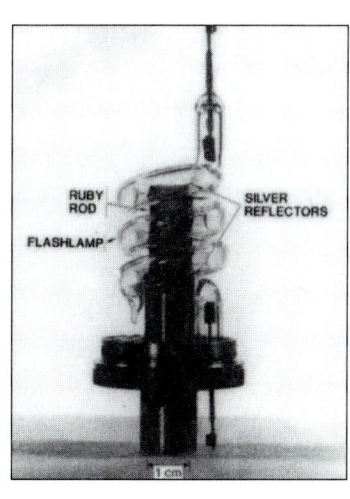

(a)

(b)

FIGURE 42.22 A schematic diagram of a laser design. The tube contains atoms, which represent the active medium. An external source of energy (e.g., optical, electrical) is needed to "pump" the atoms to excited energy states. The parallel end mirrors provide the feedback of the stimulating wave. (b) Photograph of the first ruby laser showing the flash lamp surrounding the ruby rod.   *(Courtesy of Hughes Aircraft Company)*

The mixture of helium and neon is confined to a glass tube sealed at the ends by mirrors. An oscillator connected to the tube causes electrons to sweep through it, colliding with the atoms of the gas and raising them into excited states. Neon atoms are excited to state E_3^* through this process (the asterisk * indicates that this is a metastable state) and also as a result of collisions with excited helium atoms. Stimulated emission occurs as the neon atoms make a transition to state E_2 and neighboring excited atoms are stimulated. This results in the production of coherent light at a wavelength of 632.8 nm.

Applications

Since the development of the first laser in 1960, there has been a tremendous growth in laser technology. Lasers are now available that cover wavelengths in the infrared, visible, and ultraviolet regions. Applications include surgical "welding" of detached retinas, precision surveying and length measurement, precision cutting of metals and other materials, and telephone communication along optical fibers. These and other applications are possible because of the unique characteristics of laser light. In addition to being highly monochromatic, it is also highly directional and can be sharply focused to produce regions of extremely intense light energy (with energy densities approaching those inside the laser tube).

Lasers are used in precision long-range distance measurement (range finding). It has become important, for astronomical and geophysical purposes, to measure as precisely as possible the distance from various points on the surface of the Earth to a point on the Moon's surface. To facilitate this, the Apollo astronauts set up a compact array, a 0.5-m square of reflector prisms on the Moon, which allows laser pulses directed from an Earth station to be retroreflected to the same station. Using the known speed of light and the measured round-trip travel time of a 1-ns pulse, one can determine the Earth-Moon distance, 380 000 km, to a precision of better than 10 cm.

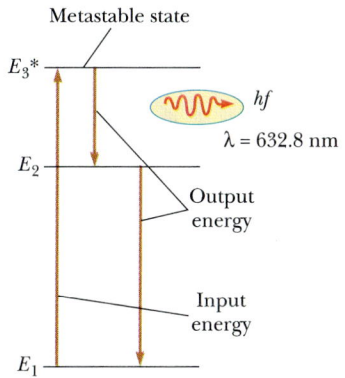

FIGURE 42.23 Energy-level diagram for the neon atom, which emits 632.8-nm photons through stimulated emission. These photons arise from the transition $E_3^* - E_2$. This is the source of coherent light in the helium-neon gas laser.

Such information would be useful, for example, in making more reliable earthquake predictions and for learning more about the motions of the Earth-Moon system. This technique requires a high-power pulsed laser for its success, since a sufficient burst of photons must return to a collecting telescope on Earth and be detected. Variants of this method are also used to measure the distance to inaccessible points on the Earth as well as in military range finders.

Novel medical applications utilize the fact that the different laser wavelengths can be absorbed in specific biological tissues. A widespread eye condition, glaucoma, is manifested by a high fluid pressure in the eye, which can lead to destruction of the optic nerve. A simple laser operation (iridectomy) can "burn" open a tiny hole in a clogged membrane, relieving the destructive pressure. Along the same lines, a serious side effect of diabetes is neovascularization, the formation of weak blood vessels, which often leak blood into extremities. When this occurs in the eye, vision deteriorates (diabetic retinopathy), leading to blindness in diabetic patients. It is now possible to direct the green light from the argon ion laser through the clear eye lens and eye fluid, focus on the retina edges, and photocoagulate the leaky vessels. These procedures have greatly reduced blindness in glaucoma and diabetes patients.

Laser surgery is now a practical reality. Infrared light at 10 μm from a carbon dioxide laser can cut through muscle tissue, primarily by heating and evaporating the water contained in cellular material. Laser power of about 100 W is required in this technique. The advantage of the "laser knife" over conventional methods is that laser radiation cuts and coagulates at the same time, leading to substantial reduction of blood loss. In addition, the technique virtually eliminates cell migration, which is very important in tumor removal. Furthermore, a laser beam can be

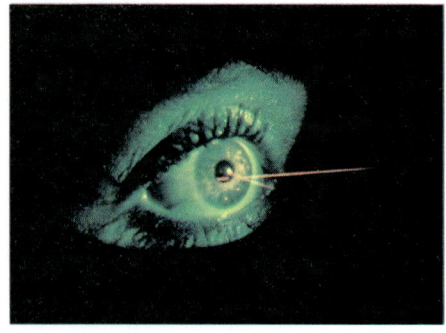

(Left) Scientists checking the performance of an experimental laser-cutting device mounted on a robot arm. The laser is being used to cut through a metal plate.   *(Philippe Plailly/SPL/ Photo Researchers, Inc.) (Right)* An argon laser passing through a cornea and lens during eye surgery. *(© Alexander Tsiaras/Science Source/Photo Researchers, Inc.)*

trapped in fine glass-fiber light-guides (endoscopes) by means of total internal reflection. The light fibers can be introduced through natural orifices, conducted around internal organs, and directed to specific interior body locations, eliminating the need for massive surgery. For example, bleeding in the gastrointestinal tract can be optically cauterized by fiber-optic endoscopes inserted through the mouth.

In biological and medical research, it is often important to isolate and collect unusual cells for study and growth. A laser cell separator exploits the fact that specific cells can be tagged with fluorescent dyes. All cells are then dropped from a tiny charged nozzle and laser-scanned for the dye tag. If triggered by the correct light-emitting tag, a small voltage applied to parallel plates deflects the falling electrically charged cell into a collection breaker. This is an efficient method for extracting the proverbial needles from the haystack.

One of the most unusual and interesting applications of the laser is in the production of three-dimensional images in a process called **holography.** Figure 42.24a shows how a hologram is made. Light from the laser is split into two parts by a half-silvered mirror at *B*. One part of the beam reflects off the object to be photographed and strikes an ordinary photographic film. The other half of the beam is diverged by lens L_2, reflects from mirrors M_1 and M_2, and finally strikes the film. The two beams overlap to form an extremely complicated interference pattern on the film. Such an interference pattern can be produced only if the phase relationship of the two waves is constant throughout the exposure of the film. This condition is met by illuminating the scene with light through a pinhole or with coherent laser radiation. The hologram records not only the intensity of the light scattered from the object (as in a conventional photograph) but also the phase difference between the reference beam and the beam scattered from the object. Because of this phase difference, an interference pattern is formed that produces an image with full three-dimensional perspective.

A hologram is best viewed by allowing coherent light to pass through the developed film as one looks back along the direction from which the beam comes. Figure 42.24b is a photograph of a hologram made using a cylindrical film. One sees a three-dimensional image of the object such that as the viewer's head is moved, the perspective changes, as for an actual object.

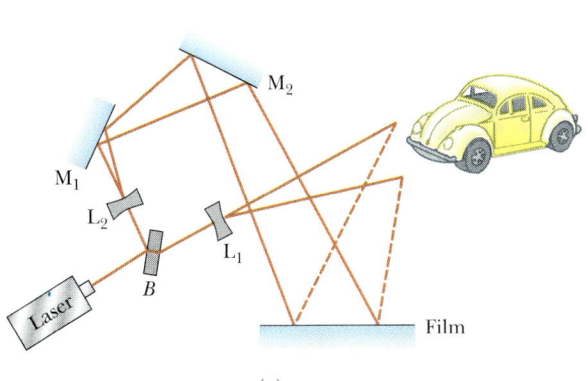

(a)

(b)

FIGURE 42.24   (a) Experimental arrangement for producing a hologram. (b) Photograph of a hologram that uses a cylindrical film.   *(Courtesy of Central Scientific Co.)*

SUMMARY

Quantum mechanics can be applied to the hydrogen atom using the potential energy function $U(r) = -k_e e^2/r$ in the Schrödinger equation. The solution to this equation yields the wave functions for the allowed states and the allowed energies:

$$E_n = -\left(\frac{k_e e^2}{2a_0}\right)\frac{1}{n^2} = -\frac{13.6}{n^2} \text{ eV} \qquad n = 1, 2, 3, \ldots \qquad (42.2)$$

The allowed energy depends only on the **principal quantum number** n. The allowed wave functions depend on three quantum numbers: n, ℓ, and m_ℓ, where ℓ is the **orbital quantum number** and m_ℓ is the **orbital magnetic quantum number**. The restrictions on the quantum numbers are

$$n = 1, 2, 3, \ldots$$

$$\ell = 0, 1, 2, \ldots, n-1$$

$$m_\ell = -\ell, -\ell+1, \ldots, \ell-1, \ell$$

All states having the same principal quantum number n form a **shell**, identified by the letters K, L, M, ... (corresponding to $n = 1, 2, 3, \ldots$). All states having the same values of n and ℓ form a **subshell**, designated by the letters, s, p, d, f, ... (corresponding to $\ell = 0, 1, 2, 3, \ldots$).

In order to completely describe a quantum state, it is necessary to include a fourth quantum number, m_s, called the **spin magnetic quantum number**. This quantum number can have only two values, $\pm\frac{1}{2}$.

An atom in a state characterized by a specific n can have the following values of *orbital angular momentum L*:

$$L = \sqrt{\ell(\ell+1)}\,\hbar \qquad (42.10)$$

where ℓ is restricted to the values $\ell = 0, 1, 2, \ldots, n-1$. The allowed values of the projection of L along the z axis are

$$L_z = m_\ell \hbar \qquad (42.11)$$

where m_ℓ is restricted to integer values lying between $-\ell$ and ℓ. Only discrete values of L_z are allowed, and these are determined by the restrictions on m_ℓ. This quantization of L_z is referred to as **space quantization**.

The electron has an intrinsic angular momentum called the **spin angular momentum**. That is, the total angular momentum of an electron in an atom can have two contributions, one arising from the spin of the electron (**S**) and one arising from the orbital motion of the electron (**L**). Electronic spin can be described by a single quantum number $s = \frac{1}{2}$. The **magnitude of the spin angular momentum** is

$$S = \frac{\sqrt{3}}{2}\hbar \qquad (42.13)$$

and the z component of **S** is

$$S_z = m_s\hbar = \pm\tfrac{1}{2}\hbar \qquad (42.14)$$

That is, the spin angular momentum is also quantized in space, as specified by the **spin magnetic quantum number** $m_s = \pm\frac{1}{2}$.

The magnetic moment $\boldsymbol{\mu}_s$ associated with the spin angular momentum of an electron is

$$\boldsymbol{\mu}_s = -\frac{e}{m}\mathbf{S} \qquad (42.15)$$

which is twice as large as the orbital magnetic moment. The z component of μ_s can have the values

$$\mu_{sz} = \pm\frac{e\hbar}{2m} \qquad (42.16)$$

The **exclusion principle** states that *no two electrons in an atom can ever be in the same quantum state.* In other words, no two electrons can have the same set of quantum numbers n, ℓ, m_ℓ, and m_s. Using this principle and the principle of minimum energy, the electronic configuration of the elements can be determined. This serves as a basis for understanding atomic structure and the chemical properties of the elements.

The allowed electronic transitions between any two levels in an atom are governed by the selection rules

$$\Delta\ell = \pm 1 \qquad \text{and} \qquad \Delta m_\ell = 0, \pm 1 \qquad (42.17)$$

X-rays are emitted by atoms when an electron undergoes a transition from an outer shell into a vacancy in an inner shell. Transitions into a vacant state in the K shell give rise to the K series of spectral lines; transitions into a vacant state in the L shell create the L series of lines, and so on. The x-ray spectrum of a metal target consists of a set of sharp characteristic lines superimposed on a broad, continuous spectrum.

QUESTIONS

1. Why are three quantum numbers needed to describe the state of a one-electron atom (neglecting spin)?
2. Compare the Bohr theory and the Schrödinger treatment of the hydrogen atom. Comment on the total energy and orbital angular momentum.
3. Why is the direction of the orbital angular momentum of an electron opposite that of its magnetic moment?
4. Why is an inhomogeneous magnetic field used in the Stern-Gerlach experiment?
5. Could the Stern-Gerlach experiment be performed with ions rather than neutral atoms? Explain.
6. Describe some experiments that support the conclusion that the spin quantum number for electrons can only have the values $\pm\frac{1}{2}$.
7. Discuss some of the consequences of the exclusion principle.
8. Why do lithium, potassium, and sodium exhibit similar chemical properties?
9. Explain why a photon must have a spin of 1.
10. An energy of approximately 21 eV is required to excite an electron in a helium atom from the $1s$ state to the $2s$ state. The same transition for the He$^+$ ion requires about twice as much energy. Explain.

11. Does the intensity of light from a laser fall off as $1/r^2$?
12. The absorption or emission spectrum of a gas consists of lines that broaden as the density of gas molecules increases. Why do you suppose this occurs?
13. How is it possible that electrons, which have a probability distribution around a nucleus, can exist in states of definite energy (e.g., $1s$, $2p$, $3d$, . . .)?
14. It is easy to understand how two electrons (one spin up, one spin down) can fill the $1s$ shell for a helium atom. How is it possible that eight more electrons can fit into the $2s$, $2p$ level to complete the $1s^2 2s^2 2p^6$ shell for a neon atom?
15. In 1914, Henry Moseley was able to determine the atomic number of an element from its characteristic x-ray spectrum. How was this possible? (*Hint:* See Figs. 42.16 and 42.17.)
16. What are the advantages of using monochromatic light to view a holographic image?
17. Why is stimulated emission so important in the operation of a laser?

PROBLEMS

Section 42.2 The Hydrogen Atom Revisited

1. In the Rutherford scattering experiment, 4.00 MeV alpha particles (^{4}He nuclei containing 2 protons and 2 neutrons) scatter off gold nuclei (containing 79 protons and 118 neutrons). If an alpha particle makes a direct head-on collision with the gold nucleus and scatters backward at 180°, determine (a) the distance of closest approach of the alpha particle to the gold nucleus and (b) the maximum force exerted on the alpha particle. Assume that the gold nucleus remains fixed throughout the entire process.

1A. In the Rutherford scattering experiment, alpha particles of energy E (^{4}He nuclei containing 2 protons and 2 neutrons) scatter off a target whose atoms have an atomic number Z. If an alpha particle makes a direct head-on collision with a target nucleus and scatters backward at 180°, determine (a) the distance of closest approach of the alpha particle to the target nucleus and (b) the maximum force exerted on the alpha particle. Assume that the target nucleus remains fixed throughout the entire process.

2. A photon is barely capable of causing a photoelectric effect when it strikes a sodium plate having a work function of 2.28 eV. Find (a) the minimum n for a hydrogen atom that can be ionized by such a photon and (b) the speed of the released electron far from the nucleus.

3. A general expression for the energy levels of one-electron atoms and ions is

$$E_n = -\left(\frac{\mu k_e^2 q_1^2 q_2^2}{2\hbar^2}\right)\frac{1}{n^2}$$

where k_e is the Coulomb constant, q_1 and q_2 are the charges of the two particles, and μ is the reduced mass, given by $\mu = m_1 m_2 / (m_1 + m_2)$. In Example 40.7, we found that the wavelength for the $n = 3$ to $n = 2$ transition of the hydrogen atom is 656.3 nm (visible red light). What are the wavelengths for this same transition in (a) positronium, which consists of an electron and a positron, and (b) singly ionized helium? (*Note:* A positron is a positively charged electron.)

4. The energy of an electron in a hydrogen atom is

$$E = \frac{p^2}{2m_e} - \frac{k_e e^2}{r}$$

According to the uncertainty principle, if the electron is localized within a distance r of the nucleus, its momentum p must be at least $\hbar r$. Use this principle

to find the minimum values of r and E. Compare the results to those of Bohr (Equations 40.24 and 40.26).

Section 42.3 The Spin Magnetic Quantum Number and
Section 42.4 The Wave Functions for Hydrogen

5. During a particular period of time, an electron in the ground state of a hydrogen atom is "observed" 1000 times at a distance $a_0/2$ from the nucleus. How many times is this electron observed at a distance $2a_0$ from the nucleus during this period of "observation"?

6. Plot the wave function $\psi_{1s}(r)$ (Eq. 42.3) and the radial probability density function $P_{1s}(r)$ (Eq. 42.7) for hydrogen. Let r range from 0 to $1.5a_0$, where a_0 is the Bohr radius.

7. The wave function for an electron in the $2p$ state of hydrogen is

$$\psi_{2p} = \frac{1}{\sqrt{3}(2a_0)^{3/2}}\frac{r}{a_0}e^{-r/2a_0}$$

What is the most likely distance from the nucleus to find an electron in the $2p$ state? (See Fig. 42.8.)

8. Show that the $1s$ wave function for an electron in hydrogen

$$\psi(r) = \frac{1}{\sqrt{\pi a_0^3}}e^{-r/a_0}$$

satisfies the radially symmetric Schrödinger equation,

$$-\frac{\hbar^2}{2m}\left(\frac{d^2\psi}{dr^2} + \frac{2}{r}\frac{d\psi}{dr}\right) - \frac{k_e e^2}{r}\psi = E\psi$$

9. If a muon (a negatively charged particle having a mass 206 times the electron's mass) is captured by a lead nucleus, $Z = 82$, the resulting system behaves like a one-electron atom. (a) What is the "Bohr radius" for a muon captured by a lead nucleus? (*Hint:* Use Equation 42.4.) (b) Using Equation 42.2 with e replaced by Ze, calculate the ground-state energy of a muon captured by a lead nucleus. (c) What is the transition energy for a muon descending from the $n = 2$ to the $n = 1$ level in a muonic lead atom?

Section 42.5 The "Other" Quantum Numbers

10. Calculate the angular momentum for an electron in (a) the $4d$ state and (b) the $6f$ state.

11. A hydrogen atom is in its 5th excited state. The atom emits a 1090-nm wavelength photon. Determine the maximum possible orbital angular momentum of the electron after emission.

□ indicates problems that have full solutions available in the Student Solutions Manual and Study Guide.

12. List the possible sets of quantum numbers for electrons in (a) the $3d$ subshell and (b) the $3p$ subshell.

13. How many sets of quantum numbers are possible for an electron for which (a) $n = 1$, (b) $n = 2$, (c) $n = 3$, (d) $n = 4$, and (e) $n = 5$? Check your results to show that they agree with the general rule that the number of sets of quantum numbers is equal to $2n^2$.

14. (a) Write out the electronic configuration for oxygen ($Z = 8$). (b) Write out the values for the set of quantum numbers n, ℓ, m_ℓ, and m_s for each electron in oxygen.

15. The ρ-meson has a charge of $-e$, a spin quantum number of 1, and a mass of 1507 times that of the electron. If the electrons in atoms were replaced by ρ-mesons, list the possible sets of quantum numbers for ρ-mesons in the $3d$ subshell.

16. Consider an atom whose M shell is completely filled. (a) Identify the atom. (b) List the number of electrons in each subshell.

17. An electron is in the N shell. Determine the maximum value of the z component of its angular momentum.

18. Determine the number of electrons that can occupy the $n = 3$ shell.

19. Find all possible values of L, L_z, and θ for an electron in a $3d$ state of hydrogen.

20. All objects, large and small, behave quantum-mechanically. (a) Estimate the quantum number ℓ for the Earth in its orbit around the Sun. (b) What energy change (in joules) would occur if the Earth made a transition to an adjacent allowed state?

21. The z component of the electron's spin magnetic moment is given by the Bohr magneton, $\mu_B = e\hbar/2m$. Show that the Bohr magneton has the numerical value of 9.27×10^{-24} J/T $= 5.79 \times 10^{-5}$ eV/T.

22. Consider an electron as a classical sphere, having the same mass density as that of a proton having a radius of 1.00×10^{-15}m. If the spin angular momentum of the electron is caused by rotation around its axis, (a) determine the speed of a point on the surface of the electron and (b) compare this speed to the speed of light. (*Hint:* $L = I\omega = \hbar/2$.)

Section 42.6 The Exclusion Principle and The Periodic Table

23. Which electronic configuration has a lower energy: $[\text{Ar}]3d^44s^2$ or $[\text{Ar}]3d^54s^1$? Identify this element and discuss Hund's rule in this case.

24. When the element 110 is discovered, what would be its probable electronic configuration?

25. Devise a table similar to that shown in Figure 42.13 for atoms containing 11 through 19 electrons. Use Hund's rule and educated guesswork.

26. (a) Scanning through Table 42.4 in order of increasing atomic number, note that the electrons fill the subshells in such a way that those subshells with the lowest values of $n + \ell$ are filled first. If two subshells have the same value of $n + \ell$, the one with the lower value of n is filled first. Using these two rules, write the order in which the subshells are filled through $n + \ell = 7$. (b) Predict the chemical valence for the elements that have atomic numbers 15, 47, and 86 and compare your predictions with the actual valences.

Section 42.7 Atomic Spectra: Visible and X-Ray

27. If you wish to produce 10-nm x-rays in the laboratory, what is the minimum voltage you must use in accelerating the electrons?

27A. If you wish to produce x-rays of wavelength λ in the laboratory, what is the minimum voltage you must use in accelerating the electrons?

28. In x-ray production, electrons are accelerated through a high voltage V and then decelerated by striking a target. Show that the shortest wavelength x-ray that can be produced is

$$\lambda_{\text{min}} = \frac{1240 \text{ nm} \cdot \text{V}}{V}$$

29. The wavelength of characteristic x-rays corresponding to the K_β line is 0.152 nm. Determine the material in the target.

30. The K_α x-ray is the one emitted when an electron undergoes a transition from the L shell ($n = 2$) to the K shell ($n = 1$). Calculate the frequency of the K_α x-ray from a nickel target ($Z = 28$).

31. Electrons are shot into a Bi target and x-rays are emitted. Determine (a) the M to L shell transitional energy for Bi and (b) the wavelength of the x-ray emitted when an electron falls from the M shell into the L shell.

32. Use the method illustrated in Example 42.7 to calculate the wavelength of the x-ray emitted from a molybdenum target ($Z = 42$) when an electron moves from the L shell ($n = 2$) to the K shell ($n = 1$).

Section 42.8 Atomic Transitions

33. The familiar yellow light from a sodium-vapor street lamp results from the $3p \rightarrow 3s$ transition in ^{11}Na. Evaluate the wavelength of this light given that the energy difference $E_{3p} - E_{3s} = 2.1$ eV.

34. The wavelength of coherent ruby laser light is 694.3 nm. What is the energy difference (in electron volts) between the upper, excited state and the lower, unexcited state?

35. A ruby laser delivers a 10-ns pulse of 1.0 MW average power. If the photons have a wavelength of 694.3 nm, how many are contained in the pulse?

35A. A laser delivers a pulse of duration t at an average power P. If the photons have a wavelength λ, how many are contained in the pulse?

ADDITIONAL PROBLEMS

36. Figure P42.36 shows the energy-level diagrams of He and Ne. An electrical voltage excites the He atom from its ground state to its excited state of 20.61 eV. The excited He atom collides with a Ne atom in its ground state and excites this atom to the state at 20.66 eV. Lasing action takes place for electron transitions from E_4 to E_3 in the Ne atoms. Show that the wavelength of this red He-Ne laser is approximately 633 nm.

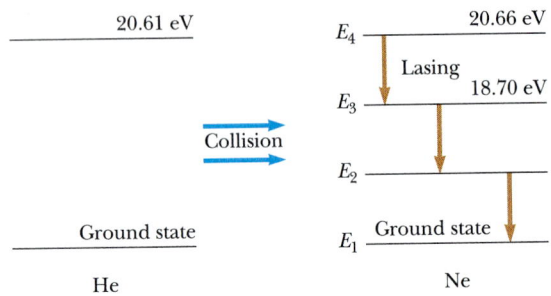

FIGURE P42.36

37. In the technique known as electron spin resonance (ESR), a sample containing unpaired electrons is placed in a magnetic field. Consider the simplest situation, that in which there is only one electron and therefore only two possible energy states, corresponding to $m_s = \pm \frac{1}{2}$. In ESR, the absorption of a photon causes the electron's spin magnetic moment to flip from a lower energy state to a higher energy state. (The lower energy state corresponds to the case where the magnetic moment $\boldsymbol{\mu}_s$ is aligned with the magnetic field, and the higher energy state corresponds to the case where $\boldsymbol{\mu}_s$ is aligned against the field.) What is the photon frequency required to excite an ESR transition in a 0.35-T magnetic field?

38. An Nd:YAG laser used in eye surgery emits a 3.0-mJ pulse in 1.0 ns, focused to a spot 30 μm in diameter on the retina. (a) Find (in SI units) the power per unit area at the retina. (This quantity is called the irradiance.) (b) What energy is delivered to an area of molecular size, say a circular area 0.60 nm in diameter?

39. A dimensionless number that often appears in atomic physics is the fine-structure constant $\alpha = k_e e^2/\hbar c$, where k_e is the Coulomb constant. (a) Obtain a numerical value for $1/\alpha$. (b) In scattering experiments, the electron size is taken to be the

classical electron radius, $r_e = k_e e^2/m_e c^2$. In terms of α, what is the ratio of the Compton wavelength (Section 40.4), $\lambda_C = h/m_e c$, to the classical electron radius? (c) In terms of α, what is the ratio of the Bohr radius, a_0, to the Compton wavelength? (d) In terms of α, what is the ratio of the Rydberg wavelength, $1/R_H$, to the Bohr radius (Section 40.6)?

40. Show that the average value of r for the $1s$ state of hydrogen has the value $3a_0/2$. (*Hint:* Use Eq. 42.7.)

41. Suppose a hydrogen atom is in the $2s$ state. Taking $r = a_0$, calculate values for (a) $\psi_{2s}(a_0)$, (b) $|\psi_{2s}(a_0)|^2$, and (c) $P_{2s}(a_0)$. (*Hint:* Use Eq. 42.8.)

42. The carbon dioxide laser is one of the most powerful developed. The energy difference between the two laser levels is 0.117 eV. Determine the frequency and wavelength of the radiation emitted by this laser. In what portion of the electromagnetic spectrum is this radiation?

43. Show that the wave function for an electron in the $2s$ state in hydrogen

$$\psi(r) = \frac{1}{4\sqrt{2\pi}}\left(\frac{1}{a_0}\right)^{3/2}\left(2 - \frac{r}{a_0}\right)e^{-r/2a_0}$$

satisfies the radially symmetric Schrödinger equation given in Problem 8.

44. For the ground state of hydrogen, what is the probability of finding the electron closer to the nucleus than the Bohr radius corresponding to $n = 1$?

45. A pulsed ruby laser emits light at 694.3 nm. For a 14-ps pulse containing 3.0 J of energy, find (a) the physical length of the pulse as it travels through space and (b) the number of photons in it. (c) If the beam has a circular cross-section of 0.60-cm diameter, find the number of photons per cubic millimeter.

45A. A pulsed laser emits light having wavelength λ. For a pulse of duration t having energy E, find (a) the physical length of the pulse as it travels through space and (b) the number of photons in it. (c) If the beam has a circular cross-section having diameter d, find the number of photons per unit volume.

46. The number N of atoms in a particular state is called the population of that state. This number depends on the energy of that state and the temperature. The population of atoms in a state of energy E_n is given by a Boltzmann distribution expression:

$$N = N_0 e^{-E_n/kT}$$

where N_0 is the population of the state as $T \rightarrow \infty$. Find (a) the ratio of populations of the states E_3^* to E_2 for the laser in Figure 42.22, assuming $T = 27°$C, and (b) the ratio of the populations of the two states in a ruby laser that produces a light beam of wavelength 694.3 nm at 4 K.

47. Consider a hydrogen atom in its ground state.
(a) Treating the orbiting electron as an effective current loop of radius a_0, derive an expression for the magnetic field at the nucleus. (*Hint:* Use the Bohr theory of hydrogen and see Example 30.3.)
(b) Find a numerical value for the magnetic field at the nucleus in this situation.

48. When electron clouds overlap, a detailed calculation of the effective charge on the nucleus may be made using quantum mechanics. For the case of the lithium atom, the effective charge on each inner electron is $-0.85e$. Use this to find (a) the effective charge on the nucleus as seen by the outer valence electron and (b) the ionization energy (compare this with 5.4 eV).

49. (a) Calculate the most probable value of r for an electron in the $2s$ state of hydrogen. (*Hint:* Let $x = r/a_0$, find an equation for x, and show that $x = 5.236$ is a solution to this equation.) (b) Show that the wave function given by Equation 42.8 is normalized.

50. All atoms have roughly the same size. (a) To show this, estimate the diameters for aluminum, molar atomic mass = 27 g/mol and density 2.70 g/cm³, and uranium, molar atomic mass = 238 g/mol and density 18.9 g/cm³. (b) What do the results imply about the wave functions for inner-shell electrons as we progress to higher and higher atomic mass atoms? (*Hint:* The molar volume is roughly proportional to $D^3 N_A$, where D is the atomic diameter and N_A is Avogadro's number.)

51. For hydrogen in the $1s$ state, what is the probability of finding the electron farther than $2.50 a_0$ from the nucleus?

52. According to classical physics, an accelerated charge e radiates at a rate

$$\frac{dE}{dt} = -\frac{1}{6\pi\epsilon_0}\frac{e^2 a^2}{c^3}$$

(a) Show that an electron in a classical hydrogen atom (see Fig. 42.3) spirals into the nucleus at a rate

$$\frac{dr}{dt} = -\frac{e^4}{12\pi^2\epsilon_0{}^2 r^2 m^2 c^3}$$

(b) Find the time it takes the electron to reach $r = 0$, starting from $r_0 = 2 \times 10^{-10}$ m.

53. Light from a certain He-Ne laser has a power output of 1.0 mW and a cross-sectional area of 10 mm². The entire beam is incident on a metal target that requires 1.5 eV to remove an electron from its surface. (a) Perform a classical calculation to determine how long it takes one atom in the metal to absorb 1.5 eV from the incident beam. (*Hint:* Assume the surface area of an atom is 1.0×10^{-20} m², and first calculate the energy incident on each atom per second.)

(b) Compare the (wrong) answer obtained in part (a) to the actual response time for photoelectric emission ($\approx 10^{-9}$ s), and discuss the reasons for the large discrepancy.

53A. Light from a laser has a power output P and a cross-sectional area A. The entire beam is incident on a metal target that requires energy ϕ to remove an electron from its surface. Perform a classical calculation to determine how long it takes one atom in the metal to absorb energy ϕ from the incident beam.

54. In interstellar space, atomic hydrogen produces the sharp spectral line called the 21-cm radiation, which astronomers find most helpful in detecting clouds of hydrogen between stars. This radiation is useful because interstellar dust that obscures visible wavelengths is transparent to these radio wavelengths. The radiation is not generated by an electron transition between energy states characterized by n. Instead, in the ground state ($n = 1$), the electron and proton spins may be parallel or antiparallel, with a resultant slight difference in these energy states. (a) Which condition has the higher energy? (b) The line is actually at 21.11 cm. What is the energy difference between the states? (c) The average lifetime in the excited state is about 10^7 y. Calculate the associated uncertainty in energy of this excited energy level.

Molecules and Solids

An artist's rendition of "buckyballs," short for the molecule buckminsterfullerene. These nearly spherical molecular structures that look like soccer balls were named for R. Buckminster Fuller, inventor of the geodesic dome. This new form of carbon, C_{60}, was discovered by astrophysicists while investigating the carbon gas that exists between stars. Scientists are actively studying the properties and potential uses of buckminsterfullerene and related molecules. For example, samples containing C_{60} doped with rubidium exhibit superconductivity at approximately 30 K.

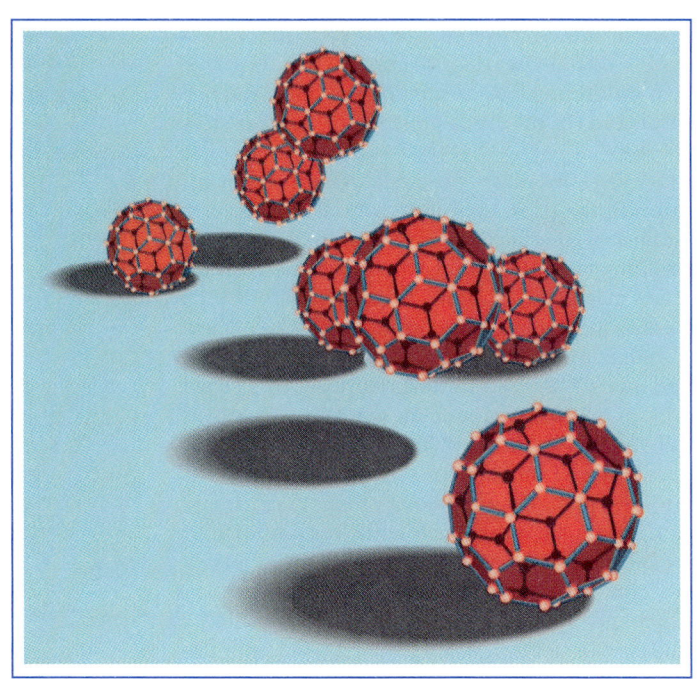

The beautiful symmetry and regularity of crystalline solids have both stimulated and allowed rapid progress to be made in the field of solid state physics in the 20th century. Although rapid theoretical progress has occurred with the most regular atomic arrangements (crystalline solids) and the most random atomic arrangements (gases), much less has been done with liquids and amorphous (irregular) solids until quite recently. The recent interest in low-cost amorphous materials has been driven by their use in manufacturing solar cells, memory elements, and fiber optic waveguides.

In this chapter, we study the aggregate of atoms known as molecules. First we describe the bonding mechanisms in molecules, the various modes of molecular excitation, and the radiation emitted or absorbed by molecules. We then take the next logical step and show how molecules combine to form solids. Then, by examining their electronic distributions, we explain the differences between insulating, metallic, and semiconducting crystals. The chapter concludes with discussions of semiconducting junctions and, the operation of several semiconductor devices.

43.1 MOLECULAR BONDS

Two atoms combine to form a molecule because of a net attractive force that exists between them whenever their separation distance is greater than their equilibrium separation distance in the molecule. Furthermore, the energy of the stable bound molecule is less than the total energy of the separated atoms.

Fundamentally, the bonding mechanisms in a molecule are primarily due to electrostatic forces between atoms (or ions). When two atoms are separated by an infinite distance, the force between them is zero, as is the electrostatic potential energy of the system they constitute. As the atoms are brought closer together, both attractive and repulsive forces act. At very large separations, the dominant forces are attractive. For small separations, repulsive forces between like charges begin to dominate.

The potential energy of a system of atoms can be positive or negative, depending on the separation between atoms. The total potential energy of the system in the absence of chemical bonding can be approximated by the expression

$$U = -\frac{A}{r^n} + \frac{B}{r^m} \tag{43.1}$$

where r is the internuclear separation, A and B are parameters associated with the attractive and repulsive forces, and n and m are small integers. Figure 43.1 sketches the total potential energy versus internuclear separation. The potential energy for large separations is negative, corresponding to a net attractive force. At the equilibrium separation, the attractive and repulsive forces just balance and the potential energy has its minimum value.

A complete description of the binding mechanisms in molecules is highly complex because it involves the mutual interactions of many particles. In this section, therefore, we discuss only some simplified models in the following order: ionic bond, covalent bond, van der Waals bond, and hydrogen bond.

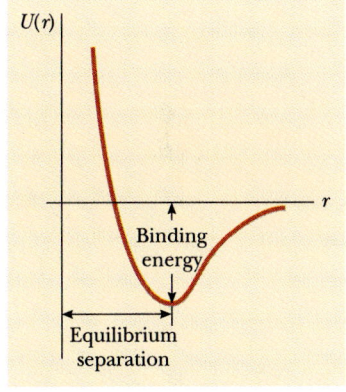

FIGURE 43.1 Total potential energy as a function of internuclear separation for a system of two atoms.

Ionic Bonds

Ionic bonds are fundamentally due to the Coulomb attraction between oppositely charged ions. A familiar example of an ionically bonded molecule is sodium chloride, NaCl, which is common table salt. Sodium, which has an electronic configuration $1s^2 2s^2 2p^6 3s$, is relatively easy to ionize, giving up its $3s$ electron to form a Na^+ ion. The energy required to ionize the atom to form Na^+ is 5.1 eV. Chlorine, which has an electronic configuration $1s^2 2s^2 2p^5$, is one electron short of the filled-shell structure of argon. Because filled-shell configurations are energetically more favorable than unfilled-shell configurations, the Cl^- ion is more stable than the neutral Cl atom. The energy released when an atom takes on an electron is called the **electron affinity**. For chlorine, the electron affinity is 3.6 eV. Therefore, an energy equal to $5.1 - 3.6 = 1.5$ eV must be provided to neutral Na and Cl atoms to change them to Na^+ and Cl^- ions at infinite separation.

Total energy versus internuclear separation is shown in Figure 43.2 for NaCl. The total energy of the molecule has a minimum value of -4.2 eV at the equilibrium separation, which is about 0.24 nm. The energy required to separate the NaCl molecule into neutral sodium and chlorine atoms, called the **dissociation energy,** is equal to 4.2 eV.

When the two ions are brought to within 0.24 nm from each other, the electrons in the filled shells begin to overlap, which results in a repulsion between the ions. When the ions are far apart, the interaction may be approximated as an

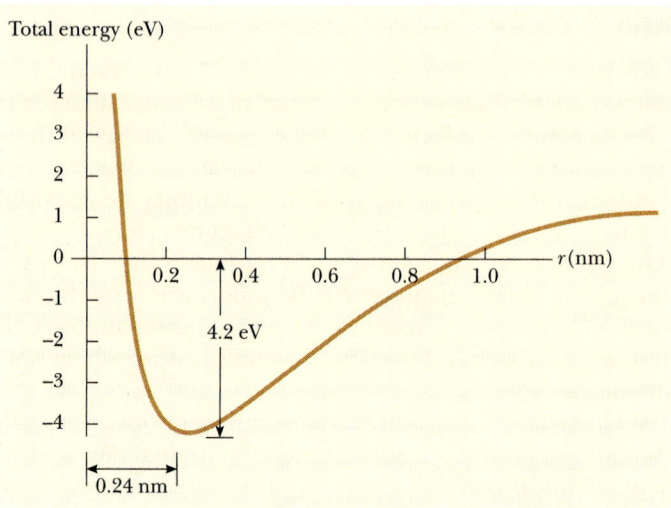

FIGURE 43.2   Total energy versus the internuclear separation for the NaCl molecule. Note that the dissociation energy is 4.2 eV.

attraction of two point charges. As they approach each other, the electron-electron repulsions, especially of core electrons, begin to dominate.

Covalent Bonds

A **covalent bond** between two atoms in a molecule can be visualized as the sharing of electrons supplied by one or both atoms. Many diatomic molecules, such as H_2, F_2, and CO owe their stability to covalent bonds. In the case of the H_2 molecule, the two electrons are equally shared between the nuclei and occupy what is called a *molecular orbital.* The electron density is large in the region between the two nuclei, so that the electrons act as the glue holding the nuclei together. The formation of the molecular orbital from the *s* orbitals of the two hydrogen atoms is shown in Figure 43.3. Because of the exclusion principle, the two electrons in the ground state of H_2 must have anti-parallel spins. Also because of the exclusion principle, if a third H atom is brought near the H_2 molecule, the third electron would have to occupy a higher energy quantum level, which is an energetically unfavorable situation. Hence, the H_3 molecule is not stable and does not form.

Stable molecules more complex than H_2 such as H_2O, CO_2, and CH_4 are also formed by covalent bonds. Consider methane, CH_4, a typical organic molecule shown schematically in the electron-sharing diagram of Figure 43.4a. In this case four covalent bonds are formed between the carbon atom and each of the hydrogen atoms. The spatial electron distribution of the four covalent bonds is shown in Figure 43.4b. The four hydrogen nuclei are at the corners of a regular tetrahedron, with the carbon nucleus at the center.

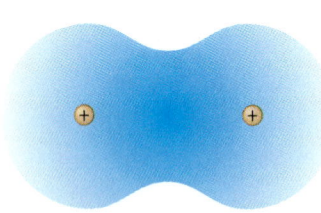

FIGURE 43.3   The covalent bond formed by the two $1s$ electrons of the H_2 molecule. The depth of blue color is proportional to the probability of finding an electron in that location.

Van der Waals Bonds

If two molecules are some distance apart, they are attracted to each other by weak electrostatic forces called **van der Waals forces.** Likewise, atoms that do not form ionic or covalent bonds are attracted to each other by van der Waals forces. For this reason, at sufficiently low temperatures where thermal excitations are negligible, gases first condense to liquids and then solidify (with the exception of helium, which does not solidify at atmospheric pressure).

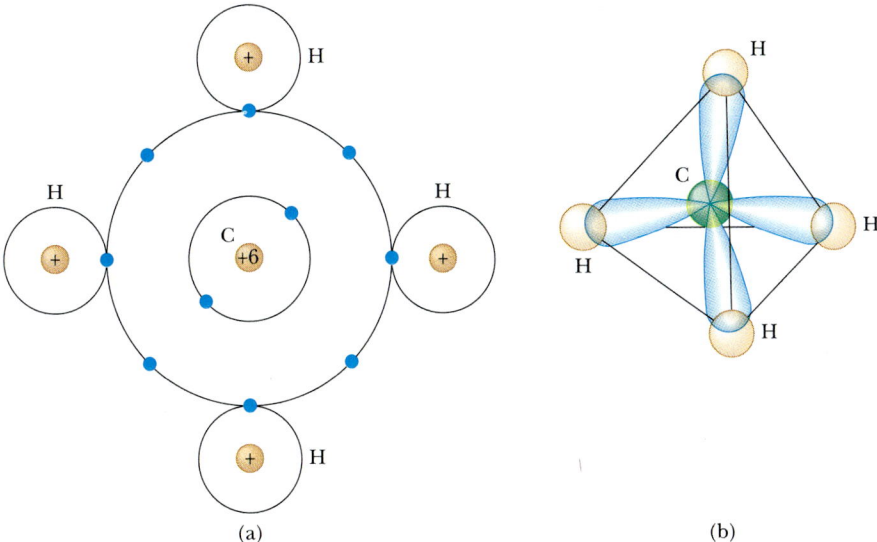

(a)   (b)

FIGURE 43.4   (a) The four covalent bonds in the CH_4 molecule. (b) A schematic representation of the angular dependence of these four covalent bonds. The carbon atom is at the center of a tetrahedron having hydrogen atoms at its corners. The electron density is greatest at the nuclei, but there is some distortion to increase the electron density between the nuclei.

There are three types of weak intermolecular forces. The first type, called the *dipole-dipole force,* is an interaction between two molecules each having a permanent electric dipole moment. For example, polar molecules such as H_2O have permanent electric dipole moments and attract other polar molecules (Fig. 43.5). In effect, one molecule interacts with the electric field produced by another molecule.

The second type is a *dipole-induced dipole force* in which a polar molecule having a permanent electric dipole moment induces a dipole moment in a nonpolar molecule.

The third type called the van der Waals or *dispersion force,* is an attractive force that occurs between two nonpolar molecules. In this case, the interaction results from the fact that, although the average dipole moment of a nonpolar molecule is zero, the average of the square of the dipole moment is nonzero because of charge fluctuations. Consequently, two nonpolar molecules near each other tend to be correlated so as to produce an attractive van der Waals force.

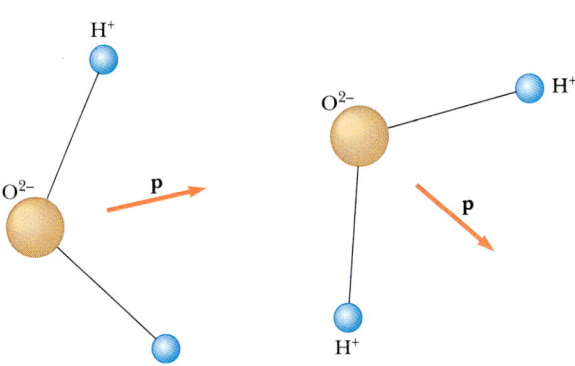

FIGURE 43.5 Water molecules have permanent electric dipole moments. The molecules attract each other because the electric field produced by one molecule interacts with and orients the moment of a nearby molecule.

The Hydrogen Bond

Because hydrogen has only one electron, it is expected to form a covalent bond with only one other atom. In some molecules, however, hydrogen forms a different type of bond, one involving more than one other atom. The bonds in such an arrangement are called **hydrogen bonds.** This is a relatively weak chemical bond, with a binding energy of about 0.1 eV. Although the hydrogen bond is weak, it is the mechanism responsible for linking giant biological molecules and polymers. For example, in the case of the famous DNA molecule that has a double helix structure, hydrogen bonds formed by sharing a proton between two atoms create linkages between the turns of the helix. The mechanism of sharing a proton is similar to the sharing of electrons in the covalent bond.

43.2   THE ENERGY AND SPECTRA OF MOLECULES

As in the case of atoms, the structure and properties of molecules can be studied by examining the radiation they emit or absorb. Before we describe these processes, it is important to first understand the various ways of exciting a molecule.

Consider a single molecule in the gaseous phase. The energy of the molecule can be divided into four categories: (1) electronic energy, due to the mutual interactions of the molecule's electrons and nuclei; (2) translational energy, due to the motion of the molecule's center of mass through space; (3) rotational energy, due to the rotation of the molecule about its center of mass; and (4) vibrational energy, due to the vibration of the molecule's constituent atoms. Thus, we can write the total energy of the molecule in the form

Excitations of a molecule

$$E = E_{el} + E_{trans} + E_{rot} + E_{vib}$$

The electronic energy of a molecule is very complex because it involves the interaction of many charged particles, but various techniques have been developed to approximate its values. The translational energy is unrelated to internal structure, hence this molecular energy is unimportant in interpreting molecular spectra.

Rotational Motion of a Molecule

Let us consider the rotation of a molecule around its center of mass, confining our discussion to the diatomic case (Fig. 43.6a), but noting that the same ideas can be extended to polyatomic molecules. A diatomic molecule has only two rotational degrees of freedom, corresponding to rotations around the y and z axes, that is, the axes perpendicular to the molecular axis.[1] If ω is the angular frequency of rotation around one of these axes, the rotational kinetic energy of the molecule can be expressed in the form

$$E_{rot} = \tfrac{1}{2}I\omega^2 \tag{43.2}$$

where I is the moment of inertia of the molecule, given by

Moment of inertia for a diatomic molecule

$$I = \left(\frac{m_1 m_2}{m_1 + m_2}\right)r^2 = \mu r^2 \tag{43.3}$$

---

[1] The excitation energy for rotations around the molecular axis is so large that such modes are not observable because nearly all the molecular mass is concentrated within nuclear dimensions of the rotation axis, giving a negligibly small moment of inertia about the internuclear line.

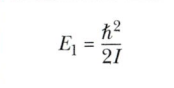

$$E_1 = \frac{\hbar^2}{2I}$$

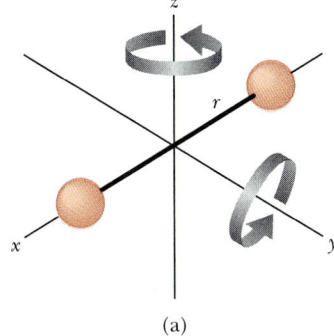

(a)

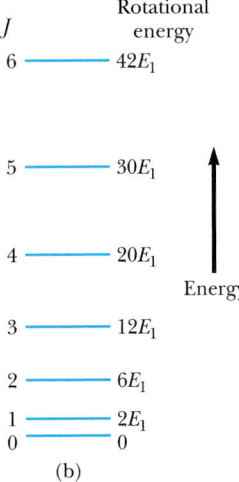

J	Rotational energy
6	$42E_1$
5	$30E_1$
4	$20E_1$
3	$12E_1$
2	$6E_1$
1	$2E_1$
0	0

Energy

(b)

FIGURE 43.6 (a) A diatomic molecule oriented along the *x* axis has two rotational degrees of freedom, corresponding to rotation around the *y* and *z* axes. (b) Allowed rotational energies of a diatomic molecule as calculated using Equation 43.6.

where m_1 and m_2 are the masses of the atoms that form the molecule, r is the atomic separation, and μ is the **reduced mass** of the molecule:

$$\mu = \frac{m_1 m_2}{m_1 + m_2} \qquad (43.4)$$

Reduced mass of a molecule

The magnitude of the angular momentum of the molecule is $I\omega$, which classically can have any value. Quantum mechanics, however, restricts angular momentum to the values

$$I\omega = \sqrt{J(J+1)}\hbar \qquad J = 0, 1, 2, \ldots \qquad (43.5)$$

Allowed values of rotational angular momentum

where J is an integer called the **rotational quantum number**. Substituting Equation 43.5 into Equation 43.2, we get an expression for the allowed values of the rotational kinetic energy:

$$E_{\text{rot}} = \tfrac{1}{2}I\omega^2 = \frac{1}{2I}(I\omega)^2 = \frac{(\sqrt{J(J+1)}\hbar)^2}{2I}$$

$$E_{\text{rot}} = \frac{\hbar^2}{2I}J(J+1) \qquad J = 0, 1, 2, \ldots \qquad (43.6)$$

Allowed values of rotational energy

Thus, we see that *the rotational energy of the molecule is quantized and depends on its moment of inertia.* The allowed rotational energies of a diatomic molecule are plotted in Figure 43.6b.

The spacings between adjacent rotational energy levels of most molecules lie in the microwave range of frequencies ($f \approx 10^{11}$ Hz) or far infrared. When a molecule absorbs a microwave photon, it jumps from a lower to a higher rotational energy level. The allowed rotational transitions of linear molecules are regulated

TABLE 43.1	**Microwave Absorption Lines Corresponding to Several Rotational Transitions of the CO Molecule**	
Rotational Transition	Wavelength of the Absorption Line (m)	Frequency of the Absorption Line (Hz)
$J = 0 \rightarrow J = 1$	2.60×10^{-3}	1.15×10^{11}
$J = 1 \rightarrow J = 2$	1.30×10^{-3}	2.30×10^{11}
$J = 2 \rightarrow J = 3$	8.77×10^{-4}	3.46×10^{11}
$J = 3 \rightarrow J = 4$	6.50×10^{-4}	4.61×10^{11}

From G. M. Barrows, *The Structure of Molecules,* New York, W. A. Benjamin, 1963.

by the selection rule $\Delta J = \pm 1$. That is, an absorption line in the microwave spectrum of a linear molecule corresponds to an energy separation equal to $E_J - E_{J-1}$. From Equation 43.6, we see that the allowed transitions are given by the condition

$$\Delta E = E_J - E_{J-1} = \frac{\hbar^2}{2I}[J(J+1) - (J-1)J]$$

Separation between adjacent rotational levels

$$= \frac{\hbar^2}{I}J = \frac{h^2}{4\pi^2 I}J \qquad (43.7)$$

where J is the rotational quantum number of the higher energy state. Because $\Delta E = hf$, where f is the frequency of the absorbed microwave photon, we see that the allowed frequency for the transition $J = 0$ to $J = 1$ is $f_1 = h/4\pi^2 I$. The frequency corresponding to the $J = 1$ to $J = 2$ transition is $2f_1$, and so on. These predictions are in excellent agreement with the observed frequencies.

The wavelengths and frequencies for the microwave absorption spectrum of the CO molecule are given in Table 43.1. From these data, the moment of inertia and bond length of the molecule can be evaluated.

EXAMPLE 43.1   **Rotation of the CO Molecule**

The $J = 0$ to $J = 1$ rotational transition of the CO molecule occurs at 1.15×10^{11} Hz. (a) Use this information to calculate the moment of inertia of the molecule.

Solution   From Equation 43.7, we see that the energy difference between the $J = 0$ and $J = 1$ rotational levels is $h^2/4\pi^2 I$. Equating this to the energy of the absorbed photon, we get

$$\frac{h^2}{4\pi^2 I} = hf$$

Solving for I gives

$$I = \frac{h}{4\pi^2 f} = \frac{6.626 \times 10^{-34}\text{ J} \cdot \text{s}}{4\pi^2(1.15 \times 10^{11}\text{ s}^{-1})}$$

$$= \boxed{1.46 \times 10^{-46}\text{ kg} \cdot \text{m}^2}$$

(b) Calculate the bond length of the molecule.

Solution   Equation 43.3 can be used to calculate the bond length, but we first need to know the value for the reduced mass μ of the CO molecule. Since $m_1 = 12$ u and $m_2 = 16$ u, the reduced mass is

$$\mu = \frac{m_1 m_2}{m_1 + m_2} = \frac{(12\text{ u})(16\text{ u})}{12\text{ u} + 16\text{ u}} = 6.86\text{ u}$$

$$= (6.86\text{ u})\left(1.66 \times 10^{-27}\frac{\text{kg}}{\text{u}}\right) = 1.14 \times 10^{-26}\text{ kg}$$

where we have used the fact that 1 u = 1.66×10^{-27} kg.

Substituting this value and the result of part (a) into Equation 43.3, and solving for r, we get

$$r = \sqrt{\frac{I}{\mu}} = \sqrt{\frac{1.46 \times 10^{-46}\text{ kg} \cdot \text{m}^2}{1.14 \times 10^{-26}\text{ kg}}}$$

$$= 1.13 \times 10^{-10}\text{ m} = \boxed{0.113\text{ nm}}$$

Vibrational Motion of Molecules

A molecule is a flexible structure in which the atoms are bonded together by what can be considered "effective springs." If disturbed, the molecule can vibrate and acquire vibrational energy. This vibrational motion and corresponding vibrational energy can be altered if the molecule is exposed to radiation of the proper frequency.

Consider the diatomic molecule shown in Figure 43.7a, where the effective spring has a force constant k. A plot of the potential energy versus atomic separation for such a molecule is sketched in Figure 43.7b, where r_0 is the equilibrium atomic separation. According to classical mechanics, the frequency of vibration for such a system is

$$f = \frac{1}{2\pi}\sqrt{\frac{k}{\mu}} \tag{43.8}$$

where again μ is the reduced mass given by Equation 43.4.

As we expect, the quantum mechanical solution to this system shows that the energy is quantized, with allowed energies

$$E_{\text{vib}} = (v + \tfrac{1}{2})\,hf \qquad v = 0, 1, 2, \ldots \tag{43.9}$$

where v is an integer called the **vibrational quantum number.** If the system is in the lowest vibrational state, for which $v = 0$, its zero-point energy is $\frac{1}{2}hf$. The accompanying vibration — the zero-point motion — is always present, even if the molecule is not excited. In the first excited state, $v = 1$ and the vibrational energy is $\frac{3}{2}hf$, and so on.

Substituting Equation 43.8 into Equation 43.9 gives the following expression for the vibrational energy:

$$E_{\text{vib}} = (v + \tfrac{1}{2})\,\frac{h}{2\pi}\sqrt{\frac{k}{\mu}} \qquad v = 0, 1, 2, \ldots \tag{43.10}$$

Allowed values of vibrational energy

The selection rule for the allowed vibrational transitions is $\Delta v = \pm 1$. From Equation 43.10, we see that the energy difference between any two successive vibrational

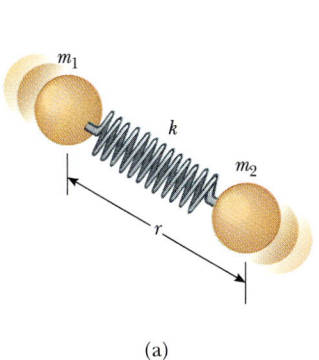

(a)

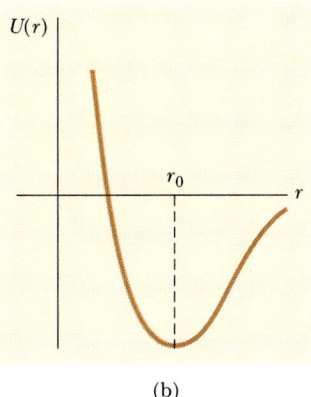

(b)

FIGURE 43.7   (a) Effective-spring model of a diatomic molecule. The fundamental vibration is along the molecular axis. (b) A plot of the potential energy of a diatomic molecule versus atomic separation, where r_0 is the equilibrium separation of the atoms.

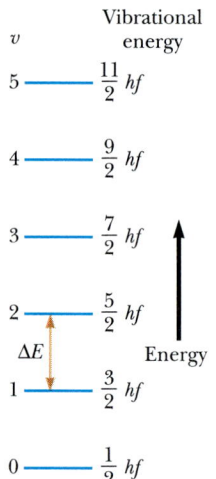

FIGURE 43.8 Allowed vibrational energies of a diatomic molecule, where *f* is the fundamental frequency of vibration, given by Equation 43.8. Note that the spacings between adjacent vibrational levels are equal.

TABLE 43.2   Fundamental Vibrational Frequencies and Effective Force Constants for Some Diatomic Molecules

Molecule	Frequency (Hz), $v = 0$ to $v = 1$	Force Constant (N/m)
HF	8.72×10^{13}	970
HCl	8.66×10^{13}	480
HBr	7.68×10^{13}	410
HI	6.69×10^{13}	320
CO	6.42×10^{13}	1860
NO	5.63×10^{13}	1530

From G. M. Barrows, *The Structure of Molecules,* New York, W. A. Benjamin, 1963.

levels is

$$\Delta E_{vib} = \frac{h}{2\pi}\sqrt{\frac{k}{\mu}} = hf \qquad (43.11)$$

The vibrational energies of a diatomic molecule are plotted in Figure 43.8. At ordinary temperatures, most molecules have vibrational energies corresponding to the $v = 0$ state because the spacing between vibrational states is large compared to $k_B T$. Transitions between vibrational levels lie in the *infrared region* of the spectrum. That is, a molecule jumps from a lower to a higher energy vibrational level by absorbing a photon having a frequency in the infrared range. The absorption frequencies corresponding to the $v = 0$ to $v = 1$ transition for several diatomic molecules are listed in Table 43.2, together with k values calculated using Equation 43.11. The stiffness of a bond can be measured by the size of the force constant k. For example, the CO molecule, which is bonded by several electrons, has a stiffer bonding than such single-bonded molecules as HCl.

EXAMPLE 43.2   Vibration of the CO Molecule

The frequency of the photon corresponding to the $v = 0$ to $v = 1$ transition for the CO molecule occurs at 6.42×10^{13} Hz. (a) Calculate the force constant k for this molecule.

Solution   From Equation 43.9, we see that the energy difference between the $v = 0$ and $v = 1$ vibrational states is

$$\Delta E = \tfrac{3}{2}hf - \tfrac{1}{2}hf = hf$$

Using Equation 43.11 and the fact that the reduced mass is $\mu = 1.14 \times 10^{-26}$ kg for the CO molecule (Example 43.1), we get

$$\frac{h}{2\pi}\sqrt{\frac{k}{\mu}} = hf$$

$$k = 4\pi^2 \mu f^2$$
$$= 4\pi^2 (1.14 \times 10^{-26}\text{ kg})(6.42 \times 10^{13}\text{ s}^{-1})^2$$

$$= 1.85 \times 10^3 \text{ N/m}$$

(b) What is the maximum amplitude of vibration for this molecule in the $v = 0$ vibrational state?

Solution   The maximum potential energy stored in the molecule is $\tfrac{1}{2}kA^2$, where A is the amplitude of vibration. Equating this to the vibrational energy given by Equation 43.10, with $v = 0$, we get

$$\tfrac{1}{2}kA^2 = \frac{h}{4\pi}\sqrt{\frac{k}{\mu}}$$

Substituting the value $k = 1.86 \times 10^3$ N/m and the value of μ from part (a), we get

$$A^2 = \frac{h}{2\pi k}\sqrt{\frac{k}{\mu}} = \frac{h}{2\pi}\sqrt{\frac{1}{k\mu}}$$

$$= \frac{6.626 \times 10^{-34}}{2\pi}\sqrt{\frac{1}{(1.86 \times 10^3)(1.14 \times 10^{-26})}}$$

$$= 2.30 \times 10^{-23}\text{ m}^2$$

so that

$$A = 4.79 \times 10^{-12} \text{ m} = 4.79 \times 10^{-3} \text{ nm}$$

Comparing this result with the bond length of 0.1128 nm, we see that the amplitude of vibration is about 4% of the bond length. Thus, we see that infrared spectroscopy provides useful information on the elastic properties (bond strengths) of molecules.

Molecular Spectra

In general, an excited molecule rotates and vibrates simultaneously. To a first approximation, these motions are independent, and so the total energy of the molecule is the sum of Equations 43.6 and 43.9:

$$E = \frac{\hbar^2}{2I}[J(J+1)] + (v + \tfrac{1}{2})hf \tag{43.12}$$

Energy levels can be calculated from this expression, and each level is indexed by two quantum numbers, J and v. From these calculations, an energy-level diagram like that shown in Figure 43.9a can then be constructed. For each allowed value of the vibrational quantum number v, there is a complete set of rotational levels corresponding to $J = 0, 1, 2, \ldots$. Note that the energy separation between successive rotational levels is much smaller than the separation between successive vibrational levels.

At ordinary temperatures, most molecules are in the $v = 0$ vibrational state but in various rotational states. When a molecule absorbs a photon, v increases by one unit while J either increases or decreases by one unit, as in Figure 43.9. Thus, the molecular absorption spectrum consists of two groups of lines: The group to the

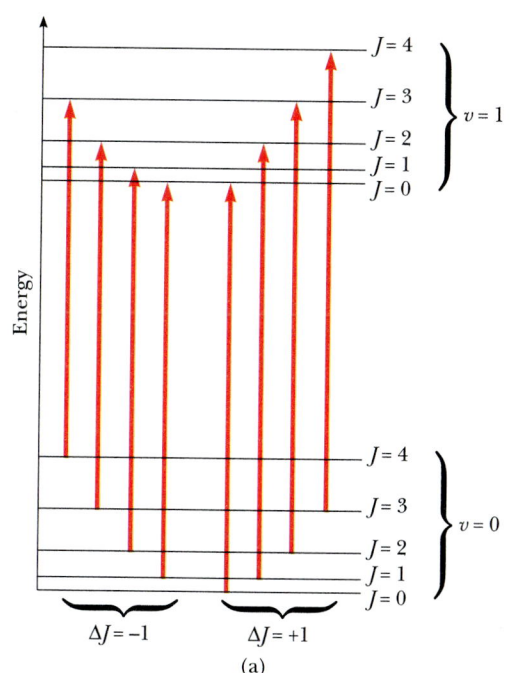

$\Delta J = -1$ $\Delta J = +1$

(a)

FIGURE 43.9 (a) Absorptive transitions between the $v = 0$ and $v = 1$ vibrational states of a diatomic molecule. The transitions obey the selection rule $\Delta J = \pm 1$, and fall into two sequences, those for which $\Delta J = +1$ and those for which $\Delta J = -1$. The transition energies are given by Equation 43.12. (b) Expected lines in the optical absorption spectrum of a molecule. The lines on the right side of center correspond to transitions in which J changes by $+1$, while the lines to the left of center correspond to transitions for which J changes by -1. These same lines appear in the emission spectrum.

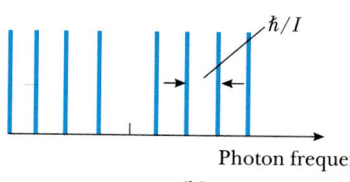

Photon frequency

(b)

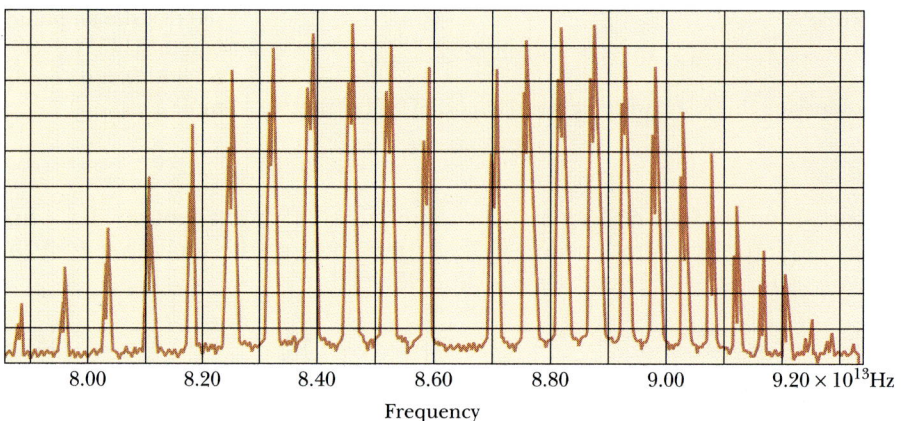

FIGURE 43.10   The absorption spectrum of the HCl molecule. Each line is split into a doublet because chlorine has two isotopes, ^{35}Cl and ^{37}Cl, which have different nuclear masses.

right of center satisfies the selection rules $\Delta J = 1$ and $\Delta v = 1$, and the group to the left of center satisfies the selection rules $\Delta J = -1$ and $\Delta v = 1$.[2]

The energies of the absorbed photons can be calculated from Equation 43.12:

$$\Delta E = hf + \frac{\hbar^2}{I}(J+1) \qquad J = 0, 1, 2, \ldots (\Delta J = +1) \qquad (43.13)$$

$$\Delta E = hf + \frac{\hbar^2}{I}J \qquad J = 1, 2, 3, \ldots (\Delta J = -1) \qquad (43.14)$$

where now J is the rotational quantum number of the *initial* state. Equation 43.13 generates the series of equally spaced lines *above* the characteristic frequency f, while Equation 43.14 generates the series *below* this frequency. Adjacent lines are separated in frequency by the fundamental unit $\hbar/I$. Figure 43.9b shows the expected frequencies in the absorption spectrum of the molecule; these same frequencies appear in the emission spectrum.

The absorption spectrum of the HCl molecule shown in Figure 43.10 follows this pattern very well and reinforces our model. However, one peculiarity is apparent: Each line is split into a doublet. This doubling occurs because of two chlorine isotopes (^{35}Cl and ^{37}Cl), have different masses, and the two HCl molecules have different values of I.

43.3   BONDING IN SOLIDS

A crystalline solid consists of a large number of atoms arranged in a regular array, forming a periodic structure. The bonding schemes for molecules described in Section 43.1 are also appropriate in describing the bonding in solids. For example, the ions in the NaCl crystal are ionically bonded, while the carbon atoms in the diamond structure form covalent bonds. The metallic bond, which is of the same general nature as ionic and covalent bonding, is responsible for the cohesion of copper, silver, sodium, and other metals.

---

[2] The selection rule $\Delta J = \pm 1$ implies that the photon (which excites a transition) is a spin-one particle with spin quantum number $s = 1$. Hence, this selection rule describes angular momentum conservation for the system molecule plus photon.

(a)   (b)

Crystalline solids. (a) A cylinder of nearly pure crystalline silicon (Si), approximately 10 inches long. Such crystals are cut into wafers and processed to make various semiconductor devices.   *(© Charles D. Winters)* (b) Crystals of natural quartz (SiO_2), one of the most common minerals on earth. Quartz crystals are used to make special lenses and prisms and in certain electronic applications.   *(Courtesy of Ward's Natural Science)*

Ionic Solids

Many crystals are formed by ionic bonding, where the dominant interaction between ions is the Coulomb interaction. Consider the NaCl crystal in Figure 43.11, where each Na^+ ion has six nearest-neighbor Cl^- ions, and each Cl^- ion has six nearest-neighbor Na^+ ions. Each Na^+ ion is attracted to its six Cl^- neighbors. The corresponding attractive potential energy is $-6k_e e^2/r$, where r is the $Na^+ - Cl^-$

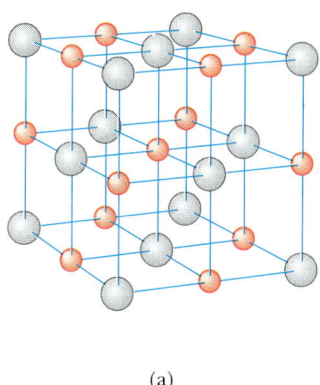
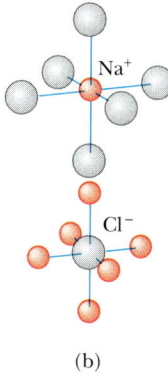

(a)   (b)

FIGURE 43.11   (a) The crystal structure of NaCl. (b) In the NaCl structure, each positive sodium ion (red spheres) is surrounded by six negative chlorine ions (gray spheres), and each chlorine ion is surrounded by six sodium ions.

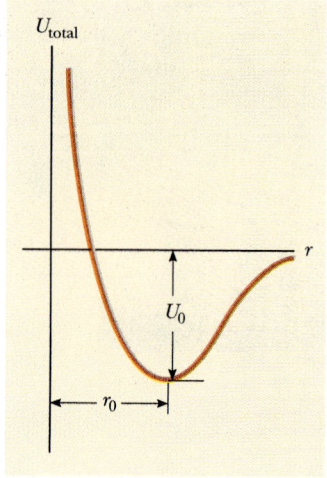

FIGURE 43.12   Total potential energy versus ion separation for an ionic solid, where U_0 is the ionic cohesive energy and r_0 is the equilibrium separation between ions.

separation. In addition, there are also 12 Na^+ ions at a distance of $\sqrt{2}r$ from the Na^+, which produce a weaker repulsive force on it. Furthermore, beyond these 12 Na^+ ions are more Cl^- ions that produce an attractive force, and so on. The net effect of all these interactions is a resultant negative electrostatic potential energy

$$U_{\text{attractive}} = -\alpha k_e \frac{e^2}{r} \tag{43.15}$$

where α is a pure number known as the **Madelung constant.** The value of α depends only on crystal structure. For example, $\alpha = 1.7476$ for the NaCl structure. When the constituent ions of a compound are brought close together, their subshells avoid overlap because electrons repel each other. This is accounted for by the potential energy term B/r^m (Section 43.1). Therefore, the total potential energy is

$$U_{\text{total}} = -\alpha k_e \frac{e^2}{r} + \frac{B}{r^m} \tag{43.16}$$

A plot of total potential energy versus ion separation is shown in Figure 43.12. The potential energy has its minimum value U_0 at the equilibrium separation, when $r = r_0$. It is left as a problem (Problem 35) to show that

$$U_0 = -\alpha k_e \frac{e^2}{r_0}\left(1 - \frac{1}{m}\right) \tag{43.17}$$

This minimum energy U_0 is called the **ionic cohesive energy** of the solid, and its absolute value represents the energy required to separate the solid into a collection of isolated positive and negative ions. Its value for NaCl is -7.84 eV per $Na^+ - Cl^-$ pair. In order to calculate the **atomic cohesive energy,** which is the binding energy relative to the neutral atoms, 5.14 eV are gained in going from Na^+ to Na, and 3.61 eV must be supplied to convert Cl^- to Cl. Thus, the atomic cohesive energy of NaCl is

$$-7.84 \text{ eV} + 5.14 \text{ eV} - 3.61 \text{ eV} = -6.31 \text{ eV}$$

Properties of ionic solids

Ionic crystals have the following general properties:

- They form relatively stable and hard crystals.
- They are poor electrical conductors because they contain no free electrons.
- They have high vaporization temperatures.
- They are transparent to visible radiation but absorb strongly in the infrared region. This occurs because the electrons form such tightly bound shells in ionic solids that visible radiation does not possess sufficient energy to promote electrons to the next allowed shell. The strong infrared absorption occurs because the vibrations of the more massive ions have a low natural frequency and experience resonant absorption in the low-energy infrared region.
- Many are quite soluble in polar liquids such as water. The water molecule, which has a permanent electric dipole moment, exerts an attractive force on the charged ions, which breaks the ionic bonds and dissolves the solid.

Covalent Crystals

Solid carbon, in the form of diamond, is a crystal whose atoms are covalently bonded. Because atomic carbon has an electron configuration $1s^2 2s^2 2p^2$, it is four electrons short of filling the $2p^6$ shell. Hence, two carbon atoms have a strong attraction for one another, with a cohesive energy of 7.37 eV.

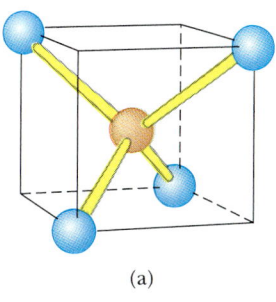

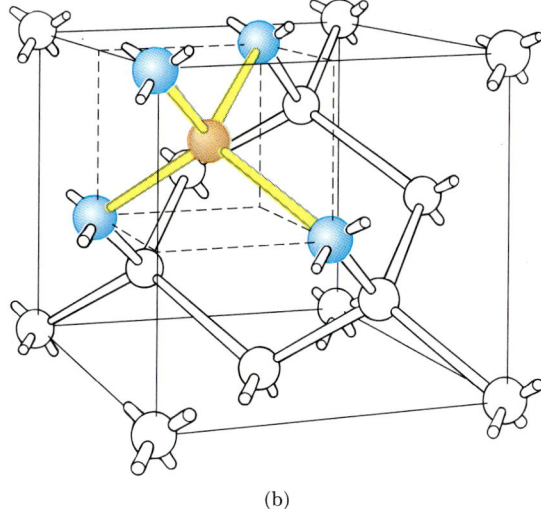

(a)

(b)

FIGURE 43.13   (a) Each carbon atom in diamond is covalently bonded to four other carbons and forms a tetrahedral structure. (b) The crystal structure of diamond, showing the tetrahedral bond arrangement.

In the diamond structure, each atom is covalently bonded to four other carbon atoms located at the four diagonally opposite corners of a cube as in Figure 43.13a. To form such a configuration of bonds, one electron of each atom must be promoted to the $1s^2 2s2p^3$ configuration, which is a half-filled p-shell and requires an energy of about 4 eV. The crystal structure of diamond is shown in Figure 43.13b. Note that each carbon atom forms covalent bonds with four nearest-neighbor atoms. The basic structure of diamond is called tetrahedral (each carbon atom is at the center of a regular tetrahedron), and the angle between the bonds is 109.5°. Other crystals such as silicon and germanium have similar structures.

The cohesive energies of some covalent solids are given in Table 43.3. The large energies account for the hardness of covalent solids. Diamond is particularly hard and has an extremely high melting point (about 4000 K). Covalently bonded solids are often very hard, have large bond energies and high melting points, and are good insulators.

TABLE 43.3	The Cohesive Energies of Some Covalent Solids
Crystal	Cohesive Energy (eV)
C (diamond)	7.37
Si	4.63
Ge	3.85
InAs	5.70
SiC	12.3
ZnS	6.32
CuCl	9.24

Metallic Solids

Metallic bonds are generally weaker than ionic or covalent bonds. The valence electrons in a metal are relatively free to move throughout the material. There are a large number of such mobile electrons in a metal, typically one or two per atom. The metal structure can be viewed as a "sea" or "gas" of nearly free electrons surrounded by a lattice of positive ions (Fig. 43.14). The binding mechanism in a metal is the attractive force between the positive ions and the electron gas.

Metals have a cohesive energy in the range of 1 to 3 eV, which is smaller than the cohesive energies of ionic or covalent solids. Since visible photons also have energies in this range, light interacts strongly with the free electrons in metals. Hence visible light is absorbed and re-emitted quite close to the surface of a metal, which accounts for the shiny nature of metallic surfaces. In addition to the high electrical conductivity of metals produced by the free electrons, the nondirectional nature of the metallic bond allows many different types of metallic atoms to be dissolved in a host metal in varying amounts. The resulting solid solutions, or

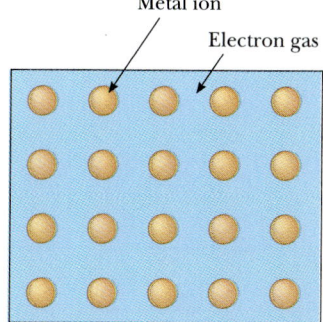

FIGURE 43.14   Schematic diagram of a metal. The blue area represents the electron gas, while the red circles represent the positive metal ion cores.

alloys, may be designed to have particular properties, such as tensile strength, ductility, electrical and thermal conductivity, and resistance to corrosion. Such properties are usually controllable and in some cases predictable.

43.4   BAND THEORY OF SOLIDS

If two identical atoms are very far apart, they do not interact and their electronic energy levels can be considered to be those of isolated atoms. Suppose the two atoms are sodium, each having a $3s$ electron that has a specific, well-defined energy. As the two sodium atoms are brought closer together, their outer orbits begin to overlap. When the interaction between them is strong enough, two different $3s$ levels are formed as shown in Figure 43.15a.

When a large number of atoms are brought together to form a solid, a similar phenomenon occurs. As the atoms are brought close together, the various atomic energy levels begin to split. This splitting in levels for six atoms in close proximity is shown in Figure 43.15b. In this case, there are six energy levels and six overlapping wave functions for the system. Since the width of an energy band arising from a particular atomic energy level is independent of the number of atoms in a solid, the energy levels are more closely spaced in the case of six atoms than in the case of two atoms. If we extend this argument to a large number of atoms (of the order of 10^{23} atoms/cm³), we obtain a large number of levels so closely spaced that they may be regarded as a continuous band of energy levels, as in Figure 43.15c. The width of an energy band depends only on nearest-neighbor interactions, whereas the number of levels within the band depends on the total number of interacting particles (hence, the number of atoms in the crystal). In the case of sodium, it is common to refer to the continuous distribution of allowed energy levels as the $3s$ band because the band originates from the $3s$ levels of individual sodium atoms. In general, a crystalline solid has a large number of allowed energy bands that arise from different atomic energy levels. Figure 43.16 shows the allowed energy bands of sodium. Note that energy gaps occur between the allowed bands; these gaps are called *forbidden energy bands* because electrons are not allowed to enter them.

If the solid contains N atoms, each energy band has N energy levels. In the case of sodium, the $1s$, $2s$, and $2p$ bands are full, as indicated by the blue-shaded areas in Figure 43.16. A level whose orbital angular momentum is ℓ can hold $2(2\ell + 1)$ electrons. The factor of 2 arises from the two possible electron spin orientations, while the factor $2\ell + 1$ corresponds to the number of possible orientations of the orbital angular momentum. The capacity of each band for a system of N atoms is $2(2\ell + 1)N$ electrons. Hence, the $1s$ and $2s$ bands each contain $2N$ electrons ($\ell =$

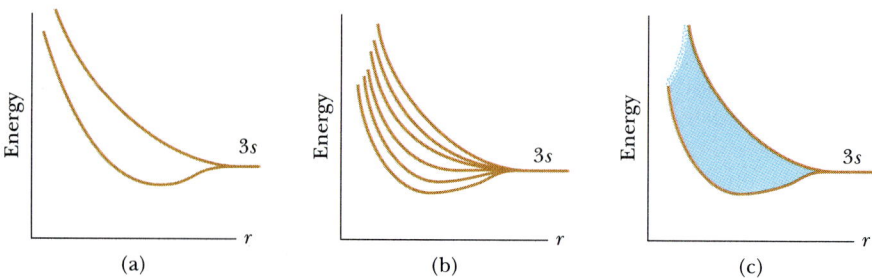

FIGURE 43.15   (a) The splitting of the $3s$ levels when two sodium atoms are brought together. (b) The splitting of the $3s$ levels when six sodium atoms are brought together. (c) The formation of a $3s$ band when a large number of sodium atoms are assembled to form a solid.

0), while the $2p$ band contains $6N$ electrons ($\ell = 1$). Because sodium has only one $3s$ valence electron and there are a total of N atoms in the solid, the $3s$ band contains only N electrons and is only half full. The $3p$ band, which is above the $3s$ band, is completely empty.

43.5   FREE-ELECTRON THEORY OF METALS

In this section, we discuss the free-electron theory of metals. In this model, we imagine that the outer-shell electrons in the metal are not strongly bound to individual atoms but are free to move through the metal.

Statistical physics can be applied to a collection of particles in an effort to relate microscopic properties to macroscopic properties. In the case of electrons, it is necessary to use quantum statistics, with the requirement that each state of the system can be occupied by only one electron. Each state is specified by a set of quantum numbers. All particles with half-integral spin, called **fermions**, must obey the Pauli exclusion principle. An electron is one example of a fermion. The probability of finding an electron in a particular state of energy E is

$$f(E) = \frac{1}{e^{(E - E_F)/k_B T} + 1} \qquad (43.18)$$

where E_F is called the **Fermi energy**, and $f(E)$ is called the **Fermi-Dirac distribution function**. A plot of $f(E)$ versus E at $T = 0$ K is shown in Figure 43.17a. Note that $f = 1$ for $E < E_F$ and $f = 0$ for $E > E_F$. That is, at 0 K, all states whose energies lie below the Fermi energy are occupied, while all states with energies greater than the Fermi energy are vacant. A plot of $f(E)$ versus E at some temperature $T > 0$ K is shown in Figure 43.17b. At $E = E_F$, the function $f(E)$ has the value $\frac{1}{2}$. The main variations in the Fermi-Dirac distribution function with temperature occur at the high energies in the vicinity of the Fermi energy. Note that only a small fraction of the levels with energies greater than the Fermi energy are occupied. Furthermore, a small fraction of the levels below the Fermi energy are empty.

In Chapter 41 we found that if a particle is confined to move in a one-dimensional box of length L, the allowed states have quantized energy levels

$$E = \frac{\hbar^2 \pi^2}{2 m L^2} n^2 \qquad n = 1, 2, 3, \ldots$$

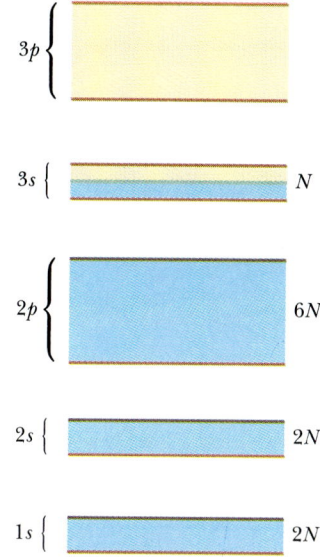

FIGURE 43.16 The energy bands of sodium. The solid contains N atoms. Note the energy gaps between the allowed bands; electrons cannot occupy states that lie in these forbidden gaps.

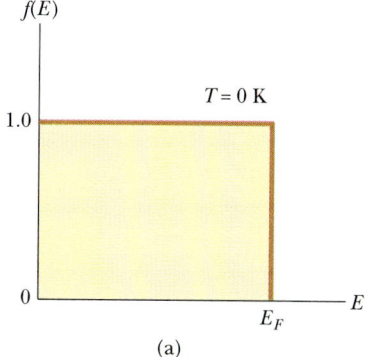

(a)

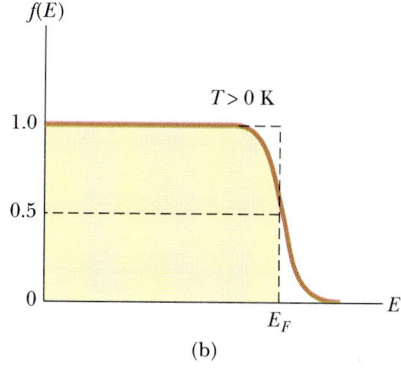

(b)

FIGURE 43.17 A plot of the Fermi-Dirac distribution function at (a) $T = 0$ K, and (b) $T > 0$ K, where E_F is the Fermi energy.

The wave functions for these allowed states are standing waves given by $\psi = A \sin(n\pi x/L)$, which satisfy the boundary condition $\psi = 0$ at $x = 0$ and $x = L$.

Now imagine an electron moving in a three-dimensional cube of edge length L and volume L^3, where the walls of the cube represent the surfaces of the metal. It can be shown (Problem 24) that the energy for such a particle is

$$E = \frac{\hbar^2\pi^2}{2mL^2}(n_x^2 + n_y^2 + n_z^2) \tag{43.19}$$

where n_x, n_y, and n_z are quantum numbers. Again, the energy levels are quantized, and each is characterized by this set of three quantum numbers (one for each degree of freedom) and the spin quantum number m_s. For example, the ground state corresponding to $n_x = n_y = n_z = 1$ has an energy equal to $3\hbar^2\pi^2/2mL^2$, and so on. In this model, we require that $\psi(x, y, z) = 0$ at the boundaries. This requirement results in solutions that are standing waves in three dimensions.

If the quantum numbers are treated as continuous variables, it is found that the number of allowed states per unit volume that have energies between E and $E + dE$ is

$$g(E)\,dE = CE^{1/2}\,dE \tag{43.20}$$

where

$$C = \frac{8\sqrt{2}\,\pi m^{3/2}}{h^3} \tag{43.21}$$

The function $g(E) = CE^{1/2}$ is called the **density-of-states function.**

In thermal equilibrium, the number of electrons per unit volume that have energy between E and $E + dE$ is equal to the product $f(E)g(E)\,dE$:

$$N(E)\,dE = C\frac{E^{1/2}\,dE}{e^{(E-E_F)/k_B T} + 1} \tag{43.22}$$

A plot of $N(E)$ versus E is given in Figure 43.18. If n is the total number of electrons per unit volume, we require that

$$n = \int_0^\infty N(E)\,dE = C\int_0^\infty \frac{E^{1/2}\,dE}{e^{(E-E_F)/k_B T} + 1} \tag{43.23}$$

This condition can be used to calculate the Fermi energy. At $T = 0$ K, the Fermi distribution function $f(E) = 1$ for $E < E_F$ and 0 for $E > E_F$. Therefore, at $T = 0$ K, Equation 43.23 becomes

$$n = C\int_0^{E_F} E^{1/2}\,dE = \tfrac{2}{3}CE_F^{3/2} \tag{43.24}$$

Substituting Equation 43.21 into Equation 43.24 and solving for E_F give

$$E_F = \frac{h^2}{2m}\left(\frac{3n}{8\pi}\right)^{2/3} \tag{43.25}$$

According to this result, E_F shows a gradual increase with increasing electron concentration. This is expected, because the electrons fill the available energy states, two electrons per state, in accordance with the Pauli exclusion principle, up to the Fermi energy.

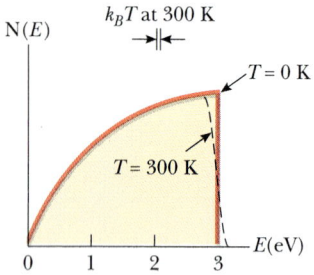

FIGURE 43.18 A plot of electron distribution versus energy in a metal at $T = 0$ K and $T = 300$ K. The Fermi energy is taken to be 3 eV.

Fermi energy at $T = 0$ K

TABLE 43.4   Calculated Values of Various Parameters for Metals at 300 K Based on the Free-Electron Theory

Metal	Electron Concentration (m^{-3})	Fermi Energy (eV)	Fermi Speed (m/s)	Fermi Temperature (K)
Li	4.70×10^{28}	4.72	1.29×10^6	5.48×10^4
Na	2.65×10^{28}	3.23	1.07×10^6	3.75×10^4
K	1.40×10^{28}	2.12	0.86×10^6	2.46×10^4
Cu	8.49×10^{28}	7.05	1.57×10^6	8.12×10^4
Ag	5.85×10^{28}	5.48	1.39×10^6	6.36×10^4
Au	5.90×10^{28}	5.53	1.39×10^6	6.41×10^4

The order of magnitude of the Fermi energy for metals is about 5 eV. Values for the Fermi energy based on the free-electron theory are given in Table 43.4, together with values for the electron speed at the Fermi level, defined by

$$\tfrac{1}{2}mv_F^2 = E_F \tag{43.26}$$

and the Fermi temperature, T_F, defined by

$$k_B T_F = E_F \tag{43.27}$$

It is left as a problem (Problem 22) to show that the average energy of a conduction electron in a metal at 0 K is

$$E_{av} = \tfrac{3}{5}E_F \tag{43.28}$$

In summary, we can consider a metal to be a system with a very large number of energy levels available to the free electrons. These electrons fill these levels in accordance with the Pauli exclusion principle, beginning with $E = 0$ and ending with E_F. At $T = 0$ K, all levels below the Fermi energy are filled, while all levels above the Fermi energy are empty. Although the levels are discrete, they are so close together that the electrons have an almost continuous distribution of energy. At 300 K, a very small fraction of the valence electrons are excited above the Fermi energy. An estimate of this fraction is given in Example 43.4.

EXAMPLE 43.3   The Fermi Energy of Gold

Each atom of gold contributes one free electron to the metal. Compute (a) the Fermi energy, (b) the Fermi speed, and (c) the Fermi temperature for gold.

Solution   (a) The concentration of free electrons in gold is 5.90×10^{28} m^{-3} (Table 43.4). Substitution of this value into Equation 43.25 gives

$$E_F = \frac{h^2}{2m}\left(\frac{3n}{8\pi}\right)^{2/3}$$

$$= \frac{(6.626 \times 10^{-34}\,\text{J}\cdot\text{s})^2}{2(9.11 \times 10^{-31}\,\text{kg})}\left(\frac{3 \times 5.90 \times 10^{28}\,\text{m}^{-3}}{8\pi}\right)^{2/3}$$

$$= 8.85 \times 10^{-19}\,\text{J} = \boxed{5.53\ \text{eV}}$$

(b) The Fermi speed is defined by the expression $\tfrac{1}{2}mv_F^2 = E_F$. Solving this for v_F gives

$$v_F = \left(\frac{2E_F}{m}\right)^{1/2} = \left(\frac{2 \times 5.85 \times 10^{-19}\,\text{J}}{9.11 \times 10^{-31}\,\text{kg}}\right)^{1/2}$$

$$= \boxed{1.39 \times 10^6\ \text{m/s}}$$

(c) The Fermi temperature is defined by Equation 43.27:

$$T_F = \frac{E_F}{k_B} = \frac{8.85 \times 10^{-19}\,\text{J}}{1.38 \times 10^{-23}\,\text{J/K}} = \boxed{6.41 \times 10^4\ \text{K}}$$

Thus, a gas of classical particles would have to be heated to approximately 64 000 K to have an average energy per particle equal to the Fermi energy at 0 K!

EXAMPLE 43.4   Excited Electrons in Copper

Calculate an approximate value for the fraction of electrons that are excited from below E_F to above E_F when copper is heated from 0 K to 300 K.

Solution Only those electrons within a range of $k_B T \approx 0.025$ eV are affected by the 300-K change in tempera-

ture. Because the Fermi energy for copper is 7.0 eV, the fraction of electrons excited from below E_F to above E_F is on the order of $k_B T / E_F$. In this case, $k_B T / E_F = 0.025/7.0 = 0.0036$ or 0.36%. Thus, we see that only a very small fraction of the electrons are affected. A more precise analysis shows that the fraction excited is $9 k_B T / 16 E_F$.

43.6   ELECTRICAL CONDUCTION IN METALS, INSULATORS, AND SEMICONDUCTORS

In Chapter 27 we found that good electrical conductors contain a high density of charge carriers, whereas the density of charge carriers in insulators is nearly zero. Semiconductors are a class of technologically important materials in which charge carrier densities are intermediate between those of insulators and those of conductors. In this section, we discuss the mechanisms of conduction in these three classes of materials. The enormous variation in electrical conductivity of these materials may be explained in terms of energy bands.

Metals

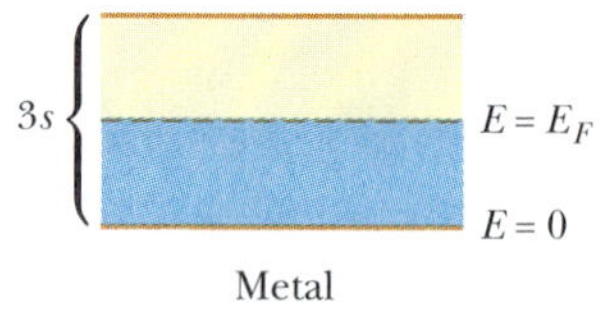

In Section 43.4, we described the energy-band picture for the ground state of metallic sodium. If energy is added to the system (say in the form of heat), electrons can move from filled states to one of many empty states.

We can obtain a better understanding of the properties of metals by considering a half-filled band, such as the $3s$ band of sodium. Figure 43.19 shows a half-filled band of a metal at $T = 0$ K, where the blue region represents levels filled with electrons. Because electrons obey Fermi-Dirac statistics, all levels below the Fermi energy are filled with electrons and all levels above E_F are empty. In the case of sodium, the Fermi energy lies in the middle of the band. At temperatures slightly greater than 0 K, some electrons are thermally excited to levels above E_F, but overall there is little change from the 0 K case. *However, if an electric field is applied to the metal, electrons having energies near the Fermi energy require only a small amount of additional energy from the applied field to reach nearby empty energy states.* Thus electrons are free to move with only a small applied field in a metal because there are many empty states available for occupancy close to the occupied energy states.

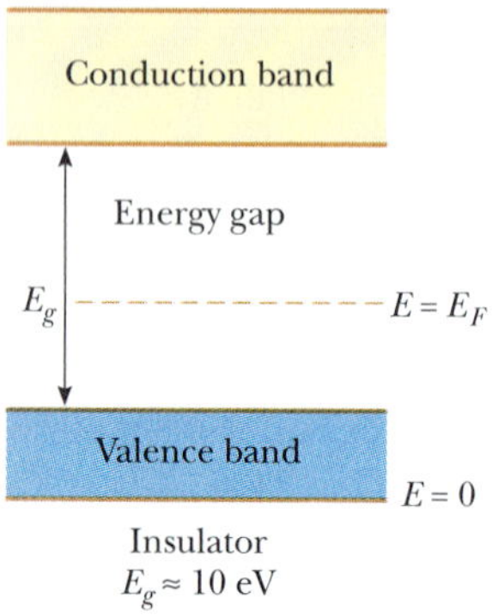

FIGURE 43.19 A half-filled band of a conductor, such as the $3s$ band of sodium. At $T = 0$ K, the Fermi energy lies in the middle of the band.

FIGURE 43.20 An insulator at $T = 0$ K has a filled valence band and an empty conduction band. The Fermi level lies somewhere between these bands.

Insulators

Now consider the two highest energy bands of a material, with the lower of these two bands filled with electrons and the higher empty at 0 K (Fig. 43.20). It is common to refer to the separation between the outermost filled and empty bands as the *energy gap*, E_g, of the material. The energy gap for an insulator is large (≈ 10 eV). The lower, filled band is called the *valence band*, and the upper empty band is the *conduction band*. The Fermi energy lies somewhere in the energy gap, as shown in Figure 43.20. At 300 K (room temperature), $k_B T = 0.025$ eV, which is much smaller than the energy gap in an insulator. At such temperatures, the Fermi-Dirac distribution predicts very few electrons thermally excited into the conduction band at normal temperatures. Thus, although an insulator has many vacant states in its conduction band, there are so few electrons occupying these states that the overall electrical conductivity is very small, resulting in a high resistivity for insulators.

Semiconductors

Materials that have an energy gap of about 1 eV are called semiconductors. Table 43.5 shows the energy gaps for some representative materials. At $T = 0$ K, all electrons are in the valence band. Thus semiconductors are poor conductors at very low temperatures. At ordinary temperatures, however, the situation is quite different. For example, the conductivity of silicon at room temperature is about 1.6×10^{-3} $(\Omega \cdot m)^{-1}$. The band structure of a semiconductor can be represented by the diagram shown in Figure 43.21. Because the Fermi level is located near the middle of the gap for a semiconductor and because E_g is small, appreciable numbers of electrons are thermally excited from the valence band to the conduction band. There are many empty nearby states in the conduction band, therefore a small applied potential can easily raise the energy of the electrons in the conduction band, resulting in a moderate current. Because thermal excitation across the narrow gap is more probable at higher temperatures, the conductivity of semiconductors increases rapidly with temperature. This contrasts sharply with the conductivity of a metal, which decreases slowly with temperature.

Charge carriers in a semiconductor can be negative and/or positive. When an electron moves from the valence band into the conduction band, it leaves behind a vacant crystal site, which is called a **hole,** in the otherwise filled valence band. This hole (electron-deficient site) appears as a positive charge, $+e$, and acts as a charge carrier in the sense that a valence electron from a nearby site can transfer into the hole. Whenever an electron does so, it creates a new hole. Thus the net effect can be viewed as the hole migrating through the material. In a pure crystal containing only one element or compound, there are equal numbers of conduction electrons and holes. Such combinations of charges are called electron-hole pairs, and a pure semiconductor that contains such pairs is called an intrinsic semiconductor (Fig. 43.22). In the presence of an electric field, the holes move in the direction of the field and the conduction electrons move opposite the field.

TABLE 43.5 **Energy Gap Values for Some Semiconductors[a]**

Crystal	E_g (eV)	
	0 K	300 K
Si	1.17	1.14
Ge	0.744	0.67
InP	1.42	1.35
GaP	2.32	2.26
GaAs	1.52	1.43
CdS	2.582	2.42
CdTe	1.607	1.45
ZnO	3.436	3.2
ZnS	3.91	3.6

[a] These data were taken from C. Kittel, *Introduction to Solid State Physics*, 5th ed., New York, John Wiley & Sons, 1976.

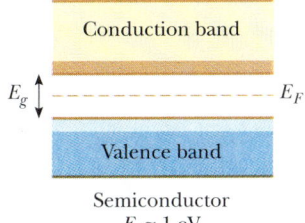

FIGURE 43.21 The band structure of a semiconductor at ordinary temperatures ($T \approx 300$ K). The energy gap is much smaller than in an insulator, and many electrons occupy states in the conduction band.

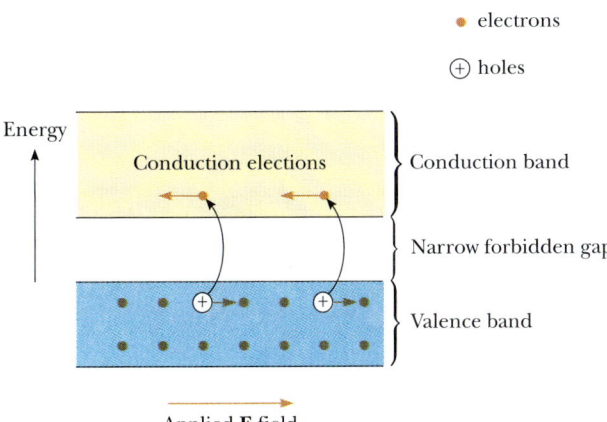

FIGURE 43.22 An intrinsic semiconductor. Note that the electrons move in the direction opposite the applied electric field and the holes move in the direction of the field.

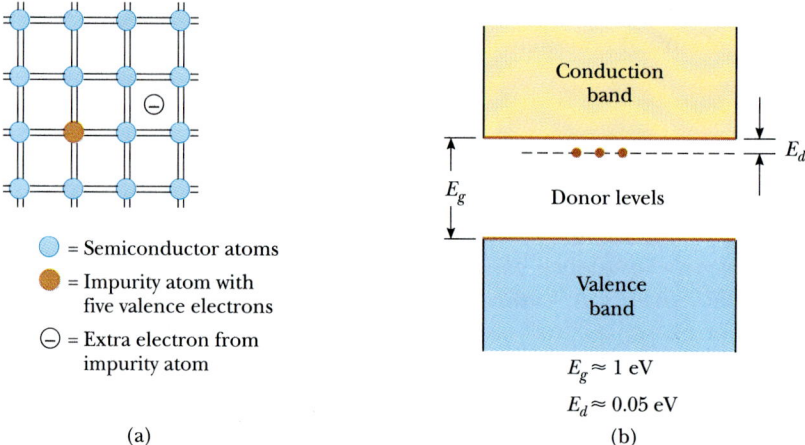

FIGURE 43.23 (a) Two-dimensional representation of a semiconductor containing a donor atom (red spot). (b) Energy-band diagram of a semiconductor in which the donor levels lie within the forbidden gap, just below the bottom of the conduction band.

Doped Semiconductors

When impurities are added to semiconductors, their band structure and resistivities are modified. The process of adding impurities, called **doping,** is important in making devices and semiconductors having well-defined regions of different conductivity. For example, when an atom containing five outer-shell electrons, such as arsenic, is added to a semiconductor, four of them participate in the covalent bonds and one is left over (Fig. 43.23a). This extra electron is nearly free and has an energy level that lies within the energy gap, just below the conduction band (Fig. 43.23b). Such a pentavalent atom in effect donates an electron to the structure and, hence, is referred to as a **donor atom.** Since the energy spacings between the donor levels and the bottom of the conduction band are very small (typically, about 0.05 eV), only a small amount of thermal energy is needed to cause an electron in these levels to move into the conduction band. (Recall that the average thermal energy of an electron at room temperature is about $k_BT \cong 0.025$ eV)

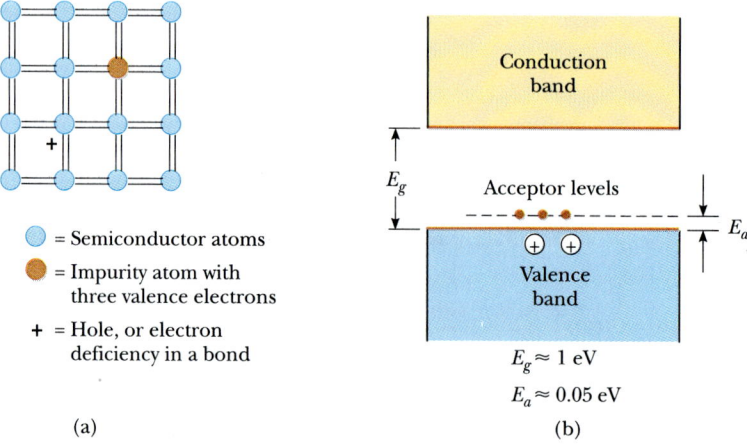

FIGURE 43.24 (a) Two-dimensional representation of a semiconductor containing an acceptor atom (red spot). (b) Energy-band diagram of a semiconductor in which the acceptor levels lie within the forbidden gap, just above the top of the valence band.

Semiconductors doped with donor atoms are called *n*-**type semiconductors** because the majority of charge carriers are electrons, the charge of which is negative.

If the semiconductor is doped with atoms containing three outer-shell electrons, such as indium and aluminum, the three form covalent bonds with neighboring atoms, leaving an electron deficiency, or hole, in the fourth bond (Fig. 43.24a). The energy levels of such impurities also lie within the energy gap, this time just above the valence band, as in Figure 43.24b. Electrons from the valence band have enough thermal energy at room temperature to fill these impurity levels, leaving behind a hole in the valence band. Because a trivalent atom in effect accepts an electron from the valence band, such impurities are referred to as **acceptors.** A semiconductor doped with trivalent (acceptor) impurities is known as a *p*-**type semiconductor** because the charge carriers are positively charged holes.

When conduction is dominated by acceptor or donor impurities, the material is called an **extrinsic semiconductor.** The typical range of doping densities for *n*- or *p*-type semiconductors is 10^{13} to 10^{19} cm^{-3}. This is to be compared with a typical semiconductor density of roughly 10^{21} atoms/cm^3.

*43.7 SEMICONDUCTOR DEVICES

The *p-n* Junction

Now let us consider what happens when a *p*-type semiconductor is joined to an *n*-type semiconductor to form a *p-n* junction. The junction consists of three distinct semiconductor regions shown in Figure 43.25a: a *p*-type region, a depletion region, and an *n*-type region. The depletion region, which extends several micrometers to either side of the center of the junction, may be visualized as arising when

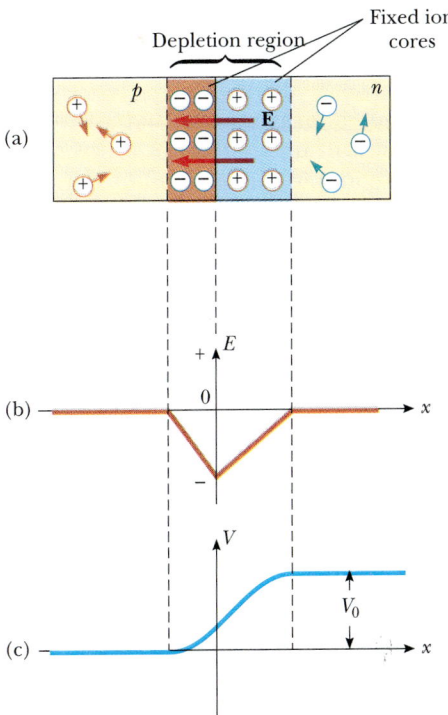

(a)

(b)

(c)

FIGURE 43.25 (a) Physical arrangement of a *p-n* junction. (b) Built-in electric field versus *x* for the *p-n* junction. (c) Built-in potential versus *x* for the *p-n* junction.

 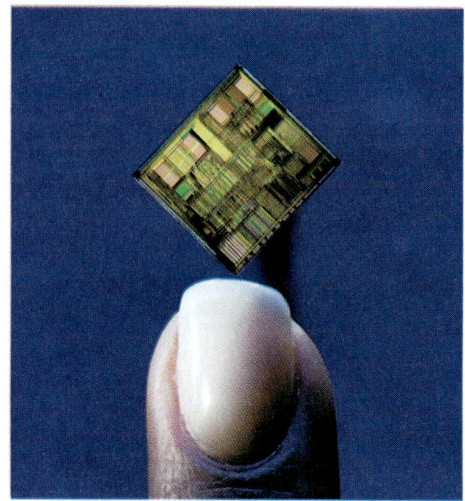

(*Left*) This high-frequency gallium arsenide diode emits pink light when biased in the forward direction.   (*Mike McNamee/SPL/Photo Researchers*) (*Right*) The high-performance Pentium™ processor chip contains approximately 3.3 million transistors, yet is about the size of a fingernail.   (*Courtesy of Intel Corporation*)

the two halves of the junction are brought together. Mobile donor electrons from the *n* side nearest the junction (blue area in Figure 43.25a) diffuse to the *p* side, leaving behind immobile positive ions. (Conversely, holes from the *p* side nearest the junction diffuse to the *n* side and leave behind a region of fixed negative ions.) Thus, the depletion region is so named because it is depleted of mobile charge carriers. It also contains a built-in electric field on the order of 10^4 to 10^6 V/cm, which serves to sweep mobile charge out of this region and keep it truly depleted. This internal electric field creates a potential barrier V_0 that prevents the further diffusion of holes and electrons across the junction and insures zero current through the junction when no external voltage is applied.

Perhaps the most notable feature of the *p-n* junction is its ability to pass current in only one direction. Such *diode* action is easiest to understand in terms of the potential diagram in Figure 43.25c. If a positive external voltage is applied to the *p* side of the junction, the overall barrier is decreased, resulting in a current that increases exponentially with increasing forward voltage, or *forward bias*. For *reverse bias* (a positive external voltage applied to the *n* side of the junction), the potential barrier is increased, resulting in a very small reverse current that quickly reaches a saturation value, I_0, as the reverse bias is increased. The current-voltage relationship for an ideal diode is

$$I = I_0(e^{qV/k_B T} - 1) \tag{43.29}$$

where q is the electronic charge, k_B is Boltzmann's constant, and T is the temperature in kelvin. Figure 43.26 shows an *I-V* plot characteristic of a real diode along with a schematic of a diode under forward bias.

The Junction Transistor

The discovery of the transistor by John Bardeen, Walter Brattain, and William Shockley in 1948 totally revolutionized the world of electronics. For this work, these three men shared a Nobel prize in 1956. By 1960, the transistor had replaced the vacuum tube in many electronic applications. The advent of the transistor

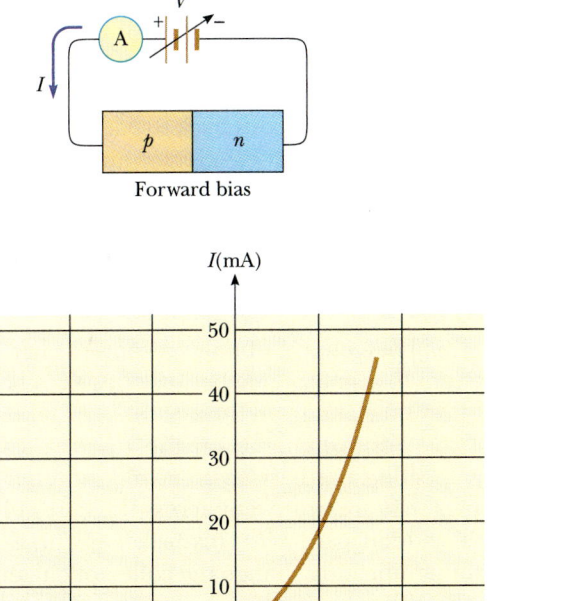

FIGURE 43.26 The characteristic curve for a real diode.

created a multibillion dollar industry that produces such popular devices as pocket radios, handheld calculators, computers, television receivers, and electronic games.

The junction transistor consists of a semiconducting material in which a very narrow *n* region is sandwiched between two *p* regions. This configuration is called a *pnp* **transistor.** Another configuration is the *npn* **transistor,** which consists of a *p* region sandwiched between two *n* regions. Because the operation of the two transistors is essentially the same, we describe only the *pnp* transistor.

The structure of the *pnp* transistor, together with its circuit symbol, is shown in Figure 43.27. The outer regions are called the **emitter** and **collector,** and the

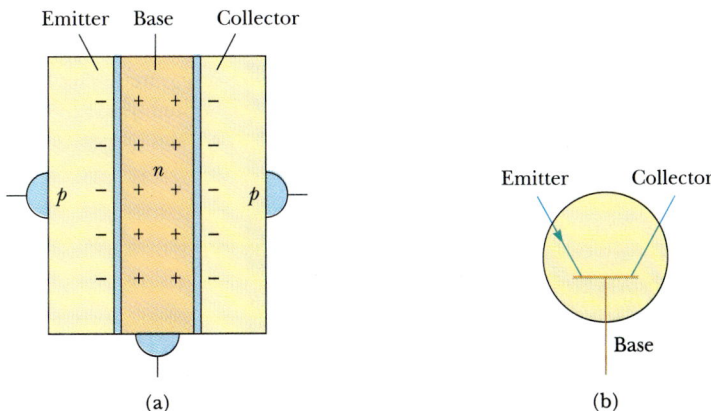

FIGURE 43.27 (a) The *pnp* transistor consists of an *n* region (base) sandwiched between two *p* regions (the emitter and the collector). (b) The circuit symbol for the *pnp* transistor.

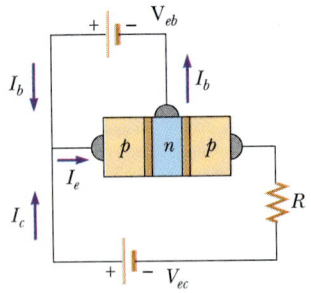

FIGURE 43.28 A bias voltage V_{eb} applied to the base as shown produces a small base current, I_b, which is used to control the collector current, I_c.

narrow central region is called the **base.** The configuration contains two junctions: the emitter-base interface and the collector-base interface.

Suppose a voltage is applied to the transistor such that the emitter is at a higher potential than the collector. (This is accomplished with battery V_{ec} in Fig. 43.28.) If we think of the transistor as two diodes back to back, we see that the emitter-base junction is forward-biased and the base-collector junction is reverse-biased.

Because the *p*-type emitter is heavily doped relative to the base, nearly all the current consists of holes moving across the emitter-base junction. Most of these holes do not recombine in the base because it is very narrow. The holes are finally accelerated across the reverse-biased base-collector junction, producing the current I_e in Figure 43.28.

Although only a small percentage of holes recombine in the base, those that do limit the emitter current to a small value because positive charge carriers accumulate in the base and prevent holes from flowing in. In order to prevent this current limitation, some of the positive charge on the base must be drawn off; this is accomplished by connecting the base to a second battery, V_{eb}, in Figure 43.28. Those positive charges that are not swept across the base-collector junction leave the base through this added pathway. This base current, I_b, is very small, but a small change in it can significantly change the collector current, I_c. If the transistor is properly biased, the collector (output) current is directly proportional to the base (input) current and the transistor acts as a current amplifier. This condition may be written as

$$I_c = \beta I_b$$

where β, the current gain, is typically in the range from 10 to 100. Thus, the transistor may be used to amplify a small, time-varying signal. The small voltage to be amplified is placed in series with the battery V_{eb} as shown in Figure 43.28. The time-dependent input signal produces a small variation in the base current, resulting in a large change in collector current and hence a large change in voltage across the output resistor.

The Integrated Circuit

The integrated circuit, invented independently by Jack Kilby at Texas Instruments in late 1958 and by Robert Noyce at Fairchild Camera and Instrument in early 1959, has been justly called "the most remarkable technology ever to hit mankind." Kilby's first device is shown in Figure 43.29. Integrated circuits have indeed

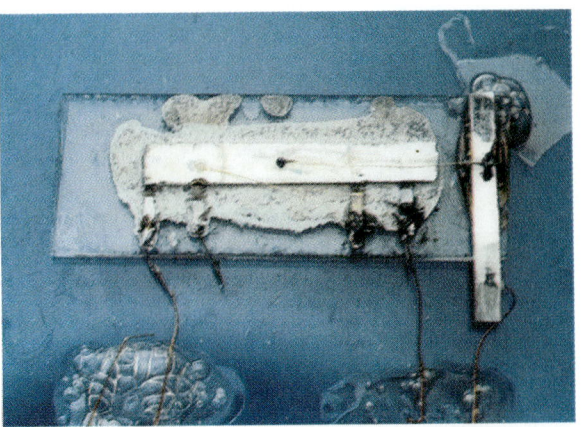

FIGURE 43.29 Jack Kilby's first integrated circuit, tested on September 12, 1958. *(Courtesy of Texas Instruments, Inc.)*

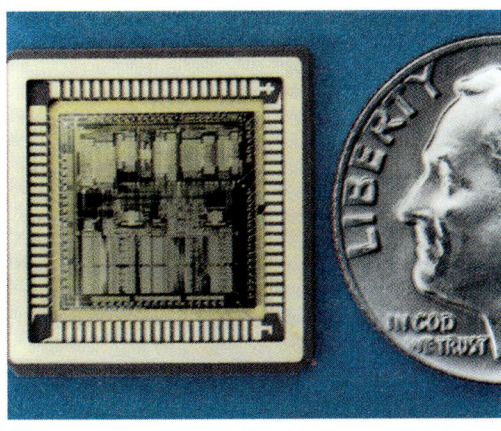

FIGURE 43.30 A 32-bit microprocessor integrated circuit containing over 200 000 components is about the size of a dime.   *(Courtesy of AT&T Bell Labs)*

started a second industrial revolution and are found at the heart of computers, watches, cameras, automobiles, aircraft, robots, space vehicles, and all sorts of communication and switching networks. In simplest terms, an integrated circuit is a collection of interconnected transistors, diodes, resistors, and capacitors fabricated on a single piece of silicon, affectionately known as a chip. State-of-the-art chips easily contain several hundred thousand components in a 1-cm^2 area (Fig. 43.30).

It is interesting that integrated circuits were invented partly in an effort to achieve circuit miniaturization and partly to solve the interconnection problem spawned by the transistor. In the era of vacuum tubes, power and size considerations of individual components set modest limits on the number of components that could be interconnected in a given circuit. With the advent of the tiny, low-power, highly reliable transistor, design limits on the number of components disappeared and were replaced with the problem of wiring together hundreds of thousands of components. The magnitude of this problem can be appreciated when we consider that second-generation computers (consisting of discrete transistors) contained several hundred thousand components requiring more than a million hand-soldered joints to be made and tested.

In addition to solving the interconnection problem, integrated circuits possess the advantages of miniaturization and fast response, two attributes critical for high-speed computers. The fast response results from the miniaturization and close packing of components, because the response time of a circuit depends on the time it takes for electrical signals traveling at about 0.3 m/ns to pass from one component to another. This time is clearly reduced by closer packing of components.

SUMMARY

Two or more atoms combine to form molecules because of a net attractive force between the atoms. The mechanisms responsible for molecular bonding can be classified as follows:

- **Ionic bonds** form in certain molecules primarily because of the Coulomb attraction between oppositely charged ions. Sodium chloride (NaCl) is one example.
- **Covalent bonds** form when the constituent atoms share electrons. For example, the two electrons of the H_2 molecule are equally shared between the nuclei.

- **Van der Waals bonds** are weak electrostatic bonds between atoms that do not form ionic or covalent bonds. They are responsible for the condensation of inert gas atoms and nonpolar molecules into the liquid phase.
- **Hydrogen bonds** are formed by the sharing of a proton between two atoms such as oxygen, nitrogen, or fluorine.

The energy of a gas molecule consists of contributions from the electronic (chemical) energy, the translation of the molecule, rotations, and vibrations. The allowed values of the rotational energy of a heteronuclear diatomic molecule are

$$E_{\text{rot}} = \frac{\hbar^2}{2I} J(J+1) \qquad J = 0, 1, 2, \ldots \tag{43.6}$$

where I is the moment of inertia of the molecule and J is an integer called the **rotational quantum number.** The selection rule for transitions between rotational levels (which usually lie in the far infrared region) is given by $\Delta J = \pm 1$.

The allowed values of the vibrational energy of a diatomic molecule are given by

$$E_{\text{vib}} = (v + \tfrac{1}{2}) \frac{h}{2\pi} \sqrt{\frac{k}{\mu}} \qquad v = 0, 1, 2, \ldots \tag{43.10}$$

where v is the vibrational quantum number, k is the force constant of the effective spring bonding the molecule, and μ is the reduced mass of the molecule. The selection rule for allowed vibrational transitions (which lie in the infrared region) is $\Delta v = \pm 1$, and the separation between adjacent levels is equal.

Bonding mechanisms in solids can be classified in a manner similar to the schemes for molecules. For example, the Na^+ and Cl^- ions in NaCl form ionic bonds, while the carbon atoms in diamond form covalent bonds. The metallic bond, which is generally weaker than covalent bonds, is characterized by a net attractive force between the positive ion cores and the mobile valence electrons.

In a crystalline solid, the energy levels of the system form a set of bands. Electrons occupy the lowest-energy states, with no more than one electron per state. There are energy gaps between the bands of allowed states.

In the free-electron theory of metals, the free electrons fill the quantized levels in accordance with the Pauli exclusion principle. The number of states per unit volume available to the conduction electrons having energies between E and $E + dE$ is

$$N(E) \, dE = C \frac{E^{1/2} \, dE}{e^{(E - E_F)/k_B T} + 1} \tag{43.22}$$

where C is a constant, and E_F is the **Fermi energy.** At $T = 0$ K, all levels below E_F are filled, all levels above E_F are empty, and

$$E_F = \frac{\hbar^2}{2m} \left(\frac{3n}{8\pi} \right)^{2/3} \tag{43.25}$$

where n is the total number of conduction electrons per unit volume. Only those electrons having energies near E_F can contribute to the electrical conductivity of the metal.

A **semiconductor** is a material having an energy gap of approximately 1 eV and a valence band that is filled at $T = 0$ K. Because of their small energy gap, a significant number of electrons can be thermally excited from the valence band into the conduction band as the temperature increases. The band structures and electrical properties of a semiconductor can be modified by adding donor atoms containing five valence electrons (such as arsenic) or acceptor atoms containing three valence electrons (such as indium). A semiconductor **doped** with donor impurity atoms is called an **n-type semiconductor,** while one doped with acceptor impurity atoms is called **p-type.** The energy levels of these impurity atoms fall within the energy gap of the material.

QUESTIONS

1. Discuss the three major forms of excitation of a molecule (other than translational motion) and the relative energies associated with the three excitations.
2. Explain the role of the Pauli exclusion principle in describing the electrical properties of metals.
3. Discuss the properties of a material that determine whether it is a good electrical insulator or a good conductor.
4. Table 43.5 shows that the energy gaps for semiconductors decrease with increasing temperature. What do you suppose accounts for this behavior?
5. The resistivity of metals increases with increasing temperature, whereas the resistivity of an intrinsic semiconductor decreases with increasing temperature. Explain.
6. Discuss the differences in the band structures of metals, insulators, and semiconductors. How does the band structure model enable you to better understand the electrical properties of these materials?
7. Discuss the models responsible for the different types of bonds that form stable molecules.
8. Discuss the electrical, physical, and optical properties of ionically bonded solids.
9. Discuss the electrical and physical properties of covalently bonded solids.
10. Discuss the electrical and physical properties of metals.

11. When a photon is absorbed by a semiconductor, an electron-hole pair is created. Give a physical explanation of this statement using the energy-band model as the basis for your description.
12. Pentavalent atoms such as arsenic are donor atoms in a semiconductor such as silicon, while trivalent atoms such as indium are acceptors. Inspect the periodic table in Appendix C, and determine what other elements might make good donors or acceptors.
13. What are the essential assumptions made in the free-electron theory of metals? How does the energy band model differ from the free-electron theory in describing the properties of metals?
14. How do the vibrational and rotational levels of heavy hydrogen (D_2) molecules compare with those of H_2 molecules?
15. Which is easier to excite in a diatomic molecule, rotational or vibrational motion?
16. The energy of visible light ranges between 1.8 and 3.2 eV. Does this explain why silicon, with an energy gap of 1.1 eV (Table 43.5), appears opaque, whereas diamond, with an energy gap of 5.5 eV, appears transparent?
17. Why is a *pnp* or *npn* sandwich (with the central region very thin) essential to transistor operation?
18. How can the analysis of the rotational spectrum of a molecule lead to an estimate of the size of that molecule?

PROBLEMS

Section 43.1 Molecular Bonds

1. A K^+ ion and a Cl^- ion are 5.0×10^{-10} m from each other. Assuming the two ions act like point charges, determine (a) the force of attraction between them and (b) the potential energy of attraction in electron volts.

2. One description of the potential energy of diatomic molecules is given by the Lenard-Jones potential,

$$U = \frac{A}{r^{12}} - \frac{B}{r^6}$$

where A and B are constants. Find, in terms of A and B, (a) the value r_0 at which the energy is a minimum

and (b) the energy E required to break up a diatomic molecule. (c) Evaluate r_0 in meters and E in electron volts for the H_2 molecule. In your calculations, take $A = 0.124 \times 10^{-120}$ eV · m^{12} and $B = 1.488 \times 10^{-60}$ eV·m^6. (Note that although this potential is still widely used, it is now known to have serious defects. For example, its behavior at both small and large values of r is greatly in error.)

Section 43.2 The Energy and Spectra of Molecules

3. The cesium-iodide (CsI) molecule has an atomic separation of 0.127 nm. (a) Determine the energy of the lowest rotational state and the frequency of the

□ indicates problems that have full solutions available in the Student Solutions Manual and Study Guide.

photon absorbed in the $J = 0$ to $J = 1$ transition. (b) What would be the fractional change in this frequency if the estimate of the atomic separation is off by 10%?

4. The CO molecule makes a transition from the $J = 1$ to $J = 2$ rotational state when it absorbs a photon of frequency 2.30×10^{11} Hz. Find the moment of inertia of this molecule.

5. Use the data in Table 43.2 to calculate the minimum amplitude of vibration for (a) the HI molecule and (b) the HF molecule. Which has the weaker bond?

6. The nuclei of the O_2 molecule are separated by 1.2×10^{-10} m. The mass of each oxygen atom in the molecule is 2.66×10^{-26} kg. (a) Determine the rotational energies of an oxygen molecule in electron volts for the levels corresponding to $J = 0$, 1, and 2. (b) The effective force constant k between the atoms in the oxygen molecule is 1177 N/m. Determine the vibrational energies (in electron volts) corresponding to $v = 0$, 1, and 2.

7. Figure P43.7 is a model of a benzene molecule. All atoms lie in a plane, and the carbon atoms form a regular hexagon, as do the hydrogen atoms. The carbon atoms are 0.110 nm apart center to center. Determine the allowed energies of rotation around an axis perpendicular to the plane of the paper through the center point O. Hydrogen and carbon atoms have masses of 1.67×10^{-27} kg and 1.99×10^{-26} kg, respectively.

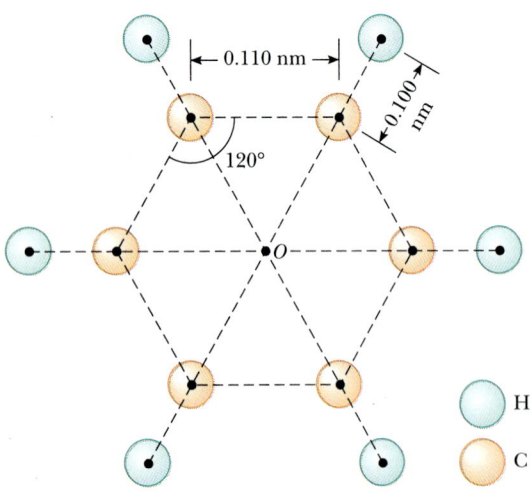

0.110 nm

0.100 nm

120°

H

C

FIGURE P43.7

8. If the CO molecule were rigid, the rotational transition into what J state could absorb the same wavelength photon as the 0 to 1 vibrational transition? (Use information given in Table 43.2.)

9. The HCl molecule is excited to its first rotational-energy level, corresponding to $J = 1$. If the distance be-

tween its nuclei is 0.1275 nm, what is the angular speed of the molecule around its center of mass?

10. (a) A diatomic molecule consists of two atoms having masses m_1 and m_2 separated by a distance r. Show that the moment of inertia around the center of mass of a diatomic molecule is given by Equation 43.3, $I = \mu r^2$. (b) Calculate the moment of inertia of NaCl around its center of mass ($r = 0.28$ nm). (c) Calculate the wavelength of radiation if a NaCl molecule undergoes a transition from the $J = 2$ state to the $J = 1$ state.

11. If the effective force constant of a vibrating HCl molecule is $k = 480$ N/m, estimate the energy difference between the ground state and the first vibrational level.

12. The rotational spectrum of the HCl molecule contains lines corresponding to wavelengths of 0.0604, 0.0690, 0.0804, 0.0964, and 0.1204 mm. What is the moment of inertia of the molecule?

13. The separation between the oxygen atoms of O_2 is 1.2×10^{-10} m. Treating the atoms as particles, determine the values of the rotational energy for the $J = 1$ and $J = 2$ states.

Section 43.3   Bonding in Solids

14. Use Equation 43.17 to calculate the ionic cohesive energy for NaCl. Take $\alpha = 1.7476$, $r_0 = 0.281$ nm, and $m = 8$.

15. The distance between the K^+ and Cl^- ions in a KCl crystal is 0.314 nm. Calculate the distance from a K^+ ion to the three closest K^+ ions.

16. Consider a one-dimensional chain of alternating positive and negative ions. Show that the potential energy of an ion in this hypothetical crystal is

$$U(r) = -k_e \alpha \frac{e^2}{r}$$

where $\alpha = 2 \ln 2$ (the Madelung constant) and r is the interionic spacing. [*Hint:* Use the series expansion for $\ln(1 + x)$.]

Section 43.4   Band Theory of Solids and
Section 43.5   Free-Electron Theory of Metals

17. Show that Equation 43.24 can be expressed as $E_F = (3.65 \times 10^{-19}) n^{2/3}$ eV, where E_F is in electron volts when n is in electrons per cubic meter.

18. The Fermi energy for silver is 5.48 eV. In addition, silver has a density of 10.6×10^3 kg/m^3 and an atomic mass of 108. Use this information to show that silver has one valence (free) electron per atom.

19. (a) Find the typical speed of a conduction electron in copper whose kinetic energy is equal to the Fermi energy, 7.05 eV. (b) How does this compare with the drift speed of 0.10 mm/s?

20. Sodium is a monovalent metal having a density of

0.971 g/cm³ and a molar mass of 23.0 g/mol. Use this information to calculate (a) the density of charge carriers, (b) the Fermi energy, and (c) the Fermi speed for sodium.

21. Calculate the energy of a conduction electron in silver at 800 K if the probability of finding the electron in that state is 0.95. The Fermi energy is 5.48 eV at this temperature.

22. Show that the average kinetic energy of a conduction electron in a metal at 0 K is $E_{av} = \frac{3}{5}E_F$. (*Hint:* In general, the average kinetic energy is

$$E_{av} = \frac{1}{n}\int EN(E)\ dE$$

where n is the density of particles, and $N(E)\ dE$ is given by Equation 43.22.)

23. Consider a cube of gold 1.00 mm on an edge. Calculate the approximate number of conduction electrons in this cube whose energies lie in the range 4.000 to 4.025 eV.

24. An electron moves in a three-dimensional box of edge length L and volume L^3. If the wave function of the particle is $\psi = A\sin(k_x x)\sin(k_y y)\sin(k_z z)$, show that its energy is

$$E = \frac{\hbar^2\pi^2}{2mL^2}(n_x^2 + n_y^2 + n_z^2)$$

where the quantum numbers (n_x, n_y, n_z) are ≥ 1. (*Hint:* The Schrödinger equation in three dimensions may be written

$$\frac{d^2\psi}{dx^2} + \frac{d^2\psi}{dy^2} + \frac{d^2\psi}{dz^2} = \frac{2m}{\hbar^2}(U - E)\psi$$

To confine the electron inside the box, take $U = 0$ inside and $U = \infty$ outside.)

25. (a) Consider a system of electrons confined to a three-dimensional box. Calculate the ratio of the number of allowed energy levels at 8.5 eV to the number at 7.0 eV. (b) Copper has a Fermi energy of 7.0 eV at 300 K. Calculate the ratio of the number of occupied levels at an energy of 8.5 eV to the number at the Fermi energy. Compare your answer with that obtained in part (a).

Section 43.6 Electrical Conduction in Metals, Insulators, and Semiconductors

26. The energy gap for silicon at 300 K is 1.14 eV. (a) Find the lowest frequency photon that will promote an electron from the valence band to the conduction band. (b) What is the wavelength of this photon?

27. Light from a hydrogen discharge tube is incident on a CdS crystal. Which spectral lines from the Balmer series are absorbed and which are transmitted?

28. A light-emitting diode (LED) made of the semicon-

ductor GaAsP emits red light ($\lambda = 650$ nm). Determine the energy band gap E_g in the semiconductor.

29. Most solar radiation has a wavelength of 10^{-6} m or less. What energy gap should the material in a solar cell have in order to absorb this radiation? Is silicon appropriate (see Table 43.5)?

*Section 43.7 Semiconductor Devices

30. For what value of the bias voltage V in Equation 43.29 does (a) $I = 9I_0$? (b) $I = -0.9I_0$? Assume $T = 300$ K.

31. The diode shown in Figure 43.26 is connected in series with a battery and a 150-Ω resistor. What battery emf is required for a current of 25 mA?

ADDITIONAL PROBLEMS

32. The effective spring constant associated with bonding in the N_2 molecule is 2297 N/m. The nitrogen atoms each have a mass of 2.32×10^{-26} kg, and the nuclei are 0.120 nm apart. Assume that the molecule is rigid and in the ground vibrational state. Calculate the J value of the rotational state that would have the same energy as the first excited vibrational state.

33. The hydrogen molecule comes apart (dissociates) when it is excited internally by 4.5 eV. Assuming that this molecule behaves like a harmonic oscillator having classical angular frequency $\omega = 8.28 \times 10^{14}$ rad/s, find the highest vibrational quantum number beneath the 4.5-eV dissociation energy.

34. Liquid helium solidifies as each atom bonds with four others, and each bond has an average energy of 1.74×10^{-23} J. Find the latent heat of fusion for helium in joules per gram. (The molar mass of He is 4 g/mol.)

35. Show that the ionic cohesive energy of an ionically bonded solid is given by Equation 43.17. (*Hint:* Start with Equation 43.16, and note that $dU/dr = 0$ at $r = r_0$.)

36. A particle of mass m moves in one-dimensional motion in a region where its potential energy is

$$U(x) = \frac{A}{x^3} - \frac{B}{x}$$

where A and B are constants with appropriate units. The general shape of this function is shown in Figure 43.12, where x replaces r. (a) Find the static equilibrium position x_0 of the particle in terms of m, A, and B. (b) Determine the depth U_0 of this potential well. (c) In moving along the x axis, what maximum force toward the negative x direction does the particle experience?

37. The Madelung constant may be found by summing an infinite alternating series of terms giving the electrostatic potential energy between a Na^+ ion and its six nearest Cl^- neighbors, its twelve next-nearest Na^+

neighbors, and so on (Fig. 43.11a). (a) From this expression, show that the first three terms of the series yield $\alpha = 2.13$ for the NaCl structure. (b) Does this series converge rapidly? Calculate the fourth term as a check.

SPREADSHEET PROBLEMS

S1. The Fermi-Dirac distribution function can be written as

$$f(E) = \frac{1}{e^{(E - E_F)/k_B T} + 1} = \frac{1}{e^{(E/E_F - 1)T_F/T} + 1}$$

Prepare a spreadsheet that will enable you to calculate and plot $f(E)$ versus E/E_F at a fixed temperature T. Examine the curves obtained for $T = 0.1T_F$, $0.2T_F$, and $0.5T_F$, where $T_F = E_F/k_B$.

S2. A certain metal at a temperature of 290 K ($k_B T = 0.025$ eV) has a Fermi energy of 5.0 eV. (a) Prepare a spreadsheet that will enable you to plot the probability function for the region 4.8 eV $\leq E \leq$ 5.2 eV and use this to estimate the area under the curve above the Fermi energy. Note that this provides the fraction of valence electrons in the conduction band. (b) What fraction of the total area is this result?

S3. Copper has a Fermi energy of 7.05 eV and a conduction electron concentration of 8.49×10^{28} m^{-3} at 300 K. Prepare a spreadsheet that plots the particle distribution function, $N(E)$, as a function of energy for a range of temperatures $T = 100$ to $T = 1300$ K. (The melting point of copper is 1356 K.) Your energy scales should range from 0 to 10 eV. Describe the behavior of the distribution function as T increases. How would the distribution change if copper is replaced by sodium? (See Table 43.4.)

CHAPTER 44

Superconductivity

Photograph of a small permanent magnet levitated above a pellet of the $Y_1Ba_2Cu_3O_{7-\delta}$ (123) superconductor cooled to liquid nitrogen temperature. *(Courtesy of IBM Research Laboratories)*

The phenomenon of superconductivity has always been very exciting, both for fundamental scientific interest and because of its many applications. The discovery of high-temperature superconductivity in certain metallic oxides in the 1980s sparked even greater excitement in the scientific and business communities. This major breakthrough is viewed by many scientists to be as important as the invention of the transistor. For this reason, it is important that all students of science and engineering understand the basic electromagnetic properties of superconductors and become aware of the scope of their current applications.

Superconductors have many unusual electromagnetic properties, and most applications take advantage of such properties. For example, once a current is produced in a superconducting ring maintained at a sufficiently low temperature, that current will persist with no measurable decay. The superconducting ring exhibits no electrical resistance to direct currents, no heating, and no losses. In addition to the property of zero resistance, certain superconductors expel applied magnetic fields so that the field is always zero everywhere inside the superconductor.

As we shall see, classical physics cannot explain the behavior and properties of superconductors. In fact, the superconducting state is now known to be a special quantum condensation of electrons. This quantum behavior has been verified through such observations as the quantization of magnetic flux produced by a superconducting ring.

In this chapter, we first give a brief historical review of superconductivity, beginning with its discovery in 1911 and ending with recent developments in high-temperature superconductivity. In describing some of the electromagnetic properties displayed by superconductors, we use simple physical arguments whenever possible. The essential features of the theory of superconductivity are reviewed with the realization that a detailed study is beyond the scope of this text. We then discuss many of the important applications of superconductivity and speculate on potential future applications.

44.1 BRIEF HISTORICAL REVIEW

The era of low temperature physics began in 1908 when the Dutch physicist Heike Kamerlingh Onnes first liquefied helium, which boils at 4.2 K. In 1911, Onnes and one of his assistants discovered the phenomenon of superconductivity while studying the resistivity of metals at low temperatures.[1] They first studied platinum and found that its resistivity, when extrapolated to 0 K, depends on its purity. Then they decided to study mercury because very pure samples could easily be prepared by distillation. Much to their surprise, the resistance of the mercury sample dropped sharply at 4.15 K to an unmeasurably small value. It was quite natural for Onnes to choose the name **superconductivity** for this new phenomenon. Figure 44.1 shows the experimental results for mercury and platinum. Note that platinum does not exhibit superconducting behavior, as indicated by its finite resistivity as T approaches 0 K. In 1913, Onnes was awarded the Nobel prize in physics for the study of matter at low temperatures and the liquefaction of helium.

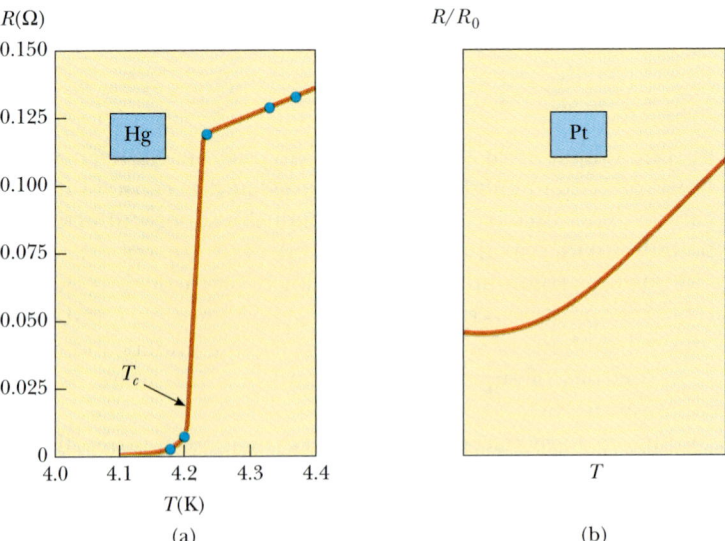

FIGURE 44.1 Plots of resistance versus temperature for (a) mercury (the original data published by Onnes) and (b) platinum. Note that the resistance of mercury follows the path of a normal metal above the critical temperature T_c and then suddenly drops to zero at the critical temperature, which is 4.15 K for mercury. In contrast, the data for platinum show a finite resistance R_0 even at very low temperatures.

[1] H. K. Onnes, *Leiden Comm.* **120b**, **122b**, **124c** (1911).

We now know that the resistivity of a superconductor is truly zero. Soon after the discovery by Onnes, many other elemental metals were found to exhibit zero resistance when their temperature was lowered below a certain characteristic temperature of the material called the **critical temperature,** T_c.

The magnetic properties of superconductors are as dramatic and as difficult to understand as their electrical properties. In 1933 W. Hans Meissner and Robert Ochsenfeld studied the magnetic behavior of superconductors and found that when certain ones are cooled below their critical temperature in the presence of a magnetic field, *the magnetic flux is expelled from the interior of the superconductor.*[2] Furthermore, these materials lost their superconducting behavior above a certain temperature-dependent **critical magnetic field,** $B_c(T)$. A phenomenological theory of superconductivity was developed in 1935 by Fritz London and Heinz London,[3] but the actual nature and origin of the superconducting state were first explained by John Bardeen, Leon N. Cooper, and J. Robert Schrieffer in 1957.[4] A central feature of this theory, commonly referred to as the BCS theory, is the formation of bound two-electron states called **Cooper pairs.** In 1962, Brian D. Josephson predicted a tunneling current between two superconductors separated by a thin (<2 mm) insulating barrier, where the current is carried by these paired electrons.[5] Shortly thereafter, Josephson's predictions were verified, and today a whole field of device physics exists based on the Josephson effect. Early in 1986, J. Georg Bednorz and Karl Alex Müller reported evidence for superconductivity in an oxide of lanthanum, barium, and copper at a temperature of approximately 30 K.[6] This was a major breakthrough in superconductivity because the highest known value of T_c at that time was about 23 K in a compound of niobium and germanium. This remarkable discovery, which marks the beginning of a new era of high-temperature superconductivity, received worldwide attention in both the scientific community and the business world. Recently, researchers have reported critical temperatures as high as 150 K in more complex metallic oxides, but the mechanisms responsible for superconductivity in these materials remain unclear.

Until the discovery of high-temperature superconductors, the use of superconductors required coolant baths of liquefied helium (rare and expensive) or liquid hydrogen (very explosive). On the other hand, superconductors with $T_c > 77$ K require only liquid nitrogen (which boils at 77 K and is comparatively inexpensive, abundant, and inert). If superconductors with T_c's above room temperature are ever found, human technology will be drastically altered.

44.2 SOME PROPERTIES OF TYPE I SUPERCONDUCTORS

Critical Temperature and Critical Magnetic Field

The critical temperatures of some superconducting elements, classified as **type I superconductors,** are given in Table 44.1. Note the absence of copper, silver, and gold, which are excellent electrical conductors at ordinary temperatures, but do not exhibit superconductivity.

[2] W. Meissner and R. Ochsenfeld, *Naturwissenschaften* **21**, 787 (1933).

[3] F. London and H. London, *Proc. Roy. Soc. (London)* A149, 71 (1935).

[4] J. Bardeen, L. N. Cooper, and J. R. Schrieffer, *Phys. Rev.* **108**, 1175 (1957).

[5] B. D. Josephson, *Phys. Letters* **1**, 251 (1962).

[6] J. G. Bednorz and K. A. Müller, *Z. Phys.* **B 64**, 189 (1986).

TABLE 44.1	Critical Temperatures and Critical Magnetic Fields (Measured at $T = 0$ K) of Some Elemental Superconductors	
Superconductor	T_c (K)	$B_c(0)$ in Tesla
Al	1.196	0.0105
Ga	1.083	0.0058
Hg	4.153	0.0411
In	3.408	0.0281
Nb	9.26	0.1991
Pb	7.193	0.0803
Sn	3.722	0.0305
Ta	4.47	0.0829
Ti	0.39	0.010
V	5.30	0.1023
W	0.015	0.000115
Zn	0.85	0.0054

In the presence of an applied magnetic field **B**, the value of T_c decreases with increasing magnetic field, as indicated in Figure 44.2 for several type I superconductors. When the magnetic field exceeds the critical field, **B**$_c$, the superconducting state is destroyed and the material behaves like a normal conductor having finite resistance.

The critical magnetic field varies with temperature according to the approximate expression

$$B_c(T) \cong B_c(0)\left[1 - \left(\frac{T}{T_c} \right)^2 \right] \qquad (44.1)$$

As you can see from this equation and Figure 44.2, the value of B_c is a maximum at 0 K. The value of $B_c(0)$ is found by determining B_c at some finite temperature and extrapolating back to 0 K, a temperature that cannot be achieved. The maximum current that can be sustained in a type I superconductor is limited by the value of the critical field.

Note that $B_c(0)$ is the maximum magnetic field required to destroy superconductivity in a given material. If the applied field exceeds $B_c(0)$, the metal ceases to

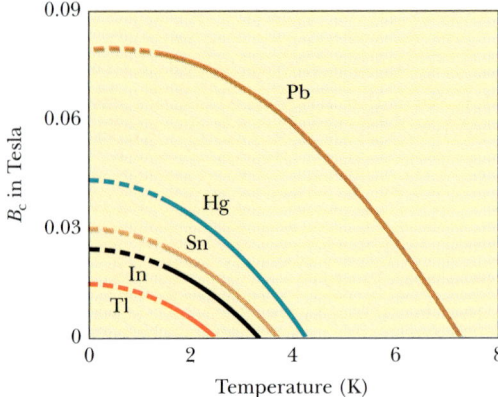

FIGURE 44.2 The critical magnetic field versus temperature for several type I superconductors. Extrapolations of these fields to 0 K give the critical fields listed in Table 44.1. For a given metal, the material is superconducting at fields and temperatures below its critical temperature and behaves like a normal conductor above that curve.

become superconducting at any temperature. Values for the critical field for type I superconductors are quite low as shown in Table 44.1. For this reason, type I superconductors cannot be used to construct high-field magnets, called superconducting magnets.

Magnetic Properties of Type I Superconductors

Simple arguments based on the laws of electricity and magnetism can be used to show that the magnetic field inside a superconductor cannot change with time. According to Ohm's law, the electric field inside a conductor is proportional to the resistance of the conductor. Thus, since $R = 0$ for a superconductor, *the electric field in its interior must be zero.* Now recall that Faraday's law of induction can be expressed as

$$\oint \mathbf{E} \cdot d\mathbf{s} = -\frac{d\Phi_B}{dt} \qquad (44.2)$$

That is, the line integral of the electric field around any closed loop is equal to the negative rate of change in the magnetic flux Φ_B through the loop. Since $\mathbf{E}$ is zero everywhere inside the superconductor, the integral over any closed path inside the superconductor is zero. Hence, $d\Phi_B/dt = 0$, which tells us that *the magnetic flux in the superconductor cannot change.* From this, we can conclude that $B (= \Phi_B/A)$ *must remain constant inside the superconductor.*

Before 1933, it was assumed that superconductivity was a manifestation of perfect conductivity. If a perfect conductor is cooled below its critical temperature in the presence of an applied magnetic field, the field should be trapped in the interior of the conductor even after the field is removed. The final state of a perfect conductor in an applied magnetic field should depend upon which occurs first, the application of the field or the cooling below the critical temperature. If the field is applied after cooling below T_c, the field should be expelled from the superconductor. On the other hand, if the field is applied before cooling, the field should not be expelled from the superconductor after cooling below T_c.

When experiments were conducted in the 1930s to examine the magnetic behavior of superconductors, the results were quite different. In 1933, Meissner and Ochsenfeld discovered that, when a metal becomes superconducting in the presence of a weak magnetic field, the field was expelled so that $\mathbf{B} = 0$ everywhere in the interior of the superconductor. Thus, the same final state $\mathbf{B} = 0$ was achieved whether the field was applied before or after the material was cooled below its critical temperature. This effect is illustrated in Figure 44.3 for a material in the shape of a long cylinder. Note that the field penetrates the cylinder when its temperature is greater than T_c (Fig. 44.3a). As the temperature is lowered below T_c, however, the field lines are spontaneously expelled from the interior of the superconductor (Fig. 44.3b). Thus, a type I superconductor is more than a perfect conductor (resistivity $\rho = 0$); it is also a perfect diamagnet ($\mathbf{B} = 0$). The phenomenon of the expulsion of magnetic fields from the interior of a superconductor is known as the **Meissner effect.** The property that $\mathbf{B} = 0$ in the interior of a type I superconductor is as fundamental as the property of zero resistance. If the applied field is sufficiently large ($\mathbf{B} > \mathbf{B}_c$), the superconducting state is destroyed and the field penetrates the sample.

Because a superconductor is a perfect diamagnet, it repels a permanent magnet. In fact, one can perform a dazzling demonstration of the Meissner effect by floating a small permanent magnet above a superconductor and achieving magnetic levitation. A dramatic photograph of magnetic levitation is shown at the

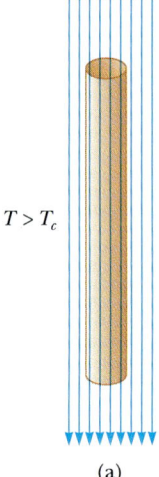

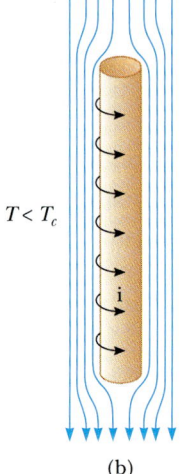

$T > T_c$ (a) $T < T_c$ (b) i

FIGURE 44.3 A type I superconductor in the form of a long cylinder in the presence of an external magnetic field. (a) At temperatures above T_c, the field lines penetrate the sample because it is in its normal state. (b) When the rod is cooled to $T < T_c$ and becomes superconducting, magnetic flux is excluded from its interior by the induction of surface currents.

opening of this chapter. The details of this demonstration are provided in Questions 17–23.

You should recall from our study of electricity that a good conductor expels static electric fields by moving charges to its surface. In effect, the surface charges produce an electric field that exactly cancels the externally applied field inside the conductor. In a similar manner, a superconductor expels magnetic fields by forming surface currents. To illustrate this point, consider again the superconductor in Figure 44.3. Let us assume that the sample is initially at a temperature $T > T_c$ as in Figure 44.3a, so that the field penetrates the cylinder. As the cylinder is cooled to a temperature $T < T_c$, the field is expelled as in Figure 44.3b. In this case, surface currents are induced on the superconductor, producing a magnetic field that exactly cancels the externally applied field inside the superconductor. As you would expect, the surface currents disappear when the external magnetic field is removed.

EXAMPLE 44.1   Critical Current in a Pb Wire

A lead wire has a radius of 3.00 mm and is at a temperature of 4.20 K. Find (a) the critical magnetic field in lead at this temperature and (b) the maximum current the wire can carry at this temperature.

Solution (a) We can use Equation 44.1 to find the critical field at any temperature if $B_c(0)$ and T_c are known. From Table 44.1, we see that the critical magnetic field of lead at 0 K is 0.0803 T and its critical temperature is 7.193 K. Hence, Equation 44.1 gives

$$B_c(4.20 \text{ K}) = (0.0803 \text{ T})\left[1 - \left(\frac{4.20}{7.193}\right)^2\right] = \boxed{0.0529 \text{ T}}$$

(b) Recall from Ampère's law (Ch. 30) that, for a wire

carrying a steady current I, the magnetic field generated at an exterior point a distance r from the wire is

$$B = \frac{\mu_0 I}{2\pi r}$$

When the current in the wire equals a certain critical current I_c, the magnetic field at the wire surface equals the critical magnetic field B_c. (Note that $B = 0$ inside because all the current is on the wire surface.) Using the expression above, and taking r equal to the radius of the wire, we find

$$I = \frac{2\pi r B}{\mu_0}$$

$$= 2\pi \frac{(3.00 \times 10^{-3} \text{ m})(0.0529 \text{ T})}{4\pi \times 10^{-7} \text{ N/A}^2} = \boxed{794 \text{ A}}$$

Magnetization

When a bulk sample is placed in an external magnetic field $\mathbf{B}_{ext}$, the sample acquires a magnetization $\mathbf{M}$ (see Section 30.9). The magnetic field $\mathbf{B}_{in}$ inside the sample is related to $\mathbf{B}_{ext}$ and $\mathbf{M}$ through the relationship $\mathbf{B}_{in} = \mathbf{B}_{ext} + \mu_0\mathbf{M}$. When the sample is in the superconducting state, $\mathbf{B}_{in} = 0$; therefore, it follows that the magnetization is

$$\mathbf{M} = -\frac{\mathbf{B}_{ext}}{\mu_0} = \chi\mathbf{B}_{ext} \qquad (44.3)$$

where χ $(= -1/\mu_0)$ in this case is again the magnetic susceptibility, as it was in Equation 30.36. That is, whenever a material is in a superconducting state, its magnetization opposes the external magnetic field and the magnetic susceptibility has its maximum negative value. Again, we see that a type I superconductor is a perfect diamagnetic.

A plot of magnetic field inside a type I superconductor versus external field (parallel to a long cylinder) at $T < T_c$ is shown in Figure 44.4a, while magnetization versus external field at some constant temperature is plotted in Figure 44.4b. Note that when $B_{ext} > B_c$, the magnetization is approximately zero.

With the discovery of the Meissner effect, Fritz and Heinz London were able to develop phenomenological equations for type I superconductors based on equilibrium thermodynamics. They could explain the critical magnetic field in terms of the energy increase of the superconducting state, an energy increase resulting from the exclusion of flux from the interior of the superconductor. According to equilibrium thermodynamics, a system prefers to be in the state having the lowest free energy. Hence, the superconducting state must have a lower free energy than the normal state. If E_s represents the energy of the superconducting state per unit volume and E_n the energy of the normal state per unit volume, then $E_s < E_n$ below T_c and the material becomes superconducting. The exclusion of a field B causes the total energy of the superconducting state to increase by an amount equal to $B^2/2\mu_0$ per unit volume. The critical field value is defined by the equation

$$E_s + \frac{B_c^2}{2\mu_0} = E_n \qquad (44.4)$$

Because the London theory also gives the temperature dependence of E_s, an exact expression for $B_c(T)$ could be obtained. Note that the field exclusion energy $B_c^2/2\mu_0$ is just the area under the curve in Figure 44.4b.

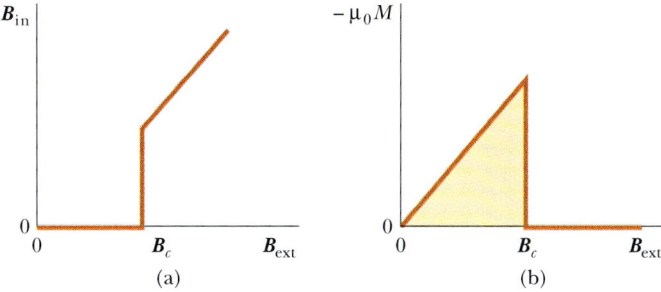

FIGURE 44.4 The magnetic field-dependent properties of a type I superconductor. (a) A plot of internal field versus external field, where $\mathbf{B}_{in} = 0$ for $\mathbf{B}_{ext} < \mathbf{B}_c$. (b) A plot of magnetization versus external field. Note that $\mathbf{M} \approx 0$ for $\mathbf{B}_{ext} > \mathbf{B}_c$.

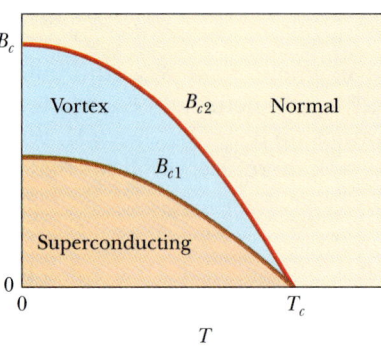

FIGURE 44.5 Critical magnetic fields as a function of temperature for a type II superconductor. Below B_{c1}, the material behaves like a type I superconductor. Above B_{c2}, the material behaves like a normal conductor. Between these two fields, the superconductor is in the vortex (mixed) state.

44.3 TYPE II SUPERCONDUCTORS

By the 1950s researchers knew there was another class of superconductors, known as type II superconductors. These materials are characterized by two critical magnetic fields, designated B_{c1} and B_{c2} in Figure 44.5. When the external magnetic field is less than the lower critical field B_{c1}, the material is entirely superconducting and there is no flux penetration, just as with type I superconductors. When the external field exceeds the upper critical field B_{c2}, the flux penetrates completely and the superconducting state is destroyed, just as for type I materials. For fields lying between B_{c1} and B_{c2}, however, the material is in a mixed state, referred to as the **vortex state.** (This name is given because of swirls of currents that are associated with this state.) While in the vortex state, the material can have zero resistance and has partial flux penetration. Vortex regions are essentially filaments of normal material that run through the sample when the external field exceeds the lower critical field, as illustrated in Figure 44.6. As the strength of the external field increases, the number of filaments increases until the field reaches the upper critical value, and the sample becomes normal.

The vortex state can be viewed as a cylindrical swirl of supercurrents surrounding a cylindrical normal metal core that allows some flux to penetrate the interior of the type II superconductor. Associated with each vortex filament is a magnetic field that is greatest at the core center and falls off exponentially outside the core. The supercurrents are the "source" of **B** for each vortex.

Critical temperatures and B_{c2} values for several type II superconductors are given in Table 44.2. The values of B_{c2} are very large when compared with the values

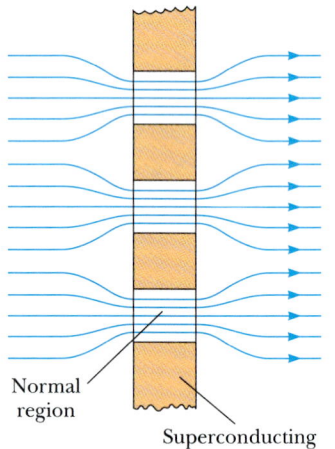

FIGURE 44.6 A schematic diagram of a type II superconductor in the vortex state. The sample contains filaments of normal (unshaded) regions through which magnetic field lines can pass. The field lines are excluded from the superconducting (shaded) regions.

TABLE 44.2 Values of the Critical Temperature and Upper Critical Magnetic Field (at $T = 0$ K) for Several Type II Superconductors

Superconductor	T_c (K)	$B_{c2}(0)$ in Tesla
Nb_3Al	18.7	32.4
Nb_3Sn	18.0	24.5
Nb_3Ge	23	38
NbN	15.7	15.3
NbTi	9.3	15
$Nb_3(AlGe)$	21	44
V_3Si	16.9	23.5
V_3Ga	14.8	20.8
PbMoS	14.4	60

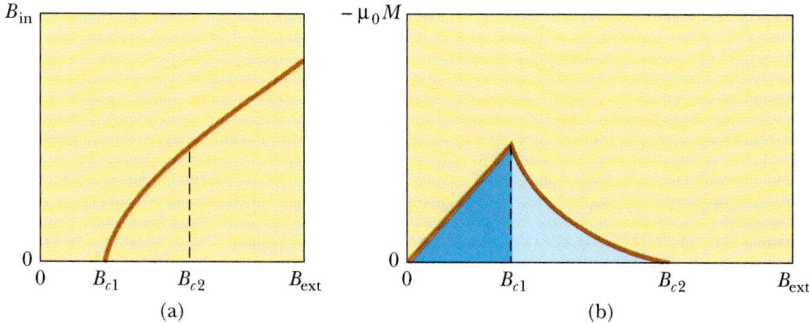

FIGURE 44.7   The magnetic behavior of a type II superconductor. (a) Plot of the internal field versus the external field. (b) Plot of the magnetization versus the external field.

of B_c for type I superconductors (Table 44.1). For this reason, type II superconductors are well suited for constructing high-field superconducting magnets. For example, using the alloy NbTi, superconducting solenoids may be wound to produce and sustain magnetic fields in the range of 5 to 10 T *with no power consumption.* Iron core electromagnets rarely exceed 2 T and consume much more power. Notice also that type II superconductors are compounds formed from elements of the transition and actinide series.

Figure 44.7a shows the internal magnetic field versus the external field for a type II superconductor, while Figure 44.7b is the corresponding magnetization versus the external field.

When a type II superconductor is in the vortex state, sufficiently large currents can cause the vortices to move perpendicularly to the current. This vortex motion corresponds to a change in flux with time and produces resistance in the material. By adding impurities or other special inclusions, one can effectively pin the vortices and prevent their motion, to produce zero resistance in the vortex state. The critical current for type II superconductors is the current that, when multiplied by the flux in the vortices, gives a Lorentz force that overcomes the pinning force.

EXAMPLE 44.2   A Superconducting Solenoid

A solenoid is to be constructed using wire made of the alloy Nb_3Al, which has an upper critical field of 32.0 T at $T = 0$ K and a critical temperature of 18.0 K. The wire has a radius of 1.00 mm, the solenoid is to be wound on a hollow cylinder of diameter 8.00 cm and length 90.0 cm, and there are to be 150 turns of wire per centimeter of length. (a) How much current is required to produce a magnetic field of 5.00 T at the center of the solenoid?

Solution   Recall from Chapter 30 that the magnetic field at the center of a tightly wound solenoid is $B = \mu_0 nI$, where n is the number of turns per unit length along the solenoid, and I is the current in the solenoid wire. Taking $B = 5.00$ T and $n = 150$ turns/cm $= 1.50 \times 10^4$ turns/m we find

$$I = \frac{B}{\mu_0 n}$$

$$= \frac{5.00 \text{ T}}{(4\pi \times 10^{-7} \text{ N/A}^2)(1.50 \times 10^4 \text{ m}^{-1})} = \boxed{265 \text{ A}}$$

(b) What maximum current can the solenoid carry if its temperature is to be maintained at 15.0 K? (Note that B near the solenoid windings is approximately equal to B on its axis.)

Solution   Using Equation 44.1, with $B_c(0) = 32.0$ T, we find $B_c = 9.78$ T at a temperature of 15.0 K. For this value of B, we find $I_{max} = 518$ A.

44.4 OTHER PROPERTIES OF SUPERCONDUCTORS

Persistent Currents

Because the dc resistance of a superconductor is zero below the critical temperature, once a current is set up in the material, it persists *without any applied voltage* (which follows from Ohm's law and the fact that $R = 0$). These **persistent currents,** sometimes called supercurrents, have been observed to last for several years with no measurable losses. In one experiment conducted by S.S. Collins in Great Britain, a current was maintained in a superconducting ring for 2.5 years, stopping only because a trucking strike delayed delivery of liquid helium necessary to maintain the ring below its critical temperature.[7]

To better understand the origin of persistent currents, consider a loop of wire made of a superconducting material. Suppose the loop is placed, in its normal state ($T > T_c$), in an external magnetic field and then the temperature is lowered below T_c so that the wire becomes superconducting as in Figure 44.8a. As with a cylinder, the flux is excluded from the interior of the wire because of the induced surface currents. However, note that *flux lines still pass through the hole in the loop.* When the external field is turned off, as in Figure 44.8b, *the flux through this hole is trapped because the magnetic flux through the loop cannot change.* The superconducting wire prevents the flux from going to zero through the appearance of a large spontaneous current induced by the collapsing external magnetic field. If the dc resistance of the superconducting wire is truly zero, this current should persist forever. Experimental results using a technique known as nuclear magnetic resonance indicate that such currents will persist for more than 10^5 years! The resistivity of a superconductor based on such measurements has been shown to be less than $10^{-26} \ \Omega \cdot m$. This reaffirms the fact that R is zero for a superconductor. (See Problem 30 for a simple but convincing demonstration of zero resistance.)

Now consider what happens if the loop is cooled to a temperature $T < T_c$ before the external field is turned on. When the field is turned on while the loop is maintained at this temperature, *flux must be excluded from the entire loop, including the*

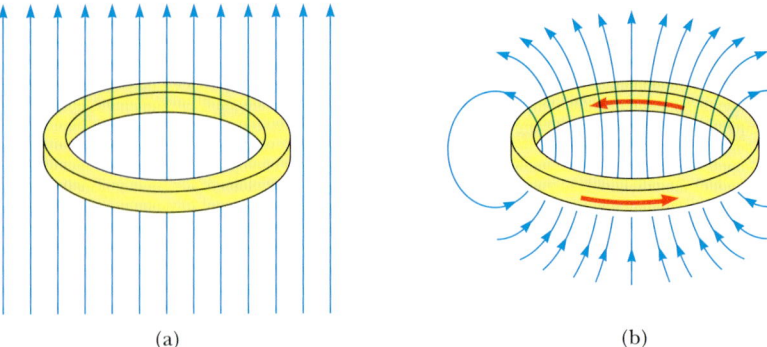

    (a)   (b)

FIGURE 44.8   (a) When a superconducting loop at $T > T_c$ is placed in an external magnetic field and the temperature is then lowered to $T < T_c$, flux passes through the hole in the loop even though it does not penetrate the interior of the material forming the loop. (b) After the external field is removed, the flux through the hole remains trapped, and an induced current appears in the material forming the loop.

[7] This charming story was provided by Steve Van Wyk.

hole, because the loop is in the superconducting state. Again, a current is induced in the loop to maintain zero flux through the loop and through the interior of the wire. In this case, the current disappears when the external field is turned off.

Flux Quantization

The phenomenon of flux exclusion by a superconductor applies only to a simply connected object; that is, one with no holes or their topological equivalent. However, when a superconducting ring is placed in a magnetic field, and the field is removed, flux lines are trapped and are maintained by a persistent circulating current as shown in Figure 44.8b. Realizing that superconductivity is fundamentally a quantum phenomenon, Fritz London suggested[8] that the trapped magnetic flux should be quantized in units of h/e. (The electronic charge e in the denominator arises because London assumed the persistent current is carried by single electrons.) Subsequent delicate measurements on very small superconducting hollow cylinders showed that the flux quantum is one-half the value postulated by London.[9] That is, the magnetic flux Φ is quantized not in units of h/e but in units of $h/2e$:

$$\Phi = \frac{nh}{2e} = n\Phi_0 \tag{44.5}$$

where n is an integer and

$$\Phi_0 = \frac{h}{2e} = 2.0679 \times 10^{-15} \text{ T} \cdot \text{m}^2 \tag{44.6}$$

is the **magnetic flux quantum.**

44.5   ELECTRONIC SPECIFIC HEAT

The thermal properties of superconductors have been extensively studied and compared with those of the same material in the normal state, and one very important measurement is specific heat. When a small amount of thermal energy is added to a normal metal, some of the energy is used to excite lattice vibrations, and the remainder is used to increase the kinetic energy of the conduction electrons. The **electronic specific heat** C is defined as the ratio of the thermal energy absorbed by the system of electrons to the increase in temperature of the system.

Figure 44.9 shows how the electronic specific heat varies with temperature for both the normal state and the superconducting state of gallium, a type I superconductor. At low temperatures, the electronic specific heat of the material in the normal state, C_n, varies with temperature as $AT + BT^3$. The linear term arises from electronic excitations, while the cubic term is due to lattice vibrations. The electronic specific heat of the material in the superconducting state, C_s, is substantially altered below the critical temperature. As the temperature is lowered starting from $T > T_c$, the specific heat first jumps to a very high value at T_c and then falls below the value for the normal state at very low temperatures. Analyses of such data show that at temperatures well below T_c, the electronic part of the specific heat is dominated by a term that varies as $\exp(-\Delta/k_B T)$, where Δ is one

[8] F. London, *Superfluids,* Vol. I, New York, John Wiley, 1954.

[9] The effect was discovered by B. S. Deaver, Jr. and W. M. Fairbank, *Phys. Rev. Letters* **7**; 43, 1961, and independently by R. Doll and M. Nabauer, *Phys. Rev. Letters* **7**; 51, 1961.

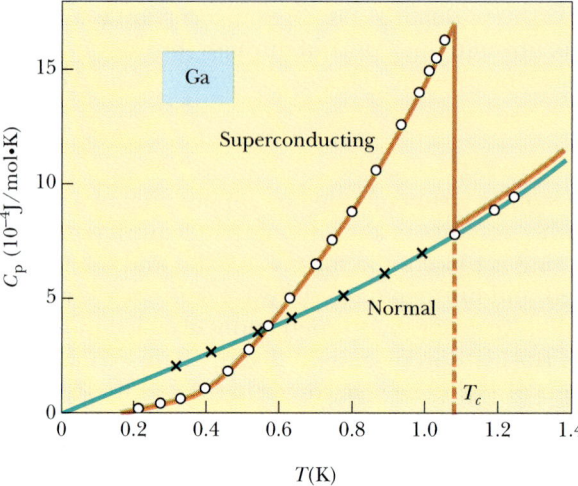

FIGURE 44.9 Electronic specific heat versus temperature for superconducting gallium (in zero applied magnetic field) and normal gallium (in a 0.020-T magnetic field). For the superconducting state, note the discontinuity that occurs at T_c and the exponential dependence on $1/T$ at low temperatures. *(Taken from N. Phillips,* Phys. Rev. *134, 385, 1964)*

half of the energy gap. This result suggests the existence of an energy gap in the energy levels available to the electrons, where the energy gap is a measure of the thermal energy necessary to move electrons from the ground to the excited states.

44.6 THE BCS THEORY

According to classical physics, part of the resistivity of a metal is due to collisions between free electrons and vibrating ions of the metal lattice and part is due to encounter between electrons and impurities/defects in the metal. Soon after the discovery of superconductivity, scientists recognized that the superconducting state could never be explained on the basis of the classical model because the electrons in a material always suffer some collisions, and therefore resistivity can never be zero. Nor could superconductivity be understood through a simple microscopic quantum-mechanical model, where an individual electron is viewed as a wave traveling through the material. Although many phenomenological theories were proposed based on the known properties of superconductors, none could explain why electrons enter the superconducting state and why electrons in this state are not scattered by impurities and lattice vibrations.

Several important developments in the 1950s led to better understanding of superconductivity. In particular, many research groups reported that the critical temperatures of isotopes of an element decreased with increasing atomic mass. This observation, called the **isotope effect,** was early evidence that lattice motion played an important role in superconductivity. For example, in the case of mercury, $T_c = 4.161$ K for the isotope ^{199}Hg, 4.153 K for ^{200}Hg, and 4.126 K for ^{204}Hg. The characteristic frequencies of the lattice vibrations are expected to change with the mass M of the vibrating atoms. In fact, the lattice vibrational frequencies are expected to be proportional to $M^{-1/2}$ [analogous to the angular frequency ω of a mass-spring system, where $\omega = (k/M)^{1/2}$]. On this basis, it became apparent that any theory of superconductivity for metals must include electron-lattice interactions.

The full microscopic theory of superconductivity presented in 1957 by Bardeen, Cooper, and Schrieffer has been successful in explaining the various features of superconductors. The details of this theory, now known as the BCS theory,

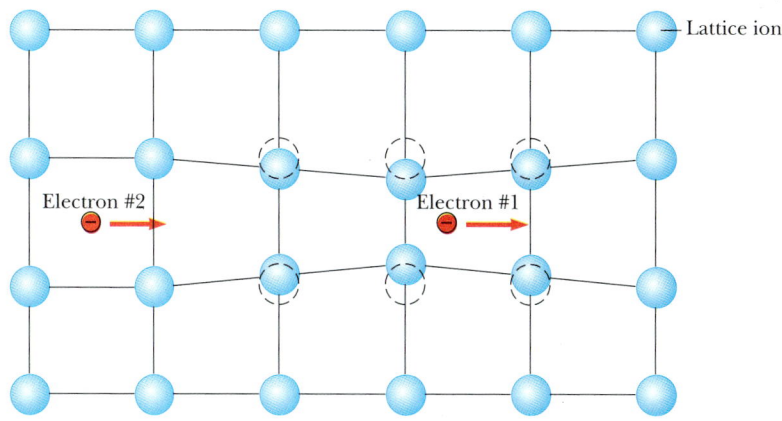

FIGURE 44.10   Basis for the attractive interaction between two electrons via lattice deformation. Electron 1 attracts the positive ions, which move inward from their equilibrium positions (dashed circles). This distorted region of the lattice has a net positive charge, and hence electron 2 is attracted to it.

are beyond the scope of this text, but we can describe some of its main features and predictions.

As mentioned in Section 44.2, the central feature of this theory is that two electrons in the superconductor are able to form a bound Cooper pair if they somehow experience an attractive interaction. This notion at first seems counter-intuitive because electrons normally repel one another because of their like charges. However, a net attraction can be obtained if the electrons interact with each other via the motion of the crystal lattice as the lattice structure is momentarily deformed by a passing electron. To illustrate this point, Figure 44.10 shows two electrons moving through the lattice. The passage of electron 1 causes nearby ions to move inward toward the electron, resulting in a slight increase in the concentration of positive charge in this region. Electron 2 (the second electron of the Cooper pair), approaching before the ions have had a chance to return to their equilibrium positions, is attracted to the distorted (positively charged) region. The net effect is a weak delayed attractive force between the two electrons resulting from the motion of the positive ions. As one researcher has beautifully put it, "The following electron surfs on the virtual lattice wake of the leading electron." In more technical terms, we can say that the attractive force between two Cooper electrons is an *electron-lattice-electron interaction*, where the crystal lattice serves as the mediator of the attractive force. Some scientists refer to this as a *phonon-mediated mechanism* because quantized lattice vibrations are called *phonons*.

A Cooper pair in a superconductor consists of two electrons having equal and opposite momenta and spin, as illustrated in Figure 44.11. Hence, in the super-

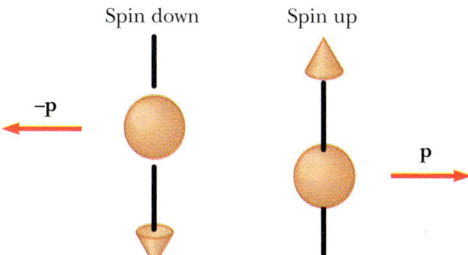

FIGURE 44.11   Schematic diagram of a Cooper pair. The electron moving to the right has a momentum p and its spin is up, while the electron moving to the left has a momentum $-p$ and its spin is down. Hence, the total momentum of the system is zero and the total spin is zero.

conducting state and in the absence of any supercurrents, *the Cooper pair forms a system having zero total momentum and zero spin.* Because Cooper pairs have zero spin, they can all be in the same state. This is in contrast to electrons, which are fermions (spin 1/2) that must obey the Pauli exclusion principle. In the BCS theory, a ground state is constructed in which *all electrons form bound pairs.* In effect, all Cooper pairs are "locked" into the *same quantum state of zero momentum.* This state of affairs can be viewed as a condensation of all electrons into the same state. It should also be noted that, because the Cooper pairs have zero spin (and hence zero angular momentum), their wave functions are spherically symmetric (like the *s*-states of the hydrogen atom). In a "semi-classical" sense the electrons are always undergoing head-on collisions, and as such, are always moving in each other's wake.

The BCS theory has been very successful in explaining the characteristic superconducting properties of zero resistance and flux expulsion. From a qualitative point of view, it can be said that to reduce the momentum of any single Cooper pair by scattering, it is necessary to simultaneously reduce the momentum of all the other pairs—in other words, all or nothing. You cannot change the velocity of one Cooper pair without changing it for all of them.[10] Lattice imperfections and lattice vibrations, which effectively scatter electrons in normal metals, have no effect on Cooper pairs! In the absence of scattering, the resistivity is zero and the current persists forever. It is rather strange, and perhaps amazing, that the mechanism of lattice vibrations responsible (in part) for the resistivity of normal metals also provides the interaction giving rise to their superconductivity. Thus, copper, silver, and gold, which exhibit small lattice scattering at room temperature, are not superconductors, whereas lead, tin, mercury, and other modest conductors have strong lattice scattering at room temperature and become superconductors at low temperatures.

As we mentioned earlier, the superconducting state is one in which the Cooper pairs act collectively rather than independently. The condensation of all pairs into the same quantum state makes the system behave like a giant quantum mechanical system that is quantized on the macroscopic level. *The condensed state of the Cooper pairs is represented by a single coherent wave function ψ that extends over the entire volume of the superconductor.*

The stability of the superconducting state is critically dependent on strong correlation between Cooper pairs. In fact, the theory explains superconducting behavior in terms of the energy levels of the "macromolecule" and the existence of an energy gap E_g between the ground and excited states of the system as in Figure 44.12a.[11] Note that, in Figure 44.12b, there is no energy gap for a normal conductor. In a normal conductor, the Fermi energy, E_F, represents the largest kinetic energy the free electrons can have at 0 K.

The energy gap in a superconductor is very small, of the order of $k_B T_c$ ($\approx 10^{-3}$ eV) at 0 K, as compared with the energy gap in semiconductors (≈ 1 eV) or the Fermi energy of a metal (≈ 5 eV). The energy gap represents the energy needed to

---

[10] Many authors choose to refer to this cooperative state of affairs as a **collective** state. As an analogy, one author wrote that the electrons in the paired state "move like mountain-climbers tied together by a rope: should one of them leave the ranks due to the irregularities of the terrain (caused by the thermal vibrations of the lattice atoms) his neighbors would pull him back."

[11] A Cooper pair is somewhat analogous to a helium atom, $^4_2$He, in that both are bosons with zero spin. It is well known that the superfluidity of liquid helium may be viewed as a condensation of bosons in the ground state. Likewise, superconductivity may be viewed as a superfluid state of Cooper pairs, all in the same quantum state.

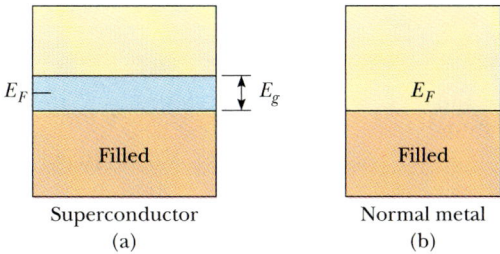

E_F — E_g

Filled

Superconductor
(a)

E_F

Filled

Normal metal
(b)

FIGURE 44.12   (a) Simplified energy-band structure for a superconductor. Note the energy gap between the lower filled states and the upper empty states. (b) The energy-band structure for a normal conductor has no energy gap. At $T = 0$ K, all states below the Fermi energy, E_F, are filled and all states above it are empty.

break up a Cooper pair. The BCS theory predicts that at $T = 0$ K,

$$E_g = 3.53 k_B T_c \tag{44.7}$$

Thus, superconductors that have large energy gaps have relatively high critical temperatures. The argument of the exponential function in the electronic heat capacity discussed in the previous section—$\exp(-\Delta/k_B T)$—contains a factor $\Delta = E_g/2$. Furthermore, the energy gap values predicted by Equation 44.7 are in good agreement with the experimental values given in Table 44.3. (The tunneling experiment used to obtain these values is described later.) As we noted earlier, the electronic heat capacity in zero magnetic field undergoes a discontinuity at the critical temperature. Furthermore, at finite temperatures, thermally excited individual electrons interact with the Cooper pairs and reduce the energy gap, which decreases continuously from a peak value at 0 K to zero at the critical temperature, as shown in Figure 44.13 for several superconductors.

TABLE 44.3   **The Energy Gap for Several Superconductors at $T = 0$ K**

Superconductor	E_g (meV)
Al	0.34
Ga	0.33
Hg	1.65
In	1.05
Pb	2.73
Sn	1.15
Ta	1.4
Zn	0.24
La	1.9
Nb	3.05

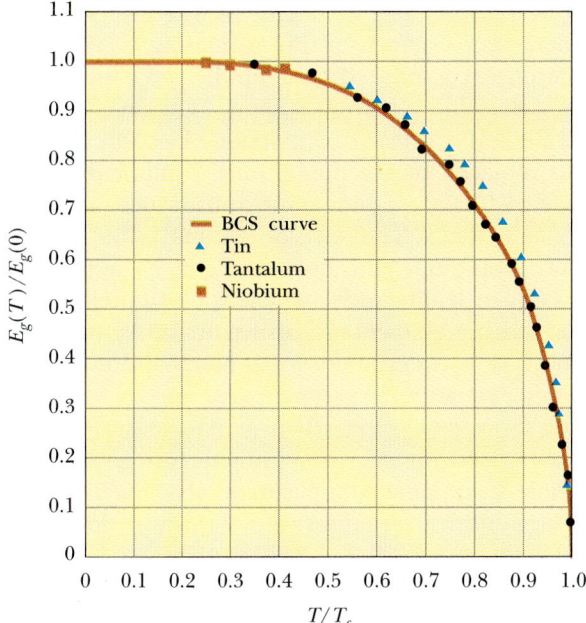

FIGURE 44.13   The points on this graph represent reduced values of the observed energy gap $E_g(T)/E_g(0)$ as a function of the reduced temperature T/T_c for tin, tantalum, and niobium. The solid curve gives the values predicted by the BCS theory. *(The data are from electron tunneling measurements by P. Townsend and J. Sutton, Phys. Ref. 28:591, 1962)*

EXAMPLE 44.3 **The Energy Gap for Lead**

Use Equation 44.7 to calculate the energy gap for lead and compare the answer with the experimental value in Table 44.3.

Solution Because $T_c = 7.193$ K for lead, Equation 44.7 gives

$$E_g = 3.53 k_B T_c = (3.53)(1.38 \times 10^{-23} \text{ J/K})(7.193 \text{ K})$$

$$= 3.50 \times 10^{-22} \text{ J} = 0.00219 \text{ eV} = 2.19 \times 10^{-3} \text{ eV}$$

The experimental value is 2.73×10^{-3} eV, corresponding to a percentage difference of about 20%.

Because the two electrons of a Cooper pair have opposite spin angular momenta, an external magnetic field raises the energy of one electron and lowers the energy of the other (see Problem 16). If the magnetic field is made strong enough, it becomes more favorable for the pair to break up into a state where both spins are pointing in the same direction to lower their energy. This value of the external field that causes the breakup corresponds to the critical field. This effect must be accounted for when dealing with type II superconductors that have very high values of the upper critical field.

44.7 ENERGY GAP MEASUREMENTS

Single Particle Tunneling

The energy gaps in superconductors can be measured very precisely in single-particle tunneling experiments (those involving normal electrons), first reported by I. Giaever in 1960.[12] As described in Section 41.9, tunneling is a phenomenon in quantum mechanics that enables a particle to penetrate through a barrier whose energy is greater than the particle energy. If two metals are separated by an insulator, the insulator normally acts as a barrier to the motion of electrons between the two metals. However, if the insulator is made sufficiently thin (less than about 2 nm), there is a small probability that electrons will tunnel from one metal to the other.

First consider two normal metals separated by a thin insulating barrier as in Figure 44.14a. If a potential difference V is applied between the two metals, electrons pass from one metal to the other and a current is set up. For small applied voltages, the current-voltage relationship is linear (the junction obeys Ohm's law). However, if one of the metals is replaced by a superconductor maintained at a temperature below T_c, as in Figure 44.14b, something quite unusual occurs. As V is increased, no current is observed until V reaches a threshold value V_t that satisfies the relationship $V_t = E_g/2e = \Delta/e$, where Δ is half the energy gap. (The factor of one half comes from the fact that we are dealing with single particle tunneling, and the energy required is one-half the binding energy of a pair, 2Δ.) That is, if $eV \geqslant 0.5 E_g$, then tunneling can occur between the normal metal and the superconductor.

Thus, single particle tunneling provides a direct experimental measurement of the energy gap. The value of Δ obtained from such experiments is in good agreement with the results of electronic heat capacity measurements. The *I-V* curve shown in Figure 44.14b shows the nonlinear relationship for this junction. Note

---

[12] I. Giaever, *Phys. Rev Letters* **5**, 147 and 464 (1960).

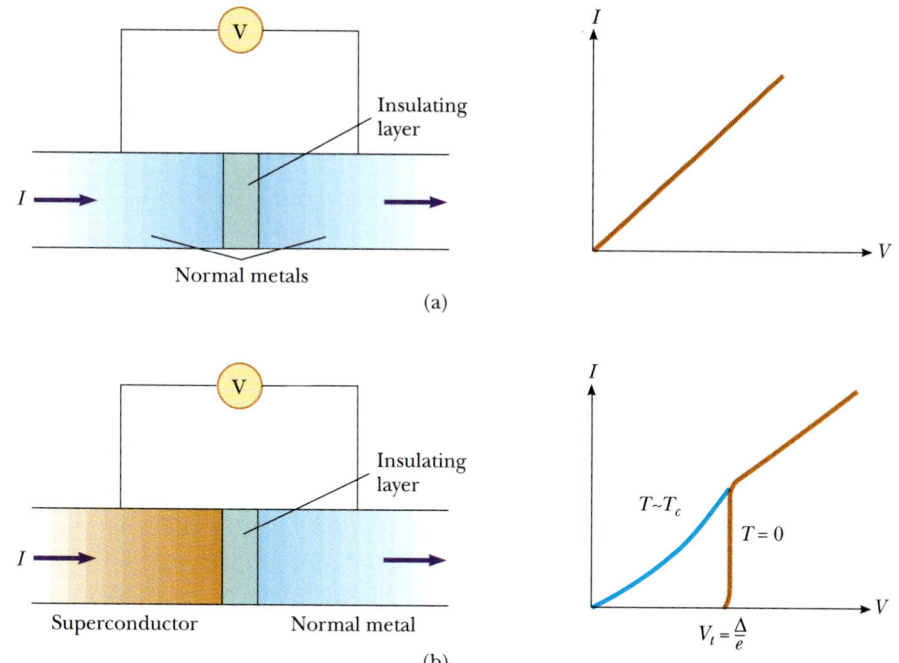

FIGURE 44.14 (a) Current-voltage relationship for electron tunneling through a thin insulator between two normal metals. The relationship is linear for small currents and voltages. (b) Current-voltage relationship for electron tunneling through a thin insulator between a superconductor and a normal metal. The relationship is nonlinear and strongly temperature-dependent. *(Adapted from N. W. Ashcroft and N. D. Mermin,* Solid State Physics, *Philadelphia, Saunders College Publishing, 1975)*

that as the temperature increases towards T_c, there is some tunneling current at voltages smaller than the energy gap threshold voltage. This is due to a combination of thermally excited electrons and a decrease in the energy gap.

Absorption of Electromagnetic Radiation

Another experiment used to measure the energy gap of superconductors is the absorption of electromagnetic radiation. In semiconductors, photons can be absorbed by the material when their energy is greater than the gap energy. Electrons in the valence band of the semiconductor absorb the incident photons, exciting the electrons across the gap into the conduction band. In a similar manner, superconductors absorb photons if the photon energy exceeds the gap energy, 2Δ. If the photon energy is less than 2Δ, no absorption occurs. When photons are absorbed by the superconductor, Cooper pairs are broken apart. Photon absorption in superconductors typically occurs in the range between microwave and infrared frequencies, as shown in the example on the next page.

44.8 JOSEPHSON TUNNELING

In Section 44.7 we described single particle tunneling from a normal metal through a thin insulating barrier into a superconductor. Now consider tunneling in two superconductors separated by a thin insulator. In 1962, Brian Josephson

EXAMPLE 44.4   The Absorption of Radiation by Lead

Find the minimum frequency of a photon that can be absorbed by lead at $T = 0$ K.

Solution From Table 44.3, we see that the energy gap for lead is 2.73×10^{-3} eV. Setting this equal to the photon energy hf and using the conversion 1 eV $= 1.60 \times 10^{-19}$ J, we find

$$hf = 2\Delta = 2.73 \times 10^{-3} \text{ eV} = 4.37 \times 10^{-22} \text{ J}$$

$$f = \frac{4.37 \times 10^{-22} \text{ J}}{6.626 \times 10^{-34} \text{ J} \cdot \text{s}} = 6.60 \times 10^{11} \text{ Hz}$$

Exercise What is the maximum wavelength of radiation that can be absorbed by lead at 0 K?

Answer $\lambda = c/f = 0.455$ mm, between the microwave region and the far infrared.

proposed that, in addition to single particle tunneling, Cooper pairs can also tunnel through such a junction. Josephson predicted that pair tunneling can occur without any resistance, producing a direct current when the applied voltage is zero and an alternating current when a dc voltage is applied across the junction.

At first, physicists were very skeptical about Josephson's proposal because it was believed that single particle tunneling would mask pair tunneling. However, when the phase coherence of the pairs is taken into account, it is found that, under the appropriate conditions, the probability for tunneling of pairs across the junction is comparable to that of single particle tunneling. In fact, when the insulating barrier separating the two superconductors is made sufficiently thin, (say, ≈ 1 nm) Josephson tunneling is as easy to observe as single particle tunneling.

In the remainder of this section, we describe two remarkable effects associated with pair tunneling.

dc Josephson Effect

Consider two superconductors separated by a thin oxide layer, typically 1 to 2 nm thick, as in Figure 44.15a. Such a configuration is known as a Josephson junction. In a given superconductor, the pairs could be represented by a wave function

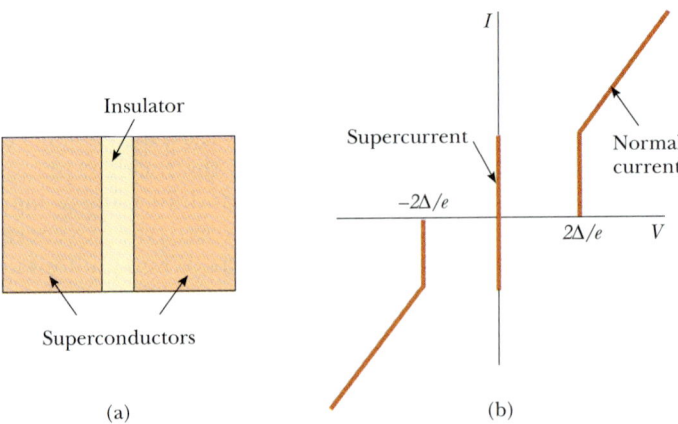

(a)   (b)

FIGURE 44.15 (a) A Josephson junction consists of two superconductors separated by a very thin insulator. Cooper pairs are able to tunnel through this insulating barrier in the absence of an applied voltage, setting up a direct current. (b) The *I-V* curve for a Josephson junction. When the bias current exceeds some threshold value, $I_{\max}$, a voltage develops and the current oscillates with a frequency $f = 2eV/h$.

$\Psi = \Psi_0 e^{i\phi}$, where ϕ is the phase and is the same for every pair. If one of the superconductors has a phase ϕ_1 and the other a phase ϕ_2, Josephson showed that at zero voltage there appears across the junction a supercurrent satisfying the relationship

$$I_s = I_{max} \sin(\phi_2 - \phi_1) = I_{max} \sin \delta \qquad (44.8)$$

where I_{max} is the maximum current across the junction under zero voltage conditions. The value of I_{max} depends on the surface area of each superconductor/oxide interface and decreases exponentially with the thickness of the oxide layer.

The first confirmation of the dc Josephson effect was made in 1963 by Rowell and Anderson. Since then, all of Josephson's other theoretical predictions have been verified. A plot of the current versus applied voltage for the Josephson junction is shown in Figure 44.15b.

ac Josephson Effect

When a dc voltage is applied across the Josephson junction, a most remarkable effect occurs: *The dc voltage generates an alternating current* given by

$$I = I_{max} \sin(\delta - 2\pi f t) \qquad (44.9)$$

where δ is a constant equal to the phase at $t = 0$, and f is the frequency of the Josephson current, given by

$$f = \frac{2eV}{h} \qquad (44.10)$$

A dc voltage of 1 μV results in a current frequency of 483.6 MHz. Precise measurements of the frequency and voltage have enabled physicists to obtain the value of e/h to unprecedented precision.

The ac Josephson effect can be demonstrated in various ways. One method is to apply a dc voltage and detect the electromagnetic radiation generated by the junction. Another method is to irradiate the junction with external radiation of frequency f'. With this method, a graph of the direct current versus voltage has steps when the voltages correspond to Josephson frequencies f that are integral multiples of the external frequency f'; that is, when $V = hf/2e = nhf'/2e$ (Fig. 44.16). Because the two sides of the junction are in different quantum states, the junction behaves like an atom undergoing transitions between these states as it

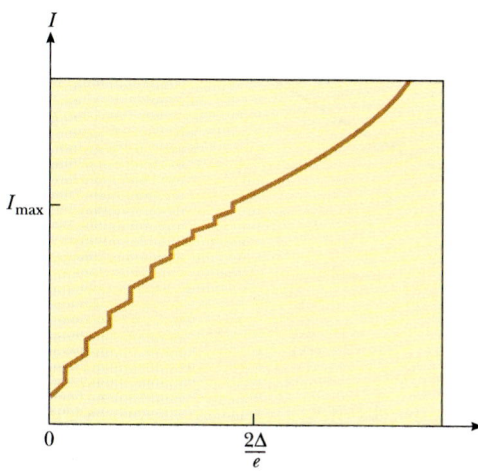

FIGURE 44.16 A plot of the dc current as a function of the bias voltage for a Josephson junction placed in an electromagnetic field. At a frequency of 10 GHz, as many as 500 steps have been observed. (In the presence of an applied electromagnetic field, there is no clear onset of single particle tunneling.)

absorbs or emits radiation. In effect, when a Cooper pair crosses the junction, a photon of frequency $f = 2eV/h$ is either emitted or absorbed by the system.

44.9   HIGH-TEMPERATURE SUPERCONDUCTIVITY

For many years, scientists have searched for materials that are superconductors at higher temperatures, and until 1986, the alloy Nb_3Ge had the highest known critical temperature, 23.2 K. There were various theoretical predictions that the maximum critical temperature for a superconductor in which the electron-lattice interaction was important would be in the neighborhood of 30 K. Early in 1986, J. Georg Bednorz and Karl Alex Müller, two scientists at IBM Research Laboratory in Zurich, made a remarkable discovery that has resulted in a revolution in the field of superconductivity. They found that an oxide of lanthanum, barium, and copper in a mixed-phase form of a ceramic became superconducting at approximately 30 K. The temperature dependence of the resistivity of their samples is shown in Figure 44.17. The superconducting phase was soon identified at other laboratories

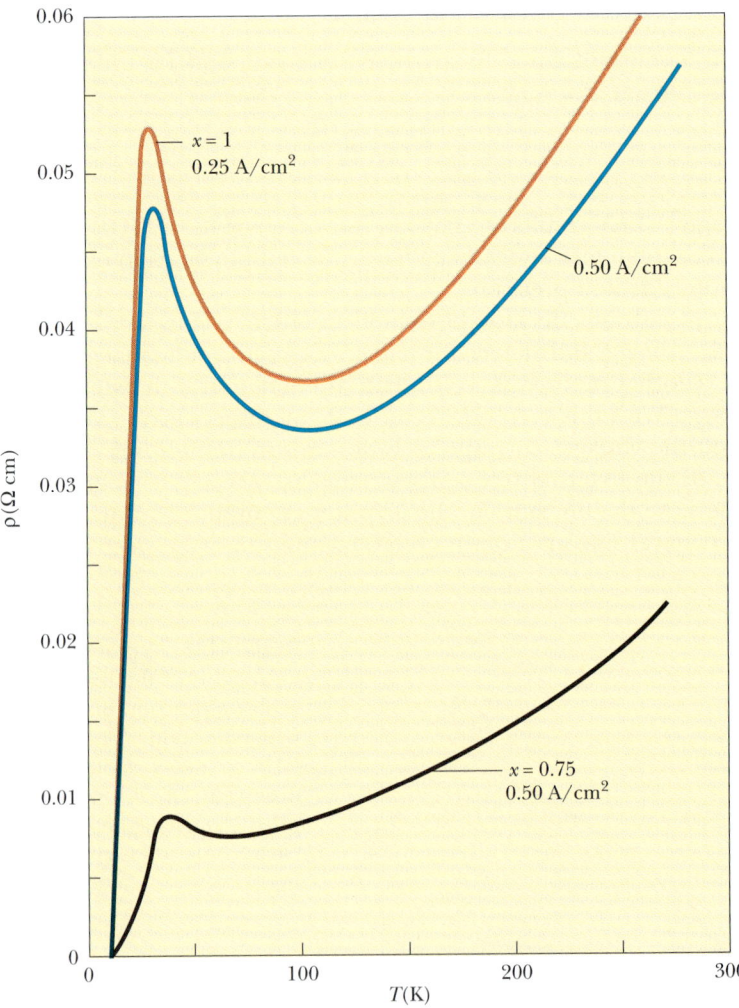

FIGURE 44.17   Temperature dependence of the resistivity of Ba-La-Cu-O for samples containing different concentrations of barium and lanthanum. The influence of current density is shown in the upper two curves. *(Taken from J. G. Bednorz and K. A. Müller, Z. Phys. B, 64:189, 1986)*

as the compound $La_{2-x}Ba_xCuO_4$, where $x \approx 0.2$. By replacing barium with strontium, researchers soon raised the value of T_c to approximately 36 K. Inspired by these developments, scientists worldwide worked feverishly to discover materials with even higher T_c values, and research in the superconducting behavior of metallic oxides accelerated at a tremendous pace. The year 1986 marked the beginning of a new era of high-temperature superconductivity. Bednorz and Müller were awarded the Nobel prize in 1987 (the fastest-ever recognition by the Nobel committee) for their most important discovery.

Early in 1987, research groups at the University of Alabama and the University of Houston announced the discovery of superconductivity near 92 K in a mixed-phase sample containing yttrium, barium, copper, and oxygen.[13] The discovery was confirmed in other laboratories around the world, and the superconducting phase was soon identified to be the compound $YBa_2Cu_3O_{7-\delta}$. A plot of resistivity versus temperature for this compound is shown in Figure 44.18. This was an important milestone in high-temperature superconductivity, because the transition temperature of this compound is above the boiling point of liquid nitrogen (77 K), a coolant that is readily available, inexpensive, and much easier to handle than liquid helium.

At an American Physical Society meeting on March 18, 1987, a special panel discussion on high-temperature superconductivity introduced the world to the newly discovered novel superconductors. This all-night session, which attracted a standing-room-only crowd of about 3000, produced great excitement in the scientific community and has been referred to as the ''Woodstock of physics.'' Realizing the possibility of routinely operating superconducting devices at liquid-nitrogen temperature and perhaps eventually at room temperature, thousands of scientists from various disciplines entered the arena of superconductivity research. The exceptional interest in these novel materials is due to at least four factors:

- The metallic oxides are relatively easy to fabricate and, hence, can be investigated at smaller laboratories and universities.
- They have very high T_c values and very high upper critical magnetic fields, estimated to be greater than 100 T in several materials (Table 44.4).
- Their properties and the mechanisms responsible for their superconducting behavior represent a great challenge to theoreticians.
- They may be of considerable technological importance.

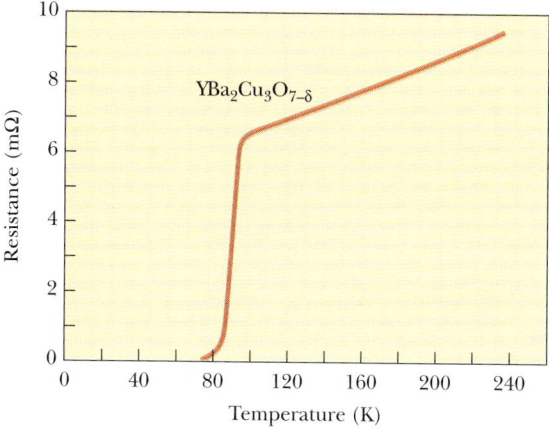

FIGURE 44.18 Temperature dependence of the resistance of a sample of $YBa_2Cu_3O_{7-\delta}$ showing the transition temperature near 92 K. *(Taken from M. W. Wu et al., Phys. Rev. Letters 58:908, 1987)*

[13] M. K. Wu, J. R. Ashburn, C. J. Torng, P. H. Hor, R. L. Meng, L. Gao, Z. J. Huang, Y. Q. Wang, and C. W. Chu, *Phys. Rev. Letters* 58:908, 1987.

TABLE 44.4	Some Properties of High-T_c Superconductors in the Form of Bulk Ceramics	
Superconductor	T_c (K)	$B_{c2}(0)$ in Tesla[a]
La-Ba-Cu-O	30	
$La_{1.85}Sr_{0.15}CuO_5$	36.2	> 36
La_2CuO_4	40	
$YBa_2Cu_3O_{7-\delta}$	92	≈ 160
$ErBa_2Cu_3O_{9-\delta}$	94	> 28
$DyBa_2Cu_3O_7$	92.5	
Bi-Sr-Ca-Cu-O	120	
Tl-Ba-Ca-Cu-O	125	
$HgBa_2Ca_2Cu_2O_{8+\delta}$	153	

[a] These are projected extrapolations based on data up to approximately 30 T.

Recently, several complex metallic oxides in the form of ceramics have been investigated, and critical temperatures above 100 K (triple-digit superconductivity) have been observed. Early in 1988, researchers reported superconductivity at approximately 120 K in a Bi-Sr-Ca-Cu-O compound and approximately 125 K in a Tl-Ba-Ca-Cu-O compound. The mercury-based cuprate compound $HgBa_2Ca_2Cu_2O_{8+\delta}$ holds the current record for the highest critical temperature,

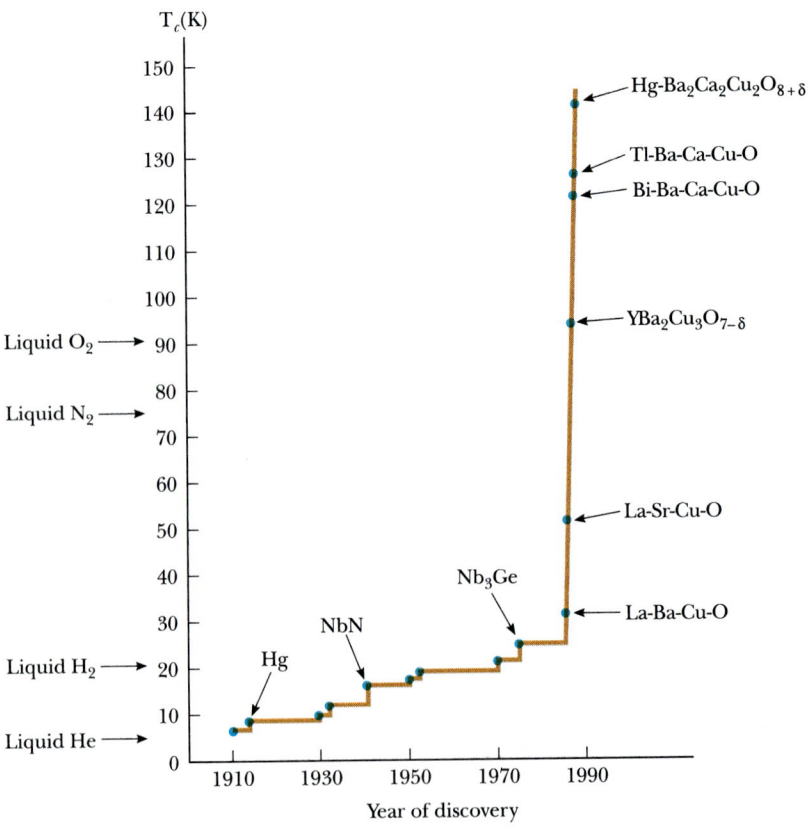

FIGURE 44.19   Evolution of the superconducting critical temperature since the discovery of the phenomenon.

$T_c = 153$ K. The increase in T_c since 1986 is highly dramatized in Figure 44.19. As you can see from this graph, the new high-T_c materials are all copper oxides of one form or another.

Mechanisms of High-T_c Superconductivity

Although the framework of the conventional BCS theory has been quite successful in explaining the behavior and properties of the "old-generation" superconductors, theoreticians are still trying to understand the nature of superconductivity in the "new-generation" high-T_c metallic oxides. The various models and mechanisms that have been proposed are far too technical to describe here, but it is interesting to note that many of the empirical observations are consistent with the predictions of the BCS theory. Evidence for this is as follows:

- Many of the copper oxides have energy gaps in the range of the predicted BCS value of $3.53kT_c$, although there are wide discrepancies in results reported by various groups.
- Pairs of charge carriers like Cooper pairs are involved in superconductivity as shown by flux quantization experiments.
- The discontinuity in the specific heat at T_c is similar to that predicted by the BCS model.

*44.10   APPLICATIONS

High-temperature superconductors may lead to many important technological advances, such as superconducting devices in every household. However, many significant materials-science problems must be overcome before such applications become reality. Perhaps the most difficult technical problem is how to mold the brittle ceramic materials into useful shapes such as wires and ribbons for large-scale applications and thin films for small electronic devices. Another major problem is the relatively low current densities measured in bulk ceramic compounds. Assuming that such problems are overcome, it is interesting to speculate on some of the future applications of these newly discovered materials.

An obvious application using the property of zero resistance to direct currents is low-loss electrical power transmission. A significant fraction of electrical power is lost as heat when current is passed through normal conductors. If power transmission lines could be made superconducting, these dc losses could be eliminated and there would be substantial savings in energy costs.

The new superconductors could have major impact in the field of electronics. Because of its switching properties, the Josephson junction can be used as a computer element. In addition, if superconducting films could be used to interconnect computer chips, chip size could be reduced and speeds would be enhanced because of this smaller size. Thus, information could be transmitted more rapidly and more chips could be contained on a circuit board with far less heat generation.

The phenomenon of magnetic levitation can be exploited in the field of transportation. In fact, a prototype train that runs on superconducting magnets has been constructed in Japan. The moving train levitates above a normal conducting metal track through eddy current repulsion. One can envision a future society of vehicles of all sorts gliding above a freeway making use of superconducting magnets. Some scientists are speculating that the first major market for levitating devices will be in the toy industry.

Another very important application of superconductivity is superconducting magnets, crucial components in particle accelerators. Currently, all high-energy

This prototype train, constructed in Japan, has superconducting magnets built into its base. A powerful magnetic field both levitates the train a few inches above the track, and propels it smoothly at speeds of 300 miles per hour or more. (*Joseph Brignolo/The Image Bank*)

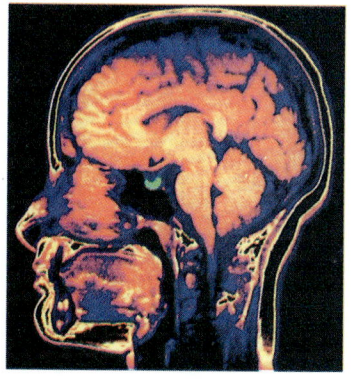

Computer-enhanced MRI of a normal human brain with the pituitary gland highlighted. (*Scott Camazine/Science Source*)

particle accelerators use liquid-helium-based superconducting technology. Again, there would be significant savings in cooling and operating costs if a liquid-nitrogen-based technology were developed.

An important application of superconducting magnets is a diagnostic tool called magnetic resonance imaging (MRI). This technique has played a prominent role in diagnostic medicine in the last few years because it uses relatively safe radio-frequency radiation to produce images of body sections, rather than x-rays. Because the technique relies on intense magnetic fields generated by superconducting magnets, the initial and operating costs of MRI systems are high. A liquid-nitrogen-cooled magnet could reduce such costs significantly.

In the field of power generation, various companies and government laboratories have worked for years in developing superconducting motors and generators. In fact, a small-scale superconducting motor using the newly discovered ceramic superconductors has already been constructed at Argonne National Laboratory in Illinois.

SUMMARY

Superconductors are materials that have zero dc resistance below a certain temperature T_c, called the **critical temperature.** A second characteristic of a type I superconductor is that it behaves as a perfect diamagnet. At temperatures below T_c, any applied magnetic flux is expelled from the interior of a type I superconductor. This phenomenon of flux expulsion is known as the **Meissner effect.** The superconductivity of a type I superconductor is destroyed when an applied magnetic field exceeds the **critical magnetic field,** B_c, which is less than 0.2 T for the elemental superconductors.

A type II superconductor is characterized by two critical fields. When an applied magnetic field is less than the lower critical field, B_{c1}, the material is entirely superconducting and there is no flux penetration. When the applied field exceeds the upper critical field B_{c2}, the superconducting state is destroyed and the flux penetrates the material completely. However, when the applied field lies between B_{c1} and B_{c2}, the material is in a **vortex state** that is a combination of superconducting regions threaded by regions of normal resistance.

Persistent currents (also called supercurrents), once set up in a superconducting ring, circulate for several years with no measurable losses and with zero applied voltage. This is a direct consequence of the fact that the dc resistance is truly zero in the superconducting state.

A central feature of the BCS theory of superconductivity for metals is the formation of a bound state called a **Cooper pair,** consisting of two electrons having equal and opposite momenta and opposite spins. The two electrons are able to form a bound state through a weak attractive interaction in which the crystal lattice of the metal serves as a mediator. In effect, one electron is weakly attracted to the other after the lattice is momentarily deformed by the first electron. In the ground state of the superconducting system, all electrons form Cooper pairs and all Cooper pairs are in the same quantum state of zero momentum. Thus, the superconducting state is represented by a single coherent wave function that extends over the entire volume of the sample.

The BCS model predicts a gap energy $E_g = 3.53 k_B T_c$, which is in contrast to a normal conductor, which has no energy gap. This energy gap represents the energy needed to break up a Cooper pair and is of the order of 1 meV for the elemental superconductors.

High-temperature superconductors have critical temperatures as high as 153 K. They are all copper oxides and the critical temperatures appear to be linked to the number of copper-oxygen layers in the structures. The new generation materials, known to be type II superconductors, have highly anisotropic resistivities and high upper critical fields. However, in the form of bulk ceramic samples, the materials have limited critical currents and are quite brittle. Although the BCS model appears to be consistent with most empirical observations on high-temperature superconductors, the basic mechanisms giving rise to the superconducting state remain unknown.

QUESTIONS

1. Describe how you would measure the two major characteristics of a superconductor.
2. Why is it not possible to explain zero resistance using a classical model of charge transport through a solid?
3. Define critical temperature, critical magnetic field, and critical current, and describe how these terms are related to each other.
4. Discuss the differences between type I and type II superconductors. Discuss their similarities.
5. What are persistent currents, and how can they be set up in a superconductor?
6. The specific heat of a superconductor in the absence of a magnetic field undergoes an anomaly at the critical temperature and decays exponentially toward zero below this temperature. What information does this behavior provide?
7. What is the isotope effect, and why does it play an important role in testing the validity of the BCS theory?
8. What are Cooper pairs? Discuss their essential properties, such as their momentum, spin, binding energy, and so forth.
9. How would you explain the fact that lattice imperfections and lattice vibrations can scatter electrons in normal metals but have no effect on Cooper pairs?
10. Discuss the origin of the energy gap of a superconductor, and how the energy-band structure of a superconductor differs from a normal conductor.
11. Define single particle tunneling and the conditions under which it can be observed. What information can be obtained from a tunneling experiment?
12. Define Josephson tunneling and the conditions under which it can be observed. What is the difference between Josephson tunneling and single particle tunneling?
13. What is meant by magnetic flux quantization, and how is it related to the concept of Cooper pairs?
14. What are the present limiting features of high-temperature superconductors as far as possible applications are concerned?
15. How would you design a superconducting solenoid? What properties must you be concerned with in your design?

16. Discuss four features of high-T_c superconductors that make them superior to the old generation of superconductors. In what way are the new superconductors inferior to the old ones?

The following questions deal with the Meissner effect. A small permanent magnet is placed on top of a high-temperature superconductor (usually $YBa_2Cu_3O_{7-\delta}$), starting at room temperature. As the superconductor is cooled with liquid nitrogen (77 K), the permanent magnet levitates as shown in Figure 44.20—a most dazzling phenomenon. Assume for simplicity that the superconductor is type I.

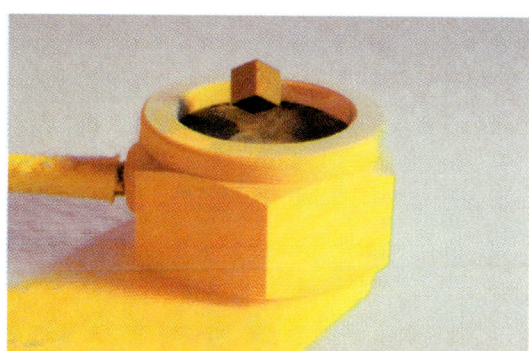

FIGURE 44.20 (Questions 17–23) Photograph of a small permanent magnet levitated above a disk of the superconductor $YBa_2Cu_3O_{7-\delta}$ that is at a temperature of 77 K. *(Courtesy of Profs. J. Dinardo and Som Tyagi, Drexel University)*

17. As a superconductor reaches its critical temperature, it expels all magnetic fields. How does this explain the levitation of the permanent magnet? What must happen to the superconductor to account for this behavior?
18. If the permanent magnet is set into rotation while levitated, it will continue to rotate for a long time. Can you think of an application that could make use of this frictionless magnetic bearing?
19. As soon as the permanent magnet has levitated, it gains potential energy. What accounts for this increase in potential energy? (This is a tricky one.)

20. When the levitated permanent magnet is pushed toward the edge of the superconductor, it will often fly off the edge. Why do you suppose this happens?
21. Why is it necessary to use a very small, but relatively strong, permanent magnet in this demonstration?
22. If the experiment is repeated by first cooling the superconductor below its critical temperature and then placing the permanent magnet on top of it, will the permanent magnet still levitate? If so, would there be any difference in its elevation compared to the previous case?
23. If a calibrated thermocouple were attached to the superconductor to measure its temperature, describe how you could obtain the critical temperature of the superconductor through the levitation effect. (*Hint:* Start the observation below T_c with a levitated permanent magnet.)

PROBLEMS

Section 44.2   Some Properties of Type I Superconductors
and Section 44.3   Type II Superconductors

1. A wire made of Nb_3Al has a radius of 2.0 mm and is maintained at 4.2 K. Using the data in Table 44.2, find (a) the upper critical field for this wire at this temperature, (b) the maximum current that can pass through the wire before its superconductivity is destroyed, and (c) the magnetic field 6.0 mm from the wire surface when the current has its maximum value.

2. A superconducting solenoid is to be designed to generate a 10-T magnetic field. (a) If the winding has 2000 turns/meter, what is the required current? (b) What force per meter is exerted on the inner windings by the magnetic field?

3. Determine the current generated in a superconducting ring of niobium metal 2.0 cm in diameter if a 0.020-T magnetic field directed perpendicular to the ring is suddenly decreased to zero. The inductance of the ring is $L = 3.1 \times 10^{-8}$ H.

4. Determine the magnetic field energy in joules that needs to be added to destroy superconductivity in 1.0 cm^3 of lead near 0 K. Use the fact that $B_c(0)$ for lead is 0.080 T.

5. An Nb wire 2.5 mm in radius is at 4.0 K. The wire is connected in series with a 100-Ω resistor and a source of emf. What emf is required to destroy the superconductivity of the wire?

6. What maximum current can be passed through a 30-cm long tightly wound solenoid made of Nb_3Ge wires of radius 1.5 mm while maintained at 4.0 K without destroying its superconductivity?

Section 44.4   Other Properties of Superconductors

7. *Persistent Currents.* In an experiment carried out by S.C. Collins between 1955 and 1958, a current was maintained in a superconducting lead ring for 2.5 years with no observed loss. If the inductance of the ring was 3.14×10^{-8} H and the sensitivity of the experiment was 1 part in 10^9, determine the maximum resistance of the ring. (*Hint:* Treat this as a decaying current in an *RL* circuit, and recall that $e^{-x} \cong 1 - x$ for small x.)

8. *Speed of Electron Flow.* Current is carried throughout the body of niobium-tin, a type II superconductor. If a niobium-tin wire of cross-section 2.0 mm^2 can carry a maximum supercurrent of 1.0×10^5 A, estimate the average speed of the superconducting electrons. (Assume the density of conducting electrons is $n_s = 5.0 \times 10^{27}/m^3$.)

9. *Diamagnetism.* When a superconducting material is placed in a magnetic field, the surface currents established make the magnetic field inside the material truly zero. (That is, the material is perfectly diamagnetic.) Suppose a circular disk, 2.0 cm in diameter, is placed in a 0.020-T magnetic field with the plane of the disk perpendicular to the field lines. Find the equivalent surface current if it all lies at the circumference of the disk.

10. *Surface Currents.* A rod of a superconducting material 2.5 cm long is placed in a 0.54-T magnetic field with its cylindrical axis along the magnetic field lines. (a) Sketch the directions of the applied field and the induced surface current and (b) estimate the magnitude of the surface current.

Section 44.6   The BCS Theory and
Section 44.7   Energy Gap Measurements

11. Calculate the energy gap for each superconductor in Table 44.1 as predicted by the BCS theory. Compare your answers with the experimental values given in Table 44.3.

12. Calculate the energy gaps for each superconductor in Table 44.2 as predicted by the BCS theory. Compare your values with those found for type I superconductors.

13. The longest wavelength photon which can excite electrons from the filled state to the empty state in Nb_3Ge is approximately 0.18 mm. Estimate the critical temperature for Nb_3Ge.

☐ indicates problems that have full solutions available in the Student Solutions Manual and Study Guide.

14. *Isotope Effect.* Because of the isotope effect, $T_c \propto M^{-\alpha}$. Use these data for mercury to determine the value of the constant α. Is your result close to what you might expect based on a simple model?

Isotope	T_c (K)
^{199}Hg	4.161
^{200}Hg	4.153
^{204}Hg	4.126

15. *Cooper Pairs.* A Cooper pair in a type I superconductor has an average particle separation of approximately 1.0×10^{-4} cm. If these two electrons can interact within a volume of this diameter, how many other Cooper pairs have their centers within the volume occupied by one pair? Use the appropriate data for lead, which has $n_s = 2.0 \times 10^{22}$ electrons/cm³.

16. *Dipole in a Magnetic Field.* The potential energy of a magnetic dipole of moment $\boldsymbol{\mu}$ in the presence of a magnetic field $\mathbf{B}$ is $U = -\boldsymbol{\mu} \cdot \mathbf{B}$. When an electron, which has spin of 1/2, is placed in a magnetic field, its magnetic moment can be aligned either with or against the field. Because the magnetic moment is negative, the higher-energy state, having energy $E_2 = \mu B$, corresponds to the case where the moment is aligned with the field, while the lower-energy state, with energy $E_1 = -\mu B$, corresponds to the case where the moment is aligned against the field. Thus, the energy separation between these two states is $\Delta E = 2\mu B$, where the magnetic moment of the electron is equal to $\mu = 5.79 \times 10^{-5}$ eV/T. (a) If a Cooper pair is subjected to a 38-T magnetic field (the critical field for Nb$_3$Ge), calculate the energy separation between the spin-up and spin-down electrons. (b) Calculate the energy gap for Nb$_3$Ge as predicted by the BCS theory at 0 K using the fact that $T_c = 23$ K. (c) How do your answers to parts (a) and (b) compare? What does this result suggest based on what you have learned about critical fields?

Section 44.8   Josephson Tunneling

17. Josephson tunneling can be used to determine accurately the e/h ratio. In a 1.500-nm thick Pb Josephson junction, the wavelength of radiation is 62.03 cm when a 1.000 μV voltage is applied across the junction. Determine e/h.

18. A Josephson junction is fabricated using indium for the superconducting layers. If the junction is biased with a dc voltage of 0.50 mV, find the frequency of the alternating current generated. (For comparison, note that the energy gap of indium at 0 K is 1.05 meV.)

19. If a magnetic flux of $1.0 \times 10^{-4}\Phi_0$ ($\frac{1}{10\,000}$ of the flux quantum) can be measured with a device called a

SQUID (Fig. P44.19), what is the smallest magnetic field change ΔB that can be detected with this device for a ring having a radius of 2.0 mm?

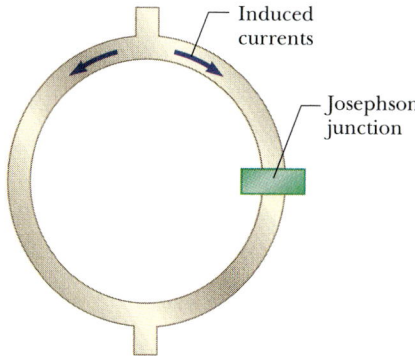

FIGURE P44.19   *Superconducting QUantum Interference Device (SQUID).*

20. A superconducting circular loop of very fine wire has a diameter of 2.0 mm and a self-inductance of 5.0 nH. The flux through the loop, Φ, is the sum of the applied flux, Φ_{app}, and the flux due to the super-current, $\Phi_{sc} = LI$, where L is the self-inductance of the loop. Because the flux through the loop is quantized, we have

$$n\Phi_0 = \Phi_{app} + \Phi_{sc} = \Phi_{app} + LI$$

where Φ_0 is the flux quantum. (a) If the applied flux is zero, what is the smallest current that can meet this quantization condition? (b) If the applied field is perpendicular to the plane of the loop and has a magnitude of 3.0×10^{-9} T, find the smallest current that circulates around the loop.

ADDITIONAL PROBLEMS

21. A solenoid of diameter 8.0 cm and length 1.0 m is wound with 2000 turns of superconducting wire. If the solenoid carries a current of 150 A, find (a) the magnetic field at the center, (b) the magnetic flux through the center cross-section, and (c) the number of flux quanta through the center.

22. *Energy Storage.* A novel method has been proposed to store electrical energy by fabricating a huge underground superconducting coil 1.0 km in diameter. This coil would carry a maximum current of 50 kA through each winding of a 150-turn Nb$_3$Sn solenoid. (a) If the inductance of this huge coil is 50 H, what is the total energy stored? (b) What is the compressive force per meter acting between two adjacent windings 0.25 m apart?

23. *Superconducting Power Transmission.* Superconductors have also been proposed for power transmission

lines. A single coaxial cable (Fig. P44.23) could carry 1.0×10^3 MW (the output of a large power plant) at 200 kV, dc, over a distance of 1000 km without loss. The superconductor would be a 2.0-cm-radius inner Nb_3Sn cable carrying the current I in one direction, while the outer surrounding superconductor of 5.0-cm radius would carry the return current I. In such a system, what is the magnetic field (a) at the surface of the inner conductor and (b) at the inner surface of the outer conductor? (c) How much energy would be stored in the space between the conductors in a 1000-km superconducting line? (d) What is the force per meter length exerted on the outer conductor?

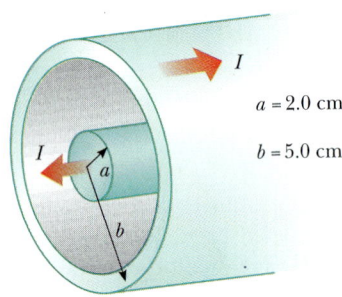

FIGURE P44.23

24. *Penetration Depth.* The penetration depth λ of a magnetic field into a superconductor, defined as the depth that the field decreases to $1/e$ of its value at the surface, is

$$\lambda = \sqrt{\frac{m_e}{\mu_0 n_s e^2}}$$

Estimate the number of superconducting electrons per cubic meter in lead at 0 K if the magnetic penetration depth near 0 K is $\lambda_0 = 4.0 \times 10^{-8}$ m.

25. *"Floating" a Wire.* Is it possible to "float" a superconducting lead wire of radius 1.0 mm in the magnetic field of the Earth? Assume the horizontal component of the Earth's magnetic field is 4.0×10^{-5} T.

26. *Magnetic Field Inside a Wire.* A type II superconducting wire of radius R carries current uniformly distributed through its cross-section. If the total current carried by the wire is I, show that the magnetic energy per unit length inside the wire is $\mu_0 I^2/16\pi$.

27. *Magnetic Levitation.* If a very small but very strong permanent magnet is lowered toward a flat-bottomed type I superconducting dish, at some point the magnet will levitate above the superconductor because the superconductor is a perfect diamagnet and expels all magnetic flux. Therefore, the superconductor acts like an identical magnet lying an equal distance below the surface (Fig. P44.27). At what height does the magnet levitate if its mass is 4.0 g and

its magnetic moment $\mu = 0.25$ A·m^2? (*Hint:* The potential energy between two dipole magnets a distance r apart is $\mu_0\mu^2/4\pi r^3$.)

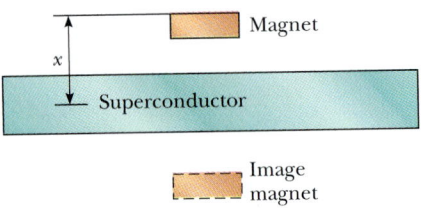

FIGURE P44.27   Magnetic levitation experiment.

28. *Designing a Superconducting Solenoid.* A superconducting solenoid was made with Nb_3Zr wire wound on a tube of diameter 10 cm. The solenoid winding consisted of a double layer of 0.50-mm-diameter wire with 984 turns (corresponding to a coil length of 25 cm). (a) Calculate the inductance, L, of the solenoid, assuming its length is large relative to its diameter. (b) The magnitude of the persistent current in this solenoid in the superconducting state has been reported to decrease by one part in 10^9 per hour. If the resistance of the solenoid is R, then the current in the circuit should decay according to $I = I_0 e^{-Rt/L}$. This resistance, although small, is due to magnetic flux migration in the superconductor. Determine an upper limit of the coil's resistance in the superconducting state. [The data were reported by J. File and R. G. Mills, *Phys. Rev. Letters* 10, 93(1963).]

29. *Entropy Difference.* The entropy difference per unit volume between the normal and superconducting states is

$$\frac{\Delta S}{V} = -\frac{\partial}{\partial T}\left(\frac{B^2}{2\mu_0}\right)$$

where $B^2/2\mu_0$ is the magnetic energy per unit volume required to destroy superconductivity. Determine the entropy difference between the normal and superconducting states in 1.0 mol of lead at 4.0 K if the critical magnetic field $B_c(0) = 0.080$ T and $T_c = 7.2$ K.

30. *A Convincing Demonstration of Zero Resistance.* A direct and relatively simple demonstration of zero dc resistance can be carried out using the four-point probe method. The probe shown in Figure P44.30 consists of a disc of $YBa_2Cu_3O_{7-\delta}$ (a high-T_c superconductor) to which four wires are attached by indium solder (or some other suitable contact material). A constant current is maintained through the sample by applying a dc voltage between points a and b, and it is measured with a dc ammeter. (The current is varied with the variable resistance R.) The potential difference V_{cd} between c and d is measured with a digital voltmeter. When the probe is immersed in liq-

TABLE 44.5	Current versus Potential Difference V_{cd} Measured in a Bulk Ceramic Sample of $YBa_2Cu_3O_{7-\delta}$ at Room Temperature[a]

I (mA)	V_{cd} (mV)
57.8	1.356
61.5	1.441
68.3	1.602
76.8	1.802
87.5	2.053
102.2	2.398
123.7	2.904
155	3.61

[a] The current was supplied by a 6-V battery in series with a variable resistor R. The values of R ranged from 10 Ω to 100 Ω. The data are from the author's laboratory.

uid nitrogen, the sample cools quickly to 77 K, which is below the critical temperature of the sample (92 K); the current remains approximately constant, but V_{cd} *drops abruptly to zero*. (a) Explain this observation on the basis of what you know about superconductors and your understanding of Ohm's law. (b) The data in Table 44.5 represent actual values of V_{cd} for different values of I taken on a sample at room temperature. Make an *I-V* plot of the data, and determine whether the sample behaves in a linear manner. From the data, obtain a value of the dc resis-

tance of the sample at room temperature. (c) At room temperature, it is found that V_{cd} = 2.234 mV for I = 100.3 mA, but after the sample is cooled to 77 K, V_{cd} = 0 and I = 98.1 mA. What do you think might explain the slight decrease in current?

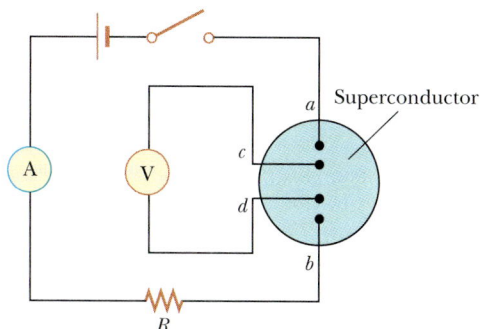

FIGURE P44.30 Circuit diagram used in the four-point probe measurement of the dc resistance of a sample. A dc digital ammeter is used to measure the current, and the potential difference between c and d is measured with a dc digital voltmeter. Note that there is no voltage source in the inner loop circuit where V_{cd} is measured.

31. An NbTi wire is made into a toroid having 20 000 turns with an inner radius of 30.0 cm and an outer radius of 40.0 cm. The toroid is kept at liquid helium temperature and carries a current of 350 A. Determine the range of the magnetic field induced inside the toroid. Is this field strong enough to destroy the superconductivity?

CHAPTER 45

Nuclear Structure

Particle tracks generated from the collision of two protons with two antiprotons. A subatomic particle known as omega meson was "born" in the collision. The tracks form in liquid hydrogen inside the bubble chamber. *(Lawrence Radiation Laboratory/Science Source/Photo Researchers, Inc.)*

In 1896, the year that marks the birth of nuclear physics, the French physicist Henri Becquerel (1852–1908) discovered radioactivity in uranium compounds. A great deal of research followed this discovery in an attempt to understand the nature of the radiation emitted by radioactive nuclei. Pioneering work by Ernest Rutherford showed that the emitted radiation was of three types, which he called alpha, beta, and gamma rays and classified according to the nature of the electric charge they possess and their ability to penetrate matter and ionize air. Later experiments showed that alpha rays are helium nuclei, beta rays are electrons, and gamma rays are high-energy photons.

In 1911, Rutherford and his students Geiger and Marsden (Section 42.1) performed a number of important scattering experiments involving alpha particles. These experiments established that the nucleus of an atom can be regarded as essentially a point mass and point charge and that most of the atomic mass is contained in the nucleus. Subsequent studies recognized the presence of a totally new type of force, the nuclear force, which is predominant at distances of less than approximately 10^{-14} m and zero for large distances. That is, the nuclear force is a short-range force.

Other milestones in the development of nuclear physics include

- The observation of nuclear reactions in 1930 by Cockroft and Walton using artificially accelerated particles
- The discovery of the neutron in 1932 by Chadwick and the conclusion that neutrons make up about half of the nucleus
- The discovery of artificial radioactivity in 1933 by Joliot and Irene Curie
- The discovery of nuclear fission in 1938 by Hahn and Strassmann
- The development of the first controlled fission reactor in 1942 by Fermi and his collaborators

In this chapter we discuss the properties and structure of the atomic nucleus. We start by describing the basic properties of nuclei, followed by a discussion of nuclear forces and binding energy, nuclear models, and the phenomenon of radioactivity. We also discuss nuclear reactions and the various processes by which nuclei decay.

45.1 SOME PROPERTIES OF NUCLEI

All nuclei are composed of two types of particles: protons and neutrons. The only exception to this is the ordinary hydrogen nucleus, which is a single proton. In describing the atomic nucleus, we must talk about the following quantities:

- The **atomic number,** Z, which equals the number of protons in the nucleus (the atomic number is sometimes called the charge number)
- The **neutron number,** N, which equals the number of neutrons in the nucleus
- The **mass number,** A, which equals the number of nucleons (neutrons plus protons) in the nucleus

In representing nuclei, it is convenient to have a symbolic way that shows how many protons and neutrons are present. The symbol used is $^A_Z X$, where X represents the chemical symbol for the element. For example, $^{56}_{26}Fe$ (iron) has a mass number of 56 and an atomic number of 26; therefore, it contains 26 protons and 30 neutrons. When no confusion is likely to arise, we omit the subscript Z because the chemical symbol can always be used to determine Z.

The nuclei of all atoms of a particular element contain the same number of protons but often contain different numbers of neutrons. Nuclei that are related in this way are called **isotopes.**

The isotopes of an element have the same Z value but different N and A values.

Isotopes

The natural abundances of isotopes can differ substantially. For example, $^{11}_6C$, $^{12}_6C$, $^{13}_6C$, and $^{14}_6C$ are four isotopes of carbon. The natural abundance of the $^{12}_6C$ isotope is about 98.9%, whereas that of the $^{13}_6C$ isotope is only about 1.1%. Some isotopes do not occur naturally but can be produced in the laboratory through nuclear reactions.

Even the simplest element, hydrogen, has isotopes: $^1_1$H, the ordinary hydrogen nucleus; $^2_1$H, deuterium; and $^3_1$H, tritium.

Charge and Mass

The proton carries a single positive charge, equal in magnitude to the charge e on the electron (where $|e| = 1.6 \times 10^{-19}$ C). The neutron is electrically neutral, as its name implies. Because the neutron has no charge, it is difficult to detect.

Nuclear masses can be measured with great precision with the help of the mass spectrometer (Section 29.5) and the analysis of nuclear reactions. The proton is approximately 1836 times as massive as the electron, and the masses of the proton and the neutron are almost equal. In Chapter 1, we defined the atomic mass unit, u, in such a way that the mass of the isotope ^{12}C is exactly 12 u, where 1 u = $1.660\ 540 \times 10^{-27}$ kg. The proton and neutron each have a mass of approximately 1 u, and the electron has a mass that is only a small fraction of an atomic mass unit:

$$\text{Mass of proton} = 1.007\ 276 \text{ u}$$

$$\text{Mass of neutron} = 1.008\ 665 \text{ u}$$

$$\text{Mass of electron} = 0.000\ 5486 \text{ u}$$

Because the rest energy of a particle is given by $E_R = mc^2$, it is often convenient to express the atomic mass unit in terms of its rest energy equivalent. For one atomic mass unit, we have

$$E_R = mc^2 = (1.660\ 540 \times 10^{-27} \text{ kg})(2.997\ 92 \times 10^8 \text{ m/s})^2$$
$$= 931.494 \text{ MeV}$$

Nuclear physicists often express mass in terms of the unit MeV/c^2, where

$$1 \text{ u} \equiv 931.494 \text{ MeV}/c^2$$

The masses of several nuclei and atoms are given in Table 45.1. The masses and some other properties of selected isotopes are provided in Appendix A.3.

TABLE 45.1   Mass of Selected Particles in Various Units

Particle	Mass		
	kg	u	MeV/c^2
Proton	$1.672\ 62 \times 10^{-27}$	1.007 276	938.28
Neutron	$1.674\ 93 \times 10^{-27}$	1.008 665	939.57
Electron	$9.109\ 39 \times 10^{-31}$	$5.485\ 79 \times 10^{-4}$	0.510 999
$^1_1$H atom	$1.673\ 53 \times 10^{-27}$	1.007 825	938.783
$^4_2$He	$6.644\ 66 \times 10^{-27}$	4.001 506	3727.38
$^{12}_6$C atom	$1.992\ 65 \times 10^{-26}$	12	11 177.9

EXAMPLE 45.1   The Atomic Mass Unit

Use Avogadro's number to show that the atomic mass unit is $1 \text{ u} = 1.66 \times 10^{-27}$ kg.

Solution   We know that exactly 12 kg of ^{12}C contains Avogadro's number of atoms. Avogadro's number, N_A, has the value 6.02×10^{23} atoms/mol.

Thus, the mass of one carbon atom is

$$\text{Mass of one } {}^{12}\text{C atom} = \frac{0.012 \text{ kg}}{6.02 \times 10^{23} \text{ atoms}}$$

$$= 1.99 \times 10^{-26} \text{ kg}$$

Since one atom of ^{12}C is defined to have a mass of 12 u, we find that

$$1 \text{ u} = \frac{1.99 \times 10^{-26} \text{ kg}}{12} = 1.66 \times 10^{-27} \text{ kg}$$

The Size and Structure of Nuclei

The size and structure of nuclei were first investigated in the scattering experiments of Rutherford, discussed in Section 42.1. In these experiments, positively charged nuclei of helium atoms (alpha particles) were directed at a thin piece of metal foil. As the alpha particles moved through the foil, they often passed near a nucleus of the metal. Because of the positive charge on both the incident particles and the nuclei, particles were deflected from their straight-line paths by the Coulomb repulsive force. In fact, some particles were even deflected backwards, through an angle of 180° from the incident direction. These particles were apparently moving directly toward a nucleus, on a head-on collision course.

Rutherford employed an energy calculation and found an expression for the distance, d, at which a particle approaching a nucleus is turned around by Coulomb repulsion. In such a head-on collision, the kinetic energy of the incoming alpha particle must be converted completely to electrical potential energy when the particle stops at the point of closest approach and turns around (Fig. 45.1). If we equate the initial kinetic energy of the alpha particle to the electrical potential energy of the system (alpha particle plus target nucleus), we have

$$\tfrac{1}{2}mv^2 = k_e \frac{q_1 q_2}{r} = k_e \frac{(2e)(Ze)}{d}$$

Solving for d, the distance of closest approach, we get

$$d = \frac{4k_e Ze^2}{mv^2}$$

From this expression, Rutherford found that the alpha particles approached nuclei to within 3.2×10^{-14} m when the foil was made of gold. Thus, the radius of the gold nucleus must be less than this value. For silver atoms, the distance of closest approach was found to be 2×10^{-14} m. From these results, Rutherford concluded that the positive charge in an atom is concentrated in a small sphere, which he called the nucleus, whose radius is no greater than about 10^{-14} m. Because such small lengths are common in nuclear physics, a convenient length unit that is used is the femtometer (fm), sometimes called the **fermi**, defined as

$$1 \text{ fm} \equiv 10^{-15} \text{ m}$$

In the early 1920s it was known that the nucleus contained Z protons and had a mass equivalent to that of A protons, where on the average $A \gtrsim 2Z$. To account for the nuclear mass, Rutherford proposed that a nucleus also contained $A-Z$ neutral particles that he called neutrons. In 1932, the British physicist James Chadwick

Ernest Rutherford (1871–1937), a physicist from New Zealand, was awarded a Nobel Prize in 1908 for discovering that atoms can be broken apart by alpha rays and for studying radioactivity. "On consideration, I realized that this scattering backward must be the result of a single collision, and when I made calculations I saw that it was impossible to get anything of that order of magnitude unless you took a system in which the greater part of the mass of the atom was concentrated in a minute nucleus. It was then that I had the idea of an atom with a minute massive center carrying a charge." *(Photo courtesy of AIP Niels Bohr Library)*

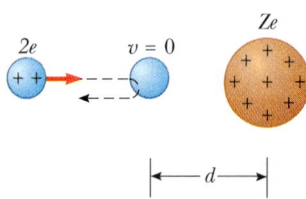

FIGURE 45.1 An alpha particle on a head-on collision course with a nucleus of charge *Ze*. Because of the Coulomb repulsion between the like charges, the alpha particle approaches to a distance *d* from the nucleus, called the distance of closest approach.

FIGURE 45.2 A nucleus can be modeled as a cluster of tightly packed spheres, where each sphere is a nucleon.

(1891–1974) discovered the neutron and was awarded the Nobel prize for this important work.

Although Rutherford's idea was that the neutron was a neutral proton-electron combination, the idea that electrons exist inside the nucleus has now been abandoned. The current view is that the neutron is not a fundamental particle and that electrons emitted in the decay of certain radioactive elements are created at the moment of decay.

Since the time of Rutherford's scattering experiments, a multitude of other experiments have shown that most nuclei are approximately spherical and have an average radius given by

Nuclear radius

$$r = r_0 A^{1/3} \tag{45.1}$$

where A is the mass number and r_0 is a constant equal to 1.2×10^{-15} m. Because the volume of a sphere is proportional to the cube of its radius, it follows from Equation 45.1 that the volume of a nucleus (assumed to be spherical) is directly proportional to A, the total number of nucleons. This suggests that *all nuclei have nearly the same density*. When the nucleons combine to form a nucleus, they combine as though they were tightly packed spheres (Fig. 45.2). This fact has led to an analogy between the nucleus and a drop of liquid, in which the density of the drop is independent of its size. We shall discuss the liquid-drop model in Section 45.4.

EXAMPLE 45.2 The Volume and Density of a Nucleus

Find (a) an approximate expression for the mass of a nucleus of mass number A, (b) an expression for the volume of this nucleus in terms of the mass number, and (c) a numerical value for its density.

Solution (a) The mass of the proton is approximately equal to that of the neutron. Thus, if the mass of one of these particles is m, the mass of the nucleus is approximately Am.

(b) Assuming the nucleus is spherical and using Equation 45.1, we find that the volume is

$$V = \tfrac{4}{3}\pi r^3 = \tfrac{4}{3}\pi r_0^3 A$$

(c) The nuclear density can be found as follows:

$$\rho_n = \frac{\text{mass}}{\text{volume}} = \frac{Am}{\tfrac{4}{3}\pi r_0^3 A} = \frac{3m}{4\pi r_0^3}$$

Taking $r_0 = 1.2 \times 10^{-15}$ m and $m = 1.67 \times 10^{-27}$ kg, we find that

$$\rho_n = \frac{3(1.67 \times 10^{-27}\ \text{kg})}{4\pi (1.2 \times 10^{-15}\ \text{m})^3} = 2.3 \times 10^{17}\ \text{kg/m}^3$$

The nuclear density is approximately 2.3×10^{14} times as great as the density of water ($\rho_{\text{water}} = 1.0 \times 10^3$ kg/m³).

Nuclear Stability

Because the nucleus is viewed as being a closely packed collection of protons and neutrons, you might be surprised that it can exist. Because like charges (the protons) in close proximity exert very large repulsive electrostatic forces on each other, these forces should cause the nucleus to fly apart. However, nuclei are stable because of the presence of the **nuclear force.** This force, which has a very short range (about 2 fm), is an attractive force that acts between all nuclear particles. The protons attract each other via the nuclear force, and at the same time they repel each other through the Coulomb force. The nuclear force also acts between pairs of neutrons and between neutrons and protons.

There are approximately 400 stable nuclei; hundreds of others have been observed, but these are unstable. A plot of *N* versus *Z* for a number of stable nuclei is given in Figure 45.3. Note that light nuclei are most stable if they contain an equal number of protons and neutrons, that is, if $N = Z$. Also note that heavy nuclei are more stable if the number of neutrons exceeds the number of protons. This can be understood by recognizing that, as the number of protons increases, the strength of the Coulomb force increases, which tends to break the nucleus apart. As a result, more neutrons are needed to keep the nucleus stable since neutrons experience only the attractive nuclear forces. Eventually, the repulsive forces between protons cannot be compensated by the addition of more neutrons. This occurs when $Z = 83$. Elements that contain more than 83 protons do not have stable nuclei.

It is interesting that most stable nuclei have even values of *A*. Furthermore, only eight nuclei have *Z* and *N* numbers that are both odd. In fact, certain values of *Z* and *N* correspond to nuclei having unusually high stability. These values of *N* and *Z*, called **magic numbers,** are

$$Z \text{ or } N = 2, 8, 20, 28, 50, 82, 126 \tag{45.2}$$

For example, the alpha particle (two protons and two neutrons), which has $Z = 2$ and $N = 2$, is very stable. The unusual stability of nuclei with progressively larger magic numbers suggests a shell structure of the nucleus similar to atomic shell structure. In Section 45.4 we briefly treat the shell model of the nucleus, which explains magic numbers.

Nuclear Spin and Magnetic Moment

In Chapter 42, we discussed the fact that an electron has an intrinsic angular momentum associated with its spin. Nuclei, like electrons, also have an intrinsic angular momentum that arises from relativistic properties. The magnitude of the nuclear angular momentum is $\sqrt{I(I + 1)}\hbar$, where *I* is a quantum number called the nuclear spin and may be an integer or half-integer. The maximum component of the angular momentum projected along the *z* axis is $I\hbar$. Figure 45.4 illustrates the possible orientations of the nuclear spin and its projections along the *z* axis for the case where $I = \frac{3}{2}$.

The nuclear angular momentum has a corresponding nuclear magnetic moment associated with it, similar to that of the electron. The magnetic moment of a nucleus is measured in terms of the **nuclear magneton** μ_n, a unit of moment defined as

$$\mu_n \equiv \frac{e\hbar}{2m_p} = 5.05 \times 10^{-27} \text{ J/T} \tag{45.3}$$

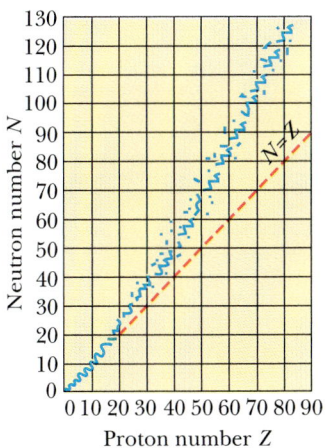

FIGURE 45.3 A plot of neutron number, *N*, versus atomic number, *Z*, for the stable nuclei (blue dots). They lie in a narrow band centered on the line of stability. The dashed line corresponds to the condition $N = Z$.

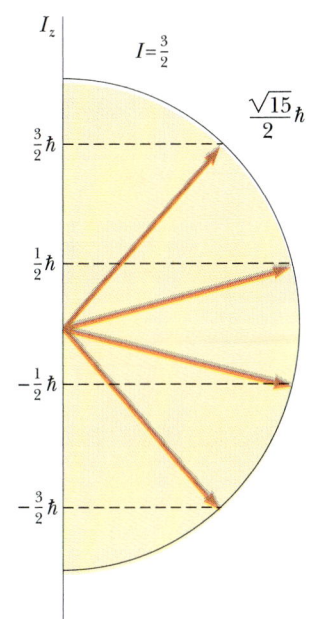

FIGURE 45.4 The possible orientations of the nuclear spin and its projections along the *z* axis for the case $I = \frac{3}{2}$.

Nuclear magneton

where m_p is the mass of the proton. This definition is analogous to that of the Bohr magneton, μ_B, which corresponds to the spin magnetic moment of a free electron (Section 42.5). Note that μ_n is smaller than μ_B by a factor of approximately 2000, due to the large difference in masses of the proton and electron.

The magnetic moment of a free proton is $2.7928\mu_n$. Unfortunately, there is no general theory of nuclear magnetism that explains this value. Another surprising point is that a neutron also has a magnetic moment, which has a value of $-1.9135\mu_n$. The minus sign indicates that this moment is opposite the spin angular momentum of the neutron.

45.2   NUCLEAR MAGNETIC RESONANCE AND MRI

Nuclear magnetic moments (as well as electronic magnetic moments) precess when placed in an external magnetic field. The frequency at which they precess, called the **Larmor precessional frequency** ω_p, is directly proportional to the mag-

Larmor frequency

netic field. This is described schematically in Figure 45.5a, where the magnetic field is along the z axis. For example, the Larmor frequency of a proton in a 1-T magnetic field is 42.577 MHz. The potential energy of a magnetic dipole moment in an external magnetic field is given by $-\boldsymbol{\mu}\cdot\mathbf{B}$. When the projection of $\boldsymbol{\mu}$ is along the field, the potential energy of the dipole moment is $-\mu B$, that is, it has its minimum value. When the projection of $\boldsymbol{\mu}$ is against the field, the potential energy is μB and it has its maximum value. These two energy states for a nucleus with a spin of $\frac{1}{2}$ are shown in Figure 45.5b.

It is possible to observe transitions between these two spin states using a tech-

Nuclear magnetic resonance

nique called **NMR,** for **nuclear magnetic resonance.** A dc magnetic field is introduced to align the magnetic moments (Fig. 45.5a), along with a second, weak, oscillating magnetic field oriented perpendicular to **B.** When the frequency of the oscillating field is adjusted to match the Larmor precessional frequency, a torque acting on the precessing moments causes them to "flip" between the two spin states. These transitions result in a net absorption of energy by the spin system, an

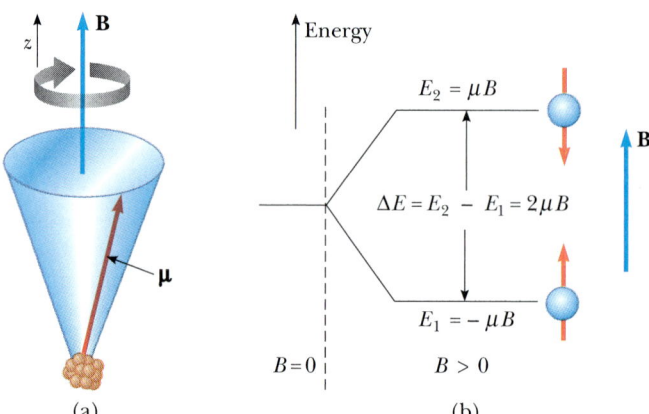

(a)   (b)

FIGURE 45.5   (a) When a nucleus is placed in an external magnetic field, **B**, the magnetic moment precesses about the magnetic field with a frequency proportional to the field. (b) A proton, whose spin is $\frac{1}{2}$, can occupy one of two energy states when placed in an external magnetic field. The lower energy state E_1 corresponds to the case where the spin is aligned with the field, and the higher energy state E_2 corresponds to the case where the spin is opposite the field. The reverse is true for electrons.

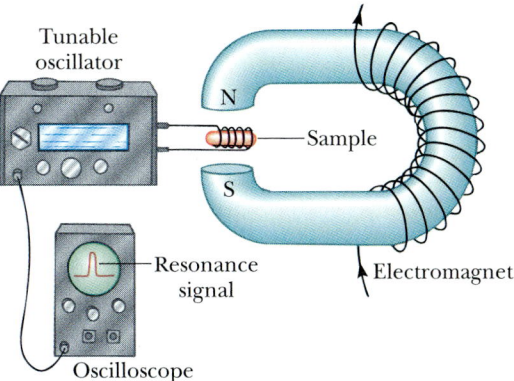

FIGURE 45.6   An experimental arrangement for nuclear magnetic resonance. The radio-frequency magnetic field of the coil, provided by the variable-frequency oscillator, must be perpendicular to the dc magnetic field. When the nuclei in the sample meet the resonance condition, the spins absorb energy from the field of the coil, and this absorption changes the Q of the circuit in which the coil is included. Most modern NMR spectrometers use superconducting magnets at fixed field strengths and operate at frequencies of approximately 200 MHz.

absorption that can be detected electronically. A diagram of the apparatus used in nuclear magnetic resonance is illustrated in Figure 45.6. The absorbed energy is supplied by the generator producing the oscillating magnetic field. Nuclear magnetic resonance and a related technique called electron spin resonance are extremely important methods for studying nuclear and atomic systems and how these systems interact with their surroundings. A typical NMR spectrum is shown in Figure 45.7.

A widely used diagnostic technique called **MRI,** for **magnetic resonance imaging,** is based on nuclear magnetic resonance. In MRI, the patient is placed inside a large solenoid that supplies a spatially varying magnetic field. Because of the gradient in the magnetic field, protons in different parts of the body precess at different frequencies, so that the resonance signal can be used to provide information

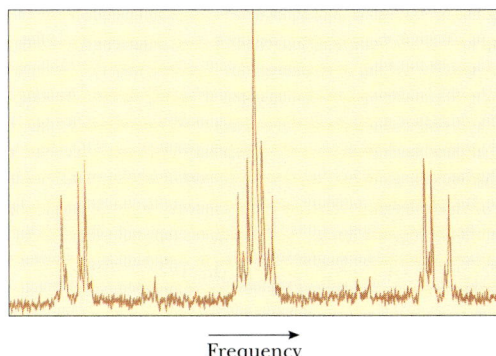

Frequency

FIGURE 45.7   NMR spectrum of ^{31}P in a bridged metallic complex containing platinum. The lines that flank the central strong peak are due to the interaction between ^{31}P and other distant ^{31}P nuclei. The outermost set of lines are due to the interaction between ^{31}P and neighboring platinum nuclei. The spectrum was recorded at a fixed magnetic field of about 4 T and the mean frequency was 200 MHz.

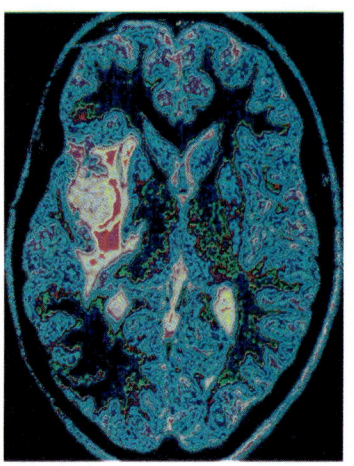

FIGURE 45.8   A computer-digitized color-enhanced MRI scan of a brain with a glioma tumor.   *(Scott Camazine/Science Source)*

on the positions of the protons. A computer is used to analyze the position information to provide data for constructing a final image. An MRI scan taken on a human head is shown in Figure 45.8. The main advantage of MRI over other imaging techniques in medical diagnostics is that it causes minimal damage to cellular structures. Photons associated with the radio-frequency signals used in MRI have energies of only about 10^{-7} eV. Because molecular bond strengths are much larger (approximately 1 eV), the radio-frequency radiation causes little cellular damage. In comparison, x-rays or γ-rays have energies ranging from 10^4 to 10^6 eV and can cause considerable cellular damage.

45.3   BINDING ENERGY AND NUCLEAR FORCES

The total mass of a nucleus is always less than the sum of the masses of its nucleons. Because mass is a measure of energy, *the total energy of the bound system (the nucleus) is less than the combined energy of the separated nucleons.* This difference in energy is called the **binding energy** of the nucleus and can be thought of as the energy that must be added to a nucleus to break it apart into its components. Therefore, in order to separate a nucleus into protons and neutrons, energy must be delivered to the system.

Conservation of energy and the Einstein mass-energy equivalence relationship show that the binding energy of any nucleus of mass M_A is

Binding energy of a nucleus

$$E_b(\text{MeV}) = (Zm_p + Nm_n - M_A) \times 931.494 \text{ MeV/u} \tag{45.4}$$

where m_p is the mass of the proton, m_n is the mass of the neutron, and the masses are all in atomic mass units. In practice, it is often more convenient to use the mass of neutral atoms (nuclear mass plus mass of electrons) in computing binding energy because mass spectrometers generally measure atomic masses.[1]

A plot of binding energy per nucleon, E_b/A, as a function of mass number for

---

EXAMPLE 45.3   The Binding Energy of the Deuteron

Calculate the binding energy of the deuteron (the nucleus of a deuterium atom), which consists of a proton and a neutron, given that the mass of the deuteron is 2.013 553 u.

Solution   From Table 45.1, we see that the proton and neutron masses are $m_p = 1.007\ 276$ u and $m_n = 1.008\ 665$ u. Therefore,

$$m_p + m_n = 2.015\ 941 \text{ u}$$

To calculate the mass difference, we subtract the deuteron mass from this value:

$$\Delta m = (m_p + m_n) - m_d$$
$$= 2.015\ 941 \text{ u} - 2.013\ 553 \text{ u}$$
$$= 0.002\ 388 \text{ u}$$

Using Equation 45.4, we find that the binding energy is

$$E_b = \Delta mc^2 = (0.002\ 388 \text{ u})(931.494 \text{ MeV/u})$$
$$= \boxed{2.224 \text{ MeV}}$$

This result tells us that, to separate a deuteron into its constituent proton and neutron, it is necessary to add 2.224 MeV of energy to the deuteron. One way of supplying the deuteron with this energy is by bombarding it with energetic particles.

If the binding energy of a nucleus were zero, the nucleus would separate into its constituent protons and neutrons without the addition of any energy; that is, it would spontaneously break apart.

---

[1] It is possible to do this because electron masses cancel in such calculations. One exception to this is the β^+ decay process.

Maria Goeppert-Mayer was born in Germany and received a PhD in physics from Göttingen University. Her thesis work, which dealt with quantum mechanical effects in atoms, was encouraged by Paul Ehrenfest. She moved to the United States in 1930 after her husband received a professorship in chemistry at Johns Hopkins University. While at home raising two children, she wrote a book with her husband on statistical mechanics. Following the publication of the book she was offered a lectureship in chemistry at Johns Hopkins University, but her presence as a woman at faculty functions was awkward.

In the late 1940s she and her husband received appointments at the University of Chicago, but her position was without pay because of a strict nepotism rule. While in Chicago, she worked with Enrico Fermi and Edward Teller. During her collaboration with Teller, she became interested in

Maria Goeppert-Mayer

| 1 9 0 6 – 1 9 7 2 |

why certain elements in the periodic table were so abundant and stable. She eventually realized that the most stable elements were characterized by particular values of atomic and neutron numbers, which she called "magic numbers." She labored over a theoretical explanation of these numbers for about one year, and finally arrived at a solution during a conversation with Fermi. This resulted in a 1950 publication in which she described the "shell" model of the nucleus. (It is interesting to note that the paper confused many Russian scientists, who translated "shell" as "grenade.") As often happens in scientific research, a similar model was simultaneously developed by Hans Jensen, a German scientist. Maria Goeppert-Mayer and Hans Jensen were awarded the Nobel prize in physics in 1963 for their extraordinary work in understanding the structure of the nucleus. When Goeppert-Mayer heard that she had won the Nobel prize, she said "Oh, how wonderful, I've always wanted to meet a king."

various stable nuclei is shown in Figure 45.9. Note that, except for the lighter nuclei, the average binding energy per nucleon is about 8 MeV. For the deuteron, the average binding energy per nucleon is $E_b/A = 2.224/2$ MeV $= 1.112$ MeV. Note that the curve in Figure 45.9 peaks in the vicinity of $A = 60$. That is, nuclei

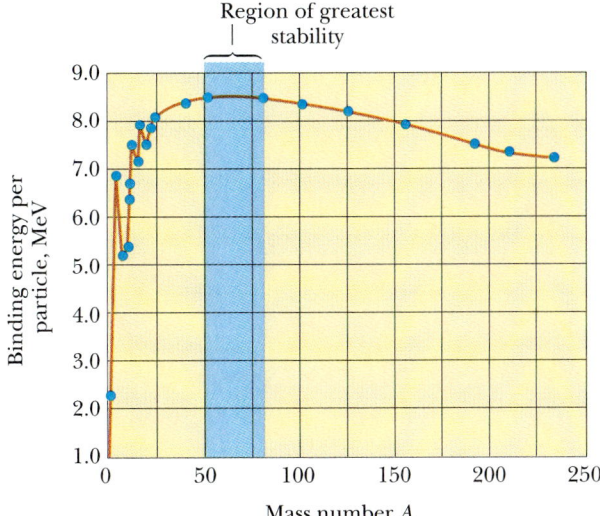

FIGURE 45.9 A plot of binding energy per nucleon versus mass number for nuclei that lie along the line of stability in Figure 45.3.

with mass numbers greater or less than 60 are not as strongly bound as those near the middle of the periodic table. The higher values of binding energy near $A = 60$ imply that energy is released when a heavy nucleus splits or fissions into two lighter nuclei. Energy is released in fission because the final state consisting of two lighter fragments is more tightly bound (lower in energy) than the original nucleus. These two important processes of fission and fusion are considered in detail in Chapter 46.

Another important feature of Figure 45.9 is that the binding energy per nucleon is approximately constant for $A > 50$. For these nuclei, the nuclear forces are said to be saturated; that is, a particular nucleon forms attractive bonds with only a limited number of other nucleons in the close-packed structure shown in Figure 45.2.

The general features of the nuclear force responsible for the binding energy of nuclei have been revealed in a wide variety of experiments and are as follow:

- The attractive nuclear force is the strongest force in nature.
- The nuclear force is a short range force that falls to zero when the separation between nucleons exceeds several fermis. The limited range of the nuclear force is evidenced by scattering experiments. The short range of the nuclear force is shown in the neutron-proton (n-p) potential energy plot of Figure 45.10a obtained by scattering neutrons from a target containing hydrogen. The depth of the potential energy is 40 to 50 MeV and contains a strong repulsive component that prevents the nucleons from approaching much closer than 0.4 fm. Another interesting feature of the nuclear force is that its size depends on the relative spin orientations of the nucleons.
- Scattering experiments and other indirect evidence show that the nuclear force is independent of the charge of the interacting nucleons. For this reason, high-speed electrons can be used to probe the properties of nuclei. The charge inde-

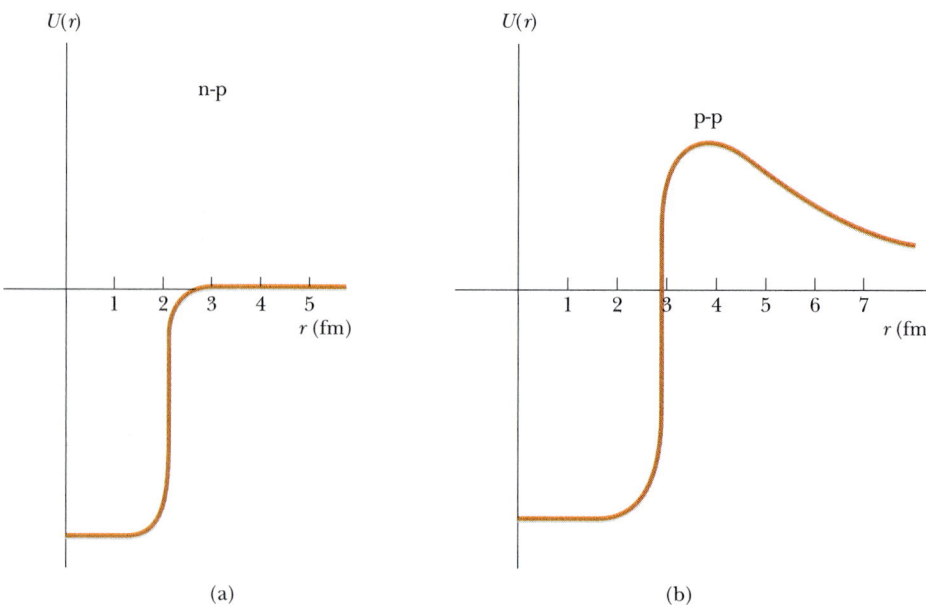

FIGURE 45.10   (a) Potential energy versus separation for the neutron-proton system. (b) Potential energy versus separation for the proton-proton system. The difference in the two curves is due mainly to the large Coulomb repulsion in the case of the proton-proton interaction.

pendence of the nuclear force also means that the only difference between the n-p and p-p interactions is that the p-p potential energy consists of a superposition of nuclear and Coulomb interactions as shown in Figure 45.10b. At distances less than 2 fm, both p-p and n-p potential energies are nearly identical, but for distances greater than this, the p-p potential energy is positive with a maximum of about 1 MeV at 4 fm.

45.4 NUCLEAR MODELS

Although the details of nuclear forces are still not well understood, several nuclear models have been proposed, and these are useful in understanding general features of nuclear experimental data and the mechanisms responsible for binding energy. The liquid-drop model accounts for nuclear binding energy, and the independent-particle model accounts for the existence of stable isotopes.

Liquid-Drop Model

The **liquid-drop model**, proposed by Bohr in 1936, treats the nucleons as if they were molecules in a drop of liquid and is closely analogous to the Thomas-Fermi model of the atom. The nucleons interact strongly with each other and undergo frequent collisions as they jiggle around within the nucleus. This jiggling motion is analogous to the thermally agitated motion of molecules in a drop of liquid.

Three major effects influence the binding energy of the nucleus in the liquid-drop model:

- **The volume effect** Earlier, we showed that the binding energy per nucleon is approximately constant, indicating that the nuclear force saturates (Fig. 45.9). Therefore, the binding energy of the nucleus is proportional to A and therefore to the nuclear volume. The contribution to the binding energy of the entire nucleus is $C_1 A$, where C_1 is an adjustable constant.
- **The surface effect** Because nucleons on the surface of the drop have fewer neighbors than those in the interior, surface nucleons reduce the binding energy by an amount proportional to the number of surface nucleons. Because the number of surface nucleons is proportional to the surface area of the nucleus, $4\pi r^2$, and $r^2 \propto A^{2/3}$ (Eq. 45.1), the surface term can be expressed as $-C_2 A^{2/3}$, where C_2 is a constant.
- **The Coulomb repulsion effect** Each proton repels every other proton in the nucleus. The corresponding potential energy per pair of interacting particles is $k_e e^2/r$, where k_e is the Coulomb constant. The total Coulomb energy represents the work required to assemble Z protons from infinity to a sphere of volume V. This energy is proportional to the number of proton pairs $Z(Z-1)/2$ and inversely proportional to the nuclear radius. Consequently, the reduction in energy that results from the Coulomb effect is $-C_3 Z(Z-1)/A^{1/3}$.

Another small effect that lowers the binding energy is significant for nuclei having a large excess of neutrons—in other words, heavy nuclei. This effect gives rise to a binding energy term of the form $-C_4 (N-Z)^2/A$.

Adding these contributions, we get as the total binding energy

$$E_b = C_1 A - C_2 A^{2/3} - C_3 \frac{Z(Z-1)}{A^{1/3}} - C_4 \frac{(N-Z)^2}{A} \qquad (45.5)$$

Semiempirical binding energy formula

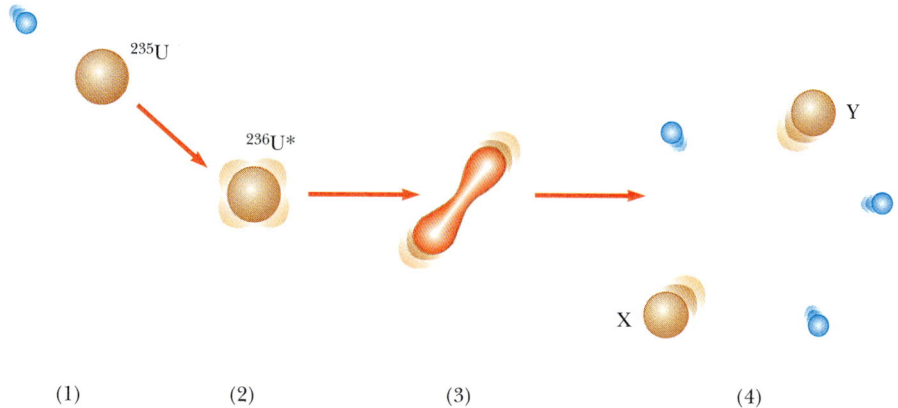

(1)   (2)   (3)   (4)

FIGURE 45.11   Steps leading to fission according to the liquid-drop model of the nucleus.

This equation, often referred to as the **semiempirical binding energy formula,** contains four constants that are adjusted to fit the expression to experimental data. For nuclei with $A \geq 15$, the constants have the values

$$C_1 = 15.7 \text{ MeV} \qquad C_2 = 17.8 \text{ MeV}$$

$$C_3 = 0.71 \text{ MeV} \qquad C_4 = 23.6 \text{ MeV}$$

Equation 45.5, together with these constants, fits the known nuclear mass values very well. However, the liquid-drop model does not account for some finer details of nuclear structure, such as stability rules and angular momentum. On the other hand, it does provide a qualitative description of nuclear fission, shown schematically in Figure 45.11.

The Independent-Particle Model

The **independent-particle model,** often called the *shell model,* is based upon the assumption that each nucleon moves in a well-defined orbit within the nucleus. This model is essentially identical to the shell model of the atom, except for the character of the force term. In this model, the nucleons exist in quantized energy states and there are few collisions between nucleons. Obviously, the assumptions of this model differ greatly from those made in the liquid-drop model.

The quantized states occupied by the nucleons can be described by a set of quantum numbers. Since both the proton and the neutron have spin $\frac{1}{2}$, Pauli's exclusion principle can be applied to describe the allowed states (as we did for electrons in Chapter 42). That is, each state can contain only two protons (or two neutrons) having *opposite* spin (Fig. 45.12). The protons have a set of allowed states, and these states differ from those of the neutron because they move in different potential wells. The proton levels are higher in energy than the neutron levels because the protons experience a superposition of Coulomb potential and nuclear potential, while the neutrons experience only a nuclear potential.

In order to understand the observed characteristics of nuclear ground states, it is necessary to include *nuclear spin-orbit* effects. Unlike the spin-orbit interaction for an electron in an atom, which is magnetic in origin, the spin-orbit effect for nucleons in a nucleus is due to the nuclear force; it is much stronger than in the atomic case, and it has opposite sign. This model is able to account for the observed magic numbers.

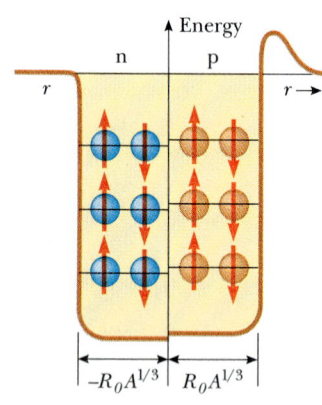

FIGURE 45.12   A square-well potential containing 12 nucleons. The red circles represent protons, and the blue circles represent neutrons. The energy levels for the protons are slightly higher than those for the neutrons because of the Coulomb potential in the protons. The difference in the levels increases as Z increases. Note that only two nucleons having opposite spin can occupy a given level, as required by the Pauli exclusion principle.

Finally, it is possible to understand why nuclei containing an even number of protons and neutrons are more stable than others. In fact, there are 160 such stable even-even isotopes. Any particular state is filled when it contains two protons (or two neutrons) having opposite spins. An extra proton or neutron can be added to the nucleus only at the expense of increasing the energy of the nucleus. This increase in energy leads to a nucleus that is less stable than the original nucleus. A careful inspection of the stable nuclei shows that the majority have a special stability when their nucleons combine in pairs, which results in a total angular momentum of zero. This accounts for the large number of high-stability nuclei (those with high binding energies) with the magic numbers given by Equation 45.2.

45.5 RADIOACTIVITY

In 1896, Henri Becquerel accidentally discovered that uranyl potassium sulfate crystals emit an invisible radiation that can darken a photographic plate when the plate is covered to exclude light. After a series of experiments, he concluded that the radiation emitted by the crystals was of a new type, one that requires no external stimulation and was so penetrating that it could darken protected photographic plates and ionize gases. This process of spontaneous emission of radiation by uranium was soon to be called **radioactivity.** Subsequent experiments by other scientists showed that other substances were more powerfully radioactive. The

The hands and numbers of this luminous watch contain minute amounts of radium salt. The radioactive decay of radium causes the watch to glow in the dark. *(Richard Megna 1990, Fundamental Photographs)*

Marie Sklodowska Curie was born in Poland shortly after the unsuccessful Polish revolt against Russia in 1863. Following high school, she worked diligently to help meet the educational expenses of her older brother and sister who had left for Paris. At the same time, she managed to save enough money for her own trip to Paris and entered the Sorbonne in 1891. Although she lived very frugally during this period (fainting once from hunger in the classroom), she graduated at the top of her class.

In 1895 she married the French physicist Pierre Curie (1859–1906), who was already known for the discovery of piezoelectricity. (A piezoelectric crystal exhibits a potential difference under pressure.) Using piezoelectric materials to measure the activity of radioactive substances, she demonstrated the radioactive nature of the elements uranium and thorium. In 1898, she and her husband discovered a new radioactive element contained in uranium ore, which they

Marie Sklodowska Curie

| 1 8 6 7 – 1 9 3 4 |

called polonium, after her native land. By the end of 1898, the Curies succeeded in isolating trace amounts of an even more radioactive element, which they named radium. In an ef-

fort to produce weighable quantities of radium, they embarked on a painstaking effort of isolating radium from pitchblende, an ore rich in uranium. After four years of purifying and repurifying tons of ore, and using their own life savings to finance their work, the Curies succeeded in preparing about 0.1 g of radium. In 1903, they were awarded the Nobel prize in physics, which they shared with Becquerel, for their studies of radioactive substances.

After her husband's death in an accident in 1906, she took over his professorship at the Sorbonne. Unfortunately, she experienced prejudice in the scientific community because she was a woman. For example, after being nominated to the French Academy of Sciences, she was refused membership, losing by one vote.

In 1911, she was awarded a second Nobel prize, this one in chemistry, for the discovery of radium and polonium. She spent the last few decades of her life supervising the Paris Institute of Radium.

most significant investigations of this type were conducted by Marie and Pierre Curie. After several years of careful and laborious chemical separation processes on tons of pitchblende, a radioactive ore, the Curies reported the discovery of two previously unknown elements, both of which were radioactive. These were named polonium and radium. Subsequent experiments, including Rutherford's famous work on alpha-particle scattering, suggested that radioactivity is the result of the decay, or disintegration, of unstable nuclei.

There are three types of radiation that can be emitted by a radioactive substance: alpha (α) decay, in which the emitted particles are ^{4}He nuclei; beta (β) decay, in which the emitted particles are either electrons or positrons; and gamma (γ) decay, in which the emitted "rays" are high-energy photons. A **positron** is a particle like the electron in all respects except that the positron has a charge of $+e$ (in other words, the positron is the antimatter twin of the electron). The symbol β^- is used to designate an electron, and β^+ designates a positron.

It is possible to distinguish these three forms of radiation using the scheme described in Figure 45.13. The radiation from a radioactive sample is directed into a region in which there is a magnetic field. The beam splits into three components, two bending in opposite directions and the third experiencing no change in direction. From this simple observation, we can conclude that the radiation of the undeflected beam carries no charge (the gamma ray), the component deflected upward corresponds to positively charged particles (alpha particles), and the component deflected downward corresponds to negatively charged particles (β^-). If the beam includes a positron (β^+), it is deflected upward.

The three types of radiation have quite different penetrating powers. Alpha particles barely penetrate a sheet of paper, beta particles can penetrate a few millimeters of aluminum, and gamma rays can penetrate several centimeters of lead.

The rate at which a particular decay process occurs in a radioactive sample is proportional to the number of radioactive nuclei present (that is, those nuclei that have not yet decayed). If N is the number of radioactive nuclei present at some instant, the rate of change of N is

$$\frac{dN}{dt} = -\lambda N \tag{45.6}$$

where λ is called the **decay constant** and is the probability of decay per nucleus per

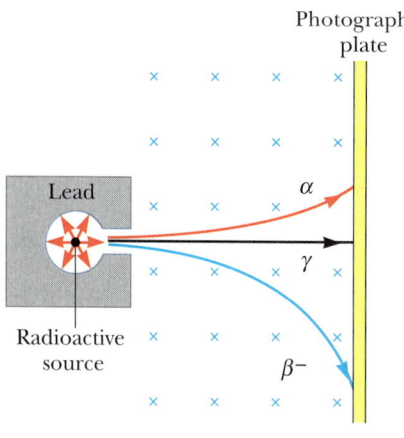

Photographic plate

FIGURE 45.13   The radiation from a radioactive source can be separated into three components by using a magnetic field to deflect the charged particles. The photographic plate at the right records the events. The gamma ray is not deflected by the magnetic field.

Lead

Radioactive source

α

γ

β^-

$\mathbf{B}_{\text{in}}$

second. The minus sign indicates that dN/dt is negative; that is, N is decreasing in time.

Equation 45.6 can be written in the form

$$\frac{dN}{N} = -\lambda\, dt$$

The solution of this equation is given by

$$N = N_0 e^{-\lambda t} \tag{45.7}$$

Exponential decay

where the constant N_0 represents the number of radioactive nuclei at $t = 0$. Equation 45.7 shows that the number of radioactive nuclei in a sample decreases exponentially with time.

The **decay rate** R, or the number of decays per second, can be obtained by differentiating Equation 45.7 with respect to time:

$$R = \left| \frac{dN}{dt} \right| = N_0 \lambda e^{-\lambda t} = R_0 e^{-\lambda t} \tag{45.8}$$

where $R_0 = N_0 \lambda$ is the decay rate at $t = 0$ and $R = \lambda N$. The decay rate of a sample is often referred to as its **activity**. Note that both N and R decrease exponentially with time. The plot of N versus t shown in Figure 45.14 illustrates the exponential decay law.

Another parameter useful in characterizing the decay of a particular nucleus is the **half-life**, $T_{1/2}$:

The **half-life** of a radioactive substance is the time it takes half of a given number of radioactive nuclei to decay.

Setting $N = N_0/2$ and $t = T_{1/2}$ in Equation 45.7 gives

$$\frac{N_0}{2} = N_0 e^{-\lambda T_{1/2}}$$

Writing this in the form $e^{\lambda T_{1/2}} = 2$ and taking the natural logarithm of both sides, we get

$$T_{1/2} = \frac{\ln 2}{\lambda} = \frac{0.693}{\lambda} \tag{45.9}$$

Half-life equation

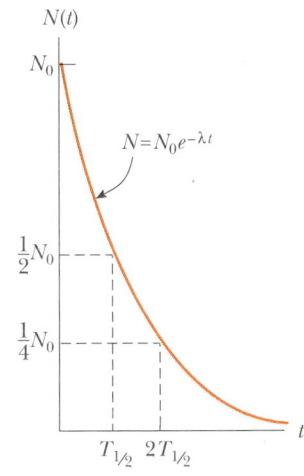

FIGURE 45.14 Plot of the exponential decay law for radioactive nuclei. The vertical axis represents the number of radioactive nuclei present at any time t, and the horizontal axis is time. The time $T_{1/2}$ is the half-life of the sample.

This is a convenient expression relating half-life to decay constant. Note that after an elapsed time of one half-life, there are $N_0/2$ radioactive nuclei remaining (by definition); after two half-lives, half of these have decayed and $N_0/4$ radioactive nuclei are left; after three half-lives, $N_0/8$ are left, and so on. In general, after n half-lives, the number of radioactive nuclei remaining is $N_0/2^n$. Thus, we see that nuclear decay is independent of the past history of a sample.

A frequently used unit of activity is the **curie** (Ci), defined as

The curie

$$1 \text{ Ci} \equiv 3.7 \times 10^{10} \text{ decays/s}$$

This value was originally selected because it is the approximate activity of 1 g of radium. The SI unit of activity is the **becquerel** (Bq):

The becquerel

$$1 \text{ Bq} \equiv 1 \text{ decay/s}$$

Therefore, 1 Ci = 3.7×10^{10} Bq. The curie is a rather large unit and the more frequently used activity units are the millicurie and the microcurie.

CONCEPTUAL EXAMPLE 45.4   How Many Nuclei Are Left?

The isotope carbon-14, $^{14}_{6}\text{C}$, is radioactive and has a half-life of 5730 years. If you start with a sample of 1000 carbon-14 nuclei, how many will still be around in 22 920 years?

Reasoning and Solution In 5730 years, half the sample will have decayed, leaving 500 carbon-14 nuclei remaining. In another 5730 years (for a total elapsed time of 11 460 years), the number will be reduced to 250 nuclei. After another 5730 years (total time 17 190 years), 125 remain. Fi-

nally, after four half-lives (22 920 years), only about 62 remain.

These numbers represent ideal circumstances. Radioactive decay is an averaging process over a very large number of atoms, and the actual outcome depends on statistics. Our original sample in this example contained only 1000 nuclei, certainly not a very large number. Thus, if we were actually to count the number remaining after one half-life for this small sample, it probably would not be 500.

EXAMPLE 45.5   The Activity of Radium

The half-life of the radioactive nucleus $^{226}_{88}\text{Ra}$ is 1.6×10^3 years. (a) What is the decay constant of $^{226}_{88}\text{Ra}$?

Solution We can calculate the decay constant λ using Equation 45.9 and the fact that

$$T_{1/2} = 1.6 \times 10^3 \text{ years}$$
$$= (1.6 \times 10^3 \text{ years})(3.16 \times 10^7 \text{ s/year})$$
$$= 5.0 \times 10^{10} \text{ s}$$

Therefore,

$$\lambda = \frac{0.693}{T_{1/2}} = \frac{0.693}{5.0 \times 10^{10} \text{ s}} = \boxed{1.4 \times 10^{-11} \text{ s}^{-1}}$$

Note that this result is also the probability that any single $^{226}_{88}\text{Ra}$ nucleus will decay in one second.

(b) If a sample contains 3.0×10^{16} such nuclei at $t = 0$, determine its activity at this time.

Solution We can calculate the activity of the sample at $t = 0$ using $R_0 = \lambda N_0$, where R_0 is the decay rate at $t = 0$ and

N_0 is the number of radioactive nuclei present at $t = 0$. Since $N_0 = 3.0 \times 10^{16}$, we have

$$R_0 = \lambda N_0 = (1.4 \times 10^{-11} \text{ s}^{-1})(3.0 \times 10^{16})$$
$$= 4.1 \times 10^5 \text{ decays/s}$$

Since 1 Ci = 3.7×10^{10} decays/s, the activity, or decay rate, at $t = 0$ is

$$R_0 = \boxed{11.1 \ \mu\text{Ci}}$$

(c) What is the decay rate after the sample is 2.0×10^3 years old.

Solution We can use Equation 45.8 as well as the fact that 2.0×10^3 year = $(2.0 \times 10^3 \text{ year})(3.15 \times 10^7 \text{ s/year})$ = 6.3×10^{10} s:

$$R = R_0 e^{-\lambda t} = (4.2 \times 10^5 \text{ decays/s}) e^{-(1.4 \times 10^{-11} \text{s}^{-1})(6.3 \times 10^{10} \text{s})}$$

$$= \boxed{1.7 \times 10^5 \text{ decays/s}}$$

EXAMPLE 45.6   The Activity of Carbon

A radioactive sample contains 3.50 μg of pure $^{11}_{6}\text{C}$, which has a half-life of 20.4 min. (a) Determine the number of nuclei in the sample at $t = 0$.

Solution The atomic mass of $^{11}_{6}\text{C}$ is approximately 11.0, and therefore 11.0 g contains Avogadro's number (6.02 × 10^{23}) of nuclei. Therefore, 3.50 μg contains N nuclei, where

$$\frac{N}{6.02 \times 10^{23} \text{ nuclei/mol}} = \frac{3.50 \times 10^{-6} \text{ g}}{11.0 \text{ g/mol}}$$

$$N = \boxed{1.92 \times 10^{17} \text{ nuclei}}$$

(b) What is the activity of the sample initially and after 8.00 h?

Solution Since $T_{1/2} = 20.4$ min $= 1224$ s, the decay constant is

$$\lambda = \frac{0.693}{T_{1/2}} = \frac{0.693}{1224 \text{ s}} = 5.66 \times 10^{-4} \text{ s}^{-1}$$

Therefore, the initial activity of the sample is

$$R_0 = \lambda N_0 = (5.66 \times 10^{-4} \text{ s}^{-1})(1.92 \times 10^{17})$$

$$= \boxed{1.09 \times 10^{14} \text{ decays/s}}$$

We can use Equation 45.8 to find the activity at any time t. For $t = 8.00$ h $= 2.88 \times 10^4$ s, we see that $\lambda t = 16.3$ and so

$$R = R_0 e^{-\lambda t} = (1.09 \times 10^{14} \text{ decays/s}) e^{-16.3}$$

$$= \boxed{8.59 \times 10^6 \text{ decays/s}}$$

A listing of activity versus time for this situation is given in Table 45.2.

Exercise Calculate the number of radioactive nuclei remaining after 8.00 h.

Answer 1.58×10^{10} nuclei.

TABLE 45.2 Activity Versus Time for the Sample Described in Example 45.6

t (h)	R (decays/s)
0	1.08×10^{14}
1	1.41×10^{13}
2	1.84×10^{12}
3	2.39×10^{11}
4	3.12×10^{10}
5	4.06×10^9
6	5.28×10^8
7	6.88×10^7
8	8.96×10^6

EXAMPLE 45.7 A Radioactive Isotope of Iodine

A sample of the isotope ^{131}I, which has a half-life of 8.04 days, has an activity of 5 mCi at the time of shipment. Upon receipt in a medical laboratory, the activity is 4.2 mCi. How much time has elapsed between the two measurements?

Solution We can make use of Equation 45.8 in the form

$$\frac{R}{R_0} = e^{-\lambda t}$$

Taking the natural logarithm of each side, we get

$$\ln\left(\frac{R}{R_0}\right) = -\lambda t$$

$$(1) \qquad t = -\frac{1}{\lambda} \ln\left(\frac{R}{R_0}\right)$$

To find λ, we use Equation 45.9:

$$(2) \qquad \lambda = \frac{0.693}{T_{1/2}} = \frac{0.693}{8.04 \text{ days}}$$

Substituting (2) into (1) gives

$$t = -\left(\frac{8.04 \text{ days}}{0.693}\right) \ln\left(\frac{4.2 \text{ mCi}}{5.0 \text{ mCi}}\right) = \boxed{2.02 \text{ days}}$$

45.6 THE DECAY PROCESSES

As we stated in the previous section, a radioactive nucleus spontaneously decays by means of one of three processes: alpha decay, beta decay, or gamma decay. Let us discuss these three processes in more detail.

Alpha Decay

If a nucleus emits an alpha particle ($^4_2$He), it loses two protons and two neutrons. Therefore, the atomic number Z decreases by 2, the mass number A decreases by 4, and the neutron number decreases by 2. The decay can be written as

$$_Z^A X \longrightarrow _{Z-2}^{A-4} Y + _2^4 He \tag{45.10}$$

where X is called the **parent nucleus** and Y the **daughter nucleus.** As examples, ^{238}U and ^{226}Ra are both alpha emitters and decay according to the schemes

$$_{92}^{238}U \longrightarrow _{90}^{234}Th + _2^4 He \tag{45.11}$$

$$_{88}^{226}Ra \longrightarrow _{86}^{222}Rn + _2^4 He \tag{45.12}$$

The half-life for ^{238}U decay is 4.47×10^9 years, and that for ^{226}Ra decay is 1.60×10^3 years. In both cases, note that the mass number of the daughter nucleus is 4 less than that of the parent nucleus. Likewise, the atomic number is reduced by 2. The differences are accounted for in the emitted alpha particle (the ^4He nucleus).

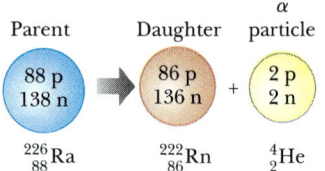

FIGURE 45.15 The alpha decay of radium. The radium nucleus is initially at rest. After the decay, both the radon nucleus and the alpha particle have kinetic energy and momentum.

The decay of ^{226}Ra is shown in Figure 45.15. When one element changes into another, as in the process of alpha decay, the process is called spontaneous decay. As a general rule, (1) the sum of the mass numbers A must be the same on both sides of the equation and (2) the sum of the atomic numbers Z must be the same on both sides of the equation. In addition, relativistic energy and momentum must be conserved. If we call M_X the mass of the parent nucleus, M_Y the mass of the daughter nucleus, and M_α the mass of the alpha particle, we can define the **disintegration energy** Q:

$$Q = (M_X - M_Y - M_\alpha)c^2 \tag{45.13}$$

Enrico Fermi, an Italian-American physicist, received his doctorate from the University of Pisa in 1922 and did postdoctorate work in Germany under Max Born. He returned to Italy in 1924 and became a professor of physics at the University of Rome in 1926. He received the Nobel prize for physics in 1938 for his work dealing with the production of transuranic radioactive elements (those more massive than uranium) by neutron bombardment.

Fermi first became interested in physics at the age of 14 after reading an old physics book in Latin. He had an excellent scholastic record and was able to recite Dante's *Divine Comedy* and much of Aristotle from memory. His great ability to solve problems in theoretical physics and his skill for simplifying very complex situations made him somewhat of an oracle. He was also a gifted experimentalist and teacher. During one of his early lecture trips to the United States, a car that he had purchased became disabled and he pulled into a nearby gas station. After Fermi repaired the car

Enrico Fermi

| 1 9 0 1 – 1 9 5 4 |

with ease, the station owner immediately offered him a job.

Fermi and his family immigrated to the United States and he became a naturalized citizen in 1944. Once in America, he accepted a position at Columbia University and later became a professor at the University of

Chicago. After the Manhattan Project was established (the project that designed and constructed the atomic bomb during World War II), Fermi was commissioned to design and build a structure (called an atomic pile) in which a self-sustained chain reaction might occur. The structure, built in the squash court of the University of Chicago, contained uranium in combination with graphite blocks to slow the neutrons to thermal speeds. Cadmium rods inserted in the pile were used to absorb neutrons and control the reaction rate. History was made at 3:45 P.M. on December 2, 1942, as the cadmium rods were slowly withdrawn and a self-sustained chain reaction was observed. Fermi's earthshaking achievement of the world's first nuclear reactor marked the beginning of the atomic age.

Fermi died of cancer in 1954 at the age of 53. One year later, the 100th element was discovered and named *fermium* in his honor.

National Accelerator Laboratory

Q is in joules when the masses are in kilograms and c is 3.00×10^8 m/s. However, when the nuclear masses are expressed in the more convenient unit u, the value of Q can be calculated in MeV using the expression

$$Q = (M_X - M_Y - M_\alpha) \times 931.494 \text{ MeV/u} \qquad (45.14)$$

The disintegration energy Q appears in the form of kinetic energy in the daughter nucleus and the alpha particle. The quantity given by Equation 45.13 is sometimes referred to as the Q value of the nuclear reaction. In the case of the ^{226}Ra decay described in Figure 45.15, if the parent nucleus decays at rest, the residual kinetic energy of the products is 4.87 MeV. Most of the kinetic energy is associated with the alpha particle because this particle is much less massive than the recoiling daughter nucleus, ^{222}Rn. That is, because momentum must be conserved, the lighter alpha particle recoils with a much higher speed than the daughter nucleus. Generally, light particles carry off most of the energy in nuclear decays.

Finally, it is interesting to note that if one assumed that ^{238}U (or other alpha emitters) decayed by emitting a proton or neutron, the mass of the decay products would exceed that of the parent nucleus, corresponding to negative Q values. These negative Q values indicate that such decays do not occur spontaneously.

EXAMPLE 45.8   The Energy Liberated When Radium Decays

The ^{226}Ra nucleus undergoes alpha decay according to Equation 45.12. Calculate the Q value for this process. Take the masses to be 226.025 406 u for ^{226}Ra, 222.017 574 u for ^{222}Rn, and 4.002 603 u for $^4_2$He, as found in Table A.3.

Solution   Using Equation 45.14, we see that

$$Q = (M_X - M_Y - M_\alpha) \times 931.494 \text{ MeV/u}$$
$$= (226.025\ 406 \text{ u} - 222.017\ 574 \text{ u}$$

$$- 4.002\ 603 \text{ u}) \times 931.494 \text{ MeV/u}$$

$$= (0.005\ 229 \text{ u}) \times \left(931.494 \frac{\text{MeV}}{\text{u}}\right) = \boxed{4.87 \text{ MeV}}$$

It is left as a problem (Problem 75) to show that the kinetic energy of the alpha particle is about 4.8 MeV, whereas the recoiling daughter nucleus has only about 0.1 MeV of kinetic energy.

We now turn to the mechanism of alpha decay. Figure 45.16 is a plot of potential energy versus distance r from the nucleus for the alpha particle nucleus system, where R is the range of the nuclear force. The curve represents the combined effects of (1) the Coulomb repulsive energy, which gives the positive peak for $r > R$, and (2) the nuclear attractive force, which causes the curve to be negative

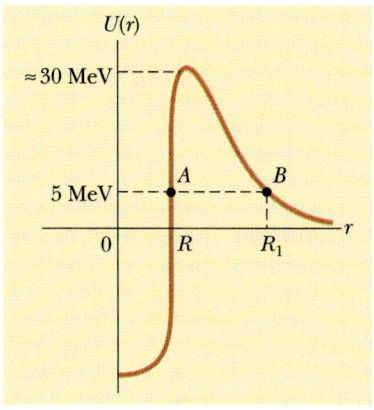

FIGURE 45.16   Potential energy versus separation for the alpha particle-nucleus system. Classically, the energy of the alpha particle is not sufficiently large to overcome the barrier, and so the particle should not be able to escape the nucleus.

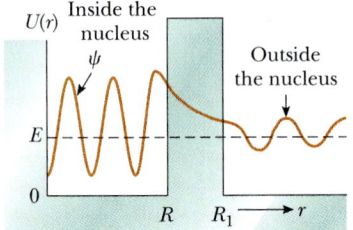

FIGURE 45.17   The nuclear potential energy is modeled as a square barrier. The energy of the alpha particle is E, which is less than the height of the barrier. According to quantum mechanics, the alpha particle has some chance of tunneling through the barrier, as indicated by the finite size of the wave function for $r > R_1$.

___

Beta decay

___

for $r < R$. As we saw in Example 45.8, the disintegration energy is about 5 MeV, which is the approximate kinetic energy of the alpha particle, represented by the lower dashed line in Figure 45.16. According to classical physics, the alpha particle is trapped in the potential well. How, then, does it ever escape from the nucleus?

The answer to this question was first provided by George Gamow in 1928 and independently by R. W. Gurney and E. U. Condon in 1929, using quantum mechanics. Briefly, the view of quantum mechanics is that there is always some probability that the particle can penetrate (or tunnel) through the barrier (Section 41.9). Recall that the probability of locating the particle depends on its wave function ψ and that the probability of tunneling is measured by $|\psi|^2$. Figure 45.17 is a sketch of the wave function for a particle of energy E meeting a square barrier of finite height, a shape that approximates the nuclear barrier. Note that the wave function exists both inside and outside the barrier. Although the amplitude of the wave function is greatly reduced on the far side of the barrier, its finite value in this region indicates a small but finite probability that the particle can penetrate the barrier. As the energy E of the particle is increased, its probability of escaping also increases. Furthermore, the probability increases as the width of the barrier is decreased.

Beta Decay

When a radioactive nucleus undergoes beta decay, the daughter nucleus has the same number of nucleons as the parent nucleus but the atomic number is changed by 1:

$$_Z^A X \longrightarrow \, _{Z+1}^A Y + \beta^- \tag{45.15}$$

$$_Z^A X \longrightarrow \, _{Z-1}^A Y + \beta^+ \tag{45.16}$$

Again, note that the nucleon number and total charge are both conserved in these decays. As we shall see later, *these processes are not described completely by these expressions.* We shall give reasons for this shortly.

Two typical beta decay processes are

$$_6^{14}C \longrightarrow \, _7^{14}N + \beta^-$$

$$_7^{12}N \longrightarrow \, _6^{12}C + \beta^+$$

Notice that in beta decay, a neutron changes to a proton (or vice versa). It is also important to note that the electron or positron in these decays is not present beforehand in the nucleus, but is created at the moment of decay from the rest energy of the decaying nucleus.

Now consider the energy of the system before and after the decay. As with alpha decay, energy must be conserved. Experimentally, it is found that the beta particles are emitted over a continuous range of energies (Fig. 45.18). These results show that beta particles having different energies are emitted. The kinetic energy of the particles must be balanced by the decrease in mass of the system, that is, the Q value. However, because all decaying nuclei have the same initial mass, *the Q value must be the same for each decay.* In view of this, why do the emitted particles have different kinetic energies? The law of conservation of energy seems to be violated! Further analysis shows that, according to the decay processes given by Equations 45.15 and 45.16, the principles of conservation of both angular momentum (spin) and linear momentum are also violated!

After a great deal of experimental and theoretical study, Pauli in 1930 proposed that a third particle must be present to carry away the "missing" energy and

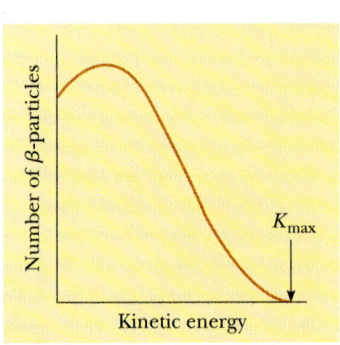

FIGURE 45.18   A typical beta decay curve. The maximum kinetic energy observed for the beta particles corresponds to the Q value for the reaction.

momentum. Fermi later named this particle the **neutrino** (little neutral one) because it had to be electrically neutral and have little or no rest mass. Although it eluded detection for many years, the neutrino (symbol ν) was finally detected experimentally in 1956. It has the following properties:

- zero electric charge
- there is increasing evidence that its mass is zero and that it travels with the speed of light
- a spin of $\frac{1}{2}$, which satisfies the law of conservation of angular momentum
- interacts very weakly with matter and is therefore very difficult to detect

Properties of the neutrino

We can now write the beta decay processes in their correct form:

$$^{14}_{6}\text{C} \longrightarrow {}^{14}_{7}\text{N} + \beta^- + \bar{\nu} \tag{45.17}$$

$$^{12}_{7}\text{N} \longrightarrow {}^{12}_{6}\text{C} + \beta^+ + \nu \tag{45.18}$$

$$\text{n} \longrightarrow \text{p} + \beta^- + \bar{\nu} \tag{45.19}$$

where the symbol $\bar{\nu}$ represents the **antineutrino,** the antiparticle to the neutrino. We shall discuss antiparticles further in Chapter 47. For now, it suffices to say that *a neutrino is emitted in positron decay and an antineutrino is emitted in electron decay.* As with the case of alpha decay, the decays listed above are analyzed by conserving energy and momentum, but relativistic expressions must be used for beta particles because their energy is large (typically 1 MeV) compared with their rest energy of 0.511 MeV.

A process that competes with β^+ decay is called **electron capture.** This occurs when a parent nucleus captures one of its own orbital electrons and emits a neutrino. The final product after decay is a nucleus whose charge is $Z - 1$:

$$^{A}_{Z}\text{X} + {}^{0}_{-1}\text{e} \longrightarrow {}^{A}_{Z-1}\text{Y} + \nu \tag{45.20}$$

Electron capture

In most cases, it is an inner K-shell electron that is captured, and this is referred to as **K capture.** One example of this process is the capture of an electron by ${}^{7}_{4}\text{Be}$ to become ${}^{7}_{3}\text{Li}$:

$$^{7}_{4}\text{Be} + {}^{0}_{-1}\text{e} \longrightarrow {}^{7}_{3}\text{Li} + \nu$$

Finally, it is instructive to mention the Q-values for the beta-decay processes. The Q-values for β^- decay and electron capture are given by $Q = (M_X - M_Y)c^2$, while the Q-values for β^+ decay are given by $Q = (M_X - M_Y - 2m_e)c^2$. These relationships are useful to determine whether or not the processes are energetically possible.

Carbon Dating

The beta decay of ^{14}C given by Equation 45.17 is commonly used to date organic samples. Cosmic rays in the upper atmosphere cause nuclear reactions that create ^{14}C. In fact, the ratio of ^{14}C to ^{12}C in the carbon dioxide molecules of our atmosphere has a constant value of approximately 1.3×10^{-12}. All living organisms have this same ratio of ^{14}C to ^{12}C because they continuously exchange carbon dioxide with their surroundings. When an organism dies, however, it no longer absorbs ^{14}C from the atmosphere, and so the $^{14}\text{C}/^{12}\text{C}$ ratio decreases as the result of the beta decay of ^{14}C, which has a half-life of 5730 years. It is therefore possible to measure the age of a material by measuring its activity per unit mass caused by the decay of ^{14}C. Using this technique, scientists have been able to identify samples of wood, charcoal, bone, and shell as having lived from 1000 to 25 000 years ago. This knowledge has helped us reconstruct the history of living organisms—including humans—during this time span.

A particularly interesting example is the dating of the Dead Sea Scrolls. This group of manuscripts was discovered by a shepherd in 1947. Translation showed them to be religious documents, including most of the books of the Old Testament. Because of their historical and religious significance, scholars wanted to know their age. Carbon dating applied to the material in which they were wrapped established their age at approximately 1950 years.

CONCEPTUAL EXAMPLE 45.9   The Age of Ice Man

In 1991, a German tourist discovered the well-preserved remains of the Ice Man trapped in a glacier in the Italian Alps. Radioactive dating of a sample of Ice Man revealed an age of 5300 years. Why did scientists date the sample using the isotope ^{14}C, rather than ^{11}C, a beta emitter with a half-life of 20.4 min?

Reasoning ^{14}C has a long half-life of 5730 years, so the fraction of ^{14}C nuclei remaining after one half-life is high enough to measure accurate changes in the sample's activity. The ^{11}C isotope, which has a very short half-life, is not useful because its activity decreases to a vanishingly small value over the age of the sample, making it impossible to detect.

If a sample to be dated is not very old, say about 50 years, then one should select the isotope of some other element whose half-life is comparable with the age of the sample. For example, if the sample contained hydrogen, one could measure the activity of ^{3}H (tritium), a beta emitter of half-life 12.3 years. As a general rule, the expected age of the sample should be long enough to measure a change in activity, but not so long that its activity cannot be detected.

The Ice Man, discovered in 1991 when an Italian glacier melted enough to expose his remains. His possessions, particularly his tools, have been able to shed light on the way people lived in the Bronze Age. Carbon-14 dating was used by archaeologists to determine the time frame of their discovery.   *(Paul Hanny/Gamma Liaison)*

EXAMPLE 45.10   Radioactive Dating

A 25.0-g piece of charcoal is found in some ruins of an ancient city. The sample shows a ^{14}C activity of 250 decays/min. How long has the tree that this charcoal came from been dead?

Solution   First, let us calculate the decay constant for ^{14}C, which has a half-life of 5730 years.

$$\lambda = \frac{0.693}{T_{1/2}} = \frac{0.693}{(5730 \text{ years})(3.16 \times 10^7 \text{ s/years})}$$
$$= 3.83 \times 10^{-12} \text{ s}^{-1}$$

The number of ^{14}C nuclei can be calculated in two steps. First, the number of ^{12}C nuclei in 25.0 g of carbon is

$$N(^{12}\text{C}) = \frac{6.02 \times 10^{23} \text{ nuclei/mol}}{12.0 \text{ g/mol}} (25.0 \text{ g})$$
$$= 1.26 \times 10^{24} \text{ nuclei}$$

Knowing that the ratio of ^{14}C to ^{12}C in the live sample was 1.3×10^{-12}, we see that the number of ^{14}C nuclei in 25.0 g *before* decay is

$$N_0(^{14}\text{C}) = (1.3 \times 10^{-12})(1.26 \times 10^{24}) = 1.6 \times 10^{12} \text{ nuclei}$$

Hence, the initial activity of the sample is

$$R_0 = N_0\lambda = (1.6 \times 10^{12} \text{ nuclei})(3.83 \times 10^{-12} \text{ s}^{-1})$$
$$= 6.13 \text{ decays/s} = 370 \text{ decays/min}$$

We can now calculate the age of the charcoal using Equation 45.8, which relates the activity R at any time t to the initial activity R_0:

$$R = R_0 e^{-\lambda t} \quad \text{or} \quad e^{-\lambda t} = \frac{R}{R_0}$$

Because it is given that $R = 250$ decays/min and because we found that $R_0 = 370$ decays/min, we can calculate t by tak-

ing the natural logarithm of both sides of the last equation:

$$-\lambda t = \ln\left(\frac{R}{R_0}\right) = \ln\left(\frac{250}{370}\right) = -0.39$$

$$t = \frac{0.39}{\lambda} = \frac{0.39}{3.84 \times 10^{-12} \text{ s}^{-1}}$$

$$= 1.0 \times 10^{11} \text{ s} = 3.2 \times 10^3 \text{ years}$$

Gamma Decay

Very often, a nucleus that undergoes radioactive decay is left in an excited energy state. The nucleus can then undergo a second decay to a lower energy state, perhaps to the ground state, by emitting a high-energy photon:

$$^A_Z X^* \longrightarrow {}^A_Z X + \gamma \tag{45.21}$$

Gamma decay

where X* indicates a nucleus in an excited state. The typical half-life of an excited nuclear state is 10^{-10} s. Photons emitted in such a de-excitation process are called gamma rays. Such photons have very high energy (in the range of 1 MeV to 1 GeV) relative to the energy of visible light (about 1 eV). Recall that the energy of photons emitted (or absorbed) by an atom equals the difference in energy between the two electronic states involved in the transition. Similarly, a gamma ray photon has an energy hf that equals the energy difference ΔE between two nuclear energy levels. When a nucleus decays by emitting a gamma ray, the nucleus doesn't change, apart from the fact that it ends up in a lower energy state.

A nucleus may reach an excited state as the result of a violent collision with another particle. However, it is more common for a nucleus to be in an excited state after it has undergone a previous alpha or beta decay. The following sequence of events represents a typical situation in which gamma decay occurs:

$$^{12}_5 B \longrightarrow {}^{12}_6 C^* + {}^0_{-1} e + \bar{\nu} \tag{45.22}$$

$$^{12}_6 C^* \longrightarrow {}^{12}_6 C + \gamma \tag{45.23}$$

Figure 45.19 shows the decay scheme for ^{12}B, which undergoes beta decay to either of two levels of ^{12}C. It can either (1) decay directly to the ground state of ^{12}C by emitting a 13.4-MeV electron or (2) undergo β^- decay to an excited state of ^{12}C* followed by gamma decay to the ground state. This latter process results in the emission of a 9.0-MeV electron and a 4.4-MeV photon.

The various pathways by which a radioactive nucleus can undergo decay are summarized in Table 45.3.

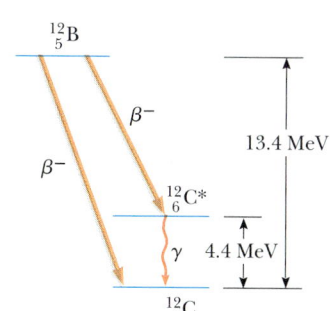

$^{12}_5 B$

β^-

β^-

13.4 MeV

$^{12}_6 C^*$

γ 4.4 MeV

$^{12}_6 C$

FIGURE 45.19   The ^{12}B nucleus undergoes β^- decay to two levels of ^{12}C. The decay to the excited level, ^{12}C*, is followed by gamma decay to the ground state.

TABLE 45.3   Various Decay Pathways

Alpha decay	$^A_Z X \rightarrow {}^{A-4}_{Z-2} X + {}^4_2 He$
Beta decay (β^-)	$^A_Z X \rightarrow {}^A_{Z+1} X + \beta^- + \bar{\nu}$
Beta decay (β^+)	$^A_Z X \rightarrow {}^A_{Z-1} X + \beta^+ + \nu$
Electron capture	$^A_Z X + {}^0_{-1} e \rightarrow {}^A_{Z-1} X + \nu$
Gamma decay	$^A_Z X^* \rightarrow {}^A_Z X + \gamma$

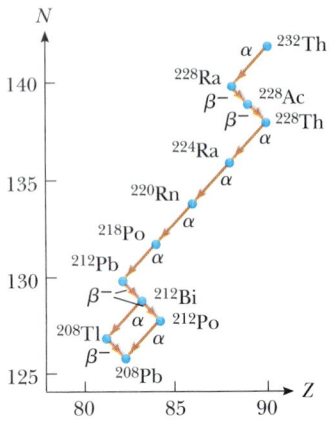

FIGURE 45.20   Successive decays for the ^{232}Th series.

45.7   NATURAL RADIOACTIVITY

Radioactive nuclei are generally classified into two groups: (1) unstable nuclei found in nature, which give rise to what is called **natural radioactivity,** and (2) nuclei produced in the laboratory through nuclear reactions, which exhibit **artificial radioactivity.**

There are three series of naturally occurring radioactive nuclei (Table 45.4). Each series starts with a specific long-lived radioactive isotope whose half-life exceeds that of any of its descendants. The three natural series begin with the isotopes ^{238}U, ^{235}U, and ^{232}Th, and the corresponding stable end products are three isotopes of lead: ^{206}Pb, ^{207}Pb, and ^{208}Pb. The fourth series in Table 45.4 begins with ^{237}Np and has as its stable end product ^{209}Bi. The element ^{237}Np is a transuranic element (one having an atomic number greater than that of uranium) not found in nature. This element has a half-life of "only" 2.14×10^6 years.

Figure 45.20 shows the successive decays for the ^{232}Th series. Note that ^{232}Th first undergoes alpha decay to ^{228}Ra. Next, ^{228}Ra undergoes two successive β decays to ^{228}Th. The series continues and finally branches when it reaches ^{212}Bi. At this point, there are two decay possibilities. The end of the decay series is the stable isotope ^{208}Pb. The entire decay sequence shown in Figure 45.20 may be characterized by a decrease in the mass number of either 4 (for alpha decays) or 0 (for beta or gamma decays).

The two uranium series are somewhat more complex than the ^{232}Th series. Also, there are several naturally occurring radioactive isotopes, such as ^{14}C and ^{40}K, that are not part of either decay series.

The existence of radioactive series in nature enables our environment to be constantly replenished with radioactive elements that would otherwise have disappeared long ago. For example, because the Solar System is approximately 5×10^9 years old, the supply of ^{226}Ra (whose half-life is only 1600 years) would have been depleted by radioactive decay long ago if it were not for the decay series that starts with ^{238}U.

45.8   NUCLEAR REACTIONS

It is possible to change the structure of nuclei by bombarding them with energetic particles. Such collisions, which change the identity of the target nuclei, are called **nuclear reactions.** Rutherford was the first to observe them in 1919, using naturally occurring radioactive sources for the bombarding particles. Since then, thousands of nuclear reactions have been observed following the development of charged-particle accelerators in the 1930s. With today's advanced technology in particle accelerators and particle detectors, it is possible to achieve particle energies of at

TABLE 45.4   The Four Radioactive Series

Series		Starting Isotope	Half-Life (years)	Stable End Product
Uranium	⎫	$^{238}_{92}$U	4.47×10^9	$^{206}_{82}$Pb
Actinium	⎬ Natural	$^{235}_{92}$U	7.04×10^8	$^{207}_{82}$Pb
Thorium	⎭	$^{232}_{90}$Th	1.41×10^{10}	$^{208}_{82}$Pb
Neptunium		$^{237}_{93}$Np	2.14×10^6	$^{209}_{83}$Bi

least 1000 GeV = 1 TeV. These high-energy particles are used to create new particles whose properties are helping to solve the mystery of the nucleus.

Consider a reaction in which a target nucleus X is bombarded by a particle a, resulting in a nucleus Y and a particle b:

$$a + X \longrightarrow Y + b \tag{45.24}$$

Nuclear reaction

Sometimes this reaction is written in the more compact form

$$X(a, b)Y$$

In the previous section, the Q value, or disintegration energy, of a radioactive decay was defined as the energy released as the result of the decay process. Likewise, we define the **reaction energy** Q associated with a nuclear reaction as *the total energy released as the result of the reaction.* More specifically, Q is defined as

$$Q = (M_a + M_X - M_Y - M_b)\, c^2 \tag{45.25}$$

Reaction energy Q

As an example, consider the reaction ^{7}Li (p, α)^{4}He, or

$$^1_1\text{H} + ^7_3\text{Li} \longrightarrow ^4_2\text{He} + ^4_2\text{He}$$

which has a Q value of 17.3 MeV. A reaction such as this, for which Q is positive, is called **exothermic**. An endothermic reaction is one in which the loss in energy of the system is balanced by an increase in the kinetic energy of the final particles. A reaction for which Q is negative is called **endothermic**. An endothermic reaction does not occur unless the bombarding particle has a kinetic energy greater than Q. The minimum energy necessary for such a reaction to occur is called the **threshold energy**.

Exothermic reaction

Endothermic reaction

Threshold energy

Nuclear reactions must obey the law of conservation of linear momentum. This assumes that the only force acting on the interacting particles is their mutual force of interaction; that is, there are no external accelerating electric fields present near the colliding particles.

If a nuclear reaction occurs in which particles a and b are identical, so that X and Y are also necessarily identical, the reaction is called a scattering event. If kinetic energy is constant as a result of the reaction (that is, if $Q = 0$), it is classified as elastic scattering. On the other hand, if $Q \neq 0$, kinetic energy is not constant and the reaction is called inelastic scattering. This terminology is identical to that used in dealing with the collision between macroscopic objects (Section 9.4).

A list of measured Q values for a number of nuclear reactions involving light nuclei is given in Table 45.5.

In addition to energy and momentum, the total charge and total number of nucleons must be conserved in any nuclear reaction. For example, consider the reaction ^{19}F(p, α)^{16}O, which has a Q value of 8.124 MeV. We can show this reaction more completely as

$$^1_1\text{H} + ^{19}_9\text{F} \longrightarrow ^{16}_8\text{O} + ^4_2\text{He}$$

We see that the total number of nucleons before the reaction (1 + 19 = 20) is equal to the total number after the reaction (16 + 4 = 20). Furthermore, the total charge ($Z = 10$) is the same before and after the reaction.

TABLE 45.5	*Q* Values for Nuclear Reactions Involving Light Nuclei	
Reaction[a]	**Measured *Q* Value (MeV)**
^{2}H(n, γ)^{3}H	6.257 ± 0.004
^{2}H(d, p)^{3}H	4.032 ± 0.004
^{6}Li(p, α)^{3}He	4.016 ± 0.005
^{6}Li(d, p)^{7}Li	5.020 ± 0.006
^{7}Li(p, n)^{7}Be	-1.645 ± 0.001
^{7}Li(p, α)^{4}He	17.337 ± 0.007
^{9}Be(n, γ)^{10}Be	6.810 ± 0.006
^{9}Be(γ, n)^{8}Be	-1.666 ± 0.002
^{9}Be(d, p)^{10}Be	4.585 ± 0.005
^{9}Be(p, α)^{6}Li	2.132 ± 0.006
^{10}B(n, α)^{7}Li	2.793 ± 0.003
^{10}B(p, α)^{7}Be	1.148 ± 0.003
^{12}C(n, γ)^{13}C	4.948 ± 0.004
^{13}C(p, n)^{13}N	-3.003 ± 0.002
^{14}N(n, p)^{14}C	0.627 ± 0.001
^{14}N(n, γ)^{15}N	10.833 ± 0.007
^{18}O(p, n)^{18}F	-2.453 ± 0.002
^{19}F(p, α)^{16}O	8.124 ± 0.007

From C. W. Li, W. Whaling, W. A. Fowler, and C. C. Lauritsen, *Physical Review* 83:512 (1951).

[a] The symbols n, p, d, α, and γ denote the neutron, proton, deuteron, alpha particle, and photon, respectively.

SUMMARY

A nuclear species can be represented by $^A_Z X$, where A is the **mass number,** which equals the total number of nucleons, and Z is the **atomic number,** which is the total number of protons. The total number of neutrons in a nucleus is the **neutron number** N, where $A = N + Z$. Elements with the same Z but different A and N values are **isotopes** of each other.

Assuming that nuclei are spherical, their radius is given by

$$r = r_0 A^{1/3} \tag{45.1}$$

where $r_0 = 1.2$ fm.

Nuclei are stable because of the **nuclear force** between nucleons. This short-range force dominates the Coulomb repulsive force at distances of less than about 2 fm and is nearly independent of charge. Light nuclei are most stable when the number of protons equals the number of neutrons. Heavy nuclei are most stable when the number of neutrons exceeds the number of protons. In addition, most stable nuclei have Z and N values that are both even.

Nuclei have an intrinsic spin angular momentum of magnitude $\sqrt{I(I + 1)}\hbar$, where I is the **nuclear spin quantum number.** The magnetic moment of a nucleus is measured in terms of the **nuclear magneton** μ_n, where

$$\mu_n \equiv \frac{e\hbar}{2m_p} = 5.05 \times 10^{-27} \, \text{J/T} \tag{45.3}$$

When a nuclear moment is placed in an external magnetic field, it precesses about the field with a frequency that is proportional to the field.

The difference between the mass of the separate nucleons and that of the compound nucleus containing these nucleons, when multiplied by c^2, gives the **binding energy** E_b of the nucleus: $E_b = \Delta mc^2$. We can calculate the binding energy of any nucleus of mass M_A using the expression

$$E_b(\text{MeV}) = (Zm_p + Nm_n - M_A) \times 931.494 \text{ MeV/u} \tag{45.4}$$

where m_p is the mass of the proton and m_n is the mass of the neutron.

The **liquid-drop model** of nuclear structure treats the nucleons as molecules in a drop of liquid. The three main contributions influencing binding energy are the volume effect, the surface effect, and the Coulomb repulsion. Summing such contributions results in the **semiempirical binding energy formula**:

$$E_b = C_1 A - C_2 A^{2/3} - C_3 \frac{Z(Z-1)}{A^{1/3}} - C_4 \frac{(N-Z)^2}{A} \tag{45.5}$$

The **independent-particle model** assumes that each nucleon moves in a well-defined quantized orbit within the nucleus. The stability of certain nuclei can be explained with this model.

A radioactive substance decays by alpha decay, beta decay, or gamma decay. An alpha particle is the ^{4}He nucleus; a beta particle is either an electron (β^-) or a positron (β^+); a gamma particle is a high-energy photon.

If a radioactive material contains N_0 radioactive nuclei at $t = 0$, the number N of nuclei remaining after a time t has elapsed is

$$N = N_0 e^{-\lambda t} \tag{45.7}$$

where λ is the **decay constant**, a number equal to the probability per second that a nucleus will decay. The **decay rate**, or **activity**, of a radioactive substance is

$$R = \left| \frac{dN}{dt} \right| = R_0 e^{-\lambda t} \tag{45.8}$$

where $R_0 = N_0 \lambda$ is the activity or number of decays per second at $t = 0$. The **half-life** $T_{1/2}$ is defined as the time it takes half of a given number of radioactive nuclei to decay, where

$$T_{1/2} = \frac{0.693}{\lambda} \tag{45.9}$$

Alpha decay can occur because some nuclei have barriers that the alpha particles can tunnel through. This process is energetically more favorable for those nuclei having a large excess of neutrons. A nucleus undergoing beta decay emits either an electron (β^-) and an antineutrino ($\bar{\nu}$) or a positron (β^+) and a neutrino (ν). In electron capture, the nucleus of an atom absorbs one of its own electrons and emits a neutrino. In gamma decay, a nucleus in an excited state decays to its ground state and emits a gamma ray (photon).

Nuclear reactions can occur when a target nucleus X is bombarded by a particle a, resulting in a nucleus Y and a particle b:

$$a + X \longrightarrow Y + b \quad \text{or} \quad X(a, b)Y \tag{45.24}$$

The energy released in such a reaction, called the **reaction energy** Q, is

$$Q = (M_a + M_X - M_Y - M_b)c^2 \tag{45.25}$$

QUESTIONS

1. Why are heavy nuclei unstable?
2. A proton precesses with a frequency ω_p in the presence of a magnetic field. If the magnetic field intensity is doubled, what happens to the precessional frequency?
3. Explain why nuclei that are well off the line of stability in Figure 45.3 tend to be unstable.
4. Why do nearly all the naturally occurring isotopes lie above the $N = Z$ line in Figure 45.3?
5. Consider two heavy nuclei X and Y having similar mass numbers. If X has the higher binding energy, which nucleus tends to be more unstable?
6. Discuss the differences between the liquid-drop model and the independent-particle model of the nucleus.
7. How many values of I_z are possible for $I = 5/2$? for $I = 3$?
8. In nuclear magnetic resonance, how does increasing the dc magnetic field change the frequency of the ac field that excites a particular transition?
9. Would the liquid-drop or independent-particle model be more appropriate to predict the behavior of a nucleus in a fission reaction? Which would be more successful in predicting the magnetic moment of a given nucleus? Which could better explain the γ-ray spectrum of an excited nucleus?
10. If a nucleus has a half-life of one year, does this mean it will be completely decayed after two years? Explain.
11. What fraction of a radioactive sample has decayed after two half-lives have elapsed?
12. Two samples of the same radioactive nuclide are prepared. Sample A has twice the initial activity of sample B. How does the half-life of A compare with the half-life of B? After each has passed through five half-lives, what is the ratio of their activities?
13. Explain why the half-lives for radioactive nuclei are essentially independent of temperature.
14. The radioactive nucleus $^{226}_{88}\text{Ra}$ has a half-life of approximately 1.6×10^3 years. In that the Solar System is about 5 billion years old, why do we still find this nucleus in nature?
15. Why is the electron involved in the reaction

$$^{14}_{6}\text{C} \longrightarrow ^{14}_{7}\text{N} + \beta^-$$

written as β^-, while the electron involved in the reaction

$$^{7}_{4}\text{Be} + \,_{-1}^{0}\text{e} \longrightarrow ^{7}_{3}\text{Li} + \nu$$

is written as $_{-1}^{0}\text{e}$?

16. A free neutron undergoes beta decay with a half-life of about 15 min. Can a free proton undergo a similar decay?
17. Explain how you can carbon-date the age of a sample.
18. What is the difference between a neutrino and a photon?
19. Does the Q in Equation 45.25 represent the quantity (final mass − initial mass) c^2, or does it represent the quantity (initial mass − final mass) c^2?
20. Use Equations 45.17 to 45.19 to explain why the neutrino must have a spin of $\frac{1}{2}$.
21. If a nucleus such as ^{226}Ra initially at rest undergoes alpha decay, which has more kinetic energy after the decay, the alpha particle or the daughter nucleus?
22. Can a nucleus emit alpha particles that have different energies? Explain.
23. Explain why many heavy nuclei undergo alpha decay but do not spontaneously emit neutrons or protons.
24. If an alpha particle and an electron have the same kinetic energy, which undergoes the greater deflection when passed through a magnetic field?
25. If film is kept in a wooden box, alpha particles from a radioactive source outside the box cannot expose the film but beta particles can. Explain.
26. Pick any beta decay process and show that the neutrino must have zero charge.
27. Suppose it could be shown that the cosmic ray intensity at the Earth's surface was much greater 10 000 years ago. How would this difference affect what we accept as valid carbon-dated values of the age of ancient samples of once-living matter?
28. Why is carbon dating unable to provide accurate estimates of very old material?
29. Element X has several isotopes. What do these isotopes have in common? How do they differ?
30. Explain the main differences between alpha, beta, and gamma rays.
31. How many protons are there in the nucleus $^{222}_{86}\text{Rn}$? How many neutrons? How many orbiting electrons are there in the neutral atom?

PROBLEMS

Table 45.6 will be useful for many of these problems. A more complete list of atomic masses is given in Table A.3 in Appendix A.

Section 45.1   Some Properties of Nuclei

1. Find the radius of (a) a nucleus of ^4_2He and (b) a nucleus of $^{238}_{92}\text{U}$. (c) What is the ratio of these radii?

□ indicates problems that have full solutions available in the Student Solutions Manual and Study Guide.

TABLE 45.6   Some Atomic Masses

Element	Atomic Mass (u)	Element	Atomic Mass (u)
$^{4}_{2}$He	4.002 603	$^{27}_{13}$Al	26.981 541
$^{7}_{3}$Li	7.016 004	$^{30}_{15}$P	29.978 310
$^{9}_{4}$Be	9.012 182	$^{40}_{20}$Ca	39.962 591
$^{10}_{5}$B	10.012 938	$^{42}_{20}$Ca	41.958 63
$^{12}_{6}$C	12.000 000	$^{43}_{20}$Ca	42.958 770
$^{13}_{6}$C	13.003 355	$^{56}_{26}$Fe	55.934 939
$^{14}_{7}$N	14.003 074	$^{64}_{30}$Zn	63.929 145
$^{15}_{7}$N	15.000 109	$^{64}_{29}$Cu	63.929 599
$^{15}_{8}$O	15.003 065	$^{93}_{41}$Nb	92.906 378
$^{17}_{8}$O	16.999 131	$^{197}_{79}$Au	196.966 560
$^{18}_{8}$O	17.999 159	$^{202}_{80}$Hg	201.970 632
$^{18}_{9}$F	18.000 937	$^{216}_{84}$Po	216.001 790
$^{20}_{10}$Ne	19.992 439	$^{220}_{86}$Rn	220.011 401
$^{23}_{11}$Na	22.989 770	$^{234}_{90}$Th	234.043 583
$^{23}_{12}$Mg	22.994 127	$^{238}_{92}$U	238.050 786

2. Construct a diagram like that of Figure 45.4 for the case when I equals (a) 5/2 and (b) 4.

3. The compressed core of a star formed in the wake of a supernova explosion can consist of pure nuclear material and is called a pulsar or neutron star. Calculate the mass of 10 cm^3 of a pulsar.

4. Singly ionized carbon is accelerated through 1000 V and passed into a mass spectrometer to determine the isotopes present (see Chapter 29). The magnetic field strength in the spectrometer is 0.200 T. (a) Determine the radii for the ^{12}C and the ^{13}C isotopes as they pass through the field. (b) Show that the ratio of radii may be written in the form

$$\frac{r_1}{r_2} = \sqrt{\frac{m_1}{m_2}}$$

and verify that your radii in part (a) agree with this.

5. From Table A.3, identify the stable nuclei that correspond to the magic numbers given by Equation 45.2.

6. Consider the hydrogen atom to be a sphere of radius equal to the Bohr radius, α_0, and calculate the approximate value of the ratio nuclear density : atomic density.

7. The atomic mass unit is exactly $\frac{1}{12}$ of the mass of an atom of ^{12}C. Before 1961, there were two units in general use:

u (physical scale) = $\frac{1}{16}$ of the mass of ^{16}O

u (chemical scale) = $\frac{1}{16}$ of the average mass of oxygen taking into account the relative isotopic abundances

Calculate the percent difference between the atomic mass unit on the present scale and the older physical scale. Use values of atomic masses from Table A.3.

8. For the stable nuclei in Table A.3, identify the number of stable nuclei that are even Z, even N; even Z, odd N; odd Z, even N; and odd Z, odd N.

9. Certain stars at the end of their lives are thought to collapse, combining their protons and electrons to form a neutron star. Such a star could be thought of as a gigantic atomic nucleus. If a star of mass equal to that of the Sun ($M = 1.99 \times 10^{30}$ kg) collapsed into neutrons ($m_n = 1.67 \times 10^{-27}$ kg), what would the radius be? (*Hint: $r = r_0A^{1/3}$.*)

10. The Larmor precessional frequency is

$$\omega_p = \frac{\Delta E}{\hbar} = \frac{2\mu B}{\hbar}$$

Calculate the radio-wave frequency at which resonance absorption will occur for (a) free neutrons in a magnetic field of 1.00 T, (b) free protons in a magnetic field of 1.00 T, and (c) free protons in the Earth's magnetic field at a location where the field strength is 50.0 μT.

11. In a Rutherford scattering experiment, alpha particles having kinetic energy of 7.7 MeV are fired toward a gold nucleus. (a) Use the formula in the text to determine the distance of closest approach between the alpha particle and gold nucleus. (b) Calculate the de Broglie wavelength for the 7.7-MeV alpha particle and compare it to the distance obtained in part (a). (c) Based on this comparison, why is it proper to treat the alpha particle as a particle and not as a wave in the Rutherford scattering experiment?

12. How much kinetic energy must an alpha particle (charge = $2 \times 1.6 \times 10^{-19}$ C) have in order to ap-

proach to within 1.0×10^{-14} m of a gold nucleus (charge = $79 \times 1.6 \times 10^{-19}$ C)?

13. (a) Use energy methods to calculate the distance of closest approach for a head-on collision between an alpha particle having an initial energy of 0.50 MeV and a gold nucleus (^{197}Au) at rest. (Assume the gold nucleus remains at rest during the collision.) (b) What minimum initial speed must the alpha particle have in order to get as close as 300 fm?

Section 45.3   Binding Energy and Nuclear Forces

14. Calculate the binding energy per nucleon for (a) ^{2}H, (b) ^{4}He, (c) ^{56}Fe, and (d) ^{238}U.

15. In Example 45.3, the binding energy of the deuteron was calculated to be 2.224 MeV. This corresponds to a value of 1.112 MeV/nucleon. What is the binding energy per nucleon for tritium, ^{3}H?

16. The energy required to construct a uniformly charged sphere of total charge Q and radius R is $U = 3k_e Q^2/5R$, where k_e is the Coulomb constant (see Problem 72). Assume that a ^{40}Ca nucleus consists of 20 protons uniformly distributed in a spherical volume. (a) How much energy is required to counter the electrostatic repulsion given by the above equation? (*Hint:* First calculate the radius of a ^{40}Ca nucleus.) (b) Calculate the binding energy of ^{40}Ca and compare it with the result in part (a). (c) Explain why the result of part (b) is larger than that of part (a).

17. The energy required to dismantle a uniform sphere of total mass M and radius R is given by $E = 3GM^2/5R$, where G is the gravitational constant. Assume that a ^{40}Ca calcium nucleus consists of 20 protons and 20 neutrons (each having a mass of 1.67×10^{-27} kg) uniformly distributed in a spherical volume. (a) How much energy is required to dismantle a ^{40}Ca nucleus? (*Hint:* First calculate the radius of a ^{40}Ca nucleus.) (b) Calculate the binding energy of ^{40}Ca and compare it with the result in part (a). (c) Based on the results of parts (a) and (b), why does the gravitational potential energy not account for the binding energy of the ^{40}Ca nucleus?

18. Calculate the binding energy of the last neutron in a $^{16}_{8}$O nucleus.

19. Two isotopes having the same mass number are known as isobars. Calculate the difference in binding energy per nucleon for the isobars $^{23}_{11}$Na and $^{23}_{12}$Mg. How do you account for the difference?

20. The peak of the stability curve occurs at ^{56}Fe. This is why iron is prevalent in the spectrum of the Sun and stars. Show that ^{56}Fe has a higher binding energy per nucleon than its neighbors ^{55}Mn and ^{59}Co.

21. Calculate the minimum energy required to remove a neutron from the $^{43}_{20}$Ca nucleus.

22. A pair of nuclei for which $Z_1 = N_2$ and $Z_2 = N_1$ are called mirror isobars (the atomic and neutron num-

bers are interchangeable). Binding energy measurements on these nuclei can be used to obtain evidence of the charge independence of nuclear forces (that is, proton-proton, proton-neutron, and neutron-neutron forces are approximately equal). Calculate the difference in binding energy for the two mirror isobars $^{15}_{8}$O and $^{15}_{7}$N.

23. The isotope $^{139}_{57}$La is stable. A radioactive isobar (see Problem 19) of this lanthanum isotope, $^{139}_{59}$Pr, is located above the line of stable nuclei in Figure 45.3 and decays by β^+ emission. Another radioactive isobar of ^{139}La, $^{139}_{55}$Cs, decays by β^- emission and is located below the line of stable nuclei in Figure 45.3. (a) Which of these three isobars has the highest neutron-to-proton ratio? (b) Which has the greatest binding energy per nucleon? (c) Which do you expect to be heavier, ^{139}Pr or ^{139}Cs?

Section 45.4   Nuclear Models

24. (a) In the liquid-drop model of nuclear structure, why does the surface-effect term $-C_2 A^{2/3}$ have a minus sign? (b) The binding energy of the nucleus increases as the volume-to-surface ratio increases. Calculate this ratio for both spherical and cubical shapes and explain which is more plausible for nuclei.

25. Using the graph in Figure 45.9, estimate how much energy is released when a nucleus of mass number 200 is split into two nuclei each of mass number 100.

26. Use Equation 45.5 and the given values for the constants C_1, C_2, C_3, and C_4 to calculate the binding energy per nucleon for the isobars $^{64}_{29}$Cu and $^{64}_{30}$Zn.

27. (a) Use Equation 45.5 to compute the binding energy for $^{56}_{26}$Fe. (b) What percentage is contributed to the binding energy by each of the four terms?

Section 45.5   Radioactivity

28. How much time elapses before 90.0% of the radioactivity of a sample of $^{72}_{33}$As disappears as measured by its activity? The half-life of $^{72}_{33}$As is 26 h.

29. A sample of radioactive material contains 10^{15} atoms and has an activity of 6.00×10^{11} Bq. What is its half-life?

30. The half-life of radioactive iodine-131 is 8 days. Find the number of ^{131}I nuclei necessary to produce a sample having an activity of 1.00 μCi.

31. A freshly prepared sample of a certain radioactive isotope has an activity of 10 mCi. After 4.0 h, its activity is 8.0 mCi. (a) Find the decay constant and half-life. (b) How many atoms of the isotope were contained in the freshly prepared sample? (c) What is the sample's activity 30 h after it is prepared?

32. Ordinary potassium from milk contains 0.0117% of the radioactive isotope ^{40}K having a half-life of 1.227×10^9 years. If you drink one glass of milk

(0.25 liter) a day, what is the activity just after you drink the milk? Assume that one liter of milk contains 2.0 g of potassium.

33. A rock sample contains traces of ^{238}U, ^{235}U, ^{232}Th, ^{208}Pb, ^{207}Pb, and ^{206}Pb. Careful analysis shows that the ratio of the amount of ^{238}U to ^{206}Pb is 1.164. (a) From this information, determine the age of the rock. (b) What should be the ratios of ^{235}U to ^{207}Pb and ^{232}Th to ^{208}Pb so that they would yield the same age for the rock? Neglect the minute amount of the intermediate decay products in the decay chains. Note that this form of multiple dating gives reliable geological dates.

34. Determine the activity of 1 g of ^{60}Co. The half-life of ^{60}Co is 5.24 years.

35. A laboratory stock solution has an initial activity due to ^{24}Na of 2.5 mCi/ml, and 10 ml is diluted (at $t_0 = 0$) to 250 ml. After 48 h, a 5-ml sample of the diluted solution is monitored with a counter. What is the measured activity?

36. A sample of radioactive material loses 10^{10} atoms per mole each second. How long does it take for 99% of all atoms to decay?

37. The radioactive isotope ^{198}Au has a half-life of 64.8 h. A sample containing this isotope has an initial activity ($t = 0$) of 40.0 μCi. Calculate the number of nuclei that decay in the time interval between $t_1 = 10.0$ h and $t_2 = 12.0$ h.

37A. A radioactive nucleus has a half-life of $T_{1/2}$. A sample containing these nuclei has an initial activity R_0. Calculate the number of nuclei that decay in the time interval between the times t_1 and t_2.

Section 45.6 The Decay Processes

38. Identify the missing nuclide (X):
    (a) $X \rightarrow {}^{65}_{28}Ni + \gamma$
    (b) ${}^{215}_{84}Po \rightarrow X + \alpha$
    (c) $X \rightarrow {}^{55}_{26}Fe + \beta^+ + \nu$
    (d) ${}^{109}_{48}Cd + X \rightarrow {}^{109}_{47}Ag + \nu$
    (e) ${}^{14}N(\alpha, X){}^{17}O$

39. Find the energy released in the alpha decay

$${}^{238}_{92}U \longrightarrow {}^{234}_{90}Th + {}^{4}_{2}He$$

You will find the following mass values useful:

$$M({}^{238}_{92}U) = 238.050\ 786\ u$$

$$M({}^{234}_{90}Th) = 234.043\ 583\ u$$

$$M({}^{4}_{2}He) = 4.002\ 603\ u$$

40. A living specimen in equilibrium with the atmosphere contains one atom of ^{14}C (half-life = 5730 years) for every 7.7×10^{11} stable carbon atoms. An archeological sample of wood (cellulose, $C_{12}H_{22}O_{11}$) contains 21.0 mg of carbon. When the sample is placed inside a beta counter whose count-

ing efficiency is 88%, 837 counts are accumulated in one week. Assuming that the cosmic-ray flux and the Earth's atmosphere have not changed appreciably since the sample was formed, find the age of the sample.

41. A ^{3}H nucleus beta decays into ^{3}He by creating an electron and an antineutrino according to the reaction

$${}^3_1H \longrightarrow {}^3_2He + \beta^- + \bar{\nu}$$

Use Table A.3 to determine the total energy released in this reaction.

42. Calculate the energy released in the alpha decay of ^{210}Po.

43. A ^{239}Pu nucleus at rest undergoes alpha decay, leaving a ^{235}U nucleus in its ground state. Determine the kinetic energy of the alpha particle.

44. Determine which decays can occur spontaneously:
    (a) ${}^{40}_{20}Ca \rightarrow {}^{0}_{1}\beta^+ + {}^{40}_{19}K$
    (b) ${}^{98}_{44}Ru \rightarrow {}^{4}_{2}He + {}^{94}_{42}Mo$
    (c) ${}^{144}_{60}Nd \rightarrow {}^{4}_{2}He + {}^{140}_{58}Ce$

45. Enter the correct isotope symbol in each open square in Figure P45.45.

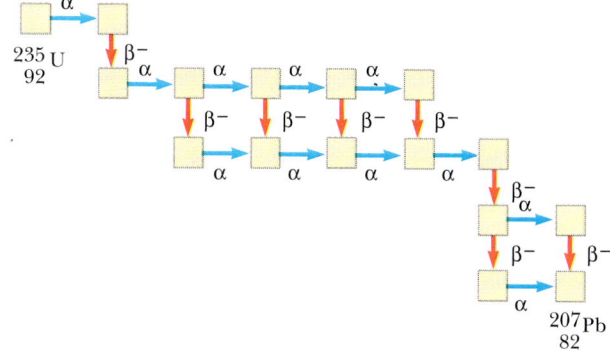

FIGURE P45.45

46. Show that β^- decay of ^{255}Md cannot occur.

47. The nucleus $^{15}_{8}$O decays by electron capture. Write (a) the basic nuclear process and (b) the decay process referring to neutral atoms. (c) Determine the energy of the neutrino. Disregard the daughter's recoil.

Section 45.8 Nuclear Reactions

48. The reaction ${}^{9}_{4}Be(\alpha, n){}^{12}_{6}C$ first observed in 1930 led to the discovery of the neutron by Chadwick. Calculate the Q value of this reaction.

49. The reaction ${}^{27}_{13}Al(\alpha, n){}^{30}_{15}P$, achieved in 1934, is the first known in which the product nucleus is radioactive. Calculate the Q value of this reaction.

50. Determine the Q value associated with the spontaneous fission of ^{236}U into the fragments ^{90}Rb and ^{143}Cs, which have mass 89.914 811 u and 142.927 220 u, re-

spectively. The masses of the other reacting particles are given in Appendix A.3.

51. (a) Determine the binding energy of the last neutron in ^{7}Li. (b) Compare this value to the Q value of the reaction, ^{6}Li(d, p)^{7}Li in Table 45.5. (c) How does the difference in the two values compare with the binding energy of the deuteron?

52. Natural gold has only one isotope, $^{197}_{79}$Au. If natural gold is irradiated by a flux of slow neutrons, β^- particles are emitted. (a) Write the appropriate reaction equations. (b) Calculate the maximum energy of the emitted beta particles. The mass of $^{198}_{80}$Hg is 197.966 75 u.

53. Using the appropriate reactions and Q values from Table 45.5, calculate the mass of ^{8}Be and ^{10}Be in atomic mass units to four decimal places.

ADDITIONAL PROBLEMS

54. After determining that the Sun has existed for hundreds of millions of years, but before the discovery of nuclear physics, scientists could not explain why the Sun has continued to burn for such a long time. [If it used a non-nuclear burning process (e.g., coal), it would have burned up in about 3000 years.] Assume that the Sun, whose mass is 1.99×10^{30} kg, consists entirely of hydrogen and that its total power output is 3.9×10^{26} W. (a) If the energy-generating mechanism of the Sun is the transforming of hydrogen into helium via the reaction,

$$4(^1_1\text{H}) \longrightarrow {}^4_2\text{He} + 2\beta^+ + 2\nu + \gamma$$

calculate the energy (in joules) given off by this reaction. (b) Determine how many hydrogen atoms are available for burning. Take the mass of one hydrogen atom to be 1.67×10^{-27} kg. (c) Assuming that the total power output remains constant, how long will it be before all the hydrogen is converted into helium, and the Sun dies? (d) Why are your results larger than the accepted lifetime of about 10 billion years?

55. Consider a radioactive sample. Determine the ratio of the number of atoms decayed during the first half of its half-life to the number of atoms decayed during the second half of its half-life.

56. (a) The first nuclear reaction was achieved in 1919 by Rutherford, who bombarded nitrogen atoms with alpha particles emitted by the isotope ^{214}Bi:

$$^4_2\text{He} + {}^{14}_7\text{N} \longrightarrow {}^{17}_8\text{O} + {}^1_1\text{H}$$

What is the Q value? (b) The first nuclear reaction utilizing particle accelerators was performed by Cockroft and Walton. Accelerated protons were used to bombard lithium nuclei:

$$^1_1\text{H} + {}^7_3\text{Li} \longrightarrow {}^4_2\text{He} + {}^4_2\text{He}$$

Because the masses of the particles involved in the reaction were well known, these results were used to obtain an early proof of the Einstein mass-energy relationship. Calculate the Q value of the reaction.

57. (a) One method of producing neutrons for experimental use bombards light nuclei with alpha particles. In one particular arrangement, alpha particles emitted by polonium are incident on beryllium nuclei:

$$^4_2\text{He} + {}^9_4\text{Be} \longrightarrow {}^{12}_6\text{C} + {}^1_0\text{n}$$

What is the Q value? (b) Neutrons are also often produced by small-particle accelerators. In one design, deuterons accelerated in a Van de Graaf generator bombard other deuterium nuclei:

$$^2_1\text{H} + {}^2_1\text{H} \longrightarrow {}^3_2\text{He} + {}^1_0\text{n}$$

Is this reaction exothermic or endothermic? Calculate its Q value.

58. The activity of a radioactive sample was measured over 12 h, with the following net count rates:

Time (h)	Counting Rate (counts/min)
1	3100
2	2450
4	1480
6	910
8	545
10	330
12	200

(a) Plot the activity curve on semilog paper. (b) Determine the disintegration constant and half-life of the radioactive nuclei in the sample. (c) What counting rate would you expect for the sample at $t = 0$? (d) Assuming the efficiency of the counting instrument to be 10%, calculate the number of radioactive atoms in the sample at $t = 0$.

59. A by-product of some fission reactors is the isotope $^{239}_{94}$Pu, an alpha emitter having a half-life of 24 000 years:

$$^{239}_{94}\text{Pu} \longrightarrow {}^{235}_{92}\text{U} + \alpha$$

Consider a sample of 1 kg of pure $^{239}_{94}$Pu at $t = 0$. Calculate (a) the number of $^{239}_{94}$Pu nuclei present at $t = 0$ and (b) the initial activity in the sample. (c) How long does the sample have to be stored if a "safe" activity level is 0.1 Bq?

60. A 25-g piece of charcoal is known to be about 25 000 years old. (a) Determine the number of decays per minute expected from this sample. (b) If the radioactive background in the counter without a sample is 20 counts/min and we assume 100% efficiency in counting, explain why 25 000 years is close to the limit of dating with this technique.

61. A large nuclear power reactor produces about 3000 MW of thermal power in its core. Three months after a reactor is shut down, the core power from radioactive by-products is 10 MW. Assuming that each emission delivers 1.0 MeV of energy to the thermal power, estimate the activity in becquerels three months after the reactor is shut down.

62. In a piece of rock from the Moon, the ^{87}Rb constant is assessed to be 1.82×10^{10} atoms per gram of material, and the ^{87}Sr content is found to be 1.07×10^{9} atoms per gram. (a) Determine the age of the rock. (b) Could the material in the rock actually be much older? What assumption is implicit in using the radioactive dating method? (The relevant decay is ^{87}Rb $\rightarrow$ ^{87}Sr $+ e^{-}$. The half-life of the decay is 4.8×10^{10} years.)

63. (a) Can ^{57}Co decay by β^{+} emission? Explain. (b) Can ^{14}C decay by β^{-} emission? Explain. (c) If either answer is yes, what is the range of kinetic energies available for the β particle?

64. (a) Why is the inverse beta decay $p \rightarrow n + \beta^{+} + \nu$ forbidden for a free proton? (b) Why is the same reaction possible if the proton is bound in a nucleus? For example, the following reaction occurs:

$$^{13}_{7}N \longrightarrow ^{13}_{6}C + \beta^{+} + \nu$$

(c) How much energy is released in the reaction given in part (b)? $[m(\beta^{+}) = 0.000\ 549\ \text{u}, M(^{13}C) = 13.003\ 355\ \text{u}, M(^{13}N) = 13.005\ 739\ \text{u}.]$

65. (a) Find the radius of the $^{12}_{6}$C nucleus. (b) Find the force of repulsion between a proton at the surface of a $^{12}_{6}$C nucleus and the remaining five protons. (c) How much work (in MeV) has to be done to overcome this electrostatic repulsion in order to put the last proton into the nucleus? (d) Repeat parts (a), (b), and (c) for $^{238}_{92}$U.

66. The ^{145}Pm nucleus decays by alpha emission. (a) Determine the daughter nuclei. (b) Using the values given in Table A.3, determine the energy released in this decay. (c) What fraction of this energy is carried away by the alpha particle when the recoil of the daughter is taken into account?

67. Consider a hydrogen atom with the electron in the $1s$ state. The magnetic field at the nucleus produced by the orbiting electron has a value of 12.5 T. (See Problem 47, Chapter 42.) The proton can have its magnetic moment aligned in either of two directions perpendicular to the plane of the electron's orbit. Because of the interaction of the proton's magnetic moment with the electron's magnetic field, there will be a difference in energy between the states with the two different orientations of the proton's magnetic moment. Find that energy difference in eV.

68. A freshly prepared sample of radioactive ^{60}Co has an initial activity of 1 Curie, that is, 3.7×10^{10} atoms decay each second. (a) If the half-life of ^{60}Co is 5.24

years, what is the total number of radioactive atoms in the sample at $t = 0$? (b) How long does it take for the activity of the sample to decrease to 0.001 Ci?

69. Carbon detonations are powerful nuclear reactions that temporarily tear apart the cores of massive stars late in their lives. These blasts are produced by carbon fusion, which requires a temperature of about 6×10^{8} K to overcome the strong Coulomb repulsion between carbon nuclei. (a) Estimate the repulsive energy barrier to fusion, using the required ignition temperature for carbon fusion. (In other words, what is the kinetic energy for a carbon nucleus at 6×10^{8} K?) (b) Calculate the energy (in MeV) released in each of these "carbon-burning" reactions:

$$^{12}C + ^{12}C \longrightarrow ^{20}Ne + ^{4}He$$

$$^{12}C + ^{12}C \longrightarrow ^{24}Mg + \gamma$$

(c) Calculate the energy (in kWh) given off when 2 kg of carbon completely fuses according to the first reaction.

70. When a material of interest is irradiated by neutrons, radioactive atoms are produced continually and some decay according to their given half-life. (a) If radioactive atoms are produced at a constant rate R and their decay is governed by the conventional radioactive decay law, show that the number of radioactive atoms accumulated after an irradiation time t is

$$N = \frac{R}{\lambda}(1 - e^{-\lambda t})$$

(b) What is the maximum number of radioactive atoms that can be produced?

71. Many radioisotopes have important industrial, medical, and research applications. One of these is ^{60}Co, which has a half-life of 5.2 years and decays by the emission of a beta particle (energy 0.31 MeV) and two gamma photons (energies 1.17 MeV and 1.33 MeV). A scientist wishes to prepare a ^{60}Co sealed source that will have an activity of at least 10 Ci after 30 months of use. (a) What is the minimum initial mass of ^{60}Co required? (b) At what rate will the source emit energy after 30 months?

72. Consider a model of the nucleus in which the positive charge (Ze) is uniformly distributed throughout a sphere of radius R. By integrating the energy density, $\frac{1}{2}\varepsilon_{0}E^{2}$, over all space, show that the electrostatic energy may be written

$$U = \frac{3Z^{2}e^{2}}{20\pi\varepsilon_{0}R}$$

73. "Free neutrons" have a characteristic half-life of 12 min. What fraction of a group of free neutrons at thermal energy (0.04 eV) will decay before traveling a distance of 10 km?

74. When, after a reaction or disturbance of any kind, a nucleus is left in an excited state, it can return to its normal (ground) state by emission of a gamma-ray photon (or several photons). This process is illustrated by Equation 45.21. The emitting nucleus must recoil in order to conserve both energy and momentum. (a) Show that the recoil energy of the nucleus is

$$E_r = \frac{(\Delta E)^2}{2Mc^2}$$

where ΔE is the difference in energy between the excited and ground states of a nucleus of mass M. (b) Calculate the recoil energy of the ^{57}Fe nucleus when it decays by gamma emission from the 14.4-keV excited state. For this calculation, take the mass to be 57 u. (*Hint:* When writing the equation for conservation of energy, use $(Mv)^2/2M$ for the kinetic energy of the recoiling nucleus. Also, assume that $hf \ll Mc^2$ and use the binomial expansion.)

75. The decay of an unstable nucleus by alpha emission is represented by Equation 45.10. The disintegration energy Q given by Equation 45.13 must be shared by the alpha particle and the daughter nucleus in order to conserve both energy and momentum in the decay process. (a) Show that Q and K_α, the kinetic energy of the alpha particle, are related by the expression

$$Q = K_\alpha\left(1 + \frac{M_\alpha}{M}\right)$$

where M is the mass of the daughter nucleus. (b) Use the result of part (a) to find the energy of the alpha particle emitted in the decay of ^{226}Ra. (See Example 45.8 for the calculation of Q.)

76. Consider the deuterium-tritium fusion reaction with the tritium nucleus at rest:

$$^2_1\text{H} + ^3_1\text{H} \longrightarrow ^4_2\text{He} + ^1_0\text{n}$$

(a) From Equation 45.1, estimate the required distance of closest approach. (b) What is the Coulomb potential energy (in eV) at this distance? (c) If the deuteron has just enough energy to reach the distance of closest approach, what is the final velocity of the deuterium and tritium nuclei in terms of the initial deuteron velocity, v_0? (*Hint:* At this point, the two nuclei have a common velocity equal to the center-of-mass velocity.) (d) Use energy methods to find the minimum initial deuteron energy required to achieve fusion. (e) Why does the fusion reaction occur at much lower deuteron energies than that calculated in part (d)?

77. The ground state of $^{93}_{43}$Tc (molar mass, 92.9102) decays by electron capture and β^+ to energy levels of the daughter (molar mass in ground state, 92.9068) at 2.44 MeV, 2.03 MeV, 1.48 MeV, and 1.35 MeV. (a) For which of these levels are electron capture and β^+ decay allowed? (b) Identify the daughter and

sketch the decay scheme, assuming all excited states de-excite by direct γ decay to the ground state.

78. The isotope $^{25}_{11}$Na decays by β^- emission (disintegration energy = 4.83 MeV; $T_{1/2} = 60.3$ s) to the ground state of $^{25}_{12}$Mg. The following γ-rays are also emitted: 0.40 MeV, 0.58 MeV, 0.98 MeV, 1.61 MeV. The 0.40-MeV γ is the first γ in a two-step cascade to the ground state of $^{25}_{12}$Mg. All other γ-rays decay from an excited state directly to the ground state. Construct the decay scheme.

79. Potassium as it occurs in nature includes a radioactive isotope, ^{40}K, which has a half-life of 1.27×10^9 years and a relative abundance of 0.0012%. These nuclei decay by two different pathways: 89% by β^- emission and 11% by β^+ emission. Calculate the total activity in becquerels associated with 1.00 kg of KCl due to β^- emission.

80. When the nuclear reaction represented by Equation 45.24 is endothermic, the disintegration energy Q is negative. For the reaction to proceed, the incoming particle must have a minimum energy called the threshold energy, E_{th}. Some fraction of the energy of the incident particle is transferred to the compound nucleus in order to conserve momentum. Therefore, E_{th} must be greater than Q. (a) Show that

$$E_{th} = -Q\left(1 + \frac{M_a}{M_X}\right)$$

(b) Calculate the threshold energy of the incident alpha particle in the reaction

$$^4_2\text{He} + ^{14}_7\text{N} \longrightarrow ^{17}_8\text{O} + ^1_1\text{H}$$

81. During the manufacture of a steel engine component, radioactive iron (^{59}Fe) is included in the total mass of 0.2 kg. The component is placed in a test engine when the activity due to this isotope is 20 μCi. After a 1000-h test period, oil is removed from the engine and found to contain enough ^{59}Fe to produce 800 disintegrations/min per liter of oil. The total volume of oil in the engine is 6.5 liters. Calculate the total mass worn from the engine component per hour of operation. (The half-life for ^{59}Fe is 45.1 days.)

82. The rate of decay of a radioactive sample is measured at 10-s intervals, beginning at $t = 0$. The following data in counts per second are obtained: 1137, 861, 653, 495, 375, 284, 215, 163. (a) Plot these data on semilog graph paper and determine the best-fit straight line. (b) From the graph, determine the half-life of the sample.

83. The rate of decay of a radioactive sample is measured at 1-min intervals starting at $t = 0$. The following data (in counts per second) are obtained: 260, 160, 101, 72, 35, 24, 13, 10, 6, 4. (a) Plot these data on semilog graph paper and sketch the best-fit straight line. De-

termine, to two significant figures, (b) the half-life for this sample and (c) the decay constant.

84. *Student Determination of the Half-Life of* ^{137}*Ba.* The radioactive barium isotope (^{137}Ba) has a relatively short half-life and can be easily extracted from a solution containing radioactive cesium (^{137}Cs). This barium isotope is commonly used in an undergraduate laboratory exercise for demonstrating the radioactive decay law. The data presented in Figure P45.84 were taken by undergraduate students using modest experimental equipment. Determine the half-life for the decay of ^{137}Ba using their data.

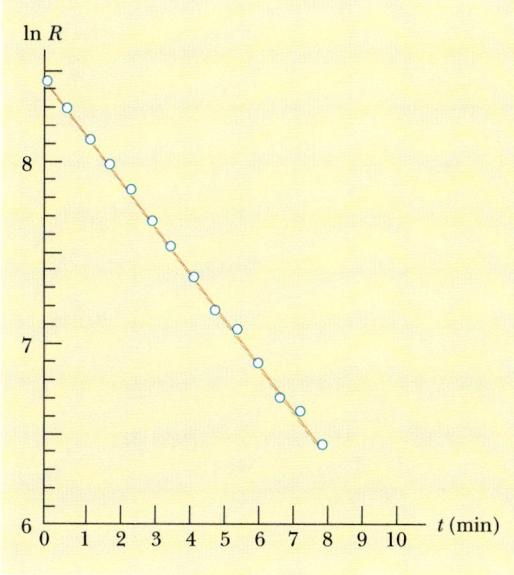

FIGURE P45.84

SPREADSHEET PROBLEMS

S1. Soon after the discovery of radioactivity, investigators realized that some radioactive nuclides decayed into nuclides that were themselves radioactive. Entire chains of radioactive nuclides would form. For example, with a half-life of 3.83 days, ^{222}Rn decays to ^{218}Po, which has a half-life of 3.05 min and decays to ^{214}Pb. For the case in which the parent and daughter both decay, the rate of decay of the parent is

$$\frac{dN_1}{dt} = -\lambda_1 N_1$$

where N_1 is the number of parent nuclei present and λ_1 is the decay constant for the parent. The rate of formation of the daughter is just

$$\frac{dN_2}{dt} = \lambda_1 N_1 \quad (formation)$$

where N_2 is the number of daughter nuclei present and λ_2 is the decay constant for the daughter. The rate of decay of the daughter is

$$\frac{dN_2}{dt} = -\lambda_2 N_2 \quad (decay)$$

The net production of the daughter is then

$$\frac{dN_2}{dt} = \lambda_1 N_1 - \lambda_2 N_2$$

Spreadsheet 45.1 calculates these rates of decay and formation of any mother-daughter pair of radioactive nuclides by integrating these coupled differential equations. (a) Let's assume that we start with 5000 atoms of the parent nuclei, which have a half-life of 5 min. The daughter nuclei further decay with a half-life of 1 min. Enter these data in the spreadsheet. How many nuclei of the parent and daughter are present after 1 min? 2 min? 30 min? (b) If the half-life of the parent is changed to 2 min and the daughter's half-life is changed to 5 min, how many nuclei of each are present at the times listed in part (a). (c) Change the parent's half-life to 1000 min and the daughter's half-life to 2 min. Discuss the number of daughter nuclei present as a function of time.

S2. Modify Spreadsheet 45.1 to take into account the decay of the granddaughter nuclei, which have a half-life of 10 min. Set the parent half-life to 5 min and the daughter half-life to 2 min. The net rate of production of the granddaughter is

$$\frac{dN_3}{dt} = \lambda_2 N_2 - \lambda_3 N_3$$

where N_3 is the number of granddaughter nuclei present and λ_3 is the decay constant for the granddaughter. How many of each species of nuclei are present at 1 min, 2 min, 5 min, 10 min, 30 min, and 60 min?

CHAPTER 46

Fission and Fusion

Interior view of the Tokamak Fusion Test Reactor (TFTR) vacuum vessel. This reactor is located at the Princeton Plasma Physics Laboratory, Princeton University, New Jersey. Graphite tiles, which protect the vessel from neutral beams, can be seen to the right along the middle of the outer wall of the chamber, and at the top and bottom of the chamber. Ion cyclotron radio-frequency launchers are seen at the left of the vessel near the shoulders of the kneeling person. *(Courtesy of Princeton Plasma Physics Laboratory)*

This chapter is concerned primarily with the two means by which energy can be derived from nuclear reactions: fission, in which a large nucleus splits, or fissions, into two smaller nuclei, and fusion, in which two small nuclei fuse to form a larger one. In either case, there is a release of energy that can then be used either destructively (bombs) or constructively (production of electric power). We also examine the interaction of radiation with matter and several devices used to detect radiation. The chapter concludes with a discussion of some industrial and biological applications of radiation.

46.1 INTERACTIONS INVOLVING NEUTRONS

In order to understand nuclear fission and the physics of nuclear reactors, we must first understand how neutrons interact with nuclei. Because of their charge neutrality, neutrons are not subject to Coulomb forces. Since neutrons interact very weakly with electrons, matter appears quite "open" to free neutrons. In general, it is found that the rate of neutron-induced reactions increases as the neutron kinetic energy decreases. Free neutrons undergo beta decay with a mean lifetime of about 10 min. On the other hand, neutrons traveling through matter are absorbed by nuclei before they decay.

A fast neutron (energy greater than about 1 MeV) traveling through matter undergoes many scattering events with the nuclei. In each such event, the neutron gives up some of its kinetic energy to a nucleus. The neutron continues to undergo collisions until its energy is of the order of the thermal energy $k_B T$, where k_B is Boltzmann's constant and T is the absolute temperature. A neutron having this amount of energy is called a **thermal neutron**. At this low energy, there is a high probability that the neutron will be captured by a nucleus, an event that is accompanied by the emission of a gamma ray. This **neutron capture** can be written

Thermal neutron

Neutron capture

$$\, _0^1 n + \, _Z^A X \longrightarrow \, _Z^{A+1} X + \gamma \qquad (46.1)$$

Although we do not indicate it here, once the neutron is captured, the nucleus $_Z^{A+1}X$ is in an excited state for a very short time before it undergoes gamma decay. Also, the product nucleus $_Z^{A+1}X$ is usually radioactive and decays by beta emission.

The neutron-capture rate depends on the nature of the target nucleus and the energy of the incident neutron. For some materials and for fast neutrons, elastic collisions are dominant. Materials for which this occurs are called **moderators** because they slow down (in other words, moderate) the originally energetic neutrons very effectively. A good moderator should be a nucleus of low mass and should not tend to capture neutrons. Boron, graphite, and water are a few examples of moderator materials.

Moderator

As we learned in Chapter 9, during an elastic collision between two particles, the maximum kinetic energy is transferred from one particle to the other when they have the same mass (Example 9.12). Consequently, a neutron loses all of its kinetic energy when it collides head-on with a proton, in analogy to the collision between a moving and a stationary billiard ball. If the collision is oblique, the neutron loses only part of its kinetic energy. For this reason, materials abundant in hydrogen atoms with their single-proton nuclei, such as paraffin and water, are good moderators.

At some point, many of the neutrons in a moderator become thermal neutrons, which means they are now in thermal equilibrium with the moderator material. Their average kinetic energy at room temperature is

$$K_{av} = \tfrac{3}{2} k_B T \approx 0.04 \text{ eV}$$

which corresponds to a neutron root mean square speed of about 2800 m/s. Thermal neutrons have a distribution of speeds, just as the molecules in a container of gas do (Chapter 21). A high-energy neutron, one whose energy is several MeV, thermalizes (that is, reaches K_{av}) in less than 1 ms when incident on a moderator. These thermal neutrons have a very high probability of undergoing neutron capture by the moderator nuclei.

46.2   NUCLEAR FISSION

As we saw in Section 45.3, nuclear fission occurs when a heavy nucleus, such as ^{235}U, splits, or fissions, into two smaller nuclei. In such a reaction, *the combined rest mass of the daughter nuclei is less than the rest mass of the parent nucleus.* Fission is initiated by the capture of a thermal neutron by a heavy nucleus and involves the energy release of about 200 MeV per fission. This energy release occurs because the smaller fission product nuclei are more tightly bound by about 1 MeV per nucleon than the original heavy nucleus.

Nuclear fission was first observed in 1938 by Otto Hahn and Fritz Strassman, following some basic studies by Fermi. After bombarding uranium ($Z = 92$) with neutrons, Hahn and Strassman discovered among the reaction products two medium-mass elements, barium and lanthanum. Shortly thereafter, Lisa Meitner and Otto Frisch explained what had happened. The uranium nucleus had split into two nearly equal fragments after absorbing a neutron. Such an occurrence was of considerable interest to physicists attempting to understand the nucleus, but it was to have even more far-reaching consequences. Measurements showed that about 200 MeV of energy were released in each fission event, and this fact was to affect the course of history.

The fission of ^{235}U by slow neutrons can be represented by the equation

Fission

$$\,^{1}_{0}n + \,^{235}_{92}U \longrightarrow \,^{236}_{92}U^* \longrightarrow X + Y + \text{neutrons} \qquad (46.2)$$

where ^{236}U* is an intermediate excited state that lasts only for about 10^{-12} s before splitting into X and Y. The resulting nuclei, X and Y, are called **fission fragments.** In any fission equation, there are many combinations of X and Y that satisfy the requirements of conservation of energy and charge. With uranium, for example, there are about 90 different daughter nuclei that can be formed.

Fission also results in the production of several neutrons, typically two or three. On the average about 2.5 neutrons are released per event. A typical fission reaction for uranium is

Fission of ^{235}U

$$\,^{1}_{0}n + \,^{235}_{92}U \longrightarrow \,^{141}_{56}Ba + \,^{92}_{36}Kr + 3(\,^{1}_{0}n) \qquad (46.3)$$

Of the 200 MeV or so released in this reaction, most goes into the kinetic energy of the heavy fragments barium and krypton.

The breakup of the uranium nucleus can be compared to what happens to a drop of water when excess energy is added to it. All the atoms in the drop have energy, but this energy is not great enough to break up the drop. However, if enough energy is added to set the drop into vibration, it elongates and compresses until the amplitude of vibration becomes large enough to cause the drop to break. In the uranium nucleus, a similar process occurs (Fig. 46.1):

- The ^{235}U nucleus captures a thermal neutron.
- This capture results in the formation of ^{236}U*, and the excess energy of this nucleus causes it to undergo violent oscillations.
- The ^{236}U* nucleus becomes highly distorted, and the force of repulsion be-

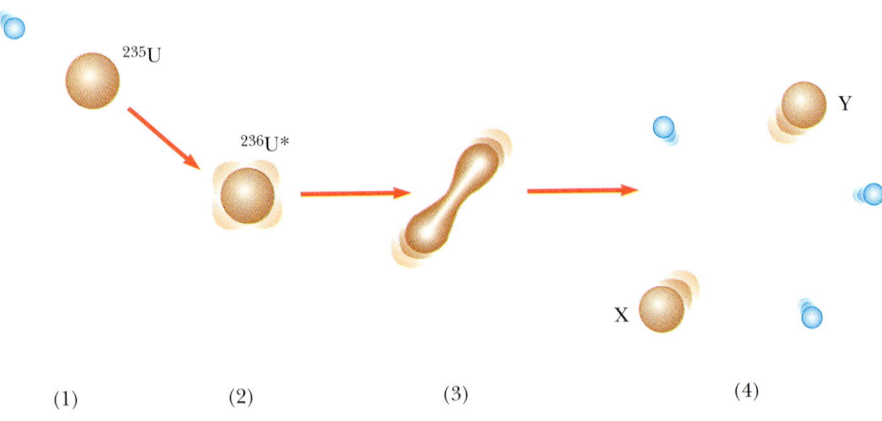

(1)   (2)   (3)   (4)

FIGURE 46.1   The stages in a nuclear fission event as described by the liquid-drop model of the nucleus.

tween protons in the two halves of the dumbbell shape tends to increase the distortion.

• The nucleus splits into two fragments, emitting several neutrons in the process.

Figure 46.2 is a graph of the distribution of fission products versus mass number A. The most probable fission events produce fragments having mass numbers $A \approx 140$ and $A \approx 95$. These fragments fall to the left of the stability line in Figure 45.3. In other words, *fragments that have a large excess of neutrons are unstable, almost instantaneously releasing two or three neutrons.* The remaining fragments are still rich in neutrons and proceed to decay to more stable nuclei through a succession of β^- decays. In the process of such decay, gamma rays are emitted by nuclei in excited states.

Let us estimate the disintegration energy Q released in a typical fission process. From Figure 45.9 we see that the binding energy per nucleon is about 7.2 MeV for heavy nuclei (those having a mass number approximately equal to 240) and about 8.2 MeV for nuclei of intermediate mass. This means that the nucleons in the fission fragments are more tightly bound and, therefore, have less mass than the nucleons in the original nucleus. This decrease in mass per nucleon appears as released energy when fission occurs. The amount of energy released is $(8.2 - 7.2)$ MeV per nucleon. Assuming a total of 240 nucleons, we find that the energy released per fission event is

$$Q = (240 \text{ nucleons}) \left(8.2 \, \frac{\text{MeV}}{\text{nucleon}} - 7.2 \, \frac{\text{MeV}}{\text{nucleon}} \right) = 240 \text{ MeV}$$

This is indeed a very large amount of energy relative to the amount released in chemical processes. For example, the energy released in the combustion of one molecule of octane used in gasoline engines is about one millionth the energy released in a single fission event!

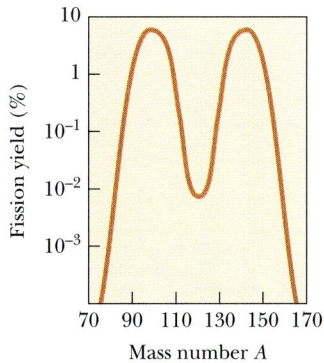

FIGURE 46.2   The distribution of fission products versus mass number for the fission of ^{235}U bombarded with slow neutrons. Note that the ordinate has a logarithmic scale.

EXAMPLE 46.1   The Fission of Uranium

In addition to the barium-lanthanum reaction observed by Meitner and Frisch and the barium-krypton reaction of Equation 46.3, two other ways ^{235}U can fission when bombarded with a neutron are (1) by forming ^{140}Xe and ^{94}Sr and (2) by forming ^{132}Sn and ^{101}Mo. In each case, neutrons are also released. Find the number of neutrons released in each event.

Solution   By balancing mass numbers and atomic numbers, we find that these reactions can be written

$$^{1}_{0}n + ^{235}_{92}U \longrightarrow ^{140}_{54}Xe + ^{94}_{38}Sr + 2(^{1}_{0}n)$$

$$^{1}_{0}n + ^{235}_{92}U \longrightarrow ^{132}_{50}Sn + ^{101}_{42}Mo + 3(^{1}_{0}n)$$

Thus, two neutrons are released in the first event and three in the second.

EXAMPLE 46.2   The Energy Released in the Fission of ^{235}U

Calculate the total energy released if 1.00 kg of ^{235}U undergoes fission, taking the disintegration energy per event to be $Q = 208$ MeV.

Solution   We need to know the number of nuclei in 1.00 kg of uranium. Since $A = 235$, the number of nuclei is

$$N = \left(\frac{6.02 \times 10^{23} \text{ nuclei/mol}}{235 \text{ g/mol}} \right) (1.00 \times 10^3 \text{ g})$$

$$= 2.56 \times 10^{24} \text{ nuclei}$$

Hence the total disintegration energy is

$$E = NQ = (2.56 \times 10^{24} \text{ nuclei}) \left(208 \, \frac{\text{MeV}}{\text{nucleus}} \right)$$

$$= 5.32 \times 10^{26} \text{ MeV}$$

Since 1 MeV is equivalent to 4.45×10^{-20} kWh, $E = 2.37 \times 10^7$ kWh. This is enough energy to keep a 100-W lightbulb burning for about 30 000 years. Thus, 1 kg of ^{235}U is a relatively large amount of fissionable material.

46.3   NUCLEAR REACTORS

In the previous section, we saw that, when ^{235}U fissions, an average of 2.5 neutrons are emitted per event. These neutrons can in turn trigger other nuclei to fission, with the possibility of a chain reaction (Fig. 46.3). Calculations show that if the chain reaction is not controlled (that is, if it does not proceed slowly), it could result in a violent explosion, with the release of an enormous amount of energy. For example, if the energy in 1 kg of ^{235}U were released, it would be equivalent to detonating about 20 000 tons of TNT! This, of course, is the principle behind the first nuclear bomb, an uncontrolled fission reaction.

A nuclear reactor is a system designed to maintain what is called a **self-sustained chain reaction.** This important process was first achieved in 1942 by Fermi at the University of Chicago, with natural uranium as the fuel. Most reactors in operation today also use uranium as fuel. Natural uranium contains only about 0.7% of the ^{235}U isotope, with the remaining 99.3% being ^{238}U. This fact is important to the operation of a reactor because ^{238}U almost never fissions. Instead, it tends to absorb neutrons, producing neptunium and plutonium. For this reason, reactor fuels must be artificially *enriched* to contain at least a few percent ^{235}U.

In order to achieve a self-sustained chain reaction, one of the neutrons emitted in ^{235}U fission, on the average, must be captured by another ^{235}U nucleus and cause it to undergo fission. A useful parameter for describing the level of reactor operation is the **reproduction constant** K, defined as *the average number of neutrons from each fission event that cause another fission event.* As we have seen, K can have a maximum value of 2.5 in the fission of uranium. However, in practice K is less than this because of several factors discussed below.

Chain reaction

Reproduction constant

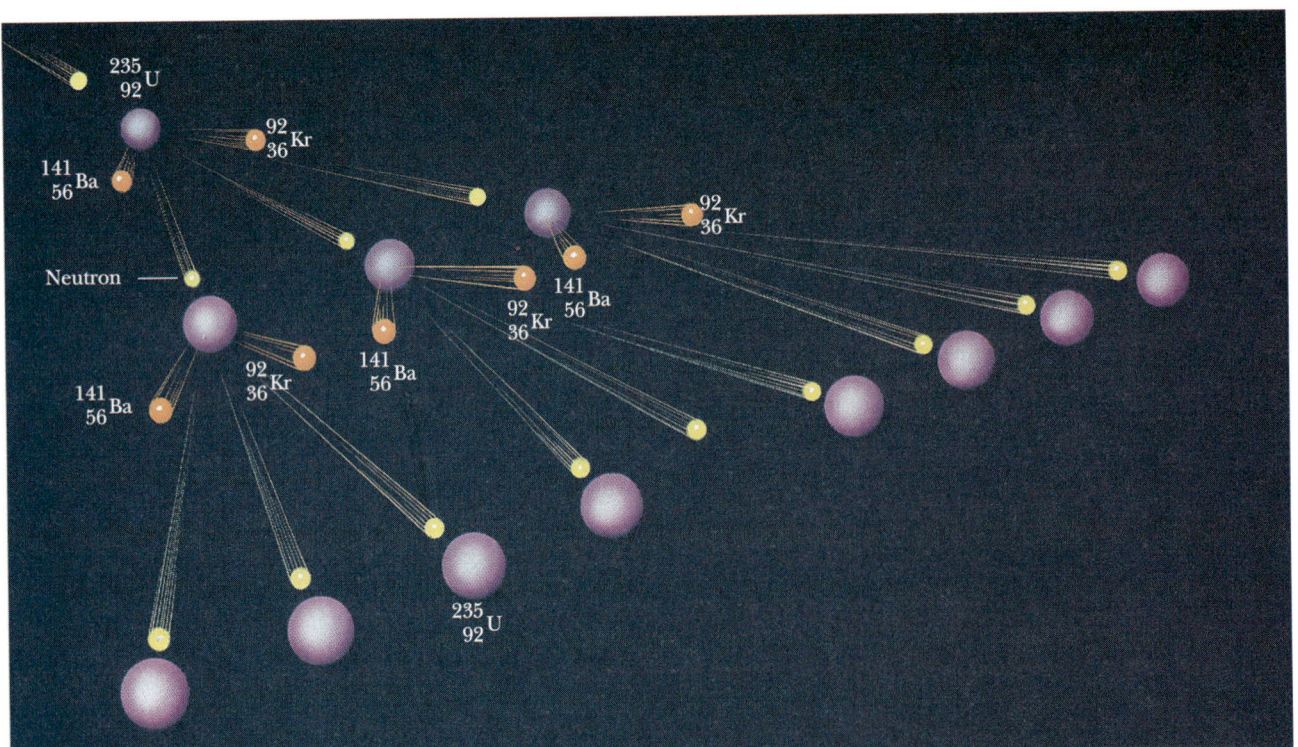

FIGURE 46.3   A nuclear chain reaction initiated by the capture of a neutron. (Many pairs of different isotopes are produced, but only one pair is shown.)

Sketch of the world's first reactor. Because of wartime secrecy, there are no photographs of the completed reactor, which was composed of layers of graphite interspersed with uranium. A self-sustained chain reaction was first achieved on December 2, 1942. Word of the success was telephoned immediately to Washington with this message: "The Italian navigator has landed in the New World and found the natives very friendly." The historic event took place in an improvised laboratory in the racquet court under the west stands of the University of Chicago's Stagg Field and the Italian navigator was Fermi. *(Courtesy of Chicago Historical Society)*

A self-sustained chain reaction is achieved when $K = 1$. Under this condition, the reactor is said to be **critical.** When $K < 1$, the reactor is subcritical and the reaction dies out. When $K \gg 1$, the reactor is said to be supercritical and a runaway reaction occurs. In a nuclear reactor used to furnish power to a utility company, it is necessary to maintain a value of K slightly greater than unity. The basic design of a nuclear reactor core is shown in Figure 46.4. The fuel elements consist of enriched uranium.

In any reactor, a fraction of the neutrons produced in fission leak out of the core before inducing other fission events. If the fraction leaking out is too large, the reactor will not operate. The percentage lost is large if the reactor is very small because leakage is a function of the ratio of surface area to volume. Therefore, a critical feature of the design of a reactor is to choose the correct surface area-to-volume ratio so that a sustained reaction can be achieved.

The neutrons released in fission events are very energetic, having kinetic energies of about 2 MeV. It is necessary to slow these neutrons to thermal energies to allow them to be captured and produce fission of other ^{235}U nuclei (Example 9.12) because the probability of neutron capture increases with decreasing energy. The energetic neutrons are slowed down by a moderator substance surrounding the fuel.

To understand how neutrons are slowed, consider a collision between a lightweight object and a very massive one. In such an event, the lightweight object rebounds from the collision with most of its original kinetic energy. However, if the collision is between objects whose masses are nearly the same, the incoming object transfers a large percentage of its kinetic energy to the target object. In the first nuclear reactor ever constructed, Fermi placed bricks of graphite (carbon) between the fuel elements. Carbon nuclei are about 12 times more massive than neutrons, but after several collisions with carbon nuclei, a neutron is slowed suffi-

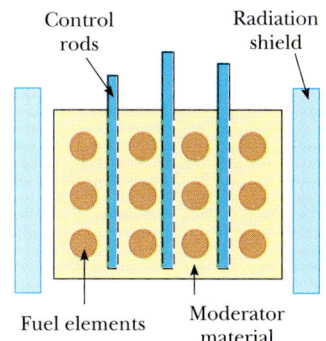

FIGURE 46.4 Cross-section of a reactor core showing the control rods, fuel elements, and moderating material surrounded by a radiation shield.

ciently to increase its likelihood of fission with ^{235}U. In this design, carbon is the moderator; most modern reactors use water as the moderator.

In the process of being slowed down, neutrons may be captured by nuclei that do not fission. The most common event of this type is neutron capture by ^{238}U, which constitutes over 90% of the uranium in the fuel elements. The probability of neutron capture by ^{238}U is very high when the neutrons have high kinetic energies and very low when they have low kinetic energies. Thus, the slowing down of the neutrons by the moderator serves the secondary purpose of making them available for reaction with ^{235}U and decreasing their chances of being captured by ^{238}U.

Control of Power Level

It is possible for a reactor to reach the critical stage ($K = 1$) after all the neutron losses described above are minimized. However, a method of control is needed to maintain a K value near unity. If K rises above this value, the heat produced in the runaway reaction would melt the reactor. To control the power level, control rods are inserted into the reactor core (see Fig. 46.4). These rods are made of materials, such as cadmium, that are very efficient in absorbing neutrons. By adjusting the number and position of these control rods in the reactor core, the K value can be varied and any power level within the design range of the reactor can be achieved.

Although there are several types of reactor systems that convert the kinetic energy of fission fragments to electrical energy, the most common reactor in use in the United States is the pressurized-water reactor. A diagram of a pressurized-water reactor is shown in Figure 46.5, and its main parts are common to all reactor designs. Fission events in the reactor core supply heat to the water contained in the primary (closed) loop and maintained at high pressure to keep it from boiling. This water also serves as the moderator. The hot water is pumped through a heat exchanger, and the heat is transferred to the water contained in the secondary

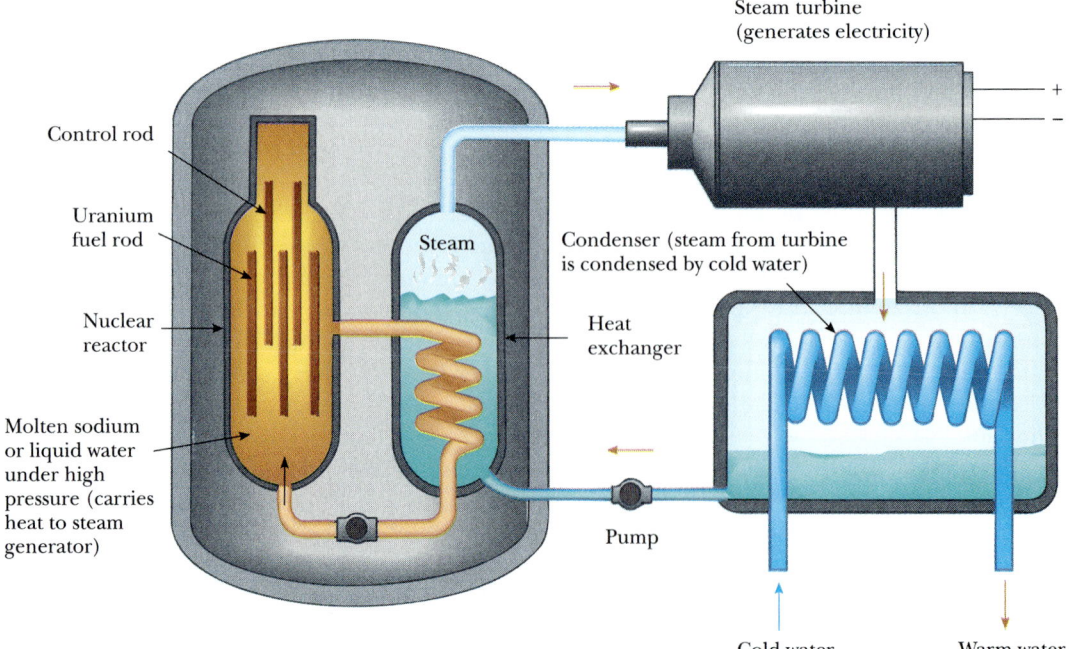

FIGURE 46.5   Main components of a pressurized-water reactor.

loop. The hot water in the secondary loop is converted to steam, which drives a turbine-generator system to create electric power. Note that the water in the secondary loop is isolated from the water in the primary loop to avoid contamination of the secondary water and steam by radioactive nuclei from the reactor core.

Safety and Waste Disposal

One of the inherent dangers in a nuclear reactor is the possibility of interrupted water flow in the secondary cooling system. If this should occur, it is conceivable that the temperature of the reactor could increase to the point where the fuel elements would melt, which would melt the bottom of the reactor and the ground below. Additionally, the large amounts of heat generated could lead to a high-pressure steam explosion. To decrease the chances of such an event, all reactors are built with a backup cooling system that takes over if the regular cooling system fails. It is important to note that a well-designed reactor cannot explode or produce melt-down (the so-called "China syndrome").

Another problem in fission reactors is the disposal of radioactive material when the core is replaced. This waste material contains long-lived, highly radioactive isotopes that must be stored over long periods of time in such a way that there is no chance of environmental contamination. At present, sealing radioactive wastes in deep salt mines seems to be the method of choice.

Another consequence of nuclear power plants is the disposal of excess heat in the cooling system. Water from a nearby river is often used to cool the reactor. This raises the river temperatures downstream from the reactor and affects the higher temperature organisms living in and adjacent to the river. Another technique to cool the reactor water is to use evaporation towers.

Even after the Chernobyl incident in Russia, the safety record of nuclear reactors is far better than other alternatives. Thus, the increased use of nuclear power, together with increased emphasis on energy conservation methods, appear to be logical components of a sensible energy policy.

46.4 NUCLEAR FUSION

In Chapter 45 we found that the binding energy for light nuclei (those having a mass number of less than 20) is much smaller than the binding energy for heavier nuclei. This suggests a process that is the reverse of fission. As we saw in Section 45.3, when two light nuclei combine to form a heavier nucleus, the process is called nuclear fusion. Because the mass of the final nucleus is less than the combined rest masses of the original nuclei, there is a loss of mass accompanied by a release of energy. The following are examples of such energy-liberating fusion reactions:

$$^1_1\text{H} + ^1_1\text{H} \longrightarrow ^2_1\text{H} + \beta^+ + \nu$$
$$^1_1\text{H} + ^2_1\text{H} \longrightarrow ^3_2\text{He} + \gamma$$

This second reaction is followed by one of the following reactions:

$$^1_1\text{H} + ^3_2\text{He} \longrightarrow ^4_2\text{He} + \beta^+ + \nu$$
$$^3_2\text{He} + ^3_2\text{He} \longrightarrow ^4_2\text{He} + ^1_1\text{H} + ^1_1\text{H}$$

These are the basic reactions in what is called the **proton-proton cycle,** believed to be one of the basic cycles by which energy is generated in the Sun and other stars that have an abundance of hydrogen. Most of the energy production takes place at the Sun's interior, where the temperature is approximately 1.5×10^7 K. As we see later, such high temperatures are required to drive these reactions, and they are therefore called **thermonuclear fusion reactions.** The hydrogen (fusion) bomb, first exploded in 1952, is an example of an uncontrolled thermonuclear fusion reaction.

All of the reactions in the proton-proton cycle are exothermic. An overall view of the proton-proton cycle is that four protons combine to form an alpha particle and two positrons, with the release of 25 MeV of energy.

Fusion Reactions

The enormous amount of energy released in fusion reactions suggests the possibility of harnessing this energy for useful purposes here on Earth. A great deal of effort is currently under way to develop a sustained and controllable thermonuclear reactor—a fusion power reactor. Controlled fusion is often called the ultimate energy source because of the availability of its fuel source: water. For example, if deuterium were used as the fuel, 0.12 g of it could be extracted from 1 gal of water at a cost of about four cents. Such rates would make the fuel costs of even an inefficient reactor almost insignificant. An additional advantage of fusion reactors is that comparatively few radioactive by-products are formed. For the proton-proton cycle described earlier in this chapter, the end product of the fusion of hydrogen nuclei is safe, nonradioactive helium. Unfortunately, a thermonuclear reactor that can deliver a net power output spread out over a reasonable time interval is not yet a reality, and many difficulties must be resolved before a successful device is constructed.

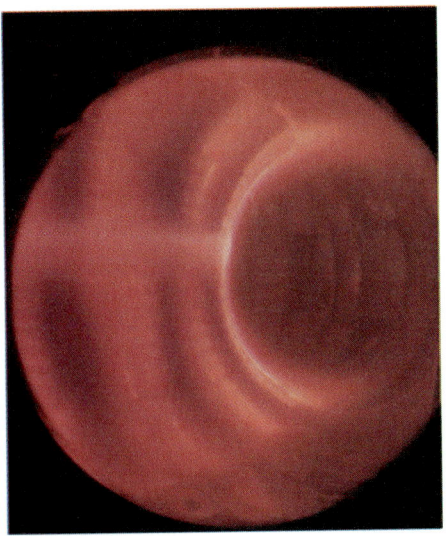

(Left) This accelerator called Saturn, one of the most powerful x-ray sources in the world, has the potential of igniting a controlled laboratory fusion reaction. It uses up to 25 trillion watts of power to produce an electron beam current of 1.25×10^7 A. The energetic electrons are converted to x-rays at the center of the machine. The glow in this photograph is due to electrical discharge in underwater switches as the accelerator is fired. *(Courtesy of Sandia National Laboratories)* *(Right)* During TFTR operation, the plasma discharge is monitored using an optical system that views the interior of the vacuum vessel. This view of a heated deuterium plasma shows a bright radiation belt in the foreground. This phenomenon occurs as the plasma density approaches the disruption limit for a given set of discharge conditions. *(Courtesy of Princeton Plasma Physics Laboratory)*

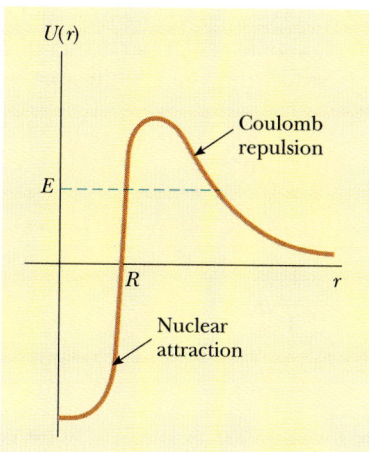

FIGURE 46.6 The potential energy as a function of separation between two deuterons. The Coulomb repulsive force is dominant at long range, whereas the nuclear attractive force is dominant at short range, where R is of the order of 1 fm. The two deuterons would require an energy E greater than the height of the barrier to undergo fusion.

We have seen that the Sun's energy is based, in part, upon a set of reactions in which hydrogen is converted to helium. Unfortunately, the proton-proton interaction is not suitable for use in a fusion reactor because the event requires very high pressures and densities. The process works in the Sun only because of the extremely high density of protons in the Sun's interior.

The fusion reactions that appear most promising for a fusion power reactor involve deuterium ($_1^2$H) and tritium ($_1^3$H):

$$_1^2\text{H} + {}_1^2\text{H} \longrightarrow {}_2^3\text{He} + {}_0^1\text{n} \qquad Q = 3.27 \text{ MeV}$$

$$_1^2\text{H} + {}_1^2\text{H} \longrightarrow {}_1^3\text{H} + {}_1^1\text{H} \qquad Q = 4.03 \text{ MeV} \qquad (46.4)$$

$$_1^2\text{H} + {}_1^3\text{H} \longrightarrow {}_2^4\text{He} + {}_0^1\text{n} \qquad Q = 17.59 \text{ MeV}$$

As noted earlier, deuterium is available in almost unlimited quantities from our lakes and oceans and is very inexpensive to extract. Tritium, however, is radioactive ($T_{1/2} = 12.3$ years) and undergoes beta decay to ^{3}He. For this reason, tritium does not occur naturally to any great extent and must be artificially produced.

One of the major problems in obtaining energy from nuclear fusion is the fact that the Coulomb repulsion force between two charged nuclei must be overcome before they can fuse. The potential energy as a function of particle separation for two deuterons (each with charge $+ e$) is shown in Figure 46.6. The potential energy is positive in the region $r > R$, where the Coulomb repulsive force dominates ($R \approx 1$ fm), and negative in the region $r < R$, where the strong nuclear force dominates. The fundamental problem then is to give the two nuclei enough kinetic energy to overcome this repulsive force. This can be accomplished by heating the fuel to extremely high temperatures (to about 10^8 K, far greater than the interior temperature of the Sun). At these high temperatures, the atoms are ionized and the system consists of a collection of electrons and nuclei commonly referred to as a plasma.

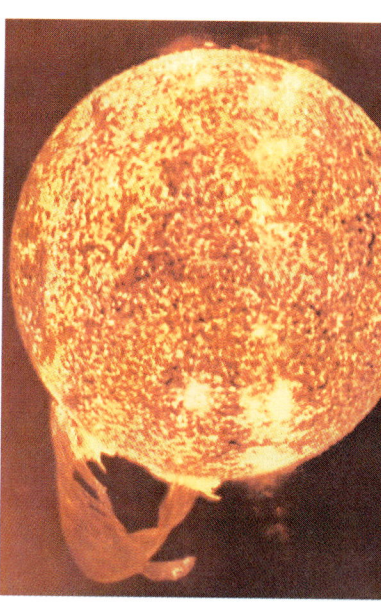

This photograph of the Sun, taken on December 19, 1973, during the third and final manned Skylab mission, shows one of the most spectacular solar flares ever recorded, spanning more than 588 000 km across the solar surface. The last picture, taken 17 hours earlier, showed this feature as a large quiescent prominence on the eastern side of the Sun. The flare gives the distinct impression of a twisted sheet of gas in the process of unwinding itself. In this photograph, the solar poles are distinguished by a relative absence of granulation and a much darker tone than the central portions of the disk. The photograph was taken in the light of ionized helium by an extreme ultraviolet instrument of the U.S. Naval Research Laboratory. *(NASA)*

EXAMPLE 46.3 The Fusion of Two Deuterons

The separation between two deuterons must be about 10^{-14} m for the attractive nuclear force to overcome the repulsive Coulomb force. (a) Calculate the height of the potential barrier due to the repulsive force.

Solution The potential energy associated with two charges separated by a distance r is

$$U = k_e \frac{q_1 q_2}{r}$$

where k_e is the Coulomb constant. For the case of two deuterons, $q_1 = q_2 = +e$, so that

$$U = k_e \frac{e^2}{r} = \left(8.99 \times 10^9 \frac{\text{N} \cdot \text{m}^2}{\text{C}^2} \right) \frac{(1.60 \times 10^{-19}\,\text{C})^2}{1.0 \times 10^{-14}\,\text{m}}$$

$$= 2.3 \times 10^{-14}\,\text{J} = \boxed{0.14\,\text{MeV}}$$

(b) Estimate the effective temperature required for a deuteron to overcome the potential barrier, assuming an energy of $\frac{3}{2} k_B T$ per deuteron (where k_B is Boltzmann's constant).

Solution   Since the total Coulomb energy of the pair of deuterons is 0.14 MeV, the Coulomb energy per deuteron is 0.07 MeV $= 1.1 \times 10^{-14}\,\text{J}$. Setting this equal to the average thermal energy per deuteron gives

$$\tfrac{3}{2} k_B T = 1.1 \times 10^{-14}\,\text{J}$$

where k_B is equal to 1.38×10^{-23} J/K. Solving for T gives

$$T = \frac{2 \times (1.1 \times 10^{-14}\,\text{J})}{3 \times (1.38 \times 10^{-23}\,\text{J/K})} = \boxed{5.3 \times 10^8\,\text{K}}$$

Example 46.3 suggests that deuterons must be heated to 6×10^8 K to achieve fusion. This estimate of the required temperature is too high, however, because the particles in the plasma have a Maxwellian speed distribution and, therefore, some fusion reactions are caused by particles in the high-energy "tail" of this distribution. Furthermore, even those particles that do not have enough energy to overcome the barrier have some probability of tunneling through the barrier. When these effects are taken into account, a temperature of "only" 4×10^8 K appears adequate to fuse the two deuterons.

The temperature at which the power generation rate exceeds the loss rate (due to mechanisms such as radiation losses) is called the **critical ignition temperature.** This temperature for the deuterium-deuterium (D-D) reaction is 4×10^8 K. According to $E \cong k_B T$, this temperature is equivalent to approximately 35 keV. It turns out that the critical ignition temperature for the deuterium-tritium (D-T) reaction is about 4.5×10^7 K, or only 4 keV. A plot of the power generated by fusion versus temperature for the two reactions is shown by the blue and red curves in Figure 46.7. The green straight line represents the power lost versus temperature via the radiation mechanism known as **bremsstrahlung.** This is the principal mechanism of energy loss, in which radiation (primarily x-rays) is emitted as the result of electron-ion collisions within the plasma. The intersections of the P_{lost} line with the P_f curves give the critical ignition temperatures.

In addition to the high temperature requirements, there are two other critical parameters that determine whether or not a thermonuclear reactor will be successful: the **ion density**, n, and **confinement time**, τ. *The confinement time is the time the interacting ions are maintained at a temperature equal to or greater than the ignition temperature.* The British physicist J. D. Lawson has shown that the ion density and confinement time must both be large enough to ensure that more fusion energy is

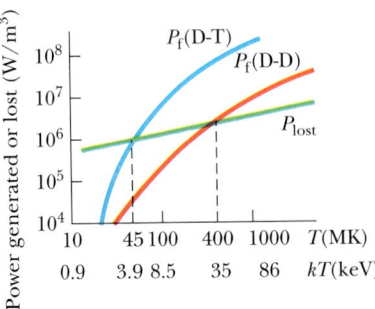

FIGURE 46.7   Power generated versus temperature for the deuterium-deuterium (red) and deuterium-tritium fusion (blue) reactions. The green curve represents the power lost versus temperature. When the generation rate P_f exceeds the loss rate P_{lost}, ignition takes place.

released than what is required to heat the plasma. In particular, **Lawson's criterion** states that a net energy output is possible under the following conditions:

$$n\tau \geqslant 10^{14} \text{ s/cm}^3 \quad \text{(D-T)}$$
$$n\tau \geqslant 10^{16} \text{ s/cm}^3 \quad \text{(D-D)}$$

(46.5)

A graph of $n\tau$ versus temperature for the D-T and D-D reactions is given in Figure 46.8.

Lawson's criterion was arrived at by comparing the energy required to heat the plasma with the energy generated by the fusion process.[1] The energy E_h required to heat the plasma is proportional to the ion density n, or $= C_1 n$. The energy generated by the fusion process is proportional to $n^2\tau$, or $E_f = C_2 n^2 \tau$. This may be understood by realizing that the fusion energy released is proportional to both the rate at which interacting ions collide (αn^2) and the confinement time, τ. Net energy is produced when the energy generated by fusion, E_f, exceeds E_h. When the constants C_1 and C_2 are calculated for different reactions, the condition that $E_f \geqslant E_h$ leads to Lawson's criterion.

In summary, the three basic requirements of a successful thermonuclear power reactor are

- The plasma temperature must be very high—about 4.5×10^7 K for the D-T reaction and 4×10^8 K for the D-D reaction.
- The ion density must be high. It is necessary to have a high density of interacting nuclei to increase the collision rate between particles.
- The confinement time of the plasma must be long. In order to meet Lawson's criterion, the product $n\tau$ must be large. For a given value of n, the probability of fusion between two particles increases as τ increases.

Current efforts are aimed at meeting Lawson's criterion at temperatures exceeding the critical ignition temperature. Although the minimum plasma densities have been achieved, the problem of confinement time is more difficult. How can a plasma be confined at 10^8 K for 1 s? The two basic techniques under investigation to confine plasmas are magnetic field confinement and inertial confinement.

Magnetic Field Confinement

Many fusion-related plasma experiments use **magnetic field confinement** to contain the charged plasma. A toroidal device called a **tokamak,** first developed in Russia, is shown in Figure 46.9. A combination of two magnetic fields is used to confine and stabilize the plasma: (1) a strong toroidal field, produced by the current in the windings, and (2) a weaker ''poloidal'' field, produced by the toroidal current. In addition to confining the plasma, the toroidal current is used to heat it. The resultant helical field lines spiral around the plasma and keep it from touching the walls of the vacuum chamber. If the plasma comes into contact with the walls, its temperature is reduced and heavy impurities sputtered from the walls ''poison'' it and lead to large power losses. One of the major breakthroughs in the 1980s was in the area of auxiliary heating to reach ignition temperatures. Experiments have shown that injecting a beam of energetic neutral particles into the

---

[1] Lawson's criterion neglects the energy needed to set up the strong magnetic field used to confine the plasma. This energy is expected to be about 20 times greater than the energy required to heat the plasma. For this reason, it is necessary to have a magnetic energy recovery system or to make use of superconducting magnets.

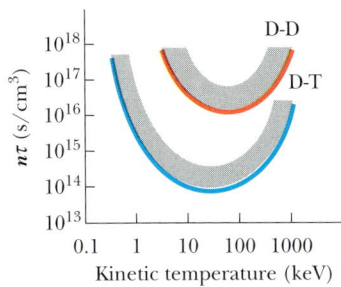

FIGURE 46.8   The Lawson number $n\tau$ versus temperature for the D-T and D-D fusion reactions.

Requirements for a fusion power reactor

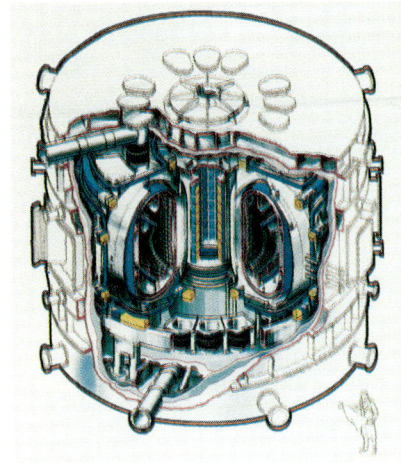

FIGURE 46.9   Schematic diagram of a tokamak used in magnetic confinement. The total magnetic field is the superposition of the toroidal field and the poloidal field.   *(Courtesy of the Plasma Physics Laboratory at Princeton)*

plasma is a very efficient method of heating the plasma to ignition temperatures (5 to 10 keV). Radio frequency heating will probably be needed for reactor-size plasmas.

The Tokamak Fusion Test Reactor (TFTR) at Princeton has reported central ion temperatures of 34.6 keV (approximately 4×10^8 K), representing a fivefold increase since 1981. During this same period, plasma confinement times have increased from 0.02 s to about 1.4 s. Values for $n\tau$ for the D-T reaction are well above 10^{13} s/cm³ and are close to the value required by Lawson's criterion. In 1991, reaction rates of 6×10^{17} D-T fusions per second were reached in the JET tokamak at Abington, England, and 1×10^{17} D-D fusions per second have been reported in the TFTR tokamak at Princeton. An international collaborative effort involving four major fusion programs is currently under way to design and build a fusion reactor called ITER (International Thermonuclear Experimental Reactor). This facility is designed to address the remaining technological and scientific issues that will establish the feasibility of fusion power.

Inertial Confinement

The second technique for confining a plasma is called **inertial confinement** and makes use of a D-T target that has a very high particle density. In this scheme, the confinement time is very short (typically 10^{-11} to 10^{-9} s), so that, because of their own inertia, the particles do not have a chance to move appreciably from their initial positions. Thus, Lawson's criterion can be satisfied by combining a high particle density with a short confinement time.

Laser fusion is the most common form of inertial confinement. A small D-T pellet, about 1 mm in diameter, is struck simultaneously by several focused, high-intensity laser beams, resulting in a large pulse of input energy that causes the surface of the fuel pellet to evaporate (Fig. 46.10). The escaping particles produce

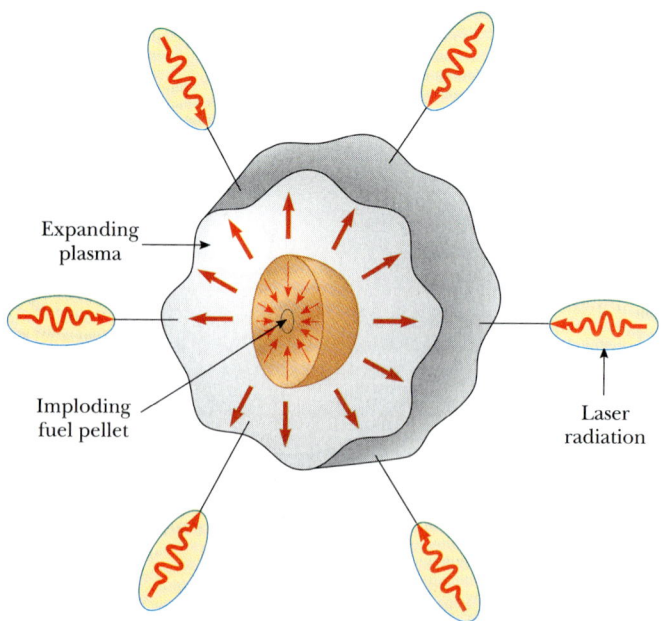

Expanding plasma

Imploding fuel pellet

Laser radiation

FIGURE 46.10 In the inertial confinement scheme, a D-T fuel pellet fuses when struck by several high-intensity laser beams simultaneously.

(a)

(b)

FIGURE 46.11   (a) The target chamber of the Nova Laser Facility at Lawrence Livermore Laboratory.   *(Courtesy of University of California Lawrence Livermore National Laboratory, Livermore, and The U.S. Department of Energy)*   (b) Spherical plastic target shells used to contain the D-T fuel shown clustered on a quarter. The shells have very smooth surfaces and are about 100-nm thick.   *(Courtesy of Los Alamos National Laboratory)*

a reaction force on the core of the pellet, resulting in a strong, inwardly moving, compressive shock wave. This shock wave increases the pressure and density of the core and produces a corresponding increase in temperature. When the temperature of the core reaches ignition temperature, fusion reactions cause the pellet to explode.

Two of the leading laser fusion laboratories in the United States are the Omega facility at the University of Rochester and the Nova facility at Lawrence Livermore National Laboratory in California. Both facilities use neodymium glass lasers; the Omega facility focuses 24 laser beams on the target and the Nova facility employs 10 beams. Figure 46.11a shows the target chamber at Nova and Figure 46.11b shows the tiny, spherical D-T pellets used. Nova is capable of injecting a power of 200 kJ into a 0.5-mm D-T pellet in 1 ns and has achieved values of $n\tau \approx 5 \times 10^{14}$ s/cm^3 and ion temperatures of 5.0 keV. These values are close to those required for D-T ignition. Progress in laser fusion experiments with time is shown in Figure 46.12. This steady progress has led the U.S. Department of Energy and other groups to plan a national facility that will involve a laser fusion device with an input energy in the 5–10 MJ range.

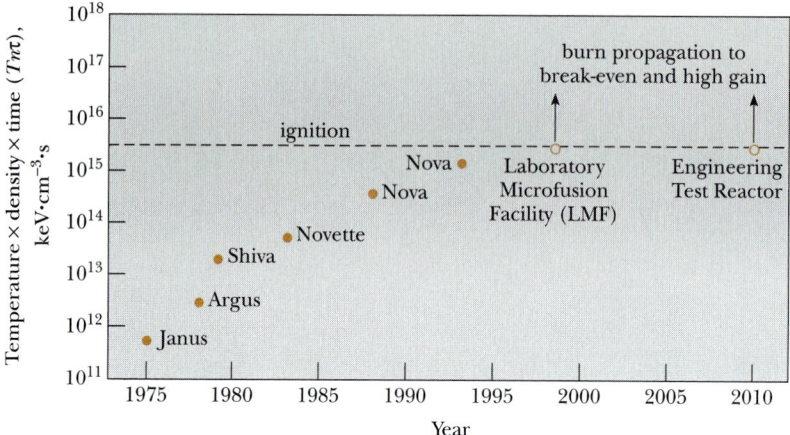

FIGURE 46.12 A plot of the product of plasma temperature, density, and confinement time for laser fusion versus year. The solid dots represent currently achieved results.

Fusion Reactor Design

In the D-T fusion reaction

$$^{2}_{1}\text{H} + ^{3}_{1}\text{H} \longrightarrow ^{4}_{2}\text{He} + ^{1}_{0}\text{n} \qquad Q = 17.59 \text{ MeV}$$

the alpha particle carries 20 percent of the energy and the neutron carries 80 percent, or about 14 MeV. While the charged alphas are primarily absorbed in the plasma and produce desired additional plasma heating, the 14 MeV neutrons pass through the plasma and must be absorbed in a surrounding blanket material to extract their large kinetic energy and generate electric power. One scheme is to use molten lithium metal as the neutron-absorbing material and to circulate the lithium in a closed heat-exchanging loop to produce steam and drive turbines as in a conventional power plant. Figure 46.13 shows a diagram of such a fusion reactor. It is estimated that a blanket of lithium about one meter thick would capture nearly 100 percent of the neutrons from the fusion of a small D-T pellet.

The capture of neutrons by lithium is described by the reaction

$$^{1}_{0}\text{n} + ^{6}_{3}\text{Li} \longrightarrow ^{3}_{1}\text{H} + ^{4}_{2}\text{He}$$

where the kinetic energies of the charged lithium and alpha particle products are converted to heat in the lithium. An extra advantage of using lithium as the energy transfer medium is the production of tritium, $^{3}_{1}\text{H}$, which may be separated from the lithium and returned as fuel to the reactor. The process is indicated in the generic fusion reactor shown in Figure 46.13.

Advantages and Problems of Fusion

If fusion power can be harnessed, it will offer several advantages over fission-generated power: (1) the low cost and abundance of the fuel (deuterium), (2) the impossibility of runaway accidents, and (3) a lesser radiation hazard than with fission. Some of the anticipated problem areas and disadvantages include (1) its as yet unestablished feasibility, (2) the very high proposed plant costs, (3) the scarcity of lithium, (4) the limited supply of the helium needed for cooling superconducting magnets used to produce strong confining fields, (5) structural damage and induced radioactivity caused by neutron bombardment, and (6) the antici-

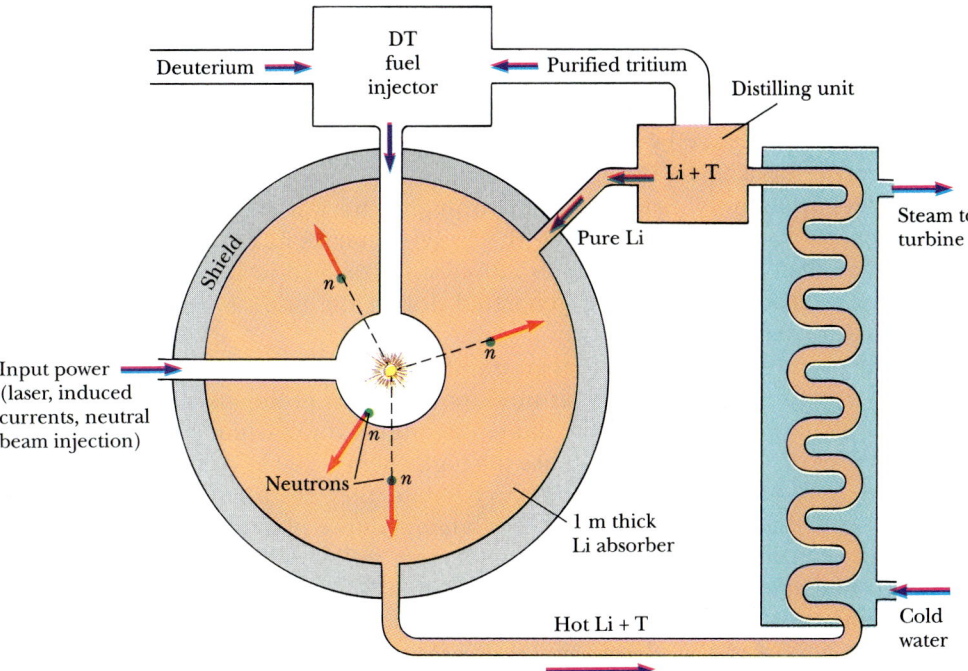

FIGURE 46.13   A generic fusion reactor.

pated high degree of thermal pollution. If these basic problems and the engineering design factors can be resolved, nuclear fusion may become a feasible source of energy by the middle of the next century.

*46.5   RECENT FUSION ENERGY DEVELOPMENTS[2]

The search for alternative energy sources has led to a highly successful international research effort to develop fusion as a practical energy source. This process is illustrated in Figure 46.14, where deuterium and tritium ions, in a high-tempera-

[2] This section was prepared by Dr. Ronald C. Davidson, Director of the Princeton Plasma Physics Laboratory and Professor of Astrophysical Physical Sciences at Princeton University.

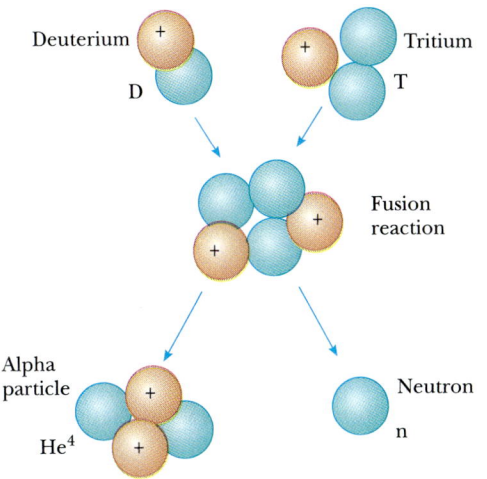

FIGURE 46.14   Schematic diagram of the deuterium (D)-tritium (T) fusion reaction. Eighty percent (80%) of the energy released is in the 14-MeV neutrons.

ture plasma, confined by a strong magnetic field, undergo fusion reactions and produce both 3.5-MeV alpha particles that are trapped in the plasma and energetic 14-MeV neutrons that carry away the heat from the reaction.

As illustrated in Figure 46.15, the power produced in laboratory fusion experiments has increased by a factor of more than 60 million since construction of the largest U.S. tokamak, the Tokamak Fusion Test Reactor at the Princeton Plasma Physics Laboratory. Scientific breakthroughs and world records in plasma performance have see-sawed among the large tokamaks in Europe, Japan, and the United States. The United States moved into the lead in December 1993, when 6.2 MW of fusion power was generated during the historic experiments on the TFTR tokamak.

Figure 46.16 shows a photograph of the TFTR device, including the high-power neutral beam injectors around the periphery of the tokamak that are used to heat and fuel the plasmas to ion temperatures approaching 4×10^8 K and ion densities of 0.6×10^{14} cm^{-3}. Once the high-energy (100 keV) neutral tritium and deuterium atoms enter the plasma chamber, they are ionized and collisionally heat the background plasma. The high-temperature plasma in TFTR is confined and insulated from contact with the chamber wall by a strong toroidal magnetic field of magnitude 5 T, which is sustained for about two seconds by external power supplies.

Properties of the plasma in TFTR are measured by an array of noninvasive diagnostic instruments, ranging from detectors that sense microwave and light emission from the plasma, to detectors that measure energetic ions escaping from

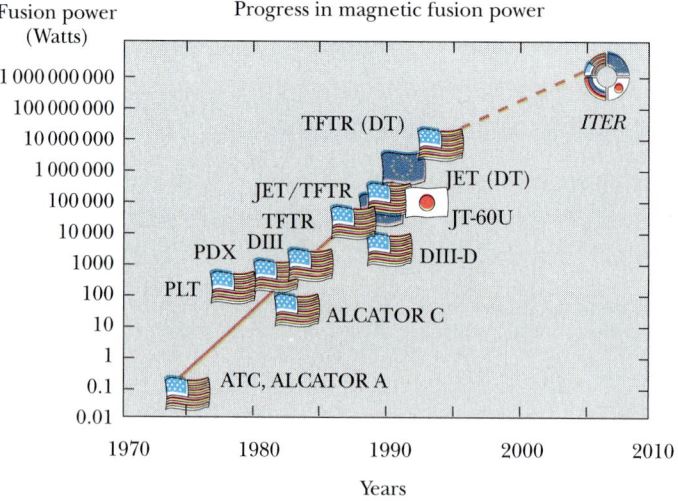

PLT	Princeton Large Torus
PDX	Princeton Divertor Experiment
JET	Joint European Torus
JT-60	Japan
ITER	International Thermonuclear Experimental Reactor
DIII & DIII-D	General Atomics Tokamak Experiments
ATC & TFTR	Princeton Plasma Physics Laboratory
ALCATOR A, C	Massachusetts Institute of Technology

FIGURE 46.15 Progress in the fusion power achieved since construction authorization of the Tokamak Fusion Test Reactor in 1976 has been a factor of 60 million.

FIGURE 46.16   Photograph of the Tokamak Fusion Test Reactor at the Princeton Plasma Physics Laboratory. The neutral beam injectors surrounding the tokamak device are capable of delivering up to 33 MW of auxiliary heating power to the plasma.

the plasma, and the 14-MeV neutrons produced in the deuterium-tritium fusion process. Typical TFTR data on fusion-power production are illustrated in Figure 46.17, where the fusion power (in the neutrons and alpha particles) is plotted versus time for three different experiments with varying concentrations of tritium and deuterium in the fuel mixture. In each case, the neutral beam injectors are "on" between $t = 0$ and $t = 0.75$ s. Note from the figure that a 50 percent tritium-50 percent deuterium mixture is the optimum for maximizing the fusion power production at 6.2 MW. Figure 46.18 is a plot of the peak ion temperature T_i in the

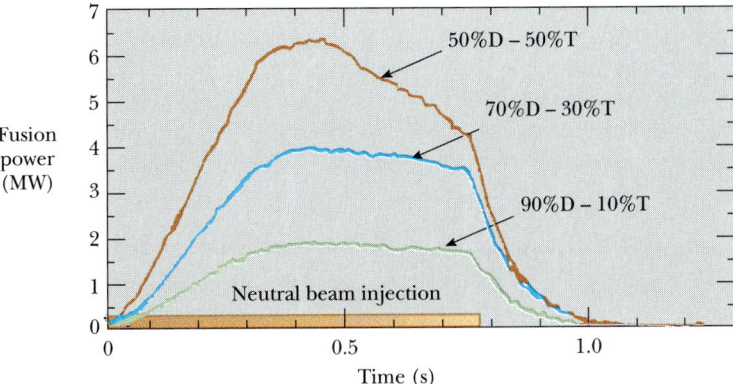

FIGURE 46.17   Plot of the total fusion power versus time obtained in the historic experiments on the Tokamak Fusion Test Reactor in December 1993.

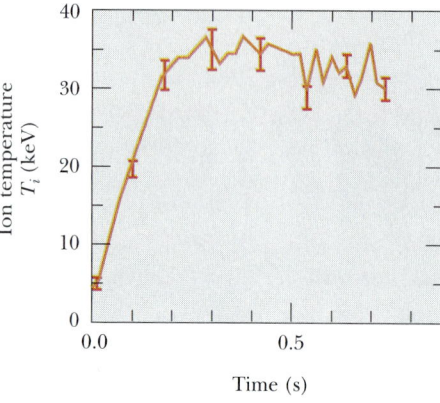

FIGURE 46.18   Plot of the peak ion temperature T_i versus time obtained in the highest fusion power experiment (6.2 MW) in Figure 46.17.

center of the plasma versus time for the same time interval as the fusion power data shown in Figure 46.17. Note that T_i increases to about 35 keV (corresponding to 3.9×10^8 K) in these experiments.

Also in the high-power experiments with deuterium-tritium plasmas, there are preliminary indications that the plasma electrons are heated collisionally by energetic 3.5-MeV alpha particles created in the fusion process. The electrons experience a temperature increase of about 0.75 keV ($\cong 0.8 \times 10^7$ K) above the baseline electron temperature of 9 keV ($\cong 10^8$ K) obtained in otherwise similar deuterium plasmas. Heating of the background plasma by the alpha particles will be very important in future-generation power reactors and experimental facilities such as ITER, which is presently being designed. The ITER device will be a tokamak device much larger in physical size than TFTR, and it is designed to produce 1000 MW of fusion power, about a factor of 100 larger than TFTR's ultimate design capability of 10 MW. In ITER, the energy content of the alpha particles will be so intense that they will heat and sustain the plasma, allowing the auxiliary heating sources to be turned off once the reaction is initiated. Such a state of "sustained burn" of a plasma is referred to as "ignition."

*46.6   RADIATION DAMAGE IN MATTER

A blue glow from Cerenkov radiation is emitted by the core of a reactor at Los Alamos National Laboratory.   *(Los Alamos National Laboratory)*

Radiation passing through matter can cause severe damage. The degree and type of damage depend upon several factors, including the type and energy of the radiation and the properties of the matter. For example, the metals used in nuclear reactor structures can be severely weakened by high fluxes of energetic neutrons, because these high fluxes often lead to metal fatigue. The damage in such situations is in the form of atomic displacements, often resulting in major alterations in the material properties. Materials can also be damaged by ionizing radiations, such as gamma rays and x-rays. For example, defects called color centers can be produced in inorganic crystals by irradiating the crystals with x-rays. One extensively studied color center has been identified as an electron trapped in a Cl^--ion vacancy.

Radiation damage in biological organisms is primarily due to ionization effects in cells. The normal operation of a cell may be disrupted when highly reactive ions or radicals are formed as the result of ionizing radiation. For example, hydrogen and hydroxyl radicals produced from water molecules can induce chemical reactions that may break bonds in proteins and other vital molecules. Furthermore, the ionizing radiation may directly affect vital molecules by removing electrons from their structure. Large doses of radiation are especially dangerous because damage to a great number of molecules in a cell may cause the cell to die. Although the death of a single cell is usually not a problem, the death of many cells may result in irreversible damage to the organism. Also, cells that do survive the radiation may become defective. These defective cells, upon dividing, can produce more defective cells and lead to cancer.

In biological systems, it is common to separate radiation damage into two categories: *somatic damage* and *genetic damage*. Somatic damage is that associated with any body cell except the reproductive cells. Such damage can lead to cancer at high dose rates or seriously alter the characteristics of specific organisms. Genetic damage affects only reproductive cells. Damage to the genes in reproductive cells can lead to defective offspring. Clearly, we must be concerned about the effect of diagnostic treatments such as x-rays and other forms of radiation exposure.

There are several units used to quantify the amount, or dose, of any radiation that interacts with a substance.

The **roentgen** (R) is that amount of ionizing radiation that produces an electric charge of $\frac{1}{3} \times 10^{-9}$ C in 1 cm^3 of air under standard conditions.

The roentgen

Equivalently, the roentgen is that amount of radiation that deposits an energy of 8.76×10^{-3} J into 1 kg of air.

For most applications, the roentgen has been replaced by the rad (which is an acronym for *radiation absorbed dose*):

One **rad** is that amount of radiation that deposits 10^{-2} J of energy into 1 kg of absorbing material.

The rad

Although the rad is a perfectly good physical unit, it is not the best unit for measuring the degree of biological damage produced by radiation because damage depends not only on the dose but also on the type of the radiation. For example, a given dose of alpha particles causes about ten times more biological damage than an equal dose of x-rays. The **RBE** (relative biological effectiveness) factor for a given type of radiation is *the number of rad of x-radiation or gamma radiation that produces the same biological damage as 1 rad of the radiation being used*. The RBE factors for different types of radiation are given in Table 46.1. The values are only approximate because they vary with particle energy and with the form of the damage. The

The RBE factor

TABLE 46.1   RBE^a for Several Types of Radiation

Radiation	RBE Factor
X-rays and gamma rays	1.0
Beta particles	1.0–1.7
Alpha particles	10–20
Slow neutrons	4–5
Fast neutrons and protons	10
Heavy ions	20

^{a} RBE = relative biological effectiveness.

RBE factor should be considered only a first-approximation guide to the actual effects of radiation.

Finally, the **rem** (radiation equivalent in man) is the product of the dose in rad and the RBE factor:

$$\text{Dose in rem} \equiv \text{dose in rad} \times \text{RBE} \tag{46.6}$$

According to this definition, 1 rem of any two radiations produces the same amount of biological damage. From Table 46.1, we see that a dose of 1 rad of fast neutrons represents an effective dose of 10 rem. On the other hand, 1 rad of gamma radiation is equivalent to a dose of 1 rem.

Low-level radiation from natural sources, such as cosmic rays and radioactive rocks and soil, delivers to each of us a dose of about 0.13 rem/year; this radiation is called *background radiation*. It is important to note that background radiation varies with geography. The upper limit of radiation dose recommended by the U.S. government (apart from background radiation) is about 0.5 rem/year. Many occupations involve much higher radiation exposures, and so an upper limit of 5 rem/year has been set for combined whole-body exposure. Higher upper limits are permissible for certain parts of the body, such as the hands and forearms. A dose of 400 to 500 rem results in a mortality rate of about 50% (which means that half the people exposed to this radiation level die). The most dangerous form of exposure is either ingestion or inhalation of radioactive isotopes, especially those elements the body retains and concentrates, such as ^{90}Sr. In some cases, a dose of 1000 rem can result from ingesting 1 mCi of radioactive material.

*46.7 RADIATION DETECTORS

Various devices have been developed for detecting radiation. These devices are used for a variety of purposes, including medical diagnoses, radioactive dating measurements, the measurement of background radiation, and the measurement of the mass, energy, and momentum of particles created in high-energy nuclear reactions.

The ion chamber

An **ion chamber** (Fig. 46.19) is a general class of detector that makes use of electron-ion pairs generated by the passage of radiation through a gas to produce an electrical signal. Two plates in the chamber are maintained at different potentials by connecting them to a voltage supply. The positive plate attracts electrons, while the negative plate attracts ions, causing a current pulse that is proportional to the number of electron-ion pairs produced when a radioactive particle enters the chamber. When the ion chamber is used both to detect the presence of a radioactive particle and to measure its energy, it is called a **proportional counter.**

Geiger counter

The **Geiger counter** (Fig. 46.20) is perhaps the most common form of ion chamber used to detect radiation. It can be considered the prototype of all counters that use the ionization of a medium as the basic detection process. It consists of a cylindrical metal tube filled with gas at low pressure and a long wire along the axis of the tube. The wire is maintained at a high positive potential (about 10^3 V) with respect to the tube. When a high-energy particle or photon enters the tube through a thin window at one end, some of the atoms of the gas are ionized. The electrons removed from these atoms are attracted toward the positive wire, and in the process they ionize other atoms in their path. This results in an avalanche of electrons that produces a current pulse at the output of the tube. After the pulse is amplified, it can either be used to trigger an electronic counter or delivered to a loudspeaker that clicks each time a particle is detected. While a

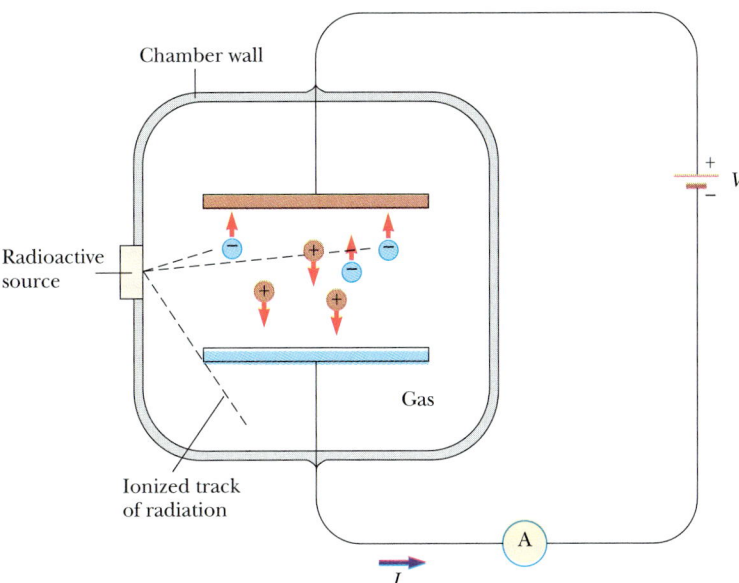

FIGURE 46.19   Simplified diagram of an ion chamber. The radioactive source creates electron-ion pairs that are collected by the charged plates. This sets up a current in the external circuit that is proportional to the particle's kinetic energy.

Geiger counter easily detects the presence of a radioactive particle, the energy lost by the particle in the counter is *not* proportional to the current pulse that it produces. Thus, a Geiger counter cannot be used to measure the energy of a radioactive particle.

A **semiconductor diode detector** is essentially a reverse-bias *p-n* junction. Recall from Chapter 43 that a *p-n* junction diode passes current readily when forward-biased and prohibits a current under reverse-bias conditions. As an energetic particle passes through the junction, electrons are excited into the conduction band and holes are formed in the valence band. The internal electric field sweeps the electrons toward the positive (*n*) side of the junction and the holes toward the negative (*p*) side. This movement of electrons and holes creates a pulse of current that is measured with an electronic counter. In a typical device, the duration of the pulse is about 10^{-8} s.

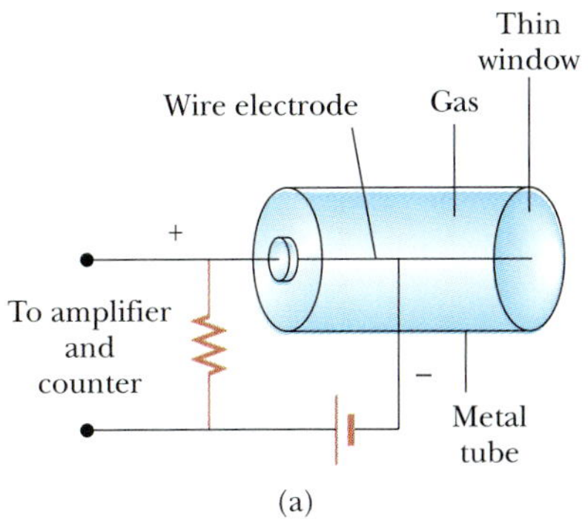

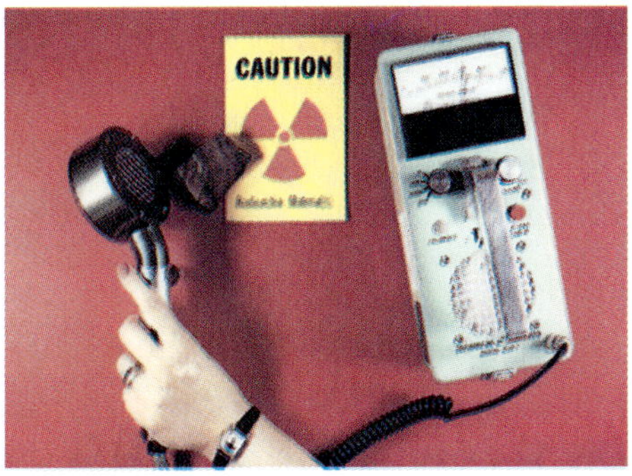

Semiconductor diode detector

FIGURE 46.20   (a) Diagram of a Geiger counter. The voltage between the central wire and the metal tube is usually about 1000 V. (b) Using a Geiger counter to measure the activity in a radioactive mineral. *(Henry Leap and Jim Lehman)*

A **scintillation counter** (Fig. 46.21) usually uses a solid or liquid material whose atoms are easily excited by radiation. These excited atoms emit visible light when they return to their ground state. Common materials used as scintillators are transparent crystals of sodium iodide and certain plastics. If such a material is attached to one end of a device called a **photomultiplier** (PM) tube, the photons emitted by the scintillator can be converted to an electric signal. The PM tube consists of numerous electrodes, called dynodes, whose potentials are increased in succession along the length of the tube as shown in Figure 46.21. The top of the tube contains a photocathode, which emits electrons by the photoelectric effect. As one of these emitted electrons strikes the first dynode, the electron has sufficient kinetic energy to eject several other electrons. When these electrons are accelerated to the second dynode, many more electrons are ejected and a multiplication process occurs. The end result is 1 million or more electrons striking the last dynode. Hence, one particle striking the scintillator produces a sizable electric pulse at the output of the PM tube, and this pulse is in turn sent to an electronic counter. Both the scintillator and diode detector are much more sensitive than a

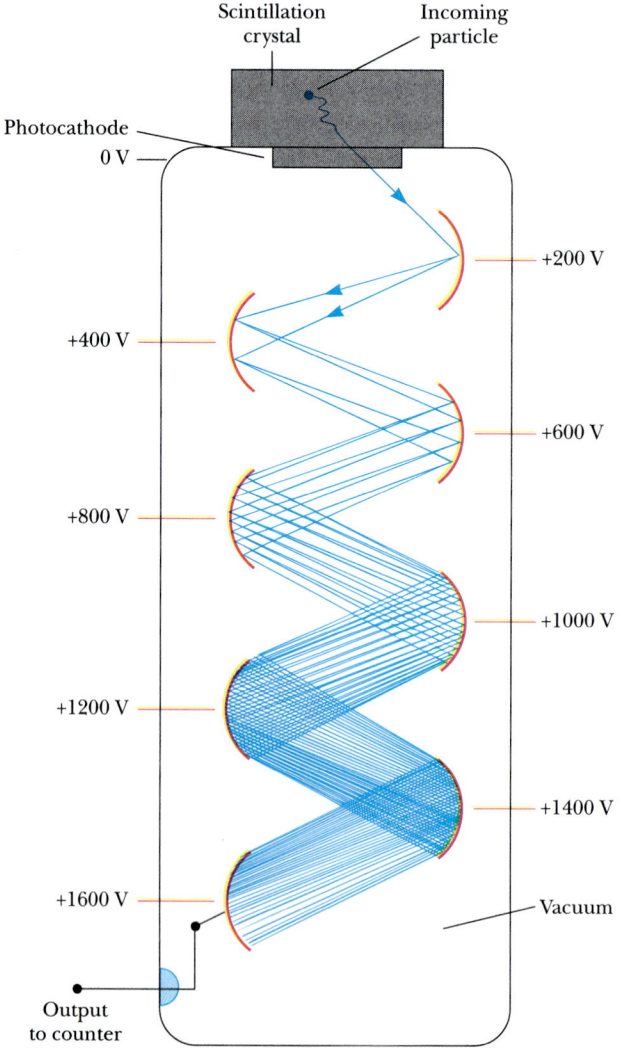

FIGURE 46.21 Diagram of a scintillation counter connected to a photomultiplier tube.

Geiger counter, mainly because of the higher density of the detecting medium. Both can also be used to measure particle energy if the particle stops in the detector.

Track detectors are various devices that are used to view the tracks of charged particles directly. High-energy particles produced in particle accelerators may have energies ranging from 10^9 to 10^{12} eV. Thus, they cannot be stopped and cannot have their energy measured with the small detectors already mentioned. Instead, the energy and momentum of these energetic particles are found from the curvature of the path of the particle in a known magnetic field, hence the necessity of track detectors. A **photographic emulsion** is the simplest example of a track detector. A charged particle ionizes the atoms in an emulsion layer. The path of the particle corresponds to a family of points at which chemical changes have occurred in the emulsion. When the emulsion is developed, the particle's track becomes visible.

A **cloud chamber** contains a gas that has been supercooled to just below its usual condensation point. An energetic particle passing through ionizes the gas along its path. These ions serve as centers for condensation of the supercooled gas. The track can be seen with the naked eye and can be photographed. A magnetic field can be applied to determine the charges of the radioactive particles, as well as their momentum and energy.

A device called a **bubble chamber,** invented in 1952 by D. Glaser, makes use of a liquid (usually liquid hydrogen) maintained near its boiling point. Ions produced by incoming charged particles leave bubble tracks, which can be photographed (Fig. 46.22). Because the density of the detecting medium of a bubble chamber is much higher than the density of the gas in a cloud chamber, the bubble chamber has a much higher sensitivity.

A **spark chamber** is a counting device that consists of an array of conducting parallel plates and is capable of recording a three-dimensional track record. Even-numbered plates are grounded, and odd-numbered plates are maintained at a high potential (about 10 kV). The spaces between the plates contain a noble gas at atmospheric pressure. When a charged particle passes through the chamber, ionization occurs in the gas, resulting in a large surge of current and a visible series of sparks along the particle path. These paths may be photographed or electronically detected and sent to a computer for path reconstruction and the determination of particle mass, momentum, and energy.

This research scientist is studying a photograph of particle tracks made in a bubble chamber at Fermilab. The curved tracks are produced by charged particles moving through the chamber in the presence of an applied magnetic field. Negatively charged particles deflect in one direction, while positively charged particles deflect in the opposite direction. *(Dan McCoy/Rainbow)*

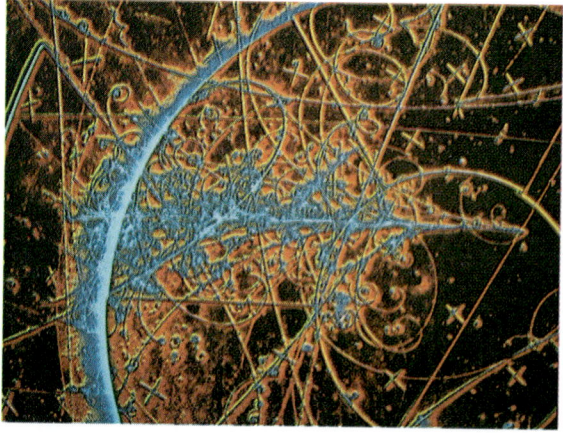

FIGURE 46.22 Artificially colored bubble-chamber photograph showing tracks of subatomic particles. *(Photo Researchers, Inc./Science Photo Library)*

Neutron detectors are more difficult to construct than charged particle detectors because neutrons do not interact electrically with atoms as they pass through matter. Fast neutrons, however, can be detected by filling an ion chamber with hydrogen gas and detecting the ionization produced by high-speed recoiling protons produced in neutron-proton collisions. Slow neutrons with energies less than about 1 MeV do not transfer sufficient energy to the protons to be detected in this way; however, slow neutrons can be detected by using an ion chamber filled with BF_3 gas. In this case, the boron nuclei disintegrate in a slow neutron capture process, emitting highly ionized alphas that are easily detected in the ion chamber.

*46.8   USES OF RADIATION

Nuclear physics applications are extremely widespread in manufacturing, medicine, and biology. Even a brief discussion of all the possibilities would fill an entire book, and to keep such a book up to date would require a number of revisions each year. In this section, we present a few of these applications and the underlying theories supporting them.

Tracing

Radioactive particles can be used to trace chemicals participating in various reactions. One of the most valuable uses of radioactive tracers is in medicine. For example, ^{131}I is an artificially produced isotope of iodine (the natural, nonradioactive isotope is ^{127}I). Iodine, a nutrient needed by the human body, is obtained largely through the intake of iodized salt and seafood. The thyroid gland plays a major role in the distribution of iodine throughout the body. To evaluate the performance of the thyroid, the patient drinks a very small amount of radioactive sodium iodide. Two hours later, the amount of iodine in the thyroid gland is determined by measuring the radiation intensity at the neck area.

A second medical application is indicated in Figure 46.23. Here, a salt containing radioactive sodium is injected into a vein in the leg. The time at which the

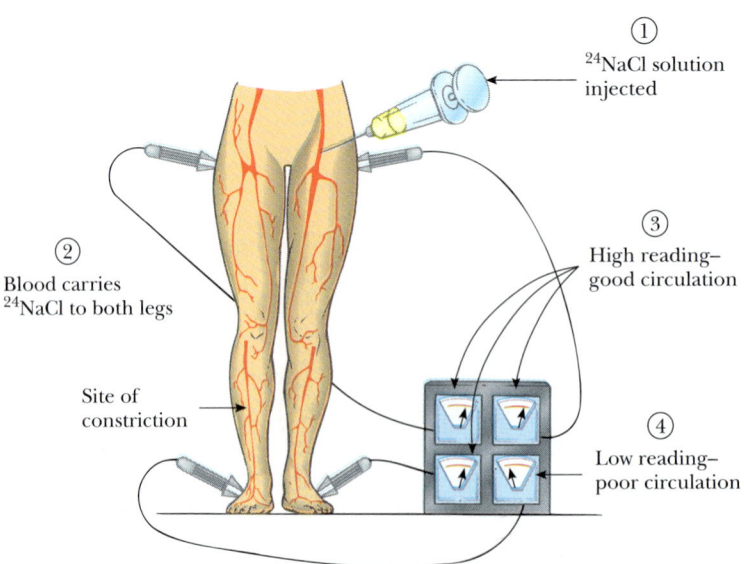

FIGURE 46.23   A tracer technique for determining the condition of the human circulatory system.

radioisotope arrives at another part of the body is detected with a radiation counter. The elapsed time is a good indication of the presence or absence of constrictions in the circulatory system.

The tracer technique is also useful in agricultural research. Suppose the best method of fertilizing a plant is to be determined. A certain element in a fertilizer, such as nitrogen, can be tagged (identified) with one of its radioactive isotopes. The fertilizer is then sprayed on one group of plants, sprinkled on the ground for a second group, and raked into the soil for a third. A Geiger counter is then used to track the nitrogen through the three groups.

Tracing techniques are as wide ranging as human ingenuity can devise. Present applications range from checking the absorption of fluorine by teeth to checking contamination of food-processing equipment by cleansers to monitoring deterioration inside an automobile engine. In the latter case, a radioactive material is used in the manufacture of the pistons, and the oil is checked for radioactivity to determine the amount of wear on the pistons.

Activation Analysis

For centuries, a standard method of identifying the elements in a sample of material has been chemical analysis, which involves testing a portion of the material for reactions with various chemicals. A second method is spectral analysis, which utilizes the fact that, when excited, each element emits its own characteristic set of electromagnetic wavelengths. These methods are now supplemented by a third technique, **neutron activation analysis.** Both chemical and spectral methods have the disadvantage that a fairly large sample of the material must be destroyed for the analysis. In addition, extremely small quantities of an element may go undetected by either method. Activation analysis has an advantage over the other two methods in both of these respects.

When the material under investigation is irradiated with neutrons, nuclei in the material absorb the neutrons and are changed to different isotopes, most of which are radioactive. For example, ^{65}Cu absorbs a neutron to become ^{66}Cu, which undergoes beta decay:

$$^{1}_{0}n + {}^{65}_{29}Cu \longrightarrow {}^{66}_{29}Cu \longrightarrow {}^{66}_{30}Zn + \beta^{-}$$

The presence of the copper can be deduced because it is known that ^{66}Cu has a half-life of 5.1 min and decays with the emission of beta particles having maximum energies of 2.63 and 1.59 MeV. Also emitted in the decay of ^{66}Cu is a 1.04-MeV gamma ray. Thus, by examining the radiation emitted by a substance after it has been exposed to neutron irradiation, extremely small traces of an element can be detected.

Neutron activation analysis is used routinely in a number of industries, but the following nonroutine example of its use is of interest. Napoleon died on the island of St. Helena in 1821, supposedly of natural causes. Over the years, suspicion has existed that his death was not all that natural. After his death, his head was shaved and locks of his hair were sold as souvenirs. In 1961, the amount of arsenic in a sample of this hair was measured by neutron activation analysis. Unusually large quantities of arsenic were found in the hair. (Activation analysis is so sensitive that very small pieces of a single hair could be analyzed.) Results showed that the arsenic was fed to him irregularly. In fact, the arsenic concentration pattern corresponded to the fluctuations in the severity of Napoleon's illness as determined from historical records.

SUMMARY

The probability that neutrons are captured as they move through matter generally increases with decreasing neutron energy. A thermal neutron (one whose energy is approximately $k_B T$) has a high probability of being captured by a nucleus:

$$\,_0^1 n + \,_Z^A X \longrightarrow \,_Z^{A+1} X + \gamma \tag{46.1}$$

Energetic neutrons are slowed down readily in materials called **moderators.** In these materials, neutrons lose their energy mainly through elastic collisions.

Nuclear fission occurs when a very heavy nucleus, such as ^{235}U, splits into two smaller fragments. Thermal neutrons can create fission in ^{235}U:

$$\,_0^1 n + \,_{92}^{235} U \longrightarrow \,_{92}^{236} U^* \longrightarrow X + Y + \text{neutrons} \tag{46.2}$$

where X and Y are the fission fragments and ^{236}U* is a compound nucleus in an excited state. On the average, 2.47 neutrons are released per fission event. The fragments and neutrons have a great deal of kinetic energy following the fission event. The fragments then undergo a series of beta and gamma decays to various stable isotopes. The energy released per fission event is about 208 MeV.

The **reproduction constant** K is the average number of neutrons released from each fission event that cause another event. In a power reactor, it is necessary to maintain $K \approx 1$. The value of K is affected by such factors as reactor geometry, the mean neutron energy, and probability of neutron capture. Proper design of the reactor geometry is necessary to minimize neutron leakage from the reactor core. Neutron energies are regulated with a moderator material to slow down energetic neutrons and therefore increase the probability of neutron capture by other ^{235}U nuclei. The power level of the reactor is adjusted with control rods made of a material that is very efficient in absorbing neutrons. The value of K can be adjusted by inserting the rods at various depths into the reactor core.

Nuclear fusion is a process in which two light nuclei fuse to form a heavier nucleus. A great deal of energy is released in such a process. The major obstacle in obtaining useful energy from fusion is the large Coulomb repulsive force between the charged nuclei at close separations. Sufficient energy must be supplied to the particles to overcome this Coulomb barrier. The temperature required to produce fusion is of the order of 10^8 K, and at this temperature all matter occurs as plasma.

In a fusion reactor, the plasma temperature must reach at least the **critical ignition temperature,** the temperature at which the power generated by the fusion reactions exceeds the power lost in the system. The most promising fusion reaction is the D-T reaction, which has a critical ignition temperature of approximately 4.5×10^7 K. Two critical parameters involved in fusion reactor design are **ion density** n and **confinement time** τ. The confinement time is the time the interacting particles must be maintained at a temperature equal to or greater than the critical ignition temperature. **Lawson's criterion** states that for the D-T reaction, $n\tau \geqslant 10^{14}$ s/cm^3.

QUESTIONS

1. Explain the function of a moderator in a fission reactor.
2. Why is water a better shield against neutrons than lead or steel?
3. Discuss the advantages and disadvantages of fission reactors from the point of view of safety, pollution, and re-

sources. Make a comparison with power generated from the burning of fossil fuels.

4. Why would a fusion reactor produce less radioactive waste than a fission reactor?
5. Lawson's criterion states that the product of ion density

and confinement time must exceed a certain number before a break-even fusion reaction can occur. Why should these two parameters determine the outcome?

6. Why is the temperature required for the D-T fusion less than that needed for the D-D fusion? Estimate the relative importance of Coulomb repulsion and nuclear attraction in each case.

7. What factors make a fusion reaction difficult to achieve?

8. Discuss the similarities and differences between fusion and fission.

9. Discuss the advantages and disadvantages of fusion power from the point of safety, pollution, and resources.

10. Discuss three major problems associated with the development of a controlled fusion reactor.

11. Describe two techniques being pursued in an effort to obtain power from nuclear fusion.

12. If two radioactive samples have the same activity measured in curies, will they necessarily create the same damage to a medium? Explain.

13. One method of treating cancer of the thyroid is to insert a small radioactive source directly into the tumor. The radiation emitted by the source can destroy cancerous cells. Very often, the radioactive isotope $^{131}_{53}I$ is injected into the bloodstream in this treatment. Why do you suppose iodine is used?

14. Why should a radiologist be extremely cautious about x-ray doses when treating pregnant women?

15. The design of a PM tube might suggest that any number of dynodes may be used to amplify a weak signal. What factors do you suppose would limit the amplification in this device?

PROBLEMS

Section 46.2 Nuclear Fission

1. Strontium-90 is a particularly dangerous fission product of ^{235}U because it is radioactive and it substitutes for calcium in bones. What other direct fission products would accompany it in the neutron-induced fission of ^{235}U? (*Note:* This reaction may release two, three, or four free neutrons.)

2. List the nuclear reactions required to produce ^{239}Pu from ^{238}U under fast neutron bombardment.

3. List the nuclear reactions required to produce ^{233}U from ^{232}Th under fast neutron bombardment.

4. (a) Find the energy released in the following fission reaction:

$$^{1}_{0}n + ^{235}_{92}U \longrightarrow ^{141}_{56}Ba + ^{92}_{35}Kr + 3(^{1}_{0}n)$$

The required masses are

$$M(^{1}_{0}n) = 1.008\ 665\ u$$

$$M(^{235}_{92}U) = 235.043\ 915\ u$$

$$M(^{141}_{56}Ba) = 140.913\ 9\ u$$

$$M(^{92}_{36}Kr) = 91.897\ 3\ u$$

(b) What fraction of the initial mass of the system is given off?

5. Find the number of ^6Li and the number of ^7Li nuclei present in 2 kg of lithium. (The natural abundance of ^6Li is 7.5 percent; the remainder is ^7Li.)

6. A typical nuclear fission power plant produces about 1000 MW of electrical power. Assume that the plant has an overall efficiency of 40 percent and that each fission produces 200 MeV of thermal energy. Calculate the mass of ^{235}U consumed each day.

7. Suppose enriched uranium containing 3.4 percent of the fissionable isotope $^{235}_{92}U$ is used as fuel for a ship. The water exerts an average frictional drag of 1.0×10^5 N on the ship. How far can the ship travel per kilogram of fuel? Assume that the energy released per fission event is 208 MeV and that the ship's engine has an efficiency of 20 percent.

Section 46.3 Nuclear Reactors

8. To minimize neutron leakage from a reactor, the surface area-to-volume ratio should be a minimum. For a given volume V, calculate this ratio for (a) a sphere, (b) a cube, and (c) a parallelepiped of dimensions $a \times a \times 2a$. (d) Which of these shapes would have minimum leakage? Which would have maximum leakage?

9. It has been estimated that there are 10^9 tons of natural uranium available at concentrations exceeding 100 parts per million, of which 0.7 percent is ^{235}U. If all the world's energy needs (7×10^{12} J/s) were to be supplied by ^{235}U fission, how long would this supply last? (This estimate of uranium supply was taken from K. S. Deffeyes and I. D. MacGregor, *Scientific American,* January 1980, p. 66.)

Section 46.4 Nuclear Fusion

10. One old prediction for the future was to have a fusion reactor supply heat to reduce garbage to elemental form, put it through a giant spectrometer, and reduce the trash to a new source of isotopically pure elements—the mine of the future. Assuming

☐ indicates problems that have full solutions available in the Student Solutions Manual and Study Guide.

an average atomic mass of 56 and an average charge of 26 (high estimate with all the organic materials), at a current of 10^6 A, how long would it take to process 1 metric ton of trash?

11. (a) If a fusion generator were built to create 3000-MW thermal power, determine the rate of fuel burning in grams per hour if the D-T reaction is used. (b) Do the same for the D-D reaction assuming that the reaction products are split evenly between (n, ^{3}He) and (p, ^{3}H).

12. Two nuclei having atomic numbers Z_1 and Z_2 approach each other with a total energy E. (a) If the minimum distance of approach for fusion to occur is $r = 10^{-14}$ m, find E in terms of Z_1 and Z_2. (b) Calculate the minimum energy for fusion for the D-D and D-T reactions (the first and third reactions in Eq. 46.4).

13. To understand why plasma containment is necessary, consider the rate at which an uncontained plasma would be lost. (a) Estimate the rms speed of deuterons in a plasma at 4×10^8 K. (b) Estimate the time such a plasma would remain in a 10-cm cube if no steps were taken to contain it.

14. Of all the hydrogen nuclei in the ocean, 0.0156% of the mass is deuterium. The oceans have a volume of 317 million cubic miles. (a) If all the deuterium in the oceans were fused to $^4_2$He, how many joules of energy would be released? (b) Present world energy consumption is approximately 7×10^{12} W. If consumption were 100 times greater, how many years would the energy calculated in part (a) last?

15. It has been pointed out that fusion reactors are safe from explosion because there is never enough energy in the plasma to do much damage. (a) In 1992, the TFTR reactor had an ion temperature of 4.0×10^8 K, a plasma density of 2.0×10^{13} cm^{-3} and a confinement time of 3.0×10^{13} s. Calculate the amount of energy stored in the plasma of the reactor. (b) How many kilograms of water could be boiled by this much energy? (The plasma volume of the TFTR reactor is about 50 m^3.)

16. In order to confine a stable plasma, the magnetic energy density in the magnetic field (Eq. 32.15) must exceed the pressure $2nk_BT$ of the plasma by a factor of at least 10. In the following, assume a confinement time $\tau = 1.0$ s. (a) Using Lawson's criterion, determine the required ion density. (b) From the ignition temperature criterion for the D-T reaction, determine the required plasma pressure. (c) Determine the magnitude of the magnetic field required to contain the plasma.

*Section 46.6 Radiation Damage in Matter

17. A building has become accidentally contaminated with radioactivity. The longest-lived material in the building is strontium-90 ($^{90}_{38}$Sr has an atomic mass

89.907 7, and its half-life is 28.8 years). If the building initially contained 1.0 kg of this substance uniformly distributed throughout the building (a very unlikely situation) and the safe level is less than 10.0 counts/min, how long will the building be unsafe?

18. A particular radioactive source produces 100 mrad of 2-MeV gamma rays per hour at a distance of 1.0 m. (a) How long could a person stand at this distance before accumulating an intolerable dose of 1 rem? (b) Assuming the radioactive source is a point source, at what distance would a person receive a dose of 10 mrad/h?

19. Assume that an x-ray technician takes an average of eight x-rays per day and receives a dose of 5 rem/year as a result. (a) Estimate the dose in rem per x-ray taken. (b) How does this result compare with low-level background radiation?

20. When gamma rays are incident on matter, the intensity of the gamma rays passing through the material varies with depth x as $I(x) = I_0 e^{-\mu x}$, where μ is the absorption coefficient and I_0 is the intensity of the radiation at the surface of the material. For 0.400-MeV gamma rays in lead, the absorption coefficient is 1.59 cm^{-1}. (a) Determine the "half-thickness" for lead—that is, the thickness of lead that would absorb half the incident gamma rays. (b) What thickness will reduce the radiation by a factor of 10^4?

21. A "clever" technician decides to heat some water for his coffee with an x-ray machine. If the machine produces 10 rad/s, how long will it take to raise the temperature of a cup of water by 50°C?

22. The danger to the body from a high dose of gamma rays is not due to the amount of energy absorbed but occurs because of the ionizing nature of the radiation. To illustrate this, calculate the rise in body temperature that would result if a "lethal" dose of 1000 rad were absorbed strictly as heat.

23. Technetium-99 is used in certain medical diagnostic procedures. If 1.0×10^{-8} g of ^{99}Tc is injected into a 60-kg patient and half of the 0.14-MeV gamma rays are absorbed in the body, determine the total radiation dose received by the patient.

24. Strontium-90 is still found in the atmosphere from the testing of atomic bombs. The decay of ^{90}Sr releases 1.1 MeV of energy into the bones of a person who has had strontium replace the calcium. If a 70-kg person receives 1.0 μg of ^{90}Sr from contaminated milk, calculate the absorbed dose rate (in J/kg) in one year. Assume the half-life of ^{90}Sr to be 29 years.

*Section 46.7 Radiation Detectors

25. In a Geiger tube, the voltage between the electrodes is typically 1.0 kV and the current pulse discharges a 5.0-pF capacitor. (a) What is the energy amplification of this device for a 0.50-MeV electron? (b) How

many electrons are avalanched by the initial electron?

25A. In a Geiger tube, the voltage between the electrodes is V and the current pulse discharges a capacitor having capacitance C. (a) What is the energy amplification of this device for an electron of energy E? (b) How many electrons are avalanched by the initial electron?

26. In a PM tube, assume seven dynodes having potentials of 100, 200, 300, . . . , 700 V. The average energy required to free an electron from the dynode surface is 10 eV. For each incident electron, how many electrons are freed (a) at the first dynode and (b) at the last dynode? (c) What is the average energy supplied by the tube for each electron? (Assume an efficiency of 100%.)

ADDITIONAL PROBLEMS

27. The atomic bomb dropped in Japan in 1945 released 1.0×10^{14} J of energy (equivalent to that from 20 000 tons of TNT). Estimate (a) the number of $^{235}_{92}$U fissioned and (b) the minimum mass of $^{235}_{92}$U needed in the bomb.

28. Compare the fractional energy loss in a typical ^{235}U fission reaction with the fractional energy loss in D-T fusion.

29. The half-life of tritium is 12 years. If the TFTR fusion reactor contains 50 m³ of tritium at a density equal to 2.0×10^{14} particles/cm³, how many curies of tritium are in the plasma? Compare this value with a fission inventory (the estimated supply of fissionable material) of 4×10^{10} Ci.

29A. The half-life of tritium is 12 years. If the TFTR fusion reactor contains tritium of volume V at a density n particles/cm³, how many curies of tritium are in the plasma?

30. Consider a nucleus at rest, which then splits into two fragments, of masses m_1 and m_2. Show that the fraction of the total kinetic energy that is carried by m_1 is

$$\frac{K_1}{K_{\text{tot}}} = \frac{m_2}{m_1 + m_2}$$

and the fraction carried by m_2 is

$$\frac{K_2}{K_{\text{tot}}} = \frac{m_1}{m_1 + m_2}$$

assuming relativistic corrections can be ignored. (*Note:* If the parent nucleus was moving before the decay, then m_1 and m_2 still divide the kinetic energy as shown, as long as all velocities are measured in the center-of-mass frame of reference, in which the total momentum of the system is zero.)

31. A 2-MeV neutron is emitted in a fission reactor. If it loses one-half its kinetic energy in each collision with a moderator atom, how many collisions must it undergo in order to achieve thermal energy (0.039 eV)?

32. A nuclear power plant operates by utilizing the heat released in nuclear fission to convert 20°C water into 400°C steam. How much water could theoretically be converted to steam by the complete fissioning of 1 g of ^{235}U at 200 MeV/fission?

32A. A nuclear power plant operates by utilizing the heat released in nuclear fission to convert water at T_c into steam at T_h. How much water could theoretically be converted to steam by the complete fissioning of N grams of ^{235}U at 200 MeV/fission?

33. About 1 of every 6500 water molecules contains one deuterium atom. (a) If all the deuterium atoms in 1 liter of water are fused in pairs according to the D-D reaction $^2\text{H} + {}^2\text{H} \rightarrow {}^3\text{He} + \text{n} + 3.3$ MeV, how many joules of energy are liberated? (b) Burning gasoline produces about 3.4×10^7 J/L. Compare the energy obtainable from the fusion of the deuterium in a liter of water with the energy liberated from the burning of a liter of gasoline.

34. The α emitter polonium-210 ($^{210}_{84}$Po) is used in a nuclear battery. Determine the initial power output of the battery if it contains 0.155 kg of ^{210}Po. Assume that the efficiency for conversion of radioactive decay energy to electric energy is 1.0 percent.

35. A certain nuclear plant generates 3065 MW of thermal power to create 1000-MW electric power. Of the waste heat, 3.0 percent is ejected into the atmosphere and the remainder is passed into a river. A state law requires that the river water be heated no more than 3.5°C when it is returned to the river. (a) Determine the amount of cooling water necessary (in kg/h and m³/h) to cool the plant. (b) If fission generates 7.80×10^{10} J/g of ^{235}U, determine the rate of fuel burning (in kg/h) of ^{235}U.

36. A very slow neutron (speed approximately equal to zero) can initiate the reaction

$$\text{n} + {}^{10}_{5}\text{B} \longrightarrow {}^{7}_{3}\text{Li} + {}^{4}_{2}\text{He}$$

If the alpha particle moves away with a speed of 9.3×10^6 m/s, calculate the kinetic energy of the lithium nucleus. Use nonrelativistic formulas.

37. Assuming that a deuteron and a triton are at rest when they fuse according to $^2\text{H} + {}^3\text{H} \rightarrow {}^4\text{He} + \text{n} + 17.6$ MeV, determine the kinetic energy acquired by the neutron.

38. A patient swallows a radiopharmaceutical tagged with phosphorus-32 ($^{32}_{15}$P), a β^- emitter. The average kinetic energy of the beta particles is 700 keV. If the initial activity is 5.22 MBq, determine the absorbed dose during a 10-day period. Assume the beta particles are completely absorbed in 10.0 g of tissue. (*Hint:* Find the number of beta particles emitted.)

39. (a) Calculate the energy (in kilowatt hours) released if 1.00 kg of ^{239}Pu undergoes complete fission and the energy released per fission event is 200 MeV. (b) Calculate the energy (in electron volts) released in the D-T fusion:

$$^2_1H + ^3_1H \longrightarrow ^4_2He + ^1_0n$$

(c) Calculate the energy (in kilowatt hours) released if 1.00 kg of deuterium undergoes fusion. (d) Calculate the energy (in kilowatt hours) released by the combustion of 1.00 kg of coal if each $C + O_2 \rightarrow CO_2$ reaction yields 4.2 eV. (e) List the advantages and disadvantages of each of these methods of energy generation.

40. The Sun radiates energy at the rate of 4.0×10^{23} kW. If the reaction

$$4(^1_1H) \longrightarrow ^4_2He + 2\beta^+ + 2\nu + \gamma$$

accounted for all the energy released, calculate (a) the number of protons fused per second and (b) the amount of energy released per second.

41. Suppose the target in a laser fusion reactor is a sphere of solid hydrogen that has a diameter of 1.5×10^{-4} m and a density of 0.20 g/cm^3. Also assume that half of the nuclei are ^{2}H and half are ^{3}H. (a) If 1.0 percent of a 200-kJ laser pulse is delivered to this sphere, what temperature does the sphere reach? (b) If all of the hydrogen "burns" according to the D-T reaction, how many joules of energy are released?

42. If the average time between successive fissions in a chain reaction in a fission reactor is 1.2 ms, by what factor will the reaction rate increase in one minute? Assume the neutron multiplication factor is 1.000 25.

43. Consider the two nuclear reactions

$$(I) \qquad A + B \longrightarrow C + E$$

$$(II) \qquad C + D \longrightarrow F + G$$

(a) Show that the net disintegration energy for these two reactions ($Q_{net} = Q_I + Q_{II}$) is identical to the disintegration energy for the reaction

$$A + B + D \longrightarrow E + F + G$$

(b) One chain of reactions in the proton-proton cycle in the Sun's interior is

$$^1_1H + ^1_1H \longrightarrow ^2_1H + \beta^+ + \nu$$

$$^1_1H + ^2_1H \longrightarrow ^3_2H + \gamma$$

$$^1_1H + ^3_2He \longrightarrow ^4_2He + \beta^+ + \nu$$

Based on part (a), what is Q_{net} for this sequence?

44. The carbon cycle, first proposed by Hans Bethe in 1939, is another cycle by which energy is released in stars and hydrogen is converted to helium. The carbon cycle requires higher temperatures than the proton-proton cycle. The series of reactions is

$$^{12}C + ^1H \longrightarrow ^{13}N + \gamma$$

$$^{13}N \longrightarrow ^{13}C + \beta^+ + \nu$$

$$^{13}C + ^1H \longrightarrow ^{14}N + \gamma$$

$$^{14}N + ^1H \longrightarrow ^{15}O + \gamma$$

$$^{15}O \longrightarrow ^{15}N + \beta^+ + \nu$$

$$^{15}N + ^1H \longrightarrow ^{12}C + ^4He$$

(a) If the proton-proton cycle requires a temperature of 1.5×10^7 K, estimate the temperature required for the first step in the carbon cycle. (b) Calculate the Q value for each step in the carbon cycle and the overall energy released. (c) Do you think the energy carried off by the neutrinos is deposited in the star? Explain.

45. When photons pass through matter, the intensity I of the beam (measured in watts per square meter) decreases exponentially according to

$$I = I_0 e^{-\mu x}$$

where I_0 is the intensity of the incident beam, and I is the intensity of the beam that just passed through a thickness x of material. The constant μ is known as the *linear absorption coefficient,* and its value depends on the absorbing material and the wavelength of the photon beam. This wavelength (or energy) dependence allows us to filter out unwanted wavelengths from a broad-spectrum x-ray beam. (a) Two x-ray beams of wavelengths λ_1 and λ_2 and equal incident intensities pass through the same metal plate. Show that the ratio of the emergent beam intensities is

$$\frac{I_2}{I_1} = e^{-[\mu_2 - \mu_1]x}$$

(b) Compute the ratio of intensities emerging from an aluminum plate 1.0 mm thick if the incident beam contains equal intensities of 50 pm and 100 pm x-rays. The values of μ for aluminum at these two wavelengths are $\mu_1 = 5.4$ cm^{-1} at 50 pm and $\mu_2 = 41.0$ cm^{-1} at 100 pm. (c) Repeat for an aluminum plate 10 mm thick.

CHAPTER 47

Particle Physics and Cosmology

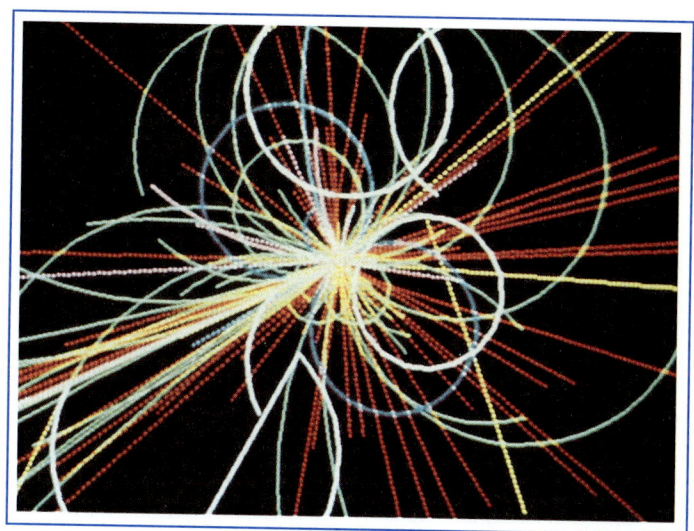

Computer-generated simulation of proton-proton collision in accelerator at Fermi National Accelerator Laboratory. *(Peter Arnold, Inc.)*

In this concluding chapter, we examine the properties and classifications of the various known subatomic particles and the fundamental interactions that govern their behavior. We also discuss the current theory of elementary particles, in which all matter is believed to be constructed from only two families of particles, quarks and leptons. Finally, we discuss how clarifications of such models might help scientists understand cosmology, which deals with the evolution of the Universe.

The word atom is from the Greek *atomos,* which means "indivisible." At one time, atoms were thought to be the indivisible constituents of matter; that is, they were regarded to be elementary particles. After 1932, physicists viewed all matter as consisting of only three constituent particles: electrons, protons, and neutrons. With the exception of the free neutron, these particles are very stable. Beginning in 1945, many new particles were discovered in experiments involving high-energy collisions between known particles. These new particles are characteristically very unstable and have very short half-lives, ranging between 10^{-6} and 10^{-23} s. So far, more than 300 of these unstable, temporary particles have been catalogued.

During the last 35 years, many powerful particle accelerators have been constructed throughout the world, making it possible to observe collisions of highly energetic particles under controlled laboratory conditions, so as to reveal the subatomic world in finer detail. Until the 1960s, physicists were bewildered by the large number and variety of subatomic particles being discovered. They wondered if the particles were like animals in a zoo, having no systematic relationship connecting them, or whether a pattern was emerging that would provide a better understanding of the elaborate structure in the subnuclear world. In the last twenty-five years, physicists have tremendously advanced our knowledge of the structure of matter by recognizing that all particles except electrons, photons, and a few related particles are made of smaller particles called quarks. Thus, protons and neutrons, for example, are not truly elementary but are systems of tightly bound quarks.

47.1   THE FUNDAMENTAL FORCES IN NATURE

As we learned in Section 6.6, all natural phenomena can be described by **four fundamental forces** between elemental particles. In order of decreasing strength, these are the strong force, the electromagnetic force, the weak force, and the gravitational force.

The **strong force,** as we mentioned in Chapter 45, represents the glue that holds nucleons together. It is very short-range and is negligible for separations greater than about 10^{-15} m (about the size of the nucleus). The **electromagnetic force,** which binds atoms and molecules together to form ordinary matter, has about 10^{-2} times the strength of the strong force. It is a long-range force that decreases in strength as the inverse square of the separation between interacting particles. The **weak force** is a short-range force that tends to produce instability in certain nuclei. It is responsible for most radioactive decay processes such as beta decay, and its strength is only about 10^{-5} times that of the strong force. Finally, the **gravitational force** is a long-range force that has a strength of only about 10^{-39} times that of the strong force (in SI units). Although this familiar interaction is the force that holds the planets, stars, and galaxies together, its effect on elementary particles is negligible. In modern physics, interactions between particles are often described in terms of the exchange of field particles, or quanta. In the case of the familiar electromagnetic interaction, for instance, the field particles are photons. In the language of modern physics, it is said that the electromagnetic force is *mediated* by photons and that photons are the quanta of the electromagnetic field. Likewise, the strong force is mediated by field particles called *gluons,* the weak force is mediated by particles called the W and Z *bosons,* and the gravitational force is mediated by quanta of the gravitational field called *gravitons.* These interactions, their ranges, and their relative strengths are summarized in Table 47.1.

TABLE 47.1   Particle Interactions

Interaction	Relative Strength	Range of Force	Mediating Field Particle
Strong	1	Short (≈ 1 fm)	Gluon
Electromagnetic	10^{-2}	∞	Photon
Weak	10^{-5}	Short ($\approx 10^{-3}$ fm)	$W^{\pm}$, Z^0 bosons
Gravitational	10^{-39}	∞	Graviton

47.2 POSITRONS AND OTHER ANTIPARTICLES

In the 1920s, the theoretical physicist Paul Adrien Maurice Dirac (1902–1984) found a relativistic quantum mechanical solution for the electron. Dirac's theory successfully explained the origin of the electron's spin and its magnetic moment. But, it had one major problem: Its relativistic wave equation required solutions corresponding to negative energy states, and if negative energy states existed, an electron in a state of positive energy would be expected to make a rapid transition to one of these states, emitting a photon in the process.

Dirac circumvented this difficulty by postulating that all negative energy states are filled. Those electrons that occupy these negative energy states are said to be in the *Dirac sea* and are not directly observable because the Pauli exclusion principle does not allow them to react to external forces. However, if one of these negative energy states is vacant, leaving a hole in the sea of filled states, the hole can react to external forces and therefore is observable. The profound implication of this theory is that, *for every particle, there is an antiparticle.* The antiparticle has the same mass as the particle, but opposite charge. For example, the electron's antiparticle (now called a *positron*) has a rest energy of 0.511 MeV and a positive charge of 1.60×10^{-19} C.

The positron was discovered by Carl Anderson in 1932, and in 1936 he was awarded the Nobel prize for his achievement. Anderson discovered the positron while examining tracks created by electron-like particles of positive charge in a cloud chamber. (These early experiments used cosmic rays—mostly energetic protons passing through interstellar space—to initiate high-energy reactions on the order of several GeV.) In order to discriminate between positive and negative charges, the cloud chamber was placed in a magnetic field, causing moving charges to follow curved paths. Anderson noted that some of the electron-like tracks deflected in a direction corresponding to a positively charged particle.

Since Anderson's initial discovery, the positron has been observed in a number of experiments. Perhaps the most common process for producing positrons is **pair production**. In this process, a gamma-ray photon with sufficiently high energy collides with a nucleus and an electron-positron pair is created. Because the total rest energy of the electron-positron pair is $2m_ec^2 = 1.02$ MeV (where m_e is the rest mass of the electron), the photon must have at least this much energy to create an electron-positron pair. Thus, electromagnetic energy in the form of a gamma ray is converted to mass in accordance with Einstein's famous relationship $E = mc^2$. Figure 47.1 shows tracks of electron-positron pairs created by 300-MeV gamma rays striking a lead sheet.

The reverse process can also occur. Under the proper conditions, an electron and positron can annihilate each other to produce two gamma-ray photons that have a combined energy of at least 1.02 MeV:

$$e^- + e^+ \longrightarrow 2\gamma$$

Very rarely, a proton and antiproton also annihilate each other to produce two gamma-ray photons.

Practically every known elementary particle has an antiparticle. Among the exceptions are the photon and the neutral pion (π^0). Following the construction of high-energy accelerators in the 1950s, many other antiparticles were discovered. These included the antiproton discovered by Emilio Segrè and Owen Chamberlain in 1955, and the antineutron discovered shortly thereafter.

Paul Adrien Maurice Dirac (1902–1984), winner of the Nobel prize for physics in 1933, is shown speaking at Yeshiva University. *(UPI/Bettmann)*

Bubble-chamber photograph of electron (green) and positron (red) tracks produced by energetic gamma rays. The highly curved tracks at the top are due to an electron-positron pair that bend in opposite direction in the magnetic field. The lower tracks are produced by more energetic electrons and a positron. *(Lawrence Berkeley Laboratory/ Science Photo Library, Photo Researchers, Inc.)*

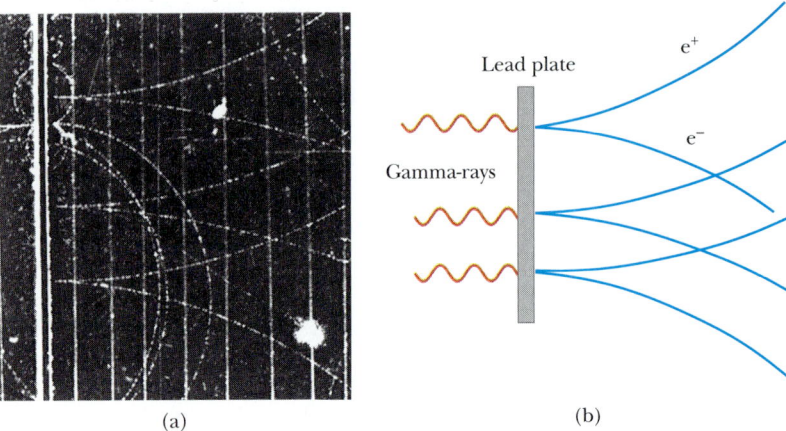

FIGURE 47.1   (a) Bubble-chamber tracks of electron-positron pairs produced by 300-MeV gamma rays striking a lead sheet.   *(Courtesy Lawrence Berkeley Laboratory, University of California)* (b) Sketch of the pertinent pair-production events. Note that the positrons deflect upward while the electrons deflect downward in an applied magnetic field directed into the page.

CONCEPTUAL EXAMPLE 47.1   **Pair Production**

When an electron and a positron meet at low speeds in free space, why are *two* 0.511-MeV gamma rays produced, rather than *one* gamma ray with an energy of 1.022 MeV?

Reasoning   Gamma rays are photons, and photons carry momentum. If only one photon were produced, momentum

would not be conserved, because the total momentum of the electron-positron system is approximately zero, whereas a single photon of energy 1.0022 MeV would have a very large momentum. On the other hand, the two gamma-ray photons that are produced travel off in opposite directions, so their total momentum is zero.

47.3   MESONS AND THE BEGINNING OF PARTICLE PHYSICS

Physicists in the mid-1930s had a fairly simple view of the structure of matter. The building blocks were the proton, the electron, and the neutron. Three other particles were known or postulated at the time: the photon, the neutrino, and the positron. These six particles were considered the fundamental constituents of matter. With this marvelously simple picture of the world, however, no one was able to provide an answer to the following important question: Since the many protons in proximity in any nucleus should strongly repel each other due to their like charges, what is the nature of the force that holds the nucleus together? Scientists recognized that this mysterious force must be much stronger than anything encountered in nature up to that time.

The first theory to explain the nature of the strong force was proposed in 1935 by the Japanese physicist Hideki Yukawa (1907–1981), an effort that later earned him the Nobel prize. To understand Yukawa's theory, it is useful to first recall that *two atoms can form a covalent chemical bond by sharing electrons.* Similarly, in the modern views of electromagnetic interactions, *charged particles interact by sharing a photon.* Yukawa used this idea to explain the strong force by proposing a new particle whose exchange between nucleons in the nucleus produces the strong force. Furthermore, he established that the range of the force is inversely proportional to the mass of this particle and predicted that the mass would be about 200 times the mass of the electron. Since the new particle would have a mass between that of the

Hideki Yukawa (1907–1981), a Japanese physicist, was awarded the Nobel prize in 1949 for predicting the existence of mesons. This photograph of Yukawa at work was taken in 1950 in his office at Columbia University.

Richard Phillips Feynman was a brilliant theoretical physicist who, together with Julian S. Schwinger and Shinichiro Tomonaga, won the 1965 Nobel prize for physics for his fundamental work in the principles of quantum electrodynamics. His many important contributions to physics include the invention of simple diagrams to represent particle interactions graphically, the theory of the weak interaction of subatomic particles, a reformulation of quantum mechanics, the theory of superfluid helium, and his contribution to physics education through the magnificent three-volume text *The Feynman Lectures on Physics.*

Feynman did his undergraduate work at MIT and received his Ph.D. in 1942 from Princeton University. During World War II, he worked on the Manhattan Project first at Princeton and then at Los Alamos, New Mexico. He joined the faculty at Cornell University in 1945 and was appointed professor of physics at California Institute of Technology in 1950, where he remained for the rest of his career.

Feynman had a passion for finding new and better ways to formulate each problem, or, as he would say, "turning it around." In the early part of his career, he was fascinated with electrodynamics and developed an intuitive view of quantum electrodynamics. Convinced that the electron could not interact with its own field, he said, "That was the beginning, and the idea seemed so obvious to me that I fell deeply in love with it . . . ." Often called the outstanding intuitionist of our age, he said in his Nobel acceptance speech, "Often, even in a physicist's sense, I did not have a demonstration of how to get all of these rules and equations, from conventional electrodynamics . . . . I never really

Richard P. Feynman

| 1 9 1 8 – 1 9 8 8 |

sat down, like Euclid did for the geometers of Greece, and made sure that you could get it all from a single set of axioms."

In 1986, Feynman was a member of the presidential commission to investigate the explosion of the space shuttle *Challenger.* In this capacity, he performed for the commission members a simple experiment that showed that one of the shuttle's O-ring seals was the likely cause of the disaster. After placing a seal in a pitcher of ice water and squeezing it with a clamp, he demonstrated that the seal failed to spring back into shape once the clamp was removed.[1]

Feynman worked in physics with a style commensurate with his personality, that is, with energy, vitality, and humor. The following quotes from some of his colleagues are characteristic of the great impact he made on the scientific community.[2]

"A brilliant, vital, and amusing

[1] Feynman's own account of this inquiry can be found in *Physics Today,* 4:26, February 1988.

neighbor, Feynman was a stimulating (if sometimes exasperating) partner in discussions of profound issues—we would exchange ideas and silly jokes in between bouts of mathematical calculation—we struck sparks off each other, and it was exhilarating." *Murray Gell-Mann*

"Reading Feynman is a joy and a delight, for in his papers, as in his talks, Feynman communicated very directly, as though the reader were watching him derive the results at the blackboard." *David Pines*

"He loved puzzles and games. In fact, he saw all the world as a sort of game, whose progress of "behavior" follows certain rules, some known, some unknown . . . . Find places or circumstances where the rules don't work, and invent new rules that do." *David L. Goodstein*

"Feynman was not a theorist's theorist, but a physicist's physicist and a teacher's teacher." *Valentine L. Telegdi*

Laurie M. Brown, one of his graduate students at Cornell, noted that Feynman, a playful showman, was "undervalued at first because of his rough manners, who in the end triumphs through native cleverness, psychological insight, common sense and the famous Feynman humor . . . . Whatever else Dick Feynman may have joked about, his love for physics approached reverence."

[2] For more on Feynman's life and contributions, see the numerous articles in a special memorial issue of *Physics Today* 42, February 1989. For a personal account of Feynman, see his popular autobiographical books, *Surely You're Joking Mr. Feynman,* New York, Bantam Books, 1985, and *What Do You Care What Other People Think,* New York, W. W. Norton & Co., 1987.

electron and that of the proton, it was called a **meson** (from the Greek *meso,* "middle").

In an effort to substantiate Yukawa's predictions, physicists began an experimental search for the meson by studying cosmic rays entering the Earth's atmosphere. In 1937, Carl Anderson and his collaborators discovered a particle of mass 106 MeV/c^2, about 207 times the mass of the electron. However, subsequent experiments showed that the particle interacted very weakly with matter and, hence, could not be the carrier of the strong force. The puzzling situation inspired several theoreticians to propose that there are two mesons with slightly different masses. This idea was confirmed in 1947 with the discovery of the pi meson (π), or simply *pion,* by Cecil Frank Powell (1903–1969) and Giuseppe P. S. Occhialini (b. 1907). The particle discovered by Anderson in 1937, the one *thought to be* a meson, is not really a meson. Instead, it takes part in the weak and electromagnetic interactions only, and it is now called the *muon.*

The pion, which is the true carrier for the strong force, comes in three varieties, corresponding to three charge states: π^+, π^-, and π^0. The π^+ and π^- particles have masses of 139.6 MeV/c^2, while the π^0 has a mass of 135.0 MeV/c^2. Pions and muons are very unstable particles. For example, the π^-, which has a mean lifetime of 2.6×10^{-8} s, first decays to a muon and an antineutrino. The muon, which has a mean lifetime of 2.2 μs, then decays into an electron, a neutrino, and an antineutrino:

$$\pi^- \longrightarrow \mu^- + \bar{\nu}$$
$$\mu^- \longrightarrow e^- + \nu + \bar{\nu} \tag{47.1}$$

The interaction between two particles can be represented in a simple diagram called a *Feynman diagram,* developed by the American physicist Richard P. Feynman (1918–1988). Figure 47.2 is such a diagram for the electromagnetic interaction between two electrons. In this simple case, a photon is the field particle that mediates the electromagnetic force between the electrons. The photon transfers energy and momentum from one electron to the other in this interaction. The photon is called a *virtual photon* because it is reabsorbed by the emitting particle, without having been detected. Virtual photons do not violate the law of conservation of energy because they have a very short lifetime, Δt, such that the uncertainty in the energy of electron plus photon, $\Delta E \approx \hbar/\Delta t$, is greater than the photon energy.

Now consider the pion exchange between a proton and a neutron via the strong force (Fig. 47.3). We can reason that the energy ΔE needed to create a pion of mass m_π is given by Einstein's equation $\Delta E = m_\pi c^2$. Again, the very existence of the pion would violate conservation of energy if it lasted for a time greater than $\Delta t \approx \hbar/\Delta E$ (from the uncertainty principle), where ΔE is the energy of the pion

Richard Feynman (1918–1988) with his son, Carl, after winning the Nobel prize for physics in 1965. The prize was shared by Feynman, Julian Schwinger, and Sin Itiro Tomonaga. *(UPI Tele-photos)*

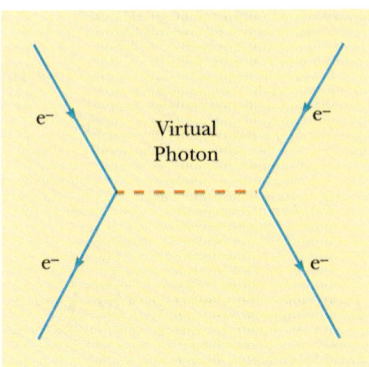

FIGURE 47.2 Feynman diagram showing how a photon mediates the electromagnetic force between two interacting electrons.

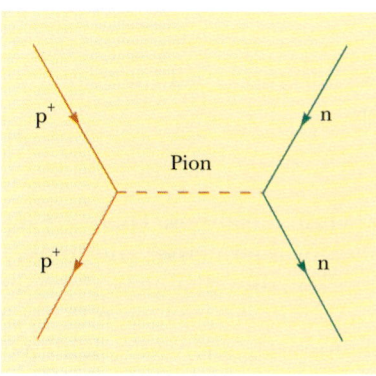

FIGURE 47.3 Feynman diagram representing a proton interacting with a neutron via the strong force. In this case, the pion mediates the strong force.

and Δt is the time it takes the pion to transfer from one nucleon to the other. Therefore,

$$\Delta t \approx \frac{\hbar}{\Delta E} = \frac{\hbar}{m_\pi c^2} \tag{47.2}$$

Because the pion cannot travel faster than the speed of light, the maximum distance d it can travel in a time Δt is $c\,\Delta t$. Using Equation 47.2 and $d = c\,\Delta t$, we find this maximum distance to be

$$d \approx \frac{\hbar}{m_\pi c} \tag{47.3}$$

From Chapter 45, we know that the range of the strong force is approximately 1.5×10^{-15} m. Using this value for d in Equation 47.3, we calculate the rest energy of the pion to be

$$m_\pi c^2 \approx \frac{\hbar c}{d} = \frac{(1.05 \times 10^{-34}\,\text{J} \cdot \text{s})(3.00 \times 10^8\,\text{m/s})}{1.5 \times 10^{-15}\,\text{m}}$$

$$= 2.1 \times 10^{-11}\,\text{J} \cong 130\,\text{MeV}$$

This corresponds to a mass of 130 MeV/c^2 (about 250 times the mass of the electron), a value in good agreement with the observed mass.

The concept we have just described is quite revolutionary. In effect, it says that a proton can change into a proton plus a pion, as long as it returns to its original state in a very short time. Physicists often say that a nucleon undergoes "fluctuations" as it emits and absorbs pions. As we have seen, these fluctuations are a consequence of a combination of quantum mechanics (through the uncertainty principle) and special relativity (through Einstein's energy-mass relationship $E = mc^2$).

This section has dealt with the particles that mediate the strong force, namely the pions, and the mediators of the electromagnetic force, photons. The graviton, which is the mediator of the gravitational force, has yet to be observed. The $W^\pm$ and Z^0 particles that mediate the weak force were discovered in 1983 by the Italian physicist Carlo Rubbia (b. 1934) and his associates using a proton-antiproton collider. Rubbia and Simon van der Meer, both at CERN, shared the 1984 Nobel prize for physics for the discovery of the $W^\pm$ and Z^0 particles and the development of the proton-antiproton collider.

47.4 CLASSIFICATION OF PARTICLES

All particles other than photons can be classified into two broad categories, hadrons and leptons, according to their interactions.

Hadrons

Particles that interact through the strong force are called **hadrons.** There are two classes of hadrons, *mesons* and *baryons,* distinguished by their masses and spins.

Mesons all have zero or integer spin (0 or 1), with masses that lie between the mass of the electron and the mass of the proton. All mesons are known to decay finally into electrons, positrons, neutrinos, and photons. The pion is the lightest of known mesons; it has a mass of about 140 MeV/c^2 and a spin of 0. Another is the K meson, having a mass of about 500 MeV/c^2 and spin 0.

Baryons, the second class of hadrons, have masses equal to or greater than the proton mass (the name *baryon* means "heavy" in Greek), and their spin is always a noninteger value (1/2 or 3/2). Protons and neutrons are baryons, as are many other particles. With the exception of the proton, all baryons decay in such a way that the end products include a proton. For example, the baryon called the Ξ hyperon first decays to the Λ^0 baryon in about 10^{-10} s. The Λ^0 then decays to a proton and a π^- in approximately 3×10^{-10} s.

Today it is believed that hadrons are composed of more elemental units called quarks. Some of the important properties of hadrons are listed in Table 47.2.

Leptons

Leptons (from the Greek *leptos* meaning "small" or "light") are a group of particles that participate in the weak interaction. All leptons have spins of 1/2. Included in this group are electrons, muons, and neutrinos, which are less massive than the lightest hadron. Although hadrons have size and structure, leptons appear to be truly elementary, with no structure (that is, point-like).

Quite unlike hadrons, the number of known leptons is small. Currently, scientists believe there are only six leptons (each having an antiparticle): the electron, the muon, and the tau and a neutrino associated with each:

$$\begin{pmatrix} e^- \\ \nu_e \end{pmatrix} \quad \begin{pmatrix} \mu^- \\ \nu_\mu \end{pmatrix} \quad \begin{pmatrix} \tau^- \\ \nu_\tau \end{pmatrix}$$

The tau lepton, discovered in 1975, has a mass equal to about twice that of the proton. The neutrino associated with the tau has not yet been observed in the laboratory.

Current evidence suggests that neutrinos travel with the speed of light and, therefore, cannot have nonzero rest mass.

47.5 CONSERVATION LAWS

In Chapter 45 we learned that conservation laws are important in understanding why certain decays and reactions occur and others do not. In general, the laws of conservation of energy, linear momentum, angular momentum, and electric charge provide us with a set of rules that all processes must follow.

A number of new conservation laws are important in the study of elementary particles. Although the two described here have no theoretical foundation, they are supported by abundant empirical evidence.

TABLE 47.2 A Table of Some Particles and Their Properties

Category	Particle Name	Symbol	Anti-particle	Rest Mass (MeV/c^2)	B	L_e	L_μ	L_τ	S	Lifetime (s)	Principal Decay Modes[a]
Photon	Photon	γ	Self	0	0	0	0	0	0	Stable	
Leptons	Electron	e^-	e^+	0.511	0	$+1$	0	0	0	Stable	
	Neutrino (e)	ν_e	$\overline{\nu}_e$	0(?)	0	$+1$	0	0	0	Stable	
	Muon	μ^-	μ^+	105.7	0	0	$+1$	0	0	2.20×10^{-6}	$e^-\overline{\nu}_e\nu_\mu$
	Neutrino (μ)	ν_μ	$\overline{\nu}_\mu$	0(?)	0	0	$+1$	0	0	Stable	
	Tau	τ^-	τ^+	1784	0	0	0	-1	0	$<4 \times 10^{-13}$	$\mu^-\overline{\nu}_\mu\nu_\tau$, $e^-\overline{\nu}_e\nu_\tau$, hadrons
	Neutrino (τ)	ν_τ	$\overline{\nu}_\tau$	0(?)	0	0	0	-1	0	Stable	
Hadrons											
Mesons	Pion	π^+	π^-	139.6	0	0	0	0	0	2.60×10^{-8}	$\mu^+\nu_\mu$
		π^0	Self	135.0	0	0	0	0	0	0.83×10^{-16}	2γ
	Kaon	K^+	K^-	493.7	0	0	0	0	$+1$	1.24×10^{-8}	$\mu^+\nu_\mu$, $\pi^+\pi^0$
		K_S^0	$\overline{K}_S^0$	497.7	0	0	0	0	$+1$	0.89×10^{-10}	$\pi^+\pi^-$, $2\pi^0$
		K_L^0	$\overline{K}_L^0$	497.7	0	0	0	0	$+1$	5.2×10^{-8}	$\pi^{\pm}e^{\mp}(\overline{\nu})_e$ $\pi^{\pm}\mu^{\mp}(\overline{\nu})_\mu$ $3\pi^0$
	Eta	η	Self	548.8	0	0	0	0	0	$<10^{-18}$	2γ, 3π
		η'	Self	958	0	0	0	0	0	2.2×10^{-21}	$\eta\pi^+\pi^-$
Baryons	Proton	p	$\overline{\text{p}}$	938.3	$+1$	0	0	0	0	Stable	
	Neutron	n	$\overline{\text{n}}$	939.6	$+1$	0	0	0	0	920	$pe^-\overline{\nu}_e$
	Lambda	Λ^0	$\overline{\Lambda}^0$	1115.6	$+1$	0	0	0	-1	2.6×10^{-10}	$p\pi^-$, $n\pi^0$
	Sigma	Σ^+	$\overline{\Sigma}^-$	1189.4	$+1$	0	0	0	-1	0.80×10^{-10}	$p\pi^0$, $n\pi^+$
		Σ^0	$\overline{\Sigma}^0$	1192.5	$+1$	0	0	0	-1	6×10^{-20}	$\Lambda^0\gamma$
		Σ^-	$\overline{\Sigma}^+$	1197.3	$+1$	0	0	0	-1	1.5×10^{-10}	$n\pi^-$
	Xi	Ξ^0	$\overline{\Xi}^0$	1315	$+1$	0	0	0	-2	2.9×10^{-10}	$\Lambda^0\pi^0$
		Ξ^-	$\overline{\Xi}^+$	1321	$+1$	0	0	0	-2	1.64×10^{-10}	$\Lambda^0\pi^-$
	Omega	Ω^-	Ω^+	1672	$+1$	0	0	0	-3	0.82×10^{-10}	$\Xi^0\pi^0$, Λ^0K^-

[a] A notation in this column such as $p\pi^-$, $n\pi^0$ means two possible decay modes. In this case, the two possible decays are $\Lambda^0 \rightarrow p + \pi^-$ or $\Lambda^0 \rightarrow n + \pi^0$.

Baryon Number

Conservation of baryon number tells us that whenever a baryon is created in a reaction or decay, an antibaryon is also created. This scheme can be quantified by assigning a baryon number $B = +1$ for all baryons, $B = -1$ for all antibaryons, and $B = 0$ for all other particles. Thus, the **law of conservation of baryon number** states that **whenever a nuclear reaction or decay occurs, the sum of the baryon numbers before the process must equal the sum of the baryon numbers after the process.**

> Conservation of baryon number

If baryon number is absolutely conserved, the proton must be absolutely stable. If it were not for the law of conservation of baryon number, the proton could decay to a positron and a neutral pion. However, such a decay has never been observed. At the present, we can say only that the proton has a half-life of at least 10^{31} years (the estimated age of the Universe is only 10^{10} years). In one recent version of a grand unified theory, physicists predicted that the proton is unstable. According to this theory, the baryon number is not absolutely conserved.

EXAMPLE 47.2 Checking Baryon Numbers

Determine whether or not the following reactions can occur based on the law of conservation of baryon number.

$$(1) \quad p + n \longrightarrow p + p + n + \bar{p}$$

$$(2) \quad p + n \longrightarrow p + p + \bar{p}$$

Solution For (1), recall that $B = +1$ for baryons and $B = -1$ for antibaryons. Hence, the left side of (1) gives a total

baryon number of $1 + 1 = 2$. The right side of (1) gives a total baryon number of $1 + 1 + 1 + (-1) = 2$. Thus, the reaction can occur provided the incoming proton has sufficient energy.

The left side of (2) gives a total baryon number of $1 + 1 = 2$. However, the right side gives $1 + 1 + (-1) = 1$. Because the baryon number is not conserved, the reaction cannot occur.

Lepton Number

Conservation of lepton number

There are three conservation laws involving lepton numbers, one for each variety of lepton. The **law of conservation of electron-lepton number** states that **the sum of the electron-lepton numbers before a reaction or decay must equal the sum of the electron-lepton numbers after the reaction or decay.**

The electron and the electron neutrino are assigned a positive lepton number $L_e = +1$; the antileptons e^+ and $\bar{\nu}_e$ are assigned a negative lepton number $L_e = -1$; and all others have $L_e = 0$. For example, consider the decay of the neutron

$$n \longrightarrow p + e^- + \bar{\nu}_e$$

Before the decay, the electron-lepton number is $L_e = 0$; after the decay it is $0 + 1 + (-1) = 0$. Thus, the electron-lepton number is conserved. It is important to recognize that the baryon number must also be conserved. This can easily be checked by noting that before the decay $B = +1$, and after the decay B is $+1 + 0 + 0 = +1$.

Similarly, when a decay involves muons, the muon-lepton number, L_μ, is conserved. The μ^- and the ν_μ are assigned positive numbers, $L_\mu = +1$; the antimuons μ^+ and $\bar{\nu}_\mu$ are assigned negative numbers, $L_\mu = -1$; and all others have $L_\mu = 0$. Finally, the tau-lepton number, L_τ, is conserved, and similar assignments can be made for the τ lepton and its neutrino.

EXAMPLE 47.3 Checking Lepton Numbers

Determine which of the following decay schemes can occur on the basis of conservation of electron-lepton number.

$$(1) \quad \mu^- \longrightarrow e^- + \bar{\nu}_e + \nu_\mu$$

$$(2) \quad \pi^+ \longrightarrow \mu^+ + \nu_\mu + \nu_e$$

Solution Because decay (1) involves both a muon and an electron, L_μ and L_e must both be conserved. Before the decay, $L_\mu = +1$ and $L_e = 0$. After the decay, $L_\mu = 0 + 0 + 1 = +1$, and $L_e = +1 - 1 + 0 = 0$. Thus, both numbers are conserved, and on this basis the decay mode is possible.

Before decay (2) occurs, $L_\mu = 0$ and $L_e = 0$. After the decay, $L_\mu = -1 + 1 + 0 = 0$, but $L_e = +1$. Thus, the decay is not possible because the electron-lepton number is not conserved.

Exercise Determine whether the decay $\mu^- \rightarrow e^- + \bar{\nu}_e$ can occur.

Answer No. The muon-lepton number is $+1$ before the decay and 0 after.

47.6 STRANGE PARTICLES AND STRANGENESS

Many particles discovered in the 1950s were produced by the nuclear interaction of pions with protons and neutrons in the atmosphere. A group of these particles, namely the kaon (K), lambda (Λ), and sigma (Σ) particles exhibited unusual properties in production and decay and, hence, were called *strange particles*.

One unusual property is that these particles were always produced in pairs. For example, when a pion collides with a proton, two neutral strange particles were produced with high probability (Fig. 47.4):

$$\pi^- + p \longrightarrow K^0 + \Lambda^0$$

On the other hand, the reaction $\pi^- + p \to K^0 + n^0$ never occurred, even though no known conservation laws would have been violated and the energy of the pion was sufficient to initiate the reaction.

The second peculiar feature of strange particles is that, although they are produced by the strong interaction at a high rate, they do not decay into particles that interact via the strong force at a very high rate. Instead, they decay very slowly, which is characteristic of the weak interaction. Their half-lives are in the range

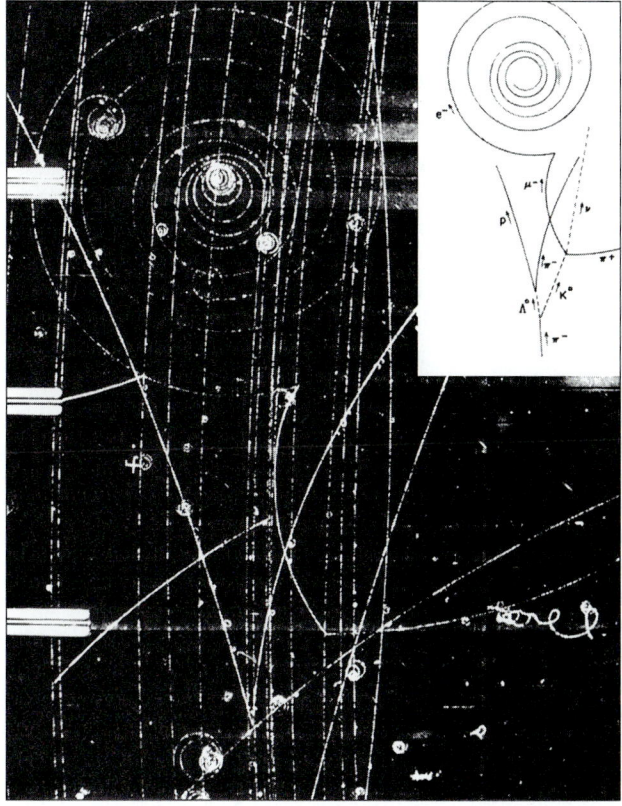

FIGURE 47.4   This bubble-chamber photograph shows many events, and the inset is a drawing of identified tracks. The strange particles Λ^0 and K^0 are formed at the bottom as the π^- interacts with a proton according to $\pi^- + p \to \Lambda^0 + K^0$. (Note that the neutral particles leave no tracks, as indicated by the dashed lines.) The Λ^0 and K^0 then decay according to $\Lambda^0 \to \pi^- + p$ and $K^0 \to \pi^+ + \mu^- + \overline{\nu}_\mu$. *(Courtesy Lawrence Berkeley Laboratory, University of California, Photographic Services)*

10^{-10} s to 10^{-8} s; most other particles that interact via the strong force have lifetimes on the order of 10^{-23} s.

To explain these unusual properties of strange particles, a law called conservation of strangeness was introduced, together with a new quantum number S called **strangeness**. The strangeness numbers for some particles are given in Table 47.2. The production of strange particles in pairs is explained by assigning $S = +1$ to one of the particles and $S = -1$ to the other. All nonstrange particles are assigned strangeness $S = 0$. The **law of conservation of strangeness** states that **whenever a nuclear reaction of decay occurs, the sum of the strangeness numbers before the process must equal the sum of the strangeness numbers after the process.**

Conservation of strangeness number

The slow decay of strange particles can be explained by assuming that the strong and electromagnetic interactions obey the law of conservation of strangeness, but the weak interaction does not. Because the decay reaction involves the loss of one strange particle, it violates strangeness conservation and, hence, proceeds slowly via the weak interaction.

EXAMPLE 47.4   Is Strangeness Conserved?

(a) Determine whether the following reaction occurs on the basis of conservation of strangeness.

$$\pi^0 + n \longrightarrow K^+ + \Sigma^-$$

Solution   The initial state has strangeness $S = 0 + 0 = 0$. Because the strangeness of the K^+ is $S = +1$ and the strangeness of the Σ^- is $S = -1$, the strangeness of the final state is $+1 - 1 = 0$. Thus, strangeness is conserved and the reaction is allowed.

(b) Show that the following reaction does not conserve strangeness.

$$\pi^- + p \longrightarrow \pi^- + \Sigma^+$$

Solution   The initial state has strangeness $S = 0 + 0 = 0$, and the final state has strangeness $S = 0 + (-1) = -1$. Thus, strangeness is not conserved.

Exercise   Show that the reaction $p + \pi^- \rightarrow K^0 + \Lambda^0$ obeys the law of conservation of strangeness.

47.7   THE EIGHTFOLD WAY

As we have seen, conserved quantities such as spin, baryon number, lepton number, and strangeness are labels we associate with particles. Many classification schemes have been proposed that group particles into families. Consider the first eight baryons listed in Table 47.2, all having a spin of $1/2$. If we plot their strangeness versus their charge using a sloping coordinate system, as in Figure 47.5a, a fascinating pattern is observed. Six of the baryons form a hexagon, while the remaining two are at its center.

Now consider the family of mesons listed in Table 47.2, having spins of zero. If we count both particles and antiparticles, there are nine such mesons. Figure 47.5b is a plot of strangeness versus charge for this family. Again, a fascinating hexagonal pattern emerges. In this case, the particles on the perimeter of the hexagon lie opposite their antiparticles, and the remaining three (which form their own antiparticles) are at its center. These and related symmetric patterns, called the **eightfold way**, were proposed independently in 1961 by Murray Gell-Mann and Yuval Ne'eman.

The groups of baryons and mesons can be displayed in many other symmetrical patterns within the framework of the eightfold way. For example, the family of

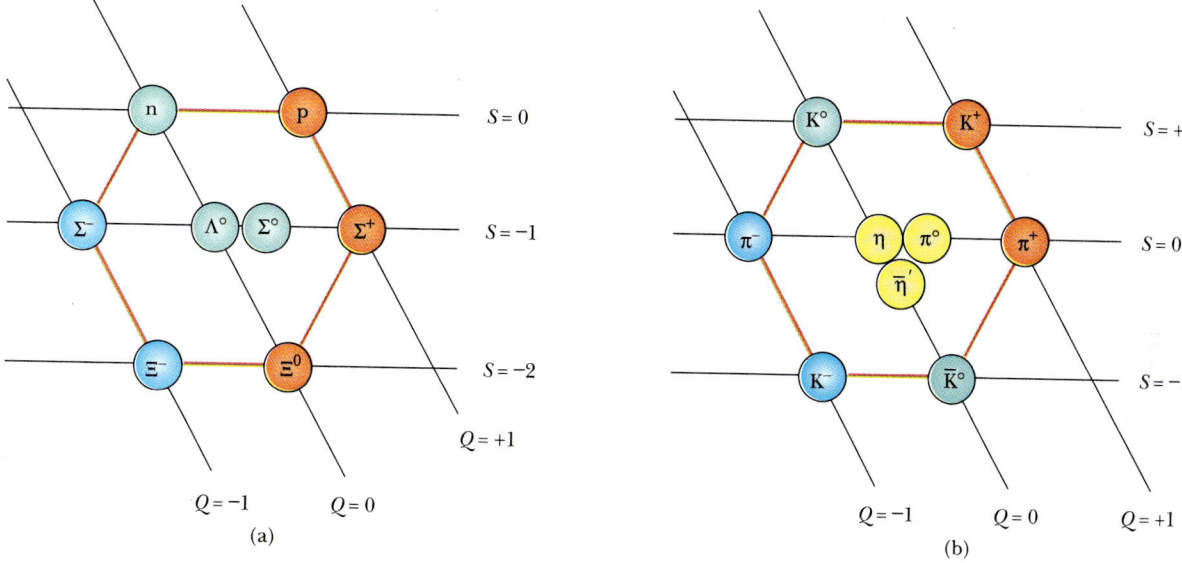

FIGURE 47.5 (a) The hexagonal eightfold-way pattern for the eight spin-1/2 baryons. This strangeness versus charge plot uses a sloping axis for the charge number Q, but a horizontal axis for the strangeness S values. (b) The eightfold-way pattern for the nine spin-zero mesons.

spin-3/2 baryons contains ten particles arranged in a pattern like the tenpins in a bowling alley. After the pattern was proposed, one of the particles was missing—it had yet to be discovered. Gell-Mann predicted that the missing particle, which he called the omega minus (Ω^-), should have a spin of 3/2, a charge of -1, a strangeness of -3, and a rest energy of about 1680 MeV. Shortly thereafter, in 1964, scientists at the Brookhaven National Laboratory found the missing particle through careful analyses of bubble chamber photographs and confirmed all its predicted properties.

The patterns of the eightfold way in particle physics have much in common with the periodic table. Whenever a vacancy (a missing particle or element) occurs in the organized patterns, experimentalists have a guide for their investigations.

47.8   QUARKS—FINALLY

As we have noted, leptons appear to be truly elementary particles because they have no measurable size or internal structure, are limited in number, and do not seem to break down into smaller units. Hadrons, on the other hand, are complex particles having size and structure. The existence of the eightfold-way patterns suggests that baryons and mesons—in other words, hadrons—have a more elemental substructure. Furthermore, we know that hadrons decay into other hadrons and are many in number. Table 47.2 lists only those hadrons that are stable against hadronic decay; hundreds of others have been discovered. These facts strongly suggest that hadrons cannot be truly elementary but have some substructure.

American physicist Murray Gell-Mann was awarded the Nobel prize in 1969 for his theoretical studies dealing with subatomic particles. *(Photo courtesy of Michael R. Dressler)*

| TABLE 47.3 | **Properties of Quarks and Antiquarks** | | | | | | | |

Quarks

Name	Symbol	Spin	Charge	Baryon Number	Strangeness	Charm	Bottomness	Topness
Up	u	$\frac{1}{2}$	$+\frac{2}{3}e$	$\frac{1}{3}$	0	0	0	0
Down	d	$\frac{1}{2}$	$-\frac{1}{3}e$	$\frac{1}{3}$	0	0	0	0
Strange	s	$\frac{1}{2}$	$-\frac{1}{3}e$	$\frac{1}{3}$	-1	0	0	0
Charmed	c	$\frac{1}{2}$	$+\frac{2}{3}e$	$\frac{1}{3}$	0	$+1$	0	0
Bottom	b	$\frac{1}{2}$	$-\frac{1}{3}e$	$\frac{1}{3}$	0	0	$+1$	0
Top	t	$\frac{1}{2}$	$+\frac{2}{3}e$	$\frac{1}{3}$	0	0	0	$+1$

Antiquarks

Name	Symbol	Spin	Charge	Baryon Number	Strangeness	Charm	Bottomness	Topness
Up	$\overline{u}$	$\frac{1}{2}$	$-\frac{2}{3}e$	$-\frac{1}{3}$	0	0	0	0
Down	$\overline{d}$	$\frac{1}{2}$	$+\frac{1}{3}e$	$-\frac{1}{3}$	0	0	0	0
Strange	$\overline{s}$	$\frac{1}{2}$	$+\frac{1}{3}e$	$-\frac{1}{3}$	$+1$	0	0	0
Charmed	$\overline{c}$	$\frac{1}{2}$	$-\frac{2}{3}e$	$-\frac{1}{3}$	0	-1	0	0
Bottom	$\overline{b}$	$\frac{1}{2}$	$+\frac{1}{3}e$	$-\frac{1}{3}$	0	0	-1	0
Top	$\overline{t}$	$\frac{1}{2}$	$-\frac{2}{3}e$	$-\frac{1}{3}$	0	0	0	-1

The Original Quark Model

In 1963 Gell-Mann and George Zweig independently proposed that hadrons have a more elemental substructure. According to their model, all hadrons are composite systems of two or three fundamental constituents called **quarks**. Gell-Mann borrowed the word *quark* from the passage "Three quarks for Muster Mark" in James Joyce's *Finnegan's Wake*. In the original quark model, there were three types of quarks designated by the symbols u, d, and s. These were given the arbitrary names *up, down,* and *sideways* (or now more commonly, *strange*).

An unusual property of quarks is that they have fractional electronic charges. The u, d, and s quarks have charges of $+2e/3$, $-e/3$, and $-e/3$, respectively. Each quark has a baryon number of $1/3$ and a spin of $1/2$. (The spin $1/2$ means that all quarks are *fermions*, defined in Section 43.5 as any particle having half-integral spin.) The u and d quarks have strangeness 0, while the s quark has strangeness -1. Other properties of quarks and antiquarks are given in Table 47.3. Associated with each quark is an antiquark of opposite charge, baryon number, and strangeness.

The composition of all hadrons known when Gell-Mann and Zweig presented their models could be completely specified by three simple rules:

- Mesons consist of one quark and one antiquark, giving them a baryon number of 0 as required.
- Baryons consist of three quarks.
- Antibaryons consist of three antiquarks.

Table 47.4 lists the quark compositions of several mesons and baryons. Note that just two of the quarks, u and d, are contained in all hadrons encountered in ordinary matter (protons and neutrons). The third quark, s, is needed only to

TABLE 47.4	**Quark Composition of Several Hadrons**
Particle	Quark Composition
Mesons	
π^+	$u\overline{d}$
π^-	$\overline{u}d$
K^+	$u\overline{s}$
K^-	$\overline{u}s$
K^0	$d\overline{s}$
Baryons	
p	uud
n	udd
Λ^0	uds
Σ^+	uus
Σ^0	uds
Σ^-	dds
Ξ^0	uss
Ξ^-	dss
Ω^-	sss

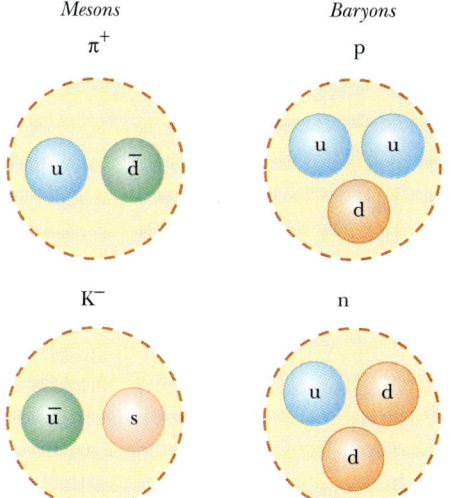

Mesons   Baryons
π⁺   p

K⁻   n

FIGURE 47.6 Quark compositions of two mesons and two baryons.

construct strange particles with a strangeness number of either +1 or −1. Figure 47.6 is a pictorial representation of the quark composition of several particles.

Charm and Other Recent Developments

Although the original quark model was highly successful in classifying particles into families, there were some discrepancies between predictions of the model and certain experimental decay rates. Consequently, a fourth quark was proposed by several physicists in 1967. They argued that if there are four leptons (as was thought at the time), then there should also be four quarks because of an underlying symmetry in nature. The fourth quark, designated by c, was given a property called **charm**. A *charmed* quark would have charge $+2e/3$, but its charm would distinguish it from the other three quarks. The new quark would have a charm of $C = +1$, its antiquark would have a charm of $C = −1$, and all other quarks would have $C = 0$, as indicated in Table 47.3. Charm, like strangeness, would be conserved in strong and electromagnetic interactions, but not in weak interactions.

In 1974, a new heavy meson called the J/Ψ particle (or simply Ψ) was discovered independently by two groups: one led by Burton Richter at the Stanford Linear Accelerator (SLAC) and the other led by Samuel Ting at the Brookhaven National Laboratory. Richter and Ting were awarded the Nobel prize in 1976 for this work. The J/Ψ particle did not fit into the three-quark model but had the properties of a combination of a charmed quark and its antiquark ($c\bar{c}$). It was much more massive than the other known mesons (~ 3100 MeV$/c^2$), and its lifetime was much longer than those that decay via the strong force. Soon, related charmed mesons were discovered corresponding to such quark combinations as $\bar{c}d$ and $c\bar{d}$, all of which have large masses and long lifetimes. In 1975 researchers at Stanford University reported strong evidence for the tau (τ) lepton, mass 1784 MeV$/c^2$. Such discoveries led to more elaborate quark models, and the proposal of two new quarks, *top* (t) and *bottom* (b). (Some physicists prefer *truth* and *beauty*.) To distinguish these quarks from the old ones, quantum numbers called *topness* and *bottomness* were assigned to these new particles and are included in Table 47.3. In

1977, researchers at the Fermi National Laboratory, under the direction of Leon Lederman, reported the discovery of a very massive new meson, Υ, whose composition is considered to be $b\bar{b}$. In March of 1995, researchers at Fermilab announced the discovery of the top quark (supposedly the last of the quarks to be found) having mass 173 GeV.

You are probably wondering whether or not such discoveries will ever end. How many "building blocks" of matter really exist? At the present, physicists believe that the fundamental particles in nature are six quarks and six leptons (together with their antiparticles). Table 47.5 lists some of the properties of these particles.

Despite many extensive experimental efforts, no isolated quark has ever been observed. Physicists now believe that quarks are permanently confined inside ordinary particles because of an exceptionally strong force that prevents them from escaping. This force, called the "color" force (discussed in Section 47.9), increases with separation distance (similar to the force of a spring). The great strength of the force between quarks has been described by one author as follows:[1]

> Quarks are slaves of their own color charge, . . . bound like prisoners of a chain gang . . . . Any locksmith can break the chain between two prisoners, but no locksmith is expert enough to break the gluon chains between quarks. Quarks remain slaves forever.

TABLE 47.5   The Fundamental Particles and Some of Their Properties

Particle	Rest Energy	Charge
Quarks		
u	360 MeV	$+\frac{2}{3}e$
d	360 MeV	$-\frac{1}{3}e$
c	1500 MeV	$+\frac{2}{3}e$
s	540 MeV	$-\frac{1}{3}e$
t	173 GeV	$+\frac{2}{3}e$
b	5 GeV	$-\frac{1}{3}e$
Leptons		
e^-	511 keV	$-e$
μ^-	107 MeV	$-e$
τ^-	1784 MeV	$-e$
ν_e	<30 eV	0
ν_μ	<0.5 MeV	0
ν_τ	<250 MeV	0

[1] Harald Fritzsch, *Quarks, The Stuff of Matter,* London, Allen Lane, 1983.

47.9   THE STANDARD MODEL

Shortly after the concept of quarks was proposed, scientists recognized that certain particles had quark compositions that were in violation of the Pauli exclusion principle. Because all quarks are fermions with spins of 1/2, they are expected to follow the exclusion principle. One example of a particle that violates the exclusion principle is the Ω^- (sss) baryon that contains three s quarks having parallel spins, giving it a total spin of 3/2. Other examples of baryons that have identical quarks with parallel spins are the Δ^{++}(uuu) and the Δ^-(ddd). To resolve this problem, it was suggested that quarks possess a new property called **color**. This property is similar in many respects to electric charge except that it occurs in three varieties (of color) called red, green, and blue. Of course, the antiquarks have the colors antired, antigreen, and antiblue. To satisfy the exclusion principle, all three quarks in a baryon must have different colors. A meson consists of a quark of one color and an antiquark of the corresponding anticolor. The result is that baryons and mesons are always colorless (or white). Furthermore, the new property of color increases the number of quarks by a factor of three.

Although the concept of color in the quark model was originally conceived to satisfy the exclusion principle, it also provided a better theory for explaining certain experimental results. For example, the modified theory correctly predicts the lifetime of the π^0 meson. The theory of how quarks interact with each other is called **quantum chromodynamics,** or QCD, to parallel quantum electrodynamics (the theory of interaction between electric charges). In QCD, the quark is said to carry a *color charge,* in analogy to electric charge. The strong force between quarks is often called the *color force.*

As stated earlier, the strong interaction between hadrons is mediated by massless particles called **gluons** (analogous to photons for the electromagnetic force). According to the theory, there are eight gluons, six of which have color charge. Because of their color charge, quarks can attract each other and form composite particles. When a quark emits or absorbs a gluon, its color changes. For example, a blue quark that emits a gluon may become a red quark, and the red quark that absorbs this gluon becomes a blue quark. The color force between quarks is analogous to the electric force between charges: Like colors repel and opposite colors attract. Therefore, two red quarks repel each other, but a red quark will be attracted to an antired quark. The attraction between quarks of opposite color to form a meson ($q\bar{q}$) is indicated in Figure 47.7a. Differently colored quarks also attract each other, but with less intensity than opposite colors of quark and antiquark. For example, a cluster of red, blue, and green quarks all attract each other to form baryons as indicated in Figure 47.7b. Thus, all baryons contain three quarks, each of which has a different color.

Recall that the weak force is believed to be mediated by the W^+, W^-, and Z^0 bosons (spin 1 particles). These particles are said to have *weak charge* just as a quark has color charge. Thus, each elementary particle can have mass, electric charge, color charge, and weak charge. Of course, one or more of these could be zero. Scientists now believe that the truly elementary particles are leptons and quarks, and the force mediators are the gluon, the photon, $W^\pm$, Z^0, and the graviton. (Note that quarks and leptons have spin 1/2 and, hence, are fermions, while the force mediators have spin 1 or higher and are bosons.)

In 1979, Sheldon Glashow, Abdus Salam, and Steven Weinberg won a Nobel prize for developing a theory that unified the electromagnetic and weak interactions. This **electroweak theory** postulates that the weak and electromagnetic inter-

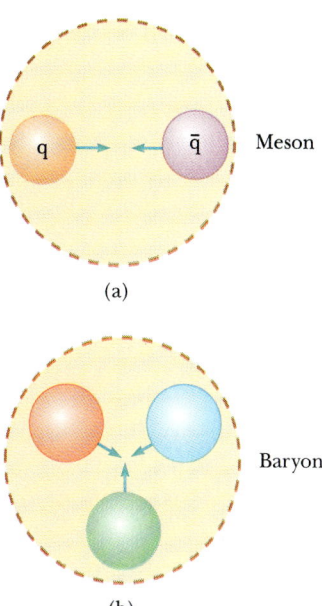

FIGURE 47.7   (a) A red quark is attracted to an antired quark. This forms a meson whose quark structure is ($q\bar{q}$). (b) Three different colored quarks attract each other to form a baryon.

actions have the same strength at very high particle energies. Thus, the two interactions are viewed as two different manifestations of a single unifying electroweak interaction. The photon and the three massive bosons ($W^{\pm}$ and Z^0) play a key role in the electroweak theory. The theory makes many concrete predictions, but perhaps the most spectacular is the prediction of the masses of the W and Z particles at about 82 GeV/c^2 and 93 GeV/c^2, respectively. The 1984 Nobel prize was awarded to Carlo Rubbia and Simon van der Meer for their work leading to the discovery of these particles at just these energies at the CERN Laboratory in Geneva, Switzerland.

The combination of the electroweak theory and QCD for the strong interaction forms what is referred to in high energy physics as the *Standard Model.* Although the details of the Standard Model are complex, its essential ingredients can be summarized with the help of Figure 47.8. The strong force, mediated by gluons, holds quarks together to form composite particles such as protons, neutrons, and mesons. Leptons participate only in the electromagnetic and weak interactions.

However, the Standard Model does not answer all questions. A major question is why the photon has no mass while the W and Z bosons do. Because of this mass difference, the electromagnetic and weak forces are quite distinct at low energies, but become similar in nature at very high energies. This behavior as one goes from low to high energies, called *symmetry breaking,* leaves open the question of the origin of particle masses. To resolve this problem, a hypothetical particle called the *Higgs boson,* which provides a mechanism for breaking the electroweak symmetry, has been proposed. The Standard Model, including the Higgs mechanism, provides a logically consistent explanation of the massive nature of the W and Z bosons. Unfortunately, the Higgs boson has not yet been found, but physicists know that its mass should be less than 1 TeV (10^{12} eV). In order to determine whether the Higgs boson exists, two quarks of at least 1 TeV of energy must collide, but calculations show that this requires injecting 40 TeV of energy within the volume of a proton.

Scientists are convinced that because of the limited energy available in conventional accelerators using fixed targets, it is necessary to build colliding-beam accelerators called **colliders.** The concept of colliders is straightforward. Particles with equal masses and kinetic energies, traveling in opposite directions in an accelera-

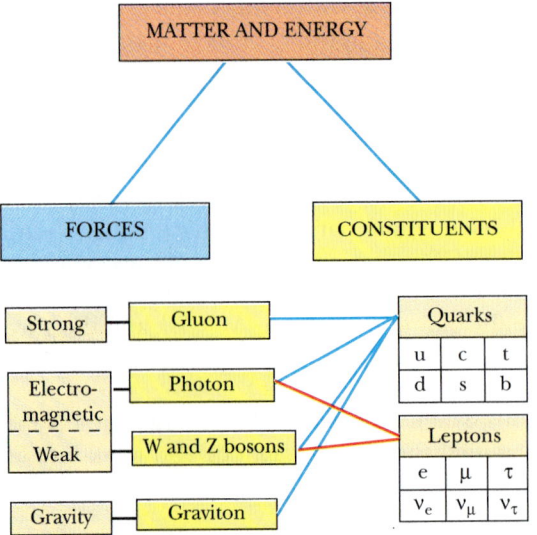

FIGURE 47.8   The Standard Model of particle physics.

tor ring, collide head-on to produce the required reaction and the formation of new particles. Because the total momentum of the interacting particles is zero, all of their kinetic energy is available for the reaction. The Large Electron-Positron collider (LEP) at CERN near Geneva, Switzerland, and the Stanford Linear Collider in California collide both electrons and protons. The Super Proton Synchrotron at CERN accelerates protons and antiprotons to energies of 270 GeV, while the world's highest-energy proton accelerator, the Tevatron, at the Fermi National Laboratory in Illinois produces protons at almost 1000 GeV (or 1 TeV). The Superconducting Super Collider (SSC), which was being built in Texas, was an accelerator designed to produce 20-TeV protons in a ring 52 mi in circumference. After much debate in Congress, and an investment of almost 2 billion dollars, the SSC project was canceled by the U.S. Department of Energy in October, 1993. CERN recently approved the development of the Large Hadron Collider (LHC), a proton-proton collider that will provide a center-of-mass energy of 14 TeV and allow an exploration of Higgs-boson physics. The accelerator will be constructed in the same 27-km circumference tunnel as CERN's Large Electron-Positron collider, and many countries are expected to participate in the project. Figure 47.9 shows the evolution of the stages of matter that scientists have been able to investigate with various types of microscopes.

A technician works on one of the particle detectors at CERN, the European center for particle physics near Geneva, Switzerland. Electrons and positrons accelerated to an energy of 50 GeV collide in a circular tunnel 2 km in circumference, located 100 m underground. *(David Parker/Science Photo Library, Photo Researchers, Inc.)*

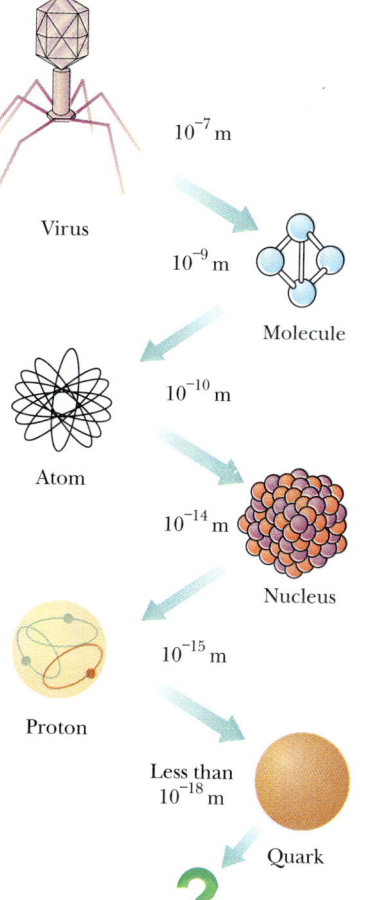

FIGURE 47.9 Looking at matter with various microscopes reveals structures ranging in size from the smallest living thing, a virus, down to a quark, which has not yet been observed as an isolated particle.

Virus

10^{-7} m

10^{-9} m

Molecule

Atom

10^{-10} m

10^{-14} m

Nucleus

Proton

10^{-15} m

Less than 10^{-18} m

Quark

?

eorge Gamow and two of his students, Ralph Alpher and Robert Herman, were the first to take the first half hour of the Universe seriously. In a mostly overlooked paper published in 1948, they made truly remarkable cosmological predictions. They correctly calculated the abundances of hydrogen and helium after the first half hour (75% H and 25% He) and predicted that radiation from the Big Bang should still be present and have an apparent temper-

George Gamow

| 1 9 0 4 – 1 9 6 8 |

ature of about 5 K. In Gamow's own words, the Universe's supply of hydrogen and helium was created very quickly, "in less time than it takes to cook a dish of duck and roast potatoes." The comment is characteristic of this interesting physicist, who is known as much for his explanation of alpha decay and theories of cosmology as for his delightful popular books, his cartoons, and his wonderful sense of humor.

47.10 THE COSMIC CONNECTION

As we have seen, the world around us is dominated by protons, electrons, neutrons, and neutrinos. Some of the other more exotic particles can be seen in cosmic rays. However, most of the new particles are produced using large, expensive machines that accelerate protons and electrons to energies in the GeV and TeV range. These energies are enormous when compared with the thermal energy in today's Universe. For example, the thermal energy $k_B T$ at the center of the Sun is only about 1 keV, but the temperature of the early Universe was high enough to reach energies higher than 1 TeV.

In this section we shall describe one of the most fascinating theories in all of science—the Big Bang theory of the creation of the Universe—and the experimental evidence that supports it. This theory of cosmology states that the Universe had a beginning and, further, that the beginning was so cataclysmic that it is impossible to look back beyond it. According to this theory, the Universe erupted from a point-like singularity about 15 to 20 billion years ago. The first few minutes after the Big Bang saw such extremes of energy that it is believed that all four interactions of physics were unified and that all matter melted down into an undifferentiated "quark soup."

The evolution of the four fundamental forces from the Big Bang to the present is shown in Figure 47.10. During the first 10^{-43} s (the ultra-hot epoch during which T $\approx 10^{32}$ K), it is presumed that the strong, electroweak, and gravitational forces were joined to form a completely unified force. In the first 10^{-32} s following the Big Bang (the hot epoch, T $\approx 10^{29}$ K), gravity broke free of this unification while the strong and electroweak forces remained as one, described by a grand unification theory. This was a period when particle energies were so great ($> 10^{16}$ GeV) that very massive particles as well as quarks, leptons, and their antiparticles existed. Then, the Universe rapidly expanded and cooled during the warm epoch when the temperatures ranged from 10^{29} to 10^{15} K, the strong and electroweak forces parted company, and the grand unification scheme was broken. As the Universe continued to cool, the electroweak force split into the weak force and the electromagnetic force about 10^{-10} s after the Big Bang.

Until about 700 000 years after the Big Bang, the Universe was dominated by radiation: Ions absorbed and re-emitted photons, thereby ensuring thermal equilibrium of radiation and matter. Energetic radiation also prevented matter from

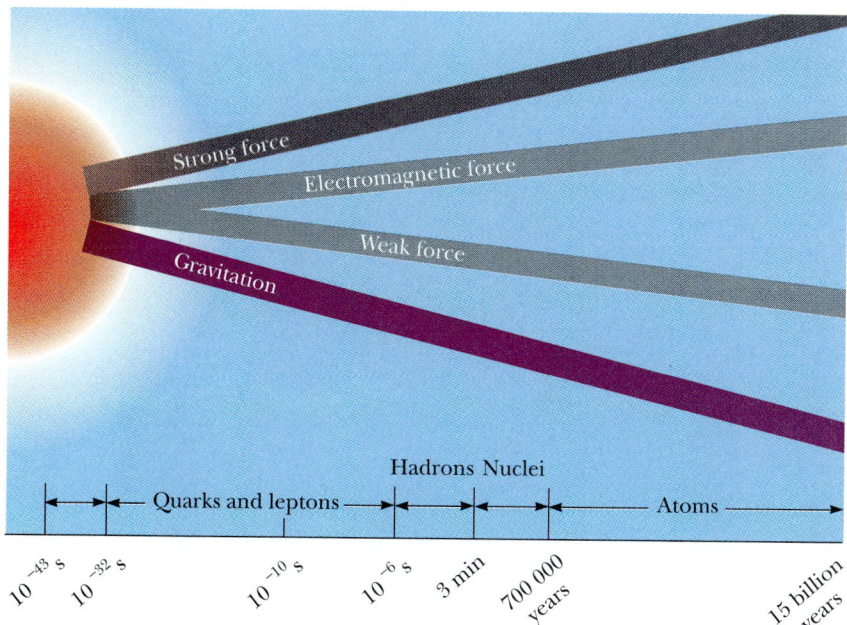

FIGURE 47.10   A brief history of the Universe from the Big Bang to the present. The four forces became distinguishable during the first microsecond. Following this, all the quarks combined to form particles that interact via the strong force. However, the leptons remained separate and exist as individually observable particles to this day.

forming clumps or even single hydrogen atoms. When the Universe was about 700 000 years old, it had expanded and cooled to about 3000 K, and protons could bind to electrons to form neutral hydrogen atoms. Because neutral atoms do not appreciably scatter photons, the Universe suddenly became transparent to photons. Radiation no longer dominated the Universe and clumps of neutral matter steadily grew—first atoms, followed by molecules, gas clouds, stars, and finally galaxies.

Observation of Radiation from the Primordial Fireball

In 1965, Arno A. Penzias and Robert W. Wilson of Bell Labs were testing a sensitive microwave receiver and made an amazing discovery. A pesky signal producing a faint background hiss was interfering with their satellite communications experiments. In spite of their valiant efforts, the signal remained. Ultimately, it became clear that they were perceiving microwave background radiation (at a wavelength of 7.35 cm) representing the leftover glow from the Big Bang.

The microwave horn that served as their receiving antenna is shown in Figure 47.11. The intensity of the detected signal remained unchanged as the antenna was pointed in different directions. The fact that the radiation had equal strengths in all directions suggested that the entire Universe was the source of this radiation. Eviction of a flock of pigeons from the 20-foot horn and cooling the microwave detector both failed to remove the "spurious" signal. Through a casual conversation, Penzias and Wilson discovered that a group at Princeton had predicted the residual radiation from the Big Bang and were planning an experiment seeking to confirm the theory. The excitement in the scientific community was high when Penzias and Wilson announced that they had already observed an excess microwave background compatible with a 3-K blackbody source.

FIGURE 47.11   Robert W. Wilson *(left)* and Arno A. Penzias *(right)* with Bell Telephone Laboratories horn-reflector antenna.   *(AT&T Bell Laboratories)*

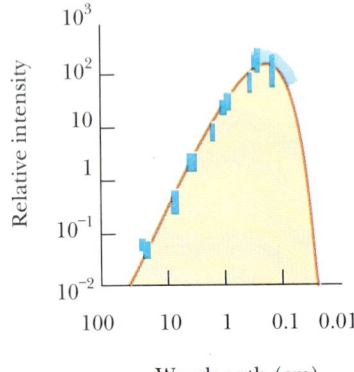

FIGURE 47.12   Radiation spectrum of the Big Bang. The blue areas are experimental results. The red line is the spectrum calculated for a black body at 2.9 K.

Because the measurements of Penzias and Wilson were taken at a single wavelength, they did not completely confirm the radiation as 3-K blackbody radiation. Subsequent experiments by other groups added intensity data at different wavelengths, as shown in Figure 47.12. The results confirm that the radiation is that of a black body at 2.9 K. This figure is, perhaps, the most clearcut evidence for the Big

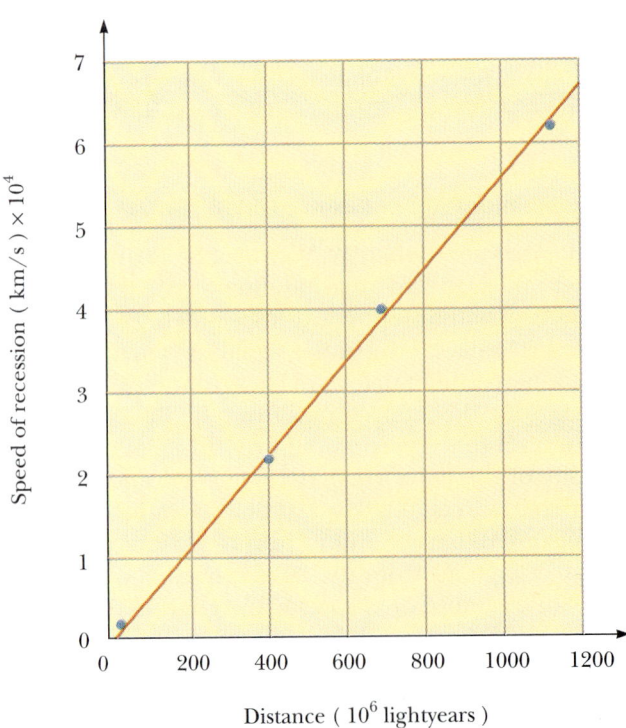

FIGURE 47.13   Hubble's law: A plot of speed of recession versus distance for four galaxies.

Bang theory. The 1978 Nobel prize for physics was awarded to Penzias and Wilson for their most important discovery.

Other Evidence for the Expanding Universe

Most of the key discoveries supporting the theory of an expanding Universe, and indirectly the Big Bang theory of cosmology, were made in the 20th century. Vesto Melvin Slipher, an American astronomer, reported that most nebulae are receding from the Earth at speeds up to several million miles per hour. Slipher was one of the first to use the methods of Doppler shifts in spectral lines to measure velocities.

In the late 1920s, Edwin P. Hubble made the bold assertion that the whole Universe is expanding. From 1928 to 1936, he and Milton Humason toiled at Mount Wilson to prove this assertion until they reached the limits of the 100-inch telescope. The results of this work and its continuation on a 200-inch telescope in the 1940s showed that the speeds of galaxies increase in direct proportion to their distance R from us (Fig. 47.13). This linear relationship, known as **Hubble's law**, may be written

$$v = HR \tag{47.4}$$

where H, called the **Hubble parameter**, has the approximate value

$$H = 17 \times 10^{-3} \text{ m/(s} \cdot \text{lightyear)}$$

EXAMPLE 47.5   Recession of a Quasar

A quasar is a star-like object that is very distant from the Earth. Its speed can be measured from Doppler shift measurements in the light it emits. A certain quasar recedes from the Earth at a speed of $0.55\,c$. How far away is it?

Solution   We can find the distance from Hubble's law:

$$R = \frac{v}{H} = \frac{(0.55)\,(3.00 \times 10^8 \text{ m/s})}{17 \times 10^{-3} \text{ m/(s} \cdot \text{lightyear)}}$$

$$= 9.7 \times 10^9 \text{ lightyears}$$

Exercise   Assuming that the quasar has moved with the speed $0.55\,c$ ever since the Big Bang, estimate the age of the Universe.

Answer $t = R/v = 1/H \approx 18$ billion years, which is in good agreement with other calculations.

Will the Universe Expand Forever?

In the 1950s and 1960s Allan R. Sandage used the 200-inch telescope at Mount Palomar to measure the speeds of galaxies at distances of up to 6 billion lightyears. These measurements showed that these very distant galaxies were moving about 10 000 km/s faster than the Hubble law predicted. According to this result, the Universe must have been expanding more rapidly 1 billion years ago and, consequently, the expansion is slowing[2] (Fig. 47.14); today, astronomers and physicists are trying to determine the rate of slowing. If the average mass density of the Universe is less than some critical density ($\rho_c \approx 3$ atoms/m^3), the galaxies will slow in their outward rush but still escape to infinity. If the average density exceeds the critical value, the expansion will eventually stop and contraction will begin, possibly leading to a superdense state and another expansion or an oscillating Universe.

---

[2] The data at large distances have large observational uncertainties and may be systematically in error from selection effects such as abnormal brightness in the most distant visible clusters.

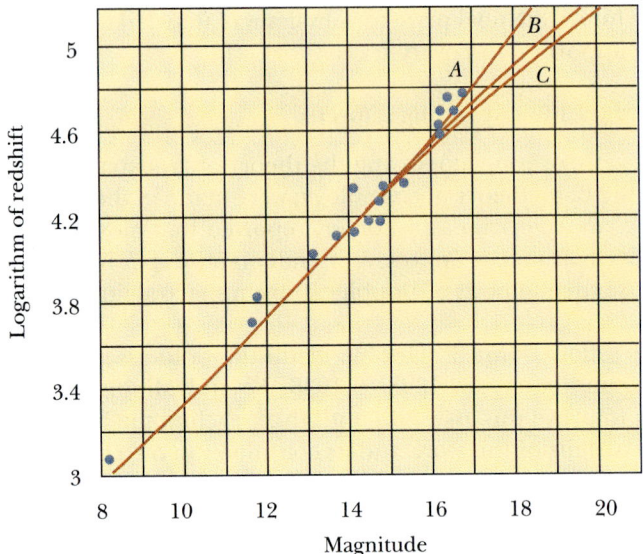

FIGURE 47.14   Red shift, or speed of recession, versus apparent magnitude of 18 faint clusters. Curve *A* is the trend suggested by the six faintest clusters of galaxies. Curve *C* corresponds to a Universe having a constant rate of expansion. If the data fall between *B* and *C*, the expansion slows but never stops. If the data fall to the left of *B*, expansion stops and contraction occurs.

EXAMPLE 47.6   The Critical Density of the Universe

Estimate the critical mass density of the Universe, ρ_c, using energy considerations.

Reasoning and Solution   Figure 47.15 shows a large section of the Universe with radius R, containing galaxies with a total mass M. A galaxy of mass m and speed v at R will just escape to infinity with zero speed if the sum of its kinetic energy and gravitational potential energy is zero. Thus,

$$E_{\text{total}} = 0 = K + U = \tfrac{1}{2}mv^2 - \frac{GmM}{R}$$

$$\tfrac{1}{2}mv^2 = \frac{Gm\tfrac{4}{3}\pi R^3 \rho_c}{R}$$

$$(1) \qquad v^2 = \frac{8\pi G}{3}R^2 \rho_c$$

Because the galaxy of mass m obeys the Hubble law, $v = HR$, (1) becomes

$$H^2 = \frac{8\pi G}{3}\rho_c$$

$$(2) \qquad \rho_c = \frac{3H^2}{8\pi G}$$

Using $H = 17 \times 10^{-3}$ m/(s·lightyear), where 1 lightyear = 9.46×10^{12} km, and $G = 6.67 \times 10^{-8}$ cm³/g·s² yields $\rho_c = 6 \times 10^{-30}$ g/cm³. As the mass of a hydrogen atom is 1.67×10^{-24} g, ρ_c corresponds to 3×10^{-6} hydrogen atoms per cm³ or 3 atoms per m³.

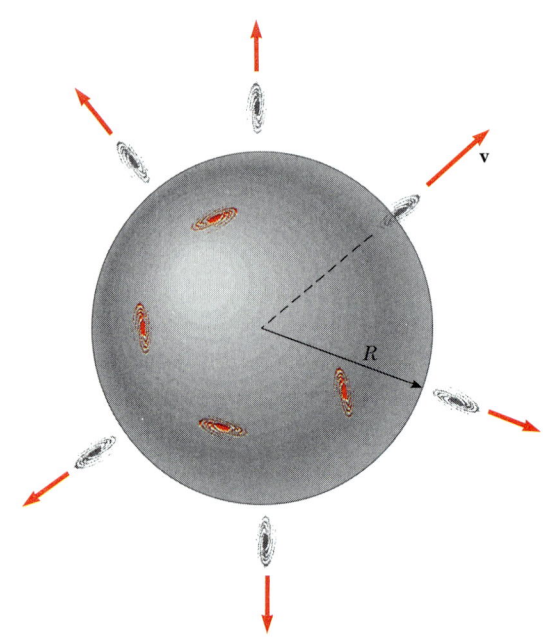

FIGURE 47.15   A galaxy escaping from a large cluster contained within radius R. Only the mass within R slows the mass m.

Missing Mass in the Universe(?)

The visible matter in galaxies averages out to 5×10^{-33} g/cm³. The radiation in the Universe has a mass equivalent of approximately 2% of the visible matter. Nonluminous matter (such as interstellar gas or black holes) may be estimated from the speeds of galaxies orbiting each other in the cluster. The higher the galaxy speeds, the more mass in the cluster. Results from measurements on the Coma cluster of galaxies, surprisingly, indicate that the amount of invisible matter is 20 to 30 times the amount present in stars and luminous gas clouds. Yet even this large invisible component, if applied to the Universe as a whole, leaves the observed mass density a factor of 10 less than ρ_c. This so-called *missing mass* (or *dark matter*) has been the subject of intense theoretical and experimental work, with exotic particles such as axions, photinos, and superstring particles suggested as candidates for the missing mass. More mundane proposals have been that the missing mass is present in certain galaxies as neutrinos. In fact, neutrinos are so abundant that a tiny neutrino rest mass on the order of 20 eV would furnish the missing mass and ''close'' the Universe.

Although we are a bit more sure about the beginning of the Universe, we are uncertain about its end. Will the Universe expand forever? Will it collapse and repeat its expansion in an endless series of oscillations? Results and answers to these questions remain inconclusive and the exciting controversy continues.

47.11 PROBLEMS AND PERSPECTIVES

While particle physicists have been exploring the realm of the very small, cosmologists have been exploring cosmic history back to the first microsecond of the Big Bang. Observation of events that occur when two particles collide in an accelerator is essential in reconstructing the early moments in cosmic history. Perhaps the key to understanding the early Universe is first to understand the world of elementary particles. Cosmologists and physicists now find they have many common goals and are joining hands to attempt to understand the physical world at its most fundamental level.

Our understanding of physics at short distances is far from complete. Particle physics is faced with many questions. Why is there so little antimatter in the Universe? Do neutrinos have a small rest mass, and if so, how do they contribute to the ''dark matter'' of the Universe? Is it possible to unify the strong and electroweak theories in a logical and consistent manner? Why do quarks and leptons form three similar but distinct families? Are muons the same as electrons (apart from their difference in mass), or do they have other subtle differences that have not been detected? Why are some particles charged and others neutral? Why do quarks carry a fractional charge? What determines the masses of the fundamental constituents? Can isolated quarks exist? The questions go on and on. Because of the rapid advances and new discoveries in the field of particle physics, by the time you read this book some of these questions will likely be resolved while others may emerge.

An important and obvious question that remains is whether leptons and quarks have a substructure. If they have a substructure, we could envision an infinite number of deeper structure levels. However, if leptons and quarks are indeed the ultimate constituents of matter, as physicists today tend to believe, we should be able to construct a final theory of the structure of matter as Einstein himself dreamed. In the view of many physicists, the end of the road is in sight, but how long it will take to reach that goal is anyone's guess.

SUMMARY

There are four fundamental forces in nature: strong (hadronic), electromagnetic, weak, and gravitational. The strong force is the force between nucleons that keeps the nucleus together. The weak force is responsible for beta decay. The electromagnetic and weak forces are now considered to be manifestations of a single force called the electroweak force. Every fundamental interaction is said to be mediated by the exchange of field particles. The electromagnetic interaction is mediated by the photon; the weak interaction is mediated by the $W^{\pm}$ and Z^0 bosons; the gravitational interaction is mediated by gravitons; the strong interaction is mediated by gluons.

An antiparticle and a particle have the same mass, but opposite charge, and other properties may have opposite values such as lepton number and baryon number. It is possible to produce particle-antiparticle pairs in nuclear reactions if the available energy is greater than $2\,mc^2$, where m is the rest mass of the particle (or antiparticle).

Particles other than photons are classified as hadrons or leptons. **Hadrons** interact through the strong force. They have size and structure and are not elementary particles. There are two types of hadrons, *baryons* and *mesons*. Mesons have baryon number zero and have either zero or integral spin. Baryons, which generally are the most massive particles, have nonzero baryon number and a spin of $1/2$ or $3/2$. The neutron and proton are examples of baryons.

Leptons have no structure or size and are considered truly elementary. They interact only through the weak and electromagnetic forces. There are six leptons: the electron e^-, the muon μ^-, the tau τ^-, and their neutrinos, ν_e, ν_μ, and ν_τ.

In all reactions and decays, quantities such as energy, linear momentum, angular momentum, electric charge, baryon number, and lepton number are strictly conserved. Certain particles have properties called **strangeness** and **charm**. These unusual properties are conserved only in those reactions and decays that occur via the strong force.

Theories in elementary particle physics have postulated that all hadrons are composed of smaller units known as **quarks**. Quarks have fractional electric charge, baryon numbers of $1/3$, and come in six "flavors": up (u), down (d), strange (s), charmed (c), top (t), and bottom (b). Each baryon contains three quarks, and each meson contains one quark and one antiquark.

According to the theory of **quantum chromodynamics,** quarks have a property called **color,** and the strong force between quarks is referred to as the **color force.**

The background microwave radiation discovered by Penzias and Wilson strongly suggests that the Universe started with a "Big Bang" about 15 billion years ago. The background radiation is equivalent to that of a black body at 3 K.

Various astronomical measurements strongly suggest that the Universe is expanding. According to **Hubble's law,** distant galaxies are receding from the Earth at a speed $v = HR$, where R is the distance from Earth to the galaxy and H is **Hubble's parameter,** $H \approx 17 \times 10^{-3}$ m/(s·lightyear).

SUGGESTED READINGS

Particle Physics

J. Bahcall, "The Solar Neutrino Problem," *Sci. American,* May 1990.

H. Brueker et al., "Tracking and Imaging Elementary Particles," *Sci. American,* August 1991.

Frank Close, *The Cosmic Onion: Quarks and the Nature of the*

Universe, The American Institute of Physics, 1986. A timely monograph on particle physics, including lively discussions of the Big Bang theory.

Harald Fritzsch, *Quarks, The Stuff of Matter,* London, Allen and Lane, 1983. An excellent introductory overview of elementary particle physics.

George Gamow, "Gravity and Antimatter," *Sci. American,* March 1961.

T. Goldman et al, "Gravity and Antimatter," *Sci. American,* March 1989.

H. Harari, "The Structure of Quarks and Leptons," *Sci. American,* April, 1983.

Leon M. Lederman, "The Value of Fundamental Science," *Sci. American,* November, 1984.

N. B. Mistry, R. A. Poling, and E. H. Thorndike, "Particles with Naked Beauty," *Sci. American,* July, 1983.

Chris Quigg, "Elementary Particles and Forces," *Sci. American,* April, 1985.

M. Riordan, "The Discovery of Quarks," *Science,* 29 May 1992.

M. J. G. Veltman, "The Higgs Boson," *Sci. American,* November 1986.

Steven Weinberg, *The Discovery of Elementary Particles,* New York, Scientific American Library, W. H. Freeman and Company, 1983. This book emphasizes the important discoveries, experiments, and intellectual exercises that reshaped physics in the 20th century.

Cosmology and the Big Bang

John D. Barrow and Joseph Silk, "The Structure of the Early Universe," *Sci. American,* April, 1980.

George Gamow, "The Evolutionary Universe," *Sci. American,* September, 1956.

David Layzer, *Constructing the Universe,* Scientific American Library, New York, W. H. Freeman and Co., 1984, Chapters 7 and 8.

David L. Meier and Rashid A. Sunyaev, "Primeval Galaxies," *Sci. American,* November, 1979.

Richard A. Muller, "The Cosmic Background Radiation and the New Aether Drift," *Sci. American,* May, 1978.

Carl Sagan and Frank Drake, "The Search for Extraterrestrial Intelligence," *Sci. American,* May, 1975.

Allan R. Sandage, "The Red-Shift," *Sci. American,* September, 1956.

QUESTIONS

1. Name the four fundamental interactions and the field particles that mediate each.
2. Describe the quark model of hadrons, including the properties of quarks.
3. What are the differences between hadrons and leptons?
4. Describe the properties of baryons and mesons and the important differences between them.
5. Particles known as resonances have very short lifetimes, of the order of 10^{-23} s. From this information, would you guess they are hadrons or leptons? Explain.
6. Kaons all decay into final states that contain no protons or neutrons. What is the baryon number of kaons?
7. The Ξ^0 particle decays by the weak interaction according to the decay mode $\Xi^0 \rightarrow \Lambda^0 + \pi^0$. Would you expect this decay to be fast or slow? Explain.
8. Identify the particle decays listed in Table 47.2 that occur by the weak interaction. Justify your answers.
9. Identify the particle decays listed in Table 47.2 that occur by the electromagnetic interaction. Justify your answers.
10. Two protons in a nucleus interact via the strong interaction. Are they also subject to the weak interaction?
11. Discuss the following conservation laws: energy, linear momentum, angular momentum, electric charge, baryon number, lepton number, and strangeness. Are all of these laws based on fundamental properties of nature? Explain.
12. An antibaryon interacts with a meson. Can a baryon be produced in such an interaction? Explain.
13. Describe the essential features of the Standard Model of particle physics.
14. How many quarks are there in (a) a baryon, (b) an antibaryon, (c) a meson, (d) an antimeson? How do you account for the fact that baryons have half-integral spins while mesons have spins of 0 or 1? (*Hint:* Quarks have spin 1/2.)
15. In the theory of quantum chromodynamics, quarks come in three colors. How would you justify the statement that "all baryons and mesons are colorless"?
16. Which baryon did Murray Gell-Mann predict in 1961? What is the quark composition of this particle?
17. What is the quark composition of the Ξ^- particle? (See Table 47.4.)
18. The W and Z bosons were first produced at CERN in 1983 (by having a beam of protons and a beam of antiprotons meet at high energy). Why was this an important discovery?
19. How did Edwin Hubble in 1928 determine that the Universe is expanding?

PROBLEMS

Section 47.2 Positrons and Other Antiparticles

1. Two photons are produced when a proton and anti-proton annihilate each other. What is the minimum frequency and corresponding wavelength of each photon?

2. A photon produces a proton-antiproton pair according to the reaction $\gamma \rightarrow p + \bar{p}$. What is the frequency of the photon? What is its wavelength?

Section 47.3 Mesons and the Beginning of Particle Physics

3. One of the mediators of the weak interaction is the Z^0 boson, mass 96 GeV/c^2. Use this information to find an approximate value for the range of the weak interaction.

4. The "mediators" of the weak force are the W and Z particles. Use the uncertainty principle to estimate the range of the weak force assuming the masses of the W and Z particles are 80 to 90 GeV/c^2.

5. A free neutron beta decays by creating a proton, electron, and antineutrino according to the reaction $n \rightarrow p + e^- + \bar{\nu}$. Assume that a free neutron beta decays by creating a proton and electron according to the reaction

$$n \longrightarrow p + e^-$$

and assume that the neutron is initially at rest in the laboratory. (a) Determine the energy released in this reaction. (b) Determine the speed of the proton and electron after the reaction. (Energy and momentum are conserved in the reaction.) (c) Are any of these particles moving at relativistic speeds? Explain.

6. A neutral pion at rest decays into two photons according to

$$\pi^0 \longrightarrow \gamma + \gamma$$

Find the energy, momentum, and frequency of each photon.

Section 47.4 Classification of Particles

7. Name one possible decay mode (see Table 47.2) for Ω^+, $\bar{K}^0$, $\bar{\Lambda}^0$, and $\bar{n}$.

Section 47.5 Conservation Laws

8. Each of the following reactions is forbidden. Determine a conservation law that is violated for each reaction.
   (a) $p + \bar{p} \rightarrow \mu^+ + e^-$
   (b) $\pi^- + p \rightarrow p + \pi^+$
   (c) $p + p \rightarrow p + \pi^+$

   (d) $p + p \rightarrow p + p + n$
   (e) $\gamma + p \rightarrow n + \pi^0$

9. If baryon number is not conserved, then one possible mechanism by which a proton can decay is

$$p \longrightarrow e^+ + \gamma$$

(a) Show that this reaction violates conservation of baryon number. (b) Assuming that this reaction occurs, and that the proton is initially at rest, determine the energy and momentum of the positron and photon after the reaction. (*Hint:* Recall that energy and momentum must be conserved in the reaction.) (c) Determine the speed of the positron after the reaction.

10. The following reactions or decays involve one or more neutrinos. Supply the missing neutrinos (ν_e, ν_μ, or ν_τ).
    (a) $\pi^- \rightarrow \mu^- + ?$ (d) $? + n \rightarrow p + e^-$
    (b) $K^+ \rightarrow \mu^+ + ?$ (e) $? + n \rightarrow p + \mu^-$
    (c) $? + p \rightarrow n + e^+$ (f) $\mu^- \rightarrow e^- + ? + ?$

11. The Λ^0 is an unstable particle that decays into a proton and a negatively charged pion. Determine the kinetic energies of the proton and pion if the Λ^0 is at rest when it decays. The rest mass of the Λ^0 is 1115.7 MeV/c^2, the rest mass of the π^- is 139.5 MeV/c^2, and the rest mass of the proton is 938.3 MeV/c^2.

12. Determine which of the reactions below can occur. For those that cannot occur, determine the conservation law (or laws) violated:
    (a) $p \rightarrow \pi^+ + \pi^0$
    (b) $p + p \rightarrow p + p + \pi^0$
    (c) $p + p \rightarrow p + \pi^+$
    (d) $\pi^+ \rightarrow \mu^+ + \nu_\mu$
    (e) $n \rightarrow p + e^- + \bar{\nu}_e$
    (f) $\pi^+ \rightarrow \mu^+ + n$

Section 47.6 Strange Particles and Strangeness

13. Determine whether or not strangeness is conserved in the following decays and reactions.
    (a) $\Lambda^0 \rightarrow p + \pi^-$
    (b) $\pi^- + p \rightarrow \Lambda^0 + K^0$
    (c) $\bar{p} + p \rightarrow \bar{\Lambda}^0 + \Lambda^0$
    (d) $\pi^- + p \rightarrow \pi^- + \Sigma^+$
    (e) $\Xi^- \rightarrow \Lambda^0 + \pi^-$
    (f) $\Xi^- \rightarrow p + \pi^-$

14. The neutral ρ meson decays by the strong interaction into two pions: $\rho^0 \rightarrow \pi^+ + \pi^-$, half-life 10^{-23} s. The neutral kaon also decays into two pions: $K^0 \rightarrow \pi^+ +$

□ indicates problems that have full solutions available in the Student Solutions Manual and Study Guide.

π^-, half-life 10^{-10} s. How do you explain the difference in half-lives?

15. For each of the following forbidden decays, determine which conservation law is violated:
    (a) $\mu^- \rightarrow e^- + \gamma$ (d) $p \rightarrow e^+ + \pi^0$
    (b) $n \rightarrow p + e^- + \nu_e$ (e) $\Xi^0 \rightarrow n + \pi^0$
    (c) $\Lambda^0 \rightarrow p + \pi^0$

Section 47.8   Quarks—Finally

16. A Σ^0 particle traveling through matter strikes a proton and a Σ^+ and a gamma ray emerges, as well as a third particle. Use the quark model of each to determine the identity of the third particle.

17. The quark compositions of the K^0 and Λ^0 particles are $d\bar{s}$ and uds, respectively. Show that the charge, baryon number, and strangeness of these particles equal the sums of these numbers for the quark constituents.

18. Neglect binding energies and estimate the masses of the u and d quarks from the masses of the proton and neutron.

19. Analyze each reaction in terms of constituent quarks:
    (a) $\pi^- + p \rightarrow K^0 + \Lambda^0$
    (b) $\pi^+ + p \rightarrow K^+ + \Sigma^+$
    (c) $K^- + p \rightarrow K^+ + K^0 + \Omega^-$
    (d) $p + p \rightarrow K^0 + p + \pi^+ + ?$
    In the last reaction, identify the mystery particle.

Section 47.10   The Cosmic Connection

20. Using Hubble's law (Eq. 47.4), estimate the wavelength of the 590 nm sodium line emitted from galaxies (a) 2×10^6 lightyears away from Earth, (b) 2×10^8 lightyears away, and (c) 2×10^9 lightyears away. *Hint:* Use the relativistic Doppler formula for wavelength λ' of light emitted from a moving source:

$$\lambda' = \lambda \sqrt{\frac{1 + v/c}{1 - v/c}}$$

21. A distant quasar is moving away from Earth at such high speed that the blue 434-nm hydrogen line is observed at 650 nm, in the red portion of the spectrum. (a) How fast is the quasar receding? (See the hint in the preceding problem.) (b) Using Hubble's law, determine the distance from Earth to this quasar.

ADDITIONAL PROBLEMS

22. The strong interaction has a range of approximately 1.4×10^{-15} m. It is thought that an elementary particle is exchanged between the protons and neutrons in the nucleus, leading to an attractive force. (a) Utilize the uncertainty principle $\Delta E \, \Delta t \geq \hbar/2$ to esti-

mate the mass of the elementary particle if it moves at nearly the speed of light. (b) Using Table 47.2, identify the particle.

23. An unstable particle, initially at rest, decays into a proton (rest energy 938.3 MeV) and a negative pion (rest energy 139.5 MeV). A uniform magnetic field of 0.250 T exists with the field perpendicular to the velocities of the created particles. The radius of curvature of each track is found to be 1.33 m. What is the rest mass of the original unstable particle?

24. What are the kinetic energies of the proton and pion resulting from the decay of a Λ^0 at rest:

$$\Lambda^0 \longrightarrow p + \pi^-$$

25. The energy flux of neutrinos from the Sun is estimated to be on the order of 0.4 W/m^2 at Earth's surface. Estimate the fractional mass loss of the Sun over 10^9 years due to the radiation of neutrinos. (The mass of the Sun is 2×10^{30} kg. The Earth-Sun distance is 1.5×10^{11} m.)

26. A Σ^0 particle at rest decays according to

$$\Sigma^0 \longrightarrow \Lambda^0 + \gamma$$

Find the gamma-ray energy.

27. If a K^0 meson at rest decays in 0.90×10^{-10} s, how far will a K^0 meson travel if it is moving at $0.96c$ through a bubble chamber?

28. A π meson at rest decays according to $\pi^- \rightarrow \mu^- + \bar{\nu}_\mu$. What is the energy carried off by the neutrino? (Assume the neutrino moves off with the speed of light.) $m_\pi c^2 = 139.5$ MeV, $m_\mu c^2 = 105.7$ MeV, $m_\nu = 0$.

29. Two protons approach each other with 70.4 MeV of kinetic energy and engage in a reaction in which a

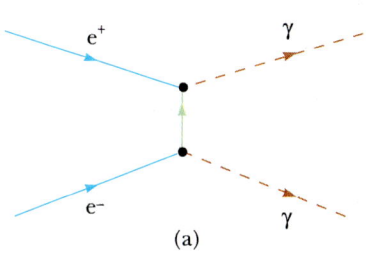

(a)

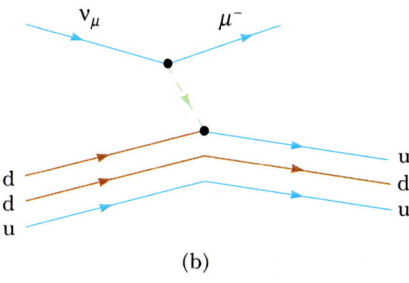

(b)

FIGURE P47.30

proton and positive pion emerge at rest. What third particle, obviously uncharged and therefore difficult to detect, must have been created?

30. What processes are described by the Feynman diagrams in Figure P47.30? What is the exchanged particle in each process?

31. Identify the mediators for the two interactions described in the Feynman diagrams shown in Figure P47.31.

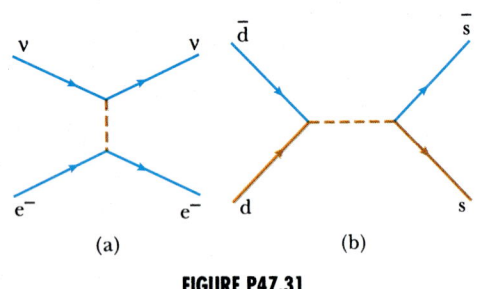

FIGURE P47.31

"Particles, particles, particles."

The Meaning of Success

To earn the respect of intelligent people and to win the affection of children;

To appreciate the beauty in nature and all that surrounds us;

To seek out and nurture the best in others;

To give the gift of yourself to others without the slightest thought of return, for it is in giving that we receive;

To have accomplished a task, whether it be saving a lost soul, healing a sick child, writing a book, or risking your life for a friend;

To have celebrated and laughed with great joy and enthusiasm and sung with exaltation;

To have hope even in times of despair, for as long as you have hope, you have life;

To love and be loved;

To be understood and to understand;

To know that even one life has breathed easier because you have lived;

This is the meaning of success.

RALPH WALDO EMERSON
and modified by Ray Serway
December 1989

TABLE A.1 Conversion Factors

Length

	m	cm	km	in.	ft	mi
1 meter	1	10^2	10^{-3}	39.37	3.281	6.214×10^{-4}
1 centimeter	10^{-2}	1	10^{-5}	0.3937	3.281×10^{-2}	6.214×10^{-6}
1 kilometer	10^3	10^5	1	3.937×10^4	3.281×10^3	0.6214
1 inch	2.540×10^{-2}	2.540	2.540×10^{-5}	1	8.333×10^{-2}	1.578×10^{-5}
1 foot	0.3048	30.48	3.048×10^{-4}	12	1	1.894×10^{-4}
1 mile	1609	1.609×10^5	1.609	6.336×10^4	5280	1

Mass

	kg	g	slug	u
1 kilogram	1	10^3	6.852×10^{-2}	6.024×10^{26}
1 gram	10^{-3}	1	6.852×10^{-5}	6.024×10^{23}
1 slug	14.59	1.459×10^4	1	8.789×10^{27}
1 atomic mass unit	1.660×10^{-27}	1.660×10^{-24}	1.137×10^{-28}	1

Time

	s	min	h	day	year
1 second	1	1.667×10^{-2}	2.778×10^{-4}	1.157×10^{-5}	3.169×10^{-8}
1 minute	60	1	1.667×10^{-2}	6.994×10^{-4}	1.901×10^{-6}
1 hour	3600	60	1	4.167×10^{-2}	1.141×10^{-4}
1 day	8.640×10^4	1440	24	1	2.738×10^{-3}
1 year	3.156×10^7	5.259×10^5	8.766×10^3	365.2	1

Speed

	m/s	cm/s	ft/s	mi/h
1 meter/second	1	10^2	3.281	2.237
1 centimeter/second	10^{-2}	1	3.281×10^{-2}	2.237×10^{-2}
1 foot/second	0.3048	30.48	1	0.6818
1 mile/hour	0.4470	44.70	1.467	1

Note: 1 mi/min = 60 mi/h = 88 ft/s.

Force

	N	dyn	lb
1 newton	1	10^5	0.2248
1 dyne	10^{-5}	1	2.248×10^{-6}
1 pound	4.448	4.448×10^5	1

TABLE A.1 *Continued*

Work, Energy, Heat

	J	erg	ft·lb
1 joule	1	10^7	0.7376
1 erg	10^{-7}	1	7.376×10^{-8}
1 ft·lb	1.356	1.356×10^7	1
1 eV	1.602×10^{-19}	1.602×10^{-12}	1.182×10^{-19}
1 cal	4.186	4.186×10^7	3.087
1 Btu	1.055×10^3	1.055×10^{10}	7.779×10^2
1 kWh	3.600×10^6	3.600×10^{13}	2.655×10^6

	eV	cal	Btu	kWh
1 joule	6.242×10^{18}	0.2389	9.481×10^{-4}	2.778×10^{-7}
1 erg	6.242×10^{11}	2.389×10^{-8}	9.481×10^{-11}	2.778×10^{-14}
1 ft·lb	8.464×10^{18}	0.3239	1.285×10^{-3}	3.766×10^{-7}
1 eV	1	3.827×10^{-20}	1.519×10^{-22}	4.450×10^{-26}
1 cal	2.613×10^{19}	1	3.968×10^{-3}	1.163×10^{-6}
1 Btu	6.585×10^{21}	2.520×10^2	1	2.930×10^{-4}
1 kWh	2.247×10^{25}	8.601×10^5	3.413×10^2	1

Pressure

	Pa	dyn/cm^2	atm
1 pascal	1	10	9.869×10^{-6}
1 dyne/centimeter^2	10^{-1}	1	9.869×10^{-7}
1 atmosphere	1.013×10^5	1.013×10^6	1
1 centimeter mercury*	1.333×10^3	1.333×10^4	1.316×10^{-2}
1 pound/inch^2	6.895×10^3	6.895×10^4	6.805×10^{-2}
1 pound/foot^2	47.88	4.788×10^2	4.725×10^{-4}

	cm Hg	lb/in.^{2}	lb/ft^2
1 newton/meter^2	7.501×10^{-4}	1.450×10^{-4}	2.089×10^{-2}
1 dyne/centimeter^2	7.501×10^{-5}	1.450×10^{-5}	2.089×10^{-3}
1 atmosphere	76	14.70	2.116×10^3
1 centimeter mercury*	1	0.1943	27.85
1 pound/inch^2	5.171	1	144
1 pound/foot^2	3.591×10^{-2}	6.944×10^{-3}	1

* At 0°C and at a location where the acceleration due to gravity has its "standard" value, 9.80665 m/s^2.

TABLE A.2 **Symbols, Dimensions, and Units of Physical Quantities**

Quantity	Common Symbol	Unit*	Dimensions†	Unit in Terms of Base SI Units
Acceleration	a	m/s²	L/T²	m/s²
Amount of substance	n	mole		mol
Angle	θ, ϕ	radian (rad)	1	
Angular acceleration	α	rad/s²	T⁻²	s⁻²
Angular frequency	ω	rad/s	T⁻¹	s⁻¹
Angular momentum	L	kg·m²/s	ML²/T	kg·m²/s
Angular velocity	ω	rad/s	T⁻¹	s⁻¹
Area	A	m²	L²	m²
Atomic number	Z			
Capacitance	C	farad (F) (= Q/V)	Q²T²/ML²	A²·s⁴/kg·m²
Charge	q, Q, e	coulomb (C)	Q	A·s
Charge density				
Line	λ	C/m	Q/L	A·s/m
Surface	σ	C/m²	Q/L²	A·s/m²
Volume	ρ	C/m³	Q/L³	A·s/m³
Conductivity	σ	1/Ω·m	Q²T/ML³	A²·s³/kg·m³
Current	I	AMPERE	Q/T	A
Current density	J	A/m²	Q/T²	A/m²
Density	ρ	kg/m³	M/L³	kg/m³
Dielectric constant	κ			
Displacement	s	METER	L	m
Distance	d, h			
Length	ℓ, L			
Electric dipole moment	p	C·m	QL	A·s·m
Electric field	E	V/m	ML/QT²	kg·m/A·s³
Electric flux	Φ	V·m	ML³/QT²	kg·m³/A·s³
Electromotive force	$\mathcal{E}$	volt (V)	ML²/QT²	kg·m²/A·s³
Energy	E, U, K	joule (J)	ML²/T²	kg·m²/s²
Entropy	S	J/K	ML²/T²·K	kg·m²/s²·K
Force	F	newton (N)	ML/T²	kg·m/s²
Frequency	f, ν	hertz (Hz)	T⁻¹	s⁻¹
Heat	Q	joule (J)	ML²/T²	kg·m²/s²
Inductance	L	henry (H)	ML²/Q²	kg·m²/A²·s²
Magnetic dipole moment	μ	N·m/T	QL²/T	A·m²
Magnetic field	B	tesla (T) (= Wb/m²)	M/QT	kg/A·s²
Magnetic flux	Φ_m	weber (Wb)	ML²/QT	kg·m²/A·s²
Mass	m, M	KILOGRAM	M	kg
Molar specific heat	C	J/mol·K		kg·m²/s²·mol·K
Moment of inertia	I	kg·m²	ML²	kg·m²
Momentum	p	kg·m/s	ML/T	kg·m/s
Period	T	s	T	s
Permeability of space	μ_0	N/A² (= H/m)	ML/Q²T	kg·m/A²·s²
Permittivity of space	ϵ_0	C²/N·m² (= F/m)	Q²T²/ML³	A²·s⁴/kg·m³
Potential (voltage)	V	volt (V) (= J/C)	ML²/QT²	kg·m²/A·s³
Power	P	watt (W) (= J/s)	ML²/T³	kg·m²/s³

continued

TABLE A.2　*Continued*

Quantity	Common Symbol	Unit*	Dimensions†	Unit in Terms of Base SI Units
Pressure	P, p	pascal (Pa) = (N/m^2)	M/LT^2	kg/m·s^2
Resistance	R	ohm (Ω)(=V/A)	ML^2/Q^2T	kg·m^2/A^2·s^3
Specific heat	c	J/kg·K	L^2/T^2·K	m^2/s^2·K
Temperature	T	KELVIN	K	K
Time	t	SECOND	T	s
Torque	$\boldsymbol{\tau}$	N·m	ML^2/T^2	kg·m^2/s^2
Speed	v	m/s	L/T	m/s
Volume	V	m^3	L^3	m^3
Wavelength	λ	m	L	m
Work	W	joule (J)(=N·m)	ML^2/T^2	kg·m^2/s^2

* The base SI units are given in upper case letters.

† The symbols M, L, T, and Q denote mass, length, time, and charge, respectively.

TABLE A.3　**Table of Atomic Masses[a]**

Z	Element	Symbol	Chemical Atomic Mass (u)	Mass Number (* Indicates Radioactive) A	Atomic Mass (u)	Percent Abundance	Half-Life (if Radioactive) $T_{1/2}$
0	(Neutron)	n		1*	1.008 665		10.4 m
1	Hydrogen	H	1.0079	1	1.007 825	99.985	
	Deuterium	D		2	2.014 102	0.015	
	Tritium	T		3*	3.016 049		12.33 y
2	Helium	He	4.00260	3	3.016 029	0.00014	
				4	4.002 602	99.99986	
				6*	6.018 886		0.81 s
3	Lithium	Li	6.941	6	6.015 121	7.5	
				7	7.016 003	92.5	
				8*	8.022 486		0.84 s
4	Beryllium	Be	9.0122	7*	7.016 928		53.3 d
				9	9.012 174	100	
				10*	10.013 534		1.5 × 10^6 y
5	Boron	B	10.81	10	10.012 936	19.9	
				11	11.009 305	80.1	
				12*	12.014 352		0.0202 s
6	Carbon	C	12.011	10*	10.016 854		19.3 s
				11*	11.011 433		20.4 m
				12	12.000 000	98.90	
				13	13.003 355	1.10	
				14*	14.003 242		5730 y
				15*	15.010 599		2.45 s
7	Nitrogen	N	14.0067	12*	12.018 613		0.0110 s
				13*	13.005 738		9.96 m
				14	14.003 074	99.63	
				15	15.000 108	0.37	
				16*	16.006 100		7.13 s
				17*	17.008 450		4.17 s

Z	Element	Symbol	Chemical Atomic Mass (u)	Mass Number (* Indicates Radioactive) A	Atomic Mass (u)	Percent Abundance	Half-Life (if Radioactive) $T_{1/2}$
8	Oxygen	O	15.9994	14*	14.008 595		70.6 s
				15*	15.003 065		122
				16	15.994 915	99.761	
				17	16.999 132	0.039	
				18	17.999 160	0.20	
9	Fluorine	F	18.99840	19*	19.003 577		26.9 s
				17*	17.002 094		64.5 s
				18*	18.000 937		109.8 m
				19	18.998 404	100	
				20*	19.999 982		11.0 s
				21*	20.999 950		4.2 s
10	Neon	Ne	20.180	18*	18.005 710		1.67 s
				19*	19.001 880		17.2 s
(10)	(Neon)			20	19.992 435	90.48	
				21	20.993 841	0.27	
				22	21.991 383	9.25	
				23*	22.994 465		37.2 s
11	Sodium	Na	22.98987	21*	20.997 650		22.5 s
				22*	21.994 434		2.61 y
				24*	23.990 961		14.96 h
12	Magnesium	Mg	24.305	23*	22.994 124		11.3 s
				24	23.985 042	78.99	
				25	24.985 838	10.00	
				26	25.982 594	11.01	
				27*	26.984 341		9.46 m
13	Aluminum	Al	26.98154	26*	25.986 892		7.4×10^5 y
				27	26.981 538	100	
				28*	27.981 910		2.24 m
14	Silicon	Si	28.086	28	27.976 927	92.23	
				29	28.976 495	4.67	
				30	29.973 770	3.10	
				31*	30.975 362		2.62 h
				32*	31.974 148		172 y
15	Phosphorus	P	30.97376	30*	29.978 307		2.50 m
				31	30.973 762	100	
				32*	31.973 908		14.26 d
				33*	32.971 725		25.3 d
16	Sulfur	S	32.066	32	31.972 071	95.02	
				33	32.971 459	0.75	
				34	33.967 867	4.21	
				35*	34.969 033		87.5 d
				36	35.967 081	0.02	
17	Chlorine	Cl	35.453	35	34.968 853	75.77	
				36*	35.968 307		3.0×10^5 y
				37	36.965 903	24.23	

continued

TABLE A.3   *Continued*

Z	Element	Symbol	Chemical Atomic Mass (u)	Mass Number (* Indicates Radioactive) A	Atomic Mass (u)	Percent Abundance	Half-Life (if Radioactive) $T_{1/2}$
18	Argon	Ar	39.948	36	35.967 547	0.337	
				37*	36.966 776		35.04 d
				38	37.962 732	0.063	
				39*	38.964 314		269 y
				40	39.962 384	99.600	
				42*	41.963 049		33 y
19	Potassium	K	39.0983	39	38.963 708	93.2581	
				40*	39.964 000	0.0117	1.28×10^9 y
				41	40.961 827	6.7302	
20	Calcium	Ca	40.08	40	39.962 591	96.941	
				41*	40.962 279		1.0×10^5 y
				42	41.958 618	0.647	
				43	42.958 767	0.135	
				44	43.955 481	2.086	
				46	45.953 687	0.004	
				48	47.952 534	0.187	
21	Scandium	Sc	44.9559	41*	40.969 250		0.596 s
				45	44.955 911	100	
22	Titanium	Ti	47.88	44*	43.959 691		49 y
				46	45.952 630	8.0	
				47	46.951 765	7.3	
				48	47.947 947	73.8	
				49	48.947 871	5.5	
(22)	(Titanium)			50	49.944 792	5.4	
23	Vanadium	V	50.9415	48*	47.952 255		15.97 d
				50*	49.947 161	0.25	1.5×10^{17} y
				51	50.943 962	99.75	
24	Chromium	Cr	51.996	48*	47.954 033		21.6 h
				50	49.946 047	4.345	
				52	51.940 511	83.79	
				53	52.940 652	9.50	
				54	53.938 883	2.365	
25	Manganese	Mn	54.93805	54*	53.940 361		312.1 d
				55	54.938 048	100	
26	Iron	Fe	55.847	54	53.939 613	5.9	
				55*	54.938 297		2.7 y
				56	55.934 940	91.72	
				57	56.935 396	2.1	
				58	57.933 278	0.28	
				60*	59.934 078		1.5×10^6 y
27	Cobalt	Co	58.93320	59	58.933 198	100	
				60*	59.933 820		5.27 y
28	Nickel	Ni	58.693	58	57.935 346	68.077	
				59*	58.934 350		7.5×10^4 y
				60	59.930 789	26.223	
				61	60.931 058	1.140	
				62	61.928 346	3.634	
				63*	62.929 670		100 y
				64	63.927 967	0.926	

TABLE A.3 *Continued*

Z	Element	Symbol	Chemical Atomic Mass (u)	Mass Number (* Indicates Radioactive) A	Atomic Mass (u)	Percent Abundance	Half-Life (if Radioactive) $T_{1/2}$
29	Copper	Cu	63.54	63	62.929 599	69.17	
				65	64.927 791	30.83	
30	Zinc	Zn	65.39	64	63.929 144	48.6	
				66	65.926 035	27.9	
				67	66.927 129	4.1	
				68	67.924 845	18.8	
				70	69.925 323	0.6	
31	Gallium	Ga	69.723	69	68.925 580	60.108	
				71	70.924 703	39.892	
32	Germanium	Ge	72.61	70	69.924 250	21.23	
				72	71.922 079	27.66	
				73	72.923 462	7.73	
				74	73.921 177	35.94	
				76	75.921 402	7.44	
33	Arsenic	As	74.9216	75	74.921 594	100	
34	Selenium	Se	78.96	74	73.922 474	0.89	
				76	75.919 212	9.36	
				77	76.919 913	7.63	
				78	77.917 307	23.78	
				79*	78.918 497		$\leqslant 6.5 \times 10^4$ y
				80	79.916 519	49.61	
				82*	81.916 697	8.73	1.4×10^{20} y
35	Bromine	Br	79.904	79	78.918 336	50.69	
				81	80.916 287	49.31	
36	Krypton	Kr	83.80	78	77.920 400	0.35	
				80	79.916 377	2.25	
				81*	80.916 589		2.1×10^5 y
(36)	(Krypton)			82	81.913 481	11.6	
				83	82.914 136	11.5	
				84	83.911 508	57.0	
				85*	84.912 531		10.76 y
				86	85.910 615	17.3	
37	Rubidium	Rb	85.468	85	84.911 793	72.17	
				87*	86.909 186	27.83	4.75×10^{10} y
38	Strontium	Sr	87.62	84	83.913 428	0.56	
				86	85.909 266	9.86	
				87	86.908 883	7.00	
				88	87.905 618	82.58	
				90*	89.907 737		29.1 y
39	Yttrium	Y	88.9058	89	88.905 847	100	
40	Zirconium	Zr	91.224	90	89.904 702	51.45	
				91	90.905 643	11.22	
				92	91.905 038	17.15	
				93*	92.906 473		1.5×10^6 y
				94	93.906 314	17.38	
				96	95.908 274	2.80	

continued

TABLE A.3　*Continued*

Z	Element	Symbol	Chemical Atomic Mass (u)	Mass Number (* Indicates Radioactive) A	Atomic Mass (u)	Percent Abundance	Half-Life (if Radioactive) $T_{1/2}$
41	Niobium	Nb	92.9064	91*	90.906 988		6.8×10^2 y
				92*	91.907 191		3.5×10^7 y
				93	92.906 376	100	
				94*	93.907 280		2×10^4 y
42	Molybdenum	Mo	95.94	92	91.906 807	14.84	
				93*	92.906 811		3.5×10^3 y
				94	93.905 085	9.25	
				95	94.905 841	15.92	
				96	95.904 678	16.68	
				97	96.906 020	9.55	
				98	97.905 407	24.13	
				100	99.907 476	9.63	
43	Technetium	Tc		97*	96.906 363		2.6×10^6 y
				98*	97.907 215		4.2×10^6 y
				99*	98.906 254		2.1×10^5 y
44	Ruthenium	Ru	101.07	96	95.907 597	5.54	
				98	97.905 287	1.86	
				99	98.905 939	12.7	
				100	99.904 219	12.6	
				101	100.905 558	17.1	
				102	101.904 348	31.6	
				104	103.905 428	18.6	
45	Rhodium	Rh	102.9055	103	102.905 502	100	
46	Palladium	Pd	106.42	102	101.905 616	1.02	
				104	103.904 033	11.14	
				105	104.905 082	22.33	
				106	105.903 481	27.33	
				107*	106.905 126		6.5×10^6 y
				108	107.903 893	26.46	
				110	109.905 158	11.72	
47	Silver	Ag	107.868	107	106.905 091	51.84	
				109	108.904 754	48.16	
48	Cadmium	Cd	112.41	106	105.906 457	1.25	
				108	107.904 183	0.89	
				109*	108.904 984		462 d
(48)	(Cadmium)			110	109.903 004	12.49	
				111	110.904 182	12.80	
				112	111.902 760	24.13	
				113*	112.904 401	12.22	9.3×10^{15} y
				114	113.903 359	28.73	
				116	115.904 755	7.49	
49	Indium	In	114.82	113	112.904 060	4.3	
				115*	114.903 876	95.7	4.4×10^{14} y
50	Tin	Sn	118.71	112	111.904 822	0.97	
				114	113.902 780	0.65	
				115	114.903 345	0.36	
				116	115.901 743	14.53	
				117	116.902 953	7.68	

TABLE A.3 *Continued*

Z	Element	Symbol	Chemical Atomic Mass (u)	Mass Number (* Indicates Radioactive) A	Atomic Mass (u)	Percent Abundance	Half-Life (if Radioactive) $T_{1/2}$
				118	117.901 605	24.22	
				119	118.903 308	8.58	
				120	119.902 197	32.59	
				121*	120.904 237		55 y
				122	121.903 439	4.63	
				124	123.905 274	5.79	
51	Antimony	Sb	121.76	121	120.903 820	57.36	
				123	122.904 215	42.64	
				125*	124.905 251		2.7 y
52	Tellurium	Te	127.60	120	119.904 040	0.095	
				122	121.903 052	2.59	
				123*	122.904 271	0.905	1.3×10^{13} y
				124	123.902 817	4.79	
				125	124.904 429	7.12	
				126	125.903 309	18.93	
				128*	127.904 463	31.70	$> 8 \times 10^{24}$ y
				130*	129.906 228	33.87	$\leq 1.25 \times 10^{21}$ y
53	Iodine	I	126.9045	127	126.904 474	100	
				129*	128.904 984		1.6×10^{7} y
54	Xenon	Xe	131.29	124	123.905 894	0.10	
				126	125.904 268	0.09	
				128	127.903 531	1.91	
				129	128.904 779	26.4	
				130	129.903 509	4.1	
				131	130.905 069	21.2	
				132	131.904 141	26.9	
				134	133.905 394	10.4	
				136*	135.907 215	8.9	$\geq 2.36 \times 10^{21}$ y
55	Cesium	Cs	132.9054	133	132.905 436	100	
				134*	133.906 703		2.1 y
				135*	134.905 891		2×10^{6} y
				137*	136.907 078		30 y
56	Barium	Ba	137.33	130	129.906 289	0.106	
				132	131.905 048	0.101	
				133*	132.905 990		10.5 y
				134	133.904 492	2.42	
				135	134.905 671	6.593	
				136	135.904 559	7.85	
				137	136.905 816	11.23	
				138	137.905 236	71.70	
57	Lanthanum	La	138.905	137*	136.906 462		6×10^{4} y
(57)	(Lanthanum)			138*	137.907 105	0.0902	1.05×10^{11} y
				139	138.906 346	99.9098	
58	Cerium	Ce	140.12	136	135.907 139	0.19	
				138	137.905 986	0.25	
				140	139.905 434	88.43	
				142*	141.909 241	11.13	$> 5 \times 10^{16}$ y
59	Praseodymium	Pr	140.9076	141	140.907 647	100	

continued

TABLE A.3 *Continued*

Z	Element	Symbol	Chemical Atomic Mass (u)	Mass Number (* Indicates Radioactive) A	Atomic Mass (u)	Percent Abundance	Half-Life (if Radioactive) $T_{1/2}$
60	Neodymium	Nd	144.24	142	141.907 718	27.13	
				143	142.909 809	12.18	
				144*	143.910 082	23.80	2.3×10^{15} y
				145	144.912 568	8.30	
				146	145.913 113	17.19	
				148	147.916 888	5.76	
				150*	149.920 887	5.64	$>1 \times 10^{18}$ y
61	Promethium	Pm		143*	142.910 928		265 d
				145*	144.912 745		17.7 y
				146*	145.914 698		5.5 y
				147*	146.915 134		2.623 y
62	Samarium	Sm	150.36	144	143.911 996	3.1	
				146*	145.913 043		1.0×10^{8} y
				147*	146.914 894	15.0	1.06×10^{11} y
				148*	147.914 819	11.3	7×10^{15} y
				149*	148.917 180	13.8	$>2 \times 10^{15}$ y
				150	149.917 273	7.4	
				151*	150.919 928		90 y
				152	151.919 728	26.7	
				154	153.922 206	22.7	
63	Europium	Eu	151.96	151	150.919 846	47.8	
				152*	151.921 740		13.5 y
				153	152.921 226	52.2	
				154*	153.922 975		8.59 y
				155*	154.922 888		4.7 y
64	Gadolinium	Gd	157.25	148*	147.918 112		75 y
				150*	149.918 657		1.8×10^{6} y
				152*	151.919 787	0.20	1.1×10^{14} y
				154	153.920 862	2.18	
				155	154.922 618	14.80	
				156	155.922 119	20.47	
				157	156.923 957	15.65	
				158	157.924 099	24.84	
				160	159.927 050	21.86	
65	Terbium	Tb	158.9253	159	158.925 345	100	
66	Dysprosium	Dy	162.50	156	155.924 277	0.06	
				158	157.924 403	0.10	
				160	159.925 193	2.34	
				161	160.926 930	18.9	
				162	161.926 796	25.5	
				163	162.928 729	24.9	
				164	163.929 172	28.2	
67	Holmium	Ho	164.9303	165	164.930 316	100	
				166*	165.932 282		1.2×10^{3} y
68	Erbium	Er	167.26	162	161.928 775	0.14	
				164	163.929 198	1.61	
				166	165.930 292	33.6	

TABLE A.3   *Continued*

Z	Element	Symbol	Chemical Atomic Mass (u)	Mass Number (* Indicates Radioactive) A	Atomic Mass (u)	Percent Abundance	Half-Life (if Radioactive) $T_{1/2}$
(68)	(Erbium)			167	166.932 047	22.95	
				168	167.932 369	27.8	
				170	169.935 462	14.9	
69	Thulium	Tm	168.9342	169	168.934 213	100	
				171*	170.936 428		1.92 y
70	Ytterbium	Yb	173.04	168	167.933 897	0.13	
				170	169.934 761	3.05	
				171	170.936 324	14.3	
				172	171.936 380	21.9	
				173	172.938 209	16.12	
				174	173.938 861	31.8	
				176	175.942 564	12.7	
71	Lutecium	Lu	174.967	173*	172.938 930		1.37 y
				175	174.940 772	97.41	
				176*	175.942 679	2.59	3.78×10^{10} y
72	Hafnium	Hf	178.49	174*	173.940 042	0.162	2.0×10^{15} y
				176	175.941 404	5.206	
				177	176.943 218	18.606	
				178	177.943 697	27.297	
				179	178.945 813	13.629	
				180	179.946 547	35.100	
73	Tantalum	Ta	180.9479	180	179.947 542	0.012	
				181	180.947 993	99.988	
74	Tungsten (Wolfram)	W	183.85	180	179.946 702	0.12	
				182	181.948 202	26.3	
				183	182.950 221	14.28	
				184	183.950 929	30.7	
				186	185.954 358	28.6	
75	Rhenium	Re	186.207	185	184.952 951	37.40	
				187*	186.955 746	62.60	4.4×10^{10} y
76	Osmium	Os	190.2	184	183.952 486	0.02	
				186*	185.953 834	1.58	2.0×10^{15} y
				187	186.955 744	1.6	
				188	187.955 832	13.3	
				189	188.958 139	16.1	
				190	189.958 439	26.4	
				192	191.961 468	41.0	
				194*	193.965 172		6.0 y
77	Iridium	Ir	192.2	191	190.960 585	37.3	
				193	192.962 916	62.7	
78	Platinum	Pt	195.08	190*	189.959 926	0.01	6.5×10^{11} y
				192	191.961 027	0.79	
				194	193.962 655	32.9	
				195	194.964 765	33.8	
				196	195.964 926	25.3	
				198	197.967 867	7.2	
79	Gold	Au	196.9665	197	196.966 543	100	

continued

TABLE A.3   *Continued*

Z	Element	Symbol	Chemical Atomic Mass (u)	Mass Number (* Indicates Radioactive) A	Atomic Mass (u)	Percent Abundance	Half-Life (if Radioactive) $T_{1/2}$
80	Mercury	Hg	200.59	196	195.965 806	0.15	
				198	197.966 743	9.97	
				199	198.968 253	16.87	
				200	199.968 299	23.10	
				201	200.970 276	13.10	
				202	201.970 617	29.86	
				204	203.973 466	6.87	
81	Thallium	Tl	204.383	203	202.972 320	29.524	
				204*	203.973 839		3.78 y
				205	204.974 400	70.476	
		(Ra E″)		206*	205.976 084		4.2 m
		(Ac C″)		207*	206.977 403		4.77 m
		(Th C″)		208*	207.981 992		3.053 m
		(Ra C″)		210*	209.990 057		1.30 m
82	Lead	Pb	207.2	202*	201.972 134		5×10^4 y
				204*	203.973 020	1.4	$\geqslant 1.4 \times 10^{17}$ y
				205*	204.974 457		1.5×10^7 y
				206	205.974 440	24.1	
				207	206.975 871	22.1	
				208	207.976 627	52.4	
		(Ra D)		210*	209.984 163		22.3 y
		(Ac B)		211*	210.988 734		36.1 m
		(Th B)		212*	211.991 872		10.64 h
		(Ra B)		214*	213.999 798		26.8 m
83	Bismuth	Bi	208.9803	207*	206.978 444		32.2 y
				208*	207.979 717		3.7×10^5 y
				209	208.980 374	100	
		(Ra E)		210*	209.984 096		5.01 d
		(Th C)		211*	210.987 254		2.14 m
				212*	211.991 259		60.6 m
		(Ra C)		214*	213.998 692		19.9 m
				215*	215.001 836		7.4 m
84	Polonium	Po		209*	208.982 405		102 y
		(Ra F)		210*	209.982 848		138.38 d
		(Ac C′)		211*	210.986 627		0.52 s
		(Th C′)		212*	211.988 842		0.30 μs
		(Ra C′)		214*	213.995 177		164 μs
		(Ac A)		215*	214.999 418		0.0018 s
		(Th A)		216*	216.001 889		0.145 s
		(Ra A)		218*	218.008 965		3.10 m
85	Astatine	At		215*	214.998 638		$\approx$ 100 μs
				218*	218.008 685		1.6 s
				219*	219.011 294		0.9 m
86	Radon	Rn					
		(An)		219*	219.009 477		3.96 s
		(Tn)		220*	220.011 369		55.6 s
		(Rn)		222*	222.017 571		3.823 d
87	Francium	Fr					
		(Ac K)		223*	223.019 733		22 m

TABLE A.3 *Continued*

Z	Element	Symbol	Chemical Atomic Mass (u)	Mass Number (* Indicates Radioactive) A	Atomic Mass (u)	Percent Abundance	Half-Life (if Radioactive) $T_{1/2}$
88	Radium	Ra					
		(Ac X)		223*	223.018 499		11.43 d
		(Th X)		224*	224.020 187		3.66 d
		(Ra)		226*	226.025 402		1600 y
		(Ms Th$_1$)		228*	228.031 064		5.75 y
89	Actinium	Ac		227*	227.027 749		21.77 y
		(Ms Th$_2$)		228*	228.031 015		6.15 h
90	Thorium	Th	232.0381				
		(Rd Ac)		227*	227.027 701		18.72 d
		(Rd Th)		228*	228.028 716		1.913 y
				229*	229.031 757		7300 y
		(Io)		230*	230.033 127		75.000 y
(90)	(Thorium)	(UY)		231*	231.036 299		25.52 h
		(Th)		232*	232.038 051	100	1.40×10^{10} y
		(UX$_1$)		234*	234.043 593		24.1 d
91	Protactinium	Pa		231*	231.035 880		32.760 y
		(Uz)		234*	234.043 300		6.7 h
92	Uranium	U	238.0289	232*	232.037 131		69 y
				233*	233.039 630		1.59×10^5 y
				234*	234.040 946	0.0055	2.45×10^5 y
		(Ac U)		235*	235.043 924	0.720	7.04×10^8 y
				236*	236.045 562		2.34×10^7 y
		(UI)		238*	238.050 784	99.2745	4.47×10^9 y
93	Neptunium	Np		235*	235.044 057		396 d
				236*	236.046 560		115,000 y
				237*	237.048 168		2.14×10^6 y
94	Plutonium	Pu		236*	236.046 033		2.87 y
				238*	238.049 555		87.7 y
				239*	239.052 157		24,120 y
				240*	240.053 808		6560 y
				241*	241.056 846		14.4 y
				242*	242.058 737		3.73×10^5 y
				244*	244.064 200		8.1×10^7 y

[a] The masses in the sixth column are atomic masses, which include the mass of *Z* electrons. Data are from the National Nuclear Data Center, Brookhaven National Laboratory, prepared by Jagdish K. Tuli, July 1990. The data are based on experimental results reported in *Nuclear Data Sheets* and *Nuclear Physics* and also from *Chart of the Nuclides,* 14th ed. Atomic masses are based on those by A. H. Wapstra, G. Audi, and R. Hoekstra. Isotopic abundances are based on those by N. E. Holden.

Mathematics Review

These appendices in mathematics are intended as a brief review of operations and methods. Early in this course, you should be totally familiar with basic algebraic techniques, analytic geometry, and trigonometry. The appendices on differential and integral calculus are more detailed and are intended for those students who have difficulty applying calculus concepts to physical situations.

B.1 SCIENTIFIC NOTATION

Many quantities that scientists deal with often have very large or very small values. For example, the speed of light is about 300 000 000 m/s and the ink required to make the dot over an i in this textbook has a mass of about 0.000 000 001 kg. Obviously, it is very cumbersome to read, write, and keep track of numbers such as these. We avoid this problem by using a method dealing with powers of the number 10:

$$10^0 = 1$$
$$10^1 = 10$$
$$10^2 = 10 \times 10 = 100$$
$$10^3 = 10 \times 10 \times 10 = 1000$$
$$10^4 = 10 \times 10 \times 10 \times 10 = 10\ 000$$
$$10^5 = 10 \times 10 \times 10 \times 10 \times 10 = 100\ 000$$

and so on. The number of zeros corresponds to the power to which 10 is raised, called the **exponent** of 10. For example, the speed of light, 300 000 000 m/s, can be expressed as 3×10^8 m/s.

In this method, some representative numbers smaller than unity are

$$10^{-1} = \frac{1}{10} = 0.1$$

$$10^{-2} = \frac{1}{10 \times 10} = 0.01$$

$$10^{-3} = \frac{1}{10 \times 10 \times 10} = 0.001$$

$$10^{-4} = \frac{1}{10 \times 10 \times 10 \times 10} = 0.0001$$

$$10^{-5} = \frac{1}{10 \times 10 \times 10 \times 10 \times 10} = 0.00001$$

In these cases, the number of places the decimal point is to the left of the digit 1 equals the value of the (negative) exponent. Numbers expressed as some power of 10 multiplied by another number between 1 and 10 are said to be in **scientific notation.** For example, the scientific notation for 5 943 000 000 is 5.943×10^9 and that for 0.0000832 is 8.32×10^{-5}.

When numbers expressed in scientific notation are being multiplied, the following general rule is very useful:

$$10^n \times 10^m = 10^{n+m} \tag{B.1}$$

where n and m can be *any* numbers (not necessarily integers). For example, $10^2 \times 10^5 = 10^7$. The rule also applies if one of the exponents is negative: $10^3 \times 10^{-8} = 10^{-5}$.

When dividing numbers expressed in scientific notation, note that

$$\frac{10^n}{10^m} = 10^n \times 10^{-m} = 10^{n-m} \tag{B.2}$$

EXERCISES

With help from the above rules, verify the answers to the following:

1. $86\ 400 = 8.64 \times 10^4$
2. $9\ 816{,}762.5 = 9.8167625 \times 10^6$
3. $0.0000000398 = 3.98 \times 10^{-8}$
4. $(4 \times 10^8)(9 \times 10^9) = 3.6 \times 10^{18}$
5. $(3 \times 10^7)(6 \times 10^{-12}) = 1.8 \times 10^{-4}$
6. $\dfrac{75 \times 10^{-11}}{5 \times 10^{-3}} = 1.5 \times 10^{-7}$
7. $\dfrac{(3 \times 10^6)(8 \times 10^{-2})}{(2 \times 10^{17})(6 \times 10^5)} = 2 \times 10^{-18}$

B.2 ALGEBRA

Some Basic Rules

When algebraic operations are performed, the laws of arithmetic apply. Symbols such as x, y, and z are usually used to represent quantities that are not specified, what are called the **unknowns.**

First, consider the equation

$$8x = 32$$

If we wish to solve for x, we can divide (or multiply) each side of the equation by the same factor without destroying the equality. In this case, if we divide both sides by 8, we have

$$\frac{8x}{8} = \frac{32}{8}$$

$$x = 4$$

Next consider the equation

$$x + 2 = 8$$

In this type of expression, we can add or subtract the same quantity from each side. If we subtract 2 from each side, we get

$$x + 2 - 2 = 8 - 2$$
$$x = 6$$

In general, if $x + a = b$, then $x = b - a$.

Now consider the equation

$$\frac{x}{5} = 9$$

If we multiply each side by 5, we are left with x on the left by itself and 45 on the right:

$$\left(\frac{x}{5}\right)(5) = 9 \times 5$$
$$x = 45$$

In all cases, *whatever operation is performed on the left side of the equality must also be performed on the right side.*

The following rules for multiplying, dividing, adding, and subtracting fractions should be recalled, where a, b, and c are three numbers:

	Rule	Example
Multiplying	$\left(\dfrac{a}{b}\right)\left(\dfrac{c}{d}\right) = \dfrac{ac}{bd}$	$\left(\dfrac{2}{3}\right)\left(\dfrac{4}{5}\right) = \dfrac{8}{15}$
Dividing	$\dfrac{(a/b)}{(c/d)} = \dfrac{ad}{bc}$	$\dfrac{2/3}{4/5} = \dfrac{(2)(5)}{(4)(3)} = \dfrac{10}{12}$
Adding	$\dfrac{a}{b} \pm \dfrac{c}{d} = \dfrac{ad \pm bc}{bd}$	$\dfrac{2}{3} - \dfrac{4}{5} = \dfrac{(2)(5) - (4)(3)}{(3)(5)} = -\dfrac{2}{15}$

EXERCISES

In the following exercises, solve for x:

Answers

1. $a = \dfrac{1}{1 + x}$ $x = \dfrac{1 - a}{a}$

2. $3x - 5 = 13$ $x = 6$

3. $ax - 5 = bx + 2$ $x = \dfrac{7}{a - b}$

4. $\dfrac{5}{2x + 6} = \dfrac{3}{4x + 8}$ $x = -\dfrac{11}{7}$

Powers

When powers of a given quantity x are multiplied, the following rule applies:

$$x^n x^m = x^{n+m} \tag{B.3}$$

For example, $x^2 x^4 = x^{2+4} = x^6$.

When dividing the powers of a given quantity, the rule is

$$\frac{x^n}{x^m} = x^{n-m} \tag{B.4}$$

For example, $x^8/x^2 = x^{8-2} = x^6$.

A power that is a fraction, such as $\frac{1}{3}$, corresponds to a root as follows:

$$x^{1/n} = \sqrt[n]{x} \tag{B.5}$$

For example, $4^{1/3} = \sqrt[3]{4} = 1.5874$. (A scientific calculator is useful for such calculations.)

Finally, any quantity x^n raised to the mth power is

$$(x^n)^m = x^{nm} \tag{B.6}$$

Table B.1 summarizes the rules of exponents.

TABLE B.1   Rules of Exponents

$$x^0 = 1$$
$$x^1 = x$$
$$x^n x^m = x^{n+m}$$
$$x^n/x^m = x^{n-m}$$
$$x^{1/n} = \sqrt[n]{x}$$
$$(x^n)^m = x^{nm}$$

EXERCISES

Verify the following:

1. $3^2 \times 3^3 = 243$
2. $x^5 x^{-8} = x^{-3}$
3. $x^{10}/x^{-5} = x^{15}$
4. $5^{1/3} = 1.709975$ (Use your calculator.)
5. $60^{1/4} = 2.783158$ (Use your calculator.)
6. $(x^4)^3 = x^{12}$

Factoring

Some useful formulas for factoring an equation are

$$ax + ay + az = a(x + y + x) \qquad \text{common factor}$$
$$a^2 + 2ab + b^2 = (a + b)^2 \qquad \text{perfect square}$$
$$a^2 - b^2 = (a + b)(a - b) \qquad \text{differences of squares}$$

Quadratic Equations

The general form of a quadratic equation is

$$ax^2 + bx + c = 0 \tag{B.7}$$

where x is the unknown quantity and a, b, and c are numerical factors referred to as **coefficients** of the equation. This equation has two roots, given by

$$x = \frac{-b \pm \sqrt{b^2 - 4ac}}{2a} \tag{B.8}$$

If $b^2 \geqslant 4ac$, the roots are real.

EXAMPLE 1

The equation $x^2 + 5x + 4 = 0$ has the following roots corresponding to the two signs of the square-root term:

$$x = \frac{-5 \pm \sqrt{5^2 - (4)(1)(4)}}{2(1)} = \frac{-5 \pm \sqrt{9}}{2} = \frac{-5 \pm 3}{2}$$

$$x_+ = \frac{-5 + 3}{2} = \boxed{-1} \qquad x_- = \frac{-5 - 3}{2} = \boxed{-4}$$

where x_+ refers to the root corresponding to the positive sign and x_- refers to the root corresponding to the negative sign.

EXERCISES

Solve the following quadratic equations:

Answers

1. $x^2 + 2x - 3 = 0$ $x_+ = 1$ $x_- = -3$
2. $2x^2 - 5x + 2 = 0$ $x_+ = 2$ $x_- = \frac{1}{2}$
3. $2x^2 - 4x - 9 = 0$ $x_+ = 1 + \sqrt{22}/2$ $x_- = 1 - \sqrt{22}/2$

Linear Equations

A linear equation has the general form

$$y = mx + b \tag{B.9}$$

where m and b are constants. This equation is referred to as being linear because the graph of y versus x is a straight line, as shown in Figure B.1. The constant b, called the **y-intercept**, represents the value of y at which the straight line intersects the y axis. The constant m is equal to the **slope** of the straight line and is also equal to the tangent of the angle that the line makes with the x axis. If any two points on the straight line are specified by the coordinates (x_1, y_1) and (x_2, y_2), as in Figure B.1, then the **slope** of the straight line can be expressed as

$$\text{Slope} = \frac{y_2 - y_1}{x_2 - x_1} = \frac{\Delta y}{\Delta x} = \tan \theta \tag{B.10}$$

Note that m and b can have either positive or negative values. If $m > 0$, the straight line has a *positive* slope, as in Figure B.1. If $m < 0$, the straight line has a *negative* slope. In Figure B.1, both m and b are positive. Three other possible situations are shown in Figure B.2.

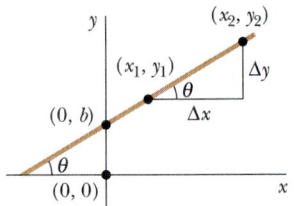

FIGURE B.1

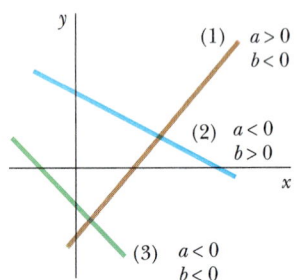

FIGURE B.2

EXERCISES

1. Draw graphs of the following straight lines:
(a) $y = 5x + 3$ (b) $y = -2x + 4$ (c) $y = -3x - 6$
2. Find the slopes of the straight lines described in Exercise 1.

Answers (a) 5   (b) -2 (c) -3

3. Find the slopes of the straight lines that pass through the following sets of points:

(a) $(0, -4)$ and $(4, 2)$, (b) $(0, 0)$ and $(2, -5)$, and (c) $(-5, 2)$ and $(4, -2)$

Answers (a) $3/2$ (b) $-5/2$ (c) $-4/9$

Solving Simultaneous Linear Equations

Consider the equation $3x + 5y = 15$, which has two unknowns, x and y. Such an equation does not have a unique solution. Instead, $(x = 0, y = 3)$, $(x = 5, y = 0)$, and $(x = 2, y = 9/5)$ are all solutions to this equation.

If a problem has two unknowns, a unique solution is possible only if we have *two* equations. In general, if a problem has n unknowns, its solution requires n equations. In order to solve two simultaneous equations involving two unknowns, x and y, we solve one of the equations for x in terms of y and substitute this expression into the other equation.

EXAMPLE 2

Solve the following two simultaneous equations:

(1) $5x + y = -8$

(2) $2x - 2y = 4$

Solution From (2), $x = y + 2$. Substitution of this into (1) gives

$$5(y + 2) + y = -8$$

$$6y = -18$$

$$y = -3$$

$$x = y + 2 = \boxed{-1}$$

Alternate Solution Multiply each term in (1) by the factor 2 and add the result to (2):

$$10x + 2y = -16$$

$$\underline{2x - 2y = 4}$$

$$12x = -12$$

$$x = -1$$

$$y = x - 2 = \boxed{-3}$$

Two linear equations containing two unknowns can also be solved by a graphical method. If the straight lines corresponding to the two equations are plotted in a conventional coordinate system, the intersection of the two lines represents the solution. For example, consider the two equations

$$x - y = 2$$

$$x - 2y = -1$$

These are plotted in Figure B.3. The intersection of the two lines has the coordinates $x = 5$, $y = 3$. This represents the solution to the equations. You should check this solution by the analytical technique discussed above.

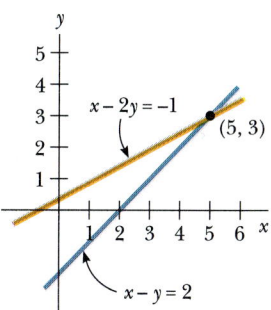

FIGURE B.3

EXERCISES

Solve the following pairs of simultaneous equations involving two unknowns:

Answers

1. $x + y = 8$ $x = 5, y = 3$
       $x - y = 2$
2. $98 - T = 10a$ $T = 65, a = 3.27$
       $T - 49 = 5a$
3. $6x + 2y = 6$ $x = 2, y = -3$
       $8x - 4y = 28$

Logarithms

Suppose that a quantity x is expressed as a power of some quantity a:

$$x = a^y \tag{B.11}$$

The number a is called the **base** number. The **logarithm** of x with respect to the base a is equal to the exponent to which the base must be raised in order to satisfy the expression $x = a^y$:

$$y = \log_a x \tag{B.12}$$

Conversely, the **antilogarithm** of y is the number x:

$$x = \text{antilog}_a y \tag{B.13}$$

In practice, the two bases most often used are base 10, called the *common* logarithm base, and base $e = 2.718 \ldots$, called the *natural* logarithm base. When common logarithms are used,

$$y = \log_{10} x \qquad (\text{or } x = 10^y) \tag{B.14}$$

When natural logarithms are used,

$$y = \ln_e x \qquad (\text{or } x = e^y) \tag{B.15}$$

For example, $\log_{10} 52 = 1.716$, so that $\text{antilog}_{10} 1.716 = 10^{1.716} = 52$. Likewise, $\ln_e 52 = 3.951$, so $\text{antiln}_e 3.951 = e^{3.951} = 52$.

In general, note that you can convert between base 10 and base e with the equality

$$\ln_e x = (2.302585) \log_{10} x \tag{B.16}$$

Finally, some useful properties of logarithms are

$$\log(ab) = \log a + \log b$$
$$\log(a/b) = \log a - \log b$$
$$\log(a^n) = n \log a$$
$$\ln e = 1$$
$$\ln e^a = a$$
$$\ln\left(\frac{1}{a}\right) = -\ln a$$

B.3   GEOMETRY

The **distance** d between two points having coordinates (x_1, y_1) and (x_2, y_2) is

$$d = \sqrt{(x_2 - x_1)^2 + (y_2 - y_1)^2} \tag{B.17}$$

Radian measure: The arc length s of a circular arc (Fig. B.4) is proportional to the radius r for a fixed value of θ (in radians):

$$s = r\theta$$
$$\theta = \frac{s}{r} \tag{B.18}$$

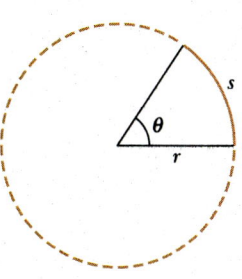

Table B.2 gives the areas and volumes for several geometric shapes used throughout this text:

FIGURE B.4

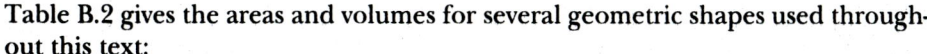

TABLE B.2	Useful Information for Geometry		
Shape	**Area or Volume**	**Shape**	**Area or Volume**

Rectangle   Area $= \ell w$

Sphere   Surface area $= 4\pi r^2$ Volume $= \dfrac{4\pi r^3}{3}$

Circle   Area $= \pi r^2$ (Circumference $= 2\pi r$)

Cylinder   Volume $= \pi r^2 \ell$

Triangle   Area $= \frac{1}{2}bh$

Rectangular box   Area $= 2(\ell h + \ell w + hw)$ Volume $= \ell wh$

The equation of a **straight line** (Fig. B.5) is

$$y = mx + b \tag{B.19}$$

where b is the y-intercept and m is the slope of the line.

The equation of a **circle** or radius R centered at the origin is

$$x^2 + y^2 = R^2 \tag{B.20}$$

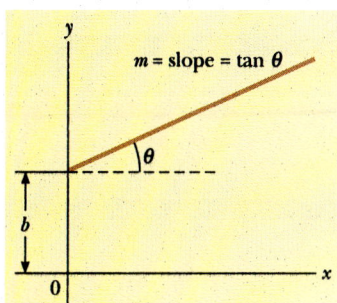

FIGURE B.5

The equation of an **ellipse** having the origin at its center (Fig. B.6) is

$$\frac{x^2}{a^2} + \frac{y^2}{b^2} = 1 \tag{B.21}$$

where a is the length of the semi-major axis (the longer one) and b is the length of the semi-minor axis (the shorter one).

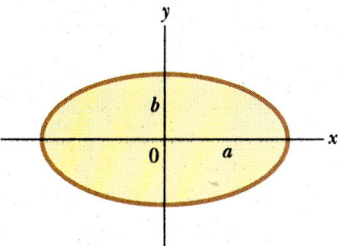

FIGURE B.6

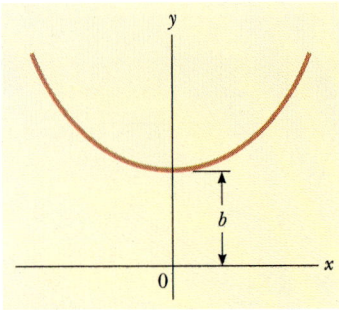

FIGURE B.7

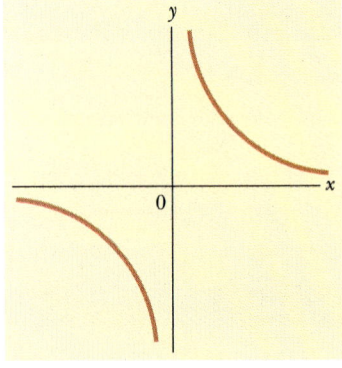

FIGURE B.8

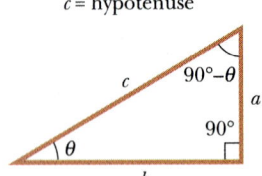

a = opposite side
b = adjacent side
c = hypotenuse

FIGURE B.9

The equation of a **parabola** the vertex of which is at $y = b$ (Fig. B.7) is

$$y = ax^2 + b \tag{B.22}$$

The equation of a **rectangular hyperbola** (Fig. B.8) is

$$xy = \text{constant} \tag{B.23}$$

B.4   TRIGONOMETRY

That portion of mathematics based on the special properties of the right triangle is called trigonometry. By definition, a right triangle is one containing at $90°$ angle. Consider the right triangle shown in Figure B.9, where side a is opposite the angle θ, side b is adjacent to the angle θ, and side c is the hypotenuse of the triangle. The three basic trigonometric functions defined by such a triangle are the sine (sin), cosine (cos), and tangent (tan) functions. In terms of the angle θ, these functions are defined by

$$\sin \theta \equiv \frac{\text{side opposite } \theta}{\text{hypotenuse}} = \frac{a}{c} \tag{B.24}$$

$$\cos \theta \equiv \frac{\text{side adjacent to } \theta}{\text{hypotenuse}} = \frac{b}{c} \tag{B.25}$$

$$\tan \theta \equiv \frac{\text{side opposite } \theta}{\text{side adjacent to } \theta} = \frac{a}{b} \tag{B.26}$$

The Pythagorean theorem provides the following relationship between the sides of a right triangle:

$$c^2 = a^2 + b^2 \tag{B.27}$$

From the above definitions and the Pythagorean theorem, it follows that

$$\sin^2 \theta + \cos^2 \theta = 1$$

$$\tan \theta = \frac{\sin \theta}{\cos \theta}$$

The cosecant, secant, and cotangent functions are defined by

$$\csc \theta \equiv \frac{1}{\sin \theta} \qquad \sec \theta \equiv \frac{1}{\cos \theta} \qquad \cot \theta \equiv \frac{1}{\tan \theta}$$

The relationship below follow directly from the right triangle shown in Figure B.9:

$$\sin \theta = \cos(90° - \theta)$$

$$\cos \theta = \sin(90° - \theta)$$

$$\cot \theta = \tan(90° - \theta)$$

Some properties of trigonometric functions are

$$\sin(-\theta) = -\sin \theta$$

$$\cos(-\theta) = \cos \theta$$

$$\tan(-\theta) = -\tan \theta$$

The following relationships apply to *any* triangle, as shown in Figure B.10:

$$\alpha + \beta + \gamma = 180°$$

Law of cosines
$$a^2 = b^2 + c^2 - 2bc \cos \alpha$$
$$b^2 = a^2 + c^2 - 2ac \cos \beta$$
$$c^2 = a^2 + b^2 - 2ab \cos \gamma$$

Law of sines
$$\frac{a}{\sin \alpha} = \frac{b}{\sin \beta} = \frac{c}{\sin \gamma}$$

Table B.3 lists a number of useful trigonometric identities.

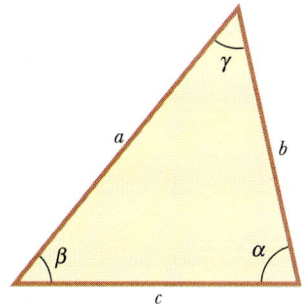

FIGURE B.10

TABLE B.3 Some Trigonometric Identities

$$\sin^2 \theta + \cos^2 \theta = 1 \qquad\qquad \csc^2 \theta = 1 + \cot^2 \theta$$

$$\sec^2 \theta = 1 + \tan^2 \theta \qquad\qquad \sin^2 \frac{\theta}{2} = \tfrac{1}{2}(1 - \cos \theta)$$

$$\sin 2\theta = 2 \sin \theta \cos \theta \qquad\qquad \cos^2 \frac{\theta}{2} = \tfrac{1}{2}(1 + \cos \theta)$$

$$\cos 2\theta = \cos^2 \theta - \sin^2 \theta \qquad\qquad 1 - \cos \theta = 2 \sin^2 \frac{\theta}{2}$$

$$\tan 2\theta = \frac{2 \tan \theta}{1 - \tan^2 \theta} \qquad\qquad \tan \frac{\theta}{2} = \sqrt{\frac{1 - \cos \theta}{1 + \cos \theta}}$$

$$\sin(A \pm B) = \sin A \cos B \pm \cos A \sin B$$
$$\cos(A \pm B) = \cos A \cos B \mp \sin A \sin B$$
$$\sin A \pm \sin B = 2 \sin[\tfrac{1}{2}(A \pm B)]\cos[\tfrac{1}{2}(A \mp B)]$$
$$\cos A + \cos B = 2 \cos[\tfrac{1}{2}(A + B)]\cos[\tfrac{1}{2}(A - B)]$$
$$\cos A - \cos B = 2 \sin[\tfrac{1}{2}(A + B)]\sin[\tfrac{1}{2}(B - A)]$$

EXAMPLE 3

Consider the right triangle in Figure B.11, in which $a = 2$, $b = 5$, and c is unknown. From the Pythagorean theorem, we have

$$c^2 = a^2 + b^2 = 2^2 + 5^2 = 4 + 25 = 29$$

$$c = \sqrt{29} = 5.39$$

To find the angle θ, note that

$$\tan \theta = \frac{a}{b} = \frac{2}{5} = 0.400$$

From a table of functions or from a calculator, we have

$$\theta = \tan^{-1}(0.400) = 21.8°$$

where $\tan^{-1}(0.400)$ is the notation for "angle whose tangent is 0.400," sometimes written as $\arctan(0.400)$.

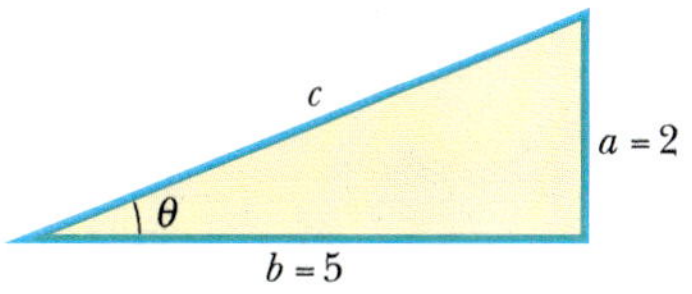

FIGURE B.11

EXERCISES

1. In Figure B.12, identify (a) the side opposite θ and (b) the side adjacent to ϕ and then find (c) $\cos \theta$, (d) $\sin \phi$, and (e) $\tan \phi$.

Answers (a) 3, (b) 3, (c) $\frac{4}{5}$, (d) $\frac{4}{5}$, and (e) $\frac{4}{3}$

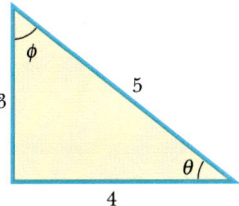

FIGURE B.12

2. In a certain right triangle, the two sides that are perpendicular to each other are 5 m and 7 m long. What is the length of the third side?

Answer 8.60 m

3. A right triangle has a hypotenuse of length 3 m, and one of its angles is 30°. What is the length of (a) the side opposite the 30° angle and (b) the side adjacent to the 30° angle?

Answers (a) 1.5 m, (b) 2.60 m

B.5 SERIES EXPANSIONS

$$(a + b)^n = a^n + \frac{n}{1!} a^{n-1}b + \frac{n(n - 1)}{2!} a^{n-2}b^2 + \cdots$$

$$(1 + x)^n = 1 + nx + \frac{n(n - 1)}{2!} x^2 + \cdots$$

$$e^x = 1 + x + \frac{x^2}{2!} + \frac{x^3}{3!} + \cdots$$

$$\ln(1 \pm x) = \pm x - \tfrac{1}{2}x^2 \pm \tfrac{1}{3}x^3 - \cdots$$

$$\left.
\begin{aligned}
\sin x &= x - \frac{x^3}{3!} + \frac{x^5}{5!} - \cdots \\[2mm]
\cos x &= 1 - \frac{x^2}{2!} + \frac{x^4}{4!} - \cdots \\[2mm]
\tan x &= x + \frac{x^3}{3} + \frac{2x^5}{15} + \cdots \quad |x| < \pi/2
\end{aligned}
\right\} \; x \text{ in radians}$$

For $x \ll 1$, the following approximations can be used:

$$(1 + x)^n \approx 1 + nx \qquad \sin x \approx x$$

$$e^x \approx 1 + x \qquad \cos x \approx 1$$

$$\ln(1 \pm x) \approx \pm x \qquad \tan x \approx x$$

B.6 DIFFERENTIAL CALCULUS

In various branches of science, it is sometimes necessary to use the basic tools of calculus, invented by Newton, to describe physical phenomena. The use of calculus is fundamental in the treatment of various problems in newtonian mechanics, electricity, and magnetism. In this section, we simply state some basic properties and "rules of thumb" that should be a useful review to the student.

First, a **function** must be specified that relates one variable to another (such as a coordinate as a function of time). Suppose one of the variables is called y (the dependent variable), the other x (the independent variable). We might have a function relationship such as

$$y(x) = ax^3 + bx^2 + cx + d$$

If a, b, c, and d are specified constants, then y can be calculated for any value of x. We usually deal with continuous functions, that is, those for which y varies "smoothly" with x.

The **derivative** of y with respect to x is defined as the limit, as Δx approaches zero, of the slopes of chords drawn between two points on the y versus x curve. Mathematically, we write this definition as

$$\frac{dy}{dx} = \lim_{\Delta x \to 0} \frac{\Delta y}{\Delta x} = \lim_{\Delta x \to 0} \frac{y(x + \Delta x) - y(x)}{\Delta x} \qquad \text{(B.28)}$$

where Δy and Δx are defined as $\Delta x = x_2 - x_1$ and $\Delta y = y_2 - y_1$ (Fig. B.13). It is important to note that dy/dx *does not* mean dy divided by dx, but is simply a notation of the limiting process of the derivative as defined by Equation B.28.

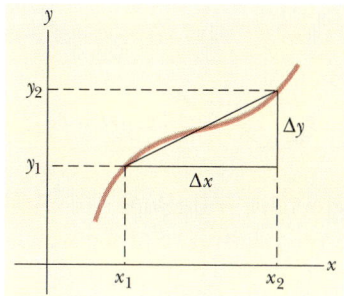

FIGURE B.13

A useful expression to remember when $y(x) = ax^n$, where a is a *constant* and n is *any* positive or negative number (integer or fraction), is

$$\frac{dy}{dx} = nax^{n-1} \qquad \text{(B.29)}$$

If $y(x)$ is a polynomial or algebraic function of x, we apply Equation B.29 to *each* term in the polynomial and take $da/dx = 0$. In Examples 4 through 7, we evaluate the derivatives of several functions.

EXAMPLE 4

Suppose $y(x)$ (that is, y as a function of x) is given by

$$y(x) = ax^3 + bx + c$$

where a and b are constants. Then it follows that

$$y(x + \Delta x) = a(x + \Delta x)^3$$
$$+ b(x + \Delta x) + c$$

$$y(x + \Delta x) = a(x^3 + 3x^2\,\Delta x + 3x\,\Delta x^2 + \Delta x^3)$$
$$+ b(x + \Delta x) + c$$

so

$$\Delta y = y(x + \Delta x) - y(x) = a(3x^2\,\Delta x + 3x\,\Delta x^2 + \Delta x^3)$$
$$+ b\,\Delta x$$

Substituting this into Equation B.28 gives

$$\frac{dy}{dx} = \lim_{\Delta x \to 0} \frac{\Delta y}{\Delta x} = \lim_{\Delta x \to 0} [3ax^2 + 3x\,\Delta x + \Delta x^2] + b$$

$$\frac{dy}{dx} = \boxed{3ax^2 + b}$$

EXAMPLE 5

$$y(x) = 8x^5 + 4x^3 + 2x + 7 \qquad\qquad \frac{dy}{dx} = 40x^4 + 12x^2 + 2$$

Solution   Applying Equation B.29 to each term independently, and remembering that d/dx (constant) $= 0$, we have

$$\frac{dy}{dx} = 8(5)x^4 + 4(3)x^2 + 2(1)x^0 + 0$$

Special Properties of the Derivative

A. **Derivative of the product of two functions**   If a function $f(x)$ is given by the product of two functions, say, $g(x)$ and $h(x)$, then the derivative of $f(x)$ is defined as

$$\frac{d}{dx}f(x) = \frac{d}{dx}[g(x)\,h(x)] = g\frac{dh}{dx} + h\frac{dg}{dx} \qquad (\text{B.30})$$

B. **Derivative of the sum of two functions**   If a function $f(x)$ is equal to the sum of two functions, then the derivative of the sum is equal to the sum of the derivatives:

$$\frac{d}{dx}f(x) = \frac{d}{dx}[g(x) + h(x)] = \frac{dg}{dx} + \frac{dh}{dx} \qquad (\text{B.31})$$

C. **Chain rule of differential calculus**   If $y = f(x)$ and $x = f(z)$, then dy/dx can be written as the product of two derivatives:

$$\frac{dy}{dx} = \frac{dy}{dz}\frac{dz}{dx} \qquad (\text{B.32})$$

D. **The second derivative**   The second derivative of y with respect to x is defined as the derivative of the function dy/dx (the derivative of the derivative). It is usually written

$$\frac{d^2y}{dx^2} = \frac{d}{dx}\left(\frac{dy}{dx}\right) \qquad (\text{B.33})$$

EXAMPLE 6

Find the derivative of $y(x) = x^3/(x+1)^2$ with respect to x.

Solution   We can rewrite this function as $y(x) = x^3(x + 1)^{-2}$ and apply Equation B.30:

$$\frac{dy}{dx} = (x+1)^{-2}\frac{d}{dx}(x^3) + x^3\frac{d}{dx}(x+1)^{-2}$$

$$= (x+1)^{-2}3x^2 + x^3(-2)(x+1)^{-3}$$

$$\frac{dy}{dx} = \frac{3x^2}{(x+1)^2} - \frac{2x^3}{(x+1)^3}$$

EXAMPLE 7

A useful formula that follows from Equation B.30 is the derivative of the quotient of two functions. Show that

$$\frac{d}{dx}\left[\frac{g(x)}{h(x)}\right] = \frac{h\dfrac{dg}{dx} - g\dfrac{dh}{dx}}{h^2}$$

Solution   We can write the quotient as gh^{-1} and then apply Equations B.29 and B.30:

$$\frac{d}{dx}\left(\frac{g}{h}\right) = \frac{d}{dx}(gh^{-1}) = g\frac{d}{dx}(h^{-1}) + h^{-1}\frac{d}{dx}(g)$$

$$= -gh^{-2}\frac{dh}{dx} + h^{-1}\frac{dg}{dx}$$

$$= \frac{h\dfrac{dg}{dx} - g\dfrac{dh}{dx}}{h^2}$$

Some of the more commonly used derivatives of functions are listed in Table B.4.

B.7   INTEGRAL CALCULUS

We think of integration as the inverse of differentiation. As an example, consider the expression

$$f(x) = \frac{dy}{dx} = 3ax^2 + b \tag{B.34}$$

which was the result of differentiating the function

$$y(x) = ax^3 + bx + c$$

in Example 4. We can write Equation B.34 as $dy = f(x)\,dx = (3ax^2 + b)\,dx$ and obtain $y(x)$ by "summing" over all values of x. Mathematically, we write this inverse operation

$$y(x) = \int f(x)\,dx$$

For the function $f(x)$ given by Equation B.34, we have

$$y(x) = \int (3ax^2 + b)\,dx = ax^3 + bx + c$$

where c is a constant of the integration. This type of integral is called an *indefinite integral* because its value depends on the choice of c.

A general **indefinite integral** $I(x)$ is defined as

$$I(x) = \int f(x)\,dx \tag{B.35}$$

where $f(x)$ is called the *integrand* and $f(x) = \dfrac{dI(x)}{dx}$.

For a *general continuous* function $f(x)$, the integral can be described as the area under the curve bounded by $f(x)$ and the x axis, between two specified values of x, say, x_1 and x_2, as in Figure B.14.

The area of the blue element is approximately $f_i\,\Delta x_i$. If we sum all these area elements from x_1 to x_2 and take the limit of this sum as $\Delta x_i \to 0$, we obtain the *true* area under the curve bounded by $f(x)$ and x, between the limits x_1 and x_2:

$$\text{Area} = \lim_{\Delta x_i \to 0} \sum_i f(x_i)\,\Delta x_i = \int_{x_1}^{x_2} f(x)\,dx \tag{B.36}$$

Integrals of the type defined by Equation B.36 are called **definite integrals**.

TABLE B.4   Derivatives for Several Functions

$$\frac{d}{dx}(a) = 0$$

$$\frac{d}{dx}(ax^n) = nax^{n-1}$$

$$\frac{d}{dx}(e^{ax}) = ae^{ax}$$

$$\frac{d}{dx}(\sin ax) = a\cos ax$$

$$\frac{d}{dx}(\cos ax) = -a\sin ax$$

$$\frac{d}{dx}(\tan ax) = a\sec^2 ax$$

$$\frac{d}{dx}(\cot ax) = -a\csc^2 ax$$

$$\frac{d}{dx}(\sec x) = \tan x\sec x$$

$$\frac{d}{dx}(\csc x) = -\cot x\csc x$$

$$\frac{d}{dx}(\ln ax) = \frac{1}{x}$$

Note: The letters a and n are constants.

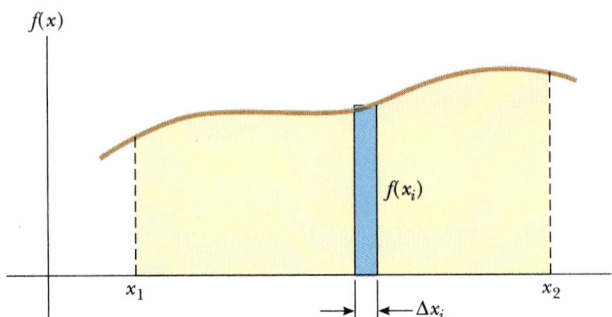

$f(x)$

$f(x_i)$

x_1 x_2

$\longrightarrow \mid \longleftarrow \Delta x_i$

FIGURE B.14

One common integral that arises in practical situations has the form

$$\int x^n \, dx = \frac{x^{n+1}}{n+1} + c \qquad (n \neq -1) \tag{B.37}$$

This result is obvious, being that differentiation of the right-hand side with respect to x gives $f(x) = x^n$ directly. If the limits of the integration are known, this integral becomes a *definite integral* and is written

$$\int_{x_1}^{x_2} x^n \, dx = \frac{x_2^{\,n+1} - x_1^{\,n+1}}{n+1} \qquad (n \neq -1) \tag{B.38}$$

EXAMPLES

1. $\displaystyle\int_0^a x^2 \, dx = \frac{x^3}{3}\bigg]_0^a = \frac{a^3}{3}$

2. $\displaystyle\int_0^b x^{3/2} \, dx = \frac{x^{5/2}}{5/2}\bigg]_0^b = \frac{2}{5}\,b^{5/2}$

3. $\displaystyle\int_3^5 x \, dx = \frac{x^2}{2}\bigg]_3^5 = \frac{5^2 - 3^2}{2} = 8$

Partial Integration

Sometimes it is useful to apply the method of *partial integration* (also called "integrating by parts") to evaluate certain integrals. The method uses the property that

$$\int u \, dv = uv - \int v \, du \tag{B.39}$$

where u and v are *carefully* chosen so as to reduce a complex integral to a simpler one. In many cases, several reductions have to be made. Consider the function

$$I(x) = \int x^2 e^x \, dx$$

This can be evaluated by integrating by parts twice. First, if we choose $u = x^2$, $v = e^x$, we get

$$\int x^2 e^x \, dx = \int x^2 \, d(e^x) = x^2 e^x - 2 \int e^x x \, dx + c_1$$

Now, in the second term, choose $u = x$, $v = e^x$, which gives

$$\int x^2 e^x \, dx = x^2 e^x - 2xe^x + 2 \int e^x \, dx + c_1$$

or

$$\int x^2 e^x \, dx = x^2 e^x - 2xe^x + 2e^x + c_2$$

The Perfect Differential

Another useful method to remember is the use of the *perfect differential*, in which we look for a change of variable such that the differential of the function is the differential of the independent variable appearing in the integrand. For example, consider the integral

$$I(x) = \int \cos^2 x \sin x \, dx$$

This becomes easy to evaluate if we rewrite the differential as $d(\cos x) = -\sin x \, dx$. The integral then becomes

$$\int \cos^2 x \sin x \, dx = -\int \cos^2 x \, d(\cos x)$$

If we now change variables, letting $y = \cos x$, we get

$$\int \cos^2 x \sin x \, dx = -\int y^2 dy = -\frac{y^3}{3} + c = -\frac{\cos^3 x}{3} + c$$

Table B.5 lists some useful indefinite integrals. Table B.6 gives Gauss's probability integral and other definite integrals. A more complete list can be found in various handbooks, such as *The Handbook of Chemistry and Physics*, CRC Press.

TABLE B.5 Some Indefinite Integrals (an arbitrary constant should be added to each of these integrals)

$$\int x^n \, dx = \frac{x^{n+1}}{n+1} \qquad (\text{provided } n \neq -1)$$

$$\int \frac{dx}{x} = \int x^{-1} \, dx = \ln x$$

$$\int \frac{dx}{a+bx} = \frac{1}{b} \ln(a+bx)$$

$$\int \frac{dx}{(a+bx)^2} = -\frac{1}{b(a+bx)}$$

$$\int \frac{dx}{a^2+x^2} = \frac{1}{a} \tan^{-1} \frac{x}{a}$$

$$\int \frac{dx}{a^2-x^2} = \frac{1}{2a} \ln \frac{a+x}{a-x} \qquad (a^2 - x^2 > 0)$$

$$\int \frac{dx}{x^2-a^2} = \frac{1}{2a} \ln \frac{x-a}{x+a} \qquad (x^2 - a^2 > 0)$$

$$\int \frac{x \, dx}{a^2 \pm x^2} = \pm \tfrac{1}{2} \ln(a^2 \pm x^2)$$

$$\int \frac{dx}{\sqrt{a^2-x^2}} = \sin^{-1} \frac{x}{a} = -\cos^{-1} \frac{x}{a} \qquad (a^2 - x^2 > 0)$$

$$\int \frac{dx}{\sqrt{x^2 \pm a^2}} = \ln(x + \sqrt{x^2 \pm a^2})$$

$$\int \frac{x \, dx}{\sqrt{a^2-x^2}} = -\sqrt{a^2 - x^2}$$

$$\int \frac{x \, dx}{\sqrt{x^2 \pm a^2}} = \sqrt{x^2 \pm a^2}$$

$$\int \sqrt{a^2-x^2} \, dx = \tfrac{1}{2}\left(x\sqrt{a^2-x^2} + a^2 \sin^{-1} \frac{x}{a} \right)$$

$$\int x\sqrt{a^2-x^2} \, dx = -\tfrac{1}{3}(a^2 - x^2)^{3/2}$$

$$\int \sqrt{x^2 \pm a^2} \, dx = \tfrac{1}{2}[x\sqrt{x^2 \pm a^2} \pm a^2 \ln(x + \sqrt{x^2 \pm a^2})]$$

$$\int x(\sqrt{x^2 \pm a^2}) \, dx = \tfrac{1}{3}(x^2 \pm a^2)^{3/2}$$

$$\int e^{ax} \, dx = \frac{1}{a} e^{ax}$$

$$\int \ln ax \, dx = (x \ln ax) - x$$

$$\int xe^{ax} \, dx = \frac{e^{ax}}{a^2}(ax - 1)$$

$$\int \frac{dx}{a+be^{cx}} = \frac{x}{a} - \frac{1}{ac} \ln(a + be^{cx})$$

$$\int \sin ax \, dx = -\frac{1}{a} \cos ax$$

$$\int \cos ax \, dx = \frac{1}{a} \sin ax$$

$$\int \tan ax \, dx = -\frac{1}{a} \ln(\cos ax) = \frac{1}{a} \ln(\sec ax)$$

$$\int \cot ax \, dx = \frac{1}{a} \ln(\sin ax)$$

$$\int \sec ax \, dx = \frac{1}{a} \ln(\sec ax + \tan ax) = \frac{1}{a} \ln\left[\tan\left(\frac{ax}{2} + \frac{\pi}{4} \right) \right]$$

$$\int \csc ax \, dx = \frac{1}{a} \ln(\csc ax - \cot ax) = \frac{1}{a} \ln\left(\tan \frac{ax}{2} \right)$$

$$\int \sin^2 ax \, dx = \frac{x}{2} - \frac{\sin 2ax}{4a}$$

$$\int \cos^2 ax \, dx = \frac{x}{2} + \frac{\sin 2ax}{4a}$$

$$\int \frac{dx}{\sin^2 ax} = -\frac{1}{a} \cot ax$$

$$\int \frac{dx}{\cos^2 ax} = \frac{1}{a} \tan ax$$

$$\int \tan^2 ax \, dx = \frac{1}{a}(\tan ax) - x$$

$$\int \cot^2 ax \, dx = -\frac{1}{a}(\cot ax) - x$$

$$\int \sin^{-1} ax \, dx = x(\sin^{-1} ax) + \frac{\sqrt{1 - a^2 x^2}}{a}$$

$$\int \cos^{-1} ax \, dx = x(\cos^{-1} ax) - \frac{\sqrt{1 - a^2 x^2}}{a}$$

$$\int \frac{dx}{(x^2+a^2)^{3/2}} = \frac{x}{a^2\sqrt{x^2 + a^2}}$$

$$\int \frac{x \, dx}{(x^2+a^2)^{3/2}} = -\frac{1}{\sqrt{x^2 + a^2}}$$

TABLE B.6   Gauss's Probability Integral and Related Integrals

$$I_0 = \int_0^\infty e^{-\alpha x^2}\, dx = \tfrac{1}{2}\sqrt{\frac{\pi}{\alpha}} \qquad \text{(Gauss's probability integral)}$$

$$I_1 = \int_0^\infty x e^{-\alpha x^2}\, dx = \frac{1}{2\alpha}$$

$$I_2 = \int_0^\infty x^2 e^{-\alpha x^2}\, dx = -\frac{dI_0}{d\alpha} = \tfrac{1}{4}\sqrt{\frac{\pi}{\alpha^3}}$$

$$I_3 = \int_0^\infty x^3 e^{-\alpha x^2}\, dx = -\frac{dI_1}{d\alpha} = \frac{1}{2\alpha^2}$$

$$I_4 = \int_0^\infty x^4 e^{-\alpha x^2}\, dx = \frac{d^2 I_0}{d\alpha^2} = \tfrac{3}{8}\sqrt{\frac{\pi}{\alpha^5}}$$

$$I_5 = \int_0^\infty x^5 e^{-\alpha x^2}\, dx = \frac{d^2 I_1}{d\alpha^2} = \frac{1}{\alpha^3}$$

$$\vdots$$

$$I_{2n} = (-1)^n \frac{d^n}{d\alpha^n} I_0$$

$$I_{2n+1} = (-1)^n \frac{d^n}{d\alpha^n} I_1$$

Periodic Table of the Elements*

Group I	Group II		Transition elements						
H 1 1.0080 $1s^1$									
Li 3 6.94 $2s^1$	**Be** 4 9.012 $2s^2$								
Na 11 22.99 $3s^1$	**Mg** 12 24.31 $3s^2$								

Symbol — **Ca** 20 — Atomic number
Atomic mass † — 40.08
$4s^2$ — Electron configuration

K 19 39.102 $4s^1$	**Ca** 20 40.08 $4s^2$	**Sc** 21 44.96 $3d^14s^2$	**Ti** 22 47.90 $3d^24s^2$	**V** 23 50.94 $3d^34s^2$	**Cr** 24 51.996 $3d^54s^1$	**Mn** 25 54.94 $3d^54s^2$	**Fe** 26 55.85 $3d^64s^2$	**Co** 27 58.93 $3d^74s^2$
Rb 37 85.47 $5s^1$	**Sr** 38 87.62 $5s^2$	**Y** 39 88.906 $4d^15s^2$	**Zr** 40 91.22 $4d^25s^2$	**Nb** 41 92.91 $4d^45s^1$	**Mo** 42 95.94 $4d^55s^1$	**Tc** 43 (99) $4d^55s^2$	**Ru** 44 101.1 $4d^75s^1$	**Rh** 45 102.91 $4d^85s^1$
Cs 55 132.91 $6s^1$	**Ba** 56 137.34 $6s^2$	57-71*	**Hf** 72 178.49 $5d^26s^2$	**Ta** 73 180.95 $5d^36s^2$	**W** 74 183.85 $5d^46s^2$	**Re** 75 186.2 $5d^56s^2$	**Os** 76 190.2 $5d^66s^2$	**Ir** 77 192.2 $5d^76s^2$
Fr 87 (223) $7s^1$	**Ra** 88 (226) $7s^2$	89-103**	**Unq** 104 (261) $6d^27s^2$	**Unp** 105 (262) $6d^37s^2$	**Unh** 106 (263)	**Uns** 107 (262)	**Uno** 108 (265)	**Une** 109 (266)

*Lanthanide series

La 57 138.91 $5d^16s^2$	**Ce** 58 140.12 $5d^14f^16s^2$	**Pr** 59 140.91 $4f^36s^2$	**Nd** 60 144.24 $4f^46s^2$	**Pm** 61 (147) $4f^56s^2$	**Sm** 62 150.4 $4f^66s^2$

**Actinide series

Ac 89 (227) $6d^17s^2$	**Th** 90 (232) $6d^27s^2$	**Pa** 91 (231) $5f^26d^17s^2$	**U** 92 (238) $5f^36d^17s^2$	**Np** 93 (239) $5f^46d^17s^2$	**Pu** 94 (239) $5f^66d^07s^2$

□ Atomic mass values given are averaged over isotopes in the percentages in which they exist in nature.
† For an unstable element, mass number of the most stable known isotope is given in parentheses.

	Group III	Group IV	Group V	Group VI	Group VII	Group 0
					H 1 1.0080 $1s^1$	**He** 2 4.0026 $1s^2$
	B 5 10.81 $2p^1$	**C** 6 12.011 $2p^2$	**N** 7 14.007 $2p^3$	**O** 8 15.999 $2p^4$	**F** 9 18.998 $2p^5$	**Ne** 10 20.18 $2p^6$
	Al 13 26.98 $3p^1$	**Si** 14 28.09 $3p^2$	**P** 15 30.97 $3p^3$	**S** 16 32.06 $3p^4$	**Cl** 17 35.453 $3p^5$	**Ar** 18 39.948 $3p^6$

Ni 28 58.71 $3d^84s^2$	**Cu** 29 63.54 $3d^{10}4s^2$	**Zn** 30 65.37 $3d^{10}4s^2$	**Ga** 31 69.72 $4p^1$	**Ge** 32 72.59 $4p^2$	**As** 33 74.92 $4p^3$	**Se** 34 78.96 $4p^4$	**Br** 35 79.91 $4p^5$	**Kr** 36 83.80 $4p^6$
Pd 46 106.4 $4d^{10}$	**Ag** 47 107.87 $4d^{10}5s^1$	**Cd** 48 112.40 $4d^{10}5s^2$	**In** 49 114.82 $5p^1$	**Sn** 50 118.69 $5p^2$	**Sb** 51 121.75 $5p^3$	**Te** 52 127.60 $5p^4$	**I** 53 126.90 $5p^5$	**Xe** 54 131.30 $5p^6$
Pt 78 195.09 $5d^96s^1$	**Au** 79 196.97 $5d^{10}6s^1$	**Hg** 80 200.59 $5d^{10}6s^2$	**Tl** 81 204.37 $6p^1$	**Pb** 82 207.2 $6p^2$	**Bi** 83 208.98 $6p^3$	**Po** 84 (210) $6p^4$	**At** 85 (218) $6p^5$	**Rn** 86 (222) $6p^6$

Eu 63 152.0 $4f^76s^2$	**Gd** 64 157.25 $5d^14f^76s^2$	**Tb** 65 158.92 $5d^14f^86s^2$	**Dy** 66 162.50 $4f^{10}6s^2$	**Ho** 67 164.93 $4f^{11}6s^2$	**Er** 68 167.26 $4f^{12}6s^2$	**Tm** 69 168.93 $4f^{13}6s^2$	**Yb** 70 173.04 $4f^{14}6s^2$	**Lu** 71 174.97 $5d^14f^{14}6s^2$
Am 95 (243) $5f^76d^07s^2$	**Cm** 96 (245) $5f^76d^17s^2$	**Bk** 97 (247) $5f^86d^17s^2$	**Cf** 98 (249) $5f^{10}6d^07s^2$	**Es** 99 (254) $5f^{11}6d^07s^2$	**Fm** 100 (253) $5f^{12}6d^07s^2$	**Md** 101 (255) $5f^{13}6d^07s^2$	**No** 102 (255) $6d^07s^2$	**Lr** 103 (257) $6d^17s^2$

SI Units

TABLE D.1   SI Base Units

Base Quantity	SI Base Unit	
	Name	Symbol
Length	Meter	m
Mass	Kilogram	kg
Time	Second	s
Electric current	Ampere	A
Temperature	Kelvin	K
Amount of substance	Mole	mol
Luminous intensity	Candela	cd

TABLE D.2   Some Derived SI Units

Quantity	Name	Symbol	Expression in Terms of Base Units	Expression in Terms of Other SI Units
Plane angle	Radian	rad	m/m	
Frequency	Hertz	Hz	s^{-1}	
Force	Newton	N	$kg \cdot m/s^2$	J/m
Pressure	Pascal	Pa	$kg/m \cdot s^2$	N/m^2
Energy: work	Joule	J	$kg \cdot m^2/s^2$	$N \cdot m$
Power	Watt	W	$kg \cdot m^2/s^3$	J/s
Electric charge	Coulomb	C	$A \cdot s$	
Electric potential (emf)	Volt	V	$kg \cdot m^2/A \cdot s^3$	W/A
Capacitance	Farad	F	$A^2 \cdot s^4/kg \cdot m^2$	C/V
Electric resistance	Ohm	Ω	$kg \cdot m^2/A^2 \cdot s^3$	V/A
Magnetic flux	Weber	Wb	$kg \cdot m^2/A \cdot s^2$	$V \cdot s$
Magnetic field intensity	Tesla	T	$kg/A \cdot s^2$	Wb/m^2
Inductance	Henry	H	$kg \cdot m^2/A^2 \cdot s^2$	Wb/A

Nobel Prizes

All Nobel Prizes in physics are listed (and marked with a P), as well as relevant Nobel Prizes in Chemistry (C). The key dates for some of the scientific work are supplied; they often antedate the prize considerably.

1901   (P) *Wilhelm Roentgen* for discovering x-rays (1895).

1902   (P) *Hendrik A. Lorentz* for predicting the Zeeman effect and *Pieter Zeeman* for discovering the Zeeman effect, the splitting of spectral lines in magnetic fields.

1903   (P) *Antoine-Henri Becquerel* for discovering radioactivity (1896) and *Pierre* and *Marie Curie* for studying radioactivity.

1904   (P) *Lord Rayleigh* for studying the density of gases and discovering argon.
(C) *William Ramsay* for discovering the inert gas elements helium, neon, xenon, and krypton, and placing them in the periodic table.

1905   (P) *Philipp Lenard* for studying cathode rays, electrons (1898–1899).

1906   (P) *J. J. Thomson* for studying electrical discharge through gases and discovering the electron (1897).

1907   (P) *Albert A. Michelson* for inventing optical instruments and measuring the speed of light (1880s).

1908   (P) *Gabriel Lippmann* for making the first color photographic plate, using interference methods (1891).
(C) *Ernest Rutherford* for discovering that atoms can be broken apart by alpha rays and for studying radioactivity.

1909   (P) *Guglielmo Marconi* and *Carl Ferdinand Braun* for developing wireless telegraphy.

1910   (P) *Johannes D. van der Waals* for studying the equation of state for gases and liquids (1881).

1911   (P) *Wilhelm Wien* for discovering Wien's law giving the peak of a blackbody spectrum (1893).
(C) *Marie Curie* for discovering radium and polonium (1898) and isolating radium.

1912   (P) *Nils Dalén* for inventing automatic gas regulators for lighthouses.

1913   (P) *Heike Kamerlingh Onnes* for the discovery of superconductivity and liquefying helium (1908).

1914   (P) *Max T. F. von Laue* for studying x-rays from their diffraction by crystals, showing that x-rays are electromagnetic waves (1912).
(C) *Theodore W. Richards* for determining the atomic weights of sixty elements, indicating the existence of isotopes.

1915   (P) *William Henry Bragg* and *William Lawrence Bragg,* his son, for studying the diffraction of x-rays in crystals.

1917   (P) *Charles Barkla* for studying atoms by x-ray scattering (1906).

1918   (P) *Max Planck* for discovering energy quanta (1900).

1919   (P) *Johannes Stark,* for discovering the Stark effect, the splitting of spectral lines in electric fields (1913).

1920   (P) *Charles-Édouard Guillaume* for discovering invar, a nickel-steel alloy with low coefficient of expansion.
(C) *Walther Nernst* for studying heat changes in chemical reactions and formulating the third law of thermodynamics (1918).

1921   (P) *Albert Einstein* for explaining the photoelectric effect and for his services to theoretical physics (1905).
(C) *Frederick Soddy* for studying the chemistry of radioactive substances and discovering isotopes (1912).

1922   (P) *Niels Bohr* for his model of the atom and its radiation (1913).
(C) *Francis W. Aston* for using the mass spectrograph to study atomic weights, thus discovering 212 of the 287 naturally occurring isotopes.

1923   (P) *Robert A. Millikan* for measuring the charge on an electron (1911) and for studying the photoelectric effect experimentally (1914).

1924   (P) *Karl M. G. Siegbahn* for his work in x-ray spectroscopy.

1925   (P) *James Franck* and *Gustav Hertz* for discovering the Franck-Hertz effect in electron-atom collisions.

1926   (P) *Jean-Baptiste Perrin* for studying Brownian motion to validate the discontinuous structure of matter and measure the size of atoms.

1927   (P) *Arthur Holly Compton* for discovering the Compton effect on x-rays, their change in wavelength when they collide with matter (1922), and *Charles T. R. Wilson* for inventing the cloud chamber, used to study charged particles (1906).

1928   (P) *Owen W. Richardson* for studying the thermionic effect and electrons emitted by hot metals (1911).

1929   (P) *Louis Victor de Broglie* for discovering the wave nature of electrons (1923).

1930   (P) *Chandrasekhara Venkata Raman* for studying Raman scattering, the scattering of light by atoms and molecules with a change in wavelength (1928).

1932   (P) *Werner Heisenberg* for creating quantum mechanics (1925).

1933   (P) *Erwin Schrödinger* and *Paul A. M. Dirac* for developing wave mechanics (1925) and relativistic quantum mechanics (1927).
(C) *Harold Urey* for discovering heavy hydrogen, deuterium (1931).

1935   (P) *James Chadwick* for discovering the neutron (1932).
(C) *Irène* and *Frédéric Joliot-Curie* for synthesizing new radioactive elements.

1936   (P) *Carl D. Anderson* for discovering the positron in particular and antimatter in general (1932) and *Victor F. Hess* for discovering cosmic rays.
(C) *Peter J. W. Debye* for studying dipole moments and diffraction of x-rays and electrons in gases.

1937   (P) *Clinton Davisson* and *George Thomson* for discovering the diffraction of electrons by crystals, confirming de Broglie's hypothesis (1927).

1938   (P) *Enrico Fermi* for producing the transuranic radioactive elements by neutron irradiation (1934–1937).

1939   (P) *Ernest O. Lawrence* for inventing the cyclotron.

1943   (P) *Otto Stern* for developing molecular-beam studies (1923), and using them to discover the magnetic moment of the proton (1933).

1944   (P) *Isidor I. Rabi* for discovering nuclear magnetic resonance in atomic and molecular beams.
(C) *Otto Hahn* for discovering nuclear fission (1938).

1945   (P) *Wolfgang Pauli* for discovering the exclusion principle (1924).

1946   (P) *Percy W. Bridgman* for studying physics at high pressures.

1947   (P) *Edward V. Appleton* for studying the ionosphere.

1948　(P) *Patrick M. S. Blackett* for studying nuclear physics with cloud-chamber photographs of cosmic-ray interactions.

1949　(P) *Hideki Yukawa* for predicting the existence of mesons (1935).

1950　(P) *Cecil F. Powell* for developing the method of studying cosmic rays with photographic emulsions and discovering new mesons.

1951　(P) *John D. Cockcroft* and *Ernest T. S. Walton* for transmuting nuclei in an accelerator (1932).

　　　(C) *Edwin M. McMillan* for producing neptunium (1940) and *Glenn T. Seaborg* for producing plutonium (1941) and further transuranic elements.

1952　(P) *Felix Bloch* and *Edward Mills Purcell* for discovering nuclear magnetic resonance in liquids and gases (1946).

1953　(P) *Frits Zernike* for inventing the phase-contrast microscope, which uses interference to provide high contrast.

1954　(P) *Max Born* for interpreting the wave function as a probability (1926) and other quantum-mechanical discoveries and *Walther Bothe* for developing the coincidence method to study subatomic particles (1930–1931), producing, in particular, the particle interpreted by Chadwick as the neutron.

1955　(P) *Willis E. Lamb, Jr.,* for discovering the Lamb shift in the hydrogen spectrum (1947) and *Polykarp Kusch* for determining the magnetic moment of the electron (1947).

1956　(P) *John Bardeen, Walter H. Brattain,* and *William Shockley* for inventing the transistor (1956).

1957　(P) *T.-D. Lee* and *C.-N. Yang* for predicting that parity is not conserved in beta decay (1956).

1958　(P) *Pavel A. Čerenkov* for discovering Čerenkov radiation (1935) and *Ilya M. Frank* and *Igor Tamm* for interpreting it (1937).

1959　(P) *Emilio G. Segrè* and *Owen Chamberlain* for discovering the antiproton (1955).

1960　(P) *Donald A. Glaser* for inventing the bubble chamber to study elementary particles (1952).

　　　(C) *Willard Libby* for developing radiocarbon dating (1947).

1961　(P) *Robert Hofstadter* for discovering internal structure in protons and neutrons and *Rudolf L. Mössbauer* for discovering the Mössbauer effect of recoilless gamma-ray emission (1957).

1962　(P) *Lev Davidovich Landau* for studying liquid helium and other condensed matter theoretically.

1963　(P) *Eugene P. Wigner* for applying symmetry principles to elementary-particle theory and *Maria Goeppert Mayer* and *J. Hans D. Jensen* for studying the shell model of nuclei (1947).

1964　(P) *Charles H. Townes, Nikolai G. Basov,* and *Alexandr M. Prokhorov* for developing masers (1951–1952) and lasers.

1965　(P) *Sin-itiro Tomonaga, Julian S. Schwinger,* and *Richard P. Feynman* for developing quantum electrodynamics (1948).

1966　(P) *Alfred Kastler* for his optical methods of studying atomic energy levels

1967　(P) *Hans Albrecht Bethe* for discovering the routes of energy production in stars (1939).

1968　(P) *Luis W. Alvarez* for discovering resonance states of elementary particles.

1969　(P) *Murray Gell-Mann* for classifying elementary particles (1963).

1970　(P) *Hannes Alfvén* for developing magnetohydrodynamic theory and *Louis Eugène Félix Néel* for discovering antiferromagnetism and ferrimagnetism (1930s).

1971   (P) *Dennis Gabor* for developing holography (1947).
       (C) *Gerhard Herzberg* for studying the structure of molecules spectroscopically.

1972   (P) *John Bardeen, Leon N. Cooper,* and *John Robert Schrieffer* for explaining superconductivity (1957).

1973   (P) *Leo Esaki* for discovering tunneling in semiconductors, *Ivar Giaever* for discovering tunneling in superconductors, and *Brian D. Josephson* for predicting the Josephson effect, which involves tunneling of paired electrons (1958–1962).

1974   (P) *Anthony Hewish* for discovering pulsars and *Martin Ryle* for developing radio interferometry.

1975   (P) *Aage N. Bohr, Ben R. Mottelson,* and *James Rainwater* for discovering why some nuclei take asymmetric shapes.

1976   (P) *Burton Richter* and *Samuel C. C. Ting* for discovering the J/psi particle, the first charmed particle (1974).

1977   (P) *John H. Van Vleck, Nevill F. Mott,* and *Philip W. Anderson* for studying solids quantum-mechanically.
       (C) *Ilya Prigogine* for extending thermodynamics to show how life could arise in the face of the second law.

1978   (P) *Arno A. Penzias* and *Robert W. Wilson* for discovering the cosmic background radiation (1965) and *Pyotr Kapitsa* for his studies of liquid helium.

1979   (P) *Sheldon L. Glashow, Abdus Salam,* and *Steven Weinberg* for developing the theory that unified the weak and electromagnetic forces (1958–1971).

1980   (P) *Val Fitch* and *James W. Cronin* for discovering CP (charge-parity) violation (1964), which possibly explains the cosmological dominance of matter over antimatter.

1981   (P) *Nicolaas Bloembergen* and *Arthur L. Schawlow* for developing laser spectroscopy and *Kai M. Siegbahn* for developing high-resolution electron spectroscopy (1958).

1982   (P) *Kenneth G. Wilson* for developing a method of constructing theories of phase transitions to analyze critical phenomena.

1983   (P) *William A. Fowler* for theoretical studies of astrophysical nucleosynthesis and *Subramanyan Chandrasekhar* for studying physical processes of importance to stellar structure and evolution, including the prediction of white dwarf stars (1930).

1984   (P) *Carlo Rubbia* for discovering the W and Z particles, verifying the electroweak unification, and *Simon van der Meer,* for developing the method of stochastic cooling of the CERN beam that allowed the discovery (1982–1983).

1985   (P) *Klaus von Klitzing* for the quantized Hall effect, relating to conductivity in the presence of a magnetic field (1980).

1986   (P) *Ernst Ruska* for inventing the electron microscope (1931), and *Gerd Binnig* and *Heinrich Rohrer* for inventing the scanning-tunneling electron microscope (1981).

1987   (P) *J. Georg Bednorz* and *Karl Alex Müller* for the discovery of high temperature superconductivity (1986).

1988   (P) *Leon M. Lederman, Melvin Schwartz,* and *Jack Steinberger* for a collaborative experiment that led to the development of a new tool for studying the weak nuclear force, which affects the radioactive decay of atoms.

1989   (P) *Norman Ramsay* (U.S.) for various techniques in atomic physics; and

Hans Dehmelt (U.S.) and *Wolfgang Paul* (Germany) for the development of techniques for trapping single charge particles.

1990 (P) *Jerome Friedman, Henry Kendall* (both U.S.), and *Richard Taylor* (Canada) for experiments important to the development of the quark model.

1991 (P) *Pierre-Gilles de Gennes* for discovering that methods developed for studying order phenomena in simple systems can be generalized to more complex forms of matter, in particular to liquid crystals and polymers.

1992 (P) *George Charpak* for developing detectors that trace the paths of evanescent subatomic particles produced in particle accelerators.

1993 (P) *Russell Hulse* and *Joseph Taylor* for discovering evidence of gravitational waves.

1994 (P) *Bertram N. Brockhouse* and *Clifford G. Shull* for pioneering work in neutron scattering.

Spreadsheet Problems

OVERVIEW

Students come to introductory physics courses with a wide variety of computing experience. Many are already accomplished programmers in one or more programming languages (BASIC, Pascal, FORTRAN, and so forth). Others have never even turned on a computer. To further complicate matters, a wide variety of hardware environments exists, although most can be classified as IBM/compatible (MS-DOS) or Macintosh environments. We have designed the end-of-chapter spreadsheet problems and the text ancillary, *Spreadsheet Investigations in Physics*, to be usable by and useful to students in all these diverse situations. Our goal is to enable students to investigate a range of physical phenomena and obtain a feel for the physics. Merely "getting the right answer" by plugging numbers into a formula and comparing the result to the answer in the back of the book is discouraged.

Spreadsheets are particularly valuable in exploratory investigations. Once you have constructed a spreadsheet, you can simply vary the parameters and see instantly how things change. Even more important is the ease with which you can construct accurate graphs of relations between physical variables. When you change a parameter, you can view the effects of the change upon the graphs simply by pressing a key. "What if" questions can be easily addressed and depicted graphically.

HOW TO USE THE TEMPLATES

The computer spreadsheet problems are arranged by level of difficulty. The least difficult problems are coded in black. For most of these problems, spreadsheets are provided on disk, and only the input parameters need to be changed. Problems of moderate difficulty, coded in blue, require additional analysis, and the provided spreadsheets must be modified to solve them. The most challenging problems are coded in magenta. For most of these, you must develop your own spreadsheets. The emphasis should be on understanding what the results mean rather than just getting an answer. For example, one spreadsheet problem explores how the distance of the horizon varies with height above the ground. You can explore why you can see farther distances when you're on top of a tall building than when you are on the ground. Why were lookouts on sailing ships placed in the crow's nest at the top of the mast?

SOFTWARE REQUIREMENTS

The spreadsheet templates are provided on a high-density (1.44 Megabyte) MS-DOS diskette using the Lotus 1-2-3 WK1 format. This format was introduced with versions 2.x of the Lotus 1-2-3 program and can be read by all subsequent versions. It can also be read directly by all the other major spreadsheet programs, including the latest Windows versions of Lotus 1-2-3, Microsoft Excel, Microsoft Works, and Novell/Wordperfect Quattro Pro as well as Microsoft Excel for the Macintosh. The program f(g) Scholar can import WK1 spreadsheets; however, some minor format changes of the templates are needed.

The Lotus WK1 format was chosen so that the templates will be usable in the widest possible variety of computing environments. Even though most spreadsheet programs operate in basically the same way, many of the latest spreadsheet programs have very powerful formatting and graphing capabilities along with many other useful features. The user of these powerful programs can exploit these capabilities to improve the appearance of their spreadsheets.

HARDWARE REQUIREMENTS

You will need a microcomputer that can run one of the spreadsheet programs in one of their many versions. Your computer should be connected to a printer that can print text and graphics. Older versions of the software will run on a 8086/8088 MS-DOS system with just a single floppy disk drive or on a 512K Macintosh with two floppy drives. Newer versions require a more powerful computer to run effectively. For example, to run Excel for Windows, Version 5.0, you must have a hard disk with about 15 megabytes of disk memory available and four to eight megabytes of RAM. Your software manuals will tell you exactly what you need to run the particular version of your spreadsheet program. However, all problems require only a minimal computer system.

There are many different software versions and many different computer configurations. You might have one floppy drive, two floppy drives, a hard disk, a local area network, and so on. The combinations are almost endless. Our best suggestion is to read your software manual and ask your instructor or computer laboratory personnel how to start your spreadsheet program.

SPREADSHEET TUTORIAL

Some students will have the required computer and spreadsheet skills to start working with the templates immediately. Other students, who have not had experience with a spreadsheet program, will need some instruction. We have written an ancillary entitled *Spreadsheet Investigations in Physics* for both these groups of students. The first part of this ancillary contains a spreadsheet tutorial that the novice student can use *independently* to gain the required spreadsheet skills. A two- or three-hour initial session with the tutorial and the computer is all that most students need to get started. Once the students have mastered the basic operations of the spreadsheet program, they should try one or two of the easier problems.

Because very few introductory physics students have studied numerical methods, we have also included a brief introduction to numerical methods in this ancillary. This section covers numerical interpolation, differentiation, integration, and the solution of simple differential equations. The student should not try to master all of this material at one time; only study the sections that are needed to solve the currently assigned problems.

The templates supplied on the distribution diskettes with *Spreadsheet Investigations in Physics* constitute an outline. You must enter the appropriate data and parameters. The parameters must be adjusted to fit the needs of your problem. Feel free to change any parameters, to expand or decrease the number of rows of output, and to change the size of increments (for example, in time or distance).

Most templates have graphs associated with them. You may have to adjust the ranges of variables plotted and the scales for the axes. See the tutorials for how to adjust the appearance of your graphs.

COPY YOUR DISTRIBUTION DISKETTES

Since you will be modifying the spreadsheets on the distribution diskettes, copy the distribution diskettes and place the originals in a safe place.

Answers to Odd-Numbered Problems

Chapter 1

1. 2.80 g/cm^3
3. 184 g
3A. $m = \frac{4}{3}\pi\rho(r_2^3 - r_1^3)$
5. (a) $4 \text{ u} = 6.64 \times 10^{-24} \text{ g}$
   (b) $56 \text{ u} = 9.30 \times 10^{-23} \text{ g}$
   (c) $207 \text{ u} = 3.44 \times 10^{-22} \text{ g}$
7. (a) 72.58 kg (b) $7.82 \times 10^{26} \text{ atoms}$
9. It is.
11. It is.
13. (b) only
15. L^3/T^3, L^3T
17. The units of G are $\text{m}^3/(\text{kg} \cdot \text{s}^2)$.
19. $1.39 \times 10^3 \text{ m}^2$
21. $8.32 \times 10^{-4} \text{ m/s}$
23. $11.4 \times 10^3 \text{ kg/m}^3$
25. (a) $6.31 \times 10^4 \text{ AU}$ (b) $1.33 \times 10^{11} \text{ AU}$
27. (a) 127 y (b) $15\ 500 \text{ times}$
29. (a) $1 \text{ mi/h} = 1.609 \text{ km/h}$ (b) 88.5 km/h
    (c) 16.1 km/h
31. $1.51 \times 10^{-4} \text{ m}$
33. $1.00 \times 10^{10} \text{ lb}$
35. 5 m
37. $5.95 \times 10^{24} \text{ kg}$
39. 2.86 cm
41. $\approx 10^6$
43. $1.79 \times 10^{-9} \text{ m}$
45. $3.84 \times 10^8 \text{ m}$
47. 34.1 m
49. $\approx 10^2$
51. (a) $(346 \pm 13) \text{ m}^2$ (b) $(66.0 \pm 1.3) \text{ m}$
53. $195.8 \text{ cm}^2 \pm 0.7\%$
55. $3, 4, 3, 2$
57. $5.2 \text{ m}^3 \pm 3\%$
59. It is not.
61. 0.449%
63. (a) 1000 kg (b) $5 \times 10^{-16} \text{ kg}$, 300 g, 0.01 g
65. (a) 10^6 (b) 10^7 (c) 10^3
S1. (a) Jud's horizon is 5.060 m and Spud's horizon is 4.382 m. It makes no difference whether you use d, s, or l. (b) On the Moon, Jud's horizon is 2.638 m and Spud's horizon is 2.285 m.
S3. (a) A plot of log T versus log L for the given data shows an approximate linear relationship. Applying the least-squares method to the data yields 1.017 647 for the slope and 0.890 686 for the intercept of the straight line that best fits the data. The data points and the best-fit line, log $T = 1.017\ 647$ log $m + 0.890\ 686$ are shown in the figure. Paying attention to significant figures, a reasonable interpretation of this result is that $n = 1.1$. In terms of T and m, our best-fit equation is $T = Cm^n$, where $C = 10^{1.1} = 12.6$.

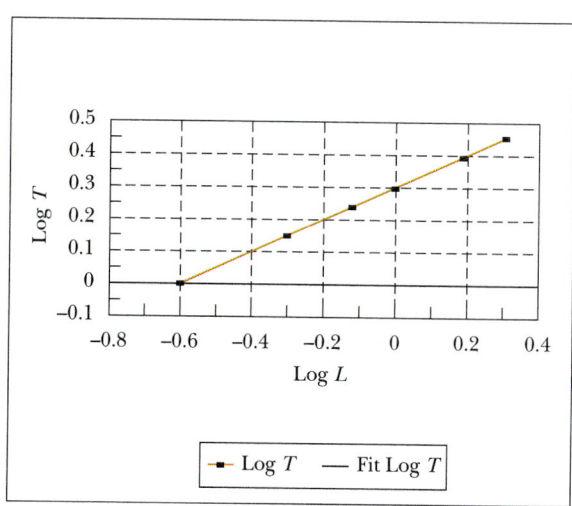

Chapter 2

1. (a) 2.3 m/s (b) 16.1 m/s (c) 11.5 m/s
3. (a) 5.0 m/s (b) 1.2 m/s (c) -2.5 m/s
   (d) -3.3 m/s (e) 0
5. (a) 3.75 m/s (b) 0
7. 50.0 km/h
9. (a) -2.4 m/s (b) -3.2 m/s (c) 4.0 s
11. (b) 1.60 m/s
13. (a) 5.0 m/s (b) -2.5 m/s (c) 0
    (d) 5.0 m/s
15. -4.00 m/s^2
17. 20.0 m/s; 6.00 m/s^2
19. (a) 20.0 m/s, 5.00 m/s (b) 262 m
21. (a) 2.00 m (b) -3.00 m/s (c) -2.00 m/s^2
23. (a) 1.3 m/s^2 (b) 2.2 m/s^2 at 3.0 s
    (c) At $t = 6.2 \text{ s}$ and for the interval $10 \text{ s} < t < 12 \text{ s}$
    (d) -2.0 m/s^2 at 8.2 s

A.43

25. -16.0 cm/s^2
27. 160 ft
29. (a) 12.7 m/s   (b) -2.30 m/s
31. (a) 20.0 s   (b) no
33. (a) 8.94 s   (b) 89.4 m/s
35. (a) 0.444 m/s^2 (b) 1.33 m/s   (c) 2.12 s
     (d) 0.943 m/s
37. (a) 45.7 s   (b) 574 m   (c) 12.6 m/s   (d) 765 s
39. (a) $x = 30t - t^2$, $v = 30 - 2t$ (b) 225 m
41. 11.4 s, 212 m
43. (a) -662 ft/s^2 (b) 649 ft
45. (a) 96 ft/s   (b) 3.08×10^3 ft/s^2
     (c) 3.12×10^{-2} s
47. (a) 10.0 m/s   (b) -4.68 m/s
49. (a) 1.53 s   (b) 11.5 m
     (c) -4.60 m/s, -9.80 m/s^2
51. (a) 29.4 m/s   (b) 44.1 m
53. 7.96 s
55. (a) 7.82 m   (b) 0.782 s
57. (a) 7.00 m/s   (b) -5.35 m/s   (c) -9.8 m/s^2
59. $a(t) = a_0 + Jt$, $v(t) = v_0 + a_0 t + \frac{1}{2} J t^2$,
     $x(t) = x_0 + v_0 t + \frac{1}{2} a_0 t^2 + \frac{1}{6} J t^3$
61. 0.509 s
63. (a) 3.00 s   (b) -15.3 m/s
     (c) -31.4 m/s, -34.8 m/s
65. (a) -6.26 m/s   (b) 6.02 m/s   (c) 1.25 s
67. 4.63 m
69. (a) 5.45 s   (b) 73.0 m
     (c) $v_{STAN} = 22.6$ m/s, $v_{KATHY} = 26.7$ m/s
71. (a) 12.5 s   (b) -2.29 m/s^2 (c) 13.1 s
73. (a) 5.28 m/s   (b) 4.83×10^{-4} m/s^2
75. $t_{AB} = t_{CD} = 2.00$ min, $t_{BC} = 1.00$ min
77. (a) 5.43 m/s^2, 3.83 m/s^2 (b) 10.9 m/s, 11.5 m/s
     (c) Maggie, 2.62 m
79. 155 s; 129 s
81. $0.577v$
S4. (a) At approximately $t = 37.0$ s after the police car
     starts, or 42.0 s after the first policeman sees the
     speeder.   (b) 74.0 m/s   (c) 1370 m

Chapter 3

1. (a) 8.60 m   (b) 4.47 m at 297°; 4.24 m at 135°
3. $(-2.75$ m, -4.76 m)
5. (2.24 m, 26.6°)
7. (2.05 m, 1.43 m)
9. 70.0 m
11. 310 km at 57° south of west
13. (a) 10.0 m   (b) 15.7 m   (c) 0
15. (a) 5.20 m at 60.0°   (b) 3.00 m at 330°
     (c) 3.00 m at 150°   (d) 5.20 m at 300°
17. 421 ft at $-2.63°$
19. 86.6 m, -50.0 m
21. 5.83 m at 59.0° to the right of first direction
23. 47.2 units at 122°

25. (b) $5i + 4j = 6.40$ at 38.7°;
     $-i + 8j = 8.06$ at 97.2°
27. 7.21 m at 56.3°
29. (a) $2i - 6j$ (b) $4i + 2j$ (c) 6.32   (d) 4.47
     (e) 288°; 26.6°
31. (a) $(-11.1$ m$)i + (6.40$ m$)j$
     (b) $(1.65$ cm$)i + (2.86$ cm$)j$
     (c) $(-18.0$ in.$)i - (12.6$ in.$)j$
33. 9.48 m at 166°
35. 390 mph at 7.37° north of east
37. $(2.60$ m$)i + (4.50$ m$)j$
39. (a) $8i + 12j - 4k$ (b) $2i + 3j - k$
     (c) $-24i - 36j + 12k$
41. (a) 5.92   (b) 19.0
43. (a) $-3i + 2j$ (b) 3.61 at 146°   (c) $3i - 6j$
47. (a) 49.5, 27.1   (b) 56.4 at 28.7°
49. 240 m at 237°
51. (10.0 m, 20.0 m)

Chapter 4

1. $(8a + 2b)i + 2cj$
3. (a) 4.87 km at 209° from east   (b) 23.3 m/s
     (c) 13.5 m/s at 209°
5. (a) $(2i + 3j)$ m/s^2
     (b) $(3t + t^2)i + (-2t + 1.5t^2)j$
7. (a) $(0.8i - 0.3j)$ m/s^2 (b) at 339°
     (c) $(360i - 72.8j)$ m; at 345°
9. (a) $v = (-5i + 0j)$ m/s, $a = (0i - 5j)$ m/s^2
     (b) $r = -5i \sin t + 4j - 5j \cos t$
         $v = -5i \cos t + 5j \sin t$
         $a = 5i \sin t + 5j \cos t$
     (c) a circle of radius 5 m centered at (0, 4 m)
11. (a) 3.34 m/s at 0°   (b) 309°
13. (a) 2.67 s   (b) 29.9 m/s   (c) $v_x = 29.9$ m/s;
         $v_y = -26.2$ m/s
15. 9.91 m/s
17. (a) clears by 0.89 m   (b) while falling 13.3 m/s
19. (a) 1.69 km/s   (b) 6490 s
21. (a) 277 km   (b) 284 s
23. 53.1°
25. 22.4° or 89.4°
27. (a) 7.90 km/s   (b) 5070 s
29. 377 m/s^2
31. (a) 1.02 km/s   (b) 2.72 mm/s^2
33. 54.4 m/s^2
35. (a) 13.0 m/s^2 (b) 5.70 m/s   (c) 7.50 m/s^2
37. 1.48 m/s^2
39. (a) $-30.8j$ m/s^2 (b) $70.4j$ m/s^2
41. (a) 26.9 m/s   (b) 67.3 m   (c) $(2i - 5j)$ m/s^2
43. 2.02×10^3 s; 21.0% longer
45. 2.50 m/s
47. 153 km/h at 11.3° north of west
49. (a) 57.7 km/h at 210°   (b) 28.9 km/h down
51. (a) 10.1 m/s^2 south at 75.7° below horizontal
     (b) 9.80 m/s^2 down
53. (a) $4.00i$ m/s   (b) (4.00 m, 6.00 m)

55. (a) 20.0 m/s, 5.00 s   (b) 31.4 m/s at 301°
    (c) 6.53 s   (d) 24.5 m away
57. 3.14 m
59. 20.0 m
61. (a) 0.60 m   (b) 0.40 m
    (c) 1.87 m/s² toward center   (d) 9.80 m/s² down
63. 4.12 m
65. (a) $\sqrt{gR}$ (b) $(\sqrt{2} - 1)R$
67. 10.8 m
69. (a) 6.80 km   (b) directly above explosion
    (c) 66.2°
71. (a) $(-7.05 \text{ cm})\mathbf{j}$ (b) $(7.61 \text{ cm})\mathbf{i} - (6.48 \text{ cm})\mathbf{j}$
    (c) $(10.0 \text{ cm})\mathbf{i} - (7.05 \text{ cm})\mathbf{j}$
73. (a) 22.2° or 67.8°   (b) 235 m   (c) 235 m
75. (a) 407 km/h at 10.6° N of E   (b) 10.8° S of E
77. (a) 5.15 s   (b) 4.85 m/s at 74.5° N of W
    (c) 19.4 m
79. (a) 43.2 m   (b) $(9.66 \text{ m/s})\mathbf{i} - (25.6 \text{ m/s})\mathbf{j}$
81. (18.8 m, −17.4 m)
83. (a) 46.5 m/s   (b) −77.6°   (c) 6.34 s
85. 2.98 km/s forward and 1.96° inward
S1. There are an infinite number of solutions to the problem; however, there is one practical limitation. We can estimate that the maximum speed that the punter can give the ball is about 20 to 30 m/s. Use a speed in this range.
S3. There are an infinite number of solutions. For example if v_0 = 22.32 m/s at 45°, then the ball clears the crossbar in 3.01 s. Or if v_0 = 24.60 m/s at 30°, then the ball clears the crossbar in 2.23 s. The time it takes the ball to clear the crossbar is immaterial, since in all likelihood time will run out before the clock is stopped.

Chapter 5

1. (a) 1/3   (b) 0.750 m/s²
3. (a) 3.00 s   (b) 20.1 m   (c) $(18.0\mathbf{i} - 9.0\mathbf{j})$ m
5. 312 N
7. $(6\mathbf{i} + 15\mathbf{j})$ N, 16.2 N
9. (a) 556 N   (b) 56.7 kg
11. (a) 4.47×10^{15} m/s² outward
    (b) 2.09×10^{-10} N inward
13. 1.72 kg, 6.97 m/s²
15. (c) Forces of brick on spring and spring on brick; of spring on table and table on spring; of table on Earth and Earth on table; of brick on Earth and of Earth on brick; of spring on Earth and of Earth on spring.
17. (a) 5.10 kN   (b) 3.62×10^3 kg
19. (a) 2.0 m/s²   (b) 170 N   (c) 2.93 m/s²
21. (a) $(2.50 \text{ N})\mathbf{i} + (5.00 \text{ N})\mathbf{j}$ (b) 5.59 N
23. 1.24 ft/s²
25. 640 N for $0 \leq t \leq 1.0$ s, 627 N at $t = 1.3$ s, 589 N at $t = 2.0$ s
27. 19N, eastward
29. 613 N
31. (b) $T_1 = 513.5$ N, $T_2 = 558.0$ N, $T_3 = 325.0$ N
33. (a) 33.9 N   (b) 39.2 N
35. (a) $g \tan \theta$ (b) 4.16 m/s²

37. (a) $F_x > 19.6$ N   (b) $F_x \leq -78.4$ N
39. survival chance better with force on larger mass
41. (a) 4.90 m/s²   (b) 3.13 m/s   (c) 1.35 m
    (d) 1.14 s   (e) no
43. (a) 706 N   (b) 814 N   (c) 706 N   (d) 648 N
45. 21.8 m/s
47. 42.2 N
49. 81.0 m/s
51. $\mu = 0.0773$
53. (a) 0.161   (b) 1.01 m/s²
55. (b) 27.2 N, 1.286 m/s²
57. (a) 1.78 m/s²   (b) 0.368   (c) 9.37 N
    (d) 2.67 m/s
59. (a) $a_1 = 2.31$ m/s² down, $a_2 = 2.31$ m/s² left, $a_3 = 2.31$ m/s² up,
    (b) $T_{\text{Left}} = 30.0$ N, $T_{\text{Right}} = 24.2$ N
61. 0.293
63. 182.5 m
65. (a) $Mg/2, Mg/2, Mg/2, 3Mg/2, Mg$ (b) $Mg/2$
67. (a) $\mu_s = \dfrac{h}{\sqrt{L^2 - h^2}}$ (b) $a = \dfrac{2L}{t^2}$

    (c) $\sin \theta = h/L$ (d) $\mu_k = \dfrac{h - \dfrac{2L^2}{gt^2}}{\sqrt{L^2 - h^2}}$
69. (a) 0.232 m/s²   (b) 9.68 N
71. (a) F forward   (b) $3F/2Mg$
    (c) $F/(M + m_1 + m_2 + m_3)$
    (d) $m_1 F/(M + m_1 + m_2 + m_3)$,
    $(m_1 + m_2)F/(M + m_1 + m_2 + m_3)$,
    $(m_1 + m_2 + m_3)F/(M + m_1 + m_2 + m_3)$
    (e) $m_2 F/(M + m_1 + m_2 + m_3)$
73. (a) friction between the two blocks   (b) 34.7 N
    (c) 0.306
75. (a) 0.408 m/s²   (b) 83.3 N
77. (a) 4.00 m   (b) 3.72 m/s
81. (a) $(-45\mathbf{i} + 15\mathbf{j})$ m/s   (b) at 162°
    (c) $(-225\mathbf{i} + 75\mathbf{j})$ m   (d) $(-227$ m, 79 m)
83. $(M + m_1 + m_2) m_2 g/m_1$
85. $T_1 = 74.5$ N, $T_2 = 34.7$ N, $\mu_k = 0.572$
87. (a) 2.20 m/s²   (b) 27.37 N
89. (a) 30.7°   (b) 0.843 N
91. 6.00 cm

Chapter 6

1. (a) 8.0 m/s   (b) 3.02 N
3. (a) 5.40 kN down   (b) 1.60 kN down
   (c) seatbelt tension plus gravity
5. $0 < v < 8.08$ m/s
7. (a) 1.52 m/s²   (b) 1.66 km/s   (c) 6820 s
9. $v \leq 14.3$ m/s
11. (a) 9.80 N   (b) 9.80 N   (c) 6.26 m/s
13. (a) static friction   (b) 0.085
15. 3.13 m/s

17. (a) 4.81 m/s   (b) 700 N up
19. (a) $(-0.163 \text{ m/s}^2)\mathbf{i} + (0.233 \text{ m/s}^2)\mathbf{j}$
    (b) 6.53 m/s   (c) $(-0.181 \text{ m/s}^2)\mathbf{i} + (0.181 \text{ m/s}^2)\mathbf{j}$
21. no
23. (a) 0.822 m/s^2 (b) 37.0 N   (e) 0.0839
25. (a) 17.0°   (b) 5.12 N
27. (a) 491 N   (b) 50.1 kg   (c) 2.00 m/s
31. (a) $3.47 \times 10^{-2} \text{ s}^{-1}$ (b) 2.50 m/s   (c) $a = -cv$
33. (a) 1.47 N·s/m   (b) 2.04×10^{-3} s
    (c) 2.94×10^{-2} N
35. (a) 8.32×10^{-8} N   (b) $9.13 \times 10^{22} \text{ m/s}^2$
37. (a) $T = (68.6 \text{ N})\mathbf{i} + (784 \text{ N})\mathbf{j}$ (b) 0.857 m/s²
39. (a) The true weight is greater than the apparent weight.
    (b) $w = w' = 735$ N at the poles; $w' = 732.4$ N at the
    equator
41. 780 N
43. 12.8 N
45. (a) 967 lb   (b) 647 lb up
47. (a) 6.67 kN   (b) 20.3 m/s

47A. (a) $mg - \dfrac{mv^2}{R}$ (b) $\sqrt{gR}$

49. (b) 2.54 s, 23.6 rev/min
51. (a) 1.58 m/s^2 (b) 455 N   (c) 329 N
    (d) 397 N upward and 9.15° inward
53. (a) 0.0132 m/s   (b) 1.03 m/s   (c) 6.87 m/s
S1. Spreadsheet 6.1 typifies the solution of second-order differential equations. In this spreadsheet, we want to solve $a = dv/dt$, where $v = dx/dt$. To find v as a function of t, we assume that a is constant over the time interval dt. Hence, we can use Euler's method to integrate $dv/dt = a$. Therefore, $v_{i+1} = v_i + a_i \Delta t$. In the Lotus 1-2-3 Spreadsheet 6.1, cells F35 and down implement this equation. To find x as a function of t, we know that v varies over the time interval, so we use Euler's modified method to integrate $dx/dt = v$. Or, $x_{i+1} = x_i + 1/2(v_{i+1} + v_i)\Delta t$. Cells F35 and down implement this equation.
S2. All objects fall faster when there is no air resistance than when there is air resistance. As the mass of the objects increases, the difference between their positions at the same time with and without air resistance becomes smaller.
S4. $F_{\min} \cong 252$ N at 31°.
S5. Try large positive speeds, that is $v_0 > 100$ m/s. You may need to increase Δt.
S7. If the terminal speeds are to be equal, then the relationship between b_1 (for $n = 1$) and b_2 (for $n = 2$) is $b_1 = b_2 \, mg$.

Chapter 7

1. 30.6 m
3. (a) 31.9 J   (b) 0   (c) 0   (d) 31.9 J
5. 5.88 kJ

7. (a) 900 J   (b) -900 J   (c) 0.383
9. (a) 137 W   (b) -137 W
11. (a) 79.4 N   (b) 1.49 kJ   (c) 1.49 kJ
11A. (a) $\mu_k \, mg/(\cos\theta + \mu_k \sin\theta)$
    (b) $\mu_k \, mg \, d \cos\theta/(\cos\theta + \mu_k \sin\theta)$
    (c) $\mu_k \, mg \, d \cos\theta/(\cos\theta + \mu_k \sin\theta)$
13. 14.0
13A. $r_1 r_2 \cos(\theta_1 - \theta_2)$
17. (a) 16.0 J   (b) 36.9°
19. $\mathbf{s} = 2\mathbf{i} + 23.5\mathbf{j}$ or $22\mathbf{i} + 8.5\mathbf{j}$
21. (a) 11.3°   (b) 156°   (c) 82.3°
23. (a) 7.50 J   (b) 15.0 J   (c) 7.50 J   (d) 30.0 J
25. (a) 575 N/m   (b) 46.0 J
25A. (a) F/d (b) $\frac{1}{2}Fd$
27. 0.299 m/s
29. (b) mgR
31. 12.0 J
31A. 3 W
33. (a) 4.10×10^{-18} J
    (b) 1.14×10^{-17} N   (c) $1.25 \times 10^{13} \text{ m/s}^2$
    (d) 240 ns
35. (a) 2.00 m/s   (b) 200 N
35A. (a) $v = \sqrt{2W/m}$ (b) $\overline{F} = W/d$
37. (a) 650 J   (b) -588 J   (c) 62.0 J
    (d) 1.76 m/s
39. 6.34 kN

41. (a) $\sqrt{\dfrac{2\,mgh}{m + M/4}}$ (b) $\sqrt{\dfrac{2\,mgh - \mu_k Mgh}{m + M/4}}$

43. 1.25 m/s
45. 2.04 m
47. (a) -168 J   (b) 184 J   (c) 500 J   (d) 148 J
    (e) 5.65 m/s
49. (a) 4.51 m   (b) no, since $f > mg \sin\theta$
51. (a) 63.9 J   (b) -35.4 J   (c) -9.51 J
    (d) 19.0 J
53. 875 W
55. (a) 7.92 hp   (b) 14.9 hp
57. (a) 7.5×10^4 J   (b) 2.50×10^4 W (33.5 hp)
    (c) 3.33×10^4 W (44.7 hp)

57A. (a) $\dfrac{1}{2} mv^2$ (b) $\dfrac{mv^2}{2t}$ (c) $\dfrac{mv^2 t_1}{t^2}$

59. 220 ft·lb/s
61. 685
63. 80.0 hp
65. (a) 1.35×10^{-2} gal   (b) 73.8   (c) 8.08 kW
67. 5.90 km/liter
69. (a) 5.37×10^{-11} J   (b) 1.33×10^{-9} J
71. 3.70 m/s
75. (a) $(2 + 24t^2 + 72t^4)$ J   (b) $a = 12t \text{ m/s}^2; F = 48t$ N
    (c) $(48t + 288t^3)$ W   (d) 1.25×10^3 J
77. 878 kN
79. (a) 4.12 m   (b) 3.35 m
81. (a) -5.60 J   (b) 0.152   (c) 2.29 rev
83. (a) $W = mgh$ (b) $\Delta K = mgh$
    (c) $K_f = mgh + mv_0^2/2$

85. 1.94 kJ

87. (b) 8.49×10^5 kg/s   (c) 7.34×10^7 m³
    (d) 1.53 km

89. 1.68 m/s

S1. (a) A plot of F as the independent variable and L as the dependent variable shows that the last four points tend to vary the most from a straight line. However, because the first point has the largest percentage deviation from a straight line, we probably are not justified in throwing any of the data points out. The slope of the best straight line obtained from the least-squares fit is 8.654 545 5 mm/N or $k = 0.116$ N/mm $= 116$ N/m. (b) The least-squares fit we used was of the form $L = aF + b$; solving for F gives $F = L/a - b/a = 0.116L - 0.561$. Hence, for $L = 105$ mm, $F = 12.7$ mm.

Chapter 8

1. (a) -196 J   (b) -196 J   (c) -196 J
   The force is conservative.

3. (b) conservative 62.7 J, nonconservative 20.7 J
   (c) $\mu = 0.330$

5. (a) 125 J   (b) 50.0 J   (c) 66.7 J
   (d) nonconservative, since W is path-dependent

7. (a) 40.0 J   (b) -40.0 J   (c) 62.5 J

9. (a) -9.00 J; No. A constant force is conservative.
   (b) 3.39 m/s   (c) 9.00 J

11. $v_A = \sqrt{3gR}$; 0.098 N downward

13. (a) $v = (gh + v_0^2)^{1/2}$ (b) $v_x = 0.6v_0$;
    $v_y = -(0.64v_0^2 + gh)^{1/2}$

15. (a) 18.5 km, 51.0 km   (b) 10.0 MJ

17. (a) 4.43 m/s   (b) 5.00 m

17A. (a) $\sqrt{2(m_1 - m_2)gh/(m_1 + m_2)}$
    (b) $2m_1h/(m_1 + m_2)$

19. (a) -160 J   (b) 73.5 J   (c) 28.8 N
    (d) 0.679

21. 489 kJ

23. (a) -4.1 MJ   (b) 9.97 m/s   (c) 50.8 m
    (d) It is better to keep the engine with the train.

25. 3.74 m/s

29. 914 N/m

31. (a) -28.0 J   (b) 0.446 m

33. 10.2 m

33A. $(kd^2/2mg) - d$

35. 0.327

37. (a) $F_r = A/r^2$

39. $\mathbf{F} = (7 - 9x^2y)\mathbf{i} - 3x^3\mathbf{j}$

41. (b) $x = 0$ (c) $v = \sqrt{0.80\, \text{J}/m}$

43. (a) $v_B = 5.94$ m/s; $v_C = 7.67$ m/s   (b) 147 J

45. (a) 1.50×10^{-10} J   (b) 1.07×10^{-9} J
    (c) 9.15×10^{-10} J

47. (a) 0.225 J   (b) 0.363 J
    (c) No. The normal force varies with position, and so the frictional force also varies.

49. $\dfrac{h}{5}(4\sin^2\theta + 1)$

51. (a) 349 J, 676 J, 741 J   (b) 174 N, 338 N, 370 N
    (c) yes

53. (a) $\Delta U = -\dfrac{ax^2}{2} - \dfrac{bx^3}{3}$ (b) $\Delta U = \dfrac{A}{\alpha}(1 - e^{\alpha x})$

55. 0.115

59. 1.24 m/s

61. (b) 7.42 m/s

63. (a) 3.19 m   (b) 2.93 m/s

65. (a) 0.400 m   (b) 4.10 m/s
    (c) It reaches the top.

67. 3.92 kJ

67A. $m_1 gd(m_2 - \mu_k m_1\cos\theta - m_1\sin\theta)/(m_1 + m_2)$

69. (a) 0.378 m   (b) 2.30 m/s   (c) 1.08 m

S1. If the particle has an initial energy less than 2471 J, it will be trapped in the potential well; for example, if $E_T = 1000$ J, it will be confined approximately to -3.15 m $\leq x \leq 4.58$ m.

S2. $x = 0$ is a point of stable equilibrium; $x = 8.33$ m is a point of unstable equilibrium.

Chapter 9

1. $(9.00\mathbf{i} - 12.0\mathbf{j})$ kg·m/s, 15.0 kg·m/s

3. 1.60 kN

5. (a) 12.0 kg·m/s   (b) 6.00 m/s   (c) 4.00 m/s

7. (a) 13.5 kg·m/s   (b) 9.00×10^3 N
   (c) 18.0×10^3 N

9. 87.5 N

11. (a) 7.50 kg·m/s   (b) 375 N

13. (a) 13.5 kg·m/s toward the pitcher
    (b) 6.75×10^3 N toward the pitcher

15. 260 N toward the left in the diagram

15A. $\dfrac{-2mv\sin\theta}{t}\mathbf{i}$

17. (a) 0.125 m/s   (b) 8 times

19. 120 m

21. (a) 1.15 m/s   (b) -0.346 m/s

23. 4.01×10^{-20} m/s

25. 301 m/s

27. (a) 20.9 m/s east   (b) 8.68 kJ into thermal energy

29. (a) 0.284, or 28.4%   (b) $K_n = 1.15 \times 10^{-13}$ J,
    $K_c = 4.54 \times 10^{-14}$ J

31. 3.75 kN; no

33. (a) 0.571 m/s   (b) 28.6 J   (c) 0.003 97

35. 91.2 m/s

37. 0.556 m

39. 497 m/s

41. $\mathbf{v} = (3.00\mathbf{i} - 1.20\mathbf{j})$ m/s

43. (a) $v_x = -9.33 \times 10^6$ m/s, $v_y = -8.33 \times 10^6$ m/s
    (b) 4.39×10^{-13} J

45. 3.01 m/s, 3.99 m/s

47. 2.50 m/s at $-60.0°$

51. -0.429 m

53. CM $= 454$ km, well within the Sun

55. $(2.54$ m, 4.75 m$)$

57. $70/6$ cm, $80/6$ cm

59. (a) $(1.40\mathbf{i} + 2.40\mathbf{j})$m/s   (b) $(7.00\mathbf{i} + 12.0\mathbf{j})$kg·m/s

61. (a) 2.10 m/s, 0.900 m/s   (b) 6.30×10^{-3} kg·m/s,
    -6.30×10^{-3} kg·m/s

61A. (a) $m_2 v_1/(m_1 + m_2)$ and $m_1 v_1/(m_1 + m_2)$
    (b) $m_1 m_2 v_1/(m_1 + m_2)$ toward the CM

63. 200 kN

65. 2150 kg

67. 0.595 m³/s

67A. $F/\rho v$

69. 291 N

71. (a) 1.80 m/s to the left   (b) 257 N to the left
    (c) larger than part b

73. (a) 4160 N   (b) 4.17 m/s

75. 32.0 kN; 7.13 MW

77. (a) 6.81 m/s   (b) 1.00 m

79. 240 s

81. $(3Mgx/L)\mathbf{j}$

83. (a) As the child walks to the right, the boat moves to the
    left, but the center of mass remains fixed.
    (b) 5.55 m from the pier   (c) Since the turtle is 7 m
    from the pier, the boy will not be able to reach the
    turtle, even with a 1 m reach.

85. (a) 100 m/s   (b) 374 J

85A. (a) $v_0 - d\sqrt{\dfrac{kM}{m^2}}$ (b) $v_0 d\sqrt{kM} - \dfrac{1}{2}kd^2\left(1 + \dfrac{M}{m}\right)$

87. $2v_0$ and 0

89. (a) 3.8 kg·m/s²   (b) 3.8 N   (c) 3.8 N
    (d) 2.8 J   (e) 1.4 J
    (f) Friction between sand and belt converts half of the
    input work into thermal energy.

S1. (a) The maximum acceleration is 100 m/s². It occurs at
    the end of the burn time of 80 s, when the rocket has its
    smallest mass. The maximum speed that the rocket
    reaches is 3.22 km/s. (b) The speed reaches half its max-
    imum after 55.3 s; if the acceleration were constant, it
    would reach half its maximum speed at 40 s (half the
    burn time), but the acceleration is always increasing dur-
    ing the burn time.

S3. (b) The disadvantages are that the ship has to withstand
    twice the acceleration and that it has only traveled half
    as far when the fuel burns out. The advantage is that it
    takes half as long to reach its final speed; hence, it trav-
    els farther in 100 s.

Chapter 10

1. (a) 4.00 rad/s²   (b) 18.0 rad

3. (a) 1.99×10^{-7} rad/s   (b) 2.66×10^{-6} rad/s

5. (a) 5.24 s   (b) 27.4 rad

7. 13.7 rad/s²

9. (a) 0.18 rad/s   (b) 8.10 m/s² toward the center of
    the track

9A. (a) v/R (b) v^2/R toward center

11. (a) 8.00 rad/s   (b) 8.00 m/s, $a_r = -64.0$ m/s²,
    $a_t = 4.00$ m/s²   (c) 9.00 rad

13. (a) 126 rad/s   (b) 3.77 m/s   (c) 1.26 km/s²
    (d) 20.1 m

15. 29.4 m/s², 9.80 m/s²

15A. $-2g\dfrac{(h - R)}{R}\mathbf{i} - g\mathbf{j}$

17. (a) 143 kg·m²   (b) 2.57×10^3 J

19. (a) 92.0 kg·m², 184 J   (b) 6.00 m/s, 4.00 m/s,
    8.00 m/s, 184 J

23. (a) $(3/2)MR^2$ (b) $(7/5)MR^2$

25. -3.55 N·m

27. 2.79%

29. (a) 0.309 m/s²   (b) $T_1 = 7.67$ N, $T_2 = 9.22$ N

29A. (a) $\dfrac{(m_2 \sin\theta - \mu_k)(m_1 + m_2 \cos\theta)}{m_1 + M/2 + m_2}g$

    (b) $T_1 = \mu_k mg + m_1 a$, $T_2 = T_1 + \frac{1}{2}Ma$

31. (a) 56.3 J   (b) 8.38 rad/s   (c) 2.35 m/s
    (d) 0.14% greater

33. (a) $2(Rg/3)^{1/2}$ (b) $4(Rg/3)^{1/2}$ (c) $(Rg)^{1/2}$

35. (a) 11.4 N, 7.57 m/s², 9.53 m/s down   (b) 9.53 m/s

37. (a) 1.03 s   (b) 10.3 rev

39. 168 N·m (clockwise)

41. (a) 4.00 J   (b) 1.60 s   (c) yes

43. (a) $\omega = \sqrt{3g/L}$ (b) $\alpha = 3g/2L$
    (c) $-\frac{3}{2}g\mathbf{i} - \frac{3}{4}g\mathbf{j}$ (d) $-\frac{3}{2}Mg\mathbf{i} + \frac{1}{4}Mg\mathbf{j}$

45. (a) $0.707R$ (b) $0.289L$ (c) $0.632R$

47. 149 rad/s

49. (a) 2.60×10^{29} J   (b) -1.65×10^{17} J/day

51. (a) 118 N, 156 N   (b) 1.19 kg·m²

51A. (a) $T_1 = m_1(a + g\sin\theta)$, $T_2 = m_2(g - a)$
    (b) $m_2 R^2 g/a - m_1 R^2 - m_2 R^2 - m_1 R^2(g/a)\sin\theta$

53. (a) -0.176 rad/s²   (b) 1.29 rev   (c) 9.26 rev

S1. The answer is not unique because the torque $\tau = FR$.

S3. Replace X^2 with $(X - H)^2$ in line 110.

Chapter 11

1. (a) 500 J   (b) 250 J   (c) 750 J

3. (a) $a_{CM} = \frac{2}{3}g\sin\theta$ (disk), $a_{CM} = \frac{1}{2}g\sin\theta$ (hoop)
    (b) $\frac{1}{3}\tan\theta$

5. 44.8 J

5A. $0.7Mv^2$

7. (a) $-17\mathbf{k}$ (b) $70.5°$

9. (a) negative z direction   (b) positive z direction

11. $45.0°$

13. $|\mathbf{F}_3| = |\mathbf{F}_1| + |\mathbf{F}_2|$, no

15. $(17.5$ kg·m²/s$)\mathbf{k}$

15A. $\frac{1}{2}(m_1 + m_2)vd$

17. $(60$ kg·m²/s$)\mathbf{k}$

19. $mvR\left[\cos\left(\dfrac{vt}{R}\right) + 1\right]\mathbf{k}$

21. $-mg\ell t\cos\theta\,\mathbf{k}$

23. (a) zero   (b) $[- mv_0^3 \sin^2 \theta \cos \theta/2g]\,\mathbf{k}$
    (c) $[- 2mv_0^3 \sin^2 \theta \cos \theta/g]\,\mathbf{k}$ (d) The downward
    force of gravity exerts a torque in the $-z$ direction.
25. (a) 0.433 kg·m²/s   (b) 1.73 kg·m²/s
27. (a) $\omega = \omega_0 I_1/(I_1 + I_2)$ (b) $I_1/(I_1 + I_2)$
29. (a) 0.360 rad/s in the counterclockwise direction
    (b) 99.9 J
31. (a) 6.05 rad/s   (b) 113 J
33. (a) $mv\ell$ down   (b) $M/(M + m)$
35. (a) 2.19×10^6 m/s   (b) 2.18×10^{-18} J
    (c) 4.13×10^{16} rad/s
37. 0.91 km/s
41. (a) The net torque around this axis is zero.
    (b) Since $\tau = 0$, $L = $ const. But initially, $L = 0$, hence it
    remains zero throughout the motion. Consequently, the
    monkey and bananas move upward with the same speed
    at any instant. The distance between the monkey and ba-
    nanas stays constant. Hence, the monkey will not reach
    the bananas.
47. 30.3 rev/s
49. (a) $v_0 r_0/r$ (b) $T = (mv_0^2 r_0^2)\,r^{-3}$

    (c) $\frac{1}{2}mv_0^2\left(\dfrac{r_0^2}{r^2} - 1\right)$ (d) 4.50 m/s, 10.1 N, 0.450 J

51. (a) $F_y = \dfrac{W}{L}\left(d - \dfrac{ah}{g}\right)$ (b) 0.306 m

    (c) $(-306\mathbf{i} + 553\mathbf{j})$ N
53. (a) 3.75×10^3 kg·m²/s   (b) 1.875 kJ
    (c) 3.75×10^3 kg·m²/s   (d) 10.0 m/s
    (e) 7.50 kJ   (f) 5.625 kJ
53A. (a) Mvd (b) Mv^2 (c) Mvd (d) $2v$
    (e) $4Mv^2$ (f) $3Mv^2$
55. $\frac{1}{3}L$
57. (c) $(8Fd/3M)^{1/2}$
61. $v_0 = [ag(16/3)(\sqrt{2} - 1)]^{1/2}$
63. F_1 clockwise torque, F_2 zero torque, F_3 and F_4 counter-
    clockwise torque
65. (a) 0.800 m/s², 0.400 m/s²
    (b) 0.600 N (top), 0.200 N (bottom)

Chapter 12

1. 10.0 N up; 6.00 N·m counterclockwise
3. $[(w_1 + w)d + w_1\ell/2]/W_2$
5. $F_W = 480$ N, $F_v = 1200$ N

5A. $\left(\dfrac{w_1}{2} + \dfrac{w_2 x}{L}\right)\left(\dfrac{d}{\sqrt{L^2 - d^2}}\right)$; $w_1 + w_2$

9. -1.50 m, -1.50 m
11. (a) 859 N   (b) 1040 N, left and upward at $36.9°$
13. 0.789
15. $F_f = 4410$ N, $F_r = 2940$ N
17. $2R/5$
19. $\frac{1}{3}$ by the left string, $\frac{2}{3}$ by the right string
21. $x = \frac{3}{4}L$
23. 4.90 mm
25. (a) 73.6 kN   (b) 2.50 mm

27. 29.2 μm

27A. $\dfrac{8m_1 m_2 gL}{\pi d^2 Y(m_1 + m_2)}$

29. (a) 3.14×10^4 N   (b) 62.8 kN
31. 1800 atm
33. $N_A = 5.98 \times 10^5$ N, $N_B = 4.80 \times 10^5$ N
35. (b) 69.8 N   (c) 0.877ℓ
37. (a) 160 N right   (b) 13.2 N right   (c) 292 N up
    (d) 192 N
39. (a) $T = w(\ell + d)/\sin \theta(2\ell + d)$ and
    (b) $R_x = w(\ell + d)\cot \theta/(2\ell + d)$; $R_y = w\ell/(2\ell + d)$
41. (a) $F_x = 268$ N, $F_y = 1300$ N   (b) 0.324

41A. (a) $\dfrac{m_1 g}{2 \tan \theta} + \dfrac{m_2 gx}{L \tan \theta}$; $(m_1 + m_2)g$

    (b) $\dfrac{m_1/2 + m_2 d/L}{(m_1 + m_2)\tan \theta}$

43. 5.08 kN, $R_x = 4.77$ kN, $R_y = 8.26$ kN
45. $T = 2.71$ kN, $R_x = 2.65$ kN, $R_y = -12.0$ N
47. (a) 20.1 cm to the left of the front edge, $\mu = 0.571$
    (b) 0.501 m

47A. (a) $\dfrac{F \cos \theta}{mg - F \sin \theta}$; $\dfrac{mgw/2 - Fh \cos \theta}{(mg - F \sin \theta)}$ (b) $\dfrac{mgw}{2F \cos \theta}$

49. (a) $W = \dfrac{w}{2}\left(\dfrac{2\mu_s \sin \theta - \cos \theta}{\cos \theta - \mu_s \sin \theta}\right)$

    (b) $R = (w + W)\sqrt{1 + \mu_s^2}$, $F = \sqrt{W^2 + \mu_s^2(w + W)^2}$
51. (a) 133 N   (b) $N_A = 429$ N, $N_B = 257$ N
    (c) $R_x = 133$ N, $R_y = -257$ N
53. 66.7 N
55. $F = \frac{3}{8}w$
57. (a) 1.67 N, 3.33 N, 1.67 N   (b) 2.36 N
59. (a) 4500 N   (b) 4.50×10^6 N/m²
    (c) This is more than sufficient to break the board.
61. $y_{cg} = 16.7$ cm

Chapter 13

1. (a) 1.50 Hz, 0.667 s   (b) 4.00 m   (c) π rad
    (d) 2.83 m
3. (b) 1.81 s   (c) no
5. (a) 13.9 cm/s, 16.0 cm/s²   (b) 16.0 cm/s, 0.262 s
    (c) 32.0 cm/s², 1.05 s
7. (b) 6π cm/s, 0.333 s   (c) $18\pi^2$ cm/s², 0.500 s
    (d) 12.0 cm
9. (a) 0.542 kg   (b) 1.81 s   (c) 1.20 m/s²
11. 40.9 N/m
13. (a) 2.40 s   (b) 0.417 Hz   (c) 2.62 rad/s
15. (a) 0.400 m/s, 1.60 m/s²
    (b) ± 0.320 m/s, -0.960 m/s²   (c) 0.232 s
17. (a) 0.750 m   (b) $x = -(0.75$ m$)\sin(2.0t)$
17A. (a) v/ω (b) $(v/\omega)\cos(\omega t + \pi/2)$
19. 2.23 m/s
21. (a) quadrupled   (b) doubled
    (c) doubled   (d) no change
23. ± 2.60 cm

25. (a) 1.55 m   (b) 6.06 s
27. (a) 3.65 s   (b) 6.41 s   (c) 4.24 s
27A. (a) $2\pi\sqrt{L/(g+a)}$ (b) $2\pi\sqrt{L/(g-a)}$
(c) $2\pi L^{1/2}(g^2 + a^2)^{-1/4}$
29. (a) 0.817 m/s   (b) 2.57 rad/s²   (c) 0.634 N
31. increases by 1.78×10^{-3} s
33. 0.944 kg·m²
35. (a) 5.00×10^{-7} kg·m²
(b) 3.16×10^{-4} N·m/rad
39. 1.00×10^{-3} s^{-1}
41. (a) 1.42 Hz   (b) 0.407 Hz
43. 318 N

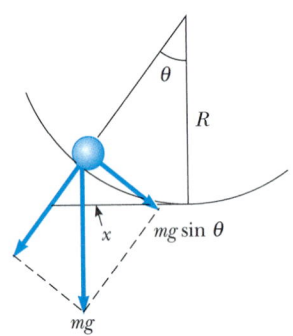

45. 1.57 s
47. (a) $E = \frac{1}{2}mv^2 + mgL(1 - \cos\theta)$ (b) $U = \frac{1}{2}m\omega^2 s^2$
49. Referring to the sketch, we have $F = -mg\sin\theta$ and
$\tan\theta = x/R$. For small displacements, $\tan\theta \approx \sin\theta$ and
$F = -(mg/R)x = -kx$ and $\omega = (k/m)^{1/2} = (g/R)^{1/2}$.

51. (a) $2Mg, T_P = Mg\left(1 + \dfrac{y}{L}\right)$

(b) $\dfrac{4\pi}{3}\sqrt{\dfrac{2L}{g}} = 2.68$ s

55. $f = \dfrac{1}{2\pi}\sqrt{\dfrac{MgL + kh^2}{ML^2}}$

57. 0.0662 m

57A. $\dfrac{\mu_s g}{4\pi^2 f^2}$

59. (a) 3.00 s   (b) 14.3 J   (c) 0.441 rad
61. 9.19×10^{13} Hz
63. (a) 15.8 rad/s   (b) 5.23 cm   (c) 1.31 cm, π
67. (a) $k = 1.74$ N/m ± 6%   (b) $k = 1.82$ N/m ± 3%;
the values of k agree   (c) $m_s = 8$ g ± 12%, in
agreement with 7.4 g
S2. Since $E_{tot} = K(t) + U(t) = \frac{1}{2}kA^2[\cos^2(\omega t + \delta) +$
$\sin^2(\omega t + \delta)] = \frac{1}{2}kA^2$, the total energy is independent of
ω and δ. It depends only on k and A.
S3. The periods increase as ϕ_0 increases. These periods are
always greater than those calculated using $T_0 = 2\pi\sqrt{L/g}$.
For example, if $\phi_0 = 45° = \pi/4$ rad, and $L = 1$ m, then
$T_0 = 2.00\ 709$ s, but the actual period is 2.08 s. (*Note:*
Due to numerical errors in the integration of the differential equations, the amplitudes may tend to increase.
If this occurs, try smaller time steps.)

Chapter 14

1. (a) 3.46×10^8 m   (b) 3.34×10^{-3} m/s²
3. (1.00 m − 61.3 nm)

5. $\dfrac{GM}{\ell^2}\left(\dfrac{2\sqrt{2}+1}{2}\right)$ toward the opposite corner

7. 35.0 N toward the Moon
9. 3.73 m/s²
11. 12.6×10^{31} kg

11A. $\dfrac{2v^3 T}{\pi G}$

13. (a) 4.39×10^{20} N   (b) 1.99×10^{20} N
(c) 3.55×10^{22} N
15. 1.90×10^{27} kg
17. Y has completed 1.30 revolutions
19. 8.98×10^7 m
21. $2GMr/(r^2 + a^2)^{3/2}$, to the left
23. 3.84×10^4 km from the Moon's center
25. 2.82×10^9 J
27. (a) 1.84×10^9 kg/m³   (b) 3.27×10^6 m/s²
(c) -2.08×10^{13} J
29. 1.66×10^4 m/s
31. 11.8 km/s
35. 1.58×10^{10} J

35A. $\dfrac{mGM_E(R_E + 2h)}{2R_E(R_E + h)} - \frac{1}{2}mv^2$

37. (a) 42.1 km/s relative to the Sun   (b) 2.20×10^{11} m
39. (a) 1.31×10^{14} N/kg   (b) 2.62×10^{12} N/kg

39A. (a) $\dfrac{GM}{(d + \ell/2)^2}$ (b) $\dfrac{GM\ell(2d+\ell)}{d^2(d+\ell)^2}$

41. (a) 7.41×10^{-10} N   (b) 1.04×10^{-8} N
(c) 5.21×10^{-9} N
43. 2.26×10^{-7}
45. 0.0572 rad/s = 32.7 rev/h
45A. $\omega = \sqrt{2g/d}$
47. 7.41×10^{-10} N

49. (a) $k = \dfrac{GmM_E}{R_E^3}, A = \dfrac{L}{2}$

(b) $\dfrac{L}{2}\left(\dfrac{GM_E}{R_E}\right)^{1/2}$, at the middle of the tunnel

(c) 1.55×10^3 m/s

51. $\dfrac{2\sqrt{2}Gm}{a^2}(-\mathbf{i})$

53. 2.99×10^3 rev/min

55. (a) $v_1 = m_2\left[\dfrac{2G}{d(m_1 + m_2)}\right]^{1/2}$

$v_2 = m_1\left[\dfrac{2G}{d(m_1 + m_2)}\right]^{1/2}$

$v_{rel} = \left[\dfrac{2G(m_1 + m_2)}{d}\right]^{1/2}$

(b) $K_1 = 1.07 \times 10^{32}$ J, $K_2 = 2.67 \times 10^{31}$ J

57. (a) 7.34×10^{22} kg   (b) 1.63×10^3 m/s
    (c) 1.32×10^{10} J
59. 119 km
61. (a) 5300 s   (b) 7.79 km/s   (c) 6.45×10^9 J
61A. (a) $2\pi(GM_E)^{1/2}(R_E + h)^{3/2}$

   (b) $\sqrt{GM_E/(R_E + h)}$ (c) $\dfrac{mGM_E(R_E + 2h)}{2R_E(R_E + h)}$

63. (a) $M/\pi R^4$ (b) $-GmM/r^2$
    (c) $-(GmM/R^4)r^2$
65. 1.48×10^{22} kg
67. (b) 981 kg/m^3
69. (b) $GMm/2R$
S5. The kinetic energy is

$$K = \tfrac{1}{2}mv^2 = \tfrac{1}{2}m(v_x^2 + v_y^2)$$

The potential energy is

$$U = \frac{GM_E m}{R_E} - \frac{GM_E m}{r}$$

where G is the universal gravitation constant, M_E is the mass of the Earth, R_E is the radius of the Earth, m is the mass of the satellite, and r is the distance from the center of the Earth to the satellite. The total energy, $K + U$, is constant. There may be a small change in the numerical value of the total energy, because of numerical errors during the integration of the differential equations.

S6. According to Kepler's laws of planetary motion the angular momentum, L, is a constant. There may be a small change in the numerical value of L, because of numerical errors during the integration of the differential equations.

Chapter 15

1. 0.111 kg
3. 3.99×10^{17} kg/m^3. Matter is mostly free space.
5. 6.24×10^6 Pa
7. 4.77×10^{17} kg/m^3

9. $P_{\text{ATM}} + \rho\sqrt{g^2 + a^2}\,(L/\sqrt{2})\cos\left(45° - \arctan\dfrac{a}{g}\right)$

11. 1.62 m
13. 77.4 cm^2
15. 0.722 mm
17. 9.12 MPa
19. 12.6 cm
21. 10.5 m; no, a little alcohol and water evaporate
23. 1.08 cm
25. 1470 N down
27. (a) 7.00 cm   (b) 2.80 kg
29. $\rho_{\text{oil}} = 1250$ kg/m^3; $\rho_{\text{sphere}} = 500$ kg/m^3
31. 0.611 kg
33. 1.07 m^2

33A. $\dfrac{m}{h(\rho_w - \rho_s)}$

35. 1430 m^3
37. (a) 17.7 m/s   (b) 1.73 mm
37A. (a) $\sqrt{2gh}$ (b) $(R/\pi)^{1/2}(8/gh)^{1/4}$
39. 0.0128 m^3/s
41. 31.6 m/s
43. (a) 28.0 m/s   (b) 392 kPa
45. Av/a
47. 103 m/s
49. $2[h(h_0 - h)]^{1/2}$
51. 5 kW at 20°C
53. (b) $\tfrac{1}{2}\rho Av^3$ if the mill could make the air stop; the same
55. 0.258 N
57. 1.91 m
59. 455 kPa
63. 8 cm/s
65. 2.01×10^6 N
69. 90.04% Zn
71. 5.02 GW
73. 4.43 m/s
75. (a) 1.25 cm   (b) 13.8 m/s
77. (a) 18.3 mm   (b) 14.3 mm   (c) 8.56 mm

Chapter 16

1. $y = \dfrac{6}{(x - 4.5t)^2 + 3}$
3. (a) longitudinal   (b) 666 s
5.

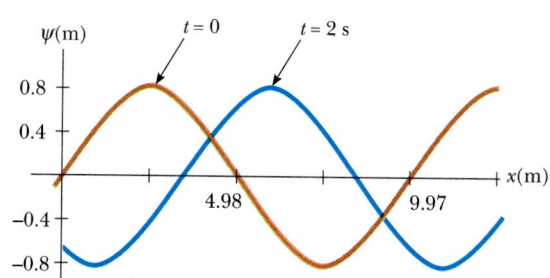

7. (a) 5.00 rad   (b) 0.858 cm
9. (a) Wave 1 travels in the $+x$ direction; wave 2 travels in the $-x$ direction.   (b) 0.75 s   (c) $x = 1.00$ m
11. 520 m/s
13. 13.5 N
15. 586 m/s
17. 0.329 s
19. (b) 0.125 s
21. 0.319 m
23. (b) $k = 18.0$ rad/m, $T = 0.0833$ s, $\omega = 75.4$ rad/s, $v = 4.20$ m/s
    (c) $y(x, t) = (0.20$ m$)\sin(18.0x + 75.4t - 0.151)$
25. $y_1 + y_2 = 11.2\sin(2.0x - 10t + 63.4°)$
27. (a) $y = (0.0800$ m$)\sin(7.85x + 6\pi t)$
    (b) $y = (0.0800$ m$)\sin(7.85x + 6\pi t - 0.785)$

29. (a) 2.15 cm   (b) 0.379 rad   (c) 541 cm/s
    (d) $y(x, t) = (2.15 \text{ cm})\cos(80\pi t + 8\pi x/3 + 0.379)$

31. $A = 2.0$ cm; $k = 2.11$ rad/m; $\lambda = 2.98$ m;
    $\omega = 3.62$ rad/s; $f = 0.576$ Hz; $v = 1.72$ m/s

33. (a) $y = (0.200 \text{ mm}) \sin[16.0x - 3140t]$
    (b) $T = 158$ N

35. 1.07 kW

35A. $2\pi^2 Mv^3 A^2/L\lambda^2$

37. (a) remains constant   (b) remains constant
    (c) remains constant   (d) p is quadrupled

39. (a) 62.5 m/s   (b) 7.85 m   (c) 7.96 Hz
    (d) 21.1 W

43. (b) $f = \frac{1}{2}(x + vt)^2$ $g = \frac{1}{2}(x - vt)^2$
    (e) $f = \frac{1}{2}\sin(x + vt)$ $g = \frac{1}{2}\sin(x - vt)$

45. (a) 3.33 m/s in the positive x direction
    (b) -5.48 cm   (c) 0.667 m, 5.00 Hz
    (d) 11.0 m/s

47. (a) 179 m/s   (b) 17.7 kW

49. (a) 39.2 N   (b) 89.2 cm   (c) 83.6 m/s

49A. (a) $2Mg$ (b) $L_0 + 2Mg/k$
    (c) $(2MgL_0/m + 4M^2g^2/km)^{1/2}$

51. (a) 5.00 m/s $+ x$ (b) 5.00 m/s $- x$
    (c) 7.5 m/s $- x$ (d) 24.0 m/s $+ x$

55. 3.86×10^{-4} (at 5.93°C)

57. $(2L/g)^{1/2}$, $L/4$

59. (a) $\dfrac{\mu\omega^3}{2k} A_0^2 e^{-2bx}$ (b) $\dfrac{\mu\omega^3}{2k} A_0^2$ (c) e^{-2bx}

Chapter 17

1. 5.56 km

3. 1430 m/s

5. 332 m/s

7. 1.988 km

9. (a) 27.2 s   (b) 25.7 s   It is shorter by 5.30%.

11. 1.55×10^{-10} m

13. 5.81 m

15. (a) 2.00 μm, 0.400 m, 54.6 m/s   (b) -0.433 μm
    (c) 1.72 mm/s

17. $(0.200 \text{ Pa}) \sin(62.8x - 2.16 \times 10^4 t)$

19. 66.0 dB

23. 100.0 m and 10.0 m

25. (a) 65.0 dB   (b) 67.8 dB   (c) 69.6 dB

27. 241 W

29. (a) 30.0 m   (b) 9.49×10^5 m

31. 50.0 km

33. 46.4°

35. 56.4°

37. 26.4 m/s

39. (a) 338 Hz   (b) 483 Hz

41. 2.82×10^8 m/s

43. (a) 56.3 s
    (b) $(56.6 \text{ km})\mathbf{i} + (20.0 \text{ km})\mathbf{j}$ from the observer

45. 130 m/s, 1.73 km

47. 80.0°

49. 1204 Hz

51. (a) 0.948°   (b) 4.40°

53. 1.34×10^4 N

55. 95.5 s

55A. $\dfrac{0.3E}{4\pi d^2 I_0} 10^{-\beta/10}$

57. (a) 55.8 m/s   (b) 2500 Hz

59. (a) 6.45   (b) 0

61. 1.60

63. The measured wavelengths depend on your monitor.
However, the wavelengths should be proportional to
those given here. For $u/v = 0.0$, $\lambda_0 = 0.75$ cm. For
$u/v = 0.5$, $\lambda_{\text{front}} = 0.37$ cm and $\lambda_{\text{back}} = 1.13$ cm.
Therefore,

$$\left|\frac{\Delta\lambda_{\text{front}}}{\lambda_0}\right| = \frac{0.37 \text{ cm} - 0.75 \text{ cm}}{0.75 \text{ cm}} = 0.5$$

and

$$\left|\frac{\Delta\lambda_{\text{back}}}{\lambda_0}\right| = \frac{1.13 \text{ cm} - 0.75 \text{ cm}}{0.75 \text{ cm}} = 0.5$$

We see that $\dfrac{\Delta\lambda}{\lambda_0} = \dfrac{u}{v}$.

Chapter 18

1. (a) 9.24 m   (b) 600 Hz

3. 0.500 s

5. (a) The path difference to A is $\lambda/2$.
    (b) $9x^2 - 16y^2 = 144$

7. at 0.0891 m, 0.303 m, 0.518 m, 0.732 m, 0.947 m, and
1.16 m

9. (a) 4.24 cm   (b) 6.00 cm   (c) 6.00 cm
    (d) $x = 0.5$ cm, 1.5 cm, 2.5 cm

11. 25.1 m, 60.0 Hz

13. (a) 2.00 cm   (b) 2.40 cm

15. (a) 0, $\pm 2\pi/3k$, $\pm 4\pi/3k$, . . .
    (b) $(\pm\pi - 2\omega t)/k$, $(\pm 3\pi - 2\omega t)/k$, $(\pm 5\pi - 2\omega t)/k$

17. (a) 60.0 cm   (b) 30.0 Hz

19. $L/4$, $L/2$

21. 0.786 Hz, 1.57 Hz, 2.36 Hz, 3.14 Hz

23. 2.80 g

23A. $m = \dfrac{Mg}{4Lf_1^2 \tan\theta}$

25. (a) $T = 163$ N   (b) 660 Hz

27. 19.976 kHz

29. 338 N

31. 20.5 kg

31A. $m_w = \rho_w A\left(L - \dfrac{v_s}{4f}\right)$

33. 50.4 cm, 84.0 cm

35. 35.8 cm, 71.7 cm

37. 349 m/s

39. (a) 531 Hz   (b) 4.25 cm

41. 328 m/s

43. (a) 350 m/s   (b) 114 cm

45. $n(206 \text{ Hz})$ and $n(84.5 \text{ Hz})$, where $n = 1, 2, 3, \ldots$
47. 1.88 kHz
49. (a) 1.59 kHz   (b) odd   (c) 1.11 kHz
51. It is.
53. (a) 1.99 Hz   (b) 3.38 m/s
55. (a) 3.33 rad   (b) 283 Hz
57. 85.7 Hz
59. $f = 50.0$ Hz; $L = 1.70$ m
61. (a) 78.9 N   (b) 211 Hz
63. $\lambda = 4.86$ m
65. 3.87 m/s *away* from the station *or* 3.78 m/s *toward* the station
67. (a) 59.9 Hz   (b) 20 cm
69. (a) 0.5   (b) $\dfrac{n^2 F}{(n+1)^2}$ (c) $\dfrac{F'}{F} = \dfrac{9}{16}$

S4. The following steps can be used to modify the spreadsheet.

   1. MOVE the entire Y_T column — one column to the right.
   2. COPY the Y_2 column — one column to the right. EDIT the heading label to Y_3.
   3. COPY the second wave input data block to the right and EDIT the labels to reflect the third wave.
   4. EDIT the Y_3 column to reflect the third wave data block addresses.
   5. EDIT the Y_T column to include Y_1, Y_2, and Y_3 in the sum.

S6. Plot $Y(t)$ versus ωt. You may want to start with three terms in the series and then add additional terms and watch how $Y(t)$ changes.

Chapter 19

1. (a) $-273.5°C$ (b) 1.27 atm, 1.74 atm
3. (a) 30.4 mm Hg   (b) 18.0 K
5. 139 K, $-134°C$ (b) 6.56 kPa
7. (a) $-320°F$ (b) 77.3 K
9. $-297°F$
11. (a) 810°F   (b) 450 K
13. (a) 90.0°C   (b) 90.0 K
15. $-40.0°C$
17. 3.27 cm
19. 1.32 cm
21. 0.548 gal
23. 217 kN
25. 1.20 cm
27. (a) 437°C   (b) 2100°C. No; they melt first.
29. (a) 99.4 cm³   (b) 0.943 cm
31. 1.08 L
33. (a) 0.176 mm   (b) 8.78 μm   (c) 93.0 mm³
35. 7.95 m³
37. 4.39 kg
39. 472 K
41. 2.28 kg
41A. $\dfrac{MP_0 V}{RT_1}\left(1 - \dfrac{T_1}{T_2}\right)$

43. 1.61 MPa = 16.1 atm
45. 594 kPa
47. 400 kPa, 448 kPa
49. 1.13
51. (a) $A = 1.85 \times 10^{-3}$ $(1/°C)$, $R_0 = 50.0$ Ω
   (b) 421 °C
53. $\alpha \Delta T \ll 1$
55. 3.55 cm

55A. $\Delta h = \dfrac{V}{A}\beta \Delta T$

57. (a) 94.97 cm   (b) 95.03 cm
59. (b) 1.33 kg/m³
63. 2.74 m
63A. $y = \frac{1}{2}\sqrt{(L + \Delta L)^2 - L^2}$, where $\Delta L = \alpha L \Delta T$
65. 30.4°C
67. (a) 18.0 m   (b) 277 kPa
69. (a) 7.06 mm   (b) 297 K
71. (a) 0.0374 mol   (b) 0.732 atm
73. (a) 6.17×10^{-3} kg/m   (b) 632 N
   (c) 580 N; 192 Hz
75. (a) $(\alpha_2 - \alpha_1)\, L\,\Delta T/(r_2 - r_1)$
   (c) It bends the other way.
S1. From the figure below, note that

$$h = \frac{L}{20}(1 - \cos\theta)$$

and

$$\frac{L}{L_0} = (1 + \alpha\,\Delta T) = \frac{\theta}{\sin\theta} = 1.000\ 055$$

Solving this transcendental equation, $\theta = 0.018\ 165$ rad and $h = 4.54$ m.

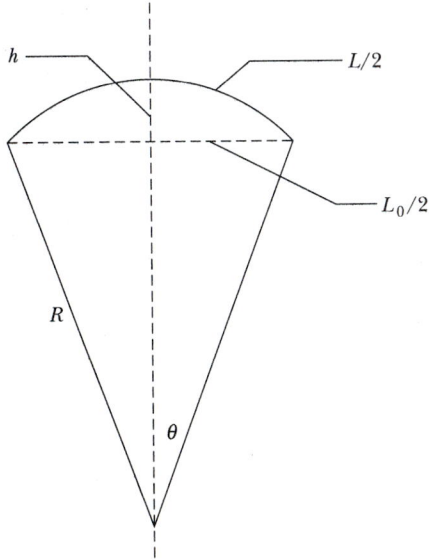

Chapter 20

1. 0.105°C
3. 10.117°C
5. 0.234 kJ/kg·°C
7. 85.3°C
9. 29.6°C
11. 34.7°C
13. 19.5 kJ
15. 50.7 ks
17. 720 cal
19. 47.1°C
21. (a) 0°C   (b) 115 g
23. 2.99 mg
25. (a) 25.760°C   (b) no
27. 0.258 g if the bullet is at 0°C
29. 810 J, 506 J, 203 J
31. 466 J
31A. $nR\,\Delta T$
33. 1.18 MJ
35. (a) 7.65 L   (b) 305 K
37. (a) -567 J   (b) 167 J
37A. (a) $W = P\,\Delta V$ (b) $\Delta U = Q - W$
39. (a) 12.0 kJ   (b) -12.0 kJ
41. (a) 23.1 kJ   (b) 23.1 kJ
43. 0.0962 g
45. (a) 7.50 kJ   (b) 900 K
47. 2.47 L
47A. $V_i = V_f e^{-W/nRT_0}$, where $W = M_w c_w (T_h - T_c)$
49. (a) 48.6 mJ   (b) 16.2 kJ   (c) 16.2 kJ
51. (a) 10.0 L·atm   (b) 0.0100 atm   (c) 7.00 kJ
53. 1.34 kW
55. 51.0°C
57. 138 MJ
59. 0.0222 W/m·°C
59A. $k = \dfrac{Pt}{A\,\Delta T}$
61. (a) 61.1 kWh   (b) $3.67
63. (a) 0.89 ft²·°F·h/BTU
    (b) 1.85 ft²·°F·h/BTU   (c) 2.08
65. 781 kg
67. 1.87 kJ
69. (a) 16.8 L   (b) 0.351 L/s
71. The bullet will partially melt.
73. 44.5°C
75. 5.26°C
79. 9.32 kW
81. 5.31 h
83. 800 J/kg·°C
S1. By varying h until $T = 45$ °C at $t = 180$ s, one finds $h = 0.00742$ cal/s·cm²·°C. Examine the associated graph for each choice of h.
S2. (b) $\Delta U = 4669$ J

Chapter 21

1. 2.43×10^5 m²/s²
3. 2.30 kmol

5. 3.32 mol
7. 8.76×10^{-21} J
9. (a) 40.1 K   (b) 6.01 km/s
11. 477 m/s
13. 109 kPa
15. 75.0 J
17. (a) 316 K   (b) 200 J

17A. (a) $dT = \dfrac{Q}{C_V} - \dfrac{Q}{C_p}$ (b) $\dfrac{QR}{C_p}$

19. (a) 3.46 kJ   (b) 2.45 kJ   (c) 1.01 kJ
21. 24.0 kJ; 68.7 kJ
23. (a) 118 kJ   (b) 6.03×10^3 kg
25. (a) 1.39 atm   (b) 366 K, 254 K
27. (a) 2.06×10^{-4} m³   (b) 560 K   (c) 8.72 K
29. (a) 0.118, so the compression ratio $V_i/V_f = 8.50$
    (b) 2.35
31. 227 K
33. 91.2 J
33A. $9\,P_0 V_0$
35. 1.51×10^{-20} J
37. 2.33×10^{-21} J
39. (a) 7.27×10^{-20} J/molecule   (b) 2.21 km/s
    (c) 3510 K
41. (a) 5.63×10^{18} m, 1.00×10^9 y
    (b) 5.63×10^{12} m, 1.00×10^9 y
43. (a) 1.028   (b) ^{35}Cl
45. (a) 2.01×10^4 K   (b) 902 K
47. (a) 3.21×10^{12} molecules
    (b) 778 km   (c) 6.42×10^{-4} s⁻¹
49. 193
51. 4.65×10^{-8} cm
55. (a) $3.65\,v$ (b) $3.99\,v$ (c) $3.00\,v$
    (d) $106\,mv^2/V$ (e) $7.98\,mv^2$
57. zero, 2.70×10^{20}
59. 0.625
63. (c) 2.0×10^3
65. (b) 5.47 km
67. (a) 10^{82} (b) 10^{12} m   (c) 10^{58} moles
69. (a) 0.510 m/s   (b) 20 ms
S1. (a) At $T = 100$ K, $f(1000)\,dv = 0.0896$ and
    $f(3000)\,dv = 0.000\,05$.
    (b) At $T = 273$ K, $f(1000)\,dv = 0.0428$ and
    $f(3000)\,dv = 0.011\,09$.
    (c) At $T = 1000$ K, $f(1000)\,dv = 0.0084$ and
    $f(3000)\,dv = 0.028\,77$.

Chapter 22

1. (a) 6.94%   (b) 335 J
3. (a) 0.333   (b) 0.667
5. (a) 1.00 kJ   (b) 0
7. (a) 0.375   (b) 600 J   (c) 2.00 kW
9. 0.330
11. (a) 5.12%   (b) 5.27 TJ
13. (a) 0.672   (b) 58.8 kW

15. 0.478°C

17. (a) 0.268   (b) 0.423

19. 453 K

21. 146 kW, 70.8 kW

23. 192 J

25. (a) 24.0 J   (b) 144 J

27. 72.2 J

27A. $W = \dfrac{\Delta T}{T_c} Q$

29. $\Delta S = -90.2$ J/K

31. 195 J/K

33. 717 J/K

35. 3.27 J/K

37. 5.76 J/K, no temperature change

39. 18.4 J/K

41. (a) 154.5 J/K   (b) 54.2 kJ

43. (a) 5.00 kW   (b) 763 W

45. (a) 4.10 kJ   (b) 14.2 kJ
    (c) 10.1 kJ   (d) 28.8%

47. (a) $2nRT_0 \ln 2$ (b) 0.273

49. (a) $10.5nRT_0$ (b) $8.5nRT_0$ (c) 0.190
    (d) 0.833

51. $nC_p \ln 3$

53. 5.97×10^4 kg/s

53A. $\dfrac{dm}{dt} = \dfrac{P}{c_w \, \Delta T}\left(\dfrac{T_h}{T_h - T_c}\right)$

57. (a) 96.9 W   (b) 1.19°C/h

59. $e = \dfrac{2(T_2 - T_1)\ln(V_2/V_1)}{3(T_2 - T_1) + 2T_2 \ln(V_2/V_1)}$

61. (b) 12.0 kJ   (c) -12.0 kJ

63. -8.26×10^5 J

Chapter 23

1. 5.14×10^5 N

3. (a) 1.59 nN away   (b) 1.24×10^{36} times larger
   (c) 8.61×10^{-11} C/kg

5. (a) 57.1 TC   (b) 3.48×10^6 N/C

7. 0.873 N at 330°

9. 40.9 N at 263°

11. 2.51×10^{-10}

13. 3.60 MN down on the top and up on the bottom of the cloud

15. (a) $(-5.58 \times 10^{-11}$ N/C$)\,\mathbf{j}$
    (b) $(1.02 \times 10^{-7}$ N/C$)\,\mathbf{j}$

17. (a) 18.8 nC   (b) 1.17×10^{11} electrons

19. (a) $(1.29 \times 10^4$ N/C$)\,\mathbf{j}$ (b) $(-3.87 \times 10^{-2}$ N$)\,\mathbf{j}$

21. (a) $k_e qx \,(R^2 + x^2)^{-3/2}$

23. (a) at the center   (b) $\left(\dfrac{\sqrt{3}\,k_e q}{a^2}\right)\mathbf{j}$

25. (a) $5.91 k_e\, q/a^2$ at 58.8°   (b) $5.91 k_e\, q^2/a^2$ at 58.8°

27. $-\pi^2 k_e\, qi/6a^2$

29. $-\left(\dfrac{k_e \lambda_0}{x_0}\right)\mathbf{i}$

---

31. (a) $(6.65 \times 10^6$ N/C$)\,\mathbf{i}$ (b) $(2.42 \times 10^7$ N/C$)\,\mathbf{i}$
    (c) $(6.40 \times 10^6$ N/C$)\,\mathbf{i}$ (d) $(6.65 \times 10^5$ N/C$)\,\mathbf{i}$

33. (a) $\dfrac{k_e Qi}{h}\,[(d^2 + R^2)^{-1/2} - ((d + h)^2 + R^2)^{-1/2}]$

    (b) $\dfrac{2k_e Qi}{R^2\, h}\,[h + (d^2 + R^2)^{1/2} - ((d + h)^2 + R^2)^{1/2}]$

35. (a) 9.35×10^7 N/C away from the center; 1.039×10^8
    N/C is 10.0% larger
    (b) 515.1 kN/C away from the center; 519.3 kN/C is
    0.8% larger

37. 7.20×10^7 N/C away from the center; 1.00×10^8 N/C
    axially away

39. $(-21.6$ MN/C$)\,\mathbf{i}$

41.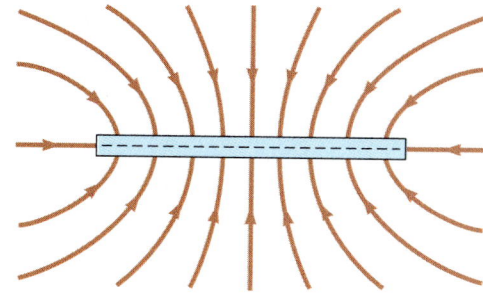

43. (a) $\dfrac{q_1}{q_2} = -1/3$ (b) q_1 is negative and q_2 is positive

45. (a) 6.14×10^{10} m/s²   (b) 19.5 μs
    (c) 11.7 m   (d) 1.20 fJ

47. 1.00×10^3 N/C in the direction of the beam

47A. K/ed parallel to $\mathbf{v}$

49. (a) $(-5.75 \times 10^{13}$ m/s²$)\,\mathbf{i}$ (b) 2.84×10^6 m/s
    (c) 49.4 ns

51. (a) 111 ns   (b) 5.67 mm
    (c) $(450$ km/s$)\,\mathbf{i} + (102$ km/s$)\,\mathbf{j}$

53. (a) 36.9°, 53.1°   (b) 167 ns, 221 ns

55. (a) 10.9 nC   (b) 5.43×10^{-3} N

55A. (a) $\dfrac{mg}{E_x \cot\theta + E_y}$ (b) $\dfrac{mg\, E_x}{E_x \cos\theta + E_y \sin\theta}$

57. (a) $\theta_1 = \theta_2$

59. 204 nC

63. (a) $-\left(\dfrac{4k_e q}{3a^2}\right)\mathbf{j}$ (b) (0, 2.00 m)

65. (a) 307 ms   (b) Yes; neglecting gravity causes a
    2.28% error.

65A. (a) $2\pi\left(\dfrac{L}{g + qE/m}\right)^{1/2}$ (b) Yes. If qE is small com-
    pared to mg, then gravity determines its period.

67. 5.27×10^{17} m/s²; 0.854 mm

71. (a) $F = \dfrac{k_e q^2}{s^2}\,(1.90)\,(\mathbf{i} + \mathbf{j} + \mathbf{k})$

    (b) $F = 3.29\,\dfrac{k_e q^2}{s^2}$ in a direction away from the vertex di-
    agonally opposite to it

S4. (a) at $d < 0.005$ m ($Y/L = 0.1$); (b) at $d > 0.124$ m ($Y/L = 2.48$); (c) same as parts (a) and (b); (d) at $d < 0.010$ m ($Y/L = 0.1$) and at $d > 0.248$ m ($Y/L = 2.48$); (e) In terms of Y/L, the answers do not change.

S5. The dipole approximation of the electric field $E = 2k_e p/x^3$ is within 20 percent of the actual value when $x > 6.2$ cm. It is within 5 percent when $x > 12.6$ cm.

Chapter 24

1. 0
3. (a) aA   (b) bA   (c) 0
5. 4.14×10^6 N/C
5A. $4\Phi/\pi d^2$
7. EhR
9. 1.87×10^3 Nm²/C
11. (a) $q/2\epsilon_0$ (b) $q/2\epsilon_0$
   (c) Plane and square look the same to the charge.
13. (a) 1.36×10^6 Nm²/C   (b) 6.78×10^5 Nm²/C
   (c) No, the same field lines go through spheres of all sizes.
13A. (a) Q/ϵ_0 (b) $Q/2\epsilon_0$
   (c) No. As the radius increases, the area increases but the field decreases to compensate.
15. -6.89×10^6 Nm²/C. The number of lines entering exceeds the number leaving by 2.91 times or more.
17. 0 if $R < d$; $2\lambda(R^2 - d^2)^{1/2}/\epsilon_0$ if $R > d$
19. 28.3 N·m²/C
21. (a) 761 nC   (b) It may have any distribution. Any and all point and smeared-out charges, positive and negative, must add algebraically to + 761 nC.
   (c) Total charge is − 761 nC.
23. (a) $\dfrac{Q}{2\epsilon_0}$ (out of the volume enclosed)

   (b) $-\dfrac{Q}{2\epsilon_0}$ (into it)
25. (a) 0   (b) 7.20×10^6 N/C away from the center
27. (a) 0.713 μC   (b) 5.7 μC
29. (a) 0   (b) $(3.66 \times 10^5$ N/C$)\hat{r}$
   (c) $(1.46 \times 10^6$ N/C$)\hat{r}$ (d) $(6.50 \times 10^5$ N/C$)\hat{r}$
31. $\mathbf{E} = (a/2\epsilon_0)\hat{r}$
33. (a) 5.14×10^4 N/C outward   (b) 646 Nm²/C
35. $\mathbf{E} = (pr/2\epsilon_0)\hat{r}$
37. 5.08×10^5 N/C up.
39. (a) 0   (b) 5.40×10^3 N/C
   (c) 540 N/C, both radially outward
41. (a) 80.0 nC/m² on each face
   (b) $(9.04 \times 10^3$ N/C$)\mathbf{k}$
   (c) $(-9.04 \times 10^3$ N/C$)\mathbf{k}$
43. (a) -99.5 μC/m²   (b) $+382$ μC/m²
43A. (a) $-q/4\pi a^2$ (b) $(Q + q)/4\pi b^2$
45. (a) 0   (b) $(8.00 \times 10^7$ N/C$)\hat{r}$ (c) 0
   (d) $(7.35 \times 10^6$ N/C$)\hat{r}$
47. (a) $-\lambda$, $+ 3\lambda$ (b) $\left(\dfrac{3\lambda}{2\pi\epsilon_0 r}\right)\hat{r}$
49. (b) $\dfrac{Q}{2\epsilon_0}$ (c) $\dfrac{Q}{\epsilon_0}$

51. (a) $\mathbf{E} = \left(\dfrac{\rho r}{3\epsilon_0}\right)\hat{r}$ for $r < a$; $\mathbf{E} = \left(\dfrac{k_e Q}{r^2}\right)\hat{r}$ for $a < r < b$;

   $\mathbf{E} = 0$ for $b < r < c$; $\mathbf{E} = \left(\dfrac{k_e Q}{r^2}\right)\hat{r}$ for $r > c$

   (b) $\sigma_1 = -\dfrac{Q}{4\pi b^2}$ inner; $\sigma_2 = +\dfrac{Q}{4\pi c^2}$ outer
53. (c) $f = \dfrac{1}{2\pi}\sqrt{\dfrac{k_e e^2}{mR^3}}$ (d) 102 pm
57. $\mathbf{g} = \left(\dfrac{GM_E r}{R_E^3}\right)\hat{r}$
59. (a) σ/ϵ_0 to the left   (b) zero
   (c) σ/ϵ_e to the right
63. $\mathbf{E} = \dfrac{\rho a}{3\epsilon_0}\mathbf{j}$

Chapter 25

1. 1.80 kV
3. (a) 152 km/s   (b) 6.50×10^6 m/s
5. (a) 2.7 keV   (b) 509 km/s
7. 6.41×10^{-19} C
9. 2.10×10^6 m/s
11. 1.35 MJ
13. 432 V; 432 eV
15. -38.9 V; the origin
17. (a) 20.0 keV, 83.8 Mm/s   (b) 7.64×10^{-23} kg·m/s
19. (a) 0.400 m/s   (b) The same
19A. (a) $\sqrt{2E\lambda d/\mu}$ (b) The same
21. 2.00 m
23. 119 nC, 2.67 m
25. 4.00 nC at $(-1, 0)$ and -5.01 nC at $(0, 2)$
27. -11.0 MV
29. (a) -386 nJ. Positive binding energy would have to be put in to separate them.   (b) 103 V
31. (a) -27.3 eV   (b) -6.81 eV   (c) 0
35. 1.74 m/s
35A. $((1 + \sqrt{2}/4) k_e q^2/Lm)^{1/2}$
37. (a) 1.00 kV $- (1.41$ kV/m$)x + (1.44$ kV$) \ln\left(\dfrac{3 \text{ m}}{3 \text{ m} - x}\right)$

   (b) $+ 633$ nJ
37A. (a) $V_0 - \dfrac{\sigma x}{2\epsilon_0} + \dfrac{\lambda}{2\pi\epsilon_0}\ln\left(\dfrac{d}{d - x}\right)$

   (b) $qV_0 - \dfrac{q\sigma d}{8\epsilon_0} + \dfrac{q\lambda}{2\pi\epsilon_0}\ln(4/3)$
39. $E_x = -5 + 6xy$ $E_y = 3x^2 - 2z^2$ $E_z = -4yz$
   7.07 N/C
41. (a) 10.0 V, -11.0 V, -32.0 V   (b) $(7.00$ N/C$)\mathbf{i}$
43. $E_x = \dfrac{3 E_0 a^3 xz}{(x^2 + y^2 + z^2)^{5/2}}$ $E_y = \dfrac{3E_0 a^3 yz}{(x^2 + y^2 + z^2)^{5/2}}$

   $E_z = E_0 + \dfrac{E_0 a^3 (2z^2 - x^2 - y^2)}{(x^2 + y^2 + z^2)^{5/2}}$ outside the sphere, and

   $E = 0$ inside.
45. $-(0.553) k_e Q/R$

47. (a) C/m² (b) $k_e \alpha \left[L - d \ln \dfrac{d+L}{d} \right]$

49. $(\sigma/2\epsilon_0)$ $(\sqrt{x^2 + b^2} - \sqrt{x^2 + a^2})$

51. 1.56×10^{12} electrons removed

53. (a) 0, 1.67 MV
    (b) 5.85×10^6 N/C away, 1.17 MV
    (c) 1.19×10^7 N/C away, 1.67 MV

55. (a) 4.50×10^7 N/C outward, 30.0 MN/C outward
    (b) 1.80 MV

57. (a) 450 kV (b) 7.50 μC

59. 5.00 μC

61. (a) 6.00 m (b) -2.00 μC

63. (a) 13.3 μC (b) 20.0 cm

65. (a) 180 kV (b) 127 kV

67. (a) $2 k_e Q d^2 (3x^2 - d^2)(x^3 - xd^2)^{-2}\mathbf{i}$
    (b) $(609 \text{ MN/C})\mathbf{i}$

69.

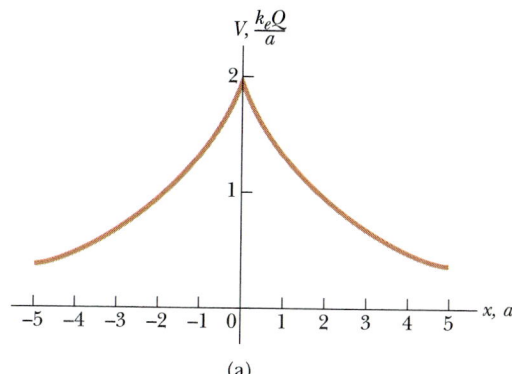

(a)

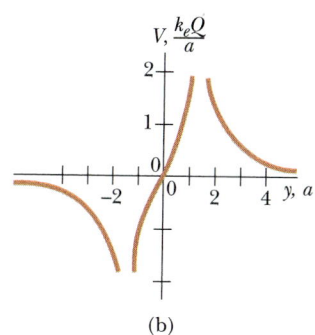

(b)

71. (a) 1.26 σ_0 (b) 1.26 E_0 (c) 1.59 V_0

73. $k_e \dfrac{Q^2}{2R}$

75. $V_2 - V_1 = -\dfrac{\lambda}{2\pi\epsilon_0} \ln\left(\dfrac{r_2}{r_1}\right)$

79. $E_y = k_e \dfrac{Q}{\ell y} \left[1 - \dfrac{y^2}{\ell^2 + y^2 + \ell\sqrt{\ell^2 + y^2}} \right]$

81. (a) $E_r = \dfrac{2k_e p \cos \theta}{r^3}$; $E_\theta = \dfrac{k_e p \sin \theta}{r^3}$; yes; no
    (b) $\mathbf{E} = \dfrac{3k_e p xy \mathbf{i}}{(x^2 + y^2)^{5/2}} + \dfrac{k_e p (2y^2 - x^2)\mathbf{j}}{(x^2 + y^2)^{5/2}}$

83. $\dfrac{3}{5}\left(\dfrac{k_e Q^2}{R}\right)$

85. $x = 0.57735$ m and $x = 0.24859$ m

Chapter 26

1. 13.3 kV

3. (a) 1.00 μF (b) 100 V

5. 684 μC

7. (a) 1.33 μC/m² (b) 13.3 pF

9. (a) 423 fF (b) 0.652

9A. (a) $4\pi\epsilon_0 (R_1 + R_2)$ (b) R_1/R_2

11. 1.52 mm

13. 4.42 μm

15. (a) 1.11×10^4 N/C toward the negative plate
    (b) 98.3 nC/m² (c) 3.74 pF (d) 74.8 pC

17. (69.1 pF) $(\pi - \theta)$

17A. $\epsilon_0 NR^2(\pi - \theta)/d$

19. 2.13×10^{16} m³

19A. $7C^3/384\pi^2\epsilon_0{}^3$

21. (a) 2.68 nF (b) 3.02 kV

23. (a) 15.6 pF (b) 256 kV

25. 66.7 nC

27. 18.0 μF

29. (a) 4.00 μF (b) 8.00 V, 4.00 V, 12.0 V, 24.0 μC, 24.0 μC, 24.0 μC

31. (a) 5.96 μF (b) 89.2 μC, 63.1 μC, 26.4 μC, 26.4 μC

33. 120 μC; 80.0 μC and 40.0 μC

35. $60R/37k_e$

37. 10

39. 83.6 μC

41. 12.9 μF

43. $\dfrac{\epsilon_0 A}{(s - d)}$

45. 90.0 mJ

47. (a) 55.9 μC (b) 4.65 V

49. 800 pJ, 5.79 mJ/m³

55. (a) 369 pC (b) 118 pF, 3.12 V (c) -45.5 nJ

55A. (a) $\epsilon_0 AV/d$ (b) $C_f = \kappa\epsilon_0 A/d$, $V_f = V/\kappa$
    (c) $-\epsilon_0 AV^2(\kappa - 1)/2d\kappa$

57. 16.7 pF, 1.62 kV

59. 1.04 m

61. $\kappa = 8.00$

63. 22.5 V

63A. 1.5 V

65. 416 pF

67. 1.00 μF and 3.00 μF

69. (b) $4\pi\epsilon_0/(a^{-1} + b^{-1})$

71. 2.33

71A. $1 + q/q_0$

73. (a) 243 μJ (b) 2.30 mJ

75. 4.29 μF

75A. $CV/(V_0 - V)$

77. 480 V

79. 0.188 m²

81. 3.00 μF

83. (b) $Q/Q_0 = \kappa$
85. 2/3
89. 19.0 kV
91. 3.00 μF

Chapter 27

1. (b) 1.05 mA
3. 400 nA
3A. $q\omega/2\pi$
5. 0.265 C
7. (a) 221 nm (b) No
9. (a) 1.50×10^5 A (b) 5.40×10^8 C
11. 13.3 μA/m^2
13. 1.32×10^{11} A/m^2
15. 1.59 Ω
17. 1.98 A
19. 1.33 Ω
19A. $R/9$
21. 1.56R
23. (a) 1.82 m (b) 280 μm
25. 6.43 A
27. (a) 3.75 kΩ (b) 536 m
29. (a) 3.15×10^{-8} $\Omega \cdot$m (b) 6.35×10^6 A/m^2
    (c) 49.9 mA (d) 6.59×10^{-4} m/s (assume 1 conduction electron per atom) (e) 0.400 V
31. 0.125
33. 20.8 Ω
35. 67.6°C
37. 26.2°C
39. 3.03×10^7 A/m^2
41. 21.2 nm
43. 0.833 W
45. 36.1%
47. (a) 0.660 kW·h (b) 3.96¢
49. (a) 133 Ω (b) 9.42 m
51. 28.9 Ω
51A. $V^2 t/mc\,(T_2 - T_1)$
53. 26.9 cents/day
55. (a) 184 W (b) 461°C
57. 2020°C
59. (a) 667 A (b) 50.0 km
63. (a) $R = \dfrac{\rho L}{\pi(r_b^2 - r_a^2)}$ (b) 37.4 MΩ
    (c) $R = \dfrac{\rho}{2\pi L} \ln\!\left(\dfrac{r_b}{r_a}\right)$ (d) 1.22 MΩ
69. Average $\rho = 1.47\ \mu\Omega \cdot$m; they agree.
S1. (a) The savings are \$254.92.

Chapter 28

1. (a) 7.67 Ω (b) 1.76 W
3. (a) 1.79 A (b) 10.4 V
5. 12.0 Ω
7. (a) 6.73 Ω (b) 1.98 Ω

9. (a) 4.59 Ω (b) 8.16%
11. 0.923 $\Omega \leqslant R \leqslant 9.0\ \Omega$
13. 1.00 kΩ
15. 55.0 Ω
17. 1.41 Ω
17A. $\sqrt{2}\,R$
19. 14.3 W, 28.5 W, 1.33 W, 4.00 W
21. (a) 0.227 A (b) 5.68 V
23. 470 Ω; 220 Ω
23A. $\frac{1}{2}R_s + (R_s^2/4 - R_s R_p)^{1/2}$ and $\frac{1}{2}R_s - (R_s^2/4 - R_s R_p)^{1/2}$
25. (a) -10.4 V (b) 141 mA, 915 mA, 774 mA
27. $\frac{11}{13}$ A, $\frac{6}{13}$ A, $\frac{17}{13}$ A
29. Starter: 171 A Battery: 0.283 A
31. 3.50 A, 2.50 A, 1.00 A
33. (a) $I_1 = \dfrac{5}{13}$ mA; $I_2 = \dfrac{40}{13}$ mA; $I_3 = \dfrac{35}{13}$ mA
    (b) 69.2 V; c
35. (a) 12.4 V (b) 9.65 V
37. (a) 909 mA (b) -1.82 V
39. 800 W, 450 W, 25.0 W, 25.0 W
41. 3.00 J
41A. $U_0/4$
43. (a) 1.50 s (b) 1.00 s
    (c) 200 μA + (100 μA)$e^{-t/(1.00\text{ s})}$
45. (a) 12.0 s (b) $i(t) = (3.00\ \mu\text{A})e^{-t/12}$
    $q(t) = (36.0\ \mu\text{C})[1 - e^{-t/12}]$
47. (a) 6.00 V (b) 8.29 μs
49. 425 mA
51. 1.60 MΩ
51A. $t/C \ln 2$
53. 16.6 kΩ
55. 0.302 Ω
57. 49.9 kΩ
59. (b) 0.0501 Ω, 0.451 Ω
61. 0.588 A
63. 60.0 Ω
65. (a) 12.5 A, 6.25 A, 8.33 A
    (b) 27.1 A; No, it would not be sufficient since the current drawn is greater than 25 A.
67. (a) 0.101 W (b) 10.1 W
69. (a) 16.7A (b) 33.3 A
    (c) The 120-V heater requires four times as much mass.
71. 6.00 Ω; 3.00 Ω
71A. $P_s/2I^2 + (P_s^2/4I^4 - P_s P_p/I^4)^{1/2}$ and
     $P_s/2I^2 - (P_s^2/4I^4 - P_s P_p/I^4)^{1/2}$
73. (a) 72.0 W (b) 72.0 W
75. (a) 40W (b) 80 V, 40 V, 40 V
77. (a) $R \leqslant 1050\ \Omega$ (b) $R \geqslant 10.0\ \Omega$
79. (a) $R \to \infty$ (b) $R \to 0$ (c) $R \to r$
81. (a) 9.93 μC (b) 3.37×10^{-8} A
    (c) 3.34×10^{-7} W (d) 3.37×10^{-7} W
83. $T = (R_A + 2R_B)C \ln 2$
85. $R = 0.521\ \Omega$, 0.260 Ω, 0.260 Ω, assuming resistors are in series with the galvanometer
87. (a) 1/3 mA for R_1, R_2 (b) 50 μC
    (c) (0.278 mA)$e^{-t/0.18\text{s}}$ (d) 0.290 s

89. (a) $1.96\ \mu C$　　(b) $53.3\ \Omega$

91. (a) $\ln\dfrac{\mathcal{E}}{V} = (0.0118\ \text{s}^{-1})\,t + 0.0882$

　　(b) $85\ \text{s} \pm 6\%$; $8.5\ \mu F \pm 6\%$

93. $48.0\ \text{W}$

93A. $1.50P$

S1. Kirchhoff's equations for this circuit can be written as

$$\mathcal{E}_1 - I_1 R_1 - I_4 R_4 = 0$$
$$\mathcal{E}_2 - I_2 R_2 - I_4 R_4 = 0$$
$$\mathcal{E}_3 - I_3 R_3 - I_4 R_4 = 0$$

and

$$I_1 + I_2 + I_3 = I_4$$

In matrix form, they are

$$
\begin{bmatrix}
R_1 & 0 & 0 & R_4 \\
0 & R_2 & 0 & R_4 \\
0 & 0 & R_3 & R_4 \\
1 & 1 & 1 & -1
\end{bmatrix}
\begin{bmatrix}
I_1 \\ I_2 \\ I_3 \\ I_4
\end{bmatrix}
=
\begin{bmatrix}
\mathcal{E}_1 \\ \mathcal{E}_2 \\ \mathcal{E}_3 \\ 0
\end{bmatrix}
$$

In Lotus 1-2-3 use the */DataMatrixInvert* and */DataMatrixMultiply* commands and in Excel use the array formulas $=MINVERSE(array)$ and $=MMULTI(array1,\ array2)$ to carry out the calculations. Using the data given in the problem, we find

$$I_1 = -1.26\ \text{A},\ I_2 = 0.87\ \text{A},\ I_3 = 1.08\ \text{A, and } I_4 = 0.69\ \text{A.}$$

Chapter 29

1. (a) West
　　(b) zero deflection　　(c) up　　(d) down
3. B_x is indeterminate; $B_y = -2.62\ \text{mT}$; $B_z = 0$
5. $48.8°$ or $131°$
7. $26.0\ \text{pN}$ west
9. $2.34 \times 10^{-18}\ \text{N}$
11. zero
13. $(-2.88\ \text{N})\,\mathbf{j}$
15. $0.245\ \text{T}$ east
17. (a) $4.73\ \text{N}$　　(b) $5.46\ \text{N}$　　(c) $4.73\ \text{N}$
19. $196\ \text{A}$ east if $\mathbf{B} = 50.0\ \mu\text{T}$ north
21. $F = 2\pi rIB \sin\theta$, up
23. $9.98\ \text{N}\cdot\text{m}$, clockwise as seen looking in the negative y-direction
23A. $NBabI \cos\theta$, clockwise as seen looking down
25. (a) $376\ \mu\text{A}$　　(b) $1.67\ \mu\text{A}$
27. (a) $3.97°$　　(b) $3.39\ \text{mN}\cdot\text{m}$
27A. (a) $\tan^{-1}(IBL/2mg)$　　(b) $\frac{1}{4} IBLd\cos\theta$
29. $1.98\ \text{cm}$
31. $65.6\ \text{mT}$
33. $r_\alpha = r_d = \sqrt{2}\,r_p$
35. $7.88\ \text{pT}$
37. $2.99\ \text{u}$; ${}_1^3\text{H}^+$ or ${}_2^3\text{He}^+$
39. $5.93 \times 10^5\ \text{N/C}$
41. $mg = 8.93 \times 10^{-30}\ \text{N}$ down, $qE = 1.60 \times 10^{-17}\ \text{N}$ up, $qvB = 4.74 \times 10^{-17}\ \text{N}$ down
43. $0.278\ \text{m}$

45. $31.2\ \text{cm}$
47. (a) $4.31 \times 10^7\ \text{rad/s}$　　(b) $5.17 \times 10^7\ \text{m/s}$
49. $70.1\ \text{mT}$
51. $3.70 \times 10^{-9}\ \text{m}^3/\text{C}$
53. $4.32 \times 10^{-5}\ \text{T}$
55. $7.37 \times 10^{28}\ \text{electrons/m}^3$
57. $128\ \text{mT}$ north at $78.7°$ below the horizon
59. (a) $(3.52\mathbf{i} - 1.60\mathbf{j}) \times 10^{-18}\ \text{N}$　　(b) $24.4°$
61. $0.588\ \text{T}$
65. $4.38 \times 10^5\ \text{Hz}$
67. $3.82 \times 10^{-25}\ \text{kg}$
69. $3.70 \times 10^{-24}\ \text{N}\cdot\text{m}$
71. (a) $(-8.00 \times 10^{-21}\ \text{kg}\cdot\text{m/s})\,\mathbf{j}$　　(b) $8.91°$

S1. The figure below shows a plot of the data and the best straight-line fit to the data. The slope is $1.58\ \mu\text{A/rad}$. The torsion constant κ is given by $\kappa = NAB \times \text{slope} = 4.74 \times 10^{-10}\ \text{N}\cdot\text{m/rad} = 8.28 \times 10^{-12}\ \text{N}\cdot\text{m/deg.}$

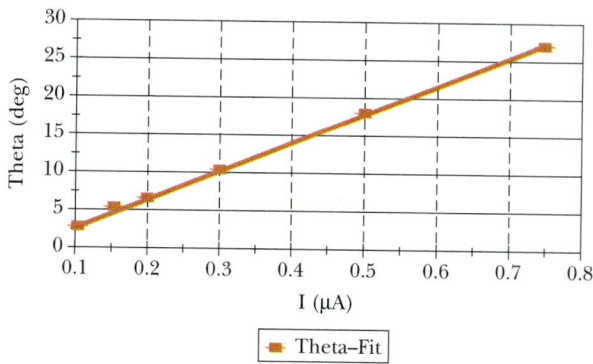

S2. This problem is somewhat similar to Problem 14.4, which dealt with gravitation and orbits, except that here the acceleration is velocity dependent. The equation of motion is

$$\mathbf{F} = q\mathbf{E} + q(\mathbf{v} \times \mathbf{B}) = m\mathbf{a} = q(E_x - v_z B_y)\mathbf{i} + qv_x B_y\,\mathbf{k}$$

The acceleration in the y direction is zero, so the y component of the velocity of the electron is constant.

Chapter 30

1. $200\ \text{nT}$
3. $31.4\ \text{cm}$
5. (a) $28.3\ \mu\text{T}$　　(b) $24.7\ \mu\text{T}$ into the page
7. $54.0\ \text{mm}$
9. $\dfrac{\mu_0 I}{4\pi x}$ into the plane of the page
11. $26.2\ \mu\text{T}$ into the plane of the page
11A. $\mu_0 I/8R$ into the plane of the page
13. $\dfrac{\mu_0 I\,(a^2 + d^2 - d\sqrt{a^2 + d^2})}{2\pi a d\,\sqrt{a^2 + d^2}}$ into the page
15. $80.0\ \mu\text{N/m}$
17. $2000\ \text{N/m}$; attractive

19. $(-27.0 \, \mu\text{N})\mathbf{i}$

21. $13.0 \, \mu\text{T}$ directed downward

21A. $\dfrac{\mu_0}{2\pi m} (400 \, I_1^2 + 69.4 \, I_2^2)^{1/2}$ at $\tan^{-1}\left(\dfrac{18.5 \, I_1 + 3.21 \, I_2}{7.69 \, I_2 - 7.69 \, I_1}\right)$ below the x-axis

23. (a) 3.98 kA   (b) ≈ 0

25. 5.03 T

27. (a) 3.60 T   (b) 1.94 T

29. Assuming constant current, make the wire long and thin. The solenoid radius does not affect the magnetic field.

31. (a) 6.34×10^{-3} N/m, inward
    (b) **F** is greatest at the outer surface.

33. 464 mT

33A. $\mu_0 \mathcal{E} \pi r/2\rho L$

35. 4.74 mT

37. (a) 3.13 mWb   (b) zero

39. (a) $BL^2/2$ (b) $BL^2/2$

41. (a) $(8.00 \, \mu\text{A})e^{-t/4}$ (b) $2.94 \, \mu\text{A}$

43. (a) 11.3×10^9 V·m/s   (b) 100 mA

45. 1.0001

47. 191 mT

49. 277 mA

51. $150 \, \mu\text{Wb}$

53. M/H

55. 1.27×10^3 turns

57. 2.02

59. (a) 9.39×10^{45} (b) 4.36×10^{20} kg

61. 675 A down

63. 81.7 A

63A. $2\pi d\mu g/\mu_0 I_A$

65. $\mathbf{B} = \dfrac{\mu_0 I}{2\pi w} \ln\left(\dfrac{b+w}{b}\right) \mathbf{k}$

67. 594 A east

69. 1.43×10^{-10} T directed away from the center

69A. $\dfrac{\mu_0 \, \omega q}{8\sqrt{2}\pi R}$

73. (a) $B = \frac{1}{3} \mu_0 b r_1^2$ (b) $B = \dfrac{\mu_0 b R^3}{3r_2}$

75. (a) 2.46 N   (b) 107.3 m/s²

77. (a) 12.0 kA-turns/m   (b) 2.07×10^{-5} T·m/A

79. $933 \, \mu\text{N}$ to the right perpendicular to the straight segments

79A. $\dfrac{\mu_0 I_1 I_2 L}{\pi R}$ to the right perpendicular to the straight segments

83. $\dfrac{\mu_0 I}{4\pi} (1 - e^{-2\pi})$ perpendicularly out of the paper

85. $20.0 \, \mu\text{T}$ toward the top of the page

87. $\dfrac{4}{3} \rho\mu_0\omega R^2$

89. $\dfrac{4}{15} \pi\rho\omega R^5$

S3. When integrating $F(r/R)$ when $r = R$, there is a singularity in the integrand at $\theta = \pi/2$ rad. The magnetic field at the wire is infinite, which we would expect even for a straight wire. We suggest you calculate the integral at $r/R = 0.0, 0.2, 0.4, 0.6, 0.8,$ and 0.9.

Chapter 31

1. 500 mV

3. 160 A

5. $+ 121$ mA

7. 61.8 mV

9. $(200 \, \mu\text{V})e^{-t/7}$

11. $Nn\pi R^2\mu_0 I_0 \alpha e^{-\alpha t} = (68.2 \text{ mV})e^{-1.6t}$ counterclockwise

11A. $Nn\pi R^2 \, \mu_0 I_0 \alpha e^{-\alpha t}$ counterclockwise

13. 272 m

15. -6.28 V

15A. $-2\pi R^2 B/t$

17. 763 mV, with the left-hand wingtip positive

19. (a) 3.00 N to the right   (b) 6.00 W

21. 2.00 mV; the west end is positive in the northern hemisphere

23. 2.83 mV

25. $145 \, \mu\text{A}$

25A. $Bd\left(\dfrac{v_2 \, R_3 - v_3 R_2}{R_1 \, R_2 + R_1 \, R_3 + R_2 \, R_3}\right)$

27. (a) to the right   (b) to the right   (c) to the right
    (d) into the plane of the paper

29. 0.742 T

31. $114 \, \mu\text{V}$ clockwise

31A. $N\mu_0 (I_1 - I_2) \pi r^2/\ell \, \Delta t$ clockwise

33. 1.80×10^{-3} N/C counterclockwise

35. (a) $(9.87 \times 10^{-3} \text{ V/m})\cos(100\pi t)$ (b) clockwise

37. (a) 1.60 A counterclockwise
    (b) $20.1 \, \mu\text{T}$ (c) up

37A. (a) $\dfrac{n\mu_0\pi r_2^2}{2R} \dfrac{\Delta I}{\Delta t}$ counterclockwise   (b) $\dfrac{n\mu_0^2\pi r_2^2}{4Rr_1} \dfrac{\Delta I}{\Delta t}$
    (c) up

39. 12.6 mV

41. (a) 7.54 kV   (b) **B** is parallel to the plane of the loop.

43. $(28.6 \text{ mV})\sin(4\pi t)$

45. (a) 0.640 N·m   (b) 241 W

47. 0.513 T

49. (a) $F = \dfrac{N^2 B^2 w^2 v}{R}$ to the left   (b) 0
    (c) $F = \dfrac{N^2 B^2 w^2 v}{R}$ to the left

51. $(-2.87\mathbf{j} + 5.75\mathbf{k}) \times 10^9$ m/s²

53. It is, with the top end in the picture positive.

57. (a) 36.0 V   (b) 600 mWb/s   (c) 35.9 V
    (d) 4.32 N·m

59. (a) 97.4 nV   (b) clockwise

61. Moving east; $458 \, \mu\text{V}$

63. It is, with the left end in the picture positive

65. $1.20 \, \mu\text{C}$

65A. Ba^2/R

67. 6.00 A

67A. b^2/aR if $t \geq b/3a$

69. (a) 900 mA   (b) 108 mN   (c) b   (d) No

71. (a) Counterclockwise   (b) $\dfrac{K\pi r^2}{R}$

73. (a) $\dfrac{\mu_0 IL}{2\pi}\ln\left(\dfrac{h+w}{h}\right)$ (b) $-4.80\ \mu V$

Chapter 32

1. 100 V
3. 1.36 μH
5. $\mathcal{E}_0/k^2 L$
7. (a) 188 μT   (b) 33.3 nWb   (c) 375 μH
   (d) field and flux
9. 21.0 μWb
11. $(18.8\ \text{V})\cos(377t)$
13. $\frac{1}{2}$
15. (a) 15.8 μH   (b) 12.6 mH
17. $(500\ \text{mA})\,(1 - e^{-10t/s}),\ 1.50\ \text{A} - (0.25\ \text{A})e^{-10t/s}$
19. 1.92 Ω
21. (a) 1.00 kΩ (b) 3.00 ms
21A. (a) $\sqrt{L/C}$ (b) $\sqrt{LC}$
23. (a) 139 ms   (b) 461 ms
25. (a) 5.66 ms   (b) 1.22 A   (c) 58.1 ms
27. (a) 113 mA   (b) 600 mA
29. (a) 0.800   (b) 0
31. (a) 1.00 A   (b) 12.0 V, 1.20 kV, 1.21 kV
    (c) 7.62 ms
33. 2.44 μJ
35. (a) 20.0 W   (b) 20.0 W   (c) 0   (d) 20.0 J
37. (a) 8.06×10^6 J/m^3 (b) 6.32 kJ
39. 44.3 nJ/m^3 in the **E** field; 995 μJ/m^3 in the **B** field
41. 30.7 μJ and 72.2 μJ
43. (a) 668 mW   (b) 34.3 mW   (c) 702 mW
45. 1.73 mH
47. 80.0 mH
49. 553 μH
51. (a) 18.0 mH   (b) 34.3 mH   (c) -9.00 mV
53. 400 mA
55. (a) -20.0 mV   (b) $-(10.0\ \text{MV/s}^2)\,t^2$
    (c) 63.2 μs
55A. (a) $-LK$ (b) $-Kt^2/2C$ (c) $2\sqrt{LC}$
57. 608 pF
59. (a) 36.0 μF   (b) 8.00 ms
61. (a) 6.03 J   (b) 0.529 J   (c) 6.56 J
63. (a) 2.51 kHz   (b) 69.9 Ω
65. (a) 4.47 krad/s   (b) 4.36 krad/s   (c) 2.53%
69. 979 mH
71. (a) 20.0 ms   (b) -37.9 V   (c) 3.03 mV
    (d) 104 mA
73. 95.6 mH
79. (a) 72.0 V; b   (c) 75.2 μs
81. $\dfrac{\mu_0}{\pi}\ln\left(\dfrac{d-a}{a}\right)$
83. 300 Ω
85. 45.6 mH
85A. $9t^2/\pi^2 C$
S1. (a) $\tau = 0.35$ ms, (b) $t_{50\%} = 0.24$ ms $= 0.686\tau$, and
    (c) $t_{90\%} = 0.80$ ms $= 2.29\tau$; $t_{99\%} = 1.6$ ms $= 4.57\tau$.

Chapter 33

3. 2.95 A, 70.7 V
5. 3.38 W
7. 14.6 Hz
9. (a) 42.4 mH   (b) 942 rad/s
11. 7.03 H
13. 5.58 A
15. 3.80 J
17. (a) $f > 41.3$ Hz   (b) $X_C < 87.5\ \Omega$
19. 2.77 nC
19A. $\sqrt{2}\ CV$
21. 100 mA
23. 0.427 A
25. (a) 78.5 Ω (b) 1.59 kΩ (c) 1.52 kΩ
    (d) 138 mA   (e) $-84.3°$
27. (a) 17.4°   (b) voltage leads the current
29. 1.88 V
29A. $R_b V_2/\sqrt{R_b^2 + 1/4\ \pi^2 f^2 C_s^2}$ where V_2 is the transformer
     secondary voltage
31. (a) 146 V   (b) 213 V   (c) 179 V   (d) 33.4 V
33. (a) 194 V   (b) current leads by 49.9°
35. 132 mm
35A. $\sqrt{800\ Ppd/\pi V^2}$ where ρ is the material's resistivity.
37. 8.00 W
39. $v(t) = (283\ \text{V})\sin(628t)$
41. (a) 16.0 Ω (b) $-12.0\ \Omega$
43. 159 Hz
45. 1.82 pF
47. (a) 124 nF   (b) 51.5 kV
49. 242 mJ
49A. $\dfrac{4\pi RCV^2\sqrt{LC}}{9L + 4CR^2}$
51. (a) 613 μF   (b) 0.756
53. 8.42 Hz
55. (a) True, if $\omega_0 = (LC)^{-1/2}$ (b) 0.107, 0.999, 0.137
57. 0.317
59. 687 V
61. (a) 9.23 V   (b) 4.55 A   (c) 42.0 W
63. 87.5 Ω
63A. $R_s = \dfrac{R_L N_1\,(N_2 V_s - N_1 V_2)}{V_2 N_2^{\,2}}$
65. 56.7 W
67. (a) 1.89 A
    (b) $V_R = 39.7$ V, $V_L = 30.1$ V, $V_C = 175$ V
    (c) $\cos\phi = 0.265$
    (d)

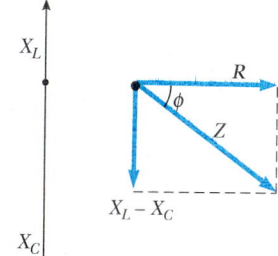

69. 99.6 mH
71. (a) 2.90 kW   (b) 5.80×10^{-3}
    (c) If the generator is limited to 4500 V, no more than
    17.5 kW could be delivered to the load, never 5000 kW.
73. (a) Circuit (a) is a high-pass filter, (b) is a low-pass filter.

   (b) $\dfrac{V_{out}}{V_{in}} = \dfrac{\sqrt{R_L^2 + X_L^2}}{\sqrt{R_L^2 + (X_L - X_C)^2}}$ for circuit (a);

   $\dfrac{V_{out}}{V_{in}} = \dfrac{X_C}{\sqrt{R_L^2 + (X_L - X_C)^2}}$ for circuit (b)

75. (a) 172 MW   (b) 17.2 kW   (c) 172 W
77. (a) 200 mA; voltage leads by 36.8°   (b) 40.0 V;
    $\phi = 0°$ (c) 20.0 V; $\phi = -90°$
    (d) 50.0 V; $\phi = +90°$
83. (a) 173 Ω   (b) 8.66 V
85. (a) 1.838 kHz

87.

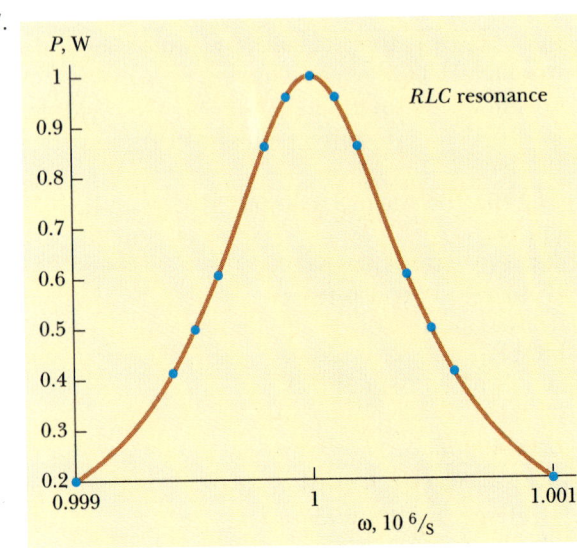

S1. (a)

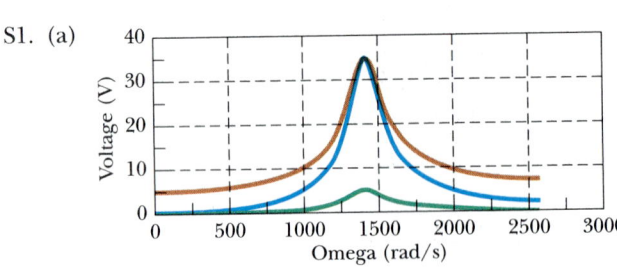

$\boxed{-V_R - V_C - V_L}$

(b) $|V_L| = |V_C|$; (c) for $\omega < \omega_0$, $V_C > V_L$; (d) for $\omega > \omega_0$, $V_L > V_C$

Chapter 34

1. 2.68×10^3 AD
5. (a) 162 N/C   (b) 130 N/C

7. 38.0 pT
9. (a) 6.00 MHz   (b) $(73.3 \text{ nT})(-\mathbf{k})$
   (c) $\mathbf{B} = (73.3 \text{ nT}) \cos(0.126x - 3.77 \times 10^7 t)(-\mathbf{k})$
11. (a) $B = 333$ nT   (b) $\lambda = 628$ nm
    (c) $f = 4.77 \times 10^{14}$ Hz
13. 5.16 m
15. 307 μW/m²
17. 66.7 kW
19. (a) 332 kW/m²   (b) 1.88 kV/m, 222 μT
21. (a) 971 V/m   (b) 16.7 pJ
23. (a) 540 N/C   (b) 2.58 μJ/m³   (c) 774 W/m²
    (d) This is 77.4% of the flux mentioned in Example 34.3
    and 57.8% of the 1340 W/m² intensity above the atmo-
    sphere. It may be cloudy at this location, or the sun may
    be setting.
25. 83.3 nPa
27. (a) 1.90 kV/m   (b) 50.0 pJ
    (c) 1.67×10^{-19} kg·m/s
29. (a) 11.3 kJ   (b) 5.65×10^{-5} kg·m/s
31. (a) 2.26 kW   (b) 4.71 kW/m²
33. 7.50 m
35. (a) $\mathcal{E}_m = 2\pi^2 r^2 f B_m \cos\theta$, where θ is the angle between
    the magnetic field and the normal to the loop. (b) Verti-
    cal, with its plane pointing toward the broadcast antenna
37. 56.2 m
37A. $2\pi cm/qB$
39. (a) 6.00 pm   (b) 7.50 cm
41. The radio audience hears it 8.41 ms sooner.
43. (a) 4.17 m to 4.55 m   (b) 3.41 to 3.66 m
    (c) 1.61 m to 1.67 m
45. 3.33×10^3 m²
47. (a) 6.67×10^{-16} T   (b) 5.31×10^{-17} W/m²
    (c) 1.67×10^{-14} W   (d) 5.56×10^{-23} N
49. 6.37×10^{-7} Pa
51. 95.1 mV/m
53. 7.50×10^{10} s = 2370 y; out the back
55. 3.00×10^{-2} degrees
57. (a) $B_0 = 583$ nT, $k = 419$ rad/m,
    $\omega = 1.26 \times 10^{11}$ rad/s; xz
    (b) 40.6 W/m² in average value   (c) 271 nPa
    (d) 406 nm/s²
59. (a) 22.6 h   (b) 30.5 s
61. (b) 1.00 MV/s
63. (a) 3.33 m; 11.1 ns; 6.67 pT

   (b) $\mathbf{E} = (2.00 \text{ mV/m}) \cos 2\pi \left(\dfrac{x}{3.33 \text{ m}} - \dfrac{t}{11.1 \text{ ns}}\right)\mathbf{j}$

   $\mathbf{B} = (6.67 \text{ pT}) \cos 2\pi \left(\dfrac{x}{3.33 \text{ m}} - \dfrac{t}{11.1 \text{ ns}}\right)\mathbf{k}$

   (c) 5.31×10^{-9} W/m²   (d) 1.77×10^{-17} J/m³
   (e) 3.54×10^{-17} Pa

Chapter 35

1. 299.5 Mm/s
3. (b) 294 Mm/s
5. 114 rad/s

7. 198 Gm

9. (a) 4.74×10^{14} Hz (b) 422 nm
   (c) 2.00×10^8 m/s

11. 70.5° from the vertical

13. 61.3°

15. 19.5°, 19.5°, 30.0°.

17. (a) 327 nm (b) 287 nm

19. 59.83°, 59.78°, 0.0422°

21. 30.0°, 19.5° at entry; 40.5°, 77.1° at exit

23. 0.171°

23A. $\sin^{-1}\left(\dfrac{\sin\theta}{n_{700}}\right) - \sin^{-1}\left(\dfrac{\sin\theta}{n_{400}}\right)$

27. 18.4°

29. 86.8°

31. 4.61°

33. 62.4°

35. (a) 24.4° (b) 37.0° (c) 49.8°

37. 1.00008

39. $\theta < 48.2°$

41. 53.6°

43. 2.27 m

45. 2.37 cm

49. 90°, 30°, No

51. 62.2%

53. $\sin^{-1}[(n^2 - 1)^{1/2} \sin\phi - \cos\phi]$ If $n \sin\phi \leqq 1, \theta = 0$

55. 82

57. 27.5°

59. 7.91°

59A. $\phi = \sin^{-1}\left[n \sin\left(90° - 2\theta + \sin^{-1}\left(\dfrac{\sin\theta}{n}\right) \right) \right]$

61. (a) 1.20 (b) 3.40 ns

S1. Using Snell's Law and plane geometry, the following relationships can be derived:

$$\sin\theta_1 = n \sin\rho_1$$
$$\rho_2 = \Phi - \rho_1$$
$$n \sin\rho_2 = \sin\theta_2$$
$$\theta = \theta_1 + \theta_2 - \Phi$$

These equations are used to find ρ_1, ρ_2, θ_2, and δ in the spreadsheet. IF statements are used in the calculation of θ_2 and δ so that error messages are not printed when ρ_2 exceeds the critical angle for total internal reflection.

Chapter 36

3. 2'11"

3A. $h/2$

5. 30 cm

7. (a) $q = 45.0$ cm, $M = -\frac{1}{2}$
   (b) $q = -60.0$ cm, $M = 3.00$
   (c) Similar to Figures 36.8 and 36.12b

9. (a) 15.0 cm (b) 60.0 cm

9A. (a) $d/2$ (b) $2d$

11. concave with radius 40.0 cm

13. (a) 2.08 m (concave)
    (b) 1.25 m in front of the object

15. (a) $q = -12.0$ cm, $M = 0.400$
    (b) $q = -15.0$ cm, $M = 0.250$
    (c) The images are erect

17. 11.8 cm above the floor

19. 1.50 cm/s

19A. v/n

21. 8.57 cm

23. 3.88 mm

25. 2

27. (a) 16.4 cm (b) 16.4 cm

29. 25.0 cm; -0.250

31. (a) 13.3 cm
    (b) A trapezoid

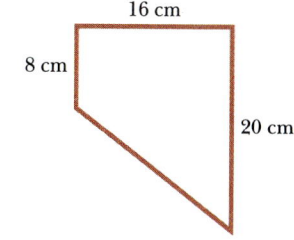

(c) 224 cm²

33. (a) -12.3 cm, to left of lens (b) $+0.615$
    (c)

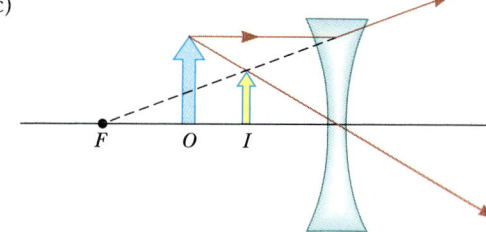

35. (a) 39.0 mm (b) 39.5 mm

37. 2.84 cm

39. $\dfrac{f}{1.41}$

41. -4 diopters, a diverging lens

43. 0.558 cm

45. 20.0 m

45A. $vf\,\Delta t/\Delta f$

47. 3.5

49. -800, image is inverted

51. -18.8

53. 2.14 cm

55. (a) 1.99 (b) 10.0 cm to the left of the lens, -2.50
    (c) inverted since the overall magnification is negative.

57. (a) 20.0 cm to the right of the second lens, -6.00
    (b) inverted (c) 6.67 cm to the right of the second lens, -2.00, inverted

59. 10.7 cm beyond the curved surface

61. 20.0 cm

61A. f

63. (a) 44.6 diopters (b) 3.03 diopters

65. $d = 8$ cm

67. 21.3 cm

67A. $h_1 \cot \left(\sin^{-1} \dfrac{nh_1}{R} - \sin^{-1} \dfrac{h_1}{R} \right) + \sqrt{R^2 - h_1^2}$

$- h_2 \cot \left(\sin^{-1} \dfrac{nh_2}{R} - \sin^{-1} \dfrac{h_2}{R} \right) - \sqrt{R^2 - h_2^2}$

69. (a) 52.5 cm   (b) 1.50 cm

71. $q' = 1.5$ m, $h' = -13.1$ mm

73. (a) -0.400 cm   (b) $q' = -3.94$ mm, $h' = 535\ \mu$m

75. (a) 30.0 cm and 120 cm   (b) 24.0 cm
    (c) real, inverted, diminished

77. real, inverted, actual size

S1. See Figure 36.14. Applying the law of sines to triangle
    OPC, one finds $R \sin \theta_1 = (r + p) \sin \alpha$. Snell's Law gives
    $n_1 \sin \theta_1 = n_2 \sin \theta_2$. Using plane geometry gives $\gamma = \theta_1 - \alpha - \theta_2$. Applying the law of sines to triangle *PIC*
    gives $q - R = R \sin \theta_2 / \sin \gamma$. These equations are used
    to calculate θ_1, θ_2, γ, and q in the spreadsheet. (a) The
    maximum angle α for which the error in the image dis-
    tance is 3 percent or less is 2.35°. (b) 2.25° (d) No. (For
    a plot of the percent error versus α, just call up the asso-
    ciated graph.)

Chapter 37

1. (a) 2.62×10^{-3} m   (b) 2.62×10^{-3} m
3. 515 nm
5. (a) 36.2°   (b) 5.08 cm   (c) 508 THz
7. 11.3 m
9. 2.50 m
9A. $c/2f$
13. 4.80×10^{-5} m
15. 423.5 nm
17. 343 nm
17A. $\lambda/2(n - 1)$
19. 641
21. (a) 2.63 rad   (b) 246 nm
23. (a) 7.95 rad   (b) 0.453
27. $10 \sin(100\pi t + 0.93)$
29. $\pi/2$
31. $360°/N$
33. No reflection maxima in the visible spectrum
35. 512 nm
37. 85.4 nm, or 256 nm, or 427 nm . . .
39. (a) Green   (b) Purple
41. 167 nm
43. $4.35\ \mu$m
45. 3.96×10^{-5} m
47. 654 dark fringes
49. (a) 5.99 m   (b) 2.99 m
49A. (a) $2\sqrt{4h^2 + d^2} - 2d$ (b) $\sqrt{4h^2 + d^2} - d$
51. 3.58°
53. 421 nm
55. 2.52 cm
57. 1.54 mm
59. 3.6×10^{-5} m

61. $x_{\text{bright}} = \dfrac{\lambda \ell \left(m + \frac{1}{2}\right)}{2hn}$, $x_{\text{dark}} = \dfrac{\lambda \ell\ m}{2hn}$

63. (b) 115 nm
65. 1.73 cm
69. (b) 266 nm
S1. (a) 9:1   (b) 1:9

Chapter 38

1. 632.8 nm
3. 560 nm
5. $\cong 10^{-3}$ rad
7. 0.230 mm
9. $\phi = \beta/2 = 1.392$ rad

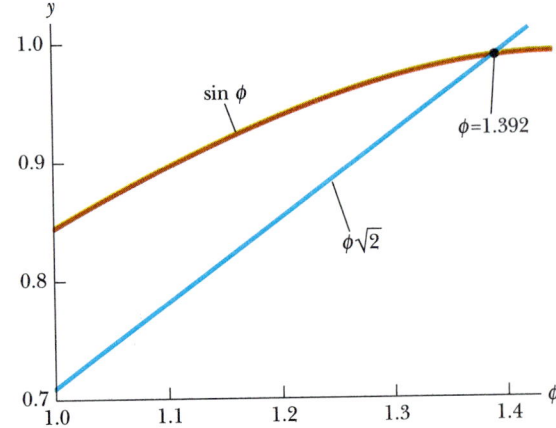

11. 1.62×10^{-2}
13. 3.09 m
15. 51.2 cm
17. 27 cm
17A. $(0.8\ \mu$m$)\ (vt/d)$
19. 11.9 km
21. 13 m
23. 105 m
27. (a) 5.7; 5 orders is the maximum
    (b) 10; 10 orders to be in the short-wavelength region.
29. 0.0683°
31. (a) 2800   (b) 4.72 μm
33. 2
35. 7.35°
37. 14.7°
39. 31.9°
41. (a) 0°, 19.9°, 42.9°   (b) 8.33 cm, 135 cm
43. (a) 0.281 nm   (b) 0.18%
45. 14.4°
47. 1.11
49. 31.2°
51. (a) 18.4°   (b) 45.0°   (c) 71.6°
53. (a) 54.7°   (b) 63.4°   (c) 71.6°
55. 60.5°

57. 2.2
59. 5.9°
61. 0.244 rad ≈ 14.0°
63. 545 nm
65. (a) 12 000, 24 000, 36 000   (b) 0.0111 nm
67. (a) 3.53×10^3 lines/cm
    (b) Eleven maxima can be observed.
69. $\frac{1}{8}$
71. (a) 6   (b) 7.50°
73. (a) 41.8°   (b) 0.593   (c) 0.262 m
S1. (a) 9:1   (b) 1:9

Chapter 39

5. (a) 60 m/s   (b) 20 m/s   (c) 44.7 m/s
5A. (a) $v_b + v$ (b) $v - v_b$ (c) $\sqrt{v^2 + v_b^2}$
7. (a) 39.2 μs   (b) Accurate to one digit. More precisely, he aged 1.78 μs less on each orbit.
9. $0.800c$
9A. $v = cL_p/\sqrt{L_p^2 + c^2 t^2}$
11. $0.436L_0$
13. $0.789c$
15. $0.696c$
17. $0.960c$
19. 42 g/cm³

19A. $\dfrac{m}{V(1 - v^2/c^2)}$

21. 1625 MeV/c
25. (a) 939.4 MeV   (b) 3.008×10^3 MeV
    (c) 2.069×10^3 MeV
27. $0.864c$
29. (a) 0.582 MeV   (b) 2.45 MeV
31. (a) 3.91×10^4 (b) $0.9999999997c$ (c) 7.66 cm
33. 4 MeV and 29 MeV
35. (a) 3.29 MeV   (b) 2.77 MeV
37. 4.2×10^9 kg/s
39. (a) $0.800c$ (b) $0.929c$

39A. (a) $\dfrac{2v}{1 + v^2/c^2}$ (b) $\dfrac{3v + v^3/c^2}{1 + 3v^2/c^2}$

41. 0.7%
43. (a) 6.67 ks   (b) 4.00 ks

43A. (a) $\dfrac{2d}{v + c}$ (b) $\dfrac{2d\sqrt{1 - v^2/c^2}}{v + c}$

47. (a) 76.0 min   (b) 52.1 min

47A. (a) $\dfrac{t_{\mathrm{II}}}{\sqrt{1 - \left(\dfrac{(v_\mathrm{I} + v_\mathrm{II})\,c}{c^2 + v_\mathrm{I} v_\mathrm{II}}\right)^2}}$ (b) $\dfrac{t_{\mathrm{II}}}{\sqrt{1 - v_\mathrm{II}^2/c^2}}$

49. 15.0 h in the spacecraft frame, 10.0 h less than in the Earth frame.
51. (a) $0.554c$, $0.866c$ (b) 0.833 m

51A. (a) $u = \dfrac{v - u'}{1 - vu'/c^2}$ where $u' = c\sqrt{1 - L'^2/L_p^2}$

    (b) $L_p\sqrt{1 - u^2/c^2}$
53. (a) 83.3 m and 62.5 m   (b) 27.0 m and 20.3 m
    (c) 6.00 ks   (d) 3.60 ks   (e) 4.80 ks
    (f) Both rockets and crew are destroyed.
55. (b) For $v \ll c$, $a = qE/m$ as in the classical description. As $v \to c$, $a \to 0$, describing how the particle can never reach the speed of light.

    (c) Perform $\displaystyle\int_0^v \left(1 - \frac{v^2}{c^2}\right)^{-3/2} dv = \int_0^t \frac{qE}{m}\, dt$ to

    obtain $v = \dfrac{qEct}{\sqrt{m^2c^2 + q^2E^2t^2}}$ and then

    $\displaystyle\int_0^x dx = \int_0^t \frac{qEtc}{\sqrt{m^2c^2 + q^2E^2t^2}}\, dt$

S1. (a)

Z	0.2	0.5	1.0	2.0
v/c	0.180	0.385	0.600	0.800

    (b) For $Z = 3.8$, $v/c = 0.870$.
S3. For $Z = 0.2$, $r = 2710$ Mly; $Z = 0.5$, $r = 5770$ Mly; $Z = 1.2$, $r = 9860$ Mly; and $Z = 3.8$, $r = 13\ 800$ Mly.

Chapter 40

1. (a) 2.57 eV   (b) 12.8 μeV
   (c) 191 neV   (d) 484 nm, 96.8 mm, 6.52 m
3. 2.27×10^{30} photons/s
5. 1.30×10^{15}/s

5A. $\dfrac{8\,\pi^2 cd^2(\lambda_2 - \lambda_1)}{(\lambda_2 + \lambda_1)^4\,(\exp(2hc/(\lambda_1 + \lambda_2)k_\mathrm{B}T) - 1)}$

7. 4.46×10^3 K
9. 9.35 μm; infrared
11. 5.18×10^3 K
15. (a) 0.350 eV   (b) 555 nm
17. 2.22 eV, 3.70 eV
19. (a) 0.571 eV   (b) 1.54 V
21. (a) only lithium   (b) 0.808 eV
23. (b) 3.87 eV   (c) 8.78 eV
25. (a) 33.0°   (b) 0.785 c

25A. With $x = E_0/mc^2$ (a) $\phi = \tan^{-1}\left(\dfrac{1}{\sqrt{1 + x}}\right)$ and

    (b) $u = \left(\dfrac{2x\sqrt{x^2 + 3x + 2}}{2x^2 + 3x + 2}\right)c$

27. 4.85 pm
29. (a) 0.843 MeV, 0.670 MeV/c
    (b) The ionized atom carries off momentum.
31. (a) 2.88×10^{-12} m   (b) 101°
33. (a) $\theta = \phi = 43.0°$
    (b) $E = 0.602$ MeV, $p = 0.602$ MeV/c
    (c) $K = 0.278$ MeV, $p = 0.602$ MeV/c

33A. With $x = E_0/mc^2$

  (a) $\theta = \phi = \cos^{-1}\left(\dfrac{1 + x}{2 + x}\right)$

  (b) $E = \left(\dfrac{x^2 + 2x}{2x + 2}\right) mc^2, \; p = \left(\dfrac{x^2 + 2x}{2x + 2}\right) mc$

  (c) $K = \dfrac{x^2 mc^2}{2x + 2}, \; p = \left(\dfrac{x^2 + 2x}{2x + 2}\right) mc$

35. 70.1°

37. (a) 200 nm $\leqslant \lambda \leqslant$ 360 nm (ultraviolet)
   (b) 800 nm $\leqslant \lambda \leqslant$ 1440 nm (infrared)

39. (a) 5   (b) No; No

41. 0.529 Å, 2.12 Å, 4.77 Å

43. (a) $E_n = -54.4 \dfrac{eV}{n^2}$, $n = 1, 2, 3, \ldots$

  (b) 54.4 eV

45. (a) 0.265 Å   (b) 0.177 Å   (c) 0.132 Å

47. (a) 3.03 eV   (b) 411 nm   (c) 7.32 × 10¹⁴ Hz

49. (a) B   (b) A   (c) B and C

51. − 6.80 eV, + 3.40 eV

53. $r_n = (1.06 \text{ Å}) n^2$, $E_n = -6.80 \text{ eV}/n^2$, $n = 1, 2, 3, \ldots$

55. 5.39 keV

57. (a) 41.4°   (b) 680 keV

59. (a) 3.12 fm   (b) − 18.9 MeV

61. (a) − 6.67 keV, − 668 eV   (b) 0.621 nm

65. (a) $E_1 = -8.178$ eV, $E_2 = -2.042$ eV,
    $E_3 = -0.904$ eV, $E_4 = -0.510$ eV
    $E_5 = -0.325$ eV
  (b) $\lambda_\alpha = 1092$ nm, $\lambda_\beta = 811$ nm, $\lambda_\gamma = 724$ nm,
    $\lambda = 609$ nm

Chapter 41

1. 4.18 × 10³⁵ photons/s

3. (a) 1.10 × 10⁻³⁴ m/s   (b) 1.36 × 10³³ s
   (c) No. The time is greater than 10¹⁵ × (age of Universe)

5. (a) 12.5 h   (b) $v = c/\sqrt{2} = 212$ Mm/s

9. $c/\sqrt{2} = 212$ Mm/s

11. 0.215 nm

13. 6.03 pm

15. At 1.00° on both sides of the central maximum

17. (a) 15 keV   (b) 124 keV

19. 1.16 Mm/s or more

21. (a) relativistic   (b) not necessarily relativistic; they
   might move as slowly as 0.021 c

23. (b) 1.84 × 10⁻¹⁶ m

25. 0.250

27. 0.513 MeV, 2.05 MeV, 4.62 MeV

29. (a) $L/2$ (b) 5.26 × 10⁻⁵   (c) 3.99 × 10⁻²
  (d) It is more likely to find the particle near $x = L/4$
   and $x = 3L/4$ than at the center, where the probability
   density is zero.

31. 0.516 MeV, 3.31 × 10⁻²⁰ kg·m/s

33. (a) 37.7 eV, 151 eV, 339 eV, 603 eV
  (b) 2.20 nm, 2.75 nm, 4.71 nm, 4.12 nm, 6.59 nm, 11.0 nm

35. 7.93 Å

35A. $L = \sqrt{\dfrac{3 \, h\lambda}{8 \, mc}}$

37. $\pm \sqrt{\dfrac{2}{L}}$

39. (a) $E = \hbar^2 / mL^2$

  (b) Requiring $\displaystyle\int_{-L}^{L} A^2 (1 - x^2/L^2)^2 \, dx = 1$ gives $A =$

  $(15/16 \, L)^{1/2}$

  (c) $47/81 = 0.580$

43. (a)

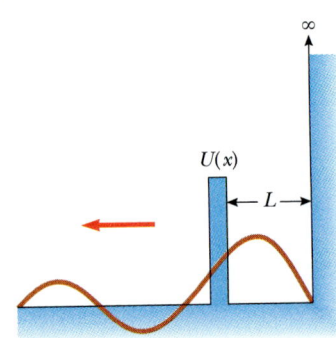

$U(x)$ $\leftarrow L \rightarrow$

  (b) $2L$

45. 1% chance of transmission, 99% chance of reflection

45A. (a) $e^{-\sqrt{8m(U-E)} \, L/\hbar}$
   (b) $1 - e^{-\sqrt{8m(U-E)} \, L/\hbar}$

47. 1.34 × 10³¹

49. 2.02 × 10⁻⁴ nm; γ-ray

51. (a) 2.82 × 10⁻³⁷ m   (b) 1.06 × 10⁻³² J
  (c) 2.87 × 10⁻³⁵% or more

53. 2.81 × 10⁻⁸

59. 0.0294

61. (a) $\Delta p \geqslant \hbar/2r$ (b) Choosing $p \approx \hbar/r$,
   $E = \dfrac{\hbar^2}{2mr^2} - \dfrac{k_e e^2}{r}$ (c) − 13.6 eV

63. (a) $\sqrt{\left(\dfrac{nhc}{2L}\right)^2 + (m_0 c^2)^2} - m_0 c^2$ $n = 1, 2, 3, \ldots$

  (b) 4.69 × 10⁻¹⁴ J; 28.6%

65. (a)

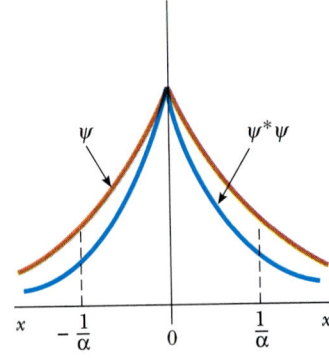

(b) $P(x \rightarrow x + dx) = \psi^\circ \psi \, dx = \begin{cases} A^2 e^{-2\alpha x} \, dx \text{ for } x > 0 \\ A^2 e^{2\alpha x} \, dx \text{ for } x < 0 \end{cases}$

(c) ψ is continuous, $\psi \rightarrow 0$ as $x \rightarrow \pm \infty$, ψ is finite at $-\infty < x < \infty$.

(d) $A = \sqrt{\alpha}$ (e) 0.632

67. (a) 1.03×10^{-3} (b) 959 pm

69. (b) $\frac{3}{2}\hbar\omega$; the first excited state

Chapter 42

1. (a) 56.8 fm   (b) 11.3 N backward
1A. (a) $r = 2Zk_e e^2/E$ (b) $E^2/2Zk_e e^2$
3. 1312 nm   (b) 164 nm
5. 797 times
7. $r = 4a_0$
9. (a) 3.13 fm   (b) -18.8 MeV   (c) 14.1 MeV
11. $\sqrt{6}\,\hbar$
13. (a) 2   (b) 8   (c) 18   (d) 32   (e) 50
15. The $3d$ subshell has $n = 3$ and $l = 2$. The particle has $s = 1$. Therefore we can have $m_\ell = -2, -1, 0, 1,$ or 2 and $m_s = -1, 0,$ or 1, leading to this table.

n	l	m_l	s	m_s
3	2	-2	1	-1
3	2	-2	1	0
3	2	-2	1	$+1$
3	2	-1	1	-1
3	2	-1	1	0
3	2	-1	1	$+1$
3	2	0	1	-1
3	2	0	1	0
3	2	0	1	$+1$
3	2	$+1$	1	-1
3	2	$+1$	1	0
3	2	$+1$	1	$+1$
3	2	$+2$	1	-1
3	2	$+2$	1	0
3	2	$+2$	1	$+1$

17. $3\hbar$
19. $L = 2.58 \times 10^{-34}$ J·s; $L_z = -2\hbar, -\hbar, 0, \hbar,$ and $2\hbar$
    $\theta = 145°, 114°, 90.0°, 65.9°,$ and $35.3°$
21. 9.27×10^{-24} J/T
23. The latter has lower energy and more unpaired spins, as a generalization of Hund's rule would suggest; chromium.
25. $1s^2 2s^2 2p^6 3s^1$ ^{11}Na
    $1s^2 2s^2 2p^6 3s^2$ ^{12}Mg
    $1s^2 2s^2 2p^6 3s^2 3p^1$ ^{13}Al
    $1s^2 2s^2 2p^6 3s^2 3p^2$ ^{14}Si
    $1s^2 2s^2 2p^6 3s^2 3p^3$ ^{15}P
    $1s^2 2s^2 2p^6 3s^2 3p^4$ ^{16}S
    $1s^2 2s^2 2p^6 3s^2 3p^5$ ^{17}Cl
    $1s^2 2s^2 2p^6 3s^2 3p^6$ ^{18}Ar
    $1s^2 2s^2 2p^6 3s^2 3p^6 4s^1$ ^{19}K

27. 124 V
27A. $V_{min} = hc/e\lambda$
29. Iron ($Z = 26$)
31. (a) 14 keV   (b) 88 pm
33. 590 nm
35. 3.50×10^{16} photons
35A. $N = \lambda Pt/hc$
37. 9.79 GHz
39. (a) 137.036   (b) $\frac{2\pi}{\alpha}$ (c) $\frac{1}{2\pi\alpha}$ (d) $\frac{4\pi}{\alpha}$
41. (a) 1.57×10^{14} m$^{-3/2}$ (b) 2.47×10^{28} m^{-3}
    (c) 8.69×10^8 m^{-1}
45. (a) 4.20 mm   (b) 1.05×10^{19} photons
    (c) 8.82×10^{16}/mm^3
45A. (a) $L = ct$ (b) $N = E\lambda/hc$
    (c) $n = 4E\lambda/\pi hc^2 d^2 t$
47. (a) $\dfrac{\mu_0 \pi m^2 e^7}{8\epsilon_0^3 h^5}$ (b) 12.5 T
49. $5.24a_0$
51. 0.125
53. (a) 0.24 s   (b) Light energy is quantized.
53A. $t = \phi A/PA'$ where A' is the face area over which one atom collects energy.

Chapter 43

1. (a) 921 pN   (b) -2.88 eV
3. (a) 40.0 μeV, 9.65 GHz   (b) 20% too large if r is 10% too small
5. (a) 0.0118 nm   (b) 0.00772 nm; HI is less stiff
7. 0, 36.9 μeV, 111 μeV, 221 μeV, 369 μeV, . . .
9. 5.69×10^{12} rad/s
11. 0.358 eV
13. 3.63×10^{-4} eV, 1.09×10^{-3} eV
15. Twelve at 0.444 nm, six at 0.628 nm, twenty-four at 0.769 nm
19. (a) 1.57 Mm/s   (b) On the order of 10^{10} times greater than the drift speed
21. 5.28 eV
23. 3.4×10^{17} electrons
25. (a) 1.10   (b) 1.55×10^{-25}
27. The red line, with photon energy 1.89 eV < 2.42 eV, is transmitted; all the other lines are absorbed.
29. 1.24 eV or less, yes
31. 4.4 V
33. $v = 7$
S2. For $T = 1$ K, N(E) increases monotonically from 0 at $E = 0$ to a maximum at the Fermi energy and then abruptly drops to zero (see Figure 43.18). As T increases the peak of the N(E) curve "rounds over" and then drops to zero less abruptly. In addition, N(E) starts to drop to zero at slightly lower E as T increases.

Chapter 44

1. (a) 30.76 T   (b) 3.07×10^5 A   (c) 7.69 T
3. 200 A
5. 2.0 kV
7. 4×10^{-25} Ω
9. 318 A
11.

Super-conductor	E_g (meV)	Super-conductor	E_g (meV)
Al	0.36	Sn	1.13
Ga	0.33	Ta	1.36
Hg	1.26	Ti	0.12
In	1.04	V	1.61
Nb	2.82	W	0.005
Pb	2.19	Zn	0.26

13. 23 K
15. 5.2×10^9 pairs
17. $e/h = 2.417 \times 10^{14}$ (V·s)$^{-1}$
19. 1.65×10^{-14} T
21. (a) 0.377 T   (b) 1.89×10^{-3} Wb
    (c) 9.15×10^{11}
23. (a) 50.0 mT   (b) 20.0 mT   (c) 2.29 MJ
    (d) 100 N outward
25. 1.75 T, No
27. 1.56 cm
29. $\Delta S = 9.83 \times 10^{-3}$ J/mol·K
31. 3.50 T to 4.67 T; since $B < B_c$ (at 4.2 K), the super-conductivity will not be destroyed.

Chapter 45

1. (a) 1.9 fm   (b) 7.44 fm   (c) 3.92
3. 2.3×10^{15} g
5. Even Z; He, O, Ca, Ni, Sn, Pb
    Even N; T, He, N, O, Cl, K, Ca, V, Cr, Sr, Y, Zr, Xe, Ba, La, Ce, Pr, Nd, Pb, Bi, Po
7. $3.18 \times 10^{-2}\%$
9. 12.7 km
11. (a) 29.5 fm   (b) 5.18 fm, small compared to the distance of closest approach
13. (a) 4.55×10^{-13} m   (b) 6.03×10^6 m/s
15. 2.85 MeV/nucleon
17. (a) 2.72×10^{-28} eV   (b) 342 MeV   (c) Nuclei are bound by the strong nuclear force, not by the gravitational force.
19. The binding energy per nucleon is greater for $^{23}_{11}$Na by 0.210 MeV. Extra electrostatic repulsion makes the magnesium nucleus less stable.
21. 7.93 MeV
23. (a) $\dfrac{A - Z}{Z}$ is greatest for $^{139}_{55}$Cs and is equal to 1.53.
    (b) ^{139}La   (c) ^{139}Cs

25. 200 MeV
27. (a) 491.3 MeV   (b) 179%, 53%, 25%, 1%
29. 1155 s
31. (a) 1.55×10^{-5} s^{-1}, 12.4 h
    (b) 2.39×10^{13} atoms   (c) 1.87 mCi
33. (a) 4.00×10^9 y if all the ^{206}Pb came from ^{238}U
    (b) 0.0199 and 4.60
35. 0.55 mCi
37. 9.46×10^9 nuclei
37A. $\dfrac{R_0 T_{1/2}}{\ln 2} \left(2^{-t_1/T_{1/2}} - 2^{-t_2/T_{1/2}}\right)$
39. 4.28 MeV
41. 18.6 keV
43. 5.16 MeV
45.

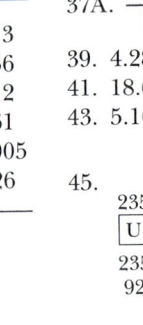

47. (a) $e^- + p \rightarrow n + \nu$ (b) ^{15}O $\rightarrow$ ^{15}N $+ \nu$
    (c) 2.75 MeV
49. -2.64 MeV
51. (a) 7.250 MeV   (b) The Q value for ^{6}Li(d, p)^7 Li is 5.020 MeV, so $E_b - Q = 2.230$ MeV, which is close to the binding energy of the deuteron.
53. 10.0135 u, 8.0053 u
55. $\sqrt{2}$
57. (a) 5.70 MeV   (b) 3.27 MeV; exothermic
59. (a) 2.52×10^{24} (b) 2.31×10^{12} Bq
    (c) 1.07 million years
61. 6×10^{19} Bq
63. (a) No.   (b) Yes.   (c) Zero to 156 keV
65. (a) 2.75 fm   (b) 152 N   (c) 2.62 MeV
    (d) $R = 7.44$ fm, $F = 379$ N, $U = 17.6$ MeV
67. 2.2×10^{-6} eV
69. (a) 7.7×10^4 eV   (b) 4.62 MeV, 13.9 MeV
    (c) 10.3×10^6 kW·h
71. (a) 12.2 mg   (b) 166 mW
73. 0.35%
75. (b) 4.79 MeV
77. (a) The 2.44 MeV state cannot be reached by β^+ decay.   (b) $^{93}_{42}$Mo
79. 1.49×10^3 Bq
81. 4.45×10^{-8} kg/h
83. (b) $T_{1/2} = 1.46$ min $\pm 6\%$ (c) $\lambda = 0.473$ min^{-1} $\pm 6\%$

S1. The following tables give the number of parent nuclei (N1) and the number of daughter nuclei (N2) present:

(a)

	N1	N2
$t = 1$ min	4349	478
$t = 2$ min	3782	648
$t = 5$ min	2488	587

(b)

	N1	N2
$t = 1$ min	3514	1391
$t = 2$ min	2470	2187
$t = 5$ min	857	2718

(c)

	N1	N2
$t = 1$ min	4997	3
$t = 2$ min	4993	5
$t = 5$ min	4983	8

Chapter 46

1. $^{142}_{54}\text{Xe}$, $^{143}_{54}\text{Xe}$, $^{144}_{54}\text{Xe}$
3. $^1_0\text{n} + {}^{232}_{90}\text{Th} \rightarrow {}^{233}_{90}\text{Th}$ $^{233}_{90}\text{Th} \rightarrow {}^{233}_{91}\text{Pa} + \beta^-$
    $^{233}_{91}\text{Pa} \rightarrow {}^{233}_{92}\text{U} + \beta^-$
5. 1.28×10^{25}, 1.61×10^{26}
7. 5800 km
9. 2664 y
11. (a) 31.9 g/h   (b) 122 g/h
13. (a) 2.22×10^6 m/s   (b) 4.5×10^{-8} s
15. (a) 2.4×10^7 J   (b) 10.6 kg
17. 1570 y
19. (a) 0.0025 rem per x-ray
    (b) 38 times the background
21. About 24 days
23. 1.14 rad

25. (a) 3.1×10^7 (b) 3.1×10^{10} electrons
25A. (a) $CV^2/2E$ (b) CV/e
27. (a) 3.0×10^{24} (b) 1.2 kg
29. 495 Ci, eight orders of magnitude less than the activity of the fission inventory
29A. 4.95×10^{-20} nV Ci
31. 26 collisions
33. (a) 1.36 GJ   (b) The fusion energy is 40 times greater than the gasoline energy.
35. (a) 4.92×10^8 kg/h $= 4.92 \times 10^5$ m^3/h
    (b) 0.141 kg/h
37. 14.1 MeV
39. (a) 22×10^6 kW·h   (b) 17.6 MeV for each D-T fusion   (c) 2.34×10^8 kW·h   (d) 9.37 kW·h
41. (a) 5.7×10^8 K   (b) 120 kJ
43. (b) 25.7 MeV
45. (b) 35.2   (c) 2.89×10^{15}

Chapter 47

1. 2.26×10^{23} Hz, 1.32×10^{-15} m
3. 2.06×10^{-18} m
5. (a) 0.782 MeV   (b) proton speed $= 0.001266c = 380$ km/s, electron speed $= 0.9185c = 276$ Mm/s
    (c) The electron is relativistic, the proton is not.
7. $\Omega^+ \rightarrow \overline{\Lambda}^0 + \text{K}^+$
    $\overline{\text{K}}^0 \rightarrow \pi^+ + \pi^-$
    $\overline{\Lambda}^0 \rightarrow \overline{\text{p}} + \Pi^+$
    $\overline{\text{n}} \rightarrow \overline{\text{p}} + \text{e}^+ + \nu_e$
9. (a) $1 \neq 0$ (b) $E_e = 469.15$ MeV, $p_e = 469.15$ MeV/c, $E_\gamma = 469.15$ MeV $p_\gamma = -469.15$ MeV/c
    (c) $0.9999994c$
11. $E_p = 5.38$ MeV $E_\pi = 32.5$ MeV
13. (b, c) conserved (a, d, e, f) not conserved
15. (a) lepton number   (b) lepton number
    (c) strangeness and charge   (d) baryon number
    (e) strangeness
19. (a) $d\overline{u} + uud \rightarrow d\overline{s} + uds$
    (b) $\overline{d}u + uud \rightarrow u\overline{s} + uus$
    (c) $\overline{u}s + uud \rightarrow \overline{u}s + d\overline{s} + sss$
    (d) $uud + uud \rightarrow d\overline{s} + uud + u\overline{d} + uds$
21. (a) $0.384c$, 38.4% of the speed of light
    (b) 6.7×10^9 lightyears
23. 1115 MeV/$c^2 = 1.98 \times 10^{-27}$ kg
25. One part in 50 million
27. 9.3 cm
29. neutron
31. (a) Z^0 boson   (b) gluon

Index

Page numbers in *italics* indicate illustrations; Page numbers followed by an ''n'' indicate footnotes. Page numbers followed by t indicate tables.

Aberrations, chromatic, *1071*, 1071–1072
  of lenses, *1070*, 1070–1072, *1071*
  spherical, 1056, *1056, 1070*, 1070–1071
Absolute pressure, 427
Absolute temperature scale, 534, 624
Absolute zero, 535, 624, 624n
Absorber, ideal, 574, 576
Absorption, selective, polarization of light waves by, *1133*, 1133–1134, *1134*
Absorption spectra, 1272
Absorption spectroscopy, 1201
ac circuits. *See* Circuit(s), alternating current.
Accelerated charge, 1221
Accelerated electron, 669
Accelerated frames, motion in, 151–154
Acceleration, 28–32, *29*
  angular, and angular velocity, 277–278
    average, 278
    instantaneous, 278, 278n, 296
    of wheel, *291*, 291–292
    torque and, *289*, 289–292
  average, 29, 44, 73, *73*
    and instantaneous, 31
  centripetal, 85, 93–94, 145, 162
  constant, 35–36
    two-dimensional motion with, 74–77, *75*
  constant angular, rotational motion with, 278–279
  dimensions of, 11t
  free-fall, 37, 39, *39*, 44, 394–395, 411
    and projectile motion, 77, 77n
    measurement of, 120–121, *121*
    variation of, with altitude, 395
  in simple harmonic motion, 362–363, *367*
  instantaneous, 29, 44, 73–74
  inverse-square law and, 396
  linear and angular, 280, *280*
  magnitude of, of connected objects with friction, 128, *128*
  mass and, 111
  of automobile, 35
    up hill, 190–191, *191*
  of center of mass, 256
  of mass, on frictionless incline, *119*, 119–120
  of moon, *396*, 396–397
  of two blocks of masses, magnitude of, 122, *122*

Acceleration *(Continued)*
  of two unequal connected masses, *121*, 121–122
  on frictionless surface, resultant force and, 113, *113*
  positive, of point charge, 668, *668*
  radial, *87*, 87–89, 93–94
  relative, 89–92, *90*
  tangential, *87*, 87–89, 280
  total, 87
  transverse, 468–469
  units of, 112t
  velocity, and position, graphical relations of, 30
Acceleration-time graph, *30*, 30–31, 32, *33*
Acceleration vector, displacement and velocity vectors, 71–74
Accelerator(s), Clinton P. Anderson Meson Physics Facility, 1189
  colliding-beam, 1430–1431
  proton-proton collision in, 1382
  Saturn, 1390
Accommodation in eye, 1074
Action forces, 114–115, *115*
Activation analysis, 1407
Adiabatic expansion, reversible, *PV* diagram for, 594–595, *595*, 608
Adiabatic free expansion, *563*, 565, 575
Adiabatic process, 565, 575
  first-law equation for, 565
  for ideal gas, 594–596
  reversible, definition of, 594
Air drag, 154
  at high speeds, 156–157, *157*
Air friction, resistive forces and, 189
Air resistance, and projectile motion, 77, 77n
Algebra, A.15–A.20
Alpha decay, 1239, 1363–1366, *1364, 1365, 1366*
Alpha particles, *1253*, 1253–1254
Ammeter, *813*, 813–814
Ampère, Andre, 833, 871
Ampère-Maxwell law, 880, *880*, 880n, 924, 925, 994
Ampère-meter 2, 842
Ampère (unit), 654, 773, 776, 789
  definition of, 870

Ampère's law, 864, 870–874, 875, 876, 890
  generalized, 924
    displacement current and, 879–881, *880*, 891
Amplitude, displacement, of sound waves, 481–482, *482*
  in simple harmonic motion, 361
  of wave, 457, *457*
  pressure, of sound waves, 481–482, *482*, 491
Analyzer, 1133
Anderson, Carl, 1415, 1418
Angle, Brewster's, 1135
  phase, 971, 983
  polarizing, 1134, 1139
  solid, 696
Angular frequency, 466
  of ac generator, 965
Angular moment, 882
Angular momentum. *See* Momentum, angular.
Angular wave number, 466
Antenna, horn-reflector, 1433, *1434*
  production of electromagnetic wave by, 1008–1011, *1010, 1011*
Antiderivative, finding of, 41
Antilogarithm, A.20
Antineutrino, 1367
Antinodes, 503, *504.* 519
Antiparticle, 1415, 1438
Antiprotons, and protons, collisions of, 1346
Antiquarks, properties of, 1426t
Archimedes, 428
Archimedes' principle, 427
  buoyant forces and, 427–431, *428, 429*, 441
Area, dimensions of, 11t
  of rectangle, 16
  of room, 16
Argon, 1269
Associative law of addition, 58, *58*
Astigmatism, 1076
Atmosphere(s), law of, 600
  molecules in, 600
Atomic clock(s), 6
  time dilation and, 1166n, 1166–1167
Atomic energy levels, *601*, 601–602
Atomic mass unit, 1349

Atomic mass(es), density and, 9–10
  table of, A.4–A.13
Atomic number, 1206, 1347, 1372
Atomic physics, 1252–1285
Atomic spectra, 1200–1202, 1272–1275
Atomic transitions, 1275, 1275–1276, 1276
Atomizer, 437, 437
Atom(s), 8, 10, 1413
  allowed energies for, 1273–1274
  Bohr's quantum model of, 1202–1208,
    1203
  donor, 1306, 1306–1307
  early models of, 1253, 1253–1255, 1254
  energy level diagram of, 1275, 1275
  excited states of, 1205, 1276
  ground state of, 221, 1205
  hydrogen. See Hydrogen atom.
  in cube, 10
  in solid, 14
  magnetic moments of, 881–883, 882, 883t
  metal, 655n
  model of, 1233–1234, 1234
  nuclear, classical model of, 1254, 1254
  nucleus of, 1254
  quantum states allowed for, 1267t, 1267–
    1268
  radiation emitted by, 1203–1204
  "seeing of," 1126
  Thomson's model of, 1253, 1253
Atwood's machine, 120–121, 121, 292, 292
Aurora Australis, 847
Aurora Borealis, 832, 846–847
Automobile(s), acceleration of, 35
    up hill, 190–191, 191
  car lift for, 425–426
  collision into wall, average force and, 241,
    241
  collision of, at intersection, 250, 250
  compact, gas consumed by, 190
  energy and, 188–191
  in U.S., gas used by, 14
  speed of, coefficient of static friction and,
    147
Avogadro's number, 541
  definition of, 10, 17

Bainbridge mass spectrometer, 848, 848
Balmer, Johann Jacob, 1202, 1205–1206
Balmer series, 1202
  for hydrogen, 1207, 1207
Band theory of solids, 1300, 1300–1301
Bardeen, John, 1308, 1319, 1328
Barometer, 427, 427
Baryon number(s), checking of, 1422
  law of conservation of, 1421
Baryons, 1420
Base, in junction transistor, 1309, 1310
Baseball player, displacement and, 24–25,
    25
Battery(ies), conducting rods connected to,
    1009, 1010

Battery(ies), (Continued)
  positive charge flows in, 786, 786, 786n
  power supplied to resistor by, 786
  terminal voltage of, 799
BCS theory, 1328–1332, 1331, 1339, 1340
Beam, uniform, horizontal, 343, 343–344
Beats, 513–515, 515
Becquerel, Henri, 1346, 1359
Becquerel (unit), 1362
Bednorz, Georg, 782, 1319, 1336, 1337
Bell, Alexander Graham, 484n
Bernoulli, Daniel, 434
Bernoulli principle, 422, 434
Bernoulli's equation, 433–435, 434
  applications of, 436, 436–437, 437
Beryllium, 1269
Beta decay, 1366, 1366–1367
Bicycle wheel, spinning, 321, 321
Big Bang, radiation spectrum of, 1434,
    1434–1435
Big Bang theory, of creation of Universe,
    161, 1432–1435, 1438
Billiard balls, collision of, 251, 251
Bimetallic strip, 539, 539
Biot, Jean Baptiste, 865
Biot-Savart law, 864, 865–869, 866, 890
Black body, 574, 576
Black hole, formation of, 1182
Blackbody radiation, 1190, 1190–1193
Body, rigid. See Rigid body(ies).
Bohr, Niels, 92, 324–325, 1190, 1202,
    1203, 1254, 1266, 1357
Bohr magneton, 883, 1266
Bohr model, quantization of angular mo-
    mentum in, 1219, 1219–1220, 1220n
Bohr radius, 1204–1205, 1208, 1257–1258
Bohr theory, assumptions of, 1203–1204
Bohr's correspondence principle, 1207–
    1208
Bohr's quantum model of atom, 1202–
    1208, 1203
Boltzmann, Ludwig, 599
Boltzmann distribution law, 599–602
Boltzmann's constant, 542, 590, 607, 1191,
    1383
Bond(s), covalent, 1288, 1288, 1289, 1311
  hydrogen, 1290, 1312
  in solids, 1296–1300
  ionic, 1287, 1311
  molecular, 1287, 1287–1290
  van der Waals, 1288, 1312
Born, Max, 1190, 1228, 1235
Boson(s), 1414
  Higgs, 1430
Boyle's law, 541
Brackett series, 1202
Bragg, W.L., 1132
Bragg's law, 1132
Brahe, Tycho, 395
Brattain, Walter, 1308
Bremsstrahlung, 1392
Brewster, Sir David, 1135

Brewster's angle, 1135
Brewster's law, 1135, 1139
Bridge(s), suspension, collapse of, 380
  Tacoma Narrows, collapse of, 380
British engineering system, 7
British thermal unit, definition of, 553
Brown, Robert, 529
Brownian motion, 529
Bubble chamber, 846, 1405, 1405
Buckminsterfullerene, 1286
"Buckyballs," 1286
Building materials, R values for, 571, 571t
Bulk modulus, 346, 347–348, 348, 349
Buoyant force(s), 155n, 427–428, 441
  and Archimedes' principle, 427–431, 428,
    429, 441

Calcite, 1136, 1136–1137
Calculus, differential, A.24–A.27
  integral, A.27–A.30
  kinematic equations derived from, 41–43
Calorie, 553
calorie (unit of heat), 553, 553n
Calorimeters, 556
Calorimetry, 556, 560
Camera, 1072, 1072–1073
  box moving past, 1164, 1164
  calculation of exposure time and, 1073
Capacitance, and geometry, 760t
  calculation of, 742–745
  definition of, 741–742, 760, 969
  equivalent, 748–749, 749
  of filled capacitors, 752, 752n
  SI unit of, 742, 760
Capacitive reactance, 970, 983
Capacitor(s), 741–742, 742, 742n, 760,
    761–762
  charged, energy stored in, 749n, 749–751,
    754–755, 755
  charging, charge versus time for, 810, 810
    current versus time for, 810, 810
  charging of, 808–810, 809
  combinations of, 745, 745–749, 746, 747
  cylindrical, 744, 744–745
  discharging, charge versus time for, 811
    current versus time for, 811
  discharging of, 811, 811
  displacement current in, 881
  electrolytic, 753–754, 754
  filled, capacitance of, 752, 752n
  high voltage, 753, 754
  in ac circuit, 969, 969–971, 975
  in RC circuit, charging of, 812, 812
    discharging of, 812–813
  instantaneous voltage drop across, 970
  maximum charge on, 809
  paper-filled, 754
  parallel-plate, 743, 743–744, 744, 758, 758
    and nonuniform electric field, 755, 755–
    756

Capacitors *(Continued)*
  effect on metal slab, 759–760, *760*
  energy stored in, 750
  partially filled, 759, *759*
  phase relationships in, 971–972, *972*
  spherical, 745, *745*
  tubular, 753, *754*
  two charged, rewiring of, 750–751, *751*
  types of, 753–756, *754, 755*
  with dielectrics, *751*, 751–756
Car lift, 425–426
Carbon, 1269
  activity of, 1362–1363, *1363*
Carbon dating, 1367–1369
Carburetor, 437
Carlson, Chester, 726
Carnot, Sadi, 620
Carnot cycle, 620, 621, *621*, 638
  PV diagram for, 621, *621*
  ratio of heats for, 622–623
Carnot-cycle heat engine, 627–628
Carnot engine, 620–624, 630
  efficiency of, 622, 623, 624
Carnot's theorem, 620, 638
Cartesian coordinate system, points in, 54, *54*
Cathode ray tube (CRT), 669–670, *670*
Cavendish, Henry, 393
Cavendish apparatus, 393, *393*
Celsius, Anders, 536n
Celsius temperature scale, 534, 536
Centripetal acceleration, 85, 93–94, 145, 162
cgs system of units, 7
Chadwick, James, 1349–1350
Chamberlain, Owen, 1415
Charge, accelerated, 1221
Charge carrier, 773
  in conductor, motion of, 783, *783*
  in electric field, motion of, 783–784, *784*
  motion of, in conductor, 774, *774*
Charge density, 777
Charge(s), conservation of, 797, 819
  electric, conserved, 650
  continuous charge distribution, 661–662
  continuous distribution, of electric field, 661–665, *662, 672*
  cylindrically symmetric distribution of, 692, *692*
  induced, 652
  linear density, 662
  negative, 650, *650*, 650–651
  nonconducting plane sheet of, 693, *693*
  on spheres, *657*, 657–658
  positive, 650
  properties of, 649–651, *651*, 671
  quantized, 651
  spherically symmetric distribution of, *690*, 690–691, *691*
  surface density, 662
  two, electric field due to, *660*, 660–661
  volume density, 662

Charge(s) *(Continued)*
  finite line of, of electric potential, 719, *719*
  maximum, 809
  point. *See* Point charge(s).
  uniform ring of, of electric field, 663–664, *664*
Charles and Gay-Lussac's law, 541
Charm, 1427–1428, 1438
Chromodynamics, quantum, 1429, 1438
Circle, equation of, A.21
Circuit(s), alternating current, 964–993, 983
  capacitors in, *969*, 969–971
  inductors in, *967*, 967–969
  power in, 975–976
  purely capacitive, 971
  purely inductive, 969
  resistors in, *965*, 965–967, *966*, 966n, 967t, 983
  sources and phasors, 965
  direct current, 797–831
  elements of, impedance values and phase angles for, 973, *973*
  filter, *979*, 979–980, *980*
  household, wiring diagram for, 817, *817*
  integrated, *1310*, 1310–1311, *1311*
  LC, 948
    charge versus time for, 950, *951*, 954
    current versus time for, 950–951, *951*, 954
    energy transfer in, 948, *949*
    oscillations in, 948–952
    angular frequency of, 950
    oscillatory, 952, *952*
    total energy in, 950, 951, *951*, 955
  multiloop, 807–808, *808*
  RC, 808–813
  RL, *941*, 941–944, *942*
    current versus time for, 944, *944*
    initial current in, 944, 954
    time constant of, *942*, 942–943, 944, *944*, 954
  RLC, *952*, 952–953, 955
    charge versus time for, 953, *953*
    overdamped, 953, *953*
  RLC series, 971–975, *972*
    analysis of, 974
    average power in, 976
    average power versus frequency for, 978, *978*
    current versus frequency for, *977*, 977–978
    impedance of, 973
    instantaneous power in, 975–976
    resonance in, 977–980
    resonating, 979
  single-loop, *806*, 806–807
  steady state, 797, 808
  time constant of, 819
Circular motion, 314, *314*
  nonuniform, *150*, 150, *151*
  of mass suspended by string, 88–89, *89*, 150–151, *151*

Circular motion *(Continued)*
  uniform, *85*, 85–87, 144–170
    and simple harmonic motion, compared, *375*, 375–377, *376*
    Newton's second law of motion and, 144–149, *145, 146*, 162
  with constant speed, 377
Classical mechanics, 1–450
Clausius, Rudolf, 628, 995, 996
Clausius statement, 618, 618n
Clocks, atomic, time dilation and, 1166n, 1166–1167
  synchronization of, 1157–1158
Cloud chamber, 1405
Coaxial cable, 946, *946*
  resistance of, 780, *780*
Coefficient(s), average, of linear expansion, 537–538, 538t, 543
  of volume expansion, 538, 543
  drag, 158–160
  Hall, 852
  of friction, 124–125, 126t
    measurement of, 127, *127*
  of kinetic friction, 125–126
    determination of, *127*, 127–128
  of performance, 627, 628
  of viscosity, *440*, 440t, 440–441
  temperature, of resistivity, 780–781
    for various materials, 777t
Coil(s), inductance of, 940, 954
  induction, 996
  induction of emf in, 910
  primary and secondary, 907
Collector, in junction transistor, 1309, *1309*
Colliders, 1430–1431
Collins, S.S., 1326
Collision frequency, 605
Collision(s), *242*, 242–243, 249
  as independent event, 784n
  average time between, 784–785
  elastic, *244*, 244–246, 262
  and inelastic, in one dimension, 243–248
  electron, in copper, 785
  inelastic, 244, 261
  linear momentum and, 235–275
  of automobiles, at intersection, 250, *250*
  of billiard balls, 251, *251*
  perfectly inelastic, 244, *244*
  proton-proton, 250–251
  slowing down of neutrons by, 247–248
  total momentum of system and, 243
  two-body, with spring, 247, *247*
  two-dimensional, *248*, 248–251
Color(s), in peacock's feathers, 1103
  quarks and, 1429, *1429*
Commutative law of addition, 57, *57*
Compact disc, diffraction grating and, 1129, *1129*
Complementarity, principle of, 1220
Compound microscope, 1078–1080, *1079*
Compressibility, 348
Compression ratio, 626

Compton, Arthur H., 1197
Compton effect, 1197–1200, *1199*, 1208, 1216
Compton scattering, at 45 degrees, 1200
  of photon by electron, *1199*, 1199–1200
Compton shift equation, 1198
  deviation of, *1199*, 1199–1200
Compton wavelength, 1198
Computers, hardware requirements and spreadsheets, A.40
  software requirements and spreadsheets, A.40–A.42
  spreadsheet problems, A.40–A.42
  spreadsheet tutorial for, A.40–A.42
Concrete, prestressed, 348, *348*
Condensation(s), heat of, 558n
  of sound waves, 479, 481
Condon, E.U., 1366
Conducting loop, 908, *908*
Conduction, charging object by, 652
  electrical. *See* Electrical conduction.
  heat, 568–571, *569*, 576, 632
  law of, 569
Conduction band, 1304
Conduction electrons, 783
Conductivity(ies), 775n, 784, 789
  and resistivity, 784, 790
  thermal, 569, 570t, 576
Conductor(s), 569
  cavity within, *722*, 722–723
  charge carrier in, motion of, 774, *774*, 783, *783*
  charged, potential of, 720–723, *721*, *722*
  current-carrying, field created by, 866, 870, *870*
    in uniform magnetic field, 838, *838*, 839, *839*, 854
    magnetic forces on, 837–841, *838*
  current in, 774, *774*, 789
  definition of, 651, 671
  grounded, 652
  in electrostatic equilibrium, 693–695, *694*, 698
  magnetic field surrounding, 866–867, *867*
  metals as, 652
  parallel, magnetic force between, *869*, 869–870
  power loss in, 786–787, 790
  resistance of, 776, 779, 790
  semicircular, force on, *840*, 840–841
  uniform, resistance of, 777
Confinement time, 1392, 1408
Conservation laws, 1420–1422
Constant-volume gas thermometer, 534, *534*
Constant-volume process, first-law equation for, 566
Contact lenses, 1077n
Continuity, equations of, 433, *433*
Convection, 572–573, *573*, 576
  forced, 572
  natural, 572

Conversion factors, A.1–A.2
Cooper, Leon, N., 1319, 1328
Cooper pair(s), 1319, *1329*, 1329–1330, 1330n, 1340
Coordinate systems, and frames of reference, 54–55
Coordinates, Galilean transformation of, *1150*, 1150–1151
Copernicus, Nicolaus, 395
Copper, excited electrons in, 1304
  Hall effect for, 853
Copper wire, drift speed of, 774–775
Cornea of eye, 1073
Corona discharge, 723
Corrals, 1216
Correspondence principle, 92
  Bohr's, 1207–1208
Cosmic rays, 845–847
Cosmology, 1432–1437
  problems and perspectives in, 1437
Coulomb, Charles, 647, 651, 653, 654
Coulomb constant, 654, 671
Coulomb repulsion effect, 1357
Coulomb (unit), definition of, 870
Coulomb's law, 649, 653–658, 655t, 662, 666, 671, 687–688, 866
  experimental proof of, 696, *696*
Coulomb's torsion balance, 651, *651*
Couple, 339, *339*
Covalent bonds, 1288, *1288*, *1289*, 1311
Critical angle, for total internal reflection, 1040
  index of refraction and, 1041, *1041*
Critical density, of Universe, 1436, *1436*
Critical ignition temperature, 1392, 1408
Critical temperature. *See* Temperature, critical.
Cross product. *See* Vector product(s).
Crystal ball, refraction and, 1063, *1063*
Crystal surfaces, tunneling microscope and, 1217
Crystalline solids, 1297, 1312
Crystals, birefringent, *1136*, 1136–1137
  covalent, 1298–1299, *1299*, 1299t
  diffraction of x-rays by, 1130n, 1130–1132, *1131*
  double-refracting, indices of refraction for, 1136t
Curie, Marie, 1359–1360
Curie, Pierre, 887, 1359–1360
Curie temperature, 887, *887*, 887t
Curie (unit), 1362
Curie's constant, 887
Curie's law, 887
Current, and motion of charged particles, 773, *773*
  and resistance, 772–796
  average, 773
  critical, in Pb wire, 1322
  definition of, 773
  density, 775, 784, 789
  direction of, 773, *773*

Current *(Continued)*
  displacement, 880, 891, 994
    capacitors in, 881
  in conductor, 774, *774*, 789
  in electron beam, *788*, 788–789
  induced, 906, 906n, *907*, 912, *912*, *915*, 915–916
  in motor, 922
  instantaneous, 773
  maximum, 809
  persistent, in superconductors, *1326*, 1326–1327, 1340
  root mean square (rms), 966, *966*, 966n, 967, 983, 984
  SI unit of, 773, 776, 789
  sinusoidal, infinite current sheet carrying, 1008
  versus time, for *RL* circuit, 944, *944*
Current loop, circular, axis of, magnetic field on, *868*, 868–869, *869*
  magnetic moment of, 842, 854
  torque on, in uniform magnetic field, *841*, 841–843, *842*
Current sheet, infinite, carrying sinusoidal current, 1008
Curvature of space-time, 1181–1182
Cyclic process, 565, 575
  irreversible, 638
  PV diagram for, 617, *617*
  reversible, 638
Cyclotron, 849–850, *850*
Cyclotron frequency, 844
Cylinder, net torque on, 288–289, *289*
  raising of, 345
  rolling, total energy of, 307–308
  rolling motion of, 307, *307*
  solid, uniform, moment of inertia of, 285, *285*

Dam, force of, 426–427, *427*
D'Arsonval galvanometer, 814, *814*
Davisson, C.J., 1220
Davisson-Germer experiment, 1220
Dead Sea Scrolls, 1368
deBroglie, Louis, 1190, 1218, 1219
deBroglie wavelength, 1218, 1218n, 1220, 1228, 1231
deBroglie waves, 1228, 1234–1235, 1243, 1255
Debye, Peter, 1197
Decay constant, 1360, 1373
Decay processes, 1363–1369, 1369t
  alpha decay, 1239, 1363–1366, *1364*, *1365*, *1366*
  beta decay, *1366*, 1366–1367
  carbon dating, 1367–1369
  gamma decay, 1369, *1369*
Decay rate, 1361, 1373
Decibel(s), 484, 484n
  sound levels in, 484, 484t
Dees, 849–850

Demagnetization, nuclear, 624n
Democritus, 8
Density-of-states function, 1302
Density(ies), and atomic mass, 9–10
  charge, 777
  critical, of Universe, 1436, *1436*
  current, 775, 784, 789
  definition of, 17
  energy, average, 1003
    in electric field, 750
    magnetic, 945
    total instantaneous, 1003
  ion, 1392, 1408
  linear charge, 662
  mass, 777
  of cube, 13
  of various substances, 9t, 422, 422t
  probability, 1228–1229, 1243–1244
    for particle in well of finite height, 1237, *1238*
  surface charge, 662, 775n
  volume charge, 662
Depth, apparent, and true depth, 1064, *1064*
  variation of pressure with, *424*, 424–427
Derivative(s), partial, 716
Descartes, Rene, 1029n, 1037
Descartes' law, 1029n
Deuterium, 1391
  and tritium, fusion reaction, *1397*, 1397–1398
Deuteron(s), binding energy of, 1178, 1354–1357
  fusion of, 1391–1392
  separation of, 1391, *1391*
Deviation, angle of, 1033–1035, *1034*
Dewar, Sir James, 574n
Dewar flask, 574, 574n
Diamagnetism, 881, 884, 887–888
Diamond, 1291
Diatomic molecule, rigid-rotor model of, 324
Dielectric constant(s), 751, 752t, 761
Dielectric strength(s), 752t, 753
Dielectric(s), atomic description of, 757–760, *758*
  capacitors with, *751*, 751–756
  definition of, 751
Differential, perfect, A.29
Differential calculus, A.24–A.27
Differential equations, 155
Diffraction, definition of, 1117, 1118, 1138
  Fraunhofer, 1118, *1119, 1120, 1121, 1123, 1123*, 1138
  Fresnel, *1118*, 1119, *1119*
  introduction of, 1117–1119, *1118*
  of x-rays, by crystals, 1130n, 1130–1132, *1131*
  pattern, circular aperture, resolution of, *1125*, 1125–1126
  single-slit, resolution of, 1124

Diffraction, *(Continued)*
  single-slit, *1119*, 1119–1124, *1120*
    intensity of, *1121*, 1121–1123
    types of, 1118
Diffraction electrons, 1221, *1222*
Diffraction grating, 1127–1130, *1128*, 1139
  compact disc and, 1129, *1129*
  intensity distribution for, 1128, *1128*
  orders of, 1129
  resolving power of, 1129–1130
Diffraction grating spectrometer, 1128, *1128*
Dimensional analysis, 10–12, 17
Diode(s), ideal, 1308
  real, 1308, 1309
  tunnel, 1239
Diopters, 1076
Dipole, electric, 761
  electric field lines for, 666, *666*
  in electric field, potential energy of, *756*, 757
  in external electric field, *756*, 756–757
  potential energy of, 756–757, 761
  torque on, 756, *756*, 761
  electric field of, 661, *661*
  electric potential of, *716*, 716–717
  oscillating electric, 1010–1011, *1011*
Dipole moment, electric, 756, 761
Dirac, Paul A.M., 1190, 1264–1265, 1415
Dirac sea, 1415
Disintegration energy, 1364
Disk, uniformly charged, electric field of, *664*, 664–665
Dispersion, 456n
Displacement, 43
  amplitude, of sound waves, 481–482, *482*
  angular, 277, *277*
  as function of time, 38, *38*
  of particle, as vector quantity, 56, *56*
  resultant, 59, *59*, 63
  velocity and speed, 24–26, *25*
  velocity as function of, 33–34
  versus time for simple harmonic motion, 361, *361, 367*
Displacement current, 880, 891, 994
  in capacitors, 881
Displacement vector, 71–72, *72*
  of projectile, 78, *79*
  velocity and acceleration vectors, 71–74
Domain walls, 885, *885*
Donor atom, *1306*, 1306–1307
Doppler, Christian Johann, 487n
Doppler effect, *487*, 487–491, *488*, 492
Dot product(s), 174, 192
  of unit vectors, 175
Double-slit experiment, 1221–1224
  Young's, 1092–1096, *1093*, 1108
Double-slit interference patterns, average intensity of, 1108
  intensity distribution of, *1096*, 1096–1098, *1098*
  phasor diagrams for, 1099, *1100*
Drag, 436

Drag coefficients, 158–160
Drift speed, 774, 784, 784n, 790
  in copper wire, 774–775
DuLong-Petit law, 598
Dyne, 112
Dyne.cm, 173

Earth, magnetic field(s) of, 888–890, *889*
Eddy currents, *922*, 922–923, *923*
Efflux, speed of, 435, *435*
Eightfold way, 1424–1425, *1425*
Einstein, Albert, 92, 454n, 454–455, 529–530, 923, 1022, 1024, 1147, 1156, 1157, 1161, 1167, 1190, 1190n, 1194, 1197, 1208
  and energy and mass, 220, 1174, 1176, 1418
  time interval measurement and, 1158
Einstein's general relativity, 1181
Einstein's light quanta, *1192*
Einstein's theory of gravitation, *1180*, 1180–1181
Einstein's theory of relativity, 4, 92–93, 1149, 1149n, 1156n, 1156–1157
Elastic, definition of, 338
Elastic limit, 347, *347*
Elastic modulus, 346, 346t, 349
Elastic potential energy, 204–205
Elasticity, in length, *346*, 346–347
  of shape, 347, *347*
  of solids, 345–349
  static equilibrium and, 337–359
  volume, 347–348, *348*
Electric, "root" of word, 647
Electric charges. *See* Charge(s), electric.
Electric current. *See* Current.
Electric field(s). *See* Field(s), electric.
Electric flux, 684–687, *685, 686*, 686n, 697
  net, through closed surface, 688, *689*
  through cube, 686–687, *687*
Electric force(s). *See* Force(s), electric.
Electric guitar, *909*, 909–910
Electric potential. *See* Potential, electric.
Electrical conduction, in doped semiconductors, *1306*, 1306–1307
  in insulators, *1304*, 1304–1305
  in metals, 1304, *1304*
  in semiconductors, 1305, *1305*, 1305t
  model for, 783–785
Electrical heater, power in, 787
Electrical instruments, 813–817
Electrical safety, 818
Electrical shock, 818
Electrically charged, 650
Electricity, and magnetism, 647–1019
  Franklin's model of, 650
  magnetism and, 833
Electromagnet, 833
  eddy current and, 922–923
Electromagnetic forces, 161, 1414

Electromagnetic induction, 907
  law of. *See* Faraday's law.
Electromagnetic radiation, absorption of, 1333
Electromagnetic waves, 455. *See* Wave(s), electromagnetic.
Electromagnetism, 648
  laws of, 905, 906
  Maxwell's theory of, 1254
Electromotive force, 787
Electron, acceleration of, 35
Electron beam, bending of, 845, *845*
  current in, *788*, 788–789
Electron capture, 1367
Electron gas, 783
Electron gun, 670, *670*
Electron microscope, 1224, *1224, 1225*
Electron-positron pairs, 1415, *1416*
Electron volt, 709
Electronic configurations of elements, 1269, 1270–1272t
Electronic transition, in hydrogen, 1206–1207
Electron(s), accelerated, 669
  bound, 1233
  charge on, 655, 671
    and mass of, 655t
  collision of, in copper, 785
  computer-simulated patterns in, 1222, *1222*
  conduction, 783
  diffraction, 1221, *1222*
  dual nature of, 1222
  excited, in copper, 1304
  free, 655, 655n
  in hydrogen, probabilities for, *1258*, 1259
  location of, *1225*, 1225–1226, 1227
  magnetic moments of, *882*, 882–883
  measurement of ratio e/m for, 849, *849*
  momentum of, 1172, 1182
  projection of, into uniform electric field, *668*, 668–669
  speed of, 1148–1149
  speedy, energy of, 1175
  spin angular momentum of, 1264n, 1265, *1265*
  spin of, 1257, *1257*, 1263–1266, 1264n
  spinning, *882*, 882–883
  stopping potential of, 1194
  superposition state of, 1222
  transmission coefficient for, 1239
  wave-like properties of, 785
  wavelength of, 1221, 1222
  x-ray scattering from, 1197, *1198*
Electrophotography, 753
Electrostatic equilibrium, conductors in, 693–695, *694*, 698
Electrostatic precipitator, 726, *726*
Electrostatics, applications of, *725*, 725–727, *726, 727, 728*
  Coulomb's law of. *See* Coulomb's law.
Electroweak forces, 161

Electroweak theory, 1429–1430
Elements, electronic configurations of, 1269, 1270–1272t
  periodic table of, A.32–A.33, 1269
Elevator motor, power delivered by, 187–188, *188*
Ellipse, equation of, A.21
Elliptical orbits, 398
emf, 787, 797–799, 818
  back, 921
  induced, 905, 908, 909
    and electric fields, *917*, 917–919
    in generator, 920
    in rotating bar, *913*, 913–914
  induction of, in coil, 910
  motional, *911*, 911–914, 925
  self-induced, 940, *940*, 947, 954
Emission, spontaneous, 1276, *1276*
  stimulated, 1276, *1276*
Emission spectra, 1272
Emissivity, 573
Emitter, in junction transistor, 1309, *1309*
Endothermic reaction, 1371
Energy, and automomobile, 188–191
  and mass, conservation of, 1177
    equivalence of, 1176–1178, 1183
  and work, 171–201
  atomic cohesive, 1298
  binding, 403, 1354, 1373
    and nuclear forces, 1354–1357, *1355, 1356*
    of deuteron, 1178, 1354–1357
    semiempirical, formula for, 1373
  carried by electromagnetic waves, 1002–1004
  conservation of, 207–210, 211, 219, 556, 797, 819
  conversion factors for, A.2
  disintegration, 1364
  electrical, and power, 786–789
    supplied to resistor, 786, 790
  equipartition of, 596–598
    theorem of, 590
  Fermi, 1301, 1302, 1303, 1312
  forms of, 171
  from wind, 437–439, *438*
  in inductor, 946
  in mass-spring collision, 214–215, *215*
  in planetary and satellite motion, 404–407
  in rotational motion, 293–295, 297
  internal, 552, 564, 565, 575
    change in, 564
    of ideal gas, 592, *592*
    total, of solid, 598
  ionic cohesive, 1298
  ionization, 221, 1205, *1205*, 1273, *1273*
  kinetic, 171–172, 181, 192, 850, 948
    allowed values of, 1232
    and work-energy theorem, *180*, 180–186, *182*, 192
    at high speeds, 191–192

Energy *(Continued)*
  average, per molecule, 590, 607
    temperature and, 590
  change in, work and, 403, 403n
  collisions and, 243–244
  for various objects, 181t
  in nuclear reactor, 1177
  loss, due to friction, 183
  mass and, 404, 404n
  of electrons, 1195
  of photoelectrons, 1194
  of simple harmonic oscillator, 368, *369*, 381
  relativistic, 191–192, 1173, 1174, *1174*, 1182
  rotational, 281–282, *282*, 308
  total, 590
    of rigid body, 307–308, 325
  liberated in decay of radium, 1365
  loss of, in resistor, 813
  mass in, 220
  mechanical, 207
    changes in, in nonconservative forces, 210–216
    conservation of, 207
    for freely falling body, 208
    total, 207, 222
  motion of standing wave and, 503–504, *504*
  of molecules, 1290–1296
  of photons, 1024, 1025
  of satellite, 221, *221*
  of simple harmonic oscillator, 368–370, *369*
  of speedy electrons, 1175
  of speedy protons, 1175–1176
  of x-rays, estimation of, 1275
  potential, 202–204, 206, 222, 708, 948
    change in, 404, 708, 728
    conservative forces and, 202, 206–207, 217
    due to point charges, electric potential and, *712*, 712–714, *713*
    elastic, 204–205, 222
    function, 206
    function for mass-spring system, 218, *218*
    gravitational, 203, *203*, 204, 222, 401–404, *402, 403*, 411
    of electric dipole, 756–757, 761
      in electric field, *756*, 757
    of simple harmonic oscillator, 368, *369*, 381
    of two point charges, 713, 729
    of well of finite height, 1236, *1236*
    terms, addition of, 403, 403n
  quantization of, 220–221, 1192, 1233
  quantized, 1024
  relativistic, 1173–1176
  released in fission of Uranium-235, 1385
  rest, 1174, 1175, 1182
  rotational, 281–283, *282*, 296
    allowed values of, 1291, *1291*
  solar, 1005

Energy *(Continued)*
thermal, 616, 618
definition of, 552
heat and, 552–554
temperature and, 558–559, 562–563, *563*
transfer of, 552, 563, 564
threshold, 1371
total, definition of, 1174
for circular orbits, 405–406, 411
of simple harmonic oscillator, 368, 381
of Universe, 219
transfer of, in *LC* circuit, 948, *949*
transmitted by sinusoidal waves on strings, *469*, 469–470
vibrational, allowed values of, 1293
zero point, 535
Energy density, average, 1003
in electric field, 750
total instantaneous, 1003
Energy diagrams, and equilibrium of system, 218–219
Energy gap, 1304
for lead, 1332
for superconductors, 1331, 1331t
measurements of, 1332–1333, *1333*
Energy-level diagram(s), 221, *221*
for hydrogen, 1272, *1272*
for oscillator, 1242, *1242*
for particle in box, 1232, *1232*
of atoms, 1275, *1275*
Energy levels, allowed, 1193, *1193*
Energy-momentum relationship, 1174–1175, 1183
Energy-time uncertainty principle, 1226, 1243
Engine(s), Carnot, 620–624, 630
efficiency of, 622
efficiency of, 619
gasoline, *625*, 625–626
Hero's, 562
internal combustion, 616
steam, 623
Stirling, 625
Entropy, 628–631, 638
and probability, 636
as measure of microscopic disorder, 636, 638
changes in, 628, 629–630, 631, 634–635, 638
in irreversible processes, 631–635
of Universe, 629, 632, 638
on microscopic scale, 635–637, 638
Entropy function, 628
Equation(s), analysis of, 12
Bernoulli's, 433–435, *434*
applications of, *436*, 436–437, *437*
Compton shift, 1198
deviation of, *1199*, 1199–1200
differential, 155
factoring of, A.17
first-law, 564, 564n
for adiabatic process, 565

Equation(s) *(Continued)*
for constant-volume process, 566
for angular momentum, 294t
for electromagnetic waves, 995–996, 998
for thin lenses, 1065, 1082
for translational motion, 294t
Galilean transformation, 1168
kinematic. *See* Kinematic equations.
Kirchhoff's loop, 942, 965
lens makers', 1065
linear, A.18–A.19
linear wave, 470–472, *471*
mass-energy equivalence, 1174
Maxwell's, 906, 923–925, 994, 997, 1001, 1013
mirror, 1057, 1082
net-force, 210
of circle, A.21
of continuity, 433, *433*
of ellipse, A.21
of motion for simple pendulum, 372, 372n
of parabola, A.22
of rectangular hyperbola, A.22
of state, 541
for ideal gas, 541–542
van der Waal's, *606*, 606–607, *607*
of straight line, A.21
quadratic, A.17–A.18
Schrodinger, 1234–1236, 1242, 1244, 1253, 1280
transformation, Galilean, 1168
Lorentz, 1167–1171, 1168n
wave, for electromagnetic waves in free space, 998
general, 998
Equilibrium, 105, 115, 337, 338, 349
electrostatic, conductors in, 693–695, *694*, 698
neutral, 219
objects in, 341
of rigid object, conditions of, *338*, 338–340, *339*
of system, energy diagrams and, 218–219
rotational, 339, 349
stable, 218
static, 337, 339
and elasticity, 337–359
rigid objects in, 341–345
thermal, 532, 543
translational, 339, 349
unstable, 219, *219*
Equilibrium position, 364
Equipartition of energy, 596–598, 607
Equipotential surface(s), 710–711, 715, *715*, 728
charged conductor in equilibrium and, 721
Equivalence principle, of Einstein, 1174, 1176, 1181
Erg, 173
Escape speed(s), 405–407, *406*, 407t
Ether, and speed of light, 1151–1152, *1152*, 1154

Euler method, 159–160, 160t
Evaporation process, 603
Excited state(s), first, of atom, 1205, *1205*
of atoms, 1276
Exclusion principle, 1253, 1266, 1281, 1312
Exit ramp, curved, design of, 147–148, *148*
Exothermic reaction, 1371
Expansion, adiabatic, reversible, *PV* diagram for, 594–595, *595*, 608
adiabatic free, *563*, 565
free, of gas, 563, 619, *619*, 633, *633*, *635*, 635–636
of ideal gas, 623
isothermal, of ideal gas, *566*, 566–567
work done during, 567
linear, average coefficient of, 537–538, 538t, 543
of railroad track, 540
thermal, of solids and liquids, 536–541, *537*, 537n
volume, average coefficient of, 538, 543
Expectation value, 1230
Exponents, rules of, A.17
Eyeglasses, 1077n
Eye(s), accommodation in, 1074
conditions of, 1074
focusing of, 1074
parts of, 1073–1074, *1074*
resolution of, 1127, *1127*

Fahrenheit, Gabriel, 536n
Fahrenheit temperature scale, 536
Farad, 742, 760
Faraday, Michael, 106, 647, 742, 833, 905, 910
Faraday's experiment, 906–908, *907*
Faraday's ice-pail experiment, 696n
Faraday's law, 905–938, 940, 947, 954, 965, 981, 994, 1321
applications of, 909–910
Farsightedness, 1074–1075, *1075*
Fermat's principle, 1042, *1042*, 1042n, 1043
Fermi, Enrico, 1364, 1383–1384, 1386, 1387
Fermi-Dirac distribution function, 1301, *1301*
Fermi energy, 1301, 1302, 1312
of gold, 1303
Fermi temperature, 1303
Fermions, 1301, 1426, 1429
Fermium, 1364
Ferromagnetism, 881, 884–887
Feynman, Richard P., 1417, 1418
Feynman diagram, 1418, *1418*, *1419*
Fiber optics, 1041, *1041*
Fictitious forces, 151–152, *152*, 162
Field-ion microscope, 727, *727*, *728*
Field(s), electric, 649–683, *658*, 658–661
amplitude, and single-slit diffraction pattern, 1121
between parallel plates of opposite charge, 711, *711*

Field(s) *(Continued)*
calculations, using Gauss's law, 697t
charge carrier in, motion of, 783–784, *784*
definition of, 658
due to charged rod, 663, *663*
due to group of charges, 659, 672
due to point charge, 659, *659*, 672
due to point source, 1004
due to solenoid, *918*, 918–919
due to two charges, *660*, 660–661
energy density in, 750
external, electric dipole in, *756*, 756–757
finding of, 672–673
induced, 918
induced emf and, *917*, 917–919
lines, *665*, 665–667, 672, 715, *715*
    around two spherical conductors, 722, *722*
    for point charge(s), *665*, 665–667, *666*, *667*
    rules for drawing, 666
nonuniform, and parallel-plate capacitor, *755*, 755–756
obtaining of, from electric potential, *715*, 715–717
of continuous charge distribution, 661–665, *662*, 672
of dipole, 661, *661*
of uniform ring of charge, 663–664, *664*
of uniformly charged disk, *664*, 664–665
potential energy of, electric dipole in, *756*, 757
projection of electron into, *668*, 668–669
radiated, 1007, *1007*
SI units of, 709
sinusoidal, 999, 1014
total, 659
uniform, motion of proton, 711–712, *712*
    potential differences in, 709–712, *710*, 728
gravitational, 203, 400–401
magnetic, 832–863, 890
    along axis of solenoid, *876*, 876–877, *877*
    created by current-carrying conductor, 866, 870, *870*
    created by current-carrying wire, *872*, 872–873, *873*
    created by infinite current sheet, 873–874, *874*
    created by toroid, 873, *873*, 890
    critical, 1319, 1340
        of type I superconductors, 1319–1321, *1320*, 1320t
    direction of, 834, *834*
    due to point source, 1004
    due to wire segment, 867–868, *868*
    energy in, 944–946, 954
    exponential decrease of, *911*, 911
    loop moving through, 916–917, *917*
    motion of charged particles in, 834–835, 836, 843–847, *844*, 854
    applications of, 847–850

Field(s) *(Continued)*
of Earth, 888–890, *889*
of solenoid, *875*, 875–876, 890, 945
on axis of circular current loop, *868*, 868–869, *869*
patterns, 834, *834*
properties of, created by electric current, 865
proton moving in, 837, *837*
radiated, 1006–1007
SI units of, 837
sinusoidal, 999, 1014
sources of, 864–904
steady, 836
strength of, 883
    magnetization and, 883–884
surrounding conductor, 866–867, *867*
uniform, current-carrying conductor in, 838, *838*, 839, *839*, 854
    force on, 838, 854
    proton moving perpendicular to, 844
    torque on current loop in, *841*, 841–843, *842*
Films, thin, interference of light waves in, *1103*, 1103–1107
    wedge-shaped, interference in, 1106–1107, *1107*
Fission, nuclear, 1383–1386, *1384*, 1408
Fitzgerald, G.F., 1156
Fixed point temperatures, 535, 535t
Fizeau, Armand H.L., 1026
Fizeau's technique for measuring speed of light, 1026, *1026*
Floating ice cube, 430, *430*
Flow, characteristics of, *431*, 431–432, 441
    irrotational, 432
    nonsteady(turbulent), 431, *432*
    steady(laminar), 431, *431*, 432
    tube of, 432, *432*
Flow rate, 433
Fluid dynamics, 422, 431–432, 441
Fluid mechanics, 421–451
Fluid(s), definition of, 421
    force exerted by, 422, *422*
    ideal, 432
    incompressible, 432
    nonviscous, 432
    pressure in, *422*, 423
        depth and, 424, *424*
        device for measuring, 422, *422*
Flux, electric, 684–687, *685*, *686*, 686n, 697
    net, through closed surface, 688, *689*
    through cube, 686–687, *687*
    expulsion, 888
    magnetic, 877, 877–878, *878*, 891
    unit of, 877
    volume, 433
Flux quantization, in superconductors, 1327
Focal length, 1057
    of two thin lenses in contact, 1069
Focal point, 1057
    image, 1066
    object, 1066

Foot, as unit of length, 5
Force law(s), 104–105
Coulomb's. *See* Coulomb's law.
Force(s), action, 114–115, *115*
applied, power delivered by, 913
as vector quantity, 56
average, in collision of automobile into wall, 241, *241*
between molecules, 588
buoyant, 155n, 427–428, 441
    and Archimedes' principle, 427–431, *428*, *429*, 441
central, 145, 162
    speed of mass and, 146
    work done by, 402
color, quarks and, 1428
concept of, 105–107
concurrent, 340, *340*
conservative, 204, 222
    and potential energy, 202, 206–207, 217
    work done by, 204
constant, horizontal, block pulled on frictionless surface by, 183–184, *184*
    measurement of, 180, *180*
    of spring, 179
    work done by, *172*, 172–174, 192
contact, 105, *106*
conversion factors for, A.1
dipole-dipole, 1289
dipole-induced, 1289
dispersion, 1289
electric, 649, 651, 653, *655*, 655–656, 671–672
    and magnetic, 836
    on proton, 660
electromagnetic, 161, 1414
electromotive, 787
electroweak, 161
equivalent, 338
external, 107–108
fictitious, 151–152, *152*, 162
    in linear motion, 152–153, *153*
field, 105–107, *106*
frictional, and power requirements for car, 189t
    block pulled on rough surface by, 184
    total, 189
fundamental, of nature, 1414, 1438
gravitational, 203, 1414
    acting on planets, 399, *399*
    and motion of particle, 410, *410*
    between extended object and particle, *407*, 407–408
    between mass and bar, 407–408, *408*
    between particles, 401–402, *402*
    between spherical mass and particle, *408*, 408–410, *409*
    between two particles, 392, *392*
    pendulum and, 371
    properties of, 393
    weight and, 394–395
gravitational weight and, *123*, 123–124
horizontal, work done using, 174, *174*

Force(s) *(Continued)*
impulse of, 239, *239*, 261
impulsive, 240, 261
in rotating system, 154, *154*
inertial, 151
Lorentz, 847, 924
magnetic, and electric, 836
between two parallel conductors, *869*, 869–870
direction of, 835, *835*, 836, *836*, 853
magnitude of, 836, 854
on current-carrying conductor, 837–841, *838*
on current segment, 874, *874*
on sliding bar, 914, *914*
perpendicular, 836
measurement of, 107, *707*
net, 105
nonconservative, 205–206, 222
normal, 115, 115n
nuclear, 161, 1351, *1351*, 1372
and energy binding, 1354–1357, *1355*, *1356*
of dam, 426–427, *427*
of friction, 124–128, *125*
of gravity, 203, 203n, 204
of nature, 106–107, 160–161, 162, 1414
on semicircular conductor, *840*, 840–841
on uniform magnetic field, 838, 854
reaction, 114–116, *115*
relativistic, 1172, 1173
resistive, 162
and air friction, 189
motion in, 154–158
proportional to speed, 155, *155*
restoring, 179
resultant, 105
acceleration of frictionless surface and, 113, *113*
location of, 656, *656*
on zero, *656*, 656–657
strong, 1414
unbalanced, 105
units of, 112, 112t
van der Waals, 1288–1289, *1289*
varying, work done by, *176*, 176–180, *178*, 192
weak, 1414
Foucault, Jean, 1024
Fourier series, 516
Fourier synthesis, 517, 518, *518*
Fourier's theorem, 516, 517
Franklin, Benjamin, 650
model of electricity of, 650
Fraunhofer diffraction, 1118, *1119, 1120, 1121*, 1123, *1123*, 1138
Free-body diagram(s), 116–117, *117*, 118, *118, 119*, 129, *130*
Free-electron theory of metals, 1301–1304, 1303t, 1312
Free expansion, adiabatic, *563*, 565
of gas, 563, 619, *619*, 633, *633*, 635–636

Free-fall, 37
definition of, 37
speed of ball in, 209, *209*
Free-fall acceleration, 37, 39, *39*, 44, 394–395, 411
and projectile motion, 77, 77n
variation of, with altitude, 395, 395t
Free space, electromagnetic waves in, wave equations for, 998
impedance of, 1003
permeability of, 865, 890
permittivity of, 654–655
Freely falling body(ies), 36–40
mechanical energy for, 208
position and velocity versus time for, 40, *40*
Freezing of water, 541
Frequency(ies), angular, 361, 466
of ac generator, 965
of motion for simple pendulum, 372
of rotating charged particle, 844, 854
average, 516
beat, 515
collision, 605
cutoff, 1194, 1195
cyclotron, 844
for mass attached to spring, 365–366
fundamental, 506
of vibrating strings, 507–508
Larmor precessional, 1352
moving train whistle, 489–490
observed, 489, 492
of electromagnetic waves, 995, 996
of matter waves, 1218–1219
of normal modes, 506
of oscillator, 502, *502*
of periodic waves, 455
of simple harmonic motion, 361
of sinusoidal waves, 466
of tuning fork, 512, *512*
of vibration, 367
precessional, 324
resonance, 379, 508, *508*, 973, 977, 984
siren, 490
with observer in motion, 488, 492
Fresnel, Augustin, 1024, 1118
Fresnel diffraction, *1118*, 1119, *1119*
Friction, as nonconservative force, 206
coefficients of, 124–125, 126t
measurement of, 127, *127*
connected masses with, magnitude of acceleration of, 128, *128*
forces of, 124–128, *125*
kinetic, coefficients of, 125–126
determination of, *127*, 127–128
force of, 124
situations involving, *182*, 182–183, *183*
loss of kinetic energy due to, 183
static, coefficient of, 124–125
speed of car and, 147
force of, 124, 129
Frictional forces, and power requirements for car, 189t
total, 189

Fringes of light waves, 1095, 1108
bright, 1120
dark, 1120–1121
distance between, 1096
order number of, 1095, 1108
Frisch, Otto, 1384
Fritzsch, Harald, 1428
Ft.lb, 173
Fuller, R. Buckminster, 1286
Fusion, latent heat of, 557–558, 558t, 559
nuclear. *See* Nuclear fusion.
Fusion reaction, 1177–1178

Galilean transformation(s), 90–91, 92, 94, 1150, 1167
of coordinates, *1150*, 1150–1151
velocity, 1151, 1169
Galileo Galilei, 1, 4, 36, 108, 1025, 1152
Gallium arsenide diode, high-frequency, 1308
Galvanometer, 814, *814, 815*, 906, *907*, 907–908
Gamma decay, 1369, *1369*
Gamma rays, 1013, 1369
Gamow, George, 1366, 1432
Gas constant, universal, 542
Gas thermometers, 533, 534, *534*, 535
Gas(es), 421
density of, 422
electron, 783
expansion of, quasi-static, *561*, 561n, 561–562, *562*, 566
free expansion of, 563, 619, *619*, 633, *633*, *635*, 635–636
adiabatic, 575
ideal, adiabatic processes for, 594–596
definition of, 541, 541n, 542
equation of state for, 541–542
free expansion of, 637
internal energy of, 592, *592*
isothermal expansion of, *566*, 566–567
law of, 541–542
macroscopic description of, 541–543, 544
molar specific heats for, 593, 607
molecular model of, 587–591, *588*
quasi-static, reversible process for, 630
specific heat of, *591*, 591–594, *592*
temperature scale of, 624
thermal energy and, 591
in equilibrium, atmospheric layer of, 599
kinetic theory of, 586–614
molar specific heats of, 593, 593t
molecules in, 542–543, 629, 636, *636*
average values of, computation of, 600–601
mean free path for, *604*, 604–605, *605*
speed distribution of, 603, *603*
moles of, 542
noble, 1269
work done by, 575
Gasoline, consumed by compact car, 190
used by cars in U.S., 14

Gasoline engine(s), *625*, 625–626
Gauge pressure, 427
Gauss, Karl Friedrich, 687
Gauss' probability integral, A.31
Gauss (unit), 837
Gaussian surface, 687, 694–695, *695*
Gauss's law, *687*, 687–689, 694, 698, 924
  application of, 698–699
    to charged insulators, 690–693
  derivation of, 696–697, *697*
  electric field calculations using, 697t
  experimental proof of, 696, *696*
  in magnetism, 878–879, *879*, 891, 924
Gauss's probability integral, A.31
Geiger, Hans, 1253, 1254, 1346
Geiger counter, 1402–1403, *1403*
GellMann, Murray, 1424, 1425, 1426
Generator(s), alternating current, 919, *919*
  angular frequency of, 965
  delivered to circuit, 975
  voltage amplitude of, 965
  and motors, 919–922
  direct current, 920–921, *921*
  emf induced in, 920
  Van de Graaff, 707, *725*, 725–726
  water-driven, 939
Geometry, A.20–A.22
  capacitance and, 760t
Germer, L.H., 1220
Giaever, I., 1332
Gilbert, William, 647, 833
Glaser, D., 1405
Glashow, Sheldon, 1429–1430
Gluons, 1414, 1429
Goeppert-Mayer, Maria, 1355
Gold, Fermi energy of, 1303
Golf ball, teeing off, *240*, 240–241
Goudsmidt, Samuel, 1257, 1264
Grain, waves, 455
Gravitation, Einstein's theory of, *1180*, 1180–1181
Gravitational constant, universal, 392, 410
  measurement of, 393–394
Gravitational field(s), 203, 400–401
Gravitational force, 1414. *See* Force(s), gravitational.
Gravitational potential energy, 203, *203*, 204, 222, 401–404, *402*, *403*, 411
Gravitational redshift, 1181
Gravitons, 1414, 1419
Gravity, center of, 254, 340–341, *341*, 349
  force of, 203, 203n, 204
  laws of, 391–420
    and motion of planets, *396*, 396–400
    Newton's laws of, 392–393, 396, 410
Grimaldi, Francesco, 1024
Ground-fault interrupter(s), 818, 909, *909*
Ground state, of atom, 1205
  of atoms, 221
  of hydrogen, 1259
Gurney, R.W., 1366
Gyroscope, motion of, 322, *323*

Hadrons, 1420, 1425, 1438
  quark composition of, 1426t, 1426–1427
Hahn, Otto, 1383–1384
Half-life of radioactive substance, 1361, 1373
Hall, Edwin, 851
Hall coefficient, 852
Hall effect, *851*, 851–853
  for copper, 853
  quantum, 853
Hall voltage, 852
Hand, weighted, *342*, 342–343
Harmonic motion, simple. *See* Motion, simple harmonic.
Harmonic series, 506–507, *507*, 519
Harmonics, of waveform, *518*
Heat, and thermal energy, 552–554, 574
  and work, 615
  conduction of, 568–571, 632
    law of, 569
  conversion factors for, A.2
  definition of, 552
  flow of, direction of, 632–633
  in thermodynamic processes, 561–563
  latent, 557n, 557–560, *558*, 575
    of fusion, 557–558, 558t, 559
    of vaporization, 558, 558n, 558t
  mechanical equivalent of, *553*, 553–554, 575
  of condensation, 558n
  of solidification, 558n
  specific, 554–557, 555n, 555t, 575
    electronic, 1327–1328, *1328*
    measurement of, 556, 556n
    molar, 555, 555t
      at constant pressure, 591
      at constant volume, 591, 607
      of gas, 593, 593t, 607
      of hydrogen, 597, *597*
      of ideal gas, *591*, 591–594, *592*
      of solids, 598, *598*
  transfer of, 560, 568–574, *569*
    irreversible, 633–634
    through slabs, 570
    to solid, 632–633
  units of, 553
Heat capacity, 554, 575
Heat engine(s), 616, *616*, 617, *617*, 637
  and second law of thermodynamics, 616–619
  Carnot-cycle, 627–628
  refrigerators as, 617–618, *618*, 628
  thermal efficiency of, 617
Heat pumps, 617, *618*, *627*, 627–628
Height, measurement of, 373
Heisenberg, Werner, 1190, 1216, 1224, 1226
Heisenberg uncertainty principle, 1224–1227
Helios, 1202
Helium, 1268, 1330n
  liquid, boiling of, 560
Helix, 844

Henry, Joseph, 647, 833, 905, 906
Henry (unit), 940, 947
Hero's engine, 562
Hertz, Heinrich, 648, 924–925, 994–995, 996–997, 1024
Higgs boson, 1430
Holography, lasers and, 1279, *1279*
Home insulation, 571–572
Hooke's law, 107, 179, 364
Hoop, uniform, moment of inertia of, 284, *284*
Horsepower, 187
Household appliances, power connections for, 817, *817*
Household wiring, 817, *817*
Hubble, Edwin P., 1435
Hubble parameter, 1435, 1438
Hubble's law, *1434*, 1435, 1438
Hund's rule, *1268*, 1269
Huygens, Christian, 1022, 1024, 1025–1026, 1036, 1037
Huygens' principle, 1036–1039, *1037*, 1043, 1119
  applied to reflection and refraction, *1038*, 1038–1039, *1039*
Hydraulic press, 425, *425*
Hydrogen, 1268
  atomic spectrum for, *1201*, 1201–1202, *1202*
  Balmer series for, 1207, *1207*
  electron charge distribution and, *1260*, 1261
  electronic transition in, 1206–1207
  ground state of, 1259
  molar specific heat of, 597, *597*
  probabilities for electron in, *1258*, 1259
  space quantization for, 1263
  wave functions for, 1257–1261, *1258*
Hydrogen atom, 657, 1252–1253, 1255–1256
  Bohr's model of, *1203*, 1203–1204, 1205, *1205*, 1227
  energies of allowed states for, 1255
  potential energy function of, 1255, *1255*
  quantum numbers for, 1256, 1256t, 1257
  radial distribution function and, 1259, *1260*
Hydrogen bonds, 1290, 1312
Hyperbola, rectangular, equation of, A.22
Hyperopia, 1074–1075, *1075*
Hysteresis, magnetic, 886
Hysteresis loops, *886*, 886–887, *887*

Ice cube, floating, 430, *430*
Ice Man, age of, 1368
Ice-pail experiment, Faraday's, 696n
Ideal absorber, 574, 576
Ideal gas. *See* Gas(es), ideal.
Ideal gas law, 541–542
Image distance, 1053
Image(s), focal point, 1066

Image(s) *(Continued)*
  formation of, 1053
  formed by converging lenses, 1068
  formed by diverging lenses, 1068
  formed by flat mirrors, 1052–1055, *1053*
  formed by refraction, *1061*, 1061–1064
  formed by spherical mirrors, *1055*, 1055–1061, *1056*
  length contraction, of box, 1164, *1164*
  of lens, location of, 1064, *1064*
  real, 1053
  virtual, 1053
Impedance, 977, 983
  in *RLC* series circuit, 973
  of free space, 1003
Impedance triangle, 973, *973*
Impulse, and linear momentum, 239–242
Impulse approximation, 240
Impulse-momentum theorem, 239, *239*, 261
Independent-particle nuclear model, 1358, *1358*, 1373
Inductance, 939–963
  and emf, calculation of, 941
  definition of, 940
  mutual, *947*, 947–948
    definition of, 947
    of two solenoids, 947–948, *948*
  of coils, 940, 954
  of solenoid, 941, 954
Induction, charging metalic object by, 652, *652*
  electromagnetic, 907
  Faraday's law of. *See* Faraday's law.
  mutual, 939, 947
Inductive reactance, 968–969, 983
Inductor(s), 939, 941
  current and voltage across, 968, *968*
  energy in, 946
  in ac circuits, *967*, 967–969
  instantaneous voltage drop across, 969
  maximum current in, 968
  phase relationships in, 971–972, *972*
Inertia, law of, 109–110
  moment(s) of. *See* Moment(s), of inertia.
Inertial field confinement, 1394, 1394–1395, *1396*
Inertial forces, 151
Inertial frames, 109–110, 129
  different, laws of mechanics in, 1149–1150, *1150*
Inertial mass, 110–111
Inertial system, definition of, 1149
Infeld, L., 454n, 454–455
Infinite current sheet, carrying sinusoidal current, 1008
  radiation from, *1006*, 1006–1008, *1007*
Infrared waves, 1012
Insulation, home, 571–572
Insulator(s), 652–653, *653*
  application of Gauss's law to, 690–693
  definition of, 651
  electrical conduction in, *1304*, 1304–1305

Integral calculus, A.27–A.30
Integral(s), definite, 42
  Gauss's probability, A.31
  indefinite, A.29–A.30
  line, 708
  path, 708
Integrated circuit, *1310*, 1310–1311, *1311*
Integration, 27, 41
  partial, A.28
Intensity(ies), average, of double-slit interference patterns, 1108
  distribution, for diffraction grating, 1128, *1128*
    of double-slit interference patterns, *1096*, 1096–1098, *1098*
  of single-slit diffraction, *1121*, 1121–1123
  of sound waves, 482–484, *483*, 486–487
  reference, 484
  relative, of maxima, 1123–1124
  wave, 1002
Interference, 460, 499
  and light waves, 1117–1118, *1118*
  conditions for, 1091–1092
  constructive, 460, *461*, 472, 500, *501*, 518
    condition for, 1095, 1097, 1103, 1104, 1108, 1109
    of light waves, 1095
  destructive, 461, *462*, 472, 500, 501, *501*, 518–519
    condition for, 1095, 1103–1104, 1108, 1109
  in soap film, 1106
  in thin films, *1103*, 1103–1107
    problems in, 1105
  in time, 513–515
  in wedge-shaped film, 1106–1107, *1107*
  of light waves, 1091–1116
  patterns of, 461, 1104
    double-slit, average intensity of, 1108
      intensity distribution of, *1096*, 1096–1098, *1098*
      phasor diagrams for, 1099, *1100*
    in water waves, *1093*
    multiple-slit, *1100*, 1101
    three-slit, *1100*, 1100–1101
  spatial, 513
  temporal, 513–515
Interferometer, Michelson, *1107*, 1107–1108, 1154, *1154*
International Thermonuclear Experimental Reactor, 1394
Interrupters, ground-fault, 818, 909, *909*
Inverse-square law, 392, 393
  and acceleration, 396
Iodine, radioactive isotope of, 1363
Ion chamber, 1402, 1403
Ion density, 1392, 1408
Ionic bonds, 1287, 1311
Ionic solids, *1297*, 1297–1298, *1298*
Ionization energy, 221, 1205, *1205*, 1273, *1273*
Ions, magnetic moments of, 883t

Irreversible process(es), 616, 616n, 619–620
  changes in entropy in, 631–635
Isentropic process, 630
Isobaric process, 566, 575
Isothermal expansion, of ideal gas, *566*, 566–567
  work done during, 567
Isothermal process, 566, 575
  PV diagram for, 606, *606*
  work done in, 566–567
Isotope effect, 1328
Isotopes, 1347–1348, 1372
Isovolumetric process, 566, 575

Jensen, Hans, 1355
JET Tokamak, 1394
Jewett, Frank Baldwin, *849*
Josephson, Brian D., 1319, 1333–1334
Josephson effect, ac, 1335–1336
  dc, 1334–1335, *1335*
Josephson junction, 1239, 1334, *1334*, 1339
Josephson tunneling, 1333–1336
Joule, 173, 553
Joule, James, 551, 552, 553
Joule heating, 787, 787n, 913
Junction, in circuit, 800–801
  Josephson, 1239, 1334, *1334*, 1339
Junction rule, Kirchhoff's, 804, 805, *805*
Junction transistor, 1308–1310, *1309*

K capture, 1367
Kamerlingh-Onnes, Heike, 782
Kaon, decay of, at rest, 238, *238*
Kelvin, definition of, 535, 543
Kelvin, Lord (William Thomson), 5n, 536n, 618
Kelvin-Planck form of second law of thermodynamics, 617, 618
Kelvin temperature scale, 534, 535, *535*, 536, 624
Kepler, Johannes, 4, 392, 395–396
Kepler's law(s), 395–396, 411
  second, 397, 411
    and conservation of angular momentum, 399–400
  third, 397–398, 411
Kilby, Jack, 1310
Kilocalorie, 553
Kilogram(s), 5, 110, 112n
  definition of, 5
Kilowatt hour, 187
Kinematic equations, 44
  choice of, 34
  derived from calculus, 41–43
  for motion in straight line under constant acceleration, 34t
  for rotational and linear motion, 279t
  rotational, 279, 296
Kinematics, 23
  rotational, 278–279

Kinetic energy. *See* Energy, kinetic.
Kinetic friction, situations involving, *182*, 182–183, *183*
Kinetic theory, of gases, 586–614
Kirchhoff's circuit law, 943–944
Kirchhoff's loop equation, 942, 965
Kirchhoff's loop rule, 967, 969
Kirchhoff's rule(s), 797, 804–808, *805*, 819
  application of, 807, *807*
  first, 816, 819
  second, 808, 811, 816, 819

Ladder, uniform, 344, *344*
Laminar flow, 431, *431*, 432
Land, E.H., 1133
Large Electron-Positron Collider, at CERN, 1431
Large Hadron Collider, 1431
Larmor precessional frequency, 1352
Laser(s), 1276–1279
  and holography, 1279, *1279*
  applications of, 1277–1279
  helium-neon gas, 1276–1277, *1277*
  in surgery, *1278*, 1278–1279
Latent heat, 575
Laue, Max von, 1130
Laue pattern, 1131
Lawrence, E.O., 849
Law(s), Ampère-Maxwell, 880, *880*, 880n, 924, 925
  Ampère's, 864, 870–874, 875, 876, 890
    generalized form of, 924
  Biot-Savart, 864, 865–869, *866*, 890
  Boltzmann distribution, 599–602
  Boyle's, 541
  Bragg's, 1132
  Brewster's, 1135, 1139
  conservation, 1420–1422
  Coulomb's. *See* Coulomb's law.
  Curie's, 887
  Descartes', 1029n
  distributive, of multiplication, 175
  Dulong-Petit, 598
  Faraday's. *See* Faraday's law.
  force, 104–105
  Galilean addition, for velocity, 1151
  Gauss's. *See* Gauss's law.
  Hooke's, 107, 179, 364
  Hubble's, *1434*, 1435, 1438
  ideal gas, 541–542
  inverse-square, 392, 393
  Kepler's, 395–396, 397–398, 411
  Kirchhoff's circuit, 943–944
  Lenz's, 908, 914n, 914–917, *915*, 925, 940
    application of, 916, *916*
  Lorentz force, 925
  Malus's, 1134, 1134n
  of atmospheres, 600
  of Charles and Gay-Lussac, 541
  of conservation of angular momentum, 318, 325

Law(s) *(Continued)*
  of conservation of linear momentum, 237
  of conservation of momentum, 1171
  of electromagnetic induction. *See* Faraday's law.
  of electromagnetism, 905, 906
  of gravity, 391–420
    and motion of planets, *396*, 396–400
    Newton's, 392–393, 396, 410
  of heat conduction, 569
  of ideal gas, 541–542
  of inertia, 109–110
  of motion, 104–143. *See also* Newton's laws of motion.
  of reflection, 1027, *1028*, 1043
  of thermodynamics. *See* Thermodynamics.
  of vector addition, 840
  Ohm's. *See* Ohm's law.
  Pascal's, 424
  power, analysis of, 12
  Rayleigh-Jeans, *1191*, 1191–1192
  Snell's. *See* Snell's law.
  Stefan's, 573
  Torricelli's, 435
  Wien's displacement, 1191
Lawson, J.D., 1392
Lawson's criterion, 1393, *1393*, 1393n, 1394, 1408
LC circuits. *See* Circuit(s), *LC.*
Lead, energy gap for, 1332
  radiation absorption by, 1334
Lederman, Leon, 1427–1428
Length contraction image, of box, 1164, *1164*
Length(s), 5, 10, 17
  contraction, *1163*, 1163–1165
  conversion factors for, A.1
  dimensional analysis of, 11
  elasticity in, *346*, 346–347
  focal, of mirror, 1057, 1082
    of two thin lenses in contact, 1069
  measured, approximate values of, 7t
  proper, 1163
  standards of, 4–8
Lens maker's equation, 1065
Lens(es), aberrations, *1070*, 1070–1072, *1071*
  biconcave, 1066, *1066*
  biconvex, 1065, *1065*, *1066*
  converging, 1082
    final image of, 1070, *1070*
    image formed by, 1068
  diverging, 1082
    image formed by, 1068
  f-number of, 1072–1073
  gravitational, 1181, *1181*
  image of, location of, 1064, *1064*
  of eye, 1073
  optical, testing of, 1105
  power of, 1076
  thin, 1064–1070, *1066*
    combination of, 1069

Lens(es) *(Continued)*
  equations for, 1065, 1082
  ray diagrams for, *1067*, 1067–1068
  shapes of, 1066, *1067*
  sign convention for, 1066, 1066t
  two in contact, focal length of, 1069
  under water, 1069
Lenz, Heinrich, 914n
Lenz's law, 908, 914n, 914–917, *915*, 925, 940
  application of, 916, *916*
Lepton number(s), checking of, 1422
  law of conservation of, 1422
Leptons, 1420, 1438
  properties of, 1428t
Leucippus, 8
Lifetime, 1227
Lift, 436–437
Light. *See also* Wave(s), light.
  and optics, 1021–1145
  dispersion of, and prisms, 1033–1036
  dual nature of, 1025, 1217, 1243
  nature and properties of, 1021–1022
  nature of, 1023–1025
  passing through slab, 1033, *1033*
  photon theory of, 1195
  photons in, 1092n
  reflection of, 1027–1028
  refraction of, *1029*, 1029–1033, *1030*
  source(s) of, coherent, 1092
    monochromatic, 1092
    wavelength of, 1096
  spectrum of, 1034–1035, *1035*
  speed of, 1151–1154, *1152*
    constancy of, 1156, 1157
    in silica, 1033
    in vacuum, 1030n, 1030–1031, 1043
    measurements of, 1025–1026, 1154
  ultraviolet, 1012–1013
  visible, 835, 1012
  wave theory of, 1217
Light ray, double-reflected, 1028, *1028*
Lightbulb, cost of operating, 788
  electrical rating of, 788
  three-way, operation of, 803–804, *804*
Lightning bolts, time interval measurement of, 1158–1159, *1159*
Limiting process, 29
Line, straight, equation of, A.21
Line integral, 708
Linear equations, A.18–A.19
Linear expansion, average coefficient of, 537–538, 538t, 543
Linear momentum. *See* Momentum, linear.
Linear motion, 314, *314*
  fictitious forces in, 152–153, *153*
Linear wave equation, 470–472, *471*
Linear waves, 459, 472
Liquid-drop nuclear model, 1357–1358, *1358*, 1373
Liquids, 422
  sound waves in, 480t, 481

Liquids (*Continued*)
  speed distribution of molecules in, 603
  thermal expansion of, *540*, 540–541
Lithium, 1268–1269
Live wire, 817, 817n
Livingston, M.S., 849
Lloyd's mirror, *1101*, 1101–1102
Load resistance, 798, 982
  power and, 799, *799*
Logarithm(s), A.20
  natural, 810
London, Fritz, 1319, 1323, 1327
London, Heinz, 1319, 1323
Loop rule, 805
"Loop-the-loop" maneuver by plane, in
    vertical circle at constant speed, 149,
    149
Lorentz, Hendrik A., 1156, 1167
Lorentz force, 847, 924
Lorentz force law, 925
Lorentz transformation, 1182
  velocity, 1169–1170, 1172, 1182
Lorentz transformation equations, 1167–
    1171, 1168n
Lyman series, 1202

Mach number, 491
Madelung constant, 1298
Magic numbers, 1351
Magnet, permanent, and superconductor,
    888
Magnetic bottle, 845, *845*
Magnetic energy density, 945
Magnetic field confinement, 1393
Magnetic fields. *See* Field(s), magnetic.
Magnetic flux, *877*, 877–878, *878*, 891
  through rectangular loop, 878, *878*
Magnetic force. *See* Force(s), magnetic.
Magnetic hysteresis, 886
Magnetic moment(s), of atoms, 881–883,
    882, 883t
  of coil, 843
  of current loop, 842, 854
  of electrons, *882*, 882–883
  of ions, 883t
  of proton, 1352
  orbital, 882, *882*
  SI units of, 842
Magnetic permeability, 884
Magnetic poles, 832, 833
Magnetic properties, of type I superconduc-
    tors, 1321–1322, *1322*
Magnetic resonance imaging, 876
  computer-enhanced, 1339
  nuclear magnetic resonance and, *1352*,
    1352–1354, *1353*, *1354*
Magnetic substances, classification of, 884–
    885
Magnetic susceptibility(ies), 884, 884t
Magnetism, electricity and, 647–1019
  Gauss's law in, 878–879, *879*, 891, 924

Magnetism (*Continued*)
  in matter, 881–888
  induction of, 833
  knowlege of, history of, 647
Magnetite, 647, 832
Magnetization, and magnetic field strength,
    883–884
  saturation, 888
Magnetization curve, 886, *886*, 887
Magnetization vector, 883
Magneton, Bohr, 883, 1266
  nuclear, 1351–1352, 1372
Magnification, angular, 1077, *1078*, 1079,
    1081
  lateral, 1053, 1079, 1082
  maximum, 1078
Magnifier, simple, *1077*, 1077–1078
Malus, E.L., 1134
Malus's law, 1134, 1134n
Manometer, open-tube, 427, *427*
Maricourt, Pierre de, 832
Marsden, Ernest, 1253, 1346
Mass density, 777
Mass-energy, conservation of, 1174
Mass-energy equivalence, 220, 1149, 1174
Mass number, 1347, 1372
Mass spectrometer, *848*, 848–849
Mass-spring collision, energy of, 214–215,
    215
  speed of mass and, 185–186
Mass-spring system, 368
  mechanical energy for, 208
  potential energy function for, 218, *218*
  simple harmonic motion for, *370*, 381
Mass(es), 17, 104
  acceleration and, 111
  acceleration of, on frictionless incline, *119*,
    119–120
  and energy, conservation of, 1177
    equivalence of, 1176–1178, 1183
  and kinetic energy, 404, 404n
  and weight, 111
  as inherent property, 111
  as scalar quantity, 111
  atomic, selected, table of, A.4–A.13
  attached to spring, *364*, 364–368, *366*, *367*
    period and frequency for, 365–366
  center of, 251–255, *252*, *253*
    acceleration of, 256
    location of, 252–253, 254, *254*
    of pair of particles, 252, *252*
    of right triangle, 255, *255*
    of rigid body, 252, *252*, 262
    of rod, *254*, 254–255
    of three particles, 254, *254*
    vector position of, 253, *253*, 262
    velocity of, 256, 262
  connected, 295, *295*, 317, *317*
  in motion, *215*, 215–216
  with friction, magnitude of acceleration
    of, 128, *128*
  conservation of, 220

Mass(es) (*Continued*)
  conversion factors for, A.1
  gravitational attraction of, 1179–1180
  gravitational force on, 394
  in free-fall, speed of, 209, *209*
  inertial, 110–111
  inertial property of, 1179–1180
  invariant, 1175
  method for lifting, 216, *216*
  missing, of Universe, 1437
  molar, 541
  moving in circular orbit, 404–405, *405*
  Newton's second law of motion applied
    to, 404
  of spring, 1178
  of Sun, 399
  of various bodies, 7t
  on curved track, *212*, 212–213
  pulled on frictionless surface, speed of,
    183–184, *184*
  pulled on rough surface by frictional
    force, 184
  rest, 1177
  rotating, 283, *283*
  SI unit of, 5–6, 110
  sliding down ramp, speed of, 212, *212*
  speed of, central force and, 146
    mass-spring system and, 185–186
  standards of, 4–8
  struck by club, range of launch of, *240*,
    240–241
  two blocks of, magnitude of acceleration
    of, 122, *122*
  two unequal connected, acceleration of,
    121, 121–122
  units of, 112, 112n, 112t
Mathematical notation, 16–17
Mathematics review, A.14–A.30
Matter, building blocks of, 8–9
  magnetism in, 881–888
  radiation damage in, 1400–1402
  stages of evolution of, investigation of, mi-
    croscopes in, 1431, *1431*
Maxwell, James Clerk, 602, 647–648, 833,
    923, 995 1022, 1147, 1151, 1254
Maxwell-Boltzmann distribution function,
    602
Maxwell's equations, 906, 923–925, 994,
    997, 1001, 1013
Mean free path, 604–605
Mean free time, 605
Mean solar day, 6
Mean solar second, 6
Measurement(s), expression of, in num-
    bers, 5n
  physics and, 3–22
Mechanical energy. *See* Energy, mechanical.
Mechanical waves, 453–527
  requirements for, 455
Mechanics, 1–450, 1148
  classical, 104
  newtonian, 1158

Mechanics *(Continued)*
  quantum. *See* Quantum mechanics.
  statistical, 596
Meissner effect, 888, 1321
Meitner, Lisa, 1384
Mendeleev, Dmitri, 1269
Mercury, superconductivity of, 1318, *1318*
Mercury barometer, 427, *427*
Mercury thermometers, 533, *533*
Mesons, 1416–1417, 1420, 1424
Metals, *1299*, 1299–1300
  as conductors, 652
  atoms of, 655n
  electrical conduction in, 1304, *1304*
  free-electron theory of, 1301–1304, 1303t, 1312
  resistivity versus temperature for, 781, *781*
  work function(s) of, 1195, 1195t, 1208
Metastable state, 1276
Meter, 5
  definitions of, 5
Michell, John, 833
Michelson, Albert A., 1107, 1154, 1155
Michelson interferometer, *1107*, 1107–1108, 1154, *1154*
Michelson-Morley experiment, *1153*, 1153n, 1153–1156, 1157
Microscope, compound, 1078–1080, *1079*
  electron, 1224, *1224, 1225*
  field-ion, 727, *727, 728*
  limiting resolution of, 1126
  scanning tunneling, *1240*, 1240–1241
Microwave horn, 1433, *1434*
Microwaves, 1011, 1012, 1012n
Millikan, Robert A., 651, 724
Millikan oil-drop experiment, *724*, 724n, 724–725
Mirror equation, 1057, 1082
Mirror(s), concave, *1055*, 1055–1057, *1056, 1057*, 1060–1061
  convex, 1058, *1058*, 1061
  diverging, 1058
  flat, images formed by, 1052–1055
  Lloyd's, *1101*, 1101–1102
  ray diagrams for, 1058–1060, *1059*
  rearview, tilting, 1054–1055, *1055*
  sign convention for, 1058t
  spherical, images formed by, *1055*, 1055–1061, *1056*
  two, multiple images formed by, 1054, *1054*
Moderator materials, 1383, 1388, 1408
Modes, normal, frequencies of, 506
  in standing waves, *505*, 505–506
  wavelengths of, 506
Modulus, bulk, 346, 347–348, *348*, 349
  elastic, 346, 346t, 349
  shear, 346, 347, *347*, 349
  Young's, 346, 349
Molar mass, 541
Molar specific heat, 555, 555t
Molecular bonds, *1287*, 1287–1290
Molecular orbital, 1288

Molecular spectra, *1295*, 1295–1296, *1296*, 1296n
Molecule(s), 1286
  average kinetic energy per, 590, 607
  carbon monoxide, microwave absorption spectrum of, 1292t
  rotation of, 1292
  collision frequency for, 605
  diatomic, possible motions of, 596, *596*
  energy and spectra of, 1290–1296
  excitations of, 1290
  forces between, 588
  in atmosphere, 600
  in gas, 629, 636, *636*
  moment of inertia of, 1290
  Newton's laws of motion and, 587
  nonpolar, 758
  oxygen, 282
  polar, 757–758
  reduced mass of, 1291
  rms speeds for, 590–591
  rotational motion of, 1290n, 1290–1291
  speeds of, distribution of, *602*, 602–604, *603*
  symmetrical, polarization and, 757, *757*
  vibrational motion of, 1293, *1293*, 1294–1295
  water, *756*, 756–757
Mole(s), 10, 541
  of gas, 542
Moment arm, 287, *288*, 288
Moment(s), angular, as quantized, 882
  magnetic. *See* Magnetic moment(s).
Moment(s) of inertia, calculation of, 283–287
  concept of, use of, 285n, 296
  of rigid objects, 286t, 296
  of uniform hoop, 284, *284*
  of uniform rigid rod, 285, *285*
  of uniform solid cylinder, 285, *285*
Momentum, and radiation pressure, 1004–1006, 1005n, *1105*
  angular, 236n
    as fundamental quantity, 324–325
    conservation of, 317–321
      Kepler's second law and, 399–400
      law of, 318, 325
    equations for, 294t
    net torque and, 322
    of particle, *312*, 312–315, 325
    of rigid body, 315, *315*, 325
    orbital, 1261, 1280
    photons and, 1273
    quantization of, in Bohr model, *1219*, 1219–1220, 1220n
    spin, 882, 1265, 1280
    torque and, 313
    total, system of particles of, 313
  concept of, 235, 236n
  conservation of, for two-particle system, *236*, 236–237, 261
  laws of, 1171

Momentum *(Continued)*
  linear, and collisions, 235–275
    and rocket propulsion, 260
    conservation of, 236–238
    definition of, 236, 261
    impulse and, 239–242
    laws of conservation of, 237
  of electron, 1172, 1182
  of recoiling pitching machine, 238, *238*
  relativistic, definition of, 1172
    Newton's laws of motion and, 1171–1173
  rotational, 236n
  total, 237, 261
    of system, in collision of automobiles, 243
    of system of particles, 256
  total linear, of system, 256
Moon, acceleration of, *396*, 396–397
  motion of, 391–392
Morley, Edward W., 1154
Moseley, Henry G.J., *1274*, 1274–1275
Motion, and scalars, 24n
  circular. *See* Circular motion.
  connected blocks in, *215*, 215–216
  horizontal, 81
  in accelerated frames, 151–154
  in elliptical orbit, 400, *400*
  in one direction, 23–52
  in plane, 76–77
  in resistive forces, 154–158
  laws of, 104–143. *See also* Newton's laws of motion.
  linear, 314, *314*
    fictitious forces in, 152–153, *153*
  of charge carrier, in conductor, 774, *774*, 783, *783*
  in electric field, 783–784, *784*
  of charged particles, electric current and, 773, *773*
  in magnetic field, 834–835, 836, 843–847, *844*, 854
    applications of, 847–850
  in uniform electric field, 667–669, 672
  of disturbance, 455
  of gyroscopes and tops, 321–324, *322, 323*
  of longitudinal pulse, 480, *480*
  of moon, 391–392
  of particle, gravitational force and, 410, *410*
  of proton in uniform electric field, 711–712, *712*
  of system of particles, 255–259
  one-dimensional, kinematic equations for, 34t
  with constant acceleration, 32–36
  oscillatory, 360–390
  periodic, 360–390
  planetary, energy in, 404–407
    Kepler's laws of, 395–396, 397–398, 411
  laws of gravity and, *396*, 396–400
  precessional, 321–322
  projectile, 77–85, 93

Motion (*Continued*)
  pure rotational, 276
  reference circle for, 376, *376*, 377
  relative, at high speeds, 92–93
  rolling, angular momentum, and torque,
      306–336
    of rigid body, *307*, 307–309, *308*
    pure, 307, *307*
  rotational, equations for, 294t
    of molecules, 1290n, 1290–1291
    with constant angular acceleration, 278–
      279
    work, power, and energy in, *293*, 293–
      295, 297
    work and energy in, 293
    work-energy relation for, 294
  satellite, energy in, 404–407
  simple harmonic, 361–364, *362*, 365
    acceleration in, 362–363, *367*
    amplitude in, 361
    and uniform circular motion, compared,
      375, 375–377, *376*
    displacement versus time for, 361, *361*,
      367
    for mass-spring system, *370*, 381
    frequency of, 361
    pendulum and, 371, *371*
    period of, 361, 380
    phase constant in, 361
    properties of, 363
    speed in, 361, 362–363, *367*
    velocity as function of position for, 369
  three types of, 23
  translational, 23
    equations for, 294t
  two-dimensional, 71–103
    with constant acceleration, 74–77, *75*
  uniform, 109
  uniform circular, *85*, 85–87
    and simple harmonic motion, compared,
      375, 375–377, *376*
    Newton's second law of motion and,
      144–149, *145*, *146*, 162
  vibrational, of molecules, 1293, *1293*,
      1294–1295
  wave, 454–478, *456*
    and vibration, 453
    simulator, 460
Motorcycle, pack leaders, relativistic, 1171,
    1171
  speeding, 1170, *1170*
Motor(s), and generators, 919–922
  definition of, 921
  induced current in, 922
Movie film, photoelectric effect and, 1196–
    1197, *1197*
Mueller, E.W., 727
Müller, Karl Alex, 782, 1319, 1336, 1337
Multiplication, distributive law of, 175
Muon(s), decay curves for, 1162, *1162*
  time dilation and, 1162, *1162*, 1164
Myopia, 1075, *1076*

National Standard Kilogram No. 20, 6
Nature, fundamental forces of, 160–161,
    162, 1414, 1438
Near point of eye, 1074
Nearsightedness, 1075, *1076*, 1076–1077
Ne'eman, Yuval, 1424
Neon, 1269
Net-force equation, 210
Neuron detectors, 1405
Neutrino, 1366–1367, 1373
  energy and momentum for, 1175
Neutron activation analysis, 1407
Neutron number, 1347, 1372
Neutron(s), 8, 9, 10
  capture, 1383
  charge and mass of, 655t
  interactions involving, 1382–1383
  slowing down of, by collisions, 247–248
  thermal, 1383
Newton, Isaac, 4, 105, 106, 108, 236, 391,
    1021, 1023, 1037, 1082, 1104, 1147
Newtonian mechanics, 1–450, 1148, 1158
Newtonian relativity, principle of, 1149–
    1154
Newton.meter, 173
Newton's laws of gravity, 392–393, 396, 410
Newton's laws of motion, 3, 4
  applications of, 116–118, 144–170
  first, 108–109, 129, 152
  molecules and, 587
  relativistic momentum and relativistic
      form of, 1171–1173
  second, 111–113, 112n, 129, 463, 471,
      472, 844
    applied to mass, 404
    damped oscillations and, 377–378
    for rotational motion, 375
    for system of particles, 256–257, *257*, 262
    in component form, 117
    linear momentum and, 236, 239, 256
    of mass attached to spring, 364
    torque and, 289–290
    uniform circular motion and, 144–149,
      145, *146*, 162
  third, *114*, 114–116, *115*, 129, 237, 242,
      392, 436, 869, 869n
Newton's rings, 1104–1105, *1105*
Newton's theory of light, 1023–1024
Newton (unit), 112
Nichrome wire, resistance of, 779
Nobel prizes, A.35–A.39
Nodes, 503, *504*, 519
Nonohmic materials, 776, 778, *778*
Normalization condition in quantum me-
    chanics, 1229–1230, 1244
Northern Lights, 832, 846–847
Notation, scientific, A.14–A.15
Nova Laser Facility, Lawrence Livermore
    Laboratory, 1395, *1395*
Noyce, Robert, 1310
Nuclear demagnetization, 624n
Nuclear fission, 1383–1386, *1384*, 1408

Nuclear force, strong, 161
  weak, 161
Nuclear force(s), 1351, *1351*, 1372
  and energy binding, 1354–1357, *1355*, *1356*
Nuclear fusion, 1389–1397, 1408
  advantages and problems of, 1396–1397
  energy of, recent developments in, *1397*,
    1397–1400
  inertial confinement in, *1394*, 1394–1395,
    1396
  magnetic field confinement in, *1393*,
    1393–1394
  proton-proton cycle of, 1389
  reactions, 1390–1393, *1392*, *1393*, *1397*,
    1397–1398
  thermonuclear reactions, 1389–1390
Nuclear fusion reactor(s). *See also specific re-
    actors.*
  design of, 1396, *1397*
Nuclear magnetic resonance, 1352, *1352*
  and magnetic resonance imaging, *1352*,
    1352–1354, *1353*, *1354*
Nuclear magneton, 1351–1352, 1372
Nuclear model(s), 1357–1359
  independent-particle, 1358, *1358*
  liquid-drop, 1357–1358, *1358*
  shell model, 1358, *1358*
Nuclear physics, 1347
Nuclear reaction(s), 1370–1371, 1372t,
    1373, 1386–1388
  self-sustained chain, 1386
Nuclear reactor(s), 1386–1389
  control of power level of, *1387*, 1388
  critical, 1387, *1387*
  first, 1387
  kinetic energy in, 1177
  pressurized-water, *1388*, 1388–1389
  reproduction constant, 1386–1387, 1408
  safety of, 1389
  waste disposal, 1389
Nuclear spin, *1351*, 1351–1352, 1372
Nuclear structure, 1346–1381
Nucleus(i), 8, 10
  atomic, 1254
  charge and mass of, 1348, 1348t
  properties of, 1347–1352
  radius of, 1350, *1350*
  size and structure of, *1349*, 1349–1350,
    1350
  stability of, 1351, *1351*
  volume and density of, 1350
Numbers, expression of measurements in,
    5n
Numerical modeling, in particle dynamics,
    158–160

Object distance, 1052–1053, *1053*
Occhialini, Giuseppe P.S., 1418
Ocean, pressure of, 426
Oersted, Hans Christian, 647, 833, 865, 870
Ohm, Georg Simon, 775, 804

Ohm (unit), 776, 969
Ohmic materials, 775–776, 778, *778*
Ohm's law, 775–776, 776n, 784, 789, 801, 973, 1321
  resistance and, 775–780
Omega facility, University of Rochester, 1395
Onnes, Heike Kamerlingh, 1318
Open-circuit voltage, 798, *798*
Optic axis, 1136
Optical illusion, of levitated professor, 1054, *1054*
Optical lenses, testing of, 1105
Optical stress analysis, 1138, *1138*
Optically active materials, 1137–1138
Optics, fiber, 1041, *1041*
  geometric, 1052–1090
    ray approximation in, *1026*, 1026–1027, *1027*, 1042
  light and, 1021–1145
Orbital, molecular, 1288
Orbits, circular, around Earth, satellite moving in, *148*, 148–149
  mass moving in, 404–405, *405*
  total energy for, 405–406, 411
  elliptical, 398
    motion in, 400, *400*
  of planets, 397, *397*, 397n, *399*
Order number, for fringes, 1095, 1108
Order-of-magnitude calculations, 13–14
Oscillating body, 363–364
Oscillation(s), damped, 377–378
  forced, *379*, 379–380
  in *LC* circuit, 948–952
  on horizontal surface, 369–370
Oscillator, critically damped, 378, *378*
  damped, 378, *378*
  frequency of, 502, *502*
  overdamped, 378
  quantized, 1193
  simple harmonic, 467–468, *468*, 468n, 1241–1242, *1242*, *1243*
    energy of, 368–370, *369*
    position of, 361, 380
  two speakers driven by, 502, *502*
  underdamped, *378*
  velocity and acceleration of, 362, 381
Oscillatory motion, 360–390
Oscilloscope, 669–671, *671*
  cathode ray tube of, 669–670, *670*
Oscilloscope simulator, 974
Otto cycle, efficiency of, 626
  PV diagram for, 625, *626*
Overtones, 506
Oxygen, magnetic field and, 864
Oxygen molecule, 282
Ozone layer, 1013

Pair production, 1415, 1416
Parabola, equation of, A.22

Parallel-axis theorem, 285–286, *287*
  application of, 287
Parallel-plate capacitors. *See* Capacitor(s), parallel-plate.
Parallelogram rule of addition, 57, *57*
Paramagnetism, 881, 884, 887
Particle dynamics, numerical modeling in, 158–160
Particle physics, 1413–1442
  Standard Model of, 1429–1431, *1430*
Particle(s), 1438
  alpha, *1253*, 1253–1254
  and extended object, gravitational force between, *407*, 407–408
  and solid sphere, gravitational force between, 408–409, *409*
  and spherical shell, gravitational force between, 408, *408*
  angular momentum of, *312*, 312–315, 325
  barrier penetration by, *1238*, 1238–1239
  charged, electromagnetic waves and, 1008–1009, *1009*
    motion of, electric current and, 773, *773*
    in magnetic field, 834–835, 836, 843–847, *844*, 854
      applications of, 847–850
    in uniform electric field, 667–669, 672
    rotating, angular frequency of, 844, 854
  classification of, 1420, 1421t
  distance traveled by, 56
  distribution of, in space, 599
  falling in vacuum, 158, *158*
  gravitational force between, 401–402, *402*
  in box, *1230*, 1230–1234, 1236, *1236*
  in well of finite height, *1236*, 1236–1237, *1238*
  interaction of, 1414, 1414t
  moving in simple harmonic motion, properties of, 363
  pair of, center of mass of, 252, *252*
  single, tunneling, 1332–1333, *1333*
  strange, and strangeness, *1423*, 1423–1424, 1438
  system of, conservation of momentum for, *236*, 236–237, 261
    motion of, 255–259
    Newton's second law of motion for, 256–257, *257*, 262
    speed of, 604
    total angular momentum of, 313
    total momentum of, 256
  systems of, 587
  three, center of mass of, 254, *254*
  three interacting, 403, *403*
  total energy and momentum of, 1175, 1182
  tunneling through barrier, *1238*, 1238–1239
  two, gravitational forces between, 392, *392*
  velocity versus time plot for, *41*, 41–42, *42*
  wave properties of, 1218–1221
Pascal, 441

Pascal, Blaise, 424
Pascal's law, 424
Paschen series, 1202
Path, mean free, 604–605
Path difference, and phase angle, 502
  and phase difference, 1121
  of light waves, 1095
Path integral, 708
Path length, 501
Pauli, Wolfgang, 1257, 1266, 1267, 1366–1367
Pauli exclusion principle, 1266, *1268*, 1312
Peacock, feathers of, colors in, 1103
Pendulum, 210, *210*, *371*, 371–375
  ballistic, 246, *246*
  conical, free-body diagram for, *146*, 146–147
  in oscillation, 380, *380*
  period of, 1163
  physical, 373–374, *374*, 381
  simple, 372, 373, 381
    period of motion for, 372–373
    period of oscillation for, 373
  torsional, 374–375, *375*
Penzias, Arno A., 1433–1435, *1434*
Performance, coefficients of, 627, 628
Period, for mass attached to spring, 365–366
  of motion, for simple pendulum, 372
  of simple harmonic motion, 361
Periodic motion, 360–390
Periodic table of elements, A.32–A.33, 1269
Periodic waves, frequency of, 455
Permeability, magnetic, 884
  of free space, 865, 890
Permittivity of free space, 654–655
Phase, of simple harmonic motion, 361
Phase angle, 971, 983
  path difference and, 502
Phase change, 557
  due to reflection, 1101–1103, *1102*
Phase constant, 467
  in simple harmonic motion, 361
Phasor diagram(s), 965, 966, 968, 972, 975, *975*, 1097, *1098*, 1099, *1100*, 1109, 1121, *1121*, *1122*, 1224, *1224*
Phasors, 965
Phipps-Taylor experiment, 1264
Phonons, 1329
Photoelectric current, 1194, *1194*
Photoelectric effect, 1024, 1025, *1194*, 1194–1195, *1195*, 1208, 1216
  applications of, *1196*, 1196–1197, *1197*
  for sodium, 1195–1196
Photoelectrons, 1194, 1208
Photographic emulsion, 1405
Photomultiplier tube, 1404, *1404*
Photon theory, 1217
Photon(s), 1024, 1192
  and electromagnetic waves, 1194–1195, 1216–1217

Photon(s) (*Continued*)
angular momentum and, 1273
Compton scattering of, *1199*, 1199–1200
energy and momentum for, 1175
energy of, 1024, 1025
high-energy, 1369
in light, 1092n
momentum of, 1218
stimulated absorption of, *1275*, 1275–1276
virtual, 1418
Phototube, 1196, *1196*
Physical quantities, symbols, dimensions, and units of, A.3–A.4
Physics, and measurement, 3–22
areas of, 1
atomic, 1252–1285
classical, 4
modern, 4
"architects" of, 1153
nuclear, 1347
particle, 1413–1442
Standard Model of, 1429–1431, *1430*
quantum, 1189
Pierce, John R., 924, 924n
Pions, 1418–1419
Pitch, 516
Pitching machine, recoiling, 238, *238*
Planck, Max, 1022, 1024, 1147, 1190, 1191, 1192, 1208
Planck's constant, 324, 882, 1024, 1272
Planck's hypothesis, 1192–1193, 1208
Plane electromagnetic waves, *997*, 997–1002, 998n, *999*, *1001*, 1001n
Plane polar coordinates, 54, *54*
Planets, data on, 397n, 397–398, 398t
gravitational force acting on, 399, *399*
motion of, energy in, 404–407
Kepler's laws of, 395–396, 397–398, 411
laws of gravity and, *396*, 396–400
orbits of, 397, *397*, 397n, *399*
Plates, in standing waves, 512–513, *513*
Platinum, superconductivity of, 1318, *1318*
Platinum resistance thermometers, 781
Point charge(s), closed surface surrounding, 688, *688*
electric field due to, 659, *659*, *691*, 691–692
electric field lines for, *665*, 665–667, *666*, *667*
electric potential of, 713
group of, electric field due to, 659, 672
outside of closed surface, 688, *688*
positive, acceleration of, 668, *668*
potential due to, 714, *714*
potential energy due to, electric potential and, *712*, 712–714, *713*
several, electric potential of, 713, *713*
two, electric potential of, *713*, 713n, 713–714
potential energy of, 713, 729
Poisson, Simeon, 1118

Polar coordinates, 55
Polar molecules, 757–758
Polarization, 652
induced, 757
of light waves, *1132*, 1132–1138, *1133, 1134, 1135, 1137*
Polarizer, 1133, *1133*
Polarizing angle, 1134, 1139
Polaroid, 1133
Poles, magnetic, 832, 833
Population inversion, 1276
Position, velocity, and acceleration, graphical relations of, 30
velocity vector as function of, 75
Position-time graph, 24, *24*, 26, *26*, 33
Position vector, 72, *72*, 74, 93
as function of time, 75
magnitude and direction of, 64, *64*
Positron, 1360, 1415
Potential, due to two point charges, 714, *714*
electric, 707–740
and potential energy due to point charges, *712*, 712–714, *713*
calculation of, 730
due to continuous charge distributions, 717–720, *718*, 729, 729t
due to uniformly charged ring, 718, *718*
obtaining electric fields from, *715*, 715–717
of dipole, *716*, 716–717
of finite line of charge, 719, *719*
of point charge, 713
of several point charges, 713, *713*
of two point charges, *713*, 713n, 713–714
of uniformly charged disk, 719, *719*
potential difference and, 708–709
SI units of, 709
Potential difference, 776
and electric potential, 708–709
in uniform electric field, 709–712, *710*, 728
Potential energy. *See* Energy, potential.
Potentiometer, 816, *816*
Pound, 112
Powell, Cecil Frank, 1418
Power, 186–188
and load resistance, 799, *799*
average, 186, 976, 978, 983–984
definition of, 186
delivered by applied force, 913
delivered by elevator motor, 187–188, *188*
delivered to wheels, 190
electrical energy and, 786–789, 790
in ac current circuit, 975–976
in algebra, A.16–A.17
in electrical heater, 787
in *RLC* circuit, 978
in rotational motion, 293–295, 297
instantaneous, 186–187, 192, 975
loss in conductor, 786–787, 790

Power (*Continued*)
of lenses, 1076
requirements for car, frictional forces for, 189t
resolving, diffraction grating and, 1129–1130
transmission of, 981–983
transmitted by sinusoidal wave, 470, 473
Power connections, for household appliances, 817, *817*
Power cord, three-pronged, 818
Power law, analysis of, 12
Powers of ten, prefixes for, 8, 8t
Poynting vector, 1002, *1002*, 1006, *1006*, 1007, 1010, 1014
Precipitator, electrostatic, 726, *726*
Pressure, 347, 422–423
absolute, 427
amplitude, of sound waves, 481–482, *482*, 491
constant, molar specific heats at, 591
conversion factors for, A.2
critical, 607
definition of, 423, 441
gauge, 427
in fluids, *422*, 423
measurement of, 427, *427*
of ocean, 426
variation of, with depth, *424*, 424–427
Primordial fireball, observation of radiation from, 1433–1435
Prince Rupert drops, 1117
Prism spectroscope, 1035, *1035*
Prism(s), and total internal reflection, *1040*, 1040–1041
dispersion of light and, 1033–1036
measurement of index of refraction using, 1036, *1036*
Probability, 637, 637t, 1229–1230, *1230*
entropy and, 636
Probability density, 1228–1229, 1243–1244
for particle in well of finite height, 1237, *1238*
Probability integral, Gauss's, A.31
Processor chip, high performance, 1308
Projectile, and target, *81*, 81–82
angle of projection and, 82–83, *83*
displacement vector of, 78, *79*
from spring-loaded popgun, speed of, 213–214, *214*
horizontal range of, and duration of flight of, 82, *82*
and maximum height of, *79*, 79–80
parabolic trajectory of, 77, *77*
range of, 80, *80*
speed of, 82
struck by club, range of, *240*, 240–241
Projectile-cylinder collision, 319, *319*
Projectile motion, 77–85, 93
Proportional counter, 1402, *1403*
Propulsion, rocket, *259*, 259–261, *260*

Proton(s), 8–9, 10
  and antiprotons, collisions of, 1346
  charge on, 655, 671
    and mass of, 655t
  electric forces on, 660
  magnetic moment of, 1352
  motion of, in uniform electric field, 711–
      712, *712*
  moving in magnetic field, 837, *837*
  moving perpendicular to uniform mag-
      netic field, 844
  -proton collision, 250–251
  speedy, energy of, 1175–1176
Ptolemy, Claudius, 395
Pythagorean theorem, A.22, 589, 1160

Quadratic equations, A.17–A.18
Quality factor, *977*, *978*, 978n, 978–979
Quantization, of electric charge, 651
  space, *1262*, 1262–1264, *1264*, 1280
Quantum chromodynamics, 1429, 1438
Quantum Hall effect, 853
Quantum mechanics, 4, 1189–1190, 1228–
    1230, 1280
Quantum number(s), 1192, 1208
  for hydrogen atom, 1256, 1256t, 1257
  nuclear spin, 1372
  orbital, 1255, 1261, 1266, 1280
  orbital magnetic, 1255, 1262–1263, 1280
  principle, 1255, 1280
  restrictions on values of, 1255
  rotational, 1291, 1312
  spin magnetic, 1256–1257, 1280
  vibrational, 1293
Quantum physics, 1189
Quantum state(s), 221, *221*, 1192
  allowed for atom, 1267t, 1267–1268
Quarks, 9, *9*, 655n, 1414, 1420, 1425–1428,
    1438
  and color force, 1428
  charmed, 1427–1428, 1438
  color and, 1429, *1429*
  free, 655n
  original model, 1426, *1427*
  properties of, 1426t, 1428t
  quantum chromodynamics amd, 1429,
      1438
Quartz crystals, 1297
Quasar, recession of, 1435

Rad, 1401
Radial acceleration, *87*, 87–89, 93–94
Radian, 277
Radian measure, A.21
Radiation, 573–574
  absorption of, by lead, 1334
  background, 1402
  blackbody, *1190*, 1190–1193
  electromagnetic, absorption of, 1333
  emitted by atom, 1203–1204

Radiation (*Continued*)
  observation of, from primordial fireball,
      1433–1435
  RBE factor for, 1401t, 1401–1402
  thermal, from human body, 1193
  uses of, 1406–1407
Radiation belts, Van Allen, 845–847, *846*
Radiation damage, genetic, 1401
  in matter, 1400–1402
  somatic, 1401
Radiation detectors, bubble chamber,
      1405, *1405*
  cloud chamber, 1405
  Geiger counter, 1402–1403, *1403*
  ion chamber, 1402, 1403
  neuron detectors, 1405
  scintillation counter, *1404*, 1404–1405
  semiconductor diode detector, 1403
  spark chamber, 1405
  track detectors, 1405
Radiation pressure, momentum and,
      1004–1006, 1005n, *1105*
Radio waves, 1011
Radioactive dating, 1368–1369
Radioactive series, 1370t
Radioactive tracing, *1406*, 1406–1407
Radioactivity, 1359–1363, *1360*, *1361*
  artificial, 1370
  natural, 1370, *1370*
Radium, activity of, 1362
  decay of, energy liberated in, 1365
Rarefactions, of sound waves, 479, 481
Ray approximation, in geometric optics,
      1026, 1026–1027, *1027*, 1042
Ray diagram(s), for mirrors, 1058–1060,
      1059
Rayleigh-Jeans law, *1191*, 1191–1192
Rayleigh's criterion, 1124–1125, *1125*,
      1138–1139
Rays, 486
  extraordinary, 1136, *1136*
  gamma, 1013
  ordinary, 1136, *1136*
  paraxial, *1055*, 1056, 1082
RBE (relative biological effectiveness), for
      radiation, 1401t, 1401–1402
RC circuit, 808–813
  capacitor in, charging of, 812, *812*
    discharging of, 812–813
Reactance, capacitive, 970, 983
  inductive, 968–969, 983
Reaction energy Q, 1371
Reaction forces, 114–116, *115*
Reaction(s), nuclear, 1386–1388
  self-sustained chain, 1386
Rectangle, area of, 16
Rectangular coordinate system, 54, *54*
Reference circle, for motion, 376, *376*, 377
Reflection, 1023
  and refraction, 1027–1033
  change in phase due to, 1101–1103, *1102*
  diffuse, 1027, *1028*

Reflection (*Continued*)
  Huygens' principle applied to, *1038*,
      1038–1039
  law of, 1027, *1028*, 1043
  of light, 1027–1028
  polarization of light waves by, 1134–1135,
      1135
  specular, 1027, *1028*
  total internal, *1039*, 1039–1041, *1040*,
      1041, 1043
  transmission of waves and, *464*, 464–465,
      465
Refracting surfaces, flat, 1063, *1063*
  sign convention for, *1061*, 1062t, 1062–
      1063
Refraction, angle of, for glass, 1032, *1032*
  double, polarization of light waves by,
      1135–1137
  Huygens' principle applied to, 1038–
      1039, *1039*
  images formed by, *1061*, 1061–1064
  index of, 1030–1032, 1031t, *1032*, 1033,
      1034, 1043
    and critical angle, 1041, *1041*
    measurement of, 1032, 1036, *1036*
  reflection and, 1027–1033
  Snell's law of. *See* Snell's law.
Refrigerators, as heat engines, 617–618,
      618, 628
Relativity, 1148–1188
  and electromagnetism, 1178–1179, *1179*
  and simultaneity, of time, 1158–1159
  Einstein's theory of, 4, 92–93, 1149,
      1149n, 1156n, 1156–1157
  general, 1179–1182
    Einstein's, 1181
  Newtonian, principle of, 1149–1154
  principle of, 1156, 1157
  special, consequences of, 1157–1167
  special theory of, consequences of, 1182
    postulates of, 1182
Rem, 1402
Reproduction constant, 1386–1387, 1408
Resistance, and Ohm's law, 775–780
  and temperature, 780–782
  current and, 772–796
  definition of, 776, 789
  equivalent, 800, 802, *802*, 818
    finding of, by symmetry arguments, 803,
        803
  load, 798, 982
    power and, 799, *799*
  of coaxial cable, 780, *780*
  of conductors, 776, 779, 790
  of nichrome wire, 779
  of uniform conductors, 777
  SI unit of, 790
Resistive forces, 162
  motion in, 154–158
  proportional to speed, 155, *155*
Resistivity(ies), 776, 777, 784, 789
  conductivity and, 784, 790

Resistivity *(Continued)*
  of various materials, 777t
  temperature coefficient of, 780–781
  temperature versus, 781, *781*
Resistor(s), 778
  color code for, 778, *778*, 778t
  current and voltage across, 966
  electrical energy supplied to, 786, 790
  energy loss in, 813
  in ac circuits, *965*, 965–967, *966*, 966n, 967t, 983
  in parallel, 800–804, *801*, 818
  in series, 799–800, *800*, 818
  phase relationships in, 971–972, *972*
  shunt, 814, *815*
  three, in parallel, *802*, 802–803
Resonance, 379, 380, *508*, 508–510, *509*, 519
  in mechanical system, 978
  in *RLC* series circuit, 978, *978*
  vibrations, 380
Resonance frequency, 379, 508, *508*, 973, 977, 984
Rest energy, 1174, 1175, 1182
Rest mass, 1177
Restoring force, 179
Retina of eye, 1074
Reversible process(es), *619*, 619–620
  for quasi-static ideal gases, 630
Richter, Burton, 1427
Right-hand rule, 278, *278*, 842, *842*
  vector product and, 310, *310*, 325
Right triangle, center of mass of, 255, *255*
Rigid body(ies), angular momentum of, 315, *315*, 325
  center of mass of, 252, *252*, 262
  conditions of equilibrium of, *338*, 338–340, *339*
  definition of, 276
  in static equilibrium, 341–345
  moments of inertia of, 286t, 296
  rolling motion of, *307*, 307–309, *308*
  rotating, angular velocity and acceleration of, 279–281, *280*, 281–282, *282*
  work and, 293, *293*
  rotation of, around fixed axis, 276–305, *277*, *315*, 315–317, 316n
  torque and, 288, *288*, 289, 289–290
  total kinetic energy of, 307–308, 325
Rings, Newton's, 1104–1105, *1105*
RL circuits. *See* Circuit(s), *RL*.
RLC circuits. *See* Circuit(s), *RLC*.
RLC series circuits. *See* Circuit(s), *RLC* series.
rms current, 966, *966*, 966n, 967, 983, 984
rms voltage, 967, 983
Robot, 1252
Rocket, escape speed of, 406
Rocket propulsion, *259*, 259–261, *260*
Rod(s), center of mass of, *254*, 254–255
  charged, electric field due to, 663, *663*

Rod(s) *(Continued)*
  conducting, connected to battery, 1009, *1010*
  in standing waves, 512–513, *513*
  rigid, uniform, moment of inertia of, 285, *285*
  rotating. *See* Rotating rod.
  swinging, 374, *374*
Roemer, Ole, 1025
Roemer's method for measuring speed of light, *1025*, 1025–1026
Roentgen, 1401
Roentgen, W., 1130
Romognosi, Gian Dominico, 833n
Root mean square (rms) current, 966, *966*, 966n, 967, 983, 984
Root mean square (rms) speed, 590, 602
  for molecules, 591
Root mean square (rms) voltage, 967, 983
Rotating masses, 283, *283*
Rotating platform, angular speed of, 320, *320*
Rotating rigid body. *See* Rigid body(ies), rotating.
Rotating rod, *290*, 290–291, 295, *295*, *316*, 316–317
Rotating system, force in, 154, *154*
Rotating turntable, 281
Rotating wheel, 279
Rotation of rigid body, around fixed axis, 276–305, *277*, *315*, 315–317, 316n
Rotational kinematics, 278–279
Rotational motion. *See* Motion, rotational.
Rowland ring, 885–886, *886*
Rubbia, Carlo, 1419
Rutherford, Ernest, 1203, 1253, 1254, 1346, 1349–1350, 1370
Rydberg constant, 1202

Salam, Abdus, 1429–1430
Sandage, Allan R., 1435
Satellite receiver-transmitter dish, 994
Satellite(s), energy of, 221, *221*
  in circular orbit around Earth, *148*, 148–149
  motion of, energy in, 404–407
  orbit of, changing of, 405
Saturn accelerator, 1390
Savart, Felix, 865
Scalar product, as commutative, 175
  definition of, 192
  of two vectors, 174–176, *175*, 175n
Scalar quantities, vector quantities and, 55–56, 64
Scalars, motion and, 24n
Scanning tunnel microscope, *1240*, 1240–1241
Scattering, polarization of light waves by, 1137, *1137*
Scattering events, 1371
Schrieffer, J. Robert, 1319, 1328

Schrodinger, Erwin, 1190, 1216, 1228, 1235
Schrodinger equation, 1234–1236, 1242, 1244, 1253, 1280
Schwinger, Julian S., 1417
Scientific notation, A.14–A.15
Scintillation counter, *1404*, 1404–1405
Scott, David, 37
Scuba diving, refraction and, 1063
Second, definition of, 6
Seesaw, 342, *342*
Segre, Emilio, 1415
Selection rule for allowed atomic transitions, 1272
Self-inductance, 940–941
Self-induction, 939
Semiconductor devices, integrated circuit, *1310*, 1310–1311, *1311*
  junction transistor, 1308–1310, *1309*
  p-n junction, *1307*, 1307–1308
Semiconductor diode detector, 1403
Semiconductor(s), 652, 1312–1313
  doped, electrical conduction in, *1306*, 1306–1307
  electrical conduction in, 1305, *1305*, 1305t
  extrinsic, 1307
  n-type, 1306–1307, 1313
  p-type, 1307, 1313
  resistivity versus temperature for, *781*, 781–782
Semiempirical formula, for energy binding, 1373
Series expansions, A.24
Shape, elasticity of, 347, *347*
Shear modulus, 346, 347, *347*, 349
Shear stress, 347
Shell nuclear model, 1358, *1358*
Shells, and subshells, atomic, 1256, 1256t, 1280
  spherical, 1258, *1258*
Shock, electrical, 818
Shock waves, *490*, 491, *491*
Shockley, William, 1308
Shunt resistor, 814, *815*
SI unit(s), 7
  base, A.34
  derived, A.34
  of average speed, 26
  of capacitance, 742, 760
  of current, 773, 776, 789
  of electric field, 709
  of force, 112
  of inductance, 940
  of magnetic field, 837
  of magnetic moment, 842
  of mass, 5–6, 110
  of potential, 709
  of resistance, 790
  of time, 6
  prefixes for, 8t
Significant figures, 15–16
Silicon, crystalline, 1297

Simple harmonic oscillator, 467–468, *468*, 468n, 1241–1242, *1242, 1243*

Simultaneity, and relativity, of time, 1158–1159

Sinusoidal waves. *See* Wave(s), sinusoidal.

Siren, sound of, frequency of, 490

Ski jumper, vertical component of velocity of, *83*, 83–84

Slide-wire potentiometer, *816*, 816–817

Slug, 112n

Snell, Willebrord, 1029

Snell's law, 1029, 1029n, 1032, 1033, 1039, 1040, 1042, 1043, 1062, 1135

Soap film, interference in, 1106

Sodium, energy bands of, 1300, *1300*

Sodium chloride, 1287, *1288*
  structure, 1131, *1131*, 1135–1136

Software requirements, for spreadsheets, A.40–A.41

Solar cells, nonreflecting coatings for, 1106, *1106*

Solar day, 6n
  mean, 6

Solar energy, 1005

Solar second, mean, 6

Solenoid(s), axis of, magnetic field along, *876*, 876–877, *877*
  electric fields due to, *918*, 918–919
  ideal, 875, *875*
  inductance of, 941, 954
  magnetic field of, *875*, 875–876, 890, 945
  superconducting, 1325
  two, mutual inductance of, 947–948, *948*

Solidification, heat of, 558n

Solid(s), 421
  amorphous, 1136
  band theory of, *1300*, 1300–1301
  birefringent, 1136
  bonding in, 1296–1300, 1312
  crystalline, 1297, 1312
  crystalline structure of, 1131, *1131*, 1135–1136
  deformation of, 345
  double-refracting, 1136
  elastic properties of, 345–349
  elasticity of, 345–349
  heat transferred to, 567–568
  ionic, *1297*, 1297–1298, *1298*
  metallic, *1299*, 1299–1300
  sound waves in, 480, 480t
  specific heat of, 598, *598*
  thermal expansion of, 536–540, *537*, 537n
  total internal energy of, 598

Sommerfeld, Arnold, 1254

Sonic boom, 491

Sound levels, in decibels, 484, 484t

Sound waves. *See* Wave(s), sound.

Space quantization, *1262*, 1262–1264, *1264*, 1280

Space-time, curvature of, 1181–1182

Spaceship(s), contraction of, 1163, *1163*, 1164

Spaceship(s) *(Continued)*
  relative velocity of, 1170, *1170*
  triangular, *1164*, 1164–1165
  voyage to Sirius, 1165

Spark chamber, 1405

Spatial interference, 513

Spatial order, 629

Spectral lines, from star, 1206
  width of, 1227

Spectrometer, diffraction grating, 1128, *1128*
  mass, *848*, 848–849

Spectroscope, prism, 1035, *1035*

Spectroscopy, absorption, 1201

Spectrum(a), absorption, 1201, *1201*, 1272
  atomic, 1200–1202, 1272–1275
  emission, 1272
  line, 1200–1201, *1201*
  molecular, *1295*, 1295–1296, *1296*, 1296n
  nuclear magnetic resonance, 1353, *1353*
  of electromagnetic waves, 1011–1013, *1012*, 1014
  of light, 1034–1035, *1035*
  of molecules, 1290–1296
  of sodium, lines of, resolving of, 1130
  of x-rays, *1274*, 1274–1275, 1281
  visible, 1201, *1201*

Speed(s), 27, 44, 73
  angular, average, 277
    instantaneous, 277, 278n, 296
  average, 26, 44, 602
  constant, circular motion with, 377
  conversion factors for, A.1
  dimensions of, 11t
  drift, 774, 784, 784n, 790
    in copper wire, 774–775
  escape, 405–407, *406*, 407t
  high, air drag at, 156–157, *157*
    kinetic energy at, 191–192
    relative motion at, 92–93
  in simple harmonic motion, 361, 362–363, *367*
  linear and angular, 280
  molecular, distribution of, *602*, 602–604, *603*
  most probable, 602
  of boat, relative to Earth, *91*, 91–92
    relative to water, 92, *92*
  of car, coefficient of static friction and, 147
  of efflux, 435, *435*
  of electromagnetic waves, 999
  of electron, 1148–1149
  of light. *See* Light, speed of.
  of mass, central force and, 146
    in free-fall, 209, *209*
    sliding down ramp, 212, *212*
  of projectile, 82
  of skier, on incline, 213, *213*
  of sound waves, 480–481, 491, 1152–1154
  of sphere, falling in oil, 156
  of wave pulse, 458
  of waves on strings, 462–464, *463*, 472

Speed(s) *(Continued)*
  resistive force proportional to, 155, *155*
  root mean square(rms), 590, 591, 602
  terminal, 155, *155*, 162
    for various objects falling through air, 157t
  transverse, 468–469

Sphere(s), charge on, *657*, 657–658
  inside spherical shell, 695, *695*
  lead, volume of, 349
  rotating, 316, *316*
    forces acting on, 150–151, *151*
  solid, gravitational force between particle and, 408–409, *409*
    rolling, free-body diagram for, 309, *309*
  speed of, falling in oil, 156
  speed of center of mass of, 308–309
  two, connected charged, 723, *723*
  uniformly charged, of electric potential, 720, *720*

Spherical aberration(s), 1056, *1056*, 1070, 1070–1071

Spherical shell, gravitational force between particle and, 408, *408*
  sphere inside, 695, *695*
  thin, electric field due to, *691*, 691–692

Spin, 865, 882
  angular momentum, 882
  nuclear, *1351*, 1351–1352, 1372
  of electron, 1257, *1257*, 1263–1266, 1264n

Spin magnetic quantum number, 1256–1257, 1280

Spreadsheet problems, A.40–A.42

Spring, collision of two-body system with, 247, *247*
  deformation of, to measure force, 107, *107*
  elastic potential energy of, 205, *205*
  force constant of, 179
    measurement of, 180, *180*
  mass attached to, *364*, 364–368, *366, 367*
  mass of, 1178
  work done by, *178*, 178–180, *179*, 204

Spring-loaded popgun, speed of projectile from, 213–214, *214*

Stable equilibrium, 218

Stanford Linear Collider, California, 1431

Star, spectral lines from, 1206

State function, 565

Stationary states, for particle in box, 1231–1232, *1232*

Statistical mechanics, 596

Steam engine, 623

Stefan's law, 573

Stern-Gerlach experiment, 1263–1264, *1264*, 1266

Stirling engine, 625

Stopping potential of electrons, 1194

Strain, 345, 349
  tensile, 346

Strain gauge, 816

Strangeness, and strange particles, *1423*, 1423–1424, 1438
  conservation of, 1424
Strangeness number, law of conservation of, 1424
Strassman, Fritz, 1383
Streamlines, *432*, 432–433, 436, *436*, 441
Stress, 345, 349
  shear, 347
  tensile, 346
  volume, 347
Submerged object, 430, *430*
Sun, 1389–1390, 1391
  mass of, 399
Super Proton Synchrotron, at CERN, 1431
Superconducting Super Collider, Texas, 1431
Superconductivity, 1317–1345
  high-temperature, *1336*, 1336–1339, *1337*, *1338*, 1338t, 1341
Superconductor(s), 782–783, 888
  critical temperatures for, 782, 782t
  energy gap for, 1331, 1331t
  flux quantization in, 1327
  high-temperature, applications of, 1339–1340
  persistent currents in, *1326*, 1326–1327
  properties of, 1317, 1340
  resistance-temperature graph for, 782, *782*
  type I, critical magnetic field of, 1319–1321, *1320*, 1320t
    critical temperature of, 1319–1321, 1320t
    magnetic properties of, 1321–1322, *1322*
    magnetization of, 1323, *1323*
    properties of, 1319–1323
  type II, *1324*, 1324t, 1324–1325, 1340
Supercurrents, in superconductors, *1326*, 1326–1327, 1340
Superposition principle, 459, 460, *461*, 472, 499, 500, 655–656, 659, 672, 1092
Susceptibility(ies), magnetic, 884, 884t
Symmetry breaking, 1430
Synchronization, of clocks, 1157–1158

Tacoma Narrow's Bridge, collapse of, 380
Tangential acceleration, *87*, 87–89
Telescope(s), 1080–1082
  reflecting, 1080, *1081*, 1081–1082
  refracting, 1080, *1080*
  resolution of, 1126–1127
Temperature, 531–550
  and zeroth law of thermodynamics, 532–533, 543
  average kinetic energy and, 590
  change in, 531–550
  coefficient, resistivity of, 780–781
  conversion of, 536
  critical, 607, 782, 1319, 1340
    for superconductor, 782, 782t
    of type I superconductors, 1319–1321, 1320t

Temperature *(Continued)*
  critical ignition, 1392, 1408
  Curie, 887, *887*, 887t
  Fermi, 1303
  fixed point, 535, 535t
  molecular interpretation of, 589–591
  resistance and, 780–782
  resistivity versus, 781, *781*
  steps for obtaining, 624
Temperature gradient, 569, 576
Temperature scale(s), 533
  absolute, 534, 624
  Celsius, 534, 536
  Fahrenheit, 536
  ideal gas, 624
  Kelvin, 534, 535, *535*, 536, 624
  thermodynamic, 535
Templates, computer spreadsheets and, A.40
Tensile strain, 346
Tensile stress, 346
Tension, 116
  free-body diagram for suspended object and, 118–119, *119*
Terminal speed, 155, *155*, 162
  for various objects falling through air, 157t
Terminal voltage, 798, 798n
Tesla, Nikola, 981
Tesla (unit), 837, 854
Tevatron, Fermi National Laboratory, 1431
Thermal conductivity(ies), 569, 570t, 576
Thermal contact, 532
Thermal efficiency, 617, 624, 637–638
  of heat engines, 617
Thermal energy, 616, 618
  definition of, 552
  heat and, 552–554, 574
  ideal gas and, 591
  temperature and, 558–559, 562–563, *563*
  transfer of, 552, 557, 563, 564, 570, 575, 628, 630–631
Thermal equilibrium, 532, 543
Thermal expansion, of solids and liquids, 536–541, *537*, 537n
Thermal neutron, 1383
Thermal radiation, from human body, 1193
Thermistor, 782
Thermodynamic temperature scales, 535
Thermodynamic variables, 542
Thermodynamics, 529–645
  first law of, 552, 563–565, 575, 615, 617
    applications of, 565–568
  second law of, 616, 617, 638
    heat engines and, 616–619
    Kelvin-Planck form of, 617, 618
  third law of, 624n
  work and heat in, *561*, 561–563, *562*
  zeroth law of, 628
    temperature and, 532–533, 543
Thermogram, 572
Thermometer(s), 533
  calibration of, 533

Thermometer(s) *(Continued)*
  constant-volume gas, 534, *534*
  gas, 533, 534, *534*, 535
  mercury, 533, *533*
  platinum resistance, 781
  problems associated with, 533, 533n
Thermonuclear power reactor, requirements for, 1393
Thermos bottle, 574
Thermostats, 539
Thin-film interference, *1103*, 1103–1107
  problems in, 1105
Thompson, Benjamin, 553
Thomson, G.P., 1220
Thomson, J.J., 849, *849*, 1203, 1253
Threshold energy, 1371
Thrust, 261
Time, 17
  conversion factors for, A.1
  dilation, 1159–1163, *1160*, *1162*, 1182
    and atomic clocks, 1166n, 1166–1167
    twins paradox and, *1165*, 1165–1166
  displacement as function of, 38, *38*
  intervals, approximate values for, 8t
  mean free, 605
  proper, 1162
  SI unit of, 6
  simultaneity and relativity of, 1158–1159
  standards of, 4–8
  velocity as function of, 32, 38, *38*
  velocity vector as function of, 74
  versus displacement for simple harmonic motion, 361, *361*, *367*
  versus velocity plot for particle, *41*, 41–42, *42*
Time constant, of *RL* circuit, 944, *944*
Time-dependent Schrodinger equation. *See* Schrodinger equation.
Ting, Samuel, 1427
Tokamak, 1393, *1393*
Tokamak Fusion Test Reactor, Princeton, 1394, *1398*, 1398–1400, *1399*, *1400*
Tomonaga, Shinichiro, 1417
Top, spinning, motion of, 321–322, *322*
Toroid, iron-filled, 885
  magnetic field created by, 873, *873*, 890
  magnetic field strength in core of, 883
Torque, 287–289, *288*, 296
  and angular acceleration, *289*, 289–292
  and angular momentum, 313
  definition of, 287, 288, 296, 842–843, 854
  net, and angular momentum, 322
    on cylinder, 288–289, *289*
  net external, 290, 296, 313, 325
  on current loop, in uniform magnetic field, *841*, 841–843, *842*
  rolling motion, and angular momentum, 306–336
  vector product and, 309–311, *310*, 325
Torricelli, Evangelista, 427
Torricelli's law, 435

Torsion balance, 1005, *1005*
  Coulomb's, 651, *651*
Tracing, radioactive, *1406*, 1406–1407
Track detectors, 1405
Trajectory, parabolic, of projectile, 77, *77*
Transducer, 479
Transformation, Galilean, 1150, 1167
  of coordinates, *1150*, 1150–1151
  Lorentz, 1182
  velocity, Galilean, 1151, 1169
    Lorentz, 1169–1170, 1172, 1182
Transformation equations, Lorentz, 1167–
    1171, 1168n
Transformer, 981–983
  ac, 981, *981*
  ideal, 981, *981, 982*
  step-down, 981
  step-up, 981, 982–983
Transistor, junction, 1308–1310, *1309*
Transmission axis, 1133, 1139
Transmission coefficient, for electron, 1239
Triangle method of addition, 57
Trigonometry, A.22–A.24
Tritium, 1391
  and deuterium, fusion reaction, *1397*,
    1397–1398
Tuning fork, frequency of, 512, *512*
Tunnel diode, 1239
Tunneling, Josephson, 1333–1336
Tunneling of particles, *1238*, 1238–1239
Turbulent flow, 431, *432*
Turning points, 218
Turntable, rotating, 281
Twins paradox, time dilation and, *1165*,
    1165–1166

Uhlenbeck, George, 1257, 1264
Ultraviolet catastrophe, 1192
Ultraviolet light, 1012–1013
Ultraviolet waves, 1012–1013
Uncertainty principle, energy-time, 1226,
    1243
  Heisenberg, 1224–1227
Unit vectors, 60–62, *61*
Units, *cgs* system of, 7
  conversion of, 13
  of acceleration, 112t
  of flux, 877
  of force, 112, 112t
  of heat, 553
  of mass, 112, 112n, 112t
  of physical quantities, A.2–A.3
  SI. *See* SI unit(s).
Universal gas constant, 542
Universe, Big Bang theory of creation of,
    161, 1432–1435, 1438
  critical density of, 1436, *1436*
  dark matter of, 1437
  entropy of, 629, 632, 638
  expansion of, 1435–1437, *1436*
  history of, 1432, *1433*

Universe *(Continued)*
  missing mass of, 1437
  total energy of, 219
Uranium-235, fission of, 1384, 1385, *1385*
  energy released in, 1385

Vacuum, particle falling in, 158, *158*
  speed of light in, 1030n, 1030–1031, 1043
Valence band, 1304
Van Allen, James, 845
Van Allen radiation belts, 845–847, *846*
Van de Graaff, Robert J., 725
Van de Graaff generator, 707, *725*, 725–726
van der Meer, Simon, 1419
van der Waals, J.D., 606
van der Waals bonds, 1288, 1312
van der Waals' equation of state, *606*, 606–
    607, *607*
van der Waals forces, 1288–1289, *1289*
Vaporization, latent heat of, 558, 558n, 558t
Vascular flutter, 437, *437*
Vector form, 655
Vector product(s), and torque, 309–311,
    310, 325
  definition of, 310
  properties of, 310
  right-hand rule and, 310, *310*, 325
  two vectors of, 311
Vector(s), 53–70
  acceleration, displacement and velocity
    vectors, 71–74
  addition of, 56–58, *57*, 62, *64*, 64–65
    component method of, *61*, 61–63
    law of, 840
  components of, *59*, 59–60, *60*, 65, *65*
  displacement, 71–72, *72*
    of projectile, 78, *79*
    velocity and acceleration vectors, 71–74
  gravitational field, 401, *401*
  magnetization, 883
  multiplication of, by scalar, 58
  negative, 58
  position, 72, *72*, 74, 93
    as function of time, 75
    magnitude and direction of, 64, *64*
  Poynting, 1002, *1002*, 1006, *1006*, 1007,
    1010, 1014
  projections of, 59–60
  properties of, 56–59
  quantities, and scalar quantities, 55–56, 64
  subtraction of, 58, *58*
  two, cross products of, 311
    equality of, 56, *56*
    scalar product of, 174–176, *175*, 175n
    sum of, 62–63
  unit, 60–62, *61*
    acceleration of particle and, 88, *88*
  velocity, 93
    as function of time, 74
Velocity, angular, and angular acceleration,
    277–278

Velocity *(Continued)*
  as function of displacement, 33–34
  as function of position for simple har-
    monic motion, 369
  as function of time, 32, 38, *38*
  as vector quantity, 56
  average, 24, 25, 27, 43, 72, *72*
    and instantaneous, 27
    calculation of, 25
  drift, 774, 784, 784n, 790
  Galilean addition law for, 1151
  instantaneous, 26, 27, 44, 73
    and speed, 26–28
    definition of, 26
  of sinusoidal waves, 466
  position, and acceleration, graphical rela-
    tions of, 30
  relative, 89–92, *90*
    of spaceship, 1170, *1170*
  tangential, 280
  transformation, Galilean, 1169
    Lorentz, 1169–1170, 1172, 1182
  versus time plot for particle, *41*, 41–42, *42*
Velocity selector, 847–848, *848*
Velocity-time graph, 29, *29, 30*, 32, 33
Velocity vector, 93
  as function of time, 74
Venturi tube, 435, *435*
Vibration, frequencies of, 367
  modes of, 453, 499
    normal modes of, 513, *514*, 519
      in standing wave, *505*, 505–506
  of string, 470
  resonant, 380
  wave motion and, 453
Virtual photons, 1418
Viscosity, 431–432, *439*, 439–441
  coefficient(s) of, *440*, 440t, 440–441
Volt, definition of, 709
  electron, 709
Voltage, Hall, 852
  open-circuit, 798, *798*
  root mean square (rms), 967, 983
  terminal, 798, 798n
    of battery, 799
Voltage amplitude, of ac generator, 965
Voltmeter, 752, 814, *814*
Volume, constant, molar specific heats at,
    591, 607
  dimensions of, 11t
  expansion, average coefficient of, 538, 543
Volume elasticity, *347*, 347–348
Volume flux, 433
Volume stress, 347
von Klitzing, Klaus, 853
Vortex state, 1324, *1324*, 1340

Wall, *R* value of, 572
Water, boiling of, 470
  freezing of, 541
  triple point of, 535

Water bed, 426
Water molecule, *756*, 756–757
Water waves, 455, 456, *457*
Watt, 187
Watt, James, 187
Wave equation(s), for electromagnetic
    waves in free space, 998
  general, 998
Wave front, 485
Wave function(s), 458, 473 472, 1228, *1229*
  allowed, 1231, 1244
  for hydrogen, 1257–1261, *1258*
  for particle in box, 1231
  for particle in well of finite height, 1237,
    1238
  for sinusoidal waves, 466, 471
  for standing waves, 503
Wave intensity, 1002
Wave number, angular, 466
Wave pulse, 456, *456*, 456n, 458–459, *459*,
    472
  speed of, 458, *463*, 463–464
Waveform(s), 516, *516*
  harmonics of, *518*
Wavelength(s), 486
  Compton, 1198
  cutoff, 1195
  deBroglie, 1218, 1218n, 1220, 1228, 1231
  definition of, 455, *455*
  for standing wave, 1231, *1231*
  of electron, 1221, 1222
  of normal modes, 506
  of rock, 1221
  "seeing ability" of wave and, 1079–1080
Wavelets, 1036, *1038*
Wave(s), amplitude of, 457, *457*
  audible, 479
  complex, *516*, 516–518
  de Broglie, 1228, 1234–1235, 1243, 1255
  electromagnetic, 455, 1000, *1000*
    detection of, 996, *996*
    energy carried by, 1002–1004
    equations for, 995–996
    frequencies of, 995, 996
    generation of, 995, 996, *996*
    in free space, wave equations for, 998
    momentum and, 1004
    phase changes of, 1102
    photons and, 1194–1195, 1216–1217
    plane, *997*, 997–1002, 998n, *999*, *1001*,
      1001n
    production of, by antenna, 1008–1011,
      1010, *1011*
    properties of, 1000, 1013–1014
    radiation pressure and, 1004–1005,
      1005, 1005n
    sinusoidal plane-polarized, *999*, 999–
      1000, 1014
    spectrum of, 1011–1013, *1012*, 1014
    speed of, 999
  grain, 455
  harmonic, traveling, 500

Wave(s) *(Continued)*
  infrared, 1012
  infrasonic, 479
  interference of, 459
  light, fringes of, 1095, 1108
    bright, 1120
    dark, 1120
      positions of, 1120–1121
    distance between, 1096
    order number of, 1095, 1108
  interference of, 1091–1116, 1117–1118,
    1118
    constructive, 1095
    in thin films, *1103*, 1103–1107
  linearly polarized, 1132, *1132*
  path difference of, 1094–1095
  plane of polarization of, 1132
  plane-polarized, 1132
  polarization of, *1132*, 1132–1138, *1133*
    by double refraction, 1135–1137
    by reflection, 1134–1135, *1135*
    by scattering, 1137, *1137*
    by selective absorption, *1133*, 1133–
      1134, *1134*
  speed of, 1151–1154, *1152*
    constancy of, 1156, 1157
  unpolarized, 1132, *1132*
  linear, 459, 472
  longitudinal, 456, *456*, 472
  matter, 1228, 1234–1235, 1243
    frequencies of, 1218–1219
  mechanical, 453–527
    requirements for, 455
  motion, 454–478, *456*
    and vibration, 453
    stimulator, 460
  nonlinear, 459
  on strings, speed of, 462–464, *463*, 472
  periodic, frequency of, 455
  phasor addition of, *1098*, 1098–1101,
    1099
  physical characteristics of, 455
  plane, 486, *486*
  radio, 1011
  shock, *490*, 491, *491*
  sinusoidal, *465*, 465–469
    frequency of, 466
    general relation for, 467
    interference of, 500
    longitudinal, 481, *481*, 491
    on strings, 467–469, *468*, 468n, 473
      energy transmitted by, *469*, 469–470
    power transmitted by, 470, 473
    traveling, 467, *467*
    velocity of, 466
    wave function for, 466, 471
  sound, 455, 456
    amplitude displacement of, 481–482, *482*
    amplitude pressure of, 481–482, *482*, 491
    harmonic, intensity of, 491
    intensity variations from source, 486–487
    interference of, 501, *501*

Wave(s) *(Continued)*
  moving train whistle, frequency of, 489–
    490
  periodic, 481–482
  pressure of, 481–482, 482, 491
    intensity of, 482–484, *483*
  siren, frequency of, 490
  speed of, 480–481, 491
    in various media, 480t, 480–481
  spherical, intensity of, 491–492
  speed of, 455
  spherical, 484–487, *485*
  standing, 499, 502–505, *504*, 519
    formation of, 505
    in air columns, 510–512, *511*, 519
    in fixed string, *505*, 505–508
    in rods and plates, 512–513, *513*
    longitudinal, 510–511, *511*
    motion of, energy and, 503–504, *504*
    wave function for, 503
    wavelengths for, 1231, *1231*
  standing states of, 1219, *1219*
  stationary, 504
  superposition of, 459
  transmission of, reflection and, *464*, 464–
    465, *465*
  transverse, 456, 472
  traveling, 456, *456*, 459–460, *461*
    harmonic, 500
    one-dimensional, 457–459
  two, combining of, 1093–1094, *1094*
  types of, 456–457
  ultrasonic, 479
  ultraviolet, 1012–1013
  visible, 1012
  water, 455, 456, *457*
    interference pattern in, *1093*
Weber per square meter, 837, 854
Weber (unit), 877
Weight, 105, 113n, 113–114, 129
  apparent, versus true weight, *123*, 123–124
  gravitational forces and, *123*, 123–124,
    394–395
  loss, and exercise, 554
  mass and, 111
Weinberg, Steven, 1429–1430
Wheatstone bridge, 815, *815*
Wheel(s), angular acceleration of, *291*,
    291–292
  power delivered to, 190
  rotating, 279
Whistle, moving train, 489–490
Wien's displacement law, 1191
Wilson, Robert W., 1433–1435, *1434*
Wind, energy from, 437–439, *438*
Wind generators, *438*, 438–439
Windmill, 439
Windshield wipers, intermittent, 811
Work, 171, 173
  and energy, 171–201
  conversion factors for, A.2
  done by central force, 402

Work *(Continued)*
  done by conservative forces, 205–206, 222
  done by constant force, *172*, 172–174, 192
  done by gas, 575
  done by spring, *178*, 178–180, *179*, 204
  done by varying force, *176*, 176–180, *178*, 192
  done in isothermal process, 566–567
  done using horizontal force, 174, *174*
  heat and, 615
  in rotational motion, *293*, 293–295, 297
  in thermodynamic processes, 561–563, *562*
  net, done by force, calculation of, from graph, 177, *177*
  units of, in systems of measurement, 173, 173t

Work-energy theorem, for rotational motion, 294
  kinetic energy and, *180*, 180–186, *182*, 192
Work function(s), of metals, 1195, 1195t, 1208

X-rays, 1013
  characteristic lines of, 1274
  diffraction of, by crystals, 1130n, 1130–1132, *1131*
  energy of, estimation of, 1275
  scattering from electrons, 1197, *1198*
  spectra of, *1274*, 1274–1275, 1281
Xerography, 726, *727*

Young, Thomas, 1022, 1024, 1092
Young's double-slit experiment, 1092–1096, *1093*, 1108
Young's modulus, 346, 348, 349
Yukawa, Hideki, 1416–1418

Zeropoint energy, 535
Zero(s), absolute, 535, 624, 624n
  significant figures and, 15
Zeroth law of thermodynamics, 628
  temperature and, 532–533, 543
Zweig, George, 1426

Standard Abbreviations and Symbols for Units

Symbol	Unit	Symbol	Unit
A	ampere	in.	inch
Å	angstrom	J	joule
u	atomic mass unit	K	kelvin
atm	atmosphere	kcal	kilocalorie
Btu	British thermal unit	kg	kilogram
C	coulomb	kmol	kilomole
°C	degree Celsius	lb	pound
cal	calorie	m	meter
deg	degree (angle)	min	minute
eV	electron volt	N	newton
°F	degree Fahrenheit	Pa	pascal
F	farad	rev	revolution
ft	foot	s	second
G	gauss	T	tesla
g	gram	V	volt
H	henry	W	watt
h	hour	Wb	weber
hp	horsepower	μm	micrometer
Hz	hertz	Ω	ohm

Mathematical Symbols Used in the Text and Their Meaning

Symbol	Meaning
$=$	is equal to
$\equiv$	is defined as
$\neq$	is not equal to
$\propto$	is proportional to
$>$	is greater than
$<$	is less than
$\gg (\ll)$	is much greater (less) than
$\approx$	is approximately equal to
Δx	the change in x
$\displaystyle\sum_{i=1}^{N} x_i$	the sum of all quantities x_i from $i = 1$ to $i = N$
$\lvert x \rvert$	the magnitude of x (always a nonnegative quantity)
$\Delta x \rightarrow 0$	Δx approaches zero
$\dfrac{dx}{dt}$	the derivative of x with respect to t
$\dfrac{\partial x}{\partial t}$	the partial derivative of x with respect to t
$\displaystyle\int$	integral